ANNUAL REVIEW OF BIOCHEMISTRY

ANNUAL REVIEW OF BIOCHEMISTRY

ESMOND E. SNELL, *Editor*
University of Texas at Austin

PAUL D. BOYER, *Associate Editor*
University of California, Los Angeles

ALTON MEISTER, *Associate Editor*
Cornell University Medical College

CHARLES C. RICHARDSON, *Associate Editor*
Harvard Medical School

VOLUME 47

1978

ANNUAL REVIEWS INC.   4139 EL CAMINO WAY   PALO ALTO, CALIFORNIA 94306

ANNUAL REVIEWS INC.
Palo Alto, California, USA

REPRINTS   The conspicuous number aligned in the margin with the title of each article in this volume is a key for use in ordering reprints. Available reprints are priced at the uniform rate of $1.00 each postpaid. The minimum acceptable reprint order is 10 reprints and/or $10.00 prepaid. A quantity discount is available.

International Standard Serial Number: 0066-4154
International Standard Book Number: 0-8243-0847-6
Library of Congress Catalog Card Number: 32-25093

Annual Reviews Inc. and the Editors of its publications assume no responsibility for the statements expressed by the contributors to this Review.

PRINTED AND BOUND IN THE UNITED STATES OF AMERICA

PREFACE

A preface, properly speaking, should introduce or provide a setting for the more detailed material that follows. But how does one provide such a preface to a volume in which 62 authors treat 33 different topics? It is impossible. One doesn't expect a preface in a monthly review journal, so why is one needed in an Annual Review? Considerations of this nature have led to the omission of a preface in recent issues of this review.

Nevertheless, it is occasionally useful to use this space to comment on matters of interest to our readers. For several years such comments revolved around one major theme: the impossibility of fulfilling the implied promise of our title in the face of the explosive increase in the literature and scope of biochemistry. We nonetheless hope to provide reasonable coverage of most areas of biochemistry over a period of years. To assist the Editorial Committee in this attempt, the categories used in our cumulative chapter title index have been updated and increased in number, and some chapters are listed under more than a single category. Certain topics in the more active categories will be reviewed almost yearly; others perhaps every three to five years. We hope this index will prove useful to the readers as well; their suggestions concerning topics for review and possible authors are always welcome.

After weighing the virtues of increased words per page against those of legibility, our aging editors have also decided to continue to use the larger type size, begun experimentally with Volume 46, in preference to the smaller type of earlier volumes.

Finally, we want to express our thanks to all concerned with production of this volume. Our pleasant and efficient Production Editor, Tate Snyder, moved on to other endeavors during the year; we welcome her replacement, Sally Lee Boyd.

THE EDITORIAL COMMITTEE

Annual Review of Biochemistry
Volume 47, 1978

CONTENTS

ANNUAL REVIEWS INC. is a nonprofit corporation established to promote the advancement of the sciences. Beginning in 1932 with the *Annual Review of Biochemistry,* the Company has pursued as its principal function the publication of high quality, reasonably priced Annual Review volumes. The volumes are organized by Editors and Editorial Committees who invite qualified authors to contribute critical articles reviewing significant developments within each major discipline.

Annual Reviews Inc. is administered by a Board of Directors whose members serve without compensation.

Annual Reviews are published in the following sciences: Anthropology, Astronomy and Astrophysics, Biochemistry, Biophysics and Bioengineering, Earth and Planetary Sciences, Ecology and Systematics, Energy, Entomology, Fluid Mechanics, Genetics, Materials Science, Medicine, Microbiology, Neuroscience, Nuclear Science, Pharmacology and Toxicology, Physical Chemistry, Physiology, Phytopathology, Plant Physiology, Psychology, and Sociology. In addition, two special volumes have been published by Annual Reviews Inc.: *History of Entomology* (1973) and *The Excitement and Fascination of Science* (1965).

Ann. Rev. Biochem. 1978. 47:1–33

EXPLORATIONS OF BACTERIAL METABOLISM

❖967

H. A. Barker

Ann. Rev. Biochem. 1978. 47:1–33

EXPLORATIONS OF BACTERIAL METABOLISM

❖967

H. A. Barker

Department of Biochemistry, University of California,
Berkeley, California 94720

CONTENTS

EARLY YEARS

I grew up mostly in California, first in Oakland till the age of 11, and then in Palo Alto till I graduated from Stanford University. My parents had been part of the western migration. My father as a young man came to California from Maine, where he had grown up on a poor farm in a rural environment that was attractive to a child, but held little promise of a good life for an

0066-4154/78/0701-0001$01.00

adult. He worked for a time as a farm hand in the San Joaquin Valley and later taught in an elementary school for a few years, until he could save a little money. In 1892 he entered Stanford with the first class, but had to drop out before graduating, for lack of funds. He returned to the public schools as a high school teacher. Later he became principal of a high school and eventually a public school administrator in several cities, including Oakland and Palo Alto, where I grew up.

My mother came to Stanford from Denver and obtained an A.B. degree in Classical Literature and an M.A. degree in Latin. She then taught languages in high school for a few years until she married. I had an older brother who was fond of literature and eventually became a professor of English. So there was nothing in my family background that predisposed me to a career in science with one possible exception. Both my father and mother were very fond of the outdoors and so each summer we spent a month or more, whenever possible, camping in the Sierras, and living a simple and quiet life in close contact with Nature. This resulted in my developing a considerable familiarity with plants and animals, and the physical environment, and perhaps even more important, developing a sense of satisfaction and accomplishment in relatively solitary activities such as fishing, hiking, and exploring new areas; this attitude was easily carried over to scientific work in a laboratory.

In high school I followed a rather standard college preparatory curriculum including mathematics, chemistry, and two foreign languages. The scholastic standards were not very high, so I had no difficulty getting adequate grades with little effort. One of my dominant interests in the last two years of high school was music. I had taken piano lessons for several years previously with only minimal results. My enthusiasm for music was greatly stimulated at this time by contacts and a developing friendship with a fellow student, Robert Vetlesen, who had unusual talents as a pianist, and at the age of 14 was already giving concerts of professional quality. As a result, I began to work hard to develop the techniques of piano playing, only to conclude after several years that my abilities in that direction are very limited. Although frustrating, this experience was beneficial in opening up to me the world of music from which I have derived much pleasure.

After graduating from high school I was fortunate to be able to spend a year (1924–1925) in Europe with my family. Most of the winter we stayed in Dresden, which at that time was a center of musical activity. I studied the piano, learned German, read classical German literature, and went to innumerable operas and concerts of every kind, usually occupying the cheapest seats. I remember that one of the highlights of the season was a musical festival honoring Richard Strauss on the occasion of his 60th birthday, during which he conducted several of his operas and ballets.

INTRODUCTION TO SCIENCE

I entered Stanford in 1925 with no idea which field I would ultimately choose as a major. Indeed, I inclined toward literary and historical subjects. Fortunately for me, much of the curriculum for the first two years was fixed, and I was required to take a course in general biology. I found much of the material both novel and interesting, and I recall that I was impressed by the enthusiasm and personalities of some of the instructors. So the following spring I decided to take a course in systematic botany from LeRoy Abrams. This turned out to be a good choice for me. The class was small and informal, and the work consisted mainly in collecting native plants in the adjacent fields and hills and learning to identify them by reference to Jepson's *Manual of Flowering Plants of California.* I soon began to appreciate the diversity of plants and the influence of environment on their distribution in nature. My knowledge in this area was later extended by taking a course in plant ecology and by accompanying a graduate student, Carl B. Wolf, on a seven-week field trip throughout the American southwest during which we collected over 5000 plants for the Stanford herbarium.

As a result of these experiences I decided, near the end of my sophomore year, that I would like to become some type of biologist. Since I had almost no background in the physical sciences, this decision meant that I had to start my real scientific education almost from the beginning. On examining the requirements for graduation in various fields, I found that I could obtain the physical sciences background I needed and fulfill the requirements for an A.B. degree most quickly by majoring in the School of Physical Sciences, which provided an introduction to mathematics, physics, chemistry, and geology. I studied all of these subjects with enthusiasm and graduated in the summer of 1929.

On entering graduate school I still had no definite idea as to which area of biology I should enter. So I decided to sample introductory courses in several areas, including plant and animal physiology, protozoology, and psychology. I found the protozoology course given by C. V. Taylor to be particularly stimulating because the class was very small and informal, which allowed close personal contact with an enthusiastic teacher. Also, because the emphasis was on microscopy, micro-manipulation, and other techniques I was able to learn something about the behavior and physiology of protozoa. The following spring I moved to the Hopkins Marine Station on the Monterey Peninsula, with a small group of premedical students, to study invertebrate zoology and embryology. The instruction was excellent and the environment was enchanting, but the most important thing that happened to me was a conversation with a fellow graduate student, Robert E. Hungate. He told me that he had been getting some instruction in microbiology from a new member of the staff, a young Dutchman by the name of C. B. van Niel, whom he had found to be a superb teacher. On Hungate's advice I decided to ask van Niel to accept me as a student in the summer quarter. There was one complication. I had planned to start on a vacation in the Sierras somewhat before the end of the summer quarter. So I asked van Niel whether he could let me start the course a week early, in order to avoid interference with my vacation plans. He was rather surprised at this request, but the following day he agreed. I was van Niel's only student that summer; he spent much time with me introducing the experiments and discussing the results. On occasion the discussions would expand into lectures lasting an hour or two, which were presented with a clarity, enthusiasm, and almost hypnotic intensity that made a deep impression on me. I quickly became convinced that microbiology was a most exciting subject. One aspect of microbiology that van Niel emphasized was the developing knowledge and theories of the biochemistry of yeast and bacterial fermentations. Most of this material was quite new to me and I soon began to realize that my knowledge of chemistry was still insufficient to understand these phenomena. This realization was responsible, in considerable part, for my later decision to change my major to chemistry.

During this summer I also assisted Taylor with some experiments on the development of starfish eggs. Taylor was committed to spend the following academic year as a visiting professor of Zoology at the University of Chicago, and he invited me to come along as his research assistant. Since this provided an opportunity to see how I would like research in protozoology and to broaden my scientific background at another institution, I accepted.

At Chicago, Taylor suggested that I investigate some aspect of the conversion of active protozoa to their resting forms, or cysts, but left me free to decide which organism to use and how to proceed. After reading the available literature and making some trials, I decided to use the ciliate Colpoda cucullus, which can be cultured readily in an infusion of hay and forms cysts abundantly under appropriate conditions. Previous studies with various ciliates had suggested that several environmental factors, including food supply, pH of the medium, accumulation of excretion products, and lowered O_2 tension, may induce encystment; but there was little solid evidence to support any of these suggestions. Making use of some things I had learned from van Niel, I was able to simplify the conditions for encystment by culturing the ciliates on a suspension of bacteria in a nonnutritive medium and show that cyst formation is almost unaffected by pH or food supply, but that it is induced by other unidentified changes in the environment associated with crowding of the ciliates. These observations formed the subject of my first scientific publication. I also investigated the nature of the factor in plant and animal infusions that induces the conversion of cysts to active ciliates, and was able to show that a number of common organic and inorganic compounds are inactive. Subsequently, Kenneth V. Thimann and I found that an acid ether extract of hay infusion contains much of the activity of the original extract. Thimann & Haagen-Smit (1) later established that the activity in ether extracts is attributable to the salts of l-malic and other organic acids.

While at the University of Chicago I also made some observations on the effect of moisture on the survival of *Colpoda* cysts exposed to high temperatures. Like bacterial spores, the cysts become more heat resistant when their moisture content is decreased. This led me to wonder whether the relation between moisture and heat resistance of living organisms could be attributed to an effect of moisture on the stability of cellular proteins. A search of the literature turned up only a few observations on this subject and so I decided this might be a suitable project to investigate later for a Ph.D. thesis. This could presumably be done in a chemistry department, where I could also increase my meager knowledge of chemistry in preparation for a career in microbiology. To do this I needed to find a sponsor who would accept me as a graduate student to work on this project. Fortunately, with the help of C. V. Taylor, I was able to persuade James W. McBain of the Stanford Chemistry Department to do so.

Before starting graduate work in Chemistry, I again spent the summer at the Hopkins Marine Station, this time as an assistant to J. P. Baumberger of the Stanford Department of Physiology. One of my duties was to serve as a teaching assistant in a small laboratory course in general physiology, and another was to investigate the toxicity of cyanide for the brine shrimp, *Artemia salina.* This remarkable creature had previously been observed to be relatively insensitive to cyanide, and I confirmed the fact that it can swim all day in a brine solution containing 50 mM KCN. However, the organism becomes sensitive to cyanide when the pH of the solution is lowered. By systematically varying the pH and the cyanide concentration we were able to conclude that the toxicity is determined primarily by the concentration of undissociated HCN. These observations were never published, but they were reported by Baumberger at a local scientific meeting.

That summer Leonor Michaelis spent several weeks at the Hopkins Marine Station as a visiting professor and I shared a small laboratory with him. Most of his time was spent either revising the manuscript of his book on oxidation-reduction potentials, or looking at the spectra of various dyes and natural pigments by means of what now seems like a rather primitive spectroscope. He also gave a few lectures on topics such as the theory and practice of electrophoresis. These were models of organization and clarity and greatly stimulated my interest in these areas of science. I bought his books on mathematics, physical chemistry for students of medicine and biology, and hydrogen ion concentration and studied them with enthusiasm. At a later time I applied for a postdoctoral position with Michaelis at the Rockefeller Institute, but nothing was available.

I spent the following two years (1931–1933) working on my Ph.D. thesis, taking the required chemistry courses and examinations, and serving as a teaching assistant in a general biology course. For my thesis research I decided to use egg albumin, since Hopkins, Sørensen, and others had developed methods for the crystallization of this protein, and it had been used in some earlier experiments on heat denaturation. I soon learned that not all the useful information about the purification of egg albumin was to be found in the scientific literature. My first attempt to prepare the crystalline protein from six dozen eggs, obtained from a local store, was unsuccessful, apparently because the eggs were too old. The only useful product was some gold cake prepared from the yolks. A second preparation, starting with newly laid eggs, gave the expected crystalline product in good yield.

My plan to study the relation between water content and the heat stability of egg albumin worked rather well at first. By the application of simple methods, I was able to get satisfactory data on the effect of relative humidity on the rate of heat denaturation and also determine the water content of native and denatured egg albumin as a function of relative humidity. However, the interpretation of the kinetic data in terms of the chemistry of denaturation was not at all clear, and I could not think of any way of further elucidating the problem. McBain had given me much good advice about methods and other technical aspects of my research, but he lacked the background in protein chemistry to be helpful at this stage. I began to doubt that I had been wise to choose my own thesis problem and to wonder whether I had the ability to carry it through. But gradually I found my way out of this gloomy mood.

McBain had several good instruments for measuring physical properties, including a polarimeter and a mercury arc light source. Since the optical rotation of protein solutions had been reported to increase during denaturation, I decided to see how this property was affected by heating egg albumin solutions. The measurements were easy to make and the observed rotations were relatively large, but my early results were confusing because different egg albumin samples that initially showed the same specific rotation gave markedly different values after being heated. Only after a period of considerable frustration did I discover that by carefully standardizing the experimental conditions and having only a single variable I could get reproducible results. This was a valuable lesson that I never forgot. The explanation of my initial difficulties was that the specific rotation of the denatured protein was determined not only by the time and temperature of heating, but also by the initial pH and the protein concentration of the solution.

POSTDOCTORAL YEARS AT PACIFIC GROVE AND DELFT

I completed my Ph.D. thesis in the depths of the Great Depression and was fortunate to get a National Research Council Fellowship for a period of two years (1933–1935) to extend my biological training at the Hopkins Marine Station. Since I had previously acquired an interest in both marine organisms and microbiology I decided to attempt to isolate some marine diatoms and dinoflagellates and learn something of their physiology and metabolism. In the course of a year I was, in fact, able to isolate pure cultures of two species of diatoms and three species of small photosynthetic dinoflagellates, and to maintain several other species of dinoflagellates in species-pure culture. The diatoms were used for studies of photosynthetic quotients by means of Warburg's manometric techniques that I learned from van Niel and from Robert Emerson, who spent the summer at the Marine Station. The main conclusion derived from my experiments was that diatoms, like green algae, produce carbohydrates, rather than fats, as major products of photosynthesis. The dinoflagellate cultures were used mainly to determine environmental conditions favorable to the growth of these little-known organisms. The optimum temperature of the photosynthetic dinoflagellates was 18°C. This required them to be grown in a refrigerated bath that was illuminated by tungsten lamps that generated considerable heat. The result was an electricity bill for culturing these organisms that strained the very modest budget of the Marine Station. Partly for this reason, I terminated my studies of photosynthetic organisms and took up an investigation of the utilization of organic substances by a colorless alga, *Prototheca zopfii,* which van Niel had brought from Delft.

I found that *Prototheca* is an essentially aerobic organism that utilizes a large number of fatty acids and alcohols, and a few sugars, as substrates for respiration, but can also convert glucose to D-lactic acid in the absence of oxygen. While investigating the ability of cell suspensions to oxidize various substances, using manometric methods, I made the unexpected observation that the quantity of oxygen consumed was much smaller than that calculated for complete oxidation. With ethanol or acetate, for example, O_2 uptake was about 50% of theoretical, and with glycerol, the value was 29%. Carbon dioxide production was also very low, which indicated that a large fraction of each substrate was converted to other products. Since relatively little organic material accumulated in the medium, a major part of each substrate must have been assimilated in some form within the cells. An analysis of all the data I had collected on O_2 uptake and CO_2 production led to the conclusion that the assimilated material, corresponding to 50–80% of the substrates, had the empirical composition of a polysaccharide. This study directed the attention of microbiologists to the quantitative importance of synthetic processes that are coupled with the aerobic degradation of organic substrates by cell suspensions of microorganisms. This so-called "oxidative assimilation," which is basically similar to the oxidative conversion of lactate to glycogen by muscle, was later shown by others to be a conspicuous feature of the aerobic metabolism of many bacteria and yeasts (2).

During my second year at the Hopkins Marine Station I had to consider what I would do when my fellowship ended. The job situation was very bleak. I applied for two positions in the Departments of Food Technology and Plant Nutrition at the University of California, but the prospects of their being funded for the following year were poor. At the same time, I answered an advertisement for a position as Junior Microbiologist in a US government laboratory, only to be informed that since I had taken only one course in microbiology I was not qualified. Fortunately, I also applied to the General Education Board of the Rockefeller Foundation for another fellowship to spend a year in the Delft Microbiology Laboratory with A. J. Kluyver, and in this I was successful. After spending a few weeks in the Sierra, my wife and I set off for Holland in July 1935.

Before leaving Pacific Grove I decided that I would like to investigate the anaerobic degradation of glutamate and the biological production of methane. These were both topics that I had learned about from van Niel. At that time a good deal was known about the bacteria responsible for several types of carbohydrate fermentation, but only a few observations, mostly with mixed cultures, had been made on the anaerobic degradation of amino acids by bacteria. van Niel thought that such studies should be done with pure cultures and that bacteria preferentially using particular amino acids could probably be obtained from soil or similar sources by the enrichment culture method. He had started anaerobic enrichment cultures using various single amino acids as energy sources and found that glutamate was a particularly good substrate. One of his students isolated a clostridium from a glutamate medium, but was unable to carry the work further; so I inherited the problem.

van Niel had done no experimental work on biological methane formation, but he had developed an ingenious hypothesis for the origin of methane, based mainly on the earlier experiments of Söhngen. The latter had shown that carbon dioxide is reduced to methane when molecular hydrogen is used as a second substrate and had also found that methane is the only hydrocarbon formed from a variety of organic substrates, irrespective of the number of carbon atoms they contain. van Niel concluded that in all these processes the organic substrate is oxidized to carbon dioxide and water, and this oxidation is coupled with the reduction of part of the carbon dioxide to methane. I decided to look for further evidence in support of this hypothesis, and to attempt to isolate cultures of methane-forming bacteria, which had not been done previously.

When I first discussed these problems with Kluyver, he was sympathetic but, I think, a little skeptical that I could make much progress on either one during the year. He suggested that while getting started on these problems I should isolate a bacterium fermenting tartaric acid and possibly other C_4-dicarboxylic acids, and investigate the chemistry of the degradation of these compounds by the method developed in the Delft laboratory, namely, quantitative determination of the fermentation products. In fact, the isolation of a tartrate-fermenting strain of *Aerobacter aerogenes,* which could also ferment fumarate and l-malate, proved to be easy, and within a few months I had data on the fermentation products. The data were interpreted to mean that the dicarboxylic acids undergo an oxidation-reduction reaction to give succinate and an oxidized product, probably oxaloacetate, that is decarboxylated to pyruvate; the latter is presumably converted to various C_1 and C_2 products characteristic of *Aerobacter aerogenes* by reactions previously observed or postulated in other systems. No effort was made to detect the postulated intermediates or enzymes. This was considered not only too difficult, but also unnecessary for the purpose of establishing the pathway of the fermentation. Since the postulated pathway was consistent with the observed yields of fermentation products and since some of the component reactions had been demonstrated previously in other biological systems, we felt safe in assuming, without further evidence, that the postulated reactions occurred in these bacteria.

I made a similar study of anaerobic glutamate degradation by first isolating a clostridium, later identified as *Clostridium tetanomorphum,* that is capable of utilizing glutamate as a major energy source, and then determining the amounts of each product formed. I finally proposed a hypothetical sequence of reactions that might account for the observed products. The latter were ammonia, carbon dioxide, hydrogen, acetate, and butyrate, and the hypothetical pathway involved a more or less simultaneous deamination and decarboxylation of glutamate to form crotonate. Crotonate could presumably undergo reduction to butyrate and a coupled oxidation, by way of β-hydroxybutyrate and acetoacetate, to acetate and hydrogen. Again, no confirmatory evidence for the postulated pathway was obtained. As I later found, the pathway is incorrect in almost every detail for glutamate degradation by *C. tetanomorphum.* However, other investigators (3–5) have shown that *Peptococcus aerogenes* and other nonsporulating bacteria degrade glutamate by a pathway similar to that originally postulated for the clostridial fermentation. Although my study of glutamate fermentation did not contribute to knowledge of intermediary metabolism, it was useful in establishing the possibility of using single amino acids as energy sources for anaerobic growth, and it eventually led to the discovery of an enzymatically active form of vitamin B_{12}.

While working on the degradation of C_4-dicarboxylic acids and glutamate, I also began to search for a way of testing van Niel's CO_2 reduction theory of methane formation with an organic substrate. Obviously what was needed was an organic compound that could be oxidized incompletely by methane-forming bacteria without producing carbon dioxide. The reduction of carbon dioxide to methane, if it occurred, could then be observed directly. A search of the literature turned up a short article by Omeliansky (6) which reported that a mineral medium containing ethanol and calcium carbonate, when inoculated with rabbit dung and incubated in the absence of O_2, undergoes a fermentation that produces gas containing mostly methane plus a little carbon dioxide. The high methane content of the gas suggested that ethanol was being oxidized only as far as acetic acid.

On the basis of this report, I started an enrichment culture for methane-producing bacteria under the conditions described by Omeliansky, but using an inoculum of sewage sludge, and soon obtained crude cultures that utilized ethanol rapidly according to the equation

$$2CH_3CH_2OH + CO_2 \rightarrow 2CH_3COOH + CH_4.$$

The cultures were also shown to oxidize butanol to butyric acid, and the latter to acetic acid, both reactions being accompanied by a disappearance of carbon dioxide and the formation of an approximately equimolar quantity of methane. These results appeared to establish the validity of the CO_2 reduction theory of methane formation for these few substrates, and with the naiveté of youth I was immediately prepared to extend this concept to methane production from all other organic compounds. This was later found to be an oversimplification.

My observations on the methane fermentation of ethanol by enrichment cultures yielded another result that was destined to have a considerable influence on my career and the development of knowledge of fatty acid metabolism in later years. When handling various ethanol-methane enrichment cultures, I became aware that some had a slightly acidic odor, attributable to acetic acid, whereas others developed a much stronger, rancid odor. Steam distillation of volatile fatty acids from cultures of the latter type yielded substantial amounts of a relatively water-insoluble liquid organic acid that was identified as n-caproic acid. This was always accompanied by butyric acid. The formation of C_4 and C_6 fatty acids in high yields from ethanol in an anaerobic environment was an unexpected discovery that I reported to Kluyver with considerable excitement. Only after a careful search of the literature did I find that in 1868 a student of Pasteur, A. Bechamp, had observed the same phenomenon and reported the isolation of 75 g of caproic acid from a culture containing 106 g of ethanol (7)!

The publication of a report on this work was delayed for a year while an industrial company, to which Kluyver served as scientific adviser, investigated the possibilities of using the process for the commercial production of caproic acid. So far as I know nothing ever came of this. Nevertheless, the company provided me with a small retainer that made it possible, the following year, to start construction of a cabin in the mountains of California which we still use each summer.

A SOIL MICROBIOLOGIST AT BERKELEY

Toward the end of the year in Delft I accepted an appointment as Instructor in Soil Microbiology and Junior Microbiologist in the Division of Plant Nutrition of the Agricultural Experiment Station, University of California. As an instructor I at first assisted C. B. Lipman in teaching a laboratory and lecture course in soil microbiology that was required of all undergraduate students in the Soil Science curriculum, and later I was given sole responsibility for the course. Since my formal training in microbiology was slight, and my knowledge of soil microbiology in particular was even smaller, I had to work hard during the first years to learn enough about the subject to teach the fundamentals and those aspects that might be of some interest to students of soils. Fortunately, the students had reasonably good backgrounds in chemistry and general biology, although I found that because of the nature of the curriculum, they were generally more interested in the inorganic and physical properties of soils than in the microbial transformations of organic compounds. Since my interest was mainly in the latter area, a few years later I developed, in collaboration with Michael Doudoroff of the Bacteriology Department, and Reese H. Vaughn and Maynard A. Joslyn of the Food Technology Department, a new course in Microbial Metabolism in the Bacteriology Department that attempted to cover the knowledge of intermediary metabolism that was rapidly developing during that period. Later, Roger Y. Stanier and Edward A. Adelberg also participated in teaching this course, which attracted graduate students from several areas of biology.

Since I had an appointment in the Agricultural Experiment Station, I was supposed to make some contribution to agricultural research. The chairman of Plant Nutrition, Dennis R. Hoagland, asked me to join in the study of a nutritional disease of fruit trees and other plants, known as "little leaf." Shortly before my appointment, Hoagland and his associates had made the important discovery that this disease is caused by a deficiency of zinc, and he was actively engaged in investigating the conditions affecting the zinc requirement. Field observations seemed to indicate that little leaf symptoms were often particularly severe in areas, such as former corrals, that had received large amounts of animal manure; this suggested that microorganisms are somehow involved in increasing the effect of zinc deficiency. Hoagland had begun to investigate this phenomenon by growing several successive crops of corn in pots of corral soil and had found that each successive crop grew more poorly, presumably because of increased zinc deficiency. Finally the condition became so severe that corn seeds would scarcely germinate. Hoagland asked me to see whether I could find any basis, microbiological or otherwise, for this phenomenon. I tried a number of experimental approaches, using sterilized and unsterilized soil, and soil reinoculated with various bacteria isolated from the original soil, but they led nowhere. Finally I made extracts of the soil to see whether they contained any material that would affect seed germination. It turned out that an extract was as poor a medium for germination as the original soil, and the explanation was that the salt concentration was just too high for corn. This terminated my experiments on corral soils. I did some other experiments on the effect of bacteria on the development and minor element nutrition of sterile plants grown in water culture but none of these produced any readily interpretable data. So with Hoagland's approval I abandoned research on bacteria-plant interrelations and devoted all my efforts to investigating simpler microbial systems.

The facilities available for microbiological research were very modest when I arrived in the Division of Plant Nutrition. They included an incubator room, a very old autoclave that did not always develop the expected temperature, a homemade oven for sterilizing glassware, a microscope, and a supply of test tubes and flasks. Most of the mechanical and electrical instruments that are now considered indispensible for research, such as centrifuges, colorimeters, respirometers, and pH meters, were lacking. Furthermore very little money was available in 1936 to purchase equipment of any sort. I well remember asking Hoagland whether I could order a $15 Seitz filter that I needed to sterilize media. He eventually approved my request but only after examining his budget to see whether we could afford it.

In part because of the limited facilities, my students and I initially concentrated on the isolation of various interesting kinds of anaerobic bacteria, which could be done with the available supplies. The bacteria included *Methanobacterium omelianskii,* the organism apparently responsible for the conversion of ethanol and carbon dioxide to acetate and methane; *Clostridium kluyveri,* responsible for the formation of butyric and caproic acids from ethanol; *Clostridium acidi-urici* and *Clostridium cylindrosporum,* which decompose uric acid and other purines; *Streptococcus allantoicus,* which degrades allantoin anaerobically; *Clostridium tetanomorphum* and *C. cochlearium,* which ferment glutamate; *Clostridium propionicum* and *Diplococcus glycinophilus,* which utilize alanine and glycine,

respectively; and *Butyribacterium rettgeri* and *Clostridium lactoaceto-
philum,* which ferment lactate in different ways. These organisms provided
many of the biochemical problems I was to investigate in later years.

The isolation of each of the above-mentioned organisms involved some
special problems, but none was as difficult as the initial isolation of *C.
kluyveri.* I have already mentioned that some enrichment cultures for
ethanol-utilizing, methane-forming bacteria produce considerable amounts
of butyric and caproic acids. Microscopic examination of such cultures
showed that they always contained a large spore-forming bacterium in
addition to a smaller bacterium (*Methanobacterium omelianskii*) that was
apparently responsible for the formation of methane. I undertook to isolate
the spore-former by serial dilution in the same medium used for the enrichment cultures but with agar. It soon became apparent that isolated
colonies of the spore-formers could not grow in this medium, since none was
found beyond the second dilution, although *M. omelianskii* grew at much
higher dilutions. As it seemed possible that the inability of the spore-former
to grow in higher dilutions might result from the absence of suitable growth
factors, I tried supplementing the medium with yeast autolysate and found
that the addition of a very high level of this material would permit it to
develop, though poorly. The problem then was to distinguish colonies of the
caproic acid-forming clostridium from the many contaminating clostridia
that thrived on yeast autolysate. This was eventually accomplished by using
a remarkably sensitive but inexpensive instrument, my nose, to detect the
presence of caproic acid in individual colonies picked with a micropipet. By
these methods, I eventually isolated a pure culture of *C. kluyveri* but was
disappointed to find that it produced little caproic acid in a yeast autolysate-
ethanol medium. Considerable additional time and effort were required to
find that the major essential nutrient derived from yeast autolysate is acetate
and the minor nutrients are carbon dioxide, biotin, and *p*-aminobenzoate.
When all these compounds were supplied, *C. kluyveri* grew readily, deriving
energy from the conversion of ethanol and acetate to butyrate, caproate,
and hydrogen (7).

EARLY EXPERIMENTS WITH RADIOACTIVE CARBON

I first became involved in experiments with radioactive carbon in 1939.
Through my colleague Zev Hassid I met Sam Ruben of the Chemistry
Department and Martin D. Kamen of the Radiation Laboratory, who had
begun to use ^{11}C in the study of photosynthesis and dark CO_2 fixation by
higher plants and algae. Ruben was the dynamic and tireless promoter of

[11]C; and he was always interested in finding new biological systems to which the isotope could be effectively applied. When I pointed out that the carbon dioxide reduction theory of methane formation from organic compounds could be tested with [11]CO_2, he was eager to collaborate.

Our experiments on the fermentation of ethanol by *M. omelianskii* confirmed the earlier conclusion that methane is derived from carbon dioxide and further demonstrated a considerable incorporation of carbon dioxide into cellular materials. An experiment on the fermentation of methanol by a *Methanosarcina* species was less convincing; although a small incorporation of carbon from carbon dioxide into both methane and cell material was observed, the results were not sufficiently quantitative to permit an unambiguous interpretation. This was a serious limitation of ^{11}C as a tracer. The 21-min half-life allowed only about 4 hr to prepare the [11]CO_2, set up the experiment, carry out the incubation, separate the products, and prepare and count the final samples. The time was generally insufficient to get more than semiquantitative data. Despite this limitation we were later able to obtain useful data on the incorporation of carbon dioxide into acetate during the fermentation of purines by *C. acidi-urici* and into the carboxyl groups of propionic and succinic acids during fermentations by propionic acid bacteria.

The more complicated experiments with ^{11}C always involved a group effort. In order to reduce the duration of an experiment to a minimum it was necessary to plan every step of the preparative and analytical procedures ahead of time, and to make a dry run to be sure that everything necessary was available and working. Since our group had the lowest priority for use of the cyclotron, the actual experiments were always done at night and frequently could not be started before 1 AM. After the incubation, everyone was busy for a while carrying out some part of the separation procedure. Then we gathered about Ruben in the early hours of the morning to watch the counting of the samples. There was always a sense of excitement and drama when the incorporation of CO_2 into some metabolic product was shown by the high speed ticking of the counter. We felt that science was really progressing!

Carbon 14 was first prepared in significant amounts by Ruben and Kamen in 1940 (8), but because of wartime restrictions and the untimely death of Ruben, the isotope did not become available for experimental purposes until 1944. At that time T. H. Norris of the Chemistry Department and I recovered the ^{14}C from several hundred liters of saturated ammonium nitrate solution that had been exposed to stray neutron radiation from the 60-inch cyclotron. This was a messy job lasting several days. It involved boiling aliquots of the solution in a 12-liter flask, passing the vapors through a condenser and over hot copper oxide, and then absorbing the CO_2 in alkali and precipitating it as $BaCO_3$. My share of the product was 1.8 g of $BaCO_3$ that had a rather low specific activity of about 1.5×10^5 cpm per mmole. This amount, small by current standards, proved to be sufficient for several fairly complicated tracer experiments on bacterial metabolism.

Although by this time I had some experience with tracer methodology, I knew virtually nothing about the technical aspects of estimating radioactivity because Ruben had previously always done the counting on a homemade counter that only he could operate. Fortunately for me, just about the time ^{14}C became available Kamen lost his position in the Radiation Laboratory because of wartime hysteria aroused by his conversation with a Russian consular official, and he was able to collaborate with me on the first tracer experiments with ^{14}C. He taught me the art of making mica window Geiger tubes and many other tricks of tracer methodology, and I in turn contributed something to his education in microbiology and biochemistry. It was a most useful and pleasant collaboration.

We first examined the role of carbon dioxide in the fermentation of glucose by *Clostridium thermoaceticum*. This bacterium had been shown to ferment glucose and xylose with the formation of over 2 moles of acetic acid per mole of sugar. The high yield of acetic acid, and the virtual absence of carbon dioxide or other one-carbon product, suggested that part of the acetic acid was formed from carbon dioxide. This hypothesis was shown to be correct by fermenting glucose in the presence of $^{14}CO_2$ and establishing that the isotope is incorporated into both carbon atoms of acetate, and that over 2 moles of carbon dioxide are actually formed and reutilized during the fermentation. Similar experiments showed that *Butyribacterium rettgeri* also uses carbon dioxide and converts it to acetic and butyric acids during the anaerobic degradation of lactate.

A somewhat more elaborate tracer experiment on the conversion of ethanol and acetate to butyrate and caproate by *C. kluyveri* provided substantial evidence that acetate, or a compound in isotopic equilibrium with acetate, is an intermediate in the conversion of ethanol to C_4 and C_6 fatty acids, and that caproic acid synthesis almost certainly involves the addition of a C_2 unit to the carboxyl carbon of butyrate rather than the reciprocal reaction (7). The latter conclusion was later confirmed by showing that ^{14}C-labeled caproic acid derived from [1-^{14}C]butyric acid and ethanol is labeled almost exclusively in the β-carbon atom.

SABBATICAL INTERLUDE

In 1941 I became eligible for my first sabbatical and was fortunate to receive a fellowship from the Guggenheim Foundation. I spent the first two months with L. F. Rettger at Yale University studying the fermentation products and cultural characteristics of various nonsporulating anaerobic bacteria, which included an organism we later called *Butyribacterium rettgeri.* The last two months were spent with W. H. Peterson at the University of Wisconsin learning methods that had been developed there for investigating bacterial nutrition and assaying for growth factors by microbiological methods. The remainder of the year was spent with Fritz Lipmann in the Surgical Laboratories of the Massachusetts General Hospital. I had been attracted to Lipmann by his studies of enzymatic pyruvate oxidation by *Lactobacillus delbrueckii,* and by his stimulating review on phosphate bond energy. When I arrived he was engaged in the isolation of the labile phosphate compound formed from pyruvate that was soon shown to be acetyl phosphate. Lipmann determined the phosphate content of the isolated product and I contributed to its characterization by estimating the acetate content.

Before working with Lipmann all my research had involved the use of living bacteria, either as growing cultures or as cell suspensions. He introduced me to methods of preparing and studying cell-free extracts, and to techniques of detecting and estimating intermediate metabolites by colorimetric and other relatively sensitive procedures. The method that Lipmann favored for making bacterial extracts consisted of simply drying cells in a vacuum desiccator over P_2O_5 and then extracting them with buffer. This seems primitive by comparison with currently available methods, but it was inexpensive and served well for a number of later studies of bacterial enzymes at Berkeley.

SUCROSE PHOSPHORYLASE

On returning to Berkeley I continued to study various bacterial fermentations, some of which have already been mentioned, and also became involved in two new lines of research: a study of enzymatic sucrose degradation and an investigation of the deterioration of dried fruit during storage.

The investigation of sucrose degradation was initiated by Michael Doudoroff. He had isolated an H_2-oxidizing bacterium that also utilized a wide range of organic substrates. An interesting peculiarity of this organism, *Pseudomonas saccharophila,* was that it oxidized sucrose more rapidly than the component monosaccharides, glucose and fructose. About the time I returned from sabbatical leave Doudoroff came to the conclusion that further analysis of this phenomenon could only be made by the use of cell extracts. At my suggestion he made some dried cell preparations and soon found that suspensions of the dried cells in a sucrose-phosphate solution caused a rapid esterification of inorganic phosphate. To identify and quantitate the products he enlisted the cooperation of Nathan O. Kaplan, who had had experience with the characterization of phosphate esters during his thesis research with David M. Greenberg, and W. Z. Hassid, who was a carbohydrate chemist. Together they demonstrated that the major enzymatic reaction is an apparently reversible conversion of sucrose and orthophosphate to fructose and glucose-1-phosphate. Because Hassid, Doudoroff, and I often had lunch together, and the conversation frequently dealt with the sucrose problem, I was gradually drawn into this research and contributed in various ways to the planning of the experiments and the isolation and characterization of sucrose and other disaccharides that can be synthesized by the phosphorylase from appropriate substrates (9). My most significant contribution to this research came as a result of an experiment that Doudoroff and I had planned to investigate the incorporation of ^{32}P into glucose-1-phosphate under various conditions. We incubated glucose-1-phosphate and ^{32}P$_i$ with sucrose or fructose expecting that the reversible enzymatic reaction would result in the formation of labeled glucose-1-phosphate. Almost as an afterthought we included a control with only glucose-1-phosphate and ^{32}P$_i$, and were surprised to find that more ^{32}P was incorporated into glucose-1-phosphate in the absence of the sugars than in their presence. In fact we did not believe the first result, and concluded that there had been a mix up of the samples. However, repetition confirmed the initial observation. We discussed the result for some time and by the next day reached the conclusion that the simplest interpretation was a reversible reaction of glucose-1-phosphate with enzyme to form a covalently bonded glucosyl enzyme and P$_i$. This soon led to the idea that sucrose was probably reacting in a similar way with the enzyme to form glucosyl enzyme and fructose. This in turn implied that the glucosyl moiety derived from sucrose could be transferred to another glucosyl acceptor such as sorbose to form glucosidosorboside in the complete absence of inorganic phosphate. Although I do not now recall the exact course of the discussion leading to these conclusions, I think that Doudoroff, who had a very agile mind, was the first to sense the probable explanation of our results. In any event, with Hassid's collaboration we were soon able to demonstrate the predicted synthesis of disaccharides by glucosyl transfer from sucrose in the absence of phosphate (10). These results established the concept that sucrose phosphorylase functions as a glucosyl-transferring enzyme, and provided substantial, though indirect, evidence for the existence of a covalent glucosyl enzyme compound, which was demonstrated many years later by Voet & Abeles (11).

RESEARCH ON DRIED FRUIT

Like many Americans in the early 1940s I felt an urge to assist in some way in the great conflict in which the nation was engaged. So in 1943 I eagerly accepted the invitation of my friend Emil M. Mrak of the Department of Food Technology to participate in a Quartermaster Corps project on methods of retarding the deterioration of dried fruit during storage, particularly since the work could be done on the campus and would not preclude other research activities. The project provided funds for an assistant; I was fortunate to select Earl R. Stadtman, a graduate of the Soil Science program who had taken my course in soil microbiology and later had assisted me in growing *Chlorella* on a large scale for Ruben. At first we knew almost nothing about the problems of preparing and storing dried fruit and soon discovered that the scientific literature dealing with these subjects was very meager. Mrak introduced us to the conventional methods of handling dried fruit, and then Stadtman and I, and later Victoria Haas, undertook a systematic study of factors influencing the deterioration of dried apricots. This required first the development of a reasonably quantitative measure of quality. Since fruit darkens progressively during storage this was accomplished by visually comparing the color of an alcoholic extract of fruit with a series of standards. We then proceeded to determine the effects of temperature, moisture, sulfur dioxide, and oxygen, and their interrelationships, on storage life, which was defined as the time required to reach an arbitrary degree of darkening (12). Several effects were revealed that had not previously been observed, or at least not adequately appreciated. Our results did not help to shorten the war, since they were not published until after its conclusion. I hope they have had some beneficial effect on the quality of commercial dried fruit, but I do not know that this is so.

CLOSTRIDIUM KLUYVERI: FATTY ACID METABOLISM AND AMINO ACID BIOSYNTHESIS

After the war Earl Stadtman decided to do his Ph.D. thesis with me and undertook to explore the enzymatic reactions participating in the energy metabolism of *C. kluyveri*. He soon found that crude extracts of dried cells are able to catalyze the anaerobic conversion of ethanol and acetate to butyrate and caproate, as well as the aerobic oxidation of ethanol and butyrate. This exciting discovery opened up the possibility of identifying the enzymatic reactions involved in the oxidation and synthesis of fatty acids. In fact the analysis of the system progressed rapidly. Stadtman found that acetyl phosphate is a product of the oxidation of both ethanol and butyrate in a phosphate buffer, and is an essential substrate for the synthesis of butyrate when hydrogen is used as a reducing agent. Other significant findings were the discovery of an acetyl-transferring enzyme (phosphotransacetylase) and an enzymatic system for using acetyl phosphate to activate other fatty acids. Later, in Lipmann's laboratory, Stadtman and his associates showed that both of these enzyme systems require CoA and catalyze the formation of acyl-CoA compounds (13, 14).

Investigation of the utilization of several C_4 compounds that had been postulated to be intermediates in the reversible conversion of butyrate to acetate and acetyl phosphate established that acetoacetate can be either reduced to β-hydroxybutyrate or cleaved to acetyl phosphate and acetate, and that vinyl acetate can undergo a dismutation forming butyrate, acetyl phosphate, and acetate. However, tracer experiments showed conclusively that neither acetoacetate nor vinyl acetate could be an intermediate in butyrate oxidation or synthesis. Since no other C_4 compound at the oxidation levels of β-hydroxybutyrate and acetoacetate was used in this system, and no intermediate accumulated in sufficient amounts to be detected by the available methods, we were forced to the conclusion that the intermediates must be relatively stable complexes of C_4 compounds with a coenzyme or other carrier. This interpretation, which I first presented in a lecture before the Harvey Society in May 1950, was developed during discussions with Stadtman, and later with Eugene P. Kennedy, who spent a year with me as a postdoctoral fellow.

The following year I was invited to give a major lecture on the formation and utilization of active acetate at the first Symposium on Phosphorus Metabolism at Johns Hopkins University. I am not sure why I was selected for this assignment, although it was probably connected with the fact that Lipmann and Ochoa, who were major contributors to this area of research, were regarded at that time as competitors, and someone thought that selection of a neutral third party would be more diplomatic. In any event, I felt a great responsibility to present a comprehensive and balanced review of the whole field, covering the work that had been done with animal as well as bacterial systems. This required a major effort; I spent several months studying the literature and trying to arrive at a unified interpretation of the often incomplete and sometimes conflicting experimental results. Finally I reached the conclusion that acyl-CoA compounds are not only primary products of the oxidation of pyruvate and acetaldehyde, and primary substrates in the synthesis of acetoacetate and citrate, as had already been demonstrated, but that they must also be intermediates in the oxidation and synthesis of butyrate (15). I proposed a pathway for butyrate oxidation to acetyl phosphate via butyryl-CoA, vinylacetyl-CoA, β-hydroxybutyryl-CoA (by implication), acetoacetyl-CoA, and acetyl-CoA that was very similar to that later demonstrated experimentally by Lynen and others.

Vinylacetyl-CoA was postulated to be the initial oxidation product of butyryl-CoA because vinyl acetate is used more readily than crotonate by extracts of *C. kluyveri*. This apparently results from the specificity of the CoA transferase in this organism. Robert Bartsch later found that *C. kluyveri* contains a special isomerase that converts vinylacetyl-CoA to crotonyl-CoA.

Another aspect of the metabolism of *C. kluyveri* that proved to be of interest was the biosynthesis of its amino acids. Since *C. kluyveri* could be grown in a medium containing ethanol, acetate, and carbon dioxide as the only carbon compounds, other than small amounts of biotin and *p*-aminobenzoate, it was apparent that the cellular amino acids must all be synthesized from C_2 compounds and carbon dioxide. Tracer experiments by Neil Tomlinson showed that about 25% of the cellular carbon was derived from carbon dioxide and 75% from acetate. Examination of the 2-, 3-, and 4-carbon amino acids derived from the proteins of bacteria grown in the presence of $^{14}CO_2$ or [1-^{14}C] acetate established that the amino acid carboxyl groups are derived from carbon dioxide and the α-carbon atoms are derived from the carboxyl carbon of acetate. The results were consistent with the interpretation that *C. kluyveri* carboxylates acetyl-CoA and pyruvate to form pyruvate and oxaloacetate and then converts these compounds into the indicated amino acids. The postulated carboxylation reactions were subsequently demonstrated in *C. kluyveri* by Stern (16).

Tomlinson also investigated the origin of the carbon atoms of glutamate in *C. kluyveri* and found, in contrast to what had been observed in other organisms, that the α-carboxyl and β-carbon atoms are derived mainly from the carboxyl carbon of acetate, and that the γ-carboxyl carbon atoms are derived mainly from carbon dioxide (17). He pointed out that these results could be accounted for by the usual reactions for the conversion of oxaloacetate and acetyl-CoA to glutamate, provided the aconitase in *C. kluyveri* had an unconventional stereospecificity resulting in the formation of a double bond in *cis*-aconitate between the central carbon atom and the methylene carbon atom derived from oxaloacetate. This change in the position of the double bond would cause a reversal of the positions of the glutamate carbon atoms derived from oxaloacetate and acetate, as compared to glutamate formed by the usual tricarboxylic acid cycle reactions. This plausible hypothesis was eventually disproved by Gottschalk, who obtained convincing evidence that the citrate synthase, rather than the aconitase of *C. kluyveri*, displays an atypical stereospecificity. He found that *C. kluyveri* contains an (R)-citrate synthase rather than the (S)-citrate synthase characteristic of most other organisms. The (R)-citrate synthase of *C. kluyveri* fully accounts for the unusual origin of the carbon atoms of glutamate. This type of citrate synthase apparently occurs in only a few anaerobic bacteria (18).

BIOCHEMISTRY OF METHANE FORMATION

Since our earlier tracer experiments with ^{11}C on the origin of methane in the fermentations of methanol and acetate had given equivocal results, when ^{14}C became available I decided to reinvestigate these problems. The immediate stimulus for this was a report by Buswell & Sollo (19) showing that little ^{14}C is incorporated into methane when unlabeled acetate is fermented in the presence of $^{14}CO_2$. This result was clearly contrary to the CO_2 reduction theory, but it did not specifically identify the source of methane carbon. I therefore encouraged Thressa Stadtman to study the fermentation of specifically labeled acetate; her results showed that virtually all the methane carbon is derived from the methyl group of acetate (20). She also established that methanogenic bacteria convert methyl alcohol to methane by a process not involving carbon dioxide reduction. In a further effort to define the chemistry of the conversion of acetate to methane Martin J. Pine investigated the fermentation of acetate labeled in the methyl group with deuterium, and found that the methyl group is incorporated as a unit into methane without loss of attached hydrogen or deuterium. The fourth hydrogen atom was shown to come from the solvent. These results appeared to exclude an oxidation-reduction of the methyl group during methane formation, although the possibility that the same hydrogen atoms are removed and returned to the methyl carbon cannot be entirely eliminated. Disregarding this possibility, the results of the various tracer experiments are consistent with a simple decarboxylation of acetate to methane and carbon dioxide. However, this still seems unlikely since it is difficult to imagine how an organism can obtain useful energy from such a process. As yet no one has succeeded in obtaining a cell-free extract with which to make a further analysis of the chemistry of the conversion of acetate to methane.

In 1956 I undertook to summarize the results of our studies on methane fermentation and to correlate this with the contributions of other groups.

This led to the proposal of a generalized pathway for the formation of methane from either acetate, methanol, or carbon dioxide, all of which are known to be used by some methane-forming bacteria (20). The main features of this pathway were the carboxylation of an unspecified carrier and the sequential reduction of the carboxyl group to a methyl group that was finally converted to methane. The methyl groups of acetate and methanol were postulated to enter this sequence by a more or less direct methyl transfer to the carrier and be either reduced to methane or oxidized to carbon dioxide by a reversal of the carbon dioxide reduction pathway, or both. This conceptual scheme seems to have been of some value to later students of methane fermentation (21).

TRANSITION FROM MICROBIOLOGY TO BIOCHEMISTRY

Since my original position at Berkeley was that of a soil microbiologist and I ended up as a biochemist, I should mention some of the stages of my metamorphosis. I remained a member of Plant Nutrition until 1950. At that time, following the death of D. R. Hoagland, its long-time chairman, five members of the faculty—Zev Hassid, Paul K. Stumpf, Eric E. Conn, Constant C. Delwiche, and I—whose interests were primarily biochemical, formed a new Department of Agricultural Biochemistry in the College of Agriculture. When the Biochemistry and Virus Laboratory was completed in 1951 we moved in along with the new Biochemistry Department and the Virus Laboratory. Although the laboratories were an improvement over those we had previously occupied, the administrative arrangements in the building were difficult for several years because of an almost constant struggle over authority and space. This situation was greatly ameliorated when Esmond Snell became chairman of the Biochemistry Department. Shortly thereafter Hassid and I transferred into that department, and the other members of Agricultural Biochemistry moved to the Davis campus of the University to establish a new, and now flourishing, Department of Biochemistry and Biophysics. In 1964 the remaining interdepartmental problems were resolved by moving the Biochemistry Department to a new building.

From 1936 to 1948 my students obtained advanced degrees in the graduate curricula of Bacteriology, Microbiology, or Agricultural Chemistry. The Biochemistry Department at Berkeley during that period was part of the Medical School; graduate degrees in biochemistry were not available to students studying with other faculty members. Since many students in other departments were doing research on biochemical problems and wished to be recognized as biochemists, there was considerable interest among both students and faculty in setting up an academic mechanism for giving degrees in biochemistry outside of the Biochemistry Department. I. L. Chaikoff of the Physiology Department and I took the lead in organizing an interdepartmental group major, called Comparative Biochemistry, to take care of this problem. A curriculum for a Ph.D. degree in Comparative Biochemistry was approved in 1948 and from then until 1958, when I joined the new Biochemistry Department, most of my students majored in this field. I took on the responsibilities of graduate student adviser in Comparative Biochemistry when the group was organized, and retained the position until my academic retirement in 1975. During this period about 75 students obtained Ph.D. degrees in Comparative Biochemistry. Subsequently many of these students have contributed substantially to the world of biochemistry; notable examples of graduates from the earlier years of this program are Elizabeth F. Neufeld, Paul A. Srere, and Earl Stadtman.

THE BR FACTOR

I have previously mentioned some experiments on *Butyribacterium rettgeri,* an anaerobic bacterium that catalyzes butyric acid fermentation of lactate and carbohydrates. In 1950 one of my students, Leo Kline, tried to grow the organism in a synthetic medium and found that it required a small amount of yeast extract in addition to the then known nutrients and growth factors. An examination of some properties of the essential material, called the BR factor, established that it was a very stable carboxylic acid, readily extractable with organic solvents from acid aqueous solutions; in addition, it occurred in several more complex forms that were not soluble in organic solvents until released by vigorous acid or alkali hydrolysis. These properties were similar to but not identical with those of some other unidentified growth factors, including a *Lactobacillus casei* factor studied by Guirard, Snell & Williams (22), and a pyruvate oxidation factor for *Streptococcus faecalis* reported by O'Kane & Gunsalus (23). At this stage, I should have contacted these investigators in order to make a closer comparison of the various preparations. Instead, after Kline had completed his thesis, I continued work in the isolation of the BR factor. I obtained about 100 pounds of *Penicillium notatum* mycelium, a good source of BR factor, prepared many gallons of autolyzate, acid hydrolyzed the material in an autoclave, built a large liquid-liquid extractor, extracted the hydrolyzate for weeks, and with the aid of an assistant, performed innumerable tedious and not always completely reproducible assays. After some additional steps, we obtained several hundred milligrams of material substantially purified but still containing a number of components both active and inactive. About this time Gunsalus visited Berkeley and in the course of conversation we found that the properties of the BR factor and the pyruvate oxidation factor were very similar. By exchanging samples we found that they were in fact identical. Since Gunsalus' preparations were considerably purer than ours, I immediately abandoned the attempt to further purify the BR factor. Subsequent observations demonstrated that lipoic acid is highly active as a growth factor for *B. rettgeri.*

The lipoate requirement of *B. rettgeri* continued to be of interest because the function of the factor appeared to be different from that in other organisms. Lipoate had been shown to function as an electron carrier in the oxidation of pyruvate. Kline and others found on the contrary that *B. rettgeri* does not require lipoate for the utilization of pyruvate, but only for the utilization of lactate. Since the products formed from lactate and pyruvate are qualitatively the same, it was concluded that lipoate probably functions as an electron carrier in the oxidation of lactate to pyruvate. Martin Flavin became interested in the role of lipoic acid in this system when he was in my laboratory, and later collaborated with C. L. Wittenberger in a study of this problem. They reached the tentative conclusion that in lactate oxidation, enzyme-bound lipoate mediates electron transfer between an unidentified electron carrier and DPN (24). Further analysis of the specific role of the lipoate-containing enzyme in the lactate-oxidizing system in *B. rettgeri* has been impeded so far by the instability of the system (25).

PURINE DEGRADATION BY CLOSTRIDIA

From 1937 to 1957 one of my major research interests was the degradation of uric acid and other purines by clostridia. I started on this project as a result of a conversation with a colleague who raised chickens. He had filled a large container with chicken droppings, which contain uric acid, and water, and was greatly impressed by the rapid rate at which the mixture developed a strong ammoniacal odor. I undertook the isolation of the responsible bacteria and had no difficulty in obtaining a number of cultures that showed a high degree of specificity for the degradation of uric acid and a few other purines. Jay V. Beck, my first graduate student, joined me in studying the physiology and nutrition of the bacteria, which we named *Clostridium acidi-urici* and *C. cylindrosporum,* and in identifying the fermentation products. We found that both organisms decompose uric acid, xanthine, and guanine readily, and hypoxanthine more slowly, with formation of acetate, carbon dioxide, and ammonia as major products; in addition, *C. cylindrosporum* forms significant amounts of glycine. Later, Norman Radin found that formate is also a fermentation product. Since both clostridium species were able to activate glycine as a reducing agent and decompose it when uric acid was simultaneously available, glycine appeared to be a normal intermediate in purine degradation. Various enzymes and metabolites known to participate in the aerobic degradation of purines were not detected in the clostridia, and consequently we concluded that the pathway of purine degradation in these bacteria is quite different from that in aerobic organisms. This conclusion was strengthened by a number of tracer experiments on the origin of the product carbon atoms. The early experiments showed that both carbon atoms of acetate and the carboxyl group of glycine are derived in part from carbon dioxide. Later experiments by Jon L. Karlsson and by Jesse C. Rabinowitz with specifically labeled purines, glycine, and formate established a similarity between the pathways of purine degradation by clostridia and of purine biosynthesis by other organisms (26). The pieces of the jigsaw puzzle of the degradative pathway were finally assembled into a coherent picture as a result of enzymatic studies initiated by Radin and carried to completion by Rabinowitz. Radin found that the first step in uric acid utilization is its reduction to xanthine, which is then converted by crude enzyme preparations to glycine, formate, carbon dioxide, and ammonia. Glycine can be oxidized to acetate, carbon dioxide, and ammonia, and serine is converted by way of pyruvate to the same products. These results, in conjunction with those of the tracer experiments, suggested that acetate is formed mainly by the sequence glycine → serine → pyruvate → acetate. Radin also obtained presumptive evidence for the formation of one or more aminoimidazoles, none of which was identical with 4-amino-5-carboxamidoimidazole, which had been implicated in purine biosynthesis. These observations suggested that the pyrimidine ring of xanthine is initially cleaved at the 1–6 bond to yield 4-ureido-5-carboxyimidazole. The formation of this intermediate was confirmed by Rabinowitz, who then proceeded to elucidate the further enzymatic steps in purine degradation, including the role of folic acid, in elegant detail (26, 27). The last contribution to this area of research from my laboratory was a study by Willard H. Bradshaw of the properties, particularly the substrate specificity, of the xanthine dehydrogenase of *C. cylindrosporum,* the enzyme responsible for the reduction of uric acid to xanthine.

SABBATICAL AT THE NATIONAL INSTITUTES OF HEALTH

As a result of my early experience with ^{14}C, I had come to rely heavily on the application of tracer methods to intact cells for the elucidation of various problems of bacterial metabolism. When the use of intact cells seemed inadequate, I occasionally encouraged my students to use cell-free extracts, but until the early 1950s we did not attempt to purify specific enzymes. The stimulus to investigate individual reactions of metabolic pathways through the use of purified enzymes was provided by Arthur Kornberg, who spent part of the summer of 1951 in my laboratory learning how to handle anaerobic bacteria. He spoke with such enthusiasm about the advantages of using purified enzymes that I decided I should get some experience in the art of enzyme isolation. The following year I spent six months with Kornberg at the National Institutes of Health. He and I shared a small laboratory and I was able to draw upon his knowledge and experience whenever it was required. I learned a great deal from him in a few months that I was later able to apply to my own research.

In Kornberg's laboratory I investigated two unrelated problems. One was the purification of the coenzyme A transferase from *C. kluyveri,* using an assay method developed by Earl Stadtman. I tried every known method of enzyme purification on this transferase, but even with Kornberg's advice I had very little success; the best preparation was purified only about fivefold with a 31% yield. However, even the methods that did not give any purification provided valuable experience, and that was what I needed. My second research problem, suggested by Kornberg, was the isolation and characterization of ATP from the sulfur-oxidizing bacterium, *Thiobacillus thiooxidans,* which had been reported to differ from the ATP of other organisms by having the phosphate groups attached to the 3', rather than the 5' position of adenosine. During the isolation of ATP I learned how useful ion exchange resins can be for separating charged molecules, and during the characterization of ATP I came to appreciate the value of enzymes as specific and convenient analytical reagents. The conclusion of our work was that the ATP of thiobacillus is the same as that of other organisms.

GLUTAMATE FERMENTATION AND B_{12} COENZYMES

I have already mentioned my early studies on glutamate degradation by *Clostridium tetanomorphum.* Further investigation of the chemistry of this process was put off for many years while I was involved in what seemed to be more exciting problems. A stimulus to return to a study of glutamate metabolism was provided indirectly by Kornberg, who isolated a histidine-degrading strain of *C. tetanomorphum* while visiting my laboratory. My student, Joseph Wachsman, investigated the early steps of histidine degradation by this organism and concluded that glutamate is an intermediate in this process, as it is in histidine degradation by aerobic organisms. He then studied the degradation of glutamate by both tracer and enzymatic methods and showed that the carbon chain is cleaved between carbon atoms 2 and 3 to form acetate from carbon atoms 1 and 2, and pyruvate from carbon atoms 5, 4, and 3 (26). The pyruvate is oxidized to carbon dioxide (carbon 5), hydrogen, and presumably acetyl-CoA (carbon atoms 4 and 3), which is mainly converted to butyrate. These results established that glutamate was being degraded by a novel pathway. A clue to the nature of the pathway was provided when Wachsman identified mesaconic acid, a branched-chain unsaturated dicarboxylic acid, as an intermediate in glutamate degradation.

$$
\begin{array}{ccc}
{}^{1}COOH & {}^{1}COOH & {}^{1}COOH \\
| & | & | \\
H{}^{2}C\text{-}NH_2 & {}^{2}CHNH_2 & {}^{2}CH \\
| & | & | \\
{}^{3}CH_2 & {}^{3}CH_3\text{-}{}^{4}CH & {}^{3}CH_3\text{-}{}^{4}C \\
| & | & | \\
{}^{4}CH & {}^{5}COOH & {}^{5}COOH \\
| & & \\
{}^{5}COOH & &
\end{array}
$$

L-Glutamate   β-Methyl aspartate   Mesaconate

The nature of the carbon skeleton rearrangement in the glutamate-mesaconate conversion was established a little later by Agnete Munch-Petersen, who converted [4-^{14}Cl] glutamate enzymatically to mesaconate and determined the position of the isotope in the product. The result proved that the bond between carbon atoms 2 and 3 of glutamate is broken, and a new bond is established between carbon atoms 2 and 4, leaving carbon atom 3 in a methyl group. A further study of the carbon chain rearrangement established that the first product formed from glutamate is the amino acid 3-methyl-L-aspartate, which is then deaminated to form mesaconate. The inter-conversion of glutamate and 3-methylaspartate proved to be the most novel and interesting step in glutamate degradation. The branched-chain amino acid was missed in the early investigations of this system because some of its properties are very similar to those of glutamate, and because the equilibria in the system are unfavorable for its accumulation in quantity. It was first detected as a product of mesaconate amination only after we found that the enzyme catalyzing its reversible conversion to glutamate can be inactivated by treatment with charcoal.

A rather detailed account of the circumstances leading to the isolation of the charcoal-absorbable cofactor for the mutase and its identification as a derivative of vitamin B_{12} has recently been published (28) and need not be repeated here. But perhaps a few comments may be of interest. In retrospect, the isolation of the corrinoid coenzymes was rather straightforward once we had reached an adequate level of understanding of the enzymatic system in which it functioned. We had a specific, sensitive, and reasonably convenient enzymatic assay; the coenzyme was relatively stable except to one environmental factor, and its physical properties were ideally suited to permit purification by ion exchange and solvent extraction techniques. Nature had put only one roadblock in our way, namely, the instability of the coenzyme to light. Our failure to recognize this property caused much difficulty and frustration during the early stages of our investigation, which lasted almost two years. Once this property was recognized, the isolation of the coenzyme could be completed in a few weeks. The critical factor for the recognition of the light effect was the development of a rapid spectrophotometric assay for the coenzyme. We should have done this much earlier, but the advantages of such an assay were not as evident at the time as they are in hindsight.

I cannot leave this topic without at least mentioning my associates, students and postdoctoral fellows, who made important contributions to the successful outcome of our work on the isolation and characterization of corrinoid coenzymes. Agnete Munch-Petersen first undertook to purify the coenzyme and established some of its ionic properties; Herbert Weissbach first recognized the coenzyme to be a corrinoid compound and contributed in many ways to the identification of its structure; Harry Hogenkamp determined the structure of the two nucleotides formed by photolysis of the coenzyme; John Toohey established optimal conditions for corrinoid coenzyme formation by *C. tetanomorphum*, and isolated and characterized several coenzyme analogs from bacteria and liver; Benjamin Volcani developed a bioautographic method for the identification of small amounts of coenzyme analogs; Jeff Ladd determined the pK_a values of the coenzymes and the effect of ionization on the absorption spectra; David Perlman of the Squibb Institute for Medical Research assisted us greatly by providing large quantities of propionic acid bacteria containing various coenzyme analogs; Axel Lezius identified the major corrinoid coenzyme in a methane-producing bacterium; Roscoe Brady demonstrated the reactions involved in the adenosylation of corrinoid compounds in *Propionibacterium shermanii;* and Robert Smyth assisted in many ways with the assay and initial isolation of the coenzymes.

While studies of the structure of the corrinoid coenzymes were progressing, we simultaneously tried to learn something about the chemistry of the mutase reaction, but with little success. Attempts to detect either free or coenzyme-bound intermediates gave negative results, so we concluded that they must have a very short life. A somewhat more significant conclusion was reached in an investigation of hydrogen transfer during the mutase reaction. Arthur Iodice found that solvent hydrogen is not incorporated into products in appreciable amounts; this supported the interpretation that hydrogen is transferred as either H^0 or H^-, but not as H^+. After Lenhert & Hodgkin (29) showed the presence of a deoxyadenosyl group in the coenzyme, Fujio Suzuki and I investigated the role of the coenzyme as a hydrogen-transferring agent by looking for a transfer of tritium from [^{3}H-methyl]3-methylaspartate to coenzyme. A significant amount of tritium was found in the coenzyme; unfortunately, the coenzyme from a control experiment without enzyme showed about half as much tritium, so the results were ambiguous. Not long thereafter Abeles and his associates clearly demonstrated that the coenzyme functions as a hydrogen-transferring agent by the use of synthetic, tritium-labeled coenzyme in the diol dehydrase reaction. We later confirmed that the coenzyme functions in the same way in the glutamate mutase reaction.

To learn more about the mode of action of glutamate mutase I felt it would be desirable to have a highly purified preparation. This turned out to be more complicated than I anticipated. Early attempts to purify the activity showed that it depends on the presence of two readily separable proteins which we called the E and S components. The relatively unstable E component, with a molecular weight of about 125,000, was purified by Suzuki; and the relatively stable S component, with a molecular weight of 17,000, was purified by Robert L. Switzer. Although we learned something about the molecular and kinetic properties of these proteins and the conditions for their interaction, we were unable to demonstrate separate functions for the two subunits, if such they be, or understand how they interact to form the catalytically active species. This remains a problem for the future.

After the discovery of the role of corrinoid coenzymes in the glutamate mutase reaction, I considered the possibility that they might also participate in the methylmalonyl mutase reaction, but never got beyond the stage of speculation. Soon afterward, several groups of investigators demonstrated that the coenzyme is indeed required for this reaction. Another process in which vitamin B$_{12}$ had been implicated by the nutritional experiments of Snell, Kitay & MacNutt (30) was the conversion of ribonucleotides to deoxyribonucleotides in *Lactobacillus leichmannii*. When Raymond Blakley came to my laboratory I encouraged him to see whether corrinoid coenzymes participate in this conversion. He was able to obtain a cell-free preparation that reduced the ribose moiety of CMP to a deoxyribose moiety and established that the reaction is strongly stimulated by corrinoid coenzymes. After returning to Canberra, Blakley purified the ribonucleotide triphosphate reductase responsible for deoxyribose formation and clarified the novel role of the coenzyme in this oxidation-reduction reaction.

LYSINE DEGRADATION BY CLOSTRIDIA AND RELATED PROBLEMS

In 1962, Olga Rochovansky came to my laboratory as a postdoctoral fellow and said she would like to investigate the anaerobic degradation of lysine while getting experience in handling anaerobic bacteria. The year before, Thressa Stadtman (31) had reported that cell-free extracts of *Clostridium sticklandii* are able to convert lysine to acetate, butyrate, and ammonia. She had identified several cofactors required for the reaction, but had been unable to detect any intermediate in lysine degradation, even when one or another of the cofactors was omitted from a reaction solution. Since it seemed possible that another organism might provide enzyme preparations more suitable for detecting intermediates, after consultation with Stadtman, Rochovansky undertook to isolate a lysine-degrading anaerobe. She succeeded in obtaining such an organism (*Clostridium* SB4) from sewage sludge and went on to show that the cofactor requirements for lysine degradation by extracts are almost the same as for *C. sticklandii.*

The search for intermediates in lysine degradation by extracts of SB4 was started by my student, Ernest A. Rimerman. We decided to begin by adding all the known cofactors except coenzyme A in the expectation that intermediates found before the CoA-dependent reaction might accumulate in larger amounts than those formed subsequently. Rimerman soon found that omission of CoA caused the accumulation of significant amounts of a heat-labile neutral compound that could be separated from other products by paper electrophoresis. This compound was identified as 3-keto-5-aminohexanoic acid, an unexpected product to be derived from lysine, which is substituted in the 2 and 6 positions. An explanation for the location of the carbonyl group was obtained by Ralph N. Costilow who was visiting my laboratory. He looked for other intermediates in lysine degradation by omitting DPN from an otherwise complete reaction solution, and by paper ionophoresis at neutral pH he detected a second basic amino acid that overlapped lysine. At a lower pH this amino acid separates readily from lysine and can be easily assayed. The new amino acid was isolated and identified as L-3,6-diaminohexanoic acid (β-lysine), a compound previously known only as a component of some polypeptide antibiotics. The formation of this amino acid indicated that the first step in lysine degradation is a migration of the amino group from the 2 to the 3 position. This was later established more firmly after purification of the responsible enzyme, L-lysine aminomutase, by Thomas P. Chirpich; he also demonstrated that the enzyme is stimulated by pyridoxal phosphate, ferrous ion, and S-adenosylmethionine. The second intermediate in anaerobic lysine degradation, and the immediate precursor of 3-keto-5-amino hexanoate, was soon found to be 3,5-diaminohexanoate. This compound was first recognized by Stadtman & Renz (32); it was independently discovered in my laboratory by Eugene E. Dekker while looking for an intermediate accumulating in the absence of corrinoid coenzyme. Thressa Stadtman and her associates at the National Institutes of Health later purified and extensively investigated the corrinoid coenzyme-

$$\underset{\substack{| \\ ^{+}NH_3}}{CH_2CH_2CH_2CH_2CHCOO^-} \quad \underset{\underset{AdoMet}{\longleftrightarrow}}{B_6-P} \quad \underset{\substack{| \qquad\quad | \\ ^{+}NH_3 \qquad ^{+}NH_3}}{CH_2CH_2CH_2CHCH_2COO^-}$$

LYSINE β-LYSINE

$$B_6-P \Big\uparrow B_{12}CoE$$

$$\underset{\substack{| \quad\; || \\ ^{+}NH_3 \;\; O}}{CH_3CHCH_2CCH_2COO^-} \quad \underset{\underset{}{\overset{NAD}{\longleftarrow}}}{} \quad \underset{\substack{| \qquad | \\ ^{+}NH_3 \;\; ^{+}NH_3}}{CH_3CHCH_2CHCH_2COO^-}$$

3-KETO, 5-AMINOHEXANOATE   3,5-DIAMINOHEXANOATE

dependent enzyme responsible for the formation of 3,5-diaminohexanoate, whereas we concentrated on the enzymes responsible for the formation and degradation of 3-keto-5-aminohexanoate.

The enzyme catalyzing the oxidative deamination of 3,5-diaminohexanoate to the 3-keto acid was purified by John J. Baker and shown to be a highly substrate-specific, but otherwise conventional, dehydrogenase. Su-Chen L. Hong and Ing-Ming Jeng found that the degradation of 3-keto-5-aminohexanoate requires the presence of acetyl-CoA, but the nature of the enzymatic reaction responsible for the degradation eluded us for some time. The acetyl-CoA requirements suggested that the degradation would follow the usual pathway for fatty acid oxidation; formation of a CoA thioester of the β-keto acid followed by a thiolytic cleavage, which in the lysine degradation system would result in the formation of 3-aminobutyryl-CoA and acetyl-CoA. However, numerous attempts to detect the postulated intermediates and products were unsuccessful.

At this point we decided to switch to another experimental approach to the problem, namely, the synthesis of the postulated 3-aminobutyryl-CoA and the test of its ability to be further degraded. Jeng soon found that extracts of our lysine-fermenting clostridium contain a highly active deaminase that converts L-3-aminobutyryl-CoA to crotonyl-CoA. The presence of this enzyme and crotonase accounted for our earlier inability to detect 3-aminobutyryl-CoA as a product of 3-keto-5-aminohexanoate degradation, but did not account for our failure to detect the other possible intermediate, 3-keto-5-aminohexanoyl-CoA. The nature of the reaction responsible for the removal of 3-keto-5-aminohexanoate in the presence of acetyl-CoA was finally determined by Takamitsu Yorifuji, who purified the responsible 3-keto-5-aminohexanoate cleavage enzyme, and found to our surprise that it catalyzes the following reaction:

$$\underset{\substack{| \\ NH_3^+}}{CH_3CHCH_2COCH_2CO_2^-} + Ac \cdot CoA \rightleftharpoons \underset{\substack{| \\ NH_3^+}}{CH_3CHCH_2CO \cdot CoA} + AcAcO^-.$$

This is a previously unrecognized type of reaction for the degradation and synthesis of β-keto acids.

Since the study of lysine degradation by clostridia had turned up several novel types of reactions, I decided to investigate analogous enzymatic reactions in two aerobic bacteria that utilize β-lysine or 3,5-diaminohexanoic acid as an energy source. Although these investigations are not yet complete, studies by Henry N. Edmunds, Su-Chen L. Hong, Gerhard Bozler, John M. Robertson, and Masahiro Ohsugi have established that the type of β-keto acid cleavage reaction discovered in clostridia also occurs in both aerobic bacteria. The β-lysine decomposing organism is of additional interest because it catalyzes both an initial acetylation of the substrate and a novel but as yet not fully defined type of deacetylation reaction at a later step in the degradation sequence.

FINAL COMMENTS

It will be obvious to the reader that the central focus of my scientific career has been the exploration of bacterial metabolism, generally the energy metabolism of anaerobic bacteria, with the objective of establishing metabolic pathways or of identifying novel enzymatic reactions. With some exceptions this has been a relatively quiet area of science, usually peripheral to the main stream of biochemical research, and therefore not subject to much competition. Consequently, I have always worked in a rather relaxed atmosphere and have been able to enjoy several weeks vacation with my family each summer in the mountains, without developing a bad conscience for neglecting my students or suffering a fear of being scooped.

Most of the research embodied in my publications, particularly in my most productive years, was done by my students and postdoctoral associates. I have been fortunate in having many bright, enthusiastic, and dedicated collaborators, several of whom regretfully could not be mentioned in this chapter because of limitations of space. Much that we have accomplished is attributable to their skill and intuition.

Literature Cited

1. Thimann, K. V., Haagen-Smit, A. J. 1937. *Nature* 140 645–46
2. Clifton, C. E. 1946. *Adv. Enzymol.* 6:269–308
3. Horler, D. F., McConnell, W. B., Westlake, D. W. S. 1966. *Can. J. Microbiol.* 12:1247–52
4. Johnson, W. M., Westlake, D. W. S. 1972. *Can. J. Microbiol.* 18:881–92
5. Buckel, W., Barker, H. A. 1974. *J. Bacteriol.* 117:1248–60
6. Omeliansky, V. L. 1916. *Ann. Inst. Pasteur Paris* 30:56–60
7. Barker, H. A. 1947. *Antonie van Leeuwenhoek J. Microbiol. Serol.* 12:167–76
8. Kamen, M. D. 1963. *J. Chem. Educ.* 40:234–42
9. Hassid, W. Z., Doudoroff, M., Barker, H. A. 1947. *Arch. Biochem.* 14:29–37
10. Doudoroff, M., Barker, H. A., Hassid, W. Z. 1947. *J. Biol. Chem.* 168:725–32
11. Voet, J. G., Abeles, R. H. 1970. *J. Biol. Chem.* 245:1020–31

12. Stadtman, E. R., Barker, H. A., Haas, V., Mrak, E. M. 1946. *Ind. Eng. Chem.* 38:541–43
13. Stadtman, E. R. 1954. *Rec. Chem. Prog.* 15 1–17
14. Stadtman, E. R. 1976. In *Reflections in Biochemistry,* ed. A. Kornberg, B. L. Horecker, L. Cornudella, J. Oro, pp. 161–72. Oxford: Pergamon
15. Barker, H. A. 1951. In *Phosphorus Metabolism,* ed. W. D. McElroy, B. Glass, 1:204–45. Baltimore: Johns Hopkins Univ. Press
16. Stern, J. R. 1965. In *Non-Heme Iron Proteins: Role of in Energy Conversion,* ed. A. San Pietro, pp. 199–209. Yellow Springs, Ohio: Antioch
17. Tomlinson, N. 1954. *J. Biol. Chem.* 209:605–9
18. Gottschalk, G., Barker, H. A. 1967. *Biochemistry* 6:1027–34
19. Buswell, A. M., Sollo, F. W. 1948. *J. Am. Chem. Soc.* 70:1778–80
20. Barker, H. A. 1956. *Ind. Eng. Chem.* 48:1438–42
21. Zeikus, J. G. 1977. *Bacteriol. Rev.* 41:514–41
22. Guirard, B. M., Snell, E. E., Williams, R. J. 1946. *Arch. Biochem.* 9:381–86
23. O'Kane, D. J., Gunsalus, I. C. 1948. *J. Bacteriol.* 56:499–506
24. Wittenberger, C. L., Flavin, M. 1963. *J. Biol. Chem.* 238:2529–36
25. Wittenberger, C. L., Haaf, A. S. 1966. *Biochim. Biophys. Acta* 122:393–405
26. Barker, H. A. 1961. In *The Bacteria* ed. I. C. Gunsalus, R. Y. Stanier, 2:151–207. New York: Academic
27. Uyeda, K., Rabinowitz, J. C. 1967. *J. Biol. Chem.* 242:24–31
28. Barker, H. A. 1976. In *Reflections on Biochemistry,* ed. A. Kornberg, B. L. Horecker, L. Cornudella, J. Oro, pp. 95–104. Oxford: Pergamon
29. Lenhert, P. G., Hodgkin, D. C. 1962. In *Vitamin B₁₂ und Intrinsic Factor, Europaisches Symposion,* ed. H. C. Heinrich, pp. 105–10. Stuttgart: Enke Verlag
30. Snell, E. E., Kitay, E., MacNutt, W. S. 1948. *J. Biol. Chem.* 175:473–74
31. Stadtman, T. C. 1963. *J. Biol. Chem.* 238:2766–73
32. Stadtman, T. C., Renz, P. 1967. *Fed. Proc.* 26:343

Ann. Rev. Biochem. 1978. 47:35–88

RETROVIRUSES[1]

❖968

J. Michael Bishop

Department of Microbiology, University of California,
San Francisco, California 94143

CONTENTS

[1]Abbreviations and nomenclature used in this review are as follows: ALSV, avian leukosis-sarcoma viruses which share a group-specific antigen; ASV, avian sarcoma virus; tdASV, transformation-defective ASV; FeLV, feline leukemia virus; MMTV, mouse mammary tumor virus; MuLV, murine leukemia virus; MuSV, murine sarcoma virus; *gag,* viral gene encoding the structural proteins of the virion core; *pol,* viral gene encoding reverse transcriptase; *env,* viral gene encoding the proteins found in the virion envelope; *src,* viral gene responsible for transformation of fibroblasts in culture and induction of sarcomas in animals; ecotropic, infectious for the species of origin; xenotropic, infectious only for hosts other than the species of origin; amphotropic, infectious for both the species of origin and other species. The nomenclature for viral proteins adopted here conforms to conventions recommended at a meeting on retroviruses at Cold Spring Harbor Laboratory, New York, in June 1977. Mature gene products are designated by their molecular weight in thousands, preceded by p (or gp for glycoproteins, pp for phosphoproteins); precursors or putative precursors by their molecular weight in thousands, preceded by Pr with a superscript to designate the viral gene encoding the precursor (e.g. Pr76^{gag}).

35

0066-4154/78/0701-0035$01.00

Tumor virology has most of its road still ahead. For what else, if not tumor virology, can lead to an unravelling of the molecular basis for the malignant behavior of cells?(1)

PERSPECTIVES AND SUMMARY

The transmission of avian leukemia by a filterable agent was reported in 1908 and Peyton Rous described the sarcoma virus that now bears his name in 1911, yet these and related viruses (known collectively as retroviruses) were largely neglected as experimental agents until the past two decades. Now, venturesome investigators have attempted to implicate retroviruses in a wide range of biological processes, including the etiology of human neoplasia, embryological development, somatic mutation, and evolution. This remarkable change in fortune is attributable in large measure to the observations that retroviruses are widely distributed among the vertebrate species as both horizontally and genetically transmitted infections, are important natural agents of oncogenesis in at least several species, and are dependent upon RNA-directed DNA synthesis for the establishment of infection.

All retroviruses share a similar architecture, a roughly homologous set of structural proteins, a unique genome that is a diploid dimer of single-stranded RNA, and a virion polymerase capable of RNA-directed DNA synthesis. Four viral genes have been identified and mapped on the viral genome; three of the genes encode proteins required for viral replication, and the fourth is responsible for a specific form of neoplastic transformation.

Retroviruses replicate through the agency of a DNA intermediate, copied from the viral genome by reverse transcriptase and integrated into the chromosomal DNA of the host cell. The assumption that integration is a prerequisite for viral gene expression has not been proven. The mechanism of integration is not known, and the sites on cellular DNA at which integration occurs have been only partially characterized.

Viral DNA serves as template for viral RNA synthesis by a host polymerase, but the location of promoter sites for initiation of RNA synthesis has not been determined. The major products of viral RNA synthesis have identical chemical polarities but divide into two distinguishable pools: RNA destined for encapsidation as viral genome, and RNA destined to serve as viral mRNA. The genesis of viral mRNAs may involve translocation of nucleotide sequences during or after transcription.

The strategy of retrovirus gene expression provides for independent expression of viral genes according to the relative need for viral gene products and meets the stricture that translation in eukaryotic cells may initiate only at 5' termini of mRNAs. Three of the four viral genes are translated from mRNAs that have the expressed gene at the 5' end; the fourth gene may be expressed by the translational "read-through" of a termination signal. All of the viral gene products required for replication are generated by cleavage of protein precursors, and some of these cleavages may be coordinated with the assembly of virions. Phosphorylation may regulate the function of several viral gene products.

Host cells can restrict the replication of retroviruses by exclusion of viral entry, or by constraints on intracellular events responsible for viral gene expression. Viral entry is blocked by the absence of cellular receptors required for penetration of the viral genome into the host cell. Intracellular restrictions can affect any of several major steps in the life cycle of the virus and testify to the intimate involvement of the host cell in viral gene expression. As a consequence of intracellular restrictions, expression of viral transforming genes can be prevented or interrupted despite the presence of the complete viral genome as integrated DNA.

The isolation of both nonconditional deletion mutants and conditional temperature-sensitive mutants of retroviruses has facilitated the analysis of viral replication and the definition of a specific viral gene responsible for neoplastic transformation. Genetic analysis of retroviruses exploits the ability of the viruses to recombine at high frequencies and to complement defective or deficient viral genes in mixed infection.

Genomes of endogenous retroviruses reside in the germ lines of many, if not all, species and segregate as normal genetic elements within these species. Expression of endogenous retroviruses is subject to control by genetic determinants of the host cell. The potential involvement of endogenous viruses in normal cellular metabolism and in natural oncogenesis remains to be properly evaluated, but it is well established that endogenous viruses can interact with exogenous viruses to generate both phenotypically mixed virus and stable genetic recombinants.

A single retrovirus gene (*src*) has been persuasively implicated in neoplastic transformation of fibroblasts. By contrast, the role of retroviruses in other forms of oncogenesis remains poorly characterized. Some retroviruses may acquire oncogenic potential by genetic recombination with viral or cellular genes in the host organism.

The mechanism of virus-induced neoplastic transformation has not been elucidated and heuristic efforts to envision a final common pathway may be misconceived. For example, the viral gene *src* elicits a pleiotropic response with individual components that can be distinguished by the use of viral mutants. The primary event (or events) that initiates the response has not been identified.

When last reviewed in this series, the biochemical analysis of retroviruses was in a primitive state. In the interim, the study of retrovirus structure and replication has advanced rapidly, galvanized mainly by the discovery of reverse transcriptase and fostered by granting agencies intent upon the hope expressed in the epigraph to this article. Fulfillment of that hope may now be within reach.

INTENTION

... either we get some kind of grip on the accumulation of thought or we continue to wallow helplessly, to starve amidst plenty. So I gamble with science and write ....(2)

A comprehensive review of retroviruses would exceed the scope of this series. I have focused on molecular structure and mechanisms as they pertain to biological phenomena. I have addressed the review more to the general reader than to the expert, yet I have tried to include sufficient detail so that controversies are clear, arguments can be followed, and the marvelous peculiarities of retroviruses can be appreciated. Knowledgeable readers will find some of the views expressed here to be biased; the opportunity to exercise personal prejudices is one of the few compensations for authorship of treatises such as this. The older literature is often summarized without citation, or is cited in the form of review articles. Many of the specific subjects discussed here have been reviewed in detail elsewhere; these reviews are cited in the subheadings. The current literature is cited selectively; I apologize in advance to my colleagues who find themselves slighted. I was at risk of starving amidst plenty.

THE TAXONOMY OF RETROVIRUSES

Four major characteristics define the family Retroviridae (3): the architecture of the virion, a diploid single-stranded RNA genome, the presence of reverse transcriptase in virions, and the requirement for a DNA intermediate in viral replication. A confusing array of taxonomic devices have been used to classify retroviruses. 1. Each strain of retrovirus is either an endogenous or an exogenous virus. The genome of an endogenous virus is encoded in the germ line DNA of a normal animal species and is perpetuated by vertical transmission through the gametes of the species (4, 5), whereas exogenous viruses persist by virtue of horizontal or epigenetic spread among members of susceptible species. 2. Retroviruses are commonly identified according to the species from which the virus has been isolated. The genome of an exogenous virus is at least partially homologous to the genome of an endogenous virus in the species of origin (6); on occasion, retroviruses fortuitously isolated from secondary hosts have subsequently been traced to their proper species of origin by application of this criterion (6, 7). Some investigators have abused this criterion in their efforts to relate putative human tumor viruses to retroviruses transmitted horizontally among other primates. 3. Three major forms of antigenicity are associated with retroviruses: group-specific antigens are shared by related viruses derived from a single host species, type-specific antigens define the most specific serological subgroups presently identified, and interspecies antigens are shared by otherwise unrelated viruses derived from different host species. 4. The host range of retroviruses provides an exceptionally useful means of classification; major examples include the subgroups of avian leukosis-sarcoma viruses (ALSV) (8), the grouping of murine leukemia viruses (MuLV) into ecotropic, xenotropic, and amphotropic classes (9), and the N/B classification of ecotropic MuLV (10). 5. Most retroviruses can be distinguished by their pathogenicity. Many retroviruses are oncogenic (Oncovirinae) and are identified according to the type of neoplasm they induce (leukemia virus, leukosis virus, sarcoma virus, mammary carcinoma virus, etc); other retroviruses are cytopathic and induce chronic degenerative diseases (Lentivirinae), or are symbiotic and evoke little or no response in the host cell (Spumavirinae).

THE ARCHITECTURE AND PROTEINS OF THE VIRION (11)

All retroviruses share a similar architecture, although three classes of virus particles have been defined by electron microscopy (12). Most retroviruses, including leukemia and sarcoma viruses, are classified as C-type particles according to certain features of their morphological maturation and mature structure. B-type particles are less common and are typified by the mammary carcinoma virus of mice. A-type particles are intracellular forms that appear in two locations: within the cytoplasm, where they may be precursors to either B-type (13) or C-type (14) particles, and within cisternae, where their nature and function are enigmatic (14a, 15). In certain cells, intracytoplasmic and intracisternal A-type particles are homologous (14a). Intracisternal A-type particles are probably not transmissible by normal routes of infection but can be established in new host cells by a specialized form of somatic cell hybridization (15).

Retroviruses generally contain six or more structural proteins encoded in the viral genome. The exact number, size, and composition of these proteins vary among different strains of virus as a function of the species from which the viruses originated (Table 1). For example, feline leukemia virus (FeLV) and RD-114 virus are unrelated strains now established in cats. However, FeLV originated in rodents, RD-114 in primates (7). The identity of the major phosphoprotein of the two viruses reflects their origins (Table 1): pp12 for FeLV, as in retroviruses still associated with rodent hosts, and pp15 for RD-114, as in the endogenous viruses of baboons. Nevertheless, homologies of size, composition, location, and function are clearly evident among the proteins of different viruses (Table 1). The major antigenicities of the virion are usually attributable to only a few of the structural proteins, but careful analysis has revealed both group-specific and type-specific determinants on most if not all viral proteins (16). Antigenic determinants can be better correlates of functional homology than is molecular size. Again, the viral phosphoproteins provide good examples. Each strain of virus contains a phosphoprotein that binds specifically to the viral genome (see below). Among mammalian viruses, these proteins have different sizes in strains derived from different species (e.g. pp12 in rodent viruses, pp15 in primate viruses; see above), yet they share certain specific antigenic determinants and hence may have evolved from a common ancestral gene (16a).

The envelope of retroviruses is derived from the plasma membrane of the host cell and may reflect the composition of that organelle in important ways. For example, a specific histocompatibility antigen and a viral glyco-protein are topographically linked on the surface of cells transformed by the Friend erythroleukemia virus (17), and virions produced by these cells contain the same histocompatibility antigen (17). The antigen and viral glycoprotein may collaborate to mediate immune rejection of the virus-producing leukemic cell by the host organism (18).

Spikes protruding from the exterior of the viral envelope are composed of glycoprotein(s) encoded in the viral genome, are often a major locus for type-specific antigenicity, are required for entry of the virus into a host cell, and are responsible for the host range of the virus (8). The spikes are generally composed of two proteins, which may be derived from a single polyprotein precursor (see below) and which remain joined by disulfide bonds (19–21). Perhaps cleavage of the precursor is required to activate the biological function of the spikes, as described recently for other enveloped animal viruses (22). The primary structure of the glycoproteins of endoge-nous and exogenous viruses of mice displays a remarkable polymorphism, which may reflect the existence of a large family of related genes whose products can also appear as differentiation antigens at the cell surface (23).

The pattern of glycosylation of spike glycoproteins is determined, at least in part, by the host cell (24). The function of the carbohydrate residues is not known; they are probably not required for infectivity (25) (although they may conceivably contribute to the precise host range of the virus), and they have no apparent effect on the antigenicity of the glycoprotein (25).

Table 1   Structural proteins of type C retroviruses[a]

Avian (ALSV)		Murine (MuLV)		Feline (FeLV)[b]		Feline (RD-114)[c]		Primate (BKD)[d]	
Protein	Location	Protein	Location	Protein	Location	Protein	Location	Protein	Location
gp85	spike	gp70	spike	gp70	spike	gp70	spike	gp70	spike
gp37[e]	spike	gp45[f]	spike						
p27	core	p30	core	p30	core[g]	p30	core[g]	p30	core[g]
pp19	RNP[h]	p15(E)[i]	?						
p15	core	p15	core	p15	core[g]	pp15	RNP	pp15	RNP
pp12	RNP	pp12	RNP	pp12	RNP(?)	p12	core[g]		
p10[j]	core	p10	RNP	p10	core[g]	p10	core[g]	p10	core[g]

[a] See text footnote 1 for nomenclature.
[b] FeLV (feline leukemia virus) is established in cats but originated from rodents (7).
[c] RD-114 is an endogenous virus of cats acquired originally by horizontal transmission from primates (7).
[d] BKD is an endogenous virus of baboons.
[e] gp37 is joined to gp85 by a disulfide bridge.
[f] gp45 is an underglycosylated form of gp70.
[g] The precise location of these proteins within the core is not known.
[h] RNP = ribonucleoprotein.
[i] p15(E) is joined to gp70 by a disulfide bridge.
[j] The existence of p10 in ALSV is in dispute.

Within the envelope is a geometrically symmetrical "core shell" composed of several virus-coded proteins, at least one of which is usually a major locus for group-specific antigenicity (11). The shell in turn encloses a helical ribonucleoprotein (26) composed of the viral genome, possibly some low molecular weight RNA (27), reverse transcriptase (28), and two small virus-specified proteins (28, 29) that bind to RNA (29, 30). The principal polypeptide constituent of the ribonucleoprotein is a highly basic protein (pp12 in ALSV, p10 in MuLV) (28) that binds to single-stranded RNA (30). Presumably, charge-shielding by this protein is essential to the assembly of the viral core. The other polypeptide in the ribonucleoprotein is generally the major phosphoprotein of the virion (pp19 in ALSV, pp12 in viruses of mice and rats, pp15 in primate viruses; see Table 1). This protein apparently binds to duplex regions in the viral RNA (30), probably at a limited number of sites (29, 30) alleged to be important in viral replication (see below). Binding occurs preferentially with homologous viral RNA (29), although other RNAs containing duplex structure can bind with lower affinities (30). The extent of phosphorylation of individual molecules of this protein varies and determines the relative affinity of the protein for viral RNA (31).

Reverse transcriptase is an essential constituent of retroviruses (32), and each virion apparently contains scores of polymerase molecules (33, 34). The structure and properties of this enzyme and its associated RNase H activity have been reviewed recently (32) and are discussed here only as they pertain to features of viral replication.

THE GENOME OF RETROVIRUSES

The genome of retroviruses is apparently diploid; two identical molecules of single-stranded RNA (mol wt 2–3 million) are joined at or near their 5' termini (35, 36). Although available evidence indicates that the subunits of the genome are linked by hydrogen bonds between complementary sequences of nucleotides (37), the exact nature of the linkage is not known, and no investigator has reported the reassembly of a native genome from its component nucleic acids. The diploidy of retroviruses is unique among the genomes of known animal viruses and may serve an essential function, such as a role in transcription of DNA from the viral genome by reverse transcriptase or facilitation of genetic recombination (see below). At the least, diploidy could account for the apparent ease with which retroviruses form heterozygotes (38, 39).

Figure 1 illustrates established features of the composition, structure, and topography of a haploid subunit of the retrovirus genome, using the genome of avian sarcoma virus (ASV) as a prototype; these general features probably pertain to all retroviruses, but only the genome of ASV has been characterized in the detail illustrated here. The 3' termini of both subunits of the dimer are polyadenylated ($\sim$ 200 residues) (36, 40), the 5' termini are "capped" by the structure 5'-m^{7}GpppGm (41, 42), and approximately 10 residues of adenosine located at specific sites within the 3' half of the genome are methylated (41, 43, 44). Polyadenylation, "capping," and a low level of internal methylation are all features of eukaryotic mRNA and, in fact, the retrovirus genome can serve as messenger for the synthesis of virus-specific proteins (see below).

A molecule of tRNA^{Trp} derived from the host cell is bound to the genome of ASV at a site 101 nucleotides distant from the 5' terminus of the genome (45); this tRNA serves as primer for the initiation of DNA synthesis by reverse transcriptase in vitro (46, 47). The identity and location of the tRNA primer can be different in retroviruses derived from other species. For example, the primer for the Moloney strain of MuLV is tRNA^{Pro} (48) located approximately 135 nucleotides from the 5' terminus of the genome (W. Haseltine, personal communication).

In addition to primer, other tRNAs derived from the host cell are less securely bound to the haploid subunit of retroviruses (49). The location of these tRNAs on the viral genome and their function (if any) in the viral life cycle are not known, but their significance has generally been discounted because many of the represented isoacceptor species are not sufficiently abundant to be present on all molecules of viral RNA (49).

The haploid subunits of the genomes of ASV (50, 51), avian myeloblastosis virus (52), and MuLV (53) are terminally redundant. For example, an identical sequence of as many as 21 nucleotides is reiterated at each terminus of the Prague-C ASV genome (Figure 1), although heterogeneity in the nucleotide sequence at the 3' terminus limits the redundancy to as few as 16 nucleotides in some molecules (51); the principal heterogeneity involves the presence or absence of the sequence C-C-A that could be added to the RNA in a reaction not requiring a template. Similar heterogeneity has been found at the 3' terminus of the genome of avian myeloblastosis virus (52).

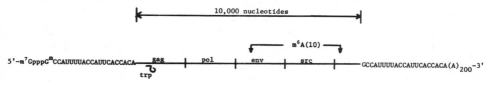

Figure 1   The genome of avian sarcoma virus. The illustration represents a composite of data obtained with a number of strains of ALSV. The redundant nucleotide sequence at the 3' and 5' termini is that found in the Prague-C strain of ASV (50, 51).

The redundant sequences in different strains of ALSV are not identical (51, 52, 54). In genetic crosses between strains of virus with different terminal redundancies, the redundant termini of a particular parental genome segregate together in recombinant progeny virus (54). This finding conforms to the suggestion that terminal redundancy serves a vital function in viral replication. The terminal redundancy in the genome of MuLV has not been fully characterized but apparently involves approximately 60 nucleotides (53).

The sequence of 119 nucleotides at the 5' terminus of the ASV genome has been determined (50, 55). Several features of this sequence may have functional significance. First, the 5' termini of two subunits can base-pair with each other and with tRNA^{Trp} in a manner that could account for the linkage of subunits into dimers (50). At present, this suggestion remains hypothetical; efforts to visualize the dimer structure of ALSV by electron microscopy have not been successful, and the subunits of other retrovirus genomes are not necessarily joined at their immediate 5' ends (56). Second, the sequence displays several features of a potential ribosome binding site (50, 55). Participation of this site in the initiation of translation has yet to be directly documented. Third, tRNA^{Trp} is positioned downstream from the putative initiator codon and could regulate the function of viral RNA by impeding translation from viral RNA in vitro. Hence, viral RNA with tRNA^{Trp} in place might serve exclusively as a template for DNA synthesis, whereas viral RNA devoid of the tRNA is a potential messenger. Fourth, intrastand base-pairing can generate extensive secondary structure [three hairpin loops with duplex stems stable at physiological ionic strength and temperature; see (55)], which could figure in the binding of either ribosomes or reverse transcriptase.

Four genes have been identified in the genome of ASV: *gag,* which encodes structural proteins of the viral core; *pol,* which encodes reverse transcriptase; *env,* which encodes the glycoprotein(s) of the viral envelope; and *src,* which is responsible for neoplastic transformation of the host cell. These genes virtually account for the coding capacity of the ASV genome (57); however, genetic variants that have most or all of *src* deleted from their genome ("transformation defective", or tdASV) can cause lymphoid leukosis (58); hence, the genome of ASV could contain an additional, unidentified genetic determinant. The genes *gag, pol,* and *env* are all required for the replication of infectious virus and are common to all virus strains that can replicate in the absence of a helper virus; by contrast, deletion of *src* has no effect on viral replication, and the selective pressures responsible for the retention of *src* through the course of countless viral generations have not been identified.

The genes of ASV have been ordered on the viral RNA by chemical analysis (Figure 1), using deletion mutants and recombinant strains to identify specific segments of the genome (59–61). Mapping of deletions by examination of heteroduplex molecules by electron microscopy has confirmed the position of *src* (62), marker rescue experiments have documented the linkage between *pol* and *env* (63), and analysis of the proteins translated from viral RNA in vitro has substantiated the position and order of *gag* and *pol* (see below). In addition, all strains of ALSV share a sequence of approximately 1000 nucleotides located at the 3' terminus of the genome and denoted "c" (for common region) (64). The function of "c" is unknown. However, its presence in the genomes of all viable strains of ALSV, and its conservation in the face of deletions in the adjacent gene *src* (64), imply that "c" is an essential constituent of the viral genome.

No other retrovirus genome has been adequately mapped, but there is provisional evidence that the gene order in MuLV/MuSV may be identical to that in ALSV. Analysis of heteroduplex molecules by electron microscopy has mapped the nucleotide sequence alleged to contain *src* close to the 3' terminus of the viral genome (65), and translation of viral RNA in vitro has provided evidence that locates *gag* and *pol* in the same positions as in the ASV genome (see below).

OTHER NUCLEIC ACIDS IN THE VIRION

Most preparations of retrovirus virions contain a small amount of high molecular weight ribosomal RNA (66), 7S RNA, which is also found in host polyribosomes (67), 5S RNA from host ribosomes (68), and a variety of host tRNAs (49). With the possible exception of the tRNAs, these RNAs are not inevitable constituents of retroviruses and are generally considered to be contaminants. However, in some strains of virus, 7S and/or 5S RNA are bound to the viral genome (49, 68); the significance of this finding is unknown. A small fraction of the tRNAs are bound to the viral genome, as discussed above, but the majority are not. The isoacceptor species of tRNA found in virus are not a random assortment of cellular tRNAs and can vary both among specific strains of virus and as a function of the host cell (69). Further details regarding the tRNAs associated with retroviruses can be found in a recent review (70). To date, none of the extragenomic RNAs of retroviruses have been implicated in either viral structure or viral replication.

On occasion, assembly of retroviruses may result in the envelopment of mRNAs. Some investigators have found globin mRNA associated with the genome of virus produced by cells synthesizing large amounts of globin

(71). In general, however, the assembly of the viral genome is a fastidious process; for example, C-type virions fail to envelope the RNA of B-type retroviruses and vice versa (72). Recent findings indicate that virions of ALSV may contain appreciable amounts of their homologous viral mRNAs complexed with the viral genome (W. Hayward, personal communication). If confirmed and extended to other strains of retroviruses, these findings may have important implications regarding viral structure and replication.

Some preparations of virus contain small amounts of DNA, most of which is probably a contaminant derived from the host cell (73). However, a small fraction of this DNA is apparently complementary to the viral genome and, when hybridized to viral RNA, can serve as primer for the initiation of DNA synthesis (74). Moreover, some of the virus-specific DNA is allegedly bound near the center of the haploid subunit in the native viral genome and may there serve as primer (74). The biological significance of these observations is presently indeterminate.

THE REPLICATIVE CYCLE

Patterns of Infection (75)

Permissive cells permit the entry and replication of virus. However, neoplastic transformation is not an inevitable consequence of viral replication. For example, both ASV and avian leukosis viruses replicate in avian fibroblasts, but only ASV transforms these cells. In general, transformation by any single strain of virus is restricted to particular target cells, whereas the host range for replication is less specific. Transformation by sarcoma viruses can occur in the absence of viral replication, but the efficiency of transformation of nonpermissive cells (e.g. mammalian fibroblast hosts for ASV) is usually quite low (10^{-4} to 10^{-6}). The varieties of mechanisms that can restrict either viral replication or cellular transformation are examined in detail below. Early studies on nonpermissive cells demonstrated that the viral genome was perpetuated in the transformed cell in the absence of viral replication and, hence, helped to set the stage for the formulation of the hypothesis that has informed all subsequent experimental analysis of retroviruses: the DNA provirus hypothesis first articulated in 1964 (76).

The Provirus Hypothesis

Studies on the physiology of retrovirus replication led to the conclusions that the synthesis of virus-specific DNA was required for the initiation of infection and that DNA served as the template for synthesis of progeny viral RNA (77). From these and other observations, Temin formulated the DNA provirus hypothesis: retroviruses transform cells and replicate through the agency of a DNA copy of the viral genome, synthesized early in infection and then integrated into the genome of the host cell in a manner analogous to that of lysogenic bacteriophage (integration of the viral genome is an essential part of transformation by DNA tumor viruses as well, but this had not been proven at the time Temin proposed the provirus hypothesis for retroviruses). Three major experimental observations have substantiated the DNA provirus hypothesis beyond any reasonable doubt: 1. virions of retroviruses contain reverse transcriptase, an enzyme that uses viral RNA as a template to synthesize DNA (32), and the function of reverse transcriptase is a prerequisite for both viral replication and virus-induced cellular transformation (78, 79); 2. the synthesis and integration of virus-specific DNA in newly infected cells has been demonstrated by molecular hybridization (80); and 3. chromosomal DNA extracted from cells infected with retroviruses can "transfect" the viral genome into permissive cells, giving rise to both viral replication and cellular transformation (81).

An Outline of Replication (82)

Figure 2 outlines the principle molecular events in the replication of retroviruses. The scheme is in part hypothetical, but the central events have been established experimentally. Transcription of parental viral genome by reverse transcriptase generates virus-specific DNA through a series of poorly characterized intermediates. The major stable products of synthesis are linear duplexes and closed circular duplexes (83–86); both forms are approximately the length of the haploid subunit of the viral genome. Circumstantial evidence has implicated the circular molecules in integration (85). Viral DNA serves as template for the synthesis of viral RNA, generating two functionally separate pools of RNA molecules (87, 88): RNA destined for encapsidation as genome, and RNA destined to serve as messenger for the synthesis of viral proteins. Available evidence suggests that integration of viral DNA is a prerequisite for the synthesis of viral RNA, but the possibility that unintegrated DNA can serve as template for RNA synthesis has not been conclusively tested. Viral protein synthesis constitutes a minor fraction of total cellular protein synthesis and has no necessary effect on cellular metabolism; retroviruses are often not cytopathic, and the host cell is transformed only if the infecting virus bears an appropriate transforming gene. Once established, virus production continues indefinitely and is usually limited only by the life span of the host cell.

Entry of the Viral Genome

The glycoproteins of the viral envelope facilitate adsorption and penetration of the virus; deletions in *env* give rise to fully assembled, noninfectious virus particles (89), and temperature-sensitive conditional mutations in *env* render virus particles noninfectious at the restrictive temperature (90). The

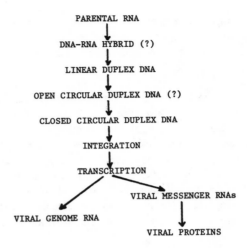

Figure 2   The replicative cycle of retroviruses. Question marks indicate hypothetical portions of the scheme. All other structures and events illustrated have been identified experimentally.

interaction between viral envelope glycoprotein and host cell receptors has been reconstructed in vitro with purified gp70 of MuLV and is highly specific (91).

The events that intervene between adsorption of the virus and the onset of viral DNA synthesis have not been elucidated, and available data are in conflict. Studies with electron microscopy suggest that the genome of ALSV migrates rapidly (within 10–60 min) to the nucleus of the infected cell (92), whereas biochemical data indicate that viral DNA synthesis commences and continues in the cytoplasm of the cell over the first 12–24 hours following infection (93).

Protein synthesis may (94) or may not (95) be required as an early event in the establishment of infection by retroviruses. Parental RNA associates with polyribosomes (94) and can be translated prior to the onset of (or in the absence of) viral DNA synthesis (96). These data were obtained with extremely high multiplicities of infection and their biological significance is presently moot: only one of the known viral gene products (*src*) does not enter the cell as a constituent of the virion, and *src* is not required for viral replication. Nevertheless, experiments with marker rescue have demonstrated that viral RNA can be expressed directly after introduction into the host cell by either microinjection (97) or application in the presence of polycations (H. Murphy, personal communication); these findings are attributable to translation from the entering RNA.

Synthesis of Viral DNA (80)

Reverse transcriptase is required for the synthesis of viral DNA in the infected cell (79). However, the mechanisms of this synthesis are poorly understood, a shortcoming largely attributable to logistical considerations. Viral DNA is not sufficiently abundant in the infected cell to permit the use of isotopic precursors for the study of DNA synthesis; consequently, investigators have relied upon molecular hybridization to analyze stable populations of viral DNA.

Synthesis of DNA by reverse transcriptase has been studied extensively in vitro with both crude and purified enzyme (32). The data obtained may be applicable to events in the infected cell because transcription of viral RNA in vitro reconstructs several salient features of viral DNA synthesis observed in vivo. In particular, the polymerase can synthesize minus strands (DNA complementary to the viral RNA) that are copies of the entire viral genome (98, 99); moreover, a small fraction of the enzymatic product is infectious (100). However, the structure of the infectious DNA synthesized in vitro has yet to be characterized, and the reaction in vitro fails to generate either full-length duplex copies of the genome or circular duplexes of any size. In general, the studies in vitro have provided useful information about the initiation and propagation of minus strands, but little has been learned to date about the synthesis of plus strands (the same polarity as the viral genome).

INITIATION AND CHAIN PROPAGATION (101)   Both the native viral genome and the separated subunits are effective templates in vitro (32), but the active form of template in the infected cell has not been identified. Synthesis of minus strands in vitro initiates on the tRNA primer located near the 5' terminus of the viral genome (101). Both tRNA[Trp] and tRNA[Pro] are used as primers by the reverse transcriptase of ALSV (102), and both tRNAs bind to the enzyme with high affinity (102, 103); this binding is alleged to direct the polymerase to the correct initiation site on the viral genome. By contrast, conventional assays cannot detect binding of tRNA primers to the reverse transcriptase of MuLV (102); binding may occur, but with an affinity too low for detection at equilibrium. In any event, the exceptional affinity between either tRNA[Trp] or tRNA[Pro] and the avian reverse transcriptase is not a prerequisite for initiation of DNA synthesis: a fragment composed of 23 nucleotides from the 3' terminus of tRNA[Trp] does not bind to the reverse transcriptase of ALSV under standard conditions, yet initiates transcription from the ASV genome with normal efficiency (B. Cordell, personal communication).

Initiation of transcription from the genome of reticuloendotheliosis viruses (avian retroviruses unrelated to ALSV) by the polymerase of these viruses may require the antecedent addition of ribonucleotides to an unidentified primer molecule (104). Addition is carried out by an enzymatic activity resident in virions (104), but the nature of this enzyme is not known. These unusual findings remain unexplained.

Propagation of minus strands in vitro proceeds from the site of initiation to the 5' end of the template (50, 55, 105, 106), then moves to the 3' terminus of the same or a separate subunit of the viral genome without interruption of polymerization (107–109) (see Figure 3). Synthesis of minus strands in the infected cell also initiates near the 5' terminus of the viral genome and then continues at the 3' terminus (110), but the primer for minus strands in vivo has not been identified.

The mechanism that transfers polymerization from the 5' to the 3' terminus of the viral genome may exploit the terminal redundancy of the genome: the DNA copy of the 5' redundancy could base-pair with the redundant sequence at the 3' terminus of the viral RNA and then serve as primer for further DNA synthesis (Figure 3). This model requires that the DNA at the 5' end of the viral RNA be separated from its template; the RNA could be removed by the RNase H activity of reverse transcriptase or displaced without hydrolysis by an unknown mechanism. Both solutions pose problems. Displacement faces a potential thermodynamic barrier because as many as 60 base pairs (e.g. in MuLV) must be disrupted. Hydrolysis by RNase H requires an attack on "capped" RNA, and the RNase H activity of reverse transcriptase is a processive exonuclease (32) whose action may be blocked by the cap structure.

Synthesis of plus strands in vivo begins prior to the completion of minus strands (110), is apparently discontinuous and must therefore require multiple initiations (110), and requires either displacement or hydrolysis of template RNA in advance of polymerization. The first segment of plus strand to be synthesized is ~250 nucleotides in length (110) and unites nucleotide sequences encoded at the 3' and 5' termini of the template RNA (designated B, T, and A in Figure 3) (110). The exact sites for initiation and termination of this segment have not been identified, but they appear to be reasonably

---

Figure 3 Proposed scheme for the synthesis of retrovirus DNA. A hypothetical scheme of viral DNA synthesis is illustrated with the genome of ASV as template. ——RNA; ∿ DNA, with arrowheads indicating direction of polymerization. 1. DNA synthesis initiates on tRNA^{Trp} primer near the 5' terminus of one of the haploid subunits of the viral genome and proceeds to the end of the subunit RNA, copying the unique nucleotide sequence A and the terminally redundant sequence T into the complementary DNA A' and T'. 2. Part or all of the RNA sequence T is removed by RNase H (or displaced without hydrolysis by an unknown mechanism). 3. The DNA sequence T' base-pairs with the RNA sequence T at the 3' end

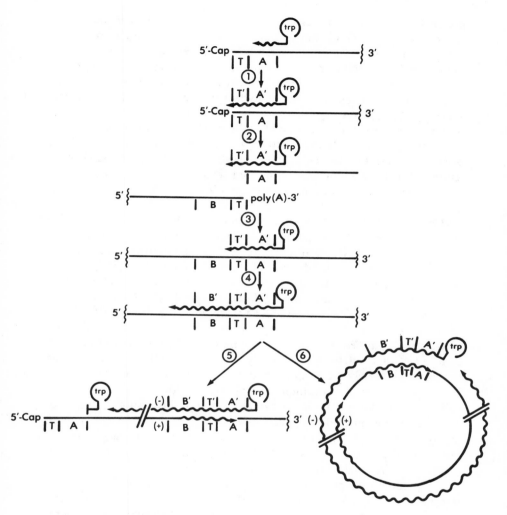

of the same or a separate subunit. 4. DNA synthesis proceeds, copying the unique nucleotide sequence B into DNA B'; the poly(A) is excluded from transcription both in vitro (343) and in vivo (110). 5. Chain propagation is illustrated after an *inter*-molecular jump of polymerization from one subunit template to another; plus-strand DNA synthesis has initiated, using viral RNA as primer and minus-strand DNA as template. If synthesis proceeds to the extreme 5' end of the subunit template, the progeny DNA will contain a second copy of A/A' and T/T'. 6. Chain propagation is illustrated after an *intra*molecular jump of polymerization that creates a circular RNA:DNA hybrid. The nucleotide sequences A, A', T, T', B, and B', and the synthesis of plus-strand DNA are all as in step 5.

precise (110). The segments synthesized subsequently are longer and hetero-geneous in length (110), but are not otherwise well characterized. The mechanism of initiation of plus strands has not been characterized, either in vitro or in vivo. One prevalent hypothesis ascribes the initiation of plus strands to the RNase H activity associated with reverse transcriptase (32). The action of this nuclease creates 3' hydroxyl termini (111); hence, hydrolysis of template RNA base-paired with minus DNA could engender ends of RNA suitable for the initiation of DNA synthesis. However, experimental efforts to implicate RNase H activity in the synthesis of plus strands have been unsuccessful to date (112). Parental viral RNA has been found covalently linked to DNA in acutely infected cells (113–115), but the validity of these observations is open to question (80).

PATHWAY OF VIRAL DNA SYNTHESIS IN THE INFECTED CELL   The synthesis of viral DNA in permissive cells infected with ASV and MuLV commences in the cytoplasm (93, 116) within one hour of infection and proceeds as far as the production of linear duplexes (117). By contrast, synthesis of viral DNA in cells infected with the cytopathic retrovirus visna occurs entirely in the nucleus (A. Haase, personal communication). The apparent rate of polymerization of ALSV DNA is exceptionally slow ($\sim$0.5 nucleotides per sec) (110) and conforms approximately to rates measured in vitro with purified reverse transcriptase (118). Synthesis of viral DNA can persist indefinitely in chronically infected cells (119, 120), but this synthesis is probably the result of continued reverse transcription from viral RNA rather than of replication of viral DNA (120).

The initial intermediate in the synthesis of viral DNA should be a base-paired hybrid composed of minus strand DNA and the RNA template; in addition, the DNA would at first be covalently linked to the primer for initiation of DNA synthesis. Hybrids with these features have not been identified in infected cells.

The predominant products of viral DNA synthesis in infected cells are linear duplex molecules (83, 84, 110, 120–123), which, in their mature form, contain minus strands roughly the length of the haploid subunit of the viral genome and plus strands interrupted perhaps 5–10 times at irregular intervals (110, 120, 121). The interruptions may consist of at least short gaps in the DNA chain (H.-J. Kung, personal communication) and could represent sites where discontinuous synthesis of plus strands was initiated. The end of the duplex that contains the 5' terminus of the minus strand probably contains nucleotide sequences corresponding to both the 5' and 3' termini of viral RNA (110). This arrangement of nucleotide sequences could be predicted from the fact that DNA synthesis initiates at the 5' terminus of the viral genome and then proceeds to the 3' terminus without interruption (see Figure 3). The remainder of the linear duplex is poorly characterized.

The results of pulse-chase experiments with density label indicate that the linear duplexes are precursors to the closed circular duplexes (117), which first appear in the nucleus 5–12 hours after infection (86). Circle formation presumably occurs in the nucleus: no circular molecules have been detected in the cytoplasm, and linear duplexes are found in the nucleus coincident with the appearance of circular DNA (117). Both the linear and circular molecules are infectious (86, 122, 123), but the specific infectivity of the linear form is appreciably greater than that of the circular DNA (122, 123); the significance and mechanism of this difference are unknown.

The contributions of host cell mechanisms to the synthesis of viral DNA have not been elucidated. With the exception of ligase to close the circular duplexes, all the required enzymatic activities may reside on reverse transcriptase. However, viral DNA synthesis is abortive when initiated in cells arrested in G_0 by deprivation of serum and is renewed only after the cells are returned to normal growth by restoration of serum to the culture medium (124). Hence, cells deprived of serum are deficient in one or more unidentified factors required for viral DNA synthesis. Normal and ASV-infected cells contain one or more proteins that can augment transcription of DNA from viral RNA and other templates by reverse transcriptase, perhaps by binding to the template and denaturing intrachain secondary structure (125). There is no evidence that this protein actually participates in viral replication.

REPRISE: CONUNDRUMS IN THE SYNTHESIS OF VIRAL DNA   Replication of viral DNA beyond the initial copy transcribed from the viral genome is unlikely; the number of copies of viral DNA synthesized in infected cells is roughly congruent with the input of infectious particles (93, 119), and no intermediates likely to be involved in the replication of viral DNA have been identified. In fact, major features of viral DNA synthesis militate against replication: part or all of the template RNA may be degraded in the process of copying the viral genome into double-stranded DNA (see above and Figure 3); reverse transcriptase cannot itself replicate DNA (32); and, in at least some instances, viral DNA synthesis is probably restricted to the cytoplasm, which may contain no replicative mechanisms other than those in mitochondria (126, 127). Since a single virus particle can establish infection (1, 4), one or two complete copies of the viral genome may suffice for integrative recombination at a suitable and probably nonrandom site on the host genome (see below).

Why is the viral genome a dimer, and why is viral DNA synthesis designed to cross from one end to another of template molecules? I find it tempting to assume that these are related aspects of viral replication and that, together, they serve one or both of two purposes: duplication of the ends of the viral genome into doubled-stranded DNA—a formal problem in the replication of any linear molecule by primer-dependent polymerization (127a)—and provision of a molecular form of DNA suitable for circularization. The additional possibility that the dimer structure facilitates recombination is discussed below. Even if this possibility proves correct, it seems likely to be an adventitious consequence of the dimer structure rather than the source of selective pressure for a dimer structure.

The mechanism that transfers polymerization from the 5' to the 3' terminus of viral RNA sacrifices one of the two copies of the redundant sequence that participate in the transfer (Figure 3); the lost copy must be recovered before or during integration in order to restore the integrity of the viral genome. I will summarize and comment on several possible means for restitution of the lost nucleotide sequence because they illuminate important unsolved problems in the synthesis of viral DNA.

1. Transfer of polymerization from one subunit to another would produce a second copy of the redundant sequence, provided that DNA synthesis does not initiate at the 5' end of the second subunit so that polymerization can proceed to the end of the template (Figure 3, step 5). This mechanism exploits (and could therefore explain) the existence of a dimeric genome but would not permit the independent expression of both subunits of a dimer and, hence, does not conform to reports that heterozygous genetic markers segregate independently in progeny virus (38, 127b). In addition, the mechanism generates a linear duplex, which is then probably converted into a circle prior to integration (see below), and it is quite possible that the process of circle formation would again sacrifice one copy of the terminal redundancy!

2. Recombination among or within molecular intermediates in viral DNA synthesis could generate competent genomes (52). Multimeric copies of the viral genome would be the most suitable forms for this mechanism, and structures of this sort have not been identified in infected cells. This mechanism might also preclude the independent expression of two components of a diploid genome.

3. The integration site in cellular DNA could contain a nucleotide sequence identical to the terminal redundancy; restitution of the lost copy would be analogous to transduction. This seems an unlikely mechanism because of the variety and size (e.g. 60 nucleotides for MuLV) of redundant sequences that are faithfully replicated in both permissive and nonpermissive cells.

4. Staggered endonucleolytic cuts that bracket a single copy of the appropriate nucleotide sequence in a circular molecule of viral DNA would generate a linear duplex with redundant ends. The strands between the cuts could separate of their own accord, be denatured by a suitable protein, or be separated as a consequence of displacement DNA synthesis by a cellular

DNA polymerase (127a). In each instance, the product would be a linear molecule whose integration would preserve the topography of the viral genome. This mechanism conforms to circumstantial evidence that circle formation is a prerequisite for integration (85).

How are the circular duplexes of viral DNA formed? Several investigators have noted that an *intra*molecular jump of polymerization as described above could produce a circular RNA : DNA hybrid (Figure 3) and have suggested that this process eventuates in the formation of circular duplex DNA. This suggestion is almost certainly a misconception, since (*a*) the circular hybrid would be disrupted when synthesis of minus and plus strands meet at the position of the tRNA primer (Figure 3); and (*b*) the precursor to circular DNA is not an RNA : DNA hybrid but a linear duplex DNA (117). Circular forms could be generated from linear duplexes (*a*) by the joining of complementary single-stranded ends as in the DNA of bacteriophage lambda (127c); (*b*) by recombination [e.g. intramolecular recombination between reiterated nucleotide sequences at or near the ends of linear molecules, as in phage P1 (127d); the pathway illustrated in Figure 3, step 5, would generate a linear duplex suitable for such recombination]; or (*c*) by ligation of blunt-ended molecules. The answer may emerge from more detailed studies on the structure of the linear and circular duplex molecules.

Why are circular duplexes formed? The question presumes that circular DNA is involved in viral replication. Granted this reasonable but unproven presumption, formation of circular molecules from linear duplexes (117) might serve any of several purposes. First, the mechanism of DNA synthesis apparently permutes the order of the viral genome in the linear duplex DNA (see above and Figure 3); circle formation provides an opportunity to eliminate the permutation. Second, the appropriate opening of a circular molecule can effect the necessary duplication of a nucleotide sequence (see above). Third, a circular form permits integration to occur by crossing over at a single site and to eventuate in a provirus that duplicates the linear order of the viral genome.

Integration of Viral DNA

Rigorous experimental documentation that viral DNA integrates into chromosomal DNA of the host has been difficult to achieve, but both centrifugation in alkali (121, 128) and analysis with restriction nucleases (personal communications from S. Hughes and R. A. Weinberg) have demonstrated apparent covalent linkage of viral DNA to high molecular weight cellular DNA; the occurrence of integration has become axiomatic to investigators of retroviruses. The role of integration in the viral life cycle is less certain. Circumstantial evidence indicates that integration is a prerequisite for viral replication (85). However, studies on nonpermissive cells indicate that integration does not mandate viral gene expression: the provirus of ASV can be present in suitable target cells without concomitant changes in cellular phenotype (129, 130), and integration of the DNA of mouse mammary tumor virus (MMTV) does not of necessity lead to synthesis of viral RNA (see below). Hence, integration is probably a necessary but not sufficient antecedent to viral gene expression. The most important variable could be the particular site in cellular DNA at which the viral genome integrates, but this supposition is at present unproven.

Nothing is known of the mechanisms that effect integration. Even the outlines of these mechanisms are difficult to anticipate: it is not known whether integrative recombination by retroviruses is legitimate (between homologous nucleotide sequences) or illegitimate (between nonhomologous nucleotide sequences), and the enzymatic mechanisms available for recombination in eukaryotic cells have not been elucidated. To date, genetic studies have failed to implicate any viral gene product in integrative recombination. Integration of viral DNA requires replication of cellular DNA (124), but the precise nature of this requirement is not known.

If integrated viral DNA is the template for synthesis of viral RNA, then at least one copy of the integrated DNA must duplicate the linear order of the viral RNA. Results of recent analyses with restriction nucleases indicate that the integration of both ASV and MMTV DNA probably meets this stricture: integrated viral DNA is joined to cellular DNA by nucleotide sequences at the 3' and 5' termini of the viral genomes (S. Hughes and G. Ringold, personal communication). This topography could result from recombination between cellular DNA and the appropriate site on a circular molecule of viral DNA; available evidence suggests but does not prove that the circular form participates in integration (85). Linear duplexes of viral DNA are also present in the nucleus at the time of integration (117) and could conceivably recombine with the host genome. As noted above, at least one end of the linear duplexes is apparently permuted with respect to the viral genome; otherwise, the structure of these molecules is too poorly characterized to permit consideration of their possible role in integration.

Studies on the nature of the integration sites in cellular DNA are presently incomplete, and available data permit only a few tentative conclusions. First, when multiple copies of provirus for either ASV or MMTV are integrated in single cells, each copy can be separately integrated; multiple copies of viral DNA are not necessarily arranged in tandem (S. Hughes, personal communication). Second, the proviruses of ALSV, MuLV, and MMTV can each integrate at multiple sites on the host genome (personal communications from S. Hughes, R. Weinberg, and G. Ringold), but the nature, specificity, and number of these sites have yet to be satisfactorily determined. Provisional evidence indicates that the number of sites may be limited (119) and that the sites can be located adjacent to different classes of cellular DNA (defined according to their frequency of reiteration) in different hosts (131). The possibility that endogenous proviruses can serve as integration sites (132) has not been adequately tested, and the more general issue of whether integrative recombination by retroviruses takes place between homologous or nonhomologous nucleotide sequences remains unresolved. Cells chronically infected with a reticuloendotheliosis virus allegedly contain provirus integrated at a unique site (133). This conclusion presumably pertains only to provirus integrated in a functional configuration because integrated viral DNA was detected by transfection rather than by molecular hybridization.

An entirely different view of integration has been proposed on the basis of data purported to demonstrate covalent linkage between parental viral RNA and cellular DNA (134). These experiments were based on the use of radioactive virus to trace the fate of parental genome in acutely infected cells, a protocol long suspect among animal virologists because of the abundance of biologically inactive particles in most virus stocks. Skepticism regarding the significance of these experimental results has been reinforced by the fact that the same protocol gave apparently erroneous results when applied to the study of host restriction of viral replication (134, 135).

Synthesis of Viral RNA (136)

The synthesis of viral RNA occurs in the nucleus of the infected cell (137) and is catalyzed by a normal cellular enzyme—probably the nucleoplasmic RNA polymerase II (138, 139); to date, no viral gene product has been implicated in the process. The template for viral RNA synthesis is DNA: synthesis is susceptible to inhibition by actinomycin D (75) and fails to occur if the production of viral DNA is prevented by either metabolic inhibitors (75) or mutations in reverse transcriptase (140). Moreover, the principal template is probably integrated DNA. Blockage of integration by either a chemical inhibitor (85) or a cellular determinant of host range (141, 142) prohibits appreciable viral RNA synthesis, and both permissive and nonpermissive cells can produce normal amounts of viral RNA in the absence of detectable unintegrated viral RNA (140). However, none of these data are conclusive; the possibility that unintegrated DNA can participate in the synthesis of viral RNA remains an open question.

Initiation of viral RNA synthesis in newly infected cells allegedly depends upon an event (or events) that occurs in the late G_2 or early M stage of the host cell cycle (143), yet infected cells may become phenotypically transformed and produce virus despite the use of periwinkle alkaloids to block cell division (144). These apparently conflicting observations could be reconciled as follows: 1. parental viral RNA can be translated early in the infectious cycle in the absence of viral DNA and RNA synthesis (96); however, this translation may generate products from only one of the four viral genes (*gag*) (96); 2. viral RNA synthesis requires an event in the cell cycle prior to metaphase, but this event progresses normally in cells blocked at metaphase by alkaloids. Synthesis of viral RNA in chronically infected cells has also been linked to events in the host cell cycle (145), but results of more recent studies with improved techniques have placed this alleged linkage in doubt (146).

The location of promoter(s) for viral RNA synthesis has not been determined. The initial product of synthesis is probably an RNA molecule not appreciably longer than the haploid subunit of the viral genome (147); hence, RNA synthesis appears to initiate within or immediately without the viral DNA template. Since the entire template must be transcribed in order to generate viable viral genomes, any promotor site located within the viral template would have to be itself transcribed. Alternatively, initiation could occur outside the viral genome at an adjacent or nearby site in cellular DNA; the initial product of viral RNA synthesis could easily contain a terminal "leader" sequence of several hundred nucleotides, which would have escaped detection by the techniques applied to date.

The mature products of viral RNA synthesis in permissive cells infected with ASV are three species of cytoplasmic RNA (Figure 4), each of which has chemical polarity identical to that of the viral genome (148, 149). The largest species has a molecular weight of 3.3×10^6 (ca. 10,000 nucleotides), a sedimentation coefficient of 38S, encodes all four viral genes, and may be identical to the viral genome. A second species has a molecular weight of 1.8×10^6 (ca. 5400 nucleotides), a sedimentation coefficient of 28S, and encodes the 3' half of the viral genome, including *env, src* and "c". The smallest species has a molecular weight of 1.2×10^6 (ca. 3600 nucleotides), a sedimentation coefficient of 21S, and encodes only *src* and "c". Since all three species are polyadenylated (148, 149) and are associated with polyribosomes (J. Lee, personal communication), all may serve as messengers for the synthesis of viral proteins. In addition, a portion of the population of 38S RNA must represent the pool of RNA which is encapsidated to form progeny virions.

Cells producing virus strains devoid of either *src* (e.g. tdASV, avian leukosis viruses, and MuLV) or *env* (e.g. replication-defective deletion mutants of ASV) contain only two species of cytoplasmic viral RNA: one species is the length of the viral genome, the other is approximately the size of the 21S RNA described above and encodes either *env* and "c" if the virus lacks *src*, or *src* and "c" if the virus lacks *env* (148, 149).

The smaller viral RNAs could be generated by any of three possible mechanisms: 1. A single precursor molecule could be processed to produce each of the viral mRNAs. As discussed below, provisional evidence for the transposition of nucleotide sequences during the genesis of viral mRNAs conforms to this possibility. The suggestion has been made that a viral structural protein may regulate the production of viral mRNAs in the infected cell: binding of pp19 to the genome of ASV apparently shields the RNA against hydrolysis by nucleases that normally cleave the genome into several smaller molecules (30). However, the precision of these cleavages has not been adequately characterized, and the role of processing in the

THE EXPRESSION OF AVIAN SARCOMA VIRUS GENES
POSTULATED SCHEME

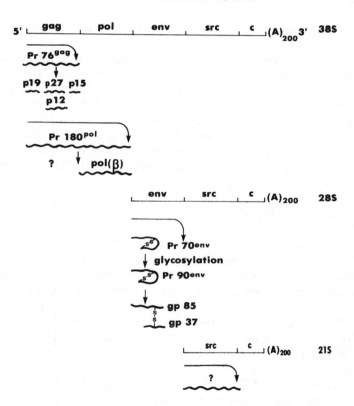

Figure 4 The strategy of retrovirus gene expression. Messenger RNAs, and primary and mature gene products are illustrated for ASV. Similar principles probably apply to other retroviruses.

genesis of viral mRNAs remains speculative. 2. Promoters within the provirus could be used to generate the individual viral mRNAs; if so, promoters for the smaller species of viral mRNAs would have to be transcribed during the synthesis of the longer species, but precedents for this do exist (150). 3. Separately integrated fragments of viral DNA might provide templates for the different species of mRNA. This seems an awkward and unnecessarily complex mechanism, but no satisfactory test for its existence has been performed.

All three species of ASV-specific cytoplasmic RNA contain nucleotide sequences at least partially homologous to a sequence of 101 nucleotides located at the extreme 5' end of the ASV genome, and this sequence is apparently not reiterated within portions of the viral genome that encode the 28S and 21S virus-specific RNAs (148). A number of separate mRNAs of adenovirus share a common 5' terminal sequence of 150–200 nucleotides, which is apparently "spliced" onto the end of the mRNAs during or after transcription (151, 152). A similar process may occur during the genesis of the 28S and 21S mRNAs of ASV. If this supposition is correct, all the mRNAs of ASV are probably generated by initiation of transcription at the same site, and "splicing" may prove to be a common feature in the production of mRNAs in eukaryotic cells. The function of the shared nucleotide sequences on viral mRNAs is not known.

The nucleus and cytoplasm of cells producing both ASV and MMTV, and the virions of ASV, contain extremely small amounts of RNA complementary to at least 40–60% of the viral genome (153). The significance of this observation is presently unknown.

Glucocorticoid Hormones and the Synthesis of MMTV RNA (154)

Glucocorticoids stimulate the production of MMTV in a variety of host cells (154). Stimulation is mediated by the same hormone-receptor complex that mediates other cellular responses to glucocorticoids (155, 156); does not require the synthesis of either cellular DNA or cellular protein (155, 157); is consequent to an increased rate of synthesis of virus-specific RNA (158, 159); and is restricted to the synthesis of MMTV RNA in cells doubly infected with MMTV and MuLV (158). It is clear that cells infected with MMTV offer an exceptionally accessible experimental system for dissecting the mechanisms by which steroid hormones modulate gene expression.

The provirus of MMTV can integrate at multiple sites on cellular DNA in newly infected cells; integration at some of these sites permits the synthesis of viral RNA, whereas integration at other sites does not (G. Ringold, personal communication). In every instance, the integrated viral DNA apparently duplicates the linear order of the viral genome (G. Ringold, personal communication). These observations raise the possibility that the cellular site of provirus integration is a determinant of viral gene expression, perhaps by providing a promoter for the synthesis of viral RNA.

Despite the apparent multiplicity of integration sites, any constitutive level of MMTV RNA synthesis is usually stimulated by glucocorticoids (157). Either the viral genome itself carried a steroid-responsive control element, or many, if not all, of the integration sites are located within regions of the host genome governed by such an element.

The Strategy of Retrovirus Gene Expression

The mechanisms that generate the final gene products of retroviruses are summarized in Figure 4, using ASV as a prototype; available data indicate that similar principles will apply to other retroviruses. Three of the four ASV genes are translated from mRNAs that contain the expressed gene at the 5' end: *gag* from 38S RNA, *env* from 28S RNA, and *src* from 21S RNA. In contrast, *pol* is probably expressed by the uninterrupted translation from *gag* and *pol* in 38S mRNA. (The mRNAs for *gag* and *pol* have been identified only according to their size; it is possible that they are chemically distinguishable from each other and from the viral genome.) This scheme conforms to previous suggestions that translation in eukaryotic cells initiates only at the 5' termini of mRNA (160), and it provides for the independent expression of viral genes according to the relative need for gene products; similar strategies have been described previously for both plant (161) and animal (162) viruses.

The primary product of translation from *gag* is a polyprotein that is subsequently cleaved to generate the individual polypeptide constituents of the viral core (163, 164). The linear order of the individual core proteins within *gag* (Figure 4) is roughly homologous between ALSV and MuLV (163, 165), although p15 may be located at the amino terminus in MuLV (166) rather than at the carboxy terminus, as in ALSV (163). Cleavage of the polyprotein fails to occur in certain host cells and is therefore probably initiated by cellular enzymes (167). However, enzymes associated with virions of both ASV (168) and MuLV (169) can cleave the precursor protein into the correct products; these findings conform to the fact that some or all of the processing of the precursor may take place in immature virions outside the cell (164, 170, 171). The origin of the virion-associated processing enzymes is in dispute; a major, virus-specified structural protein of ALSV (p15) is alleged to possess the necessary proteolytic activity (168), whereas the protease activity associated with MuLV apparently resides on a previously unrecognized protein of unknown origin (169).

In at least some instances, a fraction of the *gag* polyprotein finds its way to the surface of the cell in a glycosylated form that is neither cleaved nor incorporated into virions (172, 173). This form of the *gag* polyprotein is responsible for important antigenic determinants on the surface of leukemic cells in mice and can arise from either the expression of endogenous viral genes or the production of exogenous virus (172).

Direct evidence indicates that 38S RNA serves as messenger for the synthesis of the polyprotein encoded by *gag:* 1. Polyribosomes precipitated with antisera against one of the core proteins contain 38S virus-specific RNA (J. Lee, personal communication). A similar finding has been reported for MuLV (174), although the mRNA is somewhat smaller by virtue of the absence of *src.* 2. Pr76^{gag} can be synthesized in cell-free systems using as messenger either the genome of ASV (175, 176) or 38S RNA isolated from ASV-infected permissive cells (177); again, analogous findings have been reported for MuLV (178, 179). Synthesis of Pr76^{gag} in vitro with the genome of ASV as messenger can be inhibited by m^{7}G$^{5'}$p (180); this finding implicates the capping group at the 5' terminus of the genome in the mechanism of translation (181).

A virus-specific RNA with the size and composition expected for a *pol* messenger has not been identified (148, 149). Instead, *pol* is apparently expressed by the uninterrupted translation from *gag* and *pol* in 38S mRNA. The product of this read-through translation has a molecular weight of 180,000 (pr180^{pol}) and contains the antigenic determinants and tryptic peptides of both the *gag* and *pol* proteins (182). Pr180^{pol} has been found in virus-infected permissive (182, 183) and nonpermissive (H. Oppermann, personal communication) cells and can be synthesized in vitro with either the viral genome (178, 184) or 38S RNA isolated from infected permissive cells (A. Ullrich, personal communication) as messenger. The results of kinetic analyses indicate that Pr180^{pol} cannot be a precursor to any of the mature *gag* gene products (182), but its turnover is coordinate with the appearance of mature reverse transcriptase in virus particles (182). The polypeptides of mature reverse transcriptase may not occur within infected cells (182); rather, Pr180^{pol} is present in extracellular virions where it gradually decays in concert with the appearance of the mature polymerase (182). In summary, the available data suggest but do not prove that reverse transcriptase is generated by the processing of a polyprotein precursor, and that this maturation occurs in extracellular virus.

Read-through translation in prokaryotes results from suppression of a termination signal by a normal cellular tRNA (185). The discovery of Pr180^{pol} provides evidence that a similar process may be functionally significant in eukaryotic cells. The synthesis of Pr180^{pol} in vitro with the genome of MuLV as messenger can be substantially augmented by the use of an *amber* suppressor tRNA isolated from yeast (186). This finding identifies the signal that terminates translation from *gag,* but it does not establish the mechanism by which Pr180^{pol} is synthesized in the infected cell; *amber* codons are not usually bypassed by means other than specific suppressor tRNAs, and read-through of amber codons is therefore a very rare event in normal cells. It is also possible that the mRNAs used to synthesize Pr76^{gag} and Pr180^{pol} are not identical, but, rather, that processing mechanisms occasionally delete the termination codon for *gag*, thus generating a specific mRNA for Pr180^{pol}. In any event, the synthesis of Pr180^{pol} consequent to suppression of the *gag* terminator in vitro suggests that *pol* is located adjacent to the 3' terminus of *gag* and therefore substantiates the order of these genes as deduced from chemical analysis of recombinant viral genomes (see above).

Messenger RNAs for *env* of various retroviruses have been identified by translation in vitro (177) and in *Xenopus* oocytes (179), and by microinjection of cells infected with a deletion mutant in *env* (97). The primary product of translation from *env* is probably a protein with a mol wt of ~70,000 (Pr70^{env}), detected by short pulses of radioactive amino acids (187, 188) accumulated in infected cells in the presence of an inhibitor of glycosylation (164, 188), and translated from *env* mRNA in either a cell-free system (177) or oocytes of *Xenopus laevis* (179). Pr70^{env} may itself contain some carbohydrate residues (188), but further glycosylation generates a second form, Pr90^{env} (188–190); the apparent increase in molecular weight results from the effect of carbohydrate residues on the electrophoretic mobility of the polypeptide. Pr90^{env} is then cleaved, perhaps with concomitant further glycosylation, to produce the mature envelope glycoproteins: gp85 and gp37 of ASV, gp70 and p15(E) of MuLV. Cleavage may be coordinated with migration of the glycoprotein to the surface of the infected cell (20).

The 21S virus-specific RNA in cells infected with ASV encodes only *src* and "c," and appears likely to be the messenger for the transforming protein. Recent progress towards the identification of the product of *src* has provided evidence substantiating this supposition (see below).

On occasion, virus-specific proteins large enough to represent uninterrupted translation from the entire viral genome have been detected in infected cells (164). These proteins may contain antigenic determinants encoded in more than one viral gene, but their precise composition and their biological significance are not known. They have attracted at least passing attention because some animal virus genomes are expressed by the synthesis of a single polyprotein that contains all of the viral gene products (191).

Phosphorylation and the Function of Viral Gene Products

Phosphorylation of proteins plays an important role in the regulation of biological function (192); retroviruses offer no exception to this rule. Several gene products are phosphoproteins, and provisional evidence indicates that phosphorylation may regulate the function of at least two of these proteins. First, the ribonucleoprotein of retroviruses contains a protein (pp19 in ALSV, pp12 in MuLV) that binds to the viral RNA with great affinity and specificity (29). Phosphorylation of this protein determines its affinity for viral RNA (31) and could therefore modulate whatever function is served by the binding (e.g. regulation of the production of viral mRNA; see above). Moreover, the protein is apparently phosphorylated while still part of the polyprotein product of *gag* (193); perhaps addition of the phosphate residues influences processing of the precursor protein. Second, phosphorylation of reverse transcriptase (194), or of some constituent of its holoenzyme (195), apparently stimulates enzymatic activity (195, 196). The significance of these observations is presently indeterminate. Third, the putative product of *src* is phosphorylated (see below); this protein has been identified only recently and nothing is yet known of its function.

Retroviruses contain both phosphatase and protein kinase activities. The kinase can phosphorylate a number of viral structural proteins in vitro (197), but most of these are not normally phosphorylated in mature virus (198, 199). Hence, phosphorylation during virus production appears to be selective, and the association of phosphatase and kinase activities with virions may be adventitious.

Maturation of the Virus Particle

The viral core assembles in the cytoplasm of the infected cell and acquires an envelope by budding through the plasma membrane of the host cell. Virus particles can be assembled and enveloped without the participation of either the viral genome (200) or viral glycoproteins (201). However, maturation of the virion continues in extracellular virions: 1. The morphology of the viral core changes (26). 2. The polyprotein product of *gag* may be a major component of immature virus particles (164, 170), and cleavage of this precursor protein into the mature polypeptide constituents of the viral core may be coordinated with steps in virus maturation (170, 171). 3. The subunits of the viral genome are assembled into a stable dimer (202, 203). 4. A number of specific polypeptides of uncertain origin may be shed from the virus (204). 5. Pr180[pol] may be processed to yield the mature form of reverse transcriptase (182).

HOST RESTRICTIONS ON VIRAL REPLICATION

The host range of retroviruses can be a valuable taxonomic device and an important genetic marker. Host cells restrict viral replication by either of two means: exclusion of viral entry into the cell, and constraints on intracellular events responsible for viral gene expression.

Host Range Determined by Exclusion of Viral Entry

The susceptibility of various permissive avian cells to infection by different strains of ALSV is determined by the presence or absence of cellular receptors that permit attachment and pentration of the virus particle (8). In chickens, susceptibility to infection is a dominant trait specified by at least three separate genetic loci (8). Virus particles adsorb normally to resistant cells, but penetration and uncoating of the particles fails to occur (8). Exclusion of viral entry is also the basis for the differences in host range that distinguish the ecotropic, xenotropic, and amphotropic groups of murine leukemia viruses (205). Barriers to viral entry can be breached by artificially fusing virions into the cell (e.g. with inactivated Sendai virus) (206) or by preparing phenotypically mixed virus (often called "pseudotypes") whose envelope contains glycoproteins derived from another virus that is not excluded by the host cell in question (207).

Infection of nonpermissive mammalian cells by ASV is restricted at the cell surface (208), although the extent of restriction varies significantly among different strains of virus (208). Circumvention of the surface restriction increases the number of cells transformed by virus infection (208), but a major fraction of the infected cells still resists transformation, and no virus is produced. Hence, the expression of ASV genes in nonpermissive cells is also constrained within the host cell.

Host Range Determined by Intracellular Restrictions

Mammalian cells infected with and transformed by ASV usually produce no virus (75), although the complete viral genome is present as integrated provirus (130) and can be recovered by either fusion of the transformed cells with permissive cells (129) or transfection with DNA extracted from the transformed cells (81). Replication of ASV in mammalian cells is defective in several regards. First, the amounts of virus-specific RNA synthesized are 100- to 10,000-fold lower than those produced in permissive cells (140), and the quantities of viral gene products are also reduced (209). Second, some (albeit not all) lines of ASV-transformed mammalian cells fail to synthesize viral glycoprotein (210), and these cells are deficient in the 28S viral mRNA alleged to be the messenger for *env* (140; N. Quintrell, personal communication). Third, the polyprotein product of *gag* is not processed into core proteins (167, 211). Recently, inefficient but detectable replication of ASV in mammalian cells has been achieved by infection with phenotypically mixed stocks of MuLV (capable of independent replication in the mammalian cells) and ASV (212; J. Levy, personal communication), but the mechanism of this phenomenon has not been elucidated.

Many mammalian cells infected by ASV acquire integrated provirus for the entire viral genome without consequent neoplastic transformation (129,

130). This may be the most common outcome of ASV entry into mammalian cells, but the mechanism that prevents cellular transformation is not known.

Selective passage of a given strain of ASV through resistant cells apparently augments the efficiency with which that virus infects the same cells (208, 213). Presumably, selective pressure enriches the virus stock for genetic variants with exceptional tropism for the normally resistant cells, and this tropism is probably effected at the point of viral entry into the cell (208, 213).

A single genetic locus (Fv-1) is the major cellular determinant for susceptibility of mouse cells to infection by ecotropic MuLV (10). The host range of these viruses is defined in terms of the susceptibility of cells from two prototype strains of mice: N-tropic viruses replicate normally in cells of NIH-Swiss mice but not in cells of BALB/c mice, whereas B-tropic viruses have the opposite host range (10). Viruses with dual tropism (i.e. infectious for both N and B cells) and dual restriction (i.e. infectious for neither class of cell) can be isolated (10, 214); dual tropism is a genotypic property of the virus, whereas dual restriction results only from phenotypic mixing of viral structural proteins and has not been observed as a genotype (214). Resistance to infection is dominant (10, 215), is far from absolute (10), occurs within the cell rather than at the cell surface (216, 217), and is mediated by interaction between a cellular factor (218) and a viral protein contained within the virion (214)—possibly p30 (219). The mechanism that restricts viral replication is only partially understood; viral DNA fails to integrate (135, 141), possibly as a consequence of synthesis of faulty viral DNA (W. K. Yang, personal communication), and later events in the viral life cycle (such as synthesis of viral RNA) do not occur (142). If these observations are all correct, they implicate a structural protein of the virus in the synthesis of biologically active viral DNA.

The replication of certain strains of ALSV in permissive cells can also be restricted by an intracellular mechanism (220, 221). Provisional evidence indicates that the restrictive mechanism acts at a point subsequent to the synthesis of viral DNA (S. Hughes, personal communication).

GENETICS OF RETROVIRUSES (222)

The study of retrovirus genetics has made major contributions to the understanding of viral replication and virus-induced neoplastic transformation. In this brief summary, I review the types of viral mutants available for study, the biochemical lesions identified to date, and the genetic interactions among retroviruses. For further details, I refer the reader to the comprehensive and exceptionally comprehensible review by Vogt (222).

Nonconditional Mutants of Retroviruses

Nonconditional variants of retroviruses occur naturally and can also be induced by several forms of mutagenesis. To date, virtually all of these variants have proven to be deletion mutants that affect either the replication of the virus or virus-induced neoplastic transformation.

Nonconditional defects in viral replication account for the classical helper-dependence of many virus strains (222); the lesions can be complemented by gene product(s) specified by another virus. Replication-defective strains of ASV may have deletions in env (201, 223), pol (224), or both (201, 225). All other known sarcoma viruses are nonconditionally defective in replication. The strains of murine sarcoma virus (MuSV) studied to date bear deletions in both pol (226) and env (227); some strains are also apparently deficient in gag, other strains are not (227, 228). The deletions in MuSV are extensive and generate an exceptionally small viral genome (haploid subunit of mol wt $\sim$ 1.9 X 10^6) (65, 229).

Some strains of avian and murine leukemia viruses are also defective in replication (222; A. Rein, personal communication), but the lesions in most of these viruses have not been characterized. The genome of MC-29 virus has an extensive deletion (haploid subunit of mol wt $\sim$1.9 X 10^6; K. Bister, personal communication) that apparently affects at least parts of gag, pol, and env (230; D. Sheiness, personal communication). The genome of avian myeloblastosis virus is probably deficient in at least env (231).

Some strains of ASV segregate deletions in src at a high frequency (232), and similar deletions have allegedly been induced by mutagenesis (58). The deletions can affect either part (233) or possibly all (59, 60, 234) of the gene; all of the deletions of src mapped to date probably have common 3' termini irrespective of their total extent (233). Deletions in src abolish the ability of the virus to transform fibroblasts but have no effect on viral replication and are therefore denoted "transformation defective" (tdASV). At least some strains of tdASV can induce lymphoid leukosis in birds (58) and are therefore analogous to the naturally occurring avian leukosis viruses. Some of the partial deletions of src retain enough of the gene to generate wild-type virus when recombined with temperature-sensitive mutants of src (235) and to induce cell division in the absence of other phenotypic manifestations of neoplastic transformation (S. G. Martin, personal communication).

To date, deletions of src have been identified only in avian sarcoma viruses. Analysis of all other sarcoma viruses is complicated by their requirement for a helper virus to complement defects in replication. Nonconditional point mutations in src have been isolated from cells transformed by MuSV and then treated with a chemical mutagen (236); the mutations

were recognized because the mutagenized cells reverted to a normal phenotype and released transformation-defective MuSV when superinfected with
a helper virus.

Conditional Mutants of Retroviruses

All conditional mutants of retroviruses isolated to date are temperature-
sensitive; some affect only the transforming capacity of the virus (T class),
some affect only viral replication (R class), and some affect both transformation and replication coordinately (C class). The majority of mutants produced by present protocols for mutagenesis contain multiple lesions, which
must then be sorted out by further genetic analysis (222). Mutations affecting a viral function during the first 12 hours of infection are considered
"early," those affecting functions later in the life cycle are considered "late;"
this division is arbitrary but suitable since it probably divides the viral life
cycle into the events that occur before and during the synthesis and integration of viral DNA, and those events that occur thereafter.

T-class mutations have been identified only in *src* of ASV (22, 237) and
MuSV (22, 238); no other form of viral oncogenesis, such as leukemogenesis
or carcinogenesis, has been successfully mutagenized to date. Temperature-
sensitive lesions in *src* affect both the initiation and maintenance of virus-
induced transformation of fibroblasts (222, 237, 238). The effect of
transformation can be reversed by shifting from the restrictive to the permissive temperature (237), and this reversal requires only protein synthesis
(239) or no macromolecular synthesis at all (240). Some mutants affect all
of the phenotypic characteristics normally associated with neoplastic transformation of cells in culture; other mutants affect only a portion of the
transformed phenotype, leaving the remainder unsusceptible to the restrictive temperature (241). The capacity to dissociate various elements of the
transformed phenotype experimentally should prove helpful in studying the
mechanism of neoplastic transformation.

All R-class mutations by nature affect only the late stages of viral replication; lesions affecting the early portion of the cycle interdict viral gene
expression completely and therefore affect both viral replication and cellular transformation. Temperature-sensitive lesions in replication identified to
date include mutations that affect the processing of the polyprotein product
of *gag* (242–244), mutations that prohibit the appearance of active reverse
transcriptase in extracellular virions (245), mutations that prevent the incorporation of viral glycoprotein into mature virus particles but do not
interfere with the synthesis of viral glycoprotein in the infected cell (246),
and mutations that prevent the release of virus particles from the surface
of the infected cell (247).

Two types of C-class mutants have been identified; both involve functions required early in infection: 1. Temperature-sensitive mutations in *pol* prevent the synthesis of viral DNA at the restrictive temperature, a defect that prohibits all viral gene expression (78, 248, 249). 2. Certain mutations apparently render the viral glycoprotein nonfunctional at the restrictive temperature (222); this defect makes the infectivity of the virion abnormally labile at the restrictive temperature, blocks viral entry into the host cell, and therefore affects both viral replication and cellular transformation. Initial descriptions of these glycoprotein mutants erroneously claimed that the lesion affected only the establishment of cellular transformation (250, 251), not viral replication, and prompted speculation that an early virus function was concerned uniquely with transformation (251). Failure to detect a lesion in viral replication may underlie a more recent report of "early" T-class mutants of ASV (252).

Genetic Interactions Among Retroviruses

Mixed infections with different strains of retroviruses may result in three forms of genetic interaction: formation of heterozygous particles (38, 39), complementation of defective or deficient gene products (222), and formation of stable recombinant progeny (253, 254). Formation of heterozygotes and genetic recombination allegedly occur at exceptionally high frequencies. All complementation characterized to date is consequent to phenotypic mixing in virus particles.

Putative heterozygotes have been demonstrated with both ALSV (38) and MuLV (39) and are attributable to the fact that the viral genome is a diploid dimer. Heterozygous alleles in single particles are reputed to segregate independently in progeny virus (38, 127b), which suggests that the two subunits of the genome are independently expressed. The search for heterozygotes is subject to subtle experimental artifacts (38, 222), and the reported observations on heterozygosis remain controversial. Efforts to demonstrate heterodimer genomes (the anticipated physical correlates of heterozygosis) by electron microscopy have not succeeded to date (255), although no demonstrably heterozygous virus has yet been examined. The production of multiploid particles (containing two or more diploid genomes) offers an alternative explanation for genetic heterozygosis that deserves further consideration (247).

Recombinant retroviruses are formed by crossing over between viral genomes (256), often at multiple points in the genomes (61, 257). The mechanism of recombination has not been elucidated. Formation of heterozygotes and a subsequent second round of viral replication are allegedly prerequisites for recombination (38, 127b); this is an attractive hypothesis because topographical linkage of recombining viral genomes might account for the remarkably high frequency of recombination. However, the hypothesis remains unproven, and the possibility that recombinants can form during a single round of viral replication following mixed infection deserves further scrutiny. Provisional evidence indicates that the genetic map of ASV may be circular, i.e. that *src* is linked to *gag* during recombination (258). If correct, these findings would suggest that recombination occurs between circular molecules of nucleic acid. Unfortunately, efforts to generate a reliable genetic map by three-factor crosses have been unsuccessful to date (222).

Temperature-sensitive mutants in *src* of ASV can be sorted into four distinct groups by recombination; members of different groups can recombine to yield wild-type virus, whereas members of the same group cannot (127b). The crosses that produce wild-type virus were at first thought to represent complementation (259) and are now denoted "cooperative transformation" (127b), but it is clear that the progeny viruses are genuine recombinants (127b). Several specific crosses are either extremely infrequent or simply do not occur: recombination has not been observed between the *src* genes of different isolates of ASV (259), despite the apparent homology of these genes, as demonstrated by analysis with molecular hybridization (260); deletions in *env* generally cannot be replaced by recombination with leukosis viruses (38, 222); the deleted genes of MuSV have never been successfully reconstituted by recombination with helper virus (222); and recombination has not been demonstrated between retroviruses derived from different host species (such as ALSV and MuLV). Some of these observations could be explained by a requirement for homologous nucleotide sequences at sites of recombination.

Genetic defects in *gag, pol,* and *env* have all been complemented by mixed infection with virus capable of providing a competent gene product. *Env* is complemented by incorporation of functional glycoprotein into a phenotypically mixed virion (222). *Pol* cannot be complemented in *trans,* i.e. in mixed infections with mutant and wild-type viruses (261). Reverse transcriptase must first be incorporated into virions by phenotypic mixing in order to obtain a successful infection with the mutant genome. Apparently, the enzyme must be topographically linked to the genome it will transcribe. Temperature-sensitive mutations in *gag* have been complemented in mixed infections (243), but the precise nature of the physiological defect in these mutants is not known and the mechanism of complementation is consequently indeterminate. *Env* of a particular strain of retrovirus can be complemented by the glycoprotein of an entirely unrelated retrovirus (262, 263), or by the glycoprotein of another family of enveloped viruses [e.g. vesicular stomatitis virus (207, 264)]. Complementation of *gag* and *pol* of ALSV has been observed only between related strains of virus (262; K. Bister, personal communication), i.e. viruses with genomes that share homologous nucleotide sequences. By contrast, the multiple genetic deficiencies of MuSV can be complemented by unrelated helper viruses.

Simultaneous infection of cells with mutant virus and fragments of provirus DNA for wild-type virus can rescue the mutant allele and generate wild-type virus (63). The mechanism of marker rescue by transfection is not known, but the procedure should facilitate further characterization of viral gene functions and the consolidation of viral genetic maps.

ENDOGENOUS VIRUSES (4, 5)

The DNAs of many species carry the provirus for one or more strains of retrovirus; the genes of these endogenous viruses originated at some point in the phylogenetic histories of the species and have since evolved in concert with the remainder of the germ line DNA (265). Endogenous viruses may have been generated within the cellular genome as part of normal speciation, but during the course of phylogeny some endogenous viruses have been horizontally transmitted from one species to another and have become established in the germ line of the recipient species (7). However, infection of the germ line is not a necessary consequence of horizontal transmission. For example, viruses found in, but not endogenous to, gibbon apes and woolly monkeys apparently originated as endogenous viruses in Asiatic mice (7). At least some animal species carry two or more strains of endogenous viruses whose genomes share no detectable homology (7, 266); this could result from either horizontal transmission between ancestral species or the independent genesis of separate viral genomes in the same ancestral species.

Expression of the genes of endogenous viruses is controlled by several classes of genetic determinants: 1. The provirus itself segregates as a genetic locus that determines the capacity of the cell to produce endogenous virus and that can be traced through the progeny of genetic crosses with molecular hybridization (266, 267). 2. Cellular genetic determinants that are separate from the endogenous provirus can regulate the expression of viral genes (4, 5). In chickens, one of these determinants may be a *cis*-active regulator closely linked to the endogenous provirus (268). Other regulatory elements in chickens are probably not linked to the endogenous provirus and can specify either coordinate (269, 270) or noncoordinate (270, 271) expression of separate viral genes (such as *gag* and *env*) by regulating the relative amounts of virus-specific RNA encoding the expressed viral genes (272). The mechanism of this regulation has not been elucidated, but available evidence suggests that the amounts of mRNAs for different viral genes can be modulated independently (273; B. Baker, personal communcation). Similar genetic determinants of endogenous viral gene expression have been described in mice (5), but the mechanism of regulation has not been characterized. 3. Many retroviruses cannot reinfect cells derived from the species in which the viruses originated (7, 9). Resistance to reinfection prohibits the spread of virus and vastly reduces the likelihood that release of endogenous virus from an occasional cell can be detected. Consequently, the search for new endogenous viruses requires the use of cells from multiple species in the hope of providing a susceptible host that can amplify the virus to detectable levels (7, 9).

The expression of otherwise cryptic endogenous viruses can be induced by a variety of agents, including halogenated pyrimidines (274), chemical carcinogens (275), ionizing radiation (275), inhibitors of protein synthesis (276), and immunologically mediated events (277). Different strains of endogenous viruses may respond to different classes of inducers (9), but in no instance has the mechanism of induction been fully described. Induction by halogenated pyrimidines requires incorporation of the analogue into cellular DNA (278) during the middle of the S phase of the cell cycle (279); the incorporated analogue apparently induces or augments transcription from the endogenous provirus (279). Inhibitors of protein synthesis might block the production of an unstable repressor of transcription from viral genes (276, 280), but there is no direct evidence for this view.

Endogenous viruses constitute an important genetic backdrop to exogenous infection by retroviruses. Deletions in *env* can be complemented by phenotypic mixing with the product of an endogenous viral gene (206, 281), and the genomes of endogenous and exogenous viruses can recombine (38, 282).

RETROVIRUSES AND EVOLUTION

Many virologists now believe that viruses originated as genetic elements of normal cells that acquired autonomy during the course of evolution (283, 284). Hence, viral systems may provide clues to the existence of presently undiscovered normal cellular mechanisms. Retroviruses offer the most tangible illustration of this prospect. For example, Temin has proposed (285, 286) that RNA-directed DNA synthesis, first discovered in retroviruses, is a fundamental mechanism in normal cellular metabolism, allowing for rapid mutation and rearrangement of genes and for transfer of genetic information between cells. Occasionally, these processes may generate retroviruses that escape from the cell to assume an autonomous existence (285, 286). Hence, the genesis of retroviruses could be a continuous process in all species that draws upon cellular DNA to create new viral genes. The evidence for these views (the "protovirus hypothesis") remains circumstantial; in particular, the occurrence of RNA-directed DNA synthesis in normal cells has not been conclusively demonstrated.

The ancestral relationship between endogenous and exogenous retroviruses is uncertain. Viral genes established in germ line DNA may be highly conserved during the course of evolution (265); immunological data suggest that endogenous viruses of rodents and primates are the products of divergent evolution from a common ancestral virus (16a). By contrast, exogenous retroviruses are alleged to evolve at an uncommonly rapid rate (265), perhaps as a consequence of the exceptional error frequency of reverse transcriptase (287, 288) or of other mechanisms that could account for the remarkably high frequency of spontaneous mutations in retroviruses (213). It is presently impossible to discern whether endogenous viruses engendered exogenous viruses (e.g. as in the "protovirus hypothesis") or whether present endogenous viruses might all derive from infection of ancestral germ lines, but strong arguments have been raised against the latter possibility (285).

The ubiquity and apparent conservation of endogenous retrovirus genes have prompted the suggestion that these genes are preserved by selective pressure because they serve a useful purpose in their hosts (7, 265). One popular idea is that the products of envelope genes of endogenous viruses serve as surface markers during cellular differentiation (23), but other studies have failed to implicate the expression of endogenous viral genes in embryonic development (289).

The horizontal transmission of bacterial genes by extrachromosomal elements may have played a major role in the evolution of prokaryotes (290). Similarly, the ability of retroviruses to move from the germ line of one species to that of another has prompted the suggestion that these viruses have been a force in the evolution of vertebrates (7, 290). This proposal seems premature. According to available data, transduction by retroviruses is limited almost entirely to viral genes acquired from the genomes of endogenous viruses (38, 72).

GENETIC DETERMINANTS OF VIRUS-INDUCED NEOPLASTIC TRANSFORMATION

Retroviruses that transform fibroblasts in cell culture and induce sarcomas in animals carry a gene (*src*) that has been persuasively implicated in neoplastic transformation (222). Otherwise, the role of retroviruses in oncogenesis remains poorly characterized; for example, efforts to identify a viral gene responsible for the induction of leukemia have been unsuccessful despite the availability of assays that permit transformation of the appropriate target cell in culture (231, 291, 292).

The host cell carries genetic determinants that can appreciably influence susceptibility to specific forms of virus-induced oncogenesis. These determinants have been reviewed recently (10) and are not discussed here.

Virus-Induced Transformation of Fibroblasts: Src

Expression of *src* in fibroblasts induces a variety of phenotypic changes associated with neoplastic transformation (82). Two apparently competing hypotheses have been proposed to explain these changes: one hypothesis proposes that the product of *src* acts in the nucleus to drive the cell through repeated replications in a manner analogous to the putative action of the transforming gene product of papovaviruses (293); the other hypothesis suggests that *src* acts in the cytoplasm, possibly at the cell surface. Studies with enucleated cells indicate that *src* can act in the cytoplasm to change the morphology of the infected cell (294), but expression of the viral gene can also evoke cellular DNA synthesis (295) and mitosis (296), and the mitogenic effect of *src* has been delineated from other functions of *src* with nonconditional mutants (G. Calothy, personal communication). The response to *src* is pleiotropic; the primary event or events that initiate the response remain unidentified.

The *src* genes of ASV and of other strains of sarcoma virus can transform the same target cells, yet the nucleotide sequences that encode these genes are apparently unrelated (260). Whether the genes act by similar or different means remains to be shown. One current proposal suggests that the products of *src* genes might be analogues of the small polypeptide growth factors that stimulate cell division after binding to receptors on the cell surface (297, 298). Cells transformed by MuSV and FeSV (but not cells transformed by ASV, papovaviruses, or chemical carcinogens, nor cells infected with leukemia viruses) lack receptors for epidermal growth factor (EGF); receptor activity is lost during the first 48 hours following infection, coincident with the appearance of the transformed phenotype. Receptors for another growth factor, multiplication-stimulating activity (MSA), and for insulin and concanavalin A, are not disturbed (298). These findings led to the suggestion that the receptors for EGF are blocked by *src* gene product produced in the virus-transformed cells, and that binding of the analogue results in an enduring stimulation of cell growth. This proposal is attractive because it invokes a mechanism known to mediate pleiotropic effects resulting in cell division, but there are reasons for scepticism: the loss of receptor activity may just as likely be consequent to transformation; some of the major phenotypic consequences of transformation cannot be attributed to

EGF; and the molecular weight of EGF (6045) is only one-tenth that of the protein alleged to be encoded in *src* of ASV (see below).

A provisional identification of the protein encoded in *src* of ASV has been achieved by using antisera from mammals bearing sarcomas induced by the virus (299). The putative product of *src* is a phosphoprotein with a mol wt of ~60,000; it is present in cells transformed by ASV, but not in cells infected with a deletion mutant of *src,* and is extremely unstable in cells infected with a temperature-sensitive mutant in *src* and incubated at the restrictive temperature. An identical protein can be synthesized in vitro with either appropriate fragments of the ASV genome or 21S RNA (the putative mRNA for *src*) from ASV-infected cells as messenger (184; R. Erikson, personal communication). Surprisingly, the proteins specified by *src* of two different strains of ASV are not immunologically cross-reactive (R. Erikson, personal communication). This finding contrasts with the report that the nucleotide sequences encoding *src* in all known strains of ASV are closely related when analyzed by molecular hybridization (260), but it conforms to the failure of different strains of ASV to recombine within *src* (259).

Other investigators have attempted to identify the product of *src* by using cell-free systems to translate fragments of the ASV genome that should encode only *src* and "c" (180, 300). The two polypeptides (mol wt ~25,000 and 17,000) attributed to *src* with this approach have not been further characterized; they could be the products of premature termination, initiation within "c," or other experimental artifacts.

The transforming gene of ASV is closely related to nucleotide sequences (here denoted *sarc*) present as a few copies in the DNA of normal chickens (301), the putative host of origin of the virus. Highly conserved DNA at least partially homologous to *src* has also been found in other birds (301), and in fish, *Homo sapiens,* and other vertebrates (302). The conservation of *sarc* over large phylogenetic distances suggests that it serves a necessary function in normal cellular metabolism and, in fact, *sarc* is transcribed into a large (~6000 nucleotides) mRNA in both normal and chemically transformed avian cells (273, 303, 304). However, the amounts of *sarc* RNA in both normal and neoplastic cells are very small (1–10 copies per cell), and efforts to identify the functional significance of this RNA have been unsuccessful (303). Available evidence indicates that the nucleotide sequences of *sarc* are not linked to any of the identified endogenous retrovirus genes in the chicken genome (304, 305).

The *src* genes of various isolates of MuSV are also allegedly related to DNA in the species of origin, although the definition of nucleotide sequences encoding the transforming genes of MuSV remains circumstantial (222). The *sarc* related to the Moloney strain of MuSV (derived in mice) is present as a few copies in a highly conserved class of mammalian DNA (306, 307), whereas the *sarc* related to the Kirsten and Harvey strains of MuSV (derived in rats) may reside on the genome of a defective endogenous virus that is encoded in multiple copies in rat DNA but is not found in related species of rodents (308). Although the Kirsten and Harvey strains are separate isolates, the nucleotide sequences that constitute their putative *src* genes are indistinguishable by molecular hybridization (309).

The *sarc* of Moloney MuSV is apparently not transcribed into RNA (306). By contrast, transcription of the *sarc* of Kirsten MuSV occurs spontaneously in some lines of rat cells (308, 310), can be induced with halogenated pyrimidines in others (308), and is also detectable in both normal and neoplastic tissues from adult rat (311). The *sarc* RNA found in rat cells contains ~6000 nucleotides (310) and apparently functions as mRNA (310).

The preceding findings have led to the hypothesis that the transforming genes of ASV and MuSV were derived from the nucleotide sequences encoding *sarc* in the various species of origin of the viruses (301, 312). The mechanism by which this might have occurred is uncertain. The RNA containing *sarc* in both avian and rat cells is approximately the same size as the subunits of retrovirus genomes and might therefore be packaged during the assembly of virions in cells producing an endogenous or exogenous retrovirus. Efforts to demonstrate packaging of avian *sarc* RNA have been unsuccessful to date, but the rat *sarc* RNA is readily packaged by a variety of retroviruses that replicate in rat cells (310). Packaging of heterozygous subunits allegedly facilitates recombination between genomes of retroviruses (38, 127b); recombination between *sarc* and the genome of a retrovirus might occur by the same mechanism and generate a sarcoma virus. However, there is no evidence that either avian or rat *sarc* encodes a protein with the potential to transform cells into a neoplastic phenotype, and the nucleotide sequences of ASV *src* are appreciably diverged from those of avian *sarc* (301). In any event, the derivation of *src* genes from their putative cellular progenitors has apparently been a rare event because isolations of ASV and MuSV are extremely uncommon (313).

Certain retroviruses can transform both fibroblasts and specific hematopoietic cells, which differ for each strain of virus, e.g. phagocytic myeloid cells for MC-29 virus (292), erythroid precursors for avian erythroblastosis virus (292), and probably B lymphocytes for Abelson MuLV (313a). The mechanism of this remarkable target cell specificity is presently a mystery. The genomes of these viruses do not contain nucleotide sequences homologous to the *src* genes of conventional sarcoma viruses (314; D. Stehelin and D. Sheiness, personal communication) and must therefore have other genetic determinants for transformation of fibroblasts. The same determinants may or may not be responsible for transformation of hematopoietic cells; transformation of hematopoietic cells by Abelson virus requires specific helper viruses (see below), whereas transformation of fibroblasts by Abelson virus does not (C. Scher, personal communication).

Leukemogenesis

Leukemia viruses fall into two broad classes. Certain strains (such as myeloblastosis and MC-29 viruses of chickens, and Friend and Abelson viruses of mice) induce hematological neoplasms rapidly in vivo (within a matter of weeks), transform appropriate target cells in vitro, and are defective in one or more of the genes required for viral replication. Other strains (such as avian leukosis viruses and Moloney-MuLV) act slowly in vivo (several months required for the onset of disease), have not been found to transform target cells in vitro, and are not defective in viral replication.

The rapidly acting leukemia viruses are considered likely to bear specific genetic determinants for transformation, perhaps replacing one or more of the viral genes required for replication. However, leukemogenesis and transformation of hematopoietic cells in vitro by Abelson virus require phenotypic mixing with specific strains of helper virus (C. Scher, personal communication); the role of the helper virus in leukemogenesis is not known, and to date, a similar requirement has not been reported for other rapidly acting leukemia viruses. By contrast, the slowly acting viruses may acquire transforming genes only as a consequence of genetic interactions with other viral or cellular genes in the host organism. For example, leukemogenesis by the AKR strain of MuLV apparently ensues recombination with a second, xenotropic virus resident in the host thymus; the recombinant progeny virus (denoted MCF virus) may be the actual agent of leukemogenesis (315).

The recombination that generates MCF virus occurs within *env* (316) (although other sites of crossing over could also exist) and some investigators have speculated that the recombinant viral glycoprotein may be responsible for leukemogenesis (316, 317). This hypothesis conforms to the recent claim that the nucleotide sequences required for erythroblastosis by the rapidly acting Friend virus were acquired by recombination with the *env* gene of a xenotropic murine virus (317). These findings do not include a direct genetic identification of a viral determinant for leukemogenesis and therefore provide only circumstantial evidence regarding the possible existence and identity of such determinants.

Murine leukemia viruses can serve as cocarcinogens (318, 319), perhaps because viral replication impedes the repair of chemically induced damage in DNA (319). These findings have led to the proposal that leukemia viruses are oncogenic because they sensitize the infected cell to spontaneous or induced mutagenesis (319). However, the effects on carcinogensis and repair were studied in cells that are not normally targets for transformation by leukemia viruses, and reported efforts to demonstrate that infection with leukemia virus facilitates mutagenesis have been unsuccessful (320).

REVERSION OF THE TRANSFORMED PHENOTYPE

Nonpermissive mammalian cells transformed by ASV can segregate less transformed or even phenotypically normal variants that retain the entire viral genome (321); similar findings have been reported with permissive cells transformed by MuSV in the absence of helper virus and, hence, producing no virus (322). Reversion to a normal phenotype is stable in the ASV-infected mammalian cells, whereas some MuSV-infected revertants return to the transformed phenotype when propagated further (323). Several distinct mechanisms of reversion have been identified: 1. provirus can be deleted from the infected cell (324); 2. nonconditional mutations occur in provial *src* (236); and 3. the expression of *src* can be inhibited without loss of the functional gene (321), perhaps by attenuation of transcription from the provirus and consequent reduction in the gene dose for *src* (325). All three mechanisms have been associated with stable reversion to the normal phenotype; the mechanism responsible for unstable reversion has not been characterized. The inhibition of *src* expression in revertant cells of the third class is recessive in at least some instances, i.e. the cells can be retransformed by superinfection with the same virus as that used to elicit the original transformation (326). Paradoxically, some of the recessive revertants are resistant to transformation by other strains of tumor viruses (326). These observations obviously reflect the nature of the unknown cellular mechanism that is responsible for reversion and deserve further exploration.

ENDOGENOUS VIRAL GENES AND ONCOGENESIS

The induced expression of endogenous viral genes could provide a final common pathway for oncogenesis by various etiological agents, such as chemical and physical carcinogens, hormones, and somatic mutation (327). This "oncogene hypothesis" derived from several points of evidence: 1. retrovirus genes are transmitted in the germ lines of many, if not all, species; 2. endogenous retroviruses are often activated in concert with either spontaneous or chemically induced cellular transformation; and 3. at least one type of viral transforming gene (*src*) apparently derives from the genome of normal cells.

Admittedly, some genetically transmitted retroviruses are oncogenic in the species that bear them; well-studied examples include MuLV in strains of mice with high incidences of leukemia (267, 328), the MuLV induced during leukemogenesis by irradiation (329), and the P strain of MMTV carried by the GR strain of mice (330). However, leukemogenesis by activated endogenous MuLV may require reinfection of target cells (331) and the formation of a genetically recombinant virus (315, 316), events that cannot be implicated in oncogenesis in the absence of manifest virus. No satisfactory tests have been performed on the possible role of reinfection of target cells in carcinogensis by genetically transmitted MMTV.

The transforming gene (*src*) of murine and avian sarcoma viruses apparently derived from nucleotide sequences (*sarc*) in normal cellular DNAs (see above); hence, *sarc* is potentially an endogenous oncogene. However, efforts to implicate the expression of *sarc* in either spontaneous or chemically induced oncogenesis have been unsuccessful (303, 311), and viral particles carrying the *sarc* of rat cells in RNA do not transform cells (310).

In summary, the oncogene hypothesis in a specialized form appears to be correct: genetically transmitted retrovirus genomes can be oncogenic subsequent to full expression and release from the cell in an infectious form. But the role of endogenous viral genes in oncogenesis when virus production does not occur has yet to be properly tested.

PROSPECTS

The origins and biological significance of retroviruses remain enigmatic. Some understanding of these issues is most likely to come from further consideration of the processes that generated and conserved endogenous viral genes. Does the existence of these genes manifest an entirely novel mechanism for the genesis and translocation of new genetic information? Do these genes function in normal cellular growth and development, and, if so, how? We expect to learn more from the "molecular paleontology" of endogenous viral genes. For example, data obtained with endogenous viruses of primates have been used to propose that *Homo sapiens* originated in Asia rather than in Africa (332). Premature though this suggestion may have been, it dramatizes the precision with which the evolution of endogenous viral genomes can be traced.

Retroviruses are promising reagents for the study of developmental biology. As examples, consider the following: retrovirus genes introduced into early embryos are later expressed in only certain tissues of the adult animal (333) and, hence, can serve as readily measured markers that are responsive to the differentiated state of the host cell; transcription from a cellular gene normally active only during fetal development can be reactivated consequent to virus-induced transformation (334); and expression of a retrovirus-transforming gene (*src*) can reversibly perturb the course of cellular differentiation (335).

The cell or organism infected with an oncogenic retrovirus may have a considerable influence over its own destiny. Cellular mechanisms can reverse the established effects of a viral-transforming gene (325, 326), and host genetic determinants govern the interaction between virus and organism (10).

What of the role of retroviruses in the etiology of human neoplasia? I have avoided this controversial issue in the body of the review because the case at present is highly circumstantial and the attendant issues are not germane to my purposes here. Taken at face value, the available data suggest that *Homo sapiens* may be infected with retroviruses by horizontal transmission from other host species (336, 337) and that we harbor in our genome at least the vestige of an endogenous virus found in baboons (332, 336). There is also less persuasive evidence for at least circumstantial associations between retrovirus infection and the occurrence of human neoplasia (338–340). But we have, in the words of Burnet (341), "no decisive evidence that viral associations with malignant disease have anything more than the most limited accidental significance in the etiology of cancer," and there may be "no more place for viruses as causes of cancer than for lice as causes of poverty."

The next few years should see the molecular mechanisms of retrovirus replication fully explicated. Novel biochemical processes are in the offing, and the importance of host elements in viral replication should come into fuller view. But these certain advances do not ensure an understanding of the mechanisms of oncogenesis, and it is these mechanisms that have now come front and center in tumor virology. Granted that viruses may not be common etiological agents of malignancy, granted that the pathogenesis of malignancy in the organism is a process far removed from transformation of cells in culture; it remains true that the relative simplicity, rapidity, and precision of virus-induced transformation make it our present best hope for insight into the aberrations of the malignant cell.

> But the comfort is
> In the covenant
> We may get control
> If not of the whole
> Of at least some part
> Where not too immense,
> So by craft or art
> We can give the part
> Wholeness in a sense.(342)

RETROVIRUSES   81

ACKNOWLEDGMENTS

I thank S. Hughes, L. Levintow, G. Ringold, P. Shank, H. E. Varmus, and P. Vogt for comments on the manuscript (not all of which were well received); S. H. and H. E. V. for helpful discussions of viral DNA synthesis; H. E. V. for the intellectual companionship that has helped enliven my interest in many of the issues discussed here; numerous colleagues for permission to discuss their unpublished data; B. Cook for assistance; the National Institutes of Health and the American Cancer Society for research support; and K. I. B. for forbearance and lunch.

Literature Cited

1. Rubin, H. 1966. In *Phage and the Origins of Molecular Biology*, ed. J. Cairns, G. S. Stent, J. D. Watson, pp. 292–300. New York: Cold Spring Harbor Laboratory
2. Becker, E. 1975. In *Escape from Evil*, p. IX. New York: Free Press
3. Fenner, F. 1976. *Intervirology* 7(1–2): 61–64
4. Weiss, R. 1975. *Perspect. Virol.* 9:165–205
5. Aaronson, S. A., Stephenson, J. R. 1976. *Biochim. Biophys. Acta* 458: 323–54
6. Ruprecht, R. M., Goodman, N. C., Spiegelman, S. 1973. *Proc. Natl. Acad. Sci. USA* 70:1437–41
7. Todaro, G. J. 1975. *Am. J. Pathol.* 81:590–605
8. Weiss, R. A. 1976. *Receptors for RNA Tumor Viruses, Miles Int. Symp. Ser. No. 9*, ed. R. F. Beers, Jr., E. G. Nassett, pp. 237–51
9. Levy, J. 1978. *Curr. Top. Microbiol. Immunol.* In press
10. Lilly, F., Pincus, T. 1972. *Adv. Cancer Res.* 17:231–77
11. Bolognesi, D. P. 1974. *Adv. Virus Res.* 19:315–59
12. Dalton, A. J. 1972. *J. Natl. Cancer Inst.* 49:323–330
13. Sarkar, N. H., Whittington, E. S. 1977. *Virology* 81:91–106
14. DeGiuli, C., Hanafusa, H., Kawai, S., Dales, S., Chen, J. H., Hsu, K. C. 1975. *Proc. Natl. Acad. Sci. USA* 72:3706–10
14a. Robertson, D. L., Yau, P., Dobbertin, D. C., Sweeney, T. K., Thach, S. S., Brendler, T., Thach, R. E. 1976. *J. Virol.* 18:344–55
15. Malech, H. L., Wivel, N. A. 1976. *Cell* 9:393–91
16. Strand, M., August, J. T. 1974. *J. Virol.* 13:171–80
16a. Barbacid, M., Stephenson, J. R., Aaronson, S. A. 1977. *Cell* 10:641–48
17. Bubbers, J. E., Lilly, F. 1977. *Nature* 266:458–59
18. Blank, K. J., Freedman, H. A., Lilly, F. 1976. *Nature* 260:250–52
19. Leamnson, R. N., Halpern, M. S. 1976. *J. Virol.* 18:956–68
20. Witte, O. N., Tsukamoto-Adey, A., Weissman, I. L. 1977. *Virology* 76: 539–53
21. Mosser, A. G., Montelaro, R. C., Rueckert, R. R. 1977. *J. Virol.* 23:10–19
22. Scheid, A. 1976. In *Animal Virology, ICN-UCLA Symp. Mol. Cell. Biol.*, ed. D. Baltimore, A. Huang, C. F. Fox, 4:457–70. New York: Academic
23. Elder, J. H., Jensen, F. C., Bryant, M. L., Lerner, R. A. 1977. *Nature* 267: 23–28
24. Galehouse, D. M., Duesberg, P. H. 1976. *Virology* 70:97–104
25. Schafer, W., Fischinger, P. J., Collins, J. J., Bolognesi, D. P. 1977. *J. Virol.* 21:35–40
26. Sarkar, N. H., Nowinski, R. C., Moore, D. H. 1971. *J. Virol.* 8:564–72
27. Davis, N. L., Rueckert, R. R. 1972. *J. Virol.* 10:1010–20
28. Fleissner, E., Tress, E. 1973. *J. Virol.* 12:1612–15
29. Sen, A., Todaro, G. J. 1977. *Cell* 10:91–99
30. Leis, J. P., McGinnis, J., Green, R. W. 1977. *Virology*. In press
31. Sen, A., Sherr, C. J., Todaro, G. J. 1977. *Cell* 10:489–96
32. Verma, I. M. 1977. *Biochim. Biophys. Acta* 473:1–34
33. Panet, A., Baltimore, D., Hanafusa, T. 1975. *J. Virol.* 16:146–52
34. Krakower, J. M., Barbacid, M., Aaronson, S. A. 1977. *J. Virol.* 22:331–39
35. Beemon, K. L., Faras, A. J., Haase, A.

T., Duesberg, P. H., Maisel, J. E. 1976. *J. Virol.* 17:525–37

36. Bender, W., Davidson, N. 1976. *Cell* 7:595–607
37. Duesberg, P. H. 1970. *Curr. Top. Microbiol. Immunol.* 51:79–104
38. Weiss, R. A., Mason, W. S., Vogt, P. K. 1973. *Virology* 52:535–52
39. McCarter, J. A. 1977. *J. Virol.* 22:9–15
40. King, A. M. Q., Wells, R. D. 1976. *J. Biol. Chem.* 251:150–52
41. Furuichi, Y., Shatkin, A. J., Stavnezer, E., Bishop, J. M. 1975. *Nature* 257: 618–20
42. Rose, J. K., Haseltine, W. A., Baltimore, D. 1976. *J. Virol.* 20:324–29
43. Beemon, K. L., Keith, J. M. 1976. See Ref. 22, 4:97–105
44. Dimock, K., Stoltzfus, C. M. 1977. *Biochemistry* 16:471–78
45. Taylor, J. M., Illmensee, R. 1975. *J. Virol.* 16:553–58
46. Dahlberg, J. E., Sawyer, R. C., Taylor, J. M., Faras, A. J., Levinson, W. E., Goodman, H. M., Bishop, J. M. 1974. *J. Virol.* 13:1126–33
47. Harada, F., Sawyer, R. C., Dahlberg, J. E. 1975. *J. Biol. Chem.* 250:3487–97
48. Peters, G., Harada, F., Dahlberg, J. E., Panet, A., Haseltine, W. A., Baltimore, D. 1977. *J. Virol.* 21:1031–41
49. Sawyer, R. C., Dahlberg, J. E. 1973. *J. Virol.* 12:1226–37
50. Haseltine, W. A., Maxam, A. M., Gilbert, W. 1977. *Proc. Natl. Acad. Sci. USA* 74:989–93
51. Schwartz, D. E., Zamecnik, P. C., Weith, H. L. 1977. *Proc. Natl. Acad. Sci. USA* 74:994–98
52. Stoll, E., Billeter, M. A., Palmenberg, A., Weissman, C. 1977. *Cell.* In press
53. Coffin, J., Maxam, A., Haseltine, W. 1977. *Cell.* In press
54. Joho, R., Billeter, M. A., Weissman, C. 1977. *Virology.* In press
55. Shine, J., Czernilofsky, A. P., Friedrich, R., Goodman, H. M., Bishop, J. M. 1977. *Proc. Natl. Acad. Sci. USA* 74:1473–77
56. Kung, H.-J., Hu, S., Bender, W., Bailey, J. M., Davidson, N., Nicolson, M. O., McAllister, R. M. 1976. *Cell.* 7:609–20
57. Baltimore, D. 1974. *Cold Spring Harbor Symp. Quant. Biol.* 39:1187–1200
58. Biggs, P. M., Milne, B. S., Graf, T., Bauer, H. 1973. *J. Gen. Virol.* 18:399–403
59. Joho, R. H., Billeter, M. A., Weissmann, C. 1975. *Proc. Natl. Acad. Sci. USA* 72:4772–76
60. Wang, L.-H., Duesberg, P. H., Kawai,

S., Hanafusa, H. 1976. *Proc. Natl. Acad. Sci. USA* 73:447–51
61. Wang, L.-H., Galehouse, D., Mellon, P., Duesberg, P., Mason, W. S., Vogt, P. K. 1976. *Proc. Natl. Acad. Sci. USA* 73:3952–56
62. Junghans, R. P., Hu, S., Knight, C. A., Davidson, N. 1977. *Proc. Natl. Acad. Sci. USA* 74:477–81
63. Cooper, G. M., Castellot, S. B. 1977. *J. Virol.* 22:300–7
64. Tal, J., Kung, H.-J., Varmus, H. E., Bishop, J. M. 1977. *Virology* 79:183–87
65. Hu, S., Davidson, N., Verma, I. M. 1977. *Cell* 10:469–77
66. Bishop, J. M., Levinson, W. E., Quintrell, N., Sullivan, D., Fanshier, L., Jackson, J. 1970. *Virology* 42:182–95
67. Walker, T. A., Pace, N. R., Erikson, R. L., Erikson, E., Behz, F. 1974. *Proc. Natl. Acad. Sci. USA* 71:3390–95
68. Faras, A. J., Garapin, A. C., Levinson, W. E., Bishop, J. M., Goodman, H. M. 1973. *J. Virol.* 12:334–42
69. Wang, S., Kothari, R. M., Taylor, M. 1973. *Nature New Biol.* 242:133–35
70. Waters, L. C., Mollin, B. C. 1977. *Prog. Nucleic Acid Res. Mol. Biol.* In press
71. Ikawa, Y., Ross, J., Leder, P. 1974. *Proc. Natl. Acad. Sci. USA* 71:1154–58
72. Goldberg, R. J., Levin, R., Parks, W. P., Scolnick, E. M. 1976. *J. Virol.* 17:45–50
73. Levinson, W. E., Varmus, H. E., Garapin, A. C., Bishop, J. 1972. *Science* 175:76–78
74. Darlix, J. L., Bromley, P. A., Spahr, P. F. 1977. *J. Virol.* 22:118–29
75. Temin, H. M. 1971. *Ann. Rev. Microbiol.* 25:609–48
76. Temin, H. M. 1964. *Natl. Cancer Inst. Monogr.* 17:557–62
77. Temin, H. M. 1976. *Science* 192: 1075–80
78. Mason, W. S., Friis, R. R., Linial, M., Vogt, P. K. 1974. *Virology* 61:559–74
79. Verma, I. M., Varmus, H. E., Hunter, E. 1976. *Virology* 74:16–29
80. Weinberg, R. A. 1977. *Biochim. Biophys. Acta* 473:39–56
81. Hill, M., Hillova, J. 1976. *Adv. Cancer Res.* 23:237–98
82. Bishop, J. M., Varmus, H. E. 1975. In *Cancer: A Comprehensive Treatise,* ed. F. F. Becker, 2:3–48. New York: Plenum
83. Gianni, A. M., Smotkin, D., Weinberg, R. A. 1975. *Proc. Natl. Acad. Sci. USA* 72:447–51
84. Smotkin, D., Yoshimura, F., Weinberg, R. A. 1976. *J. Virol.* 20:621–26

85. Guntaka, R. V., Mahy, B. W. J., Bishop, J. M., Varmus, H. E. 1975. *Nature* 253:507–11
86. Guntaka, R. V., Richards, O. C., Shank, P. R., Kung, H.-J., Davidson, N., Fritsch, E., Bishop, J. M., Varmus, H. E. 1976. *J. Mol. Biol.* 106:337–57
87. Levin, J. G., Rosenak, M. J. 1976. *Proc. Natl. Acad. Sci. USA* 73:1154–58
88. Paskind, M. P., Weinberg, R. A., Baltimore, D. 1975. *Virology* 67:242–48
89. Duesberg, P. H., Kawai, S., Wang, L.-H., Vogt, P. K., Murphy, H. M., Hanafusa, H. 1975. *Proc. Natl. Acad. Sci. USA* 72:1569–73
90. Vogt, P. K., Hu, S. S. F. 1977. *Ann. Rev. Genet.* 11:203–38
91. DeLarco, J., Todaro, G. J. 1976. *Cell* 8:365–71
92. Dales, S., Hanafusa, H. 1972. *Virology* 50:440–58
93. Varmus, H. E., Guntaka, R. V., Fan, W., Heasley, S., Bishop, J. M. 1974. *Proc. Natl. Acad. Sci. USA* 71:3874–78
94. Salzberg, S., Robin, M. S., Green, M. 1977. *Virology* 76:341–51
95. Bader, J. P. 1972. *Virology* 48:485–93
96. Gallis, B. M., Eisenman, R. N., Diggelmann, H. 1976. *Virology* 74:302–13
97. Stacey, D. W., Allfrey, V. G., Hanafusa, H. 1977. *Proc. Natl. Acad. Sci. USA* 74:1614–18
98. Junghans, R. P., Duesberg, P. H., Knight, C. A. 1975. *Proc. Natl. Acad. Sci. USA* 72:4895–99
99. Rothenberg, E., Baltimore, D. 1977. *J. Virol.* 21:168–78
100. Rothenberg, E., Smotkin, D., Baltimore, D., Weinberg, R. A. 1977. *Nature* 269:122–26
101. Taylor, J. M. 1977. *Biochim. Biophys. Acta* 473:57–71
102. Haseltine, W. A., Panet, A., Smoler, D., Baltimore, D., Peters, G., Harada, F., Dahlberg, J. E. 1977. *Biochemistry.* 16:3625–32
103. Panet, A., Haseltine, W. A., Baltimore, D., Peters, G., Harada, F., Dahlberg, J. E. 1975. *Proc. Natl. Acad. Sci. USA* 72:2535–39
104. Mizutani, S., Temin, H. M. 1975. In *Fundamental Aspects of Neoplasia,* ed. A. A. Gottlieb, O. J. Plescia, D. H. L. Bishop, pp. 235–41. New York: Springer
105. Coffin, J. M., Haseltine, W. 1977. *Proc. Natl. Acad. Sci. USA* 74:1908–12
106. Collett, M. S., Dierks, P., Cahill, J. F., Faras, A. J., Parsons, J. T. 1977. *Proc. Natl. Acad. Sci. USA* 74:2389–93
107. Haseltine, W. A., Kleid, D. G., Panet, A., Rothenberg, E., Baltimore, D. 1976. *J. Mol. Biol.* 106:109–32
108. Haseltine, W., Coffin, J. M., Hageman, T. 1977. *Cell.* In press
109. Taylor, J. M., Illmensee, R., Trusal, L. R., Summers, J. 1976. See Ref. 22, 4:161–74
110. Varmus, H. E., Heasley, S., Kung, H.-J., Oppermann, H., Smith, V. C., Bishop, J. M., Shank, P. R. 1978. *J. Mol. Biol.* In press
111. Baltimore, D., Smoler, D. F. 1972. *J. Biol. Chem.* 247:7282–87
112. Collett, M. S., Faras, A. J. 1976. *J. Virol.* 17:291–95
113. Leis, J., Schincariol, A., Ishizaki, R., Hurwitz, J. 1975. *J. Virol.* 15:484–88
114. Sveda, M. M., Fields, B. N., Sveiro, R. 1976. *J. Virol.* 18:85–91
115. Takano, T., Hatanaka, M. 1975. *Proc. Natl. Acad. Sci. USA* 72:343–47
116. Lovinger, G. G., Klein, R., Ling, H. P., Gilden, R. V., Hatanaka, M. 1975. *J. Virol.* 16:824–31
117. Shank, P. R., Varmus, H. E. 1977. *J. Virol.* In press
118. Dube, D. D., Loeb, L. A. 1976. *Biochemistry* 15:3605–11
119. Khoury, A. T., Hanafusa, H. 1976. *J. Virol.* 18:383–400
120. Varmus, H. E., Shank, P. R. 1976. *J. Virol.* 18:567–73
121. Varmus, H. E., Heasley, S., Linn, J., Wheeler, K. 1976. *J. Virol.* 18:574–85
122. Smotkin, D., Gianni, A. M., Rozenblatt, S., Weinberg, R. A. 1975. *Proc. Natl. Acad. Sci. USA* 72:4910–13
123. Fritsch, E., Temin, H. M. 1977. *J. Virol.* 21:119–30
124. Varmus, H. E., Padgett, T., Heasley, S., Simon, G., Bishop, J. M. 1977. *Cell* 11:307–19
125. Hung, P. P., Lee, S. G. 1976. *Nature* 259:499–502
126. Herrick, G., Spear, B. B., Veomett, G. 1976. *Proc. Natl. Acad. Sci. USA* 73:1136–39
127. Foster, D. N., Gurney, T. Jr. 1976. *J. Biol. Chem.* 251:7893–98
127a. Watson, J. D. 1972. *Nature New Biol.* 239:197–200
127b. Wyke, J. A., Bell, J. G., Beamand, J. A. 1974. *Cold Spring Harbor Symp. Quant. Biol.* 39:897–905
127c. Yarmolinsky, M. B. 1971. In *The Bacteriophage Lambda,* ed. A. D. Hershey, pp. 97–111. New York: Cold Spring Harbor Laboratory
127d. Ikeda, H., Tomizawa, J.-I. 1968. *Cold Spring Harbor Symp. Quant. Biol.* 33:791–98

128. Markham, P. D., Baluda, M. A. 1973. *J. Virol.* 12:721–32
129. Boettiger, D. 1974. *Cell* 3:71–76
130. Varmus, H. E., Vogt, P. K., Bishop, J. M. 1973. *Proc. Natl. Acad. Sci. USA* 70:3067–71
131. Dastoor, M. N., Shoyab, M., Baluda, M. A. 1977. *J. Virol.* 21:541–47
132. Shoyab, M., Dastoor, M. N., Baluda, M. A. 1976. *Proc. Natl. Acad. Sci. USA* 73:1749–53
133. Battula, N., Temin, H. M. 1977. *Proc. Natl. Acad. Sci. USA* 74:281–85
134. Sveda, M. M., Fields, B. N., Soeiro, R. 1974. *Cell* 2:271–77
135. Sveda, M. M., Soeiro, R. 1976. *Proc. Natl. Acad. Sci. USA* 73:2356–60
136. Fan, H. 1977. *Biochim. Biophys. Acta.* In press
137. Parsons, J. T., Coffin, J. M., Haroz, R. K., Bromley, P. A., Weissmann, C. 1973. *J. Virol.* 11:761–74
138. Rymo, L., Parsons, J. T., Coffin, J. M., Weissmann, C. 1974. *Proc. Natl. Acad. Sci. USA* 71:2782–86
139. Jacquet, M., Groner, Y., Monroy, G., Hurwitz, J. 1974. *Proc. Natl. Acad. Sci. USA* 71:3045–49
140. Bishop, J. M., Deng, C. T., Mahy, B. W. J., Quintrell, N., Stavnezer, E., Varmus, H. E. 1976. See Ref. 22, 4:1–20
141. Jolicoeur, P., Baltimore, D. 1976. *Proc. Natl. Acad. Sci. USA* 73:2236–40
142. Jolicoeur, P., Baltimore, D. 1976. *Cell* 7:33–39
143. Humphries, E. H., Temin, H. M. 1974. *J. Virol.* 14:351–46
144. Bader, J. P. 1973. *Science* 180:1069–71
145. Leong, J. A., Levinson, W., Bishop, J. M. 1972. *Virology* 47:133–41
146. Humphreis, E. H., Coffin, J. M. 1976. *J. Virol.* 17:393–401
147. Fan, H. 1977. *Cell* 11:297–305
148. Weiss, S. R., Varmus, H. E., Bishop, J. M. 1977. *Cell.* In press
149. Hayward, W. S. 1977. *J. Virol.* 24:47–63
150. Kourilsky, P., Gros, F. 1974. In *Regulation of Gene Expression in Eukaryotic Cells*, ed. M. Harris, B. Thompson, pp. 19–41. Washington, DC: GPO
151. Gelinas, R. E., Roberts, R. J. 1977. *Cell* 11:533–44
152. Berget, S. M., Moore, C., Sharp, P. A. 1977. *Proc. Natl. Acad. Sci. USA* 74:3171–75
153. Stavnezer, E., Ringold, G., Varmus, H. E., Bishop, J. M. 1976. *J. Virol.* 20:343–47
154. Varmus, H. E., Ringold, G., Yamamoto, K. R. 1977. *Adv. Horm. Res.* In press
155. Ringold, G., Yamamoto, K. R., Tomkins, G. M., Bishop, J. M., Varmus, H. E. 1975. *Cell* 6:299–305
156. Young, H. A., Scolnick, E. M., Parks, W. P. 1975. *J. Biol. Chem.* 250:3337–43
157. Scolnick, E. M., Young, H. A., Parks, W. P. 1976. *Virology* 69:148–56
158. Young, H. A., Shih, T. Y., Scolnick, E. M., Parks, W. P. 1977. *J. Virol.* 21:139–46
159. Ringold, G. M., Yamamoto, K. R., Bishop, J. M., Varmus, H. E. 1977. *Proc. Natl. Acad. Sci. USA* 74:2879–83
160. Jacobson, M. F., Baltimore, D. 1968. *Proc. Natl. Acad. Sci. USA* 61:77–84
161. Hunter, T. R., Hunt, T., Knowland, J., Zimmerman, D. 1976. *Nature* 260:759–64
162. Cancedda, R., Villa-Komaroff, L., Lodish, H. F., Schlesinger, M. 1975. *Cell* 6:215–22
163. Vogt, V. M., Eisenman, E., Diggelmann, H. 1975. *J. Mol. Biol.* 96:471–93
164. Shapiro, S. Z., Strand, M., August, J. T. 1976. *J. Mol. Biol.* 107:459–77
165. Arlinghaus, R. B., Naso, R. B., Jamjoom, G. A., Arcement, L. J., Karshin, W. L. 1976. See Ref. 22, 4:689–715
166. Barbacid, M., Stephenson, J. R., Aaronson, S. A. 1976. *Nature* 262:554–58
167. Eisenman, R., Vogt, V. M., Diggelmann, H. 1974. *Cold Spring Harbor Symp. Quant. Biol.* 39:1067–75
168. Helm, K. V. D. 1977. *Proc. Natl. Acad. Sci. USA* 74:911–15
169. Yoshinaka, Y., Luftig, R. B. 1977. *Biochem. Biophys. Res. Commun.* 76:54–59
170. Yoshinaka, Y., Luftig, R. B. 1977. *Proc. Natl. Acad. Sci. USA.* 74:3446–50
171. Jamjoom, G. A., Naso, R. B., Arlinghaus, R. B. 1976. *J. Virol.* 19:1054–72
172. Snyder, H. W., Stockert, E., Fleissner, E. 1977. *J. Virol.* 23:302–14
173. Ledbetter, J., Nowinski, R. C. 1977. *J. Virol.* 23:315–22
174. Mueller-Lantzsch, N., Fan, H. 1976. *Cell* 9:579–85
175. von der Helm, K., Duesberg, P. H. 1975. *Proc. Natl. Acad. Sci. USA* 72:614–18
176. Pawson, T., Martin, G. S., Smith, A. E. 1976. *J. Virol.* 19:950–67
177. Pawson, T., Harvey, R., Smith, A. E. 1977. *Nature* 268:416–19
178. Kerr, I. M., Olshevsky, U., Lodish, H. F., Baltimore, D. 1976. *J. Virol.* 18:627–35
179. Van Zaane, D., Gielkens, A. L. J., Hesselink, W. G., Bloemers, H. P. J. 1977. *Proc. Natl. Acad. Sci. USA* 74:1855–59

180. Beemon, K., Hunter, T. 1977. *Proc. Natl. Acad. Sci. USA* 74:3302–6
181. Weber, L. A., Feman, E. R., Hickey, E. D., Williams, M. C., Baglioni, C. 1976. *J. Biol. Chem.* 251:5657–62
182. Oppermann, H., Bishop, J. M., Varmus, H. E., Levintow, L. 1977. *Cell.* In press
183. Jamjoom, G. A., Naso, R. B., Arlinghaus, R. B. 1977. *Virology* 78:11–34
184. Purchio, A. F., Erikson, E., Erikson, R. L. 1977. *Proc. Natl. Acad. Sci. USA* In press
185. Weiner, A. M., Weber, K. 1971. *Nature New Biol.* 234:206–9
186. Philipson, L., Anderson, P., Olshevsky, U., Weinberg, R., Baltimore, D. 1978. *Cell.* In press
187. Halpern, M. S., Bolognesi, D. P., Lewandowski, C. J. 1974. *Proc. Natl. Acad. Sci. USA* 71:2342–46
188. Moelling, K., Hayami, M. 1977. *J. Virol.* 22:598–607
189. Famulari, N. G., Buchhagen, D. L., Klenk, H.-D., Fleissner, E. 1976. *J. Virol.* 20:501–8
190. England, J. M., Bolognesi, D. P., Dietzschold, B., Halpern, M. S. 1977. *J. Virol.* 21:810–14
191. Shapiro, S. Z., August, J. T. 1976. *Biochim. Biophys. Acta* 458:375–96
192. Rubin, C. S., Rosen, O. M. 1975. *Ann. Rev. Biochem.* 44:831–87
193. Erikson, E., Brugge, J. S., Erikson, R. L. 1977. *Virology* 80:177–85
194. Hizi, A., Joklik, W. K. 1977. *Virology* 78:571–75
195. Tsiapalis, C. M. 1977. *Nature* 266:27–31
196. Lee, S. G., Miceli, M. V., Jungmann, R. A., Hung, P. P. 1975. *Proc. Natl. Acad. Sci. USA* 72:2945–49
197. Strand, M., August, J. T. 1971. *Nature New Biol.* 233:137–40
198. Pal, B. K., McAllister, R. M., Gardner, M. B., Roy-Burman, P. 1975. *J. Virol.* 16:123–31
199. Lai, M. M. C. 1976. *Virology* 74:287–301
200. Levin, D. J. G., Grimley, P. M., Ramseur, J. M., Berezesky, I. K. 1974. *J. Virol.* 14:152–61
201. Kawai, S., Hanafusa, H. 1973. *Proc. Natl. Acad. Sci. USA* 70:3493–97
202. Canaani, E., Helm, K. V. D., Duesberg, P. 1973. *Proc. Natl. Acad. Sci. USA* 70:401–5
203. Stoltzfus, C. M., Synder, P. N. 1975. *J. Virol.* 16:1161–70
204. Chung, K.-S., Smith, R. C., Stone, M. P., Joklik, W. K. 1972. *Virology* 50:851–64
205. Besmer, P., Baltimore, D. 1977. *J. Virol.* 21:965–73
206. Weiss, R. A. 1969. *J. Gen. Virol.* 5:511–28
207. Weiss, R. A., Boettiger, D., Murphy, H. M. 1977. *Virology* 76:808–25
208. Boettiger, D., Love, D. N., Weiss, R. A. 1975. *J. Virol.* 15:108–14
209. Stephenson, J. R., Wilsnack, R. E., Aaronson, S. A. 1973. *J. Virol.* 11:893–99
210. Kurth, R., Bosch, V., Bolognesi, D. P. 1977. *Virology* 78:511–21
211. Reynolds, F. H. Jr., Hanson, C. A., Aaronson, S. A. Stephenson, J. R. 1977. *J. Virol.* 23:74–79
212. Weiss, R. A., Wong, A. L. 1977. *Virology* 76:826–34
213. Zarling, D., Temin, H. M. 1976. *J. Virol.* 17:74–84
214. Rein, A., Kashmiri, S. V. S. 1976. *Cell* 7:373–79
215. Tennant, R. W., Myer, F. E., McGrath, L. 1974. *Int. J. Cancer* 14:504–10
216. Huang, A. S., Besmer, P., Chu, L., Baltimore, D. 1973. *J. Virol.* 12:659–62
217. Krontiris, T. G., Soeiro, R., Fields, B. N. 1973. *Proc. Natl. Acad. Sci. USA* 70:2549–53
218. Tennant, R. W., Schulter, B., Yang, W.-K., Brown, A. 1974. *Proc. Natl. Acad. Sci. USA* 71:4241–45
219. Hopkins, N., Schindler, J., Hynes, R. 1977. *J. Virol.* 21:309–18
220. Robinson, H. L. 1976. *J. Virol.* 18:856–66
221. Linial, M., Neiman, P. E. 1976. *Virology* 73:508–20
222. Vogt, P. K. 1977. In *Comprehensive Virology,* ed. H. Fraenkel-Conrat, R. Wagner. New York: Plenum. In press
223. Duesberg, P. H., Kawai, S., Wang, L.-H., Vogt, P. K., Murphy, H. M. 1975. *Proc. Natl. Acad. Sci. USA* 72:1569–73
224. Murphy, H. M. 1977. *Virology* 77:705–21
225. Hanafusa, H., Baltimore, D., Smoler, D., Watson, K. F., Yaniv, A., Spiegelman, S. 1972. *Science* 177:1188–91
226. Peebles, P. T., Gerwin, B. I., Scolnick, E. M. 1976. *Virology* 70:313–23
227. Parks, W. P., Howk, R. S., Anisowicz, A., Scolnick, E. M. 1976. *J. Virol.* 18:491–503
228. Robey, W. G., Oskarsson, M. K., Vande Woude, G. F., Naso, R. B., Arlinghaus, R. B., Haapala, D. K., Rischinger, P. J. 1977. *Cell* 10:79–89
229. Dina, D., Beemon, K., Duesberg, P. 1976. *Cell* 9:299–309
230. Bister, K., Hyman, M. J., Vogt, P. K. 1977. *Virology.* In press

231. Moscovici, C. 1975. *Curr. Top. Microbiol. Immunol.* 71:79–101
232. Vogt, P. K. 1971. *Virology* 46:939–46
233. Lai, M. M. C., Hu, S. S. F., Vogt, P. K. 1977. *Proc. Natl. Acad. Sci. USA* In press
234. Bernstein, A., MacCormick, R., Martin, G. S. 1976. *Virology* 70:206–9
235. Kawai, S., Duesberg, P., Hanafusa, H. 1977. *J. Virol.* In press
236. Greenberger, J. S., Anderson, G. R., Aaronson, S. A. 1974. *Cell* 2:279–86
237. Martin, G. S. 1970. *Nature* 227:1021–23
238. Scolnick, E. M., Stephensen, J. R., Aaronson, S. A. 1972. *J. Virol.* 10:653–57
239. Kawai, S., Hanafusa, H. 1971. *Virology* 46:470–79
240. Bader, J. P. 1972. *J. Virol.* 10:267–76
241. Becker, D., Kurth, R., Critchley, D., Friis, R., Bauer, H. 1977. *J. Virol.* 21:1042–55
242. Stephenson, J. R., Tronick, S. R., Aaronson, S. A. 1975. *Cell* 6:543–48
243. Hunter, E. M., Hayman, M. J., Rongey, R. W., Vogt, P. K. 1976. *Virology* 69:35–49
244. Rohrschneider, J. M., Diggelmann, H., Ogura, H., Friis, R. R., Bauer, H. 1976. *Virology* 75:177–87
245. Friis, R. R., Mason, W. S., Chen, C. C., Halpern, M. S. 1975. *Virology* 64:49–62
246. Mason, W. S., Yeater, C. 1977. *Virology* 77:443–56
247. Yuen, P. H., Wong, P. K. Y. 1977. *Virology* 80:260–74
248. Verma, I. M., Mason, W. S., Drost, S. D., Baltimore, D. 1974. *Nature* 251:27–31
249. Tronick, S. R., Stephenson, J. R., Verma, I. M., Aaronson, S. A. 1975. *J. Virol.* 16:1476–82
250. Wyke, J. A., Linial, M. 1973. *Virology* 53:152–61
251. Wyke, J. A. 1975. *Biochim. Biophys. Acta.* 471:91–121
252. Bookout, J. B., Sigel, M. M. 1975. *Virology* 67:474–86
253. Vogt, P. K. 1971. *Virology* 46:947–52
254. Kawai, S., Hanafusa, H. 1972. *Virology* 49:37–44
255. Shyam, D., Kung, H.-J., Bender, W., Davidson, N., Ostertag, W. 1976. *J. Virol.* 20:264–72
256. Beemon, K., Duesberg, P., Vogt, P. K. 1974. *Proc. Natl. Acad. Sci. USA* 71:4254–58
257. Joho, R. H., Stoll, E., Friis, R. R., Billeter, M. A., Weissman, C. 1976. See Ref. 22, 4:127–45
258. Hayman, M. J., Vogt, P. K. 1976. *Virology* 73:372–80
259. Wyke, J. A. 1973. *Virology* 54:28–36
260. Stehelin, D., Guntaka, R. V., Varmus, H. E., Bishop, J. M. 1976. *J. Mol. Biol.* 101:349–65
261. Linial, M., Mason, W. S. 1973. *Virology* 53:258–73
262. Sawyer, R. C., Hanafusa, H. 1977. *J. Virol.* 22:634–39
263. Vogt, P. K., Spencer, J. L., Okazaki, W., Witter, R. L., Crittenden, L. B. 1977. *Virology* 80:127–35
264. Livingston, D. M., Howard, T., Spence, C. 1976. *Virology* 70:432–39
265. Todaro, G. J., Benveniste, R. E., Callahan, R., Lieber, M. M., Sherr, C. J. 1974. *Cold Spring Harbor Symp. Quant. Biol.* 39:1159–68
266. Benveniste, R. E., Todaro, G. J. 1975. *Nature* 257:506–8
267. Chattopadhyay, S. K., Rowe, W. P., Teich, N. M., Lowy, D. R. 1975. *Proc. Natl. Acad. Sci. USA* 72:906–10
268. Cooper, G. M., Temin, H. M. 1976. *J. Virol.* 17:422–30
269. Weiss, R. A., Payne, L. N. 1971. *Virology* 45:508–15
270. Hanafusa, H., Hanafusa, T., Kawai, S., Lunginbuhl, R. E. 1974. *Virology* 58:439–48
271. Ando, T., Toyoshima, K. 1976. *Virology* 73:521–27
272. Chen, J. H., Hayward, W. S., Hanafusa, H. 1974. *J. Virol.* 14:1419–29
273. Wang, S. Y., Hayward, W. S., Hanafusa, H. 1977. *J. Virol.* In press
274. Lowy, D. R., Rowe, W. P., Teich, N., Hartley, J. W. 1971. *Science* 174:155–56
275. Weiss, R. A., Friis, R. R., Katz, E., Vott, P. K. 1971. *Virology* 46:920–38
276. Aaronson, S. A., Dunn, C. Y. 1974. *Science* 183:422–23
277. Hirsch, M. S., Phillips, S. M., Solnik, C., Black, P. H., Schwartz, R. S., Carpenter, C. B. 1972. *Proc. Natl. Acad. Sci. USA* 69:1146–50
278. Natalie, T., Lowy, D. R., Hartley, J. W., Rowe, W. P. 1973. *Virology* 51:163–73
279. Besmer, P., Smotkin, D., Haseltine, W., Fan, H., Wilson, A. T., Paskind, M., Weinberg, R., Baltimore, D. 1974. *Cold Spring Harbor Symp. Quant. Biol.* 39:1103–7
280. Aaronson, S. A., Anderson, G. R., Dunn, C. Y., Robbins, K. C. 1974. *Proc. Natl. Acad. Sci. USA* 71:3941–45
281. Hanafusa, H., Miyamoto, T., Hanafusa, T. 1970. *Proc. Natl. Acad. Sci. USA* 66:314–21
282. Stephenson, J. R., Anderson, G. R., Tronick, S. R. 1974. *Cell* 2:87–94

283. Baltimore, D. 1971. *Trans. NY Acad. Sci.* 33:327–32
284. Joklik, W. K. 1974. *Symp. Soc. Gen. Microbiol.* 24:393–20
285. Temin, H. M. 1974. *Ann. Rev. Genet.* 8:155–78
286. Temin, H. M. 1974. *Cancer Res.* 34:2835–41
287. Battula, N., Loeb, L. A. 1974. *J. Biol. Chem.* 249:4086–93
288. Mizutani, S., Temin, H. M. 1976. *Biochemistry* 15:1510–16
289. Strand, M., August, J. T., Jaenisch, R. 1977. *Virology* 76:886–90
290. Reanney, D. C. 1976. *Bacteriol. Rev.* 40:552–90
291. Rosenberg, N., Baltimore, D. 1976. See Ref. 22, 4:311–20
292. Graf, T., Royer-Pokora, B., Beug, H. 1976. See Ref. 22, 4:321–38
293. Weinberg, R. A. 1977. *Cell* 11:243–46
294. Beug, H., Peters, J. H., Graf, T. 1976. *Z. Naturforsch. Teil C* 31:766–68
295. Bell, J. G., Wyke, J. A., Macpherson, I. A. 1975. *J. Gen. Virol.* 27:127–34
296. Calothy, G., Pessac, B. 1976. *Virology* 71:336–45
297. Todaro, G. J., De Larco, J. E., Cohen, S. 1976. *Nature* 264:5581–84
298. Todaro, G. J., Larcu, J. E. D., Nissley, S. P., Rechler, M. M. 1977. *Nature* 267:525–27
299. Brugge, J. S., Erikson, R. L. 1977. *Nature* 269:346–48
300. Kamine, J., Buchanan, J. M. 1977. *Proc. Natl. Acad. Sci. USA* 74:2011–15
301. Stehelin, D., Varmus, H. E., Bishop, J. M., Vogt, P. K. 1976. *Nature* 250:170–73
302. Spector, D., Varmus, H. E., Bishop, J. M. 1978. *Proc. Natl. Acad. Sci. USA.* In press
303. Spector, D., Smith, K., Padgett, T., McCombe, P., Roulland-Dussoix, D., Moscovici, C., Varmus, H. E., Bishop, J. M. 1977. *Cell.* In press
304. Spector, D., Baker, B., Varmus, H. E., Bishop, J. M. 1977. *Cell.* In press
305. Padgett, T. G., Stubblefield, E., Varmus, H. E. 1977. *Cell* 10:649–57
306. Frankel, A. E., Fischinger, P. J. 1976. *Proc. Natl. Acad. Sci. USA* 73:3705–9
307. Frankel, A. E., Fischinger, P. J. 1977. *J. Virol.* 21:153–60
308. Scolnick, E. M., Goldberg, R. J., Williams, D. 1976. *J. Virol.* 18:559–66
309. Scolnick, E. M., Parks, W. P. 1974. *J. Virol.* 13:1211–19
310. Scolnick, E. M., Williams, D., Maryak, J., Vass, W., Goldberg, R. J., Parks, W. P. 1976. *J. Virol.* 20:570–82
311. Anderson, G. R., Robbins, K. C. 1976. *J. Virol.* 17:335–51
312. Scolnick, E. M., Rands, E., Williams, D., Parks, W. P. 1973. *J. Virol.* 12:458–63
313. Gross, L. 1970. In *Oncogenic Viruses.* Oxford/London/New York: Pergamon. 2nd ed.
313a. Premkumar, E., Potter, M., Singer, P.A., Sklar, M.D. 1975. *Cell* 6:149-59
314. Scolnick, E. M., Howk, R. S., Anisowicz, A., Peebles, P. T., Scher, C. D., Parks, W. P. 1975. *Proc. Natl. Acad. Sci. USA* 72:4650–54
315. Hartley, J. W., Wolford, N. K., Old, L. J., Rowe, W. P. 1977. *Proc. Natl. Acad. Sci. USA* 74:789–92
316. Elder, J. H., Gautsch, J. W., Jensen, F. C., Lerner, R., Hartley, J. W., Rowe, W. P. 1977. *Proc. Natl. Acad. Sci. USA* In press
317. Troxler, D., Lowy, D., Howk, R., Young, H., Scolnick, E. M. 1977. *Proc. Natl. Acad. Sci. USA* In press
318. Rasheed, S., Freeman, A. E., Gardner, M. B., Huebner, R. J. 1976. *J. Virol.* 18:776–82
319. Waters, R., Mishra, N., Bouch, N., DiMayorca, G., Regan, J. D. 1977. *Proc. Natl. Acad. Sci. USA* 74:238–42
320. Mishra, N. K., Pant, K. J., Wilson, C. M., Thomas, F. O. 1977. *Nature* 266:548–50
321. Boettiger, D. 1974. *Virology* 62:522–29
322. Stephenson, J. R., Reynolds, R. K., Aaronson, S. A. 1973. *J. Virol.* 11:218–22
323. Fischinger, P. J., Nomura, S., Tuttle-Fuller, N., Dunn, K. J. 1974. *Virology* 59:217–29
324. Frankel, A. E., Haapala, D. K., Neubauer, R. L., Fischinger, P. J. 1976. *Science* 191:1264–66
325. Deng, C.-T., Stehelin, D., Bishop, J. M., Varmus, H. E. 1977. *Virology* 76:313–30
326. Ozanne, B., Vogel, A. 1974. *J. Virol.* 14:239–48
327. Huebner, R. J., Gilden, R. V. 1971. In *RNA Viruses and Host Genomes in Oncogenesis,* ed. P. Emmelot, P. Bentvelzen, pp. 197–219. North-Holland/American Elsevier
328. Stephenson, J. R., Greenberger, J. S., Aaronson, S. A. 1974. *J. Virol.* 13:237–40
329. Decleve, A., Liebermann, M., Ihle, J. N., Kaplan, H. S. 1976. *Proc. Natl. Acad. Sci. USA* 73:4675–79
330. Bentvelzen, P. 1974. *Biochim. Biophys. Acta* 355:236–59

331. Berns, A., Jaenisch, R. 1976. *Proc. Natl. Acad. Sci. USA* 73:2448–52
332. Benveniste, R. E., Todaro, G. J. 1976. *Nature* 261:101–5
333. Jaenisch, R., Fan, H., Croker, B. 1975. *Proc. Natl. Acad. Sci. USA* 72:4008–12
334. Groudine, M., Weintraub, H. 1975. *Proc. Natl. Acad. Sci. USA* 72:4464–68
335. Holtzer, H., Biehl, J., Yeah, G., Meganathan, R., Kaji, A. 1975. *Proc. Natl. Acad. Sci. USA* 72:4051–55
336. Wong-Staal, F., Gillespie, D., Gallo, R. C. 1976. *Nature* 262:190–94
337. Kurth, R., Teich, N. M., Weiss, R., Oliver, R. T. D. 1977. *Proc. Natl. Acad. Sci. USA* 74:1237–41
338. Gallo, R. C., Gallagher, R. E., Sarngadharan, M. G., Sarin, P., Reitz, M., Miller, N., Gillespie, D. H. 1974. *Cancer* 34:1398–1405
339. Spiegelman, S., Axel, R., Baxt, W., Kufe, D., Schlom, J. 1974. *Cancer* 34:1406–20
340. Hehlmann, R. 1976. *Curr. Top. Microbiol. Immunol.* 73:141–215
341. Burnet, M. 1974. In *Intrinsic Mutagenesis: A Genetic Approach to Aging*, pp. 186–87. New York: Wiley
342. Frost, R. 1962. In *In the Clearing*, p. 56. New York: Holt, Rinehart & Winston
343. Reitz, M., Gillespie, D., Saxinger, W. C., Robert, M., Gallo, R. C. 1972. *Biochem. Biophys. Res. Commun.* 49:1216–21

Ann. Rev. Biochem. 1978. 47:89–128
Copyright © 1978 by Annual Reviews Inc. All rights reserved

HYPOTHALAMIC REGULATORY HORMONES

❖969

Andrew V. Schally, David H. Coy, and Chester A. Meyers

Endocrine and Polypeptide Laboratory, Veterans Administration Hospital, and Department of Medicine, Tulane University School of Medicine, New Orleans, Louisiana 70146

CONTENTS

89

0066-4154/78/0701-0089$01.00

PERSPECTIVES AND SUMMARY

Hypothalamic peptide hormones are those substances present in the hypothalamus, a neural structure lying at the base of the brain, which regulate the secretion of protein hormones from the anterior pituitary gland. Dramatic advances have been made in this field during the past ten years. The theory of neurohumoral control of the anterior pituitary, formulated by Harris (1) on the basis of anatomical and physiological data more than 30 years ago, suggested that hypothalamic nerve fibers of different types liberate hormonal substances into the capillaries of the primary plexus of hypophysial portal vessels in the median eminence. These substances are then carried by the portal vessels to the sinusoids of the pituitary gland where they appear to stimulate or inhibit the release of anterior pituitary hormones. However, direct evidence for the existence of specific hypothalamic neurohumors involved in the regulation of the anterior pituitary gland was lacking until 1955. That year, Saffran & Schally demonstrated the existence of a corticotropin releasing factor (CRF) (2), which opened the way to the subsequent discoveries of other hypothalamic regulatory hormones. The presence of at least nine hypothalamic regulators of the pituitary gland is now reasonably well established (3). Extracts of materials have been prepared which stimulate corticotropic, thyrotropic, and somatotropic (GH) hormones of the pituitary, and inhibit the release of growth hormone (GH), prolactin, and melanocyte stimulating hormone. The agents responsible for these biological activities are of such importance that research into their physiology and chemistry was vigorously pursued by several groups of investigators. In 1966 we isolated thyrotropin releasing hormone (TRH) from porcine hypothalami and reported the correct amino acid composition of this tripeptide (4). The work in our laboratory and that of Guillemin resulted in 1969 in the determination of structure and synthesis of porcine and ovine TRH (5–9). In 1971 we isolated luteinizing hormone- and follicle stimulating hormone-releasing hormone (LH-RH/FSH-RH), determined its structure, and synthesized it (10–14). Extensive physiological, biochemical, and immunological studies with LH-RH followed (15). Two years later, growth hormone release inhibiting hormone (GH-RIH; somatostatin) of ovine origin was isolated and structurally elucidated by Brazeau et al (16). Subsequently, we obtained this tetradecapeptide from pig hypothalami and showed that it had the same structure as the ovine hormone (17). Studies with synthetic somatostatin demonstrated that it also inhibits the secretion of thyrotropin, insulin, glucagon, gastrin, secretin, and cholecystokinin (18). Later this hormone was found in the stomach, gut, and pancreas (19), in addition to the brain. Somatostatin may play a role in the regulation of secretions not only of the pituitary but also of the pancreas, stomach, and duodenum. Thus in the last decade, the isolation, determination of structure, and synthesis of at least three hypothalamic regulatory hormones have been achieved.

Development of antisera for these hormones permitted the establishment of radioimmunological methods for their measurement (18). These methods and immunohistochemical techniques were then used to localize TRH, LH-RH, and somatostatin in various brain areas and identify the neurons which presumably synthesize, transport, and secrete these hormones (18).

Numerous analogs of these hormones have been synthesized and evaluated biologically in an attempt to determine the relationship between their structure and biological activity and to obtain peptides which would be more desirable clinically than the original hormones, including inhibitory analogs in the case of LH-RH (15, 18, 20). Several extremely potent long-acting analogs of LH-RH have been prepared, as well as some effective antagonists capable of blocking LH and FSH release and ovulation in several animal species (15). Some analogs of somatostatin have also been made which exhibit selective biological actions, such as inhibition of glucagon rather than insulin release (20, 21). Some of these substances have been evaluated clinically and are used for diagnosis and therapy (15, 18, 22). Much work is being devoted to purification of corticotropin releasing factor, prolactin releasing factor, growth hormone-releasing factor, prolactin release-inhibiting factor, melanocyte stimulating hormone (MSH)-releasing factor, and MSH-release inhibiting factor, but their chemical natures are still unknown (18). However, some of these hypothalamic hormones are likely to be identified during the next few years.

INTRODUCTION

This article reviews the most recent biochemical findings relating to each of the known hypothalamic hormones. Some earlier papers of others, as well as ourselves, are quoted as essential key references. We do not cover in detail topics such as mechanism of action, biosynthesis or control of release of hypothalamic hormones, and neurotransmitter regulation of their secretion, since they have been reviewed recently by others (20, 23). Instead, we emphasize our special interests by discussing some selected biochemical, physiological, and immunological studies on natural materials and various synthetic analogs and attempt to evaluate the medical significance of hypothalamic hormones and their derivatives.

NOMENCLATURE

The hypothalamic hypophysiotropic substances were previously called releasing factors. However, because some of these substances satisfy classical criteria for designation as hormones, we have modified their nomenclature (3, 18, 24). In this article, we use the word *hormone(s)* for those hypothalamic substances which have had their structures determined and which have been shown to be the likely physiological regulators of the secretion of the appropriate anterior pituitary hormone. Other hypothalamic regulators whose structures have not been determined are referred to as *factor(s)*, since their physiological activity cannot be correlated with a specific structure. This nomenclature (Table 1) is based on the name of the pituitary hormone, as for example, thyrotropin or growth hormone, followed by -releasing hormone, or, -release inhibiting hormone. The abbreviation -RH or -RF could represent *releasing* or *regulatory* hormone or factor, since some hypothalamic hormones appear to affect the synthesis, as well as release, of anterior pituitary hormones (3, 24). This nomenclature was chosen since it appears to be used by the majority of neuroendocrinologists, endocrinologists, clinicians, and synthetic chemists in the field. Some IUPAC-IUB approved trivial names for hypothalamic hormones are awkward, confusing, and have not been accepted by the endocrinologists, as indicated by the recent vote of the Endocrine Society and the recommendations of the publications committee of that society for its journals (25). Moreover, no abbreviations have been proposed in this IUPAC-IUB nomenclature, which would make naming of the analogs tedious. The names, and the corresponding abbreviations we proposed, are by now so well established that in the interest of historical continuity and convenience we feel that it is necessary to use them.

Table 1 Hypothalamic hormones or factors controlling the release of pituitary hormones

Full name	Abbreviation
Corticotropin (ACTH)-releasing factor	CRF
Thyrotropin (TSH)-releasing hormone	TRH
Luteinizing hormone (LH)-releasing hormone/ Follicle-stimulating hormone (FSH)-releasing hormone	LH-RH/FSH-RH
Growth hormone (GH)-release inhibiting hormone	GH-RIH; somatostatin
Growth hormone (GH)-releasing factor	GH-RF
Prolactin release-inhibiting factor	PIF
Prolactin releasing factor	PRF
Melanocyte stimulating hormone (MSH- release-inhibiting factor	MIF
Melanocyte-stimulating hormone (MSH)-releasing factor	MRF

CORTICOTROPIN RELEASING FACTOR (CRF)

Although CRF was the first hypothalamic hormone to be discovered (2, 26), it has not yet been obtained in pure form. The problems involved in the isolation of CRF, such as loss of activity during purification, have been reviewed previously (3, 24). In vitro methods using isolated anterior pituitary tissue, introduced in this field by Saffran & Schally (2), were employed to guide the early attempts at purification of CRF from posterior pituitary and hypothalamic tissue. The characteristics of CRF were those of a small peptide (27). The presence of histidine and serine in addition to the amino acids of vasopressin in a preparation of neurohypophysial CRF was detected (27). Several years later a tentative partial sequence of a CRF purified from posterior pituitary tissue was reported as

Acetyl-Ser-Tyr-Cys-Phe-His,[Asp-NH_2,Glu-(NH_2)]-Cys-(Pro,Val)-Lys-Gly-NH_2

(28). Polypeptides synthesized by coupling Ser-His or His-Ser to free amino terminal of vasopressin were reported to have an increased ratio of CRF to pressor activity (29). Vasopressin itself was persistently proposed as CRF because of its ability to liberate ACTH in vivo (probably due in part to a nonspecific stress response mediated via the hypothalamus) and even in vitro. This topic has been reviewed previously (24). However, various physiological studies, such as the demonstration of CRF in Brattleboro rats with hereditary diabetes insipidus, that do not synthesize vasopressin (30–32), show that CRF and antidiuretic activities involve different hormones. Other substances found in the central nervous system (CNS) have also been proposed during the past 30 years as physiological stimulators of ACTH release. Among these were catecholamines, histamine, acetylcholine, and other substances, but subsequent studies invalidated these claims mainly by revealing nonspecificity in the methods used for detection of CRF activity (28). Some of these substances might at best be considered as being involved at some step in the release of CRF and not ACTH (33). It was also reported (34) that serotonin stimulates ACTH release from dispersed rat pars intermedia cells in a dose-related manner. However, the pars intermedia does not secrete functionally significant quantities of ACTH (35). Moreover, the work of Jones et al (36) suggests that serotonin and acetylcholine stimulate the release of CRF and not of ACTH. In their studies, based on incubation of rat hypothalami, the response to serotonin was blocked by hexamethonium and atropine, indicating that it was mediated by a cholinergic interneuron. Noradrenaline, melatonin, and gamma-aminobutyric acid blocked CRF release (33, 36). These neurotransmitters, if involved in the regulation of CRF release, must act via different receptor mechanisms.

Synaptosomes isolated from sheep and rat hypothalami were reported to contain CRF (37). Electrical field stimulation or depolarizing concentrations (55 mM) of K^+ caused CRF release from the incubated nerve endings in the presence of Ca^{2+} (37).

Recently, several studies on the purification of CRF have been reported. Cooper et al (38) described the purification of 0.1 M HCl extract of porcine hypothalami by ultrafiltration and chromatography on Sephadex G-50. Two materials with CRF activity with mol wt of 30,000 and 1,500 and the characteristics of peptides were obtained, the former representing possibly the aggregated form (39). Jones, Gillham & Hillhouse also found two forms of CRF in media in which rat hypothalami were incubated with serotonin. The purifications consisted of gel filtration, chromatography on carboxyme-thylcellulose (CMC), and high voltage electrophoresis, but it is doubtful that this method can lead to isolation of sufficient CRF to determine its structure. Vale, Rivier & Brown (20) similarly reported two fractions with CRF activity in extracts of sheep hypothalami after separation by ion exchange chromatography. It may be pertinent to recall here that the hypothesis concerning the existence of several molecules with CRF activity was in fact put forward by us more than 14 years ago (28, 40), and the results reviewed above, along with our recent work (see below), seem to confirm it.

In our most recent studies (41), hypothalamic extracts from nearly one-half million pig hypothalami were first separated into 14 fractions by preparative gel filtration on Sephadex G-25. Significant CRF activity was found near the void volume (fraction 2, $R_f = 0.82$–0.7), in intermediate fractions (fraction 8; $R_f = 0.4$) with mol wt about 1,000, and even in the fractions 11 and 12 that had a low R_f of 0.3 and also contained catecholamines. Aliquots of the high mol wt CRF-active fractions from Sephadex with $R_f = 0.82$–0.7 were purified by column chromatography on CMC (41). CRF activity was eluted in fractions with a conductivity of 7,000–10,000 μMHOS. The CRF-active area, still contaminated with ACTH-like materials, was subjected to countercurrent distribution (CCD) in a system of

0.1% acetic acid : l-butanol : pyridine $= 11:5:3$.

The CRF active fractions, $K = 1.1$, released ACTH in vivo and in vitro in doses of 1 μg. The CRF activity of these fractions, in contrast to rat hypothalamic CRF, was not potentiated by a cofactor (42). These fractions were repurified by chromatography on SE-Sephadex to yield two areas with CRF activity (conductivities of 4,000–5,000 μMHOS and 9,600–12,000 μMHOS), detectable in doses of 0.1 μg/ml in vitro, and devoid of any inherent ACTH activity. These fractions were also active in vivo in doses of 1 μg. After rechromatography on SE-Sephadex, some CRF fractions were active in vitro in doses of 0.01 μg. Measurements of molecular weight and amino acid analysis indicated a mol wt for this CRF of about 4,000. CRF activity was completely lost after 16 hr digestion with trypsin and partially destroyed by thermolysin (41). The results indicate that CRF activity in this fraction is due to a basic polypeptide.

Fraction 8 with R_f of 0.4 from Sephadex was also repurified by chromotography on CMC (A. V. Schally, A. Arimura, T. W. Redding, K. Chihara, unpublished). CRF activity was found in well-separated acidic fractions, neutral fractions, and basic fractions. The basic fractions had the highest CRF activity, increasing ACTH release in vitro more than tenfold. However, those fractions had the highest contamination with ACTH-like peptide and attempts are being made to separate them by CCD. The same methods are being used for the purification of small mol wt CRF from Sephadex (fraction 11, 12 with $R_f = 0.3$).

Some preliminary evidence also indicates that CRF may release β-lipotropin and β-endorphin in addition to ACTH (Labrie, Schally et al, unpublished). It is difficult, if not impossible, at present to interpret the finding of multiple CRF activities, but it is possible that high mol wt fractions represent pro-CRF (a precursor of CRF), the intermediate mol wt the physiological CRF(CRH), and that the CRF activity of low mol wt fractions is due to molecules such as catecholamines or serotonin. Further work is needed for the isolation of these CRFs and determination of their physiological role. The clinical usefulness of CRF may be limited to diagnostic tests of pituitary function and to counteracting pituitary suppression in patients who have undergone therapy with adrenal cortical steroids.

THYROTROPIN RELEASING HORMONE (TRH)

Occurrence and Biological Effects

The isolation, elucidation of structure, and synthesis of porcine and ovine TRH (4–9) eliminated the skepticism concerning the existence of the "elusive" hypothalamic hormones and opened up a new field to endocrinologists, synthetic chemists, and clinicians. A variety of basic and clinical studies which were difficult or impossible to perform because of the scarcity and expense of purified materials of natural origin were now made possible. Thus, much basic and clinical data on TRH has been accumulated during the past eight years (3, 18, 20, 23). Synthetic pyroGlu-His-Pro-NH$_2$ has been shown to stimulate thyrotropin (TSH) release in all mammals studied, including mice, rats, nutria (myocaster coypus), sheep, goats, cows, humans, and birds (3, 20). However, it was inactive in the tadpole and the lungfish (43). TRH is active orally, although doses 50–100 times larger than for parenteral administration are needed. Thyroxine and triiodothyronine block the stimulatory effects of TRH by an action exerted directly on the pituitary tissue (3, 20). Subsequently, it was found that TRH will liberate prolactin in animals and humans (44–47) (see also section on PRF), but it remains to be established whether this effect is physiological. TRH can also increase the synthesis of TSH prolactin. In addition, TRH will stimulate growth hormone secretion in lactating cows (48), rats under urethane anesthesia (49), and in patients with acromegaly or renal failure (50). Recently it has been reported that TRH will stimulate colonic activity in rabbits (51). This response, which may explain side effects such as nausea and gastric cramps seen occasionally after administration of TRH, appears to originate in the CNS and is mediated by cholinergic receptors.

The interactions of TRH with centrally acting drugs have been the subject of many investigations (20, 52). The antidepressant activity of TRH in several animal models, which cannot be attributed to its effect on the pituitary-thyroid axis, as well as other neuropharmacological studies (20) suggests that TRH may act on the brain and the spinal cord (52–54). This possible role of TRH as a neurotransmitter in the CNS is supported by its presence in significant concentration in the extra-hypothalamic brain areas of various classes of vertebrates examined, including rat, chicken, snake, frog, tadpole, salmon, and lamprey (54). Networks of TRH-positive nerve terminals were also localized by immunofluorescence techniques around the motor neurons of the rat (55).

Studies on the regional and phylogenetic distribution of TRH were made possible by the generation of antibodies to TRH and the development of sensitive and specific radioimmunoassays (RIA) for this hormone (54, 56–59). Immunoreactive TRH has been detected in blood, cerebrospinal fluid, and urine of rat and man (20, 23, 57). The levels of TRH in human blood are affected by altered thyroid states (59) being less than 60 pg/ml in normal humans, below 5 pg/ml in hyperthyroidism, and 100–600 pg/ml in pituitary hypothyroidism. TRH levels can also be altered by physiological manipulations such as exposure to cold (60). Prior to its measurement in blood or urine, TRH must be extracted with alcohol or with charcoal and repurified, but even so some RIA values of TRH concentrations may have been overestimated severalfold (61).

The measurement of TRH in body fluids has also been made difficult by its rapid inactivation by proteolytic enzyme(s) first demonstrated in our laboratory (62). This inactivation can be prevented by boiling or by benzamidine, 2,3-dimercaptoethanol or 8-hydroxy quinoline plus Tween-20 (59, 62), and is probably due to a deamidase and a peptidase. Among the products of digestion of TRH with plasma or hypothalamic fragments are pyroGlu-His-Pro, proline, and proline amide (3, 18, 23, 63, 64).

The presence of degradative mechanisms has made it difficult to study the biosynthesis of TRH and conflicting findings have been reported (23, 64). However, it has been established recently that two clonal cell lines derived from rat CNS tumors can form [³H]TRH when incubated with [³H]-Pro (65).

Mechanism of Action

TRH is active in vitro in doses as low as 10 pg/ml (3). Studies on the mechanism of action suggest that the first event in TRH action probably involves selective binding to receptor sites on the external surface of the plasma membranes of TSH secreting cells. It has been determined that tritiated TRH ([³H]TRH) is specifically bound by plasma membrane receptors of bovine and murine anterior pituitary glands (66–68). The receptor affinity of analogs of TRH in most cases correlates well with the biological potency on TSH and prolactin release (66, 67), and the effects of TRH on the secretion of these two pituitary hormones may involve binding to the same receptor (67). The intracellular mechanisms which are initiated by the interaction between TRH and the membrane receptor appear to be similar to those which mediate secretory responses of other endocrine cells and to involve cyclic AMP (68, 69). The next step in the sequence of events may thus be the activation of adenylate cyclase, which is indeed associated with plasma membranes of the anterior pituitary gland (68, 69). Adenylate cyclase activity is stimulated by the addition of TRH, and derivatives of cyclic AMP can also stimulate TSH and prolactin release in vitro (69). Theophylline stimulates the release of prolactin as well as of TSH. Thus, cyclic AMP may be the mediator of the action of TRH on the pituitary cell (69, 70). However, the final criterion of Sutherland requiring the demonstration of the activation of adenylate cyclase by TRH in broken cell preparations has not yet been met (70).

Analogs of TRH

Relatively few analogs of TRH have been synthesized, as compared with LH-RH and somatostatin, perhaps because early work (71–73) soon indicated that the structural requirements of this small peptide are quite stringent. Thus, most analogs (3, 20, 23, 71–73) have little TRH activity. However, [(N^{3im}-Me-His)^2]-TRH has 3–10 times greater activity than the natural product (72) and [β-(pyrazolyl-1)-Ala^2]-TRH (74) is also more active than TRH in rats and mice. This is in contrast to [(N^{1im}-Me-His)^2]-TRH and [β-(pyrazolyl-3)-Ala^2]-TRH, which have only 0.1% and 5% activity, respectively (3, 20, 71, 74). Recently it has also been shown that [(N^{3im}-Me-His)^2]-TRH is a more potent TSH and prolactin releaser in humans than TRH (75). Since pyroGlu-Phe-Pro-NH$_2$ was reported to have about 10% activity (76), these findings indicate that both the π electron system and the basicity of the imidazole nucleus (or the isomeric pyrazole nucleus) are necessary for the full biological activity of TRH. Elongation of the polypeptide chain as in the case of pyroGlu-His-Pro-Gly-NH$_2$ (77), results in a compound with 30% of TRH activity. As stated earlier, the activities of TRH analogs for stimulating thyrotropin release and prolactin production are quite similar (67, 77) and a good correlation exists between binding affinity and these biological activities (20, 67). The latter do not depend on conformation alone, since some of these analogs with low activity such as [Leu^2]-TRH have a similar conformation to TRH in solution (78).

There do appear to be differences between the CNS and pituitary receptors for TRH as indicated by some dissociation between TRH and CNS activity, since some analogs such as homo-pyroGlu-His-Pro-NH$_2$ and homo-pyroGlu-His-Thioproline-NH$_2$ are equipotent with TRH in stimulating TSH, but have 4–15 times as much effect on the CNS (79).

Some antagonistic analogs of TRH have been reported recently (80, 81). Lybeck et al found (80) that cyclopentylcarbonylhistidine pyrrolidineamide (cpc-His-pyr) inhibited the TRH-induced TSH release in mice. However, Sievertsson showed the inhibition by cpc-His-pyr to be inconsistent, but that a related analog, cpc-Thi-pyr [Thi = β-(2-thienyl)-L-alanine] inhibited the TRH effect on the release of TSH in vitro (81).

Clinical Remarks

As one would expect from the knowledge of its physiological effects, TRH is used clinically for the differentiation between hypothalamus and pituitary hypothyroidism and for the diagnosis of mild hyper- and hypothyroidism (3, 18, 22, 82), but its proper place in the clinical armamentarium is not fully defined because it is not yet available for general use in the USA. It has been determined that in humans L-DOPA suppresses the TRH-stimulated release of TSH and prolactin (22), while a dopamine receptor agonist 2-bromo-α-ergocriptine (CB-154) inhibits only the prolactin but not the TRH response (83). Somatostatin has an opposite effect; it decreases the TSH but not the prolactin response to TRH (22).

PROLACTIN RELEASING FACTOR (PRF)

Various studies both in vivo and in vitro have revealed that the regulation of prolactin secretion by the hypothalamus is probably mediated by both a prolactin releasing factor (PRF) and a prolactin release inhibiting factor (PIF) (3, 20, 84). Both PIF and PRF activities appear to be present in the hypothalami of mammals and birds, but the available evidence indicates that PIF predominates in the former and PRF in the latter (3, 20). PIF and PRF activities seem to be located in different areas of the hypothalamus. A part of the PRF activity in the extracts of hypothalami of domestic animals is clearly due to TRH which, under specific conditions, can greatly stimulate prolactin secretion in vivo and in vitro in various mammals, including humans (3, 20, 44–47, 84) (see also section on TRH). The relative PRF activities of many analogs of TRH are comparable to their thyrotropin releasing potencies. It has also been reported recently that administration of antisera to TRH will greatly lower prolactin levels in rats (85). This finding, if confirmed, might support the concept of a physiological role for TRH in the control of prolactin secretion.

However, there is also evidence that a PRH different from TRH exists: (*a*) prolactin and TSH release induced by TRH can be dissociated under various physiological conditions (86); (*b*) clinical studies indicate that there is a lack of parallelism between TSH and prolactin levels in patients with various thyroid diseases (87). Moreover, pretreatment with 100 μg triiodothyronine inhibits the TSH but not the prolactin response to TRH in normal men (87); (*c*) work of several groups suggests that partially purified hypothalamic fractions, apparently different from TRH, can still stimulate the release of prolactin from rat pituitaries in vivo and in vitro (3, 20, 88–90). TRH is less resistant to inactivation by incubation with fresh serum than PRF and can be separated from PRF by adsorption on charcoal (87, 88). However, the work cited above (87, 88) involved relatively crude fractions of PRF. The search for PRF, in our own experience, has been made difficult by the presence in hypothalamic fractions of substances that increase the release of prolactin, such as TRH, and those which inhibit it, such as catecholamines and gamma-aminobutyric acid (91–93). These compounds can, under certain conditions, obscure or neutralize other PIF and PRF activities. When one of these Sephadex fractions was chromatographed on CMC, the PRF activity was consistently found in the elution area of TRH fractions. This could indicate that one of these PRFs is chemically similar to TRH or that all the activity is due to TRH. Estimates of TRH activity by bioassays are of no help in this case, since the TRH values would be several times too low because of the presence of somatostatin, but the RIA for TRH indicates that this PRF activity probably cannot be accounted for by TRH contamination. However, we have established that several other fractions with different physicochemical properties from TRH, as shown by different partition coefficients on CCD, basicity on CMC and SE-Sephadex, and behavior on molecular sieving on Sephadex, can also stimulate the release of prolactin from rat pituitary fragments from monolayer cell cultures (A. V. Schally, A. Arimura, T. W. Redding, unpublished).

Various drugs, among them inhibitors of catecholamine action, synthesis, or storage, such as reserpine, methyl-DOPA, α-methyl-p-tyrosine, chlorpromazine, perphenazine and other phenothiazines, haloperidol and tricyclic antidepressants, and sulpiride (a benzamide), together with others such as ether and nicotine, all lead to an increase in prolactin levels (3, 20, 23, 84). Some of these drugs act directly on the pituitary, but most of them affect the pathways or turnover of putative synaptic neurotransmitters controlling PRF and PIF secretion. Recently, evidence has been obtained that serotonin is the neurotransmitter involved in control of PRF release (94), since administration of the serotonin uptake inhibitors fluoxetine and 5-hydroxytryptophan increased the serum prolactin levels (94).

A variety of natural substances of CNS and extra-CNS origin can augment prolactin release in vivo and in some cases in vitro. Among these substances are vasopressin, estrogens, monoiodotyrosine, thyroxine and triiodothyronine, histamine, cyclic AMP, prostaglandins E_1 and E_2, and 5-hydroxytryptophan. The latter is the biosynthetic precursor of the postulated neurotransmitter of PRF. Cyclic AMP and prostaglandins can mediate in the action of PRF on the cell, but the sites and mechanisms of action of other substances are not completely understood. However, their effects can be explained by synergism, stress, or an action mediated via the CNS (84, 86–88). Similarly, the opiate peptides β-endorphin and Met-enkephalin and some of their analogs will also release prolactin, if given into the cerebral ventricle, but their effects are exerted on the CNS centers and, therefore, like the substances quoted above, they do not represent the physiological PRF, the chemistry of which remains to be elucidated.

PROLACTIN RELEASE-INHIBITING FACTOR (PIF)

Although the presence of PIF activity in hypothalamic extracts was demonstrated many years ago (for review see 3, 20, 23, 24, 84), the chemistry of PIF still remains to be elucidated. Many substances are present in the brain or hypothalamic extracts which can inhibit prolactin secretion in vitro and in vivo (3, 18, 20, 24, 90–94). Acetylcholine and serotonin have been implicated in the control of the release of prolactin, the former being inhibitory and the latter stimulatory.

It has also been demonstrated that the catecholamines influence the release of prolactin (95, 96). It was suggested that catecholamines might exert this effect by stimulating the release of PIF from the hypothalamus (97, 98). However, our recent work showed that hypothalamic catecholamines also inhibit prolactin release by a direct action on the pituitary. PIF activity present in acetic acid extracts of pig hypothalami was purified by gel filtration on Sephadex G-25, extraction with phenol, chromatography on CMC columns, countercurrent distribution, rechromatography on Sephadex G-25, and partition chromatography (91). PIF activity was detected by inhibition of prolactin release in vitro from rat pituitaries, and in vivo by infusion into a hypophysial portal vessel of the rat (99). Some of the highly purified fractions which strongly inhibited the release of prolactin in vitro and in vivo were found to contain much noradrenalin and dopamine. The magnitude of inhibition of release of prolactin was related to noradrenalin content. Synthetic noradrenalin and dopamine in doses of 10–100 ng also strongly inhibited the release of prolactin in vitro (91). When dopamine or noradrenalin were dissolved in a fresh 5% glucose solution and infused into the hypophysial vessel of the rat, prolactin secretion was significantly suppressed as compared to the glucose-infused group. The suppressive effect of catecholamines was dose dependent (99). These results (91, 99) showed that catecholamines, purified from hypothalamic tissue of synthetic origin, inhibit the release of prolactin by an action exerted directly on the pituitary gland. However, whether any catecholamine represents a physiological prolactin release inhibiting hormone remains to be determined.

It should also be noted that in favor of a role for cerebral catecholamines in the control of release of prolactin are the findings that dopamine receptor blocking agents such as perphenazine, pimozide, or haloperidol nullify catecholamine-mediated inhibition of prolactin release (20, 23, 91, 92). Apomorphine, which mimics the action of dopamine, also suppresses prolactin release, as do L-DOPA, which elevates cerebral catecholamines, and monoamine oxidase inhibitors such as iproniazid and pargyline.

Another hypothalamic substance with PIF activity, the effect of which, in contrast to catecholamines, cannot be blocked by perphenazine is gamma-aminobutyric acid (GABA) (92). GABA was isolated by us from a fraction with PIF activity obtained by chromatography on carboxymethylcellulose during the concentration of catecholamine from pig hypothalami. This fraction, chromatographically distinct from catecholamines, was purified further by six steps involving rechromatography on Sephadex G-25, CCD in two different solvent systems, free-flow electrophoresis, and chromatography on triethylaminoethyl cellulose, to yield large amounts of GABA (92). Natural and synthetic GABA inhibited prolactin release in vitro from isolated rat pituitary halves in doses as low as $0.1 \mu g/ml$. In this system, the extent of inhibition was proportional to the dose, and natural and synthetic GABA possessed identical PIF activity. Similarly, synthetic GABA suppressed prolactin release in monolayer cultures of rat pituitary cells and inhibited the TRH-stimulated prolactin secretion. The inhibition of prolactin release in vitro by GABA could not be blocked by perphenazine. GABA also had PIF activity in vivo, although large doses were needed for an effect. After serum prolactin in male or female rats was elevated by injection of monoiodotyrosine (MIT), perphenazine, chlorpromazine, haloperidol, or sulpiride, intravenous administration of GABA in doses of 1–100 mg, or an infusion, significantly decreased serum prolactin levels. Oral administration of 300 mg GABA also completely suppressed the MIT-induced elevation in prolactin levels. β-Hydroxy-GABA also significantly depressed prolactin release, but β-(p-chlorophenyl)-GABA (Lioresal®, CIBA) and four other analogs of GABA were not effective. The results indicate that GABA can inhibit prolactin release by a direct action on the pituitary gland, but again it is not known whether this effect is physiological or pharmacological, because large doses are needed to obtain an effect (92).

Other compounds with PIF activity different from catecholamines and GABA, but still unidentified, are present in pig hypothalamic extracts. One of them appears to be a polyamine and the other a polypeptide (A. V. Schally, A. Arimura, T. W. Redding, unpublished).

Somatostatin has some inhibitory effect on prolactin secretion, particularly in vitro in monolayer cultures of pituitary cells (18, 20), but its presence can be easily detected by RIA and it has been absent in some fractions with PIF activity.

L-DOPA and 2-bromo-α-ergocriptin (CB-154), an ergot alkaloid, are used in order to inhibit the release of prolactin and to suppress undesired lactation, for instance in cases of galactorrhea. It is clear that a naturally occurring compound with PIF activity could be useful clinically in the situation described above, and possibly in others, too, so that the chemistry of PIF and its mode of action on the anterior pituitary is still being actively investigated.

MELANOCYTE STIMULATING HORMONE (MSH)-RELEASING FACTOR (MRF), AND RELEASE INHIBITING FACTOR (MIF)

There is agreement now that the release of MSH from the pars intermedia of the pituitary gland is controlled by the hypothalamic stimulating factor (MRF) and inhibitory factor (MIF), the latter having a predominant role (3, 20, 23, 24, 52). However, the physiological MIF and MRF have still not been identified with certainty (52). The first prospective MIF was isolated from bovine hypothalami and identified as H-Pro-Leu-Gly-NH$_2$ in this laboratory (100, 101). The group of Walter (102) originally observed that H-Pro-Leu-Gly-NH$_2$ could be formed by incubating oxytocin with an enzyme present in hypothalamic tissue and that this tripeptide inhibits MSH release in the rat (102). The same group also demonstrated that hypothalami of rabbit and rat contain a membrane bound exopeptidase which degrades radioactively-labeled oxytocin [(9-glycinamide-1-^{14}C)-oxytocin] stepwise to produce labeled H-Pro-Leu-Gly-NH$_2$. Supernatant fractions of rabbit and rat hypothalami contain predominantly an endopeptidase which releases H-Leu-Gly-NH$_2$ from oxytocin (103). [^{3}H] and [^{14}C]-labeled H-Pro-Leu-Gly-NH$_2$ accumulated in the pineal, pituitary, kidney, liver, and adrenals; the elevated tissue to plasma ratio in the pineal, along with the identification of unchanged H-Pro-Leu-Gly-NH$_2$ there, suggests the possibility of a direct influence of the hypothalamus upon the pineal gland (104). In agreement with these findings is the recent report of Pavel et al (105) who determined that the intracarotid injection of both synthetic Pro-Leu-Gly-NH$_2$ and the corresponding purified MIF prepared from bovine hypothalami induces arginine vasotocin release into the cerebrospinal fluid of cats and significantly decreases the pineal arginine vasotocin content five minutes after the injection (105).

Various studies in animal models of Parkinsonism and depression (reviewed in 52) suggest a direct CNS action of H-Pro-Leu-Gly-NH$_2$ independent of its effect on MSH secretion. It has also been reported that H-Pro-Leu-Gly-NH$_2$ induces dopamine synthesis in brain slices (106), but our own studies failed to confirm this (52). H-Pro-Leu-Gly-NH$_2$ has been shown to be effective alone or in conjunction with L-DOPA in Parkinson's disease and clinical trials have been conducted (52).

Several analogs of MIF have been synthesized recently by the group at Ayerst Research Laboratories. These peptides include H-Pro-N-isobutyl-Gly-Gly-NH$_2$ and two stereoisomers of both H-Pro-N-Me-Leu-Gly-NH$_2$ (LL) and (LD) and H-Pro-N-Me-Leu-Ala-NH$_2$ (LLD) and (LDD) (107–109). All analogs except LL potentiated behavioral effects of L-DOPA in mice and all antagonized fluphenazine induced catalepsy in rats. In these tests, L-Pro-N-Methyl-D-Leu-Gly-NH$_2$ was the most active after parenteral or oral administration and it also affected brain catecholamine turnover (107–109).

Not all authors agree that H-Pro-Leu-Gly-NH$_2$ is in fact MIF. Other substances such as tocinoic acid, the cyclic pentapeptide ring of oxytocin, and tocinamide have been proposed as MIFs (110). These peptides were said to be active in the rat and hamster, less active in the toad and the bullfrog, and inactive in the frog (110). This report could not be confirmed by others, and the proposal was later withdrawn. Another peptide isolated by us from beef hypothalami and characterized as H-Pro-His-Phe-Arg-Gly-NH$_2$ also has some MIF activity, but it was about 1000 times less than Pro-Leu-Gly-NH$_2$ (3).

There is also evidence for an MRF. Celis et al (111) proposed that the opened N-terminal ring portion of oxytocin H-Cys-Tyr-Ile-Gln-Asn-Cys-OH constitutes MRF. However, open end or unopened cyclic pentapeptide fragments of oxytocin have not been identified in the hypothalamus.

Various drugs such as the phenothiazines (112) have also been shown to increase MSH release (112, 113) probably by depleting cerebral catecholamine content which in turn was said to lead to a decrease in MIF release (112, 113). In such a case, the release of MIF would be under adrenergic control. Evidence has also been put forward that catecholamines directly inhibit pituitary MSH secretion (114), and it was proposed that the central inhibition of MSH secretion may be mediated by dopaminergic innervation of pars intermedia cells (115). In conclusion, considerable confusion still exists as to the identity of the physiological MIF and MRF.

LUTEINIZING HORMONE-AND FOLLICLE STIMULATING HORMONE-RELEASING HORMONE

The isolation (10, 11), determination of structure (12, 13), and synthesis (14) of porcine LH-RH/FSH-RH by our group opened up vast new areas to physiological, immunological, biochemical, behavioral, immunohistological, veterinary, and clinical investigations into reproduction. In addition, the interest in possible veterinary and medical applications of LH-RH analogs caused a vastly expanded interest in their synthesis. These extensive investigations have been the subject of recent reviews by our group (15, 18, 116–119) and others (20, 23, 120), so that only the latest developments are described here. However, because this is the first review on hypothalamic hormones in this series, some earlier key findings will be touched upon for the sake of completeness and clarity. For historical details, the reader is referred to previous reviews (3, 24, 121–123).

Actions on the Pituitary

Porcine LH-RH corresponding in structure to

pyro Glu-His-Trp-Ser-Tyr-Gly-Leu-Arg-Pro-Gly-NH$_2$

was first synthesized by us (12, 14, 121–124) and then by many other groups (3, 15, 18, 20, 120). This made the new hormone readily available for a variety of studies. The structure of ovine LH-RH was later shown to be identical with the porcine hormone (125) and subsequent biochemical and immunological studies indicated that bovine, human, and rat LH-RH probably have the same structure (3, 15, 18, 126). Our original observations that natural porcine LH-RH and the synthetic decapeptide release both LH and FSH (3, 10, 11, 15, 18, 121–124) have been confirmed in and extended to a variety of animals, including rats, mice, rabbits, golden hamsters, mink, spotted skunk, impala, rock hyrax, sheep, cattle, pigs, horses, monkeys, and humans (15, 18). The fact that ovulation has been induced in most of these species with LH-RH demonstrates that this decapeptide can release enough FSH to induce follicular maturation. Increases in sex steroid levels in blood have also been reported after administration of LH-RH (15, 18). LH-RH is also active in birds such as domestic fowl and pigeons, and in some species of fish such as brown trout and carp (15, 18). This indicates also that species specificity does not occur with LH-RH, although dogfish and goldfish LH-RH may be immunologically different from the porcine and ovine hormone (15). However, it has been reported recently that the bonnet monkey is insensitive to [D-Ala^6, des-Gly^{10}]-LH-RH ethylamide (127), an analog which has been shown to be a highly effective stimulant of LH and FSH release in rats, mice, humans, sheep, and other species (15) (see also agonistic analogs of LH-RH).

The evidence that the decapeptide corresponding in structure to porcine LH-RH also has FSH-RH activity in vivo and in vitro is now indisputable. Our proposal (124) that LH-RH is also the physiological FSH-RH has not been seriously challenged, although claims have been made that other natural materials with FSH-RH activity exist (128, 129). One of these (129) was only active in vitro but not in vivo, so that it cannot be excluded that it somehow potentiated FSH release under the conditions of the test. It is also possible that these materials represent artifacts of LH-RH. Recent results indicate that LH-RH decapeptide represents the bulk of FSH-RH activity in the hypothalamus (130). Various immunological results (see below) strongly support the importance of LH-RH as the physiological FSH-releasing hormone.

Investigations of the routes of administration in rats and humans revealed that LH-RH is effective not only after intravenous, subcutaneous, intramuscular, and intracarotid injection, but also after intravaginal, intrarectal, intranasal, cutaneous (on the skin in dimethyl sulfoxide), and even oral administration (15, 18). However, the doses required for effect by extravascular routes are 100–10,000 times larger than by the parenteral.

The occasional dissociation of LH and FSH secretion can be explained in part by complex intereactions with sex steroids which have been the subject of many investigations (3, 15, 18, 20, 121–123, 131). The feedback effects of sex steroids are principally inhibitory (negative), but can also be stimulatory (positive), especially in the case of estrogen. They are exerted in part on the hypothalamus and/or another CNS center, and in part on the pituitary (3, 15, 18, 131). This direct action of sex steroids on the pituitary, which has now been unequivocally established, may involve effects on receptor-binding sites for LH-RH (15, 18, 20).

Immunological and Immunohistochemical Studies with LH-RH

A variety of immunological studies have been carried out with LH-RH. Antisera to LH-RH have been produced in rats, rabbits, guinea pigs, sheep, and humans (3, 15, 18, 132–134), and several radioimmunoassays were developed (15, 18, 126, 132, 134). Male rabbits that were actively immunized with LH-RH and had generated antibodies to it developed testicular atrophy associated with aspermatogenesis (15, 18). Passive immunization of rats with LH-RH antiserum prevented the preovulatory surge of LH and FSH, blocked ovulation (15, 18, 134), and induced hyperprolactinemia (135). In hamsters (15, 18) there was an arrest of follicular maturation, reduction in serum estradiol levels, and suppression of the LH surge and of ovulation (136). Reduction in serum FSH in addition to LH in rats after passive or active immunization with LH-RH supports the physiological role of the peptide in the regulation of FSH secretion (15, 18). LH-RH appears to be responsible for the onset of puberty (137). It is also necessary for normal implantation of fertilized ova and maintenance of pregnancy, since administration of LH-RH antisera to rats delays the former or terminates the latter, depending upon the time of injection (18, 138, 139).

The availability of antisera to LH-RH made possible various studies on localization of LH-RH by radioimmunoassays or immunocytochemical methods (15, 18, 23, 140, 141). The bulk of LH-RH appears to be localized in the median eminence and in the arcuate nucleus (15, 18, 23, 140, 141). The pathway of LH-RH containing nerve fibers in the median eminence of rats coincides with the course of the nerve fibers of the tubero-infundibular tract (33, 140, 141). The production of LH-RH in neuronal cell bodies seems to be well documented (140, 141). Immunohistological findings indicate that extrahypothalamic brain areas are also involved in the synthesis of LH-RH (141). These findings, together with (a) the significant LH-RH content of extrahypothalamic brain areas, particularly mesencephalon and organum vasculosum (b) the fact that some axons which appear to be carrying LH-RH terminate outside the median eminence, and (c) the location of LH-RH in synaptosomes, suggest that in addition to being the regulator of the release of LH and FSH, LH-RH might act as a neurotransmitter (23, 52, 142) with effects on sex drive. LH-RH has indeed been shown to excite sexual behavior in rats and to modulate the electrical activity of neurons in the CNS (142).

LH-RH has been detected by bioassay and by radioimmunoassay in the hypophysial portal blood of rats and monkeys and in peripheral circulation of women at midcycle ovulatory LH surge (15, 18, 23, 143) and in postmenopausal women (126). LH-RH, like other hormones, appears to be released in pulsatile fashion.

Biosynthesis, Degradation, Control of Secretion, and Mechanism of Action

The limited work that has been done on the biosynthesis of LH-RH was reviewed recently (15, 18, 23). LH-RH is inactivated in the blood stream and excreted by the kidney (3, 15, 18, 144). The rapid degradation of LH-RH by tissue extracts and homogenates was covered in recent reviews (3, 15, 18, 20) (see also below). The nature of putative neurotransmitters responsible for the control of secretion of LH-RH is at present the subject of some controversy, but the available evidence implicates catecholamines (20, 23).

The mechanism of action of LH-RH is also not completely clear. However, much evidence indicates that cyclic AMP (cAMP) may be the mediator of its action (69), since (*a*) cAMP, its derivatives, or agents which increase intracellular cAMP (such as prostaglandins) stimulate LH and FSH secretion in vitro (69); (*b*) LH-RH stimulates cAMP accumulation in rat anterior pituitary tissue in vitro (69); (*c*) there is a close parallel between the LH and FSH releasing activity of various stimulatory (agonistic) analogs of LH-RH and their abilities to induce accumulation of cAMP (15, 18, 20, 69, 118), and (*d*) antagonists of LH-RH inhibit this cAMP accumulation. LH-RH appears to exert its effect by activating adenyl cyclase which may lead to phosphorylation of physiologically important protein substrates (69, 118).

Analogs of LH-RH

The synthesis of many hundreds of LH-RH analogs was made possible by the use of rapid solid-phase techniques. The biological activities of these peptides have provided much insight into the role played by individual amino acids in preserving overall conformation and binding affinity to pituitary receptor sites and in triggering gonadotropin release. Since structure-function studies on LH-RH have been the subject of several recent review articles (15, 18, 20, 116–120), we limit this discussion to some of the highlights and recent developments in this field.

BIOASSAY SYSTEMS   A number of in vivo and in vitro assay systems have been developed for testing gonadotropin-releasing and anti-LH-RH effects of synthetic analogs. The most valuable of these have been stimulation and inhibition of LH and FSH release in immature male rats (146), particularly for long-acting peptides, and incubation of rat pituitary halves (2, 147) or monolayer cultures of rat anterior pituitary cells (148). A more recent and very effective method for determining potencies of the more active inhibitory analogs is the blockade of ovulation in 4-day cycling rats and in hamsters (149, 150). Since the mechanism of action of LH-RH involves interaction with pituitary plasma membrane receptors, an in vitro assay using highly purified pituitary cell membrane preparations was also developed (151) in this laboratory. We have determined that the number of binding sites for LH-RH is approximately 2.36 pM/mg of protein with a high affinity constant of 7.1×10^9 M^{-1}. We have also observed that the agonist [D-Trp^6]-LH-RH, and the antagonist [D-Phe^2, D-Trp^3, D-Phe^6]-LH-RH compete with LH-RH for its pituitary plasma membrane receptors, displacing the [^{125}I]-LH-RH more strongly than its parent hormone (E. Pedroza, J. Vilchez, and A. V. Schally, unpublished). Therefore, both stimulatory and inhibitory analogs of LH-RH apparently exert their action through the same pituitary plasma membrane receptors as those for LH-RH. The more potent and long-acting effect of such superactive analogs as [D-Trp^6]-LH-RH could be due to their higher ability to bind to the pituitary LH-RH receptors. Similarly, the blockage of ovulation by several inhibitory analogs of LH-RH, such as [D-Phe^2, D-Trp^3, D-Phe^6]-LH-RH, is explained by the same mechanism of binding to LH-RH pituitary receptors.

AGONISTS Peptides with greater activities and, therefore, greater potential therapeutic usefulness than LH-RH have resulted from changes at two positions in the decapeptide. Replacement of glycineamide in position 10 by certain alkylamine groups such as CH_3CH_2NH- and CF_3CH_2NH- result in analogs which are 3–9 times more active than LH-RH (152, 153). Replacement of glycine in position 6 by D-alanine (154) increases gonadotropin-releasing activity sixfold and this is perhaps due to the stabilization of the preferred binding conformation of LH-RH (155, 156). It has been shown (157–159) that cleavage of the Gly-Leu peptide bond of LH-RH is a major enzymatic degradative pathway using hypothalamic homogenates and that this cleavage is inhibited by the presence of D-amino acids in place of glycine; however, there is no evidence of increased plasma half-life with these analogs. The activities of D-amino acid^6 analogs increase with the size of the side chains and the D-leucine (160, 161), D-phenylalanine (162), and D-tryptophan (162) peptides increase in activity in that order to an in vivo limit in the region of 15 times the activity of LH-RH.

A logical extension of this work was the synthesis and examination of peptides containing both types of modification and, as predicted, [D-Ala^6, des-Gly-NH$_2$^{10}]-LH-RH ethylamide (163, 164), [D-Leu^6,des-Gly-NH$_2$^{10}]-LH-RH ethylamide (160, 161) and [D-Ser(Bu^t)^6,des-Gly-NH$_2$^{10}]-LH-RH ethylamide (165, 166) were found to be roughly 20–100 times more active than LH-RH in immature male rats and for induction of ovulation. Furthermore, this type of analog exhibited prolonged activity in vivo. Experiments with [^{125}I]- labeled D-Ala^6- and D-Leu^6- ethylamide peptides indicate that they do not have increased plasma half-lives and that their greater potencies and protracted gonadotropin responses are due to increases in binding affinity, and in the time the analogs remain bound to pituitary tissue (167). Surprisingly, the ethylamide analogs of [D-Phe^6]- and [D-Trp^6]-LH-

RH were found to be slightly less active than the parent peptides in the immature male rat system (162).

CLINICAL USES OF LH-RH AND ITS AGONISTS   LH-RH has been used diagnostically to determine pituitary LH and FSH reserve, therapeutically to induce ovulation in amenorrheic women, and to treat oligospermia in men (3, 15, 18, 22). The use of LH-RH and its analogs can avoid super-ovulation and the resultant multiple births that are not uncommon after administration of preparations of human menopausal gonadotropins (HMG-Pergonal®) followed by human chorionic gonadotropin (hCG). Since one injection of the superactive analogs [D-Ala^6,des-Gly-NH$_2$^{10}]-LH-RH ethylamide (163, 164), [D-Leu^6,des-Gly-NH$_2$^{10}]-LH-RH ethylamide (160, 161), [D-Ser(Bu^t)^6,des-Gly-NH$_2$^{10}]-LH-RH ethylamide (165, 166), and [D-Trp^6]-LH-RH induces protracted stimulation of release of LH and FSH lasting as long as 24 hr, they are more convenient and practical to use than LH-RH that has occasionally been given three times daily for thera-peutic purposes (15, 22). Moreover, these analogs are active not only after parenteral but also after intranasal, intravaginal, intrarectal, and oral ad-ministration if suitable doses are given (15, 18, 22).

PRE- AND POSTCOITAL CONTRACEPTIVE ACTIVITY OF LH-RH AND ITS AGONISTS   Either LH-RH (10–1,000 μg) or the superactive analogs such as [D-Ala^6,des-Gly-NH$_2$^{10}]-LH-RH ethylamide (20–100 ng) have been found to cause premature ovulation and inhibition of mating and pregnancy when administered acutely the day of diestrus or proestrus to cyclic rats (168, 169). This temporary antifertility effect was ascribed to the induction of ovulation at the physiologically "wrong time" (168). Prolonged treat-ment of immature female rats with 1–10 μg/day of superactive analogs [D-Ser(Bu^t)^6,des-Gly-NH$_2$^{10}]-LH-RH ethylamide, [D-Leu^6]-LH-RH ethyl-amide, [D-Ala^6]-LH-RH ethylamide, or [D-Trp^6]-LH-RH inhibited ovarian growth and maturation, produced dose-related delays in the vaginal open-ing, absence of cycling, and in mature rats caused cessation of cycling, atrophy of ovaries, reduction in uterine weight, and reduction in estrogen production (166, 170). There is evidence (171) that [^{125}I]-labeled [D-Leu^6, des-Gly-NH$_2$^{10}]-LH-RH ethylamide is specifically bound to ovarian recep-tors. It is therefore possible that these superactive analogs, and presumably LH-RH itself, could have a direct inhibitory influence on ovarian growth (171). Long-term administration of [D-Leu^6]-LH-RH ethylamide induced a regression of endocrine-dependent mammary tumor in rats (172). Admin-istration of [D-Ser(Bu^t)^6,des-Gly-NH$_2$^{10}]-LH-RH ethylamide in doses of 50–200 μg/kg of body weight to male rats and dogs reduced testosterone content and caused atrophy of the testes, seminal vesicles, and prostate gland (166). Recent studies (173) have shown that daily administration of as little as 8 ng of [D-Leu^6,des-Gly-NH$_2$^{10}]-LH-RH ethylamide three times a day for one week to male rats results in a 30% reduction of testicular LH/hCG and prolactin receptors with a maximal reduction of 80% at 40 ng. FSH receptor levels were not affected, but testosterone levels were reduced. These results are consistent with the marked loss of LH receptors and steroidogenic response to gonadotropins observed in the rat after daily injection of LH or hCG (174, 175). Equally dramatic is the ability of large doses of LH-RH (100–1,000 μg/day) and smaller but still pharmacological doses (1–6 μg/day) of the superactive analogs to block implantation and terminate gestation when given daily postcoitally to rats (176–179). This effect is also dose-dependent and appears to be directly related to hypersecretion of LH, functional luteolysis and/or inhibition of progesterone secretion (176, 179). In view of these paradoxical antifertility effects of large doses of LH-RH and long-acting superactive analogs, caution must be exercised in devising clinical protocols.

INHIBITORS   It is now clear that only certain modifications to positions 1, 2, and 3 appear to result in LH-RH antagonists (15, 18, 118) and it is considered that this N-terminal portion of the decapeptide constitutes its active center with respect to gonadotropin release (3, 15). The remaining C-terminal portion of the molecule is involved primarily in the binding process and in preserving overall conformation.

The first inhibitory peptide claimed to be active was [des-His$_2$]-LH-RH (180), which was found to be at best a weak inhibitor of LH-RH in the monolayer cell system. Later it was found that tryptophan in position 3 could also be altered to give mildly inhibitory peptides such as [Leu^3]-LH-RH (181). Attempts were made to increase the inhibitory activities of some of these weak inhibitors by introducing some of the superactive modifications already discussed into the same molecule. This approach was shown to be valid when we found (182) that [des-His^2,des-Gly-NH$_2$^{10}]-LH-RH ethylamide was more effective than [des-His^2]-LH-RH in the rat. [des-His^2,D-Ala^6]-LH-RH was also found (154) to be more potent. Unexpectedly, the combination inhibitory analogs containing both the D-amino acid in position 6 and the C-terminal ethylamide group were not as effective as the high activities of the parent agonist activities had suggested. Furthermore, they seemed to have generally higher residual agonist activity in vivo (117). However, as in the case of the corresponding agonist peptides, the substitution of D-Ala, D-Leu, D-Phe, or D-Trp in position 6 of a particular inhibitor increases antagonist activity dramatically in that order (118).

Another important discovery (183) in the antagonist field was that [D-Phe^2]-LH-RH was far more effective than [des-His^2]-LH-RH in vivo and in vitro. In general, the [D-Phe^2] analogs are about three times more active than the [des-His^2] analogs in vitro (184) and both more potent and longer-acting in vivo (118). At present, all of the most active inhibitory peptides are based on this modification. Analogs such as [D-Phe^2,D-Ala^6]- and [D-Phe^2,D-Leu^6]-LH-RH are capable of blocking ovulation (149, 185, 186) and the preovulatory gonadotropin surge in cyclic rats when multiple doses of 12 mg/kg or greater are given during the afternoon of proestrus. Analogs such as [D-Phe^2,D-Phe^6]- and [D-Phe^2,D-Trp^6]-LH-RH are effective in single doses of about 6 mg/kg (187, 188).

With the finding that tryptophan in [D-Phe^2,D-Phe^6]-LH-RH could be replaced by phenylalanine with no loss of inhibitory potency (187) and a beneficial reduction in residual agonist activity, many modifications to position 3 have been made resulting in the discovery of at least two even more active peptides. These are [D-Phe^2,D-Trp^3,D-Phe^6]-LH-RH (188) and [D-Phe^2,Pro^3,D-Phe^6]-LH-RH (189), both of which completely block ovulation at single doses in the region of 1 mg per rat. Recent evidence indicates that [D-Phe^2,D-Trp^3,D-Phe^6] inhibits the responses to LH-RH in humans (Gonzales-Barcena, Kastin, Coy, Nikolics, and Schally. 1977. *Lancet* 2:997–98). Work is presently in progress in this laboratory on certain dimeric forms of some of these inhibitory peptides which are up to four times more active than the best ones reported here. Thus, excellent progress is still being made in this fascinating area, which can possibly lead to development of new birth control methods.

GROWTH HORMONE-RELEASING FACTOR (GH-RF)

Physiological evidence indicates that the secretion of growth hormone (GH) from the anterior pituitary is regulated by a dual system of hypothalamic control, one inhibitory and one stimulatory (3, 18, 22, 23). The inhibition of GH secretion is mediated by somatostatin (16, 17) which is described in the next section.

The stimulatory effect on GH release of some hypothalamic fractions (190) appears to be due to a GH-RF, first detected more than 13 years ago (for review see 3, 20, 23, 24), which, under some conditions, might predominate over the inhibitory action of somatostatin because of the short half-life of the latter. It would be very desirable to obtain GH-RF clinically because of the world shortage of GH and the assumed capability of GH-RF to stimulate growth in young pituitary dwarfs and to serve as an anabolic agent free of androgenic effects (3, 18, 22). However, despite the intense effort by several groups, the nature of the physiological GH-RF is still unknown.

Several years ago we isolated a decapeptide from porcine hypothalami (191), the amino acid sequence of which (192) was similar to the amino terminal sequence of the β-chain of porcine hemoglobin. The natural deca-peptide and the corresponding synthetic material stimulated the release of bioassayable GH in the rat in certain in vitro and in vivo systems (191, 193), but did not have an effect on the levels of immunoreactive GH in rats, sheep, pigs, monkeys, and humans (192–194). Consequently, we withdrew the proposal that this decapeptide might represent GH-RF (194). Similarly, the claims of Youdaev (195) that pyroGlu-Ser-Gly-NH$_2$ might be GH-RF were dispelled when we and others (196) reported that this tripeptide neither stimulated GH release in vitro or in vivo in rats, nor increased serum GH levels in sheep or humans.

Several substances found in the hypothalamus and/or higher brain centers such as catecholamines, serotonin, TRH, vasopressin, substance P, endorphins and enkephalins (and their analogs), arginine, prostaglandins, guanosine monophosphate (cGMP), and other natural or synthetic substances such as insulin, 2-deoxyglucose, apomorphine, L-DOPA, and cAMP derivatives can, under specific conditions, stimulate the release of GH in vivo and in vitro (20, 22, 23, 197–199) (see also section on TRH). However, the effect of these substances may be explained by (*a*) the presence of nonspecific stress (e.g. that induced by vasopressin in primates, but not in rats or mice), including hypoglycemia; (*b*) the stimulation of glucose-sensitive hypothalamic receptors (insulin, 2-deoxyglucose); (*c*) their role as intermediates (messengers) in the action of GH-RF (cAMP, cGMP, prosta-glandin) (198); (*d*) the similarity of or loss of specificity to receptors, e.g. in acromegaly (TRH) (22); (*e*) their action as putative neurotransmitters involved in the stimulation of release of GH-RF (catecholamines and serotonin) or stimulation of dopamine receptors (apomorphine and L-DOPA); (*f*) action by means of the higher brain centers and other effects (enkephalins and endorphins). For a review of this complex topic, see Müller (199).

Attempts to purify GH-RF have been hampered by the presence of large amounts of somatostatin and somatostatin-like substances in hypothalamic extracts. The availability of antiserum to somatostatin has now greatly facilitated the search for GH-RF by permitting the neutralization of inher-ent growth hormone release inhibiting activity in fractions being assayed for GH-RF. Thus, using antisera to somatostatin or columns of Sepharose linked to anti-somatostatin–gamma-globulin to eliminate somatostatin and its assumed precursors, we have obtained new evidence for the existence of GH-RF and have purified several fractions with GH-RF activity by gel filtration on Sephadex, CCD, and chromatography on CMC (200). These fractions stimulated the release of growth hormone from monolayer cultures of rat anterior pituitaries at a dose of $1\mu g/ml$ and increased plasma GH in mice pretreated with chloropromazine-morphine-Nembutal (200). However, future work is needed for the isolation and structural elucidation of GH-RF.

SOMATOSTATIN

Somatostatin (growth hormone release-inhibiting hormone; GH-RIH) was isolated from ovine (16) and subsequently porcine (17) hypothalami for its ability to inhibit the release of pituitary growth hormone (GH). Synthetic somatostatin was found to produce a remarkable array of actions on diverse hormones and other substances from many tissues. These discoveries gained physiological significance when we found somatostatin-like immunoreactivity in high concentrations within those tissues affected by the hormone (19). Thus, in addition to the hypothalamus, somatostatin appears to be localized in pancreas, stomach, gut, and brain, and it inhibits the release of GH, TSH, insulin, glucagon, gastrin, secretin (and other gut hormones), pancreatic bicarbonate, and gastric acid; it also affects CNS functions. This interesting hormone has had an impact on current concepts of regulatory physiology, particularly regarding nutrient homeostasis, and it has diagnostic and therapeutic potential.

Intensive synthetic studies are in progress to determine structure-function relationships of somatostatin in various cell types, and to prepare therapeutically useful analogs. In this review only the most significant findings in each area are highlighted. Recent reviews emphasizing many aspects of somatostatin's broad range have appeared (18, 20, 22, 23).

Isolation and Structure

Somatostatin was first isolated from ovine hypothalamic extracts by successive purifications of those fractions capable of inhibiting the basal release of immunoreactive GH from cultured rat anterior pituitary cells (16). The existence of such a substance within the hypothalamus had previously been demonstrated (201). Somatostatin was determined to be a cyclic tetradecapeptide of the following sequence (202):

$$
\begin{array}{cccccccccccccc}
\text{H-Ala-Gly-Cys-Lys-Asn-Phe-Phe-Trp-Lys-Thr-Phe-Thr-Ser-Cys-OH} \\
1 \quad 2 \quad 3 \quad 4 \quad 5 \quad 6 \quad 7 \quad 8 \quad 9 \quad 10 \quad 11 \quad 12 \quad 13 \quad 14
\end{array}
$$

Subsequently, we isolated and characterized somatostatin from porcine hypothalami (17) and confirmed the sequence to be identical to somatostatin of ovine origin.

Recently we found larger, highly basic forms of somatostatin in pig hypothalami and showed that they were biologically and immunologically active (17, 203). We also recognized two types of immunoreactive somatostatin in extracts from rat pancreas, stomach, and duodenum (19). These molecules may all represent precursors of somatostatin.

Somatostatin was synthesized by several groups (204–207), thus providing sufficient quantities of the pure peptide for examination of its biological activities.

Action on Normal Tissues

PITUITARY   The inhibitory action of somatostatin on both basal (16, 18, 20) and stimulated (18, 20) secretion of GH in a variety of in vitro and in vivo assays was demonstrated in several species (18, 20), including humans (18, 20, 208–210). Both the reduced (linear) and oxidized (cyclic) forms are active (18, 22, 208, 209). In the above experiments, the elevation of in vivo GH levels prior to somatostatin infusion was induced by such varied stimuli as barbiturates (16), L-DOPA, arginine, insulin, morphine, prostaglandin E_2, exercise, sleep, electrical stimulation of hypothalamus or amygdala (20), and catecholamine infusion into the third ventricle (211). A physiological role for somatostatin in the regulation of GH secretion is supported by our observations that passive immunization in rats with anti-somatostatin elevates basal GH levels and prevents the stress-induced decrease of GH (212, 213).

Somatostatin also inhibits the TRH-induced secretion of TSH (22, 214, 215), but not of prolactin (214, 215), in vitro and in vivo, and it could play a physiological role in the regulation of TSH secretion (213, 216). The spontaneous release of prolactin in vitro is decreased by somatostatin (215).

PANCREAS   Lowered insulin levels and hypoglycemia were noted in humans and baboons, respectively (211, 217), leading rapidly to clear demonstrations in rats and dogs that both basal and stimulated secretion of insulin and glucagon were inhibited by a direct action on the α and β pancreatic islet cells (18, 20, 22). No other known pancreatic secretagogue inhibits the secretion of both insulin and glucagon in response to essentially all respective stimuli (20). That these actions may be of physiological importance is apparent from the anatomic location of somatostatin-containing cells (D-cells) between α and β islet cells (218–221) (see section on localization).

Recently, somatostatin has been shown to affect the exocrine pancreas as well, inhibiting pancreatic bicarbonate and protein secretion (222, 223).

STOMACH AND GUT   The observed hypoglycemia in the face of hypoinsulinemia mentioned above was consistent with the findings of our group

(210, 224) that the rise in blood sugar after oral administration of glucose was delayed by administration of somatostatin. These findings suggested that somatostatin may act on the gastrointestinal tract, which prompted us to examine the effect of somatostatin on gastrin (225). Both basal gastrin and that released in response to food were inhibited by somatostatin. Somatostatin inhibited gastric acid and pepsin in response to pentagastrin, as well as the gastric acid response to food and hypoglycemia in cats with gastric fistulae, indicating a direct action by somatostatin on the parietal and peptic cells (226). Thus, a direct action of somatostatin on exocrine secretion was established. Several studies have confirmed the inhibition of gastrin, gastric acid, and pepsin by somatostatin (225–228).

Subsequently, somatostatin has been found to inhibit the release of secretin (222, 223), pepsin (226), cholecystokinin/pancreozymin (223), gastric inhibitory polypeptide (229), vasoactive intestinal polypeptide (230), and motilin (230). Recently, Sacks et al (231) observed direct suppression of glucagon-stimulated glucose release from liver cells in vitro, and other direct hepatic effects have been observed in vivo (232).

KIDNEY Elevated renin levels induced by pretreatment with frusemide were reported to be suppressed by somatostatin (22, 233), and recently a diuretic effect of somatostatin was observed (234).

CENTRAL NERVOUS SYSTEM Somatostatin is widely distributed throughout the brain and spinal cord (see section on localization). Plotnikoff, Kastin, & Schally (235) were the first to systematically examine its actions on the behavior of mice to determine its CNS effects. They found that somatostatin potentiates, but does not antagonize, the behavioral effects (hyperactivity) induced by L-DOPA. There are reports that somatostatin inhibits spontaneous electrical activity of some neuronal systems (20, 236), and that it affects behavior by inducing sedation, hypothermia, and "barrel rotation"(20). In addition, somatostatin suppresses slow wave and REM sleep, and may affect appetite (20).

Actions on Pathologic Tissues

PITUITARY Somatostatin suppresses fasting GH levels in acromegalics (22, 208, 210), and ACTH in patients with Nelson's syndrome (22), but it does not affect the latter hormone in normal subjects (208). The effects on ACTH suggest differing mechanisms of action for somatostatin on secreting adenoma cells and normal pituitary cells.

The elevated TSH levels in hypothyroid patients are lowered by somatostatin infusions over several hours (22). The TSH response to TRH is also inhibited in these patients (22, 214).

PANCREAS   Pancreatic tumors secreting insulin (22) or glucagon (22, 224) are responsive to somatostatin. Interestingly, somatostatin inhibition of glucagon—but not tolbutamide—stimulated insulin secretion was observed in patients with insulinomas (20, 22), thus providing a potential diagnostic tool for detection of such tumors. The inhibitory effects of somatostatin on glucagon secretion are observed equally in normals, insulin-dependent diabetics, and acromegalics (22).

Recently, two cases of somatostatin-secreting tumors ("somato-statinomas") were reported (237, 238) in which the tumors contained excessive somatostatin, low insulin, glucagon, and other hormones, and the patients were diabetic; these findings are consistent with chronic hypersecretion of somatostatin. Removal of the tumor was in one case followed by remission of the diabetes (238).

Initial observations, that somatostatin decreased fasting blood glucose in normal subjects (239) and fasting and postprandial hyperglycemia in diabetics (240), led to proposals that glucagon plays an important role in normal regulation of blood glucose concentrations and in the pathogenesis of diabetes mellitus (239, 240). These views have been challenged by several studies (241–243), including some which have shown that (a) somatostatin interferes with carbohydrate absorption from the gut (241, 242) at doses consistent with physiologic concentrations (243); (b) prolonged infusion of somatostatin in diabetic patients results in severe hyperglycemia in spite of ongoing hypoglucagonemia (241); and (c) patients with "somato-statinomas" (237, 238) who have low insulin and glucagon levels have a diabetic response to the glucose tolerance test.

Suppression by somatostatin of gastrin secreted from a pancreatic tumor in Zollinger-Ellison syndrome was accompanied by a marked fall in gastric acid secretion (225). Similarly, somatostatin suppression of vasoactive intestinal polypeptide secretion from a pancreatic tumor in the Verner-Morrison syndrome has been reported (22).

STOMACH AND GUT   The raised plasma gastrin levels in patients with pernicious anemia were partially suppressed by somatostatin (225). One patient with bleeding ulcer was dramatically improved by suppressing gastric secretion with an infusion of somatostatin (244).

Localization

The extremely short biological half-life of somatostatin (less than 4 min) (208, 245), and the unlikelihood that the relatively high concentrations of somatostatin required by the target tissues could be present in the peripheral blood, led the earliest investigators to doubt the physiologic significance of extrapituitary effects (211, 217). We reasoned that the hypothalamus could not be the only source of somatostatin if it was involved in the physiologic regulation of gastric and pancreatic secretion (19). Our laboratory developed a highly specific radioimmunoassay for somatostatin (246) and used it to detect somatostatin-like activity in rat brain (247) [confirming results from bioassays (20)] and gastrointestinal organs (19). High concentrations of this peptide in the stomach and pancreas were found, the latter having similar concentrations to those in the arcuate nucleus of the hypothalamus, which showed the highest concentration for any hypothalamic region other than the median eminence (247). The total amount of somatostatin in either pancreas or stomach was much greater than that in the hypothalamus (19). These findings strongly suggest an important role for somatostatin in the regulation of gastrointestinal physiology in addition to its role in the pituitary.

Luft et al (218) showed somatostatin-like immunoreactivity to be present in certain islet cells of guinea pig and rat pancreas by immunohistochemical methods. It was soon shown by combined immunocytochemical and histological methods (219, 248) that pancreatic somatostatin was localized in D-cells within the islets. Although D-cells have been known for many years, their function and secretory product were hitherto uncertain. The anatomic position of islet D-cells between α and β cells suggests a possible means for direct physiologic control ("paracrine") over insulin and glucagon secretion (249).

Somatostatin has been found in other cells similar to D-cells throughout the gastrointestinal mucosa (219), in nerve fibers of the intestine and the hypothalamo-hypophyseal system (248) (see also next section), in spinal cord and fluid (20), and in a variety of vertebrate species (20). This indicates a broad phylogenetic as well as anatomic distribution.

Mechanisms of Secretion

Secretory pathways for somatostatin have so far been inferred from its localization in various neural and endocrine structures (see also previous section). Biochemical and immunohistochemical evidence of somatostatin in nerve cell bodies (248), nerve fibers, and neurosecretory granules in the hypothalamus, median eminence, and posterior pituitary suggest a double mechanism whereby somatostatin is delivered from a neuron into either portal vessels leading to the anterior pituitary (acting as a hormone) or into synapses (acting as a neurotransmitter) (248). The nature of the neurotransmitter substance(s) involved in regulating the secretion of somatostatin has not been determined. Whether or not significant quantities of somatostatin enter the peripheral blood is unknown.

Neural and endocrine elements in pancreatic islets, stomach, gut, and elsewhere contain somatostatin-like immunoreactivity, suggesting local control of secretion in these areas, but the pathways of secretion are not clear. Excellent reviews have been written on these topics (23, 141, 248).

Mechanisms of Action

The exact mechanisms by which somatostatin inhibits the release of many hormones are not known, but there is evidence that a reduction in calcium ion concentration may be involved (250–252). Increased Ca^{2+} levels have been shown to reverse the somatostatin inhibition of stimulated insulin (252) and glucagon (251) secretion. However, stimulation of insulin secretion is a biphasic event, and while both phases are inhibited by somatostatin, Ca^{2+} is only able to reverse the inhibition during the second phase, suggesting that the role of calcium in somatostatin's mechanism of action is not exclusive (250).

Efendic, Grill & Luft (253) reported that somatostatin inhibited glucose-induced insulin release by blocking the accumulation of cAMP. Although accumulation of this nucleotide was also inhibited in isolated pituitary cells (69, 254), those studies indicated somatostatin acted at a step following cAMP formation. Further, several insulin secretogogues, known to act largely by causing cAMP accumulation in the β-cell, failed to reverse the action of somatostatin on insulin release, supporting the view that somatostatin acts distally to cAMP (251).

Somatostatin analogs have been prepared which have dissociated actions on GH, insulin, glucagon, and gastric acid (see also the section on analogs), indicating that there are definite differences in the receptor recognition capabilities of various cell types and more than one discreet mechanism of action.

Somatostatin Analogs

Somatostatin is of little therapeutic value since it has many actions and a short biological half-life (22). Analogs have therefore been designed for their enhanced, selective, prolonged, or antagonistic activities. The analogs also provide insight into structure-function relationships of somatostatin in its various target tissues, and may have potential as diagnostic agents.

The use of solid-phase methodology (204–207, 255) has made possible the preparation of many analogs which have been screened for multiple activities. Several investigators have examined the systematic incorporation of D-amino acids into the somatostatin backbone (120, 256). In addition to offering resistance to degradative enzymes, and thereby prolonging activity, D-amino acid substitutions in other hormones have yielded analogs with increased or antagonistic activities (118, 154, 162). Other successful strategies in designing analogs of peptide hormones have included derivatization of various functional groups and deletion, addition, or exchange of amino acids.

ANALOGS WITH ENHANCED ACTIVITY   Rivier, Brown, & Vale (257) reported, and we later confirmed (93, 256, 258, 259), that D-Trp^8-somatostatin was 6–8 times as active as somatostatin in inhibiting GH release. We showed that this analog had a time course of inhibition identical to that of somatostatin in vivo, indicating that the increased potency was related to conformational effects rather than prolonged biological half-life (258). Multiple substitutions have been shown by us to be sometimes additive, as in [D-Ala^2,D-Trp^8]-somatostatin (256, 259), which had up to 20 times the potency of somatostatin as an inhibitor of GH release in vitro (see also the section on analogs with selective activity). However, D-Ala^2-somatostatin was only twice as active as somatostatin in that assay (120). [N-Tyr,D-Trp^8]-somatostatin had 10 times the potency of somatostatin as an inhibitor of GH release (259). Several analogs show enhanced activity, up to about 500% that of somatostatin. [Ala^2,D-Trp^8,D-Cys^{14}]-somatostatin had a potency of 400% as an inhibitor of penta gastrin-induced gastric acid secretion in cats with gastric fistulae (260).

ANALOGS WITH SELECTIVE ACTIVITY   The first demonstration that the multiple inhibitory actions of somatostatin could be dissociated came with des[Ala^1,Gly^2,Asn^5]-somatostatin, which inhibited insulin, and to a lesser extent GH, without affecting glucagon release (261). The same group later reported an analog which retained about 20% GH-release-inhibiting activity, but did not affect insulin or glucagon levels (262).

Recently, we (21, 263) and others (264) observed that D-Cys^{14}-somatostatin and [D-Trp^8,D-Cys^{14}]-somatostatin selectively inhibited GH and glucagon more than insulin release. In our hands, the latter analog had a ratio of 22:1 for the selective inhibition of glucagon over insulin, and 100:1 for that of GH over insulin (21). We have shown that the superactive analog D-Trp^8-somatostatin (see also section on analogs with enhanced activity) inhibited GH more selectively than pentagastrin-induced acid secretion (258). We recently observed that [D-Trp^8,D-Cys^{14}]-somatostatin retained the same relative degree of dissociation (5:1) between inhibition of GH and gastric acid secretion (260). Although this analog may be useful for treating diabetes since it inhibits glucagon more than insulin release (21), its absolute potency to inhibit gastric acid secretion is 2–3 times that of somatostatin, so that the clinical usefulness of this analog may be somewhat restricted. Interestingly, [D-Ala^2,D-Trp^8]-somatostatin which was 20 times as potent as somatostatin in suppressing GH release, was only twice as potent in the pantagastrin-induced gastric acid assay (260).

ANALOGS WITH PROLONGED ACTION   Despite its extremely high potency, [D-Ala^2,D-Trp^8]-somatostatin did not have prolonged activity in vivo (256), even with D-amino acids in different regions of the molecule. Inactivation of somatostatin from the N-terminus does not appear to be a major degradative pathway, since we have shown that various des[Ala^1,Gly^2]-N-Acylated[Cys^3]-somatostatin analogs, and [D-Ala^1]-somatostatin, retain full GH-release inhibiting activity, but have durations of action identical to somatostatin in rats and humans (265, 266). This view is consistent with that of Marks, Stern, & Benuck (267, 268) who demonstrated preferential cleavage by endopeptidases within the ring portion of somatostatin. Since the aromatic residues in positions 6, 7, 8, and 11 are possible sites of chymotryptic cleavage, analogs with the corresponding D-isomers were examined, but none had prolonged action in vivo, and all but the D-Trp^8 analog had low potency (256). Similarly, D-Lys^9-somatostatin had low potency (120), and D-Lys^4-somatostatin, although moderately active, was not longer-acting than somatostatin, despite the blocking of a tryptic cleavage point (265).

Other attempts to make longer-acting somatostatin analogs have been directed mostly at the disulfide bridge. Since dihydrosomatostatin (linear) was as potent as somatostatin (18, 22, 208, 209) and since [Ala^{3,14}]-somatostatin retained some biological potency (269), it seemed that the disulfide bridge might not be a requirement for activity. This was supported when Veber et al (270) prepared two nonreducible cyclic analogs of the somatostatin ring portion by replacing the sulfur atoms with methylene groups. These analogs retained 50% of somatostatin's GH-release inhibiting activity, and had full potency to inhibit pentagastrin-induced gastric secretion (270).

A 30-minute protracted action was observed for one of the nonreducible derivatives (descarboxy Cys^{14}) in the gastric secretion assay, indicating that the terminal carboxyl group is not a requirement for activity and may be a factor in the rapid inactivation of somatostatin. Little investigation of the carboxyl group has been reported since its amidation resulted in considerable loss of activity (120), yet its lack of requirement has again been substantiated recently (271) and further studies of this region are needed.

Sarantakis has reported a nonreducible cyclic analog with a shortened ring structure that significantly suppresses plasma GH for four hours after injection, but high doses were required (272).

ANALOGS FOR RIA   The initial development of a radioimmunoassay for somatostatin utilized ^{125}I-labeled Tyr^1-somatostatin as the tracer (246). This analog was fairly unstable and could not be stored longer than one week. We recently tested ^{125}I-N-Tyr-somatostatin and found it to be far superior, because it remained stable for two months at −50°C (273).

Since potent analogs of somatostatin with selective and perhaps prolonged activities can be prepared, it is likely that future analogs may be useful in the treatment of such disorders as acromegaly, diabetic ketosis and retinopathy, juvenile diabetes, insulinomas, glucagonomas, peptic ulcers, and acute pancreatitis (22).

Literature Cited

1. Green, J. D., Harris, G. W. 1947. *J. Endocrinol.* 5:136–46
2. Saffran, M., Schally, A. V. 1955. *Can. J. Biochem.* 33:408–15
3. Schally, A. V., Arimura, A., Kastin, A. J. 1973. *Science* 179:341–50
4. Schally, A. V., Bowers, C. Y., Redding, T. W., Barrett, J. F. 1966. *Biochem. Biophys. Res. Commun.* 25:165–69
5. Schally, A. V., Redding, T. W., Bowers, C. Y., Barrett, J. F. 1969. *J. Biol. Chem.* 244:4077–88
6. Bøler, J., Enzmann, F., Folkers, K., Bowers, C. Y., Schally, A. V. 1969. *Biochem. Biophys. Res. Commun.* 37: 705–10
7. Nair, R. M. G., Barrett, J. F., Bowers, C. Y., Schally, A. V. 1970. *Biochemistry* 9:1103–6
8. Burgus, R., Dunn, T. F., Desiderio, D., Guillemin, R. 1969. *C. R. Acad. Sci. Paris* 269:1870–73
9. Burgus, R., Dunn, T. F., Desiderio, D., Ward, D. N., Vale, W., Guillemin, R. 1970. *Nature* 226:321–25
10. Schally, A. V., Arimura, A., Baba, Y., Nair, R. M. G., Matsuo, H., Redding, T. W., Debeljuk, L., White, W. F. 1971. *Biochem. Biophys. Res. Commun.* 43: 393–99
11. Schally, A. V., Nair, R. M. G., Redding, T. W., Arimura, A. 1971. *J. Biol. Chem.* 246:7230–36
12. Matsuo, H., Nair, R. M. G., Arimura, A., Schally, A. V. 1971. *Biochem. Biophys. Res. Commun.* 43:1134–39
13. Baba, Y., Matsuo, H., Schally, A. V. 1971. *Biochem. Biophys. Res. Commun.* 44:459–63
14. Matsuo, H., Arimura, A., Nair, R. M. G., Schally, A. V. 1971. *Biochem. Biophys. Res. Commun.* 45:822–27
15. Schally, A. V., Kastin, A. J., Coy, D. H. 1976. *Int. J. Fertil.* 21:1–30
16. Brazeau, P., Vale, W., Burgus, R., Ling, N., Butcher, M., Rivier, J., Guillemin, R. 1973. *Science* 179:77–79
17. Schally, A. V., Dupont, A., Arimura, A., Redding, T. W., Nishi, N., Linthicum, G. L., Schlesinger, D. H. 1976. *Biochemistry* 15:509–14
18. Schally, A. V., Arimura, A. 1977. In *Clinical Neuroendocrinology,* ed. L. Martini, G. M. Besser. London: Academic. In press
19. Arimura, A., Sato, H., Dupont, A., Nishi, N., Schally, A. V. 1975. *Science* 189:1007–9
20. Vale, W., Rivier, C., Brown, M. 1977. *Ann. Rev. Physiol.* 39:473–527
21. Meyers, C., Arimura, A., Gordin, A., Fernandez-Durango, R., Coy, D. H., Schally, A. V., Drouin, J., Ferland, L., Beaulieu, M., Labrie, F. 1977. *Biochem. Biophys. Res. Commun.* 74:630–36
22. Hall, R., Gomez-Pan, A. 1976. *Adv. Clin. Chem.* 18:173–212
23. Reichlin, S., Saperstein, R., Jackson, I. M. D., Boyd, A. E. III, Patel, Y. 1976. *Ann. Rev. Physiol.* 38:389–424
24. Schally, A. V., Arimura, A., Bowers, C. Y., Kastin, A. J., Sawano, S., Redding, T. W. 1968. *Recent Prog. Horm. Res.* 24:497–588
25. US Endocrine Society. 1977. *Endocrinology* 100:14A
26. Guillemin, R., Rosenberg, B. 1955. *Endocrinology* 57:599–607
27. Schally, A. V., Saffran, M., Zimmerman, B. 1958. *Biochem. J.* 70:97–103
28. Schally, A. V., Bowers, C. Y. 1964. *Metabolism* 13:1190–1205
29. Doepfner, W., Stürmer, E., Berde, B. 1963. *Endocrinology* 72:897–902
30. Arimura, A., Saito, T., Müller, E., Bowers, C. Y., Sawano, S., Schally, A. V. 1967. *Acta Endocrinol.* 54:155–65
31. McCann, S. M., Antunes-Rodrigues, J., Nallar, R., Valtin, H. 1966. *Endocrinology* 79:1058–64
32. Krieger, D. T., Liotta, A., Brownstein, M. J. 1977. *Endocrinology* 100:227–37
33. van Loon, G. R. 1974. In *Recent Studies of Hypothalamic Function,* pp. 100–13. Basel: Karger
34. Kraicer, J., Morris, A. R. 1976. *Neuroendocrinology* 21:175–92
35. Greer, M. A., Allen, C. F., Panton, P., Allen, J. P. 1975. *Endocrinology* 96: 718–24
36. Jones, M. T., Hillhouse, E. W., Burden, J. 1976. *J. Endocrinol.* 69:1–10

37. Bennett, G. W., Edwardson, J. A. 1975. *J. Endocrinol.* 65:33–44
38. Cooper, D. M. F., Synetos, D., Christie, R. B., Schulster, D. 1976. *J. Endocrinol.* 71:171–72
39. Jones, M. T., Gillham, B., Hillhouse, E. W. 1977. *Fed. Proc.* 36:2104–9
40. Schally, A. V., Andersen, R. N., Lipscomb, H. S., Long, J. M., Guillemin, R. 1960. *Nature* 188:1192–93
41. Schally, A. V., Arimura, A., Redding, T. W., Chihara, K., Gordin, A., Huang, W. Y., Saffran, M. 1977. *Program Endocr. Soc., 59th meeting* p. 95, No. 77 (Abstr.)
42. Perlmutter, A. F., Rapino, E., Saffran, M. 1975. *Endocrinology* 97:1336–39
43. Gorbman, A., Hyder, M. 1973. *Gen. Comp. Endocrinol.* 20:588–89
44. Jacobs, L., Snyder, P., Wilber, J., Utiger, R., Daughaday, W. 1971. *J. Clin. Endocrinol. Metab.* 33:996–98
45. Tashjian, A., Barowski, N., Jensen, D. 1971. *Biochem. Biophys. Res. Commun.* 43:516–23
46. Debeljuk, L., Redding, T. W., Arimura, A., Schally, A. V. 1973. *Proc. Soc. Exp. Biol. Med.* 142:421–23
47. Bowers, C. Y., Friesen, H., Hwang, P., Guyda, H., Folkers, K. 1971. *Biochem. Biophys. Res. Commun.* 45:1033–41
48. Convey, E. M., Tucker, H. A., Smith, V. G., Zolman, J. 1973. *Endocrinology* 92:471–76
49. Kato, Y., Chihara, K., Maeda, K., Ohgo, S., Okanishi, Y., Imura, H. 1975. *Endocrinology* 96:1114–18
50. Schalch, D. S., Gonzalez-Barcena, D., Kastin, A. J., Schally, A. V., Lee, L. A. 1972. *J. Clin. Endocrinol. Metab.* 35:609–15
51. Smith, J. R., LaHann, T. R., Chestnut, R. M., Carino, M. A., Horita, A. 1977. *Science* 196:660–61
52. Kastin, A. J., Miller, L. H., Sandman, C. A., Schally, A. V., Plotnikoff, N. P. 1977. In *Essays in Neurochemistry and Neuropharmacology, Vol. 1,* ed. M. B. H. Youdim, W. Lovenberg, D. F. Sharman, J. R. Lagnado, pp. 139–76. London:Wiley
53. Martin, J. B., Renaud, L. P., Brazeau, P. 1975. *Lancet* 7931:393–95
54. Jackson, I. M. D., Reichlin, S. 1974. *Endocrinology* 95:854–62
55. Hökfelt, T., Fuxe, K., Johansson, O., Jeffcoate, S., White, N. 1975. *Neurosci. Lett.* 1:133–39
56. Bassiri, R. M., Utiger, R. D. 1972. *Endocrinology* 90:722–27
57. Oliver, C., Charvet, J. P., Codaccioni, J.
L., Vague, J., Porter, J. C. 1974. *Lancet* 1:873
58. Koch, Y., Baram, T., Fridkin, M. 1976. *FEBS Lett.* 63:295–98
59. Mitsuma, T., Hirooka, Y., Nihei, N. 1976. *Acta Endocrinol.* 83:225–35
60. Eskay, R. L., Oliver, C., Warberg, J., Porter, J. C. 1976. *Endocrinology* 98:269–77
61. Leppäluoto, J., Ling, N., Vale, W. 1976. *Int. Congr. Endocrinol. 5th, Hamburg, July 18–24, 1976,* p. 333 (Abstr.)
62. Redding, T. W., Schally, A. V. 1969. *Proc. Soc. Exp. Biol. Med.* 131:415–20
63. Griffiths, E. C., Hooper, K. C., Hutson, D., Jeffcoate, S. L., White, N. 1976. *Mol. Cell. Endocrinol.* 4:215–22
64. Bauer, K., Lipmann, F. 1976. *Endocrinology* 99:230–42
65. Grimm-Jørgensen, Y., Pfeiffer, S. E., McKelvy, J. F. 1976. *Biochem. Biophys. Res. Commun.* 70:167–73
66. Wilber, J. F., Seibel, M. J. 1973. *Endocrinology* 92:888–93
67. Hinkle, P. M., Woroch, E. L., Tashjian, A. H. Jr. 1974. *J. Biol. Chem.* 249:3085–90
68. Barden, N., Labrie, F. 1973. *J. Biol. Chem.* 248:7601–6
69. Labrie, F., Pelletier, G., Drouin, J., Belanger, A., Ferland, L., Lemay, A., Lemaire, S., Beaulieu, M. 1976. In *Basic Applications and Clinical Uses of Hypothalamic Hormones,* ed. A. L. Charro-Salgado, R. Fernandez-Durango, J. G. Lopez del Campo, pp. 100–10. Amsterdam:Excerpta Medica. 350 pp.
70. Dannies, P. S., Gautvik, K. M., Tashjian, A. H. Jr. 1976. *Endocrinology* 98:1147–59
71. Hofmann, K., Bowers, C. Y. 1970. *J. Med. Chem.* 13:1099–1101
72. Rivier, J., Vale, W., Monahan, M., Ling, N., Burgus, R. 1972. *J. Med. Chem.* 15:479–82
73. Bowers, C. Y., Weil, A., Chang, J. K., Sievertsson, H., Enzmann, F., Folkers, K. 1970. *Biochem. Biophys. Res. Commun.* 40:683–91
74. Coy, D. H., Hirotsu, Y., Redding, T. W., Coy, E. J., Schally, A. V. 1975. *J. Med. Chem.* 18:948–49
75. Sowers, J. R., Hershman, J. M., Carlson, H. E., Pekary, A. E., Reed, A. W., Nair, M. G., Baugh, C. M. 1976. *J. Clin. Endocrinol. Metab.* 43:856–60
76. Sievertsson, H., Chang, J. K., Folkers, K., Bowers, C. Y. 1972. *J. Med. Chem.* 15:219–21
77. Sievertsson, H., Castensson, S., Lindgren, O., Bowers, C. Y. 1974. *Acta Pharm. Suec.* 11:67–76

78. Bellocq, A. M., Dubien, M., Dupart, E. 1975. *Biochem. Biophys. Res. Commun.* 65:1393–99
79. Hirschmann, R. F. 1977. In *Proc. Int. Symp. Med. Chem., 5th, Paris, July 1976*, ed. J. Matthieu. In press
80. Lybeck, H., Leppäluoto, J., Virkkunen, P., Schafer, D., Carlsson, L., Mulder, J. 1973. *Neuroendocrinology* 12:366–70
81. Sievertsson, H., Castensson, S., Andersson, K., Björkman, S., Bowers, C. Y. 1975. *Biochem. Biophys. Res. Commun.* 66:1401–7
82. Hershman, J. M. 1974. *N. Engl. J. Med.* 290:886–90
83. Hirvonen, E., Ranta, T., Seppälä, M. 1976. *J. Clin. Endocrinol. Metab.* 42: 1024–30
84. Meites, J., Clemens, J. A. 1972. *Vitamins and Hormones*, ed. R. S. Harris, P. L. Munson, E. Diczfalusy, J. Glover, 30:165–221. New York: Academic. 393 pp.
85. Koch, Y., Goldhaber, G., Fireman, I., Zor, U., Shani, J., Tal, E. 1977. *Endocrinology* 100:1476–78
86. Blake, C. A. 1974. *Endocrinology* 94: 503–8
87. L'Hermite, M., Robyn, C., Golstein, J., Rothenbuchner, G., Birk, J., Loos, U., Bonnyns, M., Vanhaelst, L. 1974. *Horm. Metab. Res.* 6:190–95
88. Szabo, M., Frohman, L. A. 1976. *Endocrinology* 98:1451–59
89. Boyd, A. E., Spencer, E., Jackson, I. M. D., Reichlin, S. 1976. *Endocrinology* 99:861–71
90. Dular, R., LaBella, F., Vivian, S., Eddie, L. 1974. *Endocrinology* 94:563–67
91. Schally, A. V., Dupont, A., Arimura, A., Takahara, J., Redding, T. W., Clemens, J., Shaar, C. 1976. *Acta Endocrinol.* 82:1–14
92. Schally, A. V., Redding, T. W., Arimura, A., Dupont, A., Linthicum, G. L. 1977. *Endocrinology* 100:681–91
93. Schally, A. V., Coy, D. H., Arimura, A., Redding, T. W., Meyers, C. A., Vilchez, J., Pedroza, E., Gordin, A., Molnar, J., Kastin, A. J., Labrie, F., Hall, R., Gomez-Pan, A., Besser, G. M. 1977. In *Proc. Int. Congr. Endocrinol., 5th, Hamburg, July 18–24, 1976.* Amsterdam: Excerpta Medica. In press
94. Clemens, J. A., Sawyer, B. D., Cerimele, B. 1977. *Endocrinology* 100: 692–98
95. MacLeod, R. M. 1969. *Endocrinology* 85:916–23
96. Birge, C. A., Jacobs, L. S., Hammer, C. T., Daughaday, W. 1970. *Endocrinology* 86:120–30
97. Kamberi, I. A., Mical, R. S., Porter, J. C. 1971. *Endocrinology* 88:1012–20
98. vanMaanen, J. H., Smelik, P. G. 1967. *Acta Physiol. Pharmacol. Neerl.* 14: 519–20
99. Takahara, J., Arimura, A., Schally, A. V. 1974. *Endocrinology* 95:462–65
100. Schally, A. V., Kastin, A. J. 1966. *Endocrinology* 79:768–72
101. Nair, R. M. G., Kastin, A. J., Schally, A. V. 1971. *Biochem. Biophys. Res. Commun.* 43:1376–81
102. Celis, M. E., Taleisnik, S., Walter, R. 1971. *Proc. Natl. Acad. Sci. USA* 68: 1428–33
103. Walter, R., Griffiths, E. C., Hooper, K. C. 1973. *Brain Res.* 60:449–57
104. Redding, T. W., Kastin, A. J., Nair, R. M. G., Schally, A. V. 1973. *Neuroendocrinology* 11:92–100
105. Pavel, S., Goldstein, R., Gheorghiu, C., Calb, M. 1977. *Science* 197:179–80
106. Friedman, E., Friedman, J., Gershon, S. 1973. *Science* 182:831–32
107. Failli, A., Sestanj, K., Immer, H. U., Götz, M. 1977. *Arzneimittelforschung.* In press
108. Voith, K. 1977. *Arzneimittelforschung.* In press
109. Pugsley, T. A., Lippmann, W. 1977. *Arzneimittelforschung.* In press
110. Bower, A., Hadley, M. E., Hruby, V. J. 1971. *Biochem. Biophys. Res. Commun.* 45:1185–91
111. Celis, M. E., Taleisnik, S., Walter, R. 1971. *Biochem. Biophys. Res. Commun.* 45:564–69
112. Scott, G. T., Stillings, W. A. 1972. *Endocrinology* 90:545–51
113. Thody, A. J., Shuster, S. 1973. *J. Endocrinol.* 58:35–36
114. Bower, A., Hadley, M. E., Hruby, V. J. 1974. *Science* 184:70–72
115. Tilders, F. J. H., Mulder, A. H., Smelik, P. G. 1975. *Neuroendocrinology* 18: 125–30
116. Coy, D. H., Coy, E. J., Schally, A. V. 1975. *Res. Methods Neurochem.* 3:393–406
117. Coy, D. H., Schally, A. V., Vilchez-Martinez, J. A., Coy, E. J., Arimura, A. 1975. In *Hypothalamic Hormones*, ed. M. Motta, P. G. Crosignani, L. Martini, pp. 1–12. London: Academic. 352 pp.
118. Coy, D. H., Coy, E. J., Vilchez-Martinez, J. A., de la Cruz, A., Arimura, A., Schally, A. V. 1976. In *Hypothalamus and Endocrine Functions*, ed. F. Labrie, J. Meites, G. Pelletier, pp. 339–54. New York: Plenum. 508 pp.
119. Schally, A. V., Coy, D. H. 1977. In *Proc. NIH Peptide Releasing Horm.*

Workshop, Bethesda, Md., Nov. 1–2, 1976. pp. 99–122. New York: Plenum

120. Rivier, J., Brown, M., Rivier, C., Ling, N., Vale, W. 1977. In *Peptides 1976,* ed. A. Loffet, pp. 427–51. Brussels: Editions de L'Université de Bruxelles. 660 pp.

121. Schally, A. V., Kastin, A. J., Arimura, A. 1972. *Am. J. Obstet. Gynecol.* 114: 423–42

122. Schally, A. V., Kastin, A. J., Arimura, A. 1971. *Fertil. Steril.* 22:703–21

123. Schally, A. V., Kastin, A. J., Arimura, A. 1972. *Vitam. Horm.* 30:83–164

124. Schally, A. V., Arimura, A., Kastin, A. J., Matsuo, H., Baba, Y., Redding, T. W., Nair, R. M. G., Debeljuk, L., White, W. F. 1971. *Science* 173: 1036–38

125. Burgus, R., Butcher, M., Amoss, M., Ling, N., Monahan, M., Rivier, J., Fellows, R., Blackwell, R., Vale, W., Guillemin, R. 1972. *Proc. Natl. Acad. Sci. USA* 69:278–82

126. Mortimer, C. H., McNeilly, A. S., Rees, L. H., Lowry, P. J., Gilmore, D., Dobbie, H. G. 1976. *J. Clin. Endocrinol. Metab.* 43:882–88

127. Levitan, D., Beitins, I. Z., Milton, G., Barnes, A., McArthur, J. W. 1977. *Endocrinology* 100:918–22

128. Fawcett, C. P., Breezley, A. E., Wheaton, J. E. 1975. *Endocrinology* 96: 1311–14

129. Bowers, C. Y., Currie, B. L., Johansson, K. N. G., Folkers, K. 1973. *Biochem. Biophys. Res. Commun.* 50:20–26

130. Schally, A. V., Arimura, A., Redding, T. W., Debeljuk, L., Carter, W., Dupont, A., Vilchez-Martinez, J. A. 1976. *Endocrinology* 98:380–91

131. Aiyer, M. S., Fink, G. 1974. *J. Endocrinol.* 62:553–72

132. Kerdelhue, B., Jutisz, M., Gillessen, D., Studer, R. O. 1973. *Biochim. Biophys. Acta* 297:540–48

133. Brown, G. M., van Loon, G. R., Hummel, B. C. W., Grota, L. J., Arimura, A., Schally, A. V. 1977. *J. Clin. Endocrinol. Metab.* 44:784–90

134. Koch, Y., Chobsieng, P., Zor, U., Fridkin, M., Lindner, H. R. 1973. *Biochem. Biophys. Res. Commun.* 55:623–29

135. Kerdelhue, B., Catin, S., Kordon, C., Jutisz, M. 1976. *Endocrinology* 98: 1539–49

136. de la Cruz, A., Arimura, A., de la Cruz, K. G., Schally, A. V. 1976. *Endocrinology* 98:490–97

137. Ramirez, V. D., Sawyer, C. H. 1966. *Endocrinology* 78:958–64

138. Arimura, A., Nishi, N., Schally, A. V. 1976. *Proc. Soc. Exp. Biol. Med.* 152:71–75

139. Nishi, N., Arimura, A., de la Cruz, K. G., Schally, A. V. 1976. *Endocrinology* 98:1024–30

140. Setalo, G., Vigh, S., Schally, A. V., Arimura, A., Flerko, B. 1975. *Endocrinology* 96:135–42

141. Setalo, G., Flerko, B., Arimura, A., Schally, A. V. 1977. *Int. Rev. Cytol.* 55: In press

142. Moss, R. L. 1977. *Fed. Proc.* 36: 1978–83

143. Carmel, P. W., Araki, S., Ferin, M. 1976. *Endocrinology* 99:243–48

144. Redding, T. W., Kastin, A. J., Gonzalez-Barcena, D., Coy, D. H., Coy, E. J., Schalch, D. S., Schally, A. V. 1973. *J. Clin. Endocrinol. Metab.* 37:626–31

145. Arimura, A., Vilchez-Martinez, J. A., Coy, D. H., Coy, E. J., Hirotsu, Y., Schally, A. V. 1974. *Endocrinology* 95: 1174–77

146. Vilchez-Martinez, J. A., Coy, D. H., Coy, E. J., Arimura, A., Schally, A. V. 1976. *Fertil. Steril.* 27:628–35

147. Schally, A. V., Redding, T. W., Matsuo, H., Arimura, A. 1972. *Endocrinology* 90:1561–68

148. Vale, W., Grant, G., Amoss, M., Blackwell, R., Guillemin, R. 1972. *Endocrinology* 91:562–72

149. Corbin, A., Beattie, C. W. 1975. *Endocr. Res. Commun.* 1:1–23

150. de la Cruz, A., Coy, D. H., Schally, A. V., Coy, E. J., de la Cruz, K. G., Arimura, A. 1975. *Proc. Soc. Exp. Biol. Med.* 149:576–79

151. Pedroza-Garcia, E., Vilchez-Martinez, J. A., Hoffmann, E. O. 1977. *Program, Ann. Meet. Endocr. Soc., 59th, Chicago, June 8–10, 1977,* p. 140 (Abstr.)

152. Fujino, M., Kobayashi, S., Obayashi, M., Fukuda, T., Shinagawa, S., Yamakazi, I., Nakagami, R., White, W. F., Rippel, R. H. 1972. *Biochem. Biophys. Res. Commun.* 49:698–705

153. Coy, D. H., Vilchez-Martinez, J. A., Coy, E. J., Nishi, N., Arimura, A., Schally, A. V. 1975. *Biochemistry* 14: 1848–51

154. Monahan, M., Amoss, M. S., Anderson, H. A., Vale, W. 1973. *Biochemistry* 12:4616–20

155. Momany, F. A. 1976. *J. Am. Chem. Soc.* 98:2990–96

156. Momany, F. A. 1976. *J. Am. Chem. Soc.* 98:2996–3000

157. Koch, Y., Baram, T., Chobsieng, P., Fridkin, M. 1974. *Biochem. Biophys. Res. Commun.* 61:95–103

158. Marks, N., Stern, F. 1974. *Biochem. Biophys. Res. Commun.* 61:1458–63
159. Koch, Y., Baram, T., Hazum, E., Fridkin, M. 1977. *Biochem. Biophys. Res. Commun.* 74:488–91
160. Vilchez-Martinez, J. A., Coy, D. H., Arimura, A., Coy, E. J., Hirotsu, Y., Schally, A. V. 1974. *Biochem. Biophys. Res. Commun.* 59:1226–32
161. Fujino, M., Fukuda, T., Shinagawa, S., Kobayashi, S., Yamakazi, I., Nakayama, R., Seely, J., White, W. F., Rippel, R. H. 1974. *Biochem. Biophys. Res. Commun.* 60:406–13
162. Coy, D. H., Vilchez-Martinez, J. A., Coy, E. J., Schally, A. V. 1976. *J. Med. Chem.* 19:423–25
163. Coy, D. H., Coy, E. J., Schally, A. V., Vilchez-Martinez, J. A., Hirotsu, Y., Arimura, A. 1974. *Biochem. Biophys. Res. Commun.* 57:335–40
164. Fujino, M., Yamazaki, I., Kobayashi, S., Fukuda, T., Shinagawa, S., Nakayama, R., White, W. F., Rippel, R. H. 1974. *Biochem. Biophys. Res. Commun.* 57:1248–56
165. König, W., Sandow, J., Geiger, R. 1977. In *Peptides: Chemistry, Structure and Biology,* ed. R. Walter, J. Meienhofer. Ann Arbor: Ann Arbor Sci. Publ. In press
166. Sandow, J., von Rechenberg, W., Konig, W., Han, M., Jerabek, G., Fraser, H. 1977. In *Proc. Eur. Colloq. Hypothal. Horm., 2nd, Tübingen, July 26–28, 1976.* Hamburg: Verlag Chemie. In press
167. Reeves, J. J., Tarnavsky, G. K., Becker, S. R., Coy, D., Schally, A. V. 1977. *Endocrinology.* 101:540–47
168. Banik, U. K., Givner, M. L. 1976. *Fertil. Steril.* 27:1078–84
169. Beattie, C. W., Corbin, A. 1977. *Biol. Reprod.* 16:333–39
170. Johnson, E. S., Gendrich, R. L., White, W. F. 1976. *Fertil. Steril.* 27:853–60
171. Mayar, M. Q., Tarnavsky, G. K., Reeves, J. J. 1977. *Proc. Western Sect., Am. Soc. Anim. Sci.* 28:182–84
172. Johnson, E. S., Seely, J. H., White, W. F., DeSombre, E. R. 1976. *Science* 194:329–30
173. Auclair, C., Kelly, P. A., Labrie, F., Coy, D. H., Schally, A. V. 1977. *Biochem. Biophys. Res. Commun.* 76:855–62
174. Hsueh, A. J. W., Dufau, M. L., Catt, K. J. 1976. *Biochem. Biophys. Res. Commun.* 72:1145–52
175. Hsueh, A.J. W., Dufau, M. L., Catt, K. J. 1977. *Proc. Natl. Acad. Sci. USA.* 74:592–95
176. Humphrey, R. R., Windsor, B. L., Reel, J. R., Edgren, R. A. 1977. *Biol. Reprod.* 16:614–21
177. Corbin, A., Beattie, C. W., Yardley, J., Foell, T. J. 1976. *Endocrine Res. Commun.* 3:359–76
178. Corbin, A., Beattie, C. W., Rees, R., Yardley, J., Foell, T. J., Chai, S. Y., McGregor, H., Gorsky, V., Sarantakis, D., McKinley, W. A. 1977. *Fertil. Steril.* 28:471–75
179. Arimura, A., Pedroza, E., Vilchez, J. A., Schally, A. V. 1978. *Endocr. Res. Commun.* In press
180. Vale, W., Grant, G., Rivier, J., Monahan, M., Amoss, M., Blackwell, R., Burgus, R., Guillemin, R. 1972. *Science* 176:933–34
181. Vilchez-Martinez, J. A., Coy, D. H., Coy, E. J., Schally, A. V., Arimura, A. 1975. *Fertil. Steril.* 26:554–59
182. Coy, D. H., Vilchez-Martinez, J. A., Coy, E. J., Arimura, A., Schally, A. V. 1973. *J. Clin. Endocrinol. Metab.* 37:331–33
183. Rees, R. W. A., Foell, T. J., Chai, S., Grant, N. 1974. *J. Med. Chem.* 17: 1015–19
184. Coy, D. H., Labrie, F., Savary, M., Coy, E. J., Schally, A. V. 1976. *Mol. Cell. Endocrinol.* 5:201–8
185. Beattie, C. W., Corbin, A., Foell, T. J., Garsky, V., Rees, R. W. A., Yardley, J. 1976. *Contraception* 13:341–53
186. Yardley, J. P., Foell, T. J., Beattie, C. W., Grant, N. H. 1975. *J. Med. Chem.* 18:1244–47
187. de la Cruz, A., Coy, D. H., Vilchez-Martinez, J. A., Arimura, A., Schally, A. V. 1976. *Science* 191:195–97
188. Coy, D. H., Vilchez-Martinez, J. A., Schally, A. V. 1977. In *Peptides 1976,* ed. A. Loffet, pp. 463–69. Brussels: Editions de l'Université de Bruxelles. 660 pp.
189. Humphries, J., Wan, Y.-P., Folkers, K., Bowers, C. Y. 1976. *Biochem. Biophys. Res. Commun.* 72:939–44
190. Takahara, J., Arimura, A., Schally, A. V. 1975. *Acta Endocrinol.* 78:428–34
191. Schally, A. V., Sawano, S., Arimura, A., Barrett, J., Wakabayashi, I., Bowers, C. Y. 1969. *Endocrinology* 84: 1493–1506
192. Schally, A. V., Baba, Y., Nair, R. M. G., Bennett, E. 1971. *J. Biol. Chem.* 246:6647–50
193. Schally, A. V., Arimura, A., Wakabayashi, I., Redding, T. W., Dickerman, E., Meites, J. 1972. *Experientia* 28:205–6

194. Kastin, A. J., Schally, A. V., Gual, C., Glick, S., Arimura, A. 1972. *J. Clin. Endocrinol. Metab.* 35:326–29
195. Youdaev, N. A., Outecheva, Z. F., Novikova, T. E., Chvatchkin, Y. P., Smirnova, A. P. 1973. *Dokl. Akad. Nauk SSSR* 210:731–32
196. Schally, A. V., Redding, T. W., Takahara, J., Coy, D. H., Arimura, A. 1973. *Biochem. Biophys. Res. Commun.* 55:556–62
197. Takahara, J., Arimura, A., Schally, A. V. 1974. *Endocrinology* 95:1490–94
198. Cehovic, G., Posternak, T., Charollais, E. 1972. *Adv. Cyclic Nucleotide Res.* 1:521–24
199. Müller, E. 1973. In *Advances in Human Growth Hormone Research: Symposium, Oct. 9–12, 1973, Baltimore, Md.,* ed. S. Raiti, pp. 192–217. Washington DC:GPO. 961 pp.
200. Redding, T. W., Schally, A. V. 1977. In *Program Ann. Meet. Endoc. Soc., 59th, Chicago, June 8–10, 1977,* p. 231 Abstr.
201. Krulich, L., McCann, S. M. 1969. *Endocrinology* 85:319–24
202. Burgus, R., Ling, N., Butcher, M., Guilleman, R. 1973. *Proc. Natl. Acad. Sci. USA* 70:684–88
203. Schally, A. V., Dupont, A., Arimura, A., Redding, T. W., Linthicum, G. L. 1975. *Fed. Proc.* 34:584 (Abstr.)
204. Rivier, J., Brazeau, P., Vale, W., Ling, N., Burgus, R., Gilon, C., Yardley, J., Guillemin, R. 1973. *C. R. Acad. Sci. Paris* 276:2737–40
205. Coy, D. H., Coy, E. J., Arimura, A., Schally, A. V. 1973. *Biochem. Biophys. Res. Commun.* 54:1267–73
206. Yamashiro, D., Li, C. H. 1973. *Biochem. Biophys. Res. Commun.* 54:882–87
207. Immer, H. U., Sestanj, K., Nelson, V. R., Götz, M. 1974. *Helv. Chim. Acta* 57:730–34
208. Hall, R., Besser, G. M., Schally, A. V., Coy, D. H., Evered, D., Goldie, D. J., Kastin, A. J., McNeilly, A. S., Mortimer, C. H., Phenekos, C., Tunbridge, W. M. G., Weightman, D. 1973. *Lancet* 2:581–84
209. Siler, T. M., Vandenberg, E., Yen, S. S. C., Brazeau, P., Vale, W., Guillemin, R. 1973. *J. Clin. Endocrinol. Metab.* 37:632–34
210. Besser, G. M., Mortimer, C. H., Carr, D., Schally, A. V., Coy, D. H., Evered, D., Kastin, A. J., Tunbridge, W. M. G., Thorner, M. O., Hall, R. 1974. *Br. Med. J.* 1:352–55
211. Koerker, D. J., Ruch, W., Chideckel, E., Palmer, J., Goodner, C. J., Ensinck,

J., Gale, C. C. 1974. *Science* 184:482–84
212. Arimura, A., Smith, W. D., Schally, A. V. 1976. *Endocrinology* 98:540–43
213. Ferland, L., Labrie, F., Jobin, M., Arimura, A., Schally, A. V. 1976. *Biochem. Biophys. Res. Commun.* 68:149–56
214. Carr, D., Gomez-Pan, A., Weightman, D. R., Roy, V. C. M., Hall, R., Besser, G. M., Thorner, M. O., McNeilly, A. S., Schally, A. V., Kastin, A. J., Coy, D. H. 1975. *Br. Med. J.* 3:67–69
215. Vale, W., Rivier, C., Brazeau, P., Guilleman, R. 1974. *Endocrinology* 95:968–77
216. Arimura, A., Schally, A. V. 1976. *Endocrinology* 98:1069–72
217. Alberti, K. G. M. M., Christensen, S. E., Iversen, J., Seyer-Hansen, K., Christensen, N. J., Prange-Hansen, Aa., Lundbaek, K., Orskov, H. 1973. *Lancet* 2:1299–1301
218. Luft, R., Efendic, S., Hökfelt, T., Johansson, O., Arimura, A. 1974. *Med. Biol.* 52:428–30
219. Polak, J. M., Pearse, A. G. E., Grimelius, L., Bloom, S. R., Arimura, A. 1975. *Lancet* 1:1220–22
220. Pelletier, G., Leclerc, R., Arimura, A., Schally, A. V. 1975. *J. Histochem. Cytochem.* 23:699–701
221. Orci, L., Unger, R. H. 1975. *Lancet* 2:1243–44
222. Boden, G., Sivitz, M. C., Owen, O. E., Essa-Koumar, N., Landor, J. H. 1975. *Science* 190:163–65
223. Konturek, S. J., Tasler, J., Obtulowicz, W., Coy, D. H., Schally, A. V. 1976. *J. Clin. Invest.* 58:1–6
224. Mortimer, C. H., Carr, D., Lind, T., Bloom, S. R., Mallinson, C. N., Schally, A. V., Tunbridge, W. M. G., Yeomans, L., Coy, D. H., Kastin, A. J., Besser, G. M., Hall, R. 1974. *Lancet* 1:697–701
225. Bloom, S. R., Mortimer, C. H., Thorner, M. O., Besser, G. M., Hall, R., Gomez-Pan, A., Roy, V. M., Russell, R. C. G., Coy, D. H., Kastin, A. J., Schally, A. V. 1974. *Lancet* 2:1106–9
226. Gomez-Pan, A., Reed, J. D., Albinus, M., Shaw, B., Hall, R., Besser, G. M., Coy, D. H., Kastin, A. J., Schally, A. V. 1975. *Lancet* 1:888–90
227. Konturek, S. J., Tasler, J., Cieszkowski, M., Coy, D. H., Schally, A. V. 1976. *Gastroenterology* 70:737–41
228. Barros D'Sa, A. A. J., Bloom, S. R., Baron, J. H. 1975. *Lancet* 1:886–87
229. Pederson, R. A., Dryburgh, J. R., Brown, J. C. 1975. *Can. J. Physiol. Pharmacol.* 53:1200–05

230. Besser, G. M., Mortimer, C. H. 1976. In *Frontiers in Neuroendocrinology*, ed. L. Martini, W. F. Ganong, pp. 227–54. New York: Raven

231. Sacks, H., Waligora, K., Matthews, J., Pimstone, B. 1977. *Program Ann. Meet. Am. Diabetes Assoc., St. Louis*, A-22 (Abstr.)

232. Sacca, L., Sherwin, R. 1977. *Program Ann. Meet. Am. Diabetes Assoc., St. Louis*, A-23 (Abstr.)

233. Rosenthal, J., Escobar-Jimenez, F., Raptis, S., Pfeiffer, E. F. 1976. *Lancet* 1:772–74

234. Reid, I. A., Rose, J. C. 1977. *Endocrinology* 100:782–85

235. Plotnikoff, N. P., Kastin, A. J., Schally, A. V. 1974. *Pharmacol. Biochem. Behav.* 2:693–96

236. Renaud, L. P., Martin, J. B., Brazeau, P. 1975. *Nature* 255:233–35

237. Larsson, L.-I., Holst, J. J., Kuhl, C., Lundqvist, G., Hirsch, M. A., Ingemansson, S., Lindkaer-Jensen, S., Rehfeld, J. F., Schwartz, T. W. 1977. *Lancet* 1:666–68

238. Ganda, O. P., Weir, G. C., Soeldner, J. S., Legg, M. A., Chick, W. L., Patel, Y. C., Ebeid, A. M., Gabbay, K. H., Reichlin, S. 1977. *N. Engl. J. Med.* 296:963–67

239. Alford, F. P., Bloom, S. R., Nabarro, J. D. N., Hall, R., Besser, G. M., Coy, D. H., Kastin, A. J., Schally, A. V. 1974. *Lancet* 2:974–76

240. Gerich, J. E., Lorenzi, M., Schneider, V., Karam, J. H., Rivier, J., Guilleman, R., Forsham, P. H. 1974. *N. Engl. J. Med.* 291:544–47

241. Felig, P., Wahren, J., Sherwin, R., Hendler, R. 1976. *Diabetes* 25:1091–99

242. Wahren, J., Felig, P. 1976. *Lancet* 2:1213–16

243. Unger, R. H. 1977. *N. Engl. J. Med.* 296:998–1000

244. Mattes, P., Raptis, S., Heil, Th., Rasche, H., Scheck, R. 1975. *Horm. Metab. Res.* 7:508–11

245. Redding, T. W., Coy, E. J. 1974. *Endocrinology* 94(Suppl):A154 (Abstr.)

246. Arimura, A., Sato, H., Coy, D. H., Schally, A. V. 1975. *Proc. Soc. Exp. Biol. Med.* 148:784–89

247. Brownstein, M., Arimura, A., Sato, H., Schally, A. V., Kizer, J. S. 1975. *Endocrinology* 96:1456–61

248. Hökfelt, T., Efendic, S., Hellerstrom, C., Johansson, O., Luft, R., Arimura, A. 1975. *Acta Endocrinol.* 80(Suppl. 200):5–41

249. Unger, R., Orci, L. 1977. *Diabetes* 26:241–44

250. Curry, D. L., Bennett, L. L. 1976. *Proc. Natl. Acad. Sci. USA* 73:248–51

251. Bhathena, S. J., Perrino, P. V., Voyles, N. R., Smith, S. S., Wilkins, S. D., Coy, D. H., Schally, A. V., Recant, L. 1976. *Diabetes* 25:1031–40

252. Curry, D. L., Bennett, L. L. 1974. *Biochem. Biophys. Res. Commun.* 60:1015–19

253. Efendic, S., Grill, V., Luft, R. 1975. *FEBS Lett.* 55:131–33

254. Borgeat, P., Labrie, F., Drouin, J., Belanger, A., Immer, H., Sestanj, K., Nelson, V., Götz, M., Schally, A. V., Coy, D. H., Coy, E. J. 1974. *Biochem. Biophys. Res. Commun.* 56:1052–59

255. Merrifield, R. B. 1963. *J. Am. Chem. Soc.* 85:2149–54

256. Coy, D. H., Meyers, C., Schally, A. V., Drouin, J., Ferland, L., Beaulieu, M., Labrie, F. 1977. *Mol. Cell. Endocrinol.* In press

257. Rivier, J., Brown, M., Vale, W. 1975. *Biochem. Biophys. Res. Commun.* 65:746–51

258. Coy, D. H., Coy, E. J., Meyers, C., Drouin, J., Ferland, L., Gomez-Pan, A., Schally, A. V. 1976. *Endocrinology* 98:305A (Abstr.)

259. Schally, A. V., Coy, D. H., Arimura, A., Redding, T. W., Kastin, A. J., Vilchez-Martinez, J., Pedroza, E., Gordin, A., Molnar, J., Meyers, C. A., Labrie, F., Hall, R., Reed, D., Gomez-Pan, A., Besser, G. M. 1977. In *Proc. Int. Symp. Med. Chem., 5th, Paris, July 19–22, 1976.* Amsterdam: Elsevier. In press

260. Hirst, B. H., Reed, J. D., Shaw, B., Hunter, M., Brown, M. P., Gomez-Pan, A., Coy, D. H., Meyers, C., Schally, A. V. 1977. *J-Physiol.* In press

261. Sarantakis, D., McKinley, W. A., Juanakais, I., Clark, D., Grant, N. H. 1976. *Clin. Endocrinol.* 5s: 275–78

262. Grant, N., Clark, D., Garsky, V., Juanakais, I., McGregor, W., Sarantakis, D. 1976. *Life Sci.* 19:629–32

263. Gordin, A., Meyers, C., Arimura, A., Coy, D. H., Schally, A. V. 1977. *Acta Endocrinol.* In press

264. Brown, M., Rivier, J., Vale, W. 1977. *Science* 196:1467–69

265. Ferland, L., Labrie, F., Coy, D. H., Arimura, A., Schally, A. V. 1976. *Mol. Cell. Endocrinol.* 4:79–88

266. Evered, D. C., Gomez-Pan, A., Tunbridge, W. M. G., Hall, R., Lind, T., Besser, G. M., Mortimer, C. H., Thorner, M. O., Schally, A. V., Kastin, A. J., Coy, D. H. 1975. *Lancet* 1:1250

267. Marks, N., Stern, F. 1975. *FEBS Lett.* 55:220–24
268. Benuck, M., Marks, N. 1976. *Life Sci.* 19:1271–76
269. Sarantakis, D., McKinley, W. A., Grant, N. H. 1973. *Biochem. Biophys. Res. Commun.* 55:538–42
270. Veber, D. F., Strachan, R. G., Bergstrand, S. J., Holly, F. W., Homnick, C. F., Hirschmann, R., Torchiana, M. L., Saperstein, R. 1976. *J. Am. Chem. Soc.* 98:2367–69
271. Sarantakis, D., Teichman, J., Clark, D. E., Lien, E. L. 1977. *Biochem. Biophys. Res. Commun.* 75:143–48
272. Sarantakis, D., Teichman, J., Lien, E. L., Fenichel, R. L. 1976. *Biochem. Biophys. Res. Commun.* 73:336–42
273. Arimura, A., Coy, D. H., Chihara, M., Fernandez-Durango, R., Samols, E., Chihara, K., Meyers, C. A., Schally, A. V. 1977. In *Proc. Int. Symp. Gut Horm., Lausanne, June 18–19, 1977.* Academic. In press

Ann. Rev. Biochem. 1978. 47:129–62
Copyright © 1978 by Annual Reviews Inc. All rights reserved

BIOSYNTHESIS
OF PROCOLLAGEN

❖970

John H. Fessler and Liselotte I. Fessler

Molecular Biology Institute and Biology Department, University of California, Los Angeles, California 90024

CONTENTS

129

0066-4154/78/0701-0129$01.00

PERSPECTIVES AND SUMMARY

A number of problems in the assembly of collagen molecules and their subsequent association into fibers led to the postulate of a precursor, and the discovery of procollagen at the beginning of this decade opened the way for further insight (1). The precursor was expected to have peptide chains longer than collagen only in the amino terminal portion. However, it is now known that they have also carboxyl extensions, called carboxyl propeptide chains, that are hundreds of amino acids long, and this has raised new problems. The procollagens undergo several postribosomal alterations, and studies of the corresponding enzymes indicate a finely orchestrated series of modifications that facilitate successive assembly, secretion, and cross-linking steps from individual polypeptides to fibers. This provides opportunities for expression of genetic defects, and clinicians have successfully reopened the question of whether some connective tissue diseases may not actually be "collagen diseases" as they were once lightly alluded to.

The collagens are a family of gene products and their relationships are being defined. There are considerable homologies of amino acid sequences and these can be correlated well with the electron microscopic staining patterns of ordered aggregates of the molecules. The extra amino acid sequences of the propeptides are being determined and a significant finding is that the amino propeptides, which are cut off, contain short, well-folded collagen helices. Although we are gaining a better understanding of the structure, conformation, and successive removal of the propeptides we do not know their biosynthetic function, even though there are suggestive correlations with chain assembly and fiber growth. The detailed arrangement of molecules within a fiber remains unknown, although there are a number of models; and while the in vitro reassembly of dissolved fibers has been extensively studied we do not know the biological mechanism of fibrogenesis.

The expression of different collagen genes by cultured cells appears to be unexpectedly sensitive · to environmental conditions, and can also be changed by viral transformation. The interactions of collagens with cell surfaces and the contributions of collagens to basement membranes, which form a substratum for cell growth, are developing fields of investigation. The role of different collagens in embryonic construction of tissues is being fruitfully explored, especially with the help of powerful immunohistological techniques that allow the identification of microscopic amounts of different gene products. The concept of procollagens may need some reconsideration, as polypeptides are being found which do not behave as precursors but contain substantial collagen and noncollagen amino acid sequences.

INTRODUCTION

This review has two objectives: first, for the general reader, it reviews the problems related to the biosynthesis of a precursor protein, its modifications, and subsequent assembly into a supramolecular structure; second, for the specialist, it reviews the key points of progress. Space limitations prevent us from mentioning many fine publications, especially those in the medical, biopolymer, and enzyme fields. Thus, we endeavor to provide the reader with a key for exploration. Our survey of the literature since the last *Annual Review* article on procollagen (1) is complemented by several recent reviews (2–19).

The concept of collagens as a family of proteins is undergoing a change. The identifying criterion remains the *collagen folding* of a region of polypeptide chain to participate in a triple helix with characteristic parameters, defined by X-ray diffraction data (20). This requires that at least every third residue of this polypeptide region be glycine. It used to be thought that practically all of a collagen polypeptide was in this configuration, except for stretches of about a dozen residues at each end, which were named *telopeptides* (21), and amino acid sequencing data (9, 22) supported this concept exceptionally well. The biosynthetic precursors of these well-defined molecules are procollagens, in which each of the three constituent polypeptide chains of the collagen are covalently lengthened by *propeptides* at both the amino and carboxyl ends (23–25). These propeptides are extraordinarily large: each is of the order of magnitude of a hundred amino acids. Genetically distinct collagens exist in any one animal with defined, different amino acid sequences (4). They are known as (*a*) type I, found in bone, tendon and skin; (*b*) type II, especially found in cartilage and in the eye; (*c*) type III, associated with type I in skin, blood vessels, and smooth muscle; and (*d*) type IV, and others collagens in basement membranes. Each of these is synthesized as a procollagen [for review see type I (1); type II (26–36); type III (37–48); type IV (49, 50)]. Proteolytic removal of propeptides is rapid for procollagens I and II (25, 51–55). However, it seems to be delayed for type III (8, 40, 43) and probably does not occur at all in type IV (56–58), which is found in many basement membranes. Type IV procollagen and type IV collagen may therefore be identical, even though, by special proteolytic digestion, the type IV procollagen can be cleaved to give a residual stem that is probably entirely in the collagen fold and of very similar size to the naturally found collagen molecules of types I and II (59). Probably there are a number of other collagens which, like type IV, never lose substantial portions of polypeptide sequences that are not in a collagen fold. It is only a matter of classification whether one regards a molecule such as the component C1q of complement, with its sequenced 78-residue-long collagen triple helix (60), as a collagen or as a different protein in which the characteristic folding of collagen also occurs.

Collagen nomenclature

Although this needs to be changed we use the following system until research has settled several questions. We define, operationally, a procollagen as consisting of three pro α-chain polypeptides, which have the longest amino acid sequences that can be isolated at present with the help of inhibitors of proteases. It is very likely that the initial translation products are still longer and will be found to carry additional "signal sequences" (61) at the amino end, as has been found for several secreted proteins. Thus, just as for insulin where there is proinsulin and preproinsulin, so also there may be procollagens and preprocollagens. For a given procollagen there will be one or more physiological sequences of cleavage to some final collagen. The archetype of final collagen consists of three α polypeptide chains. Each α chain consists of a major, central portion in the collagen fold, flanked by two short telopeptide regions. The increase in length over an α chain is, in the simplest cases, one attached propeptide. If the whole amino propeptide is attached the monomeric chain is called pN-α; if the whole carboxyl propeptide is attached it is called pC-α. The corresponding trimeric molecules are pN-collagen and pC-collagen. Accidental or intentional proteolysis during isolation is likely to cleave the polypeptide chains at sites other than their physiological cleavage points; partial loss of telopeptide regions may also occur [e.g. partial loss of carboxyl telopeptides (62)]. Roman numerals indicate the genetic type of collagen chain within a species. In types II and III collagen, and probably also in type IV, each molecule consists of three identical chains. Two chains of different sequence, α1 and α2, are found in type I. The commonly found type I molecule is the heterotrimer $[(\alpha1)_2\alpha2]$ though the homotrimer $[\alpha1(I)]_3$ has been found (63–68). A material very similar to $[\alpha1(I)]_3$ has been obtained by Little & Church (69, 70) who have proposed that it represents an ontologically earlier collagen, which they named V. They suggest that other cells may synthesize this material under special culture conditions. In view of the previously mentioned uncertainties concerning proteolytic cleavage, definitive discrimination between two α chains, which are proposed to be extremely similar and yet different, is difficult. Within the limits of existing methodology Mayne et al (64) have rigorously established the existence of [α1(I)]₃, and trimers have previously been assembled in vitro from isolated α1(I) chains (71). Several new collagenous polypeptide chains have been isolated (72–76). Each peptide is given an individual name because the disruptive techniques necessary for their isolation have precluded establishment of their molecular trimeric formulas. It should also be kept in mind that a single collagenous polypeptide can fold back on itself to form a collagen triple helix. This was suggested for Ascaris cuticle collagen (77, 78) and has been considered for other invertebrate collagens (79) and for the assembly of procollagen from a very large tricistronic initial translation product (23, 80).

STRUCTURE OF PROCOLLAGENS

The concept of procollagen changed in 1974–1975. Earlier literature (1) assigned all matter found in addition to collagen to amino propeptides (also called extension peptides), which were cut off by a single enzyme, procollagen peptidase. Here we summarize evidence for the existence of both amino and carboxyl propeptides and indicate how some previous difficulties are thereby resolved. Our current view may be prejudiced by experimental factors that also influenced the persuasive evidence for the existence of only amino propeptides: namely, undetected proteolysis and use of dermatosporactic collagens. Dermatosporaxis is inherited fragility of the skin found in cattle (1), sheep (81, 82), humans (83), and cats (84). The prime defect appears to be in the amino-procollagen peptidase (1, 85–87), which results in collagens that carry extensions chemically proven to be at the amino end (1). It is, however, not established that the amino propeptides recovered from dermatosporactic and fetal skins, and used for much that follows, have not already lost some portions of the sequence that formed the beginning of the translated polypeptides. Such loss might not be limited to the sought-for "signal peptides" of secreted proteins.

Evidence for Separate Carboxyl and Amino Propeptides

Carboxyl extensions, as well as amino extensions, were found in the collagens released by cultured dermatosporactic cells (23, 88). This finding, coupled with the demonstration that procollagen is successively cut at two sites during physiological processing (25, 51–53, 89), provided the impetus to reexamine previous concepts.

TIME LABELLING STUDIES   Pulse-chase labeling studies of chick bone procollagen with Pro and Cys had not been consistent with Cys being incorporated only into amino propeptides (52), and detailed time-labeling studies of the type designed by Dintzis (90) proved that both amino acids were incorporated into a collagenase resistant fragment that was on the carboxyl side of the collagen helix, as well as into the amino propeptides (25).

VERTEBRATE COLLAGENASE EVIDENCE   Vertebrate collagenases were used in a number of ways to prove existence and locations of both carboxyl and amino propeptides. These enzymes cut the helix of completed collagens into a major portion containing the amino terminus and a minor one containing the carboxyl end (91). Vertebrate collagenases have the same action on procollagens, as was proven by time-labeling studies (24), and both the amino and carboxyl fragments have extensions (23–25). Tryptophan, which does not occur in the final collagen molecule, is found predominantly in the disulfide linked carboxyl portion of the vertebrate collagenase cleavage product of both procollagen (23–25) and of the physiological intermediate of procollagen processing, pC-collagen (previously called altered procollagen) (25, 53). Further physiological processing of pC-collagen to collagen releases this disulfide-linked carboxyl propeptide which contains Trp (25, 47, 89). Collagenase digestion of procollagens also yields a similar Trp containing disulfide-linked triple peptide (25, 89, 92, 93). When vertebrate collagenase cuts either procollagen or pN-collagen the amino fragments are not disulfide-linked; they are also larger, as judged by electrophoresis, than the corresponding amino fragments obtained from either pC-collagen or collagen (23–25). This shows that while the amino ends of procollagen and pN-collagen chains are longer than those of collagen, those of pC-collagen are the same as collagen. Furthermore, it proves that the interstrand disulfide bridges of procollagen I are located only between the carboxyl propeptides, even though the amino propeptide of pro $\alpha 1$ contains Cys.

IMMUNOLOGICAL EVIDENCE   Comprehensive analysis of the immunological data showed that two different sets of antigenic determinants are present in procollagen, and that either set can be lost from the parental molecule (94). A simple explanation is that the two sets are at opposite ends of the main collagen helix and correspond to amino and carboxyl propeptides (89). Both sets of determinants are contained only in procollagen that has been prepared with precautions against proteolysis (94–96). The amino propeptide determinants are shared by dermatosporactic collagen (97–99) and pN-$\alpha 1$ isolated from calvaria extracted with acetic acid (100). The carboxyl propeptide determinants are found in pC-collagen isolated from the media of tendon cells (93), and in the cleaved off carboxyl propeptides that are recovered from the incubation medium of calvaria (89). The carboxyl propeptide determinants have also been found in similar peptides obtained by collagenase digestion of bone (89) and cell culture (92, 101, 102) procollagen. Antibodies made against the carboxyl propeptides released into incubation media cross-react with the carboxyl fragment of procollagen cut by vertebrate collagenase (103). Cleavage with vertebrate collagenase has subsequently shown propeptides at both ends of procollagen I made by chick tendon cells (33, 34, 104) and by human fibroblasts in culture (105). Propeptides at both .ends have similarly been demonstrated in procollagen II made by chick cartilage cells (32, 34) and by chick sterna (43), and in procollagen III isolated from calf skin (107) and made by chick embryo blood vessels (43). In all these procollagens the three chains are linked by interchain disulfide bonds between the carboxyl propeptides. In addition, interchain disulfide bridges are found in procollagen III between the three amino propeptides (43, 107) and collagen chains (4). The carboxyl propeptides of the three types of chick procollagens are of approximately equal size, but the amino extensions are variable.

ELECTRON MICROSCOPIC EVIDENCE   Earlier electron microscopic examination of SLS[1] aggregates demonstrated the presence of amino extensions (1) on dermatosporactic collagen and pN-collagen that had lost the carboxyl extension due to inadequate protection against proteolysis. Carboxyl extensions were seen first in SLS aggregates of procollagen from dermatosporactic Calf cell cultures and the cleaved fragments produced from them by vertebrate collagenase (23). Narrow SLS aggregates with bushy extensions were prepared from human fibroblast cultures (108), but no clear interpretations were possible. Procollagen secreted by chick tendon cells (109) can be aggregated into SLS that have tightly packed amino extensions (15 nm) and carboxyl extensions that have fanned out, bushy ends that adhere to each other and do not pack closely. SLS aggregates of pC-collagen have only the characteristic carboxyl extensions.

Procollagen II isolated from the medium of cartilage cells (35) aggregates into SLS that closely resembles the SLS of procollagen I. Procollagen III isolated from rat skin (38) and calif skin (41) is probably not the complete precursor. SLS aggregates of the molecules demonstrate the presence of amino extensions that are not seen in collagen III. However, both procollagen III and collagen III have short (5 nm) carboxyl extensions that are removed by pepsin (38, 41).

[1]Segment Long Spacing crystallite. This is a side-by-side alignment of collagen molecules with characteristic electron microscopic appearance; its staining pattern reflects the distribution of charged side chains (see 22, 318).

Amino Propeptides

TYPE I

Amino propeptides of pNα1   The amino terminal cyanogen bromide fragment pα1-CB 0.1 of dermatosporactic pNα1(I) chains is much larger than the corresponding 19 residue fragment α1-CB 0.1 of the normal α1(I) chain (110). As shown in Figure 1 (98, 111, 112), bacterial collagenase digestion of pα1-CB 0.1 gives "compact" fragment pα1-CB 0.1, Col 1 that has practically the same amino acid composition in dermatosporactic sheep and calf (110, 111). It contains five disulfide bridges and more than one third of its residues are either Asp or Glu residues. It does not contain Hyl and has relatively few Pro, Gly, and basic residues. It contains several strong conformational antigenic determinants that are lost upon reduction and carboxymethylation and that cross-react between sheep and calf; some minor sequential order antigenic determinants are also found (98, 99). It contains both coiled and extended peptide regions (111a). Another collage-

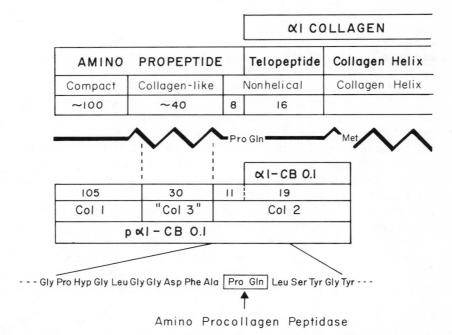

Figure 1   Diagram of the amino end of dermatosporactic sheep pNα1(I) chain together with a corresponding peptide from calf. See text for explanation. Based on references (9, 98, 111, 112).

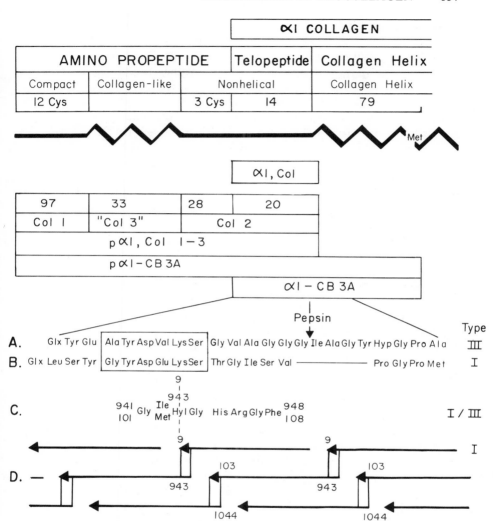

Figure 2   Diagram of the amino end of the calf pNα1(III) chain together with a comparison of cross-linking regions: (A) amino telopeptide of α1(III); (B) amino telopeptide of α1(I); (C) cross-linking sequences of α1(I) residues 941–948 (upper part of C) and residues 101–108 (lower part of C) [the same peptides occur in α1(III)]; (D) a probable staggered array of collagen I molecules in a fiber. Based on references (22, 107, 125–130).

nase-resistant peptide, pα1-CB 0.1, Col 2, has been isolated from dermatos-poractic calf pNα1 (9, 112). It contains the sequence shown in Figure 1 (112). The carboxyl end of this sequence is identified with the known amino terminus of calf α1(I) and shows that amino procollagen peptidase cleaves a Pro-Gln linkage.[2] After a short nonhelical sequence this peptide ends at its amino end with the tripeptide Gly-Pro-Hyp that is characteristic for collagen. A collagen helix rich in Gly, Pro, and Hyp extends from here to the carboxyl end of pα1-CB 0.1, Col 1 (111). The larger peptide, pα1-CB 0.1, which contains this region and pα1-CB 0.1, Col 1, readily aggregates to trimers (111a); and circular dichroism indicates that the triple helical segment is stable and melts in the range 35–45°C (111a). The antigenicity of pα1-CB 0.1 resides mostly in the compact portion pα1-CB 0.1, Col 1, and although the adjacent regions may modulate this antigenicity, these regions are themselves only weak determinants (98). A peptide correspond-ing to the amino propeptide of dermatosporactic collagen has been isolated from bovine amniotic fluid (113).

Earlier work with chick material which was probably pNα1 (extracted from calvaria by acetic acid) indicates a similar propeptide structure (100, 114, 115): the antigenic determinant is located in a Cys-containing peptide that is larger when obtained by CNBr cleavage than when recovered from a collagenase digest (100). However, amino acid analysis of the peptide obtained from chick differs from that isolated from sheep or calf (100, 110).

Amino propeptides of proα2   The amino propeptides of dermatosporactic sheep and calf pNα2 chains consist of about 60 residues and appear to have a similar design to their α1 counterparts, except that they lack the "globu-lar" amino end equivalent to pα1-CB 0.1, Col 1 (116, 117). Correspondingly they are smaller, and lack antigenic determinants (98) and Cys and Trp residues (117). However, they do have a high content of Gly, Pro, and Hyp, which suggests that they participate in a collagen triple helix. The chick amino propeptide of proα2 lacks Cys and Trp and is cleaved into small peptides by bacterial collagenase, which suggests a similar structure (J. H. & L. I. Fessler, personal communication). In contrast, a report on rat type I precursors (40) indicates that amino extensions of equal size (16,000 daltons, determined by gel filtration) are released by bacterial collagenase from pNα1 and pNα2 and these two extensions have distinct chromato-graphic properties on DEAE cellulose. This suggests that the amino exten-sion of rat pNα2 either contains a larger noncollagenous propeptide or that it is protected against cleavage by bacterial collagenase.

[2]Recent evidence (K. Kühn, private communication).

TYPE III   Substantial homologies of types III and I procollagens have been found in the structures of their amino propeptide, telopeptide, and collagen helix, and extend to potential cross-linking between adjacent chains. Type III collagen may be specially suited for making stable, thin fibrils and is the collagen of "reticulin" fibers (118, 119). It is stabilized by disulfide bridges: even after solubilization by pepsin digestion each α1(III) chain remains linked to both of its companions through two Cys residues at the carboxyl end of its collagen triple helix (120–124). Various large collagenous polypeptides have been labeled procollagen III, or collagen III, solely on the inadequate evidence that pepsin digestion leaves a disulfide-linked trimer of alpha-sized chains. The CNBr peptides of pepsin-solubilized calf and human collagen III have been characterized (4) and ordered (123, 125). The size of the α1(III) chain as it exists in an insoluble matrix is unkown. Various soluble forms have been found in rat skin (37–40), in fetal calf skin (39, 41, 42), in bovine amniotic fluid (44, 44a) in cell cultures (45–48), and in chick embryo blood vessels (43), and have been called procollagen III and collagen III. This assignment is supported by the amino acid sequence homology between fetal calf skin soluble collagens III and I (see Figure 2) (126). Ion exchange chromatography allows separation of the larger materials, called procollagen III, from the materials with alpha chain sized constituents, called collagen III (4) though a precursor-product relationship has not been established.

Amino propeptide of calf procollagen   Fetal calf amino propeptide of α1(III) Figure 2 (107) has at its amino end a 97 residue peptide (pαl, Col 1) that resembles its compact equivalent of dermatosporactic pNα1(I). It is noncollagenous and internally cross-linked by means of 12 Cys residues; Asp and Glu make up nearly one third of it, and it is immunologically active. This is followed by a 33 residue peptide (pα1, Col 3) that contains 11 Gly, 9 Hyp, and 2 Pro residues, which forms a collagen helix within the amino propeptide. Although it does not contain Cys residues it is only collagenase digestible after disulfide bridges in an adjacent peptide have been reduced. This adjacent peptide of 28 residues contains 3 Cys, is disulfide linked to the propeptides of the other two chains, is not in collagen helical fold, and connects to the amino telopeptide of α1(III); they are isolated together as the 48 residue peptide (pα1, Col 2). The main collagen helix of α1(III) continues from there, and the first Met only occurs 79 residues into this helix. CNBr treatment of the procollagen therefore cleaves off one large amino terminal peptide (pα1-CB3A), which, reading from the amino end, consists of [(pα1, Col 1)-(pα1, Col 3)-(pα1, Col 2)], [together called (pα1, Col 1–3)], followed by 79 residues of main collagen helix. The structure was deduced from differential studies with proα(III), α(III), and pepsin- and collagenase-digested materials.

Cross-linking of α1(III) and α1(I)   In Figure 2, the sequence following the blocked amino terminus of calf collagen III is given in A, and for comparison the corresponding sequence of α1(I) is shown in B. The type III collagen helix is judged to be longer by two extra triplets (126). The critical lysine residue 9 of type I, which is known to participate in cross-link formation between type I chains (22), appears as residue 8 in sequence A and is a part of closely similar sequences in both telopeptides. In type I collagen cross-links are formed between this Lys and another Lys or Hyl after one of the residues is oxidatively deaminated to the corresponding aldehyde (127). Tryptic digestion of insoluble fetal calf skin (128, 129) gave a dimer of two identical peptides held together by an aldol type of cross-link. Each peptide contained the sequence A of Figure 2, except that the Lys had become part of the cross-link. The amino telopeptide of type III collagen thus appears to cross-link with identical neighbors. Figure 2D shows a scheme of cross-linking for type I collagen fibers, which is also likely for type III. Although the detailed arrangement of molecules within a fiber is not known there is strong evidence that the type I molecules are staggered by multiples of 234 residues (22). In the arrangement shown in Figure 2D, Lys residue 9 comes to lie near to Hyl residue 943 (22, 130) and a cross-link can be formed. Figure 2C shows the sequence 941–948 of α1(I), with Ile next to the Hyl residue. The identical sequence has been isolated from near the carboxyl end of the α(III) collagen helix (123). In type I collagen the arrangement of Figure 2D also brings Lys 1044 of the carboxyl telopeptide, which is known to participate in cross-links, opposite residue 103, a hydroxylysine, in an adjacent α1(I) chain. Figure 2C shows the sequence α1(I) 101–108, which is identical to α1(I) 941–948 except for the substitution of Met for Ile. The identical sequence, including Met, has been found near the amino terminus of the α(III) collagen helical chain (123). It forms the sequence 101–108 in the known first 229 residues of human α1(III) (125), and similar sequences are found in calf α1(II) and α2(I) (22). These agreements strengthen the proposal of linking of α1(I) shown in Figure 2D and make it likely that type III collagen molecules are also arranged in this way. Electron microscopy (131) shows that while type III collagen has a distinctly different charge distribution from type I and II collagen it can also form native-type fibrils like type I.

Carboxyl Propeptides

These materials were partly described before their location at the carboxyl end of procollagen was established (1) (as noted earlier in this review).

Typically, carboxyl propeptides of procollagen I consist of three peptide chains linked by disulfide bridges, are resistant to collagenase, and have a composition very different from collagen (1, 89, 92, 95, 103). They do not have sufficient Gly for the collagen fold, lack Hyp and Hyl, are enriched in Glu, Asp, Cys, Tyr, and contain Trp, mannose, and glycosamines (89, 103, 132, 133), which are not found in collagen helices. Adjacent to the collagen helix are carboxyl telopeptides, and these remain attached after the carboxyl propeptides are physiologically cut off and accumulate during incubation of intact calvaria (25, 53, 89), of tendon cells (103), and in some cell cultures (47). There is uncertainty about several carboxyl telopeptides because they are particularly prone to accidental proteolytic degradation (62), but they are probably sufficiently large to account approximately for the difference found by some investigators in apparent molecular weights between carboxyl propeptides made by collagenase digestion of procollagen I and peptides accumulating in culture. These molecular weights are respectively, 37,000 and 33,000 as determined by $NaDodSO_4$ polyacrylamide gel electrophoresis (89). The tryptic peptide maps of these materials show good homology, but the material prepared by collagenase digestion contained additional peptides (103). The identity of the peptide of culture media with carboxyl propeptide was further established immunologically (47, 89, 103).

The native, disulfide-linked carboxyl propeptide of procollagen I has a molecular weight of 100,000, as determined by sedimentation equilibrium (103), and a sedimentation coefficient of 5.2S at both 0.5 mg/ml (103) and when much more dilute solutions of radioactively labeled propeptides were analyzed (J. H. & L. I. Fessler, unpublished). About half the Cys residues form interstrand disulfide bridges that can be reduced without denaturation and the products give sedimentation equilibria plots that show evidence for association and an average molecular weight of 43,000 $\pm$ 5,000 (103). Analyses of the reduced and alkylated carboxyl propeptides on CM-cellulose (103) and by means of $NaDodSO_4$ gel electrophoresis reveals the presence of two materials in the approximate ratio of 2:1 as judged by absorbance, Coomassie blue staining (103), and radioactivity after labeling with Trp, Leu, or Tyr (134). The electrophoretic mobilities give apparent molecular weights of 34,000 and 31,000, respectively for, the major and minor components (103). They give different tryptic peptide maps (109), though they have very similar overall amino acid and carbohydrate composition (103). The major component was proven to be the carboxyl propeptide of pCα1 and the minor component the carboxyl propeptide of pCα2. This was done by deriving them, by reduction, from separated, partly cut pC-collagens (134). The carboxyl propeptides released by bacterial collagenase from chick procollagen II (32, 54) and chick procollagen III (43) are also disulfide linked. They contain Cys and Trp, and after reduction have the same electrophoretic mobility as the carboxyl propeptide of proα1(I). The sedimentation coefficients of the denatured carboxyl propeptide of procollagen I and II are the same (J. H. & L. I. Fessler, unpublished) and gel filtration assigns similar molecular weights (32). These similarities do not imply identity and antibodies against procollagen III do not cross react with procallagens I and IV (44). The 5 nm carboxyl extension of "procollagen III" and collagen III extracted from fetal calf skin contains 56 amino acids with a high proportion of Gly, Pro, and some Hyp, which suggests the presence of some helical sequence (135). This is considerably shorter than the carboxyl propeptide isolated from chick procollagen III.

Carbohydrates On Propeptides

Mannose and glucosamine are present in the carboxyl propeptide of proα1 and proα2, and a small amount of glucosamine is found in the amino-propeptide of procollagen I secreted by chick tendon cells (132, 133). These carbohydrates are incorporated into procollagen in the presence of 2,2'-dipyridyl, which inhibits hydroxylation of lysine and thereby prevents attachment of glucose and galactose to the collagen portion. Mild alkaline hydrolysis does not release the carbohydrate from the propeptides, which indicates that these oligosaccharides are not linked to either peptidyl Ser or Thr. The carboxyl propeptide released during the physiological processing of procollagen I of chick cranial bones (89) and tendon cells (103) retains mannose and N-acetylglucosamine. Procollagen II contains mannose, glucose, and galactose as part of its collagenase resistant propeptide (30). Mannose, glucosamine, galactosamine, glucose, and galactose are linked to type IV collagen, but not fucose, mannosamine, or sialic acid. The sites of attachment are not yet established (14).

BIOSYNTHESIS

Collagen Gene Expression

Currently there are three prime areas of interest: developmental, medical, and cell biological implications of differential expression of collagen genes. Several questions recur: (a) Is there a more "primitive" collagen which can be elaborated by cells that have not reached a final, end stage of differentiation? (b) How do different cells contribute one or more collagens to a complex tissue? (c) Can one cell synthesize more than one type of collagen at the same time?

DEVELOPMENTAL CHANGES: TYPE II   Type II collagen is made during chondrogenesis (136), in development of notochord and bone (137, 138), and in limb blastema regeneration (139–141). Although type II collagen can serve as a marker of chondrocytes it is not exclusive to them. The "vitrosin" fibers of the vitreous humor of the eye are reported to consist of type II collagen and a search for the source of their formation indicated the neural retina and the vitreous body at different developmental stages (142–145). The developing eye poses a number of interesting problems in collagen fibrogenesis and one tissue, the corneal epithelium, produces two types of collagen (146).

DEVELOPMENTAL CHANGES: TYPES I AND III   In a complex tissue, such as the skin, which contains predominantly collagens I and III, increasing development and age is likely to be accompanied by decreasing content of collagen III (147–153), and skin fibroblasts are usually able to synthesize both in culture (45–47, 154). In the smooth musculature of blood vessel walls (147, 155–159) the expression of collagen I and III may be variable, and cells derived from it (160–163) usually make both collagens I and III in culture, even after cloning (74). Expression of both collagens also occurs in the highly complex lung and in cells derived from it (48, 164, 165), in peripheral nerve (166), and in peridontium (167).

MEDICAL PROBLEMS OF COLLAGEN GENE EXPRESSION   The varied elaboration of collagen I and II in intervertebral disk (168) may have medical implications. While some of the most intensely investigated aspects of collagen pathology have been genetic defects in some of the enzymes that modify procollagen after synthesis (15–17), the expression of a collagen type may also be affected, as in osteogenesis imperfecta (169, 169a) or in a lesion such as liver cirrhosis (170), hypertrophic scars (17, 155, 171), and atherosclerotic plaques (17, 159). Approaches to these problems have been to solubilize whole tissues and then either quantitate the collagens as expressed in characteristic, distinguishable CNBr peptides (67, 72, 147, 155) or to separate the still native collagens, after proteolytic solubilization, by salt precipitation, chromatography, and electrophoresis (4, 72, 172, 173, 173a). Staining histological sections with type-specific fluorescent antibodies (118, 137, 174–177) is highly informative and allows localizations at the cell level. Immunological assays have been applied to sera and synovial fluids (101, 102, 178, 179).

EXPRESSION OF COLLAGEN GENES IN CELL CULTURE   When cell cultures are derived from biopsies to test inherent capabilities and defects of collagen synthesis, a distinct problem of variable expression is encountered. A single cell can simultaneously synthesize more than one type of collagen (154, 177), as demonstrated by fluorescent antibody staining.

The collagen types synthesized even by cloned cells can change with culture conditions, and new forms of collagen may arise that have not been found so far in vivo (64). Chondrocytes appear to be particularly subject to such changes and the triggering event can be as diverse as incorporation of bromodeoxyuridine (63, 66), addition of embryo extract (180), change from monolayer to suspension culture (181–183), or variations in tonicity of the culture medium (184). Not only does the synthesis of one type, e.g. collagen II, cease and of another one, collagen I, start, but a new molecular arrangement $[\alpha 1(I)]_3$ also occurs. Detailed analysis of αII chains indicates that there are major and minor components that are probably products of separate, nonallelic genes (185). Other collagen chains, e.g. X and Y (67, 186), initially present in only trace amounts, can also increase. The types of collagen synthesized by established cell lines (187) may be changed by viral and chemical transformation (188).

Polyribosomes, Messenger RNA, and Cell-Free Translation

The genes for human type I and III collagens are on different chromosomes; thus they segregate independently in human–mouse somatic cell hybrids (189). Probes of c-DNA to collagen message are being made (190), and may help in analyzing problems of the occurrence and transcriptional control of the family of collagen genes. Research so far has been only at the cytoplasmic level, on: the location and size of polysomes making procollagen; the size of their mRNAs; in vitro translation by cell-free systems; polypeptide folding; and on posttranscriptional modifications. Cells actively synthesizing collagen have a prominent RER[3]; however, the preparation of polyribosomes from many collagen synthesizing tissues is complicated by the associated connective tissue. To overcome this difficulty, very young embryos (191–206, 208), cell cultures (193, 205, 207), and matrix free cells freshly isolated from chick embryo tendon, and sterna (209, 210), and rat chondrosarcoma (208), have been used to demonstrate the size of the polysomes active in synthesis of procollagen chains and to isolate mRNA. In tendon cells the procollagen synthesizing polysomes belong to the category of ribosomes that are tightly bound to membranes (211), as has been found for other secreted proteins.

EVIDENCE FOR MONOCISTRONIC MESSAGE   The problem of the assembly of collagen I from two pro-α1 and one pro-α2 chains caused various polycistronic mRNA hypotheses to be suggested (23, 212, 213). Although correspondingly large polysomes have been reported in association with immunologically defined procollagen (205), polysomes were later found to

[3]Rough Endoplasmic Reticulum.

aggregate at the relatively low potassium concentrations that have been used (208). Collagen synthesizing polysomes prepared from primary chick cartilage cell cultures were also found to contain a subclass of faster sedimenting polysomes that contained peptide-bound proline that could neither be hydroxylated nor cleaved by collagenase (207). Others (191) reported polysomes of the size expected for monocistronic proα messages, but polysome size alone is a poor criterion for deciding these problems. The following observations were made on chick embryo calvaria: polysomes of one size range were found to contain twice as many ribosomes that carried nascent proα1 chains than ribosomes that carried nascent proα2 chains; and the synthesis of these chains could be completed in vitro (214). Previous time-labeling studies (215) had shown that α1 and α2 chain elongation proceeds at the same rate and that each α chain is a sequentially synthesized continuous polypeptide, made in 4.8 min. Other studies (25), also of the type designed by Dintzis (90), showed that [^{35}S] Cys is incorporated into amino propeptides of procollagen I about 5.3 min before it is built into the carboxyl propeptides, and that after only 4 min of labeling the total radioactivity in the carboxyl propeptide of proα1 is twice that of proα2. All this taken together suggests that each proα1 and proα2 chain is synthesized separately as a continuous polypeptide and that the active mRNAs for the two chains are present in calvarial cells in a ratio 2:1, respectively, and that the rate of initiation, elongation, termination, and release from polysomes are all identical for the two chains. It does not exclude polycistronic messages for procollagen with separate initiation sites for the translation of each chain. Similar (90) labeling studies of collagen II (216) were used to order its CNBr peptides and gave results concordant with the above. Since the same nucleotide sequence was found to code for more than one protein in bacteriophage φX174 (217) it occurred to us that a single polynucleotide sequence might code for both chains of type I collagen according to two different reading frames. But inspection of amino acid sequences (22) around Met residues, which will be coded for by AUG, excludes use of the same mRNA.

ISOLATION OF mRNA   Messenger RNA for procollagen is estimated to be at least 4,500 nuceotides long in order to code for the amino acids of each proα chain (3), and therefore its molecular weight should be about 1.5×10^6 or greater, according to how many additional, noncoding nucleotides there are. A high G-C content is expected due to the preponderance of Pro and Gly in the protein. Procollagen mRNA is therefore expected to behave similarly to 28S rRNA and their separation has been difficult (196, 197). Oligo dT-cellulose columns (197, 200, 209, 218, 219) retain procollagen mRNA, and a poly-A sequence of about 150 nucleotides has been estimated (219, 220). Size and secondary structure complicate definitive measurements of the parameters of this mRNA by polyacrylamide gel electrophoresis and by sedimentation velocity (190, 197, 220).

CELL-FREE TRANSLATION   Messages for procollagen I, estimated to have a molecular weight of 1.6–1.8×10^6 and a sedimentation coefficient of 27–30S, were translated in the cell-free wheat germ system, yielding collagenase-sensitive protein that electrophoresed like proα1 and proα2 (195, 197, 200, 210, 220). A clear distinction between the translation products of mRNA for procollagen I and II has been demonstrated (210). Other cell-free translation systems have been prepared from reticulocytes (196) and from Krebs ascites cells (194, 218, 219). There do not seem to be species or tissue specific barriers to heterologous cell free systems (199, 204, 206). Products have been identified by electrophoretic migration, hydroxylation by proline hydroxylase, and collagenase digestibility. The multiple technical difficulties are beyond the scope of this review; rigorous control experiments are essential.

In chick calvaria, optimal synthesis of 26S, polyA-containing mRNA (221), which directs formation of collagenase sensitive polypeptides by wheat germ, is found about one day before optimal collagen synthesis (222) occurs in vivo. Synthesis of mRNA that contains presumptive collagen message also is reported to precede rRNA synthesis (223). No clear cut control of collagen synthesis, by the availability of tRNA for the amino acids prevalent in collagen, has been found (224–228). Preferential use of tRNA Gly cognate to GGU and GGC is observed with calvarial polysomes (229). An interesting indirect control was found in cartilage cells: collagen synthesis decreased when proteoglycan completion was disrupted by adding p-nitrophenyl-β-D-xylose, which independently primes growth of chondroitin chains (230).

Polypeptide Elaboration, Modification, Folding, and Trimeric Assembly

FOLDING OF NASCENT CHAINS   Although the actual configuration of nascent procollagen chains on polyribosomes is not known they can be recognized by antibodies made to collagen (203) and procollagen (205, 221). They can fold into a state in which they are both resistant to pepsin (231) and have the deuterium exchange properties expected of a collagen helix (232). It is not known how the procollagen chains enter the RER, though evidence for the aforementioned "signal peptides" (61) may come from cell-free synthesis studies with isolated message. Hydroxylation of the procollagen chains is not a prerequisite for their leaving the ribosomes (1).

HYDROXYLATION, DISULFIDE LINKAGE, AND HELICAL FOLDING
A wide range of evidence (e.g. 33, 233–238) reviewed in (1–3, 16, 17) suggests that by the time procollagen leaves the RER it is probably completely hydroxylated, glycosylated (237–240), interstrand disulfide linked, and helically folded, and probably has complex carbohydrate attached (132, 133). Furthermore, as the degree of hydroxylation and glycosylation is increased from type I to type II to type IV, so the time taken for completion of these processes is increased (2, 3, 33, 234, 241) [but also see (242)].

However, the sequence of molecular events after polypeptide synthesis is not understood. A key feature is that hydroxylation of proline residues stabilizes the helical folding of a collagen chain (1, 2), and that prolyl hydroxylases (2, 243–247) and lysyl hydroxylase (2, 248–250), as well as the lysyl glycosylases (2, 17, 251–255), require their substrates to be in unfolded, nonhelical form. It is therefore likely that proline hydroxylation and chain folding will proceed *pari passu*. Time-labeling studies (215) show that in the unperturbed system proline hydroxylation closely follows peptide synthesis. Proline hydroxylase is located within the RER, as proved by electron microscopy (2, 256, 257) and by its presence within "rough" microsomes prepared from it (258–262), so it is reasonable to envisage a coordinated progression of peptide synthesis, entry into the RER, hydroxylation, and concomitant helical folding (1–3). Unless there is refolding at a later stage this implies that the correct three chains have somehow been selected and aligned at an early stage, with the help of their amino propeptides. Disulfide linkage between the carboxyl propeptides is delayed till after synthesis of the completed pro α chains (2, 33, 52, 207, 263).

Experiments designed to check the applicability of such a scheme indicate that hydroxylation is accompanied by helix formation in tendon cells (236), but precedes it in cartilage cells (2, 3, 234), and that in both cells helix formation and disulfide interstrand linkage go hand-in-hand. Resistance to pepsin digestion was taken as a measure of helix formation (2). Unfortunately, this can only determine what proportion of the recently synthesized collagen chains are able to be in a pepsin resistant state under the conditions used for digestion: it does not distinguish the sought-after molecules that are in this state *in vivo*, from others that have the ability to be folded into this state during sample preparation. Disulfide links between chains will both stabilize the helix and facilitate the folding of a collagen helix (264, 265) and therefore there will be an experimental bias toward correlation between disulfide linkage and pepsin resistance, either because disulfide-linked trimers are less likely to become pepsin digestible or because they may become folded during sample preparation.

INHIBITION OF SECRETION   When hydroxylation is inhibited, or amino acid analogs are incorporated into the procollagen chains (1–3, 235, 257, 266–273) they are prevented from folding, accumulate in the RER, and usually cannot be secreted from the cell until either hydroxylation is carried out or the temperature is lowered so that they become folded. Thus, inhibition with 2,2' dipyridyl causes accumulation of a procollagen with an apparent T_m[4] about 15°C below body temperature. The material is hydroxylated to the normal form, with T_m slightly above body temperature, when iron is restored, and it is then processed normally (266). The important point is that it has been confirmed that the unhydroxylated procollagen consists of correctly matched trimers [(pro α1)$_2$ (pro α2)], idsulfide-linked to each other through the carboxyl propeptides (33, 266). Under different conditions other trimeric molecules were found (68). Correct matching implies correct recognition between propeptides, because the collagen main helix could not have formed at the temperature employed. If the amino propeptide of chick α2 resembles that of dermatosporactic sheep, and both lack Cys and are smaller than the α1 amino propeptides, then its interactions with the other two chains occurs primarily through its collagen-like region. But this should be destabilized by the lack of hydroxylation. In other words, chain selection is likely to have proceeded through the carboxyl propeptides, unless there is some additional large region of α2 amino propeptide that has been lost during isolation. If this is still sufficiently representative of normal, then the paradox arises that chains just entering the RER cannot participate in triple helix formation if they have to wait for matching of their carboxyl propeptides. It may be that normal selection proceeds through the amino propeptides, but that when this is prevented the carboxyl propeptides can perform this task. The action of the hydroxylases is not restricted to the point of entry of the chains and the process seems more protracted when a succession of lysyl hydroxylase and lysyl glycosylases require an unfolded form (17). The considerable dependence of hydroxylase activity on peptide length of substrate (2) fits a more complex scheme in which entering procollagen chains are prevented from premature association by attachment of hydroxylases, which may only act, however, after correct alignment between the carboxyl propeptides has occurred.

Dimers of only two pro α chains in disulfide linkage are notably sparse (33, 52) and such an intermediate assembly seems to be unstable. Interstrand disulfide linkage is interfered with when the threonine analog hydroxynorvaline is incorporated (273), or when low levels of puromycin partially inhibit completion of procollagen chains (263); it was concluded that only completely disulfide-linked trimers are secreted from cells (263).

[4]Transition temperature for the folding or unfolding of a collagen chain.

Tunicamycin, which blocks the attachment of complex carbohydrates to procollagen, does not block secretion (274), though it has various effects on collagen synthesizing cells (275). Ascorbate is required by collagen synthesizing systems but its action is complex and is not understood (2, 276-280). Cell cultures deficient in ascorbate can secrete underhydroxylated collagens, though often only breakdown products are released into the culture media. Antimycin A blocks transfer of procollagen from RER to smooth endoplasmic reticulum (233), and secretion from cells (1, 281), which indicates energy dependence for this intracellular transport. Colchicine and other antimitotic agents (1) block secretion of procollagen and cause its accumulation in large smooth vesicles (268, 282, 283).

CYTOLOGICAL EVIDENCE OF SECRETION   Although electron microscopy has proven the presence of procollagen in the RER and the Golgi Complex by using ferritin conjugated antibodies (256, 257), a unique path of secretion could not be determined (94). Special elongated smooth vesicles, called secretory vesicles, have been seen in corneal epithelium (284), in osteoblasts, and in odontoblasts (285–287). The fibrillar contents of these vesicles sometimes show striations, which suggests that they could be packages of procollagen molecules. Radioactive procollagen was found in osteoblast cell fractions that contained these vesicles, both in continuous labeling and in pulse-chase experiments (285, 288). Quantitative radioautography with [3H] proline suggests that they participate in procollagen transport (287); but this type of analysis does not clearly indicate whether the pathway of secretion from the RER is exclusively by means of the Golgi Complex (289). How procollagen leaves cells is not understood, but some form of fusion of secretory vesicle with the cell surface is envisaged (287, 290), and addition of packages of collagen molecules, in their entirety, to fibers has been suggested (290, 291). One aim is to understand how the direction of a growing fiber is determined: a correlation between osteoblast elongation and adjacent fibers has been found (292).

PROCOLLAGEN PROCESSING

Intermediates of Procollagen Cleavage

The biochemical corollary of these morphological problems is the question of when, where, and how the procollagen molecule is converted to the form in which it is found in the fiber. Pulse-chase experiments coupled with autoradiographs show that in chick embryo calvaria the amino propeptides are removed at about the time that proline-labelled protein leaves the cell (52), whereas carboxyl propeptides are cut off at a distinctly later time. This suggests that the amino propeptides are cut off near the cell surface and the carboxyl ones are removed extracellularly. Colchicine blocks both secretion and cleavage of amino propeptides. Ferritin labeled antibodies to pN-collagen, to material presumed to be pC-collagen and procollagen all reacted equally with cell fragments, therefore it was concluded that all procollagen peptidase activity is extracellular (94).

Procollagen Peptidases

Peptidases that cleave pN-collagen and procollagen are found in bone, tendon, skin, and fibroblast cell culture media and the general findings have been summarized (1). These enzymes are not serine proteases; they are not inhibited by SH reagents, but are stopped by EDTA. An amino procollagen endopeptidase from tendon has been purified (86) that cleaves pN-collagen to give amino propeptides and collagen with pGln termini (L. D. Kohn, personal communication). This enzyme is found in normal calf skin, but is missing from dermatosporactic material. Indications for more than one endopeptidase in the processing of procollagen were obtained from a number of in vitro cell culture and cell incubation experiments (43, 47, 106, 293, 294) and from the action of tunicamycin on the processing of procollagen (274). Only amino procollagen peptidase was extracted from tendon cells in culture, even though the incubation medium contained both amino and carboxyl procollagen peptidases (43). These two enzymes have been resolved and shown to be separate, specific endopeptidases (295). The physiological sequence (25, 52–54) is predominantly that first amino procollagen peptidase removes the amino propeptides and subsequently the carboxyl procollagen peptidase cleaves the carboxyl propeptide in a sequential manner (53) yielding α chains. In chick embryo blood vessels an initial cleavage of procollagen III removes a part of the propeptides, in such a way that the interstrand disulfide bridges of the amino and carboxyl ends remain intact (43). In a fibroblast culture medium a peptidase that cleaves both type III and I procollagen has been found (296). The released propeptides have been detected immunologically in serum and in amniotic fluid (93, 99, 101, 102, 297).

Fibril Formation

It has generally been supposed that the propeptides must be removed before association can occur (1, 3) though facilitation of fibril formation by propeptides has been considered (8) and investigation of the optimal temperature of fibril formation relative to body temperature suggested that a molecule other than the final collagen molecule might participate (298). Although the propeptides do prevent the formation of native type fibrils, linear associations can be seen electronmicroscopically especially in the case of pC-collagen, and these materials can be cross-linked by the action of lysyl oxidase into a series of polymeric molecules (299). Lysyl oxidase has been shown to recognize correct alignment of completed collagen molecules into native type fibers (300). Dermatosporactic pN-collagen has been shown to participate in cross-linking (87, 301). Thus it seems that thin, cross-linked fibrils can be formed without total removal of propeptides. This may be of particular importance for types III and IV procollagens.

Structure of Collagen Fibers

The potential importance of microfibrils has been particularly recognized since the discovery of diagonal step banding (302, 303, 304). However, the theoretical and experimental details are beyond the scope of this review. While it is agreed that known amino acid sequences can be arranged in a variety of ways to provide optimal opportunities for hydrophobic and electrostatic intermolecular interactions, the nature of any longer range equatorial packing order is under question (9, 11, 130, 305–317). A reference atlas for standardization of electron microscope band patterns has been published (318). Cross-linking between molecules has been reviewed (11, 127).

BASEMENT MEMBRANE COLLAGENS

Current problems of basement membranes and their collagens were the topic of a recent symposium (14), and there are earlier reviews of their biology (319) and collagen composition (1, 4, 8, 49, 50, 320–322). There are different kinds of basement membranes. Blood vessel basement membranes are of special medical interest and a wide range of large, collagenous glycopeptides have been isolated from sonicated renal glomeruli (76, 323), even when precautions were taken to prevent proteolysis (324).

Biosynthetic Systems

Procollagen-like materials have been isolated after incubating, in vitro, a number of tissues and cells that are associated with basement membranes. These are: rabbit corneal endothelium (325), rat embryonic parietal yolk sac (56, 326–329), glomeruli (57, 330–333), anterior lens capsule (56, 58, 334, 335), aortic endothelial cell cultures from calf (336, 337) and man (338), and a murine tumor (339, 340). Characteristically these materials contain collagens which cannot be classified as type I, II, or III, though these may also be produced to varying extent. They are immunologically distinct and cross-react with basement membrane collagen extracted from other species or made by other systems (331, 336, 341–343).

Structure and Heterogeneity

These polypeptides are apparently in a molecular weight range of 120–170,000 and on digestion with pepsin or similar proteases leave a collagen-helical portion approximately the length of α chains, though there may be regions that are protected from this digestion, unless the materials have been reduced (59, 344). The residual, α chain-like materials are susceptible to collagenase, contain Gly and pyrollidine residues in the expected amounts, but have a higher proportion that exists as hydroxyproline. Characteristically, the content of the unusual 3-hydroxyproline isomer is several times that of interstitial collagens (49, 247, 345, 346). The majority of hydroxylysine residues are glycosylated and carry the Glu-Gal disaccharide. Consonant with a higher overall hydroxylation it also takes longer for these collagens to be secreted, in vitro, than for interstitial collagens (3, 325, 335). Similar materials are solubilized from tissues containing basement membranes by a combination of reduction and pepsin treatment and, after partly removing interstitial collagens by gelation at 37°C, materials are left that can still be aggregated to *SLS* crystallites (75). Extraction may include use of denaturants, and CNBr yields collagenous peptides that are different from the CNBr peptides of types I, II and III but can be approximately equated to them (322). All the Cys, mannose, and glucosamine of extracted anterior lens capsule membrane collagen were found in one CNBr peptide (322). The materials split off by pepsin tend to have a noncollagen-like composition. A peptide that has some resemblance to propeptides of interstitial procollagens was found in glomerular basement membrane (347, 348). Thus a number of similarities with interstitial procollagens support the concept of a type IV collagen or procollagen.

Biosynthetic experiments have mostly been concerned with the above type of characterization. The enzyme that catalyzes formation of 3-hydroxyproline is separate from the one that catalyzes formation of 4-hydroxyproline, but these enzymes have similar requirements, e.g. unfolded substrate and presence of adjacent 4-hydroxyproline or Pro, respectively (247, 346, 349). Export has been followed by autoradiography (326, 328) and the initially secreted materials are disulfide-linked, though this process is relatively slow (57). Contrary to earlier reports (335) it is now thought that there is no subsequent cleavage (56, 58). The problem is, however, reminiscent of cell cultures making interstitial collagens: contaminating proteases can mimic procollagen peptidases, yet the correct enzymes may function more slowly than in the intact tissues. It has been suggested that the wide diversity of collagenous peptides in sonicated glomerular basement membranes could have a physiological degradative basis (76). The nature of the associations between the basement membrane collagen molecules is not known. Fine fibrils can sometimes be seen with the electron microscope and proteolytic modifications can enhance this. There probably are multiple forms of crosslinking and the overall thermal stability is high (350).

Cell Surface Interaction

Cross-linking (351) is not always confined to other collagens, and while this complicates analysis it could also be an essential part of basement membrane function. For example, acetylcholine esterase may be disulfide-linked to basement membranes through a collagenous material (352, 353) and the material called fibronectin, large external transformation sensitive (LETS) or cell attachment protein exists in basement membranes (354, 355) and has affinity for collagen (356, 357). While the possible role of collagens in cell attachment (358–360) is beyond the scope of this review, it has been found that 3-hydroxyproline-rich collagenous materials are synthesized soon after plating cells (361) and a material has been described as a special cell surface collagen (362). Fluorescent antibody studies of the distribution of collagen on fibroblast surfaces led to the suggestion of collagen forming an "exocyto-skeleton" (363).

The concept of type IV as a special procollagen consisting of three identical strands is useful, but even though immunological cross-reactions have been shown, more than one type could exist in different tissues. Clear delineation requires solubilization and chemical definition. Partial solubilization and more extensive treatment (76, 323, 364) indicate heterogeneity. New collagens, which may or may not be basement membrane collagens, have been isolated and defined following pepsin treatment of fetal membranes (72), blood vessels, and skin (73). They have distinctive CNBr peptide patterns. The sedimentation-equilibrium molecular weights of the two fetal membrane collagen chains are close to those of most α chains, though they migrate more slowly in $NaDodSO_4$ electrophoresis. Whether they exist as homo- or as hetero-triple molecules is not known (72). One of the other materials (73) has a molecular mass of only a fraction of an α chain, but this could be due to cleavage within a collagen helix; a trypsin-sensitive site exists in type III chains (365).

Invertebrate Basement Membranes

Basement membrane-like collagens may be widely distributed in the animal kingdom. Thus the mesoglea of Hydra has a collagen rich in 3-hydroxyproline, Hyl, Glu-Gal disaccharide and Gly (366). Collagen CNBr peptides of sea anemone have been characterized, and have been found to have similarities with basement membrane collagens and a summed molecular mass of an α chain (367). Intestinal basement membrane collagen of Ascaris shows heterogeneity equivalent to that found in bovine renal glomerular membranes investigated by the same techniques (368, 369).

Acknowledgments

We thank Dr. W. Salser for his insight and help in analyzing the potential for frameshift transcription, Dr. N. P. Morris and other colleagues for valuable discussion, and our sons for a lost summer. Financial support from USPHS Grant AM–13748 is gratefully acknowledged.

Literature Cited

1. Bornstein, P. 1974. *Ann. Rev. Biochem.* 43:567–603
2. Prockop, D. J., Berg, R. A., Kivirikko, K. I., Uitto, J. 1976. *Biochemistry of Collagen.* ed. G. N. Ramachandran, A. H. Reddi, pp. 163–273. New York/London: Plenum
3. Grant, M. E., Jackson, D. S. 1976. *Essays in Biochem.* 12:77–113
4. Miller, E. J. 1976. *Molec. Cell. Biochem.* 13:165–92
5. Martin, G. R., Byers, P. H., Piez, K. A. 1975. *Adv. Enzymol.* 42:167–91
6. Gallop, P. M., Paz, M. A. 1975. *Phys. Rev.* 55:418–87
7. Ross, R. 1975. *Philos. Trans. R. Soc. London* Ser. B 271:247–59
8. Veis, A., Brownell, A. G. 1975. *CRC Crit. Rev. Biochem.* 2:417–53
9. Fietzek, P. P., Kühn, K. 1976. *Int. Rev. Connect. Tissue Res.* 7:1–60
10. Furthmayr, H., Timpl, R. 1976. *Int. Rev. Connect. Tissue Res.* 7:61–99
11. Bailey, A. J., Robins, S. P. 1976. *Sci. Progress* 63:419–44
12. Grant, M. E., Harwood, R., Schofield, J. D. 1975. In *Dynamics of Connective Tissue Macromolecules,* ed. A. R. Poole, P. M. C. Burlaigh: North Holland
13. Gross, J. 1974. *Harvey Lect.* 68:351–432
14. Kefalides, N. A. ed. 1977. *Int. Symp. Basement Membranes,* New York: Academic. In press
15. Lapière, C. M., Nusgens, B. 1976. *Biochemistry of Collagen.* ed. G. N. Ramachandran, A. H. Reddi, pp. 377–447. New York/London: Plenum
16. Uitto, J., Lichtenstein, J. R. 1976. *J. Invest. Dermat.* 66:59–79
17. Kivirikko, K. I., Risteli, L. 1976. *Med. Biol.* 54:159–86
18. Cowan, M. J., Collins, J. F., Crystal, R. G. 1975. *Methods Molec. Biol.* 8:257–313
19. Reddi, A. H. 1976. *Biochemistry of Collagen,* ed. G. N. Ramachandran, A. H. Reddi, pp. 449–478. New York/London: Plenum
20. Ramachandran, G. N., Ramachandran C. 1976. *Biochemistry of Collagen,* ed. G. N. Ramachandran, A. H. Reddi, pp. 45–84. New York/London: Plenum
21. Rubin, A. L., Drake, M. P., Davison, P. F., Pfahl, D., Speakman, P. T., Schmitt, F. O. 1965. *Biochemistry* 4:181–90
22. Piez, K. A. 1976. *Biochemistry of Collagen.* ed. G. N. Ramachandran, A. H. Reddi, pp. 1–44. New York/London: Plenum
23. Tanzer, M. L., Church, R. L., Yaeger, J. A., Wampler, D. E., Park, E. D. 1974. *Proc. Natl. Acad. Sci. USA* 71:3009–13
24. Byers, P. H., Click, E. M., Harper, E., Bornstein, P. 1975. *Proc. Natl. Acad. Sci. USA* 72:3009–13
25. Fessler, L. I., Morris, N. P., Fessler, J. H. 1975. *Proc. Natl. Acad. Sci. USA* 72:4905–09
26. Dehm, P., Prockop, D. J. 1973. *Eur. J. Biochem.* 35:159–66
27. Uitto, J., Prockop, D. J. 1974. *Biochemistry* 13:4586–91
28. Müller, P. K., Jamhawi, O. 1974. *Biochim. Biophys. Acta.* 365:158–68
29. Harwood, R., Bhalla, A. K., Grant, M. E., Jackson, D. S. 1975. *Biochem. J.* 148:129–38
30. Oohira, A., Kusakabe, A., Suzuki, S. 1975. *Biochem. Biophys. Res. Commun.* 67:1086–92
31. Fessler, J. H., Greenberg, G. M. 1975. *Ann. Rheum. Dis.* 34:S40
32. Merry, A. H., Harwood, R., Woolley, D. E., Grant, M. E., Jackson, D. S.

1976. *Biochem. Biophys. Res. Commun.* 71:83–90
33. Harwood, R., Merry, A. H., Woolley, D. E., Grant, M. E., Jackson, D. S. 1977. *Biochem. J.* 161:405–18
34. Olsen, B. R., Hoffmann, H. P., Prockop, D. J. 1976. *Arch. Biochem. Biophys.* 175:341–50
35. Uitto, J., Hoffmann, H. P., Prockop, D. J. 1977. *Arch. Biochem. Biophys.* 179:654–62
36. Moro, L., Smith, B. D. 1977. *Arch. Biochem. Biophys.* In press
37. Byers, P. H., McKenney, K. H., Lichtenstein, J. R., Martin, G. R. 1974. *Biochemistry* 13:5243–48
38. Anesey, J., Scott, P. G., Veis, A., Chyatte, D. 1975. *Biochem. Biophys. Res. Commun.* 62:946–52
39. Lenaers, A., Lapière, C. M. 1975. *Biochim. Biophys. Acta* 400:121–31
40. Smith, B. D., McKenney, K. H., Lustberg, T. J. 1977. *Biochemistry* 16:2980–85
41. Timpl, R., Glanville, R. W., Nowack, H., Wiedemann, H., Fietzek, P. P., Kühn, K. 1975. *Z. Physiol. Chem.* 356:1783–92
42. Blackwell, B. A., Bensusan, H. B. 1977. *Biochem. Biophys. Res. Commun.* 75:94–101
43. Fessler, J. H., Greenberg, D. B., Fessler, L. I. 1976. In *First Int. Symp. Basement Membranes,* ed. N. A. Kefalides, New York: Academic. In press
44. Lee, G., Tate, R., Martin, G. R., Kohn, L. D. 1975. *Fed. Proc.* 34:696
44a. Foidart, J. M., Nusgens, B. 1976. *Arch. Int. Physiol.* 84:R23
45. Lichtenstein, J. R., Byers, P. H., Smith, B. D., Martin, G. R. 1975. *Biochemistry* 14:1589–94
46. Church, R. L., Tanzer, M. L. 1975. *FEBS Lett.* 53:105–9
47. Goldberg, B., Taubman, M. B., Radin, A. 1975. *Cell* 4:45–50
48. Hance, A. J., Bradley, K., Crystal, R. G. 1976. *J. Clin. Invest.* 57:102–11
49. Kefalides, N. A. 1973. *Int. Rev. Connect. Tissue Res.* 6:63–104
50. Kefalides, N. A. 1975. *J. Inv. Dermat.* 65:85–92
51. Fessler, J. H., Morris, N. P., Greenberg, G. M., Fessler, L. I. 1975. *Extracellular Matrix Influences on Gene Expression,* ed. H. C. Slavkin, R. C. Greulich, pp. 101–9. New York/San Francisco/London: Academic
52. Morris, N. P., Fessler, L. I. Weinstock, A., Fessler, J. H. 1975. *J. Biol. Chem.* 250:5719–26
53. Davidson, J. M., McEneany, L. S. G., Bornstein, P. 1975. *Biochemistry* 14:5188–94
54. Fessler, J. H., Fessler, L. I. 1976. *Fed. Proc.* 35:1355
55. Uitto, J. 1977. *Biochemistry* 16:3421–29
56. Minor, R. R., Clark, C. C., Strause, E. L., Koszalka, T. R., Brent, R. L., Kefalides, N. A. 1976. *J. Biol. Chem.* 251:1789–94
57. Williams, I. F., Harwood, R., Grant, M. E. 1976. *Biochem. Biophys. Res. Commun.* 70:200–6
58. Heathcote, J. G., Sear, C. H. J., Grant, M. E. 1977. In *First Int. Symp. Basement Membranes,* ed. N. A. Kefalides, New York: Academic. In press
59. Dehm, P., Kefalides, N. A. 1977. *Fed. Proc.* 36:680
60. Porter, R. R. 1977. *Fed. Proc.* 36:2191–96
61. Campbell, P. N., Blobel, G. 1976. *FEBS Lett.* 72:215–26
62. Rauterberg, J. 1973. *Clin. Orthop. Relat. Res.* 97:196–212
63. Mayne, R., Vail, M. S., Miller, E. J. 1975. *Proc. Natl. Acad. Sci. USA,* 72:4511–15
64. Mayne, R., Vail, M. S., Mayne, P. M., Miller, E. J. 1976. *Proc. Natl. Acad. Sci. USA* 73:1674–78
65. Narayanan, A. S., Page, R. C. 1976. *J. Biol. Chem.* 251:5464–71
66. Daniel, J. C. 1976. *Cell Differ.* 5:247–53
67. Benya, P. D., Padilla, S. R., Nimni, M. E. 1977. *Biochemistry* 16:865–72
68. Müller, P. K., Meigel, W. N., Pontz, B. F., Raisch, K. 1974. *Hoppe Seilers Z. Phys. Chem.* 355:985–96
69. Little, C. D., Church, R. L., Miller, R. A., Ruddle, F. H. 1977. *Cell* 10:287–95
70. Little, C. D., Church, R. L. 1976. *J. Cell Biol.* 70:A78
71. Tkocz, C., Kühn, K. 1969. *Eur. J. Biochem.* 7:454–62
72. Burgeson, R. E., El Adli, F. A., Kaitila, I. I., Hollister, D. W. 1976. *Proc. Natl. Acad. Sci. USA* 73:2579–83
73. Chung, E., Rhodes, R. K., Miller, E. J. 1976. *Biochem. Biophys. Res. Commun.* 71:1167–74
74. Mayne, R., Vail, M. S., Miller, E. J., Blose, S. H., Chacko, S. 1977. *Arch. Biochem. Biophys.* In press
75. Trelstad, R. L., Lawley, K. R. 1977. *Biochem. Biophys. Res. Commun.* 76:376–84
76. Sato, T., Spiro, R. G. 1976. *J. Biol. Chem.* 251:4062–70
77. Harrington, W. F., Karr, G. M. 1970. *Biochemistry* 9:3725–33

78. Evans, H. J., Sullivan, C. E., Piez, K. A. 1976. *Biochemistry* 15:1435–39
79. Kimura, S., Tanzer, M. L. 1977. *Biochemistry* 16:2554–60
80. Park, E., Tanzer, M. L., Church, R. L. 1975. *Biochem. Biophys. Res. Commun.* 63:1–10
81. Helle, O., Ness, N. N. 1972. *Acta Vet. Scand.* 3:443–5
82. Fjolstad, M., Helle, O. 1975. *J. Path.* 112:183–188
83. Lichtenstein, J. R., Martin, G. R., Kohn, L. D., Byers, P. H., McKusick, V. A. 1973. *Science* 182:298–300
84. Patterson, D. F., Minor, R. R. 1977. *Lab Investigations* In press
85. Lapière, C. M., Lenaers, A., Kohn, L. D. 1971. *Proc. Natl. Acad. Sci. USA* 68:3054–58
86. Kohn, L. D., Isersky, C., Zupnik, J., Lenaers, A., Lee, G., Lapière, C. M. 1974. *Proc. Natl. Acad. Sci. USA* 71: 40–44
87. Bailey, A. J., Lapière, C. M. 1973. *Eur. J. Biochem.* 34:91–96
88. Church, R. L., Yaeger, J. A., Tanzer, M. L. 1974. *J. Mol. Biol.* 86:785–99
89. Murphy, W. H., von der Mark, K., McEneany, L. S. G., Bornstein, P. 1975. *Biochemistry* 14:3243–50
90. Dintzis, H. M. 1961. *Proc. Natl. Acad. Sci. USA* 47:246–61
91. Gross, J. 1976. *Biochemistry of Collagen,* ed. G. N. Ramachandran, A. H. Reddi, pp. 275–317. New York/London: Plenum
92. Sherr, C. J., Taubman, M. B., Goldberg, B. 1973. *J. Biol. Chem.* 248: 7033–38
93. Dehm, P., Olsen, B. R., Prockop, D. J. 1974. *Eur. J. Biochem.* 46:107–16
94. Nist, C., von der Mark, K., Hay, E. D., Olsen, B. R., Bornstein, P., Ross, R., Dehm, P. 1975. *J. Cell Biol.* 65:75–87
95. Monson, J. M., Click, E. M., Bornstein, P. 1975. *Biochemistry* 14:4088–92
96. Park, E., Church, R. L., Tanzer, M. L. 1975. *Immunology* 28:781–90
97. Timpl, R. 1976. *Biochemistry of Collagen,* ed. G. N. Ramachandran, A. H. Reddi, pp. 319–75. New York/London: Plenum
98. Rohde, H., Becker, U., Nowack, H., Timpl, R. 1976. *Immunochem.* 13: 967–74
99. Rohde, H., Nowack, H., Becker, U., Timpl, R. 1976. *J. Immunol. Methods* 11:135–45
100. von der Mark, K., Click, E. M., Bornstein, P. 1973. *Arch. Biochem. Biophys.* 156:356–64
101. Sherr, C. J., Goldberg, B. 1973. *Science* 180:1190–92
102. Taubman, M. B., Goldberg, B., Sherr, C. J. 1974. *Science* 186, 1115–17
103. Olsen, B. R., Guzman, N. A., Engel, J., Condit, C., Aase, S. 1977. *Biochemistry* 16:3030–37
104. Uitto, J., Lichtenstein, J. R., Bauer, E. A. 1976. *Biochemistry* 15:4935–42
105. Lichtenstein, J. R., Bauer, E. A., Uitto, J. 1976. *Biochem. Biophys. Res. Commun.* 73:665–72
106. Uitto, J., Lichtenstein, J. R. 1976. *Biochem. Biophys. Res. Commun.* 71: 60–67
107. Nowack, H., Olsen, B. R., Timpl, R. 1976. *Eur. J. Biochem.* 70:205–16
108. Goldberg, B. 1974. *Cell* 1:185–92
109. Hoffmann, H. P., Olsen, B. R., Chen, H. T., Prockop, D. J. 1976. *Proc. Natl. Acad. Sci. USA* 73:4304–08
110. Furthmayr, H., Timpl, R., Stark, M., Lapière, C. M., Kühn, K. 1972. *FEBS Lett.* 28:247–50
111. Becker, U., Timpl, R., Helle, O., Prockop, D. J. 1976. *Biochemistry* 15: 2853–62
111a. Engel, J., Bruckner, P., Becker, U., Timpl, R., Rutschmann, B., 1977. *Biochemistry* 16:4026–33
112. Hörlein, D., Fietzek, P. P. 1976. *Arch. Int. Physiol.* 84:R3
113. Nowack, H., Rohde, H., Timpl, R. 1976. *Hoppe Seylers Z. Physiol. Chem.* 357:601–4
114. Bornstein, P., von der Mark, K., Wyke, A. W., Ehrlich, H. P., Monson, J. M. 1972. *J. Biol. Chem.* 247:2808–13
115. von der Mark, K., Bornstein, P. 1973. *J. Biol. Chem.* 248:2285–89
116. Becker, U., Helle, O., Timpl, R. 1977. *FEBS Lett.* 73:197–200
117. Bentz, H., Hörlein, D., Fietzek, P. P. 1976. *Arch. Int. Physiol.* 84:R3
118. Nowack, H., Gay, S., Wick, G., Becker, U., Timpl, R. 1976. *J. Immunological Methods.* 12:117–24
119. Gay, S., Fietzek, P. P., Remberger, K., Eder, M., Kühn, K. 1975. *Klin. Wochenschrift.* 53:205–8
120. Chung, E., Keele, E. M., Miller, E. J. 1974. *Biochemistry* 13:3459–64
121. Schneir, M., Miller, E. J. 1976. *Biochim. Biophys. Acta* 446:240–4
122. Rauterberg, J., Allmann, H., Henkel, W., Fietzek, P. P. 1976. *Z. Physiol. Chem.* 357:1401–7
123. Fietzek, P. P., Allmann, H., Rauterberg, J., Wachter, E. 1977. *Proc. Natl. Acad. Sci. USA* 74:84–86
124. Glanville, R. W., Allmann, H., Fietzek,

P. P. 1976. *Z. Physiol. Chem.* 357: 1663–65
125. Seyer, J. M., Kang, A. H. 1977. *Biochemistry* 16:1158–64
126. Glanville, R. W., Fietzek, P. P. 1976. *FEBS Lett.* 71:99–102
127. Tanzer, M. L. 1976. *Biochemistry of Collagen.* ed. G. N. Ramachandran, E. H. Reddi, pp. 137–62. New York/London: Plenum
128. Becker, U., Fietzek, P. P., Nowack, H., Timpl, R. 1976. *Z. Physiol. Chem.* 357:1409–15
129. Becker, U., Nowack, H., Gay, S., Timpl, R. 1976. *Immunology* 31:57–65
130. Bruns, R. R., Gross, J. 1974. *Biopolymers.* 13:931–41
131. Wiedemann, H., Chung, E., Fujii, T., Miller, E. J., Kühn, K. 1975. *Eur. J. Biochem.* 51:363–68
132. Clark, C. C., Kefalides, N. A. 1976. *Proc. Natl. Acad. Sci. USA.* 73:34–38
133. Clark, C. C., Kefalides, N. A. 1977. *J. Biol. Chem.* In press
134. Morris, N. P. 1977. PhD thesis. Univ. Calif., Los Angeles
135. Glanville, R. W., Fietzek, P. P. 1977. *Analytical Biochem.* 80:282–88
136. von der Mark, K., von der Mark, H. 1977. *J. Cell Biol.* 73:736–47
137. von der Mark, H., von der Mark, K., Gay, S. 1976. *Devel. Biol.* 48:237–49
138. von der Mark, K., von der Mark, H., Gay, S. 1976. *Devel. Biol.* 53:153–70
139. Mailman, M. L., Dresden, M. H. 1976. *Devel. Biol.* 50:378–94
140. Linsenmayer, T. F., Smith, G. N. Jr. 1976. *Devel. Biol.* 52:19–30
141. Linsenmayer, T. F., Toole, B. F. 1977. *Birth Defects* 13:19–35
142. Newsome, D. A., Linsenmayer, T. F., Trelstad, R. L. 1976. *J. Cell Biol.* 71:59–67
143. Smith, G. N. Jr., Linsenmayer, T. F., Newsome, D. A. 1976. *Proc. Natl. Acad. Sci. USA* 73:4420–23
144. Schmut, O., Reich, M. E., Hofmann, H. 1976. *Albrecht von Graefes Arch. Klin. Opthalmol.* 201:201–6
145. Swann, D. A., Caulfied, J. B., Broadhurst, J. B. 1976. *Biochim. Biophys. Acta* 427:365–70
146. Linsenmayer, T. F., Smith, G. N. Jr., Hay, E. D. 1977. *Proc. Natl. Acad. Sci. USA* 74:39–43
147. Chung, E., Miller, E. J. 1974. *Science* 183:1200–1
148. Epstein, E. H. Jr. 1974. *J. Biol. Chem.* 249:3225–31
149. Vinson, W. C., Seyer, J. M. 1974. *Biochem. Biophys. Res. Commun.* 58: 58–65
150. Shuttleworth, C. A., Forrest, L. 1975. *Eur. J. Biochem.* 55:391–95
151. Herrmann, H., von der Mark, K. 1975. *Z. Physiol. Chem.* 356:1605–12
152. Kawamoto, T., Nagai, Y. 1976. *Biochim. Biophys. Acta* 437:190–99
153. Deyl, Z., Adam, M. 1977. *Mech. Ageing Devel.* 6:25–33
154. Müller, P. K., Kühn, K. 1977. *Arznei. Forsch.* 27:199–201
155. Epstein, E. H. Jr., Munderloh, N. H. 1975. *J. Biol. Chem.* 250:9304–12
156. Trelstad, R. L. 1974. *Biochem. Biophys. Res. Commun.* 57:717–25
157. Gay, S., Balleisen, L., Remberger, K., Fietzek, P. P., Adelmann, B. C., Kühn, K. 1975. *Klin. Wochensch.* 53:899–902
158. Rauterberg, J., von Bassewitz, D. B. 1975. *Z. Physiol. Chem.* 356:95–100
159. McCullagh, K. A., Balian, G. 1975. *Nature* 258:73–75
160. Barnes, M. J., Morton, L. F., Levene, C. I. 1976. *Biochem. Biophys. Res. Commun.* 70:339–47
161. Faris, B., Salcedo, L. L., Cook, V., Johnson, L., Foster, J. A., Franzblau, C. 1976. *Biochim. Biophys. Acta* 418: 93–103
162. Layman, D. L., Epstein, E. H. Jr., Dodson, R. F., Titus, J. L. 1977. *Proc. Natl. Acad. Sci. USA* 74:671–75
163. Rauterberg, J., Allam, S., Brehmer, U., Wirth, W., Hauss, W. H. 1977. *Z. Physiol. Chem.* 358:401–7
164. Hance, A. J., Crystal, R. G. 1977. *Nature* 268:152–54
165. McLees, B. D., Schleiter, G., Pinnell, S. R. 1977. *Biochemistry* 16:185–90
166. Seyer, J. M., Kang, A. H., Whitaker, J. N. 1977. *Biochim. Biophys. Acta* 492: 415–25
167. Butler, W. T., Birkedal-Hansen, H., Beegle, W. F., Taylor, R. E., Chung, E. 1975. *J. Biol. Chem.* 250:8907–12
168. Eyre, D. R., Muir, H. 1976. *J. Biochem.* 157:267–70
169. Müller, P. K., Lemmen, C., Gay, S., Meigel, W. N. 1975. *Eur. J. Biochem.* 59:97–104
169a. Pentinnen, R. P., Lichtenstein, J. R., Martin, G. R., McKusick, V. A. 1975. *Proc. Natl. Acad. Sci. USA* 72:586–89
170. Rojkind, M., Martinez-Palomo, A. 1976. *Proc. Natl. Acad. Sci. USA* 73:539–43
171. Bailey, A. J., Bazin, S., Sims, T. J., Le Lous, M., Nicoletis, C., Delaunay, A. 1975. *Biochim. Biophys. Acta* 405: 412–21
172. Trelstad, R. L., Catanese, V. M., Rubin, D. F. 1976. *Analyt. Biochem.* 71:114–18

173. Varner, H. H. Jr., Miller, R. L. 1977. *Fed. Proc.* 36:679
173a. Sykes, B. C. 1976. *FEBS Lett.* 61: 180–85
174. Hahn, E., Timpl, R., Miller, E. J. 1974. *J. Immunol.* 113:421–23
175. Wick, G., Nowack, H., Hahn, E., Timpl, R., Miller, E. J. 1976. *J. Immunol.* 117:298–303
176. Wick, G., Kraft, D., Kokoschka, E. M., Timpl, R. 1976. *Clin. Immunol. Immunopath.* 6:182–91
177. Gay, S., Martin, G. R., Müller, P. K., Timpl, R., Kühn, K. 1976. *Proc. Natl. Acad. Sci. USA* 73:4037–40
178. Menzel, J. 1977. *J. Immunol. Methods* 15:77–95
179. Andriopoulis, N. A., Mestecky, J., Miller, E. J., Bennett, J. C. 1976. *Clin. Immunol.* 6:209–12
180. Mayne, R., Vail, M. S., Miller, E. J. 1976. *Devel. Biol.* 54:230–40
181. Deshmukh, K., Kline, W. G., Sawyer, B. D. 1976. *FEBS Lett.* 67:48–51
181a. Deshmukh, K., Kline, W. G. 1976. *Eur. J. Biochem.* 69:117–23
182. Gay, S., Müller, P. K., Lemmen, C., Remberger, K., Matzen, K., Kühn, K. 1976. *Klin. Wochensch.* 54:969–76
183. Norby, D. P., Malemud, C. J., Sokoloff, L. 1977. *Arthritis Rheum.* 20:709–16
184. Koch, F., Pawlowski, P. J., Lukens, L. N. 1977. *Arch. Biochem. Biophys.* 178: 373–80
185. Butler, W. T., Finch, J. E. Jr., Miller, E. J. 1977. *J. Biol. Chem.* 252:639–43
186. Cheung, H. S., Harvey, W., Benya, P. D., Nimni, M. E. 1976. *Biochem. Biophys. Res. Commun.* 68:1371–78
187. Goldberg, B. 1977. *Cell* 11:169–72
188. Hata, R., Peterkofsky, B. 1977. *Proc. Natl. Acad. Sci. USA* 74:2933–37
189. Raj, C. V. S., Church, R. L., Creagan, R. P., Ruddle, F. H. 1976. *J. Cell Biol.* 70:A78
190. Lehrach, H., Boedtker, H. 1976. *Z. Physiol. Chem.* 357:325
191. Lazarides, E., Lukens, L. N. 1971. *Nature New Biology.* 232:37–40
192. Diegelmann, R. F., Bernstein, L., Peterkofsky, B. 1973. *J. Biol. Chem.* 248: 6514–21
193. Kerwar, S. S., Cardinale, G. J., Kohn, L. D., Spears, C. L., Stassen, F. L. H. 1973. *Proc. Natl. Acad. Sci. USA* 70:1378–82
194. Benveniste, K., Wilczek, J., Stern, R. 1973. *Nature* 246:303–5
195. Benveniste, K., Wilczek, J., Ruggieri, A., Stern, R. 1976. *Biochemistry* 15: 830–35
196. Boedtker, H., Crkvenjakov, R. B., Last, J. A., Doty, P. 1974. *Proc. Natl. Acad. Sci. USA* 71:4208–12
197. Boedtker, H., Frischauf, A. M., Lehrach, H. 1976. *Biochemistry* 15:4765–70
198. Burns, T. M., Spears, C. L., Kerwar, S. S. 1973. *Arch. Biochem. Biophys.* 159: 880–84
199. Prichard, P. M., Staton, G. W., Cutroneo, K. R. 1974. *Arch. Biochem. Biophys.* 163:178–84
200. Neufang, O., Tiedemann, H., Balke, G. 1975. *Z. Physiol. Chem.* 356:1445–50
201. Wang, L., Andrade, H. F. Jr., Silva, S. M. F., Simoes, C. L., D'Abronzo, F. H., Brentani, R. 1975. *Prep. Biochem.* 5:45–57
202. Berman, A. E., Oborotova, T. A., Mazurov, V. I. 1975. *Biochem. SSR* 40:364–71
203. Pawlowski, P. J., Gillette, M. T., Martinelli, J., Lukens, L. N., Furthmayr, H. 1975. *J. Biol. Chem.* 250:2135–42
204. Collins, J. F., Crystal, R. G. 1975. *J. Biol. Chem.* 250:7332–42
205. Park, E., Tanzer, M. L., Church, R. L. 1975. *Biochem. Biophys. Res. Commun.* 63:1–10
206. Traut, T. W., Petruska, J. A. 1976. *Biochim. Biophys. Acta.* 418:73–80
207. Lukens, L. N. 1976. *J. Biol. Chem.* 251:3530–38
208. Diaz de Leon, L., Paglia, L., Breitkreutz, D., Stern, R. 1977. *Biochem. Biophys. Res. Commun.* 77:11–19
209. Harwood, R., Connolly, A. D., Grant, M. E., Jackson, D. S. 1974. *FEBS Lett.* 41:85–88
210. Harwood, R., Grant, M. E., Jackson, D. S. 1975. *FEBS Lett.* 57:47–50
211. Harwood, R., Durrant, B., Grant, M. E., Jackson, D. S. 1975. *Biochem. Soc. Transactions.* 3:914–15
212. Kretsinger, R. H., Manner, G., Gould, B. S., Rich, A. 1964. *Nature* 202: 438–41
213. Bankowski, E., Mitchell, W. M. 1973. *Biophys. Chem.* 1:73–86
214. Vuust, J. 1975. *Eur. J. Biochem.* 60: 41–50
215. Vuust, J., Piez, K. A. 1972. *J. Biol. Chem.* 247:856–62
216. Miller, E. J., Woodall, D. L., Vail, M. S. 1973. *J. Biol. Chem.* 248:1666–71
217. Sanger, F., Air, G. M., Barrell, B. G., Brown, N. L., Coulson, A. R., Fiddes, J. C., Hutchison, C. A. III, Slocombe, P. M., Smith, M. 1977. *Nature* 265: 687–95
218. Wang, L., Simoes, C. L., Sonohara, S., Brentani, M., Andrade, H. F. Jr., da Silva, S. M. F., Salles, J. M., Marques,

N., Brentani, R. 1975. *Nucleic Acid Research* 2:655–66
219. Salles, J. M., Sonohara, S., Brentani, R. 1976. *Mol. Biol. Rp.* 2:517–23
220. Harwood, R., Grant, M. E., Jackson, D. S. 1975. *Biochem. Soc. Transactions.* 3:916–17
221. Diaz de Leon, L., Breitkreutz, D., Pomponio, J. 1976. *Fed. Proc.* 35:1739
222. Diegelmann, R. F., Peterkofsky, B. 1972. *Devel. Biol.* 28:443–53
223. Zeichner, M., Rojkind, M. 1976. *Connective Tissue Res.* 4:169–75
224. Mäenpää, P. H., Ahonen, J. 1972. *Biochem. Biophys. Res. Commun.* 49:179–84
225. Lanks, K. W., Weinstein, I. B. 1970. *Biochem. Biophys. Res. Commun.* 40:708–15
226. Varricchio, F., Last, J. A. 1976. *Mol. Biol. Rp.* 2:465–70
227. Hussain, M. Z., Enriquez, B., Tolentino, M., Bhatnagar, R. S. 1977. *Fed. Proc.* 36:607
228. Christner, P. J., Rosenbloom, J. 1976. *Arch. Biochem. Biophys.* 172:399–409
229. Carpousis, A., Christner, P., Rosenbloom, J. 1977. *J. Biol. Chem.* 252:2447–49
230. Schwartz, N. B. 1976. *J. Cell Biol.* 70:A261
231. Brownell, A. G., Veis, A. 1976. *J. Biol. Chem.* 251:7137–43
232. Veis, A., Brownell, A. G. 1977. *Proc. Natl. Acad. Sci. USA* 74:902–5
233. Harwood, R., Grant, M. E., Jackson, D. S. 1976. *Biochem. J.* 156:81–90
234. Uitto, J., Prockop, D. J. 1974. *Biochemistry* 13:4586–91
235. Uitto, J., Hoffmann, H. P., Prockop, D. J. 1975. *Science* 190:1202–4
236. Schofield, J. D., Uitto, J., Prockop, D. J. 1974. *Biochemistry* 13:1801–6
237. Harwood, R., Grant, M. E., Jackson, D. S. 1975. *Biochem. J.* 152:291–302
238. Oikarinen, A., Anttinen, H., Kivirikko, K. I. 1976. *Biochem. J.* 156:545–51
239. Oikarinen, A., Anttinen, H., Kivirikko, K. I. 1976. *Biochem. J.* 160:639–45
240. Brownell, A. G., Veis, A. 1975. *Biochem. Biophys. Res. Commun.* 63:371–77
241. Grant, M. E., Schofield, J. D., Kefalides, N. A., Prockop, D. J. 1973. *J. Biol. Chem.* 248:7432–37
242. Schofield, J. D., Harwood, R. 1975. *Biochem. Soc. Trans.* 3:143–45
243. Tuderman, L., Kuutti, E. R., Kivirikko, K. I. 1975. *Eur. J. Biochem.* 52:9–16
244. Kuutti, E. R., Tuderman, L., Kivirikko, K. I. 1975. *Eur. J. Biochem.* 57:181–88

245. Tuderman, L., Kuutti, E. R., Kivirikko, K. I. 1975. *Eur. J. Biochem.* 60:399–405
246. Tuderman, L. 1976. *Eur. J. Biochem.* 66:615–21
247. Tryggvason, K., Risteli, J., Kivirikko, K. I. 1977. *Biochem. Biophys. Res. Commun.* 76:275–81
248. Ryhänen, L. 1976. *Biochem. Biophys. Acta* 438:71–89
249. Oppenheim, B., Englard, S. 1976. *Biochem. Biophys. Res. Commun.* 70:248–57
250. Guzman, N. A., Rojas, F. J., Cutroneo, K. R. 1976. *Arch. Biochem.* 172:449–54
251. Risteli, L., Myllyla, R., Kivirikko, K. I. 1976. *Biochem. J.* 155:145–53
252. Myllyla, R., Risteli, L., Kivirikko, K. I. 1975. *Eur. J. Biochem.* 58:517–21
253. Henkel, W., Buddecke, E. 1975. *Z. Physiol. Chem.* 356:921–28
254. Myllyla, R. 1976. *Eur. J. Biochem.* 70:225–31
255. Myllyla, R., Anttinen, H., Risteli, L., Kivirikko, K. I. 1977. *Biochem. Biophys. Acta* 480:113–21
256. Kishida, Y., Olsen, B. R., Berg, R. A., Prockop, D. J. 1975. *J. Cell Biol.* 64:331–39
257. Olsen, B. R., Berg, R. A., Kishida, Y., Prockop, D. J. 1975. *J. Cell Biol.* 64:340–55
258. Diegelmann, R. F., Bernstein, L., Peterkofsky, B. 1973. *J. Biol. Chem.* 248:6514–21
259. Cutroneo, K. R., Guzman, N. A., Sharawy, M. M. 1974. *J. Biol. Chem.* 249:5989–94
260. Harwood, R., Grant, M. E., Jackson, D. S. 1974. *Biochem. J.* 144:123–30
261. Helfre, C., Farjanel, J., Dubois, P., Fonvieille, J., Frey, J. 1976. *Biochem. Biophys. Acta.* 429:72–83
262. Peterkofsky, B., Assad, R. 1976. *J. Biol. Chem.* 251:4770–77
263. Rosenbloom, J., Endo, R., Harsch, M. 1976. *J. Biol. Chem.* 251:2070–76
264. Fessler, L. I., Rudd, C., Fessler, J. H. 1974. *J. Supramolecular Structure* 2:103–7
265. Bornstein, P. 1974. *J. Supramolecular Structure* 2:108–20
266. Fessler, L. I., Fessler, J. H. 1974. *J. Biol. Chem.* 249:7637–46
267. Uitto, J., Hoffmann, H. P., Prockop, D. J. 1976. *Arch. Biochem. Biophys.* 173:187–200
268. Ehrlich, H. P., Ross, R., Bornstein, P. 1974. *J. Cell Biol.* 62:390–405
269. Kerwar, S. S., Felix, A. M., Marcel, R. J., Tsai, I., Salvador, R. A. 1976. *J. Biol. Chem.* 251:503–9

270. Salvador, R. A., Tsai, I., Marcel, R. J., Felix, A. M., Kerwar, S. S. 1976. *Arch. Biochem. Biophys.* 174:381–92
271. Uitto, J., Hoffmann, H. P., Prockop, D. J. 1975. *Science* 190:1202–4
272. Kato, N., Yugari, Y. 1976. *Jap. J. Pharm.* 26:P80
273. Christner, P., Carpousis, A., Harsch, M., Rosenbloom, J. 1975. *J. Biol. Chem.* 250:7623–30
274. Duskin, D., Bornstein, P. 1977. *J. Biol. Chem.* 252:955–62
275. Tanzer, M. L. 1977. In *First Int. Symp. Basement Membranes,* ed. N. A. Kefalides, New York: Academic. In press
276. Levene, C. I., Bates, C. J. 1975. *Ann. NY Acad. Sci.* 258:288–306
277. Barnes, M. J. 1975. *Ann. NY Acad. Sci.* 258:264–77
278. Blanck, T. J., Peterkofsky, B. 1975. *Arch. Biochem. Biophys.* 171:259–67
279. Kao, W. W. Y., Flaks, J. G., Prockop, D. J. 1976. *Arch. Biochem. Biophys.* 173:638–48
280. Levene, C. I., Ockleford, C. D., Barber, C. L. 1977. *Virchows Archiv. (Cell Path.)* 23:325–38
281. Kruse, N. J., Bornstein, P. 1975. *J. Biol. Chem.* 250:4841–47
282. Olsen, B. R., Prockop, D. J. 1974. *Proc. Natl. Acad. Sci. USA* 71:2033–37
283. Lohmander, S., Moskalewski, S., Madsen, K., Thyberg, J., Friberg, U. 1976. *Exp. Cell Res.* 99:333–45
284. Trelstad, R. L., Coulombre, A. J. 1971. *J. Cell Biol.* 50:840–58
285. Weinstock, A., Bibb, C., Burgeson, R. E., Fessler, L. I., Fessler, J. H., 1975. *Extracellular Matrix Influences on Gene Expression,* ed. H. C. Slavkin, R. C. Greulich, pp. 321–30. New York, San Francisco & London: Academic
286. Weinstock, M. 1975. *Extracellular Matrix Influences on Gene Expression,* ed. H. C. Slavkin, R. C. Greulich, pp. 119–28. New York/San Francisco/London: Academic
287. Weinstock, M., Leblond, C. P. 1974. *J. Cell Biol.* 60:92–127
288. Burgeson, R. E., Bibb, C., Weinstock, A. 1974. *J. Cell Biol.* 63:A41
289. Ross, R., Benditt, E. P. 1965. *J. Cell Biol.* 27:83–106
290. Trelstad, R. L. 1975. *Extracellular Matrix Influences on Gene Expression,* ed. H. C. Slavkin, R. C. Greulich, pp. 331–39. New York, Academic
291. Trelstad, R. L., Hayashi, K., Gross, J. 1976. *Proc. Natl. Acad. Sci. USA* 73:4027–31
292. Jones, S. J., Boyde, A., Pawley, J. B. 1975. *Cell Tiss. Res.* 159:73–80
293. Taubman, M. B., Goldberg, B. 1976. *Arch. Biochem. Biophys.* 173:490–94
294. Tanzer, M. L., Church, R. L., Yaeger, J. A., Park, E. D. 1975. *Extracellular Matrix Influences on Gene Expression,* ed. H. C. Slavkin, R. C. Greulich, pp. 785–793. New York, San Francisco & London: Academic
295. Leung, M. K., Fessler, L. I., Greenberg, D. G., Fessler, J. H. 1977. In press
296. Nusgens, B., Lapière, C. 1976. *Arch. Int. Phys.* 84:R5
297. Taubman, M. B., Goldberg, B. 1977. In press
298. Fessler, J. H., Tandberg, W. D. 1975. *J. Supramolecular Structure* 3:17–23
299. Fessler, J. H., Doege, K. J., Siegel, R. C., Fessler, L. I. 1977. *Fed. Proc.* 36:680
300. Siegel, R. C. 1974. *Proc. Natl. Acad. Sci. USA* 71:4826–30
301. Fujii, K., Tanzer, M. L., Cooke, P. H. 1976. *J. Mol. Biol.* 106:223–27
302. Bruns, R. R., Trelstad, R. L., Gross, J. 1973. *Science* 181:269–71
303. Bruns, R. R. 1976. *J. Cell Biol.* 68:521–38
304. Miller, A. 1976. *Biochemistry of Collagen,* ed. G. N. Ramachandran, A. H. Reddi, pp. 85–136. New York, Academic
305. Fietzek, P. P., Kühn, K. 1975. *Molec. Cell Biochem.* 8:141–57
306. Traub, W. 1975. *Acta Cryst.* 31:A37
307. Piez, K. A., Torchia, D. A. 1975. *Nature* 258:87
308. Li, S., Golub, E., Katz, E. P. 1975. *J. Mol. Biol.* 98:835–39
309. Veis, A., Yuan, L. 1975. *Biopolymers* 14:895–900
310. Doyle, B. B., Hukins, D. W. L., Hulmes, D. J. S., Miller, A., Woodhead-Galloway, J. 1975. *J. Mol. Biol.* 91:79–99
311. Hukins, D. W. L., Woodhead-Galloway, J. 1976. *Biochem. Biophys. Res. Commun.* 70:413–17
312. Cunningham, L. W., Davies, H. A., Hammond, R. G. 1976. *Biopolymers* 15:483–502
313. Trus, B. L., Piez, K. A. 1976. *J. Mol. Biol.* 108:705–32
314. Piez, K. A., Trus, B. L. 1977. *J. Mol. Biol.* 110:701–4
315. Hukins, D. W. L. 1977. *Biochem. Biophys. Res. Commun.* 77:335–39
316. Lillie, J. H., MacCallum, D. K., Scaletta, L. J., Occhino, J. C. 1977. *J. Ultra Res.* 58:134–43
317. Chapman, J. A. 1974. *Connect. Tissue Res.* 2:137–50
318. Bruns, R. R., Gross, J. 1973. *Biochemistry* 12:808–15

319. Pierce, G. B. 1970. *Chemistry and Molecular Biology of Interstitial Matrix* Vol. 1, ed. E. A. Balazs, pp. 471–506. New York / San Francisco / London: Academic
320. Spiro, R. G. 1976. *Diabetologia* 12:1–14
321. Spiro, R. G. 1976. *Diabetes* 25:Suppl. 2, 909–13
322. Kefalides, N. A., Tomichek, E., Alper, R. 1975. *Extracellular Matrix Influences on Gene Expression,* ed. H. C. Slavkin, R. C. Greulich, pp. 129–36. New York / San Francisco / London: Academic
323. Hudson, B. G., Spiro, R. G. 1972. *J. Biol. Chem.* 247:4229–38
324. Freytag, J. W., Ohno, M., Hudson, B. G. 1976. *Biochem. Biophys. Res. Commun.* 72:796–802
325. Kefalides, N. A., Cameron, J. D., Tomichek, E. A., Yanoff, M. 1976. *J. Biol. Chem.* 251:730–33
326. Minor, R. R., Hoch, P. S., Koszalka, T. R., Brent, R. L., Kefalides, N. A. 1976. *Devel. Biol.* 48:344–64
327. Minor, R. R., Strause, E. L., Koszalka, T. R., Brent, R. L., Kefalides, N. A. 1976. *Devel. Biol.* 48:365–76
328. Clark, C. C., Tomichek, E. A., Koszalka, T. R., Minor, R. R., Kefalides, N. A. 1975. *J. Biol. Chem.* 250:5259–67
329. Clark, C. C., Minor, R. R., Koszalka, T. R., Brent, R. L., Kefalides, N. A. 1975. *Devel. Biol.* 46:243–61
330. Grant, M. E., Harwood, R., Williams, I. F. 1975. *Eur. J. Biochem.* 54:531–40
331. Foidart-Willems, J., Dechenne, C., Mahieu, P. 1975. *Diabetes Metab.* 1:227–34
332. Dechenne, C., Foidart-Willems, J., Mahieu, P. 1976. *J. Submicroscopic Cytol.* 8:101–19
333. Grant, M. E., Harwood, R. 1974. *Biochem. Soc. Trans.* 2:624–25
334. Grant, M. E., Schofield, J. D., Kefalides, N. A., Prockop, D. J. 1973. *J. Biol. Chem.* 248:7432–37
335. Grant, M. E., Kefalides, N. A., Prockop, D. J. 1972. *J. Biol. Chem.* 247:3539–51
336. Howard, B. V., Macarak, E. J., Gunson, D., Kefalides, N. A. 1976. *Proc. Natl. Acad. Sci. USA* 73:2361–64
337. Macarak, E. J., Howard, B. V., Kefalides, N. A. 1977. *Lab. Investigations* 36:62–7
338. Jaffe, E. A., Minick, C. R., Adelman, B., Becker, C. G., Nachman, R. 1976. *J. Exp. Med.* 144:209–25
339. Orkin, R. W., Gehron, P., McGoodwin, E. B., Martin, G. R., Valentine, T., Swarm, R. 1977. *J. Exp. Med.* 145:204–20

340. Timpl, R., Wick, G., Martin, G. R. 1977. In *First Int. Symp. Basement Membranes,* ed. N. A. Kefalides, New York: Academic In press
341. Gunson, D. E., Kefalides, N. A. 1976. *Immunology* 31:563–69
342. Gunson, D. E., Arbogast, B. W., Kefalides, N. A. 1976. *Immunology* 31:577–82
343. Arbogast, B. W., Gunson, D. E., Kefalides, N. A. 1976, *J. Immun.* 117:2181–84
344. Dehm, P., Kefalides, N. A. 1977. In *First Int. Symp. Basement Membranes,* ed. N. A. Kefalides, New York: Academic In press
345. Man, M., Adams, E. 1975. *Biochem. Biophys. Res. Commun.* 66:9–16
346. Gryder, R. M., Lamon, M., Adams, E. 1975. *J. Biol. Chem.* 250:2470–74
347. Ohno, M., Riquetti, P., Hudson, B. G. 1975. *J. Biol. Chem.* 250:7780–87
348. Bardos, P., Lanson, M., Degand, P., Gutman, N., Garrigue, M. A., Muh, J. P. 1976. *FEBS Lett.* 64:385–90
349. Risteli, J., Tryggvason, K., Kivirikko, K. I. 1977. *Eur. J. Biochem.* 73:485–92
350. Gelman, R. A., Blackwell, J., Kefalides, N. A., Tomichek, E. 1976. *Biochim. Biophys. Acta.* 427:492–96
351. Tanzer, M. L., Kefalides, N. A. 1973. *Biochem. Biophys. Res. Commun.* 51:775–80
352. Rosenberry, T. L., Richardson, J. M. 1977. *Biochemistry* 16:3550–57
353. Lwebuga-Mukasa, J. S., Lappi, S., Taylor, P. 1976. *Biochemistry* 15:1425–34
354. Linder, E., Vaheri, A., Ruoslahti, E., Wartiovaara, J. 1975. *J. Exp. Med.* 142:41–49
355. Hynes, R. O. 1976. *Biochim. Biophys. Acta.* 458:73–107
356. Engvall, E., Ruoslahti, E. 1977. *Int. J. Canc.* 20:1–5
357. Kleinman, H. K., Klebe, R. J., McGoodwin, E. B., Martin, G. R. 1976. *J. Cell Biol.* 70:A83
358. Pearlstein, E. 1976. *Nature* 262:497–500
359. Kleinman, H. K., McGoodwin, E. B., Klebe, R. J. 1976. *Biochem. Biophys. Res. Commun.* 72:426–32
360. Linsenmayer, T. F., Toole, B. P., Gross, J. 1976. *J. Cell. Biol.* 70:A59
361. Lembach, K. J., Branson, R. E., Hewgley, P. B., Cunningham, L. W. 1977. *Eur. J. Biochem.* 72:379–83
362. Lichtenstein, J. R., Bauer, E. A., Hoyt, R., Wedner, H. J. 1976. *J. Exp. Med.* 144:145–54
363. Bornstein, P., Ash, J. F. 1977. *Proc. Natl. Acad. Sci. USA* 74:2480–84

364. Daniels, J. R., Chu, G. H. 1975. *J. Biol. Chem.* 250:3531–37
365. Miller, E. J., Finch, J. E. Jr., Chung, E., Butler, W. T., Robertson, P. B. 1976. *Arch. Biochem. Biophys.* 173:631–37
366. Barzansky, B., Lenhoff, H. M. 1974. *Amer. Zool.* 14:575–81
367. Nowack, H., Nordwig, A. 1974. *Eur. J. Biochem.* 45:333–42
368. Hung, C. H., Ohno, M., Freytag, J. W., Hudson, B. G. 1977. *J. Biol. Chem.* 252:3995–4001
369. Peczon, B. D., Wegner, L. J., Hung, C. H., Hudson, B. G. 1977. *J. Biol. Chem.* 252:4002–6

Ann. Rev. Biochem. 1978. 47:163–89
Copyright © 1978 by Annual Reviews Inc. All rights reserved

MOLECULAR BIOLOGY OF BACTERIAL MEMBRANE LIPIDS

❖971

John E. Cronan, Jr.

Department of Molecular Biophysics and Biochemistry, Yale University
School of Medicine, New Haven, Connecticut 06510

CONTENTS

PERSPECTIVES AND SUMMARY

There are many advantages to using bacteria in the experimental study of the functions of lipids in biological membranes. In general, bacteria have only external membranes, with relatively simple lipid compositions. Neutral lipids are rarely found and sterol components are absent. The phospholipid components of the cell are localized in the cell membrane and are composed

163

0066–4154/78/0701–0163$01.00

of only a few species of phospholipids having relatively simple acyl group compositions. Bacteria therefore provide an ideal system for probing the functions of phospholipids in biological membranes. Workers in the field are asking several main questions. Why do cells make a variety of lipid types? Do specific membrane lipids have specific cellular functions? In view of the fact that phospholipids serve as the hydrophobic barrier in the membrane, is this the sole function of these molecules? If so, why are phospholipids made in such diversity? How is a given ratio of lipid species produced? These questions can be approached by biochemical and physiological means. However, genetic manipulation is a far more powerful approach, especially when coupled with biochemical and physiological studies. The sophistication of bacterial genetics, therefore, is the strongest reason for the use of bacteria in membrane research.

As described in this review, the depth of our knowledge of the role of lipids in bacterial membranes is uneven and in places embarrassingly scanty. Our most detailed information concerns the role of phospholipid acyl moieties in membrane function. The literature presents a clear and consistent role for acyl chains based on the dependence of the phase behavior of phospholipid molecules on their fatty acid composition. However, the contributions of polar head groups to membrane function are largely unknown.

The regulatory mechanisms that control the acyl group composition of the phospholipids have only recently been outlined by the demonstation of control sites at two different levels. The enzymes and precise sequence of the reactions involved in this regulation are not yet known. The mechanisms that control polar group composition are a mystery. The interrelationships of phospholipid synthesis with other cellular properties are not yet understood, although it appears that the nucleotide ppGpp may play a regulatory role.

Although most of the questions listed above are currently unanswered, the situation seems temporary. As described in this review and the previous review by Silbert (1), a large number of bacterial mutants defective in lipid metabolism have been isolated. The first such mutants were the unsaturated fatty acid auxotrophs of *Escherichia coli,* which have been instrumental in establishing the role of acyl chains in phospholipid functions. If the new classes of mutants defective in acyl and polar group synthesis are similarly useful, much new information should be forthcoming. In addition, the application of newly developed DNA cloning procedures to the problems of lipid metabolism has just begun. This approach should be most valuable in unraveling the regulation of lipid synthesis. Other new techniques have been developed that should allow study of the topology of lipid synthesis and movement in bacterial membranes.

The literature on these subjects deals almost exclusively with *E. coli* and this review will reflect this bias. The present review will attempt to avoid unnecessary overlap with recent ones on the same subject (2, 3) and those by Bloch & Vance (4) and by Silbert (1) in this series. Readers seeking a comparative review of bacterial lipids are referred to the review by Goldfine (5).

FUNCTIONS OF LIPIDS IN BACTERIAL MEMBRANE LIPIDS

The physiological relevance of the physical properties of bacterial membrane lipids was reviewed in 1975 by the present author and E. P. Gelmann (2). The present discussion of this subject is essentially an update of the previous review.

Physical Studies

At the time of the previous review, a reasonably consistent and straightforward picture of the dependence of the lipid phase transition on the fatty acid composition of the membrane phospholipids had emerged. These data have since been supplemented by studies using high resolution scanning calorimetry (6, 7) and proton NMR (8) that have confirmed the previous picture. However, the calorimetric studies show a pronounced asymmetry in the melting curves of membranes and lipids from an unsaturate auxotroph grown on *trans* fatty acids (6). The molecular basis of this asymmetry is as yet unclear.

The maximal amount of ordered lipid compatible with the growth of *E. coli* has been examined by three different laboratories (7–9). Each used a different physical technique and a different environmental manipulation to obtain *E. coli* cells with the minimum degree of unsaturation that will allow growth. Previously this was shown to be 15–20% of the total fatty acids (10). All three laboratories reported that about half the lipid must be in the disordered state to allow the growth of *E. coli,* which is a similar result to that obtained for *Acholeplasma laidlawii* (12).

In a complementary study, Baldassare, Rhinehart & Silbert (11) measured the membrane lipid phase transitions of cells containing a minimal amount of unsaturated fatty acid. A strain of *E. coli* blocked in two steps of fatty acid synthesis was used. When grown with *cis*-vaccenic acid, cell growth was almost normal although over 90% of the phospholipid fatty acid was *cis*-vaccenate. Under these conditions a very sharp transition was detected in both isolated membrane lipids and whole cells, at a temperature 50°C below the growth temperature. The results show that *E. coli* is able to grow with almost all of its lipid in the disordered state and suggests that the breadth of the lipid phase transition in normal *E. coli* is due to the diversity of fatty acyl chains normally found. The sharpness of the transition was similar to that found in pure synthetic lipids and indicates a large cooperative unit (13). This finding, coupled with data showing discontinuities of the partition of a spin label at temperatures far above the order-disorder transition (for review see 2), led Baldassare, Rhinehart & Silbert (11) to propose that the di-*cis*-vaccenate lipids are present in discrete and rather pure patches in the membrane. These regions should be detectable by freeze-etch electron microscopy (see below).

A number of studies using freeze-fracture electron microscopy have been interpreted in terms of the segregation of membrane proteins into patches of fluid, or disordered, lipid (for review see 2). In general, membranes frozen from temperatures above the phase transition have an approximately random distribution of intramembrane, probably protein, particles. However, in membranes frozen from temperatures below the phase transition, the particles are found in clusters boarded by large smooth areas that are devoid of particles. The smooth areas have been interpreted as patches of ordered lipid and the aggregates as membrane proteins that have been extruded from the ordered patches into patches of fluid lipid. These interpretations have recently been justified by the physical isolation of the different regions by van Heerikhuisen et al (14) and Letellier, Shechter & Moudden (15, 16). The smooth and particle-rich regions were separated by osmotic lysis of spheroplasts at temperatures below that where the lipid transition begins, followed by isopynic centrifugation. The vesicles with lower density possess fewer particles and have a higher phospholipid-to-protein ratio than the vesicles with higher density. The smooth vesicles also have a somewhat higher proportion of saturated fatty acids than do the particle-rich vesicles. However, the fractions are similar with respect to the composition of their phospholipid head groups. If osmotic lysis is performed at a temperature above that of the higher transition, the high and low density fractions are not seen. Letellier, Moudden & Shechter (16) have shown the two fractions to have characteristic phase transitions. The low density, or smooth, fraction has only the higher of the two transitions seen in intact vesicles, whereas the particle-rich fraction has only the lower transition. The low density fraction contains most of the lactate dehydrogenase activity of the spheroplast membrane but is deficient in succinic dehydrogenase and NADH oxidase (14). Sodium dodecyl sulfate polyacrylamide gels also show protein differences between the fractions (14, 15). The quantitative agreement between the data of the Dutch and French groups is quite good and there seems little doubt that these two groups have been able to isolate the products of a lipid phase separation and of the concomitant protein segregation.

Physiological Studies

CHEMOTAXIS   Lofgren & Fox (17) had reported that *E. coli* requires disordered membrane lipids for chemotaxis but not for mobility. Miller & Koshland (18) have recently reported extensive data on *E. coli* and *Salmonella typhimurium* that do not agree with the previous result (12). Miller & Koshland (18), using both the tumble-frequency and response-time chemotactic assays, found no dependence of chemotaxis on the amount of disordered membrane lipids. They attribute the results of Lofgren & Fox (17) to a loss of viability of the elaidate-supplemented cells at low temperature and the effect of this loss on the capillary assay.

DNA SYNTHESIS   Thilo & Vielmetter (19) recently reported that neither the initiation nor the propagation of DNA synthesis requires a membrane with disordered membrane lipids. These results are in direct conflict with the report of Fralick & Lark (20). Thilo & Vielmetter used a well-characterized auxotroph of *E. coli* that requires added unsaturated fatty acids for growth, whereas the previous workers had used decynoyl-*N*-acetylcysteamine, which is an inhibitor of unsaturated fatty acid synthesis. They (19) attribute the difference in results to a lack of specificity of the inhibitor and suggest that the effects observed by Fralick & Lark (20) are due to general metabolic disturbances. Indeed, other laboratories (21, 22) have found that the inhibitor affects processes other than synthesis of unsaturated fatty acids. It seems that the use of such characterized auxotrophs is to be preferred over the use of 3-decynoyl-*N*-acetylcysteamine.

LACTOSE TRANSPORT   The large amount of data, often conflicting, concerning the effect of membrane lipids on the formation and function of the lactose transport system of *E. coli* was reviewed previously (2). At that time two conclusions were drawn. First, the data in the literature did not support the original reports that the formation of the lactose transport system requires the concomitant synthesis of bulk membrane lipid or depends on the type of lipid synthesized during induction of the lactose transport protein. Also, it was concluded that the function of the lactose transport system does depend on the order-disorder state of the membrane lipid. However, due to apparent discrepancies in the literature, detailed interpretation of the data did not seem warranted. The more recent literature supports the first conclusion and somewhat clarifies the effects of the lipid state on the function of the transport system.

Tsukagoshi & Fox (24), in studying the effects of a shift in the fatty acid supplement on the induction of the lactose transport system, reported triphasic curves in Arrhenius plots of lactose transport when the fatty acid supplement was switched just prior to induction. The data were interpreted as showing a preferential association of newly formed transport protein with newly synthesized lipid (24). In the previous review (2), we considered these results to be essentially uninterpretable due to several technical points and conflicts with later data from the same laboratory. Thilo & Overath (9) have repeated these experiments and find only biphasic Arrhenius plots for cells in which the supplement was shifted, and thus no preferential association of new proteins with new lipids was seen. The previous contention (2) that the "abortive induction" phenomenon (23) can be attributed to nonphysiological conditions is also supported by Thilo & Overath (9).

Thilo, Träuble & Overath (25) have reported new experiments on the dependence of lactose transport on the order-disorder transition of the membrane lipids. Previous results from that laboratory showed a single downward break in Arrhenius plots of the rate of lactose transport versus temperature. These results suggested that, at a critical temperature, there is an abrupt discontinuous change in activation energy without a simultaneous drop in transport rate. This was difficult to understand since, for the transport rate to be constant, the change in activation enthalpy had to be exactly compensated by a change in activation entropy. Another problem, as pointed out earlier (2), was that the change in shape of the lactose transport curve was much sharper than that for the lipid phase transition. These observations led Thilo, Träuble & Overath (25) to consider the possibility that the break they observed might correlate with the upper half of the phase transition and that a second break might be found at the low temperature end of the transition. They reasoned that the second break might have been missed because of signal-to-noise problems at low transport rates. They therefore compared the rate-temperature profiles with the order-disorder transition temperature for cells of an auxotroph grown on a variety of unsaturated fatty acids. The results of these experiments showed a second upward break in the Arrhenius plots that corresponded to the low temperature end of the transition. The first change in shape occurs toward the upper end of the transition.

The results of Thilo, Träuble & Overath (25), therefore, have come to agree somewhat more closely with the results previously published by Fox and co-workers [see (2) for a review of the earlier literature], who had previously reported triphasic and, in some cases, more than triphasic Arrhenius plots for both lactose and β-glucoside transport (23, 26, 27). However, in their results (26, 27) the "break" temperatures differ from those of Thilo and co-workers (25). Also, the low temperature break is always downward in the Fox data, whereas Thilo and co-workers (25) find only upward breaks. Furthermore, the extremely abrupt discontinuities sometimes seen by Fox and co-workers (26, 27) are not evident in the data of Thilo et al (25). It seems possible that the discrepancy in the low temperature data can be attributed to a low signal-to-noise ratio in the experiments of Fox and co-workers since considerable variation is evident in the data of the various papers from this group (23, 24, 26, 27).

Despite these discrepancies, the two groups interpret their results in a similar fashion. Thilo and co-workers (25) argue that the transport proteins partition into fluid domains. Fox and co-workers (26, 27) discuss their results in terms of lipid phase separations. This difference in interpretation depends on whether the breadth of the transition depends on the low degree of cooperativity of the system or on a partial lipid-lipid phase separation. Since both mechanisms are probably involved, the difference in interpretation becomes semantic.

It should be possible to test the interpretation of the Arrhenius plots in terms of the partition of transport proteins into fluid lipid domain or phases by physical isolation of the domain by the methods of van Heerikhuisen et al (14) and Letellier et al (15, 16) referred to earlier. This technique should be most useful to test if transport protein components are partitioned into fluid parts of the membrane. The amounts of M protein, which mediates lactose transport, in each fraction could be directly measured. If the M protein is partitioned into the particle-rich fraction in the amounts indicated by the Arrhenius plots, then a rather strong case for the partition of transport proteins could be made. Such data would also strengthen the assumption that the transport data reflect the function of transport components in a lipid phase. This assumption is reasonable but is as yet unproven.

Another interesting experiment would be to examine protein segregation in protein-rich inner membranes prepared by the method of McIntyre & Bell (28). These workers found that if phospholipid synthesis was halted in a *plsB* mutant of *E. coli* (such mutants are discussed below) then both the inner and outer membranes of the cells became enriched with protein. The protein-to-phospholipid ratio of both membranes increased approximately 60%. This work demonstrated that the membrane is not normally saturated with protein and that the synthesis of membrane phospholipid is not required for the synthesis and insertion of bulk membrane protein. The protein-enriched membranes are not lethal to the cell and the enrichment is quickly normalized upon the restoration of phospholipid synthesis (29). It should be possible to see if protein segregation and lipid phase separations depend on the amount of lipid in the membrane. A currently accepted notion is that membrane proteins are surrounded by a layer, or "annulus," of lipid which does not participate in the ordered-fluid phase transition (30). If the increased protein content of these membranes proportionally increases the amount of annular lipid, then an inhibition of the ordered-fluid phase transition, lipid-lipid phase separation, and/or protein segregation might result.

MECHANISMS OF LIPID SYNTHESIS AND DEGRADATION

Recent studies on the mechanism of membrane lipid synthesis have focused on the location of phospholipids and of the synthetic enzymes within the cell, the genetics of lipid metabolism, and the enzymology of the enzymes involved. Some aspects of the latter two topics have been recently reviewed in this series (1, 4) and thus the present contribution omits most mention of the earlier literature.

Topology of Membrane Phospholipid Synthesis

Rothman & Kennedy (31) have reported that the distribution of phosphatidylethanolamine is asymmetric in *Bacillus megaterium*. In a convincing series of experiments using membrane impermeable labeling reagents, these workers showed that about one third of the phosphatidylethanolamine can be labeled rapidly by two different reagents that are specific for amine groups. The reagents were shown not to enter the cell under these conditions. When conditions that allowed slow entry of a labeling reagent were used, one third of the cellular phosphatidylethanolamine reacted quickly with the reagent, whereas the remaining phosphatidylethanolamine reacted much more slowly. In inverted membrane vesicles, two thirds of the phosphatidylethanolamine was found to be readily labeled. Since bacilli have only a single cellular membrane, these workers interpret their results as showing that the outer leaflet of the lipid bilayer, in contact with the external milieu of this bacterium, has half as much phosphatidylethanolamine as the inner leaflet. Assuming that both leaflets have equal amounts of lipid, it follows that phosphatidylglycerol, the other major lipid of this organism, must be largely in the outer leaflet of the bilayer. These results indicate that bacteria as well as eucaryotic cells (32) have membranes with an asymmetric lipid distribution.

Rothman & Kennedy (33) then coupled these labeling techniques with "pulse-chase" incorporation experiments using phospholipid precursors to approach the topology of phosphatidylethanolamine biosynthesis. They found that newly synthesized phosphatidylethanolamine molecules are resistant to modification by impermeable reagents. Upon a chase with the nonradioactive phospholipid precursor, the radioactive phosphatidylethanolamine becomes accessible to the reagents that label the external surface of the cells. Rothman & Kennedy (33) interpret the data to show that the newly synthesized phosphatidylethanolamine is first located in the inner face of the plasma membrane lipid bilayer and is later rotated, or "flip-flopped," through the lipid bilayer to become part of the lipid leaflet at the external face of the membrane. This interpretation is certainly reasonable and consistent with the data. However, these workers have not actually determined that the new lipid is in the inner face of the membrane; they have only shown that the lipid is not in the outer face. The newly synthesized lipid could also be free in the cytoplasm or bound to some nonmembrane structure.

Fatty Acid Synthesis

The first mutants to be isolated in bacterial fatty acid synthesis were those lacking the ability to synthesize unsaturated fatty acids (for review see 1, 2). Three classes of such mutants have been reported. For two classes of these auxotrophs, *fabA* and *fabB*, the enzymological defect and genetic location are known (1, 2). A single mutant thought to represent a third class, *fabC*, has been reported (34), but subsequent workers have been unable to isolate it, and report that, by genetic criteria, the *fabC* mutant is indistinguishable from a mutant with a lesion in the *fabB* gene (35). Thus, there seems to be no evidence for a third class of unsaturated auxotroph.

A mutant defective in the conversion of palmitoleate to *cis*-vaccenate was isolated previously by Gelmann & Cronan (36). The mutational lesion was known to be the lack of a fatty acid synthetic enzyme, but the specific enzyme defect was not known. For a variety of reasons (see 4) this mutant, previously called Cvc but now called *fabF*, was postulated to be deficient in β-ketoacyl-acyl carrier protein (β-ketoacyl-ACP) synthetase II. Recent results have shown that the *fabF* strain does indeed lack this enzyme (37). Furthermore, if the *fabF* mutation is introduced into a strain carrying a temperature-sensitive lesion in the *fabB* gene, the resulting strain is unable to grow at high temperature with or without an unsaturated fatty acid supplement. Since *fabB* mutants lack β-ketoacyl-ACP sythetase I, this finding argues that β-ketoacyl-ACP sythetases I and II are the major "condensing" enzymes of *E. coli* (37). The interrelationship of these two enzymes now seems clearer. Either enzyme is able to catalyze the condensation reactions of saturated fatty acid synthesis. β-Ketoacyl-ACP synthetase I is needed for the synthesis of unsaturates from the *cis* 3-decenoyl precursor to palmitoleic acid, whereas β-ketoacyl-ACP synthetase II is required only for the elongation of palmitoleate to *cis*-vaccenic acid. The reason for this curious duplication seems likely to be the temperature control of unsaturated fatty acid synthesis (see below).

Another interesting mutant is the *vtr* mutant of Broekman, cited in Silbert (1), which overproduces *cis*-vaccenic acid at all temperatures. Silbert suggested that the phenotype of this mutant could contain a lesion in the *fabD*, or malonyl transacylase, gene. However, recombinational data show that this is not the case (37). It seems likely the *vtr* mutant is an allele of the *fabF* locus.

Silbert, Pohlman & Chapman (38) have studies the enzymic defect in a second class of *E. coli* temperature-sensitive mutants defective in the synthesis of both saturated and unsaturated fatty acids. Unlike the first class of such mutants, which were defective in malonyl transacylase (1), the new class (*fabE*) has an acetyl-CoA carboxylase activity of extreme thermolability. The subunit of the enzyme responsible for this defect is not known; preliminary evidence suggests abnormalities in two of the three subunits. The *fabE* mutants have already proved useful providing cells of *E. coli* with an altered lipid composition (11).

Mutants defective in cyclopropane fatty acids have been reported recently (39). These mutants owe their phenotype to a deficiency of cyclopropane fatty acid synthase, the enzyme that transfers the methyl carbon of *S*-adenosylmethionine to the double bond of an unsaturated fatty acyl residue of a phospholipid. Although these mutants synthesize < .1% of the normal amount of cyclopropane fatty acid, they have no physiological phenotype.

Phospholipid Synthesis

A number of laboratories have isolated and characterized mutants of *E. coli* defective in membrane phospholipid synthesis. The main objective of this work is to understand the role of phospholipids, both generally and specifically, in membrane function.

Of the three known classes of mutants blocked in early steps of phospholipid synthesis, the *gpsA* mutants are the most straightforward. The *gpsA* gene codes for the enzyme that reduces dihydroxyacetone phosphate to *sn*-glycerol-3-phosphate (glycerol-P) (40, 41). Mutants defective in this gene require supplemenation with glycerol-P or glycerol for growth and phospholipid synthesis. Another class of glycerol-P auxotroph, *plsB* mutants, owe their auxotrophy to a K_m defect in glycerol-P acyltransferase, which catalyzes the first step in phospholipid synthesis (40, 42–44). These mutants are defective in their use of endogenous glycerol-P and thus require an external supply. Bell & Cronan (45) have isolated revertants of *plsB* strains that retain the Km defect in the acyltransferase and are able to grow without exogenous glycerol-P. These phenotypically suppressed strains have a lesion in the *gpsA* gene which results in an alteration of the enzyme that synthesizes glycerol-P. This enzyme, called the biosynthetic glycerol-P dehydrogenase, is normally inhibited by its product, glycerol-P. As shown with homogeneous preparations (46), the enzyme of the phenotypically suppressed *plsB* strain is about 20-fold less sensitive than the normal enzyme to inhibition by glycerol-P This lack of inhibition results in a greatly increased intracellular pool of glycerol-P that overcomes the Km defect of the glycerol-P acyltransferase in *plsB* mutants.

The third class of mutant, *plsA,* was isolated as a temperature-sensitive mutant specifically defective in phospholipid synthesis at the nonpermissive temperature. It possessed a thermolabile glycerol-P acyltransferase (47). Although genetic analysis of these mutants showed the phenotype was due to a single mutation (48), further biochemical and physiological studies showed that adenylate kinase, as well as the glycerol 3-P acyltransferase, was abnormally thermolabile (49, 50). Specific inhibition of phospholipid synthesis was observed only at the lower end of the nonpermissive temperature range; at elevated nonpermissive temperatures ATP synthesis was also inhibited (49–51). The most straightforward hypothesis to explain these complex data is that the *plsA* gene is the structural gene for adenylate kinase; its effects on phospholipid synthesis could be explained if the active glycerol-P acyltransferase were composed of two subunits, one of which is adenylate kinase. The other subunit (the product of the *plsB* gene) is probably the structural gene for acyltransferase activity (43, 52). The thermolability of phospholipid synthesis in vivo and in vitro can then be explained by interactions of the defective adenylate kinase subunits with the acyltransferase subunits.

This model has several attractive features. It explains the existing data as well as the previously unexplained effects of ATP on the acyltransferase in vitro (53, 54) and also provides a mechanism whereby phospholipid synthesis could be tied to the energy state of the cell. Atkinson, Swedes & Sedo (55) have presented compelling data that the ratio of the various adenine nucleotides rather than the supply of ATP per se determines the rate of in vivo reactions that use energy, Adenylate kinase interacts with and interconverts all three adenine nucleotides and thus seems a strong candidate for a role in linking various biosynthetic processes to the energy state of the cell. Luria, Suit & Plate (57) have reported data on protein synthesis that suggest a role for adenylate kinase in the regulation of transcription or translation.

Snider & Kennedy (52) have challenged the previous findings that *plsA* mutants have a glycerol-P acyltransferase of unusual thermolability. However, these workers did not test the conditions used previously and instead used inactivation and assay conditions that stabilize the wild-type enzyme. These conditions might, therefore, have stabilized the mutant enzyme, or allowed it to reactivate, so as to make it appear indistinguishable from the wild-type enzyme. It is well known that incubation conditions can alter the thermolability of mutant enzymes. For example, an enzyme that is abnormally temperature sensitive only in the presence of detergent has been described (58). Snider & Kennedy (52) also proposed that the preferential inhibition of phospholipid synthesis observed in *plsA* mutants at low nonpermissive temperatures could be attributed to a decrease in nucleotide concentration due to a partial inactivation of adenylate kinase. They postulated that decreased availability of nucleotides is more inhibitory to phospholipid synthesis than to nucleic acid synthesis. However, recent results seem to invalidate this explanation. Nunn et al (59) decreased the intracellular nucleotide levels by starvation of an *E. coli* adenine auxotroph for adenine (cf 55).Under these conditions, phospholipid synthesis was much less inhibited than nucleic acid or protein synthesis.

Phosphatidylethanolamine forms 75% of the total phospholipid in *E. coli*. The other lipids are phosphatidylglycerol and cardiolipin; the proportion of each depends on the growth phase (56). Three classes of mutants defective in phosphatidylethanolamine synthesis have been isolated. Hawrot & Kennedy (58) isolated a mutant with a temperature-sensitive phosphatidylserine decarboxylase that at nonpermissive temperatures was unable to synthesize phosphatidylethanolamine and thus accumulated the precursor, phosphatidylserine in amounts up to 40% of the cellular phospholipid. Mutants with a temperature-sensitive phosphatidylserine synthase have been isolated by two groups. Ohta et al (60) used an "[3H] serine suicide" technique, whereas Raetz (61) used a screening method based on enzyme analysis of heavily mutagenized single colonies. The mutants isolated by the two groups were similar by both biochemical and genetic criteria (60–63). Upon shift to a nonpermissive temperature, the synthesis of phosphatidylethanolamine ceases and large increases in anionic lipids, mostly cardiolipin, occur (60, 63). A third class of mutant, that neither synthesizes phosphatidylethanolamine nor accumulates phosphatidylserine, has been isolated (64). These mutants seem to have a normal phosphatidylserine synthase (63) and are located on the genetic map at a position distinct from the phosphatidylserine synthase mutants (64). The biochemical lesion in these mutants is not yet understood.

The most striking phenotypic characteristic of the two types of mutants defective in known steps of phosphatidylethanolamine synthesis is the absence of a rapid and severe effect on growth rate. Mutants blocked early in phospholipid synthesis, that is *gpsA* or *plsB*, grow for only about half a generation after phospholipid synthesis is halted (28, 29, 40). Since 75% of the lipid in *E. coli* is phosphatidylethanolamine, a comparable effect would be expected for mutants unable to synthesize their lipid at nonpermissive temperatures. However, both classes of these mutants grow for several generations in the absence of phosphatidylethanolamine synthesis (58, 60,

63). These results may indicate that the synthesis of phosphatidylglycerol is of more immediate importance for cell growth than for the synthesis of phosphatidylethanolamine, the major lipid of the membranes. The phenotypes of the two classes of phosphatidylethanolamine mutants are largely similar, despite the fact that the mutant defective in phosphatidylserine decarboxylase accumulates large amounts of phosphatidylserine (58). Because of the ionic character of phosphatidylserine, its incorporation into a membrane might be expected to have striking consequences on the lipid phase properties (2). The lack of such an effect is interesting and suggests that the phosphatidylserine accumulating in these mutants may not be a membrane component.

No mutants specifically defective in phosphatidylglycerol synthesis have yet been isolated, although mutants have been isolated that are almost devoid of phosphatidylglycerol phosphate synthetase activity (61). These mutants synthesize phosphatidylglycerol normally, which suggests that this enzyme may not be part of the major pathway of phosphatidylglycerol synthesis.

Pluschke, Hirota & Overath (65) have isolated a mutant of *E. coli* defective in the synthesis of cardiolipin by screening a number of temperature-sensitive mutants for alterations in phospholipid metabolism. The mutant had < 5% of the normal cardiolipin synthetase activity in vitro and contained little or no cardiolipin (i. e. <0.3% of the total phospholipid). The temperature sensitivity of the original isolate is unrelated to the defect in cardiolipin synthesis and thus cardiolipin does not seem to play an essential role in cellular membrane function.

Mutants deficient in diglyceride kinase (66) and CDP-diglyceride hydrolase (61, 63) have been isolated and briefly described by Raetz. Neither of these two membrane-bound enzymes have an assigned role in the phospholipid synthetic pathway and the mutants should prove useful in this regard. The accumulation of diglyceride by the diglyceride kinase mutant is particularly interesting (66).

The available information on eucaryotic cells indicates that very large changes in phospholipid head group composition are readily tolerated (67–69). It seems likely that the same may be true of bacteria. It seems possible that, by appropriate genetic manipulation, the simple phospholipid composition of *E. coli* could be further simplified to a membrane composed largely of phosphatidylglycerol.

Phospholipid Related Molecules

Kennedy and co-workers have isolated a novel class of oligosaccharides from *E. coli* which contain components that appear to be derived from the membrane phospholipids (70, 71). The compounds contain glucose as the sole sugar and consist of at least three species with similar molecular weights, from 4000 to 5000 (70). The glucose residues on one of the components are linked to succinic acid by an O-ester linkage, and to glycerol-P and phosphoethanolamine by phosphodiester links to C-6 of the glucose residue (71). The other fraction that has been characterized lacks the succinate and ethanolamine residues (71).

The glycerol phosphate moiety of the oligosaccharide is sn-glycerol-1-phosphate (71), and thus has the same configuration as the polar group of phosphatidylglycerol. This finding, coupled with the results of "chase" experiments, where the label lost from glycerol-labeled lipids can be recovered in the oligosaccharides, led Kennedy and co-workers to propose that phosphatidylglycerol and/or cardiolipin is the source of the oligosaccharide glycerol phosphate residues. Strong support for this pathway has recently been obtained by Schulman & Kennedy (72), who find that mutants of *E. coli* containing only traces of glucose (due to a defect in phosphoglucose isomerase) do not contain the membrane-derived oligosaccharides. In such mutants, the turnover of both phosphatidylglycerol and cardiolipin is very slow. However, when glucose is added to the medium a dramatic stimulation of turnover of the two lipids is seen. The phosphoethanolamine residue is also thought to be derived from the membrane lipids; however, this has not been demonstrated. It would be interesting to see if the mutants that lack the oligosaccharide but contain lipopolysaccharide also contain phosphoethanolamine residues.

The membrane-derived oligosaccharides found in the periplasmic space (72) are not essential for growth and are not involved in the synthesis of the diglyceride moiety of the murein lipoprotein of Braun & Hantke (73). Shulman & Kennedy (72) showed that this diglyceride moiety is derived from a long half-life lipid such as phosphatidylglycerol and/or cardiolipin, rather than a short half-life lipid such as CDP-diglyceride. The membrane-derived oligosaccharides do not seem to be required for transfer of the diglyceride moiety to the murein lipoprotein. Either cardiolipin or phosphatidylglycerol could be the source of the diglyceride moiety of the lipoprotein. It would be interesting to test if the cardiolipin-deficient mutant of Pluschke, Hirota & Overath (65) can transfer either diglyceride units to the lipoprotein, or glycerolphosphate units to the oligosaccharides.

Lampen and co-workers (74, 75) have recently demonstrated that the amino terminus of the membrane penicillinase of *Bacillus licheniformis* is phosphatidylserine. The phosphatidylserine is bound by means of the NH_2 of the serine moiety in peptide linkage to the protein. The lipid seems to be donated from a phosphatidylserine tRNA and is thought to play a role in the secretion of the exopenicillinase.

Phospholipid Degradation

The degradation of phospholipids and fatty acyl chains was reviewed by Silbert (1). The papers appearing since that review partially clarify the situation in *E. coli*. It has been shown that the mutants (*pldA*), defective in the detergent-resistant phospholipase, lack the phospholipase A activity purified by Scandella & Kornberg (76). This homogenous enzyme has now been shown to have both A_1 and A_2 activities as well as lysophospholipase activity against both 1-lyso and 2-lyso derivatives (77). The physiological role of this and the other lytic activities remains unknown (for review see 1, 78).

The β-oxidation system of *E. coli* has been shown to include a fatty acid transport system involving acyl-CoA synthetase (79). The *fadE* mutants have been shown to be defective in the electron transport flavoprotein component; and mutants in two dehydrogenases, one active on butyryl-CoA (*fadF*) and another on unsaturated acyl-CoA substrates (*fadG*), have also been described (80). Various growth abnormalities have been reported for mutants constitutive in β-oxidation (*fadR*), which suggests that the *fadR* gene product may not be a simple repressor (81). Three of the β-oxidation enzymes of *E. coli* have been shown to be present in a multienzyme complex having a molecular weight of about 300,000 (82).

Enzymology of Phospholipid Synthesis

Dowhan and co-workers have had a good deal of success in the purification of various phospholipid biosynthetic enzymes from *E. coli* (83–85). These enzymes are, with one exception, tightly bound to the inner membrane of *E. coli*. Phosphatidylserine decarboxylase and phosphatidylglycerol-P synthetase were solubilized with Triton X-100 and purified in the presence of the detergent. The latter enzyme was purified by affinity chromatography on agarose containing immobilized CDP-diglyceride (84). A similar technique, affinity elution from phosphocellulose with CDP-diglyceride, resulted in the purification of phosphatidylserine synthetase (85). A large fraction of this enzyme is found bound to ribosomes in cell extracts. This surprising finding, first reported by Raetz & Kennedy (86), was challenged by Machtiger & Fox (87). However, the total activity of phosphatidylserine synthetase observed by these latter authors is only about 10% of that observed by others; thus it seems probable that Machtiger & Fox lost or inactivated the ribosome-binding activity. Ishinaga & Kito (88) report that about 80% of the phosphatidylserine synthetase is ribosome bound; most of the remainder is soluble, although a small amount is membrane bound. The soluble enzyme is readily bound to ribosomes, membranes, or phospholipids (88, 89). When it is bound to phospholipids the K_m of the enzyme for serine decreases (89), but not when it is bound to ribosomes or membranes. The genetic evidence stated above argues strongly that *E. coli* contains only a single phosphatidylserine synthetase. The most straightforward intracellular location for this enzyme, considering its properties and its role in phospholipid synthesis, is in loose association with the inner membrane of *E. coli*. The soluble and ribosomal activities would then be artifacts resulting from cell disruption.

The enzymes involved in the early steps of phospholipid synthesis in *E. coli* have not proven as amenable to purification as the other enzymes. Perhaps the most interesting enzyme in phospholipid synthesis is the glycerol-P acyltransferase, the first enzyme of the phospholipid biosynthetic pathway (for review see 1). Two groups have attempted the purification of this enzyme (52, 90). Ishinaga, Nishihara & Kito (90) solubilized the enzyme from the cell membranes with Triton X-100 and bound this preparation to a column of immobilized 6-phosphogluconic acid. Elution of this column (the terminal 3 carbons of the immobilized ligand are structurally analogous to glycerol-P) with glycerol-P resulted in elution of 60% of the activity. This step seemed to yield a substantial purification, although quantitative data were not reported. Snider & Kennedy (52) found that treatment of membranes with low concentration of detergents resulted in a rather selective extraction of the acyltransferase with a purification of about 20-fold. Both groups found the detergent-solubilized enzyme required phospholipid for activity. Ishinaga and co-workers (90) reported that sonicated vesicles of phosphatidylglycerol were effective in the activation of the acyltransferase; however, Snider & Kennedy (52) found that mixtures of phosphatidylethanolamine and phosphatidylglycerol are more effective than phosphatidylglycerol alone. In agreement with earlier indirect data (91, 92) both groups indicate that the acyltransferase is physically associated with lipid.

The role of the glycerol-P and other acyltransferases in determining the fatty acid composition of the membrane phospholipids is the subject of a large and confusing literature. In the phospholipids of *E. coli*, saturated fatty acids are found in position 1 of the glycerol-P moiety, whereas unsaturated fatty acids are largely found in position 2 (for review see 2, 56). The mechanisms responsible for this fatty acid asymmetry are a controversial subject at present. The first reports indicated that the specific localization of fatty acids was a property of the glycerol-P acyltransferase. Vagelos and co-workers (93, 94) and Sinensky (95) reported that the product of the acylation of glycerol-P with palmityl-CoA as the substrate was 1-acyl glycerol-P, whereas unsaturated acyl-CoA substrates were incorporated as 2-acyl glycerol-P. However, several laboratories have subsequently failed to detect this specificity.

Okuyama and co-workers (96, 97) have reported that unsaturated as well as saturated CoA esters are incorporated as 1-acyl glycerol-P and thus attribute no specificity to the glycerol-P acyltransferase. Similar results have been reported by others using partially purified glycerol-P acyltransferase preparations (52, 90). Okuyama & Wakil (96) have further reported that 2-acyl glycerol-P is not acylated to phosphatidic acid by *E. coli* membrane preparations. However, all the laboratories working on this project have reported that addition of a mixture of unsaturated and saturated acyl-CoA substrates to preparations converting glycerol-P to phosphatidic acid results in the specific incorporation of saturated and unsaturated fatty acids into position 1 and 2, respectively (94, 95, 98, 99). In the detailed studies of Kito and co-workers (99, 100), the phosphatidic acid formed by acylation of glycerol-P was converted to phosphatidylglycerol (addition of CTP allows phosphatidylglycerol synthesis), and the molecular species of phosphatidylglycerol were determined. In the presence of equal concentrations of the CoA esters of palmitoleic, palmitic, and *cis*-vaccenic acids, a spectrum of molecular species of phosphatidylglycerol was formed that was very similar to the distribution found in vivo (100). If the proportion of one of these acyl-CoAs in the reaction mixture was increased, a proportional increase in the molecular species containing that fatty acid was found (99). Similar results have been reported by Okuyama et al (98).

An apparent contradiction thus exists. How can specific phospholipid species be made if acylation is not specific? The solution to this contradiction could lie in the relative rates of acylation. The incorporation of a saturated acyl-CoA into monoacyl glycerol-P has usually been reported to proceed faster than incorporation of an unsaturated acyl-CoA (93, 96, 98, 99). Conversely, addition of a second acyl group to 1-acyl glycerol-P proceeds more rapidly with an unsaturated acyl-CoA (93, 96, 97). It may be, then, that the specificity is not absolute but is a matter of rates, which in turn are dictated by the supply of acyl moieties. Since *E. coli* can change the ratio of saturated to unsaturated acyl chains (101, 102), this mechanism could operate in vivo.

However, the situation is complicated by other findings. Two groups have reported that the relative rates of incorporation of saturated as against unsaturated acyl-CoA substrates into phosphatidic acid depend on the concentration of the acceptor, glycerol-P (52, 97). It is difficult to see how this effect could have physiological relevance. The data of Bell & Cronan (45) indicate that glycerol-P levels are closely regulated in *E. coli,* and thus the glycerol-P pool should be constant in growing cells. Conversely, the growth and fatty acid composition of *E. coli* is unaffected by a large increase in the glycerol-P pool. For example, if glycerol rather than glucose is used as the sole carbon source for growth of *E. coli* K12, the intracellular glycerol-P content increases 16-fold (103). However, the phospholipids of glucose- and glycerol-grown cultures of this strain have very similar fatty acid compositions (104, 105).

These data, therefore, raise the question of in vitro artifacts. Most of the data on positional specificity were obtained using acyl-CoA substrates as acyl donors. The glycerol-P acyltransferase will accept both acyl-CoA and acyl-ACP as substrates in vitro (44, 93, 106) and genetic data indicate that the two acylation systems have at least some components in common (44). It is possible that acyl-ACP substrates, which arise either as the direct product of fatty acid synthesis or by way of acyl-ACP synthetase (see below), could be the sole donors of acyl groups in vivo. Therefore, the lengthy arguments given above may not be germane to the physiological situation. Indeed, van den Bosch & Vagelos (93) have reported differing specificities in the acylation of glycerol-P depending on whether acyl-ACP or acyl-CoA was the in vitro donor. Although that study used ACP preparations that lacked native structure (44), it raises the question of whether acyl-ACP or acyl-CoA, or both, are the physiological donors in phosphatidic acid synthesis.

The only evidence that acyl-CoA might be a donor in vivo was the report by Overath, Pauli & Schairer (107) that mutants of *E. coli* lacking acyl-CoA synthetase are unable to incorporate exogenously added fatty acids into phospholipid, whereas endogenously synthesized fatty acids are incorporated normally. From these data arose the notion that exogenous, by way of CoA, and endogenous, by way of ACP, fatty acids are transferred into phospholipid from different thioester donors (104–106). However, a subsequent report from the same laboratory weakened this idea. Klein et al (79) showed that acyl-CoA synthetase mutants were unable to transport exogenous fatty acids. This result suggested that the failure of these mutants to incorporate exogenous fatty acids into phospholipid could simply be due to a transport defect. This appears to be the case. Ray & Cronan (108) showed that intracellular fatty acids, generated by glycerol-P starvation (see below), could be incorporated into phospholipid in the absence of acyl-CoA synthetase activity.

These results led Ray & Cronan to seek an enzyme capable of the direct conversion of free fatty acid to acyl-ACP. They were able to demonstrate an enzyme catalyzing an ATP-dependent conversion of fatty acid to acyl-ACP (109). The product was shown to be acyl-ACP by chemical analysis and by transfer and elongation of the acyl group in enzymatic reactions. This enzyme is located on the inner membrane of the cell, from which it can be extracted with detergent and purified (110). The role of this enzyme is presently unknown but seems to lie in the incorporation of free fatty acids into phospholipid. These free acids could arise by turnover of phospholipid acyl groups or by cleavage of acyl-CoA substrates by the thioesterases of the cell. *E. coli* contains two thioesterases that are much more active on acyl-CoA substrates than on the homologous native or modified acyl-ACP molecules (111–113). In fact, the higher molecular weight thioesterase seems virtually unable to hydrolyze native acyl-ACP molecules (113). The thioesterases therefore seem designed to prevent the intracellular accumulation of acyl-CoA. Indeed no intracellular acyl-CoA has been detected in *E. coli,* even in cells involved in β-oxidation (79, 114). However, some indirect evidence for the presence of intracellular acyl-ACP molecules in *E. coli* has been reported (115). These considerations imply, therefore, that acyl-ACP rather than acyl-CoA is the acyl donor in phospholipid synthesis in vivo.

Why are phospholipids, which contain a saturated fatty acid residue in position 1 and an unsaturated one in position 2, the general rule in biology? This question can be clearly understood in terms of the physical properties of the lipid phase (for review see 2). Arguments based on the newly demonstrated nonequivalence of the two fatty acyl chains of phospholipids are also compelling (116, 117). However, there is no reason to expect that the large variety of molecular species observed in vivo is absolutely required. Kito and co-workers (99) have now clearly shown that phosphatidylglycerol in *E. coli* is richer in *cis*-vaccenic acid than is phosphatidylethanolamine, and have proposed that the various molecular species might play important roles in membrane physiology (118). However, *E. coli* fatty acid auxotrophs grow normally using fatty acids that greatly alter the normal pattern of molecular species (for review see 1, 2). A mutant defective in *cis*-vaccenate synthesis also grows normally (36). It seems therefore that a diversity of molecular species is not needed by *E. coli.*

The origin of the altered *cis*-vaccenate content of phosphatidylglycerol is unclear. A selection among molecular species of CDP-diglyceride by the phospholipid biosynthetic enzyme would give this result. Another possibility is that the CDP-diglyceride synthetase could select molecular species of phosphatidic acid. Both CDP and CDP-diglycerides are present in *E. coli* (119, 120) and the presence or absence of the deoxy sugar could identify molecular species. CDP-diglyceride hydrolase and/or diglyceride kinase could also be involved in the specificity of phosphatidylglycerol synthesis. It seems clear that the different fatty acid content arises by synthesis rather than by a retailoring of intact lipid molecules (121).

REGULATION OF LIPID SYNTHESIS

Several mechanisms of control for lipid synthesis have been recently outlined by genetic studies. The major result of this work is the demonstration that lipid synthesis is controlled at the levels of both fatty acid and phospholipid synthesis.

Coupling Between Fatty Acid and Phospholipid Synthesis

In *E. coli,* phospholipid and fatty acid synthesis are normally rightly coupled (56, 122, 123). However, if phospholipid synthesis and β-oxidation are blocked, fatty acid synthesis is partially uncoupled from phospholipid synthesis and a large accumulation of intracellular free fatty acid is found (123). Phospholipid synthesis is blocked by depriving a *plsB* or *gpsA* mutant of glycerol-P. However, even under these conditions, complete uncoupling is not observed (124). Nunn, Kelly & Stumfall (124) have shown that although the rate of acetate incorporation into total fatty acids is similar in a *plsB* strain that is starved for glycerol-P or that is growing normally with glycerol-P, only about 10% of the normal amount of fatty acid is synthesized. This effect seems due to a contraction of the acetate, or acetyl CoA, pool in the glycerol-P deprived cultures. Unfortunately, the acetyl CoA pool was not measured directly. The cause of this effect is unknown since glycerol-P is not a carbon source in these strains and the pools of several other metabolites are not decreased (29). Despite this difficulty, the conditions that result in a partial uncoupling of fatty acid synthesis have been quite useful in determining the site of control steps in lipid synthesis.

Several laboratories have reported that the addition of oleic acid to *E. coli* results in a decreased incorporation of acetate into phospholipid fatty acids (95, 125, 126). This effect could be due to repression, or inhibition, of fatty acid synthesis, or could more simply be the result of competition for acyl transfer between exogenously and endogenously synthesized fatty acids. Recent studies using the uncoupled system support the latter explanation (127).

Regulation of the Amount of Phospholipid in E. Coli

Phospholipid synthesis as well as RNA synthesis is under the control of the *relA* gene of *E. coli* (128–134). Amino acid starvation of stringent, that is *rel*+, strains results in a 2–3-fold decrease in the rate of phospholipid synthesis, whereas relaxed (*relA*⁻) strains synthesize phospholipid normally under these conditions (128–134). A novel nucleotide, guanosine 5'-diphosphate-3'-diphosphate (ppGpp), accumulates during amino acid starvation of *rel*+ but not *relA*⁻ strains (135), and thus seems likely to be the effector of *relA* gene control of lipid synthesis. In fact, Nunn & Cronan (133) have shown that the rate of lipid synthesis in *E. coli* is proportional to the intracellular concentration of ppGpp. When very high concentrations of ppGpp are generated in vivo, phospholipid synthesis is inhibited over 10-fold. Control of lipid synthesis by the *relA* gene thus seems to be mediated by ppGpp (133).

Nunn & Cronan (131, 132) have localized the sites of *relA* gene control in vivo. Experiments using the uncoupled system discussed above with mutants blocked in fatty acid synthesis have shown the existence of at least two sites of control, one at the level of phospholipid synthesis (131) and the other at the level of fatty acid synthesis (132). Further evidence of a control site in fatty acid synthesis has been reported by Spencer and co-workers (134).

The two enzymes that seem most likely to be controlled by ppGpp are acetyl-CoA carboxylase and glycerol-P acyltransferase since these are the first committed steps at the levels of fatty acid and phospholipid synthesis. Both enzymes have been shown to be inhibited by ppGpp. However, in both cases there is some doubt concerning the physiological relevance of the inhibition observed in vitro.

Polakis, Guchait & Lane (130) reported that ppGpp is a reversible inhibitor of the carboxyltransferase component of acetyl-CoA carboxylase. However, the inhibition of this enzyme was not linear with the ppGpp concentration, and the maximum inhibition was only 50%. The in vivo results (133) indicate that inhibition of the target enzyme should be linear with ppGpp concentration and almost complete at high concentrations.

The glycerol-P acyltransferase of *E. coli* is also inhibited by ppGpp (44, 106, 129), but only when acyl-CoA is the substrate (44, 106); no inhibition is observed if the enzyme is assayed with acyl-ACP. This curious situation is further complicated by the fact that the inhibition by ppGpp is irreversible, which conflicts with the instant reversibility observed in vivo (129, 133).

These considerations therefore make it difficult to argue that either enzyme is a site of ppGpp inhibition in vivo. The fact that inhibition is observed in reactions that involve CoA substrates raises the possibility that ppGpp inhibits these enzymes by a nonphysiological occupation of their CoA binding site. Enzymes that utilize CoA substrates are often inhibited by the 3',5'-ADP moiety of the coenzyme (136). Thus far, 3',5'-ADP and other nucleotides such as 3',5'-GDP have not been tested as inhibitors of these enzymes.

It should be noted that ppGpp also inhibits two lipid synthesizing enzymes that do not utilize CoA substrates, i.e. phosphatidylglycerol phosphate synthetase (129) and β-hydroxydecanoylthioester dehydrase (137). However, these in vitro results seem clearly artifactual, since no preferential decrease in the formation of the product of either enzyme is observed in vivo (129, 134).

The above considerations therefore indicate that data showing inhibition of lipid biosynthetic enzymes by ppGpp must be interpreted with caution and suggest that further in vivo experiments to localize the sites of ppGpp inhibition are needed. Amplification of various candidate enzymes by DNA cloning procedures seems likely to provide such information.

The genetic analysis of mutants defective in lipid synthesis has not yet provided any clues concerning the mechanism whereby the rate of lipid synthesis is controlled. The various genetic lesions in lipid synthesis are scattered almost randomly about the genetic map (138). For example, the genes that code for the two β-ketoacyl-ACP synthetases are unlinked (37). The only fatty acid biosynthetic genes thought to be linked are *fabD* and *fabF*, which code for two consecutive steps in the synthetic pathway (37).

Control of Phospholipid Head Group Composition

Little is known about the regulation of head group composition. Some alteration of head group ratios result from various culture conditions (for review see 56) and drug treatments (141, 142), but the precise mechanism of these effects is not known. The residual phospholipid synthesized in mutants blocked early in the synthetic pathway, before CDP-diglyceride, has normal ratios of head groups (29). Mutants blocked in phosphatidylserine synthesis shunt their lipid synthesis to phosphatidylglycerol and cardiolipin (60, 63), whereas mutants blocked in the conversion of phosphatidylserine to phosphatidylethanolamine accumulate the precursor lipid (58). Strains fed a phosphonate analogue of glycerol-P accumulate phosphatidylglycerol phosphonate (142). Therefore, the lack of synthesis of one phospholipid species does not seem to inhibit the synthesis of other phospholipids. Strains containing unusually high levels of phosphatidylserine synthetase do not contain an unusual amount of the serine or ethanolamine lipids (139). Analogous strains with increased levels of glycerol-P acyltransferase also show no striking effects on phospholipid synthesis (140). The results indicate that the levels of these two enzymes do not control either the rate of lipid synthesis or the distribution of head groups.

The known genes coding for the various enzymes in phospholipid head group synthesis are not linked. For example, the genes coding for two consecutive enzymes in phosphatidylethanolamine synthesis are separated by half the genetic map (138).

Regulation of Fatty Acid Chain Length

The chain length of the fatty acyl moieties of a phospholipid is a major determinant of the phase transition of the lipid (2). The specificity of the chain length in *E. coli* was thought to result from specificity of the β-ketoacyl-ACP synthetase component of fatty acid synthesis (143). However, control also seems to be exerted at the level of phosphatidic acid synthesis since the free fatty acids, accumulating during a partial uncoupling of fatty acid synthesis from phospholipid synthesis, are abnormally long (101, 123, 127). An interesting example of a bacterium that makes a bimodal distribution of fatty acid chain lengths is the *Mycobacterium smegmatis* system reviewed by Bloch (4).

Regulation of the Saturated-Unsaturated Ratio and Temperature Control

The ratio of saturated to unsaturated fatty acid synthesized by *E. coli* at a given temperature is partially determined by the levels of the β-hydroxydecanoyl thioester dehydrase (102). Recent data suggest that the levels of β-ketoacyl-ACP synthetase I and II are also important in this regard (144). A secondary site of control seems to occur at the level of phosphatidic acid synthesis (see above). However, most of the data on the regulation of the saturated-unsaturated ratio deals with the temperature control of this ratio.

Marr & Ingraham (145) first found that *E. coli* adjusts its saturated-unsaturated ratio in response to growth temperature. As the temperature of growth is decreased the proportion of *cis*-vaccenate increases greatly. Temperature control does not seem to require induction of a protein, since control is evident immediately upon temperature shift (37, 100) and is unaffected by inhibitors that block RNA and protein synthesis (37, 98).

In vivo studies demonstrate clearly that temperature regulation is yet another example of redundant control. The ratio of unsaturated to saturated species in the free fatty acid fraction, which accumulates in cells blocked in phospholipid synthesis, is altered by temperature (101). The proportion of saturated fatty acid increases greatly as the temperature is increased. This result demonstrates a direct control acting at the level of fatty acid synthesis (101). In vitro results using the soluble fatty acid synthetase are consistent with a direct temperature effect (98), but are not compelling since they are variable (146) and depend on assay conditions other than temperature (98).

A control site that acts at the level of phosphatidic acid synthesis can be considered to moderate the regulation of fatty acid synthesis. The moderation can be seen by comparing the ratios of saturated to unsaturated species, which accumulate at various temperatures in the free fatty acid and phospholipid fractions of cells starved for glycerol-P (101). The free fatty acid fraction has a higher proportion of saturated fatty acids than the phospholipid fraction. In vivo studies on the incorporation of exogenous fatty acids also show temperature control (95, 101). The site of this control seems to be the acyltransferase system, although the extent of the control observed varies widely (95, 98, 99).

Melchior & Steim (147, 148) have recently proposed a novel mechanism for temperature control based on model systems showing the differential partition of saturated and unsaturated fatty acids into ordered or fluid lipid. They found that palmitic acid was more strongly bound than oleic acid to model and *Acholeplasma* membranes at temperatures above the phase transition. From these data, they proposed that the fluidity of the membrane would determine the partition coefficient and thus the fatty acid supply of the acyltransferase. However, this interesting proposal does not appear to hold for *E. coli.* Three laboratories (95, 98, 99) have shown that the membrane-bound acyltransferase system in fluid membranes has the same specificity as the enzyme in membranes composed of fully ordered lipid.

A complex temperature control system has been demonstrated in *Bacilli* by Fulco and co-workers (for review see 149, 150). The desaturation enzyme(s) involved appears to be induced by a downward temperature shift and inactivated upon upward shift. A hyperinduction phenomenon is also seen (150). The study of the mechanism of these phenomena has thus far been hindered by the lack of an in vivo assay for the desaturatase activity.

ACKNOWLEDGMENTS

I thank Dr. Charles O. Rock for his advice on the manuscript. The unpublished experiments from the author's laboratory were supported by NIH Grants AI10186 and GM22797, and by a National Institutes of Health Research Career Development Award (GM70,411).

Literature Cited

1. Silbert, D. F. 1975. *Ann. Rev. Biochem.* 44:315–39
2. Cronan, J. E. Jr., Gelmann, E. P. 1975. *Bacteriol. Rev.* 39:232–56
3. Overath, P., Thilo, T. 1977. *MTP Int. Rev. Sci.* 1:Ser. 2. In press
4. Bloch, K., Vance, D. 1977. *Ann. Rev. Biochem.* 42:263–98
5. Goldfine, H. 1972. *Adv. Microbiol. Physiol.* 8:1–57
6. Jackson, M. B., Sturtevant, J. M. 1977. *J. Biol. Chem.* 252:4749–51
7. Jackson, M. B., Cronan, J. E. Jr. 1977. *Biochim. Biophys. Acta.* In press
8. Uehara, K., Akutsu, H., Kyogoku, Y., Akamatsu, Y. 1977. *Biochim. Biophys. Acta* 466:393–401
9. Thilo, L., Overath, P. 1976. *Biochemistry* 15:328–34
10. Cronan, J. E. Jr., Gelmann, E. P. 1973. *J. Biol. Chem.* 248:1188–95
11. Baldassare, J. J., Rhinehart, H., Silbert, D. F. 1976. *Biochemistry* 15:2986–94
12. McElhaney, R. N. 1974. *J. Supramol. Struct.* 2:617–28
13. Hinz, H.-J., Sturtevant, J. M. 1972. *J. Biol. Chem.* 247:6071–75
14. van Heerikhuisen, H., Kwah, E., van Brugzen, E. E. J., Wilhott, B. 1975. *Biochim. Biophys. Acta* 413:177–91
15. Letellier, L., Shechter, E. 1976. *J. Microsc. Biol. Cell.* 25:191–96
16. Letellier, L., Moudden, K., Shechter, E. 1977. *Proc. Natl. Acad. Sci. USA* 74:452–56
17. Lofgren, K. W., Fox, C. F. 1974. *J. Bacteriol.* 118:1181–82
18. Miller, J. B., Koshland, D. E. Jr. 1977. *J. Mol. Biol.* 111:183–201
19. Thilo, L., Vielmetter, W. 1977. *J. Bacteriol.* 128:130–43
20. Fralick, J. A., Lark, K. G. 1973. *J. Mol. Biol.* 80:459–75
21. Kass, L. R., Bloch, K. 1967. *Proc. Natl. Acad. Sci. USA* 58:1168–73
22. Nunn, W. D., Cronan, J. E. Jr. 1974. *J. Biol. Chem.* 249:724–31
23. Tsukagoshi, N., Fox, C. F. 1973. *Biochemistry* 12:2822–29
24. Tsukagoshi, N., Fox, C. F. 1973. *Biochemistry* 12:2816–22
25. Thilo, L., Träuble, H., Overath, P. 1977. *Biochemistry* 16:1283–90
26. Linden, C. D., Wright, K. L., McConnell, H. M., Fox, C. F. 1973. *Proc. Natl. Acad. Sci. USA* 70:2271–75
27. Linden, C. D., Fox, C. F. 1973. *J, Supramol. Struct.* 1:835–44
28. McIntyre, T. M., Bell, R. M. 1975. *J. Biol. Chem.* 250:9053–59

29. McIntyre, T. M., Chamberlain, B. K., Webster, R. E., Bell, R. M. 1977. *J. Biol. Chem.* 252:4487–93
30. Houslay, M. D., Warren, G. B., Birdsall, N. J. M., Metcalfe, J. C. 1975. *FBS Lett.* 51:146–51
31. Rothman, J. E., Kennedy, E. P. 1977. *J. Mol. Biol.* 110:630–48
32. Bretscher, M. S. 1973. *Science* 181: 622–29
33. Rothman, J. E., Kennedy, E. P. 1977. *Proc. Natl. Acad. Sci. USA* 74:1821–25
34. Broekmann, J. H. F. F., Hoeckstra, W. P. M. 1973. *Mol. Gen. Genet.* 124:65–67
35. Clark, D., Cronan, J. E. Jr. 1978. *J. Bacteriol.* 132:549–54
36. Gelmann, E. P., Cronan, J. E. Jr. 1972. *J. Bacteriol.* 112:381–87
37. Garwin, J., Cronan, J. E., Jr. 1978. Manuscript in preparation
38. Silbert, D. F., Pohlman, T., Chapman, A. 1976. *J. Bacteriol.* 126:1351–54
39. Taylor, F. R., Cronan, J. E. Jr. 1976. *J. Bacteriol.* 125:518–23
40. Bell, R. M. 1974. *J. Bacteriol.* 117: 1065–76
41. Cronan, J. E. Jr., Bell, R. M. 1974. *J. Bacteriol.* 118:589–605
42. Cronan, J. E. Jr., Bell, R. M. 1974. *J. Bacteriol.* 120:227–33
43. Bell, R. M. 1975. *J. Biol. Chem.* 250:7147–52
44. Ray, T. K., Cronan, J. E. Jr. 1975. *J. Biol. Chem.* 250:8422–27
45. Bell, R. M., Cronan, J. E. Jr. 1975. *J. Biol. Chem.* 250:7153–58
46. Edgar, J. R., Bell, R. M. 1977. *Fed. Proc.* 36(3):857 (Abstr. 3090)
47. Cronan, J. E. Jr., Ray, T. K., Vagelos, P. R. 1970. *Proc. Natl. Acad. Sci. USA* 65:6442–48
48. Cronan, J. E. Jr., Godson, G. N. 1972. *Mol. Gen. Genet.* 116:199–210
49. Glaser, M., Nulty, W., Vagelos, P. R. 1975. *J. Bacteriol.* 123:128–36
50. Ray, T. K., Godson, G. N., Cronan, J. E. Jr. 1976. *J. Bacteriol.* 125:136–41
51. Glaser, M., Bayer, W. H., Bell, R. M., Vagelos, P. R. 1973. *Proc. Natl. Acad. Sci. USA* 70:385–89
52. Snider, M., Kennedy, E. P. 1977. *J. Bacteriol.* 130:1072–83
53. Kito, M., Pizer, L. I. 1969. *J. Bacteriol.* 97:1321–27
54. Merlie, J. P., Pizer, L. I. 1973. *J. Bacteriol.* 116:355–66
55. Swedes, J. E., Sedo, R. J., Atkinson, D. E. 1975. *J. Biol. Chem.* 250:6930–38
56. Cronan, J. E. Jr., Vagelos, P. R. 1972. *Biochim. Biophys. Acta* 265:25–65
57. Luria, S. E., Suit, J. L., Plate, C. A.

1976. *Biochem. Biophys. Res. Commun.* 65:353–58
58. Hawrot, E., Kennedy, E. P. 1975. *Proc. Natl. Acad. Sci. USA* 72:1112–16
59. Nunn, W. D. 1977. Personal communication
60. Ohta, A., Ohonogi, K., Shibuya, I., Marrio, B. 1974. *J. Gen. Appl. Microbiol.* 20:21–32
61. Raetz, C. R. H. 1975. *Proc. Natl. Acad. Sci. USA* 72:2274–79
62. Ohta, A., Shibuya, I., Marrio, B., Ishinaga, M., Kito, M. 1974. *Biochim. Biophys. Acta* 348:449–54
63. Raetz, C. R. H. 1976. *J. Biol. Chem.* 251:3242–49
64. Cronan, J. E. Jr. 1972. *Nature New Biol.* 240:21–22
65. Pluschke, G., Hirota, Y., Overath, P. 1978. Manuscript in preparation
66. Raetz, C. R. H. 1977. *Fed. Proc.* 36(3): 638 (Abstr. 1935)
67. Horwitz, A. F. 1977. *In Growth, Nutrition, and Metabolism of Cells in Culture,* 3:110–48. New York:Academic
68. Hubbard, S. C., Brody, S. 1975. *J. Biol. Chem.* 250:7173–81
69. Sinensky, M. 1977. *J. Bacteriol.* 129: 516–24
70. van Golde, L. M. G., Schulman, H., Kennedy, E. P. 1973. *Proc. Natl. Acad. Sci. USA* 70:1368–72
71. Kennedy, E. P., Rumley, M. K., Schulman, H., van Golde, L. M. G. 1976. *J. Biol. Chem.* 251:4208–13
72. Schulman, H., Kennedy, E. P. 1977. *J. Biol. Chem.* 252:4250–55
73. Braun, V., Hantke, K. 1974. *Ann. Rev. Biochem.* 42:89–122
74. Lampen, J. O., Yamamoto, S. 1977. In *Microbiology 1977,* pp. 104–11. Washington DC: Am. Soc. Microbiol.
75. Dancer, B. N., Lampen, J. O. 1977. See Ref. 74, pp. 100–3
76. Scandella, C. J., Kornberg, A. 1971. *Biochemsitry* 10:4447–56
77. Nishijima, M., Nakaido, S., Tamori, Y., Nojima, S. 1977. *Eur. J. Biochem.* 73:115–24
78. Doi, O., Nojima, S. 1976. *J. Biochem,* 80:1247–58
79. Klein, K., Steinberg, R., Frithen, B,, Overath, P. 1971. *Eur. J. Biochem.* 19:442–50
80. Klein, K. 1973. *Acyl-CoA-Dehydrogenasen und ETF in Escherichia coli: Studien zum Fettsäureabbau.* Thesis. Univ. Köln, Köln. 180pp.
81. Vanderwinkel, E., De Vlieghere, M., Fontaine, M., Charles, D., Denanuer, F., Vandevoorde, D., De Kegel, D. 1976. *J. Bacteriol.* 127:1389–99

82. Binstock, J. F., Pramenik, A., Shultz, H. 1977. *Proc. Natl. Acad. Sci. USA* 74:492–95

83. Dowhan, W., Wickner, W. T., Kennedy, E. P. 1974. *J. Biol. Chem.* 249:3079–84

84. Hirabayashi, T., Larson, T. J., Dowhan, W. 1976. *Biochemistry* 15:5205–11

85. Larson, T. J., Dowhan, W. 1976. *Biochemistry* 15:5212–18

86. Raetz, C. R. H., Kennedy, E. P. 1972. *J. Biol. Chem.* 274:2008–12

87. Machtiger, N. A., Fox, C. F. 1973. *J. Supramol. Struct.* 1:545–54

88. Ishinaga, M., Kito, M. 1974. *Eur. J. Biochem.* 42:483–87

89. Ishinaga, M., Kata, M., Kito, M. 1974. *FEBS Lett.* 49:201–2

90. Ishinaga, M., Nishihara, M., Kito, M. 1976. *Biochim. Biophys. Acta* 450:269–72

91. Mavis, R. D., Bell, R. M., Vagelos, P. R. 1972. *J. Biol. Chem.* 274:2835–41

92. Mavis, R. D., Vagelos, P. R. 1972. *J. Biol. Chem.* 247:652–58

93. van den Bosch, H., Vagelos, P. R. 1970. *Biochim. Biophys. Acta* 218:233–48

94. Ray, T. K., Cronan, J. E. Jr., Mavis, R. D., Vagelos, P. R. 1970. *J. Biol. Chem.* 245:6442–48

95. Sinensky, M. 1971. *J. Bacteriol.* 106:449–55

96. Okuyama, H., Wakil, S. J. 1973. *J. Biol. Chem.* 248:5197–5205

97. Okuyama, H., Yamada, K., Izezawa, H., Wakil, S. J. 1976. *J. Biol. Chem.* 251:2487–92

98. Okuyama, H., Yamada, K., Kameyama, Y., Kiezawa, H., Akamatsu, Y., Nojima, S. 1977. *Biochemistry* 16:2668–73

99. Kito, M., Ishinaga, M., Nishihara, M., Kata, M., Sawada, S., Hata, T. 1975. *Eur. J. Biochem.* 54:55–63

100. Nishihara, M., Ishinaga, M., Kata, M., Kito, M. 1976. *Biochim. Biophys. Acta* 431:54–61

101. Cronan, J. E. Jr. 1975. *J. Biol. Chem.* 250:7074–77

102. Cronan, J. E. Jr. 1974. *Proc. Natl. Acad. Sci. USA* 71:3758–62

103. Lowry, O. H., Carter, J., Ward, J. B., Glaser, L. 1971. *J. Biol. Chem.* 246:6511–21

104. Silbert, D. F., Ruch, F., Vagelos, P. R. 1968. *J. Bacteriol.* 95:1658–65

105. Silbert, D. F., Cohen, M., Harder, M. E. 1972. *J. Biol. Chem.* 247:1699–1701

106. Leuking, D. R., Goldfine, H. 1975. *J. Biol. Chem.* 250:4911–17

107. Overath, P., Pauli, G., Schairer, H. U. 1969. *Eur. Biochem.* 7:554–74

108. Ray, T. K., Cronan, J. E. Jr. 1976. *Biochem. Biophys. Res. Commun.* 69:506–13

109. Ray, T. K., Cronan, J. E. Jr. 1976. *Proc. Natl. Acad. Sci. USA* 73:4374–78

110. Rock, C. O., Cronan, J. E. Jr. 1978. Unpublished data

111. Barnes, E. M. Jr., Wakil, S. J. 1968. *J. Biol. Chem.* 243:2955–62

112. Bonner, W. M., Bloch, K. 1972. *J. Biol. Chem.* 247:3123–33

113. Spencer, A., Greenspan, A., Cronan, J. E. Jr. 1978. Manuscript in preparation

114. Frerman, F. E., Bennett, W. 1973. *Arch Biochem. Biophys.* 159:434–43

115. Elovson, J., Vagelos, P. R. 1975. *Arch. Biochem. Biophys.* 168: 490–97

116. Hitchcock, P. B., Mason, R., Thomas, K. M., Shipley, G. G. 1974. *Proc. Natl. Acad. Sci. USA* 71:3036–40

117. Seelig, S., Seelig, A. 1974. *Biochem. Biophys. Res. Commun.* 57:406–13

118. Ishinaga, M., Nishihara, M., Kata, M., Kito, M. 1976. *Biochim. Biophys. Acta* 431:426–32

119. Raetz, C. R. H., Kennedy, E. P. 1973. *J. Biol. Chem.* 248:1098–1105

120. Tunaitis, E., Cronan, J. E. Jr. 1973. *Arch. Biochem. Biophys.* 155:420–27

121. Cronan, J. E. Jr., Prestegard, J. H. 1977. *Biochemistry.* 16:4738–42

122. Mindich, L. 1972. *J. Bacteriol.* 110:96–102

123. Cronan, J. E. Jr., Weisberg, L. J., Allen, R. G. 1975. *J. Biol. Chem.* 250:5835–40

124. Nunn, W. D., Kelly, D. L., Stumfall, M. Y. 1977. *J. Bacteriol.* 132:526–31

125. Silbert, D. F., Ulbright, T. M., Honegger, J. L. 1973. *Biochemistry* 12:164–71

126. Estroumza, J., Ailhaud, G. 1971. *Biochimie* 53:837–39

127. Polacco, M. L., Cronan, J. E. Jr. 1977. *J. Biol. Chem.* 252:5488–90

128. Sokawa, Y., Nakao, E., Kaziro, Y. 1968. *Biochem. Biophys. Res. Commun.* 33:108–12

129. Merlie, J. P., Pizer, L. I. 1973. *J. Bacteriol.* 116:355–66

130. Polakis, S. E., Guchhait, R. B., Lane, M. D. 1973. *J. Biol. Chem.* 248:7957–66

131. Nunn, W. D., Cronan, J. E. Jr. 1974. *J. Biol. Chem.* 249:3994–96

132. Nunn, W. D., Cronan, J. E. Jr. 1976. *J. Mol. Biol.* 102:167–72

133. Nunn, W. D., Cronan, J. E. Jr. 1976. *Biochemistry* 15:2546–50

134. Spencer, A., Muller, E., Cronan, J. E. Jr., Gross, T. A. 1977. *J. Bacteriol.* 130:114–17

135. Cashel, M. 1975. *Ann. Rev. Microbiol.* 29:301–18

136. Abiko, Y. 1975. *Metabolic Pathways* 7:1–25
137. Stein, J. P. Jr., Bloch, K. E. 1976. *Biochem. Biophys. Res. Commun.* 93:881–84
138. Bachmann, B. J., Low, K. B., Taylor, A. L. 1976. *Bacteriol. Rev.* 40:116–67
139. Raetz, C. R. H., Larson, T. J., Dowhan, W. 1977. *Proc. Natl. Acad. Sci. USA* 74:1412–16
140. Bell, R. M., Cronan, J. E. Jr. 1978. Unpublished data
141. Nunn, W. D. 1977. *Biochemistry* 16:1077–81
142. Tyhach, R. J., Engel, R., Tropp, B. E. 1976. *J. Biol. Chem.* 251:6717–21
143. Greenspan, M. D., Birge, C. H., Powell, G., Hancock, W. S., Vagelos, P. R. 1970. *Science* 170:1203–4
144. Clark, D., Garwin, J., Cronan, J. E. Jr. 1978. Unpublished data
145. Marr, A. G., Ingraham, J. L. 1962. *J. Bacteriol.* 84:1260–67
146. Cronan, J. E. Jr., Vagelos, P. R. 1969. Unpublished data
147. Melchior, D. L., Steim, J. M. 1976. *Ann. Rev. Biophys. Bioeng.* 5:205–38
148. Melchior, D. L., Steim, J. M. 1977. *Biochim. Biophys. Acta* 466:148–59
149. Fulco, A. J. 1974. *Ann. Rev. Biochem.* 43:215–41
150. Fujai, D. K., Fulco, A. J. 1977. *J. Biol. Chem.* 252:3660–70

Ann. Rev. Biochem. 1978. 47:191–216

NERVE GROWTH FACTOR[1]   ❖972

Ralph A. Bradshaw

Department of Biological Chemistry Washington University School of Medicine
St. Louis, Missouri 63110

CONTENTS

PERSPECTIVES AND SUMMARY

Since the discovery of nerve growth factor (NGF) some 30 years ago, there has been a constant interest in determining the precise nature of this substance and its true physiologic role in regulating growth, development, and maintenance in the nervous system. Although the potential importance of this understanding was clear from the outset, it has nonetheless remained an elusive goal. For the most part this has been engendered by the problems

---

[1]Abbreviations used: NGF, nerve growth factor; NSILA, nonsuppressible insulin-like activity or insulin-like growth factor; EGF, epidermal growth factor.

0066-4154/78/0701-0191$01.00

of working with neuronal tissue, which is difficult to obtain or culture in quantity, and by the problems of extrapolating in vitro results to in vivo systems. Thus, although considerable progress has been made in recent years, it is not possible at present to describe with certainty the biologically important species of NGF, its site(s) of synthesis and subsequent transport, the scope of responsive tissues, or its mechanism of action.

Fortunately, this failure to resolve the fundamental features of the chemistry and biology of NGF have not dampened the collective spirits of the increasing number of researchers interested in this problem, and important inroads have been and are being made which are summarized in this review. Perhaps the most important of these has been the realization that NGF does not represent a unique entity, charged with the singular task of maintaining the viability of selectively responsive neurons, but rather is a member of a larger group of regulating peptides and proteins that possess the same responsibility albeit to other cell types located throughout the organism. Some of these substances, such as insulin, nonsuppressible insulin-like activity (NSILA), and relaxin, actually bear structural relatedness, presumably as the result of evolution from a common precursor, while others, such as epidermal growth factor (EGF), are apparently only functionally similar. However, all of these growth factors can justifiably be classed in one group, based on the nature and time course of the effects that they elicit in their individually responsive tissues. Furthermore, these factors can properly be thought of as "maintenance" or "secondary" hormones, in contrast to "primary" hormones. As detailed in the last section of this review, the distinction between primary and secondary hormones may well hinge on an extracellular as opposed to an intracellular site of action.

Most of the knowledge about the properties of NGF are derived from studies with the male mouse submaxillary protein. However, it was first observed in the extracts of two mouse "sarcomas" and subsequently, in somewhat higher amounts, in the venoms of poisonous snakes, before the mouse submaxillary source was discovered. It has also been shown to be present, in very small amounts, in a variety of tissues, which generally correlate with sympathetic innervation. More recently this has been extended to a variety of cells in culture as well.

The molecular properties of mouse submaxillary NGF have been studied in detail. The protein occurs as a noncovalent complex of three polypeptide chains, one of which, the β-subunit, possesses all of the nerve growth-promoting activity. This subunit is composed of 2 identical polypeptide chains of 13,259 mol wt. The amino acid sequence of this unit has been determined. Partial sequence data for cobra (*Naja naja*) NGF has shown that venom NGFs also have a dimeric structure and are homologous to the mouse protein. Although studies on the biosynthesis of NGF have not definitively located the physiologically significant sites of production, other than the submaxillary gland of mice, they have established that the molecule is first constructed as prohormone of 22,000 mol wt, which is subsequently processed by proteolytic modification to the 13,000 mol wt species. The significance of this cleavage in biologic terms is not known.

The biologic effects of NGF, which have been studied in detail, are concentrated on the sympathetic and portions of the embryonic sensory nervous systems. NGF exerts a positive pleiotypic activation on responsive cells resulting in stimulation of a variety of catabolic and anabolic pathways. The polymerization of microtubules is thought to be a key step in the eventual proliferation of neurites, the morphological basis for the only reliable bioassay for the hormone. An additional important effect of the hormone is the maintenance of viability of responsive neurons in vivo and in vitro. NGF also exerts a number of effects on tissues of the central nervous systems (CNS) as well as some of nonneuronal origin. The importance of these effects, which are usually different in character from peripheral neuron responses, is not presently appreciated.

Receptors for NGF have been located in both the plasma membrane and the nucleus of peripheral neurons. The plasma membrane receptors show multiphasic specific binding of ^{125}I-NGF, similar to the behavior of insulin receptors in lymphocytes. Kinetic analyses indicate that this apparent high and low affinity binding results from negatively cooperative interactions among a single type of receptor molecules. These molecules are also readily dissolved by nonionic detergents. In contrast, nuclear receptors show only high affinity binding sites and are not solubilized by detergents. Based on considerable evidence that NGF can be internalized specifically by nerve terminals and transported to the cell bodies by retrograde axonal flow, a three step mechanism for NGF involving complexation with external membrane receptors, internalization and transport to the nucleus, and complexation with nuclear receptors to directly effect transcriptional events is proposed. In keeping with the structural and functional relationships of NGF to other maintenance hormones, such as insulin, NSILA, and relaxin, this mechanism may have broad applicability.

INTRODUCTION

Nerve growth factor has been identified in a wide variety of vertebrate tissues (1–3) but with two exceptions, all contain very low concentrations and none, as yet, has been shown to be relevant to nervous system physiology as a site of synthesis. The two sources of the factor from which milligram quantities can be obtained are adult male mouse submaxillary gland (4) and the venom of the three families of poisonous snakes, Crotalidae, Viperidae, and Elapidae (5). It should be noted that both mouse submaxillary gland and snake venom gland represent true sites of synthesis in vivo and, among other sites, may even be important elements in the network of tissues that produce NGF for the maintenance of responsive neurons, primarily those found in the sympathetic nervous system. However, the greatly increased amounts found in these sources is considerably higher than the physiologically active levels of the hormone and it seems likely that these concentrations are a manifestation of other metabolic features of the gland, which results in the increased synthesis of many compounds in addition to NGF. Thus, these high concentrations are probably unrelated to the action of NGF on the nervous system and may even play an exocrine role, possibly of a vestigial nature, in these animals.

The NGF of mouse submaxillary gland has been studied in greatest detail, due primarily to its ready availability. It seems reasonable to assume that the salient features, at least of the biologically active structure, will be common to all vertebrate NGF molecules, as judged by biologic and immunologic cross-reactivity. Thus, it may be viewed as representative of the class. However, as noted below, significant structural variations may exist, particularly at the quaternary level, between the NGF molecules isolated from different species, and those isolated from different tissues within an organism, which may be of importance in regulating one or another aspect of the synthesis, transport, or mechanism of the hormone.

This review deals primarily with mouse submaxillary gland NGF. It describes the molecular properties, as they are presently understood, and the evolutionary relationships of this hormone with other growth factors. It also summarizes the biologic properties at the organismic and cellular levels, which can be integrated into a single mechanism of action involving both extracellular and intracellular receptor interactions. Several reviews dealing more explicitly with various aspects of the chemical and biologic features of NGF are available (6–13). A more general treatment of growth factors can be found in the article by Gospodarowicz & Moran (14).

MOLECULAR PROPERTIES

Mouse Submaxillary Gland

7S NGF   Mouse submaxillary NGF is isolated from gland homogenates as a high molecular weight complex containing three types of polypeptide chains, designated α, β and γ (15, 16). The stoichiometry of the complex is $\alpha_2\beta\gamma_2$ but since the β-subunit is actually a dimer of identical chains, (17, 18) at least a twofold axis of symmetry is maintained. The molecular weight of either the α- and γ- subunits is approximately 26,000 (13, 19) while the combined molecular weight of the two polypeptides of the β-subunit is 26,518 (20). Thus, the complex has a molecular weight of approximately 130,000, which yields a sedimentation coefficient of 7S (15). This value is commonly used to differentiate this high molecular weight complex from other forms of mouse NGF. The 7S complex also contains between 1 and 2 moles zinc ion per mole (21), apparently bound to the γ-subunit. Removal of the zinc does not materially affect the stability of the complex (22).

The three unique subunits of 7S NGF are associated weakly by nonconvalent forces and can be dissociated above pH 8, below pH 5, or by simple dilution (16, 22, 23). In fact, at physiologically active concentrations, the complex is fully dissociated to its constituent subunits, suggesting that the 7S form of NGF is not important as an active entity (23). The weak association of the subunits of the 7S complex has also led Murphy et al (24) to suggest that it may not occur at all in the gland but is formed as an artifact of preparation. This conclusion is based on the observation that gland homogenates directly diluted by 10^4 and measured hydrodynamically show high molecular weight material, whereas 7S NGF under the same conditions is completely dissociated (23). However, it is unclear how such a complex would be dissociated to allow the formation of 7S NGF during the isolation procedures used and it seems rather more likely that these observations are due to a nonspecific association of the β (active)-subunit with other cellular elements after the dissociation of the naturally occurring 7S complex is induced by dilution.

The three constituent polypeptide chains of 7S NGF differ sufficiently in their isoelectric point to allow their complete separation by ion-exchange chromatography. This can be accomplished after the 7S complex has been purified to homogeneity (25) or on partially purified samples (12). In the latter case, the homogenate is first fractionated on a column of Sephadex G-100 and then chromatographed on a column of CM-cellulose eluted at pH 5.0. The β-subunit is tightly bound and is eluted at the end of the salt gradient in homogeneous form. The α-subunit, with an average pI of 4.3 passes through the column unretained and the γ-subunit, with an average pI of 5.5 (26), is eluted in the initial phase of the gradient. Since both of these subunits are contaminated only by high molecular weight material, they are readily purified to homogeneity by additional gel filtration steps. The subunits obtained by either procedure can be reassociated, essentially quantitatively, to the 7S form (12, 16).

Of the three subunits, only the β-moiety possesses nerve growth promoting activity and it is solely responsible for the maintenance of adrenergic neurons in vitro. It is unclear, therefore, what role the α- and γ-subunits play in either the associated or dissociated state. This is particularly true of the α-subunit that possesses no known biologic activity. The α-subunit, on the other hand, has been clearly established to be an arginine-specific estero-peptidase of the serine protease family (27). It is inhibited by diisopropyl-fluorophosphate and shows 35% sequence identities with bovine trypsin in the 86 of 229 residues determined (28). It is also remarkably similar to the binding protein found in the high molecular weight form of EGF, which is also derived from mouse submaxillary glands (29). Sequence analysis of limited segments of this protein have indicated about 75% identities with the NGF γ-subunit (28) and they are indistinguishable with respect to catalytic and physical properties. However, they show an absolute specificity for complex formation; i.e. the γ-subunit does not bind EGF or vice versa (29). It is possible that this specificity indicates that these enzymes may play a role in processing precursors of the hormone (see below).

2.5S or β-NGF   The active subunit of the 7S complex can be prepared by several procedures that yield similar but unique products which result from limited proteolysis that occurs during the isolation (30, 31). The most commonly used preparation is that of Bocchini & Angeletti (32) in which the active subunit is isolated from partially purified 7S NGF. This material is denoted 2.5S NGF to distinguish it from the β-NGF isolated from homogeneous 7S complex. The fundamental polypeptide unit contains 118 amino acids with amino terminal serine and carboxyl terminal arginine (20). The modifications that occur result from a foreshortening of the amino terminal portion by eight residues and the removal of the carboxyl terminal arginyl residue (20, 30). Thus, the maximally modified subunit contains only 109 amino acids. If polypeptides with and without the amino terminal octapeptide are designated A and B, respectively, and the superscripts R and T are used to denote carboxyl terminal arginine or threonine (the penultimate amino acid of the 118 residue subunit), then the ten dimeric combinations of the active subunit that are possible are:

$$A^R A^R \quad A^R A^T \quad A^T A^T \quad A^R B^R \quad A^R B^T$$

$$A^T B^R \quad A^T B^T \quad B^R B^R \quad B^R B^T \quad B^T B^T$$

Although preparations highly enriched in one or more forms have been obtained, it has not been possible to obtain a single dimer type in a homogeneous state. However, preparations containing all representations have been shown to be indistinguishable in terms of biologic and immunologic activity, (12, 30, 31) which makes this heterogeneity of only secondary interest. However, the absence of the carboxyl terminal arginyl residues does prevent β-γ interactions (30). Thus, β-NGF, treated with carboxypeptidase B to remove this residue, will not form 7S complexes, although it will still combine with the α-moiety (29). Some evidence has been obtained that partial loss of this residue results in 6S complexes where only one γ-subunit is combined with a β-subunit that contains one A^T/B^T and one A^R/B^R unit (28); the interaction presumably occurs with the latter type. The amino terminal octapeptide has been synthesized by solid-phase techniques and has been shown to be devoid of NGF activity. It was also inactive in a number of pharmacologic assays (33).

The physiologically significant form of the β-subunit may in fact be the monomer. Young et al (34), have reported a binding constant, based on hydrodynamic measurements, of 10^7 M^{-1} for the association of the two polypeptides of β-NGF, which suggests that at the ng/ml concentrations required for optimal activity only monomers are present. Shooter and co-workers (13, 35), on the other hand, using entirely different methodology, have reported values of 10^{10} to 10^{11} M^{-1} for the same constant. If correct, this value would imply that the dimeric form is the biologically active one. It has been shown that covalently cross-linked β- or 2.5S NGF (36, 37) is fully active in the bioassay, while the results of Frazier et al (38), with an insolubilized NGF prepared in 6 M guanidine HCl, suggest that the monomer also can be active. Thus, the nature of the biologically active form of β-NGF is presently unclear.

The amino acid sequence of the constituent polypeptide chain of 2.5S NGF was determined to be (39, 40):

NH_2-Ser-Ser-Thr-His-Pro-Val-Phe-His-Met-Gly-Glu-Phe-Ser-Val-Cys-Asp-
Ser-Val-Ser-Val-Trp-Val-Gly-Asp-Lys-Thr-Thr-Ala-Thr-Asn-Ile-Lys-Gly-
Lys-Glu-Val-Thr-Val-Leu-Ala-Glu-Val-Asn-Ile-Asn-Asn-Ser-Val-Phe-Arg-
Gln-Tyr-Phe-Phe-Glu-Thr-Lys-Cys-Arg-Ala-Ser-Asn-Pro-Val-Glu-Ser-Gly-
Cys-Arg-Gly-Ile-Asp-Ser-Lys-His-Trp-Asn-Ser-Tyr-Cys-Thr-Thr-Thr-His-
Thr-Phe-Val-Lys-Ala-Leu-Thr-Thr-Asp-Glu-Lys-Gln-Ala-Ala-Tyr-Arg-Phe-
Ile-Arg-Ile-Asn-Thr-Ala-Cys-Val-Cys-Val-Leu-Ser-Arg-Lys-Ala-Thr-Arg-
COOH.

The six half-cystinyl residues are paired I–IV, II–V, and III–VI and link residues 15 and 80, 58 and 108, and 68 and 110 (40). The structure was independently determined by Mobley et al (41) for β-NGF and was found to be identical to that reported by Angeletti & Bradshaw (20).

Chemical modification experiments with NGF have revealed several interesting features of the molecule but have failed to localize definitively the receptor binding site(s) or any residues that make up part of it. However, inactivation of the molecule has been achieved by conversion of two or three of the three tryptophan residues to the oxindole derivative by the action of N-bromosuccinimide (42, 43) and by complete modification of the arginine residues by 1,2-cyclohexanedione in borate buffers (44). Kinetic analyses of both reactions suggest that changes in conformation rather than conversion of specific receptor residues are the cause of the loss in biologic activity. In the case of the tryptophan modification, specific complexation of the modified hormone with dorsal root neuron receptors still occurs, albeit with a lowered affinity constant (44).

Modification of lysine residues is apparently without effect on biologic activity or receptor interactions. NGF, treated with dimethylsuberimidate to effect covalent cross-linking between the polypeptide subunits of β-NGF, was fully active biologically (36, 37). Although the number of cross-links introduced was undoubtedly considerably less, virtually all the lysine residues were at least mono-substituted. Acetylation, succinylation, and attachment of NGF to CNBr-actived Sepharose beads all produced derivatives extensively or completely modified on the ϵ-amino groups without loss of biologic activity (44).

The two tyrosine residues of NGF can also be modified without alteration in the biologic properties. Conversion of these residues to the 3-nitro-derivative by the action of tetranitromethane (43) or iodination (45) yielded modified proteins that were fully competent in their interaction with receptor molecules.

Although some information is available concerning the three-dimensional structure of NGF, particularly as it relates to the evolutionarily related family of insulins and proinsulins (see below), detailed knowledge is lacking. However, Wlodawer et al (46) have recently obtained crystals of 2.5S NGF suitable for single crystal x-ray diffraction analyses. The crystals are prepared by the hanging drop–vapor diffusion method, and have a hexagonal unit cell of $P6_122$ (or its enantiomorph $P6_522$). Appropriate heavy atom derivatives have been prepared and the determination of the complete three-dimensional structure is in progress (13).

Snake Venom

Nerve growth factor has been found to be present in the venom of the three families of poisonous snakes. Nine species of *Crotalidae,* six species of *Viperidae,* and five species of *Elapidae* have been examined to varying degrees, although homogeneous samples of NGF have only been obtained from one species of each family (11). Nonetheless, based on both structural and immunologic data, true homology exists between the various forms of snake NGF and the mouse β-NGF (47, 48).

The most detailed information is available for the NGF from cobra (*Naja naja*) venom. The homogeneous preparation of NGF, which was achieved by gel filtration, CM-cellulose column chromatography, and preparative disc gel electrophoresis, has a native molecular weight of 28,000, which was reduced by 6 M guanidine HCl to about 13,000 (49). Thus, it possesses a dimeric structure analagous to the mouse 2.5S or β-NGF. No evidence of a higher molecular weight form was obtained. Partial sequence analysis indicated that the constituent polypeptide chains are very similar or identical and are composed of 116 amino acids. A total of 73 residues were directly identified of which 61% are identical to the mouse protein. If the compositions of the undetermined regions were maximally optimized, 64% of the residues would be the same in both proteins. Although the disulfide bonds were not determined, the identical position of the half-cystinyl residues in both proteins suggests a similar pattern in the venom protein.

Cobra NGF has some properties that are distinctly different from the mouse protein. First, it is much less basic in character, possessing a pI of 6.75 as compared to 9.1 for mouse 2.5S NGF (50). Second, it does not have a carboxyl-terminal arginine residue and does not interact to form a complex with either the α- or γ-subunit of the mouse 7S complex. Finally, although it shows the same dose response curve as the mouse protein, it is only about 50% as effective in the bioassay (using embryonic chick ganglia) and it will displace only 80% of the mouse ^{125}I-NGF bound to such preparations.

In marked contrast to the results obtained for cobra NGF, which establish a clear structural relationship between the snake and mouse proteins, Pearce et al (51) have reported the isolation of a viperid NGF (*Vipera russelli*) that differs significantly from both the mouse and cobra NGFs. In particular, this NGF preparation gave molecular weight values of about 35,000 and contained 20% (w/w) carbohydrate in the form of fucose, mannose, galactose, N-acetylglucosamine, and N-acetylneuraminic acid. No evidence was provided that the molecule contained more than one polypeptide chain. Interestingly, these workers (48, 52) have also reported that the NGF from the two crotalids, *B. atrox* and *A. rhodostoma,* and another viper, *V. ammodytes ammodytes,* are glycoproteins as well, although in these cases the preparations were heterogeneous. It should be emphasized that it has been clearly established that the well characterized proteins from mouse and cobra are completely devoid of carbohydrate moieties.

The NGF from a number of other crotalid, viperid, and elapid venoms has been isolated in partially purified form (11). In general, these forms of NGF appear to be of about 25,000 mol wt and show some degree of immunologic cross-reactivity with antisera raised against venom or mouse

NGFs. This includes one study (47) in which it was shown that the NGF of *V. russelli* venom cross reacts with anticobra (*N. naja*) NGF. Thus, although the presence of carboydrate in some venom NGF preparations adds a confusing and somewhat perplexing feature to the understanding of venom NGFs in general, it seems clear that venom NGFs are homologous proteins to themselves and bear the same relationship with the NGFs of higher vertebrates.

Other Tissues

Nerve growth factor has been identified in a wide variety of normal and diseased tissues by both biologic and immunologic assay (1–3, 8). These include heart, kidney, thymus, diaphragm, uterus, vas deferens, spleen, liver, and granuloma. The adrenal medulla, in organ culture, produces NGF (53) but there is disagreement on the amount (3, 54). Young and co-workers (55–59) have reported that a variety of cells in culture, including L cells, 3T3 cells, SV40 3T3 cells, primary fibroblasts, neuroblastoma cells, melanoma cells, myoblasts, glioma cells, glioblastoma cells, and primary synovial fibroblasts all produce NGF. Of particular interest is the L-cell NGF (60). Molecular weight determinations indicate a value for this form of NGF of 160,000. However, unlike the 7S complex of mouse submaxillary gland, it is not dissociated in dilute solution. It does, however, contain a subunit similar in size and electrophoretic behavior to the polypeptide chain unit of β-NGF. The same form appears to be present both intracellularly and as an exported protein in the medium. These results clearly suggest an association of the β-NGF of these cells with a different protein(s) than the α- and γ-subunit of the submaxillary gland. However, the significance of these observations is obscured by the fact that the protein was analyzed only in crude extracts or unfractionated media. It will be necessary to obtain homogeneous samples of this form of NGF before its relationship to submaxillary gland NGF and its physiologic significance can be assessed.

EVOLUTION

As has come to be appreciated for most proteins, nerve growth factor is a member of a larger family that owes its structural relatedness to evolution from a common precursor. It is interesting that, in this case, these relationships have also been instrumental in developing and defining the family itself.

The first step in the development of the concept that growth factors constitute a protein family was the observation that NGF and proinsulin share common regions of sequence relatedness (61). When aligned at the amino termini, the regions of the proinsulin molecule that yield the B and A chains of insulin, following proteolytic activation, show 30% and 52% identities, respectively, with the corresponding regions of NGF. Furthermore, these regions are appropriately spaced to accommodate the connecting C-bridge region of proinsulin even though no significant similarity is found there. The alignment is further strengthened by the fact that the disulfide bond, NGF I–IV, corresponding to the B'-19–A-20 pair in insulin, is also conserved, which adds a potential element of three-dimensional similarity as well.

The greater length of the NGF chain (118 amino acids) as compared to that of the longest proinsulin (86 amino acids) suggests a possible additional step in the evolutionary development of NGF following the divergence created by the original gene duplication. A contiguous reduplication, which produced an internal repeat of the B region following the A chain-like region in NGF, would explain the longer length of the NGF molecule. It may also have generated repeated C and A regions as well, which were then subsequently lost by further genetic events. Alternatively, this region may be still expressed in the present structural gene product but be lost by proteolysis, perhaps by the catalytic action of the γ-subunit at some postribosomal stage. As discussed below, the latter possibility seems most likely.

Evaluation of the three-dimensional structure of NGF by physical chemical measurements and solution topographical mapping indicates that the two molecules could have similar conformations, at least in the structurally conserved regions. Generally, both molecules contain low proportions of α-helical structures (62, 63). More specific measurements (43) indicate that the two tyrosine residues of NGF and tryptophan residues 21 and 76 are in similar environments, with respect to solvent availability, as the corresponding residues in insulin as deduced from the three-dimensional structure determined by X-ray analysis (64). In contrast, Argos (65), using three-dimensional prediction methodology, has concluded that the regions of NGF most likely to be in α-helical conformations do not coincide well with the known insulin structure, which suggests that the three-dimensional structure of NGF will not, in fact, resemble that of the insulins. Resolution of this problem must await results from the present crystal studies on NGF.

Recent elucidation of the primary structure of porcine relaxin (66) and the amino terminal sequence of human NSILA (67) has established the structural relatedness of these two factors to insulin as well. Relaxin is a hormone-like substance synthesized by the granulosa cells of corpora lutea which acts on tissues of the pubic symphisis to aid in the delivery of the foetus. It may play a more general role in tissue remodeling in the nonpregnant state. As isolated relaxin is a two chain structure joined by disulfide linkages, although preliminary evidence suggests the existence of a single chain prorelaxin molecule (68). NSILA has been recognized for some time as a serum factor with biologic properties similar to those of insulin. However, it is not blocked, or suppressed, by insulin antibodies. The soluble form exists as two closely related "isohormones," designated NSILA I and II (69). It contains only a single polypeptide chain of some 75 amino acids. As shown in Figure 1, the B chain of insulin and relaxin align with the amino termini of NSILA and NGF indicating the homologous nature of the four molecules. Of particular interest is the conserved cystine residue (position B20 in insulin) found in all four molecules, and the close similarities in the region immediately following it. This region can be subdivided into two categories where insulin and NSILA show marked similarities but differ from NGF and relaxin, which share the Ser-Val-Ser-X-Trp sequence. The importance of this region as a putative portion of the receptor binding site of insulin (70) adds interest to this conservation. Similar relatedness is also found in the A chain of insulin and relaxin and the corresponding region of NGF. The NSILA sequence is not complete in this region at present.

It should also be noted that several other growth factors, such as the somatomedins, A, B, and C, multiplication stimulating factor, and fibroblast growth factor may also occupy positions in this family (71), although

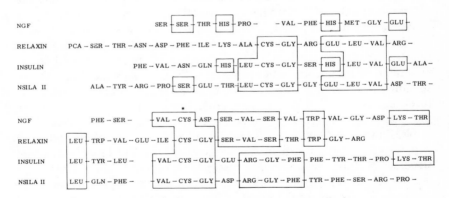

Figure 1 Sequence comparison of the amino terminal portions of mouse nerve growth factor (NGF) (20) and human nonsuppressible insulin-like activity (NSILA II) (67) with the B chains of porcine relaxin (66) and human insulin (82). Residues identical in two or more of the proteins are enclosed in boxes. Dashed spaces indicate deletions arbitrarily introduced to increase the similarity. The half cystine marked with an asterisk is bonded to the same half-cystine residue, located elsewhere in the molecule, in all four proteins.

at present insufficient structural data is available to substantiate this. Nonetheless, there exists the marked possibility that these factors share a common mechanism of action, as reflected in their ancestral development, on their respective target cells.

BIOSYNTHESIS

One of the more obscure features of the physiology of NGF is the exact in vivo site(s) of synthesis and the subsequent method of delivery to target tissues. As might be expected, studies to establish the tissue of origin have focused on the submaxillary gland of adult male mice, the richest known source of the hormone. In new born mice of either sex, the levels of hormone are low and remain so until puberty. At that time the levels in male submaxillary glands rise markedly. Such a rise can be induced in females by testosterone injections, or prevented in the male by castration (2, 72, 73). Furthermore, it has now been reasonably established that these changes are due to newly synthesized protein in situ and are not the result of increased synthesis elsewhere in the organism followed by specific uptake by the submaxillary gland (74–76). However, no other species, of either sex, contains the high level of hormone found in the male mouse nor can it be induced in them by sex hormones or by other means, neoplastic situations notwithstanding. Furthermore, there is compelling evidence to suggest that this storehouse of NGF is not utilized by the mouse, at least not to regulate the neuronal processes presently attributed to the factor. Does this mean that all mouse submaxillary gland NGF is biologically irrelevant to the mouse? The answer is probably no, although this question has not been definitively resolved. The low concentrations of hormone in the prepubescent male and the female most likely represent the "normal" concentration in that tissue. These levels are comparable to those found in other tissues that receive sympathetic innervation. Thus the submaxillary gland, as part of the network of sympathetic end organs, may be thought of as a physiologically relevant site of synthesis, because it is probably this group of organs as a whole that provides NGF to the target tissue, that is, the sympathetic nervous system.

Germane to this question are the experiments of Hendry & Iversen (77) on the effects of removal of the submaxillary gland on the circulating levels of the hormone. These authors observed that sialoadenectomy resulted in a drop in the systemic levels of the hormone, as measured by a two-site radioimmunoassay, and a subsequent return to normal levels nine weeks after the operation, without any regeneration of the excised tissue. In a similar experiment, Murphy et al (24) observed that the circulating levels of the hormone were unaffected by sialoadenectomy and concluded that the levels of systemic NGF are not controlled by the submandibular gland. Their results also differed from those of Hendry & Iverson (77) in that differences in the serum levels of male and female mice were not observed. As a result of these experiments, Murphy et al (24) suggest an exocrine role for the mouse submandibular gland with regard to NGF, which is supported by their observations that the saliva of both male and female mice contain extraordinarily high concentrations of the hormone. Wallace & Partlow (78) have reported a similar finding, but only after preferential stimulation of the gland by α-adrenergic agents. Whether salivary NGF simply represents disposed material or whether it has a direct physiologic role in the digestive process is not presently appreciated.

The relationships of submaxillary and blood NGF in the mouse is further confused by the lack of agreement on the normal levels of serum NGF. Of particular interest in this regard is the observation of Ishii & Shooter (79) that injected ^{125}I-NGF is cleared from mouse serum in vivo with a half-time of 10–30 min. Although it cannot be ruled out that this value is affected by the prior iodination of the hormone, it seems reasonable to conclude, based on the fact that this modification does not affect any other biologic or immunologic behavior, that NGF displays a rather short lifetime in serum. This raises the question whether there is *any* appreciable quantity of circulating NGF in the mouse, or any other organism. Attempts to isolate the hormone from the blood have been singularly unproductive and recent, improved analyses (H. Thoenen, personal communication) have found that normal circulating levels of the hormone are immeasurably low. The absence of serum NGF would be consistent with the proposal that NGF is a "diffusion" hormone that is not transported by the blood system. As such, submaxillary NGF, at the prepubescent or female level, would affect only sympathetic neurons innervating that organ, and its excision in the adult animal would not materially affect other elements of the peripheral nervous system, a conclusion which has already been substantiated (6). The much higher levels in the male would not alter this primary role.

By virtue of the amounts produced, the submaxillary gland is the primary producer of NGF in adult male mouse. Since it seems reasonable to assume that this increased activity is basically an amplification of the normal process, studies on the biosynthesis of NGF in that tissue should have general applications. Levi-Montalcini and Angeletti (75), using slices of adult male mouse submaxillary gland, demonstrated the incorporation of labeled amino acids into material that was immunologically precipitable with anti-NGF. They also showed a differential incorporation of labeled amino acids into the immunoprecipitate in one of the two lobes of the gland in vivo.

Berger & Shooter (80), using [^{35}S]-L-cystine, have demonstrated the incorporation of this amino acid by submaxillary glands in vitro into material that is specifically precipitated by anti-NGF. This material was shown to have a tryptic peptide profile, with respect to the migration of half-cystine containing peptides, indistinguishable from that of native NGF. Similar results were obtained in vivo.

When the labeled products were examined as a function of time (10–240 min) by SDS-gel electrophoresis, Berger & Shooter (81) observed that a 22,000 mol wt precursor appeared first, followed by the production of the 13,000 mol wt polypeptide chain of β-NGF. The amount of 22,000 mol wt material reached a maximum at 1 hr and remained constant, while the amount of β-NGF (as 13,000 mol wt units) rose rapidly. After four hours, it was present in a tenfold greater amount and after six hours the larger material had essentially disappeared. The 22,000 mol wt precursor was shown to contain all of the half-cystine peptides of β-NGF and it could be converted to the 13,000 mol wt unit by the action of either NGF γ-subunit or the EGF binding protein.

These results clearly demonstrate the biosynthesis of a proNGF species containing some 80 additional amino acids (as judged by molecular weight differences). Although it remains to be established whether these residues are removed from the amino or carboxyl terminal or both, it is interesting that the comparison of NGF and proinsulin (see above) resulted in the prediction of an additional carboxyl-terminal segment that would be susceptible to the catalytic action of the γ-subunit (61). It has not been established whether proNGF is biologically competent or if limited proteolysis is a prerequisite for the production of active hormone. It is also unknown whether the structural gene for NGF contains a region coding for a preprohormone segment. In related systems, the removal of this peptide from the amino terminal of the nascent chain ensures the proper "packaging" of the hormone for ultimate export from the cell of origin (82).

BIOLOGIC PROPERTIES

The physiologic effects of NGF have been extensively reviewed (2, 6, 7, 9) and are summarized here only as a prerequisite for mechanistic considerations.

The initial appreciation of the striking morphologic effects of NGF was actually the basis for the discovery and subsequent isolation of the hormone. In experiments involving the transplantation of sarcomas 37 and 180 into mice, it was observed that the sympathetic nervous system of the host was stimulated to innervate the tumor (83–86). The transfer of this growth-stimulating ability to an in vitro culture using check embryonic ganglia provided the basis for the bioassay still used today (86). This assay also demonstrated that two types of peripheral neurons will respond, i.e. extend neurites, in culture: the adrenergic neurons of sympathetic ganglia and the mediodorsal sensory neurons of dorsal root ganglia (6). In fact, NGF is absolutely required to maintain both cell types in vitro. In vivo, it is unsure whether or not these sensory neurons have a direct dependence on the factor. However, the requirement for NGF of sympathetic neurons is clearly maintained throughout the lifetime of the animal. Perhaps the most compelling demonstration of this fact was the experiment by Levi-Montalcini & Booker (87) which showed that injection of anti-NGF into neonatal animals causes the destruction of the sympathetic nervous system. This procedure, commonly used as a pharmacologic model (88), is termed immunosympathectomy. Positive effects can also be demonstrated by the injection of NGF into similar animals, most notably in increased levels of tyrosine hydroxylase and dopamine-β-hydroxylase, enzymes involved in the biosynthesis of catecholamines (89). Morphologic changes, such as the increase in cell volume and innervation of sympathetic end organs, also occur (6). The permanancy of these effects is, however, questionable.

The stimulation of embryonic neurons to produce neurites and the requirement of adult neurons for maintenance of viability can be translated into defined effects on the metabolic processes of these cells. In simplest terms, NGF may be viewed as a trophic agent that acts as a pleiotypic activator of various anabolic and catabolic pathways (90). Among other effects, NGF causes: increases in uridine uptake and polysome formation; protein, RNA, and lipid synthesis; and glucose utilization (6, 91, 92). It also causes microtubule (neurotubule) polymerization and in a limited way, can act as a mitogenic factor (6). As already noted, some of these effects can be translated into very specific events such as the stimulation of the two enzymes of catecholamine biosynthesis.

NGF can also act as a tropic agent to attract growing axons to their target tissues. Although it is often difficult to separate such effects from trophic actions, evidence from experiments involving transplantation of iris into the brain (93) and the unnatural position of sympathetic nerve fibers after intracerebral injection of NGF (9, 94) clearly support this facet of NGF action.

A variety of effects of NGF on cells not normally classed as responsive to NGF have been reported. In addition to effects on noradrenergic fibers in brain (95–97), it also inhibits the biosynthesis of mucopolysaccharide in chondrocytes (98) and stimulates the temporal conversion of cell surface adhesive specificity in embryonic optic tectal cells (99). It has also been reported that NGF directly affects an adenyl cyclase in the chromaffin granules of adrenal medulla (100). Three types of neoplastic tissues of neural crest origin, neuroblastoma (101), melanoma (102), and pheochromocytoma (103) have, in specific lines, been reported to bind ^{125}I-NGF. In the last case, neurite proliferation, induced by the hormone, was also observed. Finally, specific receptors for NGF have been observed in brain and a number of sympathetic end organs (104, 105). In brain, the receptors are apparently largely associated with synaptosomal elements (106, 107). The significance of any of these observed activities to the presence of specific plasma membrane receptors in unresponsive tissues cannot be readily explained. These situations may result from artifacts induced by in vitro conditions, may represent analog activities of other closely related but as yet unidentified growth factors, or may reflect true physiologic activities of the hormone not yet appreciated.

MECHANISM OF ACTION

As with other polypeptide hormones, much more is known of the molecular and biologic properties of NGF than is known of its mechanism of action. Nonetheless, several aspects have been clarified recently that offer distinct clues as to the nature of the overall process. In general, they suggest an initial interaction with a plasma membrane receptor followed by internalization and complexation with a second type of receptor, which is located in the nucleus. Since direct effects on cellular metabolism occur as the result of hormone binding to both recognitive entities, the mechanism of NGF can be viewed as biphasic with information transfer occurring at two levels. The salient evidence in support of the three main steps and a plausible model to accommodate them are summarized below.

Plasma Membrane Receptors

The first direct identification of cell surface receptors for NGF on responsive neurons was made by Frazier et al (38). In these experiments, NGF was attached to Sepharose beads that had been activated with BrCN. The coupling reaction was carried out in 6M guanidine-HCl which effectively prevented subsequent release of soluble NGF from the resin, presumably because of the formation of multiple attachment sites between the denatured protein and the carbohydrate matrix. The NGF-Sepharose conjugates were found to be active as judged by their ability to elicit neurite proliferation. Thus, it was concluded that it was not necessary for NGF to enter cells to mediate this phase of its biologic activity, although these experiments did not rule out that NGF could be internalized. It was also noted that the possibility of controlled cleavage of the protein (and/or matrix) after complexation of the conjugate with receptor could not be eliminated. If such material were directly taken into the cell it would not be detected in the diffusion controls. This limitation is germane in view of the subsequent demonstration that NGF can be internalized (108, 109), *vide infra*, and a recent report claiming that a fragment of NGF, generated by limited tryptic digestion, possesses full biologic activity (110).

The characteristics of the plasma membrane receptor for NGF have been investigated using ^{125}I-labeled hormone (45, 111, 112). In all studies, high affinity specific binding, defined as labeled NGF that can be displaced by an excess of unlabeled hormone, was observed. In one study (112), low affinity binding ($\sim 10^6 1/\text{mol}$ as opposed to $\sim 10^{10}/\text{mol}$) was observed at higher concentrations of added ^{125}I-NGF. The curvilinear Scatchard analysis of this binding data was similar to that observed for the binding of ^{125}I-insulin to cultured lymphocytes (113). The high affinity asymptote corresponded well to the concentration range required to maximally produce neurites.

Kinetic analyses of the association and dissocation of the hormone and receptor gave constants of $7.5 \times 10^6 \text{ M}^{-1} \text{ sec}^{-1}$ and $3.8 \times 10^{-4} \text{ sec}^{-1}$, at 24° C, respectively (112). Most interesting was the observation that an excess of native NGF greatly accelerated the dissociation process. This deviation from the law of mass action had been previously observed by De Meyts *et al* (113) for the dissociation of labeled insulin from lymphocytes and was attributed by these workers to negatively cooperative interactions between receptor molecules. Thus, the similarity between insulin and NGF appears to extend to molecular features of their receptors as well. Similar observations have now been made for a number of hormone receptor systems (114), which suggests negative cooperativity as a common, although not universal, feature of cell-hormone interactions. The potential advantage of this property in regulating hormonal response have been discussed elsewhere (10, 115).

The plasma membrane receptor for NGF has been solubilized by nonionic detergents from both sensory (116) and sympathetic (117) ganglia. The solubilized preparations can be assayed for specific binding using gel filtration or precipitation by polyethylene glycol to separate unbound from complexed hormone. The affinity constant of the solubilized receptor from rabbit superior cervical ganglia was found to be $5 \times 10^9 \text{liter/mol}$, in excellent agreement with the value for the high affinity binding in the membrane bound state, which indicates that the solubilization process does not appear to materially affect the receptor molecule (117).

Internalization (Retrograde Axonal Transport)

The most extensive documentation of the internalization of NGF in responsive neurons is found in the experiments of Hendry, Thoenen, and their
co-workers (108, 118–123). By injection of microgram quantities into the
anterior chamber of the eye, they demonstrated that small amounts of
hormone were specifically taken up by the sympathetic terminals innervating this organ and transported intraaxonally to the perikaryia located in the
ipsilateral superior cervical ganglia. The contralateral ganglia, which serves
as an internal control, contained significantly less radioactivity. In more
recent experiments utilizing labeled NGF of much higher specific activity
(112), this phenomenon has been confirmed using only nanogram quantities
with virtually no radioactivity reaching the contralateral ganglia (E. M.
Johnson, R. Y. Andres, and R. A. Bradshaw, unpublished experiments). In
these latter experiments, maximal accumulation was seen at 12 hr in hamsters and 16 hr in rats with the uptake system half-saturable at 15 ng of
injected hormone.

The uptake process was abolished by transection of the postganglionic
fibers and by colchicine injections, 12 hr prior to the administration of the
labeled hormone. The process was also shown to be specific for NGF.
Cytochrome c, insulin, horseradish peroxidase, ovalbumin, bovine serum
albumin, and ferritin were taken up to an extremely small extent (118, 119).
Furthermore, inactivation of NGF by oxidation of two or more tryptophan
residues, which destroys biologic activity (43), also prevented the uptake
process which suggests that only the active species can be properly recognized by the receptors responsible for the specific uptake. This was confirmed by the demonstration that excess amounts of unlabeled NGF
effectively block the uptake of iodinated hormone.

Although the nature of the uptake process at the presynaptic membrane
is unclear, it may be presumed that a pinocytotic mechanism, which results in
the vesicularization of the internalized hormone, probably occurs. This is
supported by electron microscopic autoradiography of superior cervical
ganglia after retrograde transport of ^{125}I-NGF. The transported material is
localized in vesicles attached to smooth endoplasmic reticulum and in
secondary lysosomes (123). In addition, as judged by radioactive analysis
of subcellular fractions of similarly treated superior cervical ganglia, 15-30
percent of the NGF is found in nuclear fractions (E. M. Johnson, R. Y.
Andres, and R. A. Bradshaw, unpublished experiments).

Of primary importance in the studies on retrogradily transported NGF
was the demonstration that hormone specifically transported in superior
cervical neurons causes specific increase in the levels of tyrosine hydroxy-

lase (120). The induction of this enzyme, which occurs after administration of a few micrograms of NGF to intact sympathetic neurons (89), can be demonstrated to occur with internalization of only picograms of the hormone.

Stoeckel et al (124) have also demonstrated that dorsal root neurons will transport NGF injected into the forepaw of adult rats, in 10 days. There were, however, significant differences between this process and that in the sympathetic neurons, which may be explained by the fact that these neurons appear to lose their responsiveness to NGF postembryonically (125). Norr & Varon (126) have also reported a time-linear uptake (pinocytosis) of NGF by chick embryonic dorsal root neurons.

The clear demonstration of the uptake and retrograde transport of NGF by mature neurons still leaves the question of whether these processes occur during embryonic development, before axonal extension and synapse formation is complete. The possibility remains that this internalization occurs only as a feature of mature neurons which have a special need to communicate between remote termini and their cell bodies.

Nuclear Receptors

Although Triton X-100 will readily solubilize the plasma membrane receptors of embryonic dorsal root neurons, a significant amount of specific NGF binding is still found in the insoluble pellet (116). Following subcellular fractionation, this binding was localized in the nuclear fraction. Microsomal contamination was eliminated by reconstitution experiments, and purified chromatin, prepared from these nuclei, showed identical binding properties (116).

In contrast to the plasma membrane receptors, the nuclear receptors showed only saturable high affinity binding with an association constant of approximately 5×10^9 liter/mol (116). Binding to these receptors is not inhibited by 1000-fold molar excesses of insulin, cytochrome c, and lysozyme. The highest levels are found in nuclei derived from sympathetic and sensory ganglia, with only relatively low amounts in avian erythrocytes or optic tectum cells. Interestingly, nuclear receptors for insulin have been reported in lymphocytes and liver cells with properties essentially identical to those found for NGF (127, 128).

A Mechanistic Model

Any model for the action of NGF must take into account its known biologic properties. Restricting these considerations to cells generally classified as responsive, that is sympathetic and embryonic sensory neurons, these include the uptake of metabolites, stimulation of anabolic metabolism, the polymerization of microtubules, the specific induction of enzyme synthesis, the proliferation of neurites, and the long-term maintenance of cell viability both in vitro and in vivo. These effects do not all occur on the same time scale. Metabolite uptake is relatively rapid and occurs in the first few hours of incubation, whereas neurite proliferation cannot be observed until 12 hr and does not reach maturity until 36 to 48 hr. Long term maintenance, by definition, must be judged over several days. It is interesting to note that this somewhat extended period of response is also typical of other growth factors, including insulin, and is markedly distinct from the response that is approximately proportional to receptor occupancy. It is in fact possible to subdivide hormonal agents into two general classes based on these characteristics (129). One class elicits essentially instantaneous responses, often through the generation of cyclic nucleotide second messengers, is rapidly degraded at the plasma membrane, and the duration of the effect is approximately proportional to the actual occupancy of the receptor. The second class, which have been termed maintenance or permissive hormones, show extended periods of response that last far longer than actual receptor occupancy, are degraded slowly, usually by lysosomal action, and ultimately effect protein biosynthesis. It seems clear that the primary hormones act externally exclusively through complexation with plasma membrane receptors and their associated effectors. On the other hand, growing evidence suggests that the second class, of which NGF is representative, act through a combination of events, initiated by plasma membrane receptor complexation and terminated inside the cell by combination with nuclear receptors (10). This mechanism (130) can be described by the following steps:

1. Complexation of the hormone with plasma membrane receptors, which results in metabolite uptake and the stimulation of anabolic and catabolic pathways. Subsequent events include the polymerization of tubulin to microtubules which initiates the second step.
2. Vesicularization of the hormone-receptor complex, perhaps engineered by microtubules, with the resultant internalization of the hormone.
3. Transport of the vesicle to and fusion with internal membrane structures. As a secondary phase of this process, vesicles can also be directed to lysosomes, in which case degradation of the hormone ensues.
4. Transfer of the hormone to the nucleus, followed by complexation with nuclear receptors to directly effect ongoing transcriptional events.

This mechanism finds support in many experimental observations on NGF. The presence of plasma membrane and nuclear receptors, and the capacity of neurons to internalize NGF have already been described. This mechanism also accounts for the established fact that neurite proliferation can occur without RNA synthesis but cell maintenance cannot (6). The polymerization of microtubules is probably sufficient to initiate fiber outgrowth, a process that is extensively initiated before occupation of the nuclear receptors occurs. In this regard, the observation that NGF can directly affect microtubule polymerization may indicate another role for the internalized hormone (131, 132). This model also predicts that a loss of plasma membrane receptors will occur during the transfer process. Although "down regulation," the decrease of receptors due to chronic exposure of cells to hormone has not been demonstrated for NGF, it has been clearly shown for insulin (133) and epidermal growth factor (134). It is also noteworthy that this mechanism is closely similar to that of steroid hormones (135). Although these substances do not have plasma membrane receptors they do have cytoplasmic binding proteins that serve a similar function. The subsequent binding of the steroid-receptor complex to a chromatin-bound receptor would be entirely analogous to the binding of "maintenance" hormones to their nuclear receptors. It should be noted that although this model accounts for the many facets of NGF activity, detailed information of many of the proposed steps is lacking. As with other similar hormones, such as insulin, it is unclear how the complexation with the plasma membrane receptor causes the uptake of metabolites and the resultant microtubule polymerization. It is equally unclear how the vesicle transport and transfer of the hormone to the nucleus is accomplished. Finally, the manner in which the formation of the nuclear receptor complex allows changes in RNA transcription is unknown.

In additional to these problems, many other questions remain to be answered before this mechanism can be integrated into a description of NGF action in the organism as a whole. Definitive localization of the sites of NGF synthesis must be made and the factors that regulate that synthesis determined. It will also be necessary to determine the mode of transport between the sites and the target neurons. However, the model also provides some insight into these questions. Specifically, it suggests that tissues receiving sympathetic innervation synthesize and export the hormone, which reaches the embryonic neuron by a diffusion process. Uptake occurs at the plasma membrane of the growing neuron and as neurites are extended toward this source, the uptake continues but with an additional internal transport phase. The terminal stage occurs when the growing axon has reached the end organ where the diffusion process is limited to a transsynaptic passage. In this way, only a single mechanism of action is required that accommodates the dramatic morphologic changes that occur in the target tissue during development. This mechanism also provides these neurons with a direct means for communicating between synapse and cell body, i.e. NGF becomes the "chromalytic messenger" for cells responsive to it. Although much experimental verification is required, this model provides an excellent working hypothesis for further research into the nature of NGF and other growth factors.

ACKNOWLEDGMENTS

The portions of this review dealing with studies in the author's laboratory were supported by US Public Health Service research grant NS 10229. During the preparation of this article the author was a Josiah Macy Jr. Foundation Faculty Scholar at the Howard Florey Institute of Physiology and Experimental Medicine, University of Melbourne.

Literature Cited

1. Bueker, E. D., Schenkein, I., Bane, J. L. 1960. *Cancer Res.* 20:1220–28
2. Levi-Montalcini, R., Angeletti, P. U. 1961. *Regional Neurochemistry,* ed. S. S. Kety, J. Elkes, pp. 362–76. New York: Pergamon
3. Hendry, I. A. 1972. *Biochem. J.* 128:1265–72
4. Cohen, S. 1960. *Proc. Natl. Acad. Sci. USA* 46:303–11
5. Cohen, S. 1959. *J. Biol. Chem.* 234:1129–37
6. Levi-Montalcini, R., Angeletti, P. U. 1968. *Physiol. Rev.* 48:534–69
7. Zaimis, E., ed. 1972. *Nerve Growth Factor and its Antiserum* London: Athlone. 273 pp.
8. Bradshaw, R. A., Young, M. 1976. *Biochem. Pharm.* 25:1445–49
9. Levi-Montalcini, R. 1976. *Progress in Brain Research,* ed. M. A. Corner, D. F. Swaab, 45:235–56. Amsterdam: Elsevier
10. Bradshaw, R. A., Frazier, W. A. 1977. *Current Topics in Cellular Regulation,* ed. B. L. Horecker, E. R. Stadtman, 12:1–37. New York: Academic
11. Hogue-Angeletti, R. A., Bradshaw, R. A. 1977. *Snake Venoms,* ed. C. Y. Lee. In *Handbook of Experimental Pharmacology.* Berlin: Springer-Verlag. In press
12. Jeng, I., Bradshaw, R. A. 1978. *Research Methods in Neurochemistry,* Vol. 4. ed. N. Marks, R. Rodnight. New York: Plenum. In press
13. Server, A. C., Shooter, E. M. 1978. *Adv. Prot. Chem.* In press
14. Gospodarowicz, D., Moran, J. S. 1976. *Ann. Rev. Biochem.* 45:531–58
15. Varon, S., Nomura, J., Shooter, E. M. 1967. *Biochemistry* 6:2202–9
16. Varon, S., Nomura, J., Shooter, E. M. 1968. *Biochemistry* 7:1296–1303
17. Angeletti, R. H., Bradshaw, R. A., Wade, R. D. 1971. *Biochemistry* 10:463–69
18. Greene, L. A., Varon, S., Piltch, A., Shooter, E. M. 1971. *Neurobiology* 1:37–48
19. Stach, R. W., Server, A. C., Pignatti, P.-F., Piltch, A., Shooter, E. M. 1976. *Biochemistry* 15:1455–61
20. Angeletti, R. H., Bradshaw, R. A. 1971. *Proc. Natl. Acad. Sci. USA* 68:2417–20
21. Pattison, S. E., Dunn, M. F. 1975. *Biochemistry* 14:2733–39
22. Baker, M. E. 1975. *J. Biol. Chem.* 250:1714–17
23. Pantazis, N. J., Murphy, R. A., Saide, J. D., Blanchard, M. H., Young, M. 1977. *Biochemistry* 16:1525–30
24. Murphy, R. A., Saide, J. D., Blanchard, M. H., Young, M. 1977. *Proc. Natl. Acad. Sci. USA* 74:2672–76
25. Varon, S., Nomura, J., Perez-Polo, J. R., Shooter, E. M. 1972. *Methods in Neurochemistry,* ed. R. Fried, 3:203–29. New York: Dekker
26. Varon, S., Shooter, E. M. 1970. *Biochemistry of Brain and Behavior,* ed. R. E. Bowman, S. P. Datta, pp. 41–64. New York: Plenum
27. Greene, L. A., Shooter, E. M., Varon, S. 1968. *Proc. Natl. Acad. Sci. USA* 60:1383–88
28. Silverman, R. E. 1977. *Interactions within the mouse nerve growth factor complex.* PhD thesis. Washington Univ., St. Louis, Missouri. 128 pp.
29. Server, A. C., Shooter, E. M. 1976. *J. Biol. Chem.* 251:165–73
30. Moore, J. B. Jr., Mobley, W. C., Shooter, E. M. 1974. *Biochemistry* 13:833–40
31. Mobley, W. C., Schenker, A., Shooter, E. M. 1976. *Biochemistry* 15:5543–52
32. Bocchini, V., Angeletti, P. U. 1969. *Proc. Natl. Acad. Sci. USA* 64:787–94
33. Hogue-Angeletti, R. A., Bradshaw, R. A., Marshall, G. R. 1974. *Int. J. Peptide Prot. Res.* 6:321–28
34. Young, M., Saide, J. D., Murphy, R. A., Arnason, B. G. W. 1976. *J. Biol. Chem.* 251:459–64
35. Moore, J. B. Jr., Shooter, E. M. 1975. *Neurobiology* 5:369–81

36. Stach, R. W., Shooter, E. M. 1974. *J. Biol. Chem.* 249:6668–74
37. Pulliam, M. W., Boyd, L. F., Baglan, N. C., Bradshaw, R. A. 1975. *Biochem. Biophys. Res. Commun.* 67:1281–89
38. Frazier, W. A., Boyd, L. F., Bradshaw, R. A. 1973. *Proc. Natl. Acad. Sci. USA* 70:2931–35
39. Angeletti, R. H., Mercanti, D., Bradshaw, R. A. 1973. *Biochemistry* 12:90–100
40. Angeletti, R. H., Hermodson, M. A., Bradshaw, R. A. 1973. *Biochemistry* 12:100–15
41. Mobley, W. C., Moore, J. B. Jr., Schenker, A., Shooter, E. M. 1974. *Mod. Prob. Pediat.* 13:1–12
42. Angeletti, R. H. 1970. *Biochim. Biophys. Acta* 214:478–82
43. Frazier, W. A., Hogue-Angeletti, R. A., Sherman, R., Bradshaw, R. A. 1973. *Biochemistry* 12:3281–93
44. Bradshaw, R. A., Jeng, I., Andres, R. Y., Pulliam, M. W., Silverman, R. E., Rubin, J., Jacobs, J. W. 1977. *Endocrinology: Proc. Vth Int. Cong. Endocrinol.*, ed. V. H. T. James, 2:206–12. Amsterdam: Excerpta Medica
45. Herrup, K., Shooter, E. M. 1973. *Proc. Natl. Acad. Sci. USA* 70:3884–88
46. Wlodawer, A., Hodgson, K. O., Shooter, E. M. 1975. *Proc. Natl. Acad. Sci. USA* 72:777–79
47. Angeletti, R. H. 1971. *Brain Res.* 25:424–27
48. Bailey, G. S., Banks, B. E. C., Pearce, F. L., Shipolini, R. A. 1975. *Comp. Biochem. Physiol.* 51B:429–38
49. Hogue-Angeletti, R. A., Frazier, W. A., Jacobs, J. W., Niall, H. D., Bradshaw, R. A. 1976. *Biochemistry* 15:26–34
50. Server, A. C., Herrup, K., Shooter, E. M., Hogue-Angeletti, R. A., Frazier, W. A., Bradshaw, R. A. 1976. *Biochemistry* 15:35–39
51. Pearce, F. L., Banks, B. E. C., Banthorpe, D. V., Berry, A. R., Davies, H. ff. S., Vernon, C. A. 1972. *Eur. J. Biochem.* 29:417–25
52. Glass, R. E., Banthorpe, D. V. 1975. *Biochim. Biophys. Acta* 405:23–26
53. Harper, G. P., Pearce, F. L., Vernon, C. A. 1976. *Nature* 261:251–53
54. Johnson, D. G., Gorden, P., Kopin, I. J. 1971. *J. Neurochem.* 18:2355–62
55. Oger, J., Arnason, B. G. W., Pantazis, N., Lehrich, J., Young, M. 1974. *Proc. Natl. Acad. Sci. USA* 71:1554–58
56. Young, M., Oger, J., Blanchard, M. H., Asdourian, H., Amos, H., Arnason, B. G. W. 1975. *Science* 187:361–62
57. Murphy, R. A., Pantazis, N. J., Arnason, B. G. W., Young, M. 1975. *Proc. Natl. Acad. Sci. USA* 72:1895–98
58. Arnason, B. G. W., Oger, J., Pantazis, N. J., Young, M. 1974. *J. Clin. Invest.* 53:2a
59. Saide, J. D., Murphy, R. A., Canfield, R. E., Skinner, J., Robinson, D. B., Arnason, B. G. W., Young, M. 1975. *J. Cell. Biol.* 67:376a
60. Pantazis, N. J., Blanchard, M. H., Arnason, B. G. W., Young, M. 1977. *Proc. Natl. Acad. Sci. USA* 74:1492–96
61. Frazier, W. A., Angeletti, R. H., Bradshaw, R. A. 1972. *Science* 176:482–88
62. Bradshaw, R. A., Frazier, W. A., Angeletti, R. H. 1972. *Chemistry and Biology of Peptides* ed. J. Meienhofer, pp. 423–439. Ann Arbor: Ann Arbor Sci. Publ.
63. Frank, B. H., Veros, A. J. 1968. *Biochem. Biophys. Res. Commun.* 32:155–60
64. Blundell, T. L., Cutfield, J. F., Cutfield, S. M., Dodson, E. J., Dodson, G. G., Hodgkin, D. C., Mercola, D. A., Vijayan, M. 1971. *Nature* 231:506–11
65. Argos, P. 1976. *Biochem. Biophys. Res. Commun.* 70:805–11
66. James, R., Niall, H. D., Kwok, S., Bryant-Greenwood, G. 1977. *Nature* 267:544–46
67. Rinderknecht, E., Humbel, R. E. 1976. *Proc. Natl. Acad. Sci. USA* 73:4379–81
68. Frieden, E. H., Yeh, L. A. 1976. *Fed. Proc.* 35:1628
69. Rinderknecht, E., Humbel, R. E. 1976. *Proc. Natl. Acad. Sci. USA* 73:2365–69
70. Pullen, R. A., Lindsay, D. G., Wood, S. P., Tickle, I. J., Blundell, T. L., Wollmer, A., Krail, G., Brandenburg, D., Zahn, H., Gliemann, J., Gammeltoft, S. 1976. *Nature* 259:369–73
71. Megyesi, K., Kahn, C. R., Roth, J., Neville, D. M., Nissley, S. P., Humbel, R. E., Froesch, E. R. 1975. *J. Biol. Chem.* 250:8990–96
72. Caramia, F., Angeletti, P. U., Levi-Montalcini, R. 1962. *Endocrinology* 70:915–22
73. Levi-Montalcini, R., Angeletti, P. U. 1964. *Salivary Glands and Their Secretions,* ed. L. M. Screebny, J. Meyer, pp. 129–41. Oxford: Pergamon
74. Burdman, J. A., Goldstein, M. N. 1965. *J. Exp. Zool.* 160:183–88
75. Levi-Montalcini, R., Angeletti, P. U. 1968. *Growth of the Nervous System,* ed. G. E. W. Wolstenholme, M. O'Connor, pp. 126–47. London: Churchill
76. Weis, P., Bueker, E. D. 1966. *Proc. Soc. Exptl. Biol. Med.* 121:1135–40

77. Hendry, I. A., Iversen, L. L. 1973. *Nature* 243:500–4
78. Wallace, L. J., Partlow, L. M. 1976. *Proc. Natl. Acad. Sci. USA* 73:4210–14
79. Ishii, D. N., Shooter, E. M. 1975. *J. Neurochem.* 26:843–51
80. Berger, E. A., Shooter, E. M. 1977. *Dev. Biol.* In press
81. Berger, E. A., Shooter, E. M. 1977. *Proc. Natl. Acad. Sci. USA* 74:3647–51
82. Steiner, D. F. 1977. *Diabetes* 26:322–340
83. Bueker, E. D. 1948. *Anat. Record,* 102:369–90
84. Levi-Montalcini, R., Hamburger, V. 1951. *J. Exp. Zool.* 116:321–62
85. Levi-Montalcini, R., Hamburger, V. 1953. *J. Exp. Zool.* 123:233–88
86. Levi-Montalcini, R., Meyer, H., Hamburger, V. 1954. *Cancer Res.* 14:49–57
87. Levi-Montalcini, R., Booker, B. 1960. *Proc. Natl. Acad. Sci. USA* 42:384–91
88. Steiner, G., Schönbaum, E., eds. 1972. *Immunosympathectomy* Amsterdam: Elsevier 253 pp.
89. Thoenen, H., Angeletti, P. U., Levi-Montalcini, R., Kettler, R. 1971. *Proc. Natl. Sci. USA* 68:1598–1602
90. Hershko, A., Mamont, P., Shields, R., Tomkins, G. M. 1971. *Nature New Biol.* 232:206–11
91. Partlow, L. M., Larrabee, M. G. 1971. *J. Neurochem.* 18:2101–18
92. Horii, Z. I., Varon, S. 1976. *Brain Res.* 124:121–33
93. Bjorklund, A., Bjerre, B., Stenevi, U. 1974. *Dynamics of Regeneration and Growth in Neurons* ed. K. Foxe, L. Gison, Y. Zetterman, pp. 389–409. Oxford: Pergamon
94. Levi-Montalcini, R., Chen, M. G. M., Chen, J. S. 1974. Presented at Ann. Meet. *Soc. Neurosci., 4th,* p. 305
95. Bjerre, B., Bjorklund, A., Stenevi, U. 1973. *Brain Res.* 60:161–76
96. Bjerre, B., Bjorklund, A., Stenevi, U. 1974. *Brain Res.* 74:1–18
97. Stenevi, U., Bjerre, B., Bjorklund, A., Mobley, W. 1974. *Brain Res.* 69:217–34
98. Eisenbarth, G. S., Drezner, M. K., Lebovitz, H. E. 1975. *J. Pharmacol. Exp. Ther.* 192:630–34
99. Merrell, R., Pulliam, M. W., Randono, L., Boyd, L. F., Bradshaw, R. A., Glaser, L. 1975. *Proc. Natl. Acad. Sci. USA.* 72:4270–74
100. Nikodijevic, O., Nikodijevic, B., Zinder, O., Yu, M.-Y. W., Guroff, G., Pollard, H. B. 1976. *Proc. Natl. Acad. Sci. USA.* 73:771–74
101. Revoltella, R., Bertolini, L., Pediconi,
M., Vigneti, E. 1974. *J. Exp. Med.* 140:437–51
102. Fabricant, R. N., DeLarco, J. E., Todaro, G. J. 1977. *Proc. Natl. Acad. Sci. USA.* 74:565–69
103. Tischler, A. S., Greene, L. A. 1975. *Nature* 258:341–42
104. Frazier, W. A., Boyd, L. F., Szutowicz, A., Pulliam, M. W., Bradshaw, R. A. 1974. *Biochem. Biophys. Res. Commun.* 57:1096–1103
105. Frazier, W. A., Boyd, L. F., Pulliam, M. W., Szutowicz, A., Bradshaw, R. A. 1974. *J. Biol. Chem.* 249:5918–23
106. Szutowicz, A., Frazier, W. A., Bradshaw, R. A. 1976. *J. Biol. Chem.* 251:1516–23
107. Szutowicz, A., Frazier, W. A., Bradshaw, R. A. 1976. *J. Biol. Chem.,* 251:1524–28
108. Hendry, I. A., Stoeckel, K., Thoenen, H., Iversen, L. L. 1974. *Brain Res.* 68:103–21
109. Burnham, P. A., Varon, S. 1973. *Neurobiology* 3:232–45
110. Mercanti, D., Butler, R., Revoltella, R. 1977. *Biochim. Biophys. Acta* 496:412–19
111. Banerjee, S. P., Snyder, S. H., Cuatrecasas, P., Greene, L. A. 1973. *Proc. Natl. Acad. Sci. USA.* 70:2519–23
112. Frazier, W. A., Boyd, L. F., Bradshaw, R. A. 1974. *J. Biol. Chem.* 249:5513–19
113. De Meyts, P., Roth, J., Neville, D. M. Jr., Gavin, J. R. III, Lesniak, M. A. 1973. *Biochem Biophys. Res. Comm.* 55:154–61
114. De Meyts, P. 1976. *Surface Membrane Receptors: Interface Between Cells and Their Environment* ed. R. A. Bradshaw, W. A. Frazier, R. C. Merrell, D. I. Gottlieb, R. A. Hogue-Angeletti, pp. 215–26. New York: Plenum
115. De Meyts, P., Bianco, A. R., Roth, J. 1976. *J. Biol. Chem.* 251:1877–88
116. Andres, R. Y., Jeng, I., Bradshaw, R. A. 1977. *Proc. Natl. Acad. Sci. USA.* 74:2785–89
117. Banerjee, S. P., Cuatrecasas, P., Snyder, S. H. 1976. *J. Biol. Chem.* 251:5680–85
118. Hendry, I. A., Stach, R., Herrup, K. 1974. *Brain Res.* 82:117–28
119. Stoeckel, K., Paravicini, U., Thoenen, H. 1974. *Brain Res.* 76:413–21
120. Paravicini, U., Stoeckel, K., Thoenen, H. 1975. *Brain Res.* 84:279–91
121. Stoeckel, K., Schwab, M., Thoenen, H. 1975. *Brain Res.* 99:1–16
122. Stoeckel, K., Guroff, G., Schwab, M., Thoenen, H. 1976. *Brain Res.* 109:271–84

123. Schwab, M., Thoenen, H. 1977. *Brain Res.* 122:459–74
124. Stoeckel, K., Schwab, M., Thoenen, H. 1975. *Brain Res.* 89:1–14
125. Herrup, K., Shooter, E. M. 1975. *J. Cell. Biol.* 67:118–25
126. Norr, S. C., Varon, S. 1975. *Neorobiology* 5:101–18
127. Goldfine, I. D., Smith, G. J. 1976. *Proc. Natl. Acad. Sci. USA* 73:1427–31
128. Goldfine, I. D., Smith, G. J., Wong, K. Y., Jones, A. L. 1977. *Proc. Natl. Acad. Sci. USA.*, 74:1368–72
129. Robison, G. A., Butcher, R. W., Sutherland, E. W. 1971. *Cyclic AMP.* New York: Academic. 531 pp.

130. Andres, R. Y., Bradshaw, R. A. 1977. *Biochemistry of Brain* ed. S. Kumar. Oxford: Pergamon. In press
131. Calissano, P., Cozzari, C. 1974. *Proc. Natl. Acad. Sci. USA.* 71:2131–35
132. Levi, A., Cimino, M., Mercanti, D., Chen, J. S., Calissano, P. 1975. *Biochim. Biophys. Acta.* 399:50–60
133. Gavin, J. R. III, Roth, J., Neville, D. M. Jr., De Meyts, P., Buell, D. N. 1974. *Proc. Natl. Acad. Sci. USA.* 71:84–88
134. Carpenter, G., Cohen, S. 1976. *J. Cell. Biol.* 71:159–71
135. Chan, L., O'Malley, B. W. 1976. *New Eng. J. Med.* 294:1322–28

Ann. Rev. Biochem. 1978. 47:217–49
Copyright © 1978 by Annual Reviews Inc. All rights reserved

RIBOSOME STRUCTURE

◆973

R. Brimacombe, G. Stöffler, and H. G. Wittmann

Max-Planck-Institut für Molekulare Genetik, Berlin-Dahlem, Germany

CONTENTS

PERSPECTIVES AND SUMMARY

The translation of mRNA into protein on the ribosome is a universal process that has been the object of intensive study for many years. This process, in which many complex molecules take part, occurs on ribosomes

217

0066-4154/78/0701-0217$01.00

in a number of steps that can be studied in vitro. In order to perform such an intricate function, the ribosome itself is necessarily a complex structure, consisting of many different protein and RNA molecules. Furthermore, it has become clear during the last few years that the individual functional steps in protein synthesis cannot be attributed to single ribosomal components, but rather that parts of several different components are involved at every stage. This realization has lent a new urgency to the problem of elucidating the structure of the ribosomes, for without a good understanding of the structure we cannot now expect to gain a more detailed insight into the way in which the ribosome functions.

The most widely studied species of ribosome, and the one with which this review is primarily concerned, is that of *Escherichia coli.* The obvious advantages of studying a prokaryotic ribosome are that it is easily obtainable, that it is less complicated than its eukaryotic counterparts, and that genetic studies with mutants can be carried out. Like all ribosomes, that of *E. coli* consists of two subunits. The smaller (30S) subunit contains 21 different proteins and a 16S RNA molecule, whereas the larger (50S) subunit contains 34 proteins in addition to a 5S and a 23S RNA species.

Clearly a structure of this degree of complexity cannot possibly be determined by any single experimental approach, thus research into bacterial ribosome structure has been pursued, roughly speaking, at three levels. The first of these, which has recently tended to decline in importance, is concerned with the study of general properties of the intact ribosomes. The second is the study of the structures of the isolated ribosomal components, in particular the analysis of the amino acid and nucleotide sequences of the proteins and RNA molecules. The third area of research is that of ribosomal topography, or how the various ribosomal components are arranged relative to one another in space.

In this article, we first deal with the ribosomal proteins, starting with a review of the recent rapid progress in the determination of the individual amino acid sequences. Next we consider the question of secondary and tertiary structures of the isolated proteins, before moving on to the topographical arrangement of the proteins within the ribosome. Here, a number of new techniques have been applied during the last few years, most notably immune electron microscopy, that allows the direct visualization of protein antigenic sites on the ribosome surface, and which has already yielded a great deal of information. Physicochemical techniques are also becoming increasingly important in this area, and all of these new methods have led to the important conclusion that many ribosomal proteins have elongated or irregular shapes within the ribosomal subunits. The latest developments in the more classical approaches to the topographical question, such as chemical cross-linking of protein pairs, are also discussed.

Following this, we briefly review the current status of the primary and secondary structure of the rRNA, and then describe some developments in the field of ribosome reconstitution. Here the most notable advances are the successful reconstitution of the *E. coli* 50 S subunit, and the demonstration that most if not all of the ribosomal proteins (at least from the 30S subunit) are capable of interacting with RNA. This leads naturally to a discussion of RNA topography and RNA-protein interactions, with particular reference to the recent application of RNA-protein cross-linking techniques to this problem. Finally, we include a short discussion of the evolutionary aspects of ribosome structure, by comparing the *E. coli* ribosome with ribosomes from other organisms.

RIBOSOMAL PROTEINS

Properties of Isolated Proteins

ISOLATION   The proteins from the small subunit (named S1 to S21 according to their positions on a two-dimensional electrophorerogram) as well as those from the large subunit (correspondingly named L1 to L34) have all been isolated by chromatography on cellulose ion-exchangers and by filtration through Sephadex in the presence of 6 M urea (reviewed in 1). Proteins prepared in this manner have been used mainly for immunological (2) and amino acid sequence studies (3). However, since the proteins are partly denatured by high concentrations of urea (4–6), efforts have been made during the last few years to extract and purify the ribosomal proteins under "native" conditions, that is in the complete absence of either urea or acetic acid (4, 5). Proteins isolated in this way are used for studies on their shape and other physical properties, such as secondary structure (see below).

PRIMARY STRUCTURE   The complete amino acid sequences of 33 proteins have so far been determined (Tables 1 and 2), and studies on all other *E. coli* ribosomal proteins are in progress. In all, about two thirds of the 8000 amino acids present in the *E. coli* ribosome have so far been sequenced, and this knowledge is sufficient to allow conclusions to be drawn concerning the degree of homologous structure among the ribosomal proteins. Apart from two pairs of proteins, namely L7/L12 and S20/L26, which have, respectively, almost identical or identical primary sequences (2, 3), the extent of homology is no greater than would be expected on a random basis (7). In other words the various proteins within the *E. coli* ribosome are as unrelated as completely different proteins such as haemoglobin, lysozyme, or ribonuclease.

SPATIAL STRUCTURE   Based on the amino acid sequences just described, predictions of secondary structure have been made for all those ribosomal proteins whose primary structures have been completely determined (8–11). Corroborative information concerning the secondary structure is becoming available from circular dichroism studies on proteins isolated both in the

Table 1   Properties and mutational alterations of 30S ribosomal proteins

Protein	Amino acids[a]	Molecular weight[b]	References	Shape[c]	References	References for mutants
S1				elong.	(54–56)	
S2				elong.	(57, 58)	(16, 19, 71–73)
S3				glob.	(59)	(17–19)
S4	203	23,138	(21)	elong.	(59–63)	(16–19, 74–86)
S5				elong.	(59, 64)	(17–19, 81, 85–104)
				glob.	(57)	
S6	135	15,704	(22)	sl. el.	(59)	(16, 17, 19, 22, 105)
S7				elong.	(58, 59)	(16, 17, 19)
S8	109	12,196	(23)	glob.	(56, 57, 65)	(16, 17, 19, 106–108)
S9	129	14,569	(24)			(19)
S10						(16–19, 109)
S11				elong.	(58)	(19, 104)
S12	123	13,606	(25)	elong.	(58)	(19, 84, 96, 103, 110–113)
S13	117	12,969	(26)			(17, 19, 104)
S14						(17, 19, 104)
S15	87	10,001	(27)	elong.	(58)	(16, 17, 19)
				sl. el.	(56)	
S16	82	9,191	(28)	glob.	(56)	(16, 17)
S17	83	9,573	(29)			(17, 19, 103, 114–116)
S18	74	8,897	(30)	elong.	(58)	(16, 19, 88, 104, 117)
S19	91	10,299	(31)	elong.	(66)	(17, 19, 124)
S20	86	9,554	(32)	elong.	(56, 59)	(19, 98, 104, 106, 118)
S21	70	8,369	(33)			(19)

[a] This column denotes the number of amino acids in those proteins whose sequences have been determined completely.
[b] Molecular weights are calculated from the sequence data.
[c] elong. = elongated; sl. el. = slightly elongated; glob. = globular.

Table 2 Properties and mutational alterations of 50 S ribosomal proteins[a]

Protein	Amino acids	Molecular weight	Reference	Shape	Reference	References for mutants
L1				elong.	(67)	(16–19, 86)
L2				elong.	(67)	
L3						(16–19)
L4						(19, 89, 119–122)
L5	178	20,171	(34)			(16–19, 123)
L6	176	18,832	(35)	elong.	(67)	(16–19, 124)
				sl. el.	(65)	
L7	120	12,207	(36)	elong.	(14, 15)	(17–19, 104, 125)
L8						
L9						(19, 104)
L10	164	17,736	(37, 38)			(16, 19)
L11	141	14,874	(39)	sl. el.	(58, 68)	(17–19, 126)
L12	120	12,165	(36)	elong.	(14, 15)	(17–19, 104, 125)
L13						(17, 19)
L14						(16–19)
L15	144	14,981	(40)			(16, 17, 19)
L16	136	15,296	(41)			(18, 19)
L17						(16–19)
L18	117	12,770	(42)	elong.	(58, 69)	(16, 17, 19, 127)
L19	114	13,003	(43)			(16–19, 128)
L20				elong.	(67)	
L21						(19)
L22						(16, 17, 19, 121, 122)
L23						(16–19)
L24				elong.	(70)	(16, 17, 19, 129)
L25	94	10,694	(44, 45)	elong.	(69)	(19, 104)
L26[b]	86	9,554				(19, 104)
L27	84	8,994	(46)			(19)
L28	77	8,876	(47)			(19)
L29	63	7,273	(48)			(16, 18, 19)
L30	58	6,411	(49)			(16, 18, 19, 107)
L31	62	6,971	(50)			
L32	56	6,315	(51)			(19, 104)
L33	54	6,255	(52)			(18, 19)
L34	46	5,381	(53)			

[a] See footnotes to Table 1.
[b] L26 is identical to S20.

presence or absence of urea (I. Heiland, M. Dzionara, and G. Snatzke, in preparation; J. Littlechild, J. Dijk, and C. A. Morrison, in preparation), and in the case of the protein pair L7/L12, not only the secondary but also the tertiary structure has been predicted (12). Since this protein pair has recently been crystallized (13), it will be of great interest to see how far this predicted structure agrees with that based on the current X-ray analysis. In addition, hydrodynamic methods and small-angle X-ray scattering studies both indicate that L7/L12 has an elongated shape (14, 15). A relatively large number of ribosomal proteins have been shown to be similarly elongated by these latter methods, as well as by neutron scattering and immune electron microscopy (see section on protein topography), whereas a few proteins appear to be globular (see Tables 1 and 2). Information concerning the internal structure of the proteins, in particular the interaction between hydrophobic and aromatic groups within the molecules, has been forthcoming from nuclear magnetic resonance studies of proteins isolated in both the presence and absence of urea (6). This method revealed that most proteins isolated in the presence of urea show less internal structure than their counterparts isolated under nondenaturing conditions.

MUTANTS   Many mutants with altered ribosomal proteins have recently been isolated (16–19). As listed in Tables 1 & 2, mutants are now available for almost all of the ribosomal proteins, the only exceptions being those proteins that do not show up very clearly on two-dimensional electrophorerograms. These mutants serve as very useful tools for (a) genetic studies, e.g. to establish the location and fine structure of individual ribosomal protein genes and the regulatory mechanism controlling their expression; (b) functional investigations, e.g. to ask which ribosomal functions are impaired or modified by the mutation in question; (c) protein-chemical studies, e.g. to analyze the amino acid replacements in the mutant proteins, with a view to clarifying the importance of the protein region concerned (3, 20); and (d) gaining further insight into ribosome biogenesis in vivo.

TOPOGRAPHY OF RIBOSOMAL PROTEINS

Immune Electron Microscopy

Ribosomal protein topography, or the spatial arrangement of the proteins within the subunits, is obviously of central importance in the correlation of ribosomal structure with function. Different techniques have been applied to this problem, but the most dramatic impact during the last few years has been made by immune electron microscopy (reviewed in 20). The aim of this technique is to bind a purified IgG-antibody, specific to a single ribosomal protein, to the appropriate ribosomal subunit; the bivalent antibody forms dimers between two subunits that can then be examined under the electron microscope. The location of the bound antibody on the subunit surface can be determined, and the position of an antigenic determinant of a particular protein can thus be directly visualized.

The method relies on the facts that specific IgG-antibodies are able to react with cognate proteins within the intact ribosomal subunits, and that both subunits have readily discernible shapes with recognizable morphological landmarks. The application of immune electron microscopy has advanced at an increasing pace during the last five years, and locations of all 21 proteins from the 30S subunit and 19 proteins from the 50S subunit have already been published (2, 20, 58, 61, 64, 67, 130–135). Preliminary locations have been established for the remaining 50S proteins (G. W. Tischendorf and G. Stöffler, unpublished results). Data on the localization of ribosomal components by immune electron microscopy have also been forthcoming from three other research groups (60, 66, 136–137), and there are a number of discrepancies in the data. Since most of these arise from alternative interpretations of the subunit shapes, we give a brief description of the structural features of both subunits and 70S ribosomes, as viewed under the electron microscope.

STRUCTURE OF RIBOSOMES AND SUBUNITS   30S subunits, negatively stained with uranyl acetate, reveal structures (axial ratio ~2:1) that resemble, figuratively speaking, an embryo. The particles are divided along their longer axis by a hollow or cleft ("neck") which separates the "head" from the "body" (see Figure 1). The subunits are seen in several views, some showing highly asymmetric structures, others pseudosymmetric outlines. Two lobes of unequal size protrude from the body of the subunit, and these lobes break the apparent symmetry of the subunit. Each of the distinctive views can be explained in terms of different projections of a unique 30S structure.

Electron micrographs published by the various investigators (58, 64, 138–141) show a great deal of similarity in their gross morphology, but nevertheless very different three-dimensional models have been proposed. The 30S models of Vasiliev (140) and Tischendorf et al (58, 64) are quite compatible, especially if the different methods for contrasting are taken into account. On the other hand the model of Lake (141) is more asymmetric, since it proposes a single narrow platform instead of two lobes, and is, in addition, flatter in the quasisymmetric views.

50S subunits are also seen in several forms, resembling crowns, maple leaves, or kidneys with an asymmetric notch (58, 131, 132, 138, 141–143). A statistical analysis of the distribution of the various forms has been reported (132), and again each of the views can be interpreted as a projec-

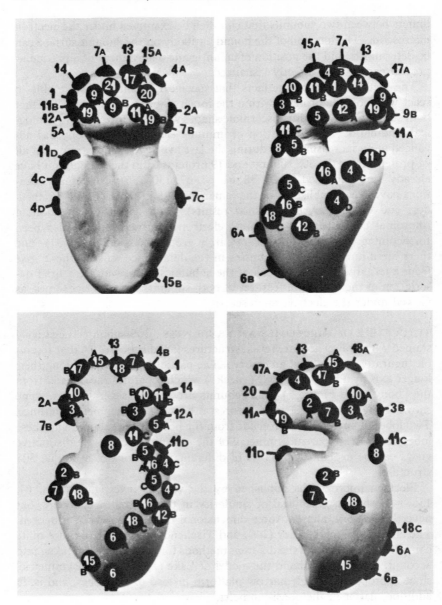

Figure 1   Three-dimensional model of the 30S subunit, with the locations of the centers of antibody binding sites for the 21 ribosomal proteins in 4 views (derived by successive rotation of the model by 90°).

tion of a unique structure, which in the 50S case is an "arm-chair-like" form with three protuberances, the central protuberance being the most prominent (see Figure 2). A similar model has been proposed by Boublik et al (136), and in both these models (132, 136) the central protuberance is in a different plane from the two lateral ones. In contrast, the model of Lake (141) places all three protuberances in the same plane. In addition, the thin appendage, which is only observed in 10–15% of the subunits by Tischendorf et al (58, 132–135), and by B. Tesche and G. W. Tischendorf (unpublished results) and which is not seen at all by Boublik (136), is interpreted by Lake (141) as one of the three main protuberances. In the 70S ribosome, the small subunit is arranged horizontally transverse with respect to the 50S subunit (58, 67), and a channel is seen between the two subunits. A similar model is described by Boublik et al (144), but again the model of Lake is in disagreement (141).

A new technique for tungsten shadowing (145) of freeze-dried ribosomal particles is proving very helpful in resolving all these differences, and the resulting resolution of approximately 6 Å has provided electron micrographs (B. Tesche and G. W. Tischendorf, unpublished results) that confirm the models proposed for the subunits by Tischendorf et al and Boublik (20, 58, 144; Figures 1, 2).

LOCATION OF ANTIGENIC SITES   It has been known for some years that the antisera that can be obtained against individual ribosomal proteins are highly specific, and that no ribosomal protein will react with any other than its cognate antiserum, with the exception of the two protein pairs L7/12 and S20/L26 [(2, 146, 147); see section above on protein structure]. As an additional precaution, the antibodies have been purified by affinity chromatography on cyanogen bromide activated agarose columns to which the cognate antigen is covalently bound (61); this ensures the removal of any contaminating antibodies.

Using such purified antibodies, Tischendorf et al (58, 61, 64) have located the antigenic sites of all twenty-one 30S proteins by immune electron microscopy, with the following results (see Figure 1). Antisera to six proteins, namely S4, S7, S11, S12, S15, and S18, revealed multiple antibody binding sites at widely separated points on the 30S subunit surface, and this can only be interpreted as an indication that the proteins concerned have highly extended conformations in situ. A further group of five proteins (S2, S3, S5, S10, S19) showed two antibody binding sites, each separated by 50–80 Å, and while these results could be compatible with slightly extended conformations [see discussion of S5 in (64)], it must be remembered that parts of the protein chain may be buried within the subunit, and consequently inaccessible to antibodies. In other words, immune electron microscopy

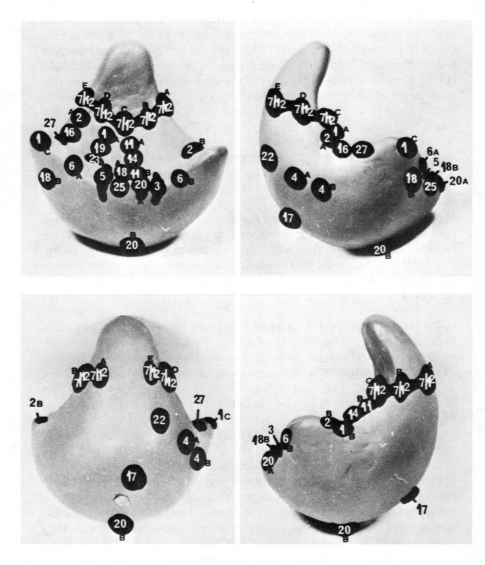

Figure 2   Three-dimensional model of the 50S subunit, indicating the centers of the antibody binding sites of 19 proteins.

gives a minimum estimate of the number of elongated proteins; in addition further antibody binding sites may well be found by using antisera raised in different animals [(see 20, 61) for more detailed discussion].

Antisera to the remaining 30S proteins showed either a single binding site (in the case of S1, S8, S13, S14, S20, and S21), or two sites located in close proximity to each other (S6, S9, S16, and S17). These proteins may be globular, but, as mentioned above, this cannot be concluded from the electron microscope data alone (cf the hydrodynamic data in Tables 1 and 2). So far, a total of 43 discernible antibody sites have been found on the 30S subunit (Figure 1). Most of these sites occur on the head of the subunit, and while there are several sites on the neck and the body, there is a notable lack of any site in the hollow between the head and the body. This could be the result of steric factors, and current experiments with monovalent antibody fragments (Fabs) may help to resolve this problem [(61), G. W. Tischendorf and G. Stöffler, unpublished results].

Six 30S proteins (i.e. S4, S5, S11, S13, S14, and S19) have been localized in a similar manner by Lake and his colleagues (60, 66, 141). The distribution of the proteins is in reasonably good agreement with the data of Tischendorf et al (see Figure 1), the main discrepancies between the two sets of data being attributable, as already discussed, to the different interpretation of the three-dimensional shape of the 30S subunit. It should also be noted here that immune electron microscopy has also been applied to the localization of a specific region of the 16S RNA (137); in this study, the 6-N,N dimethyl adenine residue near the 3'-terminus of 16S RNA was located on the "platform" of the 30S subunit, according to the model of Lake (141).

The more regular structure of the 50S subunit renders more difficult an exact placement of a given antibody site on the three-dimensional model. Nonetheless, a three-dimensional map of the locations of 19 of the thirty-four 50S proteins has been established (20, 58, 67, 130–132, 134, 135), and this is shown in Figure 2. The 50S proteins are more asymmetrically arranged than those of the 30S subunit, most of the antibody sites being clustered within the "crowned seat" region, and preliminary locations of the remaining 15 proteins are also predominantly in this region (G. W. Tischendorf & G. Stöffler, unpublished results). Six proteins (L1, L2, L6, L11, L18, and L20) each showed antibody attachment sites at points that were sufficiently remote from each other to be incompatible with globular shapes. Just as with the corresponding 30S proteins, these must therefore have elongated conformations in situ.

Proteins L7/L12 also showed multiple antibody binding sites, but since these proteins are present in the subunit in several copies (for review see 20), the existence of multiple sites alone is not indicative of an elongated conformation. The location of L7/L12 has been investigated by three different groups; Boublik et al (136) located these proteins on the two lateral protuberances, whereas Lake (148) found them on the rod-like appendage, and in contrast Tischendorf et al (58, 133–135, and see Figure 2) found multiple sites on the central protuberance. The high-resolution tungsten shadowcast technique already mentioned should help to resolve these differences.

Protein-Protein Cross-linking

The first direct measurements of neighborhoods between ribosomal proteins were made with the aid of bifunctional cross-linking reagents, and a variety of such reagents has been developed over the last few years (e.g. 149–153). In this approach, intact ribosomal subunits are treated with a suitable protein-specific reagent, the proteins are then extracted, and the components of the cross-linked protein pairs identified. The method used for this identification depends on whether or not the bifunctional reagent can be chemically cleaved following the reaction. In the case of noncleavable reagents, protein identification has mostly involved the use of Ouchterlony double diffusion tests with antisera to the individual ribosomal proteins (154), whereas in the case of cleavable reagents, the cross-linked protein pair can be split and the component proteins identified by gel electrophoresis (150, 151).

Up to now, a total of 34 pairs of cross-linked proteins from the 30S subunit have been clearly identified, and these are detailed in Table 3. Several pairs, such as S5–S8, S7–S9, S6–S18, and S13–S19, could be cross-linked with a wide variety of reagents, and further the cross-linked pairs were isolated in high yield [23–53% (151)]. It follows, despite the fact that isolated ribosomal subunits are heterogeneous and that only a small proportion of them are functionally active, that these protein pairs are almost certainly neighbors in active 30S subunits. However, this has only been shown directly in two cases, i.e. S5–S8 and S13–S19, where it was possible to reconstitute an active 30S ribosome from the cross-linked protein pair, the remaining 19 proteins, and 16S RNA (155). It should also be noted from Table 3 that the distribution of cross-linked pairs is by no means uniform; three proteins, S4, S8, S13, have each been cross-linked to eight different proteins, whereas other proteins only appear in one or two cross-linked pairs.

If the protein cross-linking data are compared with the results of immune electron microscopy (Figure 1), good agreement is found only if the elongated nature of the proteins is taken into account. The latter observation also explains some apparent contradictions with other results, notably the finding from energy-transfer experiments (see section below) that the mass centers of S14 and S19 are far apart (156, 157), despite the fact that they

Table 3 Summary of protein cross-linking data in the ribosomal subunits[a]

Protein	Neighbors	Factors
S1	S2[b], S13[b], S15 or S17[b], S18[b]	IF-2, IF-3
S2	S1[b], S3, S5, S8	IF-2
S3	S2, S4, S5, S9, S10, S12	
S4	S3, S5, S6, S8, S9, S12, S13[c], S17, S20[d]	
S5	S2, S3, S4, S8, S9, S13	
S6	S4, S8, S18[c]	
S7	S8, S9[c], S13	IF-3[g]
S8	S2, S4, S5, S6, S7, S11, S13, S15[c]	
S9	S3, S4, S5, S7[c]	
S10	S3	
S11	S8, S13	IF-2, IF-3[g]
S12	S3, S4, S13, S20, S21	IF-2, IF-3[g]
S13	S1[b], S4[c], S5, S7, S8, S11, S12, S17, S19[c], S20[d]	IF-2, IF-3
S14	S19[e]	IF-2
S15	S1[b], S8[c]	
S17[f]	S1[b], S4, S13	
S18	S1, S6[c], S21	IF-3[g]
S19	S13[c], S14[e]	IF-2, IF-3
S20	S12, S4[d], S13[d]	
S21	S12, S18	IF-3
L2	L5[h], L9, L7/L12, L10[h], L11[h], L17[h]	
L3	L5[h]	
L4	L11[h], L14[h]	
L5	L7/L12, L17[h], L23[h], L24[h], L25[h]	
L10	L11	
L11	L14[h]	
L17	L21[h], L32	
L18	L32[h]	
L22	L32[h]	

[a] For the 30S subunit, the results are cross-referenced for each protein, and data for initiation factors IF-2 and IF-3 are included. Comparisons with fluorescence measurements (see text) are indicated by appropriate footnotes. Except where indicated, the data are taken from the summaries (cited in 20, 150, 151, 156, 157). For the 50S subunit the data are not cross-referenced, and again, except where indicated, are compiled from the summaries (cited in 159, 160).

[b] A. Sommer and R. R. Traut (unpublished results).

[c] Results in agreement with fluorescence data.

[d] Results based on fluorescence data alone.

[e] Result in apparent contradiction to fluorescence data.

[f] Measurements made with a mixture of S16 and S17 are not included.

[g] K. Johnston, J. W. B. Hershey and R. R. Traut (unpublished results).

[h] J. W. Kenny and R. R. Traut (unpublished results).

can be cross-linked (Table 3). The fact that immune electron microscopy has shown S19 to be elongated resolves this disagreement, and the same argument can be used in reverse to postulate that S13 and S20 must also be elongated, although up to now only single antibody sites have been found for these proteins (compare Figure 1 and Table 3). It follows further from these considerations that attempts to build models of the protein arrangement without taking into account the elongated nature of the proteins (e.g. 158) are not very useful.

In those cases where a cross-linked pair of proteins can be isolated in high yield, it should be possible to locate the precise position of the cross-link within the respective amino acid sequences of the proteins. This is of particular importance if the proteins concerned are elongated, and in one instance the precise point of cross-linkage has already been deduced; proteins S18 and S21 were cross-linked with a sulfhydryl reagent specific for cysteine (154), and since each protein possesses only a single cysteine residue, it can be concluded that the cross-link was between Cys-10 in S18 (30) and Cys-22 in S21 (33).

A considerable number of pairs of cross-linked proteins have been observed in the 50S subunit (159, 160), and these are also included in Table 3. However, there are as yet insufficient data from other methods such as immune electron microscopy (see section above) or neutron scattering (see section below) to allow much meaningful discussion.

Neutron Scattering

The principle of the neutron scattering method, as applied to the study of ribosomal protein topography (161–164), is that 30S ribosomal subunits in which two of the 21 proteins are deuterated are prepared by in vitro reconstitution. By making neutron scattering measurements on such particles, the distance between the centers of mass of the two deuterated proteins can be deduced. One advantage of a scattering technique is that not only does it measure properties of the whole particle (and not just the particle surface), but it enables the activity of the ribosome preparation to be tested at the end of the experiment, thus reducing the danger of artifact.

Distances between the centers of mass of a number of pairs of 30S proteins have so far been measured by neutron scattering (161, 162, 164; J. A. Langer, D. M. Engelman, and P. B. Moore, personal communication) with the following results:

29 Å (S7–S9); 35 Å (S5–S8); 43 Å (S4–S5); 45 Å (S3–S9); 55 Å (S3–S5); 61 Å (S3–S4); 62 Å (S4–S8); 77 Å (S3–S8); 85 Å (S4–S9); 86 Å (S4–S7); 89 Å (S5–S9); 96 Å (S5–S6); 102 Å (S5–S7); 105 Å (S2–S5); 110 Å (S8–S9); 115 Å (S3–S7); 100–150 Å (S6–S8); 100–150 Å (S8–S10).

A further analysis of the data suggested that most, if not all, of these proteins must have axial ratios significantly deviating from unity, with S5 and S8 being the only two that appear, from the current results, to be globular. The data agree well with the results from immune electron microscopy, both with respect to the distances (Figure 1) and to the shape deductions (Tables 1 and 2). There is an apparent contradiction concerning the shape of protein S5. The electron microscope data show it to be elongated (58, 64), whereas neutron scattering experiments show it to be globular. On the other hand the short distance (35 Å) between the centers of mass of S5 and S8 is in good agreement with both electron microscopy and protein cross-linking results (see sections above), while the close proximity found between proteins S9 and S7 by neutron scattering is difficult to reconcile with the distribution of antigenic sites (cf. Figure 1).

Similar neutron scattering studies on the 50S proteins are not so far advanced; nevertheless one interprotein distance, between L7/L12 and L10, was estimated to be 100 Å (163).

Neutron scattering has also been used to determine the separation of the centers of mass of RNA and protein within the subunits. It appears that the proteins and RNA are fairly uniformly distributed in the 30S subunit (165), but there is some disagreement as to the magnitude of the displacement of the mass centers of RNA and protein in the 50S subunit (165–167). It is, however, likely that the 50S subunit has a core of high scattering density (RNA) surrounded by a low density shell of protein (166, 167), although it is not yet clear how this observation can be reconciled with the notable lack of antibody binding sites in the lower half of the 50S subunit (see section on immune electron microscopy).

Fluorescence

Another physicochemical technique that has made a significant contribution to protein topography is the study of energy transfer between fluorescently labelled protein pairs. The principle of the method is that two proteins are labelled each with a different fluorescent dye, and after reassembly in vitro into subunits the efficiency of energy transfer from one dye to the other is measured. This approach was introduced by Cantor and his colleagues (156, 157), and gives a measure of the relative distance between the centers of mass of the proteins concerned. The experiments, thus, are capable not only of detecting protein neighborhoods, but also (as with neutron scattering and immune electron microscopy) of indicating which proteins are far apart.

Several 30S proteins, i.e. S4, S6, S7, S8, S9, S13, S14, S15, S18, S19, and S20, have been labelled with fluorescent dyes (156, 157), and distance measurements have been described for 14 selected pairs. (These do not include measurements made with a mixture of S16 and S17) (156, 157). Five out of seven of these pairs, which were found to be in close proximity, have also been cross-linked (Table 3), whereas two pairs of neighbors, S4–S20 and S13–S20, have so far only been detected by the energy transfer method. Also, out of the seven protein pairs that were shown to be located at remote positions, six are absent from the cross-linking data (cf. Table 3). These pairs are S4–S15, S4–S18, S4–S19, S15–S19, S15–S18, and S19–S20; the only apparently contradictory result, namely the pair S14–S19, has already been discussed above (see section on protein-protein cross-linking). In another similar study, the distance between the binding site of the antibiotic erythromycin and proteins L7/L12 in the 50S subunit was measured (168), and the interaction of a fluorescent streptomycin derivative with *E. coli* ribosomes has also been described (169). Continuation of these experiments should contribute important information, particularly if it proves possible to attach the fluorescent label to a specific amino acid residue in the protein.

All of the techniques described above, namely immune electron microscopy, protein-protein cross-linking, neutron scattering and fluorescence measurements, provide complementary or overlapping data and should form the basis for a concerted attack on the protein topography problem.

RIBOSOMAL RNA

Primary, Secondary, and Tertiary Structure

As with the amino acid sequences of the ribosomal proteins, a large amount of information is already available concerning the nucleotide sequences of the *E. coli* ribosomal RNA species. The total sequence of 5S RNA (120 nucleotides) has been known for many years (170), and a great deal of work has been done on the sequence of the 16S RNA (1600 nucleotides); a partial sequence of this molecule was first published in 1972 (171), and more recently a modified version of the sequence has appeared (172). There are, however, still uncertainties in the ordering of a number of the ribonuclease T_1 digestion products (172), as well as a measure of disagreement as to the precise sequence of several oligonucleotides (173). Substantial progress (174–176) has also been made on the sequence of the 23S RNA molecule (3200 nucleotides).

To understand how the RNA molecules are fitted into the ribosome, it is important to have an idea of both their secondary and tertiary structure, and unfortunately there is as yet no reliable way of predicting such structures for long nucleic acid molecules. More than twenty secondary structures have so far been proposed for 5S RNA (for review see 177), and while a secondary structure has been put forward as a working hypothesis for 16S

RNA (172), and a corresponding structure for regions of 23S RNA (175, 176), there is at the moment very little evidence either for or against these structures.

Experimental investigations of the secondary structure are complicated by the finding that, at least in the case of 16S RNA, the nucleic acid conformation shows a significant dependence on the method of RNA isolation used (see section below on reconstitution). Moreover, it appears that the helical content of the ribosomal RNA is lower within the ribosomal subunits than in the isolated RNA molecules (178), and also changes upon going from a compact to an "unfolded" conformation induced by EDTA treatment, as indicated by optical rotatory dispersion measurements (reviewed in 179). Circular dichroism studies showed that the protein conformations are not altered to any significant extent by this "unfolding" process (179), but it is known that the unfolding allows random exchange of the ribosomal proteins to occur (180). Further, recent studies on the binding of isolated proteins to ribosomal RNA (see section below on reconstitution) suggest strongly that the proteins are able to influence the secondary and/or tertiary structure of the RNA, and all these results, taken together, indicate that maintenance of the correct RNA structure is vitally important for the integrity of the subunit structure as a whole.

Physicochemical studies of the overall properties of the RNA, such as those just described, inevitably only lead to general conclusions, but some studies have been made of specific secondary structural features of the ribosomal RNA within the intact subunits. These experiments involve treatment with kethoxal, a reagent that attacks nonpaired guanine residues (181), followed by identification of the precise positions, within the nucleic acid sequence, which have reacted (182). The results showed that the kethoxal-accessible sites on 16S RNA within the subunits are clustered along the RNA in a few distinct groups, but a considerable proportion of the guanine residues concerned are not in single-stranded regions of the RNA, according to the predicted secondary structure (172).

When 70S ribosomes were the target for kethoxal attack (183), the reactivity was altered as compared to that found in the 30S subunit alone. This suggested that a pronounced conformational change takes place in the 30S subunit upon 70S formation, since although some groups of kethoxal sites on the 16S RNA became less sensitive, others became more sensitive. These data corroborate the results of protein modification studies, where both lactoperoxidase-catalyzed iodination of 30S proteins (184) and N-ethyl-maleimide modification (185) led to a similar conclusion. The accessibility of guanine residues in 5S RNA within the 50S subunit has also been investigated using kethoxal (186, 187); only two sites were found to be attacked.

In addition to these secondary structural investigations, some specific tertiary features of the 16S RNA have recently been described, involving stable interactions that have been observed between regions of the RNA that are widely separated in the primary sequence (see section on RNA-protein interactions). This suggests that the topographical arrangement of the RNA within the subunits is likely to involve a variety of tertiary RNA-RNA interactions, perhaps analogous to those found in transfer RNA (188). In support of this, specific RNA-RNA cross-links can be generated between remote regions of the 16S RNA in the 30S subunit by irradiation with ultraviolet light (C. Zwieb and R. Brimacombe, unpublished results).

Functional Importance of RNA

Complementary interactions between regions of RNA are not only important in the secondary and tertiary structure of the individual rRNA molecules, but are also becoming increasingly implicated in various steps of the protein-synthetic process. In one instance, several lines of evidence suggest that the binding of tRNA to the 50S subunit involves at least in part an interaction between the TψCG loop of the tRNA and a complementary sequence on the 5S RNA molecule (for review see 177). The difficulty caused by the fact that the relevant guanine residues in the 5S RNA within the ribosome are not exposed to attack by kethoxal (186, 187) (see previous section) can be overcome by postulating appropriate conformational changes (186). While such a change has been shown to take place in the tRNA, making the TψCG sequence accessible (189, 190), the corresponding change in 5S RNA has yet to be demonstrated.

A similar case concerns the proposal made by Shine & Dalgarno (191) that the 3'-terminal sequence of 16S RNA is involved in mRNA recognition. In this instance, a dissociable oligonucleotide complex has been isolated, consisting of the 3'-terminus of the 16S RNA and an RNA fragment from a protein initiator region of R17 phage RNA (192); the two fragments contained a seven-base complementary sequence, the 3'-terminus of the 16S RNA being released from the 30S subunit as a 49-nucleotide fragment, under very mild conditions and with the help of the antibiotic colicin E3 (192, 193).

A number of other hypotheses have been put forward that involve complementary sequences of 6 to 12 base-pairs between various nucleic acid components of the protein-synthesizing system (e.g. 194–196). However, while there is some circumstantial evidence in favour of these ideas, hard facts are difficult to obtain, and some of these projected interactions may be the result of chance complementarities. Nevertheless it seems highly probable that many regions of the 16S and 23S RNA will ultimately reveal a functional importance. It is worth mentioning in this context that rRNA has been found to be the target in several affinity labelling experiments (e.g. 197–201), although it must of course be remembered that an affinity label only gives a measure of topographical neighborhood, which may or may not have functional implications.

RNA-PROTEIN INTERACTIONS

General

The three-dimensional arrangement of the ribosomal proteins (see Figures 1 and 2) is complicated enough, without the RNA. If one now adds the RNA to the picture (remembering that RNA constitutes two thirds of the mass of the particle), and concedes that the RNA is also likely to contribute significantly to the ribosomal function, then it follows that the interaction between RNA and protein becomes a vital part of, and perhaps the key to, an understanding of ribosome structure and protein-synthetic function.

Unfortunately, the study of protein-nucleic acid interactions in general has been one of the most disappointingly slow-moving fields in molecular biology. This has not been due to any lack of effort, but reflects rather the experimental difficulties and also the difficulty of formulating general rules for such interactions. It is already clear that there is a considerable variation in the ability of the individual ribosomal proteins to bind to RNA (see below), and one protein (S1) has been shown to be capable of "unwinding" double-helical RNA (202). Questions that have been asked in the past to try to establish general principles, such as "do ribosomal proteins bind to single- or double-stranded regions of the RNA?" (e.g. 203), are therefore very probably irrelevant.

Recent studies on RNA-protein interactions in the ribosome have concentrated on two aspects of the problem. The first concerns the ability of individual proteins to bind to intact rRNA, which is in turn directly connected with the problem of subunit reconstitution. The second aspect is the topographical problem, which can be considered as the question of how the RNA is "threaded through" the matrix of protein.

Reconstitution of Ribosomal Subunits

30S SUBUNIT   Early studies on the in vitro reconstitution of 30S subunits indicated that a few proteins (S4, S7, S8, S15, S17, S20, and possibly S13) (for review see 204) were able to interact specifically with 16S RNA. The remaining proteins could be added to complexes containing RNA and one or more of these "primary binding proteins" in a very specific manner. The interactions thus found between the proteins were incorporated into an "assembly map" (205), that seemed likely to reflect the topographical neighborhoods of the ribosomal proteins.

Evidence that this latter interpretation is at best an oversimplification has been provided by some new data published by Craven and his co-workers (206, 207), which rest on the simple observation that the protein-binding properties of the rRNA are dependent upon the way in which the RNA is prepared. When 16S RNA is prepared by the usual phenol extraction procedure, only the six or seven proteins named above are able to bind individually and specifically to it. In contrast, RNA prepared by an acetic acid-urea technique is able to bind an additional six or seven different proteins in a specific manner [namely S3, S5, S9, S11, S12, S13, and S18 (206, 207)]. In other words, protein binding sites can become either obscured or activated according to the conformation of the isolated RNA, and the same authors have been able to demonstrate further that one protein (S7) is able to alter the conformation of phenol-extracted 16S RNA in such a manner that, when the protein is removed, binding of S9 can subsequently take place (208). This indicates that at least a part of the effect of the primary binding proteins during assembly is the reorganization of the RNA to make other protein sites available, possibly in distant regions of the RNA.

Another recent study has demonstrated that the protein conformation is also important for RNA binding, as, by using proteins prepared under nondenaturing conditions (see section on isolated proteins above), a further new set of proteins binding to phenol-extracted RNA was found, namely S2, S5, S13, and S19 (209). The same authors found two new 23S RNA binding proteins, L11 and L15, in addition to the eleven already described (reviewed in 204, 210), and in the course of their study showed that two "native" protein-protein complexes (S13–S19 and L7/L12–L10) also were able to bind to their respective RNA molecules (211).

The conclusion from all this work is that most, if not all, of the proteins in the 30S subunit are capable of interacting directly with the 16S RNA, and that proteins and RNA can exert a strong mutual influence on their respective conformations. How far this latter point reflects the situation in vivo, where assembly takes place on precursor molecules, remains to be seen.

50S SUBUNIT   In contrast to the case of the 50S subunit from the thermophilic organism *Bacillus stearothermophilus,* which was first reconstituted several years ago (212), a successful procedure for the reconstitution of the *E. coli* 50S subunit has only recently been established (213). The procedure involves a two-step incubation, and has been shown to proceed through formation of a number of intermediate particles, that parallel, fairly closely, the corresponding in vivo intermediates. Nierhaus and his co-workers have identified those proteins that are essential for one of the early steps in this reassembly chain, which is a conformational change from a 33S to a 41S particle (214). Interestingly, all of the proteins concerned (L4, L13, L20, L22, and L24) have RNA binding sites in the 5'-proximal region of the 23S RNA [(215) also see section on RNA topography], which is a clear indication that the assembly in vitro proceeds in a general 5' to 3' direction along the rRNA.

Reconstitution experiments have also revealed a protein that is essential for assembly, but that appears to have no functional role in the completed particle. This protein is L24, which, as just mentioned, is essential for formation of the 41S intermediate particle. However, lithium chloride "core particles" lacking L24 can be prepared from 50S subunits; these core particles have a conformation corresponding to that of the 41S particle, and can be converted back into active 50S subunits in the absence of L24 (S. Spillmann and K. Nierhaus, personal communication). The fact that L24 is a very strong RNA binding protein (reviewed in 204, 210), and is also the only 50S protein that is entirely resistant to trypsin digestion within the 50S subunit (216), makes this finding doubly intriguing, and it is clearly relevant to the foregoing discussion of 30S reconstitution.

Topography of Ribosomal RNA

With the exception of the experiments already mentioned (see sections above on immune electron microscopy and neutron scattering), studies on the topography of the rRNA have been primarily concerned with the question of which specific regions of the ribosomal RNA are involved in interaction with the various individual proteins.

Here, two different approaches have been used, one involving the study of complexes between single proteins and RNA, and the other concerned with controlled nuclease digestion of the ribosomal subunits. The experiments can be further subdivided into those where an RNA-protein complex is isolated which is itself stable, and those where the complex is stabilized by a covalent cross-link between protein and RNA.

STABLE PROTEIN-RNA COMPLEXES   In the "single protein" approach, a preformed complex between protein and RNA is subjected to mild digestion with nuclease, and subsequently the RNA region "protected" by the protein can be identified. Alternatively, the binding of the protein to isolated RNA fragments of known nucleotide sequence (also produced by mild nuclease digestion) can be studied. A considerable amount of data has been accumulated using this approach, mostly by Garrett and Zimmermann and their co-workers, and the results have been reviewed in detail elsewhere (204, 210).

To summarize briefly, "binding sites" on phenol-extracted 16S RNA have been found for proteins S4, S7, S8, S13, S15, S17, and S20 (for review see 204, 210), which constitute the original set of RNA binding proteins. So far no such data are available either for acetic acid-urea extracted RNA, or for the "new" binding proteins, which are both mentioned above in the section on reconstitution of the 30S subunit (207, 209). The size and complexity of the RNA regions identified in these experiments varies considerably. Proteins S8 and S15, for example, are found associated with quite short RNA sequences, of the order of 50 nucleotides long, whereas, in contrast, S4 has a very large "binding site" comprising almost one third of the 16S molecule. It is important to realize here that the protein is not necessarily in intimate contact with the whole of its RNA site. This point is underscored by the recent finding that two regions of RNA within the S4 binding site, separated by about 100 nucleotides, are able to form a stable complex in the absence of protein (217, 218). This type of tertiary action has been mentioned above (see section on ribosomal RNA), and a similar instance has also been observed in the RNA region corresponding to protein S7 (219). Similarly, it does not follow that the whole of a protein is in contact with its RNA site, as evidenced by the finding that a fragment of protein S4 can be produced by tryptic digestion of the S4-16S RNA complex, which retains its ability to bind to the RNA (220–222).

In the case of 23S RNA, most of the available data have been obtained by making use of the fact that 23S RNA can be readily split into a 13S and an 18S fragment (223), to which individual binding proteins (224) or groups of proteins (215) can be rebound. A more precise localization has been made of the binding sites of some of the 23S RNA binding proteins (for review see 210), as well as of the three proteins, L5, L18 and L25, which bind to 5S RNA (reviewed in 177), and, very recently, detailed analyses have been published of the binding sites on 23S RNA of L1 (225) and L24 (176). All these results, for both 16S and 23S RNA, are summarized in Figure 3.

An alternative approach involving controlled hydrolysis of intact subunits has been used to gain information concerning the RNA environment of other proteins, in addition to those which are able to bind individually to phenol-extracted RNA. Here, the object is to isolate fragments of ribonucleoprotein, and to establish topographical groupings of protein and RNA by analyzing the protein and RNA moieties of the fragments. This method has been usefully applied both to the intact 30S subunit and also to a reconstituted intermediate particle, but has had only a limited success in the case of the 50S subunit (for review see 210). Again, the results are included in Figure 3. A similar type of limited digestion approach has recently been used to probe the subunit interface (228). The disadvantage

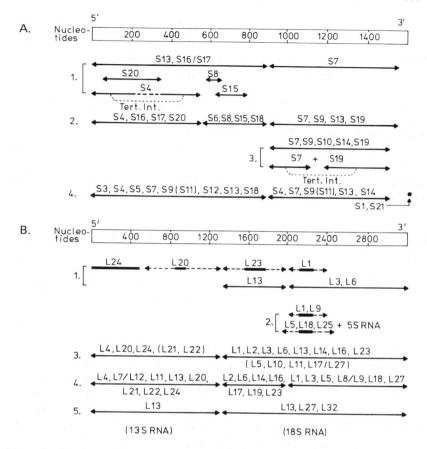

Figure 3 Summary of data on interactions of proteins with ribosomal RNA (see text for details). The scales indicate number of nucleotides, from the 5'-end.

A: 16S RNA. 1. Protein binding sites determined by nuclease digestion studies on single protein-RNA complexes (reviewed in 204, 210). The dotted line indicates RNA fragments that form a tertiary interaction in the absence of protein (see text). 2. Distribution of proteins in fragments of a reconstitution intermediate particle (204). 3. Distribution of proteins in fragments of the 30S subunit (210, 219). 4. Distribution of proteins found by RNA-protein cross-linking within the 30S subunit (229, 241).

B: 23S RNA. 1. Protein binding sites determined by nuclease digestion studies (see text for individual references). 2. Distribution of proteins in fragments of the 50S subunit (226). 3. Distribution of proteins that rebind to the 13S and 18S RNA fragments (224). Proteins in brackets require the presence of other proteins in order to bind (227). 4. Distribution of proteins by reconstitution experiments with RNA fragments (215). 5. Distribution of proteins found by RNA-protein cross-linking within the 50S subunit (241).

of all these methods is that they are limited both by the ease with which the RNA-protein complexes dissociate, and by the ease with which the specificity of the interaction is lost. Consequently, a number of laboratories have turned their attention to the covalent cross-linking of RNA to protein.

RNA-PROTEIN CROSS-LINKING   Cross-linking is a completely different type of approach to that outlined above. In the latter, a strong physical association is required to hold the RNA-protein complex together, whereas cross-linking is only a measure of topographical neighborhood. For this very reason, cross-linking studies on intact subunits are particularly interesting, as they have the potential of revealing RNA-protein contacts or neighborhoods in regions where no strong physical association exists. The complex nature of the protein arrangement (see Figure 1) would predict that many such contacts will occur for any given protein, possibly at widely separated points along the RNA chain. The cross-linking method also offers the possibility of ultimately determining the exact distance between a nucleotide and an amino acid residue.

One specialized example of RNA-protein cross-linking within the 30S subunit has been the linking of the 3'-terminus of the 16S RNA to proteins S1 and S21 (229, 230), by means of periodate oxidation of the 3'-terminal ribose followed by borohydride reduction. Initiation factor IF-3 has been cross-linked to 16S RNA by this method (231), as have also IF-1 and IF-2 (J. Hershey, personal communication). For cross-linking within the bulk of the 16S and 23S RNA molecules, more general methods are required, and some bifunctional chemical RNA-protein cross-linking reagents have recently been described (232, 233), which are beginning to yield detailed information as to which proteins can be cross-linked within the subunits (C. Oste, E. Ulmer and R. Brimacombe, unpublished results), although as yet nothing is known about the RNA regions involved in the cross-links. 1:2, 3:4 bis-epoxy butane has also been used as a cleavable RNA-protein cross-linking reagent (C. G. Kurland, personal communication).

At the present time the most widely-used method for inducing RNA-protein cross-links is by direct irradiation with ultraviolet light. This has the advantage that it is a "zero-length" cross-linking technique, in contrast to the chemical methods just mentioned (232, 233), where the reagents involved are rather long. Gorelic (234, 235) has shown that almost all the ribosomal proteins can be cross-linked to their respective RNA in the subunits by high doses of ultraviolet irradiation, but on the other hand, it has also been demonstrated that such levels of irradiation cause "unfolding" of the ribosomal subunits (236), in a manner likely to lead to protein randomization (180). Under milder conditions the reaction yields a very specific cross-linking of only S7 within the 30S subunit (236), and of L2 and L4 in the 50S (236, 237).

As with protein-protein cross-linking studies, the ultimate objective in this type of experiment should be the identification of the nucleotide and amino acid residues involved in the cross-link. A partial localization of the 16S RNA region concerned has been made in the case of the ultraviolet induced cross-linking of protein S7 to 16S RNA (219), and studies on the peptide residues involved in all three cases just mentioned (S7, L2, and L4) suggest that a single cross-link has occurred, as evidenced by the appearance in each case of a single oligonucleotide-peptide complex after trypsin and nuclease digestion (K. Möller and R. Brimacombe, unpublished results). On the other hand, when S7 and S4 were cross-linked individually to 16S RNA, by irradiation of the single protein-RNA complexes (238, 239), several peptides were found to be missing from each cross-linked protein after trypsin digestion, and this was interpreted as evidence of involvement of these peptides in the RNA-protein cross-links. Partial localizations of the RNA regions concerned in both the S4–16S RNA cross-linked complex, and a similar complex between S20 and 16S RNA, have recently been published (240).

In another study, formaldehyde has been used to cross-link proteins to RNA within the 30S and 50S subunits (241). While formaldehyde is not an ideal reagent for this purpose, as the cross-linking reaction is readily reversible, it was possible to localize partially a number of cross-linking points on the RNA. The results are included in Figure 3, and it is noteworthy that several proteins showed formaldehyde "contacts" in both halves of their respective ribosomal RNA molecules. In each case these proteins are either elongated (S4, S7; see Table 1 and Figure 1), or else have shown ambiguous binding properties in other experiments (S13, L13; see Figure 3). Initiation factor IF-3 was also shown to have contact with the 5'-proximal region of the 16S RNA (242), in addition to its ability to be cross-linked to the 3'-terminus (231). These experiments therefore support the contention made above that, within the subunits, some, and perhaps many, proteins do not possess a single RNA "binding site", but rather have a number of contacts with dispersed regions of the RNA chain. A precise analysis of such contacts will obviously play an important part in working out the topographical organization of the RNA.

COMPARISON WITH RIBOSOMES FROM OTHER ORGANISMS

So far we have only considered the *E. coli* ribosome, and it is appropriate to conclude with a comparison with ribosomes from other organisms; these can be divided into two classes, 70S and 80S ribosomes. The first class includes ribosomes from prokaryotes, i.e. bacteria and blue-green algae, and also those from chloroplasts and mitochondria, although some mitochondrial ribosomes have S values considerably different from 70S (243–244). The 80S class of ribosome comprises all eukaryotic cytoplasmic ribosomes, and two-dimensional gel electrophoresis has shown that the latter not only contain more proteins (70–80, as opposed to 50–60 in prokaryotes) but also that the average molecular weight of the proteins is greater (for review see 245). Eukaryotic ribosomes also contain larger RNA molecules. This wide variation in complexity raises the question as to how far ribosome structure has been conserved during evolution. In the case of the ribosomal proteins, information has come from immunological studies, from two-dimensional gel electrophoresis and from comparisons of the primary structures of proteins from different organisms.

The availability of antibodies specific to each of the 54 *E. coli* ribosomal proteins (146, 147), has made it possible to test for immunological cross-reaction with the corresponding proteins from other organisms (see 2, 246, 247, 248). The results of these experiments showed a strong cross-reaction with ribosomal proteins from Enterobacteriaceae (to which class *E. coli* itself belongs), but a significantly weaker degree of cross-reaction with ribosomal proteins from other classes of bacteria such as Bacillaceae (247, 248). When similar tests were made against ribosomal proteins from eukaryotic organisms, such as yeast or rat liver, there was very little cross-reaction at all, with the notable exception of antisera against *E. coli* proteins L7/L12 (245). This finding suggests that the structure of these latter proteins has been conserved during evolution, a contention which is supported by the fact that the corresponding proteins from yeast, reticulocytes, and rat liver ribosomes can be replaced by *E. coli* L7/L12 without loss of activity (245, 249).

Comparison of two-dimensional electrophorerograms revealed very similar patterns between ribosomal proteins isolated from mammals, birds, and reptiles (250, 251). The degree of similarity was much less between ribosomal proteins from amphibians and fishes on the one hand and mammals on the other, and a great diversity was observed between molluscs, crustaceans, plants, and mammals (250). Little similarity was also found when bacteria from different families, e.g. Bacillaceae and Enterobacteriaceae, were compared (247).

The most direct method for studying protein homology is clearly the analysis of the amino acid sequences of the proteins concerned. During the last few years, ribosomal proteins in pure form have been isolated from several organisms, and enough sequence information is already available to make it possible to start to compare the extent of structural homology between the *E. coli* proteins and those from other organisms. A rather high degree of amino acid sequence observation is observed in the N-terminal regions of proteins from *B. stearothermophilus* when compared with those from *E. coli* (252, 253), a result which is compatible with the functional interchangeability of the proteins from the two bacterial species (254). Homologies are also observed when the partial sequence of an acidic ribosomal protein from the halophile *Halobacter cutirubrum* (255, 256) is compared with that of *E. coli* proteins L7/L12. The acidic protein from *Halobacter* (M. Yaguchi and A. T. Matheson, personal communication) is also homologous with the corresponding protein from yeast (H. Higo, T. Itoh, and S. Osawa, personal communication) or the brine shrimp *Artemia salina* (R. Amons and W. Möller, personal communication).

On the other hand no such sequence homology has so far been detected in a comparison of the basic rat liver ribosomal proteins with those of *E. coli* or other bacteria, although two almost complete and several partial sequences are already available (A. Lin, I. G. Wool and B. Wittmann-Liebold, personal communication). This result is consistent with the low degree of immunological cross-reaction discussed above.

In the case of ribosomal RNA, less information is available. There is, however, evidence that the 16S RNA molecules from different bacteria contain large numbers of homologous sequences (257, 258), and specific fragments of RNA protected by *E. coli* protein S4, analogous to the *E. coli* S4 binding site, have been isolated from the 16S RNAs of other bacterial species (259). It is also clear that several regions of the primary sequence of prokaryotic 5S RNA have been conserved during evolution (260, 261; for review see 177), and that the 5S RNA binding proteins L5, L18, and L25 bind to the RNA in regions of conserved sequence (177). In the eukaryotic case, the larger ribosomal subunit contains a 5.8S RNA molecule in addition to a 5S molecule (reviewed in 177). Here again, the 5S sequence has been strongly conserved between various eukaryotes, and also shows some similarity to prokaryotic 5S RNA (260, 261, 177), but it has recently been shown, very interestingly, that *E. coli* proteins L18 and L25 bind to the 5.8S RNA from yeast ribosomes, and not to the 5S RNA (262). This suggests that 5.8S rather than 5S RNA in eukaryotes is the functional equivalent of prokaryotic 5S RNA.

Despite the fact that many of the ribosomal proteins have changed so extensively during evolution that virtually no structural homology exists between bacteria and mammals, the function of the ribosome has remained unaltered, even in the detailed steps of protein synthesis, i.e. initiation, elongation, and termination. This raises two interesting questions.

First, how can such a complex particle, where many proteins and several RNA molecules are in intimate contact, accommodate such drastic structural changes without a corresponding change in, or even loss of, the overall function of the ribosome? One possible answer is that the changes in primary structure of both protein and RNA have occurred in regions which are not important either structurally or functionally, and that the changes have been accommodated without gross disturbance of the overall topographical arrangement of the ribosomal components. Second, what was the evolutionary pressure that led to the greatly increased complexity of the eukaryotic ribosome, both in protein and RNA content? Perhaps the eukaryotic ribosome has extra functional roles, involving regulation of protein biosynthesis, binding to membranes, and so forth, which are unclear at present. These are both questions with which future research into ribosome structure will doubtless be concerned.

Literature Cited

1. Wittmann, H. G. 1974. In *Ribosomes,* ed. M. Nomura, A. Tissières, P. Lengyel, pp. 93–114. Cold Spring Harbor, New York: Cold Spring Harbor Laboratory. 930 pp.
2. Stöffler, G. 1974. See Ref. 1, pp. 615–67
3. Wittmann, H. G., Wittmann-Liebold, B. 1974. See Ref. 1, pp. 115–40
4. Littlechild, J. A., Malcolm, A. L. 1978. *Biochemistry.* In press
5. Dijk, J., Ackermann, I. 1978. *Biochemistry.* In press
6. Morrison, C. A., Littlechild, J., Dijk, J., Bradbury, E. M. 1977. *FEBS Lett.* 83: 348–52
7. Wittmann-Liebold, B., Dzionara, M. 1976. *FEBS Lett.* 65:281–83
8. Wittmann-Liebold, B., Robinson, S. M. L., Dzionara, M. 1977. *FEBS Lett.* 77:301–7
9. Dzionara, M., Robinson, S. M. L., Wittmann-Liebold, B. 1977. *Hoppe-Seyler's Z. Physiol. Chem.* 358:1003–19
10. Wittmann-Liebold, B., Robinson, S. M. L., Dzionara, M. 1977. *FEBS Lett.* 81:204–13
11. Dzionara, M., Robinson, S. M. L., Wittmann-Liebold, B. 1977. *J. Supramol. Struct.* In press
12. Gudkov, A. T., Behlke, J., Vtiurin, N. N., Lim, V. I. 1977. *FEBS Lett.* 82:125–29
13. Liljas, A., Kurland, C. G. 1976. *FEBS Lett.* 71:130–32
14. Wong, K. P., Paradies, H. H. 1974. *Biochem. Biophys. Res. Commun.* 61: 178–84
15. Österberg, R., Sjöberg, B., Liljas, A., Pettersson, I. 1976. *FEBS Lett.* 66: 48–51
16. Isono, K., Krauss, J., Hirota, Y. 1976. *Mol. Gen. Genet.* 149:297–302
17. Dabbs, E. R., Wittmann, H. G. 1976. *Mol. Gen. Genet.* 149:303–9
18. Isono, K., Cumberlidge, A. G., Isono,

S., Hirota, Y. 1977. *Mol. Gen. Genet.* 152:239–43
19. Dabbs, E. R. 1978. *Mol. Gen. Genet.* In press
20. Stöffler, G., Wittmann, H. G. 1977. In *Molecular Mechanism of Protein Biosynthesis,* ed. H. Weissbach, S. Pestka, pp. 117–202. New York: Academic. 720 pp.
21. Schiltz, E., Reinbolt, J. 1975. *Eur. J. Biochem.* 56:467–81
22. Hitz, H., Schäfer, D., Wittmann-Liebold, B. 1977. *Eur. J. Biochem.* 75:497–512
23. Stadler, H., Wittmann-Liebold, B. 1976. *Eur. J. Biochem.* 66:49–56
24. Chen, R., Wittmann-Liebold, B. 1975. *FEBS Lett.* 52:139–40
25. Funatsu, G., Yaguchi, M., Wittmann-Liebold, B. 1977. *FEBS Lett.* 73:12–17
26. Lindemann, H., Wittmann-Liebold, B. 1977. *Hoppe-Seyler's Z. Physiol. Chem.* 358:843–63
27. Morinaga, T., Funatsu, G., Funatsu, M., Wittmann, H. G. 1976. *FEBS Lett.* 64:307–9
28. Vandekerckhove, J., Rombauts, W., Wittmann-Liebold, B. 1977. *FEBS Lett.* 73:18–21
29. Yaguchi, M., Wittmann, H. G. 1978. *FEBS Lett.* In press
30. Yaguchi, M. 1977. *FEBS Lett.* 59: 217–20
31. Yaguchi, M., Wittmann, H. G. 1978. *FEBS Lett.* In press
32. Wittmann-Liebold, B., Marzinzig, E., Lehmann, A. 1976. *FEBS Lett.* 68: 110–14
33. Vandekerckhove, J., Rombauts, W., Peeters, P., Wittmann-Liebold, B. 1975. *Hoppe-Seyler's Z. Physiol. Chem.* 356: 1955–76
34. Chen, R., Ehrke, G. 1976. *FEBS Lett.* 69:240–45
35. Chen, R., Arfsten, U., Chen-

Schmeisser, U. 1977. *Hoppe-Seyler's Z. Physiol. Chem.* 358:531–35
36. Terhorst, C., Möller, W., Laursen, R., Wittmann-Liebold, B. 1973. *Eur. J. Biochem.* 34:138–52
37. Dovgas, N. V., Vinokurov, L. M., Velmoga, I. S., Alakhov, Y. B., Ovchinnikov, Y. A. 1976. *FEBS Lett.* 67:58–61
38. Heiland, I., Brauer, D., Wittmann-Liebold, B. 1976. *Hoppe-Seyler's Z. Physiol. Chem.* 357:1751–70
39. Dognin, J., Wittmann-Liebold, B. 1977. *FEBS Lett.* 84:342–46
40. Giorginis, S., Chen, R. 1977. *FEBS Lett.* 84:347–50
41. Brosius, J., Chen, R. 1976. *FEBS Lett.* 68:105–9
42. Brosius, J., Schiltz, E., Chen, R. 1975. *FEBS Lett.* 56:359–61
43. Brosius, J., Arfsten, U. 1978. *Biochemistry.* In press
44. Dovgas, N. V., Markova, L. F., Mednikova, T. A., Vinokurov, L. M., Alakhov, Y. B., Ovchinnikov, Y. A. 1975. *FEBS Lett.* 53:351–54
45. Bitar, K. G., Wittmann-Liebold, B. 1975. *Hoppe-Seyler's Z. Physiol. Chem.* 356:1343–52
46. Chen, R., Mende, L., Arfsten, U. 1975. *FEBS Lett.* 59:96–99
47. Wittmann-Liebold, B., Marzinzig, E. 1977. *FEBS Lett.* 81:214–17
48. Bitar, K. G. 1974. *Biochim. Biophys. Acta* 386:99–106
49. Ritter, E., Wittmann-Liebold, B. 1975. *FEBS Lett.* 60:153–55
50. Brosius, J. 1978. *Biochemistry.* In press
51. Wittmann-Liebold, B., Greuer, B., Pannenbecker, R. 1975. *Hoppe-Seyler's Z. Physiol. Chem.* 356:1977–79
52. Wittmann-Liebold, B., Pannenbecker, R. 1976. *FEBS Lett.* 68:115–18
53. Chen, R. 1976. *Hoppe-Seyler's Z. Physiol. Chem.* 357:873–86
54. Laughrea, M., Moore, P. B. 1977. *J. Mol. Biol.* 112:399–421
55. Giri, L., Subramanian, A. R. 1977. *FEBS Lett.* 81:199–203
56. Österberg, R., Sjöberg, B., Liljas, A., Garrett, R. A. Littlechild, J. 1978. *Journal Applied Cryst.* In press
57. Engelman, D. M., Moore, P. B., Schoenborn, B. P. 1975. *Proc. Natl. Acad. Sci. USA* 72:3888–92
58. Tischendorf, G. W., Zeichhardt, H., Stöffler, G. 1975. *Proc. Natl. Acad. Sci. USA* 72:4820–24
59. Rohde, M. F., O'Brien, S., Cooper, S., Aune, K. C. 1975. *Biochemistry* 14: 1079–87
60. Lake, J. A., Pendergast, M., Kahan, L.,

Nomura, M. 1974. *Proc. Natl. Acad. Sci. USA* 71:4688–92
61. Tischendorf, G. W., Stöffler, G. 1975. *Mol. Gen. Genet.* 142:193–208
62. Paradies, H. H., Franz, A. 1976. *Eur. J. Biochem.* 67:23–29
63. Österberg, R., Sjöberg, B., Garrett, R. A., Littlechild, J. 1977. *FEBS Lett.* 73:25–28
64. Tischendorf, G. W., Zeichhardt, H., Stöffler, G. 1974. *Mol. Gen. Genet.* 134:209–23
65. Giri, L., Littlechild, J., Dijk, J. 1977. *FEBS Lett.* 79:238–44
66. Lake, J. A., Kahan, L. 1975. *J. Mol. Biol.* 99:631–44
67. Tischendorf, G. W., Tesche, B., Stöffler, G. 1976. *Proc. Eur. Congr. Elect. Microscopy, 6th, Jerusalem* 2:524–26
68. Giri, L., Dijk, J., Labischinski, H., Bradaczek, H. 1978. *Biochemistry.* In press
69. Österberg, R., Sjöberg, B., Garrett, R. A. 1976. *FEBS Lett.* 65:73–76
70. Tritton, T. R., Crothers, D. M. 1976. *Biochemistry* 20:4377–85
71. Okuyama, A., Yoshikawa, M., Tanaka, N. 1974. *Biochem. Biophys. Res. Commun.* 60:1163–69
72. Yoshikawa, M., Okuyama, A., Tanaka, N. 1975. *J. Bacteriol.* 122:796–97
73. Nashimoto, H., Uchida, H. 1975. *J. Mol. Biol.* 96:443–53
74. Deusser, E., Stöffler, G., Wittmann, H. G., Apirion, D. 1970. *Mol. Gen. Genet.* 109:298–302
75. Birge, E. A., Kurland, C. G. 1970. *Mol. Gen. Genet.* 109:356–69
76. Kreider, G., Brownstein, B. L. 1971. *J. Mol. Biol.* 61:135–42
77. Zimmermann, R. A., Garvin, R. T., Gorini, L. 1971. *Proc. Natl. Acad. Sci. USA* 68:2263–67
78. Rosset, R., Vola, C., Feunteun, J., Monier, R. 1971. *FEBS Lett.* 18:127–29
79. Donner, D., Kurland, C. G. 1972. *Mol. Gen. Genet.* 115:49–53
80. Funatsu, G., Puls, W., Schiltz, E., Reinbolt, J., Wittmann, H. G. 1972. *Mol. Gen. Genet.* 115:131–39
81. Hasenbank, R., Guthrie, C., Stöffler, G., Wittmann, H. G., Rosen, L., Apirion, D. 1973. *Mol. Gen. Genet.* 127: 1–18
82. Zimmermann, R. A., Ikeya, Y., Sparling, P. F. 1973. *Proc. Natl. Acad. Sci. USA* 70:71–75
83. Olsson, M., Isaksson, L., Kurland, C. G. 1974. *Mol. Gen. Genet.* 135:191–202
84. van Acken, U. 1975. *Mol. Gen. Genet.* 140:61–68
85. Wittmann, H. G., Apirion, D. 1975. *Mol. Gen. Genet.* 141:331–41

86. Dabbs, E. R. 1977. *Mol. Gen. Genet.* 151:261–67
87. Dekio, S., Takata, R. 1969. *Mol. Gen. Genet.* 105:219–24
88. Bollen, A., Faelen, M., Lecocq, J. P., Herzog, A., Zengel, J., Kahan, L., Nomura, M. 1973. *J. Mol. Biol.* 76:463–72
89. Dekio, S., Takata, R., Osawa, S., Tanaka, K., Tamaki, M. 1970. *Mol. Gen. Genet.* 107:39–49
90. Bollen, A., Herzog, A. 1970. *FEBS Lett.* 6:69–72
91. Stöffler, G., Deusser, E., Wittmann, H. G., Apirion, D. 1971. *Mol. Gen. Genet.* 111:334–41
92. Nashimoto, H., Held, W., Kaltschmidt, E., Nomura, M. 1971. *J. Mol. Biol.* 62:121–38
93. Funatsu, G., Schiltz, E., Wittmann, H. G. 1971. *Mol. Gen. Genet.* 114:106–11
94. Funatsu, G., Nierhaus, K. H., Wittmann-Liebold, B. 1972. *J. Mol. Biol.* 64:201–9
95. Kreider, G., Brownstein, B. L. 1972. *J. Bacteriol.* 109:780–83
96. Itoh, T., Wittmann, H. G. 1973. *Mol. Gen. Genet.* 127:19–32
97. DeWilde, M., Wittmann-Liebold, B. 1973. *Mol. Gen. Genet.* 127:273–76
98. Wittmann, H. G., Stöffler, G., Piepersberg, W., Buckel, P., Ruffler, D., Böck, A. 1974. *Mol. Gen. Genet.* 134:225–36
99. Piepersberg, W., Böck, A., Yaguchi, M., Wittmann, H. G. 1975. *Mol. Gen. Genet.* 143:43–52
100. Piepersberg, W., Böck, A., Wittmann, H. G. 1975. *Mol. Gen. Genet.* 140:91–100
101. DeWilde, M., Cabezon, T., Villarroel, R., Herzog, A., Bollen, A. 1975. *Mol. Gen. Genet.* 142:19–33
102. Cabezon, T., Herzog, A., DeWilde, M., Villarroel, R., Bollen, A. 1976. *Mol. Gen. Genet.* 144:59–62
103. Yaguchi, M., Wittmann, H. G., Cabezon, T., DeWilde, M., Villarroel, R., Herzog, A., Bollen, A. 1976. *J. Mol. Biol.* 104:617–20
104. Isono, S., Isono, K. 1978. *Mol. Gen. Genet.* In press
105. Isono, K., Kitakawa, M. 1977. *Mol. Gen. Genet.* 153:115–20
106. Wittmann, H. G., Stöffler, G., Geyl, D., Böck, A. 1975. *Mol. Gen. Genet.* 141:317–29
107. Geyl, D., Böck, A., Wittmann, H. G. 1977. *Mol. Gen. Genet.* 152:331–36
108. Zubke, G., Stadler, H., Ehrlich, R., Stöffler, G., Wittmann, H. G., Apirion, D. 1977. *Mol. Gen. Genet.* 158:129–39
109. Kuwano, M., Taniguchi, H., Ono, M., Endo, H., Ohnishi, Y. 1977. *Biochem. Biophys. Res. Commun.* 75:156–62
110. Ozaki, M., Mizushima, S., Nomura, M. 1969. *Nature* 222:333–39
111. Birge, E. A., Craven, G. R., Hardy, S. J. S., Kurland, C. G., Voynow, P. 1969. *Science* 164:1285–86
112. Funatsu, G., Wittmann, H. G. 1972. *J. Mol. Biol.* 68:547–50
113. Funatsu, G., Nierhaus, K. H., Wittmann, H. G. 1972. *Biochim. Biophys. Acta* 287:282–91
114. Dekio, S. 1971. *Mol. Gen. Genet.* 113:20–30
115. Muto, A., Otaka, E., Itoh, T., Osawa, S., Wittmann, H. G. 1974. *Mol. Gen. Genet.* 140:1–5
116. Cabezon, T., Yaguchi, M., Petre, J., Herzog, A., Bollen, A. 1978. *J. Mol. Biol.* In press
117. Kahan, L., Zengel, J., Nomura, M., Bollen, A., Herzog, A. 1973. *J. Mol. Biol.* 76:473–83
118. Böck, A., Ruffler, D., Piepersberg, W., Wittmann, H. G. 1974. *Mol. Gen. Genet.* 134:325–32
119. Otaka, E., Teraoka, H., Tamaki, M., Tanaka, K., Osawa, S. 1970. *J. Mol. Biol.* 48:499–510
120. Otaka, E., Itoh, T., Osawa, S., Tanaka, K., Tamaki, M. 1971. *Mol. Gen. Genet.* 114:14–22
121. Wittmann, H. G., Stöffler, G., Apirion, D., Rosen, L., Tanaka, K., Tamaki, M., Takata, R., Dekio, S., Otaka, E., Osawa, S. 1973. *Mol. Gen. Genet.* 127:175–89
122. Pardo, D., Rosset, R. 1977. *Mol. Gen. Genet.* 156:267–71
123. Liou, Y. F., Yoshikawa, M., Tanaka, N. 1975. *Biochem. Biophys. Res. Commun.* 65:1096–1101
124. Buckel, P., Buchberger, H., Böck, A., Wittmann, H. G. 1977. *Mol. Gen. Genet.* 158:47–54
125. Watson, R. J., Parker, J., Fiil, N. P., Flaks, J. G., Friesen, J. D. 1975. *Proc. Natl. Acad. Sci. USA* 72:2765–69
126. Parker, J., Watson, R. J., Friesen, J. D., Fiil, N. P. 1976. *Mol. Gen. Genet.* 144:111–14
127. Berger, J., Geyl, D., Böck, A., Stöffler, G., Wittmann, H. G. 1975. *Mol. Gen. Genet.* 141:207–11
128. Kitakawa, M., Isono, K. 1977. *Mol. Gen. Genet.* 158:149–55
129. Cabezon, T., Herzog, A., Petre, J., Yaguchi, M., Bollen, A. 1977. *J. Mol. Biol.* 116:361–74
130. Wabl, M. 1973. PhD Thesis. Freie Universität, Berlin

131. Wabl, M. R. 1974. *J. Mol. Biol.* 84: 241–47
132. Tischendorf, G. W., Zeichhardt, H., Stöffler, G. 1974. *Mol. Gen. Genet.* 134:187–208
133. Stöffler, G., Tischendorf, G. W. 1975. In *Topics in Infectious Diseases. Drug Receptor Interactions in Antimicrobial Chemotherapy,* ed. J. Drews, F. E. Hahn, 1:117–43. Wien/New York: Springer-Verlag
134. Tischendorf, G. W., Bald, R. W., Kittler, R., Lührmann, R., Tesche, B., Stöffler, G. 1977. *Proc. FEBS Meeting, 11th, Copenhagen,* 264. (Abstr.)
135. Stöffler, G., Bald, R. W., Kittler, R., Lührmann, R., Tesche, B., Tischendorf, G. W. 1978. *FEBS Symp.* In press
136. Boublik, M., Hellmann, W., Roth, H. E. 1976. *J. Mol. Biol.* 107:479–90
137. Politz, S. M., Glitz, D. G. 1977. *Proc. Natl. Acad. Sci. USA* 74:1468–72
138. Wabl, M. R., Barends, P. J., Nanninga, N. 1973. *Cytobiologie* 7:1–9
139. Wabl, M. R., Doberer, H. G., Höglund, S., Ljung, L. 1973. *Cytobiologie* 7: 111–15
140. Vasiliev, V. D. 1974. *Acta Biol. Med. Ger.* 33:779–93
141. Lake, J. A. 1976. *J. Mol. Biol.* 105: 131–59
142. Lubin, M. 1968. *Proc. Natl. Acad. Sci. USA* 61:1454–61
143. Spiess, E. 1973. *Cytobiologie* 7:28–32
144. Boublik, M., Hellmann, W., Kleinschmidt, A. K. 1977. *Cytobiologie* 14:293–300
145. Tesche, B. 1975. *Vakuum-Technik* 4: 104–10
146. Stöffler, G., Wittmann, H. G. 1971. *Proc. Natl. Acad. Sci. USA* 68:2283–87
147. Stöffler, G., Wittmann, H. G. 1971. *J. Mol. Biol.* 62:407–9
148. Lake, J. A. 1977. *Proc. Natl. Acad. Sci. USA* 74:1903–7
149. Traut, R. R., Heimark, R. L., Sun, T. T., Hershey, J. W. B., Bollen, A. 1974. See Ref. 1, pp. 271–308
150. Sommer, A., Traut, R. R. 1976. *J. Mol. Biol.* 106:995–1015
151. Expert-Bezançon, A., Barritault, D., Milet, M., Guérin, M.-F., Hayes, D. H. 1977. *J. Mol. Biol.* 112:603–29
152. Peretz, H., Towbin, H., Elson, D. 1976. *Eur. J. Biochem.* 63:83–92
153. Kurland, C. G. 1977. In *Molecular Mechanism of Protein Biosynthesis,* ed. H. Weissbach, S. Pestka, pp. 81–113. New York: Academic. 720 pp.
154. Lutter, L. C., Zeichhardt, H., Kurland, C. G., Stöffler, G. 1972. *Mol. Gen. Genet.* 119:357–66
155. Lutter, L. C., Kurland, C. G. 1973. *Nature New Biol.* 243:15–17
156. Cantor, C. R., Huang, K., Fairclough, R. 1974. See Ref. 1, pp. 587–99
157. Huang, K., Fairclough, R. H., Cantor, C. R. 1975. *J. Mol. Biol.* 97:443–70
158. Cormick, G. G., Kretsinger, R. H. 1977. *Biochim. Biophys. Acta* 474:398–410
159. Expert-Bezançon, A., Barritault, D., Clegg, J. C. S., Milet, M., Khouvine, Y., Hayes, D. H. 1975. *FEBS Lett.* 59: 64–69
160. Kenny, J. W., Sommer, A., Traut, R. R. 1975. *J. Biol. Chem.* 250:9434–36
161. Engelman, D. M., Moore, P. B., Schoenborn, B. P. 1975. *Proc. Natl. Acad. Sci. USA* 72:3888–92
162. Engelman, D. M., Moore, P. B., Schoenborn, B. P. 1975. *Brookhaven Symp. Neutron Scattering Biol.* 27:IV20–IV37
163. Hoppe, W., May, R., Stöckel, P., Lorenz, S., Erdmann, V. A., Wittmann, H. G., Crespi, H. L., Katz, J. J., Ibel, K. 1975. *Brookhaven Symp. Neutron Scattering Biol.* 27:IV38–IV48
164. Moore, P. B., Langer, J. A., Schoenborn, B. P., Engelman, D. M. 1977. *J. Mol. Biol.* 112:199–234
165. Moore, P. B., Engelman, D. M., Schoenborn, B. P. 1974. See Ref. 1, pp. 601–13
166. Stuhrmann, H. B., Haas, J., Ibel, K., de Wolf, B., Koch, M. H. J., Parfait, R., Crichton, R. R. 1976. *Proc. Natl. Acad. Sci. USA* 73:2379–83
167. Stuhrmann, H. B., Koch, M. H. J., Parfait, R., Ibel, K., Crichton, R. R. 1977. *Proc. Natl. Acad. Sci. USA* 74:2316–20
168. Langlois, R., Lee, C. C., Cantor, C. R., Vince, R., Pestka, S. 1976. *J. Mol. Biol.* 106:297–313
169. Hall, J., Davis, J. P., Cantor, C. R. 1977. *Arch. Biochem. Biophys.* 179: 121–30
170. Brownlee, G. G., Sanger, F., Barrell, B. G. 1968. *J. Mol. Biol.* 34:379–412
171. Fellner, P., Ehresmann, C., Stiegler, P., Ebel, J. P. 1972. *Nature New Biol.* 239:1–5
172. Ehresmann, C., Stiegler, P., Mackie, G. A., Zimmermann, R. A., Ebel, J. P., Fellner, P. 1975. *Nucleic Acids Res.* 2:265–78
173. Magrum, L., Zablen, L., Stahl, D., Woese, C. 1975. *Nature* 257:423–26
174. Branlant, C., Sriwidada, J., Krol, A., Fellner, P., Ebel, J. P. 1975. *Biochimie* 57:175–225
175. Branlant, C., Korobko, V., Ebel, J. P. 1976. *Eur. J. Biochem.* 70:471–82

176. Branlant, C., Sriwidada, J., Krol, A., Ebel, J. P. 1977. *Eur. J. Biochem.* 74:155–70
177. Erdmann, V. A. 1976. *Progr. Nucleic Acid Res.* 18:45–90
178. Araco, A., Belli, M., Onori, G. 1975. *Nucleic Acids Res.* 2:373–81
179. Spitnik-Elson, P., Elson, D. 1976. *Progr. Nucleic Acid Res.* 17:77–98
180. Newton, I., Rinke, J., Brimacombe, R. 1975. *FEBS Lett.* 51:215–18
181. Litt, M. 1969. *Biochemistry* 8:3249–53
182. Noller, H. F. 1974. *Biochemistry* 13:4694–4703
183. Chapman, N. M., Noller, H. F. 1977. *J. Mol. Biol.* 109:131–49
184. Litman, D. J., Beekman, A., Cantor, C. R. 1976. *Arch. Biochem. Biophys.* 174:523–31
185. Ginzburg, I., Zamir, A. 1975. *J. Mol. Biol.* 93:465–76
186. Noller, H. F., Herr, W. 1974. *J. Mol. Biol.* 90:181–84
187. Delihas, N., Dunn, J. J., Erdmann, V. A. 1975. *FEBS Lett.* 58:76–79
188. Rich, A., Rajbhandary, U. L. 1976. *Ann. Rev. Biochem.* 45:805–60
189. Schwarz, U., Lührmann, R., Gassen, H. G. 1974. *Biochem. Biophys. Res. Commun.* 56:807–14
190. Schwarz, U., Menzel, H. M., Gassen, H. G. 1976. *Biochemistry* 15:2484–90
191. Shine, J., Dalgarno, L. 1974. *Proc. Natl. Acad. Sci. USA* 71:1342–46
192. Steitz, J. A., Jakes, K. 1975. *Proc. Natl. Acad. Sci. USA* 72:4734–38
193. Senior, B. W., Holland, I. B. 1971. *Proc. Natl. Acad. Sci. USA* 68:959–63
194. van Knippenberg, P. H. 1975. *Nucleic Acids Res.* 2:79–85
195. Herr, W., Noller, H. F., 1975. *FEBS Lett.* 53:248–52
196. Branlant, C., Sriwidada, J., Krol, A., Ebel, J. P. 1976. *Nucleic Acids Res.* 3:1671–87
197. Schwartz, I., Ofengand, J. 1974. *Proc. Natl. Acad. Sci. USA* 71:3951–55
198. Girshovich, A. S., Bochkareva, E. S., Kramarov, V. M., Ovchinnikov, Y. A. 1974. *FEBS Lett.* 45:213–17
199. Greenwell, P., Harris, R. J., Symons, R. H. 1974. *Eur. J. Biochem.* 49:539–54
200. Wagner, R., Gassen, H. G., Ehresmann, C., Stiegler, P., Ebel, J. P. 1976. *FEBS Lett.* 67:312–15
201. Yukioka, M., Hatayama, T., Omori, K. 1977. *Eur. J. Biochem.* 73:449–59
202. Szer, W., Hermoso, J. M., Boublik, M. 1976. *Biochem. Biophys. Res. Commun.* 70:957–64
203. Cotter, R. I., McPhie, P., Gratzer, W. B. 1967. *Nature* 216:864–68

204. Zimmermann, R. A. 1974. See Ref. 1, pp. 225–69
205. Held, W. A., Ballou, B., Mizushima, S., Nomura, M. 1974. *J. Biol. Chem.* 249:3103–11
206. Hochkeppel, H. K., Spicer, E., Craven, G. R. 1976. *J. Mol. Biol.* 101:155–70
207. Hochkeppel, H. K., Craven, G. R. 1977. *Mol. Gen. Genet.* 153:325–29
208. Hochkeppel, H. K., Craven, G. R. 1977. *J. Mol. Biol.* 113:623–34
209. Littlechild, J., Dijk, J., Garrett, R. A. 1977. *FEBS Lett.* 74:292–94
210. Brimacombe, R., Nierhaus, K. H., Garrett, R. A., Wittmann, H. G. 1976. *Progr. Nucleic Acid Res.* 18:1–44
211. Dijk, J., Littlechild, J., Garrett, R. A. 1977. *FEBS Lett.* 77:295–99
212. Nomura, M., Erdmann, V. A. 1970. *Nature* 228:744–48
213. Nierhaus, K. H., Dohme, F. 1974. *Proc. Natl. Acad. Sci. USA* 71:4713–17
214. Spillmann, S., Dohme, F., Nierhaus, K. H. 1977. *J. Mol. Biol.* 115:513–23
215. Chen-Schmeisser, U., Garrett, R. A. 1976. *Eur. J. Biochem.* 69:401–10
216. Crichton, R. R., Wittmann, H. G. 1971. *Mol. Gen. Genet.* 114:95–105
217. Mackie, G. A., Zimmermann, R. A. 1975. *J. Biol. Chem.* 250:4100–12
218. Ungewickell, E., Ehresmann, C., Stiegler, P., Garrett, R. A. 1975. *Nucleic Acids Res.* 2:1867–93
219. Rinke, J., Yuki, A., Brimacombe, R. 1976. *Eur. J. Biochem.* 64:77–89
220. Schulte, C., Schiltz, E., Garrett, R. A. 1975. *Nucleic Acids Res.* 2:931–42
221. Changchien, L. M., Craven, G. R. 1976. *J. Mol. Biol.* 108:381–401
222. Newberry, V., Yaguchi, M., Garrett, R. A. 1977. *Eur. J. Biochem.* 76:51–61
223. Allet, B., Spahr, P. F. 1971. *Eur. J. Biochem.* 19:250–55
224. Spierer, P., Zimmermann, R. A., Mackie, G. A. 1975. *Eur. J. Biochem.* 52:459–68
225. Branlant, C., Krol, A., Sriwidada, J., Ebel, J. P., Sloof, P., Garrett, R. A. 1976. *Eur. J. Biochem.* 70:457–69
226. Branlant, C., Krol, A., Sriwidada, J., Brimacombe, R. 1976. *Eur. J. Biochem.* 70:483–92
227. Spierer, P., Zimmermann, R. A. 1976. *J. Mol. Biol.* 103:647–53
228. Santer, M., Shane, S. 1977. *J. Bacteriol.* 130:900–10
229. Czernilofsky, A. P., Kurland, C. G., Stöffler, G. 1975. *FEBS Lett.* 58:281–84
230. Kenner, R. A. 1973. *Biochem. Biophys. Res. Commun.* 51:932–37
231. van Duin, J., Kurland, C. G., Dondon,

J., Grunberg-Manago, M. 1975. *FEBS Lett.* 59:287–90
232. Fink, G., Brimacombe, R. 1975. *Biochem. Soc. Trans.* 3:1014–15
233. Oste, C., Parfait, R., Bollen, A., Crichton, R. R. 1977. *Mol. Gen. Genet.* 152:253–57
234. Gorelic, L. 1976. *Biochemistry* 15:3579–90
235. Gorelic, L. 1976. *Biochim. Biophys. Acta* 454:185–92
236. Möller, K., Brimacombe, R. 1975. *Mol. Gen. Genet.* 141:343–55
237. Baca, O. G., Bodley, J. W. 1976. *Biochem. Biophys. Res. Commun.* 70:1091–96
238. Ehresmann, B., Reinbolt, J., Ebel, J. P. 1975. *FEBS Lett.* 58:106–11
239. Ehresmann, B., Reinbolt, J., Backendorf, C., Tritsch, D., Ebel, J. P. 1976. *FEBS Lett.* 67:316–19
240. Ehresmann, B., Backendorf, C., Ehresmann, C., Ebel, J. P. 1977. *FEBS Lett.* 78:261–66
241. Möller, K., Rinke, J., Ross, A., Buddle, G., Brimacombe, R. 1977. *Eur. J. Biochem.* 76:175–87
242. Pon, C. L., Brimacombe, R., Gualerzi, C. 1978. *Biochemistry.* 16:5681–86
243. Kroon, A. M., Agesteribbe, E., de Vries, H. 1972. In *The Mechanism of Protein Synthesis and its Regulation,* ed. L. Bosch, pp. 539–572, Amsterdam/London: North-Holland. 590 pp.
244. O'Brien, T. W., Denslow, N. D., Martin, G. R. 1974. In *Biogenesis of Mitochondria,* ed. A. M. Kroon, C. Saccone, p. 347. New York: Academic
245. Wool, I. G., Stöffler, G. 1974. See Ref. 1, pp. 417–60
246. Wittmann, H. G., Stöffler, G. 1972. In *The Mechanism of Protein Synthesis and its Regulation,* ed. L. Bosch, pp. 285–

341. Amsterdam/London: North-Holland. 590 pp.
247. Geisser, M., Tischendorf, G. W., Stöffler, G. 1973. *Mol. Gen. Genet.* 127:129–45
248. Geisser, M., Tischendorf, G. W., Stöffler, G., Wittmann, H. G. 1973. *Mol. Gen. Genet.* 127:111–28
249. Richter, G., Möller, W. 1974. In *Energy, Regulation and Biosynthesis in Molecular Biology,* ed. D. Richter, pp. 524–33. Berlin: de Gruyter. 701 pp.
250. Delaunay, J., Creusot, F., Schapira, G. 1973. *Eur. J. Biochem.* 39:305–12
251. Martini, O. H. W., Gould, H. J. 1975. *Mol. Gen. Genet.* 142:317–31
252. Yaguchi, M., Matheson, A. T., Visentin, L. P. 1974. *FEBS Lett.* 46:296–300
253. Higo, K. I., Loertscher, K. 1974. *J. Bacteriol.* 118:180–86
254. Higo, K., Held, W., Kahan, L., Nomura, M. 1973. *Proc. Natl. Acad. Sci. USA* 70:944–48
255. Visentin, L. P., Matheson, A. T., Yaguchi, M. 1974. *FEBS Lett.* 41:310–14
256. Oda, G., Strom, A. R., Visentin, L. P., Yaguchi, M. 1974. *FEBS Lett.* 43:127–30
257. Woese, C. R., Fox, G. E., Zablen, L., Uchida, T., Bonen, L., Pechman, K., Lewis, B. J., Stahl, D. 1975. *Nature* 254:83–86
258. Fischel, J. L., Krol, A., Ehresmann, C., Fellner, P., Ebel, J. P. 1975. *Biochimie* 57:885–97
259. Geisser, M., Mackie, G. A. 1976. *Eur. J. Biochem.* 70:159–70
260. Hori, H. 1975. *J. Mol. Evol.* 7:75–86
261. Hori, H. 1976. *Mol. Gen. Genet.* 145:119–23
262. Wrede, P., Erdmann, V. A. 1977. *Proc. Natl. Acad. Sci. USA* 74:2706–9

Ann. Rev. Biochem. 1978. 47:251–76
Copyright © 1978 by Annual Reviews Inc. All rights reserved

EMPIRICAL PREDICTIONS ❖974
OF PROTEIN CONFORMATION

Peter Y. Chou and Gerald D. Fasman

Graduate Department of Biochemistry, Brandeis University,
Waltham, Massachusetts 02154

CONTENTS

PERSPECTIVES AND SUMMARY

Proteins are composed of linear polypeptide chains that use the 20 different naturally occurring amino acids as their building blocks. After protein synthesis at the ribosome, the macromolecular chains fold up spontaneously to give a unique three-dimensional structure. It is the variety in the spatial architecture of proteins that allows the versatility in performing their essential roles in biological processes.

251

0066-4154/78/0701-0251$01.00

Twenty-five years ago Linderstrom-Lang (1) introduced the terms primary, secondary, and tertiary structure to describe protein architecture. The primary structure refers to the amino acid sequence of the protein. The secondary structure is the local spatial organization of the polypeptide backbone without consideration of its side-chain conformations. Hence, α-helices, β-sheets, and β-turns, which are chain reversal regions consisting of tetrapeptides, would fall within this category. The tertiary structure is the arrangement of all the atoms in space, including disulfide bridges and positions of side chains, so that all short- and long-range interactions are considered. The term quaternary structure may be used to denote the interactions between subunits such as the α- and β-chains of hemoglobins or the contact regions of insulin dimers.

While spectroscopic measurements can estimate the amount of secondary structures in a protein, as well as detect conformational changes (2–6), they cannot locate where the helical, β-sheet, and random regions are along the amino acid sequence. Only high resolution X-ray crystallography can do this precisely, although neutron diffraction analysis (7) and electron microscopy (8) also show much promise in determining protein structure to atomic resolution.

The three-dimensional structures of over 50 proteins have been elucidated by X-ray diffraction (9–14). However, compared to protein sequence determinations, of which over 700 have been completed (15), crystallographic analyses are far more expensive, painstaking, and time-consuming. Furthermore, many proteins, such as the histones, and the membrane and ribosomal proteins have not yet been crystallized, so that other avenues must be explored to yield structural information. Renaturation experiments (16–18) have shown that the information for folding the tertiary structure of a native protein is coded in its amino acid sequence. Hence, it appears likely that a close examination of the sequence data may reveal the principles of chain folding so that it may be possible to predict the three-dimensional structure of proteins.

While certain aspects of protein prediction have been treated in reviews on protein folding and sequence information (19–24), only recently has there been an exhaustive review of the literature on the empirical predictions of protein structure and their applications (25, 26). This review provides a history of work on the prediction of protein conformation, and a summary of the empirical methods currently in use. The advances in the empirical predictions of protein structure have paralleled that of the successful X-ray structural determinations of a huge variety of proteins. The earlier helix-coil predictive models have been superseded by more sophisticated empirical approaches using statistical analysis, information theory, stereochemical theory, and statistical mechanical treatments. Present predictive methods can locate helical, β-sheet, and β-turn regions in proteins with an 80% degree of accuracy. The recent discovery of the ϵ-helix in two regions of α-chymotrypsin (27) will encourage a more extensive examination into other secondary structural patterns in proteins. Hence, as the number of secondary structures increases, because of better delineation of their folding patterns in proteins, a better understanding of the various mechanisms of tertiary protein folding will also emerge. Already multistate models have been developed to predict the increasing number of different conformations in proteins. More detailed analyses of the super-secondary structures in proteins have yielded insights into the topological packing and tertiary interactions so that the prediction of the three-dimensional structure of proteins, from their amino acid sequences, is now a more feasible goal.

HISTORICAL BACKGROUND

Prediction of α-Helices

The earlier attempts at conformational prediction involved the correlation of protein secondary structure with its amino acid composition. Szent-Györgyi & Cohen (28) showed that proteins with high proline content also exhibit low helicity as determined by optical rotatory dispersion (ORD). Davies (29) found a qualitative relationship between the helical content of a protein (calculated from b_0 values of ORD) and the total percentage of those amino acids in the protein classified as helix breakers (Ser, Thr, Val, Ile, Cys) (30). Havsteen (31) also found a linear correlation between the factor $1/b_0$ and the percentages of (Ser + Thr + Pro) in the protein. Goldsack (32) confirmed that the total content of Pro, Ser, and Thr reduces the helicity in a protein, whereas a linear correlation was found between b_0 and the total residue percentage of helical forming residues (Ala + Arg + Asp + Cys + Glu + Leu + Lys). These preliminary predictive efforts relied heavily on amino acid composition and ORD data since most of the protein sequences were still unknown and the X-ray structural determination of proteins was still in its infancy. Using sequence data and chemical evidence, Scheraga (33) made a bold attempt to construct a tertiary structure of ribonuclease.

After the X-ray structure and sequence determination of myoglobin (34, 35) and hemoglobin (36, 37), Guzzo (38) pioneered the prediction of protein conformation from amino acid sequences. Assigning Pro, Asp, Glu, and His as helix terminators, nine helical regions in lysozyme were predicted of which six were in agreement with X-ray analysis. Prothero's rule (39) for helix formation required that three out of five residues be comprised of Ala, Val, Leu, or Glu. A revised model with Ala, Leu, and Glu as helix formers, and nucleation thresholds of 2/5 or 3/7, gave 75% reliability in helix prediction, although only one third of the helical residues were correctly identified (40). Periti, Quagliarotti & Liquori (41) tabulated the helical and antihelical pairs separated by up to five residues in myoglobin and hemoglobin, and used histograms as patterns of helical recognition in lysozyme. Low, Lovell & Rudko (42) used matching helical fragments of known structure to avoid the overprediction of helices in proteins. Using the results of their energy calculations, Kotelchuck & Scheraga (43) assigned residues as helix making (h) or breaking (c), and proposed that four or more h's in a row will form a helix with two c's terminating the helix. This rule correctly predicted 61% of the helices and 78% of the total residues in four proteins as helix or coil. Leberman (44) used a slightly modified model and obtained some improvements, but many helical regions remained unpredicted.

Schiffer & Edmundson (45) adopted an innovative way of locating hydrophobic arcs by the "helical wheels" method in designating helical regions. All three helices of insulin (A1–6, A12–20, B9–19) were precisely located by this model before the completion of X-ray analysis (A2–8, A13–19, B9–19) (46). It is noteworthy that Haggis (47) was the first to correctly predict the B8–18 helix of insulin by using hydrophobic arcs. Although the helical wheel model shows the view looking down an α-helix with residues at the correct angular displacement from one another, it does not convey the nature of the side chains or details of their longitudinal arrangement. Dunhill (48) tried to overcome these shortcomings by constructing helical net diagrams (a net diagram is made by slitting a cylinder open along a line parallel to its axis and laying it flat) to locate hydrophobic clusters and the distribution of charge and polar residues. However, both the helical wheel and net-diagram models encounter difficulty in representing the actual conformation of nonhelical regions such as β-sheets and chain reversal regions of globular proteins, so that their use is limited. Despite the qualitative nature of the earlier attempts, the 60–75% accuracy obtained for a two-state, helix-coil model was better than the 50% accuracy obtained by random guessing, and the precise location of some of the helices was encouraging.

Applying the Zimm-Bragg (49) σ and s parameters for helix initiation and propagation, obtained from thermal melting curves for helix-coil transitions in random copolymers of amino acids, Lewis et al (50, 51) were able to predict helices on a quantitative basis by means of helix probability profiles of denatured proteins. As the σ and s values were not available experimentally for the majority of amino acids, residues were assigned to three categories: helix makers ($s = 1.05$), helix indifferent ($s = 1.00$), and helix breakers ($s = 0.385$), with $\sigma = 5 \times 10^{-4}$ for all the residues. This formulation correctly predicted 68% of the total residues in 11 proteins.

Lewis et al (50, 51) suggested that better predictions may result when the σ and s parameters for all 20 amino acids are determined experimentally.

Wu & Kabat (52, 53) used the ϕ, ψ angles of the middle amino acid for tripeptide sequences of known protein structures and constructed a 20 X 20 table of tripeptides showing their frequencies in helical and nonhelical regions. In conjunction with the helical wheel method, this table was used to locate permissive helical regions in cytochrome c and various immuno-globulins. Applying informational theory, Robson & Pain (54) used the pairwise distribution of residues separated by 0, 1, . . . 4 apart in 11 known proteins to obtain helix-forming information. They were quite successful in predicting helical regions using only single-residue information, but some nonhelical regions were also identified as helical. Some of these over-predicted helices can be eliminated when pairwise interactions are consid-ered. However, Finkelstein & Ptitsyn (55) found no correlation between adjacent pair residues in helical and nonhelical regions and concluded that helix formation depends primarily on single residues independent of their neighbors in the polypeptide chain.

Prediction of β-Sheets

Despite the progress in protein prediction, a notable shortcoming in all the above methods is the absence of β-sheet prediction. In retrospect, it is apparent why such an important structural region was overlooked. The first proteins determined by X-ray crystallography were myoglobin and hemo-globin, both containing 80% helix and 0% β-sheet. Therefore, due to lack of statistical data on β-residues in proteins, the early predictors limited themselves to helix-coil predictions. Similarly, the theoretical studies of conformational transitions of polypeptides (56) showed the helix-forming parameters σ and s to be well defined, but no characterization was made pertaining to β-structures. The spectroscopic studies of β-sheets also lagged behind that of helices in the earlier experimental work on polypep-tides (57) so that while the helicity of proteins could be measured by b_0, attempts to estimate the percentage of β-sheets were less successful (32). Furthermore, solution studies of β-forming polypeptides were difficult due to the general insolubility of β-structures, which hindered precise spectro-photometric analysis. Hence, in the last decade, both theoretical and experi-mental factors have contributed to the sparsity of information about β-formers in proteins.

However, with increasing X-ray structural determinations of proteins it became readily apparent that β-sheets play as essential a role as α-helices in the spatial architecture of proteins. The hydrogen-bonding residues in-volved in β-sheets outnumber helices in ribonuclease (58), α-chymotrypsin (59), elastase (60), concanavalin A (61, 62), and the immunoglobulins (63,

64). Thus many of the overpredicted helices in the earlier methods were not due to long-range interactions but were simply the result of neglecting β-sheets in the predictive analysis.

The first attempt to predict β-regions in proteins was made by Ptitsyn & Finkelstein (65) using a three-state (α, β, coil) predictive algorithm. While no attempt was made to assign β-potentials to individual amino acids, they did correctly classify hydrophobic residues as strong β-formers and proline together with charged residues as β-breakers. Using pairwise interactions of residues 0, 1, . . . 6 apart, Nagano (66) developed a statistical method for locating helices, β-sheets, and loops in proteins with considerable success. Kabat & Wu (67, 68) extended their 20 X 20 tripeptide tables for locating β-sheet breaking residues. Their qualitative method predicts permissive β-regions, and sufficed to correctly locate 7 of the 12 β-regions in concanavalin A .

RECENT EMPIRICAL METHODS

Protein Conformational Parameters

As a result of their circular dichroism (CD) experiments (69), which showed Leu to be a strong helix former, Chou & Fasman (70) made a statistical analysis of 15 proteins with known X-ray structure. These studies showed that Leu occurs most frequently in the inner helical cores of proteins, and established the helix and β-sheet conformational potentials of all 20 amino acids in their hierarchical order for the first time. Extensive frequency tables of helical and β-sheet boundary residues were also published (71, 72), confirming earlier observations that negative and positive charged residues occur more often at the N- and C-terminal helix respectively (73, 74).

For the calculation of these parameters, it is essential to use high-resolution X-ray data for the delineation of the different conformational states. The conformation of a protein chain can be described by the dihedral angles ϕ and ψ which correspond to rotations about the $N-C^\alpha$ and $C^\alpha-C$ bonds. All consecutive sequences of four or more residues having ϕ, ψ angles within 40° of (–60°, –50°) are considered to be helical (right-handed) (25, 75). A consecutive sequence of three or more residues having ϕ, ψ angles within 40° of (–120°, 110°) or (–140°, 135°) are considered to be in the parallel or anti-parallel β-conformation (25, 75) even if these residues are not involved in hydrogen bonding. Residues in the protein that are not in helical or β-sheet regions are assigned to the coil conformation irrespective of the ϕ, ψ angles of the residue. The β-turns (also called β-bends, hairpin loops, reverse turns, and 3_{10} bends) consist of four consecutive residues in a protein where the polypeptide chain folds back on itself by nearly 180° (76–82).

Tetrapeptides whose C_i^α - C_{i+3}^α distances are below 7 Å and not in a helical region are characterized as β-turns (78, 82).

Using the criteria described above for the four conformational states, Chou & Fasman delineated the number of amino acids in the α, β, coil, and β-turn regions of 29 proteins (25) that contain approximately double the number of residues of the previous survey on 15 proteins (71). The average frequency for helices, β-sheets, and β-turns are respectively $\langle f_\alpha \rangle = 0.38$, $\langle f_\beta \rangle = 0.20$, and $\langle f_t \rangle = 0.32$. When the frequency of residues in the α, β, and β-turn regions are divided by their respective average frequency, their conformational parameters are obtained: $P_\alpha = f_\alpha / \langle f_\alpha \rangle$, $P_\beta = f_\beta / \langle f_\beta \rangle$, and $P_t = f_t / \langle f_t \rangle$. These conformational potentials are shown in Table 1, with α and β assignments as defined earlier (72). Although most values remained fairly constant, Met showed the most dramatic change ($P_\alpha = 1.20 \rightarrow 1.45$, $P_\beta = 1.67 \rightarrow 1.05$) due to the greater number of Met residues in the 29 proteins than in 15 proteins ($n_{\text{Met}} = 28 \rightarrow 73$). Other assignment changes included Asn, Asp, and His. It should be noted that all five charged residues are unfavorable for β formation with $P_\beta < 1.00$, while three of them (Asp, His, and Arg) are helical indifferent with $P_\alpha \sim 1.00$. On the other hand, α-breaking residues (Pro, Gly, and Asn) are strong β-turn formers with $P_t > 1.50$, while β-formers are generally found infrequently in bend regions.

Prediction Rules for α- and β-Regions

Using the conformational parameters, based on 15 proteins, Chou & Fasman formulated the following empirical rules to elucidate the secondary structural regions of proteins (72).

1. A cluster of four helical residues ($H\alpha$ or $h\alpha$) out of six residues along the protein sequence will nucleate a helix, with weak helical residues (I_α) counting as 0.5 h_α. The helical segment is extended in both directions until α-tetrapeptide breakers with $\langle P_\alpha \rangle < 1.00$ are reached (e.g. b_4, b_3i, b_2i_2, b_3h, etc.). Pro cannot occur in the inner helix or at the C-terminal helical end. Pro, Asp, Glu and His, Lys, Arg are incorporated, respectively, at the N- and C-terminal helical ends. Any segment with $\langle P_\alpha \rangle \geqslant 1.03$ as well as $\langle P_\alpha \rangle > \langle P_\beta \rangle$ is predicted as helical.
2. A cluster of three β formers ($H\beta$ or $h\beta$) out of five residues along the protein sequence will nucleate a β-sheet. The β-sheet is extended in both directions until β-tetrapeptide breakers with $\langle P_\beta \rangle < 1.00$ are reached. Any segment with $\langle P_\beta \rangle \geqslant 1.05$ as well as $\langle P_\beta \rangle > \langle P_\alpha \rangle$ is predicted as β-sheet.
3. When regions in proteins contain both α- and β-forming residues, the overlapping region is helical if $\langle P_\alpha \rangle > \langle P_\beta \rangle$, or β-sheet if $\langle P_\beta \rangle >$

Table 1 Conformational parameters for α-helical, β-sheet, and β-turn residues in 29 proteins[a]

P_α		P_β		P_t		f_i		f_{i+1}		f_{i+2}		f_{i+3}	
Glu	1.51	Val	1.70	Asn	1.56	Asn	0.161	Pro	0.301	Asn	0.191	Trp	0.167
Met	1.45	Ile	1.60	Gly	1.56	Cys	0.149	Ser	0.139	Gly	0.190	Gly	0.152
Ala	1.42	Tyr	1.47	Pro	1.52	Asp	0.147	Lys	0.115	Asp	0.179	Cys	0.128
Leu	1.21	Phe	1.38	Asp	1.46	His	0.140	Asp	0.110	Ser	0.125	Tyr	0.125
Lys	1.16	Trp	1.37	Ser	1.43	Ser	0.120	Thr	0.108	Cys	0.117	Ser	0.106
Phe	1.13	Leu	1.30	Cys	1.19	Pro	0.102	Arg	0.106	Tyr	0.114	Gln	0.098
Gln	1.11	Cys	1.19	Tyr	1.14	Gly	0.102	Gln	0.098	Arg	0.099	Lys	0.095
Trp	1.08	Gln	1.19	Lys	1.01	Thr	0.086	Gly	0.085	His	0.093	Asn	0.091
Ile	1.08	Thr	1.19	Gln	0.98	Tyr	0.082	Asn	0.083	Glu	0.077	Arg	0.085
Val	1.06	Met	1.05	Thr	0.96	Trp	0.077	Met	0.082	Lys	0.072	Asp	0.081
Asp	1.01	Arg	0.93	Trp	0.96	Gln	0.074	Ala	0.076	Thr	0.065	Thr	0.079
His	1.00	Asn	0.89	Arg	0.95	Arg	0.070	Tyr	0.065	Phe	0.065	Leu	0.070
Arg	0.98	His	0.87	His	0.95	Met	0.068	Glu	0.060	Trp	0.064	Pro	0.068
Thr	0.83	Ala	0.83	Glu	0.74	Val	0.062	Cys	0.053	Gln	0.037	Phe	0.065
Ser	0.77	Ser	0.75	Ala	0.66	Leu	0.061	Val	0.048	Leu	0.036	Glu	0.064
Cys	0.70	Gly	0.75	Met	0.60	Ala	0.060	His	0.047	Ala	0.035	Ala	0.058
Tyr	0.69	Lys	0.74	Phe	0.60	Phe	0.059	Phe	0.041	Pro	0.034	Ile	0.056
Asn	0.67	Pro	0.55	Leu	0.59	Glu	0.056	Ile	0.034	Val	0.028	Met	0.055
Pro	0.57	Asp	0.54	Val	0.50	Lys	0.055	Leu	0.025	Met	0.014	His	0.054
Gly	0.57	Glu	0.37	Ile	0.47	Ile	0.043	Trp	0.013	Ile	0.013	Val	0.053

Bracket groupings for P_α: Glu–Ala = H_α; Leu–Val = h_α; Asp–His = I_α; Arg = i_α; Thr–Cys = i_α; Tyr–Asn = b_α; Pro–Gly = B_α.

Bracket groupings for P_β: Val–Ile = H_β; Tyr–Thr = h_β; Met–His = i_β; Ala–Lys = b_β; Pro–Glu = B_β.

[a] P_α, P_β, P_t are conformational parameters of helical, β-sheet, and β-turns. f_i, f_{i+1}, f_{i+2}, and f_{i+3} are bend frequencies in the four positions of the β-turn. H_α, H_β, etc., are as defined previously (25, 72). Bend frequencies are based on 408 β-turns (25).

$\langle P_a \rangle$. The helix and β-sheet boundary frequency tables (25, 72) are also used to delineate whether the region is α or β.

Using these empirical rules, 81% of the helical residues, 85% of the β residues, and 77% of the total residues (α, β, and coil residues) in 15 proteins were correctly predicted (72). These results were clearly better than random guessing for a three-state model ($\%_N = 75\%$ vs 33%) and the earlier predictive schemes that considered only helix-coil states. The relative accuracy and simplicity of the method are the main reasons for its application to over 90 proteins (25).

Prediction of β-Turns

While the α- and β-regions in proteins have been clearly elucidated by X-ray crystallographers, chain reversal regions often are unspecified and were assumed to be part of the coil or irregular regions. Using the X-ray atomic coordinates of 29 proteins, 408 β-turns were elucidated (25) and later refined to include 459 bends (82). The positional bend frequencies did not differ appreciably using both data sets. The bend positional preferences of amino acids based on 408 bends (25) are shown in Table 1. The predominance of Pro in the second rather than the third position, Trp in the fourth but not in the second position, Cys and His in the first but not in the second position are no doubt due to stereochemical considerations that make certain residues more energetically stable at specific positions of the β-turn. Residues with the highest bend potential in all four positions are Asn, Gly, Pro, Asp, and Ser. Although the most hydrophobic residues (i.e. Val, Ile, and Leu) show the lowest bend potential, they are often found in regions just beyond the β-turns. An environmental analysis of β-turn neighboring residues shows that reverse chain folding is stabilized by anti-parallel β-sheets as well as by α-α and α-β interactions (82).

The probability of bend occurrence at residue i is calculated from $p_t = f_i \times f_i+1 \times f_i+2 \times f_i+3$ with the aid of Table 1. The average probability of β-turn occurrence is $\langle p_t \rangle = 0.55 \times 10^{-4}$. Tetrapeptides with $p_t > 0.75 \times 10^{-4}$ ($\approx 1.5 \times \langle p_t \rangle$) as well as $\langle P_t \rangle > 1.00$ and $\langle P_a \rangle < \langle P_t \rangle > \langle P_\beta \rangle$ are selected by the computer as probable bends. The percentage of bend and nonbend residues predicted correctly for 29 proteins by this computer algorithm is $\%_{t+nt} = 70\%$, while 78% of the β-turns were localized correctly within $\pm$ 2 residues (82).

Comparison of Predictive Models

Using multiple regression analysis, relationships were obtained for estimating the amount of secondary structure in a protein from its amino acid composition (83). This work improved on the earlier studies limited to consideration of helix and coil states (29, 31, 32) by taking into account combinations of residues that are formers and breakers of β-sheets and β-turns. It was shown that the frequencies (f_{aN} and f_{aC}) at the helical ends in proteins may be correlated to the experimental Zimm-Bragg helix initiation parameter σ, while the helix growth parameter, s, is similar to the P_a value or $K_a = n_a/n_c$ value (71). These discrete conformational parameters, σ' and s', for 20 amino acids were used (84, 85) instead of the tentative σ and s assignments (51) in the helix-coil transition theory, and showed improvements ($\%_a = 74\%$, $\%_N = 79\%$) over the earlier results ($\%_a = 64\%$, $\%_N = 68\%$).

Recently, the relationship between the Zimm-Bragg helix-coil transition parameters σ and s for synthetic polypeptides and helix conformational parameters for globular proteins, derived from statistical analysis and informational theory analysis, has been established on a quantitative basis (86). The cooperativity parameter σ implied by the helical lengths found in globular proteins is of the order of 10^{-1}, which is much larger than that for synthetic polypeptides. Nevertheless, the use of $\sigma = 1 \times 10^{-1}$ enables the prediction of s values calculated from stastical parameters that agree well with experimental data (86). Blagdon & Goodman (87) supplemented the central helix nucleation mechanism (71, 72) by proposing N- and C-terminal initiation of helices. Several of the α regions found by X-ray data but omitted in the Chou-Fasman prediction were located by this method (87), although it may lead to overprediction of helices if the helical nucleation rule (72) is neglected in these analyses.

The inclusion of β-sheet and β-turn conformations in the predictive rules (72) showed a marked improvement over the earlier helix-coil predictive models. Other attempts in this direction include the Bayesian approach of pattern recognition of α, β, and coil regions by the use of doublet information which gave a predictive accuracy of $\%_a = 89\%$, $\%_\beta = 88\%$, $\%_{coil} = 67\%$, and $\%_N = 78\%$ for 12 proteins (88). The nonapeptide predictive algorithm of Burgess et al (89) utilized the frequencies of occurrence of α, β, coil, and bend residues from 8 proteins and was applied to 13 proteins using a four-state model.

Robson & Pain (90–93) extended their earlier work (54, 94) on protein prediction using informational theory. Their more quantitative analysis of the directional effects of helix terminal residues (92) shows closer agreement to the helical boundary frequencies of Chou & Fasman (71). They also divided the ϕ, ψ map into eight domains, analyzed 16 possible conformational pairs in β-turns, and found that all pairs are nonrandomly distributed (93). Using a nine-state model based on the ϕ, ψ sub-domains (91) 52% of the backbone torsion angles of trypsin inhibitor were predicted to fall within 40° cells in the ϕ, ψ diagram. A similar analysis on nine other proteins produced 43% accuracy (91). While these percentages appear much smaller than two- or three-state predictions, it should be remembered that this type of analysis is greatly refined since it narrows the conformational space considerably for each residue. As such, they serve as important starting conformations for calculations of protein folding based on energy minimization (95). Another step in this direction is the five-state empirical prediction algorithm of Maxfield & Scheraga (96), which correctly assigned 56% of the residues in 20 proteins to one of five conformational states based on ϕ, ψ angles.

Based on a statistical analysis of 95 proteins of known sequence belonging to 13 families of crystallographically known conformation (66), Nagano successfully predicted 79% of the helices, 51% of the β-regions, and 71% of the loops or turns in 7 proteins not included in his original data set (97). By consideration of contiguous α-β interactions, and by observation that most parallel β-sheets are accompanied by neighboring helices, the predictability of loops was enhanced (97). Using a pseudo–free energy function $P(p,q)$, Nagano & Hasegawa (98) proposed a scheme whereby the minimization of the sum of $P(p, q)$ and energy functions (99) will lead to a plausible tertiary structure of a protein. They found that strong long-range interactions occur in regions of high β-sheet potential, and that α-β, as well as β-β, interactions are equally important in determining tertiary stability (98). This is confirmed in the β-turn environmental analysis (82), which shows 52 β-β residues versus 50 α-β and β-α residues at four positions away from the first and fourth residues of β-turns.

The zonal distribution of amino acids in the helical and nonhelical regions of 12 proteins show that Ala, Leu, Val, Ile, and Tyr occur more often in hydrophobic clustered zones, while Cys and Glu accumulate in hydrophobic-depleted zones (100). In addition, hydrophobic triplets placed in the relation 1–2–5 and 1–4–5 are found to be significant helix stabilizers. A more comprehensive stereochemical theory of protein secondary structure was formulated by Lim (101–104) using well-known architectural features of globular proteins. These organizational principles involve the compactness of form and the presence of a tightly packed hydrophobic core with a polar hydrophilic shell. Through the use of helical and anti-helical pairs and triplets at positions 1–2, 1–3, 1–4, 1–5, 1–2–5, and 1–4–5, Lim (103) developed a predictive algorithm for helices and outlined a set of rules for localizing surface and internal β-regions. When these algorithms were applied to 25 proteins of known structure, the percentage of residues predicted correctly as helical or nonhelical was $\%_{\alpha+n\alpha} = 81\%$. Likewise the accuracy for β-sheet prediction was $\%_{\beta+n\beta} = 85\%$, while $\%_N = 70\%$ was obtained for all residues using a three state model (104). The advantage of Lim's a priori theory and predictive algorithm is that it is self-sufficient and requires neither thermodynamic σ and s parameters obtained from experimental data or theoretical energy calculations (19) nor the refinement of X-ray data used in statistical and informal theory analysis (54, 66, 72). However, Lim's method tends to underpredict helices and β-regions. While the number of overpredicted and underpredicted α- and β-residues were not given in his analysis of 25 proteins (104), these numbers were cited in an earlier work based on 20 proteins (102). Hence, it is possible to calculate $\%_\alpha = 65\%$, $\%_{n\alpha} = 94\%$, $\%_{\alpha+n\alpha} = 82\%$, $Q_\alpha = 80\%$; $\%_\beta = 56\%$, $\%_{n\beta} = 94\%$, $\%_{\beta+n\beta} = 88\%$, $Q_\beta = 75\%$, and $\%_N = 72\%$. From these figures, it can be seen that Lim's method is quite conservative in predicting α- and β-regions. Since the Lim and Chou-Fasman methods tend to underpredict and overpredict α- and β-regions respectively, both schemes have been used to elucidate the conformation of plastocyanin (105), the α_1-acid glycoprotein (106), and human serum albumin (107).

The theory of protein self-organization (i.e. tertiary structure) was recently presented by Finkelstein & Ptitsyn (108–111). In paper I (108), the thermodynamic parameters that control the formation of local secondary structures (α-helices, 3_{10} helices, and β-bends) in unfolded protein chains were estimated on the basis of stereochemical analysis. These calculated σ and s parameters were shown in paper II (109) to be in good agreement with the experimental data of synthetic polypeptides. The mathematical treatment is given in paper III (110) where a modified one-dimensional Ising model for heteropolymers is used to determine the probability of formation of different secondary structures in proteins. Using the thermodynamic parameters and the matrix method, the probability profiles of helical and bend regions in the unfolded chains of 35 proteins were calculated in paper IV (111). It was concluded that fluctuating embryos of α-helices and β-bends are formed at the first stage of self-organization and are followed by β-sheet formation. However, it is the α-helices rather than β-bends that serve as blocks for initiating the tertiary folding of proteins.

A statistical mechanical treatment of protein conformation was developed by Tanaka & Scheraga in a series of six papers (112–117). In paper I, a method for evaluating statistical weights for various conformational states of amino acids, based on reported X-ray results of proteins, was described (112). In paper II, a one-dimensional three-state Ising model for specific sequence polypeptides was formulated to treat the helical, extended (β-sheet), and coil states in proteins (113). In paper III, the statistical weights derived in paper I and the theory of paper II were applied to the prediction of protein conformation. They obtained predictive accuracy in the range of $\%_\alpha = 53$ to 90%, $\%_\beta = 63$ to 88%, and $\%_N = 47\%$ to 80% for 19 proteins (114). Although the overall accuracy was not reported, this may be calculated from the total helical and β residues, which were missed in the prediction ($\Sigma a_m = 457$ and $\Sigma \beta_m = 203$) and overpredicted ($\Sigma a_0 = 333$, $\Sigma \beta_0 = 483$) to give $\%_a = 54\%$, $\%_{na} = 83\%$, $Q_a = 69\%$, $\%_\beta = 63\%$, $\%_{n\beta} = 80\%$, $Q_\beta = 72\%$, and $\%_N = 59\%$. Hence, it seems that the statistical mechanical model is more successful in correctly predicting non-helical and non-β residues than helical and β-residues in proteins. However, its lower overall accuracy of $\%_N = 59\%$ for three states when compared to other empirical methods (127) indicates that factors other than statistical weights of the 20 amino acids should be included in their model for a better correlation of protein secondary structure.

In paper IV, the statistical mechanical treatment was extended to a four-state model that included the β-bend conformation as well as the a, β, and coil states (115). Their overall result showed that 219 out of 372 β-turns were predicted correctly with 102 overpredicted bend regions. Their % bend localized (± 2 residues) was 59% and may be compared to the 78% correctly localized bends within ± 2 residues from the automated computer bend predictions (82). In paper V, a multistate model was introduced (116) using the statistical weights derived from the X-ray coordinates of 26 proteins. Using a 7×7 matrix model, the conformational probabilities of finding a residue in one of seven states was calculated for the trypsin inhibitor and flavodoxin. In paper VI, the ad hoc empirical rules of triads and tetrads used in paper III were eliminated by using the one-dimensional short-range interaction model to compute conformational-sequence probabilities for long sequences (117). This more refined model compares the predicted conformation to the a, β, or c state assigned to individual residues based on ϕ, ψ angles.

Predictive Test on Adenylate Kinase

The first objective comparison of the various protein predictive methods was made in the case of adenylate kinase, where investigators were invited to submit their predictions *prior* to any knowledge of the X-ray results (118). The overall helical predictions were found to be in good agreement with the experimental data. Chou & Fasman, and Finkelstein & Ptitsyn, localized nine of the ten helices. Barry & Friedman, as well as Lim, identified eight helices, while Nagano found seven helices. Lim, Finkelstein & Ptitsyn located the C-terminal helix 179–194 exactly, on the basis of hydrophobic clusters, whereas it was missed completely by most groups. Although this region has 9.5 helical formers out of 16 with no a breakers $(H_2 h_6 I_3 i_5)_a$, and $\langle P_a \rangle = 1.05$ according to the Chou-Fasman method, they predicted only 189–194 as helical with $\langle P_a \rangle = 1.13$, and wrongly assigned 182–188 as β-sheet due to its $\langle P_\beta \rangle = 1.29$.

The three central β strands 10–14, 90–94, and 114–118 of the five stranded parallel β-pleated sheet in adenylate kinase have been correctly located by all groups attempting β predictions. Chou & Fasman, and Nagano, overpredicted three β regions, but they were able to locate the β-strand 169–173 missed by the Lim-Finkelstein-Ptitsyn methods. Furthermore, Chou & Fasman predicted seven of the ten β-sheet boundaries in the five β regions found by X-ray analysis, to within ± 1 residue. (Five of these β-endings were identified exactly.) The various β-bend predictions show greater similarity than the α or β predictions since the bend algorithms are almost identical. Of the ten bend regions in adenylate kinase, Chou & Fasman, and Lewis, Momany & Scheraga (77), predicted seven correctly, with two overpredicted bends, while Nagano localized all ten bends with four overpredictions. It is of interest that only Nagano identified the bend 63–66 correctly, although the more refined bend frequencies based on 29 proteins (Table 1) also indicate a high probability of a β-turn at 62–65 (82).

A summary of the predictive accuracy for adenylate kinase using the various algorithms is shown in Table 2. Based on the correlation coefficient (119), Finkelstein & Ptitsyn scored the best in helix prediction ($C_\alpha = 0.62$,

Table 2 Summary of predictions of protein secondary structure[a]

Adenylate Kinase Predictors[b]	α-Helix				β-Sheet				β-Turn			
	Per-cent$_\alpha$	Per-cent$_{n\alpha}$	Q_α	C_α	Per-cent$_\beta$	Per-cent$_{n\beta}$	Q_β	C_β	Per-cent$_t$	Per-cent$_{nt}$	Q_t	C_t
Ptitsyn-Finkelstein (65)	75	87	81	0.62	54	97	76	0.58	—	—	—	—
Chou-Fasman (71, 72)	67	84	76	0.51	83	83	83	0.50	61	93	77	0.58
Nagano (66)	58	88	73	0.47	92	69	81	0.41	83	82	82	0.58
Lim (103, 104)	78	67	73	0.46	54	97	76	0.58	—	—	—	—
Barry-Friedman (118)	53	89	71	0.44	—	—	—	—	—	—	—	—
Levitt-Robson (54)	40	91	66	0.35	—	—	—	—	—	—	—	—
Kabat-Wu (67, 68)	91	37	64	0.34	96	49	72	0.30	—	—	—	—
Burgess-Ponnuswamy-Scheraga (89)	20	97	58	0.25	—	—	—	—	39	84	62	0.24
Lewis et al (50, 77)	44	79	61	0.24	—	—	—	—	72	91	81	0.62
Kotelchuck-Scheraga (43)	39	62	50	0.01	—	—	—	—	—	—	—	—
T4 Phage Lysozyme Predictors[c]												
Chou-Fasman (71, 72)	53	87	70	0.42	42	74	58	0.09	36	78	57	0.12
Schellman (26)	71	70	71	0.41	67	85	76	0.34	—	—	—	—
Burgess-Ponnuswamy-Scheraga (89)	42	93	67	0.38	0	92	46	−0.08	48	76	62	0.19
Guzzo (38)	52	81	67	0.34	—	—	—	—	—	—	—	—
Leberman (44)	45	86	65	0.32	—	—	—	—	—	—	—	—
Finkelstein-Lim-Ptitsyn (111)	64	63	63	0.26	0	93	46	−0.08	32	89	60	0.21
Ptitsyn-Finkelstein (65)	71	51	61	0.23	0	95	48	−0.06	—	—	—	—
Lim (103, 104)	60	61	61	0.21	0	93	46	−0.08	—	—	—	—
Nagano-Hasegawa (97, 98)	54	66	60	0.20	17	88	52	0.04	44	57	50	0.01
Prothero (39, 40)	62	57	59	0.19	—	—	—	—	—	—	—	—
Barry-Friedman (118)	40	73	57	0.14	50	83	66	0.22	—	—	—	—

[a] Percent$_\alpha$ = Percent of helical residues predicted correctly; Percent$_{n\alpha}$ = Percent of nonhelical residues predicted correctly. Q_α = (Percent$_\alpha$ + Percent$_{n\alpha}$)/2 is the average of the percent of helical and nonhelical residues predicted correctly. C_α = correlation coefficient between prediction and observation; C_α = 1.00 indicates perfect agreement. C_α = 0 (a prediction no better than random); C_α = −1.00 (total disagreement between prediction and observation).
[b] X-ray structure (3.0 Å): 54% α-helix, 12% β-sheet, 24% β-turn (118).
[c] X-ray structure (2.5 Å): 57% α-helix, 7% β-sheet, 15% β-turn (119).

$Q_\alpha = 81\%$) and β-sheet prediction, using the method of Lim ($C_\beta = 0.58$, $Q_\beta = 76\%$). The method of Lewis Momany & Scheraga (77) gave the best bend prediction ($C_t = 0.62$, $Q_t = 81\%$). Chou & Fasman scored the second best in all categories with $C_\alpha = 0.51$, $Q_\alpha = 76\%$ for helices, $C_\beta = 0.50$, $Q_\beta = 83\%$ for β-sheets, and $C_t = 0.58$, $Q_t = 77\%$ for β-turns. Although no scoring was kept of the inverse predictions of Kabat & Wu, it is possible to make positive predictions by considering six or more consecutive residues without helix breakers as helical, and five or more consecutive residues without β-breakers as β-sheet. While this interpretation of their rules in defining permissive but not obligatory helices was objected to (120), it is the only way to assess objectively the accuracy of their tripeptide breakers in predicting α- and β-regions. Table 2 shows that the Kabat-Wu method can correctly identify more helical ($\%_\alpha = 91\%$) and β residues ($\%_\beta = 96\%$) than the other predictive algorithms. However, the method tends to give overpredictions as evidenced from the $\%_{n\alpha} = 37\%$ and $\%_{n\beta} = 49\%$. Hence other predictive schemes should be used in conjunction with tripeptide breakers to achieve greater predictive accuracy (120).

Predictive Test on T4 Phage Lysozyme

Another assessment of the various empirical methods was the prediction of the α, β, and β-turn regions of T4 phage lysozyme made by 12 groups of investigators without prior X-ray information (119). As in the case of adenylate kinase (118), the joint prediction histograms show that better predictions are obtained near the N-terminal than the C-terminal of T4 phage lysozyme. While all the predictors identified the N-terminal helix 3–11, they all missed the short helix 137–141 (Arg-Trp-Tyr-Asn-Gln). The helical assignments of Chou & Fasman (Table 1) show that Tyr and Asn are α-breakers; Kabat & Wu also found two α-breaking tripeptides in this region. Apparently none of the predictive schemes could detect the driving forces that nucleate this small helix; its formation may be due to long-range interactions that are not yet included in current models. Matthews (119) noted that 5 of the 22 helix breakers of Kabat & Wu occur within helices and one of their 28 β-breakers occurs within a β-sheet. In addition, most of the overpredicted β-regions deduced by other predictive algorithms are also identified as "permissively sheet" by the Kabat-Wu method. Hence the test case for T4 lysozyme (119) does not support the claim that the combination of the Kabat-Wu method with others will reduce the overpredictions of α- and β-regions (120).

It is apparent from Table 2 that the secondary structural predictions for T4 lysozyme are less successful than for adenylate kinase. Based on the correlation coefficient (119), the "best" helix predictions were obtained by Chou & Fasman ($C_\alpha = 0.42$, $Q_\alpha = 70\%$) and Schellman ($C_\alpha = 0.41$,

$Q_a = 71\%$). The "best" predictions for sheets and bends were obtained by Schellman ($C_\beta = 0.34$, $Q_\beta = 76\%$) and Ptitsyn-Finkelstein ($C_t = 0.21$ $Q_t = 60\%$), respectively. It is seen that predictions for helices in T4 lysozyme are more accurate than those for β-sheets or bends, whereas comparable predictive accuracy was obtained for all three conformational states in adenylate kinase. Since the number of β-residues in T4 lysozyme is rather small (7%), the correlation statistics given for the β-sheet predictions should be treated with caution (119).

Although it is interesting to compare the relative accuracy of the different predictive schemes (Table 2), the merit of any individual method should not be judged on the basis of these two proteins alone. It may turn out that certain algorithms that yielded poor results may perform better with other proteins. It is more instructive to examine why certain methods predicted a secondary structure correctly while others did not. When such a comparison is made for a larger sampling of proteins, the strengths as well as the shortcomings of the various predictive models will become apparent. Already many of the predictive schemes used in the two test cases mentioned above have been updated. Predictors continually improve their empirical formulations by refining their conformational parameters as well as by incorporating medium- and long-range interactions based on increasing X-ray structural determinations of a greater variety of proteins.

FUTURE OUTLOOK

Improvements in Secondary Structural Predictions

Lewis & Bradbury (121) found that the attractive and repulsive electrostatic interactions of the ith residue with its neighbors $i \pm 1, 2, 3, 4, 7$ are helix breaking if there is more than one net repulsion. It was also observed that charged residues at the $i \pm 4$ positions have helix-disruptive effects, whereas helix-stabilizing effects were found with oppositely charged residues at the $i \pm 2$ and $i \pm 3$ positions (122). The electrostatic effects were used in the prediction of the helices in the histones (121, 123). A more recent prediction of the α, β and bend regions in the histones using helix-disruptive effects caused by electrostatic repulsion at $i \pm 3$ and $i \pm 4$ positions gave more compatible results between predictions and experimental data (124). Argos, Schwarz & Schwarz (125) computerized five of the predictive algorithms [Kabat & Wu (68), Nagano (66, 97), Chou & Fasman (71, 72), Burgess et al (89), and Barry & Friedman (118)], and showed that the joint probability histogram often eliminated any ambiguous predictions. Comparing the X-ray structures of 40 proteins with the combined histogram predictions, they found that helices ($Q_a \approx 72\%$) were more accurately predicted than β-sheets ($Q_\beta \approx 68\%$) or β-turns ($Q_t \approx 66\%$). They also confirmed the earlier observations (118, 119) that the overall agreement between prediction and observation within the amino terminal half of the protein is clearly superior to that for the carboxyl half, which suggests an amino terminus nucleating core. Thus, the secondary structural regions near the amino terminus are governed by short-range interactions, while long-range forces may predominate more toward the carboxyl terminus. Since long-range interactions have not been incorporated into the various predictive schemes, this would explain the poor predictions at the C-terminal portion of proteins. The joint histograms were also more successful in predicting smaller proteins than were larger ones or those proteins containing subunits where quaternary interactions exert an influence on local secondary structures. Because of these factors, caution is suggested in applying the present predictive methods and it is proposed that a perfect predictive algorithm must include a consideration of energy minimization, thermalization, and long-range interactions (125).

In predicting the α, β, and bend regions of 24 homologous ribonucleases (126), it was shown that the Chou-Fasman method (72) gives more uniform results and closer agreement with the X-ray structure than the method of Burgess, Ponnuswamy & Scheraga (89). However, an even better correlation was found with Lim's method (104), indicating that the distribution of hydrophobic residues along the primary sequence may play a greater role in the maintenance of secondary structure in the ribonucleases (126). Lenstra (127) also evaluated the secondary structure predictions in 33 proteins using four computerized methods. Of the four methods the histogram approach of Argos, Schwarz & Schwarz (125) with $C_\alpha = 0.53$, $C_\beta = 0.44$, $C_t = 0.44$ gave the best α prediction. The statistical method of Nagano (97) with $C_\alpha = 0.49$, $C_\beta = 0.53$, $C_t = 0.40$ yielded the best β prediction. The stereochemical model of Lim (103, 104) with $C_\alpha = 0.44$, $C_\beta = 0.39$ and the statistical mechanical method of Tanaka & Scheraga (114) with $C_\alpha = 0.39$, $C_\beta = 0.39$ gave less reliable results.

While the joint predictive approach may reduce certain ambiguities, it should be pointed out that the histogram method (118, 119, 125) is an aggregate of the individual predictive algorithms. To the extent that each individual method is improved, better accuracy may be expected for the joint predictions. The computerized Chou-Fasman method, which incorporates single residue, dipeptide, tripeptide, and helix and β-sheet boundary information, is a step in this direction (25). The quality of prediction for thioredoxin is $Q_\alpha = 77\%$, $Q_\beta = 89\%$, $Q_t = 83\%$ based on single-residue information. However, improvements are obtained by consideration of dipeptide, tripeptide, and conformational boundary analysis, which correctly predicts the region 86–91 as β-sheet to give $Q_\alpha = 85\%$ and $Q_\beta = 94\%$. Similarly, the predictive accuracy for superoxide dismutase ($Q_\alpha = $

47%, $Q_\beta = 76\%$, $Q_t = 74\%$) and triose phosphate isomerase ($Q_\alpha = 85\%$, $Q_\beta = 81\%$, $Q_t = 60\%$) obtained from the predictive rules (25) compares favorably with the more successful computer algorithms currently in use (127). Another innovative approach to conformational prediction was made by Kretsinger & Barry (128). By means of homologous sequences and the four EF hands of parvalbumin, which are arranged in two pairs with overall symmetry, *222*, they were able to predict a similar structure in the calcium-binding component of troponin. Hence, the inclusion of homologous sequences, knowledge of ligand binding, and symmetry factors will provide additional constraints leading to even greater accuracy in protein structural predictions.

Structural Domains and Tertiary Folding

Many globular proteins often contain two or more distinct structural domains, and it was proposed that the initiation of protein folding occurs independently in separate parts of these molecules (129). Although ϕ, ψ diagrams (130) show the distribution of residues in the α and β regions of proteins, they do not provide information on interactions between these regions. Phillips (131) was the first to use a diagonal diagram in depicting these interactions by showing the contacts between C^α atoms in ribonuclease S. This type of diagram shows the helix by a concentration of density thickening along the diagonal, and the anti-parallel β-sheet as density perpendicular to the diagonal. It also denotes centers in which side chains are in close proximity, thus nuclei in protein folding may be located. These diagonal distance maps were employed for energy calculations in refining the X-ray structures of myoglobin and lysozyme (132, 133). Rossmann & Liljas (134) extended these distance plots to the recognition of structural domains and found a domain in flavodoxin ($\beta\alpha\beta\alpha\beta$) that resembles the first two domains of lactate dehydrogenase ($\beta\alpha\beta\alpha\beta$). A common nucleotide binding domain in dehydrogenases, kinases, and flavodoxins (135), as well as a common heme binding pocket in globins and cytochrome b_5 (136) were also located. A similarity in the folding pattern of the anti-parallel β-strands between the immunoglobulin molecule and superoxide dismutase was also observed (137).

By use of a simple diagrammatic representation to depict the α and β arrangements in 31 proteins, four distinct structural patterns were observed, and it was found that secondary structures that are adjacent in the protein sequence are also often in spatial contact (138). Further analysis of diagonal distance plots revealed interacting side chains in proteins (139), as well as interesting repeating square and trapezoidal patterns that correspond to "superhelical" structures (140). A novel two-dimensional representation of hydrogen bonds between residues in parallel β-sheets was introduced, which showed the pattern of $(\beta a)_8$ super-secondary structure in triose phosphate isomerase (141). A domain with four roughly parallel a-helices was recently observed in hemerythrin, tobacco mosaic virus (TMV) protein, and tyrosyl-tRNA synthetase (142), and it was suggested that the four helical super-secondary structures found in the latter two proteins provide a common role in binding to RNA. The tertiary folding pathway of TMV protein has also been presented (143). It is based on local interactions of fluctuating helices that form a helical intermediate globule followed by transformation into one native structure by the unwinding of excess a-helices and the retention of a- and β-regions predicted from stereochemical theory (103, 104).

More recently, Richardson (144) and Sternberg and Thornton (145) independently discovered that the β-strand–helix–β-strand $(\beta a\beta)$ units in proteins fold in a right-handed sense 98% of the time. Since β-sheets are always observed to have a right-handed local twist (146) this would explain the predominance of right-handed $\beta a\beta$ units. This principle of right-handed crossover connections in β-sheets will provide a great aid in predicting protein tertiary structure (147–151).

Application of Predictive Methods

While the biological function of a protein depends on its unique three-dimensional topology, considerable information can be learned from an accurate prediction of protein secondary structure. Empirical algorithms can provide a useful starting conformation for energy minimization procedures, thereby limiting the search for the native tertiary structure. The predictive methods may also serve as guidelines for the direction of folding in low-resolution electron density maps of proteins, as in the case of TMV protein (158). The earlier predictive methods were shown to be only partially successful in locating the helices in cytochrome c (81), but later models proposed by Robson & Pain (54) and Lewis, et al (77) were found to be quite accurate in identifying the helical and bend regions of chromatium HiPIP (159). More recently, crystallographers have found the empirical rules of Chou & Fasman (72) to be consistent with their X-ray results (158, 160–163).

The conformational parameters, P_a, P_β, and P_t (Table 1) are expedient for detecting regions in proteins with potential for conformational changes due to mutations or changes in solvent conditions. The use of these parameters has suggested an $a \rightarrow \beta$ transformation in the 52–57 region of *lac* repressor mutant AP 46 (164) and in the β1–6 region of sickle cell hemoglobin (165). These predictions, as well as the proposed $\alpha \rightarrow \beta$ transition for region 19–27 in glucagon when bound to the receptor site (166), are still awaiting experimental verification. An interesting $\beta \rightarrow \alpha$ transformation in region β1–8 of 4-Zn insulin in 6% NaCl has been observed by X-ray studies (167). This is not surprising since the β1–7 region also has α-potential, but was predicted as β-sheet due to its higher β-potential $\langle P_\beta \rangle = 1.15 >$ $\langle P_\alpha \rangle = 1.07$ in agreement with X-ray data of 2-Zn insulin (72).

Secondary structural predictions are useful in the recognition of structural domains in homologous sequences. Despite sequence variations in the C-peptides of proinsulins from 10 mammalian species, in 7 proteinase inhibitors, and in 12 pancreatic ribonucleases, β-turns were predicted to be conserved, which suggests that chain reversal regions are essential for keeping the active structural domains in hormones and enzymes intact for their specific biological function (25). By use of the Chou-Fasman predictive rules (72) with the aid of homologous sequence comparisons, the coenzyme-binding domains of glutamate dehydrogenase were located (168), and conformational homologies were elucidated in growth hormones (169), immunoglobulins (170), and neurotoxins (171). Other predictive approaches that reveal structural homologies include the Zimm-Bragg formulation applied to 27 species of cytochrome c (51) and a new statistical mechanical treatment that shows similarity in conformations in 19 homologous neurotoxins (117). On the other hand, proteins that are proposed to be structurally and evolutionarily related, based on sequence alignment, have been shown to be dissimilar in conformation from predictive analysis. Thus, little structural correspondence was shown between mouse nerve growth factor and proinsulin (172) or between bromelain inhibitor and the proteinase inhibitors (25). The Chou-Fasman conformational parameters have also been used to deduce the probable polypeptide conformation on prebiotic earth (173), to test structural convergence during protein evolution (174), and to construct genealogical trees (175). Recent phytogenetic studies include rubredoxins (176) and muscular parvalbumins (177).

It is encouraging that the computed percentages of secondary structures obtained from empirical predictions are in good agreement with estimates from CD studies (72, 178–182). Utilization of the β-turn frequencies (Table 1) provides a better understanding of the conformation of many biologically active peptides (183–185). The secondary structural predictions are also helpful in building three-dimensional models with the aid of known constraints, as was done in proinsulins (186) and plastocyanin (105). Finally, empirical predictive schemes may suggest the rational design of synthetic analogs for experimental testing to see whether conservation or changes in conformation will produce alterations in hormonal or enzymatic activity. Such syntheses have already been made for ribonuclease S-peptide (187–

189), proinsulin C-peptide (181), secretin (190), and region 75–120 of human growth hormone (169). Further synthetic work of this kind should provide additional insights into structure-function relationships in proteins.

Concluding Remarks

In the historical background section, it was noted that the earliest empirical predictions of protein structure were limited to the helix-coil states. The addition of β-sheet to the predictive algorithm came at a later stage, when X-ray diffraction studies showed that the β conformation is predominant in many proteins. With increasing refinement of the X-ray data of proteins to higher resolution, residues that were originally classified as random coil were found to participate in β-turns. Hence, the prediction of helical, β-sheet, and β-turn regions allows 80% of the residues in proteins to be assigned to a particular secondary structure. The new algorithm for identifying β-turns according to the radius of curvature along the polypeptide chain using only C^α coordinates (191), and the classification of 421 β-turns into 11 bend types (82), will provide an even better understanding of these chain reversal regions of proteins. Directional effects for β-turns derived from informational theory (192), and the environmental bend positional potentials calculated from statistical analysis (82), should give predictive turn parameters that will yield better elucidation of chain reversals in proteins.

Recent advances in β-sheet prediction include the application of the one-dimensional Ising model to predict the β-regions in proteins (193, 194). Informational parameters on the β-sheet conformation have been distinguished for β residues in a pleated sheet and those that are not hydrogen bonded (192). The topological connectivities of β-sheets in 37 proteins have been systematically classified and their distribution frequencies discussed (195). More detailed analyses on the number of strands within a β-sheet, the type of connectivities between strands, and the order of the β-strands in 31 proteins have provided folding parameters to predict the strand arrangements in β-pleated sheets (196–198). Attempts have also been made to represent the $\beta a \beta$ interactions quantitatively and schematically according to strength orders and using triplet information in helix prediction so that super-secondary structures may be predicted (199, 200).

The limitations of empirical predictive methods in giving accurate information of the three-dimensional structure of a protein have been pointed out recently (95). Although it might be considered judicious to wait for a perfect predictive algorithm to be developed, the wealth of knowledge accumulated during the last decade regarding protein conformational prediction should be continually tested, refined, and applied. Though caution is necessary, courage is needed for the full use of this knowledge. The prediction of secondary structures should not be an end in itself but a means for furthering our understanding of protein conformation through synthetic analogs, corroborative experimental studies, sequence and conformational homology comparisons, refinement of X-ray analysis, and tertiary structure model building. A combination of these theoretical and experimental studies will provide greater insights into the principles of protein folding.

ACKNOWLEDGMENTS

Preparation of this review was generously supported in part by Grants from the US Public Health Service (GM 17533), National Science Foundation (PCM76-21856), and the American Cancer Society (NP-92E). This is Publication No. 1181 from the Graduate Department of Biochemistry, Brandeis University, Waltham, Massachusetts 02154. The present address of P.Y. Chow is: Department of Chemistry, Worcester Polytechnic Institute, Worcester, Massachusetts 01609.

Literature Cited

1. Linderstrom-Lang, K. U. 1952. *Lane Medical Lectures—Proteins and Enzymes,* p. 58. Stanford, California:Stanford Univ. Press, 115 pp.
2. Adler, A. J., Greenfield, N. J., Fasman, G. D. 1973. *Methods Enzymol.* 27:675–735
3. Jirgensons, B. 1973. *Optical Activity of Proteins and Other Macromolecules,* New York:Springer-Verlag. 199 pp. 2nd ed.
4. Susi, H. 1972. *Methods Enzymol.* 26:455–72
5. Tobin, M. C. 1972. *Methods Enzymol.* 26:473–97
6. Roberts, G. C. K., Jardetzky, O. 1970. *Adv. Protein Chem.* 24:447–545
7. Schoenborn, B. P. 1971. *Cold Spring Harbor Symp. Quant. Biol.* 36:569–75
8. Henderson, R., Unwin, P. N. T. 1975. *Nature* 257:28–32
9. Dickerson, R. E. 1972. *Ann. Rev. Biochem.* 41:815–42
10. Blake, C. C. F. 1972. *Prog. Biophys.* 25:83–130
11. Matthews, B. W., Bernhard, S. A. 1973. *Ann. Rev. Biophys. Bioeng.* 2:257–312
12. Liljas, A., Rossmann, M. G. 1974. *Ann. Rev. Biochem.* 43:475–507
13. Matthews, B. W. 1977. *The Proteins,* ed. H. Neurath, R. L. Hill, Vol. III. New York: Academic. 3rd ed.
14. Matthews, B. W. 1976. *Ann. Rev. Phys. Chem.* 27:493–523
15. Dayhoff, M. O. 1972. *Atlas of Protein Sequence and Structure,* Vol. V, and Suppl. 1 and 2 (1973, 1976). Silver Spring, Md: Nat. Biomed. Res. Found.
16. Anfinsen, C. B., Haber, E., Sela, M., White, F. H. Jr. 1961. *Proc. Natl. Acad. Sci. USA* 47:1309–14
17. Anfinsen, C. B. 1973. *Science* 181:223–30
18. Wetlaufer, D. B., Ristow, S. 1973. *Ann. Rev. Biochem.* 42:135–58
19. Scheraga, H. A. 1971. *Chem. Rev.* 71:195–217
20. Wu, T. T., Fitch, W. M., Margoliash, E. 1974. *Ann. Rev. Biochem.* 43:539–66
21. Anfinsen, C. B., Scheraga, H. A. 1975. *Adv. Protein Chem.* 29:205–300
22. Richards, F. M. 1977. *Ann. Rev. Biophys. Bioeng.* 6:151–76
23. Schulz, G. E. 1977. *Angew. Chem. Int. Ed. Engl.* 16:23–32
24. Chou, P. Y., Fasman, G. D. 1977. *Trends Biochem. Sci.* 2:128–31
25. Chou, P. Y., Fasman, G. D. 1978. *Adv. Enzymol.* 47. In press
26. Schellman, C. G. 1978. *CRC Crit. Rev. Biochem.* To be published
27. Srinivasan, R., Balasubramanian, R., Rajan, S. S. 1976. *Science* 194:720–22
28. Szent-Györgyi, A. G., Cohen, C. 1957. *Science* 126:697–98
29. Davies, D. R. 1964. *J. Mol. Biol.* 9:605–9
30. Blout, E. R., deLoze, C., Bloom, S. M., Fasman, G. D. 1960. *J. Am. Chem. Soc.* 82:3787–88
31. Havsteen, B. H. 1966. *J. Theor. Biol.* 10:1–10

32. Goldsack, D. E. 1969. *Biopolymers* 7:299–313
33. Scheraga, H. A. 1960. *J. Am. Chem. Soc.* 82:3847–52
34. Kendrew, J. C., Dickerson, R. E., Strandberg, B. E., Hart, R. G., Davies, D. R., Phillips, D. C., Shore, V. C. 1960. *Nature* 185:422–27
35. Edmundson, A. B. 1965. *Nature* 205:883–87
36. Perutz, M. F., Rossmann, M. G., Cullis A. F., Muirhead, H., Will, G., North, A. C. T. 1960. *Nature* 185:416–22
37. Braunitzer, G., Hilse, K., Rudloff, V., Hilschmann, N. 1964. *Adv. Protein Chem.* 19:1–71
38. Guzzo, A. V. 1965. *Biophys. J.* 5: 809–22
39. Prothero, J. W. 1966. *Biophys. J.* 6: 367–70
40. Prothero, J. W. 1968. *Biophys. J.* 8: 1236–55
41. Periti, P. F., Quagliarotti, G., Liquori, A. M. 1967. *J. Mol. Biol.* 24:313–22
42. Low, B. W., Lovell, F. M., Rudko, A. D. 1968. *Proc. Natl. Acad. Sci. USA* 60:1519–26
43. Kotelchuck, D., Scheraga, H. A. 1969. *Proc. Natl. Acad. Sci. USA* 62:14–21
44. Leberman, R. 1971. *J. Mol. Biol.* 55:23–30
45. Schiffer, M., Edmundson, A. B. 1967. *Biophys. J.* 7:121–35
46. Blundell, T., Dodson, G., Hodgkin, D., Mercola, D. 1972. *Adv. Protein Chem.* 26:279–402
47. Haggis, G. H. 1964. *Introduction to Molecular Biology*, ed. G. H. Haggis, pp. 36–75. New York:Wiley. 401 pp.
48. Dunhill, P. 1968. *Biophys. J.* 8:865–75
49. Zimm, B. H., Bragg, J. K. 1959. *J. Chem. Phys.* 31:526–35
50. Lewis, P. N., Go, N., Go, M., Kotelchuck, D., Scheraga, H. A. 1970. *Proc. Natl. Acad. Sci. USA* 65:810–15
51. Lewis, P. N., Momany, F. A., Scheraga, H. A. 1971. *Arch. Biochem. Biophys.* 144:576–83
52. Wu, T. T., Kabat, E. A. 1971. *Proc. Natl. Acad. Sci. USA* 68:1501–6
53. Wu, T. T., Kabat, E. A. 1973. *J. Mol. Biol.* 75:13–31
54. Robson, B., Pain, R. H. 1971. *J. Mol. Biol.* 58:237–59
55. Finkelstein, A. V., Ptitsyn, O. B. 1971. *J. Mol. Biol.* 62:613–24
56. Poland, D., Scheraga, H. A. 1970. *Theory of Helix-Coil Transitions in Biopolymers.* New York:Wiley. 797 pp.
57. Fasman, G. D., ed. 1967. *Poly-α-Amino Acids.* New York:Dekker. 764 pp.
58. Richards, F. M., Wyckoff, M. W. 1971. *Enzymes* 4:647–806
59. Birktoft, J. J., Blow, D. M. 1972. *J. Mol. Biol.* 68:187–240
60. Shotton, D. M., Watson, H. C. 1970. *Nature* 225:811–16
61. Reeke, G. N. Jr., Becker, J. W., Edelman, G. M. 1975. *J. Biol. Chem.* 250:1525–47
62. Hardman, K. D., Ainsworth, C. F. 1972. *Biochemistry* 11:4910–19
63. Poljak, R. J. 1975. *Nature* 256:373–76
64. Davies, D. R., Padlan, E. A., Segal, D. M. 1975. *Ann. Rev. Biochem.* 44:639–67
65. Ptitsyn, O. B., Finkelstein, A. V. 1970. *Biofizika* 15:757–67
66. Nagano, K. 1973. *J. Mol. Biol.* 75:401–20
67. Kabat, E. A., Wu, T. T. 1973. *Biopolymers* 12:751–74
68. Kabat, E. A., Wu, T. T. 1973. *Proc. Natl. Acad. Sci. USA* 70:1473–77
69. Chou, P. Y., Wells, M., Fasman, G. D. 1972. *Biochemistry* 11:3028–43
70. Chou, P. Y., Fasman, G. D. 1973. *J. Mol. Biol.* 74:263–81
71. Chou, P. Y., Fasman, G. D. 1974. *Biochemistry* 13:211–22
72. Chou, P. Y., Fasman, G. D. 1974. *Biochemistry* 13:222–45
73. Cook, D. A. 1967. *J. Mol. Biol.* 29:167–71
74. Ptitsyn, O. B. 1969. *J. Mol. Biol.* 42:501–10
75. IUPAC-IUB Commission on Biochemical Nomenclature. 1970. *Biochemistry* 9:3471–79
76. Venkatachalam, C. M. 1968. *Biopolymers* 6:1425–36
77. Lewis, P. N., Momany, F. A., Scheraga, H. A. 1971. *Proc. Natl. Acad. Sci. USA* 68:2293–97
78. Lewis, P. N., Momany, F. A., Scheraga, H. A. 1973. *Biochim. Biophys. Acta* 303:211–29
79. Kuntz, I. D. 1972. *J. Am. Chem. Soc.* 94:4009–12
80. Crawford, J. L., Lipscomb, W. N., Schellman, C. G. 1973. *Proc. Natl. Acad. Sci. USA* 70:538–42
81. Dickerson, R. E., Takano, T., Eisenberg, D., Kalli, O. B., Samson, L., Cooper, A., Margoliash, E. 1971. *J. Biol. Chem.* 246:1511–35
82. Chou, P. Y., Fasman, G. D. 1977. *J. Mol. Biol.* 115:135–75
83. Krigbaum, W. R., Knutton, S. P. 1973. *Proc. Natl. Acad. Sci. USA* 70:2809–13
84. Froimowitz, M., Fasman, G. D. 1974. *Macromolecules* 7:583–89

85. Froimowitz, M. 1977. *Macromolecules* 10:161–62
86. Suzuki, E., Robson, B. 1976. *J. Mol. Biol.* 107:357–67
87. Blagdon, D. E., Goodman, M. 1975. *Biopolymers* 14:241–45
88. Periti, P. 1974. *Estratto de Bollettino Chimico Farmaceutico* 113:187–218
89. Burgess, A. W., Ponnuswamy, P. K., Scheraga, H. A. 1974. *Isr. J. Chem.* 12:239–86
90. Robson, B. 1974. *Biochem. J.* 141:853–67
91. Robson, B., Pain, R. H. 1974. *Biochem. J.* 141:869–82
92. Robson, B., Pain, R. H. 1974. *Biochem. J.* 141:883–97
93. Robson, B., Pain, R. H. 1974. *Biochem. J.* 141:899–904
94. Robson, B., Pain, R. H. 1972. *Nature New Biol.* 238:107–8
95. Burgess, A. W., Scheraga, H. A. 1975. *Proc. Natl. Acad. Sci. USA* 72:1221–25
96. Maxfield, F. R., Scheraga, H. A. 1976. *Biochemistry* 15:5138–53
97. Nagano, K., 1974. *J. Mol. Biol.* 84:337–72
98. Nagano, K., Hasegawa, K. 1975. *J. Mol. Biol.* 94:257–81
99. Levitt, M. 1974. *J. Mol. Biol.* 82:393–420
100. Palau, J., Puigdomenech, P. 1974. *J. Mol. Biol.* 88:457–69
101. Lim, V. I. 1974. *Biofizika* 19:366–77
102. Lim, V. I. 1974. *Biofizika* 19:562–75
103. Lim, V. I. 1974. *J. Mol. Biol.* 88:857–72
104. Lim, V. I. 1974. *J. Mol. Biol.* 88:873–94
105. Wallace, D. G. 1976. *Biophys. Chem.* 4:123–30
106. Aubert, J. P., Loucheux-Lefebvre, M. H. 1976. *Arch. Biochem. Biophys.* 175:400–9
107. McLachlan, A. D., Walker, J. E. 1977. *J. Mol. Biol.* 112:543–58
108. Finkelstein, A. V., Ptitsyn, O. B. 1977. *Biopolymers* 16:469–95
109. Finkelstein, A. V., Ptitsyn, O. B., Kozitsyn, S. A. 1977. *Biopolymers* 16:497–524
110. Finkelstein, A. V. 1977. *Biopolymers* 16:525–29
111. Finkelstein, A. V., Ptitsyn, O. B. 1976. *J. Mol. Biol.* 103:15–24
112. Tanaka, S., Scheraga, H. A. 1976. *Macromolecules* 9:142–59
113. Tanaka, S., Scheraga, H. A. 1976. *Macromolecules* 9:159–67
114. Tanaka, S., Scheraga, H. A. 1976. *Macromolecules* 9:168–82
115. Tanaka, S., Scheraga, H. A. 1976. *Macromolecules* 9:812–33
116. Tanaka, S., Scheraga, H. A. 1977. *Macromolecules* 10:9–20
117. Tanaka, S., Scheraga, H. A. 1977. *Macromolecules* 10:305–16
118. Schulz, G. E., Barry, C. D., Friedman, J., Chou, P. Y., Fasman, G. D., Finkelstein, A. V., Lim, V. I., Ptitsyn, O. B., Kabat, E. A., Wu, T. T., Levitt, M., Robson, B., Nagano, K. 1974. *Nature* 250:140–42
119. Matthews, B. W. 1975. *Biochim. Biophys. Acta* 405:442–51
120. Kabat, E. A., Wu, T. T. 1974. *Proc. Natl. Acad. Sci. USA* 71:4217–20
121. Lewis, P. N., Bradbury, E. M. 1974. *Biochim. Biophys. Acta* 336:153–64
122. Maxfield, F. R., Scheraga, H. A. 1975. *Macromolecules* 8:491–93
123. Vorob'ev, V. I., Birshtein, T. M., Aleksanyan, V. I., Zalenski, A. O. 1972. *Mol. Biol. Moscow* 6:346–52. (Transl. 6:273–78)
124. Fasman, G. D., Chou, P. Y., Adler, A. J. 1976. *Biophys. J.* 16:1201–38
125. Argos, P., Schwarz, James, Schwarz, John 1976. *Biochim. Biophys. Acta* 439:261–73
126. Lenstra, J. A., Hofsteenge, J., Beintema, J. J. 1977. *J. Mol. Biol.* 109:185–93
127. Lenstra, J. A. 1977. *Biochim. Biophys. Acta* 491:333–38
128. Kretsinger, R. H., Barry, C. D. 1975. *Biochim. Biophys. Acta* 405:40–52
129. Wetlaufer, D. B. 1973. *Proc. Natl. Acad. Sci. USA* 70:697–701
130. Ramachandran, G. N., Ramakrishnan, C., Sasisekharan, V. 1963. *J. Mol. Biol.* 7:95–99
131. Phillips, D. C. 1970. In *British Biochemistry, Past and Present,* ed. T. W. Goodwin, pp. 11–28. London: Academic
132. Nishikawa, K., Ooi, T. 1972. *J. Phys. Soc. Jpn.* 32:1338–47
133. Ooi, T., Nishikawa, K. 1973. In *Conformation of Biological Molecules and Polymers,* ed. E. D. Bergman, B. Pullman, pp. 173–87. New York:Academic
134. Rossmann, M. G., Liljas, A. 1974. *J. Mol. Biol.* 85:177–81
135. Rossmann, M. G., Moras, D., Olsen, K. W. 1974. *Nature* 250:194–99
136. Rossmann, M. G., Argos, P. 1975. *J. Biol. Chem.* 250:7525–32
137. Richardson, J. S., Richardson, D. C., Thomas, K. A., Silverton, E. W., Davies, D. R. 1976. *J. Mol. Biol.* 102:221–35
138. Levitt, M., Chothia, C. 1976. *Nature* 261:552–58

139. Dunn, J. B., Klotz, I. M. 1975. *Arch. Biochem. Biophys.* 167:615–26
140. Kuntz, I. D. 1975. *J. Am. Chem. Soc.* 97:4362–66
141. Balasubramanian, R. 1977. *Int. J. Pept. Protein Res.* 9:157–60
142. Argos, P., Rossmann, M. G., Johnson, J. E. 1977. *Biochem. Biophys. Res. Commun.* 75:83–86
143. Lim, V. I., Efimov, A. V. 1976. *FEBS Lett.* 69:41–44
144. Richardson, J. S. 1976. *Proc. Natl. Acad. Sci. USA* 73:2619–23
145. Sternberg, M. J. E., Thornton, J. M. 1976. *J. Mol. Biol.* 105:367–82
146. Chothia, C. 1973. *J. Mol. Biol.* 75:295–302
147. Ralston, E., De Coen, J. L. 1974. *J. Mol. Biol.* 83:393–420
148. Tanaka, S., Scheraga, H. A. 1975. *Proc. Natl. Acad. Sci. USA* 72:3802–6
149. Ptitsyn, O. B., Rashin, A. A. 1975. *Biophys. Chem.* 3:1–20
150. Levitt, M., Warshel, A. 1975. *Nature* 253:694–98
151. Karplus, M., Weaver, D. L. 1976. *Nature* 260:404–6
152. De Haën, C., Swanson, E., Teller, D. C. 1976. *Biopolymers* 15:1825–33
153. Kuntz, I. D., Crippen, G. M., Kollman, P. A., Kimelman, D. 1976. *J. Mol. Biol.* 106:983–94
154. Tanaka, S., Scheraga, H. A. 1976. *Macromolecules* 9:945–50
155. Tanaka, S., Scheraga, H. A. 1976. *Macromolecules* 10:291–304
156. Crippen, G. M. 1977. *Macromolecules* 10:21–28
157. McCammon, J. A., Gelin, B. R., Karplus, M. 1977. *Nature* 267:585–90
158. Durham, A. C. H., Butler, P. J. G. 1975. *Eur. J. Biochem.* 53:397–404
159. Carter, C. W. Jr., Kraut, J., Freer, S. T., Xuong, N.-H., Alden, R. A., Bartsch, R. G. 1974. *J. Biol. Chem.* 249:4212–25
160. Banner, D. W., Bloomer, A. C., Petsko, G. A., Phillips, D. C., Pogson, C. I., Wilson, I. A., Corran, P. H., Furth, A. J., Milman, J. D., Offord, R. E., Priddle, J. D., Waley, S. G. 1975. *Nature* 255:609–14
161. Low, B. W., Preston, H. S., Sato, A., Rosen, L. S., Searl, J. E., Rudko, A. D., Richardson, J. S. 1976. *Proc. Natl. Acad. Sci. USA* 73:2991–94
162. Olsen, K. W., Moras, D., Rossmann, M. G., Harris, J. I. 1975. *J. Biol. Chem.* 250:9313–21
163. Mavridis, I. M., Tulinsky, A. 1976. *Biochemistry* 15:4410–17
164. Chou, P. Y., Adler, A. J., Fasman, G. D. 1975. *J. Mol. Biol.* 96:29–45
165. Chou, P. Y. 1974. *Biochem. Biophys. Res. Commun.* 61:87–94
166. Chou, P. Y., Fasman, G. D. 1975. *Biochemistry* 14:2536–41
167. Bentley, G., Dodson, E., Dodson, G., Hodgkin, D., Mercola, D. 1976. *Nature* 261:166–68
168. Wooton, J. C. 1974. *Nature* 252:542–46
169. Pena, C., Stewart, J. M., Paladini, A. C., Dellacha, J. M., Santome, J. A. 1975. In *Peptides: Chemistry, Structure, and Biology,* ed. R. Walter, J. Meienhofer, pp. 523–28. Ann Arbor: Ann Arbor Sci. Publ.
170. Low, T. L. K., Liu, Y. S. V., Putnam, F. W. 1976. *Science* 191:390–92
171. Smythies, J. R., Benington, F., Bradley, R. J., Bridges, W. F., Morin, R. D., Romine, W. O. Jr. 1975. *J. Theor. Biol.* 51:111–26
172. Argos, P. 1976. *Biochem. Biophys. Res. Commun.* 70:805–11
173. Brack, A., Orgel, L. E. 1975. *Nature* 256:383–87
174. Salemme, F. R., Miller, M. D., Jordan, S. R. 1977. *Proc. Natl. Acad. Sci. USA* 74:2820–24
175. Goodman, M., Moore, G. W. 1977. *J. Mol. Evol.* 9:In press
176. Vogel, H., Brushi, M., LeGall, J. 1977. *J. Mol. Evol.* 9:111–19
177. Goodman, M., Pechere, J. F. 1977. *J. Mol. Evol.* 9:131–58
178. Kawauchi, H., Li, C. H. 1974. *Arch. Biochem. Biophys.* 165:255–62
179. Garnier, J., Pernollet, J. C., Tertrin-Clary, C., Salesse, R., Casteing, M., Barnavon, M., de la Llosa, P., Jutisz, M. 1975. *Eur. J. Biochem.* 53:243–54
180. Munoz, P. A., Warren, J. R., Noelken, M. E. 1976. *Biochemistry* 15:4666–71
181. Vogt, H. P., Wollmer, A., Naithani, V. K., Zahn, H. 1976. *Hoppe-Seyler's Z. Physiol. Chem.* 357:107–15
182. Chen, C. H., Sonenberg, M. 1977. *Biochemistry* 16:2110–18
183. Bradbury, A. F., Smyth, D. G., Snell, C. R. 1976. *Nature* 260:165–66
184. Deber, C. M., Madison, V., Blout, E. R. 1976. *Acc. Chem. Res.* 9:106–13
185. Kopple, K. D., Go, A., Pilipauskas, D. R. 1975. *J. Am. Chem. Soc.* 97:6830–38
186. Snell, C. R., Smyth, D. G. 1975. *J. Biol. Chem.* 250:6291–95
187. Dunn, B. M., Chaiken, J. M. 1975. *J. Mol. Biol.* 95:497–511
188. Filippi, B., Borin, G., Marchiori, F. 1976. *J. Mol. Biol.* 106:315–24
189. Borin, G., Filippi, B., Moroder, L., San-

toni, C., Marchiori, F. 1977. *Int. J. Pept. Protein Res.* 10:27–38

190. Fink, M. L., Bodanszky, M. J. 1976. *J. Am. Chem. Soc.* 98:974–77
191. Rose, G. D., Seltzer, J. P. 1977. *J. Mol. Biol.* 113:153–64
192. Robson, B., Suzuki, E. 1976. *J. Mol. Biol.* 107:327–56
193. Ananthanarayanan, V. S., Bandekar, J. 1976. *Int. J. Pept. Protein Res.* 8:615–23
194. Ananthanarayanan, V. S., Bandekar, J. 1975. *Curr. Sci.* 44:609–11

195. Richardson, J. S. 1977. *Nature* 268:495–500
196. Sternberg, M. J. E., Thornton, J. M. 1977. *J. Mol. Biol.* 110:269–83
197. Sternberg, M. J. E., Thornton, J. M. 1977. *J. Mol. Biol.* 110:285–96
198. Sternberg, M. J. E., Thornton, J. M. 1977. *J. Mol. Biol.* 113:401–18
199. Nagano, K. 1977. *J. Mol. Biol.* 109:235–50
200. Nagano, K. 1977. *J. Mol. Biol.* 109:257–74

Ann. Rev. Biochem. 1978. 47:277–316
Copyright © 1978 by Annual Reviews Inc. All rights reserved

SOME ASPECTS OF EUKARYOTIC DNA REPLICATION

❖975

Rose Sheinin and Jerome Humbert

Department of Microbiology and Parasitology, University of Toronto,
Toronto, Ontario, Canada M5S 1A1

Ronald E. Pearlman

Department of Biology, York University, Toronto, Ontario, Canada M3J 1P3

CONTENTS

PERSPECTIVES AND SUMMARY

DNA replication in eukaryotes is a complex process. It has been extensively studied in vivo, using techniques such as DNA fiber radioautography and electron microscopy to examine replicating chromosomes. Intermediates of

277

replication have been identified and characterized by these methods, along with others that depend ultimately on equilibrium and velocity centrifugation analyses. The origin, rate, and direction of replication have been studied extensively.

The goals of molecular biologists go beyond visualization and description of the process of DNA replication. The hope is to understand the function of the many enzymatic and structural proteins involved, and their interaction with DNA. Our newly acquired concepts of chromatin structure have been invaluable in returning the focus of attention to the study of the replication of DNA complexed with histone and nonhistone chromosomal protein, and away from "naked" DNA. In vivo work reflects this, as do in vitro studies which are increasingly using chromatin-bound, rather than free DNA as template. Many enzymes of DNA metabolism from a number of eukaryotic sources are now well characterized. Mutant isolation and genetic analysis with both higher and lower eukaryotes will be necessary in order to assign specific roles to the many components required for replication.

Some of the major advances in understanding molecular details in DNA replication have come from work with DNA viruses and other systems which contain relatively small DNA complements, such as mitochondria, chloroplasts, and naturally occurring extrachromosomal genes from lower eukaryotes. Developing recombinant DNA technology is beginning to provide unique, cloned, eukaryotic DNA segments, which may constitute single genes or replicons. The rapidly evolving methods for DNA nucleotide sequencing (1, 2) should help open the way to full understanding of the molecular mechanisms of replication of these and other eukaryotic DNA molecules.

This review examines DNA replication at the level of the basic replication unit, within the DNA, and within the chromatin. We analyze the information which is issuing from genetic studies and in vitro systems and that will eventually enable us to put together elements of replication complexes. This approach has proved most crucial in reconstitution studies in prokaryotes that have led to the in vitro, de novo synthesis of DNA molecules with full biological activity (3). These, of course, are the goals of current work with eukaryotic organisms.

DNA replication has been studied extensively in a wide variety of eukaryotes, using many different experimental tools. It is not possible to analyze these here in depth for each organism. However, with few but very interesting exceptions, the replication of the DNA of eukaryotic cells and their DNA viruses is effected in a similar way. It therefore seems warranted to draw from all of these sources in discussing the developments in DNA replication in eukaryotes which have emerged since the subject was reviewed last for yeast (4–6), other eukaryotic microorganisms (7), the higher plants (8, 9), insects (10), and animal cells (9, 11–13).

Perhaps the most significant advances in our understanding of the molecular mechanism of semiconservative DNA synthesis will come from a consideration of this as part of the greater process of chromatin replication, which has been made easier as a result of major advances in our knowledge of chromatin structure (14, 15). Briefly, it is now known that DNA of 150–200 base pairs in length is packaged in nucleosome subunits, the precise size and conformation of which are determined by chromosomal proteins, in particular the histones. The nucleosome can be further dissected into a core particle that comprises a double-stranded helical DNA segment of about 140 base pairs, in association with an octameric aggregate of equimolar amounts of histones H2A, H2B, H3, and H4. In cells from different species the interparticle or linker DNA segment varies from very few to 60–80 base pairs. It is associated with histone H1 [or histone H5 or chromosomal protein H6 (16) in the case of nucleated avian erythrocytes and trout testis cells, respectively]. Nuclease digestion analyses suggest that the core particle-bound DNA has secondary and tertiary structure determined by 10 base-pair segments. A functional model for chromatin structure has been proposed which suggests that the core particles are composed of two isologously paired heterotetramers of one mole each of histones H2A, H2B, H3, and H4, arranged symmetrically about the interwoven helical DNA strands (17). The model is appealing because it would permit the half nucleosomes to participate in DNA replication, in chromatin replication, and in subsequent chromosome segregation without requiring major reorganization of DNA-protein associations.

Much information on DNA replication has been obtained from studies with organelles such as mitochondria and chloroplasts and from studies of processes such as repair replication, meiosis, and recombination. These are not analyzed here, not because of a lack of interesting results, but because the review cannot be all-encompassing. For the same reasons, details of experimental techniques used to obtain the results cited, are omitted. The literature has been reviewed up to and including July–August 1977.

PROTEINS OF DNA REPLICATION

Chromosomal DNA synthesis in the nucleus of eukaryotic cells proceeds in a temporally and spatially regulated pattern on many subchromosomal units of replication (5, 9-13, 18-27). Initiation of the DNA synthetic, or S phase of the cell cycle sets in train the duplication of DNA by a semiconservative mechanism (5, 9–13). The resultant daughter helices carry one parental DNA strand and one synthesized *de novo*. Once initiated, polymerization of nucleotides proceeds processively in the 5' to 3' direction along each template strand guided by many enzymes and other proteins that interact with the DNA.

In prokaryotes, genetic analysis and development of in vitro replication systems have permitted characterization and functional analysis of many components of the replication complex (3). In eukaryotes, the analagous systems are not yet well enough developed; however, many enzymes and proteins that participate in normal and repair replication of DNA in eukaryotic cells continue to be identified, isolated, and characterized. Putative replication complexes containing at least some of these proteins, and others, have been isolated from lower eukaryotic (28), plant (29), and animal cells (30, 31) (see section on in vitro systems). The nature of such complexes remains to be clarified in molecular terms. However, they undoubtedly comprise unwinding enzyme (32), DNA polymerase (33–35), DNA-ligating enzyme (36), single-strand DNA-dependent ATPase (37, 38), DNA-binding protein (32, 39, 40), endodeoxyribonuclease (41–45), exodeoxyribonuclease (46), RNase H (47–49), and/or other ribonucleases (50), RNA polymerase (51), and structural proteins of the chromatin (50). Discussion of only some of these proteins is presented in this paper.

Of the many proteins noted above, the DNA polymerases have been most extensively studied because of their clear involvement in DNA synthesis. Studies with intact cells (7, 35) and karyoplasts (52, 53) indicate that DNA polymerization is almost entirely confined to nuclei and to organelles including mitochondria, chloroplasts, and kinetoplasts. Detection of these proteins in soluble form in cytoplasmic extracts has been widely reported. However, it seems likely that these are leached from the organelles in which they exist in vivo, during cell fractionation.

Multiple polymerases have been extracted from both cytoplasm and nuclei from higher eukaryotic cells. These have been classified as DNA polymerase α, β, or γ on the basis of their molecular weight, their sensitivity to sulfhydryl inhibitors, and their reactions with DNA template and a variety of primer molecules (54). There is good evidence that essentially all of the polymerase α and β activity is associated with the nuclei (52, 55, 56). The DNA polymerase γ of rat liver cells or mycoplasma-free HeLa cells is closely related to, or identical with, the mitochondrial DNA polymerase from the homologous cell (57).

The DNA polymerase α of higher eukaryotes is a large molecular weight ($\geqslant$100,000), N-ethylmaleimide (NEM)-sensitive protein which has yet to be purified to homogeneity. It is usually found associated with other proteins of the replication complex, in particular a DNA-dependent ATPase of

60,000 daltons (37, 38), and DNA binding proteins of unknown activity (35, 58), one of which may be involved in stabilizing subunit interaction (59). It has been suggested that a subunit of 155,000 daltons may be the catalytic component that effects chain elongation (60). Only polymerase α catalyzes addition of polydeoxyribonucleotides to an RNA primer synthesized from the DNA template with *Escherichia coli* RNA polymerase (61). Recently it has been demonstrated that a modified form of DNA polymerase α, termed α_1, can be isolated from HeLa cells treated with cycloheximide (62).

DNA polymerase β is a small molecular weight enzyme of 43,000–45,000. This NEM-resistant enzyme has been purified to homogeneity from a number of eukaryotic sources and is immunologically distinct from other DNA polymerases (63, 64). DNA polymerase γ is a large molecular weight enzyme sensitive to NEM. It is present in very low amounts relative to polymerase α and β and has thus been very difficult to purify. The γ enzyme utilizes synthetic primer-templates such as poly $A:oligo\ dT_{10-18}$ more efficiently than DNA containing nicks and gaps (34, 35).

Multiple DNA polymerases have been described in some eukaryotic systems, but a single major enzyme is found in others. No low molecular weight β-like enzyme is detected in plants, protozoa, or fungi (65). This is an interesting observation but its significance in terms of the role of DNA polymerases in replication is not yet understood. Most of the enzymes from eukaryotic microorganisms have been described as polymerase α (65). A close evaluation of their properties, however, often shows extensive differences from mammalian α enzymes. It must be emphasized that although the α, β, γ classification system has been very valuable in describing higher eukaryotic DNA polymerases, this is not based on knowledge of the role of these enzymes with regard to in vivo DNA replication.

Two DNA polymerases have been isolated from the yeast *Saccharomyces cerevisiae* (66, 67). Immunological studies suggest that they are distinct. One, designated polymerase I (A), is similar, but not identical, to the polymerase α of animal cells. The yeast DNA polymerase II (B) resembles DNA polymerases II and III of prokaryotic systems, in that it carries with it a $3' \rightarrow 5'$-exonuclease activity (67). DNA synthesis on a single-stranded DNA template catalyzed by DNA polymerase I is enhanced by yeast RNA polymerases. The significance of this observation is somewhat obscure since some stimulation occurs even in the absence of ribonucleoside triphosphates (68).

In the slime mold *Dictyostelium discoideum* a single NEM-sensitive DNA polymerase of molecular weight 127,000 has been isolated (69). Two large molecular weight enzymes, designated A and B, have been isolated from *Euglena gracilis* (70, 71). Polymerase A appears to be predominantly nuclear, and the B enzyme predominantly cytoplasmic (72). Both may be oligomers of a common 3S subunit, perhaps in association with other dissimilar subunits (73), which may confer on the polymerase B its DNase and RNase H activities. Three DNA polymerases have been partially purified from macronuclei of *Paramecium* (74). These have the same molecular weight (90,000–100,000), but can be distinguished by their chromatographic and catalytic properties.

A single DNA polymerase activity, partially purified from exponentially growing *Tetrahymena pyriformis,* exhibits nuclease activity as well as other activities (75). The native molecular weight of the enzyme, as determined by gel filtration in high ionic strength conditions, is 90,000. However, SDS-polyacrylamide gel electrophoresis suggests that a 45,000-dalton subunit may be associated with the enzyme (Ganz and Pearlman, unpublished). In *Tetrahymena,* synthesis of a DNA polymerase is induced by a variety of chemical and physical agents that inhibit DNA replication (76–78). This enzyme, which may form part of a replication complex under these circumstances (28), is likely to be the same enzyme as that described above (Ganz and Pearlman, unpublished; Keiding, personal communication), and different from the mitochondrial DNA polymerase (79).

The fungus *Ustilago maydis* has at least two DNA polymerases. One, with associated 3'–5' exonuclease activity (80, 81), exhibits a molecular weight of 80,000–100,000 in high salt conditions (82). However, there is evidence for two subunit monomers of 50,000 and 55,000 molecular weight. The second DNA polymerase can be distinguished from the former because it has the ability to add polydeoxyribonucleotides to an $(rU)_{20}$ primer, hydrogen bonded to a poly dA template (G. Banks, personal communication). Of eight mutants of *U. maydis* that are defective in chromosomal DNA replication, the *pol* 1-1 strain exhibits temperature-sensitive DNA polymerase activity (83, 84), which indicates that the 100,000-dalton enzyme is essential for this process.

DNA REPLICATION UNIT (RU): SIZE AND ORGANIZATION

The concept of a unit of DNA replication initially emerged from studies of the replicon in prokaryotes (85). This term incorporates both genetic and molecular criteria to define the replication of the single DNA element of bacteria and their viruses. There is already genetic, biochemical, and biophysical evidence that indicates that the simple genomes of the viruses of eukaryotic cells, e.g. papovaviruses (86, 87), adenoviruses (88, 89), and herpesviruses (90, 91) can similarly be classified as replicons. The tools for cellular and molecular genetic analysis of eukaryotic cells are still far from ideal, which makes the identification of replicons in their complex genomes not generally feasible at present. The immediate future holds promise for some selected DNA molecules that can be isolated in clones. These include the DNAs of organelles like mitochondria (92, 93), kinetoplasts (94, 95), and chloroplasts (96, 97), particularly those of lower eukaryotes, because these can already be genetically manipulated. They embrace the naturally occurring, plasmid-like units (98) which in some instances carry the entire genome for ribosomal RNA (99–101), as well as the genomes of a number of tumor viruses that are integrated into cellular DNA (102–104) or exist as episomes (105, 106). Within this group of DNA molecules should be included the rapidly increasing number of eukaryotic DNA segments that can be experimentally cloned and studied either in prokaryotes or eukaryotes (107).

Although we cannot at present identify the many replicons of the eukaryotic genome, we are able to define replication units in structural, biochemical, and molecular terms (9, 12, 13, 108). This is a segment of the DNA double helix that is operationally defined from origin to terminus by a number of different techniques. RUs of differing sizes, usually with S values $\geqslant 60$, have been reported. Taking into account different techniques and criteria for size assessment, the evidence suggests that RUs may vary from 4 μm in length (109), i.e. 8×10^6 daltons; 1.3×10^4 base pairs, to 280 μm (110), i.e. 560×10^6 daltons; 90×10^4 base pairs, with a majority being 10–100 μm. This generalization derives from studies with yeast (4–6, 111), eukaryotic microorganisms generally (7, 12, 13, 112–116), higher plants (9, 117–122), insects (10, 123–125), fish (11), birds (11, 13), amphibia (11, 126), and a broad spectrum of mammalian cells (12, 13, 30, 109, 110, 127–147) and their viruses (148–156). A homogeneous RU of 6.93 μm has been identified within the "satellite" DNA of kangaroo rat cells (157). Similar identification of unique RUs appears to be imminent in the case of the three separate satellite DNA species of *Drosophila virilis* (124). These are related heptanucleotides that exhibit specific base-pair substitutions.

Evidence is accumulating (110, 123, 125, 126, 128, 135, 138, 140, 143, 144, 145) to support the concept (12, 13, 127) that RUs are tandemly arranged within the chromosomal DNA in clusters that can vary from very few, e.g. two (126), to many, e.g. 250 (20, 110, 123, 135, 142). The latter appears to be the case for satellite DNA (19, 124, 157), for the repeat units of the genes of ribosomal RNA (99–101), and for histones (158, 159). Such DNA clusters, the synthesis of which is coordinately regulated, may be examples of the elusive replicons of eukaryotic cells. It has been suggested that in the lower eukaryotes, e.g. yeast (4, 6), the coordinately regulated RU cluster may constitute the entire DNA thread of single, small chromosomes. In mammalian cells these clusters can be located within a single chromomere or chromosome band (10, 123–125, 160, 161), for example the late-replicating satellite DNA of the centromeric heterochromatin (19). In *Drosophila*, tandemly tied RUs comprising seven chromomeres have been identified (10, 123).

RATE OF DNA SYNTHESIS

The rate of movement of the replication complex during synthesis of DNA is very similar in yeast (4, 13, 111) and other eukaryotic microorganisms, (12, 13), higher plants (117, 121), insects (10, 123–125), and animal cells (9, 11–13, 108, 127, 128, 135, 139, 142, 143) and their viruses (153–156). This rate is 0.1–2 μm (2–40 X 10^5 daltons; 1–15 X 10^3 nucleotide residues) of DNA per minute (adjusted to 37°C). Notwithstanding the apparent uniformity of results, it is becoming clear that the rate of fork progression does vary in response to genetic controls, to temporal controls, and to the environment.

Mutant cells have been isolated among the lower and higher eukaryotes which exhibit defects in chain elongation (see section on genetic studies). DNA replication proceeds in a highly ordered and complex pattern on all chromosomes throughout S phase (18–27, 162, 163). The sequence of movement of replication complexes on chromosomal DNA appears to be species-specific (19, 22, 27, 162–165) and, unless experimentally disturbed, is repeated successively generation after generation (19, 21, 162–166). In lower eukaryotes such as yeast (4–6) and *Physarum polycephalum* (167) 80–90% of the nuclear DNA is replicated in the first 10–25% of S phase, whereas in mammalian (7, 13, 27, 129) and plant (8) cells the major fraction ($\geqslant$90%) is synthesized during mid S phase. The rate of fork progression along RUs at any given time during S phase appears to be the same (13, 127–129). However, in some mammalian (13, 129, 166) and plant cells (121) the relative rate changes throughout S, such that fork progression is more rapid when more DNA is being duplicated.

Modification of the rate of chain elongation may play a role in the fine-tuning regulation of DNA replication. It had been suggested that such modification may be a key determinant in maintaining the relatively similar duration of S phase in cells with widely varying DNA content, but this appears not to be so since the rate of DNA synthesis, as well as the size range of RUs, is quite similar in disparate eukaryotic organisms like the pea plant (121) with a haploid content of 48 pg, and mammalian cells (12, 13, 110, 127–129, 142, 143) with 5 pg of DNA per haploid nucleus (168). Even within a single phylum, the amphibia, similar observations were made in a comparative study with cells from four species with DNA content varying 40-fold (126). Since the average rate of fork progression and the average size of the RUs is similar in all eukaryotes, it seems likely that major regulation of the rate of DNA replication is effected by controlling the number of RUs initiated at any time during S (10, 12, 13).

ORIGIN AND TERMINUS OF REPLICATION UNITS

Initiation of replication is carefully controlled at the level of the RU in all organisms. The simplest model for such regulation is that of the interaction of a specific initiator protein (27, 163, 165, 169) with a genetically determined origin sequence.

In most instances, DNA synthesis is initiated at an origin within the RU, and proceeds bidirectionally by the movement of two replication forks toward the two distant termini (12, 13). There is, however, increasing evidence that unidirectional fork movement does occur during the replication of a small proportion of the DNA of insect (10, 123–125), amphibian (126), and mammalian cells (127, 128, 143, 170), and of the DNA of polyoma virus (148, 150), SV40 virus (148–150), adenovirus (151), and herpesviruses (153, 156). Proteins are implicated in regulation at this level, since unidirectional synthesis is enhanced when normal replication is impeded by inhibitors of protein synthesis (13, 128, 171)

It seems likely that characterization of specific origin sequences of chromosomal RUs will have to await their clonal isolation (107, 124, 157–159). However, tentative approaches of a more general nature have already been made (172–174). It has been suggested that initiation of synthesis on RUs requires that these have a superhelical conformation (172), perhaps with a specific origin sequence (173). In CHO cells, X-irradiated to prevent initiation (see section on inhibitors of DNA synthesis), renaturation kinetic studies have shown that the DNA most distant from the origin of the RUs is greatly enriched in repeated sequences that are distinct from satellite or "foldback" DNA (174). It was concluded that the origin sequences must therefore contain, on an average, a higher proportion of unique sequences than does the bulk chromosomal DNA.

Significant progress has been made in identifying and characterizing initiation sequences on the genetically pure RUs of the papovaviruses (87, 149, 150, 175–182), which have a double-stranded, covalently closed genome with a superhelical configuration, the adenoviruses (183–186), their associated satellite viruses (187–190), other parvoviruses (191–195), and herpesviruses (196–199). The generalization that emerges is that initiation sequences are located within larger DNA segments that contain palindromes[1] or tandem repeats of unique DNA lengths. In the case of linear DNA molecules, e.g. adenovirus, herpesvirus, and parvovirus DNA, the larger sequences are terminally redundant.

The origin of replication of the polyoma and SV40 genomes has been located by restriction enzyme analysis (87, 149, 150, 175–177). By use of enzymatic and chemical techniques a full sequence determination has been made directly on the DNA (87, 149, 150, 175, 177, 182) and on its mRNA transcript (178–181). The 73 base-pair origin segments contain an AT-rich symmetrical sequence, flanked by GC-rich regions with a twofold rotational axis of symmetry; that is, a palindrome. The latter have the properties of foldback or hairpin loop DNA (182). Essentially similar information is issuing from the sequence studies of the putative origins of replication for the other DNA viruses noted. The terminal DNA segments all appear to contain palindromic regions associated with linear foldback DNA.

The question of whether termination is determined by specific DNA sequences, or whether it is a nonevent resulting from the interaction of two processing replication complexes, has yet to be resolved. Arguments against specific termini have been based on DNA fiber radioautography studies (12, 13) and on specific investigations with SV40 and polyoma virus, the latter being the more persuasive. Thus, apparently normal termination has been observed during the replication of defective viral DNA molecules that lack the DNA segment at which synthesis ceases normally in the intact molecule, and in SV40 DNA in which the origin sequence has been enzymatically transplanted to other loci on the covalently closed DNA (87, 149, 150, 175, 177). In all instances replication was found to initiate at the origin sequence and process bidirectionally to a site 180° distant.

These studies prove very little about the presence or absence of termination sequences in eukaryotic DNA. To conclude that they do not exist may be premature in the face of the evidence for precise temporal and spatial regulation of initiation and termination on widely separated, individual, and tandemly tied RUs on the several chromosomes in eukaryotic nuclei (9–13, 19–27, 162–166, 169). In some eukaryotes unique chromosomal

---

[1]A palindrome is a word or sentence *reading* the same in both directions. This term applies to double-stranded DNA regions with an axis of twofold rotational symmetry. Only such regions will *read* (transcribe) identically from both ends of the region (99), e.g.

$$A-A-T-T-G \mid C-A-A-T-T \longrightarrow$$

$$\longleftarrow T-T-A-A-C \mid G-T-T-A-A$$

DNA sequences are made at specific stages during growth and development. Examples of these are the Z-DNA synthesized during the zygotene phase of meiosis in *Lilium* species (200, 201), specific DNA segments synthesized during larval development of the mosquito (202, 203), specific chromosomal DNA segments in larvae of *Drosophila pseudoobscura* (204), the amplification or endoreduplication of ribosomal RNA genes in *Xenopus laevis* (205), and DNA segments of polytene chromosomes of plants (206). Such regulated DNA replication argues for specificity of initiation and termination both intra- and interchromosomally. It is therefore not surprising to come upon evidence for DNA sequences that do specify termination in some viral DNA molecules that are clonally pure, and therefore amenable to direct study.

The adeno-associated virus DNA carries two apparently unique terminal nucleotide sequence repetitions (187–190) that comprise up to 4% of the total molecule. A 19 base-pair fragment that maps in the 35–65 nucleotide sequence at the 3'-end of the linear genome is known to be a terminally redundant, palindromic sequence. Analogous evidence is emerging from studies with other parvoviruses. These include the Kilham rat virus (191, 194), H1 (195) and the minute virus of mice (192, 193), the DNA of the herpesviruses (196–199), and the adenoviruses (183–186, 207–209).

It is possible that structurally and functionally similar initiation-termination sequences are present on adjoining RUs of the eukaryotic chromosomal DNA. These would not have been detected by the experiments discussed above, in which it was concluded that termination lacks specificity.

INTERMEDIATES OF DNA REPLICATION

One of the most vexing problems of DNA replication in eukaryotes has been that of the existence, kind, and number of intermediates of DNA replication (9, 12, 13). Because the DNA helix comprises two strands of opposite polarity, and because the known enzymes of DNA chain elongation exhibit a single specificity with respect to direction of deoxyribonucleotide addition (35), it was surmised that synthesis from one strand is continuous, while that from the other is discontinuous. Much of the earlier work was taken to support this hypothesis. However, the more recent in vivo studies unequivocally reveal discontinuous DNA synthesis on both template strands in yeast (4), other eukaryotic microorganisms (12, 13), higher plants (117–121), insects (10, 123, 125), and a variety of mammalian cells and their viruses (9, 12, 13, 109, 127–134, 140, 141, 143–155). Analyses with in vitro systems similarly reveal totally discontinuous synthesis of the DNA of mammalian cells (132, 210–212) and polyoma virus (213, 214). The latter data are at variance with some that indicate that the in vitro synthesis of polyoma and SV40 are semidiscontinuous (12, 215, 216).

Various other putative intermediates of DNA replication have been reported (see section on chromatin replication). The problem has been to decide which of the many nascent DNA isolates described are true intermediates of DNA replication, and which are artifacts of the experimental conditions employed to mark and extract the DNA for subsequent analysis. On balance, the in vivo studies suggest that S phase DNA replication of eukaryotic chromosomal DNA proceeds in three distinct stages. The primary event results from initiation of RUs and gives rise to single-strand DNA intermediates with a maximum size of 4–7S (6.6–7.4 $\times$ 10^4 daltons; 220–280 nucleotide residues) (9, 12, 13, 112–114, 120, 122, 125, 130, 132, 134, 135–138, 140, 142, 145, 146, 217). The secondary elongation process appears to proceed through two major size classes of intermediates. The first encompasses segments of 6–26S (9, 112–114, 118, 120, 122, 125, 130, 132, 134, 135, 138) and the second, segments of 20–100S (112–114, 118, 122, 125, 130, 134, 135, 138). The overlapping size ranges and maximal unit sizes reported by individual investigators may reflect experimental systems and analytical procedures used, or they may be determined by the specific cell types examined. The final intermediates of this series are the RUs that are arranged singly or in clusters. The third stage in DNA replication, the maturation of chromosomal DNA, results from the joining together of these RUs. (12, 112–114, 127, 128, 135, 138, 140, 142, 143).

Primary intermediates (~4S) (132, 210, 211, 218–222, 490), secondary intermediates (6–14S) (132, 210, 211, 218, 222, 223), and 20–100S intermediates (144, 210, 211, 219, 221, 490) have been detected by means of in vitro studies with both higher and lower eukaryotic cells. Similar observations have been made with extracts of mammalian cells infected with polyoma (12) and SV40 virus (12, 216, 224, 225). Formation of very large molecular weight DNA equivalent to that of nuclear chromosomes has not been observed in any in vitro system developed to date, probably because these are deficient in essential proteins.

DNA fiber radioautography, and density shift and velocity sedimentation analyses (9–13) indicate that the synthesis of the primary and secondary intermediates is indeed processive. However, the final stage of DNA replication occurs very rapidly (112–114, 127, 128) and in some cases distant in time from the earlier events (13, 112–114, 127, 128, 134, 138, 140, 142, 143). Maturation of chromosomal DNA is a quantum process, which results from the joining together of previously replicated DNA segments (13, 112–114, 128, 133–135, 138, 140, 142, 143).

One of the major puzzles of DNA replication is the intricacy and complexity of the regulation of synthesis. Various models of regulation have been put forward over the years, a majority of which have focused primarily on the enzymes and/or proteins that function in DNA replication, or on the constraints imposed by the structure of the DNA. More recently it has been proposed that processive DNA replication along the length of a single chromosomal DNA double helix is also regulated by chromatin structural organization (17, 226, 227). Such a model predicts that three classes of single-strand nascent DNA intermediates should be identifiable. The first would comprise DNA segments of length no greater than 140–280 nucleotide residues, i.e. the length of the DNA of the core particle or nucleosome subunit of the particular cell under study. The second would contain DNA molecules that are oligomers of the subunit length DNA, and could be very large. The third would be the full-length chromosomal DNA strands.

In this context it is of interest to examine the relationship between the putative intermediates of DNA replication and the DNA lengths embraced within the chromosomal subunit population. As noted in the first section, all eukaryotic chromatin is organized into core particles that carry a unit DNA of 140 base pairs. A single-strand segment of this size would be expected to have a sedimentation velocity of 3–4S and a molecular weight of 4.2×10^4. The DNA of the nucleosome subunit is longer by up to 80 nucleotide residues, depending upon the cellular species in question. Thus the newly replicated nucleosomal DNA should sediment at 4–5S and have a molecular weight of 6.6×10^4. Clearly these are the properties of the class of primary intermediates described above.

If DNA replication within the chromatin is indeed regulated by nucleosomal organization then one might expect to isolate intermediates that are oligomers of the pertinent DNA length. This is the case for putative dimeric segments of 500–800 nucleotide residues which sediment at 6–8S (15–24 $\times 10^4$ daltons), oligomeric segments of 1000–4000 nucleotide residues which sediment at 10–16, 20–26, and 30–36S ($3–12 \times 10^5$ daltons), and polymeric segments equivalent to RUs with 66,000 nucleotide residues, which sediment at 20–100S (20×10^6 daltons). Chromomeric clusters of $\geqslant 210,000$ nucleotide residues which sediment at 380 and 500S have been reported (10, 20, 160).

STUDIES WITH INHIBITORS OF DNA SYNTHESIS

Confirmatory evidence for the "three-stage" model of DNA replication noted above comes from experiments in which DNA synthesis has been deliberately interrupted. Inhibition of protein synthesis by any mechanism results in almost immediate interruption of DNA replication. This is observed in cells treated with chemicals that block the initiation or elongation steps of protein synthesis, or that produce faulty proteins by substitution of amino acids (12, 13, 112, 114, 128, 133, 143, 145, 171, 226, 228–235). It is also seen in mammalian cells that are permitted to express a temperature-sensitive defect in amino acid activation (236) or in a polypeptide of the 60S ribosomal subunit (237–239), which is an essential component of the protein-synthetic machinery.

In the case of mammalian cells, interruption of DNA replication due to inhibition of protein synthesis is manifest at the level of initiation of RUs, fork progression, and the joining together of intermediates of all size classes. Thus dependency on *de novo* polypeptide formation is seen throughout S phase. Yeast, however, can complete an already initiated S period in the presence of cycloheximide (228). In *Tetrahymena* this drug interferes with the conversion of secondary intermediates to large molecular weight DNA (235). The same does not occur if *Tetrahymena* is treated with puromycin and sparsomycin.

DNA synthesis in plasmodia of *Physarum polycephalum* proceeds through primary and secondary intermediate formation early in the S phase, and maturation of replicated units into chromosomal DNA is much delayed. The most recent studies with wild-type mold and a strain that is resistant to cycloheximide by virtue of having an altered 60S ribosomal subunit (114) have shown that the processes most sensitive to inhibition of protein synthesis are entry into S phase and maturation of chromosomal DNA. Less sensitive, but affected nonetheless, are initiation of DNA synthesis on RUs and fork progression. By judiciously choosing the time of addition of, and exposure to, cycloheximide, single-strand DNA intermediates of 4S, 6–7S, 12S, 25S, or 36–37S can be accumulated (112, 114).

Interruption of DNA replication apparently at the level of conversion of primary to secondary intermediates has been achieved with a number of chemicals. Thus mammalian cells treated with alkylating agents such as ethane methane sulfonate and N-methyl-N'-nitro-N-nitrosoguanidine, which are known to interact with specific base residues, accumulate newly replicated, single-strand segments of about 7–10S (240, 241). This is an unexpected finding, unless inhibitory cross-linking does not occur randomly, but is confined to specific DNA sequences determined by the binding of chromosomal proteins.

The antimetabolite 5'-fluoro-2'-deoxyuridine (FdUrd), which inhibits thymidylate synthetase and DNA polymerase activity (242), blocks cells early after entry into S phase (13, 22, 109, 123, 127, 130, 141, 243–245). There is some suggestion that FdUrd retards fork progression (244) and alters the sequence of DNA replication (243). This may or may not be related to the finding that FdUrd-treated cells accumulate 7–10S single-strand DNA fragments (245).

A similar phenomenon is observed in cells treated with hydroxyurea, which inhibits the ribonucleoside diphosphate reductase (246), affects chain elongation (247), and causes accumulation of newly replicated, 7S single-strand fragments (248–250). Temporary inhibition of semiconservative replication by this drug (247, 251) results in arrest of mammalian cells (135, 244, 247–252) and yeast (253, 254) early in S phase. When hydroxyurea treatment is terminated, extension of the small molecular weight DNA synthesized in the presence of drug proceeds continuously with the transient formation of molecules of approximately 1.6 and 2.4×10^8 daltons. The rate of synthesis (about 0.9 μm/min) was in excess of the normal (0.2 μm/min), perhaps reflecting the synchrony of the system, which promotes rapid joining of preformed and completed units. Once again evidence was obtained for large coordinated units of replication (135).

Three studies with mutant mammalian cells are relevant to the present discussion. Fibroblasts from human patients with Bloom's Syndrome exhibit retarded fork progression (255) and accumulate heterogeneous, single-strand DNA of maximum molecular weight smaller than that expected for an RU (256). Interruption of normal chain elongation of RUs has also been observed in cultured lymphocytes originally derived from humans with megaloblastic anemia (257) and in temperature-sensitive (*ts*) A1S9 mouse L cells incubated at nonpermissive temperature (258, 259). The latter accumulate primary and secondary intermediates of no larger than $3–5 \times 10^6$ daltons for at least 6 hr at apparently normal rate (Schwartz and Sheinin, unpublished observations), which suggests that reinitiation proceeds on already-initiated RUs, or that abnormal initiations occur.

Methotrexate, which inhibits the dihydrofolate reductase (260) and also interferes with nucleotide polymerization (261), permits human lymphoblasts to continue to replicate DNA in the semiconservative mode for about 6 hr. During this time, chain propagation proceeds at approximately the normal rate, but single-strand segments, which sediment at about 80S, accumulate (261). These are rapidly converted to normal chromosomal DNA upon removal of the drug, which suggests that methotrexate prevents maturation of newly replicated clusters of RUs. Inhibitory doses of X rays (174, 262–264) prevent initiation of synthesis on RUs in mammalian cells. X-irradiated cells do not accumulate small molecular weight DNA. DNA replication in progress continues apparently normally for about 60 min. Under these conditions segments of DNA approximating 10^9 daltons, or 25 RUs, are duplicated, which suggests that some replicons in mammalian cells may be very large indeed (263, 264).

RNA IN DNA REPLICATION

In vitro studies on the replication of prokaryotic DNA molecules (265), as well as the recognition that all known DNA polymerases are unable to initiate DNA synthesis directly on a DNA template (35), led to the search for a primer molecule. The studies with prokaryotes revealed that an RNA segment of some 40–50 nucleotide residues copied from a single-stranded

DNA template serves as primer for attachment of deoxyribonucleotides at the 3'-end. (265) Elongation and maturation of RNA-attached DNA segments to chromosomal DNA requires that the primer RNA be cleaved from the primary intermediates by a ribonuclease with a very particular specificity.

Much of the evidence for the involvement of RNA in DNA replication in eukaryotes has been reviewed previously (12, 13). Most experiments that seek to detect nascent, covalently linked RNA-DNA intermediates as fragments denser than bulk DNA, when spun to equilibrium under denaturing conditions in CsCl or Cs_2SO_4 density gradients, have been negative (12, 13, 133, 266; Chan and Sheinin, unpublished observation); however, some results have been positive (12, 13, 131, 139). Other experimental approaches using intact mammalian cells have generally proved negative (12, 13, 146).

The best evidence that RNA may act as a primer in DNA replication comes from nearest neighbor frequency analyses of nascent DNA-RNA segments synthesized in vivo by *P. polycephalum* (267) and in vitro systems derived from *S. cerevisiae* (218), from a number of mammalian cells (212, 268, 269), and from nuclei of mouse fibroblasts infected with polyoma virus (270). These indicate that 8–11 oligoribonucleotides are joined through their 3'-termini to the 5'-ends of the growing DNA strand. The 5'-end of this "initiator" RNA is not unique, but carries adenine or guanine residues. Analysis of the RNA-DNA junction reveals that all four common ribonucleotides and deoxyribonucleotides are present with about the same frequency. Other evidence suggests that each replicating fragment contains one RNA-DNA junction.

Such covalent attachment of a short oligoribonucleotide to DNA is consistent with a primer function. However, direct evidence for an initiating role for RNA in eukaryotes has not yet been presented. To date, there is little indication for specificity of such a putative initiator RNA (270). Fully competent in vitro systems for the replication of DNA from viruses such as polyoma and SV40 (225, 271–273), which have specific and now well-characterized origins of initiation (see section on the origin and terminus of replication units), should be useful in elucidating this point. Similar expectations derive from studies with *Tetrahymena* rDNA, the origin of replication of which is AT rich (274). It has now been shown that oligo A stimulates DNA synthesis in an in vitro extract from *Tetrahymena* using homologous rDNA as template (Ganz and Pearlman, unpublished observations).

Coisolation of newly made large molecular weight RNA with nascent DNA has been reported in experiments using meristematic cells of germinating *Vicia faba* seeds (275, 276), permeabilized yeast cells [where covalent attachment of RNA to DNA was demonstrated (218)], and isolated human sperm nuclei (277) (where the RNA remained tightly bound to the chromatin). The significance of these observations is not clear. They may reflect the known involvement of ribosomal RNA synthesis in initiation of S phase (see section below), or they may signal the presence of a contaminating RNA polymerase.

INITIATION OF DNA REPLICATION

Regulation of initiation of DNA replication occurs at (a), entry into the S phase, (b), initiation of synthesis on RUs during S phase, and (c), initiation of synthesis on RUs during other intervals of the growth cycle.

Regulation of traverse of the G_1/S interface during the cell division cycle of mammalian cells (27, 163, 165, 278), the higher plants (8,200), and the yeast *S. cerevisiae* (5, 6, 162, 279) appears not to result simply from initiation of early replicating RUs. Genetic, physiological, and biochemical evidence suggests that progression into S phase is determined by a regulatory protein synthesized during the G_1 or pre-DNA synthetic phase, which acts as a positive effector to de-repress DNA replication. This process undoubtedly involves activation of the protein synthetic machinery, one step of which is the formation of ribosomal RNA (162, 278, 280). This phenomenon is particularly prevalent in plant cells moving from dormancy to active growth (8, 232, 275, 276, 281, 282) and quiescent mammalian cells progressing from a G_0 or pseudo-G_1 state into cell cycle (278, 280).

Regulation of initiation of DNA replication on RUs is also effected by protein. This was first indicated by the demonstration that initiation of synthesis of primary intermediates is totally prevented by inhibition of *de novo* polypeptide formation (see the section on inhibitors of DNA synthesis). Little progress has been made in identifying specific proteins that serve as initiators of RUs on eukaryotic chromosomal DNA. Speculations abound concerning the increasing number of DNA-binding proteins that go in search of a function (32). Many of these proteins specifically stimulate DNA synthesis in vitro catalyzed by homologous DNA polymerases (39, 283–288). On the basis of similar observations made with prokaryotic DNA-binding proteins known to function in initiation of replication (3, 289), it is tempting to suggest an analogous role for these eukaryotic proteins. One class of such molecules that does have the properties of an initiator protein has received extensive study. They may collectively be described as virus-encoded proteins that are required for DNA replication of the virion genome, and in some cases, for initiation of cell DNA synthesis.

The most extensively characterized protein of this class is the T (or tumor) antigen (290) of SV40 virus, the genetic information for which is encoded in the A locus of the virus genome (86, 87, 149, 150, 291, 292).

There is now strong evidence that the T-antigen is required for initiation of viral DNA replication (86, 87, 149, 150, 293–295) and for induction of synthesis of cellular DNA (296–299). Perhaps the most elegant studies directed towards the latter point are those which showed that T-antigen synthesis and chromatin replication follow directly upon microinjection of early SV40 mRNA into quiescent monkey kidney cells (297). Recent studies suggest that induction of cellular DNA synthesis by SV40 T-antigen proceeds by an indirect mechanism that involves activation of the G_1/S interface traverse (297–300). This may be mediated by the binding of the T-antigen to cellular DNA.

The T-antigen, which is found almost exclusively in the nucleus of infected or virus-transformed cells (290), has now been purified almost to homogeneity. It exists as a multiple oligomer, the monomeric molecular weight of which is 80,000 (293, 301–303). The T-antigen is a phosphorylated (303) protein that exhibits DNA-binding specificity for double-stranded nucleic acid. It binds to the DNA isolated from cells that support SV40 replication, and from cells that restrict virus multiplication, but undergo neoplastic transformation as a result of infection (290).

The mechanism of action of T-antigen in initiation of viral DNA replication also remains to be clarified. Using virus mutants *ts* in the A gene, it has been demonstrated that incubation of infected cells at the nonpermissive temperature leads to immediate cessation of initiation without affecting ongoing replication of SV40 DNA molecules (149, 150, 290–295). These experiments suggest that the T-antigen must be bound to effect initiation, and that, once bound, it is insensitive to heat inactivation. Direct evidence that the T-antigen does bind to SV40 DNA molecules at the initiation sequence has been obtained (293, 294, 304, 305). However, binding to two other viral DNA sequences has been observed (149, 150, 175) and is perhaps related to autoregulation of synthesis of the T-antigen (306).

It seems likely that the polyoma T-antigen functions in a similar way. It too is implicated in initiation of viral DNA synthesis and in induction of cellular DNA synthesis (86, 87, 150). Although it is the first of this class of proteins to be identified, it has proved to be difficult to purify and characterize (307, 308).

Evidence has been obtained for a protein, distinct from the virus-specific T-antigen, that initiates replication of the adenovirus genome (309–317). This phosphoprotein (314, 317), which also seems to affect chain propagation (311), has a molecular weight 74,000, is encoded in the genomes of adenoviruses 2, 5, and 12 (and perhaps other related viruses), and is classified immunologically as a type-C tumor antigen (318). The adenovirus DNA-binding protein shows specificity for single-stranded virus DNA (315), and in particular for the terminal sequences of the genome identified as the sites of initiation of DNA replication (see the section on the origin and terminus of replication units). Solubilization of virus-encoded DNA-binding proteins from herpesviruses has now begun (319, 320). It seems likely that these will prove to have functions similar to those described above.

REPLICATION OF SPECIAL DNAs

One of the most promising approaches to clarifying the molecular events of DNA replication is afforded by the existence of specific nucleolar extrachromosomal genetic units. Those coding for rRNA in *Tetrahymena* (99, 321, 322) and *Physarum* (101) have already been isolated. In these, rDNA molecules are palindromes containing two genes coding for rRNA arranged as an inverted repeat. In starve-refed (323) and heat-synchronized *Tetrahymena* (324), preferential replication of the extrachromosomal nucleolar DNA has been demonstrated, which makes it possible to isolate density-labeled replicating rDNA molecules by CsCl density gradient centrifugation. Electron microscopic examination (325) of such rDNA from starve-refed *Tetrahymena* has revealed that replication proceeds bidirectionally from the center (the axis of symmetry) of the linear palindromic molecules. There is still some question as to whether or not rDNA replication occurs only during G2 of the normal vegetative cell cycle (326, 327).

The extrachromosomal rDNA of *Physarum* is synthesized throughout the intermitotic period with the exception of the first hour of the S period (328). Over-replication of this extrachromosomal nucleolar DNA (rDNA) can be demonstrated in artificially prolonged mitotic cycles (329). Although the total population of rDNA molecules doubles once per mitotic cycle, not every rDNA molecule is replicated, but some are replicated more than once (330). Electron microscopy showed formation of replication "eyes" centered over the middle portion of the rDNA, with initiation occurring about 45% and 33% from one end of the DNA. Replication is completed in about 10 minutes and proceeds bidirectionally (330). An earlier report, which describes the replication of the rDNA of *Physarum* (with a circular genome length of 3.9 μm) by a rolling circle mechanism, contradicts the above (331). Circles may be an artifact, but two different structures and two different modes of replication for the rDNA of *Physarum* cannot at present be ruled out (332).

During the sexual cycle of *Tetrahymena* a single chromosomally integrated gene coding for rRNA in the micronucleus (333) is amplified and appears as many extrachromosomal rDNA molecules in the macronucleus (334, 335). Gene amplification has been well characterized for rDNA in oocytes of *Xenopus laevis* and in many other systems (205, 336). A rolling circle mechanism is involved in at least some stage of this process (337, 338). The suggestion that RNA-dependent DNA polymerase is involved in the early stages of gene amplification, possibly in the formation of the first extrachromosomal copy, has been questioned (339). *Tetrahymena* may be a useful experimental system for the study of the early stages of gene amplification, particularly the formation of the first extrachromosomal gene copy (Pearlman et al, in preparation).

Examples of other forms of differential replication of rRNA genes in eukaryotes are known. Some of these are rDNA magnification in *Drosophila* (100, 340, 341) and yeast (342), and rDNA compensation and independent replication of rDNA in polytene chromosomes in *Drosophila* (340). Molecular details of the mechanisms of these processes are not understood, although circular rDNA molecules have been observed during rDNA magnification in *Drosophila* (341). These data support a model requiring the formation of extrachromosomal copies for rDNA magnification, a mechanism at least formally similar to rDNA amplification in *Xenopus* oocytes.

An unusual gene amplification occurs with the kinetoplast DNA of the order of protozoa, the Kinetoplastida. The term kinetoplast denotes a basophilic body situated at the base of a flagellum. It is associated with a modified region of a mitochondrion (94). The kinetoplast DNA of at least the genera *Crithidia* and *Leishmania* consists mainly of an association of approximately 27,000 covalently closed, 0.8 μm circular molecules, which are apparently held together in a definite ordered manner by topological interlocking (94, 95, 343). All of the circular molecules of one organism appear to carry the same genetic information. Light and electron microscope autoradiography, as well as buoyant density analysis, have revealed that doubling of kinetoplast DNA in a single generation results from the replication of each circular molecule rather than from repeated duplication of a small fraction of the circles. These same experiments show that replication is semiconservative. Synthesis begins at two sites separated by 180° at the periphery of the networks (94, 95, 343, 344). Long linear molecules have been observed during replication, which suggests that it does not proceed only on covalently closed minicircles.

An exciting aspect of DNA synthesis as it relates to chromosome duplication during meiosis has received intensive study (200, 201). In cells of the meiocytes of many *Lilium* species a majority of the DNA is duplicated normally during S. However, about 0.3% of the nuclear DNA undergoes delayed replication during zygotene in synchrony with chromosome pairing. This unique zygotene DNA (Z-DNA) is present on each chromosome. The Z-DNA, once replicated, remains flanked by single-stranded regions, and can be released as 32S single-strand pieces. The gaps left by Z-DNA replication are ultimately repaired by gap-filling. Using the "Black Beauty" strain of *Lilium,* which is defective in pachytene repair replication, it was demonstrated that gap-filling on either side of the Z-DNA segments occurs very late during meiosis, perhaps just before entry into the first metaphase.

These studies on Z-DNA and its replication are of particular interest for our understanding of DNA replication in eukaryotes. The nucleotide sequences that flank the Z-DNA and that remain single-stranded for as long as 5 days, are very likely to carry information for initiation and termination of replication. Their organization within the chromatin substructure is also of interest since they remain protected from degradation at all times, and from repair for an interval specified by other meiotic events.

The Z-DNA is clearly unique. Its synthesis is required for chromosome pairing. It is not united with the remainder of the newly made DNA of each chromosome until after disjunction occurs. Its continued study will undoubtedly provide information concerning the regulation of chromatin replication.

GENETIC STUDIES

The key to genetic analysis of DNA replication is the availability of mutant cells with specific, well-characterized defects in the DNA synthetic machinery. Few such eukaryotic cells have been isolated as yet. Those that are known include human cells that have been isolated from patients with genetically determined diseases (Table 1) and *ts* cells for entry into (Table 2) and traverse through (Table 3) S phase.

A majority of the available eukaryotic cells that exhibit defective DNA synthesis are mutant in a non-S-phase function (Table 2) or in proteins that affect repair of damaged DNA (Table 1). The subject of repair replication is beyond the scope of this review. However, attention is drawn to these cells

Table 1 Naturally occurring human cells mutant in DNA replication

Disease	Function affected	References
Xeroderma pigmentosum (XP)	Repair replication of DNA damaged by UV-, X-, and γ-radiation and by chemicals	345–361
Ataxia telangiectasia (AT)	Repair replication of DNA carrying specific base modifications	143, 355, 362, 363
Progeria	Repair replication of UV- and γ-radiation damage	345, 364–367
Fanconi's Anemia (FA)	Repair replication	143, 368–371
Bloom's Syndrome (BS)	Elongation of primary and secondary intermediates of DNA replication	255, 256
Megaloblastic Anemia (MA)	Elongation of primary and secondary intermediates of DNA replication	257

Table 2 Eukaryotic cell cycle mutants that block entry into S phase

Cell	Mutant designation	Cell cycle block	Function affected	References
S. cerevisiae	cdc 28	Nuclear plaque duplication	Entry into S phase	6, 279, 372
	cdc 4	Nuclear plaque separation, meiosis	Entry into S phase	6, 279, 372, 373
	cdc 7	Nuclear division	Entry into S phase	6, 279, 372
	cdc 6	Nuclear division	Entry into S phase	6, 279, 372
Chinese hamster	ts K/34C	G_1	Membrane glycoprotein synthesis	374
Chinese hamster lung (H–1)	115–46 115–47 115–53 ts–13 ts–14			237–239
	ts–41		60S ribosomal subunit synthesis	
Hamster	ts 546	Mitosis	Metaphase to anaphase transition	375, 376
	ts 665	Mitosis, cytokinesis	Prophase progression	377
	BF 113	G_1	Entry into S phase	378
Hamster-BHK 21	ts AF8	G_1	Entry into S phase	379–382
Chinese hamster (CH)	K–12	G_1	Entry into S phase	383–392
CHO	cs3–D4	G_0	Progression from G_0 to G_1	393
	11–32			394
	ts–H1		Leucyl-tRNA synthetase	236
Monkey kidney (BSC–1)	ts–13 ts–14			395–397
Mouse leukemia	ts 2	Mitosis, cytokinesis		398
(L5178Y)	ts 3			399
Mouse	ts B54	G_1	Entry into S phase	382, 400

because in some instances specific enzyme proteins have been implicated and because it is likely that some enzymes that participate in repair will also be involved in normal replication in eukaryotes (200, 422, 423).

Human cells isolated from patients with Bloom's Syndrome and megaloblastic anaemia are discussed in the section on inhibitors of DNA synthesis. Both appear to be defective in a protein involved in DNA chain propagation.

Table 3 Eukaryotic cells temperature-sensitive (*ts*) in DNA replication

Cell	Mutant designation	*ts* defect in nuclear DNA synthesis	References
Saccharomyces cerevisiae	*cdc* 2, *cdc* 8, *cdc* 21	Establishment of S phase	6, 279, 372, 401
Ustilago maydis	*ts*–220	Elongation and/or maturation	83, 84, 402–406
	ts–207, 432, 346	DNA synthesis	
	ts–20	Elongation	
	tsd 1–1	Elongation	84, 404, 407
	ts 84	Elongation	402
	pol 1–1	DNA polymerase A (I)	84, 402, 404–407
Hamster (CHO-K1)	13B11	DNA synthesis	408
Mouse BalB/C	*ts* 2	DNA synthesis	409, 410
Mouse-L	*ts* A1S9	Elongation of RU	258, 259, 411–417
	ts C1	DNA synthesis	412, 414–416, 418–421

Those *ts* cells which are mutant in a non-S-phase function that also affects DNA replication are of interest for the information they yield on the regulation of DNA replication by specific proteins, and on the interrelatedness of cell cycle events. Some of these mutant cells are discussed elsewhere in this paper. Of particular relevance here are those cells that are deficient in DNA replication per se (Table 3).

Genetic analysis of the cell division cycle (*cdc*) mutants of yeast has yielded a pathway which orders the genes relative to a hydroxyurea (HU) sensitive process that affects DNA chain elongation (372). The *cdc* 2, 8, and 21 genes, which are relatively closely linked to the HU-sensitive locus, code for products that are required throughout the entire DNA replication period. The significance of the observation that the DNA polymerase activity of *cdc* 8 cells is considerably higher than expected in cells grown at the nonpermissive temperature (424) remains to be established. The demonstration that strain *cdc* 21 (401) is a thymidylate auxotroph is very important. It should facilitate studies on DNA replication in yeast, which lacks thymidine kinase activity, and has therefore not been amenable to study by the usual dThd and BrdUrd-labeling techniques used with other cells. It should also make it easier to isolate cells that are indeed mutant in DNA synthesis.

The search for *dna*^ts mutants in *Ustilago maydis,* begun in 1970, has been very productive. Eight mutants in chromosomal DNA replication have been isolated, one of which, *pol* 1-1 (84, 404) exhibits *ts* DNA polymerase activity (see section on proteins of DNA replication). A provisional assignment of the nature of the *ts* defects of the *U. maydis* mutants isolated has been shown in Table 3. Single-strand DNA segments of 5–15S are accumulated by *ts*-20 and *tsd* 1-1 cells incubated at the nonpermissive temperature. This suggests that the block occurs at the level of formation of primary and secondary intermediates, that is, in polydeoxyribonucleotide chain elongation. The significance of the observation that *tsd* 1-1 produces excessive quantities of RHase H (407) is not clear. It is tempting to relate it to an RNA primer excision event during DNA replication (see section on RNA in DNA replication).

A tentative suggestion as to sequence assignment of gene products comes from a study of the double mutant *ts*-84 : pol 1-1. At the nonpermissive temperature, the latter cells synthesize amounts of DNA that are additive with respect to the parents carrying single mutations. Kinetic studies of DNA synthesis and development of sensitivity to ultraviolet (UV) light (402) suggest that the *ts*-84 gene product acts after *pol* 1-1. Mutants of *U. maydis* that are sensitive to UV light and are auxotrophic for pyrimidines have also been isolated (425, 426). Their study provides circumstantial evidence that the enzymes that function in the repair of UV-damaged DNA are part of the normal replication complex.

Four of the many mammalian cell isolates that show *ts* DNA synthesis appear to be mutant in a protein directly involved in DNA replication. BalB/C *ts* 2 cells carry a recessive mutation (410). The conclusion that the *ts* 2 mutation affects cellular DNA replication is strengthened by the demonstration that at the nonpermissive temperature these cells restrict the replication of polyoma DNA, a process known to be dependent upon one or more proteins of cellular DNA synthesis (409).

The *ts* A1S9 and *ts* C1 mutants of mouse L cells have received the most extensive study. Complementation analysis in intraspecific hybrids (R. Mankovitz, personal communication) indicates that two separate loci are involved. The *ts* C1 defect is corrected by information carried on the human X-chromosome (419). It has been demonstrated that the *ts* A1S9 gene codes for a protein that is involved in DNA chain elongation (258, 259) (see also section on inhibitors of DNA synthesis). Upon shift to the nonpermissive temperature (38.5°C) cells complete the S phase in progress and are subsequently blocked early in the next DNA-synthetic phase (258, 416; Sheinin, unpublished observations). The function of the *ts* C1 gene product has not yet been identified. It undergoes very rapid inactivation upon upshift to 38.5°C, and exhibits a half-life of ~3 hr. The *ts* C1 cells appear to be affected in a terminal event of chromosomal DNA synthesis (416, 420, 421). The *ts* A1S9 and *ts* C1 mutations can be distinguished by the fact that polyoma virus is replicated at 38.5°C in the former, but not the latter, cells.

It is intriguing that temperature inactivation of the *ts* A1S9 and *ts* C1 gene products, two distinct proteins of DNA replication, give rise to similar, but not identical, pleiotropic effects. Thus, when these cells are incubated at the nonpermissive temperature, nuclear semiconservative DNA replication is suppressed and repair replication is activated after an interval equivalent to that required for full cell cycling (414). Mitochondrial DNA replication proceeds apparently normally, but at a reduced rate which is coupled to that observed for nuclear DNA synthesis (412). *De novo* chromatin synthesis is interrupted once the *ts* defect is fully established, although the preformed chromatin is fully conserved (415; Sheinin and Lewis, unpublished observations). In addition there is a marked structural disorganization of the condensed chromatin (413, 416); that within the nucleoplasm becomes disaggregated, and forms small clusters, while that associated with the nucleolus and the nuclear membrane is relatively unchanged. This phenomenon may be related to the shutdown in histone formation which also occurs.

An interesting phenomenon is emerging from the study of heat inactivation of DNA synthesis and chromosomal histone formation in *ts* mutants. Thus, in those cells that appear to be dna^{ts} (*ts*A1S9 and *ts* C1), histone synthesis is tightly coupled to DNA synthesis (415; Sheinin and Lewis, unpublished observations). In contrast, in hamster CH-K12 cells, which are *ts* in a G_1 phase function required for subsequent initiation of S phase (383–392), histone biosynthesis continues at 60% of control levels after DNA synthesis has been inactivated to 14% of control levels (387). Also these cells exhibit other modifications in protein synthesis, in particular greatly enhanced synthesis of a cytoplasmic protein not yet identified.

The studies with *ts* A1S9 and *ts* C1 cells suggest that the immediate trigger of histone and chromatin replication during S phase may be the establishment of replication complexes at the origin of the earliest replicating units. They are in accord with earlier (50, 226, 427) and more recent (163, 428–430) investigations that demonstrate the regulated, coordinate synthesis of DNA, histones, and other chromosomal proteins during S phase in mammalian cells. Thus, histone synthesis is maintained for the normal lifetime of the histone mRNA (50, 427, 431), provided semiconservative replication is maintained (414). The work with *ts* A1S9 and *ts* C1 cells is in agreement with other experiments (432, 433), which show that histone synthesis is not activated by repair replication. They provide striking evidence that DNA replication must be considered within the context of chromosome structure and function.

CHROMATIN REPLICATION

The tight coupling between histone and DNA synthesis in mammalian cells has already been mentioned. It has been confirmed in studies with systems as diverse as animal cells induced to synthesize DNA by infection with certain DNA viruses (50, 427, 434, 435) or by treatment with a variety of growth factors (163), and in the lower eukaryote *S. cerevisiae* (436). Histone modification, an important factor in packaging newly replicated DNA (437, 438) occurs primarily in S phase cells (163, 427, 439) immediately upon *de novo* synthesis of the histones (440, 441). However, both phosphorylation and acetylation have been observed throughout the cell cycle. Because of the extreme diversity of nonhistone chromosomal proteins it is very difficult to assess cell cycle specificity of synthesis for this constituent of chromatin. However, some evidence indicates that certain nonhistone chromosomal proteins are synthesized in association with DNA replication (430, 439, 442). Both histones and nonhistone chromosomal proteins are fully conserved during cell growth (415, 427, 443; Sheinin and Lewis, unpublished observations).

A majority of studies on chromosome replication in eukaryotes have, until recently, focused exclusively on the nucleic acid. The revolution in our thinking about the structure and organization of nuclear chromatin has prompted a fresh look at DNA replication. Studies on nascent DNA synthesis in mammalian cells (30, 428, 429, 444–449) and in fertilized sea urchin eggs (450), have revealed that it is associated with chromatin that can be distinguished from the bulk chromatin by its density, by its protein and enzyme content (30), and by its sensitivity to digestion by endogenous and exogenous endonucleases (138, 450–452). Most recently these kinds of studies have been interpreted as indicating that the nucleosomes are less densely packed along the newly made, as opposed to the mature, chromosomal DNA. Prolonged incubation of cells prior to analysis showed that the newly replicated chromatin, both DNA and protein, acquire the properties of mature chromatin within 2–15 min, depending upon the cell under study (138, 450–452).

These studies suggested that DNA synthesis proceeds as a component of chromatin replication. A direct attack on this problem was made on the replicating chromatin of CHO cells using nucleases to explore the association of DNA replication units with nucleosomal structural units (138). Analysis of the nascent DNA revealed that it sediments at 4S and thus resembles the primary intermediates of DNA replication. This 4S nascent DNA was associated with a nuclease-resistant subnucleosomal segment.

Clearly DNA replication is initiated and proceeds within subchromosomal nucleosome units. Such experiments suggest that chromatin-carrying DNA that is in the process of replication is destabilized at the replication fork, which makes it sensitive to mechanical shear and to attack by endonucleases and proteases. The maturation process confers stability and enzyme resistance upon the growing DNA helix, presumably through interaction with newly made chromosomal proteins. Fragility of the replicating chromatin may explain the isolation of a variety of apparent intermediates of DNA replication. Double-stranded DNA carrying newly replicated material and exhibiting single-strand breaks or gaps are now routinely identified, as are small molecular weight double-stranded segments with one fully nascent strand (12, 13, 30, 131, 139, 141, 145, 217). Small DNA fragments, both of whose strands are newly synthesized, are considered to result from the annealing of complementary single strands. (30, 145).

Recent work has been directed to understanding the mechanisms by which nascent DNA is packaged with protein during chromatin replication. It is clear that histones become associated with nascent DNA within nucleosomal subunits (17, 50, 138, 226, 229, 231–234, 443–452). Most experiments indicate that newly made histones are not preferentially associated with this DNA but are randomly dispersed over the entire chromatin (233, 428, 429, 444–449). They also suggest that the newly made DNA initially acquires only preformed nucleosomes as it emerges from the replication complex. This phenomenon is particularly clear for chromatin replication in cells treated with cycloheximide or puromycin to prevent *de novo* protein synthesis (17, 226, 229, 231–234). Such cells continue to replicate their DNA for several hours, but at a reduced rate of chain elongation. The DNA-associated nucleosomes are structurally normal; however, they are less densely packed along the newly made DNA, as evidenced by the sensitivity of the chromatin to nuclease attack.

The mechanism by which parental and newly made nucleosomes are distributed to replicating eukaryotic DNA is still not clear. Conservative chromatin replication, which would predict that parental nucleosomes are retained on only one daughter strand, is clearly contradicted by the experimental evidence. Cooperative and random distribution of nucleosomes to both daughter strands has been proposed (17, 233). This model would require that equal numbers of nucleosomes, both preformed and newly formed, be distributed to both growing DNA helices. In cells treated with inhibitors of protein synthesis, the nucleosome : DNA ratio would be one half the normal value, and the nucleosomes would be of parental origin. Another model predicting nonrandom assortment of nucleosomes has also been proposed (234).

On balance the evidence favors the cooperative, random model of chromatin assembly. Support for this conclusion comes from electron microscopic studies of replicating polyoma and SV40 chromosomes (453). The nucleosome distribution on replicating viral DNA was the same as that on the unreplicated DNA portions. In infected cells treated with puromycin, the preformed nucleosomes were distributed at half the usual density along both arms of the replicating DNA. During normal viral DNA replication, newly synthesized histones were preferentially associated with replicating

DNA and assembled rapidly as nucleosomes on both daughter double helices. It is at present not clear whether these results, which differ from those cited above for cellular chromatin, represent a particular situation determined by the interaction of virus-specified proteins with the replicating DNA. Our understanding of the integrated process of chromatin replication is still far from complete.

STUDIES WITH IN VITRO SYSTEMS

Despite the extensive characterization of many enzymes and proteins implicated in DNA synthesis (see section on proteins of DNA replication), their precise role in the eukaryotic cells from which they were isolated remains for the most part unknown. Conclusions concerning the function of these have generally rested on observations of altered levels of activity in cells proceeding from a nonsynthetic state to active DNA synthesis (35). This approach is useful but not definitive. It requires supplementation by studies of replication in vitro. These permit the use of biochemical and genetic experiments for functional analysis of the components of DNA and chromatin replication in eukaryotes. The major intent is to minimally disturb the DNA-synthetic machinery and its association with the chromatin, thereby encouraging DNA replication faithful to that observed in vivo.

DNA synthesis has been studied in permeabilized whole cells (132, 211, 218, 454–463), in lysates of whole cells (211, 224, 417, 464–466), in isolated nuclei (132, 144, 214, 215, 220–223, 417, 467–490), and in subnuclear preparations (30, 216, 225, 491–496) obtained from a variety of lower and higher eukaryotes and in some instances from virus-infected cells. Studies with such in vitro systems have concentrated on elucidating the mechanism of DNA synthesis and on identifying the proteins that participate in DNA replication.

A majority of the work, some of which has been reviewed previously (12, 13), has been concerned with demonstrating that semiconservative DNA synthesis does indeed occur in vitro. This has been established for the DNA of eukaryotic cells (211, 221, 455, 459, 462, 478) and for viral DNA molecules (464). In some instances repair replication has been detected (495, 497, 498), initiated perhaps at single-strand breaks introduced during cell or template fractionation.

Most evidence suggests that de novo initiation at specific sites of origin does not occur in vitro. This conclusion depends upon the demonstration that single-strand DNA newly replicated in vitro is covalently linked to preexistent cell (132, 211, 221, 222, 478, 490) or viral DNA (224). In some instances, however, a significant portion of newly replicated DNA is not linked to the growing point preexisting in vivo (222, 461). This might reflect repair-type synthesis or breakage of a small portion of newly made strands by shearing forces. Thus a major deficiency of currently available in vitro systems is their inability to initiate DNA replication *de novo*. Most permit chain elongation even though, in general, the rate of DNA synthesis achieved in whole cell extracts is only 10–50% of that observed in homologous intact cells (132, 211, 224, 417, 462). In nuclei this rate is reduced even further to 5–30% of in vivo levels (478, 490).

Another perhaps more exciting prospect for in vitro studies rests on the possibility for dissecting out the various enzymes and proteins of DNA and chromatin replication, thereby facilitating their chemical and functional characterization. This promise is emphasized by recent studies with cell-free preparations of *ts* A1S9 (417) and *ts* C1 mouse L cells (Humbert and Sheinin, unpublished observations) which are temperature-sensitive in two distinct processes of DNA replication (see section on genetic studies). Extracts from cells grown at the nonpermissive temperature exhibit the expected *ts* phenotype when examined in vitro. In the case of *ts* A1S9 cells, the affected protein has been localized to the nucleus. Similar observations have been made with permeabilized yeast cells (218).

Studies of complexes exhibiting replicase activity have been done in mammalian (494, 496), sea urchin (499), and viral systems (486, 491, 500, 501). The nature, number, and function of the components have been suggested in some instances. Some complexes have been shown to contain a DNA polymerase α (492, 500, 501), β (500, 501), and γ (501), the stability and function of which require polyamines (487, 499), ATP (470, 499), and other less well-defined proteins (480, 492, 496, 499); some of these are synthesized during G_1 phase (492, 502). The histones, H1 in particular, can play a role in modulating the activity by altering the conformation of the replicating complex (455).

Evidence for components that participate in initiation of DNA replication has been obtained using cytoplasmic extracts of amphibian cells (271, 465), yeast (466), and mammalian cells (223). The experiments with yeast cytoplasm (466) are of interest because extracts from a number of *ts cdc* mutants (see section on genetic studies) incubated at the restrictive temperature fail to stimulate initiation. A cytoplasmic extract of SV40-infected CV-1 cells, which exhibits DNA repair synthesis, has been crudely fractionated to yield endonuclease, DNA polymerase β, and DNA ligase activities (495, 497, 498). Other effects of cytoplasmic extracts on cell (132, 221–223, 465, 466, 474, 481, 484, 485, 492, 496) and viral DNA (216, 468, 472, 479, 502) synthesis in vitro have been reported. However, no clear knowledge of the functions affected or proteins involved is available.

Replication of SV40 DNA and plasmid DNA containing *Xenopus laevis* rDNA has been followed by electron microscopic observation of the replicating molecules in cytoplasmic extracts of unfertilized *X. laevis* eggs (271).

A number of factors have been identified and a tentative suggestion for their roles in replication has been presented. The initiation (I) factor appears to be an endonuclease, possibly sequence-specific. DNA polymerase X-II with α-type activity, an RNA polymerase II, and a DNA unwinding protein have been implicated in an early event of viral DNA replication. Subsequent fork progression depends upon a DNA polymerase with γ-type activity, RNase H, and a DNA relaxation protein.

The next phase in the study of in vitro DNA replication should involve intact chromatin. This has been achieved with "minichromosomes" of SV40-infected cells (216, 225, 500). Chromatin isolated from HeLa cell nuclei (494) supports DNA synthesis on its endogenous DNA at a rate that reflects the cell cycle distribution of the cells of origin. The endogenous activity can be stimulated by added HeLa cell DNA polymerase α, β, and γ. rDNA chromatin, which can now be isolated from *Tetrahymena* (504, 505), is also capable of catalyzing endogenous DNA synthesis (Ganz and Pearlman, unpublished observations). The usefulness of isolated chromatin for the in vitro study of DNA replication is enhanced by the demonstration that protein redistribution, particularly of histones, does not occur in isolated nuclei (221).

The process of chromatin assembly has been separated from replication using SV40 DNA and a cell-free extract of *X. laevis* eggs (272). The product formed in vitro is indistinguishable from the SV40 chromosome made in vivo with respect to size, conformation, superhelical folding of the DNA, and nucleosome substructure. The assembly process proceeds from, and is dependent upon, a histone pool, even in the absence of ongoing DNA or protein synthesis. It requires a heat-sensitive factor as yet uncharacterized. Some specificity is indicated since it depends on a protein that can seal single-strand breaks produced in the double-stranded, covalently closed template when it is added to the oocyte extract (273).

Clearly much progress has been made in the last few years unravelling the complex and interrelated processes of DNA and chromatin replication. Many questions remain unanswered but the future looks bright for continued advances.

ACKNOWLEDGMENTS

The work of the authors is supported by the Medical Research Council of Canada, the National Research Council of Canada, and the National Cancer Institute of Canada. We are grateful to the many colleagues who sent us preprints and discussions of their ongoing work, those who generously contributed their editorial expertise, and most particularly our secretarial staff.

Literature Cited

1. Sanger, F., Coulson, A. R. 1975. *J. Mol. Biol.* 94:441–48
2. Maxam, A. M., Gilbert, W. 1977. *Proc. Natl. Acad. Sci. USA* 74:560–64
3. Wickner, S. 1978. *Ann. Rev. Biochem.* 47:000–00
4. Petes, T. D., Newlon, C. S., Byers, B., Fangman, W. L. 1974. *Cold Spring Harbor Symp. Quant. Biol.* 38:9–16
5. Williamson, D. H. 1974. In *Cell Cycle Controls,* ed. G. M. Padilla, I. L. Cameron, A. Zimmerman, pp. 143–52. New York: Academic
6. Carter, B. L. A. 1975. *Cell* 6:259–68
7. Prescott, D. M. 1970. *Adv. Cell Biol.* 1:57–117
8. Van't Hof, J. 1974. In *Cell Cycle Controls,* ed. G. M. Padilla, I. L. Cameron, A. Zimmerman, pp. 78–85. New York: Academic
9. Taylor, J. H. 1974. *Int. Rev. Cytol.* 37:1–20
10. Blumenthal, A. B., Kriegstein, H. J., Hogness, D. S. 1974. *Cold Spring Harbor Symp. Quant. Biol.* 38:205–24
11. Callan, H. G. 1974. *Cold Spring Harbor Symp. Quant. Biol.* 38:195–203
12. Edenberg, H. J., Huberman, J. A. 1975. *Ann. Rev. Genet.* 9:245–84
13. Hand, R. 1978. In *Cell Biology, A Comprehensive Treatise,* ed. L. Goldstein, D. M. Prescott, 7: In press
14. Kornberg, R. D. 1977. *Ann. Rev. Biochem.* 46:931–54
15. Finch, J. T., Lutter, L. C., Rhodes, D., Brown, R. S., Rushton, B., Levitt, M., Klug, A. 1977. *Nature* 269:29–36
16. Levy, B. W., Wong, N. C. W., Dixon, G. H. 1977. *Proc. Natl. Acad. Sci. USA* 74:2810–14
17. Weintraub, H., Worcel, A., Alberts, B. 1976. *Cell* 9:409–17
18. Kim, M. A., Johannsmann, R., Grzeschik, K.-H. 1975. *Cytogenet. Cell Genet.* 15:363–71
19. Back, F. 1976. *Int. Rev. Cytol.* 45:25–64
20. Willard, H. F., Latt, S. A. 1976. *Am. J. Hum. Genet.* 28:213–27
21. Bostock, C. J., Christie, S., Lauder, I. J., Hatch, F. T., Mazrimas, J. A. 1976. *J. Mol. Biol.* 108:417–33
22. Adegoke, J. A., Taylor, J. H. 1977. *Exp. Cell Res.* 104:47–54
23. Dooley, D. C., Ozer, H. L. 1977. *J. Cell Physiol.* 90:337–50
24. Nakagome, Y. 1977. *Exp. Cell Res.* 106:457–61
25. Dooley, D. C., Ozer, H. L. 1978. *Cell.* In press
26. Balazs, I., Schildkraut, C. L. 1976. *Exp. Cell Res.* 101:307–14

27. Prescott, D. M. 1976. *Adv. Genet.* 18:99–177
28. Westergaard, O., Johnson, B. 1973. *Biochem. Biophys. Res. Commun.* 55:341–49
29. Clay, W. F., Katterman, F. R. H., Bartels, P. G. 1975. *Proc. Natl. Acad. Sci. USA* 72:3134–38
30. Genta, V. M., Kaufman, D. G., Kaufmann, W. K., Gerwin, B. I. 1976. *Nature* 259:502–3
31. Yoshida, A., Cavalieri, L. F. 1977. *Biochim. Biophys. Acta* 475:42–53
32. Champoux, J. J. 1978. *Ann. Rev. Biochem.* 47:000–00
33. Holmes, A. M., Johnston, I. R. 1975. *FEBS Lett.* 60:233–43
34. Wintersberger, E. 1977. *Trends Biochem. Sci.* 2:58–61
35. Weissbach, A. 1977. *Ann. Rev. Biochem.* 46:25–47
36. Söderhäll, S., Lindahl, T. 1976. *FEBS Lett.* 67:1–8
37. Hachmann, H. J., Lezius, A. G. 1976. *Eur. J. Biochem.* 61:325–30
38. Otto, B. 1977. *FEBS Lett.* 79:175–78
39. Hoch, S. O., McVey, E. 1977. *J. Biol. Chem.* 252:1881–87
40. Otto, B. 1977. *Eur. J. Biochem.* 73:17–24
41. Mechali, M., De Recondo, A. M. 1975. *Eur. J. Biochem.* 58:461–66
42. Pedrini, A. M., Ranzani, G., Pedrali Noy, G. C. F., Spadari, S., Falaschi, A. 1976. *Eur. J. Biochem.* 70:275–83
43. Otto, B., Knippers, R. 1976. *Eur. J. Biochem.* 71:617–22
44. Lavin, M. F., Kikuchi, T., Counsilman, C., Jenkins, A., Winzor, D. J., Kidson, C. 1976. *Biochemistry* 15:2409–14
45. Wang, E.-C., Furth, J. J. 1977. *J. Biol. Chem.* 252:116–24
46. Byrnes, J. J., Downey, K. M., Black, V. L., So, A. G. 1976. *Biochemistry* 15:2817–23
47. Banks, G. R. 1974. *Eur. J. Biochem.* 47:499–507
48. Tashiro, F., Mita, T., Higashinakagawa, T. 1976. *Eur. J. Biochem.* 65:123–30
49. Modak, M. J., Marcus, S. L. 1977. *J. Virol.* 22:243–46
50. Elgin, S. C. R., Weintraub, H. 1975. *Ann. Rev. Biochem.* 44:725–74
51. Roeder, R. G. 1976. In *RNA Polymerase,* eds. R. Losick, M. Chamberlin, pp. 285–329. Cold Spring Harbor, NY: Cold Spring Harbor Lab.
52. Herrick, G., Spear, B. B., Veomett, G. 1976. *Proc. Natl. Acad. Sci. USA* 73:1136–39

53. Veomett, G., Shay, J., Hough, P. V. C., Prescott, D. M. 1976. *Methods Cell Biol.* 13:1–6
54. Weissbach, A., Baltimore, D., Bollum, F., Gallo, R., Korn, D. 1975. *Eur. J. Biochem.* 59:1–2
55. Lynch, W. E., Surrey, S., Leiberman, I. 1975. *J. Biol. Chem.* 250:8179–83
56. Foster, D. N., Gurney, T. Jr. 1976. *J. Biol. Chem.* 251:7893–98
57. Bolden, A., Noy, G. P., Weissbach, A. 1977. *J. Biol. Chem.* 252:3351–56
58. Mechali, M., De Recondo, A. M. 1978. *Nucleic Acids Res.* In press
59. Mosbaugh, D. W., Stalker, D. M., Probst, G. S., Meyer, R. R. 1977. *Biochemistry* 16:1512–18
60. Holmes, A. M., Hesslewood, I. P., Wickremasinghe, R. G., Johnston, I. R. 1977. *Biochem. Soc. Symp.* 42: In press
61. Spadari, S., Weissbach, A. 1975. *Proc. Natl. Acad. Sci. USA* 72:503–7
62. Noy, G. P., Weissbach, A. 1977. *Biochim. Biophys. Acta* 447:70–83
63. Smith, G. S., Abrell, J. W., Lewis, B. J., Gallo, R. C. 1975. *J. Biol. Chem.* 250:1702–9
64. Brun, G. M., Assairi, L. M., Chappeville, F. 1975. *J. Biol. Chem.* 250: 7320–23
65. Chang, L. M. S. 1976. *Science* 191:1183–85
66. Wintersberger, E. 1974. *Eur. J. Biochem.* 50:41–47
67. Chang, L. M. S. 1977. *J. Biol. Chem.* 252:1873–80
68. Plevani, P., Chang, L. M. S. 1977. *Proc. Natl. Acad. Sci. USA* 74:1937–41
69. Loomis, L. W., Rossomando, E. F., Chang, L. M. S. 1976. *Biochim. Biophys. Acta* 125:469–77
70. McLennan, A. G., Keir, H. M. 1975. *Biochem. J.* 151:227–38
71. McLennan, A. G., Keir, H. M. 1975. *Biochem. J.* 151:239–47
72. McLennan, A. G., Keir, H. M. 1975. *Biochim. Biophys. Acta* 407:253–62
73. McLennan, A. G., Keir, H. M. 1975. *Nucleic Acids Res.* 2:223–37
74. Tait, A., Cummings, D. J. 1975. *Biochim. Biophys. Acta* 378:282–85
75. Crerar, M., Pearlman, R. E. 1974. *J. Biol. Chem.* 249:3123–31
76. Westergaard, O., Pearlman, R. E. 1969. *Exp. Cell Res.* 54:309–13
77. Keiding, J., Westergaard, O. 1971. *Exp. Cell Res.* 64:317–22
78. Westergaard, O., Marcker, K. A., Keiding, J. 1970. *Nature* 227:708–10
79. Westergaard, O., Lindberg, B. 1972. *Eur. J. Biochem.* 28:422–31
80. Banks, G. R., Yarranton, G. T. 1976. *Eur. J. Biochem.* 62:143–50
81. Yarranton, G. T., Banks, G. R. 1977. *Eur. J. Biochem.* 77:521–27
82. Banks, G. R., Holloman, W. K., Kairis, M. V., Spanos, A., Yarranton, G. T. 1976. *Eur. J. Biochem.* 62:131–42
83. Jeggo, P. A., Unrau, P., Banks, G. R., Holliday, R. 1973. *Nature New Biol.* 242:14–16
84. Jeggo, P. A., Banks, G. R. 1975. *Mol. Gen. Genet.* 142:209–24
85. Jacob, F., Brenner, S., Cuzin, F. 1963. *Cold Spring Harbor Symp. Quant. Biol.* 28:329–48
86. *Cold Spring Harbor Symp. Quant. Biol.* 1975. 39:9–234
87. Fried, M., Griffin, B. E. 1977. *Adv. Cancer Res.* 24:67–113
88. Philipson, L., Pettersson, U., Lindberg, U. 1975. *Virol. Monogr.* Vol. 14
89. Flint, J. 1977. *Cell* 10:153–66
90. *Cold Spring Harbor Symp. Quant. Biol.* 1975. 39:657–772
91. Kaplan, A. S., ed. 1976. *Herpesviruses,* New York: Academic
92. Saccone, C., Kroon, A. M., eds. 1976. *The Genetic Function of Mitochondrial DNA.* Amsterdam: North-Holland
93. Borst, P. 1977. *Trends Biochem. Sci.* 2:31–34
94. Wolstenholme, D. R., Renger, H. C., Manning, J. E., Fouts, D. L. 1974. *J. Protozool.* 21:622–31
95. Simpson, A. M., Simpson, L. 1976. *J. Protozool.* 23:583–87
96. Bucher, Th., Neupert, W., Sebald, W., Werner, S., eds. 1977. *Genetics and Biogenesis of Chloroplasts and Mitochondria,* Amsterdam: North-Holland
97. Sager, R. 1977. *Adv. Genet.* 19:287–340
98. Livingston, D. M. 1977. *Genetics* 86:73–84
99. Engberg, J., Andersson, P., Leick, V., Collins, J. 1976. *J. Mol. Biol.* 104: 455–70
100. Stanfield, S., Helinski, D. R. 1976. *Cell* 9:333–45
101. Vogt, V. M., Braun, R. 1976. *J. Mol. Biol.* 106:567–87
102. Neer, A., Baran, N., Manor, H. 1977. *Cell* 11:65–71
103. Mayer, A. J., Ginsberg, H. S. 1977. *Proc. Natl. Acad. Sci. USA* 74:785–88
104. Evans, R. M., Shoyab, M., Drohan, W. N., Baluda, M. A. 1977. *J. Virol.* 21:942–49
105. Adams, A., Bjursell, G., Kaschka-Dierich, C., Lindahl, T. 1977. *J. Virol.* 22:373–80
106. Werner, F.-J., Bornkamm, G. W.,

Fleckenstein, B. 1977. *J. Virol.* 22:794–803

107. Sinsheimer, R. L. 1977. *Ann. Rev. Biochem.* 46:415–38
108. Roti Roti, J. L., Painter, R. B. 1977. *J. Theor. Biol.* 64:681–96
109. Taylor, J. H., Hozier, J. C. 1976. *Chromosoma* 57:341–50
110. Yurov, Y. B., Liapunova, N. A. 1977. *Chromosoma* 60:253–67
111. Petes, Th. D., Williamson, D. H. 1975. *Exp. Cell Res.* 95:103–10
112. Evans, H. H., Littman, S. R., Evans, T. E., Brewer, E. N. 1976. *J. Mol. Biol.* 104:169–84
113. Funderud, S., Haugli, F. 1975. *Nucleic Acids Res.* 2:1381–90
114. Funderud, S., Haugli, F. 1977. *Nucleic Acids Res.* 4:405–13
115. Nymann, O., Westergaard, O. 1975. *FEBS Lett.* 64:139–43
116. Johnson, B., Westergaard, O. 1976. *Eur. J. Biochem.* 62:345–52
117. Van't Hof, J. 1975. *Exp. Cell Res.* 93:95–104
118. Sakamaki, T., Fukuei, K., Takahashi, N., Tanifuji, S. 1975. *Biochim. Biophys. Acta* 295:314–21
119. Bryant, J. A. 1976. In *Molecular Aspects of Gene Expression in Plants,* ed. J. A. Bryant, pp. 1–51. New York: Academic
120. Sakamaki, T., Takahashi, N., Takaiwa, F., Tanifuji, S. 1976. *Biochim. Biophys. Acta* 447:76–81
121. Van't Hof, J. 1976. *Exp. Cell Res.* 103:395–403
122. Clay, W. F., Bartels, P. G., Katterman, F. R. H. 1976. *Proc. Natl. Acad. Sci. USA* 73:3220–23
123. Ananev, E. V., Polukarova, L. G., Yurov, Y. B. 1977. *Chromosoma* 59:259–72
124. Zakian, V. A. 1976. *J. Mol. Biol.* 108:305–31
125. Blumenthal, A. B., Clark, E. J. 1977. *Exp. Cell Res.* 105:15–26
126. Wilson, B. G. 1975. *Chromosoma* 51:213–24
127. Hand, R. 1975. *J. Cell Biol.* 64:89–97
128. Hand, R. 1975. *J. Cell Biol.* 67:761–73
129. Housman, D., Huberman, J. A. 1975. *J. Mol. Biol.* 94:173–81
130. Hozier, J. C., Taylor, J. H. 1975. *J. Mol. Biol.* 93:181–201
131. Taylor, J. H., Wu, M., Erickson, L. C., Kurek, M. P. 1975. *Chromosoma* 53:175–89
132. Tseng, B. Y., Goulian, M. 1975. *J. Mol. Biol.* 99:317–37
133. Gautschi, H. R., Clarkson, J. M. 1975. *Eur. J. Biochem.* 50:403–12

134. Friedman, C. A., Kohn, K. W., Erickson, L. C. 1974. *Biochemistry* 14:4018–4123
135. Walters, R. A., Tobey, R. A., Hildebrand, C. E. 1976. *Biochim. Biophys. Acta* 447:36–44
136. Kuebbing, D., Diaz, A. T., Werner, R. 1976. *J. Mol. Biol.* 108:55–66
137. Hershey, H. V., Werner, D. 1976. *Nature* 262:148–50
138. Hildebrand, C. E., Walters, R. A. 1976. *Biochem. Biophys. Res. Commun.* 73:157–63
139. Des Gouttes Olgiati, D., Pogo, B. G. T., Dales, S. 1976. *J. Cell Biol.* 68:557–66
140. Kowalski, J., Cheevers, W. P. 1976. *J. Mol. Biol.* 104:603–15
141. Kurek, M. P., Taylor, J. H. 1977. *Exp. Cell Res.* 104:7–14
142. Yurov, Y. B. 1977. *Cell Differ.* 6:95–104
143. Hand, R. 1977. *Human Genet.* 37:55–64
144. Krokan, H., Wist, E., Prydz, H. 1977. *Biochim. Biophys. Acta* 475:553–61
145. Planck, S. R., Mueller, G. C. 1977. *Biochemistry* 16:1808–13
146. Jering, H., Diaz, A., Werner, R. 1978. *Cell.* In press
147. Wanka, F., Brouns, R. M. G. M. E., Aelen, J. M. A., Eygensteyn, J. 1977. *Nucleic Acids Res.* 4:2083–98
148. Robberson, D. L., Crawford, L. V., Syrett, C., James, W. 1975. *J. Gen. Virol.* 26:59–69
149. Kelly, T. J., Nathans, D. 1977. *Adv. Virus. Res.* 21:85–173
150. Fareed, G. C., Davoli, D. 1977. *Ann. Rev. Biochem.* 46:471–522
151. Sussenbach, J. S., Ellens, D. J., van der Vliet, P. C., Kuijk, M. G., Steenberg, P. H., Vlak, J. M., Rozijn, T. H., Jansz, H. S. 1975. *Cold Spring Harbor Symp. Quant. Biol.* 39:539–50
152. Hirsch, I., Roubal, J., Vonka, V. 1976. *Intervirology* 7:155–75
153. Jean, J.-H., Blankenship, M. L., Ben-Porat, T. 1977. *Virology* 79:281–91
154. Ben-Porat, T., Tokazewski, S. A. 1977. *Virology* 79:292–301
155. Ben-Porat, T., Blankenship, M. L., DeMarchi, J. M., Kaplan, A. S. 1977. *J. Virol.* 22:734–41
156. Ben-Porat, T., Blankenship, M. L., Tokazewski, S. 1978. *Virology.* In press
157. Hori, T. A., Lark, K. G. 1976. *Nature* 259:504–5
158. Sures, I., Maxam, A., Cohn, R. H., Kedes, L. H. 1976. *Cell* 9:495–502
159. Birnstiel, M. L., Schaffner, W., Smith, H. O. 1977. *Nature* 266:603–7

160. Stubblefield, E. W. 1975. *Chromosoma* 53:209–21
161. Ris, H. 1975. *Ciba Found. Symp: Structure and Function of Chromosomes,* ed. D. W. Fitzsimmons, G. E. W. Wohlstenholme, pp. 7–28. Amsterdam: North-Holland
162. Burke, W., Fangman, W. L. 1975. *Cell* 5:263–69
163. Pardee, A. B. 1978. *Ann. Rev. Biochem.* 47:000–00
164. Nagl, W. 1976. *Ann. Rev. Plant Physiol.* 27:39–69
165. Shapiro, I. M., Gause, G. G. Jr., Zakharov, A. F. 1974. *Differentiation* 2:125–27
166. Klevecz, R. R., Keniston, B. A., Deaven, L. L. 1975. *Cell* 5:193–203
167. Hall, L., Turnock, G. 1976. *Eur. J. Biochem.* 62:471–77
168. Szarski, H. 1976. *Int. Rev. Cytol.* 44:93–113
169. Barlow, P. W. 1972. *Cytobios* 6:55–80
170. Lark, K. G., Consigli, R., Toliver, A. 1971. *J. Mol. Biol.* 58:873–75
171. Stimac, E., Housman, D., Huberman, J. A. 1977. *J. Mol. Biol.* 115:485–511
172. Cook, P. R., Brazell, S. S., Jost, E. 1976. *J. Cell Sci.* 22:303–24
173. Povirk, L. F., Painter, R. B. 1976. *Biochim. Biophys. Acta* 432:267–72
174. Mattern, M. R., Painter, R. B. 1977. *Biophys. J.* 19:117–24
175. Nathans, D., Smith, H. O. 1975. *Ann. Rev. Biochem.* 44:273–93
176. Shenk, T. E., Berg, P. 1976. *Proc. Natl. Acad. Sci. USA* 73:1513–17
177. Griffin, B. E., Fried, M. 1976. *Methods Cancer Res.* 12:49–86
178. Subramanian, K. N., Dhar, R., Weissman, S. M. 1977. *J. Biol. Chem.* 252:333–39
179. Subramanian, K. N., Dhar, R., Weissman, S. M., Ghosh, P. K. 1977. *J. Biol. Chem.* 252:340–54
180. Subramanian, K. N., Dhar, R., Weissman, S. M. 1977. *J. Biol. Chem.* 252:355–67
181. Dhar, R., Subramanian, K. N., Pan, J., Weissman, S. M. 1977. *J. Biol. Chem.* 252:368–76
182. Soeda, E., Miura, K., Nakaso, A., Kimura, G. 1977. *FEBS Lett.* 79: 383–89
183. Bourgaux, P., Delbecchi, L., Bourgaux-Ramoisy, D. 1976. *Virology* 72:89–98
184. Padmanabhan, R., Padmanabhan, R., Green, M. 1976. *Biochem. Biophys. Res. Commun.* 69:860–67
185. Ariga, H., Shimojo, H. 1977. *Virology* 78:415–24
186. Wu, M., Roberts, R. J., Davidson, N. 1977. *J. Virol.* 21:766–77
187. Denhardt, D. T., Eisenberg, S., Bartok, K., Carter, B. J. 1976. *J. Virol.* 18:672–84
188. Fife, K. H., Berns, K. I., Murray, K. 1977. *Virology* 78:475–87
189. Hauswirth, W. W., Berns, K. I. 1977. *Virology* 78:488–99
190. Straus, S. E., Sebring, E. D., Rose, J. A. 1976. *Proc. Natl. Acad. Sci. USA* 73:742–46
191. Gunther, M., May, P. 1976. *J. Virol.* 20:86–95
192. Tattersall, P., Ward, D. C. 1976. *Nature* 263:106–9
193. Bourguignon, G. J., Tattersall, P. J., Ward, D. C. 1976. *J. Virol.* 20:290–306
194. Salzman, L. A. 1977. *Virology* 76: 454–58
195. Rhode, S. L. III. 1977. *J. Virol.* 22:446–58
196. Roizman, B., Hayward, G. S., Jacob, R., Wadsworth, S. W., Frenkel, N., Honess, R. W., Kozak, M. 1975. In *Symposium on Herpesviruses and Oncogenesis,* ed. H. zur Hausen, F. de The, M. A. Epstein, pp. 3–38. Lyons: Int. Agency Res. Cancer
197. Delius, H., Clements, J. B. 1976. *J. Gen. Virol.* 33:125–33
198. Stevely, W. S. 1977. *J. Virol.* 22:232–34
199. Hayward, S. D., Kieff, E. 1977. *J. Virol.* 23:421–29
200. Stern, H., Hotta, Y. 1973. *Ann. Rev. Genet.* 7:37–66
201. Hotta, Y., Stern, H. 1976. *Chromosoma* 55:171–82
202. Kao, P. C., Beyer, C. F., Lang, C. A. 1976. *Biochem. J.* 154:471–80
203. Mills, B. J., Lang, C. A. 1976. *Biochem. J.* 154:481–90
204. Chatterjee, S. N., Mondal, S. N., Mukherjee, A. S. 1976. *Chromosoma* 54:117–25
205. Wellauer, P. K., Reeder, R. H. 1975. *J. Mol. Biol.* 94:151–61
206. Nagl, W. 1976. *Nature* 261:614–15
207. Weingartner, B., Winnacker, E. L., Tolun, A., Pettersson, U. 1976. *Cell* 9:259–68
208. Sussenbach, J. S., Kuijk, M. G. 1977. *Virology* 77:149–57
209. Carusi, E. A. 1977. *Virology* 76:380–94
210. Krokan, H., Cooke, L., Prydz, H. 1975. *Biochemistry* 14:4233–37
211. Gautschi, J. R., Burkhalter, M., Reinhard, P. 1977. *Biochim. Biophys. Acta* 474:512–23
212. Tseng, B. Y., Goulian, M. 1978. *Cell* 12:483–89

213. Pigiet, V., Winnacker, E. L., Eliasson, R., Reichard, P. 1973. *Nature New Biol.* 245:203–5
214. Flory, P. J. Jr. 1977. *Nucleic Acids Res.* 4:1449–64
215. Francke, B., Vogt, M. 1975. *Cell* 5:205–11
216. Edenberg, H. J., Wagar, M. A., Huberman, J. A. 1976. *Proc. Natl. Acad. Sci. USA* 73:4392–96
217. Probst, H., Hofstaetter, T., Jenke, H.-S., Gentner, P. R. 1976. *Biochim. Biophys. Acta* 442:58–65
218. Oertel, W., Goulian, M. 1977. *J. Bacteriol.* 132:233–46
219. Wist, E., Krokan, H., Prydz, H. 1976. *Biochemistry* 15:3647–52
220. Hershey, H. W., Taylor, J. H. 1975. *Exp. Cell Res.* 94:339–50
221. Seale, R. L. 1977. *Biochemistry* 16:2847–52
222. Planck, S. R., Mueller, G. C. 1977. *Biochemistry* 16:2778–82
223. Jazwinski, S. M., Wang, J. L., Edelman, G. M. 1976. *Proc. Natl. Acad. Sci. USA* 73:2231–35
224. DePamphilis, M. L., Beard, P., Berg, P. 1975. *J. Biol. Chem.* 250:4340–47
225. Su, R. T., DePamphilis, M. L. 1976. *Proc. Natl. Acad. Sci. USA* 73:3466–70
226. Weintraub, H. 1974. *Cold Spring Harbor Symp. Quant. Biol.* 38:247–56
227. Hewish, D. R. 1975. *Nucleic Acids Res.* 3:69–78
228. Golombek, J., Wolf, W., Wintersberger, E. 1974. *Mol. Gen. Genet.* 132:137–45
229. Seale, R. L. 1975. *Biochem. Biophys. Res. Commun.* 63:140–48
230. Weiner, D., Maier, G. 1975. *Eur. J. Biochem.* 54:351–58
231. Seale, R. L., Simpson, R. T. 1975. *J. Mol. Biol.* 94:479–501
232. Garcia-Herdugo, G., Gonzalez-Fernandez, A., Lopez-Saez, J. F. 1976. *Exp. Cell Res.* 104:1–6
233. Weintraub, H. 1976. *Cell* 9:419–22
234. Seale, R. L. 1976. *Cell* 9:423–29
235. Leer, J. C., Marcker, K. A., Westergaard, O. 1978. *Eur. J. Biochem.* In press
236. Thompson, L. H., Harkins, J. L., Stanners, C. P. 1973. *Proc. Natl. Acad. Sci. USA* 70:3094–98
237. Roufa, D. J., Reed, S. J. 1975. *Genetics* 80:549–66
238. Haralson, M. A., Roufa, D. J. 1975. *J. Biol. Chem.* 250:8618–23
239. Roufa, D. J., Haralson, M. A. 1975. In *DNA Synthesis and its Regulation*, ed. M. Goulian, P. Hanawalt, C. F. Fox, pp. 702–12. Menlo Park: Benjamin

240. Strauss, B., Scudiero, D. Henderson, E. 1975. *Basic Life Sci.* 5A:13–24
241. Scudiero, D., Strauss, B. 1976. *Mutat. Res.* 35:311–24
242. Kalman, T. I. 1975. *Ann. NY Acad. Sci.* 255:326–31
243. Ockey, C. H., Allen, T. D. 1975. *Exp. Cell Res.* 93:275–82
244. Ockey, C. H., Saffhill, R. 1976. *Exp. Cell Res.* 103:361–73
245. Chan, A. C., Walker, I. G. 1975. *Biochim. Biophys. Acta* 395:422–32
246. Skoog, L., Bjursell, G. 1974. *J. Biol Chem.* 249:6434–38
247. Ramseier, H. P., Burkhalter, M., Gautschi, J. R. 1977. *Exp. Cell Res.* 105:445–53
248. Fujiwara, Y. 1975. *Biophys. J.* 15:403–15
249. Walters, R. A., Tobey, R. A., Hildebrand, C. E. 1976. *Biochem. Biophys. Res. Commun.* 69:212–17
250. Martin, R. F., Radford, I., Pardee, M. 1977. *Biochem. Biophys. Res. Commun.* 74:9–15
251. Eriksen, H., Krokan, A. 1977. *Eur. J. Biochem.* 72:501–8
252. Walters, R. A., Tobey, R. A., Hildebrand, C. E. 1976. *Biochem. Biophys. Res. Commun.* 69:212–17
253. Raju, M. R., Tobey, R. A., Jett, J. H., Walters, R. A. 1975. *Radiat. Res.* 63:422–33
254. Simchen, G., Idar, D., Kassir, Y. 1976. *Mol. Gen. Genet.* 144:21–27
255. Hand, R., German, J. 1978. *Human Genet.* In press
256. Giannelli, F., Benson, P. F., Pawsey, S. A., Polani, P. E. 1977. *Nature* 265:466–69
257. Hoffbrand, A. V., Ganeshaguru, K., Hooton, J. W. L., Tripp, E. 1976. In *Clinics in Haematology*, ed. A. V. Hoffbrand, pp. 727–45. London: Saunders
258. Sheinin, R. 1976. *Cell* 7:49–57
259. Sheinin, R. 1976. *J. Virol.* 17:692–704
260. Hängii, U. J., Littlefield, J. W. 1976. *J. Biol. Chem.* 251:3075–80
261. Fridland, A., Brent, T. P. 1975. *Eur. J. Biochem.* 57:379–85
262. Makino, F., Okada, S. 1975. *Radiat. Res.* 62:37–51
263. Walters, R. A., Hildebrand, C. E. 1975. *Biochem. Biophys. Res. Commun.* 65:265–71
264. Painter, R. B., Young, B. R. 1976. *Biochim. Biophys. Acta* 418:146–53
265. Dressler, D. 1975. *Ann. Rev. Microbiol.* 29:525–59
266. Pearson, C. K., Davis, P. B., Taylor, A., Amos, N. A. 1976. *Eur. J. Biochem.* 62:451–59

267. Wagar, M. A., Huberman, J. A. 1975. *Biochim. Biophys. Acta* 383:410–20
268. Tseng, B. Y., Goulian, M. 1975. *J. Mol. Biol.* 99:339–46
269. Wagar, M. A.,Huberman, J. A. 1975. *Cell* 6:551–57
270. Reichard, P., Eliasson, R., Soderman, G. 1974. *Proc. Natl. Acad. Sci. USA* 71:4901–5
271. Benbow, R. M., Toenje, H., White, S. H., Breaux, C. B., Krauss, M. R., Ford, O. C., Laskey, R. A. 1977. In *International Cell Biology*, ed. B. Brinkley, K. Porter, pp. 453–63. New York: Rockefeller Univ. Press
272. Laskey, R. A., Mills, A. D., Morris, N. R. 1977. *Cell* 10:237–43
273. Attardi, D. G., Martini, G., Mattoccia, E., Tocchini-Valentini, G. P. 1976. *Proc. Natl. Acad. Sci. USA* 71:554–58
274. Gall, J. G. 1974. *Proc. Natl. Acad. Sci. USA* 71:3078–81
275. Jakob, K. M. 1972. *Exp. Cell Res.* 72:370–76
276. Fukuei, K., Sakamaki, T., Takahashi, N., Takaiwa, F., Tanifuji, S. 1977. *Plant Cell Physiol.* 18:173–80
277. Witkin, S. S., Bendich, A. 1977. *Exp. Cell Res.* 106:47–54
278. Novi, A. M. 1976. *Klin. Wochenschr.* 54:961–68
279. Hartwell, L. H. 1974. *Bacteriol. Rev.* 38:164–99
280. Baserga, R. 1976. *Multiplication and Division in Mammalian Cells*, Biochemistry of Disease Series, Vol. 6. New York: Dekker
281. Jakob, K. M., Bovey, F. 1969. *Exp. Cell Res.* 54:118–26
282. Mory, Y. Y., Chan, D., Sarid, S. 1972. *Plant Physiol.* 49:20–23
283. Banks, G. R., Spanos, A. 1975. *J. Mol. Biol.* 93:63–77
284. Yarranton, G. T., Moore, P. D., Spanos, A. 1976. *Mol. Gen. Genet.* 145:215–18
285. Herrick, G., Delius, H., Alberts, B. 1976. *J. Biol. Chem.* 251:2142–46
286. Otto, B., Baynes, M., Knippers, R. 1977. *Eur. J. Biochem.* 73:17–24
287. Duguet, M., Soussi, T., Rossignol, J. M., Mechali, M., De Rocondo, A. M. 1977. *FEBS Lett.* 79:160–64
288. Duguet, M., De Recondo, A. M. 1978. *J. Biol. Chem.* In press
289. Alberts, B., Barry, J., Bittner, M., Davies, M., Hiroko, H. I., Liu, C. C., Mace, D., Moran, L., Morris, C. F. 1977. In *Nucleic Acid-Protein Recognition on P&S*, ed. H. J. Vogel, New York:Academic. pp. 31–63
290. Weinberg, R. A. 1977. *Cell* 11:243–46
291. Tenen, D. G., Martin, R. G., Anderson, J., Livingston, D. M. 1977. *J. Virol.* 22:210–18
292. Rundell, K., Collins, J. K., Tegtmeyer, P., Ozer, H. L., Lai, C.-J., Nathans, D. 1977. *J. Virol.* 21:636–46
293. Jessel, D., Hudson, J., Landau, T., Tenen, D., Livingston, D. M. 1975. *Proc. Natl. Acad. Sci. USA* 72:1960–64
294. Jessel, D., Landau, T., Hudson, J., Lalor, T., Tenen, D., Livingston, D. 1976. *Cell* 8:535–45
295. Birkenmeier, E. H., May, E., Salzman, N. P. 1977. *J. Virol.* 22:702–10
296. Chou, J. Y., Martin, R. G. 1975. *J. Virol.* 15:145–50
297. Graessmann, M., Graessmann, A. 1976. *Proc. Natl. Acad. Sci. USA* 73:366–70
298. Dubbs, D. R., Kit, S. 1977. *Somatic Cell Genet.* 3:61–69
299. Segawa, K., Yamaguchi, N., Oda, K. 1977. *J. Virol.* 22:679–93
300. Dubbs, D. R., Trkula, D., Kit, S. 1978. In press
301. Tenen, D. G., Baygell, P., Livingston, D. M. 1975. *Proc. Natl. Acad. Sci. USA* 72:4351–55
302. Carroll, R. B., Smith, A. E. 1976. *Proc. Natl. Acad. Sci. USA* 73:2254–58
303. Tegtmeyer, P., Rundell, K., Collins, J. K. 1977. *J. Virol.* 21:647–57
304. Griffith, J., Dieckmann, M., Berg, P. 1975. *J. Virol.* 15:167–72
305. Reed, S. I., Ferguson, J., Davis, R. W., Stark, G. R. 1975. *Proc. Natl. Acad. Sci. USA* 72:1605–9
306. Reed, S. I., Stark, G. R., Alwine, J. C. 1976. *Proc. Natl. Acad. Sci. USA* 73:3083–87
307. Paulin, D., Cuzin, F. 1975. *J. Virol.* 15:393–97
308. Ito, Y., Spurr, N., Dulbecco, R. 1977. *Proc. Natl. Acad. Sci. USA* 74:1259–63
309. Rosenwirth, B., Shiroki, K., Levine, A. J., Shimojo, H. 1975. *Virology* 67:14–23
310. Ledinko, N. 1975. *J. Virol.* 16:807–17
311. Van der Vliet, P. C., Zandberg, J., Jansz, H. S. 1977. *Virology* 80:98–110
312. Ginsberg, H. S., Lundholm, U., Linne, T. 1977. *J. Virol.* 23:142–51
313. Sugawara, K., Gilead, Z., Green, M. 1977. *J. Virol.* 21:338–46
314. Jeng, Y.-H., Wold, W. S. M., Sugawara, K., Gilead, Z., Green, M. 1977. *J. Virol.* 22:402–11
315. Sugawara, K., Gilead, Z., Wold, W. S. M., Green, M. 1977. *J. Virol.* 22:527–39
316. Rekosh, D. M. K., Russell, W. C., Bellett, A. J. D. 1977. *Cell* 11:283–95
317. Rosenwirth, B., Anderson, J., Levine, A. J. 1976. *Virology* 69:617–25

318. Levinson, A., Levine, A. J., Anderson, S., Osborn, M., Rosenwirth, B., Weber, K. 1976. *Cell* 7:575–84
319. Purifoy, D. J. M., Powell, K. L. 1976. *J. Virol.* 19:717–31
320. Luka, J., Siegert, W., Klein, G. 1977. *J. Virol.* 22:1–8
321. Engberg, J., Nilsson, J., Pearlman, R., Leick, V. 1974. *Proc. Natl. Acad. Sci. USA* 71:894–98
322. Karrer, K. M., Gall, J. G. 1976. *J. Mol. Biol.* 104:421–53
323. Engberg, J., Mowat, D., Pearlman, R. E. 1972. *Biochim. Biophys. Acta* 272:312–20
324. Andersen, H. A., Engberg, J. 1975. *Exp. Cell Res.* 92:159–63
325. Truett, M. A., Gall, J. G. 1978. *Chromosoma.* 64:295–303
326. Charret, R. 1969. *Exp. Cell Res.* 54:353–61
327. Nilsson, J. R., Leick, V., Engberg, J. 1977. *J. Protozool.* 24:Suppl. 30A
328. Newlon, C. S., Sonenshein, G. E., Hott, C. E. 1973. *Biochemistry* 12:2338–45
329. Guttes, E. 1974. *J. Cell Sci.* 15:131–43
330. Vogt, V. M., Braun, R. 1978. In press
331. Bohnert, H.-J., Schiller, B. 1975. *Eur. J. Biochem.* 57:361–69
332. Braun, R., Hall, L., Schwärzler, M., Smith, S. S. 1977. In *Leopoldine Symposium: Cell Differentiation in Microorganisms, Plants and Animals,* ed. L. Nover, K. Mothes. Jena, German Democratic Republic: VEBG Fischer.
333. Yao, M.-C., Gall, J. G. 1977. *Cell* 12:121–32
334. Engberg, J., Pearlman, R. E. 1972. *Eur. J. Biochem.* 26:393–400
335. Yao, M.-C., Kimmel, A., Gorovsky, M. 1974. *Proc. Natl. Acad. Sci. USA* 71:3082–86
336. Tobler, H. 1975. In *Biochemistry of Animal Development,* 3:91–143. New York: Academic
337. Hourcade, D., Dressler, D., Wolfson, J. 1973. *Proc. Natl. Acad. Sci. USA* 70:2926–30
338. Rochaix, J.-D., Bird, A., Bakken, A. 1974. *J. Mol. Biol.* 87:473–87
339. Bird, A., Rogers, E., Birnstiel, M. 1973. *Nature New Biol.* 242:226–30
340. Spear, B. B. 1974. In *Ribosomes,* ed. M. Nomura, A. Tissieres, P. Lengyel, pp. 841–53. Cold Spring Harbor, N.Y. Cold Spring Harbor Lab.
341. Graziani, F., Coizzi, R., Gargano, S. 1977. *J. Mol. Biol.* 112:49–63
342. Kaback, D. B., Halvorson, H. O. 1977. *Proc. Natl. Acad. Sci. USA* 74:1177–80
343. Manning, J. E., Wolstenholme, D. R. 1976. *J. Cell. Biol.* 70:406–18
344. Simpson, L., Simpson, A. M., Wesley, R. D. 1974. *Biochim. Biophys. Acta* 349:161–72
345. Cleaver, J. E., Bootsma, D. 1975. *Ann. Rev. Genet.* 9:19–38
346. Brent, T. P. 1975. *Biochim. Biophys. Acta* 407:191–99
347. Lehmann, A. R., Kirk-Bell, S., Arlett, C. F., Paterson, M. C., Lohman, P. H. M., De Weerd-Kastelein, E. A., Bootsma, D. 1975. *Proc. Natl. Acad. Sci. USA* 72:219–23
348. Bacchetti, S., Benne, R. 1976. *Biochim. Biophys. Acta* 390:285–97
349. Ljungquist, S., Nyberg, B., Lindahl, T. 1975. *FEBS Lett.* 57:160–71
350. Sutherland, B. M., Oliver, R., Fuselier, C. O., Sutherland, J. C. 1976. *Biochemistry* 15:402–6
351. Mortelman, K., Friedberg, E. C., Slor, H., Thomas, G., Cleaver, J. E. 1976. *Proc. Natl. Acad. Sci. USA* 73:2757–61
352. Patterson, M. C., Smith, B. P., Lohman, P. H. M., Anderson, A. K. 1976. *Nature* 260:444–47
353. Patterson, M. C., Smith, B. P., Knight, P. A., Anderson, A. K. 1976. *Research Photobiology,* Vol. 7, ed. A. Castellani. New York: Plenum. pp. 207–18
354. Kuhnlein, U., Penhoet, E. E., Linn, S. 1976. *Proc. Natl. Acad. Sci. USA* 73:1169–73
355. Patterson, M. C. 1977. *Adv. Radiat. Biol.* 7:1–53
356. Goth-Goldstein, R. 1977. *Nature* 267:81–82
357. Rude, J. M., Friedberg, E. C. 1977. *Mutat. Res.* 42:433–42
358. Lehmann, A. R., Kirk-Bell, S., Arlett, C. F., Harcourt, S. A., De Weerd-Kastelein, E. A., Keijzer, W., Hald-Smith, P. 1977. *Cancer Res.* 37:904–10
359. Kuebler, J. P., Goldthwait, D. A. 1977. *Biochemistry* 16:1370–77
360. Kuhnlein, U., Lee, B., Linn, S. 1978. *Nucleic Acids Res.* 5: In press
361. Kuhnlein, U., Lee, B., Penhoet, E. E. Linn, S. 1978. *Nucleic Acids Res.* 5: In press
362. Hoar, D. I., Sargent, P. 1976. *Nature* 261:590–92
363. Lehmann, A. R., Stevens, S. 1977. *Biochim. Biophys. Acta* 474:49–60
364. Brown, W. T., Epstein, J., Little, J. B. 1976. *Exp. Cell Res.* 97:291–96
365. Rainbow, A. J., Howes, M. 1977. *Biochem. Biophys. Res. Commun.* 74:714–19
366. Rainbow, A. J., Howes, M. 1977. *Int. J. Radiat. Biol.* 31:191–95
367. Rainbow, A. J. 1977. In press

368. Fujiwara, Y., Tatsumi, M. 1975. *Biochem. Biophys. Res. Commun.* 2: 592–98
369. Auerbach, A. D., Wolman, S. R. 1976. *Nature* 261:494–96
370. Remsen, J. F., Cerutti, P. A. 1976. *Proc. Natl. Acad. Sci. USA* 73:2419–23
371. Finkelberg, R. 1977. *Studies on patients with Fanconi's anaemia* PhD thesis. Univ. Toronto, Toronto
372. Hartwell, L. H. 1976. *J. Mol. Biol.* 104:803–17
373. Simchen, G., Hirschberg, J. 1977. *Genetics* 86:57–72
374. Tenner, A., Zieg, J., Scheffler, I. E. 1977. *J. Cell Physiol.* 90:145–60
375. Wang, R. J. 1974. *Nature* 228:76–78
376. Wang, R. J. 1976. *Cell* 8:257–61
377. Wang, R. J., Yin, L. 1976. *Exp. Cell Res.* 101:331–36
378. Scheffler, I. E., Buttin, G. 1973. *J. Cell Physiol.* 81:199–216
379. Meiss, H. K., Basilico, C. 1972. *Nature* 239:66–68
380. Burstin, S. J., Meiss, H. K., Basilico, C. 1974. *J. Cell Physiol.* 84:397–407
381. Nishimoto, T., Raskas, H. J., Basilico, C. 1975. *Proc. Natl. Acad. Sci. USA* 72:328–32
382. Liskay, R. M., Meiss, H. K. 1977. *Somat. Cell Genet.* 3:343–47
383. Smith, B. J., Wigglesworth, N. M. 1973. *J. Cell Physiol.* 82:339–48
384. Roscoe, D. H., Read, M., Robinson, H. 1973. *J. Cell Physiol.* 82:325–32
385. Roscoe, D. H., Robinson, H., Carbonell, A. W. 1973. *J. Cell Physiol.* 82:333–38
386. Smith, B. J., Wigglesworth, N. M. 1974. *J. Cell Physiol.* 84:127–34
387. Rieber, M., Bacalao, J. 1974. *Cancer Res.* 34:3083–88
388. Kit, S., Jorgensen, G. N. 1976. *J. Cell Physiol.* 88:57–64
389. Dubbs, D. R., Kit, S. 1976. *Somat. Cell Genet.* 2:11–19
390. Dubbs, D. R., Kit, S. 1977. *Somat. Cell Genet.* 3:61–69
391. Marin, G., Labella, T. 1977. *J. Cell Physiol.* 90:71–78
392. Kuroki, T., Miyashita, S. Y. 1977. *J. Cell Physiol.* 90:79–90
393. Crane, M. St. J., Thomas, D. B. 1976. *Nature* 261:205–8
394. Ohlsson-Wilhelm, B. M., Freed, J. J., Perry, R. P. 1976. *J. Cell Physiol.* 89:77–88
395. Naha, P. M. 1969. *Nature* 223:1380–81
396. Naha, P. M. 1970. *Nature* 228:166–68
397. Naha, P. M. 1973. *Exp. Cell Res.* 80:467–73

398. Shiomi, T., Sato, K. 1976. *Exp. Cell Res.* 100:297–302
399. Sato, K., Shiomi, T. 1974. *Exp. Cell Res.* 88:295–302
400. Liskay, R. M. 1974. *J. Cell Physiol.* 84:49–56
401. Game, J. C. 1976. *Mol. Gen. Genet.* 146:313–15
402. Unrau, P. 1977. *Mol. Gen. Genet.* 150:13–19
403. Unrau, P. 1972. *Genet. Res. Cambridge* 19:145–55
404. Unrau, P., Olive, T. 1973. *Genetics* 74:282–83 (Suppl.)
405. Holliday, R., Dickson, J. M. 1977. *Mol. Gen. Genet.* 153:331–35
406. Unrau, P., Olive, T. 1978. In press
407. Banks, G. R. 1974. *Eur. J. Biochem.* 47:499–507
408. Marunouchi, T. 1978. *J. Cell Physiol.* In press
409. Slater, M. P., Ozer, H. L. 1976. *Cell* 7:289–95
410. Jha, K. K., Ozer, H. L. 1977. *Genetics* 86:532–34 (Suppl.)
411. Thompson, L. H., Mankovitz, R., Baker, R. M., Till, J. E., Siminovitch, L., Whitmore, G. F. 1970. *Proc. Natl. Acad. Sci. USA* 66:377–84
412. Sheinin, R., Darragh, P., Dubsky, M. 1976. *Can. J. Biochem.* 55:543–47
413. Sheinin, R., Dardick, I., Setterfield, G. 1976. *J. Cell Biol.* 70:195a
414. Sheinin, R., Guttman, S. 1977. *Biochim. Biophys. Acta* 479:105–18
415. Sheinin, R., Darragh, P., Dubsky, M. 1978. *J. Biol. Chem.* 253:In press
416. Setterfield, G., Sheinin, R., Dardick, I., Kiss, G., Dubsky, M. 1978. *J. Cell Biol.* In press
417. Humbert, J., Sheinin, R. 1978. *Can. J. Biochem.* In press
418. Thompson, L. H., Mankovitz, R., Baker, R. M., Wright, J. A., Till, J. E., Siminovitch, L., Whitmore, G. F. 1971. *J. Cell Physiol.* 78:431–40
419. Giles, R. E., Ruddle, F. H. 1976. *Genetics* 83:S26
420. Guttman, S. 1977. *The characterization of a temperature-sensitive mouse L-cell-tsCl* M.Sc. thesis. Univ. Toronto, Toronto
421. Guttman, S., Sheinin, R. 1978. *Exp. Cell Res.* In press
422. Hanawalt, P. C., Setlow, R. B., eds. 1975. *Molecular Mechanisms for Repair of DNA*, Vols. 5A, 5B. New York: Plenum
423. Grossman, L., Braun, A., Feldberg, R., Mahler, I. 1975. *Ann. Rev. Biochem.* 44:19–43

424. Wintersberger, U., Hirsch, J., Fink, A. M. 1974. *Mol. Gen. Genet.* 131:291–99
425. Moore, P. D. 1975. *Mutat. Res.* 28:355–66
426. Moore, P. D. 1975. *Mutat. Res.* 28:367–80
427. Borun, T. W. 1975. In *Results and Problems in Cell Differentiation,* ed. J. Reinert, H. Holtzer, pp. 249–90. New York: Springer
428. Jackson, V., Granner, D. K., Chalkley, R. 1975. *Proc. Natl. Acad. Sci. USA* 72:4440–44
429. Jackson, V., Granner, D., Chalkley, R. 1976. *Proc. Natl. Acad. Sci. USA* 73:2266–69
430. Vidali, G., Karn, J., Allfrey, V. G. 1975. *Proc. Natl. Acad. Sci. USA* 72:4450–54
431. Kedes, L. H. 1976. *Cell* 8:321–31
432. Stein, G. S., Park, W. D., Stein, J. L., Lieberman, M. W. 1976. *Proc. Natl. Acad. Sci. USA* 73:1466–79
433. Bases, R., Mendez, F., Neubort, S. 1976. *Int. J. Radiat. Biol.* 30:141–49
434. Liberti, P., Fischer-Fantuzzi, L., Vesco, C. 1976. *J. Mol. Biol.* 105:263–74
435. Tan, K. B. 1977. *Proc. Natl. Acad. Sci. USA* 74:2805–9
436. Moll, R., Wintersberger, E. 1976. *Proc. Natl. Acad. Sci. USA* 73:1863–67
437. Bradbury, E. M., The Biophysics Group. 1975. In *The Structure and Function of Chromatin,* ed. D. W. Fitzsimons, G. E. W. Wolstenholme, pp. 132–55. New York: Elsevier
438. Gurley, L. R., Tobey, R. A., Walters, R. A., Hildebrand, C. E., Hohman, P. G., D'Anna, J. A., Barham, S. S., Deaven, L. L. 1977. In *Cell Cycle Regulation,* ed. J. R. Jeter, I. L. Cameron, G. M. Padilla, A. M. Zimmerman. New York: Academic. In press
439. Ruiz-Carillo, A., Wangh, L. J., Allfrey, V. G. 1975. *Science* 190:117–28
440. Jackson, V., Shires, A., Chalkley, R., Granner, D. K. 1975. *J. Biol. Chem.* 250:4856–63
441. Jackson, V., Shires, A., Tanphaichitr, N., Chalkley, R. 1976. *J. Mol. Biol.* 104:471–83
442. Louie, A. J., Candido, E. P. M., Dixon, G. H. 1974. *Cold Spring Harbor Symp. Quant. Biol.* 38:803–21
443. Hancock, R. 1969. *J. Mol. Biol.* 40:457–66
444. Fakan, S., Turner, G. N., Pagano, J. S., Hancock, R. 1972. *Proc. Natl. Acad. Sci. USA* 69:2300–5
445. Tsanev, R., Russev, G. 1974. *Eur. J. Biochem.* 43:257–63
446. Hancock, R. 1974. *J. Mol. Biol.* 86:649–63
447. Seale, R. L. 1975. *Nature* 225:247–49
448. Weintraub, H. 1975. *Proc. Natl. Acad. Sci. USA* 72:1212–16
449. Seale, R. L. 1976. *Proc. Natl. Acad. Sci. USA* 73:2270–74
450. Levy, A., Jakob, K. M., Moav, B. 1975. *Nucleic Acids Res.* 2:2299–2303
451. Burgoyne, L. A., Mobbs, J. D., Marshall, A. J. 1976. *Nucleic Acids Res.* 3:3293–3304
452. Hewish, D. 1977. *Nucleic Acids Res.* 4:1881–90
453. Cremisi, C., Chestier, A., Yaniv, M. 1978. In press
454. Seki, S., Lemahieu, M., Mueller, G. C. 1975. *Biochim. Biophys. Acta* 378:333–43
455. Berger, N. A., Erickson, W. P., Weber, G. 1976. *Biochim. Biophys. Acta* 447:65–75
456. Berger, N. A., Petzold, S. J., Johnson, E. S. 1977. *Biochim. Biophys. Acta* 478:44–58
457. Sullivan, C. W., Volcani, B. E. 1976. *Exp. Cell Res.* 98:23–30
458. Yee, W. S., Decker, R. W., Brunk, C. F. 1976. *Biochim. Biophys. Acta* 447:385–90
459. Billen, D., Olson, A. C. 1976. *J. Cell Biol.* 69:732–36
460. Seki, S., Oda, T. 1977. *Biochim. Biophys. Acta* 476:24–31
461. Seki, S., Oda, T. 1977. *Cancer Res.* 37:137–44
462. Reinhard, P., Burkhalter, M., Gautschi, J. R. 1977. *Biochim. Biophys. Acta* 474:500–11
463. Le Blanc, D. J., Singer, M. F. 1976. *J. Virol.* 20:78–85
464. Hunter, T., Francke, B. 1974. *J. Virol.* 13:125–39
465. Benbow, R. M., Ford, C. C. 1975. *Proc. Natl. Acad. Sci. USA* 72:2437–41
466. Jazwinski, S. M., Edelman, G. M. 1976. *Proc. Natl. Acad. Sci. USA* 73:3933–36
467. Brown, R. L., Stubblefield, E. 1975. *Exp. Cell Res.* 93:89–94
468. Francke, B., Hunter, T. 1975. *J. Virol.* 15:97–107
469. Becker, Y., Asher, Y. 1975. *Virology* 63:209–20
470. Thompson, L. R., Mueller, G. C. 1975. *Biochim. Biophys. Acta* 414:231–41
471. Otto, B., Reichard, P. 1975. *J. Virol.* 15:259–67
472. LaColla, P., Weissbach, A. 1975. *J. Virol.* 15:305–15
473. Kolber, A. R. 1975. *J. Virol.* 15:322–31
474. Thompson, L. R., Mueller, G. C. 1975. *Biochim. Biophys. Acta* 378:344–53

475. Spaeren, U., Schrøder, K., Sudbery, C., Bjørklid, E., Prydz, H. 1975. *Biochim. Biophys. Acta* 395:413–21
476. Bolden, A., Aucker, J., Weissbach, A. 1975. *J. Virol.* 16:1584–92
477. Benz, W. C., Strominger, J. L. 1975. *Proc. Natl. Acad. Sci. USA* 72:2413–17
478. Krokan, H., Bjørklid, E., Prydz, H. 1975. *Biochemistry* 14:4227–32
479. De Pamphilis, M. L., Berg, P. 1975. *J. Biol. Chem.* 250:4348–54
480. Otnaess, A.-B., Krokan, H., Bjørklid, E., Prydz, H. 1976. *Biochim. Biophys. Acta* 454:193–206
481. Brewer, E. N. 1976. *Fed. Proc.* 35:1418
482. Shlomai, J., Asher, Y., Becker, Y. 1977. *J. Gen. Virol.* 34:223–34
483. Hooton, J. W. L., Hoffbrand, A. V. 1977. *Biochim. Biophys. Acta* 477:250–63
484. Mullbacher, A., Ralph, R. K. 1977. *Eur. J. Biochem.* 75:347–55
485. Krokan, H., Wist, E., Prydz, H. 1977. *Biochem. Biophys. Res. Commun.* 75:414–19
486. Handa, H., Shimojo, H. 1977. *Virology* 77:424–28
487. Krokan, H., Eriksen, A. 1977. *Eur. J. Biochem.* 72:501–8
488. Yamada, M., Weissbach, A. 1977. *Biochem. Biophys. Res. Commun.* 77:642–49
489. Shaw, J. E., Seebeck, T., Li, J.-L.H., Pagano, J. S. 1977. *Virology* 77:762–71
490. Funderud, S., Haugli, F. 1977. *Bio-chem. Biophys. Res. Commun.* 74:941–48
491. Yamashita, T., Arens, M., Green, M. 1977. *J. Biol. Chem.* 252:7940–46
492. Seki, S., Mueller, G. C. 1976. *Biochim. Biophys. Acta* 435:236–50
493. Burdman, J. A., Szuan, I. 1976. *J. Neurochem.* 26:1245–51
494. Knopf, K.-W., Weissbach, A. 1977. *Biochemistry* 16:3190–94
495. Girard, M., Marty, L., Cajean, C., Suarez, F. 1976. *Biochimie* 58:1101–12
496. Probst, G. S., Stalker, D. M., Mosbaugh, D. W., Meyer, R. R. 1975. *Proc. Natl. Acad. Sci. USA* 72:1171–74
497. Marty, L., Cajean, C., Suarez, F., Girard, M. 1976. *Biochimie* 58:1113–22
498. Cajean, C., Marty, L., Suarez, F., Girard, M. 1977. *Biochimie* 59:393–402
499. Murakami-Murofushi, K., Mano, Y. 1977. *Biochim. Biophys. Acta* 475:254–66
500. Wagar, M. A., Huberman, J. A., Evans, M. J. 1977. *Fed. Proc.* 36:659
501. Arens, M., Yamashita, T., Padmanabhan, R., Tsuruo, T., Green, M. 1978. In press
502. Seki, S., Mueller, G. C. 1975. *Biochim. Biophys. Acta* 378:354–62
503. Lambert, D. M., Magee, W. E. 1977. *Virology* 79:342–54
504. Leer, J. C., Nielsen, O. F., Piper, P. W., Westergaard, O. 1976. *Biochem. Biophys. Res. Commun.* 72:720–31
505. Mathis, D. J., Gorovsky, M. A. 1976. *Biochemistry* 15:750–55

Ann. Rev. Biochem. 1978. 47:317–57
Copyright © 1978 by Annual Reviews Inc. All rights reserved

STRUCTURAL AND ❖976
FUNCTIONAL PROPERTIES
OF THE ACETYLCHOLINE
RECEPTOR PROTEIN
IN ITS PURIFIED AND
MEMBRANE-BOUND STATES[1]

Thierry Heidmann and Jean-Pierre Changeux

Neurobiologie Moléculaire et Laboratoire Interactions Moléculaires
et Cellulaires associé au CNRS, Institut Pasteur, 75015 Paris, France

CONTENTS

[1]This review is dedicated to Professor David Nachmansohn.

0066-4154/78/0701-0317$01.00

SUMMARY AND PERSPECTIVES

Binding of ACh to an excitable membrane triggers the all-or-none opening of discrete channels or pores through which Na^+ and K^+ ions flow along their electrochemical gradient. The elementary functional unit that accounts for this regulation, the ACh regulator, therefore comprises at least two categories of distinct sites: 1. the ACh receptor site (AChR site), which binds ACh (and related compounds) and snake venom α-toxins, and 2. the site of ion translocation. The macromolecular units carrying each of them are referred to respectively as ACh receptor protein (AChR) and ACh ionophore (AChI) (1) [or ion conductance modulator (2)]. The ACh receptor protein has been extracted by nondenaturing detergents and purified from fish electric organ and vertebrate skeletal muscle in sizeable amounts. It is an oligomer of 250,000–300,000 mol wt that splits in the presence of SDS into a dominant component of apparent mol wt 40,000 and is labeled by an affinity reagent of the AChR site. Fractionation of *Torpedo* electric organ yields membrane fragments that contain the AChR protein as approximately 40% of their proteins, respond in vitro to ACh by a change of ion permeability, and therefore also contain the AChI. The binding properties of the membrane-bound AChR are regulated by two classes of ligands: the cholinergic agonists, which bind to the AChR site, and the local anesthetics (and related compounds), which bind to a distinct site, the local anesthetic site (or LA site). The local anesthetics block ion translocation at the level of the AChI and some evidence suggests that the LA site is carried by a membrane protein different from the AChR protein. ACh or agonist binding to the membrane-bound AChR causes fast (millisecond) and slow (from second to minute) transitions that can be monitored by various spectroscopic and relaxation methods and correspond to reversible inter-conversions of the AChR regulator between several discrete states of affinity. In the membrane at rest a low affinity state for agonists is present and at equilibrium ACh stabilizes a high affinity state that is also favored by local anesthetics. The reciprocal interactions between ACh and local anesthetics at the level of the ACh regulator present several features of the classical allosteric interactions found with regulatory enzymes. Correlations between binding and flux experiments suggest that the high affinity state for ACh corresponds to a "desensitized" state of the ACh regulator where the AChI is shut. The physiological process of activation would be viewed as the population of a state of the ACh regulator with an intermediate affinity for ACh where the AChI is open.

Perspectives of the current work on the ACh regulator include 1. the identification of the AChI through reconstitution experiments; 2. the detailed correlation in the millisecond-to-minute time range of agonist binding to the AChR site and ion translocation through the AChI; and 3. the understanding of the conformational transitions that mediate the coupling between AChR and AChI.

INTRODUCTION

As early as 1955, Nachmansohn (3) proposed that the membrane receptor for ACh is a protein and that a conformational transition is responsible for the command of the permeability change. Since then the acetylcholine receptor (AChR) has been isolated in a state that binds cholinergic ligands (4–7), purified in milligram quantities in several laboratories (for review see 8–20), and had several critical features of its functional organization recognized. It is indeed a protein and the evidence that it undergoes reversible conformational changes is now rather well documented. This review is devoted primarily to these two aspects of AChR functional organization in the case of the nicotinic receptor from fish electric organ (*Electrophorus electricus* and *Torpedo* sp.) and vertebrate neuromuscular junction.

PHARMACOLOGICAL ACTION OF ACETYLCHOLINE AND RELATED COMPOUNDS

In vivo

The physiological response to ACh monitored by electrophysiological methods at the motor end plate or the electroplaque synapse is a brief change of electrical potential across the postsynaptic membrane, which results from a transient increase of conductance (and therefore permeability)-accompanied by a net inward current carried by Na^+, K^+ (and Ca^{2+})

ions. Two main categories of currents associated with ACh release from the nerve terminal have been distinguished. The spontaneous miniature end-plate currents (mepc) are of small amplitude and triggered by a single "quantum" of ACh ($\sim$10,000 molecules). The standard end-plate currents (epc) are of much larger size and result from the evoked release by the nerve action potential of $\sim$300 quanta of ACh (i.e. 3×10^6 molecules) (for review see 21–26).

The growth phase of the epc and mepc last 0.5–1 msec (27–33) and 50–300 μsec, respectively (34, 36). They decay in an exponential manner (half-life of a few milliseconds) (29–32, 35–41) at a rate sensitive to temperature (Q_{10} from 2 to 5) (30, 32, 33, 36, 38, 42) and affected by membrane potential (31–34, 36, 37, 39, 40).

The millisecond change of ACh concentration in the synaptic cleft that causes epc and mepc has never been measured directly. Indirect electrophysiological evidence nevertheless suggests that the signal responsible for a mepc would be a transient (less than 1 msec) rise of ACh concentration up to $\sim$3 $\times$ 10^{-4} M superimposed on a background release of 10^{-8} M ACh (43, 44; see also 45).

Because of the basic difficulty in controlling the concentration of effector under conditions of physiological release, several methods have been used to assay quantitatively the electrical response to cholinergic effectors: for instance, application in bath (46–58), iontophoresis (41, 59–66), or transisomerization of a photochromic ligand into an active agonist by a brief light flash (67, 68). Each technique has serious drawbacks: in the first case, prolonged exposure may give rise to desensitization and affect the amplitude of the response (see below). In the second one, the time course and variation of absolute concentrations of the effector are not known, although methods have been developed for calibration based, for instance, on the release of droplets of ACh solution from an oil phase (44) and on the evaluation of the micropipette distance from the end plate (69–71). All these methods are limited by the difficulty in estimating maximal responses because of problems in passing enough current to clamp the subsynaptic membrane. Apparent dissociation constants (K_{app}) have nevertheless been reported for ACh in the case of frog end plate: 2.0–2.9 $\times$ 10^{-5} M (62) [for the electroplaque see (11, 57)] and for a number of depolarizing agents, such as carbamylcholine or decamethonium, which are referred to as *agonists* (see 10, 11). The shape of the dose-response curves is in general sigmoid with a Hill coefficient n (72), which varies with the nature of the agonist [for instance, $n = 2$–3 for ACh or 1.5 for decamethonium (46, 47, 51, 54–57, 62–65, 70, 73–77)] but not with the surface density of AChR sites (78).

Compounds such as *d*-tubocurarine, flaxedil, or hexamethonium block the agonist response by increasing K_{app} without changing the maximal response and are referred to as competitive antagonists (see 10, 11). In general, the pharmacological specificity is qualitatively (not necessarily quantitatively) the same for the electroplaque and the motor end plate from various groups of vertebrates and is that of a typical nicotinic AChR (79).

By contrast, noncompetitive blocking agents have little or no effect on K_{app} but decrease reversibly the amplitude of the permeability response: they include local anesthetics such as procaine, lidocaine, quotane (see 11, 80–83) and the columbian frog toxin histrionicotoxin (2, 84–88). An acidic toxin, ceruleotoxin, recently isolated from the venom of *Bungarus ceruleus* (89), has similar but less reversible effects.

The in vitro characterization of the AChR protein received a considerable impetus from the introduction of two categories of ligands that make highly specific and stable complex with the AChR site and can be radiolabeled without losing their pharmacological activity: 1. The α-toxins from the venom of a variety of snakes [e.g. *Bungarus multicinctus, Naja nigricollis, Laticauda semifasciata* (for review see 90)] are compact polypeptides with 61–74 amino acids and a flat shape revealed by X-ray crystallography (91–93); they bind noncovalently to the AChR site with K_D as low as 10^{-11} M and have a slow reversibility. 2. Affinity labeling reagents establish a covalent bond in, or near, the AChR site (94); a particularly efficient one is 4-(N-maleimido)-benzyltrimethylammonium iodide (MBTA) (see 20), which reacts with a thiol group of the AChR site after exposure to the disulfide reducing agent dithiothreitol.

It has long been known that prolonged or repetitive exposure of a muscle to ACh causes a reversible decrease of the amplitude of the response to ACh: this phenomenon is referred to as pharmacological desensitization (for review see 95). The rate of onset of desensitization increases linearly with increasing agonist concentration from minutes to seconds (57, 73, 96–99) and varies with the nature of the agonist; it is accelerated by membrane hyperpolarization (100), temperature [$Q_{10} \simeq 1.7$ at 22°C and 3 at 5°C (101)], low Na^+ (102), high Ca^{2+} (and multivalent cations), local anesthetics, and various compounds such as barbiturates and long-chain alcohols (100, 101, 103). Antagonists such as d-tubocurarine reduce the rate of onset but the effect can be overcome by increasing the agonist concentration (96).

Recovery from desensitization is independent of all the above-mentioned factors except temperature (73, 100, 101, 104) and to a large extent occurs rapidly (1–20 sec with iontophoretic test pulses). However, apparently long-lasting depression of the responses (73, 99, 100), or recoveries in the minute time range when the test pulses are applied in bath (98, 99, 105), have been observed. It is not yet clear, however, whether this slow phase is part of desensitization per se or is due to side effects such as ionic changes inside the cell (50, 75).

In Vitro

Membrane fragments isolated from the electric organ of *E. electricus* (106) or *Torpedo marmorata* (107, 108) still respond in vitro to cholinergic agonists in the absence of energy source or ionic gradient (109–114). These membrane fragments make closed vesicles or microsacs and their permeability to radioactive permeants can therefore be measured by a simple filtration assay. With an unresolved suspension of microsacs the curves of efflux do not follow a single exponential function. A recent improvement of this technique (114) came from the observation (106, 112) that the original microsac population was heterogeneous in size and nature. The functional microsacs (only 15% of the population in a crude preparation from *E. electricus*) can be separated from the nonfunctional ones by centrifugation on continuous sucrose-CsCl density gradients. With the purified population of excitable microsacs the efflux curve follows a single exponential decay and can therefore be more rigorously quantified (114).

Whatever the assay used, the agonists active on the electroplaque markedly enhance the resting efflux of $^{22}Na^+$ [in *T. marmorata* microsacs, with the apparent exception of decamethonium, which in vitro blocks the response with a significant affinity (108)]. The agonist response is competitively inhibited by nicotinic antagonists and noncompetitively by local anesthetics; α-toxins and affinity labeling reagents exert the same action in vitro as in vivo. Using rather empirical parameters to express the amplitude of the permeability change, dose-response curves can be established and K_{app} determined. For the set of ligands tested, the relative order of the affinities are the same in vitro and in vivo. Some quantitative differences are found, however, among the absolute values of K_{app} with *T. marmorata* microsacs (K_{app} are in general higher in vitro) and among the amplitudes of the maximal responses for different agonists. With heterogenous microsac populations the shape of the dose-response curves to agonists was found sigmoid in an early study (106) but hyperbolic in a more recent one (113). These differences may result from inadequate estimates of the response amplitude, from desensitization, and from alterations of membrane properties consecutive to homogenization and purification. In addition, many features of the response might be controlled by the membrane electrical potential, which is expected to be much lower with the microsacs than with the live electroplaque (106–109).

The selectivity of the AChI assayed in vitro with radioactive permeants is that expected from electrophysiological measurements: the permeability to small cations (Na^+, K^+, Ca^{2+}) increases, but not that to organic cations or to negatively charged or uncharged compounds. The ratios of the permeability increases for $^{22}Na^+$ and $^{42}K^+$ are close to one, which fits with the ratios of the conductance increases $\Delta g_{Na}/\Delta g_K$ (115, 116) recorded in vivo. Some dependence of the Na^+ and K^+ flux upon ionic environment has been noted: high external Na^+ concentration abolishes the effect of carbamylcholine on ^{41}K efflux (106) and increasing external K^+ concentration reduces by at least a factor 2 the rate of ^{22}Na efflux in the presence of carbamylcholine (112).

Finally, desensitization is observed in vitro with *T. marmorata* microsacs. Preincubation of the membrane fragments with conditioning doses of agonist (not antagonists such as *d*-tubocurarine) causes a decrease of the amplitude of the permeability response to agonist tested subsequently. The effect takes place at 0°C within minutes, is reversible, and, as expected, is enhanced by local anesthetics and Ca^{2+} (117, 118).

The microsac assay makes possible a study of the ACh regulator under well-defined ionic environments. The method has been extended with success to suspensions of myoblasts (119) or nerve cells (120).

PURIFICATION AND STRUCTURAL PROPERTIES OF AChR

Purification

The AChR protein has been extracted and purified in sizeable amounts from the electric organ of a variety of fish: *Electrophorus electricus* (14, 110, 121–126), *Torpedo marmorata* (12, 13, 111, 127–134), *T. californica* (16, 135–139), *T. nobiliana* (140), *T. ocellata* (141), *Narcine entemedor* (142), *N. brasiliensis* (136), and *Narche japonica* (143). Successful purification has also been achieved with other tissues: denervated adult skeletal muscle (17, 144–151) embryonic skeletal myotubes in culture (152), and a muscle cell line (153). Components binding snake α-neurotoxins from brain (154–157), retina (158, 159), sperm (160), and sympathetic ganglia (161, 162) have been identified and in some instances isolated, but their relevance to functional nicotinic receptors is questioned (163, 164) because of a lack of correlation between in vivo response and in vitro binding properties; thus they are not included in this review.

Recent progress in the purification of AChR was achieved by improving the subcellular fractionation of electric tissue. Standard methods have been used, mainly with homogenates of *Torpedo* electric organs, to separate membrane fragments particularly rich in AChR (about 30% of the tested proteins) from others labeled by AChE or $Na^+K^+ATPase$ (165–168). The extension of the method of affinity partitioning (169) to suspensions of AChR-rich membranes led to significant increases of specific activity (up to 4.6 μmoles α-toxin sites/g protein) (170). The procedure involves a binary-phase system formed from two water-soluble polymers: dextran and poly(ethylene) oxide conjugated to a quaternary ligand selective for the AChR. On the other hand, improvement of the standard methods by four consecutive centrifugations at low and high speed yields up to 150 mg proteins of membrane fragments with a specific activity higher than 4.0 μ moles α-toxin sites/g protein and less than 1% of that amount of acetylcholinesterase starting from 1 kg of fresh *T. marmorata* electric organ (134, 135).

SDS polyacrylamide gel electrophoresis of pure fractions of AChR-rich membranes gives distinct bands of apparent molecular weight as follows (with globular proteins as standards): 40,000, 49,000, 60,000, and 64,000 with *T. californica* (16, 136, 168) and 40,000, 43,000, 50,000, and 66,000 with *T. marmorata* (134, 135); occasionally a band of 100,000–200,000 is observed with preparations of low specific activity. The resolution of the gels used with *T. californica* might not have been good enough to separate the 40,000 and 43,000 chains. The 40,000 chain is the only one labeled by Karlin's affinity reagent MBTA (15, 134, 135) and the 43,000 chain does not comigrate with actin. The relative proportion of the different bands varies with the specific activity of the preparation. Particularly pure preparation from *T. marmorata* show only two major components of 40,000 and 43,000 and it is possible that those of higher molecular weight are either contaminants or membrane components not directly related to the ACh regulator (133, 134, 171).

The AChR is deeply integrated into the membrane and detergents are needed to release it into solution. Several nonionic detergents—Triton X-100, Tween 80, Berol, Brij 35, Emulphogen, Revex 30—and negatively charged ones, such as deoxycholate or cholate, give efficient solubilization without loss of α-toxin and cholinergic ligand bindings. Solubilization has also been achieved without detergent by extensive dialysis of *E. electricus* membrane fragments against a low ionic strength buffer followed by controlled tryptic digestion (172). In fact, for structural studies, it is recommended that detergent solubilization and subsequent purification be carried out in the presence of protease inhibitors, such as phenylmethylsulfonyl fluoride, trasylol (two serine protease inhibitors), and pepstatin (a Cathepsin D inhibitor) (145, 146). Chloroform methanol does not extract affinity-labeled AChR (8, 173), nor α-toxin-AChR complexes (173) under the conditions used by De Robertis and collaborators to extract their proteolipid (174).

The most commonly used method for purification has been affinity chromatography with a cholinergic ligand coupled covalently to a solid matrix. The ligands selected (for review see 10, 18) are snake α-toxins and a variety of quaternary ligand derivatives. Immobilized α-toxin columns are most efficient in the case of crude extracts with low specific activity, for instance from muscle tissue. On the other hand, conventional methods (111) have regained popularity since AChR-rich membrane fragments became available in significant quantities from *Torpedo* electric organ. Starting from a crude detergent extract of pure *T. marmorata* AChR-rich membranes, centrifugation is sufficient to give the AChR in a rather pure form and in milligram quantities (133, 134).

The specific activity of the best preparations ranges between 8 and 12 μmole of α-toxin sites per g of protein (Table 1). The molecular weight per α-toxin site would be close to 100,000, but cannot be taken as the exact value of the protein mass per α-toxin site because of possible interference of detergent with protein determinations, of uncertainties regarding the pharmacological activity of the radioactive ligand, and of ambiguities about the quaternary structure.

Table 1 Specific activity and chain composition of AChR-rich membranes and detergent-purified AChR

Organism	Polypeptide chains (apparent mol wt × 10⁻³)							Specific activity (μmole/g)	Ref.
AChR-rich membranes									
T. californica		41		51	60	64	(105)	4	16, 135, 168
		40		49	60	69		2.9–4.6	170
T. marmorata		40	43	(50)		(66)	(100)	4.0	133, 134
Purified AChR									
T. californica	26; 35	42						7.0	136
		40		49	60	67		—	166
		39		48	58	64		4	137
		40		48	62	66		—	138
		41		51	60	64		8.8	135
		40		48	59	67	140	10–12	139
T. nobiliana	33.5; 35.5		38.5; 43.5					12.2	140
T. ocellata		40		50	61	81		10.0	141
T. marmorata		42						9.5	111
		41.5						6	131
		46						—	175
		45		50				2.3	129
		42		50		70		—	176
		37		49		74	93; 148	(7)	132
		40		(50)		(66)		9	133, 134
Electrophorus electricus							160	11	124
		44		50				4.5	122
		43		48				6	125, 177
		42		54				7.5	110
		41		47; 53				2–4	123, 15
	32	42		49				6	126
				48; 54	60		110	3.6	178
		43		48; 54				7.5	
Rat skeletal muscle	(42)	45; 49		51	56	62	(110)	7.7–10.9	145, 146
BC₃Hl mouse muscle cell line		44		53		65; 72		2.6	153
Fetal calf myotubes		3 × 41						3.7	152

Hydrodynamic Properties

Studies of the hydrodynamic properties of detergent-extracted protein are complicated by the presence of detergents and by the existence of different states of aggregation of the molecule. Sedimentation in sucrose density gradients of detergent crude extracts or purified AChR from *Torpedo* reveals two major components: a light (L) form of standard sedimentation coefficient $8.6 \pm 0.8S$ (*T. marmorata, T. ocellata*) or $9.1 \pm 0.8S$ (*T. californica*) and a heavy (H) form of $12.5 \pm 1.3S$ (*T. marmorata, T. ocellata*) or $13.1 \pm 1.1S$ (*T. californica*) (15, 111, 175, 179–181). In some instances minor components—a very heavy (HH) form of $16.6 \pm 6.5S$ and a very light (LL) one of $5.1 \pm 0.7S$—are observed (111, 180). In *E. electricus,* L (9S) predominates under the conditions of solubilization used (182–185).

The thiol reagent *p*-chloromercuribenzoate has no effect on the L/H ratio (180), but preincubation with a disulfide reducing agent, either β-mercaptoethanol (133, 134) or dithiothreitol (139), causes an almost complete interconversion of H into L. Extraction in the presence of Ca^{2+} favors the occurrence of the L form. On the other hand, after homogenization and extraction in the presence of the alkylating reagent *N*-ethylmaleimide, most (85%) of the receptor is in the H form (139). Urea (1–2 M) shifts the balance in favor of L (134). Likely, disulfide bonds are involved in the L-H transition but need not be intermolecular. Heating at 40° C in the presence of Triton X-100 decreases H in favor of HH, but the same treatment in the absence of detergent results in a reduction of L while HH increases. At pH 10, H decreases and L does not change, but the total amount of the very light LL form increases, which suggests that H dissociates to components sedimenting as LL (180). The binding properties of H and L for cholinergic ligands and α-toxins appear essentially the same (180, 186, cf 179) although careful binding studies give K_D for ACh of $3 \pm 1 \times 10^{-9}$ M for H and of $2 \pm 1 \times 10^{-8}$ M for L (139).

Filtration on Sepharose 6B columns equilibrated with a detergent solution gives apparent Stokes radii of 6.9–7.3 nm for L (111, 182–186) and of 8.5 nm for H (111), which are not compatible with the corresponding apparent sedimentation coefficients made on the assumption that the receptor is a standard water-soluble globular protein. Centrifugation in sucrose gradients where the density is varied by replacing H_2O by D_2O (180, 184) reveals that in the presence of D_2O the receptor protein sediments more slowly relatively to standard globular proteins, which bind little or no detergent. This suggests an abnormal density of the receptor protein in detergent solution. The $\bar{v}$ calculated from amino acid composition is 0.73–0.74, typical of a globular protein. On the other hand, the apparent $\bar{v}$ calculated from the H_2O-D_2O centrifugation experiments [for a critical appraisal of the method see (187, 188)] are 0.77–0.78 (180, 184) in the presence of Triton X-100 and 0.73 (180, 184) in the presence of Na cholate. The receptor protein therefore binds significant amounts of detergent (10–20%) in aqueous solutions. Direct binding studies with radioactive cholate (125) or Triton X-100 (181) confirm this conclusion. However, a contribution of shape to the unusual hydrodynamic behavior is not excluded.

Molecular Weight and Subunit Structure

The apparent molecular weight of the various forms of the AChR protein present in detergent solution have been estimated by hydrodynamic methods (after correction for detergent binding) (180, 181, 184), SDS gel electrophoresis after crosslinking (122, 177), or osmometry (189). Most likely H is a dimer of L, HH results from the aggregation of H, and LL results from the split of L. The values reported for the molecular weight of the 9S (L) form range from 170,000 (180) to 360,000 (184) by hydrodynamic methods. Striking agreement, however, exists between the estimates given by electrophoresis in SDS after crosslinking—260,000 (122), 275,000 (after correction) (177)—and osmometry—270,000 (189). These last values look reliable and are consistent with the 8–9 nm diameter of the particles observed by electron microscopy in preparations of purified receptor from *E. electricus* (125, 190) and *T. marmorata* (12, 166, 168, 190–192). These bracelet-like particles have a characteristic electron-dense central pit and a multisubunit structure (6, or 5, subunits, 3–4 nm in diameter).

In the presence of SDS the receptor oligomer splits into subunits with apparent molecular weight by SDS gel electrophoresis ranging between 40,000 and 140,000. Table 1 gives a compilation of the available data on purified AChR from electric organ and muscle. A general agreement exists about the 40,000 band, which is dominant in almost all the preparations and, moreover, is the only one labeled by MBTA (15) [with the exception of the AChR from rat skeletal muscle where two distinct protein bands of 45,000 and 49,000 apparent mol wt react with MBTA (145)]. Four additional bands unlabeled by MBTA and of 50,000, 60,000, 65,000, and 100,000–160,000 apparent mol wt have also been systematically encountered in highly purified preparations. These bands have a different chemical composition (16), do not share antigenic determinants (19, 135), and therefore do not derive from each other, for instance by proteolytic attack (178). These bands make a stable complex with the 40,000 component under nondenaturing conditions, which is not resolved by DEAE-cellulose chromatography, isoelectric focusing in Triton X-100 and 0, 1, or 2 M urea (134). Also, antibody raised against each band isolated in SDS precipitates the whole complex (135). However, close examination of the data reveals the following: 1. The 60,000 band is not found with *T. marmorata* and *E.*

electricus receptor protein (cf 178), the 65,000 band is absent from *E. electricus* receptor, and the 140,000 band derives from the 65,000 one by intercatenary disulfide bonding (139, 193). 2. Receptor proteins from elasmobranchs, teleosts, birds, and mammals share common antigenic determinants (19, 194, 195) that, at least in two instances, seem to be carried by the 40,000 component only (196). 3. The 50,000 and 60,000 components, which comigrate on one-dimensional SDS gels with the corresponding bands of the purified AChR, are present in membrane fractions that do not contain the MBTA labeled subunit (134). 4. The ratios of the different bands (determined by densitometric recordings after staining with Coomassie Blue) are constant within a given preparation but vary from one preparation to another (133, 134). Highly purified preparations of membrane fragments containing almost exclusively the 40,000 and 43,000 components yield, after solubilization and centrifugation, an α-toxin binding protein that consists almost exclusively of the 40,000 subunit with negligible amounts (less than 10%) of the larger subunits (133, 134). 5. Analysis of the N-terminals of the 40,000 band yields only one amino acid: serine (171). These observations suggest that the receptor protein in detergent solution is an oligomer made up of only one class of polypeptide chain labeled by MBTA and of 40,000 apparent mol wt; the exact number of chains per 9S oligomer is not known but 6, or 5, is plausible (177). The significance of the other bands is not known [except for the 43,000 membrane-bound component (see below)], nor is the reason they coaggregate with the 40,000 chain upon solubilization by nondenaturing detergent. Relevant to this point, however, is that the capsid proteins of the Semliki Forest virus make stable complexes of different polypeptide chains (even with fixed stoichiometries) and detergent upon solubilization (197).

Chemical Properties

The average amino acid composition of the complex oligomer (for review see 10, 198) and of its several subunits (16) reveals an average hydrophobicity that, expressed in terms of Barrantes discriminant function Z ($Z = 0.36$–0.39), lies in between that of AChE ($Z = 0.25$) and Na^+K^+ ATPase ($Z = 0.42$) (16). Spectroscopic and chemical evidence from several laboratories (125, 129, 130, 140, 199, 200) confirm the presence of tryptophan [questioned in an earlier report (124)] in the vicinity of the acetylcholine binding site (201, 202). In agreement with early in vivo and in vitro pharmacological observations (8, 203–209) a thiol group can be selectively alkylated near (or within) the acetylcholine binding site after treatment of the purified protein by dithiothreitol (15). The content of free sulfhydryl groups in the complex oligomer ranges from 20 (12) to 34 (139) nmole/mg protein that is about 18–30% of the total cysteic acid residues; about two thirds of the free SH groups react rapidly (less than 30 min) (139). The reactivity of some SH groups in the membrane-bound AChR is influenced by cholinergic effectors and detergents (193).

AChR purified from *E. electricus* (125), denervated rat diaphragm (144), and a mouse cell line (153) reacts with concanavalin A and other plant lectins, which indicates bound α-D-gluco- and/or α-D-mannopyrannosyl residues and, at least in the case of *E. electricus,* N-acetyl-D-galactosamine. From chemical analysis, AChR from *T. californica* contains 3–5% neutral sugars: mannose, galactose, and glucose (3 : 2 : 1) and N-acetyl-D-glucosa-mine (from amino acid analysis) but no fucose (16, 176, 200). All the bands found on SDS gels stain for carbohydrate except the membrane-bound 43,000 one (134)

Isoelectric focusing of the complex oligomer in neutral detergent gives, in general, a broad peak (revealing microheterogeneity) centered around pH 5.2 [α-toxin complex (122, 185)] or 4.8 [free receptor (122, 130, 210, 211)]. Two discrete forms of pI 5.1 and 5.3 have been resolved by electrofocalization of the α-toxin AChR complex extracted from denervated diaphragm and assigned, respectively, to the junctional and extrajunctional AChR (17, 144), which otherwise do not differ by their immunological reactivity (144) or their binding specificity towards cholinergic ligands (212, 213). These two isoelectric forms are present in crude extracts of *E. electricus* membrane fragments and can be interconverted in vitro (214). The effect of NaF on this interconversion suggests that a phosphorylation-dephosphorylation of the protein takes place during this process. The corresponding enzymatic systems are indeed present in the electric organ from both *T. marmorata* and E. *electricus* (215, 216). Immunoelectrophoresis shows that AChR can be labeled in vitro with ^{32}P, but the direct demonstration that phosphoryla-tion accounts for the shift in pI is still lacking (217, 218). In addition, the 48,000 (217) or 65,000 (218) subunits appear predominantly phos-phorylated, but a phosphorylation of the 40,000 unit is also possible. The significance of this phosphorylation regarding the regulation of the permea-bility properties of the ionophore and/or the subsynaptic localization of the AChR during synaptogenesis has been discussed elsewhere (219).

A microheterogeneity of the detergent-extracted AChR and of its 40,000 subunit has been reported in other instances. In developing myotubes in vitro, the α-toxin labels only a fraction of the total content in AChR revealed by detergent extraction (220–222). These "surface" AChR have in mouse BC$_3$Hl cells an *s* value that is 0.5–0.6 X 10^{-13} sec larger than the "hidden" ones (222). SDS gel electrophoresis resolves two MBTA labeled bands of 45,000 and 49,000 apparent mol wt (145) in purified preparation from denervated rat leg muscle. High resolution two-dimensional electro-phoresis of ^{35}S methionine-labeled AChR from calf myotubes reveals that the 40,000 subunit (almost the only one observed) gives three distinct spots. These species give similar peptide maps (by cleavage of cysteine residues) and most likely differ by a single charge (152). The observed micro-heterogeneity may result from phosphorylation, glycosylation reactions, or mild proteolytic attack (or still unidentified covalent modifications) and could concern the incorporation of AChR into the membrane [for instance, split of a signal peptide (223)] or its aggregation into patches (78, 224–228).

CLASSES OF LIGAND BINDING SITES ON MEMBRANE-BOUND AND DETERGENT-PURIFIED AChR

Essential differences exist in the binding properties of the AChR in its pure and membrane-bound states possibly relevant to specific binding effects of the detergent used for solubilization, uncoupling from other membrane protein, and/or lipid stabilization of particular conformation of the protein. Extension of the data obtained with the purified protein to the membrane-bound AChR, if legitimate, should always be made with caution.

Two major categories of sites have been distinguished in AChR-rich membranes: 1. the AChR site, which binds agonists, antagonists, α-toxins, and MBTA; and 2. a site for local anesthetics (LA site).

The AChR Site

AFFINITY LABELING   Tritiated MBTA has been used efficiently to cova-lently label and quantitate AChR sites in intact cells (205, 206), in membrane fragments (229), and in solution (15, 123, 137, 230), and to identify the subunit that carries the AChR site (123, 137, 231). MBTA has a significant affinity for the unreduced (i.e. native) AChR site ($K_D = 8 \times 10^{-5}$ M) and alkylates the reduced AChR at a rate of 6×10^5 M^{-1} sec^{-1}, approximately 5000-fold faster than N-ethylmaleimide (123, 232, 233). At saturation, [^{3}H]MBTA was reported to occupy half as many sites as snake toxins (137, 230, 232). All covalent binding of MBTA is retarded by revers-ible cholinergic ligands and blocked by α-toxin (123, 206, 207, 231). Only 30% of the α-toxin binding, however, is blocked by prereaction with MBTA (230). Other potential affinity labels that attach to the AChR have been described, such as p-trimethylammonium benzene diazonium fluoroborate (234, 235), the depolarizing bromoacetylcholine (236), and trimethyloxonium fluoroborate (237). A photoaffinity label, 4-azido-2-nitrobenzyltrimethylammonium fluoroborate (138, 238, 239), reacts with the 40,000, 48,000, 62,000, and 66,000 chains of the membrane-bound and detergent-extracted AChR from *T. california* and therefore appears less specific than MBTA [*Naja n. siamensis* α-toxin prevents the alkylation of only the 40,000 chain (138)].

SNAKE VENOM α-TOXINS   Several radioactive derivatives of the α-toxins have been widely used to assay the AChR, both in vivo and in vitro, and the binding kinetics of several of them have been investigated on various membrane preparations, detergent extracts, and purified receptor (14, 16, 17, 144, 149, 164, 213, 240–250). Binding studies, carried out under conditions of excess α-toxin concentration so that all binding sites are finally occupied, give monophasic (241, 242, 249) or biphasic kinetics with two time constants different by a factor of 2–10 (16, 17, 144, 240, 246–248). Biphasicity might result from an heterogeneity of the α-toxin preparation (see 249), but the observation of monophasic kinetics under conditions of AChR excess tends to rule out this possibility (16, 242, 244). Biphasicity might also arise from an heterogeneity of the α-toxin binding sites. For instance, data obtained on membrane fragments or purified AChR from *T. californica* and skeletal muscle with [^{125}I]α-bungarotoxin and filtration techniques can be fitted on the basis of two independent binding sites with different "on" rate constants for the α-toxin (16, 246). On the other hand, different conclusions have been reached with the same toxin but with membrane fragments from *E. electricus* (240, 247, 248) and under conditions where only the amount of α-toxin bound in an "irreversible" manner (i.e. not displaced after 15 hr in the presence of excess unlabeled α-toxin or decamethonium) was measured. Biphasic kinetics for association are still observed, but the relative amplitudes (and the association rate constants) of both phases depend on the initial concentration of α-toxin, which rules out the independent sites model. A complex allosteric model, involving two subunits, preferential states, and pathways has been discussed and reported to account for the data (240, 248).

Another model has been presented for the binding of [^{3}H] Naja α-toxin to the purified AChR from *E. electricus* (241, 242). The formation of multivalent α-toxin receptor complexes would proceed from a single homogeneous population of α-toxin binding sites associated in pairs. A toxin molecule would bind to anyone of them, with a high affinity ($K_D \simeq 10^{-11}$ M), and the binding of a second α-toxin molecule would occur with a lower affinity (3×10^{-10} M) due to a steric hindrance caused by the first α-toxin molecule bound residue, which transiently impairs the development of a perfect fit. The model accounts for the accelerated dissociation of the α-toxin observed in the presence of unlabeled toxin. Additional hypotheses such as the formation of ternary complex have been proposed to account for the accelerated dissociation of the toxin receptor complex in the presence of high concentrations of small competing ligands (242, 244).

STOICHIOMETRY OF α-TOXIN AND REVERSIBLE CHOLINERGIC LIGAND BINDING   A number of radioactive or fluorescent (254–261) cholinergic ligands have been found to bind to the AChR in its membrane-bound and purified forms (for review see 10–12,14, 16, 211, 251–253), but some confusion still exists in the literature regarding the relationship of their sites to the α-toxin site and their relative stoichiometry. Simple in principle, the determination of site stoichiometries meets with several difficulties, the major one being to assess the pharmacological activity of the radioactive ligands, the α-toxin in particular, within 50% error. Moreover ACh and derivatives are known to bind, particularly in membranes, to a variety of sites—specific or not—in addition to the AChR site. An important advance in the classification of these sites was the finding that the snake α-toxins block only certain sites (106, 125, 244, 262, 263). We consider only this last category of sites to belong to the physiological AChR site *sensu stricto*.

Competitive relationships have been found for the purified and membrane-bound AChR from electric organs between antagonists such as *d*-tubocurarine or flaxedil and tritiated agonists such as decamethonium (244) or acetylcholine (125), or between [^{14}C] dimethyl-*d*-tubocurarine and carbamylcholine (251). Moreover the ratios of [^{14}C] dimethyl-*d*-tubocurarine and [^{3}H] acetylcholine binding sites is close to one (251). All these bindings are blocked by the α-toxin. Most likely, under these conditions, agonists and antagonists bind to a common (or overlapping) site(s). The abundant studies carried out on the reciprocal interactions between radioactive derivatives of α-toxins and a variety of cholinergic ligands also indicate competitive relationships (for review see 10, 16, 213, 242, 244). [Noncompetitive effects (240, 249) may be due to inadequate characterization of the α-toxin binding process itself.] Striking differences of structure exist between those toxic polypeptides and quaternary ligands. [Recent crystallographic data on the erabutoxin from *Laticauda semifasciata* suggest, however, common structural features of the three-dimensional organization (D. Tsernoglou, G. Petsko, R. Hudson, private communication).] Nevertheless, a common binding site for cholinergic ligands and snake α-toxin seems most plausible.

Data on the relative stoichiometries between α-toxin and quaternary ligands reveal a rather complex situation. Values ranging from 1 to 2 are reported for the ratio of α-toxin binding sites to radioactive ACh, decamethonium (106, 125, 186, 244, 262–264), fluorescent ligands (255–257, 259), and MBTA (137, 230, 232) binding sites. Half-of-the-site reactivity has been postulated to account for the data obtained with [^{3}H]ACh and Triton X-100 solubilized preparations of AChR from *T. california* in the presence and in the absence of nicotine and *d*-tubocurarine (265). But the possibility should be seriously considered (see next section) that the AChR site may exist both in the membrane and in solution under different states of affinity for cholinergic ligands. All of them are expected to be titrated by the α-

toxin, but only the high affinity ones would be measured with reversible cholinergic ligands, giving apparent higher ratios of α-toxin to cholinergic ligands binding sites.

Another complexity arises from the observation that with crude preparations of *E. electricus* and *Torpedo* membrane fragments, only a fraction of bound decamethonium is displaced by α-toxin and therefore considered as associated with the AChR site (106, 244, 262, 263). Part of the residual binding is actually accounted for by the interaction of decamethonium with AChE catalytic site (106, 262, 263), but approximately 50% of it persists even after blockade of AChE by anticholinesterasic agents or after AChE salt extraction (262, 263). Decamethonium could be completely displaced, however, by carbamylcholine but not by *d*-tubocurarine (106, 262, 263). The nature of the sites that account for this residual binding is still enigmatic. One possibility is that they do belong to the AChR protein even though they are not blocked by the α-toxin. The stoichiometry between decamethonium and α-toxin would become one-to-one, but only half of bound decamethonium would be displaced by the α-toxin, and this would again mimic a half-of-the-site reactivity. Another interesting possibility is that these sites do not belong to the AChR site *sensu stricto* but rather to the LA site present in these membranes (see below). This could account for the unusual dual pharmacological action of decamethonium (51, 106).

Multiple States of Affinity of the AChR Protein

In early binding studies done with membrane fragments from *T. marmorata* (266–274) it was noted that the plot of ACh binding data could not be fitted by a single hyperbola but revealed a striking heterogeneity of binding constants. Although the α-toxin test was not done, this observation has been repeated in several laboratories (for review see 14). In parallel studies (25, 245) it was found that marked changes of affinity for cholinergic ligands follow dissolution of the membrane fragments by detergents.

Reinvestigation of this problem starting from AChR-rich membranes from *T. marmorata* (186, 244) led to the following observations:

1. Binding experiments done with [^{3}H]ACh disclose after prolonged equilibration at room temperature a homogeneous population of AChR sites with high affinity. The shape of the binding curve is often sigmoid ($n = 1.3–1.5$) with a midpoint around 10 nM [positive cooperativity has also been noticed with a purified form of the AChR (275)].

2. Dissolution of the membrane fragments by Na cholate modifies the shape of the ACh binding curve and the intrinsic dissociation constants of the AChR sites without changing the total number of α-toxin sites. The Scatchard plot of the binding data can be fitted on the basis of at least three

distinct populations of sites with K_D close to 30 nM, 0.1 μM, and larger than 1 μM. The neutral detergent Triton X-100 stabilizes primarily the high affinity sites, but purification of a Triton X-100 extract on an affinity column yields a fraction that binds ACh with a low affinity (possibly as a consequence of a molecular selection).

3. The change of affinity caused by Na cholate affects more significantly the binding of agonists (by several orders of magnitude) than that of the antagonists d-tubocurarine or flaxedil (by less than a factor of 10) (125, 276). Deviation from simple binding to an homogeneous population of sites has also been noticed with agonists but not with antagonists, using the indirect α-toxin assay and *Electrophorus* purified AChR (242).

4. Removal of Na cholate by dilution regenerates the high affinity sites from the low affinity ones, but this property can be lost after prolonged storage or purification (186). A reversible interconversion between the various classes of sites takes place when detergent concentration is changed.

5. In the membrane at rest the AChR is present under a low affinity state for agonists and agonist binding causes an interconversion to the high affinity state (see following sections).

6. The differences reported between the various purified preparations of *Torpedo* and *Electrophorus* receptor protein might result, to a certain extent, either from the selection of one of the affinity states or from the stabilization of the receptor molecule in one of these states. Pharmacological differences relevant to the zoological origin might nevertheless exist.

Local Anesthetic Binding Sites in AChR-Rich Membranes

Local anesthetics and a number of toxins such as histrionicotoxin and derivatives (2, 84–88), or ceruleotoxin (89) inhibit the pharmacological response to agonists in a noncompetitive manner. The concentrations of anesthetic needed to displace cholinergic ligands from the AChR site in the membrane-bound and purified forms of AChR are much higher than those needed to block the response (89, 244, 245); it was therefore suggested that local anesthetics bind to a class of membrane sites distinct from the AChR site (244). Direct evidence for their existence in AChR-rich membranes from *T. marmorata* was provided by spectroscopic methods, using fluorescent probes such as DNS-Chol[1-(S-dimethylaminonaphthalene-1-sulfonamido)-ethane-2-trimethyl ammonium iodide] (255) and quinacrine (202, 277). It was shown that DNS-Chol, which behaves on the electroplaque as a mixed agonist and noncompetitive blocking agent, interacts both with the AChR site and with secondary sites which are revealed in the presence of agonists and associated with distinct DNS-Chol fluorescence properties (255). These secondary sites were further investigated using a pharmacologically simpler ligand, quinacrine, which behaves in vivo as a local anesthetic (202, 347). Quantitative analysis of quinacrine fluorescence intensity strongly suggests the existence of a population of saturable binding sites, with K_D close to the K_{app} for blocking the response, which can be revealed in the presence of agonists (202). The fluorescence signal emitted by DNS-Chol or quinacrine bound to these secondary sites is abolished by compounds acting as local anesthetics such as prilocaine or histrionicotoxin (HTX) (171).

Direct binding of the radioactive local anesthetic quaternary dimethisoquin to AChR-rich membranes from *T. marmorata* has been measured by ultracentrifugation techniques (251). Specific and saturable binding sites, superimposed on unspecific ones, become apparent in the presence of carbamylcholine at physiological concentrations. A preliminary estimate of one local anesthetic binding site for four ACh binding sites, has been reported (251).

A tritiated derivative of HTX—perhydrohistrionicotoxin ($H_{12}HTX$)— binds reversibly to a limited number of high affinity sites ($K_D = 0.4 \ \mu M$) present in *T. ocellata* membrane fragments (88, 362). The ratio of these sites to ACh binding sites in the membrane approaches 2:1. As expected from the pharmacological actions of $H_{12}HTX$, local anesthetics like procaine, tetracaine, and piperocaine displace [³H] $H_{12}HTX$ from its membrane sites. On the other hand, cholinergic agonists, antagonists, and α-bungarotoxin do not block [³H] $H_{12}HTX$ binding. Most likely quinacrine, [¹⁴C] quaternary dimethisoquin, and [³H] $H_{12}HTX$ bind to the same LA site present on the receptor-rich membranes but distinct from the AChR site.

Allosteric Interactions Between the AChR Site and the Local Anesthetic Binding Site.

At physiological concentrations local anesthetics cause an increase of the affinity of the membrane-bound AChR from *T. marmorata* for agonists and for some antagonists, as evidenced from the enhanced binding of radioactive and fluorescent ligands (255, 257, 261). Similarly they cause a change of shape of the ACh binding curve from a sigmoid to an hyperbola (257), an effect analogous to that reported for allosteric ligands in the case of regulatory enzymes (278, 279). Detergents, such as Triton X-100, or fatty acids at concentrations where they do not release the AChR protein into solution (280) and HTX (86) have similar effect.

Conversely, agonists modify the affinity of local anesthetics for the LA sites; indeed, as already mentioned, carbamylcholine enhances the binding of [¹⁴C] quaternary dimethisoquin to the AChR-rich membranes (251). Although it is not clear whether the enhanced fluorescence intensity observed with quinacrine in the presence of agonists and some antagonists (202, 277) is due to a change in quantum yield or to an increased binding of quinacrine to its site, the same explanation could hold. All these effects are lost upon dissolution of the membrane fragments by detergents (186, 257). The interaction between the AChR site and the LA site requires the integrity of the membrane structure.

Separation of Membrane Components Carrying the AChR Site and the LA Site

Attempts have been made to solubilize the membrane component that binds [^{3}H] H_{12}HTX using the detergent Triton X-100 (88). Filtration of the extract on a Sephadex G-200 column separates AChR (eluted in the void volume) from bound [^{3}H] H_{12}HTX which is retarded as a lower molecular weight component. However, in the presence of detergent, local anesthetics no longer affect [^{3}H] H_{12}HTX binding, which raises serious questions about its specificity and suggests nonspecific adsorption of the radioactive toxin to micelles of detergent.

Another approach has been to follow, by energy transfer from proteins, the fluorescence of quinacrine-labeled AChR-rich membranes from *T. marmorata* (171). Under these conditions a significant fraction of the signal is reduced by HTX or the local anesthetic dimethisoquin, revealing an interaction with the LA site. Dissolution of the membrane by nondenaturing detergents followed by standard fractionation yields two components: the AChR protein in a rather pure form and a high molecular weight aggregate that gives on SDS gels a single band of 43,000 apparent mol wt. This last protein species (43-K protein), freed from detergent and in the presence of quinacrine, gives a fluorescence signal that is still decreased by HTX and with the same apparent K_D as in the native membranes (8 X 10^{-7} M). On the other hand, the fluorescence signal recorded with the purified AChR protein is not sensitive to HTX. The 43-K protein therefore carries a site for HTX and local anesthetics that has been separated from the AChR protein *sensu stricto*. The demonstration that this component indeed accounts for the regulatory interactions that take place in the native membrane between agonists and local anesthetics might come from a successful reassociation between the purified components. This appears feasible since the enhanced binding of ACh by local anesthetics has been recovered after detergent elimination starting from a crude extract of AChR-rich membrane fragments (186). To some extent the interaction of these two protein components in the membrane resembles that observed between regulatory and catalytic subunits in the well-documented case of the regulatory enzyme aspartate transcarbamylase (281).

Binding of Ions

Binding of ions to specific sites is often difficult to distinguish from ionic strength effects, particularly in the case of monovalent cations. For instance, change of NaCl concentration markedly affects the binding of bisquaternary amines such as decamethonium, hexamethonium (252), fluorescent bis (3-aminopyridinium-1, 10-decane diiodide (DAP) (252, 256), and d-tubocurarine and α-toxin (244, 252, 282) to membrane-bound and purified AChR but has little effect on ACh and other monoquaternary amines (252, 283). Similar ionic strength effects have also been reported with AChE (284). Also high salt concentrations protect against the chemical modification of a subsite that might be involved in the binding of decamethonium (237).

Divalent cations act as regulatory ligands of membrane-bound AChR. Addition of millimolar concentrations of Ca^{2+} enhances the affinity for agonists in a manner that differs from that of local anesthetics since for instance it is not accompanied by a change of shape of ACh binding curve. This effect is lost upon dissolution of the AChR-rich membranes by detergent (186, 257).

The purified AChR protein from *Torpedo* and *Electrophorus* as revealed by spectroscopic measurements binds reversibly Tb^{3+} and Ca^{2+} (30–100 molecules per ACh binding site with at least three dissociation constants) (141, 285–287). Ca^{2+} inhibits the binding of ACh (285, 287) and conversely 2–6 Ca^{2+} and Tb^{3+} molecules are displaced by ACh but not by antagonists (286, 287). Subsequent addition of α-bungarotoxin leads to Ca^{2+} reuptake (287). The isolated 40,000 subunit still binds Tb^{3+} with equal affinity and, to a lesser extent, Ca^{2+} (141). Possible relationships between the divalent cations sites, the LA binding site, and/or AChI may exist but are not understood.

INTEGRATION OF AChR INTO THE CYTOPLASMIC MEMBRANE

Quantitative studies by high resolution autoradiography after labeling with radioactive α-toxins have established that the surface density of accessible AChR sites is much higher under the nerve terminals than between them [see references in (288–290)]. The revised absolute density is 50,000 ± 16,000 per μm^2 in *E. electricus* subsynaptic areas (290, 291) and 30,000 ± 7,000 per μm^2 at the crest of the folds in the neuromuscular junction (288). In muscle, outside the synapse or at the bottom of the folds the density of α-toxin sites falls within a few μm by 2–3 orders of magnitude; in *E. electricus* electroplaque an average density of 370 ± 250 sites per μm^2 is found between the synapses (290, 291).

By negative staining and/or after freeze-etching, AChR-rich (subsynaptic) membranes from *Torpedo* exhibit rosettes with the same size and shape as purified AChR (12, 166, 168, 190–192). The surface density of these 8-nm rosettes is 12,000–15,000 per μm^2, a value compatible with the density of α-toxin sites found by autoradiography if one assume several of these sites per particle; they occasionally form pseudohexagonal arrays (166, 190) with a center-to-center distance of 9–10 nm.

Analysis of the X-ray diffraction images given by the same membranes (252, 292) that have been oriented by centrifugation indicates a distance of 60 Å between dominant scatterers across the membrane instead of the 35–40 Å distance between polar heads of the classical lipid bilayer. Presumably a dense sheet of integral proteins spans the bilayer. [Indeed freeze-fracture reveals numerous "internal" particles (14, 191, 192, 293–300)] and is responsible for the large effective thickness of the membrane (90–100 Å) calculated from weight measurements. Several sharp equatorial reflections indicate a regular organization of particles within the plane of the membrane and at the distance expected for an hexagonal arrangement (292). In these membrane fragments the protein-lipid ratio is 1.5–2.0 and the buoyant density is ~ 1.2 g/cm^3 (301). The subsynaptic membrane therefore primarily consists of the dense semicrystalline assembly of ACh regulators.

In *E. electricus,* destruction of the spinal cord causes within 2–3 days a complete regression of the nerve endings, but 52 days later dense patches of α-toxin sites persist at the location of the former synaptic contacts (298, 302). The same is observed at the adult neuromuscular junction after denervation (303, 304). In the "uncovered" subsynaptic membrane the AChR shows little, if any, tendency for lateral diffusion.

This behavior contrasts with that of nonsynaptic AChR present in the cytoplasmic membrane of developing rat myotubes in culture (227). Under these in vitro conditions 90% of the surface sites labeled by a tetramethyl rhodamine derivative of α-bungarotoxin has a diffuse distribution of about 10^3 sites/μm^2 (the rest aggregates into 10–60-μm patches with 6–8 times higher density). Diffuse AChR experiences fast lateral mobility with an average effective diffusion coefficient of 10^{-11} cm^2/sec (in the patch, as in the subsynaptic membrane, no lateral mobility of AChR is observed). In developing myotubes in vivo no spontaneous patching occurs outside the area where the nerve terminal induces the aggregation of diffuse AChR into highly immobilized subsynaptic clusters (305, 306).

The rotational motion of AChR in AChR-rich membranes from *T. marmorata* can be studied by electron paramagnetic resonance spectroscopy with either noncovalent doxylpalmitoyl cholines (307, 308) or covalent (3-maleimido-2,2,5,5 tetramethyl-1-pyrrolidinyloxyl) spin label derivatives (309, 310). A strong immobilization of the acylcholines takes place when they bind to the AChR site, although significant motion is still detected (particularly when the paramagnetic probe is close to the methyl terminus of the fatty acid); this suggests that a fluid hydrophobic phase, possibly from a lipid bilayer, might be present in the vicinity of the AChR site (307, 308). Recent developments of the method of saturation transfer with covalent spin labels makes possible the measurement of slow protein motion. For instance, in fluid rod outer-segment membrane, the rotational correlation time of rhodopsin is $\tau = 20$ μsec at 20°C (310). No motion is found in this time range with the maleimide derivative of AChR ($\tau \geq 1$ msec) (309). Presumably protein-protein interactions exist in these subsynaptic membranes to restrict eventual motion in a fluid lipid bilayer.

Being an "integral" membrane protein, AChR interacts with lipids. This can be shown simply by cosedimenting in sucrose gradients a mixture of purified AChR with lipids after detergent removal (107, 311–314). Purified *Torpedo* AChR (like other membrane-bound or soluble proteins) also interacts with lipid monolayers, as shown from measurements of the surface tension or from the recovery of radioactive labels after film collection (301, 315). Incorporation into total erythrocyte lipids is enhanced at pH close to pI or by Ca^{2+} ions (315). Pure lipids are distinguished only by their hydrophobic moiety (301). AChR does not discriminate between phospholipids with different polar groups, but interacts strongly with cholesterol (and some of its analogs) and long-chain phospholipids. Interestingly, the lipid composition of the AChR-rich membranes is characterized by a high cholesterol content (weight ratio to phospholipids = 0.40) and an abundance of long-chain phospholipids (25% docosahexaenoic ester) (301).

Antibodies raised against purified AChR block the electrical response of *E. electricus* electroplaque to carbamylcholine (195, 316) and impair neuromuscular transmission [see references in (19)]. Conjugated to electron-dense ferritin molecules, they bind to the surface of AChR-rich membrane fragments from *Torpedo* (317). The AChR molecule is therefore freely exposed on the outside surface of the membrane, as expected for a pharmacological receptor. Nothing is known about the localization of the 43-K protein except that in the intact membrane it is accessible to the alkylating reagent [^{14}C]iodoacetamide (171). If little doubt exists about the identification as the AChR of the 9-mm rosettes observed after negative staining or etching, the internal particles seen after cleavage of the subsynaptic membrane may equally belong to the AChR, to the 43-K protein, or to their complex.

Several attempts have been made to reinsert the detergent-purified AChR from various sources [see references in (318)] into black lipid films, but the data presented on the recovery of a conductance response to agonist do not appear convincing. Purified AChR from *Torpedo* has also been integrated into purified liposomes and the permeability to radioactive $^{22}Na^+$ measured (199, 312). The rate of $^{22}Na^+$ influx could be accelerated by the artificial ionophores gramicidin D or valinomycin, but carbamylcholine had no effect (15, 230, 313).

A more reasonable approach is to start from purified AChR-rich membranes, which, being excitable, certainly contain the AChI. A permeability response to agonists has been repeatedly found (107; H. Sugiyama, unpublished) after reformation of closed vesicles upon elimination of detergent. The reproducibility of this result has been poor, however, which suggests that an essential step (or component) of the reconstitution was not under control. To circumvent this difficulty, reconstitution was first limited to the recovery of the affinity interactions between the AChR site and the LA sites. Reincorporation of AChR into lipid vesicles from a crude extract of AChR-rich membranes leads to a recovery (314) of the kinetic regulation of the AChR site by agonists (319; see following sections) but not on the effect of local anesthetics on this regulation (314). On the other hand, the effect of local anesthetics and Ca^{2+} on ACh binding at equilibrium was found after reassociation of a crude cholate extract of ACh-rich membranes upon elimination of the detergent by dilution (186). A better knowledge of the various components of the AChR-rich membranes, the 43-K protein in particular, is needed before functional reconstitution of the ACh regulator can be achieved.

ANALYSIS OF THE OPENING AND CLOSING OF THE CHOLINERGIC IONOPHORE IN THE PRESENCE OF AGONIST

Single-Channel Recording and Noise Analysis

By use of a microelectrode closely applied to an hypersensitive muscle fiber treated with collagenase, the voltage clamp technique, and sophisticated recording, it becomes possible to monitor currents that correspond to discrete molecular transitions taking place in the membrane in the presence of agonist. They most likely represent the opening and closing of single ACh ionophores. These statistically independent current pulses have a square shape and, under given conditions, a constant amplitude (320, 321), i.e. correspond to an all-or-none transition. In the presence of agonist the AChI would therefore continually fluctuate between open and closed conformations, the open conformation giving rise to the elementary current flow.

$$\text{closed} \underset{\alpha}{\overset{\beta}{\rightleftharpoons}} \text{open}$$

At a macroscopic level, it will result in a "noise," which can be analyzed by statistical methods (37, 65, 320–342; for review see 343) and leads to the evaluation of a relaxation time $\tau = (\alpha + \beta)^{-1}$ and of the mean AChI conductance γ. Assuming $\beta \ll \alpha$ (see 37), τ is equal to the mean AChI open time and values obtained from noise analysis are consistent with those derived from single AChI recordings (320, 321). Values of 1–3 msec for τ and 15–30 pmho for γ in the presence of acetylcholine are commonly reported. At 20°C about 10^4 Na^+ ions would flow inside the cell during the opening of one elementary AChI triggered by the binding of 1–4 ACh molecules.

The mean AChI open time varies over a tenfold range with the nature of the agonist but the amplitude of the elementary conductance change is approximately constant (320, 325, 328, 330–332). The reduced γ values reported in the case of two 3-methylammonium derivatives (330) are probably underestimates of the "true" elementary conductance, as inferred from the complex shape of the noise power spectra.

Membrane potential significantly affects τ with an e-fold change for approximately 80-mV membrane potential change (37, 320, 321, 328, 331). Lowering the temperature increases τ (Q_{10} ranges between 2 and 3 at –80 mV) (37, 65, 324, 328, 329, 331, 332) but has little effect on γ at the frog end plate between 5° and 20°C (37, 65, 329, 331, 332; cf 340). On the other hand, it has been reported with embryonic chick muscle in culture that γ is constant between 20° and 37°C, but decreases approximately 10-fold as temperature is lowered further to 5°C (341); an abrupt conductance change occurring at $\sim$ 25°C was also reported for a mouse muscle preparation (65). It was suggested that changes in the fluidity of the lipid phase could be responsible for these effects.

Extrajunctional AChI of denervated muscle fibers have 3–5 times larger mean open times than junctional AChI but comparable voltage, temperature, and agonist dependences (65, 320, 321, 324, 331, 332); slightly smaller values of γ have been reported for extrajunctional (8–18 pmho) than for junctional AChI (15–23 pmho) (54, 331, 332). Either the ACh regulator present in extrajunctional areas after denervation differs biochemically from the junctional one and/or kinetics of ion channel are influenced by the local lipid environment or by interaction between individual AChIs.

Treatment of the end plate by the disulfide reducing agent dithiothreitol leads to a slight ($\sim$ 40%) reduction of γ and a shortening of τ (329).

No detailed analysis of the effect of agonist concentration on the relaxation time derived from noise analysis is yet available; it was reported, however, using bath application of acetylcholine and carbamylcholine, that τ decreases with increasing agonist concentrations (333). Such dependence is expected if α and/or β are affected by the concentration of agonist (see below).

Voltage Jump Relaxation

Complementary information on these processes was provided by experiments using voltage jump relaxation. When membrane potential is rapidly jumped from one level to another the agonist-induced conductance relaxes to a new value with a time course $\tau = (\alpha + \beta)^{-1}$, which is an instantaneous function of the final potential (39, 40, 68, 328, 331, 333, 336–339, 344–347). Noise analysis and voltage jump relaxation done in parallel give similar values for τ and the effect of membrane potential (328, 331, 336–339).

Increasing the concentration of agonist decreases τ (39, 40, 345); the relationship between the rate constant and the agonist concentration is linear, and extrapolates for zero agonist concentration to a finite value that varies with potential and the nature of the agonist. In contrast, the constant of proportionality does not depend on membrane potential, but varies with the nature of the agonist (39, 40). The data are consistent with voltage dependence of α, small voltage dependence (if any) of β, and linear dependence of β with agonist concentration (39, 40; see also 328, 345). This last effect might imply that a bimolecular reaction between agonist and the ACh regulator either precedes or constitutes the rate-limiting step in channel opening.

Assuming without direct evidence that binding of agonist is the rate-limiting step, a forward binding rate constant for ACh of 10^7 M^{-1} sec^{-1} and a dissociation constant of 65 μM have been derived (39). The alternative model, namely that transitions of the AChI are the rate-limiting steps, implies that the relaxation rate constant levels off at high agonist concentration. Such an effect was observed with decamethonium (39, 40), but rapid desensitization or poor clamping effects may vitiate measurements in the relevant range of agonist concentration.

More complex models have been discussed in order to account also for the sigmoidicity of the dose-response curve [see discussion in (40, 345)]. Yet it appears that direct measurements of the binding processes are necessary to select among possible mechanisms.

Effect of Local Anesthetics

General anesthetics and alcohols affect the response of the end plate to agonists in a reversible but rather nonspecific manner, possibly as a consequence of changes in the fluidity and/or dielectric constant of the lipid bilayer [see references in (24)]. On the other hand, more specific interactions stand for the anesthetic action of compounds such as amine containing local anesthetics (27, 34, 321, 327, 334, 335, 339, 342, 346, 348–353), barbiturates [see references in (354)], atropine [see references in (355)], histrionicotoxin (2, 85–87), and quinacrine (347). Local anesthetics such as procaine, lidocaine, and derivatives alter the time course of the decay of the spontaneous and evoked end-plate currents and potentials, which show a fast initial and slow late component (and in some cases a third) (34, 334, 335, 348–353). Noise and voltage jump analyses also show that local anesthetics transform the simple relaxations observed in control conditions into more complex ones usually showing two (sometimes three) time constants (327, 334, 335, 339, 346).

Finally, in the presence of a lidocaine derivative, the shape of the single-channel current pulses recorded in the presence of suberyldicholine is significantly modified. The square pulses are chopped by fast blocking events whose frequency increases with increasing local anesthetic concentrations (321).

The simplest model that qualitatively and quantitatively accounts for these data assumes that these compounds bind to the open channel and block ionic transport (321, 335–339, 346, 354, 355; see also 350). Additional binding to the channel in the closed conformation has to be postulated in some cases to fit equilibrium data (346, 355).

Blocking of the open channel occurs with a voltage-dependent rate constant that does not vary significantly (at variance with backward rate constants) among the different local anesthetics and is $\sim 10^7$ M^{-1} sec^{-1} (321, 337–339, 346, 354, 355). Comparison of this value with the rate of Na^+ flux indicates that the drug binding rate is probably diffusion limited (321, 338, 346); the voltage dependence of the reaction rates further suggests that, assuming a constant membrane field, the blocked site is located at half to three quarters of its thickness within the membrane (321, 346). Although the steric blockade of ionic transport appears the most probable hypothesis, alternative schemes involving allosteric effects would fit the available data but seem nevertheless unlikely from analysis of the voltage dependences (321, 346). Models in which local anesthetics either produce (24, 353) or differentially affect (24, 39) several hypothetical populations of channels, or in which the current through each channel follows the same time course as miniature end-plate currents (327) have been rejected (28, 32, 320, 321, 327, 335, 342, 351).

DYNAMIC STUDIES IN VITRO ON THE EFFECT OF ACh ON MEMBRANE-BOUND AChR

Some kinetic features of the interaction of ACh with detergent-purified AChR were investigated, indirectly, through the effect of ACh on the relaxations of Ca^{2+} ions bound to the AChR using murexide as an indicator and the temperature jump technique (287, 356). The Ca^{2+}-murexide-AChR system exhibits a triphasic relaxation spectrum, two phases being modified in the presence of ACh. The data could be fitted assuming a bimolecular process with an "on" rate constant for the binding of ACh of 2.4 X 10^7 M^{-1} sec^{-1}, followed by an isomerization of the complex (356).

Since many of the regulatory properties of AChR in the membrane are lost upon solubilization by detergents, kinetic analyses relevant to the physiological response had to be carried out on AChR-rich membranes.

In vitro exposure of AChR-rich membrane fragments to cholinergic agonists causes a reversible change of affinity of the AChR site for the same ligands. The change of affinity was first monitored indirectly, by measuring the inhibition of α-toxin binding rates (by filtration, or ultracentrifugation in the presence of unlabeled α-toxin). An apparent 5–20-fold increase of affinity of the AChR site for agonists is observed in the minute range (213, 249, 314, 319, 363) and is accelerated by local anesthetics (314). It is not observed with antagonists such as d-tubocurarine, or flaxedil (213, 249, 319) but perhaps because of the poor time resolution of the method; in fact, preincubation of membrane preparations with "conditioning" dose of flaxedil or hexamethonium results in an increase of affinity for agonists (319). These observations have been confirmed by direct measurement of the binding of tritiated ACh, using ultrafiltration techniques (251; see also 357) and of a spin-labeled analog of decamethonium (no pharmacology reported) using ESR spectroscopy (249).

Quantitative analysis of the affinity changes of the AChR site in a large range of effector concentrations has now been achieved using fast-mixing techniques and spectroscopic recordings. A fluorescent derivative of acetylcholine, a dansyl acylcholine (260), both acts as an agonist on *E. electricus* electroplaque and displays a marked increase of fluorescence intensity when it binds to the AChR site (260, 261). Rapid mixing of AChR-rich membrane fragments with this compound reveals three distinct phases: 1. a fast (millisecond) fluorescence increase relative to high affinity AChR sites, 2. an intermediate phase relative to sites of lower affinity, and 3. a slow process, in the time range of seconds, which can be analyzed in terms of an interconversion from lower to high affinity states. From the analysis of the amplitude and rate constant of the fast phase it was inferred that a fraction of AChR sites (approximatively 20%) preexists in the membrane in a high affinity state before addition of agonists. Local anesthetics, upon preincubation with membrane fragments at physiological concentrations, increase the amplitude of the fast phase and therefore are assumed to shift a conformational equilibrium between several affinity states of the receptor in favor of the high affinity state, as do agonists even though they bind to distinct sites. Binding of the acylcholine to the AChR site in its high affinity state proceeds with a "on" rate constant close to that of diffusion control ($\sim 10^8$ M^{-1} sec^{-1}) (261).

Measurements of the intrinsic fluorescence of membrane proteins (201, 202, 358, 359) indicate that a structural change accompanies the affinity change of the AChR site (201, 359). Fast mixing of AChR-rich membrane from *T. marmorata* with agonists cause a time-dependent, saturable decrease of intrinsic fluorescence emission in the second time range. Its amplitude is proportional to the specific activity of the preparation and it is abolished by preincubation of the membranes with α-toxin. A faster signal was also detected but could not be analyzed because of limitations of the technique. Limiting time constants for the interconversion between low and high affinity states of ~5 sec in the presence of high concentration of acetylcholine and 2 sec with carbamylcholine were reported (201). No data are yet available concerning antagonists.

Despite their elegance, intrinsic protein fluorescence studies present severe drawbacks; in particular the amplitude of the changes recorded remains within 1% of the total fluorescence; extrinsic fluorescence probes have therefore been used, such as ethidium bromide (although no detailed kinetic analysis is yet available) (259) and quinacrine (202, 277).

Quinacrine acts on the electroplaque as a local anesthetic and does not compete in vitro with agonists or antagonists for the AChR site (202). Under conditions of fluorescence energy transfer from proteins to quinacrine, the equilibration process, after addition of cholinergic effectors to membrane fragments labeled with quinacrine, reveals a striking difference between agonists and antagonists (202, 277). High concentrations of agonists cause a fast increase of the fluorescence intensity (in the millisecond time range) followed by a slow decrease (minute time range) to a final intensity level different from the initial level. With antagonists, even at high concentrations, the fast increase is never observed.

The fast fluorescence signal recorded with agonists was investigated by the stopped-flow technique; the kinetics follow an exponential time course and the rate constants increase with increasing agonist concentration to reach, at high concentration, a limiting value dependent on the nature of the agonist. The data can be fitted by a scheme involving a rapid binding step for the AChR and agonist, followed by an isomerization of the complex. An alternative pathway involving rapid isomerization between two conformational states of the receptor followed by binding of the agonist to the state of highest affinity was rejected on the basis of different rate constants at high agonist concentration for different agonists. The forward isomerization rate constants differ for various agonists and range from about 100 sec^{-1} for ACh to about 1 sec^{-1} for choline; the backward isomerization constants are of the order of 1 sec^{-1} or smaller. Dissociation constants of ~70 μM for the binding of ACh and of 2 μM for the overall process (binding + isomerization reaction) were reported.

The data are accounted for by an adapted version of the model of Katz & Thesleff (73), with three interconvertible conformational states associated with different fluorescence intensity levels. Upon addition of agonist, the low affinity state, which should be present in significant amount in the membrane at rest, would be transiently interconverted into an "intermediate" state characterized by a higher fluorescence intensity level. The subsequent intensity decrease would correspond to the slow stabilization of the complex in a higher affinity state. The model accounts for the various effects of the antagonists if one further assumes that they bind preferentially either to the low or to the high affinity state but not to the intermediate state.

In conclusion, in vitro experiments unambiguously show that at least two distinct isomerization processes take place in the AChR-rich membrane, which differ in their time course: 1. "Slow" isomerization in the second time range is revealed by changes of intrinsic protein fluorescence and changes in binding properties of the AChR site for a fluorescent agonist. Although they have not yet been quantitatively analyzed, the slow affinity changes monitored in the minute range by the inhibition of α-toxin binding, by the binding of [^{3}H]ACh, or by the binding of the spin-labeled analog—as well as the slow quinacrine fluorescence changes observed with both agonists and antagonists—could also proceed with the same kinetics, but additional very slow reactions may occur. 2. "Fast" isomerization in the millisecond time range is monitored by quinacrine and observed only in the presence of agonist.

COMPARISON OF IN VITRO DATA WITH PHYSIOLOGICAL PROCESSES

It should be emphasized that the above-mentioned processes and ion permeability changes are distinct events since the former concern the AChR site or protein and the later the AChI. However, a strong correlation between the two processes should exist at the level of the pharmacological specificity of the signals and in their time courses and should account for the physiological response.

Pharmacological desensitization, which occurs upon prolonged exposure to agonist, develops within seconds and is accelerated by local anesthetics. It could therefore be correlated with the "slow" isomerization observed in vitro; according to this interpretation, the state of high affinity for the agonist would be a "desensitized" state and therefore associated with a low ionic conductance of the AChI.

In vivo, the rising phase of mepc occurs with a kinetic rate constant of 10^3 sec^{-1} or even faster and with Q_{10} values of $\sim$1.2. Closing of the open AChI takes place with a kinetic rate constant of 10^2–10^3 sec^{-1} and a Q_{10} of 2–3. The Q_{10} value of 1.3 measured for the fast isomerization process monitored by quinacrine compares well with the in vivo value found for activation and is clearly distinct from the Q_{10} value for AChI closing or desensitization.

Conversely, the value for the concentration of ACh in the synaptic cleft at its maximum [10^{-4} M (44)] fits with the dissociation constant derived from the quinacrine kinetics since it is the smallest that gives the maximal rate constant. However, the absolute values for the onward and backward isomerization rate constants are at least one to three orders of magnitude too small as compared to the in vivo values.

Activation kinetics of isolated membranes in vitro could actually be much slower than in the intact cell because of alterations of membrane properties due to isolation and purification, changes of membrane potential, and/or ionic environments, or because of the presence of bound quinacrine molecules.

Alternatively, the fluorescence signal of quinacrine could be only indirectly related to the activation process and, for instance, could result from an additional binding of quinacrine to the LA site [see preceding sections and (347)] rather than from a change of quantum yield of already bound molecules. A complete kinetic analysis should take into account the parameters of this interaction. Accordingly the isomerization rate constants for activation could be much higher than those reported. In fact, the real situation might even be more complex and involve simultaneous changes in quinacrine quantum yield and binding effects.

The definitive correlation of the affinity changes and structural transitions monitored in vitro by relaxation methods with the kinetics of activation and desensitization will rely on parallel measurements of the permeability changes of the AChR-rich membranes in vitro under well-defined conditions of membrane potential and ionic environment.

A MODEL OF THE ACh REGULATOR

In an attempt to correlate the physiological observations with the structural information presently available on the AChR protein, on the AChR and LA sites, and on the dynamic of their transitions, one may propose the following model, which is an adapted version of the model of Monod, Wyman & Changeux (279) for allosteric proteins, and of that of Katz & Thesleff (73) for desensitization (Figure 1; 14, 202, 360).

1. The *ACh regulator* is the elementary functional unit, which accounts for the regulation of cation translocation by ACh; it is made up of two distinct protein entities: the *ACh receptor* and the *ACh ionophore*. These two proteins are strongly associated in the native membrane but dissociate upon dissolution of the membrane by detergents.

2. The *ACh receptor* results from the association of a finite number (at least three, possibly more) of polypeptide chains (most likely from a single class of an apparent mol wt of 40,000) and carries several (probably the same as the number of chains) AChR sites, which bind cholinergic agonists, antagonists, and snake α-toxins.

3. The *Ach ionophore* is made up of at least one class of polypeptide chains (possibly of an apparent mol wt of 43,000) and carries both the selective ion gate, or channel, and the LA binding sites. Local anesthetics act at its level both as ligands and blocking agents of ion translocation, possibly by interacting directly with the ion gate.

4. In the membrane, AChR and AChI are coupled in a manner that resembles the association between catalytic and regulatory subunits in a regulatory enzyme such as aspartate transcarbamylase (281). Consequently, their complex, i.e. the ACh regulator, may exist under a small number of discrete conformational states that are interconvertible, preexist to ligand binding, and are therefore independent of the nature of the ligand. These states, referred to as resting R, active A, and desensitized D, respectively, differ as follows: (*a*) The AchR site has distinct affinities for cholinergic ligands: the affinity for agonists increase from R (low affinity) to A (medium affinity) to D (high affinity); on the other hand, the affinity for antagonists is lower in the A than in the R or D states. "Nonexclusive" binding (361) of a given ligand to more than one state is expected to occur. (*b*) The affinity of the LA site for local anesthetics is higher in the D and A states than in the R state. (*c*) The ion gate is open only in the A state.

5. In the membrane at rest, the R state is favored; the fraction of A state is negligible and that of the D state is small but not negligible, at least in vitro. In the presence of a given concentration of ligand, the equilibrium is shifted in favor of the state to which it binds preferentially (at the level of the AChR or LA site).

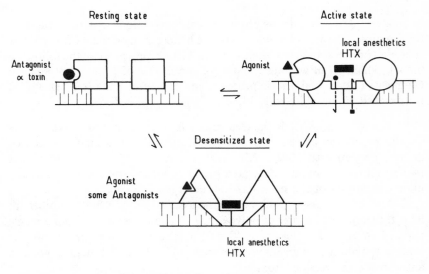

Figure 1   A model of the ACh regulator (see text).

6. Agonists will shift the equilibrium toward the D state, antagonists to either R or D state, and local anesthetics (and multivalent cations) to either the D or A state. Upon rapid change of ligand concentration, and depending on the nature of the ligand, some states can be transiently populated due to preferential pathways for the interconversions. The "activation" reaction is viewed as a transient population of the A state by ACh and the "desensitization" as the final equilibration in the D state.

This model is, in many instances, highly hypothetical. Preliminary evidence regarding the quaternary structure of the ACh regulator, which suggests that AChR site and LA site are carried by distinct protein entities, needs confirmation. The AChR and LA sites are unambiguously distinct and coupled by allosteric interactions, but only electrophysiological evidence indicates that the LA site could be part of the ionic channel and that allosteric interactions between LA and permeant ion are not excluded. Therefore there is no definitive proof yet that the 43-K protein, which carries a site for local anesthetics and HTX, is the AChI. Reconstitution experiments with purified AChR and 43-K protein should bring a direct answer to this question. Good evidence exists from both equilibrium and relaxation studies that both the AChR site and the LA site may exist under different and interconvertible states of affinity. That these structural states preexist to ligand binding is well documented in the case of the D state, which is found in significant amounts in the membrane at rest and accumulates in the presence of local anesthetics, even in the absence of agonist. The all-or-none fluctuations of the AChI observed under conditions of single-channel recording are also accounted for by discrete transitions of the ACh regulator between the R and A state. Cooperative interactions for ACh binding to the AChR observed in vitro indicate that the transition between states affects the several AChR sites present on the AChR. However, the postulate that only three states exist might be a simplification. The exact correlation between the states of affinity for agonists and for local anesthetics and the state of opening of the AChI remains to be established in detail.

ACKNOWLEDGMENTS

The experimental work carried out in the authors' laboratory was supported in part by a grant from the Muscular Dystrophy Association of America.

Literature Cited

1. Changeux, J. P., Podleski, T. R., Meunier, J. C. 1969. *J. Gen. Physiol.* 54:225S–44S
2. Albuquerque, E. X., Barnard, E. A., Chiu, T. H., Lapa, A. J., Dolly, J. O., Jausson, S. E., Daly, J., Witkop, B. 1973. *Proc. Natl. Acad. Sci. USA* 70:949–53
3. Nachmansohn, D. 1955. *Harvey Lect.* 49:57–99
4. Changeux, J. P., Kasai, M., Lee, C. Y. 1970. *Proc. Natl. Acad. Sci. USA* 67:1241–47
5. Changeux, J. P., Kasai, M., Huchet, M., Meunier, J. C. 1970. *C. R. Acad. Sci. Ser. D* 270:2864–67
6. Changeux, J. P., Meunier, J. C., Huchet, M. 1971. *Mol. Pharmacol.* 7: 538–53
7. Miledi, R., Molinoff, P., Potter, L. T. 1971. *Nature* 229:554–57
8. Karlin, A. 1974. *Life Sci.* 14:1385–1415
9. Rang, H. P. 1975. *Q. Rev. Biophys.* 7:283–399
10. Changeux, J. P. 1975. *Handbook of Psychopharmacology,* ed. L. L. Iversen, 6:235–301. New York: Plenum. 307 pp.
11. Cohen, J. B., Changeux, J. P. 1975. *Ann. Rev. Pharmacol.* 15:83–103
12. Eldefrawi, M. E., Eldefrawi, A. T., Shamoo, A. E. 1975. *Ann. NY Acad. Sci.* 264:183–202
13. Neumann, E., Bernhardt, J. 1977. *Ann. Rev. Biochem.* 46:117–41
14. Changeux, J. P., Benedetti, L., Bourgeois, J. P., Brisson, A., Cartaud, J., Devaux, P., Grünhagen, H., Moreau, M., Popot, J. L., Sobel, A., Weber, M. 1976. *Cold Spring Harbor Symp. Quant. Biol.* 40:211–30
15. Karlin, A., Weill, C., McNamee, M., Valderrama, R. 1976. *Cold Spring Harbor Symp. Quant. Biol.* 40:203–10
16. Raftery, M. A., Vandlen, R. L., Reed, K. L., Lee, T. 1976. *Cold Spring Harbor Symp. Quant. Biol.* 40:193–202
17. Brockes, J., Berg, D., Hall, Z. W. 1976. *Cold Spring Harbor Symp. Quant. Biol.* 40:253–62
18. Fulpius, B. W. 1976. *Motor Innervation of Muscle,* ed. S. Thesleff, pp. 1–29. London: Academic. 351 pp.
19. Lindstrom, J. 1976. *Receptors and Recognition,* ed. P. Cuatrecasas, M. F. Greaves, 3:1–45. London: Chapman & Hall. 166 pp.
20. Karlin, A. 1977. *Pathogenesis of Human Muscular Dystrophies,* ed. L. P. Rowland, pp. 73–84. Amsterdam/Oxford: Excerpta Med. 896 pp.
21. Katz, B. 1966. *Nerve, Muscle and Synapse.* New-York: McGraw-Hill. 193 pp.
22. Colquhoun, D. 1975. *Ann. Rev. Pharmacol.* 15:307–25
23. Magazanik, L. G. 1976. *Ann. Rev. Pharmacol.* 16:161–75
24. Gage, P. W. 1976. *Physiol. Rev.* 56:177–247
25. Steinbach, J. H., Stevens, C. F. 1976. In *Neurobiology of the Frog,* ed. R. Llinas, W. Precht, pp. 33–92. Berlin: Springer
26. Kuffler, S. W., Nicholls, J. G. 1977. *From neuron to brain.* Sunderland, Mass: Sinauer. 486 pp.
27. Deguchi, T., Narahashi, T. 1971. *J. Pharmacol. Exp. Theor.* 176:423–33
28. Kordas, M. 1969. *J. Physiol. London* 204:493–502
29. Kordas, M. 1972. *J. Physiol. London* 224:317–32
30. Kordas, M. 1972. *J. Physiol. London* 224:333–48
31. Magleby, K. L., Stevens, C. F. 1972. *J. Physiol. London* 223:151–71
32. Magleby, K. L., Stevens, C. F. 1972. *J. Physiol. London* 223:173–97
33. Takeuchi, A., Takeuchi, N. 1959. *J. Neurophysiol.* 22:395–411
34. Gage, P. W. 1968. *Nature* 218:363–65
35. Gage, P. W., McBurney, R. N. 1972. *J. Physiol. London* 226:79–94
36. Gage, P. W., McBurney, R. N. 1975. *J. Physiol. London* 244:385–408
37. Anderson, C. R., Stevens, C. F. 1973. *J. Physiol. London* 235:655–91
38. Caldwell, G. R., Gage, P. W., McBurney, R. N. 1974. *Proc. Aust. Physiol. Pharmacol. Soc.* 5:51
39. Sheridan, R. E., Lester, H. A. 1975. *Proc. Natl. Acad. Sci. USA* 72:3496–3500
40. Sheridan, R. E., Lester, H. A. 1977. *J. Gen. Physiol.* 70:187–219
41. Dionne, V. E., Stevens, C. F. 1975. *J. Physiol. London* 251:245–70
42. McBurney, R. N., Gage, P. W. 1973. *Proc. Aust. Physiol. Pharmacol. Soc.* 4:65
43. Katz, B., Miledi, R. 1977. *Proc. R. Soc. London Ser. B* 196:59–72
44. Kuffler, S. W., Yoshikami, D. 1975. *J. Physiol. London* 251:465–82
45. Lester, H. A., Koblin, D. D., Sheridan, R. E. 1976. *Neurosci. Abstr.* 2:714
46. Jenkinson, D. H. 1960. *J. Physiol. London* 152:309–24
47. Higman, H. B., Podleski, T. R., Bartels, E. 1963. *Biochim. Biophys. Acta* 75: 187–93
48. Higman, H. B., Podleski, T. R., Bartels,

E. 1964. *Biochim. Biophys. Acta* 79: 138–50
49. Mautner, H. G., Bartels, E., Webb, G. D. 1966. *Biochem. Pharmacol.* 15: 187–93
50. Karlin, A. 1967. *Proc. Natl. Acad. Sci. USA* 58:1162–67
51. Changeux, J. P., Podleski, T. R. 1968. *Proc. Natl. Acad. Sci. USA* 59:944–50
52. Karlin, A., Winnik, M. 1968. *Proc. Natl. Acad. Sci. USA* 60:668–74
53. Rang, H. P., Ritter, J. M. 1969. *Mol. Pharmacol.* 5:384–411
54. Adams, P. R. 1974. *J. Physiol. London* 241:7P–8P
55. Adams, P. R. 1975. *Pfleugers Arch.* 360:155–64
56. Adams, P. R. 1975. *Pfleugers Arch.* 360:145–53
57. Lester, H. A., Changeux, J. P., Sheridan, R. E. 1975. *J. Gen. Physiol.* 65:797–816
58. Moreau, M., Changeux, J. P. 1976. *J. Mol. Biol.* 106:457–67
59. Del Castillo, J., Katz, B. 1956. *Progr. Biophys.* 6:21–170
60. Del Castillo, J., Bartels, E., Sobrino, J. A. 1972. *Proc. Natl. Acad. Sci. USA* 69:2081–85
61. Dreyer, F., Peper, K. 1974. *Pfleugers Arch.* 348:263–72
62. Dreyer, F., Peper, K. 1975. *Nature* 254:641–43
63. Peper, K., Dreyer, F., Müller, K. D. 1976. *Cold Spring Harbor Symp. Quant. Biol.* 40:187–92
64. Müller, K. D., Dreyer, F., Peper, K. 1976. *Pfleugers Arch.* 362:R30
65. Dreyer, F., Müller, K. D., Peper, K., Sterz, R. 1976. *Pfleugers Arch.* 367: 115–22
66. Adams, P. R. 1976. *Pfleugers Arch.* 361:145–51
67. Bartels, E., Wassermann, N. H., Erlanger, B. F. 1971. *Proc. Natl. Acad. Sci. USA* 68:1820–23
68. Lester, H. A., Chang, H. W. 1977. *Nature* 266:373–74
69. McMahan, U. J., Spitzer, N. C., Peper, K. 1972. *Proc. R. Soc. London Ser. B* 181:421–30
70. Hartzell, C., Kuffler, S., Yoshikami, D. 1975. *J. Physiol. London* 251:427–63
71. Kuffler, S., Yoshikami, D. 1975. *J. Physiol. London* 251:703–30
72. Brown, W. E. L., Hill, A. V. 1922–1923. *Proc. R. Soc. London B* 94:297–334
73. Katz, B., Thesleff, S. 1957. *J. Physiol. London* 138:63–80
74. Rang, H. P. 1971. *Nature* 231:91–96
75. Jenkinson, D. H., Terrar, D. A. 1973. *Br. J. Pharmacol.* 47:363–76

76. Magleby, K. L., Terrar, D. A. 1975. *J. Physiol. London* 244:467–95
77. Karlin, A. 1967. *J. Theor. Biol.* 16: 306–20
78. Land, B. R., Podleski, T. R., Salpeter, E. E., Salpeter, M. M. 1977. *J. Physiol. London* 269:155–76
79. Dale, H. H. 1953. *Adventures in Physiology.* London: Pergamon. 652 pp.
80. Seeman, P. 1972. *Pharmacol. Rev.* 24:583–655
81. Bartels, E., Nachmansohn, D. 1965. *Biochem. Z.* 342:359–74
82. Bartels, E. 1965. *Biochim. Biophys. Acta* 109:194–203
83. Podleski, T. R., Bartels, E. 1963. *Biochim. Biophys. Acta* 75:387–96
84. Daly, J. W., Karle, J., Myers, C. W., Tokuyama, T., Waters, J. A., Witkop, B. 1971. *Proc. Natl. Acad. Sci. USA* 68:1870–75
85. Albuquerque, E. X., Kuba, K., Daly, J. 1974. *J. Pharmacol. Exp. Ther.* 189:513–24
86. Kato, G., Changeux, J. P. 1976. *Mol. Pharmacol.* 12:92–100
87. Dolly, J. O., Albuquerque, E. X., Sarvey, J. M., Mallick, B., Barnard, E. A. 1977. *Mol. Pharmacol.* 13:1–14
88. Eldefrawi, A. T., Eldefrawi, M. E., Albuquerque, E. X., Oliveira, A. C., Mansour, N., Adler, M., Daly, J. W., Brown, G. B., Burgermeister, W. B., Witkop, B. 1977. *Proc. Natl. Acad. Sci. USA* 74:2172–76
89. Bon, C., Changeux, J. P. 1977. *Eur. J. Biochem.* 74:31–51
90. Lee, C. Y. 1972. *Ann. Rev. Pharmacol.* 12:265–86
91. Tsernoglou, D., Petsko, G. 1976. *FEBS Lett.* 68:1–4
92. Low, B. W., Preston, H. S., Sato, A., Rosen, L. S., Searl, J. E., Rudko, A.D., Richardson, J. S. 1976. *Proc. Natl. Acad. Sci. USA* 73:2991–94
93. Tsernoglou, D., Petsko, G. 1977. *Proc. Natl. Acad. Sci. USA* 74: 971–74
94. Karlin, A. 1969. *J. Gen. Physiol.* 54: 245S–64S
95. Magazanik, L. G., Vyskocil, F. 1976. *Motor Innervation of Muscle,* ed. S. Thesleff, pp. 151–76. London: Academic. 351 pp.
96. Magazanik, L. G., Vyskocil, F. 1973. *Drug Receptor,* ed. H. P. Rang, pp. 105–19. London: McMillan. 321 pp.
97. Nastuk, W. L., Parsons, R. L. 1970. *J. Gen. Physiol.* 86:218–49
98. Scubon-Mulieri, B., Parsons, R. L. 1977. *J. Gen. Physiol.* 69:431–447
99. Adams, P. R. 1975. *Pfleugers Arch.* 360:135–44

100. Magazanik, L. G. Vyskocil, F. 1970. *J. Physiol. London* 210:507–18
101. Magazanik, L. G., Vyskocil, F. 1975. *J. Physiol. London* 249:285–300
102. Parsons, R. L., Schnitzler, R. M., Cohrane, D. E. 1974. *Am. J. Physiol.* 227:96–100
103. Terrar, D. A. 1974. *Br. J. Pharmacol.* 51:259–68
104. Vyskocil, F. 1975. *Pfleugers Arch.* 361:83–87
105. Rang, H. P., Ritter, J. M. 1970. *Mol. Pharmacol.* 6:357–82
106. Kasai, M., Changeux, J. P. 1971. *J. Membr. Biol.* 6:1–80
107. Hazelbauer, G. L., Changeux, J. P. 1974. *Proc. Natl. Acad. Sci. USA* 71:1479–83
108. Popot, J. L., Sugiyama, H., Changeux, J. P. 1976. *J. Mol. Biol.* 106:469–83
109. McNamee, M. G., McConnell, H. M. 1973. *Biochemistry* 12:2951–58
110. Lindstrom, J., Patrick, J. 1974. *Synaptic Transmission and Nerve Interaction*, pp. 191–216. New York: Raven
111. Potter, L. 1973. See Ref. 96, pp. 295–312
112. Hess, G. P., Andrews, J. P., Struve, G. E., Coombs, S. E. 1975. *Proc. Natl. Acad. Sci. USA* 72:4371–75
113. Hess, G. P., Andrews, J. P., Struve, G. E. 1976. *Biochem. Biophys. Res. Commun.* 69:830–37
114. Hess, G. P., Andrews, J. P. 1977. *Proc. Natl. Acad. Sci. USA* 74:482–86
115. Takeuchi, A., Takeuchi, N. 1960. *J. Physiol. London* 154:52–67
116. Lassignal, N. L., Martin, A. R. 1977. *J. Gen. Physiol.* 70:23–36
117. Popot, J. L., Sugiyama, H., Changeux, J. P. 1974. *C. R. Acad. Sci. Ser. D* 279:1721–24
118. Sugiyama, H., Popot, J. L., Changeux, J. P. 1976. *J. Mol. Biol.* 106:485–96
119. Catterall, W. A. 1975. *J. Biol. Chem.* 250:1776–81
120. Catterall, W. A. 1975. *J. Biol. Chem.* 250:4053–59
121. Olsen, R., Meunier, J. C., Changeux, J. P. 1972. *FEBS Lett.* 28:96–100
122. Biesecker, G. 1973. *Biochemistry* 12:4403–9
123. Karlin, A., Cowburn, D. 1973. *Proc. Natl. Acad. Sci. USA* 70:3636–40
124. Klett, R. P., Fulpius, B. W., Cooper, D., Smith, M., Reich, E., Possani, L. D. 1973. *J. Biol. Chem.* 248:6841–53
125. Meunier, J. C., Sealock, R., Olsen, R., Changeux, J. P. 1974. *Eur. J. Biochem.* 45:371–94
126. Chang, H. W. 1974. *Proc. Natl. Acad. Sci. USA* 71:2113–17
127. Karlsson, E., Heilbronn, E., Widlund, L. 1972. *FEBS Lett.* 21:107–11
128. Heilbronn, E., Karlsson, E., Widlund, L. 1973. In *Cholinergic Transmission of the Nerve Impulse*, pp. 151–58. Paris: INSERM
129. Heilbronn, E., Mattson, C. 1974. *J. Neurochem.* 22:315–17
130. Eldefrawi, M. E., Eldefrawi, A. T. 1973. *Arch. Biochem. Biophys.* 159:362–73
131. Changeux, J. P., Meunier, J. C., Olsen, R. W., Weber, M., Bourgeois, J. P., Popot, J. L., Cohen, J. B., Hazelbauer, G. L., Lester, H. A. 1973. See Ref. 96, pp. 273–94
132. Gordon, A., Bandini, G., Hucho, F. 1974. *FEBS Lett.* 47:204–8
133. Sobel, A., Changeux, J. P. 1977. *Biochem. Soc. Trans.* 5:511–14
134. Sobel, A., Weber, M., Changeux, J. P. 1977. *Eur. J. Biochem.* 80:215–224
135. Claudio, T., Raftery, M. A. 1977. *Arch. Biochem. Biophys.* 181:484–89
136. Schmidt, J., Raftery, M. A. 1973. *Biochemistry* 12:852–56
137. Weill, C. L., McNamee, M. G., Karlin, A. 1974. *Biochem. Biophys. Res. Commun.* 61:997–1003
138. Hucho, F., Layer, P., Kiefer, H. R., Bandini, G. 1976. *Proc. Natl. Acad. Sci. USA* 73:2624–28
139. Chang, H. W., Bock, E. 1977. *Biochemistry* 16:4513–20
140. Ong, D. E., Brady, R. N. 1974. *Biochemistry* 13:2822–27
141. Rübsamen, H., Montgomery, M., Hess, G. P., Eldefrawi, A. T., Eldefrawi, M. E. 1976. *Biochem. Biophys. Res. Commun.* 70:1020–27
142. Schmidt, J., Raftery, M. 1972. *Biochem. Biophys. Res. Commun.* 49:572–78
143. Kometani, T., Ikeda, Y., Kasai, M. 1975. *Biochim. Biophys. Acta* 413:415–24
144. Brockes, J. P., Hall, Z. W. 1975. *Biochemistry* 14:2092–99, 2100–6
145. Froehner, S., Karlin, A., Hall, Z. W. 1977, *Proc. Natl. Acad. Sci. USA* 74:4685–88
146. Froehner, S. C., Reiness, C. G., Hall, Z. W. 1977. *J. Biol. Chem.* 252:8589–96
147. Dolly, J. O., Barnard, E. 1974. *FEBS Lett.* 46:145–48
148. Dolly, J. O., Barnard, E. 1975. *FEBS Lett.* 57:267–71
149. Barnard, E., Coates, V., Dolly, J. O., Mallick, B. 1977. *Cell Biol. Int. Rep.* 1:99–106
150. Dolly, J. O., Barnard, E., Shorr, R. G. 1977. *Biochem. Soc. Trans.* 5:168–70
151. Dolly, J. O., Barnard, E. A. 1978. *Biochemistry* In press

152. Merlie, J., Changeux, J. P., Gros, F. 1978. *J. Biol. Chem.* In press
153. Boulter, J., Patrick, J. 1978. *J. Biol. Chem.* In press
154. Eterovic, V. A., Bennett, E. L. 1974. *Biochim. Biophys. Acta* 362:346–55
155. Lowy, J., McGregor, J., Rosenstone, J., Schmidt, J. 1976. *Biochemistry* 15:1522–27
156. Salvaterra, P., Mahler, H. 1976. *J. Biol. Chem.* 251:6327–34
157. Dudai, Y. 1977. *FEBS Lett.* 76:211–13
158. Yazulla, S., Schmidt, J. 1976. *Vision Res.* 16:878–80
159. Vogel, Z., Nirenberg, M. W. 1976. *Proc. Natl. Acad. Sci. USA* 73:1806–10
160. Nelson, L. 1976. *Exp. Cell Res.* 101:221–24
161. Greene, L. A., Sytkowski, A. J., Vogel, Z., Nirenberg, M. W. 1973. *Nature* 243:163–66
162. Greene, L. A. 1976. *Brain Res.* 111:135–45
163. Patrick, J., Stallcup, W. 1977. *Proc. Natl. Acad. Sci. USA* 74:4689–92
164. Patrick, J., Stallcup, W. 1977. *J. Biol. Chem.* 252:8629–33
165. Cohen, J. B., Weber, M., Huchet, M., Changeux, J. P. 1972. *FEBS Lett.* 26:43–47
166. Nickel, E., Potter, L. T. 1973. *Brain Res.* 57:508–17
167. Duguid, J. R., Raftery, M. A. 1973. *Biochemistry* 12:3593–97
168. Reed, K., Vandlen, R., Bode, J., Duguid, J., Raftery, M. A. 1975. *Arch. Biochem. Biophys.* 167:138–44
169. Flanagan, S. D., Barondes, S. H. 1975. *J. Biol. Chem.* 250:1484–89
170. Flanagan, S. D., Barondes, S. H., Taylor, P. 1976. *J. Biol. Chem.* 251:858–65
171. Sobel, A., Heidmann, T., Hofler, J., Changeux, J. P. 1978. *Proc. Natl. Acad. Sci. USA.* In press
172. Aharonov, A., Kalderon, N., Silman, I., Fuchs, S. 1975. *Immunochemistry* 12:765–71
173. Barrantes, F. J., Changeux, J. P., Lunt, G. G., Sobel, A. 1975. *Nature* 256:325–27
174. De Robertis, E. 1971. *Science* 171:963–71
175. Carroll, R., Eldefrawi, M., Edelstein, S. 1973. *Biochem. Biophys. Res. Commun.* 55:864–72
176. Mattson, C., Heilbronn, E. 1975. *J. Neurochem.* 25:899–901
177. Hucho, F., Changeux, J. P. 1973. *FEBS Lett.* 38:11–15
178. Patrick, J., Boulter, J., O'Brien, J. C. 1975. *Biochem. Biophys. Res. Commun.* 64:219–25
179. Raftery, M. A., Schmidt, J., Clark, D. G. 1972. *Arch. Biochem. Biophys.* 152:882–86
180. Gibson, R. E., O'Brien, R., Edelstein, S. J., Thompson, W. R. 1976. *Biochemistry* 15:2377–83
181. Edelstein, S., Beyer, W. B., Eldefrawi, A. T., Eldefrawi, M. E. 1975. *J. Biol. Chem.* 250:6101–6
182. Meunier, J. C., Olsen, R. W., Menez, A., Morgat, J. L., Fromageot, P., Ronseray, A. M., Boquet, P., Changeux, J. P. 1971. *C. R. Acad. Sci. Ser. D* 273:595–98
183. Meunier, J. C., Olsen, R. W., Menez, A., Fromageot, P., Boquet, P., Changeux, J. P. 1972. *Biochemistry* 11:1200–10
184. Meunier, J. C., Olsen, R. W., Changeux, J. P. 1972. *FEBS Lett.* 24:63–68
185. Raftery, M. A., Schmidt, J., Clark, D. G., Wolcott, R. G. 1971. *Biochem. Biophys. Res. Commun.* 45:1622–29
186. Sugiyama, H., Changeux, J. P. 1975. *Eur. J. Biochem.* 55:505–15
187. Tanford, C., Reynolds, J. 1976. *Biochim. Biophys. Acta* 457:133–70
188. Reynolds, J. A., Tanford, C. 1976. *Proc. Natl. Acad. Sci. USA* 73:4467–70
189. Martinez-Carrion, M., Sator, V., Raftery, M. A. 1975. *Biochem. Biophys. Res. Commun.* 65:129
190. Cartaud, J., Benedetti, L., Cohen, J. B., Meunier, J. C., Changeux, J. P. 1973. *FEBS Lett.* 33:109–13
191. Cartaud, J. 1974. *Electron Microscopy,* ed. J. V. Sanders, D. J. Goodchild, 2:284–85. Canberra: Aust. Acad. Sci.
192. Cartaud, J. 1975. *Exp. Brain Res.* 23:Suppl. p. 37
193. Suarez-Isla, B. A., Hucho, F. 1977. *FEBS Lett.* 75:65–69
194. Patrick, J., Lindstrom, J. 1973. *Science* 180:871–72
195. Sugiyama, H., Benda, P., Meunier, J. C., Changeux, J. P. 1973. *FEBS Lett.* 35:124–28
196. Valderrama, R., Weill, C. L., McNamee, M., Karlin, A. 1976. *Ann. NY Acad. Sci.* 274:108–15
197. Simons, K., Helenius, A., Garoff, H. 1973. *J. Mol. Biol.* 80:119–33
198. Barrantes, F. J. 1975. *Biochem. Biophys. Res. Commun.* 62:407–14
199. Michaelson, D., Vandlen, R., Bode, J., Moody, T., Schmidt, J., Raftery, M. A. 1974. *Arch. Biochem. Biophys.* 165:796–804
200. Moore, W. M., Holladay, L. A., Puett, D., Brady, R. N. 1974. *FEBS Lett.* 45:145–49

201. Bonner, R., Barrantes, F. J., Jovin, T. M. 1976. *Nature* 263:429–31
202. Grünhagen, H. H., Changeux, J. P. 1976. *J. Mol. Biol.* 106:497–535
203. Karlin, A., Bartels, E. 1966. *Biochim. Biophys. Acta* 126:525–35
204. Podleski, T., Meunier, J. C., Changeux, J. P. 1969. *Proc. Natl. Acad. Sci. USA* 63:1239–46
205. Karlin, A., Prives, J., Deal, W., Winnik, M. 1970. *Ciba Found. Symp.* ed. R. Porter, M. O'Connor, pp. 247–59. London: Churchill
206. Karlin, A., Prives, J., Deal, W., Winnik, M. 1971. *J. Mol. Biol.* 61:175–88
207. Karlin, A. 1973. *Fed. Proc.* 32:1847–53
208. Karlin, A., Cowburn, D., Reiter, M. 1973. See Ref. 96, pp. 193–209
209. Bartels-Bernal, E., Rosenberry, T. L., Chang, H. W. 1976. *Mol. Pharmacol.* 12:813–19
210. Heilbronn, E., Mattson, C., Elfman, L. 1974. *Proc. 9th FEBS Meet.* 37:29–37
211. Raftery, M. A., Schmidt, J., Martinez-Carrion, M., Moody, T., Vandlen, R., Duguid, J. 1973. *J. Supramol. Struct.* 1:360–67
212. Alper, R., Lowy, J., Schmidt, J. 1974. *FEBS Lett.* 48:130–32
213. Colquhoun, D., Rang, H. P. 1976. *Mol. Pharmacol.* 12:519–35
214. Teichberg, V., Changeux, J. P. 1976. *FEBS Lett.* 67:264–68
215. Teichberg, V., Changeux, J. P. 1977. *FEBS Lett.* 74:71–76
216. Gordon, A. S., Davis, C. G., Diamond, I. 1977. *Proc. Natl. Acad. Sci. USA* 74:263–67
217. Teichberg, V., Sobel, A., Changeux, J. P. 1977. *Nature* 267:540–42
218. Gordon, A. S., Davis, C. G., Milfay, D., Diamond, I. 1977. *Nature* 267:539–40
219. Teichberg, V., Changeux, J. P. 1977. *Symp. Neuromuscular Disorders*, Bath, England
220. Devreotes, P. N., Fambrough, D. M. 1975. *J. Cell Biol.* 65:335–58
221. Devreotes, P. N., Gardner, J. M., Fambrough, D. M. 1977. *Cell* 10:365–73
222. Patrick, J., McMillan, J., Watson, H., O'Brien, J. C. 1977. *J. Biol. Chem.* 252:2143–53
223. Blobel, G., Dobberstein, B. 1975. *J. Cell Biol.* 67:835–51
224. Fischbach, G., Cohen, S. A. 1973. *Dev. Biol.* 31:147–62
225. Sytkowski, A. J., Vogel, Z., Nirenberg, M. 1973. *Proc. Natl. Acad. Sci. USA* 70:270–74
226. Hartzell, H., Fambrough, D. M. 1973. *Dev. Biol.* 30:153–65

227. Axelrod, D., Ravdin, P., Koppel, D. E., Schlessinger, J., Webb, W. W., Elson, E. L., Podleski, T. R. 1976. *Proc. Natl. Acad. Sci. USA* 73:4594–98
228. Anderson, M. J., Cohen, M. W. 1974. *J. Physiol. London* 237:385–400
229. Karlin, A., Cowburn, D. A. 1974. *Neurochemistry of Cholinergic Receptors*, ed. E. De Robertis, J. Schacht, pp. 37–48. New York: Raven
230. McNamee, M. G., Weill, C. L., Karlin, A. 1975. *"Protein-Ligand Interactions,"* ed. H. Sund, G. Blauer, pp. 316–27. Berlin: de Gruyter. 486 pp.
231. Reiter, M. J., Cowburn, D. A., Prives, J. M., Karlin, A. 1972. *Proc. Natl. Acad. Sci. USA* 69:1168–72
232. Karlin, A., McNamee, M. G., Cowburn, D. A. 1976. *Anal. Biochem.* 76:442–51
233. Karlin, A. 1978. *Methods Enzymol.* In press
234. Changeux, J. P., Podleski, T., Wofsy, L. 1967. *Proc. Natl. Acad. Sci. USA* 58:2063–70
235. Mautner, H., Bartels, E. 1970. *Proc. Natl. Acad. Sci. USA* 67:74–78
236. Silman, I., Karlin, A. 1969. *Science* 164:1420–21
237. Chao, Y., Vandlen, R., Raftery, M. 1975. *Biochem. Biophys. Res. Commun.* 63:300–7
238. Ruoho, A. E., Kiefer, H., Roeder, P., Singer, S. J. 1973. *Proc. Natl. Acad. Sci. USA* 70:2567–71
239. Layer, P., Kiefer, H. R., Hucho, F. 1976. *Mol. Pharmacol.* 12:958–65
240. Bulger, J. E., Fu, J. J. L., Hindy, E. F., Silberstein, R. L., Hess, G. P. 1977. *Biochemistry* 16:684–92
241. Maelicke, A., Reich, E. 1976. *Cold Spring Harbor Symp. Quant. Biol.* 40:231–35
242. Maelicke, A., Fulpius, B. W., Klett, R. P., Reich, E. 1977. *J. Biol. Chem.* 252:4811–30
243. Menez, A., Morgat, J. L., Fromageot, P., Ronseray, A. M., Boquet, P., Changeux, J. P. 1971. *FEBS Lett.* 17:333–35
244. Weber, M., Changeux, J. P. 1974. *Mol. Pharmacol.* 10:1–14, 15–34, 35–40
245. Franklin, G. I., Potter, L. T. 1972. *FEBS Lett.* 28:101–6
246. Brockes, J. P., Hall, Z. W. 1975. *Biochemistry* 14:2092–2100
247. Bulger, J. E., Hess, G. P. 1973. *Biochem. Biophys. Res. Commun.* 54:677–84
248. Hess, G. P., Bulger, J. E., Fu, J. J. L., Hindy, E. F., Silberstein, R. J. 1975. *Biochem. Biophys. Res. Commun.* 64:1018–27

249. Weiland, G., Georgia, B., Wee, V. T., Chignell, C. F., Taylor, P. 1976. *Mol. Pharmacol.* 12:1091–1105
250. Kohanski, R., Andrews, J. P., Wins, P., Eldefrawi, M., Hess, G. P. 1977. *Anal. Biochem.* 80:531–39
251. Cohen, J. B. 1978. *Molecular Specialization and Symmetry in Membranes,* ed. A. K. Solomon. Cambridge: Harvard Univ. Park Press. In press
252. Raftery, M. A., Bode, J., Vandlen, R., Michaelson, D., Deutsch, J., Moody, T., Ross, M. J., Stroud, R. M. 1974. See Ref. 230, pp. 328–55
253. Raftery, M. A., Bode, J., Vandlen, R., Chao, Y., Deutsch, J., Duguid, J. R., Reed, K., Moody, T. 1975. *Biochemistry of Sensory Functions,* ed. L. Jaenicke, pp. 541–64. Berlin: Springer. 641 pp.
254. Weber, G., Borris, D., de Robertis, E., Barrantes, F., La Torre, J., de Carlin, M. 1971. *Mol. Pharmacol.* 7:530–37
255. Cohen, J. B., Changeux, J. P. 1973. *Biochemistry* 12:4855–64
256. Martinez-Carrion, M., Raftery, M. A. 1973. *Biochem. Biophys. Res. Commun.* 53:761–72
257. Cohen, J. B., Weber, M., Changeux, J. P. 1974. *Mol. Pharmacol.* 10:904–32
258. Barrantes, F. J., Sakmann, B., Bonner, R., Eibl, H., Jovin, T. M. 1975. *Proc. Natl. Acad. Sci. USA* 72:3097–3101
259. Raftery, M. A., Schimerlik, M. 1976. *Biochem. Biophys. Res. Commun.* 73: 607–13
260. Waksman, G., Fournié-Zaluski, M. C., Roques, B., Heidmann, T., Grünhagen, H. H., Changeux, J. P. 1977. *FEBS Lett.* 67:335–42
261. Heidmann, T., Iwatsubo, M., Changeux, J. P. 1977. *C. R. Acad. Sci.* 284:771–74 Ser. D
262. Fu, J. J. L., Donner, D. B., Hess, G. P. 1974. *Biochem. Biophys. Res. Commun.* 60:1072–80
263. Fu, J. J. L., Donner, D. B., Hess, G. P. 1977. *Biochemistry* 16:678–84
264. Moody, T., Schmidt, J., Raftery, M. A. 1973. *Biochem. Biophys. Res. Commun.* 53:761–72
265. Gibson, R. E. 1976. *Biochemistry* 15:3890–3901
266. O'Brien, R. D., Gilmour, L. P. 1969. *Proc. Natl. Acad. Sci. USA* 63:496–503
267. O'Brien, R. D., Gilmour, L. P., Eldefrawi, M. E. 1970. *Proc. Natl. Acad. Sci. USA* 65:438–45
268. Eldefrawi, M. E., Britten, A., Eldefrawi, A. T. 1971. *Science* 173:338–40
269. Eldefrawi, M. E., Eldefrawi, A. T., O'-Brien, R. D. 1971. *Proc. Natl. Acad. Sci. USA* 68:1047–50
270. Eldefrawi, M. E., Eldefrawi, A. T., O'-Brien, R. D. 1971. *Mol. Pharmacol.* 7:104–10
271. Eldefrawi, M. E., Eldefrawi, A. T. 1972. *Proc. Natl. Acad. Sci. USA* 69:1776
272. Eldefrawi, M. E., Eldefrawi, A. T., Seifert, S., O'Brien, R. D. 1972. *Arch. Biochem. Biophys.* 150:210–18
273. O'Brien, R. D., Gibson, R. E. 1974. *Arch. Biochem. Biophys.* 165:681-90
274. O'Brien, R. D., Gibson, R. E. 1975. *Arch. Biochem. Biophys.* 169:458–63
275. Eldefrawi, M. E., Eldefrawi, A. T. 1973. *Biochem. Pharmacol.* 22:3145–50
276. Meunier, J. C., Changeux, J. P. 1973. *FEBS Lett.* 32:143–48
277. Grünhagen, H. H., Iwatsubo, M., Changeux, J. P. 1977. *Eur. J. Biochem.* 80:225–42
278. Monod, J., Changeux, J. P., Jacob, F. 1963. *J. Mol. Biol.* 6:306–28
279. Monod, J., Wyman, J., Changeux, J. P. 1965. *J. Mol. Biol.* 12:88–118
280. Brisson, A., Devaux, P., Changeux, J. P. 1975. *C. R. Acad. Sci. Ser. D* 280:2153–56
281. Gerhart, J. C. 1970. *Curr. Top. Cell Regul.* 2:275–325
282. Schmidt, J., Raftery, M. A. 1974. *J. Neurochem.* 23:617–23
283. Gibson, R. E., Juni, S., O'Brien, R. D. 1977. *Arch. Biochem. Biophys.* 179: 183–88
284. Changeux, J. P. 1966. *Mol. Pharmacol.* 2:369–92
285. Eldefrawi, M. E., Eldefrawi, A. T., Penfield, L. A., O'Brien, R. D., Van Campen, D. 1975. *Life Sci.* 16:925–36
286. Rübsamen, H., Hess, G. P., Eldefrawi, A. T., Eldefrawi, M. E. 1976. *Biochem. Biophys. Res. Commun.* 68:56–63
287. Chang, H. W., Neumann, E. 1976. *Proc. Natl. Acad. Sci. USA* 73:3364–68
288. Fertuck, H. C., Salpeter, M. M. 1976. *J. Cell Biol.* 69:144–58
289. Porter, L. W., Barnard, E. A. 1976. *Ann. NY Acad. Sci.* 274:85–107
290. Bourgeois, J. P., Popot, J. L., Ryter, A., Changeux, J. P. 1978. *J. Cell Biol.* In press
291. Bourgeois, J. P., Ryter, A., Menez, P., Fromageot, P., Boquet, P., Changeux, J. P. 1972. *FEBS Lett.* 25:127–33
292. Dupont, Y., Cohen, J. B., Changeux, J. P. 1974. *FEBS Lett.* 40:130–33
293. Cartaud, J., Benedetti, L., Sobel, A., Changeux, J. P. 1978. *J. Cell Sci.* In press
294. Rash, J. E., Ellisman, M. K. 1974. *J. Cell Biol.* 63:567–86

295. Orci, L., Perrelet, A., Dunant, Y. 1974. *Proc. Natl. Acad. Sci. USA* 71:307–10
296. Heuser, J. E., Reese, J. S., Landis, D. M. D. 1974. *J. Neurocytol.* 3:109–31
297. Peper, K., Dreyer, F., Sandri, C., Akert, K., Moor, H. 1974. *Cell Tissue Res.* 149:437–55
298. Clementi, F., Conti-Tronconi, B., Peluchetti, D., Morgutti, M. 1975. *Brain Res.* 90:133–38
299. Rosenbludt, J. 1975. *J. Neurocytol.* 4:697–712
300. Heuser, J. E., Reese, T. S., Landis, D. M. 1976. *Cold Spring Harbor Symp. Quant. Biol.* 40:17–24
301. Popot, J. L., Demel, R. A., Sobel, A., Van Deenen, L., Changeux, J. P. 1978. *Eur. J. Biochem.* In press
302. Bourgeois, J. P., Popot, J. L., Ryter, A., Changeux, J. P. 1973. *Brain Res.* 62:557–63
303. Frank, E., Gautvik, K., Sommerschild, H. 1975. *Acta Physiol. Scand.* 95:66
304. Frank, E., Gautvik, K., Sommerschild, H. 1976. *Cold Spring Harbor Symp. Quant. Biol.* 40:275–82
305. Burden, S. 1977. *Dev. Biol.* 57:317–29
306. Bevan, S., Steinbach, J. H. 1977. *J. Physiol. London* 267:195–213
307. Brisson, A., Scandella, C. J., Bienvenue, A., Devaux, P., Cohen, J. B., Changeux, J. P. 1975. *Proc. Natl. Acad. Sci. USA* 72:1087–91
308. Bienvenue, A., Rousselet, A., Kato, G., Devaux, P. F. 1977. *Biochemistry* 16:841–48
309. Rousselet, A., Devaux, P. F. 1977. *Biochem. Biophys. Res. Commun.* 78:448–54
310. Baroin, A., Thomas, D. D., Osborne, B., Devaux, P. F. 1977. *Biochem. Biophys. Res. Commun.* 78:442–47
311. Changeux, J. P., Huchet, M., Cartaud, J. 1972. *C. R. Acad. Sci. Ser. D* 274:122–25
312. Michaelson, D. M., Raftery, M. A. 1974. *Proc. Natl. Acad. Sci. USA* 71:4768–72
313. McNamee, M. G., Weill, C. L., Karlin, A. 1975. *Ann NY Acad. Sci.* 264:175–82
314. Briley, M., Changeux, J. P. 1978. *Eur. J. Biochem.* In press
315. Wiedmer, T., Brodbeck, U., Zahler, P., Fulpius, B. W. 1978. *Biochim. Biophys. Acta.* In press
316. Patrick, J., Lindstrom, J., Culp, B., McMillan, J. 1973. *Proc. Natl. Acad. Sci. USA* 70:3334–38
317. Karlin, A., Holtzman, E., Valderrama, R., Damle, V., Hsu, K., Reyes, F. 1978. *J. Cell. Biol.* In press
318. Briley, M., Changeux, J. P. 1978. *Int. Rev. Neurobiol.* In press
319. Weber, M., Pfeuty-David, M. T., Changeux, J. P. 1975. *Proc. Natl. Acad. Sci. USA* 72:3443–47
320. Neher, E., Sakmann, B. 1976. *Nature* 260:799–802
321. Neher, E., Steinbach, J. H. 1978. *J. Physiol. London.* In press
322. Katz, B., Miledi, R. 1970. *Nature* 226:962–63
323. Katz, B., Miledi, R. 1971. *Nature New Biol.* 232:124–26
324. Katz, B., Miledi, R. 1972. *J. Physiol. London* 224:665–99
325. Katz, B., Miledi, R. 1973. *J. Physiol. London* 230:707–17
326. Katz, B., Miledi, R. 1973. *J. Physiol. London* 231:549–74
327. Katz, B., Miledi, R. 1975. *J. Physiol. London* 249:269–84
328. Neher, E., Sakmann, B. 1975. *Proc. Natl. Acad. Sci. USA* 72:2140–44
329. Ben-Haim, D., Dreyer, F., Peper, K. 1975. *Pfleugers Arch.* 355:19–26
330. Colquhoun, D., Dionne, V. E., Steinbach, J. H., Stevens, C. F. 1975. *Nature* 253:204–6
331. Neher, E., Sakmann, B. 1976. *J. Physiol. London* 258:705–29
332. Dreyer, F., Walther, C., Peper, K. 1976. *Pfleugers Arch.* 366:1–9
333. Sakmann, B., Adams, P. R. 1976. *Pfleugers Arch.* 365:R145
334. Ruff, R. L. 1976. *Biophys. J.* 16:433–39
335. Ruff, R. L. 1977. *J. Physiol. London* 264:89–124
336. Marty, A., Neild, T., Ascher, P. 1976. *Nature* 261:501–3
337. Ascher, P., Marty, A., Neild, T. O. 1978. *J. Physiol. London.* In press
338. Ascher, P., Marty, A., Neild, T. O. 1978. *J. Physiol. London.* In press
339. Marty, A. 1978. *J. Physiol. London.* In press
340. Feltz, A., Large, W. A., Trautmann, A. 1977. *J. Physiol. London* 269:109–30
341. Lass, Y., Fischbach, G. D. 1976. *Nature* 263:150–51
342. Dionne, V. E., Ruff, R. L. 1977. *Nature* 266:263–65
343. Neher, E., Stevens, C. F. 1977. *Ann. Rev. Biophys. Bioeng.* 6:345–81
344. Adams, P. R. 1975. *Br. J. Pharmacol.* 53:308–10
345. Adams, P. R. 1977. *J. Physiol. London* 268:271–89
346. Adams, P. R. 1977. *J. Physiol. London* 268:291–318
347. Adams, P. R., Feltz, A. 1977. *Nature* 269:609–11

348. Furukawa, T. 1957. *Jpn. J. Physiol.* 7:199–212
349. Maeno, T. 1966. *J. Physiol. London* 183:592–606
350. Steinbach, A. B. 1968. *J. Gen. Physiol.* 52:144–61, 162–80
351. Kordas, M. 1970. *J. Physiol. London* 209:689–99
352. Maeno, T., Edwards, C., Hashimura, S. 1971. *J. Neurophysiol.* 34:32–46
353. Beam, K. G. 1976. *J. Physiol. London* 258:279–300, 301–22
354. Adams, P. R. 1976. *J. Physiol. London* 260:531–52
355. Feltz, A., Large, W. A., Trautmann, A. 1977. *J. Physiol. London* 269:109–30
356. Neumann, E., Chang, H. W. 1976. *Proc. Natl. Acad. Sci. USA* 73:3994–98
357. O'Brien, R. D., Gibson, R. E. 1978. In press
358. Eldefrawi, M. E., Eldefrawi, A. T., Wilson, D. B. 1975. *Biochemistry* 14: 4304–10
359. Barrantes, F. J. 1976. *Biochem. Biophys. Res. Commun.* 72:479–88
360. Sugiyama, H., Popot, J. L., Cohen, J. B., Weber, M., Changeux, J. P. 1975. See Ref. 230, pp. 289–305
361. Rubin, M. M., Changeux, J. P. 1966. *J. Mol. Biol.* 21:265–74
362. Elliott, J., Raftery, M. A. 1977. *Biochem. Biophys. Res. Commun.* 77: 1347–53
363. Lee, T., Witzemann, V., Schimerlik, M., Raftery, M. A. 1977. *Arch. Biochem. Biophys.* 183:57–63

Ann. Rev. Biochem. 1978. 47:359–83
Copyright © 1978 by Annual Reviews Inc. All rights reserved

PROBES OF MEMBRANE STRUCTURE[1]

Hans C. Andersen

Department of Chemistry, Stanford University, Stanford, California 94305

CONTENTS

PERSPECTIVES AND SUMMARY

Biological membranes have a number of important functions in a living organism. They act as structural barriers that maintain the integrity of a cell; they are selective permeability barriers for the passage of molecules into and out of a cell or organelle; they are the site at which a number of

[1]The abbreviations used in the text: PC, 1,2-diacylphosphatidylcholine; DPPC, dipalmitoylphosphatidylcholine; NMR, nuclear magnetic resonance; ESR, electron spin resonance; TEMPO, 2,2,6,6-tetramethylpiperadine-1-oxyl; DSC, differential scanning calorimetry.

0066-4154/78/0701-0359$01.00

important enzymes act; and, in the case of nerve cells, their electrical properties are important for the transmission of information. The molecular structure of membranes is of particular concern to those wishing to understand the functioning of membranes at a molecular level.

The study of membranes has been greatly advanced by the development of model bilayer membrane systems that are structurally related to biological membranes. Bilayers are formed by dispersing phospholipids or other amphipathic lipids in water. These molecules are polar at one end and nonpolar at the other, and they spontaneously form two-dimensional bilayer structures in which the polar ends are in contact with water and the nonpolar parts are in contact only with the nonpolar parts of other molecules. These bilayers can be formed from purified lipids or from lipid mixtures extracted from biological membranes. Membrane proteins can be inserted into these bilayers to form reconstituted model membrane-protein complexes. There is a great deal of evidence that portions of biological membranes have this bilayer structure, and the most commonly used and believed working hypothesis of membrane structure is the fluid mosaic model (1), which regards biological membranes as lipid bilayers with proteins included or attached in various ways.

Many physical techniques have been widely used to study biological and model membranes. The most useful methods that give information about the molecular structure of membranes are scattering methods (namely X ray, neutron, and Raman scattering and fluorescence measurements), magnetic resonance methods, calorimetry, and freeze fracture microscopy. Each of the methods by itself provides only indirect and incomplete information, but the combined use of several of them can lead to very detailed and self-consistent pictures of membrane structure.

In this article, we discuss these probes of membrane structure. The emphasis is on the physical nature of what is measured in each experiment and the way in which the data are used to obtain molecular structural information. We give examples of the use of each technique discussed, but this article is not intended as a detailed review of the many studies in this field. We are especially concerned with the use of these techniques for the study of the model membrane systems.

MODEL MEMBRANE PREPARATIONS

Multilamellar dispersions, vesicles, and oriented multilayers are the three types of model membrane preparations that are commonly used, and in all of them the lipids are organized into bilayers. Multilamellar dispersions are obtained by dispersing phospholipids in an aqueous solution by shaking or mechanical mixing. They consist of sets of closed concentric lipid bilayers separated by aqueous regions (2). These preparations have various other names, such as "liposomes" and "Bangosomes." Vesicles are obtained by prolonged sonication of multilamellar dispersions. They consist of single bilayers that are closed upon themselves to form a spherical surface that divides the exterior aqueous solution from the interior solution (3). Oriented multilayers, in which the bilayers are flat and parallel to one another rather than curved, can be prepared in a number of ways (4–6).

Preparations of these three types have been very useful as model membranes. They can be made with various types of lipids and lipid mixtures, but for membrane-related studies one of the components is usually a phospholipid. Dipalmitoylphosphatidylcholine (DPPC) and phosphatidylcholine (PC) derived from egg yolk are two of the commonly used lipids. The external aqueous medium can be varied. Proteins and other biologically important molecules can be incorporated into them, as well as spin labeled, fluorescence labeled, or deuterium substituted molecules.

A distinctive physical characteristic of phospholipid bilayers is that they can undergo phase transitions while still maintaining the bilayer structure. [For a review of membrane phase transitions see (7).] The main transition observed in phospholipid multilamellar dispersions is called the chain melting transition. For PCs with saturated hydrocarbon chains, an additional transition, called the pretransition, is also observed at lower temperatures. The nature of these transitions, their relationship to molecular structure, and the influence of lipid composition and aqueous phase composition upon them, are topics that have been studied using many different probes of membrane structure.

TYPES OF PROBES OF MEMBRANE STRUCTURE

Scattering Methods

X-RAY SCATTERING   In X-ray scattering experiments, monochromatic X rays are beamed into a sample and the intensity of the scattered X rays is measured. When the sample of a disoriented preparation of membranes or vesicles, the scattered intensity is measured as a function of the scattering angle, θ, which is the angle between the directions of motion of the scattered rays and the incident beam. The data are usually presented as a function of $s = 2\sin\theta/\lambda$, where λ is the wavelength of the X rays. When the sample is an oriented (or partially oriented) preparation, the scattered intensity can also be measured as a function of the angles between the membrane surface and the incident beam.

The scattering is caused by variations of the electron density from one point to another in the sample. If there are significant variations in electron density over distances of length l, then there will be significant scattering for values of s which are approximately equal to $1/l$. Unlike the case of single crystal studies, X-ray scattering experiments on membrane and bilayer preparations do not give enough information to allow the precise determination of the molecular structure of the preparations. Analysis of the data always requires additional information such as that obtained from other types of experiments.

An important technique for interpreting data is to construct a physical model for the membrane and use this to calculate a mathematical model for the electron density and for the X-ray scattering. Then the model is varied to obtain agreement of the calculated with the experimental scattering intensities. The possible ambiguities associated with this method of analysis are difficult or impossible to assess. For example, Ranck et al (8) gave an example in which two very different models for a phospholipid bilayer are both consistent with the experimental X-ray scattering data. Despite the possible ambiguities associated with the interpretation of data, X-ray scattering provides some of the most useful and trustworthy information about the molecular structure of membranes.

In low angle scattering, $1/s \geqslant 10$ Å, and electron density variations over distances of the order of 10 to 200 Å determine the scattering. Thus low angle scattering can be used to measure the distance between bilayers in a preparation and the distance between the headgroup layers in a bilayer, which is usually about 50 Å. If the bilayer surface is rippled rather than planar, this affects the low angle scattering. Under favorable conditions, the electron density as a function of position in the bilayer can also be determined. In high angle scattering, $1/s \leqslant 5$ Å, and electron density variations over distances of a few Angstroms determine the scattering. In practice, high angle scattering is useful mainly for learning about the conformational state of the hydrocarbon chains and the way in which they are packed together.

The technical aspects of low angle scattering experiments have been reviewed by Akers (9). Both low and high angle X-ray scattering have been used for extensive studies of the phase diagrams and the structures of the phases of aqueous dispersions of pure phospholipids and of lipid mixtures extracted from biological systems (8, 10). The results have been reviewed by Luzzati & Tardieu (11). X-ray studies of membrane preparations from *Mycoplasma laidlawii* were performed by Engelman (12, 13), who used both high and low angle scattering to study the chain conformations and phase transition in the membrane and to estimate the membrane thickness and the electron density profile across the membrane. Wide angle scattering studies of the phase transition in *Escherichia coli* membrane preparations were performed by Esfahani et al (14).

Analysing high angle scattering from unoriented samples to learn about the packing of hydrocarbon chains is not without its ambiguities. Rand, Chapman & Larsson (15), Janiak, Small & Shipley (16), and Brady & Fein (17) studied the two solid phases of DPPC using unoriented samples. One of the solid phases exists below the pretransition temperature of about 35°C, and the other exists between 35°C and the chain melting transition temperature of about 41°C. All three groups concluded that for the lowest temperature phase the hydrocarbon chains are packed in a hexagonal array with the chains tilted from the bilayer normal. They disagreed about the structure of the higher temperature solid. The first and third groups concluded that at the pretransition there was a distinct difference in this tilt angle for the two solid phases, and the first group stated explicitly that the tilt angle is zero in the higher temperature solid phase. The second group concluded that there is a nonzero angle of tilt in both solid phases and that other structural changes, such as a change in the distortion of the hexagonal lattice and a change in the shape and symmetry of the bilayer surface, are the important ones taking place at the pretransition.

NEUTRON SCATTERING   In neutron scattering experiments, a beam of neutrons, all with nominally the same velocity, is directed upon the sample to be studied, and the intensity of the neutron scattering in various directions is measured. Whereas X rays are scattered by electrons, neutrons are scattered by the nuclei in the sample. Different nuclei, even different isotopes of the same element, differ greatly in their scattering power. It is the variation in neutron scattering density from one position to another which is responsible for neutron scattering (just as in X-ray scattering it is the variation in electron density from point to point which is important).

By selective deuteration and the use of ^2H_2O rather than ordinary water, the scattering densities of the various regions can be adjusted, since protons and deuterons have very different scattering powers. This has some important advantages for the interpretation of data, since several different experiments can be performed on samples of the same structure which differ only in the extent and nature of the deuteration.

The use of neutron scattering for the investigation of membrane structure was reviewed recently by Schoenborn (18). A collection of articles on neutron scattering can be found in the proceedings of the Symposium on Neutron Scattering for the Analysis of Biological Structures at Brookhaven National Laboratory (19). In the following paragraphs we discuss two examples that illustrate an advantage of the technique.

Zaccai et al (20) have studied the neutron scattering by bilayers of DPPC dispersed in relatively small amounts of H_2O and of ^2H_2O. Analysis of the two sets of data yielded profiles of the neutron scattering density as a function of position across the bilayer. The peaks corresponding to the polar headgroup layers of the bilayer are clearly evident. There is a lower density in the hydrocarbon region and the least density at the place where the two halves of the bilayer come into contact. Information about the location of the water in the bilayer can be obtained from the neutron scattering data by subtracting the neutron scattering density profiles obtained from the samples containing H_2O and those containing ^2H_2O. The effect of the phospholipids is canceled in this subtraction and the result is simply proportional to the density of water as a function of position in the bilayer.

Another elegant and analogous example of the use of isotopic substitution is the study of aqueous DPPC bilayers containing cholesterol (18, 21). In one experiment, the hydrocarbon chain of the cholesterol was deuterated. In another, the normal protonated form was used. From an analysis of the difference between the two results, it was possible to determine the position of the cholesterol hydrocarbon tail in the bilayer.

Neutron scattering has also been used to study biological membranes such as myelin and retinal rod outer segments (18).

RAMAN SCATTERING   In a Raman scattering experiment, a monochromatic beam of light is incident upon a sample. The light that is scattered from the material is allowed to enter a spectrometer, and the intensity of the scattered light, as a function of its frequency, is measured, keeping the frequency of the incident light fixed. The frequency difference between the incident and scattered light is of primary importance, since this corresponds to vibrational frequencies of the molecules on the sample. The quantities measured in an experiment are the frequencies, widths, and shapes of the lines or bands, and also the intensities at particular frequencies or the ratio of the intensities at two frequencies.

The first step in the interpretation of a Raman spectrum of a membrane is deciding which types of vibrations on which molecules are the cause of the various lines. Particular types of vibrations, such as carbon-carbon or carbon-hydrogen stretching vibrations, have characteristic frequencies that are approximately the same in all compounds, and so the assignment of the lines is usually a straightforward matter.

The next step in the interpretation of Raman spectra is to determine which are the important structural properties on a molecular level which influence the line positions, intensities, and shapes. There are many such factors, such as molecular conformation, molecular size, and shape, intermolecular interactions, and intramolecular interactions. Conversely, if we are interested in learning about some molecular structural property, such as the conformation of hydrocarbon chains in a membrane, we must determine which measurable features of the Raman spectrum can provide such information and how the experimental data should be interpreted.

The analysis of Raman spectra to learn molecular structural information is difficult because often many different physical effects can influence the same measurable feature of a spectrum. For example, Brown, Peticolas & Brown (22) studied the Raman spectrum of DPPC dispersions in the frequency range near 2900 cm^{-1}, where the scattering is due primarily to the CH stretching vibrations of the hydrocarbon chains. The ratio of the peak intensities of two bands at 2890 and 2850 cm^{-1} was found to decrease when the sample was heated through the chain melting phase transition. Since the chains are more mobile above the transition temperature than below it and since chain mobility is a plausible cause of intensity changes, they concluded that the ratio of these two intensities could be used to monitor changes in the mobility of the hydrocarbon chains. More detailed studies from the same laboratory (23) led to a different conclusion, namely that changes in this intensity ratio are caused by changes in the packing of hydrocarbon chains and by changes in the number of conformations that a chain can adopt.

It is difficult to validate conclusively any molecular interpretation of a Raman spectrum. Therefore, it is useful to regard the conclusions drawn from Raman spectroscopy as plausible explanations of the data rather than as necessary and unavoidable consequences of the data.

Most Raman studies of membranes have been concerned with the scattering in the carbon-carbon stretching region near 1100 cm^{-1} and the carbon-hydrogen stretching region near 2800 cm^{-1}. Lippert & Peticolas (24) and Mendelsohn (25) studied the effect of cholesterol on DPPC vesicles using the 1100 cm^{-1} region. Lippert & Peticolas (26) made assignments of some bands in the carbon-carbon stretching region of fatty acids and phospholipids and used them to study the phase transition of dioleoylphosphatidylcholine bilayers. Brown, Peticolas & Brown (22) made additional assignments for spectra of PCs. Lippert, Gorczyka & Meiklejohn (27) studied the conformation of phospholipids and proteins in red blood cell ghosts. Mendelsohn, Sunder & Bernstein (28) studied the effect of sonication on bilayers. Gaber & Peticolas (23) developed a way of extracting from Raman spectra what they consider to be quantitative measures of the hydrocarbon chain conformation and of the lateral crystal-like order between the chains.

FLUORESCENCE PROBES   In a fluorescence probe experiment, a fluorescent molecule is incorporated into a membrane preparation, and a beam of monochromatic light is directed upon the sample. The intensity of light subsequently emitted by the sample, at longer wavelength, is then measured. A fluorescent probe molecule is capable of absorbing a photon from the incident beam and of making a transition, usually to an excited singlet state. The molecule then relaxes to its lowest excited singlet state, from which it emits the photon that is detected. The details of the absorption, relaxation, and emission processes are sensitive to the molecular surroundings of the probe molecule, and this is the basis for the use of fluorescence to study the structure of membranes. Reviews by Azzi (29, 30) and by Radda (31) discuss many aspects of the fluorescence probe technique.

Many different types of fluorescent probe molecules have been used. Their fluorescence properties are usually the result of one or more aromatic hydrocarbon rings or a series of conjugated double bonds. Some of the molecules, for example 1-anilinonaphthalene-8-sulfonate, have no resemblance to the molecules normally found in biological membranes. Others, such as dansylphosphatidylethanolamine are lipids with fluorescent functional groups attached (32). A third class includes parinaric acid (9,11,13,15-octadecatetraenoic acid) and parinaroylphosphatidylcholines, which differ from biological lipids only in that they have conjugated double bonds in their hydrocarbon chains (33).

The quantities that can be measured in a fluorescence probe experiment are the intensity of the emission as a function of wavelength, the quantum yield (the ratio of the number of photons emitted to the number absorbed), the lifetime of the excited states (which is determined from the time dependence of the emission following an initial pulse of light), and the degree of depolarization of the emitted light. Experiments measuring the absorption of light as a function of wavelength are sometimes performed in conjunction with the fluorescence measurements.

These quantities are measured because they are related to the structure of the surroundings of the fluorescing molecule and because changes in the measured quantities with temperature or composition can be used as diagnostics for significant changes in membrane structure. For example, rotational motion of the probe molecule can affect the degree of depolarization, and so depolarization studies are often used to draw inferences about the "microviscosity" of a membrane.

In general, great care must be used in drawing detailed conclusions about membrane structure from fluorescence probe experiments. The types of locations and orientations of the probe molecules in a membrane must first be determined. One must be sure that the concentration of probe molecules is not high enough to affect the overall structure of the membrane. Even when the probe is present at low concentration, it is responding to its own particular surroundings which may be atypical of those parts of the membrane which are far from probe molecules. Finally, the measured property can be dependent upon a variety of physical effects, rather than just one.

An alternative way of using fluorescence probes is to regard the measured properties merely as diagnostics for observing membrane transitions. Once it has been shown that a particular property of a probe changes distinctively at a membrane phase transition, the technique can be used to investigate phase transitions even though the details of probe location in the membrane are not known. [For a review of fluorescence probe studies prior to 1975, see (29–31).]

We now list some of the more recent work in this field. Lentz, Barenholz & Thompson (34) studied phase transitions in multilamellar dispersions and vesicles of PCs using the depolarization of fluorescence of a hydrophobic probe, 1,6-diphenyl-1,3,5-hexatriene, which is thought to dissolve in the hydrocarbon interior of bilayers. They also studied two component bilayers and derived phase diagrams for various mixtures (35). Jacobson & Papahadjopoulos (36) studied the phase transitions for neutral and charged phospholipids using diphenylhexatriene and perylene as fluorescence probes. They found that the former probe, unlike the latter, was a reliable indicator for phase transitions. Sklar, Hudson & Simoni (33) used *cis* and *trans* isomers of parinaric acid and parinaroylphosphatidylcholines to measure phase transitions and phase diagrams for bilayers. Tecoma et al (37) incorporated parinaric acid biosynthetically into the membrane of *E. coli* and studied the phase transition in the membrane.

Magnetic Resonance Methods

NUCLEAR MAGNETIC RESONANCE (NMR)   In a nuclear magnetic resonance experiment, the sample to be studied is placed in a magnetic field. Those nuclei of the sample which have magnetic moments, such as protons and ^{31}P, precess in the magnetic field much as a spinning top precesses in the earth's gravitational field. There are three types of NMR experiments that are of importance for membrane studies. One is the absorption experiment, in which the frequencies at which the various nuclei precess are measured by observing the absorption of radio frequency radiation. In practice, the experiment is usually performed by keeping the frequency fixed and measuring the absorption as a function of the magnetic field strength, but it is usually easier to think about the experiment as if it were the frequency that is varied at constant field strength. The second type of experiment is a measurement of the spin-lattice relaxation time, T_1. The inverse of T_1 is a measure of the rate at which a nuclear magnet or spin can exchange energy with its surroundings. The third type of experiment is a measurement of the transverse relaxation time, T_2. The inverse of T_2 is a measure of the amplitude of fluctuations in the precessional motion of the spins caused by, among other things, the interaction of the spin with other spins in its surroundings. Each of these three measurable properties, the absorption spectrum (including line positions, shapes, and widths), the values of T_1, and the values of T_2, are affected by the structural properties of the sample on a molecular level and by molecular motions.

The most important magnetic nuclei for membrane studies are ^1H, ^2H, ^{13}C, and ^{31}P. The predominant isotopes for hydrogen and phosphorus are ^1H and ^{31}P, thus these isotopes appear in great abundance in normal biological samples. Only about 1% of the naturally occurring carbon nuclei are ^{13}C. Nevertheless, NMR is sensitive enough to make the technique useful for this isotope. Deuterium can be introduced by synthetic techniques into specific parts of molecules as a substitute for hydrogen. Also molecules containing the magnetic nucleus ^{19}F have been used as probes of membrane structure. [For reviews of NMR applied to membranes, see (38, 39).]

Factors that affect NMR absorption spectra   The various magnetic nuclei have very different precession frequencies for any particular magnetic field strength. Thus the absorption lines for different nuclei are well separated from each other.

There are three important interactions that determine NMR line shapes for any one type of nucleus. The first is the anisotropic chemical shift. The electronic cloud in the vicinity of a nucleus acts to shield the nucleus somewhat from the external magnetic field. The extent of the shielding depends on the orientation of the molecule with respect to the external magnetic field, and thus the precession frequency of the nucleus depends on this orientation. This orientation-dependent precession frequency is called the anisotropic chemical shift. The second is the magnetic dipolar interaction between spins, which broadens the absorption lines. The third, which is important only for deuterium atoms bonded to carbon atoms, is the interaction between a deuterium nucleus and the local electric field gradient in the CD bond. This interaction changes the absorption frequency of the deuterium nucleus by an amount which depends on the orientation of the CD bond with respect to the external magnetic field.

The way in which these interactions affect the spectrum depends not only upon their strength but also upon their time dependence. Suppose a nucleus in a molecule has a range of precession frequencies, $\Delta\omega$, which is caused by the fact that the molecule can be in a variety of orientations or surroundings. If the molecules cannot change their orientations or surroundings, the spectrum will contain the entire range of frequencies and a line of width $\Delta\omega$ will result. If the orientations or surroundings of the molecules can change on a time scale of the order of $1/\Delta\omega$, then a much narrower line will result. In effect, each nucleus precesses at some average frequency which is in the middle of the range of width $\Delta\omega$.

This sensitivity of the spectrum to the time dependence of the interactions is the basis for two important principles in magnetic resonance, namely proton decoupling and motional narrowing.

The dipolar coupling between nuclei is usually of much less interest in learning about membrane structure than are the other interactions. The dipolar effect of protons on the NMR spectrum of ^{2}H, ^{31}P, and ^{13}C can be decreased by irradiating the protons at their resonant frequencies with a strong radio frequency noise field. This causes the dipolar fields of the protons to fluctuate rapidly enough that their effect is reduced by averaging process. This technique, called proton decoupling, is of great usefulness in the NMR spectroscopy of all nuclei other than protons themselves.

Motional narrowing is the name given to the process by which NMR lines are narrowed by the motion of the molecules. The line broadening effect of the anisotropic chemical shift and the quadrupole coupling (for ^{2}H) can be reduced by molecular rotation, provided the rotation takes place on a rapid enough time scale (i.e. on times of the order of the reciprocal of the unnarrowed line width). The effect of dipolar coupling between spins is also reduced by molecular rotation and diffusion. In addition, overall rotation of membrane vesicles can contribute to the narrowing of lines.

Unsonicated lamellar dispersions and membrane preparations   These systems have broad NMR spectra with line widths of the order of several KHz. The lines are sharper above the phase transition temperature than they are below it.

Partially narrowed deuterium and phosphorus spectra, obtained with the aid of proton decoupling, are useful for obtaining information about the freedom of molecular rotation above the transition temperature. The theory of the motional narrowing (40, 41) shows that the extent of narrowing can be related to a quantity called the "order parameter" in the deuterium case and to a quantity called the "chemical shielding anisotropy" in the phosphorus case. These quantities contain information about certain averages of the distribution of molecular orientation with regard to the membrane surface. (This result holds only when the bilayer surface itself is not free to rotate on a timescale of 10^{-5} sec or faster.) For deuterium resonance, the molecular orientation of interest is the direction of the CD bond axis, and for ^{31}P it is the orientation of the phosphate group.

The numerical value of the order parameter or chemical shielding anisotropy can be extracted from the partially narrowed spectrum. The precise value of an order parameter or chemical shielding anisotropy ordinarily conveys little direct physical meaning. They are primarily of use for testing theories or models for molecular conformation and motion in membranes. If a model predicts values that are in disagreement with experiment, the model must be wrong. If the model gives correct predictions, it may be right, but it could still be wrong. It is possible for very different physical models to give identical order parameters and chemical shielding anisotropies. (For this reason, the term "order parameter" is an unfortunate one, since it conveys the false impression that the order parameter is a measure of some oversimplified notion of molecular order.)

In principle, proton spectra should contain some information about molecular orientations and motion. Because of the importance of the dipolar interaction between the two protons of a methylene group, the partially narrowed spectrum should be sensitive to motions that reorient the line joining the two protons. However, the spectrum is complicated by other dipolar couplings and there is some disagreement (42) about how the order parameters should be extracted from the spectrum.

Sonicated dispersions Sonicated dispersions have much narrower lines than unsonicated preparations. The types of motions that can cause such a narrowing are overall rotation of the vesicles, diffusion of molecules around the curved surface of a vesicle, and possibly an enhanced freedom of molecular motion in the curved vesicles as compared to the flatter bilayers of unsonicated dispersions. There has been some disagreement about the relative importance of these factors in determining the line widths (38, 43–45) and consequently the interpretation of the widths in terms of molecular structure is not certain.

The lines are narrow enough that small isotropic chemical split splittings between nuclei in different chemical surroundings can be observed. For example, methyl and methylene protons cause different lines in the spectrum. Measurements of T_1 and T_2, which contain information about molecular motion, can be obtained separately for each observed line.

Examples of NMR studies of membranes
1. ^{31}P and ^{2}H studies of polar head groups. Recent work done by three research groups, have provided much detailed information for testing models of the head group conformation of phospholipids in membranes. Seelig & Gally (46) studied bilayers of dipalmitoylphosphatidylethanolamine above and below the phase tradition, using both ^{31}P and ^{2}H NMR. In the deuterium work, they used molecules that were selectively deuterated at the ethanolamine carbon atoms. Their data were consistent with a model in which the head group rotates flat on the surface of the bilayer and makes rapid transitions between just two conformations. Kohler & Klein (47) measured the ^{31}P spectrum of bilayers of dipalmitoylphosphatidylethanolamine, DPPC, egg PC, and brain PC. They interpreted their findings in terms of two simple models of head group motion. Seelig, Gally & Wohlgemuth (48) and Griffin, Powers & Pershan (49) studied DPPC head group orienta-

tion. The first group worked above the phase transition temperature using unoriented samples and did both ^{31}P and ^{2}H studies using selectively deuterated samples. The second group worked below the phase transition temperature using oriented samples and performed ^{31}P experiments for a variety of angles of the magnetic field relative to the bilayer normal. Both studies were consistent with a model in which the choline head group is aligned parallel to the bilayer plane.

2. [2]H order parameter studies of hydrocarbon chains. Deuterium order parameter studies have been performed for a variety of bilayers in the fluid phase, using specifically deuterated molecules so that order parameters as a function of position along the chain could be measured. Seelig & Seelig studied deuterated DPPC (50) and deuterated 1-palmitoyl-2-oleoylphosphatidylcholine (51). Stockton et al (45) studied egg PC and egg PC-cholesterol bilayers that contained deuterated stearic acid. Stockton et al (52) incorporated specifically deuterated fatty acids biosynthetically into the membranes of *Acholeplasma laidlawii.*

The results in all cases are qualitatively similar; the order parameters are nearly constant near the head group and middle of the chains and then decrease toward the methyl ends, and the values vary with temperature, degree of unsaturation, and mole fraction of cholesterol. There have been numerous attempts (44, 50, 53, 54) to understand the physical meaning of the numerical results, to use them to obtain information about hydrocarbon chain conformation and other types of molecular motion, and to understand the relationship of these deuterium order parameters and the ESR order parameters (see below). As mentioned above, it is relatively easy to disprove models using order parameters but very difficult or impossible to prove the correctness of a model. All these analyses involve so many explicit and implicit assumptions and oversimplifications that it is difficult to know how much credence to give to the conclusions. Of special note is a recent paper by Petersen & Chan (44) which emphasizes the importance of reorientation of the chains as well as rotational isomerization in accounting for the order parameters.

3. Other studies of unsonicated systems. The study of proton spectra in unsonicated systems has led to fewer detailed and conclusive results than has the study of ^{2}H and [31]P. This is because many different types of protons contribute to a line and because of the complicating effects of dipolar coupling among all the spins. Chan and co-workers (55, 56, 44) have developed a theory for interpreting proton spectra and relaxation times. The effect of dipolar coupling on line shapes has been described differently by Ulmius et al (42) and Bloom et al (57).

The study of ^{13}C spectra in unsonicated systems has been hampered by the ineffectiveness of the usual proton decoupling techniques for removing dipolar broadening. Urbina & Waugh (58) have applied a double resonance technique to the study of DPPC dispersions. This method not only eliminates dipolar coupling, giving spectra in which the width is determined by the anisotropic chemical shift, but also gives increased sensitivity. Opella, Yesinowski & Waugh (59) applied this technique to the study of cholesterol in DPPC dispersions, using cholesterols specifically enriched with ^{13}C at two positions.

4. Sonicated vesicles. Kroon, Kainosho & Chan (60) measured protein NMR line widths and T_1 values for DPPC sonicated vesicles. They used deuterated molecules to perform isotopic dilution experiments to estimate the relative influence of intermolecular and intramolecular interactions on the results. Horwitz, Horsley & Klein (61) measured T_1 and T_2 values for sonicated egg PC vesicles. They were able to obtain individual relaxation times for several different types of protons: N-methyl protons on the choline head group and α-carbonyl, allyl, vinyl, methylene, and methyl protons on the fatty acid chains. Lee et al (62) measured proton T_1 for DPPC and egg PC vesicles and studied the effect of cholesterol. Darke et al (63) also studied the effect of cholesterol on DPPC and egg PC vesicles by measuring the changes in line shapes with composition.

$^{13}C T_1$ relaxation times give information about the correlation time for reorientation of the CH bonds in a molecule. Individual values can usually be measured for many of the carbon atoms in a molecule. Such studies have been performed, for example, by Levine et al (64) on DPPC and dioleoylphosphatidylcholine vesicles, by Godici & Landsberger (65) on egg PC vesicles, and by Barton & Gunstone (66) on a series of unsaturated PCs.

Fatty acids with magnetic nuclei have been used as probes in phospholipid vesicles. Stockton et al (45) studied ^{2}H line widths of deuterated fatty acids, and Birdsall et al (67) studied ^{19}F line widths of fluorinated fatty acids.

De Kruijff et al (68) studied the high resolution spectrum of ^{31}P in DPPC, distearoylphosphatidylcholine, and dimyristoylphosphatidylcholine vesicles as a function of temperature. Separate resonances for molecules on the inside and outside layers of the vesicle were seen.

ELECTRON SPIN RESONANCE OF SPIN LABELS   One of the most fruitful methods for studying membranes is to introduce stable free radical molecules into them and study the electron spin resonance (ESR) spectrum of the radical in the presence of an external magnetic field. The term "spin label" is often used to describe such molecules. A recent monograph (69) discusses many aspects of spin labeling.

In a spin label ESR experiment, the spin label molecule is incorporated into the membrane sample, and the sample is placed in a magnetic field. The absorption spectrum for microwave radiation is measured. (Usually, as in the case of NMR, the frequency is held fixed and the magnetic field is varied.) If oriented membrane samples are used, the spectrum can be measured as a function of the angle between the magnetic field and the direction normal to the membrane surfaces.

The most frequently used spin labels for membrane studies are nitroxides, which contain the $>$N–O group. The N atom is usually part of a 5- or 6-membered ring. This group has an unpaired electron spin located primarily in an atomic p orbital on the nitrogen atom. The axis of the orbital is normal to the plane defined by the nitrogen and oxygen atoms and the other two atoms bonded to the nitrogen.

Synthetic techniques have been developed for introducing stable nitroxide groups into many types of compounds (70). The most common types are

1. molecules that are smaller than organic molecules typically found in membranes and that are soluble in water;
2. derivatives of fatty acids, in which the nitroxide group is attached at various places;
3. derivatives of phospholipids, in which the nitroxide group is attached at various places;
4. sterol derivatives with a structure similar to cholesterol.

An example of the first type is 2,2,6,6-tetramethylpiperadine-1-oxyl (TEMPO), which is soluble in both aqueous and nonaqueous phases and which partitions between aqueous and membrane regions when added to membrane preparations. Examples of the second type are doxyl derivatives of fatty acids. (The doxyl group is a five-membered ring containing an NO group.) Examples of the third type are phosphatidylcholine molecules, one of whose acyl chains is a molecule of the second type. The 3-doxyl derivatives of cholestane-3-one and androstane-3-one-17-ol are examples of the fourth type.

Factors affecting the spectrum of a nitroxide radical   An electron in a magnetic field can absorb radiation at one particular frequency, much like a proton in a magnetic field. This would ordinarily give a single line in the absorption spectrum. For a nitroxide radical, however, the spectrum is affected by magnetic (hyperfine) interactions between the electron and the magnetic ^{14}N nucleus, which has a spin of one. The result is that there are three equally spaced lines, rather than one, corresponding to the three possible quantum mechanical states of the spin-1 nitrogen.

For a stationary molecule, the positions of the three lines depend on the orientation of the nitroxide group relative to the external field. For molecules which can rotate, the same principles of motional averaging apply to

ESR as to NMR. The shape of the spectrum is sensitive to the rate and anisotropy of molecular reorientation. In particular, the spacing between the high field and low field lines are sensitive to the molecular motion which can reorient the direction of the p orbital on the nitrogen nucleus. An important difference between NMR and ESR is that in the latter case the widths of the unbroadened lines is much larger (of the order of 10^8 Hz rather than 10^5 Hz), and hence ESR narrowing is sensitive only to motions that take place on a much shorter time scale. The various theories of line shapes for the ESR of nitroxide spin labels are reviewed in several articles in the recent monograph on spin labeling (4, 71–74).

One very common way of analyzing an ESR spectrum involves extracting from it a number called an ESR "order parameter" which bears some conceptual similarity to the deuterium and proton NMR order parameters discussed above. When this method of analysis is valid, the order parameter is a number between $-1/2$ and $+1$ which has a well-defined meaning in terms of the average ability of the spin label p orbital to rotate during times of the order of 10^{-8} sec. (Unlike the NMR case, the ESR order parameter is not a measure of the average orientation of the spin label with respect to the bilayer normal.) If a spin label has an order parameter close to unity, the physical meaning of this result is very clear, namely, that in a time of 10^{-8} sec the molecular motion does not appreciably change the direction of the p orbital containing the unpaired spin. However, a numerical value of the order parameter significantly less than unity is consistent with a wide variety of possible motions for the spin label.

Molecular motion, discussed in the previous paragraph, is the dominant factor in determining ESR spectra. There are, however, three other effects that are of importance for certain applications of spin labels. The first two are the spin-exchange interaction and the dipole-dipole interaction, which can take place if two spin labels are close enough to each other. The third is an effect of the polarity of the surroundings of a spin label upon the spectrum.

Types of analysis of nitroxide spin label spectra   In order to obtain information about the molecular structure of a membrane from ESR spin label experiments, it is first usually necessary to reduce the data to a form that is more easily interpreted. There are only a small number of ways in which this is done.

1. The spectra can be used to estimate the relative amounts of probe molecules in different environments.
2. The spectra can be used to derive order parameters or other numerical measures of probe mobility or anisotropic motion.

3. The spectra can be compared to theoretical spectra calculated from detailed models of the membrane structure and dynamics.

4. The spectra can be used to derive numerical measures of the polarity of the immediate surroundings of the probe molecule.

5. When the concentration of probe molecules is high enough, the effects of spin-exchange interaction and dipole-dipole interaction between probes on the spectra can be used to obtain information about the frequency of collisions between probe molecules and the distance between adjacent probe molecules in a membrane.

Examples of the use of ESR spin label studies The following are some of the many types of spin label studies that have been performed. [For more comprehensive and detailed reviews see (69, 75, and 76).]

1. Partitioning studies. The partitioning of TEMPO between the nonpolar interior of artificial bilayer systems and the aqueous surroundings has been used to investigate phase transitions and lateral phase separations. The technique is based on the fact that the solubility of TEMPO in a membrane interior is greater above the phase transition temperature than it is below that temperature and that the spectrum of TEMPO in water is different from the spectrum of TEMPO in a bilayer. A graph of the fraction of label which is dissolved in the membrane shows distinct changes at temperatures at which a one-component system changes its phase or a two-component system enters or leaves a two phase region of its phase diagram. TEMPO partitioning has been used to measure phase transition temperatures for bilayers composed of a single type of lipid and phase diagrams for bilayers containing two types of lipids (77, 78). One technique for studying the effect of proteins on membrane structure is to use TEMPO partitioning to study the effect of proteins on the lipid phase transition of a model membrane. This was one method used by Hesketh et al (79) in their study of a calcium transport protein. The partitioning of phospholipid spin labels between fluid and solid phases is one of the methods used to study calcium-induced lateral phase separations in phospholipid bilayers (80).

Biological membranes may have regions of varying fluidity due not only to lateral phase separation but also because protein molecules can reduce the mobility of adjacent lipids. TEMPO partitioning between the aqueous region and the fluid region of the membrane is the basis for a method for estimating the fraction of lipids in a biological membrane which are in a fluid state (81).

Partitioning methods have been used to develop spin label techniques for measuring some electrical properties of membranes. Castle & Hubbell (82) have shown that the partitioning of a charged spin label molecule, N,N-dimethyl-N-nonyl-N-tempoylammonium ion between the lipid region of phospholipid vesicles and the surrounding aqueous medium is sensitive to the electrical potential at the surface of the vesicle. They devised a procedure for estimating this surface potential from an analysis of the ESR spectrum of the label.

Cafiso & Hubbell (83) have shown that the partitioning of certain positively charged hydrophobic spin labels between a phospholipid vesicle and the surrounding aqueous medium is sensitive to the electrical potential difference between the aqueous solutions on either side of the bilayer, and have devised a method for estimating transmembrane potentials using this effect.

2. Order parameter studies and hyperfine splitting studies. A common way of presenting and analyzing ESR probe data is to derive a numerical value of the order parameter from the experimental spectrum. There are other numerical measures of probe motion which are sometimes used as alternatives to order parameters. The splittings between various hyperfine extrema (maxima and minima) in the measured spectrum or the heights of the extrema are examples of this. They are easier to obtain than the order parameters, which may require more extensive analysis of the spectrum, and for many purposes they are equally useful.

Using fatty acid spin labels and phospholipid spin labels with the doxyl group attached to various positions on the hydrocarbon chains, it is possible to insert nitroxide free radicals into a membrane at various distances from the polar headgroup region. These studies are usually performed with doxyl derivatives of fatty acids and phospholipids. For these molecules, the axis of the p orbital containing the unpaired electron is always parallel to the "backbone" of the hydrocarbon chain at the point where the doxyl group is attached. The backbone of a hydrocarbon chain is most easily defined by drawing straight lines from the midpoint of each carbon-carbon bond to the midpoint of each adjacent carbon-carbon bond. Studies using such homologous series of spin labels have yielded the intriguing result that the ESR order parameter of the spin label decreases as the spin label is moved farther from the head group of the molecule (84, 85). This variation of the order parameter has been called a "flexibility gradient" and a "fluidity gradient," has been observed in a variety of model and biological membranes (84–86), and appears to be a general property of spin labels in bilayers. (See the section above on ^2H order parameter studies of hydrocarbon chains for a brief discussion of attempts to understand the physical meaning of this effect.)

Hubbell & McConnell (84) have also used the temperature dependence of the order parameter for a spin labeled phospholipid in a DPPC bilayer as a diagnostic for measuring the phase transition temperature. The transition from a solid to a fluid bilayer leads to an increase in molecular mobility and hence to a measurable sharp decrease in the order parameter over a very short range of temperature. Van & Griffith (87) have used order parameters and anisotropy parameters of fatty acid spin labels to study the effect of an extrinsic protein, cytochrome *c,* on bilayers composed of a mixture of cardiolipin and lecithin. Changes in the splittings in spectra of spin labeled phosphatidylserine and phosphatidylcholine have been used to study Ca^{2+} induced phase transitions and lateral phase separations of bilayers containing anionic lipids (80). Schreier-Muccillo et al (88) have used stearic acid spin labels, sterol spin labels, and a stearamide spin label with a nitroxide group in the polar region of the molecule to study bilayers of DPPC and egg PC with various percentages of cholesterol added. Keith, Snipes & Chapman (89) have used two empirical motional parameters, derived from the heights of extrema in ESR spectra, to study the motion of small spin label molecules in various aqueous regions near the surface of dimyristoyl-phosphatidylcholine bilayers.

The two examples mentioned above are concerned with model membranes. An example of the use of spin labels to study biological membranes is the work of Butterfield et al (90). They used a spin label that is thought to embed itself in membranes in such a way that the nitroxide group is located near the membrane surface. They measured and compared the order parameters for this spin label embedded in erythrocytes taken from normal subjects and from patients with myotonic muscular dystrophy, Duchenne muscular dystrophy, and congenital myotonia. Relative values of the order parameters are presumably related to the relative freedom of rotational motion of molecules near the membrane surface.

3. Detailed analysis of spectral shapes. Another way to interpret ESR data is to postulate a detailed model of the dynamics of the spin labels, compute a theoretical spectrum on the basis of the model, and compare the calculated results with the experimental spectrum. The agreement between the two provides a test of whether the model is consistent or inconsistent with experiment. Many models contain some adjustable parameters that have physical meaning within the context of the model, and the parameters are varied to obtain a best fit to experiment. Examples of such models are discussed in (74). The major problem with this approach is that one must beware of interpreting the results of this type of analysis too literally, because there may be several similar, or even very different, models that agree equally well with experiment.

One example of such a study is the work of Birrell and Griffith (91). They studied oriented multilayers of DPPC containing phospholipids that had doxyl spin labels attached to the hydrocarbon chains near the head group and near the hydrocarbon tail. This work was done at room temperature, which is well below the chain melting transition temperature. They found that the spin labeled chains of both types were tilted with respect to the normal to the bilayer surface, and they were able to fit their data to a "Gaussian distribution model" for the chain backbone directions.

Another example of this type of study is the work of Gaffney & McConnell (54) who studied oriented bilayers of egg PC at room temperature. They used three different spin labeled phospholipids, with the doxyl groups at various positions along the hydrocarbon chains. A detailed fit of the spectra to a Gaussian distribution model was performed for each spin label. They concluded that the chain backbones of the spin labeled molecules were tilted with respect to the normal to the bilayer. The tilt was found to be greater when the spin label is near the head group than when it is near the methyl end of the hydrocarbon chain.

4. Other types of spin label studies of membrane structure. Certain features of the spectrum of a nitroxide spin label are affected by the polarity of its immediate surroundings. This effect is the basis of several techniques for estimating a polarity index as a function of position inside a membrane, which provides some information about the distribution of water (74, 87).

The electron-electron dipole-dipole interaction between spin labels can have a measurable effect on the spectrum if two spin label molecules are close together for a long enough period of time. The spectrum is sensitive to the distance between the spin labels. This principle was used by Marsh & Smith (92) to measure the change in the distance between nearest neighbor cholestane probes in PC-cholesterol bilayers as the composition of the bilayer was changed.

Calorimetry

Because model membranes and some biological membranes can undergo phase transitions, calorimetry techniques that are capable of measuring the latent heat associated with these transitions have been very important for the study of membranes. Standard calorimeters, which measure the total heat capacity or specific heat of a sample, are not useful for dilute phospholipid suspensions, because the bilayer heat capacity is overwhelmed by the heat capacity of the solvent. Differential scanning calorimetry (DSC), which measures the difference in heat capacity between a membrane suspension and an equal volume of the solvent, is the technique used. Recently, improvements in the design of differential scanning calorimeters have led to a new generation of instruments with high sensitivity. [For reviews of calorimetry studies of membranes, see (93, 94).]

The data obtained in a DSC experiment is a graph of the heat capacity of the sample, relative to that of a nominally equal volume of solvent, plotted as a function of temperature. If the sample absorbs an especially large amount of heat over a small enough temperature interval, the observed heat capacity will have one or more peaks that rise above the baseline. For bilayers composed of one phospholipid, a narrow peak is observed, and the quantities of particular interest are the temperature at which the heat capacity is largest and the area under the peak. They correspond to the transition temperature and latent heat of the transition, respectively. (For some PCs, two separate peaks, corresponding to two transitions, are observed.) For bilayers composed of two lipids, peaks with a more complicated shape are observed. The shapes do not have any simple physical meaning, and the goal of an analysis of the data is usually to construct a binary phase diagram from data taken over the entire range of composition for which the lipids have a bilayer structure. Such a phase diagram shows the ranges of temperature and composition over which the bilayers consist of one phase and of two phases.

It should be noted that calorimetry is, in a sense, not really a probe of membrane structure. That is, it does not provide any information about the nature of the various different structures that bilayers can have. It tells us, for example, that aqueous dispersions of bilayers of DPPC undergo a phase transition at 41°C, and that therefore it is very likely that the bilayers have a very different structure above this temperature than they do below it. However, other methods must be used to learn what these structures are and how they differ.

The following paragraph illustrates the range of membrane studies that have been performed using DSC. For more complete discussions and references, the review articles cited above should be consulted.

The transition temperatures and enthalpies for various one-component bilayers have been measured (95–97). For PCs with saturated hydrocarbon chains, two transitions are observed. The phase transition temperature of lipids with ionizable protons can be changed by changing the pH. Also, the presence of divalent cations Ca^{2+} and Mg^{2+} in the aqueous solution can affect the phase transition by interacting with the charged head groups or perhaps forming complexes. These effects have been investigated calorimetrically (36). The phase diagrams for binary mixtures of different saturated PCs (97) and of saturated PCs and saturated fatty acids (98) have been determined. The effect of cholesterol on the phase transition of PCs has been studied (99–102). Also, the lipids and membranes of *Acholeplasma laidlawii* (103) and of *E. coli* (104, 105) have been studied using DSC.

Freeze Fracture Electron Microscopy

Freeze fracture electron microscopy is a technique for taking a "photograph" of the interior surface of a membrane, which separates the two halves of the bilayer. [For a recent review of this method, see (106). Also see (107) for a discussion of the technical aspects of the method.]

The first step in such an experiment is a rapid freezing or quenching of the sample that contains the membrane. This is done with Freon or with mixtures of liquid and solid nitrogen. The freezing must be done rapidly enough to prevent any structural changes from taking place. The second step is the fracturing of the sample along one plane. It is generally accepted that membranes tend to fracture along the surface, mentioned above, that separates the two halves of a bilayer. The result then is two new frozen surfaces, and a part of each surface represents the interior surface of bilayers. In the third step, a replica of one or both surfaces is made by depositing some material, such as platinum and carbon. The replicas are stable at higher temperatures and hence they can be examined by electron microscopy. A resolution of about 30 Å can be achieved in the resulting electron micrograph.

Various types of physical features can be observed in the resulting pictures. For lipid dispersions of one- and two-component phospholipids, different textures are seen, corresponding to different phases. The physical meaning of the details of these surfaces is not as important as the fact that they can be observed and distinguished from each other. When two or more such textures appear on the same bilayer, it is clear that lateral phase separation exists in the membrane. Thus the technique is very useful for obtaining or confirming phase diagrams for lipid bilayers. For biological membranes the pictures typically show particles on a relatively smooth surface. The particles may either be homogeneously distributed or concentrated in patches. The particles are interpreted as intramembrane proteins and the smooth surfaces as lipids. Thus freeze fracture electron microscopy is also useful for obtaining information about protein distributions in model and biological membranes. [For a review of freeze fracture studies prior to 1975, see (106).]

The following are a few of the more recent freeze fracture studies. Van Dijck et al (108) and Papahadjopoulos et al (109) used freeze fracture methods in conjunction with calorimetry to study the effect of Ca^{2+} and Mg^{2+} ions upon the structure of bilayers composed of dimyristoylphosphatidylglycerol. Chapman et al (110) studied the effect of monovalent ions on the phase transitions of PC bilayers. Verkleij et al (111) studied the outer membrane of *E. coli* mutants using freeze fracture microscopy.

COMMENTS ON THE INTERPRETATION OF MEMBRANE PROBE DATA

Although many different methods for probing membrane structure are available and much work has been done using them, we have surprisingly little unambiguous knowledge about membrane structure on a molecular level. An important reason for this is that the quantities measured in these experiments, with the possible exception of X-ray and neutron scattering, are only indirectly related to structure. The arguments used to infer structural information from, for example, an NMR line width or a fluorescence depolarization ratio, or even an X-ray scattering pattern, often contain hidden structural assumptions that may be wrong. This, combined with the possibility of artifacts and systematic errors, has led to the publication of many contradictory conclusions about membrane structure. In studying the literature of this field, it is important to be aware of the physical nature of the measurements and the nature of the logic used to interpret the results obtained.

ACKNOWLEDGMENTS

This review was prepared while the author was a John Simon Guggenheim Memorial Fellow. The support of the National Science Foundation (Grant CHE 75-06634) and the National Institutes of Health (Grant GM 23085) is also gratefully acknowledged.

Literature Cited

1. Singer, S. J. 1974. *Ann. Rev. Biochem.* 43:805–33
2. Bangham, A. D. 1972. *Chem. Phys. Lipids* 8:386–92
3. Huang, C., Thompson, T. E. 1974. *Methods Enzymol.* 32B:485–89
4. Smith, I. C. P., Butler, K. W. 1976. In *Spin Labeling: Theory and Applications,* ed. L. Berliner, New York: Academic. pp. 411–51
5. Powers, L., Clark, N. A. 1975. *Proc. Natl. Acad. Sci. USA* 72:840–43
6. Powers, L., Pershan, P. S. 1977. *Biophys. J.* 20:137–52
7. Melchior, D. L., Steim, J. M. 1976. *Ann. Rev. Biophys. Bioeng.* 5:205–38
8. Ranck, J. L., Mateu, L., Sadler, D. M., Tardieu, A., Gulik-Krzywicki, T., Luzzati, V. 1974. *J. Mol. Biol.* 85:249–77
9. Akers, C. K. 1974. See Ref. 3, pp. 211–20
10. Tardieu, A., Luzzati, V., Reman, F. C. 1973. *J. Mol. Biol.* 75:711–33
11. Luzzati, V., Tardieu, A. 1974. *Ann. Rev. Phys. Chem.* 25:79–94
12. Engelman, D. M. 1970. *J. Mol. Biol.* 47:115–17
13. Engelman, D. M. 1971. *J. Mol. Biol.* 58:153–65
14. Esfahani, M., Limbrick, A. R., Knutton, S., Oka, T., Wakil, S. J. 1971. *Proc. Natl. Acad. Sci. USA* 63:3180–84
15. Rand, R. P., Chapman, D., Larsson, K. 1975. *Biophys. J.* 15:1117–24
16. Janiak, M. J., Small, D. M., Shipley, G. G. 1976. *Biochemistry* 15:4575–80
17. Brady, G. W., Fein, D. B. 1977. *Biochim. Biophys. Acta* 464:249–59
18. Schoenborn, B. P. 1976. *Biochim. Biophys. Acta* 457:41–55
19. *Neutron Scattering for the Analysis of Biological Structures. Brookhaven Symp. Biol.,* Vol. 27. To be published
20. Zaccai, G., Blasie, J. K., Schoenborn, B. P. 1975. *Proc. Natl. Acad. Sci. USA* 72:376–80
21. Blasie, J. K., Zaccai, G., Schoenborn, B. 1975. *Biophys. J.* 15:99a
22. Brown, K. G., Peticolas, W. L., Brown, E. 1973. *Biochem. Biophys. Res. Commun.* 54:358–64
23. Gaber, B. P., Peticolas, W. L. 1977. *Biochim. Biophys. Acta* 465:260–74
24. Lippert, J. L., Peticolas, W. L. 1971. *Proc. Natl. Acad. Sci. USA* 68:1572–76
25. Mendelsohn, R. 1972. *Biochim. Biophys. Acta* 290:15–21
26. Lippert, J. L., Peticolas, W. L. 1972. *Biochim. Biophys. Acta* 282:8–17
27. Lippert, J. L., Gorczyka, L. E., Meiklejohn, G. 1975. *Biochim. Biophys. Acta* 382:51–57
28. Mendelsohn, R., Sunder, S., Bernstein, H. J. 1976. *Biochim. Biophys. Acta* 419:563–69

29. Azzi, A. 1975. *Q. Rev. Biophys.* 8:237–316
30. Azzi, A. 1974. See Ref. 3, pp. 234–46
31. Radda, G. K. 1975. *Methods Membr. Biol.* 4:97–188
32. Waggoner, A. S., Stryer, L. 1970. *Proc. Natl. Acad. Sci. USA* 67:579–89
33. Sklar, L. A., Hudson, B. S., Simoni, R. D. 1977. *Biochemistry* 16:819–28
34. Lentz, B. R., Barenholz, Y., Thompson, T. E. 1976. *Biochemistry* 15:4521–28
35. Lentz, B. R., Barenholz, Y., Thompson, T. E. 1976. *Biochemistry* 15:4529–37
36. Jacobson, K., Papahadjopoulos, D. 1975. *Biochemistry* 14:152–61
37. Tecoma, E. S., Sklar, L. A., Simoni, R. D., Hudson, B. S. 1977. *Biochemistry* 16:829–35
38. Lee, A. G., Birdsall, N. J. M., Metcalfe, J. C. 1974. *Methods Membr. Biol.* 2:1–156
39. Mantsch, H. H., Saito, H., Smith, I. C. P. 1977. In *Progress in Nuclear Magnetic Resonance Spectroscopy,* ed. J. W. Emsley, J. Feeney, L. H. Sutcliffe, 11:211–72 London: Pergamon
40. Seelig, J., Niederberger, W. 1974. *J. Am. Chem. Soc.* 96:2069–72
41. Niederberger, W., Seelig, J. 1976. *J. Am. Chem. Soc.* 98:3704–6
42. Ulmius, J., Wennerström, H., Lindblom, G., Arvidson, G. 1975. *Biochim. Biophys. Acta* 389:197–202
43. Lichtenberg, D., Petersen, N. O., Girardet, J.-L., Kainosho, M., Kroon, P. A., Seiter, C. H. A., Feigenson, G. W., Chan, S. I. 1975. *Biochim. Biophys. Acta* 382:10–21
44. Petersen, N. O., Chan, S. I. 1977. *Biochemistry* 16:2657–67
45. Stockton, G. W., Polnaszek, C. F., Tulloch, A. P., Hasan, F., Smith, I. C. P. 1976. *Biochemistry* 15:954–66
46. Seelig, J., Gally, H.-U. 1976. *Biochemistry* 15:5199–5204
47. Kohler, S. J., Klein, M. P. 1977. *Biochemistry* 16:519–26
48. Seelig, J., Gally, H.-U., Wohlgemuth, R. 1977. *Biochim. Biophys. Acta* 467:109–19
49. Griffin, R. G., Powers, L., Pershan, P. S. Preprint
50. Seelig, A., Seelig, J. 1974. *Biochemistry* 13:4839–45
51. Seelig, A., Seelig, J. 1977. *Biochemistry* 16:45–50
52. Stockton, G. W., Johnson, K. G., Butler, K. W., Tulloch, A. P., Boulanger, Y., Smith, I. C. P., Davis, J. H., Bloom, M. 1977. *Nature* 269:267–68
53. Schindler, H., Seelig, J. 1975. *Biochemistry* 14:2283–87
54. Gaffney, B. J., McConnell, H. M. 1974. *J. Magn. Reson.* 16:1–28
55. Seiter, C. H. A., Chan, S. I. 1973. *J. Am. Chem. Soc.* 95:7541–53
56. Feigenson, G. W., Chan, S. I. 1974. *J. Am. Chem. Soc.* 96:1312–19
57. Bloom, M., Burnell, E. E., Valic, M. I., Weeks, G. 1975. *Chem. Phys. Lipids* 14:107–12
58. Urbina, J., Waugh, J. S. 1974. *Proc. Natl. Acad. Sci. USA* 71:5062–67
59. Opella, S. J., Yesinowski, J. P., Waugh, J. S. 1976. *Proc. Natl. Acad. Sci. USA* 73:3812–15
60. Kroon, P. A., Kainosho, M., Chan, S. I. 1976. *Biochim. Biophys. Acta* 433:282–93
61. Horwitz, A. F., Horsley, W. J., Klein, M. P. 1972. *Proc. Natl. Acad. Sci. USA* 69:590–93
62. Lee, A. G., Birdsall, N. J. M., Levine, Y. K., Metcalfe, J. C. 1972. *Biochim. Biophys. Acta* 255:43–56
63. Darke, A., Finer, E. G., Flook, A. G., Phillips, M. C. 1972. *J. Mol. Biol.* 63:265–79
64. Levine, Y. K., Birdsall, N. J. M., Lee, A. G., Metcalfe, J. C. 1972. *Biochemistry* 11:1416–21
65. Godici, P. E., Landsberger, F. R. 1974. *Biochemistry* 13:362–68
66. Barton, P. G., Gunstone, F. D. 1975. *J. Biol. Chem.* 250:4470–76
67. Birdsall, N. J. M., Lee, A. G., Levine, Y. K., Metcalfe, J. C. 1971. *Biochim. Biophys. Acta* 241:693–96
68. De Kruijff, B., Cullis, P. R., Radda, G. K. 1974. *Biochim. Biophys. Acta* 406:6–20
69. Berliner, L., ed. 1976. *Spin Labeling: Theory and Applications.* New York: Academic. 592 pp.
70. Gaffney, B. J. 1976. See Ref. 69, pp. 183–238
71. Nordio, P. L. 1976. See Ref. 69, pp. 5–52
72. Freed, J. H., 1976. See Ref. 69, pp. 53–132
73. Seelig, J. 1976. See Ref. 69, pp. 373–409
74. Griffith, O. H., Jost, P. C. 1976. See Ref. 69, pp. 453–523
75. Gaffney, B. J. 1974. See Ref. 3, pp. 161–98
76. Gaffney, B. J., Chen, S.-C. 1977. *Methods Membr. Biol.* 8:291–358
77. Shimshick, E. J., McConnell, H. M. 1973. *Biochemistry* 12:2351–60
78. Wu, S. H., McConnell, H. M. 1975. *Biochemistry* 14:847–54
79. Hesketh, T. R., Smith, G. A., Houslay, M. D., McGill, K. A., Birdsall, N. J.

M., Metcalfe, J. C., Warren, G. B. 1976. *Biochemistry* 15:4145–51
80. Ohnishi, S. 1975. *Adv. Biophys.* 8:35–82
81. McConnell, H. M., Wright, K. L., McFarland, B. G. 1972. *Biochem. Biophys. Res. Commun.* 47:273–81
82. Castle, J. D., Hubbell, W. L. 1976. *Biochemistry* 15:4818–31
83. Cafiso, D. S., Hubbell, W. L. 1978. *Biochemistry* 17: In press
84. Hubbell, W. L., McConnell, H. M. 1971. *J. Am. Chem. Soc.* 93:314–26
85. Jost, P., Libertini, L. J., Hebert, V. C., Griffith, O. H. 1971. *J. Mol. Biol.* 59:77–98
86. Rottem, S., Hubbell, W. L., Hayflick, L., McConnell, H. M. 1970. *Biochim. Biophys. Acta* 219:104–13
87. Van, S. P., Griffith, O. H. 1975. *J. Membr. Biol.* 20:155–70
88. Schreier-Muccillo, S., Marsh, D., Dugas, H., Schneider, H., Smith, I. C. P. 1973. *Chem. Phys. Lipids* 10:11–27
89. Keith, A. D., Snipes, W., Chapman, D. 1977. *Biochemistry* 16:634–41
90. Butterfield, D. A., Chestnut, D. B., Appel, S. H., Roses, A. D. 1976. *Nature* 263:159–61
91. Birrell, G. B., Griffith, O. H. 1976. *Arch. Biochem. Biophys.* 172:455–62
92. Marsh, D., Smith, I. C. P. 1973. *Biochim. Biophys. Acta* 298:133–44
93. Ladbrooke, B. D., Chapman, D. 1969. *Chem. Phys. Lipids* 3:304–56
94. Mabrey, S., Sturtevant, J. M. 1978. *Methods Membr. Biol.* 9:In press
95. Hinz, H.-J., Sturtevant, J. M. 1972. *J. Biol. Chem.* 247:6071–75
96. Chapman, D., Urbina, J., Keough, K. M. 1974. *J. Biol. Chem.* 249:2512–21
97. Mabrey, S., Sturtevant, J. M. 1976. *Proc. Natl. Acad. Sci. USA* 73:3862–66
98. Mabrey, S., Sturtevant, J. M. 1977. *Biochim. Biophys. Acta* 486:444–50
99. Ladbrooke, B. D., Williams, R. M., Chapman, D. 1968. *Biochim. Biophys. Acta* 150:333–40
100. Hinz, H.-J., Sturtevant, J. M. 1972. *J. Biol Chem.* 247:3697–3700
101. Barton, P. G. 1976. *Chem. Phys. Lipids* 16:195–200
102. Mabrey, S., Mateo, P. L., Sturtevant, J. M. 1977. *Biophys. J.* 17:82a
103. Melchior, D. L., Morowitz, H. J., Sturtevant, J. M., Tsong, T. Y. 1970. *Biochim. Biophys. Acta* 219:114–22
104. Baldassare, J. J., Rhinehart, K. B., Silbert, D. F. 1976. *Biochemistry* 15:2986–94
105. Jackson, M. B. 1976. *Fed. Proc.* 35:1531
106. Verkleij, A. J., Ververgaert, P. H. J. Th. 1975. *Ann. Rev. Phys. Chem.* 26:101–22
107. Zingsheim, H. P. 1972. *Biochim. Biophys. Acta* 265:339–66
108. Van Dijck, P. W. M., Ververgaert, P. H. J. Th., Verkleij, A. J., van Deenen, L. L. M., de Gier, J. 1975. *Biochim. Biophys. Acta* 406:465–78
109. Papahadjopoulos, D., Vail, W. J., Pangborn, W. A., Poste, G. 1976. *Biochim. Biophys. Acta* 448:265–83
110. Chapman, D., Peel, W. E., Kingston, B., Lilley, T. H. 1977. *Biochim. Biophys. Acta* 464:260–75
111. Verkleij, A., van Alphen, L., Bijvelt, J., Lugtenberg, B. 1977. *Biochim. Biophys. Acta* 466:269–82

Ann. Rev. Biochem. 1978. 47:385–417
Copyright © 1978 by Annual Reviews Inc. All rights reserved

GLYCOSAMINOGLYCANS AND THEIR BINDING TO BIOLOGICAL MACROMOLECULES

❖978

Ulf Lindahl and Magnus Höök

Department of Medical and Physiological Chemistry, Swedish University of Agricultural Sciences, College of Veterinary Medicine, The Biomedical Center, S-751 23 Uppsala, Sweden

CONTENTS

PERSPECTIVES AND SUMMARY

Glycosaminoglycans are polysaccharides that are found in animal tissues, usually in covalent association with protein. Seven types are commonly

385

0066-4154/78/0701-0385$01.00

recognized; six of these are structurally related, with a carbohydrate backbone consisting of alternating uronic acid (L-iduronic acid and/or D-glucuronic acid) and hexosamine (D-glucosamine or D-galactosamine) residues. All except one, hyaluronate, are sulfated. These polysaccharide types are distinguished by their monomer composition, by the position and configuration of their glycosidic linkages, and by the amount and location of their sulfate substituents. The hyaluronate and chondroitin sulfate species have rather simple structures whereas others, such as the dermatan sulfate, heparan sulfate and heparin species may contain a large number of different disaccharide units, arranged either in block structures or in less well-ordered, complex sequences. These structures are established during biosynthesis, by polymer-modification reactions that appear to be coordinated in a specific manner, yet operate by a partly random selection mechanism.

The glycosaminoglycans are widely distributed within the animal kingdom (1, 2). They are obviously essential for maintaining the structural integrity of many connective tissues and have in addition been assigned a number of functions relating to their characteristic physicochemical properties, for example, binding of water and microions and regulating the distribution of various macromolecules by steric exclusion (3, 4). Although no detailed structure-function relationships have been postulated these functions would not seem to warrant the structural complexity and variability observed. In the search for other functions it has become increasingly evident that the effects of binding of glycosaminoglycans to macromolecules must be considered. Such binding and its dependence on polysaccharide structure is the main topic of the present review.

A large number of macromolecules may bind to glycosaminoglycans. The majority of ligands are proteins or protein conjugates; however, glycosaminoglycans capable of interchain binding have also been implicated. In addition, a number of as yet unidentified components, for example, on cell surfaces, are believed to mediate certain observed biological effects of glycosaminoglycans. Binding between glycosaminoglycans and other macromolecules has been postulated in highly diverse functional contexts, such as the structural organization of the extracellular matrix in connective tissues, the control of hemostasis, the specific binding of plasma proteins to the blood vessel wall, and the regulation of cell behaviour and metabolism. Several binding systems have been studied in considerable detail in vitro, even though their functional significance has been unclear. Difficulties in evaluating the role of glycosaminoglycan binding in vivo may be partly ascribed to ignorance regarding the detailed topochemistry of the different polysaccharides in the mammalian organism.

Various types of binding may be discerned [for a comprehensive discussion of binding behaviour of polysaccharides see (4)]. One type, known as cooperative electrostatic binding, may involve any of a number of glycosaminoglycans; in general, these interactions are facilitated by increased charge density of the polysaccharides and by stereochemical factors that are not understood but which arise from the presence of L-iduronic acid residues. Ligands bound according to this mechanism include collagen, lipoproteins, and platelet factor 4. However, highly specific "lock and key" binding has also been described, where the polysaccharide component cannot be substituted by any other glycosaminoglycan species. One example of such binding is the interaction between heparin and antithrombin, which has been shown to involve a binding site in the heparin molecule composed of up to six disaccharide units arranged in one or more specific sequence(s). Another example is the formation of proteoglycan aggregates in cartilage by specific binding of proteoglycan core-protein to hyaluronate.

Continued analysis of binding behaviour is likely to reveal further interactions of functional importance. It may also be predicted that in these interactions the requirements for specificity regarding carbohydrate structure will vary. Binding studies should therefore be combined with more detailed characterization of glycosaminoglycan structures and, in a more extended perspective, with attempts to define the mechanisms regulating the elaboration of these structures during glycosaminoglycan biosynthesis.

STRUCTURE AND BIOSYNTHESIS OF GLYCOSAMINOGLYCANS

The different types of glycosaminoglycans commonly distinguished in vertebrate tissues are shown in Table 1. They are all polyanions that have acidic sulfate and/or carboxyl groups, and many of their functional properties have been attributed to their polyelectrolyte nature. Generally, the glycosaminoglycans do not occur as free polysaccharide chains in vivo, but as proteoglycans in which several chains are covalently linked, at the reducing terminal sugar, to a polypeptide core (however, see exceptions below). In the following, we discuss the structures of glycosaminoglycans along with the biosynthetic mechanisms leading to their formation. The presentation focuses on recent progress of relevance to the main topic of this review, that is, the binding properties of the polymers; comprehensive accounts of earlier results may be found in previous reviews on the structure (see footnote c to Table 2) and biosynthesis (8–10) of glycosaminoglycans. Keratan sulfate has been poorly investigated with regard to binding properties and is therefore not discussed in detail.

Polysaccharide Chains

DISACCHARIDE UNITS   The polymerization reactions leading to formation of glycosaminoglycans are generally believed to occur by stepwise addition of monosaccharide units from the appropriate UDP-sugars to the nonreducing ends of nascent polysaccharide chains [(8–11) however see also (13, 14) which suggests a possible role for lipid-bound disaccharide intermediates in the biosynthesis of hyaluronate and/or heparin-like polysaccharides]. With one exception, hyaluronate, the polymerization products are further modified, possibly before completion of chain elongation (15), to yield sulfated polysaccharides that often contain more than one type of disaccharide unit (references to structural data are given in Table 2).

Hyaluronate   The structure of the polymerization product is [(1→4)-β-D-glucuronosyl-(1→3)-β-D-N-acetylglucosaminyl]$_n$. No further modification takes place.

Chondroitin sulfates   The structure of the polymerization product, chondroitin, is [(1→4)-β-D-glucuronosyl-(1→3)-β-D-N-acetylgalactosaminyl]$_n$. This polymer is sulfated by sulfotransferases that utilize 3'-phosphoadenylylsulfate (PAPS) as sulfate donor and catalyze the formation of ester sulfate (O-sulfate) groups at C4 and C6 of the N-acetylgalactosamine residues (9, 10). The sulfated product usually contains about one sulfate group per disaccharide residue although considerable deviations from a 1:1 ratio have been noted (Table 1). Individual chondroitin sulfate chains may contain sulfate groups in both 4- and 6-positions but may also be exclusively 4-sulfated or 6-sulfated.

Table 1   Occurrence and properties of glycosaminoglycans[a]

Polysaccharide	Mol. wt. ($\times 10^{-3}$)	Repeating period monosaccharides	Sulfate per disaccharide unit	Other sugar components[b]	Examples of occurrence in mammalian tissues
Hyaluronate	4–8000	D-glucuronic acid D-glucosamine	0	[c]	Various connective tissues, skin, vitreous humor, synovial fluid, umbilical cord, cartilage
Chondroitin 4- and 6-sulfates	5–50	D-glucuronic acid D-galactosamine	0.1–1.3	D-galactose D-xylose	Cartilage, cornea, bone, skin, arterial wall
Dermatan sulfate	15–40	D-glucuronic acid L-iduronic acid D-galactosamine	1.0–3	D-galactose D-xylose	Skin, heart valve, tendon, arterial wall
Heparan sulfate	—	D-glucuronic acid L-iduronic acid D-glucosamine	0.4–2	D-galactose D-xylose	Lung, arterial wall, ubiquitous on cell surfaces (?)
Heparin	6–25	D-glucuronic acid L-iduronic acid D-glucosamine	1.6–3	D-galactose D-xylose	Lung, liver, skin, intestinal mucosa (mast cells)
Keratan sulfate	4–19	D-galactose D-glucosamine	0.9–1.8	D-galactosamine D-mannose L-fucose sialic acid	Cartilage, cornea, intervertebral disc

[a] See reviews (2, 5–7) for references.
[b] Including the sugar components of the polysaccharide-protein linkage region.
[c] The occurrence in hyaluronate of arabinose has been suggested but not conclusively verified.

Table 2 Structure of glycosaminoglycans

Polysaccharide	Monosaccharide Units[a]		Substituents[b]	References[c]
	A	B		
Hyaluronate	β-D-GlcUA	β-D-GlcN	R = —C(=O)CH₃	
Chondroitin sulfates	β-D-GlcUA	β-D-GalN	R = —C(=O)CH₃; R' = —H or —SO₃⁻	
Dermatan sulfate	β-D-GlcUA; α-L-IdUA	β-D-GalN	R = —C(=O)CH₃; R' = —H or —SO₃⁻	8, 20–29
Heparin, heparan sulfate	β-D-GlcUA; α-L-IdUA	α-D-GlcN	R = —SO₃⁻ or —C(=O)CH₃; R' = —H or —SO₃⁻	8, 30–39
Keratan sulfate	β-D-Gal	β-D-GlcN	R = —C(=O)CH₃; R' = —H or —SO₃⁻	

[a] The polysaccharides are depicted as linear polymers of alternating A and B monosaccharide units. For further details, see the text. Abbreviations: GlcUA, glucuronic acid; IdUA, iduronic acid; GlcN, glucosamine; GalN, galactosamine; Gal, galactose.

[b] Sulfate groups in positions other than those indicated have been reported for chondroitin sulfate [sulfated D-glucuronic acid residues (16)], for heparin [sulfate at C3 of D-glucosamine units (17)] and for keratan sulfate [sulfate at C6 of D-galactose residues (see 18 and references therein)]; the latter polymer also shows indications of branching (18).

[c] References are principally to studies on the copolymeric nature of some glycosaminoglycans (see discussion in the text). [For previously established structural properties see (2, 5–7, 9, 19)].

Dermatan sulfate   The polymerization product is presumably identical to that formed during biosynthesis of chondroitin sulfate. However, the modification of this polymer is not restricted to O-sulfaction reactions but includes also formation of L-iduronic acid residues by C5-epimerization of D-glucuronic acid units (8, 40). Considerable structural variability has been demonstrated. The D-glucuronic acid content may thus range from negligible amounts to more than 90% of the total uronic acid. Some preparations have a comparatively simple structure. For example, pig skin dermatan sulfate consists, essentially, of iduronosyl-N-acetylgalactosaminyl-4-sulfate disaccharide units. However, others are extremely complex with extensively hybridised polysaccharide backbone and, in addition, both 4- and 6-O-sulfated galactosamine residues. Oversulfated regions may be due to the occurrence of sulfated L-iduronic acid units; sulfated D-glucuronic acid has not been detected.

Heparan sulfate and heparin   The structure of the polymerization product that serves as intermediate in the biosynthesis of heparin and, presumably, heparan sulfate, is $[(1{\rightarrow}4)\text{-}\beta\text{-D-glucuronosyl-}(1{\rightarrow}4)\text{-}\alpha\text{-D-N-acetylglucosa-minyl}]_n$. Polymer-modification reactions resembling those operating in the biosynthesis of dermatan sulfate have been demonstrated (Figure 1), and in addition, two reactions, deacetylation of N-acetyl-D-glucosamine residues, followed by sulfation of the resulting free amino groups, which appear to be unique to the formation of heparin-like polysaccharides (8, 41–47). The L-iduronic acid : D-glucuronic acid ratio generally increases with increasing sulfate content; in broad terms low-sulfated, D-glucuronic acid rich polysaccharides are classified as heparan sulfate whereas high-sulfated, L-iduronic acid rich species are designated heparin. However, occasional samples have intermediate properties and it therefore seems appropriate, from a structural viewpoint at least, to consider both polymers as members of the same family of heparin-like polysaccharides.

POLYMER SEQUENCE   The spatial arrangement of the hydroxyl, sulfate, carboxyl, and acetyl side groups in a glycosaminoglycan chain is deter-

---

→

Figure 1   Sequence of polymer-modification reactions leading to formation of the predominant disaccharide unit in heparin. Some of the reactions take place in a strictly stepwise manner. N-Sulfation of the entire polysaccharide molecule thus appears to be completed before any O-sulfate groups are introduced (41). Other reactions operate in a more concerted, yet sequential, fashion; this applies for instance to the C5-epimerization and 2-O-sulfation of uronic acid residues. Due to the substrate specificities of the enzymes involved the individual polymer-modification reactions are strongly interdependent. N-Deacetylation is thus prerequisite not only to N-sulfation but also to uronosyl C5-epimerization, since only the N-sulfated polymer is recognized as substrate by the epimerase (42; I. Jacobsson, G. Bäck-

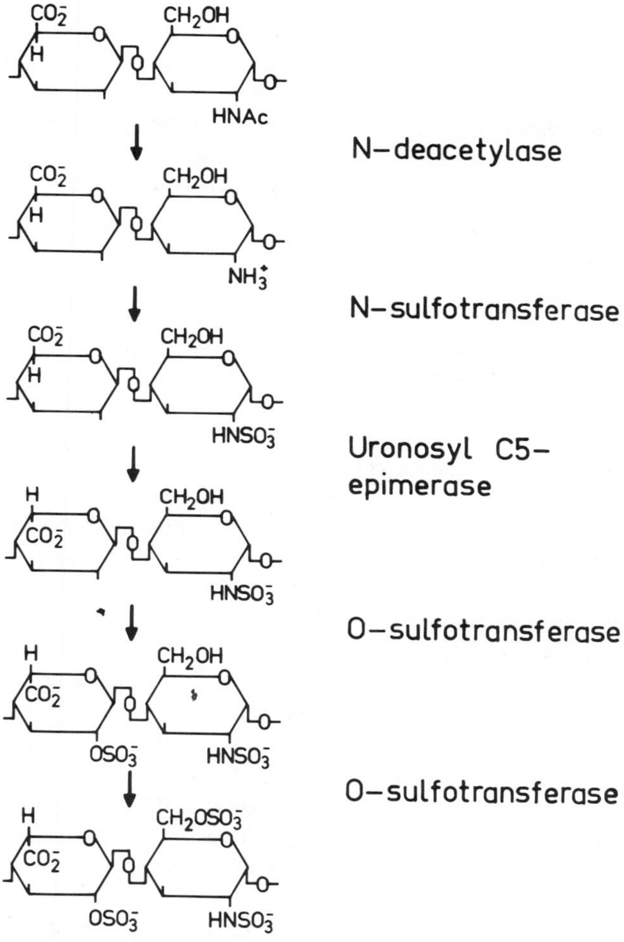

CO2-
CH2OH
H
HNAc
N-deacetylase
NH3+
N-sulfotransferase
HNSO3-
Uronosyl C5-epimerase
O-sulfotransferase
OSO3-
O-sulfotransferase
CH2OSO3-

ström, M. Höök, U. Lindahl, A. Malmström, L. Rodén, and D. S. Feingold, in manuscript). Polymer sequences that escape N-deacetylation will therefore give rise to low-sulfated, D-glucuronic acid-rich regions in the final product. Conversely, only N-deacetylated segments may proceed through the series of modification reactions and acquire high contents of sulfate and L-iduronic acid residues. However, it is important to note that disaccharide units may remain at various levels of incomplete modification. For example, the N-sulfated disaccharide units identified in heparin and heparan sulfate include not only the extensively modified, di-O-sulfated unit shown as end product of the reaction sequence, but also three different mono-O-sulfated species. These include all the possible mono-O-sulfated disaccharide permutations of the monosaccharide units shown in Table 2, with R= $-SO_3$ (I. Jacobsson, M. Höök, and U. Lindahl, unpublished observation; B. Weissman, personal communication).

mined by the types of monomer sugar residues, the position and configuration of glycosidic linkages, the positions of substituents, the conformation of monosaccharide units, and the shape of the chain. Some of these properties are established already at the stage of polymerization and are thus controlled by the specificities of glycosyltransferases (8–10), whereas others are laid down during polymer modification. Formation of different polysaccharides obviously involves different polymer-modification reactions; nevertheless, certain general principles may be discerned. In copolymeric glycosaminoglycans the L-iduronic and D-glucuronic acid residues are not randomly distributed but rather are assembled in clusters which, in turn, reflect the distribution of sulfate groups (see Table 2 for references). Copolymeric galactosaminoglycans may thus contain extended sequences of either iduronosyl-N-acetylgalactosaminyl-4-sulfate or glucuronosyl-N-acetylgalactosaminyl-6-sulfate units. Moreover, the D-glucuronic acid residues in heparin-like polysaccharides are preferentially accumulated within low-sulfated, N-acetylated regions, which predominate in heparan sulfate, whereas L-iduronic acid units prevail in high (N- and O-)-sulfated regions, which predominate in heparin. The formation of such block sequences requires coordinated action of polymer-modification enzymes and recent studies do in fact indicate how coordination may be accomplished (41, 42). During biosynthesis of heparin, polymer modification occurs in a stepwise manner (Figure 1). In each step the corresponding enzyme, or enzymes, is presented with the product from the preceding step and the modifications introduced are essential for substrate recognition by enzymes catalyzing subsequent reactions (42). When the sequential order of modification reactions and the substrate specificities of the corresponding enzymes are considered, certain characteristics of heparin-like polysaccharides, such as the codistribution of L-iduronic acid residues and sulfate groups, are readily explained (42).

On the other hand, the sequence of modification reactions, once initiated, does not necessarily proceed to completion (see legend to Figure 1). Incompletely modified disaccharide units may be expected to form less well-ordered regions, interspersed between block structures; the sequence of disaccharide units within these regions is difficult to determine, partly because of microheterogeneity but also due to lack of adequate analytical methods. Incomplete polymer modification is a major cause of structural complexity that applies not only to the heparin-like polysaccharides but also to the copolymeric galactosaminoglycans. At various steps during the process a fraction of the potential substrate units escapes modification; the mode of selection is unknown and may conceivably operate at random. The possibility of polymer modification occurring by concerted specific and random mechanisms is intriguing, particularly in view of recent findings implying the presence of specific binding sites in copolymeric glycosaminoglycans (see below).

POLYMER SHAPE   Assignments of monomer conformations in glycosaminoglycans are based on NMR (48, 49) and X-ray diffraction (50–56) studies. The data suggest that the sugar residues in glycosaminoglycans have the usual $^4C_1$ chair conformation, except the L-iduronic acid units in heparin and heparan sulfate, to which the alternative $^1C_4$ chair form has been assigned. Models with L-iduronic acid units in either chair form differ markedly with regard to polymer shape and spatial distribution of side groups (50) and it therefore seems important to ascertain whether the $^1C_4$ chair conformation applies to all L-iduronic acid residues in heparin-like polysaccharides, including those that lack O-sulfate groups at C2.

Glycosaminoglycan chains may, under certain conditions, adopt left-handed helical shapes, as demonstrated by X-ray diffraction studies on oriented polysaccharide fibers (3, 4, 50, 51). Such ordered conformations, whenever maintained in the tissues, may be expected to influence the binding behaviour of the polysaccharides in vivo. However, it is not known at present whether the helical forms ever exist under biological conditions; an ordered conformation in solution has so far only been demonstrated for hyaluronate (4).

Proteoglycans

There is now general agreement that most of the glycosaminoglycans occur in the native state covalently bound to protein (2, 3, 5, 57). The proteoglycan nature of hyaluronate remains uncertain; evidence has been presented which suggests covalent attachment of some protein to hyaluronate from synovial fluid (58). However, all available information points to hyaluronate as a single, unbranched polymer, which thus differs from conventional proteoglycans that consist of several polysaccharide chains linked to a common peptide. A typical proteoglycan of the latter type, present in cartilage, contains 5-10% protein and consists of about 100 chondroitin sulfate chains and 60 keratan sulfate chains linked to a protein core of 200,000 mol wt (see Figure 2). The core carries one substituent glycosaminoglycan chain for approximately 12 amino acid residues (5). Proteoglycans of similar kind have been found, that contain keratan sulfate but no chondroitin sulfate (59); others contain heparan sulfate (62–65), dermatan sulfate or dermatan sulfate-chondroitin sulfate copolymer (3, 5, 60, 61). The various proteoglycans differ markedly in protein content, molecular size, and number of polysaccharide chains per molecule. The polysaccharide moieties, except keratan sulfate, are all linked to protein by means of a glucuronosyl-galactosyl-galactosyl-xylosyl-O-serine sequence (5, 57).

Studies on the macromolecular properties of heparin have yielded seemingly contradictory results. Heparin is synthesized by mast cells and is stored within the cytoplasmic granules of these cells. In some tissues, such as bovine liver capsule, heparin occurs virtually exclusively as single polysaccharide chains of less than 10,000 mol wt (66). Other sources, for example, rat skin (67) and peritoneal mast cells (68), yield heparin with a molecular weight approaching 1×10^6, which apparently consists of several polysaccharide chains joined by a core structure. Yet other tissues contain both types of material (69, 70). Recent findings raise the possibility that the macromolecular heparin may indeed be a proteoglycan, with a unique polypeptide core consisting of serine and glycine in equimolar proportions (71); most of the serine residues are substituted by polysaccharide chains of mol wt $40-100 \times 10^3$. The occurrence of single-chain forms of heparin is attributed to cleavage of the polypeptide-bound chains by specific endoglycosidases at a limited number of sites, and this yields products of a molecular weight range similar to that of commercially available heparin (69–75). It is possible that heparin and heparan sulfate, though not always distinguishable in terms of carbohydrate structure (see above), differ with regard to proteoglycan type. Obviously, further work should be aimed toward elucidating the structures of the corresponding core-proteins.

BINDING PROPERTIES OF GLYCOSAMINOGLYCANS

The last decade of research has produced a rapidly growing list of biomolecules capable of binding to glycosaminoglycans. The characteristics of such interaction have sometimes been studied in detail, for example, with regard to specificity, even when the functional implications have been unclear. In other instances binding has not been demonstrated but has been inferred from functional observations; this applies particularly to the effects of glycosaminoglycans on intact cells (see below).

In the following discussion the functional aspects of glycosaminoglycan binding are emphasized whenever possible. Enzymes that have glycosaminoglycans as substrates will not be considered. The topic will be subdivided with regard to location of ligands in relation to the mammalian cell and will not be restricted to systems that involve well-characterized components. The distinction between extracellular and cell-surface ligands facilitates the presentation but may at times be artificial or arbitrary; the latter term is reserved for cell-surface structures that are generally ill-defined and are believed to mediate binding of glycosaminoglycans to the cells.

General Binding Properties

Binding of glycosaminoglycans is generally electrostatic in nature although other types of interaction may occur. In some binding systems (see below) only one particular polysaccharide will interact with the ligand. However, the majority of binding phenomena that were studied appear to involve less specific cooperative electrostatic effects. Studies with model polypeptides have revealed some general properties of such interactions. Gelman et al demonstrated by circular dichroism spectroscopy that glycosaminoglycans may drive various cationic homopolypeptides (poly-L-arginine, poly-L-lysine, poly-L-ornithine) from the random coil to the α-helix conformation (12, 76–82). The conditions employed deviated too much from the physiological milieu to permit any conclusions regarding analogous conformational effects in vivo; nevertheless, the experiments established that charge interactions with glycosaminoglycans may invoke conformational changes in proteins, a concept that has since been verified with other systems (83, 84). Interaction was facilitated by increasing the length and pK_a of the polypeptide side chains and the charge density of the polysaccharides; it was highest for heparin and lowest for hyaluronate (see also 85). However, in addition specific stereochemical factors were recognized that also seem to bear on interactions with naturally occurring ligands. Polysaccharides containing L-iduronic acid, such as dermatan sulfate and heparan sulfate, thus generally bind proteins with higher affinity than do polysaccharides such as chondroitin sulfates, that have equal charge density but contain D-glucuronic acid as the only uronic acid component (see interactions with collagen, lipoproteins, lipoprotein lipase, and platelet factor 4 below). This difference in binding behaviour remains unexplained and reinforces the need for information regarding the spatial distribution of charged groups in polysaccharide molecules under physiological conditions.

Interaction of Glycosaminoglycans with Extracellular Components

CONNECTIVE TISSUE MACROMOLECULES   The glycosaminoglycans of connective tissues, most of which occur as proteoglycans, form, along with the fibrous proteins, collagen, and elastin, the major macromolecular elements of the tissues. The relationship that these components bear to each other, and to the water and microions of the intercellular matrix (3, 4), presumably accounts for the physical properties of the tissue, and binding between the macromolecules has therefore far-reaching physiological implications. Diverse patterns of binding have been demonstrated, including not only polysaccharide-protein but also polysaccharide-polysaccharide interactions.

Interchain binding of glycosaminoglycans   A number of mainly algal poly-saccharides have the ability to form gels in which polymer strands of ordered conformation bind to each other (4). Following the initial observation of ordered conformation for a glycosaminoglycan, hyaluronate (86), a double helical structure was proposed (87, 88) but subsequently refuted (89). No further reports of inter-chain binding of glycosaminoglycans appeared until recently, when Fransson (90) described that polysaccharides (referred to as copolymeric galactosaminoglycans) that contain D-glucuronosyl- and L-iduronosyl-N-acetyl-D-galactosaminyl sulfate disaccharide units in approximately equal amounts have a marked tendency to aggregate. The interaction was demonstrated by gel chromatography and also by affinity chromatography using polysaccharide-substituted gels; it occurred at physiological ionic strength and did not require the presence of divalent cations. Also, certain other polysaccharide preparations could bind to gels substituted with copolymeric galactosaminoglycan; a comparison of binding affinities indicated that binding depends on the presence of both D-glucuronic acid- and L-iduronic acid-containing disaccharide units within the same polysaccharide chain, and furthermore, that only the 4-sulfated regions of copolymeric chains contain the binding zones. Whether these zones consist of D-glucuronic acid- and L-iduronic acid-containing disaccharide units arranged in clusters or in alternating fashion (27) remains to be established. Also copolymeric glucosaminoglycans appear capable of inter-chain binding. In a recent abstract Fransson reported that heparan sulfate preparations with L-iduronic acid/D-glucuronic acid ratios between 0.4 and 0.7 and N-acetyl/N-sulfate ratios between 0.5 and 0.7 displayed associative behavior in gel chromatography experiments (91).

The concept of inter-chain binding of glycosaminoglycans has several interesting implications. It may explain experimental anomalies such as the aberrant behaviour of copolymeric galactosaminoglycans upon gel chromatography (26) and the formation of inter-chain hemiacetal-type cross-linkages during periodate oxidation of dermatan sulfate (90, 92). Furthermore, the possibility of selective polysaccharide chain aggregation under physiological conditions offers new prospects for future research on the functional role of copolymeric glycosaminoglycans in the intercellular space and, particularly, on the cell surface.

Binding of glycosaminoglycans to collagen and elastin   The interaction between collagen and glycosaminoglycans has been studied by a variety of techniques (see also 3, 5) including electrophoresis (93), affinity chromatography (94–96), precipitation (97–100), light scattering (101–103), circular dichroism spectroscopy (95, 104), equilibrium binding (105), and agglutination of collagen-coated erythrocytes (106). In summary, the results indicate that all glycosaminoglycans except hyaluronate, which lacks sulfate groups, and keratan sulfate, which lacks carboxyl groups, bind to collagen by electrostatic interaction at physiological pH and ionic strength; that glycosaminoglycans of higher charge density and larger molecular size bind with higher affinity; and, finally, that the presence of L-iduronic acid residues appears to promote the binding process. The interaction is thus akin to the cooperative electrostatic binding of glycosaminoglycans to a number of other proteins. The number of polysaccharide chains bound to each collagen monomer varies, depending on the type of polysaccharide, between two and five (95, 103, 105); however, a dermatan sulfate of high molecular weight (41,000) could bind as many as five molecules of collagen per chain. Proteoglycans interact more strongly than the corresponding individual polysaccharide chains (95–102, 107, 108). Although this difference could be partly ascribed to the polyvalent nature of the proteoglycan molecules, recent binding experiments with isolated chondroitin sulfate proteoglycan core-protein, obtained by digestion of proteoglycan monomer with bacterial chondroitinase, point to a more complex situation. This core-protein bound strongly to collagen (96, 100), yet did not appear to inhibit binding of intact proteoglycan (99, 100). The results were explained by postulating two types of collagen-binding sites in the proteoglycan molecule, located in the core-protein and in the polysaccharide chains, respectively (99). The location of proteoglycan-binding sites in collagen molecules is unclear. Soluble collagens of types I, II, or III all bind chondroitin sulfate proteoglycan (99), although there is some indication that type II collagen may interact to a lesser extent (99, 109, 110).

While it is generally agreed that the proteoglycan-collagen interaction should influence the deposition of collagen fibres in vivo the results of model studies have been confusing and contradictory. The kinetics of fiber formation may be followed turbidimetrically by refinements (100, 111) of a technique originally described by Gross & Kirk (112). However, in applying these methods to polysaccharide-collagen mixtures glycosaminoglycans or proteoglycans had either no effect, an accelerating effect, or a decelerating effect on the gelation of collagen (95, 102, and references therein). Analyzing this situation Öbrink points out (95) that part of the variations may be ascribed to differences in timing of polysaccharide additions relative to the initial lag-phase during which nuclei are formed (111); however, the source of collagen used in the experiments is also of critical importance. In a recent study Oegema et al (100) conclude that proteoglycans have two distinct effects on collagen fibrillogenesis; they retard the assembly of collagen molecules if present early enough during the process, and they alter the final organization of the fibril, as visualized through electron microscopy, in an as yet unknown fashion.

Little information is available concerning the interaction between glycosaminoglycans and elastin. However, it was recently shown that both soluble α-elastin (113) and tropoelastin (114) precipitate in the presence of proteoglycan (mainly chondroitin sulfate) from bovine nasal cartilage, with resulting fiber formation. Circular dichroism measurements indicated a proteoglycan-induced conformational change in the α-elastin molecule, with an increased content of helical structure (115). The functional significance of these findings appears somewhat uncertain as the interactions occurred at very low pH and ionic strength. In a recent study the preferential solubilization of heparan sulfate proteoglycan on digestion of bovine aorta with elastase was taken as evidence for a specific binding between this proteoglycan and elastin (63).

Binding of hyaluronate to proteoglycan core-protein   Cartilage proteoglycans, composed of chondroitin sulfate and keratan sulfate chains linked to a common core-protein, generally occur in the tissues as aggregates of very high molecular weight, held together by binding of proteoglycan monomers to specific nonproteoglycan components (for review, see 5, 116). The first important clue to the organization of these aggregates came through the work of Hascall & Sajdera (117), who showed that the nonproteoglycan components could be separated from the monomers by density-gradient centrifugation in 4 M guanidinium chloride. One of these components, hyaluronate, was shown to have a unique role in aggregation, individual strands binding as many as 100 proteoglycan monomer molecules (5, 116, 118–123). This interaction may be demonstrated in the absence of additional components (118, 121, 124) but is apparently stabilized by two "link proteins" that form part of the native aggregates (5, 116, 117, 121, 125, 126). The link proteins are structually closely related to each other (116, 127) and are able to bind hyaluronate in the absence of proteoglycan (128). In the native aggregates hyaluronate thus presumably binds to three protein components, the two link proteins, and the hyaluronate-binding region of the proteoglycan core-protein (Figure 2). Core fragments with molecular weights in the order of 60–70,000, which contain the hyaluronate-binding region, could be isolated after mild proteolytic digestion of aggregates (121, 128, 129). Several lines of evidence indicate that the hyaluronate-binding region is located at one end of the proteoglycan core-protein molecules (5, 116, 130).

The interaction between hyaluronate and the hyaluronate-binding region has been studied in considerable detail, mostly using gel chromatography or viscometry for detection of binding. It is remarkably specific, which suggests lock and key rather than cooperative electrostatic binding. Polyelectrolytes such as dextran sulfate, chondroitin sulfate, sodium alginate, or

DNA did not interact with proteoglycan (118) nor did two isomers of hyaluronate (122; D. Heinegård, personal communication) having the structures $[(1{\rightarrow}4){-}\beta{-}\text{D-glucuronosyl-}(1{\rightarrow}3){-}\beta{-}\text{D-N-acetylgalactosaminyl}]_n$ and $[(1{\rightarrow}4){-}\beta{-}\text{D-glucuronosyl-}(1{\rightarrow}4){-}\alpha{-}\text{D-N-acetylglucosaminyl}]_n$, respectively. These structures represent chondroitin and an intermediate in heparin biosynthesis (41) and may be compared with that of hyaluronate $[(1{\rightarrow}4){-}\beta{-}\text{D-glucuronosyl-}(1{\rightarrow}3){-}\beta{-}\text{D-N-acetylglucosaminyl}]_n$. Competitive-binding experiments with hyaluronate oligosaccharides showed good binding of a decasaccharide containing a reducing N-acetylglucosamine residue, and larger oligosaccharides, but not of the homologous octasaccharide or smaller oligosaccharides (119, 122). Removal of the nonreducing terminal glucuronic acid residue from the decasaccharide did not significantly affect its binding ability (122), which indicates that a nonasaccharide sequence containing four glucuronic acid residues (or possibly an octasaccharide with glucuronic acid in reducing terminal position) is the smallest fragment capable of strong interaction with the hyaluronate-binding site on the proteoglycan core-protein (116). Modification of D-glucuronic acid residues by methyl esterification, reduction to glucose, or substitution with glycine in amide linkage abolished the binding properties of large hyaluronate oligomers. From the results of partial carboxyl modifications it could

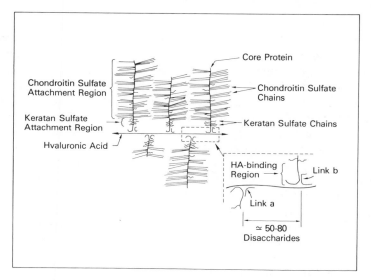

Figure 2 Schematic representation of cartilage proteoglycans and their aggregation by binding to hyaluronate (labeled hyaluronic acid or HA in the figure). Link a and Link b refer to the link proteins that stabilize the interaction between hyaluronate and the proteoglycan core-protein. (Figure provided by Dr. V. C. Hascall.)

be calculated that as many as three of the four D-glucuronic acid residues in each protein-binding site must have unsubstituted carboxyl groups (J. Christner, M. Brown, and D. D. Dziewiatkowski, personal communication). Furthermore, one or more of the N-acetylglucosamine residues must be specifically involved in binding, because chondroitin, which differs from hyaluronate only in the configuration of C4 at the hexosamine residues (see above), has no binding ability.

Reduction and alkylation of the core-protein in proteoglycan monomers prevent subsequent aggregate formation (131) and interaction with hyaluronate (132), which suggests that the conformation of the hyaluronate-binding region is critical for the binding process. Selective modification of amino acid residues in the core-protein indicated that binding to hyaluronate required intact arginine and tryptophan residues as well as ϵ-amino groups of lysine; however, fluorescence measurements suggested that tryptophan was not involved in direct subsite interactions with the hyaluronate (132). In a recent extension of these studies it was concluded that the hyaluronate-binding region contains arginine residues that are necessary for activity and, in the vicinity, lysine residues that sterically block the interaction with hyaluronate when substituted with dansyl groups but not when substituted with less bulky acetyl groups (116; D. Heinegård and V. C. Hascall, personal communication).

PLASMA PROTEINS   Several examples of interaction between glycosaminoglycans and plasma proteins are known. Whether these interactions should be credited with physiological roles depends, among other factors, on the availability of glycosaminoglycans within the blood vessel wall. The glycosaminoglycans in plasma consist essentially of chondroitin 4-sulfate (133). The arterial wall contains a variety of polysaccharides including, in addition to hyaluronate, proteoglycans of chondroitin 4- and 6-sulfates, dermatan sulfate, and heparan sulfate (21, 60, 63, 64, 134–136). Most or all of these species have also been identified in the arterial intima, and may thus be accessible to plasma components (136, 137). Polysaccharides on the endothelial cell surface [heparan sulfate (138, 139); see also below regarding binding of heparin to endothelial cells] may conceivably provide a means for the selective binding of plasma components to the vessel wall, in continued contact with the circulating blood; functional implications of such binding, although speculative, have been proposed for antithrombin (43, 140, 141) and for lipoprotein lipase (142, 143).

Heparin presents a rather special situation. It is synthesized and stored within mast cells in the close vicinity of blood vessels; although release of heparin from these cells may occur in certain situations its occurrence in plasma under normal physiological conditions is still a matter of controversy (144). The function of endogenous heparin is therefore uncertain, in spite of its conspicuous biological effects, that is, impairment of blood coagulation and acceleration of triglyceride elimination from plasma, invoked by exogenous heparin. Nevertheless, the interactions between heparin and a number of plasma proteins have been studied in considerable detail and do in fact include at least one example of specific binding (to antithrombin; see below). The interest in these interactions derives to a large extent from the extensive clinical use of heparin as an antithrombotic agent (145).

Binding of heparin to coagulation factors and to antithrombin   Heparin binds to several plasma components (146, 147), as is expected from its high negative charge density. Interactions with proteins involved in the coagulation mechanism, including factor IX (148–150), factor XI (149), and thrombin (149, 151–154), have been demonstrated and exploited in purification of these factors. Little is known of the mechanism and specificity of these interactions and there is considerable uncertainty regarding their contribution to the anticoagulant activity of heparin in vivo. In fact, all available evidence points to a more indirect mechanism of heparin action, involving a protein cofactor identified as plasma antithrombin (antithrombin III) by Abildgaard in 1968 (155).

Antithrombin is an α_2-glycoprotein composed of a single polypeptide chain with a molecular weight of about 55,000 (156, 157). It inactivates thrombin by forming a stable 1:1 stoichiometric complex, probably by interaction of an arginine residue in the inhibitor molecule with an active-site serine in the enzyme (141, 158). In the presence of heparin, complex formation is dramatically accelerated (158). Extended studies showed that not only thrombin but also factor IX_a (159, 160), factor X_a (160, 161), factor XI_a (162), and factor XII_a (163), that is, virtually all of the serine proteases in the coagulation system, with the possible exception of factor VII_a (141), were inactivated by antithrombin. In each case the rate of inactivation was greatly increased by the presence of heparin. Also plasmin, a component of the fibrinolytic mechanism, was inactivated in the same manner (164). In contrast, the inactivation of kallikrein by antithrombin was only moderately accelerated by heparin (165).

A tentative scheme of the anticoagulant mechanism of heparin is given in Figure 3. The key role of binding between heparin and antithrombin is supported by several lines of evidence (141, 153, 166). Binding of heparin to lysine residues (158) in antithrombin presumably induces a conformational change resulting in a more favorable topography of the thrombin-binding site of the latter molecule (84, 166) (I. Björk and B. Nordenman, personal communication). Furthermore, heparin apparently acts in a catalytic manner, as one molecule of heparin was shown to promote the binding of a large number of antithrombin molecules to thrombin (167). In accord with this concept heparin was found to be more strongly bound to antithrombin than to the antithrombin-thrombin complex (146, 168) (R. D. Rosenberg, personal communication).

No detailed structure-function relationships have yet been established for anticoagulant polysaccharides. Their activities are generally estimated by assays involving either clot formation in various whole-blood systems (169), or more direct measurements of antithrombin activation (140, 167) using synthetic chromogenic substrates for the appropriate serine proteases. In addition to heparin, heparan sulfate and dermatan sulfate have the ability to prevent blood coagulation, albeit at high concentrations only (140). The anticoagulant effect of dermatan sulfate does not seem to be exerted by means of antithrombin, whereas heparan sulfate activates antithrombin but is much less potent in this respect than is heparin (140). The anticoagulant activity of heparin generally increases with increasing degree of polymerization (170–175). Furthermore, N-desulfation resulted in loss of activity that could be restored by re-N-sulfation but not by N-acetylation (170, 171); modifications at the carboxyl groups of the uronic acid residues likewise impeded anticoagulant action (176). However, an altogether new dimension was added to the problem when it was shown that only about one third of the molecules in highly purified commercial heparin preparations were able

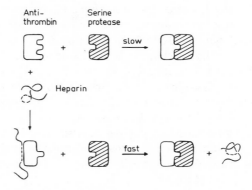

Figure 3 Schematic representation of the inactivation by antithrombin of serine proteases, for example, thrombin, participating in the coagulation mechanism. Heparin is believed to accelerate the inactivation by binding to antithrombin, thereby inducing a conformational change in the antithrombin molecule that facilitates its interaction with thrombin. (Binding of heparin to thrombin as well cannot be excluded.) The interaction requires a specific binding site (– – –) in the polysaccharide chain. For further details see the text.

to bind strongly to antithrombin (172, 177, 178). This high-affinity fraction, HA-heparin, had an anticoagulant activity of approximately 300 B.P. units/mg (178) which is about one order of magnitude higher than that of the remaining, low-affinity material, LA-heparin. The HA-heparin was found to bind to a single site in the antithrombin molecule; values for the association constant ranging between 2.3×10^6 M^{-1} (179) and about 10^8 M^{-1} (I. Björk and R. D. Rosenberg, personal communications) were obtained with molecules in free solution. LA-Heparin presumably binds to the same site, with an association constant two to three orders of magnitude lower than that recorded for the corresponding HA-heparin (I. Björk and R. D. Rosenberg, personal communications).

Oligosaccharides thought to include the antithrombin-binding site(s) of heparin were isolated after digestion of a HA-heparin-antithrombin complex with bacterial heparinase (180). No corresponding fragments could be derived from LA-heparin. The oligosaccharides had average molecular weights of about 4000, approximately corresponding to a dodecasaccharide segment of the polymer. A preliminary structural characterization failed to reveal any components other than those also present in LA-heparin. These results suggested that high-affinity binding of heparin to antithrombin requires a specific and as yet unknown sequence, or sequences, of up to six disaccharide units. The mechanisms determining the sequence of disaccharide units in heparin are only partly understood; as pointed out in the section on biosynthesis one may have to recognize, among other factors, an element of randomness. The possibility may therefore be considered that structural properties required for highly specific binding processes may be laid down in a polysaccharide molecule by partly random polymer-modification reactions.

Binding of glycosaminoglycans to platelet factor 4   The binding of platelet factor 4 (PF4) could with equal justification be discussed in the section on interaction of glycosaminoglycans with intracellular components, since PF4 normally occurs within the blood platelets, probably bound to a proteoglycan carrier (181, 182). However, PF4 has mainly been recognized for its ability to neutralize the anticoagulant effect of heparin in plasma (183) after being discharged from the platelets, and is therefore of clinical significance in relation to heparin therapy and thrombotic disease (184). The physiological role of PF4 remains an unsolved problem (182) possibly connected with the likewise unsettled questions regarding the function of endogenous heparin and its occurrence in blood (see above).

The composition of the proteoglycan-PF4 complex, as described (181), shows remarkable stoichiometry. At high ionic strength (I 0.75) the complex dissociates into the proteoglycan carrier (mol wt 59,000) and PF4 (mol wt 29,000). Characterization of the proteoglycan indicated four chains of chondroitin 4-sulfate (mol wt 12,000) in covalent linkage to a single polypeptide. The molecular weight of the carrier proteoglycan, fully saturated with PF4, was about 350,000, and it was therefore concluded that the complex occurs as a dimer, each monomer consisting of four molecules of PF4 and one molecule of proteoglycan (181). In the absence of proteoglycan, carrier PF4 is poorly soluble under physiological conditions but readily soluble at pH 3, when it dissociates into four subunits with a molecular weight of about 7000 (182).

PF4 can be displaced from its natural proteoglycan carrier by other glycosaminoglycans (181, 185); heparin has the highest, and hyaluronate the lowest affinity for the protein. Due to differences in methodology and in the sulfate contents of some of the polysaccharides tested, the relative affinities for PF4, within the series of glycosaminoglycans, cannot be directly compared with those pertaining to interactions with other ligands, for example, collagen or lipoproteins. However, the same general pattern appears to prevail; binding is promoted by high charge density as well as by the presence of L-iduronic acid residues. Displacement of PF4 from its proteoglycan carrier by heparan sulfate and/or dermatan sulfate may conceivably serve as a mechanism for the deposition of PF4 at specific sites on the vascular wall.

The amino acid composition of human PF4 indicated moderate amounts of basic amino acids; PF4 thus differs from a "cationic" protein such as protamine, which also neutralizes heparin (185). Modification of lysines by guanidination decreased heparin-neutralizing and heparin-binding activity, while modification of arginine residues had no effect (185), which suggests the involvement of lysine in a binding site for glycosaminoglycans. In support of this possibility PF4 shows a distinctive amino acid sequence at the COOH-terminal region (186) in which pairs of lysine residues are separated by pairs of leucines or isoleucines (-Lys-Lys-Ile-Ile-Lys-Lys-Leu-Leu); furthermore, a COOH-terminal peptide containing this sequence was able to reduce the heparin-induced inhibition of blood coagulation.

Binding of glycosaminoglycans to lipoproteins   The interaction between glycosaminoglycans and lipoproteins has been studied by equilibrium-binding of lipoproteins to polysaccharide-substituted agarose gel (187, 188), precipitation techniques (188,189), and fluorescence spectroscopy with pyrene-labeled lipoproteins (189) [for earlier work see (190)]. Low density lipoproteins (LDL) and very low density lipoproteins (VLDL) were found to bind glycosaminoglycans. Binding depended on electrostatic forces, increased with charge density of the polysaccharide, and could be abolished by N-acetylation of the lipoproteins. As with other interactions (see above) preferential binding of L-iduronic acid-containing species was noted (187). Divalent cations were not prerequisite to interaction but were found to stabilize the lipoprotein complexes of heparin; Mn^{2+} was more active in this respect than were either Mg^{2+} or Ca^{2+} and did in fact specifically induce the formation of insoluble complexes between heparin and HDL. The latter did not interact in the absence of metal ions (188). HDL subfractions showing a preponderance of arginine-rich apoprotein were particularly prone to complex formation (191). Following N-acetylation LDL remained amenable to complex formation with heparin in the presence of metal ions. On the basis of these and other observations the interactions between LDL and heparin were proposed to involve (a) binding between basic amino groups in the apoprotein and acidic groups in the polysaccharide to form soluble complexes; and (b) divalent cations bridging the acidic groups in heparin and the phospholipid portion of LDL, thereby producing the cross-links necessary for forming insoluble complexes (188).

No physiological role has as yet been attributed to glycosaminoglycans in lipoprotein metabolism. However, recent studies on the binding of lipoproteins to cultured cells point to interesting possibilities. LDL, the major cholesterol-carrying lipoprotein in human plasma delivers its cholesterol to cells in tissue culture by a receptor-mediated process (192). Competition experiments suggested that the same receptor may accommodate both LDL and high-density lipoproteins containing arginine-rich apoprotein (191). Binding of LDL to its cell-surface receptor could be reversed by adding heparin to the culture medium; dermatan sulfate (at higher concentration) produced a similar effect whereas chondroitin sulfate was inactive (193). These observations suggest that extracellular glycosaminoglycans may influence the binding of lipoproteins to cells, and, furthermore, raise the possibility that cell-surface glycosaminoglycans may in fact form part of the receptor structure.

A role for glycosaminoglycans, especially dermatan sulfate, in development of the atherosclerotic lesion has been postulated (194-196). During formation of the atherosclerotic plaque, lipids are deposited in the arterial wall and this process is believed to be promoted by interaction between lipoproteins and intimal glycosaminoglycans. Of the various polysaccharides known to be present in significant amounts in arterial tissue only dermatan sulfate binds to lipoproteins at physiological ionic strength (187). Furthermore, dermatan sulfate appears to be the principal glycosaminoglycan synthesized by arterial smooth muscle cells (197), which proliferate at the site of atherosclerotic lesions (195), and has indeed been found in high concentrations in human aortas with fatty streaks (198) [however, see

Radhakrishnamurthy et al (199) for comparison with monkey tissues]. In keeping with this hypothesis the alleged antiatherosclerotic effects of administered sulfated polysaccharides have been attributed to saturation of polysaccharide-binding sites on LDL molecules, which prevents their interaction with intimal glycosaminoglycans (200).

Binding of glycosaminoglycans to lipases   Intravenous injection of heparin results in an accelerated rate of removal of triglycerides from plasma, as originally described by Hahn (201). The effect has been ascribed to triglyceride-degrading enzymes, lipoprotein lipase, and hepatic lipase, which are released by heparin into the circulation from tissue sites that are presumably located at the luminal surface of capillary endothelial cells and in the liver, respectively (see 142, 202, 203). Both enzymes bind to heparin [however, see (204)] and have been purified by affinity chromatography on heparin-substituted gels (205–209); the hepatic lipase appears to bind with lower affinity than does lipoprotein lipase (210-212).

The interaction between lipoprotein lipase and heparin has been studied in some detail. Contrary to antithrombin, lipoprotein lipase binds with high affinity to all molecules in heparin preparations; the molecular species that have low affinity for antithrombin (LA-heparin), and little or no anticoagulant activity, could not be distinguished from HA-heparin (which has high affinity for antithrombin), by affinity chromatography on immobilized lipoprotein lipase. Moreover, LA-heparin efficiently released both lipoprotein lipase and hepatic lipase on intravenous injection into rats (212). The conclusion derived from these experiments, that lipoprotein lipase and antithrombin bind to different structures in the heparin molecule, was confirmed by experiments with chemically modified heparin. A partially N-desulfated and then N-acetylated heparin lacked anticoagulant activity but retained affinity for lipoprotein lipase [G. Bengtsson, T. Olivecrona, M. Höök, J. Riesenfeld, and U. Lindahl, in manuscript; see also (213)]. Furthermore, both heparan sulfate and dermatan sulfate, but not chondroitin sulfate, showed appreciable affinity for the enzyme; however, neither of these polysaccharides bound to lipoprotein lipase as strongly as did heparin nor did they release comparable amounts of enzyme in vivo.

The mechanism of lipase release in vivo is unclear. It has been suggested that a heparin-like polysaccharide which may be heparan sulfate, serving as the physiological carrier for lipoprotein lipase at the endothelial cell surface, may be displaced by heparin from a common binding site in the enzyme molecule (142). However, other possibilities appear equally plausible; heparin could thus induce a conformational change in the enzyme or interact with the cell-surface carrier structure (214); either event may lead to release of the enzyme.

Interaction of Glycosaminoglycans with Cell-Surface Components

Cultured mammalian cells contain a variety of glycosaminoglycans, including particularly heparan sulfate, associated with their surface structures (62, 139, 215-220). The physiological role of these cell-surface polysaccharides is not understood; diverse functions, such as regulation of cell growth (221), mediation of cell-to-cell communication (218), shielding of cell-surface receptors (65), and service as receptors or carriers for other macromolecules, for example, lipoproteins, lipoprotein lipase, or antithrombin (see above) have been proposed. The nature of the association between glycosaminoglycans and cells is largely unknown and probably variable since the polysaccharides may be either loosely or tightly bound to cell surfaces. Tight binding could possibly be accomplished by intercalating into the cell membrane core-proteins of proteoglycan molecules which would presumably be produced by the cell itself. However, it is clear that polysaccharides may also become associated with other cell-surface components; recent studies on the interaction of heparin with cells suggest that such binding may in fact show appreciable specificity. Heparin associates with a number of cell types, including blood platelets (222), arterial endothelium (223, 224) and liver cells (225). The occurrence of membrane-associated binding sites for heparin-like polysaccharides was suggested by the isolation of a platelet membrane fraction that has "antiheparin activity" (182), and was conclusively demonstrated in studies on the uptake of radiolabeled heparin by rat liver cells in primary culture (225). The uptake process was time- and temperature-dependent and showed saturation kinetics. It was also reversible since labeled heparin, once bound, could be displaced from the cells by the addition of excess amounts of unlabeled heparin. Similar addition of unlabeled heparin to cells containing endogenous, labeled heparan sulfate, formed during preincubation of the cells with inorganic [^{35}S]sulfate resulted in release of labeled polysaccharide into the culture medium (L. Kjellén, Å. Oldberg, and M. Höök, unpublished observation; P. Kraemer, personal communication). It thus appears that part of the heparan sulfate produced by rat liver cells occurs bound to specific sites on the cell surface. The specificity of binding was demonstrated by adding various glycosaminoglycans to cultures containing prelabeled cells; only heparin and certain heparan sulfates but none of the other polysaccharides tested, including hyaluronate, chondroitin sulfates, dermatan sulfate, and some preparations of low-sulfated heparan sulfate, were able to displace the endogenous polysaccharide.

In addition to observations of actual binding of glycosaminoglycans to cell-surface receptors a number of phenomena have been reported that indirectly implicate such binding in a functional context. Examples will be given in which a specific cellular function or behaviour has been attributed to interaction between glycosaminoglycans and structures on the cell surface.

INTERNALIZATION OF GLYCOSAMINOGLYCANS   Several reports have appeared describing the uptake and internalization of radioactively labeled glycosaminoglycans by cultured cells (226–230). These observations are of particular interest in relation to the metabolism of glycosaminoglycans in vivo, where analogous uptake may precede and initiate the degradation of extracellular polysaccharide (231). In one of the systems studied, proteoglycans of chondroitin sulfate, dermatan sulfate, and heparan sulfate were not only taken up by fibroblasts but also degraded (230). No uptake occurred with the corresponding single polysaccharide chains. The transfer of proteoglycans from culture medium to cell interior showed the kinetics expected from a receptor-mediated process; furthermore, competition experiments indicated that different proteoglycans bind to separate sites. Further studies on the mechanisms of binding and uptake must obviously include identification of the recognition markers on the proteoglycan molecules.

REGULATION OF GLYCOSAMINOGLYCAN BIOSYNTHESIS   During morphogenesis hyaluronate accumulates in the tissues at a stage characterized by extensive cell migration (232-234). Subsequent differentiation is preceded by enzymatic removal of hyaluronate and this removal appears to be essential to normal development of the tissues. Specific inhibition of differentiation expressed as lack of morphological signs of chondrogenesis, was indeed noted when hyaluronate was added to cultures of embryonic chick somite cells; the effect could not be reproduced with other biological polyanions, such as chondroitin sulfates, heparin, or nucleic acids (235, 236). Hyaluronate also inhibited the incorporation of [^{35}S]sulfate into polysaccharide, and this effect was studied further on fully differentiated chondrocytes grown in culture (237-239). Hyaluronate was found to be a very potent inhibitor of proteoglycan synthesis, with as little as $0.005\,\mu g/ml$ significantly decreasing [^{35}S]sulfate-incorporation into glycosaminoglycans. Again, other polyanions were incapable of eliciting this effect. Partially depolymerized hyaluronate, but not small oligomers such as tetra- or hexasaccharides, retained inhibitory activity (239). The effect was restricted to chondrocytes; no effect was noted on the production of sulfated polysaccharides by fibroblasts, even at relatively high concentrations of hyaluronate ($10\,\mu g/ml$). During incubation of chondrocytes with [^{14}C]hyaluronate label was taken up by the cells, and reached a plateau within about 1 hour; moreover, addition of unlabeled hyaluronate reduced this uptake. Most of the label appeared to be associated with the cell surface as it could be largely removed by mild trypsin treatment of the cells. Such treatment abolished the inhibitory effect of hyaluronate on proteoglycan synthesis (239). Taken together, these findings suggest that hyaluronate binds to specific sites on the surface of chondrocytes, and thereby inhibits the synthesis of proteoglycan by the cells. Hyaluronate did not affect chondroitin [^{35}S]sulfate synthesis when supplied to cells along with benzyl-β-D-xyloside, an exogenous initiator of chain polymerization (240). It was therefore concluded that hyaluronate acts by interfering with glycosaminoglycan chain initiation, and hence proteoglycan biosynthesis, by specifically inhibiting either the formation of normal core-protein or the xylosyl-transferase, which initiates synthesis of chondroitin sulfate chains on the protein. Irrespective of which of these alternatives prevails, the interaction between extracellular hyaluronate and the chondrocyte appears to provide an important mechanism for controling essential properties of cartilage matrices.

In contrast to hyaluronate, sulfated glycosaminoglycans were found to exert a stimulatory effect on glycosaminoglycan synthesis in both chondrocytes (241, 242) and corneal epithelial cells (243). However, comparatively high polysaccharide concentrations were required to obtain significant effects and no evidence for a receptor-mediated mechanism was obtained.

CELL AGGREGATION   Many types of mammalian cells have the ability to adhere to each other and to various natural or artificial substrates. The adhesive properties are considered essential to normal morphogenesis and differentiation and manifest themselves in markedly selective cell aggregation phenomena in vitro (244, 245). Macromolecular factors capable of promoting selective cell-adhesion have been isolated from various tissues but have not been characterized in detail; there is at present no information pointing to a general role for glycosaminoglycans in this context. Nevertheless, hyaluronate has been implicated in adhesion phenomena that may require binding of the polysaccharide to cell-surface structures. Hyaluronate was thus tentatively identified in a material deposited by cells during growth on plastic dishes; this material is believed to be somehow involved in the substrate adhesion process (246). Further, it was noted that a variety of cells in culture produced a factor, identified as hyaluronate, that could aggregate certain lymphoma cells (247, 248). Hyaluronate lost its aggregating properties on partial depolymerization and could not be substituted for by any other glycosaminoglycan; on the other hand, oligosaccharides derived from either hyaluronate or chondroitin sulfate were able to abolish aggregation induced by polymeric hyaluronate (248). A possible interpretation of these findings implies binding of both hyaluronate and chondroitin sulfate to the same cell-surface component; due to the difference in chain length only hyaluronate would be able to span the intercellular space and induce aggregation. Apparently only a limited number of cell lines possess the putative receptor, as no other cells tested, including lymphocytes and human lymphosarcoma cells, were aggregated by hyaluronate.

Interaction of Glycosaminoglycans with Intracellular Components

Glycosaminoglycans interact with many intracellular macromolecules including the enzymes involved in their biosynthesis and degradation. These interactions have mostly been demonstrated in artificial model systems and it is therefore difficult to evaluate their importance in vivo. One reason for this uncertainty is our scanty knowledge regarding the availability and distribution of glycosaminoglycans inside the animal cell. Apparently, not all of the newly-synthesized polysaccharide is secreted from the cell; a portion is diverted to an intracellular storage pool and ultimately degraded, as shown originally for fibroblasts (249). Intracellular polysaccharide could also derive from uptake of extracellular material. Some of the effects and functions attributed to intracellular glycosaminoglycans are discussed below.

GLYCOSAMINOGLYCANS AND PROTEIN BIOSYNTHESIS   Model experiments have suggested that glycosaminoglycans may affect protein synthesis on a translational as well as on a pretranslational level. Effects on nuclear function are indicated by observations that glycosaminoglycans, and particularly heparin, may induce loosening of chromatin structure (250-253), increase template availability to exogenous DNA-polymerases in isolated chromatin (250, 254), and activate endogenous DNA-polymerase in isolated nuclei (251). Some of these phenomena may conceivably be explained in terms of polysaccharides displacing DNA from complexes with histones. Effects on translation have been ascribed to binding between polysaccharide and initiation (255) or elongation (256) factors, which result in extensive inhibition of protein synthesis in cell-free systems (257, 258). Heparin, and to some extent dermatan sulfate, were the only glycosaminoglycans with significant inhibitory effect (259, 260); a partially desulfated heparin retained activity in spite of having lost its anticoagulant properties (260).

In the attempt to evaluate the biological significance of these model systems it is notable that nuclei from different kinds of cells appear to contain glycosaminoglycans in significant amounts (261-264). The presence of heparan sulfate is of particular interest since this polysaccharide could possibly assume some of the functions observed for heparin in model experiments. Circumstantial evidence in favor of a role for heparan sulfate in nuclear function has emerged from studies on the embryonic development of the sea urchin. When sea urchin embryos are grown in sulfate-deficient sea water development is not carried beyond the blastula stage. Sulfate ions are taken up by the embryo in late blastula stage (261, 265) and become incorporated into a "heparin-like polysaccharide"; this is subsequently transferred to the nucleus and apparently preferentially located in the template-active portion of the chromatin (261).

GLYCOSAMINOGLYCANS AND LYSOSOMAL ENZYMES   Addition of glycosaminoglycans such as chondroitin sulfate, dermatan sulfate, and heparin to various acid hydrolases results in either reduction or apparent increase in enzyme activity (266-269). The latter effect was considered to be due to a protective action of the polysaccharide against enzyme degradation by lysosomal proteinases (268), and this hypothesis has been elaborated into a broader concept regarding the function of lysosomal glycosaminoglycans. Studies on lysosomal enzymes in polymorphonuclear leucocytes showed that most of these enzymes interact with glycosaminoglycans in pH-dependent, reversible, electrostatic binding (266, 270). It was suggested that the enzymes in acidic primary lysosomes occur in complex with glycosaminoglycans, and are hence in a protected and relatively inactive form. Fusion of a primary lysosome with a phagosome will result in dilution of the intralysosomal acidic fluid and therefore in an increase in pH. This in turn leads to a dissociation of the complexes and a release of intact, active enzyme molecules.

GLYCOSAMINOGLYCANS IN STORAGE OR SECRETORY GRANULES   Sulfated glycosaminoglycans have been identified as constituents of various storage or secretory granules, such as the basophilic granules in mast cells (271, 272), the cortical granules of eggs from different species (273), various catecholamine-containing granules (274), the prolactin secretory granules in the pituitary gland (275), and the chromaffin granules of the adrenal medulla (276, 277). The reason for the presence of polysaccharides in many of these granules is unknown but will perhaps become clearer in view of recent considerations regarding the role of heparin in mast-cell granules (278, 279). In these granules heparin occurs as an insoluble complex with a basic, small-molecular protein (272) and this complex acts as a store for histamine. Titration data suggested that the histamine molecules are primarily bound to carboxyl groups in the protein; the protein in turn is electrostatically bound to sulfate groups in the polysaccharide. During degranulation and subsequent exposure of the heparin-protein-histamine complex to the cations (mainly Na^+) of the extracellular fluid, histamine is liberated by a simple cation-exchange mechanism. On the basis of these results a general role for glycosaminoglycan-peptide complexes in storage and release of biogenic amines was postulated (279, 280).

Literature Cited

1. Cássaro, C. M. F., Dietrich, C. P. 1977. *J. Biol. Chem.* 252:2254–61
2. Mathews, M. B. 1975. *Connective Tissue. Macromolecular Structure and Evolution.* Berlin/Heidelberg/New York: Springer. 318 pp.
3. Comper, W. D., Laurent, T. C. 1978. *Physiol. Rev.* In press
4. Rees, D. A. 1975. *MTP Int. Rev. Sci. Biochem. Ser. 1,* 5:1–41
5. Muir, H., Hardingham, T. E. 1975. *MTP Int. Rev. Sci. Biochem. Ser. 1* 5:153–222
6. Brimacombe, J. S., Webber, J. M. 1964. *Mucopolysaccharides. Chemical Structure, Distribution and Isolation.* Amsterdam: Elsevier. 181 pp.
7. Jeanloz, R. W. 1970. In *The Carbohydrates,* ed. W. Pigman, D. Horton, A. Herp, 2B:589–625. New York: Academic
8. Lindahl, U. 1976. *MTP Int. Rev. Sci. Org. Chem. Ser. 2,* 7:283–312
9. Rodén, L. 1970. In *Metabolic Conjugation and Metabolic Hydrolysis,* ed. W. H. Fishman, 2:345–442. New York: Academic. 692 pp.
10. Rodén, L., Schwartz, N. B. 1975. *MTP Int. Rev. Sci. Biochem. Ser. 1,* 5:96–152
11. Silbert, J. E., Reppucci, A. C. 1976. *J. Biol. Chem.* 251:3942–47
12. Blackwell, J., Schodt, K. P., Gelman, R. A. 1977. *Fed. Proc.* 36:98–101
13. Hopwood, J. J., Dorfman, A. 1977. *Biochem. Biophys. Res. Commun.* 75: 472–79
14. Turco, S. J., Heath, E. C. 1977. *J. Biol. Chem.* 252:2918–28
15. DeLuca, S., Richmond, M. E., Silbert, J. E. 1973. *Biochemistry* 12:3911–15
16. Suzuki, S., Saito, H., Yamagata, T., Anno, K., Seno, N., Kawai, Y., Furuhashi, T. 1968. *J. Biol. Chem.* 243:1543–50
17. Danishefsky, I., Steiner, H., Bella, A., Friedlander, A. 1969. *J. Biol. Chem.* 244:1741–45
18. Choi, H. U., Meyer, K. 1975. *Biochem. J.* 151:543–53
19. Meyer, K. 1969. *Am. J. Med.* 47: 664–72
20. Fransson, L.-Å. 1970. In *Chemistry and Molecular Biology of the Intercellular Matrix,* ed. E. A. Balazs, pp. 823–42. London: Academic. 1874 pp.
21. Fransson, L.-Å., Havsmark, B. 1970. *J. Biol. Chem.* 245:4770–83
22. Fransson, L.-Å., Malmström, A. 1971. *Eur. J. Biochem.* 18:422–30
23. Malmström, A., Fransson, L.-Å. 1971. *Eur. J. Biochem.* 18:431–35

24. Anno, K., Seno, N., Mathews, M. B., Yamagata, T., Suzuki, S. 1971. *Biochim. Biophys. Acta* 237:173–77
25. Seno, N., Akiyama, F., Anno, K. 1972. *Biochim. Biophys. Acta* 264:229–33
26. Habuchi, H., Yamagata, T., Iwata, H., Suzuki, S. 1973. *J. Biol. Chem.* 248: 6019–28
27. Fransson, L.-Å., Cöster, L., Malmström, A., Sjöberg, I. 1974. *Biochem. J.* 143:369–78
28. Fransson, L.-Å., Cöster, L., Havsmark, B., Malmström, A., Sjöberg, I. 1974. *Biochem. J.* 143:379–89
29. Cöster, L., Malmström, A., Sjöberg, I., Fransson, L.-Å. 1975. *Biochem. J.* 145:379–89
30. Cifonelli, J. A. 1968. *Carbohyd. Res.* 8:233–42
31. Helting, T., Lindahl, U. 1971. *J. Biol. Chem.* 246:5442–47
32. Cifonelli, J. A., King, J. 1972. *Carbohyd. Res.* 21:173–86
33. Cifonelli, J. A., King, J. 1973. *Biochim. Biophys. Acta* 320:331–40
34. Taylor, R. L., Shively, J. E., Conrad, H. E., Cifonelli, J. A. 1973. *Biochemistry* 12:3633–37
35. Höök, M., Lindahl, U., Iverius, P.-H. 1974. *Biochem. J.* 137:33–43
36. Hovingh, P., Linker, A. 1974. *Carbohyd. Res.* 37:181–92
37. Silva, M. E., Dietrich, C. P., Nader, H. B. 1976. *Biochim. Biophys. Acta* 437:129–41
38. Shively, J. E., Conrad, H. E. 1976. *Biochemistry* 15:3943–50
39. Linker, A., Hovingh, P. 1977. *Fed. Proc.* 36:43–46
40. Malmström, A., Fransson, L.-Å., Höök, M., Lindahl, U. 1975. *J. Biol. Chem.* 250:3419–25
41. Höök, M., Lindahl, U., Hallén, A., Bäckström, G. 1975. *J. Biol. Chem.* 250:6065–71
42. Lindahl, U., Höök, M., Bäckström, G., Jacobsson, I., Riesenfeld, J., Malmström, A., Rodén, L., Feingold, D. S. 1977. *Fed. Proc.* 36:19–23
43. Silbert, J. E., Kleinman, H. K., Silbert, C. K. 1975. *Adv. Exp. Med. Biol.* 52:51–60
44. Lindahl, U., Bäckström, G., Jansson, L., Hallén, A. 1973. *J. Biol. Chem.* 248:7234–41
45. Höök, M., Lindahl, U., Bäckström, G., Malmström, A., Fransson, L.-Å. 1974. *J. Biol. Chem.* 249:3908–15
46. Jansson, L., Höök, M., Wasteson, Å., Lindahl, U. 1975. *Biochem. J.* 149: 49–55

47. Lindahl, U., Jacobsson, I., Höök, M., Bäckström, G., Feingold, D. S. 1976. *Biochem. Biophys. Res. Commun.* 70:492–99
48. Perlin, A. S., Casu, B., Sanderson, G. R., Johnson, L. F. 1970. *Can. J. Chem.* 48:2260–68
49. Perlin, A. S. 1975. In *Proc. Int. Symp. Macromol.*, ed. E. B. Mano, pp. 337–48 Amsterdam: Elsevier
50. Atkins, E. D. T., Nieduszynski, I. A. 1977. *Fed. Proc.* 36:78–83
51. Arnott, S., Winter, W. T. 1977. *Fed. Proc.* 36:73–78
52. Atkins, E. D. T., Laurent, T. C. 1973. *Biochem. J.* 133:605–6
53. Nieduszynski, I. A., Atkins, E. D. T. 1973. *Biochem. J.* 135:729–33
54. Atkins, E. D. T., Isaac, D. H. 1973. *J. Mol. Biol.* 80:773–79
55. Atkins, E. D. T., Isaac, D. H., Nieduszynski, I. A., Phelps, C. F., Sheehan, J. K. 1974. *Polymer* 15:263–71
56. Arnott, S., Guss, J. M., Hukins, D. W. L., Mathews, M. B. 1973. *Biochem. Biophys. Res. Commun.* 54:1377–83
57. Lindahl, U., Rodén, L. 1972. In *Glycoproteins. Their Composition, Structure and Function*, ed. A. Gottschalk, pp. 491–517. Amsterdam: Elsevier. 1378 pp. 2nd ed.
58. Scher, I., Hamerman, D. 1972. *Biochem. J.* 126:1073–80
59. Axelsson, I., Heinegård, D. 1975. *Biochem. J.* 145:491–500
60. Ehrlich, K. C., Radhakrishnamurthy, B., Berenson, G. S. 1975. *Arch. Biochem. Biophys.* 171:361–69
61. Eisenstein, R., Larsson, S.-E., Kuettner, K. E., Sorgente, N., Hascall, V. C. 1975. *Atherosclerosis* 22:1–17
62. Oldberg, Å., Höök, M., Öbrink, B., Pertoft, H., Rubin, K. 1977. *Biochem. J.* 164:75–81
63. Radhakrishnamurthy, B., Ruiz, H. A., Berenson, G. S. 1977. *J. Biol. Chem.* 252:4831–41
64. Jansson, L., Lindahl, U. 1970. *Biochem. J.* 117:699–702
65. Kraemer, P. M., Smith, D. A. 1974. *Biochem. Biophys. Res. Commun.* 56:423–30
66. Jansson, L., Ögren, S., Lindahl, U. 1975. *Biochem. J.* 145:53–62
67. Horner, A. A. 1971. *J. Biol. Chem.* 246:231–39
68. Yurt, R. W., Leid, R. W., Austen, K. F., Silbert, J. E. 1977. *J. Biol. Chem.* 252:518–21
69. Ögren, S., Lindahl, U. 1971. *Biochem. J.* 125:1119–29

70. Horner, A. A. 1977. *Fed. Proc.* 36: 35–39
71. Robinson, H. C., Horner, A. A., Höök, M., Ögren, S., Lindahl, U. 1977. See Ref. 42, note added in proof
72. Ögren, S., Lindahl, U. 1975. *J. Biol. Chem.* 250:2690–97
73. Ögren, S., Lindahl, U. 1976. *Biochem. J.* 154:605–11
74. Horner, A. A. 1972. *Proc. Natl. Acad. Sci. USA* 69:3469–73
75. Horner, A. A. 1977. *Upsala J. Med. Sci.* 82:114
76. Gelman, R. A., Rippon, W. B., Blackwell, J. 1973. *Biopolymers* 12:541–58
77. Gelman, R. A., Glaser, D. N., Blackwell, J. 1973. *Biopolymers* 12:1223–32
78. Gelman, R. A., Blackwell, J. 1973. *Biopolymers* 12:1959–74
79. Gelman, R. A., Blackwell, J. 1973. *Biochim. Biophys. Acta* 297:452–55
80. Gelman, R. A., Blackwell, J. 1973. *Arch. Biochem. Biophys.* 159:427–33
81. Gelman, R. A., Blackwell, J., Mathews, M. B. 1974. *Biochem. J.* 141:445–54
82. Gelman, R. A., Blackwell, J. 1974. *Biopolymers* 13:139–56
83. Stone, A. L. 1977. *Fed. Proc.* 36:101–6
84. Villanueva, G. B., Danishefsky, I. 1977. *Biochem. Biophys. Res. Commun.* 74: 803–9
85. Scott, J. E. 1968. In *The Chemical Physiology of Mucopolysaccharides*, ed. G. Quintarelli, pp. 171–87. Boston: Little, Brown. 240 pp.
86. Atkins, E. D. T., Sheehan, J. K. 1972. *Nature New Biol.* 235:253–54
87. Atkins, E. D. T., Sheehan, J. K. 1973. *Science* 179:562–64
88. Dea, I. C. M., Moorhouse, R., Rees, D. A., Arnott, S., Guss, J. M., Balazs, E. A. 1973. *Science* 179:560–62
89. Guss, J. M., Hukins, D. W. L., Smith, P. J. C., Winter, W. T., Arnott, S., Moorhouse, R., Rees, D. A. 1975. *J. Mol. Biol.* 95:359–84
90. Fransson, L.-Å. 1976. *Biochim. Biophys. Acta* 437:106–15
91. Fransson, L.-Å. 1977. *Upsala J. Med. Sci.* 82:130
92. Fransson, L.-Å. 1974. *Carbohyd. Res.* 36:339–48
93. Mathews, M. B. 1965. *Biochem. J.* 96:710–16
94. Öbrink, B., Wasteson, Å. 1971. *Biochem. J.* 121:227–33
95. Öbrink, B. 1975. In *Structure of Fibrous Biopolymers*, ed. E. D. T. Atkins, A. Keller, pp. 81–92. London: Butterworths. 437 pp.
96. Greenwald, R. A., Schwartz, C. E.,

Cantor, J. O. 1975. *Biochem. J.* 145:601–5

97. Toole, B. P., Lowther, D. A. 1968. *Biochem. J.* 109:857–66
98. Toole, B. P., Lowther, D. A. 1968. *Arch. Biochem. Biophys.* 128:567–78
99. Toole, B. P. 1976. *J. Biol. Chem.* 251:895–97
100. Oegema, T. R., Laidlaw, J., Hascall, V. C., Dziewiatkowski, D. D. 1975. *Arch. Biochem. Biophys.* 170:698–709
101. Mathews, M. B., Decker, L. 1968. *Biochem. J.* 109:517–526
102. Öbrink, B. 1973. *Eur. J. Biochem.* 33:387–400
103. Öbrink, B., Sundelöf, L.-O. 1973. *Eur. J. Biochem.* 37:226–32
104. Gelman, R. A., Blackwell, J. 1974. *Biochim. Biophys. Acta* 342:254–61
105. Öbrink, B., Laurent, T. C., Carlsson, B. 1975. *FEBS Lett.* 56:166–69
106. Conochie, L. B., Scott, J. E., Faulk, W. P. 1975. *J. Immunol. Meth.* 7:393–98
107. Kühn, K., Bräumer, K., Zimmerman, B., Pikkarainen, J. 1970. In *Chemistry and Molecular Biology of the Intercellular Matrix*, ed. E. A. Balazs, pp. 251–73. London: Academic. 1874 pp.
108. Lowther, D. A., Toole, B. P., Herrington, A. C. 1970. In *Chemistry and Molecular Biology of the Intercellular Matrix*, ed. E. A. Balazs, pp. 1135–53. London: Academic. 1874 pp.
109. Lee-Own, V., Anderson, J. C. 1975. *Biochem. J.* 149:57–63
110. Lee-Own, V., Anderson, J. C. 1976. *Biochem. J.* 153:259–64
111. Wood, G. C., Keech, M. K. 1960. *Biochem. J.* 75:588–98
112. Gross, J., Kirk, D. 1958. *J. Biol. Chem.* 233:355–60
113. Podrazký, V., Adam, M. 1975. *Experientia* 31:523–24
114. Adam, M., Podrazký, V. 1976. *Experientia* 32:430–32
115. Podrazký, V., Štokrová, Š., Frič, I. 1975. *Connective Tissue Res.* 4:51–54
116. Hascall, V. C. 1978. *J. Supramol. Biol.* In press
117. Hascall, V. C., Sajdera, S. W. 1969. *J. Biol. Chem.* 244:2384–96
118. Hardingham, T. E., Muir, H. 1972. *Biochim. Biophys. Acta* 279:401–5
119. Hardingham, T. E., Muir, H. 1973. *Biochem. J.* 135:905–8
120. Hardingham, T. E., Muir, H. 1974. *Biochem. J.* 139:565–81
121. Hascall, V. C., Heinegård, D. 1974. *J. Biol. Chem.* 249:4232–41
122. Hascall, V. C., Heinegård, D. 1974. *J. Biol. Chem.* 249:4242–49

123. Hardingham, T. E., Muir, H. 1973. *Biochem. Soc. Trans.* 1:282–84
124. Swann, D. A., Powell, S., Broadhurst, J., Sordillo, E., Sotman, S. 1976. *Biochem. J.* 157:503–6
125. Kaiser, H., Shulman, H. J., Sandson, J. I. 1972. *Biochem. J.* 126:163–69
126. Gregory, J. D. 1973. *Biochem. J.* 133:383–86
127. Baker, J., Caterson, B. 1977. *Biochem. Biophys. Res. Commun.* 77:1–10
128. Heinegård, D., Hascall, V. C. 1974. *J. Biol. Chem.* 249:4250–56
129. Oegema, T. R., Hascall, V. C., Dziewiatkowski, D. D. 1975. *J. Biol. Chem.* 250:6151–59
130. Rosenberg, L., Hellmann, W., Kleinschmidt, A. K. 1975. *J. Biol. Chem.* 250:1877–83
131. Sajdera, S. W., Hascall, V. C. 1969. *J. Biol. Chem.* 244:77–87
132. Hardingham, T. E., Ewins, R. J. F., Muir, H. 1976. *Biochem. J.* 157:127–43
133. Calatroni, A., Donnelly, P. V., Di Ferrante, N. 1969. *J. Clin. Invest.* 48:332–43
134. Kresse, H., Heidel, H., Buddecke, E. 1971. *Eur. J. Biochem.* 22:557–62
135. Hallén, A. 1974. *Acta Universitatis Upsaliensis.* Abstracts of Uppsala Dissertations from the Faculty of Medicine. No. 204
136. Murata, K., Nakazawa, K., Hamai, A. 1975. *Atherosclerosis* 21:93–103
137. Wight, T. N., Ross, R. 1975. *J. Cell Biol.* 67:660–74
138. Buonassisi, V. 1973. *Exp. Cell Res.* 76:363–68
139. Buonassisi, V., Root, M. 1975. *Biochim. Biophys. Acta* 385:1–10
140. Teien, A. N., Abildgaard, U., Höök, M. 1976. *Thromb. Res.* 8:859–67
141. Rosenberg, R. D. 1977. *Fed. Proc.* 36:10–18
142. Olivecrona, T., Bengtsson, G., Marklund, S.-E., Lindahl, U., Höök, M. 1977. *Fed. Proc.* 36:60–65
143. Zilversmit, D. B. 1973. *Circ. Res.* 33:633–37
144. Horner, A. A. 1975. *Adv. Exp. Med. Biol.* 52:85–93
145. Wessler, S. 1975. *Adv. Exp. Med. Biol.* 52:309–22
146. Andersson, L.-O., Engman, L., Henningsson, E. 1977. *J. Immunol. Meth.* 14:271–81
147. Marciniak, E. 1974. *J. Lab. Clin. Med.* 84:344–56
148. Fujikawa, K., Thompson, A. R., Legaz, M. E., Meyer, R. G., Davie, E. W. 1973. *Biochemistry* 12:4938–45

149. Gentry, P. W., Alexander, B. 1973. *Biochem. Biophys. Res. Commun.* 50:500–9
150. Andersson, L.-O., Borg, H., Miller-Andersson, M. 1975. *Thromb. Res.* 7:451–59
151. Li, E. H. H., Orton, C., Feinman, R. D. 1974. *Biochemistry* 13:5012–17
152. Machovich, R., Blaskó, G., Pálos, L. A. 1975. *Biochem. Biophys. Acta* 379:193–200
153. Feinman, R. D., Li, E. H. H. 1977. *Fed. Proc.* 36:51–54
154. Nordenman, B., Björk, I. 1977. *Thromb. Res.* 11:799–808
155. Abildgaard, U. 1968. *Scand. J. Clin. Lab. Invest.* 21:89–91
156. Kurachi, K., Schmer, G., Hermodson, M. A., Teller, D. C., Davie, E. W. 1976. *Biochemistry* 15:368–73
157. Nordenman, B., Nyström, C., Björk, I. 1977. *Eur. J. Biochem.* 78:195–203
158. Rosenberg, R. D., Damus, P. S. 1973. *J. Biol. Chem.* 248:6490–6505
159. Rosenberg, J. S., McKenna, P. W., Rosenberg, R. D. 1975. *J. Biol. Chem.* 250:8883–88
160. Kurachi, K., Fujikawa, K., Schmer, G., Davie, E. W. 1976. *Biochemistry* 15:373–77
161. Yin, E. T., Wessler, S., Stoll, P. J. 1971. *J. Biol. Chem.* 246:3712–19
162. Damus, P. S., Hicks, M. S., Rosenberg, R. D. 1973. *Nature* 246:355–57
163. Stead, N., Kaplan, A. P., Rosenberg, R. D. 1976. *J. Biol. Chem.* 251:6481–88
164. Highsmith, R. F., Rosenberg, R. D. 1974. *J. Biol. Chem.* 249:4335–38
165. Lahiri, B., Bagdasarian, A., Mitchell, B., Talamo, R. C., Colman, R. W., Rosenberg, R. D. 1976. *Arch. Biochem. Biophys.* 175:737–47
166. Li, E. H. H., Fenton, J. W., Feinman, R. D. 1976. *Arch. Biochem. Biophys.* 175:153–59
167. Björk, I., Nordenman, B. 1976. *Eur. J. Biochem.* 68:507–11
168. Carlström, A.-S., Lledén, K., Björk, I. 1977. *Thromb. Res.* 11:785–97
169. Brozovic, M., Bangham, D. R. 1975. *Adv. Exp. Med. Biol.* 52:163–79
170. Cifonelli, J. A. 1974. *Carbohyd. Res.* 37:145–54
171. Riesenfeld, J., Höök, M., Björk, I., Lindahl, U., Ajaxon, B. 1977. *Fed. Proc.* 36:39–42
172. Andersson, L.-O., Barrowcliffe, T. W., Holmer, E., Johnson, E. A., Sims, G. E. C. 1976. *Thromb. Res.* 9:575–83
173. Laurent, T. C. 1961. *Arch. Biochem. Biophys.* 92:224–31
174. Stivala, S. S., Liberti, P. A. 1967. *Arch. Biochem. Biophys.* 122:40–54
175. McDuffie, N. M., Dietrich, C. P., Nader, H. B. 1975. *Biopolymers* 14:1473–86
176. Danishefsky, I. 1977. *Fed. Proc.* 36:33–35
177. Lam, L. H., Silbert, J. E., Rosenberg, R. D. 1976. *Biochem. Biophys. Res. Commun.* 69:570–77
178. Höök, M., Björk, I., Hopwood, J., Lindahl, U. 1976. *FEBS Lett.* 66:90–93
179. Einarsson, R., Andersson, L.-O. 1977. *Biochem. Biophys. Acta* 490:104–11
180. Hopwood, J., Höök, M., Linker, A., Lindahl, U. 1976. *FEBS Lett.* 69:51–54
181. Barber, A. J., Käser-Glanzmann, R., Jakábová, M., Lüscher, E. F. 1972. *Biochim. Biophys. Acta* 286:312–29
182. Moore, S., Pepper, D. S., Cash, J. D. 1975. *Biochem. Biophys. Acta* 379:370–84
183. Niewiarowski, S., Poplawski, A., Lipinski, B., Farbiszewski, R. 1968. *Exp. Biol. Med.* 3:121–28
184. O'Brien, J. R. 1975. *Curr. Therapeut. Res.* 18:79–90
185. Handin, R. I., Cohen, H. J. 1976. *J. Biol. Chem.* 251:4273–82
186. Deuel, T. F., Keim, P. S., Farmer, M., Heinrikson, R. L. 1977. *Proc. Natl. Acad. Sci. USA* 74:2256–58
187. Iverius, P.-H. 1972. *J. Biol. Chem.* 247:2607–13
188. Srinivasan, S. R., Radhakrishnamurthy, B., Berenson, G. S. 1975. *Arch. Biochem. Biophys.* 170:334–40
189. Nakashima, Y., Di Ferrante, N., Jackson, R. L., Pownall, H. J. 1975. *J. Biol. Chem.* 250:5386–92
190. Bernfeld, P. 1966. In *The Amino Sugars,* ed. E. A. Balazs, R. W. Jeanloz, 2B:251–66. New York: Academic. 516 pp.
191. Mahley, R. W., Innerarity, T. L. 1977. *J. Biol. Chem.* 252:3980–86
192. Brown, M. S., Faust, J. R., Goldstein, J. L. 1975. *J. Clin. Invest.* 55:783–93
193. Goldstein, J. L., Basu, S. K., Brunschede, G. Y., Brown, M. S. 1976. *Cell* 7:85–95
194. Iverius, P.-H. 1973. *Ciba Found. Symp.* 185:185–96
195. Ross, R., Harker, L. 1976. *Science* 193:1094–1100
196. Srinivasan, S. R., Dolan, P., Radhakrishnamurthy, B., Berenson, G. S. 1972. *Atherosclerosis* 16:95–104
197. Wight, T. N., Ross, R. 1975. *J. Cell Biol.* 67:675–86
198. Kumar, V., Berenson, G. S., Ruiz, H.,

Dalferes, E. R., Strong, J. P. 1967. *J. Atheroscler. Res.* 7:583–90

199. Radhakrishnamurthy, B., Eggen, D. A., Kokatnur, M., Jirge, S., Strong, J. P., Berenson, G. S. 1975. *Lab. Invest.* 33:136–40

200. Day, C. E., Powell, J. R., Levy, R. S. 1975. *Artery* 1:126–37

201. Hahn, P. F. 1943. *Science* 98:19–20

202. Robinson, D. S. 1970. *Compr. Biochem.* 18:51–116

203. Cryer, A., Davies, P., Robinson, D. S. 1975. In *Blood and Arterial Wall in Atherogenesis and Arterial Thrombosis,* ed. J. G. A. J. Hautvast, R. J. J. Hormus, F. van den Haar, pp. 102–10. Leiden: Brill

204. Fielding, P. E., Shore, V. G., Fielding, C. J. 1974. *Biochemistry* 13:4318–23

205. Egelrud, T., Olivecrona, T. 1972. *J. Biol. Chem.* 247:6212–17

206. Greten, H., Walter, B. 1973. *FEBS Lett.* 35:36–40

207. Bensadoun, A., Ehnholm, C., Steinberg, D., Brown, W. V. 1974. *J. Biol. Chem.* 249:2220–27

208. Dolphin, P. J., Rubinstein, D. 1974. *Biochem. Biophys. Res. Commun.* 57:808–14

209. Iverius, P.-H., Östlund-Lindqvist, A. M. 1976. *J. Biol. Chem.* 251:7791–95

210. Ehnholm, C., Shaw, W., Greten, H., Brown, W. V. 1975. *J. Biol. Chem.* 250:6756–61

211. Hernell, O., Egelrud, T., Olivecrona, T. 1975. *Biochem. Biophys. Acta* 381:233–41

212. Bengtsson, G., Olivecrona, T., Höök, M., Lindahl, U. 1977. *FEBS Lett.* 79:59–63

213. Danishefsky, I. 1974. *Adv. Exp. Med. Biol.* 52:105–18

214. Waite, M., Sisson, P. 1973. *J. Biol. Chem.* 248:7201–6

215. Kraemer, P. M. 1971. *Biochemistry* 10:1437–45

216. Malmström, A., Carlstedt, I., Åberg, L., Fransson, L.-Å. 1975. *Biochem. J.* 151:477–89

217. Kojima, K., Yamagata, T. 1971. *Exp. Cell Res.* 57:142–46

218. Roblin, R., Albert, S. O., Gelb, N. A., Black, P. H. 1975. *Biochemistry* 14:347–57

219. Kleinman, H. K., Silbert, J. E., Silbert, C. K. 1975. *Connect. Tissue Res.* 4:17–23

220. Akasaki, M., Kawasaki, T., Yamashina, I. 1975. *FEBS Lett.* 59:100–4

221. Ohnishi, T., Oshima, E., Ohtsuka, M. 1975. *Exp. Cell Res.* 93:136–42

222. Shanberge, J. N., Kambayashi, J., Nakagawa, M. 1976. *Thromb. Res.* 9:595–609

223. Hiebert, L. M., Jaques, L. B. 1976. *Thromb. Res.* 8:195–204

224. Hiebert, L. M., Jaques, L. B. 1976. *Artery* 2:26–37

225. Kjellén, L., Oldberg, Å., Rubin, K., Höök, M. 1977. *Biochem. Biophys. Res. Commun.* 74:126–33

226. de Oca, H. M., Dietrich, C. P. 1969. *Proc. Soc. Exp. Biol. Med.* 131:662–66

227. Saito, H., Uzman, B. G. 1970. *Exp. Cell Res.* 60:301–5

228. Saito, H., Uzman, B. G. 1971. *Exp. Cell Res.* 66:90–96

229. Saito, H., Uzman, B. G. 1971. *Exp. Cell Res.* 66:97–103

230. Kresse, H., Tekolf, W., von Figura, K., Buddecke, E. 1975. *Hoppe-Seyler's Z. Physiol. Chem.* 356:943–52

231. Wasteson, Å., Lindahl, U., Hallén, A. 1972. *Biochem. J.* 130:729–38

232. Toole, B. P., Trelstad, R. L. 1971. *Dev. Biol.* 26:28–35

233. Toole, B. P. 1972. *Dev. Biol.* 29:321–29

234. Toole, B. P., Gross, J. 1971. *Dev. Biol.* 25:57–77

235. Toole, B. P., Jackson, G., Gross, J. 1972. *Proc. Natl. Acad. Sci. USA* 69:1384–86

236. Toole, B. P. 1976. In *Neural Recognition,* ed. S. H. Barondes, pp. 275–329. New York: Plenum

237. Wiebkin, O. W., Muir, H. 1973. *FEBS Lett.* 37:42–46

238. Solursh, M., Vaerewyck, S. A., Reiter, R. S. 1974. *Dev. Biol.* 41:233–44

239. Wiebkin, O. W., Hardingham, T. E., Muir, H. 1975. In *Extracellular Matrix Influences on Gene Expression,* ed. H. C. Slavkin, R. C. Greulich, pp. 209–23. New York: Academic

240. Handley, C. J., Lowther, D. A. 1976. *Biochem. Biophys. Acta* 444:69–74

241. Nevo, Z., Dorfman, A. 1972. *Proc. Natl. Acad. Sci. USA* 69:2069–72

242. Huang, D. 1974. *J. Cell Biol.* 62:881–86

243. Meier, S., Hay, E. D. 1974. *Proc. Natl. Acad. Sci. USA* 71:2310–13

244. Moscona, A. A. 1975. In *Extracellular Matrix Influences on Gene Expression,* ed. H. C. Slavkin, R. C. Greulich. New York: Academic. 833 pp.

245. Öbrink, B., Kuhlenschmidt, M. S., Roseman, S. 1977. *Proc. Natl. Acad. Sci. USA* 74:1077–81

246. Culp, L. A. 1976. *Biochemistry* 15:4094–4104

247. Pessac, B., Defendi, V. 1972. *Science* 175:898–900

248. Wasteson, Å., Westermark, B., Lindahl, U., Pontén, J. 1973. *Int. J. Cancer* 12:169–78
249. Fratantoni, J. C., Hall, C. W., Neufeld, E. F., 1968. *Proc. Natl. Acad. Sci. USA* 60:699–706
250. Arnold, E. A., Yawn, D. H., Brown, D. G., Wyllie, R. C., Coffey, D. S. 1972. *J. Cell Biol.* 53:737–57
251. Cook, R. T., Aikawa, M. 1973. *Exp. Cell Res.* 78:257–70
252. Smith, R. M., Cook, R. T. 1977. *Biochem. Biophys. Res. Commun.* 74:1475–82
253. Saiga, H., Kinoshita, S. 1976. *Exp. Cell Res.* 102:143–52
254. Kraemer, R. J., Coffey, D. S. 1970. *Biochem. Biophys. Acta* 224:553–67
255. Waldman, A. A., Marx, G., Goldstein, J. 1975. *Proc. Natl. Acad. Sci. USA* 72:2352–56
256. Slobin, L. I. 1976. *Biochem. Biophys. Res. Commun.* 73:539–47
257. Waldman, A. A., Goldstein, J. 1973. *Biochemistry* 12:2706–11
258. Waldman, A. A., Goldstein, J. 1973. *Biochim. Biophys. Acta* 331:243–50
259. Waldman, A. A., Marx, G., Goldstein, J. 1974. *Biochim. Biophys. Acta* 343:324–29
260. Goldstein, J., Waldman, A. A., Marx, G. 1975. *Adv. Exp. Med. Biol.* 52:289–97
261. Kinoshita, S. 1974. *Exp. Cell Res.* 85:31–40
262. Bhavanandan, V. P., Davidson, E. A. 1975. *Proc. Natl. Acad. Sci. USA* 72:2032–36
263. Stein, G. S., Roberts, R. M., Davis, J. L., Head, W. J., Stein, J. L., Thrall, C.

L., Van Veen, J., Welch, D. W. 1975. *Nature* 258:639–41
264. Fromme, H. G., Buddecke, E., von Figura, K., Kresse, H. 1976. *Exp. Cell Res.* 102:445–49
265. Kinoshita, S. 1971. *Exp. Cell Res.* 64:403–11
266. Avila, J. L., Convit, J. 1976. *Biochem. J.* 160:129–36
267. Robinson, D., Stirling, J. L. 1968. *Biochem. J.* 107:321–27
268. Kint, J. A., Dacremont, G., Carton, D., Orye, E., Hooft, C. 1973. *Science* 181:352–54
269. Heijlman, J. 1974. *Biochem. Soc. Trans.* 2:638–39
270. Avila, J. L., Convit, J. 1975. *Biochem. J.* 152:57–64
271. Lagunoff, D., Phillips, M. T., Iseri, O. A., Benditt, E. P. 1964. *Lab. Invest.* 13:1331–44
272. Bergqvist, U., Samuelsson, G., Uvnäs, B. 1971. *Acta Physiol. Scand.* 83:362–72
273. Schuel, H., Kelly, J. W., Berger, E. R., Wilson, W. L. 1974. *Exp. Cell Res.* 88:24–30
274. Blaschke, E., Bergqvist, U., Uvnäs, B. 1976. *Acta Physiol. Scand.* 97:110–20
275. Giannattasio, G., Zanini, A. 1976. *Biochim. Biophys. Acta* 439:349–57
276. Margolis, R. U., Margolis, R. K. 1973. *Biochem. Pharmacol.* 22:2195–97
277. Baumgartner, H., Gibb, J. W., Hörtnagl, H., Snider, S. R., Winkler, H. 1974. *Mol. Pharmacol.* 10:678–85
278. Uvnäs, B. 1974. *Life Sci.* 14:2355–66
279. Uvnäs, B., Åborg, C.-H. 1976. *Acta Physiol. Scand.* 96:512–25
280. Uvnäs, B., Åborg, C.-H. 1977. *Acta Physiol. Scand.* 99:476–83

Ann. Rev. Biochem. 1978. 47:419–48

ERYTHROLEUKEMIC DIFFERENTIATION[1]

❖979

Paul A. Marks and Richard A. Rifkind

Departments of Medicine and of Human Genetics and Development and the Cancer Center, Columbia University, New York, 10032

CONTENTS

[1]This review was completed in August 1977.

419

0066-4154/78/0701-0419$01.00

PERSPECTIVES AND SUMMARY

Normal erythropoiesis has been a useful system for defining several aspects of the regulation of eukaryotic cell differentiation and the control of protein synthesis. However, investigations with normal erythropoietic cells have been limited by several factors, including inability to establish normal erythroid precursor cells in long term culture, limited availability of genetic variants of significance to differentiation, and difficulty in creating a differentiating population synchronized with respect to critical events during differentiation. Murine erythroleukemia cell (MELC) lines, derived from susceptible mouse spleens infected with the Friend virus complex, have been established in continuous culture. The transforming Friend virus complex includes at least two viruses: a defective spleen focus forming virus (SFFV), and a murine leukemia helper virus (MuLV). The target cell for virus infection appears to be an erythroid cell precursor, perhaps the erythropoietin-responsive cell. MELC are now extensively employed as a model system for studying mechanisms controlling cell differentiation.

MELC lines have been established that show a low but definite level of spontaneous differentiation (0.5%); when cultured with dimethylsulfoxide (Me$_2$SO) or any of a variety of other agents, they are induced to initiate erythroid differentiation at a much greater level. This program of differentiation has many similarities to normal erythropoietin-regulated differentiation, including morphogenesis, accumulation of globin mRNA, α and β globin synthesis, increase in enzymes of the heme synthetic pathway, appearance of characteristic red cell membrane proteins, and a limited capacity for cell division.

A variety of chemicals, including planar-polar compounds such as hexamethylene bisacetamide (HMBA), purines and purine derivatives, hemin, short chain fatty acids, inhibitors of DNA or RNA synthesis, as well as UV irradiation and X irradiation, are inducers of MELC differentiation. MELC do not, however, require or respond to erythropoietin. The mechanisms by which the active agents induce differentiation have been investigated but are still unresolved. The wide variety of inducing agents and the evidence implicating the cell membrane as the cellular target for inducer activity for some agents, and chromatin for others, suggest that there may be different pathways by which agents induce differentiation.

Variant cell lines have been developed that are resistent to induction by some agents, for example, Me$_2$SO, but sensitive to others, for example, HMBA. These cell lines are useful in the analysis of changes associated with induction of MELC differentiation.

Induction of MELC differentiation generally follows a latent period which varies with the genetic history of the cells, with culture conditions, and with the particular inducer. There appears to be a requirement for a cell cycle-related event for MELC to become committed to differentiate, as assayed by the capacity to express the program of erythroid differentiation after being transferred out of the inducer. The kinetics of expression of erythroid differentiation, as ascertained by globin mRNA accumulation and hemoglobin formation, appear to be stochastic. Inducers of MELC differentiation have also been shown to initiate differentiated changes in other cell systems. This supports the concept, derived from studies with MELC, that elucidation of the action of inducers in MELC may provide clues to regulatory mechanisms of more general significance to differentiation.

NORMAL ERYTHROPOIESIS: A BRIEF OVERVIEW

Regulation of the rate of erythropoiesis, that is, the rate of red blood cell formation, may be achieved by control mechanisms that exert their effects at a number of critical steps. These include 1. proliferation of the pluripotent hemopoietic stem cells; 2. commitment of hemopoietic stem cells to erythropoiesis; 3. proliferation of committed erythroid precursor cells, which include several recognized sequential stages responsive to erythropoietin; and 4. commitment of the erythroid precursor to express the program of biosynthetic and morphogenetic activities characteristic of terminal differentiation of this specialized cell lineage. These regulatory mechanisms have been reviewed extensively elsewhere (1–6) and only a brief overview of these normal processes is provided here.

Erythropoiesis constitutes one of the terminal differentiation pathways of a hemopoietic cell renewal system that is responsible for the production of at least three of the formed elements of the blood—granulocytes, megokaryocytes, and erythrocytes. The multipotency of a common hemopoietic stem cell has been amply demonstrated (1). The influence of the hemopoietic microenvironment on the developmental potential of the stem cell is supported by considerable data (7). Both the intrinsic developmental potential of the hemopoietic stem cell and the influence of the hemopoietic microenvironment appear to be under genetic control (8). Commitment of the multipotent hemopoietic stem cell to erythroid differentiation involves initiation of a series of developmental stages in the erythroid precursor cell compartment. These different stages of the erythroid precursor, as characterized in vitro, are distinguished by proliferation capacity and responsiveness to the hormone erythropoietin (9–11). The most immature precursor, termed the Day 8 Burst Forming Unit for Erythropoiesis (Day 8 BFU-E) has the greatest proliferative capabilities and the lowest responsiveness to erythropoietin among the erythropoietin-responsive precursors. The most mature precursor element, the Colony Forming Unit for Erythropoiesis (CFU-E), supports only a limited number of cell divisions, perhaps not more than 3–5, but is sensitive to low concentrations of erythropoietin. The intermediate development stage, the Day 3 BFU-E, appears to be intermediate in these properties. None of these precursor cells accumulates detectable globin mRNA or synthesizes globin (12).

From a biochemical point of view, cultures of erythroid cells respond to erythropoietin by increasing the rate of hemoglobin synthesis (2). This increase in hemoglobin synthesis is due to an increase in the numbers of differentiating erythroid cells, not to augmentation of the rate of hemoglobin synthesis by individual erythroblasts. It appears likely that the principal biological effect of erythropoietin is upon the proliferation of target erythroid precursor cells and that the differentiation of the progeny of these precursors is an inherent property of the cell lineage, which follows a probabilistic, or stochastic, model (13, 192). Stimulation of RNA synthesis is the earliest detected effect of erythropoietin on erythroid precursor cells (2). This early RNA synthetic response to the hormone includes rRNA and 4 and 5S RNA but not globin mRNA. Synthesis of globin mRNA is detected only after 6–10 hr in culture with erythropoietin (12). DNA synthesis is not required for expression of the early RNA synthetic response to the hormone but is required for the later erythropoiesis-specific biosynthetic events (15). Taken together these observations have suggested that target cell proliferation is the principal mechanism whereby erythropoiesis is regulated physiologically by erythropoietin, and that DNA synthesis is required for the expression of the differentiated function of erythroid precursor cells, namely, the transition to globin mRNA production, globin synthesis, and presumably the other differentiation-specific biosynthetic processes characteristic of the maturing erythroblast (5, 15).

Several problems severely limit experiments designed to elucidate further regulatory mechanisms in the initiation of erythropoietic differentiation, utilizing normal fetal or adult erythropoietic precursor cells in culture. First, culture conditions have not been developed that permit sustained renewal of precursor cells and prolonged erythropoiesis in vitro. Second, it has not been possible, to date, to dissociate the effects of erythropoietin on proliferation of its target cell and initiation of erythroid cell differentiation. Furthermore, it appears difficult to synchronize normal erythroid precursors with respect to critical events in differentiation. Third, the availability of developmental mutants is severely limited. For these reasons, numerous laboratories have, in recent years, been exploring the potential of the virus-transformed murine erythroleukemia cell system as a useful model for investigations into control mechanisms regulating the expression of erythroid cell differentiation.

ORIGINS OF THE MURINE ERYTHROLEUKEMIA CELL

Murine Erythroleukemia (Friend) Disease

In 1957 Charlotte Friend (16) reported recovery of a virus from the spleen cells of Swiss mice inoculated 14 months previously with filtrate from disrupted Ehrlich ascites tumor cells. Inoculation of susceptible mice with spleen cell filtrate regularly and rapidly induces massive splenomegaly and hepatomegaly, which is associated with the infiltration of these organs and the peripheral blood by immature hemopoietic elements. Initially considered to be a reticulum cell sarcoma (17, 18) associated with a reactive erythroblastosis (19), it was suggested in 1962 by Zajdela (20), on the basis of both pathological studies and radio-iron uptake, that the primary disease initiated by Friend virus complex was a neoplastic proliferation of erythropoietic elements. The primary erythropoietic nature of Friend disease was subsequently documented by histopathological studies, including electronmicroscopy (21), demonstration of erythrocytic antigens on the surface of tumor cells (22), and the isolation, by Mirand (23), of a variant of Friend virus complex that induces not only the typical manifestations of Friend disease, including splenomegaly and erythroblastosis, but a pronounced polycythemia as well. With both the Friend and the Mirand variants, the daily output of young red blood cells in infected animals is significantly augmented (24), and is associated with considerable ineffective erythropoiesis, a term for premature death of young nucleated erythroid cells (25, 26). Although the hormone erythropoietin may significantly influence the susceptibility of suitable mice to infection by Friend virus (see below), the erythroblastosis induced by the virus is independent of the titer of this hormone (27–30). Indeed, unlike normal erythropoiesis, Friend virus infected erythroid cells will even express erythropoietin-independent erythroid cell differentiation when explanted under in vitro conditions (15, 31, 32).

The case for malignant transformation of mouse erythroid cells infected by Friend leukemia virus is strong, but not conclusive. Thomson & Axelrad (33) demonstrated the malignant features of murine erythroleukemia cells (MELC) by transplantation to a murine host genetically resistant to infection by the Friend virus complex, and confirmed their unrestrained growth potential. Nevertheless, under these conditions, and using serial mouse passages, it has been demonstrated that MELC do not have indefinite transplantability unlike more typical virus transformed malignant cells (33, 34). Current evidence suggests that host environmental factors may be critical in modulating the "malignancy" of MELC in serial passage. Serial transfer to appropriately irradiated hosts (35) is sustained virtually indefinitely. Considerable evidence has accumulated indicating that host immune mechanisms modulate the expression of MELC in vivo (36–38).

Friend Virus Complex and Host Factors

The Friend virus complex has been studied with respect to its biological properties and range of host susceptibility (for review see 39–41). This work is reviewed here in summary form only. Friend virus preparations contain at least two distinct viral activities, as suggested, initially, by Mirand (42). The first of these is responsible for the rapid onset of splenomegaly, hepatomegaly, and erythroleukemia, and is assayed by its ability to induce microscopic colonies in the spleens of susceptible mice (42, 43). This virus, termed Spleen Focus-Forming Virus (SFFV), is defective and requires, for replication, a second associated viral activity or helper virus which may be any of a number of the murine leukemia virus strains (MuLV) that are responsible, in the infected animal, for the late development of a lymphocytic leukemia. MuLV may be assayed in vitro by the XC plaque assay (39). The independence of these two viral activities in the Friend virus complex has been confirmed by the isolation of each viral component; in the case of MuLV this was achieved by forced passage through rats (44, 45) and by end-point dilution (46), which is possible because of the excess of MuLV found in common strains of Friend leukemia virus complex. In the case of SFFV, Scolnick and colleagues (47) identified a BALB/c spleen homogenate that contained equal titers of SFFV and MuLV, which enabled them to establish a cell line infected by SFFV and free of replicating helper MuLV virus. Analysis of the SFFV genome, purified in this fashion, indicates that SFFV is a recombinant genome comprised of portions of the MuLV genome and sequences related to certain murine xenotropic viruses (48), which suggests analogies to the Rous sarcoma virus and the Harvey and Kirsten strains of murine sarcoma virus. It was suggested that the SFFV-specific RNA sequences which are analogous to the *sarc* sequences of Rous sarcoma virus, may have been acquired during early replications of MuLV in Swiss mice at the time of initial isolation by Charlotte Friend. Although globin mRNA sequences have been identified within Friend virus virions, the evidence indicates that these are not components of the viral genome but are adventitious passenger sequences within the viral capsid (49).

The host range of susceptibility to Friend virus complex is governed by genetic factors that control susceptibility to both the MuLV helper virus component and SFFV (39). Susceptibility and resistance may be assayed by both the in vivo and in vitro assay techniques noted above. Two independently segregating genes are responsible for the principal manifestations of host range susceptibility. The first, called Fv-1, governs relative resistance to the MuLV helper virus component. MuLV strains are defined as N-,B-, or NB-tropic, depending upon their ability to replicate in NIH Swiss (N-type) or BALB/c (B-type) strains of mice that possess the appropriate allele at the Fv-1 locus. Considerable evidence now suggests that host restriction conferred by the Fv-1 gene is not due to differences in virus receptor or cell penetration characteristics but rather to intracellular events in the viral replication cycle after absorption, penetration, and uncoating of the virus (50–52). Recent studies (53) suggest that only in the permissive host genome can provirus be integrated into cellular DNA. The second genetic host range factor is conferred by alleles at the Fv-2 locus and is specific for resistance and susceptibility to the SFFV component of Friend leukemia virus (39, 54). Mice homozygous for the recessive resistance allele at this locus are totally refractory to spleen focus formation by Friend virus. Host susceptibility is also influenced by genetic factors, genetically located in the H-2 region, which are implicated in the host immune reaction to RNA tumor viruses (39, 40), as well as by host genetic factors that regulate the availability of virus target cells (see below).

Target Cell for Friend Virus Infection

As noted above, the earliest cellular response to Friend virus infection is proliferation in hemopoietic organs, principally spleen, but also marrow and liver, of immature erythroid elements such as proerythroblasts and their progeny. Considerable evidence, albeit largely circumstantial, suggests that an erythroid precursor cell, possibly one of the several stages of erythropoietin-responsive precursors, is the target cell for Friend virus infection. Modulation of the size of the erythropoietin-responsive precursor cell population by various maneuvers including transfusion-induced polycythymia, administration of erythropoietin, hypoxia, and the combined effects of busulfan and erythropoietin, strongly implicate this cell type as the principal viral target (55–61). Nonspecific hematosuppressive agents that also suppress susceptibility to Friend virus infection (62, 63) appear to do so by reducing the number of erythropoietin-responsive precursor cells. Likewise, two genetic loci responsible for hereditary anemias, the Steel (Sl) and W alleles, confer resistance to Friend virus infection and murine erythroleukemia by virtue of their effect upon the availability of erythropoietin-responsive precursor cells (8, 64, 65).

More precise identification of the target cell for Friend virus infection has not yet been achieved. Clarke et al (66) have accomplished in vitro infection of mouse marrow cells with Friend virus, and thereby initiated the formation of erythropoietin-independent erythropoietic colonies in plasma clot cultures. No evidence of erythropoietic burst formation was obtained, which suggests that the target cell for virus infection may be the CFU-E erythroid precursor cell. Whether or not these investigators have actually achieved transformation of the target cell remains to be determined. Effects of Friend virus infection in vivo on other hemopoietic cell lineages have been reported (67–69) but these may be secondary and due to physiological adjustments of the animal to erythroleukemia. Apparent infection, by SFFV and helper virus, of a nonerythroid cell lineage has been reported (70) and transformed cells fail to display characteristic erythroid cell features. Therefore, an obligatory relationship between SFFV infection and expression of the genetic program of erythropoietic differentiation remains open to question. Conversely, although many of the maneuvers that have been employed to induce erythropoietic differentiation in MELC (see below) are accompanied by augmentation of Friend virus production, an obligate relationship between these two genetic programs, viral replication and differentiation, appears equivocal at best (71–73). Taken together, the available evidence suggests that the principal biological effect of Friend virus transformation of an erythroid precursor target cell is to make cell proliferation independent of the physiological regulator of erythroid precursor replication, erythropoietin.

Growth Characteristics of the Cell in Culture

In 1966 Friend and colleagues (74) reported the establishment of a number of cell culture lines derived from subcutaneous tumor implants of Friend virus-infected cells from DBA/2 mice. These cell cultures appear pleomorphic, in that they contain cells of immature morphology as well as all stages of nucleated erythroblast maturation. When cloned in semisolid media, individual transformed cells gave rise to both immature cells and differentiating erythroblasts (75), which suggests that the potentiality for erythroblastic differentiation is an inherent property of MELC in culture. This interpretation, that cloned MELC consist of relatively uniform populations of erythropoietic precursors, has been confirmed (76). Other MELC isolates have been established from cells growing in the ascites form (77). Most strains of MELC grow with ease in standard tissue culture media supplemented with fetal calf serum (74, 76, 78).

Most established MELC lines display a low level of spontaneous differentiation, ranging from 0.5–20%, as determined by the benzidine reaction for hemoglobin (76, 79–83). Spontaneous differentiation is a property of culture conditions as well as of cell strain. Amino acid composition of the medium, for example, can profoundly alter the rate of spontaneous differentiation (84). Attempts to propagate MELC in protein- and lipid-free media (85) have not proved generally successful in maintaining stable erythropoietic properties. As in vivo, MELC cultures display no requirement for nor response to erythropoietin (78).

AGENTS ACTIVE AS INDUCERS OF DIFFERENTIATION

Inducing Agents

The observation in 1971 that dimethylsulfoxide (Me$_2$SO) stimulates erythroid differentiation in MELC cultures (86) prompted efforts to identify other agents that are effective as inducers. These studies have resulted in the identification of a broad spectrum of agents that are active in this system (Table 1).

Mechanism of Action

Among agents particularly active as inducers are the planar-polar compounds; these agents have relatively low molecular weights (<300), with a polar, hydrophilic group and a planar, hydrophobic portion (87–89). It is apparent from the list of inducers (Table 1) that the properties of the planar-polar compounds are not shared by all agents. Short chain fatty acids, purines and purine derivatives, hemin, and inhibitors of DNA or RNA synthesis are inducers and some are quite effective. Some inducers, such as diamines (88), certain purines (90), hemin (91), and fatty acids (92, 93) are normal cellular components. In no case is there evidence that these normal cellular components exercise a normal physiological function when inducing differentiation. For example, the diacetylated diamines, such as hexamethylene bisacetamide (HMBA), are very active inducers, while diamines are relatively weak (88). This observation suggested that acetylation might be necessary for transport into the cell and that an active diamine is released by deacetylation. A series of compounds was synthesized in which the carbonyl and amino groups were reversed; these are as active as the acetylated diamines. These carboxamides cannot be deacetylated, which effectively eliminates this hypothesis (94). Purine derivatives can induce differentiation without being incorporated into RNA or DNA, and their natural catabolites are not inducers (90). Normal metabolic products of butyric acid and intermediates in heme synthesis are ineffective under conditions in which butyric acid and heme, respectively, are potent inducers (91, 92). While such data do not completely rule out a physiological role for these naturally occurring compounds, there is no persuasive evidence for such a role.

There are several lines of evidence to suggest that not all inducers act by identical mechanisms.

1. There is a considerable variety in the structure of chemical inducers, and physical agents such as X-ray and ultraviolet irradiation also induce MELC differentiation;

2. Variant cell lines have been established that are resistant to induction by some agents and sensitive to induction by others (32, 90, 95–99);

3. Certain agents, such as ouabain (100), the polar-planar compounds (87, 88, 101), and fatty acids (93, 102) have demonstrable effects on membrane function. On the other hand, agents such as X ray, UV irradiation, and very low concentrations of actinomycin (103, 104; Terada et al, unpublished observations) are inducing agents known to cause structural changes in DNA.

Table 1 Agents active as inducers of MELC differentiation[a]

Generally strong inducers	References	Generally weak inducers	References
Polar-planar compounds[b]		Polar-planar compounds	
Dimethylsulfoxide	86	2-Pyrrolidinone	87
1-Methyl-2-piperidone	87	Propionamide	87
N,N-Dimethylacetamide	87	Pyridine-N-oxide	87, 89
N-Methylpyrrolidinone	87	Piperidone	87
N-Methylacetamide	87	Pyridazine	89
N,N-Dimethylformamide	87, 101	Dimethylurea	89
N-Methylformamide	87	Antibiotics and antitumor agents	
Acetamide	87		
Triethylene glycol	87	Vincristine	198
Polymethylene bisacetamides (n = 2–8)	88	5-Fluorouracil	198
Hexamethylene bispropionamide	88	Methramycin	190
Acetamide	89	Cycloheximide	190
Tetramethyl urea	89	X irradiation	104
Antibiotics and antitumor agents		UV irradiation	103, 104
		Adriamycin	190
Bleomycin	104, 198	Cytosine arabinoside	190, 198
N-Dimethylrifampicin	198	Mitomycin C	190
Actinomycin D	103, 190	Hydroxyurea	190
Actinomycin C	103	Diamines	
Purine and purine derivatives		Cadaverine	88
Hypoxanthine	90	Fatty acids	
1-Methylhypoxanthine	90	Acetate	93
2,6-Diaminopurine	90	Propionate	93
6-Mercaptopurine	90	Butyrate	92, 93
6-Thioguanine	90	Isobutyrate	93
6-Amino-2-mercaptopurine	90	Other	
2-Acetylamino-6-mercaptopurine	90	Hemin	91
Fatty acids		Methylisobutylxanthine	unpublished observations
Butyrylcholine	104		
Other			
Ouabain	100		

[a] The relative activity as inducers assigned to the agents listed in this table are based on the best available data from our laboratory and as reported in the literature. The inducing activity of a given agent may vary with different MELC lines, culture conditions, and other, as yet unrecognized factors. It is clear (95), that agents (including Me$_2$SO) may be quite active with one cell line and totally inactive with other cell lines. An agent is arbitrarily listed as a generally strong inducer if it is reported to induce fifty percent or more of a population of MELC, at optimal concentration of the inducer. Generally weak inducers include agents that induce 5–50% of MELC at optimal concentration of the inducer.

[b] Certain compounds could be listed in more than one category. These categories are chosen for convenience in classifying the agents and not to connote possible mode of action. For example, 6-mercaptopurine could be categorized as a purine derivative or as an antitumor agent, and methyl isobutyl xanthine as a purine derivative or as a phosphodiesterase inhibitor. The categorical designation chosen was based on the experimental context in which the agent was tested and shown to be an inducer of MELC.

A systematic study of the relationship between structure and activity has been performed for a series of planar-polar compounds (88, 94).Dimerization of N-methylacetamide, by linkage through varying numbers of methylenes, results in a series of compounds such as acetylated diamines or methylene bisacetamides, that are 5–10 times more active than the simple monomer. HMBA is particularly interesting in this regard because it has been found to be the most effective compound tested in our laboratory to date by the criteria that (a) essentially the entire population of MELC is induced to differentiate, (b) a higher proportion of the total protein synthesized is hemoglobin, and (c) relatively low concentrations are effective (1–5 mM) (88). Compounds structurally related to the bisacetamides have been tested for activity, to determine the requirements for activity and the optimal chain length for each class of related compounds. Alkanes, dicarboxylic acids, and amino caroxylic acids are all inactive. The 5-carbon diamine, cadaverine, exhibits moderate activity; monocarboxylic acids are relatively active while acetamide derivatives provide maximal activity. As indicated above, reversal of amino and carbonyl groups does not alter activity. Methylene bisacetamides with unsaturated triple or double, that is cis or trans bonds are comparable in activity to the saturated compound. On the other hand, converting the 6-carbon chain into a rigid cyclohexane ring eliminates activity. Covalently linked HMBA to glutathione, which cannot penetrate beyond the plasma membrane, is inactive as inducer. Studies with radioactively labeled Me_2SO (105) and HMBA (106) suggest that these compounds must be taken up by the cells to stimulate differentiation. These results are compatible with the evidence that active planar-polar compounds must be relatively flexible, must penetrate into the cell membrane, and that an interchain length between polar groups of ∼5–8 methylenes is optimal for strong inducing activity.

There is a direct relationship between the optimal concentration for induction and the concentration at which these compounds inhibit growth and cause cell death. This is consistent for the series of effective planar-polar compounds (87, 88), as well as for purine derivatives (90) and fatty acids (93). This suggests that the mechanism of induction may involve a change in structure of a cellular component which, if too extensive, is incompatible with cell survival. Alternatively, the inducers may have toxic effects on the cell that are independent of the effect on differentiation.

Effects on Other Cell Types

A number of other cell systems have been tested with agents that induce MELC differentiation. Me_2SO induces erythroid differentiation of cell lines derived from transplantable tumors from 7,12-dimethylbenz(a)anthracene-induced erythroleukemia of rats (107). Addition of Me_2SO (108) or HMBA

(109), at concentrations comparable to those optimal for induction of MELC differentiation, to cells of mouse neuroblastoma clone NIE-115 in the confluent phase of growth, results in morphologically differentiated cultures with extensive neurite formation and the development of an excitable membrane. There are differences, however, in the patterns of response of neuroblastoma cells to the two inducing agents. Me$_2$SO has been reported to induce cilia formation in cultures of human lung cancer cells (110). Sodium butyrate produces reversible changes in the morphology, growth rate, and enzyme activities of several mammalian cell types in culture, changes that resemble induction of differentiation (7, 111). X rays have also been reported to induce morphological differentiation of neuroblastoma cells (112). It appears that some of the agents that induce MELC differentiation may also be effective as inducers of differentiation or differentiation-like changes in other transformed cell lines. Recognition of differentiation-inducing effects is, of course, entirely dependent upon an adequate assay for markers characteristic of differentiation. In any event, MELC are not unique with respect to inducibility by these agents, and may be representative of a category of transformed cells "programmed" to differentiate, but somehow blocked in their expression of that "program." There may, then, be common mechanisms regulating the commitment of programmed precursor cells, mechanisms that may be elucidated by study of chemically induced differentiation.

Effects on Oncogenicity

One of the most significant alterations in Me$_2$SO-induced MELC is a decrease in oncogenicity of the leukemic cells (86). Similar observations have been made in our own laboratory with MELC cultured with HMBA, N,N-dimethylacetamide, N-methylpyrrolidinone, and butyric acid (Epner, Fibach, Rifkind, and Marks, unpublished observations). In these studies the decrease in oncogenicity appears to be inversely related to the increase in commitment to differentiation. While in vitro treatment of MELC with inducers is clearly associated with loss in oncogenicity, in vivo studies, in which inducing agents were administered to mice with murine erythroleukemia, have been less definitive (113, 114; Reuben, Cobb, Bogden, Rifkind, and Marks, unpublished observations). A variety of polar-planar compounds have been tested in our laboratory, for therapeutic effectiveness in murine erythroleukemia. None of the compounds tested significantly increased overall survival. On the other hand, the results indicate that the time course of erythroleukemia can be altered by treatment with active inducing agents. A prolonged disease-free period is achieved, particularly with HMBA, 2-pyrrolidinone, N,N-dimethylformamide, and acetamide.

Whether eventual death is from erythroleukemia or lymphocytic leukemia due to the helper MuLV has yet to be determined. Treatment schedule and the number of mouse passages of MELC after culture affect the therapeutic response to polar-planar compounds. Since polar-planar compounds are differentiation-inducers rather than classically cytotoxic agents, the prolonged disease-free period suggests drug efficacy by a new and provocative mechanism (Reuben, Cobb, Bogden, Rifkind, and Marks, unpublished observations).

CHARACTERISTICS OF INDUCED DIFFERENTIATION

As summarized above, MELC strains have been developed that show a low level (1%) of spontaneous erythroid differentiation in culture (15). On culture with inducers (Table 1), MELC express a program of differentiation that has many morphological and biochemical similarities to that observed in normal, erythropoietin-regulated erythropoiesis (5, 15). These include chromatin condensation and other morphological changes (86), terminal cell division (90, 115), accumulation of globin mRNA (15, 72, 84, 91, 116–123), α- and β- globin chain synthesis (124, 125), increase in heme synthesis (86, 126), synthesis of characteristic erythrocyte enzymes such as catalase (127), alterations in purine metabolism (128), and the appearance of erythrocyte-specific membrane antigens (97) and other proteins, such as spectrin (129, 130). It should be emphasized that while MELC differentiation displays many characteristics of normal erythropoiesis, MELC are transformed and there are a number of aspects of MELC proliferation and differentiation that are not normal. MELC have the capacity to proliferate without erythropoietin (15, 28, 31, 32), they rarely proceed in vitro to the nonnucleated stage of differentiation characteristic of normal erythropoiesis (86, 88), and they may exhibit patterns of erythroid cell gene expression that are different from those in normal erythropoiesis (131).

The marked increase in synthesis of globin during induced differentiation provides an opportunity to examine in detail the expression of globin genes. Regulatory mechanisms in globin biosynthesis may act at any of at least four levels of control. Induction may be associated with amplification of the globin structural genes. Alternatively, a constant number of globin genes may undergo selective stimulation of transcription. The processing of globin mRNA may be altered by a decreased rate of degradation, or by facilitated transport to the cytoplasm, which leads to increased accumulation of globin mRNA available for translation. Finally, translation of globin mRNA may be selectively increased in induced cells. Each of these potential levels of control has been studied and the results are summarized below.

DNA

The mean number of globin genes per haploid MELC genome, ascertained by molecular hybridization, is 3.2–3.4 (116), a value that is consistent with observations on differentiating erythroid cells of duck (132), mouse (133), and human subjects (134). Further, the number of globin genes is identical in induced and uninduced MELC (116). Gene amplication, therefore, cannot account for induced globin synthesis.

The kinetics of DNA synthesis in induced cultures of MELC differs from that in uninduced MELC (135). In control cultures there is an initial rise in rate of pulse-labeled thymidine incorporation; a maximum value is achieved by about 10 hr and remains constant between 10 and 40 hr; it decreases to less than 10% of the peak value by 60 hr. The initial increase in rate of thymidine incorporation probably reflects entry into S-phase of cells partially synchronized in the post-logarithmic growth phase of the previous passage. The plateau level, observed between 10 and 40 hr, reflects a constant proportion of cells in S-phase due to loss of partial synchronization. The subsequent fall in DNA synthesis coincides with onset of the stationary growth phase. Although a brief initial rise is seen in cells cultured with 280 mM Me_2SO, a decrease in the rate of thymidine incorporation relative to control cells is detected by 4–6 hr. By 20 hr the rate of thymidine incorporation has fallen to its lowest value, about 25% of the rate observed in control cultures. As MELC become affected by the inducing agent, they are transiently delayed in their entry into S-phase, the period of DNA synthesis. The rate of thymidine incorporation subsequently rises, between 20 and 30 hr, to a peak value of about 75% of the highest rate in control cutlures, and it remains there until 50–60 hr. It then decreases as the cells enter the stationary growth phase and the terminal cell division characteristic of differentiation. The decrease in DNA synthesis due to transient inhibition of entry of cells into the S-phase of the cell cycle has also been documented by measuring the relative DNA content per cell, and the proportion of cells labeled during a 20 min pulse of isotopically labeled thymidine. Inhibition of DNA synthesis is associated with prolongation of G_1 phase of the cell cycle. Possible implications of these findings with regard to the mechanism of induction of MELC are discussed below.

In addition to alterations in the rate of DNA synthesis, there is evidence that structural changes in DNA occur in the course of induced MELC differentiation. DNA from MELC cultured with Me_2SO, butyric acid, or dimethylacetamide show a decrease in sedimentation rate on alkaline sucrose gradients as early as 27 hr after onset of nonsynchronous culture (103, 104). No change is detected when sedimentation is analyzed on a neutral sucrose gradient. No change in DNA sedimentation is observed with inducer-resistant MELC variants treated with the agent to which they are resistant (103). Additional evidence for alterations in DNA structure include changes in acridine orange binding, observed after 4–6 days of exposure to Me_2SO (136), and in propidium iodide binding detected by as early as 10 hr after culture with inducer (135).

The alterations detected in DNA structure may be important in the process of induction of differentiation. Alternatively, these changes in DNA could be a result of differentiation occurring subsequent to the events that are critical to the commitment of MELC to express the program of differentiation.

Nuclear Proteins

A limited number of studies have been reported on MELC chromatin structure. Peterson & McConkey (137) examined 500 proteins from untreated and Me_2SO-treated MELC and observed differences greater than 50% in only 6 nonhistone chromosomal proteins, 2 nucleoplasmic proteins, and 3 cytoplasmic proteins after treatment with inducer. A striking difference was reported in a major histone, f2a2, during the establishment of MELC in culture (138, 139). The ratio of 2 histone subfractions, $f2a2_1 : f2a2_2$, changes from a ratio of 3 to a ratio of 1 as MELC become responsive to Me_2SO-induced differentiation. Nuclei from n-butyrate-induced MELC have been reported to contain greatly increased amounts of modified forms of histone H4. These changes have been found within 24 hr of culture and are interpreted as reflecting histone acetylation (140).

A new chromosomal protein, with an apparent molecular weight of 25,000, appears during Me_2SO- or HMBA-induced differentiation (141). Nonerythroid cell lines and/or MELC resistant to induction with Me_2SO or HMBA do not accumulate this protein. Like hemoglobin, this protein is first detected at about 24 hr in culture and increases over the ensuing 48 hr of culture. It is not clear whether the several changes detected in chromatin proteins are integral to the program of erythroid differentiation or are unrelated to the mechanisms controlling differentiation.

RNA

As MELC differentiate in culture, there is a progressive decrease in RNA content which is detected by 24 hr and which affects 4S RNA and rRNA (142). A decrease in the rate of synthesis of total RNA in induced cells has been detected as early as 1.5–2 hr. An increase in the rate of accumulation of globin mRNA is detected within 6–24 hr of culture (123) depending on the inducer and culture condition. Markedly increased synthesis of globin mRNA is detected in nuclei isolated from Me_2SO-induced MELC, with as little as 5 min of nuclear transcription in vitro (143). These findings were interpreted as most consistent with transcriptional activation of globin genes upon induction of differentiation, but they do not exclude the possibility that posttranscriptional stabilization accounts for accumulation of globin mRNA.

Nuclear RNA from MELC cultured with inducers contain species of RNA molecules that have globin sequences and are larger than the mature form of globin mRNA (102, 119, 120, 144, 145). Various sizes have been assigned to these nuclear RNA species. Bastos & Aviv (119) suggest that an early transcription product is a large 27S molecule, which is cleaved into a smaller 15S species. This intermediate precursor is then converted to the 10S mRNA that accumulates in the cytoplasm. The large RNA precursor has an estimated half-life of 5 min. Curtis et al (120) identify a 15S peak in nuclear RNA with a half-life of 5 min or less, which, on the basis of hybridization to plasmid-generated α- and β-mouse globin DNA, contains β- but not α-globin sequences. They suggest that the 15S RNA is precursor for 10S cytoplasmic β-globin mRNA, while α-globin mRNA may be derived from an 11S precursor.

Examination of nuclear RNA reveals a class of polyadenylated molecules that are not found as functional mRNA molecules on MELC polyribosomes (146, 147). It is estimated that the base sequence complexity of the nuclear polyadenylated RNA is 4–5 times that of polyribosomal polyadenylated RNA. Polyribosomal polyadenylated RNA is transcribed from an estimated 1.8% of the genome, whereas nuclear polyadenylated RNA is transcribed from 7.6% of the genome (146).

Taken together, these results suggest that the biosynthesis of globin mRNA in induced MELC is a multistep process. An early event is the synthesis of large precursor nuclear RNA. This is subsequently cleaved to intermediate species and processed to 10S, functional, mRNA.

In induced MELC, globin mRNA accumulates in the cytoplasm, as assayed by cDNA : RNA hybridization or by the heterologous cell-free system for protein synthesis (72, 84, 117, 123, 125, 131, 148–150). In Me_2SO-treated cells, globin mRNA represents $\sim$0.5–1% of the total poly-(A)-containing RNA on polyribosomes (72).

The pattern of accumulation of α- and β-globin mRNA sequences differs depending on the inducer. Using specific α- and β-globin cDNAs (123), cells cultured with Me_2SO, HMBA, or butyric acid accumulate α-globin mRNA after 16, 12, and 8 hr of culture, respectively. An increase in β-globin mRNA sequences is not detected until 24 hr of culture. In cells cultured with hemin, however, both α- and β-globin mRNA are detectable by 6 hr and a constant ratio of α- to β-mRNA is maintained during induction. In fully induced cells, the α- to β-globin mRNA ratio is approximately one in cells induced by Me_2SO and HMBA; the ratio is 0.6 and 0.5

in cells induced by butyric acid and hemin, respectively. Differential expression of α- and β-globin genes during Me$_2$SO-induced differentiation of MELC has also been reported by Orkin et al (151).

Two hemoglobins are present in adult DBA/2 mice—hemoglobin major and hemoglobin minor. These hemoglobins differ in the structure of their β-globin chains (152, 153, 154, 155). Different inducers cause different patterns of accumulation of mRNA for β^{maj} relative to β^{min} (131). These observations are reviewed more fully in the discussion of cytoplasmic proteins, below.

Globins

The tryptic peptides of globin formed in induced MELC are indistinguishable from those of globin of DBA/2J adult mice (152). Globin in adult DBA/2 mice consist of α, β^{maj}, and β^{min} chains (152). The ratio of β^{maj} to β^{min} globin in normal adult DBA/2 mouse erythrocytes is approximately 4 (127, 131).

In MELC lines 745 or FSD-1 (both of DBA/2 origin) cultured for 96 hr with Me$_2$SO, up to 25% of the protein being synthesized, or approximately 10% of cellular soluble protein, is hemoglobin (127). Among several inducers, at their optimal concentrations and after 5 days of culture, HMBA induces the highest level of accumulation of hemoglobin per cell, compared with Me$_2$SO, N-methylacetamide, N,N-dimethylacetamide, and butyric acid (88).

The rate of accumulation of globin mRNA and the rate of synthesis of globin chains closely parallel each other; there is little evidence for the accumulation of untranslated globin mRNA (84). The synthesis of heme and globin are also closely coordinated, so that there is little or no detectable free hemin accumulated (113, 127).

As noted above, different agents induce different patterns of hemoglobin accumulation (127, 131, 156). DBA/2-derived MELC produced a major and a minor hemoglobin differing in the amino acid sequences of the β-chains (152–155). On the basis of available genetic and biochemical evidence (55, 153, 154), β^{maj} and β^{min} genes are closely linked and the two globins differ by a relatively few (6–9) amino acids. After 4 days in culture, cells induced by the polar-planar compounds Me$_2$SO, HMBA, or N-methylpyrrolidinone have accumulated a larger proportion of β^{maj}-containing hemoglobin (127, 131). By contrast, butyric or propionic acids induce relatively equal amounts of Hb^{maj} and Hb^{min}. The differences in rates of accumulation of these hemoglobins is reflected in the rates of synthesis of the β^{maj}- and β^{min}-globins (131). The best available evidence at this time suggests that the differences in relative amounts of β^{maj}- and β^{min}-globins cannot be attributed to translational or posttranslational regulatory mechanisms. Rather, the inducers act, directly or indirectly, to control the rates of synthesis of β^{maj}- and β^{min}-globin mRNA or the processing of these RNAs (131).

Nonglobin Proteins

Normal erythrocyte differentiation is associated with a characteristic pattern of enzymatic changes that includes an increase in activity of catalase, carbonic anhydrase, and δ-aminolevulinic acid (ALA) synthetase (15, 157, 158). MELC differentiation is also accompanied by increased activities of the enzymes involved in heme synthesis (126, 159–161) and carbonic anhydrase (127). Me_2SO-induced differentiation, for example, is associated with increased activity of ALA synthetase, ALA dehydratase, uroporphyrinogen-I-synthetase, and ferrochelatase, and an increase in heme concentration (160).

In MELC lysates there is a protein that acts to inhibit globin synthesis and is not affected by hemin (162, 163). This inhibitor has protein kinase activity that phosphorylates the small subunit of one initiation factor, (1F-E_2). Protein synthesis is inhibited by preventing initiation factor-dependent binding of methionyl-tRNA$_f$ to the 40S ribosomal subunit. There appears to be a functional similarity between this inhibitor from MELC and the hemin-responsive repressor from reticulocytes (164). The effect of inducers on the concentration or action of this repressor of globin synthesis is not known.

Plasma Membrane-Related Functions

Following exposure to inducing agents MELC undergo a variety of alterations at the plasma membrane. Some of these, which occur relatively late in the process of induced differentiation, are characteristic of normal erythropoiesis as well and, presumably, reflect the expression of the program of erythropoietic development. Other changes, occurring early in the response to inducing agents, are less readily identified as part of the normal pattern of erythroid cell differentiation and may, potentially, have significance with respect to the mechanism of action of inducing chemicals.

LATE MEMBRANE CHANGES   An increase in a variety of membrane-associated erythrocyte-specific proteins or protein-related activities have been documented during induced MELC differentiation. These include erythrocyte membrane antigens (78, 97, 165), spectrin (130), and receptors for the iron-binding serum protein, transferrin (98). Conversely, membrane activities characteristic of immature erythroid cell precursors, for example, H-2 antigenicity, may be lost during induced differentiation (166). An increase in intrinsic membrane microviscosity toward a level characteristic of the mature erythrocyte (130, 166), a late decrease in plant lectin agglutinability (129), and an overall decrease in cell volume (166) are also detected and appear to reflect the process of chemical and morphogenetic erythroid maturation. Their late occurrence, subsequent to the commitment of MELC to differentiate (see below), appears to preclude these manifestations of induction from being of significance in the events responsible for *initiation* of the process of differentiation.

EARLY MEMBRANE CHANGES   Several lines of evidence suggest that further exploration of the relationship between chemical inducers and the plasma membrane may prove fruitful for elucidation of the mechanism of action of inducing chemicals. The polar-planar class of inducing agents have been demonstrated to affect the phase transition temperature of phospholipid membranes in a fashion that correlates with their effects on differentiation (167). Likewise, the partition coefficient in octanol/water of effective compounds falls within a fairly narrow range (log P between –0.5 and –2.0) (168). In studies with doubly labeled hexamethylene bisacetamide (94), it was demonstrated that whereas the labeled methylene bridge was found with the soluble proteins, the terminal carbonyl portions of the molecule were converted to a TCA-insoluble form located principally in the plasma membrane-enriched fraction. By 10 hr of exposure to Me_2SO, MELC display a significant decrease in cell volume compared to untreated cells (169), and the magnitude of this change in volume is proportional to the number of differentiated cells seen later in culture. An additional early response to Me_2SO is an increase in plant lectin agglutinability which is tentatively attributed, by Eisen et al (129), to the development of an induced or activated cell surface protease activity.

An additional line of evidence implicating some cell surface-mediated function in the induction of differentiation of MELC is derived from the differentiation-inducing and differentiation-suppressing activities of agents with known or postulated plasma membrane activities. Among the inducing agents, ouabain, a cardiac glycoside that binds to plasma membrane sodium Na^+-K^+-Mg^{2+}- activated ATPase, is a potent inducer of erythroid differentiation in selected MELC lines. This effect is exerted in a fashion dependent upon K^+ concentration (100), which suggests that the ouabain effect is mediated by its membrane-specific disposition. Potassium sensitivity is not a feature of other known inducers and a definitive interpretation of the ouabain studies has yet to be formulated. Other agents with known or postulated membrane sites of action inhibit chemical induction of MELC. Local anesthetics, for example cocaine and tetracaine, inhibit both Me_2SO-induced hemoglobin synthesis and the pattern terminal cell divisions characteristic of MELC erythroid differentiation (168). These same agents counteract the effects of polar-planar compounds on the phase transition temperature of phospholipid membranes (167). Another class of compounds, the tumor promoters, typified by the phorbol ester 12-O-tetradecanoylphorbol-13-acetate (TPA), has been demonstrated to inhibit both spontaneous and chemically induced erythroid differentiation (170, 171). MELC must be exposed to these compounds during the first 24 hr of culture with inducers, and prior to irreversible commitment to differentiate, to achieve suppression of differentiation (171). Although a definitive site of action for the tumor-promoting phorbol esters has not been established, considerable evidence suggests an effect at plasma membrane or upon plasma membrane-related functions (172–174). It may also be noted that certain inducing agents, notably butyric acid, display effects on developmental cell systems related to cyclic nucleotide metabolism (175, 176). Taken together, the evidence suggests that further exploration of early membrane-related events during induced erythroid cell differentiation may prove valuable in the elucidation of the nature of regulatory mechanisms governing the initiation of this differentiation program.

VARIANT PHENOTYPES AND SOMATIC CELL GENETICS

The unlimited growth potential of MELC in culture recommends them for genetic studies designed to explore regulatory pathways in differentiation. To date such studies have been largely concerned with the isolation of inducer-resistant and inducer-sensitive variants, their characterization, and the use of these and related cell lines in cell hybridization studies designed to reveal genetic properties.

Isolation and Characterization of Variants

The capacity to differentiate is a uniform property of virtually all cells in a MELC population. This property has been established by several lines of evidence that include (a) repeated recloning of parent stocks, (b) the induction response of primary clones which reflects the inducibility of the parent stock (76, 177), and (c) the observation that almost 100% of some MELC lines can be induced to differentiate by active agents (88). However, variant cell lines with degrees of resistence to induction by chemical inducers can be found in MELC populations at a frequency of approximately 3 X 10^{-5} (80). Resistance variants have been isolated both by testing randomly isolated clones (79, 81, 83, 177), as well as by selection in media containing the inducer, most commonly Me$_2$SO (80, 82, 95). Under the latter conditions most MELC are induced to differentiate, and enter their terminal cell divisions. The frequency of resistance variants is significantly increased in the residual cell population capable of continued growth in the presence of Me$_2$SO. Almost all of the variant inducer-resistant cell lines, whether isolated at random or by selection, display one or both of two features that readily identify them as MELC, namely, the persistence of spontaneously differentiating erythroblasts, and the potential to differentiate when exposed to a different inducing agent (80, 83, 90, 95). Chromosome studies of several MELC variants and their parental strains have failed to identify unique karyotypes that correlate with characteristics of inducibility (83, 95, 125, 177).

Inducer-resistant variants have been characterized biochemically, largely with respect to their ability to synthesize globin and to accumulate globin RNA. In virtually every instance adequately tested, resistance variants are defective in accumulation of globin mRNA (79, 82, 95, 178). Although these observations have been adduced as evidence that the resistance defect lies at the level of globin mRNA transcription or processing, several lines of evidence suggest that the effects of inducers and of resistance variants may be expressed at a level of regulation of differentiation separate from the initiation of globin mRNA synthesis. First, induction triggers a program or series of programs (179) of differentiation including many biosynthetic and morphogenetic activities in addition to globin synthesis. Most resistance variants fail to express this program. Second, as already noted, resistance variants generally retain an appreciable level of spontaneous differentiation and the levels of spontaneous and induced differentiation vary independently (79, 80). Third, the degree of maturation of individual cells, measured as the accumulation of hemoglobin (79, 88) and globin mRNA (79), is not dependent upon the overall rate of differentiation in culture. Taken together these observations suggest, as noted by Orkin et al (79), that most inducers and resistance variants operate at a step that controls the commitment to MELC to differentiate, a step proximal to and distinct from the actual initiation of transcription of globin mRNA. The as yet unraveled complexity of these control mechanisms is exemplified by a variant cell line (Fw), characterized by Harrison et al (178), which appears to be blocked in differentiation by a heme-reversible mechanism of undetermined nature.

Somatic Cell Genetics

Most hybrids between MELC and nonerythropoietic cell lines (79, 178, 180–184) demonstrate complete extinction of both spontaneous and chemically induced erythropoietic differentiation including both globin synthesis and other features in the hybrid progeny. Inducible erythropoiesis has been demonstrated, however, in several hybrid cell lines derived from the fusion of a stable 2S (pseudo-tetraploid) MELC parent and a fibroblast cell line, which suggests that gene dosage may be important in regulating expression of the erythropoietic program of differentiation in hybrids (185). MELC-lymphoma hybrids show extinction of erythropoiesis. From these hybrids, however, subclones have been isolated in which hemoglobin production is restored. Chromosome and marker enzyme analysis demonstrates that restoration of hemoglobin synthesis is associated with loss of an X chromosome contributed by the nonerythroleukemic parent (186). These results suggest that a locus or loci on the X chromosome are capable of regulating the response of MELC to Me_2SO. Extinction of erythropoietic characteristics in MELC may also be the consequence of cytoplasmic factors contributed by the nonerythroid parent, as demonstrated by studies on the fusion of MELC with anucleated neuroblastoma cytoplasm (181). The best available evidence suggests that the cell fusion process itself is not responsible for extinction (79, 178, 180), nor is extinction due to loss of mouse globin genes from the hybrid progeny, as demonstrated by molecular hybridization (187).

Cell-cell hybridization studies have been employed to study dominance relationships among MELC strains differing with respect to levels of spontaneous or inducible differentiation (79, 178). In the studies of Harrison et al (178) hybrids that demonstrate both transdominant inducibility and recessive inducibility have been produced. Hybrids between cell strains demonstrating high and low rates of spontaneous differentiation display intermediate levels of spontaneous differentiation (79). At the present time no firm conclusions can be drawn as to the unique genetic properties of any of the available inducibility variants nor have these studies, to date, made a definitive contribution to the identification of the level of control exercised by the variant inducibility phenotypes, save for the broad interpretation that most variants appear to express their control at a level proximal to accumulation of globin mRNA (178).

RELATIONSHIP OF THE CELL CYCLE TO COMMITMENT AND EXPRESSION OF THE ERYTHROID PROGRAM

A major issue with respect to induced MELC differentiation is the nature of events determining the transition from a self-renewing, nondifferentiating population to a population of cells committed to expression of the characteristic erythroid program of development. There is evidence that the transition of MELC to the synthesis of differentiated proteins is related to an alteration in the cell cycle, which includes prolongation of G_1 and possibly a requirement for DNA synthesis (4, 15, 105, 188). For example, MELC synchronized with respect to the cell division cycle by exposure to high levels of thymidine (105), will differentiate if Me_2SO is present during at least one round of DNA synthesis. McClintock & Papaconstantinou (188), and Harrison and colleagues (4) have also provided evidence that MELC differentiation is dependent on at least one or more rounds of DNA synthesis. On the other hand, Leder et al (189) report that butyric acid induces MELC in the presence of inhibitors of DNA synthesis, hydroxyurea, or cytosine arabinoside, and in the absence of cell division. They conclude that globin gene expression does not require DNA synthesis or cell division. These conclusions are open to question because neither hydroxyurea nor cytosine arabinoside completely inhibit thymidine incorporation into DNA (15), and both hydroxyurea and cytosine arabinoside are themselves inducers of MELC differentiation (190).

After a latent period in culture with inducer, MELC may be removed into inducer-free medium and a fraction of the cells will go on to differentiate (are "committed") in the absence of inducer (105, 115, 191). The kinetics of commitment can be quantitated at the single cell level by employing inducer-free semisolid cloning media and scoring those colonies that contain benzidine-reactive cells. Each such colony represents the progeny of a single committed precursor (115, 191). The accumulation of committed MELC is linearly related to the time of exposure to inducer prior to cloning. This suggests that expression of the program of differentiation is initiated in a stochastic manner (115, 191, 192), that is, there is a constant probability of differentiation which is maintained through several generations during culture with inducer. This probability is dependent upon a number of factors including the nature and concentration of the inducer, and the genetic and culture history of the cell line (115, 191).

The evidence that a cell cycle-related event, a prolonged G_1, and/or DNA synthesis, is required for commitment of MELC to differentiate, and that the overall expression of the differentiated program displays stochastic properties, suggests the possibility that commitment and expression of differentiation, including the accumulation of mRNA for differentiation-specific proteins, represents a multistep process with different regulatory elements. There is evidence, already partly reviewed, which suggests that inducers act at a target site distinct from the globin structural genes or cis-acting regulatory genes for globin. First, culture of MELC under suboptimal conditions of inducer concentration is associated with a lower proportion of the cells committed to differentiate, but the program of differentiation in committed cells, for example, the average hemoglobin concentration per committed cell, is unaffected (88). Second, inducers have no detectable direct effect on the rate or amount of hemoglobin synthesis in cells already committed to differentiate. Third, suboptimal time in, or concentration of inducer initiates unstable differentiation, that is, mixed colonies containing both differentiating and uninduced progeny (115, 191), which reinforces the concept of a multistep process for the induction of differentiation. Finally, there are inhibitors of MELC differentiation, for example, the phorbol esters (171, 193) and 5-bromo-2'-deoxyuridine (101, 118, 194) that are effective only if added to cultures prior to commitment. They do not appear to affect the expression of globin genes once commitment has occurred.

CHROMATIN STRUCTURE AND INDUCED DIFFERENTIATION

Several lines of evidence, reviewed above, suggest that alterations in chromatin structure accompany induced MELC differentiation and might be implicated in differentiation. This evidence includes the following: (a) some active agents, such as actinomycin, and UV and X irradiation, act selectively to alter DNA structure; (b) there are demonstrable alterations in DNA and in nuclear proteins; and (c) DNA synthesis or cell division may be required for commitment to differentiate. It must be emphasized that none of the data provides direct evidence of a relationship between structural changes in chromatin and the event(s) determining MELC commitment. In each instance the observed changes may reflect effects of the inducing agent unrelated to commitment to differentiation.

There is evidence from nonerythroid systems that lends credence to a putative chromatin site for inducer action. Me_2SO (140 mM) can directly alter the physical-chemical properties of chromatin (195), by decreasing the thermostability of all base pairs and destabilizing protein-DNA interactions. Such a denaturing effect on chromatin is probably nonspecific. Nevertheless, in studies with *Escherichia coli*, Nakanishi and co-workers (196) and Travers (197) have shown that, both in vivo and in vitro, Me_2SO can initiate a specific pattern of expression of structural genes. On the basis of these considerations one may speculate that nonspecific effects of inducers on chromatin structure result in a specific program of transcription owing to the specificity of chromatin structure already established within the erythroid precursor cell.

THE MECHANISM OF INDUCED DIFFERENTIATION: AN HYPOTHESIS

MELC appear to be erythroid precursor cells transformed by the Friend virus complex. One component of this virus complex, SFFV, is responsible for uncoupling cellular replication from erythropoietin, the physiological regulator of erythroid precursor cell proliferation. By some mechanism, perhaps selection of cells with the greatest growth potential, cell lines have been established in which the normal process of erythropoietic differentiation, and concomitant terminal cell division, has become suppressed. This suppression may have been accomplished by changes in cell cycle kinetics, including shortening of the normal G_1. A wide variety of agents have been discovered that can significantly alter the probability for commitment to, and expression of, the program of erythropoietic differentiation. The diversity of these agents makes it, at present, difficult to provide a single generalizing statement as to the essential property of biologically active agents. Indeed, the present evidence suggests that different agents may have different primary sites of action, for example, at the plasma membrane or at the level of chromatin. Presumably all inducing agents affect in common certain steps required for commitment of MELC to the erythroid program of terminal differentiation. Commitment is defined as the capacity of MELC,

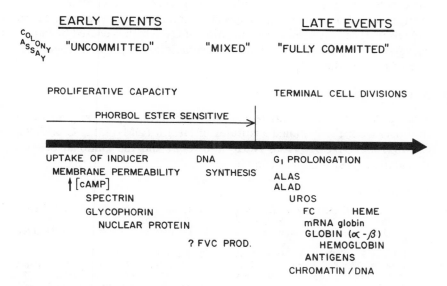

Figure 1   Hypothetical schema of induced MELC differentiation. Early events are defined as metabolic changes in MELC cultured with inducer which occur prior to commitment. Late events are defined as those occurring at or after commitment. Commitment is defined as the capacity of MELC, transferred from culture with inducer to cloning culture without inducer, to form hemoglobin and undergo terminal cell divisions. Below the thick black arrow are listed metabolic changes that appear to be induced before commitment (early events) and those that appear to be induced with or after commitment (late events). No single study has encompassed all the listed parameters, and the various parameters have been assayed in different experimental conditions and predominately in nonsynchronized cultures. Accordingly, assignment of parameters as early or late, with the probable exceptions of uptake of inducer, increased membrane permeability, hemoglobin accumulation, and terminal cell division, must be considered tentative. Abbreviations: ALAS, aminolevulinic acid synthetase; ALAD, aminolevulinic acid dehydrase; UROS, uroprotoporphyrin I synthetase; FC, ferrochelatase; FVC, Friend virus complex.

in the absence of inducer, to accumulate hemoglobin and undergo terminal cell divisions (115). Commitment involves events separate from globin mRNA accumulation and hemoglobin synthesis. Thus, commitment is affected by inducer concentration and inhibited by phorbol esters (115, 171), while hemoglobin formation in committed cells is not affected by either. Inducers cause changes in MELC that occur prior to commitment, when cells retain the capacity for extensive proliferation, and these changes are referred to as early events (Figure 1). Early events need not necessarily lead to commitment but may be involved in a sequential series of changes initiated by the inducer which culminate in commitment. The terminal or late events in expression of the erythroid program occur at or after commitment, including hemoglobin and specific membrane antigen accumulation and terminal cell division. A prolongation of G_1, which is caused by inducing agents, may be associated with the accumulation of products characteristic of the late differentiation.

ACKNOWLEDGMENTS

Work by the authors of this review was supported in part by grants and contracts from the National Institutes of Health (GM-14552, CA-13696, CA-18316, NO1-CB-4-4008, NO1-CP-6-1008) and the National Science Foundation (NSF-PCM-75-08696).

Literature Cited

1. McCulloch, E. A. 1970. In *Regulation of Hematopoiesis,* ed. A. S. Gordon, 1:133–59. New York: Appleton-Century-Crofts
2. Rifkind, R. A., Bank, A., Marks, P. A. 1974. In *The Red Blood Cell,* ed. D. McN. Surgenor, 1:51–89. New York: Academic
3. Rifkind, R. A., Marks, P. A. 1975. *Blood Cells* 1:417–28
4. Harrison, P. R. 1976. *Nature* 262: 353–56
5. Marks, P. A., Rifkind, R. A. 1972. *Science* 175:955–61
6. Rifkind, R. A., Marks, P. A., Bank, A., Terada, M., Maniatis, G. M., Reuben, R., Fibach, E. 1976. *Ann. Immunol.* 127C:887–93
7. Trentin, J. J. 1970. In *Regulation of Hematopoiesis,* ed. A. S. Gordon, 1:161–86. New York: Appleton-Century-Crofts
8. Russell, E. S. See Ref. 7, pp. 649–75
9. Stephenson, J. A., Axelrad, A. A., McLeod, D. L., Shreeve, M. 1971. *Proc. Natl. Acad. Sci. USA* 68:1542–46
10. Gregory, C. J. 1976. *J. Cell Physiol.* 89:289–302
11. Gregory, C. J., Eaves, A. C. 1977. *Blood* 49:855–64
12. Ramirez, F., Gambino, R., Maniatis, G. M., Rifkind, R. A., Marks, P. A., Bank, A. 1975. *J. Biol. Chem.* 250:6054–58
13. Till, J. E., McCulloch, E. A., Siminovitch, L. 1964. *Proc. Natl. Acad. Sci. USA* 51:29–36
14. Holtzer, H., Weintraub, H., Mayne, R., Mochan, B. 1972. *Curr. Top. Dev. Biol.* 7:229–56
15. Marks, P. A., Rifkind, R. A., Bank, A., Terada, M., Maniatis, G., Reuben, R. C., Fibach, E. 1977. In *Growth Kinetics and Biochemical Regulation of Normal and Malignant Cells,* ed. B. Drewinko, R. M. Humphrey, pp. 329–46. Baltimore: Williams & Wilkins
16. Friend, C. 1957. *J. Exp. Med.* 105: 307–18
17. Buffett, R. F., Furth, J. 1959. *Cancer Res.* 19:1063–69
18. Friend, C., Haddad, J. R. 1960. *J. Natl. Cancer Inst.* 25:1279–89

19. Metcalf, D., Furth, J., Buffett, R. F. 1959. *Cancer Res.* 19:52–58
20. Zajdela, F. 1962. *Bull. Assoc. Fr. Etude Cancer* 49:351–58
21. Ikawa, Y., Sugano, H. 1967. *Gann* 58:155–60
22. Ikawa, Y., Sugano, H., Furusawa, M. 1972. *Gann Monogr. Cancer Res.* 12: 33–45
23. Mirand, E. A. 1966. *Natl. Cancer Inst. Monogr.* 22:483–501
24. Tambourin, P. E., Gallien-Lartigue, O., Wendling, F., Huaulme, D. 1973. *Br. J. Haematol.* 24:511–24
25. Smadja-Joffe, F., Jasmin, C., Malaise, E. P., Bournoutian, C. 1973. *Int. J. Cancer* 11:300–13
26. Smadja-Joffe, F., Klein, B., Kerdiles, C., Feinendegen, L., Jasmin, C. 1976. *Cell Tissue Kinet.* 9:131–45
27. Sassa, S., Takaku, F., Nakao, K. 1968. *Blood* 31:758–65
28. Mirand, E. A., Steeves, R. A., Lang, R. D., Grace, J. T. Jr. 1968. *Proc. Soc. Exp. Biol. Med.* 128:844–49
29. McGarry, M. P., Mirand, E. A. 1973. *Exp. Hemat. Oak Ridge Tenn.* 1: 174–82
30. Horoszewicz, J. S., Leong, S. S., Carter, W. A. 1975. *J. Natl. Cancer Inst.* 54: 265–67
31. Liao, S. K., Axelrad, A. A. 1975. *Int. J. Cancer* 15:467–82
32. Hankins, W. D., Krantz, S. B. 1975. *Nature* 253:731–32
33. Thomson, S., Axelrad, A. A. 1968. *Cancer Res.* 28:2105–14
34. Steeves, R. A., Mirand, E. A., Thomson, S. 1970. In *Comparative Leukemia Research,* ed. R. M. Dutcher, pp. 624–33. Basel: Karger
35. Rossi, G. B., Friend, C. 1970. *J. Cell Physiol.* 76:159–66
36. Dietz, M., Fouchey, S. P., Longley, C., Rich, M. A., Furmanski, P. 1977. *J. Exp. Med.* 145:594–606
37. Dietz, M., Furmanski, P., Clymer, R., Rich, M. A. 1976. *J. Natl. Cancer Inst.* 57:91–95
38. Bubbers, J. E., Blank, K. J., Freedman, H. A., Lilly, F. 1977. *Scand. J. Immunol.* 6:533–39
39. Lilly, F., Pincus, T. 1973. *Adv. Cancer Res.* 17:231–77
40. Lilly, F. 1972. *J. Natl. Cancer Inst.* 49:927–34
41. Steeves, R. A. 1975. *J. Natl. Cancer Inst.* 54:289–97
42. Mirand, E. A., Steeves, R. A., Avila, L., Grace, J. T. Jr. 1968. *Proc. Soc. Exp. Biol. Med.* 127:900–4
43. Axelrad, A. A., Steeves, R. A. 1964. *Virology* 24:513–18
44. Dawson, P. J., Rose, W. M., Fieldsteel, A. H. 1966. *Br. J. Cancer* 20:114–21
45. Mirand, E. A., Grace, J. T. 1962. *Virology* 17:364–66
46. Rowson, K. E., Parr, I. B. 1970. *Int. J. Cancer* 5:96–102
47. Troxler, D. H., Parks, W. P., Vass, W. C., Scolnick, E. M. 1977. *Virology* 76:602–15
48. Troxler, D. H., Boyars, J. K., Parks, W. P., Scolnick, E. M. 1977. *J. Virol.* 22:361–72
49. Ikawa, Y., Aida, M., Saito, M. 1975. *Gann* 66:583–84
50. Sveda, M. M., Fields, B. N., Soeiro, R. 1974. *Cell* 2:271–77
51. Huang, A. S., Besmer, P., Chu, L., Baltimore, D. 1973. *J. Virol.* 12:659–62
52. Yoshikura, H. 1973. *J. Gen. Virol.* 19:321–27
53. Sveda, M. M., Soeiro, R. 1976. *Proc. Natl. Acad. Sci. USA* 73:2356–60
54. Blank, K. J., Steeves, R. A., Lilly, F. 1976. *J. Natl. Cancer Inst.* 57:925–30
55. Mirand, E. A. 1967. *Science* 156: 832–33
56. Steeves, R. A., Mirand, E. A., Thomson, S., Avila, L. 1969. *Cancer Res.* 29:1111–16
57. Tambourin, P., Wendling, F. 1971. *Nature New Biol.* 234:230–33
58. McGarry, M. P., Mirand, E. A. 1973. *Proc. Soc. Exp. Biol. Med.* 142:538–41
59. Tambourin, P. E., Wendling, F. 1975. *Nature* 256:320–22
60. Fredrickson, T., Tambourin, P., Wendling, F., Jasmin, C., Smajda, F. 1975. *J. Nat. Cancer Inst.* 55:443–46
61. Nasrallah, A. G., McGarry, M. P. 1976. *J. Natl. Cancer Inst.* 57:443–45
62. Chirigos, M. A., Marsh, R. W. 1966. *Chemotherapy* 6:489–96
63. Odaka, T. 1969. *J. Exp. Med.* 39:99–100
64. Bennett, M., Steeves, R. A., Cudkowicz, G., Mirand, E. A., Russell, L. B. 1968. *Science* 162:564–65
65. Steeves, R. A., Bennett, M., Mirand, E. A., Cudkowicz, G. 1968. *Nature* 218: 372–74
66. Clarke, B. J., Axelrad, A. A., Shreeve, M. M., McLeod, D. L. 1975. *Proc. Natl. Acad. Sci. USA* 72:3556–60
67. Brown, W. M., Axelrad, A. A. 1976. *Int. J. Cancer* 18:764–73
68. Golde, D. W., Friend, C. 1976. In *Progress in Differentiation Research,* ed. N. Muller-Berat, pp. 513–20. Amsterdam: North-Holland

69. Golde, D. W., Faille, A., Sullivan, A., Friend, C. 1976. *Cancer Res.* 36:115–19
70. Clarke, B. J., Axelrad, A. A., Housman, D. 1976. *J. Natl. Cancer Inst.* 57: 853–59
71. Dube, S. K., Pragnell, I. B., Kluge, N., Gaedicke, G., Steinheider, G., Ostertag, W. 1975. *Proc. Natl. Acad. Sci. USA* 72:1863–67
72. Pragnell, I. B., Ostertag, W., Harrison, P. R., Williamson, R., Paul, J. 1976. See Ref. 68, pp. 501–11
73. Sherton, C. C., Evans, L. H., Polonoff, E., Kabat, D. 1976. *J. Virol.* 19:118–25
74. Friend, C., Patuleia, M. C., deHarven, E. 1966. *Nat. Cancer Inst. Monogr.* 22:505–22
75. Patuleia, M. C., Friend, C. 1967. *Cancer Res.* 27:726–30
76. Singer, D., Cooper, M., Maniatis, G. M. 1974. *Proc. Natl. Acad. Sci. USA* 71:2668–70
77. Ikawa, Y., Sugano, H. 1966. *Gann* 57:641–43
78. Ikawa, Y., Ross, J., Leder, P., Gielen, J., Packman, S., Ebert, P., Hayashi, K., Sugano, H. 1973. In *Proc. Int. Symp. Princess Takomatsu Cancer Res. Fund, 4th* ed. W. Nakahara, T. Ono, T. Sugimura, H. Sugano, pp. 515–47. Tokyo: Univ. Tokyo Press
79. Orkin, S., Harosi, F. I., Leder, P. 1975. *Proc. Natl. Acad. Sci. USA* 72:98–102
80. Rovera, G., Bonaiuto, J. 1976. *Cancer Res.* 36:4057–61
81. Ikawa, Y., Aida, M., Inoue, Y. 1976. *Gann* 67:767–70
82. Ikawa, Y., Inoue, Y., Aida, M., Kameji, R., Shibata, C., Sugano, H. 1976. *Bibl. Haematol. Basel* 43:37–47
83. Preisler, H. D., Shiraishi, Y., Mori, M., Sandberg, A. A. 1976. *Cell Differ.* 5:207–16
84. Singer, D. S. 1975. *Erythropoietic differentiation in murine erythroleukemia cells.* PhD thesis. Columbia Univ., New York. 267 pp.
85. Kluge, N., Gaedicke, G., Steinheider, G., Dube, S., Ostertag, W. 1974. *Exp. Cell Res.* 88:257–62
86. Friend, C., Scher, W., Holland, J. G., Sato, T. 1971. *Proc. Natl. Acad. Sci. USA* 68:378–82
87. Tanaka, M., Levy, J., Terada, M., Breslow, R., Rifkind, R. A., Marks, P. A. 1975. *Proc. Natl. Acad. Sci. USA* 72:1003–6
88. Reuben, R. C., Wife, R. L., Breslow, R., Rifkind, R. A., Marks, P. A. 1976. *Proc. Natl. Acad. Sci. USA* 73:862–66
89. Preisler, H. D., Christoff, G., Taylor, E. 1976. *Blood* 47:363–68
90. Gusella, J., Housman, D. 1976. *Cell* 8:263–69
91. Ross, J., Sautner, D. 1976. *Cell* 8: 513–20
92. Leder, A., Leder, P. 1975. *Cell* 5: 319–22
93. Takahashi, E., Yamada, M., Saito, M., Kuboyama, M., Ogasa, K. 1975. *Gann* 66:577–80
94. Reuben, R. C., Marks, P. A., Rifkind, R. A., Terada, M., Fibach, E., Nudel, U., Gazitt, Y., Breslow, R. 1978. *Oji International Seminar on Genetic Aspects of Friend Virus and Friend Cells,* ed. Y. Ikawa. New York: Academic. In press
95. Ohta, Y., Tanaka, B., Terada, M., Miller, O. J., Bank, A., Marks, P. A., Rifkind, R. A. 1976. *Proc. Natl. Acad. Sci. USA* 73:1232–63
96. Harrison, P. R. 1978. *Br. Med. Bull.* In press
97. Furusawa, M., Ikawa, Y., Sugano, H. 1971. *Proc. Jpn. Acad.* 47:220–24
98. Hu, H.-Y. Y., Gardner, J., Aisen, P., Skoultchi, A. I. 1977. *Science* 197: 559–61
99. Malpoix, P. 1976. *Arch. Int. Physiol. Biochem.* 84:1090–91
100. Bernstein, A., Hunt, D. M., Crickley, V., Mak, T. W. 1976. *Cell* 9:375–81
101. Scher, W., Preisler, H. D., Friend, C. 1973. *J. Cell Physiol.* 81:63–70
102. Curtis, P., Weissman, C. 1976. *J. Mol. Biol.* 106:1061–75
103. Terada, M., Nudel, U., Rifkind, R. A., Marks, P. A. 1978. *Cancer Res.* In press
104. Scher, W., Friend, C. 1978. *Cancer Res.* In press
105. Levy, J., Terada, M., Rifkind, R. A., Marks, P. A. 1975. *Proc. Natl. Acad. Sci. USA* 72:28–32
106. Reuben, R. C., Rifkind, R. A., Marks, P. A. 1977. *Fed. Proc.* 36:886 (Abstr.)
107. Kluge, N., Ostertag, W., Sugiyama, T., Arndt-Jovin, D., Steinheider, G., Furusawa, M., Dube, S. K. 1976. *Proc. Natl. Acad. Sci. USA* 73:1237–40
108. Kimhi, Y., Palfrey, C., Spector, I., Barak, Y., Littauer, U. Z. 1976. *Proc. Natl. Acad. Sci. USA* 73:462–66
109. Palfrey, C., Kimhi, Y., Littauer, U. Z., Reuben, R. C., Marks, P. A. 1977. *Biochem. Biophys. Res. Commun.* 76: 937–42
110. Wenner, C. E., Hackney, J., Kimelberg, H. K., Mayhew, E. 1974. *Cancer Res.* 34:1731–37
111. Wright, J. A. 1973. *Exp. Cell Res.* 78:456–60
112. Prasad, K. N. 1971. *Nature* 234:471–73

113. Friend, C., Scher, W., Preisler, H. 1974. *Ann. NY Acad. Sci.* 241:582–88
114. Preisler, H. D., Bjornsson, S., Mori, M., Lyman, G. H. 1976. *Br. J. Cancer* 33:634–45
115. Fibach, E., Reuben, R. C., Rifkind, R. A., Marks, P. A. 1977. *Cancer Res.* 37:440–44
116. Ross, J., Gielen, J., Packman, S., Ikawa, Y., Leder, P. 1974. *J. Mol. Biol.* 87: 697–14
117. Ross, J., Ikawa, Y., Leder, P. 1972. *Proc. Natl. Acad. Sci. USA* 69:3620–23
118. Ostertag, W., Crozier, T., Kluge, N., Melderis, H., Dube, S. 1973. *Nature* 243:203–5
119. Bastos, R. N., Aviv, H. 1977. *Cell* 11:641–50
120. Curtis, P. J., Mantei, N., van den Berg, J., Weissman, C. 1977. *Proc. Natl. Acad. Sci. USA* 74:3184–88
121. Conkie, D., Affara, N., Harrison, P. R., Paul, J., Jones, K. 1974. *J. Cell Biol.* 63:414–19
122. Aviv, H., Voloch, Z., Bastos, R., Levy, S. 1976. *Cell* 8:495–503
123. Nudel, U., Salmon, J., Fibach, E., Terada, M., Rifkind, R., Marks, P. A., Bank, A. 1977. *Cell* 12:463–69
124. Boyer, S. H., Wuu, K. D., Noyes, A. N., Young, R., Scher, W., Friend, C., Preisler, H. D., Bank, A. 1972. *Blood* 40:823–35
125. Ostertag, W., Melderis, H., Steinheider, G., Kluge, N., Dube, S. K. 1972. *Nature New Biol.* 239:231–34
126. Ebert, P. S., Ikawa, Y. 1974. *Proc. Soc. Exp. Biol. Med.* 146:601–4
127. Kabat, D., Sherton, C. C., Evans, L. H., Bigley, R., Koler, R. D. 1975. *Cell* 5:331–38
128. Reem, G. H., Friend, C. 1975. *Proc. Natl. Acad. Sci. USA* 72:1630–34
129. Eisen, H., Nasi, S., Georgopoulos, C. P., Arndt-Jovin, D., Ostertag, W. 1977. *Cell* 10:689–95
130. Arndt-Jovin, D. J., Ostertag, W., Eisen, H., Klimak, F., Jovin, T. M. 1976. *J. Histochem. Cytochem.* 24:332–47
131. Nudel, U., Salmon, J., Terada, M., Bank, A., Rifkind, R. A., Marks, P. A. 1977. *Proc. Natl. Acad. Sci. USA* 74:1100–4
132. Packman, S., Aviv, H., Ross, J., Leder, P. 1972. *Biochem. Biophys. Res. Commun.* 49:813–19
133. Harrison, P. R., Birnie, G. D., Hell, A., Humphries, S., Young, B. D., Paul, J. 1974. *J. Mol. Biol.* 84:539–54
134. Ramirez, F., Natta, C., O'Donnell, J. V., Canale, V., Bailey, G., Sanguensermsri, T., Maniatis, G., Marks, P. A.,

Bank, A. 1975. *Proc. Natl. Acad. Sci. USA* 72:1550–54
135. Terada, M., Fried, J., Nudel, U., Rifkind, R. A., Marks, P. A. 1977. *Proc. Natl. Acad. Sci. USA* 74:248–52
136. Darzynkiewicz, Z., Traganos, F., Sharpless, T., Friend, C., Melamid, M. R. 1976. *Exp. Cell Res.* 99:301–9
137. Peterson, J. L., McConkey, E. H. 1976. *J. Biol. Chem.* 251:555–58
138. Blankstein, L. A., Levy, S. B. 1976. *Nature* 260:638–40
139. Lau, A. F., Ruddon, R. W. 1977. *Exp. Cell Res.* 107:35–46
140. Riggs, M. G., Whittaker, R. G., Neumann, J. R., Ingram, V. M. 1977. *Nature* 268:462–64
141. Keppel, F., Allet, B., Eisen, H. 1977. *Proc. Natl. Acad. Sci. USA* 74:653–56
142. Sherton, C. C., Kabat, D. 1976. *Dev. Biol.* 48:118–31
143. Orkin, S. H., Swerdlow, P. S. 1977. *Proc. Natl. Acad. Sci. USA* 74:2475–79
144. Ross, J. 1976. *J. Mol. Biol.* 106:403–20
145. Kwan, S-P., Wood, T. G., Lingrel, J. B. 1977. *Proc. Natl. Acad. Sci. USA* 74: 178–82
146. Minty, A., Kleiman, L., Birnie, G., Paul, J. 1977. *Biochem. Soc. Trans.* 5:679–81
147. Getz, M. J., Birnie, G. D., Young, B. D., MacPhail, E., Paul, J. 1975. *Cell* 4:121–29
148. Ostertag, W., Gaedicke, G., Kluge, N., Melderis, H., Weimann, B., Dube, S. K. 1973. *FEBS Lett.* 32:218–22
149. Gilmour, R. S., Harrison, P. R., Windass, J. D., Affara, N. A., Paul, J. 1974. *Cell Differ.* 3:9–22
150. Harrison, P. R., Gilmour, R. S., Affara, N. A., Conkie, C., Paul, J. 1974. *Cell Differ.* 3:23–30
151. Orkin, S. H., Swan, D., Leder, P. 1975. *J. Biol. Chem.* 250:8753–60
152. Hutton, J. J., Bishop, J., Schweet, R., Russell, E. S. 1962. *Proc. Natl. Acad. Sci. USA* 48:1505–13
153. Popp, R. A., Bailiff, E. G. 1973. *Biochim. Biophys. Acta* 303:61–67
154. Gilman, J. 1976. *Biochem. J.* 155: 231–41
155. Gilman, J. 1976. *Biochem. J.* 159:43–53
156. Alter, B. P., Goff, S. C. 1976. *Blood* 48:981 (Abstr.)
157. Denton, M. J., Spencer, N., Arnstein, H. R. V. 1975. *Biochem. J.* 146:205–11
158. Hunt, T. 1976. *Br. Med. Bull.* 32: 257–61
159. Friend, C., Scher, W., Rossi, G. B. 1970. In *The Biology of Large RNA Viruses,* ed. R. D. Barry, B. W. J.

Mahy, pp. 267–74. London/New York: Academic

160. Sassa, S. 1976. *J. Exp. Med.* 143:305–15

161. Sassa, S., Takaku, F., Nakao, K., Ikawa, Y., Sugano, H. 1968. *Proc. Soc. Exp. Biol. Med.* 127:527–29

162. Pinphanichakarn, P., Kramer, G., Hardesty, B. 1977. *J. Biol. Chem.* 252:2106–12

163. Cimadevilla, J. M., Hardesty, B. 1975. *Biochem. Biophys. Res. Commun.* 63: 931–37

164. Ernst, V., Levin, D. H., Rann, R. S., London, I. M. 1976. *Proc. Natl. Acad. Sci. USA* 73:1112–16

165. Ikawa, Y., Furusawa, M., Sugano, H. 1973. *Bibl. Haematol. Basel* 39:955–67

166. Arndt-Jovin, D. J., Ostertag, W., Eisen, H., Jovin, T. M. 1976. In *Modern Trends in Human Leukemia II*, ed. R. Neth, R. C. Gallo, K. Mannweiler, W. C. Moloney, pp. 137–49. Munich: Lehmanns

167. Lyman, G. H., Preisler, H. D., Papahadjopoulos, D. 1976. *Nature* 262: 360–63

168. Bernstein, A., Boyd, A. S., Crichley, V., Lamb, V. 1975. In *Biogenesis and Turnover of Membrane Macromolecules*, ed. J. S. Cook, pp. 145–59. New York: Raven

169. Loritz, F., Bernstein, A., Miller, R. O. 1977. *J. Cell. Physiol.* 90:423–38

170. Rovera, G., O'Brien, T. G., Diamond, L. 1977. *Proc. Natl. Acad. Sci. USA* 74: 2894–98

171. Yamasaki, H., Fibach, E., Nudel, U., Weinstein, I. B., Rifkind, R. A., Marks, P. A. 1977. *Proc. Natl. Acad. Sci. USA.* 74:3451–55

172. Weinstein, I. B., Wigler, M., Pietropaolo, C. 1976. In *Origins of Human Cancer* ed. H. H. Hiatt, J. D. Watson, J. A. Winsten, pp. 751–72. Cold Spring Harbor, NY: Cold Spring Harbor Lab.

173. Wigler, M., Weinstein, I. B. 1976. *Nature* 259:232–33

174. Jacobson, K., Wenner, C. E., Kemp, G., Papahadjopoulos, D. 1975. *Cancer Res.* 35:2991–95

175. Prasad, K. N., Sinha, P. K. 1976. *In Vitro* 12:125–32

176. Tallman, J. F., Smith, C. C., Henneberry, R. C. 1977. *Proc. Natl. Acad. Sci. USA* 74:873–77

177. Paul, J., Hickey, I. 1974. *Exp. Cell Res.* 87:20–30

178. Harrison, P. R., Affara, N., Conkie, D., Rutherford, T., Sommerville, J., Paul, J.

1976. In *Progress in Differentiation Research*, ed. N. Muller-Berat, pp. 135–46. Amsterdam: North-Holland

179. Eisen, H., Sassa, S., Granick, D., Keppel, F. 1978. In *Cold Spring Harbor Symp. Differentiation Normal and Neoplastic Hematopoietic Cells*, ed. B. Clarkson. Cold Spring Harbor, NY: Cold Spring Harbor Lab. In press

180. Deisseroth, A., Burk, R., Picciano, D., Minna, J., Anderson, W. F., Nienhuis, A. 1975. *Proc. Natl. Acad. Sci. USA* 72:1102–6

181. Gopalakrishnan, T. V., Thompson, E. B., Anderson, W. F. 1977. *Proc. Natl. Acad. Sci. USA* 74:1642–46

182. Miller, R. A., Ruddle, F. H. 1977. *Dev. Biol.* 56:157–73

183. Conscience, J. F., Miller, R. A., Henry, J., Ruddle, F. H. 1977. *Exp. Cell Res.* 105:401–12

184. Conscience, J. F., Ruddle, F. H., Skoultchi, A., Darlington, G. J. 1977. *Somatic Cell Genet.* 3:157–72

185. Axelrod, D. E., Gopalakrishnan, T. V., French Anderson, F. W. 1978. *Somatic Cell Genet.* In press

186. Benoff, S., Skoultchi, A. I. 1978. *Cell.* In press

187. Deisseroth, A., Velez, R., Burk, R. D., Minna, J., Anderson, W. F., Nienhuis, A. 1976. *Somatic Cell Genet.* 2:373–84

188. McClintock, P. R., Papaconstantinou, J. 1974. *Proc. Natl. Acad. Sci. USA* 71:4551–55

189. Leder, A., Orkin, S., Leder, P. 1975. *Science* 190:893:94

190. Ebert, P. S., Wars, I., Buell, D. N. 1976. *Cancer Res.* 36:1809–13

191. Gusella, J., Geller, R., Clarke, B., Weeks, V., Housman, D. 1976. *Cell* 9:221–29

192. Korn, A. P., Henkelman, R. M., Ottensmeyer, F. P., Till, J. E. 1973. *Exp. Hematol.* 1:362–75

193. Yamasaki, H., Fibach, E., Weinstein, I. B., Nudel, U., Rifkind, R. A., Marks, P. A. 1978. In press. See Ref. 94.

194. Bick, M. D., Cullen, B. R. 1976. *Somatic Cell Genet.* 2:545–58

195. Lapeyre, J- N., Bekhor, I. 1974. *J. Mol. Biol.* 89:137–62

196. Nakanishi, S., Adhya, S., Gottesman, M., Pastan, I. 1974. *Cell* 3:39–46

197. Travers, A. 1974. *Cell* 3:97–104

198. Sugano, H., Furusawa, M., Kawaguchi, T., Ikawa, Y. 1973. *Bibl. Haematol. Basel* 39:943–54

Ann. Rev. Biochem. 1978. 47:449–79
Copyright © 1978 by Annual Reviews Inc. All rights reserved

PROTEINS THAT AFFECT DNA CONFORMATION

◆980

James J. Champoux

Department of Microbiology and Immunology, School of Medicine, University of Washington, Seattle, Washington 98195

CONTENTS

449

0066-4154/78/0701-0449$01.00

PERSPECTIVES AND SUMMARY

The intracellular structure of DNA is different from the Watson-Crick double helix structure of purified DNA (1) in two major respects. First, through its interaction with cellular proteins it exists in a tightly compacted conformation. Second, information transfer during the processes of DNA replication, transcription, and recombination requires that under some conditions the DNA strands separate. Complete strand separation must involve the "melting" of the base pairs and the simultaneous unwinding of the two strands.

The determinants of the tertiary structure of DNA in chromosomes is a subject of this review. Compared to the fully extended form of purified DNA in the B structure, the DNA in a chromosome must be folded in such a way that it is compacted approximately 8000-fold in length (2). Chromosome structure has been extensively reviewed recently (3–5), therefore this article focuses attention on the recent evidence relating to the primary level of DNA organization within the chromosome. In particular we consider how the DNA duplex is folded into a nucleosome resulting in approximately a sevenfold length contraction and the formation of the unit fiber with a diameter of 100 Å.

Until recent years the problem of strand separation during DNA replication received little attention. However, it has become clear that chain elongation at the replication fork is not sufficient by itself to drive the unwinding process since, of all the DNA polymerases known, only the bacteriophage T5 DNA polymerase (6) and the *Escherichia coli* DNA polymerase I (7) are capable of separating the strands of the template DNA. In this review, we consider the following classes of proteins that may contribute to the process of strand separation in vivo: (*a*) proteins that bind specifically to single-stranded DNA (helix destabilizing proteins), or which melt duplex DNA directly (RNA polymerase); (*b*) enzymes that couple the hydrolysis of ATP to the direct unwinding of the helix; (*c*) enzymes that introduce a transient single-strand break into duplex DNA and thereby provide a swivel for helix unwinding (DNA swivelase); and (*d*) enzymes that introduce negative superhelical turns into topologically closed DNAs (DNA gyrase). These considerations obviously involve many areas in DNA biochemistry and the reader is referred to several recent and current review articles that are pertinent (8–17).

A consideration of the factors influencing DNA structure in vivo suggests that the naturally occurring superhelical turns in circular DNAs from eucaryotic sources, excluding mitochondrial DNA, may result entirely from the association of the DNA with histones to form nucleosomes, whereas superhelical turns in DNAs isolated from *E. coli* may be the result of the action of the DNA gyrase.

For the discussion below, proteins will be divided into two categories. The first category comprises those proteins that modify the conformation of DNA by binding to DNA. The second category includes those proteins that enzymatically alter DNA structure.

SUPERHELICAL TURNS

A brief description of superhelical turns and how they may arise is presented here to serve as background for the discussion to follow. We will refer to DNA as being superhelical if the axis of the DNA duplex follows a helical path in space. Examples of two kinds of regular superhelical turns, toroidal superhelices, and interwound superhelices, are illustrated in the paper by Vinograd et al (18).

We need to distinguish two basic kinds of superhelical turns depending on their origin. (a) Superhelical turns may be generated by the association of proteins or other structural elements in the cell with DNA. Such a superhelix can be formed in linear or circular DNA and is not dependent on the single strands being intact. (b) Superhelical turns may be an intrinsic property of a closed (both strands intact) circular DNA molecule or any domain of a larger DNA molecule containing intact strands bounded by barriers that somehow prevent the free winding of the helix. An example of the latter is the folded chromosome of *E. coli* that is composed of approximately 50 topologically isolated domains (19). The theory behind the occurrence of superhelical turns in a closed circular DNA also applies to these isolated domains.

A closed circular DNA is described by three parameters. The topological winding number, α, is the number of times one strand winds around the axis of the helix when the circle is forced to lie in a plane (18). This parameter, which has also been called the linking number (20, 21) is integral and invariant unless one of the strands is broken. The second parameter is the helix winding number (β) which is simply the number of helix turns in the unconstrained closed circular molecule. This number is equal to the number of turns in the nicked molecule of the same length (number of base pairs/10) providing the DNA is only moderately superhelical (22, 23). The parameter β is dependent on ionic strength, temperature, and pH (22), and is reduced by an unwinding ligand such as ethidium bromide (24).[1] If $\alpha = \beta$ then the DNA exhibits no superhelicity and is said to be relaxed. If, however, $\alpha \neq \beta$ then the circle will be superhelical. The third parameter

---

[1]The binding of each molecule of ethidium bromide to DNA is now known to unwind the helix by 26° (23, 25, 26). Early experiments measuring DNA superhelicity by titrating the turns with ethidium bromide were based on a 12° unwinding angle.

τ, the number of superhelical turns, is given by the equation $\tau = \alpha - \beta$ (18). Thus, if a molecule has an excess of topological turns ($\alpha > \beta$), τ will be positive. Likewise a deficiency of topological turns leads to negative values for τ. Virtually all naturally occurring DNA circles exhibit negative superhelical turns when purified and examined in vitro. Superhelicity is often expressed as superhelix density (σ) which is the number of superhelical turns per 10 base pairs.

There is a positive free energy associated with the negative superhelical turns in a closed circle (27–30) and this free energy predisposes such a molecule toward unwinding of the helix. Thus, a negatively superhelical molecule exhibits an early helix to coil transition when compared with its nicked counterpart during alkaline titration (18, 25). In fact, the free energy associated with the negative superhelices can assist any process requiring helix unwinding or strand separation. Below, we encounter several examples of processes that are facilitated in this way.

DNA BINDING PROTEINS THAT AFFECT CONFORMATION

Helix Destabilizing Proteins

Helix destabilizing proteins (HDP) are, by definition (14), proteins that bind tightly and preferentially to single-stranded DNA. Given this definition we can anticipate, at least theoretically, some of the properties of these proteins. (a) They should lower the equilibrium T_m for the thermally induced helix to coil transition. (b) Likewise, they may lower the optimum temperature for strand renaturation. (c) By binding to single-stranded DNA they may sterically hinder base pairing or the degradation of single strands by nucleases. (d) They may induce a conformation for the single-stranded DNA that is optimal for its function in some process, for example, as a template in replication. (e) By interacting with other proteins or enzymes they may potentiate a process involving single-stranded DNA and these proteins.

The large number of these protein molecules per cell suggests they may play a structural role in transactions involving DNA.

BACTERIOPHAGE T4 HELIX DESTABILIZING PROTEIN   The T4 gene 32 protein was the first single-strand specific DNA binding protein discovered (31, 32). This protein has been referred to as a "DNA unwinding protein" or a "DNA binding protein." Recently the term helix destabilizing protein has been suggested for this class of proteins (14). The T4 HDP is known from genetic studies to be required for DNA replication, recombination, and repair (33–35).

The protein binds only weakly to native DNA (31, 36), but exhibits a strong cooperative binding to denatured DNA (32). The cooperativity could derive from (*a*) the formation of protein oligomers (37, 38) that may possess a greatly increased binding affinity over the monomer for single strands, (*b*) the presence of protein-protein interactions such that the binding to a site on the DNA contiguous to an already bound site is much tighter than to an isolated site, (*c*) an alteration in the DNA structure adjacent to a bound protein molecule which increases the affinity of that site for the protein, or (*d*) some combination of these effects. Von Hippel and his colleagues have determined that the equilibrium constant for binding at a contiguous site is 1000 times the value for binding at an isolated site (36, 39).

The primary interaction of the protein with DNA may involve as few as two nucleotides (39) yet at saturation each protein covers between 7–10 nucleotides (32, 36). The bases are completely unstacked and the chain is more extended (4.6 Å/nucleotide) than uncomplexed single strands (2.9 Å/nucleotide) (40). However, the DNA complexed with the protein does not assume the fully extended conformation which would have a length of about 7 Å/nucleotide (32). The T4 HDP has a molecular weight of 35,000 and is asymmetric (32). From the dimensions of such a molecule, it appears that in the fully saturated DNA-protein complex, adjacent molecules are in direct contact. This suggests that protein-protein interactions may be the basis for the cooperative binding.

The affinity of the T4 HDP for single strands should lower the T_m for thermal denaturation of duplex DNA. The protein will promote the complete denaturation of poly dAT at room temperature (40°C below the normal T_m) under a variety of ionic conditions (32). However, the protein will not promote any appreciable strand separation of natural DNAs in spite of the fact that thermodynamic considerations predict that at equilibrium it should (36). This failure has been attributed to a kinetic block to the denaturation process, perhaps as a result of a requirement for an initiation loop larger than is normally present in native DNA (28, 41, 42). The failure of the protein to denature DNA may indicate that this is not one of its roles in the cell.

Although the protein does not cause extensive strand separation in vitro, there is evidence that it can bind to short regions that are rich in AT and destabilize the helix (40). Moreover, a proteolytic cleavage product of the T4 gene 32 protein (27,000 molecular weight) has been reported to have an increased capacity to separate the strands of natural DNA (43, 44). In the presence of the fixative glutaraldehyde, more extensive denaturation occurs from these sites which yields readily visible single-strand bubbles in electron micrographs. Using this technique, the early melting regions in the super-helical form of SV40 and PM2 DNA have been mapped (45, 46).

The T4 HDP will catalyze the renaturation of complementary single strands under physiological conditions (32, 47). This capacity, no doubt, reflects the removal of secondary structure from single-stranded DNA which would otherwise slow the renaturation process (48), while at the same time leaving the bases exposed.

The T4 HDP will stimulate the T4 DNA polymerase, but not *E. coli* polymerase I, when assayed with a single-stranded template (49). This stimulation is likely due to a direct interaction between the protein and the homologous polymerase as well as to an effect on the template DNA. In addition the protein will allow the T4 DNA polymerase to synthesize DNA at a single-strand break (50), presumably by facilitating the displacement of the nontemplate strand in concert with the polymerase. Finally, recent evidence indicates that the T4 HDP may interact specifically with an array of other T4 proteins involved in DNA replication and recombination (51–53).

The complex of single-stranded DNA with T4 HDP is resistant to degradation by nucleases (54, 55). Thus, one of the roles of the protein in the infected cell may be to protect exposed single strands from nucleases. The gene 32 protein regulates its own synthesis (56–58) at the level of translation. (59). This control prevents the synthesis of an excess of the protein over that needed to protect all of the single strands during infection.

E. COLI HELIX DESTABILIZING PROTEIN   A protein very similar to the T4 gene 32 protein has been isolated from uninfected *E. coli* (60–62). The *E. coli* HDP has a monomer molecular weight of 20,000 ± 2,000, but the functionally active unit is a tetramer. Unlike the T4 HDP (38) the *E. coli* protein does not appear to aggregate indefinitely in solution. The tetramer binds in a cooperative fashion to single-stranded DNA with each tetramer interacting with 30–36 nucleotides (60, 62–65). Whereas the T4 HDP causes an extension of single strands, the *E. coli* HDP causes approximately a 40% contraction in the overall length of single-stranded DNA (1.8 Å/nucleotide) (60). The protein appears to achieve this contraction by the regular folding, or perhaps supercoiling, of the single strands.

In contrast to the T4 gene 32 protein, the *E. coli* HDP will lower the T_m ($\sim 40°C$) for the helix to coil transition for natural DNA (60). However, the denaturation only occurs at low ionic strengths and in the absence of Mg^{2+}. This suggests that under physiological conditions the protein might promote renaturation of single strands rather than strand separation. In early experiments it was not possible to detect an increased rate of DNA renaturation in the presence of the *E. coli* HDP (60). More recently, it has been found that the protein will catalyze DNA renaturation in the presence of Mg^{2+} or Ca^{2+} at pH values below 6 or at physiological pH in the presence of spermidine, or spermine (65a). It has been shown that spermidine reduces the affinity of the HDP for single-stranded DNA (62), but the relationship between this observation and the effects of the polyamine on the protein catalyzed renaturation of DNA remain to be clarified.

DNA that is complexed with *E. coli* HDP is resistant to degradation by pancreatic DNase, snake venom phosphodiesterase, S1 nuclease, and *Neurospora crassa* endonuclease, but is still sensitive to micrococcal nuclease and *E. coli* exonuclease I (66). In addition, the complexes are resistant to the single-strand specific 3' to 5' nucleases associated with *E. coli* DNA polymerases I and III (67). The HDP also protects single strands from degradation by the *recBC* nuclease (68) (see section on DNA unwinding enzymes). These results suggest that the protein may be important in the cell in protecting single strands from attack by nucleases.

The *E. coli* HDP inhibits DNA synthesis on single-stranded templates by *E. coli* DNA polymerases I and III (61). However, at least for the DNA polymerase III holoenzyme,[2] the HDP is required for maximal synthesis in virtually every in vitro system that has been studied thus far (62, 70–74; see also 17). Whether this reflects an effect on the DNA template, or an interaction between the polymerase and the protein, or both, remains to be determined.

Of particular interest is the effect of the HDP on *E. coli* DNA polymerase II. DNA polymerase II is able to fill in short gaps in duplex DNA but cannot fill in long gaps. In the presence of the *E. coli* HDP the polymerase is able to use long segments of single-stranded DNA as a template (60, 61). This stimulation of polymerase II is specific since the T4 gene 32 protein is without effect. Conversely, the T4 DNA polymerase is not stimulated by the *E. coli* HDP. The *E. coli* HDP forms a specific complex with polymerase II and it is this form of the enzyme that utilizes the DNA-HDP complex as a template (67). The stimulation is due, at least in part, to the fact that in the presence of excess HDP, DNA polymerase II interacts with the DNA by a processive rather than a distributive mechanism (66, 75).

The evidence, summarized above, suggests multiple roles for the *E. coli* HDP. However, no mutants affecting the protein are known and therefore genetic evidence concerning its role in the cell is unavailable.

BACTERIOPHAGE T7 HELIX DESTABILIZING PROTEIN   Bacteriophage T7 codes for an HDP with a molecular weight between 25,000 and 31,000 which resembles the T4 gene 32 protein (76, 77). The T7 HDP aggregates with itself in the presence of Mg^{2+} (78). It will promote the

---

[2]For the distinction between *E. coli* polymerase III (or polymerase III*) and polymerase III holoenzyme see (69).

denaturation of poly dAT at 40° below the normal T_m, but there is no information on its ability to lower the T_m of natural DNA. The protein will stimulate the T7 specific DNA polymerase, but not the T4 DNA polymerase or any of the known *E. coli* DNA polymerases (76, 77). This stimulation is greatest when single-stranded templates are used and under conditions in which single strands have the greatest amount of secondary structure (for example, low temperature). The T7 HDP does not form a complex with the T7 DNA polymerase (77) which suggests that the stimulatory effect is mediated primarily at the level of the template structure. The *E. coli* HDP will stimulate the T7 DNA polymerase which offers a possible explanation for the failure to identify any T7 mutants with an altered HDP (76).

FILAMENTOUS BACTERIOPHAGE HELIX DESTABILIZING PROTEIN
Filamentous bacteriophages (M13, Fd, F1) also code for an HDP. The product of gene 5 binds specifically and cooperatively to single-stranded DNA (79–80). The protein lowers the T_m of natural DNA as well as poly dAT (79) but there is no evidence that DNA denaturation is one of the roles for the protein in the infected cell. The gene 5 protein does, however, play a crucial role in controlling the switchover from the synthesis of double-stranded progeny RF to the synthesis of single-stranded phage DNA. By complexing with newly synthesized single strands the protein apparently blocks their subsequent utilization as a template for double-strand synthesis (81–83). The protein, with a molecular weight of 10,000, forms a dimer (80, 81) and each subunit binds to approximately 4 nucleotides (79, 81). Since excess protein causes the collapse of the circular phage DNA into a rigid rod (79, 84) it has been suggested that the two ends of the DNA are bent into a hairpin configuration and that each dimer bridges two nonadjacent regions of the DNA strand (80, 84).

EUCARYOTIC HELIX DESTABILIZING PROTEINS   Proteins that bind strongly and preferentially to single-stranded DNA have been identified from a variety of eucaryotic sources, including *Lilium,* (85), mouse cells (86–88), the basidiomycetes fungus *Ustilago maydis* (89), Hela cells (90), and calf thymus (91). Such proteins have also been identified in adenovirus (92) and herpesvirus (93) infected cells. The proteins from mouse ascites cells, *U. maydis,* and calf thymus have been shown to specifically stimulate the homologous DNA polymerases (88, 89, 94). Thus, like their procaryotic counterparts, some of these proteins have been implicated in the process of DNA synthesis. Unlike the procaryotic proteins, none of the eucaryotic HDPs has been shown to bind in a cooperative fashion to single strands.

The best studied of the eucaryotic proteins are those found in calf thymus cells (91, 94, 95). At least two different species of HDPs are present in calf thymus cells, both of which stimulate the calf thymus DNA polymerase-α. Both proteins are capable of lowering the T_m of poly dAT and one of the proteins (calf HDP-I) is capable of lowering the T_m of natural DNA. (One was not tested.) Calf HDP-I did not, however, catalyze the renaturation of DNA strands. This same protein showed some affinity for double-stranded DNA, which leads to the suggestion (95) that these proteins may bind and then melt the DNA rather than bind to and expand preexisting single-strand loops in otherwise duplex DNA.

RNA Polymerase

The *E. coli* RNA polymerase is an example of a protein that binds and then melts duplex DNA (for review see 96, 97). The melting reaction can be rate limiting for initiation of RNA synthesis by the polymerase and accounts for the slow rate of chain initiation at low temperatures (98) and high ionic strengths (99). At low ionic strength the unwinding angle per bound RNA polymerase has been determined to be $\sim 240°$ (100, 101) which corresponds to the opening of about 7 base pairs. Under these conditions the polymerase need not bind to a specific promoter to melt the DNA. As expected, the extent of unwinding was found to decrease as the temperature was lowered or the ionic strength increased (101).

A closed circular DNA with negative superhelical turns is a much better template in vitro for RNA polymerase than the same DNA lacking superhelical turns (102–111). This effect is due to an increase in the rate of chain initiation and in the number of potential initiation sites rather than an increase in the rate of chain elongation. This observation is consistent with the requirement for helix opening as a step in the initiation of RNA chains by the polymerase.

Nucleosomal Proteins

The 100 Å unit fiber in eucaryotic chromatin appears to be composed of tightly packed spherical subunits called nucleosomes (5, 112, 113). Each nucleosome contains 200 ± 40 base pairs of DNA and two each of the histones H2A, H2B, H3, and H4, plus one or two H1 histone molecules. A segment of DNA approximately 140 base pairs in length, in association with histones H2A, H2B, H3 and H4, constitutes the core of the nucleosome. The DNA linker region, between nucleosome cores, is more variable in length, and is believed to be associated with the lysine rich H1 histone. H1 histone has been shown to have a greater affinity for DNA containing either positive or negative superhelical turns as compared to relaxed DNA (114–116). This observation suggests that H1 may stabilize the folding of the DNA in the linker region.

Here we are only concerned with the spatial arrangement of the DNA fiber in tightly packed nucleosomes. It is not yet possible to precisely define the overall conformation of nucleosomal DNA, but two facts must be accommodated in any model. The formation of each nucleosome results in (a) approximately a sevenfold contraction in the overall length of the DNA (112, 117) and (b) a net change in the topological linkage between the two DNA strands of about –1 ¼ (118).

The experiments of Germond et al (118) indicate that the core particle of the nucleosome is responsible for the change observed in the topological linkage. They reconstituted SV40 chromatin starting with closed circular DNA molecules that were initially relaxed (that is, $\alpha = \beta$, $\tau = 0$). After the mixture of histones that did not contain H1 had reacted with the DNA, the molecules took on the appearance of the "string of beads" structure that is characteristic of nucleosomes. In addition, the circular structures assumed a supertwisted configuration to compensate for either (a) a change that had occurred in β (so that $\tau \neq 0$), (b) the direct introduction of superhelical turns by the histones, or (c) some combination of both effects (see below). In the second case, the compensating supertwists observed in the chromatin molecules must have occurred in the sense opposite to those introduced by the histones. Neither the sign nor the number of these superhelical turns could be determined, however, by electron microscopy. In order to quantitate the topological change, the complexes were relaxed by a DNA swivelase (see below). The swivelase, by nicking and closing the DNA, removed all of the compensatory supertwists, but, of course, left unaltered the structural changes imposed upon the DNA by the histones. After relaxation, the chromatin contained 20 nucleosomes, the same number as found for native SV40 chromatin extracted from infected cells or isolated from virions (118, 119). After deproteinization the circular DNA became as negatively supertwisted as the naturally occurring form of SV40 DNA. Since native SV40 DNA has 25 ± 3 negative superhelical turns (120, 121) one can conclude that the conformational changes imposed by the histones in the nucleosome core are equivalent to the introduction of ~ 1¼ negative superhelical turns per nucleosome. Recently, the same results were obtained simply by complexing the SV40 DNA with an equimolar mixture of the histones H3 and H4 (122, 123). Therefore, these two histones may be the major determinants of the change in topological linkage that occurs during the formation of the nucleosome.

What does this tell us about the organization of the DNA in the nucleosome core? We can initially consider two limiting cases. First, assume that the association of histones affects only β, the helical winding of the

DNA. Then we can conclude that for every 200 base pairs (one nucleosome equivalent) the Watson-Crick helix is unwound by 1¼ turns. This unwinding might be evenly distributed along the DNA fiber. Alternatively one or a few regions might be locally unwound such that a total of 12 or 13 bases are unpaired. For this simple case we assume no direct introduction of superhelical turns by the histones. However, it is difficult to see how the DNA is compacted approximately sevenfold in length without some form of three-dimensional coiling. One can easily modify this model to introduce coiling by stipulating that the net contribution to the total superhelicity be zero. For example, this could be accomplished by the introduction of an equal and not necessarily integral number of positive and negative superhelical turns.

In the second limiting case, we assume that there is no effect on β, but that nucleosome formation induces 1¼ negative superhelical turns into the DNA. This is in fact the configuration of the DNA in the chromatin model suggested by Weintraub et al (113). They propose that the DNA is wrapped around the nucleosome core in a conformation approximating one left-handed toroidal superhelical turn. In solution, the free form of SV40 DNA probably exists in a configuration with interwound rather than toroidal superhelical turns (122). The free energy associated with these turns assists the process of nucleosome formation in vitro (118), but additional energy is required (122). This additional energy, which probably derives from the DNA-histone interactions, may be required for interchanging the two topologically equivalent forms of superhelical turns. We should note that this is the simplest model consistent with the data and therefore the most attractive. Alternatively, the DNA could have any combination of positive and negative superhelices such that the net contribution is approximately one negative superhelical turn per nucleosome.

It should be emphasized that neither one of these limiting cases may turn out to be correct. A large number of combinations involving both superhelical turns and changes in β are theoretically possible. Crick & Klug (124) have suggested that the bending of the DNA helix in chromatin may require "kinking." A kink requires some change in β. Crick (21) has discussed several possible models of DNA coiling plus twisting, that is, changes in β, that are compatible with the requirement that the change in topological linkage per nucleosome be −1¼.

ENZYMES THAT ALTER DNA CONFORMATION

DNA Swivelases

An enzyme with the capacity to relax a closed circular superhelical DNA was first discovered in *E. coli* and called the ω protein (125). Independently, the eucaryotic enzyme activity was discovered in nuclear extracts made from mouse embryo cells and was referred to as the DNA untwisting enzyme (126). Since then, similar activities have been identified from many sources. They have been referred to as ω proteins in Drosophila eggs (127), calf thymus (128), and *Bacillus megaterium* (129); as untwisting enzymes in rat liver (130), and yeast (130a); as DNA relaxing enzymes in KB cells (131), *Xenopus laevis* eggs (132, 133), chicken erythrocytes (134), and monkey cells (135); and as DNA nicking-closing enzymes in HeLa and mouse cells (136, 137), duck cells (122), and vaccinia virions (138). Here we will use the term DNA swivelase, first suggested by Wang (139), in a generic sense to refer to this class of enzymes. None of these names is completely satisfactory. Clearly a uniform nomenclature is required and a detailed understanding of the reaction mechanism should permit the systematic naming of these enzymes (see below).

DNA swivelases remove superhelical turns from a closed circular DNA to yield a relaxed, but still covalently closed product DNA. Since the DNA remains permanently relaxed even after the protein is removed (125, 126), these enzymes must change the topological winding number (α) rather than the helix winding number (β). The only way the topological winding can be changed is to break one strand, wind one end of that strand relative to the helix axis, and reseal the break. Thus a DNA swivelase must introduce a transient single-strand break (swivel) into duplex DNA. In one system it has now been possible to trap the nicked intermediate during the course of the enzyme catalyzed reaction (140; see below). A DNA swivelase carries out a reaction equivalent to the combined action of an endonuclease and polynucleotide ligase. However, unlike the ligases, these enzymes exhibit no requirement for an energy-donating cofactor like ATP or DPN (125, 126). This indicates that during the time the strand is broken, there must be some mechanism for conserving the energy required for resealing the break (see below). Where it has been examined, these enzymes act catalytically (130, 131, 141). Thus a given enzyme molecule must be capable of multiple cycles of nicking and sealing.

It should be emphasized that the DNA swivelases simply provide a break in the DNA which permits the strands to swivel, that is, to wind or unwind. Whether the strands wind at the break depends entirely on whether there is any strain or torsion in the molecule to drive the winding process. In the usual assays of the enzymes the free energy of superhelicity drives the relaxation of the DNA.

The enzymes can, in principle, be assayed by any procedure that discriminates relaxed from superhelical closed circular DNA. The methods that have been used include band sedimentation (125), equilibrium centrifugation in CsCl-ethidium bromide or CsCl-propidium diiodide gradients (126), direct fluorescent measurements of DNA-ethidium complexes (128, 130,

136, 137), nitrocellulose filter binding of the DNA (130, 142, 143), and agarose gel electrophoresis (131).

The DNA swivelases from *E. coli* and eucaryotic sources are different and are discussed separately.

E. COLI DNA SWIVELASE   Two different procedures for the purification of the DNA swivelase (ω protein) from *E. coli* 1100 have been published (142, 144). Both yield an enzyme that has a molecular weight of 110,000 and that appears to be a single polypeptide chain. There has been a report that the *E. coli* B enzyme contains two subunits (56,000 and 31,000) but proteolytic cleavage during purification was not ruled out (129).

The *E. coli* DNA swivelase is maximally active in the presence of 1 mM Mg^{2+} without any added salt, and is inhibited by concentrations of monovalent cations >0.30 M. The requirement for Mg^{2+} is not absolute since the enzyme shows some activity in its absence if the K^+ concentration is 50 mM (125).

The enzyme will only partially remove negative superhelical turns and will not remove positive turns (125). These two limitations may have a common basis. If binding of the enzyme is restricted to regions of DNA that are partially single-stranded, then the enzyme may only act on a substrate that is highly superhelical in the negative sense. Consistent with this interpretation is the finding that single-stranded DNA is inhibitory (125). Since tests for the removal of positive turns require the presence of ethidium bromide in the reaction mixture, an alternative explanation for the failure to remove positive turns is that the ethidium bromide is inhibitory.

The *E. coli* swivelase has been reported to abortively nick DNA in the presence of SDS, forming a complex with protein attached to the DNA (145). This behavior is similar to that exhibited by the colicin plasmid (colE1 and colE2) relaxation complexes (146, 147). By analogy with the swivelase, it would not be surprising if, eventually, conditions are found in which the proteins in the relaxation complex catalyze strand closure as well as strand breakage.

Given single-stranded fd DNA as a substrate, the *E. coli* DNA swivelase will convert the rings into forms that sediment more rapidly in alkali (148). The faster sedimenting species appear as knotted rings in electron micrographs. These topological isomers of fd DNA probably form due to the nicking-closing action of the swivelase in regions of limited secondary structure.

It has proven extremely difficult to measure swivelase activity in crude bacterial lysates due to contaminating nucleases and the presence of competing DNA. For this reason it has not been possible to rule out the presence of different enzymes which may carry out the same reaction as the *E. coli* swivelase. Recently, in fact, the product of the *nalA* gene has been purified, using an in vitro complementation assay, and found to possess a swivelase activity distinct from the well characterized *E. coli* swivelase discussed above (149) in that it relaxes both negative and positive superhelical turns, and thus could be the enzyme that removes the positive supertwists generated during DNA replication. This newly discovered enzyme activity will also abortively nick the DNA in the presence of SDS. This activity has been shown to be intimately associated with the DNA gyrase (see below). Whether it exists in the cell as a separately functioning unit remains to be determined.

EUCARYOTIC DNA SWIVELASES The eucaryotic enzymes that have been purified are composed of a single polypeptide chain with a molecular weight between 60,000 and 70,000 (130, 131). The enzymes are found in nuclei in association with chromatin (136, 143). Thus far only one enzyme activity has been identified from any source.

The eucaryotic enzymes share some features that distinguish them from the *E. coli* enzyme. They exhibit maximal activity between 0.15 M and 0.20 M monovalent cation (K^+ or Na^+) in the absence of Mg^{2+} (126–128, 131, 132, 136, 140). Recent studies with the rat liver enzyme indicate that this salt requirement reflects a relatively slow rate of dissociation of the enzyme from the DNA at low ionic strengths. Below 0.10 M K^+, the enzyme binds tightly and carries out repeated cycles of nicking and closing without dissociation (B. McConaughy and J. Champoux, in preparation). Thus, under these conditions, relaxation of all the substrate DNA only occurs at stoichiometric ratios of enzyme to DNA. In the presence of Mg^{2+}, these enzymes, like the *E. coli* enzyme, are active even at low salt (131, 136).

A second distinguishing feature of the eucaryotic enzymes is their ability to completely remove both negative and positive superhelical turns (126, 131, 136). The completely relaxed product contains the same distribution of topological isomers as one finds for DNA closed by polynucleotide ligase under the same conditions (135, 150, 151). This distribution is a reflection of the thermally induced fluctuations in helical winding in the nicked form of the DNA. Thus, there exist no constraints on the winding of the DNA at the nicks generated by the eucaryotic DNA swivelase. Treatment of one of the relaxed topological isomers with DNA swivelase regenerates the entire distribution, which demonstrates that the DNA need not be superhelical to serve as a substrate for cycles of nicking and closing by the enzyme (151).

Circular DNAs with a superhelix density intermediate between the substrate and the completely relaxed product have been observed (126, 131). Agarose gel analysis of the products from a brief exposure of SV40 DNA to the enzyme at a low temperature revealed the presence of all the possible topological intermediates in the reaction (131). This result has been interpreted to mean that the complete relaxation of a given DNA molecule may require multiple cycles of nicking and closing. Alternatively the process may be single step with a relatively long lifetime for the nicked intermediate. To explain the presence of DNA species with intermediate superhelicities one must postulate that the procedure used to stop the reaction, usually the addition of SDS, results in the premature closure of the breaks.

Since the DNA swivelase catalyzes the reversible nicking of duplex DNA, there must be some steady-state level of nicked intermediate during the course of the reaction. With highly purified rat liver enzyme at high enzyme : DNA ratios the nicked intermediate has been detected by stopping the reaction with alkali (pH > 12.5) (140). Apparently rapid denaturation of the DNA prevents closure of the nicked intermediate. The addition of KCl up to 0.50 M prior to alkaline denaturation terminates the reaction and all the breaks are sealed. This is consistent with the inability of the enzyme to relax DNA at high salt, yet indicates that the breaks made at lower salt concentrations can be sealed in the high salt. The nicked intermediate has also been trapped by rapidly adjusting the pH of the reaction mixture to 4.5 (152). Presumably the enzyme is denatured under these conditions and fails to reclose the breaks.

In order to account for the ability of the DNA swivelase to catalyze strand closure in the absence of any cofactors, it was suggested that strand breakage involves not hydrolysis of a phosphodiester bond, but transfer of one end of the broken DNA strand to a site on the enzyme (125, 126). In this way the DNA-enzyme linkage could conserve the energy for resealing the break. Recent studies with the rat liver enzyme as well as the evidence cited above for the *E. coli* enzyme suggest that this mechanism may be correct. DNA strands that have been broken by the rat liver enzyme exhibit a reduced buoyant density in alkaline CsCl which was shown to be due to the covalent attachment of protein to one of the ends of the broken strands (152). The magnitude of the buoyant shift was consistent with the attachment of a protein having a molecular weight the same as that determined for the enzyme. Since the strands containing bound enzyme could be phosphorylated by polynucleotide kinase, it was inferred that the enzyme must be attached to the DNA at the 3'-phosphate terminus.

The nature of the covalent linkage is not known. However, the stability in alkali would appear to rule out serine or threonine phosphoesters (153) as well as a mixed anhydride linkage (154, 155). Preliminary evidence indicates that the bond is stable in 3.5 M hydroxylamine at pH 4.75 (152) which suggests that a phosphoamide linkage may not be involved (153).

From these results it is apparent that the rat liver enzyme catalyzes the transfer of a 3'-polynucleotide chain from an internal position in duplex DNA back to the free end of the strand it breaks. Given this mechanism, the rat liver enzyme may probably be classified as a 3'-polydeoxynu-cleotidyltransferase, or as a DNA helix isomerase.

The reaction carried out by the *cis*A protein of ϕX174 is apparently very similar to that of a DNA swivelase. The protein catalyzes strand breakage at a specific site (156, 157), becomes attached to the 5' terminus of the broken strand (157), and is capable of recircularizing the DNA (158). Therefore, it appears that the *cis* A protein as well as the *E. coli* swivelase (145) fall into the same enzyme category as the rat liver swivelase, with the only apparent difference being the attachment of the protein at the 5' end rather than at the 3' end of the broken strand.

Conceivably, enzymes that catalyze the transfer of polynucleotide chains will also be found to play a role in recombination events that occur at specific sequences. Thus, it is tempting to hypothesize that *int* mediated recombination in λ or the translocation of a transposition or insertion sequence (159; for review see 160) may involve covalent attachment of an enzyme to the ends of the DNA at the site of breakage. In this reaction, the polynucleotide chains would be transferred to the ends of the chains in the recombinant partner rather than back to the original strand.

The rat liver enzyme has been shown to facilitate the complete renaturation of two complementary single-stranded circles (161). Normally, the renaturation of two single-stranded rings will encounter a topological block due to the requirement for the formation of the DNA helix. Rings that had been subjected to renaturation conditions and then treated with the rat liver enzyme completely renatured to form relaxed closed circular DNA molecules. In this reaction, the enzyme provided the transient swivel for the rewinding of the strands. A model is proposed suggesting how this kind of reaction may play a role in the pairing of DNA duplexes leading to recombination (15, 161).

PHYSIOLOGICAL ROLE   An attempt to identify the gene coding for the *E. coli* DNA swivelase among the mutants known to code for proteins essential in DNA replication was not successful (139). Similarly, none of the yeast cell division cycle mutants that affect DNA replication had temperature sensitive DNA swivelase activities when assayed in vitro (130a). These negative results permit no conclusion about the possible role of the enzyme in DNA replication.

It has been suggested the DNA swivelases may provide the swivel required during DNA replication to allow the DNA strands to unwind (125, 126). If this mechanism operates between the many converging replication forks in eucaryotic DNA, then there should be at least as many enzyme molecules in the cell as there are active replicons at any stage in the S period. A rat liver cell has a minimum of 70,000 enzyme molecules (130). From the size of the genome, the length of the S period, and the average rate of chain elongation in eucaryotic cells (9), one can calculate that, on the average, less than 10,000 replicons are active at the same time in S. Thus there are a sufficient number of enzyme molecules to fulfill the role of the swivel.

In eucaryotic cells, some of the enzymes involved exclusively in DNA synthesis are regulated during the cell cycle so that the highest enzyme levels are present during late G1 and/or throughout the S period (162, 163). For the same reason actively growing cells contain higher enzyme levels than resting cells. A recent report (164) claims to show modulation of the DNA swivelase activity during the cell cycle. However, no evidence for regulation of the DNA swivelase was obtained when the following situations were compared: (*a*) regenerating rat liver versus normal liver; (*b*) growing versus resting mouse 3T3 cells; (*c*) serum stimulated versus resting 3T3 cells; and (*d*) polyoma infected versus resting 3T3 cells (L. Young and J. Champoux, unpublished). The difference between these two sets of observations may be attributable to the methods used for preparing the extracts used to assay the enzyme. In the latter study total cell extracts were employed whereas in the former, isolated chromatin was extracted with salt under conditions known to effect only a partial release of the enzyme (128, 143). Thus the observed changes in enzyme levels could be attributed to differential release rather than regulation of synthesis during the cell cycle. The failure to observe a regulatory pattern for the DNA swivelase similar to that observed for other DNA enzymes suggests that the enzyme may have multiple roles in the cell.

It has been proposed that, in addition to possibly providing a swivel for DNA replication, these enzymes may also be required in transcription (139). Consider the continuous use of one strand of a DNA helix as a template for RNA synthesis under hypothetical conditions in which DNA rotation is prevented and the 5' end of the newly synthesized RNA is held fixed in space. Under these conditions the RNA chain should wrap around the DNA duplex one time for every ten nucleotides synthesized. However, electron micrographs of nascent RNA molecules show that RNA strands are not coiled around the DNA (165, 166). Therefore, either the RNA "whips" around the DNA during its synthesis, or more likely, the DNA rotates on its axis. On a given segment of DNA, transcription of different genes can occur in opposite directions at the same time. In this situation the DNA cannot rotate in one direction for one set of genes and in the opposite direction for the adjacent set unless a swivel is provided in between. It seems probable that, even for two adjacent genes transcribed in the same direction, two different rates of transcription may dictate the need for an intervening swivel.

It may be significant that vaccinia virions possess a DNA swivelase activity that appears to be distinct from the host cell enzyme (138). Since vaccinia multiplies largely in the cytoplasm, and since the DNA swivelase is found in the nucleus, it would not be too surprising if the virus coded for its own enzyme. The fact that the enzyme is found in virions has been interpreted to mean that the activity may be important for transcription early in infection (138). Alternatively, the enzyme may be involved in the maturation process or its association with the virions may be adventitious.

DNA swivelases may also be involved in DNA recombination (15, 167) and in the process of histone association with newly synthesized DNA (133; see section on nucleosomal proteins).

DNA Unwinding Enzymes

The energy for strand separation by an HDP derives from the tight binding of the protein to single-stranded DNA. The class of proteins that we will refer to as DNA unwinding enzymes apparently derive the energy for catalyzing strand separation from the hydrolysis of ATP or some other nucleoside or deoxynucleoside triphosphate.

E. COLI DNA UNWINDING ENZYME   The E. coli DNA unwinding enzyme was purified by monitoring its single-stranded DNA-dependent ATP-ase activity (168). The protein may be the largest single polypeptide chain found in E. coli with a molecular weight of 180,000. The monomer is highly asymmetric with an axial ratio of at least 11 : 1. In low salt ($<$ 50 mM KCl) the protein forms large fibrous aggregates.

The ATPase activity of the protein has an absolute requirement for single-stranded DNA (168). No activity is demonstrable with duplex DNA that is free of single strands. However, the DNA need not have free ends. The enzyme hydrolyzes ATP to yield ADP and Pi. The other ribonucleoside triphosphates are also hydrolyzed, but at a slower rate, while dATP is as effective a substrate as ATP.

The enzyme will unwind either DNA duplexes or DNA-RNA hybrids up to 2000 base pairs in length, providing they contain single-stranded tails or gaps (169, 170). Apparently the unwinding process is initiated by binding to single-stranded regions adjacent to the duplex DNA. Strand separation then ensues as the enzyme traverses the single strands and "invades" the duplex region. On a purely single-stranded DNA, the enzyme presumably migrates along the strands while consuming ATP. The enzyme is without effect on a completely duplex DNA.

The unwinding reaction requires the hydrolysis of a suitable nucleoside triphosphate, and like the ATPase activity is optimal in low salt under conditions where the enzyme is known to form aggregates. The complete unwinding of a duplex region 2000 base pairs in length requires approximately 80 enzyme molecules. This requirement is greater than what one would expect if the enzyme was only needed at the separation fork and suggests that the protein may, in addition, have to coat the single strands to prevent renaturation. Indeed, if this is not the case it is difficult to see what prevents the trailing single-stranded region from rapidly renaturing. In addition, the requirement for many enzyme molecules may reflect the need for the formation of large aggregates that cause strand separation by the sliding action of protein filaments (170). In this case, the activity of the unwinding enzyme would resemble the action of myosin in muscle.

Several mutants in genes coding for DNA replication and DNA recombination functions were tested and found to have an unaltered DNA unwinding enzyme (168). Thus, in spite of the obvious possibilities, no in vivo role has yet been established for this enzyme.

E. COLI REP PROTEIN   The *rep* function in *E. coli* was originally identified by mutants that could not replicate φX174 DNA. The defect was not in the conversion of single-stranded parental DNA to the double-stranded RF, but rather in the replication of the RF (171). The *rep* mutants appear to replicate their DNA normally although fork movement has been shown to be slower than in the wild type (172).

More recently the *rep* protein has been purified and shown to be a single-stranded DNA-dependent ATPase (173, 174, 175). In the presence of the φX174 *cis*A protein and the *E. coli* HDP the purified *rep* protein, of molecular weight 68,000, will catalyze the separation of the strands of duplex φX174 RF DNA (173). ATP is required for this reaction, apparently providing the energy for unwinding the DNA. Approximately one ATP molecule is hydrolyzed for each base pair melted.

In a separate study the *rep* protein was purified by virtue of its ATPase activity and shown to unwind the duplex region of molecules containing single-stranded tails or gaps (174, 176). Thus, like the DNA unwinding enzyme described above, the *rep* protein by itself initially attaches to single-stranded regions and melts the duplex. The unwinding reaction requires either ATP or dATP. The other ribonucleoside triphosphates are not hydrolyzed and will not serve as a cofactor in strand separation. Neither completely double-stranded DNA fragments nor duplex fd DNA containing a nick or a short gap are unwound by the enzyme. The ability of the enzyme to catalyze the separation of φX174 RF strands at a nick must be attributed to the additional effects of the *cis*A protein and/or the *E. coli* HDP.

The mechanism of strand separation by the *rep* protein is not known. However, it appears that the enzyme does not act processively at a fork, but that a supply of free protein is constantly needed (176). This suggests that melting occurs as a consequence of the continual addition of the enzyme molecules to the fork region. Protein-protein interactions as well as ATP hydrolysis probably play a role in strand separation.

E. COLI recBC NUCLEASE   The *recB* and *recC* gene products are isolated as a composite protein of molecular weight 270,000, which exhibits a bewildering array of nuclease activities in vitro (177, 178). The *recBC* nuclease, also called exonuclease V, will act as an ATP-dependent exonuclease on double- or single-stranded DNA (178). The exonuclease degrades DNA from either the 5' or the 3' end and generates oligonucleotides with an average chain length of about five. Concomitant with exonucleolytic degradation, ATP is hydrolyzed to ADP and inorganic phosphate at a rate that can exceed by an order of magnitude the number of phosphodiester bonds broken. Endonucleolytic cleavage of single-stranded DNA by the *recBC* nuclease does not require ATP, but is stimulated by its presence. Double-stranded circular DNAs are completely resistant to the action of the *recBC* nuclease (179).

From an analysis of the intermediates in the reaction, the following scheme has been suggested for the processive degradation of linear duplex DNA by the enzyme (180). The enzyme binds to both strands at the end of the duplex. The terminus of one strand remains bound to the enzyme while the other strand is first unwound and then cleaved at 100–500 nucleotide intervals. After unwinding several thousand nucleotides, the enzyme switches to the other strand and degrades it from the free end. ATP hydrolysis is thought to provide the energy for the unwinding of the DNA, but the mechanism is unknown. In the presence of excess enzyme, the exonuclease activity subsequently degrades the large single-stranded fragments to acid soluble oligonucleotides. ATP is presumably required in this reaction to promote the movement of the enzyme along the DNA. It appears that as the ATP concentration is increased the lengths of the single-stranded fragments increase (179, 180). Finally, at very high ATP concentrations, the enzyme apparently fails to switch strands and eventually ceases to unwind the DNA (181). Under these conditions, as with other substrates that are not unwound (179, 182), the ATPase activity continues.

Of particular interest is the effect of the *E. coli* HDP on the activity of the *recBC* nuclease (68). The HDP protects single-stranded DNA from exonucleolytic degradation. This means that the products from digestion of linear duplexes in the presence of HDP are long single-stranded pieces resulting from the combined action of helix unwinding and limited endonucleolytic cutting.

Clearly the intracellular environment is important in determining the nature of the reaction carried out by the *recBC* nuclease. The factors known to influence the enzyme are the ATP concentration, the presence of an HDP, the nature of the substrate DNA, and, from in vivo studies, the product of *recA* gene (183, 184). It is unlikely that free duplex ends are present in the cell. Until the in vivo substrate(s) for the enzyme is determined, the relative importance of the nucleolytic and helix unwinding activities of the enzyme to the processes of DNA repair, recombination, and possibly replication (185) remains to be determined.

A DNA unwinding enzyme that shares some properties with the *E. coli recBC* nuclease has been found in *Hemophilus influenzae* (186, 187).

BACTERIOPHAGE T7 GENE 4 PROTEIN   The gene 4 product of T7 is required for T7 DNA replication. The gene 4 protein appears to be involved both in the initiation of Okazaki fragments (188) and in the continuous elongation of DNA chains by the T7 DNA polymerase (189, 190). Recently, the protein has been purified by a complementation assay and found to consist of two species (189, 191). The smaller protein with a molecular weight of 57,000 is probably derived from the larger 66,000 molecular weight species by proteolytic processing. The purified protein exhibits two different enzymatic activities which appear to account for the roles of the protein in T7 replication.

First, the protein is an RNA polymerase that will synthesize a short oligonucleotide primer which allows the T7 DNA polymerase to utilize a single-stranded circular DNA as a template (191). The synthesis of the primer requires only ATP and CTP and the predominant product is the tetranucleotide ACCA. It is not clear whether the synthesis of the oligonucleotide requires a template sequence. With a double-stranded DNA there also appears to be some DNA chain initiation by the T7 DNA polymerase in the presence of the gene 4 protein, which suggests that priming may also occur with duplex DNA. Thus the gene 4 protein appears to have a role in the initiation of DNA chains analogous to the *dnaG* protein of *E. coli* (192).

Second, the gene 4 protein allows the T7 DNA polymerase to utilize a nicked DNA as template (190). This reaction requires the hydrolysis of deoxynucleoside triphosphates to deoxynucleoside diphosphates, but does not require the presence of the ribonucleoside triphosphates. When present, the ribonucleoside triphosphates are also hydrolyzed. Single-stranded DNA will stimulate the nucleoside triphosphatase activity of the gene 4 protein.

The tight coupling between in vitro DNA synthesis and the nucleoside triphosphatase activity suggests that the enzyme is facilitating strand separation for the T7 DNA polymerase rather than utilizing displaced single strands simply as a cofactor. Strand separation in the absence of DNA synthesis by the gene 4 protein has not been demonstrated.

It is not yet known whether each of the molecular weight species exhibits both of these enzymatic activities. It is possible that the proteolytic cleavage has some regulating or functional significance. The synthesis of an RNA primer on duplex DNA must involve the opening of the DNA helix. Thus, it is reasonable to assume that strand separation and primer synthesis are the functions of the same enzyme. During the elongation of the primer RNA by the T7 DNA polymerase, the protein may be restricted to its helix unwinding role.

DNA Gyrase

An enzyme has been isolated from *E. coli* that couples the hydrolysis of ATP to the introduction of negative superhelical turns into a closed circular DNA molecule. This activity has been termed the DNA gyrase (193). From the overall reaction one predicts that, like the DNA swivelase, the gyrase must be capable of transiently nicking the substrate DNA. However, unlike the swivelase, the gyrase must also promote the unwinding of the helix during the lifetime of the nick. There is evidence that the nicking-closing and the unwinding activities are separate functions of the enzyme which may reside on separate subunits (149, 193a). In the absence of ATP, the DNA gyrase becomes a DNA swivelase capable of relaxing both negative and positive superhelices. In addition, the drug novobiocin, and probably the similarly acting drug coumermycin, inhibits the supertwisting capability, but not the swivelase activity of the gyrase in vitro. On the other hand, the drugs nalidixic acid and oxolinic acid inhibit the overall reaction catalyzed by the gyrase. The gyrase activity from nalidixic acid-resistant mutants is insensitive to the drug in vitro. As discussed previously, the product of the *nalA* gene has been isolated, free of gyrase activity, and found to be a DNA swivelase (149). All of these results are most compatible with a model in which the DNA gyrase is composed of two different kinds of subunits. One kind of subunit is the product of the *nalA* gene that is responsible for the nicking-closing function of the enzyme. The *nalA* protein evidently can occur as a subunit of the gyrase or as a separate unit— a dimer of two identical 110,000 molecular weight subunits containing only swivelase activity. The other kind of subunit is coded for by the unlinked *cou* gene and is responsible for the ATP-dependent unwinding of the strands at the nick.

As mentioned above, ATP is required only for the unwinding component of the gyrase reaction. Since ATP is not required for the nicking-closing activity, the gyrase probably employs a mechanism similar to that used by the swivelase to conserve the energy for resealing the single-strand break.

The possibility has been considered that the gyrase unwinds the strands of the DNA, presumably in an ATP-dependent fashion, and then maintains the strands in an unwound state while the positive supertwists introduced by the unwinding are relaxed by a swivelase activity (193). According to this model, negative superhelical turns are only introduced when the protein is removed at the end of the reaction. If this is correct, the addition of a DNA swivelase to the gyrase reaction should not affect the superhelicity of the product. However, this possibility is ruled out by the observation that the *E. coli* DNA swivelase antagonizes the action of the gyrase (193).

Novobiocin and coumermycin inhibit DNA replication in vivo (194, 195). Since the drugs inhibit the DNA gyrase in vitro and since the gyrase isolated from a coumermycin resistant strain of *E. coli* is insensitive to the drugs (196), the gyrase appears to be the locus of action for these drugs. These results thus directly implicate the gyrase in the process of DNA replication. Negative superhelical turns may be required for the interaction of replication, transcription, and recombination proteins with DNA (see below) and thus the DNA gyrase may be indirectly involved in all of these processes.

As yet a DNA gyrase activity has not been detected in eucaryotic cells. When assayed in the presence of ATP under the conditions described for the *E. coli* gyrase, the rat liver DNA swivelase does not possess a gyrase type activity (B. McConaughy and J. Champoux, unpublished).

INTRACELLULAR CONSIDERATIONS

Origin of Naturally Occurring Superhelical Turns

The superhelical structure of a closed circular DNA in vitro is a direct reflection of the state of the DNA at the moment the last nick is sealed prior to or during the disruption of the cell and isolation of the DNA. If the DNA in vivo is continually subjected to cycles of nicking and sealing—by nucleases plus ligase, a DNA swivelase, or a DNA gyrase—then the superhelices reflect the steady-state condition of the DNA in the cell. Otherwise, they reflect only the state of the DNA at the moment after replication when the last break is sealed. The evidence that nonreplicating DNA is subjected to cycles of nicking and closing in vivo (197–200) is for the most part limited to systems that have been perturbed by the addition of ethidium bromide or protein synthesis inhibitors. These results, therefore, must be qualified by the possibility that these agents may be directly responsible for inducing the nicking-closing cycles. In spite of this objection the remainder of this discussion assumes continuous cycles of nicking and sealing.

We need to distinguish two possible situations regarding these cycles of nicking and closing.

1. If the nicking-closing cycles are only introduced either by nucleases plus ligase or by DNA swivelases, then the DNA helix as it exists in the cell is maintained in a "relaxed" state. This means that although the DNA might be constrained by proteins to assume a superhelical configuration, for example, histones, the DNA helix itself is essentially free of winding strain and therefore is not intrinsically superhelical (see section on superhelical turns). If this is the situation, then the superhelical structure of the DNA in vitro reflects the sum of all the structural factors operating on the DNA that distinguish the in vivo environment from the in vitro one. These factors include (a) ionic strength, temperature, and pH (22) (b) DNA unwinding proteins such as HDPs or RNA polymerase (c) DNA supertwisting proteins such as the histones, and (d) the continuous action of a DNA unwinding enzyme which results in some steady-state level of separated strands.

2. Alternatively, the nicking-closing cycles may be introduced by the DNA gyrase, in which case the extent of supertwisting is determined by this enzyme activity. If both a DNA gyrase and a swivelase are acting on a DNA, this case still applies providing the gyrase predominates. Under these conditions the extent of supertwisting reflects the steady-state balance between the two opposing activities.

The following arguments indicate that the first case applies to SV40 chromatin and by extrapolation perhaps also to all eucaryotic chromatin. After taking into account the differences in temperature and ionic strength between the in vivo environment and the conditions in vitro under which the superhelical turns are measured, one can account for all of the superhelical turns in SV40 DNA on the basis of its interaction with histones (see section on nucleosomal proteins). Moreover, treatment of the SV40 minichromosome with DNA swivelase in vitro does not cause a change in the superhelix density of the DNA measured after deproteinization (118, 201, 202). This lack of activity is not due to steric factors that block access of the enzyme to the DNA in the intact chromatin, since nicking of the chromatin DNA by the swivelase can be detected (L. Young and J. Champoux, submitted for publication). From these results it appears likely that the DNA helix in the SV40 chromatin is under no winding constraint and thus is relaxed. A few very short superhelical regions could exist, but they would have to be inaccessible to the DNA swivelase. Thus, there is no reason to suppose that the DNA in the chromatin is acted upon by a DNA gyrase or that a DNA unwinding enzyme is maintaining an unwound region in vivo. A corollary to these conclusions is that no single-stranded loops are present in the chromatin due to topological constraints imposed by the association of the DNA with proteins. Such a model was proposed by Crick (203) to provide single-stranded regions for protein recognition. These results do not, however, rule out the possibility that some regions are in a single-stranded conformation, not as a consequence of topological winding constraints, but rather due to the direct and continuous association with protein. As mentioned previously, the histones may in fact cause some partial unwinding of the helix.

The mere presence of a DNA gyrase in a cell does not necessarily mean that the DNA is maintained in a topologically underwound state, since the abundance of a potent DNA swivelase could relax the DNA as fast as the gyrase supertwists it. However, there is evidence that in *E. coli* the DNA gyrase predominates over the *E. coli* swivelase(s). The results that show this most directly involve the effects of coumermycin, the gyrase inhibitor, on the superhelical state of phage λ DNA after superinfection of a lysogen (196). (There is no replication under these conditions.) In the absence of the drug the DNA rapidly becomes superhelical, whereas in the presence of the drug the DNA remains fully relaxed. A similar inhibition of superhelix formation in crude extracts is observed for ColE1 DNA (196).

Several processes appear to require that DNA be in a superhelical conformation. The *cis*A protein of phage φX174 nicks only a superhelical substrate DNA (204). Likewise, the integrative recombination of phage λ in vitro requires that the DNA be superhelical (205). Both of these processes are inhibited by novobiocin or coumermycin, which directly implicates the gyrase in the reaction that introduces the supertwists. Recombination between a closed circular DNA and single-stranded fragments is enhanced if the DNA is superhelical (206). Finally, as mentioned previously, superhelical DNA is a better template than relaxed DNA for *E. coli* RNA polymerase. This observation may well be significant physiologically.

The observations cited above provide circumstantial evidence that in *E. coli,* circular DNAs, at least under some conditions, may be intrinsically superhelical. This condition is probably not satisfied by the process of replication since, in two instances, the immediate products of plasmid replication have been found to be relaxed closed circles (199, 200). These intermediates are quickly made superhelical, presumably by a DNA gyrase activity.

Since the *E. coli* chromosome is known to be composed of many topologically isolated domains (19), the above conclusions concerning the superhelicity of extrachromosomal elements probably also apply to the cellular chromosome.

Strand Separation and the Swivel in DNA Replication

Strand separation during DNA replication appears to be accomplished by some combination of a DNA unwinding enzyme, an HDP, chain elongation by the DNA polymerase, and preswiveling by the DNA gyrase. The relative contributions of these four factors to the overall process are not known. It seems probable that different systems will employ different combinations. Thus the gyrase would not be expected to be important in the rolling circle type replication of, for example, ϕX174, but may very well be important in strand separation during the replication of a circle by the Cairns type mechanism (207). This latter conclusion is supported by the finding that coumermycin inhibits chain elongation in the replication of colE1 DNA (196).

The energy for strand separation probably derives from the hydrolysis of ATP, or some other nucleoside triphosphate, in the reactions catalyzed by the gyrase and/or a DNA unwinding enzyme. It is possible that the DNA gyrase is involved in forcing the DNA into an underwound state which leads to the synthesis and/or stabilization of RNA primers ahead of the replicating fork. The primary role of the HDP may be to protect the exposed single strands at the replication fork. This may be especially important on the side of the fork in which the chain must be synthesized 5' to 3' in the direction opposite to fork movement.

With the discovery of the circular replicating form of the E. coli chromosome, Cairns (207) recognized the need for a swivel to allow the DNA helix to unwind. It is likely that a swivel is also needed between converging replication forks during the replication of linear eucaryotic chromosomes (9). Although it seems probable that a single-strand break serves as the swivel during DNA replication, the parental strands of circular replicons are usually intact in the isolated replicative intermediates (208–215). This observation suggests that transient swivels may be introduced into replicating DNA by a DNA swivelase or a DNA gyrase. It should be noted, however, that since the E. coli DNA swivelase cannot relax positive superhelices it cannot fulfill this role. As mentioned previously, the product of the nalA gene does relax positive superhelical turns and therefore could provide the swivel. Alternatively the transient swivels could be provided by an endonuclease in combination with polynucleotide ligase.

Given that the swivel is a transient single-strand break, one can distinguish two possible motive forces for helix unwinding at the break. In the first, ATP hydrolysis by the DNA gyrase preswivels the helix, thus aiding strand separation by "pulling" the fork along. In the second, strand separation at the fork acts to "push" against the helix which is then relaxed by a DNA swivelase. With either mechanism the process of strand separation and helix unwinding at transient nicks could theoretically continue until the last helix turn is removed from the parental strands (see also 216, 217). At this point the two newly replicated regions could physically separate with the parental strands still intact. According to this model, the termination of replication at the point where two forks converge requires no special mechanisms to preserve the continuity of the DNA chains (10, 215).

CONCLUSION

In eucaryotic cells virtually all of the DNA is complexed with proteins to form a unit fiber approximately 100 Å in diameter. Chromatin is formed by the higher order coiling of the unit fiber. In procaryotic cells, as exemplified by *E. coli*, the actual structure of the chromosome is less clear (218), but the discovery of the DNA gyrase raises the possibility that the DNA helix in the cell is maintained in an underwound state. It may be important to consider these structural features of DNA in future biochemical studies on replication, transcription, and recombination.

The recent discoveries of the DNA swivelases, the DNA gyrase, and the DNA unwinding enzymes considerably increase our knowledge of DNA biochemistry. As more is learned about these enzymes and their interaction with DNA, the prospects for understanding the details of DNA transcription, DNA recombination, and particularly DNA replication appear to be good.

ACKNOWLEDGMENTS

I thank W. Fangman, L. Young, B. McConaughy, D. Kiehn, M. Been, and J. Crosa for their critical reading of the manuscript. I also thank those people who sent me copies of their unpublished work. I am grateful to Deborah Abas for her help in preparing the manuscript. The work from this laboratory was supported by National Institutes of Health Research Grant GM 23224.

Literature Cited

1. Watson, J. D., Crick, F. H. C. 1953. *Nature* 171:736–38
2. Bak, A. L., Zeuthen, J., Crick, F. H. C. 1977. *Proc. Natl. Acad. Sci. USA* 74:1595–99
3. Kornberg, R. D. 1974. *Science* 184:868–71
4. Elgin, S. C. R., Weintraub, H. 1975. *Ann. Rev. Biochem.* 44:725–74
5. Kornberg, R. D. 1977. *Ann. Rev. Biochem.* 46:931–54
6. Fujimura, R. K., Roop, B. C. 1976. *J. Biol. Chem.* 251:2168–75
7. Kornberg, A. 1974. *DNA Synthesis*. San Francisco: Freeman. 399 pp.
8. von Hippel, P. H., McGhee, J. D. 1972. *Ann. Rev. Biochem.* 41:231–300
9. Edenberg, H. J., Huberman, J. A. 1975. *Ann. Rev. Genet.* 9:245–84
10. Gefter, M. L. 1975. *Ann. Rev. Biochem.* 44:45–78
11. Broker, T. R., Doermann, A. H. 1975. *Ann. Rev. Genet.* 9:213–44
12. Dressler, D. 1975. *Ann. Rev. Microbiol.* 29:525–59
13. Jovin, T. M. 1976. *Ann. Rev. Biochem.* 45:889–920
14. Alberts, B. M., Sternglanz, R. 1977. *Nature* 269:655–61
15. Radding, C. M. 1978. *Ann. Rev. Biochem.* 47:

16. Sheinin, R., Humbert, J. Pearlman, R. E. 1978. *Ann. Rev. Biochem.* 47:277–316

17. Wickner, S. 1978. *Ann Rev. Biochem.* 47:

18. Vinograd, J., Lebowitz, J., Watson, R. 1968. *J. Mol. Biol.* 33:173–97

19. Worcel, A., Burgi, E. 1972. *J. Mol. Biol.* 71:127–47

20. Fuller, F. B. 1971. *Proc. Natl. Acad. Sci. USA* 68:815–19

21. Crick, F. H. C. 1976. *Proc. Natl. Acad. Sci. USA* 73:2639–43

22. Wang, J. C. 1969. *J. Mol. Biol.* 43:25–39

23. Pulleyblank, D. E., Morgan, A. R. 1975. *J. Mol. Biol.* 91:1–13

24. Fuller, W., Waring, M. J. 1964. *Ber. Bunsenges. Phys. Chem.* 68:805–8

25. Wang, J. C. 1974. *J. Mol. Biol.* 89:783–801

26. Schmir, M., Révet, B. M. J., Vinograd, J. 1974. *J. Mol. Biol.* 83:35–45

27. Bauer, W., Vinograd, J. 1970. *J. Mol. Biol.* 47:419–35

28. Hsieh, T.-S., Wang, J. C. 1975. *Biochemistry* 14:527–35

29. Depew, R. E., Wang, J. C. 1975. *Proc. Natl. Acad. Sci. USA* 72:4275–79

30. Pulleyblank, D. E., Shure, M., Tang, D., Vinograd, J., Vosberg, H.-P. 1975. *Proc. Natl. Acad. Sci. USA* 72:4280–84

31. Alberts, B. M., Amodio, F. J., Jenkins, M., Gutmann, E. D., Ferris, F. L. 1968. *Cold Spring Harbor Symp. Quant. Biol.* 33:289–305

32. Alberts, B. M., Frey, L. 1970. *Nature* 227:1313–18

33. Tomizawa, J.-I., Anraku, N., Iwama, Y. 1966. *J. Mol. Biol.* 21:247–53

34. Kozinski, A. W., Felgenhauer, Z. Z. 1967. *J. Virol.* 1:1193–1202

35. Wu, J.-R., Yeh, Y.-C. 1973. *J. Virol.* 12:758–65

36. Jensen, D. E., Kelly, R. C., von Hippel, P. H. 1976. *J. Biol. Chem.* 251:7215–28

37. Carroll, R. B., Neet, K. E., Goldthwait, D. A. 1972. *Proc. Natl. Acad. Sci. USA* 69:2741–44

38. Carroll, R. B., Neet, K., Goldthwait, D. A. 1975. *J. Mol. Biol.* 91:275–91

39. Kelly, R. C., Jensen, D. E., von Hippel, P. H. 1976. *J. Biol. Chem.* 251:7240–50

40. Delius, H., Mantell, N. J., Alberts, B. M. 1972. *J. Mol. Biol.* 67:341–50

41. Gralla, J., Crothers, D. M. 1973. *J. Mol. Biol.* 78:301–19

42. Pohl, F. M. 1974. *Eur. J. Biochem.* 42:495–504

43. Hosoda, J., Takacs, B., Brack, C. 1974. *FEBS Lett.* 47:338–42

44. Moise, H., Hosoda, J. 1976. *Nature* 259:455–58

45. Morrow, J. F., Berg, P. 1973. *J. Virol.* 12:1631–32

46. Brack, C., Bickle, T. A., Yuan, R. 1975. *J. Mol. Biol.* 96:693–702

47. Wackernagel, W., Radding, C. M. 1974. *Proc. Natl. Acad. Sci. USA* 71:431–35

48. Studier, F. W. 1969. *J. Mol. Biol.* 41:199–209

49. Huberman, J. A., Kornberg, A., Alberts, B. M. 1971. *J. Mol. Biol.* 62:39–52

50. Nossal, N. G. 1974. *J. Biol. Chem.* 249:5668–76

51. Mosig, G., Breschkin, A. M. 1975. *Proc. Natl. Acad. Sci. USA* 72:1226–30

52. Breschkin, A. M., Mosig, G. 1977. *J. Mol. Biol.* 112:279–94

53. Breschkin, A. M., Mosig, G. 1977. *J. Mol. Biol.* 112:295–308

54. Huang, W. M., Lehman, I. R. 1972. *J. Biol. Chem.* 247:3139–46

55. Curtis, M. J., Alberts, B. 1976. *J. Mol. Biol.* 102:793–816

56. Krisch, H. M., Bolle, A., Epstein, R. H. 1974. *J. Mol. Biol.* 88:89–104

57. Krisch, H. M., Van Houwe, G. 1976. *J. Mol. Biol.* 108:67–81

58. Gold, L., O'Farrell, P. Z., Russel, M. 1976. *J. Biol. Chem.* 251:7251–62

59. Russel, M., Gold, L., Morrissett, H., O'Farrell, P. Z. 1976. *J. Biol. Chem.* 251:7263–70

60. Sigal, N., Delius, H., Kornberg, T., Gefter, M. L., Alberts, B. 1972. *Proc. Natl. Acad. Sci. USA* 69:3537–41

61. Molineux, I. J., Friedman, S., Gefter, M. L. 1974. *J. Biol. Chem.* 249:6090–98

62. Weiner, J. H., Bertsch, L. L., Kornberg, A. 1975. *J. Biol. Chem.* 250:1972–80

63. Molineux, I. J., Pauli, A., Gefter, M. L. 1975. *Nucleic Acids Res.* 2:1821–37

64. Ruyechan, W. T., Wetmur, J. G. 1975. *Biochemistry* 14:5529–34

65. Ruyechan, W. T., Wetmur, J. G. 1976. *Biochemistry* 15:5057–64

65a. Christiansen, C., Baldwin, R. L. 1977. *J. Mol. Biol.* 115:441–54

66. Molineux, I. J., Gefter, M. L. 1975. *J. Mol. Biol.* 98:811–25

67. Molineux, I. J., Gefter, M. L. 1974. *Proc. Natl. Acad. Sci. USA* 71:3858–62

68. MacKay, V., Linn, S. 1976. *J. Biol. Chem.* 251:3716–19

69. Wickner, W., Kornberg, A. 1974. *J. Biol. Chem.* 249:6244–49

70. Geider, K., Kornberg, A. 1974. *J. Biol. Chem.* 249:3999–4005

71. Zechel, K., Bouché, J.-P., Kornberg, A. 1975. *J. Biol. Chem.* 250:4684–89

72. Schekman, R., Weiner, J. H., Weiner, A., Kornberg, A. 1975. *J. Biol. Chem.* 250:5859–65
73. Scott, J. F., Eisenberg, S., Bertsch, L. L., Kornberg, A. 1977. *Proc. Natl. Acad. Sci. USA* 74:193–97
74. Vicuna, R., Hurwitz, J., Wallace, S., Girard, M. 1977. *J. Biol. Chem.* 252:2524–33
75. Sherman, L. A., Gefter, M. L. 1976. *J. Mol. Biol.* 103:61–76
76. Reuben, R. C., Gefter, M. L. 1973. *Proc. Natl. Acad. Sci. USA* 70:1846–50
77. Scherzinger, E., Litfin, F., Jost, E. 1973. *Mol. Gen. Genet.* 123:247–62
78. Reuben, R. C., Gefter, M. L. 1974. *J. Biol. Chem.* 249:3843–50
79. Alberts, B., Frey, L., Delius, H. 1972. *J. Mol. Biol.* 68:139–52
80. Cavalieri, S. J., Neet, K. E., Goldthwait, D. A. 1976. *J. Mol. Biol.* 102:697–711
81. Oey, J. L., Knippers, R. 1972. *J. Mol. Biol.* 68:125–38
82. Salstrom, J. S., Pratt, D. 1971. *J. Mol. Biol.* 61:489–501
83. Mazur, B. J., Model, P. 1973. *J. Mol. Biol.* 78:285–300
84. Pratt, D., Laws, P., Griffith, J. 1974. *J. Mol. Biol.* 82:425–39
85. Hotta, Y., Stern, H. 1971. *Nature New Biol.* 234:83–86
86. Salas, J., Green, H. 1971. *Nature New Biol.* 229:165–69
87. Tsai, R. L., Green, H. 1973. *J. Mol. Biol.* 73:307–16
88. Otto, B., Baynes, M., Knippers, R. 1977. *Eur. J. Biochem.* 73:17–24
89. Banks, G. R., Spanos, A. 1975. *J. Mol. Biol.* 93:63–77
90. Enomoto, T., Yamada, M. 1976. *Biochem. Biophys. Res. Commun.* 71:122–27
91. Herrick, G., Alberts, B. 1976. *J. Biol. Chem.* 251:2124–32
92. Rosenwirth, B., Shiroki, K., Levine, A. J., Shimojo, H. 1975. *Virology* 67:14–23
93. Purifoy, D. J. M., Powell, K. L. 1976. *J. Virol.* 19:717–31
94. Herrick, G., Delius, H., Alberts, B. 1976. *J. Biol. Chem.* 251:2142–46
95. Herrick, G., Alberts, B. 1976. *J. Biol. Chem.* 251:2133–41
96. Chamberlin, M. J. 1974. *Ann. Rev. Biochem.* 43:721–75
97. Chamberlin, M. J. 1976. In *RNA Polymerase,* ed. M. Chamberlin, R. Losick, pp. 159–91. New York: Cold Spring Harbor Lab. 899 pp.
98. Mangel, W. F., Chamberlin, M. J. 1974. *J. Biol. Chem.* 249:3007–13

99. Mangel, W. F., Chamberlin, M. J. 1974. *J. Biol. Chem.* 249:3002–6
100. Saucier, J.-M., Wang, J. C. 1972. *Nature New Biol.* 239:167–70
101. Wang, J. C., Jacobsen, J. H., Saucier, J.-M. 1977. *Nucleic Acids Res.* 4:1225–41
102. Westphal, H. 1971. In *The Biology of Oncogenic Viruses, Lepetit Colloq.,* ed. L. G. Silvestri, pp. 77–87. Amsterdam: North-Holland. 339 pp.
103. Hayashi, Y., Hayashi, M. 1971. *Biochemistry* 10:4212–18
104. Botchan, P., Wang, J. C., Echols, H. 1973. *Proc. Natl. Acad. Sci. USA* 70:3077–81
105. Richardson, J. P. 1974. *Biochemistry* 15:3164–69
106. Mandel, J. L., Chambon, P. 1974. *Eur. J. Biochem.* 41:367–78
107. Richardson, J. P. 1975. *J. Mol. Biol.* 91:477–87
108. Lescure, B., Oudet, P., Chambon, P., Yaniv, M. 1976. *J. Mol. Biol.* 108:83–97
109. Botchan, P. 1976. *J. Mol. Biol.* 105:161–76
110. Seeburg, P. H., Nüsslein, C., Schaller, H. 1977. *Eur. J. Biochem.* 74:107–13
111. Wang, J. C. 1974. *J. Mol. Biol.* 87:797–816
112. Oudet, P., Gross-Bellard, M., Chambon, P. 1975. *Cell* 4:281–300
113. Weintraub, H., Worcel, A., Alberts, B. 1976. *Cell* 9:409–17
114. Vogel, T., Singer, M. 1975. *J. Biol. Chem.* 250:796–98
115. Vogel, T., Singer, M. F. 1975. *Proc. Natl. Acad. Sci. USA* 72:2593–2600
116. Vogel, T., Singer, M. F. 1976. *J. Biol. Chem.* 251:2334–38
117. Finch, J. T., Noll, M., Kornberg, R. D. 1975. *Proc. Natl. Acad. Sci. USA* 72:3320–22
118. Germond, J. E., Hirt, B., Oudet, P., Gross-Bellard, M., Chambon, P. 1975. *Proc. Natl. Acad. Sci. USA* 72:1843–47
119. Griffith, J. D. 1975. *Science* 187:1202–1203
120. Keller, W. 1975. *Proc. Natl. Acad. Sci. USA* 72:4876–80
121. Shure, M., Vinograd, J. 1976. *Cell* 8:215–26
122. Camerini-Otero, R. D., Felsenfeld, G. 1977. *Nucleic Acids Res.* 4:1159–81
123. Bina-Stein, M., Simpson, R. T. 1977. *Cell* 11:609–18
124. Crick, F. H. C., Klug, A. 1975. *Nature* 255:530–33
125. Wang, J. C. 1971. *J. Mol. Biol.* 55:523–33
126. Champoux, J. J., Dulbecco, R. 1972. *Proc. Natl. Acad. Sci. USA* 69:143–46

127. Baase, W. A., Wang, J. C. 1974. *Biochemistry* 13:4299–4303
128. Pulleyblank, D. E., Morgan, A. R. 1975. *Biochemistry* 14:5205–9
129. Burrington, M. G., Morgan, A. R. 1976. *Can. J. Biochem.* 54:301–6
130. Champoux, J. J., McConaughy, B. L. 1976. *Biochemistry* 15:4638–42
130a. Durnford, J. M., Champoux, J. J. 1978. *J. Biol. Chem.* In press
131. Keller, W. 1975. *Proc. Natl. Acad. Sci. USA* 72:2550–54
132. Mattoccia, E., Attardi, D. G., Tocchini-Valentini, G. P. 1976. *Proc. Natl. Acad. Sci. USA* 73:4551–54
133. Laskey, R. A., Mills, A. D., Morris, N. R. 1977. *Cell* 10:237–43
134. Bina-Stein, M., Vogel, T., Singer, D. S., Singer, M. F. 1976. *J. Biol. Chem.* 251:7363–66
135. DeLeys, R. J., Jackson, D. A. 1976. *Nucleic Acids Res.* 3:641–52
136. Vosberg, H.-P., Grossman, L. I., Vinograd, J. 1975. *Eur. J. Biochem.* 55:79–93
137. Vosberg, H.-P., Vinograd, J. 1975. *Proc. ICN-UCLA Symp. Mol. Cell. Biol.* 3:94–120
138. Bauer, W. R., Ressner, E. C., Kates, J., Patzke, J. V. 1977. *Proc. Natl. Acad. Sci. USA* 74:1841–45
139. Wang, J. C. 1973. In *DNA Synthesis in vitro*, ed. R. D. Wells, R. B. Inman, pp. 163–74. Baltimore: University Park. 426 pp.
140. Champoux, J. J. 1976. *Proc. Natl. Acad. Sci. USA* 73:3488–91
141. Vosberg, H.-P., Vinograd, J. 1976. *Biochem. Biophys. Res. Commun.* 68: 456–64
142. Wang, J. C. 1974. *Methods Enzymol.* 29:197–203
143. Champoux, J. J., Durnford, J. M. 1975. See Ref. 137, pp. 83–93
144. Carlson, J. O., Wang, J. C. 1974. In *DNA Replication*, ed. R. R. Wickner, p. 231–37. New York: Dekker. 299 pp.
145. Depew, R. E., Liu, L. F., Wang, J. C. 1976. *Fed. Proc.* 35:1493
146. Blair, D. G., Helinski, D. R. 1975. *J. Biol. Chem.* 250:8785–89
147. Guiney, D., Helinski, D. R. 1975. *J. Biol. Chem.* 250:8796–803
148. Liu, L. F., Depew, R. E., Wang, J. C. 1976. *J. Mol. Biol.* 106:439–52
149. Sugino, A., Peebles, C. L., Kreuzer, K. N., Cozzarelli, N. R. 1977. *Proc. Natl. Acad. Sci. USA* 74:4767–71
150. Depew, R. E., Wang, J. C. 1975. *Proc. Natl. Acad. Sci. USA* 72:4275–79
151. Pulleyblank, D. E., Shure, M., Tang, D., Vinograd, J., Vosberg, H.-P. 1975.
152. Champoux, J. J. 1977. *Proc. Natl. Acad. Sci. USA* 74:3800–4
153. Shabarova, Z. A. 1970. *Prog. Nucl. Acid Res. Mol. Biol.* 10:145–82
154. Berg, P. 1956. *J. Biol. Chem.* 222: 1015–23
155. Phillips, D. R., Fife, T. H. 1969. *J. Org. Chem.* 34:2710–14
156. Henry, T. J., Knippers, R. 1974. *Proc. Natl. Acad. Sci. USA* 71:1549–53
157. Ikeda, J.-E., Yudelevich, A., Hurwitz, J. 1976. *Proc. Natl. Acad. Sci. USA* 73:2669–73
158. Eisenberg, S., Griffith, J., Kornberg, A. 1977. *Proc. Natl. Acad. Sci. USA* 74:3198–3202
159. Heffron, F., Bedinger, P., Champoux, J. J., Falkow, S. 1977. *Proc. Natl. Acad. Sci. USA* 74:702–6
160. Kleckner, N. 1977. *Cell* 11:11–23
161. Champoux, J. J. 1977. *Proc. Natl. Acad. Sci. USA* 74:5328–32
162. Watson, J. D. 1971. *Adv. Cell Biol.* 2:1–46
163. Stein, G., Baserga, R. 1972. *Adv. Cancer Res.* 15:287–330
164. Rosenberg, B. H., Ungers, G., Deutsch, J. F. 1976. *Nucleic Acids Res.* 3: 3305–11
165. Miller, O. L. Jr., Beatty, B. R. 1969. *Science* 164:955–57
166. Miller, O. L. Jr., Hamkalo, B. A., Thomas, C. A. Jr. 1970. *Science* 169:392–95
167. Meselson, M. 1972. *J. Mol. Biol.* 71:795–98
168. Abdel-Monem, M., Hoffmann-Berling, H. 1976. *Eur. J. Biochem.* 65:431–40
169. Abdel-Monem, M., Dürwald, H., Hoffmann-Berling, H. 1976. *Eur. J. Biochem.* 65:441–49
170. Abdel-Monem, M., Lauppe, H.-F., Kartenbeck, J., Dürwald, H., Hoffmann-Berling, H. 1977. *J. Mol. Biol.* 110:667–85
171. Denhardt, D. T., Dressler, D. H., Hathaway, A. 1967. *Proc. Natl. Acad. Sci. USA* 57:813–20
172. Lane, H. E. D., Denhardt, D. T. 1975. *J. Mol. Biol.* 97:99–112
173. Scott, J. F., Eisenberg, S., Bertsch, L. L., Kornberg, A. 1977. *Proc. Natl. Acad. Sci. USA* 74:193–97
174. Abdel-Monem, M., Chanal, M.-C., Hoffmann-Berling, H. 1977. *Eur. J. Biochem.* 79:33–38
175. Richet, E., Kohiyama, M. 1976. *J. Biol. Chem.* 251:808–12
176. Abdel-Monem, M., Dürwald, H., Hoffmann-Berling, H. 1977. *Eur. J. Biochem.* 79:39–45

177. Oishi, M. 1969. *Proc. Natl. Acad. Sci. USA* 64:1292–99
178. Goldmark, P. J., Linn, S. 1972. *J. Biol. Chem.* 247:1849–60
179. Karu, A. E., MacKay, V., Goldmark, P. J., Linn, S. 1973. *J. Biol. Chem.* 248:4874–84
180. MacKay, V., Linn, S. 1974. *J. Biol. Chem.* 249:4286–94
181. Eichler, D. C., Lehman, I. R. 1977. *J. Biol. Chem.* 252:499–503
182. Karu, A. E., Linn, S. 1972. *Proc. Natl. Acad. Sci. USA* 69:2855–59
183. Hertman, I. 1969. *Genet. Res.* 14:291–307
184. Willetts, N. S., Clark, A. J. 1969. *J. Bacteriol.* 100:231–39
185. Hendler, R. R., Pereira, M., Scharff, R. 1975. *Proc. Natl. Acad. Sci. USA* 72:2099–2103
186. Friedman, E. A., Smith, H. O. 1973. *Nature New Biol.* 241:54–58
187. Wilcox, K. W., Smith, H. O. 1976. *J. Biol. Chem.* 251:6127–34
188. Strätling, W., Knippers, R. 1973. *Nature* 245:195–97
189. Hinkle, D. C., Richardson, C. C. 1975. *J. Biol. Chem.* 250:5523–29
190. Kolodner, R., Richardson, C. C. 1977. *Proc. Natl. Acad. Sci. USA* 74:1525–29
191. Scherzinger, E., Lanka, E., Morelli, G., Seiffert, D., Yuki, A. 1977. *Eur. J. Biochem.* 72:543–58
192. Bouché, J.-P., Zechel, K., Kornberg, A. 1975. *J. Biol. Chem.* 250:5995–6001
193. Gellert, M., Mizuuchi, K., O'Dea, M. H., Nash, H. A. 1976. *Proc. Natl. Acad. Sci. USA* 73:3872–76
193a. Gellert, M., Mizuuchi, K., O'Dea, M. H., Itoh, T., Tomizawa, J.-I. 1977. *Proc. Natl. Acad. Sci. USA* 74:4772–76
194. Smith, D. H., Davis, B. D. 1967. *J. Bacteriol.* 93:71–79
195. Ryan, M. J. 1976. *Biochemistry* 15:3769–77
196. Gellert, M., O'Dea, M. H., Itoh, T., Tomizawa, J.-I. 1976. *Proc. Natl. Acad. Sci. USA* 73:4474–78
197. White, M., Eason, R. 1973. *Nature New Biol.* 241:46–49
198. Smith, C. A., Jordan, J. M., Vinograd, J. 1971. *J. Mol. Biol.* 59:255–72
199. Crosa, J. H., Luttropp, L. K., Falkow, S. 1976. *Nature* 261:516–19
200. Timmis, K., Cabello, F., Cohen, S. N. 1976. *Nature* 261:512–16
201. Sen, A., Levine, A. J. 1974. *Nature* 249:343–44
202. Germond, J. E., Bellard, M., Oudet, P., Chambon, P. 1976. *Nucleic Acids Res.* 3:3173–92
203. Crick, F. 1971. *Nature* 234:25–27
204. Marians, K. J., Ikeda, J.-E., Schlagman, S., Hurwitz, J. 1977. *Proc. Natl. Acad. Sci. USA* 74:1965–68
205. Mizuuchi, K., Nash, H. A. 1976. *Proc. Natl. Acad. Sci. USA* 73:3524–28
206. Holloman, W. K., Radding, C. M. 1976. *Proc. Natl. Acad. Sci. USA* 73:3910–14
207. Cairns, J. 1963. *Cold Spring Harbor Symp. Quant. Biol.* 28:44–46
208. Sebring, E. D., Kelly, T. J. Jr., Thoren, M. M., Salzman, N. P. 1971. *J. Virol.* 8:478–90
209. Jaenisch, R., Mayer, A., Levine, A. J. 1971. *Nature New Biol.* 233:72–75
210. Roman, A., Champoux, J. J., Dulbecco, R. 1974. *Virology* 57:147–60
211. Inselburg, J., Fuke, M. 1971. *Proc. Natl. Acad. Sci. USA* 68:2839–42
212. Staudenbauer, W. L. 1974. *Nucleic Acids Res.* 1:1153–64
213. Crosa, J. H., Luttropp, L. K., Falkow, S. 1976. *J. Bact.* 126:454–66
214. Cabello, F., Timmis, K., Cohen, S. N. 1976. *Nature* 259:285–90
215. Sogo, J. M., Greenstein, M., Skalka, A. 1976. *J. Mol. Biol.* 103:537–62
216. Berk, A. J., Clayton, D. A. 1974. *J. Mol. Biol.* 86:801–24
217. Sakakibara, Y., Suzuki, K., Tomizawa, J.-I. 1976. *J. Mol. Biol.* 108:569–92
218. Griffith, J. D. 1976. *Proc. Natl. Acad. Sci. USA* 73:563–67

Ann. Rev. Biochem. 1978. 47:481–532
Copyright © 1978 by Annual Reviews Inc. All rights reserved

THE OUTER MEMBRANE PROTEINS OF GRAM-NEGATIVE BACTERIA: BIOSYNTHESIS, ASSEMBLY, AND FUNCTIONS

❖981

Joseph M. DiRienzo, Kenzo Nakamura, and Masayori Inouye

Department of Biochemistry, State University of New York at Stony Brook, Stony Brook, New York 11794

CONTENTS

481

0066-4154/78/0701-0481$01.00

PERSPECTIVES AND SUMMARY

Gram-negative bacteria, like eucaryotic cells, are enclosed by an inner cytoplasmic or plasma membrane composed of a lipid bilayer structure that is interdispersed with protein. These membranes can best be described by the well-known fluid mosaic model (1). The bacterial cytoplasmic membrane houses the systems for active transport, oxidative phosphorylation, and the biosynthesis of certain macromolecules. However, Gram-negative bacteria also contain a membrane external to the cytoplasmic membrane which, although morphologically similar to the cytoplasmic membrane, contains less phospholipid, fewer proteins, and a unique carbohydrate component lipopolysaccharide (LPS). Functionally the outer membrane is quite distinct: it acts as a diffusion barrier against various compounds, for example antibiotics; it contains receptors for bacteriophages and colicins; it is involved in the process of conjugation and also cell division or, more precisely, septum formation; it contains various specific uptake systems for nutrients, such as iron, vitamins, and carbohydrates; and it contains nonspecific passive diffusion pores that allow the diffusion of low molecular weight substrates. The outer membrane also provides a protective environment for certain hydrolytic enzymes and binding proteins that reside in the periplasmic region, the area between the cytoplasmic and outer membrane, and may participate, in conjunction with the peptidoglycan, in maintaining the structural integrity of the cell.

As mentioned above, the outer membrane contains a small variety of proteins. However, these proteins appear to impart to the outer membrane many of the functions just described, and therefore, research on the outer membrane proteins is crucial to an understanding of outer membrane functions. Also, because of the small number of protein components, and their presence in rather large quantities, it is easy to isolate and characterize the outer membrane proteins. Thus, they provide an excellent system for the study of the mechanism of membrane protein synthesis, the translocation of molecules across a biological membrane, and membrane assembly processes. It is hoped that these types of investigations will also lead to an understanding of the processes that determine the differentiation of membranes.

A great deal of progress has been made in the area of outer membrane protein biosynthesis and assembly since these proteins were first separated and identified in the late 1960s and early 1970s, and today a fairly comprehensive picture of outer membrane protein biosynthesis and assembly can be formed. Most recently, it was discovered that the lipoprotein, and other major outer membrane proteins, are synthesized from precursor proteins containing an extra peptide sequence of approximately 20 amino acids at the amino terminus. These extension peptides are extremely hydrophobic, thus they have a high affinity for the cytoplasmic membrane. This property suggests that the extension peptides are the determining factors in the binding to the cytoplasmic membrane of those polyribosomes that specifically synthesize outer membrane and periplasmic proteins, and in translocating these proteins across the cytoplasmic membrane. The mechanisms of biosynthesis and translocation of outer membrane and periplasmic proteins appear to be remarkably similar to those that operate for secretory proteins in eucaryotic systems.

A vast amount of information pertaining to the assembly of proteins in the outer membrane has also been obtained. Certain outer membrane proteins have been shown (*a*) to interact to form oligomers with themselves or other neighboring proteins (*b*) to be exposed on the surface of the cell, and (*c*) to span the entire width of the outer membrane. These results, reinforced by additional studies, have demonstrated that protein pores or channels traverse the outer membrane and thus form a path for the passive diffusion of small molecules.

Extensive genetic studies on the outer membrane proteins have been providing invaluable information for elucidating their functions, the mechanisms of their biosynthesis and assembly, the mechanisms of regulation of their production, and the structural-functional relationship between them.

Another intriguing and significant approach in the study of the outer membrane proteins is the isolation and characterization of the mRNA for the lipoprotein. This is the only biologically active mRNA isolated so far from *Escherichia coli*. Its partial nucleotide sequence has been determined and this sequence verifies the presence of an extra peptide in the precursor of the lipoprotein.

This article reviews and correlates the information that has been accumulated concerning the outer membrane proteins of *E. coli* and *Salmonella typhimurium*. We discuss not only the mechanisms of the biosynthesis and assembly of the outer membrane proteins, but also their molecular properties and interactions with other components in the outer membrane. The peptidoglycan and LPS are known to be closely associated with some of the outer membrane proteins and these interactions are probably essential for the assembly of the outer membrane. Therefore it is important to know the structure and functions of these macromolecules. Comprehensive reviews have been published on the peptidoglycan (2, 3) and LPS (4–8), and there are several review articles that are pertinent to the study of the structure and biosynthesis of the *E. coli* lipoprotein (9–13).

GENERAL CHARACTERISTICS OF THE OUTER MEMBRANE

Early electron microscopic studies revealed that the cell envelope of Gram-negative bacteria is composed of two distinct membranes, the inner cytoplasmic or plasma membrane and the outer membrane, both of which demonstrate the usual double-track or bilayer appearance in thin section (14, 15). The peptidoglycan layer is located between the two membranes and this area between the cytoplasmic and the outer membranes is referred to as the periplasmic region. Both the cytoplasmic and outer membranes are 75Å thick and are 100Å apart (16). Freeze-etching studies also revealed the presence of multilayered structures in the Gram-negative cell envelope (17, 18).

Miura & Mizushima first successfully separated the cytoplasmic and outer membranes of *E. coli* by sucrose density gradient centrifugation (19). An important step in this procedure consists in preparing the crude envelope fraction from spheroplasts that were formed by treating cells with ethylenediaminetetraacetic acid (EDTA)-lysozyme. This basic membrane separation method was employed with minor modifications by other workers (20, 21). Osborn and colleagues extended the previous investigations and developed an isopycnic sucrose density gradient that separated the two membranes from *S. typhimurium* (16). Lysozyme-EDTA-prepared spheroplasts were lysed by sonication and the crude envelope fractions were centrifuged on a step gradient ranging from 30–50 percent sucrose. The advantages of this procedure over that of Miura & Mizushima (19) were that only one gradient run was required and that this method could be applied to a variety of bacterial strains. More recently these procedures were applied to the separation and preparative isolation of cytoplasmic and outer membrane from *E. coli* (22, 23).

Other purification methods were developed based on the solubility properties of the different membranes in detergents. The nonionic detergent, Triton X-100, is known as a solubilizer of the cytoplasmic membrane (24, 25). This detergent in combination with Mg^{2+} was successfully used to remove contaminating cytoplasmic membrane from crude outer membrane preparations (26). It was found that sodium lauryl sarcosinate completely solubilized the cytoplasmic but not the outer membrane (27). A procedure involving particle electrophoresis was also used to separate the two membranes of *E. coli* in order to examine the distribution of lipids (28).

The efficient separation of the cytoplasmic and outer membranes made it possible to study the characteristics of each membrane. Electron microscopy of the outer membrane from *S. typhimurium* revealed an ultrastructure of predominantly closed double-track vesicles uniform in size (0.1 μm in diameter) (16). When the outer membranes were isolated from total membrane preparations that had been exposed to salts, they displayed the form of open C-shaped and coiled structures (16, 19, 20). The outer membrane is composed of phospholipid and protein as is the cytoplasmic membrane. However, in addition to these two components, it also contains the lipopolysaccharides. There are approximately 2.5×10^6 molecules of LPS per *S. typhimurium* cell (29) and they are localized exclusively in the outer membrane (16, 29). Actually, because of the presence of the polysaccharide chains of the LPS, the density of the outer membrane is greater than that of the cytoplasmic membrane, which allows for their separation by sucrose density gradient centrifugation. The LPS occupies about 45% of the surface of the outer membrane. The lipid A portion of the LPS molecule replaces part of the phospholipid in the outer layer of the lipid bilayer. The polysaccharide chains extend toward the outside of the cell. The LPS molecules, which are closely associated with each other, are stabilized by divalent cations, predominantly Mg^{2+}. The phospholipid is localized almost exclusively in the inner layer of the outer membrane bilayer (29). In *E. coli* (28, 30) and *S. typhimurium* (16) the phospholipid composition of the outer membrane consists predominantly of phosphatidylethanolamine, phosphatidylglycerol, and cardiolipin. The rest of the outer membrane is composed of the proteins that are the topic of the remainder of this review. Reviews on the structure and function of the outer membrane (31) and cell envelope (32, 33) of Gram-negative bacteria have been published.

OUTER MEMBRANE PROTEINS

Nomenclature of the Major Proteins

In 1970, Schnaitman (34) reported that the *E. coli* outer membrane consisted of one "major outer membrane protein" band with a molecular weight of 44,000, which accounted for 70% of the total protein. However, it was soon shown by Inouye & Yee (35, 36) that this "major peak" is an artifact due to abnormal characteristics of the outer membrane proteins and that the peak actually consists of a few distinct proteins. Although it is now established that Schnaitman's "major outer membrane protein" consists of at least four different protein components, different nomenclature systems for these proteins were employed by various investigators (Table 1).

In addition to these major outer membrane proteins of higher molecular weights, another major protein, called lipoprotein (37, 38) is present. This

Table 1 A comparison of the nomenclature of major proteins of the outer membrane of *E. coli* K–12 as described by various investigators

Present paper	Inouye (36)	Bragg (48)	Schnaitman (47)	(52)	Henning (54)	(55)	Mizushima (50, 51)	Lugtenberg (53)	Uniform nomenclature[a]
Matrix protein									
Ia	Peak 4	A_1, A_2	1	1a	I	Ia	0–9	b	PG 1–2
Ib				1b		Ib	0–8	c	PG 1–3
Protein 2[b]	—	—	2		—		—	—	PG 1–1
TolG protein	Peak 7	B	3a		II*		0–10	d	HM 1–2
Protein 3b	(Peak 6)[c]	nd[d]	3b		nd[d]		(0–11)	a	HM 1–1
Lipoprotein	Peak 11	F	nd[d]		IV		0–18	nd[d]	—

[a] Proposed at Tübingen Conference, September, 1977. PG and HM represent peptidoglycan associated and heat-modifiable proteins, respectively.
[b] Protein 2 is present only in strains lysogenic for phage PA–2 (80).
[c] Parentheses around a protein band designation indicate that the correlation of these proteins to those described by other investigators is not well defined (see text).
[d] Not detected.

protein is considered to be the most abundant protein in the *E. coli* cell on the basis of molecular numbers (38) and it is discussed later. However, most of the earlier authors overlooked the existence of the lipoprotein in *E. coli* and *S. typhimurium* (34, 39–45), and even the absence of the lipoprotein was suggested (46). This error resulted chiefly from the use of sodium dodecyl sulfate (SDS)-polyacrylamide gel systems, in which proteins of smaller molecular weights could not be resolved.

The nomenclatures used for major proteins of the outer membrane of *E. coli* K-12 are summarized in Table 1. The correlation between the protein bands of different authors is based on the migration of each protein in different SDS-gel systems and the characteristics of each protein. These characteristics are described in later sections. In this article the terms "matrix proteins Ia and Ib" are used to designate the proteins that are strongly associated with the peptidoglycan; "tolG protein" designates the heat-modifiable, trypsin-sensitive protein missing in *tol*G mutants; and "lipoprotein" designates the protein of the smallest molecular weight.

"Protein 3b" of the outer membrane has not yet been well characterized. Schnaitman (47) showed that his band 3, which is resolved by the alkaline-SDS gel system of Bragg & Hou (48), is composed of two different proteins, 3a and 3b. These proteins were separated on a DEAE-ion exchange column and gave rise to different cyanogen bromide fragments (47). Protein 3a is identical to the tolG protein (49). Proteins 0–10 and 0–11 of Mizushima's group (50, 51) can be separated by a urea-SDS gel system that uses highly cross-linked gels (51), but not by Bragg and Hou's alkaline-SDS gel system (50). Protein 0–10 is identical to the tolG protein, since this protein is heat modifiable and trypsin sensitive. Protein 0–11 is probably identical to protein 3b. In contrast to protein 0–10, protein 0–11 was resistant to proteolysis and a significant portion of it was released from the cell in association with

lipopolysaccharides upon EDTA treatment (51). A recent paper by Bassford et al (52) identified band *a* of Lugtenberg et al (53) with protein 3b.
It may be questioned whether band B of Bragg & Hou (48) and band II*
of Henning's group (54, 55) consist of a single polypeptide or not, since
proteins 3a and 3b were expected to migrate as a single band in the SDS-gel
systems that were used. Recently, Henning et al (56) suggested the
heterogeneity of their band II* protein. Alternatively, because the amount
of protein 3b is known to be greatly reduced when the cells are grown at
low temperatures (52, 57, 58) such as those used by Henning et al (54, 55),
these authors might not have detected it. It is likely that peak 6 of Inouye
& Yee (36) corresponds to protein 3b, because peak 6 is one of the major
outer membrane proteins that does not correspond to the matrix protein
and is resistant to proteolysis. Peak 6 has been called Y protein and its
production has been related to DNA synthesis (59, 60).

The resolution and the order of migration of the major outer membrane
proteins greatly depends on details of the SDS-gel systems used, such as
concentration of acrylamide, extent of cross-linking, pH of the buffer, ionic
strength, voltage and current, and the way in which the sample is solubilized. SDS-polyacrylamide gel electrophoresis is an extremely useful technique for the analysis of membrane proteins, but care has to be taken in
concluding that protein bands are homogeneous, since some proteins with
unusual charges or conformations and some glycoproteins migrate anoma-
lously (61). Furthermore, the composition and amount of outer membrane
proteins may vary greatly, depending on the strain and cultural conditions
used (39, 57, 62). *E. coli* B strains have only the matrix protein Ia, K-12
strains have both Ia and Ib, while other *E. coli* strains have two proteins
comparable to those of K-12 strains that are tightly associated with the
peptidoglycan (62, 63) but differ from those of the K-12 strains in electrophoretic mobility. Both tolG protein and protein 3b are found in a variety
of *E. coli* strains (57, 62). Cultural conditions, especially the composition
of the growth medium, also affect markedly the relative amounts of the two
matrix proteins Ia and Ib, as described later.

Matrix Proteins

Matrix proteins are characterized by their tight, but noncovalent, association with peptidoglycan. They vary in composition and electrophoretic
mobility depending on the strain of origin, and can be easily isolated by an
elegant method described by Rosenbusch (64). When the whole cell envelope is solubilized in 2% SDS at 60°C these proteins are found in the
insoluble material complexed with peptidoglycan and bound lipoprotein.
The matrix proteins can be released from the peptidoglycan either by heat-

ing the complex at 100°C for 5 min in SDS (64) or by extraction at 37°C with SDS containing 0.5 M NaCl (51, 65). The former procedure apparently results in denaturation of the protein whereas the latter method does not cause denaturation (51). Two matrix proteins, Ia and Ib of strain K-12, were isolated separately either from cultures grown under different conditions (51) or from mutants lacking one of the two proteins (55, 56).

Rosenbusch (64) purified matrix protein Ia of strain B to homogeneity. Its molecular weight by several different methods was 36,500; it appears to be a single polypeptide of 336 amino acid residues and seems to lack any nonprotein moiety. The amino acid composition indicates only a moderate hydrophobicity and the polarity index, calculated according to Capaldi & Vanderkooi (66), is 45%. The isoelectric point of this protein is about 6. The first eight residues in the N-terminus sequence were determined as: NH_2-Ala-X-Tyr-Tyr-Asx-His-Lys-Glx-.

The chemical relationship between matrix proteins Ia and Ib of strain K-12 or other protein species of this group is still unknown. These proteins share various characteristics (50–52, 55–57) and can be separated in SDS-gel systems only under special conditions such as those used in the disc gel system containing 8M urea (50) or in slab gel systems (52, 53, 55) based on the method of Laemmli (67). Schmitges & Henning (55) compared the cyanogen bromide fragments of these two proteins. Among six fragments only one showed a difference in electrophoretic mobility. Since this fragment did not correspond to the C-terminal or N-terminal regions of the proteins (68), the possibility that a protein processing mechanism removes the terminal part of the polypeptide was excluded. That the matrix proteins Ia and Ib represent basically the same polypeptide is further supported by analyses of the cyanogen bromide fragments (52, 56) and by tryptic fingerprints and N-terminal amino acid sequence analyses (56). Judged from their behavior during electrofocusing (55) and on a QAE-Sephadex column (52), matrix protein Ib is more acidic than Ia, and these proteins may differ by modification of amino acid residue(s) (52, 55; C. A. Schnaitman, personal communication).

The relative amounts of the matrix proteins Ia and Ib vary greatly depending on growth conditions, especially the composition of the growth medium (39, 52, 57, 65). Cells grown in nutrient broth (52) or yeast broth (57) with no fermentable carbon source contain a decreased amount of protein Ib with a concomitant increase of protein Ia. In contrast, the amount of the protein Ia decreases with a concomitant increase of protein Ib in the cells grown in complex medium supplemented with fermentable carbon sources such as glucose and lactate, and to a lesser extent with glycerol (52). The amount of protein Ia also decreases relative to protein Ib when the growth medium is supplemented with NaCl (65, 69). The effect of NaCl is probably due to the osmolarity of the medium, since the addition of 300 mM NaCl or KCl, or 600 mM sucrose to yeast broth, caused almost the same effect (69). The physiological role of these alterations, and the mechanism by which they and the total amount of the matrix proteins are regulated, are unknown.

A striking feature of matrix proteins Ia and Ib is their extremely high content of β-structure (51, 64). This is in contrast to many other "intrinsic" membrane proteins which show high contents of α-helix. The high content of β-structure found in the native outer membrane (70) can be ascribed to the presence of these proteins and the tolG protein (51). Nakamura & Mizushima (51) showed that the β-structures of matrix proteins Ia and Ib in a K-12 strain are stable in SDS solution unless they are heated. Gross conformational change, however, occurs upon heating above 70°C in SDS solution. Circular dichroic spectra of heated samples resemble those observed for many SDS-protein complexes (71). Rosenbusch (64) showed that protein Ia of strain B does not bind SDS unless it is heated to 100°C. The stability in SDS is not an unique property of the matrix proteins. Several other outer membrane proteins, such as tolG protein (51), phospholipase A_1 (72), and tonA protein (73), are also stable in SDS solution unless they are heated.

Matrix proteins Ia and Ib migrate on SDS-gels as monomeric forms only when they are heated in SDS solution above 70°C (51, 64). Oligomeric forms of these proteins exist in unheated SDS solution in the absence of peptidoglycan (51, 74) as well as in vivo (75, 76). The matrix proteins can be released from the peptidoglycan when the protein-peptidoglycan complex is heated in SDS solution above 70°C (65). These results indicate that the gross conformational change of the matrix proteins upon heating above 70°C causes the dissociation of oligomers into subunits and the release of the proteins from the peptidoglycan. The matrix proteins, however, can be released from the peptidoglycan without conformational change by adding 0.5 M NaCl, which suggests that they are held to the peptidoglycan by ionic interactions.

There is no doubt about the interaction of the matrix proteins with peptidoglycan in vivo (64, 65). Mizushima and his co-workers (65, 77, 78) showed that the matrix proteins can specifically bind to the peptidoglycan in vitro in SDS solution. Binding is optimal at pH 8 in the presence of 5 mM Mg^{2+} and is inhibited by high concentrations of NaCl (65). Heat denatured proteins lose their ability to bind (65, 78). It is suggested that the D-diaminopimelic acid residue of the peptidoglycan plays an important role in its binding to the matrix proteins (S. Mizushima, personal communication). Yu & Mizushima (78) showed that LPS stimulates the binding of matrix proteins Ia and Ib to the peptidoglycan. Inactivation of phages by the purified matrix proteins is also stimulated by the presence of LPS (79), which suggests the interaction of the matrix proteins with LPS in vivo. The matrix protein also associates closely with another major protein, the lipoprotein, as is discussed in a later section.

Electron micrographs of the negative-stained matrix protein-peptidoglycan complex show that the matrix protein molecules are arranged as a hexagonal lattice layer with 7.7 nm repeat (64, 74). There are about 1.1×10^5 molecules of the matrix protein per cell (64) and the hexagonal lattice structure covers 60% or more of the outer surface of the peptidoglycan (74). Most probably three molecules of protein form one unit. Almost the same ultrastructure was observed with a protein layer prepared from lysozyme-spheroplasts that lacks the peptidoglycan layer (74). This indicates that protein-protein interaction is a major force in maintaining the hexagonal arrangement of the protein. The most pronounced feature of the unit hexagonal cell is a triplet of indentations with a 2 nm diameter which are readily penetrated by stain. It was suggested that these indentations represent channels that span the protein monolayer as well as the outer membrane (74). The idea that the matrix proteins span the thickness of the outer membrane is supported by the evidence that these proteins serve as receptors for phages (79–81). Recently, Kamio & Nikaido (82) showed more definitely the exposure of the matrix proteins on the outer surface of the outer membrane by the use of a nonpermeable labeling agent, CNBr-activated dextran.

Nakae showed that incorporation of the matrix protein of *S. typhimurium* (83) and *E. coli* B (84) into artificial LPS-phospholipid vesicles greatly enhances their permeability to sucrose. The vesicles reconstituted with the matrix protein showed almost the same molecular sieving properties as the intact outer membrane which excludes oligo- and polysaccharides with molecular weights higher than 900 (85). These results indicate that one function of the matrix proteins is to form the passive diffusion pores that allow the rapid diffusion of hydrophilic molecules with small molecular weights through the outer membrane. This conclusion was further supported by the analysis of mutants lacking the matrix proteins. Such mutants of *S. typhimurium* were defective in the diffusion of the β-lactam antibiotic, cephaloridine, across the outer membrane (86). Mutants of *E. coli* B selected for resistance to Cu^{2+} specifically lacked the matrix protein (87). These mutant cells exhibit a greatly reduced ability to utilize several low molecular weight metabolites such as amino acids and sugars at low concentrations. "Pleiotropic transport mutants" with decreased affinity for the uptake of substrates such as sugars, amino acids, uracil, and inorganic anions, as described by von Meyenburg (88), were also found to be missing the matrix proteins (89). Because of the nature of this protein it was designated "porin" (84).

Protein 2 described by Schnaitman (47) was found in *E. coli* K-12 strains lysogenized with a lamboid phage PA-2 (80), which utilizes the matrix protein Ib as a receptor (52, 80). In these lysogenized strains the amount of the matrix protein was greatly reduced. However, when the cells were grown in L broth containing glucose, the amount of protein was decreased with a concomitant increase in the amount of the matrix protein (39, 80). Protein 2 exhibits properties similar to those of the matrix protein, such as a similar molecular weight and a tight association with the peptidoglycan (90). Furthermore, when the *omp*B mutants that lack matrix proteins Ia and Ib were lysogenized with phage PA2 they regained the sensitivity toward colicins E2 and E3 that was lost in the parent *omp*B mutants (80). These results indicate that protein 2 is closely related to the matrix proteins. However, judging from its amino acid composition, cyanogen bromide and tryptic fragments analyses, and immunochemical analysis, protein 2 represents a completely different polypeptide than the matrix proteins (90). It was suggested that protein 2 may be coded for by a prophage gene and produced for the surface exclusion of super-infecting phage (52, 80). If this is indeed the case, interesting questions arise: what is the molecular basis of the homology between protein 2 and the matrix proteins, and how is the amount of protein 2 regulated in response to changes in the composition of the growth medium (52)? Such changes also regulate the relative amounts of matrix proteins Ia and Ib as described earlier.

TolG Protein

The tolG protein can be distinguished from the other major outer membrane proteins on SDS-gels on the basis of two characteristics. (*a*) When the whole envelope or the outer membrane is digested with trypsin or pronase, only this protein, among the major outer membrane proteins, is digested leaving a smaller polypeptide in the membrane (35, 48, 54). (*b*) The tolG protein exhibits anomalous "heat-modifiability" on SDS-gels. Its apparent molecular weight increases when the membrane is dissolved in SDS solution above 50°C (36, 47, 48, 50). The matrix protein also changes its migration upon heating, but in this case the apparent molecular weight decreases upon heating (51, 64). The reported molecular weight of the tolG protein varies. Upon heating, the apparent molecular weight increases from 27,000 to 36,000 (54), 38,000 to 48,000 (36), or 28,500 to 33,000 (91). This discrepancy is most probably due to the different SDS-gel systems employed.

The tolG protein has been purified in either the heat-modified form (50, 92) or in the nonmodified form (51, 91, 93, 94). The amino acid composition (93, 94) shows only a moderate degree of hydrophobicity. [The polarity index calculated from the amino acid composition reported by Garten et al (93) according to Capaldi & Vanderkooi (66) is 43%.] According to Garten et al (93), the minimum chemical molecular weight of 27,000 is calculated assuming that two cysteic acids arise from one mole of the protein. The sum of the apparent molecular weights of cyanogen bromide fragments also gave a value of 27,000. These results raise the possibility that the heat-modified form rather than the nonmodified form migrates atypically on SDS gels. The two cysteine residues do not seem to form an interchain disulfide bridge, since β-mercaptoethanol treatment does not change the mobility of this protein in SDS gel electrophoresis (93).

The modification of this protein by heat was studied by several investigators. Schnaitman (95) reported an increase in the intrinsic viscosity of the SDS protein complex upon heating. Results of Reithmeier & Bragg (91) suggest a more unfolded structure in the heat-modified form. In contrast to the matrix protein, however, heat modification of the tolG protein did not accompany a large increase in the binding of SDS. Thus the apparent increase in the molecular weight upon heating seems to be due to the increased asymmetry of the molecular shape rather than to increased binding of SDS. Nakamura & Mizushima (51) showed that this protein also has a high β-structure content. Like the matrix protein, this conformation was stable in SDS solution but was destroyed upon heating above 50°C.

The fact that the tolG protein is susceptible to partial proteolytic digestion upon the treatment of the outer membrane with trypsin suggests that part of this protein is exposed to the outer surface of the membrane, and this is supported by the fact that it serves as a receptor for certain phages (79, 94, 96). Furthermore, the analysis of the mutant lacking tolG protein revealed that it has an important function in F-pilus mediated conjugation (96–99). The mutant lacking tolG protein was defective as a recipient in conjugation due to a failure in pair formation (97). Purified tolG protein inhibits conjugation (94, 100) and inactivates phages K3 (94) and TuII* (79). Inhibition of conjugation and inactivation of phages by the TolG protein were greatly stimulated by the presence of LPS, which suggests the interaction of the tolG protein and LPS in vivo.

Lipoprotein

In 1969, Braun & Rehn (37) reported the existence of a lipoprotein, covalently linked to the peptidoglycan, with a molecular weight of about 7000. The complete chemical structure of this protein has been determined by Braun and his co-workers (101–104). The lipoprotein consists of 58 amino acid residues and lacks histidine, tryptophan, glycine, proline, and phenylalanine. It is linked by the ϵ-amino group of its C-terminal lysine to the carboxyl group of every tenth to twelfth meso-diaminopimelic acid residue of the peptidoglycan. The N-terminal portion of the lipoprotein consists of glycerylcysteine [S-(propane-2',3'-diol)-3-thioaminopropionic acid] to which two fatty acids are attached by two ester linkages and one fatty acid by an amide linkage. The fatty acids bound as esters are similar to the fatty acids found in the phospholipids, while the amide-linked fatty acids consist of 65% palmitate, the rest being mainly monounsaturated fatty acids (104).

Inouye and his co-workers (38, 105) independently found that the lipoprotein also exists in the E. coli membrane without covalent bonds to the peptidoglycan (that is, free instead of bound). This was achieved by analyzing an E. coli membrane fraction by SDS-gel electrophoresis. One of the major peaks, peak 11, with the smallest molecular weight of about 7500, had an extremely low content of histidine. The bound form of the lipoprotein appeared as a new peak on SDS gels only after the E. coli membrane fraction was treated with lysozyme. The bound form of the lipoprotein migrated slightly slower than the free form on SDS gels since 2–3 peptidoglycan subunits still remained bound to the lipoprotein after lysozyme treatment (106). That the protein of peak 11 (free form) has exactly the same chemical structure as the bound form of the lipoprotein, except that it does not contain any components of the peptidoglycan, was clearly demonstrated by extensive analyses of the amino acid and fatty acid composition of the protein, composition of peptidoglycan components, and of cyanogen bromide peptide fragments (105, 107).

The free as well as the bound form of the lipoprotein exists almost exclusively in the outer membrane (108, 109). There are about 2.4×10^5 molecules of the bound form per cell (37, 110), and about twice as much of the free form, i.e., about 4.8×10^5 molecules per cell (38). The total free and bound lipoprotein molecules, 7.2×10^5, makes this lipoprotein the most abundant protein, numerically, in the cell.

The amino acid sequence of the lipoprotein (102, 103) shows several striking features. First, it is strongly repetitive. At the N-terminal portion, there are three almost identical adjacent sequences. The C-terminal part of the polypeptide chain is more variable but still shows striking homology when certain sequence gaps are introduced. It was speculated that the lipoprotein gene may have evolved by repeated duplication of a gene coding originally for 15 amino acids (102). Second, the hydrophobic amino acid residues are regularly arranged in an alternating 3 to 4 pattern of repeating hydrophobic residues without exception. Since 3.6 residues make up one regular right-handed α-helical turn, all the hydrophobic residues can be aligned as two series on one face of the helical rod (10, 111). This is in good agreement with the fact that the lipoprotein has very high α-helical content (112–114). Since it also lacks proline and glycine residues there appear to be no bends in the α-helical structure. This pattern is surprisingly similar to that of the C-terminal half of tropomyosin, which also has a high α-helical content (115).

Based on the above considerations, Inouye (111) proposed a three-dimensional molecular model for assembly of the lipoprotein. In this model, an α-helix is constructed from the sequence and six such helices are arranged to form a superhelical tubular assembly with a hydrophilic interior and a hydrophobic exterior. The assembly is stabilized by seven ionic interactions between adjacent α-helices. The assembly, whose height is 76 Å, spans the full 75 Å thick outer membrane and was proposed to serve as a passive diffusion pore through the outer membrane. The three fatty acid chains linked to the N-terminal end of the lipoprotein flip back along the helices and interact with the phospholipid bilayer as well as with the hydrophobic face of the assembly. Two out of six lipoprotein molecules are convalently linked to the peptidoglycan. The conformation of the purified free lipoprotein was extensively studied by Inouye's group (107, 114). Infrared spectra indicated that it has an α-helical conformation but gave no indication of the presence of β-structure. From circular dichroic spectra, the α-helical content of the free lipoprotein was found to be as high as 88% in 0.01–0.03% SDS in the presence of 10^{-5} M Mg^{2+}.

Braun (10, 113) proposed a completely different structural model of the lipoprotein, based on calculations according to the known amino acid sequence and the conformational studies of the bound form lipoprotein. In his model, a β-loop comprised of amino acids 25–29 breaks the helical rod into two parts that are comprised of amino acids 5–24 and 30–47. Two helical rods are arranged such that they are stabilized intramolecularly either by hydrophobic or hydrophilic interactions. Three fatty acid residues stick out from the protein moiety and penetrate into the inner layer of the lipid bilayer of the outer membrane. Thus the protein part protrudes from the inside surface of the outer membrane. In this model, the bound form of the lipoprotein may serve as an anchor to connect the outer membrane with the peptidoglycan layer. However, it is not clear what the possible functions of the more plentiful free form of the lipoprotein are.

Lee et al (116) showed, by gel filtration and fluorescamine labeling of lysine residues, that the free form of the lipoprotein forms aggregates through ionic interactions in 0.1% SDS solution in vitro. When the cell wall fraction consisting of the outer membrane and the peptidoglycan was treated with cross-linking reagents, a dimer, and possibly a trimer, of the free form lipoprotein was formed (76). The in vivo formation of dimers of the lipoprotein was detected when the envelope fraction of the *E. coli* mutant (*lpp*-1), in which the lipoprotein has an extra cysteine residue (see section on genetic studies on outer membrane proteins), was isolated under oxidizing conditions and analyzed by SDS-gel electrophoresis in the absence of β-mercaptoethanol (M. DeMartini and M. Inouye, unpublished data). These results indicate that the lipoprotein molecules are closely associated with each other in vivo.

The free form of the lipoprotein was extensively purified and paracrystallized (107) and its ultrastructure, in negatively stained preparations, was examined by electron microscopy (117). The paracrystals were needle-shaped and showed highly ordered ultrastructures. Electron-transparent bands 4.7 nm in thickness are regularly spaced with a repeat distance of 22 nm. The basic unit of the paracrystals is proposed to consist of a number of lipoprotein molecules arranged with their 2.3 nm-long lipid hydrocarbon chains side by side and their 8.7 nm-long protein moieties projecting from the lipid region. In this unit the lipoprotein molecules may be assembled as a superhelical structure as proposed by Inouye (111) or as a simple coiled-coil structure.

It is not clear if the lipoprotein is exposed to the outside surface of the outer membrane, as proposed by Inouye (111). However, at least in those *E. coli* strains that have defects in LPS structure, the lipoprotein seems to react with antilipoprotein serum (10, 118). These results suggest that a part of the lipoprotein is exposed to the surface of the outer membrane.

Although the function of the lipoprotein is at present not well established, several physiological and morphological defects are observed in mutants lacking this protein (see below).

Minor Proteins

About 10–20 "minor proteins" are present in the outer membrane. The term "minor protein" may be misleading since under certain growth conditions some of these proteins are made in quantities almost as great as a "major protein." Many minor proteins, in addition to the major outer membrane proteins described above, have been identified as receptors for phages and colicins. Most of them, as well as additional proteins with no known receptor functions, are now known to have vital roles for the growth of the cell such as the uptake of nutritional substrates through the outer membrane (for reviews see 119, 120). Other minor proteins with entirely different functions have also been identified as outer membrane proteins. Table 2 lists the minor proteins of the outer membrane that have been identified so far.

The uptake of vitamin B_{12} by *E. coli* requires a specific receptor located in the outer membrane (121) that is identical to the receptor for the E type

Table 2 Minor proteins of the outer membrane of *E. coli*

Protein	Molecular Weight[a]	Function	Receptor	References
83K	83,000	(Repressed by iron)	—	119, 137, 139
feuB[b]	81,000	Fe^{3+}-enterochelin uptake	ColB	119, 134
cit	80,500	Fe^{3+}-citrate uptake	—	134
tonA[b]	78,000	Fe^{3+}-ferrichrome uptake	T1, T5, ϕ80 Albomycin, Col M	119, 127–130
cir[b]	74,000	(Repressed by iron)	ColI, ColV	137, 139, 142
bfe[b]	60,000	Vitamin B_{12} uptake	BF23, ColE1 ColE2, ColE3	122–124
lamB[b]	55,000	Maltose uptake	λ	145, 146, 146a
tsx[b]	27,000	Nucleoside uptake	T6, ColK	148, 149
Protein G	15,000		—	150
Protein D	80,000	DNA replication and cell division	—	151
Phospholipase A1	29,000	?	—	72

[a] Most of the values for the molecular weights described here are determined by SDS-gel electrophoresis. They may vary depending on the SDS-gel system that was used.

[b] These protein designations followed their gene symbols, as used by Braun (119, 120).

colicins as well as for phage BF23 (122, 123). These substances compete for the same binding sites. The receptor was identified as an outer membrane protein with a molecular weight of 60,000 (124) which is coded for by the *bfe* gene locus (125). Purified bfe protein binds colicin E3 but has lost its affinity for vitamin B_{12} (126). The addition of LPS in the presence of Triton X-100 seems to restore vitamin B_{12} binding activity (126). There are normally 200–250 receptors per cell, but mutant cells with only one or two receptors are still sensitive to bacteriophage BF23 and can transport vitamin B_{12} adequately for growth (123).

E. coli can take up iron from the medium by three independent high affinity systems: a ferrichrome-mediated, an enterochelin-mediated, and a citrate-mediated system (119, 120). *Ton*A mutants resistant to phages T5 and T1 were found to be impaired in ferrichrome iron uptake but not in enterochelin- or citrate-mediated iron uptake (127). Several competition experiments clearly showed that ferrichrome, albomycin, phages T1, T5, and ϕ80, and colicin M all share the same tonA protein as a receptor (127–129). The tonA protein, with a molecular weight of 78,000, has been characterized (128, 130). It was known that enterochelin protects sensitive cells against the killing action of colicin B by preventing colicin B adsorption (131) and that all types of colicin B tolerant mutants are impaired specifically in enterochelin-mediated iron uptake (132, 133). Hancock et al (134) showed that *feu*B mutants that are defective in ferric enterochelin uptake are unable to absorb colicin B, and are missing one of the outer membrane proteins having a molecular weight of 81,000. These results suggest that feuB protein is a receptor common for colicin B and enterochelin. Citrate-mediated systems are inducible by growth of cells on citrate. One of the outer membrane proteins, molecular weight 80,500, was found to be induced by citrate (134). Since other high affinity iron uptake systems are not inducible by citrate, this cit protein is suggested to be the receptor for the citrate iron complex. Several groups (135–139) have observed that the production of three outer membrane proteins of *E. coli* K-12 are repressed by iron in the growth medium. Two of these proteins were identified as feuB protein and the receptor protein for colicin I (cir protein) (120, 129, 139). Whether or not cir protein and a third protein, molecular weight 83,000, functions in the uptake of iron has not yet been proven. Recently, results were reported showing normal iron-enterochelin uptake in cir mutants lacking the cir protein (140, 141). The purified cir protein forms a stable complex with colicin Ia (142, 143). This protein showed a heat modifiable migration on SDS gels (143). There are about 2000–3000 colicin I receptors per cell (144).

The structural gene for phage λ receptor protein (lamB protein) is located within the *mal*B operon involved in maltose utilization. The lamB protein is required for the transport of maltose (145, 146). The *lam*B mutants were unable to grow on maltose as a carbon source at concentrations below 10 μM and a defect was found in the transport of maltose (145). When wild-type cells were treated with antibody against purified receptor protein, the transport of maltose, when at low concentrations, was greatly reduced (145). The lamB protein also may facilitate the diffusion of sugars other than maltose, such as glucose and lactose (147).

In phage T6 resistant mutants (*tsx*) an outer membrane protein with a molecular weight of about 25,000 is missing (148, 149). These mutants are also resistant to colicin K and show reduced uptake of thymidine, uridine, adenosine, and deoxyadenosine (149).

Outer membrane proteins apparently participate in the process of DNA replication and cell division. James (150) suggested that a protein G of molecular weight 15,000 has a role in the coordination of DNA replication and cell elongation. Gudas et al (151) reported that a protein of molecular weight 80,000 was synthesized at a specific time in the cell cycle shortly before DNA synthesis began and that its synthesis was inhibited by nalidixic acid. Furthermore, protein D preferentially binds to double stranded DNA in vitro. These results suggest the involvement of this protein in DNA initiation. Protein D is likely the same as the 80,000 dalton protein described by Portalier & Worcell (152), which is one of the two proteins photochemically cross-linked to DNA in vivo. However, the latter proteins were reported to be localized in the cytoplasmic membrane (152).

Phospholipase A_1 (153) is the only well-characterized, purified protein of the outer membrane so far known to have enzymatic activity (72). It may be involved in phospholipid turnover but its physiological role is not yet known. Recently, it was suggested that this enzyme is involved in the disruption of the outer membrane by EDTA (154).

GENETIC STUDIES ON OUTER MEMBRANE PROTEINS

One of the great advantages of using *E. coli* for the study of membrane proteins is its genetics. Many mutants with defective or altered outer membrane proteins are resistant to phages and colicins, resistant or supersensitive to inhibitory reagents such as antibiotics, detergents and dyes, and show impaired utilization of growth substrates or altered cellular shape. Needless to say, isolation and characterization of such mutants are extremely important for the study of biosynthesis, assembly, and function of these proteins and of the outer membrane itself. However, the observed phenotype of a mutant of a particular membrane protein does not always reflect the direct effect of that mutation. Furthermore, even if a mutant lacks a particular outer membrane protein, one should not conclude that the protein is dispensable, since it might be required for growth under different conditions. Figure 1 shows the chromosomal location so far determined of genes involved in the biosynthesis of outer membrane proteins. These genes are not located as a cluster.

Genetic studies on the matrix proteins revealed the complex aspects of their biosynthesis and assembly. The structural gene(s) for the matrix protein(s) has not been determined. As described in the preceding section, *E. coli* K-12 has two matrix proteins, Ia and Ib. Mutants lacking only protein Ib have been isolated among the strains that are resistant to phages PA-2 (*par*) (80, 90), TuIb (56, 156), or Mel (*meo*) (81). These mutants are probably identical and map at about 47.5 min on the *E. coli* linkage map (157; Figure 1). The *tol*F mutants, which are tolerant to colicins A, E2, E3, K, and L (158), are found to be specifically lacking protein Ia (52, 159). The *tol*F locus is closely related to or identical with the *cml*B locus which determines increased resistance to the antibiotics chloramphenicol and tetracycline, and maps at 20.8 min (160) on the *E. coli* K-12 chromosome linkage map (157; Figure 1). In contrast to these mutants, the *tol*IV and *tol*XIV mutants, which are tolerant to colicins E2, E3, K, and L (161) are evidently missing both proteins Ia and Ib or produce only a trace amount of protein Ia (52, 162). These mutations were mapped at the same locus, designated as *omp*B, which is located at 73.8 min on the *E. coli* K-12 chromosome linkage map (162; Figure 1). Phage resistant double mutants starting from either the *tol*F or *par* mutant were isolated by several authors (56, 159). Many of these mutants, as expected, contained neither protein Ia nor Ib. The second mutation was mapped near the *omp*B locus (56, 159). However, different types of mutants were also isolated. Bassford et al (52) selected phage PA-2 resistant mutants from a *tol*F strain. Although one such mutant exhibited the same level of tolerance to colicins as the parent *tol*F strain, it produced an almost normal amount of protein, which migrated to the position of protein Ia, and no protein Ib. The second mutation in this mutant seems to be at the *par* locus, judging from P1 transduction experiments. Henning et al (56), starting from a TuIb phage resistant strain, also isolated double mutants resistant to phage TuIa that utilize protein Ia as a receptor (79). One class of these mutants seemed to produce protein Ib but no Ia. However, these mutants remained as resistant to phage TuIb as the parent TuIb resistant strain. Two-dimensional electrophoresis and several other methods showed that the protein found in these mutants was

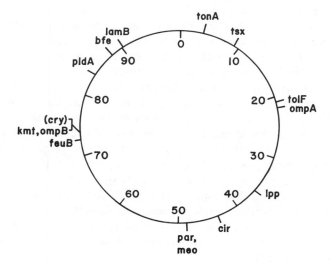

Figure 1 Chromosomal location of genes involved in the biosynthesis of outer membrane proteins. Genetic symbols are described in the text, except *pld*A for phospholipase A1 (155). Parentheses around a gene symbol indicate that the location of that marker is not well defined.

not protein Ib but exhibited all the properties of matrix protein. Its production was not due to lysogenization of a phage, and it was tentatively designated as protein Ic. Although the mutation leading to the production of protein Ic has not yet been determined, it probably does not correspond to any of the three genetic loci described above (56). These genetic results, together with the fact that the matrix proteins Ia, Ib, and Ic are chemically quite similar, suggest that they are coded by the same structural gene but are modified in different fashions. Probably the *omp*B gene is the structural gene for the precursor polypeptide and other genetic loci are involved in the modification process. However, other possibilities, such as the existence of more than one structural gene, cannot be ruled out at present. What are the functions of each matrix protein species? How are their amounts regulated in response to changes in the growth medium? These questions also remain to be answered.

As described in the preceding section, mutants lacking the matrix protein are defective in the passive diffusion of small molecular weight substances, such as sugars and amino acids. The gene locus of the "pleiotropic transport mutants" (*kmt*) of *E. coli* B/r (88), with a deficiency in the matrix protein, has been mapped at 73.7 min (89). Thus the *kmt* gene is probably identical with the *omp*B gene. The gene locus of the copper resistant mutants that cause the specific deficiency in matrix protein Ia (87) has not been reported. Quite a different type of matrix protein mutant occurs among mutants with reduced activity of the periplasmic enzymes, 3'- and 5'-nucleotidase (cryptic) (163, 164). These *Cry*⁻ mutants show greatly reduced permeability to 3'- and 5'-AMP (164). Some of the *Cry*⁻ mutants lack a major outer membrane protein that appears to be identical with the matrix protein (164). Instead, a more heavily staining band with an apparent molecular weight slightly higher than matrix protein appears in the cytoplasmic membrane fraction. Although the direct relationship between these proteins is not established, it is possible that the cytoplasmic membrane protein may represent the biosynthetic precursor(s) of the matrix protein(s). The *Cry*⁻ mutation was mapped between 72 and 79 min (163), which is the area of the *omp*B locus (73.8 min). The appearance of protein(s) with a higher molecular weight than the matrix protein was also noticed in the envelopes of some *omp*B mutants (52, 159). Recently, a putative precursor of the matrix protein containing an extra amino acid sequence was shown to be produced in toluene-treated cells (165).

Mutants missing the tolG protein were among those specifically tolerant to colicin L-JF 246 (*tol*G; 166) or resistant to phage K3 (*con;* 97) or phage TuII* (*tut;* 156). Later these mutants were shown to be identical and the gene symbols were renamed *omp*A (49, 167). This gene locus was mapped at about 21.5 min (158, 168) on the *E. coli* K-12 chromosome linkage map (127; Figure 1).

Mutants with an altered tolG protein were also found among mutants resistant to phage TuII* (93, 156) and phage K3 (96); their mutations also mapped at the *tol*G locus. When cyanogen bromide fragments of one of these altered tolG proteins were compared to those of the wild-type protein, differences were observed in only one or two fragments (169). These results indicate that the *omp*A gene is the structural gene for the tolG protein. The resistance to colicin JF 246, the resistance to various host range mutants of phage K3, and the recipient ability in conjugation were examined using a series of *E. coli omp*A gene mutants (96). Differences in the amounts or nature of the tolG protein affected these physiological processes independently which suggests differences in the role played by the tolG protein in these processes.

A mutant carrying an altered structural gene for the lipoprotein was found, fortuitously, by Suzuki et al. (170) among a collection of temperature-sensitive cell division mutants. In this mutant the lipoprotein has a free thiol group that is susceptible to modification by monoiodoacetic acid and migrates as a dimer during SDS-gel electrophoresis when the membrane fraction is dissolved in the absence of β-mercaptoethanol. Inouye et al (171) found that the arginine residue at position 57 is replaced by a cysteine residue in this mutant lipoprotein. This mutation can be explained by a single-base change from U to C and therefore, occurred in the structural gene for the lipoprotein. The mutation *(lpp–1)* was found to map at 36.5 min on the *E. coli* chromosome (170; Figure 1). It was not related to temperature-sensitive cell division and showed no special phenotype except sensitivity to mercuric compounds. Another lipoprotein mutant designated *epo,* in which both the free and bound forms of the lipoprotein are completely missing, was found by Hirota and his co-workers (172), and does not produce an active mRNA for the lipoprotein. Chromosomal DNA from the *lpo* mutant does not have the restriction fragment that can hybridize with the purified radioactive mRNA for the lipoprotein (K. Nakamura and M. Inouye, unpublished data). These data suggest that the *lpo* mutation is most likely a deletion of the structural gene *(lpp)* for the lipoprotein. The mutation is also mapped at 36.5 min on the *E. coli* chromosome (172). The *lpo* mutant seemed to grow and divide normally; however, it is extremely sensitive to EDTA and leaks considerable amounts of periplasmic enzymes into the medium. The passive transport of β-galactosides (172) and 6-aminopenicillanic acid (173) seems to be unchanged by the mutation. Mutants of *S. typhimurium* (*lky*D) which contain a decreased amount of the bound form of lipoprotein and a concomitantly increased amount of the free lipoprotein were also described (174). Originally isolated as mutants that leak periplasmic enzymes (175, 176), these mutants were found to show a defect in invagination of the outer membrane during septum formation in spite of normal ingrowth of the cytoplasmic membrane and the peptidoglycan layer. As a result the outer membrane begins to form large "blebs" over the septum region. Bleb formation was also observed in *lpo* mutants when the growth medium was low in Mg^{2+} (Y. Hirota, personal communication). Studies on *lky*D mutants in minimal medium containing a low Mg^{2+} concentration suggest that the bound lipoprotein is required for the proper invagination of the outer membrane during septum formation.

Two groups have isolated lipoprotein mutants by a suicide selection method based on the specific incorporation of radioactive amino acids into the lipoprotein during starvation for the specific amino acids that are lacking in the lipoprotein molecule (177). One such mutant isolated by Wu & Lin (178) was found to have a structurally altered lipoprotein (179) based on the following: (*a*) The mutant lipoprotein lacks the covalently linked diglyceride. (*b*) It contains an unmodified cysteine residue. (*c*) The amount of the bound form of the lipoprotein is greatly reduced. (*d*) The apparent molecular weight of the mutant lipoprotein is larger than that of the wild-type lipoprotein. (*e*) The mutant lipoprotein exists in an appreciable amount in the soluble fraction. These results, together with the fact that this mutation is mapped at or very close to the *lpp* locus (180), suggest that the mutation altered the primary structure of the lipoprotein in such a way that its modification reactions or those of its precursor, prolipoprotein, are disrupted. The determination of the altered structure may provide an important clue to solving the molecular mechanism of lipoprotein assembly in the outer membrane. Torti & Park (181) isolated a temperature-sensitive (ts) mutant in cell division that forms filamentous cells at 42°C. This mutant produced very low amounts of both bound form and free form lipoprotein at 42°C but normal amounts at permissive temperatures. Since revertants of this mutant grow normally and also make lipoprotein normally at 42°C, it was concluded that the lipoprotein may serve a vital function(s) in cellular activities. However, the finding that a mutant *lpo* lacks the lipoprotein does not favor their conclusion. The ts mutation is not in the lipoprotein structural gene since it maps at 74 min (S. Torti and J. Park, personal communication). The mutant should be interesting in terms of regulation of lipoprotein biosynthesis.

F-prime factors carrying the structural genes for several outer membrane proteins have been used to examine the gene dosage effect. Datta et al (167) could not detect any gene dosage effect for the tolG protein in strains diploid for the wild-type *omp*A gene. Also, both mutant and wild-type *omp*A alleles were expressed in heterogenotes. These results suggested that the synthesis of the tolG protein is regulated by a simple feedback mechanism. On the other hand, in the case of the vitamin B_{12} receptor, a distinct gene dosage effect has been observed in a merodiploid strain (123). Inouye's group has isolated a new F-prime factor carrying the *lpp* locus for use in examination of the gene dosage effect for the lipoprotein (R. Movva et al, submitted for publication). In the merodiploid strain, about twice as much free form lipoprotein was found as in the corresponding haploid strain; however, the amount of the bound form lipoprotein in the merodiploid strain was almost the same as in the haploid strain. These results are contrary to those obtained with the tolG protein and raise an interesting question concerning the regulation of biosynthesis and assembly of the outer membrane proteins. The report that the rate of the production of the free form lipoprotein is highest at the time of cell division (182) may also be interpreted as a result of the gene dosage effect of the *lpp* gene. This gene (36.5 min; 170) is located near the termination site of DNA replication (32 min; 157) and a round of DNA replication is completed before cell division (183). Therefore the number of copies of the *lpp* gene is doubled before cell division.

Wohlhieter et al (184) reported the expression of the *lam*B gene (phage λ receptor) of *E. coli* in *Proteus mirabilis* after conjugal transfer of the gene. Similarly, Datta et al (167) observed the functional expression of the *omp*A gene of *E. coli* as determined by phage TuII* sensitivity and SDS-gel electrophoresis in *Salmonella* and to a lesser extent in *Proteus*. These results suggest that there is common assembly mechanism of outer membrane proteins in different Gram-negative bacterial species.

It has been shown that β-galactosidase, a soluble enzyme, became inserted into the cytoplasmic membrane when the *lac*Z gene (β-galactosidase) was fused into the gene for a maltose transport protein (*mal*F) (185). It was assumed that a hybrid protein molecule is produced that is composed of an N-terminal part from the maltose transport protein and a C-terminal part from β-galactosidase. These results suggest that the N-terminal portion of the maltose transport protein is essential for its incorporation into the cytoplasmic membrane. This gene fusion technique would be especially useful in the study of the function of the peptide extension present in precursors of the outer membrane proteins, as is described later, as well as the mechanism of specific localization of the outer membrane proteins.

Molecular cloning of genes (for review see 186) is expected to provide an extremely fruitful genetic approach for the study of the biosynthesis, regulation, and assembly of outer membrane proteins. This technique has recently been applied to the study of the assembly and function of ribosomes (187) and flagella (188, 189). The gene involved in the control of the biosynthesis of capsular polysaccharide has been cloned by Berg et al (190). Among a collection of the 2000 *E. coli* strains prepared by Clarke & Carbon (191), which harbor hybrid ColE1 plasmids carrying small random segments of the *E. coli* chromosome, plasmids carrying the *fts*[+]B, E, I, M, and *par*[+]A genes were identified (Nishimura et al, submitted for publication).

Mutants in other components of the outer membrane should provide an excellent system for the study of the effect of these components on the organization, assembly, and function of the outer membrane. "Deep-rough" mutants defective in the structure of LPS show pleiotropic effects on various properties of the outer membrane, such as sensitivity to phages (192–195), permeability of various reagents (195–198a), leakage of periplasmic enzymes (199, 200), accessibility of antibody to surface antigen (118), and deficiency in conjugation (201, 202). These results suggest that the organization of the outer membrane is greatly altered in these LPS mutants. Koplow & Goldfine (43) analyzed the composition of the outer membrane of deep-rough LPS mutants of *E. coli*. The amounts of protein, relative to the other components, were greatly decreased compared to that in wild-type cells. SDS-gel electrophoresis revealed that the major outer membrane proteins were greatly diminished in these mutants. Similar results were reported by Ames et al (44) for the major outer membrane proteins of *S. typhimurium*. Randall (203) found that the phage λ receptor lamB protein is lost in LPS defective mutants. Using *E. coli* mutants with various degrees of LPS deficiency, she could correlate the loss of protein with the loss of sugar residues and phosphate from the core region of the LPS molecule. Some of the physiological defects in LPS mutants described above may be secondary effects due to the loss of the major outer membrane proteins. These results indicate that when the mutational defects in LPS structure proceed to the core region the proper assembly of the outer membrane proteins does not take place. The newly synthesized outer membrane proteins may be translocated to the outer membrane as a complex with LPS through the adhesion zones connecting the outer and cytoplasmic membranes. Alternatively, the organization or assembly of these proteins in the outer membrane in situ may be stabilized by the interaction with LPS. This possibility is supported by the fact that membrane vesicles, morphologically or functionally similar to the intact outer membrane, were reconstituted with outer membrane protein only in the presence of Mg^{2+}, LPS, and phospholipid (204, 205); and by the recent findings that purified outer membrane proteins interact with LPS in vitro (54, 78, 79, 94, 100).

When net phospholipid synthesis in a glycerol auxotrophic mutant of *E. coli* was stopped by glycerol deprivation, outer membrane proteins were synthesized normally and assembled into the outer membrane. However, the synthesis of proteins stopped when the protein : phospholipid ratio of the outer membrane increased to about 60%.

There are several mutants that cause pleiotropic effects on envelope physiology beside those already mentioned. These include cell division mutants *env* (206–209), *dna* (210–212), and *fts* (212); colicin tolerant mutants *tol* (158, 162, 213–215); and glycine tolerant mutants *qme* (216) [see (157) for map location and additional references on these mutants]. It should also be noted that changes in envelope protein composition were observed in *dna* A and B mutants (59, 212). Further analysis of these mutants in the light of the recent progress in the characterization of the outer membrane proteins should provide an insight into one of the most fascinating aspects of outer membrane research—the role of the outer membrane in DNA replication and cell division.

BIOSYNTHESIS OF THE MAJOR OUTER MEMBRANE PROTEINS

Recently, significant progress has been made in the study of the biosynthesis of the major outer membrane proteins. Most such studies have been performed using the *E. coli* lipoprotein because its unique properties have permitted use of many different approaches to this problem.

In Vivo Systems

The lipoprotein of the outer membrane lacks several of the common amino acids normally found in bacterial proteins (102). Could this protein be synthesized in cells starved for one of these amino acids? In a study using a histidine auxotroph of *E. coli* (177), approximately 95% of total protein synthesis was suppressed but 90% of the normal complement of lipoprotein was produced following a 1 hr starvation for histidine. Consequently, the only outer membrane protein made under these conditions was the lipoprotein. This afforded a reliable method for the investigation of the biosynthesis of an individual outer membrane protein in vivo without the need for mutant strains that lack one or several of the outer membrane proteins.

In an alternative approach the effects of six specific antibiotics on the biosynthesis of individual or total major outer membrane proteins were tested. Hirashima et al examined the effects of kasugamycin, tetracycline, chloramphenicol, sparsomycin, puromycin, and rifampicin on protein synthesis (217). The first five are ribosome-directed antibiotics, while rifampicin inhibits RNA synthesis. Differences in rate of the incorporation of radiolabeled amino acid into total cell envelopes or separated outer membrane proteins, in the presence and absence of the antibiotic, were used as a measure of antibiotic inhibition. In general it was found that total envelope protein synthesis was more resistant to puromycin, kasugamycin, and rifampicin than to sparsomycin, and was extremely sensitive to chloramphenicol and tetracycline. Inhibition (50%) was achieved at 500, 900, 200, 60, 4, and 0.02 μg/ml, respectively. Differential inhibitory effects were observed among the various envelope proteins with some of the antibiotics. TolG protein synthesis was only slightly inhibited by kasugamycin at a concentration where synthesis of matrix protein, protein 6, and the lipoprotein were greatly affected. The synthesis of tolG protein was again strikingly more resistant to chloramphenicol and sparsomycin than that of other major proteins; and the lipoprotein was markedly more sensitive to chloramphenicol. Although synthesis of tolG protein also had a high resistance to puromycin, lipoprotein synthesis was completely resistant to as much as 300 μg/ml of the antibiotic.

A more critical examination of the puromycin resistance of lipoprotein synthesis (217) showed that incorporation of radiolabeled fatty acid into the lipoprotein during exposure to puromycin was not reduced (217), and even histidine-starved cells synthesized lipoprotein during puromycin treatment (177). When the permeability of the cells was increased by treatment with EDTA or toluene (218), lipoprotein synthesis remained more resistant to puromycin than that of all other envelope proteins. Puromycin resistance was also tested in a cell-free polyribosome system, and immunoprecipitation of protein products with antilipoprotein serum demonstrated that lipoprotein synthesis remained resistant to the antibiotic (218). These results suggest that puromycin resistance is an intrinsic property of the lipoprotein biosynthetic machinery. The reason for this unique resistance is at present unknown; however, possible explanations are discussed later.

Cell-Free Synthesis

The specific biosynthesis of the lipoprotein was achieved using isolated polyribosomes and a soluble enzyme fraction (219). The products were analyzed by SDS-polyacrylamide gel electrophoresis and immunoprecipitation with antilipoprotein serum. The results showed that the lipoprotein was selectively synthesized on small polyribosomes, probably in the range of tri- to tetraribosomes.

It has been debated for some time whether there is a compartmentalization of ribosomes and RNA (220), that is, whether the biosynthesis of certain proteins such as outer membrane proteins and periplasmic enzymes occurs preferentially on cytoplasmic membrane-bound ribosomes. In support of this proposal it was found, by employing both chemical and immunological methods, that isolated membrane-bound polyribosomes carried 70–80% of the total nascent alkaline phosphatase (APase) peptides (221). It was suggested that polyribosomes for those proteins that had to be transported to functional sites outside the cytoplasmic membrane were tightly bound to the cytoplasmic membrane. However, the objection was raised that membrane-bound polyribosomes might be artifacts of the isolation procedures used (222). Randall & Hardy compared the ribosomal proteins of polyribosomes engaged in the synthesis of the outer membrane proteins, which are rifampicin resistant, with those made by polyribosomes engaged in total protein synthesis (223), and found no quantitative differences. Upon further examination (224) they demonstrated that membrane-bound polyribosomes selectively synthesized proteins that are secreted. These included outer membrane proteins and the maltose-binding protein of the periplasm. In addition, they demonstrated that membrane-bound polyribosome activity, in vitro, is very resistant to puromycin. The puromycin resistance of the lipoprotein biosynthesis, discussed above, may then result from the mode of binding to the cytoplasmic membrane of those polyribosomes which are specific for the synthesis of the lipoprotein. Differences in the sites or mode of binding for the various outer membrane proteins and thus the differences in sensitivity of their biosynthesis to puromycin may be due to the different affinities of individual polyribosomes for the cytoplasmic membrane.

A more defined system was developed for the in vitro study of outer membrane protein biosynthesis when the mRNA for the lipoprotein was purified approximately 250-fold from *E. coli* (225). Biosynthesis of the lipoprotein was achieved in a cell-free incubation mixture and was dependent upon the inclusion of the purified mRNA. The product was identified by immunoprecipitation and peptide mapping and had the same C-terminal peptide (residues 53 to 58), Ala-Thr-Lys-Tyr-Arg-Lys, as the lipoprotein.

Precursor Proteins

Studies with toluene-treated cells brought new insights into the understanding of the biosynthesis and assembly of the outer membrane proteins. Toluene treatment has been used to study DNA (226), RNA (227), and peptidoglycan (228) synthesis and is known to make cells permeable to macromolecules and ATP. A protein-synthesizing system, developed using toluene-treated cells (229), was totally dependent upon ATP, sensitive to tetracycline, chloramphenicol, and puromycin (229), and was found to produce membrane proteins exclusively. When these membrane proteins were treated with antilipoprotein serum, two distinct peaks were found during SDS-polyacrylamide gel electrophoresis of the immunoprecipitate; one comigrated with the in vivo lipoprotein, and the other appeared to be a new form of the lipoprotein with an apparent molecular weight of 15,000 (229, 230). This new form of lipoprotein had the same carboxyl terminal structure as the lipoprotein based on peptide mapping (230), and was therefore thought to have a peptide extension at the amino terminus. From double labeling experiments the peptide extension was determined to contain at least 18–19 extra amino acids. This new form of the lipoprotein, designated prolipoprotein, is thought to play a major role as a precursor in the biosynthesis and assembly of the lipoprotein. Thus toluene treatment appears to block processing of the prolipoprotein to the lipoprotein.

The complete amino acid sequence of the prolipoprotein was determined by analyzing the cell-free product whose synthesis was directed by the purified lipoprotein mRNA (231). This product was labeled with (^{3}H)-leucine and subjected to 46 consecutive Edman degradations. The positions of the radioactive peaks were noted and then the experiment was repeated using the cell-free product that had been labeled with other tritiated amino acids. By matching the known sequence of the lipoprotein to that of the cell-free product, the extra amino acid sequence could be determined. The prolipoprotein produced by toluene-treated cells was identical to the cell-free product. Its sequence (Figure 2) contains 20 additional amino acid residues extending from the N-terminus of the lipoprotein. Unusual properties of this extended sequence are: (a) The extended region is basic and positively charged at neutral pH because it contains two lysine but no acidic amino acid residues. (b) The region contains three glycine residues that are not present in the lipoprotein. (c) Sixty percent of the amino acid residues in the extended region are hydrophobic, in contrast to 38% in the lipoprotein. (d) The distribution of these hydrophobic amino acids along the peptide chain is completely different from their periodical distribution in the lipoprotein (111).

Do precursors for membrane proteins exist generally? Treatment of cells with toluene for a shorter time, that is, 1.5 min instead of 10 min, enhanced the incorporation of (^{35}S)-methionine into new membrane proteins of higher molecular weight (165). The production of two new membrane proteins was clearly evident, and based on results obtained by immunoprecipitation, SDS-polyacrylamide gel electrophoresis, and autoradiography it was demonstrated that these new proteins represented putative precursors of the matrix and tolG proteins with molecular weights about 2000 higher than those of the two major outer membrane proteins. Since all three major outer membrane proteins may be formed from precursors that contain about 20 extra amino acid residues, a common mechanism for their biosynthesis seems to be involved. The N-terminus of both promatrix protein and prolipoprotein is methionine (165). The distribution of leucine residues at the N-terminus region of the promatrix protein differs from that of the matrix protein, which suggests that the peptide extension is located at the N-terminal end as in the prolipoprotein. That the matrix protein consists of two distinct peptides, designated matrix proteins Ia and Ib, has already

5'-END: G-C-U-A-C-A-U-G-G-A-G-A-U-U-A-A-C-U-C-A-A-U-C-U-A-G-A-G-G-
1   10   20

S-1
┌─────────────────────────────────────┐ ┌─
1   5
MET - LYS - ALA - THR - LYS - LEU - VAL - LEU -
G-U-A-U-U-A-A-U-A-A-U-G-A-A-A-G-C-U-A-C-U-A-A-A-C-U-G-G-U-A-C-U-G-
30   40   50   60
     I-1   S-2   I-2
─────────────────────┐ ┌────────────────┐ ┌──────────────────
10   15
GLY - ALA - VAL - ILE - LEU - GLY - SER - THR - LEU - LEU - ALA -
G-G-C-G-C-G-G-U-A-A-U-C-C-U-G-G-G-U-U-C-U-A-C-U-C-U-G-
            70   80   89

20
GLY - CYS - SER - SER - ASN - ALA - LYS - ILE - ASP - GLU - LEU -
                          25   30
G-C-U-A-A-A-A-U-C-G   G-

SER - SER - ASP - VAL - GLN - THR - LEU - ASN - ALA - LYS - VAL -
                          35   40
U-C-U-U-C-U-G   G-C-U-A-A-A-G

ASP - GLU - LEU - SER - ASN - ASP - VAL - ASN - ALA - MET - ARG -
                    45   50
                    G-C-A-A-U-G

SER - ASP - VAL - GLN - ALA - ALA - LYS - ASP - ASP - ALA - ALA -
            55   60
            G-C-U-A-A-A-G

ARG - ALA - ASN - GLU - ARG - LEU - ASP - ASN - MET - ALA - THR -
65   70
                          G-A-C-A-A-C-A-U-G-G-C-U-A-C-U-

75   78
LYS - TYR - ARG - LYS
A-A-A-U-A-C-C-G

Figure 2 Nucleotide sequence of the 5' end of the mRNA for the lipoprotein and the amino acid sequence of the prolipoprotein (242; R. Pirtle, I. Pirtle, and M. Inouye, manuscript in preparation). Base sequences of oligonucleotides assigned to the amino acid sequence are also shown. Amino acid residues 1–20 represent the extended peptide region. Sections designated S-1, I-1, S-2 and I-2 signify parts of the sequence that have proposed specific functions during the translocation process (see text).

been discussed. Therefore, it has to be established whether the promatrix protein is a precursor of Ia, of Ib, or a mixture of both.

Periplasmic proteins also appear to be produced from precursors. In a cell-free system, introduced previously, an APase product was synthesized (232). It was characterized immunogenically and was able to dimerize to form an active enzyme complex exactly like the native APase protein. However, the in vitro product had a higher molecular weight and was more hydrophobic than the in vivo synthesized enzyme monomer. The larger protein was extremely hydrophobic since it bound irreversibly to decylagarose, while the native enzyme could be eluted quantitatively with 0.2 M NaCl. Unfortunately, no data are available as yet on the amino acid composition of the APase precursor.

An interesting comparison can be made between the possible mechanisms of outer membrane and periplasmic protein biosynthesis in procaryotic cells, and of secretory proteins in eucaryotic cells. Many secretory proteins in eucaryotic cells are produced from precursors that contain 16–25 extra amino acid residues at the N-terminal ends (233–239). Striking similarities exist, in at least one case, between the amino acid sequences of the prolipoprotein and the prelysozyme of chick oviduct (235).

Characteristics of Messenger RNAs for Outer Membrane Proteins

By following the biosynthesis of different outer membrane proteins after addition of rifampicin, an inhibitor of the initiation of RNA synthesis (217), these proteins were observed to have mRNAs of different stabilities. The half-lives of the mRNAs were calculated as 3.2, 4.5, 4.0, and 11.5 min for the matrix protein, protein 6, tolG protein, and lipoprotein, respectively. The existence of stable mRNAs in minicells of *E. coli* has also been reported (240). The proteins that were synthesized in minicells from the stable mRNAs were identified as the major outer membrane proteins (240, 241).

Important characteristics of the lipoprotein mRNA that permit its isolation (225) include its remarkable stability, its availability, since the lipoprotein is one of the most abundant proteins in *E. coli* (38, 110) and its small size, since the lipoprotein has a very low molecular weight (7200). The mRNA has been recently purified to approximately 85–90% (S. Wang, R. Pirtle, I. Pirtle, M. Small, and M. Inouye, manuscript in preparation), and a structural study has been started (242). For this purpose a purification procedure was developed in order to isolate (^{32}P)-labeled mRNA. The final product was identified as the lipoprotein mRNA because it comigrated with nonradioactive, functional lipoprotein mRNA and, more importantly, a T1 ribonuclease digestion of the purified mRNA produced many oligonucleotides that could be assigned to parts of the amino acid sequence of the lipoprotein (242). The base sequence of all possible 33 T1 fragments as well as fragments obtained from pancreatic ribonuclease digestion of the mRNA has been determined. Figure 2 shows oligonucleotides that have been assigned to parts of the prolipoprotein sequence. In order to determine the complete nucleotide sequence of the mRNA, partial digestions of the mRNA and several additional methods are now being examined. The following list summarizes what is currently known about the chemical structure of the lipoprotein mRNA: (*a*) The size of the mRNA is 8.2S. (*b*) It contains 360 ± 10 nucleotides. (*c*) Since the prolipoprotein contains 78 amino acid residues, 234 nucleotides are required in its code, and approximately 120 nucleotides must therefore be in nontranslatable regions of the mRNA. (*d*) The mRNA has a nontranslated region of 38 bases before the initiation codon AUG [as shown in Figure 2 (R. Pirtle, I. Pirtle, and M. Inouye, manuscript in preparation)]: One of the unique features of the 5'-end structure is that it has the same sequence of 12 bases (G-U-A-U-U-A-A-U-A-A-U-G) which contains the 80S-ribosome binding site in brome mosaic virus RNA4, a eucaryotic mRNA. (*e*) The base sequence of the 3'-end of the mRNA has been determined to be -G-C-C-A-U-U-U-U-U-U-U_{OH} (R. Pirtle, I. Pirtle, and M. Inouye, manuscript in preparation).

MODIFICATION AND ASSEMBLY

Processing and Translocation

Since the precursors of the outer membrane and periplasmic proteins are synthesized with an extra peptide sequence, cleavage of these peptides must be accomplished to obtain the native proteins and possibly to allow for their insertion into the outer membrane or the periplasmic region. Many questions remain unanswered concerning the processing of the outer membrane proteins. It is not yet known when and where processing actually takes place. The processing activity of an alkaline phosphatase precursor has been localized in the outer membrane fraction as opposed to the cytoplasmic membrane fraction (232). However, it has not been established if the corresponding events occur in an identical manner in the in vivo situation since artifactual binding of soluble enzymes to the outer membrane is known to occur (243). A model for the processing of the prolipoprotein (11, 231) proposes that processing and translocation of the protein across the cytoplasmic membrane are tightly coupled. Although the model is based on the precursor structure of the prolipoprotein, it can be readily applied to the other major outer membrane protein precursors. The extended peptide region can be divided into four separate sections based on the arrangement of the amino acids (see Figure 2). Section S-1 is hydrophilic and positively charged due to the presence of one threonine and two lysine residues. Section I-1 contains two very similar hydrophobic sequences, Leu-Val-Leu-Gly and Val-Ile-Leu-Gly, at positions 6–9 and 11–14. The remaining sections, S-2 and I-2, are hydrophilic and hydrophobic, respectively. Each section has a specific function during the translocation process. Figure 3 shows a schematic representation of the proposed translocation mechanism.

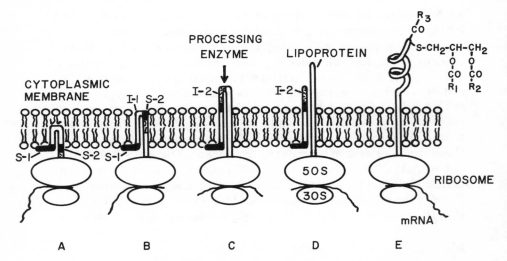

Figure 3 Proposed mechanism of translocation of the prolipoprotein across the cytoplasmic membrane. Sections of the extended peptide region of the prolipoprotein that have a specific function during translocation are designated S-1, I-1, S-2, and I-2 (see Figure 2 and text).

The positively charged section S-1 probably allows the initial attachment of the prolipoprotein and consequently the polyribosome, to the negatively charged cytoplasmic membrane through ionic interactions. As section I-1 is synthesized it is probably inserted into the membrane, as in Figure 3*A*. As the peptide elongates, section S-2 may ensure that the C-terminal end of I-1 remains on the surface of the membrane (Figure 3*B*). Section S-1 always remains on the inside surface of the cytoplasmic membrane, whereas section S-2 becomes exposed to the outside surface. At a specific stage in the translocation process the peptide extension is removed, either by an endopeptidase or an exopeptidase. The enzyme should recognize the specific amino acid sequence of section I-2 and cleave the peptide bond between the Gly and Cys residues at positions 20 and 21. Localization of the processing enzyme is important if it is to be determined when processing actually takes place. It would be also interesting to know if a specific enzyme is needed for each outer membrane and periplasmic protein.

The newly formed N-terminal end is modified immediately and, as the peptide elongates, it may begin to form the lipoprotein secondary structure thus making the translocation process irreversible (Figure 3*E*). Protein folding probably occurs outside the cytoplasmic membrane. After the protein is forced through the membrane, according to the action of the peptide extension, folding may then inhibit the protein from moving back through the cytoplasmic membrane and thus prevent the newly made protein from selective insertion into that membrane. The hydrophilic environment of the periplasmic region may induce the folding of the protein. It is interesting to point out that the three glycine residues in the peptide extension are present at positions 9, 14, and 20 which are located at the bending positions of the molecule as presented in the model in Figure 3. The fate of the peptide extension that remains in the cytoplasmic membrane is unknown. Presumably it may be rapidly digested releasing free amino acids which could then be reutilized for protein synthesis.

The model adequately explains the existence and involvement of membrane-bound polyribosomes for outer membrane protein biosynthesis. Moreover, it predicts that (a) the precursor could not be inserted into the outer membrane without being processed; (b) the prolipoprotein could not be found as an intermediate under normal growth conditions; and (c) the peptide extention would be found in the cytoplasmic membrane. Strong biochemical evidence has been reported supporting several aspects of this model (244). It was demonstrated in E. coli that nascent APase peptides move through the cytoplasmic membrane as growing peptide chains, and that polyribosomes, which are involved in the synthesis of these APase peptides, are functionally attached to the membrane and are not artifactually bound during disruption of the bacteria. These results were obtained from experiments in which acetyl (^{35}S) methionyl methylphosphate sulfone, a reagent that reacts with amino groups and does not penetrate the cytoplasmic membrane, was used to label spheroplasts. It was demonstrated, by several methods, that the label was attached to membrane bound polyribosomes by way of peptidyl-tRNA and that completed membrane and periplasmic proteins were not labeled. When the nascent peptide chains were released from the polyribosomes, 70–95% of the label was also removed. A fraction of these labeled peptides was identified, immunologically and electrophoretically, as monomers and the precursor of APase. It was suggested that energy for the irreversible translocation of the protein could be supplied by the spontaneous folding of the peptide chain outside the membrane, a process that exerts tension on the peptide; or it could be supplied by the membrane itself.

A similar model has been proposed for secretory protein processing and translocation in eucaryotic systems and is referred to as the "signal hypothesis" (245). However, there are several distinct differences between the two models, and as a comparison the reader should refer to the work performed in eucaryotic systems (245).

A discussion of the translocation of macromolecules to the outer membrane must include a consideration of Bayer's junctions, that is, localized areas of adhesion between the cytoplasmic and outer membrane in E. coli

(246). These junctions have been implicated in the translocation process of LPS (247), capsular polysaccharide (248), and the matrix protein (J. Smit and H. Nikaido, personal communication). It cannot be entirely ruled out, however, that these junctions supply a pathway for the translocation of specific phage receptor proteins (249) and F-pili protein precursors (250).

Lipid Fluidity

An important consideration in the translocation and assembly mechanisms or processes is the fluid state of the cytoplasmic and outer membranes. By using an unsaturated fatty acid auxotroph the fatty acid compositions of the membrane lipids can be dictated. By growing the cells on specific unsaturated fatty acids the fluid state of the membrane can be controlled by altering the growth temperature. It has been shown that when the membrane of *E. coli* is in the crystalline state the induction of APase is arrested (251). When the membrane is allowed to return to the noncrystalline or fluid state, induction of the enzyme can proceed. There are two possible explanations for these results: (*a*) APase monomers are synthesized but their translocation is blocked or (*b*) the actual synthesis of the monomers is arrested. Consequently, this study did not establish whether a fluid membrane is necessary for APase synthesis only, for translocation only, or for both processes. Similar experiments were performed to examine the effects of membrane fluidity on the assembly of the cytoplasmic membrane and outer membrane proteins (252). The results suggested that a fluid membrane is required for the normal assembly of the membrane proteins and that this requirement is more stringent for the outer membrane proteins than for the cytoplasmic membrane proteins. When the effects of the fluid state of the membrane on the assembly of the individual major outer membrane proteins were examined more carefully, there were remarkable differences among them (J. DiRienzo and M. Inouye, manuscript in preparation). Under conditions of a crystalline membrane state the assembly of the lipoprotein is hardly affected; however, the assembly of the matrix protein is completely inhibited. The tolG protein is synthesized, apparently, but not assembled when the membrane is in a crystalline state; it can be assembled upon subsequent shifting to conditions necessary to form a fluid membrane. The reasons for these differences are not understood at the present time. It is important to find out at which step the assembly of the matrix protein is inhibited: at translation of the mRNA, at translocation across the cytoplasmic membrane, at processing of the precursor protein, or at the insertion of the matrix protein into the outer membrane.

Studies based on the freeze fracture of *E. coli* membranes in the ordered and nonordered states may give some indication as to why there are differences in the assembly of individual outer membrane proteins. In freeze-fracture electron microscopy a change in the distribution of particles in the cytoplasmic membrane was observed at the temperature of the onset of the lateral phase separation (253). Cells frozen from above this temperature showed a net-like distribution of particles in the cytoplasmic membrane while cells frozen from a temperature below the onset of phase separation showed areas of particle aggregation and other areas devoid of particles. The extent of aggregation could be correlated with the degree of the lipid phase transition (254). Additional studies showed that the regions of the cytoplasmic membrane that were covered with particles were rich in proteins, poor in lipids, and enriched in unsaturated fatty acids (255). Those regions having no particles, that is, smooth fracture faces, were poor in protein, rich in lipids, and enriched in saturated fatty acids. Bayer demonstrated that there was a regular arrangement of equally spaced ridges in these particulate regions (256). These ridges were sometimes intersected at 90° angles by arrays of similar ridges. The ridges were composed of spherical particles 4–5 nm in diameter, and their arrangement or dissolution, depending upon the direction of the temperature shift, was extremely rapid. It was suggested that the particles probably represented proteins and that they are extruded along the ridges during membrane lipid crystallization. The outer membrane displayed these ordered arrangements to a lesser degree. Other workers verified that the cytoplasmic and outer membranes were affected differently when lipid fluidity was altered (257). In the cytoplasmic membrane 60–80% of the hydrocarbon chains take part in the transition while only 25–40% of the chains in the outer membrane become ordered. This lowered mobility of the lipid hydrocarbon chains in the outer membrane was found to be the result, either directly or indirectly, of the presence of LPS (258). When 40% of the LPS was removed the fluidity of the outer membrane was greatly increased.

Modification

The amino terminal modification process appears to be a unique feature in the biosynthetic scheme of the lipoprotein as compared to the other major outer membrane proteins. There are at least two independent reactions involved. One is the acylation of the amino group of the first cysteine residue to form an amide linkage, and the other is the addition of a diglyceride to the –SH group of the first cysteine residue. The fatty acids involved in these addition reactions have been identified (104). Since the composition of the ester-linked fatty acids is remarkably similar to the fatty acid composition of the cell membrane phospholipids it appears that the diglyceride moiety of the lipoprotein may be derived from one of the steps in phospholipid metabolism. In a study that used a glycerol-requiring strain of E. coli it was suggested that the synthesis of the protein moiety of the lipoprotein proceeds independently of the attachment of the diglyceride to the amino terminal cysteine (259). In a more detailed study employing a mutant that lacked phosphoglucose isomerase, glucose-6-phosphate dehydrogenase, and glycerophosphate dehydrogenase, it was indicated that the diglyceride moiety of the lipoprotein may be derived from the diglyceride moiety of cardiolipin (260). This conclusion was based on the results of pulse-chase experiments in which it was found that the diglyceride of the lipoprotein contained glycerol that was derived from a metabolic pool having a relatively long half-life. This would eliminate CDP-diglyceride as an immediate precursor for the lipoprotein and consequently disputes the original proposal of Hantke & Braun (104). It is not yet known whether modification of the lipoprotein precedes processing, immediately follows processing, or whether an unmodified molecule is translocated out into the periplasm or outer membrane and is then modified.

The unique structure of the lipoprotein allowed for investigation of the mechanism of biosynthesis of its diglyceride moiety with an antibiotic not normally used in the study of protein synthesis. This antibiotic, cerulenin, inhibits the enzymes involved in the fatty acid synthetase scheme (261). A preliminary examination demonstrated that biosynthesis of the free form of the lipoprotein is reduced but not completely inhibited by this antibiotic. Modification of the free-form lipoprotein did not seem to be affected (J. DiRienzo and M. Inouye, unpublished observations). This may be another indication that the diglyceride moiety of the lipoprotein molecule is derived from preexisting phospholipid. However, cerulenin inhibited the assembly of the lipoprotein into the peptidoglycan layer (262).

Assembly of the lipoprotein in the outer membrane involves conversion of part of the free form of the lipoprotein to the bound form (38). Pulse-chase experiments in vivo show that the free form is synthesized first and is then converted to the bound form. The conversion is unaffected by the inhibition of protein synthesis by amino acid starvation, by chloramphenicol treatment, or by inhibition of energy production with carbonyl cyanide m-chlorophenylhydrozone. Because the conversion reaction appears to be energy independent it was proposed that it occurs by transpeptidation. Two reactions are believed to be involved: carboxypeptidase I cleaves D-alanine from the C-terminus of the pentapeptide in the peptidoglycan (13), then carboxypeptidase II cleaves the second D-alanine with the subsequent attachment of the lipoprotein to diaminopimelic acid. While carboxypeptidase I is pencillin sensitive, carboxypeptidase II is not (263). Penicillin G (9) and penicillin FL 1060 (264) do not inhibit the conversion reaction, but high concentrations of penicillin G do cause inhibition (S. Halegoua and M. Inouye, unpublished observations). The conversion reaction could be reversible since only ~ 40% of the pulse-labeled free form is chased into the bound form after one generation, and an extended chase (lasting up to three generation times) does not increase the radioactivity in the bound form (38). Apparently the newly synthesized free form is diluted by a large preexisting pool. If, on the other hand, compartmentalization of newly synthesized free form of the lipoprotein occurs then the conversion reaction may be irreversible. Only a portion of the newly synthesized free lipoprotein may then be converted.

Bicyclomycin, an antibiotic that inhibits the synthesis of RNA and protein in growing cells of *E. coli,* when used in histidine-starved cells, inhibited the biosynthesis of both the free and bound forms of the lipoprotein (265). The synthesis of the bound form was more profoundly inhibited than that of the free form. The primary site of antibiotic action was proposed to be the conversion reaction.

The conversion reaction is thought to take place after the translocation of the free form of the lipoprotein to the outer membrane. The conversion reaction should be coordinated with peptidoglycan biosynthesis and assembly since the lipoprotein is covalently bound to the peptidoglycan. Thus, the conversion enzyme should be localized in the periplasmic region or on the interior surface of the outer membrane. Braun and co-workers have indicated that the lipoprotein is bound very slowly to newly inserted peptidoglycan subunits (264, 266). Enlargement of the peptidoglycan layer proceeds at sites where there is no bound lipoprotein. Consequently, the bound form of the lipoprotein is not inserted as preassembled lipoprotein-peptidoglycan units. Some mutants have been isolated that contain extremely low levels of the bound form of the lipoprotein (174, 179). At the time of septum formation in these cells the outer membrane cannot properly invaginate although the normal ingrowth of the cytoplasmic membrane and the peptidoglycan layer occurs (174). Presumably, the conversion enzyme(s) could be absent in these strains.

ORGANIZATION AND FUNCTION OF THE OUTER MEMBRANE

Surface Reception

Ultrastructural and biochemical data suggest that the organization of outer membrane components is highly asymmetric. Mühlradt & Golecki (267) showed that LPS molecules are exclusively localized in the outer layer of the outer membrane. Nikaido and co-workers (29, 268, 268a) suggested that the asymmetric localization of phospholipids in the outer membrane is such that most of the phospholipid molecules are localized in the inner layer.

These conclusions were based on results obtained by chemical analysis, freeze-fracture electron microscopy (29), spin-labeling (268a), and specific labeling with a reagent that cannot penetrate the outer membrane (268). Kamio & Nikaido (82) used CNBr-activated dextran, an amino group labeling reagent that does not penetrate the outer membrane, to treat mutants of *S. typhimurium* that contain an LPS with a very short polysaccharide chain, to avoid steric hindrance by carbohydrate chains. Only two or three of 18 outer membrane proteins failed to react with the reagent, which suggests that most of the proteins in the outer membrane are exposed to the external medium. Similar results were obtained using intact cells of *E. coli* B (82).

Freeze-fracture studies on *E. coli* (269, 270) and *S. typhimurium* (29) also showed the asymmetric profile of the outer membrane fracture faces. These studies revealed that the outer half of the outer membrane, that is, the concave fracture face, was densely covered with particles 8–10 nm in diameter. The convex fracture face showed only a few scattered particles. These particles probably represent outer membrane protein, judging from the following observations: (*a*) The deep rough LPS mutants of *S. typhimurium* (29) and *E. coli* (270), which lost about 50% of the major outer membrane proteins, show a decreased particle density. The degree of reduction in particle density was quantitatively correlated with the amount of the major outer membrane proteins that were lost (29). (*b*) The decrease in particle density was also observed in mutant cells that contained normal lipopolysaccharides but lacked both the matrix and tolG proteins, but not in mutant cells that lacked one or the other of these proteins (271, 272). However, in another freeze-fracture study, deep rough LPS mutants, heptoseless mutants, which had full complements of the matrix and tolG proteins, showed a reduction in outer membrane particle density (273). In wild-type cells that were treated with EDTA, a procedure that removed 50% of the LPS but no protein, a decrease in particle density also occurred. Consequently, it was concluded that these particles represent LPS aggregates that are complexed with protein and localized in the outer half of the outer membrane (273). Localization of proteins in the outer surface of the outer membrane is also indicated by the fact that most outer membrane proteins serve as receptors for various nutrients, phages, and colicins. Lipopolysaccharides also serve as receptors for many phages (4, 192). These outer membrane proteins may have evolved to facilitate the uptake of nutrients or to accomplish the conjugation process. Later, phages or colicins that utilize these proteins as receptors may have evolved (4, 120).

A striking feature of the outer membrane receptor proteins is the fact that one protein can serve as a receptor for several structurally unrelated substrates. In the case of the lamB protein, the presence of different active sites for phage λ and maltose or maltotriose was suggested (120, 145) because phage λ-resistant missense mutants transport maltose much better than nonsense mutants (145, 146), and maltose or maltotriose do not prevent the binding of phage λ (120). Similarly, high concentrations of nucleosides do not inhibit the binding of phage T6 (120). In contrast, the direct binding of ferrichrome, phages T1, T5, and ϕ80, and colicin M to the same crude preparation of receptor protein, and mutual competition for binding, has been demonstrated. Similarly, vitamin B_{12}, phage BF23, and colicin E bind to the same receptor protein and compete with each other for the binding site. The concept of narrow specificity in the structure of substrates established for enzyme reactions does not seem applicable to these outer membrane receptor proteins. Braun (119, 120) pointed out the similarity between this polyfunctional binding displayed by outer membrane receptor proteins and antigen-antibody reactions that show overlapping binding regions with different binding constants.

Although many outer membrane receptor proteins serve as common receptors for various substrates, this does not necessarily mean that they function in the same way for each substrate. Recent results, obtained by the use of genetic techniques, suggest the presence of different functional states of the bfe protein in the outer membrane (274, 275). It was shown that the bfe protein is functional for the killing of the cell by colicin E and phage BF23 only for a short period after its synthesis. The receptor protein, however, remained fully effective for the uptake of vitamin B_{12} for a long period after synthesis. Furthermore, the uptake of vitamin B_{12} follows a biphasic process. After the first phase of binding of vitamin B_{12} to a receptor there is an energy-dependent second step that requires a tonB gene function (276). However, the killing action of E colicins and phage BF 23 does not require an intact tonB gene function (213). Furthermore, colicin E-tolerant mutants still show normal sensitivity to phage BF23 and the transport of vitamin B_{12} (277). These results indicate that the mechanisms by which each of these substrates is transferred across the whole envelope, after initial binding, differ from each other.

Gene products and the function of the tonB gene are unknown at present. Besides vitamin B_{12}, uptake processes of all iron-chelator complexes require the tonB gene function (120, 276, 278, 279). Furthermore, tonB mutants are resistant to the killing actions of phages T1, ϕ80, and colicin B (120, 278). Since all these uptake processes are energy-dependent, it was postulated that the tonB gene product functions to couple the outer membrane to energized cytoplasmic membrane (280). Alternatively, the tonB gene product might be involved in maintaining the proper orientation of receptor with sites, presumably in the cytoplasmic membrane, for the subsequent uptake (276). Infection by various phages appears to take place at the adhesion zones connecting the outer and cytoplasmic membranes (250). Further analysis of the complex processes of nutrient uptake, phage DNA injection, and transmission of colicins should reveal the functional organization of the outer membrane proteins and the presumed intimate relationship between the outer and cytoplasmic membranes.

Maintenance of Structural Integrity

The structural integrity of the bacterial cell is usually attributed to the peptidoglycan or "rigid" layer (3). However, many studies have demonstrated strong interactions between the outer membrane proteins and the peptidoglycan layer. These interactions have been discussed previously in the sections dealing with the individual outer membrane proteins.

To determine how the outer membrane proteins contribute to the maintenance of cellular structure, chemical cross-linking studies were undertaken to examine protein-protein interactions. A recent review discusses the use of cross-linking reagents in studies involving membrane structure (281). An unusual finding actually led to the early cross-linking studies. Rod-shaped "ghost" membranes were isolated by treating E. coli cells successively with Triton X-100, urea, trypsin, and finally, lysozyme, and they consisted of 50–60% protein and 20–30% phospholipid, but lacked significant amounts of peptidoglycan (54, 282). The protein composition of the "ghost" membranes appeared to be predominantly composed of the major outer membrane proteins (283). In another experiment E. coli cells were treated with bifunctional cross-linking reagents and ghost membranes were isolated (284). These membranes were cross-linked over the entire surface and retained the same size and shape as the original cell. When delipidated membranes were cross-linked they were found to be composed of 80–90% protein that consisted primarily of the major outer membrane proteins. Thus Henning and co-workers originally concluded that the major outer membrane proteins, that is, the matrix and tolG proteins, are involved in the shape determination of the cell (283, 284), but this conclusion was later retracted because of the discovery of rod-shaped mutants that lacked these major proteins (156). However, these results show that there are extensive protein-protein interactions over the entire surface of the outer membrane.

The nearest-neighbor approach was used also to provide information concerning specific pairwise interaction between outer membrane proteins (75). Controlled cross-linking was achieved by using two classes of cleavable reagents—the tartaric acid derivatives and the diimido esters. Dimerization of only the matrix protein occurred. On closer examination it was found that the following were formed: dimers, trimers, and higher oligomers of the matrix protein; high molecular weight oligomers of the tolG protein; and a dimer of the lipoprotein (76). Also, the tolG protein and the free-form lipoprotein were cross-linked to the peptidoglycan and to each other. These interactions can vary causing changes in the molecular arrangement of specific areas within the outer membrane and may also contribute to the structural rigidity of the outer membrane by forming a network linked to the peptidoglycan layer.

The protein composition of the outer membrane appears to be flexible, for when one or several proteins are lost or drastically reduced due to mutation there is a compensatory increase in one or several other proteins. When matrix protein Ia was absent, the level of matrix protein Ib or both Ib and tolG protein increased so that the total amount of outer membrane protein remained constant (159). In a mutant strain lacking the matrix proteins Ia and Ib the amount of the tolG protein was approximately equal to the total amount of all three proteins in the parental strain (39). Conversely, mutants lacking the tolG protein partially compensated for this loss by synthesizing increased amounts of the matrix proteins (57). The reason for this flexibility in composition is not known. The compensation of some proteins for others may help to maintain a certain degree of rigidity of the outer membrane. On the other hand, in mutants that have defective LPS the amounts of the major outer membrane proteins decrease (43, 44, 55, 195, 201, 285), perhaps because LPS is essential for the assembly of the outer membrane proteins.

Mutants of *E. coli* have been isolated that lack both the matrix and tolG proteins (156). These mutants do not show any apparent defects in structural integrity of the cell and, consequently, they may not exhibit the functions postulated above. Triple mutants, which lack the matrix proteins, the tolG protein, and the lipoprotein have not yet been isolated.

Permeation of Substrates

Payne & Gilvarg first suggested that the cell wall acts as a diffusion barrier and thus determines the size of the molecules that can penetrate the cell (286, 287). This concept was later supported by the work of Nakae & Nikaido (85). The exclusion limit for *E. coli* and *S. typhimurium* outer membranes was found to be approximately 900 daltons for oligosaccharides. When outer membrane vesicles were reconstituted from phospholipids, LPS, and outer membrane proteins that were isolated from *S. typhimurium,* they were permeable to sucrose as well as small molecular weight oligosaccharides only when the membrane proteins were included in the reconstitution system (204). The diffusion characteristics of these reconstituted vesicles were similar to those of intact outer membrane preparations. This supported the finding that closed membrane vesicles similar to the native outer membrane were reassembled only when outer membrane proteins were included in the reconstitution mixture (205). Cytoplasmic membrane protein, when substituted in the mixture, was unable to reform membrane vesicles. It was proposed that the molecular-sieving properties of the outer membrane were due to the presence of water-filled pores (288). Reconstitution studies with selected outer membrane proteins indicated that the matrix protein produced diffusion channels in the reconstituted vesicles (83, 84). It was not possible to reconstitute lipoprotein-containing membrane vesicles that had the same diffusion properties (83, 73, 288). Mutants of *S. typhimurium* that lacked the matrix protein were deficient in the diffusion of the β-lactam antibiotic, cephaloridine, across the outer membrane (86). Confirmatory studies showed that mutants lacking the matrix protein demonstrate a greatly reduced ability to utilize low concentrations of several low molecular weight metabolites (87, 89; see section on matrix proteins). These results strongly indicate that the matrix protein is required for the formation of the passive diffusion pores.

The matrix protein is arranged in a periodic monolayer that covers most of the outer surface of the peptidoglycan (64). Detailed electron microscopy revealed that three of these protein molecules seem to be arranged in a unit cell that shows a triplet of indentations each approximately 2 nm in diameter (74). These indentations may represent the lipoprotein because (*a*) there is strong interaction between the lipoprotein and the matrix protein (289) and (*b*) the lipoprotein is dispersed evenly over the cell surface area in greater molecular numbers than the matrix protein (13). Both the free and bound forms of the lipoprotein may be involved in the interaction. The matrix protein is firmly bound to the peptidoglycan layer and cannot be dissociated even in 2% SDS at 55° C unless NaCl is present. In a mutant (*lpo*) lacking the lipoprotein (see section on genetic studies on outer membrane proteins) the matrix protein was extracted more easily from the peptidoglycan in the absence of NaCl than in wild-type cells; 80% for the *lpo* mutant as opposed to 30% for the wild type (289). When the bound form of the lipoprotein of wild-type cultures was cleaved from the peptidoglycan by trypsin, the affinity of the matrix protein for the peptidoglycan decreased to the same level as that in the *lpo* mutant strain. These results suggest that the bound form of lipoprotein plays an important role in the association of the matrix protein with the peptidoglycan (289). The bound form of the lipoprotein is not required for a simple binding of the matrix protein to the peptidoglycan in vitro (65) but is required to reconstitute membranous vesicles that resemble the outer membrane structure (77).

Based on these observations and on the properties of the matrix protein and the lipoprotein, discussed earlier, these two proteins may form an interacting complex that forms diffusion pores or channels. Pores in the outer membrane may be assembled as illustrated in Figure 4. Three molecules of the matrix protein, mainly composed of β-structures, form a hydrophilic diffusion pore with a diameter of 1.5–2.0 nm. Each of the matrix protein molecules is fixed or stabilized with a triple coiled-coil structure of the lipoprotein that is comprised of one molecule of the bound form and two of the free form. Consequently the lipoprotein may play a role in stablizing the matrix protein pore. The model is consistent with interpretations of electron micrographs of the matrix protein in which certain voids were suggested to be filled with lipoprotein molecules (74). It remains to be established that (a) the indentations shown in the electron micrographs (74) actually are filled by the lipoprotein complexes, (b) the pores exist in the electron transparent areas in the centers of the threefold symmetry of the periodic protein arrays shown in the electron micrographs (74), and (c) the lipoprotein is necessary for stabilization of diffusion pores. The extreme sensitivity to EDTA and the high requirement for Mg^{2+} observed in the *lpo* cells (172) may provide a clue to these questions. At any rate additional experimentation is necessary for a full understanding of the organization and mechanism of assembly of the outer membrane diffusion pores.

Although the matrix protein pores are nonspecific for small molecular weight, hydrophilic substrates, other specialized diffusion or uptake systems appear to be present in the outer membrane. For example, in Cu^{2+}-resistant mutant cells that lack the matrix protein, there was no reduction in the uptake of uracil, uridine (88), or thymidine (87). Furthermore, the matrix protein pores allow a rapid transmembrane diffusion of hydrophilic sub-

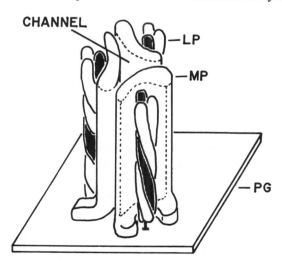

Figure 4 Assembly model of the matrix protein and lipoprotein. Abbreviations used are: LP, lipoprotein; MP, matrix protein; and PG, peptidoglycan. The bound form lipoprotein molecules are darkened.

stances with molecular weights smaller than 650, but not of saccharides with molecular weights higher than 900–1000 (85). Therefore, low molecular weight nutrients such as amino acids, sugars, and salts probably diffuse across the outer membrane through the matrix protein pores into the periplasmic space and subsequently are actively transported into the cytoplasm. On the other hand, other nutrients having molecular weights larger than the exclusion limit of the pore require their own receptors to pass through the outer membrane. Uptake of iron as a complex with ferrichrome (740 daltons) or enterochelin (746 daltons), and uptake of vitamin B_{12} (1357 daltons) depends on the presence of receptors (120). Transport of higher maltodextrins more stringently need the *lam*B protein than maltose (145, 290). Thus, the lamB protein has probably evolved for the uptake of starch and glycogen hydrolysis products (120, 145). Nucleoside transport functions of the tsx protein (149) may also have evolved for the uptake of the degradation products of nucleic acids.

Transport of maltose by way of the lamB protein, and transport of nucleosides by way of the tsx protein differ as compared with iron chelator complexes and vitamin B_{12}. The uptake of maltose and nucleosides does not require the *ton*B gene function which is required for the uptake of vitamin B_{12} and iron-chelator complexes. Mutants lacking receptor proteins show defects in the uptake of maltose and nucleosides only at low substrate concentrations (145, 149). Apparently these substrates can pass through other diffusion pores when present at higher concentrations. Similarly, mutants lacking the matrix protein cannot take up sugars or amino acids at low substrate concentrations (87, 89). These mutants can grow on glucose or lactose at concentrations > 0.2 or 1%, respectively (88). The lamB protein appears to facilitate the diffusion of sugars other than maltose, such as glucose and lactose, but not the diffusion of amino acids (147).

CONCLUSIONS

Through the works accumulated in the past few years, we are now able to construct a comprehensive outline of the sequence of events required for the biosynthesis and assembly of the outer membrane proteins. Before long, we will understand the precise molecular mechanisms of these events. Research on the outer membrane proteins is one of the most productive and progressive fields in biochemistry because the outer membrane is a simple membrane system and yet has many features in common with other biomembranes.

As we have seen from this review, the most intriguing problems are as follows:

(*a*) What are the modification processes for the precursors of the outer membrane proteins? The precise molecular mechanism of the translocation of these proteins across the cytoplasmic membrane will also provide an important clue to the mechanisms of hormone secretion and the secretion of many other proteins in animal cells.

(*b*) The mechanism regulating production of the outer membrane proteins. An interesting problem in this regard concerns the number of copies of a particular outer membrane protein and the way in which this is con-

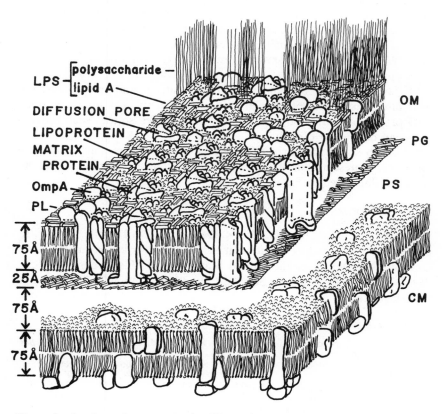

Figure 5   A schematic representation illustrating the possible molecular architecture of the *E. coli* cell envelope. Abbreviations used are: PL, phospholipid; OM, outer membrane; PG, peptidoglycan; PS, periplasmic space; and CM, cytoplasmic membrane. Polysaccharide chains in only some of the LPS molecules are shown. The designation ompA is equivalent to tolG. Note that very little phospholipid is distributed in the outer leaflet of the outer membrane.

trolled. Some outer membrane proteins are present in up to 10^5 copies; others may have only about 10^2 copies.

(c) The complete base sequence of the lipoprotein mRNA. The determination of the base sequence of the nontranslated region of the mRNA, which consists of 120 nucleotides, should provide important insights into the regulatory mechanism of translation of the mRNA for the constitutive membrane protein and the mechanism involved in the high stability of the mRNA.

(d) Genetics of the outer membrane proteins including the determination of gene location on the E. coli chromosome for structural genes, regulatory genes, and the genes required for modification.

(e) The mechanism of differentiation of membrane structures and functions. What is the mechanism that results in the formation of two distinct membranes?

(f) Structures and functions of the major outer membrane proteins and their interactions with other membrane components. The matrix protein, tolG protein, and the lipoprotein have been purified so that such interactions can be studied directly.

On the basis of the current knowledge of the E. coli cell envelope a schematic representation of the possible molecular architecture of the envelope is illustrated in Figure 5.

ACKNOWLEDGMENTS

We thank Mr. S. Halegoua, Mr. M. DeMartini, Dr. I. Pirtle, and Dr. R. Pirtle for critical reading of the manuscript. This work was supported by Grants GM19043 from the United States Public Health Service, PCM 76-07320 from the National Science Foundation, and BC-67D from the American Cancer Society.

Literature Cited

1. Singer, S. J., Nicolson, G. L. 1972. *Science* 175:720–31.
2. Ghuysen, J. M. 1968. *Bacteriol. Rev.* 32:425–64
3. Ghuysen, J. M., Shockman, G. D. 1973. In *Bacterial Membranes and Walls*, ed. L. Leive, 1:37–130. New York: Dekker. 495pp.
4. Nikaido, H. 1973. See Ref. 3, 1:131–208
5. Lüderitz, O., Staub, A. M., Westphal, O. 1966. *Bacteriol. Rev.* 30:192–255
6. Osborn, M. J. 1969. *Ann. Rev. Biochem.* 38:501–38
7. Osborn, M. J., Rick, P. D., Lehmann, V., Rupprecht, E., Singh, M. 1974. *Ann. NY Acad Sci.* 235:52–65
8. Rothfield, L., Romeo, D. 1971. *Bacteriol. Rev.* 35:14–38
9. Braun, V., Bosch, V., Hantke, K., Schaller, K. 1974. *Ann. NY Acad. Sci.* 235:66–82
10. Braun, V. 1975. *Biochim. Biophys. Acta* 415:335–77
11. Inouye, M. 1978. In *Biomembranes.* In press
12. Inouye, M., Hirashima, A., Lee, N. 1974. *Ann. NY Acad. Sci.* 235:83–90
13. Inouye, M. 1975. In *Membrane Biogenesis*, ed. A. Tzagoloff, 1:351–91. New York: Plenum. 460 pp.
14. DePetris, S. 1967. *J. Ultrastruct. Res.* 19:45–83

15. Murray, R. G. E., Steed, P., Elson, H. H. 1965. *Can. J. Microbiol.* 11:547–60
16. Osborn, M. J., Gander, J. E., Parisi, E., Carson, J. 1972. *J. Biol. Chem.* 247:3962–72
17. Nanninga, N. 1970. *J. Bacteriol.* 101:297–303
18. Bayer, M. E., Remsen, C. C. 1970. *J. Bacteriol.* 101:304–13
19. Miura, T., Mizushima, S. 1969. *Biochim. Biophys. Acta* 193:268–76
20. Schnaitman, C. A. 1970. *J. Bacteriol.* 104:890–901
21. Fox, C. F., Law, J. H., Tsukagoshi, N., Wilson, G. 1970. *Proc. Natl. Acad. Sci. USA* 67:598–605
22. Yamato, I., Anraku, Y., Hirosawa, K. 1975. *J. Biochem. Tokyo* 77:705–18
23. Mizushima, S., Yamada, H. 1975. *Biochim. Biophys. Acta* 375:44–53
24. Fox, C. F., Kennedy, E. P. 1965. *Proc. Natl. Acad. Sci. USA* 54:891–99
25. Birdsell, D. C., Cota-Robles, E. H. 1968. *Biochem. Biophys. Res. Commun.* 31:438–46
26. Schnaitman, C. A. 1971. *J. Bacteriol.* 108:545–52
27. Filip, C., Fletcher, G., Wulff, J. L., Earhart, C. F. 1973. *J. Bacteriol.* 115:717–22
28. White, D. A., Lennarz, W. J., Schnaitman, C. A. 1972. *J. Bacteriol.* 109:686–90
29. Smit, J., Kamio, Y., Nikaido, H. 1975. *J. Bacteriol.* 124:942–58
30. Cronan, J. E. Jr., Vagelos, P. R. 1972. *Biochim. Biophys. Acta* 265:25–60
31. Costerton, J. W., Ingram, J. M., Cheng, K. J. 1974. *Bacteriol. Rev.* 38:87–110
32. Salton, M. R. J. 1967. *Ann. Rev. Microbiol.* 21:417–42
33. Salton, M. R. J. 1971. *CRC Crit. Rev. Microbiol.* 1:161–97
34. Schnaitman, C. A. 1970. *J. Bacteriol.* 104:882–89
35. Inouye, M., Yee, M. 1972. *J. Bacteriol.* 112:585–92
36. Inouye, M., Yee, M. 1973. *J. Bacteriol.* 113:304–12
37. Braun, V., Rehn, K. 1969. *Eur. J. Biochem.* 10:426–38
38. Inouye, M., Shaw, J., Shen, C. 1972. *J. Biol. Chem.* 247:8154–59
39. Schnaitman, C. A. 1974. *J. Bacteriol.* 118:454–64
40. Holland, I. B., Darby, J. 1973. *FEBS Lett.* 33:106–8
41. Sekizawa, J., Fukui, S. 1973. *Biochim. Biophys. Acta* 307:104–17
42. Siccardi, A. G., Shapiro, B. M., Hirota, Y., Jacob, F. 1970. *J. Mol. Biol.* 56:475–90
43. Koplow, J., Goldfine, H. 1974. *J. Bacteriol.* 117:527–43
44. Ames, G. F. L., Spudich, E. N., Nikaido, H. 1974. *J. Bacteriol.* 117:406–16
45. Wu, H. C. 1972. *Biochim. Biophys. Acta* 290:274–89
46. Ames, G. F. L. 1974. *J. Biol. Chem.* 249:634–44
47. Schnaitman, C. A. 1974. *J. Bacteriol.* 118:442–53
48. Bragg, P. D., Hou, C. 1972. *Biochim. Biophys. Acta* 274:478–88
49. Manning, P. A., Reeves, P. 1976. *J. Bacteriol.* 127:1070–79
50. Uemura, J., Mizushima, S. 1975. *Biochim. Biophys. Acta* 413:163–76
51. Nakamura, K., Mizushima, S. 1976. *J. Biochem. Tokyo* 80:1411–22
52. Bassford, P. J. Jr., Diedrich, D. L., Schnaitman, C. A., Reeves, P. 1977. *J. Bacteriol.* 131:608–22
53. Lugtenberg, B., Meijers, J., Peters, R., Van der Hoek, P., Van Alphen, L. 1975. *FEBS Lett.* 58:254–58
54. Henning, U., Höhn, B., Sonntag, I. 1973. *Eur. J. Biochem.* 47:343–52
55. Schmitges, C. J., Henning, U. 1976. *Eur. J. Biochem.* 63:47–52
56. Henning, U., Schmidmayr, W., Hindennach, I. 1977. *Mol. Gen. Genet.* 154:293–98
57. Lugtenberg, B., Peters, R., Bernheimer, H., Berendsen, W. 1976. *Mol. Gen. Genet.* 147:251–62
58. Manning, P. A., Reeves, P. 1977. *FEMS Microbiol.* 1:275–78
59. Inouye, M., Guthrie, J. P. 1969. *Proc. Natl. Acad. Sci. USA* 64:957–61
60. Inouye, M., Pardee, A. B. 1970. *J. Biol. Chem.* 245:5813–19
61. Banker, G. A., Cotman, C. W. 1972. *J. Biol. Chem.* 247:5856–61
62. Ichihara, S., Mizushima, S. 1977. *J. Biochem. Tokyo* 81:1525–30
63. Lugtenberg, B., Bronstein, H., Van Selm, N., Peters, R. 1977. *Biochim. Biophys. Acta* 465:571–78
64. Rosenbusch, J. P. 1974. *J. Biol. Chem.* 249:8019–29
65. Hasegawa, Y., Yamada, H., Mizushima, S. 1976. *J. Biochem. Tokyo* 80:1401–9
66. Capaldi, R. A., Vanderkooi, G. 1972. *Proc. Natl. Acad. Sci. USA* 69:930–32
67. Laemmli, U. K. 1970. *Nature* 227:680–85
68. Garten, W., Hindennach, I., Henning, U. 1975. *Eur. J. Biochem.* 60:303–7
69. Van Alphen, W., Lugtenberg, B. 1977. *J. Bacteriol.* 131:623–30

70. Nakamura, K., Ostrovsky, D. J., Miyazawa, T., Mizushima, S. 1974. *Biochim. Biophys. Acta* 332:329–35
71. Reynolds, J. A., Tanford, C. 1970. *J. Biol. Chem.* 245:5161–65
72. Scandella, C. J., Kornberg, A. 1971. *Biochemistry* 10:4447–56
73. Braun, V., Schaller, K., Wolff, H. 1973. *Biochim. Biophys. Acta* 323:87–97
74. Steven, A. C., Ten Heggeler, B., Müller, R., Kistler, J., Rosenbusch, J. P. 1977. *J. Cell Biol.* 72:292–301
75. Palva, E. T., Randall, L. L. 1976. *J. Bacteriol.* 127:1558–60
76. Reithmeier, R. A. F., Bragg, P. D. 1977. *Biochim. Biophys. Acta* 466: 245–56
77. Yamada, H., Mizushima, S. 1977. *J. Biochem. Tokyo* 81:1889–99
78. Yu, F., Mizushima, S. 1977. *Biochem. Biophys. Res. Commun.* 74:1397–1402
79. Datta, D. B., Arden, B., Henning, U. 1977. *J. Bacteriol.* 131:821–29
80. Schnaitman, C. A., Smith, D., Forn de Salsas, M. 1975. *J. Virol.* 15:1121–30
81. Verhoef, C., DeGraaff, P. J., Lugtenberg, B. J. J. 1977. *Mol. Gen. Genet.* 150:103–5
82. Kamio, Y., Nikaido, H. 1977. *Biochim. Biophys. Acta* 464:589–601
83. Nakae, T. 1976. *J. Biol. Chem.* 251:2176–78
84. Nakae, T. 1976. *Biochem. Biophys. Res. Commun.* 71:877–84
85. Nakae, T., Nikaido, H. 1975. *J. Biol. Chem.* 250:7359–65
86. Nakaido, H., Song, S. A., Shaltiel, L., Nurminen, M. 1977. *Biochem. Biophys. Res. Commun.* 76:324–30
87. Lutkenhaus, J. F. 1977. *J. Bacteriol.* 131:631–37
88. von Meyenburg, K. 1971. *J. Bacteriol.* 107:878–88
89. Bavoil, P., Nikaido, H., von Meyenburg, K. 1978. *Mol. Gen. Genet.* In press
90. Diedrich, D. L., Summers, A. O., Schnaitman, C. A. 1977. *J. Bacteriol.* 131:598–607
91. Reithmeier, R. A. F., Bragg, P. D. 1977. *Arch. Biochem. Biophys.* 178: 527–34
92. Reithmeier, R. A. F., Bragg, P. D. 1974. *FEBS Lett.* 41:195–98
93. Garten, W., Hindennach, I., Henning, U. 1975. *Eur. J. Biochem.* 59:215–21
94. Van Alphen, L., Havekes, L., Lugtenberg, B. 1977. *FEBS Lett.* 75:285–90
95. Schnaitman, C. A. 1973. *Arch. Biochem. Biophys.* 157:541–52
96. Manning, P. A., Puspurs, A., Reeves, P. 1976. *J. Bacteriol.* 127:1080–84

97. Skurray, R. A., Hancock, R. E. W., Reeves, P. 1974. *J. Bacteriol.* 119: 726–35
98. Manning, P. A., Reeves, P. 1975. *J. Bacteriol.* 124:576–77
99. Manning, P. A., Reeves, P. 1977. *J. Bacteriol.* 130:540–41
100. Schweizer, M., Henning, U. 1977. *J. Bacteriol.* 129:1651–52
101. Braun, V., Sieglin, J. 1970. *Eur. J. Biochem.* 13:336–46
102. Braun, V., Bosch, V. 1972. *Proc. Natl. Acad. Sci. USA* 69:970–74
103. Braun, V., Bosch, V. 1972. *Eur. J. Biochem.* 28:51–69
104. Hantke, K., Braun, V. 1973. *Eur. J. Biochem.* 34:284–96
105. Hirashima, A., Wu, H. C., Venkateswaran, P. S., Inouye, M. 1973. *J. Biol. Chem.* 248:5654–59
106. Braun, V., Wolff, H. 1970. *Eur. J. Biochem.* 14:387–91
107. Inouye, S., Takeishi, K., Lee, N., DeMartini, M., Hirashima, A., Inouye, M. 1976. *J. Bacteriol.* 127:555–63
108. Bosch, V., Braun, V. 1973. *FEBS Lett.* 34:307–10
109. Lee, N., Inouye, M. 1974. *FEBS Lett.* 39:167–70
110. Braun, V., Rehn, K., Wolff, H. 1970. *Biochemistry* 9:5041–49
111. Inouye, M. 1974. *Proc. Natl. Acad. Sci. USA* 71:2369–2400
112. Braun, V. 1973. *J. Infect. Dis.* 128: Suppl. S, pp. 9–15
113. Braun, V., Rotering, H., Ohms, J. P., Hagenmaier, H. 1976. *Eur. J. Biochem.* 70:601–10
114. Lee, N., Cheng, E., Inouye, M. 1977. *Biochim. Biophys. Acta* 465:650–56
115. Hodges, R. S., Sodek, J., Smillie, L. B., Jurasek, L. 1972. *Cold Spring Harbor Symp. Quant. Biol.* 37:299–310
116. Lee, N., Tu, S., Inouye, M. 1977. *Biochemistry* 16:5026–30
117. DeMartini, M., Inouye, S., Inouye, M. 1976. *J. Bacteriol.* 127:564–71
118. Braun, V., Bosch, V., Klumpp, E. R., Neff, I., Mayer, H., Schlecht, S. 1976. *Eur. J. Biochem.* 62:555–66
119. Braun, V., Hancock, R. E. W., Hantke, K., Hartmann, A. 1976. *J. Supramolec. Struct.* 5:37–58
120. Braun, V., Hantke, K. 1977. In *Microbial Interactions,* ed. J. L. Reissig. London: Chapman and Hall. In press
121. White, J. C., DiGirolamo, P. M., Fu, M. L., Preston, Y. A., Bradbeer, C. 1973. *J. Biol. Chem.* 248:3978–86
122. DiMasi, D. R., White, J. C., Schnaitman, C. A., Bradbeer, C. 1973. *J. Bacteriol.* 115:506–13

123. Bradbeer, C., Woodrow, M. L., Khalifah, L. I. 1976. *J. Bacteriol.* 125: 1032–39
124. Sabet, S. F., Schnaitman, C. A. 1973. *J. Biol. Chem.* 248:1797–1806
125. Kadner, R. J., Liggins, G. L. 1973. *J. Bacteriol.* 115:514–28
126. DiMasi, D. R., Kenley, J. S., Bradbeer, C. 1977. *Ann. Meet. Am. Soc. Microbiol. 77th, New Orleans* p. 210 (Abstr.)
127. Hantke, K., Braun, V. 1975. *FEBS Lett.* 49:301–5
128. Wayne, R., Neilands, J. B. 1975. *J. Bacteriol.* 121:497–503
129. Luckey, M., Wayne, R., Neilands, J. B. 1975. *Biochem. Biophys. Res. Commun.* 65:687–93
130. Braun, V., Wolff, H. 1973. *FEBS Lett.* 34:77–80
131. Guterman, S. K. 1973. *J. Bacteriol.* 114:1217–24
132. Pugsley, A. P., Reeves, P. 1976. *J. Bacteriol.* 126:1052–62
133. Hantke, K., Braun, V. 1975. *FEBS Lett.* 59:277–81
134. Hancock, R. E. W., Hantke, K., Braun, V. 1976. *J. Bacteriol.* 127:1370–75
135. Hancock, R. E. W., Braun, V. 1976. *FEBS Lett.* 65:208–10
136. Pugsley, A. P., Reeves, P. 1976. *Biochem. Biophys. Res. Commun.* 70: 846–53
137. Ichihara, S., Mizushima, S. 1977. *J. Biochem. Tokyo* 81:749–56
138. McIntosh, M. A., Earhart, C. F. 1976. *Biochem. Biophys. Res. Commun.* 70:315–22
139. McIntosh, M. A., Earhart, C. F. 1977. *J. Bacteriol.* 131:331–39
140. Soucek, S., Konisky, J. 1977. *J. Bacteriol.* 130:1399–1401
141. Pugsley, A. P., Reeves, P. 1977. *Biochem. Biophys. Res. Commun.* 74: 903–11
142. Konisky, J., Lin, C. T. 1974. *J. Biol. Chem.* 249:835–40
143. Miguel, A., Bowles, L., Soucek, S., Konisky, J. 1977. See Ref. 126, p. 210
144. Konisky, J., Cowell, B. S. 1972. *J. Biol. Chem.* 247:6524–29
145. Szmelcman, S., Hofnung, M. 1975. *J. Bacteriol.* 124:112–18
146. Hazelbauer, G. L. 1975. *J. Bacteriol.* 124:119–26
146a. Braun, V., Krieger-Brauer, H. J. 1977. *Biochim. Biophys. Acta* 469:89–98
147. von Meyenburg, K., Nikaido, H. 1977. *Biochem. Biophys. Res. Commun.* 78: 1100–7
148. Manning, P. A., Reeves, P. 1976. *Biochem. Biophys. Res. Commun.* 71: 466–71
149. Hantke, K. 1976. *FEBS Lett.* 70: 109–12
150. James, R. 1975. *J. Bacteriol.* 124: 918–29
151. Gudas, L. J., James, R., Pardee, A. B. 1976. *J. Biol. Chem.* 251:3470–79
152. Portalier, R., Worcell, A. 1976. *Cell* 8:245–55
153. Albright, F. R., White, D. A., Lennarz, W. J. 1973. *J. Biol. Chem.* 248:3968–77
154. Hardaway, K. L., Buller, C. S. 1977. See Ref. 126, p. 209
155. Abe, M., Okamoto, N., Doi, O., Nojima, S. 1974. *J. Bacteriol.* 119: 543–46
156. Henning, U., Haller, I. 1975. *FEBS Lett.* 55:161–64
157. Backmann, B. J., Low, K. B., Taylor, A. L. 1976. *Bacteriol. Rev.* 40:116–67
158. Foulds, J., Barrett, C. 1973. *J. Bacteriol.* 116:885–92
159. Chai, T. J., Foulds, J. 1977. *J. Bacteriol.* 130:781–86
160. Foulds, J. 1976. *J. Bacteriol.* 128:604–8
161. Davies, J. K., Reeves, P. 1975. *J. Bacteriol.* 123:102–17
162. Sarma, V., Reeves, P. 1977. *J. Bacteriol.* 132:23–27
163. Beacham, I. R., Kahana, R., Levy, L., Yagil, E. 1973. *J. Bacteriol.* 116:957–64
164. Beacham, I. R., Haas, D., Yagil, E. 1977. *J. Bacteriol.* 129:1034–44
165. Sekizawa, J., Inouye, S., Halegoua, S., Inouye, M. 1977. *Biochem. Biophys. Res. Commun.* 77:1126–33
166. Chai, T., Foulds, J., 1974. *J. Mol. Biol.* 85:465–74
167. Datta, D. B., Krämer, C., Henning, U. 1976. *J. Bacteriol.* 128:834–41
168. Foulds, J. 1974. *J. Bacteriol.* 117: 1354–55
169. Henning, U., Hindennach, I., Haller, I. 1976. *FEBS Lett.* 61:46–48
170. Suzuki, H., Nishimura, Y., Iketani, H., Campisi, J., Hirashima, A., Inouye, M., Hirota, Y. 1976. *J. Bacteriol.* 127:1494–1501
171. Inouye, S., Lee, N., Inouye, M., Wu, H. C., Suzuki, H., Nishimura, Y., Iketani, H., Hirota, Y. 1977. *J. Bacteriol.* 132:308–13
172. Hirota, Y., Suzuki, H., Nishimura, Y., Yasuda, S., 1977. *Proc. Natl. Acad. Sci. USA* 74:1417–20
173. Nikaido, H., Bavoil, P., Hirota, Y. 1977. *J. Bacteriol.* 132:1045–47
174. Weigand, R. A., Vinci, K. D., Rothfield, L. I. 1976. *Proc. Natl. Acad. Sci. USA* 73:1882–86
175. Lopes, J., Gottfried, S., Rothfield, L. I. 1972. *J. Bacteriol.* 109:520–25

176. Weigand, R. A., Rothfield, L. I. 1976. *J. Bacteriol.* 125:340–45
177. Hirashima, A., Inouye, M. 1973. *Nature* 242:405–7
178. Wu, H. C., Lin, J. J.-C. 1976. *J. Bacteriol.* 126:147–56
179. Wu, H. C., Hou, C., Lin, J. J.-C., Yem, D. W. 1977. *Proc. Natl. Acad. Sci. USA* 74:1388–92
180. Yem, D. W., Wu, H. C. 1977. *J. Bacteriol.* 131:759–64
181. Torti, S. V., Park, J. T. 1976. *Nature* 263:323–26
182. James, R., Gudas, L. J. 1976. *J. Bacteriol.* 125:374–75
183. Helmstetter, C. E., Cooper, S. 1968. *J. Mol. Biol.* 31:507–18
184. Wohlhieter, J. A., Gemski, P. Jr., Baron, L. S. 1975. *Mol. Gen. Genet.* 139:93–101
185. Silhavy, T. J., Casadaban, M. J., Shuman, H. A., Beckwith, J. R. 1976. *Proc. Natl. Acad. Sci. USA* 73:3423–27
186. Sinsheimer, R. L. 1977. *Ann. Rev. Biochem.* 46:415–38
187. Morgan, E. A., Ikemura, T., Nomura, M. 1977. *Proc. Natl. Acad. Sci. USA* 74:2710–14
188. Silverman, M., Matsumura, P., Draper, R., Edwards, S., Simon, M. 1976. *Nature* 261:248–50
189. Silverman, M., Matsumura, P., Hilmen, M., Simon, M. 1977. *J. Bacteriol.* 130:877–87
190. Berz, P. E., Gayda, R., Auni, H., Zehnbauer, B., Markovitz, A. 1976. *Proc. Natl. Acad. Sci. USA* 73:697–701
191. Clarke, L., Carbon, J. 1976. *Cell* 9: 91–99
192. Lindberg, A. A. 1973. *Ann. Rev. Microbiol.* 27:205–41
193. Monner, D. A., Jonsson, S., Boman, H. G. 1971. *J. Bacteriol.* 107:420–32
194. Tamaki, S., Sato, T., Matsuhashi, M. 1971. *J. Bacteriol.* 105:968–75
195. Hancock, R. E. W., Reeves, P. 1976. *J. Bacteriol.* 127:98–108
196. Gustafsson, P., Nordström, K., Normark, S. 1973. *J. Bacteriol.* 116:893–900
197. Sanderson, K. E., MacAlister, T., Costerton, J. W., Cheng, K. J. 1974. *Can. J. Microbiol.* 20:1135–45
198. Tamaki, S., Matsuhashi, M. 1973. *J. Bacteriol.* 114:453–54
198a. Nikaido, H. 1976. *Biochim. Biophys. Acta* 433:118–132
199. Lindsay, S. A., Wheeler, B., Sanderson, K. E., Costerton, J. W., Cheng, K. J. 1973. *Can. J. Microbiol.* 19:335–43
200. Singh, A. P., Reithmeier, R. A. F. 1975. *J. Gen. Appl. Microbiol.* 21: 109–18
201. Havekes, L. M., Lugtenberg, B. J. J., Hoekstra, W. P. M. 1976. *Mol. Gen. Genet.* 190:1–8
202. Havekes, L. M., Tommassen, J., Hoekstra, W. P. M., Lugtenberg, B. J. J. 1977. *J. Bacteriol.* 129:1–8
203. Randall, L. L. 1975. *J. Bacteriol.* 123: 41–46
204. Nakae, T. 1975. *Biochem. Biophys. Res. Commun.* 64:1224–30
205. Nakamura, K., Mizushima, S. 1975. *Biochim. Biophys. Acta* 413:371–93
206. Normark, S., Boman, H. G., Matsson, E. 1969. *J. Bacteriol.* 97:1334–42
207. Normark, S. 1969. *J. Bacteriol.* 98: 1274–77
208. Rodolakis, A., Thomas, P., Starka, J. 1973. *J. Gen. Microbiol.* 75:409–16
209. Egan, A. F., Russell, R. R. B. 1973. *Genet. Res.* 21:139–52
210. Hirota, Y., Mordoh, J., Jacob, F. 1970. *J. Mol. Biol.* 53:369–87
211. Carl, P. L. 1970. *Mol. Gen. Genet.* 109:107–22
212. Hirota, Y., Ricard, M., Shapiro, B. 1971. *Biomembranes* 2:13–31
213. Davies, J. K., Reeves, P. 1975. *J. Bacteriol.* 123:96–101
214. Nagel de Zwaig, R., Luria, S. E. 1967. *J. Bacteriol.* 94:1112–23
215. Burman, L. G., Nordström, K. 1971. *J. Bacteriol.* 106:1–13
216. Wijsman, H. J. W., Pafort, H. C. 1974. *Mol. Gen. Genet.* 128:349–57
217. Hirashima, A., Childs, G., Inouye, M. 1973. *J. Mol. Biol.* 79:373–89
218. Halegoua, S., Hirashima, A., Inouye, M. 1976. *J. Bacteriol.* 126:183–91
219. Hirashima, A., Inouye, M. 1975. *Eur. J. Biochem.* 60:395–98
220. Varricchio, F. 1972. *J. Bacteriol.* 109: 1284–94
221. Canceddas, R., Schlesinger, M. J. 1974. *J. Bacteriol.* 117:290–301
222. Patterson, D., Weinstein, M., Nixon, R., Gillespie, D. 1970. *J. Bacteriol.* 101:584–91
223. Randall, L. L., Hardy, S. J. S. 1975. *Mol. Gen. Genet.* 137:151–60
224. Randall, L. L., Hardy, S. J. S. 1977. *Eur. J. Biochem.* 75:43–53
225. Hirashima, A., Wang, S., Inouye, M. 1974. *Proc. Natl. Acad. Sci. USA* 71:4149–53
226. Moses, R. E., Richardson, C. C. 1970. *Proc. Natl. Acad. Sci. USA* 67:674–81
227. Peterson, R. L., Radcliffe, C. W., Pace, N. R. 1971. *J. Bacteriol.* 107:585–88
228. Schrader, W. P., Fan, D. P. 1974. *J. Biol. Chem.* 249:4815–18

229. Halegoua, S., Hirashima, A., Sekizawa, J., Inouye, M. 1976. *Eur. J. Biochem.* 69:163–67

230. Halegoua, S., Sekizawa, J., Inouye, M. 1977. *J. Biol. Chem.* 252:2324–30

231. Inouye, S., Wang, S., Sekizawa, J., Halegoua, S., Inouye, M. 1977. *Proc. Natl. Acad. Sci. USA* 74:1004–8

232. Inouye, H., Beckwith, J. 1977. *Proc. Natl. Acad. Sci. USA* 74:1440–44

233. Yu, S., Redman, C. 1977. *Biochem. Biophys. Res. Commun.* 76:469–76

234. Shields, D., Blobel, G. 1977. *Proc. Natl. Acad. Sci. USA* 74:2059–63

235. Thibodeau, S. N., Gagnon, J., Palmiter, R. 1977. *Fed. Proc.* 36:656 (Abstr.)

236. Devillers-Thiery, A., Kindt, T., Scheele, G., Blobel, G. 1975. *Proc. Natl. Acad. Sci. USA* 72:5016–20

237. Chan, S. J., Keim, P., Steiner, D. F. 1976. *Proc. Natl. Acad. Sci. USA* 73:1964–68

238. Kemper, B., Habener, J., Ernst, M. D., Potts, J. T. Jr., Rich, A. 1976. *Biochemistry* 15:15–19

239. Burstein, Y., Schechter, I. 1977. *Proc. Natl. Acad. Sci. USA* 74:716–20

240. Levy, S. 1975. *Proc. Natl. Acad. Sci. USA* 72:2900–4

241. Garten, W., Henning, U. 1976. *FEBS Lett.* 67:303–5

242. Takeishi, K., Yasumura, M., Pirtle, R., Inouye, M. 1976. *J. Biol. Chem.* 251:6259–66

243. Osborn, M. J., Gander, J. E., Parisi, E. 1972. *J. Biol. Chem.* 247:3973–86

244. Smith, W. P., Tai, P.-C., Thompson, R. C., Davis, B. D. 1977. *Proc. Natl. Acad. Sci. USA* 74:2830–34

245. Blobel, G., Dobberstein, B. 1975. *J. Cell Biol.* 67:835–51

246. Bayer, M. E. 1968. *J. Gen. Microbiol.* 53:395–404

247. Mühlradt, P. F., Menzel, J., Golecki, J. R., Speth, V. 1973. *Eur. J. Biochem.* 35:471–81

248. Bayer, M. E., Thurow, H. 1977. *J. Bacteriol.* 130:911–36

249. Bayer, M. E. 1968. *J. Virol.* 2:346–56

250. Bayer, M. E. 1975. *Membrane Biogenesis* 1:393–404

251. Kimura, K., Izui, K. 1976. *Biochem. Biophys. Res. Commun.* 70:900–6

252. Ito, K., Sato, T., Yura, T. 1977. *Cell* 11:551–59

253. Kleeman, W., McConnell, H. M. 1974. *Biochim. Biophys. Acta* 345:220–30

254. Haest, C. W. M., Verkleij, A. J., DeGier, J., Scheek, R., Ververgaert, P. H. J., Van Deenen, L. L. M. 1974. *Biochim. Biophys. Acta* 356:17–26

255. Letellier, L., Moudden, H., Schechter, E. 1977. *Proc. Natl. Acad. Sci. USA* 74:452–56

256. Bayer, M. E., Dolack, M., Houser, E. 1977. *J. Bacteriol.* 129:1563–73

257. Overath, P., Brenner, M., Gulik-Krzywicki, T., Shechter, E., Letellier, L. 1975. *Biochim. Biophys. Acta* 389:358–69

258. Rottem, S., Leive, L. 1977. *J. Biol. Chem.* 252:2077–81

259. Lin, J. J.-C., Wu, H. C. 1976. *J. Bacteriol.* 125:892–904

260. Schulman, H., Kennedy, E. P. 1977. *J. Biol. Chem.* 252:4250–55

261. Omura, S. 1976. *Bacteriol. Rev.* 40:681–97

262. Chattopadhyay, P. K., Wu, H. C. 1977. See Ref. 126, p. 211

263. Izuki, K., Matsuhashi, M., Strominger, J. L. 1966. *Proc. Natl. Acad. Sci. USA* 55:656–63

264. Braun, V., Wolff, H. 1975. *J. Bacteriol.* 123:888–97

265. Tanaka, N., Iseki, M., Miyoshi, T., Aoki, H., Imanaka, H. 1976. *J. Antibiotics* 29:155–68

266. Braun, V., Bosch, V. 1973. *FEBS Lett.* 34:302–6

267. Mühlradt, P. F., Golecki, J. R. 1975. *Eur. J. Biochem.* 51:343–52

268. Kamio, Y., Nikaido, H. 1976. *Biochemistry* 15:2561–70

268a. Nikaido, H., Takeuchi, Y., Ohnishi, S. I., Nakae, T. 1977. *Biochim. Biophys. Acta* 465:152–64

269. Van Gool, A. P., Nanninga, N. 1973. *J. Bacteriol.* 108:474–81

270. Bayer, M. E., Koplow, J., Goldfine, H. 1975. *Proc. Natl. Acad. Sci. USA* 72:5145–49

271. Verkleij, A. J., Lugtenberg, B. J. J., Ververgaert, P. H. J. 1976. *Biochim. Biophys. Acta* 426:581–86

272. Schweizer, M., Schwartz, H., Sonntag, I., Henning, U. 1976. *Biochim. Biophys. Acta* 448:474–91

273. Verkleij, A., Van Alphen, L. V., Bijvelt, J., Lugtenberg, B. 1977. *Biochim. Biophys. Acta* 466:269–82

274. Bassford, P. J. Jr., Kadner, R. J., Schnaitman, C. A. 1977. *J. Bacteriol.* 129:265–75

275. Bassford, P. J. Jr., Schnaitman, C. A., Kadner, R. J. 1977. *J. Bacteriol.* 130:750–58

276. Bassford, P. J. Jr., Bradbeer, C., Kadner, R. J., Schnaitman, C. A. 1976. *J. Bacteriol.* 128:242–47

277. Kadner, R. J., Bassford, P. J. Jr. 1977. *J. Bacteriol.* 129:254–64

278. Wang, C. C., Newton, A. 1971. *J. Biol. Chem.* 246:2147–51
279. Frost, G. E., Rosenberg, H. 1975. *J. Bacteriol.* 124:704–12
280. Hancock, R. E. W., Braun, V. 1976. *J. Bacteriol.* 125:409–15
281. Peters, K., Richards, F. M. 1977. *Ann. Rev. Biochem.* 46:523–51
282. Henning, U., Rehn, K., Hoehn, B. 1973. *Proc. Natl. Acad. Sci. USA* 70:2033–36
283. Garten, W., Henning, U. 1974. *Eur. J. Biochem.* 47:343–52
284. Haller, I., Henning, U. 1974. *Proc. Natl. Acad. Sci. USA* 71:2018–21

285. Van Alphen, W., Lugtenberg, B., Berendsen, W. 1976. *Mol. Gen. Genet.* 147:263–69
286. Payne, J. W., Gilvarg, C. 1968. *J. Biol. Chem.* 243:6291–99
287. Payne, J. W. 1968. *J. Biol. Chem.* 243:3395–3403
288. Decad, G. M., Nikaido, H. 1976. *J. Bacteriol.* 128:325–36
289. DeMartini, M., Inouye, M. 1978. *J. Bacteriol.* In press
290. Szmelcman, S., Schwartz, M., Silhavy, T. J., Boos, W. 1976. *Eur. J. Biochem.* 65:13–19

Ann. Rev. Biochem. 1978. 47:533–606

AMINO ACID BIOSYNTHESIS AND ITS REGULATION

♦982

H. E. Umbarger

Department of Biological Sciences, Purdue University,
West Lafayette, Indiana 47907

CONTENTS

533

0066-4154/78/0701-0533$01.00

INTRODUCTION

The regulation of amino acid biosynthesis occurs at two levels: the regulation of enzyme activity or metabolite flow and the regulation of enzyme formation or gene expression. The pursuit of the first to its logical end becomes more relevant to the subject of enzymology or regulatory proteins than to amino acid biosynthesis. Pursuit of the second becomes more relevant to molecular genetics than to amino acid biosynthesis. In the nine years that have elapsed since a comprehensive review of this topic appeared in the *Annual Review of Biochemistry,* studies on a very few systems have certainly headed in these directions. For those systems where the advances have been most exciting, it has been difficult to decide what to omit as being less relevant to the selected topic. Because the goal was to consider our present understanding of the way the biosynthesis of all twenty amino acids is controlled, the decision was to limit the discussion to the physiological aspects of regulation. Where the regulatory signal has been studied in detail, the discussion is limited to describing what elements interact. The way they interact at the level of the three-dimensional structure of the protein or at the level of specific nucleotide sequences in RNA or DNA is not discussed.

The material chosen is heavily weighted toward the regulation of amino acid biosynthesis in *Escherichia coli* and *Salmonella typhimurium.* This bias could not be avoided, since these organisms have attracted most of the workers in the field. Furthermore, studies with these organisms on one pathway can be readily related to studies on other pathways or to studies on cell growth. It is hoped that the comparative coverage of regulation of biosynthesis of amino acids in other prokaryotes and the simpler eukaryotes will indicate that the *E. coli* pattern is not always the universal pattern. Only a few aspects of amino acid biosynthesis in plants have been covered. The interested reader should refer to the recent review of Mifflin & Lea (1) for a fairer treatment of this topic.

It was considered imperative to devote some space to discussing briefly the reaction sequences of the pathways themselves, since no current textbook of biochemistry contains the correct, up-to-date pathways for amino acid biosynthesis. In some cases, it seemed worthwhile to discuss some of the background material relevant to the pathways themselves and to provide some basis for assuming that biosynthesis proceeds by the pathways shown in the figures.

Much of the literature cited is quite old. This was done because the old observations are often needed to appreciate the significance of the more recent observations or because the old observations still provide the latest information on the subject and are therefore very pertinent in a comprehensive review of this kind.

THE GLUTAMATE FAMILY

The glutamate family of amino acids consists of glutamate itself and those amino acids that derive all or most of their carbon chains from glutamate: glutamine, proline, and arginine. Closely related to the glutamate family in fungi is lysine, which is derived not from glutamate itself but from α-ketoglutarate, the tricarboxylic acid cycle intermediate with which glutamate is in equilibrium. For this reason, the route to lysine in fungi is considered in this section. The route to lysine in higher plants and bacteria is considered with the other members of the aspartate family.

Glutamate and Glutamine

Because the biosynthetic pathways to glutamate and glutamine are important in the assimilation of ammonia by microorganisms and are considered in detail elsewhere in this volume (2) the topic is not reviewed here. It should be sufficient to remind the reader that, with the exception of the amidation of aspartate, all the nitrogen assimilated by most bacteria from ammonia in the medium enters metabolism as the amino group of glutamate (Reaction 1, Figure 1) or as the amide group of glutamine (Reaction 2, Figure 1). Furthermore, some microorganisms lack Reaction 1 and depend upon the animation of α-ketoglutarate at the expense of the amide group of glutamine (glutamate synthase; Reaction 3, Figure 1). Indeed, Reactions 2 and 3, which in effect provide for an ATP-linked reductive amination of

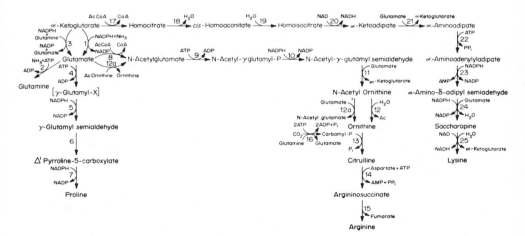

Figure 1 Biosynthesis of the glutamate family of amino acids. Also included are the steps to lysine in fungi. For lysine biosynthesis in other organisms, see Figure 3.

glutamate, are used in place of Reaction 1 in many organisms when the supply of NH_3 in the medium becomes limiting or is supplied only slowly by the deamination of other nitrogen sources. In such cases, nearly all the ammonia entering the metabolic pool does so via Reaction 2.

Whether glutamate formation proceeds via Reaction 1 or Reaction 3, the turnover of the enzyme exceeds by severalfold the rate needed to supply glutamate for protein synthesis, since the amino group finds its way directly or indirectly into all the other amino acids via transamination reactions or by the direct conversion of both the carbon and the nitrogen of glutamate into the other members of the glutamate family of amino acids: proline, arginine, and glutamine.

Proline Biosynthesis

Proline biosynthesis proceeds via a reduction of the γ-carboxyl group of glutamate, to yield glutamic-γ-semialdehyde, which spontaneously cyclizes (Reaction 6, Figure 1) to Δ^1-pyrroline-5-carboxylate. This compound is then reduced to yield proline (Reaction 7). While chemically analogous to both the formation of aspartic-β-semialdehyde from aspartate and the formation of glyceraldehyde-3-phosphate from 3-phosphoglycerate during gluconeogenesis, the formation of a phosphorylated intermediate in a separate step has not been demonstrated. Evidence for carboxyl activation (Reaction 4, Figure 1) has been obtained by Baich (3) by incubating a partially purified *E. coli* preparation with glutamate and ATP in an imidazole buffer. The reaction, a glutamate-dependent hydrolysis of ATP, was inhibited by proline. When the reaction was catalyzed by the enzyme from a strain known to lack feedback control over proline biosynthesis, the reaction was not inhibited by proline. When hydroxylamine was employed instead of imidazole, γ-glutamyl hydroxamate was formed in amounts nearly equivalent to the amount of ATP hydrolyzed. The product of the glutamokinase activity was thought to be enzyme bound and was liberated from the enzyme only with a high concentration of a nucleophilic agent, either hydroxylamine or imidazole. Whether the product was glutamyl phosphate bound to the enzyme or a glutamyl-enzyme complex was not determined.

Baich (4) has also studied in isolation the second step in proline biosynthesis (Reaction 5, Figure 1) but only in the reverse direction. The reaction was dependent on inorganic phosphate and resulted in the formation of a γ-glutamyl semialdehyde (in equilibrium with its cyclized isomer).

In *E. coli,* in which most of the experiments on proline biosynthesis have been performed, there is some question about the enzymes forming γ-glutamyl semialdehyde. The overall reaction is missing in extracts from mutants of *E. coli* with lesions in two genetic loci: *proA* and *proB*. Which reaction is missing in either of the two kinds of mutant has not been determined, and the possibility that a single enzyme (complex) catalyzes the overall reaction of carboxyl group activation and carboxyl group reduction should be considered. On the other hand, Gamper & Moses (5) have observed that extracts capable of forming γ-glutamyl semialdehyde (or Δ^1 pyrroline-5-carboxylate) are markedly sensitive to dilution, as would be expected if the two reactions were catalyzed by a complex of two dissociable proteins each exhibiting one of the two activities. Furthermore, Baich (2, 3) observed that the two activities were separated from each other by a simple fractionation step, and that only Reaction 4 was inhibited by proline.

The final step in proline biosynthesis, the reduction of Δ^1 pyrroline-5-carboxylate (Reaction 6, Figure 1), is catalyzed by a separate enzyme specified by the *proC* gene of *S. typhimurium* or *E. coli*. NADPH is the preferred hydrogen donor. The enzyme does not appear to be repressed by exogenous proline as are the *proA* and *proB* products (6).

Arginine Biosynthesis

The conversion of glutamate to arginine, like the formation of proline, involves the activation and reduction of the γ-carboxyl group of glutamate. However, to avoid the cyclization of glutamate semialdehyde, which is essential for proline biosynthesis, the glutamate amino group must be masked. This masking reaction is, indeed, the first specific step in arginine biosynthesis and, in most organisms, involves the transfer of the acetyl group of acetyl-CoA to yield N-acetylglutamate (Reaction 7, Figure 1). This reaction has been known for years to be the initial reaction in arginine biosynthesis. Unfortunately, the enzyme activity was refractory to study in extracts of *E. coli,* but the sensitivity of the reaction to end-product inhibition by arginine was very early demonstrated in vivo (7). Indeed, it was not until 1972 that Leisinger and his colleagues (8) demonstrated the in vitro synthesis of N-acetylglutamate in extracts of *Pseudomonas aeruginosa* and the inhibition of the partially purified enzyme by arginine. Later, it was shown that the *E. coli* enzyme was unstable and required several stabilizing ligands to permit its study (9).

The control of activity of the *P. aeruginosa* enzyme is more complex than mere inhibition by arginine in that the inhibition by arginine is enhanced synergistically by the products of the reaction: N-acetylglutamate and CoA (10). There is also strong inhibition of the enzyme through a synergistic effect of N-acetylglutamate and polyamines such as putrescine, cadaverine, spermidine, and spermine, the latter being the most potent. The sensitivity of the enzyme to polyamines may be related to the role of the arginine biosynthetic pathway in polyamine biosynthesis.

There is another difference between the mechanism of control of carbon flow in the arginine biosynthetic pathway of *P. aeruginosa* and that in the

E. coli pathway. This difference is related to the difference in the way the acetyl group is removed from acetyl ornithine. In *E. coli,* Enterobacteriaceae, and *Bacillus subtilis,* the function is performed by N-acetylornithinase (Reaction 12, Figure 1). In *P. aeruginosa* and a variety of other organisms, including *Brevibacterium glutamicum,* the cyanobacteria, and at least some fungi, a more energetically efficient process is used in which the acetyl group is transferred to glutamate (Reaction 12a, Figure 1). Reaction 12a thus initiates the conversion of glutamate to arginine. In such organisms the second enzyme in the pathway, N-acetylglutamokinase (Reaction 9, Figure 1), as well as the first, is subject to inhibition by arginine (11). The enzyme and its interaction with arginine has been studied in *Neurospora crassa* (12) and *Clamydomonas reinhardii* (13), in addition to *P. aeruginosa* (14). The role of N-acetylglutamate synthase (the acetyl-CoA-dependent enzyme) in such organisms is that of an anaplerotic enzyme with the function of expanding the acetylglutamate-acetylornithine pool as the cell mass increases. The role of the enzyme in maintaining this pool may be an important selective factor in the evolution of its sensitivity to the synergistic inhibition involving N-acetylglutamate (cited above).

Another aspect of the control of carbon flow in the biosynthesis of arginine is the control of carbamyl phosphate synthetase. In most, if not all, bacterial forms, a single glutamine-dependent enzyme (Reaction 16, Figure 1) forms this compound, which is a branch point intermediate in pyrimidine and arginine biosynthesis. The regulation of carbamyl phosphate synthetase activity is clearly fitted to this dual role. In *E. coli,* uridine monophosphate is an inhibitor of the enzyme. If this were the sole effector for the enzyme, it is clear that arginine biosynthesis would be prevented by an exogenous pyrimidine source. Such an occurrence is prevented, however, by ornithine, the intermediate that would accumulate if arginine biosynthesis were prevented by lack of carbamyl phosphate. Ornithine is a positive effector, which, in high enough concentrations, can reverse the uridine monophosphate effect (15). The mechanism of the interaction between the enzyme and its regulatory effectors has been extensively studied by Anderson (16).

In fungi, such as *N. crassa* and *Saccharomyces cerevisiae,* a different pattern exists, since there are two carbamyl phosphate synthetase activities (17, 18). That used for pyrimidine biosynthesis is catalyzed by the same protein that exhibits aspartate transcarbamylase activity (17, 18). The formation of carbamyl phosphate on the surface of the same protein that consumes it obviously allows for a "metabolic channelling" that would prevent the carbamyl phosphate destined for pyrimidine biosynthesis from being drawn away for citrulline biosynthesis. Only if the utilization of one of the two pools of carbamyl phosphate is impaired will that pool be able to "spill over" into the other (19, 20).

An additional factor that provides for the metabolic channelling of the intermediates in the arginine biosynthetic pathway is the localization of the arginine-specific carbamyl phosphate synthetase. This enzyme, and all the other arginine biosynthetic enzymes that lead to citrulline formation, except for acetylglutamate kinase, are present in the mitochondria or an organelle that cosediments with the mitochondria (21). In contrast, the aspartate-transcarbamylase-linked carbamyl phosphate synthetase is found in the nucleus (22). Thus, arginine is formed in the cytosol where it would be available to exert feedback control on the nonorganellar enzyme, acetyl glutamate kinase, but not on the mitochondrial, arginine-specific carbamyl phosphate synthetase. Indeed, this enzyme is not sensitive to inhibition by arginine, nor is ornithine a positive effector in *N. crassa,* but its synthesis is strongly repressed by exogenous arginine.

Another aspect of arginine metabolism that has made physical compartmentation of pools an important property in *N. crassa* is the fact that both arginine and ornithine are substrates for catabolic pathways, arginine via arginase and ornithine via a transaminase. A mechanism for sequestering pools of both amino acids is provided by the formation of a vesicle that retains the bulk of the endogenous ornithine and arginine pools (23, 24). Only when the capacity of the vesicle is exceeded in the presence of exogenous arginine does the catabolism of arginine (and ornithine, which tends to be excluded from the vesicle by high arginine) become significant.

In yeast, a unique interaction between the degradative and biosynthetic pathways for arginine is found in the inactivation of ornithine transcarbamylase upon binding to arginase (25). The inhibition of ornithine transcarbamylase that occurs upon binding of arginase is strongly dependent upon arginine as an effector. The binding of this effector occurs at a regulatory site for arginine on arginase (26). The inhibition is also dependent upon the sensitivity of ornithine transcarbamylase to substrate inhibition by ornithine itself (27). It is this substrate inhibition that appears to be enhanced by binding to arginase and particularly to arginase in the presence of arginine.

The control of the amounts of the arginine biosynthetic enzymes in *E. coli* is of considerable historical importance in the development of our present-day concepts of enzyme regulation. Not only did the term "repression" itself arise from studies of the effect of exogenous arginine on an arginine biosynthetic enzyme, but the first significant quantitative study clearly recognized as a study of the repression mechanisms was performed on an arginine biosynthetic enzyme (28, 29). Also found very early was a regulatory gene, *argR,* in which mutations led to a derepression and/or at least partial insensitivity of the arginine biosynthetic enzymes to repression by exogenous arginine (30). The arginine biosynthetic enzymes of *E. coli* are of further interest for study, since four or the corresponding structural genes are arranged in a cluster and therefore might constitute an operon, whereas the others are distributed singly on the chromosome. Not surprisingly, these enzymes do not respond coordinately as is a presumed property of enzymes specified by a single operon, but, with some exceptions in certain strains, all of the enzymes respond by repression or derepression in parallel but to different extents. Such an arrangement of genes in which the expression appears to respond to the same physiological signal has been termed a "regulon" (31). The *arg* regulon thus consists in *E. coli* strain K12 of a single cluster, genes *E, C, B,* and *H* and the remaining six genes. (The sixth gene is due to the fact that two different genes, *argF* and *argI,* specify the activity of ornithine transcarbamylase. As a consequence, there are four isozymic forms of the enzyme and the active form is a trimer). All respond to repression by arginine via the *argR* gene. (In *E. coli,* the *arg* genes are designated *argA* through *argH* in order of the specified enzymes in the biosynthetic pathway; *argI* is the exception as noted above). The *arg ECBH* cluster is not a single operon but rather a divergently expressed gene cluster with probably a common operator between genes *E* and *C.* The promoters for the *CBH* operon and for the *E* gene are probably arranged so that the leader regions (that portion of a transcript that precedes the genetic information for the protein product itself) are in part complementary to each other.

The molecular biology of the *arg* regulon has now proceeded to the point where we can be certain that *argR* does indeed specify a repressor molecule and that it acts in accord with the prediction of the Jacob-Monod model. This has been possible in the case of the arginine biosynthetic pathway as it has for several other systems by the relative ease with which it is now possible to develop either plaque-forming or defective specialized transducing phages carrying nearly any bacterial genes for which a selection mechanism exists. Such phages yield DNA that serves as template either for transcription of specific mRNAs or for the coupled transcription and translation that leads to an in vitro synthesis of the enzyme itself. A specific arginine-dependent repression by a partially purified *argR* product of the in vitro synthesis of argininosuccinate (*argH* product; Reaction 15, Figure 1) directed by a φ80d*argECBH* template was shown by Kelker et al (32). That the repression was indeed exerted at the transcriptional level was indicated by the later experiments of Cunin et al (33) using a purified RNA polymerase and the same DNA as template. They observed a repression of transcription of the *argCBH* operon dependent upon both arginine and the repressor protein. It was of further interest in view of a question (raised below) that the repressor fraction employed was free of arginyl-tRNA synthetase activity. The experiments of James and his colleagues (34, 35) have extended the evidence for the effect of repressor on transcription to *argA, argE, argF,* and *argI* in addition to the *argCBH* operon. Thus, while all the transcriptional units of the *arg* regulon have not yet been studied, it does appear that there is control at the transcriptional level that is dependent upon both arginine and the *argR* protein.

There are still some questions whether regulation by repressor and corepressor are the only arginine-specific elements affecting expression of the *arg* genes. (The question of a general metabolic control affecting the arginine as well as other amino acid biosynthetic enzymes is considered later.) Two questions that are frequently raised are whether there is any translational control and whether either arginyl-tRNA synthetase or the charging level of tRNA^{Arg} can effect repression in the *arg* regulon.

For example, Vogel et al (36) have summarized some experiments that might be indicative of a translational control. These findings were based mainly on differences between slow translation of an *arg* mRNA, formed in response to an arginine limitation, such as would occur by supplying not arginine but a slowly used precursor of arginine, and fast translations that resulted from supplying arginine itself. Slow translation resulted in a longer-lived message (and sometimes a larger number of enzyme molecules per message) than did fast translation. The longer message life might be rather trivially related to the slow movement of ribosomes beyond arginine codons and the greater number of available ribosomes when protein synthesis is slightly retarded. Both factors could result in protection of message from nuclease action for a prolonged period. The level of arginyl-tRNA would also be reduced under these conditions and, as noted below, could be involved in the control mechanism. Whether arginyl-tRNA synthetase or arginyl-tRNA is involved at any point in regulation of the *arg* regulon is currently unresolved. It was early established that none of the five isoaccepting species of tRNA^{Arg} changed in amount under repressing or derepressing conditions (37). Furthermore, Hirshfield *et al* (38) showed that mutants with *argS* lesions (giving rise to altered arginyl-tRNA synthetases) that exhibited between 2.5% and 50% of the wild-type activity did not exhibit any derepression of the *arg* regulon. It is perhaps significant that in this study the *argS* mutants were selected from strains in which there was already a physiological derepression before the selection pressure (canavanine, an arginine analog) was imposed. In contrast, Williams (39) described mutants, selected for canavanine resistance in minimal medium, that had reduced levels of arginyl-tRNA synthetase activities and nonrepressible arginine biosynthetic enzymes. Another indication that the synthetase might have a role in enzyme repression was the finding of Faanes & Rogers (40) that canavanine did not mimic arginine in its repression effect unless tRNA^{Arg} was predominantly charged with canavanine.

That there might be two distinct kinds of regulation affecting the *arg* regulon was indicated by an examination of the nature of the transcripts formed in vitro from the *argECBH* gene cluster by Krzyzek & Rogers (41). They observed, as have others before them (42–44), that during a transition from nonrepressing to repressing conditions, enzyme synthesis was reduced to almost a zero rate whereas *arg* mRNA (hybridizable to ϕ80d*argECBH* DNA) was reduced only to the steady state repressed level. This discrepancy persisted for an hour, after which enzyme synthesis increased, reaching the steady state repressed rate in about another 30 min. It was postulated that some unknown component was formed very early under the influence of arginine deprivation and that it was responsible for the complete block in arginine biosynthesis after adding arginine. Thus, the correlation between enzyme synthesis and *arg* mRNA level can be very poor with apparently a great deal of nontranslatable *arg* mRNA present during repressing conditions. Further studies revealed that under repressing conditions, the *arg* mRNA (detected by hybridization to ϕ80d*argECBH* DNA) was smaller (8S) than that formed under nonrepressing or derepressing conditions (14S) (45). It may be that the small nontranslatable transcripts represent leader sequences that have been described and have been more fully characterized as transcripts of the *trp* operon, and which are considered later. Their accumulation during the period of transient repression would then be dependent upon the unknown component described above. Whatever the nature of these short transcripts, however, they may account for the discrepancy between message levels and enzyme levels.

For arginine biosynthesis in yeast, two kinds of control mechanism have been recognized (46). One is an arginine-specific repression that is exerted over five of the eight enzymes examined. The arginine-specific repression is lost by mutations in any one of three loci, *argRI, argRII,* or *argRIII,* but all are thought to be involved in function of the same repressing complex. The second kind of control is a general amino acid control that serves to coordinate the formation of biosynthetic enzymes needed for several pathways, those leading to arginine, lysine, tryptophan, isoleucine and valine, and leucine. Thus, the biosynthetic enzymes in one pathway can be derepressed by limitation of the end product of another pathway. The two controls are independent of each other, and when both controls are removed the derepression is greater than when only one control is removed. That this general amino acid control may involve the amino acyl-tRNA synthetases or the tRNAs themselves is indicated by the finding that by limiting the charging of tRNA^{Ile} by means of a temperature-sensitive isoleucyl-tRNA synthetase, the enzymes of histidine, arginine, and isoleucine were derepressed (47). Recently Messenguy & Cooper (48) have shown that both of these controls affect gene expression at the transcriptional level.

Lysine Biosynthesis in Fungi

In most living forms, lysine is derived from aspartate and pyruvate and is considered a member of the aspartate family; the pathway in fungi is an evolutionary "blind alley" as a pathway from α-ketoglutarate. The early part of the pathway consists of the lengthening of the carbon chain of an α-ketoacid (in this case α-ketoglutarate) by one carbon (to form α-ketoadipate; Reactions 17–20, Figure 1). In so doing, the pattern followed is that which the conversion of α-ketoisovalerate is the central process in leucine cycle (49). (This pattern is encountered again in the pyruvate family, in which the conversion of α-ketoisocaproate is the central process in leucine biosynthesis.) α-Ketoadipate is converted to α-aminoadipate in a transamination reaction (Reaction 21, Figure 1). A carboxyl group activation that differs from those described earlier for the proline and arginine pathways and a reduction yield the semialdehyde (Reactions 22 and 23). The activation of α-aminoadipate is thought to involve a mixed anhydride with an adenylyl group (49). The semialdehyde receives an amino group from glutamate, not in a transamination reaction, but in a covalently linked transfer involving saccharopine as an intermediate (Reaction 24). The oxidative cleavage of saccharopine yields lysine (Reaction 25).

The control of carbon flow over the pathway occurs in part via an inhibition of the condensing enzyme by lysine in both *N. crassa* and yeast (50, 51). Yeast contains two enzymes, which are equally sensitive to lysine (51). The existence of two enzymes may account for the absence of yeast mutants lacking this enzyme. Only one of the two enzymes is appreciably repressed by exogenous lysine. Since α-aminoadipate incorporation into lysine is inhibited by lower concentrations of lysine than are required for significant inhibition of incorporation of acetate into lysine, it has been suggested that a second feedback site may exist after α-aminoadipate (52). However, it may be that the incorporation of aminoadipate was very inefficient and thus more readily antagonized by exogenous lysine. Saccharopine dehydrogenase, however, has been shown to be repressed by lysine (53).

THE SERINE FAMILY

The serine family of amino acids includes serine and the two amino acids derived from it: cysteine and glycine. The pathway serves a far greater metabolic need than the mere synthesis of the amounts of these three amino acids needed for protein. For example, most plants and microorganisms forming methionine from simple carbon sources derive the sulfur of methionine from cysteine. Again, the conversion of serine to glycine provides a significant fraction of the total one-carbon pool needed for purine, thymine, methionine, and histidine biosynthesis; the source of the remaining one-carbon needs of the cell, as is considered below, remains unknown. Finally the intact carbon chain of serine is incorporated directly into tryptophan and phospholipids, and the carbon chain of glycine is incorporated into purines and heme-containing compounds. Thus, the biosynthetic reactions draining 3-phosphoglycerate from the glycolytic pathway constitute a highly branching pathway.

Serine Biosynthesis

The biosynthesis of serine occurs via the phosphorylated pathway shown in Figure 2 (Reactions 1–3) even in organisms such as *Rhodopseudomonas capsulata,* which can convert serine to glycolytic intermediates via a non-phosphorylated pathway (54). An even more striking example is found in *Pseudomonas* AM1, which uses several one-carbon compounds as sole source of carbon and energy in a pathway that involves the formation of glycine, the conversion of glycine to serine, and the nonphosphorylytic formation of glycerate from serine. This organism grows very slowly in the absence of exogenous serine if Reaction 3 (Figure 2) is absent, and is completely dependent upon exogenous serine if the carbon source is other than a one-carbon compound (55).

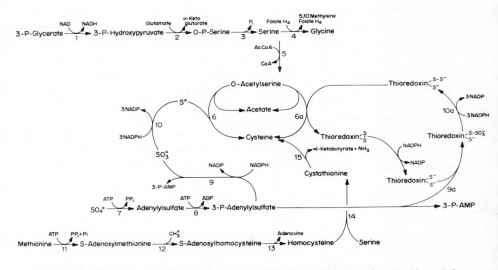

Figure 2   Biosynthesis of the serine family of amino acids. Reactions 9, 10, and 6 constitute the pathway of sulfate reduction traditionally assumed to be used for sulfate assimilation. Reactions 9a, 10a, and 6a constitute the proposed pathway involving bound intermediates. Reactions 11–15 constitute a "salvage" pathway used when methionine is the sole sulfur source or for those organisms lacking Reactions 5 and 6 (or 6a).

The control of carbon flow on the pathway to serine is achieved by the inhibition of 3-phosphoglycerate dehydrogenase by L-serine. The interaction with serine and the mechanism of the reaction catalyzed by the *E. coli* enzyme have been thoroughly studied by Pizer and co-workers (56, 57). The physiological significance of this end-product control was demonstrated by the occurrence of a mutant in which the dehydrogenase was insensitive (58). Overproduction of serine by this mutant rendered it resistant to serine hydroxamate, an analog that prevented growth by inhibiting seryl-tRNA synthetase.

Thus far, there has been no serine-specific repression of the enzymes in the pathway to serine in *E. coli* or *S. typhimurium*. It may well be that only the general, metabolic control (discussed in a later section) exists for this pathway. There is no effect on enzyme levels in *E. coli* strains containing either high or low level of seryl-tRNA synthetase activities (58, 59). Interestingly, the serine biosynthetic enzymes in a human cell line were shown to be repressed by growth of the cells in the presence of serine in the medium (60). On the other hand, the last, not the first, enzyme was inhibited by serine. Whether such an interaction could constitute an effective mechanism of control through product accumulation at the two preceding (reversible) steps remains to be demonstrated.

A hint that there is still another role for the phosphohydroxypyruvate-glutamate transaminase (Reaction 2, Figure 2) comes from the finding that *E. coli* mutants lacking this enzyme require both serine and pyridoxine (61). The basis for the double requirement, which has not been found in other organisms, has not been explained. However, it has been found that glycolaldehyde can be incorporated into pyridoxalphosphate (62). Since the uncharged molecule glycolaldehyde would be an unlikely free intermediate, there may be an "active glycolaldehyde" that is derived from serine phosphate. Alternatively, the transaminase itself may be needed for the formation of active glycolaldehyde.

Glycine Biosynthesis

The pathway from serine to glycine consists of but a single enzyme, serine hydroxymethylase (Reaction 4, Figure 2). Its activity appears not to be regulated by the product, but its formation is regulated by a complex repression mechanism involving, directly or indirectly, several products of one-carbon metabolism. That the products of single-carbon transfers might affect glycine biosynthesis might be expected if glycine itself serves as a supplementary source of one-carbon units (63). This question is discussed below.

There is disagreement over the question whether serine hydroxymethylase is derepressed by limiting methionine. Such a derepression has been observed in *metE* and *metF* mutants but not *metA* mutants of *E. coli* (64,

65), but such results have not been uniformly obtained (66, 67). The difference may be in the mode of limitation; too severe a methionine limitation would prevent the small amount of protein synthesis needed to obtain derepression. Derepression has also been obtained by purine but not thymine limitation with *S. typhimurium* (67). Repression of the enzyme has been obtained by supplementing the medium of *S. typhimurium* and of *E. coli* with the products of one-carbon metabolism (67, 68). The repression pattern was a cumulative one in the *S. typhimurium* experiments (67). This repression does not occur, however, in *metK* mutants (which have an altered S-adenosylmethionine synthetase) or in *metJ* mutants (which have derepressed methionine biosynthetic enzymes) (65, 69). In *E. coli, metK* but not *metJ* mutants themselves exhibit some derepression of serine hydroxymethylase (65, 70). Thus, serine hydroxymethylase formation appears to be controlled by the products of one-carbon metabolism, but whether the various products are acting via a common mechanism is currently not clear.

The other product formed from serine is the one-carbon unit itself in the form of 5,10-methylene tetrahydrofolate, which is, of course, essential for several biosynthetic sequences. An important question is whether other routes of de novo one-carbon unit synthesis occur and to what extent. Inspection of an idealized composition of *E. coli* (71) allows one to estimate that only about 75% of the total one-carbon-unit needs of the cell can be accounted for by formation of glycine needed for protein or for purine biosynthesis. One possibility is that the extra one-carbon unit needs are supplied by a conversion of serine to glycine, which, in turn, is cleaved to a second one-carbon unit and CO_2. Such a route would account for the formation of the one-carbon unit that is needed when serine auxotrophs are provided glycine to satisfy the serine requirement. In accord with this is the report of Newman et al (63) that the loss of the glycine cleavage pathway leads to the inability of a mutant blocked in serine biosynthesis from using glycine as a source of serine. These workers suggest that there is a third pathway for the generation of single-carbon units. The question of a third pathway is also raised by Pizer (72), who found that in the absence of serine hydroxymethylase only about 40% of the thymine methyl group came from the C-2 of glycine. The glycine cleavage pathway has been studied by Meedel & Pizer (70), who found it to be present only in cells grown with exogenous glycine. Furthermore, the cleavage pathway is dependent upon an intact serine hydroxymethylase (73). Thus, the cleavage pathway would account, along with serine hydroxymethylase, for the utilization of glycine by serine/glycine auxotrophs. However, it would not account for the additional one-carbon units that are required over and above those that are generated in the normal course of glycine biosynthesis by prototrophic cells growing on a single carbon source such as glucose. That third source remains to be elucidated.

Cysteine Biosynthesis

The pathway from serine to cysteine consists of an acetyl-CoA-dependent activation of serine to yield O-acetylserine (Reaction 5, Figure 2), which then reacts with sulfide to form cysteine (Reaction 6, Figure 2). Since sulfate is more commonly found in the environment than sulfide, the reduction of sulfate to sulfide is closely related to cysteine biosynthesis, and the enzymes that activate and reduce sulfate are specified by genes in *E. coli* and *S. typhimurium* that constitute the cysteine regulon. Genetic and biochemical correlations with mutants of both organisms have led to the idea that sulfate is reduced via the sequence shown in Reactions 7–10 (Figure 2). Three genes, *cysG, cysI,* and *cysJ,* specify the components needed for the reduction of sulfite. A sulfite reductase has been purified from *E. coli* extracts and has been shown to be a complex hemoflavoprotein of about 7×10^5 daltons with a subunit composition of $\alpha_8\beta_4$ (74). The two kinds of subunit have not yet been related to specific genes. Nevertheless, this enzyme has been thought to function in the six-electron reduction of free sulfite. Sulfite has been assumed to be formed by a phosphoadenosine sulfophosphate (PAPS) reductase, the product of the *cysH* gene. The reduction has been studied only in crude systems so far.

The possibility that assimilatory sulfate reduction may not be so simple has been raised by Tsang & Schiff (75, 76) not only for bacteria but for yeast- and chloroplast-containing organisms as well. They suggest that PAPS is not reduced directly but is transferred by PAPS-sulfotransferase to thioredoxin to yield a thiosulfonate (thioredoxin-S-SO$_3^-$), which is reduced by a thiosulfonate reductase with NADPH to yield thioredoxin-S-S$^-$. It is this covalently bound form of sulfide that serves as substrate for O-acetylserine sulfhydrylase (Reaction 6a, Figure 2) when sulfate is the sulfur source. Since thiosulfonate reductase activity probably resides in the same protein that has been identified and studied as a sulfite reductase, a nitrite reductase, and a hydroxylamine reductase activity, the mutant analysis previously performed with *E. coli* and *S. typhimurium cys* mutants is equally valid for this carrier-bound pathway. According to the Tsang and Schiff model, the reduction of free sulfite (Reaction 10, Figure 2) and incorporation of free sulfide into cysteine (Reaction 6) would thus occur only if they were present in the medium. Only in the presence of dithiols that readily form intramolecular disulfides, such as dithiothreitol, would free sulfite be liberated from PAPS.

While thioredoxin appears to be the protein used in the cell as the carrier for the sulfur undergoing reduction, it is not absolutely essential. Thioredoxin-negative mutants (*tsnC*), obtained by selecting for resistance to phage T7 (77), use the same system for this role that they use for the ribonucleoside diphosphate reductase system: glutathione, glutathione reductase, and glutaredoxin (78, 79).

The system in Chlorella and other green plants differs somewhat in that the sulfotransferase is specific for adenosine-5'-phosphosulfate, and the carrier to which the sulfur is transferred is a much smaller molecule ($\sim$ 1200 daltons) than thioredoxin (76). Furthermore, the thiosulfonate reductase of *Chlorella* is ferredoxin dependent. Whereas in *E. coli* thiosulfonate reductase and sulfite reductase appear to be the same enzyme, in *Chlorella* there is a separate sulfite reductase that functions with exogenous sulfite. The *Chlorella* pattern is found in other chloroplast-containing plants, and the *E. coli* pattern is found in yeast and other non-chloroplast-containing forms (76).

End-product inhibition in the cysteine biosynthetic pathway is of two kinds: 1. that controlling carbon flow from serine is achieved by the sensitivity of serine transacetylase to cysteine (80), 2. that controlling the consumption of energy and reducing power for sulfate reduction is achieved by the sensitivity of sulfate transport to cysteine (81).

Control of enzyme formation occurs by both repression by cysteine (or sulfide) and induction by O-acetylserine and is mediated by the *cysB* gene product in *E. coli* and *S. typhimurium* (82, 83). That the *cysB* product is indeed a protein is indicated by the isolation of *amber* and *ts* mutants (84). The *cysB* product probably contains a single kind of polypeptide chain (the apparent multicistronic nature of the *cysB* gene was probably due to intracistronic complementation). It probably binds either the inducer of the sulfate-utilizing enzymes (including the transport system), O-acetylserine, or the metabolite that represses the entire *cys* regulon (85). Repression thus "overrides" induction. Whether the actual repressing metabolite is sulfide (which would be readily converted to cysteine in the presence of O-acetylserine) or cysteine (which is converted to sulfide by the cysteine desulfhydrase present in most cells) is difficult to decide. Finally, the fact that deletion of the *cysB* gene results in loss of capacity to form all the cysteine biosynthetic enzymes indicates that the *cysB* gene serves as a positive control element, a role dependent upon O-acetylserine and antagonized by cysteine (or sulfide).

The genes of the *cys* regulon in *E. coli* and *S. typhimurium* consist of one cluster of five genes (*C-D-H-I-J*) and four genes (*A, B, E,* and *G*) that are widely separated (86, 87). Polarity studies by Laughlin (88) with the clustered genes of *S. typhimurium* revealed that three of the genes constitute an operon, *cysJIH*, with polarity and therefore transcription occurring from the *J* end of the operon; *cysC* and *cysD* are not part of that operon.

Indirect evidence for end-product control by repression has been obtained for a higher plant system, cultured tobacco cells. ATP sulfurylase is derepressed in media containing poor sulfur sources and is repressed in media containing cysteine (89). Both molybdate and selenite are toxic to tobacco cells when sulfate is the sulfur source and both cause derepression of ATP sulfurylase but apparently by different mechanisms. High concentrations of molybdate (relative to that of sulfate) compete with sulfate and prevent its activation and incorporation. Thus, in the presence of molybdate, it is limiting sulfur that causes a derepression of the enzyme. In contrast, selenite even in lower concentrations than that of sulfate is apparently incorporated in place of sulfur in some component (perhaps cysteine itself). It is thought that this in vivo–produced analog antagonizes the repressing effect of the normal sulfur-containing metabolite.

In at least some of the fungi (e.g. *Neurospora* and yeast), the uptake of sulfate and the conversion to sulfide is under the control of methionine rather than cysteine. That this can occur without interference with cysteine biosynthesis is due in part to the presence in fungi of a "reverse" transsulfuration in which the sulfur of methionine is transferred from methionine to cysteine via a sequence catalyzed by S-adenosylmethionine synthetase, S-adenosylmethionine demethylase, S-adenosylhomocysteine lyase, cystathionine-β-synthase, and γ-cystathionase (Reactions 11–15, Figure 2) (90, 91). Another factor may be the extent to which sulfide is incorporated into homocysteine via the homocysteine synthase reaction (discussed in the section on methionine biosynthesis). The incorporation of sulfide via the homocysteine synthase reaction has been proposed as the major if not the sole route for methionine biosynthesis in *S. cerevisiae* (92). The reaction is analogous to the acetyl serine sulfhydrylase reaction except that acetyl-homoserine is the substrate and homocysteine is the product. Indeed, it has been reported that the same protein in yeast serves as a sulfhydrylase for both O-acetylserine and O-acetylhomoserine (93).

In yeast, those enzymes of the sulfate assimilation pathway that have been examined are strongly derepressed when methionine is limiting and repressed when methionine is in excess (94, 95). The repression of these enzymes can be achieved by adding either S-adenosylmethionine or methionine to the medium (96). Repression by methionine involves a functional methionyl-tRNA synthetase, whereas that by S-adenosylmethionine does not. The generation of the methionine repression signal is considered in more detail later.

Neurospora and *Aspergillus nidulans* also contain a homocysteine synthase, but in both the enzyme probably plays a minor role in sulfur incorporation (90, 91). In *A. nidulans*, homocysteine provides a mechanism for incorporating sulfur into methionine (and hence cysteine) when cysteine sulfhydrylase is missing.

THE ASPARTATE FAMILY

The aspartate family consists of aspartate, asparagine, methionine, threonine, lysine (except in the fungi), and isoleucine. The carbon skeletons of lysine and isoleucine are derived in part from pyruvate but are still considered part of the aspartate family. Isoleucine biosynthesis is best considered along with the pyruvate family of amino acids since four of its biosynthetic enzymes are also needed in the valine pathway. Diaminopimelate, a lysine precursor, is needed for cell wall synthesis in bacteria but not for protein synthesis. It can therefore be considered another member of the aspartate family. As noted earlier, the methyl carbon of methionine is derived to a major extent from the β-carbon of serine and in most organisms its sulfur is from cysteine.

Since the aspartate family of amino acids is formed via a highly branched pathway, there are numerous branch points that are potential sites of regulation of carbon flow. Several patterns of control over this branching pathway have evolved in various organisms.

Aspartate Biosynthesis

The obvious way for aspartate biosynthesis to occur is via a glutamate-aspartate transaminase, which catalyzes a reaction that has long been known (Reaction 1, Figure 3). It has only recently become known, however, which enzyme did in fact perform this reaction in E. coli, which is essentially the only organism that has been studied for its variety of transaminases. It is now possible, as a result of studies in several laboratories, to give a reasonable accounting of the major transaminases in E. coli. As shall be seen, however, it is unfortunate that enzyme separations, substrate specificity studies, and genetic analyses have not been done completely in one laboratory.

These studies all must be related to the early finding of Rudman & Meister (97) of three transaminases that were designated A, B, and C. The aspartate-glutamate transaminase activity was in fraction A, which has more recently been shown to be heterogeneous, by both physical and genetic approaches. Chesne and Pelmont and their coworkers (98–101) have separated the transaminases of E. coli K12 on DEAE Sephadex columns and shown that aspartate-glutamate activity appears in two components. One component, with by far the bulk of the total activity, is attributed to transaminase A. Transaminase A also had activity for phenylalanine, tyrosine, and tryptophan but not leucine as substrate, but the affinity for at least phenylalanine among the aromatic amino acid substrates was low. A second fraction with glutamate–aromatic amino acid transaminase activity was found. A small amount of aspartate–glutamate transaminase activity was also found in this fraction but apparently was not considered significant. Chesne et al (100) attributed the activities in this fraction to two proteins: a heat-stable protein specific for phenylalanine and not repressible by tyrosine, and a heat-sensitive protein specific for phenylanine, tyrosine, and leucine and repressible by tyrosine.

These workers (100) chose to reserve the original term "transaminase A" for what is undoubtedly the product of the *aspC* gene studied by Gelfand & Rudo (102). Unfortunately, the latter workers have more recently called the tyrosine-repressible enzyme, transaminase A, the product of the *tyrB* gene which they have mapped. That the *tyrB* gene product (which would perhaps best be called "tyrosine-glutamate transaminase") may be more significant as an aspartate-glutamate transaminase than the activity profiles of Chesne et al (100) show is indicated by the observation by Gelfand & Steinberg (103) that aspartate is formed (perhaps slowly since the colony size was small) in $aspC^-$, $tyrB^+$ cells. In other words, the tyrosine-glutamate transaminase appears to have a minor role in aspartate formation, but transaminase A (the *aspC* gene product) plays the major role.

No evidence for a regulatory site for any effector molecules that would modulate the activity of proteins carrying glutamate-aspartate transaminase activity has been described, as one might deduce from the freely

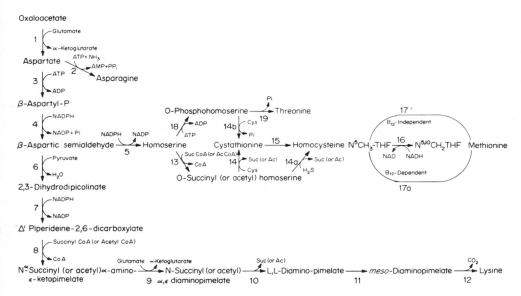

Figure 3 Biosynthesis of the aspartate family of amino acids. The steps leading from threonine to isoleucine are shown in Figure 4 along with the steps leading to valine.

reversible nature of the reaction. Evidence that aspartate in the growth medium represses bulk glutamate–aspartate transaminase activity has been reported, however (104).

Asparagine Biosynthesis

The formation of asparagine (Reaction 2, Figure 3) is a reaction that requires more energy than the formation of glutamine, since two equivalents of $\sim P$ (or high energy phosphate) are consumed. In *E. coli* and *Lactobacillus arabinosus,* the reaction is inhibited by asparagine itself. Whether this inhibition occurs by tight binding to the product site or by binding to a special regulatory site is not known (105, 106). The enzyme is repressed when asparagine is included in the medium.

The isolation of Chinese hamster ovary cell mutants containing a temperature-sensitive asparaginyl-tRNA synthase (107) has made it possible to examine the effect of limited function of asparaginyl-tRNA synthetase on asparagine synthetase formation in a mammalian cell. Arfin et al (108) found that, indeed, there was a derepression of asparagine synthetase at near the critical temperature. This is the first example of what appears to be regulation of a mammalian amino acid biosynthetic enzyme by either an amino acyl-tRNA synthetase or of the charging level of the corresponding tRNA. The same arguments can be raised in this case that were raised in earlier years when such experiments were performed with bacteria. Is it, in fact, the charging level that is important or is it that there is an altered asparaginyl-tRNA synthetase? And, finally, does the control occur at the transcriptional or the translational level?

The Common Pathway in the Formation of the Aspartate Family

The common pathway for the aspartate family of amino acids consists of three reactions that convert aspartate to the two branch point intermediates, β-aspartyl semialdehyde and homoserine. Since the regulation of these enzymes involves all four of the end products that are derived from the common pathway, they are considered together. The enzymes in *E. coli* (and *S. typhimurium*) are considered first, since they have been studied most extensively. As is shown, the *E. coli* pattern is not a universal one, however.

In *E. coli,* the formation of β-aspartyl phosphate (Reaction 3, Figure 3) is catalyzed by three different aspartokinases, which have specifically evolved to fulfil the function of threonine (and isoleucine) formation, methionine formation, and lysine formation, respectively (109). Although each aspartokinase would appear to be adapted for synthesis of β-aspartyl phosphate in an amount required for the specific end product, there appears to be no channelling of product and function in such a way. Rather, there appears to be a common pool of β-aspartyl phosphate from which all products are derived, as is demonstrated by the fact that the predominant aspartokinase varies from one strain to another (110). Furthermore, the loss of one or even two of the three aspartokinase activities does not lead to auxotrophy, i.e. a deficiency in activity by any one aspartokinase is compensated for by derepression of one or both of the remaining isozymes.

Aspartokinase I activity is associated with a second activity, homoserine dehydrogenase I (Reaction 5, Figure 3), carried on a bifunctional enzyme, which is a tetramer containing a single kind of subunit. The enzyme has been thoroughly studied physically, genetically, and physiologically in several laboratories (111–114). Some of the chemical and immunological properties of the enzyme have been recently reviewed (115).

Both activities are inhibited by threonine, but to different extents. The inhibition of aspartokinase activity is competitive with both substrates, whereas that of homoserine dehydrogenase activity is not. It is interesting that of the eight binding sites for threonine [two per subunit (116), four of these appear to be at the aspartokinase active site (114)]. When the aspartokinase activity in inactivated by titration of –SH groups, the homoserine dehydrogenase is desensitized to threonine in a parallel way (116). In contrast, when homoserine dehydrogenase is inactivated by alkylating the active site by a substrate analog, 2-amino-4-keto-5-chloropentanoate, neither aspartokinase activity nor its sensitivity to threonine is affected (117).

Although aspartokinase I–homoserine dehydrogenase I contains one kind of polypeptide chain, the chain contains two distinct functional regions (118). That carried by the amino-terminal portion, exemplified by the fragment formed by a nonsense mutant, yields a threonine-sensitive aspartokinase devoid of homoserine dehydrogenase activity. Like the native enzyme, this mutant aspartokinase exists as a tetramer. That carried by the carboxyl-terminal portion, obtained by limited proteolysis, exhibits only homoserine dehydrogenase activity. Since the aspartokinase fragment has been removed, it is insensitive to threonine; it exists as a dimer. From these and other studies, Veron et al (116) conclude that the native enzyme consists of an isologous tetrameric core of aspartokinase regions surrounded by the homoserine-dehydrogenase regions.

Aspartokinase I–homoserine dehydrogenase I is specified by the *thrA* gene in *E. coli* and *S. typhimurium* (119, 120). Like other products of the *thr* operon, it is multivalently represssed by threonine plus isoleucine (121).

In *E. coli* K12 and presumably *S. typhimurium,* aspartokinase II is also associated with a homoserine dehydrogenase activity in a bifunctional protein (122, 123). Neither activity is inhibited by methionine or any other aspartate family amino acid. It is, however, repressed by methionine. Aspartokinase II–homoserine dehydrogenase II is a dimer of identical subunits. The corresponding enzyme in *E. coli* B exhibits no homoserine dehydrogenase activity (109).

The third aspartokinase in the Enterobacteriaceae appears not to be associated with any additional activities. It was originally found to be both repressed and inhibited by lysine (124). It became apparent later that several other amino acids not directly related to lysine biosynthesis were also weakly inhibitory and in combination with lysine exerted a synergistic inhibitory effect (125). Amino acids exhibiting the synergistic effect were phenylalanine, leucine, isoleucine, valine, and methionine. An interesting consequence of the synergistic pattern of inhibition is that the addition of these "nonspecific" amino acids led to a derepression of aspartokinase III and aspartic semialdehyde dehydrogenase (126). (As shown below, the latter enzyme is most strongly derepressed by limiting lysine.) The derepression was presumably due to the synergistic effect of these amino acids and lysine on aspartokinase III, which led to a reduced lysine pool.

Aspartokinase III of *E. coli* is a dimer of identical polypeptide chains (127). Each subunit contains two nonequivalent lysine binding sites, one of low affinity, the other of high affinity (128). Binding of lysine alone is a cooperative process, but in the presence of one of the hydrophobic amino acids that act synergistically with lysine, the cooperativity disappears. The binding of lysine has been reported to promote association of the dimers into a tetrameric state (129). The binding of two of the hydrophobic amino acids, phenylalanine and leucine, has been shown to be mutually exclusive (130).

Two other properties of aspartokinase III have been reported, but their significance is unknown. One is an in vivo adenylylation reaction in which adenylate groups are covalently linked to the enzyme in a phosphodiester bond in a ratio as high as one AMP per subunit which occurs during the late stationary phase of growth (131). It is of interest that the activity of the enzyme decreases rapidly during this period (132). Very little in vitro adenylylation could be demonstrated, so it has not been possible to demonstrate a cause and effect between adenylylation and the activity loss observed in late log phase cells.

The second property is one that the lysine-sensitive aspartokinase has been shown to share with analogous enzymes in proline and arginine biosynthesis, glutamokinase and N-acetylglutamokinase, respectively. All three are susceptible to inhibition by hexose monophosphates and by AMP (133, 134). The significance of this property to proline, arginine, or lysine biosynthesis remains unclear.

The pattern of multiple enzymes found in the Enterobacteriaceae for the multifunctional step, β-aspartyl phosphate formation, is not the only way the reaction is controlled. In *B. polymyxa* (135, 136), *R. capsulata* (137), and the pseudomonads (110), there are only single aspartokinases, and their regulation is achieved by either a multivalent or a synergistic inhibition involving both lysine and threonine. In *B. polymyxa,* detailed studies revealed that in the absence of threonine, lysine alone would inhibit but only at a concentration nearly 100-fold higher than that needed with threonine (136). Similarly, in the absence of lysine, threonine was inhibitory, but a 200-fold greater concentration was required than was needed in the presence of lysine. As a consequence of the control of aspartokinase by lysine and threonine, the growth of both *R. capsulatus* and *B. polymyxa* is inhibited by a combination of lysine and threonine, and the inhibition is reversed by methionine (136, 138).

The *B. polymyxa* enzyme is composed of two subunits of about 43,000 daltons and two of about 17,000 (139). Denaturation with urea allowed separation of the subunits. Upon the removal of urea, the larger (α) subunit could be renatured to a variety of oligomeric forms that exhibited a lysine- and threonine-sensitive aspartokinase activity. However, when α subunits were renatured together with the smaller (β) subunits, a conformation resulted that was catalytically more active than that which resulted in the absence of β subunits. The sensitivity to synergistic end product inhibition was the same, however. Cross-linking studies revealed a β-α_2-β structure for the native enzyme.

In the bacilli, aspartokinase and several of the lysine biosynthetic enzymes play another role, the formation of dipicolinate, a specific spore constitutent. At the end of exponential growth, considerable aspartokinase activity is called upon for sporulation. In *B. subtilis,* there are two aspartokinases. The major one, aspartokinase I, is inhibited by diaminopimelate (140). The other enzyme, aspartokinase II, is inhibited by lysine and threonine in concert. In a derivative in which the second enzyme is derepressed, the enzyme was found to be sensitive only to lysine (141). This enzyme was found to be a tetramer with two kinds of subunits of 43,000 and 17,000 daltons. Some evidence indicated that the smaller subunit was derived from the first (142). In *Bacillus licheniformis,* the single aspartokinase is inhibited by aspartic semialdehyde, to a considerable extent by lysine alone, and to a greater extent concertedly by lysine plus threonine (143). Growth in the presence of lysine rapidly desensitizes the enzyme to lysine alone but does not affect its sensitivity to lysine plus threonine. The effect of lysine is presumably due to a rapid modification of the enzyme. On the other hand, growth with threonine enhances the sensitivity of the enzyme to inhibition by lysine alone but does not affect the sensitivity to concerted inhibition. Growth with both lysine and threonine leads to a partial repression. The activity is rapidly reduced in post log phase so that it is not possible to account for any special provision made for dipicolinate synthesis during sporulation. In contrast, the aspartokinase of *B. polymyxa* exhibits its maximal activity a few hours after the cells entered the stationary phase (144).

In those organisms in which a single aspartokinase is found, homoserine dehydrogenase activity is found on a separate protein. In general, homoserine dehydrogenase is inhibited by threonine but to various extents, so that there may be among these organisms some that contain two homoserine dehydrogenases. The taxonomic significance of the aspartokinase-homoserine dehydrogenase distribution has been discussed (110).

In yeast, the single aspartokinase is bivalently repressed by threonine plus methionine, and the single homoserine dehydrogenase is repressed only by methionine (94). In *N. crassa,* homoserine dehydrogenase is neither repressed nor inhibited by either methionine or threonine (145).

The regulation of a vascular plant homoserine dehydrogenase, that in maize, has been studied (146). The enzyme is inhibited by threonine but there is a progressive desensitization to threonine during seedling growth. Thus, homoserine dehydrogenase activity in root tips was more sensitive to threonine than that in mature cells of the root.

The remaining enzyme in the common pathway, β-aspartic semialdehyde dehydrogenase, catalyzes the NADPH-dependent reduction of β-aspartyl phosphate (Reaction 4, Figure 3). In *E. coli,* the enzyme was originally thought to be repressible by lysine alone. It now appears that the enzyme, of which there is only one kind in *E. coli,* is repressed by lysine, threonine, and methionine. The derepression obtained upon limitation of either threonine or methionine, however, is much less than that obtained upon limitation of lysine (147). The *E. coli* enzyme has been studied by Holland & Westhead (148). The enzyme in yeast has been shown to be repressible by methionine (149).

Mention should be made here of the fact that both homoserine dehydrogenase and β-aspartic semialdehyde dehydrogenase of several organisms have been shown to use either NADPH or NADH as the hydrogen donor in the biosynthetic direction. Figure 3 implies that the actual donor is NADPH, in accord with the concept that biosynthetic oxidations use NAD (or a component tightly coupled to the electron transport chain) as hydrogen acceptor, and that biosynthetic reductions employ NADPH as hydrogen donor. Support for this concept has recently been obtained by Anderson & von Meyenburg (150), who demonstrated the [NADPH]/[NADP] ratio to be much higher in *E. coli* than the [NADH]/[NAD] ratio. Thus, the tendency for the terminal electron transport system to serve as a sink to reoxidize NADH provides a channeling of reducing power into two distinct routes, one for biosynthesis and one for energy production.

Lysine Biosynthesis

With the exception of the fungi in which lysine is formed by the pathway described earlier (Figure 2), the biosynthesis of lysine occurs as a branch of the aspartate family of amino acids. The condensation of aspartic semialdehyde with pyruvate (Reaction 6, Figure 3) yields a compound that spontaneously cyclizes and is then reduced in an NADPH-linked reaction (Reaction 7, Figure 3). Ring opening for subsequent reactions is achieved by "trapping" the open chain form by succinylation as in *E. coli* (151) or by acetylation as in bacilli (152) (Reaction 8, Figure 3). After a transamination reaction (Reaction 9), the acyl group is removed to yield L,L-diaminopimelate (Reaction 10). Either this form or the *meso* form obtained by a specific racemase (Reaction 11) or both [depending upon the organism (153)] is used as a constituent of bacterial cell wall synthesis. The *meso* form is the substrate for the decarboxylase that catalyzes the final step in lysine synthesis (153) (Reaction 12).

Control of metabolite flow over the pathway from β-aspartic semialdehyde to lysine is achieved by the inhibition of the condensing enzyme by lysine (154). The case is special in bacilli since 3,6-dihydropicolinate is not only a precursor of lysine but also of dipicolinate, an important component in spores. In *B. subtilis* the enzyme is neither repressible nor inhibitable by lysine at the time of sporulation (155). In *B. cereus,* the enzyme undergoes a desensitization to lysine at the onset of dipicolinate formation (156). In *B. subtilis,* the final enzyme of the pathway, diaminopimelate decarboxylase, is inhibited by lysine (157).

The mode of regulation of formation of the lysine biosynthetic enzymes in *E. coli* has not been discovered. Mutants of *E. coli* resistant to thiosine (a lysine analog) have been isolated, and they exhibit very low activity of lysyl-tRNA synthetase (158). Such mutants would have low levels of lysyl-tRNA but do not show derepression of three lysine biosynthetic enzymes, aspartokinase III, aspartic semialdehyde dehydrogenase, and diaminopimelate decarboxylase, but there was a partial derepression of the dihydropicolinate reductase (159). An indication that lysyl-tRNA may nevertheless be involved in regulation of the lysine regulon has been obtained from examination of *hisW* and *hisT* mutants (E. Boy, F. Borne, and J. C. Patte, manuscript in preparation). These loci are involved in the modification of several tRNAs, not merely that for histidine, as is described later. A *hisW* lesion results in a slight derepression of the decarboxylase, and a *hisT* lesion leads to nonrepressibility of this enzyme. At this time it is not possible to provide a unifying interpretation encompassing these apparently contradictory observations.

Evidence that the first enzyme (aspartokinase III) may also have an effect on regulation came from the finding that the decarboxylase (but not aspartic semialdehyde dehydrogenase) is nonrepressible in the presence of an aspartokinase III that is both insensitive to lysine and derepressed owing to a *cis*-dominant mutation (O^c phenotype) (160). Evidence pointing toward effects of initial enzymes on repression patterns is cited in several later sections.

The *lys* regulon in *E. coli,* unlike the *arg* regulon, does not appear to have a common regulatory element(s) that affects the entire regulon in the same way. As discussed below, the existence of repressor proteins found for arginine as well as tryptophan synthesis in *E. coli,* so important in developing our early concepts of regulation, may be the exception rather than the rule in biosynthetic systems.

Methionine Biosynthesis

Homoserine, the terminal product in the common pathway to the aspartate family of amino acids, is the usual branch point compound for the methionine and threonine biosynthetic pathways. The interested reader should consult the comprehensive review of Flavin (91) for details about methionine biosynthesis that cannot be reviewed here. The review is particularly useful because it tells how many of our current concepts of the methionine biosynthetic pathway were discovered and some pitfalls that were encountered. The pathway shown in Figure 3 indicates that homoserine is activated by an O-acylation reaction, and in the major pathway in bacterial and in most organisms the O-acylhomoserine undergoes transsulfuration with cysteine to yield cystathionine (Reactions 13 and 14). Cystathionine is cleaved to yield homocysteine (Reaction 15), which is the acceptor for one or both of two methylating systems (Reactions 16, and 17 or 17a).

In *E. coli* and *S. typhimurium,* the acyl group activating homoserine is the succinyl group. In contrast, in fungi and the nonenteric bacteria, the acyl group is acetyl, and in higher plants the oxalyl, acetyl, or phosphoryl groups are variously used to activate homoserine (91). The utilization of a phosphoryl group as the activating group in spinach (91) and other higher plants is of additional interest, since O-phosphohomoserine is also an obligatory intermediate in threonine biosynthesis. Thus, rather than homoserine serving as the branch point intermediate, O-phosphohomoserine does under such circumstances. The requirement for succinyl-CoA in both lysine (Reaction 8, Figure 3) and methionine biosynthesis accounts for the replacement of the succinate requirement in Enterobacteriaceae mutants that lack one of the three components of the α-ketoglutarate dehydrogenase complex (161). The next intermediate in the major route, cystathionine, was not readily established as an obligatory intermediate. *S. typhimurium metB* mutants are methionine auxotrophs in which cystathionine γ-synthase is the missing enzyme. This enzyme exhibits only a limited specificity and catalyzes not only Reaction 14 but also Reaction 14a, yielding homocysteine directly (162). That Reaction 14a could not be of major importance was indicated by the methionine (or homocysteine) requirement of *metC* mutants, which lacked β-cystathionase. However, at least some of these mutants are leaky, indicating that Reaction 14a might play a minor role and allow a limited bypass of cystathionine (162).

At the methyltransferase step, there was also a problem in establishing the actual biosynthetic route. In *E. coli* and many other bacteria, genes specifying the enzymes for both reactions 17 and 17a are found. However, many bacteria are unable to make the cofactor for Reaction 17a (cobalamine or vitamin B_{12}), and must depend upon the B_{12}-independent pathway. In the absence of the latter pathway, such strains respond to either methionine or B_{12}. In those microorganisms that can form B_{12}, the B_{12}-dependent enzyme is used, usually exclusively (163).

In yeast, as indicated earlier (91, 92), Reaction 14a is probably the major mechanism for homocysteine formation. The enzyme, acetylhomoserine sulfhydrylase, is missing in yeast *met8* mutants (92). However, extracts of yeast do exhibit a weak cystathionine γ-synthase, which is also missing in *met8* mutants (164). Thus, in view of the broad specificity of the enteric enzyme, it may be that the homologous protein varies from organism to organism in the extent to which it serves as a cystathionine γ-synthase or as an acylhomoserine sulfhydrylase. In the specific case of yeast, this enzyme also appears to be the only protein with an acetylserine sulfhydrylase activity (Reaction 6, Figure 2) (93). Since this activity is less efficient than the acetylhomoserine sulfhydrylase activity, it may be that cysteine in yeast derives its sulfur both directly and via methionine through the "salvage" route (Reactions 11–14, Figure 2).

Extracts of *N. crassa* also exhibit both acetylhomoserine sulfhydrylase activity and cystathionine γ-synthase activities (165). The two activities, however, reside in different proteins. The γ-synthase is missing in *met-3* and *met-7* mutants, whereas the sulfhydrylase is not. The sulfhydrylase may play at most a minor role in methionine biosynthesis in *N. crassa,* as is indicated by the fact that *met-3* and *met-7* mutants are leaky.

In *Chlorella* and a number of the green plants, Datko et al (166) have shown that O-phosphohomoserine is converted to homocysteine directly via a sulfhydrylase reaction. This compound is also converted to homocysteine via the transsulfuration pathway as shown earlier (167). Thus, the common aspartate family pathway for green plants terminates not with homoserine but with phosphohomoserine, which is the branch point compound that can be converted to threonine or methionine.

The control of carbon flow over the pathway to methionine is achieved in bacteria by a synergistic inhibition exerted by S-adenosylmethionine and methionine (168, 169). One form of α-methylmethionine-resistant mutant has been shown to have a transsuccinylase resistant to methionine inhibition but still sensitive to S-adenosylmethionine inhibition (170). With this enzyme, the synergistic pattern of inhibition was lost. In yeast, only S-adenosylmethionine appears to be involved as an inhibitor of the corresponding transacetylase (171).

In *N. crassa,* the transacetylase is not subject to end-product control, but the second enzyme in the pathway is inhibited by S-adenosylmethionine alone (165). An obligatory positive effector for cystathionine γ-synthase is the polyglutamyl derivative of N^5-methyltetrahydrofolate, which also antagonizes the inhibitory effect of S-adenosylmethionine (172). Thus, the possibility has been considered that acetylhomoserine may have a metabolic function in addition to methionine biosynthesis.

All of the methionine biosynthetic enzymes are derepressed, although not coordinately, by growth of *E. coli* or *S. typhimurium* with limiting methionine. Derepressed mutants are readily obtained by selecting for resistance to any of several methionine analogs. Two classes of mutants have been described: those with lesions in the *metJ* gene, which may specify an aporepressor, and those with lesions in the *metK* gene, which is presumably the structural gene for S-adenosylmethionine synthetase (173–176). Because of the essential role of S-adenosylmethionine in so many biosynthetic reactions, one would expect the *metK* lesion would result only in reduced S-adenosylmethionine synthetase rather than its complete absence. Such mutants have been described for *E. coli* (174). Another class of *metK* mutants has been isolated as ethionine-resistant mutants (176). The methionine biosynthetic enzymes in this class are only slightly derepressed (in contrast to the other class), but they are subject to normal control by repression. In *S. typhimurium,* one class of *metK* mutants has no measurable S-adenosylmethionine synthetase and has constitutive levels of methionine biosynthetic enzymes (177). Another class showed normal S-adenosylmethionine synthetase activity but also had constitutive levels of methionine biosynthetic enzymes (177, 178). In still another class, the synthetase resembled that in the *E. coli* class with high K_m for methionine yet showed normal regulation of biosynthetic enzymes (177). Such observations have been interpreted as evidence for and against the possibility that S-adenosylmethionine rather than methionine is the corepressor for the *met* regulon. Another possibility is that the important regulatory element is S-adenosylmethionine synthetase itself and that when a catalytic deficiency is exhibited by *metK* mutants, it is incidental to the modification of its regulatory properties (179). The molecule would, of course, have binding sites for both methionine and S-adenosylmethionine. It may be that in the strains in which no S-adenosylmethionine synthetase is detectable, the enzyme is merely unstable. Another possibility, raised by Hobson (180), is that the enzyme is indeed missing, and that another route to S-adenosylmethionine exists.

Two methionine biosynthetic enzymes in *E. coli* are subject to a second repression signal, in addition to the *metJ*- and *metK*-dependent repression by methionine (181). The second repression signal is dependent on exogenous vitamin B_{12} and the *metH* gene (specifying the B_{12}-dependent transmethylase) and affects $N^{5,10}$-methylenetetrahydrofolate reductase and the non-B_{12}-dependent transmethylase.

In yeast, two different forms of repression are apparently mediated by methionine and by S-adenosylmethionine (96). That mediated by methionine is dependent upon a good methionyl-tRNA synthetase activity. Some evidence has been obtained by using lomofungin, a transcriptional inhibitor, to demonstrate that the methionine-mediated repression acts at the transcriptional level, whereas the S-adenosylmethionine-mediated repression acts at the translational level (182). As noted earlier when sulfur assimilation was considered, the repression by methionine and by S-adenosylmethionine affects the enzymes of sulfur assimilation as well (Reactions 7–10, Figure 2) (94, 95). Three genetic loci, *eth2, eth3,* and *eth10,* have been found to reduce the methionine-mediated repression but not that mediated by S-adenosylmethionine (95).

Threonine Biosynthesis

The remaining branch of the aspartate family involves an isomerization of homoserine, which is achieved by esterification of the hydroxyl group by phosphate (Reaction 18, Figure 3) and the removal of the phosphate group in a pyridoxalphosphate-linked β-γ elimination reaction to yield threonine (Reaction 19, Figure 3).

The final control of carbon flow into threonine is achieved by inhibition of homoserine kinase by threonine. The interaction has recently been studied by Burr et al (183), who showed that the inhibition by threonine was competitive with the substrate, homoserine. These workers have pointed out that since homoserine is present in an almost undetectable intracellular concentration, this simple kind of interaction provides an adequate regulatory loop. When the substrate concentration is relatively high, the evolution of a specific regulatory site might be expected to have occurred.

In *E. coli* and *S. typhimurium,* these two enzymes, along with aspartokinase I–homoserine dehydrogenase I, are specified by an operon of three structural genes that are under the control of a single promoter-operator locus (119, 120). The operon is bivalently repressed by threonine plus isoleucine (121). That the repression involves in some way the threonyl- and isoleucyl-tRNAs or their cognate synthetases or both is indicated by the isolation of mutants derepressed for the *thr* operon in which the threonyl- and isoleucyl-tRNA synthetases had low affinities for the amino acid substrate (184, 185). Additional evidence was provided by the finding of Nass et al (186) that the antibiotic, Borrelidin, was a specific inhibitor of threonyl-tRNA synthase. In its presence, the charging level of tRNA^{Thr} was low and there was depression of the *thr* operon. *cis*-Dominant promoter-operator mutations have been described and indicate that the *thrA* gene is nearest the operator (187, 188).

In so many studies on the regulation of amino acid biosynthesis, sites of end-product inhibition or evidence of repression or derepression are described in which it is not at all certain that the interactions observed do, in fact, play a significant role in vivo. For end-product-sensitive enzymes, a safe inference can usually be made if a mutant with an enzyme lacking the putative regulatory site is shown to overproduce the end product. Such mutations cannot always be selected, so that we are often left with some uncertainty. What is needed is some way to reconstruct all the essential components of a system and study their interactions as nearly as possible in the way they occur in vivo.

Such an attempt has recently been made by Szczesiul & Wampler (189) with the entire six enzymatic steps leading from aspartate to threonine. The enzymes from *E. coli* were separated from each other and at least partially purified. Very high enzyme concentrations and low substrate concentrations were used to simulate conditions found in the living cell. NADPH utilization was followed and the levels of intermediates were determined. The results clearly showed that the most important factor in controlling the metabolic flux was the concentration of threonine and its interaction with the aspartokinase. When the aspartokinase was insensitive to threonine, higher concentrations of threonine were reached, and the rate of NADPH utilization remained higher. Since neither aspartic semialdehyde nor homoserine accumulated under these conditions, it appeared that the other feedback sites (homoserine dehydrogenase and homoserine kinase, respectively) were less important than the aspartokinase feedback site. There was also little effect of either energy charge (190) or anabolic reducing charge (150) on metabolic flux in the systems studied. While such a system is still simple and is probably isolated from other important factors in the cell, it does demonstrate the feasibility of undertaking these important experiments. Similar attempts to study metabolic flux in the pathways to the aromatic family of amino acids have also been made and are considered later.

As mentioned in the section on methionine biosynthesis, in certain higher plants the branch point compound for methionine and threonine biosynthesis is not homoserine but O-phosphohomoserine. Interestingly, in sugar beet leaves the single threonine-specific enzyme, threonine synthase, is regulated not by threonine but by the need for the transsulfuration reaction that phosphohomoserine also undergoes (191). Threonine synthase is inhibited by cysteine (the substrate for transsulfuration) and stimulated by S-adenosyl-methionine (a product derived via transsulfuration).

Isoleucine Biosynthesis

While isoleucine is usually considered as a member of the aspartate family of amino acids, its biosynthesis is best considered along with that of the pyruvate family. Furthermore, four of the five enzymes needed for its biosynthesis are also needed for valine biosynthesis.

THE PYRUVATE FAMILY

The pyruvate family consists of the three amino acids that derive the major portion of their carbon from pyruvate: alanine, valine, and leucine. As already indicated, isoleucine, which is a member of the aspartate family but which obtains two of its six carbons from pyruvate, is most conveniently considered along with valine, since the biosynthesis of both involves primarily a common set of enzymes. Lysine, considered also a member of the aspartate family, can actually derive half of its carbon from pyruvate, depending upon which carboxyl group of *meso*-diaminopimelate is lost in the final step of lysine biosynthesis. Thus, on the average, two and one half carbons are derived from pyruvate.

Isoleucine, Valine, and Leucine Biosynthesis

The biosynthetic pathways of the three branched-chain amino acids are best considered together, not so much because of their common aliphatic character, but because their pathways are so interrelated. Similarly, the regulation of all three biosynthetic pathways is also best considered at the same time, since the regulation of one pathway is difficult to discuss in isolation. Indeed, some common elements appear to be involved in their regulation.

The biosyntheses of the five-carbon branched-chain amino acid, valine, and that of one of the six-carbon branched-chain amino acids, isoleucine, occur by a parallel set of reactions. The first step in valine biosynthesis is the condensation of pyruvate with an active acetaldehyde derived from a second molecule of pyruvate to yield α-acetolactate (Reaction 2a, Figure 4). An analogous reaction for isoleucine biosynthesis uses, rather than pyruvate as the acetaldehyde acceptor, the next higher homolog, α-ketobutyrate, yielding α-acetohydroxybutyrate (Reaction 2). With but a few exceptions, α-ketobutyrate itself is derived from the deamination of threonine (Reaction 1). The two acetohydroxy acids undergo an NADPH-dependent reduction and alkyl group migration to yield the dihydroxy acid precursors and carbon chains corresponding to those in valine and isoleucine (Reactions 3 and 3a). The α,β-dihydroxy acids undergo dehydration reactions to yield the corresponding α-keto acids (Reactions 4 and 4a), which in turn undergo transamination reactions with glutamate as the amino donor (Reactions 5 and 5a). The second transaminase for valine biosynthesis is discussed later.

The α-keto acid precursor of valine is itself a branch point compound, since it is also the substrate for the first enzyme in the leucine biosynthetic pathway (Reaction 6). The leucine pathway uses the same chain-lengthening pattern that was encountered earlier in the fungal lysine biosynthetic pathway, which is also the way oxaloacetate is converted to α-ketoglutarate via the tricarboxylic acid cycle. Thus, isopropylmalate isomerase (Reaction 7) is analogous to aconitase, and β-isopropylmalate dehydrogenase (Reac-

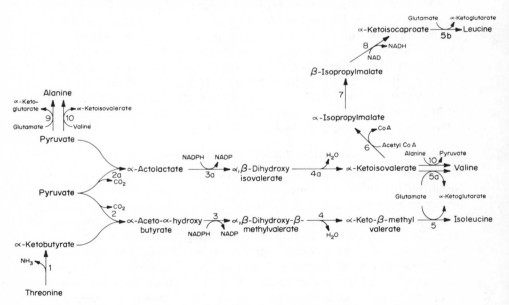

Figure 4 Biosynthesis of the pyruvate family of amino acids. The enzymes catalyzing Reactions 2, 3, and 4 also catalyze Reactions 2a, 3a, and 4a, respectively. Reactions 5, 5a, and 5b are catalyzed by the same enzyme. In addition, a second enzyme can catalyze Reaction 5b.

tion 8) is analogous to isocitrate dehydrogenase. The α-keto acid so formed is then converted to leucine by transamination (Reaction 5b).

The pathways to the branched-chain amino acids have been studied primarily in *E. coli* and *S. typhimurium* and to lesser extents in yeast and *N. crassa.* Many of these studies have been both stimulated and complicated by the many antagonisms that occur between the three branched-chain amino acids. Since the first example of this kind was reported by Gladstone (192) in 1937, with the anthrax bacillus, there have been many additional examples reported in which two or all three branched-chain amino acids are antagonizing each other at one or more points in metabolism. Since the three branched-chain amino acids (and their precursors) are analogs of each other, it is not surprising that such interactions do occur. The kinds of antagonism that have been found are interference with active transport into the cell, interference with end-product control, interference with repression control, and, perhaps, interference with incorporation into protein.

Control of carbon flow over the isoleucine biosynthetic pathway is achieved by the inhibition of threonine deaminase by isoleucine. The enzyme has been purified from yeast (193, 194), *S. typhimurium* (195), *E. coli* (196–198), *R. spheroides* (199), and *Rhodospirillum rubrum* (200). While there are differences between the enzymes from the different sources, there are several common features. For all, isoleucine is a negative effector, and valine is a positive effector that antagonizes isoleucine inhibition. The usual pattern is that the purified enzyme exhibits a hyperbolic substrate saturation curve, but in the presence of isoleucine a sigmoid saturation curve is obtained. Valine reverses the inhibition and the cooperative binding. The yeast enzyme is an exception, since the purified enzyme even in the absence of isoleucine shows cooperative binding of threonine, which is normalized by valine (193, 194). Thus it appears that the yeast enzyme spontaneously assumes the configuration that is dependent upon isoleucine in threonine deaminases from other organisms.

Recently the *S. typhimurium* enzyme has been studied, and a model for the interaction between threonine, isoleucine, and valine has been related to substrate, inhibitor, and activator sites on the enzyme (195). The enzyme exhibits a remarkable half-the-sites reactivity at two levels. Although it is a tetramer of four identical subunits, two pyridoxal phosphates and two isoleucines are bound. However, there appears to be only one functional activator site (at which valine or the substrate binds). The cooperativity in threonine binding results only from the need for threonine to bind at the activator site when isoleucine is present. If the activator site has bound valine, the requirement for the binding of threonine at the activator site is eliminated, and the substrate saturation curve is hyperbolic as it is when isoleucine is absent.

That the half-the-sites reactivity of threonine deaminase may be intimately related to the isoleucine regulatory site is indicated by the finding of a threonine deaminase in an *E. coli* B mutant with a shortened polypeptide (43,000 daltons instead of 50,000) owing to an *amber* mutation (201). This enzyme could not bind isoleucine, was present only as a dimer, and for maximal activity required the binding of pyridoxal phosphate on both subunits of the dimer. The second pyridoxal phosphate, however, was not tightly bound. Thus, the mutant exhibited a requirement for either isoleucine or pyridoxine.

Control of carbon flow over the pathway to valine occurs at the initial step in valine biosynthesis (Reaction 2a) and leads to a complication if it is an effective control. In *E. coli* strain K12, acetohydroxy acid synthase is strongly inhibited by valine. Since this enzyme also catalyzes Reaction 2, valine blocks not only valine biosynthesis but also isoleucine biosynthesis. In *S. typhimurium* and presumably most other strains of *E. coli*, there is also a valine-insensitive acetohydroxy acid synthase, so that isoleucine biosynthesis is not prevented by valine (202, 203). In such organisms, valine would be overproduced, particularly toward the end of growth. A more complete discussion of the multiple acetohydroxy acid synthases will be possible when repression of enzyme formation is considered (see below).

Most organisms exhibit acetohydroxy acid synthase activities that are only partially inhibited by valine. However, it is not always certain whether the effect is due to a single enzyme that is only partially inhibited or to multiple enzymes. In *S. typhimurium* and *E. coli* W, the question was easily answered, since the valine-insensitive and valine-sensitive enzymes exhibited different repression patterns, as is discussed later (202, 203).

Acetohydroxy acid synthases have been quite refractory to purification. An exception is that from a strain of *P. aeruginosa,* which has been purified and shown to contain two kinds of subunits of 60,000 and 15,000 daltons (204). This enzyme was inhibited by all three amino acids, and a cumulative pattern of inhibition was observed (205). The single enzyme of *B. cereus* T may have similar properties, since it is also partially inhibited by each of the three branched-chain amino acids (206).

The *P. aeruginosa* enzyme may be exceptional not only because it was readily purified but also because it did not exhibit a requirement for FAD as do most of the others from bacterial and fungal sources. The role of FAD for the enzyme is not self-evident, since the overall reaction does not involve any redox change. The interaction of FAD with what may be a similar enzyme, glyoxylate carboligase, has been more thoroughly studied by Gupta & Vennesland (207). These workers found that when the enzyme was reduced (i.e. the flavin bleached) the enzyme was inactive until it had been reactivated by oxidizing with molecular oxygen. A mutant of *P. aeruginosa* has been described in which the acetohydroxy acid synthase does exhibit a requirement for FAD (208).

The inhibition of α-isopropylmalate synthase (Reaction 6, Figure 4) by leucine provides the mechanism by which the flow of intermediates over the leucine biosynthetic pathway is controlled. The enzyme from *S. typhimurium* and yeast and its interaction with the substrates and with the inhibitor have been studied by Kohlhaw and his colleagues (209, 210). The kinetic and physical properties of a feedback-resistant enzyme from an *S. typhimurium* mutant have been compared with those of the wild-type enzyme. It appeared from these studies that the efficient inhibition of the enzyme by leucine is dependent upon a loosening of the subunits, with a strong tendency for the leucine-bound tetrameric enzyme to dissociate to monomers.

Studies on the yeast enzyme revealed an inactivation of the enzyme that was dependent upon coenzyme A (a product of Reaction 6) (211). This inactivation was antagonized by binding of leucine, the inhibitor. Whether this in vitro inactivation reaction has any in vivo counterpart is not known, but it may account for the rapid in vivo inactivation that Brown et al (212) observed when acetate-grown yeast (exhibiting elevated synthetase levels) was transferred to a glucose medium (in which growing cells exhibit a reduced isopropylmalate synthase activity). It was of interest that the coenzyme A–dependent inactivation was also found for another yeast enzyme catalyzing a similar reaction, homocitrate synthase (Reaction 17, Figure 1) (211). The enzyme from a variety of sources has been shown to have only a limited specificity; it can transfer the acetyl group not only to α-ketoisovalerate but also to pyruvate, α-ketobutyrate, α-ketovalerate, and even to α-ketoisocaproate itself. In wild-type organisms, this lack of specificity does not seem to matter. It may be that the enzyme does not encounter significant amounts of the secondary substrates. Furthermore, the K_m for α-ketoisovalerate makes it the favored substrate.

Under some circumstances, these secondary substrates can be utilized. For example, mutants of *Serratia marcescens* derepressed for the leucine biosynthetic pathway have been mutagenized to yield isoleucine auxotrophs owing to the loss of threonine deaminase (213). Such mutants give rise to isoleucine-independent strains that contain a feedback-resistant α-isopropyl malate synthase. This desensitized synthase allows the mutants to bypass the threonine deaminase step and instead form α-ketobutyrate from pyruvate via the chain-elongating mechanism of the leucine pathway. Thus pyruvate is converted to ketobutyrate, which in turn is converted to ketovalerate. As a result such abnormal amino acids as norvaline, norleucine, homoisoleucine, and homoleucine can be formed in small amounts as side reactions of the leucine biosynthetic pathway. This mechanism probably also accounts for a major pathway for isoleucine formation in *Leptospira* (214). Whether the enzymes were "borrowed" from the leucine pathway is not clear, since the overall incorporation was found to be regulated by isoleucine.

There are other mechanisms by which the isoleucine-specific step, threonine deamination, can be bypassed. In the Crookes strain of *E. coli,* glutamate mutase apparently gives rise to β-methylaspartate, which in turn is converted to β-methyloxaloacetate and decarboxylated to yield α-ketobutyrate (215). In *B. subtilis,* the threonine synthase structural gene can undergo a mutation to yield an enzyme that deaminates threonine, but only if the phosphate site is filled (216).

The transaminases effecting the final steps in branched-chain amino acid biosynthesis in the Enterobacteriaceae should also be briefly considered. The *ilv* gene cluster contains a single transaminase gene (*ilvE*) (86, 87). Loss of this gene function (transaminase B) alone leads to an absolute requirement for isoleucine, although the strains grow better with all three branched-chain amino acids. Amination of α-ketoisovalerate occurs via transaminase C, which transfers amino groups to and from alanine, valine, and α-aminobutyrate (97). It is the peculiar control of this enzyme that makes *E. coli* K12 *ilvE* mutants that grow on isoleucine alone unable to grow on isoleucine plus leucine or isoleucine plus valine; the enzyme is repressed by either valine or leucine (217). When valine is present, the formation of α-ketoisovalerate needed for leucine formation is blocked by inhibition of Reaction 2a (Figure 4) and by the repression of the transaminase C (Reaction 10). The inhibition by valine can be overcome by adding α-ketoisovalerate, which can be converted to leucine.

Questions have been raised whether there is a separate valine-glutamate transaminase that is not specified by the *ilvE* gene. However, the experiments of Monnier et al (101) and of Chesne et al (100) show clearly that the single valine-glutamate activity present in DEAE-Sephadex eluates was missing in an *ilvE* deletion strain that was capable of catalyzing Reaction 10.

The second transaminase for leucine formation is the enzyme that was referred to earlier as the tyrosine-glutamate transaminase (100, 103). Since this enzyme is repressible by tyrosine, the growth of *ilvE* mutants on isoleucine-supplemented medium is inhibited by tyrosine (218). This inhibition is overcome by leucine and not by α-ketoisovalerate. A summary of the *E. coli* transaminases is given in the following section.

Regulation of enzyme formation in the pathways to the branched-chain amino acids in the Enterobacteriaceae is complex in that no repressor protein has been identified either genetically or biochemically. It is simpler to consider the leucine biosynthetic pathway first, which has been studied most extensively in *S. typhimurium* and to a lesser extent in *E. coli.*

The structural genes for the three leucine biosynthetic enzymes are arranged in a single operon and are coordinately repressed and derepressed. (Owing to the instability of two of the three enzymes, under certain growth conditions coordinacy of enzyme synthesis may not always obtain. Thus, as an operational tool, the coordinacy of enzyme formation is not a reliable criterion for an operon-mediated control.) A control region, to the "downstream" side of the promoter site, has been identified and designated gene *leuO* (219). Whether *leuO* is the binding site for a repressor or for a positive control element is not yet known. Nevertheless, regulation of the operon involves some function of leucyl-tRNA synthetase and perhaps charged or uncharged tRNA^{Leu} itself (220). That it may involve charged or uncharged tRNA is indicated by one of the pleiotropic effects of the *hisT* lesion. This lesion prevents the formation of a pseudouridine near the anticodon region of several tRNAs including several altered tRNA isoaccepting species for leucine and leads to reduced repressibility of the *leu* gene cluster by leucine (221, 222).

In addition to mutants isolated from *S. typhimurium* and *E. coli* with lesions in either the *leuO* or the leucyl-tRNA synthetase gene (*leuS*) and that have derepressed levels of the leucine biosynthetic enzymes, derepressed mutants have also been isolated that exhibit trifluoroleucine resistance, *flrA* and *flrB* in *E. coli* B/r (223), *flrB* in *S. typhimurium* (224), or that exhibit azaleucine resistance, *azl* in *E. coli* K12 (225). The roles of these loci in the regulation of the leucine biosynthetic pathway are unknown. In some, alterations in the chromatographic profiles of the tRNA^{Leu} isoaccepting species have been observed (L. S. Williams, personal communication). Whether these alterations are due to alterations in the structural genes for the individual species, to alteration in a modification enzyme (as is true for *hisT* mutants), or to factors affecting the extent of modification of a given species of tRNA is not known. Neither is it known whether the altered profiles are the cause or an effect of the derepression. Indeed, the possibility should also be considered that the altered profile and the derepression are both coincidental effects of a common cause.

The leucine biosynthetic enzyme levels in *E. coli* have been shown to vary over a 1000-fold range (the lowest in cells grown in rich medium, the highest in cells grown in chemostat). By hybridization of mRNA with a λp*leu* phage, it was shown that enzyme level and mRNA were almost directly proportional to each other (226).

The question regarding the role of the loci known to affect the leucine biosynthetic pathway can be extended to the isoleucine and valine biosynthetic pathway, for some of the enzymes of isoleucine and valine biosynthesis are repressed by a mechanism that involves leucine as one of the multivalent repressors. Of those that do, all involve a "leucine excess" signal that is apparently generated in the same way for the *ilv* regulon as it is for the *leu* operon. Not surprisingly, therefore, wherever isoleucine or valine or both are involved in multivalent repression, their corresponding amino acyl-tRNA synthetases are also involved. The pattern of control of the *ilv* regulon, however, is not a simple one.

In *E. coli* K12, three of the enzymes, those catalyzing Reactions 1, 4 (and 4a), and 5 (and 5a, 5b), are specified by three genes in a single operon, *ilvEDA*. These three enzymes are multivalently repressed by leucine, valine, and isoleucine, and the repressed levels of activity for the three enzymes represent essentially a coordinate reduction relative to the minimal medium levels of activity (227, 228). When derepression occurs by limiting one of the three amino acids, the derepression is not coordinate. Rather, depending upon the amino acid that is limiting and upon the intensity of the restriction, the enzyme specified by the third gene in the operon is derepressed more than is that from the second gene, which is derepressed more than is the gene (*ilvE*) nearest the presumed site of transcription initiation. This noncoordinacy has been attributed to termination of transcription that can occur after the first, second, or third gene has been transcribed (228). It appears that amino acid limitation can overcome these termination sites to various degrees. It seems likely that the noncoordinate derepression observed upon amino acid limitation is dependent upon the mechanism that involves the amino acyl-tRNA synthetases or the charging status of the cognate tRNAs. Whether the coordinate repression observed upon adding excess branched-chain amino acids is also dependent on this mechanism is not known.

Repression of acetohydroxy acid synthase activity in *E. coli* K12 is different from that described above. Repression and derepression involve only leucine and valine, plus the cognate tRNAs or synthetases or both (227). However, it is now clear that there are two isozymic forms of the valine-sensitive acetohydroxy acid synthase in *E. coli* K12 (229). One is acetohydroxy acid synthase I, specified by *ilvB*, which is close to the *ilv* gene cluster; the other, acetohydroxy acid synthase III, is specified by a region linked to the *leu* operon (but not part of it) and reported to consist of two genes *ilvH* and *ilvI*. At present, the evidence that *ilvH* and *ilvI* are two separate genes is not strong, and in fact the analysis of the region that has been carried out is also compatible with a single gene that is here referred to as *ilvHI*. The difference between the two isozymes remains unclear at this time.

The remaining enzyme of the pathway, acetohydroxy isomeroreductase (catalyzing Reactions 3 and 3a) is induced by either of the two substrates (230). The induction process has been studied in vitro by means of a λ derivative carrying an *ilv-lac* fusion in which the *lac* genes are controlled by the *ilvC* control region (231). In vitro, with the DNA from this phage as the template, the synthesis of β-galactosidase is dependent upon both acetohydroxybutyrate (or, less efficiently, acetolactate) and ppGpp (232). Also needed is an unidentified component that is missing in strains bearing an *ilvDAC* deletion (J. Wild, unpublished observations). Since strains bearing smaller deletions in *ilvA* and *ilvC* are able to support β-galactosidase induction by acetohydroxybutyrate, it appears that the *ilvDAC* deletion extends into the *Y* locus, postulated several years ago but not in its presently presumed position beyond *ilvC* (233).

In those organisms for which valine is not an inhibitor of growth, such as *S. typhimurium* and *E. coli* B, there is also a second isozyme, acetohydroxy acid synthase II, specified by the *ilvG* gene (202, 203). It is multivalently controlled by isoleucine, valine, and leucine just like the enzymes of the *ilvEDA* operon. Indeed, the mapping experiments of O'Neill & Freundlich (234) place this gene between the *E* and *D* genes, so that they conclude that there is an *ilvEGDA* operon. It seems quite likely that *S. typhimurium* and *E. coli* B both lack the *ilvHI* product because acetohydroxy acid synthase–negative strains can be obtained in two steps: selection of valine-sensitive strains (*ilvG*⁻) followed by selection of isoleucine and valine auxotrophy (*ilvG*⁻,*ilvB*⁻) (W. J. Pledger, unpublished observations).

It has long been known that the K12 strain of *E. coli* can become resistant to the growth-inhibiting effect of valine by several mechanisms (235). One of these mechanisms is via the *ilvO* mutation, which originally was recognized biochemically only by derepressed levels of the activities specified by the *ilvE, D,* and *A* genes (236). It was concluded at that time the *ilvO* locus lies between *ilvA* and *ilvC* and that the mutations define the operator of an *ilvADE* operon. At that time it was not possible to account for the resistance to valine based upon the observed derepression. The experiments of Favre et al (237) showed that the most important effect of the *ilvO* mutation is to allow expression of an *ilvG* gene. The gene, as in *S. typhimurium,* specifies a valine-insensitive acetohydroxy acid synthase. Thus, in *ilvO*⁻ strains there was a third isozyme of acetohydroxy acid synthase that differed from the other two by being insensitive to valine. These workers proposed that there was in the K12 strain an *ilvADGE* operon.

A number of experimental observations, which have been summarized by Cohen & Jones (238), were not compatible with this gene arrangement. These investigators examined the location of the lesions in two presumed *ilvO* mutants from the Adelberg collection. They found them to be between *ilvE* and *rbs* (an outside marker that is linked to the *ilv* gene cluster), a view compatible with the findings of Kline et al (227) that the *ilvO* lesions were outside a deletion covering the *ilvDAC* region. Cohen & Jones (238) proposed that transcription proceeded from *E* to *A*. That this was indeed the direction of transcription was proven by the Mu-1 insertion experiments of Smith et al (228). Their experiments also revealed that *ilvG* could not be between *ilvE* and *ilvD* as it had been shown to be in *S. typhimurium* (234) and as it was reported by Favre et al (237) for an *ilvO* derivative of *E. coli* K12. The mapping data of Smith et al (228) also indicated that the *ilvO* lesion they employed was between *ilvA* and *ilvC*. The model they proposed to account for a *cis*-acting regulatory element that was "downstream" from the operon it controlled was shown to be invalid when later mapping experiments revealed (239) the gene order in the *ilv* gene cluster to be:

   rbs.........ilvG-O-E-D-A-C

The position of the *ilvB* gene, long thought to be after *ilvC* (87), is still uncertain.

This gene arrangement obviously raises the possibility of two divergent transcription units such as have been found in the *argECBH* gene cluster (240) and the *bioABFCD* gene cluster (241). However, recent hybridization experiments indicate that *ilvG* and *ilvEDA* are transcribed from the same strand of DNA (C.S. Subrahmanyam and H. E. Umbarger, unpublished observations). That the *ilvG* transcript is controlled differently from the *ilvEDA* transcript is indicated both by the noncoordinacy of the two transcription units and by the fact that, even in *ilvO⁻* strains, *ilvG* is expressed upon limitation of valine or of isoleucine but not upon limitation of leucine (239).

At present, it is difficult to offer a model that could account for the control of the four transcriptional units in the wild-type K12 *ilv* regulon or of the five transcriptional units in the *ilv* regulon in *ilvO* strains. There have been a number of proposals that some form of threonine deaminase might play such a role (either positive or negative) in the regulation of one or all genes in the *ilv* regulon (233, 242). On the other hand, both repression and derepression that appear to be normal have been found to occur in the absence of the *ilvA* product (227). The fact remains that certain *ilvA* mutants appear to exhibit an abnormal control over the *ilv* regulon. If it is assumed that these *ilvA* strains do contain only single lesions, an explanation of these effects is needed.

An "immature" form of threonine deaminase has been invoked in the Hatfield-Burns model as a repressor of the repressible *ilv* genes (243, 244). This model is based upon the binding of the branched-chain amino acyl-tRNAs to the immature but not to the mature form of threonine deaminase and the "freezing" of the enzyme in the immature form by isoleucine plus valine in concert. This model, however, could not account for the repression of the acetohydroxy acid synthases I and II in the presence of limiting isoleucine (which would cause maturation of the enzyme and removal of the putative repressor). Furthermore, apparently normal repression occurs in its absence. Perhaps a more promising lead concerning the way threonine deaminase might influence regulation is the observation of Coleman et al (245) that the branched-chain amino acyl-tRNA synthetases are unstable in the *ilvDAC* deletion strain. Additional experiments have provided strong indication that it is the lack of threonine deaminase that is responsible for the instability (246). Thus, it may be that the effect of certain *ilvA* mutations that have appeared to exert pleiotropic effects may have been due to some subtle effect involving the synthetases or even the charging levels of the tRNAs. Certainly, the in vitro synthesizing systems that are being employed should soon reveal the nature of any positive or negative elements affecting *ilv* gene expression.

The regulation of the enzymes in the pathways to all three branched-chain amino acids in *N. crassa* is positively controlled by the *leu-3* locus (247, 248). The effector molecule is the first intermediate in the leucine pathway, α-isopropylmalate. In the absence of a functional *leu-3* gene, the leucine biosynthetic enzymes are nearly fully repressed. This positive role is overcome for all the enzymes by leucine, which prevents the formation of α-isopropylmalate. In addition, threonine deaminase is repressed by isoleucine, acetohydroxy acid synthase by valine, and the isomeroreductase and dehydroxy acid dehydrase by leucine plus isoleucine.

In *S. cerevisiae*, regulation of the formation of the isoleucine and valine biosynthetic enzymes is multivalent involving all three branched-chain amino acids. An essential element for derepression by limitation for either valine or isoleucine is threonine deaminase but not for derepression by limiting leucine. It was proposed that leucyl-tRNA is the repressing element and threonine deaminase acts as a positive element that removes leucyl-tRNA by binding it (249). The leucine biosynthetic enzymes catalyzing Reactions 7 and 8 (Figure 4) are repressed by leucine, whereas that catalyzing Reaction 6 is repressed by leucine plus threonine (250).

In both *N. crassa* and yeast, some leucine and isoleucine biosynthetic enzymes are found in mitochondria or in an organelle in the mitochondrial fraction (248, 251–253). However, the distribution of the enzymes is different. In *N. crassa*, the common valine and isoleucine biosynthetic enzymes (but not threonine deaminase) are found in the mitochondria (248, 251). In yeast, all the enzymes leading from threonine and pyruvate to the keto acid precursors are in the mitochondria, as is a portion, but not all, of the branched-chain amino acid transaminase activity (252). Also in yeast, α-isopropylmalate synthase (Reaction 6, Figure 4) is in the mitochondria, but the isomerase and dehydrogenase are in the cytosol (253).

Alanine Biosynthesis

One might assume that the formation of alanine occurs via Reaction 9 (Figure 4) and that Reaction 10 (transaminase C) would at most serve a minor role. Both reactions are demonstrable in *E. coli* extracts. A leaky alanine auxotroph of *E. coli* has been described by Falkinham (254). The organism grew poorly on either D- or L-alanine, but this was attributed to additional lesions that caused poor uptake of these amino acids. Pyridoxine, leucine, or isoleucine could support growth of the mutant better. The mutant lacks an L-alanine dehydrogenase and a D-alanine oxidase activity that are probably two activities of the same enzyme. It also lacks transaminase C activity. Removal of either block abolishes the auxotrophic requirement. The organism does have an L-alanine-glutamate transaminase activity (Reaction 10), but has a reduced level of D-alanine-glutamate transaminase activity. The reason for the alanine requirement is not clear. The more important question, however, is what the relative roles of the D- and L-alanine-glutamate transaminases, transaminase C, and alanine dehydrogenase might be in alanine biosynthesis. It is of interest that the recent isotope incorporation experiments of Csonka (255) are not in accord with alanine formation simply by transamination of pyruvate. Except for the repression control over transaminase C mentioned earlier, alanine biosynthesis does not seem to be regulated.

THE AROMATIC FAMILY

The biosynthetic pathways to the aromatic amino acids have been studied in a variety of microorganisms, so that it has been possible to make comparisons of the gene-enzyme relationships in many closely as well as distantly related forms. It is striking that many examples of multifunctional enzymes are encountered in these pathways in which a single polypeptide product will catalyze two or more chemically separate reactions; sometimes, but not always, these reactions are consecutive steps in a pathway. It is also of interest that the particular combination of activities found to accompany each other on a single protein are not always the same in various organisms. This finding, along with the finding that some of the multifunctional enzymes (i.e. with "covalently aggregated" activities) aggregate with other enzymes to form greater complexes, suggests that many or even all of the aromatic amino acid biosynthetic enzymes function together in multicomponent complexes such that when some gene fusions do occur they were not selected against but rather may actually have allowed better fits into the complexes. A recent review of the comparative gene-enzyme relationships in the tryptophan biosynthetic pathway by Crawford (256) provides several examples of the varied combinations that have occurred in that pathway.

It is convenient to consider separately the common aromatic pathway that leads to chorismate from which the three aromatic amino acid pathways branch: the phenylalanine and tyrosine pathways, which in some organisms are completely separate but in others do share an additional enzyme, and the tryptophan pathway.

The Common Aromatic Pathway

Aromatic biosynthesis begins with the condensation of phosphoenolpyruvate and erythrose-4-phosphate to yield 3-deoxy-D-*arabino*-heptulosonate-7-phosphate (DAHP) (Reaction 1, Figure 5). DAHP is cyclized upon the removal of phosphate to yield 5-dehydroquinate (Reaction 2). The enzyme catalyzing this reaction requires NAD as a cofactor. Although there is no net oxidation or reduction in the overall reaction, there is an internal redox change; what had been carbon 7 of DAHP is reduced and what had been carbon 6 is oxidized (carbons 6 and 5 of 5-dehydroquinate). A removal of water and an NADPH-dependent reduction yield shikimate (Reactions 3 and 4). Shikimate is phosphorylated and condensed with another molecule of phosphoenolpyruvate (Reactions 5 and 6). A second double bond is generated upon removal of the ring phosphate to yield the branch point compound, chorismate (Reactions 5 to 7). It is from this branch point compound that the specific pathways to the three aromatic amino acids, to *p*-aminobenzoate, to menadione, to ubiquinone, and to enterochelin originate.

The control of carbon flow over the pathway to chorismate in *E. coli* and *S. typhimurium* resembles that over the common pathway in the biosynthesis of the aspartate family of amino acids. There are three DAHP synthases: one is inhibited by phenylalanine (257), one by tyrosine (258–260), and one by tryptophan (261). The same isozymic pattern has also been found in *N. crassa* (262).

In contrast to this isozymic pattern is the pattern found in *B. subtilis* (263). There is only a single DAHP synthase activity and it is carried on the same protein that carries chorismate mutase. This protein (subunit A) forms a complex with another protein (subunit B) that has almost no activity by itself but with subunit A exhibits shikimate kinase activity. Both shikimate kinase and DAHP synthase activities are inhibited by chorismate and prephenate. That the inhibition may be due to binding of chorismate or prephenate to the active site of the mutase is suggested by the fact that proteolytic cleavage of the chorismate mutase fragment from the complex was accompanied by a loss of feedback sensitivity of DAHP synthase.

Another pattern of control found in organisms with single DAHP synthases is that exhibited by the enzyme in *Corynebacterium glutamicum,* an organism that, along with many mutant derivatives, has proven so useful in the Japanese fermentation industry (264). In this organism, there is a strong synergistic inhibition of DAHP synthase by phenylalanine and tyrosine. This inhibition was increased still more by the presence of exogenous tryptophan (nearly 90%).

Still another pattern of inhibition found in organisms with single DAHP synthases is that in which the enzyme is inhibited by only a single aromatic

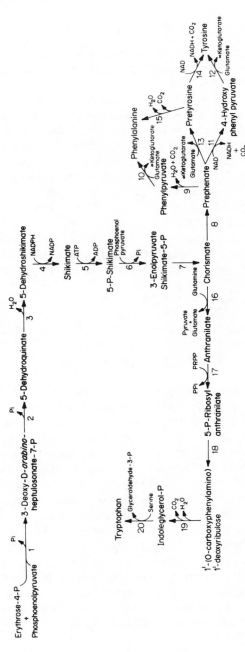

Figure 5   The biosynthesis of the aromatic amino acids.

amino acid. In *Streptomyces aureofaciens* the enzyme is inhibited by trypto-phan alone (265). The single enzyme in *P. aeruginosa* is somewhat similar in that tyrosine is the most effective inhibitor, although phenylpyruvate and tryptophan are also inhibitors (266). One might wonder whether an efficient regulation of carbon flow into the branching pathways might be achieved in such organisms. However, it may be that compensatory adjustments of metabolite flow at subsequent steps serve to prevent an excess of the DAHP synthase inhibitor from interfering with biosynthesis of the other amino acids. An attempt to account for the control over carbon flow in *P. aeruginosa* has been made by Calhous et al (267), which is considered later.

Repression of enzyme formation in the common aromatic pathway has again been studied primarily in the Enterobacteriaceae. An extensive survey of the enzymes in *E. coli* by Brown & Somerville (268) revealed that in addition to the tryptophan-specific repression of the tryptophan-sensitive DAHP synthase and the tyrosine-specific repression of the tyrosine-sensi-tive enzyme observed by others, there was a multivalent repression of the phenylalanine-sensitive enzyme by phenylalanine plus tryptophan. The re-pression of the tryptophan- and tyrosine-sensitive isozymes was dependent upon the *trpR* and *tyrR* loci, respectively. That the *tyrR* gene specifies a repressor for the *tyr* regulon was suggested by the findings of Camakaris & Pittard (269), who studied both *amber* and temperature-sensitive muta-tions in the *tyrR* gene. The repression of the phenylalanine plus trypto-phan–repressible isozyme also involves the participation of *tyrR* but not *trpR* (268, 269). No *pheR* gene has been found in *E. coli,* but in *S. ty-phimurium,* in which there is a *pheR* gene, it does not affect the multiva-lently repressed, phenylalanine-sensitive isozyme (270).

At least five of the other enzymes catalyzing the remaining steps of the common aromatic pathway are not end-product controlled. Like many other "constitutive" enzymes, there is a metabolic control that is related to growth rate. Tribe et al (271) examined this control critically and were able to relate the rate of synthesis of these enzymes to the gene dosage effect, which in turn is affected both by rate of growth and the distance of the structural gene from the replication terminus on the chromosome. Interest-ingly, this relationship enabled them to predict a probable location on the *E. coli* chromosome of the *tyrR*-controlled isozyme of shikimate kinase.

The common aromatic pathway in *N. crassa* is of particular interest, since five of the seven reactions are catalyzed by a single multienzyme complex that had long been known to be specified by the *arom* genetic region. This genetic region had previously been considered as one of the few candidates for a multicistronic operon in *N. crassa* (272). The multienzyme complex (catalyzing Reactions 2–6, Figure 5) has now been shown to be formed as a single polypeptide chain (273, 274). It had earlier been shown that the multienzyme complex not only functioned as an integral catalytic unit itself, but when combined with DAHP synthase and chorismate synthase (enzymes catalyzing Reactions 1 and 7, respectively), which were isolated as separate enzymes, and with another multienzyme complex that catalyzed three of the five tryptophan biosynthetic steps, a phenomenon termed "catalytic facilitation" resulted (275). This complex thus contained all the enzymic activities catalyzing Reactions 1-7 and 15-18 (Figure 5) except for that catalyzing Reaction 16. Thus, incubation of the two complexes and two single enzymes with erythrose-4-phosphate and phosphoenolpyruvate plus the necessary cofactors resulted in the accumulation of anthranilate. The remarkable feature of catalytic facilitation is that the rate of anthranilate production with a subsaturating amount of erythrose-4-phosphate was not increased by adding an intermediate of the pathway, shikimate. In fact, a faster rate of anthranilate formation was observed when erythrose-4-phosphate and phosphoenolpyruvate were the initial substrates than when the overall conversion was initiated with shikimate. At least part of the greater efficiency of the initial substrate over that of shikimate is due to an activation of four of the five enzymes in the complex (principally by decreasing the K_m value for their substrates) upon incubation with DAHP (276). The activation occurred even without the cofactors needed for the conversion of DAHP to the subsequent intermediates. Another feature of the activation was the reduction of the transient time (the time required for the overall catalytic function of the complex to reach its maximal rate). The shikimate kinase activity of the complex was not affected by the activation, so that this enzyme became the rate-limiting reaction in the sequence after activation had occurred.

It would be interesting to know whether these findings with *N. crassa* can be duplicated in other systems in which multienzyme complexes are active. These experiments also raise the possibility that interaction between the complex and separate enzymes was also a part of the overall catalytic facilitation. Is it possible for separate enzymes that catalyze sequential reactions to exhibit interactions in vivo that make catalysis more efficient? If so, are "nuclei" of multienzymes always needed to "seed" such interactions, or can enzymes we currently recognize as separate proteins come together for efficient catalysis? The possibility of such enzyme aggregates might well explain many of the examples of apparent "channelling" that have been reported and the near-zero level of pools of many intermediates in biosynthetic pathways of bacteria. A technology to explore this area of metabolic control may someday be developed.

Phenylalanine and Tyrosine Biosynthesis

The pathways to phenylalanine and tyrosine are considered together since they share an additional common enzyme in some organisms, although in others they are formed by the action of completely separate enzymes. The potentially common enzyme activity is chorismate mutase, which forms prephenate (Reaction 8, Figure 5). Where this serves as a branch point compound, it is the substrate for prephenate dehydratase and for prephenate dehydrogenase, the enzymes forming the α-keto acid precursors of phenylalanine and tyrosine, respectively (Reactions 9 and 11, Figure 5). The final steps (Reactions 10 and 12) are those catalyzed by one of several transaminases that are considered in a separate section.

A variation on the usual pathway to tyrosine has been found in the cyanobacteria (277). In this pathway, prephenate itself undergoes transamination to yield "pretyrosine," which is the substrate for an NAD-dependent oxidation to yield tyrosine (Reactions 13 and 14). This pathway also serves as a second route to tyrosine in *P. aeruginosa* (278). In addition to the usual prephenate dehydrogenase, this organism has a second dehydrogenase capable of converting pretyrosine to tyrosine or prephenate to 4-hydroxyphenylpyruvate (279). By means of an additional enzyme, pretyrosine dehydratase, a second route to phenylalanine is also present (Reaction 15).

In *E. coli* and *S. typhimurium,* the *tyrA* genes specify the bifunctional enzyme that converts chorismate, presumably via the enzyme-bound intermediate, prephenate, to 4-hydroxyphenylpyruvate, thereby yielding an aromatic ring (280, 281). The bifunctional enzyme chorismate mutase T–prephenate dehydrogenase is specified by the *tyrA* gene. The activity of prephenate dehydrogenase is inhibited by tyrosine, but that of chorismate mutase T is not (281).

In the same organisms, the *pheA* genes specify the other multifunctional enzyme that results in aromatization, chorismate mutase P–prephenate dehydratase (280, 282). Both activities of the purified enzymes exhibit sensitivity to phenylalanine (283, 284). Thus, in the enteric bacteria, prephenate is not a branch point compound at the terminus of a common pathway. Rather, it is an intermediate in each of two separate pathways.

In *B. subtilis,* there is but a single chorismate mutase, which, as noted earlier, is carried on the same protein that carries the single DAHP synthase (263). The mutase activity is inhibited by prephenate (285). Prephenate is thus truly a branch point compound. The pathway to tyrosine is initiated by prephenate dehydrogenase, which is inhibited by tyrosine (285). The pathway to phenylalanine is initiated by prephenate dehydratase, an enzyme that is sensitive to both phenylalanine and tryptophan (286). The tryptophan inhibition is antagonized by tyrosine. The pattern of control of carbon flow in the *B. subtilis* aromatic biosynthetic pathway, which includes the interactions described here as well as the inhibition of DAHP synthase and shikimate kinase by chorismate and by prephenate cited earlier, is called sequential feedback inhibition (286).

As mentioned earlier, the single DAHP synthase in *P. aeruginosa* is most strongly inhibited by tyrosine (266). That this does not lead to an imbalance of metabolite flow may be explained by the pattern found in the pathways to tyrosine and to phenylalanine (267, 279). A single chorismate mutase is partially inhibited by phenylalanine and strongly product-inhibited by chorismate. The conversion of prephenate to phenylpyruvate is strongly inhibited by phenylalanine and activated by tyrosine, while the flow from chorismate to tyrosine is uncontrolled by tyrosine. Thus, when the phenylalanine pool is elevated, prephenate can be converted freely to tyrosine until its supply (and that of chorismate) is quenched by the tyrosine inhibition of DAHP synthase activity. The apparently minor pathways to phenylalanine and tyrosine via pretyrosine are also at least partially subject to regulation, since the pretyrosine dehydrogenase (catalyzing Reaction 14) is inhibited by tyrosine (279).

In *E. coli,* the *tyrR* gene is involved in the repression not only of the *tyrA*-specified chorismate mutase T-prephenate dehydrogenase and the tyrosine-sensitive DAHP synthase, which are part of the same operon, but also the phenylalanine-sensitive DAHP synthase (multivalently repressed by phenylalanine plus tryptophan) and that component of transaminase A fraction that is here called the tyrosine-glutamate transaminase (287). As mentioned earlier, the *tyrR* gene probably specifies a repressor protein (269).

In *E. coli,* mutants with lesions in a phenylalanine-specific repressor gene have not been found. This could be because of the ease with which *pheO* mutations (the presumed operator gene for *pheA*) can be isolated in the search for analog-resistant mutants (282). On the other hand, an unlinked gene, *pheR,* presumably specifying a repressor, has been found in *S. typhimurium* (270). This gene affects only control of chorismate mutase P–prephenate dehydratase activity.

At present, there is no direct evidence of any involvement of the tyrosyl- and phenylalanyl-tRNA synthases in regulation of the tyrosine and phenylalanine biosynthetic enzymes. There are, however, some indications that suggest the question should remain open. For example, it was observed some years ago that the phenylalanine-sensitive DAHP synthase of *E. coli* could bind phenylalanyl tRNA, although no role for this interaction is known (288). Again, Cortese et al (222) observed that *hisT* mutants of *S. typhimurium* in which a specific pseudouridine is missing in several tRNA species are resistant to several tyrosine and phenylalanine analogs. However, an independent study revealed no alteration in the repressibility of the tyrosine-repressible enzyme in such a mutant (289). Another hint of such an involvement of the synthetase or an amino acyl-tRNA was the observation of Heinonen et al (290) that the derepression of the tyrosine biosynthetic enzyme that was due to a *tyrR* mutation was greater in a strain with a tyrosyl-tRNA synthetase with a low affinity for tyrosine than in a strain with a normal synthetase. Clearly, the question is not settled.

The Amination of Phenylpyruvate and 4-Hydroxyphenylpyruvate and the General Transaminases of E. coli

The final steps in phenylalanine and tyrosine biosynthesis in most organisms are the aminations of the two keto acids, presumably with glutamate as the amino donor. For years there has been some question concerning the specificity of the major transaminases in *E. coli,* since it was clear that the transaminase A fraction of Rudman & Meister (97) contained more than one activity (218, 291). The pertinent observations, in the absence of detailed protein fractionation, were that 1. at least part of the tyrosine-glutamate transaminase activity was repressible by tyrosine (218, 291, 292); 2. in the absence of transaminase B, the branched-chain amino acid-glutamate transaminase, the biosynthesis of leucine could occur at a significant rate, but the activity was repressed by tyrosine (218).

Even though the complete analysis has not yet been performed in any one laboratory, experiments from several laboratories have now made it possible to attribute the various activities to specific proteins and to the genes specifying those proteins. A summary of the reviewer's interpretation is given in Table 1.

The enzyme for which the term transaminase A should probably be reserved is the major aspartate-glutamate transaminase isolated by Chesne & Pelmont (98) and by Mavrides & Orr (293). It was shown to be the major mechanism for aspartate formation and to be specified by the *aspC* gene by Gelfand & Steinberg (103). Gelfand and Steinberg showed further that the *aspC* activity had an affinity only 1/300 that of the *tyrB* product for the α-keto acid precursors of phenylalanine and tyrosine. The *tyrB* product is the enzyme repressed by tyrosine. As mentioned earlier, it plays a minor role in aspartate formation and can convert α-ketoisocaproate to leucine, but is inactive with isoleucine or valine. Transaminase B is the branched-chain amino acid transaminase. It is inactive with tyrosine and aspartate but can convert phenylpyruvate to phenylalanine. The fourth major transaminase is the enzyme that makes it possible for *ilvE* mutants to grow on isoleucine alone, transaminase C. The repressibility of this enzyme by valine or leucine accounts for the requirement for leucine when *ilvE* mutants are given isoleucine plus valine and for the requirement for valine when they are given isoleucine plus leucine (207). This enzyme is not part of the *ilv* gene cluster, since a deletion of the entire cluster allows valine to be substituted for by α-ketoisovalerate. The only mutants to be shown to lack

Table 1 The general transaminases of *E. coli*[a]

Enzyme	*E. coli* gene	Amino acid substrates[b]	Repressed by	References
Transaminase A	*aspC*	aspartate glutamate phenylalanine tyrosine	constitutive	99, 100 103, 218
Tyrosine-glutamate transaminase	*tyrB*	phenylalanine tyrosine glutamate aspartate leucine	tyrosine	100, 103 218, 291 292
Transaminase B	*ilvE*	isoleucine valine leucine phenylalanine glutamate	multivalent, (by valine, isoleucine, and leucine)	101, 103 227
Transaminase C		valine alanine (α-amino butyrate)	leucine or valine	217, 254

[a] Not included are those specific transaminases that catalyze the highly specific reactions in the arginine, serine, lysine, and histidine pathways.
[b] Not included are nonphysiological amino acid substrates such as norvaline, methionine, tryptophan, etc.

transaminase C are the unique alanine auxotrophs described by Falkinham (254) that lack not only transaminase C but also L-alanine dehydrogenase. Interestingly, these mutants exhibited essentially wild-type levels of alanine-glutamate transaminase activity. The nature of their alanine requirement or the role of transaminase C in alanine biosynthesis remains unsolved.

Not included in the table is the heat-stable, phenylalanine-specific transaminase that Monnier et al (101) found in the fraction containing the tyrosine-glutamate transaminase activity, nor "enzyme B" of Mavrides & Orr (293). The latter may be the tyrosine-glutamate transaminase. The heat-stable component may be the partially inactivated tyrosine-glutamate transaminase. Complete confidence in the correlations given in Table 1 is not possible owing to the fact that the enzyme separation and the genetic analysis have not been done on the same strains. Furthermore, the complete range of substrates has not been examined with the separated enzymes nor with the strains containing only one of the four enzymes.

Tryptophan Biosynthesis

As indicated in the introduction to this section, tryptophan is formed via the same sequence of reactions in all organisms, but the distribution of activities as separate proteins is variable. The pathway is initiated by the conversion of chorismate to anthranilate in a glutamine-dependent reaction (Reaction 15, Figure 5). The phosphoribosyl moiety of phosphoribosyl pyrophosphate is transferred to anthranilate (Reaction 16). The indole ring is formed in two steps involving first an isomerization converting the ribose group to a ribulose and then a cyclization reaction to yield indole glycerol phosphate (Reactions 17 and 18). The final reaction in the pathway is always catalyzed by a single enzyme that may contain either one or two kinds of subunit and consists of the cleavage of indole glyceraldehyde-3-phosphate and condensation of the indole group with serine (Reaction 19).

The control of metabolite flow in the tryptophan pathway occurs by the inhibition of anthranilate synthase by tryptophan. The enzyme has been extensively studied in *E. coli* and *S. typhimurium,* with respect to both the structure of the protein and the interaction with substrates and inhibitors. As shown originally by Ito & Yanofsky (294) for *E. coli,* the anthranilate synthase of both organisms consists of two components. One, anthranilate synthase component I (CoI), can by itself catalyze the formation of anthranilate with ammonia as substrate. For the more efficient glutamine-dependent reaction, anthranilate synthase component II (CoII) activity is also needed. CoII activity (amido transferase activity) resides in a portion of anthranilate phosphoribosyl transferase, the enzyme catalyzing Reaction 16. Thus, anthranilate synthase is a complex of the two kinds of proteins. Grieshaber & Bauerle (295) were able to show that CoII activity alone could be obtained from certain *S. typhimurium* mutants with nonsense mutations in the phosphoribosyl transferase (*trpB*) gene. They termed the region specifying CoII as Region 1 of the gene. As a result of the complex that is formed between CoI and CoII, the binding of tryptophan to CoI inhibits not only anthranilate synthase activity but the phosphoribosyl transferase activity as well.

The CoII activity in some other organisms is associated with other proteins. For example, in *N. crassa,* it is carried on the same protein that converts phosphoribosylanthranilate in two steps to indole glycerol phosphate (296). In *B. subtilis* (297), *S. marcescens* (298), and *P. putida* (299), the amido transferase is carried on a separate protein apparently exhibiting no other activity.

CoII from *B. subtilis* is of particular interest, since it also provides the glutamine amidotransferase activity for *p*-aminobenzoate synthesis as well. One interesting mutant, with a lesion in the CoII structural gene, *trpX,* was shown to grow in a tryptophan-free medium, although slowly (300). The organisms so grown were physiologically derepressed for the tryptophan biosynthetic enzymes and were deficient in folate formation. The anthranilate synthase from this organism could use only ammonia for the formation of anthranilate. Thus it appeared that anthranilate synthase could actually function in vivo with ammonia. On the other hand, it could have been that the slow growth was in fact still due to the residual activity of the *trpX⁻* product. A more rigorous demonstration that ammonia could function in vivo was that by Zalkin & Murphy (301), who used an *E. coli* mutant in which the CoII portion of the phosphoribosyl transferase gene had been deleted. The cells grew in the absence of tryptophan when intracellular ammonia was high, but not when it was limited. In this respect the anthranilate synthase function with ammonia instead of glutamine was analogous to glutamate formation in *asm⁻ Klebsiella aerogenes* strains that lack glutamate synthase but retain glutamate dehydrogenase (302).

The consideration of the way that formation of the tryptophan biosynthetic enzyme is controlled poses a problem in a review of this kind. The study of gene expression in the *trp* operon has been as important to our current concepts of the general question of gene expression as was the study of the *lac* and *gal* operons in *E. coli* or the study of λ gene expression among viral systems. As a result, the reviewer has compromised his initial intention of considering only the kind of elements that interact in controlling amino acid biosynthesis and not how they interact at the molecular level with specific structures on DNA itself. In resisting the temptation to be even more compromising in the case of the *trp* operon, the reviewer is omitting some of the most exciting and important contributions to molecular biology that have emerged from the study of amino acid biosynthesis. It is hoped that those contributions will soon be reviewed in their appropriate context.

Genetic studies on tryptophan biosynthesis in both *E. coli* and *S. typhimurium* have demonstrated a regulatory gene (*trpR*) that was thought to specify the repressor for the *trp* operon, and a *trpO* gene was thought to be the site of the repressor interaction. [For a review of the historical background, the review of Margolin (303) should be consulted.] Physiological studies along with the genetic studies led to the conclusion that the *trp* operon did indeed respond to excess or limiting tryptophan, as the Jacob-Monod model predicted. There was nearly a coordinate control over the levels of the five enzymes except under extreme tryptophan starvation conditions and under conditions of full repression. When this occurred, the *trpA* gene product was formed at a higher rate than were the other *trp* gene products, because this protein (tryptophan synthase A protein) contains no tryptophan (304). When a mutant was used that contained tryptophan in the A protein, the preferential synthesis of this protein was eliminated (305). A deviation from coordinate repression under repressing conditions also occurs and is attributed in both *E. coli* and *S. typhimurium* to the presence of a "low level" promoter within the *trpD* gene of *E. coli* (*trpB* gene of *S. typhimurium*) that allows the unregulated expression of the last three genes in the operon.

The fortunate discovery of the $\phi 80$ phage, which forms lysogens in *E. coli* by integrating into a site very near the *trp* operon, provided a great stimulus to studies on gene expression of the *trp* operon (306). This phage readily yields specialized transducing phages carrying the *trp* operon or parts of the *trp* operon. With a collection of $\phi 80$p*trp* phages containing various lengths of the *trp* operon, it was possible to distinguish by RNA-DNA hybridization between the various parts of the *trp* operon to which *trp* mRNA was complementary (307). It was thus possible to measure not only the levels of *trp* message within the cell but also to decide from which gene(s) a given mRNA fragment had been transcribed. From such measurements, the direction of transcription, the frequency of transcription initiation, the direction of message breakdown, and the half-life of any region of the message could be determined. When correlated with kinetic studies on the appearance of the individual enzymes, several of the predictions of the genetic and physiological studies were verified and showed that much, if not all, of the control of the *trp* operon was indeed transcriptional. Among these findings were the following: 1. Transcription is initiated at the operator end and proceeds sequentially through the five genes (308, 309). 2. Translation also proceeds sequentially. 3. Repression by tryptophan is dependent on the presence of the repressor (309). 4. The onset of repression resulting from the addition of tryptophan blocks further message initiation but does not interfere with the completion of the messages initiated before adding tryptophan (308, 310). Thus, repression of the *trp* genes is also sequential. 5. Little or no transcription occurs downstream from polar mutations, although there are short stretches of message immediately beyond the polar mutation that are formed and broken down very rapidly (311, 312). 6. Degradation of the message occurs in the 5' to 3' direction (313, 314).

The availability of phages carrying *trp* DNA has also made possible in vitro studies of the interaction of repressor with operator, of the process of transcription with purified RNA polymerase, and of the coupled transcription and translation of the *trp* operon in a crude S-30 preparation of *E. coli*.

While direct binding studies between repressor and operator were not successful, the interaction was studied by measuring the quantity of mRNA made by purified RNA polymerase from the correct strand of the *trp* DNA template in the presence and absence of partially purified aporepressor and tryptophan. It was shown that, once RNA polymerase is bound to the *trp* promoter, the repressor cannot prevent transcription (315). Conversely, once repressor is bound, RNA polymerase cannot initiate transcription unless repressor is dissociated. Thus, it would appear that there is a functional overlap between the repressor binding site (operator) and polymerase binding site (promoter).

The in vitro transcription system also allowed an answer to the question of whether either tryptophanyl-tRNA or tryptophanyl-tRNA synthetase was involved in the repression mechanism. That the question was pertinent was suggested by the findings in other biosynthetic systems that derepression resulted when synthetase function was limited (i.e. when tRNA charging was low), a subject recently reviewed by Brenchley & Williams (316). There were conflicting observations since, on the one hand, strains with restricted capacity to charge tRNA^{Trp} with tryptophan did not exhibit as much repression as did normal strains (317, 318). On the other hand, 6-methyltryptophan mimicked tryptophan as a repressor but was not transferred to tRNA^{Trp}, and 7-azatryptophan, which at most represses only poorly, is readily transferred to tRNA^{Trp} (319). That complete repression of *trp* DNA transcription with purified RNA polymerase could be obtained in vitro with the *trp* repressor that had been purified free of either tryptophanyl-tRNA synthetase or tRNA^{Trp} indicated quite clearly that the operator-repressor interactions do not involve these two components. That these components do have a role in the regulation of *trp* operon expression in a quite different way has been indicated from additional studies, however. Similarly, the first enzyme in the pathway, anthranilate synthase, has also been shown to have no effect on the operator-repressor interaction (320).

The extent of tryptophan-mediated repression of *trp* operon expression in an in vitro coupled system is quite dependent upon the nature of the DNA template (321). Thus, *trp*-transducing phages with intact *trp* promoters that yielded little or no phage-promoted *trp* operon expression in the prophage state (322) yielded DNA templates that exhibited good tryptophan-dependent repression in vitro. In contrast, DNA templates derived from phages that yielded primarily "readthrough" transcription from phage promoters exhibited very little tryptophan-dependent repression of transcription. Thus, readthrough transcription is not repressible by tryptophan either in vivo or in vitro. This result is to be expected since repressor does not interfere with transcription once the polymerase is bound to the DNA. The tryptophan-dependent repression, as expected, was dependent upon both the *trpR*$^{+}$ product in the S-30 extract and in intact *trpO* region on the DNA template (321).

Quite independent of this tryptophan-dependent repression that involves the exact kind of repressor-operator interaction predicted by the Jacob-Monod model for gene expression is another tryptophan-specific control mechanism. This regulatory signal, however, affects not the initiation of transcription but the *termination* of transcription beyond a certain point (323). The existence of the site of this control was revealed by the occurrence of two mutants containing deletions of the operator-proximal *trpE* gene. Both exhibited normal *trp* operator-promoter function but did exhibit an elevated expression of the genes of the operon downstream from the deletion (*trpC, B, A*) (324). Transducing phages carrying such deletions yielded DNA templates that allowed severalfold more in vitro mRNA formation than did those with an intact *trp* operon. There were other deletion strains in which nearly the entire *trpE* gene had been deleted, but in which message formation was not affected. Thus, it appeared that there is a site between the *trpE* gene and the promoter that restricted transcription (321, 325). That this restriction of transcription was mediated by tryptophan was indicated by an in vivo increase in *trp* gene expression in *trpR⁻* strains upon severe tryptophan limitation. Tryptophan limitation caused no further increase in a strain carrying one of the unique deletions (326).

The nature of the regulatory site was revealed when the sequence of the 5' end of *trp* mRNA was determined (327). The mRNA fragment sequenced was one of about 200 nucleotides from a mutant bearing a deletion that allowed only the first 32 nucleotides of the *E* message to be made. It was found that there was a "leader" sequence about 166 nucleotides long that preceded the translation initiation site for the *E* gene product. In contrast, when *trp* mRNA was made in vitro, nearly all of the transcripts were terminated after synthesizing 145 nucleotides (328). Similar leader sequences have been found as the predominant (~90%) transcripts formed in vivo in the presence of excess tryptophan (conveniently obtained by growing *trpR⁻* cells in tryptophan-supplemented minimal medium) (329). The 3' ends of the leader sequences obtained in vivo have not been precisely identified, but it is certain that they have been terminated very near the site of termination found in vitro and are about 135–140 nucleotides long. This termination site, which is clearly controlled in some way by the level of tryptophan, has been called the attenuator, in accord with the term given to a physiologically analogous site preceding the first structural gene in the *his* operon of *S. typhimurium* (330). The DNA specifying the leader sequence has been referred to as the leader region and is designated by the genetic symbol *trpL*.

The model that has now emerged is that transcription initiation is negatively controlled by repression exerted solely by the amount of the tryptophan-repressor complex. Whether transcription terminates at the attenuator or proceeds into the structural genes is also subject to an additional tryptophan-mediated control that could be looked upon as a positive control or, more precisely, antitermination.

That purified RNA polymerase nearly always terminates transcription for a *trp* DNA template at the attenuator site (328) indicates that in vitro no additional factors are needed for this termination and that it is an intrinsic property of the polymerase and the DNA sequence in the region of termination. However, it would appear that in vivo the early termination is dependent upon the termination factor, rho, since some polarity suppressing mutations (giving an "SuA" phenotype) not only suppress the polarity generated by polar mutations but also increase wild-type *trp* operon expression severalfold (331). Strains bearing these mutations have altered rho factors (332).

The mechanism of the in vivo control over termination by tryptophan is now thought to include tryptophanyl-tRNA synthetase or the state of charging of tRNA^{Trp} itself (333–335). Antitermination is independent of the five structural genes in the *trp* operon or their products (334). At present, it would appear that the involvement of the synthetase is only in the extent of charging of tRNA^{Trp} that it can achieve. However, whether charged tRNA^{Trp} is needed, in concert with rho, to effect termination, whether uncharged tRNA^{Trp} is needed to act as an antiterminator, or whether both roles are played is not known (335). The fact that significant amounts of RNA hybridizable with *trp* DNA can be made that do not carry structural gene information may account for the findings of Lavallé & DeHauwer (336) that enzyme synthesis was completely blocked when *trp* mRNA was not. Findings of this kind in this and other systems have been used to support the idea of a translational control. The control described here might be looked upon as a post-transcription initiation control.

Finally, there are two forms of control over *trp* operon expression that reflect a more general or metabolic control. Even these, however, may have a tryptophan-related component, and the two kinds of metabolic control may be related to each other. One is a control related to the *relA* locus affecting the stringent control of ribosomal RNA synthesis upon amino acid starvation (337). The *relA⁻* strains do not show as great a derepression of *trp* operon function as do *rel⁺* strains (333). This observation implies a stimulatory effect of ppGpp on *trp* operon expression, which has indeed been demonstrated in an in vitro coupled transcription-translation system (338). One possible mechanism is that ppGpp in a very general way increases the efficiency of RNA polymerase for some promoters but not for others, but only if the promoter is already available to the polymerase (338, 339). In this way, a ppGpp accumulation due to leucine limitation would not enhance *trp* operon expression, since excess tryptophan would both repress transcription and favor termination at the attenuator. When ppGpp accumulation was due to tryptophan limitation, it would enhance the frequency of *trp* mRNA initiations over and above that resulting from removal of repression and antitermination. Another possibility is that ppGpp could instead be involved in the control at the attenuator as suggested by Morse & Morse (333).

The second kind of metabolic control is one first observed by Rose & Yanofsky (340) and serves to increase *trp* operon expression as the growth rate increases. The increased *trp* operon expression was observed both at the level of message formation and at the level of enzyme formation. Since the correlation between *trp* operon expression and growth rate were made with *trpR* cells, the control is independent of the repressor-operator interaction.

Pouwels & Pannekoek (341) have examined the nature of the *trp* transcripts at different growth rates. They found that the increased *trp* mRNA formed at faster growth rates was the result of a decrease in the frequency of termination of transcription at the attenuator site. They found that the amount of *trp* leader mRNA (or *trp* regRNA, as they called it) was essentially the same in fast growing cells as in slowly growing cells. In strains lacking the attenuator site or in strains containing a defective rho factor, the growth rate did not influence *trp* operon expression. Thus, the growth rate appears to be an additional parameter controlling the rho-dependent termination that occurs at the attenuator site. Pannekoek et al (342) had earlier demonstrated in cell extracts an antitermination factor (AT) that decreased the frequency of rho-dependent termination. Pouwels (personal communication) has more recently shown that AT is itself subject to a growth rate control much like that shown for ribosomes, amino acyl-tRNA synthetases, and the elongation factors (343, 344). Since the formation of rho factor is not affected by growth rate (345), the variation in amount of AT could account for growth rate control of *trp* operon expression.

This increase in *trp* operon expression with increase in growth rate is reversed at the higher growth rates achieved by cells growing in rich medium (340). Under such conditions, a rich medium depression occurs, resulting in levels of enzymes much lower than "maximally repressed" levels. This phenomenon is common to many biosynthetic pathways, including the nonrepressible histidine operon of *S. typhimurium* which is considered in the next section. Under conditions of fast growth rates as occur in rich media, much of the biosynthetic capacity of the cell is devoted to ribosomal synthesis. Therefore, it may be that the rich medium depression of the enzymes in many biosynthetic pathways is due in part to competition for available polymerases or even for the translation machinery. This general form of metabolic control is considered again later in the review.

The precise interrelationship between growth rate control, rich medium depression of synthesis, and ppGpp-dependent stimulation of *trp* operon expression is not completely clear. Neither is it clear by what mechanism the termination activity of rho is achieved. However, the answer is probably not possible without a clearer understanding of the mechanism by which termination of transcription itself occurs, a subject outside the scope of this review.

Needless to say, relative to what has been learned concerning the regulation of formation of the tryptophan biosynthetic enzymes in *E. coli,* little is known in other organisms. In view of the differences between the gene-enzyme relationships in various organisms, some differences between mechanisms of enzyme regulation should be expected (256). For example, in *P. putida*, phosphoribosyl anthranilate isomerase and indole glycerol phosphate synthase are specified by separate genes. The gene for the isomerase, *trpF*, is alone and separate from the *trp* gene cluster, and is apparently constitutive. The two genes for tryptophan synthase are also separate from the *trp* gene cluster and are induced by indole or indole glycerol phosphate. The regulatory molecule for this induction is the *trpA* product itself (346). The regulation of tryptophan synthase is thus autogenous. On the other hand, the regulation of the cluster of four genes is under the influence of a regulatory gene, *trpR*, but their expression is not coordinate.

It has been proposed that in *B. subtilis*, the entire tryptophan pathway is induced by chorismate (347). A single regulatory locus, *mtr*, has been described, which upon mutation leads to constitutive enzyme formation and excretion of tryptophan (348).

Again, the review of Crawford (256) should be consulted for a discussion of some of the regulatory interactions that have been observed in organisms containing gene-enzyme relationships that differ from that in *E. coli*. In some cases, there are already both genetic and physiological observations that readily suggest alternatives that would be worthwhile to explore further.

The regulation of tryptophan biosynthesis in *N. crassa* has been shown to be achieved at least in part by a kind of general amino acid control mentioned earlier in the discussion of arginine biosynthesis in yeast (46). Thus, Carsiotis & Jones (349) have shown that upon derepression of the tryptophan biosynthetic pathway with limiting tryptophan, there was a concurrent derepression of the histidine and arginine biosynthetic enzymes.

HISTIDINE BIOSYNTHESIS

Histidine is formed by a series of ten enzymic steps in which an imidazole ring is initiated on a ribose carbon chain by the transfer of carbon atom 6 and nitrogen atom 1 from a purine ring to phosphoribosylpyrophosphate. This transfer involves the formation of a condensation product between ATP and phosphoribosylpyrophosphate (Reaction 1, Figure 6). This is followed by a cleavage of a pyrophosphate group (Reaction 2). The purine ring is opened (Reaction 3) and the ribose group (originating from the original phosphoribosylpyrophosphate) undergoes isomerization to a ribulose group (Reaction 4). This is followed by an amidotransfer from glutamine and closure of the new imidazole ring and release of the purine nucleotide precursor, aminoimidazole carboxamide ribotide (Reactions 5 and 6). The imidazole compound formed is imidazole glycerol phosphate. An α-keto compound is obtained by dehydration of the glycerol phosphate moiety (Reaction 7). This compound undergoes a transamination reaction to yield a compound that upon dephosphorylation is still charged, so that it can serve as a biosynthetic intermediate (Reaction 8). Dephosphorylation (Reaction 9) yields histidinol, which is the only histidine biosynthetic intermediate that can enter the cell at an appreciable rate. Histidinol is oxidized in a two-step reaction to histidine (Reaction 10). the aldehyde, histidinal, is probably not a free intermediate, but it can serve as a substrate for the dehydrogenase.

These ten enzymic steps are catalyzed by nine enzymes that are specified in both *S. typhimurium* and *E. coli* by an operon of nine genes. The genetic analysis undertaken by Hartman and his colleagues in *S. typhimurium* constitutes a landmark in microbial genetics. The detailed genetic analysis, which was summarized a few years ago (350), made it possible to relate physiological behavior not only to specific genes but to specific regions within genes in ways possible with no other system. Another review that includes historically important observations in the analysis of the pathway and its regulation is that of Brenner & Ames (351). As a result of the extensive genetic information from histidine auxotrophs of *S. typhimurium,* most subsequent studies on histidine biosynthesis and its control have been done with this organism.

Histidine is an inhibitor of the first enzyme in the pathway, phosphoribosyl-ATP synthetase. Indeed, it was the sensitivity of this enzyme to histidine

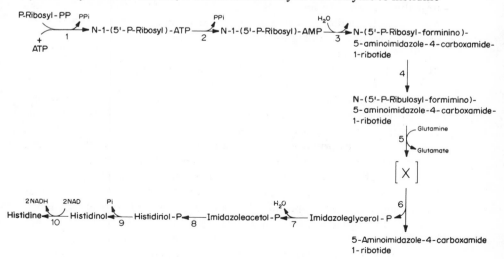

Figure 6   The biosynthesis of histidine.

that enabled Moyed & Magasanik (352) to realize that an in vitro system that produced aminoimidazolecarboxamide ribotide actually did so by a previously unknown sequence of reactions leading to histidine. The enzyme has been purified in several laboratories (353–355). It was later shown that the purification schemes employed earlier yielded an enzyme contaminated with histidase. Since the strains used would have formed only very little histidase, the contamination probably did not grossly affect the results in earlier experiments. Comparison of the wild-type enzyme with one that was insensitive to histidine revealed that, while both bound histidine, only the sensitive enzyme exhibits a conformational change upon binding of the inhibitor. The enzyme is a hexamer under conditions of a catalysis (356).

The greatest interest concerning phosphoribosyl-ATP synthetase has been its possible role in affecting the regulation of gene expression in the *his* operon. A considerable body of experimental results had led to the conclusion that this enzyme plays an important role in *his* operon expression, but as is indicated later, that role is not an essential one, nor is it understood.

Examination of mutants of *S. typhimurium* resistant to histidine analogs revealed that mutations in several loci lead to resistance (most with elevated or constitutive levels of the histidine biosynthetic enzymes). Among these loci were *hisO*, specifying the operator region (and associated structures, promoter, and attenuator; see below); *hisR*, the structural gene for tRNA^{His}; *hisS*, the structural gene for histidyl-tRNA synthetase; *hisT*, the structural gene for the enzyme that changes a uridine to a pseudouridine near the anticodon of several tRNAs including that for histidine; *hisU* and *hisW*, which are probably tRNA modification genes. [For a review of their early recognition and characterization, see (351).

The *hisR* mutants are peculiar in that they contain only about half the normal amount of tRNA^{His} (357). Thus, they could contain lesions in one of a pair of-duplicate, tandem genes for tRNA^{His}. In most of the *hisS* mutants, the histidyl-tRNA synthetase activity is reduced because of a high K_m for histidine (358). The histidine biosynthetic enzymes in such mutants are derepressed in minimal medium (and the cells may be slow growing) but approach the normal level upon the addition of histidine to the medium. The *hisT* mutations are pleiotropic in that they lead to resistance to analogs of several different amino acids. They also result in altered repression of both the *leu* and *ilv* genes (221, 222). This pleiotropy is probably because the missing enzyme is responsible for catalyzing the formation of pseudouridines in the anticodon loop of several tRNAs. The *hisU* mutants have recently been reported to be deficient in a nucleolytic activity that is involved in tRNA maturation (359). It was suggested that the *hisU* mutants may be a heterogeneous group with lesions in several closely linked loci concerned with tRNA modification.

The occurrence of these mutants made it quite clear that the signal for derepression (or the signal for repression, or both) involved tRNA^{His} and perhaps histidyl-tRNA synthetase itself. Their occurrence also made it unlikely that there was a specific repressor protein (such as the *argR* or *trpR* product). It would be possible, of course, that the repressing element played some essential role in the cell and could not be lost by mutation.

Additional evidence that the function of histidyl-tRNA synthetase had a role in repression other than merely affecting the charging level of tRNA^{His} was obtained in an examination of the effect of a second *hisS* gene in the cell (360). Elevated levels of histidyl-tRNA synthetase activity were accompanied by an elevated *his* operon expression (about twofold). It was also possible to increase the amount of histidyl-tRNA synthetase activity by a pleiotropic mutation in the *strB* locus. This increase in synthetase activity was also accompanied by an increase in *his* operon expression. It seems unlikely that the additional synthetase could serve to "trap" the repressing metabolite, histidyl-tRNA, rather it appears to be acting as a positive control element.

Evidence that still another component in the repression (or derepression) mechanism for histidine biosynthesis might be the first enzyme in the pathway came from the studies of Goldberger and his colleagues. These studies began with kinetic analyses of the onset of derepression and of the onset of repression. [For a review of these experiments, see (361). The studies of the derepression kinetics were interesting, but since they did not suggest anything about the mechanism of repression or derepression, they are not reviewed here. The studies of repression kinetics, however, indicated that the first enzyme in the histidine biosynthetic pathway played an important regulatory role in some way.

When histidine was added to a culture that was undergoing derepression, repression of the enzymes occurred "serially," provided that the strain contained the normal, histidine-sensitive phosphoribosyl-ATP synthetase (362). The serial pattern of repression was characterized by repression of the operator-proximal gene product occurring first, followed by repression of the other enzymes in the order that their structural genes occurred in the operon (362). When the feedback site on phosphoribosyl-ATP synthetase was inactive or inaccessible to histidine, repression was "concomitant" for all the enzymes. (Histidine would be prevented from interacting with the feedback site if thiazolealanine, a histidine analog, already covered the site, or if the enzyme was resistant to histidine inhibition, or if the enzyme was missing entirely owing to a nonsense mutation.) When repression was brought about by means of an analog that mimicked the repressing effect of histidine, 1,2,4-triazolealanine, the usual serial repression occurred in strains with a normal phosphoribosyl-ATP synthetase, but there was no repression in the absence of an accessible feedback site on the enzyme (363).

That this role of the first enzyme in repression of the histidine operon might be intimately related to the role of histidyl-tRNA was suggested by the fact that normal phosphoribosyl-ATP synthetase could bind histidyl-tRNA (364). In contrast, a feedback-resistant enzyme could not. The enzyme was also shown to bind to *his* operon DNA, but not if that DNA carried a small deletion in the *hisO* region (365). (The specific deletion, *hisO1242,* is referred to later.) Compatible with this capacity to bind to *his* DNA was the subsequent observation that *his*-specific transcription with RNA polymerase from a ϕ80d*his* DNA template was blocked by phosphoribosyl-ATP synthetase (366). However, the relationship of this inhibition of transcription to in vivo repression was quite unclear, since neither free histidine nor histidyl-tRNA was needed as an effector in the process. It is difficult to relate this inhibitory effect of *his* transcription by the first enzyme in the pathway to any model involving that enzyme as a mediator of an "excess histidine" signal. Rather, the free enzyme bound neither to histidyl-tRNA nor histidine would indicate histidine limitation.

Further, in accord with the view that phosphoribosyl-ATP synthetase has a direct role in regulation of the histidine operon are reports of *hisG* mutations affecting expression of the *his* operon. One mutation in the *hisG* gene, which specifies this enzyme, allows the *his* operon to be repressed by several amino acids, including aspartate (367). That aspartate causes repression in this mutant is of interest, since another *hisG* mutation led to loss of feedback sensitivity to histidine and the gain of sensitivity to aspartate.

More recently, it has been observed that ppGpp acts synergistically with low, partially inhibitory concentrations of histidine to inhibit phosphoribosyl-ATP synthetase strongly (368). This nucleotide is completely ineffective by itself, however. Furthermore, the two compounds also act in synergism to prevent formation of the complex between histidyl-tRNA and phosphoribosyl-ATP synthetase (369). The binding of histidyl-tRNA to the enzyme is not affected by histidine alone (370). In contrast, an enzyme hypersensitive to histidine does have reduced binding of histidyl-tRNA in the presence of histidine alone (371). Thus, binding of ppGpp in effect mimics the mutation by making the wild-type enzyme not only hypersensitive to histidine but also susceptible to an antagonism of histidyl-tRNA binding by histidine. This interaction with ppGpp in the presence of histidine is of interest since, as cited below, there is a strong dependence upon ppGpp in the coupled in vitro transcription-translation system with *his* DNA as the template. At first glance, it might seem contradictory that ppGpp, a positive effector for in vitro coupled synthesis, should act synergistically with histidine, a presumed negative effector, on the enzyme proposed as a major regulatory element in *his* operon function. However, it does appear that ppGpp, which accumulates upon starvation for any amino acid, can stimulate transcription of several biosynthetic operons as well as inhibit transcription of ribosomal RNA operons (338, 339). However, if histidine were not the limiting amino acid, the accumulation of ppGpp might make the histidine-mediated repression of the *his* operon even stronger if the synergistic effect on phosphoribosyl-ATP synthetase does indeed have a regulatory role. When ppGpp has accumulated because of a histidine deficit, the negative effector role of ppGpp could not occur.

It is now clear that whatever the effect of phosphoribosyl-ATP synthetase on *his* operon expression, the role is not an essential one. This conclusion is based on the observations of Scott et al (372) that deletion of essentially the entire *hisG* gene does not lead to derepression nor does it prevent the derepression that results from a mutation in *hisT*. Such experiments do not of course eliminate the possibility of an accessory role for the *hisG* product.

It should be emphasized that the histidine biosynthetic enzymes in *S. typhimurium* are different from several other biosynthetic systems in that the system is, strictly speaking, not a repressible one, i.e. the addition of histidine to a minimal medium does not repress the histidine biosynthetic enzymes (351). There is, however, the rich medium depression that has been observed in other biosynthetic systems and that was referred to in the section on tryptophan biosynthesis. The histidine-specific control, rather, is one leading to derepression. A model to account for the derepression that results from limiting histidine was proposed by Kasai (330) on the basis of in vitro transcription experiments with several DNA templates. It was found that a DNA template containing a small deletion, *hisO1242,* yielded a level of transcription about five times that of a wild-type template. In in vivo experiments, strains with this lesion mimic the derepression achieved by the *hisT1504* lesion. Kasai proposed an attenuator model in which transcripts initiated at the *his* promoter were likely to be terminated at a site on the DNA that was missing in the *hisO1242* mutants. That site he termed the attenuator. He further proposed that some positive control factor was needed to allow transcription to proceed beyond the attenuator site into the *hisG* gene. The effect of the positive control element itself in allowing readthrough rather than termination at the attenuator is apparently antagonized by charged, normal tRNA[His] but not by charged or uncharged *hisT1504* tRNA[His]. In the Kasai model, either the histidyl-tRNA synthetase or phosphoribosyl-ATP synthetase would be candidates for the positive control element, since both have binding sites for histidyl-tRNA. Some mutations in the *hisO* region are mutations that affect polymerase binding, while others primarily affect positive factor response. Therefore, the model further proposed that the positive factor recognized a site near the promoter and actually accompanied the RNA polymerase downstream to the attenuator site. This feature is an attractive one for an attenuator control model, since it would account for the lack of control at an attenuator (or polarity) site when transcription of a bacterial operon is under N gene control (by a "juggernaut" polymerase) on a transducing derivative of λ (373–375).

An in vitro demonstration of coupled transcription-translation of the *his* operon that is very much in accord with the Kasai model has been described by Artz & Broach (376). In this system, histidinol dehydrogenase was formed with a template of DNA from a λh80 transducing phage carrying either the wild-type (*hisO⁺*) *S. typhimurium his* operon or one with the *hisO1242* deletion. The enzyme formed with the *hisO1242* template greatly exceeded that with the wild type. With the *hisO1242* template, it did not matter whether the source of tRNA in the coupled reaction mixture came from a wild-type or a *hisT1504* strain. When the DNA bore the wild-type attenuator, tRNA from the *hisT1504* strain allowed twice as much enzyme formation as did tRNA from a wild-type organism.

The expression of the *his* operon is remarkably dependent upon the presence of ppGpp in the coupled reaction mixture with either a *hisO⁺* or a *hisO1242* template (377). Thus, ppGpp does not act at the level of attenuation. Since it is effective only when present during transcription, it probably acts at the level of transcription initiation.

In the Artz-Broach model, phosphoribosyl ATP synthetase is considered an accessory factor affecting repression by making the restoration of transcription more efficient. This would be in accord with the experiments of Goldberger and his colleagues (361), whose observations on the in vivo effects of an available feedback site on phosphoribosyl-ATP synthetase pointed to a transient effect of this enzyme, not on the steady state level of repression.

It is apparently not now possible to account for the mechanism by which the state of the feedback site on phosphoribosyl-ATP synthetase affects the kinetics of onset of repression. It would be of interest to know whether any of the total tRNA[His] was bound to phosphoribosyl-ATP synthetase at the time of adding histidine. If the normal, feedback-sensitive synthetase had served as a "sink" for tRNA[His], charged or uncharged, it could have perturbed the response to added histidine. Thus, there is still a possibility that the first enzyme in histidine biosynthesis has only a trivial role in control of the *his* operon, which manifests itself only during the onset of repression of cells already highly derepressed, by virtue of a metabolic interaction, not a genetic one.

Histidine biosynthesis has been studied less well in other forms. Because of the similarity of the gene arrangement in *E. coli,* it can be inferred that the findings in *S. typhimurium* would be duplicated in *E. coli.* In yeast, the genes are not arranged in operons but are separate transcriptional units. An exception might have been the *his-4* locus, which specified three enzyme activities, catalyzing Reactions 2, 3, and 10 in Figure 6 (378). Indeed, under certain conditions these activities can be separated from each other. However, it has now been demonstrated that the three activities are carried on a single polypeptide chain, and that only following proteolytic activity do they appear as separate protein fragments (R. Bigelis, personal communication).

In view of the cross-pathway regulation of the histidine biosynthetic enzymes in *N. crassa* by limiting tryptophan referred to earlier (349), it is of interest that limiting histidine also leads to derepression of the histidine, arginine, and tryptophan biosynthetic pathways (379).

GENERAL CONTROLS OVER AMINO ACID BIOSYNTHESIS

Not all the controls over amino acid biosynthesis that have been touched upon in this review have been specific controls related directly to the level of the product. For example, in the absence of specific *trpR*-mediated control of the *trp* operon, there is a growth rate control that keeps even the nonrepressed *trp* operon function proportional to the rate of growth (340). While the control mechanism may involve termination by ρ of transcription at the specific *trp* attenuator site, the extent of antitermination that occurs is dependent on the amount of AT, which in turn is growth rate controlled and may be, like ρ itself, a general control element (341, 342).

Still another type of control is the rich medium depression of synthesis (340), which occurs rather generally for amino acid biosynthetic pathways. In a rich medium, biosynthetic enzyme activities of at least several pathways are much lower than the "maximally" repressed level achieved by end-product supplementation. This level could be due in part to the commitment of either the available RNA polymerases to transcription of ribosomal RNA and ribosomal protein operons or to the available ribosomes translating ribosomal protein messages or both, depending upon the biosynthetic system being depressed.

Another element of general control in bacterial amino acid biosynthetic pathways is ppGpp. Since the transcription of several operons has been shown to be stimulated by ppGpp, one may assume that it has a general effect on many biosynthetic promoters (338, 339). While this is a general property of ppGpp, its effect is actually rendered more specific by the fact that, by itself, ppGpp does not stimulate transcription of operons unless the other factors needed for that transcription initiation are also present. Thus, ppGpp stimulates *lac* and *ilvC* gene function but not in the absence of the specific inducing metabolites (cAMP and acetohydroxybutyrate, respectively) (232).

There have been several examples reported that might be referred to as cross-pathway regulation or cross-feedback, which was anticipated 17 years ago by Monod & Jacob (380). In cross-feedback, the end product of one pathway would inhibit the first enzyme or repress the enzymes of another pathway. Among examples cited in this review are the inhibition or particularly the enhancement of inhibition of aspartokinase III by leucine and other hydrophobic amino acids (125). Rebello & Jensen (286) have described cross-feedback among the aromatic amino acids and have used the term "metabolic interlock" to describe the relationship. While such interactions might have an integrating influence to coordinate the multiple pathways of the cell, it is difficult at this time to decide whether these examples are interesting physiological freaks of nature or highly evolved general metabolic controls.

The idea of cross-pathway regulation has been more extensively encountered in yeast and *Neurospora*. As mentioned earlier, Delforge et al (46) demonstrated a general control over biosynthetic enzymes in the pathways leading to tryptophan, histidine, lysine, isoleucine and valine, leucine, and arginine. This general control was revealed not only by mutations in three regulatory loci originally identified as arginine-related regulatory loci but also by limitations in the supply of any one of the amino acid end products. Wolfner et al (381) have described two kinds of lesion that interfere with this general control. One class, affecting the *aas* locus, prevents derepression of the enzymes for biosynthesis of arginine, lysine, histidine, or tryptophan. The second class, affecting the *tra3* locus, are fully derepressed for these biosynthetic enzymes. These lesions lead to temperature sensitivity. Thus, at 36°C, growth ceases early in the G1 phase of growth. The *tra3* locus may thus be related to the sensor system that enables yeast cells to sense the availability of adequate nutrients (carbon, nitrogen, phosphorus, sulfur) before initiating cell division (382). One model involving these two control loci is one in which the *tra3* product is a negative control element preventing expression of a number of amino acid biosynthesis genes and the *aas* product is a positive control element antagonizing the *tra3* product (381). The *aas* product is inactivated by the concerted action of arginine, histidine, lysine, and tryptophan. The absence of any one of the amino acids leads to activation of the *aas* product and neutralization of the negative regulatory factor (*tra3* product).

Another important feature in the control of biosynthesis in the fungi is the existence of the kind of compartmentation that was referred to in the section on the arginine and the isoleucine and valine pathways (22, 23, 251, 252). Certainly, the separation of certain enzymes into either a cytosol or mitochondrial fraction and the enclosure of pools of products or certain intermediates into vesicles may be more widespread than is currently realized. Such mechanisms could of course be very important in controlling the regulation of branching pathways. A recent review by Davis (383) has focused on the importance of these aspects of fungal metabolism. This should continue to be a worthwhile area to explore as an alternative mode of "channelling" metabolites to that achieved by the evolution of enzyme complexes. It is this latter mode of channelling that may be more likely to occur in bacteria. It is also a field that may require some technical developments since, if many biosynthetic sequences in bacteria are catalyzed by complexes, those complexes have so far resisted detection.

CONCLUSION

At the outset of this undertaking, the reviewer had hoped to prepare a coherent overview of the vast field that is regulation of amino acid biosynthesis. It is an attempt that should have been made for the last time about five years ago. It has not been possible to give equal coverage to all twenty amino acids, for our knowledge is very uneven. Indeed, the concepts underlying the latest experimental results obtained for one pathway often differed as night from day from those obtained for another. It was also not possible to touch on the really practical developments of the Japanese amino acid fermentation industry. These developments have come from taking advantage of the way the regulation of amino acid biosynthesis can be genetically manipulated, or by bypassing the sites of end-product control. The topic was given an extensive coverage several years ago (384) that should serve to indicate a rather immediate application of the study of regulation of amino acid biosynthesis. Nevertheless, it is hoped that along with some of the more specific reviews that have been cited, this review will be of help in understanding and appreciating some of the accomplishments that have occurred and are still occurring in this field.

Literature Cited

1. Mifflin, B. J., Lea, P. J. 1977. *Ann. Rev. Plant Physiol.* 28:299–329
2. Tyler, B. 1978. *Ann. Rev. Biochem.* 47:000–000
3. Baich, A. 1969. *Biochim. Biophys. Acta* 192:462–67
4. Baich, A. 1971. *Biochim. Biophys. Acta* 244:129–34
5. Gamper, H., Moses, V. 1974. *Biochim. Biophys. Acta* 354:75–87
6. Rossi, J., Vender, J., Berg, C., Coleman, W. 1977. *J. Bacteriol.* 129:108–14
7. Vyas, S., Maas, W. K. 1963. *Arch. Biochem. Biophys.* 100:542–46
8. Haas, D., Kurer, V., Leisinger, T. 1972. *Eur. J. Biochem.* 31:290–95
9. Haas, D., Leisinger, T. 1974. *Pathol. Microbiol.* 40:140–42
10. Haas, D., Leisinger, T. 1974. *Biochem. Biophys. Res. Commun.* 60:42–47
11. Udaka, S. 1966. *J. Bacteriol.* 91:617–21
12. Cybis, J. J., Davis, R. H. 1974. *Biochem. Biophys. Res. Commun.* 60: 629–34
13. Farago, A., Denes, G. 1969. *Biochim. Biophys. Acta* 178:400–2
14. Haas, D., Leisinger, T. 1975. *Eur. J. Biochem.* 52:377–83
15. Anderson, P. M., Marvin, S. V. 1969. *Biochemistry* 9:171–78
16. Anderson, P. M. 1977. *Biochemistry* 16:587–93

17. Williams, L. G., Davis, R. H. 1970. *J. Bacteriol.* 103:335–41
18. Lue, P. F., Kaplan, J. G. 1970. *Biochim. Biophys. Acta* 220:365–72
19. Bernhard, S. A., Davis, R. H. 1971. *J. Biol. Chem.* 246:973–78
20. LaCroute, F., Pierard, A., Grenson, M., Wiame, J. M. 1965. *J. Gen. Microbiol.* 40:127–42
21. Weiss, R. L., Davis, R. H. 1973. *J. Biol. Chem.* 248:5403–8
22. Bernhardt, S. A., Davis, R. H. 1972. *Proc. Natl. Acad. Sci. USA* 69:1868–72
23. Weiss, R. L. 1973. *J. Biol. Chem.* 248:5409–13
24. Bowman, B. J., Davis, R. H. 1977. *J. Bacteriol.* 130:285–91
25. Penninckx, M., Simon, J. P., Wiame, J. M. 1974. *Eur. J. Biochem.* 49:429–42
26. Penninckx, M. 1975. *Eur. J. Biochem.* 58:533–38
27. Messenguy, F., Penninckx, M., Wiame, J. M. 1971. *Eur. J. Biochem.* 22:277–86
28. Vogel, H. J. 1953. In *Chemical Basis of Heredity,* ed. W. J. McElroy, B. Glass, pp. 276–89. Baltimore: Johns Hopkins Univ. Press. 848 pp.
29. Gorini, L., Maas, W. K. 1957. *Biochim. Biophys. Acta* 25:208–9
30. Gorini, L. 1960. *Proc. Natl. Acad. Sci. USA* 46:682–90
31. Maas, W. K., Clark, A. J. 1964. *J. Mol. Biol.* 8:359–64
32. Kelker, N. E., Maas, W. K., Yang, H.-L., Zubay, G. 1976. *Mol. Gen. Genet.* 144:17–20
33. Cunin, R., Kelker, N., Boyen, A., Yang, H.-L., Zubay, G., Glansdorff, N., Maas, W. K. 1976. *Biochem. Biophys. Res. Commun.* 69:377–82
34. Sens, D., Natter, W., James, E. 1977. *J. Bacteriol.* 130:642–55
35. Sens, D., James, E. 1975. *Biochem. Biophys. Res. Commun.* 64:169–74
36. Vogel, R. H., McLellan, W. L., Hirvonen, A. P., Vogel, H. J. 1971. In *Metabolic Pathways,* ed. H. J. Vogel, 5:463–88. New York: Academic. 576 pp.
37. Celis, T. F. R., Maas, W. K. 1971. *J. Mol. Biol.* 62:179–88
38. Hirshfield, I. N., DeDeken, R., Horn, P. C., Hopwood, D. A., Maas, W. K. 1968. *J. Mol. Biol.* 35:83–93
39. Williams, L. S. 1973. *J. Bacteriol.* 112:1419–32
40. Faanes, R., Rogers, P. 1972. *J. Bacteriol.* 112:102–13
41. Krzyzek, R. A., Rogers, P. 1976. *J. Bacteriol.* 126:348–64
42. Lavallé, R. 1970. *J. Mol. Biol.* 51:449–51
43. McLellan, W. L., Vogel, H. J. 1972. *Biochem. Biophys. Res. Commun.* 48:1027–33
44. Cunin, R., Boyen, A., Pouwels, P., Glansdorff, N., Cabeel, M. 1975. *Mol. Gen. Genet.* 140:51–60
45. Krzyzek, R. A., Rogers, P. 1976. *J. Bacteriol.* 126:365–76
46. Delforge, J., Messenguy, F., Wiame, J. M. 1975. *Eur. J. Biochem.* 57:231–39
47. Messenguy, F., Delforge, J. 1976. *Eur. J. Biochem.* 67:335–39
48. Messenguy, F., Cooper, T. G. 1977. *J. Bacteriol.* 130:1253–61
49. Bhattacharjee, J. K., Sinha, A. K. 1972. *Mol. Gen. Genet.* 115:26–30
50. Hogg, R. W., Broquist, H. P. 1968. *J. Biol. Chem.* 243:1839–45
51. Tucci, A. F., Ceci, L. N. 1972. *Arch. Biochem. Biophys.* 153:742–50
52. Tucci, A. F., Ceci, L. N. 1972. *Arch. Biochem. Biophys.* 153:751–54
53. Sinha, A. K., Kurtz, M., Bhattacharjee, J. K. 1973. *Appl. Microbiol.* 26:303–8
54. Beremand, P. D., Sojka, G. A. 1977. *J. Bacteriol.* 130:532–34
55. Harder, W., Quayle, J. R. 1971. *Biochem. J.* 121:753–62
56. Winicov, I., Pizer, L. I. 1974. *J. Biol. Chem.* 249:1348–55
57. Dubrow, R., Pizer, L. I. 1977. *J. Biol. Chem.* 252:1527–38
58. Tosa, T., Pizer, L. I. 1971. *J. Bacteriol.* 106:972–82
59. Pizer, L. I., McKitrick, J., Tosa, T. 1972. *Biochem. Biophys. Res. Commun.* 49:1351–57
60. Pizer, L. I. 1964. *J. Biol. Chem.* 239:4219–26
61. Dempsey, W. B., Itoh, H. 1970. *J. Bacteriol.* 104:658–67
62. Tani, Y., Dempsey, W. B. 1973. *J. Bacteriol.* 116:341–45
63. Newman, E. B., Miller, B., Kapoor, V. 1974. *Biochim. Biophys. Acta* 338:529–39
64. Monsouri, A., Decter, J. B., Silber, R. 1972. *J. Biol. Chem.* 247:348–52
65. Greene, R. C., Radovich, C. 1975. *J. Bacteriol.* 124:269–78
66. Taylor, R., Dickerman, H., Weissbach, H. 1966. *Arch. Biochem. Biophys.* 117:405–12
67. Stauffer, G. V., Baker, C. A., Brenchley, J. E. 1974. *J. Bacteriol.* 120:1017–25
68. Miller, B. A., Newman, E. B. 1974. *Can. J. Microbiol.* 20:41–47
69. Stauffer, G. V., Brenchley, J. E. 1977. *J. Bacteriol.* 129:740–49
70. Meedel, T. H., Pizer, L. I. 1974. *J. Bacteriol.* 118:905–10

71. Umbarger, H. E. 1977. *Biochem. Educ.* 5:67–71
72. Pizer, L. I. 1965. *J. Bacteriol.* 89:1145–50
73. Newman, E. B., Batist, G., Frazer, J., Isenberg, S., Weyman, P., Kapoor, V. 1976. *Biochim. Biophys. Acta* 421:97–105
74. Faeder, E. J., Davis, P. S., Siegel, L. M. 1974. *J. Biol. Chem.* 249:1599–1609
75. Tsang, M. L.-S., Schiff, J. A. 1975. *Plant Sci. Lett.* 4:301–7
76. Tsang, M. L.-S., Schiff, J. A. 1976. *J. Bacteriol.* 125:923–33
77. Mark, D. F., Richardson, C. C. 1976. *Proc. Natl. Acad. Sci. USA* 73:780–84
78. Holmgren, A. 1976. *Proc. Natl. Acad. Sci. USA* 73:2275–79
79. Tsang, M. L.-S., Schiff, J. A. 1977. *Fed. Proc.* 36:651
80. Kredich, N. M., Tompkins, G. M. 1966. *J. Biol. Chem.* 241:4955–65
81. Ellis, R. J. 1964. *Biochem. J.* 93:19–20P
82. Jones-Mortimer, M. C., Wheldrake, J. F., Pasternak, C. A. 1968. *Biochem. J.* 107:51–53
83. Spencer, H. T., Collins, J., Monty, K. J. 1967. *Fed. Proc.* 26:277
84. Tully, M., Yudkin, M. D. 1975. *Mol. Gen. Genet.* 136:181–83
85. Cheney, R. W. Jr., Kredich, N. M. 1975. *J. Bacteriol.* 124:1273–81
86. Sanderson, K. E. 1972. *Bacteriol. Rev.* 36:558–86
87. Bachmann, B. J., Low, K. B., Taylor, A. L. 1976. *Bacteriol. Rev.* 40:116–67
88. Laughlin, R. E. 1975. *J. Gen. Microbiol.* 86:275–82
89. Reureny, Z. 1977. *Proc. Natl. Acad. Sci. USA* 74:619–22
90. Paszewski, A., Grabski, J. 1974. *Mol. Gen. Genet.* 132:307–20
91. Flavin, M. 1975. In *Metabolic Sulfur Compounds,* Vol. VII: Metabolic Pathways, ed. D. M. Greenberg, pp. 457–503. New York: Academic. 614 pp. 3rd ed.
92. Cherest, H., Eichler, F., deRobichon-Szulmajster, H. 1969. *J. Bacteriol.* 97:328–36
93. Yamagata, S., Takeshima, K., Naiki, N. 1974. *J. Biochem. Tokyo* 75:1221–29
94. Cherest, H., Surdin-Kerjan, Y., de-Robichon-Szulmajster, H. 1971. *J. Bacteriol.* 106:758–72
95. Cherest, H., Surdin-Kerjan, Y., Antoniewski, J., deRobichon-Szulmajster, H. 1973. *J. Bacteriol.* 115:1084–93
96. Cherest, H., Surdin-Kerjan, Y., de-Robichon-Szulmajster, H. 1975. *J. Bacteriol.* 123:428–35
97. Rudman, D., Meister, A. 1953. *J. Biol. Chem.* 200:591–604
98. Chesne, S., Pelmont, J. 1973. *Biochimie* 55:237–44
99. Chesne, S., Pelmont, J. 1974. *Biochimie* 56:631–39
100. Chesne, S., Montmitonnet, A., Pelmont, J. 1975. *Biochimie* 57:1029–34
101. Monnier, N., Montmitonnet, A., Chesne, S., Pelmont, J. 1976. *Biochimie* 58:663–75
102. Gelfand, D. H., Rudo, N. 1977. *J. Bacteriol.* 130:441–44
103. Gelfand, D. H., Steinberg, R. A. 1977. *J. Bacteriol.* 130:429–40
104. Urm, E., Leisinger, T., Vogel, H. J. 1973. *Biochim. Biophys. Acta* 302:249–60
105. Cedar, H., Schwartz, J. H. 1969. *J. Biol. Chem.* 244:4112–21
106. Ravel, J. M., Norton, S. J., Humphreys, J. S., Shire, W. 1962. *J. Biol. Chem.* 237:2845–49
107. Thompson, L. H., Stanners, C. P., Siminovitch, L. 1973. *Somatic Cell Genet.* 1:187–208
108. Arfin, G. M., Simpson, D. R., Chiang, C. S., Andrulis, I. L., Hatfield, G. W. 1977. *Proc. Natl. Acad. Sci. USA* 74:2367–69
109. Patte, J. C., Lebras, G., Cohen, G. N. 1967. *Biochim. Biophys. Acta* 136:245–57
110. Cohen, G. N., Stanier, R. Y., LeBras, G. 1969. *J. Bacteriol.* 99:791–801
111. Costrejean, J. M., Guiso, N., Cowie, D. B., Cohen, G. N., Truffa-Bachi, P. 1975. *Eur. J. Biochem.* 50:431–35
112. Ogilvie, J. W., Vickers, L. P., Clark, R. B., Jones, M. M. 1975. *J. Biol. Chem.* 250:1242–60
113. Szczesiul, M., Wampler, D. E. 1976. *Biochemistry* 15:2236–44
114. Tilak, A., Wright, K., Damie, S., Takahashi, M. 1976. *Eur. J. Biochem.* 69:249–55
115. Truffa-Bachi, P., Veron, M., Cohen, G. N. 1974. *CRC Crit. Rev. Bioch.* Oct. 1974:379–415
116. Veron, M., Saari, J. C., Villar-Palasi, C., Cohen, G. N. 1973. *Eur. J. Biochem.* 38:325–35
117. Hirth, C. G., Veron, M., Villar-Palasi, C., Hurion, N., Cohen, G. N. 1975. *Eur. J. Biochem.* 50:425–30
118. Veron, M., Falcoz-Kelly, F., Cohen, G. N. 1972. *Eur. J. Biochem.* 28:520–27
119. Theze, J., Saint-Girons, I. 1974. *J. Bacteriol.* 118:990–98
120. Studdard, C. 1973. *J. Bacteriol.* 116:1–11

121. Freundlich, M. 1963. *Biochem. Biophys. Res. Commun.* 10:277–82
122. Falcoz-Kelly, F., van Rappenbosch, R., Cohen, G. N. 1969. *Eur. J. Biochem.* 8:146–52
123. Cafferata, R. L., Freundlich, M. 1969. *J. Bacteriol.* 97:193–98
124. Stadtman, E. R., Cohen, G. N., LeBras, G., deRobichon-Szulmajster, H. 1961. *J. Biol. Chem.* 236:2033–38
125. Patte, J.-C., Loviny, T., Cohen, G. N. 1965. *Biochim. Biophys. Acta* 99:523–30
126. Patte, J.-C., Zuber, M., Borne, F. 1972. *Mol. Gen. Genet.* 116:35–39
127. Richaud, C., Mazat, J. P., Gros, C., Patte, J.-C. 1974. *Eur. J. Biochem.* 40:619–29
128. Richaud, C., Mazat, J. P., Felenbok, B., Patte, J.-C. 1974. *Eur. J. Biochem.* 48:147–56
129. Funkhouser, J. D., Abraham, A., Smith, A. V., Smith, W. G. 1974. *J. Biol. Chem.* 249:5478–84
130. Shaw, J.-F., Smith, W. G. 1976. *Biochim. Biophys. Acta* 422:302–8
131. Niles, E. G., Westhead, E. W. 1973. *Biochemistry* 12:1723–29
132. Niles, E. G., Westhead, E. W. 1973. *Biochemistry* 12:1715–22
133. Baich, A. 1970. *Biochem. Biophys. Res. Commun.* 39:544–50
134. Baich, A., Hagan, S. 1970. *Biochem. Biophys. Res. Commun.* 40:445–53
135. Paulus, H., Gray, E. 1964. *J. Biol. Chem.* 239:4008–9
136. Paulus, H., Gray, E. 1967. *J. Biol. Chem.* 242:4980–86
137. Datta, P., Gest, H. 1964. *Proc. Natl. Acad. Sci. USA* 52:1004–9
138. Burlant, L., Datta, P., Gest, H. 1965. *Science* 148:1351–53
139. Biswas, C., Paulus, H. 1973. *J. Biol. Chem.* 248:2894–900
140. Rosner, A., Paulus, H. 1971. *J. Biol. Chem.* 246:2965–71
141. Moir, D., Paulus, H. 1977. *J. Biol. Chem.* 252:4648–54
142. Moir, D., Paulus, H. 1977. *J. Biol. Chem.* 252:4655–61
143. Gray, B. H., Bernlohr, R. W. 1969. *Biochim. Biophys. Acta* 178:248–61
144. Biswas, C., Gray, E., Paulus, H. 1970. *J. Biol. Chem.* 245:4900–6
145. Jenkins, M. B., Woodward, D. W. 1970. *Biochim. Biophys. Acta* 212:21–32
146. Matthews, B., Gurman, A., Bryan, J. 1975. *Plant Physiol.* 55:991–98
147. Boy, E., Patte, J.-C. 1972. *J. Bacteriol.* 112:84–92
148. Holland, M. J., Westhead, E. W. 1973. *Biochemistry* 12:2276–81
149. deRobichon-Szulmajster, H., Corrivaux, D. 1964. *Biochim. Biophys. Acta* 92:1–9
150. Anderson, K. B., von Meyenburg, K. 1977. *J. Biol. Chem.* 252:4151–56
151. Kindler, S. H., Gilvarg, C. 1960. *J. Biol. Chem.* 235:3532–35
152. Sundharadas, G., Gilvarg, C. 1967. *J. Biol. Chem.* 242:3983–84
153. Work, E. 1957. *Nature* 179:841–47
154. Yugari, Y., Gilvarg, C. 1962. *Biochim. Biophys. Acta* 62:612–14
155. Chasin, L. A., Szulmajster, J. 1967. *Biochem. Biophys. Res. Commun.* 29:648–54
156. Aronson, A. I., Henderson, E., Tincher, A. 1967. *Biochim. Biophys. Res. Commun.* 26:454–60
157. Rosner, A. 1975. *J. Bacteriol.* 121:20–28
158. Hirshfield, I. N., Zamecnik, P. C. 1972. *Biochim. Biophys. Acta* 259:330–33
159. Boy, E., Reinisch, F., Richaud, C., Patte, J.-C. 1976. *Biochimie* 58:213–18
160. Patte, J.-C. Boy, E., Borne, F. 1974. *FEBS Lett.* 43:67–70
161. Herbert, A. A., Guest, J. R. 1968. *J. Gen. Microbiol.* 53:363–81
162. Flavin, M., Slaughter, C. 1967. *Biochim. Biophys. Acta* 132:400–5
163. Salem, A. R., Foster, M. A. 1972. *Biochem. J.* 127:845–53
164. Savin, M. A., Flavin, M. 1972. *J. Bacteriol.* 112:299–303
165. Kerr, D. S., Flavin, M. 1970. *J. Biol. Chem.* 245:1842–55
166. Datko, A. H., Mudd, S. H., Giovanelli, J. 1977. *J. Biol. Chem.* 252:3436–45
167. Datko, A. H., Giovanelli, J., Mudd, S. H. 1974. *J. Biol. Chem.* 249:1139–55
168. Lee, L. W., Ravel, J. M., Shive, W. 1966. *J. Biol. Chem.* 241:5479–80
169. Wyman, A., Paulus, H. 1975. *J. Biol. Chem.* 250:3897–903
170. Lawrence, D. A. 1972. *J. Bacteriol.* 109:8–11
171. deRobichon-Szulmajster, H., Cherest, H. 1967. *Biochem. Biophys. Res. Commun.* 28:256–62
172. Selhub, J., Burton, E., Sakami, W., Flavin, M. 1971. *Proc. Natl. Acad. Sci. USA* 68:312–14
173. Chater, K. F. 1970. *J. Gen. Microbiol.* 63:95–109
174. Greene, R. C., Su, C.-H., Holloway, C. T. 1970. *Biochem. Biophys. Res. Commun.* 38:1120–26
175. Greene, R. C., Williams, R., Kung, H.-F., Spears, C., Weissbach, H. 1973. *Arch. Biochem. Biophys.* 158:249–56
176. Ahmed, A. 1973. *Mol. Gen. Genet.* 123:299–324

177. Hobson, A. C., Smith, D. A. 1973. *Mol. Gen. Genet.* 126:7–81
178. Savin, M. A., Flavin, M., Slaughter, C. 1972. *J. Bacteriol.* 111:547–56
179. Hobson, A. C. 1974. *Mol. Gen. Genet.* 131:263–73
180. Hobson, A. C. 1976. *Mol. Gen. Genet.* 144:87–95
181. Kung, H.-F., Spears, C., Greene, R. C., Weissbach, H. 1972. *Arch. Biochem. Biophys.* 150:23–31
182. Surdin-Kerjan, Y., deRobichon-Szulmajster, H. 1975. *J. Bacteriol.* 122:367–74
183. Burr, B., Walker, J., Truffa-Bachi, P., Cohen, G. N. 1976. *Eur. J. Biochem.* 62:519–26
184. Dwyer, S. B., Umbarger, H. E. 1968. *J. Bacteriol.* 95:1680–84
185. Johnson, E., Cohen, G., Saint-Girons, I. 1977. *J. Bacteriol.* 129:66–70
186. Nass, G., Poralla, K., Zähner, H. 1969. *Biochem. Biophys. Res. Commun.* 34:84–91
187. Gardner, J., Smith, O. 1975. *J. Bacteriol.* 124:161–66
188. Saint-Girons, I., Margarita, D. 1975. *J. Bacteriol.* 124:1137–41
189. Szczesiul, M., Wampler, D. E. 1976. *Biochemistry* 15:2236–44
190. Atkinson, D. E. 1971. In *Metabolic Pathways,* ed. H. J. Vogel, 5:1–21. New York: Academic. 576 pp.
191. Madison, J. T., Thompson, J. F. 1976. *Biochem. Biophys. Res. Commun.* 71:684–91
192. Gladstone, J. P. 1939. *Br. J. Exp. Pathol.* 20:189–200
193. Katsumma, T., Elsässer, S., Holzer, H. 1971. *Eur. J. Biochem.* 24:83–87
194. Ahmed, S., Bollon, A., Rogers, S., Magee, P. 1976. *Biochimie* 58:225–32
195. Decedue, C. J., Hoffler, J. G., Burns, R. O. 1975. *J. Biol. Chem.* 250:1563–70
196. Calhoun, D. H., Rimerman, R. A., Hatfield, G. W. 1973. *J. Biol. Chem.* 248:3511–16
197. Grimminger, H., Rahimi-Lavidjani, I., Lingens, F. 1973. *FEBS Lett.* 35:273–75
198. Kagan, Z. S., Dorozhko, A. I., Kovaleva, S. V., Yakovleva, L. I. 1975. *Biochim. Biophys. Acta* 403:208–20
199. Barritt, G. J., Morrison, J. F. 1972. *Biochim. Biophys. Acta* 284:508–20
200. Feldberg, R. S., Datta, P. 1971. *Eur. J. Biochem.* 21:438–46
201. Feldner, J., Grimminger, H. 1976. *J. Bacteriol.* 126:100–7
202. O'Neill, J. P., Freundlich, M. 1972. *Biochem. Biophys. Res. Commun.* 48:437–43
203. Blatt, J. M., Pledger, W. J., Umbarger, H. E. 1972. *Biochem. Biophys. Res. Commun.* 48:444–50
204. Arfin, S. M., Koziell, D. A. 1973. *Biochim. Biophys. Acta* 321:356–60
205. Arfin, S. M., Koziell, D. A. 1973. *Biochim. Biophys. Acta* 321:348–55
206. Raimond, J., Grelet, N. 1971. *Biochimie* 53:783–88
207. Gupta, N. K., Vennesland, B. 1966. *Arch. Biochem. Biophys.* 113:255–64
208. Loutit, J. S., Davis, P. R. 1970. *Biochim. Biophys. Acta* 222:222–24
209. Teng-Leary, E., Kohlhaw, G. B. 1973. *Biochemistry* 12:2980–86
210. Soper, T., Doellgart, G., Kohlhaw, G. B. 1976. *Arch. Biochem. Biophys.* 173:362–74
211. Tracy, J. W., Kohlhaw, G. B. 1975. *Proc. Natl. Acad. Sci. USA* 72:1802–6
212. Brown, H. D., Satyanarayana, T., Umbarger, H. E. 1975. *J. Bacteriol.* 121:959–69
213. Kisumi, A., Sugiura, M., Chibata, I. 1976. *J. Biochem. Tokyo* 80:333–39
214. Charon, N. W., Johnson, R. C., Peterson, D. 1974. *J. Bacteriol.* 117:203–11
215. Phillips, A. T., Nuss, J. I., Moosic, J., Foshay, C. 1972. *J. Bacteriol.* 109:714–19
216. Skarstedt, M. T., Greer, S. B. 1973. *J. Biol. Chem.* 248:1032–44
217. McGilvray, D., Umbarger, H. E. 1974. *J. Bacteriol.* 120:715–23
218. Collier, R. H., Kohlhaw, G. 1972. *J. Bacteriol.* 112:365–71
219. Calvo, J. M., Freundlich, M., Umbarger, H. E. 1969. *J. Bacteriol.* 97:1272–82
220. Alexander, R. R., Calvo, J. M., Freundlich, M. 1971. *J. Bacteriol.* 106:213–20
221. Rizzino, A. A., Bresaliar, R. S., Freundlich, M. 1974. *J. Bacteriol.* 117:449–55
222. Cortese, R., Landsberg, R., Vonder Haar, R. A., Ames, B. N. 1974. *Proc. Natl. Acad. Sci. USA* 71:1857–61
223. Kline, E. L. 1972. *J. Bacteriol.* 110:1127–34
224. Friedberg, D., Mikulka, T. W., Jones, J., Calvo, J. M. 1974. *J. Bacteriol.* 118:942–63
225. Pledger, W. J., Umbarger, H. E. 1973. *Bacteriol. Rev.* 114:183–94
226. Davis, M. G., Calvo, J. M. 1977. *J. Bacteriol.* 131:997–1007
227. Kline, E. L., Brown, C. S., Coleman, W. G. Jr., Umbarger, H. E. 1974. *Biochem. Biophys. Res. Commun.* 57:1144–51
228. Smith, J. M., Smolin, D. E., Umbarger, H. E. 1976. *Mol. Gen. Genet.* 148:111–24

229. DeFelice, M., Guardiola, J., Esposito, B., Iaccarino, M. 1974. *J. Bacteriol.* 120:1068–77
230. Ratzkin, B., Arfin, S., Umbarger, H. E. 1972. *J. Bacteriol.* 112:131–41
231. Smith, J. M., Umbarger, H. E. 1977. *J. Bacteriol.* 132:870–75
232. Wild, J., Smith, J. M., Umbarger, H. E. 1977. *J. Bacteriol.* 132:876–83
233. Pledger, W. J., Umbarger, H. E. 1973. *J. Bacteriol.* 114:195–207
234. O'Neill, J. P., Freundlich, M. 1973. *J. Bacteriol.* 116:98–106
235. Glover, S. W. 1962. *Genet. Res.* 38:448–60
236. Ramakrishnan, T., Adelberg, E. A. 1965. *J. Bacteriol.* 89:654–60
237. Favre, R., Wiates, A., Puppo, S., Iaccarino, M., Noelle, R., Freundlich, M. 1976. *Mol. Gen. Genet.* 143:243–52
238. Cohen, B. M., Jones, E. W. 1976. *Genetics* 33:201–25
239. Smith, J. M. 1977. *Genetic Organization and Regulation of the ilv Gene Cluster.* PhD thesis. Purdue Univ., West Lafayette, Indiana. 187 pp.
240. Pouwels, P. H., Cunin, R., Glansdorff, N. 1974. *J. Mol. Biol.* 83:421–24
241. Guha, A., Saturen, Y., Szybalski, W. 1971. *J. Mol. Biol.* 56:53–62
242. Levinthal, M., Levinthal, M., Williams, L. S. 1976. *J. Mol. Biol.* 102:453–65
243. Hatfield, G. W., Burns, R. G. 1970. *Proc. Natl. Acad. Sci. USA* 66:1027–35
244. Calhoun, D. H., Hatfield, G. W. 1973. *Proc. Natl. Acad. Sci. USA* 70:2757–61
245. Coleman, W. G. Jr., Kline, E. L., Brown, C. S., Williams, L. S. 1975. *J. Bacteriol.* 121:785–93
246. Williams, A. L., Whitfield, S. M., Williams, L. S. 1977. *Abstr. Ann. Mtg. Am. Soc. Microbiol., 77th, New Orleans,* p. 213
247. Polacco, J. C., Gross, S. R. 1973. *Genetics* 74:443–59
248. Olshan, A. R., Gross, S. R. 1974. *J. Bacteriol.* 118:371–84
249. Bollon, A. P., Magee, P. T. 1973. *J. Bacteriol.* 113:1333–44
250. Satyanarayana, T., Umbarger, H. E., Lindegren, G. 1968. *J. Bacteriol.* 96:2018–24
251. Cassady, W. E., Leiter, E. H., Bergquist, A., Wagner, R. P. 1972. *J. Cell Biol.* 53:66–72
252. Ryan, E. D., Kohlhaw, G. B. 1974. *J. Bacteriol.* 120:631–37
253. Ryan, E. D., Tracy, J. W., Kohlhaw, G. B. 1973. *J. Bacteriol.* 116:222–25
254. Falkinham, J. O. III. 1977. *J. Bacteriol.* 130:566–68
255. Csonka, L. N. 1977. *J. Biol. Chem.* 252:3392–98
256. Crawford, I. P. 1975. *Bacteriol. Rev.* 39:87–120
257. Simpson, R., Davidson, B. 1976. *Eur. J. Biochem.* 70:493–500
258. Dusha, I., Denes, G. 1976. *Biochim. Biophys. Acta* 438:563–73
259. Schoner, R., Herrmann, K. 1976. *J. Biol. Chem.* 251:5440–47
260. Hu, C., Sprinson, D. 1977. *J. Bacteriol.* 129:177–83
261. Camakaris, H., Pittard, J. 1974. *J. Bacteriol.* 120:590–97
262. Hoffman, P. J., Doy, C. H., Catcheside, D. E. A. 1972. *Biochim. Biophys. Acta* 268:550–61
263. Huang, L., Montoya, A., Nester, E. 1975. *J. Biol. Chem.* 250:7675–81
264. Hagino, H., Nakayama, K. 1975. *Agri. Biol. Chem.* 38:2125–34
265. Gorisch, H., Lingens, F. 1971. *Biochim. Biophys. Acta* 242:617–29
266. Jensen, R. A., Calhoun, D. H., Stenmark, S. L. 1973. *Biochim. Biophys. Acta* 293:256–68
267. Calhoun, D. H., Pierson, D. L., Jensen, R. A. 1973. *J. Bacteriol.* 113:241–51
268. Brown, K. D., Somerville, R. L. 1971. *J. Bacteriol.* 108:386–99
269. Camakaris, H., Pittard, J. 1973. *J. Bacteriol.* 115:1135–44
270. Golub, E. G., Liu, K. P., Sprinson, D. B. 1973. *J. Bacteriol.* 115:121–28
271. Tribe, D. E., Camakaris, H., Pittard, J. 1976. *J. Bacteriol.* 127:1085–97
272. Case, M. E., Giles, N. H. 1974. *Genetics* 77:613–26
273. Gaertner, F., Cole, K. 1976. *Arch. Biochem. Biophys.* 177:566–73
274. Lumsden, J., Coggins, J. R. 1977. *Biochem. J.* 161:599–607
275. Gaertner, F. H., Ericson, M. C., DeMoss, J. A. 1970. *J. Biol. Chem.* 245:595–600
276. Welch, G. R., Gaertner, F. H. 1976. *Arch. Biochem. Biophys.* 172:476–89
277. Jensen, R., Stenmark, S. L. 1975. *J. Mol. Evol.* 4:249–59
278. Stenmark-Cox, S., Jensen, R. A. 1975. *Arch. Biochem. Biophys.* 167:540–46
279. Patel, N., Pierson, D. L., Jensen, R. A. 1977. *J. Biol. Chem.* 252:5839–46
280. Dayan, J., Sprinson, D. B. 1971. *J. Bacteriol.* 108:1174–80
281. Koch, G. L. E., Shaw, D. C., Gibson, F. 1971. *Biochim. Biophys. Acta* 229:795–804
282. Im, S. W. K., Pittard, J. 1971. *J. Bacteriol.* 106:784–90
283. Dopheide, T. A. A., Crewther, P., Da-

vidson, B. E. 1972. *J. Biol. Chem.* 4447–52

284. Schmit, J. C., Artz, S. W., Zalkin, H. 1970. *J. Biol. Chem.* 245:4019–27

285. Nester, E. W., Jensen, R. A. 1966. *J. Bacteriol.* 91:1594–98

286. Rebello, J. L., Jensen, R. A. 1970. *J. Biol. Chem.* 245:3738–44

287. Im, S. W. K., Davidson, H., Pittard, J. 1971. *J. Bacteriol.* 108:400–9

288. Duda, E., Straub, M., Venetianer, P., Denes, G. 1968. *Biochem. Biophys. Res. Commun.* 32:992–97

289. Brown, B., Lax, S., Liang, L., Dabney, B., Spremulli, L., Ravel, J. 1977. *J. Bacteriol.* 129:1168–70

290. Heinonen, J., Artz, S. W., Zalkin, H. 1972. *J. Bacteriol.* 112:1254–63

291. Silbert, D. F., Jorgensen, S. E., Lin, E. C. C. 1963. *Biochim. Biophys. Acta* 73:232–40

292. Wallace, B. J., Pittard, J. 1969. *J. Bacteriol.* 97:1234–41

293. Mavrides, C., Orr, W. 1975. *J. Biol. Chem.* 250:4128–33

294. Ito, J., Yanofsky, C. 1966. *J. Biol. Chem.* 241:4112–14

295. Grieshaber, M., Bauerle, R. 1972. *Nature New Biol.* 236:232–35

296. Hulett, F. M., DeMoss, J. A. 1975. *J. Biol. Chem.* 250:6648–52

297. Kane, J. F., Holmes, W. M., Smiley, K. L. Jr., Jensen, R. A. 1973. *J. Bacteriol.* 113:224–32

298. Robb, F., Belser, W. L. 1972. *Biochim. Biophys. Acta* 285:243–52

299. Goto, Y., Zalkin, H., Keim, P., Heinrikson, R. 1976. *J. Biol. Chem.* 251:941–49

300. Kane, J. F., Holmes, W. M., Jensen, R. A. 1972. *J. Biol. Chem.* 247:1587–96

301. Zalkin, H., Murphy, T. 1975. *Biochem. Biophys. Res. Commun.* 67:1370–77

302. Brenchley, J. E., Magasanik, B. 1974. *J. Bacteriol.* 117:544–50

303. Margolin, P. 1971. In *Metabolic Pathways,* ed. H. J. Vogel, 5:389–446. New York: Academic. 576 pp.

304. Somerville, R. L., Yanofsky, C. 1964. *J. Mol. Biol.* 8:616–19

305. Brammer, W. J. 1973. *J. Gen. Microbiol.* 76:395–405

306. Matsushiro, A. 1963. *Virology* 19:475–82

307. Imamoto, F., Morikama, K., Siato, K., Mishima, S., Nishimura, T., Matsushiro, A. 1965. *J. Mol. Biol.* 13:157–68

308. Imamoto, F. 1968. *Nature* 220:31–34

309. Baker, R. F., Yanofsky, C. 1968. *Proc. Natl. Acad. Sci. USA* 60:313–20

310. Ito, J., Imamoto, F. 1968. *Nature* 220:441–44

311. Imamoto, F., Yanofsky, C. 1967. *J. Mol. Biol.* 28:25–35

312. Hiraga, S., Yanofsky, C. 1972. *J. Mol. Biol.* 72:103–10

313. Morikawa, N., Imamoto, F. 1969. *Nature* 223:37–40

314. Morse, D. E., Mosteller, R., Baker, R. F., Yanofsky, C. 1969. *Nature* 223:40–43

315. Squires, C. L., Lee, F. D., Yanofsky, C. 1975. *J. Mol. Biol.* 92:93–111

316. Brenchley, J. E., Williams, L. S. 1975. *Ann. Rev. Microbiol.* 29:251–74

317. Kano, Y., Matsushiro, A., Shimora, Y. 1968. *Mol. Gen. Genet.* 102:15–26

318. Ito, K., Hiraga, S., Yura, T. 1969. *Genetics* 61:521–38

319. Mosteller, R. D., Yanofsky, C. 1971. *J. Bacteriol.* 105:268–75

320. Hiraga, S., Yanofsky, C. 1972. *Nature New Biol.* 237:47–49

321. Zalkin, H., Yanofsky, C., Squires, C. L. 1974. *J. Biol. Chem.* 249:465–75

322. Franklin, N. 1971. In *The Bacteriophage Lambda,* ed. A. D. Hensley, pp. 621–38. New York: Cold Spring Harbor Lab. 792 pp.

323. Bertrand, K., Korn, L., Lee, F., Plott, T., Squires, C. L., Squires, C., Yanofsky, C. 1975. *Science* 189:22–26

324. Jackson, E. N., Yanofsky, C. 1973. *J. Mol. Biol.* 76:89–101

325. Rose, J. K., Squires, C. L., Yanofsky, C., Yang, H.-L. Zubay, G. 1973. *Nature New Biol.* 245:133–37

326. Morse, D. E., Morse, A. N. C. 1976. *J. Mol. Biol.* 103:209–26

327. Squires, C., Lee, F., Bertrand, K., Squires, C. L., Bronson, M. L., Yanofsky, C. 1976. *J. Mol. Biol.* 103:351–81

328. Lee, F., Squires, C. L., Squires, C., Yanofsky, C. 1976. *J. Mol. Biol.* 103:383–93

329. Bertrand, K., Squires, C., Yanofsky, C. 1976. *J. Mol. Biol.* 103:319–37

330. Kasai, T. 1974. *Nature* 249:523–27

331. Korn, L. J., Yanofsky, C. 1976. *J. Mol. Biol.* 103:395–409

332. Korn, L. J., Yanofsky, C. 1976. *J. Mol. Biol.* 106:231–41

333. Morse, D. E., Morse, A. N. C. 1976. *J. Mol. Biol.* 103:209–26

334. Bertrand, K., Yanofsky, C. 1976. *J. Mol. Biol.* 103:339–49

335. Yanofsky, C., Soll, L. 1977. *J. Mol. Biol.* 113:663–77

336. Lavallé, R., DeHauwer, G. 1970. *J. Mol. Biol.* 51:435–47

337. Edlin, G., Broda, P. 1968. *Bacteriol. Rev.* 32:206–26

338. Yang, H.-L., Zubay, G., Urm, E., Reiness, G., Cashel, M. 1974. *Proc. Natl. Acad. Sci. USA* 71:63–67
339. Reiness, G., Yang, H.-L., Zubay, G., Cashel, M. 1975. *Proc. Natl. Acad. Sci. USA* 72:2881–85
340. Rose, J. K., Yanofsky, C. 1972. *J. Mol. Biol.* 69:103–18
341. Pouwels, P. H., Pannekoek, H. 1976. *Mol. Gen. Genet.* 149:255–65
342. Pannekoek, H., Brammer, W. J., Pouwels, P. H. 1975. *Mol. Gen. Genet.* 136:199–214
343. Gausing, K. 1976. In *Control of Ribosome Synthesis,* ed. N. O. Kjeldgaard, O. Maaloe, pp. 292–303. New York: Academic. 466 pp.
344. Neidhardt, F. C., Bloch, P. L., Pedersen, S., Reeh, S. 1977. *J. Bacteriol.* 129:378–87
345 Blumenthal, R. M., Reeh, S., Pedersen, S. 1976. *Proc. Natl. Acad. Sci. USA* 73:2285–88
346. Proctor, A. R., Crawford, I. P. 1976. *J. Bacteriol.* 126:547–49
347. Kane, J. F., Jensen, R. A. 1970. *Biochem. Biophys. Res. Commun.* 38:1161–67
348. Hoch, S. O., Roth, C. W., Crawford, I. P., Nester, E. W. 1971. *J. Bacteriol.* 105:38–45
349. Carsiotis, M., Jones, R. F. 1974. *J. Bacteriol.* 119:889–92
350. Hartman, P. E., Hartman, Z., Stahl, R. C., Ames, B. N. 1971. *Adv. in Genet.* 16:1–34
351. Brenner, M., Ames, B. N. 1971. *Metab. Pathways* 5:349–87
352. Moyed, H. S., Magasanik, B. 1960. *J. Biol. Chem.* 235:149–53
353. Blasi, F., Aloj, S. M., Goldberger, R. F. 1971. *Biochemistry* 10:1409–17
354. Whitfield, H. J. Jr. 1971. *J. Biol. Chem.* 246:899–908
355. Parsons, S. M., Koshland, D. E. Jr. 1974. *J. Biol. Chem.* 249:4104–10
356. Parsons, S. M., Koshland, D. E. Jr. 1974. *J. Biol. Chem.* 249:4119–27
357. Silbert, D. F., Fink, G. R., Ames, B. N. 1966. *J. Mol. Biol.* 22:335–47
358. Roth, J. R., Anton, D. N., Hartman, P. E. 1966. *J. Mol. Biol.* 22:305–23
359. Bossi, L., Cortese, R. 1977. *Nucleic Acids Res.* 4:1945–56
360. Wyche, J. H., Ely, B., Cebula, T. A., Snead, M. C., Hartman, P. E. 1974. *J. Bacteriol.* 117:708–16
361. Goldberger, R. F., Kovach, J. S. 1972. *Current Topics Cell. Regul.* 5:285–308
362. Kovach, J. S., Berberich, M. A., Venetianer, P., Goldberger, R. F. 1969. *J. Bacteriol.* 97:1283–90
363. Kovach, J. S., Phang, J. M., Ferrence, M., Goldberger, R. F. 1969. *Proc. Natl. Acad. Sci. USA* 63:481–88
364. Smith, O. H., Meyers, M. M., Vogel, T., Deeley, R. G., Goldberger, R. F. 1974. *Nucleic Acids Res.* 1:881–88
365. Meyers, M., Blasi, F., Bruni, C. B., Deeley, R. G., Kovach, J. S., Levinthal, M., Mullinix, K. P., Vogel, T., Goldberger, R. F. 1975. *Nucleic Acids Res.* 2:2021–36
366. Blasi, F., Bruni, C. B., Avitabile, A., Deeley, R. G., Goldberger, R. F., Meyers, M. M. 1973. *Proc. Natl. Acad. Sci. USA* 70:2692–96
367. Rothman-Denes, L., Martin, R. G. 1971. *J. Bacteriol.* 106:227–37
368. Morton, D., Parsons, S. 1977. *Biochem. Biophys. Res. Commun.* 74:172–77
369. Kleeman, J. E., Parsons, S. M. 1977. *Proc. Natl. Acad. Sci. USA* 74:1535–37
370. Vogel, T., Meyers, M., Kovach, J. S., Goldberger, R. F. 1972. *J. Bacteriol.* 112:126–30
371. Sterboul, C. C., Kleeman, J. E., Parsons, S. M. 1977. *Arch. Biochem. Biophys.* 181:632–42
372. Scott, J., Roth, J., Artz, S. 1975. *Proc. Natl. Acad. Sci. USA* 72:5021–25
373. Franklin, N. C., Yanofsky, C. 1976. In *RNA Polymerase,* ed. R. Losick, M. Chamberlin, pp. 693–706. New York: Cold Spring Harbor Lab. 899 pp.
374. Adhya, S., Gottesman, M., deCrombrugghe, B., Court, D. 1976. See Ref. 373, pp. 719–30
375. Schlief, R., Greenblatt, J. 1971. *J. Mol. Biol.* 59:127–50
376. Artz, S. W., Broach, J. R. 1975. *Proc. Natl. Acad. Sci. USA* 72:3453–57
377. Stephens, J. C., Artz, S. W., Ames, B. N. 1975. *Proc. Natl. Acad. Sci. USA* 72:4389–93
378. Fink, G. R. 1966. *Genetics* 53:445–59
379. Carsiotis, M., Jones, R. F., Wesseling, A. C. 1974. *J. Bacteriol.* 119:893–98
380. Monod, J., Jacob, F. 1961. *Cold Spring Harbor Symp. Quant. Biol.* 26:389–401
381. Wolfner, M., Yep, D., Messenguy, F., Fink, G. R. 1975. *J. Mol. Biol.* 96:273–90
382. Hartwell, L. H. 1974. *Bacteriol. Rev.* 38:164–98
383. Davis, R. H. 1975. *Ann. Rev. Genet.* 9:39–65
384. Yamada, K., Kinoshita, S., Tsunoda, T., Aida, K., eds. 1972. *The Microbial Production of Amino Acids.* New York: Halsted. 548 pp.

Ann. Rev. Biochem. 1978. 47:607–34
Copyright © 1978 by Annual Reviews Inc. All rights reserved

DNA SEQUENCE ANALYSIS[1]

❖983

Ray Wu

Section of Biochemistry, Molecular, and Cell Biology, Cornell University,
Ithaca, NY 14853

CONTENTS

[1]This review covers approximately the period from March 1975 through August 1977.

607

0066-4154/78/0701-0607$01.00

PERSPECTIVES AND SUMMARY

DNA is the genetic material of nearly all organisms. Information needed to direct the synthesis of protein and RNA molecules, and to regulate genetic expressions, is encoded in DNA molecules as specific nucleotide sequences. Thus, DNA sequence analysis—determination of the primary structure of DNA molecules—is important for decoding the information stored in the genetic material. It is also a prerequisite for understanding gene organization and regulation at the molecular level.

DNA sequence analysis of a known location on a DNA molecule involves three major steps (*a*) a defined DNA segment is isolated in homogeneous form and in sufficient amount (*b*) it is labeled with a radioisotope of high specific activity at a specific position (*c*) the labeled DNA segment is partially hydrolyzed by chemical or enzymatic means to smaller fragments, and these are fractionated and their sequence identified by suitable methods.

This review is organized according to DNA sequencing techniques since the progress in this field depends heavily on the development of more rapid and versatile techniques. The initial method for DNA sequencing that was introduced nine years ago was based on the principle of repair synthesis (1). Since then, methods with increased sophistication and rapidity have been developed. Between 1968 and 1974, the rate of DNA sequence analysis has increased about fivefold; between 1974 and 1977, the rate increased another tenfold. Today a DNA sequence of 200 nucleotides can be determined within a month. By using the various methods, the DNA sequences of a number of biologically interesting molecules have been elucidated. Due to space limitations, the present review covers only the important developments of the last two years; many meritorious papers could not be included. To save space, the references but not the actual DNA sequences (approximately 15,000 base pairs) are given here. Earlier work on DNA sequence analysis which included repair synthesis and primer extension with deoxy-nucleotides or ribo-substitution, protein binding site analysis, the use of endonuclease IV, and sequencing of DNA through its RNA transcripts can be found in several excellent review articles (2–7).

ISOLATION OF HOMOGENEOUS DNA

Isolation of Intact DNA

To carry out sequence analysis the DNA molecules must be fractionated to remove impurities, including nucleases, purified to at least 95% homogeneity, and isolated in sufficient amounts. DNA molecules from bacteriophages and animal viruses are the simplest and have been given priority attention. These DNA molecules contain from 5,000–50,000 base pairs and milligram amounts can be purified relatively easily (7).

Procedures for isolation vary with the source of the DNA under consideration. Bacteriophages such as λ (1, 7), T7 (8, 9) and ϕX174 (10) are purified first to remove most of the bacterial contaminants. DNA molecules are then extracted from the bacteriophages and further purified by sucrose gradient centrifugation. DNA from simian virus 40 (SV40) is purified from monkey kidney cells by removal of most cellular DNA, followed by equilibrium centrifugation in a CsCl-ethidium bromide solution (11, 12). Polyoma DNA is isolated in a similar way (13). Mitochondrial DNA is extracted from purified mitochondria (14–16).

Digestion of DNA into Smaller Fragments

DNA molecules with even 5,000 base pairs are too long for sequence analysis unless analysis is limited to the terminal regions (9, 1, 17). The discovery of site-specific restriction endonucleases marks a major breakthrough in DNA sequence analysis (18–20). These endonucleases recognize specific tetra- to hexa-nucleotide sequences and hydrolyze DNA molecules at these specific sites. Thus, by the use of appropriate combinations of different restriction enzymes, large DNA molecules can usually be hydrolyzed into fragments of < 300 base pairs in length, each suitable for complete sequence analysis. For example, a DNA molecule of 50,000 base pairs can be hydrolyzed by a restriction endonuclease (recognizing a hexanucleotide sequence) to give ten fragments. After fractionation of the fragments by agarose gel electrophoresis, each resulting fragment can be hydrolyzed by a second restriction endonuclease (recognizing a tetranucleotide sequence) to give fragments 300 base pairs in length. If necessary, a third or fourth restriction endonuclease can be used.

Before using a restriction enzyme in DNA sequence analysis, one usually determines the number of specific DNA fragments produced by the enzyme and then arranges them into a physical map (12, 19). Even with a small

DNA such as SV40, mapping the entire 5,000 base-pair DNA into some 100 fragments has taken several years (12, 19–23). Complete sequence analysis would take much longer. Thus the time needed to carry out complete sequence analysis of a DNA of a million base pairs is prohibitive, even considering the rapid techniques for DNA sequencing. Furthermore, to sequence such long DNA, gram quantities of purified DNA are also needed. For these reasons, only selected portions of a long DNA or a complex genome have been sequenced. One of the most powerful tools to select and then amplify a portion of a complex genome is molecular cloning.

Cloning and Amplification of DNA

DNA can be broken into fragments by mechanical shearing or by specific cleavage with restriction endonucleases. Each DNA fragment can be covalently linked in vitro to a linear plasmid DNA or λ phage DNA which serves as the cloning vector. The resulting recombinant DNA species are inserted into *Escherichia coli* by transformation, and each transformed cell grown as a separate clone (24, 25). In this way, a complex genome can be divided into thousands or millions of fragments, and each fragment isolated as a clone. The essential steps involved in this method are illustrated in Figure 1. In step 1, a large DNA (X DNA) is cleaved by a restriction enzyme such as *Eco*RI (26), to produce many fragments (only four shown) with cohesive ends. These fragments are mixed with circular plasmid DNA molecules (pDNA) containing antibiotic resistant markers that are also cleaved by *Eco*RI enzyme to produce linear molecules (P_1). In step 2, the fragments are randomly associated by H-bonding through their cohesive ends, and joined covalently using DNA ligase (27, 28). Both linear and circular recombinant DNA molecules of different combinations of X and P_1 fragments are produced. In step 3, circular recombinant DNA molecules are taken up by *E. coli* during transformation, and antibiotic-resistant colonies are selected. Individual colonies are cloned and then tested for the presence of inserted DNA after digestion of the DNA with *Eco*RI endonuclease followed by gel electrophoresis. In step 4, a specific clone of recombinant plasmid, such as P_1-X_2, can be grown in quantity in the transformed *E. coli* and amplified 10^6 times or more. In step 5, the specific DNA fragment, X_2 in this example, can be obtained in quantity by cleaving the P_1-X_2 DNA with *Eco*RI endonuclease. Even if X_2 DNA is present in no more than one part in a million of the original genome, it can be isolated by the above method as long as a specific method of selection or identification is available (25). Milligram amounts of P_1-X_2 DNA can be readily obtained for sequence analysis. Using this method, the rRNA gene of *Xenopus laevis* (29), the histone gene of sea urchin (30), and the entire genomes (>90%) of *E. coli* (31), yeast (32, 33), and *Drosophila* (34) have been cloned into *E. coli* banks consisting of thousands of colonies.

A second method of cloning is based on isolating an mRNA molecule, transcribing it to a complementary DNA, and then cloning the DNA. The mRNA molecule should be relatively pure, and it is transcribed with the RNA-dependent DNA polymerase (reverse transcriptase; 35) to give the complementary DNA (cDNA). The RNA template is hydrolyzed with alkali and the single-stranded cDNA made into a double-stranded DNA with *E. coli* polymerase I (36). The double-stranded DNA is cloned after the addition of poly dA (or poly dT) tails (37, 38). The globin gene from rabbit (39–42), and the ovalbumin gene (43) have been cloned by the use of this method.

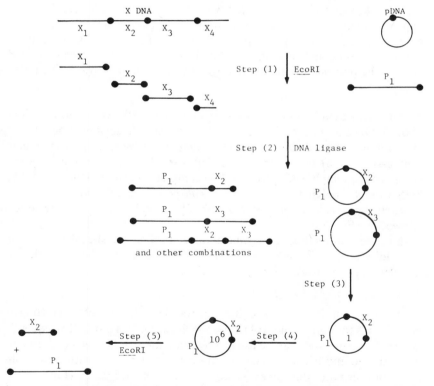

Figure 1   Cloning of DNA fragments in a plasmid. X DNA represents a DNA molecule to be cleaved and cloned. pDNA represents a plasmid DNA used as the cloning vector (24). • denotes cohesive ends, 5' d(pA-A-T-T), produced by cleavage of sensitive sites on the DNA by *Eco*RI restriction endonuclease (26).

A modification of the second method allows the use of a mixture of mRNA molecules for the production of cDNAs, double-stranded DNAs, and cloning (44, 45; see section on copying mRNA into cDNA for sequence analysis).

LABELING OF DNA AND ISOLATION OF END-LABELED FRAGMENTS

Labeling Techniques

In order to reduce the need for purified DNA for sequence analysis and to allow sequence analysis from a specific origin, the DNA must be labeled with radioisotopes of high specific activity.

5'-END LABELING   DNA labeled with ^{32}P at the 5' end is used for most sequence analysis. This labeling is catalyzed by polynucleotide kinase (46, 47) which transfers the γ-phosphate group of ATP to the 5'-OH terminus of polynucleotide chains (Fig. 2a). For DNA molecules that contain 5'-P groups, the phosphates are first removed by incubation with alkaline phosphatase.

3'-END LABELING   Two methods are available for the enzymatic labeling of the 3'-OH ends of DNA molecules. In one method, labeling of one strand of a double-stranded DNA can be accomplished by using partial repair synthesis (17, 48) to incorporate a [^{32}P]dNMP, using the 5'-protruding strand as the template (Figure 2b). For completely double-stranded DNA, both T_4 DNA polymerase (49) and E. coli polymerase I (50) can catalyze an exchange reaction between a suitable [^{32}P]dNTP and the 3' terminal nucleotide (Fig. 2c).

In the second method, deoxynucleotidyl terminal transferase is used for the addition of a [^{32}P] ribonucleotide to the 3'-end of either a single- or double-stranded DNA (51, 52; Figure 2d).

Separation of Labeled Ends

In most cases, in vitro labeling of double-stranded DNA results in labeling of both 5' ends (or both 3' ends) of the two DNA strands. To simplify sequence analysis, two methods can be used to separate the two labeled ends. In one method, the end-labeled DNA is subjected to digestion with a suitable restriction enzyme that gives a large and a small fragment (Figure 2e). Fractionation is then accomplished by electrophoresis. In the second method, the two DNA strands are separated on gel under denaturing conditions (53). Alternatively, the two strands may be separated in the presence of poly(UG) using density centrifugation in CsCl (54), or by hybridization to specific RNA molecules (55).

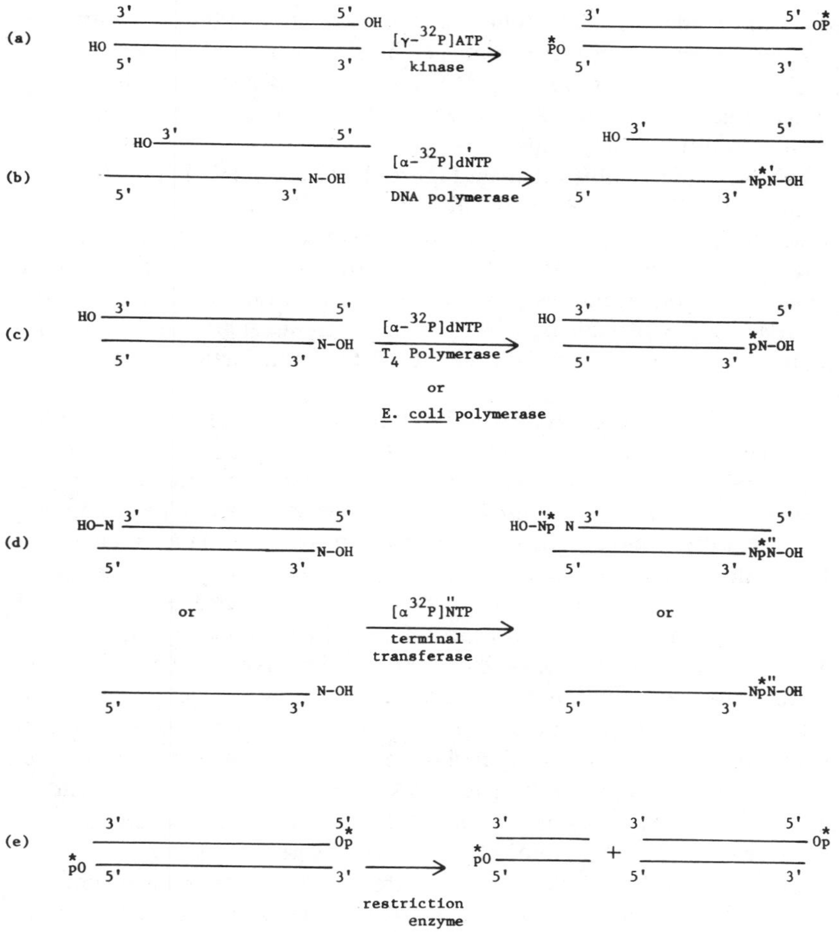

Figure 2 Methods for end-labeling of DNA molecules (a–d) and for separation of the two labeled ends (e). p̂ denotes ^{32}P; N represents any deoxynucleoside; Ñ and N̈ are specific deoxynucleosides.

THE PLUS AND MINUS METHOD
FOR SEQUENCING DNA

The Principle of the Method

A simple and rapid method for determining nucleotide sequences in single-stranded DNA has been developed by Sanger & Coulson (56). This method depends on primed synthesis with DNA polymerase to generate DNA products of different length, followed by fractionation of the single-stranded products by size on polyacrylamide gel. The success of the method depends on the possibility of generating DNA products of every length, and the ability of gel electrophoresis to separate a family of oligonucleotides differing in length by a single nucleotide.

The principle is illustrated in Figure 3 with a small hypothetical sequence, d(A–T–G–C–T–G), in a DNA chain. *E. coli* DNA polymerase I is used to extend the primer oligodeoxynucleotide by copying the template, in the presence of the four dNTPs, of which one is labeled with ^{32}P. This primer extension reaction is allowed to proceed for different lengths of time in order to generate oligonucleotides of different lengths. The extension mixture is then purified to remove the dNTPs, and further treated in one of two ways.

First, in the "minus" system, primer extension is carried out according to the partial-repair principle of Wu & Kaiser (1). The random mixture of oligodeoxynucleotides, which is still hybridized to the template DNA, is incubated with DNA polymerase I in the absence of one dNTP (using only three dNTPs). Synthesis then proceeds as far as it can on each chain and stops when points are reached where the missing dNTP is required. If dATP is the missing triphosphate (the minus-dA system), each chain will terminate at its 3' end at a position before a dA residue. Separate samples are incubated, with each one of the four dNTPs missing. The DNAs from the four incubation mixtures are then denatured and subjected to electrophoresis on polyacrylamide gel. The minus-dA lane on the gel will contain bands corresponding to positions before the dA residues in the extended chain, thus locating the positions of dAs. Similarly, the relative positions of the other nucleotides may be located and, ideally, the sequence of the DNA read off from the radioautograph of the gel.

Second, in the "plus" system, the method of Englund (49) is adopted. In the presence of a single dNTP, T_4 DNA polymerase will degrade double-stranded DNA from its 3' end, but this exonuclease action is effectively stopped at residues corresponding to the one dNTP that is present due to the incorporation of the added nucleotide. This method is applied to the random oligonucleotide mixture obtained above. Samples are incubated with T_4 DNA polymerase and one of the four dNTPs, and then fractionated by electrophoresis on a polyacrylamide gel. For example, in a plus-dA system all chains will terminate with a dA. The position of the chains will be indicated by bands on the radioautograph. Usually these bands will be one nucleotide longer than the corresponding bands in the minus-dA system (Figure 3). If there is more than one consecutive dA nucleotide, the distance between the bands in the minus-dA and plus-dA system will indicate the number of such nucleotides. Sanger & Coulson (56) can deduce the se-

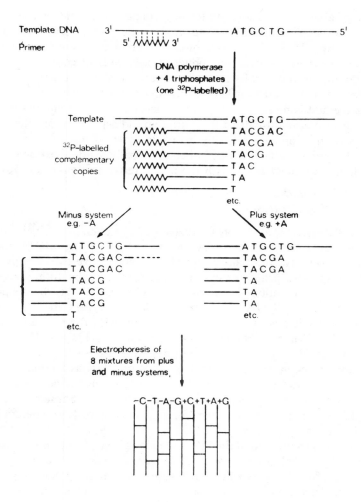

Figure 3 The principle of the plus and minus method of Sanger & Coulson (56). For sequence deduction, start from the fastest moving band (lowest band). A band in the –T lane indicates that the next residue at its 3' terminus will be a T. This is confirmed by the presence of a band in the +T lane representing an oligonucleotide one nucleotide longer. The bands with the same mobility in the +T and –A lanes thus define the dinucleotide sequence TpA. Similarly, the next largest oligonucleotides are found in the +A and –C lanes which define the dinucleotide sequence –ApC, thus establishing the sequence –TpApC.

quence of about 50–100 nucleotides in a single experiment, within a few days, by a combined use of the plus and minus systems, and by running a gel with eight lanes (Figure 3). However, there may be some ambiguities. For example, when a given mononucleotide is present consecutively the exact number of nucleotides may not be easily assigned. Furthermore, in this method, the first 15 nucleotides usually cannot be determined.

This method is most applicable to sequence analysis of single-stranded DNA, or of double-stranded DNA, after strand separation or after being made partially single-stranded by exonuclease III digestion. The primer required for this method can be either chemically synthesized or isolated from double-stranded DNA with the use of restriction enzymes (56).

Application of the Plus and Minus Method for Sequence Analysis

BACTERIOPHAGE ϕX 174 DNA   Phage ϕX 174 contains a single strand of DNA that has been shown to be the sense strand as the mRNA coding for nine known proteins. Sanger and his associates have used polypyrimidine tract analysis, primer extension coupled to ribosubstitution, and endonuclease IV digestion of uniformly labeled DNA for the sequence analysis of ϕX 174 DNA (6). During the last two years, more extensive sequence analysis has been carried out, mainly with the plus and minus method (56). Primers obtained from digesting ϕX 174 RF DNA with various restriction enzymes (Figure 4; 57) were used with both the viral (plus) and the complementary single strands of DNA as templates.

Air et al (58) determined about 50% of the sequence of gene F (Figure 4) by combining information from several sources (6). Subsequently, the entire 525 nucleotide long sequence for gene G (Figure 4) was determined with the plus and minus method (59).

Barrell et al (60) reported the DNA sequence from the promoter preceding gene D, through gene E, to gene J (Figure 4). They found that gene E amber mutations lie within the DNA sequence that codes for the gene D protein, and made an important discovery that genes D and E overlap on the DNA sequence but are translated in different reading frames. The amino acid sequence of gene D protein served as a useful check for the DNA sequence.

When the protein sequence is not available to confirm the DNA sequence, information from the plus and minus method may be supplemented by other methods, such as pyrimidine tract analysis. Brown & Smith (61) sequenced 870 nucleotides that code for about half of the gene A protein without knowing the protein sequence. In addition, over 80% of the DNA sequence that codes for the B protein was found to overlap with that coding

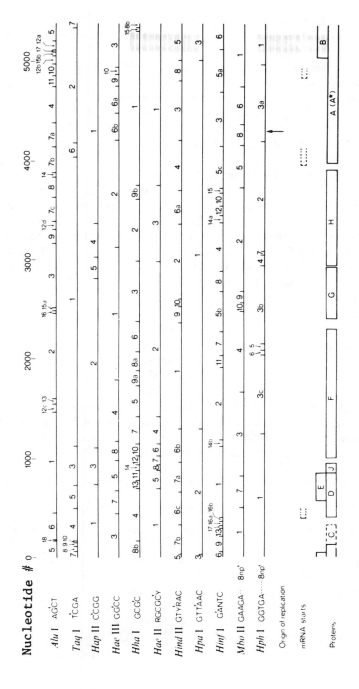

Figure 4 A physical map produced by the digestion of φX174 am3 RFI DNA with the various restriction enzymes. The zero point of the map is the *Pst* I endonuclease recognition site. [For details see reference (57).]

for the A protein. This is the second region of the ϕX 174 genome where two genes are translated from the same DNA using different reading frames (60–62).

Sequences from other parts of the ϕX 174 genome have also been reported. Robertson et al isolated four different ϕX 174 DNA fragments protected by ribosomes, and determined their sequence after synthesizing complementary RNA in vitro (63–66). Sinsheimer et al (67, 68) and Axelrod (69) reported the sequences of the 5' ends of three mRNA species synthesized in vitro and located them on the restriction enzyme map of ϕX 174 DNA.

Recently, Sanger et al (57) reported the provisional DNA sequence for the entire genome of ϕX 174, comprising 5,375 nucleotides. This impressive work served to link up or confirm all other sequences determined earlier. The sequence identifies many of the features responsible for the synthesis of the proteins of the nine genes, including initiation and termination sites for the proteins and RNAs. The unexpected finding that two pairs of ϕX 174 genes are coded by the same region of the DNA using different reading frames makes it necessary to revise the accepted concepts concerning the coding capacity of any genome.

OOCYTE 5S RIBOSOMAL DNA OF *XENOPUS LAEVIS*   The oocyte 5S rRNA genes in *Xenopus* occur in tandem clusters with repeating units of about 700 nucleotides in length (70). The primary sequence of the 5S DNA repeating unit has been deduced by means of several methods including RNA sequence analysis (71, 72) and direct DNA sequencing. The plus and minus method was employed on restriction enzyme fragments of 5S DNA, and individual repeating units cloned in bacterial plasmids (73). A 49-nucleotide-long region is believed to contain sequences important for correct initiation of transcription of the 5S DNA gene (73).

THE CHEMICAL METHOD FOR SEQUENCING DNA

The Principle of the Dimethylsulfate-Hydrazine Method

Maxam and Gilbert have developed a different, new, and rapid method for sequencing DNA (74), based on specific chemical reactions that break a terminally-labeled DNA molecule at specific nucleotides. Reaction condition is adjusted so that only one nucleotide, on the average, is reacted per DNA fragment. Subsequent cleavage of the DNA molecule at these positions will produce a family of radioactive DNA fragments extending from the same labeled end to each of the positions of that nucleotide. Four chemical reactions are used that cleave DNA preferentially at guanines (G>A), at adenines (A>G), at cytosines and thymines equally (C+T), and at cytosine alone (C). When the products of these four reactions are resolved by size, by electrophoresis on polyacrylamide gel, the DNA sequence can be read from the pattern of radioactive bands on a radioautogram (Figure 5).

This method requires that a single- or double-stranded DNA molecule be labeled at one end of one strand with ^{32}P. The labeled ^{32}P can be at the 5' end or at the 3' end as described above. A sequence of up to 100 nucleotides can be determined from the labeled end.

For the purine specific reaction, an aliquot of the DNA is treated with dimethylsulfate which methylates the guanines in DNA at the N7 position and the adenines at the N3 (75). The glycosidic bond of a methylated purine breaks on heating at neutral pH, leaving the sugar free. Alkali at 90°C will cleave the sugar from the neighboring phosphate groups. When the resulting end-labeled fragments are resolved on a gel, the radioautogram contains a pattern of dark and light bands (Figure 5; G>A lane). An adenine-enhanced cleavage can be obtained by treating the methylated DNA with acid which releases adenine preferentially (Figure 5; A>G lane).

Other aliquots are reacted with hydrazine, which cleaves cytosine and thymine, leaving a ribosylurea (76). After a partial reaction in aqueous hydrazine, the phosphate backbone of the DNA is cleaved with 0.5 M piperidine. The final gel pattern contains bands of similar intensity due to the cleavages at cytosines and thymines (Figure 5; C+T lane). However, if 2 M NaCl is included in the hydrazine reaction the reaction of thymines is suppressed. Then the piperidine breakage produces bands only from cytosines (Figure 5, C lane).

This sequencing method is applicable to both double- and single-stranded DNA; for this and other reasons it is the most versatile and useful method. In the analysis, all nucleotides are displayed, including consecutive runs of any single nucleotide. Further, 5-methylcytosine in DNA can be recognized because the methyl group interferes with the action of hydrazine (74).

Application of the Chemical Method for Sequence Analysis

A number of biologically important systems have been investigated by using the method of Maxam & Gilbert (74) for DNA sequence analysis. Several systems are briefly discussed here.

THE *LAC* OPERON   After sequence analysis of the primary *lac* operator site by the transcription method (77), Gilbert et al (78) used the chemical method for sequencing the secondary and tertiary *lac* repressor binding sites. By comparing the sequences of the three sites, the following homology was found: 5' d(G–T–G–A–G–C–G ... C–A–A). Several dG residues in this and its complementary sequence are found to be protected by repressor

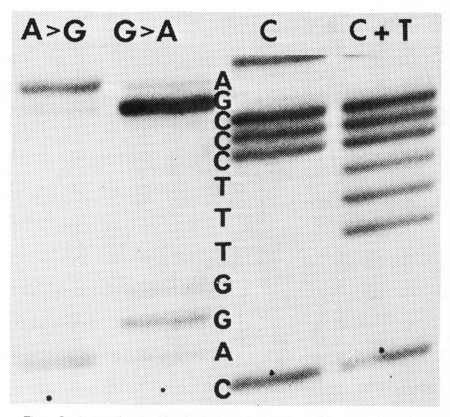

Figure 5 Autoradiogram of a sequencing gel according to Maxam & Gilbert (74). Only a portion of a gel is shown. The smallest labeled oligonucleotide moves the fastest and is shown as the bottom of the gel. The dots show the position of the bromophenol blue dye marker, between fragments 9 and 10 nucleotides long.

binding from methylation with dimethylsulfate. This suggests the importance of these nucleotides in the operator region for specific recognition by the repressor protein.

THE 5S RIBOSOMAL RNA GENE   The DNA sequence in the promoter region (the one immediately preceding) and the terminator region (the one immediately following) the yeast 5S rRNA gene, has been reported by two groups (79, 80). The DNA from yeast was either digested with *Eco*RI endonuclease, or sheared and cloned into plasmids. Clones have been selected by hybridization with labeled 5S rRNA. A restriction enzyme cleavage map of cloned *Saccharomyces cerevisiae* rDNA was made, and sequences from the ends of several restricted DNA fragments determined.

In the 103 nucleotides that precede the 5S rDNA a region with 16 A–T consecutive pairs was found. When the yeast promoter region is compared with the known bacterial promoters, no homologies are seen. After the 5S rDNA gene, a stretch of 20 (79) or 29 (80) consecutive dT was found. This may serve as a termination signal for the RNA polymerase. The sequences at the two ends of the yeast 35S rRNA gene were also determined (81), and similarity of the termination regions of the 35S and 5S rDNA gene was found.

By comparing the DNA sequence in the promoter region of yeast 5S RNA gene with that of *Xenopus* oocyte 5S RNA gene, no obvious relationship was observed.

THE SV40 GENOME   SV40 contains a circular DNA genome with approximately 5,400 base pairs. The viral DNA codes for four known proteins VP1, VP2, VP3, and T-antigen (for review see 82). The SV40 genome has been extensively mapped with a number of restriction enzymes (19), and approximately 100 cleavage sites have been located (Figure 6). With this detailed physical map available, a large percentage of the SV40 genome has been sequenced by several methods using specific DNA fragments. An appreciable amount of sequences with almost identical results have come, independently, from three different laboratories.

The major portion of the sequence was determined by Weissman et al (83–90) using RNA sequencing methods. The DNA fragment (*Hin*d-G) from restriction enzyme digestion of SV40 DNA was first used as template for the synthesis of [^{32}P] RNA by RNA polymerase. Sequence analysis was then made on the in vitro-synthesized RNA (83, 84). Subsequently, DNA sequence analysis of end-labeled DNA by the visual-inspection mobility-shift method (5, 6) was used to supplement RNA sequence information. The *Eco*RII-G fragment was found to contain the origin of DNA replication and to specify the 5' ends of early and late viral RNA synthesis (85). Together with a neighboring fragment, it includes a region 800-nucleotides long that does not encode known proteins (86). This region also contains several long palindromes and repeated sequences. More recently, the chemical method for DNA sequencing (74) was applied to supplement and confirm the sequences determined by other methods. Thus, the DNA that precedes the 5' end of mRNAs for the structural proteins VP1 (87), VP2 (88), and VP3 (89) have now been sequenced. The results showed that there is an overlap between the VP2 and VP3 genes. the mRNAs for VP2 and VP3 proteins are read in the same phase, and the initiation site for VP3 lies within the structural gene for VP2. Furthermore, there is an overlap of VP1 with the C-terminal region of VP2 and VP3, but in this case the reading frames are different (Figure 6). A similar observation has also been made by Fiers et al (unpublished). It was concluded that the gene coding for VP2

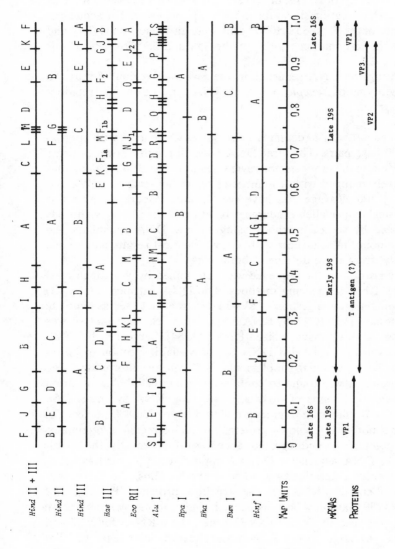

Figure 6 A physical map of the SV40 genome. The zero point of the map is the *Eco*RI endonuclease recognition site. For clarity, the circular genome is shown as a linear map opened at the *Eco*RI site. The cleavage sites, and resulting fragments for each restriction enzymes, are indicated with a separate line. Not shown here are the sites cleaved by restriction enzymes, each of which produces a single cleavage: *Bam*I at 0.15 map unit, *Hpa*II at 0.74 map unit, and *Hae*II at 0.83 map unit. [Data are from references (18–22, 82, 83, 95).]

and VP3 overlaps that coding for VP1 by 120 nucleotides. Thus, in the eucaryotic system, this part of the genome codes for the synthesis of three different proteins in two different reading frames. This finding is similar in principle, but more complex, than that found for the ϕX 174 phage (57, 69).

Another unexpected finding came from the comparison of the oligonucleotides in the late 16S mRNA with those expected in a transcript of the SV40 DNA coding for the VP1 protein (90). The results indicate that a segment of about 200 nucleotides of RNA transcribed from a distant part of SV40 DNA has become linked to the transcript of VP1 codons.

Similar findings were independently obtained with adenovirus mRNA using an electron microscopic mapping procedure. Sharp et al found that the mRNA for hexon polypeptide contained a 160-nucleotide long tail at the 5' end which was not H-bonded to the single-stranded DNA coding for the hexon mRNA (91). Chow et al observed that the 5'-terminal sequences of several adenovirus-2 mRNAs are complementary to sequences that are remote from the DNA and from which the main coding sequence of each mRNA is transcribed (92). Thus, a leader RNA sequence (100–200 nucleotides long) from an early part of the genome appears to have joined to the 5' ends of the late adenovirus mRNA molecules (91, 92, 92a).

An appreciable amount of sequence information on the SV40 genome also has come from the work of Fiers et al (93–99). The in vitro transcription method was used to sequence part of the *Hin*d H, J, and K fragments (93, 94). Subsequently, the mobility-shift method (5, 6) and the chemical method (74) were used in conjunction with, or to replace, the RNA sequencing method. The complete sequences of *Hin*d-H (94), *Hin*d-K (95), *Hin*d-L and -M (96), *Hin*d-F–*Eco*RI-2 (97) and *Hin*d-C-Hap2 (98) DNA fragments were determined. The sequence in the *Hin*d-K DNA fragment was the first reported that codes for the amino terminus sequence of the VP1 protein (95, 99).

Some sequences of the SV40 genome were obtained by Wu et al (100–103) using DNA sequencing methods without the help of RNA sequence information. The quantitative mobility-shift method (100, 101) gave the sequence of the first 10–15 nucleotides from a labeled DNA terminus, and the chemical method (74) gave the sequence of 5–100 from a labeled terminus. The sequence of the *Hin*d-M DNA fragment (102) and a highly symmetrical sequence at the *Hin*d-A/C junction (103) were the first to be reported. The sequences of 15–30 nucleotides at each of the six *Hin*dIII and seven *Hin*dII cleavage sites were determined by the quantitative mobility-shift method (Jay, Roychoudhury, Tu, and Wu, unpublished). More than 500 nucleotides located in *Hin*d-A and -B DNA fragments which code for the T-antigen, and sequences in *Hin*d-C, -L, -M, and -D have been determined by the chemical method (Yang and Wu, unpublished).

THE ATTACHMENT SITES OF λ PHAGE   One of the most thoroughly studied site-specific recombination systems is the integration and excision reaction of bacteriophage λ (104). Integration is achieved through a single reciprocal recombination event at specific λ phage *att* site POP', and the primary bacterial *att* site BOB'. Landy et al (105, 106) made clever use of restriction enzymes for the isolation and identification of restriction fragments using DNAs from four different phages that carry the four unique *att* sites POP', POB', BOP', and BOB'. Results from extensive DNA sequencing showed that the phage *att* site and bacterial *att* site have a 15-nucleotide-long sequence in common. Within this common region is a sequence, d(C–T–T–T–T–T–A), previously identified in several studies as an RNA polymerase termination site. Moreover, overlapping with this is a sequence, d(T–A–T–A–C–T–A), that matches six out of seven positions of the most common *E. coli* promoter sequence. The presumptive terminator and promoter occupy most of the common sequence of the two *att* sites. The four sequences adjacent to the common region (P arm, P' arm, B arm, and B' arm) while different, are all high in A+T contents (>70%).

OTHER METHODS FOR SEQUENCING DNA

The Quantitative Mobility-Shift Method

Oligodeoxynucleotides of different lengths can be separated by size using two-dimensional electrophoresis homochromatography (107). This system has been used for analysis of polypyrimidine sequences (108) since mobility shifts between a d(pT) and a d(pC) can be clearly distinguished. It has also been applied for sequencing DNA (109), although visual inspection of the mobility shift cannot always distinguish which one of the four nucleotides has been added. Sequence analysis needs confirmation from the complementary strand (110) or other information. A quantitative mobility-shift method was developed (100) to facilitate deduction of the correct sequence. Theoretical electrophoretic mobility shifts, calculated for the addition of any one of the four mononucleotides to a particular oligonucleotide, are compared to observed values. This method was used in conjunction with modified homochromatography mixtures (7) for sequencing phage ϕ80 DNA (111) and phage T7 DNA (112).

The quantitative mobility-shift method has recently been modified (101) for a more accurate deduction of DNA sequences. The method is especially useful for sequencing short terminally-labeled DNA such as the restriction enzyme recognition sequence. The recognition sequences of the restriction enzymes *Alu*I (23) and *Hae*II (113) have been determined after labeling different aliquots of the restricted DNA fragments at the 5' or 3' ends (see section on labeling of DNA). The sequences of approximately 6 nucleotides obtained from both the 5'-and 3'-ends served to verify the complementary sequences, and allowed unambiguous deduction of the terminal structure as even ended, or with 5'- or 3'-protruding single strands (113).

Other methods used for determining restriction enzyme recognition sequences involved analysis of 5'-end-labeled DNA either alone (114, 115, 115a), or in conjunction with 3'-end dinucleotide analyses (116, 117). While these methods often do not give sufficient overlapping sequence information, especially for DNA with 3'-protruding ends, they are adequate when other supporting data are available. Sequence analysis of recognition sites of other restriction enzymes can be found in the comprehensive review by Roberts (20).

Copying mRNA into cDNA for Sequence Analysis

A new approach has been developed in several laboratories for sequencing mRNA molecules by copying them into cDNAs, using the viral reverse transcriptase (35, 39–43) or *E. coli* DNA polymerase.

GLOBIN mRNA   Purified rabbit globin mRNA was reverse transcribed to give cDNA and this was copied to give double-stranded DNA. DNA corresponding to a large part of the globin gene was thus obtained, cloned, and its sequence analyzed (118, 119). Partial sequences of the α- and β-globin gene (118) and complete sequence of the β-globin structural gene (119) were determined with the chemical method (74). The results agree with sequence information derived from direct mRNA analysis (120) and with the amino acid sequence of the globins.

The purified human β-globin mRNA was also used as a template for the synthesis of cDNA by use of viral reverse transcriptase. Double-stranded DNA was synthesized, cleaved with restriction enzymes, terminally labeled, and sequenced by a combination of methods (121). The sequence for the translated portion of the mRNA and part of the 3'-terminal untranslated portion have been deduced. These sequences are confirmed by other published information, including that obtained by direct sequence analysis of the mRNA.

Sequence analysis of part of the 3' noncoding region of the globin mRNA molecules was carried out by Proudfoot & Brownlee (122, 123) using reverse transcription of the mRNAs. Short DNA transcripts of the mRNAs were synthesized using *E. coli* DNA polymerase I in the presence of Mn^{2+} with oligo(dT)$_{10}$ as primer. The resulting cDNA obtained was sequenced directly. A striking sequence homology, A–A–U–A–A–A, located about 20 residues away from the 3'-terminal poly(A) sequence (123) was found to be present in six different mRNA molecules: the α- and β-globin mRNAs of rabbit and human, the immunoglobulin light chain mRNA of mouse, and the ovalbumin mRNA of chicken. Recently, the complete nucleotide sequence of the 5' noncoding region of rabbit β-globin mRNA was also reported (124).

ROUS SARCOMA VIRAL RNA   The genome of the Rous sarcoma virus was partially sequenced after transcription into cDNA (125). A tRNA primer was elongated into DNA with reverse transcriptase, using the viral RNA as template. The first major DNA made was a product 101 bases from the primer to the 5' end of the template. The sequence of this major cDNA was determined using the chemical method (74). It was found that 21 bases at the extreme 5' end of the viral RNA are repeated at the 3' end of the viral genome. This sequence repetition allows the growing DNA chain to jump from the 5' end to the 3' end of the RNA template during the synthesis of cDNA in vivo.

INSULIN mRNA   A mixture of mRNA, extracted from the B cells of rat pancreas, and containing an appreciable percent of insulin mRNA, was reverse transcribed. The resulting cDNA molecules were made double-stranded and cloned (45). The DNA from a number of clones was digested by appropriate restriction enzymes, fractionated on gels, and selected fragments were sequenced by the chemical method (74). By comparing the DNA sequence with the known amino acid sequence of rat insulin, several clones containing parts of the insulin gene were identified (45). These experiments can lead to the cloning of the entire gene. It will be interesting to know whether bacteria carrying the insulin gene will synthesize the pre-proinsulin.

YEAST CYTOCHROME c mRNA   It is now possible to sequence a minor mRNA species in an unfractioned mRNA preparation. For example, the bulk mRNA from yeast contains $<$ 0.1 percent cytochrome c mRNA. However, the latter was specifically reverse-transcribed by extending a synthetic oligodeoxynucleotide primer complementary to part of the cytochrome c mRNA. A partial sequence of the resulting cDNA was then directly determined by Szostak et al (44). The cDNA can also be made double-stranded and cloned to provide an abundant source of the control regions of the cytochrome c gene for more extensive sequence analysis.

OTHER DNA SEQUENCES ANALYZED

Operator-Promoter Region

THE LAC AND THE GALACTOSE OPERON   The *lac* operator-promoter system has been reviewed recently (126–128). By sequence analysis, the nucleotide changes occuring in several *lac* promoter mutants have been determined (129). Two mutational sites were found to be symmetrically located in a sequence that may be the binding site for the catabolite gene activator protein. The same deoxyguanines in these two sites are protected by the protein from methylation by dimethylsulfate (78).

The nucleotide sequence of the operator-promoter region of the galactose operon shows extensive homology with that of the *lac* operon at the catabolite gene activator protein binding site (130).

THE BACTERIOPHAGE λ The DNA of phage λ has two operators, O_L and O_R. The partial nucleotide sequence of O_L (131) and O_R (132) has now been completed (133) using the chemical method (74). The new sequence information confirms an earlier proposal that there are three structurally similar 17-base-pair long λ repressor binding sites within each operator (133). Binding of the repressor to the λ operators blocks the methylation of certain deoxyguanine residues by dimethylsulfate within the 17-base-pair sequences (134).

The P_R and P_L promoter sites (RNA polymerase binding sites) were shown to lie within the λ operator sites (135, 136). The P_O promoter site, located near the origin of replication of λDNA, has been sequenced using a small λ-derived plasmid $\lambda dvh93$ as the source of DNA (137). Extensive sequence homologies were observed between P_O, P_R, and P_L promoters.

Recently, a 950-nucleotide-long segment of the λDNA that spans the region between the P_R promoter and the middle of the O-gene has been sequenced by using the chemical method. This region codes for the entire tof and CII proteins. The reading frame of the latter was confirmed by the sequences of two amber mutants. Moreover, a partial protein sequence coded for the O-gene can be deduced from the DNA sequence (Kössel et al, unpublished).

OTHER PROMOTER SEQUENCES   The nucleotide sequence of a strong promoter in the replicative form DNA of phage fd was determined (138). The analysis was carried out on a segment of DNA protected, by the binding of RNA polymerase, against digestion with nuclease.

The nucleotide sequence of an early T_7 phage promoter was also reported. In comparing with other promoters, Pribnow (139) observed that a closely related common sequence 5' d(T–A–T–R–A–T–G) precedes the transcription initiation site by six or seven nucleotides. However, as more promoter sequences have become known, the sequence in this region has been found to differ significantly in some cases.

The promoter sequence of the *E. coli* tyrosine tRNA gene was analyzed by using chemically synthesized oligodeoxynucleotides as primers and separated strands of $\phi80$ psuIII DNA as templates (140). After stepwise extension of the primer by repair synthesis (17), a sequence of 59 nucleotides was elucidated (140).

Large amounts of a low molecular weight RNA, VA-RNA I, are produced in adenovirus-2-infected cells, which suggests that there may be a specific initiation of transcription of this RNA. The DNA sequence that precedes the 5' end of the RNA has been analyzed by hybridization of this RNA to restricted DNA fragments of adenovirus, and sequencing the appropriate fragments (141). The DNA sequence at the "promoter" region of this adenovirus DNA bears some resemblance to promoters from procaryotes.

THE TRYPTOPHAN OPERON   The DNA sequence in the operator-promoter regions of the *trp* operon of *E. coli* (142) and *Salmonella typhimurium* (143) have been elucidated. In both organisms, the *trp* promoter-operator region was protected from *Hpa*I endonuclease cleavage by *trp* repressor or RNA polymerase (144). In *E. coli,* the sequence of 115 base pairs preceding the *trp* mRNA initiation site was deduced from both RNA and DNA sequencing results. The *trp* promoter shares similar sequences with λP_L and λP_R promoters. In *S. typhimurium* (143), the DNA sequence in the promoter region on either side of the specific *Hpa*I site was analyzed. Of the 105 base pairs analyzed, 84 are conserved in the corresponding region of the *trp* operon of *E. coli.* In both systems, the DNA sequences essential for the recognition by RNA polymerase and repressor are contained within 59 base pairs preceding the transcription start site. Within this region is a highly symmetrical 20 base-pair region containing the *Hpa*I cleavage site and the operator site.

Other Sequences Analyzed

SEQUENCES AT THE COHESIVE END REGIONS OF BACTERIOPHAGES The 3'-terminal sequences of DNA from the phages 186, P2, and 299 have been analyzed (145). Together with the known 5'-cohesive-end sequences of these phages (146–148), two pairs of symmetrical sequences were found in the cohesive ends of each phage (145). The symmetrical parts of these sequences (*cos* site) may be involved in recognition by the putative endonuclease (*ter* system). A similar *cos* site sequence exists in the phage λ (149, 150). Recently, a sequence of 50 nucleotides on both sides of the λ-*cos* site has been determined by the use of the chemical method (Donelson et al, unpublished).

THE BACTERIOPHAGE T₅ DNA   At the two 5' external even-ended termini of phage T_5, the first seven nucleotides of each strand are identical (151). In the replication of T_5 DNA, replicative concatamers are formed in which the ends of the monomers are joined to give a 14-nucleotide-long sequence with twofold rotational symmetry. This sequence may be recognized by a putative endonuclease involved in the formation of monomer-length genomes.

The phage T_5 DNA contains several site-specific nicks in one of the two strands. Several of the nicks had at least the sequence 5' d(G–C–G–C–) in common (151).

REPETITIVE MONKEY DNA   DNA fragments containing monkey DNA sequences have been isolated from defective SV40 genomes after restriction enzyme digestion (152). The sequence of one such fragment (151 base pairs), predominantly homologous to the highly reiterated class of monkey DNA, was determined by using both RNA and DNA sequencing methods (153). The nucleotide sequence is relatively free of internal repetition and thus is markedly different from the sequences reported previously for reiterated satellite and spacer regions of eucaryotic genomes.

THE HISTONE GENE   The histone H2A and H3 genes of sea urchin have been cloned in *E. coli* and their sequence partially determined (154). A sequence of 62 base pairs was found to be colinear with 20 amino acids of bovine and trout histone H2A, and a sequence of 42 base pairs colinear with 12 amino acids of bovine histone H3. The sequence information thus served to identify the cloned DNA as histone genes and to locate and orient their polarity with respect to known restriction endonuclease sites.

The histone H2B gene of sea urchin has been cloned both in a λ vector (λh22) and in a plasmid vector (155). Both the plus and minus method (56) and the chemical method (74) were used for sequence analysis. The DNA sequences coding for two thirds of the H2B histone, together with some 3' extracistronic sequences, have been determined (155). The sequence information unambiguously identified this gene and revealed its 5'-3' polarity.

CONCLUDING REMARKS

In principle, any one of the sequencing methods described above can produce unambiguous sequence information. However, in practice, the most reliable sequences are obtained either by using a second method for confirmation, or by determining the complementary sequence with the same method. For example, analysis of a double-strand DNA molecule 180 nucleotides in length can be accomplished by sequencing 100 nucleotides from each labeled 5' end and thereby overlapping ten nucleotides. However, sequence information will be more reliable if 100 nucleotides are also sequenced from each labeled 3' end. Alternatively, the DNA sequence can be checked by RNA sequence or protein sequence.

At the conclusion of the first review article on DNA sequence analysis, published in 1972 (2), I took an optimistic view by predicting that "it is expected that the primary sequence of specific regions of many biologically important DNA molecules will be elucidated within the next decade". This prediction has been fulfilled ahead of schedule, mainly due to (a) the discovery of a number of site-specific restriction enzymes for cleaving DNA into short fragments, (b) the development of two different, new, and rapid methods for DNA sequence analysis (56, 74), and (c) the development of molecular cloning techniques. These new developments have changed the direction of research in modern biology. Many complex and interesting biological problems which were intractable a few years ago are now within our reach.

ACKNOWLEDGMENTS

The author is very grateful to many colleagues for supplying reprints and manuscripts that are in press or in preparation. I thank Drs. C. P. Bahl, E. Jay, K. J. Marians, S. A. Narang, R. Roychoudhury, J. Stiles, J. Szostak, C. D. Tu, and R. Yang for participation in some of the experiments described here. Research that originated in this laboratory has been supported by grants from the American Cancer Society (VC-216), the National Cancer Institute (CA14989), and the National Science Foundation (PCM 76-05072).

Literature Cited

1. Wu, R., Kaiser, A. D. 1968. *J. Mol. Biol.* 35:523–37
2. Wu, R., Donelson, J. E., Padmanabhan, R., Hamilton, R. 1972. *Bull. Inst. Pasteur Paris* 70:203–33
3. Murray, K., Old, R. W. 1974. *Prog. Nucleic Acid Res. Mol. Biol.* 14:117–85
4. Salser, W. A. 1974. *Ann. Rev. Biochem.* 43:923–65
5. Wu, R., Bambara, R., Jay, E. 1974. *Crit. Rev. Biochem.* 2:455–512
6. Sanger, F. 1973. *ICN-UCLA Symp. Mol. Biol. 2nd,* eds. C. F. Fox, W. S. Robinson, pp. 573–99. New York: Academic
7. Wu, R., Jay, E., Roychoudhury, R. 1976. *Methods Cancer Res.* 12:87–176
8. Thomas, C. A. Jr., Abelson, J. 1966. *Proc. Nucleic Acids Res.* 1:553–61
9. Weiss, B., Richardson, C. C. 1967. *J. Mol. Biol.* 23:405–17
10. Galibert, F., Sedat, J., Ziff, E. 1974. *J. Mol. Biol.* 87:377–407
11. Hirt, B. 1967. *J. Mol. Biol.* 26:365–69
12. Danna, K. J., Nathans, D. 1971. *Proc. Natl. Acad. Sci. USA* 68:2913–17
13. Griffin, B. E., Fried, M. 1976. *Methods Cancer Res.* 12:49–86
14. Clayton, D. A., Smith, C. A., Jordan, J. M., Teplitz, M., Vinograd, J. 1968. *Nature* 220:976–79
15. Brown, W. M., Vinograd, J. 1974. *Proc. Natl. Acad. Sci. USA* 71:4617–21
16. Prunell, A., Kopecka, H., Strauss, F., Bernardi, G. 1977. *J. Mol. Biol.* 110:17–47
17. Wu, R., Taylor, E. 1971. *J. Mol. Biol.* 57:491–511
18. Smith, H. O., Wilcox, K. W. 1970. *J. Mol. Biol.* 51:379–91
19. Nathans, D., Smith, H. O. 1975. *Ann. Rev. Biochem.* 44:273–93
20. Roberts, R. J. 1976. *CRC Crit. Rev. Biochem.* 4:123–64
21. Subramanian, K. N., Pan, J., Zain, S.,

Weissman, S. M. 1974. *Nucleic Acids Res.* 1:727–52

22. Yang, R., Van de Voorde, A., Fiers, W. 1976. *Eur. J. Biochem.* 61:119–38

23. Jay, E., Wu, R. 1976. *Biochemistry* 15:3612–20

24. Cohen, S. N., Chang, A. C. Y., Boyer, H. W., Helling, R. B. 1973. *Proc. Natl. Acad. Sci. USA* 70:3240–44

25. Scott, W. A., Werner, R., eds. 1977. *Molecular Cloning of Recombinant DNA.* Miami Winter Symp., Vol. 13. New York: Academic

26. Hedgpeth, J., Goodman, H. M., Boyer, H. W. 1972. *Proc. Natl. Acad. Sci. USA* 69:3448–52

27. Gellert, M. 1967. *Proc. Natl. Acad. Sci. USA* 57:148–55

28. Weiss, B., Live, T. R., Richardson, C. C. 1968. *J. Biol. Chem.* 243:4530–42

29. Morrow, J. F., Cohen, S. N., Chang, A. C. Y., Boyer, H. W., Goodman, H. M., Helling, R. B. 1974. *Proc. Natl Acad. Sci. USA* 71:1743–47

30. Kedes, L. H., Chang, A. C. Y., Houseman, D., Cohen, S. N. 1975. *Nature* 255:533–38

31. Clarke, L., Carbon, J. 1976. *Cell* 9:91–99

32. Struhl, K., Cameron, J. R., Davis, R. W. 1976. *Proc. Natl. Acad. Sci. USA* 73:1471–75

33. Ratzkin, B., Carbon, J. 1977. *Proc. Natl. Acad. Sci. USA* 74:487–91

34. Wensink, P. C., Finnegan, D. J., Donelson, J. E., Hogness, D. S. 1974. *Cell* 3:315–25

35. Temin, H. M., Baltimore, D. 1972. *Adv. Virus Res.* 17:129–86

36. Kornberg, A. 1969. *Science* 163:1410–18

37. Jackson, D. A., Symons, R. H., Berg, P. 1972. *Proc. Natl. Acad. Sci. USA* 69:2904–9

38. Lobban, P. E., Kaiser, A. D. 1973. *J. Mol. Biol.* 78:453–71

39. Higuchi, R., Paddock, G. V., Wall, R., Salser, W. 1976. *Proc. Natl. Acad. Sci. USA* 73:3146–50

40. Maniatis, T., Kee, S. G., Efstratiadis, A., Kafatos, F. C. 1976. *Cell* 8:162–82

41. Rougeon, F., Kourilsky, P., Mach, B. 1975. *Nucleic Acids Res.* 2:2365–78

42. Rabbits, T. H. 1976. *Nature* 260:221–25

43. McReynolds, L. A., Monahan, J. J., Bendure, D. W., Woo, S. L. C., Paddock, G. V., Salser, W., Dorson, J., Moses, R. E., O'Malley, B. W. 1977. *J. Biol. Chem.* 252:1840–43

44. Szostak, J. W., Stiles, J. I., Bahl, C. P., Wu, R. 1977. *Nature* 265:61–63

45. Ullrich, A., Shine, J., Chirgwin, J., Pictet, R., Tischer, E., Rutter, W. J., Goodman, H. M. 1977. *Science* 196:1313–19

46. Richardson, C. C. 1965. *Proc. Natl. Acad. Sci. USA* 54:158–65

47. Novogrodsky, A., Hurwitz, J. 1966. *J. Biol. Chem.* 241:2923–32

48. Wu, R. 1970. *J. Mol. Biol.* 51:501–21

49. Englund, P. T. 1972. *J. Mol. Biol.* 66:209–24

50. Donelson, J. E., Wu, R. 1972. *J. Biol. Chem.* 247:4661–68

51. Roychoudhury, R., Kössel, H. 1971. *Eur. J. Biochem.* 22:310–20

52. Roychoudhury, R., Jay, E., Wu, R. 1976. *Nucleic Acids Res.* 3:863–77

53. Hayward, G. S., Smith, M. G. 1972. *J. Mol. Biol.* 63:383–95

54. Szybalski, W., Kubinski, H., Hradecna, Z., Summers, W. C. 1971. *Methods Enzymol.* 21:383–413

55. Sambrook, J., Sharp, P. A., Keller, W. 1972. *J. Mol. Biol.* 70:57–71

56. Sanger, F., Coulson, A. R. 1975. *J. Mol. Biol.* 94:441–48

57. Sanger, F., Air, G. M., Barrell, B. G., Brown, N. L., Coulson, A. R., Fiddes, J. C., Hutchison, C. A. III, Slocombe, P. M., Smith, M. 1977. *Nature* 265:687–95

58. Air, G. M., Blackburn, E. H., Coulson, A. R., Galibert, F., Sanger, F., Sedat, J. W., Ziff, E. B. 1976. *J. Mol. Biol.* 107:445–58

59. Air, G. M., Sanger, F., Coulson, A. R. 1976. *J. Mol. Biol.* 108:519–33

60. Barrell, B. G., Air, G. M., Hutchison, C. A. III. 1976. *Nature* 264:34–41

61. Brown, N. L., Smith, M. 1977. *J. Mol. Biol.* 116:1–28

62. Smith, M., Brown, N. L., Air, G. M., Barrell, B. G., Coulson, A. R., Hutchison, C. A. III, Sanger, F. 1977. *Nature* 265:702–5

63. Robertson, H. D., Barrell, B. G., Weith, H. L., Donelson, J. E. 1973. *Nature New Biol.* 241:38–40

64. Robertson, H. D. 1975. *J. Mol. Biol.* 92:363–75

65. Barrell, B. G., Weith, H. L., Donelson, J. E., Robertson, H. D. 1975. *J. Mol. Biol.* 92:377–93

66. Ravetch, J. V., Model, P., Robertson, H. D. 1977. *Nature* 265:698–702

67. Grohmann, K., Smith, L. H., Sinsheimer, R. L. 1975. *Biochemistry* 14:1951–55

68. Smith, L. H., Sinsheimer, R. L. 1976. *J. Mol. Biol.* 103:699–735

69. Axelrod, N. 1976. *J. Mol. Biol.* 108:753–79

70. Carroll, D., Brown, D. D. 1976. *Cell* 7:467–76
71. Wegnez, M., Monier, R., Denis, H. 1972. *FEBS Lett.* 25:13–20
72. Brownlee, G. G., Cartwright, E. M., Brown, D. D. 1974. *J. Mol. Biol.* 89:703–18
73. Fedoroff, N. V., Brown, D. D. 1978. *Cold Spring Harbor Symp. Quant. Biol.* In press
74. Maxam, A. M., Gilbert, W. 1977. *Proc. Natl. Acad. Sci. USA* 74:560–64
75. Lawley, P. D., Brookes, P. 1963. *Biochem. J.* 89:127–28
76. Temperli, A., Türler, H., Rüst, P., Danon, A., Chargaff, E. 1964. *Biochim. Biophys. Acta* 91:462–76
77. Gilbert, W., Maxam, A. 1973. *Proc. Natl. Acad. Sci. USA* 70:3581–84
78. Gilbert, W., Majors, J., Maxam, A. 1976. *Organization and Expression of Chromosomes*, ed. V. G. Allfrey, E. K. F. Bautz, B. J. McCarty, R. T. Schimke, A. Tissieres, pp. 167–76. Berlin: Dahlem Konferenzen
79. Valenzuela, P., Bell, G. I., Masiarz, F. R., DeGennaro, L. J., Rutter, W. J. 1977. *Nature* 267:641–43
80. Maxam, A. M., Tizard, R., Skryabin, K. G., Gilbert, W. 1977. *Nature* 267:643–45
81. Valenzuela, P., Bell, G. I., Venegas, A., Sewell, E. T., Masiarz, F. R., DeGennaro, L. J., Weinberg, F., Rutter, W. J. 1977. *J. Biol. Chem.* 252:8126–35
82. 1974. *Cold Spring Harbor Symp. Quant. Biol.* 39:53–93
83. Subramanian, K. N., Pan, J., Zain, S., Weissman, S. M. 1974. *Nucleic Acids Res.* 1:727–52
84. Dhar, R., Zain, B. S., Weissman, S. M., Pan, J., Subramanian, K. 1976. *Proc. Natl. Acad. Sci. USA* 71:371–75
85. Subramanian, K. N., Dhar, R., Weissman, S. M. 1977. *J. Biol. Chem.* 252:355–67
86. Dhar, R., Subramanian, K. N., Pan, J., Weissman, S. M. 1977. *Proc. Natl. Acad. Sci. USA* 74:827–31
87. Pan, J., Thimmappaya, B., Reddy, V. B., Weissman, S. M. 1977. *Nucleic Acids Res.* 4:2539–48
88. Dhar, R., Reddy, V. B., Weissman, S. M. 1978. *J. Biol. Chem.* 253:612–20
89. Reddy, V. B., Dhar, R., Weissman, S. M. 1978. *J. Biol. Chem.* 253:621–30
90. Celma, M. L., Dhar, R., Pan, J., Weissman, S. M. 1977. *Nucleic Acids Res.* 4:2549–60
91. Berget, S. M., Moore, C., Sharp, P. A. 1977. *Proc. Natl. Acad. Sci. USA.* 74:3171–75
92. Chow, L. T., Gelinas, R. E., Broker, T. R., Roberts, R. J. 1977. *Cell* 12:1–8
92a. Klessig, D. F. 1977. *Cell* 12:9–21
93. Fiers, W., Danna, K., Rogiers, R., Van de Voorde, A., Van Herreweghe, J., Van Heuverswyn, H., Volckaert, G., Yang, R. 1974. *Cold Spring Harbor Symp. Quant. Biol.* 39:179–86
94. Volckaert, G., Contreras, R., Soeda, E., Van de Voorde, A., Fiers, W. 1977. *J. Mol. Biol.* 110:467–510
95. Fiers, W., Rogiers, R., Soeda, E., Van de Voorde, A., Van Heuverswyn, H., Van Herreweghe, J., Volckaert, G., Yang, R. 1975. *Proc. FEBS Mtg. 10th,* pp. 17–33. Amsterdam: North Holland
96. Ysebaert, M., Thys, F., Van de Voorde, A., Fiers, W. 1977. *Nucleic Acids Res.* 3:3409–21
97. Contreras, R., Volckaert, G., Thys, F., Van de Voorde, A., Fiers, W. 1977. *Nucleic Acids Res.* 4:1001–14
98. Van Heuverswyn, H., Van de Voorde, A., Fiers, W. 1977. *Nucleic Acid Res.* 4:1015–24
99. Van de Voorde, A., Contreras, R., Rogiers, R., Fiers, W. 1976. *Cell* 9:117–20
100. Bambara, R., Jay, E., Wu, R. 1974. *Nucleic Acids Res.* 1:1503–20
101. Tu, C. D., Jay, E., Bahl, C. P., Wu, R. 1976. *Anal. Biochem.* 74:73–93
102. Tu, C. D., Roychoudhury, R., Wu, R. 1976. *Fed. Proc. Fed. Am. Soc. Exp. Biol.* 35:1595 (Abstr.)
103. Jay, E., Roychoudhury, R., Wu, R. 1976. *Biochem. Biophys. Res. Commun.* 69:678–86
104. Gottesman, M. E., Weisberg, R. A. 1971. in *The Bacteriophage Lambda,* ed. A. D. Hershey, pp. 113–38. Cold Spring Harbor, NY: Cold Spring Harbor Lab.
105. Marini, J. C., Landy, A. 1977. *Virology* 76:196–209
106. Landy, A., Ross, W. 1977. *Science* 197:1147–60
107. Brownlee, G. G., Sanger, F. 1969. *Eur. J. Biochem.* 11:395–99
108. Ling, V. 1972. *J. Mol. Biol.* 64:87–102
109. Sanger, F., Donelson, J. E., Coulson, A. R., Kössel, H., Fischer, D. 1973. *Proc. Natl. Acad. Sci. USA* 70:1209–13
110. Wu, R., Tu, C. D., Padmanabhan, R. 1973. *Biochem. Biophys. Res. Commun.* 55:1092–99
111. Bambara, R., Wu, R. 1975. *J. Biol. Chem.* 250:4607–18
112. Loewen, P. C. 1975. *Nucleic Acids Res.* 2:839–52
113. Tu, C. D., Roychoudhury, R., Wu, R.

1976. *Biochem. Biophys. Res. Commun.* 72:355–62

114. Murray, K. 1973. *Biochem. J.* 131: 569–83

115. Roberts, R. J., Myers, P. A., Morrison, A., Murray, K. 1976. *J. Mol. Biol.* 102:157–65

115a. Pirrotta, V. 1976. *Nucleic Acids Res.* 3:1747–60

116. Roberts, R. J., Myers, P. A., Morrison, A., Murray, K. 1976. *J. Mol. Biol.* 103:199–208

117. Endow, S. A., Roberts, R. J. 1977. *J. Mol. Biol.* 112:521–29

118. Browne, J. K., Paddock, G. V., Liu, A., Clarke, P., Heindell, H. C., Salser, W. 1977. *Science* 195:389–91

119. Efstratiadis, A., Kafatos, F. C., Maniatis, T. 1977. *Cell* 10:571–85

120. Paddock, G. V., Poon, R., Heindell, H. C., Isaacson, J., Salser, W. 1977. *J. Biol. Chem.* 252:3446–58

121. Marotta, C. A., Wilson, J. T., Forget, B. G., Weissman, S. M. 1977. *J. Biol. Chem.* 252:5040–53

122. Proudfoot, N. J. 1976. *J. Mol. Biol.* 107:491–525

123. Proudfoot, N. J., Brownlee, G. G. 1976. *Nature* 263:211–14

124. Baralle, F. E. 1977. *Cell* 10:549–58

125. Haseltine, W. A., Maxam, A. M., Gilbert, W. 1977. *Proc. Natl. Acad. Sci. USA* 74:989–93

126. Gilbert, W. 1976. In *RNA Polymerase,* pp. 193–205. Cold Spring Harbor, NY: Cold Spring Harbor Lab.

127. Bourgeois, S., Pfahl, M. 1976. *Adv. Protein Chem.* 30:1–99

128. Wu, R., Bahl, C. P., Narang, S. A. 1978. *Curr. Top. in Cell. Regul.* 13:In press

129. Dickson, R. C., Abelson, J., Johnson, P., Reznikoff, W. S., Barnes, W. M. 1977. *J. Mol. Biol.* 111:65–75

130. Musso, R., Di Lauro, R., Rosenberg, M., De Crombrugghe, B. D. 1977. *Proc. Natl. Acad. Sci. USA* 74:106–10

131. Maniatis, T., Ptashne, M., Barrell, B. G., Donelson, J. E. 1974. *Nature* 250:394–97

132. Pirrotta, V. 1975. *Nature* 254:114–17

133. Humayun, Z., Jeffrey, A., Ptashne, M. 1977. *J. Mol. Biol.* 112:265–77

134. Humayun, Z., Kleid, D., Ptashne, M. 1977. *Nucleic Acids Res.* 4:1595–607

135. Maurer, R., Maniatis, T., Ptashne, M. 1974. *Nature* 249:221–23

136. Walz, A., Pirrotta, V. 1975. *Nature* 254:118–21

137. Scherer, G., Hobom, G., Kössel, H. 1977. *Nature* 265:117–21

138. Schaller, H., Gray, C., Herrmann, K. 1975. *Proc. Natl. Acad. Sci. USA* 72:737–41

139. Pribnow, D. 1975. *Proc. Natl. Acad. Sci. USA* 72:784–88

140. Sekiya, T., Contreras, R., Küpper, H., Landy, A., Khorana, H. G. 1976. *J. Biol. Chem.* 251:5124–40

141. Pan, J., Celma, M. L., Weissman, S. M. 1977. *J. Biol. Chem.* 252:9047–54

142. Bennett, G. N., Schweingruber, M. E., Brown, K. D., Squires, C., Yanofsky, C. 1978. *J. Mol. Biol.* In press

143. Bennett, G. N., Brown, K. D., Yanofsky, C. 1978. *J. Mol. Biol.* In press

144. Bennett, G. N., Schweingruber, M. E., Brown, K. D., Squires, C., Yanofsky, C. 1976. *Proc. Natl. Acad. Sci. USA* 73:2351–55

145. Murray, K., Isaksson-Forsen, A. G., Challberg, M., Englund, P. T. 1977. *J. Mol. Biol.* 112:471–89

146. Padmanabhan, R., Wu, R. 1972. *J. Mol. Biol.* 65:447–67

147. Murray, K., Murray, N. E. 1973. *Nature New Biol.* 243:134–39

148. Padmanabhan, R., Wu, R., Calendar, R. 1974. *J. Biol. Chem.* 249:6197–207

149. Weigel, P. H., Englund, P. T., Murray, K., Old, R. W. 1973. *Proc. Natl. Acad. Sci. USA* 70:1151–55

150. Ghangas, G. S., Jay, E., Bambara, R., Wu, R. 1973. *Biochem. Biophys. Res. Commun.* 54:998–1007

151. Nichols, B. P., Donelson, J. E. 1977. *J. Virol.* 22:520–26

152. Rosenberg, M., Segal, S., Kuff, E. L., Singer, M. F. 1977. *Cell* 11:845–58

153. Rao, G. R. K., Singer, M. F. 1977. *J. Biol. Chem.* 252:5124–34

154. Sures, I., Maxam, A., Cohn, R. H., Kedes, L. H. 1976. *Cell* 9:495–502

155. Birnstiel, M. L., Schaffner, W., Smith, H. O. 1977. *Nature* 266:603–7

NOTE ADDED IN PROOF   A more rapid and more accurate method for determining DNA sequence is described by Sanger et al (156). This method is similar to the plus and minus method (56) but makes use of the 2', 3'-dideoxynucleoside and arabinonucleoside triphosphates (e.g. ddTTP, ddATP, ddGTP, and araCTP) as specific chain-terminating inhibitors of DNA polymerase.

Recently, the total nucleotide sequence of SV40 DNA that includes some 5230 base pairs has been completed independently by Weissman et al (personal communication) and by Fiers et al (personal communication). About 80% of the sequence codes for three late viral structural proteins, as well as two early proteins, large T and small-t antigen, which are initiated at the same position.

The nucleotide sequence of over 600 base pairs of SV40 DNA and of a human papovavirus, BK virus, has been independently analyzed and compared. About 80% sequence homology exists in the regions that code for the late mRNA leader sequence, and for the first 74 NH_2-terminal amino acids of the large T and the small t-antigen (Yang and Wu, in press). The nucleotide sequence (550 base pairs) at another region of BK virus that includes the original of replication has been determined by Dhar et al (in press).

Literature Cited Added in Proof

156. Sanger, F., Nicklen, S., Coulson, A. R. 1977. *Proc. Natl. Acad. Sci., USA* 74:5463–67.

Ann. Rev. Biochem. 1978. 47:635–653

THE PHOTOCHEMICAL ELECTRON TRANSFER REACTIONS OF PHOTOSYNTHETIC BACTERIA AND PLANTS

❖984

Robert E. Blankenship and William W. Parson

Department of Biochemistry, SJ - 70, University of Washington, Seattle, Washington 98195

CONTENTS

PERSPECTIVES AND SUMMARY

The essence of photosynthesis in chlorophyll-containing organisms is the use of light to generate oxidants and reductants (1–10). Richly pigmented membranes in these organisms act as antennas, which absorb light and funnel energy to special "reaction centers" (RCs), where the electron transfer processes begin. In the last few years, preparations of purified RCs have been isolated from a variety of bacteria, including *Rhodopseudomonas sphaeroides* (11–17), *Rps. gelatinosa* (18), *Rps. viridis* (19–22), *Rhodospirillum rubrum* (15, 16, 23–25), and *Chromatium vinosum* (26). The availability of purified RCs has made it possible to use picosecond and nanosecond spectroscopic techniques to study the primary electron transfer reaction in detail.

The primary photochemical reaction in the bacterial RC is the oxidation of a complex of bacteriochlorophyll (BChl) molecules that we call P. (P

635

0066-4154/78/0701-0635$01.00

frequently is given a subscript such as 870 to indicate the wavelength of the major absorption band of the complex.) When P is excited it transfers an electron to a quinone by way of an intermediate electron acceptor that appears to be a bacteriopheophytin (BPh). (BPh is BChl with two protons replacing the Mg.) This process occurs with a quantum yield near 100% (27), and it takes only about 2×10^{-10} sec. On a slower time scale, the oxidized BChl complex (P^+) extracts an electron from a c-type cytochrome, the reduced quinone passes an electron on to another quinone, and further carriers return electrons from the second quinone to the cytochrome, completing a cycle eventually coupled to ATP synthesis. If the secondary reactions are blocked, the primary electron transfer is fully reversible even at cryogenic temperatures.

The photosynthetic apparatus of algae and green plants is considerably more complex. Plants transfer electrons over a much wider range of redox potentials, using two photosystems in series. Photosystem II (PS II) generates a weak reductant and a strong oxidant that can oxidize H_2O to O_2. Photosystem I (PS I) generates a weak oxidant and a strong reductant that can reduce ferredoxin and subsequently NADP. The PS II reductant and PS I oxidant are coupled by an electron transport chain. This electron transport scheme is noncyclic, in contrast to the cyclic bacterial system. Like the bacterial reactions, the electron transfer reactions of plants operate with quantum yields near 100% (28).

Knowledge of the primary reactions of plants is much less detailed than that of photosynthetic bacteria. Picosecond and nanosecond absorbance measurements have not yet been reported. This is largely because purified RCs are not yet available from plants, although several preparations enriched in PS I to a level of one RC per 40 antenna chlorophylls have been obtained (29), and reports of more highly enriched particles have appeared (30, 31). Studies at cryogenic temperatures also are more difficult in plants than in bacteria, because their photochemical reactions usually are partly irreversible at low temperatures. Secondary electron transfer reactions with low activation energies may occur rapidly and prevent electrons from returning by direct back reactions. A purified RC may lack secondary electron donors and acceptors and exhibit reversible primary reactions at low temperatures. However, some highly enriched PS I preparations are inactive at low temperatures, while retaining activity at room temperature (32, 33). This could mean that the native electron acceptor is missing and that other acceptors are able to substitute at room but not at low temperature.

When little information was available about the electron acceptors, PS I frequently was likened to the bacterial system. Their electron donors are indeed very similar. However, recent work suggests that the bacterial acceptor is more like that of PS II. Parallels do exist between bacterial and plant photosynthesis, but the situation is not so simple as it once seemed.

PHOTOSYNTHETIC BACTERIA

RCs purified from most bacterial species contain three hydrophobic polypeptides, with molecular weights of approximately 20,000, 24,000, and 30,000, of which only the smaller two are essential for photochemical activity (15, 34). Those from *Rps. gelatinosa* have only two polypeptides, with molecular weights of about 24,000 and 34,000 (18). Some of the preparations contain additional polypeptides due to tightly bound cytochromes.

Purified RCs from *Rps. sphaeroides* (13, 35), *Rps. viridis* (22), and *Rds. rubrum* (25) have 4 molecules of BChl and 2 of BPh per functional unit. The pigment content has not been analyzed in the preparations from other species, but similarities in the optical absorption spectra indicate that the compositions are likely to be similar. RCs purified from strains that synthesize carotenoids also typically contain one molecule of a carotenoid (25, 36). The BChl, BPh, and the carotenoid all reside on the lightest pair of the three polypeptides (12, 15, 36). In addition, purified RCs usually contain one or more quinones, which can be either ubiquinone or menaquinone (18, 37, 38), and one atom of nonheme Fe (12). The Fe is missing in some preparations (21, 39) and it can be removed from others or replaced by Mn without destroying photochemical activity (39–41).

When the BChl complex is raised to an excited singlet state (P*), it releases an electron, which reduces another component of the RC, I. The reduced acceptor (I⁻) then transfers an electron to another acceptor, X, which appears to be a quinone. The identity of X was a puzzle for some time, because the electron spin resonance (ESR) spectrum of the reduced acceptor (X⁻) is anomalous. Rather than the sharp signal near $g = 2.00$ that is typical of semiquinones, X⁻ has a very broad ESR spectrum with a principal g factor near 1.8 (12, 41–43). The unusual spectrum evidently results from magnetic interactions between X⁻ and the nonheme Fe of the RC, because removal of the Fe causes the spectrum to become essentially identical with that of the semiquinone of ubiquinone in vitro (39, 40). Extraction of the quinone, on the other hand, causes the RCs to lose their photochemical activity (37, 44). The activity can be restored by reconstituting the RCs with ubiquinone or other related quinones. The discordant observation (45, 46) that all ubiquinone could be extracted from *C. vinosum* chromatophores without affecting the photooxidation of P has been explained by the finding that a menaquinone acts as the primary acceptor in this species (38). RCs from *Rps. viridis* also appear to contain menaquinone instead of ubiquinone (19). However, RCs and chromatophores of *Rds. rubrum* have been reported to retain activity when they contain <1 ubiquinone per P (23, 47), and this discrepancy has not been satisfactorily explained; this species apparently does not contain menaquinone (48).

Optical absorbance changes that accompany the reduction of X also support the view that X is a quinone. In RCs isolated from *Rps. sphaeroides,* the reduction of X causes an absorbance decrease at 270 nm, and increases at 320 and 450 nm (49–51). The reduction of ubiquinone to its anionic semiquinone in vitro causes similar absorbance changes.

X^- releases an electron in 10 –100 μsec, reducing a second quinone that has similar optical and ESR spectra (46, 52–56), and the Fe appears to be required for this step (R. E. Blankenship and W. W. Parson, unpublished). If RCs are excited with repeated flashes, the optical and ESR signals due to the semiquinones oscillate in amplitude with a period of two (54–56). The secondary quinone evidently becomes fully reduced after each even-numbered flash, and then rapidly transfers a pair of electrons to an external pool of quinones.

In intact chromatophores, the E_m of X decreases with increasing pH by 60 mV/pH unit up to about pH 9, depending on the bacterial species. It then becomes constant at about –180 mV (57, 58). This indicates that a group with a pK_a of about nine (presumably the semiquinone itself) takes up one H^+ as X is reduced. The effective E_m during photosynthetic electron transfer probably is near –180 mV, because protonation of X^- apparently is not fast enough to compete with the movement of an electron to the secondary quinone (57). The E_m of P is about +470 mV in most species (4); it is essentially the same in isolated RCs as it is in chromatophores, and is independent of the pH.

In the green bacteria (*Chlorobiaceae*), the E_m of the acceptor is about –550 mV (59–61), and that of P about +250 mV (60, 61). Little is known of the acceptor in these species, but it appears to resemble the acceptor in PS I of plants.

Of the four BChl molecules in the RC, two appear to interact with each other particularly closely and to share in the release of an electron in the photochemical reaction. The major evidence for this comes from the ESR and electron nuclear double resonance (ENDOR) spectra of P^+, the radical that is formed when P is oxidized. In RCs from most species, P^+ gives a narrow ESR signal at $g = 2.0025$, with a linewidth of about 9.4 G (62, 63). The linewidth is narrower, by approximately a factor of $\sqrt{2}$, than that of the cationic radical ($BChl^+$) that is formed by the oxidation of BChl in vitro (62). The ENDOR spectrum of P^+ shows splittings that are one half as large as those in the ENDOR spectrum of $BChl^+$ (63, 64). These differences can be explained by assuming that the unpaired electron in P^+ is delocalized over two molecules of BChl, so that its hyperfine interaction with each nucleus is one half as strong (62–65). In *Rps. viridis*, however, the ESR linewidth and the ENDOR splittings are not decreased by as much as they are in other species (66, 67). In this species, either P^+ does not consist of a completely symmetrical dimer of BChl molecules, or the hopping of electrons between the two molecules is relatively slow. *Rps. viridis* differs from the other species in containing BChl-*b* and BPh-*b* rather than BChl-*a* and BPh-*a*, but functionally its RCs appear to be quite similar (21).

Additional information on the interactions among the BChl molecules of the RC has come from optical absorption, CD, and linear dichroism spectra. RCs from species containing BChl-*a* have major absorption bands near 540, 600, 760, 800, and 870 nm. The 600, 800, and 870 nm bands usually are ascribed to BChl, and the 540 and 760 nm bands to BPh. When P is oxidized to P^+, the 870 nm band bleaches, the 600 nm band bleaches partially, and the 800 nm band appears to shift about 5 nm to shorter wavelengths. The 800 nm band apparently includes two overlapping absorption bands, and its shift in position seems to reflect the bleaching of one of these bands and the development of a new band at a shorter wavelength (68–73).

RCs that contain BChl-*b* and BPh-*b*, absorb light at longer wavelengths. The absorption spectrum of *Rps. viridis* RCs has five components in the near infrared, with maxima near 790, 810, 830, 850, and 960 nm (20, 22, 74). The first four of these overlap, but can be resolved at low temperatures. The 790 nm band has been ascribed to BPh, and the remainder to BChl (20). The bands at 850 and 960 nm are both associated closely with P, because both of them disappear when P is oxidized. In addition, the excitation of the RC BChl to its lowest excited singlet state (P^*) causes bleaching at both 850 and 960 nm (75). This state undoubtedly involves the BChls that make up P; it can be detected spectrophotometrically only when the photooxidation of P is blocked.

Because excitation to P^* or one-electron oxidation to P^+ causes bleaching of more than one of the near infrared absorption bands, the BChl molecules of P must interact strongly, The absorption bands would appear to be properties of the complex, rather than of two discrete molecules. Oxidation of the complex evidently disrupts the interaction, and the new absorption bands that appear at shorter wavelengths could result from the freeing of one molecule of BChl to behave essentially as a monomer (71).

Exactly how the BChl molecules of P interact is not clear. A synthetic dimer of covalently linked BChls has been found to resemble P in certain respects, but not in others (76). In wet solvents, the dimer apparently folds so that the Mg of each BChl is linked by an H_2O to the C9 keto carbonyl of the opposite BChl. The dimer undergoes photooxidation to a radical with a narrow ESR spectrum similar to that of P^+. But the absorption spectrum of the synthetic dimer has a maximum near 803 rather than 870 nm. Other similar structures have been proposed (72, 77–80), but is is not clear that any of these can account for the position of the major absorption band of the RC. Wasielewski et al (76) suggest that the special pair of BChl molecules in P interact with additional molecules of BChl or BPh, and that these interactions move the absorption to longer wavelengths. It is possible, though, that the absorption spectrum of P is strongly influenced by interactions with other components of the RC, such as electrically charged amino acids. Another possibility is that the two BChl molecules of P are arranged, not face-to-face, but end-on (74). This could give a large exciton splitting and might account for the strong absorption band at long wavelengths.

Clarke et al (81) have concluded from the zero-field splitting parameters and decay kinetics of the triplet state of P that P cannot contain a fully symmetric dimer. However, the zero-field splitting parameters of monomeric and dimeric chlorophylls are very similar (82). Studies of the triplet states of dimers with known geometries would be useful in this regard.

The participation of an intermediate electron carrier (I) between P and X was first detected by exciting RCs with short flashes at low redox potentials, when X was already reduced. Optical absorbance changes revealed the formation of a transient state (P^F) that survived for about 10 nsec (16, 75, 83, 84). State P^F appears to be the radical pair P^+I^-, and its decay under these conditions appears to involve the return of an electron from I^- to P^+. With picosecond techniques, absorbance changes reflecting the transient reduction of I can be seen also when X is functional (75, 85–87). The decay kinetics under these conditions indicate that an electron moves from I^- to X in about 200 psec. The kinetics of this step are essentially the same in *Rps. sphaeroides* and *Rps. viridis.*

In chromatophores and RCs of several species, it has been possible to trap I in its reduced state for longer periods, by prolonged illumination at low redox potentials (20, 21, 74, 75, 88–93). The RCs from *Rps. viridis, C. vinosum,* and *C. minutissimum* contain c-type cytochromes that are capable of reducing P^+ relatively rapidly. If P^+ is reduced during the 10 nsec lifetime of P^+I^-, the RC is left with I reduced. Continued illumination evidently can drive the reduction virtually to completion. Picosecond absorbance measurements show that the photooxidation of P to P^+ does not occur if I is reduced in this way (75, 94). This indicates that there probably are not additional electron carriers between P and I.

The optical absorbance changes that occur when I is reduced include bleaching in the bands due to BPh near 540 and 760 nm (790 in *Rps. viridis*) and the formation of a broad new band centered near 650 nm (680 in *Rps. viridis*) (20, 21, 74, 75, 88–92). Very similar absorbance changes accompany the reduction of BPh to its anionic radical (BPh^-) in vitro (66,

67, 95). Apparently, one of the two BPhs of the RC is a major component of I. Only one of them appears to be involved, because the 540 nm absorption bands of two BPhs can be resolved at low temperatures and the reduction of I causes only one of the bands to bleach (96).

The reduction of I also causes changes in some of the absorption bands that have been attributed to the BChl of the RC. The most dramatic of these is an apparent shift to shorter wavelengths of one of the components of the band at 800 nm (830 nm in *Rps. viridis*) (20, 21, 74, 75, 88–92). There is no bleaching in the absorption band associated with P at 870 nm (960 nm in *Rps. viridis*). In *Rps. viridis,* where the absorption bands are resolved, it is clear also that the reduction of I affects the 830 nm component, rather than the 850 nm component that is associated with P (75). The reduction also causes the disappearance of a (+) CD band near 830 nm, without affecting the (–) CD of the 850 nm component (74). This indicates that the absorption changes probably reflect a bleaching of the 830 nm band and its replacement by a new band at shorter wavelengths, rather than a simple hypsochromic shift. Bleaching and the development of a new band could occur if I consists of interacting BChl and BPh molecules, whose interactions are disrupted by the reduction of one of the molecules (75). This interpretation is analogous to the one given above for P.

The ESR spectrum of I⁻ appears to be consistent with the view that the odd electron is confined predominantly to a single molecule. The spectrum has a g factor near 2.0035 and a half-width of about 13 G (21, 67, 89–91, 97), and is similar to that of BPh⁻ in vitro (66, 67, 97). However, the anionic radical of BChl has a very similar ESR spectrum, so these observations do not prove that the molecule that is reduced in I⁻ is BPh rather than BChl (66, 67, 97). Under some conditions, the ESR spectrum of I⁻ has an additional broad doublet with a splitting of about 60 G (120 G in *Rps. viridis*), centered near $g = 2.0$ (20, 89–91). The broad signal appears to arise from magnetic coupling between I⁻ and X⁻. It evidently occurs only if the ESR signal of X⁻ is broadened by interaction with the Fe, and it is lacking if X is doubly reduced to the diamagnetic X^{2-} (93). [X normally accepts only one electron (4, 52), but prolonged illumination at low redox potentials apparently can lead to further reduction (93).]

The rate of the initial transfer of an electron from P* to I has not been measured satisfactorily. The process is too fast for the picosecond techniques that have been applied so far; one can say only that it appears to be complete in less than 10 psec (75, 85–87, 94, 98). The extraction or prior reduction of X does not seem to delay the reaction (98). Measurements of the yield of fluorescence from P* indicate that the lifetime of the excited state is on the order of 7–10 psec, in agreement with this limit (14, 99). Picosecond fluorescence lifetime measurements have given a lifetime of 15 psec for P* (100), but the excitation flash used may not have been short

enough to make these measurements definitive. If electron transfer from P*
to I is blocked because I is already reduced, P* lives for about 20 psec, just
long enough to be detectable spectrophotometrically (75). It is surprising
that the lifetime of P* is still very short, even under these conditions.

If X has been reduced or extracted from the RCs, P* still gives rise to
P^+I^- with a high quantum yield, but the radical pair cannot pass an electron
on to X. Instead, it evidently decays by the return of an electron from I^-
back to P^+. Because P^+I^- probably forms too rapidly to allow P* to undergo
intersystem crossing to a triplet state, the spins of the unpaired electrons
on P^+ and I^- initially must be opposed, but the relationship between the
spins has an opportunity to change during the 10 nsec lifetime of P^+I^-. If
the orbital overlap between P^+ and I^- is not too strong, the radical pair will
oscillate between states that are essentially singlet or essentially triplet in
character. If an electron returns from I^- to P^+ when the spins of the two
electrons are opposed, the pigments decay to their ground states. If the
return occurs with the two spins parallel, it results in the generation of a
triplet state of P. The triplet state has been detected spectrophotometrically
(and called state P^R) (16, 75, 83, 84), from its ESR spectrum (58, 101–104),
and from optically detected magnetic resonance (ODMR) (81, 105, 106).
Its quantum yield is about 15% at room temperature, but the yield becomes
close to 100% at very low temperatures (83, 96, 103). At room temperature,
the lifetime of P^R is between 5 and 50 μs in RCs from *Rps. viridis* or the
carotenoidless strains of *Rps. sphaeroides* and *Rds. rubrum* (16, 75, 83, 84).
In RCs that contain a carotenoid, P^R decays in 10–20 nsec by transferring
energy to the carotenoid and promoting it to a triplet state (16, 84). All of
the observations that have been made to date are consistent with the view
that the triplet state is not an intermediate in the normal electron transfer
reaction, but a side product that forms if electron transfer from I^- to X is
blocked. Study of the triplet is useful, nonetheless, in that it can provide
information about the radical pair state P^+I^-.

The ESR spectrum of the triplet state shows an unusual spin polarization,
i.e. a departure from thermal equilibrium among the populations of the spin
sublevels. Essentially all of the triplets are formed in the T_0 sublevel, in
which the molecule has no net magnetic moment in the direction of the
applied magnetic field (101, 102, 104). The spin polarization is inconsistent
with the normal mechanism of intramolecular intersystem crossing, but it
is readily explained on the assumption that the triplet arises from P^+I^-. One
expects the applied magnetic field to shift the energies of the T_+ and T_-
triplet sublevels in the radical pair so that only the T_0 sublevel can mix with
the singlet state (104, 107, 108). This interpretation is supported by the
observation that magnetic fields cause a decrease in the overall quantum
yield of the triplet state (107, 108).

Perhaps the most remarkable aspect of the primary photochemical reaction, is not the great speed with which an electron moves from P* to I, but the slowness with which electrons return from I⁻ to P⁺. The forward reaction is complete in less than 10 psec; the back reaction is only half-complete in 10 nsec. There are light-driven electron transfer reactions in vitro which occur at rates comparable to that of the initial reaction, but the radical pairs that are formed in these generally decay by even faster back reactions (109). The stability of P⁺I⁻ is critical if the RC is to capture the energy of light with a high quantum yield, but how the back reaction is prevented is not clear (109). One possibility is that P⁺ or I⁻ moves so that the distance between them increases immediately after the forward reaction. Another is that P and I are positioned to maximize the overlap between the donor and acceptor molecular orbitals for the forward reaction while minimizing that for the back reaction; this might be possible, because the two processes involve different orbitals on P. The amount of energy that has to be converted from electronic to vibrational modes may also be critical in determining the rate of the back reaction.

PLANT PHOTOSYSTEM I

The primary electron donor of PS I is called P700 (110). P700 is very similar to the bacterial P in E_m, ESR, and optical characteristics. When P700 is oxidized, the product, P700⁺, exhibits ESR and ENDOR spectra characteristic of a radical cation of a chlorophyll (Chl) dimer (64, 65, 111). Absorbance decreases at 700, 680, and 430 nm (110, 112) and an absorbance increase at 815 nm (113, 114) accompany the oxidation. The 815 nm absorption band is characteristic of Chl⁺ radicals (115). The CD spectrum of P700 also is consistent with a dimer (116). The E_m of P700 has been determined several times, with different results. Kok (110) measured +430 mV in acetone-extracted chloroplasts, but recent determinations have given values of +520, +493, and +375 mV (117–119). The reason for the disagreement among the measurements is unclear. Two models for the structure of the P700 dimer have been proposed recently (78, 79).

The identity of the primary electron acceptor in PS I is controversial. Malkin & Bearden reported an irreversible photooxidation of P700 at 77°K accompanied by the reduction of a species whose ESR spectrum had g factors of 1.86, 1.94, and 2.05 (120). The spectrum is indicative of an iron-sulfur protein, and the species has been named "bound ferredoxin." Quantitative measurements of the ESR signals indicate that the amount of bound ferredoxin photoreduced is stoichiometrically equal to the amount of P700 photooxidized (121). Redox titrations of PS I subchloroplast particles reveal two ferredoxin-type ESR signals (122–123). Both appear to be due to 4 Fe–4 S centers (124) and are clearly distinct from the 2 Fe–2 S soluble ferredoxin. Center A probably is bound ferredoxin; it has an E_m at pH10 of –550 mV and g factors identical to those of bound ferredoxin. Center B's role is unclear; it has an E_m at pH10 of –590 mV and g factors of 1.89, 1.92, and 2.05. The two centers interact; the $g = 1.86$ resonance of center A shifts to 1.89 when center B is reduced (123). A 4 Fe–4 S protein of 8,000 molecular weight, which may be bound ferredoxin, has been isolated and characterized to some extent (125).

Reduction of the primary acceptor also causes an optical absorbance decrease at 430 nm which is detectable at room temperature (126, 127). The component responsible for the absorbance change, P430, has an E_m of –530 mV, and has the spectral characteristics of a ferredoxin (118, 127). Its behavior parallels that of bound ferredoxin (128); the two are almost certainly the same species (4).

Redox titrations of P700 photooxidation at 77°K yield an E_m of –530 mV for the electron acceptor (118, 129, 130), which is in good agreement with the E_m of bound ferredoxin and P430. However, titrations of this sort can measure the E_m of the real primary acceptor only if the products of the photochemical reaction are stable to charge recombination on the time scale of the measurements. An earlier, more negative acceptor, which backreacts rapidly with P700$^+$, might not be detected. Measurements with continuous actinic light are particularly susceptible to this ambiguity. Under certain conditions the photooxidation of P700 at low temperatures is partially reversible (e.g. 131). The extent of reversibility seems to depend on the nature of the preparation and the experimental conditions. If bound ferredoxin were the primary acceptor it should behave similarly. However, McIntosh et al (132) found that a small amount of P700 underwent reversible photooxidation without parallel changes in the ESR signal of bound ferredoxin. An ESR signal with g factors of 1.75, 1.88, and 2.06 did appear reversibly. McIntosh et al (132) attributed the signal to a new component, X$^-$, which they interpreted as the primary acceptor, relegating bound ferredoxin to the role of a secondary acceptor. These observations have been confirmed, and reversibility of P700 photooxidation as high as 50–90% has been reported (133–135). The large extent of reversibility is important in the interpretation of the ESR signals, since it was argued that a 5–10% component of the bound ferredoxin signal would be difficult to detect (136). Recent optical measurements at low potentials suggest that there may even be two electron acceptors prior to bound ferredoxin (187).

The chemical nature of X is unknown. It has been compared to ferredoxins and to the quinone-iron complex of photosynthetic bacteria, but neither comparison seems wholly appropriate. The E_m of X has not been measured, but a value of –730 mV was estimated from a titration of P700 photooxidation at 15°K (135).

So far, X^- has been observed only under strongly reducing conditions which keep bound ferredoxin chemically reduced and therefore inactive as an electron acceptor. Demonstrating that a component such as X can be reduced under some conditions does not guarantee that it is in the preferred reaction pathway under physiological conditions. (Consider the triplet states observed in bacterial RCs at low potentials.) In enriched PS I preparations obtained so far, P700 photooxidation at low temperatures correlates with the presence of bound ferredoxin (32, 33). On the other hand, the 0.1 μs time resolution of the P430 measurements reported so far (127) would not have revealed an earlier component such as X^- if its decay kinetics were sufficiently rapid. Kinetic measurements on a faster time scale are required to establish the true sequence of reactions.

Little information exists on the progress of excited electronic states that follow excitation of P700 and lead to the relatively stable state $P700^+$-$P430^-$ (or X^-). Bleaching of the 700 nm absorption band of P700 occurs < 20 nsec after excitation (137), but this reflects mainly the loss of ground state P700 and will be rapid regardless of the mechanism of electron transfer. Measurements at 815 nm, where $P700^+$ absorbs, would be more sensitive to the mechanism; unfortunately, fast optical measurements on plant systems still are very difficult technically. However, information can be obtained by other means, especially ESR.

Radicals produced in photochemical reactions can have ESR spectra that are spin polarized. This phenomenon, called chemically induced dynamic electron polarization (CIDEP), can arise by either, or both, of two mechanisms, but the ESR spectra predicted by the two are different (138–139). In the "radical pair" mechanism, magnetic interaction between the two radicals generated by electron transfer causes spin polarization to develop. In the "triplet" mechanism, electron transfer follows intramolecular intersystem crossing from excited singlet to spin-polarized triplet states. The radical pair mechanism predicts that the two radicals will have equal and opposite polarizations (microwave emission or enhanced absorption) if the g factors of the two radicals are very different, and that both radicals will show mixed emission and absorption if the g factors are similar; the total net polarization of the system will be zero in all cases. The triplet mechanism predicts that both radicals will be polarized similarly, so that the net polarization is not zero. It is important that both radicals be observed for an unambiguous assignment of the mechanism. Unfortunately, only one radical can be seen in many cases, making interpretation difficult.

A single emissive ESR line was observed in chloroplasts under physiological conditions, and was interpreted as arising from the primary acceptor of PS I (140). Since both radicals were not observed, an unambiguous choice of mechanism was not possible, but the authors preferred the triplet mechanism. Subsequent work revealed that the signal is partly emissive and partly absorptive, and that the signal depends on the orientation of the chloroplasts with respect to the magnetic field (141). This behavior is consistent with the radical pair mechanism if the CIDEP signal arises from P700$^+$ rather than from the primary acceptor, and if the acceptor has an anisotropic g tensor and is oriented in the membrane (142). Both bound ferredoxin and X$^-$ do have anisotropic g tensors but only X$^-$ appears to have the necessary orientation in the membrane (141). The triplet mechanism seems unable to explain the observations, so emphasis has shifted to the radical pair mechanism, with electron transfer viewed as occurring from the excited singlet state of P700.

There seems to be no compelling evidence implicating triplet states in the primary reaction of PS I. Shuvalov (143) has measured luminescence from PS I particles that were chemically reduced or heat treated to inactivate P430, and has attributed it to a triplet state of Chl. Triplets also have been observed by ESR (82, 144) and by ODMR (145, 146). It is possible, however, that most of these triplets are due to antenna Chl; whether any of them orginates in the reaction centers is unclear. ESR spectra with T$_0$ polarization like that observed in bacteria are not a feature of plant systems studied to date. This is surprising if PS I photochemistry indeed operates by way of a radical pair mechanism. Fong (77) has proposed a novel mechanism of electron transfer involving long-lived triplet states, but this "upconversion" theory has little experimental support and we view it as highly speculative.

PLANT PHOTOSYSTEM II

The primary electron donor is PS II is known as P680 or Chl a$_{II}$ Excitation of PS II causes bleaching at 680 nm (690 in some preparations) (147, 148) and 435 nm (147), and an absorbance increase at 825 nm (114, 149), presumably due to the radical cation P680$^+$. A transient ESR signal similar to that of P700$^+$ also has been assigned to P680$^+$ (149–151). This is not to be confused with "ESR Signal II" which is thought to derive from a plastoquinone (152).

P680 is not nearly as well characterized as P700. This is partly because it is rereduced rapidly after photooxidation, which makes measurement of P680$^+$ difficult. In additi n, the E_m of P680 is so high that it has not yet been measurable. The E_m must be above +850 mV, because PS II can still be photoactivated at this potential (153). This is consistent with the fact that PS II serves to oxidize water to O$_2$ (E_m at pH7 of +820 mV).

The reduction of P680$^+$ has been the subject of considerable discussion. Early measurements (147) indicated a biphasic reduction, with half-times ($t_{1/2}$) of 200 μsec and 20 msec. Later, a 35 μsec component was detected (154). However, measurements at 825 nm have shown that P680$^+$ is > 80%

reduced in less than 5 μsec (155). Recently, indirect evidence for a $<$ 1 μsec component has been obtained (156). Duysens and co-workers have concluded from measurements of prompt and delayed fluorescence that P680$^+$ is reduced in $<$ 1 μsec in untreated preparations (157–158). If the system is inhibited by NH$_2$OH (157), by incubation at low pH (159–160), or by the presence of Tris (156, 160, 161), the reduction is slower ($t_{1/2}$ = 20 – 200 μsec) and may involve different electron donors. The 200 μsec reduction evidently is a back reaction with the primary acceptor. The complex reduction kinetics seen by some workers could represent reaction centers using several alternative pathways. It now appears that P680$^+$ is not stable under any conditions so far examined. Stable ESR signals have been attributed to P680$^+$ (153), but these seem likely to arise from a secondary donor rather than from P680$^+$ itself.

At 77°K, the predominant reduction path appears to be a direct back reaction with the primary acceptor (162). The $t_{1/2}$ for this is 3–5 msec (114, 150). Other donors, e.g. cytochrome b_{599} (114, 148), can reduce P680$^+$ irreversibly at low temperatures with somewhat slower kinetics. This can cause the electron acceptor to become trapped in the reduced state, so that P680 is no longer photochemically active, a situation similar to the trapping of I$^-$ in bacteria.

The primary electron acceptor in PS II is thought to be plastoquinone. Stiehl & Witt (163) first associated transient absorbance changes at 320 nm with the primary acceptor, and they identified the acceptor as plastoquinone from the spectrum of the absorbance changes. Using a PS II subchloroplast particle, van Gorkom (164) has measured a more complete spectrum which closely resembles that of the plastoquinone radical anion, and similar observations have been reported by others (165–166). Plastoquinone extraction and reconstitution experiments also support the identification of the primary acceptor as plastoquinone (167).

The absorbance increase at 320 nm, indicative of plastoquinone reduction, occurs within 1 μsec following a flash at room temperature (168), and decays with $t_{1/2}$ = 600 μsec (163). The absorbance change still occurs rapidly at low temperatures, and the majority decays with $t_{1/2}$ = 3 msec, due to the back reaction with P680$^+$; the remainder of the absorbance change at low temperatures is irreversible (166). During a train of flashes at room temperature, the amplitude of the absorbance change oscillates with a period of two (169, 170). This has been interpreted as reflecting electron transfer to a secondary acceptor which couples the one-electron photochemistry of the reaction center with the two-electron chemistry of a larger pool of quinones. Such a secondary acceptor had been proposed earlier on the basis of other experiments (171, 172). Similar oscillations in bacteria are mentioned above. In PS II, absorbance changes oscillating with a period of four are superimposed on the period two oscillations, presumably reflecting events on the oxidizing side of P680 (169, 170).

The reduction of the primary acceptor in PS II is associated also with absorbance changes near 550 nm (173, 174). However, extraction and reconstitution experiments (167, 175), and measurements at very high redox potentials (176), have revealed conditions under which PS II photochemistry can occur in the absence of these absorbance changes. This suggests that the 550 nm absorbance changes are due, not to a component which itself undergoes reduction, but rather to a molecule (most likely a carotenoid) whose absorption spectrum is shifted by a change in the redox state of the primary acceptor. The 550 nm absorbance changes have, nonetheless, been useful for redox titrations of the primary acceptor, giving an E_m at pH7 of +25 mV (174). The E_m changes by 60 mV per pH unit (177). It has been argued (178) that the effective E_m of the PS II acceptor, like that of bacteria, is more negative than the E_m at pH7, because the rate of electron transfer is fast relative to the rate at which proton uptake can occur. This would make the true E_m of the acceptor about −130 mV (178). The E_m is shifted by +70 mV by the addition of o-phenanthroline, a specific inhibitor of PS II electron transfer (177). The E_m of the bacterial electron acceptor is affected similarly by o-phenanthroline (57).

Another measure of the state of the primary acceptor is the yield of fluorescence from the antenna Chl of PS II. Duysens & Sweers (179) postulated that the RC quenched fluorescence effectively when the acceptor was oxidized but poorly when it was reduced. Fluorescence has been related to the redox state of the acceptor by many investigators subsequently, and the correlation has been found to hold in most instances (180). However, the fluorescence yield appears to depend not only on the redox state of the primary acceptor, but also on that of P680 (174) and on other factors, including the state of another quencher which is thought to be a carotenoid triplet (181, 182).

Redox titrations of fluorescence revealed two transitions, with E_m at pH7 values of −35 and −270 mV (183). More recent titrations on PS II subchloroplast particles (184) have given an extremely negative E_m at pH7 of −325 mV. The authors argue that earlier measurements reflect the reduction of secondary acceptors that are absent from the subchloroplast particles. All of the fluorescence titrations reported so far have exhibited considerable hysteresis, indicating that the systems were not in equilibrium throughout the experiments. This, coupled to the fact that the fluorescence yield is sensitive to many factors, makes interpretation of the results difficult.

Recently it has been suggested that P680 is able to transfer an electron to a second acceptor after the initial rapid reduction of P680$^+$ by the physiological donor (156, 158). This second process is not productive in O_2 evolution, and its significance is not yet clear.

In summary, the electron acceptor of PS II strikingly resembles that of photosynthetic bacteria. Both appear to be quinones that undergo reduction to anionic semiquinones. The apparent E_m and its pH dependence, the effect of o-phenanthroline, and the oscillations of period two are all very similar in the two systems. No ESR signals have been associated unambiguously with the PS II acceptor (see below), but a broad signal with a g factor of 1.8 might be anticipated on the basis of the similarities with bacteria.

Even less information is available about the mechanism of electron transfer in PS II than in PS I. McIntosh & Bolton (185) have observed CIDEP signals in deuterated algae and have interpreted them in terms of the triplet mechanism. Complex signals showing total emissive nature were seen and were interpreted as arising from two species, P680$^+$ ($g = 2.0025$) and a plastoquinone anion radical ($g = 2.0042$). The reported observation of the acceptor's signal at $g = 2.0042$ is surprising. This is a position expected for a semiquinone, and other investigators have detected ESR signals here (149, 186), but the signals seen previously appear to be due to components on the electron donor side rather than the acceptor side of PS II. If the $g = 2.0042$ resonance is indeed due to the primary acceptor in PS II, the emissive nature of both radicals strongly inplicates the triplet mechanism of CIDEP over the radical pair mechanism. If so, the similarities between PS II and bacteria are deceptive in that the basic mechanisms of electron transfer would be totally different. Further work is needed to clarify this situation.

ACKNOWLEDGMENTS

National Science Foundation (NSF) grant number GMS 74–19852 AO1 and an NSF National Needs Fellowship to (R. E. Blankenship) supported this work.

Literature Cited

1. Clayton, R. K. 1973. *Ann. Rev. Biophys. Bioeng.* 2:131–56
2. Parson, W. W. 1974. *Ann. Rev. Microbiol.* 28:41–59
3. Trebst, A. 1974. *Ann. Rev. Plant Physiol.* 25:423–58
4. Parson, W. W., Cogdell, R. J. 1975. *Biochim Biophys. Acta* 416:105–49
5. Bearden, A. J., Malkin, R. 1975. *Quart. Rev. Biophys.* 7:131–77
6. Govindjee, ed. 1975. *Bioenergetics of Photosynthesis.* New York: Academic. 689 pp.
7. Olson, J. M., Hind, G., eds. 1977. *Brookhaven Symp. Biol. Chlorophyll-Proteins, Reaction Centers, and Photosynthetic Membranes* 28:385 pp.
8. Avron, M. 1977. *Ann. Rev. Biochem.* 46:143–55
9. Ke, B. 1978. *Curr. Top. Bioenerg.* In Press
10. Clayton, R. K., Sistrom, W. R., eds. 1978. *The Photosynthetic Bacteria.* New York: Plenum. In press
11. Clayton, R. K., Wang, R. T. 1971. *Methods Enzymol.* 23:696–704
12. Feher, G. 1971. *Photochem. Photobiol.* 14:373–87
13. Reed, D. W., Peters, G. A. 1972. *J. Biol. Chem.* 247:7148–52
14. Slooten, L. 1972. *Biochim. Biophys. Acta* 256:452–66
15. Okamura, M. Y., Steiner, L. A., Feher, G. 1974. *Biochemistry* 13:1394–1403

16. Cogdell, R. J., Monger, T. G., Parson, W. W. 1975. *Biochim. Biophys. Acta* 408:189–99
17. Jolchine, G., Reiss-Husson, F. 1974. *FEBS Lett.* 40:5–8
18. Clayton, R. K., Clayton, B. J. 1977. *Proc. Ann. Meet. Am. Soc. Photobiol., 5th,* (Abstr.) FAM-D2:65
19. Peucheu, N. L., Kerber, N. L., Garcia, A. 1976. *Arch. Microbiol.* 109:301–5
20. Trosper, T. L., Benson, D. L., Thornber, J. P. 1977. *Biochim. Biophys. Acta* 460:318–30
21. Prince, R. C., Tiede, D. M., Thornber, J. P., Dutton, P. L. 1977. *Biochim. Biophys. Acta.* 462:467–90
22. Thornber, J. P., Dutton, P. L., Fajer, J., Forman, A., Holten, D., Olsen, J. M., Parson, W. W., Prince, R. C., Tiede, D. M., Windsor, M. W. 1978. *Proc. Int. Congr. Photosynth., 4th.* In Press
23. Noël, H., Van der Rest, M., Gingras, G. 1972. *Biochim. Biophys. Acta* 275: 219–30
24. Smith, W. R., Sybesma, C., Dus, K. 1972. *Biochim. Biophys. Acta* 267: 609–15
25. Van der Rest, M., Gingras, G. 1974. *J. Biol. Chem.* 249:6446–53
26. Lin, L., Thornber, J. P. 1975. *Photochem. Photobiol.* 22:37–40
27. Wraight, C. A., Clayton, R. K. 1974. *Biochim. Biophys. Acta* 333:246–60
28. Sun, A. S. K., Sauer, K. 1971. *Biochim. Biophys. Acta* 234:399–414
29. Thornber, J. P. 1975. *Ann. Rev. Plant Physiol.* 26:127–58
30. Ikegami, I., Katoh, S. 1975. *Biochim. Biophys. Acta* 376:588–92
31. Bengis, C., Nelson, N. 1977. *J. Biol. Chem.* 252:4564–69
32. Nelson, N., Bengis, C., Silver, B. L., Getz, D., Evans, M. C. W. 1975. *FEBS Lett.* 58:363–65
33. Malkin, R., Bearden, A. J., Hunter, F. A., Alberte, R. S., Thornber, J. P. 1976. *Biochim. Biophys. Acta* 430:389–94
34. Steiner, L. A., Okamura, M. Y., Lopes, A. D., Moskowitz, E., Feher, G. 1974. *Biochemistry* 13:1403–10
35. Straley, S. C., Parson, W. W., Mauzerall, D. C., Clayton, R. K. 1973. *Biochim. Biophys. Acta* 305:597–609
36. Cogdell, R. J., Parson, W. W., Kerr, M. A. 1976. *Biochim. Biophys. Acta* 430: 83–93
37. Okamura, M. Y., Isaacson, R. A., Feher, G. 1975. *Proc. Natl. Acad. Sci. USA* 72:3491–95
38. Feher, G., Okamura, M. Y. 1977. See Ref. 7, pp. 183–94
39. Loach, P. A., Hall, R. L. 1972. *Proc. Natl. Acad. Sci. USA* 69:786–90
40. Feher, G., Okamura, M. Y., McElroy, J. D. 1972. *Biochim. Biophys. Acta* 267:222–26
41. Feher, G., Issacson, R. A., McElroy, J. D., Ackerson, L. C., Okamura, M. Y. 1974. *Biochim. Biophys. Acta* 368: 135–39
42. Leigh, J. S., Dutton, P. L. 1972. *Biochem. Biophys. Res. Commun.* 46: 414–21
43. Dutton, P. L., Leigh, J. S., Reed, D. W. 1973. *Biochim. Biophys. Acta* 292: 654–64
44. Cogdell, R. J., Brune, D., Clayton, R. K. 1974. *FEBS Lett.* 45:344–47
45. Ke, B., Garcia, A. F., Vernon, L. P. 1973. *Biochim. Biophys. Acta* 292: 226–36
46. Halsey, Y. D., Parson, W. W. 1974. *Biochim. Biophys. Acta* 347:404–16
47. Morrison, L., Runquist, J., Loach, P. 1977. *Photochem. Photobiol.* 25: 73–84
48. Maroc, J., deKlerk, H., Kamen, M. D. 1968. *Biochim. Biophys. Acta* 162: 621–23
49. Clayton, R. K., Straley, S. C. 1970. *Biochem. Biophys. Res. Commun.* 39: 1114–19
50. Clayton, R. K., Straley, S. C. 1972. *Biophys. J.* 12:1221–34
51. Slooten, L. 1972. *Biochim. Biophys. Acta* 275:208–18
52. Parson, W. W. 1969. *Biochim. Biophys. Acta* 189:384–96
53. Chamorovsky, S. K., Remennikov, S. M., Kononenko, A. A., Venediktov, P. S., Rubin, A. B. 1976. *Biochim. Biophys. Acta* 430:62–70
54. Vermeglio, A. 1977. *Biochim. Biophys. Acta* 459:516–524
55. Vermeglio, A., Clayton, R. K. 1977. *Biochim. Biophys. Acta* 461:159–65
56. Wraight, C. A. 1977. *Biochim. Biophys. Acta* 459:525–31
57. Prince, R. C., Dutton, P. L. 1976. *Arch. Biochem. Biophys.* 172:329–34
58. Prince, R. C., Leigh, J. S., Dutton, P. L. 1976. *Biochim. Biophys. Acta* 440: 622–36
59. Knaff, D. B., Malkin, R. 1976. *Biochim. Biophys. Acta* 430:244–52
60. Prince, R. C., Olson, J. M. 1976. *Biochim. Biophys. Acta* 423:357–62
61. Jennings, J. V., Evans, M. C. W. 1977. *FEBS Lett.* 75:33–36
62. McElroy, J. D., Feher, G., Mauzerall, D. C. 1972. *Biochim. Biophys. Acta* 267:363–74
63. Feher, G., Hoff, A. J., Isaacson, R. A., Ackerson, L. C. 1975. *Ann. NY Acad. Sci.* 244:239–59

64. Norris, J. R., Scheer, H., Druyan, M. E., Katz, J. J. 1974. *Proc. Natl. Acad. Sci. USA* 71:4897–4900
65. Norris, J. R., Uphaus, R. A., Crespi, H. L., Katz, J. J. 1971. *Proc. Natl. Acad. Sci. USA* 68:625–28
66. Fajer, J., Davis, M. S., Brune, D. C., Spaulding, L. D., Borg, D. C., Forman, A. 1977. See Ref. 7, pp. 74–103
67. Fajer, J., Davis, M. S., Holten, J. D., Parson, W. W., Thornber, J. P., Windsor, M. W. 1978. *Proc. Int. Congr. Photosynthesis, 4th* (Abstr.) 108
68. Sauer, K., Dratz, E. A., Coyne, L. 1968. *Proc. Natl. Acad. Sci. USA* 61: 17–24
69. Reed, D. W., Ke, B. 1973. *J. Biol. Chem.* 248:3041–45
70. Philipson, K. D., Sauer, K. 1973. *Biochemistry* 12:535–39
71. Vermeglio, A., Clayton, R. K. 1976. *Biochim. Biophys. Acta* 449:500–15
72. Shuvalov, V. A., Asadov, A. A., Krakhmaleva, I. N. 1977. *FEBS Lett.* 76: 240–45
73. Mar, T., Gingras, G. 1977. *Biochim. Biophys. Acta* 460:239–46
74. Shuvalov, V. A., Krakhmaleva, I. N., Klimov, V. V. 1976. *Biochim. Biophys. Acta* 449:597–601
75. Holten, D., Windsor, M. W., Parson, W. W., Thornber, J. P. 1978. *Biochim. Biophys. Acta* 501:112–126
76. Wasielewski, M. R., Smith, U. H., Cope, B. T., Katz, J. J. 1977. *J. Am. Chem. Soc.* 99:4172–73
77. Fong, F. K. 1975. *Appl. Phys.* 6:151–66
78. Fong, F. K., Koester, V. J., Galloway, L. 1977. *J. Am. Chem. Soc.* 99:2372–75
79. Shipman, L. L., Cotton, T. M., Norris, J. R., Katz, J. J. 1976. *Proc. Natl. Acad. Sci. USA* 73:1791–94
80. Katz, J. J., Norris, J. 1977. See Ref. 7, pp. 16–54
81. Clarke, R. H., Connors, R. E., Frank, H. A. 1976. *Biochem. Biophys. Res. Commun.* 71:671–75
82. Uphaus, R. A., Norris, J. R., Katz, J. J. 1974. *Biochem. Biophys. Res. Comm.* 61:1057–63
83. Parson, W. W., Clayton, R. K., Cogdell, R. J. 1975. *Biochim. Biophys. Acta* 387:265–78
84. Parson, W. W., Monger, T. G. 1977. See Ref. 7, pp. 195–211
85. Rockley, M. G., Windsor, M. W., Cogdell, R. J., Parson, W. W. 1975. *Proc. Natl. Acad. Sci. USA* 72:2251–55
86. Kaufmann, K. J., Dutton, P. L., Netzel, T. L., Leigh, J. S., Rentzepis, P. M. 1975. *Science* 188:1301–04
87. Dutton, P. L., Kaufmann, K. J.,

Chance, B., Rentzepis, P. M. 1975. *FEBS Lett.* 60:275–80
88. Shuvalov, V. A., Klimov, V. V. 1976, *Biochim. Biophys. Acta* 440:587–99
89. Tiede, D. M., Prince, R. C., Dutton, P. L. 1976. *Biochim. Biophys. Acta* 449:447–69
90. Tiede, D. M., Prince, R. C., Reed, G. H., Dutton, P. L. 1976. *FEBS Lett.* 65:301–4
91. Dutton, P. L., Prince, R. C., Tiede, D. M., Petty, K. M., Kaufmann, K. J., Netzel, T. L., Rentzepis, P. M. 1977. See Ref. 7, pp. 213–37
92. van Grondelle, R., Romijn, J. C., Holmes, N. G. 1976. *FEBS Lett.* 72:187–92
93. Okamura, M. Y., Isaacson, R. A., Feher, G. 1977. *Biophys. J.* 17:149a (Abstr.)
94. Netzel, T. L., Rentzepis, P. M., Tiede, D. M., Prince, R. C., Dutton, P. L. 1977. *Biochim. Biophys. Acta* 460: 467–78
95. Fajer, J., Brune, D. C., Davis, M. S., Forman, A., Spaulding, L. D. 1975. *Proc. Natl. Acad. Sci. USA* 72: 4956–60
96. Clayton, R. K., Yamamoto, T. 1976. *Photochem. Photobiol.* 24:67–70
97. Feher, G., Isaacson, R. A., Okamura, M. Y. 1977. *Biophys. J.* 17:149a (Abstr.)
98. Kaufmann, K. J., Petty, K. M., Dutton, P. L., Rentzepis, P. M. 1976. *Biochem. Biophys. Res. Commun.* 70:839–45
99. Zankel, K. L., Reed, D. L., Clayton, R. K. 1968. *Proc. Natl. Acad. Sci. USA* 61:1243–49
100. Kononenko, A. A., Knox, P. P., Adamova, N. P., Paschenko, V. Z., Timofeev, K. N., Rubin, A. B. 1976. *Stud. Biophys.* 55:183–96
101. Dutton, P. L., Leigh, J. S., Seibert, M. 1972. *Biochem. Biophys. Res. Commun.* 46:406–413
102. Leigh, J. S., Dutton, P. L. 1974. *Biochim. Biophys. Acta* 357:67–77
103. Wraight, C. A., Leigh, J. S., Dutton, P. L., Clayton, R. K. 1974. *Biochim. Biophys. Acta* 333:401–8
104. Thurnauer, M. C., Katz, J. J., Norris, J. R. 1975. *Proc. Natl. Acad. Sci. USA* 72:3270–74
105. Clarke, R. H., Connors, R. E., Norris, J. R., Thurnauer, M. C. 1975. *J. Am. Chem. Soc.* 97:7178–79
106. Hoff, A. J. 1977. *Biochim. Biophys. Acta* 440:765–71
107. Blankenship, R. E., Schaafsma, T. J., Parson, W. W. 1977. *Biochim. Biophys. Acta* 461:297–305

108. Hoff, A. J., Rademaker, H., van Grondelle, R., Duysens, L. N. M. 1977. *Biochim. Biophys. Acta* 460:547–54
109. Holten, D., Windsor, M. W., Parson, W. W., Gouterman, M. 1978. *Photochem. Photobiol.* In press
110. Kok, B. 1961. *Biochim. Biophys. Acta* 48:527–33
111. Katz, J. J., Norris, J. R. Jr. 1973. *Curr. Top. Bioenerg.* 5:41–75
112. Döring, G., Bailey, J. L., Kreutz, W., Weikard, J., Witt, H. T. 1968. *Naturwissenschaften* 55:219–20
113. Inoue, Y., Ogawa, T., Shibata, K. 1973. *Biochim. Biophys. Acta* 305:483–87
114. Mathis, P., Vermeglio, A. 1975. *Biochim. Biophys. Acta* 369:371–81
115. Borg, D. C., Fajer, J., Felton, R. H., Dolphin, D. 1970. *Proc. Natl. Acad. Sci. USA* 67:813–20
116. Philipson, K. D., Sato, V. L., Sauer, K. 1972. *Biochemistry* 11:4591–95
117. Knaff, D. B., Malkin, R. 1973. *Arch. Biochem. Biophys.* 159:555–62
118. Shuvalov, V. A., Klimov, V. V., Krasnovskii, A. A. 1976. *Mol. Biol.* 10:261–72 (Engl. trans.)
119. Evans, M. C. W., Sihra, C. K., Slabas, A. R. 1977. *Biochem. J.* 162:75–85
120. Malkin, R., Bearden, A. J. 1971. *Proc. Natl. Acad. Sci. USA* 68:16–19
121. Bearden, A. J., Malkin, R. 1972. *Biochim. Biophys. Acta* 283:456–68
122. Ke, B., Hansen, R. E., Beinert, H. 1973. *Proc. Natl. Acad. Sci. USA* 70:2941–45
123. Evans, M. C. W., Reeves, S. G., Cammack, R. 1974. *FEBS Lett.* 49:111–14
124. Cammack, R., Evans, M. C. W. 1975. *Biochem. Biophys. Res. Comm.* 67:544–49
125. Malkin, R., Aparicio, P. J., Arnon, D. I. 1974. *Proc. Natl. Acad. Sci. USA* 71:2362–66
126. Hiyama, T., Ke, B. 1971. *Proc. Natl. Acad. Sci. USA* 68:1010–13
127. Ke, B. 1973. *Biochim. Biophys. Acta* 301:1–33
128. Ke, B., Beinert, H. 1973. *Biochim. Biophys. Acta* 305:689–93
129. Ke, B. 1974. in *Proc. Int. Congr. Photosynthesis, 3rd*, ed. M. Avron, pp. 373–82. Amsterdam: Elsevier. 2194 pp.
130. Lozier, R. H., Butler, W. L. 1974. *Biochim. Biophys. Acta* 333:460–64
131. Ke, B., Sugahara, K., Sahu, S. 1976. *Biochim. Biophys. Acta* 449:84–94
132. McIntosh, A. R., Chu, M., Bolton, J. R. 1975. *Biochim. Biophys. Acta* 376:308–14
133. Evans, M. C. W., Sihra, C. K., Cammack, R. 1976. *Biochem. J.* 158:71–77
134. McIntosh, A. R., Bolton, J. R. 1976. *Biochim. Biophys. Acta* 430:555–59
135. Ke, B., Dolan, E., Sugahara, K., Hawkridge, F. M., Demeter, S., Shaw, E. R. 1977. *Plant Cell Physiol. Phosynthetic Organelles*, pp. 187–99
136. Bearden, A. J., Malkin, R. 1976. *Biochim. Biophys. Acta* 430:538–47
137. Witt, K., Wolff, Ch. 1970. *Z. Naturforsch.* 25b:387–88
138. Wan, J. K. S., Elliot, A. J. 1977. *Acc. Chem. Res.* 10:161–66
139. Dobbs, A. J. 1975. *Mol. Phys.* 30:1073–84
140. Blankenship, R., McGuire, A., Sauer, K. 1975. *Proc. Natl. Acad. Sci. USA* 72:4943–47
141. Dismukes, G. C., McGuire, A., Blankenship, R., Sauer, K. 1978. *Biophys. J.* 21:In press
142. Dismukes, C., Friesner, R., Sauer, K. 1977. *Biophys. J.* 17:228a (Abstr.)
143. Shuvalov, V. A. 1976. *Biochim. Biophys. Acta* 430:113–21
144. Nissani, E., Scherz, A., Levanon, H. 1977. *Photochem. Photobiol.* 25:93–101
145. Van der Bent, S. J., Schaafsma, T. J., Goedheer, J. C. 1976. *Biochem. Biophys. Res. Commun.* 71:1147–52
146. Hoff, A. J., Govindjee, Romijn, J. C. 1977. *FEBS Lett.* 73:191–96
147. Döring, G., Renger, G., Vater, J., Witt, H. T. 1969. *Z. Naturforsch* 24b:1139–43
148. Floyd, R. A., Chance, B., Devault, D. 1971. *Biochim. Biophys. Acta* 226:103–12
149. Van Gorkom, H. J., Pulles, M. P. J., Wessels, J. S. C. 1975. *Biochim. Biophys. Acta* 408:331–39
150. Malkin, R., Bearden, A. J. 1975. *Biochim. Biophys. Acta* 396:250–59
151. Visser, J. W. M., Rijgersberg, C. P., Gast, P. 1977. *Biochim. Biophys. Acta* 460:36–46
152. Kohl, D., Wood, P. 1969. *Plant Physiol.* 44:1439–45
153. Bearden, A. J., Malkin, R. 1973. *Biochim. Biophys. Acta* 325:266–74
154. Gläser, M., Wolff, Ch., Buchwald, H.-E., Witt, H. T. 1974. *FEBS Lett.* 42:81–85
155. Mathis, P., Haveman, J., Yates, M. 1976. See Ref. 7, pp. 267–277
156. Gläser, M., Wolff, Ch., Renger, G. 1976. *Z. Naturforsch.* 31c:712–21
157. Den Haan, G. A., Duysens, L. N. M., Egberts, D. J. N. 1974. *Biochim. Biophys. Acta* 368:409–21
158. Van Best, J. A., Duysens, L. N. M. 1977. *Biochim. Biophys. Acta* 459:187–206

159. Van Gorkom, H. J., Pulles, M. P., Haveman, J., Den Haan, G. A. 1976. *Biochim. Biophys. Acta* 423:217–26
160. Haveman, J., Mathis, P. 1976. *Biochim. Biophys. Acta* 440:346–55
161. Döring, G. 1975. *Biochim. Biophys. Acta* 376:274–84
162. Butler, W. L., Visser, J. W. M., Simons, H. L. 1973. *Biochim. Biophys. Acta* 325:539–45
163. Stiehl, H. N., Witt, H. T. 1969. *Z. Naturforsch.* 24b:1588–98
164. Van Gorkom, H. J. 1974. *Biochim. Biophys. Acta* 347:439–42
165. Pulles, M. P. J., Kerkhof, P. L. M., Amesz , J. 1974. *FEBS Lett.* 47:143–45
166. Haveman, J., Mathis, P., Vermeglio, A. 1975. *FEBS Lett.* 58:259–61
167. Knaff, D. B., Malkin, R., Myron, J. C., Stoller, M. 1977. *Biochim. Biophys. Acta* 459:402–11
168. Renger, G., and Wolff, Ch. 1976. *Biochim. Biophys. Acta* 423:610–14
169. Pulles, M. P. J., Van Gorkom, H. J., Willemsen, J. G. 1976. *Biochim. Biophys. Acta* 449:536–40
170. Mathis, P., Haveman, J. 1977. *Biochim. Biophys. Acta* 461:167–81
171. Bouges-Bocquet, B. 1973. *Biochim. Biophys. Acta* 314:250–56
172. Velthuys, B. R., Amesz, J. 1974. *Biochim. Biophys. Acta* 333:85–94
173. Knaff, D. B., Arnon, D. I. 1969. *Proc. Natl. Acad. Sci. USA* 63:963–69
174. Butler, W. L. 1973. *Accts. Chem. Res.* 6:177–84
175. Cox, R. P., Bendall, D. S. 1974. *Biochim. Biophys. Acta* 347:49–59
176. Knaff, D. B., Malkin, R. 1976. *Arch. Biochem. Biophys.* 174:414–19
177. Knaff, D. B. 1975. *Biochim. Biophys. Acta* 376:583–87
178. Knaff, D. B. 1975. *FEBS Lett.* 60:331–35
179. Duysens, L. N. M., Sweers, H. E. 1963. In *Studies on Microalgae and Photosynthetic Bacteria,* ed. Jpn. Soc. Plant Physiol., pp. 353–72. Tokyo: Univ. Press
180. Papageorgiou, G. 1975. See Ref. 6, pp. 320–371
181. Zankel, K. L. 1973. *Biochim. Biophys. Acta* 325:138–48
182. Mauzerall, D. 1976. *J. Phys. Chem.* 80:2306–9
183. Cramer, W. A., Butler, W. L. 1969. *Biochim. Biophys. Acta* 172:503–10
184. Ke, B., Hawkridge, F. M., Sahu, S. 1976. *Proc. Natl. Acad. Sci. USA* 73:2211–15
185. McIntosh, A. R., Bolton, J. R. 1976. *Nature* 263:443–45
186. Blankenship, R. E., Babcock, G. T., Warden, J. T., Sauer, K. 1975. *FEBS Lett.* 51:287–93
187. Sauer, K., Mathis, P., Acker, S., Van Best, J. A. 1978. *Biochim. Biophys. Acta.* In press

Ann. Rev. Biochem. 1978. 47:655–86
Copyright © 1978 by Annual Reviews Inc. All rights reserved

THE ROLE OF ADENOSINE AND 2'-DEOXYADENOSINE IN MAMMALIAN CELLS

❖985

Irving H. Fox and William N. Kelley

The Human Purine Research Center, Departments of Internal Medicine and Biological Chemistry, University of Michigan, Ann Arbor, Michigan 48109

CONTENTS

0066-4154/78/0701-0655$01.00

PERSPECTIVES AND SUMMARY

Adenosine and 2'-deoxyadenosine are purine nucleosides that are intermediates in the pathway of purine nucleotide degradation. The multidisciplinary interest in these compounds has been stimulated by their apparent biological effects. In 1972, the first inborn error of adenosine metabolism was discovered in man; a deficiency of adenosine deaminase was found to be associated with severe immune dysfunction involving both B- and T-lymphocytes. This discovery suggested a possible relationship of this enzyme deficiency to immune dysfunction and suddenly the vast body of knowledge on the biology and biochemistry of adenosine became more visible and its relevance more widely appreciated. Despite the increasing focus on adenosine over the past five years and an even more recent interest in 2'-deoxyadenosine, effective assimilation and integration of the data available has been delayed by the unusually wide spectrum of scientific disciplines interested in this compound. In this review, we attempt to cross the boundaries of scientific disciplines to give an overview of the total body of knowledge concerning adenosine and 2'-deoxyadenosine. Of the vast literature available, however, only the studies considered most relevant to the authors' work are included. An extensive consideration of adenosine and 2'-deoxyadenosine transport and their analogues is outside the scope of this review.

Adenosine and 2'-deoxyadenosine are transported into cells by facilitated diffusion. Adenosine is formed within cells from AMP or from the degradation of S-adenosylhomocysteine. It can be removed from cells by release to the extracellular fluid, by deamination to inosine, or by reutilization to form AMP. A low concentration of adenosine tends to favor phosphorylation, while a low concentration of 2'-deoxyadenosine probably favors deamination.

Many biological properties of adenosine are identified: it is toxic to mammalian and bacterial cells, and its presence is associated with inhibition of the immune response, coronary vasodilation, delayed neurotransmission, inhibition or stimulation of hormone secretion, and changes in the metabolism of a number of tissues. Adenosine has many biochemical effects including: direct activation of adenylate cyclase, inhibition of pyrimidine biosynthesis, and diminution of phosphoribosylpyrophosphate (PP-ribose-P) synthesis; it is unclear, however, which, if any, of these properties account for its biological activity. Less information is available concerning the biological and biochemical properties of 2'-deoxyadenosine; it is known to have toxic effects and has been observed to modulate adenylate cyclase.

The study of alterations of adenosine and 2'-deoxyadenosine metabolism is important, since these abnormal states may provide clues about normal biological functions. Analysis of patients with adenosine deaminase deficiency and purine nucleoside phosphorylase deficiency has already revealed important information. There is, however, much more to learn.

Pharmacological alteration of adenosine and 2'-deoxyadenosine metabolism holds promise as one method of manipulating many types of biological events. This can occur by inhibition of adenosine or 2'-deoxyadenosine transport, modification of their metabolism, or the utilization of analogues.

It is clear that the continued study of the biology and biochemistry of adenosine, 2'-deoxyadenosine, and their analogues will be fruitful both in the understanding and the development of new ways to modify biological phenomena.

METABOLISM

The turnover of adenosine and 2'-deoxyadenosine in cells involves a complicated series of regulated reactions (Figure 1, Table 1). The activity of these biochemical pathways results in low or unmeasurable concentrations of adenosine in the cell cytosol und ͬ r normal circumstances (1–3). This compound has been detected mainly ᵢn extracellular fluid, plasma, and medium incubating or perfusing cells (1, 4–14).

Transport

Adenosine or 2'-deoxyadenosine may be generated within the cell or they may enter the cell by one of several transport mechanisms that have the characteristics of facilitated diffusion systems (15–17; Figure 1). The evidence in red cells and leukocytes suggests that there is a single carrier that mediates the uptake of all purine and pyrimidine nucleosides (17). Several carriers with different affinities for alternative substrates may exist, however, in some tissue culture cell lines (17). The K_m for adenosine transport varies from 4–42 μM in nucleated cells (17–23) and is 100 μM in erythrocytes (15). The K_m for 2'-deoxyadenosine transport varies from 2–37 μM in nucleated cells (17, 18). In murine lymphocytes there are two transport systems for adenosine—a high affinity system with a K_m for adenosine of 12 μM, and a low affinity system with a K_m of 400 μM (22).

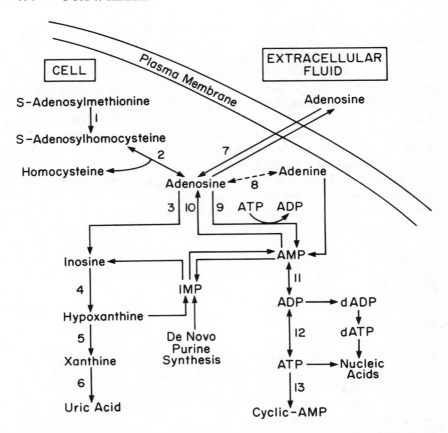

Figure 1. Metabolism of adenosine. This diagram illustrates the pathways by which adenosine is formed and degraded within cells. Adenosine may also be transported into cells or released from cells. A detailed description of these pathways is included in the section on adenosine metabolism. The metabolism of 2'-deoxyadenosine may utilize some of these pathways: *1*, S-adenosylmethionine methyltransferases; *2*, S-adenosylhomocysteine hydrolase; *3*, adenosine deaminase; *4*, purine nucleoside phosphorylase; *5 & 6*, xanthine oxidase; *7*, transport mechanisms; *8*, adenosine phosphorylase (not established); *9*, adenosine kinase; *10*, 5'-nucleotidase and non-specific phosphatase; *11*, adenylate kinase; *12*, nucleoside diphosphokinase; *13*, adenylate cyclase.

The K_m for 2'-deoxyadenosine is 230 μM for the low affinity system (22). Studies on the transport of nucleosides into mouse fibroblast cells or membrane vesicles derived from these cells suggest the existence of two distinctive mechanisms for adenosine uptake (23, 24). In intact cells, adenosine appears to enter unchanged; in membrane vesicles, however, the transport

Table 1 Enzymes of adenosine metabolism

Properties of enzymes	Adenosine deaminase	Adenosine kinase	S-Adenosylhomocysteine hydrolase	5'-Nucleotidase
E.C.	3.5.4.6	2.7.1.20	3.3.1.1	3.1.3.5
Tissue distribution	all tissues examined	all tissues examined	brain, liver, kidney, spleen, testis, heart	all tissues examined
Subcellular fraction	particulate, cytosol	cytosol; particulate (?)	probably cytosol	plasma membrane, cytosol, lysosomes
Native molecular weight	36,000; 114,000; 298,000	23,000; 40,000	—	10,000–235,000
Subunit molecular weight	36,000	40,000 (?)	—	50,000 (?)
K_m Adenosine	35–400 μM	0.4–6.0 μM	1.5 mM	—
K_m 2'-Deoxyadenosine	35–400 μM	400–1800 μM[b]	—	—
K_m Other substrates	—	ATP 25–400 μM	L-homocysteine γ 4.5 mM	nucleoside 5'-monophosphates 10–100 μM
K_i Products	inosine 60–700 μM	AMP 0.2 mM, 6-MMPR[a] 0.16 mM, ADP 0.06–1.2 mM	S-adenosylhomocysteine $I_{0.5}$ 2.5 mM	nucleosides 0.4–5.0 mM Pi 42 mM

[a] 6 methylmercaptopurine ribonucleotide.
[b] In some tissues 2'-deoxyadenosine kinase is a separate enzyme from adenosine kinase.

of adenosine leads to an increased concentration of hypoxanthine in the surrounding medium with an accumulation of ribose-1-phosphate inside the vesicles (24). This would suggest that both adenosine deaminase and purine nucleoside phosphorylase are necessary for at least this latter type of adenosine transport. Adenosine may be utilized for nucleotide synthesis or it may be degraded to inosine and purine bases. (2, 3, 6, 15, 25–36).

Formation

Phosphatases (reaction 10, Figure 1) catalyze the dephosphorylation of AMP or 2'-dAMP to form adenosine and 2'-deoxyadenosine respectively. Morphological and functional chemical studies provide convincing evidence that a form of 5'-phosphomonoesterase activity is localized in the external surface of the plasma membrane (37–43). This enzyme is important in allowing the uptake of adenosine from AMP in human lymphocytes (44). Intracellular 5'-AMP is degraded to adenosine by a specific 5'-nucleotidase (E.C.3.1.3.5) or by nonspecific phosphatases. In this pathway 5'-nucleotidase represents a low K_m enzyme, while alkaline phosphatase (E.C.3.1.3.2) is a high K_m enzyme (45, 46). Although the K_m values of some 5'-nucleoside monophosphate substrates for the 5'-nucleotidase are similar to their physiological concentrations, the nucleotide pool may normally inhibit this enzyme (46–49). In contrast, alkaline phosphatase is effectively inhibited by physiological concentrations of inorganic phosphate. In addition, the high K_m of this latter enzyme for AMP makes it unlikely that it is an important catalyst of adenosine formation under normal conditions.

The dephosphorylation of AMP has been examined in detail in Ehrlich ascites tumor cells during 2-deoxyglucose-induced purine catabolism (50, 51) and the results compared with previously published kinetic constants for 5'-nucleotidase derived from partially purified extracts of these cells (47). In this experiment, a maximum of 18% of adenosine nucleotides were dephosphorylated to adenosine (50). The concentration of AMP reached 400 μM, a value well in excess of the K_m of 5'-nucleotidase for AMP (67 μM) and equal to the K_m of alkaline phosphatase for this substrate (400 μM) (46). The concentration of ATP declined to a minimum of 400 μM, a value still well in excess of the k_i of 5'-nucleotidase for ATP (12 μM). These observations suggested that both 5'-nucleotidase and nonspecific phosphatases contributed to the formation of adenosine in this experiment. According to Henderson (51), the rate of AMP formation and its concentration in the cell, appeared, under these specific experimental conditions, to be a more important regulator of dephosphorylation than inhibition by ATP. Ineffective inhibition of alkaline phosphatase by nucleotides, and a decline of inorganic phosphate concentration, may each have contributed to the unexpected importance of the nonspecific phosphatases in this situation.

The degradation of S-adenosylhomocysteine to adenosine and homocysteine (reaction 2, Figure 1) represents another potential source of intracellular adenosine. The enzyme capable of catalyzing this hydrolysis, S-adenosylhomocysteine hydrolase (E.C.3.3.1.1), is found in yeast, plants, birds, and mammals, and is highest in liver, pancreas, and kidney (52, 53). Although the adenosylhomocysteine hydrolase equilibrium favors the synthetic reaction, the pathway in vivo is presumed to occur toward degradation since the products are easily removed by further degradation under normal conditions (54). S-adenosylhomocysteine is degraded almost completely to uric acid in rat and mouse liver (53, 55) and to inosine and hypoxanthine in brain and kidney (53). The K_m value of the enzyme is 1.5 mM for adenosine and 4.5 mM for L-homocysteine, and these compounds inhibit the hydrolytic reaction (52). The removal of S-adenosylhomocysteine is an important metabolic event because this compound is a strong inhibitor of methyltransfer reactions (56–60).

Removal

Adenosine or 2'-deoxyadenosine may be removed from cells by direct release to the extracellular environment, by synthesis to adenine nucleotides, or by degradation to purine end products. Certain tissues are known to release adenosine (4–14), and it is released in myocardium in response to oxygen lack. In this tissue, 5'-nucleotidase has been shown in some studies (61), but not others (62, 63), to be associated with membranes lining compartments open to the extracellular space. There is also evidence to suggest that in myocardium there are two nucleotide pools capable of being degraded to adenosine (64). Adenosine, inosine, and, to a greater extent, hypoxanthine, are released by lung in response to alveolar hypoxia (65). Degradative enzymes capable of releasing these compounds are located in capillary endothelial cells (65). In contrast to myocardium and lung, liver appears to release only hypoxanthine in response to hypoxia (10).

When the liver is normally oxygenated, 80% of the hypoxanthine, xanthine, and urate is removed in a single passage of blood. Under these conditions, the concentration of adenosine in hepatic venous blood is tenfold higher than its concentration in the portal vein or hepatic artery (10). This suggests that adenosine is produced and released by the liver, perhaps as a nutrient source of purines. Presumably, the adenosine released by the liver in this physiologic setting is incorporated into the circulating erythrocytes thus providing a potential source of purine substrate to other tissues (9, 10). Although adenosine release appears to be a common occurrence, there is no information concerning the mechanisms of release.

Adenosine or 2'-deoxyadenosine is actively degraded to inosine or 2'-deoxyinosine (reaction 3, Figure 1) by adenosine deaminase (E.C.3.5.4.6). The K_m for adenosine ranges from 45–150 μM in human tissues (66–70) and from 35–400 μM in other tissues (71–76). The K_m for 2'-deoxyadenosine has similar values (67, 72–74). The K_i for inosine ranges from 60–130 μM to as high as 700 μM in these enzyme preparations. While the rate of the reaction appears to be regulated by the availability and concentration of the substrate and product, the level of enzyme activity in various tissues appears to be related to the molecular form of the enzyme that predominates. [One form of the enzyme appears to be "particulate", but little more is known about it (66).] Three forms of the enzyme are soluble and interconvertible with apparent molecular weights of approximately 36,000, 114,000, and 298,000. They have been designated small, intermediate, and large, respectively. The small form of adenosine deaminase is convertible to the large form only in the presence of a protein that has an apparent molecular weight of 200,000 and has no adenosine deaminase activity (66, 67, 77). This conversion of the small form of the enzyme to a large form occurs at 4°C, exhibits a pH optimum of 5.0–8.0, and is associated with a loss of conversion activity. The small form of the enzyme predominates in tissue preparations exhibiting the higher enzyme specific activities and no detectable conversion activity, such as those from stomach and spleen (66). The large form of adenosine deaminase predominates in tissue extracts exhibiting the lower enzyme specific activities and abundant conversion activity as is evident in those from kidney and lung (66). The kinetic characteristics of the three soluble molecular species of adenosine deaminase are identical except for pH optimum, which is 5.5 for the intermediate species and 7.0–7.4 for the large and small forms. These variations of enzyme activity with conversion factor activity may be indicative of the modulating effect of this polypeptide on adenosine deaminase. The so-called "tissue specific" form of adenosine deaminase described in the literature refers to the large form of the enzyme. An optimal level of adenosine deaminase activity is essential for normal cell function in man since a deficiency of this enzyme is associated with an increased intracellular concentration of adenine nucleotides (78, 79) and an increased level of activity of this enzyme is associated with the opposite effect (80).

The removal of adenosine or 2'-deoxyadenosine can also occur through their conversion to nucleoside monophosphates (reaction 9, Figure 1) in a reaction catalyzed by adenosine kinase (E.C.2.7.1.20). This enzyme is ubiquitous but has the highest activity in liver, spleen, and lymph nodes (81). Adenosine kinase has generally been purified from the cell cytosol and has been localized in the soluble fraction in rat heart (82). Adenosine kinase may also be present, however, in the erythrocyte membrane (30). Although there is evidence to suggest that adenosine and deoxyadenosine are phosphorylated by the same enzyme (83), some experiments demonstrate the existence of a distinct enzyme for deoxyadenosine (84–86a).

Adenosine phosphorylation may be regulated in a complex manner that includes the K_m adenosine, the intracellular concentration of Mg^{2+}, cell energy charge, and the nucleotide pool. The K_m for adenosine has been reported from 0.4–6.0 μM (25, 29, 84, 87–91), values that compare with the maximum adenosine concentrations of 0.4–0.6 μM observed in canine arterial blood (4) or human coronary sinus blood (5). The K_m for 2'-deoxyadenosine is considerably higher, ranging from 400–1800 μM (84, 86a, 88). In contrast, the K_m for ATP is well below its normal intracellular concentration (29, 89, 92) which suggests that this compound may not be an important physiological regulator of adenosine phosphorylation under normal circumstances. An increasing ATP "charge", however, has a stimulating effect on adenosine phosphorylation (89). It is possible that the latter effect may be related to the decreasing concentration of ADP, since the K_i of adenosine kinase for ADP is in the range of its known intracellular concentration (89, 92). Phosphorylation of adenosine may occur in the absence of a divalent cation, but reports on the effects of Mg^{2+}, Mg^{2+}-ATP, or ATP alone are conflicting (29, 87–89, 93); optimum ATP/Mg concentrations are reported all the way from a value of 10 to a value of 1. Mg ATP may be a poorer substrate than ATP alone since the velocity of the reaction is reduced with Mg^{2+}, and an increasing $S_{0.5}$ for ATP results (89). In these studies, there is a sigmoidal response of adenosine kinase to increasing ATP concentrations with 1–2 mM Mg^{2+}, but a hyperbolic response with no

Mg^{2+} (89). Mg^{2+} may be important in the active site of the adenosine kinase (94).

It is possible that adenosine can be removed by a direct reversible phosphorolysis to adenine (reaction 8, Figure 1) in a reaction similar to that catalyzed by purine nucleoside phosphorylase (E.C.2.4.4.1). While crude tissue preparations exhibit such a pathway for adenosine (33, 89, 95, 96) and 2'-deoxyadenosine (96), highly purified preparations of purine nucleoside phosphorylase have no such activity (97, 98). A distinct adenosine phosphorylase separate from purine nucleoside phosphorylase has been isolated from *Bacillus subtilis* (99).

Regulation

When adenosine enters the cell, several factors determine whether it is to be converted to nucleotides or to inosine. The utilization of adenosine is dependent upon the relative activities of adenosine deaminase and adenosine kinase (25, 29, 30, 35, 96). The K_m values for adenosine deaminase are roughly an order of magnitude higher than those reported for adenosine kinase. A low concentration of adenosine, therefore, will tend to favor reutilization of adenosine by phosphorylation to AMP. In human erythrocyte ghosts most adenosine is converted into adenine nucleotides when adenosine concentration is below 3 μM, whereas at higher concentrations degradation of adenosine to inosine and hypoxanthine is favored (30). In Ehrlich ascites tumor cells, adenosine was rephosphorylated during normal cellular metabolism. Adenosine itself does not accumulate unless an adenosine deaminase inhibitor, coformycin, is present (50). While adenosine rephosphorylation is an important route of adenosine metabolism when ATP concentration is high, as it usually is, this pathway is decreased with a severe depletion of ATP (50). Nonetheless, adenosine kinase is an important enzyme in the maintenance of intracellular ATP levels, since inhibition of the enzyme leads to a decrease in the intracellular content of ATP.

While less information is available on the metabolism of 2'-deoxyadenosine, deamination appears to be a more important route of deoxyadenosine metabolism than phosphorylation in mouse brain, heart, kidney, and liver as well as in mouse and human erythrocytes (96).

Adenosine and 2'-deoxyadenosine metabolism may be altered during cell proliferation. Adenosine deaminase activity increases in regenerating mouse liver (100). Transformation of monocytes to macrophages is accompanied by a two- to ninefold increase in the levels of adenosine deaminase activity (101). Adenosine deaminase activity increases up to fourfold in phytohemagglutinin-stimulated lymphocytes at 24–60 hr (102) and decreases to 44% of resting values at 72 hr (35), while adenosine kinase does not change (35). In stimulated or unstimulated lymphocytes, phosphorylation occurs when adenosine is present at a concentration below 5 μM or 0.5 μM, respectively, without any change of the adenosine concentration which produces half-maximal rates of phosphorylation. In still another study, stimulation of human peripheral blood lymphocytes by phytohemagglutinin and concanavalin A results in diminution or loss of the tissue-specific enzyme of adenosine deaminase (103). Despite normal or decreased ADA activity in the latter experiment, there is a 12-fold elevation in the rate of adenosine deamination and a 6-fold increase in the phosphorylation of adenosine. The authors suggest that there is either an increased formation of adenosine or an increased entry of adenosine into these stimulated lymphocytes. Nucleoside transport rates have, indeed, been found to increase during cell proliferation (17). Inhibition of adenosine deaminase activity with or without the addition of adenosine or 2'-deoxyadenosine prevents the proliferation of monocytes and lymphocytes (35, 86a, 101, 102, 104–106). Thus adenosine metabolism modulated by adenosine deaminase activity, may be important during cell proliferation.

Specific adenosine binding proteins have been observed (107). The nature of these proteins or their significance to adenosine metabolism is not clear.

SPECIAL EFFECTS

Biochemical Effects

Adenosine has many biochemical effects (Table 2). In many instances, however, it is not possible to correlate a specific chemical action of adenosine with a specific biological activity of this compound. Much less information is available concerning the biochemical effects of 2'-deoxyadenosine.

CYLIC AMP   The intracellular concentration of cyclic AMP is altered by adenosine in many tissues. Adenosine, as well as adenosine analogues increase the cyclic AMP concentration in lymphocytes or mouse thymocytes (108–110), brain slices or cell lines derived from mouse neuroblastoma, astrocytoma, glioma (111–119), vagus nerve (120), pig skin (121), bone cells (122), platelets (123), ventricular myocardium (124), and adrenal and leydig tumor cells (125). A similar increase in the concentration of cyclic AMP is observed in mouse thymocytes (109) and ventricular myocardium (124) using concentrations of 2'-deoxyadenosine that are approximately 20-fold higher than the concentrations of adenosine used. In contrast, in mouse neuroblastoma cells (115, 118) or guinea pig brain slices (116) 2'-deoxyadenosine not only did not increase cyclic AMP levels, but it blocked the stimulatory effect of adenosine or prostaglandin E on cyclic AMP levels.

Table 2 Effects of adenosine and 2′-deoxyadenosine

Biochemical effects	Biological effects
Increases cyclic AMP (108–135)[a]	Causes cell toxicity (83, 106, 138, 139,
Blocks pyrimidine biosynthesis	142, 152, 153, 158–161, 171)[a]
(136–143)	Causes immunosuppression (101, 102,
Inhibits phosphoribosylpyrophosphate	104–106, 108, 162–165)[a]
synthesis (144–148)	Alters cell morphology (125, 173–176)
Increases S-adenosylhomocysteine (148a)	Acts as a vasodilator (1, 4, 13, 177–176)
Miscellaneous	Stimulates hormone secretion (125,
Inhibits nucleic acid synthesis (102,	234–240)[a]
104, 105, 136, 149–152)[a]	Metabolic effects
Inhibits thiamine synthesis (152)	Inhibits lipolysis (126–134)[a]
Reduces urea synthesis (154)	Potentiates histamine release (241)
Inhibits protein kinases (155, 156)[a]	Increases liver glycogen (242)
Increases ATP levels (78, 79, 154)	

[a] A similar effect has been documented for 2′-deoxyadenosine.

Adenosine decreases cyclic AMP concentration in fat cells (126–134), rat brain, kidney cortex, liver, and Ehrlich ascites tumor cells (135). Adenosine and 2′-deoxyadenosine were equally potent in decreasing cyclic AMP concentrations in fat cells (126). The basis for this apparent variability in response to adenosine and 2′-deoxyadenosine among tissues is not clear. Differences could be related to the concentration of adenosine used in these experiments since this ranges from less than 1 μM to as high as 5000 μM.

While elevated levels of cyclic AMP could occur as a direct result of the metabolic transformation of adenosine to cyclic AMP, a more important mechanism relates to the activation of plasma membrane adenylate cyclase by adenosine. The stimulation of adenylate cyclase by adenosine and some of its analogs appears to be related to their stability in the glycosidic high anticonformation (135a). The reduced potency of 2′-deoxyadenosine as an activator of adenylate cyclase may therefore be related to its instability in the high anticonformation. Theophylline inhibits adenosine stimulation of adenylate cyclase in many tissues (110, 114, 116–125) due to direct competition by theophylline for adenosine binding to the enzyme.

In many systems, adenosine and cyclic AMP have similar activities. Indeed, in some of these circumstances, cyclic AMP is most likely degraded to adenosine by plasma membrane ectoenzymes (125, 136, 137) and the effects attributed to cyclic AMP may actually be due to adenosine.

PYRIMIDINE SYNTHESIS Several observations suggest that adenosine inhibits pyrimidine synthesis de novo. The addition of adenosine to cultured mammalian cells leads to (*a*) accumulation of orotic acid (137–139) (*b*) depletion of the pyrimidine nucleotide pool (136–140), and (c) expansion of the adenine nucleotide pool (136–140). Recent studies have localized this inhibitory effect of adenosine to a reduction in the activity of orotate phosphoribosyltransferase (141), the enzyme that catalyzes the conversion of orotic acid to orotidine 5'-monophosphate. Similar alterations have been observed in mammalian cells and in bacteria incubated with adenine (136, 142, 143). When lymphoblastogenesis is studied using incorporation of radioactive uridine or thymidine, adenosine produces an apparent increase in the uptake of these compounds initially because of the decreased size of the pyrimidine nucleotide pools (104, 136).

PP-RIBOSE-P METABOLISM   The effects of adenosine and adenine on pyrimidine synthesis described above suggest that PP-ribose-P metabolism may also be altered in the presence of adenosine. PP-ribose-P is a rate-limiting substrate for orotate phosphoribosyltransferase (141). It is possible, therefore, that inhibition of pyrimidine synthesis is secondary to a decrease in the intracellular concentration of PP-ribose-P. Adenosine reduces the intracellular concentration of PP-ribose-P in human erythrocytes (144, 145), human lymphoblasts (35, 137), and *Escherichia coli* (146, 147). This is due to a decrease in the synthesis of PP-ribose-P when inorganic phosphate is limiting (144, 145). Since PP-ribose-P is an essential substrate not only for pyrimidine biosynthesis de novo, but also for purine biosynthesis de novo, purine salvage pathways, and pyridine nucleotide synthesis (148), a decrease in the intracellular concentration of this compound could have profound metabolic effects.

S-ADENOSYLHOMOCYSTEINE METABOLISM   As mentioned earlier, an increase in the concentration of adenosine may reverse or inhibit S-adenosylhomocysteine hydrolase resulting in the accumulation of S-adenosylhomocysteine, a potent inhibitor of methyltransferase reactions (54–60). Indeed adenosine concentrations above 15 μM cause significant elevations in intracellular levels of S-adenosylhomocysteine and S-adenosylmethionine as well as inhibition of DNA methylation in vivo in S 49 cells (148a).

MISCELLANEOUS EFFECTS   Adenosine appears to have a number of other biochemical effects. There is evidence to show that adenosine, deoxyadenosine, and other purine nucleosides may inhibit DNA, RNA, or protein synthesis in tumor cells (149–151). The inhibition of mitogen-mediated lymphoblastogenesis, measured by the decreased incorporation of radioactive thymidine into nucleic acid (102, 104, 105, 136, 152), is also a reflection of decreased nucleic acid synthesis. In bacteria, adenosine inhibits the synthesis of thiamine (152) as well as the synthesis of GMP, which is catalyzed by XMP aminase (153). Adenosine also induces an elevation of adenine nucleotides in mammalian tissues (78, 79, 154). The following additional effects of adenosine have been observed in one study but not in others (154): reduction of gluconeogenesis from lactate, pyruvate, or gluta-mine; inhibition of the accelerating effects of oleate and dibutyryl cyclic AMP on gluconeogenesis; and a diminution of urea synthesis from NH_4Cl under optimum conditions. Adenosine, 2'-deoxyadenosine, or their ana-logues, have also been observed to competitively inhibit protein kinases (155, 156); the significance of this effect is unclear.

Adenosine may have biological activity without being converted to IMP or AMP. This may be due to the existence of an adenosine receptor. Adeno-sine has vasodilator properties after being covalently linked to an oligosac-charide, which prevents its entry into coronary smooth muscle cells (157). This suggests that adenosine activity is due to its effect at the surface of the coronary myocyte. Many of the other biological and biochemical effects of adenosine described imply the existence of a receptor at the cell surface. More studies will be necessary to define the existence and nature of this adenosine receptor as a distinct entity separate from plasma membrane adenylate cyclase. The occurrence of adenosine binding protein has been observed (107), but its relevance to the effects of adenosine is unclear.

Biological Properties

CELL TOXICITY AND IMMUNOSUPPRESSION   Adenosine has been im-plicated in a multiplicity of biological phenomena (Table 2). One of the most dramatic effects of this purine nucleoside is its toxicity to cultured mammalian cells and to bacteria at concentrations ranging from 1–1000 μM. Bacteriostatic effects of adenine or adenosine have been observed in *Aerobacter aerogenes, E. coli, B. subtilis,* and *Micrococcus sodenesis* (143, 152, 153). Inhibition of growth with adenosine has occurred in cultured human diploid fibroblasts, human lymphoblasts, lymphosarcoma T-cells, and Chinese hamster ovary cells (106, 138, 139, 158–161). Derivatives of purine nucleosides, which are 2'-deoxyribonucleosides, also appear to be toxic (86a, 158–158b).

The toxicity of adenosine and 2'-deoxyadenosine is enhanced with inhibi-tion of adenosine deaminase. For example, the inhibition of lymphocyte-mediated cytolysis, mitogen-mediated lymphoblastogenesis, and monocyte to macrophage transformation resulting from added adenosine, is po-tentiated by the presence of an adenosine deaminase inhibitor (101, 102, 104–106, 108). In addition, the inhibition of mitogen-mediated lympho-blastogenesis by 2'-deoxyadenosine is potentiated by inhibition of adeno-sine deaminase (96a). The inhibitors of adenosine deaminase [erythro-9-(2-

hydroxy-3-nonyl) adenine hydrochloride, coformycin or 2'-deoxycofor-mycin], may cause decreased mitogen-mediated lymphoblastogenesis, inhibition of monocyte to macrophage transformation, prolonged survival of allografts in mice, and failure to reject an allograft tumor (101, 102, 104, 162, 163) even without the addition of adenosine. The combination of 2'-deoxyadenosine and 2'-deoxycoformycin was highly toxic to mouse fibroblasts (158a). The increase of adenosine deaminase activity during immune cell transformation, described above, also provides indirect evidence for the importance of adenosine and 2'-deoxyadenosine degradation (35, 101, 102).

It is becoming increasingly apparent that these toxic effects of adenosine or 2'-deoxyadenosine may be related to inhibition of the immune response. Lymphocytes appear to be particularly sensitive to the inhibitory effect of 2'-deoxyadenosine as well as 2'-deoxyinosine and 2'-deoxyguanosine (86a). A low concentration of coformycin (1 μM) has no effect on T- or B-cell function in vitro, but has a profound inhibitory effect on precursor lymphocytes (164). Maturation of precursor cells into T-cells is very sensitive to coformycin, while maturation of precursor cells into immunoglobulin-secreting cells is not impaired. This suggests that developing T-cells are more sensitive than B-cells to adenosine toxicity. It is unclear whether the increased sensitivity of chronic lymphocytic leukemia lymphocytes to adenosine (165) might be related to this apparent selective T-cell toxicity.

Intensive research studies have attempted to delineate the essential biochemical basis for adenosine toxicity and immunosuppression. The observations that the toxic effects of this compound are associated with a block of pyrimidine synthesis (136–142) and that toxicity is completely or partially reversed with uridine suggest that inhibition of pyrimidine synthesis may be responsible for the toxic effects of adenosine (104, 137–139). Although adenosine toxicity is often accompanied by inhibition of pyrimidine synthesis, this correlation does not prove a cause and effect relationship. In fact, uridine does not correct the toxic effects of adenosine in fibroblasts deficient in adenosine deaminase despite normal incorporation of adenosine into adenine and guanine nucleotides (166). This may be due to the reduced ability of adenosine deaminase deficient cells to convert adenosine to hypoxanthine. Although most intracellular hypoxanthine originates from IMP rather than adenosine (50) the normal production of hypoxanthine may provide an alternative substrate for nucleotide synthesis when high concentrations of adenosine are present. The complete reversal of the toxic effects of 6-methylmercaptopurine riboside, an analogue of adenosine, by both uridine and hypoxanthine together tends to support this possibility (167). The utilization of hypoxanthine as a substrate for purine nucleotide synthesis could conceivably be relevant to immune cell suppression, since hypoxanthine is a major component of transfer factor (168). These observations suggest a possible block of both purine and pyrimidine synthesis during adenosine toxicity.

Other biochemical effects of adenosine may potentially be responsible for the toxic effects of this compound. The decreased intracellular content of PP-ribose-P produced by adenosine could itself theoretically account for the toxic effects of this compound. Further studies will be necessary to establish the relevance of disordered PP-ribose-P metabolism to adenosine toxicity.

Since increased cyclic AMP levels are known to inhibit immune responsiveness (169) and actually inhibit immune cytolysis (108, 110), an accumulation of cyclic AMP has been proposed to underly the immunological suppressive effects and toxicity of adenosine. Adenosine, however, continues to be toxic to T-cells, even when specific mutations are induced in the cells that block the adenosine-mediated increase in cyclic AMP (106), which implies that alteration of cyclic AMP levels is not critical for adenosine cytoxicity.

Recent studies have demonstrated that toxicity, not reversible with uridine, occurs when adenosine concentrations exceed 15 μM in S 49 mouse lymphoma cells. In this setting, the toxicity appears to be secondary to inhibition of S-adenosylmethionine-mediated methylation reactions by the adenosine metabolite, S-adenosylhomocysteine (148a). This mechanism of adenosine toxicity is supported by the observations that 5'-deoxy-5'-isohomocystine, an analogue of S-adenosylhomocysteine (170), inhibits mitogen-induced blastogenesis of human and rabbit lymphocytes as well as oncogenic transformation of chicken fibroblasts. It is not clear at present if other mechanisms or biochemical effects of adenosine described above, including of nucleic acid synthesis (102, 104, 105, 136, 149–151) and the increase of adenine nucleotide concentration (78, 79, 154), may also contribute, in some alternative manner, to the toxicity of this compound.

Less information is available concerning the mechanism of 2'-deoxyadenosine toxicity. Incubation of stimulated lymphocytes with 0.1–1 mM 2'-deoxyadenosine causes a dramatic elevation of intracellular dATP levels and a diminution of dTTP, dCTP, and dGTP concentrations as blastogenesis is inhibited (158). The inhibitory effects of 2'-deoxyadenosine are reversible with deoxycytidine but not uridine (86a). It is possible that an imbalance of deoxyribonucleotide precursors for DNA synthesis or inhibition of ribonucleotide reductase by accumulating dATP may account for these toxic effects.

Evaluation of cells resistant to adenosine toxicity has provided further information relevant to this question. A deficiency of adenosine kinase has been associated with resistance to adenosine toxicity (83, 138, 160). This would imply that under some circumstances the conversion of adenosine to adenine nucleotide derivatives is necessary for adenosine toxicity. Adenosine resistance, however, has also been observed with normal adenosine kinase activity (160), and sensitivity to adenosine has occurred with a deficiency of adenosine kinase (161). These latter observations suggest that, in contrast to other studies, adenosine kinase activity may not be essential for the toxicity of this compound. Although there remains the possibility that kinetic alterations of this enzyme may explain these discrepencies, other mechanisms for adenosine resistance exist. Resistance to adenosine toxicity in a mouse cell line resistant to azauridine suggests that an alteration of pyrimidine synthesis may account for this abnormality (171). Cells with a mutation of PP-ribose-P synthetase may possibly be resistant to adenosine, since such cells are able to grow in the presence of the adenosine analogue, 6-methylmercaptopurine ribonucleoside (172). These observations provide further potential mechanisms to account for adenosine toxicity. It is conceivable that there is no single mechanism for cytotoxicity, but rather the summation of a number different mechanisms result in the abnormality observed. More work will be necessary to define which activity of adenosine is relevant to its cytotoxic effect.

MORPHOLOGICAL CHANGES   Alterations in cell morphology or cell surface properties have been identified during incubation with adenosine. Cells incubated with adenosine at concentrations ranging from 1 μM–4000 μM for one hour become refractile and develop spindly processes (173). These morphological changes are reversed by colchicine and occur without changes in the intracellular concentration of cyclic AMP. Cultured adrenal tumor cells become rounded when steroidogenesis is stimulated by adenosine (125). Adenosine or its analogs cause the nucleolus of chick embryo fibroblasts to lose material and unravel, over a period of several hours, into beaded strands termed nucleolar necklaces (174, 175). This phenomenon is presumed to result from the inhibition of mRNA and ribosome synthesis (174, 175). Finally, the addition of adenosine plus thymidine stimulate growth, increase concanavalin A agglutinability, and increase the synthesis of a plasminogen activator in baby hamster kidney cells (176). It is not clear if some of these morphological changes may be a cause or a result of the toxicity of adenosine.

VASOACTIVE EFFECTS   Adenosine has been implicated in the local regulation of coronary blood flow. The intracoronary injection of adenosine and its nucleotide derivatives cause vasodilation (1, 4, 177–181). The concentration of adenosine is increased in coronary sinus blood of the dog during myocardial hypoxia (1, 4, 13, 179, 180) and in humans with angina pectoris (5). Finally, maximal coronary vasodilation results from the adenosine concentrations that are achieved during hypoxia (182). Further support for the thesis that adenosine is a regulator of coronary blood flow comes from the striking parallelism observed between coronary flow and adenosine production when oxygen tension is decreasing (183). This effect of adenosine appears to be unrelated to adrenergic function since vasoactive properties of catecholamines are unaltered by dipyridamole, a drug known to potentiate the adenosine effect (184, 185). Other purine nucleotide derivatives such as ATP, ADP, and AMP have also been found to have coronary vasomotor effects (181), although it is unclear whether these compounds are effective without further metabolism to adenosine. Although the possibility that adenosine is a major determinant in the regulation of coronary blood flow is attractive and consistant with much of the data, it is clear that other local controls of coronary flow, such as prostaglandins, also exist (186, 187).

It is not known how adenosine causes vasodilation. The vasodilator properties of adenosine may be related to its attachment to an adenosine receptor on the surface of the coronary artery smooth muscle cells, since vasodilation occurs with adenosine covalently linked to an oligosaccharide (157). The effect of adenosine on vascular smooth muscle may be mediated by cyclic AMP (188) since it and adenosine produce muscle relaxation (187–192). Adenosine at vasodilating concentrations, however, does not appear to release cyclic AMP (192, 193), and no increase in coronary venous cyclic AMP is detected during angina pectoris (5). It is also unclear how other myocardial actions of adenosine may be related to these mechanisms. Adenosine reduces myocardial oxygen consumption (194). Although this effect appears to be separate from coronary vasodilation or negative inotropic or chronotropic actions of adenosine (194), another study has shown that the effect of adenosine on myocardial oxygen consumption, free carbon dioxide, and hydrogen ion release are homodirectional with effects on myocardial dynamics (195).

Adenosine is capable of dilating cerebral (196, 197) and skeletal muscle arterioles (7, 8, 198), and constricting renal arterioles (199, 200). In rat brain adenosine concentrations increase with decreased arterial blood pressure or a decreased arterial partial pressure of oxygen (201). Thus, adenosine may also play a role in the regulation of cerebral blood flow in response to hypoxia.

NEURON FUNCTION   The existence of purinergic nerves utilizing purines as neurotransmitters has been proposed and popularized by Burnstock (202). A role for nervous system influence on tissue adenine nucleotide levels was suggested initially by the profound transient diminution of ATP following the denervation of the vas deferens (203). Adenosine and ATP have depressive activities in the peripheral nervous system. These effects include slowing of the hart rate and impaired atrioventricular conduction (204–207), relaxation of smooth muscle of vascular (189, 190), intestinal (191, 208–212), or tracheal (211, 213, 214) origin, and inhibition of noradrenalin release by nerve stimulation of kidney, isolated canine subcutaneous adipose tissue, and vas deferens (215, 216). Direct stimulation of vascular tissue (217) or sympathetic nerves (14) causes the release of adenosine or its metabolites. There is a relaxant effect by 2'-deoxyadenosine on vascular smooth muscle (217a), but not on intestinal smooth muscle (191, 217a).

Adenosine, 2'-deoxyadenosine, and other adenosine derivatives and analogues have a potent depressive action on neurons of the cerebral cortex (218–222), an effect that is accompanied by an elevation of the concentration of ATP and phosphocreatine (218). Adenosine deaminase inhibitors, hexobendine and papaverine, compounds that inhibit adenosine uptake, potentiate the adenosine effect; the effect is antagonized by theophylline and caffeine. Cyclic AMP, adenosine, or adenine also antagonize the analgesic action of morphine in mice (222).

A physiological role for adenosine has been suggested by studies of its release from the nervous system. Cerebral tissues, stimulated following incubation with radioactive adenine, release radioactive compounds that include adenosine and smaller quantities of inosine, hypoxanthine, and adenine nucleotides (12, 223–226). The rate of formation of adenosine from ATP in synaptosomes may be dependent upon 5'-nucleotidase activity rather than the activity of ATPases and adenylate kinase (227). Electrical stimulation leading to release of radioactivity requires calcium (223, 228) and is partially inhibited by tetrodotoxin (223). The observations that calcium ions are frequently necessary for secretory processes, including neurotransmitter release, and that tetrodotoxin inhibits neural conduction dependent on sodium ion entry, provide further support for the hypothesis that neurotransmission may be related to the release of adenosine and its metabolites.

Injection of tritiated adenosine into different parts of the central nervous system is followed by movement of radioactive label in retrograde and anterograde directions along axons (229–232). In these studies, the finding of radioactivity within cells of the next order neuron indicates release of label into the intercellular space and subsequent uptake by adjacent terminals. These observations suggest that adenosine or its derivatives are released either as primary transmitters or as additional transmitters with complementary function. This property of adenosine and its metabolites may prove valuable for neuroanatomical studies. In the context of the potential neurotransmitter function of adenosine, it is important to recognize that the detection of a purine compound after nerve stimulation, or detection of an apparent neurogenic effect of a purine derivative, does not necessarily constitute evidence that this compound is a neurotransmitter. It is possible that the purine is merely associated with an unknown transmitter substance (233) or modifies the release or effect of such a substance.

HORMONE SECRETION   While adenosine appears to have profound effects on hormone secretion, its effect on steroidogenesis has been studied most extensively. Parenteral administration of adenosine in the rat is associated with an increase in plasma corticosterone levels (234). Steroidogenesis by adrenal tumor cells or Leydig tumor cells in culture is stimulated by adenosine, 2'-deoxyadenosine, 5'-deoxyadenosine, 2-chloroadenosine, ATP, adenyl-5'-yl imidodiphosphate, adenosine 5'(β λ-methylene)-triphosphate, ADP, AMP, NAD, and FAD; but not by adenine or other nucleoside triphosphates (125). Adenosine is 15 times more potent than 2'-deoxyadenosine. This effect of adenosine is potentiated by dipyridamole or erythro-9-[3-(nonane-2-ol)] adenine and inhibited by the addition of crystalline adenosine deaminase or theophylline. Effects of ATP, NAD, and cyclic AMP are inhibited by adenosine deaminase which suggests that their effects might also be mediated by adenosine. The effect of adenosine on steroidogenesis is not limited to endocrine tissues since administration of adenosine or some of its metabolites to starved rats increases the activity of liver microsomal 3-hydroxy-3-methylglutaryl-CoA reductase as well as the incorporation of radioactive acetate into sterol in liver slices (235).

The effect of adenosine on hormone secretion is clearly not limited to steroidogenesis. Adenosine or inosine, but not uridine, stimulate insulin secretion from pancreatic islets (236, 237), possibly by being metabolized and furnishing glycolytic intermediates which function as analogs of glucose (238). Glucagon secretion is also stimulated by adenosine, cyclic AMP, AMP, ADP, NADP, and NADPH (239). Adenosine and guanosine stimulate the incorporation of radioactive leucine into prolactin and block the inhibitory effects of apomorphine on prolactin synthesis and release (239). These nucleosides also stimulate growth hormone release. Erythropoiesis in plethoric mice is activated by adenosine, AMP, cyclic AMP, and dibutyryl cyclic AMP, but not by cytidine, its nucleotides, or cyclic GMP (240). This stimulation appears to be due to erythropoietin since it is prevented by antierythropoietin.

MISCELLANEOUS EFFECTS   Adenosine has an effect on the metabolism of several tissues. It may regulate fat cell metabolism by antagonizing cyclic AMP accumulation and inhibiting lipolysis in a manner similar to the regulation by prostaglandin E and nicotinic acid (126–134). The effect of 2'-deoxyadenosine was similar to that of adenosine (126). The addition of purified adenosine deaminase to fat cells potentiates cyclic AMP accumulation caused by norepinephrine, increases lipolysis, and abolishes the antilipolytic activity of insulin (131, 132). These effects of adenosine deaminase on cyclic AMP are abolished by coformycin (132). Other effects of adenosine include a threefold potentiation of mast cell histamine release (241), an eightfold increase in the active form of rat liver glycogen synthetase activity one hour after intraperitoneal injection (242), stimulation of hepatic gluconeogenesis (243), the reversal of methotrexate cytotoxicity in mouse bone marrow cells when used in combination with thymidine (244), and a potentiation of potassium loss in energy-depleted human erythrocytes (245). In the latter study 2'-deoxyadenosine reduced potassium loss from erythrocytes, an effect opposite to that observed with adenosine (245).

ALTERATIONS OF METABOLISM

The existence of spontaneous abnormalities of adenosine and 2'-deoxy-adenosine metabolism in man that are associated with specific abnormalities in biological function has served to illustrate the importance of adenosine. The detailed study of these disorders may allow a further understanding of the biochemistry and biology of adenosine.

Adenosine Deaminase Deficiency

In 1972, the deficiency of adenosine deaminase was found in two unrelated patients with severe combined immunodeficiency disease (246). The disease is genetic in origin and has an autosomal mode of inheritance. Homozygous deficient subjects show a failure to thrive, have infections, diarrhea, malabsorption, and candidiasis with atrophic tonsils, adenoids, and thymus. Both cell and antibody mediated immunity are severely impaired (247). X rays demonstrate evidence of osteoporosis, small or absent thymus gland, and a number of bony abnormalities. Pathological evaluation reveals extreme thymic involution. Evidence for an abnormality of purine catabolism is provided by the elevated plasma adenosine concentration and an increased quantity of adenine in the urine, plasma, and erythrocytes (248). ATP formation is increased in incubated erythrocytes (78) and lymphocytes (249), and ATP concentration in lymphocytes is ten times greater than the normal value (79). Therapy of this disorder has included bone marrow transplantation and transfusion of normal erythrocytes (79). The favorable response to enzyme replacement suggests that a direct etiological relationship exists between enzyme deficiency and immune dysfunction. Abnormalities of immune function may result from the accumulation of adenosine and/or 2'-deoxyadenosine and the resultant toxic effect on the immune system by activation of adenylate cyclase, inhibition of PP-ribose-P synthesis, blockade of pyrimidine synthesis, elevation of intracellular S-adenosyl-homocysteine, or other mechanisms described above.

Adenosine deaminase has been studied in patients with severe combined immunodeficiency disease. The enzyme from one patient has 0.5% of normal activity in splenic tissue (250). This mutant form of adenosine deaminase has an S_{20w} of 7.4 S and a molecular weight of 115,000, which corresponds to the intermediate form of the normal enzyme. The substrate specificity, pH optimum, K_M for adenosine (130 μM), or K_i for inosine (1.6 mM) of the adenosine deaminase from this patient, are comparable to the corresponding molecular form from normal subjects (66). These observations suggest that the structure of adenosine deaminase from this patient is altered. Hemolysate from another patient inhibits adenosine deaminase partially purified from a normal individual which suggests the production of an adenosine deaminase inhibitor in this adenosine deaminase deficient subject (251). Cultured skin fibroblasts from two different patients reveal < 1% and 10% of normal adenosine deaminase activity, respectively (252). In the latter case, the pH optimum and the K_m are similar to the normal values; the mutant enzyme is resistant to heat inactivation and has a greater negative charge than the normal enzyme. The fibroblasts from four other adenosine deaminase deficient patients have about 20% of normal enzyme activity, a normal K_m for adenosine of 65 μM, increased electrophoretic mobility, increased heat stability, and a molecular weight of $\sim$ 260,000 (253). Immunochemical studies of adenosine deaminase from patients with combined immunodeficiency disease show one enzyme with an amount of immunoreactive enzyme that is greater than the catalytically active enzyme (254), and three enzymes with an equivalent amount of immunoreactive and catalytically active enzyme (254, 255). Although both large and small forms of adenosine deaminase in these examples are deficient, conversion factor activity is normal (256, 257). Each of these studies provides further support for the possibility that there is genetic heterogeneity of the mutations involving the structural gene coding for adenosine deaminase.

Increased Adenosine Deaminase Activity

A 45- to 70-fold increase of erythrocyte adenosine deaminase activity has been observed in kindred with hereditary hemolytic anemia (80). Patients with this dominantly inherited entity have a mild anemia and decrease of erythrocyte adenine nucleotides to less than 50% of that of comparable reticulocyte-rich blood. The decreased erythrocyte adenine nucleotide concentrations, which appear to be responsible for the hemolytic anemia, may result from the diminished reutilization of adenosine to adenine nucleotides as a result of excessive destruction of adenosine by elevated adenosine deaminase levels.

Purine Nucleoside Phosphorylase Deficiency

Purine nucleoside phosphorylase deficiency in man leads to a disturbance of T-cell function (258). The discovery of this new disease has reemphasized the etiological relationship between disorders of purine catabolism and immune function. The first three patients described with this disorder had hypouricemia, hypouricosuria, and an excessive urinary excretion of inosine, guanosine, deoxyinosine, and deoxyguanosine (259, 260). In one patient on a low purine diet, the serum uric acid was 1.0 mg/dl, the urinary uric acid 11 mg/24 hr, and the urinary nucleosides 1520 mg/24 hr (259). These patients are, therefore, overproducers of purines despite the hypouricemia. This overproduction of purines is accompanied by an elevation in the concentration of PP-ribose-P in erythrocytes (259) to a level similar to that observed in patients with a complete deficiecy of hypoxanthine-guanine phosphoribosyltransferase (148).

Recently, two brothers were observed with a less severe abnormality of T-cell dysfunction, a serum uric acid of 2.5 mg/dl and 0.45% of normal erythrocyte purine nucleoside phosphorylase activity (261). The purine nucleoside phosphorylase from these two patients has a tenfold higher K_m for inosine, a reduced ability of inosine to protect against thermal inactivation, and an altered electrical charge; these figures provide strong evidence for a structurally-altered enzyme (262), when compared to those for the normal enzyme. While there is evidence for a block of purine nucleotide degradation in vivo and in vitro, no block of pyrimidine biosynthesis de novo is apparent (261). In these two brothers, the normal induction of T-cell maturation in bone marrow precursors by human thymic epithelium-conditioned medium or thymosin suggests normal T-cell generation and intact thymic epithelial function (263). The selective impairment of certain T-cell functions in these patients may be explained by a shortened survival of nucleoside phosphorylase-deficient T-lymphocytes in vitro and in vivo (263). However, the exact metabolic basis for this altered lymphocyte function remains unclear.

A deficiency of purine nucleoside phosphorylase leads to inosine accumulation. Increased intracellular inosine concentrations may then act as a product inhibitor of adenosine deaminase (106). Thus, the deficiency of purine nucleoside phosphorylase may lead to a functional deficiency of adenosine deaminase.

PHARMACOLOGICAL ALTERATIONS OF ADENOSINE OR 2'-DEOXYADENOSINE METABOLISM

The multiple biological and biochemical properties of adenosine or 2'-deoxyadenosine allow for pharmacological interventions to inhibit, potentiate, or mimic the activity of these compounds. The possible interventions may be extensive and include coronary vasodilation, central nervous system activation, inhibition of specific components of immune function, or modification of hormone synthesis or release.

Inhibition of Transport

A number of substances chemically unrelated to nucleosides are potent inhibitors of nucleoside transport, with kinetic characteristics similar to those of competitive inhibitors of enzymatic activity (16, 17). Dipyridamole inhibits nucleoside uptake in heart, erythrocytes, hepatoma cells, and chick fibroblasts. This appears to be a nonspecific effect, since the transport of other types of compounds is also inhibited. Colchicine inhibits the uptake of nucleosides by an effect separate from its activity on microtubular protein. Phenethyalcohol, dimethyl sulfoxide, inorganic and organic mercurials, streptovaricin D, cytochalasin B, theophylline, papaverine, and prostaglandins influence the transport of nucleosides in general. Some of these compounds probably act directly by interacting with membrane components and indirectly by modifing carrier activity.

A variety of nucleosides and nucleoside analogues modify nucleoside uptake (16, 17). Since nucleosides generally share a common membrane carrier, they act in general as reversible competitive inhibitors of transport. A group of nucleoside analogs, purine 6-thioethers, or 2-amino purine ribonucleoside derivatives, are specific inhibitors of the uptake of nucleosides. One such compound, nitrobenzylthioinosine, is itself absorbed by two processes (a) a diffusional mechanism proportional to concentration and (b) a saturable binding to specific membrane sites. This compound has been used to study nucleoside transport since it inhibits only the saturable component of nucleoside uptake (16). Although the data described above apply to general aspects of nucleoside transport they are applicable to inhibition of adenosine and 2'-deoxyadenosine transport. While it is clear that a wide variety of compounds, including other nucleosides, can interfere with adenosine or 2'-deoxyadenosine transport into cells, a major question is whether inhibition of transport has any relevance to the biochemical effects of adenosine or 2'-deoxyadenosine.

Dipyridamole inhibits adenosine uptake into human erythrocytes or erythrocyte ghosts (15, 30, 264, 265), human platelets (266), cultured chick heart cells (6), perfused rabbit heart (267), or a number of canine tissues (268). In guinea pig cortical slices dipyridamole does not inhibit the uptake of 2'-deoxyadenosine in contrast to its effect on adenosine (114). Reduction of adenosine uptake potentiates the effect of adenosine as a vasodilator (184, 268–271), as an inhibitor of platelet aggregation (266), as a stimulant of intracellular cyclic AMP (118, 124), and as a relaxer of smooth muscle (212–214). A number of compounds have pharmacological properties resembling dipyridamole (264, 266, 268–270, 272–274). Theophylline also inhibits transport (16, 17, 275) and modifies some biological activity of adenosine. However, unlike dipyridamole, theophylline tends to antagonize the biological properties of adenosine. It is believed that theophylline competes with adenosine for membrane binding sites. Theophylline inhibits the adenosine-induced dilatation of cerebral (196) and coronary (157) arterioles, the constriction of renal arterioles (200), the relaxation of intestinal (192) and tracheal (216) smooth muscle, the depression of cerebral cortex (220, 221), the reduction of myocardial oxygen consumption (194), the stimulation of steroidogenesis (125), the inhibition of release of acetylcholine from cholinergic nerves (211, 276), and the potentiation of mast cell mediator release (241). The adenosine stimulation of adenylate cyclase is inhibited by theophylline in central nervous system cells (114, 116–119), vagus nerve (120), immune cells (110), pig skin (121), bone cells (122), platelets (123), adrenal cells (125), or ventricular myocardium (124).

Modification of Metabolism

Modification of the metabolism of adenosine or 2'-deoxyadenosine has been directed toward increasing or decreasing their conversion to inosine and 2'-deoxyinosine respectively. Inhibition of adenosine deaminase by erythro-9-(2-hydroxy-3-nonyl) adenine hydrochloride, coformycin, or 2'-deoxycoformycin, potentiates the toxic effects or immune suppressive effects of adenosine (101, 102, 104, 138, 139, 154–161). Adenosine deaminase inhibitors also potentiate the adenosine-related depression of spontaneous firing in cortical neurons (221), the adenosine-related inhibition of lipolysis (131, 132), or adenosine activation of steroidogenesis (125). In contrast, increased adenosine destruction is caused by the addition of crystalline adenosine deaminase. The latter effect is associated with lack of stimulation of steroidogenesis in adrenal tumor cells (125) and increased lipolysis and cyclic AMP concentrations in fat cells (131, 132).

Adenosine Analogues

The study of the pharmacology of adenosine or 2'-deoxyadenosine analogues is in a period of exponential growth and has been extensively reviewed recently (277). Experiments have shown that nucleoside analogues have the potential to be used in a multiplicity of areas that might include antimicrobial effects, antitumor effects, cardiovascular effects, immunosuppression or immunostimulation, platelet aggregation, hypolipidemic effects, and use as an insecticide. The major types of structural modifications used to produce analogues include interchange of carbon and nitrogen of the purine ring, changes of substituent groups attached to the purine ring, alteration of the carbohydrate moiety, or change of the site of attachment of the carbohydrate moiety to the ring (278). The simultaneous use of more than one nucleoside analogue can lead to a synergistic effect when both compounds act on the same pathway or one compound inhibits the metabolic conversion of the other and causes it to accumulate (278).

Adenosine or 2'-deoxyadenosine analogues can potentially undergo the many metabolic interconversions of adenosine or 2'-deoxyadenosine themselves. Bloch (278) has described the antibiotic, tubercidin (7-deazaadenosine), as an example. This nucleoside is cytotoxic to microorganisms and to mammalian cells. It is phosphorylated to the nucleoside triphosphate derivative and can be incorporated into nucleic acid or NAD in place of the adenine derivative. There are multiple potential sites for toxicity of this compound; the most important effect, however, appears to be an inhibition of cellular utilization of glucose by the interference of phosphofructokinase by its triphosphate derivative. Many other adenosine and 2'-deoxyadenosine analogues have growth-inhibiting properties (158a, 279–295). These compounds are activated by conversion to nucleoside phosphate derivatives (158a, 281, 285–287, 289, 291–293), have synergistic effects when used in combination (158a, 279, 281, 287, 289, 291–294), and some have been shown to inhibit pyrimidine synthesis de novo as well as purine pathways (289, 290).

There are other biological properties of adenosine analogues. The halogenated derivatives seem to be particularly active compounds. Adenosine analogues depress the firing rate of cortical neurons (221), stimulate steroidogenesis by adrenal tumor cells and by Leydig tumor cells in culture (125), are highly effective antilipolytic agents (296), have antiviral activity (297), and inhibit the lipolytic effects of adenosine deaminase in fat cells (132). Relaxation of smooth muscle of intestine or blood vessels is also induced by adenosine analogues (189, 217a, 298–301). Finally, analogues of adenosine, especially 2-halogenated derivatives, activate adenylate cyclase in a large number of tissues (114–116, 120, 122, 125), but inhibit the activation of adenylate cyclase by norepinephrine in fat cells (126).

THE FUTURE

The biology and biochemistry related to adenosine and 2'-deoxyadenosine have become a major subject of investigation in many different areas of the biological sciences. Each new observation has posed a further question. Information about adenosine and 2'-deoxyadenosine metabolism in different tissues is limited. Further work is necessary to unravel the mechanisms of adenosine release from tissue and the changes in adenosine and 2'-deoxyadenosine metabolism in proliferating cells. Which tissues are dependent upon adenosine release from the liver as a purine source? Are problems of lymphocyte nutrition the basis for immunodeficiency in disorders of adenosine metabolism?

Although adenosine and 2'-deoxyadenosine have many biological properties in the experimental situation, the relevance of these to normal homeostasis is unclear. Do these nucleosides regulate lymphocyte function? Is adenosine a neurotransmitter? Is adenosine a physiological regulator of adenylate cyclase? Beyond these types of questions are those that ask which biochemical effects of adenosine and 2'-deoxyadenosine are responsible for which biological activities of these compounds. It may be that careful study of the biochemical and biological disorders associated with the alterations of adenosine and deoxyadenosine metabolism may provide some important clues about these relationships.

The pharmacological potential of adenosine and 2'-deoxyadenosine analogues seems endless. The development of specific analogues or related compounds for a defined effect will await a better understanding of adenosine biology and biochemistry. The use of such potent, specific agents in the management of human diseases has already occurred, but promises to develop extensively in the future.

Finally, the wide array of biological and biochemical effects of adenosine must pose more fundamental questions. Is adenosine a cell messenger or regulator in much the same way as cyclic nucleotides? The fact that this compound modifies the activity of adenylate cyclase in many different issues may support this concept. It is possible that adenosine is an intercellular messenger, while cyclic AMP is an intracellular messenger.

ACKNOWLEDGMENTS

This work was supported in part by grants from the United States Public Health Service AM 19045, AM 19674 and 5MO1RR42, American Heart Association 77-849, and the National Foundation 1-393. We wish to thank Jumana Judeh for her careful typing of the manuscript.

NOTE ADDED IN PROOF   Recent studies demonstrate that dATP and dADP concentrations are increased in erythrocytes from patients with adenosine deaminase deficiency (302, 303). This data indicates that the previously reported elevated concentrations of ADP and ATP in erythrocytes (79, 304) may be largely deoxynucleotide derivatives. Thus the immune deficiency associated with adenosine deaminase deficiency could be related to the toxic effect of deoxynucleotide accumulation. A similar mechanism may account for the immunodeficiency state in purine nucleoside phosphorylase deficiency. Since deoxyguanosine kinase and deoxyinosine kinase are localized to lymphoid tissues in man (86a), deoxynucleotide derivatives could accumulate almost selectively in lymphocytes from patients with purine nucleoside phosphorylase deficiency. The observation that 2'-deoxyguanosine inhibits mitogen-mediated lymphoblastogenesis (86a) provides further support for this potential mechanism of immunodeficiency.

Literature Cited

1. Berne, R. M., Rubio, R., Dobson, J. G. Jr., Curnish, R. R. 1971. *Circ. Res.* 28–29: Suppl. 1, pp. I115–19
2. Liu, M. S., Feinberg, H. 1971. *Am. J. Physiol.* 220:1242–48
3. Mustafa, S. J., Rubio, R., Berne, R. M. 1975. *Am. J. Physiol.* 228:62–67
4. Rubio, R., Berne, R. M., Katori, M. 1969. *Am. J. Physiol.* 216:56–62
5. Fox, A. C., Reed, G. E., Glassman, E., Kaltman, A. J., Silk, B. B. 1974. *J. Clin. Invest.* 53:1447–57
6. Mustafa, S. J., Berne, R. M., Rubio, R. 1975. *Am. J. Physiol.* 228:1474–78
7. Dobson, J. G. Jr., Rubio, R., Berne, R. M. 1971. *Circ. Res.* 29:375–84
8. Bockman, E. L., Berne, R. M., Rubio, R. 1976. *Am. J. Physiol.* 230:1531–37
9. Lerner, M. H., Lowy, B. A. 1974. *J. Biol. Chem.* 249:959–66
10. Pritchard, J. B., O'Connor, N., Oliver, J. M., Berlin, R. D. 1975. *Am. J. Physiol.* 229:967–72
11. Sim, M. K., Maguire, M. H. 1972. *Circ. Res.* 31:779–88
12. Pull, I., McIlwain, H. 1972. *Biochem. J.* 130:975–81
13. Schrader, J., Haddy, F. J., Gerlach, E. 1975. *Recent Adv. Stud. Card. Struct. Metab.* 7:171–75
14. Fredholm, B. B. 1976. *Acta Physiol. Scand.* 96:422–30
15. Roos, H., Pfleger, K. 1972. *Mol. Pharmacol.* 8:417–25
16. Paterson, A. R. P., Kim, S. C., Bernard, O., Cass, C. E. 1975. *Ann. NY Acad. Sci.* 255:402–10
17. Berlin, R. D., Oliver, J. M. 1975. *Int. Rev. Cytol.* 42:287–336
18. Steck, T. L., Nakata, Y., Bader, J. P. 1969. *Biochim. Biophys. Acta* 190:237–49
19. Mizel, S. B., Wilson, L. 1972. *Biochemistry* 11:2573–78
20. Lemkin, J. A., Hare, J. D. 1973. *Biochim. Biophys. Acta* 318:113–22
21. Quinlan, D. C., Hochstadt, J. 1974. *Proc. Natl. Acad. Sci. USA* 71:5000–3
22. Strauss, P. R., Sheehan, J. M., Kashket, E. R. 1976. *J. Exp. Med.* 144:1009–21
23. Li, C., Hochstadt, J. 1976. *J. Biol. Chem.* 251:1175–80
24. Li, C., Hochstadt, J. 1976. *J. Biol. Chem.* 251:1181–87
25. Olsson, R. A., Snow, J. A., Gentry, M. K., Frick, G. P. 1972. *Circ. Res.* 31:767–68
26. Wiedmeier, V. T., Rubio, R., Berne, R. M. 1972. *J. Mol. Cell. Cardiol.* 4:445–52
27. Wiedmeier, V. T., Rubio, R., Berne, R. M. 1972. *Am. J. Physiol.* 223:51–54
28. Namm, D. H. 1973. *Circ. Res.* 33:686–95
29. Meyskens, F. L., Williams, H. E. 1971. *Biochim. Biophys. Acta* 240:170–79
30. Schrader, J., Berne, R. M., Rubio, R. 1972. *Am. J. Physiol.* 223:159–66
31. Parks, R. E. Jr., Brown, P. R. 1973. *Biochemistry* 12:3294–302
32. Zachara, B. 1975. *Vox. Sang.* 28:453–55
33. Zimmerman, T. P., Gersten, N., Miech, R. P. 1970. *Proc. Am. Assoc. Cancer Res.* 11:87
34. Rapaport, E., Zamecnik, P. C. 1976. *Proc. Natl. Acad. Sci. USA* 73:3122–25
35. Snyder, F. F., Mendelsohn, J., Seegmiller, J. E. 1976. *J. Clin. Invest.* 58:654–66
36. Peer, L. J., Winkler, H., Snider, S. R., Gigg, J. W., Gaumgartner, H. 1976. *Biochem. Pharmacol.* 25:311–15
37. Parker, J. C. 1970. *Am. J. Physiol.* 218:1568–74
38. Widnell, C. C. 1972. *J. Cell. Biol.* 52:542–58
39. DePierre, J. W., Karnovsky, M. L. 1974. *Science* 183:1096–98
40. Farquhar, M. G., Bergeron, J. J. M., Palade, G. E. 1974. *J. Cell. Biol.* 60:8–25
41. Brownlee, S. T., Heath, E. C. 1975. *Arch. Biochem. Biophys.* 166:1–7
42. Stefanovic, V., Mandel, P., Rosenberg, A. 1975. *J. Biol. Chem.* 250:7081–83
43. Woo, Y. T., Manery, J. F. 1975. *Biochim. Biophys. Acta* 397:144–52
44. Fleit, H., Conklyn, M., Stebbins, R. D., Silber, R. 1975. *J. Biol. Chem.* 250:8889–92
45. Fox, I. H., Marchant, P. J. 1976. *Can. J. Biochem.* 54:462–69
46. Fox, I. H., Marchant, P. J. 1976. *Can. J. Biochem.* 54:1055–60
47. Murray, A. W., Friedrichs, B. 1969. *Biochem. J.* 111:83–89
48. Drummond, G. I., Yamamoto, M. 1971. In *The Enzymes,* ed. P. D. Boyer, Vol. 4, p. 337. New York: Academic 3rd ed.
49. Fox, I. H. 1978. In *Uric Acid,* ed. W. N. Kelley, I. M. Weiner. Springer. In press
50. Lomax, C. A., Henderson, J. F. 1973. *Can. Res. Dev.* 33:2825–29
51. Lomax, C. A., Bagnara, A. S., Henderson, J. F. 1975. *Can. J. Biochem.* 53:231–41
52. Walker, R. D., Duerre, J. A. 1975. *Can. J. Biochem.* 53:312–19

53. Schatz, R. A., Vunnam, C. R., Sellinger, O. Z. 1977. *Life Sci.* 20:375–83
54. De la Haba, G., Cantoni, G. L. 1959. *J. Biol. Chem.* 234:603–8
55. Cortese, R., Perfetto, E., Arcari, P., Prota, G., Salvatore, F. 1974. *Int. J. Biochem.* 5:535–45
56. Hurwitz, J., Gold, M., Anders, J. 1964. *J. Biol. Chem.* 239:3474–82
57. Zappia, V., Zydek-Cwick, C. R., Schlenk, F. 1969. *J. Biol. Chem.* 244:4499–509
58. Akamatsu, Y., Law, J. H. 1970. *J. Biol. Chem.* 245:709–13
59. Deguchi, T., Barchas, J. 1971. *J. Biol. Chem.* 246:3175–81
60. Pegg, A. E. 1971. *FEBS Lett.* 16:13–6
61. Rubio, R., Berne, R. M., Dobson, J. G. Jr. 1973. *Am. J. Physiol.* 225:938–53
62. Nakatsu, K., Drummond, G. I. 1972. *Am. J. Physiol.* 223:1119–27
63. Borgers, M., Schaper, J., Schaper, W. 1971. *J. Mol. Cell. Cardiol.* 3:287–96
64. Schrader, J., Gerlach, E. 1976. *Pfluegers Arch.* 367:129–35
65. Metzer, R. M. Jr., Rubio, R., Berne, R. M. 1975. *Am. J. Physiol.* 229:1625–31
66. Van der Weyden, M. B., Kelley, W. N. 1976. *J. Biol. Chem.* 251:5448–56
67. Akedo, H., Nishihara, H., Shinkai, K., Komatsu, K., Ishikawa, S. 1972. *Biochim. Biophys. Acta* 276:257–71
68. Osborne, W. R. A., Spencer, N. 1973. *Biochem. J.* 133:117–23
69. Trams, E. G., Lauter, C. J. 1975. *Biochem. J.* 152:681–17
70. Daddona, P., Kelley, W. N. 1977. *J. Biol. Chem.* 252:110–15
71. Brady, T. G., O'Connell, W. 1962. *Biochim. Biophys. Acta* 62:216–29
72. Coddington, A. 1965. *Biochim. Biophys. Acta* 99:442–51
73. Pfrogner, N. 1967. *Arch. Biochem. Biophys.* 119:147–54
74. Sim, M. K., Maguire, M. H. 1971. *Eur. J. Biochem.* 23:17–29
75. De Boeck, S., Rymen, T., Stockx, J. 1975. *Eur. J. Biochem.* 52:191–95
76. Piggott, C. O., Brady, T. G. 1976. *Biochim. Biophys. Acta* 429:600–7
77. Nishihara, H., Ishikawa, S., Shinkai, K., Akedo, H. 1972. *Biochim. Biophys. Acta* 302:429–42
78. Agarwal, R. P., Crabtree, G. W., Parks, R. E. Jr., Nelson, J. A., Keightley, R., Parkman, R., Rosen, F. S., Stern, R. C., Polmar, S. H. 1976. *J. Clin. Invest.* 57:1025–35
79. Polmar, S. H., Stern, R. C., Schwartz, A. L., Wetzler, E. M., Chase, P. A., Hirschhorn, R. 1976. *N. Engl. J. Med.* 295:1337–43

80. Valentine, W. N., Paglia, D. E., Tartaglia, A. P., Gilsanz, F. 1977. *Science* 195:783–85
81. Krenitsky, T. A., Miller, R. L., Fyfe, J. A. 1974. *Biochem. Pharmacol.* 23:170–2
82. De Jong, J. W. 1973. *Fed. Proc.* 32:660
83. Lomax, C. A., Henderson, J. F. 1972. *Can. J. Biochem.* 50:423–27
84. Streeter, D. G., Simon, L. N., Robins, R. K., Miller, J. P. 1974. *Biochemistry* 13:4543–49
85. Krygier, V., Momparler, L. 1971. *J. Biol. Chem.* 246:2745–51
86. Krygier, V., Momparler, L. 1971. *J. Biol. Chem.* 246:2752–57
86a. Carson, D. A., Kaye, J., Seegmiller, J. E. 1977. *Proc. Natl. Acad. Sci. USA.* 74:5677–81
87. Schnebli, H. P., Hill, D. L., Bennett, L. Jr. 1967. *J. Biol. Chem.* 242:1997–2004
88. Lindberg, B., Klenow, H., Hansen, K. 1967. *J. Biol. Chem.* 242:350–56
89. Murray, A. W. 1968. *Biochem. J.* 106:549–55
90. Divekar, A. Y., Hakala, M. T. 1971. *Mol. Pharmacol.* 7:663–73
91. De Jong, J. W., Kalkman, C. 1973. *Biochim. Biophys. Acta* 320:388–96
92. Henderson, J. F., Mikoshiba, A., Chu, S. Y., Caldwell, I. C. 1972. *J. Biol. Chem.* 247:1972–75
93. Caputto, R. 1951. *J. Biol. Chem.* 189:801–14
94. Kaneti, J. 1973. *J. Theor. Biol.* 38:169–79
95. Divekar, A. Y. 1976. *Biochim. Biophys. Acta* 422:15–28
96. Snyder, F. F., Henderson, J. F. 1973. *J. Biol. Chem.* 248:5899–904
97. Kim, B. K., Cha, S., Parks, R. E. Jr. 1968. *J. Biol. Chem.* 243:1771–76
98. Krenitsky, T. A., Elion, G. B., Henderson, A. M., Hitchings, G. H. 1968. *J. Biol. Chem.* 243:2876–81
99. Senesi, S., Falcone, G., Mura, U., Sgarrella, F., Ipata, P. L. 1976. *FEBS Lett.* 64:353–57
100. Rothman, I. K., Silber, R., Klein, K. M., Becker, F. F. 1971. *Biochim. Biophys. Acta* 228:307–12
101. Fischer, D., Van der Weyden, M. B., Snyderman, R., Kelley, W. N. 1976. *J. Clin. Invest.* 58:399–407
102. Hovi, T., Smyth, J. F., Allison, A. C., Williams, S. C. 1976. *Clin. Exp. Immunol.* 23:395–403
103. Hirschhorn, R., Levytska, V. 1974. *Cell. Immunol.* 12:387–95
104. Carson, D. A., Seegmiller, J. E. 1976. *J. Clin. Invest.* 57:274–82

105. Fox, I. H., Keystone, E. C., Gladman, D. D., Moore, M., Cane, D. 1975. *Immun. Commun.* 4:419–27
106. Ullman, B., Cohen, A., Martin, D. W. Jr. 1976. *Cell* 9:205–11
107. Hsu, H. H. T. 1975. *Proc. Soc. Exp. Biol. Med.* 149:698–701
108. Wolberg, G., Zimmerman, T. P., Hiemstra, K., Winston, M., Chu, L. 1975. *Science* 187:957–59
109. Zenser, T. V. 1975. *Biochim. Biophys. Acta* 404:202–13
110. Zimmerman, T. P., Rideout, J. L., Wolberg, G., Duncan, G. S., Elion, G. B. 1976. *J. Biol. Chem.* 251:6757–66
111. Schrier, B. K., Gilman, A. G. 1973. *Fed. Proc.* 32:680
112. Schultz, J., Daly, J. W. 1973. *J. Biol. Chem.* 248:860–66
113. Clark, R. B., Gross, R. 1974. *J. Biol. Chem.* 249:5296–303
114. Mah, H. D., Daly, J. W. 1975. *Biochim. Biophys. Acta* 404:49–56
115. Blume, A. J., Foster, C. J. 1975. *J. Biol. Chem.* 250:5003–8
116. Mah, H. D., Daly, J. W. 1976. *Pharmacol. Res. Commun.* 8:65–79
117. Clark, R. B., Seney, M. N. 1976. *J. Biol. Chem.* 251:4239–46
118. Green, R. D., Stanberry, L. R. 1977. *Biochem. Pharmacol.* 26:37–43
119. Sattin, A., Rall, T. W., Zanella, J. 1975. *J. Pharmacol. Exp. Ther.* 192:22–32
120. Roch, P., Salamin, A. 1976. *Experientia* 32:1419
121. Iizuka, H., Adachi, K., Halprin, K. M., Levine, V. 1976. *Biochim. Biophys. Acta* 444:685–93
122. Peck, W. A., Carpenter, J. G., Schuster, R. J. 1976. *Endocrinology* 99:901–9
123. Haslam, R. J., Rosson, G. M. 1975. *Mol. Pharmacol.* 11:528–44
124. Huang, M., Drummond, G. I. 1976. *Biochem. Pharmacol.* 25:2713–19
125. Wolff, J., Cook, G. H. 1977. *J. Biol. Chem.* 252:687–93
126. Fain, J. N., Pointer, R. H., Ward, W. F. 1972. *J. Biol. Chem.* 247:6866–72
127. Fain, J. N. 1973. *Mol. Pharmacol.* 9:595–604
128. Schwabe, U., Ebert, R., Erbler, H. C. 1973. *Arch. Pharmacol.* 276:133–48
129. Ebert, R., Schwabe, U. 1973. *Arch. Pharmacol.* 278:247–59
130. Schwabe, U., Schonhofer, P. S., Ebert, R. 1974. *Eur. J. Biochem.* 46:537–45
131. Schwabe, U., Ebert, R., Erbler, H. C. 1975. *Adv. Cyclic Nucleotide Res.* 5:569–84
132. Fain, J. N., Wieser, P. B. 1975. *J. Biol. Chem.* 250:1027–34
133. Hjemdahl, P., Fredholm, B. B. 1976. *Acta Physiol. Scand.* 96:170–79
134. Turpin, B. P., Duckworth, W. C., Solomon, S. S. 1976. *Clin. Res.* 24:30A
135. McKenzie, S. G., Bar, H. P. 1973. *Can. J. Physiol. Pharmacol.* 51:190–96
135a. Miles, D. L., Miles, D. W., Eyring, H. 1977. *Proc. Natl. Acad. Sci. USA* 74:2194–98
136. Hilz, H., Kaukel, E. 1973. *Mol. Cell. Biochem.* 1:229–39
137. Snyder, F. F., Seegmiller, J. E. 1976. *FEBS Lett.* 66:102–6
138. Green, H., Chan, T. S. 1973. *Science* 182:836–37
139. Ischii, K., Green, H. 1973. *J. Cell. Sci.* 13:429–39
140. Kaukel, E., Fuhrmann, U., Hilz, H. 1972. *Biochem. Biophys. Res. Commun.* 48:1516–24
141. Planet, G., Fox, I. H. 1977. *Clin. Res.* 25:323A
142. Aronow, L. 1961. *Biochim. Biophys. Acta* 47:184–85
143. Hosono, R., Kuno, S. 1974. *J. Biochem.* 75:215–20
144. Planet, G., Fox, I. H. 1975. *Proc. Can. Fed. Biol. Soc.* 18:163
145. Planet, G., Fox, I. H. 1976. *J. Biol. Chem.* 251:5839–44
146. Bagnara, A. S., Finch, L. R. 1973. *Eur. J. Biochem.* 36:422–27
147. Bagnara, A. S., Finch, L. R. 1974. *Eur. J. Biochem.* 41:421–30
148. Fox, I. H., Kelley, W. N. 1971. *Ann. Int. Med.* 74:424–33
148a. Kredich, N. M., Martin, D. W. 1977. *Cell.* 12:931–8
149. Vornovitskaya, G. I., Ioannesyants, I. A., Borzenko, B. G., Shapot, V. S. 1975. *Vopr. Med. Khim.* 21:192–94
150. Bynum, J. W., Volkin, E. 1976. *J. Cell. Physiol.* 88:197–206
151. Foltinova, I., Kuzela, S. 1976. *Neoplasma* 23:223–26
152. Shobe, C. R., Campbell, J. N. 1973. *Can. J. Microbiol.* 19:1275–84
153. Kida, M., Kawashima, F., Imada, A., Nogami, I., Suhara, I., Yoneda, M. 1969. *J. Biochem.* 66:487–92
154. Lund, P., Cornell, N. W., Krebs, H. A. 1975. *Biochem. J.* 152:593–99
155. Hirsch, J., Martelo, O. J. 1976. *Life Sci.* 19:85–90
156. Kariya, T., Field, J. B. 1976. *Biochim. Biophys. Acta* 451:41–47
157. Olsson, R. A., Charles, M. D., Davis, J., Khouri, E. M., Patterson, R. E. 1976. *Circ. Res.* 39:93–98
158. Tattersall, M. H. N., Ganeshaguru, K., Hoffbrand, A. V. 1975. *Biochem. Pharmacol.* 24:1495–98

158a. Lapi, L., Cohen, S. S. 1977. *Biochem. Pharmacol.* 26:71–76

158b. Scott, F. W., Henderson, J. F. 1977. *Can. Fed. Biol. Soc.* 20:721

159. Krooth, R. S. 1964. *Cold Spring Harbor Symp. Quant. Biol.* 29:189–211

160. McBurney, M. W., Whitmore, G. F. 1975. *J. Cell. Physiol.* 85:87–100

161. Hershfield, M. S., Snyder, F. F., Seegmiller, J. E. 1977. *Science* 197:1284–87

162. Chassin, M. M., Chirigos, M. A., Johns, D. G., Adamson, R. H. 1977. *N. Engl. J. Med.* 296:1232

163. Lum, C. T., Sutherland, D. E. R., Najarian, J. S. 1977. *N. Engl. J. Med.* 296:819

164. Ballet, J. J., Insel, R., Merler, E., Rosen, F. S. 1976. *J. Exp. Med.* 143:1271–76

165. Bajaj, S., Insel, J., Quagliata, F., Hirschhorn, R., Silber, R. 1977. *Clin. Res.* 25:405A

166. Benke, P. J., Dittmar, D. 1976. *Pediatr. Res.* 10:642–46

167. Grindey, G. B., Lowe, J. K., Divekar, A. Y., Hadala, M. T. 1976. *Can. Res.* 36:379–83

168. Tomar, R. H., Knight, R., Stern, M. 1976. *J. Allergy Clin. Immunol.* 58:190–97

169. Bourne, H. R., Lichtenstein, L. M., Melmon, K. L., Henney, C. S., Weinstein, Y., Shearer, G. M. 1974. *Science* 184:19–28

170. Bona, C., Robert-Gero, M., Lederer, E. 1976. *Biochem. Biophys. Res. Commun.* 70:622–29

171. Hashmi, S., May, S. R., Krooth, R. S., Miller, O. J. 1975. *J. Cell. Physiol.* 86:191–200

172. Green, C. D., Martin, D. W. Jr. 1973. *Proc. Natl. Acad. Sci. USA* 70:3698–702

173. Yin, H. H., Berlin, R. D. 1975. *J. Cell. Physiol.* 85:627–34

174. Granick, D. 1975. *J. Cell. Biol.* 65:389–417

175. Granick, D. 1975. *J. Cell. Biol.* 65:418–27

176. Taylor, J. C., Hill, D. W., Rogolsky, M. 1975. *Exp. Cell. Res.* 90:468–71

177. Winbury, M. M., Papierski, D. H., Hemmer, M. L., Hambourger, W. E. 1953. *J. Pharmacol. Exp. Ther.* 109:255–60

178. Wolf, M. W., Berne, R. M. 1956. *Circ. Res.* 4:343–48

179. Berne, R. M. 1963. *Am. J. Physiol.* 204:317–22

180. Rubio, R., Berne, R. M. 1969. *Circ. Res.* 25:407–15

181. Moir, T. W., Downs, T. D. 1972. *Am. J. Physiol.* 222:1386–90

182. Cobb, F. R., Bache, R. J., Greenfield, J. C. Jr. 1974. *J. Clin. Invest.* 53:1618–25

183. Rubio, R., Wiedmeier, V. T., Berne, R. M. 1974. *J. Mol. Cell. Cardiol.* 6:561–66

184. Parrat, J. R., Wadsworth, R. M. 1972. *Br. J. Pharmacol.* 46:585–93

185. Moir, T. W., Downs, T. D. 1973. *Proc. Soc. Exp. Biol. Med.* 144:517–22

186. Morris, J. J., Peter, R. H. 1971. In *Cardiac & Vascular Diseases,* ed. H. L. Conn, Jr., O. Horowitz, Vol. 1, pp. 123–38. Philadelphia: Lea & Febiger

187. Afonso, S. 1974. *J. Physiol.* 241:299–308

188. Poch, G., Kukovetz, W. R. 1972. *Adv. in Cyclic Nucleotide Res.* 1:195–211

189. Kalsner, S. 1975. *Br. J. Pharmacol.* 55:439–45

190. Seidel, C. L., Schnarr, R. L., Sparks, H. V. 1975. *Am. J. Physiol.* 229:265–69

191. McKenzie, S. G., Frew, R., Bar, H. 1977. *Eur. J. Pharmacol.* 41:183–92

192. Herlihy, J. T., Bockman, E. L., Berne, R. M., Rubio, R. 1976. *Am. J. Physiol.* 230:1239–43

193. McKenzie, S. G., Frew, R., Bar, H. 1977. *Eur. J. Pharmacol.* 41:193–203

194. Gross, G. J., Warltier, D. C., Hardman, H. F. 1976. *J. Pharmacol. Exp. Ther.* 196:445–54

195. Weissel, M., Brugger, G., Raberger, G., Kraupp, O. 1974. *Pharmacology* 12:120–28

196. Oberdorster, G., Lang, R., Zimmer, R. 1975. *Eur. J. Pharmacol.* 30:197–204

197. Kozniewska, E., Trzebski, A., Zielinski, A. 1976. *J. Physiol.* 256:96P–97P

198. Tominaga, S., Watanabe, K., Nakamura, T. 1975. *Tohoku J. Exp. Med.* 115:185–95

199. Osswald, H., Schmitz, H. J., Heidenreich, O. 1975. *Pflugers Arch.* 357:323–33

200. Osswald, H. 1975. *Arch. Pharmacol.* 288:79–86

201. Rubio, R., Berne, R. M., Bockman, E. L., Curnish, R. R. 1975. *Am. J. Physiol.* 228:1896–902

202. Burnstock, G. 1972. *Pharmacol. Rev.* 24:509–81

203. Rowe, J. N., Van Dyke, K., Westfall, D. P., Stitzel, R. E. 1976. *Pharmacology* 14:193–204

204. Urthaler, F., James, T. N. 1972. *J. Lab. Clin. Med.* 79:96–105

205. Hopkins, S. V. 1973. *Biochem. Pharmacol.* 22:335–39

206. Chiba, S., Kubota, K., Hashimoto, K. 1973. *Eur. J. Pharmacol.* 21:281–85

207. Shrader, R., Rubio, R., Berne, R. M. 1975. *J. Mol. Cell. Cardiol.* 7:427–33

208. Weston, A. H. 1973. *Br. J. Pharmacol.* 48:302–8
209. Satchell, D. G., Burnstock, G. 1975. *Eur. J. Pharmacol.* 32:324–28
210. Satchell, D. G., Maguire, M. H. 1975. *J. Pharmacol. Exp. Ther.* 195:540–48
211. Sawynok, J., Jhamandas, K. H. 1976. *J. Pharmacol. Exp. Ther.* 197:379–90
212. Spedding, M., Weetman, D. F. 1976. *Br. J. Pharmacol.* 57:305–10
213. Farmer, J. B., Farrar, D. G. 1976. *J. Pharm. Pharmacol.* 28:748–52
214. Coleman, R. A. 1976. *Br. J. Pharmacol.* 57:51–57
215. Hedqvist, P., Fredholm, B. B. 1976. *Naunyn-Schmiedeberg's Arch. Pharmacol.* 293:217–23
216. Clanachan, A. S., Paton, D. M. 1977. *Br. J. Pharmacol.* 59:534
217. Su, C. 1975. *J. Pharmacol. Exp. Ther.* 195:159–66
217a. Leslie, S. W., Borowitz, J. L., Miya, T. S. 1973. *J. Pharm. Sci.* 62:1449–52
218. Maitre, M., Ciesielski, L., Lehmann, A., Kempf, E., Mandel, P. 1974. *Biochem. Pharmacol.* 23:2807–16
219. Kostopoulos, G. K., Limacher, J. J., Phillis, J. W. 1975. *Brain Res.* 88:162–65
220. Phillis, J. W., Kostopoulos, G. K. 1975. *Life Sci.* 17:1085–94
221. Phillis, J. W., Edstrom, J. P. 1976. *Life Sci.* 19:1041–54
222. Gourley, D. R. H., Beckner, S. K. 1973. *Proc. Soc. Exp. Biol. Med.* 144:774–78
223. Pull, I., McIlwain, H. 1973. *Biochem. J.* 136:893–901
224. Huang, M., Daly, J. W. 1974. *J. Neurochem.* 23:393–404
225. Sulakhe, P. V., Phillis, J. W. 1975. *Life Sci.* 17:551–56
226. Lewin, E., Bleck, V. 1976. *Neurochem. Res.* 1:429–35
227. Kluge, H., Zahlten, W., Hartmann, W., Wieczorek, V., Ring, U. 1975. *Acta Biol. Med. Ger.* 34:27–36
228. Kluge, H., Fischer, H. D., Schwarzenfeld, I. V., Haubenreiser, J., Zahlten, W., Hartmann, W. 1975. *Acta Biol. Med. Ger.* 34:1279–82
229. Schubert, P., Kreutzberg, G. W. 1975. *Brain Res.* 85:317–19
230. Hunt, S. P., Kunzle, H. 1976. *Brain Res.* 112:127–32
231. Rose, G., Schubert, P. 1977. *Brain Res.* 121:353–57
232. Kruger, L., Saporta, S. 1977. *Brain Res.* 122:132–36
233. Hulme, M. E., Weston, A. H. 1974. *Br. J. Pharmacol.* 50:609–11
234. Formento, M. L., Borsa, M., Zoni, G.

1975. *Pharmacol. Res. Commun.* 7:247–57
235. Rao, G. S., George, R., Ramasarma, T. 1975. *Biochem. J.* 154:639–45
236. Feldman, J. M., Jackson, T. B. 1974. *Endocrinology* 94:388–94
237. Capito, K., Hedeskov, C. J. 1976. *Biochem. J.* 158:335–40
238. Weir, G. C., Knowlton, S. D., Martin, D. B. 1975. *Endocrinology* 97:932–36
239. Hill, M. K., MacLeod, R. M., Orcutt, P. 1976. *Endocrinology* 99:1612–17
240. Schooley, J. D., Mahlmann, L. 1975. *Proc. Soc. Exp. Biol. Med.* 150:215–19
241. Sullivan, T. J., Marquardt, D. L., Parker, C. W. 1976. *Clin. Res.* 24:544A
242. Chagoya de Sanchez, V., Brunner, Z., Sanchez, M. E., Lopez, C., Pina, E. 1974. *Arch. Biochem. Biophys.* 160:145–50
243. Haeckel, R. 1977. *Adv. Exp. Med. Biol.* 76A:488–99
244. Pinedo, H. M., Zaharko, D. S., Bull, J. M., Chabner, B. A. 1976. *Can. Res.* 36:4418–24
245. Cotterrell, D. 1976. *J. Physiol.* 256:131P–32P
246. Giblett, E. R., Anderson, J. E., Cohen, F., Pollara, B., Meuwissen, H. J. 1972. *Lancet* 2:1067–69
247. Meuwissen, H. J., Pickering, R. J., Pollara, B., Porter, I. H. 1975. *Combined Immunodeficiency Disease and Adenosine Deaminase Deficiency: A Molecular Defect.* New York: Academic
248. Mills, G. C., Schmalsteig, F. C., Trimmer, K. B., Goldman, A. S., Goldblum, R. M. 1976. *Proc. Natl. Acad. Sci. USA* 73:2867–71
249. Raivio, K. O., Schwartz, A. L., Stern, R. C., Polmar, S. H. 1977. *Adv. Exp. Med. Biol.* 76A:456–62
250. Van der Weyden, M. B., Buckley, R. H., Kelley, W. N. 1974. *Biochem. Biophys. Res. Commun.* 57:590–95
251. Trotta, P. P., Smithwick, E. M., Balis, M. E. 1976. *Proc. Natl. Acad. Sci. USA* 73:104–8
252. Chen, S. H., Scott, C. R., Swedberg, K. P. 1975. *Am. J. Hum. Genet.* 27:46–52
253. Hirschhorn, R., Beratis, N., Rosen, F. S. 1976. *Proc. Natl. Acad. Sci. USA* 73:213–17
254. Carson, D. A., Goldblum, R., Keightley, R., Seegmiller, J. E. 1977. *Adv. Exp. Med. Biol.* 76A:463–70
255. Daddona, P. E., Van der Weyden, M. B., Kelley, W. N. 1976. *Clin. Res.* 24:575A
256. Hirschhorn, R., Levytska, V., Pollara, B., Meuwissen, H. J. 1973. *Nature New Biol.* 246:200–2

257. Hirschhorn, R. 1975. *J. Clin. Invest.* 55:661–67
258. Giblett, E. R., Ammann, A. J., Wara, D. W., Diamond, L. K. 1975. *Lancet* 1:1010–13
259. Cohen, A., Doyle, D., Martin, D. W. Jr., Ammann, A. J. 1976. *N. Engl. J. Med.* 295:1449–54
260. Siegenbeek Van Heukelom, L. H., Akkerman, J. W. N., Staal, G. E. J., De Bruyn, C. H. M. M., Stoop, J. W., Zegers, B. J. M., De Bree, P. K. Wadman, S. K. 1977. *Clin. Chim. Acta* 74:271–79
261. Edwards, N. L., Gelfand, E. W., Biggar, W. D., Fox, I. H. 1978. *J. Lab. Clin. Med.* In press
262. Fox, I. H., Andres, C. M., Gelfand, E. W., Biggar, W. D. 1977. *Science* 197:1084–86
263. Gelfand, E. W., Dosch, H., Biggar, W. D., Fox, I. H. 1978. *J. Clin. Invest.* In press
264. Van Belle, H. 1970. *Eur. J. Pharmacol.* 11:241–48
265. Kolassa, N., Pfleger, K. 1975. *Biochem. Pharmacol.* 24:154–56
266. Philp, R. B., Francey, I., McElroy, F. 1973. *Thromb. Res.* 3:35–50
267. Liu, M. S., Feinberg, H. 1973. *Biochem. Pharmacol.* 22:1118–21
268. Afonso, S., O'Brien, G. S. 1971. *Arch. Int. Pharmacodyn. Ther.* 194:181–96
269. Sano, N., Satoh, S., Hashimoto, K. 1972. *Jpn. J. Pharmacol.* 22:857–65
270. Paoloni, H. J., Wilcken, D. E. L. 1972. *J. Pharmacol. Exp. Ther.* 183:137–45
271. Degenring, F. H., Curnish, R. R., Rubio, R., Berne, R. M. 1976. *J. Mol. Cell. Cardiol.* 8:877–88
272. Afonso, S., O'Brien, G. S., Crumpton, C. W. 1968. *Circ. Res.* 22:43–48
273. Van Belle, H. 1970. *Eur. J. Pharmacol.* 10:290–92
274. Hansing, C. E., Folts, J. D., Afonso, S., Rowe, G. G. 1972. *J. Pharmacol. Exp. Ther.* 181:498–511
275. Woo, Y. T., Manery, J. F., Dryden, E. E. 1974. *Can. J. Physiol. Pharmacol.* 52:1063–73
276. Vizi, E. S., Knoll, J. 1976. *Neuroscience* 1:391–98
277. Bloch, A. 1975. *Ann. NY Acad. Sci.* 255:1–610
278. Bloch, A. 1975. *Ann. NY Acad. Sci.* 255:576–96
279. Nelson, J. A., Parks, R. E. Jr. 1972. *Can. Res.* 32:2034–41
280. Hare, J. D., Hacker, B. 1972. *Physiol. Chem. Physics* 4:275–85
281. Scholar, E. M., Brown, P. R., Parks, R. E. Jr. 1972. *Can. Res.* 32:259–69
282. Warnich, C. T., Patterson, A. R. P. 1973. *Can Res.* 33:1711–15
283. Shantz, G. D., Smith, C. M., Fontenelle, L. J., Lau, H. K. F., Henderson, J. F. 1973. *Can. Res.* 33:2867–71
284. Tritsch, G. L. 1973. *Can. Res.* 33:310–12
285. Smith, C. M., Snyder, F. F., Fontenelle, L. J., Henderson, J. F. 1974. *Biochem. Pharmacol.* 23:2023–25
286. Zimmerman, T. P., Chu, L. C., Bugge, C. J. L., Nelson, D. J., Miller, R. L., Elion, G. B. 1974. *Biochem. Pharmacol.* 23:2737–49
287. Plunkett, W., Cohen, S. S. 1975. *Can. Res.* 35:1547–54
288. Hakala, M. T., Slocum, H. K., Gryko, G. J. 1975. *J. Cell. Physiol.* 86:281–91
289. Bennett, L. L. Jr., Allan, P. W. 1976. *Can. Res.* 36:3917–23
290. Grindey, G. B., Lowe, J. K., Divekar, A. Y., Hakala, M. T. 1976. *Can. Res.* 36:379–83
291. Johns, D. G., Adamson, R. H. 1976. *Biochem. Pharmacol.* 25:1441–44
292. Plunkett, W., Cohen, S. S. 1977. *Ann. NY Acad. Sci.* 284:91–102
293. Rose, L. M., Brockman, R. W. 1977. *J. Chromatogr.* 133:335–43
294. Sloan, B. J., Kielty, J. K., Miller, F. A. 1977. *Ann. NY Acad. Sci.* 284:60–80
295. Bennett, L. L. Jr., Allan, P. W., Carpenter, J. W., Hill, D. L. 1976. *Biochem. Pharmacol.* 25:517–21
296. Schillinger, E., Loge, O. 1974. *Biochem. Pharmacol.* 23:2283–89
297. Giziewica, J., De Clercq, E., Luczak, M., Shugar, D. 1975. *Biochem. Pharmacol.* 24:1813–17
298. Hellmann, D. B., Pitt, B. 1976. *Am. J. Physiol.* 231:1495–500
299. Paoletti, R., Berti, F., Spano, P. F., Michal, G., Weimann, G., Nelboeck, M. 1973. *Pharmacol. Res. Commun.* 5:87–100
300. Imai, S., Otorii, T., Takeda, K., Katano, Y., Horii, D. 1975. *Jap. Heart J.* 16:421–32
301. Prasad, R. N., Fung, A., Tietje, K., Stein, H. H., Brondyk, H. D. 1976. *J. Med. Chem.* 19:1180–86

REFERENCES ADDED IN PROOF:

302. Coleman, M. S., Donofrio, J., Hutton, J. J., Daoud, A., Lampkin, B., Dyminski, J. 1977. *Blood* 50:Suppl. 1, p. 292
303. Cohen, A., Hirschhorn, R., Horowitz, S., Rubinstein, A., Polmar, F., Hong, R., Martin, D. 1978. *Proc. Natl. Acad. Sci. USA* 75:472–76
304. Schmalstieg, F. C., Nelson, J. A., Mills, G. C., Monahan, T. M., Goldman, A. S., Goldblum, M. D. 1977. *J. Pediatr.* 91:48–51

Ann. Rev. Biochem. 1978. 47:687–713

SPHINGOLIPIDOSES

❖986

Roscoe O. Brady

Developmental and Metabolic Neurology Branch, National Institute of Neurological and Communicative Disorders and Stroke, National Institutes of Health, Bethesda, Maryland 20014

CONTENTS

PERSPECTIVES AND SUMMARY

The sphingolipidoses comprise a group of 11 inherited disorders of metabolism that are characterized by pathological accumulations of a specific class of lipids in various organs and tissues of afflicted individuals. Most of these metabolic disturbances are fatal in the first few years of life, although an

0066-4154/78/0701-0687$01.00

increasing number of patients are being reported with longer life spans. However, the majority of patients seen by practicing physicians are those with adult (Type 1) Gaucher's disease and Fabry's disease. The central nervous system is apparently spared from the lipid accumulation in these conditions, although large amounts may be found in the systemic organs.

Remarkable advances have occurred in this area during the past 12 years. The enzymatic defects in each of the 11 disorders have been demonstrated. This information has been widely used for the development of diagnostic enzyme assays that use easily accessible materials such as leukocytes, serum, cultured skin fibroblasts, and tears. These tests have been refined so that healthy heterozygous carriers may also be identified. Most importantly, these assays now permit the prenatal diagnosis of any of the sphingolipid storage diseases. These procedures are now universally employed for the control of these metabolic diseases.

More recent investigations deal with improvement and acceleration of diagnostic procedures, probing the molecular nature of the genetic defects and the development of effective therapy for these diseases. Highest on the list of possibilities with regard to the latter aspect is replacement of the enzymes that are deficient in these disorders. Initial findings in Gaucher's disease and Fabry's disease are encouraging and indicate that extensive clinical trials should be undertaken. Furthermore, a major breakthrough appears imminent in conditions where the central nervous system is involved. It has recently been demonstrated that it is possible to temporarily alter the blood-brain barrier so that exogenous enzymes may reach the brain. Much additional investigation is required to determine whether this procedure will become a realistic therapeutic modality.

INTRODUCTION

I restrict this review to a description of the present status of research in the sphingolipid storage diseases and an indication of further developments that might be anticipated in this area in the near future. This approach seems indicated by the large number of reviews that have recently been devoted to this subject (e.g. 1–5). Therefore, I intend to provide only a brief survey of the chemistry of the involved lipids, the nature of the enzymatic defects, and diagnostic principles so that investigators unfamiliar with the field will be able to follow the lines I wish to develop. I shall focus on 1. newer diagnostic concepts that have not received extensive coverage because of the recency of their introduction, 2. studies now emerging on the molecular nature of the lipid storage diseases, 3. an assessment of the prospects for enzyme replacement therapy, and 4. novel approaches for understanding and treating these disorders.

CHEMISTRY

All of the lipids discussed in this chapter are derived from the 18-carbon amino alcohol sphingosine [4t-sphingenine, *trans*(2*S*,3*R*)-2 amino-1,3 dihydroxy 4-octadecene]. Lesser amounts of the C_{20} homologue of sphingosine also occur in some tissues. The chemistry of these long chain bases has been described in detail (6). Fatty acids whose chain length ranges from 16 to 26 carbon atoms are joined by an amide bond to the nitrogen atom on carbon 2 of sphingosine. This combination of sphingosine and fatty acid is called ceramide (Cer), and it is the common hydrophobic portion of all sphingolipids. Various substituents are linked to the hydroxyl group on carbon 1 of the sphingosine moiety of ceramide. Thus, sphingomyelin, a major lipid of many cell membranes, consists of ceramide and phosphocholine linked by a phosphodiester bond (Figure 1, Row 3). Glycosphingolipids are comprised of ceramide to which monosaccharides or oligosaccharides are joined by glycosidic bonds (Figure 1, Rows 2, 4–11). These substances have been divided into 1. neutral sphingoglycolipids, 2. gangliosides, which, in addition to hexose and hexosamine, contain one or more molecules of N-acetylneuraminic acid (NeuAc), and 3. sulfated glycolipids that contain sulfuric acid esterified to the oxygen on carbon 3 of galactosphingolipids. The nomenclature of the sphingolipids has recently undergone extensive revision (7, 8), and this terminology is employed wherever appropriate.

BIOSYNTHESIS OF SPHINGOLIPIDS

The formation of the sphingolipids has been investigated extensively. It is generally agreed that their assembly occurs through the function of enzyme complexes acting in a concerted, sequential fashion (9). The specific reactions involved have been cited in a previous discussion of this subject (10). Since, with one exception (11), all of the metabolic abnormalities concerning sphingolipids are due to deficiencies of lipid hydrolases, I concentrate on the catabolism of sphingolipids in this communication.

BIODEGRADATION OF SPHINGOLIPIDS

Generic Reactions

The catabolism of sphingolipids occurs through the stepwise hydrolytic cleavage of the various component molecules beginning with the terminal hydrophilic portion of the molecule. For example, the catabolism of ganglioside G_{M1} [Galβ1→3GalNAcβ1→4Gal(3←2αNeuAc)β1→4Glcβ1→1'Cer], a major brain ganglioside, is initiated through the action of a β-galactosidase that cleaves the terminal molecule of galactose to yield ganglioside G_{M2} [GalNAcβ1→4Gal(3←2αNeuAc)β1→4Glcβ1→1'Cer].

DISEASE	SIGNS AND SYMPTOMS
FARBER'S DISEASE	HOARSENESS, DERMATITIS SKELETAL DEFORMATION, MENTAL RETARDATION
GAUCHER'S DISEASE	SPLEEN AND LIVER ENLARGEMENT EROSION OF LONG BONES AND PELVIS, MENTAL RETARDATION ONLY IN INFANTILE FORM
NIEMANN-PICK DISEASE	LIVER AND SPLEEN ENLARGEMENT MENTAL RETARDATION, ABOUT 30 PERCENT WITH RED SPOT IN RETINA
KRABBE'S DISEASE (GLOBOID LEUKODYSTROPHY)	MENTAL RETARDATION, ALMOST TOTAL ABSENCE OF MYELIN, GLOBOID BODIES IN WHITE MATTER OF BRAIN
METACHROMATIC LEUKODYSTROPHY	MENTAL RETARDATION, PSYCHOLOGICAL DISTURBANCES IN ADULT FORM, NERVES STAIN YELLOW-BROWN WITH CRESYL VIOLET DYE
CERAMIDE LACTOSIDE LIPIDOSIS	SLOWLY PROGRESSING BRAIN DAMAGE, LIVER AND SPLEEN ENLARGEMENT
FABRY'S DISEASE	REDDISH-PURPLE SKIN RASH, KIDNEY FAILURE, PAIN IN LOWER EXTREMITIES
TAY-SACHS DISEASE	MENTAL RETARDATION, RED SPOT IN RETINA, BLINDNESS, MUSCULAR WEAKNESS
TAY-SACHS VARIANT	SAME AS TAY-SACHS DISEASE BUT PROGRESSING MORE RAPIDLY
GENERALIZED GANGLIOSIDOSIS	MENTAL RETARDATION, LIVER ENLARGEMENT, SKELETAL DEFORMITIES, ABOUT 50 PERCENT WITH RED SPOT IN RETINA
FUCOSIDOSIS	CEREBRAL DEGENERATION, MUSCLE SPASTICITY, THICK SKIN

Figure 1 Principal manifestations, stored lipids, and metabolic defects in the sphingolipidoses.

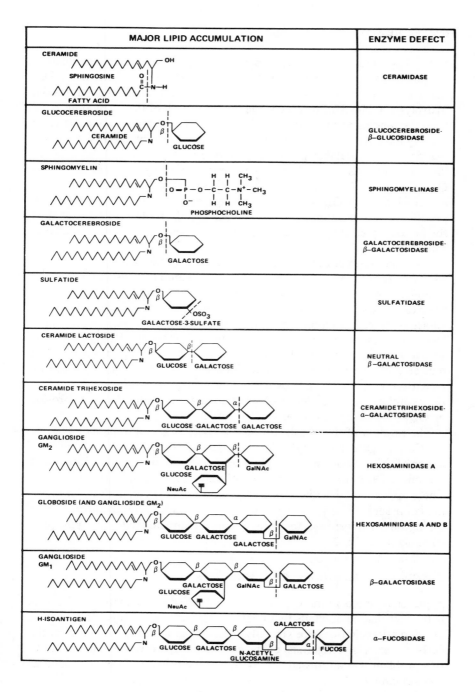

Figure 1   (continued).

Since G_{M2} is branched in its terminal portion, two paths for its degradation are potentially available: namely, through the action of an hexosaminidase that cleaves the molecule of GalNAc or a neuraminidase that catalyzes the hydrolysis of NeuAc. In spite of the fact that both of these reactions have been demonstrated in vitro (12, 13), and the respective products of their activity have been found in vivo (14), the relative contribution of these two reactions to the catabolism of G_{M2} in vivo is still unsettled. Since more G_{M2} accumulates in the brain of patients with Tay-Sachs disease than G_{A2} (GalNAcβ1→4Galβ1→4Glcβ1→1'Cer) (15), it is presumed that the principal route of G_{M2} catabolism involves the cleavage of GalNAc; the product of this reaction, G_{M3} (NeuAcα2→3Galβ1→4β1→1'Cer), is then further degraded through the action of a neuraminidase to ceramidelactoside (Figure 1, Row 6).

The catabolism of ceramidelactoside requires participation of a β-galactosidase. There are two β-galactosidases in mammalian tissues whose pH optima are 4.2 and 4.8, respectively. Either of these enzymes can catalyze the cleavage of galactose from ceramidelactoside in vitro, depending on the choice of detergent employed. The more acidic enzyme also catalyzes the cleavage of galactose from galactocerebroside (Galβ1→1'Cer) (16). The second β-galactosidase preferentially catalyzes the cleavage of the terminal molecule of galactose of G_{M1} in addition to ceramidelactoside (17). More recently, two β-galactosidases with more neutral pH optima that catalyze the hydrolysis of ceramidelactoside but are inactive with galactocerebroside or G_{M1} (18) have been isolated from liver.

The product of the β-galactosidase activity is glucocerebroside (Glcβ1→1'Cer) (Figure 1, Row 2). This compound is catabolized through the action of the enzyme known as glucocerebrosidase to yield free glucose and ceramide. Ceramide is, in turn, hydrolyzed to sphingosine and fatty acid. There appear to be acidic (19) and neutral (20, 21) ceramidases in mammalian tissues. Finally, sphingosine itself is degraded after the formation of sphingosine-1-phosphate to yield a long chain aldehyde and phosphoethanolamine (22, 23).

Detergents, Activators, and Specifiers

A principal and continuing problem encountered by sphingolipid chemists is the fact that compounds that contain ceramide linked to various mono-, di-, and trisaccharides are virtually insoluble in aqueous solutions. In the early stages of investigations of the enzymatic hydrolysis of sphingolipids, this problem was addressed simply by the inclusion of detergents such as sodium cholate (19) or Cutscum (isooctylphenoxypolyoxyethanol) (24) in the reaction mixtures. Even the hydrolysis of water-soluble glycolipids such as G_{M1} is markedly enhanced by the addition of a variety of detergents (25). The artificiality of such mixtures was clearly recognized, and several groups of investigators have devoted much effort to investigating the possibility that naturally occurring detergents or activators may be found in vivo. An early indication of the existence of such a substance was provided by Mehl & Jatzkewitz (26), and subsequent studies in the latter author's laboratory have provided considerable clarification of the nature of the material. It was shown that the hydrolysis of sulfatide by cerebroside sulfatase (EC 3.1.6.8) is enhanced by the addition of a protein with a molecular weight of approximately 21,500 (27). Later studies revealed that the activator was localized in lysosomes (28) [as are most of the sphingolipid hydrolases (29)] and that the activator binds to sulfatide in a 1 to 1 complex, which indicates that it acts on the substrate rather than on the enzyme (30).

Another important contribution along this line was the demonstration by Ho & O'Brien that spleens of patients with Gaucher's disease contain a heat-stable factor that enhances the activity of glucocerebrosidase (EC 3.2.1.45) (31, 32). Subsequently, a strong binding between the enzyme and its activator was reported, but which occurred only in the presence of phospholipids (33, 34). The "effector" is a glycoprotein with a molecular weight of 20,000 (33; cf 27). However, the intracellular localization of this material is said to be in the cytosol (31). The activator was subsequently purified from Gaucher spleen tissue, and it was shown to be relatively acidic glycoprotein comprised of 16% carbohydrate with a molecular weight of 11,000 (35). The specificity of the factor in Gaucher spleen has recently been questioned since the activity of human liver G_{M1}-β-galactosidase is activated tenfold by the addition of this material (36). A separate, apparently nonglycoprotein activator has been identified in normal human liver (35).

The third entry into this area was the report by Li et al that liver contains a heat-stable nondialyzable factor that promotes the catabolism of G_{M2} by purified hexosaminidase A (EC 3.2.1.52) (37). This substance also promoted the hydrolysis of other glycolipids such as G_{M1} and ceramidetrihexoside (Figure 1, Row 7) (38; cf also 39). Note that the anomeric configuration of terminal digalactoside in the latter compound is α, whereas in glucocerebroside and G_{M2} it is β. Kinetic data indicated an apparent stoichiometric reaction between G_{M1} and the activator (40). This activator is also a heat-stable glycoprotein. More recent evidence indicates that the hexosaminidase activator has been separated from materials that promote the hydrolysis of other sphingolipids (S.-C. Li, communicated September 30, 1977). There appear to be some notable discrepancies between the modes of action of these various activators, since the addition of the factor purified by Ho and co-workers causes an augmentation of the hydrolysis of such artificial substrates as 4-methylumbelliferyl-β-D-glucopyranoside as well as glucocerebroside, whereas the factors dealt with by the other investigators affect only the hydrolysis of the natural sphingolipids and do not influence the activity of the enzymes on fluorogenic glycosides (e.g. 41). Recently, Hechtman (41) and Hechtman & LeBlanc (42) have further characterized the hexosaminidase-activating protein from human liver. In contrast to the previously described activators, the G_{M2}-β-hexosaminidase factor whose molecular weight was 36,000 was not found to be a glycoprotein, and heating it for 5 min at 100° destroyed 80–90% of the activity (41). These authors have suggested that a ternary complex might be formed between substrate, hexosaminidase A, and activating protein. A further important observation was the fact that the hydrolysis of G_{A2} (see above), which is readily catalyzed by hexosaminidase A, was unaffected by the addition of activating factor.

Activators have also been described that promote the enzymatic hydrolysis of sphingomyelin (39, 43). Thus, the overwhelming evidence points to the presence of relatively heat-stable factors that enhance the catalytic activity of various sphingolipid hydrolases in normal and pathologic tissues. However, several aspects of these studies have raised doubts in the minds of some investigators concerning the physiological roles of such substances and whether they exist as separate entities within cells. Perhaps the most distressing aspect of these studies is the relatively small actual enhancement of catalytic activity conferred to hexosaminidase A (37, 41). For example, the ratio of activity with artificial fluorogenic glucosaminides vs G_{M2} decreases from 600 : 1 in unfractionated lysosomes to 1×10^6 : 1 with purified hexosaminidase A (44). The maximum reported stimulation of G_{M2} hydrolysis with the addition of activators to purified enzyme (41, 42) falls well below the activity observed with unfractionated lysosomes. Both we (45) and others (35) have concluded that the "activator" of glucocerebrosidase may not be physiologically important in the normal degration of glucocerebroside.

What, then, is the nature of all of the factors, and if they are not important, how does sphingolipid catabolism really occur in vivo? It is not possible to answer these questions categorically at this time, but the following concepts might be adduced as bases for further investigation. It is to be noted that in almost all cases a thermal inactivation step is employed in the isolation of the "activator" (27, 34, 38, 40, 43) that consists of immersing the homogenized tissue in a boiling water bath for 5 min or incubation for 30 min at 60° (42). The molecular size of the activators varies considerably. They are usually glycoproteins. Glucocerebrosidase (46, 47), and other lysosomal lipid hydrolases have been shown to be glycoproteins (48–52). Perhaps some of the so-called activators are fragments of the lysosomal enzymes or lysosome membranes produced by heating, proteolysis, or the action of endoglycosidases (53–55) during the preparation of these substances (cf 42). This speculation seems consistent with the following observations:

1. The activator could not be demonstrated in untreated homogenates of tissue but only after dialysis at acidic pH (42) (enhancement of proteolysis) and then incubation at elevated temperature (endoglycosidase activity).
2. Most of these factors, like the enzymes themselves, are glycoproteins.
3. For the most part, they have been shown to be localized in lysosomes, as are the enzymes.

How then, can one account for the large discrepancy in the catabolism of sphingolipids such as G_{M2}, which do not require detergents for enzymatic degradation if lysosomes are employed (13), whereas the addition of detergents is mandatory with purified hexosaminidase A, and even then, very low activity is seen with the natural lipid substrate (44)? It seems to this reviewer that one should consider the possibility that sphingolipid catabolism may well occur in a manner analogous to the concerted series of reactions postulated (9) and later shown to be correct for the biosynthesis of G_{M1} (56). Sphingolipid hydrolases may also be aligned in an ordered fashion on lysosomal membranes. It seems likely that hydrolysis of sphingolipids in this manner would be much more efficient than by random access of various substrates to enzymes freely admixed within a lysosomal sac. One may extend this concept to the visualization of sets of enzymes concerned with the complete degradation of materials such as globoside (Figure 1, Row 9) from senescent red blood cells or ganglioside G_{D1a} (NeuAcα2$\rightarrow$3Galβ1$\rightarrow$3GalNAcβ1$\rightarrow$4Gal(3$\leftarrow$2αNeuAc)β1$\rightarrow$4Glcβ1$\rightarrow$1'Cer) a major ganglioside in the brain (15), thyroid (57), and testis (58). Some support for this concept may be seen in the experiments of Dawson, Sloolmiller, and Radin, who found that the degradation of complex lipids such as G_{M2} or ceramidelactoside was unimpaired in the presence of an inhibitor of β-glucosidase, a step considered to be obligatory for the complete catabolism of these substances (59).

ABNORMAL ENZYMOLOGY OF THE SPHINGOLIPIDOSES

The specific metabolic defect in each of the lipid storage diseases has been unequivocally demonstrated and amply confirmed for all of these disorders with the exception of lactosylceramidosis (Figure 1, Row 6). A deficiency of a neutral β-galactosidase has recently been reported in this disorder (60) that, if correct, should dispel the controversy that has arisen concerning this condition (61). The various enzyme defects have been repeatedly reviewed (e.g. 1–5, 62–65). Rather than recapitulating this information in the present contribution, I have summarized them in schematic form in Figure 1. For present purposes, I should like to discuss several recent contributions concerning the molecular basis of these disorders. It was anticipated some time ago that the genesis of these diseases lay in genetic mutations affecting the primary structure of the various enzymes since all patients with adult (Type I) and juvenile (Type III) Gaucher's disease had clear evidence of residual glucocerebrosidase activity in their tissues (66). Furthermore, the K_m for glucocerebroside in one of the pathological spleens was found to be an order of magnitude larger than that in control specimens. The concept of an inactive or less than normally efficient protein has been clearly demonstrated in metachromatic leukodystrophy (67) and in G_{M1} gangliosidosis (68, 69). In a liver preparation obtained from a patient with the latter disorder, the K_m for G_{M1} was found to be five times greater than normal, whereas the K_m for 4-methylumbelliferyl-β-D-galactopyranoside was twice that in controls (70). The presence of material that cross-reacts with antibodies raised against the normal enzyme (CRM) has also been reported in a patient with Sandhoff-Jatzkewitz disease (Figure 1, Row 9) (71). The most perplexing concern at this time is the failure to demonstrate CRM in urine or liver from patients with Fabry's disease (72, 73). The importance of this point is encountered again in the section dealing with the therapy of lipid storage diseases.

Remarkable progress has been made in the past few years towards the solution of problems concerned with altered biochemical genetics in the sphingolipidoses by use of the technique of cell hybridization. Cultured skin fibroblasts have been particularly useful for unraveling the isoenzyme abnormalities in Tay-Sachs disease and Sandhoff-Jatzkewitz disease. In the former condition, only hexosaminidase isoenzyme A is lacking (74), whereas in the latter, both major mammalian hexosaminidases A and B are deficient (75). Uncertainties concerning the interrelationship between these isoenzymes have been resolved by the demonstration that the deficiency of hexosaminidase A in these respective disorders is due to different mutations and that restoration of the activity can be obtained through complementation (76–79). These studies have also provided strong support for the concept that hexosaminidase A is composed of two nonidentical subunits $[(\alpha\beta)_n]$ and hexosaminidase B is $(\beta\beta)_n$ (80, 81). Previously suggested models concerning the interrelationship of these enzymes, including the "conformational isomer" hypothesis (44), must now be discarded. Hybridization studies have also been extremely useful for identifying the chromosomes that carry the genetic code for enzymes. Thus, the cistron for the α subunits of hexosaminidase A has been assigned to chromosome 15 (82, 83) and that for β subunits to chromosome 5 (83–85). The gene locus for α-L-fucosidase has been assigned to chromosome 1 (86).

DIAGNOSTIC TESTS

Detection of Homozygotes and Heterozygotes

A major concern of clinical biochemists at this time is the development of facile diagnostic enzyme assay procedures. This aspect is particularly important since it represents the principal application, for practical community benefit, of the discoveries of the enzymatic defects in these disorders. Thus, the use of readily available patient materials such as leukocytes (87), cultured skin fibroblasts (88), serum (89), urine (90), tears (91, 92), or hair follicles (93) and the establishing of optimal conditions for enzyme assay systems provides important diagnostic advantages over the histologic or quantitative lipid analytical procedures that were previously required. Because the enzymes in most of these specimens are stable to freezing, the samples may be shipped from any part of the world to an appropriate facility for assay. Furthermore, automated procedures have been developed for many of these determinations (94–96).

The original tests along these lines were based on the use of natural sphingolipids that were usually labeled in the hydrophilic portion cleaved by the enzyme. The product was separated by extracting unreacted substrate with a nonpolar solvent or by coprecipitation with denatured protein. It soon became apparent that artificial chromogenic or fluorogenic glycosides could be used in many instances instead of costly radioactive materials. Although difficulties have been encountered regarding the consistency of the correlation of the activity between the natural and artificial substrates, conditions are gradually being developed that permit their use with confidence (96, 97). Until very recently, the diagnosis of Niemann-Pick disease and Krabbe's disease (cf Figure 1) depended on the use of labeled lipids. Unsubstituted nitrophenyl derivatives of phosphocholine or galactose did not provide the discrimination required. By inspecting the structure of the accumulating molecules in the region of the site of enzymatic cleavage, it was considered likely that the introduction of a long chain aliphatic amide ortho to the phenolic hydroxyl of p-nitrophenyl derivatives of phosphocholine or β-D-galactopyranoside might yield a compound that resembles the structure of sphingomyelin and galactocerebroside sufficiently that these materials would be specifically cleaved by the respective hydrolases (98). Accordingly the sphingomyelin analogue was synthesized (99) (Figure 2-III). It was found to be a completely reliable substrate for the diagnosis of Niemann-Pick disease and for the detection of heterozygous carriers of this disorder (100). The corresponding analogue of galactocerebroside was then prepared (Figure 2-IV). This substance has been shown to be similarly useful for the detection of victims and carriers of Krabbe's disease (101). Two precautions were required with the latter material. The first was the necessity of pulverizing the reagent so that water-clear solutions were obtained in the presence of detergent such as sodium taurodeoxycholate (Compound III is freely water soluble). The second was the necessity of carrying out the determinations of enzyme activity by using subcellular particles of cultured skin fibroblasts or leukocytes. Unfractionated homogenates of these cells contain additional galactosidase activity that, unless it is removed prior to assay, lessens the usefulness of the analogue. By use of this procedure, the analogue has been found to be reliable for the diagnosis of Krabbe's disease and detection of carriers in whom leukocytes as well as fibroblasts are sources of the enzyme (A. E. Gal, R. O. Brady, K. Suzuki, W. G. Johnson, unpublished observations). The centrifugation step is not required when homogenates of brain or liver tissue are used (101). Next the β-D-glucopyranosyl derivative of 2-hexadecanoylamino-4-nitrophenol was synthesized (Figure 2-V). This compound may be used for the diagnosis of Gaucher's disease (102).

At the present time, the most difficult diagnostic tasks concern the correct identification of heterozygous carriers of genetic disorders. With care, enzyme assays with leukocytes and, especially, cultured skin fibroblasts, have been reliable for the detection of heterozygotes (e.g. 98, 100, 101). However, in the largest screening program mounted so far for the identification of carriers of Tay-Sachs disease, a number of serious mistakes have occurred.

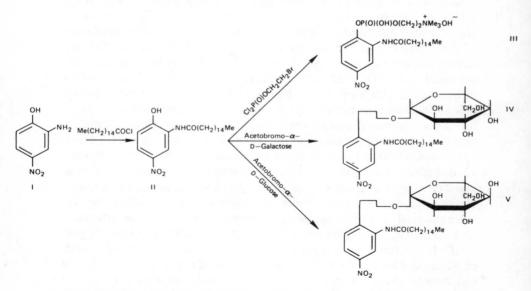

Figure 2 Chromogenic analogues of sphingomyelin (III), galactocerebroside (IV), and glucocerebroside (V) for the diagnosis of Niemann-Pick disease, Krabbe's disease, and Gaucher's disease, respectively.

The test is usually based on the differentiation of hexosaminidase isoenzymes by heat-inactivation of hexosaminidase A (89). Difficulties have been encountered in the uncritical use of this procedure and intensive efforts have been undertaken by the organization primarily concerned with this effort to establish and maintain quality controls (103).

Prenatal Diagnosis

At this time, the maximum practical use of the ability to diagnose these and other heritable disorders of metabolism lies in accurate tests carried out on cells derived from fetuses at risk for the various diseases. Except for Fabry's disease, all of the established lipid storage diseases are transmitted as autosomal recessive disorders. Both parents must be carriers to have an affected child. When both parents are heterozygotes, statistically, there is one chance in four that the child will be affected with each pregnancy. Two children will be carriers, and one will not be involved at all. The mode of transmission of ceramide lactosidosis has not been determined, although the index case, which is presumed to be a homozygote, was a female, and therefore it, too, is likely to be an autosomal recessive condition. In Fabry's disease, only the female need be a carrier to have an affected son. Half of her sons will be hemizygous (affected) and half of her daughters will be carriers. Characteristically, when monitoring pregnancies at risk, a transabdominal amniocentesis is performed around the 14th gestational week, although on occasion this procedure has been carried out as early as the 11th week (104). Approximately 20 ml of amniotic fluid are withdrawn. Most of the suspended cells that have desquamated from the skin or mucous membranes of the fetus are nonviable. However, the surviving cells can be propagated in tissue culture, and usually after 3 or 4 weeks a sufficient number of cells is available for enzyme assays. The reliability of this procedure is now established for most of the sphingolipidoses (105), and this technique is in wide usage throughout the world.

Two important aspects of this procedure are receiving considerable attention from biochemists at the moment. The first is the perfection of chromogenic and fluorogenic substrates for antenatal diagnosis. The reliability of the chromogenic analogue of sphingomyelin (Figure 2-III) has been established (4), and it is anticipated that the galactocerebroside analogue (Figure 2-IV) will also prove satisfactory. A difficult situation appears to have arisen in Tay-Sachs disease. Apparently healthy normal individuals have been detected who lack hexosaminidase A isoenzyme in their serum or leukocytes (106, 107), and another individual has been reported without demonstrable hexosaminidase isoenzymes A or B (108) as measured with 4-methylumbelliferyl-β-D-N-acetylglucosaminide. It has been shown that the individuals in the former group have 50% of the normal level of G_{M2}-

cleaving activity in their leukocytes (109) and are thus properly classified as Tay-Sachs heterozygotes. More recent studies using a variety of techniques have revealed that hexosaminidase A isoenzyme accounted for 3.5–6.9% of total hexosaminidase activity in cultured skin fibroblasts derived from these individuals (110). Because this level of hexosaminidase A is still lower than that seen in patients with juvenile G_{M2} gangliosidosis (111), these individuals would be classified as affected. Therefore, we (109) and others (111) have stressed the importance of using labeled G_{M2} as substrate in order to prevent incorrect diagnosis of individuals with unusual hexosaminidase mutations. Procedures for the preparation of G_{M2} labeled in the hexosaminyl moiety are readily available (111–113). The facile enzymatic conversion of G_{M1}, which is available commercially, to G_{M2} for use as starting material in the procedure described in References 111 and 112 has been published (114).

Accelerated Prenatal Testing

Although the first prenatal diagnosis of Tay-Sachs disease was made by direct determination of hexosaminidase A activity in amniotic fluid (115), many investigators are uncomfortable using this source of material (116, 117) and prefer to use cells that can be more confidently considered to be of fetal origin. It appeared logical to try to develop more rapid procedures for antenatal diagnosis by improving the sensitivity of enzyme analyses that utilize amniotic cells of demonstrated viability (118, 119) or after only a few passages in tissue culture (120, 121). The latter technique has now been demonstrated to be useful for the diagnosis of a number of lipid storage diseases (122).

A number of alternative procedures are also under consideration. These include the isolation of fetal cells from maternal blood, fetal blood sampling, fetal tissue biopsy, and somatic cell hybridization (122, 123). The latter procedure entails the formation of hybrids between deficient parenchymal cells and cultured amniotic fluid cells. Since the number of heterokaryons obtained is likely to be low, efficient microanalytical techniques such as those now being developed will probably be required.

Another potentially interesting approach to the diagnosis of heritable disorders has been proposed by Hösli (123). He observed that the inclusion of an unmetabolizable substance such as sphingomyelin in the tissue culture medium bathing cells from patients with Niemann-Pick disease causes a selective increase in alkaline phosphatase activity in the cells. Similar augmentation of this enzyme was obtained by adding glycogen to cells from patients with Pompe's disease that are deficient in α-glucosidase and by the addition of Tamm-Horsfall urinary glycoprotein to fibroblast cultures derived from patients with cystic fibrosis (124). Although these assays depend upon an increase in the activity of an ancillary enzyme that is not involved in the initial catabolic reaction for the biodegradation of these substances and thus would appear to be subject to variation, the diagnostic potential of this observation appears interesting. Another obvious implication is the use of this procedure to help identify unknown metabolic defects.

THERAPY

Organ Allografts

Following the decade of rapid progress in identifying enzymatic defects, the development of diagnostic procedures, and the antenatal detection of fetuses afflicted with such disorders, investigators have begun enzyme replacement therapy trials for individuals afflicted with the lipid storage diseases (125). Much effort is now devoted to devising strategies for the treatment of lipid storage diseases as well as other heritable metabolic disorders. High on the original list of therapeutic possibilities was the potential use of organ allografts, even though they were considered less desirable than replacement with purified enzymes. Because of the extreme difficulty encountered in obtaining requisite quantities of pure enzymes, organ transplantation was undertaken in several centers. In the first instance, a spleen was grafted into a patient with juvenile (Type III) Gaucher's disease (126). There was no significant improvement in the recipient's condition, and severe immunological complications were encountered (127). Another patient with infantile (Type II) Gaucher's disease received a kidney allograft (128). Here again, minimal benefit was observed. On the other hand, patients with Fabry's disease, who often suffer from severe renal difficulties in their late forties because of the accumulation of ceramide-trihexoside (CTH) in this organ, have had noticeable improvement of their symptoms of uremia (129, 130). Although the transplanted kidneys do not accumulate CTH, most investigators now believe that its accumulation elsewhere in the body is not reduced by kidney allografts (131–133). More recently, a preliminary report has appeared describing apparent improvement in the epileptic seizure pattern in a patient with Type A Niemann-Pick disease after liver transplantation (134). The patient has since died, and a full report of the effects of the procedure is awaited.

Enzyme Replacement

TAY-SACHS DISEASE   While the effects of organ grafting were being investigated, work was begun in several laboratories to devise enzyme isolation procedures that would permit investigations of the effectiveness of exogenous enzymes as specific therapeutic agents. These efforts received encouragement from the report of Porter et al (135), who demonstrated that the addition of arylsulfatase A to the culture medium containing cells from patients with metachromatic leukodystrophy (Figure 1, Row 5) restored the uptake and catabolism of labeled sulfatide to that in normal cells. In order to minimize the possibility of sensitizing recipients to exogenous proteins, it was considered advisable to utilize human materials as the source of enzymes. The first enzyme that became available in sufficient purity was hexosaminidase A isolated from human urine. This enzyme was infused intravenously into a patient with Sandhoff-Jatzkewitz disease (Figure 1, Row 9). A normal level of hexosaminidase A was temporarily established in the serum by this procedure (136). However, biopsies of the brain indicated that the enzyme has not crossed the blood-brain barrier, a not unexpected finding. Furthermore, there was no increase of hexosaminidase A activity in the spinal fluid or urine. The half-life of the enzyme in the circulation was approximately 7.5 min. Most of it appeared to have been taken up by the liver. When hexosaminidase A activity in a pre-infusion liver biopsy was compared with that in a specimen obtained 45 min after injection, an unexpectedly large increase in hexosaminidase A activity was observed. There was 4.8 times more total hexosaminidase A in the liver than had actually been infused. The 2.4-fold augmentation reported previously (137) is probably an incorrect figure. This value was based on total hexosaminidase A infused in equal doses on two successive days. Later experiments revealed that all exogenous enzyme disappears from the liver of recipients within 24 hr. Enhancement of catalytic activity was also seen in a subsequent investigation of enzyme replacement in Fabry's disease (138; cf 139).

Another important discovery made in the study with hexosaminidase was that the level of globoside (Figure 1 Row 9), which is increased several fold over normal in the blood of patients with Sandhoff-Jatzkewitz disease, had decreased 43% by 4 hr after injection of hexosaminidase A. Although the ability of hexosaminidase A to catalyze the cleavage of globoside is well known (44, 140), the important lesson in this experiment stemmed from the kinetics of the rapid uptake of the enzyme and the subsequent removal of globoside from the circulation. These observations indicate that hexosaminidase A exerted its catalytic effect after it had been taken up by tissues such as the liver. The demonstrated acidic pH optimum for this enzyme (44) suggests that it had been incorporated into lysosomes and carried out its catalytic function within these organelles (13).

FABRY'S DISEASE   The earliest attempt at enzyme replacement in Fabry's disease was carried out by Mapes and co-workers, who infused whole plasma into two Fabry hemizygotes (141). These authors reported a slow decrease of plasma CTH, which reached 50% of the preinfusion level

10 days following infusion. In addition a 22- to 35-fold augmentation of plasma ceramidetrihexosidase activity was observed. In a subsequent study, no decrease in circulating CTH was observed following infusion of fresh plasma (138), and other investigators have failed to verify an augmentation of circulating α-galactosidase activity [142, 143; see, however, (144)]. On the other hand, a decrease of plasma CTH did occur in a patient with Fabry's disease when infused with a leukocyte-enriched plasma suspension (138). However, suitable preparations of this type are difficult to obtain for long-range therapy because of the potential hazards of mismatching the cells. Accordingly, it seemed reasonable to determine the effect of purified ceramidetrihexosidase in patients with Fabry's disease. The enzyme was obtained from human placental tissue (145). Following infusion of the enzyme, a significant decrease of plasma CTH was observed in two patients with this disorder (138). The decrease occurred several hours after the enzyme was injected, and it was proportional to the amount of enzyme administered. There was no increase of ceramidelactoside, the immediate product of the reaction in the plasma. When one considers that placental ceramidetrihexosidase is virtually inactive at the pH of blood, these findings strongly indicate that this enzyme also exerted its catalytic activity extracirculatorily.

Several other observations made in the course of this investigation deserve comment. The first is the fact that a 4-fold enhancement of α-galactosidase activity was seen in a liver biopsy specimen obtained from one of the patients 1 hr after injection of enzyme. It was postulated that this augmentation of catalytic activity was due to an association of a monomer of active placental ceramidetrihexosidase with catalytically inactive subunits in the patient's tissue forming a heteropolymeric enzyme with restored catalytic configuration (137; see 146, 147). However, the fact that CRM has not been demonstrated in tissues of patients with Fabry's disease (73) lessens the attractiveness of this hypothesis at this time. In spite of this possible discrepancy in interpretation, certain deductions concerning the potential usefulness of enzyme replacement therapy in Fabry's disease seem warranted. Most of the CTH that accumulates in the blood vessels, kidneys, and other organs in these patients appear to be derived from neutral glycolipids present in the stroma of senescent erythrocytes. CTH presumably enters the circulation from sites of red cell catabolism such as the spleen and liver. Therefore, a sustained reduction in the quantity of CTH in the blood might be expected to exert a beneficial effect on the vascular and renal problems in these individuals. Second, both patients who received the placental enzyme were skin-tested a year following infusion for possible reaction to the exogenous protein. There was no indication whatsoever that these men had been sensitized to the placental preparation (4). Therefore, the long-range investigation of enzyme replacement therapy in Fabry's disease seems clearly indicated at this time. The principal difficulty at the moment is the lack of a large-scale procedure for the production of enzyme of acceptable purity for human administration. Investigations are underway in several laboratories to overcome this hindrance (50, 148).

GAUCHER'S DISEASE   Even more impressive findings have been obtained in investigations of enzyme replacement in Gaucher's disease. In the studies of Tay-Sachs disease and Fabry's disease a critical unanswered question was whether the administration of exogenous enzyme would reduce the quantity of accumulated material in the tissues of patients with sphingolipidoses. This problem has been resolved affirmatively in the study in Gaucher's disease. When two patients were infused with glucocerebrosidase isolated from human placental tissue (149), a 26% reduction in the quantity of accumulated glucocerebroside was observed in both recipients (150). Later, in a third patient who received the enzyme, only 8% of the accumulated glucocerebroside was cleared from the liver (151). However, 10 to 20 times more lipid has accumulated in this patient's liver than in the first two recipients. A constant relationship appeared with regard to the quantity of glucocerebroside catabolized and the amount of enzyme injected in all three trials. For each unit of enzyme infused, approximately 0.4 nmole of lipid was cleared from the liver (151). It should be noted that these patients had previously had their spleens removed. In a more recent study involving two patients whose spleens had not been excised, this proportionality did not obtain. In these individuals, liver glucocerebroside decreased 12 and 14%, respectively, following injection of glucocerebrosidase (R. O. Brady, J. A. Barranger, F. S. Furbish, A. E. Gal, P. G. Pentchev, in preparation). Further studies of the effects of splenectomy on enzyme replacement therapy in Gaucher's disease are underway.

Another impressive finding in enzyme replacement in Gaucher's disease was the return of the elevated quantity of glucocerebroside in the blood, in four of the five recipients, to the normal level by 72 hr after infusion of enzyme. This reduction persisted over a period of months (152). In the second recipient, it was still 20% lower than the preinfusion value more than 2.5 years later (4). This long-lasting effect provides much hope for enzyme replacement in Gaucher's disease. No decrease in blood glucocerebroside was observed in the third recipient. This finding was not surprising since glucocerebroside in the blood is apparently in equilibrium with that in tissues (153) and only a small reduction occurred percentagewise in the liver in this recipient. The duration of the decrease of circulating glucocerebroside in the two recipients whose spleens had not been removed is being followed.

Other groups are also engaged in enzyme replacement in Gaucher's disease, and reports of these activities are beginning to appear (47, 154). Probably the most encouraging observation is that of Belchetz and co-workers, who observed that the size of the liver decreased 3 cm in a patient who received intermittent infusions of glucocerebrosidase over a 13-month period (154). These investigators have also presented evidence for improved function of the patient's reticuloendothelial system following administration of enzyme. If this observation can be confirmed, it will provide further impetus for enzyme replacement trials in Gaucher's disease, since these cells have long been considered to be primarily involved in the storage of glucocerebroside (e.g. 155). Belchetz and co-workers encapsulated glucocerebrosidase in liposomes consisting of egg phosphatidyl choline and cholesterol or phosphatidyl choline, cholesterol, and phosphatidic acid. The recipient experienced considerable pain after the injections, which was attributed to the use of "unsonicated" liposomes. In our hands, solutions of purified glucocerebrosidase stabilized with human serum albumin caused no untoward effects. However, the optimal method of administering the enzyme is not yet established and this aspect is presently under investigation (47, 150, 156).

In summary, evidence is now accumulating that the administration of purified glucocerebrosidase exerts a beneficial effect in Gaucher's disease and that these results are of impressive duration. Since a high-yield large-scale procedure is now available for the isolation of human placental glucocerebrosidase (47), investigators are now in a position to carefully assess the clinical effectiveness of this therapeutic modality. The findings in these critical investigations will have much impact on the future of this form of treatment for heritable metabolic disorders.

Further Strategies

Despite the optimism surrounding the possibilities of enzyme replacement for patients with Gaucher's disease and Fabry's disease, who, in fact, comprise the majority of patients with lipid storage diseases encountered clinically, similar success has not occurred in disorders where the central nervous system is damaged. At this time, reliance on genetic counseling is required to provide relief to the families in which these conditions occur. A major research interest continues to be the possibility of devising a safe procedure for delivering enzymes to the brain. Two recent developments, which are still in the preliminary stage, indicate that this feat may be possible. The first was the demonstration that α-mannosidase, which is normally present in relatively low activity in the brain, can enter the central nervous system if the blood-brain barrier is temporarily altered by injecting hyperosmotic solutions of mannitol into the external carotid artery (157).

The level of mannosidase activity in the brain increased from 1250 to 2467 units per gram wet weight, an amount equivalent to the normal physiological complement. Thus, this procedure should permit full restoration of enzymatic activity in a totally deficient situation.

The second series deals with altering the blood-brain barrier by exposing animals to intermittent periods of hyperbaric oxygen (158). Both techniques have inherent potential dangers, and investigators must be ever mindful of the possibility of developing alternative procedures. Furthermore, there are several other troublesome aspects of these investigations that must be carefully considered. The first is the concern whether neuronal cells will take up exogenous enzymes such as hexosaminidase. We have examined the effect of adding purified human placental hexosaminidase to the tissue culture medium bathing neuroblastoma cells, and, so far, we have not been able to demonstrate uptake of this enzyme by these cells (J. W. Kusiak, C. J. Lauter, R. O. Brady, unpublished observations). Current research is directed towards modifying the carbohydrate portion of the enzyme to try to obtain a preparation that will be endocytosed (cf 159). The second reservation, which applies primarily to hexosaminidase A, is the extraordinarily low activity of the purified enzyme with G_{M2} (37, 44). A primary concern is whether this enzyme preparation will exert catalytic activity with the natural lipid in vivo. An early experiment along this line failed to show restoration of G_{M2} catabolizing activity when purified hexosaminidase A was added to brain lysosomes obtained from a patient with Sandhoff-Jatzkewitz disease (160).

On the other hand, hope that this approach may eventually be useful was raised by Feder's demonstration of an apparent exchange of glucuronidase between brain cells in vivo (161). Another important experiment along this line was reported by Reuser and co-workers, who co-cultivated human skin fibroblasts from normal individuals and patients with Sandhoff-Jatzkewitz disease, G_{M1}-gangliosidosis, and glycogenosis II (162). These investigators obtained evidence for the intercellular exchange of hexosaminidase, but not β-galactosidase or α-glucosidase. It remains to be determined whether the carbohydrate composition of the latter enzymes precludes their uptake, or if there is a defect in recognition sites on the surface of the deficient cells. If the latter condition obtains, enzyme replacement therapy would appear to be remote for disorders of this type.

A number of other concepts have been advanced as potential approaches for the treatment of inherited metabolic disorders (66, 163–165). Chief among these is the possibility of allosterically modifying the patient's mutated enzyme so that sufficient catalytic activity will be restored. The enhancement of enzymatic activity described in the sections on the therapy of Tay-Sachs disease and Fabry's disease seems to indicate that such an approach may be reasonable. If this approach is attempted, one must not engage in a trade-off through the use of a nonmetabolizable allosteric modifier, and careful toxicity studies will be required.

In therapeutic trials such as those outlined above, the prospect of using animal models of human disorders is becoming exceedingly important. Sporadic reports of mutations in animals that resemble human diseases have appeared over the years (166); the latest is a cat with total hexosaminidase deficiency (167). These analogues will be extremely helpful if they can be reproducibly obtained, as in the case of the canine form of Krabbe's disease (168). A particularly important point in this regard is the lingering uncertainty of how much enzyme will be required to restore normal function in deficient humans. For example, no data are currently available on enzyme levels in organs and tissues of human heterozygotes. Since it has been known for years that the activity of sphingolipid hydrolases can be induced by presentation of augmented quantities of material to be metabolized (169), it has been assumed that homozygotes exist under conditions of maximal induction. Perhaps a partial answer to this question is now available. The level of galactocerebrosidase activity in the spleen and liver of neonatal dogs heterozygous for canine globoid leukodystrophy (Krabbe's disease) was only slightly less than that in normals, whereas it was only 50% of that in controls in the kidney and brain. Although the activity of this enzyme in brain increased with age even in the homozygous dogs, the absence of central nervous system damage in the neonatal heterozygotes may indicate that half normal enzymatic activity may be compatible with unimpaired function and that this figure might be used as a rough index of the amount of exogenous enzyme that will be required by patients with sphingolipidoses. However, it must be emphasized that this speculation is tenuous at best since there are discrepancies between the canine and human forms of this disorder (170). Certainly, the level of enzyme activities in tissues of heterozygotes for other disorders will have to be determined to substantiate the reasonableness of this deduction. Since automated enzyme assays are readily available for most of these conditions (see section on diagnosis), the minimal effort and funds required to screen surplus domestic animals for their potential carrier status is warranted, and this project should be undertaken forthwith.

A more sophisticated approach may be required for the development of useful conventional laboratory animal models such as rats or mice. An important step in this direction has recently been achieved with the demonstration that injection of AY-9944 (*trans*-1,4 bis[2-chlorobenzylaminomethyl]cyclohexane dihydrochloride) into neonatal rats causes a marked reduction of sphingomyelinase activity in the tissues of these animals (171). Furthermore, there was a 292% increase in sphingomyelin with fatty acids less than 20 carbons in length in the liver of animals treated for 17 days with this reagent. This species of sphingomyelin is particularly increased in patients with Niemann-Pick disease (172). Since the compound had no effect on the catalytic activity of sphingomyelinase in vitro, and there was no evidence of the formation of an inhibitor of sphingomyelinase in vivo, it appears that the synthesis of this enzyme was impaired. These circumstances provide a useful, acute, animal model of Niemann-Pick disease. This development is particularly welcome since even though sphingomyelinase has been purified to apparent homogeneity (173), the low yield of pure enzyme precludes studies of enzyme replacement in patients with Niemann-Pick disease at this time.

More recently, another promising approach to the production of animal models of lipid storage diseases has appeared. These analogues are based on the demonstration of Mintz & Illmensee of the totipotency of teratocarcinoma cells, which, when fused with mouse blastocysts, result in the production of normal animals capable of reproduction (174, 175). In order to utilize this principle, a tissue culture medium has been developed that apparently selectively permits the growth of glucocerebrosidase-less mutated cells (R. O. Brady, A. E. Gal, P. G. Pentchev, J. K. Steusing, in preparation). The combination of these techniques should permit the development of a strain of mice with a condition analogous to Gaucher's disease in humans. Once demonstrated, this principle can be extended to the development of other animal models of sphingolipidoses. Their availability will materially accelerate research in the lipid storage diseases.

CONCLUDING REMARKS

Current investigations in the lipid storage diseases may be subdivided into several areas. The first is the practical application of the discoveries of the enzymatic defects in these disorders, which were made over the past 12 years through delivery and improvement of diagnostic and carrier-screening procedures on an ever-widening scale. These tests and the accurate genetic counseling made possible by them have helped thousands of heterozygous couples to have healthy normal families, a truly impressive feat over a relatively short period of time. The second aspect concerns the examinations that are beginning to be made of the nature of the mutations that occur in these disorders. Although most of these conditions appear to be the result of alterations of the structure of enzymes, the possibility of a regulatory mutation cannot be excluded in some of these conditions such as Fabry's disease. The third line of investigation deals with the studies on enzyme replacement therapy in these disorders. The development of a large-scale procedure for the isolation of the requisite enzyme has been reported only for Gaucher's disease. Much additional work will be required before this goal is realized in the other disorders. However, the encouraging results obtained in enzyme replacement in Gaucher's disease and Fabry's disease mandate that these efforts be continued. Finally, additional procedures for the selective removal of accumulating lipids from the tissues of patients with sphingolipidoses now in the planning stage will be assayed in the future.

Literature Cited

1. Neufeld, E. F., Lim, L., Shapiro, L. J. 1975. *Ann. Rev. Biochem.* 44:357–76
2. Desnick, R. J., Thorpe, S. R., Fiddler, M. B. 1976. *Physiol. Rev.* 56:57–99
3. Sandhoff, K. 1977. *Angew. Chem. Int. Ed. Engl.* 16:273–85
4. Brady, R. O. 1977. *Metab. Clin. Exp.* 26:329–45
5. Pentchev, P. G., Barranger, J. A. 1978. *J. Lipid Res.* In press
6. Stoffel, W. 1973. *Chem. Phys. Lipids* 11:318–34
7. IUPAC-IUB Commission on Biochemical Nomenclature. 1977. *Lipids* 12:455–68
8. Sweeley, C. C., Siddiqui, B. 1977. In *Biochemistry of Mammalian Glycoproteins and Glycolipids,* ed. W. Pigman, M. I. Horowitz, pp. 459–540. New York: Academic
9. Roseman, S. 1970. *Chem. Phys. Lipids* 5:270–87
10. Fishman, P. H., Brady, R. O. 1976. *Science* 194:906–15
11. Fishman, P. H., Max, S. R., Tallman, J. F., Brady, R. O., Maclaren, N. K., Cornblath, M. 1975. *Science* 187:68–70
12. Kolodny, E. H., Kanfer, J. N., Quirk, J. M., Brady, R. O. 1971. *J. Biol. Chem.* 246:1426–31
13. Tallman, J. F., Brady, R. O. 1972. *J. Biol. Chem.* 247:7570–75
14. Sandhoff, K., Harzer, K., Wässle, W., Jatzkewitz, H. 1971. *J. Neurochem.* 18:2469–89
15. Svennerholm, L. 1962. *Biochem. Biophys. Res. Commun.* 9:436–41
16. Wenger, D. A. 1974. *Chem. Phys. Lipids* 13:327–39
17. Tanaka, H., Suzuki, K. 1975. *J. Biol. Chem.* 250:2324–32
18. Ben-Joseph, Y., Shapira, E., Edelman, D., Burton, B. K., Nadler, H. L. 1977. *Arch. Biochem. Biophys.* 184:373–79
19. Gatt, S. 1963. *J. Biol. Chem.* 238: PC3131–33
20. Nilsson, A. 1969. *Biochim. Biophys. Acta* 176:339–47
21. Sugita, M., Williams, M., Dulaney, J. T., Moser, H. W. 1975. *Biochim. Biophys. Acta* 398:125–31
22. Stoffel, W., Sticht, G., LeKim, D. 1968. *Hoppe-Seyler's Z. Physiol. Chem.* 349:1745–48
23. Keenan, R. W., Maxam, A. 1969. *Biochim. Biophys. Acta* 176:348–56
24. Brady, R. O., Kanfer, J. N., Shapiro, D. 1965. *J. Biol. Chem.* 240:39–43
25. Norden, A. G. W., O'Brien, J. S. 1973. *Arch. Biochem. Biophys.* 159: 383–92
26. Mehl, E., Jatzkewitz, H. 1968. *Biochim. Biophys. Acta* 151:619–27
27. Fischer, G., Jatzkewitz, H. 1975. *Hoppe-Seyler's Z. Physiol. Chem.* 356: 605–13
28. Mraz, W., Fischer, G., Jatzkewitz, H. 1976. *Hoppe-Seyler's Z. Physiol. Chem.* 357:1181–91
29. Weinreb, N. J., Brady, R. O., Tappel, A. L. 1968. *Biochim. Biophys. Acta* 159:141–46
30. Fischer, G., Jatzkewitz, H. 1977. *Biochim. Biophys. Acta* 481:561–72
31. Ho, M. W., O'Brien, J. S. 1971. *Proc. Natl. Acad. Sci. USA* 68:2810–13
32. Ho, M. W., O'Brien, J. S., Radin, N. S., Erickson, J. S. 1973. *Biochem. J.* 131:173–76
33. Ho, M. W., Light, N. D. 1973. *Biochem. J.* 136:821–23
34. Ho, M. W., Rigby, M. 1975. *Biochim. Biophys. Acta* 397:267–73
35. Peters, S. P., Coyle, P., Coffee, C. J., Glew, R. H., Kuhlenschmidt, M. S., Rosenfeld, L., Lee, Y. C. 1977. *J. Biol. Chem.* 252:563–73

36. Peters, S. P., Coffee, C. J., Glew, R. H., Lee, R. E., Wenger, D. A., Li, S.-C., Li, Y.-T. 1977. *Arch. Biochem. Biophys.* 183:290–97

37. Li, Y.-T., Mazzotta, M. Y., Wan, C.-C., Orth, R., Li, S.-C. 1973. *J. Biol. Chem.* 248:7512–15

38. Li, S.-C., Wan, C.-C., Mazzotta, M. Y., Li, Y.-T. 1974. *Carbohydr. Res.* 34:189–93

39. Mraz, W., Fischer, G., Jatzkewitz, H. 1976. *FEBS Lett.* 67:104–9

40. Li, S.-C., Li, Y.-T. 1976. *J. Biol. Chem.* 251:1159–63

41. Hechtman, P. 1977. *Can. J. Biochem.* 55:315–24

42. Hechtman, P., LeBlanc, D. 1977. *Biochem. J.* 167:693–701

43. Baraton, G., Revol, A. 1977. *Clin. Chim. Acta* 76:339–43

44. Tallman, J. F., Brady, R. O., Quirk, J. M., Villalba, M., Gal, A. E., 1974. *J. Biol. Chem.* 249:3489–99

45. Pentchev, P. G., Brady, R. O. 1973. *Biochim. Biophys. Acta* 297:491–96

46. Dale, G. L., Beutler, E. 1976. *Proc. Natl. Acad. Sci. USA* 73:4672–74

47. Furbish, F. S., Blair, H. E., Shiloach, J., Pentchev, P. G., Brady, R. O. 1977. *Proc. Natl. Acad. Sci. USA* 74:3560–63

48. Goldstone, A., Konecny, P., Koenig, H. 1971. *FEBS Lett.* 13:68–72

49. Beutler, E., Guinto, E., Kuhl, W. 1975. *J. Lab. Clin. Med.* 85:672–77

50. Kusiak, J. W., Quirk, J. M., Brady, R. O. 1978. *J. Biol. Chem.* 253:184–90

51. Graham, F. R. B., Roy, A. B. 1973. *Biochim. Fiophys. Acta* 329:88–92

52. Balasubra nanian, K. A., Bachhawat, B. K. 1975. *Biochim. Biophys. Acta* 403:113–21

53. Tarentino, A. L., Maley, F. 1974. *J. Biol. Chem.* 249:811–817

54. Nishigaki, M., Muramatsu, T., Kobata, A. 1974. *Biochem. Biophys. Res. Commun.* 59:638–45

55. Tsay, G. C., Dawson, G., Sung, S.-S. J. 1976. *J. Biol. Chem.* 251:5852–59

56. Cumar, F. A., Fishman, P. H., Brady, R. O. 1971. *J. Biol. Chem.* 246:5075–5084

57. Mullin, B. R., Fishman, P. H., Lee, G., Aloj, S. M., Ledley, F. M., Winand, R. J., Kohn, L. D., Brady, R. O. 1976. *Proc. Natl. Acad. Sci. USA* 73:842–46

58. Lee, G., Aloj, S. M., Kohn, L. D. 1977. *Biochem. Biophys. Res. Commun.* 77:434–41

59. Dawson, G., Stoolmiller, A. C., Radin, N. S. 1974. *J. Biol. Chem.* 249:4638–46

60. Burton, B. K., Ben-Joseph, Y., Nadler, H. L. 1977. *Lactosyl Ceramidosis: a Deficiency of Neutral β-Galactosidase in Liver and Cultivated Fibroblasts.* Presented at Ann. Meet. Am. Soc. Hum. Genet., *29th,* San Diego

61. Wenger, D. A., Sattler, M., Clark, C., Tanaka, H., Suzuki, K., Dawson, G. 1975. *Science* 188:1310–12

62. Brady, R. O. 1973. *Adv. Enzymol.* 38:293–315

63. Brady, R. O. 1974. *Chem. Phys. Lipids* 13:271–82

64. Brady, R. O. 1976. *Arch. Neurol.* 33:145–51

65. Brady, R. O. 1976. In *Basic Neurochemistry,* ed. G. J. Siegel, R. W. Albers, R. Katzman, B. W. Agranoff, pp. 556–68. Boston: Little, Brown. 825 pp. 2nd ed.

66. Brady, R. O. 1966. *N. Engl. J. Med.* 275:312–18

67. Stumpf, D., Neuwelt, E., Austin, J., Kohler, P. 1971. *Arch. Neurol.* 25:427–31

68. Meisler, M., Rattazzi, M. C. 1974. *Am. J. Hum. Genet.* 26:683–91

69. O'Brien, J. S., Norden, A. G. W. 1977. *Am. J. Hum. Genet.* 29:184–90

70. Norden, A. G. W., O'Brien, J. S. 1975. *Proc. Natl. Acad. Sci. USA* 72:240–44

71. Srivastava, S. K., Beutler, E. 1974. *J. Biol. Chem.* 249:2054–57

72. Rietra, P. J. G. M., Molenaar, J. L., Hamers, M. N., Tager, J. M., Borst, P. 1974. *Eur. J. Biochem.* 46:89–98

73. Schram, A. W., Hamers, M. N., Brouwer-Kelder, B., Donker-Koopman, W. E., Tager, J. M. 1977. *Biochim. Biophys. Acta* 482:125–37

74. Okada, S., O'Brien, J. S. 1969. *Science* 165:698–700

75. Sandhoff, K., Andreae, U., Jatzkewitz, H. 1968. *Life Sci.* 7:278–85

76. Thomas, G. H., Taylor, H. A. J., Miller, C. S., Axelman, J., Migeon, B. R. 1974. *Nature* 250:580–82

77. Galjaard, H., Hoogeveen, A., deWit-Verbeek, H. A., Reuser, A. J. J., Keijzer, W., Westerveld, A., Bootsma, D. 1974. *Exp. Cell. Res.* 87:444–48

78. Ropers, H. H., Grzeschik, K. H., Bühler, E. 1975. *Humangenetik* 26:117–21

79. Rattazzi, M. C., Brown, J. A., Davidson, R. G., Shows, T. B. 1976. *Am. J. Hum. Genet.* 28:143–54

80. Beutler, E., Kuhl, W. 1975. *Nature* 258:262–64

81. Srivastava, S. K., Wiktorowicz, J. E., Awasthi, Y. C. 1976. *Proc. Natl. Acad. Sci. USA* 73:2833–37

82. Kucherlapati, R. P., Ruddle, F. H. 1976. *Cytogenet. Cell Genet.* 16:181–83

83. Hoeksema, H. L., Reuser, A. J. J., Hoogeveen, A., Westerveld, A., Braidman, I., Robinson, D. 1977. *Am. J. Hum. Genet.* 29:14–23

84. Gilbert, F., Kucherlapati, R., Creagan, R. P., Murnane, M. J., Darlington, G. J., Ruddle, F. H. 1975. *Proc. Natl. Acad. Sci. USA* 72:263–67

85. Lalley, P. A., Brown, J. A., Shows, T. B. 1976. *Cytogenet. Cell Genet.* 16:188–91

86. Turner, V. S., Turner, B. M., Kucherlapati, R. P., Ruddle, F. H., Hirschhorn K. 1976. *Proc. 3rd Int. Workshop Hum. Gene Mapp.,* Baltimore, 1975 pp. 238–40

87. Kampine, J. P., Brady, R. O., Kanfer, J. N., Feld, M., Shapiro, D. 1967. *Science* 155:86–88

88. Sloan, H. R., Uhlendorf, B. W., Kanfer, J. N., Brady, R. O., Fredrickson, D. S. 1969. *Biochem. Biophys. Res. Commun.* 34:582–88

89. O'Brien, J. S., Okada, S., Chen, A., Fillerup, D. L. 1970. *N. Engl. J. Med.* 283:15–20

90. Navon, R., Padeh, B. 1972. *J. Pediatr.* 80:1026–30

91. Carmody, P. J., Rattazzi, M. C., Davidson, R. G. 1973. *N. Engl. J. Med.* 289:1072–74

92. Singer, J. D., Cotlier, E., Kummer, R. 1973. *Lancet* 2:1116

93. Grim, T., Wienker, T. F., Ropers, H. H. 1976. *Hum. Genet.* 32:329–34

94. Lowden, J. A., Skomorowski, M. A., Henderson, F., Kaback, M. 1973. *Clin. Chem. Winston-Salem NC* 19:1345–49

95. Saifer, A., Parkhurst, G. W., Amoroso, J. 1975. *Clin. Chem.* 21:334–42

96. Kolodny, E. H., Mumford, R. A. 1976. *Clin. Chim. Acta* 70:247–57

97. Peters, S. P., Lee, R. E., Glew, R. H. 1975. *Clin. Chim. Acta* 60:391–96

98. Brady, R. O., Johnson, W. G., Uhlendorf, B. W. 1971. *Am. J. Med.* 51:423–31

99. Gal, A. E., Fash, F. J. 1976. *Chem. Phys. Lipids* 16:71–79

100. Gal, A. E., Brady, R. O., Hibbert, S. R., Pentchev, P. G. 1975. *N. Engl. J. Med.* 293:632–36

101. Gal, A. E., Brady, R. O., Pentchev, P. G., Furbish, F. S., Suzuki, K., Tanaka, H., Schneider, E. L. 1977. *Clin. Chim. Acta* 77:53–59

102. Gal, A. E., Pentchev, P. G., Fash, F. J. 1976. *Proc. Soc. Exp. Biol. Med.* 153:363–66

103. Kaback, M. M., Rimoin, D. L., O'Brien, J. S. 1977. *Tay-Sachs Disease: Screening and Prevention.* New York: Alan R. Liss. 450 pp.

104. Schneider, E. L., Ellis, W. G., Brady, R. O., McCulloch, J. R., Epstein, C. J. 1972. *J. Pediatr.* 81:1134–39

105. Milunsky, A. 1975. *The Prevention of Genetic Disease and Mental Retardation,* p. 239. Philadelphia: Saunders. 506 pp.

106. Navon, R., Padeh, B., Adam, A. 1973. *Am. J. Hum. Genet.* 25:287–93

107. Vidgoff, J., Buist, N. R. M., O'Brien, J. S. 1973. *Am. J. Hum. Genet.* 25:372–81

108. Dreyfus, J.-C., Poenaru, L., Svennerholm, L. 1975. *N. Engl. J. Med.* 292:61–63

109. Tallman, J. F., Brady, R. O., Navon, R., Padeh, B. 1974. *Nature* 252:254–55

110. Navon, R., Geiger, B., Ben-Joseph, Y., Rattazzi, M. C. 1976. *Am. J. Hum. Genet.* 28:339–49

111. O'Brien, J. S., Norden, A. G. W., Miller, A. L., Frost, R. G., Kelley, T. E. 1977. *Clin. Genet.* 11:171–83

112. Bach, G., Suzuki, K. 1975. *J. Biol. Chem.* 250:1328–32

113. Tallman, J. F., Kolodny, E. H., Brady, R. O. 1975. *Methods Enzymol.* 35:541–48

114. Kolodny, E. H., Brady, R. O., Quirk, J. M., Kanfer, J. N. 1970. *J. Lipid Res.* 11:144–49

115. Schneck, L., Valenti, C., Amsterdam, D., Friedland, J., Adachi, M., Volk, B. W. 1970. *Lancet* 1:582–83

116. Brady, R. O. 1974. In *Clinical Biochemistry: Principles and Methods,* ed. H. C. Curtius, M. Roth, pp. 1277–91. New York: de Gruyter. 1677 pp.

117. Rattazzi, M. C., Davidson, R. G. 1977. *Pediatr. Res.* 11:1072–77

118. Moss, M. L., Harris, W. W. 1972. *Enzyme Assays in Isolated Amniotic Cells.* Presented at Symp. Recent Dev. Res. Methods Instrum., Bethesda, Md. p. 13

119. Brady, R. O. 1973. *Sci. Am.* 229:88–97

120. Galjaard, H., Niermeijer, M. F., Hahnemann, N., Mohr, J., Sorenson, S. A. 1974. *Clin. Genet.* 5:368–77

121. Hösli, P. 1974. In *Birth Defects,* ed. A. Motulsky, W. Lenz, pp. 226–33. Amsterdam: Excerpta Med. 373 pp.

122. Galjaard, H., Hoogeveen, E., van der Veer, E., Kleijer, W. J. 1977. In *Human Genetics,* ed. S. Armendares, R. Lisker, 5:194–206. Amsterdam: Excerpta Med.

123. Hösli, P. 1976. *Adv. Exp. Med. Biol.* 68:1–13

124. Hösli, P., Erickson, R. P., Vogt, E. 1976. *Biochem. Biophys. Res. Commun.* 73:209–16

125. Brady, R. O., Gal, A. E., Pentchev, P. G. 1974. *Life Sci.* 15:1235–48
126. Groth, C. G., Hagenfeldt, L., Dreborg, S., Löfstrom, B., Öckerman, P. A., Samuelsson, K., Svennerholm, L., Werner, B., Westberg, G. 1971. *Lancet* 1:1260–64
127. Groth, C. G., Bergström, K., Collste, L., Egberg, N., Högman, C., Holm, G., Möller, E. 1972. *Clin. Exp. Immunol.* 10:359–65
128. Desnick, S. J., Desnick, R. J., Brady, R. O., Pentchev, P. G., Simmons, R. L., Najarian, J. S., Swaiman, K., Sharp, H. L., Krivit, W. 1973. *Birth Defects Orig. Artic. Ser.* 9:109–19
129. Desnick, R. J., Simmons, R. L., Allen, K. Y., Woods, J. E., Anderson, C. F., Najarian, J. S., Krivit, W. 1972. *Surgery* 72:203–11
130. Philippart, M., Franklin, S. S., Gordon, A. 1972. *Ann. Intern. Med.* 77:195–200
131. Clarke, J. T. R., Guttmann, R. D., Wolfe, L. S., Beaudoin, J. G., Morehouse, D. D. 1972. *N. Engl. J. Med.* 287:1215–18
132. Spence, M. W., MacKinnon, K. E., Burgess, J. K., d'Entremont, D. M., Belitsky, P., Lannon, S. G., MacDonald, A. S. 1976. *Ann. Intern. Med.* 84:13–16
133. Van den Bergh, F. A. J. T. M., Rietra, P. J. G. M., Kolk-Vegter, A. J., Bosch, E., Tager, J. M. 1976. *Acta Med. Scand.* 200:249–56
134. Delvin, E., Glorieux, F., Daloze, P., Gorman, J., Bloch, P. 1974. *Am. J. Hum. Genet.* 26:25 (Abstr.)
135. Porter, M. T., Fluharty, A. L., Kihara, H. 1971. *Science* 172:1263–65
136. Johnson, W. G., Desnick, R. J., Long, D. M., Sharp, H. L., Krivit, W., Brady, B., Brady, R. O. 1973. *Birth Defects Orig. Artic. Ser.* 9:120–24
137. Brady, R. O., Pentchev, P. G., Gal, A. E. 1975. *Fed. Proc.* 34:1310–15
138. Brady, R. O., Tallman, J. F., Johnson, W. G., Gal, A. E., Leahy, W. R., Quirk, J. M., Dekaban, A. S. 1973. *N. Engl. J. Med.* 289:9–14
139. Bakay, B., Nyhan, W. L. 1972. *Proc. Natl. Acad. Sci. USA* 69:2523–27
140. Sandhoff, K., Wässle, W. 1971. *Hoppe-Seyler's Z. Physiol. Chem.* 352:1119–33
141. Mapes, C. A., Anderson, R. L., Sweeley, C. C., Desnick, R. J., Krivit, W. 1970. *Science* 169:987–89
142. Christensen, E. 1974. *N. Engl. J. Med.* 290:630–31
143. Patrick, A. D. 1974. In *Enzyme Therapy in Lysosomal Storage Diseases*, ed. J. M. Tager, G. J. M. Hoogwinkel, W. T. Daems, p. 300. Amsterdam: North Holland. 301 pp.
144. Desnick, R. J. 1974. See Ref. 143, p. 300
145. Johnson, W. G., Brady, R. O. 1972. *Methods Enzymol.* 28:849–56
146. Rotman, B., Celada, F. 1968. *Proc. Natl. Acad. Sci. USA* 60:660–67
147. Melchers, F., Messer, W. 1970. *Eur. J. Biochem.* 17:267–72
148. Mayes, J. S., Beutler, E. 1977. *Biochim. Biophys. Acta* 484:408–16
149. Pentchev, P. G., Brady, R. O., Hibbert, S. R., Gal, A. E., Shapiro, D. 1973. *J. Biol. Chem.* 248:5256–61
150. Brady, R. O., Pentchev, P. G., Gal, A. E., Hibbert, S. R., Dekaban, A. S. 1974. *N. Engl. J. Med.* 291:989–93
151. Brady, R. O., Pentchev, P. G., Gal, A. E., Hibbert, S. R., Quirk, J. M., Mook, G. E., Kusiak, J. W., Tallman, J. F., Dekaban, A. S. 1976. *Adv. Exp. Med. Biol.* 68:523–32
152. Pentchev, P. G., Brady, R. O., Gal, A. E., Hibbert, S. R. 1975. *J. Mol. Med.* 1:73–78
153. Dawson, G., Sweeley, C. C. 1970. *J. Biol. Chem.* 245:410–16
154. Belchetz, P. E., Braidman, I. P., Crawley, J. C. W., Gregoriadis, G. 1977. *Lancet* 2:116–17
155. Brady, R. O., King, F. M. 1973. In *Lysosomes and Storage Diseases*, ed. H. G. Hers, F. Van Hoof, pp. 381–94. New York: Academic. 666 pp.
156. Ihler, G. M., Glew, R. H., Schnure, F. W. 1973. *Proc. Natl. Acad. Sci. USA.* 70:2663–66
157. Barranger, J. A., Pentchev, P. G., Rapoport, S. I., Brady, R. O. 1977. *Ann. Neurol.* 1:496 (Abstr.)
158. Rattazzi, M. C., McCullough, R. A., Lanse, S. B., Jacobs, E. A. 1977. Fifth Int. Conf. Birth Defects, Montreal, Aug. 21–27. 73 pp. (Abstr.)
159. Kaplan, A., Achord, D. T., Sly, W. S. 1977. *Proc. Natl. Acad. Sci. USA* 74:2026–30
160. Tallman, J. F., Johnson, W. G., Brady, R. O. 1972. *J. Clin. Invest.* 51:2339–45
161. Feder, N. 1976. *Nature* 263:67–69
162. Reuser, A., Halley, D., deWit, E., Hoogeveen, A., van der Kamp, M., Mulder, M., Galjaard, H. 1976. *Biochem. Biophys. Res. Commun.* 69:311–18
163. Brady, R. O. 1973. *Angew. Chem. Int. Ed. Engl.* 12:1–11
164. Brady, R. O. 1976. *Science* 193:733–39
165. Peters, S. P., Lee, R. E., Glew, R. H. 1977. *Medicine (Baltimore)* 56:425–42
166. Baker, H. J., Mole, J., Lindsey, J. R.,

Creel, R. M. 1976. *Fed. Proc.* 35:1193–1201

167. Cork, L. C., Munnell, J. F., Lorenz, M. D., Murphy, J. V., Baker, H. J., Rattazzi, M. C. 1977. *Science* 196:1014–17

168. Fletcher, T. F., Suzuki, K., Martin, F. B. 1977. *Neurology* 27:758–66

169. Kampine, J. P., Kanfer, J. N., Gal, A. E., Bradley, R. M., Brady, R. O. 1967. *Biochim. Biophys. Acta* 137:135–39

170. Suzuki, Y., Austin, J., Armstrong, D., Suzuki, K., Schlenker, J., Fletcher, T. 1970. *Exp. Neurol.* 29:65–75

171. Sakuragawa, N., Sakuragawa, M.,

Kuwabara, T., Pentchev, P. G., Barranger, J. A., Brady, R. O. 1977. *Science* 196:317–19

172. Brunngraber, E. G., Berran, B., Zambotti, V. 1973. *Clin. Chim. Acta* 48:173–81

173. Pentchev, P. G., Brady, R. O., Gal, A. E., Hibbert, S. R. 1977. *Biochim. Biophys. Acta* 488:312–21

174. Mintz, B., Illmensee, K. 1975. *Proc. Natl. Acad. Sci. USA* 72:3585–89

175. Illmensee, K., Mintz, B. 1976. *Proc. Natl. Acad. Sci. USA* 73:549–53

Ann. Rev. Biochem. 1978. 47:715–50

ANIMAL CELL CYCLE ❖987

Arthur B. Pardee, Robert Dubrow, Joyce L. Hamlin, and Rolf F. Kletzien

Department of Basic Sciences, Sidney Farber Cancer Institute, and Department of Pharmacology, Harvard Medical School, Boston, Massachusetts 02115

CONTENTS

0066-4154/78/0701-0715$01.00

PERSPECTIVES AND SUMMARY

The observation that animal cells duplicate their DNA during a discrete interval in interphase allowed the cell cycle to be divided into four classical phases: G_1, S (DNA synthetic period), G_2, and M (mitosis). G_1 is the gap period between mitosis and the initiation of DNA synthesis; G_2 is the period between S and M. For most growing cell lines in tissue culture the interval between divisions is 10–30 hr.

Variation in cell cycle times among different cell types is mainly due to variation in the length of G_1, with the duration of S (6–8 hr) + G_2 (2–6 hr) + M (1 hr) being relatively constant. In addition, there is much variability in the length of G_1 among individual cells in a single population. This variability has been explained by phenotypic variation in cells at birth. An alternative model proposes that the nature of the cell cycle is such that cells exit from G_1 into S with a constant probability per unit time, thus creating an inherent G_1 variability. Differences in generation times among populations would be accounted for by differences in the magnitude of this transition probability.

Animal cells can also exist in a nongrowing quiescent state during which they do not divide for long periods. Under most circumstances, normal cells that have ceased to grow have the G_1 content of DNA. Whether these cells have left the cell cycle to enter a qualitatively distinct G_0 state or are arrested in a prolonged G_1 is a subject of debate.

The crucial control events for the regulation of growth seem to reside in G_1. Evidence has accumulated for the existence of a restriction or commitment point in mid- to late G_1, at which time a cell decides whether to initiate DNA synthesis and undergo division or to cease proliferation. Environmental conditions influence this decision; suboptimal growth conditions shift normal cells into quiescence. Transformed cells can lose this restriction-point control in whole or in part.

In order to perform biochemical studies on the cell cycle, it is generally necessary to obtain a population of cells that is synchronous with respect to the cell cycle. This can be done by selectively detaching mitotic cells from the growth surface and replating them, by blocking cells at a specific point in the cycle with a drug and then releasing them, by using serum or amino acid limitation or growth to confluence to shift cells into quiescence and then stimulating them to grow by the addition of complete medium, or by various combinations of these methods.

When quiescent cells are stimulated to divide, an array of biochemical changes occur at various times before the initiation of DNA synthesis. This set of metabolically unrelated biochemical reactions that fluctuate coordinately with changes in cell growth has been termed the "pleiotypic response" (1). It has been a difficult task to distinguish which of the many observed changes during the G_0 to S or the M to S transition are either necessary or sufficient for entry into S. We do not know whether the animal cell cycle is a linear sequence of dependent events or whether it consists of several relatively independent biochemical pathways.

Two approaches to the study of causal relationships in this process are the isolation of temperature-sensitive mutants that are blocked at a specific point in the cell cycle, and the study of the effects of drugs that inhibit specific biochemical processes. The study of differences between normal and transformed cells also suggests which biochemical parameters are important in growth regulation, as does the study of the factors that stimulate quiescent cells to proliferate.

The literature on the cell cycle has been well covered through about 1974. Special attentton is drawn to reviews by Baserga (2), Prescott (3, 4), and Mitchison (5). We confine our review mainly to articles appearing between 1975 and mid-1977. We stress the biochemical aspects, including events that may soon be open to biochemical study, and we deal mainly with the cycle of mammalian lines in culture, since most data have been obtained with these cells. Information on lymphocytes, chick cells, and lower eukaryotes is not reviewed extensively here, nor do we consider the numerous studies in which normal and transformed cells are compared (6, 7).

G_0 AND G_1

Normal cells make a choice between proliferation or quiescence when they reach G_1. According to one view, quiescent cells withdraw from the cell cycle into a qualitatively distinct G_0 state. Definite kinetic and biochemical differences between G_0 and G_1 cells support this alternative (reviewed in 2, 4). G_0 cells take longer to reach S than do G_1 cells progressing to S from mitosis (e.g. 8, 9). WI-38 human fibroblasts kept quiescent for prolonged periods entered a progressively deeper G_0 state characterized by reduced transcriptional activity. From this state the cells were increasingly difficult to rescue (10).

The alternative view holds that quiescent cells slowly traverse the cell cycle with a greatly extended G_1 period (11, 12). Up to 96% of quiescent chick embryo cells become labelled after 120 hr of continuous exposure to ^{3}H-thymidine (12). No kinetic evidence was found for an "out-of-cycle" G_0 compartment in 3T3, 3T6, or SV3T3 cells (13).

These two views can be reconciled by the probabilistic cell cycle model (14). According to this model, cells after mitosis enter an "A" state, in which their activity is not directed towards replication. The probability per unit time that any cell in a population will leave the "A" state and enter the "B" phase is constant. The "B" phase consists of S, G_2, and M, and may also include a portion of the conventional G_1. The transition probability is determined by the cell type and by environmental conditions. Changes in proliferation rate are accounted for by changes in transition probability.

Kinetic data on the age distribution of intermitotic times of exponentially growing populations, on differences in intermitotic times of siblings, and on the rate of entry of serum-stimulated G_0 cells into S were consistent with this model (13–17), as were kinetic studies on changes in environmental conditions such as serum concentration (18, 19), hormones, (15, 20), and cell density (13, 15).

In this framework, a "G_0" population of cells may be characterized as a population with a low transition probability, with most quiescent cells residing in the "A" state. Rapidly proliferating cells have a high transition probability and spend relatively little time, on the average, in the "A" state. A biochemically distinct "A" state with low transition probability would be equivalent to the proposed distinct G_0 state. However, quiescent cells have not withdrawn into a special "out-of-cycle" state. They have a finité probability of leaving the "A" state and undergoing cell division. Regardless of the mechanism, we will refer to cells that have ceased rapid proliferation, and that have the G_1 content of DNA, as being in the G_0 state.

Restriction Point

Normal cells respond to suboptimal growth conditions by entering the G_0 state. Diverse conditions could result in the same quiescent state, and cells could make this commitment to enter G_0 or to continue to proliferate at a single point in G_1 called the restriction point (21). It has been proposed that some transformed cells are defective in their restriction-point control (9, 21–32), while other transformed lines retain the ability to enter a resting state under some conditions (23, 27, 28, 33–36).

Conditions that shift cells into G_0 include high cell density (e.g. 35–40); serum limitation (9, 12, 23, 25); limitation of some amino acids (9, 41–45) or of other nutrients, such as phosphate, glucose (41, 42), or lipids and biotin (32); and the presence of certain drugs (24, 29, 30, 31, 46, 47).

The mechanism of density-dependent inhibition of growth is not yet well understood. Factors proposed include medium depletion (48–51), limitation of available growth surface (48, 49), the formation of a diffusion boundary layer close to the cell surface (52, 53), direct contact interactions (36, 40, 54), a diffusible inhibitory factor released by the cell (40), and the amount of cell surface area exposed to the growth medium (35). Density inhibition of cell movement is apparently not causally related to density inhibition of growth; cytochalasin B inhibited serum stimulated movement of quiescent 3T3 cells without inhibiting the initiation of DNA synthesis (55). Epidermal growth factor bound as well to dense as to sparse human glial cells, indicating that a reduced binding of this growth factor cannot explain density-dependent inhibition (56). For many cell types, density dependent inhibition is not an all-or-none phenomenon that occurs suddenly at confluence, but rather begins well before confluence and proceeds gradually (35, 36, 40).

Limitation of all the amino acids in the medium simultaneously (41), or of isoleucine (43), histidine plus glutamine (42), or arginine (44, 45), has been reported to arrest cells in G_0. The arginine block in CHO cells was not nearly as stringent as the isoleucine block, however (57). The metabolic effects of amino acid deprivation could be mediated by the appearance of uncharged tRNA (58).

Drugs that shift proliferating cells into quiescence include streptovitacin A, caffeine (24), succinylated concanavalin A (47), picolinic acid (30), and methylglyoxal bis(guanylhydrazone) (31). Glucocorticoids arrested primary chick embryo fibroblasts (29) as well as a variant 3T3 cell line (46) in G_0. Rat serum very-low-density lipoprotein prevented quiescent fetal rat hepatocytes from entering the cell division cycle (59), and interferon suppressed the transition from G_0 to S in 3T3 cells (60). Finally, monolayer 3T3 cells were stopped in G_0 (61) when they were placed in suspension culture.

There is some controversy over whether cells arrested in G_0 by different mechanisms are actually in identical physiological states. BHK cells blocked by serum, isoleucine, or glutamine deprivation and then released into complete medium all required the same length of time to initiate DNA synthesis (21). However, the interval between G_0 and S in Chinese hamster lung cells depended on the condition used to arrest growth (9). Transfer of cells from one blocking condition to another never resulted in the initiation of DNA synthesis, suggesting that they are blocked in the same state (9, 21). However, experiments involving a temperature-sensitive BHK cell cycle mutant suggested that serum deprivation arrests these cells at a different position in G_0 than does isoleucine deprivation (62). Cell motility and thymidine and deoxyglucose transport were different in isoleucine- and serum-deprived Nil-8 hamster cells. Thus the two quiescent cell populations were not identical (63).

Evidence has accumulated in favor of a restriction or commitment point between G_0 and S. Presumably, once the biochemical event associated with this point has occurred, a cell is irreversibly committed to initiate DNA synthesis and undergo cell division. A growth stimulus, such as serum, can be added to quiescent cells and then removed at various times. After the restriction point, cells will go on to initiate DNA synthesis in the absence of the growth stimulus. For confluent chick embryo fibroblasts, this commitment point was located in mid-G_1, four hours prior to S, when serum was used as the growth stimulus (37). Combination of serum removal with lowering of the pH of the culture medium (12), or use of the purified growth factor MSA (multiplication stimulating activity) (64), placed the restriction point at or near the G_1/S boundary. In human glial cells the commitment point was found to be in late G_1, five hours prior to S, whether the growth stimulus was serum, platelet-rich serum, or purified epidermal growth factor (23, 38). The same conclusion was reached using 3T3 cells, based upon a five-hour lag in the change in rate of entry into S upon a shift from a high to a low serum concentration (19). Apparent commitment after serum removal may reflect the action of residual serum (12, 64).

Drugs that preferentially inhibit the commitment event can be added at various times after growth stimulation. The restriction point is defined as the point at which the drug is no longer effective in preventing cells from entering S. The location of the restriction point was estimated to be midway through the transition between G_0 and S in BHK cells based upon drug studies with caffeine, 5-fluorouracil, and puromycin aminonucleoside (24). Studies with low levels of actinomycin D [see (2) for earlier work] identified a commitment point in Chinese hamster cells midway between mitosis and S (8). Evidence has been presented for a commitment point in the G_1 period of the preceding cell cycle, which was sensitive to thymidine and FUdR in murine mastocytoma cells (65).

The biochemical nature of the restriction event is unknown. It may be either probabilistic (14) or deterministic. A probabilistic event could be the result of an oscillatory phenomenon with a certain amount of noise (66). The kinetics of changes in the rate of entry into S during serum shiftup and shiftdown indicate that serum could control at least two distinct processes that determine the transition probability (19). A deterministic event could be the attainment of a crucial rate in a process such as net protein accumulation (67), or a crucial amount of a specific protein or RNA (68–70). Whether the attainment of a critical cell mass is related to the commitment event has been a subject of controversy (4).

Cell Cycle Mutants

Mutants that are temperature-sensitive for growth and that are blocked at specific points in the cell cycle, should enable specific biochemical control events in the cell cycle to be identified. Several mutants are blocked in G_1 at the nonpermissive temperature (reviewed in 4). Balb/3T3 cell lines blocked at several different points in G_1 have been isolated (71). A Chinese hamster mutant blocked in G_1 at 40°C was found to be defective in glycoprotein synthesis at 40°C (72).

A cold-sensitive CHO mutant was arrested in G_0 at 33°C (73). At 39°C the cells had a fibroblastic morphology, whereas at 33°C they were round and epithelioid. Colchicine-resistant CHO clones had a cold-sensitive block in G_1 (74). Partial revertants for growth displayed increased colchicine sensitivity, suggesting that reduced colchicine permeability and the G_1 block were the result of the same mutation.

Transformed cells can be viewed as variants that have lost aspects of growth regulation. An important objective is to identify biochemical control points that have become defective in transformed cells.

Growth Factors

Serum, a complex mixture of substances, has been used routinely in cell culture medium to provide necessary growth factors. An ultimate goal is to replace the use of serum with defined, purified growth factors. A variety of polypeptide hormones are mitogenic in at least some systems. Growth factors have been reviewed recently by Gospodarowicz & Moran (75). We restrict our discussion to relevant work that has appeared subsequent to their review.

The principal growth factor in serum for primate dermal fibroblasts, arterial smooth muscle cells, and glial cells is derived from platelets (76, 77). A basic polypeptide growth factor (13,000 mol wt) for Balb/c-3T3 cells purified from human serum was heat stable and could be extracted from human platelets by heating at 100°C for two minutes (78, 79).

A factor that was released into the medium by SV40-transformed BHK cells, and that stimulated both growth and migration of 3T3 cells, has been purified (80, 81). It was a basic protein with a molecular weight of 18,000. A factor in rat plasma that stimulated confluent 3T3 cells to initiate DNA synthesis was separated from a factor that promoted cell multiplication (82). A basic polypeptide growth factor (11,000–13,000 mol wt) for chondrocytes and 3T3 cells has recently been partially purified from cartilage (83). Unlike fibroblast growth factor, which requires glucocorticoids to exert its maximal effect, the cartilage-derived factor alone stimulated DNA synthesis in 3T3 cells maximally. The initiation of cell proliferation by prostaglandin F_2 was markedly potentiated by insulin and inhibited by hydrocortisone (84, 85).

A carboxymethylcellulose column depleted serum of factors required by an ovarian cell line (86). Either the eluate from the column or pituitary extract restored growth. A rat pituitary line grew in a defined serum-free medium supplemented with transferrin, triiodothyronine, thyrotropin-releasing hormone, parathyroid hormone, and a partially purified somatomedin preparation (87). BHK and HeLa cells grew in serum-free medium supplemented with 25 hormones (87).

Serum, glucocorticoids, epidermal growth factor, insulin, glucagon, inosine, and arginine have been reported to be necessary for the growth of cultured hepatocytes (44, 88, 89). A high molecular weight serum factor stimulated both growth and arginine uptake (4).

Polypeptide hormones interact with plasma membrane receptors. MSA, insulin, nonsuppressible insulin-like activity, proinsulin, and somatomedin A all bind to the MSA receptor of chick embryo fibroblasts (90, 91). This binding was correlated with the stimulation of DNA synthesis in quiescent cells. The MSA receptor was shown to be distinct from the insulin receptor.

Two types of insulin binding sites exist in chick embryo fibroblasts—one of high affinity and low capacity and one oi low affinity and high capacity (92). The stimulation of 2-deoxyglucose uptake and DNA synthesis in quiescent cultures by insulin correlated with the level of occupancy of the low affinity sites. These may be equivalent to the MSA receptor.

Insulin receptors in quiescent Balb/3T3 cells were two- to nine-fold higher than in growing cells (93). The number of glucocorticoid receptors per HeLa cell doubled between late G_1 and S (94). Glucocorticoid receptors in mitogen-stimulated lymphocytes increased (95).

Quiescent cells can be stimulated to proliferate by a mild protease treatment (96). This stimulation was not due to the removal of the major surface protein (250,000 mol wt) of chick embryo fibroblasts (97, 98). Proteases potentiated the action of other growth factors, including serum, perhaps by increasing the efficiency of their utilization (99–101). Secretion of plasminogen activator, a cellular protease, diminished about ten-fold as 3T3 cells became density-inhibited (102), although intracellular levels did not change.

Divalent Cations

Ca^{2+} induced the transition from G_0 to S in Balb/3T3 and rat heart cells (103, 104). Ca^{2+} may be necessary in late G_1. Quiescent Balb/3T3 cells stimulated with serum in Ca^{2+}-deficient medium did not initiate DNA synthesis. However, the addition of Ca^{2+} as late as 10 hr after serum addition enabled DNA synthesis to proceed as in control cultures (105). Thus Ca^{2+} may play a key role in the control of cell growth. Ca^{2+} and cyclic nucleotides interact in the control of lymphocyte mitogenesis (106).

Mg^{2+} deprivation caused chick embryo fibroblasts to become quiescent (107). The intracellular availability of Mg^{2+}, as determined by the amount bound to cellular membrane systems, has been proposed as a key regulator of cell growth. Mg^{2+} would regulate those metabolic pathways in which the rate-limiting steps are transphosphorylation reactions. Subtoxic concentrations of certain divalent metal ions such as Zn^{2+}, Cd^{2+}, and Hg^{2+} stimulated DNA synthesis in chick embryo cells (108).

Cell Surface

The relationship between cell growth and the cell surface has been reviewed (109). Several recent studies suggest ways in which the cell surface may be involved in density-dependent inhibition of growth. A surface membrane enriched fraction from 3T3 cells inhibited the entry of sparsely growing 3T3 cells into S (110). The membrane fraction might produce the same effect as adjacent cells at confluence. A glycoprotein associated with the membrane of density-inhibited hamster melanocytes was released into the medium. This glycoprotein restored density-inhibition of growth to malignant melanocytes (111). Similarly, acid mucopolysaccharides isolated from rat liver cell coats restored density-inhibition of growth to Yoshida ascites hepatoma cells (112).

It has been proposed that glycosaminoglycans play a role in density-dependent growth inhibition. Confluent 3T3 cells contained increased amounts of chondroitin sulfate and heparin sulfate and decreased amounts of hyaluronic acid at the cell surface compared with sparse cultures (113–115). When stimulated to divide, chick embryo fibroblasts exhibited increased hyaluronic acid production (116). Hyaluronic acid and heparin sulfate may have opposite roles in cell growth. Sulfated polysaccharides blocked the transition from G_0 to S in BHK cells (117).

Differences in glycolipid synthesis and organization in the cell cycle have been reviewed (118). Density-dependent enhancement of glycolipid synthesis is the most prominent change. Growth inhibition by glycosylation of surface acceptors via the surface glycosyltransferases, for adjacent cells, has been proposed (119). However, an exogenous galactosyltransferase activity markedly increased after serum stimulation of BHK cells (120).

There is some controversy over whether plasma membrane fluidity is important in growth regulation. Random distribution of intramembranous particles was found in both sparse and confluent 3T3 cells (121, 122), in contrast to an earlier report of an intensely clustered distribution in confluent cultures (123). Measurements of the degree of microviscosity of membrane lipids indicated that sparse cultures had higher fluidity than confluent cultures (124).

It has been proposed that surface modulating assemblies, consisting of cell surface glycoprotein receptors and a submembranous array of microtubules, microfilaments, and associated contractile and membrane proteins, control surface receptor mobility and regulate growth signals from the cell surface to the interior (125). In support of this hypothesis, colchicine and other microtubule disrupting agents blocked the transition from G_0 to S in serum-stimulated chick embryo fibroblasts (126), serum-stimulated neuroblastoma cells (127), and lectin-stimulated lymphocytes (128). Concanavalin A, in doses that inhibit surface receptor mobility, also inhibited entry into S (126). The increased phosphatidylinositol turnover that is an early event in concanavalin A–induced lymphocyte mitogenesis was inhibited by colchicine (129). Other early events—increased protein and RNA synthesis, lymphotoxin synthesis, and increased phospholipid turnover— were not inhibited by colchicine (130, 131). These findings were interpreted as being incompatible with the concept that microtubule-disruptive agents prevented lymphocyte activation (130). However, colchicine may inhibit events necessary for entry into S without inhibiting all early events following stimulation. Microfilaments or related cytochalasin-sensitive structures may be involved in lymphocyte activation (132).

Transport

The rate of transport of small-molecular-weight nutrients across the cell membrane has been proposed as a primary regulator in switching cells between quiescence and proliferation (2, 133). Transport activity would determine the availability inside the cell of one or more critical nutrients, which in turn would determine the growth state of the cell.

Consistent with this proposal, quiescent cells have reduced rates of uptake of phosphate, uridine, hexoses, and some amino acids. When quiescent cells are stimulated to grow, uptake of phosphate, uridine, and hexoses begins to increase within minutes. Transport changes are specific and do not represent a generalized alteration in membrane permeability. For example, the transport of adenosine and of many amino acids does not change with the growth state of the cell. The kinetics and mechanism of the transport changes differ among nutrients (reviewed in 134, 135).

In more recent studies, the addition of serum to quiescent 3T3 cells stimulated ouabain-sensitive (Na^+-pump mediated) $^{86}Rb^+$ (a K^+ tracer) influx two- to four-fold within 10 min (136, 137). The addition of ouabain at the time of serum addition blocked entry into S, suggesting that the increase in K^+ transport was a necessary event for the transition from G_0 to S. Quiescent 3T3 cells had a reduced activity of (Na^+-K^+)-ATPase (138).

Two distinct components of sugar transport in BHK cells were reported —a fast component associated with growing cells, and a slow component associated with quiescent cells (139). It has been proposed that the transition of a given cell from one state to the other is a discrete and sudden event.

Reduced rates of transport in the A, but not the L, system for neutral amino acids were associated with quiescence (35, 140, 141). In isolated membrane preparations, Na^+-gradient-dependent transport of α-aminoisobutyric acid was lower in confluent compared with growing Balb/3T3 cells (141, 142).

Putrescine transport was greatly reduced in quiescent compared with growing human fibroblasts (143).

The question of whether changes in uridine and glucose uptake are actually changes in membrane transport or are changes in the intracellular rate of phosphorylation of these compounds has been addressed recently. There are conflicting data with regard to uridine uptake. Several lines of evidence, including kinetic and inhibitor studies and measurements of phosphorylated and nonphosphorylated pools, indicated that phosphorylation was the rate-limiting step in uridine uptake by quiescent 3T3 cells, and that serum stimulated the rate of phosphorylation and not transport (144). However, membrane vesicles from confluent 3T3 cells transported uridine at a rate three times lower than that of vesicles from growing cells, demonstrating that the lowered rate of uridine uptake in quiescent cells was due to a membrane alteration and not to changes in subsequent metabolic processes (145).

It was reported that serum-deprived 3T3 cells exhibited a decreased rate of phosphorylation of 2-deoxyglucose, but not of transport (135). However, the methodology used in this study has been questioned (147). 3-O-Methylglucose, a nonmetabolizable glucose analogue, was used to demonstrate that quiescent 3T3 cells exhibited reduced rates of hexose transport (147).

Several recent studies have helped to elucidate the relationship between transport and growth control. Studies on revertant lines of transformed 3T3 cells indicated that reduced transport was linked with normal growth arrest in G_0, and not with random growth arrest throughout the cycle (147).

Changes in phosphate and glucose uptake were shown not to be causally linked with shifts between quiescence and proliferation (148–151). Doses of glucocorticoids that stimulated quiescent 3T3 cells to proliferate did not cause increases in phosphate or hexose transport (148, 150). In addition, the uptake rate of glucose or phosphate was lowered by adjusting the concentration of these substrates in the medium such that the uptake rate was below that of quiescent cells. This had no effect on the growth of proliferating cells or on the stimulation of quiescent cells (148, 149, 151). Serum stimulation of 3T3 cells grown to quiescence in 1% serum actually resulted in a decrease in phosphate transport (148). The addition of fetuin or low concentrations of serum to quiescent 3T3 cells produced increased hexose uptake without subsequent proliferation (150).

Dosages of cytochalasin B that inhibited glucose transport did not inhibit the initiation of DNA synthesis in serum-stimulated 3T3 cells (55). Hexose transport did not decrease in density-inhibited androgen responsive mouse mammary tumor cells (35). α-Aminoisobutyric acid transport, which decreased with density in these cells, did not increase upon addition of testosterone, which stimulated the growth of density-inhibited cells. This indicated that an increase in the activity of the "A" neutral amino acid transport system was not necessary for proliferation.

The content of cellular amino acids increased two- to three-fold in quiescent cells (140), perhaps as a consequence of reduced protein synthesis. Therefore, it is unlikely that the level of intracellular amino acids regulates growth.

A recent study in which quiescent 3T3 cells were stimulated with a purified growth factor indicated that early transport increases were an integral aspect of growth stimulation (81). However, no experiment has yet demonstrated that transport is a primary regulator of cell growth, and evidence is accumulating that contradicts this hypothesis. A more likely possibility is that decreases in transport activity associated with quiescence are only feedback-type adaptations to a reduced requirement for nutrients.

Protein Synthesis and Degradation

Protein synthesis increases when quiescent cells are stimulated (152, 153), probably secondarily to increases in RNA content. However, the effect of serum on protein synthesis in Vero African green monkey kidney cells may be mediated through a translational control mechanism (152).

The synthesis of a protein(s) may be involved in the restriction point event. The delay in the initiation of DNA synthesis after treatment of CHO cells in mid-G_1 with cycloheximide was greater than the length of the cycloheximide treatment (68). A labile protein with a three-hour half-life, which must build up to a critical concentration for commitment to DNA synthesis to occur, was consistent with these results. Synthesis of an 80,000 mol wt cytoplasmic protein in CHO cells began in early G_1 and peaked in late G_1; it was not synthesized during S (69). Newly synthesized proteins in both the cytoplasm and nucleus were different in growing and quiescent hamster embryo fibroblasts (154). The synthesis of a non-DNA-binding protein was low in quiescent hamster cells and increased four-fold during the first few hours after the cells were stimulated (155). Extracts from proliferating cell lines can initiate some DNA synthesis in nuclei isolated from resting tissues (156). The activity was tentatively associated with one or more proteins of greater than 50,000 mol wt.

It has been reported that slowly turning over proteins degrade more rapidly in resting than in growing cells, whereas for rapidly turning over proteins there was no difference (67, 157, 158). In another report, however, rapidly turning over proteins were degraded faster in stationary cultures of L cells than in growing cultures (159), and slowly turning over proteins were degraded at the same rate. The degradation rates of actin and tubulin, two slowly turning over proteins, were about twice as high in quiescent as in growing 3T3 cells (160).

The degradation rate of bulk slowly turning over membrane protein was the same in growing and density-inhibited monkey epithelial cells (161). However, differences in the relative rates of turnover of individual membrane proteins were observed (162).

The mechanism underlying changes in protein degradation rate is unclear. Since the activity of the lysosomal enzyme cathepsin D was twice as high in resting as in growing L cells (159), this enzyme might be responsible for the increased degradation of rapidly turning over proteins in resting L cells. Changes in degradation rates upon the addition or removal of serum occur rapidly (157, 163). Protein synthesis may be required to maintain the increased protein degradation in quiescent cells; cycloheximide blocked both processes (158, 163).

There was a close inverse correlation between the rate of protein degradation and the amount of DNA synthesis induced following stimulation of quiescent rat embryo fibroblasts with varying serum concentrations (157). Rates of de novo protein synthesis and protein degradation varied with the proliferation rate of mouse mammary tumor cells modulated by cell density and testosterone (67); the net rate of protein accumulation was always directly proportional to proliferation rate. It was proposed that parameter determines proliferation rate by determining the concentration in the cell of a critical protein.

RNA

Ribosomal RNA (rRNA) and transfer RNA (tRNA) increase when quiescent cells are stimulated to proliferate (2, 4). Increased synthesis (164) can account entirely for the increase in these RNA species, although evidence has been presented to show that rRNA processing pathways may also change (165).

Most eukaryotic messenger RNA (mRNA) is covalently attached to polyadenylic acid (poly A) at the 3' end of the molecule. This has provided a convenient tool by which mRNA levels can be determined in quiescent and proliferating cells. Several laboratories have examined mRNA (poly A+) production in cells passing from quiescence to proliferation. While they all agree that there is more mRNA on the ribosomes, which accounts for increased protein synthesis when cells are stimulated to proliferate, they disagree about its origin. Penman, Green, and collaborators have reported that 3T3 and 3T6 cells that are stimulated to proliferate had two to four times more cytoplasmic mRNA than quiescent cells (166–168). The increased mRNA content of proliferating cells appeared not to be the result of increased transcription since there was no change in the rate of synthesis of heterogenous nuclear RNA (HNRNA), the putative precursor of mRNA. An alteration in the efficiency of a post-transcriptional step could be responsible for the increased mRNA. Further work, employing several different means of labelling the RNA and measuring the poly A+ RNA content, has probed the efficiency of transfer of mRNA from the nucleus to the cytoplasm (166–168). These studies support the hypothesis that increased efficiency of mRNA transfer from the nucleus to the cytoplasm causes increased cytoplasmic mRNA in proliferating cells. The post-transcriptional site altered by growth stimulation has not been identified. However, the kinetics of mRNA processing (time required for the production of a complete, processed transcript and for its appearance in the cytoplasm) were the same in quiescent and proliferating cells (167). Levis et al (169) have recently extended these studies to an epithelial cell line (CHO). Proliferating CHO cells contain two to three times more RNA than quiescent cells, and the increased mRNA appeared to be the principal factor controlling the rate of protein synthesis.

Rudland (170) and Bandman & Gurney (171) have studied mRNA content of quiescent and proliferating Balb/c–3T3 cells, and both groups reported that there was at most a 20–30% increase in cytoplasmic mRNA in proliferating cells. Thus, the slight increase could not account for the two- to three-fold increase in protein synthesis. Both groups (170, 171) found that quiescent cells exhibited enhanced extrapolysomal poly A+ RNA, presumed to be messenger ribonucleoprotein (mRNP) particles. Stimulating quiescent cells by adding serum to the culture resulted in reordering the mRNP to the polyribosomes. For example, only 15% of the cytoplasmic poly A+ RNA was in polysomes in quiescent cells; this increased to 80% by 4–6 hr following growth stimulation (172). The percentage in mRNP decreased from 85% in quiescent cells to 20% in this same period. Thus, the increase in polysomal poly A+ RNA sufficiently accounted for the increase in protein synthesis. It was proposed that a translational mechanism of control causes increased polysomal mRNA observed in proliferating cells. Differences in mRNA half-life appear not to contribute to the controversy since half-lives of mRNA in quiescent and proliferating cells appear to be the same (172, 173).

Proliferating cells contain more polysomal mRNA, regardless of the mechanism by which the mRNA got there. Is mRNA in proliferating cells coding for proteins different from or similar to those in quiescent cells? The complexity, relative abundance classes, and homology of cytoplasmic poly A+ RNA in quiescent and proliferating 3T6 cells have been studied (174). Both types of cells had a low-complexity RNa class of about 2000 sequences and a high-complexity class of about 6400. Only a slight difference in the relative abundance was noted (high complexity contained 22% of total mRNA in proliferating cells versus 28% in quiescent). The homology between mRNAs from the two growth states was tested by preparing complementary DNA (cDNA), with mRNA from quiescent cells as template, and hybridizing the cDNA to proliferating-cell mRNA. The reciprocal cross was also performed. Surprisingly, more than 90% of the sequences were found in common. At most, 3% of the mRNA in quiescent cells was not found in proliferating cells, while about the same percentage of the mRNA in proliferating cells was not found in quiescent cells. If this percentage difference were concentrated in the high-complexity class, it could amount to a difference in as many as 1400 sequences, while if it were concentrated in the low-complexity class, the difference would represent 400 sequences (174). Thus, while 90% of the mRNA from quiescent and proliferating cells apparently codes for the same proteins, many sequences do exist that could code for proteins unique to and possibly responsible for either quiescence or proliferation.

Messenger RNA sequences in quiescent and proliferating mouse embryo cells in culture (AKR-2B) have also been examined, and similar results were obtained (175). Greater than 90% of increased polysomal poly A+ RNA in cells stimulated to proliferate was also found in quiescent cells. There were some poly A+ RNA species unique to proliferating cells. Yet all poly A+ RNA species in quiescent cells were also found in proliferating cells.

Cyclic Nucleotides and Associated Enzymes

The role of cyclic nucleotides in growth regulation has been reviewed extensively (176–178) in recent years. We concentrate here on reports published since 1975. Studies undertaken prior to 1975 commonly concluded that cyclic AMP (cAMP) acted somehow to arrest cell growth, while cyclic GMP (cGMP) acted to stimulate growth (176–178). It was proposed that cGMP opposed the effects of cAMP (179). Evidence for this hypothesis with respect to cell growth was based on the following: (a) cAMP levels dropped when cells were stimulated to proliferate, while cGMP levels rose (179–182); (b) fibroblast growth factor was reported to activate guanylate cyclase and raise cGMP levels (183); (c) SV40-transformed 3T3 cells had higher cGMP levels than untransformed 3T3 cells (182).

Recent work does not support this hypothesis. For example, studies have been conducted involving guanylate cyclase activity, cGMP levels, and cGMP phosphodiesterase activity in Balb/c–3T3 and normal rat kidney (NRK) cells, as well as with cell lines transformed by a chemical carcinogen and by RNA and DNA tumor viruses (184). Guanylate cyclase activity was found to parallel adenylate cyclase activity; both increased as cells ceased to grow, and both decreased in transformed cells (184, 185). Cyclic GMP concentrations were found to be lower in transformed cells than in untransformed cells. Cyclic GMP phosphodiesterase activity did not correlate with transformation; rather, it paralleled cAMP phosphodiesterase activity. In addition, no stimulation of guanylate cyclase by fibroblast growth factor was observed. It has recently been reported (186) that experiments could not be repeated in which a transient increase in cGMP in the G_1 phase of the cell cycle was observed. However, the importance of the cAMP/cGMP ratio to growth regulation has been emphasized (186), and this ratio does change as a function of the cell cycle. Other reports have indicated that cAMP fluctuates as a function of the cell cycle (187, 188) or growth state (178). These findings are thus consistent with the idea that cyclic AMP regulates specific stages in the cycle, particularly in growth arrest in G_1. What role cGMP may play, if any, is unclear.

Adenylate cyclase and cAMP phosphodiesterase are responsible for cAMP synthesis and breakdown. Increased activity of adenylate cyclase has been shown to correlate with increased levels of cAMP as cell growth was arrested (176–178) and cyclase activity fluctuated with the cell cycle (189, 190). As cell growth was arrested, cyclase activity remained high while phosphodiesterase activity leveled off with a resultant increase in cAMP (191). Serum stimulation of quiescent BHK fibroblasts caused rapid stimulation of cAMP phosphodiesterase (192), which can account for the observed decrease in cAMP as these cells were stimulated to grow.

It has been proposed that most if not all cAMP effects in mammalian cells are mediated by cAMP-dependent protein kinases (193), although evidence exists that cAMP and a cAMP-binding protein do interact directly with chromatin (194). Cyclic AMP-dependent protein kinase is composed of two catalytic subunits and two regulatory subunits. Two types (type I and type II) have been identified in mammalian tissues. The catalytic subunits are identical while the regulatory subunits differ. Both type I and type II protein kinase were found in CHO cells and their activities fluctuated during the cell cycle (195). Type I activity was high in mitosis, low in G_1, and constant throughout the rest of the cycle. Type II activity increased near the end of G_1 and began to decrease in mid-S phase. A type I protein kinase was found in SV40-transformed 3T3 cells that was not found in untransformed 3T3 cells (196).

Fluctuations observed in the enzymes involved in cAMP metabolism and the cAMP-dependent protein kinase are consistent with, but do not prove, the hypothesis that cAMP plays some role in the regulation of cell growth. Tomkins and collaborators showed in a mutant lymphosarcoma cell line (S49) that cAMP is involved in growth arrest but not in progress through the cell cycle (197–199). Adding B_2cAMP specifically stopped wild type S49 cells in the G_1 phase of the cell cycle, and cell death followed in 48–72 hr. B_2cAMP-resistant cells were selected for by maintenance in B_2cAMP-containing medium. Detailed characterization of the cAMP-binding proteins and cAMP-dependent protein kinase activity revealed that both were diminished in three mutant lines (199). The biological response (cell growth arrest, percent cells in G_1) of the mutants to exogenous B_2cAMP directly correlated with the amount of protein kinase associated with the mutant. When maintained in the presence or absence of B_2cAMP, mutant cell lines without cAMP-dependent protein kinase exhibited cell cycle times identical to those of wild type S49 in the absence of B_2cAMP. Thus these period fluctuations of cAMP levels were not required for, nor did they determine, progress through the cycle, at least in S49 cells. In addition, cAMP-dependent protein kinases mediated cAMP regulation of cell growth arrest in S49 cells.

A recent report (200) has established a role for cAMP-dependent protein kinase in inhibiting meiotic cell division of *Xenopus* oocytes induced by progesterone. In this system the incubation of oocytes with progesterone in the first meiotic prophase causes synchronous progression to the second meiotic metaphase. This progesterone-stimulated cell division sequence was found to be inhibited by microinjection of a highly purified catalytic subunit of protein kinase. Without progesterone, cell division could be induced directly by microinjection of the purified regulatory subunit of protein kinase. Inhibition of progesterone-induced cell division by microinjection of the catalytic subunit was noted only during the first hour of progesterone treatment, indicating that only early events in meiotic cell division were affected. Thus the catalytic subunit of protein kinase is necessary and sufficient to block meiotic cell division in *Xenopus* oocytes. Direct proof that cAMP-dependent protein kinase mediates growth inhibition in mammalian cells may be obtainable by a similar approach.

Polyamines and Ornithine Decarboxylase

Increases in polyamines (putrescine, spermine, and spermidine) and in the key enzyme regulating their synthesis (ornithine decarboxylase) occur very early when cells are stimulated to grow. Increased levels of spermine and spermidine may be required for optimal DNA synthesis in some cells (201); they are correlated with passage through G_1, and entry into S (202, 203). Hormones that stimulate cell growth (204), chemicals that promote tumor formation (205), and infection with tumor viruses (206) have been reported to induce ornithine decarboxylase activity.

Recently, inhibitors of ornithine decarboxylase have been employed to assess the importance of polyamines to cell growth. α-Methyl ornithine blocks proliferation of hepatoma cells in culture (207). The entry of 3T3 and WI-38 from G_0 into S induced by serum stimulation was blocked by ornithine decarboxylase inhibitors (203). Methylglyoxal bis(guanylhydrazone), an inhibitor of S-adenosylmethionine decarboxylase, when added to log growing cultures of rat embryo fibroblasts, arrested them in G_0 (31). Adding the inhibitor to quiescent cells stimulated to proliferate by the addition of serum did not block entry into the first S phase but did block entry into the subsequent one. Thus a certain minimum level of polyamines is necessary for progression through the cell cycle.

Clark & Duffy (208) studied the role of polyamines in the growth stimulation of density inhibited 3T3 cells by 12 different mitogenic agents. While activation of polyamine synthesis and increase in cell number showed similar dose response curves for serum stimulation of growth, there was a poor correlation among the growth promoting ability of the other mitogens and increases in polyamines or ornithine decarboxylase activity. Increases in polyamine synthesis did not induce cell division in quiescent 3T3 cells. Dexamethasone or cortisol both induced cell division in quiescent 3T3 without increased synthesis of polyamines. Thus it appears that increased polyamine is neither necessary nor sufficient for growth.

Nuclear Changes

Changes in chromatin structure and function early in the transition from G_0 to S have been reviewed (209). These changes include increased chromatin template activity, positive ellipticity in the 250–300 nm region of the circular dichroism spectrum, and binding of intercalating dyes. Recent studies have demonstrated progressive dispersion of interphase chromatin as the cell advances toward S (210, 211). Chromatin changes may represent gene activation or preparation for subsequent DNA synthesis.

The DNA in human diploid fibroblasts became susceptible to S_1 nuclease (single strand–specific) as early as 4 hr after serum stimulation of resting cultures, even though measurable DNA synthesis did not commence until 9 hr after addition of serum (212). Although this could be interpreted as a preparation for ensuing DNA synthesis, it may be a manifestation of the earliest stages of the S period. This interpretation would agree with data which suggest that DNA synthesis at a low rate begins much earlier in the cell cycle (i.e. in what is termed "G_1") than previously thought (213).

Nonhistone chromosomal proteins may be responsible for chromatin changes. The synthesis of nonhistone chromosomal proteins increases early after stimulation of quiescent cells or after mitosis and reaches a maximum in late G_1 (214, 215). It has been suggested that their synthesis may be coordinated, based on preferential inhibition of their synthesis by TPCK and pactamycin (216). Differences in the amounts of specific nonhistone chromosomal proteins have also been observed between resting and growing cells (217, 218).

Nonhistone chromosomal proteins are observed to increase on chromatin during G_1. The question of their binding sites arises, since the genome has not as yet doubled. A model has been proposed involving one high- and several low-affinity sites for regulatory proteins; the low-affinity sites soak up the proteins when they are made in G_1, but these are redistributed to the newly made high-affinity sites during S (219).

Nonhistone chromosomal proteins begin to be phosphorylated just after stimulation of serum-starved BHK cells, with the maximum rate occurring late in G_1 (220). This observation is supplemented by studies on cyclic AMP-dependent protein kinases in WI38 cells (221), in CHO cells (195), and in bovine lymphocytes (222), in which it was shown that the activity of at least some kinases was highest in late G_1 or early in S.

Synthesis of specific DNA-binding proteins differs between resting and proliferating cells (223–226). The synthesis of several DNA-binding proteins was low in resting cells and increased following stimulation, prior to the initiation of DNA synthesis (155, 223, 226, 227). The synthesis of one protein was dramatically elevated in quiescent 3T6 cells (223). After stimulation its synthesis declined to zero before the onset of S. Several DNA-binding proteins increased between M and S in mitotically selected cells (227, 228).

Which of the many changes discussed above relates to preparation for entry into the S period remains to be seen. It should be borne in mind that viable genetic variants of Chinese hamster cells have been found that have no measurable G_1 (229). Furthermore, DNA synthesis in isolated early G_1 nuclei can be provoked by trypsin treatment (230). Presumably, under these circumstances, any required preparations for S can be accomplished in a very short time.

THE S PERIOD

In most eukaryotic cells in culture, the S period is 6–8 hr long, but it ranges in vivo from 10 min in *Schizosaccharomyces pombe* (231) to as long as 35 hr in mouse ear skin (232). In eukaryotic organisms the genome is distributed among several chromosomes. Each DNA fiber is divided into many replicating units, termed *replicons* (233, 234); each replicon has a centered origin from which growing forks proceed outward in both directions, apparently fusing with the forks of adjacent replicons. It is not known whether the origin of replication is a unique nucleotide sequence, nor whether the terminus of replication is fixed. The rate of fork travel is estimated to be between 0.5 and 2 μm min^{-1} (234, 235), but may vary during the S period (236, 237, 238). The spacing between origins varies in size from 7 to 100 μm (234, 239). Several lines of evidence indicate that not all replicons function simultaneously, and that cytologically defined loci replicate at characteristic times within the S period (reviewed in 4). However, in the cleaving embryo in several species, the DNA is duplicated approximately

100 times faster than in adult somatic cells of the same species by a mechanism involving larger numbers of small replicons, possibly all functioning simultaneously (240, 241).

Duplication of the eukaryotic genome is further complicated by the fact that the DNA is tightly associated with at least an equal weight of protein (242). Histone and nonhistone chromosomal proteins undoubtedly serve a host of structural and regulatory roles, and must be duplicated at some time during the cell cycle. It is also clear that the replication machinery must be able to dislodge or loosen these proteins in order to replicate the DNA with which they interact. In this context it has recently been proposed that the "nu body" structure of the nucleo-protein complex (242) may determine the sites at which DNA synthesis initiates on chromatin (243, 244).

Initiation of DNA Replication at an Origin of a Replicon

The origin of DNA replication in SV40 has recently been sequenced and has been shown to be an inverted repetitive sequence (palindrome) that can theoretically generate either one or two hairpin loops per strand due to intra-chain base pairing (245). Adenovirus 2 (246) and the small adeno-associated virus AAV (247) have also been shown to contain palindromic sequences at both ends of the DNA duplex, and DNA synthesis presumably initiates at these ends. The parvoviruses also contain inverted repetitive sequences at their origins, and a model has been suggested in which animal cells and their viruses may both utilize palindromes for initiation of DNA synthesis (248).

There is, in fact, evidence for the existence of inverted repetitive sequences in eukaryotic genomes. The DNA cross-linking agent Psoralen has been used to show that 6% of the human genome can form hairpin loops with a heterogenous but average length of 190 base pairs per inverted repeat sequence (249). These sequences were found to occur in all families of sequence complexity by $C_0 t$ analysis and were further shown to be distributed over the length of all chromosomes. However, inverted repeat sequences may not exist as hairpin loops in situ, but rather may exist as linear structures (250).

The hairpin loop is attractive for control purposes not only because it can be recognized by proteins as a discontinuity in the double helical structure, but also because of the single-strandedness of a few residues at the center of the palindrome, through a lack of base pairing at the turn. These could conceivably serve as template for an RNA or DNA primer, as a site for nuclease cleavage, or as a high-affinity site for unwinding proteins. Any of these modes could serve to initiate DNA synthesis at a particular site.

DNA fiber autoradiographic studies suggest that the origins of replication are fixed, but that the time of replication of a given replicon in the S period is not precise (251). Moreover, late-replicating bovine satellite DNA from the kangaroo rat appears in fiber autoradiographs to contain replicons whose size varies little around a mean value of 7 μm or multiples thereof (239). Since satellite DNA is known to be a highly repeating sequence, the size of the replicon may relate somehow to the size of the repeating nucleotide sequence.

In eukaryotic cells, the actual event that triggers initiation at an origin of replication is a mystery. An early event could be the generation of a nick, providing ends for DNA or RNA polymerase. In this connection, several laboratories have reported endonucleases that attack both single- and double-stranded DNA (252, 253). In calf thymus the enzyme produced 7S double-stranded DNA pieces which, after denaturation, could act as both primer and template for DNA polymerase. Interestingly, these pieces contain hairpin loops (252). An endonuclease has been found in 3T3 and polyoma-infected 3T3 cells that cleaves double-stranded polyoma DNA into five discrete pieces and thereby demonstrates its specificity (254).

Whatever the mechanism of initiation at a replicon, it must explain time-ordered synthesis of replicons within the S period. Because of the well-known dependence of DNA synthesis on concomitant protein synthesis, several investigators have asked whether inhibition of protein synthesis prevents initiation exclusively (as opposed to elongation), as it does in bacteria (255) and in yeast (256). In DNA fiber autoradiographic studies, Hand (257) observed a decrease in the rate of fork movement, a lowered number of initiation events, and an increase in unidirectional replication. Gautschi (258) and Evans et al (259), on the other hand, found that the rate of travel of the replication fork, but not the number of initiations, is lowered by cycloheximide or puromycin treatment. In yeast it is quite clear that once the cell has entered the S period, protein synthesis is not required to complete S, even though time-ordered synthesis of genetic markers occurs in this system (256).

Several groups have reported the isolation of DNA-membrane complexes from eukaryotic cells. In pulse-chase experiments with radioactive thymidine, the DNA seemed to move through the membrane complex with the kinetics expected if DNA were initiated and synthesized on the membrane (260–263). The DNA polymerase in a membrane complex isolated from normal human lymphoid cells appeared to be different from those hitherto described (see next section), and DNA synthesis in this system was sensitive to RNAse (260). A striking observation is that administration of inhibitors of sterol synthesis to mouse L cells immediately arrested DNA synthesis, but not protein or RNA synthesis (264). DNA synthesis was restored to near-normal levels by the addition of mevalonate. However, other groups find no evidence for the initiation of DNA synthesis at the nuclear membrane in autoradiographic studies on thin sections examined with the electron microscope (265, 266).

General Character of DNA Elongation in Eukaryotes

Once the cell has selected a locus for DNA synthesis through some unknown initiation event, it now proceeds to polymerize DNA in both directions. As in prokaryotes, eukaryotic DNA polymerases function only in the 5' to 3' direction (267), suggesting that discontinuous synthesis must occur on at least one strand of the double helix. In Chinese hamster ovary cells and in mouse cells, 30 sec pulses of radioactive thymidine resulted in the formation of 4S double-stranded DNA pieces, which were chased to 20–60S pieces within 2–8 min (268, 269). By 2 hr the label was found in high-molecular-weight DNA. These entities presumably correspond to Okazaki fragments (270), which are then ligated into replicon-sized pieces which, in turn, coalesce to eventually form chromosome-sized DNA fibers. In *Drosophila* the Okazaki fragment may result from the ligation of even smaller 40–50 b.p. fragments (271). In vitro, several groups have also observed the formation of the 2–4S pieces in short thymidine pulses, followed by conversion to intermediate (20–100S) and, finally, to very large (greater than 10^6 daltons) fragments (272–275). Since all label in short pulses is found in 4S pieces by most laboratories, the implication is that DNA synthesis is discontinuous (i.e. proceeds by Okazaki fragments) on both strands. However, in an in vitro DNA synthesizing system prepared from Physarum homogenates, one strand may be replicated continuously in the 5' to 3' direction, without the intermediate formation of Okazaki fragments (276). In an in vitro system from HeLa cells, the 4S piece may be ligated to a larger (14 to 30S) DNA fragment instead of to other 4S pieces (277). The latter observation can be explained if the ligase in HeLa cells is extremely efficient. The above data should be considered in the light of recent work suggesting that the conditions maintained during radiolabeled thymidine pulses markedly affect the apparent size of the DNA fragments obtained (278).

Several laboratories have obtained evidence for the involvement of RNA primers in eukaryotic DNA synthesis (reviewed in 279). Labeled uridine is found in the 4S Okazaki pieces after very brief pulses (280), and the density of a fraction of 4S fragments on $CsCl_2$ gradients is high enough to indicate the presence of both RNA and DNA in these fragments (281). However, uridine is converted to deoxycytidine, at least in Erlich ascites cells, and no evidence for RNA primers was found in this system (282). Adventitious binding of RNA to DNA in $CsCl_2$ gradients may also occur (283, 284).

Machinery Utilized at All Replicating Loci

The proteins and enzymes responsible for the polymerization of DNA during the S period in eukaryotic cells have not been identified, but several activities are good candidates for involvement in the process.

UNWINDING AND UNTWISTING PROTEINS   Proteins that unwind the DNA double helixes have been found in calf thymus (285), in mouse ascites cells (286), and in HeLa cells (287). Several subspecies of unwinders were found in calf thymus. All bind preferentially to single-stranded DNA, but also have a low affinity for double-stranded DNA. The latter property suggests that unwinding proteins could be utilized as initiator proteins, since DNA may be unavailable to RNA or DNA polymerases unless the interior of the helix is exposed. Alternatively, if these proteins bind exclusively to single-stranded DNA in situ, several possibilities exist for localized, relatively specific binding sites: (a) low melting, A-T-rich regions that "breathe" periodically into transient, single-stranded structures; (b) preexisting single-stranded structures such as have been found next to polypyrimidine clusters in sea urchins (288) and the single-stranded character of G_1 DNA (212); (c) the single-stranded knobs at the end of hairpin loops; and (d) localized stretches of single-stranded DNA next to an initiation locus already partially destabilized by a presumptive initiation protein.

The calf thymus unwinding proteins were found to stimulate DNA synthesis in vitro and, interestingly, do not facilitate reannealing of complementary DNA sequences, as does the T4 gene 32 protein (285). The mouse ascites cell unwinding protein has a molecular weight of 30,000–35,000, and has been shown to stimulate polymerase I preferentially when bound to a single-stranded template. The ascites unwinding protein can be phosphorylated in vitro with a chromatin-associated protein kinase, whereupon it loses its stimulating effect on the polymerase. The HeLa cell protein was shown to have a molecular weight of 34,000 and to constitute 3% of the cell protein. This is not a surprising figure if these proteins are required in nearly stoichiometric amounts with the replicating DNA, as is the gene 32 protein in T4 (289). The HeLa cell protein is more prominent in growing cells than in confluent cultures, suggesting its involvement with DNA synthesis.

In order to relieve the torque that must develop in a double helix during replication, the need for a swivel enzyme has been postulated both in pro- and eukaryotes. Swivel or untwisting enzymes have been reported in rat liver (290–292), normal human lymphocytes (293), mature chick erythrocytes (294), human fibroblasts (295), and Xenopus laevis oocytes (296). Unlike unwinding proteins, which are required in stoichiometric amounts and which leave both phosphodiester backbones of the helix intact, swivel enzymes are catalytic and produce a nick in one strand, allowing free rotation and relaxation of supercoiling (see 297 for discussion). Presumably, the same enzyme then reseals the nick. The rat liver enzyme is a single polypeptide of molecular weight 65,000, and has been shown to produce a nicked intermediate (291). The Xenopus enzyme was capable of removing both positive and negative supercoils (296). The activity of the lymphocyte enzyme is highest during the S period (293).

DNA POLYMERASES   The DNA polymerases found in animal cells have recently been reviewed (267, 298). Of interest are a low-molecular-weight nuclear enzyme (usually referred to as polymerase β or II, $\sim$ 3.5S) and a higher-molecular-weight enzyme usually found in both cytoplasm and nucleus (polymerase α or I, ca. 6–8S). Recent evidence indicates that the latter enzyme may be localized almost exclusively in the nucleus (299, 300). Both enzymes have been studied in the following cell types in recent years: HeLa (301, 302); BHK (303, 304); Chinese hamster fibroblasts (305); rat ascites hepatoma (306); mouse liver and testes (307); human WI38 (308); human lymphocytes (309); chick erythrocytes (310); and in mouse SVT2 and human KB cells (300). Both enzymes are template-dependent and are able to extend DNA chains on 3'-hydroxyl primers. Neither enzyme has the proofreading, 3' to 5' exonucleolytic activity of *E. coli* polymerases (297), but a third enzyme identified in human erythroid hyperplastic bone marrow has this function, and manifests template- and primer-dependent DNA polymerization (311).

Although it has not been established which, if either, of these two polymerases is the DNA replicase in animal cells, the large polymerase I seems the more likely candidate, based on its behavior during different growth states. In human lymphocytes stimulated to divide with phytohemagglutinin, polymerase I activity increased concomitantly with DNA synthesis, while polymerase II increased only later (309). In HeLa cells, polymerase I increased ten-fold during G_1 but then decreased after the S period was completed (302). Polymerase II did not change during this interval. In mice, polymerase I increased during liver regeneration (307), and the ratio of polymerase I to II changed with the state of BHK cells, being highest in log cultures, and lowest during quiescence (304). In addition, when WI38 cells were infected with Cytomegalovirus, host DNA synthesis was induced, and polymerase I increased, with no change observed in polymerase II (308). The situation is not so clear in erythrocyte development. Both polymerases were high in blast cells, but as the cells ceased division, polymerase I decreased gradually while polymerase II decreased abruptly (310).

LIGASES   The conversion of 4S pieces to larger DNA fragments during S in eukaryotes suggests the existence of ligases, and enzymes analogous to ligating enzymes in microorganisms have been found. In several cell types there were two distinct enzymes, both of which require ATP; both were found in the nucleus and cytoplasm (312). The high-molecular-weight

DNA ligase I was prevalent during liver regeneration and in rapidly prolife-rating tissues (313), while the low-molecular-weight ligase II remained relatively constant under a variety of growth conditions.

Histone Proteins and the S Period

The synthesis of histones occurs predominantly during the S period and is apparently tightly coupled to DNA synthesis: Compounds that inhibit DNA synthesis, such as hydroxyurea and high levels of thymidine, block histone production (314). Hydroxyurea and cytosine arabinoside provoke a disappearance of histone messenger RNA from the polyribosomes with a half-life of 15–30 min (315, 316), which is a much shorter half-life than that measured in the presence of actinomycin D during the S period (317). The loss of histone mRNA during hydroxyurea inhibition of DNA synthe-sis apparently can be prevented with cycloheximide, implying the synthesis of some kind of regulatory protein induced by shutdown of DNA synthesis (315).

It has been suggested that histones may be involved in some way with DNA chain elongation (318), and evidence was presented that newly made DNA was covered within one minute with newly made protein (presumably histone). Tsanev & Russev came to a similar conclusion in experiments in which radiolabeled proteins (mostly histones) were found to associate pref-erentially with DNA labeled during the same time interval (319). However, Jackson et al have come to the conclusion that newly replicated DNA is covered with both old and new histones, and that newly made histones are distributed at random in the genome (320, 321). Within 10 min of its replication, DNA is packaged into nuclease-resistant particles that are indistinguishable from those of nonreplicating chromatin (268, 322). It is worthwhile to note that although yeast cells make histones during S, they can complete the S period without concomitant protein (ergo, histone) synthesis (323). Furthermore, during *Xenopus* development histone synthe-sis is apparently not coordinated either temporally or quantitatively with DNA synthesis (324). If histones are required at a replicating fork, then both of these latter findings would have to be explained by the presence of sufficient pools to cover new DNA.

Balhorn et al (325) have evidence that both old and new histones (particu-larly H1) are phosphorylated during the S period; Gurley et al (326, 327) maintain that H1 is phosphorylated in late G_1 and in S. Hydroxyurea did not prevent phosphorylation of H1 (327), suggesting that histones elabo-rated in previous S periods are modified during S.

It is of interest to note that after UV damage, in which 100–140 base pairs are presumably repaired by a polymerase, histone synthesis does not seem to be stimulated (328). Another finding worth noting is that low levels of histone (but not polylysine) added to isolated HeLa cell nuclei stimulated DNA synthesis, while high levels of both inhibited polymerization (329).

In Vitro DNA Synthesizing Systems

Many questions concerning eukaryotic DNA synthesis, particularly those relating to the enzymology, will be answered by using in vitro systems. Current attempts to establish such systems fall into four categories: (a) cells permeabilized to macromolecules (e.g. deoxynucleotide triphosphates) by Tween 80 (330), lysolecithin (331), Brij 58 (275), toluene (332), hypotonic treatment (333, 334), and cold shock (335); (b) nuclei isolated in *aqueous* solvents (156, 272, 274, 336) and in *organic* solvents (273); (c) extracts prepared from nuclei (337), from whole *Xenopus* eggs (338), or from whole cell lysates (272); and (d) isolated chromatin (339).

Most of these systems carry out at least the formation and ligation of Okazaki pieces into replicon-sized DNA fragments. Only two of these reports claim to have achieved some initiation in vitro, as opposed to the continuation of replicons initiated in vivo (156, 336). Benbow & Ford (336) stimulated nuclei from nondividing frog liver cells to replicate DNA by exposure to the cytoplasm from eggs of the same species. Jazwinski, Wang & Edelman stimulated DNA synthesis in nuclei isolated from resting lymphocytes by using extracts from growing cells (156). The "eye" forms characteristic of bidirectional replication were observed in electron micrographs of the DNA isolated from these nuclei. The extracts of monkey cells (337, 339) and of *Xenopus* eggs (338) were used to support the replication of SV40, which presumably utilizes host cell replication machinery. In (337) and (339), covalently closed SV40 chromosomes were formed from the SV40 resident in the cell when the extracts were prepared.

Replication of Mitochondrial DNA

Mitochondrial DNA (mtDNA) is apparently replicated at all times during the eukaryotic cell cycle (340). Two reports suggest that the synthesis of mitochondrial and nuclear DNA are independently regulated. When resting 3T3 cells were stimulated to divide, cycloheximide prevented nuclear but not mitochondrial DNA replication (341). Ad2 virus infection eventually turned off host but not mitochondrial DNA synthesis (342). Once the circular chromosome in a mitochondrion has replicated, it apparently has as great a chance of being replicated again as does the DNA in a mitochondrion that has not been recently replicated (340).

Mitochondrial DNA in HeLa cells is apparently attached through its origin to a membrane component (343). Thus, it appears that mtDNA may be replicated in a manner like that of the bacterial chromosome.

G₂, M, AND DIVISION

The cell enters G_2 after completion of the S period. Damage to DNA may prevent progress toward mitosis (344). In G_2 cells there may be an inhibitor of DNA synthesis associated loosely with the nucleus: When pre-washed G_2 HeLa cell nuclei were injected into *Xenopus* eggs, they synthesized DNA, whereas unwashed G_2 nuclei did not (345). This contrasts with earlier studies in which fusion between G_2 and S phase HeLa cells did not suppress DNA synthesis in the S phase nucleus (346).

There is evidence for a weak control point in G_2, since various conditions have been reported to arrest cells there; for example, dibutyryl cAMP arrested V79 cells (347). Many drugs used in cancer chemotherapy arrest cells in G_2, but are generally not reversible (348–351). Hence these drugs cannot be utilized to promote G_2-specific synchrony.

Preparations for mitosis probably occur in most cell types during G_2. Protein synthesis is required in order to advance toward mitosis (but see 352). A twenty-fold increase in the reaction of phenylglyoxal with cell-surface arginine has been observed during G_2 (353), as well as a five-fold increase in oligoadenosine diphosphoribose (354). As the cell rounds up in preparation for division, there is a reorganization of actin filaments (355). Tubulin, a component of the mitotic apparatus, is made during both late S and in G_2 (356). Agents that combine with microtubules, such as maytansine (357) and colchicine, arrest cells in metaphase. Tubulin phosphorylation was higher in M and in S than in G_1 or G_2 cells, and a tubulin-associated protein kinase was maximal during M (358). The lysine-rich histones of HTC cells were also extensively phosphorylated during mitosis, at sites different from those phosphorylated during interphase. Phosphate hydrolysis from histones was three times higher at M and was extremely rapid as the cells entered G_1 (359).

The event that initiates mitosis is unknown. The initiation of mitosis may depend upon the level of inhibitory cytoplasmic factors; fused S/G_2 cells had the same G_2 duration as fused S/S cells; fused early-G_2/late-G_2 cells had an intermediate G_2 duration (360). However, fusion of mitotic cells with interphase cells provokes premature chromosome condensation in the interphase chromatin, suggesting the presence of cytoplasmic inducers during mitosis (361). Premature chromosome condensation of growing human lymphocytes is reported to occur in the presence of allogenic or xenogenic mitotic cells without fusion, suggesting an activator of mitosis that can escape into the medium (362).

Cell division (cytokinesis) rapidly follows mitosis. It is inhibited by cytochalasin B, yielding binucleate cells (360). A quite similar result was obtained with a temperature-sensitive mutant of Chinese hamster cells at the restrictive temperature, even though metabolism proceeded relatively normally in these cells at the higher temperature (363).

Following cytokinesis the daughter cells seem largely to go their separate ways (see correlations of G_1 discussed above). But there is some evidence that the motions of the daughter cells are in mirror-symmetric directions, suggesting internal structures whose symmetric orientation at division can be retained (364).

CONCLUSION

Research on the animal cell cycle has entered a new phase during the last few years. The identification and purification of growth factors have made possible studies on the interaction of these factors with cell membrane receptors and will eventually lead to elucidation of the steps between absorption of growth factors and their ultimate effects on activating processes within the cell. These investigations will fuse with those on membranal and submembranal structure and function, as well as with studies on the role of intracellular "second messengers" such as cyclic nucleotides. The G_0 and G_1 parts of the cell cycle are still largely "gaps" with regard to our knowledge of biochemical events which are either necessary or sufficient for the initiation of DNA synthesis. That the transition from G_0 to S requires prolonged exposure to growth factors argues against the sufficiency of an early triggering event. Serum has been the substance most used to stimulate quiescent cells. The great complexity of serum leaves open the possibility that some of the serum-induced biochemical changes may be unrelated to the initiation of DNA synthesis. The use of purified growth factors to stimulate cells will help to define more clearly the biochemical changes directly related to growth stimulation. Attempts are underway to identify proteins, RNAs, smaller molecules, and structures that are directly involved in progress through the cycle and to determine the ways in which their synthesis is regulated.

Ideas about the nature of quiescence have been in flux, due in part to the introduction of the probabilistic model and the restriction point hypothesis. The proposal that normal cells enter quiescence if they cannot accomplish a specific regulatory event in G_1, and that transformed cells have escaped in whole or in part from the need to accomplish this event, or can accomplish it more easily, provides the basis for further experimentation on the nature of cancer. Elucidation of the timing and the biochemical nature of the restriction event is on the horizon.

Changes in the nucleus and chromatin occur throughout the cell cycle. Their causal relationships to other events in the cell cycle and to progress through the cycle remain to be discovered. We will have to ask questions regarding the importance of the transcription of RNA and its subsequent processing and transport into the cytoplasm, nuclear preparation for DNA synthesis during G_1, the event directly responsible for initiation of DNA synthesis, and the orderly progression of replication of different parts of the genome.

G_2 is also almost devoid of landmarks. Inhibitors are known to be more effective at certain times in G_2 than at other times in the cycle.

Our ability to obtain cell cycle mutants is still limited. Each new mutant, however, should lead to further information about necessary cycle steps. In this regard, the more developed studies on yeast cell cycle mutants should suggest regulatory mechanisms that might also be operative in higher cells.

ACKNOWLEDGMENTS

We are indebted to David Schneider for help in preparing this review. This investigation was aided by Grants CA-19864-01 and CA-19949-02 awarded by the National Cancer Institute, DHEW. Rolf Kletzien was a postdoctoral Fellow of the Jane Coffin Childs Foundation. Joyce Hamlin was a Fellow of The Medical Foundation, Inc.

Literature Cited

1. Hershko, A., Mamont, P., Shields, R., Tomkins, G. M. 1971. *Nature New Biol.* 232:206–11
2. Baserga, R. 1976. *Multiplication and Division in Mammalian Cells.* New York: Marcel Dekker. 239 pp.
3. Prescott, D. M. 1976. *Adv. Genet.* 18:99–177
4. Prescott, D. M. 1976. *Reproduction of Eukaryotic Cells.* New York: Academic. 177 pp.
5. Mitchison, J. M. 1971. *The Biology of the Cell Cycle.* Cambridge: The University Press. 313 pp.
6. Levine, A. J. 1976. *Cancer Res.* 36: 4278–331
7. Nicolson, G. L. 1976. *Biochim. Biophys. Acta* 458:1–72
8. Epifanova, O. I., Abuladze, M. K., Zosimovskaya, A. I. 1975. *Exp. Cell Res.* 92:23–30
9. Martin, R. G., Stein, S. 1976. *Proc. Natl. Acad. Sci. USA* 73:1655–59
10. Rossini, M., Lin, J. C., Baserga, R. 1976. *J. Cell. Physiol.* 88:1–11
11. Dell'Orco, R. T., Crissman, H. A., Steinkamp, J. A., Kraemer, P. M. 1975. *Exp. Cell Res.* 92:271–74
12. Rubin, H., Steiner, R. 1975. *J. Cell. Physiol.* 85:261–70
13. Shields, R., Martin, J. A. 1977. *J. Cell. Physiol.* 91:345–56
14. Smith, J. A., Martin, L. 1973. *Proc. Natl. Acad. Sci. USA* 70:1263–67
15. Robinson, J. H., Smith, J. A., Totty, N. F., Riddle, P. N. 1976. *Nature* 262:298–300
16. Shields, R. 1977. *Nature* 267:704–7
17. Minor, P. D., Smith, J. A. 1974. *Nature* 248:241–43
18. Brooks, R. F. 1975. *J. Cell. Physiol.* 86:369–78
19. Brooks, R. F. 1976. *Nature* 260:248–50
20. Jimenez de Asua, L., O'Farrell, M., Bennett, D., Clingan, D., Rudland, P. 1977. *Nature* 265:151–53
21. Pardee, A. B. 1974. *Proc. Natl. Acad. Sci. USA* 71:1286–90
22. Baker, M. E. 1976. *Biochem. Biophys. Res. Commun.* 68:1059–65
23. Lindgren, A., Westermark, B. 1976. *Exp. Cell Res.* 99:357–62
24. Pardee, A. B., James, L. J. 1975. *Proc. Natl. Acad. Sci. USA* 72:4994–98
25. Bartholomew, J. C., Yokota, H., Ross, P. 1976. *J. Cell. Physiol.* 88:277–86
26. Schiaffonati, L., Baserga, R. 1977. *Cancer Res.* 37:541–45
27. Holley, R. W., Baldwin, J. H., Kiernan,

J. A., Messmer, T. O. 1976. *Proc. Natl. Acad. Sci. USA* 73:3229–32

28. Lindgren, A., Westermark, B. 1977. *Exp. Cell Res.* 104:293–99
29. Fodge, D. W., Rubin, H. 1975. *Nature* 257:804–6
30. Fernandez-Pol, J. A., Bono, V. H., Johnson, G. S. 1977. *Proc. Natl. Acad. Sci. USA* 74:2889–93
31. Rupniak, H. T., Paul, D. 1977. In *Advances in Polyamine Research*, ed. R. A. Campbell. New York: Raven Press. In press
32. Hatten, M. E., Horowitz, A. F., Burger, M. M. 1977. *Exp. Cell Res.* 107:31–34
33. Gill, G. N., Weidman, E. R. 1977. *J. Cell. Physiol.* 92:65–76
34. Baker, M. E. 1975. *Exp. Cell Res.* 95:121–26
35. Robinson, J. H., Smith, J. A. 1976. *J. Cell Physiol.* 89:111–22
36. Skehan, P. 1976. *Exp. Cell Res.* 97:184–92
37. Temin, H. M. 1971. *J. Cell. Physiol.* 78:161–70
38. Lindgren, A., Westermark, B., Pontin, J. 1975. *Exp. Cell Res.* 95:311–19
39. Bartholomew, J. C., Neff, N. T., Ross, P. A. 1976. *J. Cell. Physiol.* 89:251–58
40. Canagaratna, M. C. P., Riley, P. A. 1975. *J. Cell. Physiol.* 85:271–82
41. Holley, R. W., Kiernan, J. A. 1974. *Proc. Natl. Acad. Sci. USA* 71:2942–45
42. Kamely, D., Rudland, P. 1976. *Nature* 260:51–53
43. Tobey, R. A. 1973. In *Methods in Cell Biology*, ed. D. M. Prescott, VI:67–112. New York: Academic
44. Paul, D., Walter, S. 1975. *J. Cell. Physiol.* 85:113–24
45. Popescu, N. C., Casto, B. C., DiPaolo, J. A. 1975. *J. Cell. Physiol.* 86:599–604
46. Armelin, M. C. S., Armelin, H. A. 1977. *Nature* 265:148–51
47. Mannino, R. J., Burger, M. M. 1975. *Nature* 256:19–22
48. Dulbecco, R., Elkington, J. 1973. *Nature* 246:197–99
49. Thrash, C. R., Cunningham, D. D. 1975. *J. Cell. Physiol.* 86:301–10
50. Fodge, D. W., Rubin, H. 1975. *J. Cell. Physiol.* 85:635–42
51. Hassell, J., Engelhardt, D. L. 1977. *Exp. Cell Res.* 107:159–67
52. Stoker, M. G. P. 1973. *Nature* 246:200–3
53. Froelich, J. E., Anastassiades, T. P. 1975. *J. Cell. Physiol.* 86:567–80
54. Skehan, P., Friedman, S. J. 1976. *Exp. Cell Res.* 101:315–22
55. Brownstein, B. L., Rozengurt, E., Ji-

menez de Asua, L., Stoker, M. 1975. *J. Cell. Physiol.* 85:579–85
56. Westermark, B. 1977. *Proc. Natl. Acad. Sci. USA* 74:1619–21
57. Weissfeld, A. S., Rouse, H. 1977. *J. Cell Biol.* 73:200–205
58. Grummt, F., Grummt, I. 1976. *Eur. J. Biochem.* 64:307–12
59. Leffert, H. L., Weinstein, D. B. 1976. *J. Cell Biol.* 70:20–32
60. Sokawa, Y., Watanabe, Y., Watanabe, Y., Kawade, Y. 1977. *Nature* 268:236–38
61. Otsuka, H., Moskowitz, M. 1976. *J. Cell. Physiol.* 87:213–20
62. Burstin, S. J., Meiss, H. K., Basilico, C. 1974. *J. Cell. Physiol.* 84:397–408
63. Kohn, A. 1975. *Exp. Cell Res.* 94:15–22
64. Bolen, J. B., Smith, G. L. 1977. *J. Cell. Physiol.* 91:441–48
65. Thomas, D. B., Lingwood, C. A. 1975. *Cell* 5:37–44
66. Klevecz, R. R. 1976. *Proc. Natl. Acad. Sci. USA* 73:4012–16
67. Robinson, J. H., Smith, J. A., Dee, L. A. 1976. *Exp. Cell Res.* 102:117–26
68. Schneiderman, M. H., Dewey, W. C., Highfield, D. P. 1971. *Exp. Cell Res.* 67:147–55
69. Ley, K. D. 1975. *J. Cell Biol.* 66:95–101
70. Rao, P. N., Prasad, S. S., Wilson, B. A. 1977. *Proc. Natl. Acad. Sci. USA* 74:2869–73
71. Naha, P. M., Meyer, A. L., Hewitt, K. 1975. *Nature* 258:49–53
72. Tenner, A., Zieg, J., Scheffler, I. E. 1977. *J. Cell. Physiol.* 90:145–60
73. Crane, M. St. J., Thomas, D. B. 1976. *Nature* 261:205–8
74. Ling, V. 1977. *J. Cell. Physiol.* 91:209–24
75. Gospodarowicz, D., Moran, J. S. 1976. *Ann. Rev. Biochem.* 45:531–58
76. Rutherford, R. B., Ross, R. 1976. *J. Cell Biol.* 69:196–203
77. Westermark, B., Wasteson, A. 1976. *Exp. Cell Res.* 98:170–74
78. Antoniades, H. N., Stathakos, D., Scher, C. D. 1975. *Proc. Natl. Acad. Sci. USA* 72:2635–39
79. Antoniades, H. N., Scher, C. D. 1977. *Proc. Natl. Acad. Sci. USA* 74:1973–77
80. Burk, R. R. 1976. *Exp. Cell Res.* 101:293–98
81. Bourne, H. R., Rozengurt, E. 1976. *Proc. Natl. Acad. Sci. USA* 73:4555–59
82. Wolf, V., Kohler, N., Roehm, C., Lipton, A. 1975. *Exp. Cell Res.* 92:63–69
83. Klagsbrun, M., Langer, R., Levenson, R., Smith, S., Lillehei, C. 1977. *Exp. Cell Res.* 105:99–108

84. Jimenez de Asua, L., Clingan, D., Rudland, P. S. 1975. *Proc. Natl. Acad. Sci. USA* 72:2724–28
85. Jimenez de Asua, L., Carr, B., Clingan, D., Rudland, P. 1977. *Nature* 265:450–52
86. Nishikawa, K., Armelin, H. A., Sato, G. 1975. *Proc. Natl. Acad. Sci. USA* 72:483–87
87. Hayashi, I., Sato, G. H. 1976. *Nature* 259:132–34
88. Richman, R. A., Claus, T. H., Pilkis, S. J., Friedman, D. L. 1976. *Proc. Natl. Acad. Sci. USA* 73:3589–93
89. Leffert, H. L., Moran, T., Boorstein, R., Koch, K. S. 1977. *Nature* 267:58–61
90. Rechler, M. M., Podskalny, J. M. 1976. *Nature* 259:134–36
91. Rechler, M. M., Podskalny, J. M., Nissley, S. P. 1977. *J. Biol. Chem.* 252:3898–910
92. Raizada, M. K., Perdue, J. F. 1976. *J. Biol. Chem.* 251:6445–55
93. Thomopoulos, P., Roth, J., Lovelace, E., Pastan, I. 1976. *Cell* 8:417–23
94. Cidlowski, J. A., Michaels, G. A. 1977. *Nature* 266:643–45
95. Smith, K. A., Crabtree, G. R., Kennedy, S. J., Munck, A. U. 1977. *Nature* 267:523–25
96. Reich, E., Rifkin, D., Shaw, E., eds. 1975. *Proteases and Biological Control.* New York: Cold Spring Harbor Laboratory
97. Teng, N. N. H., Cheng, L. B. 1975. *Proc. Natl. Acad. Sci. USA* 72:413–17
98. Blumberg, P. M., Robbins, P. W. 1975. *Cell* 6:137–47
99. Noonan, K. D. 1976. *Nature* 259:573–76
100. Zetter, B. R., Sun, T.-t., Chen, L. B., Buchanan, J. M. 1977. *J. Cell. Physiol.* 92:233–40
101. Brown, M., Kiehn, D. 1977. *Proc. Natl. Acad. Sci. USA* 74:2874–78
102. Chou, I.-N., O'Donnell, S. P., Black, P. H., Roblin, R. O. 1977. *J. Cell. Physiol.* 91:31–37
103. Dulbecco, R., Elkington, J. 1975. *Proc. Natl. Acad. Sci. USA* 72:1584–88
104. Swierenga, S. H. H., MacManus, J. P., Whitfield, J. F. 1976. *In Vitro* 12:31–36
105. Boynton, A. L., Whitfield, J. F., Isaacs, R. J. 1976. *In Vitro* 12:120–23
106. Whitfield, J. F., MacManus, J. P., Rixon, R. H., Boynton, A. L., Youdale, T., Swierenga, S. 1976. *In Vitro* 12:1–18
107. Rubin, H. 1975. *Proc. Natl. Acad. Sci. USA* 72:3551–55
108. Rubin, H. 1975. *Proc. Natl. Acad. Sci. USA* 72:1676–80

109. Pardee, A. B. 1975. *Biochim. Biophys. Acta* 417:153–72
110. Whittenberger, B., Glaser, L. 1977. *Proc. Natl. Acad. Sci. USA* 74:2251–55
111. Lipkin, G., Krecht, M. E. 1976. *Exp. Cell Res.* 102:341–48
112. Ohnishi, T., Ohshima, E., Ohtsuka, M. 1975. *Exp. Cell Res.* 93:136–42
113. Underhill, C. B., Keller, J. M. 1976. *J. Cell. Physiol.* 89:53–64
114. Cohn, R. H., Cassiman, J. J., Bernfield, M. R. 1976. *J. Cell Biol.* 71:280–94
115. Vannucchi, S., Chiarugi, V. P. 1977. *J. Cell. Physiol.* 90:503–10
116. Moscatelli, D., Rubin, H. 1975. *Nature* 254:65–66
117. Clarke, G. D., Smith, C. 1973. *J. Cell. Physiol.* 81:125–32
118. Hakomori, S. I. 1975. *Biochim. Biophys. Acta* 417:55–89
119. Shur, B. D., Roth, S. 1975. *Biochim. Biophys. Acta* 415:473–512
120. La Mont, J. T., Gammon, M. T., Isselbacher, K. J. 1977. *Proc. Natl. Acad. Sci. USA* 74:1086–90
121. Pinto da Silva, P., Martinez-Palomo, A. 1975. *Proc. Natl. Acad. Sci. USA* 72:572–76
122. Micklem, K. J., Abra, R. M., Knutton, S., Graham, J. M., Pasternak, C. A. 1976. *Biochem. J.* 154:561–66
123. Scott, R. E., Furcht, L. T., Kersey, J. H. 1973. *Proc. Natl. Acad. Sci. USA* 70:3631–35
124. Inbar, M., Yuli, I. Raz, A. 1977. *Exp. Cell Res.* 105:325–35
125. Edelman, G. M. 1976. *Science* 192:218–26
126. McClain, D. A., Eustachio, P. D., Edelman, G. M. 1977. *Proc. Natl. Acad. Sci. USA* 74:666–70
127. Baker, M. E. 1976. *Nature* 262:785–86
128. Gunther, G. R., Wang, J. L., Edelman, G. M. 1976. *Exp. Cell Res.* 98:15–22
129. Schellenberg, R. R., Gillespie, E. 1977. *Nature* 265:741–42
130. Resch, K., Bouillon, D., Gemsa, D., Averdunk, R. 1977. *Nature* 265:349–51
131. Sherline, P., Mundy, G. R. 1977. *J. Cell Biol.* 74:371–76
132. Greene, W. C., Parker, C. M., Parker, C. H. W. 1976. *Exp. Cell Res.* 103:109–17
133. Holley, R. 1972. *Proc. Natl. Acad. Sci. USA* 69:2840–41
134. Pardee, A. B. 1976. In *International Workshop on Cell Surfaces and Malignancy,* pp. 117–24. Bethesda, Md.: National Institutes of Health.
135. Plagemann, P. G. W., Richey, D. P. 1974. *Biochim. Biophys. Acta* 344:263–306

136. Rozengurt, E., Heppel, L. A. 1975. *Proc. Natl. Acad. Sci. USA* 72:4992–95
137. Tupper, J. T., Zorgniotti, F., Mills, B. 1977. *J. Cell. Physiol.* 91:429–40
138. Elligsen, J. D., Thompson, J. E., Frey, H. E., Kruuv, J. 1974. *Exp. Cell Res.* 87:233–40
139. Eilam, Y., Vinkler, C. 1976. *Biochim. Biophys. Acta* 433:393–403
140. Oxender, D. L., Lee, M., Cecchini, G. 1977. *J. Biol. Chem.* 252:2680–83
141. Parnes, J. R., Garvey, T. Q., Isselbacher, K. J. 1976. *J. Cell. Physiol.* 89:789–94
142. Lever, J. A. 1976. *J. Cell. Physiol.* 89:779–88
143. Pohjanpelto, P. 1976. *J. Cell Biol.* 68:512–20
144. Rozengurt, E., Stein, W. D., Wigglesworth, N. M. 1977. *Nature* 267:442–44
145. Quinlan, D. C., Hochstadt, J. 1974. *Proc. Natl. Acad. Sci. USA* 71:5000–5003
146. Hassell, J. A., Colby, C., Romano, A. H. 1975. *J. Cell. Physiol.* 86:37–46
147. Dubrow, R., Pardee, A. B., Pollack, R. 1978. *J. Cell. Physiol.* In press
148. Greenberg, D. B., Barsh, G. S., Ho, T.-S., Cunningham, D. D. 1977. *J. Cell. Physiol.* 90:193–210
149. Barsh, G. S., Greenberg, D. B., Cunningham, D. D. 1977. *J. Cell. Physiol.* 92:115–28
150. Thrash, C. R., Cunningham, D. D. 1974. *Nature* 252:45–47
151. Naiditch, W. P., Cunningham, D. D. 1977. *J. Cell. Physiol.* 92:319–32
152. Hassell, J. A., Engelhardt, D. L. 1976. *Biochemistry* 15:1375–81
153. Mostafapour, M. K., Green, H. 1975. *J. Cell. Physiol.* 86:313–20
154. Becker, H., Stanners, C. P. 1972. *J. Cell. Physiol.* 80:51–62
155. Melero, J. A., Salas, M. L., Salas, J. 1976. *Eur. J. Biochem.* 67:341–48
156. Jazwinski, S. M., Wang, J. L., Edelman, G. M. 1976. *Proc. Natl. Acad. Sci. USA* 73:2231–35
157. Warburton, M. J., Poole, B. 1977. *Proc. Natl. Acad. Sci. USA* 74:2427–31
158. Epstein, D., Elias-Bishko, S., Hershko, A. 1975. *Biochemistry* 14:5199–204
159. Tanaka, K., Ichihara, A. 1976. *Exp. Cell Res.* 99:1–6
160. Fine, R. E., Taylor, L. 1976. *Exp. Cell Res.* 102:162–68
161. Kaplan, J., Moskowitz, M. 1975. *Biochim. Biophys. Acta* 389:290–305
162. Kaplan, J., Moskowitz, M. 1975. *Biochim. Biophys. Acta* 389:306–13
163. Gunn, J. M., Ballard, F. J., Hanson, R. W. 1976. *J. Biol. Chem.* 251:3586–93
164. Mauck, J. C., Green, H. 1973. *Proc. Natl. Acad. Sci. USA* 70:2819–23
165. Purtell, M. J., Anthony, D. D. 1975. *Proc. Natl. Acad. Sci. USA* 72:3315–19
166. Johnson, L. F., Williams, J. G., Abelson, H. T., Green, H., Penman, S. 1975. *Cell* 4:69–75
167. Johnson, L. F., Levis, R., Abelson, H. T., Green, H., Penman, S. 1976. *J. Cell Biol.* 71:933–38
168. Johnson, L. F., Penman, S., Green, H. 1976. *J. Cell. Physiol.* 87:141–46
169. Levis, R., McReynolds, L., Penman, S. 1977. *J. Cell. Physiol.* 90:485–502
170. Rudland, P. S. 1974. *Proc. Natl. Acad. Sci. USA* 71:750–54
171. Bandman, E., Gurney, T. 1975. *Exp. Cell Res.* 90:159–68
172. Rudland, P. S., Weil, S., Hunter, A. R. 1975. *J. Mol. Biol.* 96:745–66
173. Abelson, H. T., Johnson, L. F., Penman, S., Green, H. 1974. *Cell* 1:161–67
174. Williams, T. G., Penman, S. 1975. *Cell* 6:197–206
175. Getz, M. J., Elder, P. K., Benz, E. W., Stephens, R. E., Moses, H. L. 1976. *Cell* 7:255–65
176. Pastan, I., Johnson, G. S. 1974. *Adv. Cancer Res.* 19:303–36
177. Chlapowski, F. J., Kelly, L. A., Butcher, R. W. 1975. *Adv. Cyclic Nucleotide Res.* 6:245–310
178. Pastan, I. H., Johnson, G. S., Anderson, W. B. 1975. *Ann. Rev. Biochem.* 44:491–522
179. Goldberg, N. D., Haddox, M. K., Dunham, E., Lopez, C., Hadden, J. W. 1976. In *Control of Proliferation in Animal Cells*, ed. B. Clarkson, R. Baserga, p. 609. New York: Cold Spring Harbor Laboratory. 1029 pp.
180. Seifert, W. E., Rudland, P. S. 1974. *Nature* 250:138–40
181. Seifert, W. E., Rudland, P. S. 1974. *Proc. Natl. Acad. Sci. USA* 71:4920–24
182. Moens, W., Vokaer, A., Kram, R. 1975. *Proc. Natl. Acad. Sci. USA* 72:1063–67
183. Rudland, P. S., Gospodarowicz, D., Seifert, W. 1974. *Nature* 250:741–43
184. Nesbitt, J. A., Anderson, W. B., Miller, Z., Pastan, I., Russell, T. R., Gospodarowicz, D. 1976. *J. Biol. Chem.* 251:2344–51
185. Miller, Z., Lovelace, E., Gallo, M., Pastan, I. 1975. *Science* 190:1213–15
186. Zeilig, C. E., Goldberg, N. D. 1977. *Proc. Natl. Acad. Sci. USA* 74:1052–54
187. Costa, M., Gerner, E. W., Russell, D. H. 1976. *Biochim. Biophys. Acta* 425:246
188. Burger, M. M., Bombik, B. M., Breck-

enridge, B. M., Sheppard, J. R. 1972. *Nature New Biol.* 239:161–64

189. Makman, M. H., Klein, M. I. 1972. *Proc. Natl. Acad. Sci. USA* 69:456–61
190. Millis, A. J. T., Forrest, G. A., Pious, D. A. 1974. *Exp. Cell Res.* 83:335–42
191. Anderson, W. B., Russell, T. R., Carckman, R. A., Pastan, I. 1973. *Proc. Natl. Acad. Sci. USA* 70:3802–8
192. Pledger, W. J., Thompson, W. J., Strada, S. J. 1975. *Nature* 256:729–31
193. Kuo, J. F., Greengard, P. 1969. *Proc. Natl. Acad. Sci. USA* 64:1349–55
194. Kallas, J. 1977. *Nature* 265:705–9
195. Costa, M., Gerner, E. W., Russell, D. H. 1976. *J. Biol. Chem.* 251:3313–20
196. Gharret, A. J., Malkinson, A. M., Sheppard, J. R. 1976. *Nature* 264:673–76
197. Daniel, V., Litwack, G., Tomkins, G. M. 1973. *Proc. Natl. Acad. Sci. USA* 70:76–81
198. Coffino, P., Gray, J. W., Tomkins, G. M. 1975. *Proc. Natl. Acad. Sci. USA* 72:878–82
199. Insel, D. A., Bourne, H. R., Coffino, P., Tomkins, G. M. 1975. *Science* 190:896–99
200. Maller, J. L., Krebs, E. G. 1977. *J. Biol. Chem.* 252:1712–19
201. Fillingame, R. H., Jorstaed, G. M., Morris, D. R. 1975. *Proc. Natl. Acad. Sci. USA* 72:4042–45
202. Heby, O., Marton, L. J., Zardij, L., Russell, D. H., Baserga, R. 1975. *Exp. Cell Res.* 90:8–14
203. Boynton, A. L., Whitfield, J. F., Isaacs, R. J. 1976. *J. Cell Physiol.* 89:481–88
204. Nissley, S. P., Passamani, J., Short, P. 1976. *J. Cell. Physiol.* 89:393–98
205. Yuspa, S. H., Lichti, U., Ben, T., Patterson, E., Hennings, H., Slaga, T. J., Colburn, N., Kelsey, W. 1976. *Nature* 262:402–4
206. Gazdar, A. F., Stull, H. B., Kilton, L. J., Bachrach, U. 1976. *Nature* 262:696–98
207. Momont, P. S., Bohlen, P., McCann, P. P., Bey, P., Schuber, F., Tardif, C. 1976. *Proc. Natl. Acad. Sci. USA* 73:1626–30
208. Clark, J. L., Duffey, P. 1976. *Arch. Biochem. Biophys.* 172:551–57
209. Baserga, R., Nicolini, C. 1976. *Biochim. Biophys. Acta* 458:109–34
210. Rao, P. N., Wilson, B., Puck, T. T. 1977. *J. Cell. Physiol.* 91:131–42
211. Nicolini, C., Giaretti, W., De Saive, C., Kendall, F. 1977. *Exp. Cell Res.* 106:119–25
212. Collins, J. M. 1977. *J. Biol. Chem.* 252:141–47
213. Klevecz, R. R., Keniston, B. A., Deaven, L. L. 1975. *Cell* 5:195–203

214. Baserga, R. 1974. *Life Sci.* 15:1057–71
215. Gerner, E. W., Meyn, R. E., Humphrey, R. M. 1976. *J. Cell. Physiol.* 87:277–88
216. Vidali, G., Karn, J., Allfrey, V. G. 1975. *Proc. Natl. Acad. Sci. USA* 72:4450–54
217. Yeoman, L. C., Taylor, C. W., Jordan, J. J., Busch, H. 1975. *Exp. Cell Res.* 91:207–15
218. Yeoman, L. C., Seeber, S., Taylor, C. W., Fernbach, D. J., Falletta, J. M., Jordan, J. J., Busch, H. 1976. *Exp. Cell Res.* 100:47–55
219. Pall, M. L. 1974. *Differentiation* 2:363–65
220. DeMorales, M. M., Blat, C., Harel, L. 1974. *Exp. Cell Res.* 86:111–19
221. Bombik, B. M., Baserga, R. 1976. *Biochim. Biophys. Acta* 442:343–57
222. Sens, W., Unsold, H. J., Knippers, R. 1976. *Eur. J. Biochem.* 65:263–69
223. Salas, J., Green, H. 1971. *Nature New Biol.* 229:165–69
224. Tsai, R. L., Green, H. 1973. *J. Mol. Biol.* 73:307–16
225. Stein, G. H. 1975. *Exp. Cell Res.* 90:237–48
226. Melero, J. A., Salas, M. L., Salas, J., Macpherson, I. A. 1975. *J. Biol. Chem.* 250:3683–89
227. Choe, B. K., Rose, N. R. 1974. *Exp. Cell Res.* 83:261–70; 271–80
228. Fox, T. O., Pardee, A. B. 1971. *J. Biol. Chem.* 246:6159–65
229. Liskay, R. M. 1977. *Proc. Natl. Acad. Sci. USA* 74:1622–25
230. Brown, R. L., Clark, R. W., Chiu, J.-F., Stubblefield, E. 1977. *Exp. Cell Res.* 104:207–13
231. Bostock, C. J. 1970. *Exp. Cell Res.* 60:16–26
232. Blenkinsopp, W. K. 1968. *Exp. Cell Res.* 50:265–76
233. Cairns, J. 1966. *J. Mol. Biol.* 15:372–73
234. Huberman, J. A., Riggs, A. D. 1968. *J. Mol. Biol.* 32:327–41
235. Painter, R. B., Schaeffer, A. W. 1969. *J. Mol. Biol.* 45:467–79
236. Painter, R. B., Schaeffer, A. W. 1971. *J. Mol. Biol.* 58:289–95
237. Housman, D., Huberman, J. A. 1975. *J. Mol. Biol.* 94:173–81
238. Van't Hof, J. 1976. *Exp. Cell Res.* 103:395–403
239. Hori, T. A., Lark, K. G. 1976. *Nature* 259:504–5
240. Callan, H. G. 1973. *Cold Spring Harbor Symp. Quant. Biol.* 38:195–204
241. Blumenthal, A. B., Kriegstein, H. J., Hogness, D. S. 1973. *Cold Spring Harbor Symp. Quant. Biol.* 38:205–24

242. Elgin, S. C. R., Weintraub, H. 1975. *Ann. Rev. Biochem.* 44:725–74
243. Hewish, D. R. 1976. *Nucleic Acids Res.* 3:69–78
244. Rosenberg, B. H. 1976. *Biochem. Biophys. Res. Commun.* 72:1384–91
245. Jay, E., Roychoudhury, R., Wu, R. 1976. *Biochem. Biophys. Res. Commun.* 69:678–86
246. Padmanabhan, R., Padmanabhan, R., Green, M. 1976. *Biochem. Biophys. Res. Commun.* 69:860–67
247. Straus, S. E., Sebring, E. D., Rose, J. A. 1976. *Proc. Natl. Acad. Sci. USA* 73:742–46
248. Tattersall, P. J., Ward, D. C. 1976. *Nature* 263:106–9
249. Dott, P. J., Chuang, C. R., Saunders, G. F. 1976. *Biochemistry* 15:4120–25
250. Cech, T. R., Pardue, M. L. 1976. *Proc. Natl. Acad. Sci. USA* 73:2644–48
251. Amaldi, F., Buongiorno-Nardelli, M., Carnevalli, C., Leoni, L., Mariotti, D., Pomponi, M. 1973. *Exp. Cell Res.* 80:79–87
252. Wang, E. C., Henner, D., Furth, J. J. 1975. *Biochem. Biophys. Res. Commun.* 65:1177–83
253. Urbanczyk, J., Studinski, G. P. 1974. *Biochem. Biophys. Res. Commun.* 59:616–22
254. McGuire, M. S., Center, M. S., Consigli, R. A. 1976. *J. Biol. Chem.* 251:7746–52
255. Maaloe, O., Hanawalt, P. C. 1961. *J. Mol. Biol.* 3:144–55
256. Hereford, L. M., Hartwell, L. H. 1974. *J. Mol. Biol.* 84:445–61
257. Hand, R. 1975. *J. Cell Biol.* 67:761–73
258. Gautschi, J. R. 1974. *J. Mol. Biol.* 84:223–29
259. Evans, H. H., Littmann, S. R., Evans, T. E., Brewer, E. N. 1976. *J. Mol. Biol.* 104:169–84
260. Cavalieri, L. F., Carroll, E. 1975. *Biochem. Biophys. Res. Commun.* 67:1360–69
261. Clay, W. F., Katterman, F. R. H., Bartels, P. G. 1975. *Proc. Natl. Acad. Sci. USA* 72:3134–38
262. Infante, A. A., Firshein, W., Hobart, P., Murray, L. 1976. *Biochemistry* 15:4810–17
263. Genta, V. M., Kaufman, D. G., Kaufmann, W. K. 1976. *Nature* 259:502–3
264. Kandutsch, A. A., Chen, H. W. 1977. *J. Biol. Chem.* 252:409–15
265. Sparvoli, E., Galli, M. G., Mosca, A., Paris, G. 1976. *Exp. Cell Res.* 97:74–82
266. Fakan, S., Hancock, R. 1974. *Exp. Cell Res.* 83:95–102
267. Weissbach, A. 1977. *Ann. Rev. Biochem.* 46:25–47
268. Hildebrand, C. E., Walters, R. A. 1977. *Biochem. Biophys. Res. Commun.* 73:157–63
269. Gautschi, J. R., Clarkson, J. M. 1975. *Eur. J. Biochem.* 50:403–12
270. Okazaki, T., Okazaki, R. 1969. *Proc. Natl. Acad. Sci. USA* 64:1242–48
271. Blumenthal, A. B., Clark, E. J. 1977. *Exp. Cell Res.* 105:15–26
272. Tseng, B. Y., Goulian, M. 1975. *J. Mol. Biol.* 99:317–37
273. Spaeren, U., Schroder, K., Sudbery, C., Bjorklid, E., Prydz, H. 1975. *Biochim. Biophys. Acta* 395:413–21
274. Krokan, H., Bjorklid, E., Prydz, H. 1975. *Biochemistry* 14:4227–32
275. Reinhard, P., Burkhalter, M., Gautschi, J. R. 1977. *Biochim. Biophys. Acta* 474:500–11
276. Brewer, E. N. 1975. *Biochim. Biophys. Acta* 402:363–71
277. Krokan, H., Cooke, L., Prydz, H. 1975. *Biochemistry* 14:4233–37
278. Kuebbing, D., Diaz, A. T., Werner, R. 1976. *J. Mol. Biol.* 108:55–66
279. Edenberg, H. J., Huberman, J. A. 1975. *Ann. Rev. Genet.* 9:245–84
280. Waqar, M. A., Huberman, J. A. 1975. *Cell* 6:551–57
281. Taylor, J. H., Wu, M., Erickson, L. C., Kurek, M. P. 1975. *Chromosoma* 53:175–89
282. Probst, H., Gentner, P. R., Hofstätter, T., Jenke, S. 1974. *Biochim. Biophys. Acta* 340:361–73
283. Mendelsohn, J., Castagnola, J. M., Goulian, M. 1975. *Biochim. Biophys. Acta* 407:283–91
284. Pearson, C. K., Davis, P. B., Taylor, A., Amos, N. A. 1976. *Eur. J. Biochem.* 62:451–59
285. Herrick, G., Alberts, B. 1976. *J. Biol. Chem.* 251:2124–32
286. Otto, B., Baynes, M., Knippers, R. 1977. *Eur. J. Biochem.* 73:17–24
287. Enomoto, T., Yamata, M. 1976. *Biochem. Biophys. Res. Commun.* 71:122–27
288. Case, S. T., Baker, R. F. 1975. *J. Mol. Biol.* 98:69–92
289. Sinha, N. K., Snustad, D. P. 1971. *J. Mol. Biol.* 62:267–71
290. Champoux, J. J., McConaughy, B. L. 1976. *Biochemistry* 15:4638–42
291. Champoux, J. J. 1976. *Proc. Natl. Acad. Sci. USA* 73:3488–91
292. Thomas, T. L., Patel, G. L. 1976. *Proc. Natl. Acad. Sci. USA* 73:4364–68
293. Rosenberg, B. H., Ungers, G., Deutsch,

J. F. 1976. *Nucleic Acids Res.* 3: 3305–11

294. Bina-Stein, M., Vogel, T., Singer, D. S., Singer, M. 1976. *J. Biol. Chem.* 251:7363–66

295. Keller, W. 1975. *Proc. Natl. Acad. Sci. USA* 72:2550–54

296. Mattoccia, E., Attardi, D. G., Tocchini-Valentini, G. P. 1976. *Proc. Natl. Acad. Sci. USA* 73:4551–54

297. Kornberg, A. 1974. *DNA Synthesis* San Francisco: W. H. Freeman. 399 pp.

298. Weissbach, A. 1975. *Cell* 5:101–8

299. Herrick, G., Spear, B. B., Veomett, G. 1976. *Proc. Natl. Acad. Sci. USA* 73:1136–39

300. Foster, D. N., Gurney, T. Jr. 1976. *J. Biol. Chem.* 251:7893–98

301. Spadari, S., Weissbach, A. 1975. *Proc. Natl. Acad. Sci. USA* 72:503–7

302. Chiu, R. W., Baril, E. F. 1975. *J. Biol. Chem.* 250:7951–57

303. Craig, R. C., Keir, H. M. 1975. *Biochem. J.* 145:215–24

304. Craig, R. C., Costello, P. A., Keir, H. M. 1975. *Biochem. J.* 145:233–40

305. Roufa, D. J., Moses, R. E., Reed, S. J. 1975. *Arch. Biochem. Biophys.* 167: 547–59

306. Tsuruo, T., Hirayama, K., Ukita, T. 1975. *Biochim. Biophys. Acta* 383: 274–81

307. Hecht, N. B. 1975. *Biochim. Biophys. Acta* 383:388–98

308. Hurai, K., Watanabe, Y. 1976. *Biochim. Biophys. Acta* 447:328–39

309. Bertazzoni, U., Stefanini, M., Noy, G. P., Giulotto, E., Nuzzo, F., Falaschi, A., Spadari, S. 1976. *Proc. Natl. Acad. Sci. USA* 73:785–89

310. Wang, H. F., Popenoe, E. A. 1977. *Biochim. Biophys. Acta* 474:98–108

311. Byrnes, J. J., Downey, K. M., Black, V. L., So, A. G. 1976. *Biochemistry* 15:2817–23

312. Söderhäll, S., Lindahl, T. 1975. *J. Biol. Chem.* 250:8438–44

313. Söderhäll, S. 1976. *Nature* 260:640–42

314. Spalding, J., Kajiwara, K., Mueller, G. C. 1966. *Proc. Natl. Acad. Sci. USA* 56:1535–42

315. Gallwitz, D. 1975. *Nature* 258:247–48

316. Borun, T. W., Gabrielli, F., Ajiro, K., Zweidler, A., Baglioni, C. 1975. *Cell* 4:59–67

317. Breindl, M., Gallwitz, D. 1974. *Eur. J. Biochem.* 45:91–97

318. Weintraub, H. 1972. *Nature* 240: 449–53

319. Tsanev, R., Russev, G. 1974. *Eur. J. Biochem.* 43:257–63

320. Jackson, V., Granner, D. K., Chalkley, R. 1975. *Proc. Natl. Acad. Sci. USA* 72:4440–44

321. Jackson, V., Granner, D., Chalkley, R. 1976. *Proc. Natl. Acad. Sci. USA* 73:2266–69

322. Seale, R. L. 1975. *Nature* 255:247-49

323. Moll, R., Wintersberger, E. 1976. *Proc. Natl. Acad. Sci. USA* 73:1863–67

324. Adamson, E. D., Woodland, H. R. 1974. *J. Mol. Biol.* 88:263–85

325. Balhorn, R., Jackson, V., Granner, D., Chalkley, R. 1975. *Biochemistry* 14: 2504–11

326. Gurley, L. R., Walters, R. A., Tobey, R. A. 1975. *J. Biol. Chem.* 250:3936–44

327. Gurley, L. R., Walters, R. A., Tobey, R. A. 1974. *Arch. Biochim. Biophys.* 164:469–77

328. Stein, G. S., Park, W. D., Stein, J. L., Lieberman, M. W. 1976. *Proc. Natl. Acad. Sci. USA* 73:1466–70

329. Hahn, E. C. 1974. *Biochem. Biophys. Res. Commun.* 57:635–40

330. Billen, D., Olson, A. C. 1976. *J. Cell Biol.* 69:732–36

331. Miller, M. R., Castellot, J. J., Pardee, A. B. 1978. *Biochemistry* In press

332. Hilderman, R. H., Goldblatt, P. J., Deutscher, M. P. 1975. *J. Biol. Chem.* 250:4796–801

333. Seki, S., Oda, T. 1977. *Cancer Res.* 37:137–44

334. Seki, S., Lemahieu, M., Mueller, G. C. 1975. *Biochim. Biophys. Acta* 378: 333–43

335. Berger, N. A., Johnson, E. S. 1976. *Biochim. Biophys. Acta* 425:1–17

336. Benbow, R. M., Ford, C. C. 1975. *Proc. Natl. Acad. Sci. USA* 72:2437–41

337. Su, R. T., DePamphilis, M. L. 1976. *Proc. Natl. Acad. Sci. USA* 73:3466–70

338. Attardi, D. G., Martini, G., Mattoccia, E., Tocchini-Valentini, G. P. 1976. *Proc. Natl. Acad. Sci. USA* 73:554–58

339. Edenberg, H. J., Waqar, M. A., Huberman, J. A. 1976. *Proc. Natl. Acad. Sci. USA* 73:4392–96

340. Bogenhagen, D., Clayton, D. A. 1976. In *Genetics and Biogenesis of Chloroplasts and Mitochondria*, ed. T. Bucher, W. Neupert, W. Sebald, S. Werner, pp. 597–604. Amsterdam: North Holland, 895 pp.

341. Fischer-Fantuzzi, L., Marin, G., Vesco, C. 1975. *Eur. J. Biochem.* 60:505–11

342. Fisher, P. B., Horwitz, M. S. 1977. *J. Virol.* 22:340–45

343. Albring, M., Griffith, J., Attardi, G. 1977. *Proc. Natl. Acad. Sci. USA* 74:1348–52

344. Rao, A. P., Rao, P. N. 1976. *J. Natl. Cancer Inst.* 57:1139–43
345. De Roeper, A., Smith, J. A., Watt, R. A., Barry, J. M. 1977. *Nature* 265:469–70
346. Rao, P. N., Johnson, R. T. 1970. *Nature* 225:159–64
347. Stambrook, P. J., Velez, C. 1976. *Exp. Cell Res.* 99:57-62
348. Krishan, A., Frei, E. III. 1976. *Cancer Res.* 36:143–50
349. Barlogie, B., Drewinko, B., Schumann, J., Freireich, E. J. 1976. *Cancer Res.* 36:1182–87
350. Tobey, R. A. 1975. *Nature* 254:245–47
351. Barranco, S. C., Novak, J. K., Humphrey, R. M. 1975. *Cancer Res.* 35:1194–1204
352. Wheatley, D. N., Henderson, J. Y. 1974. *Nature* 247:281–83
353. Stein, S. M., Berestecky, J. M. 1975. *J. Cell. Physiol.* 85:243–49
354. Kidwell, W. R. 1975. *J. Biochem. Tokyo*

77:6P–7P
355. Sanger, J. W. 1975. *Proc. Natl. Acad. Sci. USA* 72:1913–16
356. Snyder, J. A., McIntosh, J. R. 1976. *Ann. Rev. Biochem.* 45:699–720
357. DeFilippes, W., Bono, V. H., Dion, R. L., Johns, D. G. 1975. *Biochem. Pharmacol.* 24:1735–38
358. Piras, R., Piras, M. M. 1975. *Proc. Natl. Acad. Sci. USA* 72:1161–65
359. Balhorn, R., Jackson, V., Granner, D., Chalkley, R. 1975. *Biochemistry* 14:2504–11
360. Rao, P. N., Smith, M. L. 1976. *Exp. Cell Res.* 103:213–18
361. Johnson, R. T., Rao, P. N. 1970. *Nature* 226:717–22
362. Stroud, A. N., Nathan, R., Harami, S. 1975. *In Vitro* 11:61–68
363. Hatzfeld, J., Buttin, G. 1975. *Cell* 5:123–29
364. Albrecht-Buehler, G. 1977. *J. Cell Biol.* 72:595–603

Ann. Rev. Biochem. 1978. 47:751–77
Copyright © 1978 by Annual Reviews, Inc. All rights reserved

THE PLASMA LIPOPROTEINS: STRUCTURE AND METABOLISM

♦988

Louis C. Smith, Henry J. Pownall, and Antonio M. Gotto Jr.

Department of Medicine, Baylor College of Medicine
and The Methodist Hospital, Houston, Texas 77030

CONTENTS

0066-4154/78/0701-0751$01.00

PERSPECTIVES AND SUMMARY

In plasma, lipids are integral components of several macromolecular lipid-protein complexes, termed lipoproteins, which have characteristic sizes, densities, and compositions. All lipoproteins contain protein components, called apoproteins, and polar lipids in a surface film surrounding a neutral lipid core. The apoproteins range in molecular weight from 5,700 to 75,000 and are distributed among lipoproteins of different density classes; the primary amino acid sequence of five of the eight major apoproteins is known.

Plasma lipoproteins function to transport lipids in a water-soluble form. Specifically, chylomicrons carry dietary triglyceride from the intestine to nonhepatic tissues for utilization or storage; very low density lipoproteins (VLDL) contain triglyceride made primarily in the liver; the low density lipoproteins (LDL) derive from VLDL catabolism, and the high density lipoproteins (HDL), made in the liver, contain the bulk of the plasma cholesterol. The enzymes, lipoprotein lipase and lecithin cholesterol acyl-transferase, modify the structure of lipoproteins by catalyzing the hydrolysis of triglyceride and by forming cholesteryl ester from cholesterol and phosphatidylcholine, respectively. Furthermore, LDL may regulate de novo cholesterol synthesis in nonhepatic tissues; HDL may promote cholesterol transport from peripheral tissues to the liver.

Physical studies of the apoproteins have shown that they have a low energy of stabilization. These findings suggest that hydrophobic residues are exposed to the aqueous phase, and that these exposed residues may be involved in hydrophobic self-association and in lipid binding. All of the apoproteins sequenced to date contain amphipathic helical regions that are thought to be essential for apoprotein-phospholipid interaction. In LDL, but not in HDL or VLDL, sufficient amounts of cholesteryl ester exist as a separate phase that undergoes cooperative melting. The microviscosity of the apolar regions of lipoprotein increases with increasing apoprotein content.

Studies of lipoprotein biogenesis have shown that rat intestine synthesizes apoA-I as a component of chylomicrons. The hepatic origin of the "arginine-rich" protein has been demonstrated. At the subcellular level, an mRNA for a major apoprotein of chicken VLDL has been translated in vitro. The catabolism of triglyceride-rich lipoproteins is substantially different in humans than it is in rats, the experimental animal principally studied. The primary structure of apoC-II has been determined. The minimal sequence of apoC-II necessary for activation of lipoprotein lipase is contained in the COOH-terminal half of the apoprotein and involves two different portions of the activator. Triglyceride hydrolysis is maximal at a molar ratio of enzyme to apoC-II of 1:1. Lipid transfer between lipoproteins can occur by way of a monomolecular species in the aqueous solution.

In future years, the dynamics of lipid:apoprotein interaction and transport will be correlated with the structure of apoproteins and lipids. Delineation of the chemical structure of the lipoprotein receptors on cell membranes, and the events associated with intracellular cholesterol synthesis, will lead to correlation of the in vitro studies of lipoprotein metabolism with normal and pathological phenomena.

INTRODUCTION

The plasma lipoproteins are lipid-protein complexes that transport lipids in the circulation and regulate lipid synthesis and catabolism. Recent studies of lipoproteins, published since this topic was last reviewed by Morrisett et al (1), have substantially expanded the body of information about the composition and structural properties of lipoproteins and have provided in vitro evidence that correlates lipid structure and function. Herein, we consider recent developments in (a) apoprotein structure and (b) lipoprotein metabolism. Our review focuses on the distribution, composition, structure, and metabolism of the plasma lipoproteins in various species, with the major emphasis on man. We shall use the *A, B, C,* etc nomenclature for identifying the apoproteins (2). We have compiled and related the different nomenclatures for lipoproteins and apoproteins in Table 1. Many other aspects of lipoprotein structure and metabolism are covered in recent reviews (1, 3–7, 49, 154, 155).

Table 1   Composition of human plasma lipoproteins

Properties	Chylomicrons	VLDL	LDL	HDL
Major apoproteins	ApoA-I ApoB[a] ApoC	ApoB ApoC-I ApoC-II ApoC-III ApoE	ApoB	ApoA-I ApoA-II
Minor apoproteins	ApoA-II ApoE[b] PRP[c]	ApoA-I ApoA-II ApoD[d]	ApoC	ApoC-I ApoC-II ApoC-III ApoD ApoE

[a] Also termed apoLDL.
[b] Also termed arginine-rich protein (52).
[c] Proline-rich protein (10).
[d] Also termed "thin-line" protein (9) and apoA-III (9).

APOPROTEINS

Natural Distribution

The distribution of the major human plasma apoproteins is well-known (1). Isolation and characterization of a number of apoproteins of lesser abundance have been reported. McConathy & Alaupovic (8) have studied the properties of apoD and LP-D, a distinct class or subclass of lipoproteins isolated from HDL_3. They obtained apoD by successive treatment of HDL_3 with neuraminidase, chromatography of the product on concanavalin A-Sepharose 4B, elution of the retained LP-D with 0.2 M methyl-α-D-glucopyranoside, and final chromatography on hydroxyapatite-cellulose. An alternative procedure combines chromatography of HDL_3 on an immunosorber containing anti-apoD antibodies followed by hydroxyapatite chromatography. With either procedure, they have judged that the purified LP-D is homogeneous, based on a single, symmetrical boundary in the analytical ultracentrifuge, a single band on gel electrophoresis, and unique antigenic properties. The composition is about 70% protein and 30% lipid, the former being a single glycoprotein called apoD with a molecular weight of 22,100. Kostner uses a similar procedure but finds traces of apoA-I in all of the purified fractions and therefore he assigns this protein to the A-family and designates the purified component as apoA-III rather than apoD (9). The same protein has been given the trivial name "thin-line" apoprotein because of its characteristic thin precipitin line near the antigen well when tested against antibodies (9).

Sata et al (10) have described the isolation of a proline-rich protein (PRP) from human plasma. After removal of most of the lipoproteins from plasma by centrifugation at d = 1.21 g/ml, the infranatant fraction is bound to Intralipid. After recentrifugation the resulting supernatant is chromatographed on a 4% agarose gel and the eluted Intralipid-protein complex delipidated and chromatographed on 4% agarose and in 6 M urea on DEAE-cellulose. The purified protein aggregate, of $>10^6$ daltons, is rich in proline (8.9 mole %) and forms a single subunit of molecular weight 74,000 in polyacrylamide gel electrophoresis in sodium dodecyl sulfate. PRP is found in plasma only in a lipid-free state and in chylomicrons. Its plasma concentration varies between 12 and 41 mg/dl. Its importance in lipid metabolism is not known and we consider it an open question whether it should be classified with the plasma apoproteins.

Structure and Properties

The primary structure of apoA-I, apoA-II, apoC-I, and apoC-III have been reviewed previously (1). Little progress has been reported on the structure of apoB and there are widely varying reports as to its molecular weight.

Jackson et al (11) have completed the primary structure of apoC-II (Figure 1). The calculated molecular weight for the 78 amino acids is 8837. Three of the four prolines occur within the first 12 residues; these "helix-breaking" amino acids would prevent helix formation in this part of the protein. Except for an additional proline at residue 42, the remainder of the apoprotein can readily form an α-helical structure. The presence of four adjacent and three 1→4 pairs of oppositely charged amino acid residues in apoC-II may be important in the stabilization of its amphipathic helical structure within VLDL (1).

Using spectroscopic methods, Gwynne et al (12) have observed a reversible thermal unfolding of apoA-I between 43 and 75°C; this transition is absent in HDL where the apoprotein presumably is stabilized by lipid-protein interactions. Tall et al (13, 14) have studied the structural stability of apoA-I when challenged by various chemical and thermal changes. They identify a reversible two-state thermal transition between 43 and 71°C that has an enthalpy of protein unfolding of 64 kcal/mole. From the unusually low free energy of stabilization of apoA-I at 37°C (−2.4 kcal/mole), they suggest that native apoA-I has a loosely folded tertiary structure in which

Thr -Glu -Gln -Pro -Gln - Gln - Asp -Glu - Met - Pro -Ser -Pro - Thr -Phe- Leu -
           5   10   15

Thr - Glu - Val - Lys - Glu - Trp - Leu - Ser - Ser - Tyr - Gln - Ser - Ala - Lys - Thr -
           20   25   30

Ala - Ala - Gln - Asn - Leu - Tyr - Glu - Lys - Thr - Tyr - Ieu - Pro - Ala - Val - Asp -
           35   40   45

Glu - Lys - Leu - Arg - Asp - Leu - Tyr - Ser - Lys - Ser - Thr - Ala - Ala - Met - Ser -
           50   55   60

Thr - Tyr - Thr - Gly - Ile - Phe - Thr - Asp - Gln - Val - Leu - Ser - Val - Leu - Lys -
           65   70   75

Gly - Glu - Glu
        78

Figure 1 Amino acid sequence of apoC-II.

a large number of hydrophobic areas of the protein are exposed to water. Reynolds (15) has arrived at a similar conclusion based upon the fact that low concentrations of guanidine-HCl unfold apoA-I and apoA-II. The exposed hydrophobic areas of the protein may be important in the phospholipid-binding and self-associative properties of apoproteins. Relevant to this concept are the monolayer experiments of Phillips et al (16) which show that apoA-I, in contrast to other hydrophobic proteins such as casein, forms a loose helical structure at the water-lipid interface. McLachlan (17) has hypothesized that the helical pattern in apoA-I is a basic repeating unit of 22 amino acids that appears eight times. He has further suggested that residues 47–240 have appeared by gene duplication from an ancestral unit of 22 amino acids.

Stone & Reynolds (18) have stated that apoA-I dimerizes with an association constant of 1.3×10^4 M^{-1}. Vitello & Scanu (19) have found a monomer-dimer-tetramer-octamer model for apoA-I and their results are similar to those of Stone & Reynolds. Vitello & Scanu also find that the self-association is relatively insensitive to changes in pH and ionic strength and therefore discount electrostatic effects.

The self-association of apoA-II (18, 20, 21), of reduced apoA-II (22), and of reduced carboxymethylated (RCM) apoA-II (23) has also been studied. Stone & Reynolds (18) find only noncovalent dimers of apoA-II between 0.4 and 1.0 mg/ml. Gwynne et al (20), on the other hand, observe an apoA-II dimer that maximally associates at 25°C and has an association constant $\sim 3 \times 10^4$ M^{-1}. In contrast, Vitello & Scanu (21) report successive monomer-dimer-trimer equilibria for the self-association of apoA-II at concentrations between 0.8 and 1.5 mg/ml. They obtain similar results with reduced apoA-II (22) and with RCM apoA-II (23). These latter results raise the question as to whether the disulfide linkage is necessary for the functional integrity of apoA-II, an impression enforced by the fact that apoA-II from other species has serine substituted for the cysteine at residue six (24).

In the self-association studies of apoA-II, the fluorescence depolarization and mean residue ellipticity of the monomeric and self-associated apoprotein are substantially different (20, 22, 23). With increasing protein concentration, the negative mean residue ellipticity at 222 nm and the polarization of apoA-II fluorescence both increase in magnitude. These findings suggest that oligomeric apoA-II has a different secondary structure than the monomer. The qualitative differences in the extent of self-association of apoA-I and apoA-II as measured in various laboratories may be due to the buffer systems used. From the studies we have reviewed, we conclude that apoA-I and apoA-II do self-associate and that lipid-protein interactions must compete effectively with protein-protein interactions to form a stable lipid-protein complex.

LIPOPROTEIN STRUCTURE

High Density Lipoprotein (HDL)

Investigations of the composition, structure, and properties of intact lipoprotein have refined the earlier models for HDL. The recent studies have
focused on the microscopic structure of lipoproteins, and physicochemical
methods have, understandably, been an important component of these investigations.

The high field nuclear (proton) magnetic resonance (NMR) spectra of
whole HDL have revealed several distinctive features. In porcine HDL_3,
Hauser (25) has observed that the chemical shifts of the phosphatidylcholine (PC) protons in the polar head group and first two $-CH_2-$units of the
acyl chain are different from those of pure egg PC; the remainder of the
protons have chemical shifts similar to egg PC. He attributes this result to
lipid-protein interaction between the phospholipid polar head group and
apoprotein within HDL_3. Analysis of the linewidth data suggests that the
phospholipids are not completely immobilized within HDL (28). Based on
paramagnetic ion-induced shifts, practically all of the phospholipid polar
head groups are on the surface of HDL (26, 27), a concept supported by
the kinetics of phospholipase A_2 HDL hydrolysis (29). Immunochemical
and cross-linking experiments also show that a substantial percentage of the
protein is located at the HDL surface. Schonfeld et al (30) detect only 10%
of the ApoA-I in HDL by RIA. By contrast, Mao et al (31) obtain values
for apoA-II in HDL that exceed 100% of the apoA-II measured by chemical analysis. The apoproteins of HDL are readily accessible to the bifunctional cross-linking reagent, 1,5-difluoro-2,4-dinitrobenzene (32, 33).
ApoA-I can be cross-linked with apoA-II, but no cross-linking products are
found containing the same apoproteins. From these results, Grow & Fried
(32, 33) suggest that within HDL, apoA-I and apoA-II have specific topographical distributions in which all of the apoA-II molecules are adjacent
to an apoA-I molecule but with little or no self-association of either apoA-I
or apoA-II. It is difficult to relate this result to the published molar ratio
of apoA-I to apoA-II of 2 : 1 (1, 3, 34) and 1 : 1 (35). Grow & Fried's study
raises the possibility that within HDL there may be a specific stoichiometry
and topographic relationship of apoA-I and apoA-II and that the association of apoA-I with apoA-II could contribute to this relationship.

The microscopic properties of intact HDL have been investigated by
fluorescence depolarization of the probe, 1,6-diphenyl-1,3,5-hexatriene (36),
by pyrene excimer fluorescence (37), and by partitioning of the ESR probe,
2,2,6,6-tetramethylpiperidine-1-oxyl (TEMPO) (37). The depolarization
data indicate that the lipid region of HDL is more viscous than that of lipids
extracted from HDL (36). TEMPO, which partitions preferentially into

fluid regions of phospholipids, does not partition into HDL as well as it does into lipids extracted from HDL (37). Similarly, the rate of diffusion of pyrene in HDL is much lower than that observed in fluid phospholipids. These findings suggest that the apoproteins may restrict the mobility of small molecules in the lipid domain of HDL.

Tall et al (38) have analyzed the thermal behavior of HDL by differential scanning calorimetry (DSC), by low angle X-ray diffraction, and by polarized light microscopy. They find no thermal transitions in HDL between 0 and 60°C, whereas the extracted lipids of HDL exhibit well-defined thermal transitions that coincide with the melting of cholesteryl esters. The absence of a cholesteryl ester transition in HDL suggests that this lipid is confined to domains containing less than the requisite number of molecules required for cooperative melting. When HDL was heated above 71°C, a broad endothermic peak corresponding to the irreversible release of apoA-I was observed (38). In contrast, no release of apoA-II occurs until HDL is heated above 90°C. One interpretation of these results is that apoA-II is more tightly bound to HDL than is apoA-I.

On the basis of these studies, several statements can be made about the structure of HDL. First, the phospholipid polar head groups are probably located at the surface of HDL. Second, the apoproteins are also at or near the surface, although photoaffinity labeling indicates that penetration of hydrophobic segments into the particle core may occur (39). Third, the interior of the HDL particle is highly viscous and contains its component lipids in domains smaller than those required for cooperative melting.

Low Density Lipoprotein (LDL)

Small and co-workers have studied the thermal properties of LDL by a variety of techniques (40–43). Calorimetric and X-ray scattering data reveal a broad reversible transition between 20 and 45°C (40, 41), that represents a liquid crystalline → liquid phase transition involving cholesteryl esters. This assignment requires that the cholesteryl esters of LDL exist in a separate domain containing enough esters to exhibit cooperative melting. Studies of mixtures of triglycerides and cholesteryl esters and of LDL having different triglyceride content show that an increase in the triglyceride to cholesteryl ester ratio decreases the thermal transition temperature. It is probable that a significant quantity of the triglyceride of intact LDL is confined to the same lipid domain as cholesteryl ester. ^{13}C-NMR studies of LDL over the temperature range between 20 and 45°C showed that there is increased mobility of the steroid moiety of the cholesteryl esters as LDL is raised above its phase transition (42). The thermally-induced changes in the molecular structure of LDL cholesteryl esters previously observed by small angle X-ray scattering have been reevaluated by Atkinson et al (43), who infer that the cholesteryl esters are radially distributed in the core of LDL in an arrangement similar to that of the cholesteryl ester, cholesteryl myristate (CM). The similarity of the intense $1/36$ Å^{-1} scattering from LDL and CM at 10°C is evidence that the LDL cholesteryl esters are in a smectic layered organization in which pairs of ester molecules are translated parallel to their molecular long axis; the pairs of opposed steroid groups with interlocked C_{18} and C_{19} angular methyl groups observed in the crystal structure are retained in this model of the smectic phase. Mateu et al (44) report a reversible low temperature thermal transition ($<$0°C) in LDL in which the $1/36$ Å^{-1} scattering is absent; the significance of this finding remains to be shown.

Jonas (36) has found, on the basis of fluorescence depolarization of DPH, that, of the human plasma lipoproteins, LDL has the most viscous (6 Poise) apolar region. The microviscosity of the extracted lipids of LDL is lower (2.4 Poise) than that of LDL, perhaps because the lipid-protein interactions restrict the movement of DPH. The thermal data show no discontinuities characteristic of a phase transition, a finding at variance with directly measured thermal transitions (40, 41). This discrepancy may be due to the localization of the DPH to regions of LDL that do not undergo a thermal transition.

The topographical distributions of phospholipids in LDL have been studied by NMR. On the basis of the quenching of the ^{31}P signal of human LDL by a Mn^{2+}/EDTA complex, Henderson et al (26, 27) assign 50% of the phospholipid polar head groups to the surface of the particle. Finer et al (28), in studies of porcine LDL by ^{1}H-NMR at 220 MHz, find that two thirds of the choline groups have a sharp resonance with the remaining one third too immobilized to give a signal in the absence of detergent; all of the resonances of the choline groups in porcine LDL are accessible to the shift reagent, sodium ferricyanide. Finer et al conclude that the phospholipids of LDL are in two different chemical environments and propose a model in which a core of protein is surrounded by an inverted monolayer containing one third of the phospholipids having their head groups tightly packed and in contact with 15% of the protein. The outer surface of the particle is a phospholipid monolayer containing the rest of the phospholipids and apoB. The neutral lipids are assigned to a central layer with some interdigitation with opposing phospholipid acyl chains on the outside and in the protein core. The resulting model may be represented as a trilayer structure.

In contrast, Shen et al (45) suggest, on the basis of the size and composition of lipoproteins, that all lipoprotein structures can be unified into a simple model in which the neutral lipids are in a central core and the proteins and phospholipids are in a monolayer on the particle surface. In their model cholesterol is located just beneath the protein layer. The location of the protein on the surface is also consistent with the findings of Harmony & Cordes (46) that the glycoprotein component of LDL is exposed to the extent that it reacts with concanavalin A.

Although adequate support for a model of LDL is not yet available, several conclusions can be made in light of recent studies. First, LDL contain cholesteryl esters in separate domains in which some LDL triglycerides are probably solubilized. A part of the cholesteryl esters is in a domain such that it participates in a reversible thermal transition. Second, much of the LDL protein is at or near the surface in association with a phospholipid monolayer. Some of the phospholipid is relatively immobilized but the mechanism by which this immobilization occurs is unclear. Third, cholesterol is probably partitioned between the surface monolayer and the neutral core with a partition coefficient that is dependent on the composition of these two regions. These three generalizations apply only to where the *major* fraction of these components reside. The location of minor amounts of lipids or a hydrophobic segment of the apoB in other regions of the particle cannot be excluded. It is important to keep in mind that the components are very likely in a state of dynamic equilibrium.

Very Low Density Lipoprotein (VLDL)

Little is known concerning the structure of VLDL. Deckelbaum et al (47) have noted that cholesteryl esters in VLDL are probably interspersed throughout the rest of the neutral lipids since no transition due to melting of a separate cholesteryl ester domain can be detected by DSC; separate solubility studies show that cholesteryl esters are soluble in triglyceride.

Lipoprotein-X (LP-X)

Patsch et al (48) have reported a detailed structural analysis of LP-X, found in the plasma of patients with advanced obstructive liver disease. Three different fractions, LP-X$_1$ (d = 1.038 g/ml), LP-X$_2$ (d = 1.049 g/ml), and LP-X$_3$ (d = 1.058 g/ml) can be isolated by zonal ultracentrifugation. All three populations are rich in phospholipids (65%) and free cholesterol (25%) and are relatively poor in triglycerides (~5%). LP-X $_{1,2,3}$, respectively, exhibit flotation rates of 17.3, 9.7, and 3.2 Svedbergs, and Stokes radii of 339, 343, and 249 Å. In all three particles, the protein constituents contain a large fraction of α-helical structure (41–65%) and the fluidities of the lipid regions of the particles are very low. All three contain serum albumin and C-peptides but only LP-X$_{1,2}$ contain apoA-I and apoE.

LIPOPROTEIN BIOGENESIS

The assembly and secretion of the plasma lipoproteins occurs only in the liver and the intestine. The liver secretes "nascent" VLDL, containing triglycerides of endogenous origin, and nascent HDL. The intestine synthesizes and secretes dietary triglyceride, transported largely as chylomicrons and, to a lesser extent, as VLDL. Once secreted, the nascent lipoproteins undergo rapid modification in the plasma by physical transfer of lipid and apoprotein components and by enzymatic modification by lecithin cholesterol acyltransferase (LCAT) and lipoprotein lipase (LPL). By the action of these enzymes, HDL and LDL, the most abundant and long-lived lipoproteins in the plasma, are formed. Molecular details of the synthesis and assembly of lipoprotein components, the secretion of intact lipoproteins, and the mechanisms affecting hormonal and dietary regulation of these processes are largely unknown. Lipoprotein biogenesis has been comprehensively reviewed; we use the earlier reviews (3, 49) as a starting point.

Chylomicrons

ApoA, apoB, and apoC are the major protein constituents of chylomicrons in the lymph (50, 51). ApoE (52) and the proline-rich protein are also present (10). ApoB is synthesized by the intestine, whereas apoC and apoE are probably acquired by transfer from HDL (3, 53). Glickman & Green (54) have shown by [^{3}H]leucine incorporation that during lipid absorption, the rat intestine actively synthesizes apoA-I, most of which (85%) is present in lipoproteins with d < 1.006, as determined by quantitative immunoelectrophoresis. Some apoA-I of lymph HDL is produced in the intestine, but filtration of plasma HDL has precluded quantification of the intestinal contribution. The coordinate synthesis of apoprotein and lipid components of chylomicrons has not been studied systematically. Such studies are necessary to clarify the regulatory mechanisms that control the incorporation of dietary cholesterol into chylomicrons, which likely involves coenzyme A-dependent esterification of cholesterol (55).

Very Low Density Lipoprotein (VLDL)

An mRNA for a major apoprotein of chicken VLDL, apoVLDL-II (56), has been partially purified and translated in a heterologous cell-free system (57). Chan et al have shown that estrogen stimulation of VLDL synthesis in the cockerel produces an increased amount of translatable mRNA for apoVLDL-II. The number of specific nuclear binding sites for estrogen rapidly increases. This change is associated with enhanced activities of RNA polymerase I and II and with an increase in the number of RNA synthesis initation sites (58). In this system, the control of VLDL synthesis by estrogen resides, in part, at the level of gene transcription. Synthesis of immunoprecipitable VLDL and the stimulation by estrogen have also been demonstrated in nonproliferating chicken liver in cell cultures (59). Isolated rat hepatocytes in suspension also synthesize a material which, by immunological, flotation, and electrophoretic criteria, appears to be VLDL (60).

The site within the hepatic cell at which VLDL is synthesized is the subject of dispute (61, 62). Alexander et al (62) have suggested that both triglycerides and phospholipids are made in the smooth endoplasmic reticulum, independently of apoprotein synthesis in the rough endoplasmic reticulum. According to Nestruck & Rubinstein (63), apoB combines with the lipid components in the first stage of VLDL assembly. In their view, apoC proteins are acquired subsequently during or after secretion into the space of Disse. Glucosamine incorporation occurs during passage through the Golgi (64). VLDL of intestinal origin contains apoA, while hepatic VLDL reportedly does not (3). The basis for this difference, which exists in spite of apoprotein exchange, has not been studied.

Low Density Lipoprotein (LDL)

Plasma LDL is produced primarily by the action of lipoprotein lipase on triglyceride-rich lipoproteins and is discussed later in this review as well as elsewhere (49, 5).

High Density Lipoprotein (HDL)

Hamilton et al (65) have isolated from the perfused rat liver newly secreted, nascent HDL that are disc-shaped in negatively stained electron micrographs, and differ from plasma HDL by a low content of esterified cholesterol (4.3 versus 23.8%) and by apoprotein composition; the amount of apoE is much greater than that of apoA-I (66). However, the discs are still highly active as a substrate of LCAT (65).

LIPOPROTEIN CATABOLISM

Catabolism of plasma lipoproteins proceeds by three distinct processes: (a) physical processes of transfer and exchange of lipid and apoprotein components; (b) enzymatic changes in composition involving LCAT and LPL, and (c) receptor-mediated cellular uptake and passive endocytosis.

The dynamics and the contributions of each process (Table 2) under various nutritional and physiological states are poorly defined, and complicated further by the appreciable differences in lipoprotein composition and metabolism in various species studied. With new information about lipoprotein structure and catabolic mechanisms, rapid progress in the understanding of lipoprotein metabolism should ensue.

Table 2 Lipoprotein synthesis and metabolism

Lipoprotein	Location	Process	Major components
Chylomicron	Intestine	assembly, secretion	apoA, apoB, phospholipid, cholesterol, cholesteryl ester, triglyceride
	Lymph	transfer	apoC, apoE[a]
	Plasma	transfer	apoC, cholesterol
	Endothelial cell	hydrolysis[b]	triglyceride, phospholipid
		transfer	fatty acid, cholesterol, phospholipid, apoC, apoA, apoE[a]
VLDL	Liver	assembly, secretion	apoB, apoC, cholesterol, phospholipid, triglyceride, cholesteryl ester
	Plasma	transfer	apoC[a], apoE[a], cholesteryl ester
	Endothelial cell	hydrolysis[a]	triglyceride, phsopholipid
		transfer	fatty acid, cholesterol, phospholipid, apoC, apoE[a]
LDL	Plasma	formation	from VLDL and chylomicrons
		exchange	cholesterol, phospholipid
	Peripheral tissues, liver[a]	uptake, degradation, regulation	cholesterol, cholesteryl ester
HDL	Liver	assembly, secretion	apoA, apoE, phospholipid, cholesterol
	Plasma	formation[a]	from chylomicrons[a]
		acyltransfer[c]	phosphatidylcholine, cholesterol
		transfer	cholesteryl ester, apoC, apoE[a]
		exchange	apoC, cholesterol, phospholipid
	Liver, peripheral tissues[a]	uptake, degradation	cholesterol, cholesteryl ester

[a] Inferred but not determined.
[b] Lipoprotein lipase (triacylglycerol hydrolyase, apoC-II activated).
[c] Lecithin cholesterol acyltransferase.

Exchange and Transfer

It has been known for many years that phospholipid and cholesterol exchange takes place among lipoproteins and various blood components. However, the mechanisms of these transfers are poorly understood. Published values for the rates of transfer of many lipids are of questionable validity since the times required to isolate the lipoproteins for analysis were much greater than the half-times of exchange. The use of high concentrations of salt (67) for ultracentrifugal separation of reactants, and of heparin-MnCl$_2$ to precipitate lipoproteins (68) creates further ambiguity concerning the interpretation of the experiments, since the effects of these procedures on lipoprotein structure are uncertain. For further discussion the reader is referred to Bruckdorfer (69) and Bell (70).

Pyrene transfer between HDL has provided a model of lipid exchange (71). When pyrene-labeled HDL are mixed with unlabeled lipoproteins, pyrene excimer fluorescence decreases with a half-time of approximately 3 msec. The fact that changes in the concentration of either pyrene-labeled or unlabeled HDL do not affect the half-time of pyrene exchange is compelling evidence that the limiting step is the dissociation of the lipid into the aqueous solvent. Similar phenomena have been shown with pyrene-containing lipids, 10-(3-pyrenyl)-decanoic acid (72), 1-oleyl-2-[4(3-pyrenyl)butanoyl]glycerol, a fluorescent alkylacylglycerol (73), and 3'-pyrenemethyl-23,24-*dinor*-5-cholen-22-oate-3β-ol, a cholesterol analog (74). Exchange of these lipids between lipoproteins appears to be a physiochemical process that may not involve apoproteins. ^{1}H-NMR evidence excludes fusion of phosphatidylcholine vesicles as a mechanism of cholesterol transfer (75).

The mechanism of phospholipid exchange between lipoproteins has not been studied (70). However, in an artificial, protein-free system, Martin & MacDonald (76) find that transfer between phosphatidylcholine vesicles occurs by way of a monomolecular species in the aqueous solution. A similar mechanism has been demonstrated for lysophosphatidylcholine, from true or micellar solution, which transfers into LDL or HDL across a dialysis membrane (77). Proteins that catalyze the exchange of phospholipids between various organelle membranes in vitro have been described (78). A similar function for apoproteins or other plasma proteins in promoting lipid transfer or exchange between lipoproteins remains an interesting but untested possibility. The report that a high molecular weight plasma protein stimulates cholesteryl ester exchange between VLDL and LDL (79) also merits further attention. Both rapid exchange and net transfer of apoC between chylomicrons or VLDL and HDL (49, 80–83) occur in vivo and in vitro, but no studies of mechanism or kinetics have been conducted. It is not known whether lipids are involved in these transfers of apoproteins.

The physiological significance of lipid exchange remains to be determined. The equilibrium of lipid components of lipoproteins with the plasma compartment may regulate physical properties such as fluidity of the lipoprotein and membrane surfaces which in turn may influence enzymatic degradation. In addition, cholesterol and other lipids may be removed by transfer apart from endocytosis of the carrier lipoprotein. An important task of future studies will be to define the contribution of the physical, enzymatic, and endocytotic processes of lipid removal from plasma in various tissues.

Enzymatic Modification

LECITHIN CHOLESTEROL ACYLTRANSFERASE (LCAT)   LCAT is an extracellular enzyme of hepatic (84) and intestinal origin (85) that circulates in the plasma and catalyzes the transfer of the C-2' fatty acid from phosphatidylcholine to cholesterol (86). LCAT is activated by apoA-I (87) and by apoC-I (88). Activation by apoA-I is greatest with unsaturated PC, and exceeds by fourfold the stimulation by apoC-I, which is equally effective with both saturated and unsaturated PC (88). One report claims that apoD activates LCAT (89). The transfer of apoE from HDL to VLDL occurs concomitantly with LCAT action (90). The relationship of these apoproteins to LCAT catalysis awaits clarification by quantitative studies with LCAT preparations which have no contaminating apoproteins. The reaction catalyzed by LCAT is poorly understood because there is no stable homogeneous enzyme preparation; various attempts yielded only a partially purified enzyme (89, 91, 92). In a complex set of procedures, Albers et al (93) prepared an enzyme of high purity, the stability of which was not reported. They removed contaminating apoD with an immobilized antibody to this protein. The apparent molecular weight of the purified LCAT is 68,000.

One approach with great potential for studying apoprotein activation of LCAT is that of solid-phase peptide synthesis of activators. ApoC-I, synthesized independently by Sparrow and associates (94) and by Harding et al (95), spontaneously forms a lipid-protein complex when mixed with synthetic PC (94). Synthetic apoC-I activates LCAT indistinguishably from the naturally occurring apoprotein. A synthetic fragment composed of residues 17–57 is as active as the entire apoprotein in enhancing LCAT activity, while synthetic fragment 24–57 is only 60% as active (96). These findings suggest that the residues 17–57 contain the determinants for LCAT activation. Thus by peptide synthesis of native and model apoproteins, functionally important regions of apoproteins can be localized.

The exact role of LCAT in lipoprotein metabolism remains to be defined. The presence of apoA-I in intestinal chylomicrons (50, 54) suggests that LCAT has an important role in the processing of the HDL components— PC, cholesterol, and apoA-I—originating from chylomicrons during catabolism (98). A lipoprotein deficient in cholesteryl ester appears in the HDL density range in the postprandial plasma of LCAT-deficient patients (97). LCAT may remove cholesterol and PC in excess of the amount assimilated by cellular membranes. LCAT action may follow transfer of substrates and apoA to preexisting HDL. Alternatively, by analogy with the studies in perfused liver systems, there may be coordinate release of a nascent lipoprotein from which HDL (66) is formed by LCAT action. Characterization and thorough kinetic studies of the various lipoprotein substrates will lead to an understanding of the physiological role of this enzyme. The transfer of cholesteryl ester from HDL to VLDL has been reported (99); the quantitative importance is not known. The major pathway for removal of cholesteryl ester formed by LCAT action appears to be uptake of the intact HDL in the liver (100–102).

LIPOPROTEIN LIPASE   The hydrolysis of di- and triglyceride in chylomicrons and VLDL in extrahepatic tissues is accomplished by an exoenzyme, lipoprotein lipase, at the luminal surface of the endothelial cell (103, 104). For maximal activity, this enzyme requires apoC-II, a component of the surface film of the lipoprotein substrate (105, 106). Interaction of LPL with apoC-II produces an extremely stable complex. The calculated equilibrium dissociation constant for the enzyme:apoC-II complex is $<10^{-8}$ M (107) from one laboratory and 3×10^{-10} M from another (108). The rate of hydrolysis increases as a function of apoC-II concentration and is maximal at a molar ratio of enzyme to apoC-II of $1:1$ (107). This stoichiometry has also been deduced by Fielding & Fielding (108) on the basis of kinetic experiments. An LPL activated by apoC-I (109–111) has been reported but not yet confirmed in other laboratories.

LPL is one of several lipases released into the plasma by the intravenous injection of heparin. The nature of the binding of LPL to the endothelial cell surface is not known but may involve glycosaminoglycans (103). Olivecrona et al (112) have speculated that LPL is electrostatically bound to heparan sulfate polymers that extend 20–50 nm from the endothelial cell surface. The changes in the levels of activity of LPL in various tissues such as adipose, mammary, heart, and skeletal muscles, support the view that synthesis of the enzyme occurs in cells that respond to nutritional and endocrine stimulation (113–116). The mechanism of LPL transport from the site of synthesis to the endothelial cell surface remains unknown, as does the molecular basis by which enzyme synthesis is regulated. Synthesis of LPL by the endothelial cell appears unlikely but has not been excluded.

Properties of LPL from various sources are summarized in Table 3. Specific activities have little meaning since the assay conditions, substrates, and purity of the enzymes are not comparable. LPL from all sources are glycoproteins with similar carbohydrate content. They show characteristic inhibition by protamine and high salt; the latter was shown to be anion-specific (125). Heparin is not present in the two purified rat postheparin plasma enzymes (119), but has either no effect (118), or stimulates (107, 119, 123), depending on the assay system and enzyme preparation used. The basis for these different effects of heparin is unknown. The enzyme from bovine milk has immunological determinants in common with LPL of human postheparin plasma (126, 127).

The proportions of apoC-I, apoC-II, and apoC-III in VLDL are somewhat different in hypertriglyceridemia, compared to normal VLDL (128). They also change with size (129) and in response to diet (130). The potential regulation of LPL activity by apoproteins has been explored in a number of laboratories in efforts to find the underlying defects in hypertriglyceridemia. Fielding & Fielding found (108) that apoE and apoD do not affect the reaction catalyzed by LPL. The inhibitory effects of apoC-I and apoC-III in experiments in vitro (131, 107, 123) have been related to the nature of the artificial substrates, as well as the ratios of apoC-II, inhibitory apoprotein, and substrate concentrations (132). The apoC proteins in plasma are in equilibrium between HDL and the triglyceride-rich lipoproteins. Since the relative ratio of apoC in isolated plasma lipoproteins is so constant (80, 81, 83, 133), the inhibitory effects of apoC-I and apoC-III may be artifacts of the in vitro assays. Whether these in vitro inhibitory effects have physiological correlates remains unanswered.

Determination of the amino acid sequence of apoC-II (Figure 1) by Jackson et al (11) has allowed Kinnunen et al (134) to identify the minimum sequence necessary for activation of bovine milk LPL. Cyanogen bromide fragments of apoC-II corresponding to residues 1–9 and 10–59 have virtually no ability to activate the enzyme. However, the COOH-terminal cyanogen bromide fragment and the synthetic peptide corresponding to residues 60–78 increases hydrolysis three- to fourfold in an experiment in which the same concentration of apoC-II increases hydrolysis by ninefold. By contrast, synthetic peptide fragment 55–78 gives an activation within experi-

Table 3   Properties of lipoprotein lipase

Source	Molecular weight	Specific activity (mmol of fatty acid/ mg of protein/hr)	Reference
Rat heart	37,500	4	117, 118
Rat adipose tissue	69,250	10–14	119
Rat heart	34,000	0.6	107
Bovine milk	62–66,000	24–28	120
	48–51,000	37	121
	55,000	29	122
Porcine adipose tissue	60–62,000	3.4	123
Porcine heart	73,000	2.4	124

mental error (±10%) of that produced by apoC-II. The synthetic peptide containing residues 66–78 does not activate; removal of the three COOH-terminal residues, Gly-Glu-Glu, from CNBr fragment 60–78 also abolishes the activation. These studies suggest that the maximal activation of LPL by apoC-II requires a minimal sequence contained within residues 55–78. It is speculated that these two COOH-terminal glutamic acid residues are sites for ionic interaction between LPL and apoC-II, that residues 55–65 are involved in the enhancement of enzymatic catalysis, and that the NH_2-terminal 49 residues interact with phospholipids.

LPL is specific for primary esters but has no absolute stereo-specificity. Fatty acids at position one of sn-triacylglycerol (135), 1(3)-monoacylglycerol (136), and enantiomeric alkyldiacylglycerols (137) are released preferentially over those at position three. In the case of sn-triacylglycerols, the ratio is 3:1. With emulsions of sn-1,2- or 2,3-diacylglycerols (138), the preference is less than that of sn-triacylglycerols and depends on the precise fatty acid composition. No specificity is observed with enantiomeric dialkylacylglycerols; the primary ester bonds were hydrolyzed at the same rate (139). These findings suggest that the ester carbonyl at position two may orient the substrate with respect to the catalytically active site of the enzyme. No detectable specificity is associated with double bonds in unsaturated triglycerides (140).

Scow & Egelrud (141) have shown that hydrolysis of [³H]palmitate and [³²P]phosphate-labeled chylomicron phosphatidylcholine, by bovine milk LPL and bee venom phospholipase A₂, produces 2- and 1-monoacylphosphatidylcholine, respectively, in agreement with stereochemical preference demonstrated for di- and triglycerides. LPL from rat postheparin plasma hydrolyzes phosphatidylcholine and phosphatidylethanolamine at comparable rates—about 2–3% of the rate of triglyceride hydrolysis (119).

HEPATIC LIPASE   The triglyceride hydrolases in postheparin plasma of different tissue origins are distinguished primarily by the different conditions for optimal in vitro assays and by immunological criteria. Addition of heparin to a liver perfusion medium rapidly releases a lipolytic enzyme into the perfusate (142–145). The hepatic enzyme is not activated by plasma apoproteins (145) and is not inhibited by protamine sulfate or by high ionic strength (146, 147). The heparin-releasable hepatic triglyceride hydrolyase has the same stereospecificity (142) as LPL. The two enzymes have different immunological characteristics (148–151). Hepatic lipase binds to concanavalin A-sepharose and is probably a glycoprotein (152). The function of the hepatic lipase in vivo is not known. It could have a role in the further catabolism of the triglyceride-depleted lipoproteins produced by LPL in extrahepatic tissues (49, 153).

Catabolism of Chylomicrons and VLDL

Catabolism of chylomicrons and VLDL proceeds in two phases. Initially, the action of LPL on chylomicrons and VLDL at the endothelial surface removes triglyceride from the apolar core to leave a remnant from the chylomicrons and an intermediate density lipoprotein (IDL) from the VLDL. These products are smaller in size, are relatively enriched in cholesteryl ester, sphingomyelin, and apoB, but are poor in apoC and apoE. ApoB is the major apoprotein. In the second phase of catabolism, the chylomicron remnant is taken up and further degraded by the liver and other tissues. IDL are subsequently converted to LDL; the site and mechanism in vivo are unknown. The lipids released by lysosomal hydrolysis presumably equilibrate with the intracellular biosynthetic pools. These processes are discussed briefly in the last section, below, and in detail in other reviews (154, 155).

The catabolism of triglyceride-rich lipoproteins is significantly different in humans than in rats, the experimental animal principally studied. Sigurdsson et al (156) found in humans that most, if not all, VLDL are converted to LDL. By contrast, in rats, most of the VLDL apoB is removed by the liver and only 3% appears in LDL (157, 49).

In humans, VLDL containing triglycerides of either endogenous or exogenous origin cannot be distinguished by present methods of isolation. [^{14}C]Retinyl acetate, which is entirely of dietary origin, has been found in VLDL, which was thought to be of endogenous origin in hypertriglyceridemic subjects (158). Whether this finding is valid for individuals with normal plasma lipids has not been determined. For purposes of this review, the chylomicrons and VLDL are discussed together, because there is insufficient information to do otherwise. However, based on the observed differences in apoprotein content, the early phases of the metabolism of lipoproteins containing exogenous and endogenous triglyceride may prove to be significantly different. Much of the earlier literature has been reviewed in depth by Eisenberg & Levy (49) and by Jackson et al (3).

Quantitative information about changes in lipid and apoprotein composition during lipoprotein metabolism in vitro and in vivo is meager. Methods of measuring quantitatively all of the apoproteins are not available. Densitometric scanning of apoproteins separated by electrophoresis in polyacrylamide gels under denaturing conditions (159) or by isoelectric focusing (160, 161) is currently used, despite variable dye binding by some of the apoproteins (129, 159). Immunoassays are available for some but not all apoproteins but these may yield variable results. The usual methods of isolation of lipoproteins require lengthy ultracentrifugations in high salt concentration (162, 163), which causes dissociation and irreversible aggregation of some apoproteins (164, 165). The lipid values are often incompletely reported in studies of lipoprotein metabolism. Phospholipids are usually analyzed only as lipid phosphorous. After saponification, triglyceride is determined as glycerol. Total cholesterol includes both the free and esterified forms. Information about fatty acids, the minor phospholipid components, and isomers of monoacylphospholipids and of mono- and diglycerides, is lost in these determinations. Consequently, the precise composition of the circulating lipoproteins is not known with assurance. Additional information of the physical properties and distribution of the various lipid and apoproteins will be necessary to go beyond our present understanding of the interrelationship of lipoprotein structure and metabolism.

The interaction of LPL with triglyceride-rich substrates has been studied by Higgins & Fielding (166) to determine if there is preferential removal of larger, intact lipoproteins in the perfused rat heart. Clinical observations show that the larger triglyceride-rich particles are cleared more rapidly from the circulation. In perfused rat heart, intact triglyceride-rich lipoproteins compete with the remnant lipoproteins at the capillary site of hydrolysis in proportion to their abundance in the perfusion medium. The relative rate of hydrolysis of remnant lipoproteins, compared to the intact particles, decreases only slightly (25%) until about two thirds of the original triglycerides are removed, at which point the rate drops rapidly and falls to less than 10% of the initial rate as the triglyceride depletion approaches 95%. The reasons for the slower rate of triglyceride hydrolysis with decreasing triglyceride content and smaller size are unknown. Loss of apoC-II and an increased proportion of cholesterol are possible contributory factors.

Fielding & Higgins (167) have concluded that the membrane-bound LPL in the perfused rat heart and the heparin-released soluble myocardial LPL interact and hydrolyze the lipoprotein substrates at equal rates. Lymph chylomicrons, relative to plasma chylomicrons, contain fourfold less apoC-II (83) but do not differ significantly when the rates of fatty acid release are compared. These results may reflect the rate-limiting release of the fatty acid product at the site of hydrolysis since ^{125}I VLDL degradation in vivo is accelerated by about tenfold by heparin-induced lipolysis (168).

As chylomicrons and VLDL are hydrolyzed by LPL, apoC and apoE are transferred to high density plasma fractions in vitro and in vivo (81, 133, 169). Eisenberg (153) has noted that the mass of apoB in all VLDL and LDL is the same and is not related to size, weight, or content of other lipid and apoprotein constituents of the lipoprotein. This finding supports a direct precursor-product relationship between VLDL and LDL. Analysis of the smaller lipoproteins produced by LPL in postheparin plasma from supradiaphragmatic rats, which contain only LPL, shows that apoB, cholesteryl ester, and possibly sphingomyelin (80) are retained completely in the product lipoproteins. Qualitatively, the triglyceride-rich lipoproteins and the remnants contain the same apoprotein constituents (133, 169). The decrease in apoC content appears to be related to the changes in composition of the lipoprotein surface film during lipolysis, since the dissociation of apoC from VLDL proceeds at a rate proportional to that of fatty acid release, and in the absence of lipoprotein acceptors for apoC (170).

The events associated with the release of partially degraded chylomicrons or VLDL from the site of hydrolysis are unknown. Felts et al (171) have proposed that the specific recognition of the chylomicrons remnant by the liver depends on the presence of the LPL in the remnant, although only about one in ten chylomicron remnants contain enzymatically active LPL.

Molecular details of transfer of products of triglyceride hydrolysis from the plasma compartment to the tissue are not known. Scow et al (104) have proposed that the surface film of the chylomicron can fuse with the external leaflet of the endothelial cell membrane. Hypothetically, the transfer would involve lateral diffusion of acylglycerol and fatty acid into the endothelial cell through an interfacial continuum. The transfer of lipids as individual molecules through solution to the endothelial cell probably occurs. The existence and relative contribution of these proposed mechanisms remain to be investigated. Since the partition coefficient of the fatty acid (172) between the aqueous compartment and the endothelial cell surface greatly favors the distribution of the fatty acid in the cell membrane, lateral diffusion may contribute significantly to transport of fatty acid across the endothelial cell by way of the transendothelial channels (173).

Uptake and Degradation by Cultured Cells

Plasma lipoproteins, produced through the enzymatic action of LCAT and lipoprotein lipase and modified by the physical processes of transfer and exchange, are ultimately removed by endocytosis in a variety of cell types. The following overview is drawn largely from studies by Goldstein & Brown of LDL interaction with cultured human fibroblasts, summarized in recent reviews (154, 155). In human nonpathological states, the extent of uptake of circulating LDL and HDL appears to be determined by requirements of the specific cell type for the cholesterol and cholesteryl ester component of these lipoproteins. The uptake is balanced with cellular synthesis to satisfy the requirements for membrane structure and for synthesis of steroid hormones. In the pathological state, dysfunction in regulation of the uptake of lipoproteins and consequently of intracellular cholesterol synthesis may occur.

Specific binding sites for LDL present in normal cells are reported to be absent or defective in fibroblasts from subjects with homozygous familial hypercholesterolemia (154, 155). Interaction of LDL with a specific receptor initiates a series of complex processes involving endocytosis, lysosomal degradation of the internalized lipoprotein, concomitant activation of cholesterol esterifying enzymes within the cell, and ultimately, by a mechanism presently unknown, the suppression of the HMG-CoA reductase activity in these cell lines. These phenomena have also been demonstrated in lymphocytes and in arterial smooth muscle cells.

The existence of specific LDL receptors in the adrenal tissue and in the endothelial cell has been inferred. Current evidence does not support their occurrence in the liver. Breslow et al (174) have recently shown that in isolated hepatocytes the HMG-CoA reductase activity is unaffected by the level of LDL. Cholesterol synthesis in rat liver is regulated by chylomicron remnants (175, 176), and other circulating serum lipoproteins are very much less effective in their inhibition of hepatic cholesterol synthesis (177). Whether the LDL is degraded exclusively in extrahepatic tissues has not been firmly established but peripheral tissues at least appear to be the major site. Catabolism of plasma LDL in swine is increased by functional hepatectomy (178). Administration of 4-aminopyrazole-3,4-d-pyrimidine decreases serum cholesterol levels in the rat to very low levels (175, 176). The reduction in plasma cholesterol is accompanied by a 2- to 16-fold increase in the rate of sterol synthesis in extrahepatic tissues. This supports the view that serum lipoproteins, primarily LDL, can regulate sterol synthesis in extrahepatic tissue but that the synthesis is usually suppressed under physiological conditions.

The nature of the specific receptor for LDL is not known. Electron microscopy has shown that LDL is catabolized and internalized in lysosomal vesicles but there is little information about the events associated with endocytosis (154, 155). Furthermore, how these studies done in cultured cells are related to similar cells under physiological conditions in vivo is not known. Whether the mechanism demostrated in vitro is physiologically important or functions as a regulatory mechanism in vivo remains to be demonstrated. The production of the major biochemical and morphological features of atherosclerosis in aortic smooth cells in vitro by an altered LDL (179) is provocative in its inference that the disease in vivo occurs by the accumulation of altered lipoproteins that bypass the normal pathways of uptake and regulation. Reduction of the binding, internalization, and degradation of LDL by HDL has been shown in cultured cells (180) and has been postulated as the apparent mechanism by which HDL decreases the atherogenic process (181). Correlation of these in vitro studies with in vivo phenomena portends an exciting future in the field of lipoprotein metabolism.

ACKNOWLEDGMENTS

This review was completed August 1, 1977. The authors wish to express appreciation to their colleagues Drs. Richard L. Jackson, Paavo K. J. Kinnunen, Sandra H. Gianturco, Alberico Catapano, Yin J. Kao, Simon T. Mao, and William A. Bradley for their helpful criticism of this review. The authors are indebted to Ms. Barbara Allen for her assistance in the preparation of the manuscript. Work from the authors' laboratory in this review was supported in part by United States Public Health Service Grants HL-17269, HL-15648, HL-19459, RR-00350, and LRC Contract 71-2156, and by the Robert A. Welch Foundation Q-343 and the American Heart Association. H. J. Pownall is an established investigator of the American Heart Association.

Literature Cited

1. Morrisett, J. D., Jackson, R. L., Gotto, A. M. Jr. 1975. *Ann. Rev. Biochem.* 44:183–207
2. Alaupovic, P., Lee, D. M., McConathy, W. J. 1972. *Biochim. Biophys. Acta* 260:689–707
3. Jackson, R. L., Morrisett, J. D., Gotto, A. M. Jr. 1976. *Physiol. Rev.* 56:259–316
4. Scanu, A. M., Edelstein, C., Keim, P. 1975. In *The Plasma Proteins*, ed. F. W. Putnam, 7:317–91. New York: Academic. 481 pp.
5. Day, C. E., Levy, R. S., eds. 1976. *Low Density Lipoproteins*. New York: Plenum. 445 pp.
6. Gennis, R. B., Jonas, A. 1977. *Ann. Rev. Biophys. Bioeng.* 6:195–238
7. Morrisett, J. D., Jackson, R. L., Gotto, A. M. Jr. 1977. *Biochim. Biophys. Acta* 472:93–133
8. McConathy, W. J., Alaupovic, P. 1976. *Biochemistry* 15:515–20
9. Kostner, G. 1974. *Biochim. Biophys. Acta* 336:383–95
10. Sata, T., Havel, R. J., Kotite, L., Kane, J. P. 1976. *Proc. Natl. Acad. Sci. USA* 73:1063–67
11. Jackson, R. L., Baker, H. N., Gilliam, E. B., Gotto, A. M. Jr. 1975. *Proc. Natl. Acad. Sci. USA* 74:1942–45
12. Gwynne, J., Brewer, H. B. Jr., Edelhoch, H. 1975. *J. Biol. Chem.* 250:2269–74
13. Tall, A. R., Shipley, G. G., Small, D. M. 1976. *J. Biol. Chem.* 251:3749–55
14. Tall, A. R., Small, D. M., Shipley, G. G., Lees, R. S. 1975. *Proc. Natl. Acad. Sci. USA* 72:4940–42
15. Reynolds, J. A. 1976. *J. Biol. Chem.* 251:6013–15
16. Phillips, M. C., Hauser, H., Leslie, R. B., Oldani, D. 1975. *Biochim. Biophys. Acta* 406:402–14
17. McLachlan, A. D. 1977. *Nature* 267:465–66
18. Stone, W. L., Reynolds, J. A. 1975. *J. Biol. Chem.* 250:8045–48
19. Vitello, L. B., Scanu, A. M. 1976. *J. Biol. Chem.* 251:1131–36
20. Gwynne, J., Palumbo, G., Osborne, J. C. Jr., Brewer, H. B. Jr., Edelhoch, H. 1975. *Arch. Biochem. Biophys.* 170:204–12
21. Vitello, L. B., Scanu, A. M. 1976. *Biochemistry* 15:1161–65
22. Osborne, J. C. Jr., Palumbo, G., Brewer, H. B. Jr., Edelhoch, H. 1975. *Biochemistry* 14:3741–46
23. Osborne, J. C. Jr., Palumbo, G., Brewer, H. B. Jr., Edelhoch, H. 1976. *Biochemistry* 15:317–20
24. Edelstein, C., Noyes, C., Keim, P., Heinrikson, R. L., Fellows, R. E., Scanu, A. M. 1976. *Biochemistry* 15:1262–68
25. Hauser, H. 1975. *FEBS Lett.* 60:71–75
26. Henderson, T. O., Kruski, A. W., Davis, L. G., Glonek, T., Scanu, A. M. 1975. *Biochemistry* 14:1915–20
27. Glonek, T., Henderson, T. O., Kruski, A. W., Scanu, A. M. 1974. *Biochim. Biophys. Acta* 348:155–61
28. Finer, E. G., Henry, R., Leslie, R. B., Robertson, R. N. 1975. *Biochim. Biophys. Acta* 380:320–37
29. Pattnaik, N. M., Kezdy, F. J., Scanu, A. M. 1976. *J. Biol. Chem.* 251:1984–90

30. Schonfeld, G., Bradshaw, R. A., Chen, J.-S. 1976. *J. Biol. Chem.* 251:3921–26
31. Mao, S. J. T., Gotto, A. M. Jr., Jackson, R. L. 1975. *Biochemistry* 14:4127–31
32. Grow, T. E., Fried, M. 1975. *Biochem. Biophys. Res. Commun.* 66:352–56
33. Grow, T. E., Fried, M. 1977. *Biochem. Biophys. Res. Commun.* 75:117–24
34. Friedberg, S. J., Reynolds, J. A. 1976. *J. Biol. Chem.* 251:4005–9
35. Curry, M. D., Alaupovic, P., Suenram, C. A. 1976. *Clin. Chem.* 22:315–22
36. Jonas, A. 1977. *Biochim. Biophys. Acta* 486:10–22
37. Morrisett, J. D., Pownall, H. J., Jackson, R. L., Segura, R., Gotto, A. M. Jr., Taunton, O. D. 1977. In *Polyunsaturated Fatty Acids,* 8:139–61. Champaign, Illinois: American Oil Chemists Society. 258 pp.
38. Tall, A. R., Deckelbaum, R. J., Small, D. M., Shipley, G. G. 1977. *Biochim. Biophys. Acta* 487:145–53
39. Stoffel, W., Därr, W., Salm, K.-P. 1977. *Hoppe-Seylers Z. Physiol. Chem.* 358: 453–62
40. Deckelbaum, R. J., Shipley, G. G., Small, D. M., Lees, R. S., George, P. K. 1975. *Science* 190:392–94
41. Deckelbaum, R. J., Shipley, G. G., Small, D. M. 1977. *J. Biol. Chem.* 252:744–54
42. Sears, B., Deckelbaum, R. J., Janiak, M. J., Shipley, G. G., Small, D. M. 1976. *Biochemistry* 15:4151–57
43. Atkinson, D., Deckelbaum, R. J., Small, D. M., Shipley, G. G. 1977. *Proc. Natl. Acad. Sci. USA* 74:1042–46
44. Mateu, L., Kirchhausen, T., Camejo, G. 1977. *Biochim. Biophys. Acta* 487: 243–45
45. Shen, B. W., Scanu, A. M., Kézdy, F. J. 1977. *Proc. Natl. Acad. Sci. USA* 74:837–40
46. Harmony, J. A. K., Cordes, E. H. 1975. *J. Biol. Chem.* 250:8614–17
47. Deckelbaum, R. J., Tall, A. R., Small, D. M. 1977. *J. Lipid Res.* 18:164–68
48. Patsch, J. R., Aune, K. C., Gotto, A. M. Jr., Morrisett, J. D. 1977. *J. Biol. Chem.* 252:2113–20
49. Eisenberg, S., Levy, R. I. 1975. *Adv. Lipid Res.* 13:1–89
50. Kostner, G., Holasek, A. 1972. *Biochemistry* 11:1217–23
51. Fainaru, M., Havel, R. J., Felker, T. E. 1976. *Biochim. Biophys. Acta* 446:56–68
52. Havel, R. J., Kane, J. P. 1973. *Proc. Natl. Acad. Sci. USA* 70:2015–19
53. Roheim, P. S., Edelstein, D., Pinter, G. G. 1976. *Proc. Natl. Acad. Sci. USA* 73:1757–60
54. Glickman, R. M., Green, P. H. R. 1977. *Proc. Natl. Acad. Sci. USA* 74:2569–73
55. Haugen, R., Norum, K. R. 1976. *Scan. J. Gastroenterol.* 11:615–21
56. Jackson, R. L., Lin, H. Y., Chan, L., Means, A. R. 1977. *J. Biol. Chem.* 252:250–53
57. Chan, L., Jackson, R. L., O'Malley, B. W., Means, A. R. 1976. *J. Clin. Invest.* 58:368–79
58. Jackson, R. L., Chan, L., Snow, L. D., Means, A. R. 1978. *Fed. Proc.* In press
59. Tarlow, E. M., Watkins, P. A., Reed, R. E., Miller, R. S., Zwergel, E. E., Lane, M. D. 1977. *J. Cell. Biol.* 73:332–53
60. Jeejeebhoy, K. N., Ho, J., Breckenridge, C., Bruce-Robertson, A., Steiner, G., Jeejeebhoy, J. 1975. *Biochem. Biophys. Res. Commun.* 66:1147–53
61. Glaumann, H., Berstrand, A., Ericsson, J. L. E. 1975. *J. Cell. Biol.* 64:356–77
62. Alexander, C. A., Hamilton, R. L., Havel, R. J. 1976. *J. Cell. Biol.* 69:241–63
63. Nestruck, A. C., Rubinstein, D. 1976. *Can. J. Biochem.* 54:617–28
64. Dolphin, P. J., Rubinstein, D. 1976. *Can. J. Biochem.* 55:83–90
65. Hamilton, R. L., Williams, M. C., Fielding, C. J., Havel, R. J. 1976. *J. Clin. Invest.* 58:667–80
66. Marsh, J. B. 1976. *J. Lipid Res.* 17: 85–90
67. Rubenstein, B., Rubinstein, D. 1972. *J. Lipid Res.* 13:317–20
68. Illingworth, D. R., Portman, O. W. 1972. *Biochim. Biophys. Acta* 280: 281–89
69. Bruckdorfer, K. R., Graham, J. M. 1976. In *Biological Membranes,* ed. D. Chapman, 3:103–52. New York: Wiley. 362 pp.
70. Bell, F. P. 1976. See Ref. 5, pp. 111–13
71. Charlton, S. C., Olson, J. S., Hong, K. Y., Pownall, H. J., Louie, D. D., Smith, L. C. 1976. *J. Biol. Chem.* 251:7952–55
72. Sengupta, P., Sackmann, E., Kuhnle, W., Scholz, H. P. 1976. *Biochim. Biophys. Acta* 436:869–78
73. Charlton, S. C., Hong, K. Y., Smith, L. C. 1976. *Circulation* 54(II):210 (Abstr.)
74. Kao, Y. J., Charlton, S. C., Smith, L. C. 1977. *Fed. Proc.* 56:936 (Abstr.)
75. Haran, N., Shporer, M. 1977. *Biochim. Biophys. Acta* 465:11–18
76. Martin, H. J., MacDonald, R. C. 1976. *Biochemistry* 15:321–27
77. Portman, O. W., Illingworth, D. R. 1973. *Biochim. Biophys. Acta* 326:34–42
78. Wirtz, K. W. A. 1974. *Biochim. Biophys. Acta* 344:95–117

79. Zilversmit, D. B., Hughes, L. B., Balmer, J. 1975. *Biochim. Biophys. Acta* 409:393–98
80. Eisenberg, S., Schurr, D. 1976. *J. Lipid Res.* 17:578–87
81. Havel, R. J., Kane, J. P., Kashyap, M. L. 1973. *J. Clin. Invest.* 52:32–38
82. Fidge, N., Poulis, P. 1975. *J. Lipid Res.* 16:367–78
83. Fielding, C. J., Fielding, P. E. 1976. *J. Lipid Res.* 17:419–23
84. Nordby, G., Berg, T., Nilsson, M., Norum, K. R. 1976. *Biochim. Biophys. Acta* 450:69–77
85. Clark, S. B., Norum, K. R. 1977. *J. Lipid Res.* 18:293–300
86. Glomset, J. A., Norum, K. R. 1973. *Adv. Lipid Res.* 11:1–65
87. Fielding, C. J., Shore, V. G., Fielding, P. E. 1972. *Biochem. Biophys. Res. Commun.* 46:1493–98
88. Soutar, A. K., Garner, C. W., Baker, H. N., Sparrow, J. T., Jackson, R. L., Gotto, A. M. Jr., Smith, L. C. 1975. *Biochemistry* 14:3057–64
89. Kostner, G. 1974. *Scand. J. Clin. Lab. Invest.* 33:Suppl. 137, pp. 19–21
90. Norum, K. R., Glomset, J. A., Nichols, A. V., Forte, T., Albers, J. J., King, W. C., Mitchell, C. D., Applegate, K. R., Gong, E. L., Cabana, V., Gjone, E. 1975. *Scand. J. Clin. Invest.* 35:Suppl. 142, pp. 31–55
91. Varma, K. G., Soloff, L. A. 1976. *Biochem. J.* 155:583–88
92. Varma, K. G., Nowotny, A. H., Soloff, L. A. 1977. *Biochim. Biophys. Acta* 486:378–84
93. Albers, J. J., Cabana, V. G., Stahl, Y. D. B. 1976. *Biochemistry* 15:1084–87
94. Sigler, G. F., Soutar, A. K., Smith, L. C., Gotto, A. M. Jr., Sparrow, J. T. 1976. *Proc. Natl. Acad. Sci. USA* 73:1422–26
95. Harding, D. R. K., Battersby, J. E., Husbands, D. R., Hancock, W. S. 1976. *J. Am. Chem. Soc.* 98:2664–65
96. Sigler, G. F., Soutar, A. K., Smith, L. C., Gotto, A. M. Jr., Sparrow, J. T. 1976. *Circulation* 54(II):165 (Abstr.)
97. Glomset, J. A., Norum, K. R., Nichols, A. V., King, W. C., Mitchell, C. D., Applegate, K. R., Gong, E. L., Gjone, E. 1975. *Scand. J. Clin. Lab. Invest.* 35:Suppl. 142, pp. 3–30
98. LaRosa, J. C., Levy, R. I., Brown, W. V., Fredrickson, D. S. 1971. *Am. J. Physiol.* 220:785–91
99. Nichols, A. V., Smith, L. 1965. *J. Lipid Res.* 6:206–10
100. Van Berkel, T. J. C., Koster, J. F., Hulsmann, W. C. 1977. *Biochim. Biophys. Acta* 486:586–89
101. Nakai, T., Otto, P. S., Kennedy, D. L., Whayne, T. F. 1976. *J. Biol. Chem.* 251:4914–21
102. Drevon, C. A., Berg, T., Norum, K. R. 1977. *Biochim. Biophys. Acta* 487:122–36
103. Robinson, D. S. 1970. *Comp. Biochem.* 18:51–116
104. Scow, R. O., Blanchette-Mackie, E. J., Smith, L. C. 1976. *Circ. Res.* 39:149–62
105. LaRosa, J. C., Levy, R. I., Herbert, P., Lux, S. E., Fredrickson, D. S. 1970. *Biochem. Biophys. Res. Commun.* 41:57–62
106. Havel, R. J., Shore, V. G., Shore, B., Bier, D. M. 1970. *Circ. Res.* 27:595–600
107. Chung, J., Scanu, A. M. 1977. *J. Biol. Chem.* 252:4202–9
108. Fielding, C. J., Fielding, P. E. 1977. In *Cholesterol Metabolism and Lipolytic Enzymes,* ed. J. Polonovski, pp. 165–72. New York: Masson USA. 211 pp.
109. Ganesan, D., Bass, H. B. 1975. *FEBS Lett.* 53:1–4
110. Ganesan, D., Bass, H. B. 1975. *Artery* 2:143–52
111. Ganesan, D., Bradford, R. H., Ganesan, C., Alaupovic, P., Bass, H. B. 1975. *J. Appl. Physiol.* 39:1022–33
112. Olivecrona, T., Bengtsson, G., Marklund, S.-E., Lindahl, U., Höök, M. 1977. *Fed. Proc.* 36:60–65
113. Scow, R. O. 1977. *Fed. Proc.* 36:182–85
114. Linder, C., Chernick, S. S., Fleck, R. T., Scow, R. O. 1976. *Am. J. Physiol.* 231:860–64
115. Tan, M. H., Sata, T., Havel, R. J. 1977. *J. Lipid Res.* 18:363–70
116. Chajek, T., Stein, O., Stein, Y. 1977. *Biochem. Biophys. Acta* 488:140–44
117. Fielding, C. J. 1976. *Biochemistry* 15:879–84
118. Fielding, P. E., Shore, V. G., Fielding, C. J. 1977. *Biochemistry* 16:1896–1900
119. Fielding, P. E., Shore, V. G., Fielding, C. J. 1974. *Biochemistry* 13:4318–22
120. Egelrud, T., Olivecrona, T. 1972. *J. Biol. Chem.* 247:6212–17
121. Iverius, P.-H., Östlund-Lindqvist, A.-M. 1976. *J. Biol. Chem.* 251:7791–95
122. Kinnunen, P. K. J. 1977. *Medical Biology* 55:187–91
123. Bensadoun, A., Ehnholm, C., Steinberg, D., Brown, W. V. 1974. *J. Biol. Chem.* 249:2220–27
124. Ehnholm, C., Kinnunen, P. K. J., Huttunen, J. K., Nikkilä, E. A., Ohta, M. 1975. *Biochem. J.* 149:649–55

125. Fielding, C. J., Fielding, P. E. 1976. *J. Lipid Res.* 17:248–56
126. Hernell, O., Egelrud, T., Olivecrona, T. 1975. *Biochim. Biophys. Acta* 381:233–41
127. Huttunen, J. K., Ehnholm, C., Kinnunen, P. K. J., Nikkilä, E. A. 1975. *Clin. Chim. Acta* 63:335–47
128. Carlson, L. A., Ballantyne, D. 1976. *Atherosclerosis* 23:563–68
129. Kane, J. P., Sata, T., Hamilton, R. L., Havel, R. J. 1975. *J. Clin. Invest.* 56:1622–34
130. Schonfeld, G., Weidman, S. W., Witztum, J. L., Bowen, R. M. 1976. *Metabolism* 25:261–75
131. Ganesan, D., Bass, H. B., McConathy, W. J., Alaupovic, P. 1976. *Metabolism* 25:1189–95
132. Östlund-Lindqvist, A.-M., Iverius, P.-H. 1975. *Biochem. Biophys. Res. Commun.* 65:1447–55
133. Eisenberg, S., Rachmilewitz, D. 1975. *J. Lipid Res.* 16:341–51
134. Kinnunen, P. K. J., Jackson, R. L., Smith, L. C., Gotto, A. M. Jr., Sparrow, J. T. 1977. *Proc. Natl. Acad. Sci. USA* 74:4848–51
135. Morley, N. H., Kuksis, A., Buchnea, D. 1974. *Lipids* 9:481–88
136. Twu, J.-S., Nilsson-Ehle, P., Schotz, M. C. 1976. *Biochemistry* 15:1904–9
137. Paltauf, F., Esfandi, F., Holasek, A. 1974. *FEBS Lett.* 49:119–23
138. Morley, N. H., Kuksis, A., Buchnea, D., Myther, J. J. 1975. *J. Biol. Chem.* 250:3414–18
139. Paltauf, F., Wagner, E. 1976. *Biochim. Biophys. Acta* 431:359–62
140. Morley, N. H., Kuksis, A. 1977. *Biochim. Biophys. Acta* 487:332–42
141. Scow, R. O., Egelrud, T. 1976. *Biochim. Biophys. Acta* 431:538–49
142. Akesson, B., Gronowitz, S., Herslof, B. 1976. *FEBS Lett.* 71:241–44
143. Pykalisto, O. J., Vogel, W. C., Bierman, E. L. 1974. *Biochim. Biophys. Acta* 369:254–63
144. Jansen, H., Hülsmann, W. C. 1974. *Biochim. Biophys. Acta* 369:387–96
145. Greten, H., Walter, B., Brown, W. V. 1972. *FEBS Lett.* 27:306–10
146. Assmann, G., Krauss, R. M., Fredrickson, D. S., Levy, R. I. 1973. *J. Biol. Chem.* 248:1992–99
147. LaRosa, J. C., Levy, R. I., Windmueller, H. G., Fredrickson, D. S. 1972. *J. Lipid Res.* 13:356–63
148. Huttunen, J. K., Ehnholm, C., Kinnunen, P. K. J., Nikkilä, E. A. 1975. *Clin. Chim. Acta* 63:335–47
149. Klose, G., De Grella, R., Greten, H. 1976. *Atherosclerosis* 25:175–82
150. Ehnholm, C., Huttunen, J. K., Kinnunen, P. K. J., Miettinen, T. A., Nikkilä, E. A. 1975. *N. Engl. J. Med.* 292:1314–17
151. Jansen, H., Hulsmann, W. C. 1975. *Biochim. Biophys. Acta* 398:337–46
152. Ehnholm, C., Shaw, W., Greten, H., Brown, W. V. 1975. *J. Biol. Chem.* 250:6756–61
153. Eisenberg, S. 1976. See Ref. 5, pp. 73–92
154. Goldstein, J. L., Brown, M. S. 1977. *Ann. Rev. Biochem.* 46:897–930
155. Goldstein, J. L., Brown, M. S. 1976. *Curr. Top. Cell. Regul.* 11:147–81
156. Sigurdsson, G., Nicoll, A., Lewis, B. 1975. *J. Clin. Invest.* 56:1481–90
157. Faergeman, O., Bierman, E. L., Havel, R. J. 1975. *J. Clin. Invest.* 55:1210–18
158. Hazzard, W. R., Bierman, E. L. 1976. *Metabolism* 25:777–801
159. Kane, J. P. 1973. *Anal. Biochem.* 53:350–64
160. Gidez, L. I., Swaney, J. B., Murane, S. 1976. *J. Lipid Res.* 18:59–68
161. Swaney, J. B., Gidez, L. I. 1977. *J. Lipid Res.* 18:69–76
162. Havel, R. J., Eder, H. A., Bragdon, J. H. 1955. *J. Clin. Invest.* 34:1345–53
163. Patsch, J. R., Sailer, S., Kostner, G., Sandhofer, F., Holasek, A., Braunsteiner, H. 1974. *J. Lipid Res.* 15:356–66
164. Fainaru, M., Havel, R. J., Imaizumi, K. 1977. *Biochem. Med.* 17:347–55
165. Herbert, P. N., Forte, T. M., Shulman, R. S., La Piana, M. J., Gong, E. L., Levy, R. I., Fredrickson, D. S., Nichols, A. V. 1975. *Prep. Biochem.* 5:93–129
166. Higgins, J. M., Fielding, C. J. 1975. *Biochemistry* 14:2288–93
167. Fielding, C. J., Higgins, J. M. 1974. *Biochemistry* 13:4324–30
168. Eisenberg, S., Bilheimer, D. W., Levy, R. I., Lindgren, F. T. 1973. *Biochim. Biophys. Acta* 326:361–77
169. Mjøs, O. D., Faergeman, O., Hamilton, R. L., Havel, R. J. 1975. *J. Clin. Invest.* 56:603–15
170. Glangeaud, M. C., Eisenberg, S., Olivecrona, T. 1977. *Biochim. Biophys. Acta* 486:23–35
171. Felts, J. M., Itakura, H., Crane, R. T. 1975. *Biochem. Biophys. Res. Commun.* 66:1467–75
172. Sallee, V. L. 1974. *J. Lipid Res.* 15:56–64

173. Simionescu, N., Simionescu, M., Palade, G. E. 1975. *J. Cell. Biol.* 64:586–607
174. Breslow, J. L., Lothrop, D. A., Clowes, A. W., Lux, S. E. 1977. *J. Biol. Chem.* 252:2726–33
175. Andersen, J. M., Dietschy, J. M. 1977. *J. Biol. Chem.* 252:3646–51
176. Andersen, J. M., Dietschy, J. M. 1977. *J. Biol. Chem.* 252:3652–59
177. Nervi, F. O., Dietschy, J. M. 1975. *J. Biol. Chem.* 250:8704–11

178. Sniderman, A. D., Carew, T. E., Chandler, J. G., Steinberg, D. 1974. *Science* 183:526–28
179. Goldstein, J. L., Anderson, R. G. W., Buja, L. M., Basu, S. K., Brown, M. S. 1977. *J. Clin. Invest.* 59:1196–1202
180. Miller, N. E., Weinstein, D. B., Carew, T. E., Koschinsky, T., Steinberg, D. 1977. *J. Clin. Invest.* 60:78–88
181. Carew, T. E., Koschinsky, T., Hayes, S. B., Steinberg, D. 1976. *Lancet* 1:1315–17

Ann. Rev. Biochem. 1978. 47:779–817

INSECT PLASMA PROTEINS

❖989

G. R. Wyatt

Department of Biology, Queen's University, Kingston,
Ontario K7L 3N6, Canada

M. L. Pan

Department of Zoology, University of Tennessee, Knoxville, Tennessee 37916

CONTENTS

0066-4154/78/0701-0779$01.00

INTRODUCTION

Although the insects are by far a more numerous and diverse group of animals than the vertebrates, biochemical knowledge of them is, by comparison, rudimentary. Recently, there has been some improvement in this situation. Interest in understanding how insects function has increased with the realization of their potential for research in developmental biology and molecular evolution, and also to support the development of new methods of insect control. Yet we are only just ready for a first review of the biochemistry of insect plasma proteins. The first quantitative study of the individual plasma proteins of an insect during development was the pioneering immunochemical analysis on the cecropia silkmoth, *Hyalophora cecropia*, by Telfer & Williams in 1953 (1). The first recognition of function of a specific insect hemolymph protein (other than hemoglobin, which occurs only in a limited group, and certain enzymes) was Telfer's (2) study of the "female protein," now known as vitellogenin. Purification and characterization of a functionally recognized insect plasma protein, other than hemoglobin, was first achieved for the diglyceride transport lipoprotein, by Chino and co-workers in 1969 (3). Many publications in this area go little beyond the presentation of electrophoretic patterns of unidentified components.

In a recent summary of this topic, Chen (4) expressed the need for "an unequivocal classification of insect hemolymph proteins based on accurate physico-chemical data." Our discussion of these proteins is arranged by function rather than by structure, but emphasizes their physicochemical properties wherever information is available. We also consider what comparisons can be made with the blood proteins of vertebrates, and what is known about the control of synthesis of hemolymph proteins in insects.

A few words on the insect circulatory system may be helpful to some readers (5). The hemolymph is the only circulating fluid, and fills the body cavity, or hemocoel. It is separated from the cellular tissues by only a thin, permeable connective tissue membrane, and is maintained in circulation by a tubular, dorsal heart, that is sometimes assisted by accessory pulsatile organs in the limbs. Hemolymph comprises about 10%–40% of the body's volume and researchers have naturally tended to select insects such as the larger Lepidoptera, from which relatively large amounts can be obtained. The hemocytes are of several types with roles that include phagocytosis, encapsulation of parasites, and wound healing; they do not transport oxygen. The plasma is a biochemically rich solution, very different from vertebrate plasma, with widely varying proportions of inorganic ions, high levels of free amino acids, usually a rather high level of trehalose, and sometimes substantial amounts of other solutes such as organic phosphates, citrate, glycerol, and peptides (6–9). In intimate association with the hemolymph is a cellular tissue, the fat body, which combines many of the roles of the liver and adipose tissue of vertebrates. It is responsible for intermediary metabolism, the synthesis of plasma proteins, and the storage of reserves for metamorphosis, oogenesis, and physical activity (10–12).

TECHNIQUES

Since much of the research with insect material is done by workers whose training is not primarily biochemical, and insects present some special problems, some comments on methodology may be useful.

Preparation of Hemolymph Plasma

Hemolymph may be collected by dropping from a cut in the cuticle, or, when the volume is small, directly into a capillary pipette. To prevent coagulation (which occurs only in some species and stages; see below), prior chilling of the insect and glassware, and rapid dilution into saline, are generally effective, although citrate, EDTA, and other additives may sometimes help (13, 14). the action of polyphenol oxidase, which causes darkening and eventual formation of melanic precipitate, may be inhibited by addition of a few crystals of phenylthiourea (15). Hemocytes, although often few in number, should be removed by centrifugation before analysis of the plasma.

Electrophoresis

The vast majority of recent studies on insect plasma proteins have used electrophoretic techniques. Polyacrylamide gel, with advantages of high sensitivity and resolving power, is generally the most efficient medium. Because of the large sizes of some of the insect proteins, 4.5–6% acrylamide generally gives better resolution than the 7–7.5% commonly used for human serum; this effect has been illustrated (16, 17). Improved separation of proteins of diverse molecular sizes may be achieved by employing discontinuous gradient gels, as in a recent study of lepidopteran plasma proteins (18), which also illustrates the advantages of slab over disc gels for comparing many samples. Still greater resolution may be obtained by two-dimensional electrophoresis, the second dimension preferably using agarose in which the rate of migration depends solely on net charge (19). Some insect proteins require elevated ionic strength in the buffer in order to maintain solubility; in this case it is advisable to perform electrophoresis in the cold because of the increased current flow.

The identification of the separated proteins is often difficult, and it must be emphasized that migration rate alone is not a sufficient criterion. Some proteins, particularly those serving in transport, may have different electro-phoretic mobilities depending on their functional status; this has been observed with the lipid-carrier protein of the cecropia silkworm (W. H. Telfer and M. L. Pan, in preparation).

Immunochemical Methods

Immunochemical techniques are invaluable in the identification and quanti-tation of proteins and in their isolation for measurement of incorporated radioactivity. Most insect plasma proteins are good antigens in rabbits. While effective immunization protocols may have to be worked out for particular antigens, the procedures recommended for bovine serum albumin (20) are generally suitable. A simple technique for isolation of an antigen is to elute the band of interest from acrylamide gel after electrophoresis. After homogenization of a gel section, large proteins which may not elute readily can be injected together with the polyacrylamide, which can then serve an adjuvant role in the rabbit (21).

For qualitative analysis of antigens, the simplest technique is the double diffusion method of Ouchterlony (19). For demonstrating identity and cross-reaction studies of proteins from different species, analysis should be performed with hyperimmune sera against proteins of every species under investigation; otherwise erroneous conclusions may be reached. Double immunodiffusion can also be used to determine the diffusion coefficients of antigens (22).

Immunoelectrophoresis (23) has been widely used for qualitative analy-sis. Different buffer salts, which profoundly influence the separation of human serum proteins (24), can also affect insect material (25). For the presentation of results, superimposing a stained electrophoretogram per-mits ready correlation of the positions of proteins and precipitation lines (26–28). Electrophoresis in polyacrylamide gel, followed by immunodiffu-sion in agar, is an excellent method for identification of the separated proteins (29); lengths of siliconized glass tubing of outside diameter equal to that of the acrylamide gels can serve as molds to make depressions in the agar layer for subsequent insertion of the acrylamide gels.

Quantitative determination of individual antigens from a mixture can be performed by means of Oudin's single immunodiffusion method, which was first used for insect plasma proteins by Telfer & Williams (1). The density of the precipitation bands can also serve as a measure of the cross reactivity of antigens (27). Another quantitative technique of great sensitivity is the so-called rocket immunoelectrophoresis of Laurell (30). In addition to mea-suring the amount of antigen, this technique permits assay of incorporated radioactivity in the precipitation zone. It, and its variant, crossed immuno-electrophoresis (31, 32), deserve wider application to insect material.

In studying protein synthesis, specific products are commonly isolated for counting of incorporated radioactivity in the form of washed immuno-precipitates. Specific antisera must be prepared either by the use of highly purified antigens or by appropriate absorption. [For example, antiserum prepared against egg yolk protein is absorbed with hemolymph from male insects (2).] Immunoprecipitates may be isolated and washed either by centrifugation or filtration; a suggestion that the latter procedure may be invalidated by high blanks (33) has been discounted in a recent study (34).

Purification and Storage of Proteins

Ion exchange cellulose chromatography is widely used in the separation of insect plasma proteins. Yet, when pupal cecropia silkmoth hemolymph was chromatographed on DEAE cellulose in a Tris-citrate buffer (35) a consid-erable amount of yellow color appeared in the final Triton-X wash of the column, which suggests loss of carotenoid from lipoproteins. Lowered lipid content has been observed in avian yolk lipoprotein isolated from TEAE cellulose columns (36). A safer method for the isolation of some lipoproteins may be density gradient centrifugation (37), which has been applied to silkmoth plasma proteins (38, 39). Metrizamide, a recently developed iodinated density gradient material (40), may be valuable for the isolation of insect lipoproteins. The selection of buffer salts may be important; for example, separation of cecropia plasma proteins during sucrose gradient centrifugation was poorer in Tris than in other buffers (M. L. Pan and R. A. Wallace, in preparation).

Though proteins are generally best stored cold, there are exceptions; Crowle (19) has documented unfavorable conditions for several vertebrate proteins. Freezing causes irreversible precipitation of some proteins, partic-ularly lipoproteins, including avian lipovitellins. Cecropia lipoprotein I, after repeated freezing and thawing, gives a fuzzy precipitation zone in immunodiffusion (W. H. Telfer, personal communication). Such proteins are best stored as precipitates in heavy metal ion-free ammonium sulfate solution.

PROTEIN PATTERNS

Data on changes in the total hemolymph protein content during the devel-opment of various insects have been summarized previously (4, 8, 41), and some recent references are listed in Table 1. Typically, the protein level rises during each instar and falls at each molt. In early instars, the level is low (about 1–2 g/100 ml), but in the last larval stage it rises much more steeply, to 6–10 g/100 ml in some Lepidoptera, or as high as 20 g/100 ml in blowflies. There may then be a fall before the larval-pupal ecdysis, a transient rise in the pupa, and a fall during development of the adult (e.g. 42, 43). Hemimetabolous insects show similar but less extreme developmental changes. During vitellogenic and previtellogenic stages, the hemolymph of females may contain substantially more protein than that of males of a species (18), and in the adult there may be fluctuations correlated with reproductive cycles (44).

The number of protein components detected in insect plasma is limited only by the resolving power and sensitivity of the technique used. Recent analyses by polyacrylamide gel electrophoresis have described from about 10–30 bands. Usually, however, only a few of these are predominant in quantity. The numerous papers presenting the plasma protein patterns of different insects and describing changes in these during development and in response to various stimuli cannot be reviewed in detail. Table 1 may serve as a guide to the published information on different insects. Especially thorough developmental studies, based on quantitative analyses of electrophoretic patterns, include those of Patel (45) on the cecropia silkmoth, Cölln (42) on the meal moth, *Anagasta kuehniella,* and Schmidt and coworkers (46–51) on several species of ants. A number of authors have recorded changes in plasma protein levels and patterns during parasitism, and bacterial and virus infection. In virus infection, a decline in hemolymph protein level has repeatedly been noted, probably due to preemption of the fat body, the chief site of plasma protein synthesis, by virus multiplication. We have not tried to cover the reports on effects of insecticides and chemosterilants on plasma protein levels and patterns. In general, however, few conclusions can be drawn from studies in which the protein components were not identified.

Those proteins that have been at least partially purified from insect hemolymph, and to some extent characterized, receive emphasis in the discussion to follow. These are listed in Table 2.

VITELLOGENINS

In 1954, Telfer (2) demonstrated, in the cecropia silkmoth, a female-specific antigen that was transferred during oogenesis from the hemolymph to the eggs. Such yolk precursor proteins, or vitellogenins (206), have since been found to occur generally in the hemolymph of adult female insects. They are deposited in the yolk in a form which is appropriately termed vitellin, since there may be modification from the hemolymph precursor (298). Such modification may affect only the lipid moiety (238) and in most cases has not been examined. The yolk protein itself has also been, rather confusingly, called vitellogenin, or lipovitellin, the latter being a term borrowed from vertebrate systems. Since the vitellins reflect the properties of their precursor vitellogenins and are often easier to prepare in quantity, we shall include them in this discussion, although they are not strictly plasma proteins. For purification of the vitellins, extraction in high salt buffer followed by precipitation at low ionic strength has often been useful (282). Accumulation of vitellogenin in hemolymph, which facilitates its isolation, can sometimes be brought about by ovariectomy (28).

The properties of insect vitellogenins and vitellins that have been isolated are summarized in Table 2. The best characterized vitellogenin is that of the saturniid silkmoth, *Philosamia cynthia,* which was isolated from pupal hemolymph as "diglyceride carrier lipoprotein II" (3) and was recently identified as the vitellogenin of this species (237). In the electron microscope, it is seen as a globular protein of diameter 12.6 nm. The yolk proteins of several Orthoptera and Lepidoptera all have molecular weights of 500,000–600,000. In the cockroach, *Leucophaea maderae,* aggregation from a 14S (mol wt 560,000) to a 28S (mol wt 1,600,000) form takes place in the egg (282). Mosquito (*Aedes aegypti*) vitellin has a molecular weight of 270,000, and values of about 300,000 are also observed for the major yolk proteins (not yet purified) of two other Diptera, *Drosophila melanogaster,* and a blowfly, *Calliphora erythrocephala* (D. Harnish, personal communication).

Table 1 Systematic summary of literature on insect plasma proteins

Insect order and species	Method[a]	Observations	Reference
Coleoptera			
Dendroctonus pseudotsugae	EP	plasma and ovarian proteins	52
Dermestes frischi	EP	pattern during oogenesis	52a
	EP	parasitism effect	53
Leptinotarsa decemlineata	EP	pattern, species difference	54
	EC, IE	vitellogenin synthesis, photoperiod effect	55
	EC, IE	plasma protein and vitellogenesis	56
	EP	specific diapausing plasma proteins	57
	Ch, EP	leptinotarsin, a toxic protein, properties	58
	Enz	esterases, properties, ontogeny	392, 393
L. juncta	EP	pattern, species difference	54
Phyllophaga fusca	ES	pattern	59
Tenebrio molitor	EP	pattern, glyco-, and lipoproteins	60
	IE	antigenic changes, ontogeny	61
	IE	vitellogenic protein, synthesis, timing	62
	EC	vitellogenic protein from fat body	63
Trogoderma granarium		patterns, diapause, and control	64
Popillia japonica	ES, EC, Ch, Ce	bacterial infection effects	65–67
Dermaptera			
Anisolabis littorea	ES	pattern	68
Diptera			
Aedes aegypti	EP, ID	vitellogenin synthesis; vitellin purification	70
	IP	vitellogenin synthesis, control	71–73
	EP	parasitism effect	74

Table 1 (Continued)

Insect order and species	Method[a]	Observations	Reference
Calliphora erythrocephala	EP	protein synthesis by fat body	75–77
	EP	plasma and fat body proteins, ontogeny	78
	Cm	calliphorin, purification and properties	79, 80
	ID, IE	calliphorin, occurrence, synthesis by fat body	81, 82
	Enz	prophenoloxidase activation	83
Calliphora stygia	EP, Cm	pattern, concentration, synthesis, uptake by fat body, control	84–88
	EP, ID	concentration, origin, fate, properties	89
C. vicina	Cm	calliphorin synthesis, mRNA translation	90, 91
Chironomus (8 spp.)	ES	hemoglobin polymorphism	92, 373–375
C. tentans	Ce, Ch	hemoglobin characterization	93
	EC, EP, ES	hemoglobin polymorphism	376, 377
	Cm	protein concentration changes	94
C. thummi	Ch	protein uptake by salivary glands	95
	Cm, Ch	hemoglobin crystallization, properties	341, 378
		hemoglobin synthesis in fat body	129, 389
Culex pipiens	EP	patterns: autogenous, and anautogenous forms	96
Cuterebra sp.	ES	pattern	59
Drosophila grisea	EP	vitellogenin, JH effect on uptake	97
D. hydei	Ch	ecdysone binding	98
	EP	vitellogenin synthesis by fat body	99
D. macroptera	EP	vitellogenin in diapause; JH effect on uptake	97
D. melanogaster	ES, EP	ontogeny, genetics; two-dimensional gel analysis	100, 101
	Cm	turnover	102
	EP, ES	metabolism in wild and lethal mutant	103
		fat body protein granule and plasma protein	104
	EP, IE, Ch	vitellogenin synthesis; vitellin purification	105
	EP, ID	vitellogenin, characterization, synthesis, uptake	106
	EP	JH and vitellogenin uptake in ap^4 mutant	107, 108
	Cm	calliphorin-like protein mRNA translation	91
	EP	vitellogenin identification, metabolism	109
	EC, EP	larval-specific proteins, ontogeny	110
D. mulleri	EP	vitellogenin synthesis by fat body	99
D. pseudoobscura	EP	changes during development	111
D. virilis	EP	vitellogenin characterization, synthesis	106
	EP	vitellogenin synthesis by fat body	99
Drosophila (8 spp.)	EP	vitellin SDS gel analysis	106
Lucilia cuprina	EP	genetics of lucilin	112
Musca domestica	EC	protein synthesis	113
	EP	nutrition and plasma proteins	114
	ID, EA	origin of vitellogenin	115
	EP, Cm	proteins and oogenesis	116
	EA	lipoprotein phospholipid, diet effect	117
	EP, ES	esterases	118, 119
	EP	chemosterilant effect	120, 121
Phormia regina	EP	pattern and quantitative changes	122, 123
	EA	allatectomy effect	124
Rhynchosciara americana	EP	plasma protein used for cocoon	125
	EP	carotenoid metabolism disorder	126
Sarcophaga barbata	EP, Enz	tyrosinase: activity and distribution	127
S. bullata	IE	vitellogenin synthesis after allatectomy	128
Ephemeroptera			
Hexagenia verunata	EP	parasitism effect	69
Hemiptera			
Dysdercus cingulatus	Cm	concentration changes	130
Oncopeltus fasciatus	ID, IE, EP	pattern, concentration, JH effect	131–134
	EP	vitellogenin identification, JH effect	135
	EP, ID	vitellogenin, antigenic study	136
Eurygaster sp.	EP	parasitism effect	137

Table 1 (Continued)

Insect order and species	Method[a]	Observations	Reference
Pyrrhocoris apterus	Cm	concentration changes, hormonal control	138
	EC	ontogeny, concentration changes	139
Rhodnius prolixus	EC, ES	synthesis by fat body, yolk formation	140–142
	ID	allatectomy and yolk protein in plasma	143–145
	EP	vitellogenin identification and hormonal control	146
Triatoma infestans	EP, ID	detection of two vitellogenins	147
T. protracta	EP, IE	vitellogenin identification and hormonal control	148
Hymenoptera			
Apis mellifica	ES	patterns: castes and ontogeny	149
	EP, ES, ID, IE	pattern, ontogeny	150–153
	EP	esterase, dehydrogenase zymograms	394
	ID, IE	vitellogenin in queens and workers	154
	EC	vitellogenin occurrence, synthesis, control	155–159
Arge pectoralis	ES	pattern	59
Bombus terrestris	EP	pattern, vitellogenin, parasite effect	160
Formica polyctena	EPa, ES	pattern	46, 47, 51, 148a
F. pratensis	EP	pattern, changes during metamorphosis	48, 50, 51
	EP	dehydrogenase, esterase zymograms	395
F. rufa	EP	pattern, changes during metamorphosis	47, 49–51
Neodiprion (4 spp.)	ES	pattern	59
Pristophora erichsonii	ES	parasitism effect	161
Vespula arenaria	ES	pattern	59
Isoptera			
Macrotermes subhyalinus	EP	synthesis and release by queen fat body	162
Lepidoptera			
Antheraea mylitta	EP	pattern, vitellogenin detection	18
A. pernyi	Enz	prophenoloxidase activation	163
A. polyphemus	EP	lipoproteins, JH binding	164
	EP	JH effect; vitellogenin detection	165
	EP	pattern; vitellogenin detection	18
Archips cerasivorana	ES	pattern	59
Anagasta (Ephestia) kuehniella			
	EC	pattern, ontogeny	166
	EC	parasitism effects	167
	EP	changes in basic proteins	168
	EC	quantitative pattern, ontogeny, ligation, and ecdysone effects	42
Bombyx mori	Cm	plasma protein synthesis by fat body	169
	EP	pattern, ontogeny	170
	EP	inheritance of "albumin"	171
	EP	vitellogenin detection	172
	EP, Ch, ID, IE	vitellogenin occurrence and synthesis	173
	EP	plasma proteins after ovariectomy; *sm* mutant	174, 175
	EP	prophenoloxidase characterization	176
	ES	esterase zymograms	177
	EP, Ch	lysozyme characterization	178
	EP	virus effects	179
Callosamia promethea	EP	pattern, vitellogenin detection	14
Calpodes ethlius	EP, ID	fat body sequesters proteins into granules	180, 181
	Enz	soluble acid phosphatases	182
Danaus plexippus	ID, IE, IP	vitellogenin synthesis timing, control	26, 183
	IE, IP	vitellogenin synthesis not affected by ovary	184
Diatraea grandiosella	EP	ontogeny, relationship of fat body and plasma	185, 186
Eacles imperialis	ES	pattern	59
Galleria mellonella	ES, EP	pattern	59, 60, 187
	Ch, ES	esterases, "cooling protein", "ligature protein"	188
	Ch, EF, EP	pattern, molecular weights, pI	189
	EP, ID	fat body sequesters plasma proteins	190
	Ch, EP	lysozyme characterization, properties	178, 191
	EP	virus, protozoan infection effects	192

Table 1 (Continued)

Insect order and species	Method[a]	Observations	Reference
Heliothis 3 spp.	EC	pattern, ontogeny	193
Heliothis 2 spp.	EP	parasitism effects	194, 195
H. zea	EP	esterases	196
Hyalophora cecropia	ID	ontogeny of six plasma proteins	1
	ES, ID	enzymes, ontogeny, cross reactivity	197–200
	EP	pattern; vitellogenin detection	18
	Cm	protein synthesis: diapause, injury	201
	Cm, EP, Ch	protein synthesis: injury, perfusion	16, 45, 202, 203
	EP	protein synthesis: plasma and cuticle	204
	ID	discovery of vitellogenin	2
	ID	protein uptake by oocytes	205
	IP	vitellogenin synthesis in fat body	206
	IP, EP	vitellogenin synthesis, no JH effect	207, 208
	Ch	vitellogenin isolation, characterization	35
	IE, Ch, Cm	vitellogenin species comparison	28
	Ce	lipoproteins, characterization, synthesis	38, 209
	EP	lipoproteins, JH transport	164
	Ch	lipoproteins, cholesterol transport	210
	EP, ES, Enz	deoxyribonuclease multienzyme complexes	211
	ID, EP, Ch	carboxylesterases	212
	EP, ID, Ch	storage proteins, characterization	213
H. gloveri	EP	pattern; vitellogenin detection	18
	Ce	lipoproteins, vitellogenin, characterization	39
	EP	JH binding	164
	EP	esterases, JH induction	214
	EP, ID, Ch	carboxylesterases	212
Hyphantria cunea	ES	pattern	59
Malacosoma americana	ES	pattern, ontogeny, lipoproteins, chromo-proteins	59
	ID, ES, Ch	ontogeny, distribution	215–217
	EP	protein synthesis, JH analog effect	218
Mamestra brassicae	EP	ontogeny; virus effect	219
Manduca sexta	EP	pattern	220
	EP	ontogeny, comparative study	18
	Cm	protein concentration, diet effect	221
	EP, Ch	protein transport into cuticle	222, 223
	EP	vitellogenin identification	224
	Ch	JH-binding protein, isolation, properties	225–227
	EP, IE, IP, Ch	JH-binding protein synthesis by fat body	228
	Ch, Enz	JH esterase purification, role, ontogeny	229–231, 396
Nymphalis antiopa	EP	pattern, vitellogenin identification	232
Ostrinia nubilalis	EP	pattern, enzymes, relations to fat body	233, 234
P. cynthia	ES, ID	ontogeny, enzymes cross reactivity	197–200
	Ch, EP	lipoproteins: purification, properties	3, 235, 236
	Ch, EP	vitellogenin characterization	237, 238
	Ce, Ch	ecdysone transport	239
	EP	ontogeny, vitellogenin detection	18
Philosamia ricini	Cm	concentration changes	240
Pieris brassicae	ES	pattern, ontogeny, esterase	241, 242
	EP	pattern, zymogram, diet effect	243, 244
	EP	fat body stores two plasma proteins	245
P. rapae	ES	pattern	59
Plodia interpunctella	EP	pattern, histochemistry, JH-binding protein	246–249
	EP	lipoproteins: isolation and properties	250
Protoparce quinquemaculata	ES, ID	pattern, ontogeny	59, 251
Pseudaletia unipuncta	EP	virus infection effect	252
Rothschildia orizaba	ES, ID	ontogeny, distribution	215–217
Sitotroga cerealella	EP	protein uptake by fat body	253
Spodoptera mauritia acronyctoides	EP	ontogeny, virus infection effect	254
Schizura concinna	ES	pattern	59
Trichoplusia ni	EP	parasitism effect	255
	EA, ID, EP	virus effects	256, 257

Table 1 (Continued)

Insect order and species	Method[a]	Observations	Reference
Neuroptera			
Sialis sp.	EP	parasitism effect	69
Odonata			
Uropetala carovei	EP	pattern, zymogram	259
Orthoptera			
Acheta domesticus	EP	pattern, histochemistry; conjugated proteins	60, 260
	Cm	concentration changes	261, 262
Anacridium aegyptium	EP	parasitism effect, vitellogenin detection	263
Blatta orientalis	Ce	larval-specific protein and molting	27, 264
Blattella germanica	EP, ID, IE, Ch	vitellogenin identification, properties	265–267
	Ce, IE	vitellogenin and plasma proteins, properties	28
	IE, IP	vitellogenin synthesis: effects of JH, ovary	184
	Ce, ID, IE	larval-specific protein, cross reactions	27
Byrsotria fumigata	ID	vitellogenin production, uptake, control timing	268, 269
Cophopodisma pyrenaea	EP	pattern and speciation	270
Diploptera punctata		protein granules in fat body	271
Leucophaea maderae	EP	vitellogenin synthesis and control	272–274
	Ce, IE, IP	vitellogenin synthesis, JH control, polysomes	128, 275–279
	Ce, EP, IE, IP	vitellogenin, vitellin, isolation, properties	280–286
	Enz	dopa and tyrosine decarboxylases	397
Locusta migratoria	Cm	concentration, synthesis, hemolymph volume	287–289
	ES	protein associated with molting	290
	EP	lipoprotein and flight	291
	EP	pattern, ontogeny, oogenesis, control	292
	EA, Cm	vitellogenin, synthesis, timing, control	293, 294
	EP, ID	vitellogenin identification, quantitation	295, 296
	Ce, EP, IE, IP	vitellogenin, vitellin, purification, properties	297–300
	Ce, Cm	lipoprotein characterization	301–303
	Ce, EF	lipoprotein, JH-binding, properties	304
	EC, Enz	JH carboxyesterase	305
	EP, Ch	yolk protein utilization	306
	EP	synthesis	307
Melanoplus bivittatus	ES	pattern	59
Nauphoeta cinerea	EP	vitellogenin detection	308
	ID	vitellogenin induction in nymph by JH	309
	EP, IP	vitellogenin, JH, and oocyte growth	310, 311
	EP, Enz	amine oxidases	312
Periplaneta americana	EPa	conjugated proteins, pI	313, 314
	ES	pattern	59
	EP	protein levels, nerve severance, tumors	315, 316
	EC	pattern, vitellogenin detection	317
	EP, Ch	pattern changes during vitellogenesis	318, 319
	IP	vitellogenin synthesis in fat body; timing	206
	EP, ID, IE	vitellogenin, synthesis, uptake, specificity	29, 320–324
	EP	vitellogenin, detection, JH analog effect	325
	EP	ovary implantation effect in males	326
	EP	pattern during molting cycle	327
	Cm	ion binding	328
	EP	lipoprotein, lipid transport	329
	Enz	lipase	330
	EP, Enz	amine oxidases	312
	EP	α-glucosidase activity	331
	Enz	dopa and tyrosine decarboxylases	398
Schistocerca gregaria	EPa	hormonal control	332
	Ce, EP	pattern, characterization of components	333
	EF, IP	vitellogenin detection	334
	EP	parasitism effect	335
S. vaga	IE	vitellogenin detection	128
Stenobothrus lineatus	EP	pattern and speciation	270

[a] Methods: Ce, centrifuge; Ch, chromatography; Cm, chemical analysis; EA, electrophoresis in agarose; EC, electrophoresis in cellulose acetate; EP, electrophoresis in polyacrylamide; EPa, electrophoresis in filter paper; ES, electrophoresis in starch gel; Enz, enzyme assay; ID, immunodiffusion; IE, immunoelectrophoresis; IP, immunoprecipitation.

Table 2 Physical and chemical properties of some insect plasma proteins[a]

Protein	Insect source	$S_{20,w}$[b]	Molecular weight × 10^{-3} Native	Subunits[c]	Lipid (%)	Carbo-hydrate (%)	Amino acid analysis?	Other characteristics	References
Vitellogenins (vg) and vitellins (vn)									
vg	Blattella germanica	16.8	659	100	15.7	4.5	yes	$\bar{v}$, diameter, f/f_o, pI = 5.0	28
vg	B. germanica	–	–	52	7.6	8.0	–	–	267
vg, vn	Leucophaea maderae	14.0	560	57, 87, 96, 118	7.0	8.6	yes	lipid, sugar composition, P content	282–286
vg, vn	Locusta migratoria	17.1	550	52–140	9.6	13.3	yes	$\bar{v}$, sugar composition	298, 299
vn	L. migratoria	16.3	530	55–130	–	11.0	yes	$D_{20,w}$, f/f_o, E, pI = 6.9	300
vn	L. migratoria	–	–	–	–	14.0	–	sugar composition, amino acid analysis of peptide	297
vn	Bombyx mori	–	540	50,200	–	–	–	–	336
vg, vn	Hyalophora cecropia	15.9	500	43,120	10	1	yes	$\bar{v}$, diameter, $D_{20,w}$, f/f_o, pI = 5.7, lipid composition	28, 35, 38
vg	H. gloveri	16.0	–	–	–	–	–	lipid composition	39
vg, vn	Philosamia cynthia	13.3	500	55, 230	10	2.5	yes	$\bar{v}$, diameter, lipid, sugar composition	3, 237, 238
vn	Aedes aegypti	–	270	–	10.5	–	–	–	70
vn	Drosophila melanogaster	–	98[d]	48.6	–	–	–	–	105

Table 2 (Continued)

Lipid transport proteins									
Yellow protein	*Locusta migratoria*	6.9	340	—	31.5	—	yes	—	302
HDL	*Hyalophora cecropia, H. gloveri*	8.7	189^d	—	48	—	—	v̄, lipid content, composition	38, 39
DGLP-I	*Philosamia cynthia*	9.4	700	—	44	—	yes	Diameter, lipid content	3, 235
Juvenile hormone binding protein									
	Manduca sexta	—	28	—	—	—	yes	pI = 4.95	226
Storage proteins									
Protein 1	*Bombyx mori*	—	485	85	—	—	—	—	337
Protein 2	*B. mori*	—	500	85	—	—	—	—	337
Protein 1	*Hyalophora cecropia*	—	480	85	—	—	yes	—	213
Protein 2	*H. cecropia*	—	530	89	—	—	yes	—	213
Calliphorin	*Calliphora erythrocephala*	19.4	528	87	—	0.5	yes	$D_{20,w}$ f/f$_o$, E_{280}/E_{250}	80
Protein C	*C. stygia*	—	250	83	—	—	yes	—	89
Lucilin	*Lucilia cuprina*	—	250	83	—	—	yes	—	112
Larval protein 1	*Drosophila melanogaster*	—	450–480	75–81	—	—	yes	—	368

Table 2 (Continued)

Protein	Insect source	$S_{20,w}$ [b]	Molecular weight × 10^{-3}		Lipid (%)	Carbo-hydrate (%)	Amino acid analysis?	Other characteristics	References
			Native	Subunits [c]					
Hemoglobins									
	Chironomus plumosus	2	31.4	—	—	—	—	pI = 5.40	338
	C. plumosus	2.7 / 1.7[e]	32 / 14.6[e]	—	—	—	—	$D_{20,w}$	339
	C. tentans	1.7	15.9	—	—	—	—	—	93
	C. thummi	—	16, 15, 15.9	—	—	—	—	—	340 339 341
Enzymes									
Lysozyme	Bombyx mori	—	16.5	—	—	—	—	—	178
Lysozyme	Galleria mellonella	—	14.7	—	—	—	yes	—	178
Prophenol-oxidase	Bombyx mori	6.6	80	40	—	—	yes	pI = 4.98, Cu = 0.15%	176
Prophenol-oxidase	Calliphora erythrocephala	9.4	115	77, 70, 49	—	—	—	—	342
JH esterase	Manduca sexta	4.98	67	—	—	—	—	D, pI = 5.3, 5.4, 5.5	230

Table 2 (Continued)

Antibacterial protein									
P-5	*Hyalophora cecropia*	—	96	24	—	—	—	—	343
Toxin									
Leptino-tarsin	*Leptinotarsa decemlineata*	—	50	—	—	—	—	—	58
Chromoprotein									
Insecti-cyanin	*Manduca sexta*	—	72	23	—	yes	$pI = 6.1$	—	344
Other plasma proteins									
Bg I	*Blattella germanica*	19	476	80	0	yes	$\bar{v}$, diameter, $D_{20,w}$, f/f_o, $pI = 5.5$	—	28
Bg II	*B. germanica*	5.4	511	—	53.2	—	$\bar{v}$, diameter, $D_{20,w}$, f/f_o, $pI = 6.9$	—	28
Lipopro-tein	*Plodia interpunctella*	—	70	—	—	yes	—	—	250
Protein B	*Calliphora stygia*	—	240	81	—	yes	—	—	89

[a] Only proteins that were at least partially purified before characterization are included.
[b] Extrapolation to zero concentration was not done in all cases.
[c] Not all of the polypeptides observed for vitellogenins are primary subunits; see text.
[d] Probably inaccurate; see text.
[e] pH-dependent.

The apparent subunits seen in SDS-polyacrylamide gel electrophoresis vary considerably, although a component of about 50,000 daltons is observed in almost all species. Although in some species the subunits appear to occur in simple proportions in the molecule (238, 285, 286), their nature must be assessed critically. In *Locusta migratoria,* they do not occur in simple molar proportions, and pulse labeling in fat body with ^{3}H-leucine has shown initial synthesis of components of about 250,000 daltons, followed by intracellular processing, presumably by proteolytic cleavage, to produce peptides with the molecular weights found in the secreted protein (298, 299). That some, at least, of the polypeptides in the locust and a cockroach are secondary in origin, is further suggested by the different proportions found in vitellin compared with vitellogenin (285, 286, 298). In *D. melanogaster,* three subunits of closely similar size (about 50,000 daltons) are found (106, 109). Identification of the primary units of translation and their relationships to the final polypeptides in the yolk proteins of different insect groups is badly needed.

In certain cockroaches and in the bug, *Triatoma infestans,* two electrophoretically separable vitellogenins are found in the hemolymph, but their relationship to one another is not yet known (29, 147, 345). The milkweed bug, *Oncopeltus fasciatus,* is unusual in that the hemolymph in diapause contains an antigenically incomplete vitellogenin which, after diapause, is modified or joined by another molecule to produce the complete protein; on entry into the occyte this is further modified, with altered electrophoretic mobility, to form vitellin (136).

The vitellogenins and vitellins are lipoglycoproteins. In addition to phospholipids, *P. cynthia* vitellogenin contains diacylglycerol, cholesterol, and some carotenoids (which account for the yellow color), while the vitellin from eggs also contains triacylglycerol and altered proportions of the other lipids (3, 238). In some species, a small amount of nonlipid P may be present (283, 299, 300), but highly phosphorylated yolk proteins, like the phosvitin of amphibia and birds, have not been found in insects. The carbohydrate component varies widely in amount but, wherever examined, contains mannose and glucosamine (238, 282, 297, 299). In a glycopeptide from locust vitellin, the mannose is apparently linked to asparagine (297).

A number of insect vitellogenins and vitellins show a common pattern in their amino acid composition, characterized by high aspartic and glutamic acids (or asparagine and glutamine) which remarkably resembles the amino acid composition of some lipovitellins of birds and amphibia, while differing from that of some nonvitellogenic insect plasma proteins (28, 237). However, immunochemical tests show no cross reaction far outside of the generic level among cockroaches (28) or outside of the family level in saturniid silkmoths (2, 200). This suggests marked divergence in amino acid sequence, despite similarity in net composition.

Insect vitellogenins are synthesized in the fat body and appear in the hemolymph usually only in the adult, close to the time of yolk deposition during oogenesis (references in Table 1). The timing of synthesis has been demonstrated by incorporation of isotopic amino acids into immunochemically isolated products in a number of insect species. The silkmoths are exceptional in that vitellogenin is synthesized first in the pharate pupa shortly after the cocoon is spun (173, 207). In the honeybee, vitellogenin occurs in workers as well as queens, and changes in concentration under various conditions have been plotted in detail (155, 156). Although vitellogenins are regarded, and often defined, as specific to the female, Telfer (2), on the basis of careful immunochemical tests, reported traces (about one thousandth of the female levels) in hemolymph of male cecropia pupae; and synthesis in male *Drosophila*, after implantation of ovaries, has recently been reported (106). Because of its bearing on the question of gene control, further information on the sex limitation of vitellogenin synthesis will be of great interest.

Vitellogenin synthesis in the fat body of many species is induced by juvenile hormone, secreted from the corpora allata under appropriate physiological or environmental stimuli (reviewed in 346, 347). This system is being used for the biochemical study of juvenile hormone action (285, 298, 348). The fat body must first acquire competence to respond, since vitellogenin synthesis can be induced precociously in the last and penultimate, but not earlier, larval instars of two cockroach species, and in the pupal, but not the larval, stage of the monarch butterfly (184, 309). In the silkmoths, however, eggs can develop in the absence of corpora allata (for review see 349), and the lack of any influence of juvenile hormone on vitellogenin production in cecropia has recently been established (208). The mosquito, *Aedes aegypti* is exceptional among insects examined in that vitellogenin synthesis in the fat body is induced by ecdysone which is produced in the ovary after a blood meal (73, 350); this relationship is analogous to that in amphibia and birds, where liver vitellogenin synthesis is induced by estrogen from the ovary (351).

In the uptake of vitellogenin from the hemolymph, the ovaries are selective both among the plasma proteins (352) and taxonomically. After interspecific transplantations of ovaries in cockroaches, vitellogenin sequestration took place only when implant and host were of the same family (322), and experiments with a cockroach and a silkmoth showed that oocytes of each took up only their own kind of radioactively labeled vitellogenin (28). In some insects at least, juvenile hormone is required for vitellogenin uptake by ovaries (135, 353). In the *Drosophila* mutant *apterous*[4], little or no yolk is formed, although vitellogenin is present in the hemolymph, and treatment with juvenile hormone caused the oocytes to sequester yolk (107, 108). The *sm* mutant of the mulberry silkworm, *Bombyx mori,* has a similar phenotype, but no influence of juvenile hormone on uptake has been demonstrated (175).

The structural biochemistry, control of synthesis, and control of uptake of the insect vitellogenins and vitellins are currently active subjects of research.

LIPOPROTEINS AND LIPID TRANSPORT

Insect plasma contains several lipoprotein components (Table 1). Both for the cecropia silkmoth and for locusts, for example, six bands staining for lipid have been observed after polyacrylamide gel electrophoresis (164, 304). Certain of these components have recently been isolated and characterized and their roles studied, particularly in lipid transport, where, in contrast to analogous processes in vertebrates, diglycerides have an especial importance (354).

From hemolymph of cynthia silkmoth pupae, Chino et al (3, 235) purified to homogeneity two lipoproteins designated as diglyceride-transport lipoproteins (DGLP-I and DGLP-II). DGLP-II has since been identified as a vitellogenin (237). DGLP-I, renamed lipoprotein I (354), has the capacity to accept diglycerides from the fat body for transport in the hemolymph to sites of need, including ovaries and, presumably, muscle. Cynthia lipoprotein I (Table 2) has a molecular weight of 700,000 and is globular with a diameter of 13.5 nm, as shown by the electron microscope (3). The subunit composition has not been reported. It contains about 44% lipid, of which more than half is diglyceride; the remainder consists of cholesterol, phospholipids, a trace of triglyceride, and some carotenoids, which impart a yellow color (3, 355). It is a glycoprotein, with about 1% of mannose (238). From the related silkmoths *H. cecropia* and *H. gloveri,* by density gradient flotation, Thomas & Gilbert (38, 39) isolated three fractions designated as low density, high density, and very high density lipoproteins (LDL, HDL, and VHDL). Although not homogeneous electrophoretically, VHDL appears to correspond principally to vitellogenin and HDL to lipoprotein I (354); the LDL fraction, containing 76% lipid, has not been further studied.

A yellow diglyceride-carrying lipoprotein from locust hemolymph, isolated by sucrose gradient centrifugation, has a molecular weight of 340,000 (Table 2; 302). Two preparations contained 26% and 36% of lipid (which possibly reflects different states of loading), consisting predominantly of diglycerides and phospholipids.

During development of cecropia silkworms, the HDL and VHDL fractions are reported to be not detectable in larval instars I to IV, and to build up during the fifth instar to maxima in the pharate pupa and pupa (209). Incorporation of labeled palmitate and amino acids was obtained in fat body from silkworms and locusts during incubation in vitro, which shows this tissue to be a site of synthesis (209, 301). Inhibition of protein synthesis, however, did not affect the uptake of lipid by the lipoprotein, which indicates that the apoprotein can be loaded without requirement for de novo protein synthesis. In *P. cynthia,* lipoprotein I isolated from eggs contained the identical apoprotein but a much lower content of lipid than that obtained from hemolymph, which demonstrates that it can unload lipid in the oocytes (238).

The loading of diglyceride-carrying lipoproteins, however, appears to be more complex than simple association. In the silkmoth pupa, diglyceride transfer depends on a specific interaction between the fat body and the protein, an interaction that is inhibited by respiratory poisons and does not take place when the isolated lipoprotein is incubated with labeled diglyceride (3, 356). Elevation and altered distribution of lipids in hemolymph electrophoretic components are observed in locusts after flight, which stimulates release of lipid from the fat body into the hemolymph through action of an adipokinetic hormone (291). A quantitative analysis of effects of the hormone by gel exclusion chromatography shows shift of diglyceride into a new chromatographic fraction, accompanied by depletion of two protein components, which suggests the formation of a complex (303). In another study on locusts, the addition of a second protein fraction enhanced the transfer of diglyceride from the fat body to the yellow lipoprotein (302).

In the locusts, lipid freshly absorbed from the gut is also found in the hemolymph, chiefly as lipoprotein-bound diglycerides (357). In the cockroach, however, freshly absorbed lipid is found chiefly in the form of triglycerides bound to a protein fraction of low electrophoretic mobility (329).

HORMONE-BINDING PROTEINS AND POLYPEPTIDE HORMONES

Transport of Juvenile Hormone

The juvenile hormones (JH) of insects comprise three closely related homosesquiterpenoid esters which, before metamorphosis, are responsible for maintaining the larval state of development. After metamorphosis, in many insects, they assume a second role in stimulating reproductive maturation (346, 347). Although sufficiently water soluble to provide the concentrations found in hemolymph (225, 358), they circulate chiefly in company with certain hemolymph proteins.

The association of ^{3}H-JH, either after injection or after incubation with hemolymph, with one of the major lipoproteins has been reported (164, 304, 359, 360). In silkmoths, most label was in the lipoprotein I (or HDL) fraction, which is involved in lipid transport (164). In *Locusta* also, a major yellow lipoprotein was chiefly labeled (304). The affinity of such lipoproteins for JH is relatively weak, however [$K_d = 4 \times 10^{-5}$ M for the Colorado potato beetle, *Leptinotarsa* (360)] and nonspecific. JH acid, tripalmitin, and synthetic JH analogues are also bound (358, 360).

A distinct, low molecular weight, higher affinity, specific JH-binding protein was subsequently identified in tobacco hornworm (*Manduca sexta*) hemolymph (225) and then purified (along with a second, closely related protein component) from this source (Table 2; 226). This is not a lipoprotein, but possesses a single specific binding site with $K_d = 4 \times 10^{-7}$ M for JH-I. The relative affinities of homologues indicate that the site has hydrophobic character (358). Neither the inactive JH acid and diol derivatives, nor several hormonally active synthetic JH analogues are significantly bound. Similar low molecular weight JH-specific binding proteins have been identified in larval hemolymph of the Indian meal moth (247) and in a number of other insects belonging to three orders (358). A recent study of larval and adult Colorado potato beetles, however, could detect JH binding only in the lipoprotein fraction (360).

Synthesis of JH-binding protein has been demonstrated in isolated *Manduca* larval fat body (228). The biological role of these proteins appears to be protection of JH against inactivation by hemolymph esterases. Bound JH-I is protected from attack by the nonspecific esterases which are generally present in insect hemolymph, but remains sensitive to a specific JH esterase that appears in the hemolymph of *Manduca* during the second half of the last larval instar; this is the stage when elimination of JH permits metamorphosis, and the interaction of JH with binding protein and esterases is believed to contribute to the regulation of JH titer (229, 230). The presence of binding protein synergized the biological action of JH on imaginal disks in vitro (229) and diminished the degradation of the hormone by epidermal and fat body tissue (227, 248). Binding protein, however, does not appear to be necessary for tissue recognition of JH (248), which is consistent with the observation that the protein lacks affinity for some biologically active JH analogues. The lipoproteins mentioned above bind JH only at relatively high concentrations, do not protect against esterase activity, and may have little biological role with regard to this hormone (360).

Transport of Ecdysone

Ecdysones are the steroidal growth, molting, and differentiation hormones of insects (346, 347). By analogy with the steroid-binding globulins of vertebrate plasma, the existence of ecdysone-binding proteins in insect hemolymph has seemed likely, although ecdysones are sufficiently water soluble for transport in the free form. In hemolymph of the bug, *Pyrrhocoris* (139), the blowfly, *Calliphora* (361), and *Drosophila* (98), some association of ^{3}H-α-ecdysone has been shown with macromolecular components separated by Sephadex chromatography and other procedures. However, the association was lost during dialysis, and no binding protein has been isolated, or binding constant measured. Labeled ecdysone did not accompany the major lipoproteins of cynthia silkmoths during purification (239). These experiments were all conducted with α-ecdysone, which was available at the necessary specific activity, and not with the biologically more active derivative β-ecdysone; however, both forms are normally transported in hemolymph. Recently, in locusts, the apparently specific association of a small proportion of ^{3}H-β-ecdysone with a macromolecular component has been reported (362). The nature and significance of this interaction require further study.

Polypeptide Hormones

Insects use many nonlipoidal hormones, probably polypeptide or protein in nature, in the control of metabolism, homeostasis, development, and behavior (346, 347). We do not review these, since their concentrations in hemolymph are low and they have generally been detected only by bioassay. It is worth recording, however, the recent report of complete purification of the adipokinetic hormone of locusts, identified structurally as a decapeptide (363, 364). Another interesting development, indicating remarkable evolutionary stability, is the demonstration that the hyperglycemic and hypoglycemic hormones of the tobacco hornworm react with antiserum against mammalian glucagon and insulin, respectively (365). In addition, hornworm neuroendocrine extract contains a component that reacts with mammalian gastrin antiserum (366).

STORAGE PROTEINS

An important class of proteins that play a major role in insect metamorphosis was discovered through electron microscopic studies on the blowfly, *C. erythrocephala* (81). Larval extracts revealed abundant protein particles about 10 nm in diameter that were then isolated by centrifugation and given the name calliphorin. Purified calliphorin has a molecular weight of 528,000 and dissociates in guanidine hydrochloride or at pH values above 6.5 into 87,000 dalton subunits, which indicates hexameric structure (80). Very similar proteins have subsequently been isolated in trimeric form (mol wt about 250,000) from *C. stygia* (protein C; 89) and the Australian sheep blowfly, *Lucilia cuprina* (lucilin; 112) (Table 2). They are characterized by exceptionally high contents of tyrosine and phenylalanine, and carry about 0.5% carbohydrate and a small amount of lipid (80, 89). Wild populations of *L. cuprina* show numerous phenotypic subunit patterns, suggesting both gene duplications and polymorphism at each locus (112). Genetic crosses indicate at least 12 structural loci, some of which map on the second chromosome.

In fully grown blowfly larvae, at the conclusion of feeding, calliphorin is extraordinarily abundant, making up 75–80% of the total hemolymph protein, some 60% of the total buffer-soluble protein of the animal, or about 7 mg per individual of 120 mg body weight (81, 89). The protein is synthesized in the larval fat body and secreted into the hemolymph, where it accumulates (82, 85). At the conclusion of the feeding period, synthesis stops and reabsorption takes place from the hemolymph into the fat body concurrently with the appearance of dense protein granules in the fat body cells (87); at adult emergence, only 0.03 mg of soluble plasma calliphorin per individual remains. When ^{14}C-labeled calliphorin was administered to late larvae, $^{14}CO_2$ was released during adult development and adult proteins were labeled (89). It is inferred that calliphorin is a storage protein synthesized during the feeding stage and used as a source of amino acids and energy for the synthesis of new proteins during metamorphosis (41).

Another *Calliphora* hemolymph protein (protein II or protein B; Table 2) of similar molecular weight and subunit structure, but distinct in its antigenic properties and developmental timing of accumulation, remains unknown in function (81, 89).

The relationship between specific hemolymph proteins and fat body granules has been established in a recent study on the cecropia silkmoth (213). Dense granules, 1.5–3 μm in size, isolated from pupal fat body by centrifugation and then dissolved in buffer, yielded proteins 1 and 2 (Table 2), with native and subunit molecular weights close to those of calliphorin; their amino acid compositions resemble that of calliphorin but are lower in tyrosine and phenylalanine. The changing contents of these proteins in hemolymph and fat body were in accord with synthesis and secretion by the fat body, followed by reabsorption, as suggested for *Calliphora*. The fat body granules showed periodic fine structure indicating crystalline array of the proteins. The proteins and granules occur in greater amounts in the female than in the male. Recent work with the silkworm, *B. mori*, shows two similar proteins (Table 2), one of which occurs exclusively in the female (337). This probably corresponds to a previously reported larval female-specific hemolymph protein of *B. mori* (172).

Protein storage in hemolymph and fat body is evidently a widespread or universal phenomenon in insects. Proteins showing complete or partial antigenic identity with calliphorin were demonstrated in several flies, and particles of similar size but giving no immunological cross reaction were found in some unrelated insects (81). A comparable protein, cross-reacting with calliphorin, has been identified in *Drosophila* and called drosophilin or larval serum protein 1 (91, 110, 367, 368). The presence of abundant dense protein granules ("proteinaceous spheres" and other terms) in fat body cells of many insects during metamorphosis has been recognized since the late nineteenth century, although their origin was unknown (reviewed in 11, 369). The participation of protein sequestered from hemolymph in the formation of fat body granules was first indicated by fine-structural studies and quantitative analyses on the skipper, *Calpodes ethlius* (370). The generality of this process is supported by evidence for selective uptake of protein from hemolymph into fat body in many species (42, 87, 190, 213, 216, 245, 251, 288).

While protein storage is best established during metamorphosis, there is some evidence for a similar phenomenon in other developmental stages. In locusts (290) and cockroaches (27) a larval-specific hemolymph protein rises in concentration before each molt and falls sharply during or after ecdysis. The cockroach protein is of high molecular weight (16S) and its decline in titer, just when cuticle synthesis is most rapid, has led to the suggestion of a role in nutrient storage (27). In an adult viviparous cockroach, also, fat body protein granules form and disappear during the reproductive cycle in a manner suggesting a storage role, though in this case no relation to hemolymph protein has been shown (271).

The developmentally programmed synthesis, secretion, and uptake of massive amounts of specific storage proteins demonstrate impressive regulation of fat body cell activity. Recent experiments with *B. mori* indicate that storage protein synthesis may depend on declining juvenile hormone titer in the last larval instar (S. Tojo, personal communication). The formation of fat body protein granules in *Calpodes* is prevented by ligature of larvae behind the thorax and is induced by ecdysone (180, 181). In a cockroach, accumulation of larval-specific hemolymph protein is correlated with ecdysone-induced molting (264). In *Drosophila*, ecdysone stimulates, but is apparently not necessary for, granule formation (104); in *Calliphora*, however, the uptake of hemolymph proteins was not stimulated by ecdysone (88). Thus, the control of protein uptake and granule formation remains unclear. As a step toward analysis of subcellular mechanisms, mRNA has been prepared from *Calliphora* and *Drosophila* larval fat body and translated, in the wheat germ cell-free system, into calliphorin-like proteins, identified by immunochemical reaction and molecular weight (91, 371).

HEMOGLOBINS

Hemoglobin occurs in only a few insects: within specialized tracheal cells in horse bot fly larvae and some aquatic bugs, and free in the hemolymph plasma in the larvae of certain midges (Chironomidae) (5). The hemoglobin of *Chironomus* has an exceptionally high affinity for oxygen ($p_{\frac{1}{2}} = 0.7$ mm Hg) (372), and it apparently provides some storage during brief periods of oxygen lack in pond bottom burrows. In fully grown larvae, hemoglobin may account for 20–40% (341) or 90% (373) of the total hemolymph protein (neither paper gives data). By electrophoretic methods, 8–14 hemoglobin variants can be resolved from the hemolymph of individual *Chironomus* larvae, the pattern differing between species, geographic races, and developmental stages (Table 1). This observation has stimulated biochemical and genetic study of *Chironomus* hemoglobins.

In *C. plumosus* and *C. thummi,* both monomeric (mol wt ~ 16,000) and dimeric forms of hemoglobin occur (Table 2), dissociation being favored at elevated pH, although the situation in vivo is not certain (339, 341, 378–380). In *C. tentans* and some other species, however, only the monomers are found (Table 2; 93). A hemoglobin from *C. thummi* has been isolated in crystalline form (341) and sequenced (381, 382; note corrections from the preliminary report), and the tertiary structure has been determined by Fourier analysis at 2.5 Å resolution (383). Only 25% of the amino acids correspond to the sequence of lamprey hemoglobin, and 20% to that of whale myoglobin, yet the tertiary structure is closely similar to whale myoglobin. There are also extensive sequence differences among several hemoglobin forms from *C. thummi* (384).

Since multiple monomeric hemoglobins are found in individual insects, multiple genes must be involved, and 12–14 loci have been suggested (375). All are located on chromosome 3 and presumably arose through gene duplication (376, 385, 386). An interesting regulatory mutation greatly increases the quantity of one hemoglobin form and decreases that of another (387, 388). During development, hemoglobin first appears in instar II; in the last larval instar (IV) the quantity increases and the pattern of forms changes (373, 375). The fat body is the major site of synthesis (129, 389). In metamorphosis, hemoglobin is largely destroyed before emergence of the adult, and the midgut has been invoked as a site of degradation (390, 391). Mechanisms for developmental regulation of the synthesis and degradation of this protein must therefore exist. As larval-specific proteins, most abundant before metamorphosis, the *Chironomus* hemoglobins resemble the storage proteins (see above), and Thomson (41) suggests that this was their primary role, heme having associated with them secondarily to function in respiration. This would circumvent the problem of their evolutionary origin in isolated members of an animal group; yet the similarity of molecular size and conformation strongly suggests an origin in common with the vertebrate hemoglobins.

ENZYMES

Many enzyme activities found in insect hemolymph have been listed in previous reviews (6–8), and a few recent reports are added in Table 1, although we have not attempted a complete listing. For some, notably esterases and dehydrogenases, isozymes and developmental changes have been described (e.g. 198, 395). Very few enzymes have been purified and characterized from insect plasma, and evidence is usually lacking also for their sources and biological roles. Trehalase has been found in plasma from blowflies (399) and the American cockroach (331), but a recent study of its distribution in cockroach hemolymph (400) has concluded that when the hemocytes are removed, under appropriate isotonic conditions, they contain all of the enzyme, none being found in the plasma. It will be of interest to determine whether this finding may be extended to other enzymes reported in insect plasma.

Lysozyme in hemolymph is discussed below in relation to its role in defense against bacteria. We select for further discussion here the phenoloxidase system, which is characteristic of insect hemolymph and has been studied in some detail.

Phenoloxidases

The presence of phenoloxidase (tyrosinase) in insect hemolymph has long been known through the darkening and eventual melanic precipitation that occurs when hemolymph is exposed to air (6). Phenoloxidase is involved in the production of agents for tanning of the insect cuticule, and has been extensively studied in relation to the biochemistry and endocrine control of this process (401, 402). The enzyme system is found in the cuticle as well as in the hemolymph, and the participation of the plasma enzyme in the tanning process, assumed by some authors (403), has been questioned by others (401, 404). Hemolymph phenoloxidase may be important in defense against internal parasites through melanic encapsulation (405), but its total biological role is not yet clear. The phenoloxidases of insects represent a complex system which still presents many problems despite extensive investigation; only a few recent contributions can be mentioned here.

Recent work on hemolymph phenoloxidase in Lepidoptera has centered on its activation from a proenzyme. In hemolymph of the oak silkmoth, *Antheraea pernyi,* the kinetics of activation (163) and the inhibitory influence of salts (406) have been studied, and evidence presented that tyrosine is an endogenous substrate (407). The action and inhibition of the *Spodoptera* enzyme have been described (408). Intensive studies with the silkworm, *B. mori,* have led to purification of the hemolymph prophenoloxidase as a copper protein with a monomeric molecular weight of 40,000 and a tendency to associate to dimers and higher aggregates (Table 2; 176). The content of prophenoloxidase in silkworm hemolymph rises at the time of pupation to the remarkably high level of about 1 mg/ml (M. Ashida, personal communication). The proenzyme can be activated by chymotrypsin (409) and is naturally activated, with the release of a peptide, by an activator protein from the cuticle (410, 411, 412).

The phenoloxidase system of Diptera appears to be more complex. In *Calliphora,* several electrophoretically separable forms of the enzyme occur in hemolymph (342, 413). The purified proenzyme (Table 2) has a molecular weight of 115,000 and, although aggregation occurs after activation, a previously reported much higher molecular weight (414) is attributed to contamination with calliphorin. In *Sarcophaga,* the activation of hemolymph proenzyme by an activator from the cuticle has been described, as well as conversion to the active form by activator in the hemolymph at the time of puparium formation (127, 415). During incubation in vitro, whole hemolymph, but not plasma, increased its prophenoloxidase content; thus, the hemocytes have been suggested as a source of the enzyme (415). In *Drosophila,* a complex, multicomponent, cascade-type activation system, with differences among mutants, has been described (416, 417); although this work was done with preparations from whole insects, hemolymph phenoloxidase is probably chiefly involved.

Some characteristics of substrate specificity and activation of cockroach hemolymph phenoloxidase have also been reported (418).

PLASMA PROTEINS IN DEFENSE

Responses to Wounding

The hemolymph of many, but not all, insects coagulates rapidly upon escape from the body. The process has been studied chiefly by microscopic techniques and apparently involves a class of fragile hemocytes as well as proteins from the plasma (for review see 419). Although chelating agents are sometimes effective in preventing coagulation (13, 14), and the presence of vertebrate-type clotting factors has been claimed (420), there is no clear evidence for a thrombin-fibrinogen mechanism. Changes in lipoproteins have been reported to accompany coagulation of cockroach hemolymph (314), and coagulation of locust hemolymph involves the disappearance of one electrophoretic component representing about 20% (in contrast to an earlier report; 421) of the total plasma protein (422). Surprisingly, the biochemistry of insect hemolymph clotting seems not to have been studied further.

After injury to the integument of diapausing silkmoth pupae, protein synthesis is stimulated in several tissues, the effect on synthesis of hemolymph proteins being especially great (201, 202, 423). Although the physiology of this response is not understood, it is presumably related to recovery from the wound (346).

Inducible Antibacterial Factors

Since early in this century it has been known that insects inoculated with living nonpathogenic or killed pathogenic bacteria can acquire resistance to a subsequent challenge by bacterial pathogens (reviewed in 424–426). Most experiments have used Lepidoptera, such as the wax moth (*Galleria mellonella*) or silkworms, but other insects, including milkweed bugs (427), locusts (428), bees (429), and *Drosophila* (430) have also been successfully immunized. Some bacteria make more effective vaccines than others, and cell-free preparations such as *Pseudomonas aeruginosa* cell-surface lipopolysaccharide (431), or the polysaccharide derived therefrom (432), are also active. The response is much more rapid and less specific than the antibody response of vertebrates, and insects certainly do not produce immunoglobulins, although occasional unsubstantiated references to immunoglubulins in insects persist (260, 433).

While phagocytosis by hemocytes is important in the insect defense system, the induced resistance depends chiefly on soluble bactericidal factors in the plasma. Several earlier studies described factors that were dialyzable and stable to heat, acid, and alkali (427, 434–436). None of these was obtained pure, however, and from the varied properties reported it is difficult to draw conclusions as to their chemical nature. Multiple factors may be involved. Normal hemolymph often has some antibacterial activity (437), and hemagglutinating activity of unknown signifiance has also been demonstrated (438–441).

Recently, the participation of hemolymph plasma proteins in the antibacterial reaction has been studied. Kawarabata (442) vaccinated silkworm pupae with live *Escherichia coli* or heat- or formalin-killed *P. aeruginosa* and within 12 hr obtained immunity to the latter bacterium, which is normally highly pathogenic. Hemolymph from immunized pupae was bactericidal in vitro and this activity was nondialyzable and heat sensitive; a trypsin-sensitive active component was isolated by Sephadex G-100 chromatography. Boman and co-workers (443) have studied the inducible reaction in the hemolymph of saturniid silkmoth pupae (*H. cecropia* and related species). For immunization, living *Enterobacter cloacae* was most effective. The response, measured as the rate of killing of *E. coli* by diluted hemolymph in vitro, was maximal after 3 days and lasted more than a week. The induction of antibacterial activity was blocked by actinomycin D or cycloheximide and was accompanied by the synthesis of at least 8 hemolymph polypeptides (444). [Other reports that actinomycin failed to block insects' responses to bacteria may have depended on too low doses (178, 445).] After fractional precipitation with ammonium sulfate, recombination of three fractions was necessary for restoration of maximal bactericidal activity. One protein (P5) has been purified and characterized (Table 2) and shown to enhance the activity of other fractions, although it was inactive alone (343). *E. coli* mutants lacking some carbohydrate components of the cell surface lipopolysaccharide showed increased sensitivity to killing, and the lipopolysaccharide or the derived lipid A inhibited the bactericidal reaction (443, 446); thus, this complex is proposed as one site of attack. Similarities to the mammalian complement system have been pointed out (444).

Bacillus thuringiensis, an insect pathogen, produces inhibitors that interfere with the action of the hemolymph bactericidal factors and may contribute to its virulence (447). A bactericidal substance induced after infection or injury in the hemolymph of fleshfly (*Sarcophaga*) larvae appears to be a protein of low molecular weight (448).

Insect hemolymph often contains lysozyme (426). The enzymes from *Bombyx* and *Galleria* have been purified (Table 2) and some kinetic differences from egg white lysozyme, as well as a higher specific activity, have been found (178, 191). After injection of killed bacteria, hemolymph lysozyme activity may rise greatly (178, 449, 450), and a primary role of lysozyme in antibacterial defense has been suggested (450). Quantitatively, however, the lysozyme activity does not parallel the bacterial resistance (442, 449) and the hemolymph lysozyme has been separated from other bactericidal factors (437, 442, 444), so that it appears to be only one component of the antibacterial system.

These studies indicate a complex, multicomponent, inducible hemolymph protein system involved in defense against bacterial infection. Continued isolation and characterization of the active components is required. In diapausing silkmoth pupae, the spectrum of induced proteins appears to be identical after immunization with different bacteria, and overlaps, but shows differences from, that resulting from simple wounding (444). However, kinetic and inhibition studies indicate different mechanisms in the killing of different bacteria (451). The role of the nonprotein factors found by earlier workers is not clear, but since the assays were generally not quantitative and recombined fractions were not tested, it is easy to see how active proteins could be overlooked.

The nature of insect resistance to viral infection is poorly understood, but one laboratory (442) has reported a viral inhibitory factor, apparently protein in nature, in hemolymph of nuclear polyhedrosis-infected silkworms.

OTHER PLASMA PROTEINS

Chromoproteins

The leaf-green color of the hemolymph and integument of many insects, originally thought to be chlorophyll, has long been recognized as due to yellow carotenoid protein (see section on lipoproteins) together with a blue bile pigment-containing chromoprotein (452, 453). Recently, from hemolymph of *Manduca sexta* larvae, the blue protein has been purified to homogeneity, crystallized, named insecticyanin, and studied in detail (Table 2; 344). The prosthetic group is probably the unusual isomer biliverdin IXγ. Insecticyanins from hemolymph and integument of *Manduca* are immunologically identical; among species of the family Sphingidae, reactions of partial identity are shown, while there is no precipitin cross reaction with hemolymph from other families of Lepidoptera.

A Toxin

A toxic component, named "leptinotarsin," of the hemolymph of the Colorado potato beetle, *Leptinotarsa decemlineata,* has been purified as a protein of molecular weight 50,000 (Table 2), although the activity is stable to several proteases (58). Leptinotarsin is highly toxic when injected into other insects or mice, yet harmless by feeding, and has no obvious natural function.

CONCLUSIONS AND PROSPECTS

Knowledge of the plasma proteins of insects has advanced greatly during the past few years. Yet it is still not possible, for any insect species, to attribute properties and functions even to all of the *major* protein components that are readily demonstrable by zonal electrophoresis.

Several classes of insect plasma proteins can now be recognized and compared functionally with some of the plasma proteins of vertebrates. Certain insect lipoproteins, like those of vertebrates, have roles in the transport of lipids, yet the special importance of diglycerides appears to be peculiar to the insects. No chylomicra as vectors for triglycerides have been detected in insects, and there is no evidence for any serum albumin, which could carry fatty acids. The vitellogenins, serving as precursors for major yolk proteins, perhaps exhibit the closest resemblance in the two animal groups, and here the possibility of evolutionary homology must be considered. Hemoglobin occurs as a plasma protein in only one group of insects, and its probable homology with vertebrate hemoglobins has been questioned. The storage proteins appear to be special adaptations to insect

molting, metamorphosis, and cyclic reproduction, and have no known analogue in the vertebrates. For defense against infection, the insects have no
immunoglobulins, but do possess antibacterial proteins that have been compared to components of complement, though this needs the support of
physicochemical characterization. Certain insect polypeptide hormones do
appear to be homologous, on immunochemical evidence, with mammalian
hormones.

Some of the insect plasma proteins provide favorable material for the
study of biochemical evolution. The storage proteins and the vitellogenins
(or vitellins) occur generally among insects in amounts sufficient for isolation and study. The few available immunochemical comparisons indicate
much greater rates of divergence among these proteins than among the
enzymes and polypeptide hormones. This is in accordance with the low
selective pressure to be expected for proteins whose primary role is nutrient
storage. Systematic comparative studies including functional, immunochemical, and physicochemical characterization would be of great interest.

The insect plasma proteins can also serve in the analysis of gene regulation. Insect metamorphosis, it has been suggested, represents the successive
expression of three gene sets—larval, pupal, and adult (454), while the
evidence has also been interpreted as supporting no switchover of gene sets
(4; p. 96). A recent review of this question, based chiefly on the hemolymph
proteins of higher Diptera, which exhibit the most complete metamorphosis
among all insects, concludes that two gene sets, larval and adult, but not
pupal, can be recognized (41). Our own impression is that the control of
synthesis of individual proteins varies so greatly that it can be questioned
whether regulatory gene sets, which may exist, have yet been recognized at
all. Thus, storage proteins, which are generally accumulated before metamorphosis and thus regarded as larval, may, in some insects, be produced
in the adult as reserves for oogenesis. In the cecropia silkmoth, storage
protein and vitellogenin are both produced in the fat body at the end of the
larval stage, but with distinctly different timing which indicates independent regulation. The immediate triggers for changes in protein synthesis are
often endocrine, but this only brings into focus the problem of developmental changes in gene programming. Thus, in many adult, female insects,
synthesis of vitellogenin in the fat body is induced by the presence of
juvenile hormone, and in the mosquito it is induced by ecdysone. Yet,
vitellogenin was not made in the larval fat body under exposure to the same
hormones. These problems are currently under investigation. The question
of the biological stability and turnover of insect hemolymph proteins, which
must also contribute to control of their steady-state levels, has scarcely yet
been touched (102, 455).

ACKNOWLEDGEMENTS

We thank all those who have provided unpublished information. The preparation of this review was assisted by grants from the United States National Institutes of Health (HD07951) and the National Research Council of Canada to G. R. Wyatt and the Faculty Research Fund of the University of Tennessee to M. L. Pan.

Literature Cited

1. Telfer, W. H., Williams, C. M. 1953. *J. Gen. Physiol.* 36:389–413
2. Telfer, W. H. 1954. *J. Gen. Physiol.* 37:539–58
3. Chino, H., Murakami, S., Harashima, K. 1969. *Biochim. Biophys. Acta* 176: 1–26
4. Chen, P. S. 1971. *Biochemical Aspects of Insect Development,* pp. 66–77, 94–100. Basel:Karger. 230 pp.
5. Chapman, R. F. 1969. *The Insects Structure & Function,* pp. 659–91. New York: Am. Elsevier. 819 pp.
6. Wyatt, G. R. 1961. *Ann. Rev. Entomol.* 6:75–102
7. Jeuniaux, C., Florkin, M. 1974. In *The Physiology of Insecta,* ed. M. Rockstein 5:255–307. New York: Academic. 648 pp. 2nd ed.
8. Jeuniaux, C. 1971. In *Chemical Zoology,* ed. M. Florkin, B. Scheer, 6:63–118. New York: Academic. 484 pp.
9. Bodnaryk, R. P., Levenbook, L. 1968. *Biochem. J.* 110:771–73
10. Kilby, B. A. 1963. *Adv. Insect Physiol.* 1:111–74
11. Price, G. M. 1973. *Biol. Rev. Camb. Phil. Soc.* 48:333–75
12. Wyatt, G. R. 1975. *Verh. Dtsch. Zool. Ges.* 1975:209–26
13. Beard, R. L. 1950. *Physiol. Zool.* 23:47–57
14. Gregoire, C. 1953. *Arch. Int. Physiol.* 61:237–39
15. Schneiderman, H. A. 1967. In *Methods in Developmental Biology,* ed. F. H. Wilt, N. K. Wessells, pp. 753–66. New York: T. Y. Crowell. 813 pp.
16. Patel, N. G., Schneiderman, H. A. 1969. *J. Insect Physiol.* 15:643–60
17. Smilowitz, Z. 1973. *Ann. Entomol. Soc. Am.* 66:93–99
18. Whitmore, E., Gilbert, L. I. 1974. *Comp. Biochem. Physiol.* 47B:63–78
19. Crowle, A. J. 1973. *Immunodiffusion, New York:* Academic. 545 pp. 2nd ed.
20. Leskowitz, S., Waksman, B. H. 1960. *J. Immunol.* 84:58–72
21. Weintraub, M., Raymond, S. 1963. *Science* 142:1677–78
22. Allison, A. C., Humphrey, J. H. 1960. *Immunology* 3:95–106
23. Grabar, P., Williams, C. A. 1955. *Biochim. Biophys. Acta* 17:67–74
24. Crowle, A. J. 1961. *Immunodiffusion,* New York: Academic. 333 pp. 1st ed.
25. Barlow, J. S. 1962. *Anal. Biochem.* 3:206–11
26. Pan, M. L., Wyatt, G. R. 1971. *Science* 174:503–5
27. Kunkel, J. G., Lawler, D. M. 1974. *Comp. Biochem. Physiol.* 47B:697–710
28. Kunkel, J. G., Pan, M. L. 1976. *J. Insect Physiol.* 22:809–18
29. Bell, W. J. 1970. *J. Insect Physiol.* 16: 291–99
30. Laurell, C. B. 1966. *Anal. Biochem.* 15:45–52
31. Laurell, C. B. 1965. *Anal. Biochem.* 10:358–61
32. Norrild, B., Bjerrum, O. J., Vestergaard, B. F. 1977. *Anal. Biochem.* 81:432–41
33. Borovsky, D., VanHandel, E. 1977. *J. Insect Physiol.* 23:655–58
34. Hagedorn, H. H. Kunkel, J. G., Wheelock, G. 1978. *J. Insect Physiol.* In press
35. Pan, M. L., Wallace, R. A. 1974. *Am. Zool.* 14:1239–42
36. Wallace, R. A. 1965. *Anal. Biochem.* 11:297–311
37. Scanu, A. M., Wisdom, C. 1972. *Ann. Rev. Biochem.* 41:703–30
38. Thomas, K. K., Gilbert, L. I. 1968. *Arch. Biochem. Biophys.* 127:512–21
39. Thomas, K. K., Gilbert, L. I. 1969. *Physiol. Chem. Phys.* 1:293–311
40. Birnie, G. S., Rickwood, D., Hell, A. 1973. *Biochim. Biophys. Acta* 331: 283–94
41. Thomson, J. A. 1975. *Adv. Insect Physiol.* 11:321–98
42. Cölln, K. 1973. *Wilhelm Roux' Arch. Entwicklungsmech. Org.* 172:231–57
43. Kinnear, J. F.. Thomson, J. A. 1975. *Insect Biochem.* 5:531–52
44. Scheurer, R., Leuthold, R. 1969. *J. Insect Physiol.* 15:1067–77

45. Patel, N. G. 1971. *Insect Biochem.* 1:391–427
46. Schmidt, G. H. 1965. *J. Insect Physiol.* 11:71–77
47. Schmidt, G. H., Hess, U. 1973. *Comp. Biochem. Physiol.* 46B:15–35
48. Schmidt, G. H., Wirth, B. 1974. *J. Insect Physiol.* 20:1421–66
49. Schmidt, G. H., Schwankl, W. 1975. *Comp. Biochem. Physiol.* 52B:365–80
50. Schmidt, G. H., Wirth, B., Schwankl, W. 1976. *Insect Biochem.* 6:697–301
51. Schmidt, G. H., Schwankl, W., Wirth, B. 1876. *Insect Biochem.* 6:391–98
52. Sahota, T. S. 1970. *Can. J. Zool.* 48:1307–12
52a. Kuethe, H. W. 1972. *J. Insect. Physiol.* 18:2411–23
53. Thong, C. H. S., Webster, J. M. 1975. *J. Invert. Pathol.* 26:91–98
54. Parker, R. 1971. *J. Insect Physiol.* 17:1689–98
55. DeLoof, A., De Wilde, J. 1970. *J. Insect Physiol.* 16:157–69
56. DeLoof, A., De Wilde, J. 1970. *J. Insect Physiol.* 16:1455–66
57. DeLoof, A. 1972. *J. Insect Physiol.* 18:1039–47
58. Hsiao, T. H., Fraenkel, G. 1969. *Toxicon* 7:119–30
59. Whittaker, J. R., West, A. S. 1962. *Can. J. Zool.* 40:655–71
60. Wang, C. M., Patton, R. L. 1968. *J. Insect Physiol.* 14:1069–75
61. Butler, J. E., Leone, C. A. 1966. *Comp. Biochem. Physiol.* 19:699–711
62. Laverdure, A. 1972. *J. Insect Physiol.* 18:1369–85
63. Pemrick, S. M., Butz, A. 1970. *J. Insect Physiol.* 16:1443–54
64. Cohen, E. 1972. *Insect Biochem.* 2:161–66
65. Bennett, G. A., Shotwell, O. L., Hall, H. H. 1968. *J. Invert. Path.* 11:112–18
66. Bennett, G. A., Shotwell, O. L. 1970. *J. Invert. Path.* 15:157–64
67. Bennett, G. A., Shotwell, O. L. 1973. *Biotechnol. Bioeng.* 15:1023–37
68. Leader, J. P., Bedford, J. J. 1972. *J. Insect Physiol.* 18:2229–35
69. Chambers, S. P., Hall, J. E., Hitt, S. Z. 1975. *J. Invert. Pathol.* 25:171–78
70. Hagedorn, H. H., Judson, C. L. 1972. *J. Exp. Zool.* 182:367–78
71. Hagedorn, H. H., Fallon, A. M. 1973. *Nature* 244:103–5
72. Hagedorn, H. H., Fallon, A. M., Laufer, H. 1973. *Develop. Biol.* 31:285–94
73. Fallon, A. M., Hagedorn, H. H., Wyatt, G. R., Laufer, H. 1974. *J. Insect Physiol.* 20:1815–23
74. Andreadis, T. G., Hall, D. W. 1976. *Exp. Parasitol.* 39:252–61
75. Price, G. M. 1966. *J. Insect Physiol.* 12:731–40
76. Price, G. M. 1967. *J. Insect Physiol.* 13:69–79
77. Price, G. M., Bosman, T. W. 1966. *J. Insect Physiol.* 12:741–45
78. Julien, P., Corrivault, G. W., Perrott, J. M. 1977. *Insect Biochem.* 7:109–14
79. Munn, E. A., Feinstein, A., Greville, G. D. 1967. *Biochem. J.* 102:5P–6P
80. Munn, E. A., Feinstein, A., Greville, G. D. 1971. *Biochem. J.* 124:367–74
81. Munn, E. A., Greville, G. D. 1969. *J. Insect Physiol.* 15:1935–50
82. Munn, E. A., Price, G. M., Greville, G. D. 1969. *J. Insect Physiol.* 15:1601–05
83. Thomson, J. A., Sin, Y. T. 1970. *J. Insect Physiol.* 16:2063–74
84. Kinnear, J. F., Martin, M. D., Thomson, J. A., Neufeld, G. J. 1968. *Aust. J. Biol. Sci.* 21:1033–45
85. Kinnear, J. F., Martin, M. D., Thomson, J. A. 1971. *Aust. J. Biol. Sci.* 24:275–89
86. Martin, M. D., Kinnear, J. F., Thomson, J. A. 1969. *Aust. J. Biol. Sci.* 22:935–45
87. Martin, M. D., Kinnear, J. F., Thomson, J. A. 1971. *Aust. J. Biol. Sci.* 24:291–99
88. Thomson, J. A., Kinnear, J. F., Martin, M-D., Horn, D. H. S. 1971. *Life Sci.* 10:203–11
89. Kinnear, J. F., Thomson, J. A. 1975. *Insect Biochem.* 5:531–52
90. Sekeris, C. E., Scheller, K. 1977. *Dev. Biol.* 59:12–23
91. Sekeris, C. E., Perassi, R., Arnemann, J., Ullrich, J., Scheller, K. 1977. *Insect Biochem.* 7:5–10
92. Thompson, P. E., English, D. S. 1966. *Science* 152:75–76
93. Thompson, P., Bleecker, W., English, D. S. 1968. *J. Biol. Chem.* 243:4463–67
94. Firling, C. E. 1977. *J. Insect Physiol.* 23:17–22
95. Schin, K., Laufer, H. 1974. *J. Insect Physiol.* 20:405–11
96. Chen, P. S. 1967. *Nature* 215:316–17
97. Kambysellis, M. P., Heed, W. B. 1974. *J. Insect Physiol.* 20:1779–86
98. Butterworth, F. M., Berendes, H. D. 1974. *J. Insect Physiol.* 20:2195–2204
99. Gelti-Douka, H., Gingeras, T. R., Kambysellis, M. P. 1974. *J. Exp. Zool.* 187:167–72

100. Duke, E. J. 1966. *Genet. Res.* (*Camb.*) 7:287–94
101. Duke, E. J., Pantelouris, E. M. 1963. *Comp. Biochem. Physiol.* 10:351–55
102. Boyd, J. B., Mitchell, H. K. 1966. *Arch. Biochem. Biophys.* 117:310–19
103. Ruegg, M. K. 1968. *Z. Vgl. Physiol.* 60:275–307
104. Thomasson, W. A., Mitchell, H. K. 1972. *J. Insect Physiol.* 18:1885–99
105. Gavin, J., Williamson, J. H. 1976. *J. Insect Physiol.* 22:1457–64
106. Kambysellis, M. 1977. *Am. Zool.* 17:535–49
107. Postlethwait, J. H., Weiser, K. 1973. *Nature* 244:284–85
108. Gavin, J. A., Williamson, J. H. 1976. *J. Insect Physiol.* 22:1737–42
109. Bownes, M., Hames, B. D. 1977. *J. Exp. Zool.* 200:149–56
110. Roberts, D. B., Wolfe, J., Akam, M. E. 1977. *J. Insect Physiol.* 23:871–78
111. Pasteur, N., Kastritsis, C. D. 1972. *Experientia* 28:215–16
112. Thomson, J. A., Radok, K. R., Shaw, D. C., Whitten, M. J., Foster, G. G., Birt, L. M. 1976. *Biochem. Genet.* 14:145–60
113. Petzelt, C., Bier, K. 1970. *Wilhelm Roux' Arch. Entwicklungsmech. Org.* 164:359–66
114. Bodnaryk, R. P., Morrison, P. E. 1966. *J. Insect Physiol.* 12:963–76
115. Bodnaryk, R. P., Morrison, P. E. 1968. *J. Insect Physiol.* 14:1141–46
116. Hall, T. J., Sanders, S. M., Cummings, M. R. 1976. *Insect Biochem.* 6:13–18
117. Dwivedy, A. K., Bridges, R. G. 1973. *J. Insect Physiol.* 19:559–76
118. Menzel, D. B., Craig, R., Hoskins, W. M. 1963. *J. Insect Physiol.* 9:479–93
119. Collins, W. J., Forgash, A. J. 1968. *J. Insect Physiol.* 14:1515–23
120. Gadallah, A., Kilgore, W. W. 1972. *J. Econ. Entomol.* 62:393–96
121. Painter, R. R., Kilgore, W. W. 1972. *J. Econ. Entomol.* 65:23–27
122. Chen, P. S., Levenbook, L. 1966. *J. Insect Physiol.* 12:1595–1609
123. Chen, P. S., Levenbook, L. 1966. *J. Insect Physiol.* 12:1611–27
124. Mjeni, A. M., Morrison, P. E. 1973. *Can. J. Zool.* 51:1069–79
125. De Bianchi, A. G., Terra, W. R. 1976. *J. Insect Physiol.* 22:535–40
126. Terra, W. R., De Bianchi, A. G., Paesde Mello, M., Basile, R. 1976. *Experientia* 32:433–34
127. Hughes, L., Price, G. 1975. *J. Insect Physiol.* 21:1373–84
128. Engelmann, F., Hill, L., Wilkens, J. L. 1971. *J. Insect Physiol.* 17:2179–91
129. Bergtrom, G., Laufer, H., Rogers, R. 1976. *J. Cell Biol.* 69:264–74
130. Jalaja, M., Prabhu, V. K. K. 1971. *Experientia* 27:639–40
131. Terando, M. L., Feir, D. 1966. *Comp. Biochem. Physiol.* 18:163–68
132. Terando, M. L., Feir, D. 1967. *Comp. Biochem. Physiol.* 20:431–36
133. Terando, M. L., Feir, D. 1967. *Comp. Biochem. Physiol.* 21:31–35
134. Bassi, S. D., Feir, D. 1971. *Insect Biochem.* 1:433–38
135. Kelly, T. J., Davenport, R. 1976. *J. Insect Physiol.* 22:1381–93
136. Kelly, T. J., Telfer, W. H. 1978. *Develop. Biol.* 61:58–69
137. Belyaeva, T. G., Stepanyan, E. B. 1975. *Zool. Zh.* 54:998–1003
138. Slama, K. 1964. *J. Insect Physiol.* 10:773–82
139. Emmerich, H. 1970. *J. Insect Physiol.* 16:725–49
140. Coles, G. C. 1965. *J. Insect Physiol.* 11:1317–23
141. Coles, G. C. 1965. *J. Insect Physiol.* 11:1325–30
142. Coles, G. C. 1965. *J. Exp. Biol.* 43:425–31
143. Pratt, G. E., Davey, K. G. 1972. *J. Exp. Biol.* 56:201–14
144. Pratt, G. E., Davey, K. G. 1972. *J. Exp. Biol.* 56:215–21
145. Pratt, G. E., Davey, K. G. 1972. *J. Exp. Biol.* 56:223–37
146. Baehr, J. C. 1974. *Gen. Comp. Endocrinol.* 22:146–53
147. Perassi, R. 1973. *J. Insect Physiol.* 19:663–72
148. Mundall, E., Engelmann, F. 1977. *J. Insect Physiol.* 23:825–36
148a. Brunnert, H. 1967. *Z. Naturforsch.* 22:336–39
149. Liu, T. P., Dixon, S. E. 1965. *Can. J. Zool.* 43:873–79
150. Lensky, Y. 1971. *Comp. Biochem. Physiol.* 38B:129–39
151. Lensky, Y. 1971. *Comp. Biochem. Physiol.* 39B:325–41
152. Engels, W., Fahrenhorst, H. 1974. *Wilhelm Roux' Arch. Entwicklungsmech. Org.* 174:285–96
153. Gilliam, M. 1972. *Experientia* 28:341
154. Rutz, W., Lüscher, M. 1974. *J. Insect Physiol.* 20:897–909
155. Engels, W. 1972. *Wilhelm Roux' Arch. Entwicklungsmech. Org.* 171:55–86
156. Engels, W. 1974. *Am. Zool.* 14:1229–37
157. Engels, W., Goncalves, L. S., Engels, E. 1976. *J. Apic. Res.* 15:3–10
158. Engels, W., Fahrenhorst, H. 1975. *Wilhelm Roux' Arch. Dev. Biol.* 178:79–88

159. Engels, W., Ramamurty, P. S. 1976. *J. Insect Physiol.* 22:1427–32
160. Roseler, I., Roseler, P.-F. 1973. *J. Insect Physiol.* 19:1741–52
161. Barlow, J. S. 1962. *J. Insect Path.* 4:274–75
162. Wyss-Huber, M., Lüscher, M. 1975. *J. Insect Physiol.* 21:1697–1704
163. Evans, J. J. T. 1967. *J. Insect Physiol.* 13:1699–1711
164. Whitmore, E., Gilbert, L. I. 1972. *J. Insect Physiol.* 18:1153–67
165. Blumenfeld, M., Schneiderman, H. A. 1968. *Biol. Bull. (Wood's Hole)* 135:466–75
166. Egelhaaf, A. 1965. *Z. Vererbungsl.* 97:150–56
167. Fisher, R. C., Ganesalingam, V. K. 1970. *Nature* 227:191–92
168. Imberski, R. B., Gertson, P. N. 1974. *Insect Biochem.* 4:341–44
169. Shigematsu, H. 1958. *Nature* 182:880–82
170. Nakasone, S., Kobayashi, M. 1965. *Nippon Sanshigaku Zasshi* 34:257–62
171. Gamo, T. 1968. *J. Genet.* 43:271–77
172. Doira, H. 1968. *Sci. Bull. Fac. Agric., Kyushu Univ.* 23:205–14
173. Ono, S., Nagayama, H., Shimura, K. 1975. *Insect Biochem.* 5:313–29
174. Doira, H., Kawaguchi, Y. 1972. *J. Fac. Agric. Kyushu Univ.* 17:119–27
175. Kawaguchi, Y., Doira, H. 1973. *J. Insect Physiol.* 19:2076–96
176. Ashida, M. 1971. *Arch. Biochem. Biophys.* 144:749–62
177. Kai, H., Hasegawa, K. 1972. *J. Insect Physiol.* 18:133–42
178. Powning, R. F., Davidson, W. J. 1973. *Comp. Biochem. Physiol.* 45B:669–86
179. Matsuzaki, K., Yoshida, M. 1975. *Nippon Sanshigaku Zasshi* 44:229–30
180. Collins, J. V. 1969. *J. Insect Physiol.* 15:341–52
181. Collins, J. V. 1974. *Can. J. Zool.* 52:639–42
182. Collins, J. V. 1975. *Can. J. Zool.* 53:480–89
183. Pan, M. L., Wyatt, G. R. 1976. *Dev. Biol.* 54:127–34
184. Kunkel, J. G., Pan, M. L. 1977. *Am. Zool.* 17:914
185. Chippendale, G. M. 1970. *J. Insect Physiol.* 16:1057–68
186. Chippendale, G. M. 1970. *J. Insect Physiol.* 16:1909–20
187. Croizier, G., Odier, F. 1974. *Experientia* 30:189–90
188. Marek, M. 1969. *Z. Naturforsch.* 24B:732–40
189. Lysenko, O. 1972. *J. Invert. Pathol.* 19:335–41
190. Collins, J. V., Downe, A. E. R. 1970. *J. Insect Physiol.* 16:1697–1708
191. Powning, R. F., Davidson, W. J. 1976. *Comp. Biochem. Physiol.* 55B:221–28
192. Weiser, J., Lysenko, O. 1972. *Acta Entomol. Bohemoslovaca* 69:97–100
193. Vinson, S. B., Lewis, W. J. 1969. *Comp. Biochem. Physiol.* 28:215–20
194. Vinson, S. B., Barras, D. J. 1970. *J. Insect Physiol.* 16:1329–38
195. Barras, D. J., Kisner, R. L. 1972. *Comp. Biochem. Physiol.* 43B:941–47
196. Sell, D. K., Whitt, G. S., Metcalf, R. L., Lee, L.-P. K. 1974. *Can. Entomol.* 106:701–9
197. Laufer, H. 1960. *Ann. NY Acad. Sci.* 89:490–515
198. Laufer, H. 1961. *Ann. NY Acad. Sci.* 94:825–35
199. Laufer, H. 1963. *Ann. NY Acad. Sci.* 103:1137–54
200. Laufer, H. 1964. In *Taxonomic Biochemistry and Serology*, ed. C. A. Leone, pp. 171–189. New York: Ronald. 728 pp.
201. Telfer, W. H., Williams, C. M. 1960. *J. Insect. Physiol.* 5:61–72
202. Skinner, D. M. 1963. *Biol. Bull. Wood's Hole* 125:165–76
203. Ruh, M. F., Willis, J. H., Hollowell, M. P. 1972. *J. Insect Physiol.* 18:151–60
204. Ruh, M. F., Willis, J. H. 1974. *J. Insect Physiol.* 20:1277–85
205. Telfer, W. H. 1960. *Biol. Bull. Wood's Hole* 118:338–51
206. Pan, M. L., Bell, W. J., Telfer, W. H. 1969. *Science* 165:393–94
207. Pan, M. L. 1971. *J. Insect Physiol.* 17:677–89
208. Pan, M. L. 1977. *Biol. Bull Wood's Hole.* 153:336–45
209. Thomas, K. K. 1972. *Insect Biochem.* 2:107–18
210. Chino, H., Gilbert, L. I. 1971. *Insect Biochem.* 1:337–47
211. Riechers, L. A., Meyers, F. W., Berry, S. J. 1969. *J. Insect Physiol.* 15:743–53
212. Whitmore, D., Gilbert, L. I., Ittycheriah, P. I. 1974. *Mol. Cell. Endocrinol.* 1:37–54
213. Tojo, S., Betchaku, T., Ziccardi, V. J., Wyatt, G. R. 1978. *J. Cell Biol.* In press
214. Whitmore, D., Whitmore, E., Gilbert, L. I. 1972. *Proc. Natl. Acad. Sci. USA* 69:1592–95
215. Loughton, B. G. 1965. *J. Insect Physiol.* 11:1651–61
216. Loughton, B. G., West, A. S. 1965. *J. Insect Physiol.* 11:919–32
217. Loughton, B. G., West, A. S. 1968. *Can. J. Zool.* 46:625–28

218. Mansingh, A., Steele, R. W. 1975. *J. Insect Physiol.* 21:733–40
219. Van Der Geest, L. P. S., Wassink, H. J. M. 1969. *J. Invert. Pathol.* 14:419–20
220. Greene, J. R., Dahlman, D. L. 1973. *J. Insect Physiol.* 19:1241–50
221. Dahlman, D. L. 1969. *J. Insect Physiol.* 15:2075–84
222. Koeppe, J. K., Gilbert, L. I. 1973. *J. Insect Physiol.* 19:615–24
223. Koeppe, J. K., Gilbert, L. I. 1974. *J. Insect Physiol.* 20:981–92
224. Nijhout, M. M., Riddiford, L. M. 1974. *Biol. Bull. Wood's Hole* 146:377–92
225. Kramer, K. J., Sanburg, L. L., Kezdy, F. J., Law, J. H. 1974. *Proc. Natl. Acad. Sci. USA* 71:493–97
226. Kramer, K. J., Dunn, P. E., Peterson, R. C., Seballos, H. L., Sanburg, L. L., Law, J. H. 1976. *J. Biol. Chem.* 251:4979–85
227. Hammock, B., Nowock, J., Goodman, W., Stamoudis, V., Gilbert, L. I. 1975. *Mol. Cell Endocrinol.* 3:167–84
228. Nowock, J., Goodman, W., Bollenbacher, W. E., Gilbert, L. I. 1975. *Gen. Comp. Endocrinol.* 27:230–39
229. Sanburg, L. L., Kramer, K. J., Kezdy, F. J., Law, J. H., Oberlander, H. 1975. *Nature* 253:266–67
230. Sanburg, L. L., Kramer, K. J., Kezdy, F. J., Law, J. H. 1975. *J. Insect Physiol.* 21:873–87
231. Vince, R. K., Gilbert, L. I. 1977. *Insect Biochem.* 7:115–20
232. Herman, W. S., Bennett, D. C. 1975. *J. Comp. Physiol.* B99:331–38
233. Chippendale, G. M., Beck, S. D. 1966. *J. Insect Physiol.* 12:1629–38
234. Chippendale, G. M., Beck, S. D. 1967. *J. Insect Physiol.* 13:995–1006
235. Chino, H., Sudo, A., Harashima, K. 1967. *Biochim. Biophys. Acta* 144:177–79
236. Chino, H., Downer, R. G. H., Takahashi, K. 1977. *Biochim. Biophys. Acta* 487:508–16
237. Chino, H., Yamagata, M., Takahashi, K. 1976. *Biochim. Biophys. Acta* 441:349–53
238. Chino, H., Yamagata, M., Sato, S. 1977. *Insect Biochem.* 7:125–31
239. Chino, H., Gilbert, L. I., Siddall, J. B., Hafferl, W. 1970. *J. Insect Physiol.* 16:2033–40
240. Pant, R., Agrawal, H. C. 1965. *Biochem. J.* 96:824–28
241. Lamy, M. 1964. *Proc.-Verb. Soc. Sci. Phys. Bordeaux* 25:242–46
242. Clements, A. N. 1967. *J. Insect Physiol.* 13:1021–30
243. Van Der Geest, L. P. S. 1968. *J. Insect Physiol.* 14:537–42
244. Van Der Geest, L. P. S., Borgsteede, F. H. M. 1969. *J. Insect Physiol.* 15:1687–93
245. Chippendale, G. M., Kilby, B. A. 1969. *J. Insect Physiol.* 15:905–26
246. Pentz, S., Kling, H. 1972. *J. Insect Physiol.* 18:2277–83
247. Ferkovich, S. M., Silhacek, D. L., Rutter, R. R. 1975. *Insect Biochem.* 5:141–50
248. Ferkovich, S. M., Silhacek, D. L., Rutter, R. R. 1976. In *The Juvenile Hormones,* ed. L. I. Gilbert, pp. 342–53. New York: Plenum. 572 pp.
249. Ferkovich, S. M., Oberlander, H., Rutter, R. R. 1977. *J. Insect Physiol.* 23:297–302
250. Kling, H., Pentz, S. 1975. *Comp. Biochem. Physiol.* 50B:103–4
251. Hudson, A. 1966. *Can. J. Zool.* 44:541–55
252. Tanada, Y., Watanabe, H. 1971. *J. Invert. Pathol.* 17:127–29
253. Chippendale, G. M. 1971. *Insect Biochem.* 1:122–24
254. Takei, G. H., Tamashiro, M. 1975. *J. Invert. Pathol.* 26:147–58
255. Smilowitz, Z. 1973. *Ann. Entomol. Soc. Am.* 66:93–99
256. Young, S. Y., Scott, H. A. 1970. *J. Invert. Pathol.* 16:57–62
257. Young, S. Y., Lovell, J. S. 1971. *J. Invert. Pathol.* 17:410–18
259. Bedford, J. J., Leader, J. P. 1975. *J. Entomol. Ser. A: Physiol. Behav.* 50:1–7
260. Kuo, C. C., Patton, R. L. 1975. *Insect Biochem.* 5:519–29
261. Nowosielski, J. W., Patton, R. L. 1965. *J. Insect Physiol.* 11:263–70
262. Woodring, J. P., Clifford, C. W., Roe, R. M., Mercier, R. R. 1977. *J. Insect Physiol.* 23:559–67
263. Girardie, J. 1977. *J. Insect Physiol.* 23:569–78
264. Kunkel, J. G. 1975. *Comp. Biochem. Physiol.* 51B:177–80
265. Tanaka, A. 1973. *Dev. Growth & Differ.* 15:153–68
266. Tanaka, A., Ishizaki, H. 1974. *Dev. Growth & Differ.* 16:247–55
267. Oie, M., Takahashi, S. Y., Ishizaki, H. 1975. *Dev. Growth & Differ.* 17:237–46
268. Barth, R. H. Jr., Bell, W. J. 1970. *Biol. Bull. Wood's Hole* 139:447–60
269. Bell, W. J., Barth, R. H. 1970. *J. Insect Physiol.* 16:2303–13
270. Marty, R., Zalta, J. P. 1968. *J. Insect Physiol.* 14:861–68
271. Stay, B., Clark, J. K. 1971. *J. Insect Physiol.* 17:1747–62

272. Scheurer, R. 1969. *J. Insect Physiol.* 15:1411–19
273. Scheurer, R. 1969. *J. Insect Physiol.* 15:1673–82
274. Wyss-Huber, M., Lüscher, M. 1972. *J. Insect Physiol.* 18:689–710
275. Brookes, V. J. 1969. *Develop. Biol.* 20:459–71
276. Brookes, V. J. 1976. *J. Insect Physiol.* 22:1649–57
277. Engelmann, F., Penney, D. 1966. *Gen. Comp. Endocrinol.* 7:314–25
278. Engelmann, F. 1971. *Arch. Biochem. Biophys.* 145:439–47
279. Engelmann, F. 1974. *Insect Biochem.* 4:345–54
280. Brookes, V. J., Dejmal, R. K. 1968. *Science* 160:999–1001
281. Dejmal, R. K., Brookes, V. J. 1968. *J. Insect Physiol.* 14:371–81
282. Dejmal, R. K., Brookes, V. J. 1972. *J. Biol. Chem.* 247:869–74
283. Engelmann, F., Friedel, T. 1974. *Life Sci.* 14:587–94
284. Engelmann, F., Friedel, T., Ladduwahetty, M. 1976. *Insect Biochem.* 6:211–20
285. Koeppe, J., Ofengand, J. 1976. See Ref. 248, pp. 486–504
286. Koeppe, J., Ofengand, J. 1976. *Arch. Biochem. Biophys.* 173:100–13
287. Tobe, S. S., Loughton, B. G. 1967. *Can. J. Zool.* 45:975–84
288. Tobe, S. S., Loughton, B. G. 1969. *J. Insect Physiol.* 15:1659–72
289. Turner, A. E., Loughton, B. G. 1975. *Insect Biochem.* 5:791–804
290. McCormick, F. W., Scott, A. 1966. *Experientia* 22:228–29
291. Mayer, R. J., Candy, D. J. 1967. *Nature* 215:987
292. Bentz, F., Girardie, A., Cazal, M. 1970. *J. Insect Physiol.* 16:2257–70
293. Minks, A. K. 1967. *Arch. Neerl Zool.* 17:175–258
294. Minks, A. K. 1971. *Endocrinol. Exp. (Bratislava)* 5:73–77
295. Bar-Zev, A., Wajc, E., Cohen, E., Sapir, L., Applebaum, S. W., Emmerich, H. 1975. *J. Insect Physiol.* 21:1257–63
296. Bakker-Grunwald, T., Applebaum, S. W. 1977. *J. Insect Physiol.* 23:259–63
297. Yamasaki, K. 1975. *Insect Biochem.* 4:411–22
298. Chen, T. T., Couble, P., DeLucca, F. L., Wyatt, G. R. 1976. In *The Juvenile Hormones,* ed. L. I. Gilbert, pp. 505–29 New York: Plenum. 572 pp.
299. Chen, T. T., Strahlendorf, P. W., Wyatt, G. R. 1978. *J. Biol. Chem.* In press
300. Gellissen, G., Wajc, E., Cohen, E., Emmerich, H., Applebaum, S. W., Flossdorf, J. 1976. *J. Comp. Physiol. B.* 108:287–301
301. Peled, Y., Tietz, A. 1973. *Biochim. Biophys. Acta* 296:499–509
302. Peled, Y., Tietz, A. 1975. *Insect Biochem.* 5:61–72
303. Mwangi, R. W., Goldsworthy, G. J. 1977. *J. Comp. Physiol. B* 114:177–90
304. Emmerich, H., Hartmann, R. 1973. *J. Insect Physiol.* 19:1663–75
305. Pratt, G. E. 1975. *Insect Biochem.* 5:595–607
306. McGregor, D. A., Loughton, B. G. 1974. *Can. J. Zool* 52:907–17
307. Philips, D. R., Loughton, B. G. 1976. *Comp. Biochem. Physiol.* 55B:215–20
308. Adiyodi, K. G. 1967. *J. Insect Physiol.* 13:1189–95
309. Lanzrein, B. 1974. *J. Insect Physiol.* 20:1871–85
310. Wilhelm, R., Lüscher, M. 1974. *J. Insect Physiol.* 20:1887–97
311. Buhlmann, G. 1976. *J. Insect Physiol.* 22:1101–10
312. Shambaugh, G. F., Zeller, L. H., Bylica, S. M. 1974. *Insect Biochem.* 4:185–96
313. Siakotos, A. N. 1960. *J. Gen. Physiol.* 43:999–1013
314. Siakotos, A. N. 1960. *J. Gen. Physiol.* 43:1015–30
315. Prabhu, V. K. K., Hema, P. 1969. *Experientia* 25:1115
316. Hema, P., Prabhu, V. K. K. 1970. *J. Insect Physiol.* 16:1165–70
317. Thomas, K. K., Nation, J. L. 1966. *Biol. Bull. Wood's Hole* 130:254–64
318. Nielsen, D. J., Mills, R. R. 1968. *J. Insect Physiol.* 14:163–70
319. Krolak, J. M., Clore, J. N., Petrovitch, E., Mills, R. R. 1977. *J. Insect Physiol.* 23:381–85
320. Bell, W. J. 1969. *J. Insect Physiol.* 15:1279–90
321. Bell, W. J. 1971. *J. Insect Physiol.* 17:1099–1111
322. Bell, W. J. 1972. *J. Exp. Zool.* 181:41–48
323. Bell, W. J. 1972. *J. Insect Physiol.* 18:851–55
324. Bell, W. J., Bohm, M. K. 1975. *Biol. Rev.* 50:373–96
325. Prabhu, V. K. K., Nayar, K. K. 1972. *J. Insect Physiol.* 18:1435–40
326. Prabhu, V. K. K., Hema, P. 1970. *J. Insect Physiol.* 16:147–56
327. Fox, F. R., Mills, R. R. 1969. *Comp. Biochem. Physiol.* 29:1187–95
328. Weidler, D. J., Sieck, G. C. 1977. *Comp. Biochem. Physiol.* 56A:11–14
329. Reisser-Bollade, D. 1976. *Insect Biochem.* 6:241–46

330. Downer, R. G. H., Steele, J. E. 1973. *J. Insect Physiol.* 19:523–32
331. Matthews, J. R., Downer, R. G. H., Morrison, P. E. 1976. *J. Insect Physiol.* 22:157–63
332. Hill, L. 1962. *J. Insect Physiol.* 8:609–19
333. Kulkarni, A. P., Mehrotra, K. N. 1970. *J. Insect Physiol.* 16:2181–99
334. Dufour, D., Taskar, S. P., Perron, J. M. 1970. *J. Insect Physiol.* 16:1369–77
335. Gordon, R., Webster, J. M., Hislop, T. G. 1973. *Comp. Biochem. Physiol.* 46B:575–93
336. Takesue, S. In preparation
337. Tojo, S., Kimura, S. In preparation
338. Svedberg, T., Eriksson-Quensel, I. B. 1934. *J. Am. Chem. Soc.* 56:1700–6
339. Behlke, J., Scheler, W. 1967. *Eur. J. Biochem.* 3:153–57
340. Huber, R., Formanek, H., Braun, V., Braunitzer, G., Hoppe, W. 1964. *Ber. Bunsenges. Phys. Chem.* 68:818–19
341. Braun, V. V., Formanek, H., Braunitzer, G. 1968. *Z. Physiol. Chem.* 349:45–53
342. Munn, E. A., Bufton, S. F. 1973. *Eur. J. Biochem.* 35:3–10
343. Pye, A. E., Boman, H. G. 1977. *Infect. Immun.* 17:408–14
344. Cherbas, P. T. 1973. *Biochemical Studies of Insecticyanin.* PhD thesis. Harvard Univ. Cambridge, Mass. 258 pp.
345. Kunkel, J. G. 1970. Cited in Ref. 346. p. 452
346. Wyatt, G. R. 1972. In *Biochemical Actions of Hormones,* ed. G. Litwack, Vol. 2, pp. 385–490. New York: Academic. 542 pp.
347. Doane, W. W. 1973. In *Developmental Systems: Insects,* ed. S. J. Counce, C. H. Waddington, Vol. 2, pp. 291–497. London/New York: Academic. 615 pp.
348. Engelmann, F. 1976. See Ref. 248, pp. 470–85
349. Wigglesworth, V. B. 1964. *Adv. Insect Physiol.* 2:243–332
350. Hagedorn, H. H., O'Connor, J. D., Fuchs, M. S., Sage, B., Schlaeger, D. A., Bohm, M. K. 1975. *Proc. Natl. Acad. Sci. USA* 72:3255–59
351. Tata, J. R. 1976. *Cell* 9:1–4
351. Telfer, W. H. 1965. *Ann. Rev. Entomol.* 10:161–84
353. Bell, W. J., Barth, R. H. 1971. *Nature* 230:220–21
354. Gilbert, L. I., Chino, H. 1974. *J. Lipid Res.* 15:439–56
355. Harashima, K. 1970. *Int. J. Biochem.* 1:523–31
356. Chino, H., Gilbert, L. I. 1965. *Biochim. Biophys. Acta* 98:94–110

357. Weintraub, H., Tietz, A. 1973. *Biochim. Biophys. Acta* 306:31–41
358. Kramer, K. J., Dunn, P. E., Peterson, R. C., Law, J. H. 1976. See Ref. 248, pp. 327–41
359. Trautmann, K. H. 1972. *Z. Naturforsch.* 27B:263–73
360. Kramer, S. J., De Kort, C. A. D. 1977. *Insect Biochem.* In press
361. Thamer, G., Karlson, P. 1972. *Z. Naturforsch.* 27B:1191–95
362. Feyereisen, R. 1977. *Experientia* 33:1111–13
363. Stone, J., Mordue, W., Bately, K., Morris, H. 1976. *Nature* 263:207–11
364. Cheeseman, P., Goldsworthy, G. J., Mordue, W. 1977. *Life Sci.* 21:231–36
365. Tager, H. S., Makese, J., Kramer, K., Speirs, R., Childs, C. 1976. *Biochem. J.* 156:515–20
366. Kramer, K. J., Speirs, R. D., Childs, C. N. 1977. *Gen. Comp. Endocrinol.* 32:423–26
367. Roberts, D. B. 1971. *Nature* 233:394–97
368. Wolfe, J., Akam, M. E., Roberts, D. B. 1977. *Eur. J. Biochem.* 79:47–53
369. Wigglesworth, V. B. 1972. *The Principles of Insect Physiology,* p. 443. London: Chapman & Hall, 7th ed. 827 pp.
370. Locke, M., Collins, J. V. 1968. *J. Cell Biol.* 36:453–83
371. Kemp, D. J., Thomson, J. A., Peacock, W. J., Higgins, T. J. V. 1977. *Biochem. Genet.* In press
372. Amiconi, G., Antonini, E., Bounori, M., Formanek, H., Huber, R. 1972. *Eur. J. Biochem.* 31:52–58
373. Manwell, C. 1966. *J. Embryol. Exp. Morphol.* 16:259–70
374. Braunitzer, G., Braun, V. 1966. *Z. Physiol. Chem.* 346:303–5
375. English, D. S. 1969. *J. Embryol. Exp. Morphol.* 22:465–76
376. Thompson, P., English, D. S., Bleecker, W. 1969. *Genetics* 63:183–92
377. Braunitzer, G., Braig, S., Buse, G., Plagens, U. 1971. *Limnologica (Berlin),* 8, 1:119–24
378. Braun, V., Crichton, R. R., Braunitzer, G. 1968. *Z. Physiol. Chem.* 349:197–210
379. Braunitzer, G., Buse, G., Braig, S. 1968. *Z. Physiol. Chem.* 349:263–64
380. Braunitzer, G. 1968. *Z. Physiol. Chem.* 349:1789–91
381. Buse, V. G., Braig, S., Braunitzer, G. 1969. *Z. Physiol. Chem.* 350:1686–90
382. Braunitzer, G., Buse, G., Gersonde, K. 1974. In *Molecular Oxygen in Biology,* ed. O. Hayaishi, pp. 183–218. Amsterdam: North-Holland, 367 pp.

383. Huber, R., Epp, O., Steigemann, W., Formanek, H. 1971. *Eur. J. Biochem.* 19:42–50

384. Braunitzer, G. 1970. *Z. Physiol. Chem.* 351:1289–90

385. Tichy, H. 1970. *Chromosoma* 29:131–88

386. Tichy, H. 1975. *J. Mol. Evol.* 6:39–50

387. Thompson, P. E., Patel, G. 1972. *Genetics* 70:275–90

388. Thompson, P., Horning, M. J. 1973. *Biochem. Genet.* 8:309–19

389. Laufer, H., Bergtrom, G., Rogers, R. 1976. In *Invertebrate Tissue Culture*, ed. E. Kurstak, K. Maramorosch, pp. 227–40 New York: Academic. 398 pp.

390. Laufer, H., Poluhowich, J. 1971. *Limnologica* 8:125–26

391. Schin, K., Poluhowich, J., Gamo, T., Laufer, H. 1974. *J. Insect Physiol.* 20:561–71

392. Kramer, S. J., De Kort, C. A. D. 1976. *Life Sci.* 19:211–18

393. Kramer, S. J., De Kort, C. A. D. 1976. *Mol. Cell. Endocrinol.* 4:43–53

394. Gilliam, M., Jackson, K. 1972. *Comp. Biochem. Physiol.* 42B:423–27

395. Schmidt, G. H., Wirth, B., Schwankl, W. 1976. *Insect Biochem.* 6:297–301

396. Weirich, G., Wren, J., Siddall, J. B. 1973. *Insect Biochem.* 3:397–407

397. Wirtz, R. A., Hopkins, T. L. 1977. *Insect Biochem.* 7:45–49

398. Hopkins, T. L., Wirtz, R. A. 1976. *J. Insect Physiol.* 22:1167–71

399. Friedman, S. 1960. *Arch. Biochem. Biophys.* 87:252–58

400. Katagiri, C., Downer, R. G., Chino, H. 1977. *Insect Biochem.* 7:351–53

401. Cottrell, C. B. 1964. *Adv. Insect Physiol.* 2:175–218

402. Brunet, P. C. J. 1965. In *Aspects of Insect Biochemistry*, ed. T. W. Goodwin, pp. 49–77. London: Academic. 108 pp.

403. Karlson, P., Mergenhagen, D., Sekeris, C. E. 1964. *Z. Physiol. Chem.* 338:42–50

404. Hackman, R. H., Goldberg, M. 1967. *J. Insect Physiol.* 13:531–44

405. Nappi, A. J. 1975. In *Invertebrate Immunity*, ed. K. Maramorosch, R. E. Shope, pp. 293–326. New York: Academic. 365 pp.

406. Evans, J. J. T. 1968. *J. Insect Physiol.* 14:107–19

407. Evans, J. J. T. 1968. *J. Insect Physiol.* 14:277–91

408. Ishaaya, J. 1972. *Insect Biochem.* 2:409–16

409. Ohnishi, E., Dohke, K., Ashida, M. 1970. *Arch. Biochem. Biophys.* 139:143–48

410. Ashida, M., Ohnishi, E. 1967. *Arch. Biochem. Biophys.* 122:411–16

411. Dohke, K. 1973. *Arch. Biochem. Biophys.* 157:210–21

412. Ashida, M., Dohke, K., Ohnishi, E. 1974. *Biochem. Biophys. Res. Commun.* 57:1089–95

413. Sin, Y. T., Thomson, J. A. 1971. *Insect Biochem.* 1:56–62

414. Karlson, P., Liebau, H. 1961. *Z. Physiol. Chem.* 326:135–43

415. Hughes, L., Price, G. M. 1976. *J. Insect Physiol.* 22:1005–11

416. Mitchell, H. K., Weber, U. M., Schaar, G. 1967. *Genetics* 57:357–68

417. Seybold, W. D., Meltzer, P. S., Mitchell, H. K. 1975. *Biochem. Genetics* 13:85–108

418. Preston, J. W., Taylor, R. L. 1970. *J. Insect Physiol.* 16:1729–44

419. Gregoire, C. 1974. In *Physiology of Insecta*, ed. M. Rockstein, 5:309–60. New York: Academic. 2nd ed. 692 pp.

420. Gilliam, M., Shimanuki, H. 1970. *Experientia* 26:908–9

421. Bowen, T. J., Kilby, B. A. 1953. *Arch. Intern. Physiol.* 61:413–16

422. Brehelin, M. 1971. *C. R. Acad. Sci. Paris* D273:1598–1601

423. Berry, S. J., Krishnakumaran, A., Schneiderman, H. A. 1964. *Science* 146:938–40

424. Briggs, J. D. 1964. See Ref. 419, 3:259–83

425. Chadwick, J. S. 1967. *Fed. Proc.* 26:1675–79

426. Whitcomb, R. F., Shapiro, M., Granados, R. R. 1974. See Ref. 419, 5:447–536

427. Gingrich, R. E. 1964. *J. Insect Physiol.* 10:179–94

428. Hoffmann, D. 1972. *C. R. Acad. Sci., Paris* D274:1109–12

429. Gilliam, M., Jeter, W. J. 1970. *J. Invert. Pathol.* 16:69–70

430. Boman, H. G., Nilsson, I., Rasmuson, B. 1972. *Nature* 237:232–35

431. Chadwick, J. S., Vilk, E. 1969. *J. Invert. Pathol.* 13:410–15

432. Chadwick, J. S. 1971. *J. Invert. Pathol.* 17:299–300

433. Marek, M. 1970. *Comp. Biochem. Physiol.* 35:737–43

434. Briggs, J. D. 1958. *J. Exp. Zool.* 138:155–88

435. Stephens, J. M., Marshall, J. H. 1962. *Can. J. Microbiol.* 8:719–25

436. Hink, W. F., Briggs, J. D. 1968. *J. Insect Physiol.* 14:1025–34

437. Kinoshita, T., Inoue, K. 1977. *Infect. Immun.* 16:32–36

438. Feir, D., Walz, M. A. 1964. *Ann Ent. Soc. Am.* 57:388
439. Scott, M. T. 1971. *Arch. Zool. Exp. Gen.* 112:73–80
440. Scott, M. T. 1972. *J. Invert. Pathol.* 19:66–71
441. Anderson, R. S., Day, N. K. B., Good, R. A. 1972. *Infect. Immun.* 5:55–59
442. Kawarabata, T. 1970. *Sci. Bull. Fac. Agr. Kyushu Univ.* 24:231–54
443. Boman, H. G., Nilsson-Faye, I., Paul, K., Rasmuson, T. Jr. 1974. *Infect. Immun.* 10:136–45
444. Faye, I., Pye, A., Rasmuson, T., Boman, H. G., Boman, I. A. 1975. *Infect. Immun.* 12:1426–38
445. Kamp, H. 1968. *Z. Vgl. Physiol.* 58:441–64
446. Boman, H. G., Nilsson-Faye, I., Rasmuson, T. 1974. In *Lipmann Symposium: Energy, Biosynthesis and Regulation*, ed. D. Richter, pp. 103–14. Berlin: W. de Gruyter
447. Edlund, T., Siden, I., Boman, H. G. 1976. *Infect. Immun.* 14:934–41
448. Natori, S. 1977. *J. Insect Physiol.* 23:1169–73
449. Chadwick, J. S. 1970. *J. Invert. Path.* 15:455–56
450. Mohrig, W., Messner, B. 1968. *Biol. Zentralbl.* 87:439–70
451. Rasmuson, T., Boman, H. G. 1978. In *Developmental Immunobiology*, ed. B. Solomon, J. Horton, J. Amsterdam: Elsevier-North Holland. pp. 83–90
452. Hackman, R. H. 1952. *Arch. Biochem. biophys.* 41:166–74
453. Cromartie, R. I. T. 1959. *Ann. Rev. Entomol.* 4:59–76
454. Williams, C. M., Kafatos, F. C. 1971. *Bull. Soc. Ent. Suisse* 44:151–62
455. Tobe, S. S, Davey, K. G. 1975. *Can. J. Zool.* 53:614–29

Ann. Rev. Biochem. 1978. 47:819–46
Copyright © 1978 by Annual Reviews Inc. All rights reserved

FLUORESCENCE ENERGY TRANSFER AS A SPECTROSCOPIC RULER[1]

Lubert Stryer

Department of Structural Biology, Sherman Fairchild Center,
Stanford University School of Medicine, Stanford, California 94305

CONTENTS

[1]Energy transfer was previously reviewed in this series by Steinberg (5). For recent reviews of different facets of energy transfer, see the articles by Schiller (6) and by Fairclough & Cantor (7). Energy transfer in photosynthesis (8, 140), bioluminescence (9–12), and insect photoreception (141) are not considered here because of space limitations.

0066-4154/78/0701-0819$01.00

INTRODUCTION AND PERSPECTIVE

Electronic excitation energy can be efficiently transferred between a fluorescent energy donor and a suitable energy acceptor over distances as large as 70 Å. In 1948, Förster (1) proposed a theory for this dipole-dipole energy transfer process which postulated that the rate of transfer depends on the inverse sixth power of the distance between the donor and acceptor. This predicted distance dependence was verified by fluorescence studies of donor-acceptor pairs separated by a known distance in well-defined model systems (2–4). Stryer & Haugland (3) then suggested that energy transfer could be used as a spectroscopic ruler in the 10–60 Å range to reveal proximity relationships in biological macromolecules.

This review begins with a brief summary of Förster's theory and of experimental tests of its validity. It then shows that the orientation factor introduces relatively little uncertainty in estimating distances from the transfer efficiency. The real challenge, discussed in the subsequent section, is to specifically label the macromolecule of interest with a suitable donor and acceptor. The repertoire of fluorescent labeling reagents and of fluorescent analogs of prosthetic groups and other biomolecules is enlarging at a rapid rate. The use of enzymes to catalyze the insertion of probes is another promising approach to specific labeling. The section on experimental strategy also includes some comments on the measurement of transfer efficiency. Applications of energy transfer as a spectroscopic ruler are considered in the third part of this review. More than 25 uses of energy transfer to elucidate proximity relationships in a wide variety of biological macromolecules and assemblies have been reported (Table 1). A number of these applications are discussed to illustrate the scope of the energy transfer method and the kind of information about structure and dynamics that can be derived from its use. This review ends with some conclusions and comments about future directions.

PRINCIPLES

Förster's Theory

According to Förster's theory[2] (1, 51), the rate of energy transfer k_T and the efficiency E are given by

$$k_T = r^{-6} K^2 J n^{-4} k_F \times 8.71 \times 10^{23} \text{ sec}^{-1},$$

$$E = r^{-6}/(r^{-6} + R_0^{-6}),$$

---

[2]An excellent translation by R. S. Knox of Förster's classic paper can be obtained at cost from the Department of Physics and Astronomy, University of Rochester, Rochester, New York 14627.

where R_0, the distance at which the transfer efficiency is 50%, is

$$R_0 = (J K^2 Q_0 \, n^{-4})^{1/6} \times 9.7 \times 10^3 \text{ Å}.$$

The geometric variables in the above expressions are: r, the distance between the centers of the donor and acceptor chromophores; and K^2, the orientation factor for a dipole-dipole interaction. The spectroscopic variables are: J, the spectral overlap integral; n, the refractive index of the medium between the donor and acceptor; k_F, the rate constant for fluores-

Table 1   Applications of energy transfer as a spectroscopic ruler

Biomolecule	References
Transfer RNA	
Yeast phenylalanine tRNA	12
E. coli fMet tRNA	13
E. coli glutamate tRNA	13
Oligopeptides	
Adrenocorticotropic hormone	14
Enkephalin analog	15
Oligomers of 2-hydroxyethyl-L-glutamine	16, 17
Metalloproteins	
Transferrin	18, 19
Carboxypeptidase A	20
Thermolysin	21
Receptors and transporters	
Rhodopsin	22–24
Galactose receptor	25
Immunoglobulin G	26–28
Ca^{2+}-ATPase	29
Gramicidin A	30
Multisubunit enzymes and assemblies	
Ribosomes	31, 32
RNA polymerase	33, 34
Pyruvate dehydrogenase complex	35–38
Cytochrome oxidase	39
Chloroplast coupling factor	40, 41
Bacterial luciferase	42
Aspartate transcarbamylase	43
Microtubules	44
Myosin	45
Other macromolecules	
Aspartokinase	46
Protease inhibitors	47–49
Erythrocyte membranes	50

cence emission by the energy donor; and Q_0, the quantum yield of fluorescence of the energy donor in the absence of acceptor. Efficient transfer requires that the energy donor and acceptor be in resonance, which means that the fluorescence emission spectrum of the donor must overlap the absorption spectrum of the acceptor, as measured by the spectral overlap integral J (in cm^3 M^{-1})

$$J = \frac{\int F(\lambda)\,\epsilon(\lambda)\,\lambda^4\,d\lambda}{\int F(\lambda)\,d\lambda},$$

where $F(\lambda)$ is the fluorescence intensity (in arbitrary units) of the energy donor at wavelength λ (in cm), and $\epsilon(\lambda)$ is the extinction coefficient (in cm^{-1} M^{-1}) of the energy acceptor. The medium between the energy donor and acceptor has a relatively small effect on the transfer process (n^{-4}) provided that it is transparent over the range of wavelengths at which the donor emits.

Tests of Förster's Theory in Model Systems

The validity of Förster's theory of dipole-dipole energy transfer has been established by studies of well-defined model systems. Latt, Cheung & Blout (2) introduced the use of molecular sticks to separate an energy donor and acceptor by a known distance. Their observed transfer efficiencies of about 50% for donors and acceptors attached to the ends of a 20 Å molecular stick, formed by the fusion of two steroids, were in good agreement with values predicted by Förster's equations. The dependence of energy transfer on distance was explicitly determined by Stryer & Haugland (3), who used oligomers of poly-L-proline ($n = 1$–12) to separate a naphthalene energy donor from a dansyl energy acceptor by distances ranging from 12–46 Å. The observed transfer efficiencies, ranging from 100%–16%, varied with the inverse 5.9 ± 0.3 power of the distance between the donor and acceptor, in excellent agreement with the r^{-6} dependence predicted by Förster's theory. An r^{-6} dependence was also found by Kuhn and co-workers (4) in their molecular sheet experiments, which used cyanine dyes separated by multilayers of fatty acids of known dimensions. Another important aspect of Förster's theory has been verified in a well-defined model system. Haugland, Yguerabide & Stryer (52) showed that the transfer rate k_T, as measured by nanosecond fluorimetry, is in fact proportional to J, the spectral overlap integral, when the magnitude of J was varied over a 40-fold range.

Orientation Factor

Dipole-dipole energy transfer has an angular dependence expressed by K^2, the orientation factor, which is given by

$$K^2 = (\cos\alpha - 3\cos\beta\cos\gamma)^2,$$

where α is the angle between the donor and acceptor transition moments, β is the angle between the donor moment and the line joining the centers of the donor and acceptor, and γ is the angle between the acceptor moment and the line joining the centers of the donor and acceptor. K^2 can have a value between 0 and 4. If both the donor and acceptor rotate freely in a time that is short compared to the excited state lifetime of the donor, then K^2 is equal to two thirds. The question arises as to how to treat K^2 if this condition is not met. At first thought, the potential range of 0–4 for K^2 would seem to vitiate the usefulness of energy transfer in estimating distances in biological macromolecules. As discussed below, this is not so because of the effect on K^2 of partial rotational freedom, the existence of electronic transitions polarized along more than one direction, and the nature of its probability distribution function.

From the observed transfer efficiency, an *apparent distance r'* can be calculated assuming that K^2 = two thirds. The actual distance r is then related to the apparent distance r' by

$$r = \alpha r' = (1.5 \, K^2)^{1/6} \, r',$$

where α is the sixth root of the ratio of the actual value of K^2 to the assumed value of two thirds. Since K^2 can be between 0 and 4, the potential range of α is 0–1.35. In practice, the range of α is much smaller. First, let us consider the effect of rotational mobility, which has been analyzed in detail by Dale & Eisinger (53–55). A nanosecond fluorescence polarization experiment should be carried out to determine the extent of rotational mobility of the energy donor (and the energy acceptor, if it is fluorescent). If the emission anisotropy of the donor, in the absence of any rotational motion, is A_0, and the initial value observed at a time short compared to the excited lifetime is A_1, then the donor can be treated as though it were rotating very rapidly within the volume of a cone having a semiangle θ, which Kawato, Kinosita & Ikegami (56) have shown is given by

$$A_1/A_0 = \cos^2 \theta \, (1 + \cos\theta)^2 \, /4.$$

The absolute limits on α for a donor rotating very rapidly within a cone of semiangle θ are

$$[0.75 \, (1 - \cos^2 \theta)]^{1/6} \leqslant \alpha \leqslant (6 \cos^2 \theta)^{1/6}.$$

The dependence of α on the cone semiangle θ is shown in Figure 1. Consider an energy donor with $A_0 = 0.35$ and $A_1 = 0.23$, which corresponds to $\theta = 30$ degrees, a typical value. For this donor, α can range between 0.68 and 1.32. If the apparent distance r' is 40 Å, then the actual distance r must be between 27 and 52.8 Å. The likely range of r can be further narrowed by considering the probability distribution function of K^2 (57, 58; and L. Stryer, to be published). There is a 10% probability that α will be above the upper dashed line, and a 10% probability that α will be below the lower dashed line in Figure 1. Thus, there is an 80% probability that α will assume a value in the shaded region of Figure 1. For $\theta = 30°$, $p = 0.8$ that α is between 0.74 and 1.14, which means that $p = 0.8$ that r is between 29.6 and 45.6 Å for $r' = 40$ Å. It should be noted that this analysis assumes that the energy acceptor has no rotational freedom and that the donor and acceptor transition moments are entirely along unique directions. In practice, the range of α is *narrower* than given in Figure 1 because these conditions are not strictly met in real systems. The uncertainty in α can be further diminished by using several different donor-acceptor pairs at the same site. If their apparent distances agree closely, then α almost certainly has a value near 1. In fact, apparent distances estimated

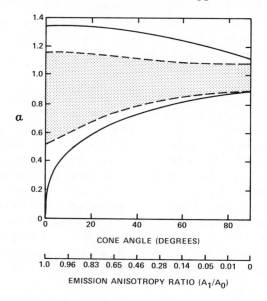

Figure 1 Effect of rotational mobility on the uncertainty in the estimated distance. The ratio α of the actual distance to the apparent distance is plotted as a function of the semiangle of the cone in which the donor rotates rapidly. The cone semiangle (upper scale on the x-axis) is derived from the observed emission anisotropy ratio (lower scale). The solid lines denote the absolute limits of α, whereas the lower and upper dashed lines give the 10% and 90% values of the probability distribution function for α. Thus, there is an 80% probability that α has a value in the shaded region. This calculation assumes that the energy acceptor has no rotational freedom and that the donor and acceptor transitions are polarized along unique directions. In practice, the range of α is narrower than shown here because these assumptions are not met.

from the observed transfer efficiencies for two or more donor-acceptor pairs at the same site have agreed closely thus far: e.g. 72, 75, and 77 Å in rhodopsin (22); 51 and 53 Å in cytochrome oxidase (39); and 27 and 30 Å in bacterial luciferase (42).

For chromophores that contain degenerate transitions polarized along orthogonal directions, the range of α is even smaller than discussed above. For example, metalloporphyrins have degenerate transitions that are polarized along perpendicular directions in the plane of the porphyrin (59), as noted in the energy transfer study of cytochrome oxidase (39). Even more favorable for reducing the range of α are chromophoric metal ions such as Co^{2+} and the fluorescent lanthanides Tb^{3+} and Eu^{3+}, which have electronic transitions along all three perpendicular directions (18–21, 60). If such a spectroscopically isotropic ion is used as an energy donor or energy acceptor, the range of α is 0.89–1.12 (20). If these ions are used as both the donor and acceptor, then α is equal to 1 (21).

In summary, the uncertainty in distance introduced by the orientation factor is relatively small—usually less than 20%. The problem is readily handled by measuring the extent of rotational mobility of the donor or acceptor to ensure that it is adequate ($\theta > \sim 30°$) and by using more than one donor-acceptor pair. For spectroscopically isotropic ions, there is almost no uncertainty in the orientation factor, even if they are rotationally immobile.

EXPERIMENTAL STRATEGY

The use of energy transfer as a spectroscopic ruler depends on having a fluorescent energy donor at a specific site and a suitable energy acceptor at another one. The absorption spectrum of the energy acceptor must overlap the emission spectrum of the energy donor. Furthermore, the spectral characteristics of the donor and acceptor must be sufficiently distinctive so that the number of photons absorbed and emitted by each can readily be determined. The aim of this section of the review is to aid the prospective user of the energy transfer technique in finding suitable donor-acceptor pairs. Fluorescent biomolecules and analogs are considered first, followed by a discussion of covalent fluorescent labeling reagents and of enzyme-catalyzed insertion reactions.

Fluorescent Biomolecules and Analogs

The first step in identifying a donor-acceptor pair is to ascertain whether the macromolecule has an intrinsic chromophore that can serve as a donor or acceptor. A single tryptophan residue in a protein, as in the galactose-binding protein from *Salmonella typhimurium* (25), is an excellent donor because its emission, centered at about 330 nm, overlaps the absorption of many potential energy acceptors. The fluorescent Y base at the anticodon of yeast phenylalanine transfer RNA (12) is another gift of nature to the fluorescence spectroscopist. Several species should be considered in searching for an intrinsic chromophore of a single kind, as exemplified by human serum albumin, which has one tryptophan, whereas bovine serum albumin has two. Fluorescent coenzymes and substrates such as reduced nicotinamide adenine nucleotide (61), oxidized flavin mononucleotide (62), and pyridoxal phosphate (63) are also attractive potential donors and acceptors. The heme group in heme proteins (64) is invariably a good acceptor because it has an absorption band extending over the entire visible region, but it is not fluorescent and hence cannot serve as a donor. Similarly, the retinal group in rhodopsin (22) is a suitable energy acceptor that has the additional advantage that it can be erased by light.

The use of fluorescent analogs (Table 2 and Figure 2) should also be considered. For example, protoporphyrin and its zinc and tin derivatives are fluorescent analogs of heme (39, 71, 72). Thiochrome diphosphate, which is fluorescent, can be used in place of thiamine diphosphate, which is not (35, 73). The fluorescent lanthanides Tb^{3+} and Eu^{3+} can often substitute for Ca^{2+} (21, 81–84). Zn^{2+} can be replaced by Co^{2+}, which can serve as an energy acceptor because of its absorption bands in the visible (20). Formycin (69, 70, 85), 2-aminopurine riboside (69, 70, 85), ethenoadenosine (65–67), and benzoadenosine (68) are fluorescent analogs of adenosine (Figure 2). Fluorescent etheno derivatives of ATP and other adenine nucleotides have been synthesized and studied in a variety of proteins by Leonard and co-workers (65–67). These fluorescent analogs can often substitute for cyclic AMP, AMP, ADP, ATP, NAD$^+$, FAD, and CoA and serve as energy

Table 2 Fluorescent analogs of biomolecules

Biomolecule	Fluorescent analog
ATP, NAD$^+$, FAD, CoA, and other adenine nucleotides	ethenoadenosine derivatives (65–67) benzoadenosine derivatives (68) formycin derivatives (69, 70, 85)
Heme	protoporphyrin (39, 71) zinc and tin protoporphyrin (71, 72)
Thiamine diphosphate	thiochrome diphosphate (35, 73)
Fatty acids	α- and β-parinaric acid (74, 75)
Phospholipids	phospholipids containing parinaric acid (76) dialkyl cyanine dyes (4, 77) fluorescent derivatives of phosphatidyl ethanolamine (78–80)
Ca^{2+}	Tb^{3+} and Eu^{3+} (21, 81–84)

donors. Another important recent development is the use of conjugated polyene fatty acids as fluorescent membrane probes. Sklar, Hudson & Simoni (74) have shown that the α and β isomers of parinaric acid (Figure 2) are excellent fluorescent analogs of unsaturated and saturated fatty acids, respectively. The spectroscopic characteristics of these fluorescent fatty acids and of parinaroyl phosphatidyl choline, and their use in detecting phase transitions in model membrane systems have been described in detail (74–76).

Finally, fluorescent derivatives of substrates, inhibitors, and other ligands can be advantageously used in energy transfer studies. For example, dansyl galactosides bind to the *lac* carrier protein in membrane vesicles from *Escherichia coli* (86). Anthroylouabain binds specifically to the cardiac glycoside inhibitory site on the Na^+-K^+ ATPase (87). A fluorescent derivative of erythromycin, an inhibitor of translocation in protein synthesis, was used in energy transfer studies of the 50S ribosome (32). Dansyl and fluorescein haptens served as the donor and acceptor in energy transfer studies of immunoglobulin G (26).

Fluorescent Labeling Reagents

Another approach to obtaining donor-acceptor pairs for energy transfer studies is to covalently insert suitable fluorescent groups. The repertoire of fluorescent labeling reagents has increased markedly in recent years, as reviewed by Kanaoka (88). The most useful types of reactions for introducing fluorescent groups are summarized in Figure 3. It is important to note that this list is not exhaustive, nor should it be assumed that a class of

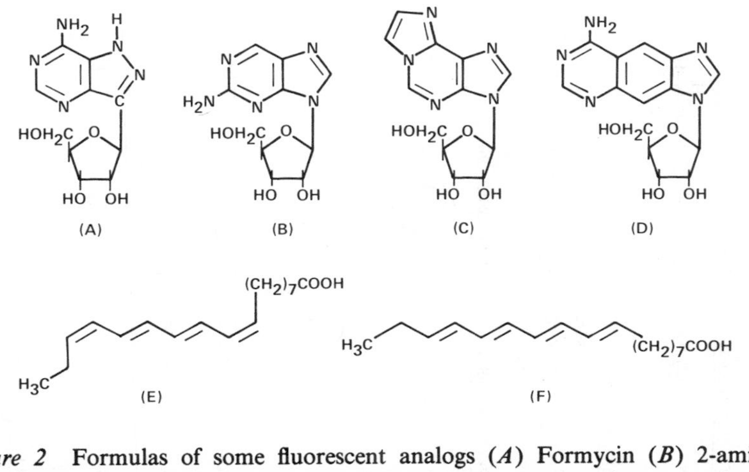

Figure 2   Formulas of some fluorescent analogs (*A*) Formycin (*B*) 2-aminopurineribonucleoside, (*C*) ethenoadenosine, and (*D*) benzoadenosine are analogs of adenosine. α and β parinaric acid (*E*) and (*F*) are analogs of fatty acids.

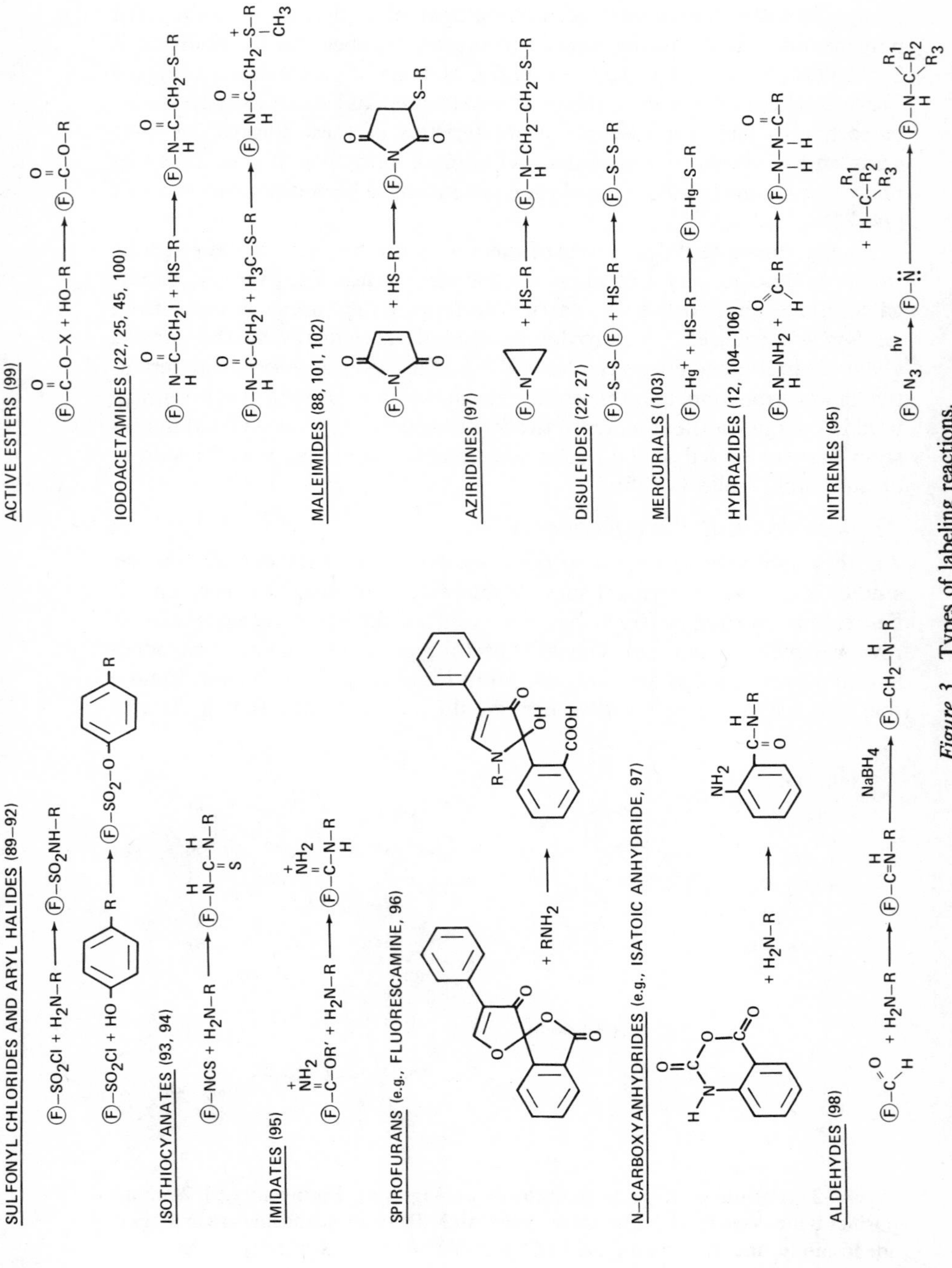

Figure 3 Types of labeling reactions.

compounds only undergoes the reaction cited in the figure. For example, maleimides are good reagents for labeling sulfhydryls, but they do react under certain conditions with other groups on proteins, such as amino groups. The structural formulas of several useful labeling reagents are given in Figure 4.

A single sulfhydryl group in a protein (or a particularly reactive one if there are several) is a choice target for the covalent attachment of a fluorescent probe. Fluorescent derivatives of iodoacetamide, such as N-(iodoacetamidoethyl)-1-aminonaphthalene-5-sulfonate (IAENS, Figure 4), which was introduced by Hudson & Weber (100), should be among the first to be tried. These compounds can also be used to label methionine residues in proteins devoid of cysteine. Maleimides (e.g. benzoxazolylphenyl-maleimide, Figure 4), aziridines, and mercurials are also useful labeling reagents for sulfhydryl residues. Yet another means of labeling sulfhydryls is to carry out a disulfide-sulfhydryl interchange reaction with fluorescent disulfides such as didansyl cystine.

Several classes of compounds, such as sulfonyl chlorides, isothiocyanates, and N-carboxyanhydrides, react with amino groups (Figure 3) but one

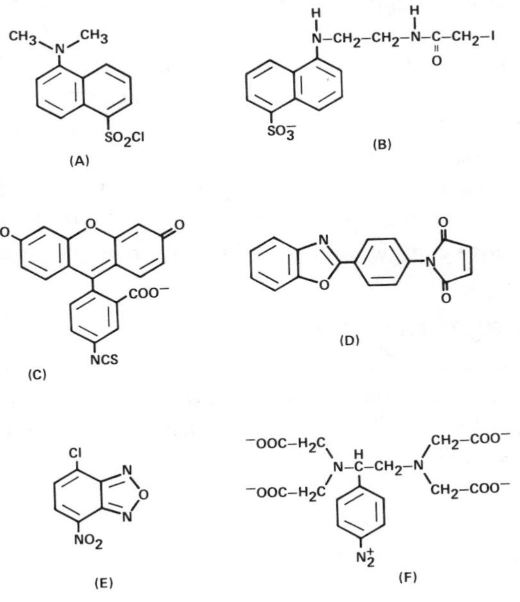

Figure 4   Formulas of some fluorescent labeling reagents (*A*) Dansyl chloride (*B*) N-(iodoacetamidoethyl)-1-amino-naphthalene-5-sulfonate (*C*) fluorescein isothiocyanate (*D*) N-[p-(2-benzoxazolyl) phenyl]-maleimide (*E*) 7-chloro-4-nitrobenzo-2-oxa-1,3-diazole (*F*) 1-(p-benzenediazonium)-ethylenodinitrilotetraacetic acid.

should not expect labeling at a unique site in view of the large number of amino groups on most proteins. Dansyl chloride, fluorescein isothiocyanate, and chloronitrobenzoxadiazole (Figure 4) are valuable labeling reagents for amino groups. Fluorescamine (Figure 3), introduced by Udenfriend and co-workers (96), is especially interesting because its reaction products with primary amines are highly fluorescent, whereas fluorescamine itself and its degradation products are nonfluorescent.

The specificity and reactivity of the active sites of enzymes can sometimes be exploited. For example, p-nitrophenyl anthranilate reacts specifically with serine 195 of α-chymotrypsin to form a stable, fluorescent acyl-enzyme complex (99). Acetylcholinesterase is rapidly phosphorylated by pyrene-butanolethylphosphorofluoridate to give a fluorescent phosphoryl enzyme (107). Fluorescent affinity labeling reagents for a variety of proteins are now being developed. Some of these reagents are aryl azides that can be converted photochemically into highly reactive nitrenes. A cyclic AMP-dependent protein kinase has been specifically labeled with an azidoetheno derivative of cyclic AMP (143).

Terminal sugar residues in glycoproteins, nucleic acids, and polysaccharides can be labeled by oxidizing them with periodate and coupling fluorescent hydrazides to the resulting aldehydes (108). For example, this approach has been used to label the 3'-CCA end of transfer RNA (12).

A noteworthy recent development is the synthesis of bifunctional chelating agents that are structurally similar to EDTA, such as 1-(p-benzenediazonium)-ethylenedinitrilotetraacetic acid (Figure 4) (109). Leung & Meares (60) have used these compounds to attach stable chelates of terbium and europium to proteins. As mentioned previously, these trivalent lanthanides are attractive energy donors because of the nearly isotropic character of their emission.

Enzyme-Catalyzed Insertion

A promising approach for the specific labeling of proteins is the use of enzymes to catalyze the insertion of a fluorescent group. Transglutaminase from guinea pig liver (110) catalyzes the insertion of primary amines into glutamine sidechains in peptides and proteins:

$$R-NH_2 + R'-\overset{O}{\overset{\|}{C}}-NH_2 \longrightarrow R'-\overset{O}{\overset{\|}{C}}-NH-R + NH_3 .$$

Dutton & Singer (111) have used transglutaminase to cross-link membrane proteins and to attach dansyl cadaverine to the Ca^{2+}-ATPase in sarcoplasmic reticulum membranes. Rhodopsin in retinal disc membranes has also been specifically labeled using transglutaminase (24, 125; J. S. Pober, V.

Iwanij, E. Reich, and L. Stryer, manuscript in preparation). One mole of dansyl cadaverine was incorporated per mole of rhodopsin. The site of labeling is the protease-sensitive region between the F1 and F2 units of rhodopsin. Energy transfer studies using dansyl cadaverine as the donor and 11-*cis* retinal as the acceptor were carried out.

The carbohydrate units of rhodopsin were labeled by Shaper & Stryer (106) using two enzymes. Galactose was specifically inserted into the carbohydrate units of rhodopsin by incubating retinal disc membranes with UDP-galactose and galactosyl transferase. This enzyme catalyzes the formation of a 1,4-glycosidic bond between galactose and N-acetylglucosamine (112). These modified membranes were treated with galactose oxidase to generate an aldehyde at the C-6 position of the inserted galactose units and then reacted with dansyl hydrazide to yield a fluorescent hydrazone. This procedure for the specific attachment of a spectroscopic probe should be generally applicable to membrane glycoproteins. It will be interesting to learn whether fluorescent derivatives of UDP-galactose can serve as substrates for galactosyl transferase. If they can, then it would be feasible to specifically label the N-acetylglucosamine terminus of oligosaccharides in a single step. The use of other sugar transferases should also be explored.

The biosynthetic incorporation of a fluorescent substrate is another means of specifically labeling complex macromolecules and assemblies. A dansylated UDP-pentapeptide can serve as a substrate in the synthesis of bacterial cell walls (113). The incorporated dansyl group is responsive to changes in the membrane environment during the synthesis of the nascent peptidoglycan. A fluorescent fatty acid was incorporated into the membrane phospholipids of an unsaturated fatty acid auxotroph of *E. coli* (114). The fluorescence properties of the incorporated parinaric acid were sensitive to phase transitions in the bacterial cell membrane.

Measurement of Transfer Efficiency

The efficiency of energy transfer E can be determined in three different ways:

1. From the *excited state lifetime of the energy donor* in the absence (τ_0) and presence (τ_T) of the energy acceptor.

$$E = 1 - (T_T/T_0)$$

2. From the *quantum yield* (*or relative fluorescence intensity*) *of the energy donor* in the absence (Q_0) and presence (Q_T) of the energy acceptor.

$$E = 1 - (Q_T/Q_0)$$

3. From the *excitation spectrum of the energy acceptor,* if it is sufficiently fluorescent (or from the action spectrum of a photochemical change). Let

G (λ) be the magnitude of the corrected excitation spectrum (or action spectrum) of the energy acceptor excited at wavelength λ. The extinction coefficients of the energy donor and acceptor at that wavelength are $\epsilon_D(\lambda)$ and $\epsilon_A(\lambda)$, respectively. G is measured at two wavelengths: at λ_1, where the donor has no absorption, and at λ_2, where the extinction coefficient of the donor is large compared to that of the acceptor. The transfer efficiency is then given by

$$E = [G(\lambda_2)/G(\lambda_1) - \epsilon_A(\lambda_2)/\epsilon_A(\lambda_1)] \times [\epsilon_A(\lambda_1)/\epsilon_D(\lambda_2)].$$

Under certain conditions, the transfer efficiency can also be determined from the emission kinetics of the energy acceptor (115).

Transfer efficiencies determined by methods 1 and 2 assume that the insertion of the acceptor does not alter the quantum yield of the donor except by the energy transfer process itself. In other words, the local environment of the donor must be the same in the absence and presence of acceptor. This assumption can be tested by measuring the emission spectrum and rotational correlation time of the donor in the presence and absence of acceptor. If these parameters are not constant, methods 1 and 2 cannot be applied. Method 3 can be used even if the local environment of the donor is different in the presence of an acceptor, provided that the absorption spectra of the donor and acceptor moieties are known. It is desirable to measure the transfer efficiency by two of these methods, which should agree if the donor-acceptor pairs in the sample are homogeneous (22).

SELECTED APPLICATIONS OF ENERGY TRANSFER

Rhodopsin

Fluorescence energy transfer studies provided the first indication that rhodopsin is an elongated molecule. Wu & Stryer (22) specifically labeled bovine rhodopsin in disc membranes or in digitonin solution at three sites. Site A, a sulfhydryl group, was alkylated by fluorescent derivatives of iodoacetamide, such as IAENS (Figure 2). Site B, a different sulfhydryl group, was labeled with fluorescent disulfides such as didansyl-L-cystine. Acridine derivatives were tightly bound to site C by noncovalent interactions. The labeled rhodopsins retained their 500 nm absorption band and were regenerable after bleaching. The transfer efficiencies determined from the quantum yield and excited-state lifetime of the donor showed that the apparent distances between 11-*cis* retinal and sites A, B, and C are 75, 55, and 48 Å, respectively. Different donors at the same time gave nearly the same apparent distances (e.g. 73, 75, and 77 Å for site A), which suggests that the apparent distances closely approximate the actual ones. Sites A, B, and C are relatively close to each other ($\sim$ 30 Å). The 75 Å estimated distance between site A and 11-*cis* retinal in digitonin solution (22) and in disc membranes (23) was interesting because it suggested that rhodopsin has an elongated shape. A 40,000 dalton sphere would have a diameter of only 45 Å. Wu & Stryer (22) concluded that rhodopsin is sufficiently long to traverse the disc membrane if suitably oriented and that it could serve as a light-controlled gate. Subsequent neutron and X-ray scattering studies (116, 117) have shown that rhodopsin in detergent solution is in fact at least 75 Å long. Several lines of evidence suggest that rhodopsin traverses the disc membrane (118), and it has recently been found that light increases the number of ionic channels in reconstituted bilayers containing rhodopsin (119).

Pober, Iwanij, Reich & Stryer (24, 139, 142) have used transglutaminase to specifically insert dansyl cadaverine into rhodopsin in its native membrane environment. The site of labeling was ascertained by taking advantage of the finding (120) that rhodopsin is cleaved by a variety of proteolytic enzymes into two fragments of different size. F1 has a molecular weight of about 26,000 and contains the carbohydrate units, whereas F2 has a molecular weight of about 12,000 and contains the site A sulfhydryl group and the N-retinyl group formed by borohydride reduction. In fact, the site labeled by transglutaminase is at the junction of the F1 and F2 units. The transfer efficiency from dansyl cadaverine at this site to 11-*cis* retinal in rhodopsin in disc membranes was 31%, which gives an apparent distance of 61 Å for an R_0 of 53 Å, and provides additional evidence that rhodopsin has an elongated shape.

Galactose Receptor Protein

The binding of a specific ligand by a sensory receptor is expected to produce a conformational change that can be transmitted to the next component of the sensory system. The fluorescence studies of Zukin, Hartig & Koshland (25) provide clear-cut evidence for such a propagated conformational change in the galactose binding protein of *S. typhimurium* when it interacts with galactose. This 33,000 dalton receptor for galactose in the chemotactic response contains a single tryptophan and a single methionine residue. The emission maximum of this tryptophan residue shifts 5 nm to the blue when the receptor binds galactose. A similar shift occurs in the related galactose binding protein from *E. coli* (121), which contains more than one tryptophan. The presence of a single methionine in the *S. typhimurium* protein made it possible to answer a key question: Is the conformational change reported by the altered tryptophan emission localized to the immediate vicinity of the galactose binding site or is it propagated to a distant region of the protein? To answer this question, a second fluorescent group was inserted into the protein by alkylating the single methionine with a 1000-fold excess of 5-iodoacetamidofluorescein at pH 7 for 24 hr. The binding of galactose also changed the fluorescence emission spectrum of the acetamidofluorescein linked to methionine, shifting its maximum 4 nm to the blue and decreasing its intensity by 10%. The spatial extent of the conformational change induced by the binding of galactose to one site on the receptor was then ascertained by using fluorescence energy transfer to estimate the distance between the single tryptophan residue and the acetamidofluorescein group on the single methionine residue. The transfer efficiency of 22%, determined from the excitation spectrum of the acetamidofluorescein fluorescence, showed that the apparent distance between these sites is 41 Å. Fluorescence polarization measurements placed limits on K^2, corresponding to a distance range of 32–52 Å. Since galactose is only 5 Å in length, it is evident that the conformational change sensed by fluorescent groups at least 30 Å apart must be delocalized. A model based on these and other fluorescence experiments is shown in Figure 5. Zukin, Hartig & Koshland (25) have noted that environmentally sensitive reporter groups located far from the active site should be generally useful in detecting propagated conformational changes in biological macromolecules. The distinctive role of fluorescence energy transfer in this experimental approach is that it reveals the minimal spatial extent of the conformational change elicited by the binding of ligand.

Metalloproteins

Fluorescent trivalent lanthanide ions can serve as probes of the structure of metal-binding proteins, as was first shown by Luk (18) in his studies of

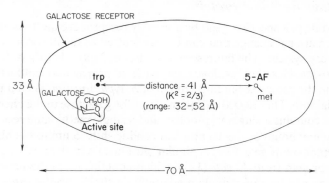

Figure 5 Schematic diagram of the *Salmonella* galactose receptor, showing that the labeled methionine residue is about 41 Å from the tryptophan residue adjacent to the galactose binding site. The binding of galactose produces a delocalized conformational change. From Zukin, Hartig & Koshland (25).

transferrin. This 77,000 dalton protein contains two binding sites for Fe^{3+} that can be replaced by Tb^{3+}. When excited at 295 nm, the fluorescence intensity of Tb^{3+} in Tb^{3+}-transferrin is about 10^5 greater than that of the free ion because of short-range energy transfer from ionized tyrosines in the immediate vicinity of the bound Tb^{3+}, which can then transfer its excitation energy to Fe^{3+} with an R_0 of 27 Å. Luk observed no change in the excited state lifetime of Tb^{3+} when Fe^{3+} was presumed to be bound to the other site on the same molecule, and concluded that these sites are at least 43 Å apart. Recent advances in the understanding of the binding of metal ions to transferrin stimulated Meares & Ledbetter (19) to reinvestigate this system. They prepared transferrin demonstrated to contain Tb^{3+} at one site and Fe^{3+} at the other, and found a transfer efficiency of 64%, indicating that these sites are 25 ± 2 Å apart. There is virtually no uncertainty in K^2 because the donor and acceptor transitions are nearly isotropic.

Energy transfer has been used by Horrocks, Holmquist & Vallee (21) to measure the distance between two ions in a metalloenzyme. Thermolysin, a heat-stable protease, contains a zinc at its active site and four calcium ions at other locations. One of the Ca^{2+} was replaced by Tb^{3+}, and the active-site Zn^{2+} by Co^{2+}, which has an absorption spectrum with a maximal extinction coefficient of about 80 cm^{-1} M^{-1} that overlaps the Tb^{3+} emission. The bound Tb^{3+} was excited by energy transfer from tryptophan. In turn, Tb^{3+} transferred its excitation energy to Co^{2+} with an efficiency of 89%. The R_0 distance for this donor-acceptor pair is not precisely known because of uncertainty as to the quantum yield of Tb^{3+} fluorescence, but a plausible estimate of 0.51 gives an R_0 of 19.6 Å. This estimated R_0 of 19.6 Å, and the observed transfer efficiency of 89%, indicate that the distance between the Tb^{3+} and Co^{2+} is 13.7 Å, in excellent agreement with the distance known from X-ray crystallographic studies (122). Again, K^2 is known to be two thirds because of the isotropic character of both the energy donor and acceptor.

The usefulness of Co^{2+} as an energy acceptor in place of Zn^{2+}, which is silent in absorption and fluorescence spectroscopy, had previously been shown by Latt, Auld & Vallee in their energy transfer studies of cobalt carboxypeptidase A (20). The energy donor was the fluorescent dansyl group at the amino terminus of a series of peptide substrates. The R_0 value for transfer from dansyl to Co^{2+} is 16 Å. Stopped-flow fluorescence measurements of these enzyme-substrate complexes showed that the distance from the dansyl group to Co^{2+} ranged from less than 7 Å in Dansyl-Gly-L-Trp to 12 Å in Dansyl-(Gly)$_3$-L-Trp, which suggests that these substrates assume an extended conformation when bound to the active site of carboxypeptidase A.

Cytochrome c Oxidase

Cytochrome c oxidase, the terminal member of the respiratory chain, contains seven polypeptide chains ranging in molecular weight from 4,500–40,000. The three largest subunits (I, II, and III) of this membrane-bound assembly are encoded by mitochondrial DNA and are highly hydrophobic, whereas the four small ones (IV–VII) are encoded by nuclear DNA and are relatively hydrophilic. Dockter, Steinemann & Schatz (39) have used energy transfer to establish proximity relationships between three sites of purified cytochrome c oxidase from yeast: 1. a highly reactive sulfhydryl group on subunit II (34,000 daltons), which was alkylated by the 1,5 or 2,6 isomer of N-(iodoacetamidoethyl)-1-amino-naphthalene-5-sulfonate; 2. endogenous heme a; and 3. cytochrome c, or its fluorescent porphyrin analog, covalently attached to subunit III (23,000 daltons) by a disulfide bond. The sulfhydryl group of cytochrome c was reacted with 5,5'-dithiobis-(2-nitrobenzoate) to give thionitrobenzoate-activated cytochrome c, which in turn reacted with the accessible sulfhydryl group of subunit III.

Five distances in the labeled cytochrome c oxidase complex were measured. Energy transfer from the two isomers of acetamidoethylaminonaphthalene sulfonate attached to the sulfhydryl group on subunit II to heme a indicated that these sites are about 52 Å apart. The heme of cytochrome c is closer to this sulfhydryl group on subunit II, about 35 Å away, as shown by transfer from the same fluorescent energy donors. The fluorescent porphyrin analog of cytochrome c was used as the energy donor to determine the distance between the heme group of cytochrome c and endogenous heme a, which are about 25 Å apart. There was little uncertainty about the orientation factor in these energy transfer experiments because the fluorescent donors attached to subunit II have considerable rotational mobility and the heme acceptor has degenerate x- and y-polarized absorption bands. Furthermore, the distances estimated using the 1,5 and 2,6 isomers of the energy donor were in excellent agreement. A model of the cytochrome c-cytochrome c oxidase complex depicting these proximity relationships is shown in Figure 6. The 25 Å separation between the heme of bound cytochrome c and the nearest heme a group of the oxidase raises the intriguing question as to how electron transfer occurs over such a large distance. Dockter, Steinemann & Schatz (39) also cite their interesting finding that the distance between the subunit II sulfhydryl group and heme a is about 30% less in a mutant cytochrome c oxidase complex containing a 2000 dalton shorter subunit II than it is in the wild type. It is evident that the energy transfer approach provides valuable structural information that can complement three-dimensional image reconstruction studies of cytochrome c oxidase crystals (123).

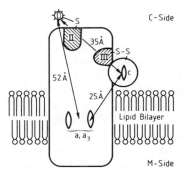

Figure 6   Schematic diagram of the yeast cytochrome *c* oxidase complex, showing the three labeled sites and the estimated distances between them. From Dockter, Steinemann & Schatz (39).

Pyruvate Dehydrogenase Complex

Proximity relationships in the pyruvate dehydrogenase complex of *E. coli* have been investigated by Hammes and co-workers (35–38). This multienzyme assembly of 60 polypeptide chains has a molecular weight of about 4.6 million and a diameter of about 300 Å, as shown by Reed (124). The complex consists of three types of enzymes: the pyruvate dehydrogenase component (E_1), which contains thiamine pyrophosphate and catalyzes the decarboxylation of pyruvate; dihydrolipoyl transacetylase (E_2), which contains lipoamide and catalyzes the oxidation of the two-carbon intermediate and its transfer to CoA; and dihydrolipoyl dehydrogenase (E_3), which contains FAD and catalyzes the regeneration of the oxidized form of lipoamide. E_1, a 192,000 dalton dimer, contains two sulfhydryl groups that can be labeled by a variety of maleimides without loss of enzymatic activity (38). The fluorescence properties of these sulfhydryl labels are markedly altered by the binding of substrate or of allosteric effectors such as acetyl CoA. The binding of these molecules results in a conformational change that exposes the fluorescent group on the sulfhydryl residue to a more polar environment and increases its rotational mobility. The distance between this sulfhydryl group and the active site of E_1 was estimated by using thiochrome diphosphate (73), a highly fluorescent analog of thiamine diphosphate. 1-Anilinonaphthalene-8-sulfonate (ANS) binds to the CoA regulatory site on E_1, and so it was possible also to estimate the distance between the reactive sulfhydryl and the CoA site. Papadakis & Hammes (38) found that these distances are more than 40 Å. Hence, the fluorescent labels on the reactive sulfhydryls of E_1 sense conformational changes that extend over large distances from the catalytic and regulatory sites of E_1.

Energy transfer between fluorescent groups on different enzymes of this multienzyme complex was also investigated. Moe, Lerner & Hammes (36) found that the efficiency of energy transfer from thiochrome diphosphate at the active site of E_1 to FAD at the active site of E_3 is 8%, which gives an apparent distance of about 45 Å. Shepherd & Hammes (37) observed no energy transfer from ANS at the acetyl CoA regulatory site on E_1 to FAD on E_3, which indicates that these sites are at least 58 Å apart. These distances between the active sites of E_1 and E_3 have interesting implications for the catalytic mechanism of the pyruvate dehydrogenase complex. The activated intermediates are thought to be transferred from the active site of E_1 to E_2 to E_3 by the lipoamide prosthetic group, which acts as a flexible arm (124). The simplest version of this hypothesis requires that the maximum distance between active sites be 28 Å, twice the length of the 14 Å lipoamide arm. However, the energy transfer studies showed that the active sites of E_1 and E_3 are at least 45 Å apart, which points to alternative mechanisms such as substrate diffusion within the complex, or acyl transfer between adjacent lipomide residues (36). Another possibility is that the partial reactions catalyzed by the component enzymes drive substantial conformational changes that translocate the reactive intermediate on lipoamide from one active site to the next.

Ribosomes

One of the most challenging problems studied thus far by fluorescence energy transfer techniques is the 30S ribosome of *E. coli*, which contains 21 proteins and a 16S RNA, and has a long axis of about 190 Å. Huang, Fairclough & Cantor (31) have prepared 30S ribosomes containing a fluorescent energy donor attached to one of the 21 proteins and an energy acceptor linked to another. Purified individual ribosomal proteins were reacted with N-(iodoacetamidoethyl)-1-aminonaphthalene-5-sulfonate to insert the energy donor, or with fluorescein isothiocyanate to insert the energy acceptor. This donor-acceptor pair was chosen because its R_0 distance of about 42 Å is one of the largest attainable. 30S ribosomes containing pairs of labeled protein were then reassembled. As might be expected, labeled proteins containing fewer fluorescent groups than the average of the population were preferentially incorporated into the reassembled structure. Typically, ribosomes containing an average of 0.5 molecules of donor or acceptor were reconstituted and shown to be effective in synthesizing polyphenylalanine. The observed transfer efficiencies were analyzed in terms of two limiting models. One model assumes that the proteins are labeled at single, unique sites, whereas the other model, which had been developed in detail by Gennis & Cantor (49), assumes that the surface of the protein is randomly labeled. Proximity relationships derived from these analyses were similar and led to the following conclusions:

1. Four pairs of proteins are so close that extensive protein-protein contact is likely. In fact, it is known that two of these pairs (S7-S9 and S6-S18) can be cross-linked (115).
2. Six other pairs of proteins are sufficiently close to be nearest neighbors in the 30S particle, although some RNA probably separates them.
3. The ten other pairs of labeled proteins are far enough apart that other proteins or extensive regions of RNA lie between them.
4. Many protein pairs that are linked in assembly do not make extensive contact with each other, which suggests that they are linked by the action of RNA.

It will be interesting to compare the proximity relationships emerging from energy transfer studies with those being elucidated by complementary approaches such as neutron scattering (125), electron microscopy using antibodies (126, 127), and chemical cross-linking (128).

The distance between two functionally important regions of the 50S ribosome from *E. coli* has been determined by Langlois et al (32). The protein L7, which is required for elongation factor G-dependent GTPase activity, was selectively removed (along with L12) from 50S particles, labeled with IAENS, and added back to form a functional 50S ribosome containing a fluorescent energy donor. The energy acceptor was a tightly bound fluorescent analog of erythromycin synthesized from fluorescein isothiocyanate and erythromycylamine. Erythromycin inhibits peptide bond formation and translocation and is thought to bind near the peptidyl transferase center of the 50S particle. The energy transfer efficiency showed that the distance between these sites is about 70 Å, which is interesting because it shows that two sites involved in translocation are far apart. Thus, translocation is probably mediated by conformational changes that extend over an appreciable portion of the 50S ribosome.

Transfer RNA

Energy transfer studies of proximity relationships in transfer RNA were first investigated by Beardsley & Cantor (12), who took advantage of the naturally occurring fluorescent base at the anticodon of yeast phenylalanine tRNA. The energy acceptors in their experiments were fluorescent hydrazides attached to aldehyde groups generated by periodate-oxidation of the 3'-terminal adenosine. The transfer efficiency indicated that the anticodon is more than 40 Å away from the amino acid attachment site. Subsequent X-ray crystallographic studies of this tRNA showed that this distance is about 75 Å (129, 130).

The relationship between the tertiary structure of tRNA in the crystalline state and in solution has been explored in depth by Yang & Söll (13) using energy transfer techniques. They prepared five species of tRNA (either fMet or glutamate tRNA from *E. coli*) labeled with acridine, dansyl, anthranila-mide, or coumarin derivatives at two of the following sites (Figure 7): (*a*) the 5' end, (*b*) 4-thiouridine, (*c*) dihydrouridine, (*d*) 2-thiouridine at the anticodon, (*e*) pseudouridine, and (*f*) 3' terminal adenosine. The distances estimated from the observed transfer efficiencies are 24 Å for *a* to *f*, 38 Å for *b* to *f*, 36 Å for *c* to *e*, > 65 Å for *d* to *f*, and 55 Å for *e* to *f*. The corresponding distances in yeast tRNA-determined crystallographically are 25, 41, 23, 74, and 53 Å respectively. The distances estimated from the transfer efficiencies agree closely with those determined crystallograph-ically, except for *c* to *e*, which is 23 Å in the crystal and estimated to be 36 Å in solution. From these promixity relationships, Yang & Söll con-cluded that tRNA in solution adopts a clover-leaf conformation similar to the one in the crystal, the amino acid acceptor end is far from the pseudouri-dine loop, the anticodon is far from the amino acid acceptor end, and the CCA end is extended. It should now be feasible to use energy transfer to detect conformational transitions in transfer RNA as it interacts with its cognate synthetase and with the ribosome.

Membrane Channels and Pumps

Energy transfer can provide information about the lateral distribution of lipids and proteins in membranes (29, 30, 131). In particular it can reveal whether a particular peptide or protein is monomeric or oligomeric in its membrane environment, as shown by Veatch & Stryer (30) in their study of the molecularity of the gramicidin A transmembrane channel. Tyrosine

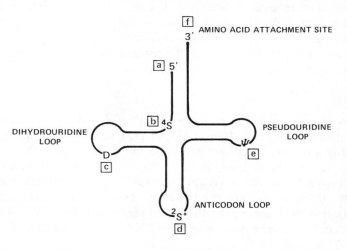

Figure 7 Cloverleaf diagram of transfer RNA showing the six labeled sites in the energy transfer studies of Yang & Söll (13).

11 of gramicidin C was labeled with a dansyl group, which served as the prospective energy donor, or with an azo dye, which served as the energy acceptor. The relative quantum yield Q/Q_0 of the dansyl fluorescence in liposomes containing both gramicidins was measured as a function of the mole fraction a of the gramicidin bearing the energy acceptor. The surface density of gramicidin was kept below 1 per 1000 phospholipids to avoid energy transfer between nonassociated gramicidins. For a monomer, $Q/Q_0 = 1$, independent of a, whereas for oligomers, Q/Q_0 decreases with a. For the simplest dimer model, $Q/Q_0 = 1-Ea$, where E is the transfer efficiency within a dimer. The observed dependence of Q/Q_0 versus a for the labeled gramicidins best fits a dimer model with E = 75%, which indicates that the donor and acceptor chromophores in a dimer are about 33 Å apart. These experiments complement other fluorescence (132) and conductance (30, 133–135) studies that show the presence of two molecules of gramicidin in a channel.

A similar approach was used by Vanderkooi et al (29) to investigate the Ca^{2+}-ATPase system in reconstituted vesicles. Purified ATPase from sarcoplasmic reticulum was labeled with a fluorescent energy donor (IAENS) or an energy acceptor (iodoacetamidofluorescein). Energy transfer was observed in reconstituted vesicles containing donor-labeled and acceptor-labeled ATPase molecules. The efficiency of energy transfer was unaffected by a tenfold dilution of the lipid phase with egg phosphatidylcholine. In contrast, the addition of a tenfold molar excess of unlabeled ATPase abolished the energy transfer which indicates that energy transfer occurred within a complex of two or more ATPase molecules rather than between randomly distributed monomers.

Translational Dynamics of Flexible Polypeptide Chains

In the applications discussed thus far, energy transfer has been used to estimate distances that do not change appreciably in terms of nanoseconds. Energy transfer is also well-suited to measuring changes in distance occurring during the excited-state lifetime of the energy donor. Haas, Katchalski-Katzir & Steinberg (17) have investigated the translational dynamics of the ends of oligopeptides by measuring energy transfer between chromophores attached to the amino and carboxyl termini. The underlying principle is that energy transfer is enhanced by translational diffusion because a donor-acceptor pair that is too far apart for efficient transfer at the instant of excitation may diffuse toward each other during the excited-state lifetime so that the distance between them becomes short enough for efficient transfer (136, 137). Oligomers of 2-hydroxyethyl-L-glutamine (n = 4–9 residues), containing a naphthalene energy donor at one end and a dansyl energy acceptor at the other, were studied by nanosecond fluorimetry. The mean end-to-end distances of this series of oligomers ranged from 15.2 Å for n = 4, to 21.6 Å for n = 9, which are comparable to the R_0 value of 22 Å (16). Convincing evidence is presented that the orientation factor does not pose a problem in determining the distribution of end-to-end distances (17), contrary to an earlier criticism (138). The nanosecond emission kinetics of the naphthalene group was measured in solvents of varying viscosity, ranging from 1–900 centipoise in going from trifluoroethanol to glycerol, and from 0.58–9.6 cp in going from methanol to 50% methanol in glycerol. The naphthalene-excited state lifetime in the absence of acceptor was 63.5 nsec. In the pentamer containing the energy acceptor, the average excited state lifetime of the donor ranged from 14.3 nsec in pure glycerol to 6.4 nsec in trifluoroethanol. Similar decreases in the average excited state lifetime were observed in the other oligomers as the solvent viscosity was decreased, because the rate of intramolecular Brownian motion is higher in the more fluid media. Values of the diffusion coefficient D ranging from about 5×10^{-9}–10^{-7} cm^2 sec^{-1} were derived from the nanosecond emission kinetics of the donor. These values are about an order of magnitude lower than expected for the unattached chromophores which shows that the polypeptide backbone possesses appreciable internal friction that resists the Brownian motion of the ends. The second major conclusion is that the magnitude of the internal friction, estimated from the extrapolated value of the diffusion coefficient at zero viscosity, is higher for the shorter chain molecules.

PROSPECTS AND CONCLUSIONS

Fluorescence energy transfer has been effectively used as a spectroscopic ruler in the 10–80 Å range in a wide variety of biological macromolecules and assemblies. The distinctive features of this approach are its *low spatial resolution, high temporal resolution, high sensitivity,* and *applicability in complex systems.* The spatial resolution of the energy transfer technique is limited by the appreciable size of most donors and acceptors and by the uncertainty in the orientation factor. Recent calculations and experiments indicate that the error in the estimated distance introduced by the orientation factor is unlikely to be greater than 20%. Thus, energy transfer can readily reveal whether two sites are separated by 30 Å or 60 Å, for example, whereas the technique is not well-suited to discriminating between distances of say 30 and 35 Å. Proximity information at this level of resolution can have important biological implications, such as showing that a protein is long enough to span a membrane or that a conformational change in a receptor is delocalized. A second feature of energy transfer spectroscopy is its high temporal resolution. Transient intermediates having lifetimes of milliseconds, and sometimes even of microseconds, are within the scope of the technique. It should be feasible to detect changes in distance of a few angstroms accompanying dynamic processes such as enzymatic catalysis, muscle contraction, signal transduction, and active transport. Furthermore, translational motions of lipids and proteins in membranes should be evident in energy transfer experiments if the product of the diffusion coefficient of the donor-acceptor pair and the excited-state lifetime of the donor is comparable to or greater than R_0^2. An additional advantage of energy transfer spectroscopy is its high sensitivity. A nanomole of sample usually suffices. The sensitivity of fluorescence techniques is being further enhanced by advances in instrumentation, as exemplified by photon-counting fluorimeters and by the use of synchrotron radiation as an intense pulsed light source. Finally, it should be noted that the energy transfer technique can be used to investigate proximity relationships in complex systems such as the plasma membrane of intact cells. The challenge with complex systems is to specifically label them with a suitable donor and acceptor. Significant progress is being made in this regard through the design of fluorescent analogs, substrates, and inhibitors and by the use of highly specific enzyme-catalyzed insertion reactions. Some of the most interesting energy transfer experiments in the years ahead will be carried out using fluorescence microscopes and fluorescence-activated cell sorters.

ACKNOWLEDGMENTS

The work carried out in the author's laboratory was supported by research grants from the National Institute of General Medical Sciences (GM-24032) and the National Eye Institute (EY-02005).

Literature Cited

1. Förster, T. 1948. *Ann. Physik.* 2:55–75
2. Latt, S., Cheung, H. T., Blout, E. R. 1965. *J. Am. Chem. Soc.* 87:995–1003
3. Stryer, L., Haugland, R. P. 1967. *Proc. Natl. Acad. Sci. USA* 58:719–26
4. Bucher, H., Drexhage, K. H., Fleck, M., Kuhn, H., Mobius, D., Schafer, F. P., Sondermann, J., Sperling, W., Tillmann, P., Wiegand, J. 1967. *Mol. Cryst.* 2:199–230
5. Steinberg, I. Z. 1971. *Ann. Rev. Biochem.* 40:83–114
6. Schiller, P. W. 1975. In *Biochemical Fluorescence: Concepts*, ed. R. F. Chen, H. Edelhoch, 1:285–303. New York: Dekker. 408 pp.
7. Fairclough, R. H., Cantor, C. R. 1978. *Methods Enzymol.* In press
8. Govinjee, N. F. N. 1975. *Bioenergetics of Photosynthesis.* New York: Academic. 999 pp.
9. Morin, J. G., Hastings, J. W. 1971. *J. Cell. Physiol.* 77:313–18
10. Morise, H., Shimomura, O., Johnson, F. H., Winant, J. 1974. *Biochemistry* 13:2656–62
11. Ruby, E. G., Nealson, K. H. 1977. *Science* 196:432–34
12. Beardsley, K., Cantor, C. R. 1970. *Proc. Natl. Acad. Sci. USA* 65:39–46
13. Yang, C. H., Söll, D. 1974. *Proc. Natl. Acad. Sci. USA* 71:2838–42
14. Schiller, P. W. 1971. *Proc. Natl. Acad. Sci. USA* 69:975–79
15. Schiller, P. W., Chun, F. Y., Lis, M. 1977. *Biochemistry* 16:1831–38
16. Haas, E., Wilchek, M., Katchalski-Katzir, E., Steinberg, I. Z. 1975. *Proc. Natl. Acad. Sci. USA* 72:1807–11
17. Haas, E., Katchalski-Katzir, E., Steinberg, I. Z. 1978. *Biopolymers.* In press

18. Luk, C. K. 1971. *Biochemistry* 10: 2838–43
19. Meares, C. F., Ledbetter, J. E. 1978. *Biochemistry.* In press
20. Latt, S. A., Auld, D. S., Vallee, B. L. 1970. *Proc. Natl. Acad. Sci. USA* 67:1383–89
21. Horrocks, W. D. Jr., Holmquist, B., Vallee, B. L. 1975. *Proc. Natl. Acad. Sci. USA* 72:4763–68
22. Wu, C.-W., Stryer, L. 1972. *Proc. Natl. Acad. Sci. USA* 69:1104–8
23. Steinemann, A., Wu, C.-W., Stryer, L. 1973. *J. Supramol. Struct.* 1:348–53
24. Pober, J. S. 1976. *Limited fragmentation of specifically-labeled rhodopsins.* PhD thesis. Yale Univ., New Haven. 158 pp.
25. Zukin, R. S., Hartig, P. R., Koshland, D. E. Jr. 1977. *Proc. Natl. Acad. Sci. USA* 74:1932–36
26. Werner, T. C., Bunting, J. R., Cathou, R. E. 1972. *Proc. Natl. Acad. Sci. USA* 69:795–99
27. Bunting, J. R., Cathou, R. E. 1973. *J. Mol. Biol.* 77:223–35
28. Bunting, J. R., Cathou, R. E. 1974. *J. Mol. Biol.* 87:329–38
29. Vanderkooi, J. M., Ierkomas, A., Nakamura, H., Martonosi, A. 1977. *Biochemistry* 16:1262–67
30. Veatch, W., Stryer, L. 1977. *J. Mol. Biol.* 113:89–102
31. Huang, K. H., Fairclough, R. H., Cantor, C. R. 1975. *J. Mol. Biol.* 97:443–70
32. Langlois, R., Lee, C. C., Cantor, C. R., Vince, R., Pestka, S. 1976. *J. Mol. Biol.* 106:297–313
33. Wu, F. Y.-H., Wu, C.-W. 1974. *Biochemistry* 13:2562–66
34. Wu, C.-W., Yarbrough, L. R., Wu, F. Y.-H., Hillel, Z. 1976. *Biochemistry* 15:2097–104
35. Moe, O. A., Hammes, G. G. 1974. *Biochemistry* 13:2547–52
36. Moe, O. A., Lerner, D. A., Hammes, G. G. 1974. *Biochemistry* 13:2552–57
37. Shepherd, G. B., Hammes, G. G. 1976. *Biochemistry* 15:311–17
38. Papadakis, N., Hammes, G. G. 1977. *Biochemistry* 16:1890–96
39. Dockter, M. E., Steinemann, A., Schatz, G. 1977. *J. Biol. Chem.* In press
40. Cantley, L. C., Hammes, G. G. 1975. *Biochemistry* 14:2976–81
41. Cantley, L. C., Hammes, G. G. 1976. *Biochemistry* 15:9–14
42. Tu, S.-C., Wu, C.-W., Hastings, J. W. 1977. *Biochemistry.* In press
43. Matsumoto, S., Hammes, G. G. 1975. *Biochemistry* 14:214–24
44. Becker, J. S., Oliver, J. M., Berlin, R. D. 1975. *Nature* 254:152–54
45. Haugland, R. P. 1975. *J. Supramol. Struct.* 3:338–47
46. Wright, K., Takahashi, M. 1977. *Biochemistry* 16:1548–54
47. Gennis, L. S., Cantor, C. R. 1976. *J. Biol. Chem.* 251:769–75
48. Gennis, L. S., Gennis, R. B., Cantor, C. R. 1972. *Biochemistry* 11:2517–24
49. Gennis, R. B., Cantor, C. R. 1972. *Biochemistry* 11:2509–17
50. Peters, R. 1971. *Biochim. Biophys. Acta* 233:465–68
51. Forster, T. 1966. In *Modern Quantum Chemistry,* ed. O. Sinanoglu, Section III B:93–137. New York: Academic. 999 pp.
52. Haugland, R. P., Yguerabide, J., Stryer, L. 1969. *Proc. Natl. Acad. Sci. USA* 63:23–30
53. Dale, R. E., Eisinger, J. 1974. *Biopolymers* 13:1573–605
54. Blumberg, W. E., Dale, R. E., Eisinger, J., Zukerman, D. M. 1974. *Biopolymers* 13:1607–20
55. Dale, R. E., Eisinger, J. 1975. See Ref. 6, pp. 115–284
56. Kawato, S., Kinosita, K., Ikegami, A. 1977. *Biochemistry* 16:2319–24
57. Jones, R. 1970. *Nanosecond fluorimetry.* PhD thesis. Stanford Univ. 300 pp.
58. Hillel, Z., Wu, C.-W. 1976. *Biochemistry* 15:2105–13
59. Gouterman, M. 1961. *J. Mol. Spectrosc.* 6:138–63
60. Leung, C. S., Meares, C. F. 1977. *Biochem. Biophys. Res. Commun.* 75:149–55
61. Velick, S. F. 1958. *J. Biol. Chem.* 233:1455–67
62. Velick, S. F. 1961. In *Light and Life,* ed. W. D. McElroy, B. Glass, pp. 108–43. Baltimore: John Hopkins. 924 pp.
63. Goldberg, M. E., York, S., Stryer, L. 1968. *Biochemistry* 7:3662–67
64. Weber, G., Teale, F. W. J. 1959. *Discuss. Faraday Soc.* 27:134–39
65. Secrist, J. A., Barrio, J. R., Leonard, N. J. 1972. *Science* 175:646–47
66. Leonard, N. J., Tolman, G. L. 1975. *Ann. NY Acad. Sci.* 255:43–58
67. Luisi, P. L., Baici, A., Bonner, F. J., Aboderin, A. A. 1975. *Biochemistry* 14:362–68
68. Scopes, D. I. C., Barrio, J. R., Leonard, J. H. 1977. *Science* 195:296–98
69. Ward, D. C., Reich, E., Stryer, L. 1969. *J. Biol. Chem.* 244:1228–37
70. Ward, D. C., Reich, E. 1972. *J. Biol. Chem.* 247:705–19

71. Leonard, J. J., Yonetani, T., Callis, J. B. 1974. *Biochemistry* 13:1460–64
72. Vanderkooi, J. M., Adar, F., Erecinska, M. 1976. *Eur. J. Biochem.* 64:381–87
73. Wittorf, J. H., Gubler, C. J. 1970. *Eur. J. Biochem.* 14:53–60
74. Sklar, L. A., Hudson, B. S., Simoni, R. D. 1975. *Proc. Natl. Acad. Sci. USA* 72:1649–53
75. Sklar, L. A., Hudson, B. S., Petersen, M., Diamond, J. 1977. *Biochemistry* 16:813–19
76. Sklar, L. A., Hudson, B. S., Simoni, R. D. 1977. *Biochemistry* 16:819–28
77. Yguerabide, J., Stryer, L. 1971. *Proc. Natl. Acad. Sci. USA* 68:1217–21
78. Waggoner, A. S., Stryer, L. 1970. *Proc. Natl. Acad. Sci. USA* 67:579–89
79. Wu, E.-S., Jacobson, K., Papahad-jopoulos, D. 1977. *Biochemistry* 16:3936–41
80. Schlessinger, J., Koppel, D. E., Axelrod, D., Jacobson, K., Webb, W. W., Elson, E. L. 1976. *Proc. Natl. Acad. Sci. USA* 73:2409–13
81. Epstein, M., Reuben, J., Levitzki, A. 1977. *Biochemistry* 16:2449–57
82. Horrocks, W. D. Jr., Schmidt, G. F., Sudnick, D. R., Kittrell, C., Bernheim, R. A. 1977. *J. Am. Chem. Soc.* 99:2378–80
83. Furie, B. C., Furie, B. 1975. *J. Biol. Chem.* 250:601–8
84. Brittain, H. G., Richardson, F. S., Martin, R. B. 1977. *J. Am. Chem. Soc.* 75:149–55
85. Ward, D. C., Horn, T., Reich, E. 1972. *J. Biol. Chem.* 247:4014–20
86. Schuldiner, S., Weil, R., Robertson, D. E., Kaback, H. R. 1977. *Proc. Natl. Acad. Sci. USA* 74:1851–54
87. Fortes, P. A. G. 1977. *Biochemistry* 16:531–40
88. Kanaoka, Y. 1977. *Angew. Chem. Int. Ed. Engl.* 16:137–47
89. Weber, G. 1952. *Biochem. J.* 51:155–67
90. Veatch, W. R., Blout, E. R. 1976. *Biochemistry* 15:3026–30
91. Ghosh, P. B., Whitehouse, M. W. 1968. *Biochem. J.* 108:155–56
92. Birkett, D. J., Price, N. C., Radda, G. K., Salmon, A. G. 1970. *FEBS Lett.* 6:346–48
93. Huang, K., Cantor, C. R. 1972. *J. Mol. Biol.* 67:265–75
94. Cherry, R. J., Cogoli, A., Oppliger, M., Schneider, G., Semenza, G. 1976. *Biochemistry* 15:3653–56
95. Peters, K., Richards, F. M. 1977. *Ann. Rev. Biochem.* 46:523–51
96. Udenfriend, S., Stein, S., Böhlen, P., Dairman, W., Leimgruber, W., Weigele, M. 1972. *Science* 178:871–72
97. Stryer, L. 1970. In *Ciba Foundation Symp. on Mol. Prop. of Drug Receptors,* ed. R. Porter, M. O'Connor, pp. 133–53. London: Churchill. 999 pp.
98. Rifkin, D. B., Compans, R. W., Reich, E. 1972. *J. Biol. Chem.* 247:6432–37
99. Haugland, R. P., Stryer, L. 1967. In *Conformation of Biopolymers,* ed. G. N. Ramachandran, 1:321–35. New York: Academic. 999 pp.
100. Hudson, E. N., Weber, G. 1973. *Biochemistry* 12:4154–61
101. Weltman, J. K., Szaro, R. P., Frackelton, A. R. Jr., Dowben, R. M., Bunting, J. R., Cathou, R. E. 1973. *J. Biol. Chem.* 248:3173–77
102. Wu, C.-W., Yarbrough, L. R., Wu, F. Y.-H. 1976. *Biochemistry* 15:2863–68
103. Leavis, P. C., Lehrer, S. S. 1974. *Biochemistry* 13:3042–48
104. Renthal, R., Steinemann, A., Stryer, L. 1973. *Exp. Eye Res.* 17:511–15
105. Weber, P., Hof, L. 1975. *Biochem. Biophys. Res. Commun.* 65:1298–302
106. Shaper, J., Stryer, L. 1977. *J. Supramol. Struct.* 6:291–99
107. Berman, H. A., Taylor, P. 1977. *Fed. Proc.* 36:2892 (Abstr.)
108. Reines, S. A., Cantor, C. R. 1974. *Nucleic Acids Res.* 1:767–86
109. Sundberg, M. W., Meares, C. F., Goodwin, D. A., Diamanti, C. I. 1974. *J. Med. Chem.* 17:1304–307
110. Connellan, J. M., Chung, S. I., Whetzel, N. K., Bradley, L. M., Folk, J. E. 1971. *J. Biol. Chem.* 246:1093–98
111. Dutton, A., Singer, S. J. 1975. *Proc. Natl. Acad. Sci. USA* 72:2568–71
112. Brew, K., Vanaman, T. C., Hill, R. L. 1968. *Proc. Natl. Acad. Sci. USA* 58:491–97
113. Weppner, W. A., Neuhaus, F. C. 1977. *J. Biol. Chem.* 252:2296–2303
114. Terroma, E. S., Sklar, L. A., Simoni, R. D., Hudson, B. S. 1977. *Biochemistry* 16:829–35
115. Chen, R. F., Edelhoch, H., eds. 1975. See Ref. 6
116. Yeager, M. J. 1975. *Brookhaven Symp. Biol.* 27:3–36
117. Sardet, C., Tardieu, A., Luzzati, V. 1976. *J. Mol. Biol.* 105:383–407
118. Fung, B. 1977. *Molecular structure of rhodopsin: proteolytic cleavage and enzymatic iodination in disc membranes and reconstituted membranes.* PhD thesis. Univ. of Calif., Berkeley. 299 pp.
119. Montal, M., Darszon, A., Trissl, H. W. 1977. *Nature* 267:221–25

120. Pober, J., Stryer, L. 1975. *J. Mol. Biol.* 95:477–81
121. Boos, W., Gordon, A. S. 1971. *J. Biol. Chem.* 246:621–28
122. Matthews, B. W., Weaver, L. H. 1974. *Biochemistry* 13:1719–25
123. Henderson, R., Capaldi, R. A., Leigh, J. S. 1977. *J. Mol. Biol.* 112:631–48
124. Reed, L. J. 1974. *Accounts Chem. Res.* 7:40–46
125. Engelman, D. M., Moore, P. B., Schoenborn, B. P. 1975. *Proc. Natl. Acad. Sci. USA* 72:3888–92
126. Lake, J. A. 1976. *J. Mol. Biol.* 105:131–60
127. Stoffler, G., Wittmann, H. G. 1977. In *Protein Synth.* ed. H. Weissback, S. Pestka, pp. 9–99. New York: Academic. 721 pp.
128. Lutter, L. C., Bode, U., Kurland, C. G., Stoffler, G. 1974. *Mol. Gen. Genet.* 129:167–76
129. Suddath, F. L., Quigley, G. J., McPherson, A., Sneden, D., Kim, J. J., Kim, S. H., Rich, A. 1974. *Science* 248: 20–24
130. Robertus, J. D., Ladner, J. E., Finch, J. T., Rhodes, D., Brown, R. S., Clark, B. F. C., Klug, A. 1974. *Nature* 250: 546–51
131. Fernandez, S. M., Berlin, R. D. 1976. *Nature* 264:411–15
132. Veatch, W. R., Mathies, R., Eisenberg, M., Stryer, L. 1975. *J. Mol. Biol.* 99:75–92
133. Tosteson, D. C., Andreoli, T. E., Tieffenberg, M., Cook, P. 1968. *J. Gen. Physiol.* 51:373–84 Suppl.
134. Bamberg, E., Lauger, P. 1973. *J. Membr. Biol.* 11:177–94
135. Zingsheim, H. P., Neher, E. 1974. *Biophys. Chem.* 2:197–207
136. Elkana, Y., Feitelson, J., Katchalski, E. 1968. *J. Chem. Phys.* 48:2399–404
137. Steinberg, I. Z., Katchalski, E. 1968. *J. Chem. Phys.* 48:2404–10
138. Dale, R. E., Eisinger, J. 1976. *Proc. Natl. Acad. Sci. USA* 73:271–73
139. Iwanij, V. 1978. *Eur. J. Biochem.* In press
140. Glazer, A. N. 1978. *Mol. Cell. Biochem.* In press
141. Kirschfeld, K., Franceschini, N. 1977. *Nature* 269:386–90
142. Pober, J. S., Iwanij, V., Reich, E. R., Stryer, L. 1978. *Biochemistry.* In press
143. Dreyfuss, G., Schwartz, K., Blout, E. R., Barrio, J. R., Liu, F. T., Leonard, N. J. 1978. *Proc. Natl. Acad. Sci. USA.* In press

Ann. Rev. Biochem. 1978. 47:847–80
Copyright © 1978 by Annual Reviews Inc. All rights reserved

GENETIC RECOMBINATION: STRAND TRANSFER AND MISMATCH REPAIR

❖991

Charles M. Radding

Departments of Medicine and Molecular Biophysics & Biochemistry, Yale University School of Medicine, New Haven, Connecticut 06510

CONTENTS

847

PERSPECTIVES AND SUMMARY

The Biological Picture

TYPES OF RECOMBINATION   Genetic recombination includes a variety of processes that produce new linkage relationships of genes or parts of genes (1). We can subdivide the field into 1. *general genetic recombination,* which takes place between homologous chromosomes, more or less anywhere along their length, and 2. recombination that does not require extensive homology. The latter category comprises (*a*)*site-specific recombination,* which depends upon the existence of specific sites in one or both molecules and which includes interactions of viral genomes and insertion sequences with chromosomes of prokaryotes and eukaryotes (2–7), and (*b*) less well defined instances of recombination that appear to require neither extensive homology nor special sites (8–10). This review deals only with general recombination.

EUKARYOTES   We get the clearest picture of recombination from observations on meiosis in certain fungi. The reason is simply that we can recover all of the products of meiosis in four or eight cells. These cells are the direct descendents of a single zygote in which homologous chromosomes paired once and exchanged genetic information. Although its details are complex, the pattern of recombination in the fungi can be represented by the simple diagram in Figure 1 (11–13). Exchanges between nearby genetic markers are often nonreciprocal: one copy of a marker is lost and is replaced by the corresponding marker of the other parent, a replacement that is called *gene conversion.* This kind of recombination is also included under the rubric *intragenic recombination.* When an exchange involves close markers, distant flanking markers either do not recombine, or they recombine reciprocally, a frequent event that falls in the category of *reciprocal crossing-over* or *intergenic recombination.* According to one view of this pattern, the region of nonreciprocity is the actual site of chemical interaction between two recombining molecules of DNA: The chemical interaction involves some intermediate, the resolution of which determines whether the flanking arms of DNA emerge in their original relationships, or in new recombinant relationships (Figure 1). Observations made on recombination in *Drosophila melanogaster* suggest that this pattern is not limited to the fungi but may be a general one in eukaryotes (14–16).

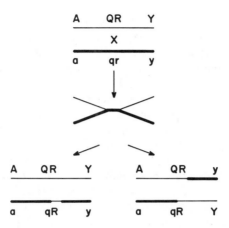

Figure 1 Characteristic patterns of meiotic gene conversion and segregation in fungi: The recombination of nearby markers is often nonreciprocal. When close markers exchange, distant flanking markers either do not recombine (*bottom left*) or they recombine reciprocally (*bottom right*). One way to interpret this pattern is indicated by the middle diagram, namely that the region of nonreciprocal exchange is the actual site of chemical interaction.

PROKARYOTES In prokaryotes the recombination of distant markers sometimes appears to be nonreciprocal (17–20). However, in prokaryotes we cannot study the equivalent of meiotic recombination in which homologous pairing happens only once. Instead, we observe populations of progeny or molecules that usually have replicated and recombined many times. In phage crosses we also see selective effects of the maturation and packaging of DNA (21, 22). These circumstances obscure the outcome of individual exchanges. On the other hand, in a number of instances reciprocal recombination has been observed in *Escherichia coli* (17, 23–28).

The Molecular Basis of General Recombination

Breakage and reunion of DNA play a major role in recombination both in eukaryotes and prokaryotes (23, 29–32). Genetic and biophysical experiments in prokaryotes have shown that reunion is accomplished via the formation of a *heteroduplex joint,* a molecular splice in which one strand comes from each parent [See examples of heteroduplex joints in Figure 2; see also (18, 23, 33, 34).] When mutant sites are included in the heteroduplex joint, one or more bases are mismatched. As long as the mismatch persists, the DNA will be heterozygous. In the fungi, in the first mitotic division after meiosis a wild type and a mutant sometimes segregate, apparently from a single copy of DNA (11, 12). Roughly half of such *postmeiotic segregations* are associated with recombination of flanking markers. These observations on postmeiotic segregation suggest that the heteroduplex joint serves the same purpose in eukaryotes as in prokaryotes.

Evidence also supports the view (35–39) that the formation and repair of heteroduplex joints are responsible for nonreciprocal recombination of nearby markers (cf Figure 1). When heteroduplex DNA made in vitro is introduced into prokaryotes, some pairs of mismatched bases are corrected by a process called *mismatch repair,* which is conceived as the excision of mismatched bases from one strand and repair by copying of the intact strand derived from the other parent. Observations on some fungi are consistent with the idea that mismatch repair is one major source of gene conversion. The genetic effects of mismatch repair arise from the resynthesis of DNA, which changes the number of strands in which a marker is represented. It follows that any resynthesis of DNA associated with heteroduplex joints can have a similar effect, for example, synthesis involved in the formation of a joint (see Figure 2 III).

Thus, according to our current concepts, the prominent features of recombination in prokaryotes and eukaryotes can be related to the formation and processing of heteroduplex joints. Observations on the relevant chemistry can be divided into those on *strand transfer* and those on *mismatch repair.*

INITIATION OF STRAND TRANSFER

The importance of the heteroduplex joint tells us that recombination involves the separation of strands and their pairing in new hybrid combinations. Recent research has brought us fresh insight into the biochemistry of strand separation and pairing, as well as progress in understanding how coincident breaks are made in homologous regions of two molecules.

Separation of Strands

HELIX DESTABILIZING PROTEINS (HDPs)   In *E. coli* infected by phage T4, Alberts & Frey (40) discovered the first of a class of widespread proteins (41–48) which, by virtue of their preferential binding to single-stranded DNA, are capable in principle of destabilizing double-stranded DNA and thus promoting its denaturation (49).[1] The prototype of this class, the phage

[1]These proteins have been termed variously, "DNA binding," "unwinding," and "melting" proteins. Since there is reason to reject each of these terms, I follow the recent suggestion of Alberts & Sternglanz (49a) by using the term "helix destabilizing protein" (HDP).

T4 gene 32 protein, has been implicated in replication, recombination, and repair (50) of T4 by studies of gene 32 mutants (51). At low ionic strength, the T4 gene 32 protein promotes the denaturation of poly d(A-T) but not of natural DNA (40, 52), although some observations suggest that a cleaved form of the protein may lower the T_m of T4 DNA (53–55). The HDPs of *E. coli* and calf thymus promote denaturation of natural DNA at concentrations of salt that are well below those likely to prevail in vivo (45, 56).

Apart from promoting denaturation directly, which is probably not their physiological role, HDPs may contribute to strand separation in several other ways; the first is by protecting single-stranded DNA from endonucleolytic and exonucleolytic degradation, and the second is by promoting strand displacement in replication (see section on strand displacement, below). The HDP of *E. coli* protected single-stranded DNA from degradation by a number of DNases including the *recBC* DNase of *E. coli,* pancreatic DNase, snake venom phosphodiesterase, the endonuclease of *Neurospora crassa,* endonuclease S_1 of *Aspergillus oryzae,* and the exonuclease associated with phage T4 DNA polymerase (41, 57). The *E. coli* HDP did not protect single-stranded DNA against digestion by micrococcal nuclease, *E. coli* exonuclease I, or the exonucleolytic activity of *E. coli* DNA polymerase II (41). The HDP of *Ustilago maydis* protected single-stranded DNA from digestion by *U. maydis* DNase I (48). Curiously the *U. maydis* HDP was even more efficient in protecting double-stranded DNA from digestion by *E. coli* exonuclease III. In vitro the gene 32 protein protected poly dT from hydrolysis by the exonucleolytic activity of T4 polymerase (58), and protected single-stranded DNA from degradation by pancreatic DNase (59). In vivo, 32 protein prevented degradation of single-stranded regions of phage T4 DNA (50, 59, 60), which may be a sufficient explanation for the failure of gene 32 mutants to form joint molecules (61). Breschkin & Mosig (62) observed phenotypic differences among different gene 32 mutants that led them to propose that 32 protein is part of a multienzyme complex. Parenthetically, the importance of the HDPs for the integrity of single-stranded DNA and the apparent role of single-stranded DNA in regulating the synthesis of 32 protein (63, 64) support the inference that little free single-stranded DNA exists in cells.

DNA-DEPENDENT ATPases   A second mechanism for separating strands of DNA uses the energy of ATP. Certain ATP-dependent DNases of *Haemophilus influenzae* and *E. coli* provided the first example (65–68). These enzymes act exonucleolytically on double-stranded and single-stranded DNA. The enzyme from *E. coli* also cleaves single-stranded DNA endonucleolytically. Acting on double-stranded DNA, the enzymes from *H. influenzae* and *E. coli* initially produce long single-stranded tails and single-stranded fragments far in excess of acid-soluble nucleotides, which implies that strand separation involves melting of long stretches of DNA rather than the uncovering of one strand by exonucleolytic digestion of its complement. MacKay & Linn (57) found that the ratio of melting activity to DNase activity of the *E. coli* enzyme, the *recBC* DNase, could be increased by the presence of the *E. coli* HDP, which protects the single strands from further degradation. Wilcox & Smith (66) clearly showed that the melting activity of the enzyme from *H. influenzae* is catalytic; a single molecule of enzyme separates many thousands of base pairs. The putative role of these enzymes in recombination is discussed below (See section on selected topics).

More recently, Hoffman-Berling and his colleagues discovered in *E. coli* a DNA-dependent ATPase that binds to single strands and lacks any nucleolytic activity, but is capable of denaturing double-stranded DNA that is adjacent to a single-stranded region (69–71). One substrate, for example, was phage fd DNA with about 1000–2000 nucleotides in a double-stranded region and the rest, 4000–5000 nucleotides, as single-stranded DNA. For maximal unwinding, some 80 molecules of enzyme were required per molecule of partially double-stranded fd DNA. The enzyme acted processively: after it had bound to substrate, the addition of excess single-stranded DNA did not inhibit unwinding, which must involve multiple steps. The enzyme did not bind to double-stranded DNA nor hydrolyze ATP in its presence. The enzyme unwound neither short double-stranded fragments with two unpaired nucleotides at each end, nor double-stranded fd DNA with nicks or short gaps. While the ATP-dependent DNase of *H. influenzae* melted 150 base pairs per second (66), the enzyme discovered by Abdel-Monem et al (70) melted only about 400 per minute. The enzyme, a fibrous protein of molecular weight 180,000, is the largest soluble peptide in *E. coli*. Studies of mutants failed to relate the enzyme to the activity of *recA, recB, recC, lex, uvrA, rep, dnaB,* or *dnaE*.

Kornberg and his colleagues (72) recently discovered that the *E. coli rep* protein, which is required for replication of phages ϕX174 and P2 (73), hydrolyzes ATP in the presence of single-stranded DNA. Moreover, *rep* protein will separate the strands of RFI in a mixture that also contains ATP, Mg^{2+}, *E. coli* HDP, and ϕX174 gene A protein, the latter responsible for introducing a nick in the RFI (See also section on strand displacement, below.) Hoffman-Berling and collaborators (personal communication) have purified from *E. coli* an ATPase of 75,000 mol wt that will unwind DNA and which may be related to *rep* and the ATPase isolated by Richet & Kohiyama (74). Hoffman-Berling et al (personal communication) also found that the phage T4 ATPase described by Ebisuzaki et al (75, 76) will separate strands of DNA.

The mechanism by which these enzymes use the energy of ATP to unwind DNA is mysterious and fascinating. Abdel-Monem et al proposed that the ATPase they discovered (180,000 mol wt) binds to a single-stranded region and uses the energy of ATP to move along one strand into an adjacent region of double-stranded DNA, unwinding the latter by a wedge-like action (71). Consistent with that view, the enzyme behaves as if it can move only in one direction, 5' to 3', along the single strand to which it bound initially [cited in (71)]. The ATP-dependent DNase of *E. coli* needs ATP not only for melting but even to cleave the first few nucleotides (77). The amount of ATP used in vitro is large. Wilcox & Smith (66) estimated that 17 moles of ATP are hydrolyzed per mole of base pair melted by the DNase of *H. influenzae*. The large DNA-dependent ATPase of Abdel-Monem et al (69, 70) may use even an order of magnitude more ATP per base pair melted.

Gellert et al (78, 79) recently discovered another kind of ATPase, which unwinds DNA and is required for site-specific recombination in vitro. (See section on strand uptake and the role of superhelical DNA in recombination below). An observation of Das et al may relate another kind of ATPase activity to recombination (80). Lowery-Goldhammer & Richardson (81) discovered that the transcription factor rho is a poly(C)-dependent ATPase. Das et al (80) isolated a thermosensitive rho mutant and showed that rho protein purified from the mutant has thermosensitive ATPase activity. A cross of phage λ in the mutant strain at high temperature showed a seventy fold reduction in recombination.

STRAND DISPLACEMENT ASSOCIATED WITH NEW DNA SYNTHESIS
Strand displacement by new synthesis is an integral feature of replication forks, in which case the displaced strand itself soon becomes the template for new synthesis (82, 83). When new synthesis starts at a nick in vitro, the displaced strand has a free end and may remain unreplicated under some circumstances (Figure 2a). The first such example was provided by DNA polymerase I of *E. coli*, which, starting at a nick, catalyzes the limited displacement of the strand ahead of the enzyme (84–86). Fujimura & Roop (87) showed that phage T5 DNA polymerase promotes strand displacement and reviewed indirect evidence that other polymerases do the same. Kolodner & Richardson (87a) found that the phage T7 gene 4 protein, which enables T7 DNA polymerase to start at a nick, is a DNA-dependent nucleoside 5'-triphosphatase. They postulated that gene 4 protein promotes strand displacement by unwinding DNA.

HDPs from different sources stimulate DNA synthesis in vitro partly by specific interactions with polymerases (43, 56, 88–92). Nossal (93) found that the T4 gene 32 protein enabled the T4 polymerase to start new DNA

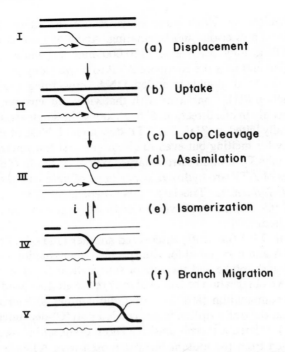

Figure 2 Mechanisms of strand transfer, depicted in a hypothetical scheme that is an elaboration of part of the model of Meselson & Radding (104). With the exception of isomerization, all of these mechanisms have been observed in vitro. (*a*) Displacement. New synthesis (*wavy lines*) displaces a strand with a free end. (*b*) Uptake. The free end of the displaced strand is taken up by a homologous double-stranded molecule in a reaction that is driven by the energy of superhelix formation. (*c*) Loop cleavage. Strand uptake produces a D loop in which a strand of the recipient molecule becomes a ready target for endonucleases while the two helices remain connected by a single phosphodiester bond (162; cf Figure 3c), which presents a much smaller target. (*d*) Assimilation. After cleavage of the loop, the strand displaced by new synthesis can be assimilated by exonucleolytic digestion at the 5' terminus symbolized by the open circle. Heteroduplex DNA forms in only one of the two molecules. (*e*) Isomerization. A one-strand crossover (III) becomes a two-strand crossover (IV) by isomerization (as in Figure 4) or by branch migration (as in Figure 3c, d). Isomerization puts the arms flanking the strand crossover in a recombinant configuration; branch migration leaves the flanking arms in the parental configuration (Figure 3d). (*f*) Branch migration. The migration of a two-strand crossover causes symmetrical exchange of strands and symmetrical formation of heteroduplex DNA.

For clarity, nicks have been left open at their original sites, but these might be closed by ligase at any time. Cleavage of the connecting strand or strands terminates an exchange, and mismatched bases in the heteroduplex regions are subject to excision and repair.

synthesis at a nick, presumably because 32 protein promoted strand displacement. Alberts et al (83) observed more extensive displacement of single strands by a system based on six purified proteins of T4 replication. A striking example of strand displacement as a result of new DNA synthesis is that carried out jointly by polymerase III, *rep* protein, and HDP of *E. coli,* plus the gene A protein of ϕX174 (72). Eisenberg, Griffith & Kornberg (94) showed that the HDP (which they called DBP) promotes the reaction in part by coating the displaced single strand. From studies of the replication of phage f1 in vivo Mazur & Model (95) proposed a similar role for the f1 gene 5 protein.

Pairing of Strands

HELIX DESTABILIZING PROTEINS (HDPs)   As discussed above, the HDPs acting alone promote strand separation only at low ionic strength. At higher concentrations of counterion the T4 gene 32 protein accelerates the renaturation of DNA, a kinetic effect that Alberts & Frey attributed to the disruption of intrastrand base pairs, and extension of the otherwise folded single strands (40). Wackernagel & Radding (96) showed that gene 32 protein acting in vitro could make biologically active joint molecules of λ DNA from half molecules with single-stranded ends. The HDPs of *Lilium* and *Ustilago* also promote renaturation in vitro (46, 48). Recent observations by Christiansen & Baldwin (97) on the helix destabilizing protein of uninfected *E. coli* suggest that yet another member of this class is capable of playing a role in strand pairing. Whereas previous studies had revealed no activity of the *E. coli* HDP in annealing, Christiansen & Baldwin found a 7000-fold acceleration of reannealing at 37°C in the presence of spermine or spermidine at concentrations that may reasonably be considered physiologic (97). The promotion of both enzymic strand displacement (72, 83, 94; see also above) and pairing of homologous single strands is consistent with a role of HDPs in creating the first strand crossover in recombination (Figure 2a, b).

STRAND UPTAKE AND THE ROLE OF SUPERHELICAL DNA IN RECOMBINATION   According to a popular idea, recombination starts with the invasion of a double-stranded molecule by the free end of a single strand, leading to the formation of a triple-stranded joint in which one strand of the recipient molecule is subsequently cleaved (98–105, also Figure 2). This idea has been applied to the initiation of nonreciprocal recombination, as in bacterial transformation and conjugation (100, 101, 103), as well as to the initiation of reciprocal crossing-over in eukaryotes (98, 99, 102, 104). A possible physicochemical basis for the formation of a triple-stranded joint was discovered by Holloman et al (106) who reported that superhelical DNA of ϕX174 (RFI) spontaneously takes up homologous single-stranded fragments in vitro. Beattie et al (107) and Wiegand et al (108) showed that the product of the reaction is a molecule that contains a displacement loop (D loop), in which the donor strand is paired to the homologous strand of the recipient, and the displaced strand of the recipient forms a single-stranded loop. The length of the heteroduplex region was approximately that predicted from the superhelix density of the recipient DNA. Beattie et al observed little difference in the rate of uptake of specific single-stranded fragments that varied in length and nucleotide sequence. Strand uptake by superhelical DNA in vitro is a slow reaction that is probably limited in rate by a step that is required for homologous pairing, namely the unstacking of a small number of base pairs (107). In vivo, this step might be promoted by a protein.

The intercalating agent ethidium bromide inhibits strand uptake by superhelical DNA in vitro (106). Seto & Tomasz recently reported that intercalating agents inhibit transformation of *Diplococcus pneumoniae* (109). The temperature dependence of a step that occurs just before integration of transforming DNA in *D. pneumoniae* (110, 111) is strikingly similar to that of strand uptake in vitro (107).

When Holloman & Radding (112) transfected spheroplasts with mutant double-stranded circular DNA plus fragments of wild-type single-stranded DNA from ϕX174, they observed efficient, nonselective rescue of the wild-type marker. This recombination depended upon the use of spheroplasts from *recA$^+$* cells as well as superhelical DNA. By contrast, when joint molecules of mutant RFI and wild-type fragments were first made by strand uptake in vitro, the production of wild-type progeny did not require the *recA$^+$* function. These experiments implicate *recA$^+$* in the formation of joint molecules, provided that a single pathway produces recombinants both from joint molecules and from a mixture of RFI plus fragments. The experiment also shows that superhelicity can play a role in at least one kind of general recombination in vivo. Earlier evidence of the biological significance of superhelicity was provided by Puga & Tessman (113), who observed that RFI DNA was transcribed in vivo 4 to 5 times more efficiently than RFII.

Mizuuchi & Nash (114) and Gellert et al (78) found that site-specific recombination of phage λ in vitro requires superhelical DNA. This observation led them to discover an enzyme they called DNA gyrase. In a reaction that requires ATP, this enzyme can reduce the number of times that the two strands of the helix are topologically interwound, thereby making negatively superhelical DNA from relaxed closed circular DNA. Novobiocin and coumermycin, which inhibit replication in vivo and in vitro, inhibit the gyrase. Gellert et al partially purified gyrase from a strain of *E. coli* that was resistant to coumermycin, and showed that the enzyme itself was resistant to inhibition by coumermycin (79). The need for gyrase in vitro for site-specific recombination (78) and for replication of ϕX174 RF (115) can be bypassed by the use of superhelical DNA as substrate. In the case of replication in vitro, Marians et al showed further that superhelicity promoted the binding of a specific essential protein, the gene A protein (115).

In relation to linear or nicked circular DNA in solution, most closed circular molecules extracted from cells have a deficit in the number of right-hand turns of the helix, a deficit that is remedied by the addition of superhelical turns. A manifestation of this deficit is free energy that promotes the binding of any ligand that unwinds the helix and consequently removes the superhelical turns (116, 117). The list of such ligands, which is long, includes proteins and nucleic acids (107, 116–118; see also above). Because of its augmented affinity for a variety of ligands, it is unlikely that superhelical DNA in vivo exists in the free state that we study in vitro. Rather, the energy that is required to form the superhelix must determine a hierarchy of binding affinities which differs from that of relaxed DNA (107). By inference, I conclude that enhanced affinities both for single strands and certain proteins may be part of the mechanism by which superhelicity promotes recombination.

Superhelicity is not limited to small circular molecules of DNA. Chromosomal DNA from a variety of sources, both prokaryotic and eukaryotic, behaves as if it were organized in superhelical loops (119–126), which supports the hypothesis that superhelicity may be of general significance for recombination (106). The constraints that make a linear molecule of DNA behave like a series of closed loops may be related to intracellular packing and folding (122, 127). Since special care must be taken to isolate the folded chromosome with which the properties of superhelicity are associated (119, 120) the failure to isolate DNA in a superhelical form is not evidence that it lacks superhelicity in vivo. Specifically, the reisolation of linear phage DNA from infected cells does not justify the conclusion that this DNA was not superhelical in vivo.

Recognition vs Breakage

SPECIAL SITES   In eukaryotes there appear to be special sites at which recombination begins (11, 12, 128). These special sites have been inferred from the polarization of gene conversion (129), which is most readily explained by assuming that the formation of heteroduplex DNA begins at a site on one side of a gene and extends into the gene for variable distances (11, 12). In crosses between different strains of *Neurospora,* Angel, Austin & Catcheside identified a site, *cog,* near the *his* 3 locus, that increases recombination locally (130, 131). Certain observations on *cog* can be interpreted to mean that chemical events that precede transfer of a strand can take place at a distance from the site where strands are ultimately exchanged (11, 12, 132; see also below).

Stahl and his collaborators (133) and Henderson & Weil (134) demonstrated the existence of sites at which recombination in coliphage λ is enhanced. While these sites, called *chi* sites, are not normally found in λ, they can be produced by mutation, and they may be found as often as once every 5–15 genes when pieces of *E. coli* DNA are inserted in λ DNA (Malone, Chattoraj, Faulds, Stahl, and Stahl, personal communication). The effect of chi sites has been examined only in variants of λ that lack their own system of recombination (Red) and have another mutation, *gam⁻*, a combination that confers a selective advantage upon recombinant molecules for reasons that seem to be related to maturation of DNA (135–137). Like the *cog* site (see above), the *chi* sites have a property that is interesting from a mechanistic viewpoint. In addition to stimulating recombination in their own vicinity, *chi* sites stimulate recombination at a distance, even when they are located in a large region of nonhomology. This observation suggests that chemical events that lead to recombination can start at a distance from the site at which homologous pairing begins. For example, such events might include binding of proteins, nick translation, strand displacement (82), or exonucleolytic digestion.

Sobell (138, 139) and Wagner & Radman (140) postulated that inverted repetitions or palindromes constitute special sites of cleavage that lead to the initiation of strand exchange.

PAIR AND BREAK vs BREAK AND PAIR   While reunion in general genetic recombination is accomplished ultimately by the pairing of homologous nucleotide sequences, we do not know whether recognition precedes or follows breakage. McGavin reported that a four-stranded helix can be built that entails hydrogen bonding of like base pairs A:T with A:T, and G:C with G:C, and about 10 bases per turn of the four-stranded helix (141). Unless long molecules of DNA are free to rotate about each other, such pairing of two double helices would be limited to short regions, which, however, might be adequate to establish mutual recognition of homologous double-stranded molecules.

Similarly, because DNA is a plectonemic coil in which the two strands are wrapped around each other about once every ten base pairs, most workers have assumed that at least one strand must acquire a free end if it is to form a duplex with another strand. Cross & Lieb (142) and Moore (143) proposed, however, that single strands from different helices might pair prior to any breakage. Some recent striking experiments show that a free end is not essential to form a duplex from single strands. Kirkegaard and Wang (personal communication) and Champoux (144) discovered that double-stranded closed circular DNA was formed when circular molecules of complementary single-stranded DNA were incubated with ω protein of *E. coli* (145) or the comparable enzyme from rat liver (146). Kirkegaard & Wang used DNA from φX174; Champoux used DNA from SV40. Both enzymes acted much as they do upon negatively superhelical DNA. In this case they produced short-lived interruptions that permitted circular single strands to wind about each other about 500 times. The enthalpy of helix formation presumably drives the reaction.

We know, however, that interruptions in duplex DNA such as ends, nicks, and gaps stimulate recombination (51, 147–151), and that single-stranded DNA appears during transformation (152–154) and conjugation (155). Such observations have led to the view that breakage by any means can generate single strands which pair with homologous strands in other molecules (99). According to one version of that idea, initiation involves the interaction of the free end of a single strand with a double-stranded homolog, followed by enzymic cleavage of a specific strand of the recipient (98–105; Figure 2c). The breakage that produces a donor strand may be random, but the cutting of a recipient molecule is determined by the structure of the first intermediate formed by homologous pairing. A possible stereochemical basis for such an induced cutting was envisaged in the formation of a single-stranded loop when superhelical DNA takes up a homologous strand (106; Figure 2c). Wiegand et al (108) showed that the S_1 nuclease of *A. oryzae* and the *RecBC* DNase of *E. coli* readily cleaved the displaced single-stranded loop of the recipient molecule. The *RecBC* DNase cleaved the displaced loop as rapidly as it cleaved a circular single strand 18 times as long, which suggests that the enzyme specifically recognized some feature of the D loop in addition to single-strandedness. After the loop was digested by S_1 nuclease, polynucleotide ligase covalently linked one or the other end of a sizable fraction of the donor strands to the recipient molecule.

Howard-Flanders and his colleagues studied the cutting of circular DNA that was induced by the presence of repairable lesions in homologous molecules, which they called "cutting-in-trans." Ross & Howard-Flanders (156, 157) showed that mixed infection of *E. coli* with damaged and undamaged phage λ caused cleavage of the undamaged DNA in a reaction that required homology as well as the function of $uvrA^+$ and $recA^+$. The observations showed that cutting was not the consequence of a general induction of enzymes involved in repair. Cassuto, Mursalim & Howard-Flanders (158) demonstrated cutting-in-trans of φX174 RFI DNA in extracts. The reaction in vitro required homology of the damaged and undamaged DNA, as well as extract from an induced $uvrB^+$ strain that carried λp $recA^+$ as a prophage (159; see section on RecA in E. coli, below). Following the protocol of Cassuto et al, Holloman, Cunningham & Beattie (personal communication) found that homologous single-stranded fragments also stimulated the cutting of ϕX174 or G4 RFI in crude extracts. Cutting-in-trans results from the interaction of homologous molecules, one that very likely is related to recombination. It is possible that the interaction is like strand uptake by superhelical DNA and that cutting-in-trans is the cleavage of a D loop (see above). These observations (156–159) may open the way for a direct enzymological approach to recombination.

PROPAGATION OF STRAND TRANSFER

Once a cross connection has been established between homologous molecules, two kinds of concerted mechanism may increase the length of heteroduplex regions: either (a) asymmetrically, by transferring a strand from one molecule to another, or (b) symmetrically, by exchanging strands between molecules. The concept of asymmetric strand transfer is based on indirect genetic and biochemical data. The concept of symmetric strand exchange has been established more directly by observations made in vitro. Both concepts help to account for nonreciprocal intragenic recombination and reciprocal intergenic recombination (cf Figures 1 and 2).

Assimilation and Asymmetric Strand Transfer

To study the role of λ exonuclease in recombination, Cassuto et al (160, 161) constructed several substrates consisting of double-stranded DNA with single-stranded branches. By digesting the strand that had a 5' terminus, λ exonuclease caused the complete assimilation into the helix of the single-stranded branch that had a 3' terminus (160, 161). This reaction, which Cassuto & Radding dubbed strand-assimilation, led to the concept of a concerted transfer of a strand by displacement plus assimilation, displacement promoted by new synthesis, and assimilation promoted by exonucleolytic degradation (104, 162, 163; Figure 2 III). Such a concerted reaction might be catalyzed by a single enzyme or complex. As a model we can cite DNA polymerase I of E. coli, which promotes all of the requisite functions: synthesis, strand displacement, and 5' → 3' exonucleolytic digestion (82). We return later to this idea, which has been useful in rationalizing recombination as it is seen in eukaryotes (104). Details of strand assimilation have been reviewed before (163). In principle strand assimilation may be promoted by exonucleases that act either on double-stranded or single-stranded DNA (163, 164).

The recent discovery of DNA gyrase (78, 79) suggests another mechanism that might promote incorporation of a single strand into a helix. Acting on topologically constrained DNA, gyrase should be able to promote extensive uptake of a single strand by increasing the free energy of the superhelix (107; see also section on strand uptake and the role of superhelical DNA in recombination, above).

Branch Migration and Symmetric Exchange of Strands

Branch migration, the exchange of like strands at a branch in DNA, was first recognized by Lee, Davis & Davidson (165). In micrographs of renatured DNA from a permuted and terminally redundant genome, they saw single-stranded branches of the type illustrated in Figure 3a. The branch point in one strand adopted the various positions permitted by homologous base pairing. Broker & Lehman used the concept of branch migration to interpret their micrographs of intermediates made by recombination of

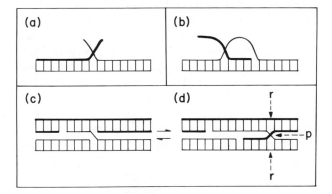

Figure 3   Branch migration. (*a*) A single-stranded branch. The branch point can occur anywhere along the extent of the branch (165). (*b*) A displacement loop (D loop). Experiments have shown that the loop can wind into the helix and displace the strand represented by the heavily shaded line. (*c*) A one-strand crossover in a diagram that illustrates the connection of two helices by a single phosphodiester bond. Since the two helices are juxtaposed and there is no net change in the number of paired bases, the one-strand crossover (*c*) may become a two-strand crossover (*d*) by branch migration, but this reaction has not been demonstrated experimentally. (*d*) A two-strand crossover, whose migration causes symmetrical exchange of strands and symmetrical formation of heteroduplex DNA. The labels *r* and *p* mark alternate points of cleavage proposed by Holliday (168) to account for genetically different progeny. Cutting two strands at *r* would produce molecules with a recombinant configuration of the arms flanking the two-strand connection, and cutting at *p* would produce the parental configuration (see section on ideas on reciprocal crossing-over).

phage T4 under conditions of limited DNA synthesis (166, 167). They used the terms *single-strand branch migration* to describe the exchange of strands in structures like that shown in Figure 3a, and *double-strand branch migration* to describe the exchange of strands in structures like that shown in Figure 3d. The latter is the recombination intermediate postulated by Holliday (37, 168).

Recent research has demonstrated that the two-strand crossover postulated by Holliday is an intermediate in prokaryotic recombination. Doniger, Warner & Tessman (169) proposed that two circular molecules of DNA linked by a two-strand crossover might be an intermediate in recombination of phages S13 and ϕX174, following which Thompson et al (170) demonstrated the existence of the predicted figure-8 structures. By examination of DNA isolated from cells mixedly infected by a deletion mutant and an *amber* mutant of ϕX174, Benbow et al (105) showed that many figure-8's were recombinant, and they observed about a tenfold reduction in the frequency of figure-8 molecules in $recA^-$ strains. They also found figure-8 molecules when they prepared RF DNA in $recA^-$ cells and incubated it in an extract of $recA^+$ cells. No figure-8 molecules were isolated when the extract was made from $recA^-$ cells (105). By using conditions that cause partial denaturation, Valenzuela & Inman (171) visualized junctions in λ DNA with the fine structure expected for the two-strand crossover and showed that molecules were joined at homologous sites. While some of the structures they observed may have arisen as a result of replication, others seemed likely to be the products of recombination (171). Potter & Dressler (172) produced striking micrographs of two-strand crossovers between molecules of colicin plasmid DNA. By cutting each monomeric circle in the figure-8 with a restriction endonuclease they showed that the crossover connected homologous sites. In $recA^-$ strains there were at least fiftyfold fewer figure 8's (172). From a strain of *Bacillus subtilis,* Köhnlein & Hutchinson isolated a four-stranded form of DNA, which they suggested might consist of two helices held together by a two-strand crossover (173).

The two-strand crossover is particularly interesting and important because it can migrate, thereby making heteroduplex DNA symmetrically in the joined molecules. Because branch migration entails no net change in the number of paired bases, investigators have supposed that it requires little energy and is driven by random diffusion (165, 174). According to that view, the kinetics should resemble those of a random walk in one dimension. To study double-strand branch migration, Thompson, Camien & Warner (175) used a restriction endonuclease as described above to convert figure-8 molecules of phage G4 DNA into X-shaped structures. Migration of the crossover to the end of the X form should produce two linear monomers. Thompson et al (175) measured the rate of disappearance of X forms. By simulating a random walk in a computer they derived a function that relates the fraction of molecules dissociated to the number of steps in the walk, a step in this case being a shift of the crossover by one pair of nucleotides. They calculated that the time required per step at 37°C was about 170 μ sec. Since the rate was only 0.3% of that calculated for DNA of the same size by the theoretical treatment of Meselson (174), Thompson et al concluded that something other than rotary diffusion is rate-limiting in double-strand branch migration. They calculated a probability of 1 in 3 that a branch would be found more than 850 nucleotide pairs away from its starting point in 8 min, which may or may not be adequate to account for the formation in vivo of heteroduplex regions of the order of 1000 nucleotides long (13, 167, 176, 177). Double-strand branch migration might be enzymically driven in vivo. Sigal & Alberts pointed out that enzymically driven movement of a one-strand crossover (Figure 2 III) would necessarily move a two-strand crossover located in the same molecule (162). In addition to replication, the action of enzymes like gyrase or DNA-dependent ATPases might induce the axial rotation that would drive branch migration.

Thompson, Sussman, and Warner (personal communication) have found that double-strand branch migration in vitro is blocked by rather gross nonhomologies. In crosses of phage λ under conditions of stringently limited replication, Sodegren and Fox (personal communication), studied the effect of nonhomology on the transfer of a density label. The density of the selected recombinants was consistent with failure of a strand exchange to proceed beyond the bracketing regions of nonhomology.

Robberson & Clayton (178) attributed the loss of naturally occurring D loops from nicked mitochondrial DNA to displacement of the noncovalently bonded strand by branch migration (Figure 3b). To study the kinetics of single-strand branch migration, Radding et al (179) used D loops made in vitro from superhelical ϕX174 DNA and defined single-stranded fragments. The dissociation of D loops exhibited biphasic kinetics, which was attributed in part to the topology of the D loop. Branch migration of a D loop was inhibited by salt. Under physiological conditions, 0.15 M NaCl and 37°C, more than half the D loops survived for more than a minute. The intrinsic stability of D loops, even in relaxed DNA is consistent with their postulated role in the initiation of recombination (106, 112). In vivo the lifetime of a D loop is likely to be governed by the presence of HDPs and nucleases.

Ideas on Reciprocal Crossing-Over

When Sigal & Alberts (162) built the molecular model of a two-strand crossover they found that the proper angles and distances were maintained without unstacking any bases on either side of the exchange (Figure 3d).

In other words, a crossover need not be a single strand but only a phosphodiester bond. To account for genetic data (cf Figure 1) Holliday had suggested that a two-strand crossover might be resolved by cutting in either of two ways to yield respectively molecules with the parental or recombinant configuration of the flanking arms (168; Figure 3d). Examination of models indicates that these alternative cleavages are not sterically equivalent unless bases are unstacked and the flanking arms are rotated 180° (104, 172, 180). Sigal & Alberts proposed instead that joined molecules might isomerize freely about the two-strand crossover, one isomer having the parental configuration, the other having the recombinant configuration of the flanking arms. Cutting of the connecting strands would then produce both kinds of progeny. Sobell (180) demonstrated with models that the isomerization suggested by Sigal & Alberts involved 180° of rotation not around one axis but around each of two axes. Such an isomerization is complicated but not inconceivable given the flexibility of DNA in solution (104).

Meselson & Radding (104) proposed a simpler isomerization of a structure in which only one strand connects two helices. In that case isomerization involves only the rotation of two arms of the helix around an axis that is between and parallel to them (Figure 4). Isomerization about the one-strand crossover (Figure 2 III) produces a two-strand crossover (Figure 2 IV and Figure 4) and puts the flanking arms in the recombinant orientation. Alternatively, branch migration might convert the one-strand crossover into a two-strand crossover while the flanking arms remain in their original relationship [Figure 3c, d and Figure 2 in (104)]. Since two helices connected by a phosphodiester bond are juxtaposed (162), branch migration of a one-strand crossover might move either a 5' or 3' terminal nucleotide from one helix to the other. Once produced, either by isomerization or by branch migration, the two-strand crossover can migrate, as discussed above, making heteroduplex DNA symmetrically in both partners.

Genetic data (181–183) indicate that in the fungi some exchanges involve the formation of heteroduplex DNA in only one of two recombining molecules, as if only one strand were being transferred (Figure 2 III). As indicated above, the model proposed by Meselson & Radding (104) reconciles the asymmetric formation of heteroduplex DNA with the reciprocal exchange of flanking markers (see Figures 2 and 4). If recombination begins by forming heteroduplex DNA in one molecule, and only later forms heteroduplex DNA in both molecules, exchanges near the site of initiation should differ genetically from distal exchanges. One can detect some of the possible genetic consequences of forming heteroduplex DNA symmetrically in fungal recombination. For example, assuming that each gene is represented once in a haploid cell, the postmeiotic segregation of a marker from two cells in a tetrad is evidence that the marker was in heteroduplex DNA in both cells. Hence the marker was in heteroduplex regions in two molecules of DNA in the zygote that yielded the four haploid cells. The observed outcome is called an aberrant 4:4 segregation. By this and an independent criterion, Paquette and Rossignol recently found evidence in *Ascobolus immersus* for the idea that recombination begins by the transfer of one strand and later tends to involve the mutual exchange of strands (personal communication).

Hotchkiss (99) has extensively reviewed models of recombination, including some that formally resemble the Meselson-Radding model. Related ideas were published by Benbow, Zuccarelli & Sinsheimer (105). Different views have been put forward recently by Moore (143) and Chipchase (184).

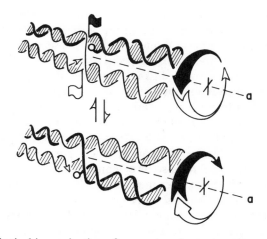

Figure 4 Hypothetical isomerization of a one-strand crossover to produce a two-strand crossover (104). The top diagram represents DNA molecules linked by the crossover of one strand. The structure is identical to that illustrated in Figures 2 III and 3c. The arrowhead represents a 3' end, and the open circle represents a 5' end. Isomerization, which can be demonstrated with molecular models, proceeds by rotation of the two arms to the right of the flags, about an axis, *a,* that is between the arms and parallel to them. The heavily shaded arm would rotate above the plane of the paper and the lightly shaded arm would rotate behind the plane of the paper. A phosphodiester bond at the point signalled by the dark flag will become the cross-connection closer to the viewer, and a bond signalled by the light flag will become the other cross-connection. The isomerization is reversible unless branch migration supervenes as in Figure 2f, or the nick is sealed where the 3' end (*arrowhead*) abuts on the 5' end (*open circle*).

MISMATCH REPAIR

Early observations on meiotic gene conversion and bacterial transformation provide the backdrop for more recent experiments, including those on transfection by heteroduplex DNA, which verify the hypothesis of mismatch repair: namely that mismatched bases in heteroduplex regions are corrected by a process that resembles excision-repair of UV lesions. Less success has attended efforts to study the enzymology.

Meiotic Gene Conversion

Meiotic gene conversion and postmeiotic segregation have been described in the section, perspectives and summary. These phenomena, and conversion-like events in prokaryotes (see below), have led to the idea that mismatched bases in heteroduplex joints are sometimes corrected by a kind of excision-repair (11, 36). There are two kinds of observation on fungal recombination that strongly support this idea. First, Leblon (185, 186) and Yu-Sun, Wickramaratne & Whitehouse (187) observed in *Ascobolus immersus* and *Sordaria brevicollis*, respectively, that mutations produced by different mutagens showed distinct characteristics with regard to (*a*) the direction of a conversion, mutant to wild type or vice versa, and (*b*) the frequency of postmeiotic segregation, the latter reflecting the formation of heteroduplex DNA without any correction of mismatches. Second, nearby mutations are often converted together. If mismatch-repair produces conversion, and the absence of repair produces postmeiotic segregation, then juxtaposition on one chromosome of a mutant that converts frequently with one that still segregates after meiosis should cause both to convert. Such was the case in *A. immersus* (188) and *Saccharomyces cerevisiae* (Fogel and Mortimer, personal communication). Related observations on *Schizosaccharomyces pombe* were also made by Gutz (189). These observations suggest that different pairs of mismatched bases are recognized differently and that recognition provokes a process of correction that affects neighboring regions of DNA. There are, however, observations that cannot be reconciled easily with a process that discriminates one mismatch from another. An enzyme system that recognizes some mismatches better than others might also be expected to show some bias in the conversion of any given mismatch from wild type to mutant or vice versa, as is true in *A. immersus,* for example. In *S. cerevisiae,* however, all mutations that are converted, including deletions, tend to convert from wild type to mutant as often as the reverse (13, 190–192).

Preliminary observations in vitro support the naive expectation that passive branch migration cannot form heteroduplex DNA containing a deletion or a sizable region of nonhomology (see section on branch migration, above). Hence the mechanism that converts deletion mutants (185, 190, 191) is especially interesting. If the deletion temporarily becomes part of a heteroduplex joint (see 51), the mechanism for including the deletion may involve either an actively driven form of strand transfer or a process more like the reannealing of strands in solution.

While this discussion is focused on mismatch repair, the section perspectives and summary introduced the more general idea that any resynthesis associated with the heteroduplex joint is a potential source of gene conversion. For example, the formation of heteroduplex DNA in only one of two recombining molecules implies that resynthesis must occur sooner or later (Figure 2 III). Asymmetric formation of heteroduplex DNA probably contributes to conversion (see section on ideas on reciprocal crossing-over); but it is not a sufficient explanation because asymmetric formation by itself would lead to postmeiotic segregation, which is not the major manifestation of intragenic recombination. There may also be repair synthesis, independent of mismatches, that contributes to conversion (35, 39, 193). Evidence of new DNA synthesis specifically associated with recombination comes from a prokaryote. In the recombination of bacteriophage λ, under conditions of limited replication, new DNA synthesis is greatest in the region of genetic exchange (194, 195).

Transformation of D. pneumoniae

In prokaryotes, various manifestations of recombination may be related to mismatch repair, including negative interference (196–199) and marker effects (183, 200–205). The transformation of *D. pneumoniae* is one of the most extensively studied cases of conversion-like phenomena in prokaryotes. Transformation involves the insertion of a strand of donor DNA into a recipient molecule to form a region that is heteroduplex and heterozygous (34). Some markers transform efficiently (HE) and others inefficiently (LE) (206). For several markers, the number of transformants approaches the number of donor genomes that are taken up by the cells (207, 208). Ephrussi-Taylor & Gray (205) observed that juxtaposition of an LE marker and an HE marker on the same strand of DNA caused both to transform inefficiently. They proposed that LE markers produce mismatches that are subject to destruction or excision repair. Lacks (209) isolated mutants of pneumococcus, called *hex⁻*, which cause all markers to transform with high efficiency, as if the system that recognizes some mismatched bases were lost. Tiraby & Fox (210) reasoned that if *hex⁺* eliminates certain heterozygotes, *hex⁻* strains might show an increased rate of mutagenesis because of failure to recognize transient mismatches that become mutations. Indeed they found a four- to thirtyfold increase in the frequency of spontaneous mutations in *hex⁻* strains. Roger (211) found that opposite strands were not equally effective in transformation which is consistent with specific recognition of different mismatches. Shoemaker & Guild (212) showed that the kinetics of destruction of LE markers is consistent with their elimination after heteroduplex DNA has been formed.

For any base pair there are six possible mismatches, and on the average one would expect correction to eliminate the donor marker in about half of all cases. Since LE markers transform 1/5 to 1/20 as efficiently as HE markers (213) the mechanism that recognizes mismatches in *D. pneumoniae* must either correct the donor strand specifically or destroy the genome carrying the mismatch (213, 214).

While the observations on meiotic gene conversion and transformation of *pneumococcus* discussed above suggest that enzyme systems recognize and correct mismatched base pairs by a kind of excision repair, there are clearly aspects of specificity that we do not understand.

Transfection by Heteroduplex DNA

To study mismatch repair more directly, a number of investigators have transfected bacteria with heteroduplex DNA made in vitro. Heteroduplex molecules have been prepared in two ways; (*a*) by denaturing and reannealing DNA of different genotypes, which produces a mixture of the parental types and the two types of heteroduplex molecules, L_a/H_b and L_b/H_a, or (*b*) by separating strands L and H and reannealing them in the two new combinations which produces homogeneous preparations of L_a/H_b or L_b/H_a with mutant and wild-type markers located on identified strands. As the following accounts show, such experiments prove to be complex and difficult to interpret.

Early examples of apparent correction of heteroduplex DNA in *E. coli* were provided by Doerfler & Hogness (215) and Baas & Jansz (216) for phages λ and ϕX174 respectively. Enea, Vovis & Zinder (217) observed a similar instance of apparent correction of heteroduplex DNA of phage f1. Spatz & Trautner (198) made a more extensive study involving the transfection of *B. subtilis* by defined heteroduplex molecules of phage SPP1 DNA. They measured the frequencies of bursts containing only wild-type phage, only mutant phage, or both. All heteroduplexes yielded some pure bursts, as if some mechanism of correction were operative. However, if correction was related to recognition of the mismatch, such specificity was masked by other effects, including (*a*) preferential correction of one or the other strand for many markers, and (*b*) an apparent correlation between map position and the frequency of correction. Transfection by phage SPP1 DNA requires recombination, which may be related to the complexity of these observations. However, the observations of Schlaeger & Spatz (218) argued against invoking nonspecific excision and resynthesis associated with recombination to account for the observed mismatch correction.

By crosses between density-labeled heavy and light λ phages under conditions that stringently blocked replication, White & Fox (196, 197) studied the relationship between strand transfer and double exchanges between close markers, i.e. high negative interference. In a cross of the type $+b+ \times a+c$, they found that wild-type recombinants, $+ + +$, seldom resulted just from the exchange of a little bit of DNA between a and c. Under the special conditions imposed by the restriction of replication they found evidence of extensive formation of heteroduplex DNA and they proposed that the wild-type recombinants usually resulted from mismatch repair. When they transfected cells with specific heteroduplex molecules made in vitro they were able to simulate the pattern of recombination that they observed in crosses.

Wildenberg & Meselson (177) prepared heteroduplex DNA of phage λ by denaturing and reannealing pairwise combinations of eight different *amber* mutations. They transfected *E. coli* in the presence of helper phage. To eliminate "conventional recombination" (i.e. strand transfer) all phage were int^- red^- and the bacteria were $recA^-$. Wild-type progeny were produced at frequencies up to 10%. The calculated frequencies of repair of mismatches at single sites varied from 3% to 20% for different markers. Wagner & Meselson (22) employed the same method, extended by the use of homogeneous preparations of heteroduplex DNA with mismatched bases at four sites. About 40% of bursts from single cells contained phage of two genotypes, in pairs that could have arisen by mismatch repair. Surprisingly, 60% of bursts were pure bursts of one or another of 16 possible genotypes, as if all sites had been repaired (see also 219). Alternatively, Wagner & Meselson reasoned that these pure bursts might have resulted from the random failure of one strand to give rise to progeny, a failure or loss that followed some correction of the original heteroduplex. Comparison of the frequencies of the 16 possible genotypes among pure bursts and mixed bursts showed that the observed outcome could have resulted from random loss of a strand rather than from repair of every mismatch. These observations suggest that a pure burst by itself is an equivocal criterion of mismatch repair, a conclusion that is in accord with the observation of other kinds of bias in the transmission of genetic information to phage progeny (18, 21, 216, 217). Mismatches at various sites were not repaired independently as shown clearly by the behavior of three markers within a region of ~4000 base pairs (22). Of cases characterized by correction of the outside markers, more than half involved repair of all three mismatches on the same strand, which is consistent with repair by excision of tracts of nucleotides. Estimates of the extent of such tracts were less than 2000 nucleotides in one study (177) and 3000 nucleotides in a second study (22). The data of Wagner & Meselson included a correlation between the relative frequencies of certain genotypes issuing from reciprocal heteroduplexes, L_a/H_b and

L_b/H_a, and those expected if excision started at a mismatch and proceeded in the direction, 5' to 3'.

Nevers & Spatz examined the effect of various mutations on the frequency of pure bursts from cells transfected by heteroduplex λ DNA with a mismatch in the cI gene (220). They observed a two- to threefold decrease of pure bursts in *uvrD* and *uvrE* mutants of *E. coli.* Since *uvrE⁻* has a mutator phenotype (221, 222), Nevers & Spatz (220) likened their observations to those on the effect of *hex⁻* on transformation and mutability in *Pneumococcus* (210).

Lai & Nathans (223, 224) prepared heteroduplex DNA from numerous pairs of noncomplementing mutants of SV40. Upon transfection of African green monkey cells, many of these heteroduplexes produced plaques. Comparision of heteroduplexes, L_a/H_b plus L_b/H_a, from mutants that blocked replication with heteroduplexes from mutants that blocked some other step suggested that formation of plaques depended on something other than the ability of the original heteroduplex molecule to replicate (224). Miller et al (225) transfected mouse embryo cells with heteroduplexes of polyoma DNA that was genetically marked at four sites, and analyzed 18 plaque isolates of virus. More than half were recombinants. In both of these studies, conventional recombination or strand transfer may have played a role by exchanges either (*a*) between the reciprocal heteroduplexes L_a/H_b and L_b/H_a, which together have all of the wild-type information, or (*b*) between the progeny produced by replication. Because of the apparently high frequency at which recombinants appeared, the investigators interpreted the observations as evidence of mismatch-repair.

These observations on transfection by heteroduplex DNA justify several conclusions. The transmission to progeny of genetic information from one or the other strand of heteroduplex phage DNA is biased by means that we do not understand (22, 198), a conclusion that calls for caution in interpreting all experiments on transfection by heteroduplex DNA. At the same time, the evidence shows that in *E. coli* enzymes appear to recognize and correct mismatched bases by a process akin to excision repair. For the reason stated, the data on transfection of mammalian cells by heteroduplex DNA are more ambiguous, but are consistent with the occurrence of mismatch-repair.

Enzymology of Mismatch Repair

In addition to restriction endonuclease (226) at least five endonucleases exist in *E. coli,* several of which recognize structural abnormalities in double-stranded DNA (227–233). While none has been shown to recognize mismatched base pairs, such a role appears not to have been excluded either (229). According to the prevailing view, mismatch-repair is like the excision-repair of UV lesions (234), but evidence from several organisms suggests that the respective pathways of repair are not identical (209, 220, 235–237).

In the smut fungus *Ustilago maydis,* Holloman & Holliday (238) found that a combination of mutants at two loci, *nuc-1* and *nuc-2* (239) reduced the total DNase activity in extracts to 5–25% of the wild-type level. The doubly mutant strains also showed reduced gene conversion (238, 239). In a small number of crosses, the respective deficiencies in recombination and DNase activity segregated together. However, the relationships of these two mutations to recombination and DNase activities are complex. The defect in recombination resulted from two genetic defects, but one mutation, *nuc-2,* reduced the activity of at least two DNases. Holloman & Holliday (238, 240) purified and characterized one of the nucleases, whose activity was reduced in the double mutant. The enzyme, called DNase I, readily digests single-stranded DNA or RNA, producing oligonucleotides and mononucleotides. Acting on double-stranded DNA, the enzyme introduces nicks (240), whose number can be increased by prior UV irradiation (241). To study the possible role of DNase I in mismatch repair, Ahmad, Holloman & Holliday used heteroduplex DNA of phage SPP1 as a substrate and measured the ability of the DNA to transfect *B. subtilis* (242). The enzyme inactivated only a fraction of the molecules of native, homoduplex, or heteroduplex DNA respectively. In each case the remaining fraction was resistant to further attack. The fraction of heteroduplex DNA that was inactivated was three- to sevenfold greater than the fraction of homoduplex DNA.

Endonucleases that specifically digest disordered polynucleotide chains have been isolated from other fungi (243–246), mammalian cells (247, 248), and *B. subtilis* (249). Shenk et al (250) observed that the S_1 nuclease from *Aspergillus oryzae* can recognize a mismatch of apparently a single base pair. An amount of S_1 nuclease that was large in relation to its activity on single-stranded DNA cleaved only a small fraction of heteroduplex molecules. Dodgson & Wells (251) studied the action of S_1 nuclease on defined mismatches in which one to six internal residues of dA or dG were unpaired in the oligonucleotide dG-dC. While S_1 nuclease cleaved the larger mismatches more readily than the smaller ones, all were poor substrates.

SELECTED TOPICS

RecA in E. coli

Some mutations in the *recA* gene of *E. coli* reduce recombination by 3 to 6 orders of magnitude (252), by which property *recA* was discovered (253); but mutations in *recA* are pleiotropic (252), and some mutations that may be in the *recA* gene do not strongly affect recombination (254, 255). The pleiotropy of *recA* mutants includes 1. increased sensitivity to UV light, ionizing radiation, and alkylating agents, 2. decreased inducibility of lambdoid phages by UV light or certain chemical agents, 3. decreased mutability by some agents, 4. increased breakdown of DNA after irradiation by UV light, and 5. changes in the regulation of cell division. In addition, the function of *recA* is essential for an inducible system of repair that is associated with increased mutagenesis (256–258). The inducible system of repair is affected by *lexA* and *tsl,* mutations at 90 min on the map; as well as by *recA, tif, zab,* and *lexB,* all in the vicinity of 58 min on the map (254, 255, 259; Clark, Templin, and Czonka, personal communication).

Given the pleiotropy of *recA* mutations, there are obviously many ways in which they might affect recombination, including structural alterations in the DNA (252, 260). Recent research has given new impetus to efforts to discover the molecular basis of the functions governed by *recA*.

From *E. coli,* McEntee & Epstein (159, 261) isolated a transducing variant, phage λ *precA,* which carries the *recA* gene. The gene was not shut off by the λ repressor since it was expressed in a lysogen. When λ *precA* infected a noninducible lysogen that had been irradiated with UV to reduce the synthesis of host proteins, most of the radioactivity incorporated into protein was in a single electrophoretic band corresponding to a monomer with a molecular weight of 43,000 (261). With other isolates of λ *precA* carrying mutations in the *recA* region, McEntee observed qualitative changes in the protein that show it to be the product of *recA* (262). Isolated by gel filtration, the *recA* protein behaved like a tetramer.

Inouye (263), Gudas & Pardee (264, 265), and Sedgwick (266) discovered a protein, called protein X, that is part of the inducible pathway of repair defined by the mutations described above: *lexA, tsl, recA, tif, zab,* and *lexB.* McEntee (262) found that mutations that alter the *recA* protein made by λ *precA* also alter the inducible X protein; Gudas & Mount (267) showed that *recA1* alters the electrophoretic mobility of protein X; Little & Kleid (267a) observed that limited proteolysis of purified protein X and *recA* protein yielded similar sets of peptides; and Emmerson & West (267b) found that *tif-1* alters the isoelectric point of protein X. These experiments show that protein X and *recA* protein are the same, and that *tif-1* is a mutation in the *recA* gene, findings that considerably clarify the regulatory interactions of *recA* (262, 267–267b). Gudas & Pardee (265) observed that protein X binds to single-stranded DNA. Recently Roberts and Roberts (personal communication) purified an activity that inactivates the repressor of phage λ, during which the repressor is cleaved (267c). The purified protein appears to be identical to the *recA* protein. These observations promise to elucidate how the *recA* gene works its many effects.

To recapitulate information about the effect of *recA* on molecular recombination: Benbow et al (105) and Potter & Dressler (172) found that *recA*⁺ is required for the formation of figure-8 recombinant molecules (see section on branch migration and symmetric exchange of strands, above); Cassuto et al (158) observed cutting-in-trans when they used extracts of a thermally induced lysogen of λ *precA* (see section on recognition vs breakage, above); Holloman & Radding (112) observed that the need of *recA*⁺ for recombination of superhelical DNA and homologous fragments in spheroplasts was bypassed when joint molecules were used for transfection (see section on strand uptake and the role of superhelical DNA in recombination, above). These observations point to an early role of *recA*⁺ function in recombination.

ATP-Dependent DNases[2]

Clark (252) has reviewed the literature up to 1973 on the role in recombination of the widespread class of ATP-dependent DNases, the exemplar of which is the enzyme encoded by the *recB* and *recC* genes of *E. coli*. Mutants in various species of bacteria that lack this enzyme are less viable than wild-type cells (268–270), and in some kinds of cross show little or no defect in recombination (26, 27, 269, 271–275). In crosses of noncomplementing mutants of β-galactosidase, Birge & Low (276) observed nearly normal levels of β-galactosidase in *recB*⁻ zygotes, but less than 1% of viable recombinants, an observation that was confirmed by Bergmans, Hoekstra & Zuidweg (277). According to that observation, some form of transcribable recombinant DNA was formed in the absence of *recBC* DNase, which therefore might be required to complete some late step in recombination and/or to assure the survival of the zygote. When Vovis (278) and Kooistra & Venema (270) studied mutants of *D. pneumonia* and *H. influenzae* that lack ATP-dependent DNase they found that transformation proceeded normally to a late stage in the formation of recombinant molecules. Wilcox & Smith nonselectively isolated a number of mutants of the ATP-dependent DNase in *H. influenzae* and studied their effects on transformation (269). When they took into account the nonviable cells formed in cultures of such mutants, they estimated that the frequency of transformation was reduced only fourfold at the most. Because of all of these observations, we cannot decide whether in cells that lack ATP-dependent DNase the preferential loss of zygotes creates a spurious defect in recombination or whether a defect in resolving intermediates in recombination results in the preferential loss of zygotes.

[2]Although there are other ATP-dependent DNases (226), I use the term here only for the subclass of enzymes that are like the *recBC* DNase of *E. coli*.

The multiple activities of the ATP-dependent DNases were described above (see section on separation of strands: ATPases). In search of information about the significance of these activities in vivo, Kushner (279) partially purified the enzyme from thermosensitive *recB* and *recC* mutants of *E. coli.* Of the three nucleolytic activities, only the exonucleolytic activity on double-stranded DNA was thermosensitive in vitro. From experiments on transfection of spheroplasts of *rec+* and *recB−* cells by different forms of DNA, Benzinger, Enquist & Skalka (280) also concluded that the exonucleolytic activity on double-stranded DNA is the most prominent manifestation of *recBC* DNase in vivo.

Recombination of Phage T7 DNA in vitro

Sadowski & Vetter (281) observed recombination and packaging of phage T7 DNA in extracts of infected cells. Because most of the required gene products played a role in the maturation and packaging of DNA, Sadowski (282) could not tell which products also acted in recombination. To eliminate packaging from the observations, Vlachopoulou & Sadowski (283) examined the fate of [^{32}P]5-bromodeoxyuridine T7 DNA that was added to extracts of cells infected by phage of light density. Large pieces of heavy DNA appeared to join covalently to light DNA. Experiments with thermosensitive mutants showed that active T7 exonuclease (284) had to be present during the incubation of exogenous DNA in the extracts (283). Joining was also blocked if extracts were prepared from cells infected by T7 mutants that cannot replicate. These included mutants in gene 1.3 (ligase), gene 3 (endonuclease), gene 4 (DNA synthesis), and gene 5 (DNA polymerase).

CONCLUSION

The view of recombination presented here is the same as that summarized by Hotchkiss (99): Recombination probably begins at the ends of single strands which form a bridge of one or two strands. After some lateral migration these bridges can be cut to release helices with either new or old arrays of outside markers. Resynthesis of DNA associated with the formation or repair of heteroduplex joints will sometimes produce unequal recoveries of genetic markers.

ACKNOWLEDGMENTS

This research was sponsored by grant No. ACS NP 90 from the American Cancer Society and grant No. USPHS CA 16038 from the National Cancer Institute. This manuscript was prepared in part while I was a Visiting Miller

Professor at the University of California, Berkeley. I am grateful to Dr. A. J. Clark for many stimulating discussions and to the Department of Molecular Biology, and the Miller Institute for their hospitality. My colleagues, Richard Cunningham and Kenneth Beattie, made many helpful suggestions.

Literature Cited

1. Clark, A. J. 1971. *Ann. Rev. Microbiol.* 25:437–64
2. Echols, H. 1971. *Ann. Rev. Biochem.* 40:827–54
3. Gottesman, M. E., Weisberg, R. A. 1971. *The Bacteriophage λ,* ed. A. D. Hershey, pp. 113–38. Cold Spring Harbor. 792 pp.
4. Starlinger, P., Saedler, H. 1976. *Curr. Top. Microbiol. Immunol.* 75:111–53
5. Cohen, S. N. 1976. *Nature* 263:731–38
6. Landy, A., Ross, W. 1977. *Science* 197:1147–60
7. Bukhari, A. I. 1976. *Ann. Rev. Genet.* 10:389–412
8. Murray, N. E., Murray, K. 1974. *Nature* 251:476–81
9. Lai, C. J., Nathans, D. 1974. *J. Mol. Biol.* 89:179–93
10. Mertz, J. E., Berg, P. 1974. *Proc. Natl. Acad. Sci. USA* 71:4879–83
11. Whitehouse, H. L. K. 1973. *Towards an Understanding of the Mechanism of Heredity.* New York: St. Martin's Press. 528 pp.
12. Fincham, J. R. S., Radford, A., Day, P. R. 1978. *Fungal Genetics.* Oxford: Blackwell Sci. Publ. 4th ed. In press
13. Hurst, D. D., Fogel, S., Mortimer, R. K. 1972. *Proc. Natl. Acad. Sci. USA* 69:101–5
14. Carlson, P. S. 1971. *Genet. Res. Cambridge* 17:53–81
15. Chovnick, A., Ballantyne, G. H., Holm, D. G. 1971. *Genetics* 69:179–209
16. Chovnick, A. 1973. *Genetics* 75:123–31
17. Sarthy, P. V., Meselson, M. 1976. *Proc. Natl. Acad. Sci. USA* 73:4613–17
18. Enea, V., Zinder, N. D. 1976. *J. Mol. Biol.* 101:25–38
19. Berg, D. E., Gallant, J. A. 1971. *Genetics* 68:457–72
20. Hershey, A. D., Rotman, R. 1949. *Genetics* 34:44–71
21. Ross, D. G., Freifelder, D. 1976. *Virology* 74:414–25
22. Wagner, R. Jr., Meselson, M. 1976. *Proc. Natl. Acad. Sci. USA* 73:4135–39
23. Meselson, M. 1967. *Heritage from Mendel,* ed. A. Brink, pp. 81–104. Madison: Univ. Wisc. Press
24. Herman, R. K. 1968. *J. Bacteriol.* 96:173–79
25. Wackernagel, W., Radding, C. M. 1974. In *Mechanisms in Recombination,* ed. R. F. Grell, pp. 111–22. New York: Plenum
26. Hobom, G., Hogness, D. S. 1974. *J. Mol. Biol.* 88:65–87
27. Bedbrook, J. R., Ausubel, F. M. 1976. *Cell* 9:707–16
28. Potter, H., Dressler, D. 1977. *Proc. Natl. Acad. Sci. USA* 74:4168–72
29. Taylor, J. H. 1965. *J. Cell Biol.* 25:57–67
30. Oppenheim, A. B., Riley, M. 1967. *J. Mol. Biol.* 28:503–11
31. Kellenberger-Gujer, G., Weisberg, R. A. 1971. *The Bacteriophage λ,* ed. A. D. Hershey pp. 407–15. Cold Spring Harbor, NY: Cold Spring Harbor Lab. 792 pp.
32. Siddiqi, O., Fox, M.S. 1973. *J. Mol. Biol.* 77:101–23
33. Tomizawa, J. 1967. *J. Cell Physiol.* 70:Supp. 1, pp. 201–14
34. Gurney, T. Jr., Fox, M. S. 1968. *J. Mol. Biol.* 32:83–100
35. Taylor J. H., Haut, W. F., Tung, J. 1962. *Proc. Natl. Acad. Sci. USA* 48:190–98
36. Holliday, R. 1962. *Genet. Res. Cambridge* 3:472–486
37. Holliday, R. 1974. *Genetics* 78:273–87
38. Whitehouse, H. L. K. 1965. *Genetics Today,* Proc. 11th Int. Cong. Genet., The Hague, Sept. 1963, Vol. 2, ed. S. J. Geerts, pp. 87–88. Oxford: Pergamon. 494 pp.
39. Meselson, M. 1964. *J. Mol. Biol.* 9:734–45
40. Alberts, B. M., Frey, L. 1970. *Nature* 277:1313–18
41. Molineux, I. J., Gefter, M. L. 1975. *J. Mol. Biol.* 98:811–25
42. Reuben, R. C., Gefter, M. L. 1974. *J. Biol. Chem.* 249:3843–50
43. Scherzinger, E., Litfin, F., Jost, E. 1973. *Mol. Gen. Genet.* 123:247–62
44. Oey, J. L., Knippers, R. 1972. *J. Mol. Biol.* 68:125–38
45. Herrick, G., Alberts, B. 1976. *J. Biol. Chem.* 251:2133–41

46. Hotta, Y., Stern, H. 1971. *Dev. Biol.* 26:87–99
47. Hotta, Y., Stern, H. 1971. *Nature New Biol.* 234:83–86
48. Banks, G. R., Spanos, A. 1975. *J. Mol. Biol.* 93:63–77
49. Jensen, D. E., Von Hippel, P. H. 1976. *J. Biol. Chem.* 251:7198–7214
49a. Alberts, B. Sternglanz, R. 1977. *Nature* 269:655–61
50. Wu, J.-R., Yeh, Y.-C. 1973. *J. Virol.* 12:758–65
51. Broker, T. R., Doermann, A. H. 1975. *Ann. Rev. Genet.* 9:213–44
52. Jensen, D. E., Kelly, R. C., Von Hippel, P. H. 1976. *J. Biol. Chem.* 251:7215–28
53. Hosoda, J., Takacs, B., Brack, C. 1974. *FEBS Lett.* 47:338–42
54. Moise, H., Hosoda, J. 1976. *Nature* 259:455–58
55. Anderson, R. A., Coleman, J. E. 1975. *Biochemistry* 14:5485–91
56. Sigal, N. Delius, H., Kornberg, T., Gefter, M. L., Alberts, B. 1972. *Proc. Natl. Acad. Sci. USA* 69:3537–41
57. MacKay, V., Linn, S. 1976. *J. Biol. Chem.* 251:3716–19
58. Huang, W. M., Lehman, I. R. 1972. *J. Biol. Chem.* 247:3139–46
59. Curtis, M. J., Alberts, B. 1976. *J. Mol. Biol.* 102:793–816
60. Kozinski, A., Felgenhauer, Z. Z. 1967. *J. Virol.* 1:1193–1202
61. Tomizawa, J., Anraku, N., Iwama, Y. 1966. *J. Mol. Biol.* 21:247–53
62. Breschkin, A. M., Mosig, G. 1977. *J. Mol. Biol.* 112:279–94
63. Gold, L., O'Farrell, P. Z., Russel, M. 1976. *J. Biol. Chem.* 251:7251-62
64. Krisch, H. M., Bolle, A., Epstein, R. H. 1974. *J. Mol. Biol.* 88:89–104
65. Friedman, E. A., Smith, H. O. 1973. *Nature New Biol.* 241:54–58
66. Wilcox, K. W., Smith, H. O. 1976. *J. Biol. Chem.* 251:6127–34
67. Karu, A. E., MacKay, V., Goldmark, P. J., Linn, S. 1973. *J. Biol. Chem.* 248:4874–84
68. MacKay, V., Linn, S. 1974. *J. Biol. Chem.* 249:4286–94
69. Abdel-Monem, M., Hoffmann-Berling, H. 1976. *Eur. J. Biochem.* 65:431–40
70. Abdel-Monem, M., Dürwald, H., Hoffmann-Berling, H. 1976. *Eur. J. Biochem.* 65:441–49
71. Abdel-Monem, M., Lauppe, H. F., Kartenbeck, J., Dürwald, H., Hoffmann-Berling, H. 1977. *Eur. J. Biochem.* 110:667–85
72. Scott, J. F., Eisenberg, S., Bertsch, L. L., Kornberg, A. 1977. *Proc. Natl. Acad. Sci. USA* 74:193–98
73. Denhardt, D. T. 1975. *CRC Crit. Rev. Microbiol.* 161–223
74. Richet, E., Kohiyama, M. 1976. *J. Biol. Chem.* 251:808–12
75. Ebisuzaki, K., Behme, M. T., Senior, C., Shannon, D., Dunn, D. 1972. *Proc. Natl. Acad. Sci. USA* 69:515–19
76. Purkey, R. M., Ebisuzaki, K. 1977. *Eur. J. Biochem.* 75:303–10
77. Eichler, D. C., Lehman, I. R. 1977. *J. Biol. Chem.* 252:499–503
78. Gellert, M., Mizuuchi, K., O'Dea, M. H., Nash, H. A. 1976. *Proc. Natl. Acad. Sci. USA* 73:3872–76
79. Gellert, M., O'Dea, M. H., Itoh, T., Tomizawa, J. 1976. *Proc. Natl. Acad. Sci. USA* 73:4474–78
80. Das, A., Court, D., Adhya, S. 1976. *Proc. Natl. Acad. Sci. USA* 73:1959–63
81. Lowery-Goldhammer, C., Richardson, J. P. 1974. *Proc. Natl. Acad. Sci. USA* 71:2003–7
82. Kornberg, A. 1974. *DNA Synthesis.* San Francisco: Freeman. 399 pp.
83. Alberts, B., Barry, J., Bittner, M., Davies, M., Hama-Inaba, H., Liu, C-C., Mace, D., Moran, L., Morris, C. F., Piperno, J., Sinha, N. K. 1977. In *Nucleic Acid—Protein Recognition,* ed. H. J. Vogel, pp. 31–63. New York: Academic
84. Masamune, Y., Richardson, C. C. 1971. *J. Biol. Chem.* 246:2692–2701
85. Kelly, R. B., Cozzarelli, N. R., Deutscher, M. P., Lehman, I. R., Kornberg, A. 1970. *J. Biol. Chem.* 245:39–45
86. Dumas, L. B., Darby, G., Sinsheimer, R. L. 1971. *Biochim. Biophys. Acta* 228:407–22
87. Fujimura, R. K., Roop, B. C. 1976. *J. Biol. Chem.* 251:2168–75
87a. Kolodner, R., Richardson, C. C. 1977. *Proc. Natl. Acad. Sci. USA* 74:1525–29
88. Molineux, I. J., Gefter, M. L. 1974. *Proc. Natl. Acad. Sci. USA* 71:3858–62
89. Molineux, I. J., Friedman, S., Gefter, M. L. 1974. *J. Biol. Chem.* 249:6090–98
90. Reuben, R. C., Gefter, M. L. 1973. *Proc. Natl. Acad. Sci. USA* 70:1846–50
91. Sherman, L. A., Gefter, M. L. 1976. *J. Mol. Biol.* 103:61–76
92. Huberman, J. A., Kornberg, A., Alberts, B. M. 1971. *J. Mol. Biol.* 62:39–52
93. Nossal, N. G. 1974. *J. Biol. Chem.* 249:5668–76
94. Eisenberg, S., Griffith, J., Kornberg, A. 1977. *Proc. Natl. Acad. Sci. USA* 74:3198–3202
95. Mazur, B. J., Model, P. 1973. *J. Mol. Biol.* 78:285–300

96. Wackernagel, W., Radding, C. M. 1974. *Proc. Natl. Acad. Sci. USA* 71:431–35
97. Christiansen, C. H., Baldwin, R. L. 1977. *J. Mol. Biol.* 115:441–54
98. Hotchkiss, R. D. 1971. *Adv. Genet.* 16:325–48
99. Hotchkiss, R. D. 1974. *Ann. Rev. Microbiol.* 28:445–68
100. Fox, M. 1966. *Macromolecular Metabolism,* pp. 183–96. New York: Little, Brown. 366 pp.
101. Lacks, S. 1966. *Genetics* 53:207–35
102. Paszewski, A. 1970. *Genet. Res. Cambridge* 15:55–64
103. Kunicki-Goldfinger, W. J. H. 1971. In *Recent Advances in Microbiology,* 10th Int. Cong. Microbiol., ed. A. Pérez-Miravete, D. Peláez, pp. 291–303. Mexico: Editorial Muñoz, S. A.
104. Meselson, M. S., Radding, C. M. 1975. *Proc. Natl. Acad. Sci. USA* 72:358–61
105. Benbow, R. M., Zuccarelli, A. J., Sinsheimer, R. L. 1975. *Proc. Natl. Acad. Sci. USA* 72:235–39
106. Holloman, W. K., Wiegand, R., Hoessli, C., Radding, C. M. 1975. *Proc. Natl. Acad. Sci. USA* 72:2394–98
107. Beattie, K. L., Wiegand, R. C., Radding, C. M. 1977. *J. Mol. Biol.* 116:783–803
108. Wiegand, R. C., Beattie, K. L., Holloman, W. K., Radding, C. M. 1977. *J. Mol. Biol.* 116:805–24
109. Seto, H., Tomasz, A. 1977. *Proc. Natl. Acad. Sci. USA* 74:296–99
110. Collins, C. J., Guild, W. R. 1972. *J. Bacteriol.* 109:266–75
111. Shoemaker, N. B., Guild, W. R. 1972. *Proc. Natl. Acad. Sci. USA* 69:3331–35
112. Holloman, W. K., Radding, C. M. 1976. *Proc. Natl. Acad. Sci. USA* 73:3910–14
113. Puga, A., Tessman, I. 1973. *J. Mol. Biol.* 75:83–97
114. Mizuuchi, K., Nash, H. A. 1976. *Proc. Natl. Acad. Sci. USA* 73:3524–28
115. Marians, K. J., Ikeda, J. E., Schlagman, S., Hurwitz, J. 1977. *Proc. Natl. Acad. Sci. USA* 74:1965–68
116. Bauer, W., Vinograd, J. 1970. *J. Mol. Biol.* 47:419–35
117. Bauer, W., Vinograd, J. 1974. *Basic Principles in Nucleic Acid Chemistry,* ed. P.O.P. Ts'o 2:265–303. New York: Academic. 519 pp.
118. Wang, J. C. 1974. *J. Mol. Biol.* 87:797–816
119. Worcel, A., Burgi, E. 1972. *J. Mol. Biol.* 71:127–47
120. Pettijohn, D. E., Hecht, R. 1973. *Cold Spring Harbor Symp. Quant. Biol.* 38:31–41
121. Griffith, J. D. 1975. *Science* 187:1202–3
122. Germond, J. E., Hirt, B., Oudet, P., Gross-Bellard, M., Chambon, P. 1975. *Proc. Natl. Acad. Sci. USA* 72:1843–47
123. Cook, P. R., Brazell, I. A. 1975 *J. Cell. Sci.* 19:261–79
124. Ide, T., Nakane, M., Anzai, K., Andoh, T. 1975. *Nature* 258:445–47
125. Benyajati, C., Worcel, A. 1976. *Cell* 9:393–407
126. Piñon, R., Salts, Y. 1977. *Proc. Natl. Acad. Sci. USA* 74:2850–54
127. Champoux, J. J. 1978. *Ann. Rev. Biochem.* 47:449–79
128. Hastings, P. J., Whitehouse, H. L. K. 1964. *Nature* 201:1052–54
129. Lissouba, P., Rizet, G. 1960. *Compt. Rend.* 250:3408–10
130. Angel, T., Austin, B., Catcheside, D. G. 1970. *Aust. J. Biol. Sci.* 23:1229–40
131. Catcheside, D. G. 1974. *Ann Rev. Genet.* 8:279–300
132. Catcheside, D. G., Angel, T. 1974. *Aust. J. Biol. Sci.* 27:219–29
133. Stahl, F. W., Crasemann, J. M., Stahl, M. M. 1975. *J. Mol. Biol.* 94:203–12
134. Henderson, D., Weil, J. 1975. *Genetics* 79:143–74
135. Lam, S. T., Stahl, M. M., McMilin, K. D., Stahl, F. W. 1974. *Genetics* 77:425–33
136. Enquist, L. W., Skalka, A. 1973. *J. Mol. Biol.* 75:185–212
137. Stahl, F. W., Chung, S., Crasemann, J., Faulds, D., Haemer, J., Lam, S., Malone, R. E., McMilin, K. D., Nozu, Y., Siegel, J., Strathern, J., Stahl, M. 1973. *Virus Research,* ed. C. F. Fox, W. S. Robinson, pp. 487–503. New York: Academic. 599 pp.
138. Sobell, H. M. 1972. *Proc. Natl. Acad. Sci. USA* 69:2483–87
139. Sobell, H. M. 1975. *Proc. Natl. Acad. Sci. USA* 72:279–83
140. Wagner, R. E. Jr., Radman, M. 1975. *Proc. Natl. Acad. Sci. USA* 72:3619–22
141. McGavin, S. 1971. *J. Mol. Biol.* 55:293–98
142. Cross, R. A., Lieb, M. 1967. *Genetics* 57:549–60
143. Moore, D. 1974. *J. Theor. Biol.* 43:167–86
144. Champoux, J. J. 1977. *Proc. Natl. Acad. Sci. USA.* 74:5328–32
145. Wang, J. C. 1971. *J. Mol. Biol.* 53:523–33
146. Champoux, J. J., Dulbecco, R. 1972. *Proc. Natl. Acad. Sci. USA* 69:143–46
147. Miller, R. C. Jr. 1975. *Ann. Rev. Microbiol.* 29:355–76

148. Howard-Flanders, P. 1975. *Molecular Mechanisms for Repair of DNA,* ed. P. C. Hanawalt, R. B. Setlow, pp. 265–74. New York: Plenum
149. Benbow, R. M., Zuccarelli, A. J., Sinsheimer, R. L. 1974 *J. Mol. Biol.* 88:629–51
150. Konrad, E. B. 1977. *J. Bacteriol.* 130:167–72
151. Michalke, W. 1967. *Mol. Gen. Genet.* 99:12–33
152. Lacks, S. 1962. *J. Mol. Biol.* 5:119–31
153. Dubnau, D., Cirigliano, C. 1972. *J. Mol. Biol.* 64:9–29
154. LeClerc, J. E., Setlow, J. K. 1975. *J. Bacteriol.* 122:1091–1102
155. Bialkowska-Hobrzańska, H., Kunicki-Goldfinger, W. J. H. 1977. *Mol. Gen. Genet.* 151:319–26
156. Ross, P., Howard-Flanders, P. 1977. *J. Mol. Biol.* 117:137–58
157. Ross, P., Howard-Flanders, P. 1977. *J. Mol. Biol.* 117:159–74
158. Cassuto, E., Mursalim, J., Howard-Flanders, P. 1978. *Proc. Natl. Acad. Sci. USA.* 75:620–24
159. McEntee, K., Epstein, W. 1977. *Virology* 77:306–18
160. Cassuto, E., Radding, C. M. 1971. *Nature New Biol.* 229:13–16; 230:128
161. Cassuto, E. Lash, T., Sriprakash, K. S., Radding, C. M. 1971. *Proc. Natl. Acad. Sci. USA* 68:1639–43
162. Sigal, N., Alberts, B. 1972. *J. Mol. Biol.* 71:789–93
163. Radding, C. M. 1973. *Ann. Rev. Genet.* 7:87–111
164. Sriprakash, K. S., Lundh, N., Moo-On-Huh, M., Radding, C. M. 1975. *J. Biol. Chem.* 250:5438–45
165. Lee, C. W., Davis, R. W., Davidson, N. 1970. *J. Mol. Biol.* 48:1–22
166. Broker, T. R., Lehman, I. R. 1971. *J. Mol. Biol.* 60:131–49
167. Broker, T. R. 1973. *J. Mol. Biol.* 81:1–16
168. Holliday, R. 1964. *Genet. Res. Cambridge* 5:282–304
169. Doniger, J., Warner, R. C., Tessman, I. 1973. *Nature New Biol.* 242:9–12
170. Thompson, B. J., Escarmis, C., Parker, B., Slater, W. C., Doniger, J., Tessman, I., Warner, R. C. 1975. *J. Mol. Biol.* 91:409–19
171. Valenzuela, M. S., Inman, R. B. 1975. *Proc. Natl. Acad. Sci. USA* 72:3024–28
172. Potter, H., Dressler, D. 1976. *Proc. Natl. Acad. Sci. USA* 73:3000–4
173. Köhnlein, W., Hutchinson, F. 1976. *Mol. Gen. Genet.* 144:323–31
174. Meselson, M. 1972. *J. Mol. Biol.* 71:795–98
175. Thompson, B. J., Camien, M. N., Warner, R. C. 1976. *Proc. Natl. Acad. Sci. USA* 73:2299–2303
176. Tsujimoto, Y., Ogawa, H. 1977. *J. Mol. Biol.* 109:423–36
177. Wildenberg, J., Meselson, M. 1975. *Proc. Natl. Acad. Sci. USA* 72:2202–6
178. Robberson, D. L., Clayton, D. A. 1973. *J. Biol. Chem.* 248:4512–14
179. Radding, C. M., Beattie, K. L., Holloman, W. K., Wiegand, R. C. 1977. *J. Mol. Biol.* 116:825–39
180. Sobell, H. M. 1974. In *Mechanisms in Recombination,* ed. R. F. Grell, pp. 433–38. New York: Plenum
181. Fogel, S., Mortimer, R. K. 1974. *Genetics* 77: Suppl. 22
182. Stadler, D. R., Towe, A. M. 1971. *Genetics* 68:401–13
183. Stadler, D. R. 1973. *Ann. Rev. Genet.* 7:113–27
184. Chipchase, M. 1976. *J. Theor. Biol.* 57:249–79
185. Leblon, G. 1972. *Mol. Gen. Genet.* 115:36–48
186. Leblon, G. 1972. *Mol. Gen. Genet.* 116:322–35
187. Yu-Sun, C. C., Wickramaratne, M. R. T., Whitehouse, H. L. K. 1977. *Genet. Res. Cambridge* 29:65–81
188. Leblon, G., Rossignol, J.-L. 1973. *Mol. Gen. Genet.* 122:165–82
189. Gutz, H. 1971. *Genetics* 69:317–37
190. Fink, G. R. 1974. In *Mechanisms in Recombination,* ed. R. F. Grell, pp. 287–93. New York: Plenum
191. Fink, G. R., Styles, C. A. 1974. *Genetics* 77:231–44
192. Lawrence, C. W., Sherman, F., Jackson, M., Gilmore, R. A. 1975. *Genetics* 81:615–29
193. Meselson, M. 1965. In *Ideas in Modern Biology,* ed. J. A. Moore, pp. 1–16. Garden City, NY: Natural History Press
194. Siegel, J. 1974. *J. Mol. Biol.* 88:619–28
195. Stahl, F. W., McMilin, K. D., Stahl, M. M., Nozu, Y. 1972. *Proc. Natl. Acad. Sci. USA* 69:3598–3601
196. White, R. L., Fox, M. S. 1974. *Proc. Natl. Acad. Sci. USA* 71:1544–48
197. White, R. L., Fox, M. S. 1975. *Genetics* 31:33–50
198. Spatz, H. C., Trautner, T. A. 1970. *Mol. Gen. Genet.* 109:84–106
199. Crawford, I. P., Preiss, J. 1972. *J. Mol. Biol.* 71:717–33
200. Norkin, L. C. 1970. *J. Mol. Biol.* 51:633–55
201. Stadler, D., Kariya, B. 1973. *Genetics* 75:423–39
202. Bresler, S. E., Kreneva, R. A., Kushev, V. V. 1971. *Mol. Gen. Genet.* 113:204–13

203. Berger, H., Pardoll, D. 1976. *J. Virol.* 20:441–45
204. Benz, W. C., Berger, H. 1973. *Genetics* 73:1–11
205. Ephrussi-Taylor, H., Gray, T. C. 1966. *J. Gen. Physiol.* 49:211–31
206. Ephrussi-Taylor, H., Sicard, A. M., Kamen, R. 1965. *Genetics* 51:455–75
207. Fox, M. S. 1957. *Biochim. Biophys. Acta* 26:83–85
208. Lerman, L. S., Tolmach, L. J. 1957. *Biochim. Biophys. Acta* 26:68–82
209. Lacks, S. 1970. *J. Bacteriol.* 101:373–83
210. Tiraby, J.-G., Fox, M. S. 1973. *Proc. Natl. Acad. Sci. USA* 70:3541–45
211. Roger, M. 1977. *J. Bacteriol.* 129:298–304
212. Shoemaker, N. B., Guild, W. R. 1974. *Mol. Gen. Genet.* 128:283–90
213. Tiraby, J.-G., Fox, M. S. 1974. *Genetics* 77:449–58
214. Guild, W. R., Shoemaker, N. B. 1976. *J. Bacteriol.* 125:125–35
215. Doerfler, W., Hogness, D. S. 1968. *J. Mol. Biol.* 33:661–78
216. Baas, P. D., Jansz, H. S. 1972. *J. Mol. Biol.* 63:557–68
217. Enea, V., Vovis, G. F., Zinder, N. D. 1975. *J. Mol. Biol.* 96:495–509
218. Schlaeger, E. J., Spatz, H. C. 1974. *Mol. Gen. Genet.* 130:165–75
219. White, R. L., Fox, M. S. 1975. *Mol. Gen. Genet.* 141:163–71
220. Nevers, P., Spatz, H.-C. 1975. *Mol. Gen. Genet.* 139:233–43
221. Siegel, E. C. 1973. *J. Bacteriol.* 113:145–60
222. Mattern, I. E. 1971. In *1st Eur. Biophys. Cong.*, ed. E. Broda, A. Locker, H. Springer-Lederer, pp. 237–40. Vienna: Wiener Med. Akad.
223. Lai, C. J., Nathans, D. 1975. *Virology* 66:70–81
224. Nathans, D., Lai, C. J. 1975. *Cellular Modifications of Transfecting SV40 DNA.* Pasadena, Calif: Elsevier
225. Miller, L. K., Cooke, B. E., Fried, M. 1976. *Proc. Natl. Acad. Sci. USA* 73:3073–77
226. Nathans, D., Smith, H. O. 1975. *Ann. Rev. Biochem.* 44:273–93
227. Gates, F. T. III, Linn, S. 1977. *J. Biol. Chem.* 252:1647–53
228. Gates, F. T. III, Linn, S. 1977. *J. Biol. Chem.* 252:2802–7
229. Radman, M. 1976. *J. Biol. Chem.* 251:1438–45
230. Ljungquist, S., Lindahl, T., Howard-Flanders, P. 1976. *J. Bacteriol.* 126:646–53
231. Ljungquist, S. 1977. *J. Biol. Chem.* 252:2808–14
232. Braun, A., Grossman, L. 1974. *Proc.* *Natl. Acad. Sci. USA* 71:1838–42
233. Verly, W. G., Rassart, E. 1975. *J. Biol. Chem.* 250:8214–19
234. Grossman, L., Braun, A., Feldberg, R., Mahler, I. 1975. *Ann. Rev. Biochem.* 44:19–43
235. Holliday, R., Dickson, J. M. 1977. *Mol. Gen. Genet.* 153:331–35
236. DiCaprio, L., Hastings, P. J. 1976. *Mutation Res.* 37:137–40
237. Guild, W. R., Shoemaker, N. B. 1974. *Mol. Gen. Genet.* 128:291–300
238. Holloman, W. K., Holliday, R. 1973. *J. Biol. Chem.* 248:8107–13
239. Badman, R. 1972. *Genet. Res. Cambridge* 20:213–29
240. Holloman, W. K. 1973. *J. Biol. Chem.* 248:8114–19
241. Holliday, R. Holloman, W. K., Banks, G. R., Unrau, P., Pugh, J. E. 1974. In *Mechanisms in Recombination*, ed. R. F. Grell, pp. 239–62. New York: Plenum
242. Ahmad, A., Holloman, W. K., Holliday, R. 1975. *Nature* 258:54–56
243. Shishido, K., Ando, T. 1975. *Biochim. Biophys. Acta* 390:125–32
244. Wiegand, R. C., Godson, G. N., Radding, C. M. 1975. *J. Biol. Chem.* 250:8848–55
245. Kedzierski, W., Laskowski, M. Sr. 1973. *J. Biol. Chem.* 248:1277–80
246. Linn, S., Lehman, I. R. 1965. *J. Biol. Chem.* 240:1294–1304
247. Otto, B., Knippers, R. 1976. *Eur. J. Biochem.* 71:617–22
248. Pedrini, A. M., Ranzani, G., Pedrali Noy, G. C. F., Spadari, S., Falaschi, A. 1976. *Eur. J. Biochem.* 70:275–83
249. Ciarrocchi, G., Fortunato, A., Cobianchi, F., Falaschi, A. 1976. *Eur. J. Biochem.* 61:487–92
250. Shenk, T. E., Rhodes, C., Rigby, P. W. J., Berg, P. 1975. *Proc. Natl. Acad. Sci. USA* 72:989–93
251. Dodgson, J. B., Wells, R. D. 1977. *Biochemistry* 16:2374–79
252. Clark, A. J. 1973. *Ann. Rev. Genet.* 7:67–86
253. Clark, A. J., Margulies, A. D. 1965. *Proc. Natl. Acad. Sci. USA* 53:451–59
254. Castellazzi, M., Morand, P., George, J., Buttin, G. 1977. *Mol. Gen. Genet.* 153:297–310
255. Morand, P., Blanco, M., Devoret, R. 1977. *J. Bacteriol.* 131:572–82
256. Miura, A., Tomizawa, J.-I. 1968. *Mol. Gen. Genet.* 103:1–10
257. George, J., Devoret, R., Radman, M. 1974. *Proc. Natl. Acad. Sci. USA* 71:144–47
258. Witkin, E. M. 1976. *Bacteriol. Rev.* 40:869–907

259. Morand, P., Goze, A., Devoret, R. 1977. *Mol. Gen. Genet.* 157:69–82

260. Tomizawa, J., Ogawa, H. 1968. *Cold Spring Harbor Symp. Quant. Biol.* 33:243–50

261. McEntee, K., Hesse, J. E., Epstein, W. 1976. *Proc. Natl. Acad. Sci. USA* 73:3979–83

262. McEntee, K. 1977. *Proc. Natl. Acad. Sci. USA* 74:5275–79

263. Inouye, M. 1971. *J. Bacteriol.* 106:539–42

264. Gudas, L. J., Pardee, A. B. 1975. *Proc. Natl. Acad. Sci. USA* 72:2330–34

265. Gudas, L. J., Pardee, A. B. 1976. *J. Mol. Biol.* 101:459–77

266. Sedgwick, S. G. 1975. *Nature* 255:349–50

267. Gudas, L., Mount, D. W. 1977. *Proc. Natl. Acad. Sci. USA* 74:5280–84

267a. Little, J. W., Kleid, D. 1977. *J. Biol. Chem.* 252:6251–52

267b. Emmerson, P. T., West, S. C. 1977. *Mol. Gen. Genet.* 155:77–85

267c. Roberts, J. W., Roberts, C. W., Mount, D. W. 1977. *Proc. Natl. Acad. Sci. USA* 74:2283–87

268. Capaldo, F. N., Ramsey, G., Barbour, S. D. 1974. *J. Bacteriol.* 118:242–49

269. Wilcox, K. W., Smith, H. O. 1975. *J. Bacteriol.* 122:443–53

270. Kooistra, J., Venema, G. 1976. *J. Bacteriol.* 128:549–56

271. Hall, J. D., Howard-Flanders, P. 1972. *J. Bacteriol.* 110:578–84

272. Lin, P. F., Bardwell, E., Howard-Flanders, P. 1977. *Proc. Natl. Acad. Sci. USA* 74:291–95

273. Tessman, I. 1968. *Science* 161:481–82

274. Benbow, R. M., Hutchison, C. A. III, Fabricant, J. D., Sinsheimer, R. L. 1971. *J. Virol.* 7:549–58

275. Kobayashi, I., Ikeda, H. 1977. *Mol. Gen. Genet.* 153:237–45

276. Birge, E. A., Low, K. B. 1974. *J. Mol. Biol.* 83:447–57

277. Bergmans, H. E. N., Hoekstra, W. P. M., Zuidweg, E. M. 1975. *Mol. Gen. Genet.* 137:1–10

278. Vovis, G. F. 1973. *J. Bacteriol.* 113:718–23

279. Kushner, S. R. 1974. *J. Bacteriol.* 120:1219–22

280. Benzinger, R., Enquist, L. W., Skalka, A. 1975. *J. Virol.* 15:861–71

281. Sadowski, P. D., Vetter, D. 1976. *Proc. Natl. Acad. Sci. USA* 73:692–96

282. Sadowski, P. D. 1977. *Virology* 78:192–202

283. Vlachopoulou, P. J., Sadowski, P. D. 1977. *Virology* 78:203–15

284. Kerr, C., Sadowski, P. D. 1972. *J. Biol. Chem.* 247:311–18

Ann. Rev. Biochem. 1978. 47:881–931
Copyright © 1978 by Annual Reviews Inc. All rights reserved

CHEMICAL APPROACHES TO THE STUDY OF ENZYMES CATALYZING REDOX TRANSFORMATIONS[1]

❖992

Christopher Walsh

Departments of Chemistry and Biology, Massachusetts Institute of Technology, Cambridge, Massachusetts 02139

CONTENTS

[1]Literature review was terminated on August 15, 1977.

PERSPECTIVES AND SUMMARY

Enzymatic oxidation-reduction sequences lie at the heart of cellular energy metabolism since the energy released in oxidation of reduced organic or inorganic compounds is captured with varying efficiencies in useful forms such as ATP, membrane potentials, or reduced coenzymes. Given this physiological role, mechanistic scrutiny of the various categories of enzymes catalyzing these electron transfer processes has been actively pursued. When electrons are removed from a molecule undergoing oxidation they must be passed to a cosubstrate undergoing concomitant reduction. Examination of the functional groups in proteins indicates no obvious electron sinks for transient electron storage and almost without exception, oxidoreductases require conjugated organic cofactors or redox active transition metals such as Cu, Fe, and Mo as conduits to facilitate low energy passage of electrons out of substrate into product. The conjugated cofactors include nicotinamide coenzymes for simple dehydrogenations that are probably two electron transfer processes. Flavins, pterins, and heme coenzymes can be used in enzymes where one electron transfers occur, and all three, in their reduced forms, can pass electrons to O_2 enzymically.

Just as water is a ubiquitous nucleophile for cellular group transfer reactions, molecular oxygen is a common electron acceptor in aerobic organisms. O_2 can experience enzyme-mediated four electron reduction to water, two electron reduction to H_2O_2, or it can be reductively activated for transfer of one of its oxygen atoms (monooxygenation) or both oxygen atoms (dioxygenation). Oxygenation can be used for biosynthetic purposes, for example, catecholamine or corticosteroid formation, for degradation of aromatic rings for example, tyrosine or tryptophan catabolism, or for metabolic activation of various drugs and carcinogens in detoxification sequences. Chemical studies on oxygen-utilizing enzymes have made significant progress on the mechanisms of all these processes in recent years.

Alternatives to conjugated organic cofactors in enzymatically mediated electron transfers are redox active metal ions. While much effort continues to go into biophysical studies, the scope and mechanism of copper, iron, and molybdenum biochemistry, and the beginnings of selenium involvement, are being unraveled by bioorganic and bioinorganic approaches.

Many organic substrates are probably oxidized in heterolytic (ionic) bond cleavage processes. Homolytic (radical) mechanisms may be more prevalent among inorganic molecules. With the probable exception of NAD-dependent dehydrogenations it is likely that C–H bonds in oxidizable substrates are usually broken to produce H^+ and carbanions, electron-rich species that are then readily oxidized both by one and two electron processes. A particularly clear cut sequence of oxidations by way of stabilized carbanionic intermediates is that of pyridoxal-P dependent reactions at the α, β, and γ-carbons of amino acids.

This article attempts to review recent experimental work related to the issues raised in the preceding paragraphs and to illustrate the power of chemical approaches to the unraveling of reaction mechanisms of the various categories of redox enzymes. General mechanistic criteria and methodological tools are evaluated in the context of specific enzymatic reactions. Discussion focuses on partitioning data, isotope effects, use of substrate and coenzyme analogues including mechanism-based inactivators, spectroscopic detection of intermediates, and their structural characterization, among others. The aim is to tie in diverse enzyme phenomenology, particularly where O_2 is cosubstrate for some reductive chemistry, with chemically sound or at least permissible patterns of electron flow. Interwoven with these threads is an attempt to codify various biological redox transformations and to gain some predictive ability concerning the kind of enzyme-cofactor combination that may be used to solve a given chemical problem in redox metabolism.

INTRODUCTION

This chapter reviews some recent experiments involved in chemical approaches to the study of enzyme catalysts. In a companion chapter in this volume I. A. Rose and M. Wimmer present criteria for the detection and identification of intermediates in enzymatic reactions, with focus on enzymes catalyzing phosphoryl, carboxyl, or methyl transfers. To complement their chapter the focus of this chapter is on enzymes that carry out oxidation/reduction reactions on their substrates. The chemical criteria for analysis of reaction mechanisms apply equally well, of course, to the topics of both these chapters. Coverage of simple NAD-dependent dehydrogenases is excluded because a separate chapter in this area was originally scheduled for the present volume. This chapter presents selected, analytical coverage of a few topics from the recent redox enzyme literature rather than a descriptive encyclopedic listing of all published papers on the subject. The first half deals with chemical studies on redox enzymes, the second half with recent work on pyridoxal-P-dependent enzymes that also can be analyzed in terms of redox catalysis. Biophysical studies, also critical in unraveling redox enzyme mechanisms, are not covered for lack of space.

Most organic molecules undergo oxidation by loss of two electrons in enzymatic reactions, a few lose four electrons. When one substrate molecule is oxidized, another must be reduced: the electrons removed from the molecule undergoing oxidation must be transferred to some acceptor. Molecular oxygen is a common electron acceptor in aerobic systems, and undergoes two electron reduction to H_2O_2 or four electron reduction to water. Some inorganic molecules such as sulfite or dinitrogen can accept six electrons while bound at an enzyme active site, and undergo smooth reduction to 2 moles of sulfide or ammonia respectively.

Enzymes that promote the electron transfers between substrate molecules accelerate reaction velocities by lowering activation energies. A prevalent mechanism, used almost without exception for redox enzymes, is the obligate requirement for an organic coenzyme molecule to act as electron acceptor, or donor in the reverse direction, or a transition metal to act as a conduit for low energy passage of electrons out of one substrate molecule to an acceptor molecule. The organic coenzymes for redox catalysis are, most frequently, nicotinamides and flavins, and less frequently, pterins. Redox active transition metals include iron, copper, and molybdenum. Recent experiments on the ways in which the chemistry open to either type of these cofactors is utilized by the biological system will be examined. Reductive oxygen metabolism involves not only the production of H_2O_2 or H_2O but also the controlled activation of O_2 as an oxygen transfer agent, both for hydroxylations (monooxygenation) and for dioxygenase catalysis.

The coenzyme pyridoxal-phosphate is admirably designed for the stabilization of carbanionic intermediates during enzymatic catalysis, with consequent rate accelerations. A fruitful mechanistic analysis of participation in redox reactions is provided by examination of reactions at the α-, β-, or γ-carbons of the reacting amino acid substrates.

REDOX ENZYMES

Flavin-Dependent Oxidoreductases

The two major redox coenzymes in simple enzymatic dehydrogenations are nicotinamide-based coenzymes and flavin coenzymes. There is some consensus that a hydride ion is transferred between NADH and substrate. At present there is no similar mechanistic consensus about flavoenzyme oxidative catalysis. While dihydronicotinamides are inert to oxidation by O_2 on a useful biological time scale, dihydroflavins are not, which reflects the probable restriction of nicotinamide coenzymes to two electron chemistry whereas the flavins can participate in both one and two electron chemistry in enzymatic processes.

One can categorize flavoenzymes in two ways and Table 1 [modified from (1, 2)] does this on the basis of the type of functional group in the substrate that is oxidized. Table 2 arranges categories according to the nature of the electron acceptor which reoxidizes the reduced flavin cofactor in the second half reaction.

Table 1 Substrate classes for flavoenzyme catalyses

Reduced form	Oxidized form	Example
1. $-\overset{\overset{\displaystyle H}{\vert}}{\underset{\underset{\displaystyle OH}{\vert}}{C}}-$	$-\overset{\overset{\displaystyle }{\Vert}}{\underset{\underset{\displaystyle O}{}}{C}}-$	E. coli membrane D-lactate dehydrogenase; glucose oxidase
2. $-\overset{\overset{\displaystyle H}{\vert}}{\underset{\underset{\displaystyle NH_2}{\vert}}{C}}-$	$\overset{\overset{\displaystyle }{\Vert}}{\underset{\underset{\displaystyle O}{}}{C}} + NH_4^+$	D- and L-amino acid oxidases; amine oxidases
3. $-CH-CH-\underset{\underset{\displaystyle O}{\Vert}}{C}-$	$-\overset{\vert}{C}=\overset{\vert}{C}-\underset{\underset{\displaystyle O}{\Vert}}{C}-$	Succinate dehydrogenase; acyl CoA dehydrogenases; α-glycerophosphate dehydrogenases
4. NADH	NAD$^+$	NADH dehydrogenases; transhydrogenases; dihydroorotate dehydrogenases
5. SH SH	S——S	Lipoamide dehydrogenase; glutathione reductase
6. SO_3^{2-}	S^{2-}	Sulfite reductase

Category 1 of Table 2, which comprises the flavoprotein dehydrogenases, is physiologically a key one in biological energy metabolism. The functional distinction between flavoenzyme dehydrogenases and oxidases is that in the flavin-dehydrogenase holoenzyme complexes, the intrinsic high reactivity of reduced flavins with O_2 has been *suppressed*. O_2 is not a rapid physiological reoxidant: either O_2 is excluded from the active site or it cannot achieve proper geometry when bound for initiation of electron transfer. Instead, the physiological electron acceptors for flavoprotein dehydrogenases are molecules such as quinones, for example, coenzyme Q and Vitamin K, heme-iron-containing cytochromes, such as, cytochrome b_5 and cytochrome P_{450}, or nonheme iron-sulfur cluster proteins such as adrenodoxin, in the adrenal steroid hydroxylation sequences, or putidaredoxin, in the *pseudomonad* camphor hydroxylation multicomponent system.

FLAVOPROTEIN OXIDASES AND DEHYDROGENASES   The majority of known flavoprotein dehydrogenases—Escherichia coli D-lactate dehydrogenase (3), D-alanine dehydrogenase (4), succinate dehydrogenase (5), acyl CoA dehydrogenase (6), NADH dehydrogenase (7), and dihydroorotate dehydrogenase (8), are membrane bound in keeping with their role as

Table 2 Classification of flavoenzymes by electron acceptors

Category	Comment
1. Dehydrogenases	Do not use O_2 as acceptor. Acceptors are often $1e^-$ acceptors such as quinones and cytochromes
2. Oxidases	O_2 is the electron acceptor; reduced by $2e^-$ to H_2O_2
3. Oxidase-decarboxylases	O_2 the acceptor; reduced by $4e^-$ to H_2O
4. Hydroxylases	O_2 the acceptor; one atom ends up in H_2O, one in hydroxylated product
5. Metalloflavoenzymes	Require a bound transition metal ion for catalysis; Fe^{II} or Fe^{III} and Mo^{VI}; may be hydrogenases or oxidases
6. Flavodoxins	$1e^-$ transfer proteins; semiquinone form clearly involved

electron carriers in energy-yielding membrane-associated redox chains. This localization poses purification problems and has slowed mechanistic study (8a).

Mechanisms of electron transfer between substrate and flavin coenzyme
The soluble flavoprotein oxidases and monooxygenases have been examined in much more detail. Among the favorite oxidases for mechanistic study is pig kidney D-amino acid oxidase. It may be a model for other flavoenzymes oxidizing amino acids or hydroxy acids.

Previous studies with β-chloroalanine (9, 10), which partitions between oxidation to β-chloropyruvate and NH_4^+, or α,β-elimination to pyruvate, Cl^-, and NH_4^+, had been interpreted in support of an α-carbanionic intermediate or transition state at an early stage in catalysis. Direct evidence that the oxidase could process carbanions was provided by elegant studies showing that nitroalkane anions and O_2 were converted catalytically to nitrite, H_2O_2, and the corresponding aldehyde (11). In a study with β-haloalanines using β-fluoro-, β-chloro-, or β-bromoalanine (16), the partitioning ratio between oxidation to β-halo-α-imino acid and α,β-elimination of HX to amino acrylate, which ketonizes and hydrolyzes to pyruvate, was controlled by the leaving group tendency of the halogen. Thus β-bromoalanine gave only pyruvate, Br^-, and NH_4^+, β-chloroalanine gave mixtures, and β-fluoroalanine yielded β-fluoropyruvate exclusively, which supports the idea that the partitioning intermediate is the α-carbanion (I) (12, 13, 14).

X = F, Cl, Br

I

A major unresolved question is how electrons are subsequently transferred from such a carbanionic species to the flavin coenzyme. Arguments in favor of covalent adducts (11, 13) and charge transfer complexes (15) had been advanced, as well as the idea that the redox process may occur one electron at a time to produce the substrate radical and the flavin semiquinone as fleeting species before passage of the second electron (14, 16, 17). A covalent flavin N^5-amino acetonitrile adduct can be isolated when cyanide ion is added to D-amino acid oxidase processing preformed nitroethane anion (11). In catalytic turnover presumably a product N^5-imine is trapped as shown by equation 1. The imine progenitor to aldehyde product, could arise from expulsion of the nitro group as nitrite ion. Cyanide trapping leads to irreversible inactivation. Porter and Bright have argued that such an adduct (11, 13) may represent an intermediate for a normal amino acid

1.

substrate. Model N^5-adducts are known (18, 19), including the N^5-sulfite adduct (20) which could be an intermediate in bacterial AMP-sulfate reductase catalysis (21), but Bruice has argued that such adducts may be the result of nonproductive side equilibria and not be kinetically competent reaction intermediates in model reductions of carbonyl compounds by dihydroflavins (14). Recently Alston, Mela & Bright (22) have observed that the preformed carbanion of 3-nitropropionate irreversibly inactivates succinate dehydrogenase activity in mitochondria and may be the basis for the toxic properties of this plant metabolite when ingested by livestock. They argue that 3-nitropropionate is isoelectronic with succinate, and that the inactivator carbanion may add to N^5 of the FAD cofactor and then lose an adjacent hydrogen to yield a stable eneamine adduct as the modified enzyme (equation 2). Structural verification of this supposition will be difficult

2.

due to the covalent linkage of FAD in the membraneous succinate dehydrogenase to N_3 of a histidyl residue of the protein. In this connection,

Kenney, Singer, Edmondson, and their colleagues have recently analyzed the structure of some other covalently linked flavoenzymes (23, 24). The linkage is also N_3-his in D-6-hydroxy nicotine oxidase and bacterial sarcosine dehydrogenase, but N_1-his in thiamine dehydrogenase from a soil bacterium (25), in β-cyclopiazonate oxidocyclase from a Penicillium mold (26), and in L-gulono-γ-lactone oxidase, a liver microsomal enzyme performing the last step in ascorbate biosynthesis (27). No mechanistic imperatives for covalent versus noncovalent attachment of FAD have been provided yet, but the mechanism of covalent loading through the C_8methylene group of the coenzyme may be accessible with in vitro preparations of the D-6-hydroxynicotine oxidase (28).

The N_5 locus of flavins is not the only site postulated for covalent adduct formation. Model data (29, 30) have shown that thiols can add to carbon 4a of oxidized flavins and such a species may form during dihydrolipoamide dehydrogenase catalysis. The FAD of that monocarboxymethylated enzyme can still be partially reduced by 1 mM NAD with formation of a new bond at 384 nm ($\epsilon = 8.7$ mM^{-1}). Thorpe & Williams (31) suggest that on binding of NAD the 4a adduct between an enzyme cysteinyl group and FAD is stabilized and that this may reflect that the adduct is a catalytically significant species in disulfide reduction. [Without derivitizing the other active site-SH as S-CH$_2$COO$^-$ the adduct (II) would not accumulate.]

II

We note later in this review that the 4a bridgehead carbon is also the site of interaction with molecular oxygen.

Returning to the β-halo-α-amino acids and D-amino acid oxidase, we note that the α,β-elimination reaction stereochemistry has been studied with 2-[^{3}H]-2R,3R-2-amino-3-chlorobutyrate (D-erythro) and the 2-[^{3}H]-2R,3S-diastereomer (D-threo) (32). The α-^{3}H is removed as a proton and, after Cl$^-$ elimination, is specifically added back to the resulting bound eneamine, part of the time, for a net intramolecular transfer to C_3 of the developing 2-imino-butyrate. For the D-erythroisomer this triton transfer occurs 7.3% of the time and 2.6% of the time for the D-threo. The resultant 2-imino-3-[^{3}H]-butyrate was analyzed for chirality by decarboxylation with H$_2$O$_2$ to 2-[^{3}H]-propionate, conversion to propionyl CoA, and carboxylation by propionyl CoA carboxylase, which is known to remove only the 2R proton on carboxylation. Surprisingly both D-erythro- and D-threo-

2-[^{3}H]-2-amino-3-chlorobutyrates yield 2-imino-2-R-[^{3}H]-ketobutyrate as product. The steric course of HCl loss to net *syn* for D-erythro and net *anti* for D-threo is a remarkable changeover. Both observed paths lead to a common trans-eneamino acid bound at the active sites as the initial elimination product and this transition state to trans-eneamine may be lower than to cis-eneamine from either substrate (32). There is a similar syn/anti elimination course for threonine and allothreonine deamination by sheep liver threonine deaminase (33) and the same argument for a common trans-eneamino acid initial bound product could hold there. Among other things, these data show that the α-hydrogen of substrate is held as a proton at the D-amino acid oxidase active site, sequestered from interaction with bulk solvent. This kinetic inaccessibility to bulk water also holds for physiological amino acids as the elegant double stopped flow experiments of Bright, Voet, & Porter with D-2-[^{2}H]-phenylalanine have demonstrated (34).

We noted above that Bruice has argued that substrate carbanions may not form covalent enzyme-flavin adducts but rather may oxidize subsequently by radical paths (14, 16). In fact Bright & Porter had suggested that when nitroalkane anions are mixed with glucose oxidase from *Aspergillus niger,* some flavin semiquinone accumulated (35). Chan & Bruice have now used 5-hydroxy carbonyl substrates such as III as glucose analogues and the

III   IV

nitroxide IV as an $^3O_2$ analogue with the enzyme (36). They find carbanion formation, then large amounts of E-FADH· formation both in coenzyme reduction and reoxidation. Indeed the E-flavin semiquinone could be generated quantitatively by incubation of enz-FAD, glucose, and nitroxide briefly followed by gel filtration on a Sephadex G-25 column. They suggest that similar one electron redox chemistry occurs normally and Williams and Bruice have calculated for various model reactions that activation energies for radical pairs are acceptably within activation energies for the overall reactions (16). Porter and Bright do not support this as a necessary mechanism with glucose oxidase (35).

Mechanism-based inactivators of flavoenzymes   Over the past five years a variety of mechanism-based inactivators, or suicide substrates, have been designed to inactivate flavin-linked enzymes, many of them based on the premise that carbanions are intermediates in flavoenzyme catalysis and that generation of a carbanion in a substrate analogue might activate some chemically latent functional group (often an acetylene) for rearrangement to an electrophilic, conjugated allene (37–39). Various criteria for mechanistic evaluation of enzymatic suicide substrates have been elaborated elsewhere (37–39). The nitroalkane anions possess certain properties of this type of inactivator. The utility of a suicide substrate can be several-fold. If a specific kind of chemical intermediate must be formed to uncover the latently reactive group and if inactivation is observed, then the initial mechanistic postulate is supported. If the enzyme nucleophile that has added can be identified in the covalent, inactive adduct, then that active site nucleophile may be important in catalysis. In passing it is noted that such compounds may offer maximal in vivo specificity as rationally designed drugs (40), a point that is reiterated in the pyridoxal section.

With acetylenic substrate analogues for *Mycobacterial* L-hydroxy acid oxidase (41, 42) and for mitochondrial monoamine oxidase (43), namely 2-hydroxy-3-butynoate and N,N-dimethylpropynylamine, which satisfy kinetic and chemical criteria (e.g., first order loss of activity, and stoichiometric active site labeling) for suicide substrates, the structure of the covalent adducts have now been identified by Fourier transform NMR, degradation, and preparation of model compounds. In both instances the inactivator reacts with the flavin coenzyme not with the apoprotein. Hydroxybutynoate (44) yields the C_{4a},N_5 cyclic adduct V (equation 1) while the propynylamine (45) yields an acylic flavocyanine linked at N_5 of the cofactor [also see (46, 47) for model studies].

$$V \qquad\qquad VI \qquad\qquad 3.$$

The structures of these two adducts confirm that C_{4a} and N_5 are loci proximal to and reactive with these suicide substrates although in the hydroxy acid oxidase N^5 has attacked the oxidizable center while in monoamine oxidase that atom has attacked the terminal carbon of the inactivator. Reaction with the flavin coenzyme exclusively is an intriguing observation. The compound pargyline VII also inactivates monoamine oxidase by com-

VII

bination (structurally uncharacterized) with the covalently bound FAD of the enzyme (48).

In six distinct 2-hydroxy acid oxidizing flavoenzymes, 2-hydroxy-3-butynoate is an irreversible inactivator (49), modifying the coenzyme and not the apoprotein. The adduct may be a cyclic C_6, N_5 structure in the *Peptidostreptococcus elsdenii* D-lactate dehydrogenase (S. Ghisla, personal communication). In all of those enzymes hydroxybutynoate functions as a substrate, partitioning between turnover to product and enzyme inactivation by covalent modification. This partition ratio ($k_3 : k_4$) (equation 4) is an important number to determine (38) for a suicide substrate since it may

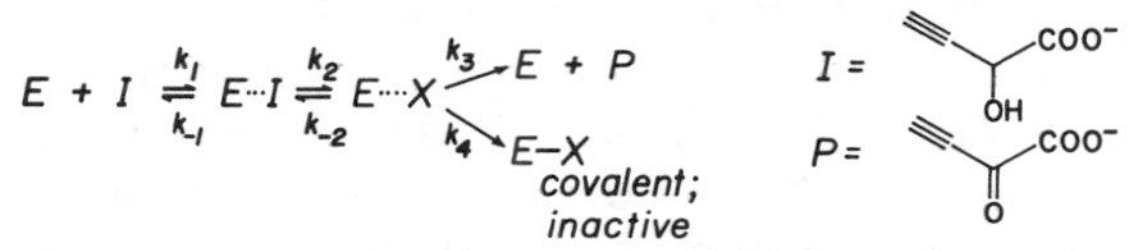

govern in vivo specificity, especially if the product P is a reactive electrophile released into solution, as is 2-keto-3-butynoate in this case. For several of these hydroxy acid oxidases, about 100 turnovers occur before inactivation. Given that 2-keto-3-butynoate is reactive, it could qualify as inactivator before release from the active site, and undergo attack by the eneamine linkage (C_{4a}) of the dihydroflavin (there is no nucleophile in oxidized flavin). Alternatively an initial acetylenic carbanion may form. It has some resonance contribution from the allenic anion form IX, which would admittedly

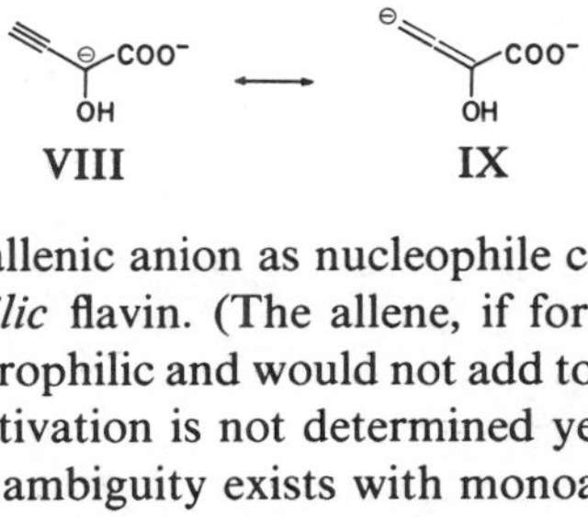

VIII   IX

be very unstable. The allenic anion as nucleophile could now attack C_{4a} of the oxidized *electrophilic* flavin. (The allene, if formed, cannot be protonated for then it is electrophilic and would not add to oxidized flavin.) Thus, the mechanism of inactivation is not determined yet for the hydroxy acid oxidases and the same ambiguity exists with monoamine oxidase. The use of 2-hydroxy-3-butenoate (vinylglycolate) X was an attempt to resolve this ambiguity (49, 50). It can inactivate only by oxidation to bound product and Michael attack by dihydroflavin. In every case vinylglycolate is readily oxidized but does not cause inactivation by flavin modification; while supporting the allenic anion hypothesis these data are not ironclad since

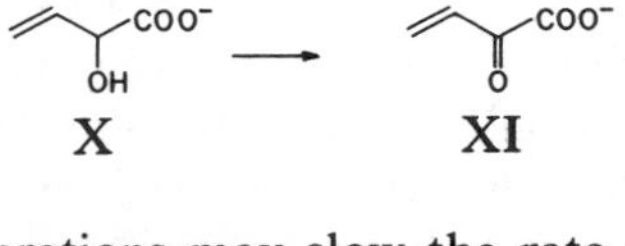

X   XI

stereoelectronic considerations may slow the rate of dihydroflavin to the conjugated olefin compared to the conjugated acetylenic keto acid, and permit selective release of XI before it undergoes attack (42, 49). (However, no flavin adducts form even after ~10,000 turnovers in several enzymes (49).) It is worth noting that the hydroxybutynoate may be mechanistically specific for flavin-linked hydroxy acid dehydrogenases and not for similar NAD-linked enzymes. Hydroxybutynoate and vinylglycolate have been utilized with *E. coli* membrane vesicles containing a flavoprotein D-lactate dehydrogenase to generate the electrophilic keto acids that are specific inactivators of the PEP-dependent phosphotransferase system responsible for hexose active transport in many bacteria (51).

In 1975 Horiike et al (52) reported that acetylenic D-propargylglycine (2-amino-4-pentynoate) XII, (a natural product as the L-isomer (53)) inactivated D-amino acid oxidase by covalent modification of the apoprotein, not the coenzyme, and generated a new chromophore at 317 nm. Loss of activity occurred after approximately 600 turnovers. In a study of inactivation mechanism Marcotte & Walsh (54) reported that L-amino acid oxidase turns over L-propargylglycine without inactivation and accumulates chromophoric products that differ depending on the buffer nucleophile; the same products were produced by D-amino acid oxidase before loss of activity which suggested that the initial oxidation product, 2-imino-4-pentynoate rearranges nonenzymically to 2-imino-3,4-pentadienoate, which is a conjugated, electrophilic allenic keto acid, XIII (equation 5). This is the likely killing species which can also be trapped by buffer nucleophiles such as butylamine. The butylamine adduct absorbed at 317 nm ($\epsilon = 15,000$).

XIII

5.

With 2-[^{14}C] propargylglycine, inactivated D-amino acid oxidase had 1–2 labels per subunit and the 317 nm absorbance was calculated as $\epsilon \approx 15,000$, which suggests the addition of an ϵ-amino group of a lysine residue to the inactivator still bound at or near the active site (54).

In fact, modified enzyme was not completely inactive but rather showed specificity different from that shown by native enzyme toward various D-amino acids (55). The remaining enzyme activity, ~20% with D-alanine, and ~50% with D-phenylalanine, was insensitive to more propargylglycine. The modified enzyme bound competitive inhibitors benzoate and anthranilate 50 to 100-fold less well and product release was no longer the rate-determining step, all of which suggests substrate binding was impaired. Coenzyme binding was unaffected. Thus propargylglycine is not a suicide substrate; the enzyme only "wounds" itself quite specifically and the mechanism is distinct from the acetylenic inactivators noted in preceding paragraphs.

In a similar category are experiments on active site chlorination of the enzyme by N-chloro-D-leucine (56). This substrate gives up two chlorines per active site by the first two molecules processed. The half time for decline of activity is one second at 1 mM compound as determined by stopped flow studies. Again the apoprotein is altered and the coenzyme is left intact. Porter and Bright suggest that chlorination of active site tyrosine (one or two residues) has occurred (56). As with D-propargylglycine the modified enzyme retains some activity; with D-alanine V_{max} is 20% that of enzyme. For D-phenylalanine and D-leucine it is 3.3% and 4.5% respectively. Careful presteady state kinetic analyses prove that the enzyme chlorination specifically slows the flavin reduction step, and lowers that rate 1000-fold selectively. Now flavin reduction is rate-determining rather than release of product as in native enzyme. There is specificity in the N-chloro amino acid. N-chloro-isoleucine, -norvaline, and -norleucine cause modification while N-chloro-D-ala and -D-valine do not. The actual chlorinating species could be molecules of bound N-chloroimminuim acid product before release from the active site (the N-chloroamino acids themselves show a half-time of ~ 1 hr at 25°C).

In contrast to these mechanism-based enzyme modifications that are active site-directed, or proximal to active site-directed, and leave some catalytic activity, L-vinylglycine(2-amino-3-butenoate) completely inactivates L-amino acid oxidase from snake venom, partitioning to 2000 product molecules per inactivating event (54). The apoprotein is modified, not the FAD. Finally preliminary data (57) show that both N-allyl- and N-propargylglycine inactivate sarcosine oxidase, presumably by initial oxidation to the imino acid products and Michael attack by an apoenzyme nucleophile. Undoubtedly more work will appear on mechanism-based flavoenzyme inactivators.

Flavin coenzyme analogues   Along with substrate analogues that can react by chemical pathways distinct from physiological substrates and thus shed light on the nature of intermediates, flavin coenzyme analogues offer potential for dissecting out phases of the complex redox catalysis. In dihydro-flavins, N_1, N_5, and C_{4a} (XIV) are key parts of the electron sink. With

XIV   XV

the synthesis of 5-carba-5-deazariboflavin XV reported in 1970 (58), it became possible to evaluate whether this carbon analogue would be functional in catalysis. After initial studies on binding to an apoflavodoxin (59) and model studies on reduction of simple deazaisoalloxazines by n-propylmcotinamides (60), enzymatic studies were conducted on 5-deaza-FMN and 5-deaza-FAD (61–67), prepared at the coenzyme levels by chemical and enzymatic conversions of 5-deazariboflavin. The 5-carba-analogue is reduced sluggishly ($\sim 10^{-2}$- to 10^{-5}-fold the rates of FMN or FAD) by a variety of flavoprotein dehydrogenases and oxidases, and yields a 1,5-dihydro derivative with direct hydrogen transfer in each case: N-methyl-glutamate synthase (61); D-amino acid oxidase (64); glycolate oxidase (67); L-lactate oxidase (decarboxylating) (63); *Benekea harveyii* NADH flavin oxidoreductase (62, 66). This lack of exchange with solvent reflects the fact that substrate-derived hydrogen transfers to a nonexchange-able C_5 locus (65), which proves the site of electron entry into the analogue, and shows that in 1,5-dihydrodeazaflavins, the central ring is no longer a readily autoxidizable dihydropyrazine but a dihydropyridine as in NADH.

Redox potential determination confirmed that the redox potential of the 5-deazaflavin/5-deazaflavin–H_2 couple is –320 mv, very close to NADH (65). (Massey & Stankovich found a value of –278 mv (68)), and about 100 mv more negative than normal flavins. This thermodynamic barrier to reduction could contribute to the kinetic sluggishness of this analogue. In keeping with these properties dihydro-5-deazaflavins are some 10^3–10^4 less reactive with O_2 than dihydroflavins ($t \frac{1}{2} < 1$ sec). (NADH is also essentially inert to rapid autooxidation by O_2.) A functional consequence is that 5-deazaflavin reconstituted oxidases undergo coenzyme reduction but not reoxidation; the analogue functions stoichiometrically not catalytically. Blankenhorn (69) has determined that the redox potential for 5-deazaflavin semiquinone formation is –650 mv, a value so low he suggests that this analogue will be restricted to kinetically rapid two electron chemistry only and may be a probe for that aspect of flavoenzyme chemistry. In one case, light EDTA reduction of 5-deaza-FAD-D-amino acid oxidase, the one electron reduced analogue appears sufficiently stabilized in the particular protein environment to form, but it is not, apparently, catalytically significant (70). C_5 of 1,5-dihydro-5-deazaflavins is prochiral in the same way as

6.

C_4 of NADH and in principal could allow classification of flavoenzymes by relative stereochemistry (for example, "A" side, "B" side) as performed for nicotinamides a couple of decades ago (equation 6). While enzymatic experiments show chiral recognition, rapid disproportionation of mixtures of oxidized and reduced 5-deazaflavins, which results in racemization of chirality at C_5, has to date thwarted those goals (65). Recently, Massey & Hemmerich and colleagues (71) have taken advantage of the remarkably strong reducing power of the 5-deazaflavin radical and have used it as a reductant for other flavoenzymes, iron-sulfur enzymes, and even catalase. The thermodynamic problem of radical accessibility is solved by generating it photochemically.

One use for which 5-deazaFAD was uniquely suited was evaluation of the role of FAD in the enzyme glyoxalate carboligase which catalyzes the nonoxidative decarboxylation condensation of two molecules of glyoxalate, in a thiamin-PP dependent reaction, to tartronate semialdehyde and CO_2 (equation 7). No obvious redox role for FAD was ascribable by previous

7.

workers. When inactive apoglyoxalate carboligase was reconstituted with 5-deazaFAD, 98% of the V_{max} of native FAD enzyme was achieved (72), which strongly suggests the role of the slowly redox active analogue ($10^{-2}–10^{-5}$ normal FAD) as an effector molecule only. It is likely that the flavin found in the enzyme hydroxynitrilase (73) and acetolactate synthase (74) may have similar nonredox regulatory roles.

Another key flavin coenzyme analogue, 1-carba-1-deazariboflavin, which has the nitrogen N_1 at the other end of the election sink replaced by carbon, has been synthesized by Ashton et al (75) at Merck and is blue-black (λ_{max} 535, $\epsilon = 6800$) rather than yellow. On reduction (Equation 8), chemical or enzymatic, the major tautomer is not 1,5-dihydro but rather

8.

the 2-enol, and above pH 5.6 is the stabilized enolate anion (76). Unlike the sluggish, reduced 5-deazaflavins, dihydro-1-deazaflavins are rapidly autoxidized ($t\frac{1}{2} <$ one second) by O_2, are catalytically competent in turnover with flavoenzymes, and undergo both one and two electron redox chemistry (77). The E'_0 for the 1-deazaflavin couple is -280 mv, between the 5-deaza- and riboflavin ($E'_0 = -200$ mv) systems. With the *B. harveyii* NADH oxidoreductase, V_{max} is 2.5% that of riboflavin. Experiments with 4-R-[^{2}H]-NADH show that the hydrogen transfer step is still rate-determining and the V_{max} reduction may be a reflection of control of turnover by the flavin redox potential in this system. On the other hand when 1-deazaFAD-reconstituted D-amino acid oxidase is assayed the V_{max} for several amino acids *equals* the V_{max} for FAD-holoenzyme. This is explicable (77). The slow step in catalysis for both forms of enzyme is release of imino acid product. The rate-determining *physical* step masks any effect the more negative redox potential of 1-deazaFAD has on slowing down the *chemical* step of coenzyme analogue reduction. With 1-deazaFAD-glucose, oxidase V_{max} is 10% the V_{max} of holoenzyme and [1-^{2}H]glucose shows a V_{max} isotope effect of 8.5. The V_{max} ratios of any FAD-enzyme/1-deazaFAD enzyme pair may be an index of whether the elementary step of coenzyme reduction controls V_{max} totally or partially (77).

Some other coenzyme analogues are of recent interest. 8-HydroxyFAD has been isolated from the electron transferring flavoprotein (the electron acceptor for the acyl CoA dehydrogenase) of *Peptidostreptococcus elsdenii* (78). This orange flavin (XVI) shows a spectroscopic pK_a of 4.8, reflecting ionization of the 8-hydroxyl and electronic rearrangement to the para quinoid form of the anion. The 8-hydroxy FAD was inactive with

XVI

apo-D-amino acid oxidase but at the 8-hydroxyFMN level was active as coenzyme for an apoflavodoxin (78). With the *B. harveyii* NADH oxidoreductase, lowering the pH towards the pK_a of 4.8 generates active coenzyme in keeping with conversion back to the "flavin-like" electron density pattern in the ring system (79). A 6-hydroxy FAD has been found in fractional amounts in the same *P. elsdenii* protein (79) and also in glycolate oxidase from pig liver (80). Both hydroxy forms may be oxidative artifacts either in vivo or in vitro and not functional coenzyme forms. Additional coenzyme analogues recently prepared (75) and tested with the oxidoreductase (77) are 1,5-dideazariboflavin which is essentially inert in turnover, 3-

deazariboflavin, which is the enolate anion XVII at neutral pH, riboflavin N-oxide, which eliminates H_2O and oxidizes back to riboflavin on enzymatic or chemical reduction, and roseoflavin, XVIII, a natural red flavin

XVII   XVIII

antimetabolite from a streptomyces (81). Roseoflavin is inert to the *B. harveyii* oxidoreductase and probably to other flavoenzymes, quite possibly because of a mechanism analogous to the 8-hydroxyflavin, which releases electron density into the ring from the dimethylamino group and thereby decreases the ease of reduction (77). Lastly, the synthesis of 8-azidoFAD has been reported (82). The compound shows 66% the V_{max} of native glucose oxidase after reconstitution with apoenzyme 6–24 hr, but is inactive with apo-D-amino acid oxidase. No photolytic studies on the azido moiety were reported.

Before proceeding to flavin-linked monooxygenases two other sets of recent flavoenzyme papers merit note. The first deals with yeast old yellow enzyme, the first flavoprotein discovered. Abramowitz and Massey reported a one step affinity column protocol to homogeneity (83). Old yellow enzyme has long been known to purify in green form, with tightly bound p-hydroxybenzaldehyde (84) as ligand. Structural variations of potential ligands reveal that an aromatic thiol or phenol is a minimal ionizable structure. Proof that the enzyme-ligand complex represents a charge transfer complex (15) with electron rich ligands as donor and oxidized flavin as acceptor was accumulated by correlation of λ_{max} (the energy for the transition) of the new spectroscopic transition in the complex with the Hammet σ_p parameter for a series of para substituted phenols. The corollary experiment of varying the charge transfer acceptor by way of a series of flavin analogues, including 5-deazaflavin, with different redox potentials, completed the study elegantly (85).

Two bacterial enzymes recently discovered in Huenneken's laboratory (86) termed the R and F flavoproteins serve as the physiological reducing system for the coenzyme B_{12}-dependent methionine synthase of *E. coli.* A priming problem exists for isolated methionine synthase in that the associated aquo B_{12} must be converted to methyl B_{12} as the methyl donor; this may involve two sequential one electron reductions of cobalt III in resting B_{12} coenzyme to supernucleophilic cobalt I (87). Huennekens (86) suggests that NADH is the ultimate electron source (equation 9). (This functional multienzyme sequence has analogy to the multicomponent cytochrome

$$NADH \xrightleftharpoons{2e^-} \underset{\text{FAD}}{\text{protein R}} \xrightleftharpoons{1e^-} \underset{\text{FMN}}{\text{protein F}} \longrightarrow \underset{\text{methionine synthase}}{(B_{12}\text{-Co}^{III}}$$ 9.

P_{450} systems noted later.) The R protein (27,000 mol. wt) accepts two electrons from NADH, and gives up one electron at a time to the FMN-F protein. That is, they behave as a flavodoxin reductase and flavodoxin respectively. The semiquinone form of the F protein is blue, oxygen inert, and shows a redox potential of –0.29v. The two electron E'_0 for the R protein is –0.30v. Huennekens notes that the $Co^{II}B_{12} \rightarrow Co^IB_{12}$ $E^{0'}$ is –0.70v, or +9 kcal/mole endergonic from the semiquinone form of the F-protein. If the Co^I species forms at all, only a small amount can be obtained. (The overall equilibrium may be favored by rapid demethylation of S-adenosylmethionine by Co^I, which is exergonic by –17 kcal/mole.)

The *E. coli* and *Apobacter aerogenes* glutamate synthase is an iron-sulfur FMN and FAD flavoenzyme that has the indicated stoichiometry:

glutamine + α-ketoglutarate + NADPH → NADP + 2 glutamates.

Apparently glutaminase activity releases "nascent ammonia" at the active site to form an imine with bound α-ketoglutarate. This ketimine is reduced by one of the dihydroflavins that has received electrons from the NADH. With [4S- ^{3}H]NADPH, tritium ends up in solvent water undergoing exchange, presumably while at N_5 of the dihydroflavin (88). In the absence of glutamine, ammonia can serve as amino donor (also with the iron-free, flavin-free apoenzyme). In this case 4-S-[^{3}H]-NADPH yields L-[α-^{3}H]glutamate, which indicates some direct hydrogen transfer, and this suggests bypass of the flavin and iron/sulfur cluster redox elements. Antiglutamate dehydrogenase antibodies (89) had been employed previously to attempt to rule out trace contamination by that enzyme.

Flavin-Linked Monooxygenases

This subject has been reviewed extensively (90, 91) and only salient new developments are mentioned here. The reaction of dihydroflavins with O_2, either for 2e$^-$ reduction to H_2O_2, or for hydroxylation processes, has been suggested, for some years now, to involve initial one electron transfer to flavin semiquinone plus superoxide ion, followed by radical recombination to yield a 4a flavin hydroperoxide, **XIX**. If the enzyme is an oxidase, the

XIX

peroxide falls apart to oxidized flavin and hydrogen peroxide readily. If the enzyme is a monooxygenase, it can act as an oxygen transfer agent or a precursor to one. Firm evidence for the 4a-hydroperoxy flavins has been accrued by Kemal & Bruice who prepared and crystallized N_5-alkylated-4a-peroxy adducts by adding H_2O_2 to the oxidized 5-alkyl cations (equa-

10.

tion 10) in a reaction akin to pseudobase formation by water attack at 4a (92). The N^5 ethyl adduct is light green with λ_{max} at 370 nm ($\epsilon = 8000–10,000$). This chromophore is identical to one seen in presteady state kinetic studies by Massey and colleagues on several flavin-linked monooxygenases. In a careful study on a pseudomonad p-hydroxybenzoate hydroxylase (93), they actually detected three spectroscopically distinct intermediates; the first appears to be the 4a peroxide, the last the 4a-OH pseudobase, and the second, they suggest, may be a ring opened flavin. They put forth the mechanism depicted in equation 11, noting that only activated aromatic

11.

rings, for example, phenols, are substrates for the flavin-dependent mono-oxygenases, and that hydroxylation proceed ortho to yield 3,4-dihydroxy-benzoate. The initial hydroxylated product would be the unstable cyclohexadienone tautomer of the aromatic product. In support of the idea that a flavin 4a-peroxy adduct could be an oxygen transfer agent, Kemal et al (94) demonstrated that for oxidation of thioxane to its sulfoxide, the model N^5-alkyl-4a-OOH flavins were 10^4 to 10^5-fold better oxidants than nonpolarized alkyl hydroperoxides. Ghisla et al have placed N^5-ethyl-4a-hydroxy FAD on apo p-hydroxybenzoate hydroxylase and mimicked the spectrum of the postulated 4a-OH intermediate (95).

While most flavin-linked hydroxylases act on aromatic substrates, bacterial luciferase is an exception, oxidizing a long chain aldehyde to an acid with blue light emission (equation 12). The required $FMNH_2$ is provided

$$FMNH_2 + O_2 + RCHO \xrightarrow{luciferase} FMN + H_2O + RCOO^-$$

12.

by the NADH oxidoreductase alluded to earlier which has been obtained in homogeneous form from marine bacteria (96). Elegant cryoenzymological (97, 98) studies of Hastings, Balny & Bouzou indicated that the 4a-OOH-FMN · · · enzyme intermediate could be accumulated at low temperatures and was a competent intermediate on addition of light and aldehyde. Kemal and Bruice's model 4a-OOH adduct (92) has an identical UV-visible spectrum to that intermediate; further, their adduct produces chemiluminescence on addition of aldehydes. The light output is slow, reaching a maximum with formaldehyde at 88 min as against ∼1 sec for bioluminescence. (Is this, perhaps, a case of enzymic acceleration?). When N^5-methyl and N^5-ethyl flavins were mixed with H_2O_2 and luciferase, disappointingly no bioluminescence occurred, perhaps because the N^5-alkyl substituent interferes with some enzyme residue that normally binds to N^5 of FMN (94). With 1-[^{2}H]-aldehydes Kemal and Bruice saw kinetic isotope effects of ∼1.6 on light production, which bears some analogy to luciferase. They suggested that a mixed peroxide between flavin and aldehyde was progenitor to the light emitting species in a Baeyer-Villiger type sequence (XX). However, stopped flow kinetic studies with deuterated aldehydes

XX

(1-[^{2}H]-decanal or dodecanal) and luciferase show that a step with a $k_H/k_D \approx 5$ precedes light emission (P. Shannon, R. Spencer, R. Presswood, J. Hastings, C. Walsh, unpublished observations). The 1.6-fold k_H/k_D on light emission represents an incomplete expression of the intrinsic isotope effect (99) by partial suppression due to a slow step interposed between C–H(^{2}H) bond breakage and light emission. These data rule out the Baeyer-Villiger mechanism, at least that part concerning generation of the light emitting species concerted with C_1–H cleavage in the aldehyde. A proposal put forth by Ingraham and colleagues (100), namely a symmetry forbidden 2+2 cycloreversion of a cyclic adduct, is consistent with the enzymic data. They note that a nonenzymic thermal cycloreversion of equation 13 may be a model for how the 4a flavin hydroperoxide could be subsequently

13.

processed by luciferase. Note that attack on the aldehyde is pictured to occur from nucleophilic N_5 in the peroxy adduct, and the aldehyde C–H

14.

bond breaks by hydride shift (equation 14). The 1,3-oxazetidine would be the species undergoing 2+2 cycloreversion, driven by relief of angle strain and increase in product resonance energy. Since this 2+2 cleavage is thermally forbidden by orbital symmetry to give the ground state of products when proceeding by such a four center transition state, this process would necessarily give excited state products. The exact nature of the emitting excited state of FMN, or some derivative, remains uncertain (101).

An interesting alternate pathway for oxygen transfer occurs with the *p*-hydroxybenzoate hydroxylase when 4-thiobenzoate is used as substrate analogue in place of 4-hydroxybenzoate. The product is not the 3-hydroxy species but the disulfide (equation 15) (102). Stopped flow kinetics and

15.

spectrophotometry show the 4a–OOH–FAD is formed but is apparently captured preferentially by the nucleophilic sulfur atom at C_4 of the thiobenzoate instead of the carbon center at C_3. Preferential hydroxylation on sulfur would yield the sulfenic acid, a highly reactive grouping which in the presence of another molecule of 4-thiobenzoate would yield the observed

16.

disulfide by the indicated displacement of the –OH group (equation 16). The idea of sulfenates as oxygenated sulfur derivatives in metabolism has received attention recently by Allison (103) who has reviewed evidence that glyceraldehyde-3-P dehydrogenase on mild oxidation has cys_{149} converted to the sulfenate with loss of aldehyde dehydrogenase activity and gain in acyl phosphatase activity. He has also suggested that sulfenates can form in papain and even that a sulfenate may be a catalytically important group in plasma amine oxidase action. Finally, sulfenates may be formed physiologically by the only well-characterized flavin monooxygenase known in mammalian systems (104). This monooxygenase was initially isolated as an activity from pig liver microsomes which oxidizes (and oxygenates) secondary and tertiary amine groups in drug molecules to the corresponding secondary hydroxylamines and N-oxides respectively. It also participates in N-demethylation of such molecules as benzphetamine, by a mechanism quite distinct from that achieved by the hepatic cytochrome P_{450} (105). In addition, this enzyme catalyzes oxygen transfer to sulfur atoms (for example, in thiourethylenes), and Ziegler has recently suggested that its major cellular role may be in the direction of protein folding by the control of disulfide bond formation rates in proteins such as RNAse. It would achieve this by functionalizing specific cysteine residues as cysteine sulfenates which would then be attacked by other nearby cysteines in the protein to form disulfides, as illustrated above with 4-thiobenzoate (104). This would be an efficient chemical route for enzymatic acceleration of oxidation of thiols to disulfides.

There remain two final comments about flavin-linked monooxygenases using NADH (NADPH) as reductant. (*a*) Kaplan (106) has recently pointed out that six such hydroxylases all remove the pro R hydrogen from C_4 of the dihydronicotinamide and has predicted that this will be a consistent stereochemical result, reflecting active site evolution to recognize only the *re* face of the substituted pyridine ring of NADH(NADPH). The complementary stereochemical question of whether flavoenzymes show preferentially the *re* or *si* face of the isoalloxazine ring system to incoming oxidants or reductants is unanswered. (*b*) Babior and Kipnes (107) have shown that the NADPH-dependent superoxide producing activity of human neutrophils (108) requires FAD and suggest that this enzyme, used for destruction of invading microorganisms, may have an active site designed to let O_2^- get away before the second electron can be transferred to form H_2O_2.

Complex Flavoproteins

There are many recent reviews (109, 110) containing a wealth of information on physical properties which will not be discussed here. Within the past year several complex flavoenzymes, that is, ones containing some additional cofactor or chromophore, have been purified, although little mechanistic data were reported. The *E. coli* membrane NADH dehydrogenase, which is ~2% of the membrane protein, was solubilized with triton and purified to homogeneity. Curiously, the purified enzyme has no bound flavin; it requires exogenous FAD at a K_m of 4 μM (7). A formate dehydrogenase from *Pseudomonas oxalyticus* is an FMN and iron-sulfur cluster enzyme (111), in contrast to the *E. coli* (112) membrane enzyme from anaerobically grown cells, that contain molybdenum, heme iron/iron-sulfur clusters, and selenium (probably as selenocysteine, T. Stadtman, personal communication). *Alcaligenes eutrophus* contains a hydrogenase carrying out reversible activation of molecular hydrogen to $2H^+$ and $2e^-$, probably by way of a hydride-proton mechanism. This enzyme has now been purified to homogeneity, aerobically, as an inactive, oxidized form. Unlike the more studied *Clostridium pasteurianum* hydrogenase (109), this inactive hydrogenase can be reductively activated so that it will carry out the reaction of equation 17. The *A. eutrophus* hydrogenase is an FMN and iron-sulfur

$$NAD^+ + H_2 \underset{enz}{\rightleftharpoons} NADH + H^+ \qquad 17.$$

cluster enzyme (113), again in contrast to the *C. pasteuriarum* hydrogenase which has three four iron-sulfur clusters as the only redox element. The FMN is the likely $2e^-$ acceptor of electrons from NADH, which is not active with the clostridial hydrogenase. Two papers appeared in 1976 on the interconversion of the molybdoflavoiron-sulfur enzymes, xanthine dehydrogenase (NAD as acceptor) and xathine oxidase (O_2 as acceptor). Waud and Rajagopalan (114) noted that the avian liver dehydrogenase acquires oxidase activity by incubation at 37°, by selective proteolysis, or by treatment with dithiobisnitrobenzoate, all conditions which could alter enzyme conformation and flavin redox properties. Only in the dehydrogenase form does the flavin semiquinone signal appear on an anaerobic reduction with xanthine; in the oxidase form only $FADH_2$ accumulates. The suggestion was made that selective stabilization of the FAD semiquinone was correlated with NAD reactivity, a suggestion with possible implication for one electron transfer. Barber et al (115) pointed out that only $FADH_2$ and not the FADH·semiquinone is reactive with O_2, which would explain the lack of oxidase activity in the dehydrogenase form of the enzyme on kinetic grounds. The oxidase form has an $FAD/FADH_2$ redox couple 130 mv more positive (-230 mV vs -360 mV) than the dehydrogenase form, and is observed to have 40-fold more $FADH_2$ available to react with O_2, which correlates with an O_2 turnover number of 16.2 sec^{-1} as against 0.40 sec^{-1} for the dehydrogenase.

Cytochrome P_{450}-Dependent Monooxygenases

Cytochrome P_{450}-linked monooxygenases are multienzyme complexes of two types; a two enzyme variant (NADH-oxidizing flavoenzyme and P_{450}) and a three enzyme type [NADH(NADPH)-oxidizing flavoenzyme, iron-sulfur cluster electron transfer protein, and P_{450}]. The inducible xenobiotic processing liver monooxygenases exemplify the first type; the camphor 5-exo hydroxylation system of *P. putida* is the paradigm for the second variety. A number of the three enzyme variants were found in 1976, including a complex from *Bacillus megaterium* to hydroxylate 3-oxo-Δ^4-steroids at the 15β locus (116), and systems from chick kidney responsible for hydroxylation of 25-hydroxy vitamin D_3 either at carbon 24 or, a distinct system, at carbon 1 to yield either dihydroxy compound or the 1,24,25-trihydroxy vitamin D_3 (117, 118). Preliminary solubilization of a monoterpene hydroxylase, which converts geraniol or nerol to the 10-hydroxy isomeric products from *Vinca rosea*, has been reported (119).

Three other multienzyme complexes related to the above categories have also been examined. One is a para-methoxybenzoate demethylase from

$$H_3CO\langle\bigcirc\rangle COO^- + O_2 + NADH \longrightarrow HO\langle\bigcirc\rangle COO^- + _H\overset{O}{\underset{}{\overset{\|}{C}}}_H + NAD^+ + H_2O \qquad 18.$$

P. putida containing two components; an iron-sulfur cluster and FMN-dependent NADH reductase, and a 2Fe/2S cluster dimeric protein with monooxygenase activity, which is the *first known monooxygenase* with the iron/sulfur cluster as oxygen transfer agent (120). The reaction (equation 18) presumably involves hydroxylation of the methyl group and subsequent decomposition of the hydroxymethyl aryl ether to the observed products. The second is a methane monooxygenase from *Methylosinus trichosporium* which converts CH_4 to CH_3OH by way of a three component system, one of which is a copper protein (121). Third, in the biosynthesis of cholesterol, cholestane-Δ^7-ene-3-β-ol is converted to the $\Delta^{5,7}$-diene-3-β-ol as a proximal precursor. The reaction, known to require O_2 and NADPH for some time, has stoichiometric analogy to the microsomal stearyl CoA desaturase complex known to use a flavoprotein NADH reductase, cytochrome b_5, and a structurally uncharacterized iron-containing desaturase (122). Now Reddy et al (123, 124) have shown that antibody to cytochrome b_5 inhibits diene formation (equation 19) in microsomes by 73%, which argues for

$$\text{HO}\cdots + O_2 + NADH \longrightarrow NAD^+ + 2H_2O + \underset{5,7\text{-}diene}{\text{HO}\cdots} \longrightarrow \underset{cholesterol}{\text{HO}\cdots} \qquad 19.$$

b_5 participation. While it is known that the 5α and 6α hydrogens are lost in an overall *syn* elimination, the nature of the intermediates, including a 6-α-hydroxy species, and how dioxygen becomes reduced smoothly by four electrons, remains unclear in both this and the stearyl CoA desaturase case.

A number of laboratories have reported purification of several inducible hepatic cytochrome P_{450} (P_{448}) isozymes to homogeneity (125, 126) with attendant characterization of overlapping specificity. One intriguing report involves purification, by octylamine-sepharose chromatography (127), of the hitherto difficultly accessible P_{450} component from adrenal cell mito-chondria responsible for cholesterol side chain cleavage-activity. On recon-stitution with the NADPH adrenodoxin reductase and the Fe/S-containing adrenodoxin, side-chain cleavage activity, but not 11-β-hydroxylase activ-ity (carried out by the previously purified adrenal P_{450} 11-β-hydroxylase), was restored. This should open the way to further study of this remarkable first step in steroid hormone biosynthesis, where cholesterol is fragmented to pregnenelone and isocapraldehyde, a C_7 fragment, in a reaction consum-ing *three* moles NADH, *three* moles O_2, and oxidizing the initial substrate by *six electrons.* The process appears to involve a 20α,22-R-diol inter-

mediate which then cleaves across the C_{20-22} bond (equation 20). It has been suggested that some $Fe^V=O$, a "ferryl" species, is involved sequentially in 2 electron oxidation, first at C_{22}, then at C_{20}, and finally at oxidative cleav-

age of the $C_{20}-C_{22}$ bond (equation 21) (128). An interestingly distinct fragmentation pattern occurs when a 20-aryl analogue of cholesterol is incubated with adrenal mitochondrial acetone powder. Now cleavage at the $C_{17}-C_{20}$ bond occurs to yield acetophenone and a 17-methyl-18-norandrostadienol (equation 22) (129). (No such C_{17}–C_{20} cleavage bond occurs with cholesterol itself.) The analogue also is cleaved in the normal mode to produce pregnenelone, the physiological product, and phenol. The abnormal cleavage could arise from rearrangement of an initial carbonium ion, which could form the more stable tertiary carbonium ion followed

23.

by collapse with loss of a proton from C_{17} (equation 23) (129). Unproved as yet is whether the P_{450} complexes the alcohol as shown, and whether the carbonium ion mechanism is ever followed with cholesterol as substrate.

A variety of kinetic and physical studies have appeared on the tight interaction between the first two components of the adrenal mitochondrial hydroxylation complex, namely adrenoredoxin reductase and adrenodoxin (K_{diss}=10^{-9} M) (130–132). Complexation lowers adrenodoxin's midpoint potential by 40 mV (–291 → –331 mV) while leaving the two electron potential of the reductase unchanged at –295 mV (133). Also, the first component of the inducible two enzyme liver system, NADPH liver P_{450} reductase, has now been purified to homogeneity (134–136) by affinity methods after detergent solubilization, and this preparation will reduce P_{450} in vitro *unlike* previously prepared proteolyzed forms of this unusual FMN and FAD-containing reductase.

The nature of the activated P_{450}-oxygen transfer species continues to be studied intensively (137–140) and several reports have surfaced that various organic peroxides can stimulate specific substrate hydroxylations such as that of cyclohexane and steroids, and O- or N-dealkylations mediated by P_{450} enzymes. These may occur by direct generation of transient feryl ion derivatives of P_{450}(Fe^{IV}–O^-), analogous to intermediates proposed for catalase and peroxidase (141–143). Stopped flow kinetic studies on purified liver microsomal P_{450} with benzphetamine as substrate show that added cytochrome b_5 can act as an effector molecule, increasing the rate of formation of oxidized P_{450} without detectable redox change in the b_5 cytochrome (144); other studies suggest reduced b_5 can donate electrons to P_{450} (145, 146).

Since the proximal products of P_{450}-mediated monooxygenation of aromatic substrates are arene oxides, solution chemistry of arene oxides as reviewed by Bruice & Bruice (147) is germane to this metabolism. In analyzing the breakdown of these aromatic epoxides they noted that the much bruited reactivity difference between K region, which is not fully aromatic, and non-K region epoxides of aromatic polycyclics is *not* because the K region epoxide is "necessarily a better alkylating agent," but because that epoxide survives for a longer period in aqueous solution and this allows it greater opportunity to react.

The enzyme epoxide hydrase, responsible for ring opening of arene oxides to the dihydrodiols for detoxification, is available in homogeneous form (148) and substrate specificity has been analyzed (149). Actually, action of epoxide hydrase can put the organism at greater rather than lesser risk in certain cases. For example, benzo-[a]-pyrene is initially oxygenated to the 7,8-epoxide and then opened to the 7,8-*trans*-dihydrodiol by the liver microsomal epoxide hydrase. This 7,8-diol is again substrate for the P_{450} aryl hydrocarbon hydroxylase and is epoxidized stereospecifically at the 9,10-locus. This 7,8-diol 9,10-epoxide may be the proximal carcinogen of benzo-[a]-pyrene metabolism in covalent reaction with DNA (150). It is worth passing note that while simple aromatics such as anthracene or

$7,8-diol-\beta-9,10-epoxide$

XXI

naphthalene are processed enzymically to *trans*-1,2-dihydrodiols in animal cells by sequential P_{450} monooxygenase then epoxide hydrase action, they are processed to the *cis*-1,2-dihydrodiols in bacteria (151, 152) by action of single dioxygenase enzymes (equation 24).

24.

An additional family of enzymes involved in detoxifying, nucleophilic capture of arene oxides or other electrophilic xenobiotics are the glutathione S-tranferases which were purified recently by Jacoby and colleagues (153) and studied for mechanism (154) also against organic nitrates and thiocyanates. 1-Chloro-2-nitrobenzenes substituted with groups at the 4-position show linear behavior in turnover with σ^- constants confirming nucleophilic catalysis with a strong dependence on the electrophilicity of the site undergoing attack by the glutathione thiol. There is kinetic evidence for covalent intermediate formation with some substrates (154), but only modest rate enhancements occur in the best of cases over corresponding nonenzymic thiol attacks, which suggests that the major accelerative role of the enzyme is a hydrophobic milieu for enhancing nucleophilicity of bound glutathione's thiolate anion.

A variety of substrates deuterated at aliphatic hydroxylatable sites have been tested with liver microsomal P_{450} enzymes to see if C–H bond cleavage limits V_{max} but no consistent sense has developed except that isotope effects on V_{max}, when detectable, are often as low as 2. Recently, Hjelmeland et al (155) assigned a $k_H/k_D = 11$ for the *intrinsic* isotope effect [see Northrop (99)] in benzylic hydroxylation of 1,3-diphenylpropane (1-dideutero) by liver microsomes (equation 25). Product alcohol was analyzed by mass

$$\underset{\phi}{\overset{HO \; H \; D \; D}{\diagdown}}\underset{\phi}{\diagup} \xleftarrow{\; k_H \;} \underset{\phi}{\overset{H \; H \; D \; D}{\diagdown}}\underset{\phi}{\diagup} \xrightarrow{\; k_D \;} \underset{\phi}{\overset{H \; H \; D \; OH}{\diagdown}}\underset{\phi}{\diagup} \qquad 25.$$

spectrometry for mono- or dideuterium content. No V_{max} rate effects were reported, but the high isotope effect on the elementary step was likened to the k_H/k_D values $\geqslant 6$ for chromyl chloride-mediated benzylic oxidations as a P_{450} model.

In general, for aromatic substrates metabolized to phenols, deuterium isotope effects in metabolism are not observed, and it has been pointed out that none are expected if the arene oxide pathway is involved (155a). This expectation rests on the fact that in arene oxide breakdown the rate-determining step is initial fission of the epoxide C-O bond to the carbonium ion which then rapidly produces phenol, directly or by way of cyclohexadienone tautomer. Nor is there any C-H cleavage involved in arene oxide formation so that no expression of isotope is anticipated. An exception, experimentally, is the conversion of methyl phenyl sulfide to the metahydroxylated sulfide (equation 26) (155a). These data suggest a route to

$$\underset{S-CH_3}{\overset{H(D)}{\diagup}} \xrightarrow{microsomes} \underset{S-CH_3}{\overset{OH}{\diagup}} \qquad k_H/k_D = 1.60 \qquad 26.$$

the metaphenol independent of an arene oxide intermediate, possibly by way of direct insertion of oxygen into an aromatic C-H bond.

Pterin-Dependent Monooxygenases

Pterin-linked monooxygenases are used to hydroxylate unactivated aromatic rings, in contrast to the substrates for flavin-dependent hydroxylases, in phenylalanine, tyrosine, and tryptophan. The well-studied hepatic phenylalanine hydroxylase is partially phosphorylated when isolated (0.3 moles/subunit) and further incubation with ATP and a cAMP-dependent protein kinase produces 1.0 mole phosphoryl groups/subunit with a 2.6 fold increase in V_{max} (156). Kaufman and colleagues (159) have now also purified bovine adrenal tyrosine hydroxylase to ~90% homogeneity after initial chymotryptic solubilization, and found that both phosphorylation and phosphatidylserine addition activate the enzyme. Phenylalanine hydroxylase is activated by lysophosphatidylcholine, which produces a 20- to 50-fold increase in V_{max} (156). Two microbial pterin-linked enzymes for parahydroxylation of benzoate (157) or L-mandelate (158) respectively have been isolated, both requiring tetrahydropterin and Fe^{II} for activity in concert with known requirements for the three aromatic amino acid hydroxylases above.

Given the structural similarity of flavins and pterin ring systems and the functional analogy of tetrahydropterins to dihydroflavins, there has been speculation that a pterin hydroperoxide (XXII) might be the active oxygenating species, and Kaufman (159) have presented preliminary data

XXII

for a spectroscopically detectable intermediate in phenylalanine hydroxylation that could conceivably be such an alkyl hydroperoxide. Yet, tetrahydroflavins are considerably more sluggish in reaction with O_2 than dihydroflavins (90), and it is possible that the ubiquitously required ferrous iron atom interacts with and activates oxygen, in these reactions possibly as Fe^{III}-superoxide. The reduced pterin could be a primary reductant for the iron; confirmation will require further study.

Two enzymes in microbial oxidative one-carbon metabolism may also require pterin coenzymes. Methanol dehydrogenase of *Pseudomonas C* takes methanol to formaldehyde and then to formate using phenazine methosulfate as electron acceptor (160). The purified enzyme has absorption peaks with λ_{max} at 282 and 345 nm, ruling out a flavin but not a pterin cofactor. Similarly Steenkamp & Mallinson (161) have studied a trimethylamine dehydrogenase (forming dimethylamine and formaldehyde) from a facultative methyltroph. The enzyme contains a four iron-sulfur cluster, demonstrated by the chromophore extrusion technique (162), and an unidentified yellow cofactor covalently attached to the protein. This may be the first example of a covalent pterin linkage to the protein (163). However, it now seems that the chromophore is a novel FMN derivative substituted at carbon 6 by a cysteinyl sulfur (T. P. Singer, personal communication).

Copper-Dependent Oxidases and Oxidoreductases

Copper-containing enzymes generally use O_2 as obligate electron acceptor as a probable consequence of the high midpoint potential, Cu^{II}/Cu^{I}, of $+0.3 - +0.6$ V, which excludes essentially all other known biological electron acceptors. Galactose oxidase, plasma amine oxidase, and lysyl oxidase, discussed below, are nonblue copper oxidases that oxidize substrates by two electrons while reducing O_2 by two electrons to H_2O_2. Ascorbate oxidase, brilliantly blue, oxidizes two ascorbates by a total of four electrons and takes O_2 down to $2H_2O$ molecules. Dopamine-β-hydroxylase, a copper-linked monooxygenase, also has nonblue copper and transfers one atom of oxygen into product. The structural basis for blue forms of copper may be distorted ligand geometry and the presence of at least two histidine and one cysteinyl groups as ligands (163a).

In distinction to the FAD-linked monoamine oxidase of mitochondrial membranes are oxidases that work on monoamines or diamines (164) and have Cu^{II} at the active site. A generalized stoichiometry (equation 27) indicates that the imine is the initial product, suffering nonenzymatic

$$O_2 + RCH_2NH_3^+ \xrightarrow[H_2O_2]{ENZ\text{-}Cu^{++}} RCH=NH_2^+ \xrightarrow[H_2O]{non\text{-}enz} RCHO + NH_4^+ \qquad 27.$$

hydrolysis. A variety of studies (164) have suggested that these amine oxidases have an additional prosthetic group, inactivated by carbonyl reagents, which has been claimed to be pyridoxal-P. The most recent basis for this claim is the broad absorption peak, around 500 nm, that arises on treatment of enzyme from pig kidney with cycloserine (165). Hamilton (1) has previously written speculative mechanisms for a unique O_2-mediated reoxidation of enzyme-bound pyridoxamine-P at the end of turnover back to pyridoxal-P but there are no known precedents. Other workers have disputed the idea of pyridoxal-P and suggested that a sulfenic acid grouping (103) or unknown electron sink is present. The release of product as an imine is a serious problem for any pyridoxal-P based mechanism.

Two substrate analogues have provided evidence for substrate-derived carbanions as early intermediates before the redox step. The amine oxidase from bovine plasma converts β-chlorophenylethylamine to phenylacetaldehyde and chloride ion in an O_2-independent catalytic process, most simply construed as abstraction of α-H as a proton and elimination of the β-chloro group as chloride ion, with formation of an eneamine product that tautomerizes to the imine and hydrolyzes (37). In contrast, β-bromoethylamine is a suicide substrate, possibly by way of oxidative conversion to the β-bromoimine, an α-halocarbonyl, which alkylates the enzyme in an S_N^2 process before release from the active site (166). With [14C]-aminoacetonitrile, the enzyme catalyzes its own inactivation with covalent modification, probably by initial carbanion formation. This carbanion has ketenimine character and as such is a reactive electrophile after protonation on nitrogen (equation 28) (167). This scheme has obvious mechanistic application to

$$H_3C-\underset{\underset{N}{\overset{\overset{H}{|}}{\underset{\|\|}{C}}}{\overset{}{C}}-NH_3^+ \rightleftharpoons H_3C-\underset{\underset{N}{\overset{\ominus}{\underset{\|\|}{C}}}}{\overset{}{C}}-NH_3^+ \longleftrightarrow H_3C-\underset{\underset{N}{\overset{}{\underset{\ominus}{\|}}}}{\overset{}{C}}-NH_3^+ \rightleftharpoons H_3C-\underset{\underset{NH}{X:}}{\overset{}{C}}-NH_3^+ \longrightarrow H_3C-\underset{\underset{NH_3^+}{X-C}}{\overset{}{C}}-NH_3^+ \qquad 28.$$

alkylated enzyme

how the natural plant compound β-aminopropionitrile effects structural defects in collagen and elastin of grazing animals. Normal cross-linking of these structural proteins is initiated by oxidation of the ϵ-amino group of certain lysine side chains to an aldehyde that can then condense with amino groups on other collagen strands. The lysyl oxidase has recently been purified by collagen affinity chromatography and is indeed a Cu^{II}-dependent amine oxidase (168). Thus β-aminopropionitrile may be processed and cause inactivation by a similar carbanion to electrophilic ketenimine rearrangement (37–39). A final comment on the plasma benzylamine oxidase is that oxidation of 1-[$^2H_2$]-benzylamine proceeds with no isotope effect on V_{max} but a V_{max}/K_m effect of 3.1, suggesting that the effect on the elementary C–H(C–D) bond cleavage step is suppressed by some subsequent slow step in catalysis which controls V_{max} (169). A direct examination of the redox step by stopped flow kinetic analysis showed a k_H/k_D of 2.8 on enzyme chromophore reduction (169).

The enzyme galactose oxidase converts the C_6 primary alcohol of galactose (and other analogues) to an aldehyde, while O_2 goes to H_2O_2. Since the aldehyde can be reduced nonenzymatically with NaB^3H_4, this enzyme has been used to label galactosyl groups of membrane glycoproteins that are accessible to the external medium. The role and oxidation state of the enzyme-bound copper have been quite controversial (170–173). While isolated enzyme has up to 70% of its copper as esr-detectable Cu^{II}, Hamilton's group has seen activation by ferricyanide with disappearance of the esr signal. They favor the idea of one electron oxidation to Cu^{III}, a unique involvement of trivalent copper in biological catalysis: thus Cu^{II}-enzyme is held to be inactive. Hamilton suggests galactose is oxidized by rate-determining (174) hydrogen removal at C_6 with two electron passage to convert the putative Cu^{III} directly to Cu^I. The reoxidation of Cu^I by molecular oxygen could proceed in single electron steps by way of a $Cu^{II}-O_2^-\cdot$ intermediate. It is observed that a molecule of superoxide ion leaks out on the average once in 2–5×10^3 turnovers (173, 175). This would leave the enzyme in an inactive Cu^{II} oxidation state, to be reactivated oxidatively on addition of ferricyanide ion.

Dopamine-β-hydrolase is a copper-dependent monooxygenase catalyzing hydroxylation at the prochiral carbon 2 with retention of configuration (176, 177). Fumarate is an activator affecting substrate K_m values but not V_{max}; the olefinic diacid reduces the V_{max} deuterium isotope effect in 2-[H$_2$]-phenylethylamine hydroxylation from 5 to 2 (177). Klinman (178; personal communication) has now found that with 2-[^{3}H]-dopamine, the tritium isotope selection [a V_{max}/K_m effect (99)] varies also with O$_2$ concentration, from a value of 15 at zero oxygen (by extrapolation) to a low value of 4 at infinite O$_2$. This variation is interpreted in support of a random addition of dopamine or O$_2$ to enzyme: if O$_2$ added first the kinetic isotope effect would not be influenced by O$_2$. If O$_2$ added second, the V/K tritium effect would decrease to 1.0 as O$_2$ reached infinite concentration. Continuing the tritium isotope analysis, Klinman finds that when fumarate is present, the V/K tritium effect does indeed go to 1.0 as O$_2 \to \infty$. These data suggest that fumarate activation involves preferential flux of reactants through the sequence where dopamine binding preceeds O$_2$ binding, and represents a fruitful use of enzymatic isotope effect studies.

Dioxygenases

The bulk of the dioxygenases purified and analyzed contain iron as co-factor in a structurally uncharacterized environment (i.e., not iron-sulfur clusters). A few contain heme iron. Hayaishi (179) among others has classified dioxygenases as intramolecular if both oxygen atoms are incorporated into one product and intermolecular if incorporated into two separate product molecules. Most of the intermolecular dioxygenases use α-ketoglutarate as cosubstrate and oxidatively decarboxylate it to succinate during substrate hydroxylation (180). Persuccinates have been suggested as oxygenating intermediates (181) but are inactive when directly tested with prolyl hydroxylase (180). Mechanistic studies have been slow but the γ-butyrobetaine hydroxylase (equation 29) from *Pseudomonas SP AK1* has now

$$\gamma\text{-butyrobetaine} + O_2 + \alpha\text{-ketoglutarate} \xrightarrow{\text{Enz-Fe}^{II}} \text{carnitine} + \text{succinate} + CO_2 \qquad 29.$$

been purified to homogeneity in large quantity (181). It requires added ferrous irons for activity as do other members of this class such as the recently purified lysyl 5-hydroxylase (182). Suggestions of Fe^{II}–O$_2 \leftrightarrow$ Fe^{III}–O$_2^-$ as initial activated oxygen reagent are widely accepted. Indeed, Hayaishi and colleagues (183) have shown that the heme-iron-containing indolenine 2,3-dioxygenase (ring cleaving) purifies as a mixture of Fe^{III} and Fe^{II} species. Both ^{18}O$_2^-$ and ^{18}O$_2$ were utilized for ring cleavage of tryptophan to formylkyneurenine, the O$_2$ presumably reacting with Fe^{II}-enzyme and the superoxide with Fe^{III}-enzyme.

Recent reports (184, 185) have described a heme enzyme with oxygenase activity toward the side chain of indolyl-3-alkanes (e.g., tryptophan) which is distinct from the ring-cleaving tryptophan 2,3-dioxygenase and indoleamine 2,3-dioxygenase (183). The mechanism of this enzyme is unclear. It appears to convert tryptophanamide to 1-amino-2-hydroxy-3-indole (a hydroxylation, decarboxylation sequence) (184). L-Tryptophan is processed to NH_3, CO_2, and an unstable compound, tentatively identified as 3-indolylglycoladehyde based on derivitization as the stable 2-(3-indolyl)-quinoxaline (equation 30) (185). Most recently it was noted that tryptophanyl

30.

peptides are converted to α,β-dehydrotryptophanyl residues, possibly by hydroxylation at the β-carbon and dehydration (186).

Marked progress has been made with the purification of the enzyme complex responsible for converting arachiidonate or $\Delta^{8,11,14}$-eicosatrienoate into prostaglandins E_2 and E_1 respectively, the prostaglandin cyclooxygenase. Hayaishi et al (187) purified cyclooxygenase from bovine vesicular gland microsomes and found two fractions, I and II. Fraction I is a single band on native acrylamide gels, has a molecular weight of $\sim 3.5 \times 10^5$ and requires added heme for activity. (It perhaps reconstitutes a heme dioxygenase.) Fraction 1 converts eicosatrienoate to the unstable prostaglandin E_1, a precursor to PGE_1 which has a cyclic endoperoxide group and a 15α-hydroperoxy group. If tryptophan is added in addition to heme, Enzyme I will carry out a peroxidase-type cleavage of the α-hydroperoxy group of PGG_1 to a hydroxyl group, generating PGH_1. This unstable hydroxy endoperoxide is a substrate for fraction II, prostaglandin endoperoxide isomerase, (188) which uses glutathione as a specific cofactor for endoperoxide opening to prostaglandin E_1 the major biosynthetic prostaglandin metabolite from the acylic precursor (equation 31). A related

31.

enzyme complex has also been purified from sheep vesicular glands and yields only one type of subunit (of molecular weight 70,000) by sodium dodecyl sulfate gel electrophoresis. When [^{3}H]-aspirin, a known cyclooxygenase inactivator (189), is used, the subunit is labeled covalently. Van Dorp and colleagues (190) feel that heme iron is likely to be a loosely dissociable cofactor for both the dioxygenase and peroxidase activities. The sheep enzyme appears to be a dimeric glycoprotein representing a major protein component of vesicular gland microsomes (190).

The cyclooxygenase catalyzes a double dioxygenation sequence, producing a five-membered cyclic endoperoxide and an acyclic allylic hydroxperoxide, but the mechanism may involve an allylic hydroperoxylation in each instance, followed by cyclization in one instance. The redox chemistry could occur in two electron steps or by way of radical, one electron processes. One could write a scheme (as in equation 32) involving proton

$$32.$$

abstraction, or attack could be initiated on the double bond by coordinated superoxide with hydrogen atom expulsion. In support of radical intermediates are consistent observations that cyclooxygenases undergo self-destruction during turnover within 30–60 sec. Egan, Paxton & Kuehl (191) have found that reductive processing of the PGG$_2$ intermediate generates radicals that inactivate the enzyme. Phenol and methional, good radical scavengers (of OH· for example), increase both initial rates and extents of catalytic reaction before inactivation and promote formation of PGH$_2$ at the expense of PGG$_2$ (starting initially from the tetraene-arachidonate). Preliminary ESR studies show radicals forming during turnover in the absence of radical scavengers; about 5000 turnovers occur before an inactivating event. The PGG$_2$-derived oxygen radical, adventitious breakdown of a highly reactive chemical intermediate, may be reacting with a disulfide at the enzyme active site. The allylic hydroperoxylation mechanism for dioxygen introduction (homolytically or heterolytically) probably applies to lipoxygenase catalysis also. Kinetic studies with soybean lipoxygenase have been plagued by lags and limited extent of reaction. While radical inactivations could also be occurring here, Gibbian & Galway have noted (192) that at 10^{-9} M concentrations, lipoxygenase adsorbs physically to cuvette walls and so inactivates. Also, addition of the product allylic hydroperoxide abolishes the initial lag which suggests time-dependent activation of enzyme occurs after a few turnovers. Then as product builds up, competitive inhibition sets in and slows rates. The activation by ROOH could be some change in active site iron-redox state, e.g., $Fe^{II} \rightarrow Fe^{III}$. The tetraacetylenic analogue of arachidonate $\Delta^{5,8,11,14}$-eicosatetraynoate is an inhibitor of cyclooxygenase, possibly irreversibly (193). The $\Delta^{5,8,11}$-triynoate shows preferential inhibition of lipooxygenase (194). Inactivation may occur by C_{13}-H or C_{10}-H abstraction, respectively, with formation of conjugated allenes as potential electrophiles, though there is no evidence yet.

A final dioxygenase example is that of 2-nitropropane from the yeast *Hansunela mraki* (195). Two molecules of nitropropane are converted to two molecules of acetone and nitrite at the expense of one O_2 (equation 33).

$$2 \underset{NO_2}{\bigvee} + O_2 \xrightarrow{E \cdots \overset{\text{III}}{Fe} \atop E \cdots FAD} 2 \underset{O}{\bigvee} + 2 NO_2^- \qquad 33.$$

The homogeneous enzyme contains FAD and a mole of Fe^{III} but no labile sulfur. $^{18}O_2$ is partly incorporated into product acetone, which suggests that this is a unique variant of an intermolecular dioxygenase. No mechanistic proposals have been presented.

Miscellaneous Redox Reactions

The reactions discussed in this section are diverse and related only in the sense that oxidation-reduction steps are the key ones in the overall transformations. S-adenosylhomocysteine lyase is a hydrolase that cleaves substrate to adenosine and homocysteine, which represents a formal displacement of the homocysteinyl moiety by water. To probe the catalytic mechanism Palmer & Abeles (196) purified the enzyme and demonstrated a mole of tightly bound NAD; addition of adenosine produces E-NADH and presumably 3'-ketoadenosine since the holoenzyme carries out rapid exchange of the 4'-hydrogen of adenosine with water probably by way of the enolate anion. This clue suggests hydrolysis of SAH involves oxidation, α,β-elimination to the conjugated olefinic 3'-keto-4',5'-dehydro-5'-deoxy adenosine, Michael addition of water to the olefinic terminus, and re-reduction of the 3'-keto by E-NADH; a case of redox chemistry below the surface (equation 34).

$$34.$$

Sulfite oxidase is a molybdenum and heme-containing enzyme in mammalian mitochondria that catalyzes the terminal two electron oxidation step in sulfur metabolism, sulfite to sulfate, while O_2 is reduced to H_2O_2. Recently the enzyme from both chicken liver (197) and rat liver (198) have been shown to undergo controlled nicking by trypsin to a small fragment, 9,500–11,000, containing the heme, and a larger fragment, $\sim$50,000, containing the molybdenum. Preliminary structural studies show the heme fragment is very similar to cytochrome b_5 and to a heme fragment of the yeast flavohemoprotein cytochrome b_2, which suggests evolutionary relationships. Both fragments of rat liver sulfite oxidase have been crystallized and the Mo^{VI} fragment can still oxidize sulfite but must use ferricyanide as electron acceptor. A dividend of the fact that the enzyme seems to show two distinct domains is that the molybdenum peptide can now be examined spectroscopically in a protein for the first time; in all known intact molybdo-enzymes some iron chromophore obviates this possibility. The Mo^{VI} shows absorbance at 350 nm and a broad band from 450–600 nm, both of low intensity and sulfur ligands, have been suggested. Rajagopalan (198) hypothesizes that a primordial gene for a b type cytochrome may have been adapted for specific function in the three present day enzymes. Fusion to a molybdenum polypeptide fragment produces sulfite oxidase, to an FMN fragment yields yeast cytochrome b_2 with lactate dehydrogenase activity, and to a hydrophobic tail yields the microsomal cytochrome b_5.

An interesting attribute of ribulose diphosphate carboxylase is its recently uncovered monooxygenase activity, processing ribulose diphosphate and O_2 to 3-phosphoglycerate and 2-phosphoglycolate, an alternate sequence of possible physiological significance (199). While spinach carboxylase/monooxygenase was claimed to have copper[II] (200), this has been denied recently (201); the lack of redox active metal would make this a rare oxygenase but the substrate-derived enediolate intermediate is an easily oxidized anion which could react with O_2 by a radical path. The simple epoxide glycidate is claimed to inactivate 82% of spinach enzyme's oxygenase activity without affecting carboxylase activity (202); this dichotomy is also preserved with iodoacetamide and may augur selective involvement of an enzyme thiol or, maybe, a thiolate radical, for the oxygenase reaction. It could be involved in oxygen activation whereas carboxylation ought to be a simpler two electron process.

In some analogy are the luciferases of cypridinia, coelenterates, and fireflys. They function as monooxygenases (and C–C bond cleavage enzymes) without oxidative organic cofactors or redox active metals (101). The *Renilla pyriformis* luciferase has recently been purified 12,000-fold to homogeneity and fits this category in carrying out the reaction of equation 35 (203) by way of the oxyluciferin monoanion. McCapra (101)

35.

has suggested that chemiluminescent reactions of acridan esters are good models for these forms of bioluminescence and that the central features are carbanion formation, one electron transfer to O_2, peroxyanion formation by way of a second electron transfer, and possible closure to a cyclic-4-membered dioxetane which then decomposes to the excited state of the ketone, and then emits light on return to ground state (equation 36). He

$$R-\overset{|}{\underset{|}{C}}H \longrightarrow R-\overset{|}{\underset{|}{C}}{}^{\ominus} \xrightarrow{O_2} R-\overset{|}{\underset{|}{C}}{}^{\oplus} + \cdot O_2^{\ominus} \longrightarrow R-\overset{|}{\underset{|}{C}}-O-O^{\ominus} \xrightarrow{?} dioxetane \longrightarrow \left[R-\overset{O}{\underset{}{C}}\right]^{*} \longrightarrow h\nu \qquad 36.$$

has suggested that these processes look like directed autoxidations and the only role of the luciferases may be catalyzed removal of the acidic hydrogen of a specific luciferin to yield the carbanion which may be easily oxidized, possibly in one electron steps. He also argues that enzyme-initiated bioluminescence may yield singlet states of excited products directly and exclusively, which would account for the very high quantum yields observed. Precisely this chemical susceptibility of electron-rich carbanions to electron removal (oxidation) has been exploited by Christen and his colleagues for some time (204). Using hexacyanoferrate as oxidant for carbanionic, or eneamine, intermediates generated by fructose diphosphate aldolase either from muscle (class I) or yeast (class II), they observe time-dependent inactivation in the presence of substrates. Inactivation follows first order kinetics and this criterion is consonant with oxidative modification of an enzyme functional group by a substrate radical anion formed by one electron oxidation of the normal carbanionic intermediate (205). When transaldolase is used with fructose-6-phosphate and 2 equivalents of ferricyanide, glyceraldehyde-3-P and hydroxypyruvaldehyde are produced. Unlike the normal cleavage product dihydroxyacetone, the oxidized product is released from the enzyme which leads to oxidative turnover at 0.65% the V_{max} of the normal transfer reaction. With this enzyme, also, there is a slow loss of activity which may be due to the radical anion species reacting with enzyme (206).

Another type of carboxylase with some as yet uncharacterized oxygen utilizing capacity is the vitamin K-dependent activity responsible for γ-carboxyglutamyl formation in proteins such as prothrombin and factor X (207). The activity has been solubilized from microsomes (208) and uses the pentapeptide phe-leu-glu-glu-val as substrate for $^{14}HCO_3$ incorporation (209), which facilitates assay. It seems that ATP, CoASH, and biotin are not involved; the dihydro form (hydroquinone) of vitamin K is involved, and O_2 is required, since no carboxylation occurs anaerobically. The nature of products from O_2 is unknown, nor is it clear how CO_2 or the γ-methylene group of a glutamyl residue are activated, though for the latter presumably it is as a carbanion equivalent. A possible clue (210) is that inhibition of this in vitro maturation phase of prothrombin synthesis and the O_2-dependent formation of vitamin K-2,3-epoxide (XXIII) are inhibited in parallel by

vitamin K oxide

XXIII

anticoagulants. One can write a scheme, entirely speculative and without facts, that might tie in various facets (equation 37); alternatively the vita-

37.

min K hydroquinone could be a CO_2 carrier like biotin, or it could merely be an electron donor with O_2 as electron acceptor (it should then go to H_2O_2) with the mysterious chemistry occurring somewhere in between.

A speculative review (210a) has focused attention on the concept that a variety of antineoplastic agents, such as mitomycin C, adriamycin, or camptothecin function as alkylating agents only after undergoing in vivo reductions. Both mitomycins and the anthracyclines, exemplified by adriamycin, have quinones groupings which on reduction to hydroquinones can undergo facile eliminations of adjacent substituents and thereby generate quinone methide equivalents and uncovered electrophiles which can undergo Michael addition by protein or nucleic acid nucleophiles. Thus for adriamycin a possible sequence (210a) is as shown in equation 38. Continuing the

38.

argument, camptothecin, a potent antitumor agent, could conceivably be reduced enzymically to the dihydropyridine and thus activated for E ring

opening and subsequent Michael attack (equation 39) (210a). These speculations await experimental test.

$$39.$$

PYRIDOXAL PHOSPHATE-DEPENDENT ENZYMES

The chemical role of pyridoxal-P is to stabilize carbanionic intermediates
during enzyme-catalyzed reactions at the α- β- or γ-carbon of specific
amino acid substrates by acting as a delocalizing electron sink for the extra
electron density. As stated in earlier sections of this review, these processes
are both carbanion chemistry and enzyme-induced redox chemistry where
the proton and itinerant electron pair are separated early in the reaction.
The redox outcome at the α, β, or γ-carbons of substrate depends on the
kinetic specificity for reprotonation shown by a given enzyme, and this in
turn is a consequence of steric disposition and effective concentration of
BH^+ groups at a particular enzyme's active site. While the basic rules are
worked out for pyridoxal-P enzymes in general, some recent mechanistic
insights have been derived from four areas: (a) model studies and coenzyme
analogues; (b) NMR studies on rates of specific proton exchanges with
solvent hydrogens, and the stereochemistry of those exchanges; (c) maps of
the stereochemical outcome in β-replacement and β-elimination reactions
where X-CH$_2$-group is converted to a CH$_3$-group; (d) development and
testing of suicide substrates. These four areas will be examined separately.

Model Studies and Coenzyme Analogues

Although the C_4-carbonyl of pyridoxal-P is essential for catalytic activity,
the 4-vinyl and 4-acetylenic analogues have recently been prepared and
tested. The vinyl analogue not surprisingly inhibits pyridoxal kinase and the
FMN-linked pyridoxamine-P oxidase of rat liver (211); it also inhibits
growth of mouse sarcoma 180 and mammary adenocarcinoma TA3. The
4-ethynyl PLP analogue irreversibly inactivates apoaspartate transaminase
(212) and possibly undergoes addition by the active site lysine ϵ-amino
group to yield an initial eneamine that could tautomerize to the fully
conjugated p-quinoid species shown XXIV. The 5-trans vinyl carboxylate

XXIV

analogue of **PLP** also slowly inactivates apoaspartate transaminase with concomitant change in absorbance peak from 385–417 nm, a covalently bound species. Model studies with 1,3-diamino compounds (e.g., α,γ-diaminobutyrate) produce compounds with a λ_{max} of 410–417 nm which are thought to be hexahydropyrimidine derivatives as shown, and mimics for enzyme inactivation. The vinyl carboxylate side chain acts as electron sink for loss of the 4-hydrogen (equation 40) (213).

$$40.$$

The key intermediate in pyridoxal-P dependent β-eliminations of a good leaving group, or its eventual replacement by an incoming nucleophile in a variety of reactions (e.g., serine dehydrase), is an α-aminoacry-late-Schiff base, and λ_{max} values of 455–470 nm have been ascribed to them in various enzymes (214). A model study validates this idea (215). When L-tryptophan is mixed with N-methyl pyridoxal and Al^{III} in aqueous base, a 514-nm absorbing species, presumably the p-quinoid form of the chelated, stabilized α-carbonion **XXV**, forms within 15 min; during 8 hr this decays to a 467 nm species that is the aminoacrylate chelate after loss of the β-indole

XXV

substituent. Addition of parachlorothiophenol produces the s-chlorophenyl derivative of cysteine. For amino acids without a good leaving group at the β-carbon, the 467 nm species does not form.

NMR Proton Exchange Studies

Proton magnetic resonance studies have recently been performed with several pyridoxal-P enzymes. Babu & Johnston (216) noted that L-alanine transaminase catalyzes exchange of all carbon-bound hydrogens of alanine, the α-H and the three β-methyl hydrogens, which suggests capacity to form β-carbanionic intermediates. This capacity was exploited with L-propargylglycine; β-carbanion-formation on this analogue should uncover the latent acetylene by proparglylic rearrangement and does in fact lead to active site-directed stoichiometric inactivation (217). Both Cooper (218) and Jenkins (219) have quantitated exchange rates at the α-H and β-Hs of L-alanine and find the β-H exchanges as fast as α-H, and that α-H, β-D intermediates are observed with enzyme in D_2O. Jenkins suggested that α-H removal is not rate-limiting but Cooper found 2-[^{2}H]-alanine shows a 2.3-fold kinetic isotope effect. Both suggest there is appreciable storage of the substrate-derived α-H at the active site sequestered from solvent, possibly at the 4'-carbon of PLP, during β-carbanion formation and β-hydrogen exchange, but they differ as to whether one base or two bases are involved in the process.

We note that β-carbanion formation is not required for transaminase catalysis and is not seen with L-aspartate transaminase (216–219). Further, tryptophanase (220), B. subtilis alanine racemase (216), and pyridoxamine pyruvate transaminase exchange only the α-hydrogens of alanine. On the other hand, PLP-enzymes which carry out chemistry at the γ-carbon of amino acids must make β-carbanion equivalents, and indeed γ-cystathionase (221) and methionine-γ-lyase (K. Soda, personal communication) exchange all the hydrogens of alanine. When γ-cystathionase processes α-aminobutyrate, both prochiral β-hydrogens are exchanged at equal rates; (in contrast cystathionine-γ-synthase of Salmonella (222) is stereospecific in β-hydrogen exchange with that C_4 amino acid.) When the next longer homologue, norvaline, is used, a 24:1 selectivity for one β-H exchange occurs. Presumably at the γ-cystathionase active site the ethyl group of aminobutyrate can rotate freely but the propyl group of norvaline cannot (221).

A final NMR experiment deals with transamination of β,β-difluoro-oxalacetate by aspartate transaminase. Although a very slow substrate (10^{-5} the rate of oxalacetate), this nonenolizable keto acid is converted to a stable difluoroaspartate as monitored by ^{19}F-NMR (223, 224). This stability contrasts with previous work on mono-β-fluoro oxalacetate which apparently transaminates to β-fluoroaspartate which loses HF rapidly to yield oxalacetate. Loss of HF from difluoroaspartate may be retarded inductively.

Stereochemistry of Reactions at the β-Carbon

A number of stereochemical studies have focused on the fate of the α-aminoacrylate-schiff base in β-eliminations and β-replacements. Since β-replacements must occur at the active site, steric control is expected and observed. Tryptophan synthase converts serine labeled chirally at the β-carbon (e.g., 2R–CHTOH or 2S–CHTOH) to tryptophan with net retention of configuration (225). The indole β-carbanion attacks the same face of the aminoacrylate double bond as the –OH departed from (equation 41).

$$\text{RX} \xrightarrow[\beta\text{-replacement}]{\text{RX}^-} \qquad \xrightarrow{\beta\text{-elimination}} \qquad + \text{NH}_4^+ \qquad 41.$$

Similarly, using comparably β-chiral tritiated O-acetyl-L-serines, Floss and colleagues demonstrated that O-acetylserinesulfhydrylase adds H_2S to form cysteine again with retention; only one face of the bound aminoacrylate is accessible to solvent. In each instance both 3R and 3S isomers of serine were used and gave complementary results (226). For β-elimination enzymes the timing of enamine to imine tautomerization has been less clear. Aminoacrylate is presumably formed but it might be protonated and tautomerized chirally to pyruvate imine before release from the active site if these enzymes have a BH^+ kinetically and sterically accessible, or the enamine could protonate achirally free in solution. In fact, D-serine dehydrase (acting on β-chiral serines (227) or on D-threonine (228), L-serine dehydrase (229), tyrosine phenol lyase (230, 231), and tryptophanase (232) all produce pyruvate which is chiral when the experiment is done to introduce all three isotopes of hydrogen in the methyl group; in each instance retention of configuration is observed, confirming the shielding of bound aminoacrylate by enzyme groups so that only front side protonation occurs stereospecifically and before release. When sheep liver threonine deaminase (33), actually not a PLP enzyme but with bound α-ketobutyrate as cofactor, eliminates H_2O from L-threonine (2S, 3R) or L-allothreonine (2S, 3S) in D_2O, the α-ketobutyrate product is predominantly $3R–D_1$ in *both* cases. This convergent stereochemistry is reminiscent of the stereochemical outcome from erythro and threo β-chloroaminobutyrate eliminations with D-amino acid oxidase, noted earlier in this article, and is again consistent with net *trans* and net syn elimination (or vice versa) to yield a common, possibly transoid, enamine which is reprotonated chirally (32). David & Kapke speculate that various PLP-linked enzymes in D_2O will convert L-amino acids to $R–3–D_1–2$–keto acids and D-amino acids to $S–3–D_1$–keto acids (33).

What happens with a β-replacement enzyme, tryptophan synthase, which acts chirally in β-replacements, when it is allowed to carry out a β-elimination on 2S,3R, or 2S,3S-[3-T]-serine? It yields *achiral* 3-[^{1}H,^{2}H,^{3}H]-pyruvate in D$_2$O (232). There is no need for a β-replacement enzyme to have a BH$^+$ group to protonate aminoacrylate; indeed it would not want to use up the eneamine that way. Consequently, on adventitious hydrolysis, the eneamine is released into solution without enzymically mediated tautomerization to the more stable imine. Recent studies with serine hydroxymethylase (233) show that as the chiral-CHTOH group of serine is converted to the methylene group of 5,10-methylene tetrahydrofolate a partial loss of chirality occurs (a 76:24 split rather than the expected 100:0 ratio is observed) which suggests that free CH$_2$ = O is produced as an intermediate and, on the average, 24 of 100 molecules at the active site are free long enough to rotate and thus scramble chirality.

Suicide Substrates

Within the last three years a large number of suicide substrates, mechanism-based inactivators, have been designed and tested with specific pyridoxal enzymes, and some results are tabulated in Table 3. The inactivators fall into two general categories: 1. olefinic or acetylenic amino acids, where the isolated unsaturated group becomes activated either by carbanion formation adjacent to it (propargylglycine) or by two electron oxidation to a bound conjugated ketimine (vinylglycine), and 2. the β-haloalanines, mono, di or trihalogenated, which can undergo enzyme-mediated loss of the elements of HX to produce an aminoacrylate schiff base. In both cases the activated electrophiles are attacked by enzyme nucleophiles at or near the active site. The efficiency of inactivation in any given instance can be determined, as it has been for some entries in the table, from the partitioning ratio, which is the average number of turnovers per inactivation event (38). This reflects the competition between hydrolytic release of potential alkylator and its rate of adventitious capture by an enzyme nucleophile. It is noted that the aminoacrylate intermediate is a normal one for β-replacement and β-elimination enzymes but is not formed normally by transaminases [e.g. D-amino acid transaminase (234), L-alanine (217), L-aspartate transaminases (235)], decarboxylases [β-aspartate decarboxylase (236)], or racemases [*E. coli* alanine racemase (237)]. These latter types have not had to evolve active site geometries where bases that could attack such intermediates are held away from attack; thus, when presented with the unexpected electrophile, they occasionally make a nucleophilic mistake. This happens once every 1500 catalytic cycles with the *B. subtilis* D-amino acid transaminase (237a), and once every 800 cycles with E. coli alanine racemase (237).

Four points about specific inhibitors are worth making:

1. Vinylglycine inactivates some transaminases as a consequence of cat-

Table 3 Suicide substrates for pyridoxal-P-linked enzymes

Category	Compound	Number of catalytic turnovers/inactivation	References
I. α-Carbon reactions			
A. Transaminases			
1. L-Aspartate transaminase	L-chloroalanine	nd[a]	235
	amino-methoxybutenoate	nd	240
	L-propargylglycine	nd	248
	L-vinylglycine	1	250, 234
2. L-Alanine transaminase	L-propargylglycine	2	217
3. D-Amino acid transaminase	D-chloroalanine	1,500	237a
	D-vinylglycine	400–800	234
4. GABA transaminase	ethanolamine-O-sulfate	nd	249
	gabaculine	nd	242–244
	γ-vinyl GABA, γ-acetylenic GABA	nd	40
5. Lysine ε-transaminase	4,5-lysyne	40	unpublished results, P. Shannon & C. Walsh
	4,5-cis-lysene	1,700	unpublished results, P. Shannon & C. Walsh
	4,5-trans-lysene	160	unpublished results, P. Shannon & C. Walsh
B. Decarboxylases			
1. Histidine decarboxylase	L-histidine methyl ester	nd	251
2. Ornithine decarboxylase	1,4-diamino butyne	nd	252
3. β-Aspartate decarboxylase	L-chloroalanine	nd	236
C. Racemases			
1. E. coli alanine racemase	D,L-chloroalanine	~800	237
	D,L-fluoroalanine	~800	237
	D-O-acetyl serine, D-O-carbamyl serine	~800	237
	D-cycloserine	nd	237
II. β-Carbon reactions			
1. Threonine deaminase	L-serine, L-chloroalanine	~10,000	253
2. Tryptophanase	trifluoroalanine	nd	247
3. β-Cystathionase	trifluoroalanine	nd	247
	rhizobitoxine	nd	39
4. Tryptophan synthetase	trifluoroalanine	nd	247
	dichloroalanine	nd	247
	cyanoglycine	nd	254
III. γ-Carbon reactions			
1. γ-Cystathionase	L-propargylglycine	nd	255, 256
	L-trifluoroalanine	nd	247
2. Cystathionine-γ-synthetase	L-propargylglycine	nd	239
3. Methionine-γ-lyase	L-propargylglycine	nd	unpublished results, M. Johnston, C. Walsh, & K. Soda

[a] nd = not determined.

alyzed *1,3-azaallylic* isomerizations to the conjugated imino acid, an uncovered electrophile. Other PLP enzymes, such as threonine deaminase (238), γ-cystathionase (W. Washtien and C. Walsh, unpublished), and cystathionine-γ-synthase make the same initial α-carbanion but carry out *1,3-allylic* isomerization to bound aminocrotonate, which is harmlessly released and isomerized to the more stable α-ketobutyrate and ammonia.

2. 2-Amino-3-methoxy-trans-3-butenoate, a substituted vinylglycine, turns over with aspartate transaminase and then inactivates, producing an enzyme adduct with λ_{max} at 350 nm that is distinct from the vinylglycine adduct. Rando and colleagues suggest that the initial michael adduct then loses the elements of methanol to yield the new chromophore (equation 42) (240).

42.

3. The compound gabaculine (241), 5-amino-1,3-cyclohexadienylcarboxylate, isolated from a streptomyces, is a powerful and specific mechanism-based inactivator of GABA transaminase from microorganisms or mouse brain (242); $K_I = 5.8 \times 10^{-7}$ M, ($k_{inact} = 1.35 \times 10^{-3}$ sec^{-1}) some 10^3-fold lower than the K_m for GABA (γ-aminobutyrate). It is likely that gabaculine is transaminated and then the bound imine can lose an adjacent acidic hydrogen, thereby aromatize, and produce a meta-anthranilate derivative of PLP which is resistant to ready hydrolysis (equation 43)

43.

(243). Gabaculine treatment elevates GABA levels in mouse brain (244). A particularly promising suicide substrate for GABA transaminase is

XXVI

γ-acetylenic GABA (XXVI) (40) which is highly effective in vivo in brain GABA transaminase blockage and looks like a promising clinical candidate as an antiepileptic drug (245, 246).

4. Trifluoroalanine inactivates a variety of PLP-enzymes carrying out β- or γ-carbon chemistry where monohaloalanines do, not possibly because Michael addition to β-diffuoroaminoacrylate-Schiff base intermediates is enhanced inductively (247). When 1-[^{14}C]-trifluoroalanine inactivated γ-cystathionase, two labels per tetramer (half-site reactivity?) were fixed, and the chromophore λ_{max} was now at 490 and 518 nm indicating a stable paraquinoid tautomer. Since the inactivation rate profile was mirrored by

loss of all three fluorine atoms as F⁻, this new chromophore may be an aminomalonate derivate, and decarboxylation, expected to be facile from such a structure, proceeded on denaturation, consistent with equation 44 (247).

$$44.$$

One can expect more developments on mechanism-based inactivators of pyridoxal enzymes based on functional group activation achieved by way of carbanion chemistry of intermediates.

ACKNOWLEDGMENTS

Research cited from the author's laboratory has been supported by grants from the National Institutes of Health, the National Science Foundation, and the American Heart Association; as well as an Alfred P. Sloan Fellowship (1975–1977) and a Camille and Henry Dreyfus Teacher-Scholar Award (1976–1980).

Literature Cited

1. Hamilton, G. A. 1971. In *Prog. Bioorg. Chem.* 1:83–159
2. Walsh, C. 1978. *Enzymatic Reaction Mechanisms.* San Francisco: Freeman. In press
3. Kohn, L. D., Kaback, H. R. 1973. *J. Biol. Chem.* 248:7012–7017
4. Kaczorowski, G., Shaw, L., Laura, R., Walsh, C. 1975. *J. Biol. Chem.* 250:8921–30
5. Singer, T. P., Gutman, M., Massey, V. 1973. In *Iron-Sulfur Proteins,* ed. W. Lovenberg, 1:225–300 New York: Academic
6. Hall, C. L., Heijkenskjold, L., Bartfai, T., Ernster, L., Kamin, H. 1975. *Arch. Biochem. Biophys.* 177:402–14
7. Dancey, G. F., Levine, A. E., Shapiro, B. M. 1976. *J. Biol. Chem.* 251:5911–20
8. Forman, H. J., Kennedy, J. 1977. *J. Biol. Chem.* 252:3379–81
8a. Hatefi, Y., Stigall, D. 1976. In *The Enzymes,* ed. P. Boyer, 13:175–295. 3rd ed.
9. Walsh, C. T., Schonbrunn, A., Abeles, R. H. 1971. *J. Biol. Chem.* 246:6855–66
10. Walsh, C. T., Krodel, E., Massey, V., Abeles, R. H. 1973. *J. Biol. Chem.* 248:1946–55
11. Porter, D. J. T., Voet, J. G., Bright, H. J. 1973. *J. Biol. Chem.* 248:4400–11
12. Dang, T. Y., Cheung, Y., Walsh, C. 1976. *Biochem. Biophys. Res. Commun.* 72:960–68
13. Porter, D. J. T., Bright, H. J. 1975. See Ref. 8a, 12:421–505
14. Bruice, T. C. 1976. *Ann. Rev. Biochem.* 45:331–73
15. Massey, V., Ghisla, S. 1974. *Ann. NY Acad. Sci.* 227:446–65
16. Williams, R. F., Bruice, T. C. 1976. *J. Am. Chem. Soc.* 98:7752–68
17. Kemal, C., Bruice, T. C. 1976. *J. Am. Chem. Soc.* 98:3955–64
18. Bruice, T. C. 1976. *Prog. Bioorg. Chem.* 4:1–88
19. Hemmerich, P. 1978 *Adv. Chem. Soc.* In press
20. Massey, V., Muller, F., Feldberg, R., Schuman, M., Sullivan, P. A., Howell, L. G., Mayhew, S. G., Matthews, R. G., Foust, G. P. 1969. *J. Biol. Chem.* 244:3999–4008
21. Bramlett, R. N., Peck, H. D. 1975. *J. Biol. Chem.* 250–2979–86

22. Alston, T., Mela, L., Bright, H. J. 1977. *Proc. Natl. Acad. Sci. USA.* 74:3767–71
23. Singer, T. P., Edmondson, D. E. 1974. *FEBS Lett.* 42:1–14
24. Kearney, E. B., Kenney, W. C. 1974. *Horizons Biochem. Biophys.* 1:62–96
25. Kenney, W. C., Edmondson, D. E., Seng, R. L. 1976. *J. Biol. Chem.* 251:5386–90
26. Edmondson, D. E., Kenney, W. C., Singer, T. P. 1976. *Biochemistry* 15:2937–45
27. Kenney, W. C., Edmondson, D. E., Singer, T. P., Nakagawa, H., Asano, A., Sato, R. 1976. *Biochem. Biophys. Res. Commun.* 71:1194–200
28. Decker, K. 1976. *Trends Biochem. Sci.* 1:184–85
29. Yokoe, I., Bruice, T. C. 1975. *J. Am. Chem. Soc.* 97:450–51
30. Loechler, E. L., Bruice, T. C. 1975. *J. Am. Chem. Soc.* 97:3235–36
31. Thorpe, C., Williams, C. H. 1976. *J. Biol. Chem.* 251:7726–28
32. Cheung, Y. F., Walsh, C. 1976. *Biochemistry* 15:2432–41
33. Kapke, G., Davis, L. 1976. *Biochemistry* 15:3745–49
34. Porter, D. J. T., Voet, J. G., Bright, H. J. 1977. *J. Biol. Chem.* 252:4464–73
35. Porter, D. J. T., Bright, H. J. 1977. *J. Biol. Chem.* 252:4361–70
36. Chan, T. W., Bruice, T. C. 1977. *J. Am. Chem. Soc.* 99:2387–89
37. Abeles, R., Maycock, A. 1976. *Acc. Chem. Res.* 9:313
38. Walsh, C. 1977. *Horizons Biochem. Biophys.* 3:36–81
39. Rando, R. 1974. *Science* 185:320; 1975. *Acc. Chem. Res.* 8:281
40. Jung, M. J., Metcalf, B. W. 1975. *Biochem. Biophys. Res. Commun.* 67:301–6; Lippert, B., Metcalf, B. W., Jung, M. J., Casara, P. 1977. *Eur. J. Biochem.* 74:441–45
41. Ghisla, S., Ogata, H., Massey, V., Schonbrunn, A., Abeles, R. H., Walsh, C. T. 1976. *Biochemistry* 15:1791–97
42. Schonbrunn, A., Abeles, R. H., Walsh, C. T., Ghisla, S., Ogata, H., Massey, V. 1976. *Biochemistry* 15:1798–807
43. Maycock, A. L., Abeles, R. H., Salach, J. I., Singer, T. P. 1976. *Biochemistry* 15:114–25
44. Walsh, C. T., Schonbrunn, A., Lockridge, O., Massey, V., Abeles, R. H. 1972. *J. Biol. Chem.* 247:6004–6
45. Maycock, A. L., Abeles, R. H., Salach, J. I., Singer, T. P. 1976. In *Flavins and Flavoproteins*, ed. T. Singer. pp. 218–29. New York: Elsevier

46. Gartner, B., Hemmerich, P., Zeller, E. A. 1976. *Eur. J. Biochem.* 63:211–21
47. Maycock, A. L. 1975. *J. Am. Chem. Soc.* 97:2270–72
48. Chuang, H. Y. K., Patek, D. R., Hellerman, L. 1974. *J. Biol. Chem.* 249:2381–84
49. Cromartie, T. H., Walsh, C. T. 1975. *Biochemistry* 14:3482–90
50. Walsh, C. T., Abeles, R. H., Kaback, H. R. 1972. *J. Biol. Chem.* 247:7858; Walsh, C. T., Kaback, H. R. 1973. *J. Biol. Chem.* 248:5456–62
51. Snyder, M. A., Kaczorowski, G. J., Barnes, E. M., Walsh, C. 1976. *J. Bacteriol.* 127:671–73
52. Horiike, K., Nishina, Y., Miyake, Y., Yamano, T. 1975. *J. Biochem. Tokyo* 78:57–63
53. Scannell, J. P., Pruess, D. L., Demny, T. C., Weiss, F., Williams, T., Stempel, A. 1971. *J. Antibiot.* 24:239
54. Marcotte, P., Walsh, C. 1976. *Biochemistry* 15:3070–76
55. Marcotte, P., Walsh, C. T. 1978. *Biochemistry*. Submitted for publication
56. Porter, D. J. T., Bright, H. J. 1976. *J. Biol. Chem.* 251:6150–53
57. Kraus, J. L., Belleau, B. 1975. *Can. J. Chem.* 53:3141–44
58. O'Brien, D. E., Weinstock, L. T., Cheng, C. C. 1970. *J. Heterocycl. Chem.* 7:99–107
59. Edmondson, D. E., Barman, B., Tollin, G. 1972. *Biochemistry* 11:1133–38
60. Shinkai, S., Bruice, T. C. 1973. *J. Am. Chem. Soc.* 95:7526–28
61. Jorns, M. S., Hersh, L. B. 1974. *J. Am. Chem. Soc.* 96:4012–14
62. Fisher, J., Walsh, C. 1974. *J. Am. Chem. Soc.* 96:4345–46
63. Averill, B. A., Schonbrunn, A., Abeles, R. H., Weinstock, L. T., Cheng, C. C., Fisher, J., Spencer, R., Walsh, C. 1975. *J. Biol. Chem.* 250:1603–5
64. Jorns, M. S., Hersh, L. B. 1975. *J. Biol. Chem.* 250:3620–28
65. Spencer, R., Fisher, J., Walsh, C. 1976. *Biochemistry* 15:1043–53
66. Fisher, J., Spencer, R., Walsh, C. 1976. *Biochemistry* 15:1054–64
67. Jorns, M. S., Hersh, L. B. 1976. *J. Biol. Chem.* 251:4872–81
68. Stankovich, M. T., Massey, V. 1976. *Biochim. Biophys. Acta* 452:335–44
69. Blankenhorn, G. 1976. *Eur. J. Biochem.* 67:67–80
70. Hersh, L. B., Jorns, M. S., Peterson, J., Currie, M. 1976. *J. Am. Chem. Soc.* 98:865–67
71. Massey, V., Hemmerich, P. 1978. *Biochemistry.* 17:1–15

72. Cromartie, T. H., Walsh, C. T. 1976. *J. Biol. Chem.* 251:329–33
73. Becker, W., Pfeil, E. 1966. *Biochem. Z.* 346:301
74. Stormer, F. C., Umbarger, H. E. 1964. *Biochem. Biophys. Res. Commun.* 17:587–92
75. Ashton, W., Brown, D., Brown, J., Graham, D., Rogers, E. 1977. *Tetrahedron Lett.* 30:2551–54
76. Spencer, R., Fisher, J., Walsh, C. 1977. *Biochemistry* 16:3586–94
77. Spencer, R., Fisher, J., Walsh, C. 1977. *Biochemistry* 16:3594–3602
78. Ghisla, S., Mayhew, S. G. 1976. *Eur. J. Biochem.* 63:373–90
79. Walsh, C., Fisher, J., Spencer, R., Graham, D., Ashton, W., Brown, J., Brown, R., Rogers, E. 1978. *Biochemistry.* In press
80. Mayhew, S. G., Whitfield, C. D., Ghisla, S., Schuman-Jorns, M. 1974. *Eur. J. Biochem.* 44:579
81. Otani, S., Takatsu, M., Nakano, M., Kasai, S., Miura, R., Matsui, K. 1974. *J. Antibiot.* 27:88–89
82. Koberstein, R. 1976. *Eur. J. Biochem.* 67:223–29
83. Abramovitz, A. S., Massey, V. 1976. *J. Biol. Chem.* 251:5321–26
84. Matthews, R. G., Massey, V., Sweeley, C. C. 1976. *J. Biol. Chem.* 250:9294–98
85. Abramovitz, A. S., Massey, V. 1976. *J. Biol. Chem.* 251:5327–36
86. Fujii, K., Galivan, J. H., Huennekens, F. M. 1977. *Arch. Biochem. Biophys.* 178:662–70
87. Schrauzer, G. N., Deutsch, E., Windgassen, R. J. 1968. *J. Am. Chem. Soc.* 90:2441
88. Geary, L. E., Meister, A. 1977. *J. Biol. Chem.* 252:3501–8
89. Mantsala, P., Zalkin, H. 1976. *J. Biol. Chem.* 251:3300–5
90. Massey, V., Hemmerich, P. 1976. See Ref. 8a, 12:191
91. Flashner, M., Massey, V. 1974. In *Molecular Mechanisms of Oxygen Activation,* ed. O. Hayaishi, p. 245. New York: Academic
92. Kemal, C., Bruice, T. C. 1976. *Proc. Natl. Acad. Sci. USA* 73:995–99
93. Entsch, B., Ballou, D. P., Massey, V. 1976. *J. Biol. Chem.* 251:2550–63
94. Kemal, C., Chan, T. W., Bruice, T. C. 1977. *Proc. Natl. Acad. Sci. USA* 74:405–9
95. Ghisla, S., Entsch, B., Massey, V., Husain, M. 1977. *Eur. J. Biochem.* 76:149–56
96. Jablonski, E., DeLuca, M. 1977. *Biochemistry* 16:2932–36
97. Hastings, J. W., Balny, C. 1975. *J. Biol. Chem.* 250:7288–93; 1976. *Biochemistry* 14:4719
98. Douzou, P. 1977. *Adv. Enzymol.* 45:157–272
99. Northrop, D. B. 1975. *Biochemistry* 14:2644–51
100. Lowe, J. N., Ingraham, L. L., Alspach, J., Rasmussen, R. 1976. *Biochem. Biophys. Res. Commun.* 73:465–69
101. McCapra, F. 1976. *Acc. Chem. Res.* 9:201–8
102. Entsch, B., Ballou, D. P., Husain, M., Massey, V. 1976. *J. Biol. Chem.* 251:7367–79
103. Allison, W. S. 1976. *Acc. Chem. Res.* 9:293–99
104. Poulsen, L. L., Ziegler, D. M. 1976. *Fed. Proc.* 35:1653.
105. Prough, R. A., Ziegler, D. M. 1977. *Arch. Biochem. Biophys.* 180:363–73
106. You, K. S., Arnold, L. J., Kaplan, N. O. 1977. *Arch. Biochem. Biophys.* 180:550–54
107. Babior, B. M., Kipnes, R. S. 1978. *Blood.* In press
108. Babior, B. M., Curnutte, J. T., McMurrich, B. J. 1976. *J. Clin. Invest.* 58:989–96
109. Lovenberg, W., ed. *Iron-Sulfur Proteins,* Vols. 1–3. New York: Academic
110. Palmer, G. 1975. See Ref. 13, p. 1
111. Ruschig, U., Muller, U., Willnow, P., Hopner, T. 1976. *Eur. J. Biochem.* 70:325–30
112. Enoch, H. G., Lester, R. L. 1975. *J. Biol. Chem.* 250:6693–705
113. Schneider, K., Schlegel, H. G. 1976. *Biochim. Biophys. Acta* 452:66–80
114. Waud, W. R., Rajagopalan, K. V. 1976. *Arch. Biochem. Biophys.* 172:365–79
115. Barber, M. J., Bray, R. C., Cammack, R., Coughlan, M. P. 1977. *Biochem. J.* 163:279–89
116. Berg, A., Gustafsson, J. A., Ingelman-Sundberg, M., Carlstrom, K. 1976. *J. Biol. Chem.* 251:2831–38
117. Pedersen, J. I., Ghazarian, J. G., Orme-Johnson, N. R., DeLuca, H. F. 1976. *J. Biol. Chem.* 251:3933–41
118. Madhok, T. C., Schnoes, H. K., DeLuca, H. F. 1977. *Biochemistry* 16:2142
119. Madyastha, K. M., Meehan, T. D., Coscia, C. J. 1976. *Biochemistry* 15:1097–102
120. Bernhardt, F. H., Nastainczyk, W., Seydewitz, V. 1977. *Eur. J. Biochem.* 72:107–15
121. Tonge, G. M., Harrison, D. E., Higgins, I. J. 1977. *Biochem. J.* 161:333

122. Strittmatter, P. 1974. *Proc. Natl. Acad. Sci. USA* 71:4565–69
123. Reddy, V. V. R., Kupfer, D., Caspi, E. 1977. *J. Biol. Chem.* 252:2797–801
124. Reddy, V. V. R., Caspi, E. 1976. *Eur. J. Biochem.* 69:577–82
125. Haugen, D. A., Coon, M. J. 1976. *J. Biol. Chem.* 251:7929–39
126. Huang, M. T., West, S. B., Lu, A. Y. H. 1976. *J. Biol. Chem.* 251:4659–65
127. Wang, H. P., Kimura, T. 1976. *J. Biol. Chem.* 251:6068–74
128. Vanlier, J. E., Rousseau, J. 1976. *FEBS Lett.* 70:23–27
129. Hochberg, R. B., McDonald, P. D., Ponticorvo, L., Lieberman, S. 1976. *J. Biol. Chem.* 251:7336–42
130. Chu, J. W., Kimura, T. 1973. *J. Biol. Chem.* 248:5183–87
131. Lambeth, J. D., Kamin, H. 1976. *J. Biol. Chem.* 251:4299–306
132. Lambeth, J. D., Kamin, H. 1977. *J. Biol. Chem.* 252:2908–17
133. Lambeth, J. D., McCaslin, D. R., Kamin, H. 1975. *J. Biol. Chem.* 251:7545–50
134. Yasukochic, Y., Masters, B. S. S. 1976. *J. Biol. Chem.* 251:5337–44
135. Knapp, J. A., Dignam, J. D., Strobel, H. W. 1977. *J. Biol. Chem.* 252:437–43
136. Dignam, J. D., Strobel, H. W. 1977. *Biochemistry* 16:1116–23
137. Jollow, D., Kocsis, J., Snyder, R., Valinio, H., eds. 1977. *Biological Reactive Intermediates.* New York: Plenum
138. Yasunobu, T., Mower, H., Hayaishi, O., eds. 1976. *Iron and Copper Proteins.* New York: Plenum. 593 pp.
139. Jerina, D., ed. 1977. *Drug Metabolism Concepts.* ACS Symp. Ser., Vol. 44. 196 pp.
140. Freudenthal, R., Jones, P., eds. 1977. *Carcinogenesis: A Comprehensive Survey,* Vol II. New York: Raven, 450 pp.
141. Hrycay, E. G., Gustafsson, J. A., Ingelman-Sundberg, M., Ernster, L. 1976. *Eur. J. Biochem.* 61:43–52
142. Nordblom, G. D., White, R. E., Coon, M. J. 1976. *Arch. Biochem. Biophys.* 175:524–33
143. Lichtenberger, F., Nastainczyk, W., Ullrich, V. 1976. *Biochem. Biophys. Res. Commun.* 70:939–46
144. Guengerich, F. P., Ballou, D. P., Coon, M. J. 1976. *Biochem. Biophys. Res. Commun.* 70:951–56
145. Imai, Y., Sato, R. 1977. *Biochem. Biophys. Res. Commun.* 75:420–26
146. Kamataki, T., Kitagawa, H. 1977. *Biochem. Biophys. Res. Commun.* 76:1007–13
147. Bruice, T. C., Bruice, P. Y. 1976. *Acc. Chem. Res.* 9:378–84
148. Lu, A. Y. H., Ryan, D., Jerina, D. M., Daly, J. W., Levin, W. 1975. *J. Biol. Chem.* 250:8283–88
149. Bentley, P., Schmassmann, H., Sims, P., Oesch, F. 1976. *Eur. J. Biochem.* 69:97–103
150. Yang, S. K., McCourt, D. W., Roller, P. P., Gelboin, H. V. 1976. *Proc. Natl. Acad. Sci. USA* 73:2594–98
151. Jeffrey, A. M., Yeh, H. J. C., Jerina, D. M., Patel, T. R., Davey, J. F., Gibson, D. T. 1975. *Biochemistry* 14:575
152. Jerina, D. M., Selander, H., Yagi, H., Wells, M. C., Davey, J. F., Mahadevan, V., Gibson, D. T. 1976. *J. Am. Chem. Soc.* 98:5988–96
153. Habig, W. H., Pabst, M. J., Jakoby, W. B. 1974. *J. Biol. Chem.* 249:7130–39
154. Keen, J. H., Habig, W. H., Jakoby, W. B. 1976. *J. Biol. Chem.* 251:6183–88
155. Hjelmeland, L. M., Aronow, L., Trudell, J. R. 1977. *Biochem. Biophys. Res. Commun.* 76:541–49
155a. Tomaszewski, J. E., Jerina, D. M., Daly, J. W. 1975. *Biochemistry* 14:2024–31
156. Abita, J. P., Milstien, S., Chang, N., Kaufman, S. 1976. *J. Biol. Chem.* 251:5310–14
157. Reddy, C. C., Vaidyanathan, C. 1976. *Arch. Biochem. Biophys.* 177:488–98
158. Bhat, S. G., Vaidyanathan, C. S. 1976. *Arch. Biochem. Biophys.* 176:314–23
159. Kaufman, S. 1976. See Ref. 138, pp. 94–96
160. Goldberg, I. 1976. *Eur. J. Biochem.* 63:233–40
161. Steenkamp, D. J., Mallinson, J. 1976. *Biochem. Biophys. Acta* 429:705–19
162. Gillum, W. O., Mortenson, L. E., Chen, J. S., Holm, R. H. 1977. *J. Am. Chem. Soc.* 99:584–95
163. Steenkamp, D. J., Singer, T. P. 1976. *Biochem. Biophys. Res. Commun.* 71:1289–95
163a. Solomon, E. I., Hare, J. W., Gray, H. B. 1976. *Proc. Natl. Acad. Sci. USA* 73:1389–93
164. Malstrom, B., Andreasson, L. E., Reinhammar, B. 1975. See Ref. 13, p. 507
165. Finazzi Agro, A., Guerrieri, P., Costa, M. T., Mondovi, B. 1977. *Eur. J. Biochem.* 74:435–40
166. Neumann, R., Hevey, R., Abeles, R. H. 1975. *J. Biol. Chem.* 250:6362–67
167. Maycock, A. L., Suva, R. H., Abeles, R. H. 1975. *J. Am. Chem. Soc.* 97:5613–14
168. Siegel, R. C., Fu, J. C. C. 1976. *J. Biol. Chem.* 251:5779–85

169. Olsson, B., Olsson, J., Petterson, G. 1976. *Eur. J. Biochem.* 64:327–31
170. Hamilton, G. A., Libby, R. D., Hartzell, C. R. 1973. *Biochem. Biophys. Res. Commun.* 55:333–40
171. Giordano, R. S., Bereman, R. D., Kosman, D. J., Ettinger, M. J. 1974. *J. Am. Chem. Soc.* 96:1023.
172. Cleveland, L., Coffman, R. E., Coon, P., Davis, L. 1975. *Biochemistry* 14:1108–15
173. Dyrkacz, G. R., Libby, R. D., Hamilton, G. A. 1976. *J. Am. Chem. Soc.* 98:626–28
174. Maradufu, A., Cree, G. M., Perlin, A. S. 1971. *Can. J. Chem.* 49:3429
175. Hamilton, G. A., Dyrkacz, G. R., Libby, R. D. 1976. See Ref. 138, p. 489.
176. Taylor, K. B. 1974. *J. Biol. Chem.* 249:454–58
177. Bachan, L., Storm, C. B., Wheeler, J. W., Kaufman, S. 1974. *J. Am. Chem. Soc.* 96:6799–800
178. Klinman, J. P. 1977. *Fed. Proc.* 36:2082
179. Hayaishi, O., ed. 1974. See Ref. 91, p. 1–29
180. Abbott, M., Udenfriend, S. 1975. See Ref. 91, p. 168–215
181. Lindstedt, G., Lindstedt, S., Nordin, I. 1977. *Biochemistry* 16:2181–88
182. Turpeenniemi, T. M., Puistola, U., Anttinen, H., Kivirikko, K. 1977. *Biochem. Biophys. Acta* 483:215–19
183. Hayaishi, O., Hirata, F., Ohnishi, T., Henry, J. P., Rosenthal, I., Katoh, A. 1977. *J. Biol. Chem.* 252:3548–50
184. Roberts, J., Rosenfeld, H. J. 1977. *J. Biol. Chem.* 252:2640–47
185. Takai, K. Ushiro, H., Noda, Y., Narumiya, S., Tokuyama, T., Hayaishi, O. 1977. *J. Biol. Chem.* 252:2648–56
186. Noda, Y., Takai, K., Tokuyama, T., Narumiya, S., Ushiro, H., Hayaishi, O. 1977. *J. Biol. Chem.* 252:4413–15
187. Miyamoto, T., Ogino, N., Yamamoto, S., Hayaishi, O. 1976. *J. Biol. Chem.* 251:2629–36
188. Ogino, N., Miyamoto, T., Yamamoto, S., Hayaishi, O. 1977. *J. Biol. Chem.* 252:890–95
189. Hemler, M., Lands, W. E. M., Smith, W. L. 1976. *J. Biol. Chem.* 251:5575–79
190. Van der Ouderaa, F. J., Buytenhek, M., Nugteren, D. H., Van Dorp, D. A. 1977. *Biochim. Biophys. Acta* 487:315–31
191. Egan, R. W., Paxton, J., Kuehl, F. A. 1976. *J. Biol. Chem.* 251:7329–35
192. Gibian, M. J., Galaway, R. A. 1976. *Biochemistry* 15:4209–14
193. Downing, D. T., Ahern, D. G., Bachta, M. 1970. *Biochem. Biophys. Res. Commun.* 40:218
194. Hammarstrom, S. 1977. *Biochim. Biophys. Acta* 487:517–19
195. Kido, T., Soda, K., Suzuki, T., Asada, K. 1976. *J. Biol. Chem.* 251:6994–7000
196. Palmer, J. L., Abeles, R. H. 1976. *J. Biol. Chem.* 251:5817–19
197. Guiard, B., Lederer, F. 1977. *Eur. J. Biochem.* 74:181–90
198. Johnson, J. L., Rajagopalan, K. V. 1977. *J. Biol. Chem.* 252:2017–25
199. Andrews, T. J., Lorimer, G. H., Tolbert, N. E. 1973. *Biochemistry* 12:11–18
200. Siegel, M., Wishnick, M., Lane, M. D. 1977. See Ref. 8a, 6:169
201. Lorimer, G. H., Andrews, T. J., Tolbert, N. E. 1973. *Biochemistry* 12:18–23
202. Wildner, G. F., Henkel, J. 1976. *Biochem. Biophys. Res. Commun.* 69:268–75
203. Matthews, J. C., Hori, K., Cormier, M. J. 1977. *Biochemistry* 16:85–91
204. Healy, M. J., Christen, P. 1972. *J. Am. Chem. Soc.* 94:7911; 1973. *Biochemistry* 12:35
205. Christen, P., Cogoli-Greuter, M., Healy, M. J., Lubini, D. 1976. *Eur. J. Biochem.* 63:223–31
206. Christen, P., Gasser, A. 1976. *J. Biol. Chem.* 251:4220–23
207. Stenflo, J., Suttie, J. W. 1977. *Ann. Rev. Biochem.* 46:157–72
208. Mack, D., Suen, E. T., Girardot, J. M., Miller, J. A., Delaney, A., Johnson, B. C. 1976. *J. Biol. Chem.* 251:3269–76
209. Suttie, J. W., Hageman, J. M., Lehrman, S. R., Rich, D. H. 1976. *J. Biol. Chem.* 251:5827–30
210. Bell, R. G., Stark, P. 1976. *Biochem. Biophys. Res. Commun.* 72:619–25
210a. Moore, H. W. 1977. *Science* 197:527–32
211. Korytnyk, W., Hakala, M. T., Potti, P. G. G., Angelino, N., Chang, S. C. 1976. *Biochemistry* 15:5458–66
212. Yang, I. Y., Harris, C. M., Metzler, D. E., Korytnyk, W., Lachmann, B., Potti, P. P. G. 1975. *J. Biol. Chem.* 250:2947–55
213. Miura, R., Metzler, D. E. 1976. *Biochemistry* 15:283–90
214. Snell, E. E., DiMari, S. J. 1970. See Ref. 8a, 2:335
215. Karube, Y., Matsushima, Y. 1976. *J. Am. Chem. Soc.* 98:3725–26
216. Babu, U. M., Johnston, R. B. 1974. *Biochem. Biophys. Res. Commun.* 58:460–66
217. Marcotte, P., Walsh, C. 1975. *Biochem. Biophys. Res. Commun.* 62:677

218. Cooper, A. J. L. 1976. *J. Biol. Chem.* 251:1088–96
219. Golichowski, A., Harruff, R. C., Jenkins, W. T. 1977. *Arch. Biochem. Biophys.* 178:459–67
220. Snell, E. E. 1975. See Ref. 98, 42:287
221. Washtien, W., Cooper, A. J. L., Abeles, R. H. 1977. *Biochemistry* 16:460–287
222. Guggenheim, S., Flavin, M. 1969. *J. Biol. Chem.* 244:6217
223. Briley, P. A., Eisenthal, R., Harrison, R., Smith, G. D. 1977. *Biochem. J.* 161:383–87
224. Briley, P. A., Eisenthal, R., Harrison, R., Smith, G. D. 1977. *Biochem. J.* 163:325–31
225. Skye, G. E., Potts, R., Floss, H. G. 1974. *J. Am. Chem. Soc.* 96:1593–95
226. Floss, H. G., Schleicher, E., Potts, R. 1976. *J. Biol. Chem.* 251:5478–82
227. Cheung, Y. F., Walsh, C. 1976. *J. Am. Chem. Soc.* 98:3397–98
228. Yang, I. Y., Huang, Y. Z., Snell, E. E. 1975. *Fed. Proc.* 34:496
229. Yang, I. Y., Huang, Y. Z., Snell, E. E. 1975. See Ref. 228, footnote 2
230. Fuganti, C., Ghiringhelli, D., Giangrasso, D., Grasselli, P. 1974. *Chem. Commun.* 726
231. Sawada, S., Kumagai, H., Yamada, H., Hill, R. K. 1975. *J. Am. Chem. Soc.* 97:4334–37
232. Schleicher, E., Mascaro, K., Potts, R., Mann, D. R., Floss, H. G. 1976. *J. Am. Chem. Soc.* 98:1043–44
233. Tatum, C. M., Benkovic, P. A., Benkovic, S. J., Potts, R., Schleicher, E., Floss, H. G. 1977. *Biochemistry* 16:1093–102
234. Soper, T. S., Manning, J. M., Marcotte, P. A., Walsh, C. T. 1977. *J. Biol. Chem.* 252:1571–75
235. Morino, Y., Okamoto, M. 1973. *Biochem. Biophys. Res. Commun.* 50:1061–67
236. Relyea, N. M., Tate, S. S., Meister, A. 1974. *J. Biol. Chem.* 249:1519–24
237. Wang, E., Walsh, C. 1978. *Biochemistry.* 17: In press
237a. Soper, T. S., Jones, W. M., Lerner, B., Trop, M., Manning, J. M. 1977. *J. Biol. Chem.* 252:3170–75
238. Kapke, G., Davis, L. 1975. *Biochem. Biophys. Res. Commun.* 65:765–69
239. Deleted in proof
240. Rando, R. R., Relyea, N., Cheng, L. 1976. *J. Biol. Chem.* 251:3306–12
241. Kobayashi, K., Miyazawa, S., Endo, A. 1977. *FEBS Lett.* 76:207–10; 1976. *Tetrahedron Lett.* 7:537–40
242. Rando, R. R., Bangerter, F. W. 1976. *J. Am. Chem. Soc.* 98:6762–64
243. Rando, R. R., Bangerter, F. W. 1977. *J. Am. Chem. Soc.* 99:5141–45
244. Rando, R. R., Bangerter, F. W. 1977. *Biochem. Biophys. Res. Commun.* 76:1276–81
245. Schechter, P. J., Tranier, Y., Jung, M. J., Lippert, B., Bohlen, P. 1977. *Eur. J. Pharmacol.* In press
246. Schechter, P. J., Tranier, Y. 1978. *Psychopharmacologia.* In press; 1977. *Pharmacol. Biochem. Behav.* In press
247. Silverman, R. B., Abeles, R. H. 1976. *Biochemistry* 15:4718–23
248. Tanase, S., Morino, Y. 1976. *Biochem. Biophys. Res. Commun.* 68:1301–8
249. Fowler, L. J., John, R. A. 1972. *Biochem. J.* 130:569
250. Rando, R. R. 1974. *Biochemistry* 13:3859–63
251. Lane, R. S., Manning, J. M., Snell, E. E. 1976. *Biochemistry* 15:4180–85
252. Relyea, N. M., Rando, R. R. 1975. *Biochem. Biophys. Res. Commun.* 67:392–402
253. McLemore, W. O., Davis, L., Metzler, D. E. 1968. *J. Biol. Chem.* 243:441
254. Miles, E. W. 1975. *Biochem. Biophys. Res. Commun.* 64:248–55
255. Abeles, R. H., Walsh, C. T. 1973. *J. Am. Chem. Soc.* 95:6124–25
256. Washtien, W., Abeles, R. H. 1977. *Biochemistry* 16:2485–91

Ann. Rev. Biochem. 1978. 47:933–65
Copyright © 1978 by Annual Reviews Inc. All rights reserved

CELLULAR TRANSPORT
MECHANISMS

❖993

David B. Wilson

Section of Biochemistry, Molecular, and Cell Biology, Wing Hall,
Cornell University, Ithaca, New York 14853

CONTENTS

933

0066-4154/78/0701-0933$01.00

PERSPECTIVES AND SUMMARY

In this review the assumption is made that living cells are surrounded by lipid bilayer membranes that are inherently impermeable to most hydrophilic compounds. It is also assumed that hydrophilic molecules are transported into cells by specific transport systems that are present in the membrane and are made up of one or more proteins. These transport systems either equilibrate substrates across the membrane (facilitated diffusion) or use energy to concentrate substrates (active transport).

There have been two important recent technical advances in transport research. One is the development of methods for making closed membrane vesicles that retain the activity of certain transport systems (1, 2). In some cases it is possible to make vesicles that either have the overall orientation present in the original cell or that have an inverted orientation (3–5). Vesicles have the advantage that they are free of endogenous energy sources and do not metabolize most substrates. But they have the disadvantage that certain transport systems are not active in them, and some membrane proteins partially alter their orientation during vesicle formation (6). The second advance is the development of a number of procedures for incorporating membrane proteins into phospholipid vesicles so as to reconstitute transport activity (7). These techniques not only provide assays for transport proteins but also provide simple systems for the study of purified transport proteins.

There have also been two important conceptual advances in this field. One is a realization that transport models involving mobile or rotating carriers are not applicable to proteins containing exposed hydrophilic residues (8). It is likely that at least one component of each transport system spans the membrane—probably as an oligomer—to form a hydrophilic pore through the membrane. In addition, special mechanisms are probably required to insert transport proteins into the membrane with the correct orientation. The other advance is a realization that many active transport systems function as either proton or sodium symports (9); that is, the movement of substrate through the membrane is always accompanied by the parallel movement of the appropriate ion. In many cases these systems are electrogenic, so that energy for transport comes not only from the ion gradient, but also from the membrane potential.

Only work on bacterial and mammalian systems is discussed in this review. A detailed review of plant transport appeared in References 10–12, and this subject is regularly discussed in the *Annual Review of Plant Physiology*. At the present time it appears that most bacterial systems carry out active transport and fall into one of three classes: group translocation systems, membrane-bound transport systems, and binding protein transport systems. Within the class of membrane-bound transport systems there appear to be some important differences between various systems. Mammalian transport systems appear to be fairly evenly divided between facilitative diffusion systems and active transport systems. Most mammalian active transport systems are either coupled directly to ATP hydrolysis or are sodium symports, although it is quite possible that other mechanisms will be found as more systems are thoroughly characterized.

INTRODUCTION

The purpose of this article is to review the information about the mechanisms by which molecules are moved through membranes and concentrated by living cells. It is impossible to give a complete account of all the work in this extensive field, so this article is limited to a discussion of general topics and a limited number of specific systems chosen to illustrate the different transport mechanisms. Despite their importance, proton transport systems will not be discussed, since work in this field is well reviewed (13–16). Even with this limitation, the literature is so extensive that no attempt is made to cite every paper. Instead, in many cases, recent reviews are cited that can be consulted for the references to specific articles.

General Topics

There is a fundamental controversy in the transport field. Most workers believe that living cells are surrounded by a membrane composed of a lipid bilayer containing proteins, some of which function to specifically transport hydrophilic molecules through the membrane. However, a small group maintains that the cell is permeable to most molecules and that the process called transport is due either to direct binding of molecules to cytoplasmic compounds or to altered properties of the cell water, which result from the high concentration of charged macromolecules in the cytoplasm (17, 18). Some of the arguments on both sides were reviewed in an article in *Science* (19) and in letters commenting on it (20).

Multiple Systems

A major difficulty that must be dealt with in transport studies is the fact that most substrates are transported by several systems. This phenomena is seen most strikingly in bacteria—in *Escherichia coli* at least five systems transport galactose (21)—but it is also observed in higher organisms (22).

The multiple K_m values seen for the uptake of a given substrate can be due either to several independent systems or to a single system with multiple binding sites. The leucine transport system provides a good illustration of the problem. A multiple site model has been proposed to explain the complex kinetics observed for leucine transport in *E. coli* (23). There are at least three K_m values for transport, so that a model with a single carrier having three binding sites was proposed and shown to fit the kinetic data slightly better than a model using two independent systems. However, a model containing three independent systems, which is the one proposed by other workers (24–26), would give the best fit to the kinetic data. Unfortunately, when there are more than two K_ms of uptake, it is probably not possible to decide the controversy by kinetic data alone because of the many degrees of freedom in the models and the errors of transport measurements. Some criteria distinguishing between independent systems and a single allosteric system have been discussed (27).

The fact that mutants can be isolated in separate genes that each inactivate a single system, without affecting the other systems, is evidence that favors three independent leucine transport systems (28). In addition, two of the systems are associated with periplasmic binding proteins and use ATP to drive active transport, whereas the third system is driven by the proton motive force (25). Finally, there is evidence showing differences in the regulation of the different systems (29). Thus, the data for leucine transport indicate that there are three distinct leucine transport systems, although there may be some interactions between them.

Glutamate transport in *E. coli* is another system showing complex kinetics in which variations of a single system model have been presented (30, 31). However, a recent study provides strong evidence that there are three different glutamate transport systems (32). These different systems were distinguished by the use of inhibitors and by the isolation both of mutations that inactivate certain systems and mutations that increase the level of certain systems significantly. The three systems are: a binding protein associated system, a sodium-stimulated membrane-bound system, and a nonsodium-stimulated membrane-bound system. Each system has a unique substrate specificity; moreover, the properties of each mutant are consistent with the systems being independent.

There are at least two systems where the evidence is consistent with an allosteric model. One is a potassium transport system in *Neurospora* (33), and the other is an aromatic amino acid transport system in *Bacillus subtilis* (34). Even for these systems the evidence is still indirect, so that further work is required to insure that a single carrier is responsible for the dual K_ms seen for these systems.

Properties of Membrane Proteins

The synthesis and properties of membrane proteins, with special emphasis on their significance for transport systems, are the subject of a recent article (8). While direct evidence for some of the conclusions of this article is still limited, the concepts appear very reasonable and should be considered by all workers in this field. The basic conclusion is that because almost all proteins contain both hydrophobic and hydrophilic regions, they cannot migrate through the membrane in their native state. Therefore, proteins that are either outside the plasma membrane, span the membrane, or are present on the side opposite to that on which protein biosynthesis occurs, will be synthesized on membrane-bound ribosomes. Many of these proteins will be inserted into or moved through the membrane as they are synthesized. A hydrophobic segment is present at the N terminus of some excreted proteins that is removed after the protein has passed through the membrane (35). Furthermore, bacterial periplasmic proteins and outer membrane proteins, which are moved through the inner membrane, are synthesized on membrane-bound ribosomes (36). Post-transcriptional processing such as glycosylation may also function to insert molecules into the membrane. However, there is no direct evidence for this process at the present time.

Multiple Mechanisms

A final complexity in the transport field is the existence of quite different mechanisms that are used by living cells to carry out active transport. Specific examples of the different classes of transport systems and their properties are discussed in the rest of this article.

BACTERIAL TRANSPORT SYSTEMS

Introduction

There are at least three distinct classes of active transport systems in bacteria (37). One class couples the transport of molecules across the cell membrane with chemical modification, a process called group translocation. A second class transports molecules against a concentration gradient with no change in the transported molecule. All of the proteins required by members of this class are firmly bound to the cell membrane and are retained in isolated membrane vesicles. Members of this class are called membrane-bound transport systems. About 40% of all known E. coli transport systems belong to this class, including transport system for sugars, ions, and amino acids. The third class also concentrates molecules without modification, but transport systems in this class require at least one component that is lost during the formation of membrane vesicles. This component is a soluble protein that specifically binds the substrates of the transport system. Therefore, members of this class are called binding protein transport systems, and they include systems that transport ions, sugars, amino acids, and vitamins. Another 40% of the transport systems of *E. coli* are in this class.

It is not clear what factors determine whether a given substrate is transported by a membrane-bound or a binding protein transport system. In fact, some substrates are transported by members of both classes. The one generalization that can be made is that binding protein systems appear to have somewhat lower K_m values than membrane-bound systems transporting the same or similar substrates.

There is strong evidence that each class utilizes a different mechanism to couple energy to the accumulation of substrates. The group translocation systems utilize the chemical energy of the modification reaction to accumulate substrates. The membrane-bound systems utilize the proton-motive force across the membrane to drive active transport, whereas the binding protein transport systems utilize ATP or a related compound to drive active transport.

In the case of the membrane-bound systems, the proton motive force can be used either directly by a proton symport mechanism or indirectly by symport or antiport with some other ion. The most commonly used indirect mechanism is sodium symport. The sodium gradients used by bacteria to drive transport appear to be produced by proton-driven sodium pumps (38). This description of the energy coupling process is a simplification of a very complex and controversial subject. Since there are a number of excellent reviews (9, 39–41) dealing with this question, it is not discussed further here. Neither are bacterial group translocation systems, as they have been well covered in recent reviews (42–45).

The uptake of nonmetabolizable substrates by all transport systems eventually reaches a steady state. The steady state for most bacterial transport systems belonging to either membrane-bound or binding protein classes results from a dynamic equilibrium between an entry reaction and an exit reaction. One exception is phosphate transport in *E. coli* (46), which is similar to amino acid transport in yeast in showing little or no exit, but instead the uptake reaction is inhibited by the pool of substrate in the cell (47).

Several reports suggest a relationship between membrane-bound and binding protein transport systems. Thus, in an ATPase mutant of *E. coli* where ATP levels are maintained by glycolysis colicin K inhibits the uptake of both glutamine, which is transported by a binding protein system, and proline, which is transported by a membrane-bound system (48). However, as glutamine transport is inhibited to a lesser extent and at a slower rate than is proline transport, the two systems may be inhibited in different ways. In another study a mutation was reported to inactivate both classes of transport systems (49). However, no energy source was provided during the transport measurements, so that the mutation may have affected energy metabolism, not transport. In a more recent study, an energy source was provided in the transport assay, but only membrane-bound transport systems were tested (50). A mutant that resembles the one just discussed was shown to inactivate only membrane-bound transport systems when an energy source was provided during uptake measurements (51).

Finally, it was reported that the extent of stimulation of a large number of amino acid transport systems by an increase in the energy supply was directly proportional to the hydrophobicity of the amino acid (52). This relationship is true for members of both classes of transport systems. The basis for this phenomenon is not known, but the relationship is quite clear and was tested with five members of each class.

In summary, the latest evidence does not indicate a clear link between membrane bound and binding protein transport systems.

Membrane-Bound Transport Systems

LACTOSE PERMEASE   E. coli lactose permease is the most extensively studied bacterial membrane-bound transport system, both because it was the first transport system to be identified and because it has a very high V_{max}, about 200 nmol min^{-1} mg^{-1} protein. In addition, nonmetabolizable substrates such as thiomethyl galactoside are available for studying this system. Only one protein is required for lactose transport, since every mutation that inactivates the transport system is found in the Y gene of the lactose operon (53). This is true also for mutants that are unable to concentrate lactose permease substrates, but that can move such substrates through the membrane at a normal rate (54, 55).

The product of the Y gene (the M protein) has been isolated in an inactive form by labeling it with radioactive N-ethylmaleimide after protecting it with substrate during a preincubation with unlabeled N-ethylmaleimide (56). It is a membrane protein with a molecular weight of 30,000 (57). Lactose permease is inducible, and the M protein appears to represent about 4% of the protein in the membrane fraction, or 0.3% of the total protein in induced cells (57–59). Lactose transport activity has been reconstituted into membrane vesicles prepared from a Y^- mutant by use of an aprotic solvent extract from induced wild-type membranes (60). However, the extract was quite unstable, and the protein responsible for transport activity was not purified.

Lactose permease is a proton symport system; that is, the movement of lactose through the membrane is coupled to the movement of one or more protons. The movement of lactose into energy-deficient cells removes protons from the medium with a stoichiometry of 1 proton per lactose molecule (61), and the imposition of a proton gradient on whole cells drives lactose transport (62). In addition, the lactose-permease–dependent hydrolysis of o-nitrophenyl galactoside (ONPG) in starved cells is greatly stimulated by uncouplers that allow the protons moved into the cell by the lactose carrier to leak out, thus dissipating the membrane potential produced by their inward movement (63). Exit of lactose from the cell moves protons outside, and this proton movement can be used to drive the transport of proline (64). The lactose-permease–dependent hydrolysis of ONPG by thoroughly starved cells in the absence of uncouplers requires energy with an estimated stoichiometry of 1 proton per ONPG transported (65). However, the stoichiometry of lactose transport in membrane vesicles is pH-dependent, being 1 at pH 5.5 and 2.1 at pH 7.5. This result suggests that a group with a pK of 6.8 is responsible for this change (66). Finally, the energy requirements for forming and maintaining a lactose gradient have been determined in whole cells by microcalorimetry (67). These studies found a value that was consistent with a stoichiometry of 1 proton per lactose transported when the inward flux of lactose was less than 100 nmole/min/mg. In addition, the maintenance of the steady state required only a small amount of energy because most of the exiting lactose appeared to take a proton outside, and it could then be used to re-transport another lactose molecule. At higher rates of lactose uptake there was a 10-fold increase in the energy consumption. It is unlikely that this 10-fold increase represents energy required directly for transport. It seems more likely that either the increased energy consumption resulting from the higher rate of transport or the higher internal concentration of lactose present in these experiments activates some other energy-requiring process.

The complexity of lactose permease is shown by the following studies. First of all, there is trans-stimulation in which the presence of a substrate on one side of the membrane stimulates the movement of the same or other substrates across the membrane from the other side (68, 69). This phenomenon is not observed during active transport but can be observed in exit studies and counter-flow experiments (70). In addition, efflux of accumulated substrate can cause the transient accumulation of another substrate that is present outside the cell. This process, called counter flow, can occur at a rate close to the V_{max} of active transport in cells that are unable to carry out active transport because of the presence of energy poisons (70, 71). While some of the accumulation may result from the proton gradient produced by the exiting substrate, part of the accumulation results from some other process, since uncouplers cause only a partial inhibition of counter flow. A clear difference between counter flow and active transport lies in their responses to the external pH. Counter flow is inhibited as the pH is raised from 5 to 8, while active transport has a broad pH optimum in this range (70).

Other observations that indicate that the system is complex come from studies on *E. coli* membrane vesicles that have the same overall membrane orientation as the whole cell, although the orientation of a number of proteins may be randomized during their preparation (6, 72, 73). A series of dansyl galactosides in which the dansyl group and the galactose residue are joined by from 0 to 6 methylene groups was used in this work. These molecules bind specifically to the lactose carrier and are not transported across the membrane (74). Their binding can be readily measured as their fluorescence is greatly altered upon binding. There is a low (about 10% of the maximum) but significant binding (measured with radioactive nitrophenyl galactoside) in the absence of any energy source (59). This binding is not inhibited by uncouplers and does not increase at low pH, which indicates that maximum binding requires more than the binding of a proton to the carrier. This binding is probably the same as that studied previously in a sonicated membrane preparation (75). Maximal binding of dansyl galactoside occurs when a respiratory substrate is added or a membrane potential is applied, and this binding is inhibited by uncouplers (58). Maximum binding also occurs when the vesicles are preloaded with substrates, and efflux occurs in the presence of dansyl galactoside. This binding is not inhibited by uncouplers and is probably the same phenomenon as the uncoupler-resistant counter flow mentioned previously (70). The K_d values for the energized and nonenergized binding are about the same. A study of thiodigalactoside binding to membrane vesicles found a high level of binding in the presence of azide, about 50% of the binding seen for energized vesicles (76). The ΔH of binding was very high, -21 kcal/mole (76). The reason for the high level of energy-independent binding seen in these experiments is not known.

To explain the existence of energy-dependent and energy-independent binding, it was proposed that the carrier can exist in two states in the absence of energy: one with high affinity, and one with very low or no affinity for substrates. In addition, it was proposed that these two states are in equilibrium and that energization shifts the equilibrium to the high-affinity form (73). However, the data yield an equilibrium constant of 10 for the interconversion of these two forms which requires only a small energy difference between the two forms, whereas direct measurements indicate that large amounts of energy are required to give maximal binding (58, 77). Therefore, although there are two types of carriers in the membrane, they probably are not in equilibrium. One species of carrier shows high-affinity binding at all times, whereas the other form binds with high affinity only when the appropriate gradient is present. Since only one gene product is required for lactose permease, the two species of the carrier must result either from a chemical modification of part of the carrier or from different carrier molecules interacting with two types of membrane lipids or proteins.

There is evidence in whole cells for a carrier species that promotes facilitated diffusion in the absence of energy as long as an uncoupler is present to dissipate the potential produced by the protons carried in by the carrier (63). The energized carrier has a V_{max} of 180 μmol g^{-1} min^{-1} and a K_m of 0.7 mM for ONPG transport, while the deenergized system shows two types of transport, one that has a K_m of about 1 mM and a V_{max} of 40 μmol/g/min and a second that shows a linear increase with increasing ONPG concentration. The saturable component probably represents activity of the carrier species responsible for the non–energy-requiring binding observed in vesicles. This activity represents 23% of the total activity, compared with 10% found by the binding experiments. The many differences between the two experiments may explain this discrepancy, or, alternatively, the non–energy-requiring species may be preferentially inactivated during vesicle formation. Since no ONPG transport occurs in starved cells in the absence of uncoupler, both forms of the carrier must function as proton symporters, and therefore in energized cells both forms of the carrier would carry out active transport. The presence of the nonsaturable component seen in starved cells probably indicates that the energy-requiring component can bind substrate but with a low affinity. This would explain how a lactose gradient could convert the low-affinity form into the high-affinity form. If the energy-requiring state of the carrier did not bind lactose at all, it would be difficult to understand how it could respond to a lactose gradient.

Direct evidence from two different laboratories indicates that both species of the lactose carrier have the same binding properties on each side of the membrane and that active transport can occur in either direction, depending on the direction of the proton motive force (i.e. the carrier is symetrical). One study used *E. coli* membrane vesicles prepared by breaking the cells with a French pressure cell. Such vesicles have a membrane orientation opposite that of the intact cell (4). These vesicles can concentrate lactose in response to either a membrane potential (positive outside) or a pH gradient (alkaline inside). Efflux of accumulated lactose can be driven by ATP or by respiration, both of which produce a membrane potential and a pH gradient opposite to that required for uptake (78). Furthermore, the binding of dansyl galactoside to these vesicles (77) responds to a pH gradient, a membrane potential, or lactose efflux in the same way as do normal vesicles. The other study compared the efflux, counter flow, and uptake of lactose driven by an artifical potassium gradient in membrane vesicles that have normal orientation and the same vesicles after brief sonication, which was shown to invert the membrane (79). In each case the two classes of vesicles gave identical results, except that the normal vesicle preparation appeared to contain some leaky vesicles.

A major study (80) concluded that energy did not effect the uptake of lactose but converted exit from a saturable reaction with a K_m close to the uptake reaction to a first order reaction. A later study (81) confirmed the last finding, but showed that the V_{max} of the uptake reaction was also increased by the presence of energy, as it was in the vesicle studies and the studies on ONPG transport discussed previously. The fact that energization of the membrane lowers the affinity of the exit reaction indicates that the presence of a membrane potential, or a proton gradient across the membrane, converts the site of the high-affinity form of the carrier present on the inside surface of the membrane to a low-affinity form.

Dansyl galactosides of varying chain length also have been used to try to determine the distance of the lactose binding site from the external water phase by measuring the ability of a hydrophilic quencher to inhibit fluorescence of the bound galactosides as well as the changes in their fluorescence (82). These experiments suggest that the immediate environment of the binding site is hydrophobic and that the external aqueous phase is about 6 or 7 Å from the binding site on the carrier.

In conclusion, almost all studies on this system are consistent with a model in which the lactose carrier is present in two different forms. One, representing about 20% of the carrier molecules, has a high affinity for substrates in the absence of energy; the other has a low affinity for substrates unless it is activated by either a pH gradient, a membrane potential, or a lactose gradient. In the absence of energy both forms of the carrier are symmetric in that they have identical properties on each side of the membrane. The presence of either a pH gradient or a membrane potential converts both forms into an asymmetric carrier having a high-affinity site on one side of the membrane and a low-affinity site on the other. This asymmetry can occur in either direction, depending on the direction of the proton motive force across the membrane. Transport of lactose through the membrane by either form of the carrier is always associated with the movement of one or of two protons across the membrane, depending on the external pH.

THE *E. COLI* DICARBOXYLIC ACID TRANSPORT SYSTEM   The dicarboxylic acid transport system of *E. coli* is another well-studied membrane-bound transport system. This system transports succinate, malate, and fumarate and is inactivated by mutations in three different genes. One of these genes codes for a protein that binds lactate in addition to the above three compounds. This protein has been purified to homogeneity and appears to function to transport the compounds that it binds through the outer membrane (83). The evidence for this is as follows. Mutants that lack this protein do not transport these compounds, whereas membrane vesicles prepared from the mutant strain transport them normally. In addition, N-ethylmaleimide inhibits transport in whole cells but not in membrane vesicles, and this compound inhibits the binding activity of the outer membrane protein.

Proteins with a similar function apparently are required for glycerol phosphate transport (84) and glucose-6-phosphate transport (85) in whole cells. This suggests that the outer membrane may be impermeable to large, negatively charged ions. Although these proteins can be released by osmotic shock, they are clearly not the same as the periplasmic binding proteins associated with binding protein transport systems. They have higher K_ds and, more importantly, the transport systems with which they are associated function normally in membrane vesicles that lack these proteins, whereas the true binding protein transport systems are inactive in membrane vesicles.

The products of the other two genes that are required for succinate transport in both whole cells and vesicles have been isolated and purified to apparent homogeneity from a lubrol extract of washed membranes by affinity chromatography (86). These proteins bind succinate, malate, and fumarate, but do not bind lactate. Labeling and crosslinking studies indicate that both these proteins appear to span the membrane and are close to each other in the membrane. Furthermore, it appears that the binding site of protein SBP_2, which has a K_d for succinate of 2 μM, is present on the external side of the inner membrane, whereas the site of SB_1, which has a K_d for succinate of 23 μM, is on the inside surface of the inner membrane. Vesicles treated with a crosslinking reagent under conditions in which 90% of the two membrane proteins are coupled together still show 70% of the rate of uptake seen in uncrosslinked vesicles. Treatment with phospholipase D, which specifically hydrolyses phosphoserine, inhibits 80% of the succinate uptake in the vesicles while causing only a small inhibition of proline transport (87).

The dicarboxylic acid transport system is energized by a proton gradient (88, 89). Unlike other membrane-bound transport systems, succinate uptake was reported to be inhibited by an ATPase mutation (87). But other workers (90) did not see this effect, so it may result from some secondary property of the particular ATPase mutant used in the first study.

Since lactose permease requires only one membrane protein whereas this system requires two membrane proteins, we may conclude that even two systems using the same energy source show differences in their transport machinery. In summary, the dicarboxylic acid system is currently the most thoroughly characterized membrane-bound transport system, despite the fact that no successful reconstitution of succinate transport has been reported.

THE HALOBIUM GLUTAMATE TRANSPORT SYSTEM   Another membrane-bound system that appears to differ in some ways from lactose permease and the dicarboxylic acid transport system is the glutamate transport system of *Halobacterium halobium*. This organism has been the subject of many recent studies because it contains a novel light-driven proton pump, bacterial rhodopsin, in its membrane (16). This organism lives in a high salt environment and contains a proton, sodium antiport system (91); thus it has a large sodium gradient across the membrane. Membrane vesicles having a normal orientation can be prepared that retain an active light-driven proton pump and most transport activities. The transport of amino acids by these vesicles appears to be driven by both the sodium gradient and the membrane potential; however, the transport of glutamate, which has been studied most extensively, appears to be driven by a sodium gradient alone. Efflux of all amino acids can be driven by an inverted sodium gradient and it appears that sodium is cotransported during the exit reaction (92).

A protein that binds glutamate has been isolated by cholate extraction of *Halobacterium* membranes (93). It has a molecular weight of about 50,000 and has been purified about 15-fold. When this protein is incorporated into soybean phospholipid vesicles it catalyzes equilibration of glutamate across the membrane but does not carry out active transport. Surprisingly, the equilibration does not require sodium. This result indicates either that another component is required for active transport or that the protein has been partially modified during isolation.

A number of other bacterial uptake systems have been shown to be driven by sodium gradients: melibose (94) and glutamate (95–97) in *E. coli,* proline in *Mycobacterium phlei* (98), and melibose in *Salmonella* (99, 100).

THE PS3 ALANINE TRANSPORT SYSTEM   Another unusual organism that has been used in transport studies is the thermophilic organism, PS3. Alanine transport was successfully reconstituted into phospholipid vesicles by use of a cholate extract of PS3 membranes (101). The reconstituted vesicles could carry out active transport of alanine driven by a membrane potential created by the valinomycin-catalyzed efflux of potassium. The alanine carrier was purified about 13-fold, which suggests that a single protein may be responsible for the transport activity, but until a pure preparation of the carrier protein is obtained, this conclusion remains tentative.

PROLINE TRANSPORT IN E. COLI AND M. PHLEI   A proline transport protein has been extracted from *E. coli* membranes with acidic *n*-butanol and then chromatographed on Sephadex LH-20, which gave a four-fold purification (102). The initial extract contained only 2% of the membrane protein. Both the extract and the LH-20 material were reconstituted with *E. coli* phospholipids, and these vesicles catalyzed the transient accumulation of proline driven by valinomycin-induced potassium efflux. The vesicles reconstituted with the butanol extract were reported to transport glutamate and cysteine, in addition to proline. These isolated proteins were quite unstable, so that further purification may be difficult.

An *M. phlei* protein that binds proline has been extracted from membranes by use of cholate and Triton X-100 (103). This protein was purified by several steps, including isoelectric focusing. There were two peaks of binding activity: one that bound several amino acids and one that bound only proline. The purified fractions were tested for transport activity by reconstitution into detergent-extracted *M. phlei* vesicles and into liposomes. In each case there was no activity for the peak showing multiple binding activity, but the other peak did give activity that was specific for proline.

VITAMIN TRANSPORT IN *LACTOBACILLUS CASEI*   *L. casei* carries out the active transport of folate compounds apparently by a single system. A membrane protein that binds folate has been extracted from *L. casei* membranes with 5% Triton X-100 and purified 100-fold (104). The purified protein, which moves as a single band of mol wt 25,000 on gel electrophoresis in the presence of SDS, binds 0.85 mole of folate per mole of protein. The amino acid composition indicates that it is an extremely hydrophobic protein. Several mutants have been isolated that are deficient in folate transport; in these strains the binding activity is altered (105). In addition, the specificity and pH optimum of the binding activity and transport reaction are similar. These studies indicate that the isolated protein is the carrier for the folate transport system.

A similar *L. casei* protein that binds thiamine has been isolated and purified to homogeneity (105). This protein is also extremely hydrophobic, and the properties of the binding reaction and the transport system are similar. Even though there is no genetic evidence that this protein functions in the transport of thiamine, the strong indirect evidence mentioned above makes it likely that it is the thiamine carrier.

Binding Protein Transport Systems

INTRODUCTION   Binding proteins are present in the periplasmic space, the area between the inner and outer membrane of the cell (106). Proteins in this region can be selectively removed from the cell by a gentle procedure called osmotic shock (85). Despite the fact that the specificity and binding constants for each binding protein are nearly identical to the specificity and K_ms for the associated transport system, the binding proteins do not appear to be the membrane carriers for these systems, since they are hydrophilic proteins that do not interact strongly with membranes. In addition, strains with mutations in the structural gene for galactose-binding protein still retain the ability to transport substrates of the associated transport system with the same V_{max} as wild-type cells but with a greatly increased K_m (about 1000-fold) (107, 108).

Mutations in the binding protein drastically reduce the ability of the systems to concentrate substrates which indicates that binding proteins are essential for the physiological function of all the associated transport systems (109–113). It is unlikely that the binding proteins function simply to allow substrates to pass through the outer membrane, since the properties of proteins with this function are quite different from those of the binding proteins discussed here. One class of outer membrane proteins that function to move large molecules through the outer membrane (proteins required for maltose, vitamin B_{12}, and ferrichrome-mediated iron transport) are lipophilic molecules that are tightly bound in the outer membrane (114). A second class that appear to move charged molecules through the outer membrane (proteins associated with succinate, a-glycerol phosphate and possibly glucose-6-phosphate transport) were discussed in the section on succinate transport.

There have been many studies of binding proteins as they are readily isolated and purified, and these have been recently reviewed (13, 115). The complete amino acid sequence of the arabinose-binding protein has been determined (116) as well as its three-dimensional structure at a resolution of 2.8 Å (117). The structure of this protein can be described as an ellipsoid with dimensions 70 Å by 35 Å by 35 Å, and it contains two globular domains separated by a cleft. The galactose-binding protein is structurally related to the arabinose-binding protein, since the two molecules show immunological cross reactivity (118).

A number of reports over the past ten years claim to successfully reconstitute binding-protein transport by the addition of purified binding protein to shocked cells or vesicles. However, none of these studies so far has been successfully repeated, and in many cases the reported activity probably resulted from the binding of substrate to the added binding protein.

For all systems studied so far, the product of at least one gene in addition to the binding protein is required for a functional transport system (109, 111–119). This gene product has not been identified for any system, despite extensive efforts. Strong indirect evidence that the *Salmonella* histidine binding protein interacts directly with the product of the second gene, required for the high-affinity histidine transport system, has been provided by genetic experiments (120). Certain mutations in the histidine binding protein structural gene were found to be suppressed by mutations in the second required gene and vice versa. This finding provides important additional evidence that the binding protein participates directly in the transport process rather than playing an auxiliary role.

THE *E. COLI* β-METHYL GALACTOSIDE TRANSPORT SYSTEM   The high-affinity galactose transport system, called the β-methyl galactoside transport system, which is associated with the galactose binding protein has been extensively studied. Although galactose itself is transported by at least four other transport systems, β-methyl galactoside and β glycerol galactoside, are transported only by the β-methyl galactoside transport system in cells that lack lactose permease and TMG II permease.

It was suggested that this system used proton motive force to drive active transport (121), but more recent studies have shown that, like all other binding-protein transport systems that have been studied, the system uses ATP to drive active transport (122). This conclusion is supported by the absence of any proton influx associated with β-methyl galactoside transport in whole cells (123, 124).

All of the known mutations that inactivate this system map in a region called the *mgl* locus (119, 125, 126). This locus contains three complementation groups, one of which—the *mgl* B cistron—is the structural gene for the galactose-binding protein (110, 127). There is some evidence that the two non–binding-protein cistrons, *mgl* A and *mgl* C, are actually a single cistron showing intra-cistronic complementation. In the first place, eight out of thirty-two mutants in the A or C loci do not complement any other A or C mutation (127). This seems a very high proportion of double mutations, but is a reasonable proportion of noncomplementing mutants. In the second place the phenotypes of A and C mutants are identical in a number of studies (108, 127, 128). If the two cistrons are really one, then the A, C product must be at least a dimer in its functional state to allow the observed complementation.

There have been many studies of the galactose-binding protein (129–131), some of which have produced conflicting results. The most recent experiments (132) confirm the original studies (129, 130) and show that this protein has a single substrate binding site like all other binding proteins that have been studied. In addition to its role in transport, the galactose-binding protein is the receptor for galactose chemotaxis (133).

The properties of transport and chemotaxis in a number of binding-protein mutants were examined, and all mutants that altered the K_m of transport also altered the K_m of galactose chemotaxis (127). Some mutants were found that had normal transport but had lost galactose chemotaxis, and one mutant was found that had lost the β-methyl galactoside transport system but had normal chemotaxis. These results indicate that chemotaxis and transport utilize the same substrate binding site, but that both require other regions of the molecule. Furthermore, the regions required for chemotaxis and transport cannot be exactly the same. The galactose-binding protein probably interacts with the product of the *trg* gene, signaling the galactose concentration to the cell, since mutants in the *trg* gene do not show galactose or ribose chemotaxis but do respond normally to other chemotaxis substrates (127).

A functional difference between strains carrying *mgl* B mutations and those carrying *mgl* A or C mutations has been observed (107, 108). The *mgl* B mutant strains can grow on high concentrations of β-methyl galactoside and retain a β-methyl galactoside transport activity with the same V_{max} as the wild-type strain but with a 1000-fold higher K_m, while the *mgl* A, C mutant strains do not grow on β-methyl galactoside and have no transport activity. It was suggested that the *mgl* A, C gene codes for the membrane carrier of the system and that the binding protein functions to reduce the apparent K_m of the carrier for substrates. However, a study of the β-methyl galactoside exit reaction suggests that the system may require a third component (128). Uptake of the nonmetabolizable substrate β-methyl galactoside reaches a plateau that was shown to result from a dynamic equilibrium between the entry reaction and an exit reaction. The exit reaction followed first order kinetics even when the internal substrate concentration was 0.1 M. Exit was greatly stimulated in energized cells. This stimulation required ATP and did not occur with proton motive force alone. The exit reaction did not require the product of any *mgl* cistron. Equilibration of β-methyl galactoside into cells that lack the *mgl* products either because of mutations or from growth on the repressing substrate, glucose, occurred readily and was stimulated by an energy source. Furthermore, the equilibration showed specificity, since two substrate analogues, thio-methyl galactoside and β-ethyl galactoside, which are not substrates of the β-methyl galactoside transport system did not equilibrate into the cells. Since equilibration occurs normally in *mgl* A, C mutants, it seems possible that the carrier involved in the exit reaction also functions in the entry reaction. If so, the uptake reaction, studied in *mgl* B mutant strains, would probably result from the residual activity of the mutant binding proteins, since none of the *mgl* B cistron mutants contain deletions or chain-termination mutations (108). If these suggestions are correct, the function of the *mgl* A, C product could be to catalyze the energy-coupling reaction.

A possible model that integrates these results is given below. The binding protein that carries substrate would bind to a site on the inner membrane that includes the exit carrier and the *mgl* AC gene product. ATP would be utilized in a reaction catalyzed by the *mgl* AC product to cause a conformational change in the binding protein so that it loses its affinity for substrate. The released substrate would pass through the inner membrane into the cell via the exit carrier, while the binding protein would dissociate and return to its normal conformation ready to repeat the cycle.

The use of energy to alter the affinity of the binding protein for its substrates appears to be required in any model of binding-protein transport systems, because the affinity of these proteins for their substrates is so high. This means that a carrier that could accept substrates from a binding protein would have an extremely high affinity and so would be detected by binding studies with whole membranes. Such a protein has not been found for any binding-protein system. Furthermore, the calculated rate of dissociation of substrate from the amount of glutamine-binding protein present in *E. coli* is only about 10% of the rate of glutamine transport (134). So little is currently known about the membrane proteins involved in binding-protein transport systems that any model is highly speculative. It is clear that the mechanism of these systems will differ in many ways from that of the class of membrane-bound systems.

Anomalous Transport Systems

There are two transport systems in *E. coli* that do not fall clearly into any one of the three classes. These are the cystine-specific (135) and the lysine-specific transport systems (136). These two systems are not inhibited by osmotic shock when the cells are grown in minimal medium, but are inhibited when the cells are grown in a medium containing yeast extract and bactotryptone. Furthermore, shock fluid from the cells grown in rich medium binds lysine and cystine; however, these activities are unstable and could not be purified further, so they are not well characterized. Both these amino acids are transported by vesicles (137), but the relationship of the systems in the vesicles to those seen in the whole cell has not been clearly established.

In the *E. coli* maltose transport system, which is clearly a binding-protein system, some results are not completely consistent with ATP being the energy source (138). However, in the cited study no controls were presented to show that arsenate had inhibited the ATP pool, so the possibility remains that ATP is the energy source as it is for all other binding-protein transport systems that have been studied.

The nature of the energy source for three potassium transport systems of *E. coli* has been studied (139). One system associated with a binding protein used ATP. The second was a membrane-bound system driven by proton motive force. The third system appeared to require both ATP and proton motive force; since it was not inhibited by osmotic shock and was not active in membrane vesicles it may represent a new class of transport system. The dual energy requirement may result from use of one type of energy to drive active transport, and another to regulate transport. This system, which is the major potassium transport system in *E. coli,* clearly warrants more study.

MAMMALIAN TRANSPORT SYSTEMS

Red Cell Systems

INTRODUCTION   Human red cells have been used to study a class of transport systems that move substrates through the membrane but do not concentrate them. Because of their simplicity these systems have been the subject of extensive kinetic studies in an attempt to see if they can be explained by a simple carrier model (140). A mechanism has been proposed that involves the binding of substrate to a single site followed by its movement across the membrane. The substrate then dissociates from a second site on the other side, and the carrier is altered so that the original site is available for another cycle. A thorough kinetic study of uridine transport indicates that this system does fit the model for a simple asymmetric carrier (141).

HEXOSE TRANSPORT SYSTEM   The hexose transport system of red cells has been the subject of many studies, some of which are not consistent with a simple carrier model (142–144). A recent study concludes that there are two carriers in the membrane. The major one is responsible for about 85% of the transport and appears to satisfy the kinetic criteria for a simple asymmetric carrier (145). Because of its relatively low activity, kinetic properties of the second carrier have not been determined accurately.

Both kinetic data and studies with a number of inhibitors indicate that the carrier has different properties on the two sides of the membrane (146). Moreover, use of modified sugars showed that compounds with large substituents at the 4 or 6 position bind and inhibit at the outside surface of the membrane, but not on the inside, while molecules with large substituents at the 1 position given the opposite result (147).

This system has been the subject of a number of studies designed to determine which of the membrane proteins identified by gel electrophoresis (146) functions in sugar transport. These studies were of two types: In one, membranes were reacted with labeled inhibitors and the labeled protein was identified; in the other, membranes were extracted to remove as many proteins as possible while retaining glucose transport. Most of these studies were interpreted to indicate that a protein present in band 3 functioned in glucose transport. However, one of these studies, which used an impermeant maleimide, indicated that a protein present in zone 4.5 had this function (148). Another method used to identify the glucose carrier was the reconstitution of glucose transport into phospholipid vesicles by Triton extracts of red cell membranes (149). This assay was then used to purify the protein responsible for glucose transport, and it was shown to be present in zone 4.5 (150). The reconstituted transport was stereospecific and was inhibited by several inhibitors of red cell glucose transport. The purified protein had a specific activity 15-fold higher than the initial Triton extract and a mol wt of 55,000. It is a glycoprotein and was estimated to represent about 2% of the protein in the red cell membrane. The purified carrier moved as a very broad but symmetrical band during SDS gel electrophoresis, while zone 4.5 is broad and irregular, which suggests that it contains several components. Another group also used reconstitution to show that the glucose carrier is present in zone 4.5 (151). The turnover number of the glucose carrier is almost 500 sec^{-1}, which is about average for transport systems. A study on the transport of polyols of different sizes by this system has been interpreted as supporting a gated pore type model (152), which will be discussed further in the section on the mitochondrial adenine nucleotide transporter. This seems to be a reasonable mechanism for this type of carrier.

ANION TRANSPORT SYSTEM   A third extensively studied red cell system is the anion transport system. Its properties have been recently reviewed (153), and only the work on the nature of the carrier itself is discussed here. This system has a very high turnover number, 7×10^5 ions $site^{-1}$ sec^{-1} (154). A number of amino reagents inhibit anion transport; the most specific reagents are the disulfonated stilbene derivatives, SITS and DIDS. Use of labeled derivatives showed that more than 90% of the label present in membrane proteins was in band 3 (155). This protein is one of the major red cell membrane proteins and has been shown to span the membrane (156). There has been a study of the position of different parts of the band 3 protein within the membrane (157). The C terminus is on the outside surface, and the N terminus is on the inside surface. Carbohydrate is attached toward the C-terminal end of this protein. When the outside surface of the membrane is treated with trypsin, a single cut produces a 63,000 mol wt fragment containing the original N terminus and a 30,000 mol wt fragment. Extensive digestion of the inside surface leaves only a 17,000 mol wt fragment of the protein in the membrane.

A number of other irreversible inhibitors also bind to to this same protein (158–160). The photo affinity compound N-(4-azido-2-nitrophenyl)-2-amino ethyl sulfonate, which is a substrate of the anion carrier, irreversibly inactivates this system in the light and is mainly bound to the band 3 protein (161).

Further evidence that band 3 is the anion carrier is the fact that red cell membranes depleted of most proteins except band 3, retain anion transport. The anion transport in these membranes has the same properties as that present in intact red cells (162). Finally band 3 has been incorporated into phospholipid vesicles that then were able to transport anions. Vesicle transport was not inhibited by several compounds that normally inhibit the anion transport system. However, when band 3 was isolated from membranes in which anion transport had been inhibited by an irreversible inhibitor, it did not reconstitute anion transport when it was incorporated into vesicles (162). Although the reconstitution results alone are not definitive, the results of all the different experiments indicate that band 3 functions in anion transport.

Two-dimensional gel electrophoresis of purified band 3 gave several components. The major protein represented 75% of the total protein and appears to be the anion carrier (163, 164). This protein probably is present in the membrane as a dimer (165, 166). Furthermore, reaction of one polypeptide per dimer pair with DIDS inhibits anion transport completely.

Mitochondrial Transport Systems

ADENINE NUCLEOTIDE CARRIER   The movement of ADP into mitochondria and of ATP out of mitochondria is catalyzed by an electrogenic exchange carrier. This system appears to be the rate-limiting step in the overall process of mitochondrial ATP production, despite the fact that the carrier makes up 6% of the mitochondrial membrane protein (167). This system has been extensively studied, and there is strong evidence that it functions by a two-state gated pore mechanism. In this mechanism ADP binds to a form of the carrier present on the outside surface, and the carrier then undergoes a conformational change that allows the ADP to pass through the pore into the mitochondria. In the new conformation the binding site is now accessible from the inside and binds ATP. The carrier now reverts to its original conformation, allowing the ATP to pass through the pore to the cytoplasm and altering the binding site so it opens to the outside.

The studies that support this model have been reviewed in two articles (168, 169) that will be summarized. These studies were made possible by the existence of two compounds, atractyloside and bongkrekate, which bind tightly to the carrier, causing competitive inhibition. Atractyloside appears to bind predominantly to the form of the carrier present on the outside surface of the mitochondria; bongkrekate is a lipid soluble molecule that appears to bind predominantly to the site present on the inside surface of the mitochondria. Studies of ADP and ATP binding to whole mitochondria and to submitochondrial particles in which the membrane is inverted indicate that the carrier binding sites present on the two sides of the membrane are quite different. The outside site has a high affinity for ADP and a low affinity for ATP; the inside site has the opposite specificity. This same conclusion has been reached in studies of the binding of spin-labeled inhibitors of the exchange carrier to mitochondria and submitochondrial particles (170, 171). These studies were interpreted to show that the carrier molecule does not span the membrane, at least when the inhibitors are bound; however, the evidence presented does not provide any strong support for this conclusion, and it is inconsistent with the other data for this system.

The carrier has been isolated and purified to homogeneity using the binding of labeled atractyloside or bongkrekate as an assay. The same molecule was isolated in each case, but the properties of the two carrier preparations were very different. The purified carrier has a subunit mol wt of 29,000 and appears to be present as a dimer in the membrane and in the detergent extract. Antibodies prepared against the two forms of the carrier did not cross-react, but the two forms could be interconverted by the addition of the appropriate inhibitor or adenine nucleotide. The binding of nucleotides to the carrier changes the shape of the mitochondrial membrane to such an extent that the change can be observed by light scattering or electron microscopy. This provides further evidence that binding of nucleotides to the carrier causes a large conformational change.

Another assay that has been used to isolate the carrier is the reconstitution of activity into phospholipid vesicles (172). The carrier was purified about 50-fold; the largest protein present in the preparation had a mol wt of 29,000. The 50-fold purification was higher than would be predicted if 6% of the protein is the carrier but may result from partial inactivation during the extraction step. The properties of the reconstituted system were similar to those found in whole mitochondria.

PHOSPHATE TRANSPORTER   Mitochondria have two different pathways for transporting phosphate. One exchanges phosphate for dicarboxylic acids and has a V_{max} of 15 nmole/min/mg; the other pathway exchanges phosphate for OH and has a V_{max} of 200 nmole/min/mg. Both systems have K_m values for phosphate near 1.5 mM (173). The carrier for the latter system has been extracted from mitochondria with octylglucoside and reconstituted into phospholipid vesicles (174). The reconstitution of activity was used as an assay to obtain about a six-fold purification of the carrier. The reconstituted phosphate/OH⁻ exchange reaction was shown to be electrogenic. The orientation of the reconstituted carrier resembed that of mitochondria. It is inhibited by both N-ethylmaleimide and N-benzylmaleimide, in contrast to the carrier in submitochondrial particles, which is inhibited only by N-benzylmaleimide.

When both the purified phosphate carrier and the adenine nucleotide carrier were reconstituted into the same vesicles, phosphate stimulated ADP/ATP exchange (175). This result supports the proposal that these two electrogenic carriers move charges in opposite directions in order to maintain electroneutrality during oxidative phosphorylation in mitochondria. Evidence that this system may function by a gated pore type mechanism is the observation that the treatment of mitochondria, loaded with phosphate, with an SH reagent that inhibits the phosphate carrier causes the loss of an amount of phosphate that is approximately equal to the number of phosphate carrier molecules present in the mitochondria (176). A suggested explanation for this finding is that the inhibitor binds to the carrier site only on the outside of the mitochondria, and normally the carrier is present on the inside, bound to phosphate; thus when the carrier moves a phosphate across the membrane it becomes inhibited, releasing the phosphate bound to it into the medium.

Sodium-Potassium ATPase

The sodium-potassium ATPase creates and maintains the sodium and potassium gradients present in the cells of most higher organisms and also maintains osmotic equilibrium. The gradients that it produces, higher potassium and lower sodium inside the cell, are used for excitation by nerve and muscle cells and to drive active transport of certain compounds in many cells. Many articles review the properties of this enzyme. Some of the more recent ones are listed in References 177–181.

The enzyme has been purified to homogeneity by quite different procedures, and in each case the purified enzyme contained two polypeptides. One has a mol wt of 95,000 and contains both an aspartic residue that becomes phosphorylated during the transport reaction and the ouabain binding site. A tripeptide containing the phosphorylation site has been isolated by pronase digestion of ^{32}P-labeled enzyme and has the sequence $\left(\begin{smallmatrix} thr \\ ser \end{smallmatrix}\right)$-asp-lys. The same labeled tripeptide was isolated after pronase digestion of the Ca^{2+} ATPase of the sarcoplasmic recticulum (182).

The second polypeptide has a mol wt of 50,000 and is a glycoprotein. Antibodies prepared against this subunit partially inhibit the ATPase activity of the native enzyme. Crosslinking studies on purified ATPase show that the large and small subunit are adjacent and that the large subunit is present as a dimer (183). Other workers who studied the enzyme both in intact membranes and in its purified form found evidence for dimer formation with both the large and small subunits in both forms of the enzyme but did not find crosslinking between the large and the small subunit (184).

There is some disagreement about the ratio of small to large subunits in the enzyme, with values of 2, 1, and 1/2 being reported (179, 183, 185). The last value is the most recent and gives a mol wt of 250,000 for the native enzyme, which agrees with the value reported from radiation inactivation studies. There appears to be one ouabain binding site and either one or two phosphorylation sites per 250,000 mol wt (186). These values are consistent with kinetic studies that indicate that the enzyme shows 1/2 site reactivity. About 2 moles of sodium are bound with high affinity per 250,000 grams of protein and this binding is inhibited by ouabain (187). Since the ouabain binding site is present on the outer surface of the membrane, and the phosphorylation site is present on the inside surface, the large subunit must span the membrane.

The purified enzyme has been reconstituted into phospholipid vesicles, which then carry out sodium and potassium transport. The properties of the transport reaction in the reconstituted system are quite similar to those of the native system (179, 188).

The basic reaction catalyzed by this enzyme is the movement of $3Na^+$ from the inside of the cell to the external medium and the movement of $2K^+$ in the opposite direction. During this reaction a molecule of ATP is split to ADP and inorganic phosphate, providing the energy to move the ions against concentration gradients. The reaction is reversible if the ion gradients are large. There is evidence that the reaction occurs in a number of steps. In one proposed mechanism the first step is the binding of $3Na^+$ and ATP to the enzyme, which is followed by the hydrolysis of ATP to form phosphorylated enzyme and ADP. The next step is a conformational change in the enzyme that allows the Na^+ to pass through the membrane. Then, $2K^+$ ions bind to the enzyme and the phosphorylated group is hydrolyzed to give phosphate. In the final step the enzyme returns to its original conformation, which allows the K^+ to enter the cell. Mg^{2+} is required for the overall reaction. The last step may require the binding of ATP to a second low-affinity site on the enzyme (189). Kinetic arguments have been raised against a sequential model, but the direct evidence in favor of it seems much stronger (190).

The K_m for Na^+ on the inside surface of the membrane is about 1/100 the K_m for Na^+ on the outside surface, whereas the K_m values for K^+ on the two surfaces show the opposite relationship. There is also evidence that the Na^+-binding sites are different from the K^+-binding sites (180). Tryptic digestion of the small subunit produced Na^+-dependent ionophoric activity as tested by a black lipid membrane, but this finding could not always be reproduced (191). Potassium-dependent ionophoric activity was present in the peptides produced by cyanogen bromide cleavage of the whole enzyme. Other indirect evidence that Na^+ and K^+ binding is associated with the small subunit is the fact that Na^+ and K^+ partially protect this subunit against tryptic digestion (192).

In different experiments, which used much lower levels of trypsin, digestion of the intact enzyme in the presence of Na^+ gave a different set of products than when K^+ was present during digestion. This suggests that the two ions are bound to different conformational states of the enzyme (193). Additional evidence for two conformational states is the formation of two classes of antibody against the pure enzyme that appear to react with different conformational states, although the evidence for this is indirect (194).

The binding of Mg^{2+} and P_i to the enzyme was studied by microcalorimetry. The ΔH values for binding of these compounds were –49 kcal mol^{-1} and –42 kcal/mole respectively (195). These large values indicate that the binding of either substrate results in a conformational change in the enzyme. The binding of either molecule may cause the same conformational change, since the binding of both molecules gave a ΔH of –41 kcal mole^{-1}. The binding of these compounds altered the reactivity of the enzyme to both sodium borohydride and to 7-chloro-4-nitrobenzo-oxa-1-diazole, although the changes were not large. Finally it has been shown that cardiac glycosides bind 100 times more tightly to the phosphorylated form of the enzyme than to the nonphosphorylated form (183).

Kinetic studies indicate that in the presence of K^+, ATP, and free Mg^{2+} the enzyme can undergo a conformational change to an inactive form and that this change may occur in vivo (196). The overall properties of this enzyme are consistent with the two-state gated pore model described for the adenine nucleotide carrier, with the additional feature that the interconversion of the two forms is driven by ATP hydrolysis so that active transport can occur.

Sarcoplasmic Reticulum Ca^{2+} ATPase

The sarcoplasmic reticulum is a vesicular membranous component of muscle cells that controls muscle contraction by changing the internal Ca^{2+} concentration. In the resting state, Ca^{2+} is inside the reticulum vesicles and stimulation causes its release. The cell is returned to the resting state by the transport of Ca^{2+} into the vesicle. About 65% of the protein in this membrane is a single polypeptide, with a mol wt of 100,000, which catalyzes the ATP-driven movement of Ca^{2+} across the membrane. The properties of this enzyme have been the subject of several recent reviews (197–199).

The stoichiometry of the reaction is clearly 2Ca^{2+}-transported per ATP hydrolyzed and probably 1Mg^{2+} moved in the opposite direction, although this is not as well defined. The enzyme is composed of the 100,000 mol wt polypeptide and variable amounts of a low molecular weight proteolipid. The enzyme can be phosphorylated by ATP in the presence of Ca^{2+}. About 1 mole of phosphate is incorporated per mole of enzyme. The β carboxyl group of an aspartic residue accepts the phosphate group, and a peptide containing this residue has been sequenced (200). The enzyme can also be phosphorylated by inorganic phosphate. If a calcium gradient is present, this phosphate residue can be donated to ADP in a reversal of the normal reaction. Under these conditions only one mole of phosphate per two enzyme molecules is incorporated when inorganic phosphate is the donor.

ATP synthesis also occurs with purified enzyme that is not present in sealed vesicles (201). In these experiments the binding of Ca^{2+} ions probably provides the energy for the synthesis of ATP from ADP and P_i. Only about 0.14 mole of ATP were formed per mole of enzyme in these experiments. The purified enzyme binds Ca^{2+}, but the exact amount and the affinities of the sites vary between different studies. Rapid kinetic experiments indicate that phosphorylation of the enzyme causes after a short lag, the release of bound Ca^{2+} (202).

Isoelectric focusing separates the pure enzyme into 6 species, all of which have the original 100,000 mol wt (203). This separation is probably the result of either chemical modification or differential binding of charged molecules to a single class of polypeptide chains, since all the ATPase molecules can be phosphorylated during the transport reaction. In addition, limited tryptic digestion of sarcoplasmic reticulum vesicles cleaves all the enzyme molecules to give species of 55,000 and 45,000 mol wt. This cleavage does not inhibit either ATPase or transport activity. The larger fragment, which contains the phosphorylation site, is mostly present on the outside surface of the membrane, since it becomes heavily labeled during iodination of sarcoplasmic reticulum vesicles. The small fragment appears to be buried inside the membrane and has a considerably more hydrophobic amino acid composition than the larger fragment. More extensive digestion cleaves the 55,000 fragment to species of 30,000 and 20,000 mol wt. The site of phosphorylation is in the 30,000 mol wt fragment.

The enzyme has Ca^{2+}-dependent ionophoric activity when it is either succinylated or digested with trypsin (191). The ionophoric activity of a limited tryptic digest was associated with a 20,000 mol wt fragment that was derived from the 55,000 mol wt fragment discussed before. When the 20,000 mol wt fragment was cleaved with cyanogen bromide, the products retained ionophoric activity. The 45,000 mol wt fragment produced by a single cut in the enzyme also had ionophoric activity, but it did not show any ion dependence. The proteolipid fraction also has a nonspecific ionophoric activity.

A direct role for the proteolipid in Ca^{2+} transport was proposed, based on reconstitution experiments. Preparations of the ATPase deficient in the proteolipid gave good ATPase activity after reconstitution, but had low Ca^{2+} transport (204) activity. The ratio of Ca^{2+} transport to ATPase activity was restored to its normal value by the addition of purified proteolipid. Furthermore, when the proteolipid was incorporated into phospholipid vesicles, they became permeable to Ca^{2+}. It is not known whether the proteolipid functions directly to allow Ca^{2+} to pass through the membrane or indirectly by changing the membrane in a nonspecific fashion. The proteolipid was characterized in an earlier study, but no further chemical study has been reported, in contrast to the proteolipid associated with proton-translocating ATPase, which has been thoroughly characterized (205).

There have been extensive studies on the phospholipid requirements of the Ca^{2+} ATPase (206). The minimum amount of phospholipid that still preserves activity is 30 molecules per ATPase molecule. The specificity of the lipid requirement for ATPase activity is different from the specificity for Ca^{2+} transport.

The Ca^{2+} ATPase also appears to function in Ca^{2+} efflux. This efflux is extremely rapid and does not appear to be a reversal of the uptake reaction. It has been proposed that the efflux occurs through a pore formed by the aggregation of 4 ATPase molecules (207). A number of results suggest that ATPase molecules do aggregate, but this model is not yet proven.

Sodium Symport Systems

There is considerable evidence that many mammalian active transport systems are electrogenic sodium symports. Two such systems are the neutral amino acid system (208–210) and a sugar transport system present in the brush border membrane of intestinal and kidney cells (211, 212). The properties of the sugar transport systems of kidney cells have been reviewed recently (213). Some of the strongest evidence for a sodium symport mechanism comes from studies using membrane vesicles (211, 214, 215). There is still controversy about this mechanism, especially in the case of the neutral amino acid transport system for which it has been suggested that transport may also be directly coupled to the oxidation of DPNH (216). Much of the controversy arises because of the difficulties involved in determining the membrane potentials present in the cell under different conditions. Overall, the recent data strongly supports a sodium symport model. An amino acid binding protein that functions in amino acid transport has been extracted from ascites cell membranes and partially purified (217). This protein can reconstitute amino acid uptake into phospholipid vesicles. The reconstituted transport requires a sodium gradient and has essentially the same specificity as does the neutral amino acid transport system of whole cells.

Sodium-dependent glucose transport has also been reconstituted into phospholipid vesicles by use of a Triton extract of brush border membranes (218). The protein or proteins responsible for this have not yet been purified to homogeneity, but work is continuing on this project.

Binding Proteins

Proteins that bind calcium and are linked by indirect evidence to calcium transport have been isolated from the intestinal mucosa of a number of species (for review see 219, 220). These proteins are not hydrophobic, and probably do not function as membrane carriers. If they function at the external surface of the membrane, as has been suggested, some mechanism is required to keep them from diffusing away from the cell. If they are intracellular proteins, a nontransport role would seem more probable, and the fact that the primary structure of one of these proteins is similar to a large number of intracellular Ca^{2+}-binding proteins that do not function in transport makes this a possibility (221).

Two proteins that function in vitamin B_{12} transport, transcobalamine II and intrinsic factor, are clearly present outside the cell and function to bring vitamin B_{12} to the membrane (222, 223). The available evidence, although still indirect, suggests that the B_{12} protein complex is moved into the cell by pinocytosis. Mutants altering each protein are known and appear to inactivate B_{12} transport.

Intestinal Hydrolyases

The mammalian intestinal brush border membranes contain a number of hydrolytic enzymes and transport systems. These function in the digestion and uptake of metabolites from the lumen. Most of each hydrolyase molecule is external to the membrane, but a small segment of about 10,000 mol wt at the N terminal end of each molecule is integrated into the membrane (224, 225). Studies on an intestinal amino peptidase have shown that part of the hydrophobic tail is present on the inside surface of the membrane, which indicates that this molecule spans the membrane (226).

It has been suggested that sucrase-isomaltase not only catalyzes the hydrolysis of its substrates but, in addition, transports some of the products across the membrane (227, 228). Most of the hydrolysis products are transported by the Na^+-dependent transport systems present in the membrane; however, some enter the cell by another mechanism. This transport only occurs with products resulting from the activity of the enzyme. There was an early report that purified sucrase-isomaltase, reconstituted into black lipid membranes, transported some of the glucose derived from sucrose (229). However, this work utilized enzyme that lacked the hydrophobic segment and has not been repeated. In more recent studies that used the intact enzyme and liposomes, there was no transport by the reconstituted enzyme (230). It appears at present that while there is evidence suggesting a limited transport by intestinal hydrolyases, it is not of major physiological significance. The process is complex and still not understood.

The γ-Glutamyl Cycle

γ-Glutamyl transferase is present in the membrane fraction of most cells and catalyzes the reaction:

Amino acid + Glutathione → γ-Glutamyl-Amino acid + Cys-Gly.

It has been proposed that this and several other enzymes function in the transport of amino acids (231) by a cyclic set of reactions. Evidence for this proposal is the fact that the transferase is bound to the membrane and is particularly active in tissues that transport amino acids at a high rate. The symptoms of patients with a defect in the enzyme γ-glutamyl-cysteine synthetase, which catalyzes one of the proposed set of reactions required for transport by γ-glutamyl transferase, are consistent with this model but do not prove it. However, a fibroblast cell line has been isolated that has less than 0.5% of the wild-type activity of this enzyme, and amino acid transport in this strain is indistinguishable from that in the wild-type strain (232). This result indicates that if γ-glutamyl transferase functions in amino acid transport, it must be responsible for only a minor component of the total transport by fibroblasts.

ACKNOWLEDGMENTS

I would like to thank all the people who sent reprints and preprints of their work. In addition I would particularly like to thank my wife Nancy for her help in editing this article.

Literature Cited

1. Kaback, H. R. 1971. *Methods Enzymol.* 22:99–120
2. Sacktor, B. 1976. *Curr. Top. Bioenerg.* 6:39–81
3. Steck, T. L., Kant, J. A. 1974. *Methods Enzymol.* 31:172–79
4. Hertzberg, E. L., Hinkle, P. C. 1974. *Biochem. Biophys. Res. Commun.* 58:178–84
5. Bhattacharyya, P., Barnes, E. M. Jr. 1976. *J. Biol. Chem.* 251:5614–19
6. Adler, L. W., Rosen, B. P. 1977. *J. Bacteriol.* 129:959–65
7. Racker, E. 1978. *Methods Enzymol.* In press
8. Singer, S. J. 1977. *J. Supramol. Struct.* 6:313–23
9. Harold, F. M. 1976. *Curr. Top. Bioenerg.* 6:83–149
10. Zimmerman, M. H., Milburn, J. A. eds. 1975. *Encyclopedia of Plant Physiology,* 1:00–00. New York: Springer
11. Lüttge, U., Pitman, M. G., eds. 1976. *Encyclopedia of Plant Physiology,* 2:00–00. New York: Springer
12. Heber, U., Stocking, C. R., eds. 1976. See Ref. 11, 3:00–00
13. Wilson, D. B., Smith, J. B. 1978. In *Bacterial Transport,* ed. B. Rosen, Chap. 10. New York: Dekker. In press
14. Kozlov, I. A., Skulachev, V. P. 1977. *Biochim. Biophys. Acta* 463:29–89
15. Papa, S. 1976. *Biochim. Biophys. Acta* 456:39–84
16. Bogolmolni, R. A., Baker, R. A., Lozier, R. H., Stoechenius, W. 1978. *Biochim. Biophys. Acta.* In press
17. Ling, G. N. 1977. *Mol. Cell. Biochem.* 15:159–72
18. Damadian, R. 1973. *Crit. Rev. Microbiol.* 2:377–422
19. Kilata, G. B. 1976. *Science* 192:1220–21
20. 1976. *Science* 193:528–29
21. Wilson, D. B. 1974. *J. Biol. Chem.* 249:553–58
22. Oxender, D. L. 1972. *Ann. Rev. Biochem.* 41:777–814
23. Awazu, S., Amanuma, H., Morikawa, A., Anraku, Y. 1975. *J. Biochem. Tokyo* 78:1047–56
24. Rahmanian, M., Claus, D. R., Oxender, D. L. 1973. *J. Bacteriol.* 116:1258–66
25. Wood, J. M. 1975. *J. Biol. Chem.* 250:4477–85
26. Guardiola, J., DeFelice, M., Klopotowski, T., Iaccarino, M. 1974. *J. Bacteriol.* 117:382–405
27. Borst-Pauwels, G. W. F. H. 1976. *J. Theor. Biol.* 56:191–204
28. Anderson, J. J., Oxender, D. L. 1977. *J. Bacteriol.* 130:384–92
29. Anderson, J. J., Quay, S. C., Oxender, D. L. 1976. *J. Bacteriol.* 126:80–90

30. Kahane, S., Marcus, M., Metzer, E., Halpern, Y. S. 1976. *J. Bacteriol.* 125:770–75
31. Willis, R. C., Furlong, C. E. 1975. *J. Biol. Chem.* 250:2581–86
32. Schellenberg, G. D., Furlong, C. E. 1977. *J. Biol. Chem.* 252:9055–64
33. Slayman, C. W. 1970. *Biochim. Biophys. Acta* 211:502–12
34. Glover, G. I., D'Ambrosio, S. M., Jensen, R. A. 1975. *Proc. Natl. Acad. Sci. USA* 72:814–18
35. Inouye, H., Beckwith, J. 1977. *Proc. Natl. Acad. Sci. USA* 74:1440–44
36. Randall, L. L., Hardy, S. J. S. 1977. *Eur. J. Biochem.* 75:43–54
37. Kepes, A. 1973. *Biochimie* 55:693–702
38. Harold, F. M., Papineau, D. 1972. *J. Membr. Biol.* 8:45–62
39. Boos, W. 1974. *Ann. Rev. Biochem.* 43:123–46
40. Simoni, R. D., Postma, P. W. 1975. *Ann. Rev. Biochem.* 44:523–54
41. Rosen, B. P., Kashket, E. R. 1978. See Ref. 13, Chap. 12
42. Kundig, W. 1976. In *The Enzymes of Biological Membranes*, ed. A. Martonosi, 3:31–55. New York: Plenum
43. Hays, J. 1978. See Ref. 13, Chap. 2
44. Moczydlowski, E., Saier, M. H. Jr. 1978. See Ref. 13, Chap. 3
45. Cordaro, C. 1976. *Ann. Rev. Genet.* 10:341–59
46. Medveczky, N., Rosenberg, H. 1971. *Biochem. Biophys. Acta* 241:494–506
47. Crabeel, M., Grenson, M. 1970. *Eur. J. Biochem.* 14:197–204
48. Plate, C. A., Suit, J. L., Jetten, A. M., Luria, S. E. 1974. *J. Biol. Chem.* 249:8138–43
49. Lieberman, M. A., Hong, J. S. 1974. *Proc. Natl. Acad. Sci. USA* 71:4395–99
50. Lieberman, M. A., Simon, M., Hong, J. S. 1977. *J. Biol. Chem.* 252:4056–67
51. Plate, C. A. 1976. *J. Bacteriol.* 125:467–74
52. Sprott, G. D., Wood, J. M., Martin, W. G., Schneider, H. 1977. *Biochem. Biophys. Res. Commun.* 76:1099–1106
53. Kennedy, E. P. 1970. In *The Lactose Operon*, ed. J. R. Beckwith, D. Zipser, pp. 49–92. New York: Cold Spring Harbor Lab.
54. Wilson, T. H., Kashket, E. R., Kusch, M. 1972. In *The Molecular Basis of Biological Transport*, ed. J. F. Wiessner, F. Huijing, pp. 219–47. New York: Academic
55. West, I. C., Wilson, T. H. 1973. *Biochem. Biophys. Res. Commun.* 50:551–58
56. Fox, C. F., Kennedy, E. P. 1965. *Proc. Natl. Acad. Sci. USA* 54:891–99
57. Jones, T. H. D., Kennedy, E. P. 1969. *J. Biol. Chem.* 244:5981–87
58. Schuldiner, S., Kerwar, G. K., Weil, R., Kaback, H. R. 1975. *J. Biol. Chem.* 250:1361–70
59. Rudnick, G., Schuldiner, S., Kaback, H. R. 1976. *Biochem.* 15:5126–31
60. Altendorf, K., Muller, C. R., Sandermann, H. Jr. 1977. *Eur. J. Biochem.* 73:545–51
61. West, I. C., Mitchell, P. 1973. *Biochem. J.* 132:587–92
62. Flagg, J. L., Wilson, T. H. 1976. *J. Bacteriol.* 125:1235–36
63. Cecchini, G., Koch, A. L. 1975. *J. Bacteriol.* 12:187–95
64. Flagg, J. L., Wilson, T. H. 1978. *Membrane Biochem.* In press
65. Purdy, D. R., Koch, A. L. 1976. *J. Bacteriol.* 127:1188–96
66. Ramos, S., Kaback, H. R. 1977. *Biochemistry.* 16:4271–75
67. Long, R. A., Martin, W. G., Schneider, H. 1977. *J. Bacteriol.* 130:1159–74
68. Koch, A. L. 1964. *Biochim. Biophys. Acta* 79:177–200
69. Robbie, J. P., Wilson, T. H. 1969. *Biochim. Biophys. Acta* 173:234–44
70. Bentaboulet, M., Kepes, A. 1977. *Biochim. Biophys. Acta.* 471:125–34
71. Wong, P. T. S., Wilson, T. H. 1970. *Biochim. Biophys. Acta* 196:336–50
72. Kaback, H. R. 1974. *Science* 186:882–92
73. Kaback, H. R., Ramos, S., Robertson, D. E., Stroobant, P., Tokuda, H. 1978. *J. Supramol. Struct.* In press
74. Schuldiner, S., Kung, H. F., Kaback, H. R., Weil, R. 1975. *J. Biol. Chem.* 250:3679–82
75. Kennedy, E. P., Rumley, M. K., Armstrong, J. B. 1974. *J. Biol. Chem.* 249:33–37
76. Belaich, A., Simopietri, P., Belaich, J. P. 1976. *J. Biol. Chem.* 251:6735–38
77. Lancaster, J. R. Jr., Hinkle, P. C. 1977. *J. Biol. Chem.* 252:7657–61
78. Lancaster, J. R. Jr., Hinkle, P. C. 1977. *J. Biol. Chem.* 252:7662–66
79. Teather, R. M., Hamelin, O., Schwartz, H., Overath, P. 1977. *Biochim. Biophys. Acta* 467:386–95
80. Winkler, H. H., Wilson, T. H. 1966. *J. Biol. Chem.* 241:2200–11
81. Lancaster, J. R., Hill, R. J., Struve, W. G. 1975. *Biochim. Biophys. Acta* 401:285–98
82. Schuldiner, S., Weil, R., Robertson, D. E., Kaback, H. R. 1977. *Proc. Natl. Acad. Sci. USA.* 74:1851–54

83. Lo, T. C. Y., Sanwal, B. D. 1975. *J. Biol. Chem.* 250:1600–2
84. Silhavy, T. J., Hartig-Beecken, I., Boos, W. 1976. *J. Bacteriol.* 126:951–58
85. Heppel, L. A. 1969. *J. Gen. Physiol.* 54:95–113
86. Lo, T. C. Y., Sanwal, B. D. 1975. *Biochem. Biophys. Res. Commun.* 63:278–85
87. Lo, T. C. Y. 1978. *J. Supramol. Struct.* In press
88. Ramos, S., Kaback, H. R. 1977. *Biochemistry* 16:854–59
89. Gutowski, S. J., Rosenberg, H. 1975. *Biochem. J.* 152:647–54
90. Kang, S. Y., Cowell, J. L., Wilson, D. B., Heppel, L. A. 1975. *Am. Soc. Microbiol. Abstr.* K130
91. Lanyi, J. K., MacDonald, R. E. 1976. *Biochemistry* 15:4608–14
92. Lanyi, J. K., Rentahal, R., MacDonald, R. E. 1976. *Biochemistry* 15:1603–10
93. Lanyi, J. K. 1977. *J. Supramol. Struct.* 6:169–77
94. Tsuchiya, T., Raven, J., Wilson, T. H. 1977. *Biochem. Biophys. Res. Commun.* 76:26–31
95. Halpern, Y. S., Barash, H., Dover, S., Druck, K. 1973. *J. Bacteriol.* 114:53–58
96. Miner, K. M., Frank, L. 1974. *J. Bacteriol.* 117:1093–98
97. Tsuchiya, T., Hasan, S. M., Raven, J. 1977. *J. Bacteriol.* 131:848–53
98. Hirata, H., Kosmakos, F. C., Brodie, A. F. 1974. *J. Biol. Chem.* 249:6965–70
99. Stock, J., Roseman, S. 1971. *Biochem. Biophys. Res. Commun.* 44:132–138
100. Tokuda, H., Kaback, H. R. 1977. *Biochemistry* 16:2130–36
101. Hirata, H., Sone, N., Yoshida, M., Kagawa, Y. 1976. *Biochem. Biophys. Res. Commun.* 69:665–71
102. Amanuma, H., Motojima, K., Yamaguchi, A., Anraku, Y. 1977. *Biochem. Biophys. Res. Commun.* 74:366–73
103. Lee, S.-H., Cohen, N. S., Jacobs, A. J., Bodie, A. F. 1978. *J. Supramol. Struct.* 7: In press
104. Henderson, G. B., Zevely, E. M., Huennekens, F. M. 1977. *J. Biol. Chem.* 252:3760–65
105. Henderson, G. B., Zevely, E. M., Kadner, R. J., Huennekens, F. M. 1977 *J. Supramol. Struct.* 6:239–47
106. Heppel, L. A. 1971. In *Structure and Function of Biological Membranes*, ed. L. I. Rothfield, pp. 223–47. New York: Academic
107. Robbins, A. R., Rotman, B. 1975. *Proc. Natl. Acad. Sci. USA* 72:423–27
108. Robbins, A. R., Guzman, R., Rotman, B. 1976. *J. Biol. Chem.* 251:3112–16
109. Ames, G. F. L., Lever, J. E. 1972. *J. Biol. Chem.* 247:4309–16
110. Boos, W. 1972. *J. Biol. Chem.* 247:5414–24
111. Kellermann, O., Szmelcman, S. 1974. *Eur. J. Biochem.* 47:139–49
112. Aksamit, R. R., Koshland, D. E. Jr. 1974. *Biochemistry* 13:4473–78
113. Rahmanian, M., Claus, D. R., Oxender, D. L. 1973. *J. Bacteriol.* 116:1258–66
114. Kadner, R. J., Bassford, P. 1978. See Ref. 13, Chap. 8
115. Oxender, D. L., Quay, S. C. 1976. *Methods Membr. Biol.* 6:183–242
116. Hogg, R. W., Hermodson, M. A. 1977. *J. Biol. Chem.* 252:5135–41
117. Quiocho, F. A., Gilliland, G. L., Phillips, G. N. Jr. 1977. *J. Biol. Chem.* 252:5142–49
118. Parsons, R. G., Hogg, R. W. 1974. *J. Biol. Chem.* 249:3608–14
119. Ordal, G. W., Adler, J. 1974. *J. Bacteriol.* 117:509–16
120. Ames, G. F. L., Spudich, E. N. 1976. *Proc. Natl. Acad. Sci. USA* 73:1877–81
121. Parnes, J. R., Boos, W. 1973. *J. Biol. Chem.* 248:4429–35
122. Wilson, D. B. 1974. *J. Bacteriol.* 120:866–71
123. Henderson, P. J. F., Giddons, R. A., Jones-Mortimer, M. C. 1977. *Biochem. J.* 162:309–20
124. Singh, A. P., Bragg, P. D. 1977. *J. Supramol. Struct.* Suppl. 1:667
125. Ganesan, A. K., Rotman, B. 1968. *J. Mol. Biol.* 16:42–50
126. Boos, W., Sarvas, M. 1970. *Eur. J. Biochem.* 13:526–33
127. Ordal, G. W., Adler, J. 1974. *J. Bacteriol.* 117:517–26
128. Wilson, D. B. 1976. *J. Bacteriol.* 126:1156–65
129. Anraku, Y. 1968. *J. Biol. Chem.* 243:3116–4123
130. Anraku, Y. 1968. *J. Biol. Chem.* 243:3123–27
131. Boos, W. 1974. *Curr. Top. Membr. Transp.* 5:51
132. Zukin, R. S., Strange, P. G., Heavey, L. R., Koshland, D. E. Jr. 1977. *Biochemistry* 16:381–86
133. Hazelbauer, G. L., Adler, J. 1971. *Nature New Biol.* 230:101–4
134. Weiner, J. H., Heppel, L. A. 1971. *J. Biol. Chem.* 246:6933–41
135. Berger, E. A., Heppel, L. A. 1972. *J. Biol. Chem.* 247:7684–94
136. Rosen, B. P. 1973. *J. Biol. Chem.* 246:3653–62
137. Lombardi, F. J., Kaback, H. R. 1972. *J. Biol. Chem.* 247:7844–57

138. Ferenci, T., Boos, W., Schwartz, M., Szmelcman, S. 1977. *Eur. J. Biochem.* 75:187–93
139. Rhoades, D. B., Waters, F. B., Epstein, W. 1976. *J. Gen. Physiol.* 67:325–41
140. Lieb, W. R., Stein, W. P. 1974. *Biochim. Biophys. Acta* 373:178–96
141. Cabantchik, Z. I., Ginsburg, H. 1977. *J. Gen. Physiol.* 69:75–96
142. LeFevre, P. G. 1975. *Ann. NY Acad. Sci.* 264:398–413
143. Jung, C. Y. 1975. In *The Red Blood Cell,* ed. D. Surgenor, Vol. 2, Chap. 16. New York: Academic. 2nd ed.
144. Eilam, Y. 1975. *Biochim. Biophys. Acta.* 401:349–63
145. Ginsburg, H. 1978. *Biochim. Biophys. Acta.* 506:119–35
146. Steck, T. L. 1974. *J. Cell Biol.* 62:1–19
147. Barnett, J. E. G., Holman, C. D., Chalkley, R. A., Munday, K. A. 1975. *Biochem. J.* 145:417–29
148. Batt, E. R., Abbott, R. E., Schachter, D. 1976. *J. Biol. Chem.* 251:7184–90
149. Kasahara, M., Hinkle, P. C. 1976. *Proc. Natl. Acad. Sci. USA* 73:396–400
150. Kasahara, M., Hinkle, P. C. 1977. *J. Biol. Chem.* 252:7384–90
151. Kahlenberg, A., Zala, C. A. 1978. *J. Supramol. Struct.* In press
152. Bowman, R. J., Levitt, D. G. 1977. *Biochim. Biophys. Acta* 466:68–83
153. Fortes, P. A. G. 1977. In *Membrane Transport in Red Cells,* ed. J. C. Eldory, V. L. Lew, pp. 175–95. New York: Academic
154. Sachs, J. R., Knauf, P. A., Dunham, P. B. 1975. See Ref. 143, 2:613
155. Cabantchik, Z. I., Rothstein, A. 1974. *J. Membrane Biol.* 15:207–27
156. Bretscher, M. S. 1971. *J. Mol. Biol.* 59:351–57
157. Steck, T. L., Ramos, B., Strapazon, E. 1976. *Biochemistry* 15:1154–60
158. Cabantchik, Z. I., Balshin, M., Breuer, W., Rothstein, A. 1975. *J. Biol. Chem.* 250:5130–36
159. Zaki, L., Fasold, H., Schuhmann, B., Passow, H. 1975. *J. Cell Physiol.* 86:471–94
160. Rothstein, A., Cabantchik, Z. I., Knauf, P. A. 1976. *Fed. Proc.* 35:3–10
161. Wolosin, J. M., Cabantchik, Z. I., Ginsburg, H. 1977. *J. Biol. Chem.* 252: 2419–27
162. Ross, A., McConnell, H. M. 1977. *Biochem. Biophys. Res. Commun.* 74: 1318–25
163. Conrad, M. J., Penniston, J. T. 1976. *J. Biol. Chem.* 251:253–55
164. Vimr, E. R., Carter, J. R. Jr. 1976. *Biochem. Biophys. Res. Commun.* 73: 779–84
165. Yu, J., Steck, T. L. 1975. *J. Biol. Chem.* 250:9176–84
166. Wang, K., Richards, F. M. 1974. *J. Biol. Chem.* 249:8005–18
167. Akerboom, T. P. M., Bookelman, H., Tager, J. M. 1977. *FEBS Lett.* 74: 50–54
168. Klingenberg, M. 1976. See Ref. 42, 3:383–439
169. Klingenberg, M. 1977. In *Biochemistry of Membrane Transport,* ed. G. Semeza, E. Carafoli, pp. 567–79. New York: Springer
170. Devaux, P. F., Bienvenue, A., Lauquin, G., Brisson, A. D., Vignais, P. M., Vignais, P. V. 1975. *Biochemistry* 14: 1272–80
171. Lauquin, G. J. M., Devaux, P. F., Bienvenue, A., Villiers, C., Vignais, P. V. 1977. *Biochemistry* 16:1202–8
172. Schertzer, H. G., Racker, E. 1976. *J. Biol. Chem.* 251:2446–52
173. Coty, W. A., Pedersen, P. L. 1974. *J. Biol. Chem.* 249:2593–98
174. Banerjee, R. K., Shertzer, H. G., Kanner, B. I., Racker, E. 1977. *Biochem. Biophys. Res. Commun.* 75:772–78
175. Schertzer, H. G., Kanner, B. I., Banerjee, R. K., Racker, E. 1977. *Biochem. Biophys. Res. Commun.* 75:779–84
176. Guerin, M., Guerin, B. 1975. *FEBS Lett.* 50:210–13
177. Jorgensen, P. L. 1975. *Q. Rev. Biophys.* 7:239–74
178. Albers, R. W. 1976. See Ref. 42, 3:283–302
179. Hokin, L. E. 1977. See Ref. 169, pp. 374–88
180. Post, R. L. 1977. See Ref. 169, pp. 352–62
181. Korenbrot, J. I. 1977. *Ann. Rev. Physiol.* 39:19–49
182. Bastide, F., Meissner, G., Fleischer, S., Post, R. L. 1973. *J. Biol. Chem.* 248:8385–91
183. Kyte, J. 1972. *J. Biol. Chem.* 247: 7642–49
184. Giotta, G. J. 1976. *J. Biol. Chem.* 251:1247–52
185. Jorgensen, P. L. 1974. *Biochim. Biophys. Acta* 356:53–67
186. Hopkins, B. E., Wagner, J. H., Smith, T. W. 1976. *J. Biol. Chem.* 251:4365–71
187. Kanike, K., Lindenmayer, G. E., Wallick, E. T., Lane, L. K., Schwartz, A. 1976. *J. Biol. Chem.* 251:4794–95
188. Anner, B. M., Lane, L. K., Schwartz, A., Pitts, B. J. R. 1977. *Biochim. Biophys. Acta* 467:340–45

189. Henderson, G. R., Askari, A. 1977. *Arch. Biochem. Biophys.* 182:221–26
190. Garrahan, P. J. Garay, R. P. 1974. *Ann. NY Acad. Sci.* 242:445–58
191. Shamoo, A. E., Goldstein, D. A. 1977. *Biochim. Biophys. Acta* 472:13–53
192. Churchill, L., Hokin, L. E. 1976. *Biochim. Biophys. Acta* 434:258–64
193. Jorgensen, P. L. 1977. *Biochim. Biophys. Acta* 466:97–108
194. McCans, J. L., Lane, L. K., Lindenmayer, G. E., Butler, V. P. Jr., Schwartz, A. 1974. *Proc. Natl. Acad. Sci. USA* 71:2449–52
195. Kuriki, Y., Halsey, J., Biltonen, R., Racker, E. 1976. *Biochemistry* 15:4956–61
196. Cantley, L. C. Jr., Josephson, L. 1976. *Biochemistry* 15:5280–87
197. MacLennan, D. H., Holland, P. C. 1976. See Ref. 42, 3:221–59
198. Hasslebach, W., Beil, F. U. 1977. See Ref. 169, pp. 461–28
199. Korenbrot, J. I. 1977. *Ann. Rev. Physiol.* 39:19–49
200. Allan, G., Green, N. M. 1976. *FEBS Lett.* 63:188–92
201. Knowles, A. F., Racker, E. 1975. *J. Biol. Chem.* 250:1949–50
202. Ikemoto, N. 1976. *J. Biol. Chem.* 251:7275–77
203. Madeira, V. M. C. 1977. *Biochim. Biophys. Acta* 464:583–88
204. Racker, E., Eytan, E. 1975. *J. Biol. Chem.* 250:7533–34
205. Fillingame, R. H. 1976. *J. Biol. Chem.* 251:6630–37
206. Warren, G. B., Houslay, M. D., Metcalfe, J. C., Birdsall, N.J. 1975. *Nature* 255:684–87
207. Martonosi, A., Nakamura, H. L., Jilka, R. L., Vanderkooi, J. M. 1977. See Ref. 169, pp. 401–15
208. Heinz, E., Geck, P., Pietrzyk, C., Burckhardt, G., Pfeiffer, B. 1977. *J. Supramol. Struct.* 6:125–33
209. Eddy, A. A., Philo, R., Earnshaw, P., Brocklehurst, R. 1977. See Ref. 169, pp. 250–60
210. Christensen, H. N., deCespedes, C., Handlogten, M. E., Ronquist, G. 1973. *Biochim. Biophys. Acta* 300:487–522
211. Hopfer, U., Sigrist-Nelson, K., Murer, H. 1975. *Ann. NY Acad Sci.* 264:414–27
212. Kimmich, G. A., Carter-Su, C., Randles, J. 1978. *Am. J. Physiol.* In press
213. Silverman, M. 1976. *Biochim. Biophys. Acta* 457:303–51
214. Kinne, R., Murer, H., Kinne-Saffran, E., Thees, M., Sachs, G. 1975. *J. Membr. Biol.* 21:373–95
215. Lever, J. E. 1977. *J. Supramol. Struct.* 6:103–24
216. Garcia-Sancho, J., Sanchez, A., Handlogten, M. E., Christensen, H. N. 1977. *Proc. Natl. Acad. Sci. USA* 74:1488–91
217. Cecchini, G., Payne, G. S., Oxender, D. L. 1978. *J. Supramol. Struct.* 7: In press
218. Crane, R. K., Malathi, P., Preiser, H. 1977. See Ref. 169, pp. 261–68
219. Wasserman, R. H., Fullmer, C. S., Taylor, A. N. 1978. In *Vitamin D,* ed. D. E. M. Lawson, London: Academic. In press
220. Carafoli, E., Crompton, M. 1976. See Ref. 42, 3:203–7
221. Baker, W. C., Ketchem, L. K., Dayhoff, M. O. 1977. In *Calcium Binding Proteins and Calcium Function,* ed. R. H. Wasserman. New York: Elsevier. pp. 73–75
222. DiGirolamo, P. M., Huennekens, F. M. 1975. *Arch. Biochem. Biophys.* 168:386–93
223. Fiedler-Nagy, C., Rowley, G. R., Coffey, J. W., Miller, O. N. 1976. *Br. J. Haematol.* 31:311–21
224. Semenza, G. 1977. See Ref. 42, 3:349–76
225. Maroux, S., Louvard, D. 1976. *Biochim. Biophys. Acta* 419:189–95
226. Louvard, D., Semeriva, M., Maroux, S. 1976. *J. Mol. Biol.* 106:1023–35
227. Ramaswamy, K., Malathi, P., Caspary, W. F., Crane, R. K. 1974. *Biochim. Biophys. Acta* 345:39–48
228. Ramaswamy, K., Malathi, P., Crane, R. K. 1976. *Biochem. Biophys. Res. Commun.* 162–68
229. Storelli, C., Wogeli, H., Semenza, G. 1972. *FEBS Lett.* 24:287–92
230. Brunner, J., Hauser, J., Semenza, G., Wakcer, H. 1977. See Ref. 169, pp. 105–13
231. Meister, A., Tate, S. S. 1976. *Ann. Rev. Biochem.* 45:560–604
232. Pellefigue, F., Butler, J. D., Spielberg, S. P., Hollenberg, M. D., Goodman, S. I., Schulman, J. D. 1976. *Biochem. Biochem. Biophys. Res. Commun.* 73:997–1002

Ann. Rev. Biochem. 1978. 47:967–96
The US Government has the right to retain a nonexclusive,
royalty-free license in and to any copyright covering this paper

CONTROL OF TRANSCRIPTION TERMINATION

❖994

Sankar Adhya and Max Gottesman

Laboratory of Molecular Biology, National Cancer Institute,
National Institutes of Health, Bethesda, Maryland 20014

CONTENTS

PERSPECTIVES AND SUMMARY

The chromosome of prokaryotic organisms is organized into transcriptional units. Such organization requires signals to initiate transcription at one end of a unit and terminate it at the other. Operons are transcriptional units

967

0066-4154/78/0701-0967$01.00

regulated at the level of transcription initiation, usually by an interplay of activator factors, repressor molecules, and RNA polymerase at the operator-promoter loci of the unit (1; reviewed in 2). Operons may be polycistronic, i. e. may encode more than one polypeptide. Individual polypeptides are formed because polycistronic messenger RNA contains translation punctuation signals. Recently it has come to be appreciated that transcription punctuation signals exist not only at the ends of operons but also within them (3, 4). The presence within operons of signals that terminate transcription, called terminators (*t*), is not predicted by the operon theory. Furthermore, that the activity of these terminators can be regulated reveals a dimension of gene control that had not been previously perceived.

A review of the literature of transcription termination in *Escherichia coli* and its bacteriophages has been published (5). Our article concentrates on concepts of transcription termination based on recent exciting findings in the field: (*a*) The code for transcription termination may reside in the secondary structure of nucleic acids. (*b*) The activity of terminators, like the activity of promoters, can control gene expression. Some terminators have evolved along with the systems that regulate them. (*c*) Polarity in gene expression is the consequence of premature transcription termination at discrete sites by termination factor Rho. (*d*) The positive control function, pN, of phage λ regulates distal gene expression by antagonizing termination at specific terminators. Some of these terminators appear to be Rho-independent, suggesting that pN is a transcription antitermination factor, rather than an anti-Rho.

WHAT IS TRANSCRIPTION TERMINATION?

The formation of an RNA transcript begins with the attachment of an RNA polymerase holoenzyme molecule to DNA at a promoter region (2, 6). Transcription begins near the site of attachment. The chain elongates during the movement of RNA polymerase along the DNA, polymerizing template-encoded ribonucleotides at a rate of 20–50 nucleotides per second at 37°C. Transcription termination is usually thought of as three separate events (7): (*a*) cessation of movement of RNA polymerase and of RNA chain growth at a terminator; (*b*) release of the completed RNA chain from the enzyme; and (*c*) dissociation of RNA polymerase from the DNA template.

These reactions have been studied in vitro under a variety of experimental conditions. Specific, unbound RNA transcripts can be synthesized on bacteriophage DNA templates in purified systems containing RNA polymerase and the four ribonucleoside triphosphates (reviewed in 5). The simplified system probably yields free RNA polymerase at the end of the reaction. It is not clear, however, wheteher the occurrence of steps (*b*) and (*c*) is as written, is reversed, or is simultaneous. The transcripts formed in these systems can also be isolated from infected cells [Lozeron et al quoted in (8)], which suggests that the in vitro situation can be faithful to that in vivo.

The termination of other defined RNA molecules requires an additional protein factor, Rho (3). The products of Rho-mediated transcription termination are free RNA transcripts. The fate of RNA polymerase after termination at a Rho-dependent termination site is not known, since Rho is active on most templates only at low ionic strengths, where the RNA product prevents RNA polymerase from acting catalytically (6). Whether the inactive RNA polymerase is free or is bound to DNA following termination is not known. Rho-dependent transcripts can also be isolated from whole cells (9–11).

RHO-INDEPENDENT TERMINATION

A number of DNA templates will cause termination at defined sites when incubated with RNA polymerase and ribonucleoside triphosphates. These include the DNAs of the *trp* operon of *E. coli* or of *Salmonella,* and the DNAs of the coliphages T7, T3, ϕX174 and the lambdoid phages. Transcription of λ DNA in the simplified system yields at least two unbound RNA molecules, both of which correspond to in vivo transcripts (12, 13). One of these, the 6S transcript, is 193 nucleotides long and derives from DNA between the early and late gene regions of the phage (13a) (see Figure 5). Although a physiological role for this transcript has not been demonstrated, it has been suggested that it may serve as a primer for the transcription of λ late mRNA (13b, 14, 14a). According to this model, the λ function, gpQ, suppresses the termination of the transcript, permitting its elongation into the late-gene region. The sequence of the 6S RNA was established early, thanks to the fact that it is an abundant species when λ DNA is transcribed in high (0.18–0.30 M KCl) salt concentrations. Of possible significance is the observation that the 6S RNA can encode a polypeptide of 42 amino acids; whether the transcript ever serves in vivo as mRNA is not known. The second species, the 4S or "oop" RNA, is 77 nucleotides long and derives from the λ early region [(15) E. Schwarz, G. Scherer, G. Hobom, and H. Kössel, in preparation.)] (see Figure 1). Its role in phage development is obscure at the moment.

The 3'-OH end of the 4S and 6S transcripts is frequently found to bear adenylate residues that are not template-encoded (16, 17). The significance of this adenylation, which occurs both in vivo and in purified transcription mixtures, is unknown; the adenylation of these transcripts would prevent their serving as primer molecules for longer transcripts.

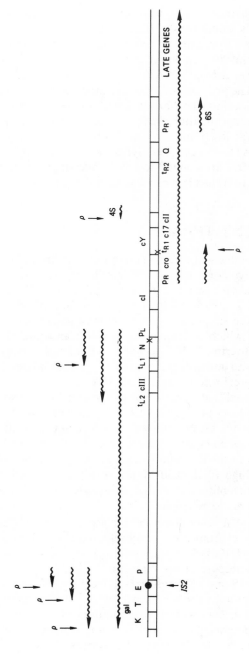

Figure 1 A partial genetic map of prophage λ and the neighboring *gal* operon. The *gal* operon has three structural genes *K*, *T*, and *E*. The ● in *E* signifies an IS2 element. The lambda promoters, terminators, and genes shown in the Figure are discussed in the text. The wavy lines show the origin and end of various transcripts, the direction of the arrows representing the orientation of the transcripts. The RHO (*ρ*)-dependent terminators are shown by vertical arrows. The unmarked transcripts are Rho-independent. The two "X" marks show the location of the two pN recognition elements, *nut*_L and *nut*_R, in the *P*_L and *P*_R operons, respectively. The map is drawn approximately to scale.

Although both the 6S and 4S transcripts are formed in the absence of Rho, the addition of the termination factor affects both their yields and sequences (12, 17a). The synthesis of 6S RNA is stimulated about 20%; 4S RNA synthesis increases some 500% in the presence of Rho. The exact termination point of the transcripts is, in part, dependent upon the inclusion of Rho in the incubation mixture. In the absence of Rho, both transcripts terminate in U_6A-OH; in its presence, some of the 6S RNA and most of the 4S RNA terminates after the incorporation of an additional, template-encoded uridylate residue, as U_6AU-OH.

The stimulation of 4S RNA synthesis by Rho appears to be caused by increased transcription of the DNA encoding the 4S transcript (12). Rho may accelerate the release of the 4S RNA and RNA polymerase after completion of the transcript. Because of the small size of the DNA region involved, a long-dwelling RNA polymerase molecule could effectively block the synthesis of a second 4S transcript. Interference by bound RNA polymerase of RNA synthesis by a second polymerase molecule has been noted when ϕX174 and λ DNA are transcribed at high RNA polymerase concentrations (18; M. Rosenberg, personal communication).

Both T3 and T7 DNA templates yield defined transcripts when incubated with *E. coli* RNA polymerase in the absence of Rho (7, 19, 20; P. Sarkar, D. Valenzuela, and U. Maitra, personal communication). Transcription of T7 DNA produces a single species of RNA covering the early region, whereas the transcription of T3 DNA results in two species of early RNA, the larger resulting from inefficient termination of the other.

The *trp* leader RNA is a bacterial transcript corresponding to the region of the *trp* operon between the promoter and the first structural gene (21–23). The transcript has been isolated from intact cells and can be synthesized in vitro in the absence of Rho. The terminator at the 3'-OH end of the *trp* leader—or attenuator—is highly efficient in vitro; approximately 95% of the transcription originating at the *trp* promoter terminates at the attenuator. Unlike the λ 4S and λ 6S transcripts, the *trp* leader RNA remains bound within the RNA polymerase-DNA complex at the end of the reaction (23a.) The *trp* leader, like the 6S RNA of λ, encodes a potential polypeptide. The control of *trp* leader synthesis in vivo is considered in a later section.

When the *gal* operon is transcribed in the absence of cyclic AMP and the cyclic AMP binding protein, transcription initiates five basepairs "upstream" from the initiation site used in the presence of these factors, and terminates after the synthesis of a hexanucleotide (24; R. DiLauro, personal communication). It is not known whether the hexanucleotide is free or is bound to RNA polymerase.

Transcription termination at defined sites can also be induced by reducing the concentration of ribonucleoside triphosphates, although such substrate starvation does not lead to RNA chain release. In the case of *gal*, transcription from the cyclic AMP–independent promoter terminates with the synthesis of a trinucleotide (R. DiLauro, personal communication). Transcription of the *lac* operon at low ribonucleoside triphosphate levels yields two abbreviated RNA molecules, 6 and 17 nucleotides long, both of which initiate at the *lac* promoter (25). The second transcript terminates 5 nucleotides beyond a GC-rich sequence. The importance of the GC-rich sequence in this termination event is suggested by analysis of mutations lying in or near the sequence. Of four mutations that substitute an AT pair for a GC pair in this region, 3 reduce or eliminate the termination. A mutation that adds an additional GC pair to the GC-rich region increases the frequency of termination at this locus. It has been suggested that during elongation DNA—RNA hybrids form that must be denatured before polymerase can proceed (25). Elongation arrest may occur near GC-rich sequences because rGC:dGC hybrid duplexes are more stable than dGC:dGC duplexes. However, it is far from clear that substantial stretches of RNA-DNA hybrid do, in fact, form. Furthermore, while several natural termination sites are preceeded by a GC-rich region, others are not [e.g. in *gal* or in t_{R1} of λ (see section below)]. The significance of these observations is not known.

RHO-DEPENDENT TERMINATION

While certain termination signals are capable, with different degrees of efficiency, of terminating transcription in the presence of RNA polymerase alone, others absolutely require the *E. coli* termination protein, Rho (see 5). Although termination does not occur at these sites in the absence of Rho, RNA polymerase appears to pause at these loci during elongation. On T4 and T7 DNA, RNA polymerase pauses at sites in the early gene region where termination would occur in the presence of Rho (26–28). Similarly, transcription initiating at the λ p_R promoter pauses at the Rho-dependent termination site t_{R1} (3, 29). In the absence of Rho, RNA polymerase pauses at t_{R1} for about 60 sec at 37°C before transcribing the DNA beyond the terminator (see Table 1).

A relationship between RNA polymerase pausing and Rho-mediated termination is indicated by an analysis of a group of λt_{R1} mutations that affect both processes (see the following section).

A Rho-dependent signal, t_{L1}, is found approximately 400 nucleotide pairs to the left of λ p_L(3). While both t_{L1} and t_{R1} are efficient terminators in vitro—termination at t_{R1} is about 90% efficient—the strength of these

Table 1 Efficiency of wild-type and mutant lambda t_{R1} terminators in vitro

DNA[a]	Percentage readthrough after sec				
	25	40	55	300	300 (+ Rho)
λ^+	0	4	14	100	20
λ $cin1$	—	—	—	100	9
λ $cin1$ $cnc1$	88	92	—	94	95

[a] Transcription mixtures contain RNA polymerase, λ DNA, and Rho where indicated. Rifampicin was added at initiation to prevent more than a single round of transcription. The α-^{32}P GMP labeled, p_R- initiated rightward mRNA was isolated by hybridization and the RNase T_1 oligonucleotides analyzed. Percentage readthrough is calculated as the ratio of a post t_{R1} oligonucleotide (residues 326–340) to a pre-t_{R1} oligonucleotide (residues 258–280) (see Figure 3). [From (29)]

termination signals is different in vivo; t_{L1} remains highly efficient, whereas transcription from p_R reads through t_{R1} 40–50% of the time (9) (see Table 2). It is not known whether this difference reflects sequence specificity, or whether it is caused by the other factors that influence transcription termination (see the section on transcription termination in vivo, below). The sequence of t_{R1} is shown in Figure 2; the sequence of t_{L1} has not yet been determined.

Rho-dependent termination signals are also found in bacterial DNA and in the 1200 base-pair DNA insertion element, *IS2* (4). By in vitro assay, these signals fall into two classes: those active at low Rho concentrations, e.g. the termination signals at the end of the *gal* operon and within IS2; and a second class, represented by termination signals within operons such as *gal* and *lac,* that require high levels of Rho for activity.

Table 2 Efficiency of wild-type and mutant lambda t_{R1} terminators in vivo

Infecting phage[a]	Percentage readthrough 2.5–5.0 min after infection	
	Host	
	Wild-type	*rhots15*
λN^-	48	100
λN^- $cin1$	<15	95
λN^- $cin1$ $cnc1$	90	100
λN^+	95	—
λN^+ $cin1$	45	85
λN^+ $cin1$ $cnc1$	100	100

[a] Cells were infected at 39°C with the indicated phage at a multiplicity of 5. ^{32}P-orthophosphate was added 2.5 min after infection and the cells were harvested at 5 min. The RNA was extracted and analyzed as in Table 1. (D. Court, D. Wulff, M. Rosenberg, C. Brady, in preparation)

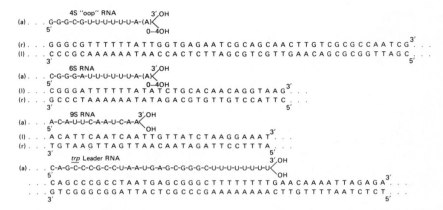

Figure 2 Nucleotide sequences beyond the sites of transcription termination. Shown are three lambda and one *E. coli* nucleotide sequences: a λ4S "oop" RNA that is stimulated by Rho (15, 17); a λ6S RNA that is Rho-independent (13, 17a); a λ9S RNA that is Rho-dependent (29); and an *E. coli trp* leader RNA that is Rho-independent in vitro but partially Rho-dependent in vivo (23). (a) is the 3'-terminal sequence of the RNAs; (1) and (r) represent the l-strands and r-strands of λ DNA, respectively.

THE SEQUENCES OF SOME TRANSCRIPTION TERMINATION SIGNALS

A number of Rho-independent transcription termination signals, and the Rho-dependent t_{R1} signal have been sequenced (Figures 1–3). An analysis of these sequences reveals the following:

1. No relationship can be seen in the DNA regions beyond the termination sites, which suggests these sequences play little role in transcription termination (Figure 2).
2. The primary sequences of the Rho-independent termination signals show a strong commonality. In each case, termination occurs downstream from a GC-rich region within a run of uridine residues. The Rho-dependent termination signal, t_{R1}, however, has neither the GC-rich region nor the string of uridine residues.
3. The most striking feature of all the terminators that have been sequenced is the region of dyad symmetry displayed just proximal to the termination point. For each terminator, a Gierer structure (30) at the level of DNA and a stem-loop structure in the RNA near its 3'-OH end can be formed. The size and the potential stability of the stems vary from terminator to terminator. The stem of the Rho-independent λ 6S is quite

extensive and rich in GC base-pairs. The stem at the *trp* attenuator is smaller, but is likewise GC-rich. The Rho-dependent λ *cro* transcript displays the shortest and least stable stem (Figure 3).

This symmetry may explain why RNA polymerase ceases elongation at these sites. (*a*) DNA Gierer structures may form during transcription which pose a barrier to further elongation (27). (*b*) Alternatively, the stem-loop structures of RNA in the region of the termination signal may bind to RNA polymerase, impeding further movement (29). At Rho-dependent sites, these structures would be less stable; RNA polymerase would only pause, and not terminate.

In an attempt to decide among the various possibilities, the analog ITP was used in place of GTP in a purified transcription system (22). Since rI:rC or rI:dC bonds are weaker than the corresponding rG:rC or rG:dC complexes, RNA or RNA-DNA hybrid structures containing rI should be less stable than structures containing rG. ITP does appear to eliminate termination at the *trp* attenuator, which suggests that RNA is involved in transcription termination. Whether the weakening of an RNA-DNA duplex or of an RNA-RNA structure is responsible for the ITP effect was not resolved by this experiment.

Mutations that alter the stem region of the *cro* transcript have been isolated and shown to affect transcription termination at t_{R1} (31, 29). These are of two types. The first has a single representative, *cin*1. This mutation adds an additional A:U base-pair to the stem; functionally it increases both the length of the RNA polymerase pause and the efficiency of termination at t_{R1}, although it does not make the terminator Rho-independent (Table 1). Although the characteristics of *cin*1 are consistent with the idea that longer stems increase the strength of the terminator, the mutation also alters a heptamer sequence found at the base of t_{R1} (TATGGTG) which is related to similar sequences often found in close association with promoter regions. It is possible that affinity between RNA polymerase and the heptamer sequence may contribute to the pause at t_{R1}. The wild-type sequence (TATGGTG) resembles the heptamer at the *lac* promoter (TATGTTG): The mutated sequence (TATAGTG) more closely resembles the heptamer found with *lacP*s (TATATTG), a cyclic AMP–independent, and presumably stronger, variant of the *lac* promoter. *cin*1 may thus act by increasing the affinity of t_{R1} for RNA polymerase. Although the mutant t_{R1} does not gain a transcription initiating capability, it is possible that free RNA polymerase may bind to this region, since high polymerase concentrations appear to block transcription at a point some 40–50 nucleotides upstream from the site of the *cin*1 mutation (M. Rosenberg, personal communication).

A second type of mutation, *cnc*, was derived from *cin*1 and has not yet been separated from it (31). The *cnc* mutations disrupt the stem and eliminate both the pause and the Rho-mediated termination (29; see Table 1, Figures 3C and 4).

Of interest is the fact that nine of eleven nucleotides in a sequence encompassing the loop and part of the stem of t_{R1} are capable of hydrogen-bonding with nucleotides (forming a dyad symmetry) in a sequence 35 bases beyond the terminator. Lambda mutations of the *cY* type reside in this distal sequence (29). A secondary structure involving the t_{R1} and *cY* regions would prevent the formation of the t_{R1} secondary structure implicated in termination. It is believed that the λ functions gpcII and gpcIII increase termination at t_{R1} by interacting with the *cY* region. The possibility that these functions act by promoting one form of RNA secondary structure at the expense of another is intriguing.

Although point mutations in the *trp* attenuator are not yet available, several DNA deletions in this region have been isolated and characterized (23). One deletion, *del*147, starts within the *trpC* gene and terminates between the *trp* leader and *trpE,* no more than 11 base-pairs beyond the

A.

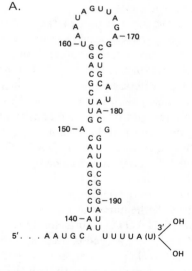

Figure 3   (above and opposite) The 3'-OH ends of three transcripts synthesized in vitro and in vivo are shown in their potential stem-loop configurations. A (*above*) The λ6S transcript (13); B. (*opposite, top*) The *trp* attenuator, shown in two possible and conflicting structures (22); C. (*opposite, bottom*) The λ *cro* transcript, with the mutation *cin*1, which adds an AU base-pair to the stem and *cnc*1 and *cnc*8, which disrupt the stem structure (29).

B.

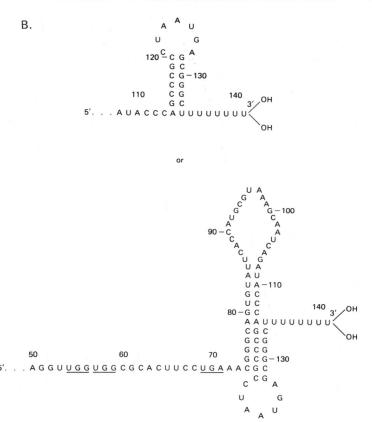

or

C.

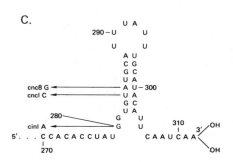

furthest site of *trp* leader termination in vivo. This deletion has no functional consequences either in vitro or in whole cells. Deletion *del*142 extends from *trpD* into the stem region including nucleotide 126: attenuator function is entirely eliminated. A third deletion, *del*1419, with one endpoint in *trpC* and the other halfway within the string of uridine residues at the attenuator, eliminates transcription termination in vitro but only slightly reduces attenuation in vivo. The loss of in vitro attenuation is not reversed by the addition of Rho.

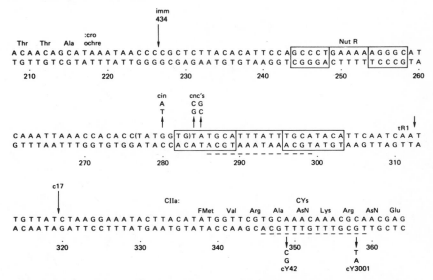

Figure 4   The t_{R1} Region of Bacteriophage Lambda (29). The lambda gene *cro*, ending with the ochre codon UAA at position 219. The region of nonhomology with phage 434 extends to position 226 and is followed by the recognition sequence for pN on the right, *nutR*. The *nutR* sequence displays dyad symmetry; it differs from the p*N* recognition element on the left, *nutL*, by an AT to GC change at position 252. Shown in brackets at positions 277 to 283 is the heptamer found at the base of a large region of dyad symmetry. Related heptamer sequences are often associated with promoter regions. Mutations affecting rightward transcription termination are located at position 280 (*cin*1) 284 (*cnc*1) and 285 (*cnc*8). The arrow at position 312 indicates the most distal termination point of the 9S transcript. The block of nucleotides between positions 286 and 299 shows symmetry with the block between positions 346 and 359. The *c*17 mutation, which creates a new start point for rightward transcription beyond t_{R1}, is formed by a duplication of nucleotides in positions 310–319. The *cII* gene, whose product is required for leftward transcription of the t_{R1} region, begins at position 338; some mutations that prevent the action of p*cII*, *cY*42 (position 349), and *cY*3001 (position 358) are located in the *cII* gene itself.

Both the *trp* leader and the λ *cro* RNA show sequence heterogeneity at their 3'-OH termini (23, 29). The *trp* leader terminates at CU_7-OH or CU_8-OH in vitro. Sequence heterogeneity in vivo is even broader, with *trp* leader transcripts terminating at CU_4-OH through CU_8-OH. The *cro* transcript terminates at AU-OH, AUC-OH, AUCA-OH or $AUCA_2$-OH both in vivo and in vitro. Whether this heterogeneity has physiological meaning is not clear.

More than one secondary structure is possible at the 3'-OH terminus of the *trp* leader (22). Of the 12 nucleotides in positions 108 to 119, 10 can hydrogen-bond with nucleotides in positions 74 to 85. Such an interaction would prevent the stem-loop structure shown in Figure 3B from forming, since nucleotides in positions 114 to 119 are components of the stem. Perhaps the regulation of the *trp* operon involves an equilibrium between these two potential structures.

THE MECHANISM OF ACTION OF RHO

Rho protein was first isolated in 1969 by Roberts and has been purified to homogeneity (3). Rho has a subunit molecular weight of about 50,000 and is free of nucleolytic activity. It acts catalytically to promote the release of RNA chains from a transcribing complex at specific sites on the DNA template (32). For this activity, Rho need not be present at the time of transcription initiation (32). Two lines of evidence suggest that Rho and RNA polymerase interact. First, while Rho will catalyze RNA chain termination on T3 and T7 DNA being transcribed by *E. coli* RNA polymerase, it is inactive when T3 or T7 RNA polymerase synthesizes RNA on these templates (32). Second, a mutant Rho protein that is inactive with wild-type RNA polymerase will act in conjunction with a specifically altered RNA polymerase (A. Das and C. Merril, personal communication). The later polymerase does not terminate without Rho.

While originally detected as a transcription termination factor on a λ DNA template, Rho catalyzes two activities: transcription termination, and an RNA-dependent phosphohydrolysis of nucleoside triphosphates (NTPase) (33). A variety of tests demonstrate that these activities are carried out by the same protein, although the monomeric and tetrameric forms of the enzyme appear to be active as NTPases, whereas a tetrameric or higher order oligomer of Rho is required for transcription termination (34, 35). The NTPase activity of Rho is necessary for transcription termination:

The β-γ imido analogues of the ribonucleoside triphosphates can be incorporated into RNA by RNA polymerase, but cannot be cleaved by Rho. When these analogues are used as substrates for RNA synthesis, Rho is inactive in terminating transcription; the addition of a small amount of a normal ribonucleoside triphosphate restores termination activity (36, 37).

Both the NTPase and the termination activities of Rho require the persistence of the RNA product (38; R. Crouch, personal communication). If a single round of RNA synthesis is carried out in the presence of pancreatic and T1 ribonucleases, the ability of Rho to terminate, measured as an inhibition of pyrophosphate release from the ribonucleoside triphosphate precursors, is eliminated. It should be noted, however, that very high concentrations of nuclease must be added to see this effect; perhaps the RNA must be degraded rapidly after formation to block completely Rho's NTPase activity.

Rho displays little specificity with respect to the nucleoside triphosphate substrate (39; R. Crouch, personal communication). ATP has a somewhat lower Km than the other ribonucleotides; dGTP and 3'-dGTP are both efficiently hydrolyzed.

The polynucleotide requirement for NTPase activity is more specific (40; R. Crouch, personal communication). Artificial polymers show large differences in their capacity to support ATP hydrolysis. The most efficient polymers are polyC, polyUC, and polyAUC. Whereas polyU$_{20}$C is as active as polyC, polyU is only 5% as active. PolyA and polyAC are inactive. Polydeoxynucleotides such as polydT, polydC, polydA, and polydA : dT do not support Rho-catalyzed ATP hydrolysis; the 2'-O methyl–substituted derivatives of polyC, polyU, and polyA are likewise inert in the reaction.

These data suggest that the primary RNA sequence recognized by Rho may not be very complex; a region rich in pyrimidines containing a minimum of 1 cytidine residue per 20 nucleotides appears sufficient to promote Rho NTPase.

That the secondary structure of RNA may influence Rho activity is suggested by the observation that polyI, when annealed to polyC, inactivates that polymer as a template for NTPase (40). Similarly, the highly structured RNAs of bacteriophage MS2 and Qβ do not stimulate Rho hydrolysis of ATP (R. Crouch, personal communication). The requirement for relatively unstructured RNA may explain the influence of ionic conditions on Rho activity. In a polyC-supported reaction, Rho NTPase is insensitive to high KCl concentrations (37). RNA synthesized from fd DNA templates also supports extensive NTPase activity over a broad range of salt concentrations. In contrast, NTPase activity promoted by T7 RNA is inefficient and highly salt-sensitive; at a concentration of 0.15 M KCl, Rho ATPase activity is completely inhibited. As expected, transcription termination activity on T7 DNA templates follows the same KCl inhibition curve as the ATPase activity. A likely explanation for these observations is that T7 RNA has more secondary structure than polyC or fd RNA and that this structure, which is inhibitory to Rho action, is stabilized by high ionic concentrations.

A model linking the two activities of Rho has been proposed (40; see also the following section). In this scheme, Rho attaches to nascent RNA and moves along the RNA chain in the 5' to 3' direction, toward RNA polymerase. The hydrolysis of ATP or some other nucleoside triphosphate provides the energy for this movement. Eventually, Rho contacts a polymerase that has paused at a Rho-dependent termination site and reacts with it, resulting in the release of the nascent RNA chain.

It is possible that a transcribing complex arrested anywhere during elongation is a substrate for Rho action. Thus, the RNA molecules that terminate randomly under conditions of substrate starvation, or the *trp* leader RNA, are released from their complex with RNA polymerase under the influence of Rho (J. Richardson, personal communication; T. Platt, personal communication). It is possible that a Rho-RNA complex is an intermediate in such dissociation.

It is clear from the model that factors interfering with Rho attachment to, or movement on, RNA will inhibit transcription termination, e.g. RNA secondary structure or the presence of ribosomes on the RNA template. The inhibitory effect of RNA secondary structure on Rho activity has been discussed above. The potential interference between ribosomes and Rho activity forms the basis of a theory of polarity and will be discussed in the following section.

The notion that Rho moves on RNA chains is supported by several experiments. There is a positive correlation between the size of an RNA polymer and its capacity to support Rho NTPase activity (40; R. Crouch, personal communication). PolyC polymers smaller than C_{11} are inactive, while C_{15} is about 20% and C_{45} about 50% as active as very long polyC molecules. Assuming that the rate-limiting step in the NTPase reaction is the attachment or detachment of Rho from the polynucleotide, and that ATP hydrolysis correlates with the movement of the enzyme from its point of attachment to the 3' end of the polynucleotide, it is clear that small polymers would be less efficient cofactors of NTPase than large polymers.

A recent experiment provides direct evidence for the 5'-3' displacement of Rho along an RNA chain (G. Galluppi and J. P. Richardson, personal communication; 41). Rho was incubated with polyC in the presence and absence of ATP, and the complex then treated with polynucleotide phosphorylase, a 3' exonuclease. Although ATP does not stimulate the binding of Rho to polyC, it dramatically increases the resistance of the polyC-Rho complex to the 3' phosphorolytic activity of the phosphorylase. The simplest model is that ATP stimulates the movement of Rho towards the 3'-OH end of polyC and thus protects the polymer against degradation.

TRANSCRIPTION TERMINATION IN VIVO: A MODEL

We have indicated above that observations made in vitro on transcription termination can accurately reflect the in vivo situation. For example, the λ *cro* transcript synthesized in vitro is identical to the *cro* transcript that can be isolated from whole cells (D. Court, D. Wulff, C. Brady, and M. Rosenberg, in preparation). Mutations in t_{R1} that affect *cro* termination in vivo can have comparable effects in the test tube (see Table 1). Similarly, the *trp* leader has been isolated both from intact cells and from purified transcription mixtures (22). Nevertheless, in a variety of cases there are discrepancies between the way transcription terminates in vitro and in vivo. The frequency of, or the requirements for, termination differ in the two systems in the cases of *trp*, λ, T3, T7, and for termination within bacterial operons. In some instances, the in vitro system apparently lacks components that regulate termination.

Polarity: Transcription Termination with Operons

Polar mutations are characterized by their capacity to inactivate both the cistron in which they are located and cistrons promoter-distal to the site of the mutation (1, 42, 43). These mutations are of two types. (*a*) Some nonsense (translation-termination) mutations, but not all, are polar. In general, nonsense mutations located near the aminoterminal portion of a cistron tend to be more polar than mutations near the carboxyl terminal end (44). (*b*) Insertion elements, such as IS1 (800 base-pairs) and IS2 (1200 base-pairs) are highly polar (45, 46; see also 47). The polarity of IS2 insertions is independent of the site of insertion of the element within a cistron (54). In vitro experiments indicate that IS2 carries a transcription-termination signal responsive to very low concentrations of Rho (4). IS1 and nonsense mutations do not introduce transcription termination signals at the mutated site.

The Model

Several years ago, we proposed a model for Rho-mediated transcription termination (48–51). The model also explains polarity, i.e. transcription termination within operons. A discussion of the various aspects of the model, including supporting evidence, are presented here (Figure 5).

For Rho to terminate transcription, the factor must have free access to RNA. When an operon is being expressed, the nascent mRNA is normally covered with ribosomes; these ribosomes block the attachment of Rho to the RNA, or its movement along the chain. Ribosomes are released from mRNA when polypeptide chain elongation ceases. This release occurs some distance beyond the chain-terminating nonsense codon, and involves elements that are not yet well characterized, either biochemically or genetically. When the RNA is no longer covered with ribosomes, Rho is able to proceed along the chain towards an RNA polymerase molecule paused at a Rho-dependent termination signal. Between Rho and the transcribing

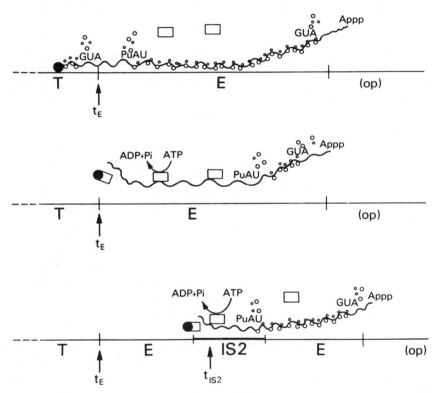

Figure 5   The model of transcription termination is shown for the *gal* operon. The straight lines represent the DNA; the wavy lines, the messenger RNA; the filled circles, the RNA polymerase; the boxes, the Rho molecules; and the open circles, the ribosomes. The vertical arrows indicate the location of termination signals. AUG and UAPu represent, respectively, initiation and termination codons for translation. Details of the model are explained in the text. Top: Wild-type *gal* operon, in which the Rho-dependent transcription termination signal between *E* and *T* is suppressed by translation of the *E* gene. Middle: A nonsense mutation early in *E* causes transcription termination at the end of the *E* gene. Bottom: Transcription termination following translation termination within the IS2 element (shown in thick line) in *E*.

complex a reaction then takes place that leads to release of a completed RNA molecule. This situation, we presume, prevails at the ends of operons; a region of untranslated RNA about 40 nucleotides long is known to exist at the end of the *trp* operon (A. Wu, T. Platt, personal communication).

When a nonsense codon is followed shortly by a restart codon, such as AUG or GUG, ribosome release does not occur, and polypeptide chain synthesis recommences. The model proposes that this takes place in the intercistronic regions of operons characterized by equal expression of all cistrons (Figure 5, top). Other operons, for example, the early gene region of T7, show a "natural polarity," i.e. promoter-proximal cistrons are transcribed more frequently than promoter-distal ones (27, 28, 52, 53). Intercistronic distances in these operons may be sufficiently long to permit occasional ribosome detachment. Rho would then attach to the unprotected mRNA, and transcription would terminate at the first Rho-dependent termination site within the operon.

A nonsense codon introduced into a cistron by a mutation will be polar depending upon its orientation with respect to restart codons, ribosome attachment sites, and transcription termination signals in the operon (Figure 5, middle). If the mutation is followed immediately by a restart codon, translation will resume and Rho activity will be inhibited. The mutation will not be polar. If there is no transcription termination signal between the nonsense codon and the next cistron, the mutation likewise will be nonpolar, since free ribosomes will attach at the ribosome binding site of this cistron. Finally, if the nonsense mutation is close to a Rho termination signal, sufficient RNA may not be generated by the mutation to permit Rho action. These considerations may explain why a curve plotting degree of polarity vs position of a nonsense mutation within a cistron is a gradient with peaks and valleys (54, 55).

Rho-dependent transcription-termination signals have now been detected by in vitro assays in several operons, including *gal, lac,* and *trp* (4, 56). In *gal,* a Rho site is located at or near the end of *galE;* in *lac,* a site lies in *lacZ,* about 1000 base-pairs from the promoter. Some of these sites may be active in vivo only when preceded by a nonsense mutation.

The in vivo transcription of *trp* operons bearing polar nonsense mutations has been studied extensively (57–59). Consistent with the model, transcription continues beyond the nonsense codon and terminates at a defined point within the operon. These sites of termination represent, presumably, the Rho-dependent termination signals within *trp* (56).

The polarity of IS1 insertion elements is also the consequence of Rho-mediated transcription termination. IS1 elements, which are highly polar at any insertion site in a cistron or in any orientation with respect to an operon, do not themselves carry Rho termination signals (4). It was proposed that IS1 elements carry nonsense codons at both ends and generate long, untranslated stretches of RNA (49). Termination of transcription occurs at the Rho-dependent signals within the operons. This hypothesis has been proved by the demonstration that the 768 base-pair long IS1 element carries nonsense codons in all three reading frames within the first 100 base-pairs in both orientations (60).

IS2 elements differ from IS1 elements in two important respects: They are polar only in one orientation (61); and they carry Rho-sensitive termination signals (4). We assume that translation and transcription terminate within the IS2 element. Some aspects of insertion polarity still remain to be tested. Although it is known that certain transfer RNA suppressors, e.g. suIII, do not reduce the polarity of IS1 or IS2 insertion elements (62), the gamut of tRNA suppressors has not been assayed. Additionally, while it has been demonstrated that an insertion element reduces the transcription of an operon, the site of transcription termination has not been carefully defined. However, consistent with the central idea that the polarity of these elements, like the polarity of nonsense mutations, is the result of Rho-mediated transcription termination, insertion polarity as well as nonsense polarity is absent in a Rho-defective mutant [(51) see also the following section].

Attenuation of *trp* is analogous to natural polarity; it may be modulated by the same factors that affect transcription termination in general, as well as by factors unique to the *trp* operon (23, 63). Although attenuation is at least 95% efficient in vitro, attenuation in vivo ranges from 85% to about 25%, depending on the growth conditions and the genetic makeup of the strain. Attenuation is maximal when cells are grown in excess tryptophan. This may be the result of high levels of charged tryptophanyl tRNA, since mutations in *trpT* or *trpS*, which reduce the level of charged $tRNA^{trp}$, inhibit attenuation, suggesting that attenuation is affected by translation, presumably of the *trp* leader sequence. Although the *trp* leader does encode a polypeptide, its putative product has yet to be isolated from whole cells. Nevertheless, a deletion fusing the *trp* promoter and the 5' end of the *trp* leader to *lacI* results in the production of an abnormal *lac* repressor, with several *trp* leader-encoded aminoacids at its amino-terminal end (U. Schmeissner and J. Miller, personal communication). This is strong evidence for translation of the leader sequence. While the precise mechanism of *trp* regulation is obscure, it has been suggested that the translation of two tandem *trp* codons in the leader is the key to attenuation. Were this the case, *trp* attenuation, unlike mutational polarity, would be *positively* correlated with translation efficiency. The control of *trp* attenuation can, however, be demonstrated only in a $relA^+$ strain, suggesting that the entire regulation may be mediated in an unknown way by ppGpp (63, 64; T. Platt, personal communication).

Like mutational polarity, *trp* attenuation appears to be at least partially relieved in vivo by a *rho* mutation (65–67).

Transcription of certain bacterial operons and of portions of the T4 chromosome is prevented by treating *E. coli* with the protein synthesis inhibitor, chloramphenicol (68, 69). That this is a form of polarity is shown by the fact that transcription is restored if a *rho* mutation is introduced into the cells (69; E. Brody, personal communication).

MUTATIONS AFFECTING TRANSCRIPTION TERMINATION

Mutations Located at the Site of Termination

Point mutations in t_{R1} of λ can either improve termination (*cin* 1) or eliminate it altogether (*cnc*). A mutant IS2 element that does not show its characteristic polar effect when present in λ or *gal* has been isolated (70). The mutation is likely to affect either the translational stop signal or the Rho-dependent transcription-termination site in IS2; it has not been tested in vitro. Deletions of the *trp* attenuator have been selected by demanding increased *trp* enzyme synthesis (23). Other mutations, some of which are known to be deletions, remove termination signals between operons, resulting in operon fusion (71).

Mutations Not Linked to the Affected Termination Site

RNA POLYMERASE MUTATIONS   Two mutations, *nitB* and *rif*501, which lie in the *rpoB* gene (encoding the RNA polymerase β subunit), reduce termination at the λ t_{L1} site (72, 73). The *nitB* mutant is temperature-sensitive for growth. It is not known what other termination events are affected by these mutations, or whether mutations altering the other subunits of RNA polymerase might also affect transcription termination.

RHO MUTATIONS   A variety of selective techniques have been used to isolate many mutations that affect transcription termination and that co-transduce with the *ilv* locus by phage P1 (65, 71, 72, 74–76). It is generally assumed that these mutations all reside in the structural gene for Rho protein, *rho,* since most appear to alter Rho-mediated transcription termination and/or Rho NTPase activity in vitro (72, 76–78). It should be noted, however, that this has not been demonstrated in all cases [for example, *suA1* has no characteristic in vitro behavior (79)] nor has extensive complementation analysis been performed; it thus remains possible that these mutations represent more than one cistron.

The mutations show allele specificity in their ability to suppress termination at different sites. In part this may be due to the "leakiness" of most *rho* mutations, which display substantial residual Rho activity in vitro. The leaky mutations, isolated as suppressors of nonsense polarity, also suppress IS1 polarity, but do not suppress the polarity of IS2 elements (50). This phenotype fits well with the model of polarity presented above, since low Rho concentrations suffice to terminate transcription in IS2 insertions. In contrast, termination at the natural intraoperonic sites, which we believe is the cause of nonsense polarity, requires large amounts of Rho in vitro. We assume that these *rho* mutants do not produce enough active Rho to terminate transcription at these sites.

A *rho* mutation isolated as a suppressor of IS2 polarity *rho*ts15, as expected, suppresses nonsense and IS1 polarity as well (51, 76). Very little in vitro transcription termination activity can be demonstrated by Rho isolated from the *rho*ts15 strain.

Other aspects of the *rho* allele specificity are not obviously due to mutational leakiness. Two *rho* mutations, *psuA*120 and *psu*2, suppress nonsense polarity to about the same extent, yet only the latter affects transcription termination in phage λ in vivo (50). Conversely, *ilv*-linked mutations (*sun*), isolated as suppressing transcription termination in λ, are incapable of suppressing nonsense polarity (75). *trp* attenuation, which is Rho-independent in vitro, is suppressed in vivo by some, but not all, *rho* mutations (65, 67). The level of *trp* enzymes in derepressed cells is increased 3–4-fold in the *rho* mutants *psu*2 and *rho*ts15; the *rho* mutation *psu*103, on the other hand, has no effect on *trp* attenuation.

The fact that most *rho* mutations are leaky is explained in two ways. (*a*) Most *rho* mutants show increased levels of Rho protein, although the specific activity of the enzyme is reduced (80). The normal level of Rho is 0.1–0.15% of the cell protein, while the temperature-sensitive *rho* mutant *nitA*ts702 has a Rho concentration of 0.3% at 32°C, the permissive temperature, and 0.6% at the nonpermissive temperature, 42°C (81). It has been proposed that the increase in Rho concentration in mutant cells reflects a Rho-dependent attenuation of transcription of the *rho* gene (80). This compensatory mechanism for maintaining constant levels of Rho activity makes it difficult to isolate truly Rho-deficient strains. (*b*) Several *rho* mutants are conditionally lethal for growth (76, 81). Both *rho*ts15 and *nitA*ts702 fail to grow at 42°. The amber *rho* mutant, *nitA*am112, isolated in a *supD*ts background, is also temperature-sensitive for growth. Mutations of this type might be excluded by some selection techniques.

The question of the lethality of *rho* mutations is not fully resolved. The isolation of *rpoB* mutations which suppress the lethality of *nitA*am112 is consistent with the idea that failure to terminate transcription causes the death of these Rho defective cells (83). However, the *rho*ts15 mutant displays a pleiotropic phenotype, which may not be directly related to defects in transcription termination (86). The mutant is also defective in membrane ATPase (Unc), recombination and UV-repair (possibly Rec*BC*), and ability to grow phage P2 (Rep). Like *rho,* these defects appear to involve the loss of an ATPase activity, suggesting that the product of the *rho* gene participates in a variety of cellular reactions, other than transcription termination, which utilize ATP. Derivatives of *rho*ts15 have been isolated which grow at 42°, but in which IS2 insertions remain non-polar. Conversely suppressors of *rho*ts15 which restore IS2 polarity but which do not correct the *ts* phenotype have been isolated. This suggests that some process other than termination, affected by the *rho* mutation, is responsible for the conditional lethal phenotype.

As indicated above, some mutations in the *rpoB* locus restore transcription termination to the *rho*ts15 or the *nitA*am112 mutant (A. Das and C. Merril personal communication; 83). It has been suggested that the altered RNA polymerase in these mutants is capable of terminating transcription at Rho-dependent sites in the absence of Rho factor (83). Alternatively, the RNA polymerase may have the capacity to interact with the altered Rho in *rho*ts15 or with the amber fragment of *nitA*am112. The latter has a molecular weight of 44,000 daltons and retains immunological activity (81). Preliminary *in vitro* complementation studies suggest that the latter alternative may be the case for the *rpoB* mutations that suppress *rho*ts15 (A. Das and C. Merril, personal communication).

Transcription termination at the ends of the *trp* and *bio* operons appears to be Rho-mediated; the *tsu1* and *rho*ts15 mutations permit extensive readthrough beyond the promoter-distal ends of these operons into the respective neighboring operons (67, 71).

The expression of a number of operons appears to be normal in the *rho*ts15 mutant strain (67; M. Winkler, personal communication). The expression of the *lac* and *his* operons is not increased in *rho*ts15 compared to the wild-type parent. The latter operon is thought to have an attenuator in *Salmonella* (84, 85); if this is so in *E. coli* as well, it is Rho-independent or active with the Rho protein of a variety of *rho* mutant strains. Transcription termination at the end of the λ 6S RNA, as well as in the λ *N*–operon, is also observed in the *rho*ts15 mutant (A. Oppenheim, D. Court, M. Rosenberg, and M. Gottesman, unpublished results). As in the case of the *his* operon, it is not possible to distinguish between the two possibilities: These may represent true, Rho-independent termination sites or, alternatively, sites at which the mutant Rho protein can still act.

A *rho* mutation can, in principle, affect a particular step in the transcription-termination reaction. Thus the *nitA*ts702 and *nitA*ts18 mutations have normal polyC-dependent ATPase activity, but show altered ATPase activity on nascent RNA templates (82). The *rho*ts15 mutant, which displays almost no transcription termination activity in vitro, will carry out polyC-dependent ATP hydrolysis at low, but not at high temperatures (76). Some altered properties of the mutant Rho enzymes may be artifactual, since it has been demonstrated that enzymes isolated from several *nitA* mutants are abnormally sensitive to proteolysis in extracts (82).

ANTITERMINATION

The regulation of transcription termination was first demonstrated and remains best understood in the case of bacteriophage λ. For this purpose, λ synthesizes a function, the product of its *N* gene, whose role is to prevent transcription termination at sites on the phage chromosome.

After infection or prophage induction of lambda, phage genes are expressed sequentially, beginning with the immediate-early genes *N* and *cro*, which are transcribed, respectively, from the promoters p_L and p_R (9-11). Transcription of the early genes (e.g. *cIII* and *cII* lying beyond *N* and *cro* is prevented or reduced by the Rho-dependent termination signals, t_{L1} and t_{R1}. In an *N* mutant, or in the presence of chloramphenicol, the early genes are never efficiently transcribed (87). Normally, however, the accumulation of pN results in the extension of the p_L and p_R transcripts beyond the terminators and into the early-gene region of the phage; pN acts as a positive control function for the expression of early genes by suppressing transcription termination (9, 88). That pN is a protein is evident from the isolation of nonsense (*N7*, *N53* and *N217*) as well as temperature-sensitive (*N*ts8) mutations in the *N* gene (89, 90, 91). A 13,500 dalton protein (detected by SDS-gel electrophoresis of ^{35}S-labeled proteins synthesized after lambda infection of UV-irradiated *E. coli*) is thought to be the product of gene *N*; the protein is missing after infection with a lambda *N* amber mutant but present with a *N* missense (91a).

pN Acts at Many Sites

Gene *N* product can suppress the termination of transcription at a variety of termination signals. Transcription originating at p_L, under the influence of pN, extends not only beyond t_{L1}, but beyond other termination sites thought to lie in the region between p_L and *att* (A. Oppenheim, personal communication; W. Szybalski, personal communication; 67). At least one of these sites, t_{L2}, appears to be active in the *rho*ts15 mutant (67) (Figure 1). If this site(s) is truly Rho-independent, then pN would act as an *antitermination* factor, rather than as an anti-Rho factor, as has been suggested (3). If prophage excision is blocked, p_L-initiated transcription extends through the prophage-bacterial junction and into the bacterial chromosome at least as far as the *gal* operon, some 21 kilobases away from p_L (48). This

p_L-promoted transcription of *gal* results in the "escape synthesis" of *gal* enzymes. Transcription of *trp* and *lac* due to the suppression of termination by pN of p_L-initiated transcription has also been demonstrated in λ transducing phage (92–94; I. Herskowitz, personal communication). Similarly, pN-mediated antitermination can produce readthrough transcription of *gal* from a prophage p_R promoter (O. Reyes and C. Dambly, unpublished data). Under the influence of pN, transcription fails to terminate at the ends of operons, e.g. *bioA* and *trp* (48, 95).

The transcriptional barriers in operons exposed by IS1 insertions (48) or polar nonsense mutations (48, 92, 94), as well as the transcription termination signal introduced by IS2 (48), are all overcome by λ *N*–function. Thus the polarity caused by all the known types of polar mutations is suppressed by pN when the affected operons are expressed from a phage promoter.

Of interest is the fact that the λ chromosome carries a site(s) to the left of the *b*2 region resistant to the action of pN (49). The site is suppressed by the *rho*ts15 mutation (M. Gottesman, in preparation). Little is known about the site(s), but it probably functions to prevent p_L-initiated transcription from reading through into the antisense strand of the late-gene region.

Recognition Specificity of pN

An interesting property of pN is its specificity of recognition. For example, lambda pN will suppress polarity in *gal* expression when the operon is transcribed from a lambda promoter, but will not suppress the polarity of the same mutation when transcription originates at the *gal* promoter. Clearly some recognition site(s) required *in cis* for pN mediated antitermination lies in the phage chromosome (97, 48). These recognition sites for pN, designated *nut* for N utilization (95a) [also called *ren* (49)] are distinct from the sites of action of pN (t_{R1}, t_{L1} etc). Their locations on the phage chromosome have been deduced from studies of the effects of a variety of prophage deletions on *gal* escape synthesis (48, C. Dambly personal communication), and from analyses of genetic hybrids formed in crosses between a number of related temperate bacteriophages with diverse *N*-like gene functions (97). Hybrid phage consisting mostly of λ DNA have been constructed by crossing lambda with coliphages 434 (98) or 21 (99), or with *Salmonella* phage P22 (100). The resultant hybrids, with the exception of λ*imm*[434] which carries the *N* gene of λ, contain heterologous and functionally distinct *N* genes. Although the *N* genes of λ, λ*imm*[21] or λ*imm*[22] perform analogous roles in phage growth and are capable of antiterminating transcription at identical sites, each displays specificity of recognition (97, 101–104). Lambda can supply pN to λ*imm*[434] but not to λ*imm*[21] and λ*imm*[22]. Similarly, λ*imm*[21] cannot complement λ*N*⁻, λ*imm*[434] *N*⁻

or λimm^{22} N^- mutants. Finally, under rigorous conditions, pN of λimm^{22} will only antiterminate transcription initiating from λimm^{22} promoters (104). The non-complementing hybrids thus must carry non-identical *nut* sites corresponding to their different *N*-functions, and these sites must lie in the limited regions of non-homology between the various phage. However, since λimm^{434} and lambda *can* exchange *N*-functions, their respective *nut* sites would appear to lie in the regions of homology between these two phages. These arguments, combined with the analysis of the effects of prophage deletions on *gal* escape synthesis (48), would lead to the prediction that *nut*L should lie just to the left of the p_L promoter and *nut*R just to the right of gene *cro*.

Two sequences of 17 base pairs, identical except for one nucleotide and positioned in the same orientation with respect to the direction of transcription, lie in the predicted locations on the λ map (29). The sequence presumed to be *nut*L is located between the start point of p_L-promoted transcription and gene *N*; the second, *nut*R, lies between the carboxyl terminus of gene *cro* and t_{R1} (see Figures 1 and 4). The sequences show dyad symmetry. Phage mutants, λnutL$^-$, have been isolated (104a) and by sequence analysis their alterations have been shown to lie within the 17 base pair sequence (M. Rosenberg, J. Salstrom and D. Court, personal communication). The mutant phage produce pN which acts to antiterminate transcription initiating at p_R, but functions beyond t_{L1} are not expressed, indicating that pN can no longer act to antiterminate transcription arising from p_L.

Stoichiometry of pN Requirement

By measuring the amount of *N*-dependent *gal* or *trp* escape-synthesis, a quantitative estimate of pN activity can be obtained. When *gal* escape levels are determined in a series of *N*am lysogens carrying tRNA suppressors of varying efficiencies, it is found that the amount of escape is directly proportional to efficiency of amber suppression (S. Adhya and M. Gottesman, in preparation). Similarly, when the amount of pN is limited by infecting with a $\lambda trp N$ts mutant at intermediate temperatures, the expression of *trp* enzymes is inversely related to the temperature (92). These experiments indicate that pN is not made in excess, contradicting a previous report based on a more difficult assay (96).

The efficiency of pN antitermination is less than complete. Thus, the amount of *gal* enzyme escape increases as the *gal* operon is brought closer to p_L by a series of deletions removing the intervening genetic material (49). This is consistent with the notion that several transcription-termination signals lie between p_L and *gal* and that the antitermination efficiency of pN is less than 100% at each signal. Furthermore, the presence of an IS2 element between p_L and *gal* in two otherwise identical lysogens reduces the amount of *gal* escape approximately 50% (S. Adhya and M. Gottesman, in preparation).

Involvement of Host Factors

A number of host mutations that alter the pN activity of bacteriophage have been isolated. Mutations in either of two genes, *nusA* or *nusB* (*gro*), whose products are as yet unidentified, prevent the growth of λ or λ*imm*[21] at elevated temperatures; the phage appear phenotypically N^- in the mutant hosts (105–108). Although λ*imm*[22] does grow in these strains, *gal* escape synthesis from the p_L promoter of λ*imm*[22], λ, or λ*imm*[21], which is dependent upon pN, is drastically reduced in the *nusA* mutant (104). The effect of the *nusB* mutations on *gal* escape synthesis has not yet been tested.

Some mutations in *rpoB* that confer resistance to rifampicin can also affect pN activity. Such a mutation, which by itself has no effect on λ growth, will, in combination with $nusA^-$ or $nusB^-$, severely restrict the activity of λ pN even at low temperatures (109, 110). Another class of mutation, *ron*, also located in *rpoB*, inhibits the growth of λ that carry an otherwise silent mutation in their *N* gene (*mar*); wild-type λ grow well in *ron* hosts (111). Other mutant λ capable of growing in *nus* hosts have been isolated; some of these carry mutations in their *N* genes (D. Friedman, personal communication).

Bypass of pN Requirement

A priori, two kinds of mutations should permit the growth of λ N^- phage: mutations in the host that alter or eliminate Rho; and mutations in the phage that destroy or circumvent termination signals.

Some *rho* mutants do, in fact, permit the growth of λ N^- or the expression of certain phage functions dependent upon pN (50, 72, 76, 112). This suggests that the vital phage functions are not separated from their promoters by Rho-independent termination signals. We have already mentioned that *gal* escape synthesis in a λ N^- lysogen does not occur in a *rho*ts15 mutant; the presumed Rho-independent site, located between 67.7 and 70.8 λ-fractional lengths, does not affect an essential region of the phage chromosome. Lambda Q^- mutants fail to grow in Rho-deficient strains (50, 76). It has been suggested that gpQ antiterminates the Rho-independent 6S transcript, permitting expression of the λ late genes (13b, 14).

The analysis of the λ mutants that overcome a pN defect is not entirely straightforward. The mutant λ*nin*5 is apparently deleted for t_{R2} (113); λ*nin*5 grows well on nus^- hosts and λ N^- *nin*5 forms small plaques on a wild-type *E. coli* (105). A point mutation *byp*, located just to the right of the *nin*5 deletion, permits λ growth on *nus*⁻ strains, but does not by itself suppress an N^- mutation. An additional mutation, *c17*, which is a new promoter to the right of, and bypassing, the termination signal t_{R1}, is required for a λ*N-byp* to plate (114, 115) (see Figure 1).

Models of Antitermination

How pN acts to prevent termination at transcription-termination sites is not yet known. A partial purification of the protein has been accomplished using crude coupled transcription-translation systems. Two such systems have been reported: one in which pN permits transcription originating at p_R to extend into gene *R*, with the resultant synthesis of pR, an endolysin (116); and a second, where pN-dependent readthrough from p_L into *trp* on a λ*trp* DNA template is measured (117). As yet, no purified transcription assay system responding to pN has been established.

Because the control of transcription termination is complex, pN could act on any one of a number of factors to produce antitermination. In addition, it should be recalled that the recognition site for pN is physically separate from its sites of action. Does pN remain at the *nut* sites, acting at a distance to antagonize termination; or does it move from the recognition sites to the distal termination signals?

Of the several possible mechanisms by which pN could work, the following seem the most plausible:

1. Interaction with RNA polymerase: In this model, an interaction takes place between pN and RNA polymerase at the *nut* region of the phage DNA. The RNA polymerase is altered in such a way that it fails to terminate at distal transcription-termination sites (48, 92, 97). From the genetic evidence, which suggests an interplay between the β subunit of RNA polymerase and pN, it is conceivable that pN may attach to RNA polymerase, via the β subunit, and travel with the enzyme during the elongation (109–111).

2. Interaction with ribosomes: Translation termination and ribosome release appear to be prerequisites for Rho-mediated transcription termination. It is possible that the primary action of pN is to antagonize ribosome discharge, which would have the secondary effect of blocking transcription termination (49). This model leaves unexplained the apparent ability of pN to antiterminate transcription at Rho-independent sites.

3. Interaction with RNA: pN could bind to λ RNA at the *nut* loci and affect the secondary structure of the nascent RNA, preventing transcription termination. This might be accomplished by the movement of the nascent RNA, under the action of pN, into a cellular domain where con-

straints influence RNA secondary-structure formation. Relevant to this idea is the report that λ DNA, when transcribed extensively, is found attached to a fast-sedimenting cellular component, and that this attachment requires pN (118).

In order to distinguish among the several models of pN action, further genetic and biochemical analysis is necessary. Identification of the products of *nusA* and *nusB* genes could shed light on the mechanism of pN-mediated antitermination. Finally, the establishment of an in vitro transcription system in which pN functions could resolve the many questions concerning this interesting phenomenon.

ACKNOWLEDGEMENTS

We appreciate very much the communication to us of unpublished results, and critical reading of this manuscript by R. Crouch, D. Court, A. Das, A. Oppenheim, C. Merril, M. Rosenberg, D. Friedman, C. Dambly, D. Wulff, L. Guarente, J. Beckwith, J. Richardson, F. Blattner, J. Salstrom, T. Platt, G. Hobom, J. Miller, U. Maitra, C. Yanofsky, and M. Imai. We thank Connie Rafferty for typing the manuscript.

We specifically acknowledge the contribution of B. de Crombrugghe, A. Das, D. Court, and O. Reyes in the formulation of our model for transcription termination described here.

Literature Cited

1. Jacob, F., Monod, J. 1961. *Cold Spring Harbor Symp. Quant. Biol.* 26:193–211
2. Chamberlin, M. J. 1976. In *RNA Polymerase,* ed. R. Losick, M. Chamberlin, pp. 17–67. New York: Cold Spring Harbor Laboratory
3. Roberts, J. W. 1969. *Nature* 224:1168–74
4. de Crombrugghe, B., Adhya, S., Gottesman, M., Pastan, I. 1973. *Nature New Biol.* 241:260–64
5. Roberts, J. 1976. See Ref. 2, pp. 247–71
6. Krakow, J. S., Rhodes, G., Jovin, T. M. 1976. See Ref. 2, pp. 127–57
7. Maitra, U., Lockwood, A. H., Dubnoff, J. S., Guha, A. 1971. *Cold Spring Harbor Symp. Quant. Biol.* 35:143–56
8. Szybalski, W. 1974. In *Control of Transcription,* ed. B. B. Biswas, R. K. Mandal, A. Stevens, W. E. Cohn, pp. 201–12. New York: Plenum Press
9. Lozeron, H. A., Dahlberg, J. E., Szybalski, W. 1976. *Virology* 71:262–77
10. Kourilsky, P., Bourguignon, M.-F., Gros, F. 1971. In *The Bacteriophage λ,* ed. A. D. Hershey, pp. 647–66. New York: Cold Spring Harbor Laboratory
11. Heinemann, S. F., Spiegelman, W. G. 1971. *Cold Spring Harbor Symp. Quant. Biol.* 35:315–18
12. Howard, B., de Crombrugghe, B., Rosenberg, M. 1977. *Nucleic Acids Res.* 4:827–42
13. Lebowitz, P., Weissman, S. M., Radding, C. M. 1971. *J. Biol. Chem.* 246:5120–39
13a. Sklar, J., Yot, P., Weissman, S. M. 1975. *Proc. Natl. Acad. Sci. USA* 72:1817–21
13b. Blattner, F. R., Dahlberg, J. E. 1972. *Nature New Biol.* 237:227–36
14. Roberts, J. 1975. *Proc. Natl. Acad. Sci. USA* 72:3300–304
14a. Sklar, J., Weissman, S. M. 1976. *Fed. Proc.* 35:1537
15. Dahlberg, J. E., Blattner, F. R. 1973. In *Virus Research,* ed. C. F. Fox, W. S. Robinson, p. 533. New York: Academic
16. Smith, G. R., Hedgpeth, J. 1975. *J. Biol. Chem.* 250:4818–21
17. Rosenberg, M., Weissman, S. 1975. *J. Biol. Chem.* 250:4755–64

17a. Rosenberg, M., de Crombrugghe, B., Musso, R. 1976. *Proc. Natl. Acad. Sci. USA* 73:717–21
18. Axelrod, N. 1976. *J. Mol. Biol.* 108:753–70
19. Millette, R. L., Trotter, C. D., Herrlich, P., Schweiger, M. 1971. *Cold Spring Harbor Symp. Quant. Biol.* 35:135–42
20. Chakraborty, P. R., Bandyopadhyay, P., Huang, H. H., Maitra, U. 1974. *J. Biol. Chem.* 249:6901–9
21. Bertrand, K., Korn, L. J., Lee, F., Platt, T., Squires, C. L., Squires, C., Yanofsky, C. 1975. *Science* 189:22–26
22. Lee, F., Yanofsky, C. 1977. *Proc. Natl. Acad. Sci. USA* 74:4365–69
23. Bertrand, K., Korn, L. J., Lee, F., Yanofsky, C. 1978. *J. Mol. Biol.* In press
23a. Yanofsky, C. 1976. In *Molecular Mechanisms in the Control of Gene Expression*, ed. D. Nierlich, W. J. Rutter, C. F. Fox. ICN–UCLA Symposia on Molecular and Cellular Biology. pp. 75–87
24. Musso, R., diLauro, R., Adhya, S., de Crombrugghe, B. 1978. *Cell* 12:847–54
25. Gilbert, W. 1976. See Ref. 2, pp. 193–205
26. Darlix, J. L., Fromageot, P. 1972. *Biochimie* 54:47–54
27. Darlix, J. L. 1974. *Biochimie* 56:693–703
28. Darlix, J. L., Horaist, M. 1975. *Nature* 256:288–92
29. Rosenberg, M., Court, D., Wulff, D. L., Shimatake, H., Brady, C. 1978. *Nature.* In press
30. Gierer, A. 1966. *Nature* 212:1480–81
31. McDermit, M., Pierce, M., Staley, D., Shimaji, M., Shaw, R., Wulff, D. L. 1976. *Genetics* 82:417–22
32. Goldberg, A. R., Hurwitz, J. 1972. *J. Biol. Chem.* 247:5637–45
33. Lowery-Goldhammer, C., Richardson, J. P. 1974. *Proc. Natl. Acad. Sci. USA* 71:2003–7
34. Minkley, E. G. Jr. 1973. *J. Mol. Biol.* 78:577–80
35. Oda, T., Takanami, M. 1974. *J. Mol. Biol.* 83:289–304
36. Howard, B., de Crombrugghe, B. 1976. *J. Biol. Chem.* 251:2520–24
37. Galluppi, G., Lowery, C., Richardson, J. P. 1976. See Ref. 2, pp. 657–65
38. Darlix, J. L. 1973. *Eur. J. Biochem.* 35:517–26
39. Lowery, C., Richardson, J. P. 1977. *J. Biol. Chem.* 252:1375–80
40. Lowery, C., Richardson, J. P. 1977. *J. Biol. Chem.* 252:1381–85
41. Richardson, J. P. 1978. *Proc. FEBS Mtg. 11th.* In press
42. Franklin, N. C., Luria, S. E. 1961. *Virology* 15:299–311
43. Ames, B. N., Hartman, P. E. 1963. *Cold Spring Harbor Symp. Quant. Biol.* 28:349–65
44. Newton, W. A., Beckwith, J. R., Zipser, D., Brenner, S. 1965. *J. Mol. Biol.* 14:290–96
45. Shapiro, J. 1969. *J. Mol. Biol.* 40:93–105
46. Jordan, E., Saedler, H., Starlinger, P. 1968. *Mol. Gen. Genet.* 102:353–63
47. Shapiro, J. A., Adhya, S. L., Bukhari, A. I. 1977. *DNA Insertion Elements, Plasmids and Episomes.* New York: Cold Spring Harbor Laboratory. pp. 3–11
48. Adhya, S., Gottesman, M., de Crombrugghe, B. 1974. *Proc. Natl. Acad. Sci. USA* 71:2534–38
49. Adhya, S., Gottesman, M., de Crombrugghe, B., Court, D. 1976. See Ref. 2, pp. 719–30
50. Reyes, O., Gottesman, M., Adhya, S. 1976. *J. Bacteriol.* 126:1108–12
51. Das, A., Court, D., Gottesman, M., Adhya, S. 1977. In *DNA Insertion Elements, Plasmids and Episomes*, eds. A. I. Bukhari, J. A. Shapiro, S. L. Adhya. Cold Spring Harbor Laboratory. pp. 93–97
52. Minkley, E. G., Pribnow, D. 1973. *J. Mol. Biol.* 77:255–77
53. Hercules, K., Jovanovich, S., Sauerbier, W. 1976. *J. Virol.* 17:642–58
54. Shapiro, J. A., Adhya, S. L. 1969. *Genetics* 62:249–64
55. Zipser, D., Zabell, S., Rothman, J., Grodzicker, T., Wenk, M., Novitski, M. 1970. *J. Mol. Biol.* 49:251–54
56. Shimizu, N., Hayashi, M. 1974. *J. Mol. Biol.* 84:315–35
57. Morse, D. E., Yanofsky, C. 1969. *Nature* 224:328–31
58. Hiraga, S., Yanofsky, C. 1972. *J. Mol. Biol.* 72:103–10
59. Franklin, N. C., Yanofsky, C. 1976. See Ref. 2, pp. 693–706
60. Ohtsubo, H., Ohtsubo, E. 1978. *Proc. Natl. Acad. Sci. USA.* 75:615–619
61. Fiandt, M., Szybalski, W., Malamy, M. H. 1972. *Mol. Gen. Genet.* 119:223–31
62. Adhya, S. L., Shapiro, J. A. 1969. *Genetics* 62:231–47
63. Morse, D. E. Morse, A. N. C. 1976. *J. Mol. Biol.* 103:209–26
64. Pouwels, P. H., Pannekoek, H. 1976. *Mol. Gen. Genet.* 149:255–65
65. Korn, L. J., Yanofsky, C. 1976. *J. Mol. Biol.* 103:395–409
66. Korn, L. J., Yanofsky, C. 1976. *J. Mol. Biol.* 106:231–41

67. Gottesman, M., Adhya, S., Court, D., Das, A. 1978. *Proc. FEBS Mtg. 11th.* In press
68. Salser, W., Bolle, A., Epstein, R. H. 1970. *J. Mol. Biol.* 49:271–95
69. Varmus, H. E., Perlman, R. L., Pastan, I. 1971. *Nature New Biol.* 230:41–44
70. Tomich, P. K., Friedman, D. I. 1978. See Ref. 51, pp. 99–107
71. Guarente, L. P., Mitchell, D. H., Beckwith, J. 1977. *J. Mol. Biol.* 112:423–36
72. Inoko, H., Imai, M. 1976. *Mol. Gen. Genet.* 143:211–21
73. Lecocq, J. P., Dambly, C. 1973. *Arch. Int. Biochem. Physiol.* 31:383–84
74. Beckwith, J. 1963. *Biochim. Biophys. Acta* 76:162–64
75. Brunel, F., Davison, J. 1975. *Mol. Gen. Genet.* 136:167–80
76. Das, A., Court, D., Adhya, S. 1976. *Proc. Natl. Acad. Sci. USA* 73:1959–63
77. Richardson, J. P., Grimley, C., Lowery, C. 1975. *Proc. Natl. Acad. Sci. USA* 72:1725–28
78. Ratner, D. 1976. See Ref. 2, pp. 645–55
79. Dambly, C., Court, D., Brachet, P. 1976. *Mol. Gen. Genet.* 148:175–82
80. Ratner, D. 1976. *Nature* 259:151–53
81. Imai, M., Shigesada, K. 1978. *J. Mol. Biol.* In press
82. Shigesada, K., Imai, M. 1978. *J. Mol. Biol.* In press
83. Guarente, L., Beckwith, J. 1978. *Proc. Natl. Acad. Sci. USA.* 75:294–97
84. Artz, S. W., Broach, J. R. 1975. *Proc. Natl. Acad. Sci. USA.* 72:3453–57
85. Winkler, M. E., Roth, D. J., Hartman, P. E. 1978. In press
86. Das, A., Court, D., Adhya, S. 1978. In *Molecular Basis of Host Virus Interactions,* ed. M. Chakravarty. New Jersey: Science Press. In press
87. Skalka, A., Butler, B., Echols, H. 1967. *Proc. Natl. Acad. Sci. USA* 58:576–83
88. Portier, M.-M., Marcaud, L., Cohen, A., Gros, F. 1972. *Mol. Gen. Genet.* 117:72–81
89. Campbell, A. 1961. *Virology* 14:22–32
90. Thomas, R., Leurs, C., Dambly, C., Parmentier, D., Lambert, L., Brachet, P., Lefebvre, N., Mousset, S., Porcheret, J., Szpirer, J., Vauters, D. V. 1967. *Mutat. Res.* 47:735–41
91. Brown, A., Arber, W. 1964. *Virology* 24:237–39
91a. Shaw, J., Jones, B., Pearson, M., *Proc. Natl. Acad. Sci. USA.* In press

92. Franklin, N. C. 1974. *J. Mol. Biol.* 89:33–60
93. Mercereau-Puijalon, O., Kourilsky, P. 1976. *J. Mol. Biol.* 108:733–51
94. Segawa, T., Imamoto, F. 1974. *J. Mol. Biol.* 87:741–45
95. Segawa, T., Imamoto, F. 1976. *Virology* 70:181–84
95a. Salstrom, J. S., Szybalski, W. 1976. *Fed. Proc.* 35:1538
96. Radding, C. M., Echols, H. 1968. *Proc. Natl. Acad. Sci. USA* 60:707–12
97. Friedman, D. I., Wilgus, G. S., Mural, R. J. 1973. *J. Mol. Biol.* 81:505–16
98. Kaiser, A. D., Jacob, F. 1957. *Virology* 4:509–21
99. Liedke-Kulke, M., Kaiser, A. D. 1967. *Virology* 32:465–74
100. Botstein, D., Herskowitz, I. 1974. *Nature* 251:584–89
101. Friedman, D. I., Ponce-Campos, R. 1975. *J. Mol. Biol.* 98:537–49
102. Friedman, D. I., Jolly, C. A., Mural, R. J., Ponce-Campos, R., Baumann, M. F. 1976. *Virology* 71:61–73
103. Couturier, M., Dambly, C. 1969. *C. R. Acad. Sci.* 270:428–30
104. Hilliker, S., Gottesman, M., Adhya, S. 1978 *Virology.* In press
104a. Salstrom, J. S., Szybalski, W. 1978. *J. Mol. Biol.* In press
105. Friedman, D. I., Jolly, C. T., Mural, R. J. 1973. *Virology* 51:216–26
106. Friedman, D. I., Baron, L. S. 1974. *Virology* 58:141–48
107. Friedman, D. I., Baumann, M., Baron, L. S. 1976. *Virology* 73:119–27
108. Keppel, F., Georgopoulos, C. P., Eisen, H. 1974. *Biochimie* 56:1503–9
109. Baumann, M. F., Friedman, D. I. 1976. *Virology* 73:128–38
110. Sternberg, N. 1976. *Virology* 73:139–54
111. Ghysen, A., Pironio, M. 1972. *J. Mol. Biol.* 65:259–72
112. Belfort, M., Oppenheim, A. B. 1976. *Mol. Gen. Genet.* 148:171–73
113. Court, D., Campbell, A. 1972. *J. Virol.* 9:938–45
114. Butler, B., Echols, H. 1970. *Virology* 40:212–22
115. Hopkins, N. 1970. *Virology* 40:223–29
116. Greenblatt, J. 1972. *Proc. Natl. Acad. Sci. USA* 69:3606–10
117. Dottin, R. P., Pearson, M. L. 1973. *Proc. Natl. Acad. Sci. USA* 70: 1078–82
118. Hallick, L., Boyce, R. P., Echols, H. 1969. *Nature* 223:1239–41

Ann. Rev. Biochem. 1978. 47:997–1029
Copyright © 1978 by Annual Reviews Inc. All rights reserved

PROSTAGLANDINS AND THROMBOXANES

❖995

B. Samuelsson, M. Goldyne, E. Granström, M. Hamberg, S. Hammarström, and C. Malmsten

Department of Chemistry, Karolinska Institutet, S-104 01 Stockholm, Sweden

CONTENTS

PERSPECTIVES AND SUMMARY

The prostaglandins are cyclopentane derivatives formed from polyunsaturated fatty acids by most mammalian tissues and by tissues of lower vertebrates and certain invertebrates. The prostaglandins modulate a number of

997

0066-4154/78/0701-0997$01.00

hormonal, neurohormonal, or other stimuli and they have a variety of other pharmacological actions. However, their physiological role is known only to a limited extent.

A new and rapid development of the field has taken place since the previous review (1). Although experimental evidence for the existence of an endoperoxide structure in the biosynthesis of prostaglandins was obtained in 1965 it was not until 1973 that the first unstable endoperoxide, PGH_2, was isolated. The endoperoxides were found to have unique biological effects particularly on platelets and vascular and respiratory smooth muscle. These effects could not be explained by their conversion into the previously known primary prostaglandins. The recognition of a platelet aggregating factor derived from arachidonic acid, which was different from the endoperoxides and the known prostaglandins, led to the discovery of a new group of compounds, thromboxanes. An intermediary derivative, thromboxane A_2, is highly unstable ($t_{1/2}$=30–40 sec) and possesses pronounced biological activity. It was also found to constitute the major part of a biological activity formed, for example, during anaphylaxis in guinea pig lung, and earlier referred to as rabbit aorta contracting substance. Thromboxanes have now been identified in a variety of tissues including platelets, leucocytes, spleen, inflammatory granuloma, brain, and kidney. They have a physiological role in normal hemostasis and are of considerable pathophysiological interest in thrombo-embolic diseases and anaphylactic reactions. There is intensive, ongoing research in this area involving biochemistry, physiology, and pharmacology.

The availability of the endoperoxides also paved the way for the recognition of another prostaglandin, PGI_2, alternatively referred to as prostacyclin. In a search for thromboxane-forming enzymes using biological methods, it was found that the endoperoxide was converted into material with vasodilator and antiaggregating properties that cause effects opposite to those caused by the thromboxanes.

Enzymological studies on the formation of prostaglandins have also been facilitated by the isolation of the endoperoxides, and substantial progress is being made in this area concerning enzymes involved in both the formation and transformation of endoperoxides. The diversification of the field at a biochemical level which has taken place in the past few years will probably lead to a better understanding of the physiological role of prostaglandins and thromboxanes in the control of cellular processes.

BIOSYNTHESIS

Enzymes

Figure 1 illustrates the enzymology of prostaglandin and thromboxane biosynthesis and shows the structures of substrate, intermediates, and stable products. The prostaglandin synthase system from bovine vesicular glands was first solubilized and resolved into two components by DEAE cellulose chromatography (2). One enzyme converted precursor acids to prostaglandin endoperoxides (PGG and PGH) whereas the other one converted PGH to PGE. The enzymes have been designated prostaglandin endoperoxide synthase (EC 1.14.99.1;*1*, Figure 1) and prostaglandin endoperoxide E isomerase (EC 5.3.99.3;*3*, Figure 1), respectively.

Prostaglandin endoperoxide synthase has been purified to homogeneity from bovine (3) and sheep vesicular glands (4, 5). The first successful purification (3) was based on Tween 20 solubilization from microsomes, DEAE cellulose chromatography, and isoelectric focusing, which gives a 750-fold increase of specific activity. This material, which was pure as judged by polyacrylamide gel electrophoresis under nondenaturing conditions, catalyzed the conversion of 8,11,14-eicosatrienoic acid to PGG_1 in the presence of heme. Hemoglobin, myoglobin, and catalase could be used instead of heme whereas hematoheme and cytochrome c were ineffective. In the presence of heme and thryptophan (or heme and one of several other

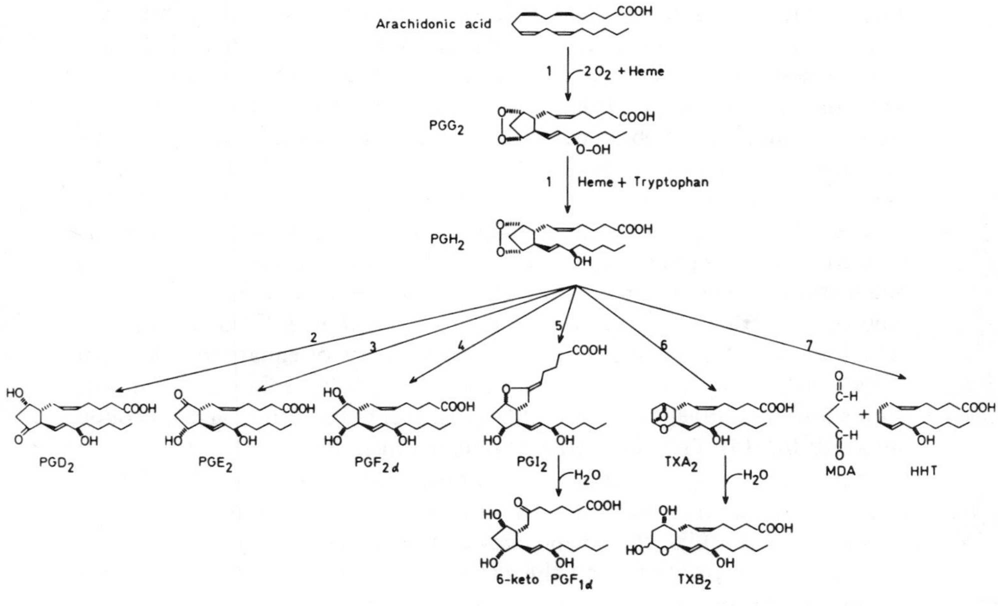

Figure 1. Enzymes and cofactors of prostaglandin and thromboxane biosynthesis. 1. Prostaglandin endoperoxide synthase (EC 1.14.99.1) (3–5).2. Serum albumin (19, 20), glutathione-S-transferase (19). 3. Prostaglandin endoperoxide E isomerase (EC 5.3.99.3) (13). 4. Prostaglandin endoperoxide reductase (guinea pig uterus) (22). 5. Prostaglandin endoperoxide I isomerase (6,9 oxocyclase, ref. (76). 6. Prostaglandin endoperoxide: thromboxane A isomerase (16). 7. Not designated.

aromatic compounds) pure prostaglandin endoperoxide synthase also catalyzed a cleavage of the hydroperoxy group of PGG_1. It is possible that the role of the aromatic compound is to supply electrons for the latter reaction. The specific activities for the two reactions catalyzed by the enzyme (cyclooxygenase and peroxidase) were 2.5 and 3.0 μmol/min mg, respectively. The pH optimum was about 8 and the isoelectric point 7.2. Prostaglandin endoperoxide synthase from sheep vesicular glands has been purified by: acid precipitation, Tween 40 solubilization, ammonium sulfate fractionation, DEAE cellulose chromatography, affinity chromatography, isoelectric focusing (4) or $NaClO_4$ extraction, Tween 20 solubilization, molecular sieving, and isoelectric focusing (5). The results obtained differ in: degree of purification (230- and 64-fold), specific activity (23 and 8–11 μmol/min mg), purity ($\sim$ 95% and 100%) and isoelectric points (6.5 and 7.0) (4, 5). The purified enzyme contained only small amounts of hemin but readily bound up to 4 moles of this cofactor for each 70,000 dalton subunit (see below) (4). Binding of 1 mole of hemin per 70,000 daltons was sufficient for maximal enzyme activity. The hemin content of sheep vesicular gland microsomes corresponded to 2 moles/mole of synthase (5) but this heme was apparently lost during the purification of the enzyme. A high content of nonheme iron was reported in one case (4), whereas the other preparation contained neither copper, iron, nor zinc in stoichiometric amounts (5). The high iron values were obtained with reaggregated enzyme and it is possible that iron was trapped or nonspecifically bound (cf also 5). The molecular weight of prostaglandin endoperoxide synthase has been determined at 124,000 daltons from the Stokes radius (53Å) and sedimentation coefficient (7.4 S) (with correction for bound detergent), and at 129,000 daltons by sedimentation equilibrium analysis (5). Sodium dodecyl sulfate polyacrylamide gel electrophoresis showed a single polypeptide of 72,000 daltons (5), which indicates that the enzyme has two subunits of equal size. Only one N-terminal amino acid (alanine) was present and the amino acid composition showed relatively high contents of hydrophobic amino acids, proline, and histidine (5). The pure ovine prostaglandin endoperoxide synthase also contained 12 moles per mole of mannose and 5 moles per mole of N-acetylmannoseamine, was free of phospholipids, but contained 69% w/w of Tween 20 (5). Like the bovine enzyme it catalyzed both cyclooxygenase and peroxidase reactions and the ratio of these activities did not change during purification (5). Self-destruction of the prostaglandin synthase system which had been reported earlier by, for example, Lands et al (see 1) was also noted with pure prostaglandin endoperoxide synthase (3, 5). Both the cyclooxygenase and the peroxidase activities decreased after 15–30 sec of catalysis, apparently because an inactivator was formed during the reaction (3). Hydroquinone prevented the auto-destruction and protected the enzyme from inactivation by H_2O_2 (5) which suggests that the inactivator is a hydroperoxy derivative of the substrate fatty acid (5). PGG has also been proposed as a mediator of the self-destruction (6). The formation of an activator during prostaglandin biosynthesis has been reported (7).

Chemiluminescence, possibly caused by singlet oxygen decay (8) and additional evidence has been reported for the involvement of singlet oxygen in prostaglandin biosynthesis (9). Furthermore, cooxygenation of organic compounds occurred, for example: (a) oxyphenbutazone to 4-hydroxyoxyphenbutazone (10), (b) 1,3-diphenylisobenzofuran to o-dibenzoylbenzene, (c) luminol to aminophtalate, and (d) benzopyrene to a hydroxybenzopyrene (11). a-Lipoic acid, which inhibited cooxygenation, increased the K_m and V_{max} of prostaglandin endoperoxide synthase (12) and thus inhibited the enzyme at low substrate concentration.

Prostaglandin endoperoxide E isomerase has been partially purified by Tween 20 solubilization, DEAE cellulose chromatography, and hydrophobic chromatography on ω-aminobutyl Sepharose 4B (13). The purification was 26-fold from microsomes. The enzyme had a half-life of only 30 min at 25° but could be protected from inactivation by glutathione, dithiothreitol, 2-mercaptoethanol, and cysteine. However, glutathione not only stabilized the enzyme but also served as a cofactor for the isomerization of PGH to PGE. The latter effect was specific for glutathione. Glutathione was not oxidized during the reaction. Both PGG and PGH were substrates for the enzyme but the rate of PGG isomerization was only half of that of PGH. The isomerase did not catalyze a peroxidase reaction since the product formed from PGG was 15-hydroperoxy PGE. Moreover, this product was not converted to PGE by pure prostaglandin endoperoxide synthase which converts PGG to PGH (see above). This indicates the following pathway for PGE biosynthesis: $20:n \rightarrow PGG_{n-2} \rightarrow PGH_{n-2} \rightarrow PGE_{n-2}$ ($n = 3$–5). An alternative pathway (14) in which 15-hydroperoxy PGE_{n-2} rather than PGH_{n-2} is an intermediate seems unlikely.

An enzyme that converts prostaglandin (PG) endoperoxides to thromboxanes has been detected in platelets from several species (15–17) (see also paragraph on thromboxanes below). It may be referred to as prostaglandin endoperoxide thromboxane A isomerase or thromboxane synthase (6, Figure 1). A particulate fraction sedimenting between 12,000 and 100,000 X g had the highest concentration of both prostaglandin endoperoxide synthase and thromboxane synthase (16, 17). The fraction consists mainly of the so called dense tubular system of platelets. Plasma membranes, dense bodies, a-granules (16), and the soluble fraction (16, 17) did not have detectable levels of these enzymes. The human platelet particulate fraction was treated with Triton X-100 or Tween 20 and the solubilized material was fractionated by DEAE cellulose chromatography (16). Two fractions (I and

III) were required for thromboxane B_2 biosynthesis from arachidonic acid. Fraction I converted arachidonic acid to PG endoperoxide but did not alter PGH_2. Fraction III converted PGH_2 to thromboxane B_2 but left arachidonic acid unaltered. The purification scheme thus resolved two enzymes required to convert unsaturated fatty acids to thromboxanes, namely prostaglandin endoperoxide synthase and thromboxane synthase.

Two or three soluble proteins catalyze the conversion of prostaglandin endoperoxides to PGDs (2, Figure 1). One such protein (mol wt 36,000–42,000) was detected in rat lung, stomach, and intestine (18). A similar protein from sheep lung has been purified to homogeneity and identified as a glutathione-S-transferase (19). The isomerase reaction was apparently independent of the transferase reaction but required glutathione as a cofactor (19). The ratio of PGD to PGE formed during nonenzymatic degradation of PGH in water is 0.20:0.25. Addition of serum albumin ($\geqslant$3 mg/ml) raised the PGD to PGE ratio to between 0.3 and 4.5 depending on the species of origin of the albumin (20). Boiling, but not treatment with N-ethylmaleimide, greatly reduced the isomerase activity of albumin (20). Similarly, fatty acids inhibited the conversion of PGH to PGD, which suggests that the fatty acid binding site of albumin needs to be unoccupied (19, 20). Prostaglandin D_2 formation has been observed in platelet rich plasma (21). The conversion of a potent platelet aggregating agent (PGH_2) to an inhibitor of platelet aggregation (PGD_2) by serum albumins, may have a protective role against unwanted platelet aggregation (cf the paragraph on PGI_2 below).

Microsomes from guinea pig uteri possessed prostaglandin endoperoxide F_α reductase activity [4, Figure 1 (22)]. The reducing factor had several characteristics of an enzyme, i.e. saturability at high substrate concentrations, inhibition by p-hydroxymercuribenzoate, ability to be solubilized by Cutscum (but not by Tween 20) and subsequent elution with the void volume of a Sephadex G-75 column. The factor was, however, very heat resistant; some 80% of the activity remained after 10 min at 100°C. Neither reduced glutathione nor NADPH stimulated PGH to PGF_α conversion in the presence of uterine microsomes. The factor was not detected in guinea pig lung, liver, or kidney.

Prostaglandin A isomerase (23) and prostaglandin C isomerase (24) have been partially purified from rabbit serum. These enzymes catalyze the conversions of PGA to PGC and of PGC to PGB, respectively, but have not yet been found in tissues or in human serum.

Inhibitors

Experiments with purified prostaglandin-forming enzymes showed that nonsteroid antiinflammatory drugs such as aspirin and indomethacin inhibited prostaglandin endoperoxide synthase but not prostaglandin endoperoxide E isomerase (2) and that only the cyclooxygenase and not the peroxidase reaction was inhibited (3). Elegant work using acetyl-^{3}H aspirin furthermore showed that aspirin is an active site acetylating agent (25, 26). Only catalytically active enzyme appeared to be acetylated (27). Indomethacin, unlike aspirin, did not covalently modify prostaglandin endoperoxide synthase (28). A free carboxyl group and a halogen atom are common structural features of several "time dependent" inhibitors of prostaglandin synthesis (29). Antiinflammatory drugs also inhibited prostaglandin metabolizing enzymes (30).

Inhibition of PG biosynthesis by some new acetylenic acids has been reported (31). Among eicosatriynoic acids, the 8,11,14-isomer was more potent than the 7,10,13-isomer which in turn was more potent than the 5,8,11-isomer. Although a poor inhibitor of prostaglandin endoperoxide synthase, 5,8,11-eicosatriynoic acid was a relatively good inhibitor of the ω-8 lipoxygenase which forms 12L-hydroperoxy-5,8,10,14-eicosatetraenoic acid (HPETE) in platelets (32).

Thiol analogs of $PGF_{2\alpha}$ and PGE_2 ($ID_{50} \sim 1$ μM) and other thiols, notably 2-mercaptoethanol ($ID_{50} = 10 \mu$M) inhibited the cyclooxygenase but not the peroxidase activity of prostaglandin endoperoxide synthase (33). Other inhibitors of prostaglandin biosynthesis include saturated fatty acids (34), 12-methyl-8,11,14-3icosatrienoic acid (35), and antioxidants such as cresols and xylenols, which are potent inhibitors of prostaglandin synthesis in 3T3 fibroblasts and platelets (36). Adenosine 3',5'-monophosphate (cAMP) has been reported to inhibit PG endoperoxide biosynthesis in human platelets (37, 38). The site of action is not clear; inhibition of prostaglandin endoperoxide synthase (37) or of phospholipase A_2 (38) has been suggested. Inhibition of PG formation by corticosteroids was first reported in skin (39) and subsequently in other tissues and cells. Evidence that the inhibitory effect is exerted at the level of prostaglandin release as opposed to the level of synthesis has been presented for adipose tissue (40, 41). This mechanism of inhibition does not, however, operate in other cells (42–45). A more general mechanism of action seems to be inhibition of precursor fatty acid release from phospholipids (46–49), first suggested by Gryglewski et al (42).

Several inhibitors of thromboxane synthase have been reported (16, 17, 50–54) namely: sodium p-benzyl-4 [1-oxo-2-(4-chlorobenzyl)-3-phenylpropyl] phenylphosphonate [N-0164 (50, 54)], 9,11-epoxymethanoprostanoic acid (17), 9,11-epoxymethano-15-hydroxyprosta-5,13-dienoic acid (17, 54), imidazole (51, 52, 54), nordihydroguaiaretic acid, 12L-hydroperoxy-5,8,10,14-eicosatetraenoic acid (HPETE) (16), 9,11-azo-15-hydroxyprosta-5,13-dienoic acid, and 2-isopropyl-3-nicotinylindole (L-8027) (54). A reported inhibition of thromboxane synthase by benzydamine (53) may be nonspecific (54). It has been reported (55) that 15-hydroperoxy-5,8,11,13-eicosatetraenoic acid is an inhibitor of prostaglandin endoperoxide I isomerase $(IC_{50} = 1.3 \ \mu M)$. $9\beta,15(S)$-Dihydroxy-prosta-5(cis), 13-($trans$)-dienoic acid and related compounds are potent competitive inhibitors of PGA-isomerase (56). A comprehensive review, "Inhibition of Prostaglandin Synthesis," has been published (57).

Thromboxanes

Three major products were formed from arachidonic acid in human platelets (58): 12L-hydroxy-5,8,10-heptadecatrienoic acid (HHT) and the hemiacetal of 8-(1-hydroxy-3-oxopropyl)-9,12L-dihydroxy-5,10-heptadecadienoic acid (PHD), which were derived from PGG_2, and 12-L-hydroxy-5,8,10,14-eicosatetraenoic acid (HETE) which was formed by an independent pathway, involving a new lipoxygenase. Platelet aggregation, induced by thrombin, was accompanied by the release of large amounts of these products whereas the amounts of $PGG_2 + PGH_2$, PGE_2, and $PGF_{2\alpha}$ were small (59). This suggested a new concept of PG action, in which PG endoperoxides rather than stable prostaglandins were mediators. In studies on the mechanism of formation of PHD from PGG_2, an unstable intermediate was trapped by the addition of methanol, ethanol, or sodium azide (60). This yielded analogues of PHD in which the hemiacetal hydroxyl group was replaced by methoxy- , ethoxy- , and azido-functions, respectively. Additional experiments with $^{18}O_2$ and CH_3O^2H permitted the assignment of an oxane:oxetane structure for the intermediate. The unstable compound was named thromboxane A_2 because of its origin from thrombocytes and its structure (Figure 1). By analogy the stable degradation product, PHD, was renamed thromboxane B_2. Thromboxane A_2 is a potent inducer of platelet aggregation, platelet release reaction (60, 61), and smooth muscle contractions. Rabbit aorta contracting substance (RCS), originally believed to be a PG endoperoxide, was shown to consist of thromboxane A_2 and to a lesser degree PG endoperoxide (60, 62). Thromboxane formation has been demonstrated in platelets (58–60), lung (63–65), spleen (64), polymorphonuclear leucocytes (66–68), brain (69), and inflammatory granuloma (70). The formation of rabbit aorta contracting and platelet aggregating factors from 8,11,14-eicosatrienoic, arachidonic, and 5,8,11,14,17-eicosapentaenoic acids has been investigated (71). The results indicated that although PGH_1, PGH_2, and PGH_3 were formed by sheep vesicular glands, only PGH_2 and PGH_3 were converted to thromboxanes by platelet microsomes. Furthermore, 5,8,11,14,17-eicosapentaenoic acid, and the endoperoxides and thromboxanes derived from it did not induce platelet aggregation. Conversion of 8,11,14-eicosatrienoic acid to thromboxane B_1 in platelets has however been reported (72). A factor which releases arachidonic acid in lung, thus generating RCS, was found in lung perfusates during anaphylaxis (73). The RCS-releasing factor appeared to be a peptide containing less than 10 amino acid residues.

Prostaglandin I_2

Arachidonic acid relaxed coronary arteries whereas PGE_2 and $PGF_{2\alpha}$ caused contraction (74). The relaxing effect was prevented by indomethacin (74) and mimicked by PGH_2 (75). A factor which both relaxed blood vessels and inhibited platelet aggregation was formed from PGG_2 and PGH_2 by aortic microsomes (76, 77). The factor, originally referred to as PGX and now named PGI_2, was more potent than PGE_1 or PGD_2 in preventing platelet aggregation and was unstable in aqueous solution ($t_{1/2}$ <10 min at 37°C) (77). It is thus distinguished from the primary prostaglandins. The enzymic formation of PGI_2 was inhibited by 15-hydroperoxy-5,8,11,13-eicosatetraenoic acid (55). It has been suggested that PGI_2 formation by arterial walls may explain their ability to resist platelet adhesion (76) and that a balance between PGI_2 formation by this tissue and prostaglandin endoperoxide and thromboxane formation by platelets may be crucial to the control of thrombus formation in blood vessels (77). It was also shown that arteries could utilize prostaglandin endoperoxides formed by platelets for PGI_2 biosynthesis (78). The isolation and structural elucidation of PGI_2 were recently reported (79). The product, formed from PGI_2 by decomposition in aqueous solution, was identified as 6-keto-$PGF_{1\alpha}$ by comparison with synthetic material. Previously, conversion of arachidonic acid (80) and PGH_2 (81) to 6-keto-$PGF_{1\alpha}$ by forestomach homogenates had been reported. The formation of 6-keto $PGF_{1\alpha}$ from PGI_2 suggested that PGI_2 might be 6,9α-epoxy-11α,15(S)-dihydroxyprosta-5,13-dienoic acid a proposed intermediate in the formation of 6-keto-$PGF_{1\alpha}$ (83). By comparisons of biological and chemical properties of PGI_2 and synthetic 6,9α-epoxy-11 α,15(S)dihydroxyprosta-5,13-dienoic acid this was confirmed. Two other groups of investigators have independently arrived at the same results (84, 85). It has been shown that PGI_2 mediated the relaxing effect of arachidonic acid and PGH_2 on coronary artery (86). PGI_2 raised cAMP levels in platelets (87, 88) and was 10- and 30-fold more potent in this respect than PGD_2 and PGE_1, respectively (88). 6-Keto-$PGF_{1\alpha}$ was at least 1000-fold less potent than PGI_2 in raising cAMP levels (88). The formation of PGI_2 and/or 6-keto $PGF_{1\alpha}$ has been demonstrated in aorta (76, 77), mesenteric arteries (77), coronary arteries (86), ductus arteriosus (89), fetal arteries (90), heart (91, 92), stomach (77, 80, 81, 83, 92), lung (92), uterus (93), seminal vesicles (55, 94), and inflammatory granuloma (95). It has also been shown that PGI_2 is formed by 3T3 fibroblasts and that it mediates a stimulatory effect by arachidonic acid or PGG_2 on cAMP levels (96).

Prostaglandin Biosynthesis in Cell Culture

Many animal cells in culture release prostaglandins into the growth medium (97–109). The cells can be stimulated to produce high PG levels by serum (98, 102, 107), thrombin (102), bradykinin (102, 108, 109), angiotensin II (108, 109), arginine vasopressin (108), cAMP (48), histamine (109), serotonin (109), and mechanical manipulation (102). All of these agents appear to act by increasing the release of precursor fatty acids from cellular lipids (103, 108, 109). Serum probably also stimulates prostaglandin production by providing free arachidonic acid (98, 102, 107). The stimulation of PG synthesis by serum, thrombin, and bradykinin was prevented by actinomycin D and cycloheximide (104) and by uncouplers of oxidative phosphorylation (105). Although many cells produce low PG concentrations in the absence of stimulatory agents, certain virus-transformed cells have a much higher rate of basal PG biosynthesis. This was first shown using baby hamster kidney fibroblasts transformed by polyoma virus (97), and subsequently by using 3T3 mouse embryo fibroblasts transformed by the same virus (98) or SV40 (100). These cells produced higher PGE_2 and $PGF_{2\alpha}$ levels than the parent nontransformed cells. Elevated levels of PGE_1 and $PGF_{1\alpha}$ (99), PGA + PGB (100), and 13,14-dihydro-15-keto-PGF_α (106) have also been detected in polyoma (99) and SV40-transformed 3T3 cells (100, 106). A different clone of SV40-transformed 3T3 cells did not produce elevated PGE_2 and $PGF_{2\alpha}$ levels (98).

Miscellaneous

Prostaglandin endoperoxide and thromboxane formation has been demonstrated using isolated human umbilical artery (110) and may have a role in the closure of the vessel at birth. PGG_2 and PGH_2 were mainly converted to PGE_2 and $PGF_{2\alpha}$ by kidney (111) and brain (112). The conversion to PGE_2 was prevented by previous boiling of the homogenates, whereas the conversion to $PGF_{2\alpha}$ was not. Ratios of PGE_2 to $PGF_{2\alpha}$ of 0.5–1.9 were obtained with different regions of kidney and brain. Human seminal vesicles, but not human prostate or testis, converted 8,11,14-eicosatrienoic acid to PGE_1 (113). In contrast to mammalian prostaglandin biosynthesis, the formation of PGA_2 by the coral *plexaura homomalla* involved neither PGE_2 nor prostaglandin endoperoxides as intermediates (114). Human platelets converted 8,11,14-eicosatrienoic acid to four hydroxy acids (72). Two were analogues of HETE and HHT (see above under thromboxanes) whereas the other compounds were C_{20} acids with hydroxy groups at the 8,11,12- and 8,9,12-positions, and double bonds at the 9,14- and 10,14-positions, respectively. Rabbit polymorphonuclear leucocytes efficiently converted arachidonic acid to 5L-hydroxy-6,8,11,14-eicosatetraenoic acid, and 8,11,14-eicosatrienoic acid to 8L-hydroxy-9,11,14-eicosatrienoic acid (115).

ASSAY METHODS

Multiple Ion Analysis

A survey of the use of multiple ion analysis as a tool for quantitative determination of PGE_2 and $PGF_{2\alpha}$ and their metabolites in plasma and urine appeared in the previous review (1). The detection and isolation of the endoperoxides (18, 116, 117) as well as the isolation of thromboxane B_2, 12-L-hydroxy-5,8,10-heptadecatrienoic acid and 12-L-hydroxy-5,8,10,14-eicosatetraenoic acid following incubation of arachidonic acid with, for example, blood platelets and lung (58, 63, 118) made it necessary to develop further methods for quantitative determination by multiple ion analysis. In addition, methods for determination of arachidonic acid (119), PGA_2 (120), PGD_2 (89, 92, 121), PGE_1 (122a, 122b, 99), 13,14-dihydro-PGE_2 (122, 123), and 6-keto-$PGF_{1\alpha}$ (89, 92) have been described.

The method used for determination of PG endoperoxide [sum of PGG_2 and PGH_2 (62)] was based on the fact that the endoperoxides spontaneously isomerize in aqueous medium into a mixture of the corresponding PGE and PGD compounds ($t_{1/2}$ at 37°C, 4–5 min) and the fact that they are reducible into $PGH_{2\alpha}$ by, for example, $SnCl_2$, an agent that does not reduce the keto-prostaglandins. In practice an aqueous medium containing PGG_2 and PGH_2 was immediately split into two equal parts. One part was treated with 15 volumes of ethanol and the other with 15 volumes of ethanol containing 0.5% $SnCl_2$. The amounts of $PGF_{2\alpha}$ in these samples were determined (124) and the difference was taken as a measure of the amount of PG endoperoxides initially present. By this method release of PGG_2 and/or PGH_2 from human platelets during aggregation (59, 62, 117) and from guinea pig lung during infusion of arachidonic acid (62) was demonstrated.

Quantitative determination of thromboxane B_2 (TXB_2) was carried out using octadeuterated TXB_2 as internal standard (59). The intensity of the ions at m/e 256 (protium form) and m/e 260 (deuterium form; TMSiO-$C^2H=C^2H-CH_2-C^2H=C^2H-(CH_2)_3-COOCH_3^+$.) present in the mass spectra of the trimethylsilyl ether derivative of the methyl esters were monitored against time. This method demonstrated an extensive release of TXB_2 from human platelets treated with thrombin, collagen, and other aggregating agents (59, 125). Whole homogenates of guinea pig lung and spleen incubated for 30 min were found to contain large amounts of TXB_2 (7.1 and 3.0 $\mu g/g$ of tissue, respectively) (64). Large amounts of TXB_2 were released from intact perfused guinea pig lung following administration of arachidonic acid (63) and upon anaphylaxis (126). Furthermore release of TXB_2 from isolated human umbilical artery (110) and the presence of TXB_2 in human burn blister fluid (127) were demonstrated.

12L-Hydroxy-5,8,10,14-eicosatetraenoic acid (HETE) was determined using the 5,6,8,9,11,12,14,15-octadeuterated derivative as internal standard (59). Human platelets were found to release HETE when treated with a number of aggregating agents (59, 125). Whole homogenates of guinea pig lung and spleen incubated for 30 min contained very large amounts of HETE (15.1 and 29.5 $\mu g/g$ of tissue, respectively) (64). A release of HETE from intact perfused guinea pig lung following administration of arachidonic acid was also demonstrated (63). HETE was not detectable in uninvolved epidermis of patients suffering from psoriasis; however, involved epidermis contained 4.1 $\mu g/g$ of tissue (119).

Arachidonic acid was determined by multiple ion analysis using 5,6,-8,9,11,12,14,15-$^2H_8$ arachidonic acid as internal standard (119). Psoriatic epidermis was found to contain 36.3 $\mu g/g$ of nonesterified arachidonic acid as compared to 1.4 $\mu g/g$ in uninvolved epidermis (119). A method for quantitative determination of PGA_2 was recently developed (120). The internal standard used was [3,3,4,4-$^2H_4$]PGA_2, and the trimethylsilyl ether derivative of the methyl ester was analyzed. PGA_2 could not be detected in human blood plasma and thus was less than 10 pg/ml which indicates that the PGA_2-like material detected in plasma by radioimmunological techniques (128) was different from PGA_2.

Radioimmunoassay

During the past few years, the number of developed PG radioimmunoassays (RIAs) and published studies employing such assays has expanded rapidly. While the earlier radioimmunoassays mainly were aimed at primary prostaglandins, more recent methods have been developed for a large number of other prostaglandins—PG metabolites, PG analogues, thromboxanes, and thromboxane derivatives. Furthermore, the measurements are not limited to blood plasma/serum, or urine, which was common earlier, but have been extended to a vast number of biological fluids, to tissue analyses, and to analyses of PG production by isolated cells.

The majority of published studies are however still concerned with PG levels in blood plasma or serum. A few years ago it was demonstrated that measured blood levels of primary prostaglandins do not reliably reflect the true endogenous levels of these compounds (see 1 and references therein). When studying the levels of circulating prostaglandins, the major plasma metabolites should be focused on instead, that is, the 15-keto-13,14-dihydro compounds. This view has now been widely accepted, and a considerable number of radioimmunoassays for 15-keto-13,14-dihydro-PGF$_{2\alpha}$ and 15-keto-13,14-dihydro-PGE$_2$ have been published (129–142). Due to the serious sources of error that appear when monitoring primary prostaglandins in the peripheral circulation, this kind of study is now fairly seldom performed. Most scientists prefer other approaches, examples of which can be taken from the reproductive physiology. The role of PGF$_{2\alpha}$ in luteolysis has, for instance, been studied by (a) measurement of the compound as such in the venous drainage close to the uterus and/or the ovary (143–155); or (b) as a metabolite, either 15-keto-13,14-dihydro-PGF$_{2\alpha}$, in the peripheral circulation (156–158) or 5α,7α-dihydroxy-11-ketotetranorprostanoic acid in urine (159), or by (c) measuring the capacity of the uterus to synthesize prostaglandins (22). Others have measured PGF$_{2\alpha}$ in uterine fluid or jet wash specimens (160, 161).

Due to the possibility of rapid fluctuations in the PG production, neither blood nor tissue analyses are entirely reliable if samples are not collected frequently enough (see e.g. 158). When dealing with small animals or with human subjects, it may not be possible to obtain a sufficiently large number of samples. In such studies, analysis of the daily excretion of a urinary metabolite may provide a better solution to the problem. Several laboratories have developed radioimmunoassays for urinary PGF metabolites and applied their methods to various kinds of studies (159, 162–164). When the basal daily excretion of the major PGF metabolite in the human was measured, all these assays gave results comparable to those earlier obtained by a gas chromatographic-mass spectrometric(GC-MS) method (165). The radioimmunoassays were employed both in the human and other species for the determination of PG production/excretion under the following conditions: after treatment with PG synthetase inhibitors (159); after injection of furosemide (166); after administration of PGF$_{2\alpha}$ (159, 163, 164, 167); during estrous cycles, pregnancy, parturition, and estrogen treatment (159).

In general, both the sensitivity and the specificity of PG radioimmunoassays have increased during recent years. Recently produced antibodies often have higher avidities and specificities, which may to some extent be explained by the use of more suitable carrier molecules for preparation of the conjugagates (168–170), better coupling methods (171), the use of other species for antibody production (172, 173), or other reasons. The use of labeled ligands of better quality also contributes considerably to the increased sensitivity. An example is the use of ^{125}I labeled PG derivatives, which can be prepared with very high specific activities (138, 162, 164, 174).

On the other hand, the accuracy of many radioimmunoassays unfortunately seems to be very low when they are applied to biologic material. Sometimes enormous discrepancies are found when RIA results are compared with those obtained by the more specific GC-MS methods. As an example, a number of radioimmunoassays developed for the measurement of PGA compounds in plasma have resulted in "normal" levels around 1000–2000 pg/ml in the human (128, 175–180); in contrast, mass spectrometric methods for PGA_2 demonstrated that the compound can hardly be detected at all in human plasma (120, 181). [Limit of detection in the most sensitive assay is 5 pg/ml (120); reported basal levels of PGA_2 in healthy adults < 10 pg/ml.] That these enormous discrepancies are caused by the presence of 1000–2000pg PGA_1 per ml is not likely. Besides, the only RIA that selectively measured PGA_1 gave the lowest levels, only around 20 pg/ml (182). Neither can the extremely high PGA levels obtained by RIA be explained by cross reactions with other prostaglandins, as unphysiological levels of these generally will be required to give such results. When interpreting RIA results, it must be kept in mind that the antigen-antibody binding can be inhibited by a multitude of factors that are completely unrelated to the compound intended to be measured (183); if this is not considered, such inhibition will erroneously be interpreted as being caused by high amounts of prostaglandin in the sample.

It was recently demonstrated that the prostaglandin endoperoxides are converted to only a minor extent into the stable prostaglandins PGE_2 and $PGF_{2\alpha}$ in certain tissues and cells, for example, platelets and lung tissue (58, 63). Instead, the major part of the endoperoxides is transformed into the highly unstable TXA_2, which is subsequently converted into the stable TXB_2 (60). Thus, in order to get a reliable picture of biochemical events in these cells, it is necessary to monitor the thromboxanes instead of the prostaglandins.

Several different types of assays for TXB_2 have recently been developed, including an RIA (184). This method allows the determination of TXB_2 in a large number of minute samples and is thus suitable for detailed kinetic studies of, for example, the events during platelet aggregation. It was demonstrated that large amounts of TXB_2 were formed during aggregation induced by addition of collagen or arachidonic acid (184, 185). Recently, a similar radioimmunoassay was published which was used to study events during cardiac anaphylaxis (186).

For many kinds of studies, however, it is not sufficient to measure only TXB_2, but also desirable to quantitate the unstable precursor, TXA_2. A simple way to solve this problem is to convert TXA_2 into a stable derivative, mono-O-methyl TXB_2, with excess amounts of methanol (60) and to monitor this derivative. This solution was employed in a recent publication describing an RIA for mono-O-methyl TXB_2 (185), and the data obtained by this RIA reliably reflected the TXA_2 formation, as shown by the fact that the compound could be detected only relatively early during platelet aggregation (cf 60, 187). Furthermore, the half-life of TXA_2 in aqueous medium at pH 7.4 was found to be 32.5 ± 2.5 sec at 37°C; this agrees well with earlier published data (60). The RIA has also been employed for the study of the properties of TXA_2 and it was demonstrated, for example, that the compound is stabilized considerably by the presence of albumin (188).

Thromboxane measurements have so far been carried out only in vitro, either in platelet experiments or in perfusion studies of various organs. It would be highly desirable to be able to measure thromboxanes also in vivo, but this seems to be very difficult. Studies on the metabolic fate of TXB_2 (189, 190) indicate that it is a poor substrate for 15-hydroxyprostanoate dehydrogenase. TXB_2 is in fact the dominating compound in the circulation after injection (189). As a great artifactual formation of the compound can be expected during sampling, measured "peripheral plasma levels" of TXB_2 will no doubt be very high and will certainly not reflect the true endogenous circulating amounts of the compound. The problem is similar to that pertaining to PG measurements in the circulation [see above and (1)]. However, in the case of PG assays, the problem has been overcome by monitoring the 15-keto-13,14-dihydro metabolites instead, which are not formed artifactually during blood sampling and reliably reflect endogenous levels. In the case of assay of thromboxanes in vivo, the problem of artifactual formation is even greater (cf 59, 184). No major circulating metabolite of the compound seems to exist, as is the case with the prostaglandins. A possible approach is the measurement of the major metabolite, dinor-TXB_2 in urine (189, 190).

PROSTAGLANDINS AND CYCLIC NUCLEOTIDES

In view of the large number of publications since the previous review (1) it is not possible to provide a comprehensive survey of the literature. The interested reader may consult several recent reviews dealing solely with this topic (191–193). Prostaglandins and cyclic nucleotides in platelets are discussed in the section on platelets.

Prostaglandins and cAMP Biosynthesis

Specific binding of PGE to a variety of cells or tissues (see section on receptors) has been correlated with the accumulation of adenosine 3',5'-monophosphate(cAMP) (194), or with the activation of adenylate cyclase (195–200) both of which are independent of β-receptor participation (201–206). An additional correlation has been documented in rat myometrium between treatment with PGE, and increased saturation of an intracellular cAMP receptor protein (207). The inability of PGF (206, 208) to significantly affect adenylate cyclase or to increase cAMP levels, except at high concentrations ($\geqslant 10^{-5}$ M), has been repeatedly documented (199, 202, 203, 209–213). The activation of adenylate cyclase by PGE varies depending on the cell in question. In rat liver, the adenylate cyclase of a sinusoidal endothelial cell responded to PGE (214, 215), whereas the cyclase of the Kupffer cells, although responding to fluoride ion, GTP, and catecholamines, failed to respond to prostaglandins (214). Different isolation techniques may account for the conflicting data regarding the presence of a PGE sensitive adenylate cyclase in the hepatic parenchym cells (214, 215). In human adipose tissue, 5 μM PGE_1 caused a 15-fold increase in the level of cAMP of adipocytes as compared to a 95-fold increase in fibroblasts from the same tissue (216). Lymphocyte subpopulations from mice demonstrated variable responses to 10^{-6} M PGE_1, which correlated with their anatomical origin; thymic lymphocytes were most responsive (36-fold increase in cAMP level), followed by splenic lymphoid cells (6-fold), lymph node cells (3-fold), and peripheral blood lymphocytes (1.2-fold) (217).

The study of different populations of cells within a given organ has also helped to explain some seemingly paradoxical observations. In cell-free preparations from guinea pig gastric mucosa, PGE and histamine had opposing effects on HCl secretion, but both stimulated adenylate cyclase. These observations were reconciled by observations that membrane preparations from the gastric antrum had a PGE responsive, histamine unresponsive, adenylate cyclase, whereas membranes from the gastric fundus had a PGE and histamine responsive adenylate cyclase. Thus, the opposing effects of PGE and histamine may result from stimulation of adenylate cyclases in different cells that respond with antagonistic functions (218).

PGE_1 and PGE_2 activated adenylate cyclase in plasma membranes from normal liver but not in that from hepatomas (199). Similarly, leukemic lymphocytes were less responsive to PG stimulation than normal lymphocytes (206). Normal rat kidney (NRK) cells, but not NRK cells transformed by various murine sarcoma viruses, had a PGE responsive adenylate cyclase (198, 219). In contrast, mouse neuroblastomas (221, 222), virus-induced maloney sarcomas (220), and NRK cells transformed by the Schmidt-Ruppin strain of avian sarcoma virus did respond to PGE_1 stimulation (198). PGE_1 and PGE_2 were equally effective in raising levels of cAMP in BALB/c 3T3 mouse fibroblasts transformed by polyoma virus, and it was shown that endogenously produced PGE_2 markedly raised the cellular levels of cAMP (223). The endogenous PGE_1 levels were tenfold lower than those of PGE_2 and consequently had less effect on cAMP levels (99). Regular 3T3 cells responded to PGE_1 at concentrations equivalent to those endogenously produced, but failed to respond to PGE_2.[1] It is noteworthy that age- (224) and cell cycle- (225) dependent alterations in the cAMP response to prostaglandins have been documented. The possible

[1]Goldyne, M. E., Lindgren, J. Å., Claesson, H.-E. Hammarström, S. Data to be published.

effects on cell growth of these changes in PGE-cAMP responsiveness have been covered by other reviews (226–228).

PGE stimulation of adenylate cyclase in purified cell membranes from platelets (229), liver (199, 203, 209, 211), spleen (212), thyroid (231, 232), pancreatic islets (230), and fibroblasts (219) is enhanced by GTP. PGE_1 combined with GTP produced an increase in the V_{max} of the adenylate cyclase in pancreatic islets (230). Neither PGE_1 nor GTP affected phosphodiesterase. In rat liver (199), GTP increased receptor affinity for PGE_1 without altering receptor concentration. In addition, GTP plus PGE_1 decreased the K_m and increased the V_{max} of adenylate cyclase.

The recently discovered PGI_2 (see section on biosynthesis) increased the levels of cAMP in platelets (87, 88, 233). In Balb/c 3T3 mouse fibroblasts (96), arachidonic acid elevated the cellular levels of cAMP. Although the effect was blocked by indomethacin it could not be explained by conversion of arachidonic acid to PGD_2, PGE_2, or $PGF_{2\alpha}$. Arachidonic acid was converted by the cells to PGE_2 and PGI_2 and the latter product appeared to mediate the effect on cAMP. A similar situation may exist in C3H mouse mammary tumors (234) and human synovial cells (47) which responded to arachidonic acid in a manner similar to that of 3T3 cells.

cAMP and Prostaglandin Biosynthesis

Stimulation by cAMP of PG biosynthesis has been demonstrated in cultured mammalian cells (48, 235), Graafian follicles (236), thyroid cells (237, 238), adrenal cortex (239), and adipocytes (249). In BALB/c 3T3 mouse fibroblasts (48), cAMP caused the release of arachidonic acid and 8,11,14-eicosatrienoic acid, but no appreciable release of stearic, oleic, linoleic, or linolenic acids. cAMP activated a triglyceride lipase which liberated arachidonic acid from pig thyroid (238). In contrast, cAMP inhibited PG formation in platelets (37, 38) (see also under inhibitors in the biosynthesis section).

The ability of certain PGEs to activate adenylate cyclase and that of cAMP, in turn, to stimulate PGE biosynthesis suggest a positive feedback system. A stop signal in this system might be provided by the phenomenon of agonist specific desensitization of adenylate cyclase which has been described in peritoneal macrophages (202), synoviocytes (241, 242), astrocytoma cells (205, 243, 244), epidermis (246), fibroblasts (245), and Graafian follicles (247). In all cases, the cAMP response to PGE decreased in proportion to time of PG exposure. In two studies (205, 244) a decreased cAMP response to epinephrine was also observed. Possible mechanism(s) of the desensitization which have been suggested (243, 247, 248) include phosphodiesterase activation, synthesis of inhibitory proteins, alteration or loss of agonist binding sites, and altered interaction of the agonist binding site with adenylate cyclase.

cAMP-Independent Effects of Prostaglandins

Although many PGE effects appear to be mediated by cAMP, some PGE induced alterations in cell functions are not accompanied by alterations in cAMP levels. Examples of this include prostaglandin E_2, $F_{1\alpha}$, and $F_{2\alpha}$ inhibition of amino acid uptake into human diploid WI 38 fibroblasts (248), and PGE_2 stimulated hematopoietic stem cell proliferation (249).

Prostaglandins and cGMP

Since the original observation that $PGF_{2\alpha}$ raised the level of guanosine 3':5'-monophosphate (cGMP) in the rat uterus (250), additional studies have documented the ability of PGFs to elevate cGMP in liver (251), venous smooth muscle (252, 253), 3T3 fibroblasts (254), and brain (255). PGEs appear to stimulate cGMP accumulation only when used at high concentrations (7–10 μM) (251, 254, 256). The possibility that PG endoperoxides may also stimulate the production of cGMP has been reviewed (257) (see also section on platelets) and recent evidence indicates that PGG_2 activates the soluble guanylate cyclase from platelets (259).

Prostaglandins and the "Yin-Yang" Hypothesis

The "Yin-Yang" hypothesis (260, 261) proposes that cAMP and cGMP are the opposing arms of a bidirectional intracellular control system. It has found support in studies on: the contraction of vascular smooth muscle (252, 253) and myocardium (262); the release of lysozomal enzymes (263); leukocyte chemotaxis (264); growth regulation in cell culture (227, 228, 265); protein phosphorylation (266); and the regulation of certain functions in various organs (267, 268). PGEs and PGFs also have opposing effects on certain tissue and cell responses and these have been correlated with their respective capacities to stimulate the production of cAMP and cGMP. Thus PGE_1/cAMP relax and $PGF_{2\alpha}$/cGMP contract venous smooth muscle (252, 253) whereas PGE/cAMP retard (264, 269) and $PGF_{2\alpha}$/cGMP enhance leukocyte chemotaxis (264, 270). The abilities of PG endoperoxides (258), PGI_2 (87, 88, 96, 233), and thromboxanes to affect cyclic nucleotide levels creates a potentially more complex regulatory system. In addition, there are studies in which PGE_1 and PGE_2 have antagonistic effects (271, 272) that present a problem in the context of the "Yin-Yang" hypothesis.

Prostaglandins and Hormone-Cyclic Nucleotide Interactions

The effects of luteinizing hormone (LH) on cAMP in isolated Graafian follicles from estrous rabbits (236) and intact rat ovaries (273, 274) suggest that, even though LH can raise PG levels, PG biosynthesis is not required for the LH-induced elevation of cAMP. Rather, the elevation of cAMP in response to LH appears to trigger PG synthesis (236). Likewise, stimulation of cAMP production by LH-releasing hormone (LHRH) in rat hemipituitaries and the subsequent release of LH appear to be independent of endogenous PG formation (275).

In contrast, prolactin induced elevation of cGMP in mouse mammary gland explants may be mediated by $PGF_{2\alpha}$ (276) whereas the ability of bradykinin to elevate cAMP in human synovial fibroblasts may be PGE mediated (47).

A recent study and a review on the thyroid gland (277, 278) underline the complexity of potential interactions between prostaglandins, hormones, and cyclic nucleotides. Thyroid stimulating hormone (TSH) appears to directly activate phospholipase A_2 and thereby induce PG synthesis (277). In addition TSH raises cAMP which activates a lipase that further stimulates PG synthesis (277).

RECEPTORS

Specific binding of prostaglandins E, F, and A has been demonstrated in several tissues. Although this binding has often not been correlated with a biochemical or physiological response, the term PG receptors has been used to describe saturable binding of high affinity, which does not metabolically alter.

Prostaglandin E Receptors

The first report on a PG receptor, as such, came from experiments with rat adipocytes (194, 279). This receptor was specific for PGE_1 (K_D 3–8 nM), and various other prostaglandins had lower affinities, in proportion to their relative potencies, to stimulate ovarian adenylate cyclase (194, 279). PGE specific binding has subsequently been demonstrated in stomach and thyroid gland (see 1), adrenal gland (195), corpus luteum (280, 281), fibroblasts (196, 198, 200, 282), liver (1, 199), lymphoma cells (200), neuroblastoma cells (196), platelets (283), thymocytes (1, 284), and uterus (285–288). K_D values for PGE binding in some of these tissues were one to two orders of magnitude lower than PGE concentrations required for half maximal stimulation of adenylate cyclase in the same tissues. [Binding and enzyme assays were performed in different laboratories (289)]. This emphasizes the importance of measuring PG effects and PG binding in the same preparation. A few such studies have been performed. Good agreement between binding data and adenylate cyclase activation was thus reported for cultured L cells, neuroblastoma, and lymphoma cells (196, 200) and for adrenal glands (195). It is interesting to note that several cell lines that are unresponsive to PG stimulation lack detectable PGE binding (196, 198,

200, 282). These studies provide evidence that receptors are involved in PG stimulation of adenylate cyclase. Heterogeneity of PGE receptors with respect to ligand affinity has been reported by several authors (199, 280, 288). High affinity sites (K_D 1.3–17.6 nM) and low affinity sites (K_D 10–136 nM) were detected. The biological significance of the heterogeneity is unknown. It has been suggested that the low affinity sites may be more important than the high affinity sites for adenylate cyclase activation (280). Binding specificity studies have shown that different tissues have different types of PGE receptors. Thus corpus luteum (280) and thymocyte (284) receptors bound PGE_1 and PGE_2 with equal affinity. Myometrium receptors from hamster (287) and monkey (288) had higher affinity for PGE_2 than for PGE_1. Finally PGE receptors from L cells, neuroblastoma cells and lymphoma cells (196, 200) had higher affinity for PGE_1 than for PGE_2. In all tissues prostaglandins A, B, D, and F and prostaglandin E metabolites had considerably lower affinities than prostaglandin E. CA^{2+} and Mg^{2+} ions as well as testosterone and dihydrotestosterone increased PGE_1 binding in corpus luteum membranes (280). Progesterone increased and estradiol decreased PGE binding in uterus (288), and GTP, dGTP, and GMP P(N)P increased PGE binding in liver (199).

Prostaglandin F Receptors

A physiological role of $PGF_{2\alpha}$ as a luteolytic hormone in nonprimate mammals has been established (for review see 290). In these animals $PGF_{2\alpha}$ is released by the uterus, and transferred from the utero-ovarian vein to the ovarian artery by a countercurrent mechanism. This causes inhibition of progesterone secretion and regression of the corpus luteum. Specific binding of $PGF_{2\alpha}$ was first demonstrated in corpora lutea from sheep (291). In 1974, binding with similar characteristics was reported for bovine corpora lutea (292). Under equilibrium conditions (at 23°C, K_D of 50–100 nM, and receptor site concentrations of 1.3–1.6 pmol/mg) microsomal protein was obtained by Scatchard plot analyses. Several structural modifications of $PGF_{2\alpha}$, for example, reduction of the carboxyl to a hydroxymethyl group, hydrogenation of the Δ^5 cis double bond, or oxidation of either of the hydroxyl groups at C–9 or C–15, increased the dissociation constant 40–180-fold. In contrast, hydrogenation of the Δ^{13} trans double bond or oxidation of the hydroxyl group at C–11, increased the K_D less (4–9-fold). Decreasing the length of the carboxyl side chain by one methylene group increased the K_D 52-fold, whereas adding two methylene groups to the methyl side chain approximately doubled the K_D. The affinities of several prostaglandins and PG analogs for the receptor in bovine corpus luteum agreed reasonably well with the luteolytic potencies of these compounds in vivo in hamsters (293). This indicated that the receptor is involved in PG-induced luteolysis. A biochemical response of $PGF_{2\alpha}$ related to luteolysis has not been demonstrated in cell free preparations from corpora lutea. Thus, it has not been possible to correlate binding to the receptor with PG effects. Subcellular distribution studies showed that the $PGF_{2\alpha}$ receptor in bovine corpus luteum was localized in plasma membranes (294). Preformed $[^3H]PGF_{2\alpha}$-receptor complex has been solubilized by sodium deoxycholate or Triton X-100 (295, 296). Analyses on Sepharose 6B columns and by density gradient centrifugations in media containing H_2O and D_2O permitted a molecular weight determination of the receptor from its Stoke's radius (63 Å) and sedimentation coefficient (4.8S) with correction for bound detergent [26% w/w, 107,000 daltons (296)]. Some of these investigations on the $PFG_{2\alpha}$ receptor in corpora lutea were recently summarized (297). Several reports from other laboratories have confirmed the occurrence of $PGF_{2\alpha}$ receptors in bovine (298–307) and equine (308) corpora lutea. High (K_D 1.6–5.1 nM) and "low affinity" binding sites (K_D 18–24 nM) were found (198, 303). The former sites required calcium ions in order to be detected (300) and the two classes of binding sites were not caused by negative cooperativity (304). Various proteolytic and lipolytic enzymes as well as amino acid–modifying reagents inhibited $PGF_{2\alpha}$ binding to corpus luteum membranes (301–303). At physiological concentrations progesterone also decreased $PGF_{2\alpha}$ binding (305). A comparison between binding affinities and luteolytic potencies of prostaglandin analogues in heifers showed some correlation (307).

PLATELET FUNCTIONS

Formation of PGE_2 (310, 311) and PGD_2 (21) has been shown during platelet aggregation in vitro induced by thrombin, collagen, and epinephrine. A number of workers have also demonstrated that incubation of intact platelets or platelet homogenates with arachidonic acid causes formation of PGE_2 and $PGF_{2\alpha}$ (311). Arachidonic acid also causes a rapid and irreversible aggregation of human platelets (312, 313).

Intravenous infusion of arachidonic acid causes sudden death in rabbits with accumulation of platelets in heart blood and pulmonary circulation (314). Furthermore, injection of arachidonic acid into the carotid artery of rats produces a stroke syndrome with obstruction of the cerebral microcirculation by platelet aggregates (315). Arachidonic acid was also found to be considerably more potent in inducing thrombocytopenia in guinea pigs than lauric acid, which might indicate that thromboembolism induced by arachidonic acid differs from that induced by intravenous injection of other fatty acids (316).

Arachidonic acid does not induce platelet aggregation or release of ADP or serotonin in the presence of PG synthesis inhibitors (312), which indicates that PG formation is essential for platelet response that is induced by arachidonic acid (312). However, the stable prostaglandins formed from arachidonic acid do not explain aggregation induced by arachidonic acid, since none of these compounds can induce the release response per se, although PGE_2 in small concentrations potentiates the release reaction induced by an active agent (317).

The formation of the two unstable endoperoxide intermediates PGG_2 and PGH_2 in PG biosynthesis during platelet aggregation (59, 117, 125) has been demonstrated. These compounds also induce rapid and irreversible aggregation in vitro (59). The PG endoperoxides correspond to the labile aggregation stimulating substance (LASS) (318) formed during platelet aggregation induced by arachidonic acid. They also correspond to the potential inflammatory mediators found under the same conditions (313), and to the proposed intermediate in PG biosynthesis during platelet release (31, 319). The compounds induce release of ADP and serotonin from platelets (125) without liberation of cytoplasmic enzymes (320). They also cause ultrastructural changes corresponding to other aggregating compounds (321, 322). The endoperoxide intermediates are rapidly transformed to the stable prostaglandins PGE_2 and $PGF_{2\alpha}$ in platelet suspensions (59) and to some extent to PGD_2 (323). The formation of PGD_2 from PGH_2 has been reported to be considerably increased in the presence of plasma proteins (20). In addition to the prostanoic structures, the endoperoxides form thromboxane B_2 (58). The finding of an intermediate with a half-life shorter than PGG_2 but more potent in inducing platelet aggregation and the release reaction led to the discovery of thromboxane A_2 (60–62), and formation of this intermediate has been demonstrated during aggregation and release by different agents (185).

A physiological role for PG endoperoxides and thromboxanes in platelets has been demonstrated (125) and different cases with prolonged bleeding time due to platelet cyclo-oxygenase deficiency have been studied (125, 325, 326).

A condition similar to cyclo-oxygenase deficiency in the platelets is obtained after treatment with PG synthesis inhibitors such as aspirin and indomethacin (327–330). Ultrastructural studies of platelets after incubation with aspirin or indomethacin demonstrate that the contractile wave, corresponding to the release reaction, is inhibited when platelets are stimulated with different aggregators (331, 332). However, the release reaction is not inhibited when induced by PG endoperoxides (125), thromboxane A_2 (61), or a stable $9\alpha,11\alpha$-azo analog of the endoperoxides (333).

The finding that aggregation and release induced by thrombin, was not completely inhibited by PG synthesis inhibitors (125, 334) might indicate the existence of a complementary mechanism of aggregation and release (334).

The question as to whether both endoperoxides and thromboxane A_2 are potent in inducing platelet response is still under debate, and evidence that thromboxane formation is not essential has been presented (71). However, it was found that the platelets from a patient with prolonged bleeding-time did not aggregate normally when incubated with PGG_2 but responded normally in the presence of platelets with thromboxane synthetase, and this might suggest the existence of a "thromboxane synthetase deficiency" (326). Different inhibitors of thromboxane synthetase have also been used to separate the biological effect of the endoperoxides and TXA_2. Thus L8027 has been reported to exert its inhibiting effect on platelet function by means of its effect on thromboxane synthetase (335, 336) (see also under inhibitors in the section on biosynthesis).

The interrelationship between prostaglandins and cyclic nucleotides in human platelets is another controversial subject. Since a number of inhibitors of platelet aggregation increase the level of cAMP while agents that produce or augment aggregation reduce the level of cAMP it was proposed that platelet aggregation is favored by a decrease in cAMP and inhibited by an increase in cAMP (271, 317). It was also proposed that the platelet aggregation and release reaction induced by the PG endoperoxides was due to a decrease of cAMP, and that the essential role of PGG_2 during aggregation induced by different agents was its ability to decrease the level of cAMP (271). Support for this hypothesis was obtained from studies that showed an inhibiting effect of PGG_2 upon adenylate cyclase activity (337).

PGE_1 increased cAMP and caused inhibition of platelet aggregation (338). PGE_2 in small concentrations increased platelet aggregation induced by other agents and also caused a decrease of cAMP, whereas higher concentration of PGE_2 had the same effect as PGE_1 (338, 339). PGD_2 is a very potent inhibitor of platelet aggregation (340, 341) and the compound is also a potent stimulator of human platelet adenylate cyclase (342). The 13,14-dehydro analogue of PGI_2 is still more potent as an antiaggregating and adenylate cyclase stimulating compound (85) which might indicate that PGI_2 physiologically exerts its antiaggregating effect (77, 87, 92, 343–345) by this mechanism.

Interestingly, morphine antagonizes the antiaggregating effect of PGE_1 (346) as does PGE_2, but neither morphine nor PGE_2 can induce platelet aggregation per se (217, 346). This finding indicates that a decrease of cAMP does not cause aggregation and release by itself but rather modifies the response to some other factor (320).

The observation that both thrombin (347) and ADP (320) cause decreased platelet cAMP without prostaglandin synthesis is contradictory to the hypothesis that prostaglandin biosynthesis is a prerequisite for the decrease of the cAMP level (271). Furthermore, it was found that the decrease of cAMP with PGG_2, might be explained by the rapid release of ADP, which in turn causes a secondary decrease of cAMP (320).

Incubation of a platelet suspension with dibutyryl-cAMP or an endogenous increase of cAMP causes a decrease in the formation of biologically active compounds (348) and PGG_2 (37) from arachidonic acid and collagen, which indicates that cAMP to some extent may modify prostaglandin formation. However, the finding that PGE_2 potentiation of platelet aggregation induced by the endoperoxides (LASS) is normal after aspirin ingestion indicates that the effect of cAMP, is, to some extent, independent of PG synthesis (349).

In addition to an effect which in some way involves PG formation cAMP also has a general antiaggregating and antiadhesive effect which does not involve prostaglandins. Thus ADP causes aggregation, but no release, in the presence of a PG synthesis inhibitor (125), whereas a marked elevation of the cAMP level also inhibits this aggregation (317, 350, 351). Recently it was demonstrated that cAMP inhibits the availability of arachidonate to PG synthetase in human platelet suspensions (38) which opens another possible level of interaction between cAMP and prostaglandins.

Cyclic GMP has been reported to have the opposite effect of cAMP in human platelets. Thus compounds known to cause aggregation of human platelets give an increased level of cGMP (352, 353, 354, 355, 356) while inhibitors of platelet aggregation, like aspirin, have the opposite effect (352). However, other inhibitors of platelet function, like cytochalasin B, do not interfere with cGMP formation (355). It has been reported from some workers that cGMP potentiates platelet response to other agents (357) while other workers did not find any significant effect of cGMP upon the release reaction or prostaglandin formation (37, 320).

Prostaglandin endoperoxides (333) and the stable $9\alpha,11\alpha$-azo-analogue (333) induce a rapid release of ^{14}C-serotonin and ADP. The question of whether the platelet aggregation induced by these compounds is due to release ADP or whether the endoperoxides and the azo-analogue cause platelet aggregation per se has not been clearly answered, although the observation that the aggregation, at least to some extent, is due to liberated ADP seems logical (320). The platelet aggregation induced by the endoperoxide G_2 is also markedly decreased in platelet disorders with decreased content of ADP [e.g. in Hermansky-Pudlak Syndrome (358)] or in disorders characterized by a decreased response to ADP [e.g. in Thrombasthenia Glanzmann (358)]. Evidence against the hypothesis that PG endoperoxide induced aggregation is due to ADP has been reported from studies with rabbit platelets that had been depleted of ADP by thrombin treatment (359).

No significant decrease of cyclo oxygenase of thromboxane synthase activity was found in platelets from patients with Bernard-Soulier Syndrome, Thrombasthenia Glanzmann, or Hermansky-Pudlak Syndrome

(358), whereas a decreased formation of prostaglandins was reported in a patient with Storage Pool Disease (360). However, it was not demonstrated whether this is a real decrease of the cyclo oxygenase activity or whether it reflects the decreased potentiation of ADP (320, 358).

The mechanism of action of PG endoperoxides and thromboxanes in the platelets is still unknown. They are formed in the dense tubular system (361) that is in close contact with the annular bundle of microtubuli. These are beneath the cell-membrane that is involved in the contractile process of the release mechanism (362).

Like PG endoperoxides the divalent ionophore A 23187 induces platelet aggregation and release which is not inhibited by aspirin or indomethacin (363, 364, 365, 366), whereas a marked increase of cAMP inhibits the response (364). There is evidence that the calcium required for the contractile process (367) is of intracellular origin (368, 369). The process, from surface stimulation by an aggregating agent to liberation of intracellular calcium, seems to involve an essential step involving formation of PG endoperoxides and thromboxanes which seem to act as intracellular messengers (370).

It is of interest to correlate circulatory defects with prostaglandin synthesis because of the role of prostaglandins and thromboxanes in normal platelet function (125), and the involvement of platelets in hemostasis and thrombosis (309).

Consequently, the findings that in thrombosis there is an increase of PG endoperoxides (371) and collagen-stimulated formation of PG endoperoxides (372) are both remarkable.

Also interesting is the release of PG-like substances in coronary venous blood flow following myocardial infarction (373).

A logical therapeutic approach to prevent thrombus formation that involves platelets is to inhibit the platelet activity in some way (374–376). Compounds that increase the cAMP level might be active in this aspect (317, 377, 378) as well as compounds that inhibit the PG formation (377, 379–381). Among these substances dihomo-γ-linoleic acid has drawn special attention since its mode of action is a combination of both mechanisms (382–385). Further inhibitors are described in the section on biosynthesis under *Inhibitors*. It was recently demonstrated that the liberation of arachidonic acid from the phospholipid pool might be an important modulator of platelet function which might be influenced by pharmacologically active compounds (125, 334, 358, 386–388).

However, after the finding that PG endoperoxides are transformed by the vessel wall to the potent antiaggregating PGI_2 (77, 87, 92, 343–345) the advantage of using prostaglandin synthesis inhibition to prevent thrombus formation is doubtful. A specific inhibitor of thromboxane synthetase might be a more efficient pharmacological tool for this purpose.

Literature Cited

1. Samuelsson, B., Granström, E., Gréen, K., Hamberg, M., Hammarström, S. 1975. *Ann Rev. Biochem.* 44:669
2. Miyamoto, T., Yamamoto, S., Hayaishi, O. 1974. *Proc. Natl. Acad. Sci. USA* 71:3645
3. Miyamoto, T., Ogino, N., Yamamoto, S., Hayaishi, O. 1976. *J. Biol. Chem.* 251:2629
4. Hemler, M., Lands, W. E. M., Smith, W. L. 1976. *J. Biol. Chem.* 251:5575
5. van der Ouderaa, F. J., Buytenhek, M., Nugteren, D. H., van Dorp, D.A. 1977. *Biochim. Biophys. Acta.* 487:315
6. Egan, R. W., Paxton, J., Kuehl, F. A. 1976. *J. Biol. Chem.* 251:7329
7. Cook, H. W., Lands, W. E. M. 1975. *Biochem. Biophys. Res. Commun.* 65:464
8. Marnett, L. J., Wlodawer, P., Samuelsson, B. 1974. *Biochem. Biophys. Res. Commun.* 60:1286
9. Rahimtula, A., O'Brien, P. J. 1976. *Biochem. Biophys. Res. Commun.* 70:893
10. Portoghese, P. S., Svanborg, K., Samuelsson, B. 1975. *Biochem. Biophys. Res. Commun.* 63:748
11. Marnett, L. J., Wlodawer, P., Samuelsson, B. 1975. *J. Biol. Chem.* 250:8510
12. Marnett, L. J., Wilcox, C. L. 1977. *Biochim. Biophys. Acta* 487:222
13. Ogino, N., Miyamoto, T., Yamamoto, S., Hayaishi, O. 1977. *J. Biol. Chem.* 252:890
14. Raz, A., Schwartzman, M., Kenig-Wakshal, R. 1976. *Eur. J. Biochem.* 70:89
15. Needleman, P., Moncada, S., Bunting, S., Vane, J. R., Hamberg, M., Samuelsson, B. 1976. *Nature* 261:558
16. Hammarström, S., Falardeau, P. 1977. *Proc. Natl. Acad. Sci. USA* 74:3691
17. Sun, F. F. 1977. *Biochem. Biophys. Res. Commun.* 74:1432
18. Nugteren, D. H., Hazelhof, E. 1973. *Biochim. Biophys. Acta* 326:448
19. Christ-Hazelhof, E., Nugteren, D. H., van Dorp, D. A. 1976. *Biochim. Biophys. Acta* 450:450
20. Hamberg, M., Fredholm, B. B. 1976. *Biochim. Biophys. Acta* 431:189
21. Oelz, O., Oelz, R., Knapp, H. R., Sweetman, B. J., Oates, J. A. 1977. *Prostaglandins* 13:225
22. Wlodawer, P., Kindahl, H., Hamberg, M. 1976. *Biochim. Biophys. Acta* 431:603
23. Polet, H., Levine, L. 1975. *Arch. Biochem. Biophys.* 168:96
24. Polet, H., Levine, L. 1975. *J. Biol. Chem.* 250:351

25. Roth, G. J., Majerus, P. W. 1975. *J. Clin. Invest.* 56:624
26. Roth, G. J., Stanford, N., Majerus, P. W. 1975. *Proc. Natl. Acad. Sci. USA* 72:3073
27. Rome, L. H., Lands, W. E. M., Roth, G. J., Majerus, P. W. 1976. *Prostaglandins* 11:23
28. Stanford, N., Roth, G. J., Shen, T. Y., Majerus, P. W. 1977. *Prostaglandins* 13:669
29. Rome, L. H., Lands, W. E. M. 1975. *Proc. Natl. Acad. Sci. USA* 72:4863
30. Pace-Asciak, C., Cole, S. 1975. *Experentia* 31:143
31. Goetz, J. M., Sprecher, H., Cornwell, D. G., Panganamala, R. V. 1976. *Prostaglandins* 12:187
32. Hammarström, S. 1977. *Biochim. Biophys. Acta* 487:517
33. Ohki, S., Ogino, N., Yamamoto, S., Hayaishi, O., Yamamoto, H., Miyake, H., Hayashi, M. 1977. *Proc. Natl. Acad. Sci. USA* 74:144
34. Robak, J., Dembinska-Kieć, A., Gryglewski, R. 1975. *Biochem. Pharmacol.* 24:2057
35. Hoi Do, U., Sprecher, H. 1976. *J. Lipid Res.* 17:424
36. Lindgren, J. Å., Claesson, H. E., Hammarström, S. 1977. *Prostaglandins* 13:1093
37. Malmsten, C., Granström, E., Samuelsson, B. 1976. *Biochem. Biophys. Res. Commun.* 68:569
38. Minkes, M., Stanford, N., Chi, M. M-Y., Roth, G. J., Raz, A., Needleman, P., Majerus, P. W. 1977. *J. Clin. Invest.* 59:449
39. Greaves, M. W., McDonald-Gibson, W. 1972. *Br. J. Pharmacol.* 46:172
40. Lewis, G. P., Piper, P. J. 1975. *Nature* 254:308
41. Chang, J., Lewis, G. P., Piper, P. J. 1977. *Br. J. Pharmacol.* 59:425
42. Gryglewski, R. J., Panczenko, B., Korbut, R., Grodzinska, L., Ocetkiewicz, A. 1975. *Prostaglandins* 10:343
43. Kantrowitz, F., Robinson, D. R., McGuire, M. B., Levine, L. 1975. *Nature* 258:737
44. Tashjian, A. H., Voelkel, E. F., McDonough, J., Levine, L. 1975. *Nature* 258:739
45. Floman, Y., Zor, U. 1976. *Prostaglandins* 12:403
46. Hong, S.-C. L., Levine, L. 1976. *Proc. Natl. Acad. Sci. USA* 73:1730
47. Newcombe, D. S., Fahey, J. V., Ishikawa, Y. 1977. *Prostaglandins* 13:235

48. Lindgren, J. Å., Claesson, H. E., Hammarström, S. 1978. *Adv. Prostaglandin Thromboxane Res.,* 3:167–174
49. Hammarström, S., Hamberg, M., Duell, E. A., Stawiski, M. A., Anderson, T. F., Voorhees, J. J. 1977. *Science* 197:994
50. Kulkarni, P. S., Eakins, K. E. 1976. *Prostaglandins* 12:465
51. Raz, A., Needleman, P. et al 1977. *Prostaglandins* 13:611
52. Needleman, P., Raz, A., Ferrendelli, J. A., Minkes, M. 1977. *Proc. Natl. Acad. Sci. USA* 74:1716
53. Moncada, S., Needleman, P., Bunting, S., Vane, J. R. 1976. *Prostaglandins* 12:323
54. Diczfalusy, U., Hammarström, S. 1977. *FEBS Lett.* 82:107
55. Moncada, S., Gryglewski, R. J., Bunting, S., Vane, J. R. 1976. *Prostaglandins* 12:715
56. Jones, R. L. 1974. *Biochem. J.* 139:381
57. Lands, W. E. M., Rome, L. H. 1976. In *Prostaglandins: Chemical and Biochemical Aspects,* ed. S. Karim, pp. 87–137
58. Hamberg, M., Samuelsson, B. 1974. *Proc. Natl. Acad. Sci. USA* 71:3400
59. Hamberg, M., Svensson, J., Samuelsson, B. 1974. *Proc. Natl. Acad. Sci. USA* 71:3824
60. Hamberg, M., Svensson, J., Samuelsson, B. 1975. *Proc. Natl. Acad. Sci. USA* 72:2994
61. Svensson, J., Hamberg, M., Samuelsson, B. 1976. *Acta Physiol. Scand.* 98:285
62. Svensson, J., Hamberg, M., Samuelsson, B. 1975. *Acta Physiol. Scand.* 94:222
63. Hamberg, M., Samuelsson, B. 1974. *Biochem. Biophys. Res. Commun.* 61:942
64. Hamberg, M. 1976. *Biochim. Biophys. Acta* 431:651
65. Dawson, W., Boot, J. R., Cockerill, A. F., Mallen, D. N. B., Osborne, D. J. 1976. *Nature* 262:699
66. Samuelsson, B. 1976. *Adv. in Prostaglandin Thromboxane Res.* 1:1–6
67. Goldstein, I. M., Malmsten, C. L., Kaplan, H. B., Kindahl, H., Samuelsson, B., Weissmann, G. 1977. *Clin. Res.* 21 (Abstr.)
68. Higgs, G. A., Bunting, S., Moncada, S., Vane, J. R. 1976. *Prostaglandins* 12:749
69. Wolfe, L. S., Rostworowski, K., Marion, J. 1976. *Biochem. Biophys. Res. Commun.* 70:907
70. Chang, W. C., Murota, S., Tsurufuji, S. 1977. *Prostaglandins* 13:17

71. Needleman, P., Minkes, M., Raz, A. 1976. *Science* 193:163
72. Falardeau, P., Hamberg, M., Samuelsson, B. 1976. *Biochim. Biophys. Acta* 441:193
73. Nijkamp, F. P., Flower, R. J., Moncada, S., Vane, J. R. 1976. *Nature* 263:479
74. Kulkarni, P. S., Roberts, R., Needleman, P., 1976. *Prostaglandins* 12:337
75. Needleman, P., Kulkarni, P. S., Raz, A. 1977. *Science* 195:409
76. Moncada, S., Gryglewski, R., Bunting, S., Vane, J. R. 1976. *Prostaglandins* 263:663
77. Gryglewski, R. J., Bunting, S., Moncada, S., Flower, R. J., Vane, J. R. 1976. *Prostaglandins* 12:685
78. Bunting, S., Gryglewski, R., Moncada, S., Vane, J. R. 1976. *Prostaglandins* 12:897
79. Johnson, R. A., Morton, D. R., Kinner, J. H., Gorman, R. R., McGuire, J. C., Sun, F. F. 1976. *Prostaglandins* 12(6):915
80. Pace-Asciak, C. 1976. *Experientia* 32:291
81. Pace-Asciak, C. 1976. *J. Am. Chem. Soc.* 98:2348
82. Pace-Asciak, C., Wolfe, L. S. 1971. *Biochemistry* 10:3657
83. Pace-Asciak, C. R., Nashat, M. 1977. *Biochim. Biophys. Acta* 487:495
84. Corey, E. J., Keck, G. E., Székely, I. 1977. *J. Am. Chem. Soc.* 99:2006
85. Fried, J., Barton, J. 1977. *Proc. Natl. Acad. Sci. USA* 74:2199
86. Dusting, G. J., Moncada, S., Vane, J. R. 1977. *Prostaglandins* 13:3
87. Gorman, R. R., Bunting, S., Miller, O. V. 1977. *Prostaglandins* 13:377
88. Tateson, J. E., Moncada, S., Vane, J. R. 1977. *Prostaglandins* 13:389
89. Pace-Asciak, C., Rangaraj, G. 1977. *Biochim. Biophys. Acta* 486:583
90. Powell, W. S., Solomon, S. 1977. *Biochem. Biophys. Res. Commun.* 75:815
91. Isakson, P. C., Raz, A., Denny, S. E., Pure, E., Needleman, P. 1977. *Proc. Natl. Acad. Sci. USA* 74:101
92. Pace-Asciak, C., Rangaraj, G. 1977. *Biochim. Biophys. Acta* 486:579
93. Fenwick, L., Jones, R. L., Naylor, B., Poyser, N. L., Wilson, N. H. 1977. *Br. J. Pharmacol.* 59:191
94. Chang, W. C., Murota, S. 1977. *Biochim. Biophys. Acta* 486:136
95. Chang, W. C., Murota, S., Matsuo, M., Tsurufuji, S. 1976. *Biochem. Biophys. Res. Commun.* 72:1259
96. Claesson, H. E., Lindgren, J. Å., Hammarström, S. 1977. *FEBS Lett.* 81:415

97. Hammarström, S., Samuelsson, B., Bjursell, G. 1973. *Nature New Biol.* 243:50

98. Hammarström, S. 1977. *Eur. J. Biochem.* 74:7

99. Goldyne, M., Lindgren, J. Å., Claesson, H. E., Hammarström, S. 1978. Submitted for publication

100. Ritzi, E. M., Stylos, W. A. 1976. *J. Natl. Cancer Inst.* 56:529

101. Ishikawa, Y., Ciosek, C. P. Jr., Fahey, J. V., Newcombe, D. S. 1976. *J. Cyclic Nucleotide Res.* 2:115

102. Hong, S.-C. L., Polsky-Cynkin, R., Levine, L. 1976. *J. Biol. Chem.* 251:776

103. Hong, S. L., Levine, L. 1976. *J. Biol. Chem.* 251:5814

104. Pong, S., Hong, S.-C. L., Levine, L. 1977. *J. Biol. Chem.* 252:1408

105. Pong, S., Levine, L. 1977. *Prostaglandins* 13:65

106. Ritzi, E. M., Boto, W. O., Stylos, W. A. 1975. *Biochem. Biophys. Res. Commun.* 63:179

107. Dunn, M. J., Staley, R. S. 1976. *Prostaglandins* 12:37

108. Zusman, R. M., Keiser, H. R. 1977. *J. Biol. Chem.* 252:2069

109. Alexander, R. W., Gimbrone, M. A. 1976. *Proc. Natl. Acad. Sci. USA* 73:1617

110. Tuvemo, T., Strandberg, K., Hamberg, M., Samuelsson, B. 1976. *Acta Physiol. Scand.* 96:145

111. Pace-Asciak, C., Nashat, M. 1975. *Biochim. Biophys. Acta* 388:243

112. Pace-Asciak, C., Nashat, M. 1976. *J. Neurochem.* 27:551

113. Hamberg, M. 1976. *Lipids* 11:249

114. Corey, E. J., Ensley, H. E., Hamberg, M., Samuelsson, B. 1975. *J. Chem. Soc. Chem. Commun.* 8:277

115. Borgeat, P., Hamberg, M., Samuelsson, B. 1976. *J. Biol. Chem.* 251:7816

116. Hamberg, M. Samuelsson, B. 1973. *Proc. Natl. Acad. Sci. USA* 70:899

117. Hamberg, M., Svensson, J., Wakabayashi, T., Samuelsson, B. 1974. *Proc. Natl. Acad. Sci. USA* 71:345

118. Nugteren, D. H. 1975. *Biochim. Biophys. Acta* 380:299

119. Hammarström, S., Hamberg, M., Samuelsson, B., Duell, E. A., Stawiski, M., Voorhees, J. J. 1975. *Proc. Natl. Acad. Sci. USA* 72:5130

120. Gréen, K., Steffenrud, S. 1976. *Anal. Biochem.* 76:606

121. Knapp, H. R., Oelz, O., Oates, J. A. 1977. *Fed. Proc.* 36:1020 (Abstr. No. 3944)

122. Whorton, A. R., Hubbard, W., Sweetman, B., Carr, K., Frolich, J., Bux-

baum, D. M., Oates, J. 1977. *Fed. Proc.* 36:308 (Abstr. No. 195)

122a. Goldyne, M. E., Hammarström, S. 1978. *Anal. Biochem.* In press

123. Samuelsson, B., Gréen, K. 1974. *Biochem. Med.* 11:298

124. Gréen, K., Granström, E., Samuelsson, B., Axen, U. 1973. *Anal. Biochem.* 54:434

125. Malmsten, C., Hamberg, M., Svensson, J., Samuelsson, B. 1975. *Proc. Natl. Acad. Sci. USA* 72:1446

126. Hamberg, M., Svensson, J., Hedqvist, P., Strandberg, K., Samuelsson, B., 1976. *Adv. Prostaglandins Thromboxane Res.* 1:495

127. Hamberg, M., Jonsson, C.-E. 1977. To be published

128. Pletka, P., Hickler, R. B. 1974. *Prostaglandins* 7:107

129. Granström, E., Samuelsson, B. 1972. *FEBS Lett.* 26:211

130. Gréen, K., Granström, E. 1973. In *Prostaglandins in Fertility Control,* ed. S. Bergström, Vol. 3. Stockholm: Karolinska Ins. pp. 55–61

131. Levine, L., Gutierrez-Cernosek, R. M. 1973. *Prostaglandins* 3:785

132. Stylos, W. A., Burstein, S., Rosenfeld, J., Ritzi, E. M., Watson, D. J. 1974. *Prostaglandins* 4:553

133. Cornette, J. C., Harrison, K. L., Kirton, K. T. 1974. *Prostaglandins* 5:155

134. Fairclough, R. J., Payne, E. 1975. *Prostaglandins* 10:266

135. Granström, E., Kindahl, H. 1976. *Adv. Prostaglandin Thromboxane Res.* 1:81

136. Mitchell, M. D., Flint, A. P., Turnbull, A. C. 1976. *Prostaglandins* 11:319

137. Liebig, R., Bernauer, W., Peskar, B. A. 1974. *Naunyn-Schmiedeberg's Arch. Pharmacol.* 284:276

138. Sors, H., Maclouf, J., Pradelles, P., Dray, F. 1977. *Biochim. Biophys. Acta* 486:553

139. Haning, R. V., Kieliszek, F. X., Alberino, S. P., Speroff, L. 1977. *Prostaglandins* 13:455

140. Fitzpatrick, F. A., Wynalda, M. A. 1976. *Anal. Biochem.* 73:198

141. Peskar, B. A., Holland, A., Peskar, B. M. 1974. *FEBS Lett.* 43:45

142. Tashjian, A. H., Voelkel, E. F., Levine, L. 1977. *Biochem. Biophys. Res. Commun.* 74:199

143. Scaramuzzi, R. J., Baird, D. T., Boyle, H. P., Land, R. B., Wheeler, A. G. 1977. *J. Reprod. Fertil.* 49:157

144. Sharma, S. C., Fitzpatrick, R. J. 1974. *Prostaglandins* 6:97

145. Newcomb, R., Booth, W. D., Rowson, L. E. 1977. *J. Reprod. Fertil.* 49:17

146. Pexton, J. E., Weems, C. W., Inskeep, E. K. 1975. *Prostaglandins* 9:501
147. Douglas, R. H., Ginther, O. J. 1977. *Prostaglandins* 11:251
148. Barcikowski, B., Carlsson, J. C., Wilson, L., McCracken, J. A. 1974. *Endocrinology* 95:1340
149. Roberts, J. S., Barcikowski, B., Wilson, L., Skarnes, R. C., McCracken, J. A. 1975. *J. Steroid Biochem.* 6:1091
150. Currie, W. B., Cox, R. I., Thorburn, G. D. 1976. *Prostaglandins* 5:1093
151. Gleeson, A. R., Thorburn, G. D., Cox, R. I. 1974. *Prostaglandins* 5:521
152. Castracane, V. D., Jordan, V. C. 1975. *Biol. Reprod.* 13:587
153. Shemesh, M., Hansel, W. 1975. *Proc. Soc. Exp. Biol. Med.* 148:243
154. Shemesh, M., Hansel, W. 1975. *Proc. Soc. Exp. Biol. Med.* 148:123
155. Shaikh, A. A., Naqvi, R. H., Saksena, S. K. 1977. *Prostaglandins* 13:311
156. Peterson, A. J., Fairclough, R. J., Payne, E., Smith, J. F. 1975. *Prostaglandins* 10:675
157. Kindahl, H., Edqvist, L.-E., Bane, A., Granström, E. 1976. *Acta Endocrinol. Copenhagen* 82:134
158. Kindahl, H., Edqvist, L.-E., Granström, E., Bane, A. 1976. *Prostaglandins* 11:871
159. Granström, E., Kindahl, H. 1976. *Prostaglandins* 12:759
160. Demers, L. M., Yoshinaga, K., Greep, R. O. 1974. *Prostaglandins* 5:513
161. Demers, L. M., Halbert, D. R., Jones, D. E., Fontana, J. 1975. *Prostaglandins* 10:1057
162. Ohki, S., Hanyu, T., Imaki, K., Nakazawa, N., Hirata, F. 1974. *Prostaglandins* 6:137
163. Cornette, J. C., Kirton, K. T., Schneider, W. P., Sun, F. F., Johnson, R. A., Nidy, E. G. 1975. *Prostaglandins* 9:323
164. Ohki, S., Imaki, K., Hirata, F., Hanyu, T., Nakazawa, N. 1975. *Prostaglandins* 10:549
165. Hamberg, M. 1973. *Anal. Biochem.* 55:368
166. Abe, K., Otsuka, Y., Yasujima, M., Ciba, S., Seino, M., Irokawa, N., Yoshinaga, K., Hirata, F., Ohki, S., Nakazawa, N., Hanyu, T. 1976. *Prostaglandins* 12:843
167. Ohki, S., Nishigaki, Y., Imaki, K., Kurono, M., Hirata, F., Hanyu, T., Nakazawa, N. 1976. *Prostaglandins* 12:181
168. Raz, A., Schwartzman, M., Kenig-Wakshal, R., Perel, E. 1975. *Eur. J. Biochem.* 53:145
169. Stylos, W., Howard, L., Ritzi, E., Skarnes, R. 1974. *Prostaglandins* 6:1
170. Christensen, P., Leyssac, P. P. 1976. *Prostaglandins* 11:399
171. Axen, U. 1974. *Prostaglandins* 5:45
172. Bauminger, S. 1976. *J. Immunol. Methods* 13:253
173. Maclouf, J., Andrieu, J. M., Dray, F. 1975. *FEBS Lett.* 56:273
174. Maclouf, J., Pradel, M., Pradelles, P., Dray, F. 1976. *Biochim. Biophys. Acta* 431:139
175. Zusman, R., Caldwell, B., Speroff, L., Behrman, H. 1972. *Prostaglandins* 2:41
176. Jaffe, B., Behrman, H., Parker, C. 1973. *J. Clin. Invest.* 52:398
177. Lee, J. 1973. *Prostaglandins* 3:241
178. Zusman, R., Spector, D., Caldwell, B., Speroff, L., Schneider, G., Mulrow, P. 1973. *J. Clin. Invest.* 52:1093
179. Van Orden, D. E., Whalen, J., Farley, D. B. 1975. *Gynecol. Invest.* 6:27
180. Brouhard, B. H., Lagrone, L., Travis, L. B., Cunningham, R. J. 1977. *Clin. Chim. Acta* 75:287
181. Frölich, J. C., Sweetman, B., Carr, K., Hollifield, J., Oates, J. 1975. *Prostaglandins* 10:185
182. Zia, P., Golub, M., Horton, R. 1975. *J. Clin. Endocrinol. Metab.* 41:245
183. Kirkham, K., Hunter, W., eds. 1971. *Radioimmunoassay Methods,* Edinburgh: Churchill Livingstone
184. Granström, E., Kindahl, H., Samuelsson, B. 1976. *Anal. Lett.* 9:611
185. Granström, E., Kindahl, H., Samuelsson, B. 1976. *Prostaglandins* 12:929
186. Anhut, H., Bernauer, W., Peskar, B. A. 1977. *Eur. J. Pharmacol.* 44:85
187. Samuelsson, B., Hamberg, M., Malmsten, C., Svensson, J. 1976. *Adv. Prostaglandin Thromboxane Res.* 2:737
188. Folco, G., Granström, E., Kindahl, H. 1977. *FEBS Lett.* 82:321
189. Kindahl, H. 1977. *Prostaglandins* 13:619
190. Roberts, L. J., Sweetman, B. J., Morgan, J. L., Payne, N. A., Oates, J. A. 1977. *Prostaglandins* 13:631
191. Kuehl, F. A. Jr. 1974. *Prostaglandins* 5:325
192. Kuehl, F. A. Jr., Oien, H. G., Ham, E. A. 1974. In *Prostaglandin Synthetase Inhibitors,* ed. H. J. Robinson, J. R. Vane, p. 53. New York: Raven
193. Kuehl, F. A. Jr., Cirillo, V. J., Oien, H. G. 1976. In *Prostaglandins: Chemical and Biochemical Aspects,* ed. S. M. Karim, p. 192. M.T.P.
194. Oien, H. G., Mandel, L. R., Humes, J. L., Taub, D., Hoffsommer, R. D.,

Kuehl, F. A. Jr. 1975. *Prostaglandins* 9:985

195. Dazord, A., Morera, A. M., Bertrand, J., Saez, J. M. 1974. *Endocrinology* 95:352

196. Brunton, L. L., Wiklund, R. A., Van Arsdale, P. M., Gilman, A. G. 1976. *J. Biol. Chem.* 251:3037

197. Maguire, M. E., Brunton, L. L., Wiklund, R. A., Anderson, H. J., Van Arsdale, P. M., Gilman, A. G. 1976. *Recent Progr. Horm. Res.* 32:633

198. Davies, P. J. A. 1976. *Fed. Proc.* 35:1438

199. Okamura, N., Terayama, H. 1977. *Biochim. Biophys. Acta* 465:54

200. Brunton, L. L., Maguire, M. E., Anderson, H. J., Gilman, A. G. 1977. *J. Biol. Chem.* 252:1293

201. Kelly, L. A., Butcher, R. W. 1974. *J. Biol. Chem.* 249:3098

202. Remold-O'Donnel, E. 1974. *J. Biol. Chem.* 249:3615

203. Tomasi, V., Ferretti, E. 1975. *Mol. Cell. Endocrinol.* 2:22

204. Adachi, K., Yoshikawa, K., Halprin, K. M., Levine, V. 1975. *Br. J. Dermatol.* 92:381

205. Su, Y. F., Cubeddu, X. L., Perkins, J. P. 1976. *J. Cyclic Nucleotide Res.* 2:257

206. Polgar, P., Vera, J. C., Rutenburg, A. M. 1977. *Proc. Soc. Exp. Biol. Med.* 154:493

207. Harlson, S., Do Khac, L., Vesin, M. F. 1976. *Mol. Cell. Endocrinol.* 6:17

208. Borgeat, P., Labrie, F., Garneau, P. 1974. *Can. J. Biochem.* 53:455

209. Sweat, F. W., Wincek, T. J. 1973. *Biochem. Biophys. Res. Commun.* 55:522

210. Burstein, S., Gagnon, G., Hunter, S. A., Maudsley, D. V. 1976. *Prostaglandins* 11:85

211. Yamashita, L., Sweat, F. W. 1976. *Biochem. Biophys. Res. Commun.* 70:438

212. Wincek, T. J., Sweat, F. W. 1976. *Biochem. Biophys. Acta* 437:571

213. Fredholm, B. B., Hamberg, M. 1976. *Prostaglandins* 11:507

214. Wincek, T. J., Hupka, A. L., Sweat, F. W. 1975. *J. Biol. Chem.* 250:8863

215. Ferretti, e., Biorrdi, C., Tomasi, V. 1976. *FEBS Lett.* 69:70

216. Dixon-Shanies, D., Knittle, J. L. 1976. *Biochem. Biophys. Res. Commun.* 68:982

217. Bach, M. A. 1975. *J. Clin. Invest.* 55:1074

218. Wollin, A., Code, C. F., Dousa, T. P. 1976. *J. Clin. Invest.* 57:1543

219. Anderson, W. B., Gallo, M., Pastan, I. 1974. *J. Biol. Chem.* 22:7041

220. Humes, J. L., Strausser, H. R. 1974. *Prostaglandins* 5:183

221. Penit, J., Huot, J., Jard, S. 1976. *J. Neurochem.* 26:265

222. Blume, A. J., Foster, C. J. 1976. *J. Neurochem.* 26:305

223. Claesson, H. E., Lindgren, J. A., Hammarström, S. 1977. *Eur. J. Biochem.* 74:13

224. Haslam, R. J., Goldstein, S. 1974. *Biochem. J.* 144:253

225. Penit, J., Cantan, B., Huot, J., Jard, S. 1977. *Proc. Natl. Acad. Sci. USA* 74:1575

226. MacManus, J. P., Whitfield, J. F. 1974. *Prostaglandins* 6:475

227. Pastan, I. H., Johnson, G. S., Anderson, W. B. 1975. *Ann. Rev. Biochem.* 44:491

228. Chlapowski, F. J., Kelly, L. A., Butcher, R. W. 1975. *Adv. Cyclic Nucleotide Res.* 6:245

229. Krishna, G., Harwood, J. P., Barker, A. J., Jamieson, G. A. 1972. *J. Biol. Chem.* 247:2253

230. Johnson, D. G., Thompson, W. J., Williams, R. H. 1974. *Biochemistry* 13:1920

231. Wolff, J., Cook, G. H. 1973. *J. Biol. Chem.* 248:350

232. Sato, S., Takashi, Y., Furihata, R., Makinchi, M. 1974. *Biochim, Biophys. Acta* 332:166

233. Best, L. C., Martin, T. J., Russell, R. G. G., Preston, F. E. 1977. *Nature* 267:850

234. Burstein, S., Gagnon, G., Hunter, S. A., Maudsley, D. V. 1977. *Prostaglandins* 13:41

235. Hamprecht, B., Jaffe, B. M., Philpott, G. W. 1973. *FEBS Lett.* 36:193

236. Marsh, J. M., Yang, N. S. T., Lemaire, W. J. 1974. *Prostaglandins* 7:269

237. Haye, B., Champion, S., Jacquemin, C. 1974. *FEBS Lett.* 41:89

238. Burke, G., Chang, L. L., Szabo, M. 1973. *Science* 180:872

239. Laychock, S. G., Warner, W., Rubin, R. P. 1977. *Endocrinology* 100:74

240. Dalton, C., Hope, W. C. 1974. *Prostaglandins* 6:227

241. Newcombe, D. S., Ciosek, C. P. Jr., Ishikawa, Y., Fahey, J. V. 1975. *Proc. Natl. Acad. Sci. USA* 72:3124

242. Ciosek, C. P. Jr., Fahey, J. V., Ishikawa, Y., Newcombe, D. S. 1975. *J. Cyclic Nucleotide Res.* 1:229

243. Su, Y. F., Johnson, G. L., Cubeddu, X. L., Leichtling, B. H., Ortmann, R., Perkins, J. P. 1976. *J. Cyclic Nucleotide Res.* 2:271

244. Leichtling, B. H., Drotar, A. M., Ortmann, R., Perkins, J. P. 1976. *J. cyclic Nucleotide Res.* 2:89

245. Frankling, T. J. 1973. *Nature* 246:146
246. Adachi, K., Lizuka, H., Halprin, K. M., Levine, V. 1977. *Biochim. Biophys. Acta* 497:428
247. Lamprecht, S. A., Zor, U., Salomon, Y., Koch, Y., Ahren, K., Lindner, H. R. 1977. *J. Cyclic Nucleotide Res.* 3:69
248. Polgar, P., Taylor, L. 1977. *Biochem. J.* 162:1
249. Fehér, I., Gidáli, J. 1974. *Nature* 247:550
250. Kuehl, F. A. Jr., Cirillo, V. J., Ham, E. A., Humes, J. L. 1973. *Adv. Biosci.* 9:155
251. Muira, Y., Fukui, N. 1976. In *Advances in Enzyme Regulation,* ed. G. Welser, p. 14. Oxford: Pergamon
252. Dunham, E. W., Haddox, M. K., Goldberg, N. D. 1974. *Proc. Natl. Acad. Sci. USA* 71:815
253. Kadowitz, P. J., Joiner, P. D., Hyman, A. L., George, W. J. 1975. *J. Pharmacol. Exp. Ther.* 192:677
254. De Asua, L. J., Clingan, D., Rudland, P. S. 1975. *Proc. Natl. Acad. Sci. USA* 72:2724
255. Nahorski, S. R., Pratt, C. N. F. W., Rogers, K. J. 1976. *Br. J. Pharm.* (Abstr.) 445 pp.
256. Matsuzawa, H., Nirenberg, M. 1975. *Proc. Natl. Acad. Sci. USA* 72:3472
257. Gorman, R. R. 1975. *J. Cyclic Nucleotide Res.* 1:1
258. Glass, D. B., Gerrard, J. M., Townsend, D. W., Carr, D. W., White, J. G., Goldberg, N. D. 1977. *J. Cyclic Nucleotide Res.* 3:37
259. Goldberg, N. D. 1976. *Meet. Rep.: Int. Titisee Conf., J. Cyclic Nucleotide Res.* 2:189
260. Goldberg, N. D., Haddox, M. K., Hartle, D. K., Hadden, J. W. 1973. *Proc. Int. Congr. Pharmacol., 5th 1972 San Francisco,* 5:146. Basel: Karger
261. Goldberg, N. D., O'Dea, R. F., Haddox, M. K. 1973. *Adv. Cyclic Nucleotide Res.* 3:155
262. Nawrath, H. 1976. *Nature* 262:509
263. Weissmann, G., Goldstein, I., Hoffstein, S. 1976. *Adv. Prostaglandin Thromboxane Res.* 2:803
264. Hill, H. R., Estensen, R. D., Quie, P. G., Hogan, N. A., Goldberg, N. D. 1975. *Metabolism* 24:447
265. Seifert, W. E., Rudland, P. S. 1974. *Nature* 248:138
266. Sandoval, I. V., Cuatrecasas, P. 1976. *Nature* 248:511
267. Kuehl, F. A. Jr., Ham, E. A., Zanetti, M. E., Sanford, C. H., Nicol, S. E., Goldberg, N. D. 1974. *Proc. Natl. Acad. Sci. USA* 71:1866
268. Sapag-Hagar, M., Greenbaum, A. L. 1974. *FEBS Lett.* 46:180
269. Rivlein, I., Rosenblatt, J., Becker, E. L. 1975. *J. Immunol.* 115:1126
270. Diaz-Perez, J. L., Goldyne, M. E., Winkelmann, R. K. 1976. *J. Invest. Dermatol.* 66:149
271. Salzman, E. W. 1976. *Adv. Prostaglandin Thromboxane Res.* 2:767
272. Manku, M. S., Mtabaji, J. P., Horrobin, D. F. 1977. *Prostaglandins* 13:701
273. Zor, U., Bauminger, S., Lamprecht, S. A., Koch, Y., Chobsieng, P., Lindner, H. R. 1973. *Prostaglandins* 4:499
274. Grönquist, L., Perkler, T., Harén, K. 1974. *Prostaglandins* 6:303
275. Naor, Z., Koch, Y., Bauminger, S., Zor, U. 1975. *Prostaglandins* 9:211
276. Rillema, J. A. 1975. *Nature* 253:466
277. Spaulding, S. W., Burrow, G. N. 1975. *Endocrinology* 96:1018
278. Mashiter, K., Field, J. B. 1974. *Fed. Proc.* 33:78
279. Kuehl, F. A. Jr., Humes, J. L. 1972. *Proc. Natl. Acad. Sci. USA* 69:480
280. Rao, C. V. 1974. *J. Biol. Chem.* 249:7203
281. Rao, C. V. 1975. *Prostaglandins* 9:569
282. Brunton, L. L., Gilman, A. G. 1977. *Fed. Proc.* 36:455 (Abstr. No. 1371)
283. McDonald, J. W. D., Stuart, R. K. 1975. *Clin. Res.* 22:399A
284. Grunnet, I., Bojesen, E. 1976. *Biochim. Biophys. Acta* 419:365
285. Johnson, M., Jessup, R., Ramwell, P. W. 1974. *Prostaglandins* 6:433
286. Wakeling, A. E., Wyngarden, L. J. 1974. *Endocrinology* 95:55
287. Kimball, F. A., Wyngarden, L. J. 1975. *Prostaglandins* 9:413
288. Kimball, F. A., Kirton, K. T., Wyngarden, L. J. 1975. *Prostaglandins* 10:853
289. Rao, C. V. 1975. *Prostaglandins* 9:579
290. Horton, E. W., Poyser, N. L. 1976. *Physiol. Rev.* 56:595
291. Powell, W. S., Hammarström, S., Samuelsson, B. 1974. *Eur. J. Biochem.* 41:103
292. Powell, W. S., Hammarström, S., Samuelsson, B. 1975. *Eur. J. Biochem.* 56:73
293. Powell, W. S., Hammarström, S., Samuelsson, B., Miller, W. L., Sun, F. F., Fried, J. J., Lin, C. H., Jarabak, J. 1975. *Eur. J. Biochem,* 59:271
294. Powell, W. S., Hammarström, S., Samuelsson, B. 1976. *Eur. J. Biochem.* 61:605
295. Hammarström, S., Kyldén, U., Powell, W. S., Samuelsson, B. 1975. *FEBS Lett.* 50:306

296. Kyldén, U., Hammarström, S. 1978. Submitted for publication
297. Hammarström, S., Powell, W. S., Kyldén, U., Samuelsson, B. 1976. *Adv. Prostaglandin Thromboxane Res.* 1:235
298. Rao, C. V. 1975. *Biochem. Biophys. Res. Commun.* 64:416
299. Kimball, F. A., Lauderdale, J. W. 1975. *Prostaglandins* 10:313
300. Rao, C. V. 1975. *Biochem. Biophys. Res. Commun.* 67:1242
301. Rao, C. V. 1976. *Biochim. Biophys. Acta* 436:170
302. Rao, C. V. 1976. *Fed. Proc.* 35:458
303. Rao, C. V. 1976. *Mol. Cell. Endocrinol.* 6:1
304. Rao, C. V. 1976. *Life Sci.* 18:499
305. Rao, C. V. 1976. *Steroids* 27:831
306. Mitra, S., Rao, C. V. 1977. *Fed. Proc.* 36:375
307. Kimball, F. A., Lauderdale, J. W., Nelson, N. A., Jackson, R. W. 1976. *Prostaglandins* 12:985
308. Kimball, F. A., Wyngarden, L. J. 1977. *Prostaglandins* 13:553
309. Day, H. J. 1975. *Ser. Haematol.* 8:23
310. Smith, J. B., Ingerman, C., Kocsis, J. J., Silver, M. J. 1973. *J. Clin. Invest.* 52:965
311. Smith, J. B., Ingerman, C., Kocsis, J. J., Silver, M. J. 1974. *Thromb. Res.* 4 (Suppl.) 1:49
312. Silver, M. J., Smith, J. B., Ingerman, C., Kocsis, J. J. 1973. *Prostaglandins*, p. 863
313. Vargaftig, B. B., Zirinis, P. 1973. *Nature New Biol.* 244:114
314. Silver, M. J., Hoch, W., Kocsis, J. J., Ingerman, C. M., Smith, J. B. 1974. *Science* 183:1085
315. Furlow, T. W. Jr., Bass, N. H. 1975. *Science* 187:658
316. Sughara, T., Takahashi, T., Yamaya, S., Ohsaka, A. 1976. *Jpn. J. Med. Sci. Biol.* 29:255
317. Salzman, E. 1972. *N. Engl. J. Med.* 286:358
318. Willis, A. L., Vane, F. M., Kuhn, D. C., Scott, C. G., Petrin, M. 1974. *Prostaglandins* 8:453
319. Smith, J. B., Ingerman, C., Kocsis, J. J., silver, M. J. 1974. *J. Clin. Invest.* 53:1468
320. Claesson, H.-E., Malmsten, C. 1977. *Eur. J. Biochem.* 76:277
321. Gerrard, J. M., White, J. G. 1975. *Am. J. Pathol.* 80:189
322. Gerrard, J. M., Townsend, D., Stoddard, S., Witkop, C. J. Jr., White, J. G. 1977. *Am. J. Pathol.* 86:99
323. Smith, J. B., Ingerman, C. M., Silver, M. J. 1976. *Thromb. Res.* 9:413
324. Gordon, J. L., MacIntyre, D. E., McMillan, R. M. 1976. *Eur. J. Pharmacol.* 58:299
325. Dechavanne, M., Lagarde, M., Bryon, P. A. 1976. *Nouv. Rev. Fr. Hematol.* 16:421
326. Weiss, H. J., Larges, B. A. 1977. *Lancet* 8014:760
327. Hamberg, M., Svensson, J., Samuelsson, B. 1974. *Lancet* 2:223
328. DeGaetano, G., Donati, M. B., Grathini, S. 1975. *Thromb. Diath. Haemorrh.* 34:285
329. Willis, A. L. 1974. *Science* 183:325
330. Jafari, E., Saleem, A., Shaikh, B. S., Demers, L. M. 1976. *Prostaglandins* 12:829
331. Gerrard, J. M., White, J. G. 1976. *Am. J. Pathol.* 82:513
332. Lewis, J. C., Hagedorn, A. B. 1976. *Thromb. Res.* 9:647
333. Corey, E. J., Nicolaou, K. C., Machida, Y., Malmsten, C. L., Samuelsson, B. 1975. *Proc. Natl. Acad. Sci. USA* 72:3355
334. Vargaftig, B. B. 1977. *J. Pharm. Pharmacol.* 29:222
335. Gryglewski, R. J. 1977. *Naunyn Schmiedebergs Arch. Pharmacol.* 297: 585
336. Gryglewski, R. J., Zmuda, A., Dembinska-Kiec, A., Kreichow, E. 1977. *Pharm. Res. Commun.* 9:109
337. Miller, O. V., Gorman, R. R. *Cyclic Nucl. Res.* 2:79
338. Salzman, E. W. 1973. p. 331. Paris: Institute National de la Sante et de la Recherche Medicale (INSERM)
339. Bruno, J. J., Taylor, L. A., Droller, M. J. 1974. *Nature* 251:721
340. Smith, J. B., Silver, M. J., Ingerman, C. M., Kocsis, J. J. 1974. *Thromb. Res.* 5:291
341. Nishizawa, E. E., Miller, W. L., Gorman, R. R., Bondy, G. L., Svensson, J., Hamberg, M. 1975. *Prostaglandins* 9:109
342. Mills, D. C. B., MacFarlane, D. E. 1974. *Thromb. Res.* 5:401
343. Moncada, S., Higgs, E. A., Vane, J. R. 1977. *Lancet* 1(8001):18
344. Gryglewski, R., Moncada, S., Bunting, S. 1976. *Prostaglandins* 12:897
345. Dusting, G. J., Lattiner, N., Moncada, S., Vane, J. R. 1977. *Br. J. Pharmacol.* 59:443
346. Gryglewski, R. J., Szczeklik, A., Bierow, K. 1975. *Nature* 256:56
347. Lagarde, M., Dechavanne, M. 1977. *Biomed. Express* 27:110
348. Vargaftig, B. B., Chignard, M. 1975. *Agents Actions* 5:137

349. Weiss, H. J., Willis, A. L., Kuhn, D., Brand, H. 1976. *Br. J. Haematol.* 32:257

350. Wang, T. Y., Hussey, C. V., Garanics, J. C. 1977. *Am. J. Clin. Pathol.* 67:362

351. Ryo, R. 1976. *Acta Haematol. Jpn.* 39:85

352. Haslam, R. J., McClenaghan, M. D. 1974. *Biochem. J.* 138:317

353. Jakobs, K. H., Bohme, E., Mocikat, C. 1974. *Naunyn Schmiedebergs Arch. Pharmacol.* 282:R40 (Suppl.)

354. Glass, D. B., Frey, W., Carr, D. W., Goldberg, N. D. 1977. *J. Biol. Chem.* 252:1279

355. Haslam, R. J., Davidson, M. M., McClenaghan, M. D. 1975. *Nature* 253:455

356. Chiang, T. M., Beachey, E. H., Kang, A. H. 1975. *J. Biol. Chem.* 250:6916

357. Chiang, T. M., Dixit, S. N., Kang, A. H. 1976. *J. Lab. Clin. Med.* 88:215

358. Malmsten, C., Kindahl, H., Samuelsson, B., Levy-Toledano, S., Tobelem, G., Caen, J. P. 1977. *Br. J. Haematol.* 35:511

359. Kinlough-Rathbone, R. L., Mustard, J. F., Packham, M. A. 1976. *Science* 192:1011

360. Willis, A. L., Weiss, H. J. 1973. *Prostaglandins* 4:783

361. Gerrard, J. M., White, J. G., Rao, G. H. R., Townsend, D. W. 1976. *Amer. J. Pathol.* 83:283

362. Behnke, O. 1976. *Anat. Rec.* 158:121

363. White, J. G., Rav, G. H., Gerrard, J. M. 1974. *Am. J. Pathol.* 77:135

364. Feinstein, M. B., Fraser, C. 1975. *J. Gen. Physiol.* 66:561

365. Mitani, M., Umetsu, T., Yamanish, T., Otake, N. 1977. *J. Antibiot.* 30:239

366. Costa, J. L., Detwiler, T. C., Feinman, R. D., Murphy, D. L., Patlak, C. S., Pettigrew, K. D. 1977. *J. Physiol.* 264:297

367. Polland, T. D., Fujiwara, K. Handin, R., Weiss, G. 1977. *Ann. NY Acad. Sci.* 283:218

368. LeBreton, G. C., Dinerstein, R. J., Roth, L. J., Feinberg, H. 1976. *Biochem. Biophys. Res. Commun.* 71:362

369. Massini, P., Luscher, E. F. 1976. *Biochim. Biophys. Acta* 436:652

370. Silver, M. J., Smith, J. B. 1975. *Life Sci.* 16:1635

371. Lagarde, M., Dechavanne, M. 1977. *Lancet* 8002:88

372. Lagarde, M., Dechavanne, M. 1977. *Biomed. Express* 27:119

373. Kraemer, R. J., Phernetton, T. M., Folts, J. D. 1976. *J. Pharmacol. Exp. Ther.* 199:611

374. DeClerck, F., Goossens, J., Vermylen, J., Hornstra, G., Reneman, R. S. 1976. *Arch. Int. Phamacodyn. Ther.* 222:233

375. Weiss, H. J. 1976. *Am. Heart J.* 92:86

376. Morse, E. E. 1977. *Ann. Clin. Lab. Sci.* 7:68

377. Holmsen, H. 1975. *Ser. Haematol.* 8:50

378. Hornstra, G., Vergroesen, A. J. 1976. *Acta Biol. Med. Ger.* 35:1065

379. Flower, R. J. 1974. *Pharmacol. Rev.* 26:33

380. Rosenberg, J. C., Sell, T. L. 1975. *Arch. Surg.* 110:980

381. Fujitani, B., Tsuboi, T., Takeno, K., Yoshida, K., Shimizu, M. 1976. *Thromb. Haemostasis* 36:401

382. Rose, J. C., Johnson, M., Ramwell, P. W., Kot, P. A. 1975. *Proc. Soc. Exp. Biol. Med.* 148:1252

383. Willis, A. L., Coma, K., Kuhn, D. C., Paulsrud, J. 1974. *Prostaglandins* 8:509

384. Sim, A. K., McCraw, A. P. 1977. *Thromb. Res.* 10:385

385. Farrow, J. W., Willis, A. L. 1975. *Br. J. Pharmacol.* 55:316P.

386. Picket, W. C., Jesse, R. L., Cohen, P. 1976. *Biochem. J.* 160:405

387. Vincent, J. E., Zijlstra, F. J. 1976. *Prostaglandins* 12:971

388. Schoene, N. W., Jacono, J. M. 1976. *Adv. Prostaglandin Thromboxane Res.* 2:763

Ann. Rev. Biochem. 1978. 47:1031–78
Copyright © 1978 by Annual Reviews. All rights reserved

MECHANISMS OF ENZYME-CATALYZED GROUP TRANSFER REACTIONS

❖996

Mary J. Wimmer and Irwin A. Rose

The Institute for Cancer Research, Philadelphia, Pennsylvania 19111

CONTENTS

PERSPECTIVES AND SUMMARY

The importance of the coupling of two catalyzed reactions on a single enzyme is seen in the pyruvate kinase reaction, which is a mixture of two fundamental semi-independent processes: phosphoryl transfer and keto-enol tautomerism (equation 1).

0066-4154/78/0701-1031$01.00

The occurrence of the two steps in a single enzyme can be rationalized as a means of increasing the efficiency of both steps. The phosphoryl transfer step and the ketonization step are tightly coupled as shown by the finding that enolization does not occur unless K^+, Mg^{2+}, and ATP or an anionic analog of ATP are present (1). Although stereospecific Mg^{2+}-dependent decarboxylation of oxalacetate is also catalyzed by this enzyme, the enol-pyruvate-enzyme complex that is formed, does not couple to phosphorylation with ATP (2). This is not surprising in view of the lack of requirement for K^+ and ATP (or P_i) for the decarboxylation. In spite of the tight coupling in the pyruvate + ATP reaction, a mechanism did not evolve that bypassed the enolpyruvate intermediate (Equation 2):

Presumably resonance stabilization of the anion $(O{=}\dot{C}{=}C)^-$ is important in lowering the energy barrier for coupling the two reactions and the extended transition state required for concerted action is not.

Several other examples of 1,3 substitutions have been studied which involve rearrangement of the kind: $C{=}C{\sim}_C{\sim}H \rightarrow HC{\sim}^C{\equiv}C$. In these cases, the observations of 1,3-proton transfer (3–6) imply the existence of an intermediate carbanion and rule out a concerted process. Evidently the stepwise proton transfer process can be very efficient as seen by the particular example, $\Delta^{4,6}$-3-keto steroid isomerase of bacteria, the V_{max} of which is very great, $\sim 10^5 s^{-1}$.

Other examples of combinations of reactions on a single enzyme are seen in the biotin-dependent reactions and acyl and phosphoryl transfer where similar questions of stepwise or concerted mechanisms arise. In the case of the biotin enzymes the frequent distribution of function among separate subunits may be a good example of the modular combination of functions into single structures in other enzyme classes. The question of concerted versus stepwise coupling of separate functions in biotin-dependent carboxyl transfer remains uncertain. Here the concerted mechanism requires the interaction of species that are, in many cases, bound to different subunits. In the cases of proton, acyl, and phosphoryl transfer much progress is being made in distinguishing between concerted and stepwise mechanisms of transfer. The present chapter briefly describes useful methods for studying this question and provides several examples where clear-cut results have been obtained.

METHODS OF STUDY

A number of recent papers have attempted to use the absence of predicted behavior (negative evidence) as support for an alternate mechanism of enzyme catalysis. This approach threatens the logical coherence of our science. We discuss several examples in the text, among them the claim that an intermediate has been ruled out by the failure to observe partial reactions by isotope exchange in the presence of possible synergistic analogues, and the proposal that the absence of discrimination in H_2O/D_2O indicates that the medium is the direct source of H^+ for product formation. With current interest in the use of isotope effects, we are seeing cases in which the absence of an isotope effect is taken to support one mechanism as opposed to another. A misdemeanor not quite in this class is the all too frequent claim that MgATP is the species of ATP that reacts with an enzyme because a $1/v$ vs $1/MgATP$ plot is linear. In fact, if the formation of the ternary complex of enzyme, ATP, and Mg^{2+} is rapid relative to a subsequent step of the reaction, it is not possible to obtain any information about the path of its formation from steady state kinetics. Such evidence, per se, would not allow one to say that the Mg and ATP were associated with each other in a complex on the enzyme. It has been proposed that the *failure* of substitution-inert chromium complexes of ADP and ATP to react in the pyruvate kinase reaction "indicates" that, in the normal reaction with Mg and ATP, the metal-to-γP bond must be broken in the transition state. We still suffer from the problem of overinterpreting stereochemical results. In using steady state kinetic data, we do not take seriously the quantitation of minor paths of substrate addition, speaking of ordered or random as absolute. Therefore we precede our discussion of group transfer processes with a commentary on some currently used methods. Some important kinetic methods, such as the use of the transition state analogue concept, will not be considered because of space limitation. This method has been recently reviewed (7–9).

Techniques that provide enzyme structural information, such as X-ray crystallography, NMR and ESR spectroscopy, and active-site directed chemical labeling, are insufficient to define the chemical mechanism of a reaction. It is not clear whether certain catalytic amino acid residues are "better" than others for a particular type of chemistry. The acid and alkaline phosphatases proceed by way of phosphohistidine and phosphoserine E-P intermediates, respectively, and the same is true for phosphoglycerate and phosphoglucomutases. The acylenzyme intermediates in reactions of most proteases are known to involve either serine or cysteine residues.

Stereochemical Principles and Methods

Recent attempts to delineate rules for a mechanistic interpretation of reaction stereospecificity as it occurs on an enzyme (10, 11), are summarized. Racemic products from a reaction are rare but important indicators of mechanism. They occur when enols are the real enzyme reaction products or when torsionally symmetric compounds occur as intermediates. Inversion can result from a stepwise displacement or a concerted displacement. Any interpretation based on reactant and product stereochemistry assumes that a directed rearrangement of an intermediate did not occur; this is the *minimum motion assumption*. A case in which this assumption fails is the citrate to isocitrate conversion by aconitase (12). On the basis of overall stereochemistry one would predict that transaconitate would be the intermediate, not *cis*-aconitate. In cases of prochirality at a carbon (Caabe) or achirality (Caabb or abC=Cba) the reaction stereochemistry can have no metabolic significance. As an example, the citrate formed by condensation of acetyl CoA and oxalacetate is the same for all of the four possible stereochemistries of the condensation, except with regard to isotope positioning. When members of a class of such cryptic reactions have a common stereochemistry, this may have mechanistic significance, especially when the enzymes of the class do not show common features and the reactants seem metabolically and structurally distant in each example. Such classes might be: the aldolases, the sugar isomerases, the biotin carboxytransferring enzymes, the hydroxylases, and the Claisen cleavages such as citratase. When reactions are known to occur by a succession of stereospecific steps, especially in the so-called "ping-pong" mode as for the aldolases, the retention of stereochemistry may reflect provision of an active site with a minimum number of functional groups:

A model for keto-deoxygluconic-6-P aldolase using $-CO_2^-$ as $-:B^-$ has been discussed by Meloche & Glusker (13). Attempts to show isotope transfer in such reactions have not been reported. Midelfort and Rose (unpublished, 1974) tried to do so using fructose-1,6-P_2 in dimethyl sulfoxide containing a trace of TOH. Aldolase in normal water was added but no evidence was obtained for significant transfer of tritium into the dihydroxyacetone-P. A negative result in this isotope partition experiment is insufficient to rule out a single base mechanism, however, if the off rate of tritium is rapid at any stage of the dilution or transfer. The "one base" mechanism, which seems most likely for the sugar isomerases (14) may give the kinetic advantage of avoiding rate limiting proton dissociations in recycling the product form of the enzyme to the substrate form. Intramolecular proton transfers are known to be much faster than ionizations of moderate acids (10^{13} vs $10^3 s^{-1}$). Likewise the principle of "minimum number" of catalytic residues that do the same job avoids redundancy in the active site, especially with acids and bases.

A structural feature of enzymes may also be important in explaining the repetition of reaction stereochemistry within the class of addition/elimination reactions where the occurrence of the anti (trans-) mode of addition to double bonds is common. These reactions are often thought to proceed by a stepwise sequence, so that the stereoelectronic requirement that concerted reactions show antistereochemistry should not be a factor. It is proposed that the placement of donors of H^+ and OH^- may best be made across a cleft containing the unsaturated substrate. Active site clefts have been observed in many crystallographic studies of enzymes and may be a device for increasing the area of surface contact with the substrate(s) as well as separating groups with opposite functions.

"Stereochemical equivalence" has long been a method for recognizing particular intermediates in chemical reactions as in the classic example of Bender's recognition of a tetrahedral adduct in ester hydrolysis (15). It took Ogsten to point out that this would not work with enzymes that were assymetric reagents and therefore able to distinguish identical a groups in a Caabc site (16), as expected from the formation of diastereomeric transition states. Clearly, however, an enzyme cannot distinguish among the identical groups on Caaab, except in the special case of a = H and then only by means of isotope discrimination.

An important property of torsionally symmetrical groups is that they may leave evidence of their formation as intermediates by acting as pools for the mixing of isotope in a reaction. Thus it is common to exclude $-NH_2$ as the acceptor of a proton from a substrate by the failure to dilute isotope in a hydrogen transfer reaction (17), or conversely to suspect $-NH_2$ as the base when extent of transfer is small (18). Formation of a $-CH_3$ group was excluded in the activation of [Z,3T]phosphoenolpyruvate for condensation with erythrose-P because stereospecificity was not lost (19) and was postulated in the coenzyme B_{12} reactions as the basis for labeling of the 5' methylene group from tritiated substrate (20, 21). A list of groups that are torsiosymmetric in solution and possibly on an enzyme includes: $-CH_3$, $-C^+H_2$, $-C(CH_3)_2^+$, $-CH_2^-$, $-CO_2^-$, $-SO_2^-$, $-NH_3^+$, $-NH_2$, $-PO_3^{2-}$, and metaphosphate, PO_3^-. A period of progress may have been stimulated recently by studies in which the torsional symmetry of $-PO_3^{2-}$ was used to demonstrate phosphoryl bond cleavage in ATP (see section on positional exchange). Likewise the avoidance of torsional equivalence in phosphate monoesters by the use of thiophosphosphoryl analogues, has been used with great success by Eckstein and coworkers to study the stereochemical course of enzyme-catalyzed phosphoryl transfer reactions (22, 23).

Nondestructive and Destructive Approaches that Give Rate Constants

Kinetic constants for discrete reaction steps are not easily obtained. It is not generally stated but needs to be remembered that such expressions as k_{cat} and V_{max}/E_t only represent lower limits to the real reaction rate constants and that only a small portion of E_t may participate in a rate limiting step even in the absence of dead-end complexes if the equilibria prior to the slow step are unfavorable. The problem, of course, is that rates, by which we generally mean steady state rates, are products of concentrations and rate constants, and for reversible steps the observed rate is only a difference between rates. The separation of rate constants and intermediate concentrations seems necessary if effects of pH, isotope substitution, temperature, medium, second substrates, effectors and so forth, on rates can be properly evaluated. Some examples of useful methods have recently appeared: Rao, Buttlaire & Cohn (24) have conducted a study of the arginine kinase reaction at equilibrium using ^{31}P NMR at catalytic and stoichiometric enzyme concentrations. Because three out of four of the reactants produced resonances that were easily distinguished, the equilibrium distribution could be evaluated directly from the integrated signal areas without perturbing the system to terminate the reaction. Exchange rates between substrates and products were calculated from the change in linewidths upon addition of EDTA to stop the reaction. The fact that the number and areas of the ^{31}P resonances could be fully accounted for by the four reactants argues against the existence of a phosphorylated intermediate at a significant concentration on the enzyme at equilibrium. The equilibrium constant *on the enzyme* was found to favor P-arginine by a factor of 15 relative to the catalytic equilibrium constant, and the rate of substrate interconversion was nearly an order of magnitude faster than the overall reaction in the forward direction, which implies that steps other than the chemical transformation itself are rate limiting in arginine kinase catalysis.

Because different steps in an overall reaction usually have different free energies of activation, studies at very low temperatures may serve to stabilize transient intermediates long enough to permit fairly extensive physicochemical characterization (25–28), including the possibility of X-ray crystallographic analysis (29). Accumulation of intermediates can be achieved in concentrations sufficient to permit rate studies between consecutive steps upon slight warming. Controlling the ionic environment and demonstrating that the enzyme and catalytic pathway are unchanged by subzero temperatures and aqueous organic solvents are necessary, however, to ensure that these types of studies relate to the reaction under normal conditions. Douzou et al (25) initiated this approach using spectrophotometric and ORD analytical techniques to observe intermediate complexes I and II of the horseradish peroxidase-catalyzed reaction, and evidence for multistep reactions has now been reported for several proteases and glycosidases (29).

The demonstration of burst kinetics is often ascribed to formation of an intermediate and hence presumed to disfavor a concerted displacement. Subzero temperatures have been used to decrease the steady state rate of β-galactosidase and β-glucosidase so that the burst production of the aglycone, o-nitrophenol, could be seen (30, 31). This method has some limitations when used alone as evidence for an intermediate. It is possible that a direct displacement reaction occurred but that the rate-limiting step is either departure of product or recycling of the enzyme from the product conformation to the conformation required for action with the substrate. This latter case is interesting because exact stoichiometry would also be expected.

Another method used for measuring rate constants from rates and concentration of intermediates is a destructive method. To measure the concentration of intermediates in the muscle aldolase reaction in the steady state (32), fructose-1,6-P_2 was generated slowly and continuously in the presence of a large excess of aldolase such that all of the substrate would be bound. The reaction was run with a high activity of product-trapping enzyme so that whenever the solution was quenched in acid all of the fructose-1,6-P_2, dihydroxyacetone-P and glyceraldehyde-3-P that were assayed represented species on the enzyme that decayed in acid to these stable forms. It was concluded that the apparent k_{off} rate constant for glyceraldehyde-3-P was about 20 times the steady-state rate and probably much faster than the aldol condensation with bound dihydroxyacetone-P.

In any destruction analysis experiment such as this, special methods of quenching may be needed to determine those different bound species that have become indistinguishable by the process of the quenching. Thus in the aldolase experiment dihydroxyacetone-P will have come from the carbinolamine, the Schiff's base, the enamine, and from bound dihydroxyacetone-P itself. Loss of stereospecificity in the quenching of enols and enamines provides a way of distinguishing them from the keto forms on an enzyme. A similar problem in the chemical anomerization of sugars has been solved by chemical quenching to form trimethyl silyl ethers and subsequent identification of the stable derivatives. The problem with the enzyme is more difficult than this because the enzyme is itself a catalyst. During the quench process an unstable intermediate may form a linkage to the enzyme. In this case therefore, not only is the intermediate lost but its covalency to the enzyme would be misinterpreted. The existence of a hemiacetal ester in the sucrose phosphorylase mechanism was indicated by studies of Voet & Abeles (33) in which [^{14}C]glucose-1-P with enzyme was destroyed in acid or sodium dodecyl sulfate solution. Almost an equivalent of ^{14}C, without phosphate, was covalently linked to the denatured protein. Methanolysis of the denatured enzyme in acid produced α-methyl glucoside. Therefore it was concluded that the bond to a carboxyl group of the enzyme had been formed with inversion of the original α-glucosyl linkage of the substrate. In spite of these indications of a direct S_N2 displacement of P_i by enzyme, the authors keep open the possibility of a carbonium ion mechanism in which the intermediate collapsed to a specific ester with the protein upon denaturation. Ray & Long (34) have been concerned with how best to mix an enzyme incubation with a quenching solution to stop a reaction well within a single turnover time so that the composition of species in the steady state can be obtained.

Chemical quenching may also be of use in identifying intermediates such as carbanions and carbonium ions. When an equilibrium mixture formed from large amounts of fumarase and T-fumarate was quenched in hot methanol, I. A. Rose and E. L. O'Connell (unpublished) were not able to detect significant amounts of either 2-methoxy succinate (expected from a carbonium ion intermediate) or [RS,3T]malate (expected from the random protonation of a carbanion). Both were present at $< 10^{-3}$ of the concentration of bound substrates. However, the ratio of malate to fumarate on the enzyme was found displaced from the solution equilibrium observed with a small amount of enzyme; from this distribution of fumarate and malate in the quench mixture it was evident that the reaction was stopped before intermediates had disappeared.

The measurement of rate constants for release of substrate from its functional binary complex with an enzyme is important, not only because it represents a necessary step in the reverse direction but also because if it is slow relative to k_{cat}, the forward reaction may be said to be diffusion-limited. The enzyme may be thought of as having evolved to a state of "perfection" such that no improvement in rate can be achieved by increasing catalytic efficiency. Evolution may continue to increase the V_{max} although this avenue has limited appeal because the cell must operate at ever higher concentration of metabolites, which may cause difficulties for the cell. The thesis of Albery & Knowles (35), that triose-P isomerase has evolved to perfection, depends on their assumption that this enzyme normally acts in the cell well below saturation and thereby acts in a way that cannot be speeded up by increasing the catalytic rate. This thesis of perfection and its evolutionary interpretation requires that the enzyme never be rate-determining in metabolism, otherwise the concentration of substrate would increase to give a reaction-limited, not a diffusion-limited rate. Little reliable information exists about these matters because the "concentrations" of the substrates as determined are probably not true values and compartment sizes are only known by assumption in most cases. However, in the case of gluconeogenesis in the intact rat liver, primary isotope effect studies (36) show that triose-P isomerase is a rate-limiting step for utilization of glycerol, and, by extension from earlier labeling patterns (37), this is also true for the utilization of lactate for glycogen synthesis. In this case, the expected increase of substrate beyond the capacity of the isomerase could be corrected by increase in V_{max}, which is only about $3000s^{-1}$, and sufficiently slow to allow the escape of most of the hydrogen that is being transferred in the reaction.

A method for determining the apparent off-rate constants of substrate from binary and ternary complexes with enzyme and other substrates evolved in the laboratories of Bar Tana & Rose (38), out of an effort to clarify the isotope trapping technique introduced by Meister's group (39) to demonstrate covalent intermediates. One substrate, containing isotope, is mixed with enzyme and then rapidly mixed with unlabeled substrate and second substrate or cofactor, whichever is necessary to complete the mixture for reaction with the enzyme. The reaction is stopped as soon as possible after at least 5 half-lives. The extent to which isotope at the original specific activity appears in the product as a function of the concentration of the trapping species, second substrate, or cofactor, can be used to calculate the rate of dissociation of S from a functional state on the enzyme. The method provides evidence for proper binding in the binary complex, at least to the extent that bound isotope can be trapped. Incomplete conversion into product of the substrate bound in the "pulse" can arise for two reasons: part of the bound substrate may be bound incorrectly to enzyme and cannot be righted without dissociation, and/or substrate may dissociate from the *ternary* complex too rapidly to be trapped as product completely. However, if some trapping occurs then dissociation from the ternary complex cannot be rapid compared with k_{cat}. Defining $K_{1/2}$ as the concentration of second component required to reach half maximum trapping it is possible to derive some simple equations (38, 40) relating $K_{1/2}$ and the desorption rate of substrate from the binary complex. Furthermore, if incorrect binding is assumed not to be the explanation for the failure to trap all of the substrate that was bound, one can estimate the upper limit to the desorption rate of substrate from the ternary substrates complex.

"Parallel Line Kinetics" and Isotope Exchange

The most common case in which kinetic studies contribute to a clear decision between mechanisms is where parallel plots of $1/v$ vs $1/S_1$ at different S_2 are obtained. The significance of this is that an irreversible step precedes the addition of S_2 and this step is often the release of first product, P_1. This thereby signifies the presence of an intermediate derived from ES_1. It is necessary to substantiate this interpretation with isotope exchange studies ($S_1 \rightleftharpoons P_1$ in the absence of S_2) because parallel line kinetics can be obtained in other circumstances (41–44).

Many mechanisms have been "established" with the aid of the isotope exchange technique. However, in those cases in which the observed rate is not kinetically competent, further investigation is required before definite conclusions are drawn. Exchange rates may be lowered by substrate or product inhibition, and this can only be evaluated by varying conditions to obtain the extrapolated maximum velocity. Evidence for an intermediate by the use of isotope exchange may be masked if product dissociation is slow or does not occur in the absence of one or more components of the overall reaction. On the other hand, a slow exchange may represent a side reaction or a contaminating enzyme activity, neither of which are relevant to the catalytic mechanism in question. (The ATP:ADP exchanges that occur in the presence of hexokinase (45) or chicken liver pyruvate carboxylase (46) seem to illustrate this point.) Too often the possibility that a contaminating substrate may be responsible for the exchange is completely ignored. A simple test of this is to determine the effect on the exchange rate of adding substrate. In the case of a true intermediate capable of forming without addition of the reaction-completing substrate the addition of that substrate should lower the rate of isotope exchange, not increase it as if the observed exchange were due to contamination.

It is important to recall that isotope exchange will not corroborate the presence of an intermediate that is suspected from steady state kinetics when two products are formed in producing the intermediate: $E + S_1 \rightleftharpoons ES_1 \rightleftharpoons EX + P_1 + P_2$. Because both products will be present in an amount equal to EX, the study of exchange between one of the products and S_1 is unlikely to give significant rates. This is true whatever the order of release of products. Once a product is released at zero concentration, not only is the path for back exchange of the other cut off, but also, in the absence of S_2, no pathway exists to regenerate E from EX to permit successive turnovers. As an example, most carboxylases containing the biotin prosthetic group catalyze Mg^{2+} and HCO_3^--dependent $ATP:P_i$ and $ATP:ADP$ exchanges (see section on acyl transfer). However, the former requires the presence of ADP, and the latter, P_i (47). The absence of partial reactions is meaningless, although E-carboxybiotin is known to be an intermediate, because of the formation of two products.

Perhaps the chief reason that a stepwise mechanism does not give parallel line kinetics, and also fails the exchange reaction test in the absence of the last substrate, is that product release somehow requires the presence of the complete system. In such cases, investigation of intramolecular positional exchange in a reisolated substrate may provide otherwise hidden mechanistic information.

Intramolecular Positional Exchange

An early study of catalytic mechanism by positional exchange was made by Gold & Osber (48) who attempted to demonstrate that, although no glucose-1-P:P_i exchange is observed, the glycogen phosphorylase reaction proceeds via a glucosyl-enzyme or carbonium ion mechanism. Their approach involved incubation of the enzyme with glucose-1-P, labeled with ^{18}O in the C_1–O–P position, in the absence of glucosyl acceptor (Equation 4)

If a carbonium ion or glucosyl-enzyme intermediate were formed, and if P_i can rotate freely on the enzyme, then exchange of the C–O–P oxygen with the external PO_3 oxygens of reisolated glucose-1-P would occur. The rate of this exchange was determined to be only 0.3% of the overall rate of reaction with enzyme + glucose-1-P. However, in a later study by Kokesh & Kakuda (49), the presence of a glucosyl-acceptor analog was found to enhance the oxygen scrambling rate, demonstrating that cleavage precedes transfer to the glucosyl acceptor, but requires some type of activation by the second substrate.

Midelfort & Rose (50) developed a method to detect reversible cleavage of bound ATP to bound $ADP \cdot X\text{-}PO_3$ which does not require that ADP dissociate from the enzyme. The ATP $\beta\gamma$-bridge oxygen was made highly enriched in oxygen-18. When this ATP is cleaved to bound ADP, the isotope becomes scrambled among the $\beta\text{-}PO_3$ oxygens of ADP if rotation about the $P_\alpha O\text{-}P_\beta$ bond can occur at the catalytic site (Equation 5).

$$\left[\begin{array}{c} AMP_\alpha - O - \overset{\overset{O}{|}}{\underset{\underset{O}{|}}{P}_\beta} - \bullet - \overset{\bullet}{\underset{\underset{\bullet}{|}}{P}_\gamma} - \bullet \end{array}\right] \rightleftharpoons \left[\begin{array}{c} AMP - O \overset{O}{\overset{}{\underset{O}{\diagdown}}} P \overset{\bullet}{\underset{}{\diagdown}} \cdot X - PO_3 \end{array}\right]$$

$$\left[\begin{array}{c} AMP - O - \overset{\bullet}{\underset{\underset{O}{|}}{P}} - O - \overset{\bullet}{\underset{\underset{\bullet}{|}}{P}} - \bullet \end{array}\right] \rightleftharpoons \left[\begin{array}{c} AMP - O - P \overset{\bullet}{\diagup} \overset{}{\diagdown} O \cdot X - PO_3 \\ \diagdown O \end{array}\right] \qquad 5$$

Reformed ATP will exhibit β-nonbridge ^{18}O enrichment due to positional exchange of the $\beta\gamma$-bridge and β-nonbridge oxygens. If the β–PO_3 group of bound ADP is symmetrical during the reversible cleavage, the exchange rate will equal the reversible cleavage rate of ATP, enabling the kinetic competency of an $[ADP \cdot X\text{-}PO_3]$ intermediate to be evaluated. By this method, the kinetic feasibility of a γ-glutamyl-P intermediate was demonstrated in the glutamine synthetase reaction; this will be discussed later in the section on acyl transfer.

If ATP dissociation is slow relative to reversible P_β–^{18}O-P_γ bond cleavage, multiple cleavages will occur in each molecule of ATP that reacts. If, as might be expected, the β-PO_3 oxygens of ADP become equivalent during the initial cycle, multiple cleavages will not be detected by the $\beta\gamma$-bridge : β-nonbridge exchange technique, and the observed positional exchange rate will represent a minimum value. Evidence for multiple cycles may be obtained, however, if the reversible cleavage reaction results in a "change" in the ATP γ-phosphoryl group as would occur if P_β–O-P_γ bond scission is brought about by hydrolysis. Wimmer & Rose (51) used this approach to study the mechanism of ATP : H_2O oxygen exchange in the photophosphorylation system of chloroplasts (see section on phosphoryl transfer).

Assuming rotational freedom the minimum rate of positional isotope PIX required by the overall reaction kinetics of an enzyme is $(V_-)(V_+)/[(V_-) + (V_+)]$ where V_+ and V_- are V_{max}s in each direction.[1]

[1]This can be seen as follows: using saturating S to determine the maximum positional exchange rate, V_x, with its statistical correction, and saturating S and P in V_+ and V_- studies;

$$E + S \underset{2}{\overset{1}{\rightleftharpoons}} ES \underset{4}{\overset{3}{\rightleftharpoons}} E^x_y \underset{6}{\overset{5}{\rightleftharpoons}} EP \rightleftharpoons E + P$$

One obtains for the exchange:

$$\frac{E_t}{V_x} = \frac{k_2 + k_3 + k_4}{k_2 k_4} + \frac{1}{k_3}$$

The rate constant k_3 represents the contribution that V_+ makes to V_x and it is

$$\geqslant V_+/E_t \text{ or } \frac{1}{k_3} \leqslant \frac{E_t}{V_+}$$

(continued on next page)

Isotope Effects

Important books (52–55) and articles (56, 57) on isotope effects of special interest to biochemists have appeared recently. They reflect the recognition of the unique contribution that these approaches afford in enzyme reaction studies. Only a brief survey of important concepts and approaches can be ventured at this point.

An important conceptual step is taken with the recognition that isotope discrimination in a competitive situation corresponds to a comparison of competitive kinetics as a ratio of V/K effects (58, 59). V/K corresponds to the apparent second order rate constant at $S \ll K_m$. That is, if discrimination between two competing isotopic, or otherwise different, substrate species is studied, the ratio of utilization is given by the ratio of V_m/K_m of the two species if they had been measured separately. The character of V/K is such that it does not have any of the kinetic terms of the reaction pathway that are beyond the first irreversible step of the reaction. Instead it includes all steps up to and including the rate limiting step. From the expression for V/K one can sort out k_{on} and k_{off} of the substrate by equation 6:

$$V/K = \frac{k_{on} \cdot k_{cat}}{k_{off} + k_{cat}} \qquad\qquad 6.$$

where k_{on} and k_{off} are *not* lumped constants but single rate constants for the productive contact and release of substrate. The k_{off} may therefore differ from the lumped constants determined by steady state or isotope trapping experiments, and both may differ from the combined constants that make up the dissociation constant of the ES complex. Equation 6 describes the

---

For the reverse reaction at saturating P there may be intermediate forms between EP and E_y^x. These will lower V_- relative to V_x which will have all the enzyme distributed between the two forms ES and E_y^x. If such forms do not exist one obtains:

$$\frac{E_t}{V_-} = \frac{k_2 + k_3 + k_4}{k_2 k_4}$$

where K_m of P is $(k_2 + k_3)/k_4$. If they do exist

$$\frac{k_2 + k_3 + k_4}{k_2 k_4} < \frac{E_t}{V_-}$$

Therefore

$$\frac{E_t}{V_x} = < \frac{E_t}{V_-} + \leq \frac{E_t}{V_+}$$

from which the initial equation is derived as a lower limit.

partition of the first properly contacted species of substrate between forward and reverse flow. When two substrate species are competing, the ratio of partition functions of first contacted species gives the ratio of the two species to give products which, when compared to the initial ratio gives the discrimination effect. If k_{cat} for one of these substrates is altered by isotope substitution, the ratio of the V/K values predicts the competitive discrimination that will be observed (40, 59). The k_{cat} in this expression includes all rate constants of V_{max} except those that follow the first irreversible step. For a linear sequence such as occurs with muscle aldolase:

$$DG + E \underset{2}{\overset{1}{\rightleftharpoons}} E^{DG} \underset{4}{\overset{3}{\rightleftharpoons}} E^D_G \overset{5}{\longrightarrow} E^D \overset{7}{\longrightarrow} E + D,$$

$$\frac{V_m}{K_m} = \frac{k_1 k_3 k_5 / k_4}{k_2 + k_3 k_5 / k_4}$$

whereas

$$\frac{V_{max}}{E_t} = \frac{k_{cat} \cdot k_7}{k_{cat} + k_7}$$

For the isotope effect in bond cleavage to be expressed in V_{max} requires that $k_7 \gg k_{cat}$, whereas to be expressed in V/K requires that $k_2 \gg k_{cat}$. When V_{max}'s are compared from separate studies in order to obtain the isotope effect for the sequence between the ground state of ES and last product formation, it is of great importance to exclude inhibitors that might be introduced as impurities in the substrate preparations. Both isotopic species should be prepared in parallel. Ground state distortion ($E_{DG} \rightleftharpoons E'_{DG}$) as a mechanism of catalysis will be included in the k_{cat} term of both V/K and V and therefore, if it were characterized by an isotope effect, there would be no way to separate it from the isotope effect of the bond breaking step by steady state kinetic studies.

If other non–isotope–sensitive steps are partly rate-determining for either V_{max} or V/K, their contribution will have to be evaluated in order to obtain the intrinsic isotope effects on these terms. Such corrections are generally most difficult to calculate so that only if rather large isotope effects have been found, i.e. the critical step is kinetically isolated, can the measured effect be used with some confidence to compare with calculations for different transition states than might be proposed. In the case of hydrogen isotope effects where data with two isotope substitutions are possible, the extent of the correction can be evaluated and the intrinsic effect calculated, in principle. This procedure has been standardized by Northrop (59). In cases of moderate isotope effects this procedure requires that suitably precise data be obtained (60). It is worth spending time with altered reaction conditions (temperature or pH) or poorer substrates, attempting to improve the isolation of the bond breaking step before proceeding to collect final data that might require correction for the contribution of other steps.

Having obtained the isotope effect for the bond breaking step it will be desirable eventually for enzymologists to fit various reasonable transition state structures to determine those that are compatible. When this is done with more than one isotope the range of acceptable geometries is rapidly narrowed. To calculate the isotope effect corresponding to a particular model with variations in bond lengths and angles the current practice is to use spectroscopically derived force constants in a program devised by Wolfsberg & Stern (61, 62). A good example of this approach is the study of Burton et al on the hydrolysis of tertiary butyl chloride (63). This study might be a good model for glycoside transfer reactions that go by a carbonium ion mechanism. Without making the full valence force field calculation, assuming the transition state to approximate a known structure, the transition state/reactant equilibrium (on which absolute rate theory rests) may be approximated by the equilibrium of known compounds for which an isotope effect can be obtained, and thus it is often possible to describe the mechanism in general terms. Tables of fractionation factors and the definitions necessary to calculate reaction equilibria can be found (56, 59, 64). The fractionation factor ϕ_{RH} represents the ratio RD/RH that would result in a hypothetical H(D) exchange equilibrium with some very abundant reference compound that is a 50:50 mixture of the two isotopic species. The reference compound that has been used by organic chemists is acetylene because most exchange equilibria with acetylene would give fractionation factors > 1. Biochemists are accumulating data with water as the reference because there are many equilibria that can be measured relative to water. To relate the two conventions the acetylene related fractionation factors are multiplied by 0.62 to obtain the water related values when H and D are compared. Tritium fractionation factors can be derived with the Swain equation, $k_H/k_T = (k_H/k_D)^{1.44}$.

To complicate matters however, very large primary isotope effects may result from quantum mechanical tunneling in a proton transfer reaction (65, 66) if the proton is not solvated in the ground or transition states. Large effects can also come from a sequence of C–H bond activating reactions in which more than one makes a contribution to the rate (67). A distinction between the normal isotope effect and tunneling can be made by the requirement that tunneling effects and zero point energy effects have different activation energies. At lower temperatures the tunneling effect, requiring less activation energy, will tend to flatten out an Arrhenius plot more so for H than D because of the larger contribution of tunneling to H. The extrapolated slopes will cross and give $A^D_{calc}/A^H_{calc} > 2$ (66). Examples of large primary isotope effects in enzyme reactions are: the coenzyme B_{12} dehydrase, $k_H/k_T > 50$ (68, 69), and glucose oxidase, $k_H/k_D = 10\text{--}15$ (70).

Isotope effects can result from binding interactions of enzymes and ligands if there is a change in bond character in the complex. This has been used to suggest ground state activation of NADH by alcohol dehydrogenase (71), but this evidence has not been confirmed (72). Lewis & Wolfenden have used secondary isotope effects to demonstrate that aldehydes that are good inhibitors of papain show lower K_i values with 1–D substitution (73). This is taken as evidence that the inhibitor is either the tetrahedral gemdiol or an aldehyde addition compound with the enzyme.

Schimerlik, Rife & Cleland have developed an equilibrium perturbation method to determine kinetic isotope effects (74, 75). Equilibrium concentrations of reactants and products are mixed with a heavy isotope substitute at a primary or secondary position in one of the reactants only. There will be an imbalance in the forward and reverse rates if an isotope effect occurs. By using a spectral parameter of the reaction, the "overshoot" of the equilibrium is followed and seen to return to equilibrium with time. The isotope effect is calculated from aspects of the rate curve and is capable of sufficient sensitivity to measure heavy atom isotope effects. This method owes its unusual sensitivity to the fact that it measures the difference between the two rates directly rather than the rates themselves. The isotope effect is a V/K effect as "competition" between normal isotope in one direction and heavy isotope in the opposite direction causes the temporary shift from equilibrium. Depending on the reactant pair involved and the order of product release, the equilibrium perturbation effect can be identical to the usual V/K effect, or can include additional kinetic terms (75). This is because products as well as substrates are present in the equilibrium measurements, and several levels of reactants may have to be tried to obtain the maximum observable isotope effect.

GROUP TRANSFER REACTIONS

The phrase "group transfer" excludes few of the enzymes that we know, if H^+ is included. In general the questions we are concerned with are: is transfer between donor and acceptor direct or by way of an intermediate? and does the enzyme function as a transient acceptor and donor? A recently published summary (76) has given the impression that stepwise catalysis is a preferred if not exclusive reaction pathway. It seems reasonable to suppose that the evolution of "complex" reactions such as pyruvate kinase and glutamine synthetase occurred by joining simpler steps within protein units forming more complex structures. In this way a bound intermediate would either always be found or at least be expected in some examples of a reaction class.[2] Transamination between amino acids is simple in the sense that a single chemistry applies to both half reactions whereas transamination between glutamine and glucosamine-6-P by glucosamine-6-P synthase is probably chemically complex in a sense similar to pyruvate kinase, glutamine synthetase, carbamyl P synthetase, sucrose phosphorylase, and many others, in which mirror image steps are not involved in completing the transfer and therefore additional catalytic components must be required.

Proton Transfer

The glyoxalase reaction, previously considered to be analogous to an internal Cannizarro reaction (hydride transfer), has been shown by Hall et al (79) to be a proton abstraction mechanism. The reaction occurs primarily with intramolecular transfer but also with entry of a proton as shown in D_2O. Earlier studies of a similar nature failed to appreciate the importance of small amounts of exchange labeling as an indication of a proton transfer mechanism (Equation 7)

7.

Evidently a basic group on the 2 si face removes the proton from the tetrahedral C-1 of the glutathione adduct. A thiohemiacetal is involved in the oxidation of aldehydes by glyceraldehyde-P dehydrogenase (80) and of the formaldehyde glutathione adduct by formaldehyde dehydrogenase (81). These are not base catalyzed abstractions. The difference is, of course, the presence of the C_2-carbonyl in the methylglyoxal that facilitates enolization of an aldehydic proton even from a carbon that does not generally show H^+ exchange.

Polarization of an α-carbonyl by an electrophile on the enzyme would facilitate enolization. Evidence for increased positive character at the carbonyl carbon has been obtained with triose P isomerase by showing that reduction by BH_4^- is approximately seven times faster than in solution (82).

---

[2]Examples of reaction classes where differences in mechanism within a class may suggest a progression in evolution are not infrequent: phosphoglycerate mutase of plants and animals differ greatly. In the latter both enzyme and glycerate-2,3-P_2 are used as carriers of P whereas some more direct paths of transfer may apply for the plant enzyme (77). Although phosphoryl transfer between ATP and acceptors clearly involves E-P in the case of nucleoside diphosphokinase (78), evidence for E-P in most other kinase reactions is negative.

Reduction of pyruvate by borohydride was shown by Kosicki & Westheimer to be catalyzed by pyruvate kinase (83) and is strictly stereospecific (2) with a catalytic rate about 10^5 times greater than the rate of pyruvate reduction in neutral solution.

A continuing problem is how a C–H bond is activated to a carbanion when neighboring to a carboxyl group and when there is no evidence for thioester formation in a prior step. Mechanisms of proton abstraction, such as electrophilic substitution or elimination, are much less attractive than enol or enamine formation. A concerted electrophilic carboxylation by carboxybiotin of substrates such as pyruvate that are, in other reactions, given to enolization, is considered below. The Bruices (84) have shown that an elimination mechanism for proton activation can be imposed on oxalacetate by tertiary amines (equation 8).

$$ 8. $$

The comparable role for glutathione as a coenzyme of glutathione S-transferase (85) in the enolization of 3-keto steroid would be as shown in equation 9.

$$ 9. $$

Tobias & Kezdi (86) considered the possibility of an elimination with a good leaving –OR group

as a proton labilizing mechanism but found no support for this. It is clear from ^{18}O studies by Dinovo & Boyer (87) with fumarase that no additional ^{18}O enters the reaction as required if the C–4 carboxyl of malate were hydrated to activate the 3–H for an elimination process. ^{18}O should be found in the carboxyl group of the products due to torsional symmetry of the adduct if this were the case.

An elimination mechanism for production of the enediol in sugar isomerases has been considered in the past. It is known that the gemdiol form of glyceraldehyde-3-P is not a substrate for triose-P isomerase (88). However for the enzymes that isomerize pentoses and hexoses it is known that the cyclic forms are responsible for the reaction (14). One could therefore easily write an elimination mechanism that would solve both the ring opening and enolization problem in one step. Stereochemical studies would be consistent with an anti elimination in all cases. Evidence against such a mechanism is the observation that 5-deoxyglucose-6-P, which cannot be an analogue of the abundant pyranose form of glucose-6-P but exists in large part in the aldehyde form, is a very good substrate for glucose-6-P isomerase (C. F. Midelfort, unpublished). Further support for a two step mechanism with this enzyme is its ability to anomerize mannose-6-P very well (89) whereas mannose-6-P is not isomerized significantly (14).

The ability to operate on the open chain forms as intermediates requires that this isomerase bind them tightly to displace the unfavorable equilibria of ring opening [$\sim 10^{-5}$ for glucose in water (90)], and yet have the freedom for torsion at C_1–C_2 and C_2–C_3 needed for anomerization. This requires a strong interaction with the newly formed 5–OH. The alternate explanation for catalysis of mutarotation by this enzyme through dissociation of the open chain form is ruled out by demonstrating that catalysis of isomerization occurs with both α- and β-forms of the substrate.

Monder and co-workers (91–93) have discovered a versatile enzyme (system) in the path of formation of C_{21}-acid steroids. Transfer of tritium from C_{21} of 11-deoxycorticosterone to C_{20} of the corresponding C_{21} acid was observed during catalysis by 21-hydroxysteroid dehydrogenase, and is very suggestive of an enediol intermediate in the path to aldehyde (equation 10).

10.

In addition the enzyme interconverts the C_{20} epimers of the aldehydes that are intermediates toward oxidation. This combination of isomerization and epimerization may make it difficult to determine if cis- or trans-geometry is required in the presumed enediol intermediate.

It is of interest to ask if there are any data in support of a true enediol or enediolate ion intermediate in the isomerases. Is there a true conjugate acid formed from a basic group of the enzyme and the substrate derived proton? An alternate explanation by Harris & Feather (94) for the intramolecular transfer seen in the sugar isomerases is based on the nonenzymatic, acid-catalyzed transfer reaction, [1-T]fructose → [2-T]glucose. In 2N H$_2$SO$_4$ at 100°C, glucose that was formed had $\sim$ 79% of the specific activity at C–2 of the starting fructose. However, more hydride transfer (equation 11) was shown in aqueous acid than is seen in many isomerase reactions, which suggests that these make use of base catalysis (equation 12).

$$11.$$

Gleason & Barker (95) have obtained evidence for intramolecular transfer (0.3%) in 0.8N KOH for [2-T]ribose → [1-T]arabinose. This is less transfer than seen in the sugar isomerases but suggests that base catalysis may be able to explain transfer without the implication that the enzyme contains the conjugate acid. These and other observations suggested that the abstracted tritium may transfer within a complex with the developing carbanion in a conducted tour mechanism or base catalysis (equation 12).

$$12.$$

This mechanism is perhaps to be preferred to the acid-catalyzed hydride mechanism which cannot explain proton exchange.

The conducted tour mechanism could predict the large range of exchange/transfer ratios, 200/1 to 1/5000, that have been reported for various isomerases. The proton would be partitioned in equilibria among the three central intermediates. It would be able to exchange only from the conjugate acid-enediol form so that the exchange rate could be very slow if this form were not favored relative to the shared-proton states. If proton transfer to the base were rate limiting for exchange, the base could be considered a counter-ion for the complex and the conjugate acid would have only the superficial role of permitting proton exchange with the medium. In this case the observation of greater than 33% transfer would not rule out $-NH_2$ as the base. The observation of extensive exchange would imply that the counter-ion role of the base is insignificant compared to its role as a general base.

Referring to Equation 12, it is possible that separate routes of proton exchange could occur if the carbanions were not rapidly interconverted in step 2, that is if formation of EBH and its exchange with the medium precede the formation of the enediol. This would be recognized if tritium from water partitions differently into substrate and product when measured in the two directions of reaction, i.e.: $(S_T/P_T)_S \rightarrow P > (S_T/P_T)_P \rightarrow S$. In the three cases that have been examined, glucose-P (96), triose-P (97, 98), and glucosamine-P (99) isomerases, the ratios are very close to equal.

The standard technique for establishing a carbanion intermediate mechanism in an enzyme such as aldolase is to observe proton exchange in the absence of a second substrate. This is often not observed when it might be expected, perhaps because the enzyme-bound proton does not readily exchange. A technique introduced and explored by Walsh and Abeles et al (100, 101), to provide evidence for proton activation, depends on the use of a pseudosubstrate which, upon proton abstraction, becomes unstable and decomposes to a stable product by elimination of an adjacent or vinylagous halogen. In this way these authors were able to support a carbanion mechanism of substrate activation for flavin reduction. This approach, which is discussed in the chapter by Walsh in this volume, is also considered in the section on carboxyl transfer below.

The use of suicide reagents, recently reviewed by Abeles & Maycock (102) and by Rando (103), is usually based on an enzyme's capability for proton abstraction from a substrate analogue. The proton removal activates an innocuous compound to an active electrophile capable of forming a covalent bond with the enzyme that activated it. It is of interest that nature has used the suicide activation in the design of specific toxins such as penicillin (104).

Proof that proton transfers occur by virtue of action of the enzyme as a base is difficult to obtain except in special cases where transfer between stable positions of substrate and product can be detected. However, Yamada & O'Leary (105) have recently proposed that a lack of isotopic discrimination in forming a C–H(D) bond in H_2O/D_2O mixed medium can be taken as evidence for participation of EBH in the path between H_2O and product. This would indeed be a useful method, but the failure to show discrimination can have other explanations. Such a conclusion would indeed be justified if a discrimination were seen at low substrate concentration but not high concentration in a one substrate reaction (such as fumarase or aldolase). This could occur only if H^+ entered the reaction prior to substrate:

$$E \underset{1}{\rightleftharpoons} E^H \underset{2}{\rightleftharpoons} E^H_S \overset{3}{\longrightarrow} E + PH$$

Discrimination should disappear if the base has no hydrogens (not $-NH_2$ or $-ZnOH$) because high substrate would prevent the reverse of step 2. In the case that the base contains one or more protons, the discrimination may not disappear but will decrease from a partially intermolecular to a fully intramolecular effect. Discrimination will not vary with S if a group from the medium is the proton donor or if the proton reacts with the enzyme after the substrate. In the case that E^H_S is randomly reached a difference in the discrimination ratio with [S] implies action of enzyme as proton donor. In the case of two-substrate reactions a change in discrimination with substrate S_2 concentration will not be relevant if protonation of S_1 has preceded the step of S_2 addition. Therefore it seems that only under the special condition of ordered reaction—proton addition followed by first substrate at subsaturating concentration of other substrates—does a decrease in discrimination indicate the role of the enzyme in general acid catalysis.

Methyl Transfer

Initial rate studies give sequential kinetics for methyl transfer enzymes that use S-adenosylmethionine (107, 108) and hence provide no suggestion of an $E-CH_3$ intermediate. In order to determine the stereochemistry of methyl transfer, Mascaro et al (109) fed *Streptomyces grisceres* on methionine containing an isotopically chiral methyl-group. The antibiotic indolmycin was recovered from the medium and stereochemistry of its $-CH_3$ group was compared with that of the methionine. It is presumed that methyl transfer occurs from S-adenosyl methionine as indicated by equation 13.

13.

It was shown that the methyl transfer occurred with overall inversion (109). Thus, an S_N2 transition state including donor and receptor is likely.

The stereochemistry of the substitution at C–3 of indolpyruvate was determined by feeding 3R or 3S tritiated tryptophan also labeled with ^{14}C (110). The T/C^{14} ratio in the indolmycin was unchanged from that of the 3S compound and only 2% of the 3R tryptophan which indicates that activation occurred at the 3R position. Therefore, replacement of the hydrogen was with net retention, strongly indicating a stepwise process. Hörnemann & Floss (110) raise the question of whether the methyl transferase causes the enolization of the indolpyruvate itself or requires a separate enzyme. It may be of interest to note that, in the recent study by Rétey et al (111) of the stereochemistry of phenylpyruvate enolase, the H_R proton at C_3 is also specifically labilized.

An important study of the mechanism of $-CH_3$ transfer catalyzed by catechol-O-methyl transferase using secondary isotope effects was made by Hegazi et al (112). The enzyme was given either CH_3- or CD_3-S-adenosyl methionine and dihydroxyacetophenone as acceptor. Using exemplary caution to avoid introducing artifacts differentially, they found that the data establish a large inverse isotope effect, $V_H/V_D = 0.86 \pm 0.04$. This helped to rule out an enzyme methyl-carbonium intermediate ($sp^3 \rightarrow sp^2$, $V_H/V_D > 1$), and favors an S_N2 transition state as suggested from the stereochemistry of indolmycin synthesis (109).

On the other hand the intramolecular, vicinyl, $-CH_3$ migration in cholesterol synthesis has been shown by Phillips & Clifford (113) to go with retention. A double displacement mechanism was not considered by these authors although enzyme acting as an attacking group is often proposed in isoprenyl reactions. Rather a cyclic carbonium transition state is proposed in analogy to the pinacol rearrangement.

Acyl Transfer

Nonenzymatic and enzymatic reactions involving formation of oxygen esters, thiol esters, and amides from carboxylic acids generally require two steps: an activation process that transforms one carboxyl oxygen into a good leaving group, and a subsequent nucleophilic substitution by –OR, –SR, or –NHR at the carboxyl carbon. Enzyme-catalyzed activation and transacylation are often accomplished by a single enzyme with activation by adenylylation or phosphorylation incident to the cleavage of the P_α–OP_β or $P_\beta O$–P_γ bond of ATP. One of the key pieces of evidence for this initial step of the mechanism is the appearance of one of the original carboxyl oxygens in the AMP or P_i formed upon transacylation. The nucleophilic substitution step most likely proceeds by way of a tetrahedral intermediate formed upon addition of the nucleophile before departure of the leaving group.

It is not apparent why different ATP-utilizing acyl transfer enzymes have evolved, specifically employing one of the two activation mechanisms. In most adenylylation processes, partial exchange reactions have been found to occur, i.e. $ATP:PP_i$ exchange in the absence of acyl acceptor, while in cases of phosphorylation by ATP, partial reactions are not usually observed. Perhaps dissociation of PP_i in the former is favored by the potential for metabolic coupling of the reaction, for example to pyrophosphatase, thus increasing the concentration of enzyme in the acyl adenylate-complexed form. In most bacteria the ATP-dependent conversion of acetate to acetyl-CoA can occur by two pathways differing with respect to the carboxyl activation mechanism. One route involves two enzymes with free acetyl-P as the coupling intermediate, and the other is catalyzed by a single enzyme by way of formation of bound acetyl-AMP (equation 14).

The acetyl-P pathway does not occur in most higher organisms, although the overall equilibrium constants for the upper and lower pathways are similar in magnitude. Interestingly, an amoebal acetyl-CoA synthetase has been discovered recently which appears to catalyze the reaction: acetate + ATP + CoA $\rightleftharpoons$ acetyl-CoA + ADP + P_i (114). One might propose a mechanism for this enzyme which includes a bound acetyl-P intermediate, thus combining the activities of acetate kinase and phosphotransacetylase within a single enzyme. Perhaps, for efficient conversion of acetate to acetyl CoA, the acetyl CoA synthetase route has evolved because of the potential, via PP_i metabolism, to drive the reaction through utilization of both high energy bonds of ATP.

ACTIVATION BY ADENYLYLATION   Evidence for the participation of aminoacyl adenylates in the reactions catalyzed by aminoacyl-tRNA synthetases is abundant (115–119). Amino acid carboxyl activation is evidenced by the appearance of one carboxyl oxygen in product AMP. A parallel-line kinetic pattern is displayed which indicates that an irreversible step occurs after ATP and amino acid bind, but before tRNA reacts. This has been taken to suggest formation of amino acyl-AMP and dissociation of PP_i during the initial steps of the reaction. Consistent with this mechanism, most of the enzymes catalyze amino acid-dependent ATP:PP$_i$ exchange 10-100 times faster than the overall reactions; aminoacyl adenylates can be used as substrates, reacting either with PP$_i$ to form ATP or with tRNA to form aminoacyl-tRNA; active complexes of Enz-aminoacyl adenylate can be isolated; reaction in the presence of hydroxylamine results in formation of aminoacyl hydroxamates; and a kinetically significant AMP-dependent aminoacyl-tRNA:tRNA exchange is inhibited by PP$_i$ in a manner suggesting competition between tRNA and PP$_i$ for a common intermediate.

Loftfield (115) argues against the two step mechanism, pointing out that not all synthetases catalyze a tRNA-independent ATP:PP$_i$ exchange reaction. He favors an alternate, concerted pathway, in which Enz-aminoacyl-AMP would be looked upon as a dead-end complex. A pulse chase double-labeling experiment was designed by Lovgren et al (120) in which a limiting amount of tRNA was supplied to a mixture of free enzyme and enzyme-bound [^{14}C]aminoacyl adenylate complex in a medium containing Mg^{2+}, ATP, and tritiated free amino acid. If PP$_i$ was allowed to accumulate, almost all of the aminoacyl tRNA formed appeared to originate from the free, tritiated amino acid (120). However, Kim et al (121) found that in the presence of pyrophosphatase to remove PP$_i$, the rate of transfer of amino acid to tRNA from the preformed enzyme complex was consistently faster than transfer from free amino acid. It was concluded that the aminoacyl-tRNA synthetase reaction proceeds by way of an aminoacyl adenylate intermediate and that the results obtained in the absence of pyrophosphatase were complicated by a rapid PP$_i$-activated exchange between bound aminoacyl adenylate and free amino acid.

An interesting alternate method of peptide biosynthesis, exemplified by peptide antibiotic synthesis in *Bacillus brevis* (122, 123), involves a multienzyme complex-directed process that is independent of nucleic acids and that is formally analogous to the fatty acid synthetase system. The ribosomal and antibiotic synthesizing systems appear to have in common the activation of the amino acid carboxyl group by formation of aminoacyl adenylates. While in the former system transacylation occurs to the 3' (or 2' ?)-hydroxyl of tRNA, the antibiotic synthesizing complex uses the activated amino acids to form of specific thioester linkages with the enzyme itself.

During studies of the argininosuccinate synthetase reaction, substrate activation was evident from the transfer of oxygen to AMP from the ureido group of citrulline, although no citrulline-dependent ATP:PP$_i$ exchange was observed, as would be expected if AMP-citrulline were an intermediate in the reaction (124). In fact no partial reactions were found with any substrate:product pair that could not be explained by a tightly-bound argininosuccinate contaminant. The possibility of a concerted mechanism was therefore considered. As an alternative to exchange studies which require that substrate and product dissociate during the partial reactions, a pulse-labeling experiment was carried out by preincubation of enzyme with [^{14}C]citrulline and ATP followed by addition of excess unlabeled citrulline and aspartate (125). The 60% trapping observed could simply reflect slow citrulline dissociation in the presence of ATP and not necessarily formation of a covalent intermediate. However, studies performed using enzyme at substrate levels demonstrated citrulline-dependent cleavage of ATP to PP_i stoichiometric with the amount of enzyme present, suggesting formation of citrulline adenylate which remains tightly bound to the enzyme in the absence of aspartate. PP_i formed must also remain bound to explain the lack of $ATP:PP_i$ exchange in the presence of citrulline (125).

ACTIVATION BY PHOSPHORYLATION   Evidence for the intermediate formation of γ-glutamyl phosphate in the glutamine synthetase reaction (ATP + glutamate + $NH_3 \rightleftharpoons$ ADP + P_i + glutamine) has been reviewed by Meister (126). The marked inhibition of the enzyme by one of the four isomers of methionine sulfoximine is dependent on the presence of ATP and a divalent metal and is associated with cleavage of ATP to ADP (127). Meister and co-workers (128) observed that a phosphorylated derivative of methionine sulfoximine is produced slowly in this "suicide" reaction and is stoichiometric and tightly complexed with the enzyme. The structure of this inhibitor form is analogous to both intermediates postulated to occur during the normal synthetase reaction:

$$
\begin{array}{ccc}
{}^=O_3P-O-\overset{\displaystyle O}{\underset{\displaystyle \,}{C}}{\Large \diagup} & {}^=O_3P-N=\overset{\displaystyle CH_3}{\underset{\displaystyle \,}{S}}=O & {}^=O_3P-O-\overset{\displaystyle NH_3}{\underset{\displaystyle \,}{C}}-O^- \\
| & | & | \\
CH_2 & CH_2 & CH_2 \\
| & | & | \\
CH_2 & CH_2 & CH_2 \\
| & | & | \\
CHNH_3{}^+ & CHNH_3{}^+ & CHNH_3 \\
| & | & | \\
CO_2{}^- & CO_2{}^- & CO_2{}^-
\end{array}
$$

γ- Glutamyl Phosphate	Methionine Sulfoximine Phosphate	Postulated Tetrahedral Intermediate

However, the absence of $ATP:ADP$ exchange in the presence of glutamate alone and the lack of kinetic evidence left some doubt as to the existence of γ-glutamyl-P along the glutamine synthetase reaction pathway. Midelfort & Rose (50) have recently provided evidence against a concerted substitution mechanism by showing that in the presence of glutamate and the absence of NH_3, ATP participates in a phosphoryl transfer as rapidly as required for the process to be a component of the overall reaction. They observed $\beta\gamma$-bridge:β-nonbridge ^{18}O positional exchange in ATP reisolated after incubation under the above conditions, which indicates that this ATP at one time existed in the form of ADP on the enzyme (see the section on methods of study). No exchange of ATP γ-PO$_3$ oxygens with glutamate oxygens occurred indicating that P$_i$ had not been formed and returned to ATP in the absence of ammonia. The rate of glutamate-dependent reversible ATP cleavage was at least as fast as the maximum velocity in the slower, reverse direction, and was inhibited by approximately 50% in the presence of saturating NH$_4$Cl. The fact that total inhibition by NH$_3$ did not occur must mean that bound ATP that has been reversibly cleaved can return to free ATP from complexes that contain NH$_3$, i.e. NH$_3$ can be bound to forms of the enzyme between E$_{free}$ and E$\cdot^{ADP}_{Glu-P}$. The absence of γ-PO$_3$: γ-glutamate oxygen exchange in the presence of NH$_3$ (C. F. Midelfort, personal communication) indicates that, assuming P$_i$ is torsionally symmetrical on the enzyme, the return of bound P$_i$ to ATP is kinetically less favorable than product dissociation.

A two-step reaction sequence is the generally accepted mechanism for ATP-dependent transfer of CO$_2^-$ from bicarbonate to acceptor (such as pyruvate in pyruvate carboxylase) catalyzed by the biotin-dependent carboxylases (47). Evidence for the participation of Enz-biotin as an intermediate carboxyl carrier has come from initial velocity and product inhibition kinetics (129–131); resolution of *E. coli* acetyl CoA carboxylase into the three interdependent subunits biotin carboxylase, carboxyl carrier protein (containing biotin), and carboxyl transferase (132); and observations of the catalyzed carboxylation, at the 1'-N position, of both free and prosthetic d-biotin in the absence of carboxyl acceptor, with catalyzed transfer to the acceptor when it is added (47). The occurrence of ATP:ADP, ATP:P$_i$, and acceptor:carboxylated acceptor exchange reactions, which require only the components of one or the other partial reaction, is often cited as a major line of evidence for the existence of an Enz-biotin-CO$_2^-$ intermediate along the main reaction pathway. However, in general, the ATP:ADP and ATP:P$_i$ exchanges proceed too slowly to be used as a main source of mechanistic information. Both of these reactions depend on Mg^{2+} and HCO$_3^-$ and, respectively, on P$_i$ and ADP, but as discussed in the section on methods of study, this requirement for both products should be expected. The reason for the apparent lack of kinetic competence may reflect product inhibition, as discussed by Kaziro et al (133), or the slow dissociation of ADP and P$_i$ in the absence of carboxyl acceptor substrate.

The chemical mechanism of carboxybiotin formation from ATP and bicarbonate remains a subject of controversy. Three main schools of thought exist (47, 134), all consistent with the fact that bicarbonate is the substrate species and that during the carboxylase reaction, one bicarbonate oxygen can be found in P$_i$ for every –CO$_2^-$ transferred (133, 135).

1. The concerted mechanism was first postulated by Lynen et al (136) and is based primarily on oxygen tracer studies, the lack of exchange reactions associated with the first partial reaction unless all components are present, and the identification of the biotin carboxylation product as 1'-N-carboxybiotin (47).

2. A two-step mechanism proposed by Kaziro et al (133) with carboxyphosphate as intermediate is supported by the finding (137) that the isolatable biotin carboxylase subunit of acetyl CoA carboxylase catalyzes ATP synthesis from ADP and carbamyl phosphate. Although biotin is required for ATP synthesis from carbamyl-P, the prosthetic group does not appear to participate directly in the phosphotransferase reaction (138).

3. Model studies with ureido phosphate analogues provide the basis for Kluger & Adawadkar's hypothesis of an O-phosphobiotin intermediate in the biotin carboxylation (134). However, modified forms of biotin which would prevent phosphorylation or carboxylation of the ureido ring do not abolish ATP synthesis from carbamyl-P and ADP by the carboxylase subunit of acetyl CoA synthetase (138).

Sauers et al (139) have questioned the concerted mechanism for production of carboxybiotin from bicarbonate, which lacks reactivity toward nucleophilic attack relative to the carbon dioxide molecule itself. They propose that the breakdown of carboxyphosphate formed from ATP and bicarbonate serves to deliver CO_2 at high concentration to the active site for reaction with biotin. One might question why the enzyme would not choose to use reactive carboxyphosphate directly rather than go through the trigonal-linear-trigonal atomic rearrangements required for participation of bound CO_2.

The reaction catalyzed by carbamyl phosphate synthetase is essentially irreversible and has been postulated to proceed by the following mechanism (140, 141):

$$E + ATP + HCO_3^- \rightleftharpoons E \cdot O_2COPO_3 + ADP$$
$$E \cdot O_2COPO_3 + NH_3 \rightleftharpoons E \cdot O_2CNH_2 + P_i$$
$$E \cdot O_2CNH_2 + ATP \rightleftharpoons E + H_2NCO_2PO_3 + ADP$$

Overall:

$$2\,ATP + HCO_3^- + NH_3 \rightleftharpoons 2\,ADP + P_i + H_2NCO_2PO_3$$

This enzyme differs from the carbamate kinase of many microorganisms which catalyzes only the third step in a readily reversible fashion (142). In the development of carbamyl phosphate synthetase, a "carbamate kinase" function is coupled with a "carbamate synthetase" activity on the same enzyme, thereby increasing the concentration of $E \cdot O_2CNH_2$ for more efficient metabolism of ammonia. Carbamyl phosphate synthetase does catalyze the reverse of the third step, but done in phosphate buffer, the reaction produces only one mole of ATP and NH_3 per mole of carbamyl phosphate used (143). If this reaction indeed represents reversal along the main mechanistic pathway, then it can be explained if carbamate desorption and hydrolysis proceeds much more readily than enzyme-catalyzed phosphorolysis; alternatively, if P_i participates chemically in the reaction, hydrolysis of a carboxyphosphate intermediate may occur faster than phosphoryl transfer to ADP.

In general (144, 145), ATP:ADP and ATP:P_i exchanges are not catalyzed by CPSase in the absence of NH_3 which has been shown to be the amine substrate form (146). In fact, extensive kinetic studies (147, 148) indicate that ADP and P_i dissociate after carbamyl-P in a mechanism reported to be ordered sequentially. Powers et al (149) have observed that P_1P_5-bis(adenosine-5')-pentaphosphate is a strong inhibitor of CPSase implicating two ATP binding sites, one presumably involved with HCO_3^- activation, the other with carbamate phosphorylation. Consistent with the kinetic mechanism would be the demonstration that both ATP molecules were bound to the enzyme simultaneously during the reaction. CPSase catalyzes bicarbonate-dependent cleavage of ATP in the absence of NH_3 at a rate 10–20% the rate of carbamyl-P production, possibly limited by the slow dissociation of ADP and P_i in the absence of NH_3 (143).

Evidence seems to be accumulating concerning carboxyphosphates as the means of HCO_3^- activation by ATP. In an early study Jones & Spector (150) determined in the case of the overall reaction that one oxygen from HCO_3^- is found in the P_i formed, implicating the direct involvement of ATP in the activation process. This could be the result of a concerted process of carbamate formation, or of a reaction involving carboxyphosphate, possibly including bound CO_2. Pulse chase studies have been conducted by Anderson & Meister (141) which suggest either production of a bound intermediate from HCO_3^- and the γ-PO_3 of ATP, or slow HCO_3^- and ATP dissociation relative to conversion to carbamate, ADP, and P_i. Expanding this approach, Rubio & Grisolia (146) found that in the absence of NH_3, an amount of bound "carboxyphosphate" could be accumulated to a stoichiometry of 4.1 moles of intermediate per mole of enzyme. However, CPSase has been reported to exist in a monomer-dimer equilibrium, with one active site per monomer (151), and this fact raises a question as to the meaning of the Rubio & Grisolia stoichiometry of 4.1. Their proposed enzyme-bound intermediate decomposed upon high dilution with a half-time of approximately 0.7 sec at 25°, which is an order of magnitude slower than the estimated nonenzymatic breakdown rate of carboxy-P ($t_{1/2} \leqslant 0.069$

sec (139)) and must reflect stabilization by the enzyme. Powers & Meister (152) have reported trapping of a carboxyphosphate intermediate in the *E. coli* carbamyl phosphate synthetase system by reduction in dimethylsulfoxide with KBH_4 or by methylation with diazomethane. The stoichiometry (1.14 nmol with 3.6 nmol enzyme, assuming monomer form) may reflect the inherent instability of the intermediate as well as its steady state enzyme-bound level during the ATPase reaction in the absence of NH_3. Hard evidence for the kinetically competent formation of a carboxyphosphate intermediate in the CPSase mechanism of action remains scant at the present time.

During the carboxylation of phosphoenolpyruvate (PEP) to produce oxalacetate catalyzed by PEP carboxylase (PEPC), bicarbonate has been reported to be the substrate (153, 154), except in the case of the maize enzyme (155), and the reaction is essentially irreversible (153). This is in contrast to the PEP $\rightarrow$ oxalacetate conversions catalyzed by PEP carboxykinase (PEPCK) and PEP carboxytransphosphorylase (PEPCTrP) which use the CO_2 form as substrate (135) and are readily reversible because the high energy bond of PEP is conserved (156) (equation 15).

$$\text{PEP} + \text{HCO}_3 \xrightarrow{\text{PEPC}} \text{Oxalacetate} + \text{OPO}_3$$
$$\text{PEP} + \text{CO}_2 + \text{IDP} \underset{\text{}}{\overset{\text{PEPCK}}{\rightleftharpoons}} \text{Oxalacetate} + \text{ITP} \qquad 15.$$
$$\text{PEP} + \text{CO}_2 + \text{P}_i \underset{\text{}}{\overset{\text{PEPCTrP}}{\rightleftharpoons}} \text{Oxalacetate} + \text{PP}_i$$

A concerted mechanism has been postulated for all three reactions (153, 156–158). However, that pyruvate formation from oxalacetate is catalyzed by PEPCK and PEPCTrP suggests participation of an enolpyruvate intermediate in these two reactions (159, 160). Interestingly, pyruvate is formed nonstereospecifically while oxalacetate is formed stereospecifically by PEPCTrP (161) which suggests that in the absence of CO_2, enolpyruvate dissociates and is protonated nonenzymatically. Consistent with this is the inability of PEPCTrP to catalyze detritriation of [3-^{3}H]pyruvate (161). In the PEPCTrP case, a pentacoordinated pyrophosphoenolpyruvate intermediate, which could decompose to enolpyruvate and PP_i in the absence of CO_2 has also been considered (162).

An important piece of evidence concerning the mechanism of the PEPC reaction is the report that approximately one bicarbonate oxygen is found in the P_i formed for every two in oxalacetate (153). The carboxyl activation involved has been proposed to occur concomitant with carboxylation of PEP because of the failure to observe any partial reactions catalyzed by the enzyme. However, this concerted mechanism involves a cyclic pentacoordinated-P transition state in which adjacent rather than in-line concerted displacement is required. This would seem to contradict the "preference rules" of 5-coordinate phosphate chemistry (163) and may require a pseudorotation step before CO-P bond scission (see section on phosphoryl transfer). A two-step mechanism is also consistent with the data if intermediates remain enzyme-bound, and may be formulated to involve activation of HCO_3^- to bound carboxyphosphate by phosphoryl transfer from PEP, concomitant with formation of enolpyruvate.

ATP-INDEPENDENT TRANSACYLATION   Succinyl CoA : 3-ketoacid coenzyme-A transferase (CoA transferase) catalyzes a carboxyl activation/transacylation reaction that does not involve ATP cleavage because the acyl acceptor itself contains the group responsible for the activation (equation 16).

$$\text{Acetoacetate} + \text{succinyl CoA} \rightleftharpoons \text{Acetoacetyl CoA} + \text{succinate} \qquad \textbf{16.}$$

A ping pong mechanism is displayed for the overall reaction (164) and a covalent Enz-SCoA intermediate has been demonstrated to exist in the form of a γ-glutamyl thioester (164–166). The formation and reaction of

$$\begin{array}{c} \text{O} \\ \parallel \\ \text{Enz} -\text{C}-\text{S}-\text{CoA} \end{array}$$

thus has the same mechanistic alternatives as the overall process.

Oxygen-18 tracer experiments by Boyer et al (167, 168) established that a carboxyl oxygen of the acid substrate is transferred to the carboxyl of the acid product with no dilution by the medium, but with intermediate transfer to a group on the enzyme. The overall reaction has been proposed to occur by either a concerted pathway, or a stepwise mechanism involving an anhydride intermediate (167, 169) (equation 17).

(The latter would be an example of carboxyl activation by acylation.) No exchange of free CoASH into either acyl-CoA substrate is catalyzed by the enzyme, so that if the stepwise mechanism is correct, $CoAS^-$ is not free to dissociate (170). Substituent effect studies (170) have been suggested to favor the stepwise mechanism, but to date, no further evidence in support of the postulated anhydride intermediate has been reported. In fact, the substituent effect data can also be interpreted in terms of preequilibrium effects in a reaction in which, after thiol ester formation is complete, a conformational change or product dissociation step is slow.

The enzyme-catalyzed hydrolysis of peptides and proteins can be viewed as consisting of two transacylation reactions involving acid and/or base catalysis, but requiring no activation step for the leaving groups to depart. The generally accepted mechanism of acyl transfer for the serine proteases, likely applicable to the sulfhydryl peptidases as well, involves tetrahedral and acyl enzyme intermediates. Evidence for this mechanistic pathway is well-documented, and discussed by Kraut in the previous Annual Review (171).

Henken & Abeles (172) have provided an extensive and critical study of phosphotransacetylase [Ac-P + HSCoA $\rightleftharpoons$ AcSCoA + P$_i$] in an effort to establish if Acyl-E is an intermediate. Sequential kinetics were observed. Although slow partial reactions were seen, these were traced to contamination of the enzyme or substrate with the reaction-completing substrate. Isotope trapping was seen in large measure in both directions indicating that functional species were formed on the enzyme in the pulse. It may be questioned whether the trapping resulted from the formation of $E_{\cdot P_i}^{-Ac}$ or $E_{\cdot CoA}^{-Ac}$ in equilibrium with the Michaelis complexes or if all of the trapping resulted from slow dissociation of these complexes. The major evidence against a sequential/stepwise mechanism was failure to obtain labeled enzyme when the enzyme was passed through Sephadex G-25 after incubation with [^{14}C]-acetyl-P or [^{14}C]-acetyl-CoA. However, only if acetyl transfer to the enzyme and dissociation of first product had occurred would the label be fixed to the extent that separation of enzyme from small molecules on the column would prevent reversal of the labeling step; and then the ability to find labeled enzyme would depend on its stability to hydrolysis. That CoA is not released from E-acetyl CoA was shown independently by a measure of the acetyl CoA hydrolysis rate of the enzyme, which was very slow. The inability to show partial reactions probably makes the Sephadex experiment redundant. Henken & Abeles conclude from their data that an acetyl-enzyme intermediate can be ruled out for the phosphotransacetylase. They take issue with the proposition of Spector (76) that negative evidence cannot disprove the existence of covalent catalysis.

Wimmer and Rose (unpublished 1976) have sought additional evidence for an acetyl-enzyme intermediate by positional oxygen exchange in ^{18}O-labeled acetyl-P (equation 18).

18.

Positional exchange could not be detected at the rate required for a mechanism in which acyl enzyme with bound, but torsionally-symmetrical, P_i is formed in the absence of CoA.

CO_2 Transfer

Recent discussions of the reaction mechanism of carboxy transfer from biotin-CO_2 to acceptor have appeared (47, 139, 173, 174). The concerted electrophilic substitution mechanism, equation 19, is the most specific proposal.

19.

It was suggested by Rétey & Lynen (175) in an attempt to rationalize retention stereochemistry of propionyl CoA carboxylase (176). Retention has also been shown for pyruvate carboxylase (4) and for transcarboxylase with both propionyl CoA and pyruvate (177), and acetyl CoA carboxylase (178). The cyclic mechanism was also proposed by Mildvan et al (179, 180) to explain the requirement that biotin of pyruvate carboxylase must be in the carboxylated form in order to achieve the release of tritium from [3-T]pyruvate. A similar observation in propionyl CoA carboxylase (176) and transcarboxylase (177) for propionyl CoA might be interpreted as support for a concerted reaction in which the biotin carbonyl serves as the base for proton abstraction. The requirement for carboxylation of biotin to obtain proton abstraction from substrate is rather compelling in view of the oft stressed structural independence of the carboxyl carrying and transferring portions of these enzymes (181–184) occurring on different subunits in some cases. A typical enolization mechanism should not require that the biotin be carboxylated, although such arguments are always suspect. The retention of stereochemistry is only indicative of the cyclic mechanism. However, it points to the importance of doing the stereochemistry of several other enzymes of this class because one example of inversion would be very damaging to the idea of a concerted mechanism.

The concerted mechanism helps to explain a problem that has been pointed out many times—that biotin itself has none of the properties one would expect of a good acid or base. The active species for accepting the carboxyl would best be the enol form. But protonation of the carbonyl requires 10 N HCl (185) and removal of CO_2 from the amide link has a half-time of about 4 hr at 25° (186) because of the poor electron withdrawing qualities of the urea oxygen. In the concerted reaction these energy barriers are overcome by the exact positioning of groups so that the reaction has the barrier of a tautomerization instead. Evidence for the importance of the cyclic structural element for changing the strength of hydrogen bonds is cited by Stallings (187) in presenting a scheme for the carboxylation of biotin by hydrogen bonding of bicarbonate to the urea part of the bicyclic ring. Significant lengthening of the C–O and shortening of the C–N bonds relative to bond lengths in urea and biotin are seen in crystals of urea-phosphoric acid (188) and uroniun nitrate (189).

An argument for the cyclic mechanism is the observation that in the transcarboxylase reaction with [3^3H]pyruvate about 5% of the tritium that was activated was found in the propionyl CoA that was isolated (190). The role of biotin as perceived in this mechanism is to serve as a common base first to remove the proton from pyruvate as it is carboxylated and then to add the proton to the methylmalonyl CoA as it is being decarboxylated. On the other hand, it is possible that some group on the biotin peptide could act as a common base to explain the tritium transfer. The sequence reported for the biotin-containing peptide of transcarboxylase is glu-ala-met-biocytin-met-glu-ile (184). One of the glutamates could possibly serve as a base. It is interesting, in comparing this region with pyruvate carboxylase of higher animals, to note that the second glutamate is conserved (184). The first one is also present in the peptide from *E. coli* acetyl CoA carboxylase. The underlined region is conserved in all of these enzymes (184).

To explain the intermolecular tritium transfer one could propose that a small stagnant pool of water is common to two active site bases, one on each carboxyl acceptor subunit in the transcarboxylase. This picture seems contrary to the one generally given of the biotin communicating over a large distance to the two acceptor sites (47). That transcarboxylase may be an exception to this picture is suggested by Fung et al (191) who showed that the distance between spin label of CoA nitroxide and ^{13}C-pyruvate, places the CoA site and the methyl carbon of pyruvate within 8 Å. If, as seems likely, the biotin was placed in this small space there is little room for the ten molecules of water that would be required to dilute the tritium to the point of 5% transfer. This interpretation of the distance measurements has been questioned (183) on the basis that perhaps at the high pyruvate concentration used in these studies there was occupancy of pyruvate on the acyl CoA subunit as well. This possibility was derived from the kinetic studies of Northrop (129) showing that oxalacetate inhibits competitively with respect to propionyl-CoA.

Two other mechanistic proposals use enolization as a first step toward carboxylation followed by attack by the resulting carbanion on the electrophilic carbon center which is either on carboxybiotin or CO_2. In the former case a tetrahedral intermediate is formed that decays unimolecularly to the isourea form of the biotin and the carboxylated product (Equation 20).

$$\text{20.}$$

The arguments for classical enolization as an independent step are not sufficient to establish the point. The absence of proton exchange and the occurrence of retention stereochemistry in three cases is unsubstantial negative evidence. Using propionyl CoA carboxylase, Stubbe & Abeles (192) have observed that $[\beta\text{-F}]$propionyl CoA does not undergo liberation of F^- unless ATP and HCO_3^- are present. However these authors did observe the liberation of F^- in the complete system which suggests either that $[\beta\text{-F}]$methylmalonyl CoA was unstable, or that the presumed "enol" is decomposed in the complex

$$\text{E}\begin{bmatrix} \text{BiCO}_2^- \\ H^+ \\ \text{FCH}_2\text{CH}^-\text{COSCoA} \end{bmatrix}$$

The important observation was made that F^- was liberated approximately seven times faster than expected from the production of ADP. This suggests that the complex from which the presumed acrylyl CoA is formed is able to accept a new molecule of $[\beta\text{-F}]$propionyl CoA for "enolization" many times before liberation of ADP and P_i (equation 21)

$$\text{21.}$$

The conclusion is that excess fluoride liberation would be difficult to explain from the electrophilic substitution mechanism. On the other hand, it is possible that the enolization-coupled liberation of F^- is only a reflection of the greater instability of an already unstable compound when bound to the active site of the enzyme. In this case the binding of ATP and HCO_3^- or the carboxylation of biotin could satisfy secondary roles in the suggested mechanism.

The distinction between the two enolization mechanisms is in the mode of carboxyl transfer. The formation of a tetrahedral addition compound with the carboxybiotin followed by cleavage to product and enol biotin perceives the biotin moiety, in an isourea form, to be comparable as a leaving group to the enol species of the substrate; this is required by the easy reversibility of CO_2 transfer shown by the exchange of oxalacetate and pyruvate (193). The two steps in the tetrahedral mechanism are favored by different selective protonations, first of the carboxylate group in forming the intermediate, and second of the biotin carbonyl. The bound-CO_2 mechanism requires protonation of the biotin carbonyl of the biotincarboxylate. Sauers et al have argued for a CO_2 intermediate by analogy with the decomposition of $ROCO_2^-$ and

$$\text{R-N-C} \begin{array}{c} H \\ | \end{array} \begin{array}{c} O \\ \diagup \diagup \\ \diagdown \\ OAr \end{array}$$

(194) where addition-cleavage mechanisms are not seen, but rather elimination pathways followed by addition of water to $O=C=O$ or $RN=C=O$. The evolution of biotin for this purpose is perhaps difficult to explain in that carboxyphosphate, the presumed product of the first step of the carboxylase reactions, seems to be a more reactive species than CO_2 (139) whereas, judging from stability in solution, carboxybiotin is the most stable of the three [$t_{1/2}$ for hydrolysis are: $\leqslant$ .07s (139), $\sim$ 700s (139) and 6000s (174), respectively at pH 7.0.] Therefore carboxyphosphate would lead to much higher steady state accumulation of CO_2, in an aqueous solution, than carboxybiotin.

Such considerations, without reference to the enzymes, are probably very misleading. For example, it is known with many biotin enzymes that decarboxylation of the enzyme in the absence of acceptor molecules is considerably faster with the native than with the denatured carboxybiotin enzyme which suggests that the native protein environment promotes processes that may be favorable for carboxyl transfer (47). In this case it seemed reasonable to inquire into the nature of the decarboxylation product. The rationale was that the observation of bicarbonate in the enzyme system would support a nucleophilic attack by water or $^-$OH and hence perhaps a reactive nucleophile such as enolpyruvate or enolpropionyl CoA in the normal systems. Indeed bicarbonate was observed to be the major product of the decarboxylation of $[\beta^{14}C]$methylmalonyl CoA by transcarboxylase and a very significant one with propionyl CoA carboxylase (D. Kuo and I. A. Rose, unpublished 1977). However, in the avidin-sensitive reaction by methylmalonyl CoA decarboxylase, CO_2 and not bicarbonate, was produced. The argument for bound CO_2 as an intermediate, first in its transfer to biotin and then in product formation, puts carboxybiotin on a side path and gives no real role for the biotin in the decarboxylase. Thus the formation of bicarbonate was seen only in the enzymes that participate in carboxyl transfer.

Phosphoryl Transfer

MECHANISMS Phosphoryl transfer reactions are believed to proceed by one of three different mechanisms: by way of meta-P, by direct displacement, or by a pentacoordinated-P intermediate. Evidence for the occurrence of each pathway, based primarily on nonenzymatic studies, has recently been reviewed by Benkovic & Schray (195) and by Westheimer (196).

The monomeric metaphosphate route is analogous to the carbonium ion mechanism for nucleophilic substitution on carbon except for the resonance-stabilized negative charge on the intermediate. The reaction takes place in two steps, the first of which is the rate-limiting formation of reactive PO_3^- upon departure of a protonated leaving group, followed by rapid addition of a nucleophile to complete the substitution reaction. This mechanism has become the generally accepted route of nucleophilic substitution for phosphate monoester monoanions and unsubstituted acyl phosphates. A question arises concerning the stereochemistry expected in the meta-P pathway. Non-enzymatically one would predict racemization due to the planarity of the PO_3^-, but to what extent is difficult to say. In hydrolysis reactions, for example, acid catalysis will aid in leaving group departure, but the high reactivity of meta-P may well prevent symmetrical solvolysis if diffusion away of the leaving group is slow relative to nucleophile capture. In an enzyme-directed process, in which the attacking nucleophile most likely approaches from one side of the plane, racemization could still take place because of the inherent symmetry of the PO_3^- group. The degree to which racemization occurs would depend on the relative competing rates of positional exchange of oxygens, by torsion about one P–O bond, and nucleophilic attack on phosphorous (equation 22).

22.

Leaving group protonation and meta-P stabilization by the enzyme may increase the chance for racemization as well as improve the selectivity of this reactive intermediate.

The concerted displacement mechanism is believed to be operative in reactions involving substitution at phosphorus in acyclic di- and tri-esters. A transition state is postulated in which the five phosphorous ligands are in trigonal bipyramidal geometry, and as in S_N2 substitutions at saturated carbon, the attacking and leaving groups are in line with each other and partially bonded. The stereochemical course for the concerted reaction is necessarily one of inversion.

The pentacoordinated phosphate intermediate mechanism follows a two-step sequence of addition then elimination. The intermediate phosphate is coordinated with five ligands in a trigonal bipyramidal arrangement. Through studies with various organophosphorous esters (197) and phosphorous halides (198), the chemistry of such phosphorous intermediates has been observed to follow a set of preference rules as to the positioning and reactivity of the five ligands (163): (*a*) the geometry is defined by two apical and three equatorial, or basal, ligands; (*b*) groups can leave from apical positions only; (*c*) the more electronegative groups tend to occupy apical positions while the more electropositive groups prefer the equatorial positions; (*d*) a five-membered ring spans one apical and one equatorial position; (*e*) positional exchange can occur between the apical and basal groups by a process of bond bending and stretching called pseudorotation (Equation 23).

$$\qquad\qquad\qquad\qquad\qquad\qquad\qquad\qquad 23.$$

The stereochemistry of phosphoryl transfer by a pentacoordinate phosphorous pathway depends on the direction of nucleophilic addition and leaving group departure. Because groups enter and depart from the apical positions, the location of the departing group after formation of the intermediate dictates whether or not pseudorotation will be required to position the leaving group apical. That is, in what is referred to as "in-line" attack of the nucleophile, the leaving group is in a position to leave immediately and the reaction stereochemistry would be indistinguishable from an S_N2 reaction, i.e. inversion. In the "adjacent" scheme, the leaving group initially occupies a basal position and therefore at least one pseudorotation step is required. Product stereochemistry depends on the relative rates of pseudorotation and bond cleavage. If cleavage is the more rapid, retention of configuration will be observed, whereas positional exchange much faster than cleavage could result in racemization (199). In an enzyme-catalyzed reaction, the tightness of binding and disposition of enzyme residues that bind to the phosphorous ligands or provide acid or base catalysis will play an important role in determining the ultimate stereochemistry of the reaction. Racemization seems unlikely unless the pentacoordinated species is the one that actually departs from the enzyme and rearranges to the eventual product without steric restrictions.

NUCLEOTIDE TRANSFER In enzyme-catalyzed reactions involving dinucleotide phosphate cleavage or synthesis, knowledge of the stereochemical pathway may allow mechanistic conclusions to be drawn, and the stereochemistry can be determined by substitution of one phosphorous oxygen with sulfur. The dinucleotide or cyclic mononucleotide then possesses a chiral phosphorous center, as do the hydrolysis products if $H^{18}OH$ is used. An elegant set of studies on the stereochemical mechanism of action of ribonuclease-A was conducted by Eckstein and co-workers (200, 201). The hydrolytic reaction catalyzed by the enzyme occurs in two steps with intermediate formation of the cyclic 2',3'-nucleotide. Both in-line and adjacent pentacoordinate-P mechanisms had been postulated for each partial reaction (202). Through the use of one isomer of uridine 2',3'-cyclic phosphorothioate and nonenzymatic ring opening or closure reactions of known stereochemistry, both steps were found to occur by an in-line displacement. (Knowledge of the absolute configuration of the cyclic phosphorothioate was not necessary to elucidate the stereochemical mechanism in this case, but in later examples this will not be true). If base catalysis of nucleophilic addition and acid catalysis of leaving group departure are provided by groups at the active site, this mechanism requires that two such groups be involved and that they recycle during each turnover. That the ribonuclease mechanism is in-line displacement leaves unresolved the existence of a pentacoordinated phosphate as a distinct intermediate in the reaction.

The diasteriomers of adenosine 5'-(0-1-thiotriphosphate), ATPαS, and of adenosine 5'-(0-2-thiotriphosphate), ATPβS, have been shown to react differentially with several kinases (22), and this has enabled their resolution into isomers designated A and B. The absolute configurations remain to be determined, but use has been made of the ATPαS isomers to initiate studies of the stereochemisty of adenylyl transfer reactions involving displacements at the α phosphorous.

Eckstein and co-workers have recently investigated the mechanisms of *E. coli* DNA-dependent RNA-polymerase and yeast tRNA nucleotidyl-transferase (203,204). Both of these enzymes catalyze 3',5'-internucleotide bond formation upon attack of a nucleotide 3'-hydroxyl at the α phosphate of ATP, with elimination of pyrophosphate. Interestingly, each enzyme has been found to use the A isomer of ATPαS as a substrate, and each produces a phosphorothioate diester bond having the R configuration. The results indicate that no racemization has taken place in either case, ruling out a pentacoordinate-P intermediate which rapidly pseudorotates relative to pyrophosphate departure. Whether inversion or retention has occurred will be solved with elucidation of the absolute configuration of ATPαS.

HYDROLYSIS TO P_i Phosphatase reactions are often known to yield P_i which contains more water oxygens than the single one expected from the cleavage step. In the case of myosin ATPase, this excess exchange has been proposed to occur prior to formation of ADP by a stepwise pathway involving a pentacoordinated intermediate formed upon addition of H_2O to the ATP γ-PO_3 (205). If elimination of one of the original γ-PO_3 oxygens occurs more rapidly than elimination of ADP, there will be oxygen exchange. Alternatively, multiple hydrolytic cleavages of the $P_\beta O$–P_γ bond of ATP before release of ADP and P_i will result in excess water oxygen incorporation into P_i if the oxygens of bound P_i are equivalent (206, 206a). The pentacoordinated-P and reversible hydrolysis mechanisms have also been proposed to explain the ATP:H_2O oxygen exchanges that are catalyzed by the mitochondrial and chloroplast ATP synthetase systems (207–209).

Distinction between the two pathways, and therefore a test for the occurrence of a pentacoordinate-P intermediate during the overall hydrolysis or synthesis reactions, would be possible if the rates of exchange and of $P_\beta O$–P_γ bond cleavage in ATP could be independently compared. If, upon cleavage of $\beta\gamma$-bridge ^{18}O-labeled ATP, the β-phosphoryl of bound ADP is torsionally symmetrical, $\beta\gamma$-bridge:β-nonbridge positional oxygen exchange in reisolated ATP provides a measure of the cleavage reaction (50). Wimmer & Rose (51) were thus able to ask if ATP molecules that participated in γ-PO_3:H_2O exchange in the presence of illuminated chloroplasts also underwent cleavage. By analyzing the γ-phosphoryl group in the form of trimethyl phosphate (50), these authors found that ATP molecules that experienced water oxygen exchange did so nearly to equilibrium before returning to the medium. Of these same molecules, approximately 80% had also undergone reversible cleavage. Because of the high extent of γ-PO_3:H_2O exchange, the relative rates of cleavage and exchange cannot yet be determined. [Interestingly, during net synthesis of ATP from ADP and P_i in the same system, the rate of ATP dissociation is nearly an order of magnitude faster than in the absence of net phosphorylation, as indicated by a marked decrease in the extent of γ-PO_3:H_2O exchange in synthesized ATP (51).] However, the results indicate that, although net hydrolysis of ATP to ADP and P_i does not occur in intact chloroplasts (210), reversible hydrolysis on the ATP synthetase does occur and thus may be responsible for the ATP : H_2O exchange.

Eargle et al (211) conducted a study of the P_i : H_2O oxygen exchange catalyzed by *E. coli* alkaline phosphatase, analyzing P_i in the form of the tris (trimethylsilyl) derivative which, like the trimethylphosphate method of analysis, allows determination of the extent of exchange in a given molecule. The results indicated that multiple oxygen exchange was a rare event, i.e. the release of P_i was very rapid in comparison to the hydrolysis rate.

E–P OR NO E–P   The choice of a concerted, meta-P, or pentacoordinate-P mechanism applies to all phosphoryl transfer reactions. The path from substrate to product may be direct, or may proceed via a phosphorylated intermediate. The question concerning the presence or absence of an E–P in reactions catalyzed by the kinases has been addressed for some time, and as pointed out by Morrison & Heyde (212), controversy still exists. The kinetic mechanisms for nearly all kinases are of the sequential type; in most cases, ATP : ADP partial exchange reactions are found to proceed much slower than required for kinetic competence, and few phosphoryl enzymes have been isolated. However, as discussed at the beginning of this review, masking of evidence for a covalent enzyme-X in kinetic studies may occur, emphasizing the need for a different approach to the answer of the intermediate question.

Nucleoside diphosphokinase is one exception to the usual kinase pattern in that the enzyme exhibits ping pong kinetics, and can be isolated in an active phosphorylated form (78, 213, 214). Acetate kinase has been suggested to proceed via an E–P despite its sequential mechanism (215–217). A phosphorylated form of this enzyme is produced from either phosphoryl donor substrate, and the phosphate moiety can be transferred to both acceptor substrates (218). Acetyl-P : acetate exchange catalyzed by this enzyme requires the presence of bound MgADP while ATP : ADP exchange is reported to proceed without cosubstrate (216, 217). However, the rates of the observed partial reactions relative to the overall reaction appear to be very slow (216, 217, 219). In the case of phosphoglycerate kinase, Johnson et al (77) have recently presented convincing evidence that the "phosphoryl-enzyme" form of this enzyme is actually a tight complex of E· 1,3-bisphosphoglycerate.

Hexokinase undergoes transformation to an inactive phosphoenzyme upon incubation with ATP and reactivation in the presence of ADP (220, 221); both processes require D-xylose. From the observation that approximately 300 catalytic cycles may occur before the lethal modification, the possibility was considered that inactivation is effected by rearrangement of an active E-P. That the slow ATP : ADP exchange catalyzed by this kinase appears to be due to a side reaction (45) argues against the existence of E-P along the main catalytic pathway.

In a study of the stereochemistry of phosphoryl transfer by kinases of the glycolytic pathway, Knowles's group (J. R. Knowles, personal communication) prepared one of the isomers of glycerate-2,3-cyclic phosphorothioate which was converted to glycerate-2-0-[^{18}O]thiophosphate by hydrolysis in H^{18}OH and subsequently to glycerate-3-0-[^{18}O]thiophosphate by reactions of enolase, pyruvate kinase, hexokinase, and other glycolytic enzymes. After ring closure, the original cyclic phosphorothioate isomer was reisolated and found to contain ^{18}O, indicating that its configuration was opposite that of the starting material. Because the initial ring opening and the final ring closure proceed by the same in-line inversion mechanism, overall inversion of configuration of the cyclic phosphorothioate dictates that the transfer of the thiophosphoryl group from the two to the three position of glycerate-O-thiophosphate in the long route via ATP must have occurred with retention. Thus, the reactions catalyzed by pyruvate kinase and hexokinase necessarily proceed by the same stereochemical mechanism. The fact that no racemization was observed argues against the existence of a freely rotating metaphosphate intermediate in either of the phosphoryl transfer reactions. A means to elucidate the stereochemical pathway of the two kinases is not yet available. However, if an inversion mechanism is ultimately found, the occurrence of a phosphoenzyme intermediate in either kinase reaction would be disfavored.

An E-P intermediate is well-established in the intermolecular phosphoryl transfers of phosphogluco- and phosphoglycerate mutases (222). The distinct synthase enzymes that result in net synthesis of glucose-1,6-P$_2$ and glycerate-2,3-P$_2$ also involve formation of E-P, with glycerate 1,3-P$_2$ as the phosphoryl donor in both cases. Phosphoglycerate mutase and glycerate-2,3-P$_2$ synthase are closely related enzymes in many physical features and in the fact of a histidine-P intermediate in both cases (223). For both glucose bisphosphate enzymes, glucose-1,6-P$_2$ gives rise to E-P, and, although glycerate-1,3-P$_2$ seems not to interact significantly with phosphoglucomutase, both form serine-P intermediates (222, 224) which suggests a common origin for these enzymes as well. The possibility of a meta-P intermediate in phosphoryl transfer to the synthase enzymes could be examined by isotope scrambling because glycerate-1,3-P$_2$ is not regenerated from E-P and glycerate-3-P with either synthase (equation 24).

24.

Under conditions where half of the glycerate-1,3-P_2 was consumed in net reaction, no scrambling was seen (224).

Evidence to support the obligatory role of E-P in glucose-1,6-P_2 synthase was obtained by rapid quench procedures (224). The rates of phosphorylation of the enzyme at saturating glycerate-1,3-P_2 and transfer of ^{32}P from E-P to glucose-1-P were measured separately. They exceeded by a small margin the net reaction rate with glucose-1-P, the best acceptor. Since the overall reaction obeys ping pong kinetics, it is expected that the individual rates should be consistent with the overall rate, i.e. no effect of acceptor binding should be felt on the formation of E-P.

A pyrophosphoryl as well as a phosphoryl enzyme intermediate has been found to occur during the formation of PEP catalyzed by pyruvate, phosphate dikinase of *Propionibacteria* (225, 226). The mechanism of the reaction is supported by isotopic exchange and tracer studies (equation 25).

$$\text{Enz} + \text{ATP} \rightleftharpoons \text{Enz-P}_\beta P_\gamma + \text{AMP}$$
$$\text{Enz-P}_\beta P_\gamma + P_i \rightleftharpoons \text{Enz-P}_\beta + P_\gamma P_i \qquad 25.$$
$$\text{Enz-P}_\beta + \text{pyruvate} \rightleftharpoons \text{PEP}_\beta + \text{Enz}$$

Three independent substrate sites are indicated, one for each substrate/product pair, i.e. ATP, AMP; P_i, PP_i; Pyruvate, PEP. Both pyrophosphoryl and phosphoryl enzyme have been isolated and shown to be catalytically active.

In *E. coli,* a similar synthesis of PEP from pyruvate occurs except that the reaction does not require inorganic phosphate and has been postulated to proceed according to the scheme of equation 26 (227):

$$\text{Enz} + \text{ATP} \rightleftharpoons \text{Enz-P}_\beta P_\gamma + \text{AMP}$$
$$\text{Enz-P}_\beta P_\gamma + H_2O \rightleftharpoons \text{Enz-P}_\beta + P_{\gamma i} \qquad 26.$$
$$\text{Enz-P}_\beta + \text{pyruvate} \rightleftharpoons \text{PEP}_\beta + \text{Enz}$$

Active phosphorylated enzyme has been isolated (228), but no report on the finding of a stable pyrophosphoryl enzyme has yet appeared. Berman et al (229) ruled out a PEP-ADP intermediate by $H^{18}OH$ studies showing that one water oxygen appeared in each P_i formed, with none in AMP or PEP. Their results are consistent not only with a pyrophosphoryl enzyme intermediate, but also with pyrophosphoryl enolpyruvate, while the exchange studies are consistent with the former and not the latter mechanism (227). A look at $P_i : H_2O$ oxygen exchange with E-P + P_i proved alone to be inconclusive in that AMP was required for the rate to be comparable to the ATP : P_i exchange rate (230). The data are consistent with a reaction sequence in which AMP remains bound to the pyrophosphoryl enzyme until P_i is formed, i.e. both $P_\alpha O-P_\beta$ and $P_\beta O-P_\gamma$ bond cleavages occur before dissociation of products AMP and P_i (230).

CONCLUDING REMARKS

Enzymology has seemed to come a long way from the early days when kinetic studies were done to establish by the "saturation" phenomenon a physical interaction between enzyme and substrate. As we see, efforts to characterize the interactions necessary for catalysis do not lead to a single general principle but rather the assurance that chemical principles are at work. Different type reactions therefore present different problems which genetic evolution has brought to the present "perfect state" wherein the true transition state of kinetic importance often describes the conformational changes for release of products. This, and the unfortunate occurrence of sticky protons, sticky AMP and ADP, etc has created problems for many of our standard methods of identifying reaction intermediates, where they exist. Therefore the future for enzyme reaction chemistry will depend, in many cases, on being able to devise new methods to see the chemical steps. Kinetic isotope effects makes a strong claim to our attention. Nonkinetic approaches—the study of the reaction at equilibrium by "seeing" the intermediates, either by spectral, crystallographic, or chemical methods—are strong approaches that emphasize the intermediate. In this chapter we hope we have put some of our present knowledge into a useful perspective.

ACKNOWLEDGMENTS

This review was supported by National Institutes of Health Grants GM-20940, CA-06927, and RR-05539; National Institutes of Health Fellowship GM-05499 to M. J. Wimmer; and an appropriation from the Commonwealth of Pennsylvania. The authors would like to thank Dr. J. P. Klinman for her helpful discussion concerning this review.

Literature Cited

1. Rose, I. A. 1960. *J. Biol. Chem.* 235: 1170–77
2. Creighton, D. J., Rose, I. A. 1976. *J. Biol. Chem.* 251:61–68
3. Wang, S.-F., Kawahara, F. S., Talalay, P. 1963. *J. Biol. Chem.* 238:576–85
4. Ayling, J. E., Dunathan, H. C., Snell, E. E. 1968. *Biochemistry* 7:4537–42
5. Klinman, J. P., Rose, I. A. 1971. *Biochemistry* 10:2259–66
6. Hashimoto, H., Günther, H., Simon, H. 1972. *FEBS Lett.* 33:81–83
7. Wolfenden, R. 1976. *Ann. Rev. Biophys. Bioeng.* 5:271–306
8. Wolfenden, R. 1972. *Acc. Chem. Res.* 5:10–18
9. Lienhard, G. E. 1972. *Science* 180: 149–54
10. Rose, I. A. 1972. *Crit. Rev. Biochem.* 1:33–57
11. Hanson, K. R., Rose, I. A. 1975. *Acct. Chem. Res.* 8:1–10
12. Rose, I. A., Hanson, K. R. 1976. *Applications of Biochemical Systems in Organic Chemistry, Part II.* pp. 507–53 New York: Wiley
13. Meloche, H. P., Glusker, J. P. 1973. *Science* 181:350–52
14. Rose, I. A. 1975. *Advan. Enzymol.* 43:491–517
15. Bender, M. L., Heck, H. d'A. 1967. *J. Am. Chem. Soc.* 89:1211–20
16. Ogston, A. G. 1948. *Nature* 162:963
17. Rose, I. A., O'Connell, E. L. 1961. *J. Biol. Chem.* 236:3086–92

18. Dunathan, H. C. 1971. *Advan. Enzymol.* 35:71–134
19. Floss, H. G., Onderka, D. K., Carroll, M. 1972. *J. Biol. Chem.* 247:736–44
20. Rétey, J., Arigoni, D. 1966. *Experientia* 22:783–84
21. Frey, P. A., Essenberg, M. K., Abeles, R. H. 1967. *J. Biol. Chem.* 242:5369–77
22. Eckstein, F., Goody, C. S. 1976. *Biochemistry* 15:1685–91
23. Eckstein, F. 1970. *J. Am. Chem. Soc.* 92:4718–23
24. Rao, B. D. N., Buttlaire, D. H., Cohn, M. 1976. *J. Biol. Chem.* 251:6981–86
25. Douzou, P., Sireix, R., Travers, F. 1970. *Proc. Nat. Acad. Sci. USA* 66:787–92
26. Douzou, P. 1971. *Biochimie* 53:1135–45
27. Fink, A. L. 1976. *J. Theor. Biol.* 61:419–45
28. Fink, A. L. 1977. *Accts. Chem. Res.* 10:233–39
29. Makinen, M. W., Fink, A. L. 1977. *Ann. Rev. Biophys. Bioeng.* 6:301–43
30. Fink, A. L., Good, N. E. 1974. *Biochem. Biophys. Res. Commun.* 58:126–31
31. Fink, A. L., Angelides, K. J. 1975. *Biochem. Biophys. Res. Commun.* 64:701–08
32. Rose, I. A., O'Connell, E. L. 1977. *J. Biol. Chem.* 252:479–82
33. Voet, J. G., Abeles, R. H. 1970. *J. Biol. Chem.* 245:1020–31
34. Ray, W. J., Jr., Long, J. W. 1976. *Biochemistry* 15:3990–93
35. Albery, W. J., Knowles, J. R. 1976. *Biochemistry* 15:5631–40
36. Rose, I. A., Kellermeyer, R., Sternholm, R., Wood, H. G. 1962. *J. Biol. Chem.* 237:3325–31
37. Schambye, P., Wood, H. G. 1954. *Radioisotope Conf., Oxford, Eng.* 1:346–50
38. Rose, I. A., O'Connell, E. L., Litwin, S., Bar Tana, J. 1974. *J. Biol. Chem.* 249:5163–68
39. Krishnaswamy, P. R., Pamiljans, V., Meister, A. 1962. *J. Biol. Chem.* 237:2932–40
40. Cleland, W. W. 1975. *Biochemistry* 14:3220–24
41. Ning, J., Purich, D. L., Fromm, H. J. 1969. *J. Biol. Chem.* 244:2840–46
42. Ogasawara, N., Gander, J. E., Henderson, L. M. 1966. *J. Biol. Chem.* 241:613–19
43. Litwack, G., Cleland, W. W. 1968. *Biochemistry* 7:2072–79
44. Cleland, W. W. 1970. *The Enzymes* 2:1–65
45. Solomon, F., Rose, A. 1971. *Arch. Biochem. Biophys.* 147:349–50
46. Scrutton, M. C., Utter, M. F. 1965. *J. Biol. Chem.* 240:3714–23
47. Moss, J., Lane, M. D. 1971. *Advan. Enzymol.* 35:321–442
48. Gold, A. M., Osber, M. P. 1972. *Arch. Biochem. Biophys.* 153:784–87
49. Kokesh, F. C., Kakuda, Y. 1977. *Biochemistry* 16:2467–73
50. Midelfort, C. F., Rose, I. A. 1976. *J. Biol. Chem.* 251:5881–87
51. Wimmer, M. J., Rose, I. A. 1977. *J. Biol. Chem.* 252:6769–75
52. Collins, C. J., Bowman, N. S., eds. 1970 *Isotope Effects in Chemical Reactions,* New York: Van Nostrand-Reinhold
53. Caldin, E., Gold, V., eds. 1975. *Proton Transfer Reactions.* New York: Wiley
54. Cleland, W. W., O'Leary, M. H., Northrop, D. B., eds. 1977. *Isotope Effects on Enzyme Catalyzed Reactions.* Baltimore: University Park Press
55. Schowen, R. L., Gandour, R., eds. *Transition States of Biochemical Processes.* New York: Plenum In Press
56. Schowen, R. L. 1972. *Progr. Phy. Org. Chem.* 9:275–332
57. Klinman, J. P. 1978. *Advan. Enzymol.* 46:413
58. Simon, H., Palm, D. 1966. *Angew Chemie* 5:920–33
59. Northrop, D. B. 1975. *Biochemistry* 14:2644–51
60. Albery, J. W., Knowles, J. R. 1977. *J. Am. Chem. Soc.* 99:637–38
61. Wolfsberg, M., Stern, M. J. 1964. *Pure Appl. Chem.* 8:225–42
62. Wolfsberg, M. 1972. *Accts. Chem. Res.* 5:225–33
63. Burton, G. W., Sims, L. B., Wilson, J. C., Fry, A. 1977. *J. Am. Chem. Soc.* 99:3371–79
64. Hartshorn, S. R., Shiner, V. J., Jr. 1972. *J. Am. Chem. Soc.* 94:9002–12
65. Lewis, E. S. 1975. in *Proton Transfer Reactions,* ed. E. Caldin, V. Gold. pp. 317–38 New York: Wiley
66. Caldin, E. F. 1969. *Chem. Rev.* 69:135–56
67. Kosower, E. M. 1977. *J. Phys. Chem.* 81:807–08
68. Essenberg, M. K., Frey, P. A., Abeles, R. H. 1971. *J. Am. Chem. Soc.* 93:1242–51
69. Weisblat, D. A., Babior, B. M. 1971. *J. Biol. Chem.* 246:6064–71
70. Bright, H. J., Gibson, Q. H. 1967. *J. Biol. Chem.* 242:944–1003
71. Bush, K., Mahler, H. R., Shiner, V. J. 1971. *Science* 172:478–80

72. deJuan, E., Taylor, K. B. 1976. *Biochemistry* 15:2523–27
73. Lewis, C. A., Wolfenden, R. 1978. *Biochemistry* 16:4890–95
74. Schimerlik, M. I., Rife, J. E., Cleland, W. W. 1975. *Biochemistry* 14:5347–54
75. Cleland, W. W. 1977. See Ref. 54, pp. 153–75
76. Spector, L. B. 1973. *Bioorganic Chem.* 2:311–21
77. Johnson, P. E., Maister, S. G., Semeriva, M., Young, J. M., Knowles, J. R. 1976. *Biochemistry* 15:2893–2901
78. Mourad, N., Park, R. E. 1966. *J. Biol. Chem.* 241:271–78
79. Hall, S. S., Doweyko, A. M., Jordan, F. 1976. *J. Am. Chem. Soc.* 98:7460–61
80. Krimsky, I., Racker, E. 1955. *Science* 122:319–21
81. Rose, Z. B., Racker, E. 1962. *J. Biol. Chem.* 237:3279–82
82. Webb, M. R., Knowles, J. R. 1974. *Biochem. J.* 141:589–92
83. Kosicki, G. W., Westheimer, F. H. 1968. *Biochemistry* 7:4303–10
84. Bruice, P. Y., Bruice, T. C. 1974. *J. Am. Chem. Soc.* 96:5523–32
85. Benson, A. M., Talalay, P., Keen, J. H., Jakoby, W. B. 1977. *Proc. Nat. Acad. Sci. USA* 74:158–62
86. Tobias, P. S., Kezdy, F. J. 1969. *J. Am. Chem. Soc.* 91:5171–73
87. Dinovo, E. C., Boyer, P. D. 1971. *J. Biol. Chem.* 246:4586–93
88. Trentham, D. R., MacMurray, C. H., Pogson, C. I. 1969. *Biochem. J.* 114:19–23
89. Rose, I. A., O'Connell, E. L., Schray, K. J. 1973. *J. Biol. Chem.* 248:2232–34
90. Los, J. M., Simpson, L. B., Wiesner, K. 1956. *J. Am. Chem. Soc.* 78:1564–68
91. Martin, K. O., Monder, C. 1976. *Biochemistry* 15:576–81
92. Lee, H. J., Monder, C. 1977. *Biochemistry* 16:3810–14
93. Martin, K. O., Oh, S.-W., Lee, H. J., Monder, C. 1977. *Biochemistry* 16:3803–09
94. Harris, D. W., Feather, M. S. 1975. *J. Am. Chem. Soc.* 97:178–81
95. Gleason, W. B., Barker, R. 1971. *Can. J. Chem.* 49:1433–40
96. Rose, I. A. 1962. *Brookhaven Symp. Biol.* 15:293–309
97. Maister, S. G., Pett, C. P., Albery, W. J., Knowles, J. R. 1976. *Biochemistry* 15:5607–12
98. Fletcher, S. J., Herlihy, J. M., Albery, W. J., Knowles, J. R. 1976. *Biochemistry* 15:5612–17
99. Midelfort, C. F., Rose, I. A. 1977. *Biochemistry* 16:1590–96
100. Walsh, C., Schonbrunn, A., Abeles, R. 1971. *J. Biol. Chem.* 246:6855–66
101. Walsh, C., Krodel, E., Massey, V., Abeles, R. 1973. *J. Biol. Chem.* 248:1946–54
102. Abeles, R. H., Maycock, A. L. 1976. *Accts. Chem. Res.* 9:313–19
103. Rando, R. R. 1974. *Science* 185:320–24
104. Rando, R. R. 1974. *Accts. Chem. Res.* 8:281–88
105. Yamada, H., O'Leary, M. H. 1976. *J. Am. Chem. Soc.* 99:1660–61
106. Rose, I. A. 1977. See Ref. 54, p. 149
107. Frohe, L., Schwabe, K. P. 1970. *Biochem. Biophys. Acta* 220:469–76
108. Coward, J. K., Slisz, E. P., Wu, F. Y.-H. 1973. *Biochemistry* 12:2291–97
109. Mascaro, J. Jr., Horhammer, R., Eisenstein, S., Sellers, L. K., Mascaro, K., Floss, H. G. 1977. *J. Am. Chem. Soc.* 99:273–74
110. Hörnemann, L. Z., Floss, H. G. 1975. *Biochem. Physiol. Pflanzen* 168:19–25
111. Rétey, J., Bartl, K., Ripp, E., Hull, W. E. 1977. *Eur. J. Biochem.* 72:251–57
112. Hegazi, M. F., Borchardt, R. T., Schowen, R. L. 1976. *J. Am. Chem. Soc.* 98:3048–49
113. Phillips, G. T., Clifford, K. H. 1976. *Eur. J. Biochem.* 61:271–86
114. Reeves, R. E., Warren, L. G., Susskind, B., Lo, H. 1977. *J. Biol. Chem.* 252:726–31
115. Loftfield, R. B. 1972. *Prog. Nucl. Acid Res. Mol. Biol.* 12:87–128
116. Soll, D., Schimmel, P. R. 1974. *The Enzymes* 10:489–538
117. Boyer, P. D., Koeppe, D. J., Luchsinger, W. W. 1956. *J. Am. Chem. Soc.* 78:356–57
118. Myers, G., Blank, H. U., Soll, P. 1971. *J. Biol. Chem.* 246:4955–64
119. Midelfort, C. F., Chakraburtty, K., Steinschneider, A., Mehler, A. H. 1975. *J. Biol. Chem.* 250:3866–78
120. Lovgren, T. N. E. Heinonen, J., Loftfield, R. B. 1975. *J. Biol. Chem.* 250:3854–60
121. Kim, J. P., Chakraburtty, K., Mehler, A. H. 1977. *J. Biol. Chem.* 252:2698–2701
122. Lipmann, F., Gevers, W., Kleinkauf, H., Roskoski, R. 1971. *Advan. Enzymol.* 35:1–34
123. Laland, S. G., Zimmer, T. 1973. *Essays Biochem.* 9:31–57
124. Rochovansky, O., Ratner, S. 1961. *J. Biol. Chem.* 236:2254–60
125. Rochovansky, O., Ratner, S. 1967. *J. Biol. Chem.* 242:3839–49
126. Meister, A. 1974. *The Enzymes* 10:699–754

127. Weisbrod, R. E., Meister, A. 1973. *J. Biol. Chem.* 248:3997–4002
128. Ronzio, R. A., Rowe, W. B., Meister, A. 1969. *Biochemistry* 3:1066–75
129. Northrop, D. B. 1969. *J. Biol. Chem.* 244:5808–19
130. McClure, W. R., Lardy, H. A., Wagner, M., Cleland, W. W. 1971. *J. Biol. Chem.* 246:3579–83
131. Barden, R. E., Fung, C., Utter, M. F., Scrutton, M. C. 1972. *J. Biol. Chem.* 247:1323–33
132. Lane, M. D., Moss, J., Polakis, S. E. 1974. *Curr. Topics Cell. Regul.* 8: 139–95
133. Kaziro, Y., Hass, L. F., Boyer, P. D., Ochoa, S. 1962. *J. Biol. Chem.* 237: 1460–68
134. Kluger, R., Adawadkar, P. D. 1976. *J. Am. Chem. Soc.* 98:3741–42
135. Cooper, T. G., Tchen, T. T., Wood, H. G., Benedict, C. R. 1968. *J. Biol. Chem.* 243:3857–63
136. Lynen, F., Knappe, J., Lorch, E., Juetting, G., Ringelmann, E., Lachance, J. 1961. *Biochem. Z.* 335:123–67
137. Polakis, S. E., Guchhait, R. B., Lane, M. D. 1972. *J. Biol. Chem.* 247: 1335–37
138. Polakis, S. E., Guchhait, R. B., Zwergel, E. E., Lane, M. D., Cooper, T. G. 1974. *J. Biol. Chem.* 249:6657–67
139. Sauers, C. K., Jencks, W. P., Groh, S. 1975. *J. Am. Chem. Soc.* 97:5546–5553
140. Jones, M. E. 1965. *Ann. Rev. Biochem.* 34:381–418
141. Anderson, P. M., Meister, A. 1965. *Biochemistry* 4:3803–09
142. Raijman, L., Jones, M. E. 1973. *The Enzymes* 9:97–119
143. Anderson, P. M., Meister, A. 1966. *Biochemistry* 5:3157–63
144. Guthohrlein, G., Knappe, J. 1969. *Eur. J. Biochem.* 8:207–14
145. Marshall, M., Metzenberg, R. L., Cohen, P. P. 1958. *J. Biol. Chem.* 233: 102–05
146. Rubio, V., Grisolia, S. 1977. *Biochemistry* 16:321–29
147. Elliott, K. R. F., Tipton, K. F. 1974. *Biochem. J.* 141:807–16
148. Elliott, K. R. F., Tipton, K. F. 1974. *Biochem. J.* 141:817–24
149. Powers, S. G., Griffith, D. W., Meister, A. 1977. *J. Biol. Chem.* 252:3558–60
150. Jones, M. E., Spector, L. B. 1960. *J. Biol. Chem.* 235:2897–2901
151. Clarke, S. 1976. *J. Biol. Chem.* 251: 950–61
152. Powers, S. G., Meister, A. 1976. *Proc. Nat. Acad. Sci. USA* 73:3020–24
153. Marayama, H., Easterday, R. L., Chang, H. C., Lane, M. D. 1966. *J. Biol. Chem.* 241:2405–12
154. Cooper, T. G., Wood, H. G. 1971. *J. Biol. Chem.* 246:5488–90
155. Waygood, E. R., Mache, R., Tan, C. K. 1969. *Can. J. Botany* 47:1455
156. Utter, M. F., Kolenbrander, H. M. 1972. *The Enzymes* 6:117–68
157. Miller, R. S., Lane, M. D. 1968. *J. Biol. Chem.* 243:6041–49
158. Wood, H. G., O'Brien, W. E., Michaels, G. 1977. *Advan. Enzymol.* 45:85–155
159. Utter, M. F., Kurahashi, K. 1954. *J. Biol. Chem.* 207:821–41
160. Davis, J. J., Willard, J. M., Wood, H. G. 1969. *Biochemistry* 8:3127–36
161. Willard, J. M., Rose, I. A. 1973. *Biochemistry* 12:5241–46
162. O'Brien, W. E., Singleton, R., Wood, H. G. 1973. *Biochemistry* 12:5247–53
163. Westheimer, F. H. 1968. *Accts. Chem. Res.* 1:70–8
164. Hersh, L. B., Jencks, W. P. 1967. *J. Biol. Chem.* 242:3468–80
165. Hersh, L. B., Jencks, W. P. 1967. *J. Biol. Chem.* 242:3481–86
166. Solomon, F., Jencks, W. P. 1969. *J. Biol. Chem.* 244:1079–81
167. Falcone, A. B., Boyer, P. D. 1959. *Arch. Biochem. Biophys.* 83:337–44
168. Benson, R. W., Boyer, P. D. 1969. *J. Biol. Chem.* 244:2366–71
169. Jencks, W. P. 1962. *The Enzymes* 6:373–85
170. White, H., Jencks, W. P. 1976. *J. Biol. Chem.* 251:1688–99
171. Kraut, J. 1977. *Ann. Rev. Biochem.* 46:331–58
172. Henken, J., Abeles, R. H. 1976. *Biochemistry* 15:3472–79
173. Bruice, T. C. 1976. *Ann. Rev. Biochem.* 45:331–73
174. Knappe, J. 1970. *Ann. Rev. Biochem.* 39:757–76
175. Retey, J., Lynen, F. 1965. *Biochem. Z.* 342:256–71
176. Prescott, D. J., Rabinowitz, J. L. 1968. *J. Biol. Chem.* 243:1551–57
177. Cheung, Y. F., Fung, C. H., Walsh, C. 1975. *Biochemistry* 14:2981–86
178. Sedgwick, B., Cornforth, J. W., French, S. J., Gray, R. T., Kelstrup, E., Willadsen, R. 1977. *Eur. J. Biochem.* 75:481–95
179. Mildvan, A. S., Scrutton, M. C., Utter, M. F. 1966. *J. Biol. Chem.* 241:3488–98
180. Mildvan, A. S., Scrutton, M. C. 1967. *Biochemistry* 6:2978–94
181. Alberts, A. W., Vagelos, P. R. 1972. *The Enzymes* 6:37–82

182. Lane, M. D., Polakis, S. E., Moss, J. 1975. *Subunit Enzymes* 2:181–221
183. Wood, H. G., Zwolinski, G. K. 1976. *Critical Rev. Biochem.* 4:47–102
184. Wood, H. G., Barden, R. E. 1977. *Ann. Rev. Biochem.* 46:385–413
185. Caplow, M. 1969. *Biochemistry* 8: 2656–58
186. Caplow, M., Yager, M. 1967. *J. Am. Chem. Soc.* 89:4513–21
187. Stallings, W. C. 1977. *Arch. Biochem. Biophys.* 183:189–99
188. Kostansek, E. C., Busing, W. R. 1972. *Acta Cryst.* B28:2454–59
189. Worsham, J. E., Busing, W. R. 1969. *Acta Cryst.* B25:572–78
190. Rose, I. A., O'Connell, E. L., Solomon, F. 1976. *J. Biol. Chem.* 251:902–4
191. Fung, C. H., Mildvan, A. S., Leigh, J. S. Jr. 1974. *Biochemistry* 13:1160–69
192. Stubbe, J., Abeles, R. H. 1977. *J. Biol. Chem.* 252:8338–40
193. Scrutton, M. C., Keech, D. B., Utter, M. F. 1965. *J. Biol. Chem.* 240:574–81
194. Williams, A. 1972. *J. Chem. Soc. Perkin Trans.* 2:808–12
195. Benkovic, S. J., Schray, K. J. 1973. *The Enzymes* 8:201–38
196. Westheimer, F. H. 1977. *Pure and Appl. Chem.* 49:1059–67
197. Haake, P. C., Westheimer, F. H. 1961. *J. Am. Chem. Soc.* 83:1102–9
198. Muetterties, E. L., Schunn, R. A. 1966. *Quart. Rev. Chem. Soc.* 20:245
199. Gallagher, M. J., Jenkins, I. D. 1968. *Topics in Stereochemistry* 3:1–96
200. Usher, D. A., Richardson, D. I., Eckstein, F. 1970. *Nature* 228:663–65
201. Usher, D. A., Erenrich, E. S., Eckstein, F. 1972. *Proc. Nat. Acad. Sci. USA* 69:115–18
202. Usher, D. A. 1969. *Proc. Nat. Acad. Sci. USA* 661–67
203. Eckstein, F., Armstrong, V. W., Sternbach, H. 1976. *Proc. Nat. Acad. Sci. USA* 73:2987–90
204. Eckstein, F., Sternbach, H., von der Haar, F. 1977. *Biochemistry* 16: 3429–32 /
205. Young, J. H., McLick, J., Korman, E. F. 1974. *Nature* 249:474–76
206. Bagshaw, C. R., Trentham, D. R., Wolcott, R. G., Boyer, P. D. 1975. *Proc. Nat. Acad. Sci. USA* 72:2592–96
206a. Boyer, P. D., deMeis, L., Carvalho, M. G. C., Hackney, D. D. 1977. *Biochemistry* 16:136–40
207. Korman, E. F., McLick, J. 1972. *Bioenergetics* 3:147–58
208. Rosing, J., Kayalar, C., Boyer, P. D. 1977. *J. Biol. Chem.* 252:2478–85
209. Shavit, N., Skye, G. E., Boyer, P. D. 1967. *J. Biol. Chem.* 242:5125–30
210. Avron, M., Jagendorf, A. T. 1959. *J. Biol. Chem.* 234:967–72
211. Eargle, D. H., Licko, V., and Kenyon, G. L. 1977. *Anal. Biochem.* 81:186–95
212. Morrison, J. F., Heyde, E. 1972. *Ann. Rev. Biochem.* 41:29–54
213. Pedersen, P. L. 1968. *J. Biol. Chem.* 243:3205–11
214. Garces, E., Cleland, W. W. 1969. *Biochemistry* 8:633–40
215. Janson, C. A., Cleland, W. W. 1974. *J. Biol. Chem.* 249:2567–71
216. Anthony, R. S., Spector, L. B. 1971. *J. Biol. Chem.* 246:6129–35
217. Akarstedt, M. T., Silverstein, E. 1976. *J. Biol. Chem.* 251:6775–83
218. Anthony, R. S., Spector, L. B. 1970. *J. Biol. Chem.* 245:6739–41
219. Rose, I. A., Grunberg-Manago, M., Korey, S. R., Ochoa, S. 1954. *J. Biol. Chem.* 211:737–56
220. Dela Fuente, G. 1970. *Eur. J. Biochem.* 16:240–43
221. Cheung, L., Inagami, T., Colowick, S. P. 1973. *Fed. Proc.* 32:667
222. Ray, W. J. Jr., Peck, E. J. Jr. 1972. *The Enzymes* 6:407–77
223. Rose, Z. B., Dube, S. 1976. *J. Biol. Chem.* 251:4817–22
224. Wong, L.-J., Rose, I. A. 1976. *J. Biol. Chem.* 251:5431–39
225. Evans, J. H., Wood, H. G. 1968. *Proc. Nat. Acad. Sci. USA* 61:1448–53
226. Milner, Y., Wood, H. G. 1972. *Proc. Nat. Acad. Sci. USA* 69:2463–68
227. Cooper, R. A., Kornberg, H. L. 1967. *Biochim. Biophys. Acta* 141:211–13
228. Cooper, R. A., Kornberg, H. L. 1967. *Biochem. J.* 105:49c–50c
229. Berman, K., Itada, N., Cohn, M. 1967. *Biochim. Biophys. Acta* 141:214–16
230. Berman, K., Cohn, M. 1970. *J. Biol. Chem.* 245:5319–25

Ann. Rev. Biochem. 1978. 47:1079–1126
Copyright © 1978 by Annual Reviews Inc. All rights reserved

POST-TRANSCRIPTIONAL AND TRANSLATIONAL CONTROLS OF GENE EXPRESSION IN EUKARYOTES[1]

❖997

M. Revel and Y. Groner

Department of Virology, Weizmann Institute of Science, Rehovoth, Israel

CONTENTS

[1]A list of abbreviations appears on the following page.

0066-4154/78/0701-1079$01.00

Abbreviations:

VSV   vesicular stomatitis virus
EMC   Encephalomyocarditis virus
STNV   satellite tobacco necrosis virus
CPV   cytoplasmic polyhydrosis virus
RSV   Rous sarcoma virus
SFV   semliki forest virus
5' m^7G 7-methyl guanosine at 5' end
    of RNA
AUG   adenylyl-uridylyl-guanylyl
    trinucleoside diphosphate
pUp   3',5' uridine diphosphate
ppX   5' nucleoside diphosphate end of
    RNA
m^7pG 7-methyl 5' guanosine mono-
    phosphate
m^6A 6-methyl adenine
DRB   dichloro   ribofuranosyl   ben-
    zimidazol
AMPPNP   5' adenylyl imidodiphos-
    phate
IF2,IF3   prokaryotic initiation factor 2
    or 3
eIF-2   eukaryotic initiation factor 2
eIF-4B, eIF-5   eukaryotic   initiation
    factor 4B,5 etc

GSSG   oxidized glutathione
met-tRNA$_f$ methionyl transfer RNA
fmet-tRNA$_f$ formyl methionyl trans-
    fer RNA
mRNP   messenger ribonucleoproteins
premRNAs   mRNA precursors
hnRNA   heterogenous nuclear RNA
$m^7G^5ppp^5Xm$ 7-methyl   guanosine-
    (5')triphosphate(5')2–0
    methylated nucleoside
MS2   *E. coli* RNA-bacteriophage MS2
tc   translational control
HCR   hemin controlled repressor
HRI   hemin regulated inhibitor
Ts   prokaryotic elongation factor Ts
I_{Hb} initiation factor stimulating hemo-
    globin mRNA translation
DAI   dsRNA activated inhibitor
RWF   ribosomal wash fluid
SMWI   small molecular weight inhibi-
    tor
EF-2   eukaryotic elongation factor 2

PERSPECTIVES AND SUMMARY

In eukaryotic cells, genetic information flows from nucleus to cytoplasm as RNA transcripts of the DNA. These transcripts, called pre-mRNA, are combined with proteins and undergo processing and modifications to become translatable mRNAs. Processing includes cleavage of large transcripts, capping and methylation of the 5' end, internal methylation, and addition of poly A to the 3' end. An exciting recent discovery has been that different segments of pre-mRNA can be spliced and rejoined to form the final nucleotide message that will be translated into proteins. Extensive control of gene expression may occur post-transcriptionally at the level of mRNA maturation.

The binding of mRNA to ribosomes in the cytoplasm has been shown to involve the nucleotide sequence close to the initiation AUG codon of the message, and in many cases, recognition of the 5' capping structure. Seven initiation factors are involved in binding the initiator methionyl transfer RNA (met-tRNA$_f$) and the mRNA onto the ribosome. The role of structural ribosomal protein and ribosomal RNA in mRNA recognition is not yet clear in eukaryotes, but in bacteria a direct mRNA:rRNA base pairing is known to take place during initiation. Proteins that modify this interaction can determine the preferential selection of one mRNA over another, a process often called mRNA discrimination. Various mRNAs seem to have different affinities for the ribosome and compete with each other for initiation.

Untranslated mRNAs are found as ribonucleoproteins (mRNP) in the nucleus and sometimes in the cytoplasm of many cells, which indicates that some translational control of gene expression operates in addition to transcriptional control. Some untranslated RNA can also be rapidly degraded. Very few mechanisms of translational control have been elucidated, but a few cases have been studied in great detail during the past years and are reviewed here. In reticulocytes, initiation of globin synthesis depends on the presence of hemin, and is inhibited by oxidized glutathione. A complex translational inhibitory system, involving the activation of a protein kinase phosphorylating initiation factor eIF-2, was discovered. Protein synthesis may in this way be coupled to the energy metabolism of the cell. Double-stranded RNA triggers another similar system that inhibits initiation. These controls operate to a varying extent in different cells, and may be induced or repressed, causing increase or decrease in initiation factor activity and changes in the rate at which various mRNA are translated.

Interferon seems to exert its antiviral effect, at least in part, by controlling translation of viral mRNAs. Treatment of cells by interferon induces a translational inhibitory system that can be activated by double-stranded

RNA (dsRNA). As in reticulocytes there is activation of a protein kinase, independent of cyclic nucleotides, which phosphorylates several proteins, among them eIF-2, and inhibits initiation. Double-stranded RNA activates an enzyme that converts ATP into a potent nucleotidic inhibitor of translation. Both initiation and elongation are subject to this regulation. Elongation seems blocked because certain minor species of transfer RNA (tRNA) are not utilized properly by ribosomes from interferon-treated cells.

In virus-infected cells, viral mRNAs are, in general, translated more efficiently than cellular mRNAs, and there is a shutoff of cellular protein synthesis in the infected cell. Understanding how interferon is able to produce the opposite effect, that is, the continuance of host protein synthesis while viral mRNAs are inhibited, could answer many questions in translational control.

The biochemical mechanisms of RNA and protein synthesis have been extensively reviewed recently (1–4). This discussion reviews only the latest information on the post-transcriptional and translational processes that may control genetic expression.

POST-TRANSCRIPTIONAL MODIFICATIONS OF mRNA AND THEIR ROLE IN TRANSLATION

In eukaryotic cells, formation of translatable mRNA is a complex process. The primary DNA transcripts, pre-mRNAs, synthesized in the nucleus, undergo post-transcriptional modifications. These include: cleavage of the large RNA precursors, 3'-end polyadenylation, 5'-end capping and methylation, internal methylation, and splicing together of different segments of pre-mRNA into one mRNA chain.

Pre-mRNA Cleavage and RNA Splicing

It is now accepted that many cellular and viral mRNA molecules are derived from high molecular weight precursors (5). Studies with animal viruses, in particular adenovirus 2 (6–8), and more recently globin-mRNA (9–12) have clearly shown large pre-mRNAs, synthesized in the nucleus. The various aspects of pre-mRNA and RNA processing are covered in comprehensive reviews published during the last few years (13–22). This discussion is restricted to the recently discovered phenomenon of RNA splicing, and in particular to processes where control of gene expression may take place.

Late after adenovirus 2 infection, the nucleus contains high molecular weight viral specific RNAs that are complementary to the "r" strand of viral DNA. Experiments by Darnell and co-workers (23, 24), indicated that transcription of this large viral RNA is initiated at a specific site near position 15 on the adenovirus genome. The long (>20,000 nucleotides) primary transcript which is complementary to about 85% of the adeno-r-strand, is converted to adenovirus messengers (1,000–4,000 nucleotides) by a series of post-transcriptional modifications. From several studies (25–28) it became evident that late adenovirus mRNAs contain 5'-capped leader sequences of about 150 nucleotides which are transcribed from DNA sequences located at distant sites of the adenovirus chromosome. For example, Berget, Moore & Sharp (25) reported that the adenovirus hexon mRNA consists of four segments—the "body," which is the major part of the mRNA, and three smaller sequences, transcribed from three different chromosomal sites, and spliced together to form the 5' leader of the molecule. Similar observations were made by others for several adenovirus late mRNAs (27, 28, 37).

In simian virus 40 (SV40) a similar situation occurs. A common capped leader sequence at the 5' terminus of both the 19S and 16S mRNAs is transcribed from a DNA segment not adjacent to the coding portion of these mRNAs (29–33).

How and why such chimeric RNA molecules are formed, what function they serve, and at what stage splicing occurs are still open questions. Various models have been suggested (25–33) to explain the biogenesis of these mRNA molecules. They fall into two major classes. The first proposes that such mRNA molecules are produced by processing of a larger precursor RNA molecule. Two or more regions of a large capped RNA transcript are brought together by looping out the RNA sequences to be deleted. The juxtaposed segments are then covalently joined by intramolecular ligation. This hypothesis requires specific mechanisms and enzymes for cleavage and resealing of RNA ends (34, 35). The data concerning transcriptions of adenovirus (23, 24), are consistent with the "RNA looping" model and the existence of a large RNA precursor. The second class of models proposes that mRNA molecules are transcribed from a preformed modified DNA template folded in such a way that RNA polymerase will glide and transcribe nonadjacent sequences.

The joining of distant RNA sequences is probably a general process (504–506) and may explain a common structural feature of mRNA. Because heterogeneous nuclear (hnRNA) and mRNA have a common 3' poly(A) linked segment (13), part of the mRNA is probably excised from the 3' region of the hnRNA. However, the finding that caps at the 5' ends of hnRNA constitute precursors of mRNA caps (36, 38, 117), together with the observations that capping is coupled to initiation of transcription and occurs on triphosphate ends (39–42), raised the possibility that mRNA segments are located at the 5' termini of large precursor molecules. The RNA splicing mechanism may explain how a 5'-capped region of long precursor RNA molecule can be joined to the poly(A)-containing coding region located at the 3' side of the same molecule.

Splicing determines the final nucleotide sequence of the mRNA and as such appears as a most important step where control could occur. The precise role that capped leader sequences play, is still open to question and is discussed below. In general the role of nontranslated sequences, as found at the 5' and 3' end of mRNA, is still in question (Figure 1).

The 3'-Poly(A) Segment in mRNA Translation and Stability

Most eukaryotic mRNAs and viral mRNAs contain at the 3' end a poly(A) tail of 50–200 residues (16, 43). The poly(A) segment is added post-transcriptionally; polyadenylation can take place both before and after processing of hnRNA, in the nucleus as well as in the cytoplasm (44–50). Poly(A) is absent from some classes of cellular messengers (53–58) such as histone mRNAs (51, 52), and from some viral mRNAs such as reovirus (58a) and many plant viruses (59). Is the poly(A) segment essential for mRNA function in the cell? Many studies suggest that it is not required for translation in cell-free extracts. Histone mRNAs are actively translated in Krebs ascites systems (60–62) and deadenylated rabbit globin mRNA is translated in wheat embryo, Krebs ascites, and rat liver extracts with the same efficiency as intact globin mRNA (63). Translation of L-cell RNA, in wheat germ system, is not markedly reduced after removal of the poly(A) segments by a 3' exonuclease (64). Similarly ribonuclease IV, which is a poly-(A)-specific ribonuclease from chick oviduct (65), has no influence on the translational capacity of oviduct cell-free extracts (66). Formation of duplex between poly(U) and the poly(A) segment of mRNA does not impair translation (67). In poliovirus, the poly(A) is necessary for viral RNA to be infectious, but not for its in vitro translation (68).

Nevertheless, a role for poly(A) in mRNA stability may have been demonstrated by Soreq et al (69) and Williamson et al (70). Globin mRNA,

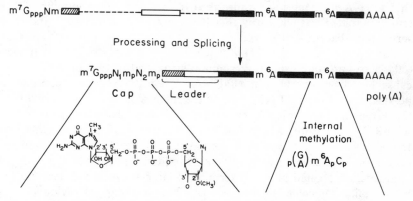

Figure 1 Post-transcriptional modifications of eukaryotic cellular and viral mRNAs.

from which the poly(A) was enzymatically removed with polynucleotide phosphorylase, directs globin synthesis in Krebs ascites cell-free system at the same initial rate as does poly(A)-containing mRNA. The deadenylated mRNA became, however, less active after longer periods of incubation. Furthermore, injection of globin mRNA into frog oocytes, which allows the study of translation for several days, showed that after an initial translation at similar rates, the poly(A)-free mRNA became considerably less active than intact globin RNA (71). This effect could be attributed to the greater stability of polyadenylated mRNA (72). Direct measurements of globin mRNA content showed that 56 hr after injection about 85% of the poly(A)-free mRNA was degraded, while native poly(A)-containing mRNA was preserved. The minimal length of the poly(A) segment required for stability was found to be about 30 adenylate residues (73) and it was shown that the deadenylated mRNA is broken down in the frog oocyte by a translational-dependent mechanism (74). Readdition of a poly(A) segment to deadenylated mRNA restored its functional stability in the frog oocytes (75).

It is not clear to what extent the oocyte system allows conclusions about the role of poly(A) in the cell from which the mRNA is taken. For example, histone mRNAs in the cytoplasm of most cells are as stable as poly(A)-containing mRNAs, provided DNA synthesis is active (76). In frog oocytes, however, histone mRNAs are as unstable as deadenylated globin mRNA, and 20 hr after injection no histone synthesis is detected (77). Enzymatic polyadenylation of histone mRNA increases its half-life in oocytes to more than 48 hr (78). The oocytes could therefore degrade the deadenylated mRNAs more actively than do other cells. On the other hand, histone mRNAs are degraded rapidly in cells when no DNA synthesis occurs, and this may explain their behaviour in oocytes.

The role of poly(A) in mRNA stability has still to be more firmly established. Deadenylated mengo virus RNA, which is translated in frog oocytes, showed no loss of activity even 60 hr after injection (79). Finally, the conclusion that poly(A) is not needed for translation per se has been challenged: deadenylated ovalbumin mRNA was shown to be less efficiently reinitiated than intact mRNA (80).

In conclusion, most of the data available show that the 3' poly(A) has an important function in mRNA stability. As discussed below, specific proteins may bind to the poly(A) segment (217, 262–264) and protect it from nuclease degradation. The 5' cap seems also to protect mRNA from exonuclease degradation.

5' Capping and Methylation in mRNA Translation

Many eukaryotic, viral mRNAs and nuclear pre-mRNAs, contain methylated nucleotides. At the 5' termini, a capping structure of the form $m^7G^{5'}ppp^{5'}XmYm \ldots$ is generally found, in which the terminal 7-methyl-guanosine is linked by a 5',5'-pyrophosphate bridge to the adjacent methylated nucleotide (for review see 81). In addition, methylated adenosine residues (N^6-methyladenosine) are found at internal positions of some eukaryotic cellular and viral mRNAs. Methylation and capping introduce several possibilities for post-transcriptional regulation of genetic information (Figure 1).

The functional importance of 5'-terminal m^7G in caps for eukaryotic mRNA translation was first demonstrated by Shatkin and co-workers (82, 83). mRNAs lacking 5' m^7G were prepared either by synthesizing reovirus and vesicular stomatitis virus (VSV) mRNAs in vitro under conditions where methylation is inhibited, or by removing chemically m^7G from reovirus and globin mRNAs by β-elimination. These messengers when tested in a wheat germ cell-free system, demonstrated that efficient translation depends upon the presence of 5' terminal m^7G in the caps (82, 83). Similar evidence was obtained for other cellular mRNAs (84) in wheat germ system, and for viral and cellular mRNAs in cell-free translation systems from *Artemia salina* (85). In vivo, Rose & Lodish (86) observed that in VSV-infected cells, mRNAs with 5' triphosphate ends are not attached to ribosomes, whereas all the VSV polysomal mRNAs are capped. Questions were, however, raised on whether the dependence of mRNAs translation on 5' caps is a general phenomena in all eukaryotic cells and for all mRNAs. Indeed, in cell-free systems of rabbit reticulocytes the impairment of reovirus, VSV, and globin mRNAs translation by the removal of the 5' m^7G was not dramatic (87, 88, 89) and β-eliminated brome mosaic virus (BMV) RNA-4 showed only a twofold reduction of template activity in the wheat germ system (90). The decapped VSV and BMV mRNA retained their ability to code for authentic viral proteins. In addition, not all eukaryotic mRNAs are capped. Poliovirus RNA isolated from the polyribosomes of infected cells is not capped; its terminus is pUp. .(91–93), and it also does not have the protein bound to the 5' end of virion RNA (91a, 92a). Encephalomyocarditis (EMC) RNA (94) and plant satellite tobacco necrosis virus (STNV) RNA (95–97) similarly do not have a 5' m^7G; yet all these mRNAs are translated efficiently in eukaryotic cell-free systems (4, 84, 98–100).

Decapping of mRNA with T4 polynucleotide kinase (103) did not decrease its activity (102); in this case 5'-monophosphate termini (pX. .) were formed. It was suggested (89, 102) that chemical decapping by β-elimination could have side effects resulting in inactivation of mRNA. This was ruled out (104) by recapping and remethylating β-eliminated RNA with purified vaccinia guanylyltransferase and guanine-7-methyltransferase, which restored ribosome binding and translation of the RNA; capping without methylation did not restore activity. Furthermore, decapping with two enzymes, tobacco and potato pyrophosphatase, to yield 5'-diphosphate termini (ppX . . .) without mRNA cleavage (106, 107, 109), produced a 90% loss of mRNA activity (105, 108). The different result with T4 nucleotide kinase (102) could reflect a better translation of RNA with 5'-monophosphate ends than other forms of uncapped RNAs, or it may result from inefficient removal of 5'm^{7}GDP by the T4 enzyme (118).

A likely explanation for these conflicting observations on cap requirement is that with certain mRNAs, in which the initiator AUG is far from the 5' terminus, and with certain ribosomal systems, which have a high affinity for the mRNA binding sequence, the cap structure is not as important as with other mRNAs or ribosomes. In addition, the conditions of in vitro translation, such as the sources of initiation factors (496), potassium concentration (101), and temperature (497), can alter the stringency of cap requirement.

The 5' Cap Structure in mRNA Stability

The 5' cap structure has at least two functions that explain its role in mRNA translation. The m^{7}G is recognized during initiation, as discussed in detail in a later section. But the cap itself also protects the mRNA from 5'-exonucleolytic degradation. Reovirus mRNAs with blocked 5' termini are more stable when injected to oocytes or translated in cell-free extracts than unblocked mRNAs (110). Unmethylated G^5'ppp^5'X . . . terminated RNA, although poorly translated, is as stable as m^{7}G^5'ppp^5'X . . . RNA. RNAs of cytoplasmic polyhydrosis virus (CPV), from which the 5' m^{7}pG was removed by tobacco pyrophosphatase (106, 107) were also less stable than intact RNA in wheat germ extracts (105). In reticulocyte lysates, however, there is no difference in the stability of capped and unblocked mRNAs (110).

Enzymatic removal of caps in cells may serve as negative translational control. Cap-hydrolyzing enzymes have been detected in extracts of HeLa cells (111, 112) tobacco cultured cells (106), and potato (109). The tobacco pyrophosphatase specifically removes m^{7}pG from capped mRNAs without scission in other parts of the RNA molecule (106, 107). On the other hand, the HeLa enzyme, m^{7}G-pyropase, cleaves only free caps or caps attached to small oligonucleotides, but does not attack caps in mRNA. The presumed function of this enzyme is to eliminate caps from short 5' oligonucleotides that have been generated by mRNA degradation, and thus prevent them from inhibiting translation (111, 112). The 5' terminal cap has the same turnover as the rest of the mRNA (93, 113, 114). Thus, in order to reconcile this observation with a possible in vivo function of decapping in mRNA stability, one has to assume that following removal of caps there is a rapid degradation of the whole mRNA molecule.

Cap removal blocks translation and allows degradation of nontranslated mRNAs. This process contrasts, therefore, with poly(A) removal which causes degradation of mRNAs still being actively translated.

Internal Methylation of mRNA

Methylation of adenosine residues has been reported at internal positions of eukaryotic and viral mRNAs (115–124), although the exact locations of these residues have yet to be established. They are not part of the poly(A) segment (116, 119). In HeLa cells (115), L cells (38), B77 Avian sarcoma virus RNA (120), and SV40 late mRNA (168), m^6A occurs mainly in two sequences—Apm^6ApC and Gpm^6Apc. This remarkable degree of sequence specificity argues for an important biological function. In addition, the m^6A-containing sequences are conserved during processing of hnRNA (38). The functional significance of m^6A is still unknown. Since various viral and cellular mRNAs lack m^6A (125–131) it seems that this modified residue is not essential for mRNA translation. An interesting possibility is that such groups constitute recognition sites for RNA processing (44) and/or splicing enzymes. Alternatively the m^6A could have a function in translation; in association with termination signals they may act to block translation of a specific region of mRNA.

To conclude, a complete picture of the function of mRNA methylation and of the 5' capping structure is not yet available, but it appears that these post-transcriptional modifications control translation both by a positive effect—increasing mRNA affinity for ribosomes, and by a negative effect—preventing breakdown of mRNA. Details of the role of methylated caps in ribosome binding are discussed in the next section.

RECOGNITION OF mRNA DURING INITIATION

Present models of mRNA recognition by ribosomes are often based on the information available from prokaryotic systems (2, 132, 133, 160). In bacterial mRNAs, the site that binds to ribosomes contains (*a*) the initiator codon AUG to which f-met-tRNA$_f^2$ base pairs, and (*b*) an initiation signal composed of several purine nucleotides preceding the AUG codon on its 5' side by about 10 nucleotides; this signal base pairs to a complementary sequence near the 3' end of 16S ribosomal RNA. The number of base pairs formed varies from one mRNA cistron to another. In the ribosome, the structure of the 3' end of 16S ribosomal RNAs (rRNAs), and the proteins located around it, influence the binding of mRNA and f-met-tRNA$_f$ during

---

[2] f-met-tRNA$_f$ is formyl methionine transfer RNA that has been formylated by N^{10}-formyltetrahydrofolic acid and a transformylase.

initiation. Some of these proteins are initiation factors, such as IF3 and IF2, and some are ribosomal proteins, such as S1 (interference factor i-α) and protein S12 (the streptomycin protein). These proteins can produce changes in rRNA or mRNA configuration and affect the stability of the interaction between them, determining thereby the affinity of ribosomes to a given cistron (2, 133, 134).

It is also likely that in eukaryotes both RNA-RNA and protein-RNA interactions play a role in mRNA recognition and discrimination by ribosomes. But structural studies on eukaryotic ribosomes are not yet at a stage where models of the precision obtained in *Escherichia coli* can be formulated. Furthermore, it is already clear that there are several important differences in eukaryotic mRNAs which have to be taken into account for any model of initiation. First, the cap structure, absent in prokaryotes, is positively recognized; second, the nucleotide sequences to which ribosome bind appear quite variable; third, eukaryotic mRNAs are functionally monocistronic. Finally, the number of translational initiation factors and ribosomal proteins is much larger in mammalian cells than in bacteria.

Recognition of 5' m⁷G in Caps During Initiation

The functional importance of 5' m^7G in caps is manifested at early stages of protein synthesis initiation. Ribosome binding experiments with reovirus mRNA indicate that mRNA molecules containing 5' terminal m^7G are preferentially bound to wheat germ 40S ribosomal subunit (135). Furthermore, cap analogs such as m^7pG, m^7GTP, and $m^7GpppXm$ have been shown to specifically inhibit in vitro ribosomal binding and translation of several viral and cellular capped mRNAs in wheat germ, reticulocyte, *A. salina,* and also L-cell extracts (99, 100, 136–139, 189). In uncapped messengers like STNV RNA, EMC RNA, T4 mRNA, and synthetic SV40 cRNA, the absence of 5' terminal m^7G does not preclude translation, but these mRNAs are insensitive to inhibition by cap analogs. Recently, it was observed that U1 nuclear RNA, which has the 5' terminal sequences $m_3^{2,2,7}Gp(p)pAmpUm$ . . . (140, 141) binds to wheat germ ribosomes: U1 RNA is not translated in wheat germ extracts, but markedly inhibits mRNA translation in this system (142). On the other hand, a short 5' oligonucleotide $m^7GpppGmpCpUp(Xp)_3Gp$, derived by T1 RNase from reovirus mRNA (135), or the oligomer $m^7GpppGp$ from BMV (143) fail to bind to wheat germ ribosomes, indicating that the cap by itself is not sufficient to form a stable initiation complex.

A methyl group at the 7th position of the guanosine and a 5' phosphate are both essential for exerting the inhibitory effect of cap analogues (100, 101). The positive charge imposed by the methyl group at the 5' terminal m^7G could be required merely for maintaining a specific conformation at the 5' end of the mRNA, which facilitates binding of ribosome to the adjacent initiation site. Nuclear magnetic resonance analysis of cap analogs (144), and studies with a variety of chemical derivatives of m^7GDP (144a) suggest that electrostatic interaction between the positively charged m^7G and the negatively charged phosphate leads to a rigid preferred conformation at the 5' end of the mRNA. But the finding that cap analogs inhibit translation of capped messengers by blocking the formation of the mRNA-ribosome initiation complex, indicates that the cap is positively recognized during initiation. A protein fraction that recognizes the 5'-terminal m^7G in mRNA has been detected in high-salt wash of *A. salina* ribosomes (145). Shafritz and co-workers studied the mechanism of inhibition of protein synthesis by m^7pG in fractionated reticulocyte system using purified globin mRNA and initiation factors (138; for nomenclature see Table 1). They discovered that m^7pG specifically inhibits the interaction of capped mRNA with purified reticulocyte initiation factor eIF-4B. Although purified eIF-2 and eIF-5 can also form complexes with mRNA, only eIF-4B-mRNA interaction is inhibited by the cap analog m^7pG. They concluded that eIF-4B is most likely the putative cap binding protein. Other structural features or sequences within the mRNA are also recognized by eIF-4B since (*a*) the amount of cap analogs required for inhibition of either protein synthesis (99, 100, 136, 137), or eIF-4B-mRNA interaction (138) greatly exceeds (> 1000-fold) the amount of mRNA present in the system and

Table 1   Eukaryotic initiation factors[a]

Nomenclature		Molecular weight		
New	Old	Subunits	Native	Activity
eIF–1	IF–E1, –	15,000	15,000	40S complex formation
eIF–2	IF–E2, IF–MP	55,000 50,000 35,000	125,000	met–tRNA$_f$ binding GTP binding
eIF–3	IF–E3, IF–M5	Many	⩾500,000	Dissociation, mRNA binding
eIF–4A	IF–E4, IF–M4	50,000	50,000	Natural mRNA binding
eIF–4B	IF–E6, IF–M3	(80,000)?	(80,000)	mRNA (cap) recognition
eIF–4C	IF–E7, IF–M2Bβ	19,000	17,000	Stabilization
eIF–4D	–, IF–M2Bα	17,000	15,000	Subunit joining, elongation?
eIF–5	IF–E5, IF–M2A	150,000	125,000	80S formation, GTPase
eIF–2A	–, IF–M1	65,000	65,000	tRNA binding to 40S subunit

[a] Adapted from (196).

(b) eIF-4B binds naturally uncapped EMC virus RNA (138), STNV RNA, and β-eliminated globin mRNA (146); this binding is not sensitive to cap analogs. Enzymatic addition of capping structure to either STNV RNA or β-eliminated globin mRNA, renders their interaction with eIF-4B sensitive to cap analogs (146). Due to the low affinity of eIF-4B to m^7pG, no direct demonstration of their interaction was established. This could possibly be achieved by affinity labeling of eIF-4B with m^7pG derivatives.

Consistent with the model that the 5' cap structure plays an important role in mRNA-ribosome interactions, is the observation (147) that attachment of reovirus mRNAs to 40S ribosomal subunits protects the 5' $m^7GpppGm$. . . region against RNase. However, when 80S ribosome-protected fragments were isolated it was found (147) that in many cases the 5' cap was no longer protected. Hence the formation of initiation complex involves, besides recognition of the 5' cap, interactions with other structural components forming binding sites on mRNA, as well as the participation of several initiation proteins (see below).

The base composition of the RNA fragment adjacent to the cap affects the relative affinity of the fragment for ribosomes (87, 148). Besides the m^7G, 2'-O-methylation of the penultimate nucleotide (X) in the cap (m^7GpppX) enhances the extent of ribosome binding to cap-containing synthetic polymers (87). Vaccinia mRNAs with A caps, namely mRNAs containing adenosine residue in the penultimate position of the cap (m^7GpppA . . .), bind with a higher affinity to wheat germ or reticulocyte ribosomes than to mRNA with G caps (m^7GpppG . . .) (104). It is possible that the higher binding efficiency of A caps containing vaccinia mRNAs is due to some other structural feature, such as a common leader sequence (see above). Nevertheless, it is interesting that in addition to vaccinia (149) both adenovirus (122, 123) and SV40 (124) late mRNAs have been shown to contain A caps.

Sequences of Ribosome Binding Sites in mRNA

Nucleotide sequences of more than ten eukaryotic initiation regions are now available (Table 2; 150–159) and surprisingly, there are very few common features among them. Each has an AUG codon preceded by a nontranslated region and a 5' terminal cap, but the initiator AUG codon of rabbit β-globin mRNA is located 54 nucleotides from the 5' terminus (156), that of RSV (154) 82 nucleotides from it, and in BMV (151) the AUG is 10 nucleotides from the 5' cap. Thus, the distance between the cap and the AUG is quite variable in eukaryotic mRNAs. The 5' portion of RSV (154) and β-globin (156) mRNAs contain extensive potential secondary structure; however, not many regions of secondary structure can be drawn from any of the three protected sites of VSV (157). In the six reovirus initiation sites, the common features are a G residue at the 3' side of the AUG codon, and the sequence CUA adjacent to the 5' cap (153). However, these are not found in other known mRNAs.

It has been proposed (160) that in eukaryotes, as in bacteria, (161) a short sequence of mRNA at the 5' side of the initiator AUG, is base paired during initiation with complementary regions near the 3' end of the 18S ribosomal RNA, whose eight terminal nucleotides, ^{3}'AUUACUAG^5'. . ., are probably conserved throughout the entire eukaryotic kingdom (162–164). It was also suggested (165), that interaction between 18S rRNA and globin mRNA may be important for mRNA translation. However, the available 5' terminal sequences of eukaryotic mRNAs show very little significant complementarity with the above mentioned 18S rRNA sequence, and the potential interactions are predicted to be far less stable than those demonstrated with prokaryotic messengers (160). Furthermore, some of the suggested interactions (166) occur around the initiation AUG codon in such a way that the met-tRNA$_f$ and the 18S rRNA would compete for the AUG sequence. It is possible that additional sequences from the 3' end of 18S rRNA will be found to interact with mRNA, but at present the evidence for rRNA–mRNA base pairing in eukaryotes is weak. The limited sequence homology (5' cap terminus and AUG codon) among the known eukaryotic initiation sites, led Kozak & Shatkin (150, 152) to propose a "minimal recognition" mechanism for initiation of eukaryotic mRNAs. This involves merely the presence of an AUG codon close to the 5' terminus of the mRNA which is usually, but not necessarily, capped.

The Functionally Monocistronic Behavior of Eukaryotic mRNAs

In 1968, Jacobson and Baltimore suggested (167) that in animals, in contrast to bacteria, mRNA does not contain multiple initiation sites. Several cellular mRNAs have been shown to code for only one protein (156, 169–170). In poliovirus and EMC virus RNA, which code for as much as 10 proteins, translation proceeds from one initiation site to give a long polypeptide chain, which is then cleaved to form the final proteins (171–175). Recently, two different initiation sites were detected in vitro on poliovirus RNA (176), and many mRNAs were shown to contain several initiation sites. In all cases, however, only one of them is active in vitro. The active initiation sites are the ones closest to the 5' end of the messenger (for review see 177). The internal, inactive initiation site is activated when smaller mRNA species, containing the "silent" initiation site at the 5' end, are generated either by cleavage of the larger mRNA or by differential transcription.

In BMV (178) one of the viral mRNAs, RNA-3 has two initiation sites, but only one of them is active in vitro. There is an RNA-4 that contains part of the sequences of RNA-3, and in which the second initiation site of coat protein, is active. In sindbis virus and semliki forest virus (SFV), virion 42S RNA contains one active initiation site and directs in vitro synthesis of nonstructural proteins (179, 180). Late in infection a smaller 26S mRNA species appears which is derived from the 3' end of the 42S RNA (181–183), and directs the synthesis of three viral structural proteins (179, 181, 184). It is not known whether the silent initiation sites of the 42S are active in vivo. A similar situation was found in tobacco mosaic virus (TMV) where a small 9S mRNA, localized at the 3' end of the virion RNA, codes for the coat protein in cell-free systems.

A related situation was reported for SV40 and polyoma viruses (185–187). Late SV40 mRNAs consist of two RNA sedimenting as 19S and 16S. Both mRNAs are capped (188, 124) and methylated (190). The 16S mRNA shares common nucleotide sequence with the 3' region of the 19S species (191–193). Prives et al, have found (185, 186) that cell-free translation of late SV40 19S mRNA results in the production of viral protein 2 (VP-2), while the 16S mRNA codes for the major capsid protein VP-1. They have suggested that during 19S RNA translation, the initiation site for VP-1 is inaccessible and can be translated only after the cleavage of the 19S RNA to 16S, which is then recapped.

Whether or not internal initiation sites are inactive because they are buried in secondary structure, like the masked synthetase initiation site of RNA phages (132), is not known. Nevertheless, this process controls viral protein synthesis, for by generating small RNA, a specific region of the viral genome that codes for structural proteins can be amplified (177).

ROLE OF RIBOSOMAL PROTEIN FACTORS IN mRNA RECOGNITION AND TRANSLATION

Initiation Factors involved in mRNA Binding

Among the seven mammalian initiation factors for protein synthesis (194–196) that have been identified and purified from rabbit reticulocytes (Table 1) at least three have been implicated in binding and recognition of natural mRNA as opposed to synthetic polynucleotides. Originally, a role for eIF-3, a very large multimeric factor with molecular weight in excess of 500,000 which is found on free native 40S ribosomes (197, 198), was proposed on the basis of studies with hemoglobin mRNA (199–201) and myosin mRNA (202). In addition, the role of two smaller monomeric factors, eIF-4A (50,000 mol wt) and eIF-4B (70–80,000 mol wt), has been clearly established for globin mRNA binding to the 40S ribosomal initiation complex (195, 200, 201, 203). A strict requirement for eIF4A is seen with EMC or mengo viral RNA (195, 204).

Because in early studies eIF-3 was isolated together with eIF-4B (195, 200), and because one of eIF-3 polypeptides may correspond to eIF-4A (203), the genuine function of eIF-3 in natural mRNA recognition is questionable. eIF-3 is required (205) to bind and stabilize the ternary complex [met-tRNA$_f$.eIF-2.GTP] (206) to 40S ribosomes. Binding of the ternary complex can precede mRNA attachment (205, 207, 208), and eIF-3 may stimulate mRNA binding indirectly as does eIF-2 (201). It is likely, however, that as in prokaryotes (2), mRNA and initiator tRNA bind independently; when both are present, the complex formed is more stable. Recently, a partial requirement for AUG codons in met-tRNA$_f$-binding to 40S ribosomes with eIF-2 and eIF-3 was demonstrated (196, 209). Omission of eIF-3 has indeed a more profound effect on mRNA binding than on the binding of met-tRNA$_f$ (201); eIF-3 should therefore be considered as an mRNA binding factor (Figure 2).

Hydrolysis of ATP is required for a yet unknown step in hemoglobin mRNA binding to ribosomes; mRNA binding factors eIF-4A, eIF-4B, and ATP are not needed when AUG is used instead of mRNA to form 80S ribosomal initiation complexes (195, 196, 201). Other factors that may have some role in natural mRNA binding include eIF-1 (15,000 mol wt), possibly preventing ribosomal subunit joining when mRNA is absent (195, 210) and eIF-4C (17,000 mol wt), that is partially required for hemoglobin mRNA binding (201).

Another method for determining which initiation factors play a role in mRNA binding involves measurement, the absence of ribosomes, of mRNA

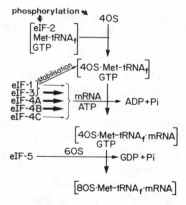

Figure 2   Initiation of mammalian protein synthesis [adapted from (201)].

binding to the proteins adsorbed on nitrocellulose filters. The met-tRNA$_f$ binding factor eIF-2 (205; Table 1) has in addition a high affinity for several mRNAs, but also for poly(A) (211, 138), for phage RNA or dsRNA (212). eIF-2 is a minor protein of the mRNP particles (211). Protein M1 (eIF-2A), which stimulates a GTP-independent initiator tRNA binding to 40S ribosomes (213, 214), also strongly binds mRNA (215). M1 is probably not involved in initiation (208, 216), but may have some regulatory function. RNA binding by eIF-5, the ribosomal subunit joining factor (195), was also reported (138), but may be related to its affinity for rRNA, rather than for mRNA. The clearest mRNA binding function was demonstrated for eIF-4B: binding of several mRNAs by this protein is inhibited by the cap analog m^{7}pG which suggests that what is recognized includes the 5' terminal cap (138). The cap is not the only part of mRNA recognized by eIF-4B, since it very strongly binds EMC viral RNA, a noncapped mRNA (94, 138); but with this RNA, cap analogs had no effect. Direct binding may not be synonymous to a role in mRNA-ribosome binding. Indeed, eIF-3 and eIF-4A, surprisingly, do not bind RNA by the direct filter test. These factors may exert their effects by modulating the mRNA-ribosome interaction.

The role of ribosomal protein is poorly known as yet. Little is known also of the function of the proteins bound to mRNA in polysomal mRNP particles (217–219) and their relationship to initiation factors (195, 211, 220–222).

Initiation Factors in mRNA Selection and Discrimination

To evaluate the role of initiation factors in mRNA selection, their relative effects on the translation of different mRNAs were compared. Such studies led to the conclusion that different mRNAs need different amounts of given initiation factors to be optimally translated and this constitutes a basis for mRNA discrimination effects.

EMC and mengo viral RNAs need more reticulocyte eIF-4A than does globin mRNA (195). This explains the discriminatory properties of a factor from Krebs ascites cells, IF$_{EMC}$, reported by Wigle & Smith (223) and more recently by Blair et al (204). These authors compared 12 different mRNAs, and showed that eIF-4A will stimulate Krebs cell ribosomes to translate much more EMC or mengo RNA than globin mRNA or oviduct mRNA. When tested with ribosomes that completely lack initiation factors, eIF-4 is needed for globin mRNA as well (195). The discriminatory effect results only from the different amounts needed. The situation appears similar to that of *E. coli* ribosomes, where different amounts of IF-3 and S1(i-α) are required for different cistrons on MS2 RNA (2, 133, 224).

In contrast to IF$_{EMC}$(eIF-4A), another factor I$_{Hb}$ has a discriminatory effect that favors the translation of globin mRNA over that of mengo or

EMC RNAs. This factor was purified from reticulocytes (225, 134) by its ability to differentially stimulate globin mRNA translation, but not mengo RNA, in Krebs ascites cell-free extracts. In a plasmocytoma cell-free system, Golini et al (226) showed that this discriminatory factor is similar to eIF-4B. When EMC RNA is added, in vitro or by preinfecting the cells with EMC virus, the plasmocytoma mRNAs, as globin mRNA, are competed-out from the translation system, and this competition is relieved by adding eIF-4B. There could be a direct competition between the two mRNAs for binding this protein. A discriminatory effect of I_{Hb} in favor of globin mRNA was, however, seen either when mengo RNA is added as competitor (227), or when the two mRNAs are added independently (225), which suggests that the effect on globin mRNA is not only the result of sequestration of the factor by the viral RNA. The discriminatory effect of eIF-4B (or I_{Hb}), which is a cap binding factor (138), could be related to the fact that globin mRNA is capped while mengo and EMC RNA are not capped. Capped mRNA will not bind efficiently to ribosomes unless the cap binding factor is present in sufficient amount. With low amounts of eIF-4B, the viral RNAs would be more efficient than globin mRNA in binding to ribosomes.

Competition between mRNAs for a limiting amount of factor or ribosomes may help explain the ability of I_{Hb} (or eIF-4B) to preferentially stimulate α-globin mRNA translation over that of β-globin mRNA in extracts of Krebs cells; this effect was originally shown by Nudel et al (225). In reticulocytes, increasing the amount of unfractionated $(\alpha+\beta)$-globin mRNA decreases the ratio of α-/β-globin synthesis (228–231, 169). Recently, Kabat & Chappell (232), reported that eIF-4B, and to a lesser extent eIF-4A, relieve the competition between β- and α-globin mRNAs in a purified ribosomal system from reticulocytes. This was interpreted as due to a higher affinity of β-globin mRNA for the factor (232, 215), leaving little of it for α chains. An increase in factor's concentration allows α-globin mRNA to be translated as well or better than β-globin mRNA. Another interpretation of these data is that more factor is needed for optimal α-globin mRNA translation than for β-globin mRNA, and this produces the discriminatory effect as seen above for globin versus mengo mRNAs. Cap recognition by eIF-4B may be one of the parameters that influence the affinity for α- and β-globin mRNAs: m^7pG was shown to inhibit more α- than β-globin mRNA translation (139, 233). It is possible that α-globin mRNA depends more on cap recognition, while the β-globin mRNA binds more to eIF-4B and ribosomes through the long initiation sequence following the cap (see Table 2). Uncapped mRNA would also bind strongly through this sequence. Discrimination between uncapped mengo RNA and capped globin mRNA was also induced by cap analogs (100).

Can these effects be called mRNA discrimination? Variations in the relative translation of different mRNAs, due to changes in mRNA concentration (229, 232), protein synthesis inhibitors (235, 236), and salt concentrations in vitro (229) and in vivo (234, 237) have been used as arguments against the existence of mRNA discrimination factors. Lodish (4, 238, 498) showed that any change in the rate of initiation will alter the translation ratio of two mRNAs if (a) they are competing with each other for limiting initiation capacity, and (b) they bind to ribosomes at different efficiencies. In this model the weakest mRNA for α-globin, which is less efficiently initiated than β-globin (235), will be favored if overall initiation is increased. The opposite will occur if overall initiation is inhibited, for example, by a hypertonic shock (237). It seems, however, that the effects of eIF-4B (or I_{Hb}) cannot be explained simply by this model since threefold increases in the α-/β-globin ratio are observed under conditions where the total globin produced is not increased (232) or even decreased, as was found in a wheat germ cell-free system (134). Furthermore, in Krebs ascites systems, I_{Hb} was shown to give the same increase in α-/β-globin translation ratio at all mRNA concentrations and at all salt concentrations (134).

Table 2 Initiation regions of eukaryotic mRNAs

mRNA		Sequences	References
Brome mosaic virus RNA-4[a]		m^{7}GpppGUAUUAAUAAUGUCGA	(151)
Rous sarcoma[a] virus		m^{7}GpppGmCCAUUUUACCAUUCACCACAUUGGGUGUGCACCUGGGUUGAU GGUUGGACCGUCGAUUCCCUAACGAUUGCGAACACCUGAAUGAAGCA GAAGGCUUCA	(154)
Rabbit β-globin[a]		m^{7}Gpppm^{6}AmCACUUGCUUUUGACACAACUGUGUUUACUUGCAAUCCCC CAAAACAGACAGAAUGGUGC	(155, 156)
Reovirus mRNAs[b]	S54	m^{7}GpppGmCUAUUUUG(CCUCUUCC,C)AGACGUUGUCGCAAUGGAGGU GUGCUUGCCCAACG	(150, 152)
	S45	m^{7}GpppGmCUAAAGUCACGCCUGUCGUCGUCACUAUGGCUUCCUCA CUCAG	(153)
	S46	m^{7}GpppGmCUAUUCGCUGGUCAGUUAUGGCU(CGCUGC)GCGUUCCU AUUCAAG	(153)
	m52	m^{7}GpppGmCUAAUCUGCUGACCGUUACUCUGCAAAGAUGGGGAA CG(CU,CUUC)CUAUCG	(153)
	m44	m^{7}GpppGmCUAAAGUGACCGUGGUCAUGGCUUCAUUCAAGGGAUU CUCCG	(153)
	m30	m^{7}GpppGmCUAUUCGCGGUCAUGGCUUACAUCGCAG	(153)
Vesicular stomatitis mRNAs	N[c]	m^{7}Gppp(m^6)AmACAGUAAUCAAAUGUCUGUUACAGUCAAG	(157)
	NS[c]	m^{7}Gppp(m^6)AmACAGAUAUCAUGGAUAAUCUCACAAAG	(157)
	G[d]	m^{7}Gppp(m^{6}A)AmAC....UUUCCUUGACACUAUGAAGUGCCUUUUGU ACUUAG	(157)

[a] Sequence done without ribosome protection.
[b] Sequence protected by 40S ribosomes.
[c] Sequence protected by 80S ribosomes.
[d] Sequence protected by 80S ribosomes but cap was not protected.

Finally, the opposite effects of IF_{EMC} (eIF-4A) and I_{Hb} (eIF-4B), on mengo RNA, EMC RNA on the one hand and globin mRNA translation on the other (even in the absence of mRNA competition, as above) could not be easily explained by overall changes in initiation.

We proposed (2) that efficiency of a mRNA, that is, what makes it a strong or a weak template, is not constant, but can be modified by the addition of more or less of a given initiation factor. It is probable that the efficiency of mRNA recognition by the ribosomal machinery, is not just due to the structure of the RNA or to binding by a specific factor, but to the overall kinetic parameters of the initiation system considered. For example, there does not have to be a constant one to one relationship between the number of eIF-4 molecules needed and the number of globin or EMC RNA molecules, but changes in the concentrations of the factors, when comparing cell-free systems from different origins or when adding or removing exogenous factors, would actually modify the relative affinity of different mRNAs for the ribosomal machinery. The question of whether such factors can be called mRNA-discriminating is then a semantic one. The important point is that ribosome can be made to discriminate between mRNAs. Actually, components other than protein factors could modify mRNA affinity, as suggested by recent work on reticulocyte lysates treated with micrococcal nuclease (499), and studies on tcRNA (see p. 1100).

Other Reports on mRNA-Specific Initiation

If some mRNA-specificity can be explained by differences in the amount of a given initiation factor required for optimal translation, it seems more difficult to explain the effects of partially purified "eIF-3" fractions from different tissues reported by Heywood et al (202, 239, 240). In wheat germ or in reticulocyte cell-free systems, myosin mRNA translation was reported to be stimulated by muscle "eIF-3," but not by its reticulocyte counterpart; globin mRNA was not stimulated by muscle factor. Similar specificities were found for myosin versus myoglobin mRNA (202) and tubulin mRNA (241). It is again possible that optimal translation of myosin mRNA requires a different concentration of a given factor than does globin mRNA. The "eIF-3" preparations from different tissues may differ in the amount of individual factors they contain: the "eIF-3" preparations used sediment as 18-20S complexes (239) and could still represent complexes of initiation factors, such as those described recently (242), which contain at least eIF-2, eIF-3, eIF-4C, eIF-4D, and eIF-5. On the other hand, purified eIF-3 still contains 9 polypeptides and may have interchangeable or modifiable subunits responsible for mRNA specificity (201).

Comparison of purified initiation factors from liver, reticulocytes, or other cells has, until now, shown no tissue specificity (195, 198, 200, 243–246), and almost all mRNAs are translatable to some extent in heterologous systems (98). The efficiency may, however, be sometimes very low in heterologous systems, for example, globin in a protozoal extract (247). Nevertheless, at present, the discriminatory effects of initiation factors seem due rather to quantitative differences in their requirement, than to qualitative specificities. The really relevant question is still whether mRNAs may remain untranslated in a tissue because of lack of a given initiation or translation factor.

REGULATION OF mRNA TRANSLATION IN CELLS

The Existence of Untranslated mRNAs as mRNP Particles

In the cytoplasm of eukaryotic cells, defined mRNAs coding for proteins that can be made by these cells, may be found as free messenger ribonucleoprotein particles (mRNP) nonassociated with polyribosomes (248–253; for review see 98). In duck immature erythrocytes, globin mRNA exists as free 20S mRNPs, in addition to the 15S mRNPs that can be recovered by EDTA treatment of polyribosomes (254). The 15S mRNP is presumably the translated form of globin mRNP. Active mRNA, which directs the synthesis of globin, can be recovered by deproteinization from both types of mRNPs, which suggested that one of the proteins of the 20S mRNP prevents translation (254). Differences in the protein patterns of 15S and 20S mRNPs were found with SDS-polyacrylamide gel electrophoresis; differences in proteins were also observed in mRNPs containing different mRNAs in these cells (255). The protein composition of mRNPs is complex; in addition to three major and several minor structural proteins (217–219, 221, 250, 256–259), initiation factors may be present (98, 211, 222, 260, 261), but contamination by other cytoplasmic protein complexes (242) is always hard to exclude. An interesting observation (255, 494) is the absence, in free 20S mRNPs, of the specific poly(A) binding protein of 75,000 mol wt that is found in all polyribosomal mRNPs (217, 262–264). This suggested that the poly(A) tail of untranslated mRNA may be shielded (255), possibly stabilizing the mRNA in this way (see p. 1084; 71, 265).

A model implying a role for poly(A) in untranslated mRNPs that may exist in muscle cells (253, 266) was proposed by Heywood et al (267), to explain the translational inhibition caused by translational control RNA (tcRNA) (202, 268). This oligonucleotide of about 20–30 residues is extracted from free mRNPs. It forms a complex with myosin mRNA, and it is suggested that by an oligo(U) sequence, it base pairs with the 3'-poly(A) tail of mRNA (268) and is also complementary to a 5' region of the mRNA (267). Because of the double-stranded structure formed, the tcRNA-mRNA complex would not be translatable. A dsRNA melting factor was described by Ilan & Ilan (269), that could act by unwinding mRNA, as does factor S1 in *E. coli* (270). Uridine-rich tcRNA may be formed by nucleolytic cleavage during processing of a longer pre-mRNA molecule (267, 271). A shorter tcRNA, that is noninhibitory or even stimulatory, is found in active polyribosomal mRNP (267, 268, 272). Specificity of tcRNA for mRNA was also suggested (267, 268).

An equivalent of tcRNA was found by Ochoa's group in *A. salina* cell sap (273). It is a 20 nucleotide-long RNA that is rich in U and C, resistant to RNase T1, but sensitive to RNase A. It inhibits aminoacyl tRNA binding to ribosomes without mRNA specificity, and even blocks translation of synthetic polynucleotides. It does not require the poly(A) tail of mRNA. An activator RNA, working presumably by base pairing to tcRNA, may form by the action of RNase A. Dormant *A. salina* cysts contain mRNA that is not translated (274–276) until development resumes after hydration (277); a role for tcRNA in inhibiting protein synthesis during cryptobiosis is possible.

Unfertilized sea urchin eggs also contain mRNA that is not translated (278–281), and contain a dominant inhibitor of aminoacyl tRNA binding until the eggs are fertilized (282). Developing sea urchin embryo ribosomes have less inhibitor (283). The inhibitor was not well characterized, but thought to be a protein. In *Blastocladiella,* an oligonucleotide inhibitor was found on the ribosomes (284), but its structure is very different from tcRNA since it contains dimethylguanosine. Cap analogs (99, 100, 136–139) and a nucleotidic inhibitor formed in interferon-treated cells (see later section) are other examples of translational inhibitors that should be distinguished from tcRNA.

Developmental Controls of Protein Synthesis

Changes in initiation factor activities have been claimed to accompany developmental changes in cells. Fertilization of sea urchin eggs increases protein synthesis which allows use of stored histone maternal nRNAs (278–283). Hydration of *A. salina* cysts causes a 20-fold increase in eIF-2 level, which then allows natural mRNAs to be translated (273, 285). Both activation and synthesis of eIF-2 take place. Several years ago (287) changes in initiation factor activities during muscle differentiation (286) were reported, but this remains to be confirmed. The regulation of mRNA translation during insect development by initiation factors and tRNA isoacceptors has been reviewed at length (288, 289). In slime mold, *Dictyostelium,* differentiation leads to decreased protein synthesis initiation (4), and selective reduction in the translation of some mRNAs (495).

Regulation at the post-transcriptional level during hormonal enzyme induction had been originally proposed by Tomkins (290), based on the observation that actinomycin D can increase the amount of tyrosine transaminase induced by steroids in cultured hepatoma cells. Although it is clear that transcription of new mRNAs is the primary effect of hormone induction (291), the problem of superinduction by temporary addition of inhibitors of RNA and protein synthesis (290) remains unsolved. A model based on mRNA competition (see p. 1096) was proposed by Palmiter & Schimke (292). A dialyzable factor induced by cortisol, which stimulates tyrosine transaminase synthesis in an in vitro protein synthesis system, was reported (293). Oestradiol and other hormones may increase translational activity (294, 295), and synthesis of vitellogenin has been studied in this respect (296, 297). In the induction of interferon by poly I:C in human cultured cells, superinduction by cycloheximide and actinomycin D (298) or dichloro ribofuranosyl benzimidazol (DRB) (299) produces a ten-fold increase of interferon synthesis. In this case, the data available show that not only more interferon mRNA seems to be made but its stability may be increased (300, 301). Whatever superinduction may be, cycloheximide or actinomycin D were shown to affect initiation of protein synthesis (302), but also have complicated effects on cellular metabolism (303).

Control of histone synthesis during the cell cycle has recently been studied in greater detail. Histone synthesis stops rapidly when DNA synthesis phases out at the end of the S phase, or when inhibitors of DNA synthesis, such as cytosine arabinoside, are added (304, 305). Even during S phase, when histone synthesis is active, some histone mRNA is found in free mRNPs not associated with the polyribosomes (306). Upon arrest of DNA synthesis, the histone mRNA released from polyribosomes does not, however, accumulate as free mRNP in the cytoplasm, but undergoes rapid degradation (305–307). This mRNA degradation requires protein synthesis (308), which may explain the effects of cycloheximide in superinduction mentioned above. Melli et al (309) demonstrated recently, with a sea urchin histone DNA probe cloned in λ-h22 coliphage, that histone mRNA sequences are produced throughout the cell cycle in the nucleus, but are found in the cytoplasm only during S phase. It is assumed that a specific and regulated degradation of histone mRNA takes place in the cytoplasm when DNA synthesis is not active. Accumulation of histones in the cytoplasm could trigger this process (305) but feedback translational inhibition by the product has never been clearly established (310). Another cell-cycle related translational control is the inhibition of protein synthesis during mitosis (311).

Untranslated globin mRNA sequences have also been found in some variants of Friend erythroblasts, in Friend cell x lymphoma cell hybrids (312), and in avian erythroblasts (313). Evidence for some post-transcriptional control during Friend cell differentiation by hemin could be demonstrated (312, 314). Hemin availability may regulate globin synthesis during erythroid development by a mechanism similar to that found in vitro in reticulocyte lysates (see next section).

MECHANISMS OF TRANSLATIONAL CONTROL: STUDIES IN RETICULOCYTES

Mammalian reticulocytes have been a system of choice for cell-free studies on the regulation of protein synthesis because over 90% of the protein made by these nucleated cells consists of equal amounts of α- and β-hemoglobin chains, actively synthesized on unequal amounts of preformed mRNAs (for review see 4). During maturation, reticulocytes lose their ability to initiate globin chains (315), and eventually degrade all mRNA and ribosomes. Micrococcal nuclease has recently been used to prepare mRNA dependent reticulocyte lysates (316). Protein synthesis in rabbit reticulocytes is controlled by the availability of iron or of hemin; in the absence of hemin, reticulocyte polyribosomes disaggregate as a result of a block in reinitiation, a phenomenon which has been extensively described and reviewed (317–322), but much excitement has been generated lately by the partial elucidation of the mechanism of hemin action in vitro.

Hemin Control of Initiation in Reticulocytes: the "Hemin Controlled Repressor (HCR)"

When reticulocyte lysates are prepared and hemin is not added, protein synthesis proceeds for only a few minutes and then stops abruptly (319). This is preceded by the loss of[met-tRNA$_f$.40S] ribosomal initiation complexes (323). With added hemin, the translational rate is maintained for hours. A translational inhibitor forms in hemin deficient cell sap from a latent proinhibitor (324–326). The inhibitor is at first reversible by hemin addition but later becomes irreversible (320, 327, 328). Formation of inhibitor is affected by temperature (320, 329), and triggered by N-ethyl maleimide (326). Partial isolation by Gross & Rabinowitz (330), indicated that the preformed inhibitor catalytically blocks translation in hemin-supplemented lysates, with a typical lag period. Since HCR reduces the amount of initiator tRNA bound to 40S ribosomal subunits (323, 331), and since the met-tRNA$_f$ binding factor eIF-2 (205, 206, 332, 333; see Table 1) can restore initiation in a hemin-deprived lysate and overcome HCR's effect (212, 334–337), it was concluded that HCR produces a gradual inactivation of the functional pool of eIF-2 during the lag period (338). Further support for this model has been obtained recently by the isolation of irreversible HCR, or HRI, in purified form (339), and the demonstration that HCR is a protein kinase whose only, or main, substrate is the smallest subunit (35,000 mol wt) of the trimeric initiation factor eIF-2 (340–342), as originally suggested by Balkow, Hunt & Jackson (343).

Irreversibly activated HCR, formed in rabbit reticulocyte cell sap by prolonged incubation or by N-ethyl maleimide treatment, was purified several thousand fold (339, 340), and appears as a 6S protein, with a molecular weight estimated at about 140,000 by gel filtration. Higher molecular weight of 300,000 had been reported for crude HCR (330). By SDS-polyacrylamide gel electrophoresis, HCR appears to be a 96,000 mol wt polypeptide, which autophosphorylates (340). Purified HCR gives two peaks by isoelectric focusing (pI values 5.56 and 6.36). Two types of HCR were also observed by Kramer et al (341, 344); one phosphorylates eIF-2, the other a 40S ribosomal protein of 30,000 mol wt. But since this protein is also a substrate to the cAMP dependent kinase (342), artifacts are possible. A reticulocyte ribosomal protein S13 (or S6 in other cells) is multiphosphorylated in vivo under various conditions but does not show a clear relationship to translational control (345, 346).

The protein kinase activity of HCR is cAMP independent (339–341) and incorporates 1–2 phosphate atoms per molecule of pure eIF-2 (347). Purified HCR phosphorylates the same 1–2 sites in the eIF-2-35K polypeptide as do cruder preparations of hemin-deprived lysates (340). It should be noted that one of the larger subunits of eIF-2 (50,000 mol wt) is phosphorylated by another enzyme that is considered unrelated to the hemin-regulated translational inhibitor (348, 349), except in one report (350). The 50K subunit of eIF-2 is the only one phosphorylated in intact reticulocytes, together with one subunit of eIF-3 and eIF-4B (351).

The inhibitory effect of purified HCR on initiation can be overcome by adding eIF-2 in molar excess over the 40S ribosome present (339, 340, 347). The more HCR present, the more eIF-2 is needed (352). ATP hydrolysis is required for HCR to inhibit binding of met-tRNA$_f$ to 40S ribosomes (340, 343, 355); GTP can also function (344). HCR does act catalytically (330); one molecule of HCR is enough to inactivate a system containing 100–1000 ribosomes and eIF-2 molecules, since 0.1 μg of HCR (0.7 pmoles) inhibit 1 ml of lysate in which 200 pmoles ribosomes and eIF-2 are estimated to be present (339, 340). The claim that the eIF-2-35K protein kinase activity is the translational inhibitor is based on copurification and on the requirement for ATP, or at least GTP, hydrolysis. Phosphorylation of eIF-2-35K protein by HCR can also be inhibited by a series of purines, 3'-5' cAMP, 2 amino purine (but not hydroxy purine or cGMP), and by caffeine or theophyline, all at unphysiological millimolar concentrations. These nucleotides inhibit HCR action on initiation and reverse the effects of hemin deprivation (353–355). There is good correlation between the efficiencies of these compounds to inhibit eIF-2 phosphorylation and to relieve the translational inhibitor (340), but their precise mode of action is unknown; purines inhibit several protein kinases (356). Many sugars are also known to reverse the effect of hemin-deprivation on translation in reticulocyte lysates (353, 357), but they seem to inhibit HCR formation rather than its action once formed. High levels of GTP reverse the effect of HCR (343, 355), presumably by overcoming the defect in eIF-2 activity, since GTP participates directly with eIF-2 in the formation of the[met-tRNA$_f$.40S]ribosomal complex (205, 206, 332, 333; Figure 2)

Phosphorylation of eIF-2 and Inhibition of Translation

Important questions remain to be resolved. First, it is not clear how eIF-2 phosphorylation inhibits formation of the [met-tRNA$_f$.40S]ribosomal complex. With highly purified preparations of initiation factors and ribosomes, extensively phosphorylated eIF-2 was found to be as efficient as unphosphorylated eIF-2 factor in promoting the formation of 40S and 80S initiation complexes and in restoring the rate of protein synthesis in HCR-inhibited systems (347, 357). It is important to note that the effect of eIF-2 in these assays is stochiometric; one globin chain is initiated per eIF-2 molecule added. This suggests that what may be inhibited is the recycling of eIF-2 between ribosomes (338). The release of eIF-2 itself following formation of the 80S ribosomal initiation complex, is not inhibited by phosphorylation (347), but if one of the subunits of eIF-2 plays a role in the release of the GDP and the rebinding of GTP (similar to that played by factor Ts in prokaryotes in *E. coli*), inhibition of this function might impair recycling.

In crude, or less purified systems, HCR was shown to inhibit formation of the [eIF-2.met-tRNA$_f$.GTP.40S] ribosomal complex (340, 344). Formation of the ternary complex[eIF2.met-tRNA$_f$.GTP]was not inhibited in these experiments [However, one report involving *A. salina* eIF-2 factor did report such inhibition (358)]. Since inhibition is not observed with pure eIF-2 and HCR, it is possible that other factors are also modified by HCR. Phosphorylation of a 40S ribosomal protein may be one of HCR activities (344), and additional translation factors, beside eIF-2, can reverse the effect of HCR (359, 360). eIF-2 may be only the most apparent substrate of HCR's protein kinase activity. Another hypothesis proposed (331, 361, 222) is that alteration of some eIF-2 function in ribosome subunit joining triggers the deacylation of met-tRNA$_f$ bound to 40S ribosomes. This deacylation would be the real cause of inhibition and would occur only when other factors, such as the specific met-tRNA$_f$ deacylase (214, 362–365), are present. Formylation of met-tRNA$_f$ restores initiation in hemin-deprived lysates (366) and prevents the deacylase action (363). eIF-2 and GTP also prevent deacylation (363, 365). Clearly, much work remains to be done to elucidate the relationship between eIF-2 phosphorylation by HCR and the loss of [met-tRNA$_f$.40S]ribosomal complexes. Eventually, it may be found that inhibition is not directly related to phosphorylation.

Another important unresolved question concerns the mechanism of hemin action. Is the purified HCR the same inhibitor as that formed from proinhibitor during hemin deprivation? Their mode of action appears similar (352) and less eIF-2 kinase activity was found in heminated reticulocyte cell sap than in cell sap where the proinhibitor had been activated (340). However, identification of the proinhibitor, and demonstration of its conversion to eIF-2-kinase by a process involving hemin deprivation, remains to be done. Datta et al (358, 367) have proposed that the proinhibitor is an inactive eIF-2-kinase which can be activated by phosphorylation with a cAMP-dependent protein kinase from muscle, which is a model similar to activation of phosphorylase kinase in muscle (368). Several cAMP dependent and independent protein kinases with specificity for ribosomal components have been isolated from reticulocytes (369, 370). Hemin may inhibit activation of an eIF-2 kinase-kinase by cAMP (367), inhibit activation of proinhibitor (327, 339), or inhibit directly the activity of several protein kinases, (350, 357, 371). Hemin may also stimulate a phosphoprotein phosphatase (372); if a cycle of eIF-2 phosphorylation and dephosphorylation takes place normally during initiation, hemin deprivation may interfere with this cycle. This is, however, not in line with the fact that heminated lysate shows no phosphorylated eIF-2, as judged from two-dimensional gel electrophoresis (340). Finally, hemin may interact with initiation factors themselves, as suggested by Raffel et al (373).

Inducement of Translational Inhibitor by Oxidized Glutathione (GSSG)

The addition of diamide, which oxidizes glutathione, to reticulocytes, or of GSSG itself to lysates, induces an inhibitor that blocks initiation with the typical HCR-like biphasic kinetics and the loss of [met-tRNA$_f$.40S] ribosomal complexes (374, 375, 322, 335). GSSG-induced inhibition is reversed by eIF-2, requires ATP, and is accompanied by phosphorylation of the 35K subunit of eIF-2 (335, 357). Prevention of inhibition by a wide range of sugars suggested that the level of NADPH may be important (353), but NADPH itself has no protective effects on translation (357). Nevertheless, activation of HCR by –SH reagents (326, 340) suggests that oxidation may be one mechanism by which proinhibitor can be activated.

Biological Significance of Hemin and GSSG Initiation Control

Stimulation of initiation by hemin is not restricted to erythroid cells or to the coordination of hemin and globin synthesis (376–380). Translational inhibitors, with eIF-2 protein kinase activity, have been isolated from rat liver (381) and undifferentiated Friend leukemia cells (382). Similar activities have been detected in Ehrlich ascites cells (383), HeLa cells (379), and mature erythrocytes (384).

A wide variety of environmental factors affect the rate of protein synthesis initiation in mammalian cells: glucose starvation, amino acid starvation, serum deprivation, high temperatures, mitosis, hypertonicity, hormones, and mitogens (for review see 385), but only a few cases have been examined in detail.

Glucose starvation, anaerobiosis, and inhibition of oxidative phosphorylation, all lead to ATP-deprivation, and induce in reticulocytes a translational inhibitor very similar to that due to hemin deprivation or GSSG, reversed by thiols and sugars (386, 353). Amino acid starvation of growing cells inhibits initiation of protein synthesis (met-tRNA$_f$-binding) (387, 388), possibly also as a result of a loss of ATP and GTP (389); there is no synthesis of ppGpp (390). The biological significance of these controls of initiation is hard to assess at present, but a reasonable hypothesis is that protein synthesis is coupled to the energy metabolism of the cell. Hemin could be involved because of its role in cytochrome synthesis, and serve as a link between energy metabolism and mRNA translation.

The question of whether these controls affect overall protein synthesis, or can produce changes in the translation of specific mRNAs, has to be considered in the light of the model of Lodish (4, 238) in which strong mRNAs will be favored over weak mRNAs by reducing overall initiation rates. Experimental proofs of such discriminatory effects was, for example, obtained using amino acid starvation (236) or hypertonicity (237). As a rule, the strongest mRNAs, producing the major protein products of the system considered, were more resistant to the treatment. Furthermore, viral mRNAs had in general, an advantage over host mRNAs (234), which suggests that translational controls may operate in virus-infected cells. Several types of virus-related translational controls are discussed in the next sections.

Induced Inhibition of Translation in Reticulocytes by Double-Stranded RNA

Double-stranded RNA (dsRNA) appears to be an important biological signal (391). It can be produced by symmetrical transcription of DNA in normal cells or during virus infection. Its major source is probably the replication of viral RNA genomes. Hunt and Ehrenfeld's discovery of a dsRNA-induced inhibition of globin synthesis (392) was in fact prompted by the idea that poliovirus replicative RNA may be the agent that causes shut off of host protein synthesis during virus infection. It turned out that shutoff does not require RNA replication, as is discussed in a later section. The effect of dsRNA remains one of the most intriguing control mechanisms in protein synthesis.

In reticulocyte lysates, dsRNAs such as reovirus RNA, polio or EMC RNA replicative form, dsRNA from *Penicillium chrysogenum,* or synthetic poly I:C, inhibit initiation of translation by a mechanism similar, but not identical, to that of HCR described above. Single-stranded RNA, transfer RNA, and DNA have no such effects. dsRNA activates a translational inhibitor and an eIF-2-Kinase (340); its effects are largely reversed by adding eIF-2 (335) in line with the decrease in [met-tRNA$_f$.40S] ribosomal complexes (393). The inhibitor works catalytically: very small amounts of dsRNA (1–50 ng/ml) inactivate a large excess of ribosomes (394); surprisingly high concentrations of dsRNA (1–20 μg/ml) neither inhibit initiation nor cause phosphorylation of eIF-2 (340). As in the case of HCR, inhibition requires ATP, but is prevented by GTP and by purines, present as cAMP or 2-aminopurine (343, 354).

In contrast to HCR, the dsRNA-induced inhibitor and eIF-2-protein kinase are activated by dsRNA on the ribosomes themselves without the need for cell sap (340, 357). Another difference is that in these crude ribosomes, a 67,000 mol wt protein (protein 67) becomes phosphorylated when dsRNA is added, as well as the 35K subunit of eIF-2. With more purified ribosomes, dsRNA does not inhibit (335). It was proposed that dsRNA activates a specific protein kinase—DAI or dsRNA activated inhibitor (340). This could be due to direct dsRNA binding (*a*) to a proinhibitor (*b*) to eIF-2 [as shown by Kaempfer (212)] which would make it available to the protein kinase, or (*c*) by an indirect effect of dsRNA triggering an activation system for the protein kinase (see also section on interferon below). High levels of dsRNA or 2-amino-purine prevent the activation of the inhibitor and phosphorylation of protein 67, but do not inhibit phosphorylation of eIF-2 after the inhibitor has been preactivated with ATP at low dsRNA levels (340). Phosphorylation of protein 67 may therefore be related to activation of DAI. But since DAI from reticulocytes has not yet been purified, it is not possible to say whether it is a single enzyme that autophosphorylates or whether it represents a chain of reactions. Other proteins may also be phosphorylated when dsRNA is added, but be hard to detect in crude ribosome preparations (357). There are probably additional dsRNA-induced reactions such as synthesis of a nucleotidic inhibitor (424, 500; also see section on interferon) and it remains to be proven, here as in the case of HCR, whether the eIF-2-kinase is the real inhibitor.

In other cells, such as Krebs ascites cells, dsRNA can inhibit translation of mRNAs, including EMC RNA and hemoglobin mRNA, but only at concentrations of dsRNA at least 40-fold higher than those needed in reticulocyte lysates (395). This may be due in part to a dsRNA nuclease similar to *E. coli* RNase III, but low sensitivity to dsRNA seems to be a general property of mammalian cells growing in tissue culture (396). Normal cells survive dsRNA levels of serveral hundred μg/ml, but cells preexposed to interferon, which is an antiviral glycoprotein whose synthesis is itself induced at the transcriptional level by dsRNA (300, 391), are much more sensitive to the toxic effects of dsRNA (397). Furthermore, cell-free extracts from interferon-treated mouse cells show inhibition of mRNA translation at dsRNA concentrations of a few nanograms per milliliter (398, 399), as do reticulocytes. It is possible that reticulocytes, which are non-growing terminally differentiated cells, have acquired dsRNA sensitivity that can be induced in growing cells by interferon. In this respect, it is of interest that interferon produces a limited inhibition of cell growth (400).

THE INTERFERON SYSTEM: AN EXAMPLE OF INDUCED TRANSLATION CONTROL

Exposure of sensitive cells to interferon induces an antiviral state, which at least in part results from an inhibition of viral mRNA translation (401–408). Host protein synthesis is much less affected, if at all, by interferon (408–409). Although interferon probably has additional effects in cells, it provides the first good example of an inducible translational control system.

Extracts from interferon-treated uninfected cells contain a dominant inhibitor of translation (410, 411, 412). Formation of this inhibitor, which is not interferon itself, correlates well with the development of the antiviral state; in the intact cells both take several hours to appear and are blocked by actinomycin D (413). The amount of inhibitor increases with the dose of interferon. Amino acid starvation (414) and virus infection (415) can strongly enhance the interferon-induced translational inhibition; the virus effect was attributed to dsRNA (398).

The translational inhibitor present in lysates, high-salt ribosomal extracts, or cell sap from interferon-treated cultures is potentiated if incubated with ATP and dsRNA prior to addition to protein synthesis systems of either untreated L cells (416–418) or reticulocytes (419). With ribosomal extracts, activation was also seen with ATP alone, but not with the imido analog (AMPPNP), and the lag period before the onset of inhibition could be much reduced by this activation (420). During activation with dsRNA and ATP, several processes requiring ATP hydrolysis seem to take place: phosphorylation of certain ribosome associated proteins (418–423), activation of an eIF-2-kinase (419, 420), synthesis of a small nucleotidic inhibitor of protein synthesis (424, 425, 420), and activation of a nuclease (427).

Protein Phosphorylation

The dsRNA-dependent phosphorylation of a 67,000 mol wt polypeptide (protein 67) is a most apparent difference between interferon-treated and control cell fractions (418–422), and can serve as a marker for interferon's action; it appears only under conditions where the antiviral state is induced. The 35K subunit of eIF-2 is also phosphorylated more actively by ribosomal extracts from interferon-treated cells (418, 420) together with 2–3 other polypeptides, as yet unidentified.

The interferon-induced, dsRNA-dependent eIF-2 kinase activity, measured with exogenous eIF-2 substrate, copurifies with protein 67 phosphorylation activity (420). Phosphorylation of the 35K subunit of eIF-2 remains dsRNA-dependent in purified systems. On the other hand, the dsRNA dependency of protein 67 phosphorylation decreases during purification (418): dsRNA stimulates the reaction rate, but without dsRNA, phosphorylation of protein 67 proceeds more slowly. A phosphoprotein phosphatase, which acts rather specifically on protein 67 and on eIF-2, was shown (420, 426) to cause the strict dsRNA-dependency in crude extracts: only when dsRNA is added does the rate of protein 67 phosphorylation exceed phosphatase activity. By the use of poly(I:C)-Sepharose, it was shown that dsRNA binds protein 67, eIF-2 and the other phosphorylated substrates; one of dsRNA effects could be to increase the substrates affinity for the protein kinase (420, 426). It appears, however, that dsRNA is required by the protein kinase system in at least two steps: (*a*) phosphorylation of substrates, such as eIF-2 and (*b*) activation of the protein kinase system itself.

What are the components of the interferon-induced protein kinase system and how is it activated? Interferon-treated cell extracts differ from controls, by the presence of a protein kinase activity characterized by its specificity for arginine-rich histones H_3, H_4 (420). This enzyme, PK-i, is separable from the other protein kinases of mouse ribosomes (428), is not cAMP or cGMP-dependent, but can be activated by preincubation with dsRNA and ATP (420, 422, 426). High concentrations of dsRNA (20 μg/ml) inhibit the activation of PK-i and eIF-2 kinase, but once activated with low dsRNA (up to 0.4 μg/ml) phosphorylation of histones or eIF-2 is resistant to high dsRNA. It is not clear yet how PK-i is related to eIF-2 kinase and protein 67 phosphorylation, or if these are different enzymes. As suggested for reticulocytes (340), protein 67 phosphorylation which takes place during activation, may be linked to the eIF-2 kinase activity. Since protein 67 autophosphorylates, it has not been possible to identify yet a protein 67-kinase. Fractions of PK-i, devoid of eIF-2-kinase, have been observed (420). Finally, a PK-i activation factor, called A, was obtained from cell sap of interferon-treated cells (420). The data available suggest a chain of reactions that lead to activation of a protein kinase phosphorylating eIF-2 and possibly other translational factors (420).

The Translational Inhibitory Complex

Purification of the translational inhibitor from interferon-treated cell extracts confirms that it is a complex of several activities (420). On the one hand fractions containing PK-i, protein 67 phosphorylation activity, or PK-i activation factor A, inhibit mRNA translation. As in reticulocytes, it is likely that some of the translational inhibition is related to phosphorylation. On the other hand part of the interferon-induced translational inhibitor was found to be due to a separate dsRNA and ATP-dependent inhibitory activity (420, 424). This activity was first reported by Kerr and his group, who showed (422, 424, 425) that incubation of interferon-treated cell fractions with dsRNA and ATP leads to the formation of a heat-stable small molecular weight inhibitor of translation. This nucleotidic compound (425) is synthesized from ATP by an enzyme that can be separated from protein kinase PK-i and from protein 67 (420). The enzyme is a large thermolabile protein, which requires dsRNA and ATP to inhibit translation; its activity is low in control extracts. It can bind to poly(I:C)-Sepharose, from which, upon incubation with ATP, the inhibitory nucleotide is released (424). The nucleotide inhibits protein synthesis in crude systems at nanomolar concentrations, but with a lag period similar to that seen without preincubation (424). It is possible that the inhibitory nucleotide does not inhibit by itself, but triggers other reactions.

Very recently, the inhibitory material has been identified as an oligonucleotide pppA(2')p(5')A(2')p(5')A . . . (501). Purified enzyme E, from interferon-treated cells (420) catalyzes the formation of such 2'-5' dinucleoside monophosphate bonds (426). The oligomeric form of the nucleotide suggests a slow polymerization reaction. It may be noteworthy that(2'),(5')-adenosine diphosphate was also shown recently to inhibit translation (502).

Are the different components of the interferon-induced translational inhibitor related? Several of them bind to dsRNA and may form a complex on a double-stranded region of mRNA or rRNA. One attractive hypothesis would be that the heat-stable nucleotide inhibitor is related to the dsRNA-ATP-dependent activation of PK-i or eIF-2-kinase. On the other hand the substance could inhibit protein synthesis by a mechanism totally unrelated to protein phosphorylation. The step in translation that is inhibited by the oligonucleotide has not yet been identified.

The apparent complexity of the translational inhibitor induced by interferon may help to explain the multiple alterations of the protein synthesis machinery which have been reported in this system.

Steps of the Translational Process Inhibited in Interferon-Treated Cell Extracts

Kinetic studies in cell-free system show that mRNA translation is inhibited after interferon treatment both at the polypeptide elongation and initiation steps (399, 416, 429, 430). When translation is studied without addition of dsRNA, the interferon-induced block appears first in elongation. Initiation proceeds for some time and then stops (399). When dsRNA is added, initiation is blocked earlier, but the elongation defect is still present (416).

Analysis of the elongation defect showed that the polypeptide chains formed are shorter and incomplete (399, 431). Ribosomes are blocked along the mRNA and addition of excess amounts of tRNA overcomes the interferon-induced elongation defect (399, 431, 432, 433). Minor species of leucyl tRNA were found to be required to restore mengo RNA translation in mouse L cell extracts; these species could not be replaced by the major leu-tRNAs, even those recognizing the same codon (434). Different tRNAs were active for globin mRNA and mengo RNA in this system (432, 434). Poly(U,C) translation was also inhibited and could be restored by a yeast leu-tRNA (435, 436). In Friend cells extracts, a lysyl-tRNA was reported to restore EMC RNA translation (437, 438), which suggests that different cells may have different rate-limiting tRNA species. L cells are very poor in minor leu-tRNAs, while the same species are more abundant in Krebs cells and rabbit liver (431, 434). In another work, bacterial, yeast, and rat-liver tRNA were found to be less active than Krebs cell tRNA (430).

The initiation block, which develops more slowly, is accompanied by a loss of [met-tRNA$_f$.40S]ribosomal complexes, while mRNA still binds to ribosomes (399). Addition of formylated met-tRNA$_f$ overcomes the initiation block (399, 429). The effect of dsRNA appears to be essentially an increase in the rate of inhibition, which in its absence takes place more slowly. In the presence of dsRNA, or after activation of the translational inhibitors by dsRNA and ATP, the restoring effect of tRNA disappears (416, 420), as would be expected if the initiation block were then to predominate over the elongation block. On the other hand, the interferon-induced translational inhibitors, assayed without dsRNA, are reversed by tRNA (420).

An important question is whether there are two independent mechanisms of translational inhibition—one dsRNA dependent and one dsRNA independent (and tRNA reversible); or whether there is only a single system. A unifying hypothesis could be that inhibition results from alteration of ribosomal translation factors involved in tRNA binding during initiation and elongation. This alteration could result from phosphorylation of eIF-2, which is a met-tRNA$_f$ binding factor (Table 1). One of its subunits, or another phosphorylated protein (such as protein 67) could have a function in tRNA binding during elongation. The altered ribosomal machinery may be unable to use tRNA species present in too low concentrations.

Deacylation of aminoacyl tRNA by an interferon-induced activity has also been proposed; total leu-tRNA seems more rapidly deacylated in interferon-treated cell extracts (439). A more curious deacylase which attacks the aminoacylated form of plant TMV RNA was also described in interferon-treated cell extracts (440). When leu-tRNA isoacceptors are examined, no specific change in the species active to restore protein synthesis is seen (434, 441). These species are still found in interferon-treated cells (431). It cannot be concluded, therefore, that the tRNA requirement results from a change in the ribosomal machinery or a change in tRNA. A further question is whether there is a relationship between the tRNA effect and the fact that amino acid starvation inhibits initiation (387) and potentiates the interferon effect (442). Finally, tRNA may have a role unrelated to amino acid transfer (438).

Discrimination Between Host and Viral mRNAs

Discrimination is apparent in cells infected by viruses such as reovirus (409) and SV40 which do not shut off host protein synthesis. In SV40-infected cells (408), interferon given *after* infection results in a clear inhibition of viral mRNA translation; the SV40 RNAs are made, have a normal size, but are not in polyribosomes. At the same time, host proteins are still actively synthesized. This mRNA discrimination is, however, lost when the cells are homogenized to produce cell-free systems (420), and an inhibition of both viral and host mRNAs becomes apparent. Until now the discrimination seen in intact cells was not reproduced in the cell-free systems (409, 410). In one case, discrimination was reported (412) but the viral RNA may have been contaminated by dsRNA. Overall inhibition of initiation by eIF-2 phosphorylation would be expected (4, 237) to favor translation of the stronger viral mRNAs; interferon activity produces the opposite effect in the cells. The tRNA effect could be the basis for discrimination, since the tRNA population of the cell may be tailored for host mRNAs, giving more chances that viral mRNAs would become limited by a given tRNA. Differences in the tRNA populations between cells of the same organism suggested an adaptation of tRNAs to the type of mRNAs translated (443–446). Furthermore, in vitro discrimination between mRNAs was seen with different tRNA species in interferon-treated cell extracts (434, 435, 438). Competition between host and viral mRNAs could take place at the elongation step, and, as predicted by Lodish's model (4), this could favor the weakest, or host RNA.

The interferon-induced translational regulation is probably still more complex, and cannot be explained by simple models. Discrimination against viral mRNA may require coordinate induction of several activities. Besides the protein kinase and nucleotidic inhibitors discussed above, a dsRNA-ATP-dependent endonuclease activity was found by Lengyel's group in interferon-treated extracts (427, 447, 448). This activity could cleave preferentially certain viral RNAs, such as reovirus and VSV RNAs. An interferon-induced membrane nuclease has been described in chick cells (449). A membrane bound nuclease that cuts 28S rRNA and inhibits translation was purified from reticulocytes (450, 451). A small nucleolytic enzyme that inhibits translation was recently identified in ribosome extracts from interferon-treated L cells (503). The degradation of viral mRNAs may be involved in the effect of interferon against some viruses, but there are exceptions: in the case of SV40, when interferon is given late after infection, viral mRNA translation stops but the untranslated viral mRNAs are not degraded (408). The same applies to vaccinia (402).

Another activity found in interferon-treated cells, is an inhibitor of mRNA methylation and capping (452, 453). Since not all viral mRNAs are capped [picorna viral RNA as EMC, polio, and mengo are uncapped (94)], this inhibition cannot explain all the effects of interferon. Many viruses contain their own methylase and capping enzymes, and it seems that the inhibitor acts on the host methylases (454). Increased RNA methylation by interferon has been observed (455). In SV40-infected interferon-treated cells, the methylation of viral RNA is also increased (507).

The complexity of the interferon-induced translational regulation may be necessary to allow the discrimination between host and viral protein synthesis, which seems to be the basis of interferon's antiviral action.

SHUTOFF OF HOST PROTEIN SYNTHESIS BY VIRUSES

Viral infection usually leads to an inhibition of host RNA and protein synthesis (456). In some cases, as in tumor viruses, inhibition is delayed and occurs during the late part of the lytic cycle, while early after infection host enzymes may even be induced. With many viruses, however, there is a rapid shutoff of host protein synthesis early after infection. This occurs with Sindbis, Newcastle disease, vesicular stomatitis, adeno, herpes, and vaccinia viruses, for example; but one of the best studied case is picornavirus infection (177, 456–459). Poliovirus infection of HeLa cells, or mengo or EMC infection of mouse cells, results in a rapid decrease in total protein synthesis which resumes when viral proteins are synthesized. During the early phase host mRNA translation is inhibited at the initiation step (460, 461). The host mRNAs are not degraded (460–462) and continue to be synthesized. Polyadenylation (463), capping, and methylation (464) of host mRNA are not affected. Poliovirus infection can also shut off translation of other viral mRNAs when VSV- (465) or herpes simplex- (466) infected cells are superinfected by poliovirus. The VSV RNA that are not translated continue to be synthesized, polyadenylated, capped, and methylated normally and are not degraded (467).

The mechanism of the specific translational discrimination that allows picornavirus RNA to synthesize proteins, while other mRNAs are not translated, has not been elucidated. It now seems unlikely that dsRNA (392) produced during infection is responsible for the host translation shutoff. Both host and viral mRNAs are sensitive to dsRNA and, furthermore, shutoff is observed at times where very little dsRNA is yet produced (468), or with polymerase mutants unable to form replicative RNA (469). The involvement of dsRNA both in host protein shutoff and in interferon-induced viral protein synthesis inhibition (see above), would also seem paradoxical. Another model proposed that the stronger viral RNA outcompetes host mRNAs from ribosomal initiation (470, 237). In vitro competition of EMC RNA against cellular mRNAs, which is relieved by eIF-4B, was demonstrated (226). A decreased initiation rate induced in cells by hypertonicity or other agents leads to preferential viral protein synthesis in infected cells that resembles shutoff (234, 237, 471), even in the case of SV40 where shutoff does not normally take place (472). Competition between viral and cellular mRNAs may play some role in the fine regulation of translation, but does not explain the picornavirus early shutoff of host protein synthesis which occurs before there are large amounts of viral RNA in the cells (461, 469). Complete shutoff occurs after poliovirus infection even in the presence of guanidine, which prevents viral RNA replication (460, 473).

Genetic evidence indicates that a poliovirus capsid protein is required for host shutoff (469), and acts early after infection without need for viral protein synthesis, if the multiplicity of infection is high enough. Since, in several experiments, cell-free systems prepared from picornavirus-infected cultures were active to translate exogenous cellular or viral (nonpicornal) mRNAs (470, 474), it was proposed that the viral coat protein could act on the cell membrane (177, 459). Experiments of Carrasco & Smith (475) indicate that properties of the cell membrane are altered, leading to an influx of sodium ions in the cell. Because various mRNAs have different salt optimums for translation, the addition of sodium ions allows viral mRNAs to be translated optimally, while cellular mRNAs are inhibited. The effect of salt in this case would be selective and at variance from the model of Koch et al (234, 237) in which salt decreases overall initiation. It seems clear that exposure of cells to high salt increases shutoff, while low salt prevents it. Since, however, several agents besides salt (234, 476), influence protein synthesis in this way, it is not clear whether ions act directly on ribosome activity or induce indirectly a cellular mechanism of translational control.

Indeed, recent data suggest a poliovirus-induced translational control. Poliovirus capsids added to reticulocyte lysates trigger an inhibition of globin synthesis with stochiometry and kinetic-suggesting activation of an inhibitory system (477). Furthermore, extracts obtained from HeLa cells, 2 hr after poliovirus infection, when shutoff is completed, do show an impaired protein synthesis capacity in vitro (461). In these experiments, Kaufmann et al measured endogenous initiation on cellular polysomes and showed that the reduced activity is due to the factors removable from ribosomes by high-salt wash. It seems possible that a poliovirus component triggers in the cell a mechanism that inactivates a translational initiation factor. As for hemin or interferon control, purification of the system is necessary to establish the mechanism involved.

Other shut-off systems, such as vaccinia virus (478–481) or VSV (471, 482, 483) have been studied but have not yielded more definitive information than the picornaviruses. It may be noted that vaccinia virus induces the phosphorylation of two specific ribosomal proteins not phosphorylated in uninfected cells (484), possibly by a capsid enzyme (485). But the effect of such ribosomal protein modifications is unclear (346, 368, 369). Other effects of viruses on protein synthesis, such as the helper effect of SV40 on adenovirus in monkey cells (486, 487) are still not well established or understood.

FINAL REMARKS

As our knowledge of the mechanisms of mRNA translation in eukaryotes progresses, the processes that control translation of genetic information into proteins will become clearer. This review is focused on a few steps and systems about which much information has been obtained in the past years. Many other studies on protein synthesis, which were not included, may have a direct relationship to translational control. Several chemicals and antibiotics act on initiation; toxins, such as diphteria toxin, ADP-ribosy-late EF-2, and others, such as abrin, ricin, and many others, inhibit elongation by specific mechanisms; these were recently reviewed (488, 489). The effects of carcinogens, hormones, and viruses on tRNA populations (490) could be important controls of protein synthesis. Intracellular compartmentalization of protein synthesis and amino acyl tRNAs is still intriguing (491). Translational processing of proteins on membranes of the endoplasmic reticulum (492, 493) have become an important field of investigation.

Data on controls of polypeptide chain termination, and the possible role of suppressor tRNAs in read-through of nonsense codons, are beginning to appear (177). Clearly, much remains to be done; eukaryotic cells seem to have a complicated biochemical machinery that enables them to process, store, and translate genetic information in a controlled manner. The new discoveries of the past years reviewed here, should, however, give moderate optimism that cell regulation can be understood, if one does not stick too rigidly to simplistic models.

ACKNOWLEDGMENTS

We are grateful to our colleagues who graciously communicated manuscripts before publication. The expert secretarial assistance of Mrs. Peggy Wynick is gratefully acknowledged. Work from our laboratory was supported by National Cancer Institute Contract NO1 CP33220, the US Public Health Service, and by Gesellschaft für Strahlen und Umweltforschung (GSF), München, Germany, and by the United States-Israel Binational Science Foundation (BSF), Jerusalem, Israel.

Literature Cited

1. Chambon, P. 1975. *Ann. Rev. Biochem.* 44:613–38
2. Revel, M. 1977. In *Molecular Mechanism of Protein Biosynthesis,* ed. H. Weissbach, S. Petska, pp. 254–321. New York: Academic
3. Weissbach, H., Ochoa, S. 1976. *Ann. Rev. Biochem.* 45:191–216
4. Lodish, H. F. 1976. *Ann. Rev. Biochem.* 45:39–72
5. Scherrer, K. 1974. In *Control of Gene Expression. Advances in Experimental Medicine and Biology,* Vol. 44, ed. A. Kohn, A. Shatkay, pp. 169–219, New York: Plenum
6. Parsons, J. T., Gardner, J., Green, M. 1971. *Proc. Natl. Acad. Sci. USA* 68:557–60
7. Wall, R., Phillipson, L., Darnell, J. E. 1972. *Virology* 50:27–34
8. McGuire, P. M., Swart, C., Hodge, L. D. 1972. *Proc. Natl. Acad. Sci. USA.* 69:1578–82
9. Ross, J. 1976. *J. Mol. Biol.* 106:403–20
10. Curtis, P. J., Weissman, C. 1976. *J. Mol. Biol.* 106:1061–75
11. Kwan, S., Wood, T. G., Lingrel, J. B. 1977. *Proc. Natl. Acad. Sci. USA* 74:178–82
11a. Strair, R. K., Skoultchi, A. I., Shafritz, D. A. 1977. *Cell* 12:131–42
12. Bastos, R. N., Aviv, H. 1977. *Cell* 11:641–50
13. Darnell, J. E., Jelinek, W. R., Molloy, G. R. 1973. *Science* 181:1215–21
14. Weinberg, R. A. 1973. *Ann. Rev. Biochem.* 42:329–54
15. Davidson, E. H., Britten, R. J. 1973. *Q. Rev. Biol.* 48:565–613
16. Brawerman, G. 1974. *Ann. Rev. Biochem.* 43:621–42
17. Greenberg, J. R. 1975. *J. Cell Biol.* 64:269–88
18. Lewin, B. 1975. *Cell* 4:11–20
19. Lewin, B. 1975. *Cell* 4:77–93
20. Perry, R. P. 1976. *Ann. Rev. Biochem.* 45:605–29
21. Molloy, G., Puckett, L. 1976. *Prog. Biophys. Molec. Biol.* 31:1–38
22. Chan, L., Harris, S. E., Rosen, J. M., Means, A. R., O'Malley, B. W. 1977. *Life Sci.* 20:1–16
23. Weber, J., Jelinek, W., Darnell, J. E. 1977. *Cell* 10:611–16
24. Goldberg, S., Weber, J., Darnell, J. E. 1977. *Cell* 10:617–21
25. Berget, S. M., Moore, C., Sharp, P. A. 1977. *Proc. Natl. Acad. Sci. USA* 74:3171–75
26. Gelinas, R. E., Roberts, R. J. 1977. *Cell* 11:533–44
27. Chow, L., Gelinas, R. E., Broker, J., Roberts, R. J. 1977. *Cell* 11:819–36
28. Klessig, D. F. 1977. *Cell* 12:9–21
29. Aloni, Y., Dhar, R., Laub, O., Horo-

witz, M., Khoury, G. 1977. *Proc. Natl. Acad. Sci. USA* 74:3686–90
30. Celma, M., Dhar, R., Pan, J., Weissman, S. M. 1977. *Nucleic Acids Res.* 4:2549–59
31. Lavi, S., Groner, Y. 1977. *Proc. Natl. Acad. Sci. USA* 74:5323–27
32. Hsu, M.-T., Ford, S. 1977. *Proc. Natl. Acad. Sci. USA* 74:4982–85
33. Haegeman, G., Fiers, W. 1978. *Nature.* In press
34. Ohtsuki, K., Groner, Y., Hurwitz, J. 1977. *J. Biol. Chem.* 252:483–91
35. Silber, R., Malathi, V. G., Hurwitz, J. 1972. *Proc. Natl. Acad. Sci. USA* 69:3009–13
36. Perry, R. P., Kelley, D. E. 1976. *Cell* 8:433–42
37. Chow, L. T., Gelinas, R. E., Borker, T. R., Roberts, R. J. 1977. *Cell* 12:1–8
38. Schibler, U., Kelley, D. E., Perry, R. P. 1978. *J. Mol. Biol.* In press
39. Groner, Y., Hurwitz, J. 1975. *Proc. Natl. Acad. Sci. USA* 72:2930–34
40. Winicov, I., Perry, R. P. 1976. *Biochemistry* 15:5039–46
41. Wei, C.-M., Moss, B. 1977. *Proc. Natl. Acad. Sci. USA* 74:3758–61
42. Groner, Y., Gilboa, E., Aviv, H. 1978. *Biochemistry.* In press
43. Brawerman, G. 1975. *Prog. Nucleic Acid Res. Mol. Biol.* 17:117–48
44. Derman, E., Darnell, J. E. 1974. *Cell* 3:255–64
45. Mendecki, J., Lee, S. Y., Brawerman, G. 1972. *Biochemistry* 11:792–98
46. Perry, R. P., Kelley, D. E., LaTorre, J. 1974. *J. Mol. Biol.* 82:315–31
47. Slater, I., Gillespie, D., Slater, D. W. 1973. *Proc. Natl. Acad. Sci. USA* 70:406–11
48. Wilt, F. H. 1973. *Proc. Natl. Acad. Sci. USA* 70:2345–49
49. Diez, J., Brawerman, G. 1974. *Proc. Natl. Acad. Sci. USA* 71:4091–95
50. Brawerman, G., Diez, J. 1975. *Cell* 5:271–80
51. Adesnik, M., Darnell, J. E. 1972. *J. Mol. Biol.* 67:397–406
52. Greenberg, J. R., Perry, R. P. 1972. *J. Mol. Biol.* 72:91–98
53. Milcarek, C., Price, R., Penman, S. 1974. *Cell* 3:1–10
54. Nemer, M., Graham, M., Dubroff, L. M. 1974. *J. Mol. Biol.* 89:735–54
55. Fromson, D., Duchastel, A. 1975. *Biochim. Biophys. Acta.* 378:394–404
56. Nemer, M., Dubroff, L. M., Graham, M. 1975. *Cell* 6:171–78
57. Greenberg, J. R. 1975. *J. Cell Biol.* 67:144a (Abstr.)

58. Sonenshein, G. E., Geoghegan, T. E., Brawerman, G. 1976. *Proc. Natl. Acad. Sci. USA* 73:3088–92
58a. Stoltzfus, C. M., Shatkin, A. J., Banerjee, A. K. 1973. *J. Biol. Chem.* 248:7993–98
59. Semancik, J. S. 1974. *Virology* 62:288–91
60. Jacobs-Lorena, M., Baglioni, C., Borun, T. W. 1972. *Proc. Natl. Acad. Sci. USA* 69:2095–99
61. Gross, K. W., Jacobs-Lorena, M., Baglioni, C., Gross, P. R. 1973. *Proc. Natl. Acad. Sci. USA* 70:2614–18
62. Breindl, M., Gallwitz, D. 1973. *Eur. J. Biochem.* 32:381–91
63. Sippel, A. E., Stavrianopoulos, J. G., Schultz, G., Feigelson, P. 1974. *Proc. Natl. Acad. Sci. USA* 71:4635–39
64. Bard, E., Efron, D., Marcus, A., Perry, R. P. 1974. *Cell* 1:101–11
65. Muller, W. E. G. 1976. *Eur. J. Biochem.* 70:241–48
66. Muller, W. E. G., Seibert, G., Steffen, R., Zahn, R. K. 1976. *Eur. J. Biochem.* 70:249–58
67. Munoz, R. F., Darnell, J. E. 1974. *Cell* 2:247–52
68. Spector, D. H., Baltimore, D. 1974. *Proc. Natl. Acad. Sci. USA* 71:2983–87
69. Soreq, H., Nudel, U., Salomon, R., Revel, M., Littauer, U. Z. 1974. *J. Mol. Biol.* 88:233–45
70. Williamson, R., Crossley, J., Humphries, S. 1974. *Biochemistry* 13:703–7
71. Huez, G., Marbaix, G., Hubert, E., Leclercq, M., Nudel, U., Soreq, H., Salomon, R., Lebleu, B., Revel, M., Littauer, U. 1974. *Proc. Natl. Acad. Sci. USA* 71:3143–46
72. Marbaix, G., Huez, G., Burny, A., Cleuter, Y., Hubert, E., Leclercq, M., Chantrenne, H., Soreq, H., Nudel, U., Littauer, U. Z. 1975. *Proc. Natl. Acad. Sci. USA* 72:3065–67
73. Nudel, U., Soreq, H., Littauer, U. Z., Marbaix, G., Huez, G., Leclercq, M., Hubert, E., Chantrenne, H. 1976. *Eur. J. Biochem.* 64:115–21
74. Huez, G., Marbaix, G., Burny, A., Hubert, E., Leclercq, M., Cleuter, Y., Chantrenne, H., Soreq, H., Littauer, U. Z. 1977. *Nature* 266:473–74
75. Huez, G., Marbaix, G., Hubert, E., Cleuter, Y., Leclercq, M., Chantrenne, H., Soreq, H., Nudel, U., Littauer, U. Z. 1975. *Eur. J. Biochem.* 59:589–92
76. Perry, R. P., Kelley, D. E. 1973. *J. Mol. Biol.* 79:681–96
77. Huez, G., Marbaix, G., Weinberg, E., Gallwitz, D., Hubert, E., Cleuter, Y. 1977. *Biochem. Soc. Trans.* 5:936

78. Chantrenne, H. 1977. *Nature* 269: 202
79. Soreq, H., Littauer, U. Z. Unpublished results
80. Doel, M. T., Carey, N. H. 1976. *Cell* 8:51–58
81. Shatkin, A. J. 1976. *Cell* 9:645–53
82. Both, G. W., Banerjee, A. K., Shatkin, A. J. 1975. *Proc. Natl. Acad. Sci. USA* 72:1189–93
83. Muthukrishnan, S., Both, G. W., Furuichi, Y., Shatkin, A. J. 1975. *Nature* 255:33–37
84. Kemper, B. 1976. *Nature* 262:321–23
85. Muthukrishnan, S., Filipowicz, W., Sierra, J. M., Both, G. W., Shatkin, A. J., Ochoa, S. 1975. *J. Biol. Chem.* 250:9336–41
86. Rose, J. K. 1975. *J. Biol. Chem.* 250: 8098–8104
87. Muthukrishnan, S., Morgan, M., Banerjee, A. K., Shatkin, A. J. 1976. *Biochemistry* 15:5761–68
88. Rose, J. K., Lodish, H. F. 1976. *Nature* 262:32–37
89. Lodish, H. F., Rose, J. K. 1977. *J. Biol. Chem.* 252:1181–88
90. Shih, D. S., Dasgupta, R., Kaesberg, P. 1976. *J. Virol.* 19:637–42
91. Nomoto, A., Lee, Y. F., Wimmer, E. 1976. *Proc. Natl. Acad. Sci. USA* 73: 375–80
91a. Lee, Y. F., Nomoto, A., Detjen, B. M., Wimmer, E. 1977. *Proc. Natl. Acad. Sci. USA* 74:59–63
92. Hewlett, M. J., Rose, J. K., Baltimore, D. 1976. *Proc. Natl. Acad. Sci. USA* 73:327–30
92a. Flanegan, J. B., Pettersson, R. F., Ambros, V., Hewlett, M. J., Baltimore, D. 1977. *Proc. Natl. Acad. Sci. USA* 74: 961–65
93. Fernandez-Munoz, R., Darnell, J. E. 1976. *J. Virol.* 126:719–26
94. Frisby, D., Eaton, M., Fellner, P. 1976. *Nucleic Acids Res.* 3:2771–87
95. Wimmer, E., Chang, A. Y., Clark, J. M. Jr., Reichmann, M. E. 1968. *J. Mol. Biol.* 38:59–73
96. Horst, J., Fraenkel-Conrat, H., Mandeles, S. 1971. *Biochemistry* 10:4748–52
97. Moss, B. 1977. *Biochem. Biophys. Res. Comm.* 74:374–83
98. Shafritz, D. A. 1977. In *Molecular Mechanism of Protein Synthesis,* ed. H. Weissbach, S. Petska, pp. 555–601, New York: Academic
99. Roman, R., Brooker, J. D., Seal, S. N., Marcus, A. 1976. *Nature* 260: 359–60
100. Canaani, D., Revel, M., Groner, Y. 1976. *FEBS Lett.* 64:326–31
101. Weber, L. A., Hickey, E. D., Nuss, D. L., Baglioni, C. 1977. *Proc. Natl. Acad. Sci. USA* 74:3254–58
102. Abraham, K. A., Pihl, A. 1977. *Eur. J. Biochem.* 77:589–93
103. Abraham, K. A., Lillehaug, J. R. 1976. *FEBS Lett.* 71:49–52
104. Muthukrishnan, S., Moss, B., Cooper, J. A., Maxwell, E. S. 1977. *J. Biol. Chem.* In press
105. Shimotohno, K., Kodama, Y., Hashimoto, J., Miura, K. 1977. *Proc. Natl. Acad. Sci. USA* 74:2734–38
106. Shinshi, H., Miwa, M., Sugimura, T., Shimotohno, K., Miura, K. 1976. *FEBS Lett.* 65:254–57
107. Ohno, T., Okada, Y., Shimotohno, K., Miura, K., Shinshi, H., Miwa, M., Sugimura, T. 1976. *FEBS Lett.* 67: 209–13
108. Zan-Kowalczewska, M., Bretner, M., Sierakowska, H., Szczesna, E., Filipowicz, W., Shatkin, A. J. 1977. *Nucleic Acids Res.* 4:3065–81
109. Kole, R., Sierakowska, H., Shugar, D. 1976. *Biochim. Biophys. Acta* 438: 540–50
110. Furuichi, Y., La Fiandra, A., Shatkin, A. J. 1977. *Nature* 266:235–39
111. Nuss, D. L., Furuichi, Y., Koch, G., Shatkin, A. J. 1975. *Cell* 6:21–27
112. Nuss, D. L., Furuichi, Y. 1977. *J. Biol. Chem.* 252:2815–21
113. Perry, R. P., Kelley, D. E. 1976 *Cell* 8:433–42
114. Ouellette, A. J., Reed, S. L., Malt, R. A. 1976. *Proc. Natl. Acad. Sci. USA* 73:2609–13
115. Wei, C.-M., Gershowitz, A., Moss, B. 1975. *Cell* 4:379–86
116. Perry, R. P., Kelley, D. E., Friderici, K., Rottman, F. 1975. *Cell* 4:387–94
117. Salditt-Georgieff, M., Jelinek, W., Darnell, J. E., Furuichi, Y., Morgan, M., Shatkin, A. 1976. *Cell* 7:227–37
118. Efstratiadis, A., et al 1977. *Nucleic Acids Res.* 4:4165–74
119. Desrosiers, R. C., Friderici, K. H., Rottman, F. M. 1975. *Biochemistry* 14:4367–74
120. Dimock, K., Stoltzfus, C. M. 1977. *Biochemistry* 16:471–78
121. Lavi, U., Fernandez-Munoz, R., Darnell, J. E. 1977. *Nucleic Acids Res.* 4:63–69
122. Moss, B., Koczot, F. 1976. *J. Virol.* 17:385–92
123. Sommer, S., Salditt-Georgieff, M., Bachenheimer, S., Darnell, J. E., Furuichi, Y., Morgan, M., Shatkin, A. J 1976. *Nucleic Acids Res.* 3:749–65
124. Groner, Y., Carmi, P., Aloni, Y. 1977. *Nucleic Acids Res.* 4:3959–68

125. Moyer, S. A., Abraham, G., Adler, R., Banerjee, A. K. 1975. *Cell* 5: 59–67
126. Furuichi, Y., Miura, K. 1975. *Nature* 253:374–75
127. Wei, C.-M., Moss, B. 1975. *Proc. Natl. Acad. Sci. USA* 72:318–22
128. Perry, R. P., Scherrer, K. 1975. *FEBS Lett.* 57:73–78
129. Yang, N.-S., Manning, R. F., Gage, L. P. 1976. *Cell* 7:339–47
130. Surrey, S., Nemer, M. 1976. *Cell* 9:589–95
131. Moss, B., Gershowitz, A., Weber, L. A., Baglioni, C. 1977. *Cell* 10:113–20
132. Steitz, J. A. 1975. In *RNA Bacteriophages*, ed. N. Zinder, pp. 319–52. Cold Spring Harbor, New York: Cold Spring Harbor Lab.
133. Steitz, J. A. 1977. In *Biological Regulation and Control*, ed. R. Goldberger. New York: Plenum. In press
134. Revel, M., Goldberg, G., Nudel, U., Zilberstein, A., Dudock, B., Canaani, D., Groner, Y. 1977. In *Cell Differentiation in Microorganisms, Plants and Animals*, ed. R. Nover, K. Mothes, Leopoldina Symp., Jena: Fisher. pp. 158–81
135. Both, G. W., Furuichi, Y., Muthukrishnan, S., Shatkin, A. J. 1975. *Cell* 6:185–95
136. Hickey, E. D., Weber, L. A., Baglioni, C. 1976. *Proc. Natl. Acad. Sci. USA* 73:19–23
137. Weber, L. A., Feman, E. R., Hickey, E. D., Williams, M. C., Baglioni, C. 1976. *J. Biol. Chem.* 251:5657–62
138. Shafritz, D. A., Weinstein, J. A., Safer, B., Merrick, W. C., Weber, L. A., Hickey, E. D., Baglioni, C. 1976. *Nature* 261:291–94
139. Suzuki, H. 1976. *FEBS Lett.* 72:309–13
140. Reddy, R., Ro-Choi, T. S., Henning, D., Busch, H. 1974. *J. Biol. Chem.* 249:6486–94
141. Ro-Choi, T. S., Reddy, R., Choi, Y. C., Raj, N. B., Henning, D. 1974. *Fed. Proc.* 33: 1548
142. Rao, S. M., Blackstone, M., Busch, H. 1977. *Biochemistry.* 16:2756–62
143. Dasgupta, R., Harada, F., Kaesberg, P. 1976. *J. Virol.* 18:260–67
144. Hickey, E. D., Weber, L. A., Baglioni, C., Kim, C. H., Sarma, R. H. 1977. *J. Mol. Biol.* 109:173–83
144a. Adams, B. L., Morgan, M., Muthukrishnan, S., Hecht, S. M., Shatkin, A. J. *J. Biol. Chem.* In press
145. Filipowicz, W., Furuichi, Y., Sierra, J. M., Muthukrishnan, S., Shatkin, A. J., Ochoa, S. 1976. *Proc. Natl. Acad. Sci. USA* 73:1559–63
146. Shafritz, D. A., Padilla, M., Canaani, D., Groner, Y., Weinstein, J. A., Bar-Joseph, M., Merrick, W. C. 1978. *J. Biol. Chem.* In press
147. Kozak, M., Shatkin, A. J. 1976. *J. Biol. Chem.* 251:4259–66
148. Both, G. W., Furuichi, Y., Muthukrishnan, S., Shatkin, A. J. 1976. *J. Mol. Biol.* 104:637–58
149. Boone, R., Moss, B. 1977. *Virology* 79:67–80
150. Kozak, M., Shatkin, A. J. 1977. *J. Mol. Biol.* 112:75–96
151. Dasgupta, R., Shih, D. S., Seris, C., Kaesberg, P. 1975. *Nature* 256:624–28
152. Kozak, M., Shatkin, A. J. 1977. *J. Biol. Chem.* 252:6895–6908
153. Kozak, M. 1978. *Nature.* In press
154. Haseltine, W. A., Maxam, A. M., Gilbert, W. 1977. *Proc. Natl. Acad. Sci. USA* 74:989–93
155. Baralle, F. E. 1977. *Cell* 10:549–58
156. Efstratiadis, A., Kafatos, F. C., Maniatis, T. 1977. *Cell* 10:571–85
157. Rose, J. K. 1977. *Proc. Natl. Acad. Sci. USA* 74:3672–76
158. Shine, J., Czernilofsky, A. P., Friedrich, R., Goodman, H. M., Bishop, J. M. 1977. *Proc. Natl. Acad. Sci. USA.* In press
159. Van de Voorde, A., Contreras, R., Rogiers, R., Fiers, W. 1976. *Cell* 9:117–20
160. Steitz, J. A., Jakes, K. 1975. *Proc. Natl. Acad. Sci. USA* 72:4734–38
161. Shine, J., Dalgarno, L. 1975. *Nature* 254:34–38
162. Dalgarno, L. Shine, J. 1973. *Nature New Biol.* 245:261–62
163. Eladari, M. E., Galibert, F. 1975. *Eur. J. Biochem.* 55:247–55
164. Sprague, K. U., Kramer, R. A., Jackson, M. B. 1975. *Nucleic Acids Res.* 2:2111–18
165. Kabat, D. 1975. *J. Biol. Chem.* 250:6085–92
166. Legon, S. 1976. *J. Mol. Biol.* 106:37–53
167. Jacobson, M. F., Baltimore, D. 1968. *Proc. Natl. Acad. Sci. USA* 61:77–84
168. Canaani, D., Groner, Y. Unpublished
169. Palmiter, R. D. 1974. *J. Biol. Chem.* 249:6779–87
170. Swan, D., Aviv, H., Leder, P. 1972. *Proc. Natl. Acad. Sci. USA* 69:1967–71
171. Smith, A. E. 1973. *Eur. J. Biochem.* 33:301–13
172. Oberg, B. F., Shatkin, A. J. 1972. *Proc. Natl. Acad. Sci. USA* 69:3589–93
173. Villa-Komaroff, L., Guttman, N., Baltimore, D., Lodish, H. F. 1975. *Proc. Natl. Acad. Sci. USA* 72:4157–61
174. Esteban, M., Kerr, I. M. 1974. *Eur. J. Biochem.* 45:567–76

175. Jacobson, M. F., Asso, J., Baltimore, D. 1970. *J. Mol. Biol.* 49:657–69
176. Celma, M. L., Ehrenfeld, E. 1975. *J. Mol. Biol.* 98:761–80
177. Smith, A. E., Carrasco, L. 1977. *MTP Int. Rev. Sci., Biochem. Ser. II,* 7: ed. H. V. R. Arnstein, London: Butterworth. In press
178. Shih, D. S., Kaesberg, P. 1973. *Proc. Natl. Acad. Sci. USA* 70:1799–1803
179. Cancedda, R., Schlesinger, M. J. 1974. *Proc. Natl. Acad. Sci. USA* 71:1843–47
180. Simmons, D. T., Strauss, J. H. 1974. *J. Mol. Biol.* 86:397–409
181. Glanville, N., Ranki, M., Morser, J., Kääriainen, L., Smith, A. E. 1976. *Proc. Natl. Acad. Sci. USA* 73:3059–63
182. Rosemond, H., Sreevalsan, T. 1973. *J. Virol.* 11:399–415
183. Kennedy, S. I. T. 1972. *Biochem. Biophys. Res. Commun.* 48:1254–58
184. Cancedda, R., Swanson, R., Schlesinger, M. J. 1974. *J. Virol.* 14:664–71
185. Prives, C. L., Aviv, H., Gilboa, E., Revel, M., Winocour, E. 1974. *Cold Spring Harbor Symp. Quant. Biol.* 39:309–16
186. Prives, C. L., Gluzman, Y., Gilboa, E., Revel, M., Winocour, E. 1976. *Papers Presented 1975 Tumor Virus Meeting on SV40 Polyoma and Adenoviruses.* Cold Spring Harbor. p. 21
187. Smith, A. E., Kamen, R. I., Mangel, W. F., Shure, H., Wheeler, T. 1976. *Cell* 9:481–87
188. Lavi, S., Shatkin, A. J. 1975. *Proc. Natl. Acad. Sci. USA* 72:2012–16
189. Groner, Y., Grosfeld, H., Littauer, U. Z. 1976. *Eur. J. Biochem.* 71:281–93
190. Aloni, Y. 1975. *FEBS Lett.* 54:363–67
191. Weinberg, R. A., Warnaar, S. O., Winocour, E. 1972. *J. Virol.* 10:193–202
192. Aloni, Y. 1974. *Cold Spring Harbor Symp. Quant. Biol.* 39:165–78
193. Weinberg, R. A., Ben-Ishai, Z., Newbold, S. E. 1974. *J. Virol.* 13:1263–73
194. Anderson, W. F., Bosch, L., Cohn, W. E., Lodish, H., Merrick, W. C., Weissbach, H., Wittmann, H. G., Wool, I. G. 1977. *FEBS Lett.* 76:1–10
195. Staehelin, T., Trachsel, H., Erni, B., Boschetti, A., Schreier, M. H. 1975. *Proc. FEBS Mtg., 10th,* Vol. 39, pp. 309–23. North Holland/American Elsevier
196. Merrick, W. C., Peterson, D. T., Safer, B., Lloyd, M., Kemper, W. M. 1977. *Proc. FEBS Meet., 11th,* 43:17–26. Oxford:Pergamon
197. Sundkvist, I. C., Staehelin, T. 1975. *J. Mol. Biol.* 99:401–18
198. Benne, R., Hershey, J. W. B. 1976. *Proc. Natl. Acad. Sci. USA* 73:3005–9
199. Schreier, M. H., Staehelin, T. 1973. In *Regulation of Transcription and Translation in Eukaryotes Mosbach Colloquium, 24th,* ed. E. K. F. Bautz, P. Karlson, H. Kerten, pp. 335–48, Heidelberg: Springer
200. Safer, B., Adams, S. L., Kemper, W. M., Berry, K. W., Lloyd, M., Merrick, W. C. 1976. *Proc. Natl. Acad. Sci. USA* 73:2584–88
201. Trachsel, H., Erni, B., Schreier, M., Staehelin, M. 1977. *J. Mol. Biol.* 116: 755–67
202. Heywood, S. M., Kennedy, D. S., Bester, A. J. 1974. *Proc. Natl. Acad. Sci. USA* 71:2428–34
203. Benne, R., Luedi, M., Hershey, J. W. B. 1977. *J. Biol. Chem.* 252:5798–803
204. Blair, G. E., Dahl, H. H. M., Truelsen, E., Lelong, J. C. 1977. *Nature* 265:651–53
205. Schreier, M., Staehelin, T. 1973. *Nature New Biol.* 242:35–37
206. Gupta, N. K., Woodley, C. L., Chen, Y. C., Bose, K. K. 1973. *J. Biol. Chem.* 248:4500–11
207. Darnbrough, C., Legon, S., Hunt, T., Jackson, R. J. 1973. *J. Mol. Biol.* 76:379–403
208. Adams, S. L., Safer, B., Anderson, W. F., Merrick, W. C. 1975. *J. Biol. Chem.* 250:9083–89
209. Majumdar, A., Roy, R., Das, A., Dasgupta, A., Gupta, N. K. 1977. *Biochem. Biophys. Res. Commun.* 78:161–69
210. Benne, R., Hershey, J. W. B. 1978. *J. Biol. Chem.* 253
211. Hellerman, J. G., Shafritz, D. A. 1975. *Proc. Natl. Acad. Sci. USA* 72:1021–25
212. Kaempfer, R. 1974. *Biochem. Biophys. Res. Commun.* 61:591–97
213. Merrick, W. C., Anderson, W. F. 1975. *J. Biol. Chem.* 250:1107–1206
214. McCuiston, J., Parker, R., Moldave, K. 1976. *Arch. Biochem. Biophys.* 172: 387–98
215. Shafritz, D. A. Personal communication
216. Filipowicz, W., Sierra, J. M., Nombela, C., Ochoa, S., Merrick, W. C., Anderson, W. F. 1976. *Proc. Natl. Acad. Sci. USA* 73:44–48
217. Blobel, G. 1973. *Proc. Natl. Acad. Sci. USA* 70:924–28
218. Morel, C. M., Gander, E. S., Herzberg, M., Dubochet, J., Scherrer, K. 1973. *Eur. J. Biochem.* 36:455–64
219. Lebleu, B., Marbaix, G., Huez, G., Temmerman, J., Burny, A., Chan-

trenne, H. 1971. *Eur. J. Biochem.* 19:264–69

220. Nudel, U., Lebleu, B., Zehavi-Willner, T., Revel, M. 1973. *Eur. J. Biochem.* 33:314–22

221. Grubman, M. J., Shafritz, D. A. 1977. *Virol.* 81:1–16

222. Safer, B., Anderson, W. F. 1978. In *Critical Reviews in Biochemistry,* Vol. 5, ed. G. D. Fassman. Cleveland: CRC. In press

223. Wigle, D. T., Smith, A. E. 1973. *Nature New Biol.* 242:136–40

224. Steitz, J. A., Wahba, A. J., Laughrea, M., Moore, P. B. 1977. *Nucleic Acids Res.* 4:1–7

225. Nudel, U., Lebleu, B., Revel, M. 1973. *Proc. Natl. Acad. Sci. USA* 70:2139–44

226. Golini, F., Thach, S. S., Birge, C. H., Safer, B., Merrick, W. C., Thach, R. E. 1976. *Proc. Natl. Acad. Sci. USA* 73:3040–44

227. Lebleu, B., Nudel, U., Falcoff, E., Prives, C., Revel, M. 1972. *FEBS Lett.* 25:97–103

228. Hall, N. D., Arnstein, H. R. V. 1973. *FEBS Lett.* 35:45–50

229. McKeehan, W. D. 1974. *J. Biol. Chem.* 249:6517–26

230. Beuzard, Y., London, I. M. 1974. *Proc. Natl. Acad. Sci. USA* 71:2863–66

231. Temple, G., Lodish, H. F. 1975. *Biochem. Biophys. Res. Commun.* 63: 971–79

232. Kabat, D., Chappell, M. R. 1977. *J. Biol. Chem.* 252:2684–90

233. Suzuki, H. 1977. *FEBS Lett.* 79:11–14

234. Koch, G., Oppermann, H., Bilello, P., Koch, F., Nuss, D. 1976. In *Modern Trends in Human Leukemia II.* ed. R. Neth, R. Gallo, K. Mannweiler, W. C. Moloney, Vol. 19, pp. 541–55. München: Lehmans

235. Lodish, H. F., Jacobsen, M. 1972. *J. Biol. Chem.* 247:3622–29

236. Sonenshein, G. E., Brawerman, G. 1976. *Biochemistry* 15:5497–501

237. Nuss, D. L., Oppermann, H., Koch, G. 1975. *Proc. Natl. Acad. Sci. USA* 72: 1258–62

238. Lodish, H. F. 1974. *Nature* 251:385–88

239. Kennedy, D. S., Heywood, S. M. 1976. *FEBS Lett.* 72:314–18

240. Heywood, S. M., Kennedy, D. S. 1976. *Prog. Nucleic Acid Res. Mol. Biol.* 19:477–84

241. Gilmore-Hebert, M. A., Heywood, S. M. 1976. *Biochim. Biophys. Acta* 454:55–66

242. Voorma, H. O., Amesz, H., Goumans, H., Van der Mast, C., Thomas, A. 1977.

FEBS Meet. 11th, Copenhagen. pp. A 2–5–252–2 (Abstr.)

243. Schreier, M. H., Staehelin, T. 1973. *Proc. Natl. Acad. Sci. USA* 70:462–65.

244. Picciano, D. J., Prichard, P. M., Merrick, W. C., Shafritz, D. A., Graf, H., Crystal, R. G., Anderson, W. F. 1973. *J. Biol. Chem.* 248:204–14

245. Thompson, H. A., Sadnik, I., Scheinbuks, J., Moldave, K. 1977. *Biochemistry* 16:2221–30

246. Strycharz, W. A., Ranki, M., Dahl, H. H. M. 1974. *Eur. J. Biochem.* 48:303–10

247. Ilan, J., Ilan, J. 1976. *J. Biol. Chem.* 251:5718–25

248. Spirin, A. 1969. *Eur. J. Biochem.* 10:20–35

249. Brawerman, G. 1974. *Ann. Rev. Biochem.* 43:621–42

250. Gander, E. S., Stewart, A. G., Morel, C. M., Scherrer, K. 1973. *Eur. J. Biochem.* 38:443–52

251. Williamson, R. 1973. *FEBS Lett.* 37:1–6

252. Jacobs-Lorena, M., Baglioni, C. 1972. *Proc. Natl. Acad. Sci. USA* 69:1425–28

253. Heywood, S. M., Kennedy, D. S., Bester, A. J. 1975. *FEBS Lett.* 53:69–72

254. Civelli, O., Vincent, A., Buri, J. F., Scherrer, K. 1976. *FEBS Lett.* 72:71–76

255. Vincent, A., Civelli, O., Buri, J. F., Scherrer, K. 1977. *FEBS Lett.* 77:281–86

256. Bryan, R. N., Hayashi, W. 1973. *Nature New Biol.* 244:271–74

257. Liautard, J. P., Setyono, B., Spindler, E., Kohler, K. 1976. *Biochim. Biophys. Acta* 425:373–83

258. Kumar, A., Pederson, T. 1975. *J. Mol. Biol.* 96:353–65

259. Barrieux, A., Ingraham, H. A., Nystul, S., Rosenfeld, M. G. 1976. *Biochemistry* 15:3523–28

260. Barrieux, A., Rosenfeld, M. G. 1977. *J. Biol. Chem.* 252:392–98

261. Ilan, J., Ilan, J. 1973. *Nature New Biol.* 241:176–80

262. Kwan, S. W., Brawerman, G. 1972. *Proc. Natl. Acad. Sci. USA* 69:3247–51

263. Jeffery, R. J., Brawerman, G. 1975. *Biochemistry* 14:3445–51

264. Schwartz, H., Darnell, J. E. 1976. *J. Mol. Biol.* 104:833–51

265. Bergman, I. E., Brawerman, G. 1977. *Biochemistry* 16:259–64

266. Buckingham, M. E., Cohen, A., Gros, F. 1976. *J. Mol. Biol.* 103:611–26

267. Bester, A. J., Kennedy, D. S., Heywood, S. M. 1975. *Proc. Natl. Acad. Sci. USA* 72:1523–27

268. Heywood, S. M., Kennedy, D. S., Bester, A. J. 1975. *Eur. J. Biochem.* 58:587–93
269. Ilan, J., Ilan, J. 1977. *Proc. Natl. Acad. Sci. USA* 74:2325–29
270. Kolb, A., Hermoso, J. M., Thomas, J. O., Szer, W. 1977. *Proc. Natl. Acad. Sci. USA* 74:2379–83
271. Molloy, G. R., Thomas, W. L., Darnell, J. E. 1972. *Proc. Natl. Acad. Sci. USA* 69:3684–88
272. Bogdanovsky, D., Hermann, W., Schapira, G. 1973. *Biochem. Biophys. Res. Commun.* 54:25–32
273. Lee-Huang, S., Sierra, J. M., Naranjo, R., Filipowicz, W., Ochoa, S. 1977. *Arch. Biochem. Biophys.* 180:276–87
274. Nilsson, M. O., Hultin, T. 1975. *FEBS Lett.* 52:269–72
275. Grossfeld, H., Littauer, U. Z. 1975. *Biochem. Biophys. Res. Commun.* 67:176–81
276. Sierra, J. M., Filipowicz, W., Ochoa, S. 1976. *Biochem. Biophys. Res. Commun.* 69:181–89
277. Huang, F. L., Warner, A. H. 1974. *Arch. Biochem. Biophys.* 163:716–27
278. Gross, P. R., Gross, K. W., Skoultchi, A. I., Ruderman, J. V. 1973. In *Protein Synthesis in Reproductive Tissue, Karolinska Symp., 6th*, pp. 244–62. Stockholm: Karolinska Institutet
279. Ruderman, J. V., Gross, P. R. 1975. *Dev. Biol.* 36:286–98
280. Gabrielli, F., Baglioni, C. 1977. *Nature* 269:529–31
281. Gross, K. W., Jacobs-Lorena, M., Baglioni, C., Gross, P. R. 1973. *Proc. Natl. Acad. Sci. USA* 70:2614–18
282. Gambino, R., Metafora, S., Felicetti, L., Raisman, J. 1973 *Biochim. Biophys. Acta* 312:377–91
283. Hille, M. B. 1974. *Nature* 249:556–58
284. Adelman, T. G., Lovett, J. S. 1974. *Biochim. Biophys. Acta* 335:236–45
285. Filipowicz, W., Sierra, J. M., Ochoa, S. 1975. *Proc. Natl. Acad. Sci. USA* 72:3947–51
286. Yaffe, D., Yablonka, Z., Kessler, G. Dym, H. 1975. *Proc. FEBS Meet. 10th*, ed. G. Bernardi, F. Gros, 38:313–25 Amsterdam: North Holland
287. Heywood, S. M., Kennedy, D. S. 1974. *Dev. Biol.* 38:390–93
288. Ilan, J., Ilan, J. 1971. *Dev. Biol.* 25:280–92
289. Ilan, J., Ilan, J. 1975. *Dev. Biol.* 42:64–74
290. Tomkins, G. M., Levinson, B. B., Baxter, J. D., Dethlefsen, L. 1972. *Nature New Biol.* 239:9–14
291. Towle, H. C., Tsai, M. J., Hirose, M., Tsai, S. Y., Schwartz, R. J., Parker, M. G., O'Malley, B. W. 1976. In *The Molecular Biology of Hormone Action*, pp. 107–36. New York: Academic
292. Palmiter, R. D., Schimke, R. T. 1973. *J. Biol. Chem.* 248:1502–12
293. Beck, G., Beck, J. P., Defer, N. 1976. *Hoppe Zeylers Z. Physiol. Chem.* 357:493
294. Clemens, M. J., Tata, J. R. 1973. *Eur. J. Biochem.* 33:71–80
295. Palmiter, R. D. 1975. *Cell* 4:189–97
296. Maenpaa, P. H. 1976. *Biochem. Biophys. Res. Commun.* 72:347–54
297. Tata, J. R. 1976. *Cell* 9:1–14
298. Vilcek, J., Havell, E. A. 1973. *Proc. Natl. Acad. Sci. USA* 70:3909–14
299. Sehgal, P. B., Tamm, I., Vilcek, J. 1976. *Virology* 70:532–41
300. Raj, N. B. K., Pitha, P. 1977. *Proc. Natl. Acad. Sci. USA* 74:1483–87
301. Cavalieri, R. L., Havell, E. A., Vilcek, J., Petska, S. 1977. *Proc. Natl. Acad. Sci. USA* 74:4415–19
302. Goldstein, E. S., Reichman, M., Penman, S. 1974. *Proc. Natl. Acad. Sci. USA* 71:4752–56
303. Steinberg, R. A., Levinson, B. B., Tomkins, G. M. 1975. *Cell* 5:29–35
304. Robbins, E., Borun, T. W. 1967. *Proc. Natl. Acad. Sci. USA* 57:409–16
305. Butler, W. B., Mueller, G. C. 1973. *Biochim Biophys. Acta* 294:481–96
306. Borun, T. W., Gabrielli, F., Ajiro, K., Zweidler, A., Baglioni, C. 1975. *Cell* 4:59–67
307. Perry, R. P., Kelley, D. E. 1973. *J. Mol. Biol.* 79:681–96
308. Gallwitz, D. 1975. *Nature* 257:247–48
309. Melli, M., Spinelli, G., Arnold, E. 1977. *Cell* 12:167–74
310. Stevens, R. H., Williamson, A. R. 1975. *Proc. Natl. Acad. Sci. USA* 72:4679–84
311. Fan, H., Penman, S. 1970. *J. Mol. Biol.* 50:655–70
312. Harrison, P. R. 1976. *Nature* 262:353–56
313. Therwath, A., Scherrer, K. 1974. *Experientia* 30:710
314. Harrison, P. R., Affara, N., Conkie, D., Rutherford, J., Sommerville, J., Paul, J. 1976. In *Progress in Differentiation Research* ed. N. Muller-Berat pp. 135–46. Amsterdam: North Holland
315. Herzberg, M., Revel, M., Danon, D. 1969. *Eur. J. Biochem.* 11:148–53
316. Pelham, H. R. B., Jackson, R. J. 1976. *Eur. J. Biochem.* 67:247:56
317. Rabinovitz, M., Freedman, M. L., Fisher, J. M., Maxwell, C. R. 1969.

Cold Spring Harbor Symp. Quant. Biol. 34:565–78

318. Adamson, S. D., Herbert, E., Kemp, S. F. 1969. *J. Mol. Biol.* 42:247–58

319. Zucker, W. V., Schulman, H. M. 1968. *Proc. Natl. Acad. Sci. USA* 59:582–87

320. Hunt, T., Vanderhoff, G., London, I. M. 1972. *J. Mol. Biol.* 66:471–81

321. Rabinowitz, M. 1974. *Ann. NY Acad. Sci.* 241:322–33

322. London, I. M., Clemens, M. J., Ranu, R. S., Levin, D. H., Cherbas, L. F., Ernst, V. 1976. *Fed. Proc.* 35:2218–22

323. Legon, S., Jackson, R. J., Hunt, T. 1973. *Nature New Biol.* 241:150–52

324. Adamson, S. D., Yan, M. P., Herbert, E., Zucker, W. V. 1972. *J. Mol. Biol.* 63:247–64

325. Maxwell, C. R., Kamper, C. S., Rabinovitz, M. 1971. *J. Mol. Biol.* 58:317–27

326. Gross, M., Rabinovitz, M. 1972. *Biochim. Biophys. Acta* 287:340–52

327. Gross, M., Rabinovitz, M. 1972. *Proc. Natl. Acad. Sci. USA* 69:1565–68

328. Gross, M. 1974. *Biochim. Biophys. Acta* 366:319–32

329. Gross, M., Rabinovitz, M. 1973. *Biochim. Biophys. Acta* 299:472–79

330. Gross, M., Rabinovitz, M. 1973. *Biochim. Biophys. Res. Commun.* 50:832–38

331. Balkow, K., Mizuno, S., Fisher, J. M., Rabinovitz, M. 1973. *Biochim. Biophys. Acta* 324:397–409

332. Safer, B., Anderson, W. F., Merrick, W. C. 1975. *J. Biol. Chem.* 250:9067–75

333. Benne, R., Wong, C., Luedi, M., Hershey, J. W. B. 1976. *J. Biol. Chem.* 251:7675–81

334. Clemens, M. J., Henshaw, E. C., Rahamimoff, H., London, I. M. 1974. *Proc. Natl. Acad. Sci. USA* 71:2946–50

335. Clemens, M. J., Safer, B., Merrick, W. C., Anderson, W. F., London, I. M. 1975. *Proc. Natl. Acad. Sci. USA* 72:1286–90

336. Clemens, M. J. 1976. *Eur. J. Biochem.* 66:413–22

337. Hardesty, B., Kramer, G., Cimadevilla, J. M., Pinphanichakarn, P., Konecki, D. 1976. In *Modern Trends in Human Leukemia II*, ed. R. Neth, pp. 67–80. Munich: Lehmans

338. Cherbas, L., London, I. M. 1976. *Proc. Natl. Acad. Sci. USA* 73:3506–10

339. Ranu, R. S., London, I. M. 1976. *Proc. Natl. Acad. Sci. USA* 73:4349–53

340. Farrell, P. J., Balkow, K., Hunt, T., Jackson, R. J. 1977. *Cell* 11:187–200

341. Kramer, G., Cimadevilla, J. M., Hardesty, B. 1976. *Proc. Natl. Acad. Sci. USA* 73:3078–82

342. Levin, D. H., Ranu, R. S., Ernst, V., London, I. M. 1976. *Proc. Natl. Acad. Sci. USA* 73:3112–16

343. Balkow, K., Hunt, T., Jackson, R. J. 1975. *Biochem. Biophys. Res. Commun.* 67:366–75

344. Kramer, G., Henderson, A. B., Pinphanichakarn, P., Wallis, M. H., Hardesty, B. 1977. *Proc. Natl. Acad. Sci.* 74:1445–49

345. Traugh, J. A., Porter, G. G. 1976. *Biochemistry* 15:610–16

346. Gressner, A. M., Wool, I. G. 1974. *J. Biol. Chem.* 249:6917–25

347. Safer, B., Peterson, D., Merrick, W. C. 1977. In *Translation Synthetic Natural Polynucleotides*, ed. A. B. Legocki, pp. 24–31

348. Issinger, O. G., Benne, R., Hershey, J. W. B., Traut, R. R. 1976. *J. Biol. Chem.* 251:6471–74

349. Traugh, J. A., Tahara, S. M., Sharp, S. B., Safer, B., Merrick, W. C. 1976. *Nature* 263:163–65

350. Lenz, J. R., Baglioni, C. 1977. *Nature* 266:191–93

351. Benne, R., Edman, J., Traut, R. R., Hershey, J. W. B. 1978. *Proc. Natl. Acad. Sci. USA* 75:108–12

352. Ranu, R. S., Levin, D. H., Delaunay, J., Ernst, V., London, I. M. 1976. *Proc. Natl. Acad. Sci. USA* 73:2720–24

353. Giloh, H., Schochot, L., Mager, J. 1975. *Biochim. Biophys. Acta* 414:309–20

354. Legon, S., Brayley, A., Hunt, T., Jackson, R. J. 1974. *Biochem. Biophys. Res. Commun.* 56:745–52

355. Ernst, V., Levin, D. H., Ranu, R. S., London, I. M. 1976. *Proc. Natl. Acad. Sci. USA* 73:1112–16

356. Iwai, H., Inamasu, M., Takeyama, S. 1972. *Biochem. Biophys. Res. Commun.* 46:824–30

357. Levin, D., Ranu, R. S., Ernst, V., Trachsel, H., London, I. M. 1977. See Ref. 196. 43:27–36

358. Datta, A., De Haro, C., Sierra, J. M., Ochoa, S. 1977. *Proc. Natl. Acad. Sci. USA* 74:1463–67

359. Gross, M. 1976. *Biochim. Biophys. Acta* 447:445–59

360. Gross, M. 1977. *Arch. Biochem. Biophys.* 180:121–30

361. Balkow, K., Mizuno, S., Rabinovitz, M. 1973. *Biochem. Biophys. Res. Commun.* 54:315–23

362. Morrisey, J., Hardesty, B. 1972. *Arch. Biochem. Biophys.* 152:385–97

363. Gupta, N. K., Aerni, R. J. 1973. *Biochem. Biophys. Res. Commun.* 51:907–12

364. Cimadevilla, J. M., Morrisey, J., Hardesty, B. 1974. *J. Mol. Biol.* 83:437–46
365. Nygard, O., Hultin, T. 1977. *Eur. J. Biochem.* 72:537–42
366. Cahn, F., Lubin, M. 1975. *Mol. Biol. Rep.* 2:49–58
367. Datta, A., De Haro, C., Sierra, J. M., Ochoa, S. 1977. *Proc. Natl. Acad. Sci. USA* 74:3326–29
368. Rubin, C. S., Rosen, O. M. 1975. *Ann. Rev. Biochem.* 44:831–87
369. Traugh, J. A., Sharp, S. B. 1977. *J. Biol. Chem.* 252:3738–44
370. Levin, D. H., Ranu, R. S., Ernst, V., Fifer, M. A., London, I. M. 1975. *Proc. Natl. Acad. Sci. USA* 72:4849–53
371. Hirsch, J. D., Martelo, O. J. 1976. *Biochem. Biophys. Res. Commun.* 71:926–32
372. Mumby, M., Traugh, J. A. 1977. *Fed. Proc.* 36:2666
373. Raffel, C., Stein, S., Kaempfer, R. 1974. *Proc. Natl. Acad. Sci. USA* 71:4020–24
374. Kosower, N. S., Vanderhoff, G. A., Benerofe, B., Hunt, T., Kosower, E. M. 1971. *Biochem. Biophys. Res. Commun.* 45:816–21
375. Kosower, N. S., Vanderhoff, G. A., Kosower, E. M. 1972. *Biochim. Biophys. Acta* 272:623–37
376. Mathews, M. B. 1972. *Biochim. Biophys. Acta* 272:108–18
377. Beuzard, Y., Rodvien, R., London, I. M. 1973. *Proc. Natl. Acad. Sci. USA* 70:1022–26
378. Palmiter, R. 1973. *J. Biol. Chem.* 248:2095–2106
379. Weber, L. A., Feman, E. R., Baglioni, C. 1975. *Biochemistry* 14:5315–21
380. Lodish, H. F., Desalu, O. 1973. *J. Biol. Chem.* 248:3520–27
381. Delaunay, J. C., Ranu, R. S., Levin, D. H., Ernst, V., London, I. M. 1977. *Proc. Natl. Acad. Sci. USA* 74:2264–69
382. Pinphanichakarn, P., Kramer, G., Hardesty, B. 1977. *J. Biol. Chem.* 252:2106–12
383. Clemens, M. J., Pain, V. M., Henshaw, E. C., London, I. M. 1976. *Biochem. Biophys. Res. Commun.* 72:768–75
384. Freedman, M. L., Geraghty, M., Rosman, J. 1974. *J. Biol. Chem.* 249:7290–94
385. Jackson, R. J. 1975. MTP *Int. Rev. Sci. Biochem. Ser. I* 7:89–135
386. Giloh, H., Mager, J. 1975. *Biochim. Biophys. Acta* 414:293–308
387. Vaughan, M. H., Hansen, B. S. 1973. *J. Biol. Chem.* 248:7087–96
388. Pain, V. M., Henshaw, E. C. 1975. *Eur. J. Biochem.* 57:335–42
389. Grummt, F., Grummt, I. 1976. *Eur. J. Biochem.* 64:307–12
390. Thammana, P., Buerk, R. R., Gordon, J. 1976. *FEBS Lett.* 68:187–90
391. Carter, W. A., DeClercq E. 1974. *Science* 186:1172–78
392. Ehrenfeld, E., Hunt, T. 1971. *Proc. Natl. Acad. Sci. USA* 68:1075–78
393. Darnbrough, C., Hunt, T., Jackson, R. J. 1972. *Biochem. Biophys. Res. Commun.* 48:1556–64
394. Hunter, T., Hunt, T., Jackson, R. J., Robertson, H. D. 1975. *J. Biol. Chem.* 250:409–17
395. Robertson, H. D., Mathews, M. B. 1973. *Proc. Natl. Acad. Sci. USA* 70:225–29
396. Graziadei, W. D. III, Lengyel, P. 1972. *Biochem. Biophys. Res. Commun.* 46:1816–23
397. Stewart, W. E., De Clercq, E., De Somer, P. 1973. *J. Gen. Virol.* 18:237–46
398. Kerr, I. M., Brown, R. E., Ball, L. A. 1974. *Nature* 250:57–59
399. Content, J., Lebleu, B., Nudel, U., Zilberstein, A., Berissi, H., Revel, M. 1975. *Eur. J. Biochem.* 54:1–10
400. Stewart, W. E., Gresser, I., Tovey, M. G., Bandu, M. T., LeGoff, S. L. 1976. *Nature* 262:300–2
401. Levy, H. B., Carter, W. A. 1968. *J. Mol. Biol.* 31:561–77
402. Jungwirth, C., Horak, I., Bodo, G., Lindner, J., Schultze, B. 1972. *Virology* 48:59–70
403. Metz, D. H. 1975. *Cell* 6:429–39
404. Metz, D. H., Esteban, M., Danielescu, G. 1975. *J. Gen. Virol.* 27:197–209
405. Osterhoff, J., Jager, M., Jungwirth, C., Bodo, G. 1976. *Eur. J. Biochem.* 69:535–43
406. Wiebe, M. E., Joklik, W. K. 1975. *Virology* 66:229–40
407. Repik, P., Flamand, A., Bishop, D. H. L. 1974. *J. Virol.* 14:1169–78
408. Yakobson, E., Prives, C., Hartman, J. R., Winocour, E., Revel, M. 1977. *Cell* 12:73–81
409. Gupta, S. L., Graziadei, W. D. III, Weideli, H., Sopori, M. L., Lengyel, P. 1974. *Virology* 57:49–63
410. Falcoff, E., Falcoff, R., Lebleu, B., Revel, M. 1973. *J. Virol.* 12:421–30
411. Gupta, S. L., Sopori, M. L., Lengyel, P. 1973. *Biochem. Biophys. Res. Commun.* 54:777–83
412. Samuel, C. E., Joklik, W. K. 1974. *Virology* 58:476–91
413. Falcoff, E., Falcoff, R., Lebleu, B., Revel, M. 1972. *Nature New Biol.* 240:145–47

414. Kerr, I. M., Friedmann, R. M., Esteban, M., Brown, R. E., Ball, L. A., Metz, D. H., Risby, D., Tovell, D. R., Sonnabend, J. A. 1973. *Adv. Biosci.* 11:109–23

415. Friedman, R. M., Metz, D. H., Esteban, R. M., Tovell, D. R., Ball, L. A., Kerr, I. M. 1972. *J. Virol.* 10:1184–98

416. Kerr, I. M., Brown, R. E., Clemens, M. J., Gilbert, C. S. 1976. *Eur. J. Biochem.* 69:551–59

417. Roberts, W. K., Clemens, M. J., Kerr, I. M. 1976. *Proc. Natl. Acad. Sci. USA* 73:3136–40

418. Zilberstein, A., Federman, P., Shulman, L., Revel, M. 1976. *FEBS Lett.* 68:119–24

419. Cooper, J. A., Farrell, P. J. 1977. *Biochem. Biophys. Res. Commun.* 77:124–31

420. Revel, M., Gilboa, E., Kimchi, A., Schmidt, A., Shulman, L., Yakobson, E., Zilberstein, A. 1977. Proc. FEBS Meet 11th, Vol. 43, pp. 47–58. Oxford: Pergamon

421. Lebleu, B., Sen, G. C., Shaila, S., Cabrer, B., Lengyel, P. 1976. *Proc. Natl. Acad. Sci. USA* 73:3107–11

422. Roberts, W. K., Hovanessian, A., Brown, R. E., Clemens, M. J., Kerr, I. M. 1976. *Nature* 264:477–80

423. Samuel, C. E., Farris, D. A., Eppstein, D. A. 1977. *Virology.* 83:56–71

424. Hovanessian, A. G., Brown, R. E., Kerr, I. M. 1977. *Nature* 268:537–40

425. Kerr, I. M., Brown, R. E., Hovanessian, A. G. 1977. *Nature* 268:540–43

426. Zilberstein, A., Kimchi, A., Schmidt, A., Revel, M. 1978. *Proc. Natl. Acad. Sci. USA.* In press

427. Sen, G. C., Lebleu, B., Brown, G. E., Kawakita, M., Slattery, E., Lengyel, P. 1976. *Nature* 264:370–73

428. Schmitt, M., Kempf, J., Quirin-Stricker, C. 1977. *Biochim. Biophys. Acta* 481:438–49

429. Kerr, I. M., Friedman, R. M., Brown, R. E., Ball, L. A., Brown, J. C. 1974. *J. Virol.* 13:9–21

430. Samuel, C. E. 1976. *Virology* 75:166–76

431. Revel, M., Content, J., Zilberstein, A., Nudel, U., Berissi, H., Dudock, B. 1975. In *In Vitro Transcription and Translation of Viral Genomes Colloques INSERM Paris* 47:397–406

432. Content, J., Lebleu, B., Zilberstein, A., Berissi, H., Revel, M. 1974. *FEBS Lett.* 41:125–30

433. Gupta, S. L., Sopori, M. L., Lengyel, P. 1974. *Biochem. Biophys. Res. Commun.* 57:763–70

434. Zilberstein, A., Dudock, B., Berissi, H.,

Revel, M. 1976. *J. Mol. Biol.* 108:43–54

435. Falcoff, R., Lebleu, B., Sanceau, J., Weissenbach, J., Dirheimer, G., Ebel, J. P., Falcoff, E. 1976. *Biochem. Biophys. Res. Commun.* 68:1323–31

436. Weissenbach, J., Dirheimer, G., Falcoff, R., Sanceau, J., Falcoff, E. 1977. *FEBS Lett.* 82:71–76

437. Hiller, G., Winkler, I., Viehhauser, G., Jungwirth, C., Bodo, G., Dube, S., Ostertag, W. 1976. *Virology* 69:360–63

438. Mayr, U., Bermayer, H. P., Weidinger, G., Jungwirth, C., Gross, H. J., Bodo, G. 1977. *Eur. J. Biochem.* 76:541–51

439. Sen, G. C., Gupta, S. L., Brown, G. E., Lebleu, B., Rebello, M. A., Lengyel, P. 1976. *J. Virol.* 17:191–203

440. Sela, I., Grossberg, S. E., Sedmak, J. J., Mehler, A. H. 1976. *Science* 194:527–29

441. Colby, C., Penhoet, E. E., Samuel, C. E. 1976. *Virology* 74:262–64

442. Revel, M. 1975. *ICRS J. Med. Sci.* 3:609

443. Garel, J. P. 1974. *J. Theor. Biol.* 43:211–25

444. Smith, D. W. E. 1975. *Science* 190:529–35

445. LeMeur, M., Gerlinger, P., Ebel, J. P. 1976. *Eur. J. Biochem.* 67:519–26

446. Hatfield, D., Matthews, C. R., Caicuts, M. 1978. *Eur. J. Biochem.* In press

447. Brown, G. E., Lebleu, B., Kawakita, M., Shaila, S., Sen, G. C., Lengyel, P. 1976. *Biochem. Biophys. Res. Commun.* 69:114–22

448. Ratner, L., Sen, G. C., Brown, G. E., Lebleu, B., Kawakita, M., Cabrer, B., Slattery, E., Lengyel, P. 1977. *Europ. J. Biochem.* 79:565–77

449. Marcus, P. I., Terry, T. M., Levine, S. 1975. *Proc. Natl. Acad. Sci. USA* 72:182–87

450. Wreschner, D., Herzberg, M. 1976. *Eur. J. Bioch.* 64:399–404

451. Wreschner, D., Melloul, D., Herzberg, M. 1977. *FEBS Lett.* 77:83–86

452. Sen, G. C., Lebleu, B., Brown, G. E., Rebello, M. A., Furuichi, Y., Morgan, M., Shatkin, A. J., Lengyel, P. 1975. *Biochem. Biophys. Res. Commun.* 65:427–34

453. Sen, G. C., Shaila, S., Lebleu, B., Brown, G. E., Desrosiers, R. C., Lengyel, P. 1977. *J. Virol.* 21:69–83

454. Desrosiers, R. C., Lengyel, P. 1977. *Fed. Proc.* 36:812

455. Rozee, K. R., Katz, L. J., McFarlane, E. S. 1969. *Can. J. Microbiol.* 15:969–71

456. Bablanian, R. 1975. *Prog. Med. Virol.* 19:40–83

457. Baltimore, D. 1969. In *The Biochemistry of Viruses* ed. H. Levy pp.101–76. New York: Marcel Dekker
458. Levintow, L. 1974. In *Comprehensive Virology*, ed. H. Fraenkel-Conrat, R. Wagner, 2:109–69. New York: Plenum
459. Carrasco, L. 1977. *FEBS Lett.* 76: 11–15
460. Leibowitz, R., Penman, S. 1971. *J. Virol.* 8:661–68
461. Kaufmann, Y., Goldstein, E., Penman, S. 1976. *Proc. Natl. Acad. Sci. USA* 73:1834–38
462. Colby, D. S., Finnerty, V., Lucas-Lenard, J. 1974. *J. Virol.* 13:858–69
463. Koshel, K. 1974. *J. Virol.* 13:1061–66
464. Fernandez-Munoz, R., Darnell, J. E. 1976. *J. Virol.* 126:719–26
465. Doyle, M., Holland, J. J. 1972. *J. Virol.* 9:22–28
466. Saxton, R. E., Stevens, J. G. 1972. *Virology* 48:207–20
467. Ehrenfeld, E., Lund, H. 1977. *Virology* 80:297–308
468. Celma, M. L., Ehrenfeld, E. 1974. *Proc. Natl. Acad. Sci. USA* 71:2440–44
469. Steiner-Pryor, A., Cooper, P. D. 1973. *J. Gen. Virol.* 21:215–25
470. Lawrence, C., Thach, R. E. 1974. *J. Virol.* 14:598–610
471. Nuss, D. L., Koch, G. 1976. *J. Virol.* 19:572–78
472. Oppermann, H., Koch, G. 1976. *Arch. Virol.* 52:123–34
473. Bablanian, R., Eggers, H. J., Tamm, I. 1965. *Virology* 26:100–13
474. Abreu, S. L., Lucas-Lenard, J. 1976. *J. Virol.* 18:182–94
475. Carrasco, L., Smith, A. E. 1976. *Nature* 264:807–9
476. Pong, S. S., Nuss, D. L., Koch, G. 1975. *J. Biol. Chem.* 250:240–45
477. Racevkis, J., Kerwar, S. S., Koch, G. 1976. *J. Gen. Virol.* 31:135–38
478. Moss, B. 1968. *J. Virol.* 2:1028–37
479. Esteban, M., Metz, D. H. 1973. *J. Gen. Virol.* 19:201–16
480. Rosemond-Hornbeak, H., Moss, B. 1975. *J. Virol.* 16:34–42
481. Oppermann, H., Koch, G. 1976. *J. Gen. Virol.* 32:261–73
482. Mudd, J. A., Summers, D. F. 1970. *Virology* 42:328–40
483. Wertz, G. W., Youngner, J. S. 1972. *J. Virol.* 9:85–89
484. Kaerlin, M., Horak, I. 1976. *Nature* 259:150–52
485. Kleinmann, J. H., Moss, B. 1975. *J. Biol. Chem.* 250:2430–37
486. Fox, R. I., Baum, S. G. 1974. *Virology* 60:45–53
487. Nakajima, K., Ishitsuka, H., Oda, K. I. 1974. *Nature* 252:649–53
488. Petska, S. 1977. In *Molecular Mechanism of Protein Biosynthesis,* ed. H. Weissbach, S. Petska, pp. 468–536. New York: Academic
489. Vazquez, D. 1976. In *Reflection on Biochemistry,* ed. A. Kornberg pp. 347–56 Oxford: Pergamon
490. Littauer, U. Z., Inouye, H. 1973. *Ann. Rev. Biochem.* 42:439–70
491. Ussery, M. A., Tanaka, W. K., Hardesty, B. 1977. *Eur. J. Biochem.* 72:491–500
492. Blobel, G., Dobberstein, B. 1975. *J. Cell. Biol.* 67:835–51
493. Katz, T., Rothman, J., Lingappa, V., Blobel, G., Lodish, H. 1977. *Proc. Natl. Acad. Sci. USA* 74:3278–82
494. Van Venrooij, W. J., Van Eekelen, C. A. G., Jansen, R. T. P., Princen, J. M. G. 1977. *Nature* 270:189–91
495. Alton, T. H., Lodish, H. F. 1977. *Cell* 12:301–10
496. Held, W. A., West, K., Gallagher, T. 1977. *J. Biol. Chem.* 252:8489–97
497. Weber, L. A., Hickey, E. E., Baglioni, C. 1978. *J. Biol. Chem.* 253:178–83
498. Lodish, H. F., Froshauer, S. 1977. *J. Biol. Chem.* 252:8804–11
499. Stewart, A. G., Lloyd, M., Arnstein, H. R. V. 1977. *Eur. J. Biochem.* 80: 453–59
500. Hovanessian, A. G., Kerr, I. M. 1978. *Eur. J. Biochem.* In press
501. Kerr, I. M., Brown, R. E. 1978. *Proc. Natl. Acad. Sci. USA* 75:256–60
502. Buchwald, I., Hackett, P. B., Egberts, E., Traub, P. 1977. *Mol. Biol. Rep.* 3:315–21
503. Eppstein, D. A. and Samuel, C. E. 1977. *Biochem. Biophys. Res. Commun.* 79:145–53
504. Breathnach, R., Mandel, J. L., Chambon, P. 1977. *Nature* 270:314–19
505. Jeffreys, A. J., Flavell, R. A. 1977. *Cell* 12:1097–1108
506. Tilgham, S. M., Tiemeier, D. C., Seidman, J. G., Peterlin, B. M., Sullivan, M., Maizel, J. V., Leder, P. 1978. *Proc. Natl. Acad. Sci. USA* 75:725–29
507. Yakobson, E. Kahana, H. Revel, M. Groner, Y. 1978. *Ann. Meet. Israel Bioch. Soc.* p. 169 (Abstr.)

Ann. Rev. Biochem. 1978. 47:1127–62

REGULATION OF THE ASSIMILATION OF NITROGEN COMPOUNDS

❖998

Bonnie Tyler

Department of Biology, Massachusetts Institute of Technology,
Cambridge, Massachusetts 02139

CONTENTS

1127

0066-4154/78/0701-1127$01.00

PERSPECTIVES AND SUMMARY

In all biological systems the assimilation of nitrogen into macromolecules is essential for growth. The metabolic pathways of nitrogen metabolism can be divided into two classes: the assimilatory pathways necessary for the utilization of nitrogen from compounds available in the medium, and the biosynthetic pathways leading to the production of the nitrogen-containing compounds of the cell. The specific steps in these pathways vary with the organism, but in virtually all cells glutamate and glutamine serve as the nitrogen donors for biosynthetic reactions. Thus a knowledge of the formation of glutamate and glutamine from various nitrogen sources is crucial to our understanding of cell growth. Since the article by H. E. Umbarger in this volume deals with many of the biosynthetic pathways of nitrogen metabolism, this review deals exclusively with those pathways involved in the formation of glutamate or glutamine.

Several reviews have dealt with some of the reactions leading to the production of glutamate and glutamine in both prokaryotic (1–4) and eukaryotic (5–7) cells. These articles have discussed the factors regulating the activity of some enzymes responsible for the formation of these compounds, but have said relatively little about the regulation of synthesis of these enzymes. However, during the last few years studies on enteric bacteria have led to the notion that a general control element mediates the formation of a wide array of enzymes involved in the assimilation of nitrogen into glutamate and glutamine. This regulatory protein is the enzyme responsible for glutamine biosynthesis, glutamine synthetase (GS). It appears that this enzyme interacts with DNA to activate transcription of certain genes. The evidence indicates that in these organisms GS is responsible not only for the biosynthesis of glutamine but also for the assimilation of nitrogen into glutamate from: (a) NH_3, the preferred source of nitrogen, (b) a wide variety of organic compounds, (c) NO_3^- and N_2. In addition, it appears that the rate of GS formation is determined by a mechanism involving autogenous regulation. Consequently GS regulates nitrogen assimilation at several different levels through a cascade effect analogous to the one responsible for the regulation of the enzymatic activity of this complex protein. Recent experiments indicate that nitrogen fixation by other organisms may be affected by GS. Thus it may be that, in a sense, GS is ultimately responsible for all nitrogen assimilation since the nitrogen cycle begins with N_2.

In order to discuss adequately this recent literature on the regulation of assimilation of nitrogen into glutamate and glutamine, this review is limited to bacterial nitrogen metabolism and deals primarily with experiments conducted in the enteric bacteria. Some of this material has been reviewed previously (8, 9).

INTRODUCTION

Enteric bacteria growing in a defined mineral medium utilize glucose as the preferred source of carbon, and ammonia as the preferred source of nitrogen. Under these growth conditions the nitrogenous groups for intermediary metabolism derive from glutamate and glutamine as illustrated in Figure 1. Glutamate is the primary amino group donor for the amino acids while the amide group of glutamine is the direct nitrogen donor for certain steps in the biosynthesis of some amino acids, purines, pyrimidines, and certain other key compounds. The question then arises, how is nitrogen incorporated into glutamate and glutamine and how is this assimilation of nitrogen regulated?

Glutamine is synthesized in only one way, by the addition of ammonia to glutamate in a reaction catalyzed by glutamine synthetase (L-glutamate: ammonia ligase, EC 6.3.1.2). However, in the enteric bacteria, glutamate can be produced by a variety of reactions that can be divided into three main classes according to the immediate origin of the nitrogen and carbon atoms. Briefly stated these are:

1. from ammonia and 2-oxoglutarate, either by a reaction catalyzed by glutamate dehydrogenase (L-glutamate: NADP$^+$ oxidoreductase, EC 1.4.1.4) or as the result of a coupled reaction catalyzed by glutamine synthetase and glutamate synthase (L-glutamate: NADP$^+$ oxidoreductase, E.C. 1.4.13) (see Figure 1),
2. from the carbon and nitrogen atoms of another amino acid as a direct product of degradation of that amino acid,
3. from the amino groups of another amino acid and 2-oxoglutarate by a transamination reaction.

Some background information on the enzymes catalyzing the production of glutamate from ammonia is essential for understanding the regulation of the other two classes of glutamate-producing reactions. These enzymes are glutamate dehydrogenase, glutamate synthase, and glutamine synthetase.

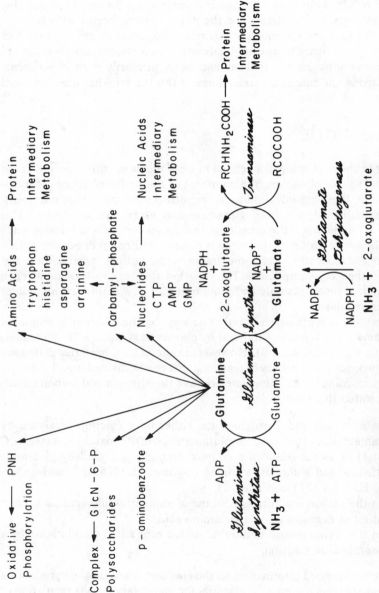

Figure 1   Pathways of ammonia assimilation in the enteric bacteria for the production of glutamate and glutamine, and some of the roles of these compounds in intermediary metabolism.

FORMATION OF GLUTAMATE AND GLUTAMINE FROM AMMONIA

The Role of Glutamate Dehydrogenase (GDH)

BIOCHEMISTRY OF GDH   Glutamate dehydrogenases catalyze the reductive amination of 2-oxoglutarate by ammonia in a reversible reaction utilizing either NADPH or NADH. They occur in a wide variety of organisms (reviewed in 6, 7). In microorganisms NAD-dependent GDHs appear to serve a catabolic function, while the enzymes utilizing NADPH serve primarily for the biosynthesis of glutamate. The enteric bacteria contain only an NADPH-dependent GDH. The GDHs of *Escherichia coli* and *Salmonella typhimurium* are of similar size [molecular weight 300,000 (10) and 280,000 (11) respectively]; the *E. coli* enzyme apparently consists of six identical subunits (10, 12). As with most of the bacterial enzymes (7), the GDHs of these organisms (10–14) have a relatively high K_m, on the order of 1 mM, for both 2-oxoglutarate and ammonia, which suggests that for cells growing under conditions of ammonia limitation, this enzyme is relatively ineffective in nitrogen assimilation.

PHYSIOLOGY OF GDH PRODUCTION   It is unlikely that GDH plays any catabolic role in *E. coli* since growth in the presence of glutamate results in repression of this enzyme (15–17). Such regulation might be predicted if GDH serves only a biosynthetic function and would not be expected if the enzyme played any role in the degradation of glutamate. Mutants (16) lacking GDH will grow slowly on glutamate as a source of carbon. In addition, studies (119) on *E. coli* strains unable to catabolize glutamate (see below) suggest that GDH is not involved in the major pathway of glutamate metabolism. It is of interest that the GDH of *Salmonella* is not repressed by glutamate (18). However the relative magnitude of the K_ms of this enzyme for 2-oxoglutarate and glutamate strongly suggest that the *Salmonella* GDH only serves in a biosynthetic capacity. Formation of GDH is repressed in nitrogen-starved cells of *Klebsiella aerogenes* (19) and *K. pneumoniae* (21, 99) but not in similar cultures of *E. coli* (22) or *S. typhimurium* (18). Such regulation is discussed later. The molecular basis for GDH repression by glutamate has not been studied.

GENETICS OF GDH FORMATION   Mutants of *E. coli* (16, 23), *K. aerogenes* (24) or *S. typhimurium* (25)) lacking GDH do not require glutamate for growth, an observation consistent with the existence of an alternative pathway for glutamate biosynthesis (see Figure 1). Derivatives of these *gdh* mutants, which have an additional lesion in a gene for glutamate synthase (*gltB*), are glutamate auxotrophs, while strains lacking only glutamate synthase activity will grow on glucose-minimal medium containing high concentrations of ammonia. Thus it appears that enteric bacteria possess only two pathways for synthesizing glutamate from ammonia. Strains containing both of these mutations (*gdh, gltB*) have been used to locate (27) the *gdh* gene at 26 min on the Bachmann linkage map of *E. coli* (71). The gene order is *gdh–trp–pyrF* on the chromosome of *K. aerogenes* (100) and *purB–gdh–trp* on the *E. coli* chromosome (27).

Strains lacking GDH activity have no discernible phenotype; such mutants grow as well as wild-type strains on both high and low concentrations of ammonia and all other nitrogen sources tested (24, 27). Apparently the role of GDH in ammonia assimilation can be completely replaced by the glutamate synthase-GS pathway. The question then arises whether glutamate synthase activity is equally dispensable to the cell.

The Role of Glutamate Synthase (GOGAT)

BIOCHEMISTRY OF GOGAT   Glutamate synthase is abbreviated GO-GAT from its previous trivial name: glutamine amide-2-oxoglutarate aminotransferase (oxidoreductase, NADP). It is one of many enzymes that catalyze the transfer of the amide group of glutamine in various biosynthetic reactions. In this case, glutamine reacts with 2-oxoglutarate to form two molecules of glutamate (see Figure 1). This reaction was discovered in 1970 by Tempest and co-workers (28) while attempting to explain how nitrogen-limited cultures of *Aerobacter aerogenes* are able to grow in a glucose-ammonia-mineral medium where the cells are essentially devoid of GDH. Subsequent work has shown that GOGAT exists not only in the enteric bacteria (14, 18, 19, 23, 29) but also in every prokaryote examined (4, 29), as well as in some strains of yeast (30, 31). Before the discovery of GOGAT it was not known how ammonia is assimilated by some bacteria (e.g. *Erwinia carotovora* and some *Bacilli*) that lack GDH activity. Aspartase (32) and alanine dehydrogenase (33) had previously been considered as possible catalysts for reactions leading to ammonia assimilation in these organisms.

Glutamate synthase has been purified from *E. coli* (13, 14, 34, 35) and from *K. aerogenes* (35, 36). In both cases, this iron-sulfide flavoprotein consists of a dimer of two unequal subunits; the smaller polypeptides from the two species of bacteria have essentially the same molecular weight (53,000) but the larger subunit from *K. aerogenes* is significantly heavier (mol wt 175,000) than that from *E. coli* (mol wt 135,000). The subunit composition of intracellular GOGAT is not clear. One group reports that enzyme purified from *E. coli* has a molecular weight of 800,000 (14), indicating the association of four dimers. Others find full GOGAT activity in a single dimer isolated from *K. aerogenes* (36).

It was reported (14) that, in contrast to most glutamine amidotransferases, this enzyme cannot utilize NH_3 as a substrate. More recently others (13, 34–36) have found NH_3-dependent activity in highly purified preparations of GOGAT. However, the mechanism of the GOGAT-catalyzed reaction, utilizing either glutamine or NH_3 as a substrate, differs in several ways from that of other glutamine amidotransferases. Other glutamine amidotransferases bind glutamine to the light subunit and then transfer the amide nitrogen to the heavy subunit where the reaction occurs. The heavy subunit also catalyzes the amidation from ammonia. However, in the case of GOGAT, the glutamine-binding site is located on the heavy subunit while ammonia binds to the smaller subunit (36). Mantsala & Zalkin (73) reported that the ammonia-dependent reaction can be carried out by the small subunit alone. The mechanism of the ammonia-dependent GOGAT-catalyzed reaction differs from that of other glutamine amidotransferases (13, 35) and is essentially the same whether GOGAT or GDH catalyzes the reaction. Consequently it is possible, despite arguments to the contrary (13, 34), that the enzyme preparations used for these studies (13, 34–36) were contaminated with trace amounts of GDH. Presumably this controversy could be resolved by purifying GOGAT from a *gdh* mutant devoid of GDH.

PHYSIOLOGY OF GOGAT PRODUCTION   Relatively little is known about the regulation of formation of GOGAT. The few studies reported have all used slightly different growth conditions. Thus, one group working with *K. aerogenes* found higher levels of GOGAT in ammonia-limited cultures than in those grown under conditions of excess ammonia (19). In contrast, other workers (38) using a different strain of *K. aerogenes,* found that the level of GOGAT was lower in cells grown with growth-rate limiting concentrations of ammonia or glutamine than in cells grown with excess ammonia. Cultures of *K. pneumoniae* grown in glucose-minimal medium with either excess ammonia or molecular nitrogen as the growth rate–limiting nitrogen source contain equivalent GOGAT levels (29). In addition, cells of either *K. aerogenes* or *E. coli* grown with glutamate as the sole nitrogen source contain low levels of GOGAT (19, 39). In an extensive study of *S. typhimurium,* Brenchley et al (18) found high levels of GOGAT in wild-type cells grown on glucose-minimal medium with excess ammonia regardless of the presence of glutamate and glutamine. However, low activity was found in cultures grown on nutrient broth or on glucose-ammonia medium supplemented with (*a*) aspartate or (*b*) a combination of amino acids that inhibit GS activity (40) or (*c*) growth rate–limiting supplies of glutamate or glutamine. In addition, cells grown with glutamate as the sole nitrogen source had very low GOGAT activity.

These observations could be explained by a model in which either high intracellular concentrations of glutamate, the end product of the GOGAT-catalyzed reaction, or very low intracellular levels of glutamine or ammonia, result in a lack of synthesis of this enzyme. This model fits the data if we assume that in *S. typhimurium,* as in *E. coli* (41, 42), glutamate enters the cell relatively slowly. In this case the intracellular pool of glutamate would only be high when cultures are grown in nutrient broth or in minimal medium supplemented with aspartate, because these media contain compounds that are rapidly transported and metabolized to glutamate. Secondly, one must also assume that growth limitation apparently caused by the limited availability of glutamate actually results from low intracellular levels of glutamine.

Based on this analysis, it is possible that other observations on alterations in GOGAT synthesis may actually result from variations in the internal pools of glutamate and/or glutamine. Prusiner et al (43) found that addition of 5 mM cAMP to *cya*⁻ mutants of *E. coli* decreased the level of GOGAT about twofold, but also altered the growth rate. A more dramatic change in GOGAT levels, a fivefold increase, was observed by LaPointe et al (44) when cells of *E. coli* carrying a thermosensitive mutation in the glutamyl-tRNA synthetase were grown at a partially restrictive temperature where the growth rate was reduced. However, in this experiment the level of GS also increased 50-fold, which might alter the intracellular level of glutamine. Although it is not possible to make definitive statements about GO-GAT regulation from any of these experiments, it is clear that synthesis of GOGAT is regulated. However, many questions remain to be answered. Is there a specific repressor or activator regulating transcription of the *gltB* gene? Is synthesis of GOGAT affected by the same general control element (GS) which appears to regulate the formation of many other enzymes involved in nitrogen metabolism? What is the role of glutamate in the regulation of GOGAT?

GENETICS OF GOGAT FORMATION   Strains altered in GOGAT formation have been described in *K. pneumoniae* (29), *K. aerogenes* (24), *E. coli* (23, 27), and *S. typhimurium* (45). As discussed above, these mutants do not require glutamate for growth on mineral salts medium containing glucose and excess ammonia. However, in contrast to *gdh* mutants, strains lacking GOGAT activity have a distinct phenotype. These mutants are unable to grow on glucose-minimal medium containing a low concentration of ammonia or a variety of other nitrogenous compounds as the sole source of nitrogen. Therefore they were originally described as Asm⁻ strains (29) for ammonia assimilation negative. Table 1 summarizes the growth patterns of wild-type strains and Asm⁻ mutants growing with glucose as a source of carbon and various nitrogen sources.[1] These data clearly indicate the relative roles of the two pathways (see Figure 1) for ammonia assimilation in enteric bacterial. When cells are grown in the presence of glucose and excess ammonia, glutamate is formed by way of GDH, a reaction not requiring a large investment of energy as does the coupled GS-GOGAT pathway. Under these conditions relatively low GS and GOGAT activities suffice to provide the glutamate and glutamine necessary for protein synthesis and for biosynthetic reactions (Figure 1). However, when the available supply of ammonia is limited, the GDH reaction does not function efficiently and ammonia is assimilated by the ATP-driven GS-GOGAT pathway. Thus, one might expect the specific activity of both GS and GOGAT to be higher in nitrogen-limited cultures than in those grown with excess ammonia. Such increased activity is observed with GS (see next section) but no dramatic difference is seen in GOGAT levels (4, 18, 29, 38, 39). Some data (18, 38, 39) even show an inverse relationship between GS and GOGAT levels.

A gene (*gltB*) involved in GOGAT biosynthesis in *E. coli* was originally reported to be linked to the *malP,Q* locus (23). More recent studies (27) have relocated this gene in another segment of the *E. coli* chromosome; it is closely linked to *argG* and is, in fact, identical to the locus previously called *aspB* (46). Since GOGAT consists of two nonidentical subunits, one might expect the Asm⁻ phenotype to result from mutations in two different structural genes, and possibly also from alterations in one or more regulatory proteins. Therefore it is of interest that for a substantial number of independent Asm⁻ mutants of *E. coli* (27), *K. aerogenes* (47), and *K. pneumoniae* (48) the loss of GOGAT activity is due to lesions tightly linked to *argG* (27). There have been no reports of Asm⁻ strains lacking GOGAT activity due to a mutation unlinked to *argG*, a result that might suggest that the genes for both subunits are linked to *argG*, or that only mutations in one gene, that for the large subunit, give an Asm⁻ phenotype.

The Role of Glutamine Synthetase (GS)

BIOCHEMISTRY OF GS   Glutamine synthetase catalyzes the ATP-dependent production of glutamine from ammonia and glutamate as shown in

---

[1]Some of the nitrogen compounds listed in Table 1 (e.g. histidine, arginine and proline) can serve as both a carbon and nitrogen source. For these compounds cAMP will serve to activate the synthesis of the degradative enzymes. Asm⁻ strains are not altered in the ability to synthesize these degradative enzymes under conditions where cAMP levels are high.

Table 1 Utilization of compounds as the source of nitrogen in the presence of glucose[a]

Nitrogen compound	Wild-type organism		Asm⁻ mutant[b]	Asm⁻ GlnC mutant[c]
	E. coli	Klebsiella		
L-Alanine	+[d]	+	−[d]	−
L-Arginine	+	+	−	+
L-Asparagine	+	+	+	+
L-Aspartate	+	+	+	+
γ-Aminobutyrate	+	+	−	+
L-Citrulline	−	+	nd[d]	nd
L-Cysteine	−	+	nd	nd
L-Glutamate	+	+	+	+
L-Glutamine	+	+	+	+
Glycine	+	+	−	−
L-Histidine	−	+	−	+
L-Leucine	−	−	nd	nd
L-Lysine	−	−	nd	nd
L-Methionine	+	nd	−	−
L-Ornithine	+	+	−	+
L-Proline	+	+	−	+
L-Putrescine	−	+	−	+
L-Pyroglutamate	+	nd	−	+
D-Serine	+	+	+	+
L-Serine	+	+	−	−
L-Threonine	−[e]	+	nd	nd
L-Tryptophan	−	+	−	+
Adenine	nd	+	−	nd
Agmatine	−	+	−	+
Allantoin	nd	+	−	nd
Cytidine	+	+	−	nd
Thymine	+	nd	nd	nd
Uracil	+	+	−	nd
Urea	−	+	−	−
Uric Acid	nd	+	−	nd
Xanthine	nd	+	−	nd
NH₄⁺, 30 mM	+	+	+	+
NH₄⁺, 1 mM	+	+	−[f]	−
Nitrate, sodium	−	+	−	−
Nitrite, sodium	−	+	−	−
N₂	−	+[g]	−	−

[a] These data are taken from references 8, 9, 27, 29, 38, 158, 159 and from personal communication with B. Magasanik. The data for *K. pneumoniae* (29) and *K. aerogenes* (8, 9, 38, B. Magasanik, personal communication) are listed together since these two organisms differ primarily with respect to the *nif* genes. Glucose was present at 0.4%.

[b] These data are for Asm⁻ mutants of *Klebsiella* and *E. coli*. No differences are observed in the phenotypes of Asm⁻ mutants of these organisms except for the ability to utilize 1 mM ammonia.

[c] These data give the growth characteristics for an Asm⁻ GlnC mutant of the wild-type organism which grows on the indicated nitrogen compound.

[d] Symbols are as follows: + indicates growth, − indicates no growth, and nd indicates no data available.

[e] Cells of *E. coli* will not grow with threonine as the sole source of nitrogen, unless leucine is present to induce the degradative threonine dehydrogenase (136).

[f] Asm⁻ mutants of *Klebsiella* do not grow with 1 mM ammonia as the sole source of nitrogen, but Asm⁻ mutants of *E. coli* do grow with this concentration of ammonia in the growth medium (27, 38). In fact, *E. coli* Asm⁻ mutants only fail to grow with ammonia as the sole nitrogen source when the concentration of ammonia is reduced to 75 μM. This difference in the phenotype of Asm⁻ mutants of *E. coli* and *K. aerogenes* presumably reflects the difference in the levels of GDH on nitrogen-limiting growth media.

[g] Data are for cells of *K. pneumoniae* (29).

Figures 1 and 2. Stadtman and his co-workers have rigorously studied the biophysical and enzymatic properties of the protein purified from *E. coli* W (reviewed in 49). The enzyme has a molecular weight of 600,000 and contains twelve identical subunits. The subunits are arranged in two hexagonal units layered with a 4.5 nm spacing between them. In addition, the divalent cation Mg^{2+} or Mn^{2+} is required for stability.

The elegant studies by Stadman & Ginsburg and Holzer et al (reviewed in 49, 50) have shown that the enzymatic activity of GS is regulated by three different mechanisms: 1. by the interconversion of a relaxed (inactive) and taut (active) form in response to variations in concentrations of divalent cations, 2. by cumulative feedback inhibition by various endproducts of glutamine metabolism (Figure 1), and 3. by covalent alterations of the enzyme by the reversible adenylylation of a specific tyrosyl residue on each subunit (Figure 2). Maximum biosynthetic activity is obtained when the enzyme is completely unadenylylated (E_0) and decreases over a wide range as the degree of adenylylation (adenylylation state) increases. However, the fully adenylylated (E_{12}) biosynthetically inactive enzyme retains glutamyl transferase activity and thus can be measured easily. This ATP-dependent modification apparently occurs in a variety of Gram-negative bacteria. The adenylylated form of the enzyme is more sensitive than the unmodified form to feedback inhibition by products of glutamine metabolism (see Figure 2).

The attachment and removal of the covalently bound AMP residues from GS is accomplished by the action of an adenylyltransferase (ATase), a monomer having a molecular weight of 130,000 (49) (see Figure 2). There are two factors that determine whether adenylylation or deadenylylation predominates: the levels of several metabolites and the interaction of ATase with a small regulatory protein, P_{II}. The levels of 2-oxoglutarate and glutamine have antagonistic effects on the activity of ATase. When the relative concentrations of 2-oxoglutarate is high, the deadenylylation activity of ATase is increased; when the relative level of glutamine is high, the adenylylation reaction is favored. In addition, ATP stimulates the ATase to deadenylylate. The small regulatory protein (P_{II}) that interacts with ATase is a tetramer composed of four identical subunits (mol wt $\sim$ 11,000;41). It exists in two states. Unmodified P_{II} (P_{IIA}) stimulates the adenylylation of GS by ATase and uridylylated P_{II}(P_{IID}) enhances the deadenylylation reaction. The modified form of P_{II} is the result of a covalent attachment of a UMP residue to P_{IIA} by the action of a uridylyltransferase (UTase). The activity of this enzyme is also modulated by the concentrations of 2-oxoglutarate and glutamine as indicated in Figure 2. The uridylyl residue is removed by a uridylyl-removing enzyme (UR) which may be the same protein as the UTase. The adenylylation state of the GS protein is affected by the relative level of glutamine and 2-oxoglutarate not only in experiments with cell extracts (52) but also in whole cells (39).

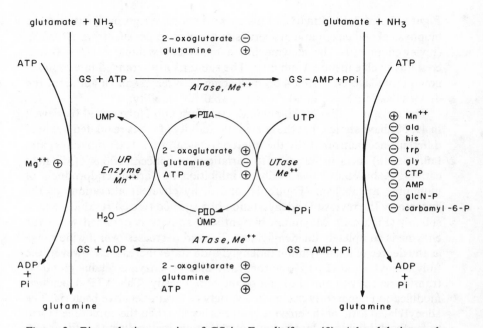

Figure 2 Biosynthetic capacity of GS in *E. coli* (from 49). Adenylylation and deadenylylation of GS results from the interaction of P_{IIA} or P_{IID} with ATase and GS as described in the text. These reactions are stimulated, $\oplus$, or inhibited, $\ominus$, by the levels of various metabolites and by Mg^{2+} or Mn^{2+}. The interconversion of P_{IIA} and P_{IID} by UTase and UR enzyme responds to the levels of these same metabolites. The UTase is also activated by either Mg^{2+} or Mn^{2+} while the UR enzyme is activated only by Mn^{2+}. Nonadenylylated GS (E_0), which has maximum biosynthetic activity, requires Mg^{2+} to catalyze the formation of glutamine. As the adenylylation state of the enzyme increases, the Mg^{2+}-dependent biosynthetic capacity decreases and the enzyme exhibits some biosynthetic activity when Mn^{2+} replaces Mg^{2+}. Fully adenylylated GS (E_{12}) will catalyze the biosynthetic reaction at a low rate in the presence of Mn^{2+} but not Mg^{2+}. Adenylylated GS is more sensitive to inhibition $\ominus$ by various endproducts of glutamine metabolism than is the nonadenylylated form of the enzyme.

Experiments by Tronick and co-workers (53) with antisera prepared against GS from *E. coli* (GS_E) suggested that the GS from all Gram-negative bacteria are antigenically related and covalently modified, possibly by adenylylation. However, the antisera failed to react with GS from Gram-positive bacteria, with the exception of *Streptomyces rutgersensis*. More detailed studies on the GS of *K. aerogenes* (GS_K) by Magasanik and collaborators have shown that the enzymatic activity of GS_K is also regulated by ATase, UTase, and a P_{II} protein (53, 54, 55). The enzyme from *K.*

aerogenes appears identical to that of *E. coli* when viewed with the electron microscope (55, 56, 57). Comparison of GS derived from *E. coli* with that from *K. aerogenes* has revealed small differences in various biochemical parameters (pH optima, divalent cation dependence), which differentiate the two proteins (58). In addition, the subunits of GS_E and GS_K have slightly different electrophoretic mobilities on SDS polyacrylamide gels: GS_K migrates more slowly (59) which might imply that the subunit of GS_K is slightly heavier than that of GS_E.

The enzymatic activity of the glutamine synthetase of *S. typhimurium* (GS_S), like that of *E. coli* and *K. aerogenes,* also appears to be modulated by adenylylation (18, 49, 53, 60, 61). The size of the GS_S subunit is similar to that of GS_E (59, 61) as are several biochemical parameters including the isoactivity pH (60). Extensive biophysical studies indicate that (61) purified GS_S is very similar to GS_E except in proton relaxation rates and electron paramagnetic resonance (62).

PHYSIOLOGY OF GS PRODUCTION   The GS in enteric bacteria grown in the presence of excess ammonia tends to be more highly adenylylated and consequently less biosynthetically active than that found in nitrogen-limited cultures (22, 39, 47, 50, 58, 60, 63, 66, 73). The absolute level of GS protein in these cells is inversely related to the availability of nitrogen in glucose-minimal medium (18, 20, 39, 45, 47, 50, 60, 64, 65, 67, 68). In a nitrogen-limited culture, the level of GS is five- to tenfold higher than that observed in cells grown with excess ammonia, and represents several percent of the total cell protein. However, the absolute intracellular level of GS can vary over 100-fold in other growth media. Thus cells grown in nutrient broth (18, 66), in glucose minimal medium supplemented with certain amino acids (18, 67), or in minimal medium with certain other carbon sources (68) contain extremely low levels of highly adenylylated GS. Bender & Magasanik (68) have suggested that the extremely low rate of GS synthesis in cells grown in such medium does not result simply from repression by ammonia but rather reflects the ratio of glutamine to 2-oxoglutarate. This notion follows from the proposal (47, 54, 55) of Magasanik and co-workers that the absolute level of GS in wild-type cells reflects the adenylylation state of the enzyme. The evidence for this view will be discussed below. Earlier experiments by Senior (39) using cells from ammonia-limited chemostat cultures strongly suggested that the adenylylation state of GS in *E. coli* W reflects the intracellular ratio of 2-oxoglutarate to glutamine, as predicted from the elegant work of Stadtman and co-workers (1, 49, 52). However the data of Senior (39) show significant variations in the intracellular level of GS under conditions where no changes occur in the adenylylation state of the enzyme.

GENETICS OF GS FORMATION   Mutations affecting the synthesis of GS have been isolated in *K. aerogenes* (38, 47, 54, 55, 64, 69), *S. typhimurium* (45, 73, 74, 75), *K. pneumoniae* (20, 21, 48, 76, 99), and *E. coli* K-12 (27, 69, 70). Genetic studies have revealed that these mutations are at a number of distinct loci around the chromosome: *glnA, glnB, glnD, glnE, glnF.*

Mutations at the glnA locus   The *glnA* locus has been located at 85 min (27, 69, 70) on the Bachmann linkage map of E. coli (71). In *E. coli* (70), *K. aerogenes* (72) and *S. typhimurium* (73), there exist lesions at this site resulting in a temperature-dependent Gln⁻ phenotype. The GS from these mutants is more thermolabile than that isolated from wild-type cells; hence it is likely that this site is the structural gene for the GS subunit. In all the enterics, the *glnA* gene is linked to the *rha* locus (27, 69, 70, 74; R. Goldberg, unpublished observations). In *K. aerogenes* the order of markers is *rbs-glnA-rha-metB-ilv* with 91% cotransduction by phage P1 between the *glnA* gene (*glnA*$_K$) and *rbs* but no detectable linkage of *glnA*$_K$ with *ilv* (69). Recent unpublished studies by G. Pahel, P. Rushner, and B. Tyler on the *glnA* gene of E. coli (*glnA*$_E$) have revised the order of genes in this region of the *E. coli* chromosome. The new order is *rbs-ilv-fad-polA-glnA-rha-pfk-tpi-metB* with at least 65% cotransduction by phage P1 between *polA* and *glnA*$_E$ but no cotransduction between *ilv* and *glnA*.

There are three types of mutations tightly linked to the *glnA* gene. The first class (*glnA*⁻) results in a Gln⁻ phenotype and the apparent absence of GS polypeptide (21, 27, 69, 70, 72–74). The second class of mutations results in the constitutive synthesis of high levels of enzymatically active GS (GlnC phenotype) (27, 38, 69, 74) under all growth conditions. A third class of mutations [Gln(AC)⁻ phenotype] (21, 27, 69), allows the production, on glucose-ammonia medium, of high levels of enzymatically inactive GS protein that can be detected immunologically, or on polyacrylamide gels, or by other regulatory characteristics of the GS protein (see below). Fine scale mapping by three point crosses of the lesions in *glnA*⁻, *glnC,* and *gln(AC)*⁻ mutants in *K. aerogenes* has suggested that these three classes of mutations may occur within the same gene (47, 69). Two mutations resulting in the GlnC phenotype apparently lie between mutations that cause the production of enzymatically inactive GS antigen. However, fine-scale deletion mapping is needed to prove this point unequivocally.

Mutations at the glnB site   The only mutation clearly altering the *glnB* locus has been isolated in *K. aerogenes* (38, 64, 69). Analogous mutations may exist in *S. typhimurium* (74). The gene of *K. aerogenes* was located (69) by complementation with an *E. coli* F' episome with the region between 52 and 56 min of the Bachmann map (71). This implies that an analogous gene exists in this region of the *E. coli* chromosome. Mapping by phage Pl (69) revealed that this locus is cotransducible with *tyrA* at a low frequency. Phenotypically, this mutation results in glutamine-dependent growth. The level of GS in a *glnB* strain is about one tenth of that found in wild-type strains and the enzyme is fully adenylylated. Biochemical complementation with purified proteins of the adenylylation system of *K. aerogenes* strongly suggests that the *glnB* mutation results in an altered P_{II} protein no longer able to stimulate the deadenylylation of GS by the ATase (54). Strains carrying *glnB* lesions revert at a very high frequency to yield strains having the GlnC phenotype (constitutive GS production) due to mutations at either *glnA* and *glnE* (see below).

Mutations at the glnD locus   Lesions in the *glnD* gene cause a reduction in the amount of GS in the cell and increased adenylylation of the GS, particularly in cells grown with excess nitrogen. The phenotype of *glnD* mutants depends on both the strain and the organism. In *K. aerogenes* (F. Foor, personal communication), and *K. pneumoniae* (48), lesions in *glnD* cause a partial glutamine auxotrophy in most strain backgrounds. In strains of *E. coli* (48) and *S. typhimurium* (75), a mutation in *glnD* results in an absolute growth requirement for glutamine. Nitrogen-limited cultures of *K. aerogenes glnD* mutants contain high levels of relatively nonadenylylated enzyme. However the level of GS in *glnD* strains of *E. coli* is extremely low under all growth conditions (48). Consequently, it is of great interest that in all these organisms the *glnD* mutants are devoid of the P_{II} modifying enzyme, UTase (48, 75, F. Foor, unpublished observations). The *glnD* gene is near the *leu* operon on the chromosome of all four organisms (48, 75, S. Streicher, unpublished observations). In *E. coli* the *glnD* gene is 98% cotransducible with the *dapD* locus by phage P1 (73, F. F. Bloom and B. Tyler, unpublished observations).

Mutations at glnE yield the GlnC phenotype   There exists one mutation, unlinked to *glnA,* that results in a GlnC phenotype. This mutation was found in *K. aerogenes* (54, 55). The lesion, at the *glnE* locus, is very close, but separable from the *glnB* gene (55). Foor, Janssen & Magasanik (54) have shown that the ATase in this *glnE* mutant is altered so that it always stimulates the deadenylylation of GS. Thus, in contrast to the situation with *glnA*-linked mutations that result in the GlnC character and highly adenylylated GS, this *glnE* mutation results in high intracellular levels of GS under all growth conditions but the GS is only slightly adenylylated.

Mutations at the glnF site   Gln⁻ mutants of *S. typhimurium* (75), *E. coli* (27), *K. aerogenes* (47), and *K. pneumoniae* (48) also result from lesions in the *glnF* gene which is cotransducible with the *argE* gene in *S. ty-phimurium,* and with the corresponding gene, *argG,* in the other enterics. In *E. coli* (G. Pahel and B. Tyler, unpublished observations) but not in *K. aerogenes* (47) the *glnF* and *gltB* genes are very tightly linked. In most cases, strains altered in the *glnF* gene contain less than 10% of the GS enzymatic activity found in the wild-type strains grown on glucose-ammonia medium. Garcia and coworkers (75) showed that the *glnF* strains of *S. typhimurium* do not contain any protein immunologically similar to GS and are not altered in the enzymes of the adenylylation system. The identity and function of the *glnF*-product are not known.

FORMATION OF GLUTAMATE AND GLUTAMINE FROM OTHER SOURCES OF NITROGEN

Although glucose and ammonium salts are the preferred sources of carbon and nitrogen for enteric bacteria growing in a mineral salts medium, these organisms have the ability to utilize the nitrogen from a large variety of organic compounds and, in some cases, from NO_3^- and molecular nitrogen. Glutamate is then formed from these compounds by one of the three pathways outlined in the introduction. Glutamine synthesis from these alternative nitrogen sources depends not only on the amount of biosynthetically active GS but also on the intracellular concentration of ammonia generated from these nitrogen compounds. Some of these nitrogen compounds can also serve as sources of carbon and energy for the cell. In either case the utilization of these nitrogen compounds generally requires proteins not present in cells growing in glucose-ammonia medium. The synthesis of these proteins is often subject to regulation involving the interaction of a small effector molecule with a regulatory protein specific for a particular degradative pathway, as was originally elucidated for the *lac* operon (77). However, relief of this specific control system is usually not a sufficient condition for maximum synthesis of the proteins necessary for the utilization of these nitrogen compounds. In general, the presence of another regulatory protein, a positive control element, is required. The critical point is that this positive control element is not pathway-specific; rather, one particular protein can increase transcription from a large variety of genes so long as the pathway-specific regulatory protein is in the correct conformation to allow synthesis to occur. In some cases synthesis of the enzymes necessary for the assimilation of a nitrogen compound requires only the more general positive control elements (see Table 2).

It is of particular interest that in wild-type cells the regulation of degradation of some nitrogen compounds can require different positive control elements depending on the physiological state of the cell. The best example

Table 2 Regulation of assimilation of some nitrogen compounds

Nitrogen compound	Organism	Regulation by			Reference
		Induction	CAP-cAMP	GS	
Arginine	K[a]	+[b]	+	+	95
	E[c]	nd	nd	+	27
Asparagine	K	−[d]	−	+	97
	E	−	−	−	117, 118
γ-Aminobutyrate	E	+[e]	+	+	27, 160–162
Histidine	K	+	+	+	38, 64, 78, 79 80–83
N$_2$	K	−	−	+	76, 99, 163–164
Ornithine	K	+	+	+	—[e]
	E	nd[f]	nd	+	27
Proline	K	+	+	+	81
	E	+	+	+	27
Putrescine	K	+	+	+	9[f]
D-Serine	E	+	+	−	125–127
Tryptophan	K	−	−	+	9, 98
	E	+	+	−	65, 129–130
Urea	K	−	−	+	96

[a] indicates *Klebsiella*.
[b] indicates regulation.
[c] indicates *E. coli*.
[d] indicates no regulation.
[e] B. Friedrich, personal communication.
[f] no data available.

of this situation is seen in the regulation of histidine catabolism. As with a number of organic compounds, histidine can serve as both a source of carbon and a source of nitrogen. One positive control element is involved in the synthesis of the *hut* (histidine-utilization) enzymes when the cell has need for energy, while another positive control element is generally utilized for the activation of transcription of the *hut* genes when the cell senses a need for nitrogen. In cells growing on a good source of carbon and using histidine as a source of nitrogen, the synthesis of the *hut* enzymes is activated by GS protein. Furthermore, it now appears that the regulation of assimilation of nitrogen from various compounds can be divided into two classes:

1. regulation that is mediated by GS.
2. regulation that is independent of GS.

The evidence for this view, which originates from studies on mutants lacking GOGAT activity, is reviewed below after a discussion of the data indicating that GS serves to activate *hut*-gene transcription.

Regulation of Assimilation of Nitrogen Compounds by Glutamine Synthetase

THE HUT OPERON IS THE MODEL SYSTEM FOR GS REGULATION
Much of the evidence indicating that GS can regulate transcription came from studies on the *hut* genes of *K. aerogenes*. Synthesis of the *hut* enzymes is subject to regulation by a specific repressor (78, 79) and by more general control elements (38, 64, 80–83). Since histidine is degraded to glutamate and can serve as a source of carbon and energy, transcription of the *hut* operons is subject to catabolite repression (80–83). Recent studies have suggested that adenosine 3',5'-monophosphate (cAMP) and a cAMP binding protein (CAP) will activate transcription of all catabolite-sensitive operons (reviewed in 87). It is thought that the level of cAMP inside the cell is inversely related to the pool of catabolites in the cell. Thus the intracellular concentration of cAMP senses the cell's need for energy. Consequently it is not surprising that Neidhardt & Magasanik (88) reported that *K. aerogenes* produced very low levels of histidase, the first enzyme in the histidine degradative pathway, in a medium containing histidine, excess ammonia and glucose, the preferred source of carbon and energy. However, they also observed (88) that *K. aerogenes* produced high levels of histidase during nitrogen-limited growth in medium containing glucose, histidine, and no ammonia, a condition which should result in a large pool of catabolites inside the cell. Clearly these results are at variance with the cAMP model of catabolite repression, according to which restriction of metabolism by nitrogen limitation should result in a low intracellular level of cAMP.

Therefore, Prival & Magasanik (81) reinvestigated this phenomenon of histidase "escape" from catabolite repression under conditions of nitrogen limitation using a mutant of *K. aerogenes* unable to synthesize cAMP. They found that histidase and β-galactosidase, a catabolite-sensitive enzyme not involved in glutamate or NH₃ formation, were controlled in a similar manner as long as the nitrogen source did not limit the growth rate. Neither histidase nor β-galactosidase was produced in the absence of exogenous cAMP when a high level of ammonia was present in the medium. However, when this strain was grown on a nitrogen source that limits the growth rate, histidase, but not β-galactosidase, was formed at a high rate in the absence of exogenous cAMP. Similar results were obtained with proline oxidase, which like histidase is involved in the generation of glutamate from an amino acid. From these experiments, it was concluded (81) that nitrogen-starved cells contain a product capable of stimulating the synthesis of histidase and proline oxidase in the absence of cAMP. Since the level of histidase correlates directly with the level of GS in glucose-grown cells of

K. aerogenes, Magasanik and co-workers suggested that this regulatory product is GS (38, 64).

This correlation may be summarized as follows. In wild-type cells growing with glucose as a source of carbon, the level of histidase is low in ammonia-rich medium where the level of GS is low, and the level of histidase is high in nitrogen-poor medium where the level of GS is high (38, 64). Mutants unable to synthesize high levels of GS do not synthesize the Hut enzymes when grown on a growth-rate limiting nitrogen source (27, 38, 64). Mutants constitutive for the synthesis of GS have high levels of histidase even when grown in glucose-minimal medium with excess ammonia (38, 64, 89).

It is most important that this correlation between histidase levels and GS protein levels holds for a strain of *K. aerogenes* that synthesizes high levels of enzymatically inactive glutamine synthetase antigen due to a mutation in *glnA,* the structural gene for GS [Gln(AC)⁻ phenotype] (89). From this result it is apparent that the GS protein is involved, rather than a product of the reaction catalyzed by GS. When an episome carrying the *glnA* region of *E. coli* was introduced into the *gln(AC)*⁻ mutant of *K. aerogenes* (89) the GS antigen produced by the *K. aerogenes* chromosome was repressed and the level of histidine was again very low under conditions of nitrogen excess. All these observations support the hypothesis that GS participates as a positive control element in the regulation of the *hut* operon.

The hypothesis was tested directly using a highly purified system (82) in which transcription of *hut*-specific message was dependent on the presence of certain factors. Addition of cAMP and CAP together caused an eightfold increase in transcription of *hut* specific message. Nonadenylylated GS (the form present in nitrogen-limited cells) stimulated *hut* transcription to a similar extent as did the CAP and cAMP combination, but adenylylated GS (the form present in cells grown with excess nitrogen) did not cause a significant stimulation of transcription. Similarly, nonadenylylated GS treated with EDTA, which is known to cause a conformational change in the protein, failed to stimulate transcription. Further control experiments showed that nonadenylylated glutamine synthetase did not stimulate transcription of λDNA or DNA of the *gal* operon, another catabolite-sensitive operon not involved in nitrogen-assimilation. This evidence strongly supports the proposal that nonadenylylated GS is an activator of transcription of the *hut* operons and that adenylylation of GS causes a conformational alteration so that the protein can no longer function as a positive effector of transcription.

If GS protein is directly responsible for activating *hut* transcription, then GS must interact with DNA. In fact two lines of evidence indicate that GS binds to DNA. Biochemical experiments by Streicher & Tyler (90) strongly suggest that GS complexes to DNA in crude extracts of the enteric bacteria. They found that all the GS can be sedimented from these extracts under conditions where purified GS does not sediment. This sedimentation is dependent upon the integrity of the DNA. Recent experiments by Burgess (91) have shown that DNA-binding proteins, such as RNA polymerase, can be selectively precipitated from crude extracts by the addition of polyethylene glycol under conditions that precipitate DNA. Streicher & Tyler (90) have shown that such conditions also result in precipitation of all the GS from crude extracts but do not precipitate purified GS. It is of interest that some of the enzymes of the GS-adenylylation system purify with this GS-DNA complex (B. Tyler, S. G. Rhee, and S. Streicher, unpublished observations). The GS obtained by polyethylene glycol precipitation can be purified away from the DNA and other proteins by a procedure that yields pure GS in good yields. This GS, as well as that obtained by more standard methods (40, 66, 82), appears to bind to DNA when viewed under the electron microscope (57).

Although the evidence indicates that the GS protein has affinity for DNA, none of these experiments reveal any sequence specificity in the formation of the GS-DNA complex. These results are similar to those originally found with the CAP protein (92, 93); in this case specific binding was only observed with particular small DNA fragments carrying the *lac* promoter (94). Similar experiments may be necessary to demonstrate specific binding of GS to the *hut* promoter. However, the highly purified GS used in the *hut* transcription studies (82) did contain traces of other polypeptides which could include the GS transcription factor (60) discussed below. It may be that this factor is necessary for sequence-specific binding of GS to DNA.

GS-MEDIATED REGULATION OF SYNTHESIS OF OTHER SYSTEMS The experiments presented in the previous section constitute the evidence that GS serves as a positive control element for *hut*-gene transcription. The question then arises whether GS serves as a general regulatory protein that activates the synthesis of other enzymes involved in nitrogen assimilation in the enteric bacteria. In attempts to answer this question, experiments were conducted on wild-type and mutant strains of *K. aerogenes* (9, 38, 81, 96–100) and *K. pneumoniae* (76, 99) to determine whether the level of GS also correlated with the rate of synthesis of other enzymes involved in assimilation of nitrogen. This approach has involved assaying GS and the enzyme of interest in various strains grown under different conditions:

1. in nitrogen-limited cultures of wild-type cells which should contain high levels of GS,
2. in nitrogen-limited cultures of *glnA* mutants which do not produce any GS polypeptide,
3. in cultures of *glnC* mutants grown with glucose and excess ammonia, a situation where these mutants produce high levels of GS.

The correlation between GS levels and enzyme levels observed in these studies have led to the conclusion that, in *Klebsiella,* GS can activate the synthesis of enzymes responsible for the utilization of proline (81), arginine (95), putrescine (9), urea (96), asparagine (97), tryptophan (98), and molecular nitrogen (76, 99). Since an inverse relationship was observed between the intracellular level of GS and GDH (38, 89, 100) it was also proposed that GS represses the synthesis of GDH (38). Some of these enzymes are regulated like those of the Hut system by specific induction and by cAMP as well as by GS; the synthesis of others does not require induction, does not respond to cAMP, and is regulated exclusively by GS (see Table 2).

This GS-mediated regulation of enzyme synthesis appears to require a certain "critical level" of GS. Thus, when cultures of wild-type cells are shifted from a condition of nitrogen excess to one of nitrogen limitation, the rate of *hut* enzyme synthesis does not immediately increase in parallel with the increased rate of GS synthesis; rather GS must accumulate to a certain level before the rate of *hut* enzyme synthesis increases. This "threshold" level of GS can be seen in the early work on GS-mediated transcription of the *hut* (64) and *gdh* (38) genes. More careful measurements in *K. aerogenes* (F. Foor and B. Magasanik, unpublished observations), *E. coli,* and *K. pneumoniae* (F. R. Bloom and B. Tyler, unpublished observations) show that approximately the same specific activity of GS is required for activation of *hut*-gene expression in all three organisms. In addition, this GS must be in a low adenylylation state; both cell-free experiments (82) and studies with whole cells (96) have shown that highly adenylylated GS will not serve to activate gene expression.

The question arises whether the same intracellular concentration of GS is required for activation of transcription from all promoters or whether this critical level of GS varies with different promoters as in the case of cAMP and CAP (101). Therefore in order to establish that the synthesis of an enzyme is never subject to regulation by GS, it is important to show that the absence of GS-mediated enzyme synthesis is not due to the lack of a critical level of GS. One should examine the regulation of the enzyme synthesis in GlnC strains carrying mutations that are known to overcome ammonia repression of other enzymes. However, the inability of a GlnC mutant to overcome ammonia repression of enzyme synthesis cannot argue definitively for the absence of GS regulation of that enzyme; not all *glnA*-linked mutations resulting in equivalent high levels of GS allow the synthesis of all GS-regulated enzymes (27). Apparently, some minor changes in the structure of GS, which do not dramatically alter the catalytic activity of the enzyme, can affect the affinity of the protein for some DNA sequences but not for others. On the other hand, if the synthesis of an enzyme can be activated by GS, there should exist some *glnA*-linked mutations that result in the GlnC phenotype and overcome ammonia repression of the synthesis of that enzyme. The absence of such mutations can be viewed as suggestive evidence that GS does not regulate the synthesis of that enzyme. This reasoning has led to a genetic approach for rapidly screening for GS regulation of nitrogen assimilation (27). The design of these experiments follows from observations on Asm⁻ strains of *Klebsiella* (38).

In the presence of glucose, Asm⁻ mutants will grow with glutamate or glutamine as the sole source of nitrogen, but are unable to use as a source of nitrogen a number of compounds used by the wild-type strain (see Table 1). For some of these compounds (e.g. N_2, NO_3^-, glycine, and serine), all the nitrogen is metabolized to ammonia; consequently these compounds will not serve as a nitrogen source for an Asm⁻ strain if the intracellular level of ammonia from their metabolism is too low for the formation of glutamate by the GDH-catalyzed reaction. However, degradation of several of these compounds (e.g. histidine and proline) is known to yield glutamate which can serve as the sole source of nitrogen for Asm⁻ strains. The failure of Asm⁻ strains to grow in glucose medium with histidine or proline as a growth-rate limiting source of nitrogen can possibly be explained by the fact that the level of the *hut* and *put* (proline utilization) enzymes in these cells is not sufficient to produce enough glutamate for growth. (In fact the level of GS tends to be lower in nitrogen-limited cultures of Asm⁻ mutants than in cultures of wild-type cells grown under the same conditions, and there is little GS-activated enzyme synthesis in an Asm⁻ strain (38).)

In *K. aerogenes,* revertants of an Asm⁻ strain lacking GOGAT activity were selected for the ability to use histidine as a nitrogen source in the presence of glucose (38). The revertants fell into two classes: (*a*) those that regained GOGAT activity and therefore were able to utilize as nitrogen sources all the compounds used by the wild-type strain; and (*b*) those that produced GS constitutively (GlnC phenotype) and could utilize histidine, tryptophan, and arginine, but not glycine or low concentrations of ammonia as a source of nitrogen in this medium. Thus the existence of GlnC mutants among revertants of an Asm⁻ strain selected for the ability to utilize a compound as a nitrogen source can serve as a diagnostic tool indicating

GS-mediated regulation of the utilization of that compound. However, if GS regulates the assimilation of a compound (e.g. N_2, NO_3^-) from which the nitrogen only is converted to ammonia, then this compound may not be utilized by any GlnC Asm⁻ strains since GOGAT activity may be essential for the assimilation of the ammonia into glutamate.

This genetic approach has been used to investigate the controversial subject of whether GS regulates the expression of genes native to *E. coli* (27), a question of interest since cells of *S. typhimurium* are deficient in GS-activated gene expression (60, 64, 81, 102). Previous work had established that the GS of *E. coli* can activate transcription of the *hut* genes of *K. aerogenes* (65, 89) and of *S. typhimurium* (60) and showed that the *nif* genes, which require GS for transcription in the host organism, *K. pneumoniae* (76, 99), are expressed in *E. coli* (103, 104). On the other hand, the GDH of *E. coli* is not regulated by GS (89), but this may be due to the structure of the GDH promoter (27). However, studies on the regulation of nitrite reductase and proline utilization led Newman & Cole to propose (105) that in *E. coli,* as in *S. typhimurium,* GS fails to regulate the synthesis of native enzymes. Therefore it is significant that recent genetic and physiological studies (27) clearly contradict the conclusions of Newman & Cole; apparently in both *E. coli* and *Klebsiella* GS regulates the expression of genes involved in the assimilation of nitrogen from a number of amino acids (see Tables 1 and 2). Previous studies had shown that the activity of a glutamine transport system correlated with the level of GS in wild-type cells of *E. coli* (106). The data of Pahel, Zelenetz & Tyler (27) also extend the evidence that this system is regulated by GS.

THE GS-TRANSCRIPTION FACTOR   Cells of *S. typhimurium* can grow on a variety of compounds as the sole source of carbon and nitrogen (107). However, little is known about the regulation of synthesis of the enzymes necessary for the degradation of these compounds, except for the Hut (80, 81, 102, 109–113) and Put (81, 108) systems. In these cases, the enzymes are subject to induction and catabolite repression as in *K. aerogenes* (81) and *E. coli* (27). However, in cells of *S. typhimurium* the native GS cannot substitute for CAP and cAMP to activate transcription of the *hut*(s) and *put*(s) genes (60, 64, 81, 102). This lack of GS-activation of transcription is not attributable to insufficient levels of nonadenylylated GS in the cytoplasm of *S. typhimurium* (60). The situation is not due to an alteration in the *hut* promoter since GS activates the synthesis of histidase from the *hut*(s) gene when these genes are present in the cytoplasm of *K. aerogenes* (81) or *E. coli* (60) or in a purified transcription system (82). Moreover, the recent studies by Bloom, Streicher & Tyler (60) have shown that transcription of the *hut*(s) genes is activated by the GS of *S. typhimurium* in the cytoplasm of *E. coli* or *K. aerogenes.* The authors conclude that GS is necessary but not sufficient for activation of transcription of the *hut* genes; apparently another factor must also be present. This factor is active in both *K. aerogenes* and *E. coli* but missing or altered in *S. typhimurium.* The present evidence, summarized below, favors the notion that *S. typhimurium* does contain an altered GS transcription factor.

In cultures of *S. typhimurium* grown under different conditions, the expression of *glnA,* as reflected in levels of GS, is similar to that in *K. aerogenes,* where it has been suggested that GS regulates *glnA* expression. In addition, the *nif* genes of *K. pneumoniae,* believed to be regulated by GS (76, 99), are expressed and normally regulated in *S. typhimurium* LT-7 (104). Small variations in the activity of a glutamine transport system in *S. typhimurium* LT-2 may reflect GS activation of transcription of the structural gene for a glutamine permease (115, 116). These may be examples of GS-mediated regulation that do not require the transcription factor. However it is more likely that *S. typhimurium* contains a GS transcription factor that interacts with the promoters of the *glnA* and *nif* genes but not with the promoters of the *hut* genes.

Obvious candidates for the GS transcription factor are ribonucleic acid polymerase, one of the enzymes of the GS adenylylation system, or the product of the *glnF* gene.

GS-Independent Regulation of Assimilation of Nitrogen Compounds

The genetic and physiological studies outlined above have failed to provide evidence for GS-mediated regulation of assimilation of certain nitrogen compounds that support growth of wild-type strains but are not utilized by Asm⁻ mutants; however no definitive studies have been published on the regulation of assimilation of most of these compounds (e.g. alanine, glycine, methionine, L-serine) (see Table 1). The catabolism of tryptophan (65, 128–130), serine (131, 132), glycine (133), and threonine (2, 134–136) has been studied in *E. coli* K-12. Some enzymes responsible for the degradation of these amino acids are subject to induction and catabolite repression. There is no clear evidence that GS plays a role in the regulation of assimilation of any of these nitrogen compounds, although in some cases (132, 136) the level of the enzymes does increase slightly when cells are grown in ammonia-free medium. It is of interest that utilization of tryptophan (65, 98, 128–130) or asparagine (97, 118, 119) is not regulated by GS in *E. coli* K-12 but is stimulated by GS in *K. aerogenes.*

Relatively little is known about the regulation of assimilation of most nitrogen compounds utilized by Asm⁻ strains. Thus definitive studies are needed on the regulation of the transport system and aspartase activity required for aspartate utilization. The only published studies on the regulation of glutamate utilization (except for those discussed in the section on GDH physiology) deal with the glutamate permease (41, 42) which is controlled by a regulatory gene. However, no published studies have investigated whether more general control elements are needed for synthesis of this transport system. The nature of the regulation of the enzymes (aspartase and aminotransferase) required for glutamate metabolism is not known although mutants lacking these proteins have been isolated (119, 120).

Glutamine can serve as a nitrogen source if a transport system and glutaminase or GOGAT activity are present. For glutamine to serve as a source of carbon, the enzymes of glutamate metabolism are also needed. The regulatory characteristics of glutaminase enzyme activity have been the subject of reviews (121, 122), and some reports have described the regulation of formation of the two glutaminases present in *E. coli* (123, 124). The level of one of these enzymes is independent of growth conditions while the level of the other is low in nitrogen-starved cells, responds to induction by glutamine, and appears to be inversely related to the level of cAMP in the cell (123). Two permeases capable of transporting glutamine into cells of *S. typhimurium* have been identified. As mentioned earlier, GS apparently regulates the activity of one of these transport systems (106, 115, 116); little is known about the regulation of the other system except that it also transports methionine.

D-Serine is converted to pyruvate and ammonia by D-serine deaminase in a detoxifying reaction; consequently it is not surprising that this degradative enzyme is synthesized in Asm⁻ strains (Table 1). Synthesis of this enzyme appears to require positive activation of the *dsdA* gene by a regulatory protein complexed to D-serine (125). Bloom & McFall (126) have shown that synthesis of D-serine deaminase does not increase during nitrogen-limited growth of strains altered in the *dsdC* gene, which codes for the pathway-specific positive activator. These observations imply that GS does not affect transcription of the *dsdA* gene. However, maximal expression of this gene requires cAMP, although this requirement can be partially overcome by high concentrations of D-serine (127). Thus it appears that the cell views D-serine deaminase not only as a detoxifying enzyme but also as an enzyme of carbon metabolism. Therefore, it is of interest that utilization of this nitrogen compound does not appear to be regulated by GS. Authors studying GS-activated synthesis of D-serine deaminase used a mutant strain (*dsdC⁻*) in which this synthesis is more sensitive to catabolite repression than in wild-type strains (126). Apparently, this *dsdC* product does not activate transcription in the absence of D-serine as well as the wild-type *dsdC* product does in its presence. Consequently these mutants may be affected in their response to GS activation of the *dsdA* gene in the absence of D-serine, and it would be reasonable to compare regulation of D-serine deaminase production in wild-type cells and GlnC mutants containing wild-type *dsd* genes.

REGULATION OF SYNTHESIS OF GLUTAMINE SYNTHETASE

The Synthesis of GS is Subject to Autogenous Regulation

From the experiments presented above it is clear that the enzymatic and regulatory activities of GS are critical for the regulation of assimilation of nitrogen compounds. The question arises, how is the intracellular level of GS regulated?

As pointed out earlier, the absolute level of GS enzyme activity can vary 50- to 100-fold. Weglenski & Tyler showed (137) that this variation in GS protein reflects the intracellular concentration of *glnA*-mRNA; thus the regulation of GS synthesis occurs primarily at the level of transcription. Their experiments suggest that the promoter of the *glnA* gene is located at the *rha* proximal end of the gene.

Genetic and physiological experiments have examined the nature of this regulation. Three point crosses were used to order mutations in the *glnA* region of the chromosome (47, 69). The data suggest autogenous regulation; two mutations leading to the GlnC phenotype (constitutive formation of GS protein) lie between mutations that result in the production of enzymatically inactive GS protein (47, 72, 89). However, since these data are complicated by both positive and negative interference, they do not exclude the possibility that all the mutations resulting in the GlnC phenotype might be contiguous and lie in a neighboring gene that codes for a regulatory protein rather than the GS polypeptide. This possibility was tested (89, 138) by examining the dominance and complementation patterns of various *glnA*-linked mutations in *K. aerogenes*. The data show that the wild-type *glnA*$^+$ allele decreases expression of a mutation resulting in the GlnC character in haploid cells. This observation is predicted if the GlnC phenotype results from a loss of repressor function, and is not expected if the GlnC phenotype is due to an increase in a positive activator for *glnA* transcription. However, the data are not decisive on this point due to the possibility of subunit mixing (139). Regardless of whether the control is positive or negative the complementation pattern of mutations resulting in the GlnC or GlnA$^-$ phenotype is consistent with the view that the mutations are in the same gene (138). Finally, it was found that alterations in the enzymes of the GS adenylylation system affect the rate of production of GS in a manner inverse to their effects on the adenylylation state of the enzyme (54, 55, 69). From these studies Magasanik and co-workers suggested that adenylylated GS is the inhibitor of *glnA* transcription (54). According to this model (47, 54, 55, 68) adenylylation of GS results in a conformational change that allows the protein to function as a repressor. Consequently, the rate of GS synthesis is determined by the intracellular concentration of 2-oxoglutarate and glutamine which determine the adenylylation state of the GS protein. However, it is also proposed (68) that the two classes of *glnA*-linked regulatory mutations (GlnC and Gln(AC)⁻ affect the formation of GS for different reasons. In GlnC strains, which produce high levels of GS regardless of the growth conditions, the GS is frozen in the nonrepressing conformation, whatever the state of adenylylation. In Gln(AC)⁻ strains, where very low levels of GS are produced under certain growth conditions, the alterations in regulation of GS formation are due only to abnormally low levels of intracellular glutamine. These notions about the state of the GS protein could be tested directly in a cell-free transcription system.

IS *glnA* TRANSCRIPTION REGULATED BY POSITIVE OR NEGATIVE CONTROL?   Other aspects of this model have been examined in whole cells. Weglenski & Tyler (137) measured the level of *glnA*-specific mRNA in wild-type and mutant cells of *K. aerogenes* and found a positive correlation between the intracellular level of GS and the level of this RNA; they were unable to detect *glnA*-mRNA in strains devoid of GS protein. These results are not predicted by a model where transcription of *glnA* is regulated simply by repression mediated through the GS protein, although they do support the notion of autogenous control. They suggest that regulation of *glnA*-gene expression involves activation of transcription by GS but do not exclude the possibility that a negative control element also affects *glnA*-transcription. Recent studies (see below) have lead Gaillardin & Magasanik to propose (47) that GS serves as both a positive and negative effector of *glnA* transcription.

THE ROLE OF THE *glnF* GENE IN THE SYNTHESIS OF GS   Both genetic and physiological evidence indicate that some additional factor is involved in GS regulation of *glnA* transcription. This notion follows logically from the observations by Bloom, Streicher & Tyler (60) that another factor is required for GS activation of *hut*-gene transcription. In addition, experiments by Bartnik & Tyler (140) showed that the level of GS did not increase in proportion to the number of copies of the *glnA* gene in nitrogen-limited cultures that apparently contained only nonadenylylated GS; these observations also indicate that some factor other than GS limits *glnA* expression. The identity of this factor is not known but a likely candidate is the product of the *glnF* gene. Its function is not known but it appears to act through the GS protein. Gaillardin & Magasanik (47) have shown that the Gln⁻

phenotype of *glnF* mutants can be reversed by recessive mutations (called *glnA*[R]) which apparently are located within the *glnA* gene. From these data, the Gln[+] phenotype of these mutants cannot be due to a mutation in the promoter region of the *glnA* gene as has been suggested for revertants of *glnF* mutants of *Salmonella* (75). However the *glnA*[R] mutations strongly suggest that the *glnF* gene product plays a role in the autogenous regulation of GS synthesis.

In cultures grown on severely repressing medium, the level of GS was higher in GlnA[R] GlnF[−] strains than in the GlnA[+] GlnF[−] parent or wild-type cells (47). However, *glnA*[R] mutants, whether *glnF*[−] or *glnF*[+], were unable to derepress GS synthesis in response to nitrogen limitation. It is proposed that GS formation in these mutants represents a "basal" level that is independent of repression or activation. In addition it is suggested: (*a*) that any decrease in GS synthesis from this basal level is independent of the *glnF*-product and results from repression of *glnA*-transcription by GS, (*b*) that any increase in GS levels from basal level is attributable to *glnF* product–mediated activation of *glnA* transcription by nonadenylylated GS; and (*c*) that in the absence of the *glnF* product, GS always represses the transcription of the *glnA* gene. Gaillardin & Magasanik further suggest (47) that the interaction of GS with adenylyltransferase and/or P$_{II}$ may be necessary for the *glnF* product to interact with GS. This notion reconciles these studies on GS formation with those of Streicher & Tyler (90) on GS-DNA complexes and their affinity for the enzymes of the adenylylation system.

It has not been established that *glnF* mutants are altered in *glnA* transcription. It might be that the product of the *glnF* gene is not directly involved in regulating GS synthesis, but that it codes for an enzyme involved in maintaining some other regulatory element, for example a tRNA or ppGpp (see below). However, Bloom and Tyler (unpublished) have isolated a temperature-sensitive *glnF* mutant of *K. pneumoniae* in which GS synthesis decreases immediately upon shifting to the nonpermissive temperature. These results suggest that regulation of GS formation is directly affected by the *glnF* product rather than by some molecule resulting from its action. Clearly further experiments are needed to elucidate the role of the *glnF* gene in GS synthesis.

The Effect of Energy Limitation on the Synthesis of Glutamine Synthetase

GlnF mutants of *E. coli* K-12 but not of *K. aerogenes* or *K. pneumoniae* are conditional glutamine auxotrophs. In eight independent isolates, *E. coli* strains altered in *glnF* required glutamine to grow on glucose-minimal medium containing either excess ammonia or a growth-limiting source of nitrogen. However, the growth of these strains was independent of glutamine if the supply of carbon and/or energy is limited. The level of GS protein increased at least eightfold when *glnF⁻* mutants were subjected to energy limitation (G. Pahel, A. D. Zelenetz, and B. Tyler, unpublished). This "energy effect" is independent of the adenylylation system and autogenous regulation, since it was still observed in a strain carrying mutations in both the *glnF* and *glnD* genes. (The *glnD* mutation results in highly adenylylated GS under all growth conditions.) An analysis of the data of Senior (39) reveal this effect of energy starvation on the level of GS in chemostat-grown cultures of *E. coli* W. Thus it appears that in *E. coli, glnA* transcription has an additional level of regulation that is mediated by an "energy factor" the identity of which is not presently known.

One possible mechanism for stimulation of GS synthesis by energy limitation is through the CAP protein and cAMP. Prusiner, Miller & Valentine (43) examined the level of GS in a *cya⁻* strain of *E. coli* K-12 growing with excess ammonia in the presence or absence of exogenous cAMP. They observed a doubling of GS activity and suggested that GS is regulated by CAP and cAMP. This effect is small compared to that observed for more classical catabolite-sensitive operon (87). Since cAMP has a wide range of physiological effects on *E. coli* K-12 [e.g. changes in membrane functions such as active transport and salt sensitivity (141–144)] it is possible that the observations of Prusiner et al (43) are an indirect consequence of cAMP enhancing the energy effect through some factor other than the CAP protein. An examination of the energy effect in a strain lacking the CAP protein should resolve this point.

Two other modified nucleotides respond to conditions of energy downshift in *E. coli:* ppGpp (145, 146) and phantom spot (147). However, the changes in these nucleotides are somewhat transient so they are unlikely to play a role in the energy effect. Perhaps the best way to identify the "energy factor" is to obtain from GlnF⁻ strains, new mutants that have either lost the energy effect or increased this effect sufficiently to become Gln⁺ on glucose-ammonia medium.

Are Attenuation and/or ppGpp Involved in the Regulation of Glutamine Synthetase?

The regulation of a number of biosynthetic operons for example, *his* and *trp*, apparently involves the amino acyl-tRNA and/or the amino acyl tRNA synthetase that are specific for that operon. In addition ppGpp has been implicated in regulation of synthesis in a number of biosynthetic enzymes. Present models suggest that these factors regulate initiation of transcription of a leader sequence and/or rho-mediated termination of some mRNA chains at an attenuator site on the leader sequence proximal to the structural genes (148–151). Since GS is a biosynthetic enzyme it is reasonable to ask if attenuation and/or ppGpp are involved in the regulation of GS synthesis. The evidence thus far is inconclusive.

STUDIES OF MUTANTS ALTERED IN tRNA FORMATION   LaPointe and co-workers (44) have examined GS synthesis in an *E. coli* mutant containing a temperature-sensitive mutation in the structural gene for glutamyl-tRNA synthetase (*gltX*). Cells of this strain contain 50-fold more GS when grown at the somewhat nonpermissive temperature of 37°C. However, these effects may be indirect: the increased level of glutamate in the mutant grown at 37°C might alter the concentration of 2-oxoglutarate and/or glutamine which then affects the synthesis of GS. In addition the level of glutamate synthase is elevated in the mutant grown at the the higher temperature. The higher level of this enzyme might result in a depletion of the glutamine pool and the subsequent depression of GS synthesis. A number of other mutations that specifically affect the glutamyl-tRNA synthetase, *gltE* and *gltM* (152, 153) have been isolated and mapped. The synthesis of GS has not been examined in these mutants. Since *gltE* and *gltM* mutants are conditionally dependent on streptomycin, it is of interest that strains resistant to spectinomycin contain slightly higher levels of GS under all growth conditions (A. D. Zelenetz, G. Pahel, and B. Tyler, unpublished). A strain of *E. coli* altered in the glutaminyl-tRNA synthetase is not affected in the synthesis of GS (154). In *S. typhimurium*, *hisW* mutants which are altered in the synthesis of some tRNAs contain 50% less GDH than wild-type strains but are not affected in the regulation of GS synthesis (155). Thus there is no conclusive evidence for the involvement of tRNA in the regulation of GS.

A Working Model for the Regulation of Glutamine Synthetase Formation

The key aspects of a working model for the regulation of GS formation are as follows: (*a*) Regulation of GS formation occurs primarily at the level of transcription. (*b*) GS synthesis is subject to autogenous regulation. Activation of *glnA* transcription is effected by nonadenylylated GS through the action of the *glnF* product and, possibly, some of the enzymes of the adenylylation system. Repression of *glnA* transcription is effected by GS and does not require the *glnF* product. A more direct role of the adenylylation system enzymes in *glnA* repression is not ruled out. The identity of the *glnF* product is not known. (*c*) An energy factor acts independently of autogenous regulation to increase transcription of the *glnA* gene but its identity is not known. (*d*) Attenuation of *glnA* transcription influences the rate of GS formation. There is scant evidence for the involvement of attenuation in the regulation of *glnA* transcription but it is likely that such regulation occurs.

Several observations do not fit this model. Specifically, several experiments (39; F. Foor, personal communication) show little correlation between the adenylylation state and the level of GS, and indicate that the adenylylation state of the GS is not the only factor regulating the level of GS. Possibly the activity of the *glnF* product is subject to regulation. These observations also raise other questions. What is the adenylylation state of each GS molecule in a cell? Is *glnA* transcription regulated by the ratio or absolute level of two populations of nonadenylylated and completely adenylylated GS, or is the rate of transcription a function of the average adenylylation state of each GS molecule?

Deletion mapping of mutations in the *glnA*-region is needed to confirm the notion of autogenous regulation of GS synthesis. Further studies are needed to expand our knowledge of the energy effect and to allow an evaluation of the role of attenuation in GS formation. For these experiments the approach described by Brenchley and co-workers (45, 156) in which they selected mutants resistant to the glutamine analogue, methionine sulfoximine, may prove extremely valuable. In order to analyze fully the various types of regulation of GS it is important to know how the target sites for the effectors of *glnA* transcription are arranged in the controlling region of the gene. It is not known whether GS binds to the same site on the DNA when it functions as an activator and as a repressor. In fact, it is not known whether GS-mediated activation of transcription involves the direct interaction of GS with DNA. It might be that the *glnF* gene product binds to the *glnA* promoter and to GS in such a way that the GS molecule never actually interacts with DNA. Alternatively the *glnF* product may simply enhance the direct binding of GS to DNA. However, the question then arises whether the *glnF* product binds to DNA or only to GS. In addition, it is not known whether the energy factor is acting at the same site(s) as the GS. There is not yet any evidence for the existence of a leader region or attenuator site in the *glnA* gene. Clearly some knowledge of the relative physical positions of these control sites is necessary for a complete understanding of the regulation of *glnA* transcription. To get this information it will be necessary to select many more mutants near the structural gene for GS. Finally, experiments dealing with the binding of GS to DNA and with the regulation of *glnA* transcription in a cell-free system should provide the most definitive test of the working model presented above.

CONCLUDING REMARKS

Clearly there is now considerable information on the regulation of assimilation of nitrogen compounds in the enteric bacteria. Much of this information originates from the proposal that GS activates the formation of enzymes involved in nitrogen metabolism. However, many questions remain. What is the mechanism of regulation of *glnA* transcription and the nature of the GS transcription factor? Is the *glnF* product, which is necessary for GS-mediated activation of *glnA,* also involved in GS activation of transcription of other genes? Does GS regulate the assimilation of nitrogen compounds other than amino acids? There is some indication, for example, (165) that pyrimidine catabolism in *E. coli* may be affected by GS. It is obviously important to demonstrate directly, in a cell-free transcription system, the role of GS and GS transcription factors on the expression of other genes.

There remain other general questions about nitrogen metabolism in the enteric bacteria. For example, are there factors other than GS that serve as general control elements for nitrogen metabolism? Broach and co-workers (166) have described mutants of *S. typhimurium* affected in nitrogen metabolism due to alterations in a gene for which the function is not known. The authors suggest that these mutants are affected in an ammonia transport system, but other explanations are quite possible. It is also important to understand more about the regulation of GOGAT formation.

Finally the question arises whether GS regulates gene expression in organisms other than the enteric bacteria. One report implicates GS in the regulation of nitrogen fixation in *Rhizobium* (167). It appears (168) that the synthesis of GS in *Bacillus subtilis* may be subject to autogenous regulation since mutations in the structural gene for GS alter the production of the enzyme. This possibility is particularly intriguing since the GS of this organism differs significantly from that of the enteric bacteria; the enzymatic activity is not regulated by adenylylation but may be affected by sulfhydryl group modification (169). In addition, some studies (170; A. L. Sonenshein, personal communication) suggest that GS may be involved in the regulation of sporulation which results from nitrogen limitation. Thus GS may regulate transcription of genes involved in nitrogen metabolism in a wide variety of bacteria.

It is not yet clear if there is any general control of nitrogen assimilation in any organisms other than the enteric bacteria. If future experiments show that such regulation exists, it will be of interest to determine the nature of the general control elements involved in the assimilation of nitrogen compounds.

ACKNOWLEDGMENTS

I am grateful to Forrest Foor, Gregory Pahel, Stanley Streicher, and Boris Magasanik for many fruitful discussions; and to Forrest Foor and Marjorie Brandriss for their critical comments on the original manuscript of this review. I would like to thank Hilda Harris for her expert secretarial help, without which this review could not have been written. Original research from the author's laboratory was funded by Public Health Service Research Grant GM22527 from the National Institute of General Medical Sciences.

Literature Cited

1. Stadtman, E. R., Ginsburg, A., Ciardi, J. E., Yeh, J., Hennig, S. B., Shapiro, B. M. 1970. *Adv. Enzyme Regul.* 8:99–118
2. Umbarger, H. E. 1973. *Adv. Enzymol.* 37:349–95
3. Wheelis, M. L. 1975. *Ann. Rev. Microbiol.* 29:505–24
4. Tempest, D. W., Meers, J. L., Brown, C. M. 1973. In *The Enzymes of Glutamine Metabolism,* ed. S. Prusiner, E. R. Stadtman, pp. 167–82. New York: Academic. 615 pp.
5. Meister, A. 1968. *Adv. Enzymol.* 31: 183–218
6. Goldin, B. R., Frieden, C. 1971. *Curr. Top. Cell Regul.* 4:77–117
7. Hillar, M. 1974. *Bioenergetics* 6:89–124
8. Magasanik, B. 1976. *Prog. Nucleic Acid Res.* 17:99–115
9. Magasanik, B. 1977. *Trends Biochem. Sci.* 2:9–12
10. Sakamoto, N., Kotre, A. M., Savageau, M. A. 1975. *J. Bacteriol.* 124:775–83
11. Coulton, J. W., Kapoor, M. 1973. *Can. J. Microbiol.* 19:427–38
12. Coulton, J. W., Kapoor, M. 1973. *Can. J. Microbiol.* 19:439–50
13. Mantsala, P., Zalkin, H. 1976. *J. Biol. Chem.* 251:3300–5
14. Miller, E. R., Stadtman, E. R. 1972. *J. Biol. Chem.* 247:7409–19
15. Halpern, Y. S., Umbarger, H. E. 1960. *J. Bacteriol.* 80:285–88
16. Vender, J., Rickenberg, H. V. 1964. *Biochim. Biophys. Acta* 90:218–20
17. Varricchio, F. 1969. *Biochim. Biophys. Acta* 177:560–64
18. Brenchley, J. E., Baker, C. A., Patil, L. G. 1975. *J. Bacteriol.* 124:182–89
19. Meers, J. L., Tempest, D. W., Brown, C. M. 1970. *J. Gen. Microbiol.* 64: 187–94
20. Kondorosi, A., Svab, Z., Kiss, G. B., Dixon, R. A. 1977. *Mol. Gen. Genet.* 151:221–26
21. Shanmugan, K. T., Chan, I., Morandi, C. 1975. *Biochim. Biophys. Acta* 408: 101–11
22. Mecke, D., Holzer, H. 1966. *Biochim. Biophys. Acta* 122:341–51
23. Berberich, M. A. 1972. *Biochem. Biophys. Res. Commun.* 47:1498–1503
24. Brenchley, J. E., Magasanik, B. 1974. *J. Bacteriol.* 117:544–50
25. Patil, L., Kuchta, J., Brenchley, J. E. 1977. *Ann. Meet. Am. Soc. Microbiol.* 77th, Abstr. K91, p. 201
26. Foor, F., Janssen, K. A., Streicher, S. L., Magasanik, B. 1975. *Fed. Proc.* 34:514 (Abstr.)
27. Pahel, G., Zelenetz, A. D., Tyler, B. 1977. *J. Bacteriol.* In press
28. Tempest, D. W., Meers, J. L., Brown, C. M., 1970. *Biochem. J.* 117:405–07
29. Nagatani, H., Shimizu, M., Valentine, R. C. 1971. *Arch. Mikrobiol.* 79:164–75
30. Brown, R., Burn, S., Johnson, J. 1973. *Nature New Biol.* 246:115–16
31. Roon, R. J., Even, H. L., Larimore, F. 1974. *J. Bacteriol.* 118:89–95
32. Freeze, E., Park, S. W., Cashel, M. 1964. *Proc. Natl. Acad. Sci. USA* 51:1164–72
33. Hong, M. M., Shen, S. C., Braunstein, A. E. 1959. *Biochim. Biophys. Acta* 36:288–89
34. Mantsala, P., Zalkin, H. 1976. *J. Biol. Chem.* 251:3294–99
35. Geary, L. E., Meister, A. 1977. *J. Biol. Chem.* 252:3501–8
36. Trotta, P. P., Platzer, K. E. B., Haschemeyer, R. H., Meister, A. 1974. *Proc. Natl. Acad. Sci. USA* 71:4607–11
37. Mantsala, P., Zalkin, H. 1976. *J. Bacteriol.* 126:539–41
38. Brenchley, J. E., Prival, M. J., Magasanik, B. 1973. *J. Biol. Chem.* 248: 6122–28
39. Senior, P. J. 1975. *J. Bacteriol.* 123: 407–18
40. Woolfolk, C. A., Stadtman, E. R. 1964. *Biochem. Biophys. Res. Commun.* 17: 313–19
41. Halpern, Y. S., Lupo, M. 1965. *J. Bacteriol.* 90:1288–95
42. Marcus, M., Halpern, Y. S. 1969. *J. Bacteriol.* 97:1118–28
43. Prusiner, S., Miller, R. E., Valentine, R. C. 1972. *Proc. Natl. Acad. Sci. USA* 69: 2922–26
44. LaPointe, J., Delcuve, G., Duplain, L. 1975. *J. Bacteriol.* 123:843–50
45. Steimer-Veale, K., Brenchley, J. E. 1974. *J. Bacteriol.* 119:848–56

46. Reiner, A. M. 1969. *J. Bacteriol.* 97: 1431–36
47. Gaillardin, C. M., Magasanik, B. 1978. *J. Bacteriol.* 133: In press
48. Streicher, S. L., Bloom, F. R., Foor, F., Levin, M., Tyler, B. 1977. See Ref. 25, K88, p. 200
49. Ginsburg, A., Stadtman, E. R. 1973. In *The Enzymes of Glutamine Metabolism*, ed. S. Prusiner, E. R. Stadtman, pp. 9–44. New York: Academic Press 615 pp.
50. Wohlhueter, R. M., Schutt, H., Holzer, H. 1973. See Ref. 4, pp.45–61
51. Adler, S. P., Purich, D., Stadtman, E. R. 1975. *J. Biol. Chem.* 250:6264–72
52. Segal, A., Brown, M. S., Stadtman, E. R. 1974. *Arch. Biochem. Biophys.* 161:319–27
53. Tronick, S. R., Ciardi, J. E., Stadtman, E. R. 1973. *J. Bacteriol.* 115:858–68
54. Foor, F., Janssen, K. A., Magasanik, B. 1975. *Proc. Natl. Acad. Sci. USA* 72:4844–48
55. Janssen, K. A., Magasanik, B. 1977. *J. Bacteriol.* 129:993–1000
56. Valentine, R. C., Shapiro, B. M., Stadtman, E. R. 1968. *Biochemistry* 7: 2143–52
57. Eisenberg, D., Burton, Z., Blumenberg, M., Magasanik, B., Streicher, S. L., Tyler, B. 1976. In *Molecular Mechanisms in the Control of Gene Expression*, ed. D. P. Nierlich, W. J. Rutter, pp. 171–76. New York:Academic
58. Bender, R. A., Janssen, K. A., Resnick, A. D., Blumenberg, M., Foor, F., Magasanik, B. 1977. *J. Bacteriol.* 129: 1001–9
59. Bender, R. A., Streicher, S. L. 1976. *Ann. Meet. Am. Soc. Microbiol.* 76th K144:160 (Abstr.)
60. Bloom, F. R., Streicher, S. L., Tyler, B. 1977. *J. Bacteriol.* 130:983–90
61. Balakrishnan, M. S., Villafranca, J. J., Brenchley, J. E. 1977. *Arch. Biochem. Biophys.* 181:603–15
62. Villafranca, J. J., Ash, D. E., Wedler, F. C. 1976. *Biochemistry* 15:544–53
63. Holzer, H., Schutt, H., Masek, Z., Mecke, D. 1968. *Proc. Natl. Acad. Sci. USA* 60:721–24
64. Prival, M. J., Brenchley, J. E., Magasanik, B. 1973. *J. Biol. Chem.* 248: 4334–44
65. Goldberg, R. B., Bloom, F. R., Magasanik, B. 1976. *J. Bacteriol.* 127:114–19
66. Woolfolk, C. A., Shapiro, B., Stadtman, E. R. 1966. *Arch. Biochem. Biophys.* 116:177–92
67. Shanmugan, K. T., Morandi, C. 1976. *Biochim. Biophys. Acta* 437:322–32

68. Bender, R. A., Magasanik, B. 1977. *J. Bacteriol.* 132:100–5
69. Streicher, S. L., Bender, R. A., Magasanik, B. 1975. *J. Bacteriol.* 121:320–31
70. Mayer, E. P., Smith, O. H., Fredricks, W. W., McKinney, M. A. 1975. *Molec. Gen. Genet.* 137:131–42
71. Bachmann, B. J., Low, A. B., Taylor, A. L. 1976. *Bacteriol. Rev.* 40:116–67
72. DeLeo, A. B., Magasanik, B. 1975. *J. Bacteriol.* 121:313–19
73. Kustu, S. G., McKereghan, K. 1975. *J. Bacteriol.* 122:1006–16
74. Funange, V. L., Brenchley, J. E. 1977. *Genetics* 86:513–26
75. Garcia, E., Bancroft, S., Rhee, S. G., Kustu, S. 1977. *Proc. Natl. Acad. Sci. USA* 74:1662–66
76. Tubb, R. S. 1974. *Nature* 251:481–84
77. Jacob, F., Monod, J. 1961. *J. Mol. Biol.* 3:318–56
78. Gerson, S. L., Magasanik, B. 1975. *J. Bacteriol.* 124:1269–75
79. Goldberg, R., Magasanik, B. 1975. *J. Bacteriol.* 122:1025–31
80. Parada, J. L., Magasanik, B. 1975. *J. Bacteriol.* 124:1263–68
81. Prival, M. J., Magasanik, B. 1971. *J. Biol. Chem.* 246:6288–96
82. Tyler, B., DeLeo, A. B., Magasanik, B. 1974. *Proc. Natl. Acad. Sci. USA* 71: 225–29
83. Magasanik, B., Lund, P., Neidhardt, F. C., Schwartz, D. T. 1965. *J. Biol. Chem.* 240:4320–24
84. Magasanik, B., Bowser, H. R. 1955. *J. Biol. Chem.* 213:571–80
85. Revel, H. R. B., Magasanik, B. 1958. *J. Biol. Chem.* 233:930–35
86. Lund, P., Magasanik, B. 1965. *J. Biol. Chem.* 240:4316–19
87. Pastan, I., Adhya, S. 1976. *Bacteriol. Rev.* 40:527–51
88. Neidhardt, F. C., Magasanik, B. 1957. *J. Bacteriol.* 73:253–59
89. Streicher, S. L., DeLeo, A. B., Magasanik, B. 1976. *J. Bacteriol.* 127:184–92
90. Streicher, S. L., Tyler, B. 1976. *Fed. Proc.* 35:1471 (Abstr.)
91. Gross, C., Engbaek, F., Flammang, T., Burgess, R. 1976. *J. Bacteriol.* 128: 382–89
92. Nissley, P., Anderson, W., Gallo, M., Pastan, I., Perlman, R. 1972. *J. Biol. Chem.* 247:4264–69
93. Riggs, A., Reines, G., Zubay, G. 1971. *Proc. Natl. Acad. Sci. USA* 68:1222–25
94. Majors, J. 1975. *Nature* 256:672–74
95. Friedrich, B., Magasanik, B. 1978. *J. Bacteriol.* 133:680–85
. 96. Friedrich, B., Magasanik, B. 1977. *J. Bacteriol.* 131:446–52

97. Resnick, A. D., Magasanik, B. 1976. *J. Biol. Chem.* 251:2722–28
98. Paris, C. G. 1977. *Studies on the metabolism of tryptophan in K. aerogenes: enzymology and regulation.* PhD thesis. MIT, Boston, 180 pp.
99. Streicher, S. L., Shanmugan, K. T., Ausubel, F., Morandi, C., Goldberg, R. B. 1974. *J. Bacteriol.* 120:815–21
100. Bender, R. A., Macaluso, A., Magasanik, B. 1976. *J. Bacteriol.* 127:141–48
101. Lis, J. T., Schlief, R. 1973. *J. Mol. Biol.* 79:149–62
102. Brill, W., Magasanik, B. 1969. *J. Biol. Chem.* 244:5392–402
103. Dixon, R. A., Postgate, J. R. 1972. *Nature* 237:102–3
104. Cannon, F. C., Dixon, R. A., Postgate, J. R. 1976. *J. Gen. Microbiol.* 93:111–25
105. Newman, B. M., Cole, J. A. 1977. *J. Gen. Microbiol.* 98:369–77
106. Willis, R. C., Iwata, K. K., Furlong, C. E. 1975. *J. Bacteriol.* 122:1032–37
107. Gutnick, D., Calbo, J. M., Klopotowski, T., Ames, B. N. 1969. *J. Bacteriol.* 100:215–19
108. Dendinger, S., Brill, W. J. 1970. *J. Bacteriol.* 103:144–52
109. Smith, G. R., Magasanik, B. 1971. *J. Biol. Chem.* 246:3330–41
110. Meiss, H. K., Brill, W. J., Magasanik, B. 1969. *J. Biol. Chem.* 144:5382–91
111. Smith, G. R., Halpern, Y. S., Magasanik, B. 1971. *J. Biol. Chem.* 246:3320–29
112. Smith, G. R., Magasanik, B. 1971. *Proc. Natl. Acad. Sci. USA* 68:1493–97
113. Hagen, D. C., Magasanik, B. 1973. *Proc. Natl. Acad. Sci. USA* 70:808–12
114. Postgate, J. R., Krishnapillai, V. 1977. *J. Gen. Microbiol.* 98:379–85
115. Betteridge, P. R., Ayling, P. D. 1976. *J. Gen. Microbiol.* 95:324–34
116. Betteridge, P. R., Ayling, P. D. 1975. *Molec. Gen. Genet.* 138:41–52
117. Cedar, H., Schwartz, J. H. 1968. *J. Bacteriol.* 96:2043–48
118. Wriston, J. C., Yellin, T. O. 1973. *Adv. Enzymol.* 39:185–248
119. Marcus, M., Halpern, Y. S. 1969. *Biochim. Biophys. Acta* 177:314–20
120. Gelfand, D. H., Steinberg, R. A. 1977. *J. Bacteriol.* 130:429–40
121. Prusiner, S. 1973. See Ref. 4, pp. 293–316
122. Hartman, S. C. 1973. See Ref. 4, pp. 319–30
123. Prusiner, S. 1975. *J. Bacteriol.* 123:992–99
124. Varricchio, F. 1972. *Arch. Mikrobiol.* 81:234–38

125. Bloom, F. R., McFall, E., Young, M. C., Carothers, A. M. 1975. *J. Bacteriol.* 121:1092–1101
126. Bloom, F. R., McFall, E. 1975. *J. Bacteriol.* 121:1078–84
127. McFall, E. 1973. *J. Bacteriol.* 113:781–85
128. Gartner, T. K., Riley, M. 1965. *J. Bacteriol.* 89:313–18
129. Gartner, T. K., Riley, M. 1965. *J. Bacteriol.* 89:319–25
130. Botsford, J. L. 1975. *J. Bacteriol.* 124:380–90
131. Pardee, A. B., Prestidge, L. S. 1955. *J. Bacteriol.* 70:667–74
132. Isenberg, S., Newman, E. B. 1974. *J. Bacteriol.* 118:53–58
133. Newman, E. B., Batist, G., Fraser, J., Isenberg, S., Weyman, P., Kapoor, V. 1976. *Biochim. Biophys. Acta.* 421:97–105
134. Shizuta, Y., Hayaishi, O. 1970. *J. Biol. Chem.* 245:5416–23
135. Newman, E. B., Kapoor, V., Potter, R. 1976. *J. Bacteriol.* 126:1245–49
136. Potter, R., Kapoor, V., Newman, E. B. 1977. *J. Bacteriol.* 132:385–91
137. Weglenski, P., Tyler, B. 1977. *J. Bacteriol.* 129:880–87
138. Bender, R. A., Magasanik, B. 1977. *J. Bacteriol.* 132:106–12
139. Beckwith, J., Rossow, P. 1974. *Ann. Rev. Genet.* 8:1–13
140. Bartnik, E., Tyler, B. 1977. See Ref. 25, K90, p. 201
141. Hempfling, W. P., Beeman, D. K. 1971. *Biochem. Biophys. Res. Commun.* 45:929–30
142. Ezzell, J. W., Dobrogosz, W. J. 1975. *J. Bacteriol.* 124:815–24
143. Kumar, S. 1976. *J. Bacteriol.* 125:545–55
144. Dills, S. S., Dobrogosz, W. J. 1977. *J. Bacteriol.* 131:854–65
145. Hansen, M., Pato, M. L., Molin, S., Fiil, N., von Meyenburg, K. 1975. *J. Bacteriol.* 122:585–91
146. Braedt, G., Gallant, J. 1977. *J. Bacteriol.* 129:564–66
147. Gallant, J., Shell, L., Bittner, R. 1976. *Cell* 7:75–84
148. Bertrand, K., Korn, L., Lee, F., Platt, T., Squires, C. L., Squires, C., Yanofsky, C. 1975. *Science* 189:22–26
149. Artz, S. W., Broach, J. R. 1975. *Proc. Natl. Acad. Sci. USA* 72:3453–57
150. Stephens, J. C., Artz, S. W., Ames, B. N. 1975. *Proc. Natl. Acad. Sci. USA* 72:4389–93
151. Smith, J. M., Smolin, D. E., Umbarger, H. E. 1976. *Molec. Gen. Genet.* 148:111–24

152. Murgola, E. J., Adelberg, E. A. 1970. *J. Bacteriol.* 103:20–26
153. Murgola, E. J., Adelberg, E. A. 1970. *J. Bacteriol.* 103:178–83
154. Korner, A., Magee, B. B., Liska, B., Low, K. B., Adelberg, E. A., Soll, D. 1974. *J. Bacteriol.* 120:154–58
155. Puskas, R. S., Reid, T. J. III, Brenchley, J. E. 1977. See Ref. 25, K86, p. 200
156. Brenchley, J. 1973. *J. Bacteriol.* 114:666–73
157. Brenchley, J. E., Patil, L. G. 1976. See Ref. 59, K147, p. 161
158. O'Donovan, G. A., Neuhard, J. 1970. *Bacteriol. Revs.* 34:278–343
159. Wang, T. P., Sable, H. Z., Lampen, J. O. 1950. *J. Biol. Chem.* 184:17–28
160. Dover, S., Halpern, Y. S. 1972. *J. Bacteriol.* 109:835–48
161. Dover, S., Halpern, Y. S. 1972. *J. Bacteriol.* 110:165–70
162. Dover, S., Halpern, Y. S. 1974. *J. Bacteriol.* 117:494–501
163. Prejko, R. A., Wilson, P. W. 1970. *Can. J. Microbiol.* 16:681–85
164. Streicher, S. L., Gurney, E. G., Valentine, R. C. 1972. *Nature* 239:495–99
165. Ban, J., Vitale, L., Kos, E. 1972. *J. Gen. Microbiol.* 73:267–72
166. Broach, J., Neumann, C., Kustu, S. 1976. *J. Bacteriol.* 128:86–98
167. Ludwig, R. A., Signer, E. R. 1977. *Nature* 267:245–48
168. Dean, D. R., Hoch, J. A., Aronson, A. I. 1977. *J. Bacteriol.* 131:981–85
169. Deuel, T. F. 1971. *J. Biol. Chem.* 246:599–605
170. Reysset, G., Aubert, J.-P. 1975. *Biochem. Biophys. Res. Commun.* 65:1237–41

Ann. Rev. Biochem. 1978. 47:1163–91
Copyright © 1978 by Annual Reviews Inc. All rights reserved

DNA REPLICATION PROTEINS ❖999
OF *ESCHERICHIA COLI*

Sue Hengren Wickner

Laboratory of Molecular Biology, National Cancer Institute, National Institutes of Health, Bethesda, Maryland 20014

CONTENTS

PERSPECTIVES AND SUMMARY

Implicit in the double-helical structure of DNA proposed by Watson & Crick was the means for its replication: each strand could serve as a template for the synthesis of a complementary strand (1). However, the molecu-

0066-4154/78/0701-1163$01.00

lar mechanisms by which DNA is replicated have been elusive. One reason for this elusiveness is that the process involves many more proteins than initially imagined. Biochemical studies of replication have been difficult not only due to the complex nature of the process, but also due to the absence of a stable DNA replication structure analogous to the ribosome for protein synthesis. Attempts to purify intact DNA replication forks with their associated proteins have thus far failed. On the other hand, much progress has been made in isolating individual protein components, characterizing these proteins, reconstituting DNA replicating systems, and elucidating the mechanisms by which they catalyze DNA replication.

The isolation of DNA replication proteins has been approached in three general ways. One method has been to isolate enzymatic activities that are logically involved in replication, such as DNA polymerases, DNA ligases, and DNA binding proteins. To demonstrate the role in vivo for a protein purified in this way, a mutant that has an altered protein must be isolated and the physiological effects of the defect studied. A second method has been to isolate essential DNA replication proteins by in vitro complementation assays. If a particular gene product is essential for a DNA synthesizing reaction, then extracts prepared from cells carrying a thermolabile defect in that gene will be inactive for DNA synthesis at high temperature. The corresponding wild-type DNA replication protein can be purified from wild-type cells by isolating the component that is able to restore DNA synthetic activity to the mutant extract at high temperature. Presumably the most heat labile protein in the extract is the temperature-sensitive DNA replication protein, and protein that restores activity is the product of the wild-type allele. Proof that the correct protein has been purified is obtained by purifying the protein from the temperature-sensitive mutant and demonstrating that it is more thermolabile than the wild-type protein for activity in vitro. While this method allows the isolation of the products of genes known to be important in replication, proteins whose genes have not yet been identified by mutation cannot be studied by this method. A third method has been to fractionate crude extracts and reconstitute multicomponent DNA replication pathways with purified protein fractions. This method allows the isolation of proteins that are not defined by temperature-sensitive DNA replication mutants and whose enzymatic activities are unknown. The combined use of these three approaches has allowed the isolation of many DNA replication proteins and the demonstration of their biological roles.

Various aspects of the biochemistry and genetics of procaryotic DNA replication have been reviewed in recent years (2–9). This review summarizes some multicomponent DNA replicating pathways that use defined DNA templates, and some of the proteins shown to be required for *Es-*

cherichia coli chromosome replication in vivo. The systems that will be emphasized are those that replicate small single-stranded and double-stranded circular DNA molecules, including fd (M13), G4, ST-1, ϕX174 single-stranded phage DNA, ColE1 double-stranded plasmid DNA, and ϕX174 RF I[1] double-stranded phage DNA. These systems replicate DNA by several enzymatic pathways; they differ in the mechanisms by which they initiate and elongate DNA chains. This review does not summarize other procaryotic in vitro DNA replicating systems such as T4 (10–14), T7 (15–26e), λ (27–29), and M13 RF (30, 31) (for review see 8).

ESCHERICHIA COLI CHROMOSOME REPLICATION

In Vivo DNA Replication

The process of *E. coli* DNA replication has been characterized extensively in vivo (reviewed in 2–5). In brief, the *E. coli* chromosome is a double-stranded circular DNA molecule of molecular weight 2.5×10^9. Replication starts at a unique site on the genome and proceeds bidirectionally by a semiconservative mechanism. One and perhaps both strands are synthesized discontinuously; the replication intermediates are about 1000 nucleotides long and are referred to as (10S) Okazaki fragments. Okazaki and co-workers have presented evidence that there are rNMP residues covalently attached to the 5' ends of the fragments (32–36), but this work has not yet been confirmed (37). In addition, recent studies suggest that some of the newly synthesized fragments do not result from de novo initiation. Mutants that have reduced levels of dUTPase accumulate short (4S) Okazaki-like fragments. In these mutants, dUMP is first misincorporated into DNA and then excised by a repair mechanism (38, 39). The transient nicks introduced during repair are not easily distinguished from those due to discontinuous synthesis. It is not yet known what percentage of the newly synthesized fragments seen in the presence of normal levels of dUTPase are the result of repair of misincorporated bases.

Genetic studies of *E. coli* have defined many functions essential for chromosome replication (reviewed in 9, 40, 41). These have been identified by the isolation of mutants thermosensitive or cryosensitive for DNA replication and growth. These mutants are phenotypically of two classes. Mutants blocked in the initiation, or termination, of chromosome replication do not cease DNA synthesis immediately upon shift to the nonpermissive temperature, but finish replicating the chromosomes already initiated at the time of the temperature shift. Genes whose functions are required only for initiation of rounds include *dnaA, dnaI,* and *dnaP* (9). Mutants blocked in

---

[1]ϕX174 RF I is double-stranded covalently closed circular DNA.

chromosome elongation cease DNA synthesis immediately upon shift to the nonpermissive temperature; elongation genes include *dnaB, dnaC(D),*[2] *dnaG, dnaZ,* and *polC* (initially referred to as *dnaE*) (9). While some *dnaB* and *dnaC(D)* mutants are blocked in elongation (40, 42, 45), other *dnaB* and *dnaC(D)* mutants have the properties of initiation mutants (46, 47). The *dnaG, dnaZ,* and *polC* elongation gene products may or may not also be required for chromosome initiation. Since the process of chromosome elongation involves Okazaki fragment synthesis, some of the DNA elongation functions may be involved in the de novo initiation of these fragments. The *lig* and *polA*ex genes are required for the joining of Okazaki fragments (9).

Inhibitors of DNA synthesis have defined other proteins involved in replication (reviewed in 48). Two antibiotics, chloramphenicol and rifampin, inhibit initiation of chromosomal DNA synthesis, but not elongation, which indicates that protein synthesis is required (49, 50). Rifampin inhibits initiation of DNA synthesis in synchronized cultures at a time when initiation is no longer sensitive to chloramphenicol, which suggests that transcription itself is required for initiation (51, 52). In addition, genetic evidence suggests that RNA polymerase and the *dnaA* gene product interact (53). Novobiocin (and its analogue, coumermycin) and nalidixic acid (and its analogue, oxolinic acid) inhibit DNA elongation in *E. coli* (48). The target proteins for these drugs are two protein components of DNA gyrase (discussed on page 1183; 54–56). Arabinosyl nucleotides inhibit DNA synthesis, presumably through their effect on DNA polymerase III (48). Edeine also inhibits DNA synthesis; however, the target protein for this drug has not yet been identified (48).

Genetic and drug inhibitor studies have shown that a minimun of 13 functions are required for *E. coli* DNA replication. These genes and the proteins coded by them are summarized in Table 1 and are discussed in this review. One reason for the probable incompleteness of this list may be that the *dna* mutants have been selected for their inability to synthesize DNA at the nonpermissive temperature while still being able to synthesize RNA and protein. Thus, proteins involved in cellular processes other than DNA synthesis have not been identified as DNA replication gene functions.

In Vitro DNA Replication

E. coli cells made permeable to small molecules have been useful for studying replication in vitro. Cells treated with ether, toluene, or sucrose continue propagating the replication forks initiated in vivo (reviewed in 3, 4). DNA synthesis in these systems is semiconservative, discontinuous, and

---

[2]*dnaC* and *dnaD* genes were initially thought to be distinct (42). They have since been shown genetically (43) and biochemically (44) to be identical.

Table 1 *E. coli* functions required for DNA replication[a]

Locus	Characterization of protein	References
Required for initiation of chromosome replication		
dnaA	?	
dnaB	~250,000 dalton active protein; 48,000 dalton subunit; ~10 molecules/cell; rNTPase stimulated by single-stranded DNA	64–67
dnaC(D)	~25,000 dalton active protein; interacts physically and functionally with *dnaB* protein	44, 68
dnaI	?	
dnaP	?	
rpoB	RNA polymerase β subunit	
Required for chromosome elongation		
dnaB	see above	
dnaC(D)	see above	
dnaG	~65,000 dalton active protein; priming protein which synthesizes ribo- deoxy- and mixed oligonucleotides; ~10 molecules/cell	69–72
dnaZ[b]	~125,000 dalton active protein; interacts with DNA EF III; functions in DNA elongation with DNA EF I, DNA EF III, and DNA polymerase III	74, 75
polC (*dnaE*)	DNA polymerase III; ~180,000 dalton active protein; ~10 molecules/cell	76–80
lig	DNA ligase	81
*polA*ex	5′-3′ exonuclease of DNA polymerase I	82
cou	component of DNA gyrase sensitive to novobiocin and coumermycin	54
nalA	component of DNA gyrase sensitive to nalidixic acid and oxolinic acid	55, 56
rep[c]	~67,000 dalton active protein; involved in ϕX174 RFI replication	83, 84
Required for in vitro DNA replication		
?	DNA binding protein; ~80,000 dalton active protein; ~18,500 dalton subunit; binds to single-stranded DNA, not to double-stranded DNA	2–5, 85
?	DNA elongation factor I; copolymerase III*; β component of DNA polymerase III holoenzyme; involved in DNA elongation of primed single-stranded DNA	74, 75, 86
?	DNA elongation factor III; γ or δ component of DNA polymerase III holoenzyme; ~65,000 dalton active protein; involved in DNA elongation of primed single-stranded DNA; forms a complex with dnaZ protein	74, 75, 86
?	Replication factors X, Y, Z; factors i, n; required for the transfer of dnaB protein to ϕX174 DNA covered with DNA binding protein in conjunction with dnaC protein; factors Y and Z bind to single-stranded DNA; factor Y is a single-stranded DNA-dependent ATPase (dATPase); native factor Y ~55,000 daltons; native factor X ~45,000 daltons	87, 88, 89

[a] The original *dnaH* mutation (57) is actually two mutations, one resulting in an unusually high thymine requirement and the other being a *dnaA* allele (58). The *dnaS* mutation (38) is a defect in dUTPase (39) and the *dnaF* mutation is a defect in ribonucleotide reductase (59); these two genes are no longer referred to as *dna* genes. Other loci have recently been described that may define new functions required for *E. coli* replication: *dnaJ* and *dnaK* (60–62); *dnaL* and *dnaM* (63).

[b] The *dnaZ* mutation was originally referred to as *dnaH* (73).

[c] The *rep* gene product is required for phage ϕX174 growth. It is not an essential function for *E. coli* growth, although *rep* cells contain more growing forks per chromosome than wild-type cells (164, 165).

ATP-dependent. It is inhibited by drugs that inhibit DNA synthesis in vivo: nalidixic acid, oxolinic acid, novobiocin, coumermycin, edeine, and arabinosyl nucleotides. The *dnaB, dnaG, dnaZ,* and *polC* gene products are required. Synthesis is not inhibited by rifampicin and *dnaA* is not required, presumably because permeabilized cells do not initiate chromosome replication. Permeabilized cell systems have limited application to the biochemical study of replication since the cells are not fully permeable to macromolecules and the endogenous DNA template used is complex and undefined.

Gentle lysates of concentrated cells on cellophane membranes replicate chromosomal DNA; synthesis by this system has properties identical to the permeabilized cell system (reviewed in 3, 4). The advantage of this system over the permeable cell systems is that the effects of macromolecules on replication can be studied. The stimulation of DNA synthesis of lysates of *dna* temperature-sensitive cells by protein fractions from wild-type cells has provided complementation assays for the purification of *polC* and *dnaG* gene products (78, 90). Extracts prepared similarly from phage-infected or plasmid-carrying cells replicate the phage and plasmid DNA. Extracts of uninfected cells convert exogenously added ϕX174 DNA to duplex DNA (91). For the study of *E. coli* chromosome replication, the lysed cell system retains the disadvantage of utilizing an endogenous DNA template and not initiating rounds of replication. Two soluble protein systems that replicate exogenously added *E. coli* DNA in the form of folded chromosomes have been described (92, 93), but the biochemistry has not been extensively studied.

Soluble membrane-free systems have been developed that replicate exogenously added phage and plasmid DNA (3, 4, 94). In general, these systems are prepared by gently lysing concentrated cell suspensions and removing cellular DNA and membranes by centrifugation at high speed. Extracts made in this way have been used to replicate single-stranded phage DNA [fd (M13), ϕX174, and G4 (ST-1)], double-stranded phage DNA (ϕX174 RF I and T7) and plasmid DNA (ColE1 and R100). These DNA molecules are replicated in *E. coli* and in vitro by several enzymatic pathways using various combinations of host proteins and phage proteins. Some of the gene functions required for the in vivo and in vitro replication of several phage and plasmid DNAs are summarized in Table 2.

SINGLE-STRANDED CIRCULAR DNA-DEPENDENT DNA SYNTHESIS

The membrane-free systems that convert single-stranded circular phage DNA to double-stranded DNA have been studied most extensively. Their main advantage is that some steps in DNA replication can be studied

Table 2 Protein requirements of some procaryotic DNA replicating systems

	E. coli in vivo	fd DNA in vitro	G4 DNA in vitro	φX174 DNA in vitro	φX174 RF I DNA in vitro	ColE1 DNA in vitro	λ phage in vivo	T7 phage in vivo	T4 phage in vivo
E. coli functions									
dnaA	+[a]	−[b]	−	−	nt[c]	−	−	−	−
dnaI, dnaP	+	−	−	−	nt	nt	−	−	−
dnaB	+	−	−	+	+	+	+	−	−
dnaC(D)	+	−	−	+	+	+	−	−	−
dnaG	+	−	+	+	+	nt	+	−	−
dnaZ	+	+	+	+	+	+	+	−	−
polC (DNA pol III)	+	+	+	+	+	+	+	−	−
rpoB (RNA pol)	+	+	−	−	−	+	+	−	−
nalA (DNA gyrase)	+	−	−	−	+	+	nt	nt	−
cou (DNA gyrase)	+	−	−	−	+	+	nt	+	−
lig (DNA ligase)	+	−	−	−	nt	. +	+	−	−
polA (DNA pol I)	−	−	−	−	−	+	−	−	−
rep	−	−	−	−	+	nt	−	−	−
tsnB, tsnC	−	−	−	−	nt	nt	−	+	−
grpC (dnaJ, dnaK)	+	−	−	−	nt	nt	+	−	−
grpD, grpE	−	−	−	−	nt	nt	+	−	−
DNA binding protein	nt	+	+	+	nt	nt	nt	nt	nt
DNA EF I, β protein, copol III*	nt	+	+	+	nt	nt	nt	nt	nt
DNA EF III	nt	+	+	+	nt	nt	nt	nt	nt
Replication factors X, Y, Z; i, n	nt	−	−	+	nt	nt	nt	nt	nt
Phage functions	−	−	−	−	+	−	+	+	+

[a] Needed for replication.
[b] Not needed for replication.
[c] Not tested.

independently of reactions that are unique to double-stranded DNA. It is presumed that when an in vitro DNA replicating system shares similar properties and uses some of the same proteins as in vivo replication, the mechanism of the in vitro reaction will resemble the in vivo reaction. These phage replicating systems have been fractionated into many protein components. With these purified proteins and defined DNA templates, it has been possible to dissect the overall process into partial reactions: (a) priming of DNA synthesis and (b) elongation of primed DNA.

The *E. coli* DNA binding protein, first described by Sigal et al (85) (reviewed in 2–4, 95–97) is required for both of these reactions and is involved in all of the reconstituted single-stranded DNA-dependent replicating systems discussed below. Functions of DNA binding protein in in vitro replication include *(a)* binding to single-stranded DNA as well as maintaining the DNA in the proper conformation for primer synthesis, (*b*) melting small duplex DNA regions which might otherwise block elongation, and (*c*) affecting activities of replication proteins by direct protein-protein interactions. It has been shown that the complex of DNA binding protein with single-stranded DNA affects the activities of some proteins that recognize single-stranded DNA. For example, the protein-DNA complex cannot be transcribed by RNA polymerase (97, 98) and cannot be degraded by some nucleases (85, 99). It has also been found that DNA binding protein interacts physically with other proteins such as exonuclease I, DNA polymerase II, and T7 DNA polymerase, as shown by Molineux & Gefter (99). While in vitro results suggest that DNA binding protein may be important in DNA replication, as yet no mutants defective in DNA binding protein have been isolated. Thus its involvement in *E. coli* replication has not been proven.

Other *E. coli* DNA binding proteins, referred to as HU (100), HD (101), H_1 (102), H_2 (102), and D (103) have also been described. While they may or may not be distinct from each other, they are distinct from *E. coli* DNA binding protein; no mutants have been identified to establish their role in DNA replication.

Priming of Single-Stranded DNA Synthesis

Synthesis from single-stranded circular phage DNA templates requires de novo initiation of DNA synthesis. Since all three of the *E. coli* DNA polymerases require 3' OH termini for DNA elongation, a priming event must precede DNA elongation (2). The term priming is defined operationally as an event that allows subsequent DNA elongation. So far, three enzymatic pathways have been discovered by which de novo oligonucleotide primer synthesis occurs on single-stranded circular phage DNA templates. Each is unique in its nucleotide and protein requirements and its DNA specificity: (*a*) RNA polymerase catalyzes RNA priming of phage fd (and M13) DNA synthesis. (*b*) *dnaG* protein catalyzes oligonucleotide priming of phage G4 and ST-1 DNA synthesis when these DNAs are covered with DNA binding protein. (*c*) *dnaG* protein also catalyzes priming of phage ϕX174 DNA synthesis but only after prepriming reaction involving *dnaB* and *dnaC(D)* proteins, DNA binding protein, and two or three other proteins. These three priming pathways are discussed below; the DNA elongation of primed templates is discussed in the section on DNA elongation.

PRIMING OF DNA SYNTHESIS CATALYZED BY RNA POLYMERASE
In crude extracts of uninfected *E. coli*, the conversion of phage fd (M13) single-stranded DNA to duplex DNA requires four rNTP's and four

dNTPs (94); the products synthesized are nearly full-length linear DNA molecules that originate from a unique region on the fd DNA (94, 104). DNA synthesis requires RNA polymerase, since the reaction is rifampicin-sensitive and inhibited by antibody to RNA polymerase; *dnaA, dnaB, dnaC(D)*, and *dnaG* gene products are not required (94, 105, 106; Table 2).

Three contradictory reports exist on the protein requirements for the reconstitution of the fd DNA synthesizing reaction: (*a*) DNA binding protein and RNA polymerase are required for priming (107). (*b*) DNA binding protein and a protein fraction referred to as RNA polymerase III, which contains RNA polymerase and other protein components, are required (108). (*c*) RNA polymerase, DNA binding protein, RNase H, and two purified protein protein fractions, referred to as discriminatory proteins α and β, are required (109, 110). In all of these reports, RNA polymerase functions in fd DNA synthesis by initiating RNA chains de novo and thus providing primer ends for DNA elongation (Figure 1). The region of initiation of DNA synthesis is unique with all of these in vitro reaction conditions (104, 107, 109). The region of initiation of RNA primer synthesis is also unique and sequence data suggest that it is a duplex structured region (107, 107a, K. Geider, E. Beck, H. Schaller, personal communication).

PRIMING OF DNA SYNTHESIS CATALYZED BY *dnaG* PROTEIN   In crude extracts, phage G4 and ST-1 DNA synthesis requires *dnaG* protein as demonstrated by the thermolability of DNA synthesis in extracts prepared from *dnaG* thermolabile cells (70). It does not require RNA polymerase, *dnaB* protein, or *dnaC(D)* protein (111).

The G4 and ST-1 DNA-synthesizing reactions have been reconstituted with purified *dnaG* protein, *E. coli* DNA binding protein, and DNA elongation components (70, 72, 111, 112; Table 2). The *dnaG* protein was originally purified by in vitro complementation assays by measuring either

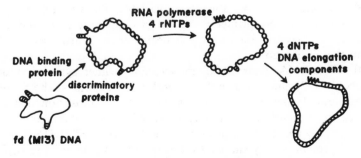

Figure 1   Mechanism of priming of fd (M13) phage DNA.

E. coli chromosome synthesis in the cellophane disc system (90) or φX174, G4, or ST-1 DNA-dependent DNA synthesis in the soluble extract system (69, 70). The protein isolated was in fact the *dnaG* gene product because the activity of the protein purified from a *dnaG* thermolabile mutant was thermolabile in vitro (69). It was suggested that the function of the *dnaG* protein in the conversion of G4 and ST-1 single-stranded DNA to duplex DNA was to synthesize an RNA primer for DNA elongation (70, 111). This has been confirmed and it has been further shown that *dnaG* protein synthesizes deoxyribooligonucleotides as well as oligonucleotides containing both rNMP and dNMP residues (72). As final proof that the *dnaG* protein is involved in oligonucleotide synthesis, primer synthesis was shown to be thermolabile when thermolabile *dnaG* protein was used (72). The mechanism of priming of G4 and ST-1 by *dnaG* protein is suggested by the isolation of various intermediates in the reaction (70–72; Figure 2). First, *dnaG* protein binds reversibly to specific sites present in G4 and ST-1 DNA. The DNA must be covered with *E. coli* DNA binding protein for *dnaG* protein to bind; about one molecule of *dnaG* protein binds per circle of DNA. Second, *dnaG* protein catalyzes the synthesis of ribo- deoxyribo- or mixed ribo- and deoxyribooligonucleotides depending on the nucleotides added to the reaction.

With G4 DNA, deoxyribooligonucleotide primer synthesis requires ADP (or ATP at higher concentrations) in addition to dNTPs (72). This ADP requirement is not yet understood but ADP seems to be incorporated at the 5' end of the oligonucleotide and AMP appears to be incorporated internally. The region of initiation of G4 DNA synthesis is unique in vivo (113) and in vitro (112). The DNA sequence in the region of initiation (J. Simms and D. Dressler, personal communication; G. N. Godson and J. Fiddes, personal communication) contains the complementary sequence of the primer made in vitro (7lb). while the in vitro synthesized primer may be as long as 30 nucleotides (71b), shorter oligonucleotides also support DNA elongation (72, S. Wickner, unpublished results). The 5' ends of the oligonucleotides are unique but the 3' ends are variable. The structure suggested by the sequence of the primer is a stem of eight base pairs and a loop containing five bases. Primer synthesis begins at the fourth base from the stem in the 5' direction.

With ST-1 DNA, deoxyribooligonucleotide primer synthesis requires only dNTPs (72) and can be initiated with a dA residue at the 5' end (S. Wickner, unpublished result). This is the first demonstration that DNA synthesis can be initiated by the synthesis of a deoxyribooligonucleotide primer. The sequence of the primer synthesized from ST-1 DNA in vitro by *dnaG* protein is very similar to that of the G4 primer. The structure suggested by the sequence is identical to that suggested by the G4 primer

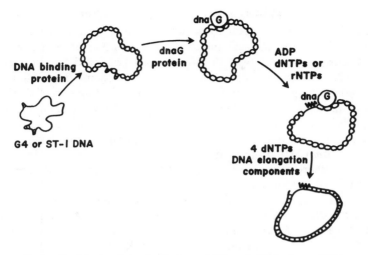

Figure 2   Mechanism of priming of G4 and ST-1 phage DNA.

sequence; that is, a stem of eight base pairs and a loop of five bases (J. Simms, D. Dressler, D. Capon, M. Gefter, S. Wickner, unpublished results).

PRIMING BY *dnaG* PROTEIN FOLLOWING REACTIONS INVOLVING *dnaB* AND *dnaC(D)* PROTEINS IN CONJUNCTION WITH OTHER *E. coli* PROTEINS   In crude extracts, phage ϕX174 DNA-dependent DNA synthesis requires only ATP of the rNTPs and four dNTPs. It also requires *dnaB, dnaC(D)*, and *dnaG* gene products, but not RNA polymerase or *dnaA* gene product (106, 105).[3] The *dnaB* and *dnaC(D)* proteins have been purified using the in vitro ϕX174 DNA-dependent complementation assays and have been studied individually. The homogeneous *dnaB* protein is a ribonucleoside triphosphatase that is stimulated by single-stranded DNA and not by double-stranded DNA or RNA (65). rNTPs, but not dNTPs, are hydrolyzed, and the products are inorganic phosphate and rNDPs (65). The role of this rNTPase in DNA synthesis is not yet known. Purified *dnaC(D)* protein, in the presence of ATP, forms a physical complex with dnaB protein and inhibits its DNA-independent ribonucleoside triphosphase activity (68).

As yet there is no data to prove that *dnaB* and *dnaC(D)* proteins interact in vivo. However, as mentioned earlier, in vivo, some *dnaB* and *dnaC(D)*

[3]It is now clear that *dnaA* gene product is not required (64), although earlier reports suggested that it was required (106).

mutants appear to be defective in initiation of rounds of chromosome replication, and others appear to be defective in elongation. Extracts prepared from both types of *dnaB* and *dnaC(D)* mutants catalyze thermolabile φX174 DNA synthesis (44, 46, 64). Several possibilities exist: (*a*) the phenotypic designations of initiation of chromosome replication and chromosome elongation are not strictly valid, (*b*) the function of *dnaB* and *dnaC(D)* proteins in φX174 DNA synthesis resembles the function of these proteins in chromosome initiation, (in which case, it must be assumed that elongation type mutants are also defective in initiation), and (*c*) in vivo the *dnaB* and *dnaC(D)* proteins defective in one function are still able to perform another function, but in vitro the altered proteins are unable to function at all due to their lability.

The φX174 DNA–synthesizing reaction has been reconstituted with purified *dnaB, dnaC(D)*, and *dnaG* proteins, *E. coli* DNA binding protein, two or three other *E. coli* proteins defined only by their requirement in this reaction and referred to as replication factors X, Y, and Z (88) or factors i and n (87), and DNA elongation components (Table 2). Replication factor Y is a single-stranded DNA-dependent ATPase or dATPase; the products of the reaction are inorganic phosphate and ADP or dADP (89). Under some assay conditions, this ATPase activity is specifically stimulated by φX174 DNA but not by fd or ST-1, which do not require this protein for their replication.

The φX174 DNA–synthesizing reaction has been divided into partial reactions (68, 88, 114–117). The various intermediates have been isolated by agarose gel filtration (114–117; S. Wickner, unpublished results; Figure 3). DNA binding protein and replication factors Y and Z bind to the DNA. The *dnaB* protein is transferred to the protein-covered DNA in a reaction requiring ATP, *dnaC(D)* protein, and replication factor X. The protein-DNA complex contains *dnaB* protein, DNA binding protein, and replication factors Y and Z; it does not contain *dnaC(D)* protein or factor X. Presumably, *dnaG* protein, and the protein-DNA complex synthesize an oligonucleotide primer as they do with G4 DNA. It has not been possible to demonstrate in vivo or in vitro that the origin of φX174 complementary–strand DNA synthesis is unique using methods that have shown that the origin of conversion of fd and G4 DNA to duplex DNA is unique (104, 117, 118).

TEMPLATE SPECIFICITY OF SINGLE-STRANDED DNA REPLICATING SYSTEMS   Each of the three enzymatic pathways by which single-stranded circular DNA can be primed is DNA template–specific. Very likely this specificity resides in the recognition of DNA structural sites by proteins.

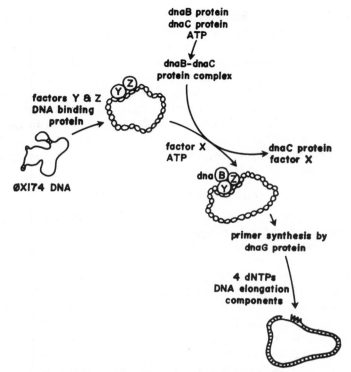

Figure 3   Mechanism of priming of φX174 phage DNA.

While crude extracts catalyze rifampicin-sensitive priming of fd DNA synthesis, they are unable to catalyze rifampicin-sensitive priming of G4, ST-1, and φX174 DNA synthesis (94, 111, 105). Most likely, when fd DNA is covered with DNA binding protein and the other proteins required for primer synthesis, RNA polymerase is able to catalyze the synthesis of RNA from a specific region. Presumably, when G4, ST-1, or φX174 DNA is covered with these proteins, there are no exposed sites for RNA polymerase to initiate primer synthesis (107–110).

The priming reaction catalyzed by *dnaG* protein is also template specific; it is specific for G4 and ST-1 DNA and cannot act on fd or φX174 DNA. The *dnaG* protein is unable to bind to fd or φX174 DNA covered by DNA binding protein or to catalyze oligonucleotide synthesis on these DNAs (72). Presumably, fd and φX174 DNA do not have binding sites that can be recognized by *dnaG* protein.

Similarly, φX174 DNA is primed by a template specific reaction in which *dnaB* protein is transferred to the DNA prior to primer synthesis, in a reaction involving DNA binding protein, *dnaC(D)* protein, and two or three other proteins. This transfer reaction is specific for φX174 DNA; with identical reaction conditions, *dnaB* protein cannot be transferred to fd DNA (S. Wickner, unpublished results). The φX174-DNA protein complex serves as a substrate for *dnaG* protein activity. If the function of *dnaG* protein in φX174 DNA synthesis is similar to its function in G4 DNA synthesis, then perhaps the other proteins required for φX174 DNA synthesis, but not for G4 synthesis, are needed to form a site on the DNA that can be recognized by *dnaG* protein.

DNA Elongation of Primed Single-Stranded DNA

Once a primer is synthesized on single-stranded circular phage DNA or on any single-stranded DNA, elongation can occur in vitro by any one of several mechanisms: (*a*) DNA polymerase I alone can elongate DNA (2), (*b*) with some assay conditions, DNA polymerase II will elongate primed DNA that is covered with DNA binding protein (119), and (*c*) with other conditions, DNA polymerase II or III will elongate primed DNA in combination with *dnaZ* protein and two other *E. coli* proteins referred to as DNA elongation factors I and III (DNA EF I and DNA EF III) (74, 75). The synthetic activities of DNA polymerase I and II are probably not essential *E. coli* functions as demonstrated by the isolation of viable *polA* (polymerase I⁻) (120) and *polB* (polymerase II⁻) (121) mutants that are deficient in these enzymes. Thus mechanisms (*a*) and (*b*) may not be involved in replication. Mechanism (*c*) probably shares the most similarities with the in vivo mechanism of DNA elongation and is discussed in detail here.

DNA elongation in vitro and *E. coli* DNA replication require the *polC* gene product. This gene product is a component of DNA polymerase III, since DNA polymerase III activity purified from *polC* thermosensitive strains is thermolabile in vitro (76, 77). The fundamental properties of DNA polymerases I, II, and III have been reviewed previously (2–4). Purified DNA polymerase III catalyzes the incorporation of 10–50 nucleotides per gap with double-stranded DNA templates containing small single-stranded gaps; it is inactive with long single-stranded DNA templates (122). DNA polymerase III has an associated 3'-5' exonuclease activity which hydrolyzes single-stranded DNA producing 5'-mononucleotides (80). It also has an unusual 5'-3' exonuclease activity (80) that requires a 5' single-stranded end, but will proceed into an adjacent double-stranded region. The subunit structure of DNA polymerase III is still unresolved. There is agreement that purified DNA polymerase III contains a 140,000 dalton subunit (78, 79, 86, S. Sugrue and J. Hurwitz, personal communication). If the enzyme contains other essential subunits, then other genes required for DNA polymerase III activity remain to be identified. No properties of DNA polymerase III have yet been found that explain why this polymerase is more ideally suited for *E. coli* replicative synthesis than DNA polymerases I or II. Possibly its requirement for *E. coli* chromosome synthesis reflects its interaction with other replication proteins.

DNA elongation in vitro and *E. coli* DNA replication also require the dnaZ protein, which was purified using an in vitro complementation assay. This assay measured the stimulation of fd, ST-1, and φX174 DNA-dependent DNA synthesis in crude extracts of dnaZ thermolabile cells by fractions from wild-type dnaZ protein fractions (74). The dnaZ protein functions in vitro in the elongation of primed single-stranded DNA in combination with DNA polymerase II or III and DNA EF I and DNA EF III. DNA EF I and DNA EF III were purified using as an assay their requirement for DNA synthesis in conjunction with DNA polymerase III and dnaZ protein (74, 75). These proteins and the reaction catalyzed by them have been studied extensively recently (74, 75, 86, 123–127) and the mechanism by which they catalyze DNA elongation has been suggested by the physical isolation of protein-protein and protein-DNA complexes preceding dNMP incorporation (75; Figure 4): (*a*) DNA EF III and dnaZ protein form a protein complex independent of DNA polymerase III, DNA EF I, primed template, and nucleotides. (*b*) Together they catalyze the transfer of DNA EF I to a primed DNA template. This transfer reaction requires ATP or dATP and both primer and template; it does not require DNA polymerase III. The DNA EF I-primed template complex does not contain dnaZ protein, DNA EF III, or ATP. The function of ATP or dATP in this reaction is unknown. (*c*) DNA polymerase III binds to the DNA EF I-primed DNA complex, but not to primed templates alone. The binding of DNA polymerase III proceeds in the absence of ATP (or dATP), dnaZ protein, and DNA EF III and without further addition of DNA EF I. (*d*) The complex of DNA polymerase III, DNA EF I, and primed template catalyzes DNA synthesis upon addition of dNTPs without addition of ATP as a cofactor. Thus, the DNA EF I-primed DNA complex facilitates or stabilizes the binding of DNA polymerase III to the primer. This may be the explanation for the stimulation by these three proteins of dNMP incorporation by DNA polymerase III. It is not known yet if DNA EF I is associated with DNA polymerase III during chain propagation.

This mechanism of DNA elongation suggests that dnaZ protein and DNA EF III are not tightly associated with the DNA template. However, recent experiments have shown that the *E. coli* DNA binding protein increases the affinity of dnaZ protein and DNA EF III for single-stranded DNA such that complexes of these three proteins with any single-stranded

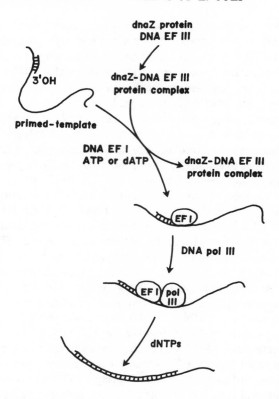

Figure 4   Mechanism of DNA elongation catalyzed by DNA polymerase III, *dnaZ* protein, DNA EF I, and DNA EF III.

DNA can be physically isolated (S. Wickner, unpublished results). Binding of dnaZ protein and DNA EF III to DNA covered with DNA binding protein occurs in the absence of ATP or dATP cofactor, primer ends, DNA EF I, and DNA polymerase III. The binding of DNA EF I to the complex still requires ATP or dATP and primer ends; the binding of DNA polymerase III still requires that DNA EF I be bound first.

Initial contradictory reports of the protein structure and enzymatic activities of the elongation apparatus have been reconciled in part. In one laboratory, Hurwitz et al (126) observed that the reaction could be catalyzed by DNA polymerase III and a protein fraction referred to as stimulatory protein. With further work the stimulatory fraction was separated into two fractions, referred to as DNA elongation factors I and II (127). Later, factor II was fractionated into two components, one identified as the *dnaZ* gene product and the other referred to as DNA elongation factor III (74). It is still possible that more proteins will be found to be required for DNA elongation. In another laboratory, W. Wickner et al observed that DNA elongation required two protein fractions, referred to as DNA polymerase III* and copolymerase III* (123), and that the reaction could be catalyzed by a single protein preparation called DNA polymerase III holoenzyme (125). These reports claimed that copolymerase III* was a 77,000 dalton polypeptide, polymerase III* a tetramer of 90,000 dalton polypeptides, and polymerase III a holoenzyme containing two copolymerase III* subunits and two polymerase III* subunits. It is now accepted that this DNA elongation reaction requires at least four different proteins (74, 86). DNA polymerase III holoenzyme has been recently reported to consist of at least four subunits, referred to as α (140,000 daltons; presumably a component of DNA polymerase III), β (40,000 daltons, presumably copolymerase III*), γ (52,000 daltons), and δ (32,000 daltons) (86). Either γ or δ presumably is the product of the *dnaZ* gene. Operationally, the combination of dnaZ protein, DNA EF III, DNA EF I, and DNA polymerase III carries out the same reaction as DNA polymerase III holoenzyme. The availability of dnaZ protein, DNA polymerase III, DNA EF I, and DNA EF III in active form, and purified free of each other, has allowed the dissection of the reaction. Whether the components of the DNA elongation apparatus can best be described as subunits of a holoenzyme or individual proteins is impossible to say without more protein structure studies. In this connection, *E. coli* DNA polymerase II and *Bacillus subtilis* DNA polymerase III are able to function in vitro in conjunction with dnaZ protein, DNA EF I, and DNA EF III (74, 126–128). It may be that the lack of specificity of dnaZ protein, DNA EF I, and DNA EF III for DNA polymerase III in vitro means that this DNA elongation reaction is not the essential function of DNA polymerase III in *E. coli*.

In summary, there are three enzymatic pathways by which *E. coli* proteins can initiate single-stranded DNA-dependent DNA synthesis, and several mechanisms by which *E. coli* proteins can elongate primed single-stranded DNA in vitro. We should be careful not to draw strong analogies between *E. coli* chromosome replication in vivo and single-stranded phage DNA synthesis in vitro. Very likely, each of the pathways for replicating DNA in vitro is in some ways similar to host DNA replication and in other ways distinct. The priming of fd DNA is similar to initiation of *E. coli* chromosome replication in that both use RNA polymerase. However, *E. coli* chromosome initiation requires in addition at least dnaA, dnaB, dnaC(D), dnaI, and dnaP proteins. Elongation of the *E. coli* chromosome involves de novo initiation of the DNA chain which is growing overall in the 3' to 5' direction. The initiation of ϕX174 DNA synthesis resembles Okazaki fragment synthesis in that both require dnaB, dnaC(D), dnaG, dnaZ, and polC proteins and neither requires *dnaA, dnai,* or *dnaP* gene products or RNA polymerase. The replication of G4 DNA is more puzzling. With ϕX174 DNA, dnaG protein requires four or five additional proteins to make a primer; with G4 DNA these other proteins are not required. It is possible that (*a*) some Okazaki fragments are synthesized by the G4 pathway and others by the ϕX174 pathway or (*b*) *E. coli* DNA does not contain the required structural sites to allow priming by dnaG protein alone.

DOUBLE-STRANDED CIRCULAR SUPERCOILED DNA-DEPENDENT DNA SYNTHESIS

Double-stranded DNA replication presents some particular problems that are different from those of single-stranded DNA replication: (*a*) initiation of double-stranded DNA synthesis, (*b*) separation of the two strands of the double helix, and (*c*) overall 3'-5' DNA elongation from one strand and simultaneous 5'-3' synthesis from the other. ColE1 and ϕX174 RF I DNA replicating systems will be discussed below. Both of these supercoiled double-stranded circular DNAs require some of the *E. coli* replication proteins for their own replication and thus these systems are useful for studying the mechanisms involved in double-stranded DNA replication.

ColE1 DNA Replication In Vitro

ColE1 DNA, like the *E. coli* chromosome, is supercoiled, closed, circular, duplex DNA. Its in vivo and in vitro replication have been reviewed recently (8, 129). Most of the molecules in vivo replicate semi-conservatively in Cairns-type theta structures. DNA synthesis initiates at a fixed point and proceeds unidirectionally. Plasmid replication, in contrast to *E. coli* replication, is insensitive to inhibition of protein synthesis. In fact, in the presence of chloramphenicol, the number of copies of ColE1 DNA per cell increases from about 30 to 1000. Initiation of ColE1 replication, like initiation of *E. coli* replication, requires RNA polymerase. Presumably, with ColE1 DNA, RNA serves as a primer for initiation of DNA synthesis. ColE1 DNA synthesis has been reported to be inhibited in *dnaC(D)* and *dnaG* strains at nonpermissive temperature but not in *dnaB* strains; contradictory results have been reported for *dnaA* and *polC*. Replication requires DNA polymerase I and/or its 3'-5' exonuclease since ColE1 cannot be maintained in *polA* cells. In addition, bacterial mutants that are temperature sensitive for the replication of ColE1 have been isolated and some have been found to map in *polA*.

Gently lysed extracts of wild-type *E. coli* or *E. coli* carrying the ColE1 plasmid catalyze ColE1 DNA synthesis in a fashion similar to the in vivo mechanism. The reaction depends on all four rNTPs and four dNTPs and is rifampicin sensitive (130–133). Plasmid coded functions are not required for in vitro replication, although they may participate when they are present. In vitro, replication is semiconservative and the DNA products are closed double-stranded circular molecules of normal superhelicity (130). Electron micrographs of reaction products show that replication begins at a unique region and proceeds unidirectionally (134). Both strands of the replication loops are double-stranded, indicating that replication occurs nearly simultaneously on both parental strands. The two daughter molecules generally separate before replication is complete, which results in two open circular molecules with gaps at the origin-terminus region (135). The gaps are then filled and two supercoiled molecules are formed. Occasionally catenated molecules consisting of two interlocked closed circular monomeric units are formed. The DNA products are active in transfection and can also serve as templates for further replication (130).

INITIATION OF COLE1 REPLICATION   Partially replicated ColE1 molecules, referred to as early replicative intermediates, accumulate under particular in vitro conditions such as 5% glycerol (131). These molecules contain unique replication loops in the region of the ColE1 origin of replication. This was determined by measuring the distance from the single Eco RI restriction endonuclease cleavage site to the branch point of the loop and comparing the in vivo and in vitro replicating molecules (129, 133). Replication loops synthesized in vitro contain one or two double-stranded branches, consisting of 6S DNA fragments (134). When only one fragment is present, it is the L-strand (136). These DNA fragments are several hundred nucleotides long (134). The early replicative intermediates are true intermediates in DNA replication because they can be chased into complete molecules. This can be accomplished, for example, by lowering the glycerol concentration in the reaction mixture (131, 136). While it is not clear why early ColE1 replicative intermediates accumulate in vitro, their existence has allowed the study of early steps in replication.

Synthesis of the L-strand fragment requires RNA polymerase since the reaction requires 4 rNTPs and is sensitive to rifampicin (130, 131). RNA synthesis precedes DNA synthesis; ColE1 DNA, incubated with rNTPs in a first reaction with an extract from which nucleotides and small molecules have been removed, supports rifampin-resistant DNA synthesis in a second reaction upon the addition of dNTPs. Synthesis of early replicative intermediates also requires DNA polymerase I and/or its 3'–5' exonuclease activity (132).

The point of initiation of the RNA primer of ColE1 DNA replication in vitro has not yet been determined. While all four rNTPs are required for

DNA synthesis, purified 6S L-strand early replicative intermediates retain only one or two rNMP residues at their 5' ends (138, 139). This suggests that the RNA primer, but not the DNA, must be efficiently removed, possibly by RNase H (140–142) or some similar enzyme. The transition from RNA to DNA occurs in one of three nucleotides in a DNA region consisting of five dAMP residues (138, 139). Both the 3' and 5' sides of this region are rich in dGMP and dCMP residues. This in vitro origin is distinct from the site on ColE1 DNA of the protein-DNA complex referred to as the relaxation complex (143).[4]

In summary, L-strand ColE1 DNA synthesis appears to be primed by RNA polymerase and elongated for some distance by DNA polymerase I. However, mixtures of purified RNA polymerase and DNA polymerase I, plus or minus DNA binding protein, do not alone catalyze 6S L-strand DNA synthesis, which suggests that other *E. coli* proteins are required (J. Tomizawa and T. Itoh, personal communication). H-strand ColE1 DNA synthesis appears to be primed by another enzymatic pathway (136), possibly one of those used to initiate G4 or ϕX174 single-stranded DNA-dependent synthesis.

DNA ELONGATION OF COLE1 EARLY REPLICATIVE INTERMEDIATES   Electron micrographs of partially replicated molecules synthesized in vitro show that replication of the early replicative intermediates continues by unidirectional expansion of the replication loop (134). Since synthesis only occurs by 5'–3' extension, then at least the H strand of the early replicative intermediates is elongated by a discontinuous mechanism. While small fragments have been seen in vivo and in toluenized cells (129), the in vitro system has not yet been exploited in the study of Okazaki fragment synthesis and processing.

Some of the protein requirements for elongation of the early replicative intermediates have been examined (129; Table 2). Synthesis does not require RNA polymerase since elongation is rifampin resistant (131). It does require dnaB and dnaC(D) proteins; extracts of *dnaB* and *dnaC(D)* temperature-sensitive strains do not synthesize full-length DNA but can be stimulated to do so by the addition of purified dnaB and dnaC(D) proteins (T. Itoh and J. Tomizawa, personal communication). The *dnaZ* and *polC* gene products also appear to be involved (132, 146). The requirements for other *dna* gene products have not been studied.

[4]It has been proposed that replication was initiated from the site of the relaxation complex (144). However, this site is approximately 300 nucleotides in the direction of overall DNA replication from the origin identified by the in vitro experiments described above (129) and is not essential for the maintenance of the ColE1 plasmid (145).

ColE1 DNA-dependent synthesis requires a recently described enzyme, DNA gyrase. DNA gyrase was first identified in the course of studies on integrative recombination of phage λ DNA in vitro (147–149). Recombination required a supercoiled substrate, but covalently closed double-stranded circular DNA (relaxed DNA) could be used if first incubated with ATP and an extract of *E. coli*. These results suggested that there was an *E. coli* enzyme capable of converting relaxed closed circular DNA to a supercoiled form. The enzyme has been purified and catalyzes ATP-dependent negative supercoiling of all relaxed closed circular DNAs tested (149). Two components of DNA gyrase have been identified, the *nalA* and *cou* gene products. As mentioned earlier, these two gene products are also required for *E. coli* chromosome replication. The ATP-dependent supercoiling activity of wild-type DNA gyrase is sensitive to both novobiocin (or coumermycin) (54) and nalidixic acid (or oxolinic acid) (55, 56); enzyme purified from strains resistant to novobiocin or nalidixic acid is resistant to novobiocin or nalidixic acid, respectively.

The requirement for DNA gyrase for ColE1 DNA synthesis was shown by the sensitivity of synthesis to novobiocin (54) and oxolinic acid (55) in extracts of drug-sensitive strains but not in drug-resistant strains. A requirement for DNA gyrase in the elongation of the ColE1 early replicative intermediates was suggested by the observation that the intermediates are not elongated in extracts in the presence of coumermycin and rifampin (J. Tomizawa and T. Itoh, personal communication).

In addition to functioning in integrative recombination of phage λ, ColE1 replication, and *E. coli* replication, DNA gyrase is also involved in ϕX174 RF I replication (56, 150, 151). The nalA protein component of gyrase has been purified by complementation in the ϕX174 RF I DNA–dependent DNA-synthesizing reaction (discussed below), and is free of detectable DNA gyrase activity (56). It is not clear if the nalA protein functions solely through its participation as a component of DNA gyrase or has an independent physiological role.

Several other properties of DNA gyrase have been reported. In the absence of ATP, DNA gyrase catalyzes relaxation of either positively or negatively supercoiled DNA (55, 56). This relaxing activity is sensitive to nalidixic acid (or oxolinic acid) but not to novobiocin; relaxing activity of gyrase prepared from a nalidixic acid-resistant strain is resistant to this drug (55, 56). These results suggest that relaxing activity resides in the nalA protein component of DNA gyrase. Another activity of gyrase is site-specific double-stranded DNA cleavage induced by treatment of the gyrase-DNA complex with oxolinic acid, followed by treatment with sodium dodecyl sulfate; nalidixic acid–resistant gyrase does not induce cleavage with these conditions (55, 56). Novobiocin does not induce this reaction or inhibit breakage induced by nalidixic acid. It has been observed that chromosomal DNA isolated from *E. coli* treated with nalidixic acid has a reduced single-strand molecular weight (151a). Since the relaxing or super-coiling of DNA mechanistically requires nicking and closing, the nalidixic acid–induced nuclease activity may reflect the trapping of such an intermediate.

There is another *E. coli* enzyme that catalyzes a nicking-closing reaction, referred to as ω protein (152, 153). Unlike DNA gyrase, it is not coded by any of the *E. coli* genes known to be involved in replication. In addition, it can only relax negatively supercoiled DNA and is not capable of relaxing the positive supercoils generated during replication.

Cairns (154) first proposed that DNA replication required a swivel activity based on his discovery of the circular structure of the *E. coli* chromosome. DNA gyrase is ideally suited for this function. Its activity could drive the replication fork by unwinding the parental strands ahead of the fork and could reduce the positive supercoil strain generated by the unwinding of parental strands as they are replicated. These potential functions of DNA gyrase in the replication of double-stranded, covalently closed, circular supercoiled DNA suggest that DNA synthesis may occur on the separated strands at replication forks. Thus, the mechanisms of DNA priming and elongation dependent on single-stranded DNA described above may resemble those operating to replicate double-stranded DNA templates after the strands of the double helix have been separated.

TERMINATION OF COLE1 REPLICATION   Termination may require no special protein functions other than those required for replication, as proposed by Sakakibara, Suzuki & Tomizawa (155). Simply, replication may continue to completion; the parental strands in the unreplicated portion could rotate by the action of a nicking-closing enzyme in the direction of negative superhelicity until they could no longer be held together. DNA gyrase would be the most likely enzyme to carry out this process. If the parental strands were completely unwound when the final nicking-closing reaction occurred, then two daughter molecules would be formed. If the parental strands were still wound around each other, but no longer held together, a catenane of two interlocked open circular molecules would be formed. Replication would be completed with the filling of the gaps and the introduction of supercoils. This model is more attractive in its simplicity than an alternative one that requires synthesis of a terminal redundancy at the final stage of replication to ensure circularity of progeny molecules (3).

φX174 RF I DNA Replication In Vitro

After φX174 single-stranded viral DNA is converted to duplex DNA, the remaining gap is filled, perhaps by DNA polymerase I. The nick is sealed by DNA ligase and negative supercoils are introduced, possibly by DNA gyrase. These covalently closed supercoiled molecules (referred to as RF I) serve as templates for the production of progeny RFI molecules. It is not clear how this occurs in vivo. One possible mechanism is that the growing point moves around the circular genome from origin to terminus with each strand of the duplex serving as a template for the synthesis of its complement, as described above for ColE1. An alternative mechanism (rolling circle model) is that the growing point extends the 3' end of a nick and displaces the 5' end from its original site at the origin. A complementary strand is then synthesized from the displaced strand. At present in vivo data are consistent with either model (reviewed in 6, 8, 156, 157).

The ϕX174 gene A protein is the only phage function required for RF I replication (8, 156). In vivo studies showed that this function is involved in introducing a site-specific nick in the viral strand of the RF I molecule (158). A gene A protein was first purified by a double-label procedure and was found to be a viral strand–specific endonuclease specific for ϕX174 RF I and ϕX174 single-stranded DNA (159). Other DNAs were not nicked by the protein.

More recently, ϕX174 RF I replication has been studied in crude extracts of gently lysed E. coli (150, 160–162). By a complementation assay for ϕX174 RF I DNA–dependent DNA synthesis, the gene A activity has been purified (84, 162). In addition to confirming earlier results, the gene A protein was shown to specifically nick ϕX174 supercoiled RF I and not relaxed RF I (162). By electron microscopy of restriction nuclease-treated complexes of ϕX174 RF I DNA and gene A protein, the gene A protein was seen specifically bound to a site in the A cistron of the RF molecules (84) and the nick made was in the A cistron (162). In vivo the gene A function is cis dominant (8, 156). That is, a wild-type A gene does not support replication of the A⁻ genome in the same cell. However, in vitro, purified gene A protein can nick ϕX174 RF I molecules other than those from which it was transcribed and translated.

The purified gene A protein also catalyzes the relaxation of negatively supercoiled ϕX174 RF I DNA, which indicates that the protein is able to carry out a nicking-closing reaction (162). Treatment of the relaxed RF I molecules with phenol or protease produces nicked molecules. Most likely, the gene A protein remains covalently attached to the 5' end of the nick, since these molecules were susceptible to exonuclease III hydrolysis but not to attack by the 5'–3' exonuclease of DNA polymerase I. The existence of a nicking-closing activity associated with gene A protein, the attachment of the protein at the 5' end of the nicked molecule, and the requirement for gene A function for progeny viral strand synthesis (163), suggest that the reversal of the gene A nicking action may be responsible for the circularization of progeny viral single-stranded DNA.

In vivo, as well as in vitro, ϕX174 RF I replication requires the product of the *E. coli rep* gene. This gene was identified as a function required for ϕX174 phage growth (164). Other phage, including G4, ST-1, M13, and P2, also require this function for growth. While the *rep* function does not appear to be essential for *E. coli* growth, the number of replicating forks per chromosome increases in *rep* mutants, which suggests that the forks move slower than normal (165). In this mutant, ϕX174 single-stranded DNA is converted to RF I, but then replication stops (164). The rep protein has been purified using as an assay ϕX174 RF I DNA–dependent complementation (83, 84, 150). It has been suggested that the purified rep protein catalyzes ATP-dependent DNA strand separation in conjunction with ϕX174 gene A protein and DNA binding protein (83, 84). There is evidence that rep protein–complementing activity and single-stranded DNA-dependent ATPase activity purify together; however, it has not yet been shown that rep protein–complementing activity and the DNA unwinding activity purify together.

Other *E. coli* proteins, besides the rep protein, may be able to carry out strand separation during replication. One such protein, discussed above, is DNA gyrase. Another is an ATP-dependent DNA unwinding protein isolated from *E. coli* and referred to as DNA helicase (166–168). This enzyme separates the strands of RNA-DNA or DNA-DNA duplexes. Like the rep protein preparations, helicase contains DNA-stimulated ATPase activity. It was proposed that the enzyme binds to single-stranded regions of DNA and then moves processively into double-stranded regions separating the strands. Unlike DNA gyrase, helicase has not been implicated in *E. coli* chromosome replication.

ϕX174 RF I replication in vitro also requires dnaB, dnaC(D), dnaG, and dnaZ proteins (150; C. Sumida-Yasumoto and J. Hurwitz, personal communication). These results are consistent with those obtained from in vivo studies measuring phage production after ϕX174 RF I DNA transfection (8, 156, 157, 169, 170). Replication in vitro also requires DNA gyrase but not RNA polymerase since the reaction is sensitive to both novobiocin and nalidixic acid but insensitive to rifampicin (149). By using a similar approach to that used to study single-stranded DNA-dependent systems, it should be possible to reconstitute the ϕX174 RF I DNA replicating system with purified components and to dissect the reaction into partial reactions. ϕX174 RF I dependent synthesis of single-stranded DNA has been reconstituted by the combination of gene A protein, rep protein, DNA binding protein, and DNA polymerase III holoenzyme (84). Unlike the in vivo conversion of ϕX174 RF I to progeny RF I, the products are not RF I, and unlike the in vivo conversion of RF I to single-stranded circles, phage functions are not required. Attempts to reconstitute ϕX174 RF I replication with mixtures of purified ϕX174 gene A, rep, dnaB, dnaC(D) and dnaG proteins, DNA gyrase, factors required for ϕX174 single-stranded DNA synthesis, DNA binding protein, and DNA elongation components have so far been unsuccessful (C. Sumida-Yasumoto and J. Hurwitz, personal communication). This suggests that other *E. coli* proteins are required for this reaction.

CONCLUSIONS

In vivo and in vitro studies of procaryotic DNA replication have utilized both genetic and biochemical procedures and have produced much information about this complex basic metabolic pathway. The fractionation of entire DNA replicating systems into their protein components, the reconstitution of DNA replicating systems with purified proteins and defined DNA templates, and the separation of the overall replication reactions into partial reactions has prepared the way for future studies of the mechanisms by which replication proteins participate in DNA replication.

In contrast to the synthesis of RNA and proteins, the synthesis of DNA is catalyzed by several different enzyme pathways even within one organism. For example, only host functions are required for the conversion of fd, G4, and ϕX174 phage DNAs to duplex DNA, and for the replication of ColE1 plasmid DNA, while only phage functions are required for the replication of T4 phage DNA. T7 and λ phage DNA, fd, G4, and ϕX174 RF I DNA, and F plasmid DNA are replicated by various combinations of *E. coli* proteins and phage or plasmid proteins. This diversity of enzymatic pathways is reflected in the specificity of these pathways for their DNA templates. As summarized above, the DNA sequence and/or structure of fd, G4, and ϕX174 DNA determines which enzymatic pathway will catalyze the initiation of DNA synthesis. Since diversity may be the rule rather than the exception, the study of a variety of different systems may be of more value in the case of DNA replication than in other biochemical processes.

Although a great deal is being learned about how DNA replication is accomplished, almost nothing is known yet about the control of replication. Future work must include genetic and biochemical studies in this area.

ACKNOWLEDGMENTS

I wish to thank M. Gellert, H. Nash, N. Nossal, Max Gottesman, A. Das, J. I. Tomizawa, and D. Botstein for their help.

Literature Cited

1. Watson, J. D., Crick, F. H. C. 1953. *Nature* 171:737–38
2. Kornberg, A. 1974. *DNA Synthesis,* San Francisco: Freeman, 399 pp.
3. Gefter, M. L. 1975. *Ann. Rev. Biochem.* 44:45–78
4. Geider, K. 1976. *Curr. Top. Microbiol. Immunol.* 74:55–112
5. Alberts, B. M., Sternglanz, R. 1977. *Nature* 269:655–61
6. Dressler, D. 1975. *Ann. Rev. Microbiol.* 29:525–59
7. Jovin, T. M. 1976. *Ann. Rev. Biochem.* 45:889–920
8. Lewin, B. 1977. *Gene Expression,* Vol. 3, New York: J. Wiley and Sons. 925 pp.
9. Wechsler, J. A. 1978. In *DNA Synthesis: Present and Future,* ed. I. Molineux, M. Kohiyama, pp. 49–70. New York: Plenum
10. Barry, J., Alberts, B. M. 1972. *Proc. Natl. Acad. Sci. USA* 69:2717–21
11. Morris, C. F., Sinha, N. K., Alberts, B. M. 1975. *Proc. Natl. Acad. Sci. USA* 72:4800–4
12. Alberts, B., Barry, J., Bittner, M., Davies, M., Hama-Inaba, H., Liu, C. C., Mace, D., Moran, L., Morris, C. F., Piperno, J., Sinha, N. 1977. In *Nucleic Acid-Protein Recognition,* ed. H. J. Vogel, pp. 33–63, New York: Academic
13. Imae, Y., Okazaki, R. 1976. *J. Virol.* 19:435–45
14. Imae, Y., Shinozaki, K., Okazaki, R. 1976. *J. Virol.* 19:765–74
15. Hinkle, D. C., Richardson, C. C. 1974. *J. Biol. Chem.* 250:2974–84
16. Scherzinger, E., Litfin, F. 1974. *Mol. Gen. Genet.* 135:73–86
17. Scherzinger, E., Seiffert, D. 1975. *Mol. Gen. Genet.* 141:213–32
18. Scherzinger, E., Klotz, G. 1975. *Mol. Gen. Genet.* 141:233–49
19. Modrich, P., Richardson, C. C. 1975. *J. Biol. Chem.* 250:5508–14
20. Modrich, P., Richardson, C. C. 1975. *J. Biol. Chem.* 250:5515–22
21. Hinkle, D. C., Richardson, C. C. 1975. *J. Biol. Chem.* 250:5523–29
22. Masker, W. E., Richardson, C. C. 1976. *J. Mol. Biol.* 100:557–67
23. Masker, W. E., Richardson, C. C. 1976. *J. Mol. Biol.* 100:543–56
24. Mark, D. F., Richardson, C. C. 1976. *Proc. Natl. Acad. Sci. USA* 73:780–84
25. Scherzinger, E., Lanka, E., Morelli, G., Seiffert, D., Yuki, A. 1977. *Eur. J. Biochem.* 72:543–58
26. Kolodner, R., Richardson, C. C. 1977. *Proc. Natl. Acad. Sci. USA* 74:1525–29
26a. Scherzinger, E., Lanka, E., Hillenbrand, G. 1978. *Nucleic Acid Res.* 4:4151–63
26b. Stratling, W., Knippers, R. 1973. *Nature* 245:195–97
26c. Deleted in proof
26d. Kolodner, R., Masamune, Y., LeClerc, J. E., Richardson, C. C. 1978. *J. Biol. Chem.* 253:566–73
26e. Kolodner, R., Richardson, C. C. 1978. *J. Biol. Chem.* 253:574–84
27. Klein, A., Powling, A. 1972. *Nature New Biol.* 239:71–73
27a. Shizuya, H., Richardson, C., 1974. *Proc. Natl. Acad. Sci. USA* 71:1758–62
28. Klein, A., Bremer, B., Kluding, H., Symmons, P. 1978. *Eur. J. Biochem.* 83:59–66
29. Skalka, A. 1977. *Curr. Top. Microbiol. Immunol.* 78:201–37
30. Staudenbauer, W. L. 1974. *Eur. J. Biochem.* 47:353–63
31. Kessler-Liebscher, B., Staudenbauer, W. L. 1976. *Eur. J. Biochem.* 70:523–29
32. Hirose, S., Okazaki, R., Tamanoi, F. 1973. *J. Mol. Biol.* 77:501–17
33. Okazaki, R., Hirose, S., Okazaki, T., Ogawa, T., Kurosawa, Y. 1975. *Biochem. Biophys. Res. Commun.* 62:1018–24
34. Kurosawa, Y., Ogawa, T., Hirose, S., Okazaki, T., Okazaki, R. 1975. *J. Mol. Biol.* 96:653–64
35. Okazaki, R., Okazaki, T., Hirose, S., Sugino, A., Ogawa, T., Kurosawa, Y., Shinozaki, K., Tamanoi, F., Saki, T., Machida, Y., Fujiyama, A., Kohara, Y. 1975. In *DNA Synthesis and Its Regulation,* ed. M. Goulian, P. Hanawalt, C. F. Fox, pp. 832–62. Menlo Park: W. A. Benjamin
36. Ogawa, T., Hirose, S., Okazaki, T., Okazaki, R. 1977. *J. Mol. Biol.* 112:121–40
37. Uyemura, D., Eichler, D. C., Lehman, I. R. 1976. *J. Biol. Chem.* 251:4085–89
38. Konrad, E. B., Lehman, I. R. 1975. *Proc. Natl. Acad. Sci. USA* 72:2150–54
39. Tye, B.-K., Nyman, P. O., Lehman, I. R., Hochhauser, S., Weiss, B. 1977. *Proc. Natl. Acad. Sci. USA* 74:154–57
40. Wechsler, J. A., Gross, J. D. 1971. *Mol. Gen. Genet.* 113:273–84
41. Gross, J. D. 1972. *Curr. Top. Microbiol. Immunol.* 57:39–74
42. Carl, P. 1970. *Mol. Gen. Genet.* 109:107–22
43. Wechsler, J. A. 1973. In *DNA Synthesis In Vitro,* ed. R. D. Wells, R. Inman, pp. 375–82. Baltimore: University Park
44. Wickner, S., Berkower, I., Wright, M.,

Hurwitz, J. 1973. *Proc. Natl. Acad. Sci. USA* 70:2369–73
45. Wechsler, J. A. 1975. *J. Bacteriol.* 121:594–99
46. Zyskind, J. W., Smith, D. W. 1977. *J. Bacteriol.* 129:1476–86
47. Schubach, W. H., Witmer, J. D., Davern, C. I. 1973. *J. Mol. Biol.* 74:205–21
48. Cozzarelli, N. R. 1977. *Ann. Rev. Biochem.* 46:641–68
49. Lark, K. G., Renger, H. 1969. *J. Mol. Biol.* 42:221–36
50. Ward, C. B., Glaser, D. A. 1969. *Proc. Natl. Acad. Sci. USA* 64:905–12
51. Lark, K. G. 1972. *J. Mol. Biol.* 64: 47–60
52. Messer, W. 1972. *J. Bacteriol.* 112:7–12
53. Bagdasarian, M. M., Izakowska, M., Bagdasarian, M. 1977. *J. Bacteriol.* 130:577–82
54. Gellert, M., O'Dea, M. H., Itoh, T., Tomizawa, J.-I. 1976. *Proc. Natl. Acad. Sci. USA* 73:4474–78
55. Gellert, M., Mizuuchi, K., O'Dea, M. H., Itoh, T., Tomizawa, J.-I. 1977. *Proc. Natl. Acad. Sci. USA.* 74:4772–76
56. Sugino, A., Peebles, C., Kreuzer, K., Cozzarelli, N. R. 1977. *Proc. Natl. Acad. Sci. USA.* 74:4767–71
57. Sakai, H., Hashimoto, S., Komano, T. 1974. *J. Bacteriol.* 119:811–20
58. Derstine, P. L., Dumas, L. B. 1976. *J. Bacteriol.* 128:801–9
59. Fuchs, J. A., Karlstrom, H. O., Warner, H. R., Reichard, P. 1972 *Nature New Biol.* 238:69–71
60. Saito, H., Uchida, H. 1977. *J. Mol. Biol.* 113:1–25
61. Sunshine, M., Feiss, M., Stuart, J., Yochem, J. 1977. *Mol. Gen. Genet.* 151: 27–34
62. Georgopoulos, C. P. 1977. *Mol. Gen. Genet.* 151:35–39
63. Sevastopoulos, C. G., Wehr, C. T., Glaser, D. A. 1977. *Proc. Natl. Acad. Sci. USA* 74:3485–89
64. Wright, M., Wickner, S., Hurwitz, J. 1973. *Proc. Natl. Acad. Sci. USA* 70:3120–24
65. Wickner, S., Wright, M., Hurwitz, J. 1974. *Proc. Natl. Acad. Sci. USA* 71:783–87
66. Lanka, E., Geschke, B., Schuster, H. 1978. *Proc. Natl. Acad. Sci. USA* 75: 799–803
67. Ueda, K., McMacken, R., Kornberg, A. 1978. *J. Biol. Chem.* 253:261–69
68. Wickner, S., Hurwitz, J. 1975. *Proc. Natl. Acad. Sci. USA* 72:921–25
69. Wickner, S., Wright, M., Hurwitz, J. 1973. *Proc. Natl. Acad. Sci. USA* 70:1613–18
70. Bouche, J. P., Zechel, K., Kornberg, A. 1975. *J. Biol. Chem.* 250:5995–6001
71. Rowen, L., Kornberg, A. 1978. *J. Biol. Chem.* 253:758–64
71a. Bouche, J., Rowen, L., Kornberg, A. 1978. *J. Biol. Chem.* 253:765–69
71b. Rowen, L., Kornberg, A. 1978. *J. Biol. Chem.* 253:770–74
72. Wickner, S. 1977. *Proc. Natl. Acad. Sci. USA* 74:2815–19
73. Filip, C. C., Allen, J. S., Gustafson, R. A., Allen, R. G., Walker, J. R. 1974. *J. Bacteriol.* 119:443–49
74. Wickner, S., Hurwitz, J. 1976. *Proc. Natl. Acad. Sci. USA* 73:1053–57
75. Wickner, S. 1976. *Proc. Natl. Acad. Sci. USA* 73:3511–15
76. Gefter, M. L., Hirota, Y., Kornberg, T., Wechsler, J. A., Barnoux, C. 1971. *Proc. Natl. Acad. Sci. USA* 68:3150–53
77. Nusslein, V., Otto, B., Bonhoeffer, F., Schaller, H. 1971. *Nature New Biol.* 234:285–86
78. Otto, B., Bonhoeffer, F., Schaller, H. 1973. *Eur. J. Biochem.* 34:440–47
79. Livingston, D. M., Hinkle, D. C., Richardson, C. C. 1975. *J. Biol. Chem.* 250:461–69
80. Livingston, D. M., Richardson, C. C. 1975. *J. Biol. Chem.* 250:470–78
81. Lehman, I. R. 1974. *Science* 186:790–97
82. Lehman, I. R., Uyemura, D. G. 1976. *Science* 193:963–69
83. Scott, J. F., Eisenberg, S., Bertsch, L. L., Kornberg, A. 1977. *Proc. Natl. Acad. Sci. USA* 74:193–97
84. Eisenberg, S., Griffith, J., Kornberg, A. 1977. *Proc. Natl. Acad. Sci. USA* 74: 3198–202
85. Sigal, N., Delius, H., Kornberg, T., Gefter, M. L., Alberts, B. 1972. *Proc. Natl. Acad. Sci. USA* 69:3537–41
86. McHenry, C., Kornberg, A. 1977. *J. Biol. Chem.* 252:6478–84
87. Schekman, R., Weiner, J. H., Weiner, A., Kornberg, A. 1975. *J. Biol. Chem.* 250:5859–65
88. Wickner, S., Hurwitz, J. 1974. *Proc. Natl. Acad. Sci. USA* 71:4120–24
89. Wickner, S., Hurwitz, J. 1975. *Proc. Natl. Acad. Sci. USA* 72:3342–46
90. Nusslein, V., Bonhoeffer, F., Klein, A., Otto, B. 1973. In *DNA Synthesis In Vitro*, ed. R. Wells, R. Inman, pp. 185–94. Baltimore:University Park
91. Olivera, B., Bonhoeffer, F. 1972. *Proc. Natl. Acad. Sci. USA* 69:25–29
92. Kornberg, T., Lockwood, A., Worcel, A. 1974. *Proc. Natl. Acad. Sci. USA* 71:3189–93

93. Nusslein, V., Henke, S., Johnston, L. H. 1976. *Mol. Gen. Genet.* 145:183–90
94. Wickner, W., Brutlag, D., Schekman, R., Kornberg, A. 1972. *Proc. Natl. Acad. Sci. USA* 69:965–69
95. Weiner, J. H., Bertsch, L. L., Kornberg, A. 1975. *J. Biol. Chem.* 250:1972–80
96. Molineux, I. J., Gefter, M. L. 1974. *Proc. Natl. Acad. Sci. USA* 71:3858–62
97. Molineux, I. J., Friedman, S., Gefter, M. L. 1974. *J. Biol. Chem.* 249:6090–98
98. Geider, K., Kornberg, A. 1974. *J. Biol. Chem.* 249:3999–4005
99. Molineux, I. J., Gefter, M. L. 1975. *J. Mol. Biol.* 98:811–25
100. Rouviere-Yaniv, J., Gros, F. 1975. *Proc. Natl. Acad. Sci. USA* 72:3428–32
101. Berthold, V., Geider, K. 1976. *Eur. J. Biochem.* 71:443–49
102. Cukier-Kahn, R., Jacquet, M., Gros, F. 1972. *Proc. Natl. Acad. Sci. USA* 69:3643–47
103. Ghosh, S., Echols, H. 1972. *Proc. Natl. Acad. Sci. USA* 69:3660–64
104. Tabak, H. F., Griffith, J., Geider, K., Schaller, H., Kornberg, A. 1974. *J. Biol. Chem.* 249:3049–54
105. Wickner, R. B., Wright, M., Wickner, S., Hurwitz, J. 1972. *Proc. Natl. Acad. Sci. USA* 69:3233–37
106. Schekman, R., Wickner, W. T., Westergaard, O., Brutlag, D., Geider, K., Bertsch, L. L., Kornberg, A. 1972. *Proc. Natl. Acad. Sci. USA* 69:2691–95
107. Schaller, H., Uhlmann, A., Geider, K. 1976. *Proc. Natl. Acad. Sci. USA* 73:49–53
107a. Gray, C., Sommer, R., Polke, C., Beck, E., Schaller, H. 1977. *Proc. Natl. Acad. Sci. USA* 75:50–53
108. Wickner, W., Kornberg, A. 1974. *Proc. Natl. Acad. Sci. USA* 71:4425–28
109. Vicuna, R., Hurwitz, J., Wallace, S., Girard, M. 1977. *J. Biol. Chem.* 252:2524–33
110. Vicuna, R., Ikeda, J. E., Hurwitz, J. 1977. *J. Biol. Chem.* 252:2534–44
111. Schekman, R., Weiner, A., Kornberg, A. 1974. *Science* 186:987–93
112. Zechel, K., Bouche, J. P., Kornberg, A. 1975. *J. Biol. Chem.* 250:4684–89
113. Martin, D. M., Godson, G. N. 1977. *J. Mol. Biol.* 117:321–35
114. Wickner, S., Hurwitz, J. 1975. See Ref. 35, pp. 227–38
115. Ray, R., Capon, D., Gefter, M. L. 1976. *Biochem. Biophys. Res. Commun.* 70:506–12
116. Weiner, J. H., McMacken, R., Kornberg, A. 1976. *Proc. Natl. Acad. Sci. USA* 73:752–56
117. McMacken, R., Ueda, K., Kornberg, A. 1977. *Proc. Natl. Acad. Sci. USA* 74:4190–94
118. Eisenberg, S., Harbers, B., Hours, C., Denhardt, D. T. 1975. *J. Mol. Biol.* 99:107–23
119. Sherman, L. A., Gefter, M. L. 1976. *J. Mol. Biol.* 103:61–76
120. De Lucia, P., Cairns, J. 1969. *Nature* 224:1164–66
121. Campbell, J. L., Soll, L., Richardson, C. C. 1972. *Proc. Natl. Acad. Sci. USA* 69:2090–94
122. Kornberg, T., Gefter, M. L. 1971. *Proc. Natl. Acad. Sci. USA* 68:761–64
123. Wickner, W., Schekman, R., Geider, K., Kornberg, A. 1973. *Proc. Natl. Acad. Sci. USA* 70:1764–67
124. Wickner, W., Kornberg, A. 1973. *Proc. Natl. Acad. Sci. USA* 70:3679–83
125. Wickner, W., Kornberg, A. 1974. *J. Biol. Chem.* 249:6244–49
126. Hurwitz, J., Wickner, S., Wright, M. 1973. *Biochem. Biophys. Res. Commun.* 51:257–67
127. Hurwitz, J., Wickner, S. 1974. *Proc. Natl. Acad. Sci. USA* 71:6–10
128. Low, R. L., Rashbaum, S. A., Cozzarelli, N. R. 1976. *J. Biol. Chem.* 251:522–26
129. Tomizawa, J. I. 1978. See Ref. 9. pp. 797–826
130. Sakakibara, Y., Tomizawa, J.-I. 1974. *Proc. Natl. Acad. Sci. USA* 71:802–6
131. Sakakibara, Y., Tomizawa, J.-I. 1974. *Proc. Natl. Acad. Sci. USA* 71:1403–7
132. Staudenbauer, W. L. 1976. *Mol. Gen. Genet.* 145:273–80
133. Tomizawa, J.-I., Sakakibara, Y., Kakefuda, T. 1975. *Proc. Natl. Acad. Sci. USA* 72:1050–54
134. Tomizawa, J.-I., Sakakibara, Y., Kakefuda, T. 1974. *Proc. Natl. Acad. Sci. USA* 71:2260–64
135. Sakakibara, Y., Tomizawa, J.-I. 1974. *Proc. Natl. Acad. Sci. USA* 71:4935–39
136. Tomizawa, J.-I. 1975. *Nature* 257:253–54
137. Staudenbauer, W. L. 1976. *Mol. Gen. Genet.* 149:151–58
138. Tomizawa, J.-I., Ohmori, H., Bird, R. E. 1977. *Proc. Natl. Acad. Sci. USA* 74:1865–69
139. Bird, R. E., Tomizawa, J.-I. 1978. *J. Mol. Biol.* In press
140. Berkower, I., Leis, J., Hurwitz, J. 1973. *J. Biol. Chem.* 248:5914–21
141. Henry, C. M., Ferdinand, F.-J., Knippers, R. 1973. *Biochem. Biophys. Res. Commun.* 50:603–11
142. Miller, H. I., Riggs, A. D., Gill, G. N. 1973. *J. Biol. Chem.* 248:2621–24

143. Clewell, D. B., Helinski, D. R. 1969. *Proc. Natl. Acad. Sci. USA* 62:1159–63
144. Helinski, D., Lovett, M., Williams, P., Katz, L., Collins, J., Kupersztoch-Portonoy, Y., Sato, S., Leavitt, R., Sparks, R., Hershfield, V., Guiney, D., Blair, D. 1975. See Ref. 35. pp. 514–36
145. Inselburg, J. 1977. *J. Bacteriol.* 132:332–40
146. Staudenbauer, W. L. 1978. See Ref. 9. pp. 827–38
147. Nash, H. A. 1975. *Proc. Natl. Acad. Sci. USA* 72:1072–76
148. Mizuuchi, K., Nash, H. A. 1976. *Proc. Natl. Acad. Sci. USA* 73:3524–28
149. Gellert, M., Mizuuchi, K., O'Dea, M. H., Nash, H. A. 1976. *Proc. Natl. Acad. Sci. USA* 73:3872–76
150. Sumida-Yasumoto, C., Yudelevich, A., Hurwitz, J. 1976. *Proc. Natl. Acad. Sci. USA* 73:1887–91
151. Marians, K. J., Ikeda, J. E., Schlagman, S., Hurwitz, J. 1977. *Proc. Natl. Acad. Sci. USA* 74:1965–68
151a. Pisetsky, D., Berkower, I., Wickner, R., Hurwitz, J. 1972. *J. Mol. Biol.* 71:557–71
152. Wang, J. C. 1971. *J. Mol. Biol.* 55:523–33
153. Wang, J. C. 1978. See Ref. 9. pp. 347–66
154. Cairns, J. 1963. *J. Mol. Biol.* 6:208–13
155. Sakakibara, Y., Suzuki, K., Tomizawa, J. I. 1976. *J. Mol. Biol.* 108:569–82
156. Denhardt, D. T. 1975. *Crit. Rev. Microbiol.* 4:161–223
157. Dumas, L. 1978. In *The Single-Stranded DNA Phages,* ed. D. Dressler, D. Denhardt, D. Ray. Cold Spring Harbor, NY: Cold Spring Harbor Lab.
158. Franke, B., Ray, D. 1971. *J. Mol. Biol.* 61:565–86
159. Henry, T. J., Knippers, R. 1974. *Proc. Natl. Acad. Sci. USA* 71:1549–53
160. Eisenberg, S., Scott, J. F., Kornberg, A. 1976. *Proc. Natl. Acad. Sci. USA* 73:1594–97
161. Eisenberg, S., Scott, J. F., Kornberg, A. 1976. *Proc. Natl. Acad. Sci. USA* 73:3151–55
162. Ikeda, J. E., Yudelevich, A., Hurwitz, J. 1976. *Proc. Natl. Acad. Sci. USA* 73:2669–73
163. Fujisawa, H., Hayashi, M. 1976. *J. Virol.* 19:416–24
164. Lane, H. E., Denhardt, D. T. 1975. *J. Mol. Biol.* 97:99–112
165. Denhardt, D. T., Dressler, D. H., Hathaway, A. 1967. *Proc. Natl. Acad. Sci. USA* 57:813–18
166. Abdel-Monem, M., Hoffmann-Berling, H. 1976. *Eur. J. Biochem.* 65:431–40
167. Abdel-Monem, M., Durwald, H., Hoffmann-Berling, H. 1976. *Eur. J. Biochem.* 65:441–49
168. Abdel-Monem, M., Lauppe, H., Kartenbeck, J., Durward, H., Hoffmann-Berling, H. 1977. *J. Mol. Biol.* 110:667–85
169. Taketo, A. 1973. *Mol. Gen. Genet.* 122:15–22
170. Taketo, A. 1975. *Mol. Gen. Genet.* 139:285–91

AUTHOR INDEX

SUBJECT INDEX

CUMULATIVE INDEXES

CONTRIBUTING AUTHORS, VOLUMES 43-47

CHAPTER TITLES, VOLUMES 43-47